DATE DUE

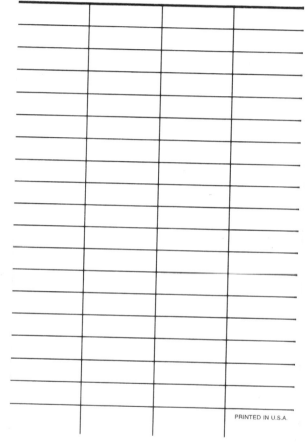

PRINTED IN U.S.A.

OC '87

THE OXFORD COMPANION TO GERMAN LITERATURE

BY

HENRY AND MARY GARLAND

SECOND EDITION

BY

MARY GARLAND

OXFORD NEW YORK
OXFORD UNIVERSITY PRESS
1986

Oxford University Press, Walton Street, Oxford OX2 6DP

Oxford New York Toronto
Delhi Bombay Calcutta Madras Karachi
Petaling Jaya Singapore Hong Kong Tokyo
Nairobi Dar es Salaam Cape Town
Melbourne Auckland

and associated companies in
Beirut Berlin Ibadan Nicosia

Oxford is a trade mark of Oxford University Press

First edition 1976
Second edition 1986

British Library Cataloguing in Publication Data
Garland, Henry
The Oxford companion to German literature.—
2nd ed.
1. German literature—Dictionaries
I. Title   II. Garland, Mary
830'.03'21   PT41
ISBN 0-19-866139-8

Printed in Great Britain
at the University Printing House, Oxford
by David Stanford,
Printer to the University

PREFACE TO THE SECOND EDITION

THE FIRST edition of the *Oxford Companion to German Literature*, which was the joint work of my husband and myself, proved to be of service to readers from all walks of life. It stimulated a lively response, and we were very grateful to all who wrote to us with suggestions and queries.

In this new edition the structure of the first edition has been preserved intact. Many suggestions from readers have been incorporated, and where necessary entries updated. New author entries treat representative writers from all the German-speaking countries. Here it has been my aim to provide concise information about those trends which have left their mark since the 1960s, and which reflect the political and cultural climate of the period.

I should like to thank my colleagues in the Department of German in the University of Exeter, William P. Hanson, Isabel S. A. Howe, John R. P. McKenzie, Dr Gerald Opie, and Jochen Rohlfs, for having so readily answered my queries and read individual entries. I am particularly grateful to Professor W. Edgar Yates for his stimulating advice on many points of detail and for his kind support when I was working under great pressure of time. I am also indebted to a great number of colleagues, in this country and abroad, who reviewed the first edition and made valuable suggestions. Waltraud Hagen's *Handbuch der Editionen* (1979, Berlin/GDR) with its detailed information on a large selection of critical and 'historisch-kritische' editions is one of the less well-known lexicographical works which have proved to be of great help. I also wish to acknowledge with deep gratitude the assistance which I have received from the Deutsches Literaturarchiv, Marbach, and in particular from its librarians Frau Margot Pehle and Frau Monika Seibold, who have generously devoted their time and expertise to the final checking of bibliographical details. Needless to say, responsibility for adjustments and additions to the first edition rests with me. I am indebted to Miss Gisela Fischer for her help in preparing the typescript and xerox copies, to the staff of the University Library for their unfailing kindness, and to Dr Sara Dewhirst for her help and suggestions at the final proof stage. And once more it has been my pleasure to work with Bruce L. Phillips and the staff of the Oxford University Press. I am especially grateful to Betty Palmer for her untiring assistance during the final stages of this edition.

May 1986   M.G.

PREFACE TO THE FIRST EDITION

A CHARACTERISTIC feature of the Oxford Companions to Literature is the wide variety of entries listed under a single alphabet and linked by a system of cross-references. *The Oxford Companion to German Literature* was conceived in this tradition, and although exigencies of space finally made it necessary to drop such entries as conspicuous characters in literary works, we have sought to produce a Companion to the historical and cultural background to German literature as well as to the writers and works themselves. The Companion spans the period from *c*. 800 to the early 1970s, and our aim has been to cover in a reasonably representative way every period of the literature of each German-speaking country. It is our hope that the frequent cross-references will serve to draw attention to the interconnection between literature and all aspects of history, as well as helping the reader to find specific items of information.

When, in 1965, we accepted the invitation to undertake the Companion we realized that the task would be a demanding one, and it is only the fact that we were both of us engaged on this enterprise that has enabled the work to be completed within ten years. Illness took its toll and greatly augmented the burden of the unafflicted partner. We have, of course, had occasion to put questions to helpful and patient colleagues to whom we here express our gratitude. At the same time in a work of this size it is, we fear, inevitable that some errors will occur. Elaborate checks have been made, but as the late Percy Scholes neatly put it in the original preface (1938) to *The Oxford Companion to Music*: 'a certain proportion of errors corrected during the day creep back during the night'. We owe it to all those to whose advice we are indebted to stress that the responsibility for the statements printed and for the decision on what to include rests with us.

We particularly acknowledge the help given by Dr. L. Peter Morris of the Department of History of the University of Exeter, who checked the historical entries, and by Dr. W. Donald Hudson of the Department of Philosophy, who gave us similar assistance with the entries on philosophers. In the German Department Miss Hilda Swinburne and Mr. Keith Dickson read and made comments on some of the medieval and twentieth-century drama entries respectively. Professor W. Edgar Yates kindly offered to read a substantial number of galleys when we were in the grip of a tight time-table and we are grateful for his suggestions. Miss Gisela Fischer, through much of the Companion's history, gave it her wholehearted interest and typed long stretches of it. In our reading and checking of proofs against copy we had some valuable assistance from Miss Diana Guthrie. We also thank with special pleasure the staff of the University Library for the patient help they have given while we gathered our material; and we are indebted to the section

Information of Inter Nationes, Bonn–Bad Godesberg, and to the Deutsches Literaturarchiv, Marbach, for their assistance.

We have worked in the closest collaboration with Mr. Bruce L. Phillips of the Clarendon Press and we wish to express our sincerest thanks for the encouragement, help, and advice we have received from the advisers and staff of the Press and for the courtesy and understanding which they have invariably displayed.

Lastly, we cannot forbear to add how much the sense of mutual assistance and reliable collaboration has served to keep each of us going to the end of a very long road.

H.G.
M.G.

November 1975

NOTE

THE entries are arranged in alphabetical order; A and Ä, O and Ö, U and Ü are in each case treated as if identical. Titles beginning with the definite article in the nominative (*der, die,* or *das*) are placed under their second word (e.g. *Zauberberg, Der*); in other cases (e.g. *Des Sängers Fluch*) and with the indefinite article (e.g. *Ein Bruderzwist in Habsburg*), the article determines the order of appearance.

Medieval authors are arranged in order according to their first name, whereas authors later than the sixteenth century are listed according to their true surname. Thus Walther von der Vogelweide appears under W, Heinrich von Kleist under K. In order to assist the reader to follow up a cross-reference indicated by 'q.v.', first names which determine alphabetical positions are indicated by small capitals (e.g. HUGO von Trimberg). This also applies to the pseudonymous Romantic writer Jean Paul, who is placed under J.

Titles of works beginning with *Sanct, Sankt, Sante,* or *St.* are all listed under *Sankt.*

Place of birth precedes date of birth and place of death follows date of death after the name.

As is customary in German, 'SS' replaces the double letter ß in those names and words which are printed in capitals.

For the Christian names Thomas and Theodor the German abbreviation Th. has been used. The long-established form E. T. A. Hoffmann has, however, been retained.

A

Aachen, German city close to the Belgo-German frontier, known to English and French historians as Aix-la-Chapelle. Of Roman origin, it became the capital of Charlemagne (see KARL I, DER GROSSE), who is buried in the 8th-c. chapel which forms the central and oldest part of the cathedral. From 813 to 1531 the German Kings (see DEUTSCHER KÖNIG) were crowned in this cathedral, and the city in this period was the scene of seventeen Diets of the Empire. Two peace treaties were signed there: the first marked the close of the War of Devolution (1668), the second (Aachener Friede) ended the War of the Austrian Succession (1748, see ÖSTERREICHISCHER ERBFOLGEKRIEG). During the Napoleonic period Aachen was from 1801 to 1815 a French possession.

AAL, JOHANNES (Bremgarten, Aargau, *c.* 1500–51, Solothurn), a Roman Catholic, left Switzerland for religious reasons and studied in Freiburg/Breisgau. In 1533 the Roman Catholic Church was reinstated in Solothurn, and Aal was appointed a canon. He is the author of a play on John the Baptist (*Tragoedia Johannis des Täufers*, 1549, reissued 1929). It is designed for performance on two successive days.

ABBT, THOMAS (Ulm, 1738–66, Bückeburg), son of a wig-maker, studied at Halle and became in 1760 a professor of philosophy at the University of Frankfurt on the Oder. In the following year he was appointed professor of mathematics at Rinteln, after which he became director of schools in Bückeburg, the capital of the little state of Schaumburg-Lippe, ruled by a notably enlightened prince, Graf Wilhelm (q.v.). Abbt, who reached manhood at the time of the Seven Years War (see SIEBENJÄHRIGER KRIEG), was stirred by the tenacity of Prussia and the courage and genius of Friedrich II (q.v.). His admiration is reflected in the modernized stoicism of his tract *Vom Tode für das Vaterland* (1761). His *Vom Verdienst* (1765) is a didactic work setting out a scale of virtue which gives first place to the man of action and soldier, though he does not neglect gentler qualities. It enjoyed a considerable contemporary success. Abbt contributed to the *Literaturbriefe* (q.v.), signing his articles with the initial B. Among his friends were the well-known originators of 'Enlightenment', F. Nicolai and M. Mendels-sohn (qq.v.). Nicolai published Abbt's *Vermischte Werke* (6 vols., 1768–81).

Abderiten, Geschichte der, a satire by C. M. Wieland (q.v.), first published in part in the periodical *Der teutsche Merkur* in 1774, and complete in book form in 1780. It has the subtitle *Eine sehr wahrscheinliche Geschichte*. Wieland was occupied in writing it at intervals from 1773 to 1779. The book is a satire on the self-satisfied parochial life of German small towns in Wieland's day. Its setting and disguise is the ancient Greek town of Abdera in Thrace, the inhabitants of which (with the exception of Democritus) were noted for their narrow-mindedness.

The book consists of five parts. The first, *Demokritus unter den Abderiten*, tells of Democritus's return to Abdera and the doubts which the Abderites entertain about his sanity. In the second, *Hippocrates in Abdera*, the great physician Hippocrates is asked to investigate Democritus's sanity. He pronounces Democritus sane and the inhabitants mad. The third part, *Euripides in Abdera*, is largely literary satire. In the fourth, *Der Prozeß um des Esels Schatten*, a silly dispute between an ass driver and the hirer of the ass is allowed to develop into a legal case of the first magnitude. Finally, in *Die Frösche der Latona*, the Abderites allow themselves to be driven from home, because they do not take steps against the sacred frogs.

Abdias, a Novelle by A. Stifter (q.v.) written in 1842 and first published in 1843. Stifter prepared the final version in 1845 for publication in vol. 4 of the *Studien*. The story is divided into three sections under the headings 'Esther', 'Deborah', and 'Ditha', the names of Abdias's mother, wife, and daughter. The first two show episodes in the life of Abdias in an ancient Jewish settlement, which is situated among Roman ruins in a remote part of the African desert. Aron, his father, as befits a man of wealth, offers Esther and her son a sheltered life of luxury, but when the boy grows up, he sends him away to acquire the skills with which to accumulate riches as well as the knowledge of the hazards awaiting him in foreign lands.

After fifteen years Abdias returns, and his marriage to the beautiful Deborah brings him reward, but an illness deprives him of his handsome looks and of Deborah's love. Abdias finds compensation in his increasing wealth and power. During a battle in the desert he proves his worth as a commander, and justifies his right to high authority; but the flights of ambition are checked when he returns home to experience revenge and humiliation at the hands of an enemy, Melek. He is branded as a traitor to the settlement, which has suffered destruction and plunder during the attack. Deserted and humiliated by all, Abdias discovers Deborah in her once luxurious apartment. During the terror of the attack she has not only given birth to a daughter but has also found a new spirit of love for Abdias. Having satisfied himself that the enemy has not discovered his hidden wealth, Abdias now believes he sees at last the finger of destiny. With humility and wisdom he arranges for the basic needs of mother and child, but Deborah dies from the effects of childbirth.

In the third and longest section Abdias devotes his life and wealth to Ditha. As soon as she is strong enough, they secretly depart for Europe. In a quiet Austrian valley he makes her a new home. When she is 4 years old he discovers that she is blind, and in order to secure her future he resumes his trade and accumulates riches. When Ditha is 11 years old a flash of lightning gives her sight. Abdias enjoys the reward of her love and sight and cultivates the valley in which Ditha grows up. She is 16 when a flash of lightning kills her. Abdias's world is shattered, although he lives on for thirty years and more. Stifter himself discusses in his introduction the problematical nature of human destiny, which Abdias's unusual story unfolds, without passing judgement.

Abecedarium nordmannicum, a runic alphabet of sixteen signs with a simple explanatory text in alliterative verse. This is written in Old Saxon with traces of Anglo-Saxon. The alphabet is entered in a 9th-c. MS. at St. Gall containing grammatical sections from the *Etymologiae* of Isidor (q.v.).

ABEL, JAKOB FRIEDRICH (Vaihingen, 1751–1829, Schorndorf), one of Schiller's teachers, was appointed professor of philosophy at the Militär-Akademie at the Solitude in 1772. It was he who first introduced Schiller (q.v.) to Shakespeare's plays. In 1790 he was appointed to a chair of philosophy at Tübingen, and in 1823 he became general superintendent in Stuttgart. The episode underlying Schiller's story *Der Verbrecher aus verlorener Ehre* (q.v., 1786) was experienced by his father, and Abel himself published an account of it in *Sammlung und Erklärung merkwürdiger Erscheinungen aus dem menschlichen Leben* (1787).

ABEL, KASPAR (Hindenburg, Brandenburg, 1667–1763, Westdorf nr. Aschersleben), a schoolmaster and pastor, made translations from Ovid and Boileau and wrote verse satires (*Auserlesene satirische Gedichte,* 1714).

ABELE VON UND ZU LILIENBERG, MATTHIAS (Vienna, c. 1616–77, Vienna), a clerk of the court in Vienna, later in Krems-Und-Stein, wrote anecdotal books on cases before the courts (*Metamorphosis telae judiciariae, Das ist Seltzame Gerichtshändel,* 1651; *Vivat Unordnung,* 1669). He also published a *Sterbebüchlein* (1650).

Abenteuer der Sylvester-Nacht, a story by E. T. A. Hoffmann (q.v.), written in 1815 and published in the same year in vol. 4 of *Fantasiestücke in Callots Manier.* (see RAHMEN). The narrator encounters his former love Julia at an evening party in Berlin. He rushes out without hat or coat into the bitter winter night and calls at a tavern to warm himself. There he meets the shadowless Peter Schlemihl (see PETER SCHLEMIHLS WUNDERSAME GESCHICHTE) and another man, who has lost his reflection. The story of how this man ceded his reflection to the seductive courtesan Giulietta (the affinity with Julia is intentional) is then told under the title 'Die Geschichte vom verlornen Spiegelbild'. This story is the source of Act II of Offenbach's opera *Les Contes de Hoffmann.*

Abenteuerliche Simplicissimus Teutsch, Der, a novel by J. J. C. von Grimmelshausen (q.v.), first published pseudonymously in 1669 (dated 1668) in an edition containing 5 Books. Only a few copies of this edition have survived, and the best known extant form is the second edition of 1669, to which a sixth book was added in the same year: *Continuatio des abentheurlichen Simplicissimi oder Schluß desselben.* Grimmelshausen himself designed the well-known copper engraving on the title-page of the edition of 1683–4; dominated by a grotesque monster displaying an open book, it is a hieroglyphic composition alluding to the novel's multifarious motifs. The edition of 1671, containing minor alterations, is the version which is usually reprinted. The long sub-title of 92 words, which sets forth the book's programme, contains the true name of Simplicissimus, Melchior Sternfels von Fuchshaim, as well as the pseudonym of the author, German Schleifheim von Sulsfort; like other pseudonyms used by Grimmelshausen, they are anagrams of his full name, and his authorship was not discovered until the 19th c.

The background of *Simplicissimus* is the Thirty Years War (see DREISSIGJÄHRIGER KRIEG). Of the six books the last (*Continuatio*) is apparently an afterthought prompted by the immense success of the work. The action is best recounted book by book. Bk. 1: the hero, a child, whose innocence is symbolized by his name, grows up in the Spessart on the farm of his putative father. Plundering troops raid the farm and torture or rape the inhabitants. The boy flees and is sheltered by a hermit, who gives him the name Simplicius, educates him, and instructs him in religion. On the death of the hermit, Simplicius goes to Hanau. Bk. 2: in Hanau he survives an attempt to drive him insane. He is carried off by Croat soldiers from whom he escapes, and then, falling in with imperial troops, he forms a friendship with a young man called Herzbruder. Bk. 3: he becomes an efficient and daring soldier, and his exploits with his comrade Springinsfeld gain him the nickname Der Jäger von Soest, where he is stationed. Eventually he falls into Swedish captivity and has various amorous adventures, one of which leads to a forced marriage. Bk. 4: Simplicius sets out for Cologne, where he has money, but encounters difficulties in obtaining it. He next accompanies two noblemen to Paris, where he has many nocturnal exploits with fine ladies. On his return he falls ill with smallpox, and, finding himself alone and penniless on his recovery, makes his way to Germany as a doctor. Bk. 5: having met Herzbruder again, Simplicius goes with him on a pilgrimage to Einsiedeln. Herzbruder dies, Simplicius finds that his own wife has died, and makes a second, unsuccessful marriage with a country girl. He learns that he is of noble birth and that his true father was the hermit who succoured him after his escape from the marauding soldiers. His name proves to be that given on the title-page, Melchior Sternfels von Fuchshaim. After further adventures, including a visit to Moscow, he becomes convinced of the vanity of earthly things and becomes a hermit. Bk. 6: the world tempts him again; after various dangers and sufferings he is wrecked on an island in the South Atlantic, and there resumes his life as a hermit; rejecting a chance of repatriation, he ends his life there.

The novel is a work on at least four planes. It is an absorbing, racily told adventure story; at the same time it recounts the development and maturing of a character; it comments with dry irony on human affairs; and finally, it presents a view of life, a recognition of the vanity of worldly things and a resigned and half-humorous acceptance of what fortune brings. The final book, moreover, is the first Robinsonade (q.v.), or 'Robinson Crusoe story', in German literature. The incidental character of the Landstörzerin Courasche (Bk. 5 ch. 6, but there called only 'Landstörzerin'), to whom Grimmelshausen later devoted the novel *Trutz-Simplex*

(q.v.), is the original of Brecht's Mutter Courage in *Mutter Courage und ihre Kinder* (q.v.). Of all the massive literature of the 17th c., *Simplicissimus* is one work which remains outstandingly alive. It forms the basis of *Des Simplicius Simplicissimus Jugend* (1948) by Karl Amadeus Hartmann (b. 1905).

Abenteuerroman, a kind of novel intended usually for entertainment, but sometimes serving more serious purposes. The Abenteuerroman can be traced back to some of the verse tales of the Middle Ages, such as *Ruodlieb, Salman und Morolf, König Rother,* or *Herzog Ernst* (qq.v.) and manifests itself later in Arthurian romance. The tradition is continued in prose tales of the 16th c. of which the chap-book *Fortunatus* (q.v.) and *Der Goldfaden* by G. Wickram (q.v.) are examples. The adventure novel rises to its greatest height in Grimmelshausen's *Der abenteuerliche Simplicissimus* (q.v.). Also worthy of mention are Grimmelshausen's other 'Simplician novels', the works of J. Beer (q.v.), and, in the early 18th c., *Die Insel Felsenburg* (q.v., 1731–43) by J. G. Schnabel (q.v.). In the late 18th c. wildly romantic adventure novels (Schauerromane and Ritterromane), which correspond to the English 'Gothick' novel, occur in great profusion. The Romantic period itself brought attractive variations of the genre in Eichendorff's novels. Exciting adventure, as suitable for boys as for men, is offered in the novels of F. Gerstäcker and, later, of Karl May (qq.v.), nearly all set in America. In the 20th c. the Abenteuerroman occurs either frankly as entertainment or in satirical or ironic form (e.g. Thomas Mann's *Bekenntnisse des Hochstaplers Felix Krull,* q.v.). A special form of the Abenteuerroman is the Schelmenroman (q.v.), a category which comprises some of the works and authors mentioned above, viz. Grimmelshausen, Beer, and Mann's *Krull.*

Abenteurer und die Sängerin, Der, a verse play by H. von Hofmannsthal (q.v.), included in *Theater in Versen* (1899), and described as 'Ein Gedicht in zwei Aufzügen'. It is based on the *Mémoires* of Casanova (q.v.), whom the ageing adventurer who goes by the name Baron Weidenstamm may be said to represent. He revisits Vittoria, whom he had briefly loved many years before, and by whom he has a son. In the intervening years Vittoria has become a great singer, and she bears no resentment against 'Weidenstamm' for the fickleness of his love, for the experience of former years awakened her great gift for music.

'Aber abseits, wer ist's?', first line of the text of the *Alto Rhapsody* by J. Brahms (q.v.). It is taken from Goethe's *Harzreise im Winter* (q.v.), ll. 29–50.

Abfall der Niederlande, see GESCHICHTE DES ABFALLS DER VEREINIGTEN NIEDERLANDE VON DER SPANISCHEN REGIERUNG.

Abgekühlte, Der, a poem by H. Heine (q.v.) included in the section *Lamentationen* of the *Romanzero* (q.v.). No. 9 of the subdivision *Lazarus,* it begins 'Und ist man tot, so muß man lang/Im Grabe liegen'. It is a poem of nostalgia in the face of death.

Abgesang, see MINNESANG.

Abitur, DAS (Reife- or Abschlußprüfung), the German matriculation which qualifies candidates for university study. It was introduced in 1788 as part of the school reforms implemented by Freiherr K. A. von Zedlitz (1731–93), who served as minister under Friedrich II (q.v.) of Prussia and his successor. In 1787 he founded the Prussian grammar school (Oberschulkollegium); this as well as the Abitur was revised by W. von Humboldt (q.v.).

The right of entry which the Abitur traditionally confers was modified by some Länder of the Federal Republic (see BUNDESREPUBLIK DEUTSCHLAND) in 1970.

Ablaßkram, the sale of indulgences, an ecclesiastical practice which had existed for centuries before Luther, but increased enormously in the late 15th c., degenerating into a mere means of raising revenue. Luther's opposition to the flagrant exploitation of the sale of indulgences, as practised on behalf of the Archbishop of Mainz, resulted in his first step towards his breach with the Papacy (see LUTHER, M. and THESES, 95). The Fastnachtspiel (q.v.) *Der Ablaßkrämer* by N. Manuel (q.v.), one of the most outstanding plays of the Reformation, ridicules the misuse of this practice.

Abor und das Meerweib, title given to a fragment of 136 lines from a lost Middle High German verse romance written *c.* 1300. It narrates an episode in which the hero, Abor, is succoured by a mermaid, who gives him a garment of invulnerability and a herb which enables him to understand the speech of beasts and birds. The author is unknown.

ABRAHAM A SANTA CLARA (Kreenheinstetten, Hegau, Württemberg, 1644–1709, Vienna), ecclesiastical name of the Augustinian friar, Johann Ulrich Megerle. An innkeeper's

son, he was educated at the Jesuit college at Ingolstadt and then at the Benedictine university at Salzburg. He entered his Order in 1662, was ordained in 1668, and was sent to Vienna as penitential preacher in 1672. His popular mode of preaching attracted attention in high circles and he was given the style Imperial Preacher (Kaiserlicher Prediger). He was subprior in Vienna in 1677, prior in 1680, was transferred to Graz in 1683, and from 1695 occupied a high position in his Order in Vienna.

Abraham applied the gifts that were apparent in his preaching to his moral writings. His style bubbles over with homely and caustic wit, with proverbs and common sayings aptly or drastically applied, and with an extraordinary exuberance of vocabulary. This lively writing, which is reflected in the vivid unconventionality of his titles, is applied to the serious purpose of moral reform, for his works hold up a mirror to the folly and vice of the world, and especially to Vienna, which he knew and served.

Abraham was an indefatigable and inexhaustible publicist, and only the titles of the principal works are given here. The first of his moral satirical works is *Mercks Wienn* (1679), which vehemently derides and castigates the foibles and sins of the Viennese, a message underlined by the epidemic of plague raging at the time and the threat of Turkish invasion. *Auff, auff ihr Christen* (1683) urges unity and strength against the Turkish siege of Vienna. *Judas der Ertz-Schelm* (1686) purports to be a biography of Judas Iscariot, based on legendary accretions, but its narrative form is constantly interrupted and distorted by Abraham's irrepressible inventiveness, and by his addiction to illustrative episodes and anecdotes. A series of moral satires followed at the end of the century, with *Heilsames Gemischgemasch* (1704), *Huy und Pfuy der Welt* (1704), and *Wohlgefüllter Weinkeller* (1710). A hundred male follies are derided in *Centifolium stultorum* (1709), and an equal number of feminine ones in *Mala galina* (1713). Abraham's spontaneous and kaleidoscopic humour caused him at one time to be regarded as a literary buffoon, but his moral purpose is clear and persistent, and he is one of the few writers of the century who succeeded in turning baroque exuberance to homely and popular account.

A false image was also unintentionally created by Schiller, who modelled the sermon of his eccentric Capuchin in *Wallensteins Tod* (see WALLENSTEIN) on some of Abraham's more outrageous puns and locutions.

Abraham's *Werke* (3 vols.), ed. K. Bertsche, appeared 1943–5, and selections as *Auslese* (6 vols.), ed. H. Strigl, 1904–6, and *Auswahl,* ed. W. Höllerer (q.v.), 1959.

Abreise, a poem written in unrhymed trochaic verse by E. Mörike (q.v.) in 1846. Its subject, lightly and deftly touched, is the departure of a coach from an inn, severing the ties of a new love, and leaving behind a sadness which lasts no longer than a summer shower.

Abrogans, customary designation of a late Latin dictionary of synonyms, the German translation of which (*Deutscher Abrogans*) is the oldest document written in German. It is so called because 'abrogans' is the first Latin word listed. Dating from the second half of the 8th c. (probably between 764 and 783), it was written in the chapter school of Freising, Bavaria. The original MS., in which Latin and German were interlinear, is lost; the three existing MSS. are all Alemannic: (1) *Pariser Glossen*, (2) *Keronisches Glossar,* so called because it was formerly ascribed to a hypothetical monk named Kero or Gero (St. Gall), (3) *Reichenauer Glossar.* All three are believed to be the work of monks in Reichenau or Murbach (Alsace). A later Bavarian version, known as the *Samanunga,* was written c. 790 in Regensburg and was formerly designated *Pseudohrabanisches Glossar.* (See GLOSSEN.)

ABSCHATZ, HANS ASSMANN, FREIHERR VON (Würbitz, Silesia, 1646–99, Liegnitz), studied at Strasburg and Leyden, afterwards making a grand tour in France and Italy. On his return he lived on his estates, and from 1675 held office in the Duchy of Liegnitz. He translated Guarini's pastoral play *Il pastor fido* in 1678 and wrote elegant playful erotic poetry which, in form, was indebted to Opitz and HOFFMANN von Hoffmannswaldau (qq.v.). These poems, most of which were written in his youth, were published posthumously in *Poetische Übersetzungen und Gedichte* (1704).

Abschied, a well-known poem by J. von Eichendorff (q.v.), which bears the superscription 'Im Walde bei Lubowitz'. Its first line runs 'O Täler weit, o Höhen', and it is a farewell to the forests of the poet's home. It appears, untitled, in the novel *Ahnung und Gegenwart* (q.v.). Two other poems by Eichendorff have the same title: the sonnet 'Laß Leben, nicht so wild die Locken wehen!' and the song 'Abendlich schon rauscht der Wald'.

Abschied, Der, an ode written by F. Hölderlin (q.v.) after his parting from Susette Gontard (Diotima). Its subject is the sense of desolation. It is in Asclepiadic verse and exists in three versions. (See ASKLEPIADISCHE STROPHE.)

ABSOLON is the author of a lost Middle High German epic poem dealing with the exploits and death of the Emperor Barbarossa (see FRIEDRICH I, KAISER). He is mentioned by RUDOLF von Ems (q.v.) in *Alexander* and *Willehalm von Orlens* and his poem was written between 1190 and c. 1230, an instance of a virtually contemporary subject. Absolon came of a family resident in the Lake Constance district.

Absurda Comica oder Herr Peter Squentz, a comedy (Schimpff-Spiel) by A. Gryphius (q.v.), published in 1657 or 1658. Its plot is an adaptation of the Bottom–Peter Quince (Squentz) scenes in Shakespeare's *A Midsummer Night's Dream*; it is likely that Gryphius knew this episode indirectly, through a comedy, *Peter Squenz,* by Daniel Schwenter, a professor of Altdorf University. Gryphius's play contains a parody of Meistergesang (q.v.), complete, in the original edition, with music.

Abu Telfan oder Die Heimkehr vom Mondgebirge, a novel written by W. Raabe (q.v.) in 1865–7, and published in 1868. Raabe has given it a pessimistic motto, said to be of Turkish origin, 'Wenn Ihr wüßtet, was ich weiß, sprach Mahomed, so würdet Ihr viel weinen und wenig lachen.' The novel begins with the return to Germany of Leonhard Hagebucher, who has been ten years in Africa in captivity, from which he has been saved by an ivory dealer, Cornelius van der Mook. Paradoxically, his long confinement has turned him into a freer man than his compatriots who have remained in Bumsdorf and Nippenburg. Hagebucher's free outlook and his exotic history quickly alienate the narrow lower middle class, from which he derives, and he finds refuge with another outcast, Cousin Wassertreter, who had suffered imprisonment for his political opinions in 1817 and has been cold-shouldered by society ever since. Wassertreter's philosophy is to laugh at the foibles of a detestable world which cannot be expected to change.

The story develops as a commentary on the political and social structure of Germany, often savagely exposing the shortcomings and vices of the rulers and the ruled and simultaneously illuminating their narrowness and conservatism as a source of stability and strength. The shortcomings of the Bürgertum are clearest in Hagebucher's prejudiced and self-satisfied family, the vices of the aristocracy in the titled and protected seducer and swindler, Baron von Glimmern. Hagebucher is brought into contact with three people whose lives have been virtually destroyed by crimes which society declines to expose. First is Frau Claudine Fehleysen, whose husband has been driven to death and her son to flight by the vindictive intrigues of von Glimmern. Frau Claudine waits in patience and in hope at the Katzenmühle for the return of her son Viktor, who was formerly betrothed to Nikola von Einstein. Nikola is the second acquaintance of Hagebucher to have been involved in disaster caused by von Glimmern, whom she is presently driven to marry; the third is Lieutenant Kind, whose daughter has been seduced by von Glimmern. Kind spends his life preparing the case against Glimmern, pursues him when he flees to London, and kills him in a duel in which he, too, perishes. Frau Claudine has a fleeting encounter with the long-lost Viktor, now masquerading as the Dutch ivory hunter Van der Mook, but Viktor, avoiding the widowed Nikola von Glimmern, goes to the United States and falls in the Civil War. Frau Claudine and Nikola live together at the Katzenmühle. Hagebucher gives the two women what help he can.

The richest part of the book lies in the gallery of eccentric and humane characters, peripheral to the conventional world, above all Vetter Wassertreter, and marginal figures such as Herr von Bumsdorf, Professor Reihenschlager, or Täubrich-Pascha. Raabe seems more at home with positive human values than with the conscious pessimism or the rather stagey action supporting the political denunciation.

Achilleis, an uncompleted epic poem in hexameters by Goethe. Eight cantos were projected, of which only the first (of 651 lines) was written. It was to have for its subject the death of Achilles, filling the gap between *Iliad* and *Odyssey.* The starting-point of the extant canto is the scene before Troy as the flames of Hector's funeral pyre die down. Written in 1799 at the height of Goethe's classical enthusiasm, it has wonderful passages such as the opening ('Hoch zu Flammen entbrannte die mächtige Lohe noch einmal' etc.) but also appreciable mythological *longueurs* and many mechanically constructed hexameters.

ACHLEITNER, FRIEDRICH (Schalchen, Austria, 1930–), collaborated as a member of the Wiener Gruppe (q.v.) with H. C. Artmann and G. Rühm (qq.v.) in the *avant-garde* volume of dialect poems *hosn, rosn, baa* (1959). A selection of this edition appeared together with the texts *schwer schwarz* (1960) and *der rote reiter—drei geschichten* (1967) in the collection *prosa. konstellationen. montagen. dialektgedichte. studien.* (1970). *quadrat-roman & andere quadratsachen* appeared in 1973, *quadrat-studie* in 1974, and *friedrich achleitner + gerhard rühm/super record 50+50,* ed. H. Bäcker, in 1980. See KONKRETE POESIE.

ACHLER, ELSBETH, see KÜGELIN, KONRAD.

'**Ach, wie ist's möglich dann**', first line of an 18th-c. Thuringian folk-song. It occurs as an inserted song in *Eginhardt und Emma* (1817) by Helmina von Chézy (q.v.). The melody to which it is now sung is a folk-tune, adapted by Friedrich Krücken in 1827.

ACKERMANN, HANS, a citizen of Zwickau in the 16th c., wrote in the spirit of the Reformation the biblical plays *Der verlorene Sohn* (1536) and *Tobias* (1539).

Ackermann aus Böhmen, Der, a dialogue by Johann von Saaz (q.v.) written *c.* 1400 and first printed in 1460. It comprises a formal dispute between the 'Ploughman' (who ploughs with a pen, in other words a clerk) and Death, who has deprived him of his young wife. It is believed to have originated in personal experience, the wife of the author having died in childbirth in 1399 or 1400. The Ploughman and Death speak alternately, the former making his complaint, which Death answers, in the manner of a legal process. After sixteen interchanges between the parties, God delivers judgement in favour of Death, yet praises the spirit of the complainant. The work closes with the Ploughman's prayer to God for the soul of his dead wife. The formal pattern is shot through with deep feeling, and the often rhetorical and exuberant prose has real poetic power. Medieval in essence, it shows signs of awakening humanism.

Acolastus, see GNAPHEUS, W. and BEHEMB, M.

Acta Eruditorum, the first German learned periodical, founded in Leipzig in 1682 by Professor Otto Mencke (1644–1707). Including the *Nova Acta Eruditorum* it comprised 117 volumes. It ceased publication after 100 years in 1782.

Adams erstes Erwachen und erste seelige Nächte, a short prose idyll by F. Müller (q.v.) published in 1778. The speaker is Adam, who in exile from Eden recounts his first awakening in Paradise and his life up to the creation of Eve.

Adamslegende, title given to a Middle High German poem of nearly 4,000 lines, the author of which gives his name as Lutwin. He was probably a cleric and a native of Austria, and wrote the poem early in the 14th c. It deals with the life of Adam and Eve after their expulsion from the Garden, recounting their attempts to conciliate God. Adam receives an olive branch and the assurance of salvation when it bears fruit. The fruit is Christ, for of this tree the Cross will be made. The source of the poem is a Latin *Vita Adae et Evae.*

ADAMUS, FRANZ, pseudonym of Ferdinand Bronner (Auschwitz, 1867–1948, Goisern, Salzkammergut), who wrote a Naturalistic trilogy of plays, *Jahrhundertwende* (1900), of which only the first, *Familie Wawroch*, made appreciable impact. It strongly recalls G. Hauptmann's *Die Weber* (q.v.).

A.D.B., abbreviation for *Allgemeine Deutsche Biographie* (q.v.).

Adebar, the stork, used especially of the bird in its legendary relationship to birth. The word is of Low German origin.

Adelaide, a poem written by Friedrich Matthisson (q.v.) and published in his *Gedichte*. It was set to music by Beethoven in 1795. The first line runs 'Einsam wandelt dein Freund im Frühlingsgarten'.

ADELBRECHT, PRIESTER, see JOHANNES BAPTISTA.

Adelheid, a character in Goethe's play *Götz von Berlichingen* (q.v.), whose full name is Adelheid von Walldorf. She is the embodiment of alluring sensuality and uninhibited sexual appetite and Goethe declares in *Dichtung und Wahrheit* (q.v.) that he was himself, like Pygmalion, in love with his own creation.

Adel und die Revolution, Der, a fragment of autobiography by Eichendorff (q.v.). It was first published in *Aus dem literarischen Nachlasse* (1866).

Adel und Untergang, a volume of verse by J. Weinheber (q.v.), published in 1934. It combines deep feeling with considerable virtuosity. Among its contents are *Antike Strophen*, a cycle divided into three sections, 'Gesang vom Manne', 'Die Oden', and 'Gesang vom Weibe', a cycle of 'Variationen auf eine hölderlinische Ode', which quotes as its theme the ode 'An die Parzen' (q.v.), and a further cycle called *Heroische Trilogie*, the first and third parts of which consist of sonnets, while the second is in *terza rima*. The section entitled *Blumenstrauß* is composed of attractive flower poems.

ADELUNG, JOHANN CHRISTOPH (Spantekow nr. Anklam, 1732–1806, Dresden), lexicographer, was a schoolmaster in Erfurt and later a librarian in Dresden. He compiled an important dictionary, *Versuch eines vollständigen grammatisch-kritischen Wörterbuchs der hochdeutschen Mundart* (5 vols., 1774–86), which is still of value for 18th-c. vocabulary and usage. His grammatical publications include *Deutsche Sprachlehre*

(1781), written for the Prussian Ministry of Education, *Umständliches Lehrgebäude der deutschen Sprache* (1782), and *Über den Stil* (3 vols., 1785–6).

Of general cultural interest are *Versuch einer Geschichte der Kultur des menschlichen Geistes* (1782) and *Geschichte der menschlichen Narrheit* (7 vols., 1785–9).

ADENAUER, KONRAD (Cologne, 1876–1967, Rhöndorf nr. Bonn), German statesman, entered the administration of Cologne in 1906 and was appointed Oberbürgermeister in 1917. His political affinities were with the Roman Catholic Zentrum (q.v.). His refusal to co-operate with the National Socialists enabled him to re-enter political life after 1945. He became the leader of the Christian Democratic Union, and in 1949 was the first chancellor of the Federal Republic (see BUNDESREPUBLIK DEUTSCHLAND), retaining the office until 1963. He supported a policy of hostility towards Russia, and of conciliation towards France.

Ad equum errehet, superscription of an Old High German spell to cure lameness in a horse. The spell, which is preserved on the same page of a 12th-c. MS. in Paris as the spell *Contra caducum morbum* (q.v.), is set in three folk-song-like verses.

Adjutantenritte und andere Gedichte, by Detlev von Liliencron (q.v.), published in 1883. It contained 97 poems, including *Tod in Ähren* (q.v.), *Die Attacke*, and *Wer weiß, wo?* (q.v.), and concluded with a prose sketch entitled *Adjutantenritte*, based on experiences of 1870–1. This collection of poetry and prose achieved an unexpected success with the younger generation of realistic writers, and is one of the landmarks in the growth of Naturalism (see NATURALISMUS).

ADLER, ALFRED (Vienna, 1870–1937, Aberdeen), was a pupil of S. Freud (q.v.) from 1902. In 1907 he began to develop his own system of Tiefenpsychologie (i.e. psychology of the subconscious) under the name Individualpsychologie, and his divergent views led in 1911 to a permanent breach with Freud. Adler discards Freud's views on the all-important central position of sex and sees the urge for power, prestige, and domination as the principal factor in human behaviour; this urge is the subject's attempt at compensation for a sense of inferiority towards his environment. The term inferiority complex (Minderwertigkeitskomplex) was originated by Adler. After 1933 he lived in exile.

Adler's works include *Über den nervösen Charakter* (1912) and *Menschenkenntnis* (1927).

7

ADLER, Viktor or Victor (Prague, 1852–1918, Vienna), of Jewish descent, began as a doctor in the slums of Vienna and was drawn to politics by the misery he saw there. After a short association with Georg von Schönerer's German National Party he turned to the Social Democrats, founding the *Arbeiter-Zeitung* in 1889, which he edited until 1918. In 1899 he was the leading spirit in the formulation of the Brünner Programm (q.v.) and from 1905 was a deputy in parliament and the Social Democratic Party leader. Adler was tireless in his exposure of exploitation of the working classes and in consequence served various terms of imprisonment. In later years he associated his party with the Crown, which led to the Social Democrats being derided as 'Kaisersozialisten'; and he supported the State throughout the 1914–18 War, participating in the suppression of a strike in January 1918. In October 1918 he accepted government office. He died at his desk on 11 November, the last day of the old regime and the first of the new.

Adlerorden, name of two Prussian orders of chivalry.

(1) Der schwarze Adlerorden was the highest Prussian order. It was founded in 1701 by Friedrich I (q.v.) of Prussia. Recipients, if not already of noble birth, were invariably ennobled.

(2) Der rote Adlerorden, which came next in the Prussian hierarchy of orders, began as 'ordre de la sincérité' founded by the heir apparent of Bayreuth, Georg Wilhelm. It became a Prussian order in 1792 after the transfer of the Margravate of Bayreuth to the Prussian Crown. Both orders became extinct in 1919.

Ad me ipsum, an autobiographical document contained in the *Aufzeichnungen* of H. von Hofmannsthal (q.v.).

ADOLF VON NASSAU (Worms, c. 1255–98, Göllheim, nr. Worms), was elected German King (see Deutscher König) in 1292. A man with little financial backing, he was elected with the strong support of the Electoral Archbishop of Cologne, who intended to use Adolf to extend his own power. Adolf's determination to pursue a policy of his own led the electoral princes to combine against him, and in June 1298 they took the unprecedented step of deposing him. Determined to resist by force of arms, Adolf was defeated and killed at the battle of Göllheim on 2 July 1298.

ADOLPH, Karl (Vienna, 1869–1931, Vienna), a workman, wrote novels of working-class life in Vienna (*Haus Nr. 37*, 1908; *Schackerl*, 1912; *Töchter*, 1914).

ADORNO, Theodor (Frankfurt/Main, 1903–69, Brig, Switzerland), whose name was originally Wiesengrund, began a university career at Frankfurt in 1930, emigrated in 1933 to Oxford and then to the U.S.A., where he became a friend of Th. Mann (q.v.), and returned in 1950 to a chair at Frankfurt. His writings are partly philosophical and partly musicological. Philosophically he is primarily a critic of phenomenology, Existentialism, and neo-positivism. He wrote on Wagner (*Versuch über Wagner*, 1952) and linked music with sociology in his *Einleitung in die Musiksoziologie* (1962). *Eingriffe* (1963), which invokes the name of Karl Kraus (q.v.), contains nine essays directed against the misuse of language. *Gesammelte Schriften*, ed. R. Tiedemann, 16 vols., appeared 1970–80. Vol. 2, *Noten zur Literatur*, contains mostly previously published articles on modern European writers and movements.

Adriatische Rosemund, Die, a novel by P. von Zesen (q.v.).

Ad signandum domum contra diabolum, superscription of an Old High German spell to bless a house in protection against the devil. It consists of two lines of alliterative verse.

ADSO VON TOUL, see Antichrist.

Adultera, L', a short novel by Th. Fontane (q.v.), published in *Nord und Süd* in 1880 and in book form in 1882. It is the first of his Berlin novels. The wealthy financier van der Straaten married when he was 42 an attractive 17-year-old of good family, Melanie de Caparoux. The story begins ten years later, after two daughters have been born. Van der Straaten spoils his wife in every way, but a streak of vulgarity, which he deliberately cultivates, grates on Melanie's more sensitive nature. Half innocently and half recklessly he subjects the marriage to a strain by inviting a young business associate, Ebenezer Rubehn, to stay in his house. Melanie and Rubehn slide gradually into an intimate relationship and decide to go off together. Van der Straaten learns of their intention and tries unsuccessfully to persuade Melanie, who is pregnant by Rubehn, to stay. The lovers leave for Rome, and after the divorce are married there. The child is born in Venice, whereupon they return to Berlin, where Melanie hopes to resume her social life. But she finds herself ostracized, even by her own young daughters of the first marriage. Meanwhile Rubehn's family business collapses. The couple have to work for their living and Melanie, finding a new happiness in partnership, ceases to regard 'Society' as essential to life. Van der Straaten sends a typically eccentric token of reconciliation. The title refers to Tintoretto's Venetian picture of the woman taken in adultery, which van der Straaten, by a curious quirk, has had copied for Melanie at the beginning of the novel. At the end he sends, as his peace offering, a miniature of it in a locket.

Aegidius, see TRIERER ÄGIDIUS.

A.E.I.O.U., abbreviation for *Austriae est imperare orbi universo* (it falls to Austria to rule over the whole globe), motto of the Habsburg Emperor Friedrich III (q.v.). An alternative Latin version is *Austria erit in orbe ultima*, and the German form is: 'Alles Erdreich ist Österreich untertan'.

AELST, PAUL VAN DER (*fl.* Deventer, Holland, *c.* 1600), a printer, published a collection of poems, including translations into High German from the French, Italian, Latin, and Dutch, as well as High German originals (*Blum und Ausbund allerhand auserlesener weltlicher züchtiger Lieder und Reime*, 1602). Many of the poems in this anthology are taken from Forster's *Frische Teutsche Liedlein* (see FORSTER, GEORG). He also made a High German translation of a Dutch version of the folk tale *Von den vier Heimonskindern*, published at Cologne in 1604 (see HAIMONS-KINDER).

AEMILIA, GRÄFIN VON SCHWARZBURG-RUDOLSTADT (Rudolstadt 1637–1706 Rudol-stadt), *née* Barby, sister-in-law of Gräfin Ludaemilia (q.v.), was the authoress of some Protestant hymns.

Affekt, technical term for 'emotion' in 18th-c. literary criticism; the word is first recorded in the 16th c.

Agadir Incident, the most conspicuous in-cident in the international crisis of 1911. Tension over French troop movements in Morocco, which represented a possible threat to German interests there, led the German government to make a demonstration of force by sending to Agadir the small gunboat S.M.S. *Panther*. The crisis was peacefully resolved by slight French concessions in Central Africa, but a consequence of the episode was a closer Anglo-French under-standing. The incident is sometimes termed 'der Panthersprung'.

Agathon, see GESCHICHTE DES AGATHON.

Aglaja, a literary annual founded in Vienna in 1815 by J. Sonnleithner (q.v.) and later edited by J. Schreyvogel (q.v.). It lasted until 1832. Among its contributors were Charlotte Birch-Pfeiffer, I. F. Castelli, Helmine von Chézy,

J. L. Deinhardstein, F. de la Motte FOUQUÉ, F. Grillparzer (first publication of *Das Kloster bei Sendomir*, q.v.), J. von Hammer-Purgstall, E. von Houwald, C. Pichler, F. Rückert, F. Schlegel, J. G. Seidl, Z. Werner, and J. C. von Zedlitz (qq.v.).

Agnes, a poem written by E. Mörike (q.v.) in 1831, often known by its first line 'Rosenzeit! Wie schnell vorbei'. It was dedicated to Luise Rau (q.v.).

Agnes Bernauer, a five-act tragedy in prose, 'Ein deutsches Trauerspiel', by F. Hebbel (q.v.). Written in 1851, it was first performed under the direction of F. Dingelstedt (q.v.) in the Munich Hoftheater in March 1852. It was published in 1855. The action is based on an episode in Bavarian history (1428–36), the secret marriage of Albrecht (see ALBRECHT III), heir to the Wittelsbach Duke Ernst (see ERNST I) of Bavaria/Munich, which ended in the arrest of his wife, the Augsburg barber's daughter Agnes Bernauer. She was tried, condemned to death for witchcraft, and drowned in the Danube at Straubing in 1435.

Hebbel adapted the historical figures to suit his purpose. Agnes rejects the courtship of Theobald, the barber's assistant, against the advice of her father, Caspar Bernauer, who urges her to marry within her class. Albrecht's friends, Graf Törring, Nothafft von Wernberg, and Rolf von Frauenhoven, likewise try to prevent the young duke from rashly marrying the barber's daughter, whom he meets during his visit, and who is known as the 'angel of Augsburg' because of her exceptional beauty. Caspar blesses the couple, accepting Agnes's conviction that her love is an expression of God's will. Duke Ernst, fearing for the safety of his dynasty, disinherits Albrecht at a tourna-ment in Regensburg in favour of his nephew Adolf. At the castle of Straubing Albrecht, as he is about to go to a tournament at Ingolstadt, learns of Adolf's death. Hardly has he left the castle than Agnes is arrested. Offered the choice of renouncing Albrecht or suffering death, she chooses the latter and is drowned in the Danube. Duke Ernst accepts full responsibility for her execution which he has ordered for no reason other than to safeguard his dynasty, for Adolf's death has left his son Albrecht as the only pos-sible heir. Ernst's chancellor von Preising confirms that Agnes has to die because she is 'beautiful and virtuous'. On hearing the news of Agnes's death, Albrecht revolts against his father, who is taken prisoner by Albrecht's men. But the Emperor's herald arrives to pronounce ban and excommunication upon the rebel duke. In the final confrontation between father and son, Ernst recognizes Agnes, in death, as duchess, her low station having prevented him from doing so during her life. He abdicates, and passes the ducal staff to his son. Albrecht's acceptance reaffirms the dynasty.

By executing one innocent of any crime Duke Ernst appears inhuman, although Hebbel, under the impact of the 1848 revolutions, finds motivation for the political theme of the play. He also intended Agnes's tragedy to demonstrate his conception of tragic guilt inherent in the beauty and purity which constitute Agnes's unwitting 'hubris'. Hebbel also saw his play as a modernization of Sophocles's *Antigone*, in that Agnes, like Antigone, represented the 'absolute right', which was in conflict with the 'relative right', embodied in the state and represented by King Creon and Duke Ernst respectively. In such a conflict the representatives of secular power must inevitably incur personal guilt in the administration of the law. Duke Ernst emerges fully aware of the ethical conflict which destroys his humanity.

Critical comments on the political theme have not ceased since the first performance of the play, which has nevertheless maintained its position as one of Hebbel's major tragedies. For other treatments of the subject see BERNAUER, AGNES.

Agnes Bernauerin, a tragedy by J. A. von Törring (q.v.), published in 1780. It is described as 'Ein vaterländisches Trauerspiel' and dedicated 'Meinem Vaterlande Bayern'. It deals with an episode in Bavarian history (see BERNAUER, AGNES). It begins immediately after Duke Albrecht's secret marriage to the burgher's daughter Agnes in 1432, shows her imprisonment and execution by drowning on stage, and ends with a reconciliation on patriotic grounds between Albrecht and his father Duke Ernst. It is written in a prose that has poetic pretensions.

Agramer Hochverratsprozeß, a treason trial initiated on false evidence by the Hungarian government, and supported by the Austrian government in 1909 in order, in drawing attention to Pan-Serbian ambitions, to provide a pretext for the annexation of Bosnia and Herzegovina. The trial of the 53 accused collapsed in face of the exposures of corruption made by T. G. Masaryk (later first president of Czechoslovakia).

AGRICOLA, JOHANNES (Eisleben, 1494–1566, Berlin), one of the early Protestants, studied at Wittenberg and was a friend of Luther and Melanchthon (qq.v.), with whom his irascible temperament later brought him into conflict. In middle life headmaster of the school at Eisleben

(1526), he became in 1540 court chaplain to the Elector Joachim II of Brandenburg (q.v.). Johannes Agricola wrote a controversial play condemning J. Hus (*Tragedia Johannis Huß*, 1537). He published two collections of proverbs, the Low German *Dre hundert Sprikwörde* (1528), which appeared in High German a year later, and *Sibenhundert und fünfftzig Deutsche Sprichwörter* (1548), reprinted 1971 (*Die Sprichwörtersammlungen,* 2 vols., ed. S. L. Gilman). He also wrote hymns, of which 'Ich ruf zu dir, Herr Jesu Christ' is the best known.

AGRICOLA, MARTIN (Schwiebus, 1486–1556 Magdeburg), a German musician and cantor in Magdeburg, wrote a collection of hymns entitled *Deutsche Musica* (1560). He also wrote on theory, and his verse *Musica instrumentalis, deudsch* (1528 or 1529), with examples in musical notation, is a valuable document of musical history.

AGRICOLA, RUDOLF (Holland, 1443–85, Heidelberg), the Latin name adopted by Roelof Huysman, who became one of the leading spirits of the early German Renaissance. Agricola's works, written in Latin, are slender, but his learning, his command of languages, and his attractive personality gained him considerable influence and many friends, including Erasmus and Melanchthon (qq.v.). He became a professor at Heidelberg in 1483.

Ägyptische Helena, Die, an opera in two acts composed by R. Strauss (q.v.) with a libretto provided by H. von Hofmannsthal (q.v.). It was first performed on 6 June 1928 at Dresden five weeks before Hofmannsthal's death. A revised version had its first performance on 14 August 1933 at Salzburg. The plot, a free adaptation of Euripides' *Helena*, deals with the ancient myth, introducing into the story of Helen of Troy and the vengeful Menelaus (Menelas in the play) an Egyptian princess and sorceress, Aitra, who assuages the husband's anger and restores conjugal harmony.

Ahnen, Die, a work of fiction in six volumes and seven 'sections' (Abteilungen) by G. Freytag (q.v.), published 1873–81. It takes the form of seven short, linked novels, tracing the history of a German family through various periods of history. Freytag conceived the plan while accompanying the Crown Prince (see FRIEDRICH III) in the field in 1870 and dedicated it to the Crown Princess (later Kaiserin Friedrich, q.v.).

The first volume contains two sections, *Ingo* set in the 4th c. A.D. at the time of the migration of peoples, and *Ingraban* set in the 8th at the time of St. Boniface, who plays a part in the story. In II a descendant of Ingo in the 11th c. appears as a supporter of Emperor Heinrich II. III (*Die Brüder vom deutschen Hause*), set in the 13th c., touches on the crusades and on the eastern campaigns of the Knights of the Teutonic Order (see DEUTSCHER ORDEN). *Marcus König* (IV) treats of the decline of this Order and introduces in the final chapter the figure of Martin Luther. V (*Die Geschwister*) covers a long period from the end of the Thirty Years War (see DREISSIGJÄHRIGER KRIEG) to the reign of FRIEDRICH WILHELM I (q.v.), and VI (*Aus einer kleinen Stadt*) is focused on the Napoleonic Wars (q.v.) with a symbolic opposition of French colonel and Prussian doctor as rivals in love. An epilogue carries the story up to the end of the 1848 Revolution (see REVOLUTIONEN 1848–9). *Aus einer kleinen Stadt* has frequently been reprinted as a separate novel.

Ahnfrau, Die, a five-act tragedy in trochaic verse by F. Grillparzer (q.v.). It is his first published play. The plot, adapting a French anecdote of the robber Jules Mandrin and an anonymous ghost novel (*Die blutende Gestalt mit Dolch und Lampe oder die Beschwörung im Schlosse Stern bei Prag*), was originally conceived for a story, but J. Schreyvogel (q.v.) encouraged Grillparzer to dramatize it (1816). Grillparzer revised the play with Schreyvogel's help and it was successfully performed on 31 January 1817 in the Theater an der Wien. Schreyvogel defended the play against unfavourable criticism in a preface ('Vorbericht') to the first edition (1817).

The action, accomplishing the extinction of the house of Graf Zdenko von Borotin, is mainly supported by the ghost of the adulterous ancestress ('Ahnfrau'), whose appearance to the characters signifies their doom, which she can foresee but not prevent. The sequence of events reveals the love of Borotin's daughter Bertha for Jaromir, who turns out to be her brother. Although believed to have died as a boy, Jaromir has been brought up by the robber Boleslav, who returns to reveal the true relationship between the characters as Borotin is about to die from an injury inflicted by Jaromir when trying to escape arrest as a robber. Bertha, upon hearing that she has loved her brother, and that the son has unwittingly killed his father, dies from shock, while the 'Ahnfrau' saves Jaromir from arrest and condemnation by killing him with her kiss.

Ahnung und Gegenwart, a novel written by J. von Eichendorff (q.v.) in 1811–12 and published by F. de la Motte FOUQUÉ (q.v.) in 1815. It is divided into three books. In the first the young Graf Friedrich takes leave of his university friends and sets out for his native Austria.

He has many random adventures and encounters. He meets the beautiful Rosa, and he and she are mutually attracted. Going on alone, he is overtaken by darkness and seeks shelter in a remote tumbledown mill. In the night he is attacked, but, with the help of an unknown girl, he wards off his assailants. From this point he is followed by a beautiful and melancholy boy named Erwin. Friedrich, who was wounded in the nocturnal encounter, is taken into the mansion of Graf Leontin, where he recovers. Leontin is a gay, capricious character and he has a sister, who proves to be the Rosa to whom Friedrich is already attracted. Faber, an eccentric man of letters, who appears at intervals in the story, is also an inhabitant of the mansion. The friends set out on a romantic journey without destination, but Rosa leaves with a Gräfin Romana for the 'Residenz' (Vienna). Friedrich and Leontin continue their aimless journey, staying for a time at the house of Herr von A., where they meet various eccentrics. Herr von A. has a daughter, Julie, who falls in love with Leontin. On learning that there is talk of a marriage between him and Julie, Leontin takes flight and Friedrich follows.

In Bk. II Friedrich arrives at the 'Residenz'. He meets Rosa at a ball, and is disappointed at her frivolity; the social life of the capital seems to him shallow and meaningless. He calls at Gräfin Romana's mansion, spends the night there and is repelled by her sensuality. He meets the heir apparent (Erbprinz), a superficially attractive and successful philanderer. Friedrich next sets out with Leontin and Erwin for the Rhine. Two hunters whom they meet turn out later to be Rosa and Romana in disguise. The heir apparent presently carries Rosa off.

Bk. III begins with war, the defeat of the nation, and the beginning of a new life. Friedrich joins the patriotic forces in the mountains and distinguishes himself in the fighting. He meets Gräfin Romana again, who is irresistibly drawn to him. He repels her, whereupon she shoots herself. After the military defeat Friedrich is proscribed and wanders alone through the mountains. He meets the boy Erwin, who proves to be the girl who had once come to his aid at the mill and now takes the name Erwine. She dies and Friedrich, joined by Leontin, wanders on, discovering on the way his long-lost and half-crazy brother Rudolf, who proves to be Erwine's father. Friedrich resolves to turn his back on the world and enter a monastery, Leontin marries Julie and departs for the New World. Faber returns to the old life. As Friedrich leaves the monastery chapel, a veiled lady, who is Rosa, falls insensible to the ground.

Ahnung und Gegenwart is plentifully bestrewn with poems by Eichendorff, including some of his best-known songs, such as 'O Täler weit, o Höhen' (*Abschied*, q.v.), 'In einem kühlen Grunde' (*Das zerbrochene Ringlein*, q.v.), 'Es weiß und rät es doch keiner' (*Die Stille*, q.v.), and 'Vergangen ist der lichte Tag' (*Nachtlied*, q.v.). It contains, especially in Bks. II and III, much criticism of contemporary society. It has all the Romantic ingredients of landscape and atmosphere and at the same time affirms seriousness of purpose and religious faith. An obvious indebtedness to Goethe's *Wilhelm Meisters Lehrjahre* (q.v.) does not detract from its refreshing originality.

AICHINGER, Ilse (Vienna, 1921–), grew up in Linz and Vienna and, being of partly Jewish descent, suffered harsh treatment during the period of National Socialist rule in Austria. She gave up medical studies in 1948 in order to write, and, after a period in a publisher's office, devoted herself entirely to literature. She married the poet Günter Eich (q.v.) in 1953. Her novel *Die größere Hoffnung* (1948) is a symbolic tragedy set in the time of Hitlerian persecution. *Rede unter dem Galgen* (1952) is the Austrian title of a collection of stories better known in Germany and abroad by the title *Der Gefesselte* (1953). She has also written radio plays, including *Besuch im Pfarrhause*, which also contains dialogues (1961), and has published a selection of stories and verse (*Wo ich wohne*, 1963), a volume of stories (*Eliza, Eliza*, 1965), and a group of radio plays (*Auckland*, 1969). The collection of 21 texts with the radio play *Gare Maritime* (1976) entitled *schlechte Wörter* (1976), is a sequel to *Eliza, Eliza*. It displays increasing reluctance to present a 'story'. *Meine Sprache und ich. Erzählungen* (1978) contains all prose written by the late 1970s. The title story is an outstanding example of her overriding preoccupation with the resonance between language and inner experience (the term 'Fremdwörter', stripped of its etymological connotations and applied as a key word for verbal and consequently existential alienation, illustrates the technique). *Verschenkter Rat. Gedichte* (1978) contains thematically arranged poems written between 1955 and 1978. A pioneer in the lapidary dialogue form, Ilse Aichinger published in 1980 an extended edition of *Zu keiner Stunde. Szenen und Dialoge* (1957). Of the German and Austrian honours awarded to her, the Trakl-Prize of 1979 reflects a degree of affinity with G. Trakl (q.v.).

Ainune, title given to a fragment (314 lines) of a Middle High German verse romance written *c.* 1230 or soon after. The principal characters are a King and a Queen ('Ainune', i.e. Oenone, whom the king woos). The fragment consists of courtly dialogue. The author is unknown.

Akademien, in the sense of corporations of scholars, scientists, and men of learning such as the Royal Society in Great Britain, were founded in Germany in the 18th c. The most distinguished was the Preußische Akademie der Wissenschaften, Royal (Königl.) until 1918; it was founded in 1700 by statute of Friedrich I in accordance with plans prepared by Leibniz (q.v.), and inaugurated in 1711 in Berlin. It at first bore the title 'Societät der Wissenschaften'. It was concerned with the natural sciences, mathematics, philosophy, and history. Among its distinguished members may be mentioned W. and A. von Humboldt, Schleiermacher, Th. Mommsen, R. Niebuhr, and Ranke (qq.v.). Of its many important publications the *Monumenta Germaniae historica* are the most famous. In 1946 it was renamed Deutsche Akademie zu Berlin with headquarters in East Berlin, and is one of the few organizations to have a membership drawn from both Germanies.

Similar institutions were founded in Bavaria in 1759, in Saxony in 1846, in Austria in 1847, in Baden in 1909; particularly notable was the Königl. Gesellschaft der Wissenschaften zu Göttingen, founded in 1751 by the professor of medicine, A. von Haller (q.v.), which published the *Göttinger Gelehrte Anzeigen*. A Preußische Akademie der Künste, though mooted by Ranke in the 19th c., was not founded until the 20th c., and the 'Sektion Dichtkunst' was not incorporated until 1926. This was dissolved in May 1933 after stormy sessions in March. The President H. Mann, F. Werfel, L. Frank, L. Fulda, G. Kaiser, B. Kellermann, A. Mombert, R. Pannwitz, A. Paquet, R. Schickele, F. von Unruh, and J. Wassermann (qq.v.) were expelled. Th. Mann, A. Döblin, and Ricarda Huch (qq.v.) resigned. A new National Socialist Akademie der Dichtung was announced in May 1933. The National Socialists wanted Stefan George (q.v.) as president, but he refused. The original members were W. Beumelburg. H. F. Blunck, P. Dörfler, P. Ernst, F. Griese, Hans Grimm, H. Johst, E. G. Kolbenheyer, Agnes Miegel, Borries von Münchhausen, W. Schäfer, E. Strauß, and W. Vesper (qq.v.). Ernst Jünger and Hans Carossa (qq.v.) refused an invitation to join the Akademie.

Akten des Vogelsangs, Die, a novel by W. Raabe (q.v.), written in 1893–5 and published in 1895. A retrospective 'frame' novel (see Rahmen), it begins with the receipt by the civil servant Karl Krumhardt of the news of the death of his boyhood friend, Velten Andres. He learns from Velten's former sweetheart Helene Trotzendorf (the name is symbolical of her wilful, obstinate character) the manner of his friend's life and death. Velten, having failed to win the woman he loved, who married a wealthy American speculator, has squandered his great intellectual gifts. Years later he was succoured in his last illness in Berlin by Helene, returned from America. The 'Vogelsang' is the once rural suburb of the small city in which Velten, Helene, and Karl Krumhardt grew up.

Aktion, Die, a political and (until 1918) literary journal, published from 1911 to 1932, until its later years as a weekly. Its editor was Franz Pfemfert (q.v.) and among its contributors were H. Ball, J. R. Becher, G. Benn, F. Blei, G. Britting, M. Brod, A. Ehrenstein, C. Einstein, I. Goll, G. Heym, H. Kasack, E. Lasker-Schüler, E. Stadler, C. Sternheim, F. Werfel, and C. Zuckmayer (qq.v). (See EXPRESSIONISMUS.) A reprint, 4 vols. (1911–14), ed. P. Raabe, appeared in 1961.

Akzente, *Zeitschrift für Literatur,* an annual periodical, edited from 1954, the year of its inception, to 1967 by W. Höllerer and H. Bender (qq.v.) and later by H. Bender and M. Krüger. Originally described as *Zeitschrift für Dichtung,* it is West Germany's most important literary journal promoting contemporary literary debate as well as original fiction and lyric poetry.

Alamodezeit, derived from 'à la mode', refers to one aspect of the 17th c., and especially to the second half of it, viz. the imitation in literature, language, clothing, and manners, of French models. The most conspicuous instances in literature are to be found in the Second Silesian School (see SCHLESICHE SCHULEN). The 'Alamode' tendency aroused contemporary opposition, notably from Moscherosch and Logau (qq.v.).

Alarcos, a verse tragedy written by Friedrich Schlegel (q.v.) and performed in Weimar in 1802 while Goethe was Director of the Court Theatre. It is termed 'Ein Trauerspiel in zwei Aufzügen' and is set in Spain. Count Alarcos has married Doña Clara, but then finds the Infanta wishes to marry him. He kills his wife, only to find that the Infanta has died of grief. He thereupon kills himself. The Weimar audience understandably proved unable to take the play seriously and was rebuked by Goethe, who stood up in his box and sternly ordered, 'Man lache nicht!' Its principal interest lies in its variety of irregular verse, both ancient and modern. Trimeters predominate, there are passages of blank verse, 'heroic couplets', and even a sonnet.

ALARICH I (*c.* 370–*c.* 410), King of the West Goths, invaded Italy in 401 and fought an indecisive battle at Pollentia (402) against the Roman general, Stilicho. In 410 he occupied and plundered Rome, and afterwards set out to conquer North Africa. He died on the way at Cosenza and was buried in the bed of the River Busento. He is the subject of several dramas and novels by minor writers, of Platen's poem *Das Grab im Busento* (two versions, see PLATEN), and of Dahn's novel *Stilicho* (1901, see DAHN, F.).

Albanus, title given to an anonymous Middle High German poem probably written at the end of the 12th c.; it recounts the legend of St. Alban, which has a close resemblance to the story of Oedipus. Alban, the son of an incestuous union between father and daughter, is exposed as an infant, and is found and brought up by the King of Hungary. As a grown man he returns and marries, unwittingly, his mother–sister. The two sinners expiate their guilt in penitence. The story is also told in the *Ehebüchlein* of ALBRECHT von Eyb (q.v.).

ALBER, see TUNDALUS.

ALBÉRIC DE BESANÇON, see ALEXANDERLIED.

Alberich, name of a dwarf who plays a part in Germanic legend. Alberich appears in the *Nibelungenlied* (q.v.) in aventiure III, where he seeks to avenge Nibelung and Schilbung, is vanquished by Siegfried, and is forced to surrender the Tarnkappe (q.v.), and in aventiure VIII, where he provides Siegfried with forces for the wooing of Brünhild in Iceland. Alberich also appears in the Dietrich legend (see DIETRICHSAGE) and in *Ortnit* (q.v.), where he is the hero's father. In Wagner's *Der Ring des Nibelungen* (q.v.) Alberich, who is Hagen's father, steals the gold from the Rhine daughters, only to lose it to Wotan.

ALBERT, HEINRICH (Lobenstein, Thuringia, 1604–51, Königsberg), German composer, who was a pupil and son-in-law of Heinrich Schütz (q.v.). He spent the greater part of his life as cathedral organist in Königsberg. Albert was also a lyric poet; his poems express for the most part intimations of mortality. He composed the melody for *Ännchen von Tharau* (q.v.), and it is thought by some that he also wrote the words. He was prominent among the Königsberg poets (see KÖNIGSBERGER DICHTERSCHULE).

ALBERTI, KONRAD, pseudonym of Konrad Sittenfeld (Breslau, 1862–1918, Berlin), a journalist on the *Berliner Morgenpost,* who began with writings on literature (*Herr L'Arronge und das deutsche Theater*, 1884; *Gustav Freytags Leben und Schaffen*, 1884; *Bettina von Arnim*, 1885; *Ludwig Börne*, 1886). He is the author of collections of Novellen, including *Riesen und Zwerge* (1886), *Plebs* (1887), *Federspiel* (1890), and many novels, some of which, in their sensational treatment of social questions, caused a temporary stir. Among them were *Wer ist der Stärkere?* (1888), *Die Alten und die Jungen* (1889), and *Das Recht auf Liebe* (1890).

ALBERTINUS, Ägidius (Deventer, Holland, *c.* 1560–1620, Munich), possibly of German descent in spite of his Dutch birth, was educated in a Jesuit school and in 1593 was in the service first of Duke Wilhelm der Fromme of Bavaria and then of Maximilian I (q.v.), Wilhelm's son and successor. From 1604 he was court librarian, from 1618 secretary of council (Ratssekretär). Albertinus was a staunch and unswerving Catholic, who wrote prolifically in support of the Counter-Reformation (see GEGENREFORMATION). Many of his abundant moral and religious writings are translations from the Spanish, especially of Antonio de Guevara.

Among the more important of his works are his *Hausspolicey* (1602), *Lucifers Königreich und Seelengejaid* (1616, reprinted 1884), which treats of Lucifer's 'hunting preserves', and *Der Hirnschleifer* (1618), the moral interpretation of a set of engravings. With his translation of *Guzmán de Alfarache* by Mateo Alemán (*Der Landstörtzer: Oder Gusman von Alfarache oder Picaro genannt*, 1615) Albertinus introduced the picaresque novel (see SCHELMENROMAN) into German literature.

ALBERTUS MAGNUS (Lauingen, 1200–80, Cologne), by birth a Graf von Bollstädt, became a Dominican monk in 1223 and taught at various schools of the Order, becoming a doctor in Paris and later teaching in Cologne. He was made Provincial of his Order in 1254, and was bishop of Regensburg from 1260 to 1262. Albertus was a scholar of immense knowledge and tireless industry; his paraphrase of the works of Aristotle furthered the Aristotelian philosophy of the Middle Ages. His exceptional expertise in the natural sciences led to imputations of magic. His writings are entirely in Latin. St. Thomas Aquinas was his pupil, and in later life Albertus defended Aquinas's reputation. Beatified in 1622, Albertus was canonized in 1931. He is sometimes termed Albrecht von Köln.

ALBERUS, Erasmus (Sprendlingen, Hesse, *c.* 1500–53, Neubrandenburg), met Luther and Melanchthon (qq.v.) in Wittenberg and became an ardent supporter of the Reformation. His polemical verse satire *Der Barfüßer Mönche Eulenspiegel und Alkoran* (1542) attacks the adherents of the traditional faith. His chief literary merit is as a writer of fables. *Das Buch von der Tugend und Weisheit* (1550, reprinted 1892) contains his version of 49 of Aesop's fables, some of which had appeared as early as 1534. He expands the narratives of the fables, emphasizes the human characteristics of the animals, and localizes the action in various districts of Germany. The tone is notably more gentle than that of many of his contemporaries.

Albigenser, Die, an epic poem by N. Lenau (q.v.), published in 1842. Written in rhyming iambic verse (*c.* 3,500 lines), it treats of the Albigensian heresy and its suppression under Pope Innocent III. It is in the form of a cycle of poems of varying length, described by Lenau as 'Freie Dichtungen'.

ALBINUS, Johann Georg (Weißenfels, 1624–79, Naumburg), a minor baroque poet and pastor at St. Othmar's Church, Naumburg. He composed florid poetry (*Trauriger Cypressenkranz*, 1653) and made verse translations in similar style (e.g. *Himmelflammende Seelenlust*, 1675, from H. Hugon's *Pia Desideria*). Some of the plainer poetry from his *Geistliche und Weltliche Gedichte* (1659) survives in hymn-books, notably 'Straf mich nicht in Deinem Zorn' and 'Welt, ade, ich bin Dein müde'.

ALBRECHT I, Deutscher König (1255–1308, Brugg on the Aare), was the eldest son of Rudolf I (q.v.) of Habsburg. In 1292 he failed to be elected German King, the Electors preferring the less powerful Adolf von Nassau (q.v.), but he attained his ambition six years later, after Adolf's death. His election was not recognized by the Pope until 1303. Albrecht sought to extend his power by force of arms, but was assassinated in 1308 by his nephew Johann Parricida, an episode which is recounted in Schiller's *Wilhelm Tell* (q.v.). He is buried in Speyer Cathedral.

ALBRECHT II (1397–1439, Neszmély, Hungary), son of Duke Albrecht IV of Austria, succeeded to the dukedom in 1404 as Albrecht V and was elected German King (see DEUTSCHER KÖNIG) in 1438 in succession to Sigismund (q.v.). He succeeded Sigismund also as ruler in Bohemia and Hungary. The advance of the Turks into south-eastern Europe drew Albrecht to the defence of southern Hungary. While waiting for reinforcements after the loss of Semendria to the Turks, he fell ill of a fever and died on the return journey to Vienna after a reign as German King of just over a year. With him the house of Habsburg returned after 132 years to the imperial throne.

ALBRECHT III, DER FROMME (Munich, 1401–60, Munich), son of Duke Ernst I von Bayern/München (q.v.), succeeded his father in 1438. He was brought up at the court of his aunt, Queen Sophia of Bohemia. For dynastic reasons he was betrothed to Elizabeth of Württemberg, who, however, married another. In 1432 (or 1433) he secretly married the commoner Agnes Bernauer (q.v.),·whom his father caused to be drowned in the Danube in 1435 after Albrecht had repeatedly refused to abandon her. It needed the intervention of the Emperor Sigismund (q.v.) to reconcile father and son. In 1436 Albrecht married Anna of Brunswick. He was a well-educated and well-liked ruler, an effective reformer of the Bavarian monastic system, and a patron of the arts. He is the Albrecht of Hebbel's tragedy *Agnes Bernauer* (q.v.).

ALBRECHT IV, HERZOG VON BAYERN (Munich, 1447–1508, Munich), also known as Albrecht der Weise, acceded to the dukedom on his father's death in 1460, reigning for the next seven years jointly with his brother. In his forty-eight years of rule he materially strengthened the dynasty. He was receptive to the arts and literature and was the patron for whom Ulrich Füetrer (q.v.) wrote *Das Buch der Abenteuer.*

ALBRECHT names himself as the author of the 13th-c. verse romance known as *Der jüngere Titurel* (see ALBRECHT VON SCHARFENBERG and TITUREL). His identity is uncertain.

ALBRECHT, JOHANN FRIEDRICH ERNST (Stade, 1752–1814, Hamburg), a physician who abandoned his profession to devote himself to literature, is best known for a prose adaptation of Schiller's *Don Carlos* (q.v.). He became a bookseller in Prague in 1792 and in 1802 a theatre director in Hamburg. He was the husband of the actress Sophie Albrecht (q.v.).

ALBRECHT, SOPHIE (1757–1840) née Baumer, actress-wife of J. F. E. Albrecht (q.v.), played the part of Luise in the first production of Schiller's play *Kabale und Liebe* (q.v.). On their first acquaintance in 1784 Schiller became for a short time her devoted admirer.

ALBRECHT, MARSCHALL VON RAPPERSWIL, a great Swiss nobleman and minor Middle High German poet of the end of the 13th c., whose three extant Minnelieder show an indebtedness to KONRAD von Würzburg, GOTTFRIED von Neifen, and ULRICH von Winterstetten (qq.v.).

ALBRECHT VON EYB (nr. Ansbach, 1420–75, Eichstätt) translated three Latin comedies (*Menaechmi* and *Bacchides* of Plautus, and *Philogenia* of U. Pisani), adapting them to the conditions of German life and, notably, substituting homely German for Latin names. His manual of marriage (*Ehebüchlein*, 1472, also largely composed of translations, some drawn from the *Decameron*) achieved a popularity which lasted into the 16th c. He acquired his enthusiasm for neo-humanism during early study at Padua, Bologna, and Pavia; in later life he was a canon of Bamberg and Eichstätt. His German works were published as *Deutsche Schriften*, ed. M. Herrmann, 2 vols., 1890.

ALBRECHT VON HALBERSTADT, Middle High German poet, author of a verse translation of Ovid's *Metamorphoses*, written probably in 1190, possibly in 1210 (a clumsy reference to date in the poem is ambiguous). Albrecht wrote it at Jechaburg near Sondershausen for the Thuringian court of Landgraf Hermann (q.v.). Unlike most of his contemporaries, he has the merit of translating from the Latin original instead of a French intermediary, though the task taxes his poetic gifts. The poem is known through two fragments of a MS. (*c.* 400 lines) and a considerably adapted version published more than three centuries later by J. Wickram (q.v.) in 1545.

ALBRECHT VON JOHANNSDORF, a Middle High German Minnesänger, was a minor nobleman in the service of Bishop Wolfger of Passau (q.v.). Documents referring to him indicate that his lifespan included the period 1185–1209. He may have known WALTHER von der Vogelweide (q.v.), and is believed to have participated in a crusade. His handful of Minnelieder conform outwardly to the pattern in which the man subordinates himself to the woman above him; and he is responsible for the classical formulation of the educative value of Minnedienst: 'daz ir deste werde sît und dâ bî hôchgemuot' (daß Euer Wert wächst und zugleich Euer hoher Mut). His poetry reflects a balanced and harmonious personality. His integrity and warmth of heart are most evident in the poems referring to departure for the crusade.

ALBRECHT VON KEMNATEN or KEMENATEN, author of the Middle High German poem *Goldemar*, is believed to have lived in what is now Switzerland, but nothing is known of him with certainty except his authorship of this poem, which was written in the second half of the 13th c. Only the first nine stanzas survive, but indirect evidence provides an indication of the story. Dietrich, with the help of giants, liberates a beautiful maiden held captive in a mountain by the dwarf king Goldemar. The poem adapts the heroic legend of Dietrich (see DIETRICHSAGE) to courtly presentation.

Albrecht was formerly held to be the author of three other Middle High German poems, the *Eckenlied*, *Sigenot*, and *Virginal* (qq.v.). He is possibly the originator of the verse form common to all four, a twelve-line stanza, known as the Eckenstrophe (q.v.).

ALBRECHT VON KÖLN, see ALBERTUS MAGNUS.

ALBRECHT VON SCHARFENBERG, Middle High German poet of the late 13th c., author of two epic poems, *Merlin* and *Seifrid von Ardemont*, known only through adaptations in *Das Buch der Abenteuer* by Ulrich Füetrer (q.v.). He has also often been regarded as the Albrecht who wrote *Der jüngere Titurel* (see TITUREL).

ALCUIN (also ALKUIN, ALCHUINE, etc.) (York, 735–804, Tours), the great ecclesiastic, is important in the history of Germany and German literature for his service to Charlemagne (see KARL I, DER GROSSE). Alcuin, who adopted the Latin name Albinus Flaccus, was educated at York. While in Italy in 780 he met Charlemagne, who invited him to become his adviser and honoured him with his friendship. From 781 until his death, except for two brief intervals in England, Alcuin's services were at the disposal of Charlemagne, who had him installed as abbot of the great Abbey of St. Martin at Tours (793). Alcuin himself instructed Charlemagne in astronomy, rhetoric, and dialectics, and he was responsible for carrying out Charlemagne's policy of developing knowledge of Latin and Roman civilization throughout his empire. To this end Alcuin supervised the education of the clergy, who provided the educated stratum of society from which emerged officials and educators. His works, written in Latin, include poems, grammatical and rhetorical dialogues, and theological treatises. A number of Latin letters to Charlemagne survive, as also does a Middle High German translation of his *Tractatus de virtutibus et vitiis*, dating from the 12th c.

Alemannen, a Germanic race, deriving from the Semnones. The Alemannen are first heard of in the 3rd c. A.D., when they were resident on the upper Main. In the time of the migration of peoples (see VÖLKERWANDERUNG) they moved south and east, into France, Switzerland, and Italy, finally settling in northern Switzerland and present-day Baden-Württemberg. In time the term Alemannen was limited to those inhabiting Switzerland. The inhabitants of Baden and Württemberg (Swabia), though of identical race, came to be known as Swabians (see SCHWABEN). The adjective 'alemannisch' occurs in literature in relation to dialect and also in a title, J. P. Hebel's *Alemannische Gedichte* (q.v.).

Alemannische Gedichte, a collection of dialect poetry published in 1803 by J. P. Hebel (q.v.). The dialect is the Swabian of Hebel's southern Black Forest home in the Wiesental, and the first poem, *Die Wiese*, is named after its river. The original edition contained 32 poems, and 12 more were subsequently added (1820). The volume is aimed, according to the original title-page, at 'Freunde ländlicher Natur und Sitten'. The subjects are simple ones, country life and talk, e.g. *Die Marktweiber in der Stadt, Die Mutter am Christabend, Wächterruf, Sonntagsfrühe*. Some of the poems are in the sophisticated form of hexameters (*Die Wiese, Der Karfunkel, Der Statthalter von Schopfheim, Das Habermus*). The collection is a landmark in dialect poetry (see DIALEKTDICHTUNG).

Alemannische Tochter Syon, see LAMPRECHT VON REGENSBURG and MÖNCH VON HEILSBRONN.

Alexander, an unfinished Middle High German poem by RUDOLF von Ems (q.v.). Though a fragment, it contains more than 21,000 lines and progresses as far as Alexander's expedition to the Oxus, approximately half the projected work. It gives a picture of an ideal ruler, brave, just, generous, and humane. Like *Willehalm von Orlens* (q.v.) it contains a catalogue of contemporary poets.

Alexanderlied des Pfaffen Lamprecht (see LAMPRECHT, PFAFFE), an Early High German poem recounting the life of Alexander the Great. It is avowedly a translation of a French poem by Albéric de Besançon (*Alberich von Bisinzo*), of which only a fragment is extant. After a brief introduction it portrays Alexander's youth and his wars against Darius and Porus, and then proceeds to his legendary adventures in the East and visit to Paradise. It ends with his death and a general exhortation to repentance and piety.

The *Alexanderlied* is composed of two barely compatible elements. On the one hand Alexander is portrayed as an admirable hero with knightly virtues, on the other his story serves as a demonstration of the futility of worldly ambition and the vanity of all things temporal. The explanation of this contradiction is to be found, at least in part, in the source of the story. The

Macedonian king is portrayed in terms of the Greek story known as the Pseudo-Callisthenes, and the medieval poem reproduces unintentionally its humane values. The standpoint of the medieval ascetic is embodied in the introduction and in an episode at the conclusion, which is a 'memento mori'. It is not impossible, however, that the clerical author was also influenced by the knightly ideals which were then in the making.

The history of the poem is complex. Pfaffe Lamprecht, believed to be a priest of Trier, wrote only one-fifth of the entire work, probably between 1130 and 1150. This part is included in the Vorauer Handschrift (q.v.). The fullest version was contained in a MS. destroyed in the bombardment of Strasburg in 1870. It is believed that this represented adaptation and continuation by another hand. A third version at Basel is a crude abridgement, probably made in the 15th c. from a lost MS. of the 13th c. The *Alexanderlied* has little poetic value, but it is important as one of the earliest substantial medieval poems dealing with a worldly subject. It is available in an edition by F. Maurer, 1940. All three texts were edited by K. Kinzel, 1885.

Alexander und Antiloie, an anonymous Middle High German poem of some 500 lines, recounting the exposure and castigation of Alexander's unfaithful counsellors by the dwarf Antiloie. The poem, a kind of verse Schwank (q.v.), was probably written *c.* 1260–70.

Alexandriner, German term for the French 'Alexandrin', a line of verse consisting of 12 syllables (or 13 if the last is unstressed), divided by a caesura after the sixth, and rhyming in pairs. It was introduced into German literature in the 17th c. by Lobwasser, Melissus, and Weckherlin (qq.v.) and established by Opitz (q.v.). Up to the middle of the 18th c. it was the accepted verse form for the drama and the epic, and was also frequently used in other poetic forms. Its use declined in the later 18th c., surviving longest in Austria. In 1831 Goethe revived the Alexandrine for satirical and symbolical purposes in *Faust* (q.v.) *Pt. II* (Act IV, *Des Gegenkaisers Zelt*).

ALEXIS, WILLIBALD, pseudonym of the novelist Georg Wilhelm Häring (Breslau, 1798–1871, Arnstadt) and the name by which he is generally known. Alexis, who claimed to be of French Huguenot descent, lost his father in 1802. In 1806 his family moved to Berlin, where he had his schooling, and in 1815, while still a schoolboy, he participated in the Waterloo campaign as a volunteer serving in the Colberg Grenadiers. In 1817 he entered Berlin University to study law, migrating in 1818 to Breslau University. In 1820 he entered the Prussian civil service as a junior legal official.

At this time he began to write, his earliest efforts including an epic (he called it 'ein scherzhaft idyllisches Epos') *Die Treibjagd* (1822), a historical Novelle *Die Schlacht bei Torgau* (1823), and some horror stories. His first historical novel *Walladmor* (1824) appeared under false colours as 'nach dem Englischen des Walter Scott', and three years later he published *Schloß Avalon* (1827) with a similar attribution. In 1824 he had given up his career to devote himself to literature and journalism. He edited the *Berliner Konversationsblatt* (later combined with *Der Freimütige*), from 1827 to 1829 with F. Förster, and from then until 1835 he was its sole editor.

Alexis dabbled in drama with *Ännchen von Tharau* (1829) and with the comedies *Exzellenz* (1848) and *Der Salzdirektor* (1851), but the form most congenial to him was the historical novel, which he had first attempted in the shadow of Scott. *Cabanis*, the first of these novels to be published under his pseudonym, Alexis, appeared in 1832, and, though he temporarily turned to the novel of contemporary life with *Haus Düsterweg* (1835) and *Zwölf Nächte* (1838), it was with a series of novels touching the history of Prussia that he made his name with the public; and it is for these patriotic stories ('vaterländische Romane', he called them) that he is now (albeit faintly) remembered. *Der Roland von Berlin* (q.v.) appeared in 1840, *Der falsche Woldemar* (q.v.) in 1842, followed by his best-known work *Die Hosen des Herrn von Bredow* (q.v.) in 1846–8. Its sequel, *Der Wärwolf* (q.v.), followed in 1848, *Ruhe ist die erste Bürgerpflicht* (q.v.), a story of the days preceding Jena, in 1852 and the sequel to this, *Isegrimm* (q.v.), in 1854. *Dorothe* (q.v., in modern spelling *Dorothee*), a story set in the late 17th c., was published in 1856.

Later in that year Alexis had a serious stroke, which virtually brought his writing to an end; and though he survived for fifteen years, his declining life was passed in increasing and distressing infirmity. Alexis's novels do not deserve the almost total neglect into which they have fallen. They are based on serious and careful historical study, their characters are robustly and acutely drawn, the dialogue is lively and varied, and they are diversified with shrewd touches of humour; moreover their tone is liberal and their nationalism never chauvinistic or vainglorious. Their eclipse is probably due to their formal unoriginality.

Alexis wrote several travel works (*Herbstreise durch Scandinavien*, 1828; *Wanderungen im Süden*, 1828; *Wiener Bilder*, 1833; and *Schattenrisse aus Süddeutschland*, 1834). Together with J. E. Hitzig (q.v.), he helped to edit the collection of accounts of crimes published in 30 vols. between 1842 and 1862 as *Der neue Pitaval* (see PITAVAL). Alexis's *Gesammelte Werke* appeared in 18 vols., 1861–6, in 20 vols., 1874, *Vaterländische Romane* (10 vols.), ed. L. Lorenz and A. Bartels, 1912–25.

Alexis und Dora, a poem in elegiac metre written by Goethe in May 1796 and first published in Schiller's *Almanach für das Jahr 1797*. Its subject is the onset of love and the pain of parting, compressed into one experience. The Greek youth Alexis, on the point of departure by ship, is seized with love for Dora, whom he must instantly leave. Except for a short introduction and a shorter postlude, the 158-line poem is a monologue spoken on shipboard by Alexis as the shore recedes. It is one of the purest products of Goethe's classicism. Schiller wrote of it, 'so voll Einfalt ist sie [sc. die Elegie], bei einer unergründlichen Tiefe der Empfindung'.

Alldeutscher Verband (the Pan-German League) was founded in 1891 as a result of the outcry against the Anglo-German treaty by which Germany conceded its interests in Zanzibar in exchange for Heligoland. The mood which created it was compounded of a desire for colonies as tokens of status in the modern world, an energetic nationalism, and a brisk xenophobia. Up to 1894 it was known as Der Allgemeine Deutsche Verband.

The League developed into a pressure group which sought to orientate German foreign policy (1) towards the acquisition of colonies, (2) towards the maintenance of German national consciousness in 'Auslandsdeutsche', especially in North and South America, and (3) towards the support of German interests by open or barely veiled force. Under the presidency of H. Class (q.v.) the League attacked German foreign policy in the years 1908–14 for its supposed weakness. During the 1914–18 War it strongly supported the annexation of Belgium. After 1918 it was active in the agitation against the Treaty of Versailles, and condoned the political murder of W. Rathenau (q.v., 1922). It was dissolved under the National Socialists in 1939.

'Alle menschliche Gebrechen/Sühnet reine Menschlichkeit', closing lines of an eight-line dedicatory poem written by Goethe on the flyleaf of a copy of *Iphigenie auf Tauris* (q.v.), which he gave to the Berlin (and former Weimar) actor Georg Wilhelm Krüger (1791–1841) on 31 March 1827. It is often quoted as an apt poetic summing-up of *Iphigenie* by its author. The superscription, which the poem bears in Goethe's

collected works, is *Dem Schauspieler Krüger mit einem Exemplar der 'Iphigenie'*.

'Alles schweige, jeder neige', first line of a student song sung as accompaniment to the 'Landesvater' (q.v.). It dates from 1781, though the tune 'Landesvater, Schutz und Rater' is older.

Allgemeine Deutsche Bibliothek, a prominent journal in support of the Aufklärung (q.v.), which appeared from 1765 to 1805 and was edited by F. Nicolai (q.v.). In 1793 the title was amended to *Neue Allgemeine Deutsche Bibliothek*.

Allgemeine Deutsche Biographie (A.D.B.), comprehensive biographical work of reference, equivalent to the *Dictionary of National Biography*. Edited by the historical committee of the Bavarian Academy of Sciences (Königlich Bayrische Akademie der Wissenschaften), it comprises 56 vols. and an index and was published in the years 1875–1912. A successor, also published by the Bavarian Academy of Sciences, the *Neue Deutsche Biographie*, began to appear in 1953.

Allgemeine Zeitung, a newspaper founded in 1798 by J. F. Cotta (q.v.) in Stuttgart and transferred to Augsburg in 1810 and to Munich in 1882. In 1908 it was converted into a periodical. In its heyday (up to c. 1850) it had a number of well-known contributors, including Heine and F. Dingelstedt (qq.v.).

Alliteration occurs in German as an adornment of rhetorical or poetic speech, e.g.

... sie kann nicht vor euch her wie sonst
Die Fahne tragen—schwere Bande fesseln sie,
Doch frei aus ihrem Kerker schwingt die Seele
Sich auf den Flügeln ihres Kriegsgesangs.

      Schiller, *Die Jungfrau von Orleans*.

It is also abundant in proverbial expressions (Haus und Hof, über Stock und Stein). Such uses are related to its role in the earlier poetry of the Old High German period. Its function then was analogous to that of rhyme as a means of cementing verse and isolating it from prose and it is, in fact, described as Stabreim in German. Old High German alliterative verse normally contained three similar initial sounds in each line, two in the first half and one early in the second. An example is the line

    heuwun harmlicco   huitte scilti

from the *Hildebrandslied* (q.v.). The use of Stabreim died out in the 9th c. In the 19th c. Fouqué (q.v.) revived it for songs in *Der Held des Nordens* (1810); and Richard Wagner used

Stabreim in *Der Ring des Nibelungen* (q.v. 1852–4) and other works.

ALLMERS, HERMANN (Rechtenflet, Bremen, 1821–1902, Rechtenflet), a minor poet and dramatist, was deeply rooted in his Friesian homeland and devoted himself to its preservation and to the encouragement of Heimatkunst (q.v.). A well-to-do farmer and mayor of his native village, he travelled widely in youth. His best work is in the field of environmental description (*Marschenbuch*, 1858; *Die Pflege des Volksgesangs im deutschen Nordwesten*, 1878; *Dörnberg*, 1887). In later years he came to be regarded as a G.O.M. of letters in north-west Germany. His poem *Feldeinsamkeit* was set by Brahms (q.v.). His *Sämtliche Werke* appeared in 6 vols., 1891–6.

Almanach de Gotha, see GOTHAISCHE GENEALOGISCHE TASCHENBÜCHER.

Almansor, a tragedy in blank verse by H. Heine (q.v.), first published in *Tragödien, nebst einem lyrischen Intermezzo* in 1823. It is set in Spain at a time when the Moorish power is yielding to Spanish Christian expansion. Many of the Moors have been converted, including the beautiful Zuleima, now known as Donna Clara, who is betrothed to Don Enrique, whom she does not love. Her former lover, the Moor Almansor, carries her off, but is overtaken by pursuing Spaniards. With Zuleima in his arms he throws himself to death from a high rock. There are no acts, but the play falls into eight unnumbered divisions.

Alpen, Die, a descriptive poem by A. von Haller (q.v.). It consists of 50 rhyming stanzas, each of ten Alexandrines (see ALEXANDRINER). It contrasts the rural peace and innocent virtue of the inhabitants of the Alps with the vices, misery, and corruption of civilized humanity. In its condemnation of sophistication and praise of simplicity, it anticipates Rousseau, and its description of the Alps opens up new poetic territory. Haller visited the Alps in 1728, when he was 19, wrote the poem in the following winter, and published it in 1732. The present first stanza was added in the second edition in 1734.

Lessing quoted two stanzas (40 and 41) in *Laokoon* (q.v.), using them to condemn the principle of descriptive poetry, though he conceded that the passage is 'ein Meisterstück in seiner Art'.

Alpenkönig und der Menschenfeind, Der, a play by F. Raimund (q.v.), written in 1828 and first performed at the Theater in der Leopoldstadt, Vienna, in October 1828. Described as a Romantisch-komisches Original-Zauberspiel, it was published in vol. 1 of Raimund's *Sämtliche Werke* (1837). The principal character Rappelkopf is a misanthrope, who lives in seclusion, tormenting his womenfolk and his servants by his suspicious nature. When their situation is at its worst the 'Alpenkönig' Astragalus appears and promises to cure Rappelkopf's misanthropy. Rappelkopf is translated into the bodily form of his brother-in-law, thereby discovering that his earlier suspicions about his family are unfounded. The 'Alpenkönig' takes Rappelkopf's own form and horrifies the metamorphosed Rappelkopf by his misanthropic and neurotic behaviour. The cure is successful. Rappelkopf is restored to his own shape and recovers his mental balance, and the harmony of the family is complete.

The play is in prose with considerable tracts of verse, a development which is symptomatic of Raimund's increasing ambition to write serious drama. There are numerous arias and songs, including the famous sextet, 'So leb denn wohl, du stilles Haus'.

Alpharts Tod, an anonymous Middle High German poem written *c.* 1250. It is preserved in a defective MS. of the 15th c., which contains 467 eight-line stanzas of the form known as Hildebrandston (q.v.). The story is an offshoot of the Dietrich legend (see DIETRICHSAGE). Alphart, a warrior of Dietrich and a nephew of Hildebrand, rides out against Ermanarich's forces, subdues the warrior Witege, but is then treacherously attacked simultaneously by Witege and Heime and falls after a heroic resistance. The poem ends with Dietrich's victory at Bern (Verona). The standpoint of the poem is chivalric; Ermanarich's warriors are condemned for their shameful disregard of the code of knightly conduct.

ALRAM VON GRESTEN, is the name, otherwise unknown, to which three Minnelieder of the 13th c. are ascribed in one MS.; the same poems appear elsewhere with different attribution. The name also appears in the form Waltram.

Alsace-Lorraine, see ELSASS-LOTHRINGEN.

Alsfelder Passionsspiel (Alsfeld, Hesse), title given to a late medieval German religious play of more than 8,000 lines. It covers events from the death of John the Baptist to the Ascension and the dispersal of the disciples to preach the gospel. It makes much of the wiles of the evil powers embodied in a host of devils, led by Lucifer, with Satan as his chief lieutenant. The play is intended to take up three days and was first performed in 1501.

'Als Noah aus dem Kasten war', first line of a well-known drinking song by A. Kopisch (q.v.).

Also sprach Zarathustra, a long rhapsodic work, mostly in prose, by F. Nietzsche (q.v.), the publication of which was spread over the years 1883–92. Of the four parts, the first and second were written respectively in Rapallo and Sils Maria in 1883 and published in the same year. Pt. III was written at Nice in 1884 and published in that year; Pt. IV was written at Mentone and Nice in the winter 1884–5 and privately printed in 1885. The complete work was published in 1892, three years after Nietzsche's mental collapse.

Also sprach Zarathustra is dedicated in a subtitle to 'everyone and no one' ('Ein Buch für alle und Keinen'). Nietzsche's Zarathustra is not the Persian fire-god Zoroaster, but a cloak for Nietzsche himself in the role of prophet and seer. Nietzsche's philosophical works all have an emotional tinge, but *Zarathustra* discards altogether what is normally understood as philosophy, and instead expresses a 'Weltanschauung' in a series of declamatory and oracular sermons, most of which end with the formula 'Also sprach Zarathustra'. Nietzsche denounces the civilization of his day, attacking those who look nostalgically to the past or believe in a future life ('Von den Hinterweltlern'), condemning asceticism ('Von den Verächtern des Leibes'), castigating pity ('Von den Predigern des Todes'), and glorifying war ('Vom Krieg und Kriegsvolk'). Pt. I closes with the summation 'Tot sind alle Götter: nun wollen wir, daß der Übermensch lebe.' *Also sprach Zarathustra* develops no coherent argument, but operates by reiteration and intensification, so that the later parts in the main repeat the contents of the first, though in different tones. Nietzsche denounces Christianity and democracy because they tend to equalize strong and weak, and rejects the virtues of tolerance, meekness, humility, and pity. He replaces these outdated ideas by the concept of the Superman who possesses the urge for domination and power and the strength to achieve his aims. It is consonant with Nietzsche's worship of strength that he is contemptuous of the emancipation of women; and *Zarathustra* contains the famous utterance, 'Du gehst zu Frauen? Vergiß die Peitsche nicht!', quoted as the words of the wise Old Woman (das alte Weiblein). At the end of Pt. II Zarathustra faces the superior strength of Death, and in the section 'Vom Gesicht und Rätsel' of Pt. III he finds an answer to the fear of nothingness in the idea of perpetual recurrence (ewige Wiederkehr), a cyclical view of history repeating itself *ad infinitum*. The third and fourth parts are

not only more poetic in tone, as in 'Die sieben Siegel (Oder das Ja- und Amenlied)', they contain four poems, one of which occurs twice in slightly different forms. As 'Das trunkene Lied 12' (Pt. IV) it is called 'Zarathustras Rundgesang', beginning 'O Mensch, gib Acht!', and this was used by G. Mahler (q.v.) in the 4th movement of his 3rd Symphony.

Also sprach Zarathustra was much abused on moral grounds when it first appeared, but in the early years of the 20th c. it came to be accepted by the *avant-garde*, as well as by some lonely souls, as of virtually biblical authority. It was in this light that R. Strauss (q.v.) composed his tone poem *Also sprach Zarathustra* (1896), and that Delius was inspired to write his 'Mass of Life' (*Eine Messe des Lebens*, 1909), the text of which consists of passages from Nietzsche's book. A later age has tended to look askance at its authoritarian outlook and euphoric excess.

Alster, two adjoining lakes in the centre of Hamburg, the small Binnenalster and the much larger Außenalster.

ALT, Georg or Simon, town clerk of Nürnberg in the late 15th c., translated the Latin chronicle of world history by H. Schedel (q.v.) into German in 1493.

ALTDORFER, Albrecht (Regensburg, *c.* 1480–1538, Regensburg), German painter; he composed fantastic landscapes with a pronounced sense of atmosphere as background for religious or historical scenes. Among his most notable works are the *Alexanderschlacht* in Munich (Alte Pinakothek) and the series of religious pictures in the Abbey of St. Florian near Linz (Austria).

Altdorf University, founded in 1623 and dissolved in 1809, served as university for Nürnberg, which from 1504 to 1806 included Altdorf, a few miles south-east, within its territory. The university buildings are occupied by the Innere Mission (see PROTESTANTISMUS).

ALTE DESSAUER, DER, popular appellation of LEOPOLD, Fürst von Anhalt-Dessau (q.v.).

ALTE FRITZ, DER, sobriquet for Friedrich der Große, King Friedrich II (q.v.) of Prussia.

Alte Mann mit der jungen Frau, Der, a play by J. N. Nestroy (q.v.), written probably in 1848 or 1849, but not then performed. Nestroy incorporated some parts of it in *Mein Freund,* a comedy performed in 1851. *Der alte Mann mit der jungen Frau* is a more serious piece than the usual run of Nestroy's plays. The young wife,

Regine Kern, flirts with noble neighbours in the absence of her husband, and is on the point of committing herself further when she is discovered by him. She begs him to let her stay with him, but he insists on emigrating to Australia, where she may rejoin him, if she still wishes, in a year's time. Kern's integrity and generosity are demonstrated in a sub-plot in which he assists an escaped political prisoner. This element would have incurred the hostility of the censorship. The play was first published in Nestroy's *Gesammelte Werke*, 1890–1.

Alte Mutter und Kaiser Friedrich, Die, a comic verse tale of the 13th c. in which a spendthrift son escapes his mother's accusations at the Emperor's court by pretending that another person is her son.

ALTENBERG, PETER, pseudonym of Richard Engländer (Vienna, 1859–1919, Vienna). Altenberg spent his whole life in Vienna, for a time as a bookseller, but mostly as a familiar figure of Viennese night life. The Viennese café was his natural terrain, and to maintain his indolent way of life he had frequently to seek assistance from others. His work consists mainly of impressionistic sketches, sophisticated essays, and semi-autobiographical writings. His style is sensitive, over-refined, even effete. The principal titles are *Wie ich es sehe* (1896), *Ashantee* (1897), *Was der Tag mir zuträgt* (1901), *Prodromos* (1906), *Neues Altes* (1911), *Vita ipsa* (1918), and *Mein Lebensabend* (1919). *Ausgewählte Werke* (2 vols.), ed. D. Simon, appeared in 1979.

Alte Nester, a novel by W. Raabe (q.v.), written 1877–9 and published in 1880. It deals with the childhood and subsequent history of four characters, Ewald Sixtus and his sister Eva, Irene Everstein, and Fritz Langreuter, who is the narrator of the story. The earlier part is retrospective. The four characters grew up together, the Sixtuses in the forester's house and Irene, the squire's daughter, in Schloß Werden. An important aspect of the story is the conflict between a sentimentally recollected idyll of childhood and the necessity to achieve maturity and face reality. Ewald and Irene, though mutually attracted, clash and drift apart. Fritz Langreuter fails, through his self-absorption, to perceive Eva's affection for him. The problems of the characters are resolved through the personality and insight of the simple yet balanced character of the eccentric Vetter Just, who brings Ewald and Irene to a realization of their true desires. They abandon their tug of war, marry, and go to live in Ireland. Just marries Eva, and Fritz persists in the benevolent bachelordom for which he is clearly suited.

Alten Weibes List, a Middle High German erotic poem of the 13th c. Husband and wife meet at the house of a procuress, so both are guilty of infidelity. Though the wife beats the husband, the pair are reconciled. The poem, which falsely purports to be by KONRAD von Würzburg (q.v.), is of unknown authorship.

Alternden, Die, title of a cycle of Novellen by A. Schnitzler (q.v.) which was published in the collected works in 1924 and includes *Frau Beate und ihr Sohn, Doktor Gräsler, Badearzt,* and *Casanovas Heimfahrt* (qq.v.).

Altershausen, an unfinished novel by W. Raabe (q.v.), begun in 1899 and later abandoned. It was published posthumously in 1912. The famous professor of medicine Geheimrat Friedrich Feyerabend sets out on his retirement for his native town and there, in contact with two friends of his childhood, reviews his early life in a new perspective.

Älteste Urkunde des Menschengeschlechts, Die, a theological work by J. G. Herder (q.v.), published 1774–6 (2 vols.). A bold and arbitrary interpretation of the story of the Creation, it bears the sub-title *Eine nach Jahrhunderten enthüllte heilige Schrift.*

Alte Turmhahn, Der, an idyll (Idylle) by E. Mörike (q.v.), begun in 1840, completed and published in 1852. It consists of 293 lines of Knittelverse (q.v.), and Mörike, directing his irony upon himself, sets the idyll in his own parish 'Zu Cleversulzbach im Unterland'. The old weathercock, taken down after 113 years, is rescued from the smithy by the pastor and set up in his study. The cock tells of his old and new life and, in doing so, reflects the idyllic life of the village parsonage.

Alt-Heidelberg, dramatization by W. Meyer-Förster (q.v.) of his own sentimental novel *Karl Heinrich.* It had for many decades an immense success in Germany, and as a film was shown all over the world.

'Alt-Heidelberg, du Feine', first line of a poem in praise of Heidelberg by J. V. von Scheffel (q.v.). It occurs in Canto 2 of the long poem *Der Trompeter von Säkkingen* (q.v.). The melody is by Zimmermann.

Althochdeutsch, Old High German, often abbreviated in English to OHG. See GERMAN LANGUAGE, HISTORY OF.

Altkatholiken, designation of the Roman Catholics who refused to accept the dogma of

Papal Infallibility promulgated in 1870, and who preferred in consequence to adhere to an independent church. Prominent among them were university theologians, including I. von Döllinger (q.v.). The 'Altkatholiken' (known in Switzerland as 'Christkatholiken') were recognized in most German states in the 1870s, but not until 1918 in Bavaria and Austria. They continue to exist as a small minority. *Der Pfarrer von Kirchfeld* (1871) by L. Anzengruber (qq.v.) reflects the stresses which led to the breaking-away of the 'Altkatholiken' from Rome.

Altmark, the flat plain lying to the west of Berlin which was the original nucleus of the margravate from which grew the Electorate of Brandenburg and later the Prussian state. Its principal town was Stendal.

Altsächsische Genesis, title given to fragments of an epic poem of the 9th c. written in Old Saxon alliterative verse (Stabreim). The view that its author also wrote the *Heliand* (q.v.) has been discarded. Its subject is the Creation and early biblical history; the fragments deal with Adam and Eve, Cain and Abel, Abraham, and the destruction of Sodom, and the source is the Vulgate. The author treats the story freely, inventing dialogue and inserting the story of the fall of the rebellious angels. The MS. is in the Vatican.

ALTSWERT, Meister (Altschwert), the name assumed by an Alsatian poet of the late 14th c., whose four poems (published in 1850) are allegories of ideal love, linked with satirical portrayal of existing morals. They are *Das alte Schwert, Der Kittel, Der Tugenden Schatz,* and *Der Spiegel.* The first and fourth are relatively short (*c.* 300 lines), the other two have between 1,500 and 2,000 lines.

ALVERDES, Paul (Strasburg, 1897–1979, Munich), an officer's son, spent his childhood in several German regions including the Rhineland at Düsseldorf. In 1914 he volunteered, and in 1915 received a severe throat wound, spending months in hospital. This experience underlies his best-known work, the Novelle *Die Pfeiferstube* (q.v., 1929). After studying at Jena and Munich universities he launched out in 1922 as a free-lance writer and journalist in Munich. From 1934 to 1944 he edited with K. B. von Mechow (q.v.) the monthly *Das Innere Reich.*

His works include poems (*Die Nördlichen,* 1922), plays (*Die feindlichen Brüder,* 1923), radio plays (notably *Die Freiwilligen,* 1934, dealing with volunteers at the front at Langemarck in the autumn of 1914), and a number of Novellen (*Novellen,* 1923; *Reinhold oder die Verwandelten,*

including *Der Kriegsfreiwillige Reinhold* and *Reinhold im Dienst,* 1931; *Das Zwiegesicht,* 1937; *Jette im Wald,* 1942; *Grimbarts Haus,* 1949; *Legende vom Christ-Esel,* 1953; *Die Hirtin auf dem Felde,* 1954).

ALXINGER, Johann Baptist von (Vienna, 1755–97, Vienna), became in 1796 secretary of the Burgtheater (q.v.). A representative of the Aufklärung (q.v.), he pleaded in verse and journalism for freedom of thought. His *Gedichte* (1780 and 1784) were collected in 2 vols. in 1788. His principal works are two verse romances with medieval settings in which he embodied the ideals of the Freemasons. *Doolin von Mainz* (1787) has as its background the world of Charlemagne; *Bliomberis* (1791) is an Arthurian romance. His *Sämtliche Werke* (10 vols.) appeared in 1812.

Amadis de Gaule, a novel usually known by the French title, but of Spanish or Portuguese origin, and published by Montalvo in Spanish *c.* 1508 (1 vol. comprising 4 Bks.). The French translator Herberay des Essarts expanded it into 8 Bks. (1527). The book was so popular that it was repeatedly extended. A German version began to appear in 1569, attained 13 Bks. by 1583, and finally reached 24 in 1595. It was published by S. Feyerabend (q.v.). J. Fischart (q.v.) translated Bk. 6. The extravagant mixture of heroic, amorous, and miraculous adventures satisfied a need, and it retained some of its popularity right through the 17th c., and even in the 18th c. the theme persisted. C. M. Wieland (q.v.) completed his long poem *Der neue Amadis* in 1771, and a short early poem by Goethe bears the same title; it is likely that this title alludes to Wieland's poem, and may be an addition by F. H. Jacobi (q.v.), who published it in 1775.

Ambraser Handschrift, a MS. collection of medieval epic poems made between 1504 and 1516 for the Emperor Maximilian I (q.v.). Among its contents are various 'Dietrichepen' (see Dietrichsage), Hartmann's *Erec* (q.v.), the *Frauenbuch* of Ulrich von Lichtenstein (q.v.), and, in addition, the only surviving text of *Kudrun* (q.v.). Until 1806 it was kept in the Imperial Schloß Ambras or Amras, near Innsbruck, and since that date has been in the Hofbibliothek, now Nationalbibliothek, in Vienna.

Amerika, see Der Verschollene.

Amfortas, the wounded Grail King in R. Wagner's *Parsifal* (q.v.). In Wolfram von Eschenbach's *Parzival* (q.v.) the name appears as Anfortas.

22

Am grünen Strand der Spree, a novel by H. Scholz (q.v.).

Amis, see STRICKER.

Amor Dei, a historical novel by E. G. Kolbenheyer (q.v.), published in 1908. Its central figure is the philosopher Spinoza (q.v.).

Amphitryon. Ein Lustspiel nach Molière, a three-act blank-verse comedy by H. von Kleist (q.v.), completed and published in 1807. As the sub-title indicates, this comedy is an adaptation of Molière's *Amphitryon* (1668), which in turn is based on Plautus' *Amphitruo*—the first of the numerous plays on the legend of Amphitryon, king of Tiryns, and of Alcmene, whom he married in Thebes after Zeus (Jupiter) had visited her during Amphitryon's absence in battle. Alcmene became mother of twin sons, of whom Heracles was the son of Zeus.

Although Kleist may at one time have contemplated providing a German version of the French comedy, he radically changed his course and made his divergence from Molière the most significant feature of the play. The theme of changed identity is not only duplicated, but moves in two separate spheres of experience. Kleist exploits Jupiter's impersonation of Alkmene's husband Amphitryon, 'Feldherr' of the Thebans, to expose the twofold dilemma of the god's failure to win Alkmene's love, save through his disguise, and of Alkmene's failure to recognize the sublime deceit, which reflects on her relationship both to the god and to her husband. The metaphysical and psychological issues emerging from innate error and confusion of the emotions take human experience into the sphere of tragedy. A contrast is provided by the grotesquely comic experience of Sosias, the servant, now, of two masters, who confronts his own double in the disguised Mercury, while Charis, Sosias' wife, is placed in the same situation as Alkmene.

For a later treatment of the theme of Amphitryon see HACKS, P.

Am Quell der Donau, a hymnic ode by F. Hölderlin (q.v.). It is prefaced by an introductory passage which is incomplete.

Amsel, Die, a story by R. Musil (q.v.), published in 1927 as the last of his short narratives. A first-person narrator introduces two characters, mathematically denoted as *Aeins* and *Azwei*. At the height of his maturity, *Azwei* relates three episodes of his life to his former schoolfriend (and sceptical *alter ego*). The first recounts his escape from married life in Berlin into a dream world of mystic yearning evoked by the melodious song

of the 'divine' bird, the nightingale, which the blackbird, *die Amsel* of the title, mimics. The second episode recalls an incident on the Tyrolean front when he mistakenly believed himself to be the target of a lethal dart (Fliegerpfeil) dropped by an aircraft in 1915. In the third, rather cryptic, episode the death of his parents leads him back to his boyhood and to the discovery of his old blackbird cage in which he places another blackbird as if it were the original one. There is no further allusion to the nightingale which had appeared to him as a symbol of spiritual transcendence. The simple ending merely underlines the experience of an insoluble dilemma.

Am Tor des Himmels, a Novelle by G. von LE FORT (q.v.), published in 1954. The central story consists of the so-called 'Galileisches Dokument', in which a student of Galileo relates the events leading to Galileo's appearance before the Inquisition in Rome. Galileo himself, referred to not by name but as 'der Meister', remains in the background of the action, which centres on three characters: the German student, who is the narrator of the central story representing the 'document', Diana, and the Cardinal, her uncle, who provide the opportunity to discuss the problems which science presented to the Church in challenging the validity of the Ptolemaic system. The student and Diana move from Padua to Rome shortly before Galileo's trial. Here the student is invited to work in the Cardinal's observatory. The Cardinal himself believes in the truth of Galileo's theories, but fears that its acknowledgement by the Church will lead to disbelief and confusion and thus threaten the authority of the Church. His fears appear to be justified, for his niece Diana has already abandoned her faith in God and dedicated herself to 'the master'. In pronouncing the anathema, the Cardinal accordingly denies the truth. In the author's interpretation of the historic recantation, Galileo responds by similarly masking the truth. Galileo's student, believing throughout that the new discoveries should not affect true faith, returns to his native Germany.

The frame introducing the 'document' (see RAHMEN) is set in and after the 1939–45 War, thus making it possible to examine the relevance of the problem to modern times. An air raid destroys the document as well as the faith of the narrator, a distant relation of the student. After the war her friend, a young scientist, returns to continue his research, convinced, like the student, that the progress of science cannot be halted. Yet it is he who resumes the search for God, and thus revives the abandoned conviction of the 17th-c. student that faith should grow independently of the laws of causality. The symbolical title has also a specific reference to the observatory in Padua.

Amulett, Das, the first Novelle of C. F. Meyer (q.v.), published in 1873, and based on *La Chronique du règne de Charles IX* by P. Mérimée. The story, which comprises ten short chapters, is told in the first person by Hans Schadau. On 14 March 1611 he visits Renat Boccard, the aged father of Schadau's long-dead friend Wilhelm Boccard. The visit and the sight of a locket worn by Wilhelm arouse Schadau's memories of the young man's death nearly forty years before, and he recalls his own early life and his departure for Paris in 1572 to join the defenders of the Calvinist cause against Catholic oppression. Schadau meets Gasparde, a niece of Admiral Coligny. He becomes the admiral's secretary, and marries Gasparde. In the massacre of St. Bartholomew Coligny is among those killed. Schadau is helped by a Catholic friend, Wilhelm Boccard, who carries about with him a locket with a picture of the Blessed Virgin of Einsiedeln, the 'Amulett' of the title. This amulet was once lent to Schadau for a duel, and saved his life by stopping a sword thrust. Together the two friends hasten to protect Gasparde from the massacre. They arrive to find her surrounded by a mob, whom she keeps at bay with a pistol. The two men come to her assistance and hurry away. As they flee, Boccard is shot, in spite of the protective power of the locket, which he is wearing. Schadau and his wife escape and reach the Swiss border. Meyer's interest in this adventurous tale is concentrated on the tangled religious conflicts of the time, of which the amulet is a symbol, as it saves the life of a heretic and fails to protect a devout Roman Catholic.

Anakreons Grab, a short poem written by Goethe probably in 1785 and first published in *Goethes Schriften*, 1789. It is in elegiac metre and though only six lines long conveys an atmosphere of idyllic harmony and gentle sadness. It has been set to music by Hugo Wolf (q.v.).

Anakreontiker, German poets mostly writing in the middle of the 18th c., whose poems were avowedly based on 'Anacreon', i.e. on 60 *Anakreonteia*, largely pseudo-Anacreontic odes, published in France by Henri Estienne (1531–98) in 1554. Though stylistically often scarcely distinguishable, the best poems treat the themes of love, wine, nature, and friendship with elegance, charm, and wit. They are playful products of the imagination, sometimes sung at convivial gatherings in Horace's spirit ('carpe diem'). They have been classified under the literary rococo (see ROKOKO). Their virtues are formal and a reaction against prevailing conventions. The principal German Anacreontic poets were Hagedorn, Gleim, J. N. Götz, Uz, Ramler, and Zachariä (qq.v.). Various other poets, including G. E. Lessing and Goethe (qq.v.), wrote poems in this manner.

Anatol, a cycle of seven one-act plays written by A. Schnitzler (q.v.) 1889–90, and published in 1893. Though some of the plays were performed separately, the first production of the cycle in German took place in Vienna in December 1910. A performance in Czech had been given in Prague in 1893.

The seven plays present incidents in the amours of the young philanderer and man-about-town Anatol. In *Die Frage an das Schicksal* he hypnotizes Cora in the hope of hearing from her the truth about her fidelity; but as soon as she is unconscious he flinches from putting the question. *Weihnachtseinkäufe* consists in a short conversation in the street on Christmas Eve between Anatol and Frau Gabriele, a well-to-do acquaintance; when she has gone he realizes that he has neglected an opportunity. *Episode,* the first of the plays to be written, sees him bitterly disillusioned when Bianca, a circus-girl, with whom he has once had an *affaire,* fails to recognize him on her return to Vienna. *Denksteine* reveals to Anatol the mercenary nature of his mistress Emilie. In *Abschiedssouper* he gives what he intends to be a farewell supper for the ballerina Annie, of whom he has tired, only to find, to his unjustified indignation, that she has come with a similar intention. The dialogue of *Agonie,* in its dissimulation and hypocrisy, shows the love between Anatol and Frau Else degenerating into boredom. The last, *Anatols Hochzeitsmorgen,* sees Anatol deceiving his bride with Ilona, a former mistress, the night before his wedding. Max, a bachelor friend, is present in *Die Frage an das Schicksal, Episode, Abschiedssouper,* and *Anatols Hochzeitsmorgen.* The other three plays are duologues. The fluent dialogue of the cycle is by turns witty and pungent, the atmosphere evocative. Erotic psychology is convincingly, though cynically, analysed. The plays are accompanied by a verse prologue (*Einleitung*) by Schnitzler's friend H. von Hofmannsthal (q.v.). An eighth play, *Anatols Größenwahn,* not published in the author's lifetime, was performed in 1932.

An Belinden, a poem written by Goethe at the time of his engagement to Lili Schönemann (q.v.). It expresses his wonderment at the love which holds him in an uncongenial social environment. Written in 1775, it was published in

Jacobi's *Iris* in the same year. See also NEUE LIEBE, NEUES LEBEN and LILIS PARK.

An den christlichen Adel deutscher Nation von des christlichen Standes Besserung, one of Luther's three important theological tracts of the year 1520. It consists of three parts: (1) an attack on the privileges of the clergy and the powers assumed by the Pope; (2) a denunciation of the wealth and temporal interests of the Pope; (3) a general exposure of ecclesiastical abuses.

Andenken, a poem by Friedrich Matthisson (q.v.), published in his *Gedichte* and set to music by Beethoven (q.v.) in 1809. The first line runs 'Ich denke dein, wenn durch den Hain . . .'

An den Mond, a poem by Goethe which exists in two versions, one written between 1776 and 1778, the other and better known one probably between 1785 and 1789. Its conception belongs to the first years of Goethe's friendship with Charlotte von Stein (q.v.). The poem was first published (in its second version) in *Goethes Schriften*, 1789. The first version became known on the publication of Goethe's letters to Frau von Stein in 1848. The first line of the poem, in which nature and love intertwine, runs 'Füllest wieder Busch und Tal'.

'An der Saale hellem Strande', first line of the poem *Rudelsburg*, written in 1826 by Franz Kugler (q.v.). It is sung to a tune composed four years before by F. E. Fesca.

ANDERSCH, ALFRED (Munich, 1914–80, Berzona), an officer's son, was a member of the youth organization of the Communist Party, and in 1933 spent six months in Dachau concentration camp. He served in the 1939–45 War from 1940, deserted in 1944 on the Italian front, and became a U.S. prisoner of war. Andersch's development from Marxism and French existentialism towards a materialistic European outlook is evident in his essays, stories, novels, and radio plays from 1945 onwards, in which year he became a prominent spokesman for the place and responsibilities of his generation in post-war Germany.

With H. W. Richter (q.v.) he became joint editor of the periodical *Der Ruf*, which was prohibited in 1947. He was co-founder of Gruppe 47 (q.v.) and contributed to the *Frankfurter Hefte* (ed. Eugen Kogon and Walter Dirks) and *Neue Zeitung* (under Erich Kästner, q.v.). In 1949 he published a representative selection of essays by international authors under the title *Europäische Avantgarde*. In 1955 he started another periodical, *Texte und Zeichen*, in which a great variety of intellectuals, writers,

and poets (e.g. Th. Adorno, Celan, Enzensberger, Heißenbüttel, Grass, Golo Mann, Arno Schmidt, W. Jens, M. Walser, qq.v.) are represented.

His story *In der Nacht der Giraffe* became one of his successful radio plays. His autobiographical work *Die Kirschen der Freiheit* argues in provocative form the theme of desertion, unusual in German post-war literature. He did not succeed in publishing it until 1952. This theme is also contained in *Sansibar oder Der letzte Grund* (q.v., 1957) and in the adventures of Franziska Lukas in his novel *Die Rote* (1960, in revised form 1972). In *Efraim* (1967) Andersch treats the motif of the Wandering Jew in the context of intellectual, social, and national insecurity which arises from his own sense of homelessness and ideological disillusionment. The mythological ingredient is meant to support the message of its 'European' author, who in 1958 took up residence outside Germany. His travel books combine irony in lighter vein with serious topical economic and environmental problems. *Wanderungen im Norden* (1962) go back to travels in 1953. *Aus einem römischen Winter. Reisebilder* was published in 1966, and *Hohe Breitengrade*, based on travels to Spitzbergen, in 1969. A further volume of stories, *Mein Verschwinden in Providence*, appeared in 1971, the novel *Winterspelt* in 1974, *Der Vater eines Mörders* in 1980, and the travel book *Flucht in Etrurien* in 1981. Essays include *Öffentlicher Brief an einen sowjetischen Schriftsteller, das Überholte betreffend* (1977). The volume *Bericht, Roman, Erzählungen* was published in 1965, *empört euch der himmel ist blau*. *Gedichte und Nachdichtungen 1946–1977* in 1977, *Hörspiele* and *Neue Hörspiele* in 1973 and 1979 respectively, and *Sämtliche Erzählungen* in 1983.

ANDERSEN, LALE, see LILLI MARLEEN.

An die Engel, a poem by H. Heine (q.v.), included in the section *Lamentationen* of the *Romanzero* (q.v.). No. 15 in the subdivision *Lazarus*, it begins 'Das ist der böse Thanatos'. Its subject is the compassion of the dying man for his wife.

An die ferne Geliebte, a song cycle (Liederkreis) by Beethoven (q.v., op. 98), written in 1816 and dedicated to Fürst Joseph von Lobkowitz. It contains six songs by Alois Jeitteles (q.v.), linked by a continuous and prominent piano accompaniment. The songs are I. 'Auf dem Hügel sitz' ich', II. 'Wo die Berge so blau', III. 'Leichte Segler in den Höhen', IV. 'Diese Wolken in den Höhen', V. 'Es kehret der Maien', VI. 'Nimm sie hin denn, diese Lieder'.

An die Freude, Schiller's 'Ode to Joy', written in 1785 in the first flush of new happiness with his Leipzig friends. It uses cosmic images to elevate joy into something universal. The original version, published in *Die Thalia* in 1786, consisted of nine strophes with nine choruses, but Schiller later deleted the last strophe and chorus. Beethoven (q.v.) used the first three strophes and the first, third, and fourth choruses as the text for the last movement of his Ninth Symphony.

An die Geliebte, a sonnet written in 1830 by E. Mörike (q.v.). It is linked with his love for Luise Rau (q.v.).

An die Parzen, an early Alcaic ode written in 1798 by F. Hölderlin (q.v.) beginning: 'Nur Einen Sommer gönnt, ihr Gewaltigen!' It is an impassioned plea to the Fates to grant him time for his poetic genius to ripen and express itself.

An die preußische Armee, see ODE AN DIE PREUSSISCHE ARMEE.

Andorra, a play in the Expressionist manner by M. Frisch (q.v.), written between 1958 and 1961 and published in 1961. It was first performed in the Schauspielhaus, Zurich, in November 1961. It is described as a 'Stück in zwölf Bildern'. The twelve 'pictures' are linked by comments made in a witness box down stage ('Vordergrund'). The young man Andri, adopted son of the schoolmaster (der Lehrer), is believed to be a Jew; in the end he proves not to be, but the attitude of his environment has imprinted indelibly upon him the supposed Jewish characteristics, and he perishes as a Jew when a fascist foreign power (die Schwarzen) invades Andorra. Barblin, the girl he loves, is also maltreated because she is regarded as a 'Jew's whore'. Andri is successively seen in contact with the coarse-grained soldier Peider (who rapes Barblin), with the prejudiced population (Doktor, Wirt, Tischlermeister, Geselle), and with the Church (Pater). All, in varying degrees, incur guilt in their conduct to him. Not least guilty is the schoolmaster who, with good intention, invents the lie about Andri. When 'die Schwarzen' invade, no resistance is offered and none raises a finger to help Andri. Each appears in turn in the witness box and pleads good intention or ignorance.

The play is designed as a parable on the theme of collective guilt and is addressed as much to the future as to the past, as much to the spectator nations, such as Switzerland, as to the persecutors.

ANDREAE, JOHANN VALENTIN (Herrenberg, Württemberg, 1586–1654, Stuttgart), religious administrator and mystic, studied physical and mathematical sciences as well as theology at Tübingen and, as a young man, attended meetings of the Rosicrucian Order (see ROSENKREUZER). Between 1607 and 1614 he visited the Rhineland (including Heidelberg and Strasburg), France and Switzerland (notably Geneva), and Rome. On these journeys he took particular note of political and social organization as well as of religious and scientific matters. In 1614 he became curate of Vaihingen; encountering suspicion and hostility in the parish, he devoted himself to religious writings. In 1630 he was made headmaster of the school at Calw, undertaking an enlightened reform of religious teaching. After nine years at Calw he was appointed to the Consistory Council in Stuttgart and in 1650 became abbot of Bebenhausen (not a monastic appointment in Protestant Württemberg). He rendered useful service in ecclesiastical legislation and organization and was the founder of the Tübinger Stift (q.v.). He also became in 1646 a member of the Fruchtbringende Gesellschaft (q.v.), in which he was known as 'der Mürbe'.

His writing, much of which is in Latin, is mainly devotional or expository prose. He inclined to mysticism and Pietism (see PIETISMUS). Apart from plays (*Esther,* 1602; *Hyacinth,* 1603; *Turbo,* 1616) he wrote the following works in Latin: *Mythologia christiana* (1618), *Theophilus* (1649), and *Vita ab ipso conscripta* (not published until 1849, though a German translation appeared in 1789). Most notable is the Utopian novel *Reipublicae Christianapolitanae descriptio* (1619, ed. by R. van Dülmen as *Christianopolis. 1619. Originaltext und Übertragung nach D. S. Georgi 1741,* 1972), in which life in an imaginary state is based on the imitation of Christ. Among the German works are *Geistliche Gemälde* (1612), *Vom letzten und edelsten Beruf des wahren Dieners Gottes* (1615), *Chymische Hochzeit des Christiani Rosenkreuz* (1616), *Geistliche Kurtzweil* (1619), and *Adelicher Zucht-Ehrenspiegel* (1623).

Andreas Hofer, a patriotic poem by J. Mosen (q.v.). It begins 'Zu Mantua in Banden'.

Andreas Hofer, der Sandwirt von Passeier, a tragedy (Trauerspiel) in five acts by K. L. Immermann (q.v.). It appeared originally as *Ein Trauerspiel in Tyrol* in 1828 and was revised and published under its present title in 1834. It is written chiefly in blank verse with a few scenes in prose. (See HOFER, ANDREAS.)

ANDREAS-SALOMÉ, LOU (St. Petersburg, 1861–1937, Göttingen), daughter of a Russian general (von Salomé) of French descent, was one of the earliest women students and read theology at Zurich University. In 1882 she formed an attachment to Nietzsche (q.v.) and in 1887 married F. C. Andreas, a professor of Göttingen University. An 'Ibsenite', she wrote a perceptive book on his characters (*Ibsens Frauengestalten*, 1892). Of her numerous stories *Ruth* (1895), the cycle *Im Zwischenland* (1902), *Das Haus* (1919), and *Rodinka* (1922) deserve mention. She was one of the many highly placed and intelligent ladies who devoted themselves to Rilke (q.v.), whom she met in 1897 and accompanied to Russia 1899 and 1900. Shortly before the 1914–18 War she turned to psychoanalysis and was prominent in Freud's circle. The subtlety and sensitivity of her writing reflect her acute intelligence as well as her intensely intellectual temperament.

ANDREAS VON REGENSBURG, an Augustinian monk ordained priest in 1405, was the author of a *Chronik von den Fürsten zu Baiern.*

ANDRES, STEFAN (Breitwies nr. Trier, 1906–70, Rome) was at first a candidate for the Roman Catholic priesthood, but abandoned this intention and turned to university study at Cologne, Jena, and Berlin, and settled in 1937 in Positano, southern Italy, in order to protect his wife from racial persecution. He lived at Unkel near Königswinter on the right bank of the Rhine before moving to Rome in 1961.

His first work of fiction was *Das Märchen im Liebfrauendom* (1928) and his first novel, *Bruder Lucifer*, appeared in 1932. He wrote many Novellen, of which *El Greco malt den Großinquisitor* (q.v., 1936) and *Wir sind Utopia* (1943) are particularly noteworthy. Collections of Novellen include *Moselländische Novellen* (1937), *Das Grab des Neides* (1940), *Das Wirtshaus zur weiten Welt* (1943), and *Das goldene Gitter* (1943). His novels include *Eberhard im Kontrapunkt* (1933), *Die unsichtbare Mauer* (1934), *Der Mann von Asteri* (1939), *Der gefrorene Dionysus* (1943), *Die Hochzeit der Feinde* (1947), *Ritter der Gerechtigkeit* (1948), *Der Knabe im Brunnen* (1953), *Die Reise nach Portiuncula* (1954), *Der Mann im Fisch* (1963), and an ambitious philosophical trilogy *Die Sintflut* (q.v., 1949–59).

Andres wrote with verve and assurance. The existence of evil, the diversity of human motive, and the problems of guilt and expiation dominate his fiction, in which he shows a masterly handling of tension. He is also the author of plays and published several collections of poetry, of which *Der Granatapfel* (1950) deserves mention. It includes *Der Skorpion*, a poem which he valued highly. Collected *Gedichte* appeared in 1976. Other posthumous publications include the novel *Die Versuchung des Synesio* (1971), *Die große Lüge* (1973), *Das Fest der*

Fischer (1973), speeches and essays, *Der Dichter in dieser Zeit* (1974), and correspondence, *Lieber Freund—lieber Denunziant* (1977).

ANDRIAN-WERBURG, LEOPOLD, FREIHERR VON (Vienna, 1875–1951, Fribourg, Switzerland), of distinguished aristocratic family and, on his mother's side, of Jewish descent, which included the composer Meyerbeer, contributed poems to Stefan George's *Blätter für die Kunst* (q.v., 1894–1901). He also published in 1895 a Novelle, *Der Garten der Erkenntnis*. He entered the Austrian foreign service in 1899, was chargé d'affaires in Athens in 1908, and during the war was appointed governor-general of Poland. After the war he published *Die Ständeordnung des Alls. Rationales Weltbild eines katholischen Dichters* (1930) and *Österreich im Prisma der Idee* (1937). When Austria was annexed by the Third Reich in 1938 he emigrated to Switzerland.

An Ebert, an ode in unrhymed verse written by F. G. Klopstock (q.v.) in 1748. In it the poet, revelling in grief, imagines the death of all his friends and of his future wife. The addressee is J. A. Ebert (q.v.).

Anegenge (i.e. 'Anfang'), an Early Middle High German theological poem treating of redemption and salvation, which are seen to be God's plan from the beginning. The title is in the original. The substance of Genesis and the Gospels is subordinated to dogmatic interpretation. The Austrian cleric, who wrote it c. 1160, was unskilled in abstract thought, and the view has been advanced that his work is a versification of lecture notes. This poem is sometimes known as *das jüngere Anegenge*, to distinguish it from the *Ezzolied* (q.v.), which also discourses 'von dem anegenge'.

An eine Äolsharfe, an ode written by E. Mörike (q.v.) in 1837. The sad fleeting strains of the Aeolian harp recall past grief and reflect the movement of the poet's heart. Written in free unrhymed verse, it is preceded by a quotation from Horace (*Carmina* II, ix, 9–12).

An einem Wintermorgen, vor Sonnenaufgang, a poem written in 1825 by E. Mörike (q.v.).

An Fanny, an ode in 11 unrhymed strophes written by F. G. Klopstock (q.v.) in 1748. 'Fanny' (q.v.) conceals the identity of Marie Sophie Schmidt, whom Klopstock loved in vain.

Anfortas, in Wolfram von Eschenbach's *Parzival* (q.v.) the Grail King who suffers from a perpetually open wound, the punishment for sensual love, a sin against the Grail. He can be cured only by Parzival's compassionate question. Anfortas proves to be Parzival's maternal uncle and brother of Trevrizent. As Amfortas (q.v.) he appears in R. Wagner's *Parsifal* (q.v.).

Angela Borgia, the last Novelle written by C. F. Meyer (q.v.), published in the *Deutsche Rundschau* in 1891, and based mainly on *Lucrezia Borgia* by Ferdinand Gregorovius (q.v.). It is divided into 12 chapters.

The two Borgia ladies, Lukrezia and Angela, are of equal importance in the story. In contrast to Lukrezia, Angela represents innocence and purity coupled with strength of character. The first chapters are devoted to the splendour of Lukrezia's entry into Ferrara as the bride of Duke Alfonso, the eldest of the four Este brothers, of whom the other three are Ippolito, Ferrante, and Giulio. With her new husband, Alfonso, Lukrezia hopes to begin a new life, untarnished by the Borgia heritage of evil, which Angela, who has enjoyed the sheltered life of a convent, has escaped. Lukrezia breaks her virtuous resolve, while Angela, exposed to the same worldliness and intrigue, preserves her nobility of character.

However, the action also develops from the natures of the four Este brothers: Alfonso is magnanimous, Ippolito power-loving and addicted to intrigue; Don Ferrante is guided by malice, whereas Don Giulio, though free from evil intention, has succumbed to crime through reckless lust, and languishes in prison. On Lukrezia's entry into Ferrara an amnesty frees Giulio. When he kneels before the future duchess, Angela, who accompanies Lukrezia, is both fascinated and frightened by the beauty of his eyes, which express his love of the sun and of life. Meanwhile the Judge of Ferrara, Herkules Strozzi, succumbs to Lukrezia's beauty, and Ippolito, in spite of his clerical office, lusts for Angela. As Angela remains under the spell of Giulio's eyes, Ippolito has Giulio blinded and blames Angela for his crime. While Alfonso refrains from punishing Ippolito, Ferrante plans to murder them both, and turns to Giulio for support. He finds it, though only after he learns of Angela's sense of guilt at his blindness. The plot fails, and both brothers are convicted by Strozzi of high treason.

As they are led to the scaffold, Ippolito pleads for their lives, and Alfonso commutes the sentence to imprisonment. But Ferrante, dominated by a fear of life, poisons himself. Giulio accepts mercy and imprisonment, and he and Angela are secretly married. Meanwhile, Lukrezia's notorious brother Don Cesare Borgia has been released from prison in Spain and appeals to Lukrezia to send Strozzi to him. Yet before Lukrezia can, by her support of her brother, betray Italy, Alfonso, alert to the danger, leaves for Bologna and appoints his wife as regent and Ippolito as her adviser. Although Lukrezia tries to sacrifice Strozzi for her brother's cause, her intrigues are doomed as Cesare is killed in a siege. Upon the Duke's return Lukrezia confesses her disloyalty. To conceal the plot from the public, the Duke has Strozzi murdered and then accords him a state funeral. During the following five years Lukrezia's love and loyalty remain constant, and she bears Alfonso two sons. Angela's marriage is discovered and is approved by Lukrezia and the Duke. A message arrives from the dying Ippolito, who has resigned his office and retired to Milan, pleading for Don Giulio's freedom, to which the Duke unhesitatingly agrees. This Renaissance tale of violence thus reaches an unexpectedly conciliatory and balanced ending.

ANGELUS SILESIUS, real name Johann Scheffler (Breslau, 1624–77, Breslau), used the pseudonym (meaning Silesian messenger) for his poetic works, for which he is best known. He studied medicine at Strasburg, Leyden, and Padua, and in 1648 entered the service of Duke Sylvius Nimrod of Württemberg and Oels as court physician. In 1653 he returned to his native Breslau, discarded the Lutheran faith in which he had been raised, and became a Roman Catholic. Ordained in 1661, he devoted himself to furthering the Counter-Reformation (see GEGENREFORMATION); his numerous polemical tracts were collected in *Ecclesiologia* (1677), which he published under his real name.

In 1657 appeared his *Geistreiche Sinn- und Schlußreime* in five, and his *Heilige Seelen-Lust oder geistliche Lieder, der in ihren Jesum verliebten Psyche* in four books. The latter is composed of strophic hymns, some of which derive from his Protestant youth, as well as mystical songs of Roman Catholic character. The former became his most famous work after being reissued, retitled, and augmented by a sixth book; it appeared in 1675 as *Der cherubinische Wandersmann,* containing upwards of 1,600 brief poems, most of which are mystical and paradoxical epigrams of one or two pairs of Alexandrines (e.g. 'Dann wird das Blei zu Gold, dann fällt der Zufall hin,/Wann ich mit Gott durch Gott in Gott verwandelt bin.').

Angelus Silesius, who was influenced by Böhme as well as by Abraham von Franckenberg and by Daniel von Czepko (qq.v.), is a poet who pregnantly shapes the thought of others, rather than an original thinker. In his own day his most popular work was a book of edification, *Sinnliche Beschreibung der vier letzten Dinge* (1675). An edition of his works by G. Ellinger (2 vols.)

includes some of the polemical writings. The poetic works, *Sämtliche poetische Werke* (3 vols.), ed. H. L. Held, appeared 1949–52.

ANGELY, Louis (Leipzig, 1787–1835, Berlin), a successful comic actor and theatre director in Berlin, was an adroit and prolific fabricator of successful comedies, most of which have never been published. The most successful were *Sieben Mädchen in Uniform* (1825), *Das Fest der Handwerker* (1828), and *Die Reise auf gemeinschaftliche Kosten* (1834). Angely was also a well-known restaurateur.

Angst des Tormanns beim Elfmeter, Die, a story by P. Handke (q.v.), published in 1970. The goalkeeper facing the penalty kick is intended as an image of a man who does not himself initiate action but reacts to what he encounters. Handke makes the image plausible by casting the principal figure, Josef Bloch, as a once well-known goalkeeper. Bloch is sacked from his job as a fitter, drinks heavily, spends the night with a girl and kills her in the morning on a sudden inexplicable impulse. The murder is presented as something that *happens to Bloch*. In his flight Bloch resembles the goalkeeper who does not know from which direction a shot may come. He is seen throughout as undergoing a process of self-alienation, which Handke seeks to express by his dry matter-of-fact narrative prose. A film version was directed by Wim Wenders (1972); English title: *The Goalkeeper's Fear of the Penalty*, American title: *The Goalie's Anxiety at the Penalty Kick*.

Anilin, a technological novel by K. A. Schenzinger (q.v.).

An mein Volk, proclamation signed on 17 March 1813 at Breslau by Friedrich Wilhelm III (q.v.) of Prussia, summoning the Prussian people to war against the French. It marks the beginning of the War of Liberation (see NAPOLEONIC WARS). The text was drafted by Theodor von Hippel, a civil servant who was a nephew of the poet and novelist Theodor Gottlieb von Hippel (q.v.).

Anmerkungen übers Theater, a tract published anonymously in 1774 and at first attributed by some to Goethe (q.v.). The author was J. M. R. Lenz (q.v.), and the work is the script of a lecture read in 1772 to the Société de Philosophie et de Belles-Lettres of Strasburg. It is an unsystematic essay which ranges with wit and assurance over the whole history of drama. Lenz recognizes the validity of Aristotle's *Poetics* for the Greece of his day but does not consider the Stagyrite's views to be binding for a later age. Though he mocks at French interpretations of Aristotle, his own slight criticisms of the philosopher are expressed with respect. Lenz's main thesis is that character should be the basis of tragedy and action the foundation of comedy. Lenz speaks enthusiastically of Shakespeare, praising especially the historical plays. Other English authors who win his approval are Samuel Richardson and Laurence Sterne. Included as an appendix to the lecture is a prose translation of *Love's Labour's Lost* under the title *Amor vincit omnia*. The English title is not mentioned. The translation is also anonymous, but, like the tract, it is by Lenz.

'Anmutig Thal! du immergrüner Hain!', first line of Goethe's poem *Ilmenau* (q.v.).

ANNA AMALIA, HERZOGIN (Wolfenbüttel, 1739–1807, Weimar), a Brunswick princess, was married to Duke Ernst August Konstantin of Saxe-Weimar in 1756. On his death in 1758, she was left with two infant sons and acted as regent for the elder, Karl August (q.v.), until 1775. A woman with lively intellectual interests, she ruled responsibly and humanely and sought to educate her elder son in the principles of good government. To this end she appointed C. M. Wieland (q.v.), whose political novel *Der goldene Spiegel* (q.v.) had impressed her, as tutor to Karl August in 1772. After her withdrawal from affairs in 1775 she maintained her literary and musical interests. A well-known water-colour by G. M. Kraus (q.v.) depicts her evening circle *c.* 1795, including Goethe, Herder, and H. Meyer (qq.v.).

Annalen, an alternative title for Goethe's *Tag- und Jahreshefte* (q.v.).

Ännchen von Tharau, a well-known 17th-c. poem, often sung and commonly regarded as a Volkslied (q.v.). It was long believed to be by S. Dach (q.v.), but doubt has been cast on this attribution, some considering that it is by H. Albert (q.v.). It is a nuptial song intended, and perhaps commissioned, for the wedding of Annke, daughter of the pastor of Tharau in East Prussia. The original is in dialect, and the High German version normally sung was made by Herder (q.v.) in 1778. The tune is by Silcher (q.v.).

Anne Bäbi Jowäger, abbreviated title of Gotthelf's novel *Wie Anne Bäbi Jowäger haushaltet und wie es ihm mit dem Doktern geht* (q.v.).

Annette, a collection of 19 poems written by Goethe (q.v.) in Leipzig, and copied by his friend Behrisch (q.v.) in 1767. The MS. was discovered in 1894, first published in vol. 37 of the *Große Weimarer Ausgabe* and in facsimile in 1923. The Annette of the title and addressee of some of the poems is Anna Katharina Schönkopf (q.v.). The whole collection is in Goethe's earliest poetic style, showing rococo and Anacreontic affinities.

Annke van Tharow, original Low German form of *Ännchen von Tharau* (q.v.).

ANNO, St. (*c.* 1010–75, Siegburg), Archbishop of Cologne, canonized in 1183. Anno was for a time a political power in the Empire, but lost influence after 1063. His lifetime coincides with the struggle for power between Emperor and Papacy. He himself was in conflict with Adalbert, Archbishop of Bremen. See ANNOLIED.

Annolied, a poem in praise of Archbishop Anno (q.v.) of Cologne, written towards the end of the 11th c., not long after Anno's death (1075), in order to support his claim for canonization. The poem is curiously constructed, prefacing Anno's biography with a summary of universal history. No MS. survives and it is known only through a printed edition published by Martin Opitz (q.v.) in 1639 (repr. 1946).

ANSCHÜTZ, HEINRICH (Luckau, 1785–1865, Vienna), famous as an actor in Vienna at the Burgtheater (q.v.) from 1821. Anschütz, who had previously played in various German cities, was a man of good education. His intelligence and vocal quality made him an outstanding performer in the principal older roles (Falstaff, Lear, Wallenstein, and Miller in *Kabale und Liebe*). His autobiography, *Heinrich Anschütz, Erinnerungen aus dessen Leben und Wirken,* was published posthumously in 1866.

An Schwager Kronos, a poem written by Goethe (q.v.) in October 1774 and first printed in *Schriften* 1789. In the MS., written for Frau von Stein (q.v.) in 1777, the date and place of origin is given: 'In der Postchaise den 10. Oktober 1774'. The coachman (Schwager) of the title is Time (Chronos), for the 'K' is merely an error which has been perpetuated. The poet sees his life as a journey, with Time as coachman and a welcome at the end from the princes of the Underworld. In its combination of exaltation and humour, it is akin to *Wanderers Sturmlied* (q.v.). The original differs appreciably from the version of 1789 and is now usually preferred.

Ansichten eines Clowns, a novel by H. Böll (q.v.), published in 1963. It is told in the first person and mainly in retrospect by Hans Schnier, the son of well-to-do parents, who opts

out of his middle-class environment and becomes a professional clown. He does so from disgust at the pretensions and lack of integrity of the world of his upbringing. Yet in choosing to be an outsider he also chooses solitude, for which he is not fitted, and under the stress of this situation he breaks down. The book is intended as a bitter indictment of the new Germany.

Anthologie auf das Jahr 1782, collection of poems published by Schiller (q.v.) in Stuttgart in the autumn of 1781. Some friends contributed, but the majority of the poems are by Schiller himself. The most important are the tormented, sensual odes addressed to 'Laura'.

Antichrist, in Judaic tradition, a false and hostile figure, appearing before the Messiah, became in the 8th c. the subject of a legend which was formulated by Adso of Toul in his Latin *Libellus de Antichristo* in the 10th c. According to Adso, Antichrist will appear when the various kingdoms have seceded from the Roman Empire (see DEUTSCHES REICH, ALTES). When the Emperor comes to Rome and lays down his insignia in token of the end of the Empire, then shall Antichrist reign for three and a half years until he is struck down by God.

The figure of Antichrist is mentioned in *Muspilli* (q.v.) in the 9th c., and is the subject of a poem by Frau Ava (q.v.) in the 12th c. The most considerable work connected with Antichrist is the medieval Latin play *Ludus de Antichristo* (q.v.), written in Bavaria in the 12th c. He appears again in a satirical Fastnachtspiel of the 15th c. (*Des Entchrist Vasnacht,* q.v.).

Anti-Goeze, a series of 11 polemical pamphlets addressed by G. E. Lessing (q.v.) in 1778 to Pastor J. M. Goeze (q.v.) of Hamburg. Its origin lay in the publication by Lessing in 1777 of some rationalistic comments on the New Testament by H. S. Reimarus (q.v.), which provoked bitter attacks on Lessing by Goeze. *Anti-Goeze* is one of the most virulent polemical works of modern times, and its violence frustrates its defence of independent thought and tolerant judgement. The controversy was stopped by the intervention of Lessing's employer, the Duke of Brunswick, in July 1778. *Nathan der Weise* (q.v.), written by Lessing a few months after the ban, is sometimes termed the 'twelfth *Anti-Goeze* pamphlet'.

Antigone, a free translation in five acts by F. Hölderlin (q.v.) of Sophocles' tragedy. Kreon is portrayed as the maker of his own destiny and laments his folly: 'aus fremdem/Irrsal nicht, sondern selber hat er gefehlt' judges the chorus. The didactic element rather than Kreon's longing for destruction dominates the ending. In his *Anmerkungen zur Antigone* Hölderlin defines his aim as 'den Geist der Zeit ... festzuhalten und zu fühlen, wenn er einmal begriffen ist'. B. Brecht (q.v.) adapted his version (see ANTIGONE DES SOPHOKLES, DIE).

Antigone des Sophokles, Die, the stage adaptation with prologue and *Vorspiel*, but with no division into acts, by B. Brecht (q.v.) of Hölderlin's translation of *Antigone* (q.v.). It was first performed at Chur in Switzerland in 1948 without the *Vorspiel*, which Brecht revised in 1951.

It is set during the last days of the fighting in Berlin in April 1945. Two sisters return home from an air-raid shelter and discover that their soldier brother has been there during their absence. They discover, too, that the SS have been quick to hang him as a deserter. Brecht uses the contrasting reactions of the two girls, whose own lives are suddenly at stake should their identity as the sisters of a deserter be established, as the moral crux of the play. The one is too frightened to seek out the body and denies to the SS man the identity of the hanged man. But the other goes out to cut her brother down from the butcher's hook not only in the faint hope that she may revive him, but also because she loves him. The prologue invites the audience to look for further parallels to the action in their own age.

Small touches promote the understanding of the political message. Thus Kreon is not addressed as 'King', but as 'Mein Führer'. Basically Brecht pursues the same aim as Hölderlin, who had adapted the Theban tragedy for a patriotic end: the ancient setting was to promote the spectator's detachment and to recognize 'den Geist der Zeit' in the way formulated by Hölderlin in his *Anmerkungen zur Antigone*. Brecht retains long stretches of Hölderlin's verse and does not allow his alterations to disrupt the rhythmic form. His *Antigonemodell* of 1948 comments further on the production. Brecht's *Antigone-Legende* represents a variant in epic form, using 'Brückenverse' in flexible hexameters.

Antimachiavel, L', a political treatise by Friedrich II (q.v.) of Prussia, written in 1739–40 and published anonymously at The Hague in 1740. It purports to refute *Il principe* by the Florentine Niccolo Machiavelli (1464–1527). Friedrich argues against tyranny and aggression and claims that the ruler must consider himself to be the first servant of the state ('le premier domestique de l'état'), balancing good and evil in man by his justice and skill. He concurs with Machiavelli in justifying the ruler's right to break treaties and to provoke war if he thereby prevents a situation which would endanger his country's safety, thus prophetically justifying his own aggression in 1740 and 1756 (see SCHLESISCHE KRIEGE and SIEBENJÄHRIGER KRIEG).

Antiqua, see FRAKTUR.

Antonie, see RAT KRESPEL.

Anton Reiser, a novel by K. P. Moritz (q.v.). The first volume appeared in 1785, the second and third in 1786, and the fourth and last in 1790. *Anton Reiser* bears the sub-title 'Ein autobiographischer Roman', and the book closely follows the first twenty-one years of the author's life (as he recalls them), and operates largely by introspection. Reiser has an unhappy childhood with a background of parental discord, is apprenticed to a pietistic hatmaker, proves unsatisfactory and is sent back to school. He runs away to go on the stage, but fails as an actor. The book leaves his problems unsolved. Moritz's intense psychological interest and perception explain the attraction which the book has exercised in the present century. He is absorbed by the significance of apparently trivial experiences ('wie dasjenige im Fortgange des Lebens sehr wichtig werden kann, was anfänglich klein und unbedeutend schien').

Anton Reiser, fünfter Teil (1794) is by a friend, J. F. Klischnig, and consists of recollections of Moritz in the years 1783–93.

ANTON ULRICH, HERZOG VON BRAUNSCHWEIG (Hitzacker, 1633–1714, Salzdahlum), second son of Duke Ernst August, was educated by J. G. Schottel and Siegmund von Birken (qq.v.), travelled extensively, and assisted his elder brother in the business of government, being admitted in 1685 to equal powers with him. His pro-French and anti-Hanoverian policy in the period leading up to the War of the Spanish Succession culminated in his eviction from Brunswick. In 1710 he was converted to Roman Catholicism, while safeguarding the rights of his Protestant subjects.

Anton Ulrich, who as 'Der Siegprangende' became in 1659 a member of Die Fruchtbringende Gesellschaft (q.v.), began by writing Protestant odes (*Christfürstliches Davids-Harfen-Spiel*, 1667) and composed the words for a number of court operas and ballets. (*Bühnendichtungen*, 2 vols., ed. B. L. Spahr, 1982). Two long novels convey his sense of aristocratic mission through a complex web of high-born characters. *Die Durchleuchtige Syrerin Aramena* (5 vols., 1669–73, ed. B. L. Spahr, Pt. I–IV, 1975, Pt. V, 1983), set in patriarchal times, presents the numerous characters (including 17 pairs of lovers) as a mirror of the state: Aramena's marriage provides a symbol of good government. The whole is a reflection of an exclusively aristocratic ideal. The basis of *Die römische Octavia* (6 vols., 1677–1707) is similar. Enacted in the time of Nero, it portrays the vicissitudes besetting 27 pairs of lovers, all carefully and hierarchically differentiated. Opinion about the merits of Anton Ulrich's elaborately constructed and heavily weighted novels is sharply divided. Leibniz (q.v.), who admired *Aramena*, may have found a mathematical attraction in the complex construction.

ANTSCHEL, PAUL, see CELAN, P.

ANZENGRUBER, LUDWIG (Vienna, 1839–89, Vienna), orphaned in 1844, grew up in straitened circumstances and in 1856 became an assistant in a bookshop. In 1860 he went on the stage, appearing first in Wiener Neustadt, then in Krems, and later in Steyr. All three theatres were in dire straits, and Anzengruber, whose talent was meagre, sank to touring in Austria and Hungary with the lowest kind of travelling troupe. He found in these depressing years the energy to write, achieving in 1864 a brief success with the play *Der Versuchte* (since lost). A Volksstück, *Glacéhandschuh und Schurzfell*, written in the same year and preserved though never acted, treats the clash of social class. Other early plays are lost. Anzengruber returned in 1868 to Vienna, and in 1869 he was glad to find a small post at police headquarters. In November 1870 his rural Volksstück in dialect, *Der Pfarrer von Kirchfeld* (q.v.), was acted with immense success, making him famous overnight. Its dialect successors, *Der Meineidbauer* (q.v., 1871), *Die Kreuzelschreiber* (q.v., 1872), and *Der G'wissenswurm* (1874), strengthened Anzengruber's reputation, but a falling-off became apparent in the next few years with the peasant farces *Doppelselbstmord* (1876) and *'s Jungferngift* (1878). In 1873 *Elfriede* (q.v.), a Schauspiel in High German dealing with marriage, had failed in the Burgtheater. Two other High German plays had peasant settings, *Hand und Herz* (1873) and *Der ledige Hof* (1877). Anzengruber turned to Viennese Lokalstücke and achieved with *Das vierte Gebot* (q.v., 1877) another resounding success, which was not maintained by its successors *Alte Wiener* (1878), *Ausm g'wohnten Gleis* (1879), and *Heimg'funden!* (1885). He then reverted to the rural setting with *Stahl und Stein* (1886) and *Der Fleck auf der Ehr* (1889), an adaptation of his Novelle *Wissen macht Herzweh*. Anzengruber wrote two novels, *Der Schandfleck* (q.v., 1877) and *Der Sternsteinhof* (2 vols., 1885), and a number of stories published in the collections *Bekannte von der Straße* (1881), *Feldrain und Waldweg* (1882), *Allerhand Humore* (1883), and *Wolken und*

Sunn'schein (1888). One volume of stories (*Letzte Dorfgänge*, 1894) and a Volksstück appeared posthumously (*Brave Leut' vom Grund*, 1892). His *Sämtliche Werke* appeared in 17 vols., ed. R. Latzke and O. Rommel, 1920–2.

Apart from his novels, Anzengruber's best work was done in the unpretentious and provincial Bauernstück, with its inclusion of songs and use of dialect speech. His best plays include a village philosopher such as Wurzelsepp (*Der Pfarrer von Kirchfeld*) or Steinklopferhans (*Die Kreuzelschreiber*). The Wiener Lokalstücke, apart from the powerful tragedy *Das vierte Gebot*, are secondary to the rural plays, but have a certain nostalgic charm. Though the narrow restrictions of the Volksstück have been thought to have hampered Anzengruber's development, he proved unable to write freely in High German forms.

APEL, JOHANN AUGUST (Leipzig, 1771–1816, Leipzig), became a lawyer and later a librarian in Leipzig. With F. A. Schulze (q.v.), he published a large collection of ghost stories (*Gespensterbuch*, 1810–17), one of which is the basis of Weber's opera *Der Freischütz* (q.v.). His early plays (*Polyidos*, 1805; *Die Aitolier*, 1806; *Kallirhöe*, 1806) were in the classical manner.

APEL, PAUL (Berlin, 1872–1946, Berlin), an actor and later a radio personality, wrote several plays, of which one, *Hans Sonnenstößers Höllenfahrt* (1911), achieved a nation-wide success.

Apis und Este, see *Throne stürzen, Die.*

Apollonius von Tyrland, see HEINRICH VON NEUSTADT.

APOSTATA, pseudonym of M. Harden (q.v.).

Apostel, Der, a short story by G. Hauptmann (q.v.), published in 1890. It is a fantasy of a man who, in his love of peace, desires to be like Christ. The themes and style of the story, which is a sketch for Hauptmann's novel *Der Narr in Christo Emanuel Quint* (q.v.), recall Büchner's Novelle *Lenz* (q.v.).

Apotheker Heinrich, a once widely read novel by H. Heiberg (q.v.), published in 1885. Heinrich, a well-to-do pharmacist, marries at 48 Dora, the 18-year-old daughter of a friend. Dora, sensitive and sentimental, has already fallen in love with a cousin. The loveless marriage with Heinrich wears her down until she escapes by drowning herself. The story is set in a town on the Baltic coast of Schleswig.

Appenzeller Reimchronik, a Middle High German account in verse of the war of liberation (Appenzeller Krieg) which the peasants of Appenzell successfully waged between 1401 and 1404 against the Austrian Abbot of St. Gall. The unknown author, whose literary gifts were slender, has adhered faithfully to the facts.

Aquis submersus, a Novelle by Theodor Storm (q.v.), first published in the *Deutsche Rundschau* in 1876. It is the first of Storm's 'Chroniknovellen', and is set in a frame (see RAHMEN) linking the present with a distant past. This is unfolded in two chronologically separated parts of a fictitious 17th-c. MS., for which Storm uses an appropriate archaic style. An 'Erinnerungsnovelle', it contains the recollections of Johannes, a German painter, who, for reasons of class, is prevented from marrying Katharina, the woman he loves. After the death of Katharina's sympathetic and enlightened father, her brother, Junker Wulf, tries to marry her against her will to Junker von der Risch. Wulf is portrayed as a ruthlessly autocratic character, symbolized by his ferocious dogs, to which he eventually falls victim. Wulf's persecution drives the lovers into union. In revenge Wulf fires at Johannes when he asks for Katharina's hand, and marries the girl off to a pastor, who accepts a pregnant bride in return for a living. She bears a son, and years after, when Johannes and Katharina unexpectedly come together again, the boy is drowned while they embrace. At the pastor's request Johannes paints the dead child's portrait and adds the letters c. p. a. s. (culpa patris aquis submersus) in token of his repentance. Although the story implies social criticism of the land-owning Junker class, it is primarily a tragedy of passion. A sense of fate contributes a subdued atmosphere, conveying the vanity of earthly happiness. This is underlined by the brief final section of the 'frame', which indicates that both the painter and his work are long forgotten.

Arabella, an opera, described as a 'lyrische Komödie', composed by R. Strauss (q.v.) with a libretto provided by H. von Hofmannsthal (q.v.). It is based on a Novelle by Hofmannsthal entitled *Lucidor,* and was first performed on 1 July 1933 at Dresden. Set in Vienna about 1860, it is a comedy of misunderstanding. Graf Waldner, his wife Adelaide, and their two daughters, Arabella and Zdenka, are in desperate financial straits because the count has gambled away the family fortune and continues to pile up debts. They are able to maintain appearances only by dressing Zdenka inexpensively as a boy, who, to the outside world, is known as Zdenko. Arabella is deaf to the pleas

of her suitor Matteo, an officer, who, however, receives moving love letters apparently from her, which in reality are written by Zdenka/Zdenko, who is deeply in love with Matteo.

Graf Waldner, in order to escape from his financial embarrassments, proposes to marry Arabella to Mandryka, a rich brother officer of his regimental days, and sends him a portrait of the girl. Instead of the expected old gentleman, a handsome young man of the same name appears and becomes a suitor for Arabella's hand. The original Mandryka has died, and this is his nephew and heir. Young Mandryka and Arabella fall in love, and the finances of the Waldner family appear to be restored. One difficulty remains: the love and jealousy of Matteo. At a ball he receives an assignation, accompanied by a key, apparently from Arabella. He keeps the assignation, which takes place in darkness. Mandryka suspects Arabella's fidelity, and a storm of anger looms over the family. At this moment Zdenka appears as the young girl she really is. It is she who has made and kept the assignation. By her confession she wins Matteo for herself and reconciles Mandryka and Arabella.

Aramena, see ANTON ULRICH, HERZOG VON BRAUNSCHWEIG.

ARBERG, GRAF PETER VON, see PETER VON ARBERG, GRAF.

Arc de Triomphe, a novel by E. M. Remarque (q.v.).

ARCHENHOLZ, JOHANN WILHELM VON (Langfuhr, Danzig, 1743–1812, Hamburg), a German historian, was first an officer, serving in the later campaigns of the Seven Years War (see SIEBENJÄHRIGER KRIEG), sustaining wounds and retiring in 1763 as a captain. After travelling extensively in Europe, Archenholz settled in 1780 in Magdeburg where he set up as a publicist, founding the periodical *Litteratur- und Völkerkunde,* which ran from 1782 to 1791. His *England und Italien* (1785) presents a balanced account of countries he had visited. He became known as a historian through the 20 vols. of *Annalen der britischen Geschichte* (1789–98) and his *Geschichte des siebenjährigen Krieges* (1787). From 1792 until his death he lived near Hamburg, editing the learned journal *Minerva.*

Archibald Douglas, a ballad written by Th. Fontane (q.v.) in 1854. Douglas overcomes the hostility of King James V to his family by his transparent and profound love of his homeland. The ballad has been set by Karl Loewe (q.v.), op. 28, 1857.

Archipelagus, Der, a poem by F. Hölderlin (q.v.) written in 1800. Though often termed an elegy because of its tone, it is not written in elegiac verse but in 296 classical hexameters. It laments the loss of the greatness of Greece and looks forward to a revival of Hellenic greatness in western Europe.

ARCHIPOETA, pseudonym of an unknown Latin poet of the 12th c. 'The Archpoet' was a German of lower nobility, a *ministeriale*, protected by Reinald von Dassel, Chancellor to the Emperor Friedrich I (q.v.). He is best known for a poem in praise of Friedrich and for his ironical confession *Estuans intrinsecus ira vehementi*, which contains the well-known song 'Mihi est propositum in taberna mori' (q.v.). It has been supposed that he was born *c.* 1130 and that his surviving poems were written between 1159 and 1167.

Ardinghello, und die glückseeligen Inseln, a novel published in 2 vols. in 1787 by J. J. W. Heinse (q.v.). A sub-title, *Eine italiänische Geschichte aus dem sechszehnten Jahrhundert* was omitted in the second and later editions. The preface (Vorbericht) is dated 1785. The novel stems from Heinse's years in Italy (1780–3), his enthusiasm for Renaissance painting, and his sensual attraction to women.

Ardinghello (in reality Prospero Frescobaldi, a Florentine painter living in exile in Venice) represents an ideal of natural man, in whom courage, energy, and sexual prepotency are united. In Venice he loves and is loved by the high-born Cäcilia. Her betrothed, suspecting the truth, hires assassins, but it is he who is murdered by Ardinghello, who then flees to Genoa. Here he rescues two ladies, Fulvia and Lucinde, from a galley which is carrying them off to Moorish slavery. Without forgetting Cäcilia, he becomes the lover of both, though his preference is for Lucinde. With another noble lady, Bianca, he goes to Rome, participates in a Bacchanalian orgy, and is smitten with the charms of the courtesan Fiordimona. In Rome he commits a second murder in the pursuit of his amours. Having made Italy rather too hot to hold him, he flees abroad and is enabled, with the help of a Moorish prince, who is under an obligation to him, to acquire the islands of Paros and Naxos. Here Ardinghello establishes a Utopia of love with a new religion of Nature, and is joined not only by new loves, but also by Cäcilia and Fiordimona. In Heinse's own words: 'So schwang die Liebe in allerhöchster Freiheit ihre Flügel'.

In spite of this abundance of incident, the story, which is chiefly told through letters and conversations, contains many descriptions of Italian paintings and lengthy discussions of art.

ARENDT, Erich (Neu-Ruppin, 1903–84, East Berlin), a contributor to *Der Sturm* and *Rote Signale* (qq.v., 1931), emigrated in 1933, returning to East Berlin in 1950. He is the author of several volumes of poetry, including *Trug doch die Nacht den Albatros* (1951), *Bergwindballade* (1952), *Gesang der sieben Inseln* (1957), and *Unter den Hufen des Windes* (1966, ed. V. Klotz). *Aus fünf Jahrzehnten. Gedichte* (1968) includes the political poetry written during Arendt's exile. His poetry, at one time influenced by his friend P. Celan (q.v.), reaches a climax in *Ägäis* (1967), the fruit of his travels to the Aegean. The poems were edited by G. Wolf as *Starrend von Zeit und Helle. Gedichte der Ägäis* (1980, DDR; 1981, BRD). *Zeitsaum* appeared in 1978, *entgrenzen* and *Das zweifingrige Lachen. Ausgewählte Gedichte 1921–1980* in 1981.

Among the poets he chose for translation (many in collaboration with Katja Hayek-Arendt) are the Chilean Pablo Neruda, the Spanish poets Rafael Alberti, Vincente Aleixandre, Nicolás Guillén, and Miguel Hernández, and Walt Whitman.

Argenis (1621), a Latin novel by John Barclay (q.v.).

Argonauten, Die, see GOLDENE VLIESS, DAS.

Ariadne auf Naxos, a one-act opera with a prelude (Vorspiel) composed by R. Strauss (q.v.) with a libretto provided by H. von Hofmannsthal (q.v.). The one-act opera (without Vorspiel) originally concluded, as an opera within a play, a version of Molière's *Le Bourgeois Gentilhomme* (in a German translation as *Der Bürger als Edelmann*). The single act of *Ariadne* presents the virtually static situation in which Ariadne laments her desertion by Theseus and awaits death, but Hofmannsthal gives it life and movement by introducing the figures of the *commedia dell'arte*, Zerbinetta, Harlekin, Scaramuccio, Truffaldin, and Brighella, who seek to console Ariadne and divert her thoughts. Ariadne remains disconsolate until the arrival of Bacchus renews her interest in life and love.

Der Bürger als Edelmann with the original *Ariadne auf Naxos* was first performed on 25 October 1912 in Stuttgart, and proved a failure. Hofmannsthal persuaded Strauss to detach *Ariadne* from the French comedy, and wrote for it a prelude, which shows the preparations behind stage for the entertainment and the dismay of the music master and the composer when they learn that their beautiful performance of *Ariadne* is to be combined with the antics of Italian comedians. In this revised form it was first performed on 4 October 1916, and ever since has held its place in the repertoire.

Arier denotes originally a member of the Indo-Iranian group of peoples which included the Medes, Persians, and Indians. In ancient Mesopotamia they constituted a ruling class. The word is also used to denote Indo-Europeans. In the 19th c. the linguistic and historical term was extended and virtually transferred to political use, coming to signify the Germanic peoples as predestined to be a master race, and acquiring a special anti-Semitic sense. *Arier* (adj. *arisch*), meaning non-Jewish, was regularly employed in official language under National Socialism.

ARIGO, pseudonym of the German translator of Tommaso Leoni's *Fiori di vertù* (1468) and of Boccaccio's *Decameron* (q.v., 1472). These translations, which were formerly attributed to H. Steinhöwel (q.v.), are now believed to be the work of Heinrich Schlüsselfelder, a patrician of Nürnberg.

Aristoteles und Fillis (Phyllis), an anonymous short Middle High German verse tale written in the late 13th c. Aristotle thwarts the love of his pupil, the boy Alexander of Macedon, for Phyllis, who in revenge inflames the desire of the philosopher. She persuades him to be saddled and bridled, and ridden by her in an orchard, where he is observed and mocked by the court. In disgust he withdraws and writes a book denouncing the wiles of women. The author models his style on Gottfried von Straßburg, from whose *Tristan* (q.v.) he lifts a number of lines.

ARMAND, pseudonym of F. A. Strubberg (q.v.).

Armee hinter Stacheldraht, Die, a war book by E. E. Dwinger (q.v.), published in 1929. Sub-titled *Sibirisches Tagebuch* it is, like many war books of the 1930s, on the borderline between novel and autobiography. It tells of the author's wartime experiences in captivity after being wounded while serving as a dragoon cadet on the eastern front in 1915. It is a story of intense and partly unnecessary physical suffering in hospital, followed by years of hardship and deprivation, first in southern Russia, then in Siberia. The Russian revolution brings hope of return, but the camp is taken over by the White Russians and repatriation is frustrated. It ends as the author resolves to escape. The book was a best seller. It is followed by two related novels, *Zwischen Weiß und Rot* and *Wir rufen Deutschland* (qq.v.).

Arme Heinrich, Der, a Middle High German poem, some 1,500 lines in length, by HARTMANN von Aue (q.v.). It was written *c.* 1195, after *Gregorius* and before *Iwein* (qq.v.), and is believed to mark Hartmann's emergence from a crisis to a harmonious state of mind. Its source is unknown. Heinrich is a nobleman of courtly demeanour and humane temper who in the midst of prosperity and happiness is stricken with leprosy. From a doctor at Salerno he learns that the only cure is the heart's blood of a marriageable virgin, willingly given. The daughter of the tenant farmer with whom the leper has chosen to lodge offers herself as a willing sacrifice, and insists, in spite of resistance by Heinrich and her parents. At Salerno she is bound naked on the operation table but Heinrich, hearing the sharpening of the surgical knife and perceiving her innocent beauty through a chink, bids the surgeon desist, to the distress of the girl. Heinrich, who has been moved by 'eine niuwe güete' (eine neue Güte), is, by the grace of God, cured of his leprosy, and the couple are married.

The poem, half legend, half fairy-tale, is one of the most attractive works of the Middle Ages in its combination of humanity and piety. The story has been more recently, and less successfully, treated by G. Hauptmann (q.v.) in his play *Der arme Heinrich* (1902).

Armeleutdichtung denotes plays (Armeleut-stücke), poetry, and novels having settings of poverty, which were popular in the early days of Naturalism (see NATURALISMUS). A. Holz, J. Schlaf, and G. Hauptmann (qq.v.) provide conspicuous examples in drama, Holz in poetry, and M. Kretzer (qq.v.) in the novel.

ARME MANN IM TOGGENBURG, DER, self-descriptive pseudonym of U. Bräker (q.v.).

Armen, Die, a satirical novel by H. Mann (q.v.), published in 1917, and dealing with the *Bürger* in relation to the proletariat. Though it is the second novel in Mann's 'Kaiserreich-Trilogie', it anticipated the complete publication of the first, *Der Untertan* (q.v.), by one year. The first edition was illustrated by Käthe Kollwitz (q.v.), the directness and integrity of whose compassionate drawings contrast with the acidity of Mann's oblique satire. Some of the characters of the first novel reappear, notably Diederich Heßling himself, now a prosperous factory owner exploiting his poverty-stricken workers. The plot is concerned with the attempt of a workman, Balrich, to prove that he, not Heßling, is the legal owner, and of Heßling's manœuvrings to frustrate this true claim. Eventually the house of Balrich's supporter, Professor Klinkorum, is burned down, and the flames consume the document supporting

Balrich's case. It is 1914, war breaks out, and Balrich is called up. *Die Armen* is a weaker novel than *Der Untertan*. *Der Kopf* (q.v.) completes the trilogy.

Armenbibel, see BIBLIA PAUPERUM.

Armer Konrad, designation of an association of peasants in 1514 in Württemberg directed against the extortions and misgovernment of Duke ULRICH, Herzog von Württemberg (q.v.).

Arme Spielmann, Der, a Novelle by F. Grillparzer (q.v.), begun in 1831 and published as 'Erzählung' in the Almanach *Iris* (q.v.) in 1848. The figure of the Spielmann is based on an old violinist whom Grillparzer had frequently heard while dining in the inn Zum Jägerhorn in the Dorotheergasse in Vienna; he presumably lost his life during the floods early in 1830.

In the opening section of the frame (see RAHMEN) the narrator introduces the Spielmann, who arouses his curiosity when he plays at the annual church festival in Brigittenau, a Viennese suburb. His conscientious but execrable playing makes him the poorest of street-musicians, and yet the narrator hears him speak with great sensitivity of the ennobling function of art. In the inner story the narrator listens to the life story of Jakob, the Spielmann, in his home, a single room which he shares with two artisans. The son of an influential civil servant, he had proved a misfit in the eyes of his ambitious father and, as it turned out, in practical life and society. Innocent and unsuspecting to the point of foolishness, he had allowed himself to be deprived of his heritage and possible marriage to Barbara, a grocer's daughter, who became instead the wife of a butcher. With her he lost the only sympathetic and helpful person in his life. Henceforth his violin became his only companion as he fiddled in the streets for a lean livelihood, and at home for God's pleasure and his own. He taught Barbara's son, Jakob, the song which had once awakened his love for her. When the narrator, in the final section of the frame, comes back to see him, the Spielmann is about to be buried. From his landlady he learns how he had met his death after rescuing children from the floods. Barbara becomes the loyal custodian of the old, worn violin which the narrator tries to purchase.

The confrontation of the narrator with Jakob (his 'Original') reflects the ambivalence of artistic striving with an ingenious mixture of sympathy and irony. The 'Spielmann' succeeds in preserving inner peace and harmony in his self-imposed isolation from an indifferent, materialistic, and coarse-grained society.

Arme Vetter, Der, an Expressionist play by E. Barlach (q.v.), completed just before the 1914–18 War, published in 1918 and, in 1919, with 36 lithographs. It was first performed in 1919 (Hamburger Kammerspiele).

On an Easter Sunday Hans Iver strolls along the banks of the Elbe. Unlike others who have come here from near-by Hamburg, Iver seeks not a temporary, but the final, escape from reality and intends to sacrifice himself to the spirit of the Resurrection. He shoots himself and is carried by passers-by to a near-by inn. Here his physical condition is not taken seriously. Among those he has already met in the open are Siebenmark and his fiancée Lena Isenbarn, who detects in Iver's suffering and spiritual longing her own identity, while Siebenmark vainly tries to break through the barriers of his own materialistic world. When Iver is exposed to the mockery of the inn guests, headed by the grotesque 'Frau Venus' (a male vet), and Lena stands up for him, Siebenmark makes a last attempt to secure Lena by tempting Iver to compromise with his world. He follows Iver into the night to settle his account with him by a sum of money (and thus in terms of his reality), but is rejected. Left to himself Iver opens his wound and bleeds to death. His body is discovered and brought into the barn where Siebenmark meets Lena, and puts before her the choice of him or the dead man. She chooses Iver, identifying herself with his message. The original title of the play was *Osterleute*. The final title underlines the symbolic relationship between Siebenmark and Iver.

ARMINIUS (*c.* 18 or 16 B.C.–A.D. 19 or 21), known in later ages as Hermann der Cherusker, was for a time an officer in the Roman service. On returning to his homeland in north Germany he led, as chief of the Cherussi, a revolt against the Roman forces and in A.D. 9 annihilated a Roman army under Varus in the Teutoburg Forest (see TEUTOBURGER WALD). Though defeated by Germanicus in A.D. 16, Arminius was able to maintain himself, and after Germanicus' withdrawal he re-established his power. He was assassinated by rivals in A.D. 21. His wife Thusnelda was captured by Germanicus and confined in Rome. Arminius' victory is commemorated by the Hermannsdenkmal.

As Hermann der Cherusker, Arminius has been celebrated in literature as a national hero since the middle of the 18th c. The more important works which have him as their subject are: Ulrich von Hutten's dialogue *Arminius* (1529); Lohenstein's novel *Grossmütiger Feldherr Arminius samt seiner durchlauchtigsten Thusnelda* (1689); J. E. Schlegel's play *Hermann* (1743);

Klopstock's poem *Hermann und Thusnelda* (1752) and his plays *Hermanns Schlacht* (1769), *Hermann und die Fürsten* (1784), and *Hermanns Tod* (1787); H. v. Kleist's *Die Hermannsschlacht* (1809); F. de la Motte Fouqué's play *Hermann* (1818); Grabbe's *Die Hermannsschlacht* (1835); and Otto Ludwig's fragment *Arminius* (1851).

Armut, Reichtum, Schuld und Buße der Gräfin Dolores, a novel published in 1810 by L. J. von Arnim (q.v.). Dolores, daughter of a count who has fled the country, lives with her sister in poverty in a ruinous castle. She marries the wealthy Graf Karl and leads with him a spoilt and discontented life. She deceives him but afterwards repents and is forgiven. Her father, returning from India with new-won wealth, becomes minister at court and persuades the ruling princess to visit Dolores and her husband, who are now living in Sicily. The princess falls in love with Graf Karl, who nevertheless remains faithful to Dolores. Dolores, however, still troubled in conscience by her own earlier infidelity, believes that she has lost his love. She dies of grief, and the princess poisons herself.

ARNDT, ERNST MORITZ (Rügen, 1769–1860, Bonn), born a Swedish subject, since Rügen belonged to the Swedish Crown, was the son of a bailiff, later turned farmer, and was at first educated at home. In 1786 he was sent to Stralsund grammar school and in 1791 he matriculated at Greifswald University, studying theology and history. His studies were interrupted by two years spent at home on Rügen (1794–6) and by a European tour embracing Austria, Italy, and France as well as many German states, most of which he carried out on foot (1798–9), for he possessed a strong physique and attached great importance to exercise. His first publication of note was an account of this tour (*Reise durch einen Teil Deutschlands, Italiens und Frankreichs in d. J. 1798 u. 1799*, 1801–3). He graduated in 1800 and immediately became a lecturer at Greifswald. In the same year he married, but lost his wife in childbirth in 1801. He married a second time in 1817. He visited Sweden in 1804 and again published an account of his journey (*Reise durch Schweden im Jahre 1804*, 1806). In 1806 he was appointed a professor and was immediately given leave to carry out historical research in Stockholm (1806–9). In the year of his arrival serfdom was abolished in Rügen and Swedish Pomerania, a measure which was attributable to Arndt's treatise *Versuch einer Geschichte der Leibeigenschaft in Pommern* (1803).

The political events of 1805–6, which established Napoleon's hegemony in central Europe, deeply affected Arndt, who conceived an intense desire to see a European awakening and rising, which quickly developed into the aim of a universal German crusade against Napoleon. His first important work on this theme is *Geist der Zeit* (1806), which was followed by a second volume in 1809 and a third in 1813. In 1810 Arndt visited Berlin under a false name in order to confer with anti-French personalities and arrived at the conviction, which he retained for life, that the salvation of Germany could only be achieved and retained by the predominance of Prussia. As the breach between France and Russia became inevitable, Arndt moved to Breslau and then, at the request of Baron vom Stein (q.v.), went to St. Petersburg, where he became Stein's principal anti-French propagandist, pouring out a succession of tracts, articles, and broadsheets with such titles as *Die Glocke der Stunde, Kurzer Katechismus für deutsche Soldaten, An die Preußen*, and *Was bedeutet Landsturm und Landwehr?* Arndt's single-minded fanaticism and his energetic, direct prose style made him particularly apt for this role. He had already written poetry (*Gedichte*, 1803), but he now appeared as a political poet animated by an Old Testament-like faith, and writing vigorous poems praising military virtues, hatred of the French enemy, and death for the fatherland. Published in *Lieder für Teutsche* and *Bannergesänge und Wehrlieder* (both 1813), these were among the most effective patriotic poems of the War of Liberation (see NAPOLEONIC WARS).

Arndt continued his propagandist activity throughout the war, but his views on German unity and constitutional government were suspect in high quarters. He soon became suspicious of the Congress of Vienna and openly attacked it. He had identified himself with Prussia and hoped as a reward to secure appointment to a chair at the new University of Bonn. This was granted to him in 1818, but in 1820 he was suspended on suspicion of subversive activity and, though no charge was preferred, he was not reinstated until the accession of Friedrich Wilhelm IV (q.v.), in 1840. He published his memoirs, *Erinnerungen aus dem äußeren Leben*, in 1840. In 1848 Arndt was elected to the new German parliament at Frankfurt (see FRANKFURTER NATIONALVERSAMMLUNG), but he lost interest when the Prussian king declined the imperial crown, and resigned in 1849. He continued to lecture at Bonn until 1854. His last publication of importance was his recollections of Stein, *Meine Wanderungen und Wandelungen mit dem Reichsfreiherrn Heinrich Carl Friedrich vom Stein* (1858, reissued, ed. W. Steffens, 1957).

Arndt's best writing is to be found in the clarity and vividness of his travel books and retrospective works. The undoubtedly sincere combination of religion and ruthless bellicosity in his patriotic songs is understandable in the historical circumstances of their genesis.

Arndt's poems were collected as *Gedichte* (2 vols., 1860), to which was added *Spät erblüht* in 1889. Select editions of his works were published in 1908 (16 vols., ed. H. Meisner and R. Geerds) and in 1913 (12 vols., ed. A. Leffson and W. Steffens).

ARNDT, JOHANN (Ballenstedt, Anhalt, 1555–1621, Celle), religious writer, a pastor's son who, after studying at Helmstedt, Wittenberg, Strasburg, and Basel, was appointed pastor at Badeborn near Quedlinburg. He was dismissed by his Calvinistic sovereign duke because of his Lutheran leanings. Subsequently he held cures in Quedlinburg, Brunswick, Halberstadt, and Eisleben, and in 1611 took up his final pastorate in Celle. He consistently endeavoured to instil living feeling into the rigid system of Lutheranism, and his mystic tendencies suggest an affinity with later Pietism (see PIETISMUS). His principal work is *Vom wahren Christentum*, which appeared in four books between 1606 and 1609. His *Paradiesgärtlein* (1612) is a book of edification and his *Postille* (1615) a book of devotions for home use. He also wrote a commentary on Luther's Catechism (*Luthers Katechismus*, 1617). His name occurs also as Arnd.

ARNIM, ACHIM VON, see ARNIM, L. J. VON.

ARNIM, BETTINA VON (Frankfurt/Main, 1785–1859, Berlin), *née* Bettina (Elisabeth) Brentano, was a daughter of Maximiliane Brentano, *née* La Roche, and granddaughter of Sophie von La Roche (q.v.). Clemens Brentano (q.v.) was her brother. Her mother died in 1793 and her father in 1797. From 1793 to 1798 she was educated at a convent and then went to live with her grandmother. In 1806 she visited Frankfurt and became friendly with Goethe's mother, whose recollections of her son she noted down. She first met Goethe in 1807 and remained in close contact with him until 1811 when, provoked by Bettina's behaviour to his wife, he formally severed all connection. In the same year she married her brother's friend Achim von Arnim (see ARNIM, L. J. VON), whom she had met in 1808. After his death in 1831 she settled in her Berlin home, frequenting literary circles which included Friedrich Jacobi, L. Tieck, F. Schleiermacher, the Grimm brothers, and the Humboldt brothers (qq.v.).

Her first book, *Goethes Briefwechsel mit einem Kinde* (q.v., 1835), was less a documentary collection than a free and imaginative rehandling of a correspondence. Her second publication, *Die Günderode* (1840), is a work of similar

character, based on the letters of her friend Karoline von Günderode (q.v.), whom she first met when staying with F. K. von Savigny (q.v.) in Marburg in the early 1800s. A third work involving the free adaptation of letters was devoted to her brother (*Clemens Brentanos Frühlingskranz*, 1844), and a fourth (*Ilius Pamphilius und die Ambrosia*, 1848) relates to a minor writer, Philipp E. Nathusius (1815–72), and the period of *c.* 1835–40. As she not only cut, compressed, and invented, but also destroyed originals, she is not popular with scholars, but these hybrid works have charm and freshness of feeling, and reveal Bettina's warmth of heart. As a widow she took an interest in social questions and politics, taking a progressive view of the emancipation of women and leaning towards socialism. *Dies Buch gehört dem König* (q.v., 1843) is a persuasive tract of generalized social reform backed with a section of individual examples. A sequel, *Gespräch mit Dämonen*, appeared in 1852. *Sämtliche Schriften* appeared in her lifetime (11 vols., 1853). *Das Armenbuch* remained unfinished (ed. W. Vortriede, 1969). W. Oehlke edited *Sämtliche Werke*, 7 vols., 1920–2; *Werke und Briefe* were edited by G. Konrad (5 vols., 1959–63).

ARNIM, GRAF HARRY VON (Moitzelsitz, Pomerania, 1824–81, Nice), Prussian diplomat who became involved in a clash with Bismarck (q.v.). Arnim was Minister in Lisbon in 1862, in 1864 at Munich, and in the same year at the Vatican. In 1872 he became ambassador in Paris. At odds with Bismarck over the latter's French policy, Arnim believed that he could hold his own through direct contact with the Prussian court. In 1874 he was recalled and placed on the retired list. A few months later he was arrested on a charge of retaining official documents. Condemned to a year's imprisonment, he fled abroad and attacked Bismarck's policy in a pamphlet which made use of official secrets (*Pro nihilo!*, 1876). For this he was condemned *in absentia* to five years' penal servitude. In 1880 he was granted a safe conduct in order to return to Germany to seek a new trial, but died before he could undertake the journey. The 'Arnim-Affäre' was one of the great scandals in the new Germany of the 1870s. It is referred to in Chapter 7 of Fontane's *Irrungen Wirrungen* (q.v.). See also HOLSTEIN, F. VON.

ARNIM, LUDWIG JOACHIM VON (Berlin, 1781–1831, Wiepersdorf), commonly known as Achim von Arnim, came of a well-known, distinguished, and wealthy Prussian noble family. His mother died at his birth, and he was brought up in Berlin by his grandmother. He had his schooling at the Joachimsthal Gymnasium, and then studied natural sciences and law at Halle (1798–9) and Göttingen (1800–1), where he formed a lasting friendship with Clemens Brentano (q.v.). In 1801 he set out on the grand tour, including in a three-year journey visits to Switzerland, Italy, France, England, and the Netherlands. His first works, the novels *Hollins Liebesleben* (1802) and *Ariels Offenbarungen* (1804), were derivative. After his return he settled in Heidelberg with Brentano, and the two friends devoted themselves to the collection of German folk-songs which, when published in 1805–8 under the title *Des Knaben Wunderhorn* (q.v.), was to make them acknowledged leaders of the Heidelberg Romantic school (see ROMANTIK). At the outbreak of war in 1806, Arnim published a volume of patriotic poetry entitled *Kriegslieder*. He also edited in Heidelberg the periodical *Zeitung für Einsiedler* (q.v.), which appeared in book form as *Tröst-Einsamkeit* (1808). He settled in Berlin in 1809, writing a collection of Novellen (*Der Wintergarten*, 1809), a play, *Halle und Jerusalem* (q.v., 1811), and the novel *Armut, Reichtum, Schuld und Buße der Gräfin Dolores* (q.v., 1810). It is in this work that he first appears as a truly independent author. Arnim was a Prussian patriot, and *Nachtfeier* (1810) is a mourning cantata written on the occasion of the death of Queen Luise (q.v.) of Prussia. In 1811 he married Bettina Brentano (see ARNIM, BETTINA VON), the gifted sister of his friend Clemens. When Prussia mobilized against Napoleon after the disastrous Russian campaign, Arnim volunteered for service, but the unit in which he was commissioned as a captain was almost immediately disbanded. Meanwhile he had published *Vier Novellen* (1812), of which the most important is *Isabella von Egypten* (q.v.). In the critical years 1813 and 1814 he supported the war effort with the proceeds of a volume of plays (*Schaubühne*, q.v., 1813) and with patriotic journalism. Disappointed with the political consequences of the war, he withdrew to his considerable estates at Wiepersdorf south-east of Berlin, and spent the remainder of his life managing them, as well as writing novels and plays. His unfinished novel *Die Kronenwächter* (q.v.) appeared in 1817, his play *Die Gleichen* (q.v.) in 1819, and he also wrote the Novellen *Der tolle Invalide auf dem Fort Ratonneau* (q.v., 1818) and *Die Majoratsherren* (1822). Possessing great facility, Arnim remained something of a dilettante all his life. A cultivated and urbane gentleman, he had a wide circle of friends among the men of letters of his day, including J. J. von Görres, F. K. von Savigny, L. Tieck, the brothers Grimm, A. von Chamisso, Adam Müller, and J. Kerner (qq.v.). He died suddenly of a stroke. Arnim's *Sämtliche Werke* (22

vols.), ed. W. Grimm, appeared 1839–56; *Sämtliche Romane und Erzählungen* (3 vols.), ed. W. Migge, 1962–5.

ARNOLD, ÊWART (*êwart* is a MHG word for 'priest'), a cleric who gives his name as the author of a Middle High German poem recounting the legend of St. Juliana. She withstands torments, binds the devil who has tried to tempt her in her prison to carnal sin, and is finally decapitated. Traditionally allotted to the first half of the 12th c., the poem was possibly written somewhat later. Êwart Arnold is no longer believed to be identical with the Priester ARNOLD (q.v.) who wrote *Von der Siebenzahl*.

ARNOLD, GOTTFRIED (Annaberg, Saxony, 1666–1714, Perleberg), Pietist and mystic, discovered while at Wittenberg University the great gulf between the Church and primitive Christianity and devoted his life to the fostering of a more truly Christian outlook. At Dresden, where he was a private tutor, he became friendly with P. J. Spener (q.v.) and, settling in Quedlinburg, sought to spread his new gospel through a circle of friends. His first work, dealing with primitive Christianity (*Die erste Liebe, d.i. wahre Abbildung der ersten Christen*), was published in 1696. For a short time in 1697–8 he was a professor of history at Gießen University, but soon returned to Quedlinburg and began the monumental work which made his name a household word for nearly a century. The *Unparteiische Kirchen- und Ketzerhistorie* (1699–1715) treats ecclesiastical history as a conflict between the tendency of the Church to petrify into rigidity and the efforts of heretics in every age to renew it. Arnold's ecclesiastical unorthodoxy exposed him to persistent attack, and in 1702 he secured a refuge in Prussia, where he was appointed historiographer royal. In his last years he wrote devotional books (*Episteln*, 1704; *Evangelien*, 1706; *Wahre Abbildung des inwendigen Christentums*, 1709). Arnold's numerous hymns, which were not collected in his lifetime, were especially popular in the pietistic community. The best known of them is 'So führst Du doch recht selig, Herr, die Deinen'.

ARNOLD, JOHANN GEORG DANIEL (Strasburg, 1780–1829, Strasburg), held a clerical appointment in his native city during the Revolutionary period (1795–8), studied afterwards in Göttingen and became a professor of law at Koblenz, migrating in 1811 to Strasburg. He wrote a comedy in Alsatian dialect, *Der Pfingstmontag* (1816).

ARNOLD, MÜLLER JOHANN, see ARNOLDSCHER PROZESS.

ARNOLD, PRIESTER, an Austrian priest of the 12th c. who composed a religious poem on the Holy Ghost, based on the mystical significance of the number seven (*Von der Siebenzahl zum Lobe des heiligen Geistes, c.* 1130). He is no longer believed to be identical with the Êwart ARNOLD (q.v.), who wrote a life of St. Juliana (*Die heilige Juliane*).

Arnoldscher Prozeß, a famous lawsuit in Prussia under Frederick the Great (see FRIEDRICH II, KÖNIG). Johann Arnold of Pommerzig was a miller, who, it was alleged, failed to pay his rent. He was taken to court in 1778, lost his suit and, through distraint, his mill. The King intervened in 1779 in the belief that an injustice had been done, quashed the judgement, reinstated the miller, and imprisoned the judges. The courts upheld the judges, but the King intervened a second time in support of the miller, dismissing most of the judges and sentencing them to a term of detention (Festungshaft). The case is a famous example of 'Kabinettsjustiz', of the overruling of the judiciary by the sovereign of a despotic state. The judges were later rehabilitated under Frederick's successor.

Arnsteiner Mariengebet, a Middle High German poem, which is an early document of the cult of the Virgin Mary. It begins with a dogmatic section on the Virgin Birth, and continues with a prayer of praise of Mary and intercession for repentant sinners. It was written in the second half of the 12th c. It has been suggested that its author was Guda, wife of Graf Ludwig von Arnstein, who founded the monastery at Arnstein on the River Lahn near Nassau.

ARP, HANS (Strasburg, 1887–1966, Basel), also Jean Arp, was a painter and graphic artist, a sculptor in stone and wood, and a poet who wrote French as well as German poetry. He studied in Weimar and Paris, and in 1911 joined the group of artists Der blaue Reiter (q.v.). His early verse is conventional. In 1916 he was one of the founders of Dada (see DADAISMUS) in Zurich. His later poetry is composed without heed to grammatical connections, proceeding in a series of dream-like word associations, the absurdity of which corresponds to Arp's conviction of the meaninglessness of the universe.

Arp's poetic publications (which often contain variants of poems already published) include *der vogel selbdritt, die wolkenpumpe* (both 1920), *Der Pyramidenrock* (1924), *Konfigurationen, Weißt du schwarzt du* (both 1930), *Des taches dans le vide* (1937), *Wortträume und schwarze sterne* (1953, contains a selection of his work from 1911 on), *Worte mit und ohne Anker* (1957), *Mondsand*

(1960), *Sinnende Flammen,* and *Vers le blanc infini* (both 1961). *Gesammelte Gedichte* (3 vols.) appeared 1963–84. *Unsern täglichen Traum* (1955) is autobiographical.

Arria und Messalina, a verse tragedy by A. Wilbrandt (q.v.).

Arthurian Romance, see ARTUS, KÖNIG.

ARTMANN, HANS CARL (St. Achatz am Walde, Austria, 1921–), usually referred to as H. C. Artmann, is an Austrian *avant-garde* writer and was from 1952 to 1958 a member of the Wiener Gruppe (q.v.). He is an experimenter with words, interested in montage and 'concrete poetry' (see KONKRETE POESIE) and in exploring the potentialities of baroque parody, of the grotesque, and of Viennese dialect. He published *reime verse formeln* (1954), *XXV epigrammata, in teuschen alexandrinern gesetzt* (1956), and *bei überreichung seines herzens* (1958). His dialect experiment *med ana schwoazzn dintn. gedichtar aus bradnsee* (1958) attracted much attention. He also collaborated with F. Achleitner and G. Rühm (qq.v.) in the volume of dialect poetry *hosn rosn baa* (1959). *Verbarium* followed in 1966, *allerleirausch* in 1967, and *ein lilienweißer brief aus lincolnshire. gedichte aus 21 jahren* in 1969. Artmann dislikes genres and it is difficult to classify his works, but *kein pfeffer für czermak* (1954) and *die mißglückte luftreise* (1955) have the designation 'kasperlstück', *Grünverschlossene Botschaft* (1967) is described as '90 Träume' and *tök ph' rong süleng* (1967) is termed 'roman-fragment'. His stories include *Die Anfangsbuchstaben der Flagge* (1968) and *Frankenstein in Sussex* (1969). Artmann applied the diversity of his style to his translations of works from various languages; their authors include Calderón, Lope de Vega, Molière, Beaumarchais, Marivaux, Goldoni, and T. H. White. *Grammatik der Rosen. Gesammelte Prosa* (3 vols.), ed. K. Reichert, appeared in 1979; the collection includes *Die Jagd nach Dr. U. oder Ein einsamer Spiegel in dem sich der Tag reflektiert* (1977) and *Nachrichten aus Nord und Süd* (1978). *Sämtliche persische Qvatrainen* (from the years 1967 and 1968) and a volume of stories, *Die Wanderer,* appeared in 1978, and the volume *Die Sonne war ein grünes Ei* and the prose of *Von der Erschaffung der Welt und ihren Dingen* in 1982.

Artmann spent a number of years travelling in various European countries, lived for a time in Sweden and in West Berlin, and in 1972 settled in Salzburg.

Artus, König, the legendary King Arthur of the Round Table. The historical Arthur was a

British military leader in war in the 5th c. or 6th c. His name later became a focus of Celtic legend, first fully developed in France in the 12th c. in the *matière de Bretagne*, which has its most conspicuous embodiment in the work of Chrétien de Troyes. Arthurian legend becomes the symbol of an ideal world, a frankly unreal but consistent panorama, which developed and exercised the highest courtly and knightly qualities. Arthurian romance begins in Germany in the late 12th c. with Hartmann von Aue's *Erec*, which is quickly followed by other poems, including a further romance by Hartmann (*Iwein*), Ulrich von Zazikhoven's *Lanzelet*, Wolfram von Eschenbach's *Parzival*, and Wirnt von Grafenberg's *Wigalois* (qq.v.). Gottfried von Straßburg's *Tristan* (q.v.) is not strictly Arthurian, but the world it portrays has close analogies with the Arthurian picture. With the waning of the short-lived courtly ideal in the first half of the 13th c., the Arthurian world loses its *raison d'être* and fades from German literature. In the 19th c. the quasi-Arthurian story of Tristan was revived by R. Wagner (q.v.) and early in the 20th c. E. Stucken (q.v.) wrote a series of plays on themes drawn from the legend.

Artushof, Der, a story by E. T. A. Hoffmann (q.v.), written in 1815 and published in 1816 in Brockhaus's *Urania*. In 1819 it was included in vol. 1 of *Die Serapionsbrüder*. The story begins in the hall known as the Artushof (King Arthur's court) in Danzig. Its theme is the discovery by a young businessman that his real calling is art. He breaks off his engagement to a potentially good housewife (Christina), and pursues in vain an ideal of womanly beauty (Felizitas), rejecting for her the passionate and loving Dorina. In the end he realizes that Felizitas is a symbol of art, not a woman to be possessed, and leaves for Italy to marry Dorina.

ARX, CAESAR VON (Basel, 1895–1949, Nieder-Erlinsbach nr. Aarau), became a professional theatrical producer, successively in Basel, Leipzig (1919–23), and Zurich (1923–5). After 1925 he lived in the village of Erlinsbach, devoting his time to the writing of plays. These include festival plays (Festspiele) for Solothurn (1922), Lucerne (1928), the Bernese Oberland (1926), and the Swiss Confederation (1941), and the plays *Die Geschichte vom General Johann August Suter* (1929), *Der Verrat von Novara* (1934), *Das Drama vom verlorenen Sohn* (1934; an adaptation of a play by H. Salat, q.v., 1535), *Der heilige Held* (1936, on Nikolaus von der Flüe, q.v.), *Brüder in Christo* (1947), and the comedy *Vogel, friß oder stirb* (1932). He befriended G. Kaiser (q.v.) while the latter was a refugee in Switzerland. He died by his own hand.

ASAM, COSMAS DAMIAN (Benediktbeuren, 1686–1739, Munich) and ÄGID (also EGID) QUIRIN (Tegernsee, 1692–1750, Mannheim), brothers who were notable architects and outstanding decorators of Bavarian churches from 1717. Among their best-known achievements are the abbey churches of Weltenburg, Aldersbach, Rohr, and Osterhofen. The brothers built at their own expense the exquisite church of St. Johann Nepomuk in Munich (1733–46, commonly known as the Asam-Kirche).

Aschenbrödel (Cinderella), one of the *Kinder- und Hausmärchen* (q.v., 1812–23) of the brothers Grimm (q.v.). The story has since been treated by Platen (q.v.), in *Der gläserne Pantoffel, Eine heroische Komödie* (1826) written in a mixture of prose and verse; by Grabbe (q.v.), in *Aschenbrödel, Dramatisches Märchen* (1830); and by Benedix (q.v.), in *Aschenbrödel* (1868).

Aschenmann, Der, the song 'So mancher steigt herum' in Raimund's *Das Mädchen aus der Feenwelt* (q.v.).

Asiatische Banise, Die, a once famous novel by Heinrich Anselm von ZIEGLER und Kliphausen (q.v.).

Askanier, a German princely house which at different times ruled in Brandenburg, Saxony, Lauenburg, and Anhalt. In Brandenburg the rule of the Askanier ended in 1319, but the family provided the ruling house in Anhalt until 1918.

Asklepiadische Strophe, a four-line strophe, in which the central portion of the first two lines is made up of two (sometimes three) choriambs (see CHORIAMBUS). The name is derived from the Greek poet Asclepiades (*c.* 300 B.C.). Klopstock occasionally uses the Asclepiadic stanza, e.g. in *Die Frühlingsfeier* (q.v.). The principal German exponent is Hölderlin (q.v.), who is the author of the following characteristic example:

Trennen/wollten wir uns?/wähnten es gut/und klug?
Da wirs/taten, warum/schröckte, wie, Mord,/die Tat?
Ach! wir kennen uns wenig,
Denn es waltet ein Gott in uns.
        (*Der Abschied.*)

ASMODI, HERBERT (Heilbronn, 1923–), made a name with his comedies *Pardon wird nicht gegeben* (1958), *Nachsaison* (1959), *Die Menschenfresser* (1961), and *Mohrenwäsche* (1964).

ASMUS, pseudonym used by M. Claudius (q.v.).

Aspern, a village in Lower Austria which since 1905 has been included in Vienna. It is on the left bank of the Danube, c. 5 miles from the city centre. Between Aspern and Eßling took place on 21–2 May 1809 the battle of Aspern in which the Austrians under the Archduke Charles (see KARL, ERZHERZOG) defeated the French under Napoleon. Both sides suffered exceptionally heavy losses. This was Napoleon's first defeat; its effect was quickly annulled by his victory on 5–6 July 1809 at Deutsch-Wagram (see NAPOLEONIC WARS).

ASSIG, HANS VON (Breslau, 1650–94, Schwiebus), after studying at Leipzig, served for some years with the Swedish navy, seeing action in war against the Danes. Towards the end of his life he held office in Schwiebus. He wrote occasional and religious poetry, which was published posthumously in 1719.

ASSING, LUDMILLA (Hamburg, 1821–80, Florence), was a niece of Varnhagen von Ense (q.v.), whose posthumous papers, letters, diaries, etc., she published between 1859 and 1870. She also wrote biographies of Luise, Gräfin von Ahlefeldt (1857) and Sophie von La Roche (q.v., 1859). She lived in Berlin from 1842 for many years and met or corresponded with some of the most notable figures of her time, including A. von Humboldt, F. Hebbel, G. Keller, and Fürst Pückler-Muskau (qq.v.). In 1861, because of an offence against the censorship, she was sentenced to imprisonment, which she avoided by settling in Florence.

ASSMANN, HANS, FREIHERR VON, see ABSCHATZ, HANS ASSMANN, FREIHERR VON.

Ästhetische Feldzüge, a series of lectures delivered at Kiel University in 1833 by L. Wienbarg (q.v.) and published in 1834. They came to be regarded as the programme of the Young German movement (see JUNGES DEUTSCHLAND).

Athenäum, Das, a literary periodical edited by F. and A. W. Schlegel (qq.v.), which appeared twice yearly from 1798 to 1800. It was the organ of the group of early Romantic writers (see ROMANTIK) which included the Schlegels, Novalis, and Schleiermacher (qq.v.), all of whom wrote for *Das Athenäum.* Particularly notable contributions are Novalis's *Blütenstaub* and *Hymnen an die Nacht,* and F. Schlegel's *Ideen, Fragmente,* and *Charakteristik der Meisterischen Lehrjahre von Goethe.*

Athis und Prophilias, an anonymous Middle High German poem existing only in fragments of a 13th-c. MS.; it was written, probably in Thuringia, about 1215. It is based on a French poem, *L'Estoire d'Athènes,* the theme of which is a story of friendship, probably oriental in origin, but transferred to the classical world (Athens and Rome). Prophilias loves Athis's wife so deeply and with such danger to his health that Athis yields her up to him. Athis is suspected of murder, and Prophilias sacrifices himself for him. The true murderers are discovered and Prophilias freed. Finally, Athis, after various vicissitudes, is rewarded with the love and hand of Prophilias's sister Gayte. The story is prolonged by a series of battles and loves concerning other characters. The German version is remarkable for its elegant, mannered style.

Atriden-Tetralogie, Die, a cycle of four plays by G. Hauptmann (q.v.), first published in this collected form posthumously in 1949. The individual plays are: (1) *Iphigenie in Aulis* (1943); (2) *Agamemnons Tod* (1948); (3) *Elektra* (1948); (4) *Iphigenie in Delphi* (1941). They are Hauptmann's response to the 1939–45 War, interpreting the course of events as the work of a blind and malevolent Fate, inspiring an equal malevolence in men. It has been suggested that 'Deutschland' could in many places be substituted for 'Hellas'.

Atta Troll, title of a satirical epic written by H. Heine (q.v.) in 1841 after his return from a holiday in the Pyrenees and first published in part in *Zeitung für die elegante Welt* (q.v.) in 1843. It appeared in full and final form as a separate publication in 1847. It is sub-titled 'ein Sommernachtstraum'. Written in rhymeless, trochaic, four-line stanzas and comprising 27 'chapters' (Kaput), it begins with an account of the escape of Atta Troll, a dancing bear, from his owner. The crude and unkempt Atta Troll is the object of satire directed against the German radicals of the 1830s and the bear himself castigates the failings of the human race. The poem then recounts the hunt for Atta Troll, but diverges for a time into a dream vision (Sommernachtstraum), a kind of phantasmagoria, in which three forms of the *femme fatale* appear to the hunter. In the end Atta Troll is shot and 'honoured' in an epitaph that parodies the literary style of Ludwig I, König von Bayern (q.v.):

Atta Troll, Tendenzbär; sittlich
Religiös; als Gatte brünstig;
Durch Verführtsein von dem Zeitgeist,
Waldursprünglich Sanskülotte;

Sehr schlecht tanzend, doch Gesinnung
Tragend in der zottgen Hochbrust;
Manchmal auch gestunken habend;
Kein Talent, doch ein Charakter!

ATTILA, King of the Huns, who died suddenly
in 453 (see ILDIKO), succeeded his uncle in 433.
In 451 he invaded western Europe, swept
across Germany and deep into France until
defeated near Châlons-sur-Marne by an army of
Goths, Burgundians, Franks, and Romans under
Aëtius. In 452 he invaded Italy. The impression
made in Germany by this brief but devastating
incursion was so profound that Attila (as Etzel)
became a legendary figure.

ATZE, GERHARD, a knight at the court of
HERMANN, Landgraf von Thüringen (q.v.), who
is authenticated in a document of 1196.
WALTHER von der Vogelweide (q.v.) wrote two
poems against Atze, who had shot a horse belonging to him.

*Auch eine Philosophie der Geschichte zur
Bildung der Menschheit,* a provocative essay
written by J. G. Herder (q.v.) in Bückeburg and
published in 1774. It is largely negative, denouncing the Aufklärung (q.v.) ideal of progress and lamenting the decline from original
nature, which represents the true order of
things.

Auch Einer, an eccentric novel by F. Th.
Vischer (q.v.), published in 2 vols. in 1879. The
(at first) anonymous hero A.E. (Auch Einer)
meets the narrator by chance on a Swiss tour.
The opening section of the first volume is
devoted to his angry and unending struggle
against the crosses imposed by the physical conditions of existence ('die Tücke des Objekts').
The second section is an inserted story, *Der
Besuch: Eine Pfahldorfgeschichte,* written by A.E.
and constituting, in the guise of a story about a
lake village, a satirical commentary on contemporary civilization. In the second volume the
narrator identifies A.E. as Albert Einhart and
learns of his violent death in a courageous but
quixotic act. The last section purports to be
A.E.'s diary and discloses the tragic story of his
love, revealing him, in Vischer's phrase, as a
man who has 'Kaliber', and comments now
humorously, now bitterly on human affairs. This
last part undoubtedly contains much of
Vischer himself.

Auch ich war in Arkadien, a short 'Phantasie'
by J. von Eichendorff (q.v.), published in 1834.
It is a political satire, made from a conservative
standpoint, ridiculing the radical trends of the

time. There is a central allusion to the Hambacher Fest (q.v.).

AUERBACH, BERTHOLD (Nordstetten, Black
Forest, 1812–82, Cannes), whose real name was
Moyses Baruch, came of a poor Jewish family
and was originally intended to be a rabbi. He
turned away, however, from Jewish theology to
study law and philosophy at Tübingen and
Munich universities. From the latter he was sent
down as a 'demagogue' (he was a member of the
Burschenschaft, q.v.) and in 1837 he was imprisoned for two months in the fortress of
Hohenasperg in Württemberg. After completing
his studies in Heidelberg he became associated
with the Young German movement (see JUNGES
DEUTSCHLAND), wrote a tract *Das Judentum und
die neueste Literatur* (1836) directed against W.
Menzel, the arch-enemy of Young Germany,
and collaborated in A. Lewald's magazine
Europa (1838–40). His first novels were *Spinoza*
(2 vols., 1837) and *Dichter und Kaufmann* (2 vols.,
1840), both historical. *Spinoza* was the expression of Auerbach's admiration for the
philosopher; *Dichter und Kaufmann* treats, with
much detail of Jewish life, the obscure 18th-c.
German-Jewish poet Ephraim Kuh (q.v.).
Auerbach established himself with *Schwarzwälder Dorfgeschichten* (q.v., 4 vols., 1843–53),
twenty stories in which the simplicity of rural
life is, by implication, elevated above the complexities and insincerities of urban existence, or
else the two contrasting environments are
brought into confrontation, as in *Die Frau
Professorin* (q.v., 1846). These stories caught on
with the public, and Auerbach, relieved of the
pressure of poverty, lived in various cities, including Dresden (1850–9), settling for the rest of
his life in Berlin. The volume *Schrift und Volk.
Grundzüge der volkstümlichen Literatur, angeschlossen
an eine Charakteristik J. P. Hebels* (1846) makes
plain his support of folk poetry and literature on
political and social grounds. In 1850 appeared
his tragedy *Andre Hofer* (see HOFER, A.). *Neues
Leben* (3 vols., 1852), a novel in which the
aristocrat becomes a village schoolmaster,
further reflects Auerbach's political and social
views. *Barfüßele* (q.v., 1856), *Joseph im Schnee*
(q.v., 1860), and *Edelweiß* (q.v., 1861), in spite
of sentimentality, are among his most attractive
stories with Black Forest settings. *Auf der Höhe*
(q.v., 3 vols., 1865) shows political conflict resolved in democratic harmony, and *Das Landhaus am Rhein* (q.v., 5 vols., 1869), with its modern
picture of a 'Napoleon of commerce', contains
some of Auerbach's most acute and convincing
portrayals of character. The political novel
Waldfried (3 vols.) appeared in 1874 and his last
novel, *Der Forstmeister* (2 vols.), in 1879. Auerbach's fiction has mostly been dismissed as

sentimental or didactic, but he possessed a real grasp of character and an ability to evoke the urban and Black Forest backgrounds which he knew so well. *Sämtliche Schwarzwälder Dorfgeschichten* appeared in 1871 in 8 vols., and in 1884 in 10 vols. *Gesammelte Schriften* (20 vols.) appeared in the 1850s (completed 1858), and *Werke* (22 vols.), 1863–4. A selection, ed. A. Bettelheim (15 vols.) appeared in 1913. *Briefe an seinen Freund Jakob Auerbach* (2 vols.) was published in 1884 with an introduction by F. Spielhagen (q.v.). Extant unpublished writings appeared in 1893 as *Dramatische Eindrücke*.

Auerbachs Keller, a noted tavern in Leipzig, dating from 1530. Its name derives from its first owner Heinrich Stromer, who was commonly known as Heinrich Auerbach, after his birthplace. In the 17th c. Auerbachs Keller became associated with the Faust legend, becoming the scene of the episode in which Faust rides up the stairs astride a barrel. Goethe's version of this scene in *Faust I* (q.v.) is headed *Auerbachs Keller*.

AUERNHEIMER, RAOUL (Vienna, 1876–1948, Oakland, U.S.A.), was an Austrian civil servant and reserve officer, who became a dramatic critic and 'cultural essayist' (Feuilletonist) for the *Neue Freie Presse* in Vienna. He wrote Novellen and a number of comedies, including *Die große Leidenschaft* (1905), *Der gute König* (1908), *Das Paar nach der Mode* (1913), and *Casanova in Wien* (1924). In 1938 he was for a short time in a concentration camp but found refuge in America. His novel *Metternich* was first published in English in 1940, then in German in 1947. He published a book on the dramatist F. Grillparzer (q.v.) in 1948.

AUERSPERG, ANTON ALEXANDER, GRAF VON, see GRÜN, ANASTASIUS.

Auerstedt, a village in the former Prussian Provinz Saxony, has given its name to the battle fought there on 14 October 1806 between the French and a Prussian army under Duke Karl von Braunschweig (see BRAUNSCHWEIG). The Duke was mortally wounded and his army totally defeated. This disaster, coinciding with that at Jena (q.v.) on the same day, ended effective Prussian resistance to Napoleon for six years (see NAPOLEONIC WARS).

Auf das Grab von Schillers Mutter, a poem consisting of six distichs written by E. Mörike (q.v.) in 1835. Schiller's sister Luise was the wife of Pastor Franck, Mörike's predecessor in Cleversulzbach, and Frau Schiller, the poet's

mother, died there in 1802 and was buried in the churchyard.

Auf dem See, a poem written by Goethe during his Swiss journey in the early summer of 1775. The occasion was a trip by boat on the Lake of Zurich on 15 June, and the poem refers to Goethe's love for Lili Schönemann (q.v.) and his attempt to escape from it. It was published in 1789 in a version slightly different from the original form and opening with the lines 'Und frische Nahrung, neues Blut/Saug' ich aus freier Welt'.

Auf dem Staatshof, a Novelle by Th. Storm (q.v.), written in 1858 and published in *Argo* in 1859. It describes through its narrator, Marx, the eclipse of a family of landowners, the van der Bodens, whose last descendant, Anne Lene, inherits the impoverished Staatshof. Her fiancé, a fortune-hunting nobleman, abandons her, and at a party not long after she falls through the rotten floorboards of a garden pavilion and drowns in the stream which flows beneath it. No one knows whether her death was accidental or intentional. After her death the Staatshof is demolished and a new house built.

Auf dem Wasser zu singen, a poem written by F. L. von Stolberg (q.v.) in 1782 shortly after his marriage. A rhythmic barcarolle in three stanzas, it was set to music by Franz Schubert (q.v.) in 1822.

Auf den Marmorklippen, an allegorical fantasy by E. Jünger (q.v.) in the form of a short novel. It was published in 1939. The peaceful state of Marina is insidiously attacked and overwhelmed by the brutal destructive forces of the elemental 'Oberförster'. See INNERE EMIGRATION.

Auf der Höhe, a three-volume novel by B. Auerbach (q.v.), published in 1865. The symbolical title alludes to the rural inhabitants of the hills and also to court life on the heights of human society, which are the two environments involved in the novel. The young queen of an unnamed German state is about to bear a child, and a woodcutter's wife from the hills, Walpurga, is persuaded to become the wet nurse. Walpurga's *naïveté* and integrity are contrasted with the sophistication and complex formalism of the court. The King, a handsome young man of good intentions and strong character, is attracted to a lady-in-waiting, the young Countess Irmgard, who is friendly towards Walpurga. Court gossip brings this *affaire* to the notice of Irmgard's father, the Liberal Count Eberhard von Wildenort, and causes him to die of a stroke. Irmgard takes flight and is believed drowned, but is actually succoured and sheltered by Walpurga, whose husband, thanks to her earnings at court, has bought a farm (the Freihof) in the hills. Here Irmgard lives in simplicity and seclusion, beset by remorse. The King's marriage, though maintained in outward appearance, has become a nullity. He becomes involved, through his autocratic temperament, in a constitutional crisis, but subdues his natural inclinations and appoints a Liberal minister in accordance with the elections. On her death-bed Irmgard is reconciled with the Queen and a reconciliation then follows between the royal spouses. *Auf der Höhe* shows acute perception of character and alert analysis of social factors. It is arranged in eight books, the seventh of which is the diary kept by Irmgard during her four years of repentance.

The figure of Walpurga was caricatured in *Walpurga die taufrische Amme*, one of the parodies in *Nach berühmten Mustern* by F. Mauthner (q.v.).

Auf der Universität, a Novelle by Th. Storm (q.v.), written in 1862 and published in 1863. Philipp, the narrator, relates in eight sections his memories of the low-born daughter of a French tailor and a German mother, Lenore (Lore) Beauregard; the heading of the fifth section provides the story's title. Lore's exotic southern grace and beauty attract the attention of the sons of a select upper-middle-class circle of a class-conscious small town at a dancing class, at which Lore stands in for a missing partner. Through this experience she becomes torn between hatred and longing for a social class from which her parents are excluded. Her engagement to Christoph, a sturdy, good-natured carpenter, opens the prospect of a settled future. But when she believes a false report of Christoph's unfaithfulness in a distant town, she abandons her self-respect and gives herself up to a life of pleasure and luxury with a wealthy but irresponsible student known as the Raugraf. On receiving an invitation from Christoph to marry him, she drowns herself in the lake near the students' dance hall.

Auf ein altes Bild, a poem written by E. Mörike (q.v.) in 1837. The poet contemplates a picture in which the infant Christ and the Virgin appear in the foreground, while further back is seen the tree from which the Cross will be fashioned. The poem, of only six lines, has been set to music by Hugo Wolf (q.v.).

Auf eine Christblume, two linked poems by E. Mörike (q.v.). Mörike found the flower (Christmas rose, *Helleborus niger*) growing on a grave in Neuenstadt cemetery in October 1841.

Auf eine Lampe, a poem in ten lines of antique trimeters, written by E. Mörike (q.v.) in 1846. It affirms the intrinsic value of a work of art. Its last line, 'Was aber schön ist, selig scheint es in ihm selbst', gave rise in 1951 to a critical controversy, in which, among others, E. Staiger and M. Heidegger (q.v.) were involved.

Auf einer Wanderung, a poem in irregular rhyming verse written by E. Mörike (q.v.) in 1845. It conveys with great intensity the experience of listening to a beautiful singing voice, by which the whole world seems to the poet momentarily transfigured. It was set to music by Hugo Wolf (q.v.). An earlier version, much longer and greatly inferior, bore the title *Auf zwei Sängerinnen* (1841). Mörike also used the title *Auf einer Wanderung* for a poem of different character written in 1856–8.

AUFFENBERG, JOSEF, FREIHERR VON (Freiburg/Breisgau, 1798–1857, Freiburg), a descendant of a great noble family, made an unsuccessful attempt in 1815 to reach Greece in order to fight for the Greek cause. He then joined the Austrian army for a short time, transferring to the Grand Ducal Guard in Baden. In 1817 he was encouraged by J. Schreyvogel (q.v.) to write for the stage and wrote in the next twelve years or so a number of verse tragedies after the manner of Schiller. They include *Die Flibustier*, *Wallace*, *Der Admiral Coligny*, *Die Syrakuser*, *König Erich*, *Die Verbannten* (all 1819–21) and *Pizarro*, *Die Spartaner*, *Viktorin* (all 1823). He also published in 1827 three plays based on novels by Walter Scott, *Fergus MacIvor* (*Waverley*), *Der Löwe von Kurdistan* (*The Talisman*), and *Ludwig XI in Peronne* (*Quentin Durward*). His novel *Die Furie von Toledo* appeared in 1832, and a semi-autobiographical work, based on a visit to Spain in 1832, *Humoristische Pilgerfahrt nach Granada und Cordova*, was published in 1835. In 1839 he was appointed court chamberlain at Karlsruhe. His complete works (*Sämtliche Werke*, 1843–5) occupy 23 vols.

'**Auf Flügeln des Gesanges**', first line of an untitled poem by H. Heine (q.v.) included in the *Buch der Lieder* (q.v.) as No. IX of 'Lyrisches Intermezzo'. It has been set to music by F. Mendelssohn (q.v.).

Aufgeregten, Die, a play begun by Goethe in 1791 and left unfinished. It is described as 'Politisches Drama in fünf Akten', but the fifth act is given only in scenario form. It was published in 1817. The play, conceived as a comedy, represents the impact of the French Revolution on the tensions of a German village.

Aufgesang, see MINNESANG.

Auf halbem Wege, a political novel by E. E. Dwinger (q.v.), published in 1939. It deals with the Kapp-Putsch (q.v.) and the Ruhr revolt of 1920 from a pronounced right-wing standpoint. Prominent personalities, including President Ebert, Noske, Kapp (qq.v.), Jagow (1865–1941), and General von Lüttwitz (1859–1942), are introduced as characters, as well as figures from the earlier post-war novel *Die letzten Reiter* (q.v.).

Aufhaltsame Aufstieg des Arturo Ui, Der, a play (Parabelstück) with 15 scenes, prologue (two versions), and epilogue by B. Brecht (q.v.), written in 1941, for the most part in blank verse. It was published posthumously in *Stücke*, vol. 9. It was not finally revised by Brecht, and further publications have taken account of amendments found among his papers at a later stage. The action takes place in Chicago and Cicero, and portrays the methods by which the chief gangster Arturo Ui and his accomplices Ernesto Roma, Emanuela Giri, and Givola rise to power in an economic crisis. The point of the transparent parable of Hitler's rise to power is made more explicit in the added *Zeittafel* covering the years 1929–38, from the 1929 world crisis up to the annexation of Austria. The episodes include Hitler's dealings with Hindenburg (q.v.), Dogsborough of the play, his rise to power, for which he prepares by taking lessons in speech and gesture from an actor, the burning of the wholesalers' warehouse, an allusion to the Reichstag, and the trial of the supposed fire raiser, Fish. The play also takes in the murder of one of the gangsters, Ernesto Roma, and his companions, a reference to the assassination of E. Röhm (q.v.) on 30 June 1934. The parody of the garden scene of *Faust I* in scene 12 depicts Hitler's negotiations with the Austrian Chancellor E. Dollfuß (q.v.), Ignatius Dullfeet in the play. Dullfeet represents Frau Marthe, his wife Betty is Gretchen, Givola (Goebbels, q.v.) in his flower garden corresponds to Mephisto, and Ui (Hitler) to Faust himself. The sudden death and state funeral of Dullfeet precede the final scene, which demonstrates Ui's 'free' election. He thus returns in triumph from Chicago to Cicero, which he had left as a poor man, and now re-enters with his army. German readers will have gathered well before the last couple of lines that the name of Ui is derived from 'Pfui!', an untranslatable expression of disgust.

Aufklärung, 'Enlightenment', term applied to the German phases and aspects of the European movement of rationalism and humanitarianism which extended from the middle of the 17th c. to the beginning of the 19th c. Its systematic philosophical ancestor is Descartes, its first political thinker Hobbes, and its first psychologist Locke. The American Declaration of Independence and the French Revolution are its most obvious political manifestations. The scientific aspect of this age is less conspicuous in Germany.

The Aufklärung in Germany is usually said to begin late in the 17th c. with Christian Thomasius (q.v.), a professor at Leipzig and later at Halle, who viewed his task as the broad propagation of an attitude of common-sense rationalism and humanitarianism. To this end he lectured in German, not in the traditional Latin, and in this spirit he strongly opposed the widespread witch trials and burnings of the age. The first German philosopher of the Aufklärung was Leibniz (q.v.), but its earliest influential thinker was Christian Wolff (q.v.), who propounded an easily understandable philosophical system of popularized rationalism. The climax of the philosophical Aufklärung is the work of Immanuel Kant (q.v.), who also provided its most famous definition: 'der Ausgang des Menschen aus seiner selbstverschuldeten Unmündigkeit . . . *Sapere aude!* Habe Mut, dich deines eigenen Verstandes zu bedienen! ist daher der Wahlspruch der Aufklärung' (*Beantwortung der Frage: Was ist Aufklärung?*). The spread of 'Enlightenment' coincided with a rise in the cultural and economic importance of the middle classes (Bürgertum), who provided its principal support and its most characteristic exemplars. It is particularly concerned with social questions within the framework of this class, notably with the position of women.

The first phase of the Aufklärung in literature (1700–50) is marked by the broadly didactic poetry of B. H. Brockes (q.v.) and the efforts of J. C. Gottsched (q.v.) to establish a practical system of poetry and drama based on a common-sense application of French principles. The narrowness of Gottsched's views is overcome by the broader view of the Swiss theorists J. J. Bodmer and J. J. Breitinger (qq.v.). With C. F. Gellert (q.v.) the age became conscious of emotion, praising controlled feelings, which commonly expressed themselves in sentimentality (Empfindsamkeit, q.v.), and rejecting uncontrolled passion.

The highest phase of the literary Aufklärung was reached in the years 1750–80. One side of it is expressed in the urbane and civilized writings of C. M. Wieland (q.v.), the other in the forthright intellectual clarity of G. E. Lessing (q.v.), who swept away aesthetic and ethical prejudices, creating a new German drama in *Emilia Galotti* and, in *Nathan der Weise* (qq.v.), powerfully advocating religious tolerance.

The Sturm und Drang (q.v.) and the new Classicism of Goethe and Schiller (qq.v.) reacted against the Aufklärung, but both are considerably indebted to its ethical and aesthetic advances.

The pedestrian aspects of the Aufklärung, as represented in the inflexible dullness of the ageing F. Nicolai (q.v.), became prominent at the end of the 18th c. and its late manifestations were ridiculed by Heine in *Die Harzreise* (q.v., 1827). Many of its fundamental, ethical, and social ideas, together with the literary realism which it half-consciously fostered, however, remain as an essential part of the heritage of the 19th c. and 20th c.

Auf meines Kindes Tod, a cycle of three poems by J. von Eichendorff (q.v.), occasioned by the death of his 2-year-old youngest daughter in 1832. The best known of the three is the poem beginning 'Von fern die Uhren schlagen'.

Auf Miedings Tod, an obituary poem written by Goethe in February and March 1782 and printed in the same year. The occasion was the death of the court carpenter, Johann Martin Mieding, who had acted as stage manager and general handyman for the amateur theatrical performances of the Weimar court.

Aufruf an mein Volk, see AN MEIN VOLK.

Aufstand der Fischer von St. Barbara, Der, a story by Anna Seghers (q.v.), published in 1928. It deals with the revolt of a community of Breton fishermen against their employers because of low wages. After lives have been lost and damage done, the rising collapses because the fishermen at other ports fail to maintain solidarity.

Aufstieg und Fall der Stadt Mahagonny, an opera with 20 scenes by B. Brecht and K. Weill (qq.v.), written 1927–9, published in 1929 and first produced at Leipzig on 9 March 1930. Partly in prose and partly in free verse, it is based on Brecht's earlier Songspiel (*sic*) *Mahagonny* (1927) and is, as stated in *Versuche,* vol. 2, 'ein Versuch in der epischen Oper: eine Sittenschilderung'. Three rogues (Willy der Prokurist, Dreieinigkeitsmoses, and Frau Leokadja Begbick) found the get-rich-quick city of enjoyment Mahagonny ('die Goldstadt'). It does not, however, prove a success until the woodcutter Paul Ackermann from Alaska proposes a ban on all prohibitions ('Vor allem aber achtet scharf/Daß man hier alles dürfen darf'), whereupon prosperity comes to Mahagonny, where money is the only criterion. Paul Ackermann becomes penniless, and, since this is of all crimes the worst, he is executed for failing to pay for whisky he has consumed. The city burns and the inhabitants chant their songs of pleasure and,

above all, money. It is right, so we are shown, that Mahagonny should be destroyed, for money brings no freedom. It is also inevitable that the city, which has been spared by the hurricane, is destroyed by the more powerful corruption of human nature among the inhabitants of the city.

Aufzeichnungen des Malte Laurids Brigge, Die, a novel by R. M. Rilke (q.v.) written in 1904, 1908–9 and published in 1910. The eponymous hero, a 28-year-old Danish poet of noble origin, lives in poverty in Paris, where he relates in a series of episodes his experience, past and present. Deliberately choosing formlessness and diverging from the traditional form of the novel, Brigge's authorship and approach form the unity of the novel, which is influenced by Rilke's review of the posthumous writings of the Norwegian poet S. Obstfelder and the Danish writer J. P. Jacobsen. The novel is written in poetic style depicting, in fleeting impressions of Parisian scenes, symbols of suffering, sickness, poverty, ugliness, horror, and fear (e.g. a blind trader, pregnant women, neglected babies, bleak hospital corridors, the remaining wall of a demolished house). The poor, the beggars, are to Malte outcasts, refuse of fate, like himself. Sickness intensifies his imagination, re-creating moments of horror in the past (Christine's ghost, the play with masks) and probing the many faces of death. Childhood reminiscences, occupying large tracts of the novel, open with the death of Malte's grandfather, Christoph Detlev Brigge, and culminate in a surgical stab into the heart of his dead father following a request of the deceased. Death meets him in the streets and in books depicting historical figures (Grischa Otrepjow and Charles the Bold). Malte's love focuses on Abolone, a younger sister of Countess Sibylle, his dead mother. He cherishes an ideal of woman which he finds already in Sappho's lament of love, the essence of which he sees in a woman's excess of self-surrender. An elderly friend, translator of fragments of Sappho's poetry, comprehends the unity of the ancient classical age. In this perspective he understands a young girl, who expects from the consummation of love the increasing loneliness of the lovers. The idea that love enhances separateness rather than togetherness occurs in further variation in the song of a young woman in Venice and in the free adaptation of the parable of the Prodigal Son with which Brigge's notes break off.

Aufzug der Romanze, Der, see KAISER OKTAVIANUS.

Augsburg, historic German city included in Bavaria since 1806. Of Roman origin, it developed from the military *colonia* Augusta Vindelicorum, founded in 15 B.C. In 1276 it became a Free Imperial City, losing this status only with the dissolution of the Empire (1806). It owed its immense prosperity in the later Middle Ages to its situation on the trade route between Italy and northern Europe. The powerful Augsburg merchant family of Fugger (q.v.) was responsible for the Fuggerei, an early experiment in social welfare. In 1530 Augsburg was the scene of the Diet in which the differences between Catholics and Protestants were debated, and the Protestant document prepared for this occasion is known as the Augsburgische Konfession (q.v.); and in 1555 the Religious Peace (see AUGSBURGER RELIGIONSFRIEDE) was signed there. Agnes Bernauer (q.v.), the subject of plays by Hebbel (q.v.), and others, was the daughter of a barber surgeon of Augsburg, which is also the birthplace of B. Brecht (q.v.).

Augsburger Chronik, see MEISTERLIN, SIGMUND and ZINK, BURCHARD.

Augsburger Gebet, title by which an Old High German prayer is known. It consists of four lines of rhyming verse and is a translation of a Latin prayer. It was probably written at the end of the 9th c. or early in the 10th c. The MS., formerly in Augsburg, is now in Munich.

Augsburger Kreidekreis, Der, a story by B. Brecht (q.v.) set in Augsburg during the Thirty Years War, and included in the collection *Kalendergeschichten* (q.v.). It was written before 1940 and published in Moscow in 1941. In spite of the difference in place and time, it forms a basis for the play *Der kaukasische Kreidekreis* (q.v.).

Augsburger Liederbuch, a collection of folksongs made by one Erhart Oeglin and printed in 1512.

Augsburger Passionsspiel, a passion play, probably of the 14th c., which is one of the sources of the *Oberammergauer Passionsspiel* (q.v.).

Augsburger Reichstage. The most important of the Imperial Diets held at Augsburg were that of 1530 (see AUGSBURGISCHE KONFESSION) and that of 1555 (see AUGSBURGER RELIGIONSFRIEDE).

Augsburger Religionsfriede, a settlement of the religious problems of 16th-c. Germany, promulgated as an imperial law on 25 September 1555. Its principal provisions were: (1) confirmation of existing rights of territorial sovereignty irrespective of religion; (2) that the religion of a sovereign should determine the religion of his subjects (this decision, summed up in the legal Latin tag *cuius regio, eius religio*, is the best-known clause of the Peace); (3) that subjects refusing to conform must emigrate; (4) that sovereigns who changed their religion should forfeit their throne. See also AUGSBURGISCHE KONFESSION and LANDESKIRCHEN.

Augsburgische Konfession, document submitted to the Diet of Augsburg in 1530 as a joint statement of the beliefs held by the Protestant princes and certain cities. It was drawn up in both German and Latin (*Confessio Augustana*), was printed in the same year and widely circulated. An authentic text was established by Melanchthon (q.v.), who, however, modified subsequent editions, notably in 1540, when an unsuccessful attempt was made to accommodate the Calvinists. From 1560 onwards bitter controversy raged between the adherents of the original and later versions.

The *Augsburgische Konfession* was recognized in 1555 in the Religious Peace of Augsburg (see AUGSBURGER RELIGIONSFRIEDE) as the doctrinal basis of the Protestant Church in Germany.

AUGUST II, DER STARKE, King of Poland and, as Friedrich August I, Kurfürst von Sachsen (Dresden, 1670–1733, Warsaw), was the second son of the Elector Johann Georg III, and became elector in 1694 and king of Poland in 1697. He had temporarily to cede the crown to Stanislas Leszczyński. In his residences in Dresden and Warsaw August sought to emulate Louis XIV. Meißen (Dresden) porcelain was invented in his reign by Tschirnhaus und Böttger. He left only one legitimate son (see AUGUST III), but had numerous illegitimate children, one of whom, Moritz von Sachsen (see SAXE, MAURICE DE, MARÉCHAL), achieved military fame in French service. The adjective 'strong' referred, not to his character or policy, but to his immense physical prowess.

AUGUST III, King of Poland and, as Friedrich August II, Kurfürst von Sachsen (Dresden, 1696–1763, Dresden), was elected in 1733 king of Poland. He was the only legitimate son of Augustus the Strong (see AUGUST II). Married to a daughter of the Emperor Joseph I (q.v.), he had, like his father, become a Roman Catholic. In the Polish War of Succession (1733–5) he was successful against his rival claimant Stanislas Leszczyński. In the Austrian War of Succession (see ÖSTERREICHISCHER ERBFOLGEKRIEG) and in the first Silesian Wars (see SCHLESISCHE KRIEGE) he supported Austria against Prussia without profiting by it, having to capitulate soon after the opening of the Seven Years War (see SIEBENJÄHRIGER KRIEG). As Saxony was occupied by Friedrich II (q.v.) of Prussia, the luckless Augustus fled to Poland.

AUGUSTENBURG, FRIEDRICH, HERZOG VON (Augustenburg, 1829–80, Wiesbaden), a claimant to the ducal throne of Schleswig and Holstein, when it became vacant by the death of King Frederick VII of Denmark without male issue. Augustenburg's claim was obstructed by Bismarck (q.v.), and the duchies were included in Prussia as the new Provinz Schleswig-Holstein in 1866. The full style of the Duke was Herzog zu Schleswig-Holstein-Sonderburg-Augustenburg. See also SCHLESWIG-HOLSTEINISCHE FRAGE and FRIEDRICH CHRISTIAN, HERZOG VON AUGUSTENBURG.

AUGUSTIN, ERNST (Hirschberg, Silesia, 1927–), qualified as a doctor, and later travelled in the Indian sub-continent. He is the author of three novels, which have been compared by some to Kafka's fiction: *Der Kopf* (1962), *Das Badehaus* (1963), *Mamma* (1970). The novel *Raumlicht: Der Fall Evelyne B.* appeared in 1976, *Eastend* in 1982.

AUGUSTIN VON HAMERSTETEN, a 15th-c. knight who was at first in the service of the Emperor and then, after 1490, in that of the Elector of Saxony at Torgau. He wrote a prose romance with the title *Die Hystori vom Hirs mit den guldin ghurn* (Hirsch mit dem goldenen Gehörn) *und der Fürstin vom pronnen*. It is a story of a knight, whom a princess persuades to go to the Holy Land. In due course he returns and is received into her favour.

Aurea Gemma, see LUCIDARIUS.

Aus dem bürgerlichen Heldenleben, a cycle of social comedies by C. Sternheim (q.v.).

Aus dem Leben eines Taugenichts, a story (Novelle) by J. von Eichendorff (q.v.), published in 1826. The first chapter had previously been published in 1823 in the periodical *Deutsche Blätter.* The idle 'good-for-nothing' of the title is sent packing by his father and goes light-heartedly wherever chance takes him. He becomes a gardener, then a toll-collector at a great house (Schloß) near Vienna. He falls in love with a lady at the house, whom he takes to be a countess; believing her to be socially beyond him, he decamps, finds himself in Italy, and has strange adventures at a castle. He learns that his lady loves him, so makes his way back to the great house and there discovers to his delight that she is no countess, but the porter's niece, so that he himself is an eligible, and soon an accepted, suitor. All the time, without his ever realizing it, he has played a part in the romance and elopement of the real countess. The plot of the story matters little; mood and atmosphere are all-important. The Taugenichts (we never learn his name, and indeed so prosaic an attribute would be irrelevant) goes his way easily and light-heartedly through a blissful world of unspoiled landscape where it never rains and every night is moonlit.

The Novelle is saved from sentimentality by the self-irony of the Taugenichts (who narrates his own story) and by the gentle irony of his creator. In the work are inserted six poems and a number of detached stanzas. The poems include such well-known examples as 'Wem Gott will rechte Gunst erweisen' (*Der frohe Wandersmann*), 'Wohin ich geh und schaue' (*Der Gärtner*), 'Wer in die Fremde will wandern' (*Heimweh*), 'Die treuen Berg' stehn auf der Wacht' (*An der Grenze*), and 'Schweigt der Menschen laute Lust' (*Der Abend*, qq.v.). They are all given in the text without title. The romantic figure of the Taugenichts has become so well known that he is now virtually a figure of German folklore.

Aus dem Regen in die Traufe, a Novelle by O. Ludwig (q.v.), first published in 1857 with *Die Heiteretei* under the group title *Die Heiteretei und ihr Widerspiel.* The starting-point of the story is the 'Gründermarkt' which also opens *Die Heiteretei,* and the heroine of the earlier story is in the background of *Aus dem Regen in die Traufe.* The tailor Hannes Bügel lives with his mother, a disciplinarian who still canes him in spite of his thirty years. They are helped in the house by a sweet-natured girl, Sannel, who loves Hannes. In order to escape from his mother's tutelage Hannes gives a written promise of marriage to a young woman ('die Schwarze'), who installs herself in the house and bullies its inhabitants; and so Hannes finds himself fallen 'out of the frying-pan into the fire' (aus dem Regen in die Traufe). A good-natured journeyman retrieves the situation, opening Hannes's eyes to the attractions of Sannel, and luring 'die Schwarze' away from Hannes without committing himself. Hannes marries Sannel, and Frau Bügel amiably abdicates her authority.

Aus dem Tagebuch einer Schnecke, a novel by G. Grass (q.v.).

'Aus den Gruben, hier im Graben', first line of the child's song in the *Novelle* (q.v.) by Goethe. It has been suggested that the first part of the line alludes to Daniel's situation in the lion's den, and the second to the 'Burggraben' of the old castle in the *Novelle*.

Aus der Chronika eines fahrenden Schülers, an uncompleted narrative written by C. Brentano (q.v.) in 1802 and published in 1818 in F. Förster's *Sängerfahrt.* The stimulus to write it came from perusal of the *Limburger Chronik* (q.v.), and Brentano has successfully caught the naïve chronicle tone. His story purports to be taken from an ancient MS., begun in 1338 by 'Johannes der Schreiber', a name derived from his source. The story begins in Strasburg, where Johannes has been taken into service by an aged knight, Veltlin von Türlingen. The opening pages, containing passages praising the beauty of Strasburg Minster, form a frame (see RAHMEN) to the story. Veltlin asks Johannes to read the story of his life, which is entitled 'Chronika des fahrenden Schülers Johannes Laurenburger, von Polsnich an der Lahn'. It begins with childhood recollections, which include the knowledge of his knightly birth and the humble circumstances in which he lived with his pious widowed mother. This phase of the story proves to be a frame within the original frame, for Johannes now gives his mother's story told in the first person, 'Was mir meine selige Mutter, die schöne Laurenburger Els, in dem Häuslein meines seligen Großvaters, des Voglers Kilian, auf der Hirzentreu von sich und dem lieben Großvater erzählt hat.' Els tells of the early death of her mother, at which point the story breaks off. The book contains a number of inserted songs, of which the best known is, 'Es sang vor langen Jahren/Wohl auch die Nachtigall'.

A revised and slightly longer version of the *Chronika* was found in Brentano's papers and published in 1880 in *Stimmen von Maria Laach.*

Aus der Gesellschaft, a drama (Schauspiel) in four acts by E. von Bauernfeld (q.v.), first performed in the Burgtheater, Vienna, in 1867. The commoner Magda Werner, brought up in an aristocratic household, is loved by the Minister of State, Fürst Lübbenau. His relatives oppose the match as a *mésalliance*, and Magda's own pride makes her reluctant. But the Fürst has his way, and the more important relatives are reconciled. The Viennese of Bauernfeld's day regarded the play as an important social document.

Aus der Gesellschaft, title given in 1844 to a series of six novels by Gräfin I. von Hahn-Hahn (q.v.). It was originally the title of the first of these, which was retitled *Ilda Schönholm* in 1844.

Aus einer kleinen Stadt, the sixth and last section of G. Freytag's long chronicle novel *Die*

Ahnen (q.v.). It can be read, and has been published, as a separate novel.

Aus engen Wänden, original title of *Bötjer Basch* (q.v.), a Novelle by Th. Storm (q.v.).

Ausgabe letzter Hand, general term for a definitive edition of collected works supervised by the author but chiefly employed with reference to Goethe, whose *Goethes Werke. Vollständige Ausgabe letzter Hand* appeared in 40 volumes between 1827 and 1830. The publisher was Cotta of Stuttgart. A supplementary series of 20 volumes (41–60) followed in the years 1832–42 with the title *Goethes Nachgelassene Werke.* Its editors were J. P. Eckermann and F. W. Riemer (qq.v.). Important works which were published for the first time in the *Ausgabe letzter Hand* include the poems *Trilogie der Leidenschaft* (q.v., 1827), the complete *Wilhelm Meisters Wanderjahre* (q.v., 1829), the *Zweiter römischer Aufenthalt* (1829), the *Tag- und Jahreshefte* (q.v., 1830), and (in the supplementary series) *Faust II* (q.v., 1832), the first versions of *Götz von Berlichingen* (q.v., 1833) and of *Iphigenie auf Tauris* (q.v., 1842), the fourth part of *Dichtung und Wahrheit* (q.v., 1843), and the *Paralipomena zu Faust* (1842). The edition exists in two formats, 12° and 8°; the contents of both forms are identical.

Ausgleich, Österreichisch-ungarischer, an agreement made in 1867 radically revising the relationship between Austria and Hungary. It was made necessary by the defeat of Austria in the German War of 1866 (see DEUTSCHER KRIEG), which excluded Austria from Germany and encouraged Hungarian particularism. The Ausgleich established a dual state, the two components of which dealt independently with their internal affairs, while foreign policy, military matters, and the related finances were dealt with by joint ministries. The Ausgleich constituted the state Österreich-Ungarn. The arrangement functioned adequately for a time, but from 1897 on there was constant friction over the financial terms, which had to be renegotiated every ten years.

Aus meinem Leben, see DICHTUNG UND WAHRHEIT.

Ausnahme und die Regel, Die, a Lehrstück (q.v.) in nine episodes by B. Brecht (q.v.), written in 1929–30 for schools and published in *Versuche,* vol. 10. It is in prose, but has a prologue, six songs, and an epilogue in verse. It was first produced in 1938 in Palestine. In 1948 Paul Dessau (q.v.) set the poems to music. The action takes place in Mongolia in the early 20th c. It portrays a merchant who attempts to cross the desert on his way to the city of Urga ahead of his rivals in quest of profit from the oil resources. He dismisses his guide, whom he distrusts, because he lacks ruthlessness, and enters the perilous desert in the sole company of his porter, a coolie; here they run into trouble. The coolie, disregarding his own sufferings, places his master's needs before his own and offers him some water. The suspicious merchant misunderstands the coolie's good intentions and shoots him. The coolie's widow claims justice in court, where the merchant is accused of murder. But since he is a rich and therefore a 'good' man he is acquitted, and his pretence that he has acted in self-defence is accepted. The exception to the rule, that a poor man's intentions were good, remains unrecognized. A prologue introduces the theme and invites the school to follow the action with detachment. The epilogue ends with an appeal for action against the misuse of justice, which is the rule, in defence of humanity, the exception.

Austerlitz, a battle, fought on 2 December 1805 between the French armies under Napoleon and the combined Austrian and Russian armies. The battlefield lay to the west of Austerlitz, a small town in Moravia, then Austrian territory. The forces engaged amounted to 65,000 French and 82,000 allies. The French outmanœuvred and outfought their opponents, who lost in casualties and prisoners 30 per cent of their strength. The battle ended the campaign and led to the Treaty of Preßburg (1805), by which Austria suffered substantial territorial loss, and ultimately to the dissolution of the Holy Roman Empire in 1806. (See NAPOLEONIC WARS and DEUTSCHES REICH, ALTES.)

Austrasien or **Austrien,** in the early Middle Ages denoted the eastern half of the Frankish Empire of the Merovingians (see MEROWINGER) from 430 to 751, covering roughly present-day Rhineland, Hesse, Württemberg (later Bavaria also), as well as Holland, Belgium, and northeastern France. Its chief towns were Reims and Metz.

Austria as it is or *Sketches of Continental Courts by an Eye Witness,* a book published anonymously in 1828. Its author was discovered in 1864 to be C. Sealsfield (q.v.). It is not a satirical work, but describes the customs and the mode of life in Austria and other German states, especially those of the courts and aristocracy, and is straightforwardly and sharply critical of corruption, of the secret police, and of the policies of Metternich (q.v.). A corrupt version in German was published in 1834 with the title *Seufzer aus Österreich und seinen Provinzen.* An accurate translation (*Österreich, wie es ist*) was published in

1919. Sealsfield wrote the book in German and the translation is supposed to have been made in London for the publisher. The original MS. was not preserved and all German and French versions published are translations from the English.

Austrian Succession, War of the, see ÖSTER-REICHISCHER ERBFOLGEKRIEG.

Austrien, see AUSTRASIEN.

Auswahl aus des Teufels Papieren, an early novel by JEAN Paul (q.v.).

Automate, Die, a story by E. T. A. Hoffmann (q.v.), written and published in 1814 in the *Zeitung für die Elegante Welt* (q.v.). A fragment had previously appeared in the *Allgemeine Musikalische Zeitung* (1814). The story was subsequently included in *Die Serapionsbrüder* (vol. 2, 1819). It combines the account of an ingeniously designed speaking figure with the story of a young man's love for a woman possessed of a beautiful expressive and highly trained voice. This element in the story relates to Hoffmann's love for Julia Marc.

AVA, FRAU (d. 1127), a medieval poetess, believed to have been an anchoress in the neighbourhood of Melk. In earlier years a married woman, she discloses at the end of her life of Jesus that her two sons, both clerics, helped her with her writings. Apart from the nun Roswitha (q.v.), who, however, wrote exclusively in Latin, she is the first known German woman writer.

Frau Ava's poems, intended by her to constitute a single work, comprise *Das Leben Jesu* and two shorter works, *Der Antichrist* and *Das jüngste Gericht,* together with a brief autobiographical conclusion. All these are contained in the Vorauer Handschrift (q.v.); the later Görlitz MS. includes also a poem on John the Baptist. The poems, which are believed to have been written towards the end of her life, are broad, easy-paced narrations. The poetry, naïve and unskilled, reveals her as a simple, pious, warm-hearted woman.

AVANCINI, NIKOLAUS (Brez nr. Trento, 1611–86, Rome), a nobleman from South Tyrol, entered the Society of Jesus in 1627, was for a time in Graz, Agram, and Laibach, and then, from 1640 to 1665, was a professor in Vienna, to which he returned in 1666 after a short spell of duty in Passau. In 1682 he was moved to the Society's headquarters in Rome. Avancini was a notable exponent of Jesuit drama (see JESUITENDRAMA), writing a large number of Latin plays, many of them spectacular allegorical works written in praise of Habsburg during Avancini's residence in Vienna. The most important of these laudatory dramas is *Pietas victrix sive Flavius Constantinus Magnus de Maxentio tyranno victor* (1659).

AVENARIUS, FERDINAND (Berlin, 1856–1923, Kampen, Sylt), founded the journal *Der Kunstwart* (q.v.) in 1887 and devoted himself to the education of literary taste. In 1903 he extended his activity to the plastic arts with the foundation of the Dürerbund. He was a prolific, but undistinguished lyric poet and the editor of an influential anthology of poetry (*Hausbuch deutscher Lyrik*, 1903). *Faust. Ein Spiel* (1919) was intended as a modern substitute for Pt. II of Goethe's *Faust* (q.v.).

AVENTINUS, whose real name was Johannes Turmaier (Abensberg, 1477–1534, Regensburg), studied in Vienna under Celtis (q.v.) and in 1508 entered the service of the Bavarian court, becoming court historiographer in 1517. He wrote textbooks of language (1512) and music (1516). His chief work is a history of Bavaria written first in Latin (*Annales ducum Boiariae*, 1511), then in German (*Bayerische Chronik*, 1533). Aventinus also produced in 1519 the first map of Bavaria.

Aventiure in Middle High German literature is a type of encounter characteristic of chivalric life in the courtly romance; it is the acceptance of risk in equal combat in the hope of augmented honour and renown. It is succinctly expressed by Hartmann von Aue in *Iwein* (ll. 527–37):

,âventiure? waz ist daz?'
,daz wil ich dir bescheiden baz:
nû sich, wie ich gewâfent bin.
ich heize ein rîter und hân den sin,
daz ich suochende rîte
einen man, der mit mir strîte,
der gewâfent sî als ich.
daz prîset in, ersleht er mich.
gesige ich aber im an,
so hât man mich vür einen man
und wirde werder, danne ich sî.'

(,Aventiure? was ist das?' 'Das will ich dir weiter mitteilen. Nun sieh, wie ich gewappnet bin. Ich heisse ein Ritter und habe den Sinn, dass ich reite, um einen Mann zu suchen, der mit mir streiten will und der wie ich gewappnet ist. Erschlägt er mich, so erhöht das ihn. Siege ich aber über ihn, so hält man mich für einen Mann; und so werde ich würdiger, als ich es jetzt bin.')

Aventiure is personified in Wolfram von Eschenbach's *Parzival* (q.v.). The word is used, as a deliberate archaism, by V. von Scheffel

(q.v.) in *Frau Aventiure. Lieder aus Heinrich von Ofterdingens Zeit*, 1863.

In the *Nibelungenlied* (q.v.) and other medieval works 'aventiure' serves to denote the separate cantos.

Avisensänger, see BÄNKELSÄNGER.

AYRENHOFF, CORNELIUS HERMANN VON (Vienna, 1733–1819, Vienna), entered the Austrian army as an ensign (Fähnrich) in 1751, attaining in 1794 the rank of field-marshal. Ayrenhoff began to write plays in his thirties and became a prolific author of neo-classical tragedies in the manner of Gottsched (*Aurelius*, 1766; *Hermann und Thusnelde*, 1767) and of comedies in the French style. One of these, *Der Postzug* (1769), was singled out for praise by Friedrich II (q.v.) of Prussia in *De la littérature allemande* (q.v.). His *Sämtliche Werke* appeared in 6 vols., 1803–7.

AYRER, JAKOB (Nürnberg, 1543–1605, Nürnberg), a prolific dramatist, lived in Bamberg from 1570 to 1593, when he returned to Nürnberg and was active as a notary and as public prosecutor (Stadtprokurator). He is the author of 106 plays, written between 1592 and 1602, of which 69 survive. The chief repository of his dramatic work is the posthumous *Opus theatricum* (1618, ed. A. von Keller, 1865, repr. 1973), in which 66 of his plays are printed. They are divided in approximately equal numbers into serious and comic (see FASTNACHTSPIEL). Ayrer, who drew on German sources and the Italian *novelle*, so popular in the 16th c., is said to have been influenced by travelling troupes of English actors (see ENGLISCHE KOMÖDIANTEN). His earliest play is *Die Erbauung der Stadt Bamberg*, written *c*. 1570. Among other titles are *Tragedia von Erbauung der Stadt Rom, Tragedia von Keiser Otten dem dritten, Hug Dieterich, Ortnit*, and *Wolfdieterich, Julius Redivivus* (a free translation of Frischlin's Latin play), *Comedia vom König Edwarto, Comedia Vom König in Zypern, Comedia von der schönen Phaenica und Graf Tymbri von Golison aus Arragonien*, which has the same source and action as the main plot of *Much Ado About Nothing*, and *Comedia von der schönen Sidea*, a play of magic and enchantment with an action resembling the episode of Ferdinand and Miranda in *The Tempest*. He also wrote Singspiele (see SINGSPIEL) and, in his earlier years, a chronicle of Bamberg. His plays are all in verse, in the form known as Knittelverse (q.v.). He is the first German dramatist to make extensive use of stage directions. Ayrer's resemblances to Shakespeare are superficial, his style is without distinction, and his characters do not carry conviction.

B

BAADER, FRANZ XAVER (also Benedikt) (Munich, 1765–1841, Munich), qualified as a mining engineer in Freiberg (Saxony), where he met A. von Humboldt (q.v.). A Roman Catholic, he acquired a keen interest in religious and philosophical ideas and felt that religion should exercise a greater influence on politics, a view expressed in his tract *Über das durch die französische Revolution beigeführte Bedürfniß einer neuen und innigen Verbindung der Religion und Politik* (1815). A visit to Russia, intended to promote this aim, which resembles that of the Holy Alliance (see HEILIGE ALLIANZ), ended in disillusionment. In 1826 he became a professor at Munich University. He leaned increasingly towards an unorthodox theosophical mysticism and disapproved of papal absolutism. He was influenced by Jakob Böhme (q.v.), and himself influenced Schelling (q.v.), who in turn transmitted his influence to the Romantic movement (see ROMANTIK).

Baal, the first major play by B. Brecht (q.v.), written in 1918, published in 1922, and first performed at the Altes Theater, Leipzig, and the Deutsches Theater, Berlin, in 1923. For both performances the second version was used. The first two versions bear the motto 'Baal frißt! Baal tanzt!! Baal verklärt sich!!!'. A fragment of 1930 is entitled *Der böse Baal der asoziale,* meaning that Baal is asocial because he lives in an asocial world. Brecht composed the songs himself. The first version is a parody of *Der Einsame,* a play about C. D. Grabbe (q.v.) by H. Johst (q.v.). Three further versions, of 1918, 1919, and 1926 (entitled *Lebenslauf des Mannes Baal*), were published in 1966 as *Baal. Drei Fassungen.*

The characterization of the poet of 'immortal' songs, the 'Lyriker' Baal, who meets a lonely death in the woods after a career of carousing, seduction, and murder, is deliberately exaggerated to stress Brecht's contempt of any form of idealism or authoritarianism. The *Vorspruch* (*Schriften zum Theater* 2) prepares for the 'abnormality' of the 20th-c. Baal, the 'passive genius' (a parody of Johst's portrayal of Grabbe as reflecting the traditional concept of genius).

The play opens with the *Choral vom großen Baal.* He is called after the pagan god by virtue of his vices and animal instincts. He exploits 'das große Weib Welt', but does not allow himself to be exploited. The structure of the 22 scenes, the crisp dialogue, as well as the songs,

show that in this play Brecht, despite his indebtedness to Expressionism, laid the foundations for his epic theatre (see EPISCHES THEATER). In commenting on the play twenty years later (*Stücke* 1) Brecht concedes that it lacks wisdom; but the theme was still attractive enough for him to project a libretto for an opera, which would show the impossibility of killing man's longing for happiness. This idea was to be given concrete form by the failure of the executioner to kill the 'fat little god of happiness' who had been condemned to death for trying, with his disciples, to make the people happy after a long war.

Babenberg, name of two noble families, whose relationship is uncertain. The older Babenberg counts occupied a castle in Bamberg and succumbed in 906 to the Konradiner, with whom they had been involved since 902 in a conflict over supremacy in Franconia (*Babenberger Fehde*).

The younger Babenbergs (*Neubabenberger*) were margraves of the Eastern Mark (present-day Austria) from 976. Through the Emperor Friedrich I (q.v.) they were made dukes of Austria in 1156. The line became extinct in 1246.

BABO, JOSEPH MARIUS VON (Ehrenbreitstein nr. Koblenz, 1756–1822, Munich), became a protégé of the Elector Karl Theodor of the Palatinate and Bavaria, who employed him in Mannheim and afterwards in Munich. Babo was director of the Munich Nationaltheater, as well as a professor of aesthetics and official censor of books. He is the author of a number of plays, including a military drama, *Arno* (1776), the comedy *Winterquartier in Amerika* (1778), and *Die Römer in Teutschland* (1780), a patriotic play of early German history. His best-known work was the Bavarian tragedy *Otto von Wittelsbach, Pfalzgraf in Bayern* (1782); his almost equally successful comedy *Bürgerglück* (1792) was in the sentimental manner of middle-class drama popular towards the end of the 18th c.

Babylonische Gefangenschaft, a penitential sermon in verse, preserved at Maria-Saal, Austria, and probably written *c.* 1140. The Babylonian captivity of the title symbolizes everlasting death, which is to be avoided by

53

repentance. The sermon is based on a numerical parallel: 70 years of captivity, but for the repentant Christian only the 70 penitential days from Septuagesima to Easter. Some lines are missing at the beginning, and possibly also at the end.

BACH, JOHANN SEBASTIAN (Eisenach, 1685–1750, Leipzig), the most famous member of a family that produced six generations of musicians. Johann Sebastian's father Ambrosius was Stadtmusikus of Eisenach, and his two uncles, his paternal grandfather, and his great-uncle were all organists, and, in addition, one of his great-grandfathers was also a musician. Orphaned at 10, he was brought up at Ohrdruf and then at Lüneburg. He occupied various musical posts in Weimar, Arnstadt, and Mühlstadt between 1703 and 1717. From then until 1723 he was court musical director (Hofkapellmeister) at Köthen (or Cöthen) in Anhalt. He was appointed choirmaster at St. Thomas's Church, Leipzig, and city music director in 1723, a post he occupied until his death.

J. S. Bach wrote numerous orchestral works, suites, and concertos, including the six Brandenburg Concertos, as well as violin and keyboard concertos. He composed *Das wohltemperierte Klavier*, generally known as 'The Forty-eight Preludes and Fugues', the Goldberg Variations, and much organ music. His vocal works include more than 200 cantatas for ecclesiastical use, two secular works (*Kaffeekantate* and *Bauernkantate*), and the great choral masterpieces, the B minor Mass (*Hohe Messe in H-moll*), the Christmas and Easter Oratorios, and the St. Matthew and St. John Passions. For many of the choral works a great part of the text was biblical and the cantatas, passions, and oratorios use hymns current in Bach's time. The texts for the recitatives and arias in some of these works were written by Picander, i.e. C. F. Henrici (q.v., *Weihnachtsoratorium* and *Matthäus-Passion*), though Bach also made use of poetry by Brockes (q.v., *Johannis-Passion*). The poet of the *Oster-Oratorium* is not known. Picander was responsible for the words of the Coffee and Peasant Cantatas.

Bach's eldest son, Wilhelm Friedemann (1710–84), primarily an instrumental composer, was chiefly active in Dresden and Halle, but also in Leipzig, Brunswick, and Berlin. He is the subject of the novel *Friedemann Bach* (1858) by A. E. Brachvogel (q.v.).

The second son, Carl Philipp Emanuel (1714–88), who wrote vocal as well as instrumental music, setting works by Klopstock and Gellert (qq.v.), was in the service of Friedrich II (q.v.) of Prussia, and from his later residence in Hamburg 1767–88 is known as 'der Hamburger Bach'.

Johann Christoph Friedrich (1732–95) worked chiefly in Bückeburg. His songs include settings of poems by Herder (q.v.).

'The London Bach', Johann Christian (1734–82), after two years as cathedral organist at Milan, settled in London in 1762 as master of music to Queen Charlotte, consort of George III.

Bacharach, a small town on the left bank of the Rhine between Majnz and Koblenz. Its fame as a vine-growing centre is indicated by the traditional rhyme:

'Zu Bacharach am Rhein, zu Klingenberg am Stein,
Zu Hochheim an dem Main, da gibt's die besten Wein'.'

In Brentano's Lorelei poem, *Die Lore Lay*, Bacharach is the original home of the beautiful witch who is associated with the Lorelei rock *c.* 6 miles downstream on the opposite bank (see LORELEI).

BACHERACHT, THERESE VON, see STRUWE, THERESE VON.

BACHERL, FRANZ (Waldmünchen, 1808–69, Columbus, Nebraska), a Bavarian schoolmaster, claimed that the anonymous tragedy *Der Fechter von Ravenna*, successfully performed in 1854, was a slightly altered copy of his own tragedy *Die Cherusker in Rom*, submitted to H. Laube (q.v.), and rejected by him. The resulting controversy was temporarily checked by the disclosure by F. Halm (q.v.) in a declaration of 27 March 1856 that he was the author. Bacherl's play was published in 1856. In 1857 he resigned his post in order to devote himself (unsuccessfully) to writing. He emigrated to America in 1867.

BACHLER, WOLFGANG (Augsburg, 1925–), journalist and broadcaster in Germany and foreign correspondent in Paris, is the author of a novel (*Der nächtliche Gast*, 1950) and of collections of poetry (*Die Zisterne*, 1950; *Lichtwechsel*, 2 vols., 1955–61; *Türklingel*, 1962; *Türen aus Rauch*, 1963).

BACHMANN, INGEBORG (Klagenfurt, 1926–73, Rome), spent her girlhood in Carinthia, from the age of 12 to 18 under war conditions. She studied at Graz, Innsbruck, and Vienna universities, and came to public notice as a writer through reading her poems at a meeting of the Gruppe 47 (q.v.) in 1952. She published two volumes of poetry, *Die gestundete Zeit* (1953), which contains the poem *Große Landschaft bei*

Wien, and *Anrufung des großen Bären* (1956). A volume of stories, *Das dreißigste Jahr*, appeared in 1961, containing, in addition to the title-work, *Alles, Unter Mördern und Irren, Ein Schritt nach Gomorrha, Ein Wildermuth*, and *Undine geht*; a separate story, *Jugend in einer österreichischen Stadt* (1961), was included in a new edition of *Das dreißigste Jahr* in 1966. In 1959–60 she was invited to lecture on poetry at Frankfurt University. Extracts from these lectures are included with essays, old and new poems, and some of the stories in the volume *Ingeborg Bachman: Gedichte, Erzählungen, Hörspiel, Essays* (1964). The *Hörspiel* in this title is *Der gute Gott von Manhattan* (1958); her earlier radio plays were *Herrenhaus* (1954) and *Die Zikaden* (1955). She collaborated with the composer H. W. Henze (q.v.), writing the libretti for the operas *Der Prinz von Homburg* (1960) and *Der junge Lord* (1965). The former is based on H. von Kleist's play *Prinz Friedrich von Homburg* (q.v.), the latter on a story by W. Hauff (q.v.), *Der Scheik von Alexandrien und seine Sklaven*. She is the author of the novel *Malina* (1971), and of *Simultan* (1972) which contains five stories. In 1964 she received the Büchner Prize. Her early writing, in both her poetry and her narrative work, is bleak, even harsh in imagery and phrasing, but in its stringency and poetry especially possesses an austere beauty. She died of burns sustained in her apartment. Four volumes of *Werke*, ed. C. Koschel and I. von Weidenbaum, were published in 1978. Vol. 1 contains *Gedichte, Hörspiele, Libretti, Übersetzungen*, vol. 2 *Erzählungen*, vol. 3 *Todesarten: Malina und unvollendete Romane*, vol. 4 *Essays, Reden, Vermischte Schriften, Anhang*. An edition of *Sämtliche Erzählungen* appeared in 1980.

In the 1960s Bachmann had planned a cycle of novels that was to include *Malina*; two other projects, one of which was to be entitled *Todesarten*, remained unfinished. The editors of *Werke* published the unfinished cycle as *Todesarten*; the longer fragment apeared as *Der Fall Franza*, the other as *Requiem für Fanny Goldmann*. The disturbing experience of women in their relationships with men, a major theme, is analysed in pervasive monologistic form, in which myth and dream support the portrayal of different layers of consciousness. Recognized by Max Frisch (q.v., in *Montauk*), she influenced the feminist cause of women writers, notably Christa Wolf and Irmtraud Morgner (qq.v.).

BACHMANN, ZACHARIUS, see RIVANDER, ZACHARIAS.

BACZKO, LUDWIG FRANZ JOSEF VON (Lyck, East Prussia, 1756–1823, Königsberg), went blind at the age of 21 while a student. He became a lecturer in history at the School of Artillery in Königsberg, and in his later years was director of the Institute for the Blind there. Baczko wrote poems, historical novels and stories (including *Ehrentisch oder Erzählungen aus den Ritterzeiten*, 1793), plays, and a posthumously published autobiography (*Geschichte meines Lebens*, 1824). His most considerable work was a history of Prussia (*Geschichte Preußens*, 1792–1800).

Baden, former German margravate (1100–1805), Grand Duchy (1805–1918), and Land of the Weimar Republic (q.v.). For changes after 1945 see WÜRTTEMBERG.

Baden-Baden is, with Wiesbaden, the best-known German spa. Its double name denotes both city and state, distinguishing it from other Badens. Its reputation as a fashionable resort dates from the first decade of the 19th c. It was much visited by royalty and was the scene in 1860 of the Badener Fürstentag, a meeting between Napoleon III, King Wilhelm I (q.v.) of Prussia, and various German princes.

Badener Lehrstück vom Einverständnis, Das, a play (Lehrstück, q.v.) in the form of an oratorio by B. Brecht (q.v.) with music by Hindemith (q.v.), written in 1929 for the Baden-Baden Music Festival. It consists of 11 short sections with headings, and debates whether four survivors calling for help after a plane crash deserve to be rescued. All are denied the right to live, and the theme of rescue is treated in terms of ideological conversion to socialist collectivism represented by the chorus (gelernter Chor). The three engineers conform, and are honoured for abandoning their individual identity while confronting death. But the pilot Charles Nungesser (named after a French pilot of the 1914–18 War) defies the chorus, and is committed to nothingness (das Nichts). The theme of aviation, linking the play with *Der Ozeanflug* (q.v.), reaches its climax in the ninth section entitled *Ruhm und Enteignung*. Brecht states in *Versuche*, vol. 2, that the didactic element is unsatisfactory, since too much importance is attached to death.

Baedeker, the famous publishers of guide-books, originated as a firm of printers and booksellers in Bielefeld at the beginning of the 18th c. Its founder was Dietrich Baedeker (1680–1716). The present firm was established as 'G. D. Baedeker' in Essen in 1798. The publication of guide-books was begun under Karl Baedeker with a guide to Koblenz (1829), which was followed by a successful guide to the Rhine (*Rheinreise von Mainz bis Cöln* by J. A. Klein q.v., 1832). In 1846 the first guides in other languages were issued. The firm moved to Leipzig in 1872, and after 1945 to Stuttgart.

BAGGESEN, JENS IMMANUEL (Korsör, 1764–1820, Hamburg), Danish writer, whose mother tongue was German. From 1811 to 1814 he was a professor at Kiel. Baggesen wrote sentimental and satirical poems, but is chiefly known for the part he played in drawing the attention of the Danish nobles (see FRIEDRICH, HERZOG VON AUGUSTENBURG and SCHIMMELMANN, H. E., GRAF VON) to Schiller's plight during his illness, the consequence of which was the handsome pension granted to Schiller in December 1791.

Bahnwärter Thiel, a Novelle by G. Hauptmann (q.v.), published in the periodical *Die Gesellschaft* in 1888, and designated 'eine novellistische Studie'. In linesman Thiel, Hauptmann portrays the dual existence of a man of humble position but of extraordinary physique and spiritual awareness. Thiel's first wife, Minna, has died in childbirth, and, to provide for the needs of his son Tobias, so Thiel assures his pastor, he marries again. In his second marriage to the robust Lene, a former dairymaid, Thiel's life soon becomes divided into his life at home, where he gradually loses authority, and his life in his isolated railway hut in the woods. Here he maintains spiritual communion with Minna. But when Lene has a son of her own to nurse, Tobias becomes the victim of the ruthlessly brutal side of her nature, against which Thiel fails to assert himself, even when he unexpectedly witnesses Tobias's helpless suffering. The first two sections of the Novelle lead up to this event. The third, the last and longest section, opens with the effect of Tobias's 'passion', as Thiel sees it, upon his relationship with Minna. She appears to him that night in an agonizing dream, walking along the railway line away from him; he feels that she has claimed her son, through whom Thiel has betrayed her love. The following night his mind is further tormented when Lene tells him that she intends to accompany him with the children to his hut to plant potatoes on a strip of land he has been allowed to cultivate. Only Tobias's joy at going to his father's hut restores Thiel's peace of mind, and in the sunlit forest next morning he feels for a while that he is again in complete harmony with Minna's spirit. Lene's visit brings disaster. Tobias wanders on to the line, and Thiel sees him run over by a passing train. The following day Lene and her baby are found dead, covered with their own blood. When Thiel, the suspected murderer, is found, he is sitting on the railway line fondling Tobias's cap. He ends in a mental hospital. Although written during Hauptmann's early Naturalistic phase the work includes mystic elements, and employs symbolic language.

BAHR, HERMANN (Linz/Danube, 1863–1934, Munich), studied in Vienna, Czernowitz, and Berlin, where, in the 1880s, he met A. Holz and M. Kretzer, the Naturalist writers (qq.v.). He worked for a time as a journalist and publisher's reader in Berlin, and then became a dramatic critic in Vienna (1892). Bahr moved frequently, living again for a time in Berlin, then in Salzburg, and once more in Vienna, until in 1922 he settled finally in Munich. In 1909 he married the opera singer Anna Mildenburg. For a short time in 1918 he was senior dramatic adviser (Dramaturg) at the Burgtheater (q.v.). Towards the end of his life Bahr went out of his mind.

Always a step ahead of the latest movement, Bahr wrote, as early as 1891, *Die Überwindung des Naturalismus,* a book of essays rejecting Naturalism and calling for a 'Literatur der Nerven'. His numerous plays, many in Viennese dialect, include *Die neuen Menschen* (1887), *Die große Sünde* (1889), *Die Mutter* (1891), *Aus der Vorstadt* (1893, with C. Karlweis, q.v.), *Der Franzl* (1901, on F. Stelzhamer, q.v.), *Die häusliche Frau* (1893), *Das Tschaperl* (1898), *Der Meister* (1904), *Das Konzert* (1909), *Wienerinnen* (1911), *Die Kinder* (1911), *Das Prinzip* (1912), *Das Phantom* (1912), *Der Querulant* (1914), *Die Stimme* (1916), *Unmensch* (1919), *Spielerei* (1919), *Ehelei* (1920), the title of which alludes to Schnitzler's *Liebelei* (q.v.), and *Altweibersommer* (1924). Of these *Das Tschaperl* and *Das Konzert* are the best known. Among his novels are *Theater* (1897), *Drut* (1909), later entitled *Die Hexe Drut,* and three works of a cycle intended to contain 12 volumes setting forth the shape and temper of the age: '*O Mensch*' (1910), *Himmelfahrt* (1916), and *Die rotte Korahs* (1918). *Wiener Theater* (1899) and *Rezensionen* (1903) contain some of his theatre criticism. His autobiography (*Selbstbildnis*) appeared in 1923. Bahr proclaimed his reconversion to the Roman Catholic faith in 1916. Select editions, *Meister und Meisterbriefe um Hermann Bahr,* ed. J. Gregor, appeared in 1947 and *Zur Überwindung des Naturalismus. Theoretische Schriften 1887–1904,* ed. G. Wunberg, in 1968.

BALDE, JAKOB (Ensisheim, Alsace, 1604–68, Neuburg/Danube, Bavaria), a neo-Latin poet of distinction, was educated at Molsheim near Strasburg. In the turmoil of war the school migrated to Germany, and Balde completed his education between 1622 and 1626 at Ingolstadt University and Jesuit colleges. He entered the Society of Jesus in 1624 and from 1626 to 1628 taught in a Jesuit school at Munich. In 1630 he became a professor at Innsbruck; eight years later he was appointed court preacher at Munich, and at the same time was entrusted with the education of the two Bavarian princes. Balde resigned his office in 1646 because of delicate health, but continued until 1648 as court historiographer. He was for a time at Landshut and then at Amberg, and spent the last years of his life as court preacher at Neuburg. Balde's German poetry is insignificant, but his Latin poems gained him a great reputation, and he was frequently compared with Horace, sometimes to the Roman poet's disadvantage. His work, in which nature poetry (including pastoral and hunting poems) and ecstatic odes in adoration of the Virgin Mary are especially notable, includes *Batrachomyomachia* (1637) (an adaptation of the *Froschmeuseler*, see ROLLENHAGEN, G.), *Jephtias* (drama, 1637), *Poema de vanitate mundi* (1638), *Lyricorum libri IV* (1643), *Medicinae gloria* (1651), *Poemata* (1660), and *Urania victrix* (1663). His works were first collected in 1729 (*Opera omnia*).

BALDEMAR VON PETERWEIL, see FRANKFURTER SPIELE.

'Bald gras' ich am Neckar, bald gras' ich am Rhein', first line of a folk-song, which first appeared in print in *Des Knaben Wunderhorn* (q.v.), 1808. It has been set to music by Gustav Mahler (q.v.).

BALDNER, LEONHARD (Strasburg, 1612–94, Strasburg), a fisherman and later a fishery official, spent many years making careful studies chiefly of birds and fish, but also of mammals, amphibia, and insects. The results, which included much detailed observation of behaviour, were recorded in a MS. which was destroyed by fire at Strasburg. A handsome MS. copy of Baldner's *Vogel-Fisch- und Thierbuch* was made in 1666 for the Elector Karl Ludwig of Hesse-Heidelberg. It is accurately illustrated in colour by the Strasburg painter Johann Walther der Ältere. A facsimile of this copy, which is now at Kassel, was published in 1973–4, and is the only printed edition. A MS. of Baldner's work was used by the English naturalist John Ray (1627–1705) in preparing his *Ornithologia* (Latin 1676, English 1678).

BALDUIN VON TRIER (1285–1354), archbishop of Trier from 1307, was prominent in German politics, contriving the election of two emperors, Heinrich VII (q.v.) in 1308 and Ludwig IV (q.v.) in 1314. For a time he supported imperial anti-papal policy and was a leading spirit in the Rhenser Kurfürstentag (q.v.). He was later reconciled with the Papacy

and in 1346 was a party to the election of Karl IV (q.v.). He was one of the most powerful German prelates of his day, maintaining claims to the archbishopric of Mainz, and the dioceses of Speyer and Worms, while retaining the archiepiscopal see of Trier.

BALL, HUGO (Pirmasens, 1886–1927, Sant' Abbondio, Ticino), after beginning a commercial career, studied philosophy at Munich, Heidelberg, and Basel universities and then became a theatrical producer. In Munich he was associated with F. Wedekind (q.v.) in the court theatre, the Kammerspiele. A pacifist, he emigrated to Switzerland in 1915 and in 1916 founded the Cabaret Voltaire (see DADAISMUS). After the war he became a Roman Catholic convert. In his Klanggedichte and speech-song (Sprechgesang) he aimed at a distancing effect to language associated with war and the bourgeois society. In print he employed the technique of collage in the form of an arbitrary arrangement of different types (*Karawane*, 1917). He is the author of the plays *Die Nase des Michelangelo* (1911) and *Der Henker von Brescia* (1914), of the novel *Flametti* (1918), and of a biography of his friend H. Hesse (q.v., 1927). His political polemics of the post-war years include *Zur Kritik der deutschen Intelligenz* (1919). His diary, *Die Flucht aus der Zeit. Tagebücher 1912–1921* (1927) appeared again in 1946, *Briefe 1911–1927* in 1957, and *Gesammelte Gedichte*, ed. A. Schütt-Hennings, in 1970. Ball was married to Emmy Ball-Hennings (q.v.).

Ballade, term used in German literature since the time of the Sturm und Drang (q.v.) for narrative poems in stanzas, especially those with a dramatic climax. Bishop Percy's *Reliques of Ancient English Poetry* (1765) was a major influence on the growth of the modern German ballad. Many older examples of ballad poetry are to be found among the folk-songs of the later Middle Ages and the 16th c., and these are nowadays designated 'Volksballaden'. The cult of the ballad in the 18th c. owed much to Herder's enthusiasm for the collection of folksongs. Notable early modern ballads are *Lenore* by G. A. Bürger (q.v.) and Goethe's *Erlkönig* and *Der König in Thule* (qq.v.). The ballad was cultivated in a more intellectual and conscious fashion by Goethe and Schiller, particularly in their so-called Balladenjahr 1797. The Romantic movement (see ROMANTIK) revived the ballad as an aspect of the folk-song. Older examples were popularized by *Des Knaben Wunderhorn* (q.v.) and new ones written by Chamisso, Brentano, Eichendorff, and Uhland (qq.v.), among others. Many poets of the 19th c.

wrote ballads and, among many that are mediocre, those of Th. Fontane (q.v.) stand out. The principal 20th-c. writer of ballads is Börries von Münchhausen (q.v.). The word has occasionally been used as a title, as in *Ballade des äußeren Lebens* by H. von Hofmannsthal (q.v.). Ballad-type poems of a more satirical and even scurrilous type have also been cultivated in this century, e.g. by B. Brecht (q.v.) and by cabaret performers such as F. Wedekind (q.v.). It has become increasingly ideological in orientation since the 1960s.

Balladenalmanach, the customary designation for *Der Musenalmanach* for 1798, in which Goethe and Schiller published a number of ballads. Goethe's contributions included *Der Schatzgräber, Die Braut von Korinth, Der Gott und die Bajadere,* and *Der Zauberlehrling* (qq.v.). Of Schiller's ballads *Der Ring des Polykrates, Die Kraniche des Ibykus, Der Taucher* (qq.v.), *Ritter Toggenburg, Der Handschuh* (q.v.), and *Der Gang nach dem Eisenhammer* appeared in the *Balladenalmanach.* The volume was published towards the end of 1797.

Ballade vom verschütteten Leben, a symbolical religious poem by R. Hagelstange (q.v.), published in 1952. It tells of six German soldiers buried in an underground foodstore from 1945 to 1951. Two survive. The poem is concerned with spiritual values. The underlying story was published in the press as true, but is apocryphal.

Ballhausplatz, square in Vienna, close to the Hofburg. The Foreign Ministry of the Austro-Hungarian Empire was situated in the Ballhausplatz, which was often used metonymously to designate the Ministry.

BALL-HENNINGS, Emmy (Flensburg, 1885–1948, nr. Lugano), *née* Cordsen and of humble origin, married one Hennings and made a successful career as a *diseuse.* In 1920 she married H. Ball (q.v.), with whom she co-operated in the Cabaret Voltaire in Zurich. She published poems (*Die letzte Freude,* 1913; *Helle Nacht,* 1920; *Der Kranz,* 1939), the fairy-tales *Märchen am Kamin* (1943), a novel *Das Brandmal* (1920), and a number of stories.

BALLHORN or Balhorn, Johann (Lübeck, 1528–1603, Lübeck), a printer of Lübeck whose name has given a word to the German language. 'Verballhornen' is to debase, distort, or bowdlerize a printed work. He appears to have been harshly treated by tradition, for there is no evidence to show that he modified the works he printed any more than was customary in his day.

Balmung, in the German Nibelungensage (q.v.), the name of the sword which Siegfried acquires from the treasure of the Nibelungs. In the *Nibelungenlied* (q.v.) Kriemhild uses it to slay Gunther and Hagen.

BALTICUS, Martin (Munich, 1532–1600), rector of the grammar school at Ulm, wrote biblical plays for school performance in both Latin and German. They include *Adelphopolae,* 1556 (which appeared in German as *Josephus,* 1579) and *Drama Danielis,* 1558.

Bamberger Glaube und Beichte, see Beichtformeln.

Bamberger Reiter, a late 13th-c. equestrian statue in the interior of Bamberg Cathedral. This serene figure of a medieval nobleman has been variously identified as the Emperor Constantine, St. Stephen, St. George, Konrad III (q.v.), and one of the Magi, but none of these attributions has been generally accepted.

Bambi, an animal story by F. Salten (q.v.).

BAMM, Peter, pseudonym of Curt Emmrich (Hochneukirch, 1897–1975, Zurich), a fluent and versatile essayist, a surgeon by profession. He volunteered for service in the 1914–18 War, and afterwards studied medicine at Munich, Göttingen, and Freiburg universities, travelled extensively as a ship's surgeon and then devoted himself to literature, returning to medicine as an army surgeon (Stabsarzt) in the 1939–45 War. In the 1950s he again travelled in the eastern Mediterranean. Most of Bamm's essays first appeared in the literary pages of newspapers. His publications include *Die kleine Weltlaterne* (1935), *Der i-Punkt* (1937), *Der Hahnenschwanz* (1939), *Ex ovo* (1948), *Die unsichtbare Flagge* (1952, recollections of the war), *Frühe Stätten der Christenheit* (1955), *Wiege unserer Welt* (1956), *Welten des Glaubens* (1959), *An den Küsten des Lichts* (1961), and *Anarchie mit Liebe* (1962).

Bänkelsänger, a street singer of the 18th c. and 19th c. who sang ballads recounting crimes and catastrophes such as fires, shipwrecks, and mine disasters, and then sold broadsheets containing the ballad with crude illustrations. The word first occurs in 1709 and refers to the elevated position of the singer standing on a bench or platform. The Bänkelsänger, who was sometimes also called Marktsänger, is a descendant of the Avisensänger or Zeitungssänger, who disseminated news in similar fashion. The sensational ballads retailed by the Bänkelsänger are known as Moritaten (see Moritat). The form underwent sophisticated exploitation in cabaret in the late 19th c.

Barbara oder die Frömmigkeit, novel by F. Werfel (q.v.), published in 1929. The central character, Ferdinand R., a ship's surgeon, conjures up the story by recalling his past life. It ranges back to his childhood, to his parents, who have always remained remote from him, to military and monastic schooling, the war, prerevolutionary days in Vienna, his medical studies, and the inflation of the early 1920s. Through all this the one element of security given him is the selfless devotion of the old family servant Barbara, who, on his last visit to her, has given him her life's savings in gold. As the liner steams through the night, he commits the gold to the sea in order to save this precious memorial of Barbara from the contamination of the world.

BARBAROSSA, the Italian nickname of the Emperor Friedrich I (q.v.). It is also the title of a poem by F. Rückert (q.v.), written in 1813, which embodies the legend of Barbarossa's readiness to return in time of need.

BARCLAY, John (Pont-à-Mousson, 1582–1621, Rome), of mixed Scottish and French descent, is the author of a Latin political novel, *Argenis* (1621), which achieved European popularity and was translated into German by Martin Opitz (q.v.) in 1626–31.

BARDACH, Hans, Edler von Chlumberg, wrote under the pseudonym Hans Chlumberg (q.v.).

Bardendichtung, a rhapsodic and nationalistic style of poetry cultivated in Germany in the 1760s and 1770s, nourished in part by a vague conception of the role of the early medieval bard and by Ossian (q.v.). The form was short-lived. See Gerstenberg, Klopstock, Denis, also Bardiet.

Bardiet, a word invented by F. G. Klopstock (q.v.) as a generic term for his patriotic historical plays *Hermanns Schlacht* (1769), *Hermann und die Fürsten* (1784), and *Hermanns Tod* (1787). Klopstock adapted it from the Latin *barditus* used by Tacitus in the *Germania III* ('Sunt illis haec quoque carmina, quorum relatu, quem barditum vocant, accendunt animos, futuraeque pugnae fortunam ipso cantu augurantur').

Bärenhäuter, popular term for an indolent fellow, in use since the 16th c. A character called 'Der Bärenhäuter' is the subject of a popular fairy-tale, told by Grimmelshausen (in *Der erste Bärenhäuter*, 1670), L. J. von Arnim (in *Isabella*

von Egypten), and C. Brentano (in *Geschichte und Ursprung des ersten Bärenhäuters*) (qq.v.). The 'Bärenhäuter' makes a pact with the devil, undertaking not to wash, pare his nails, or cut his hair for seven years; having successfully complied with the conditions he is released, and rewarded with a beautiful bride.

Barfüßele, a story by B. Auerbach (q.v.), published in 1856 and subsequently included in *Sämtliche Schwarzwälder Dorfgeschichten* (1884). It has a contemporary rural setting. The orphan Amrei, known in derision as 'Barfüßele', is brought up kindly by a witch-like old woman, 'die schwarze Marann'. As goose-girl, Amrei occupies the humblest position in the village. Later she is taken into service by a local farmer's wife, the Rodelbauerin, and at a distant wedding, to which she is invited out of kindness, she sees an obviously well-to-do young farmer's son, who is taken with her, as she is attracted to him. She sees no more of him until he turns up at the Rodelbauers' farm, sent by his family to look for a wife. The farmer attempts to arrange a match for his ill-tempered sister Rosel, who ruins her chances by mishandling Amrei. The visitor recognizes her as the girl he met at the wedding and determines to marry her. After a row with the Rodelbauers he carries Amrei off to his parents, who are at first put out at the prospect of a portionless daughter-in-law, but are soon won over by Amrei's transparent honesty and goodness. The marriage is a happy and prosperous one.

Barfüßer, term applied until the 18th c. to the Franciscan Order, whose members went barefoot.

Barlaam und Josaphat, a Middle High German poem of some 16,000 lines by Rudolf von Ems (q.v.). It is his second surviving work and is believed to represent a reaction against the worldliness of his early opinions and lost early poetry. It tells of a noble heathen prince (Josaphat) who is converted by a Christian teacher (Barlaam) and achieves peace in communion with God and in renunciation of the lusts of the flesh. Its origin is the legend of Buddha, which reached Europe in the course of the crusades; and the immediate source was a Latin version, which Rudolf received from Bishop Wido of Cappel, near Zurich, between 1220 and 1223. The poem has been construed as Rudolf's rejection of the outlook of Gottfried von Straßburg (q.v.), the master on whom he had modelled himself.

Two other Middle High German versions are known, the *Laubacher Barlaam* (q.v.) of the late

12th c. and a late 13th-c. fragment. F. Pfeiffer's edn. of 1843 was reprinted in 1975.

BARLACH, ERNST (Wedel, Holstein, 1870–1938, Rostock), a richly gifted personality, who became sculptor, graphic artist, dramatist, and novelist, studied art in Hamburg, Dresden, and Paris. He lived in various towns, including Wedel and Berlin, settling in Güstrow, Mecklenburg, in 1910. A visit to Russia in 1906 and another to Florence in 1909 were decisive artistic experiences. In the 1920s, when Expressionism was in fashion, many honours were conferred upon him. He was made a member of the Berlin Academy in 1919 and of the Munich Academy in 1925, was awarded the Kleist Prize (for literature) in 1924 and the highest Prussian order Pour le mérite in 1933. Under the National Socialist regime Barlach's art was regarded as degenerate (see ENTARTETE KUNST) and his works were destroyed or impounded, and until his death he was subject to continuing attacks and provocations. He is buried in Ratzeburg. His sculptures include monuments in stone and bronze (at Cologne, Kiel, Lübeck, and Magdeburg) and numerous carvings in wood, one of which (*Lesender Klosterschüler*) plays a central part in the story *Sansibar oder Der letzte Grund* by A. Andersch (q.v.). Some of his lithographs are illustrations to his own literary works.

Barlach made a number of early experiments in writing, but his first works of importance were written when he was over 40. Between 1912 and 1929 he published seven Expressionistic plays; *Der tote Tag* (q.v., 1912) and *Der arme Vetter* (q.v., 1918) are both concerned with the relationship of father to son, a personal preoccupation linked with the birth of Barlach's illegitimate son in 1906. The other dramatic works, *Die echten Sedemunds* (1920), *Der Findling* (1922), *Die Sündflut* (1924), *Der blaue Boll* (1926), and *Die gute Zeit* (1929, qq.v.) frequently touch the same theme, but also represent the individual wrestling with the ties of the material world in search of God. The dramatic fragment *Der Graf von Ratzeburg* (q.v., 1951) presents this theme in particularly complex symbolism. Barlach's two novels (*Der gestohlene Mond* and *Seespeck*, both 1948) are unfinished.

The collected works, *Das dichterische Werk* (3 vols.), ed. K. Lazarowicz and F. Droß, appeared 1956–9 (*Die Dramen*, *Die Prosa I*, and *Die Prosa II*); they were followed by correspondence (2 vols.), ed. F. Droß, *Briefe aus den Jahren 1888–1924* and *Briefe aus den Jahren 1925–1938* (1969).

Barometermacher auf der Zauberinsel, Der, a farce by F. Raimund (q.v.), written in 1823 and first performed at the Theater in der Leopoldstadt, Vienna, in December 1823. It was published posthumously in vol. 3 of Raimund's *Sämtliche Werke* (1837). It is described as 'Zauberposse mit Gesang in zwei Aufzügen'. The fairy Rosalinde decides to allocate her gifts by chance, and the barometer-maker Quecksilber, living on the magic island, proves to be the recipient. He is given a wand which turns all things to gold, a horn to summon irresistible armies, and a sash which transports him whither he wishes. He admires Princess Zoraide, who relieves him of his magic gifts. Quecksilber then finds magic figs which give people large noses. With its aid he wins back the gifts, marries the Princess's maid, and leaves the Princess with an outsize nose. It is a pantomime with obvious humour and no sentimentality; it contains songs, arias, choruses, and a duet. Quecksilber speaks broad Viennese.

Baron von B., Der, a short anecdotal story by E. T. A. Hoffmann (q.v.), first published in 1819 in the *Allgemeine Musik-Zeitung*. Hoffmann included it in vol. 3 of his collection *Die Serapionsbrüder* (1820). It describes a baron of immense musical knowledge and taste, who falsely believes himself to be a great violinist. It illustrates the gulf which can separate the critical and the creative.

Baroque (Barock) was once a term of denigration applied to supposedly extravagant and grotesque examples of 17th-c. and 18th-c. architecture and even to absurdity or whimsicality in the abstract (e.g. 'a baroque idea' or 'conceit'). The word first acquired a respectable connotation through the art historian Jakob Burckhardt (q.v.). It was given a more precise definition in the work of Heinrich Wölfflin (q.v.), who at the turn of the century established baroque as a style in the arts of painting, architecture, and sculpture, in which dynamic energy, flux, swaying movement, and dramatic contrast, all declining to be fixed in space, replace the harmonious completeness and immobility of the classical, from which some of its elements are nevertheless derived. In the second decade of the 20th c. the concept 'Barock' was extended to German literature, notably by F. Strich, H. Cysarz, and E. Ermatinger.

The exponents of this new conception of 'Barockdichtung' considered what had previously been regarded as bombast and a degenerate taste for excessive ornament to be a conscious and purposefully designed style, which created a new technique of accumulation, synonymity, antithesis, and parallelism to express the flickering chiaroscuro of a world in which grandeur and suffering, pomp and death coexisted in the closest proximity. The most conspicuous figure of baroque literature is

Andreas Gryphius (q.v.), both as tragic dramatist and as poet. The high-flown rhetoric of Gryphius appears in augmented form in Lohenstein and Hofmannswaldau (qq.v.). The pregnant syntactical formalism of the style is evident in the mystical poetry of ANGELUS Silesius and Quirinus Kuhlmann (qq.v.). Among other exemplars, Friedrich von Spee, Weckherlin, Fleming, Hock, Zesen, and J. U. König (qq.v.) should be mentioned; in the later phases a satirical baroque prose is developed by C. Reuter and ABRAHAM a Santa Clara (qq.v.). Among the more conspicuous forms are dramas of political or religious violence (Gryphius and Lohenstein), funerary odes, and sonnets for the New Year reviewing the personal misfortunes of the old and the probable disasters of that about to dawn. Titles such as *Schlußgesang von der Flüchtigkeit und Nichtigkeit des menschlichen Lebens* (Zesen) are not surprising in an age which knew the Thirty Years War, the repeated ravages of the plague, and the gruesome and protracted execution of capital sentences. But readers should guard against too personal and intimate an interpretation of this poetry, which is frequently ceremonial and highly formal. Moreover, though it has perhaps been less emphasized, there is much baroque erotic poetry, some of it of a pastoral character.

The truly baroque features are not conspicuous in all the writers of the 17th c. The hymn writers in particular, of whom P. Gerhardt (q.v.) is the outstanding figure, belong to a popular tradition which can trace its descent from Luther; and the great novelist of the century, Grimmelshausen (q.v.), is perhaps the least baroque figure of all.

The temporal delimitation of the baroque age has been made difficult by the attempt to find some consonance between style in the various arts. It has often seemed convenient, but has in practice proved unfortunate, to treat 'Barock' as synonymous with the 17th c. The literary symptoms seem most clearly observable in the middle and later part of the century, though one school of critics opts for an earlier period (1580–1640). Baroque music is primarily associated with H. Schütz, with J. S. Bach (qq.v.), and with Handel running therefore well into the 18th c. The architecture, too, belongs to the late 17th c. and to much of the following century up to 1750. The problem is complicated by the concept of 'rococo' (see ROKOKO) and the difficulty of finding any point at which a frontier line can be drawn between baroque and rococo.

BARTELS, ADOLF (Wesselburen, 1862–1945, Weimar), a prolific historian of literature, was for a short time a journalist, after which he established himself in Weimar, where he

founded the Deutscher Schillerbund in 1907. His histories of literature include *Geschichte der deutschen Literatur*, 1901–2, and *Einführung in die Weltliteratur*, 1912–13; he also compiled a bibliographical handbook (*Handbuch zur Geschichte der deutschen Literatur*, 1906) and a lavishly illustrated historical study of the German peasant (*Der Bauer in der deutschen Vergangenheit*, 1900). He wrote historical novels (of which *Die Dithmarscher*, 1898, is the best known), the play *Martin Luther* (1903), and numerous poems. He was a tireless publicist, propagating his strongly nationalistic and violently anti-Semitic opinions. He was, as might be expected, an ardent advocate of Heimatkunst (q.v.) and came into his own under the National Socialists (see NSDAP).

BARTH, EMIL (Haan, 1900–58, Düsseldorf), a minor poet and novelist, worked as a printer until 1924, when he gave up printing for writing. He lived for a time in Munich, but spent the last twenty-five years of his life in Düsseldorf, except for a short interlude in Xanten. His restrained conventional poetry, much of it on subjects drawn from nature, is contained in the collections *Totenfeier* (1928), *Ex voto* (1933), *Xantener Hymnen* (1948), *Tigermuschel* (1946), and *Meerzauber* (1961). Barth also wrote the novels *Das verlorene Haus* (1936), *Der Wandelstern* (1939), *Das Lorbeerufer* (1943), and *Enkel des Odysseus* (1951).

BARTH, KARL (Basel, 1886–1968, Basel), ordained a pastor in the Reformed Church in 1911, adopted pacifist views under the stress of the 1914–18 War. His tract *Der Römerbrief* (1919) led to his appointment as professor in Göttingen in 1921. In 1925 he moved to Münster, and in 1930 to Bonn University. A confirmed opponent of National Socialism, he was dismissed in 1935. Between 1933 and 1935 he was prominent in the formation of the Bekennende Kirche (q.v.), and in 1935 accepted a chair at Basel University. He opposed Hitler (q.v.) in speech and writings and, at the same time, criticized the western democracies. After 1945 he returned to the standpoint that the Church should refrain from political engagement. Barth was the leading spirit behind the Barmen Declaration of 1934 which formulated the principles of the Bekennende Kirche, which are based on the belief in the sole revelation of God through Christ. His thorough reassessment of the Reformation owes some of its influence to his command of language. His theological dialectics are expressed in *Die christliche Dogmatik im Entwurf I* (1927), and, in considerably revised form, in a series of works beginning with *Die kirchliche Dogmatik*, vol. 1, 1932.

Until the mid-1920s Barth collaborated with Emil Brunner (1889–1966), professor of theology at Zürich University from 1924 to 1953 and well known in the U.S.A. Both were critical of the views of F. Schleiermacher (q.v.). Barth's essay *Nein! Antwort an Emil Brunner*, emanating from a doctrinal dispute, appeared in 1934.

BARTH, KASPAR VON (Küstrin, 1587–1658, Leipzig), a child prodigy in poetry, later studied at Wittenberg and Halle, wrote Latin poetry and composed the first Christian epic in German (*Der deutsche Phönix*, 1626).

BARTHEL, LUDWIG FRIEDRICH (Marktbreit, 1898–1962, Munich), a minor poet, served in both world wars and between the wars was an archivist in Munich. He began with hymnic verse in the manner of Hölderlin (q.v.), wrote political poetry and fiction in the Nazi period, and turned after 1945 to poetry of elegiac tone. His principal poetic publications are: *Gedichte der Landschaft, Gedichte der Versöhnung* (both 1932), *Dem inneren Vaterlande* (1933), *Komme, o Tag!* (1937), *Dom aller Deutschen* (1938), *Inmitten* (1939), *Liebe, du große Gefährtin* (1944), *Kelter des Friedens* (1952), *In die Weite* (1957), *Die Auferstandenen* (1958), *Das Frühlingsgedicht* (1960), and *Sonne, Nebel, Finsternis* (1961). *Das Leben ruft* (1935) contains stories; three Novellen were published separately: *Das Mädchen Phöbe* (1940), *Runkula* (1954), and *Hol über* (1961). He also wrote a novel, *Die goldenen Spiele* (1936).

BARTHEL, MAX (Dresden, 1893–1975, Waldbröl), a mason's son, wrote anti-war and revolutionary poetry. His socialist poetry was collected in *Verse aus den Argonnen* (1916) and *Arbeiterseele* (1920). He later became a National Socialist, publishing poetry (*Danksagung*, 1938), stories, and novels, including *Das Land auf den Bergen* (1939), *Die Straße der ewigen Sehnsucht* (1941), and *Das Haus an der Landstraße* (1942). *Das Haus, in dem wir wohnen. Neue Gedichte* appeared in 1963 and *Roter Mohn. Lieder und Gedichte* in 1964. The volume *Schulter an Schulter* (1934) was produced in collaboration with H. Lersch and K. Bröger (qq.v.).

BARTSCH, RUDOLF HANS (Graz, 1873–1952, Graz), the son of an army officer, spent his working life in the War Office archives in Vienna, retiring to Graz with the rank of captain in 1911. He was a prolific and popular writer of novels and stories, mostly touching on social problems and often tinged with sentimentality. Bartsch outlived his once considerable reputation. His first novel, *Als Österreich zerfiel . . . 1848* (1905), was reissued in 1913 under a new title—*Der letzte Student.* His novel of life in Graz, *Zwölf aus*

der Steiermark (1908), was the first of his works to make a distinct impact. *Elisabeth Kött* (1909) has an actress as its central figure. *Schwammerl* (1912), one of Bartsch's best-known works, is a sentimental version of the life of Schubert. Other novels include *Das deutsche Leid* (1911), *Die Geschichte von der Hannerl und ihren Liebhabern* (1913), *Frau Utta und der Jäger* (1914), *Lukas Rubesam* (1917), *Ein Landstreicher* (1921), *Seine Jüdin oder Jakob Böhmes Schusterkugel* (1921), *Grenzen der Menschheit*, a trilogy of which the parts are entitled *Der Königsgedanke, Der Satansgedanke*, and *Erlösung* (1924), *Die Salige* (1924), *Die Verliebten und ihre Stadt* (1927), *Der große und der kleine Klaus* (1931), *Das Lächeln der Marie Antoinette* (1932), *Lumpazivagabundus* (1936), and *Wenn Majestäten lieben* (1949). Bartsch's best work is probably found in his Novellen, of which he published the following collections: *Vom sterbenden Rokoko* (1909), *Bittersüße Liebesgeschichten* (1910), and *Unerfüllte Geschichten* (1916), which contains the story *Beethovens Weg zum Glück.*

BARUCH, LÖB, see BÖRNE, L.

BARUCH, MOYSES, real name of the novelist better known by his pseudonym Berthold Auerbach (q.v.).

BASEDOW, JOHANN BERNHARD (Hamburg, 1723–90, Magdeburg), made a name for himself in the second half of the 18th c. with writings on education. He instituted in Dessau under the patronage of Prince Leopold Friedrich von Dessau the Philanthropinum (1774), which incorporated his principles. He is one of the great educational reformers of the Aufklärung (q.v.), insisting on the teaching of the mother tongue, physical education, and the establishment of links between school work and the world outside.

His works include *Praktische Philosophie für alle Stände* (1758), *Theoretisches System der gesunden Vernunft* (1765), *Methodischer Unterricht in der überzeugenden Erkenntnis der biblischen Religion* (1764), *Vorstellung an Menschenfreunde und vermögende Männer* (1768), *Methodenbuch für Väter und Mütter der Familien und Völker* (1770), and *Elementarwerk* (1774), which last is his fundamental treatise on education.

Goethe records in Bk. 14 of *Dichtung und Wahrheit* (q.v.) the visit Basedow paid to him in Frankfurt in 1774, quoting the last four lines of the poem *Zwischen Lavater und Basedow*, which he wrote in 1774.

Basel, Peace of (Baseler Friede), signed on 5 April 1795 by Prussia and France (see REVOLUTIONSKRIEGE).

Baseler Konfession, statement of the twelve articles of belief of the Reformed Church (see REFORMIERTE KIRCHE) drawn up in 1534 and adopted at Mülhausen in 1537; it was valid in Basel until 1872.

Baseler Konzil, ecclesiastical council summoned in 1431 by Pope Martin V to consider measures of reform. Martin V died before the Council opened. The radical nature of the reforms proposed caused a conflict between Council and Pope Eugenius IV, which led to the deposition of the latter and the dissolution of the Council, which, however, continued to sit until 1449, moving in its last year to Lausanne. These quarrels led to the Great Schism of 1439–49. The reforms proposed by the Council, though temporarily adopted in some countries, were in the end abandoned.

BASIL, OTTO (Vienna, 1901–), a former *avant-garde* writer, living chiefly in Vienna, but also in Munich, published poetry: *Zynische Sonette* (1919), *Sonette an einen Freund* (1925), *Freund des Orients* (1940), *Sternbild der Waage* (1945), *Apokalyptischer Vers* (1947). He is the author of a novel, *Der Umkreis* (1933), and of a study of Nestroy (q.v., 1967).

BASSERMANN, ALBERT (Mannheim, 1867–1952, Zurich), famous German actor of the late 19th c. and early 20th c. Bassermann was a member of the Meiningen troupe (see MEININGER) from 1890 to 1895 and from 1899 was in Berlin. He made a name for himself in Ibsen, but was also successful in classical roles such as Mephistopheles, Egmont, Philipp II, and Wilhelm Tell.

BASSOMPIERRE, FRANÇOIS DE (1579–1646), created marshal of France for services to Louis XIII during the Huguenot rising of 1621, began his career at the court of Henri IV in 1598. His activities as a soldier and diplomat came to an abrupt end when Richelieu, suspecting his implication in a plot, had him arrested and imprisoned in the Bastille in 1631. He was not released until after Richelieu's death in 1643. While in prison he wrote his memoirs, which were published in Cologne in 1665 under the title *Mémoires du Maréchal de Bassompierre contenant l'histoire de sa vie*. Goethe incorporated two of his stories in an almost literal translation in his *Unterhaltungen deutscher Ausgewanderten* (q.v.). One of these, the night spent by the Marshal with a woman of low station, served as the basis for H. von Hofmannsthal's Novelle *Das Erlebnis des Marschalls von Bassompierre* (q.v., 1900). The other, also dealing with an illicit love, provided the central event in a Novelle by E. Strauß (q.v.) entitled *Der Schleier* (q.v., 1920).

Batrachomyomachia, a Latin translation of the *Froschmeuseler* by G. Rollenhagen (q.v.), made by J. Balde (q.v.).

BAUDISSIN, WOLF HEINRICH, GRAF VON (Copenhagen, 1789–1878, Dresden), a civil servant, gained a reputation as a translator. He contributed 13 plays to the Schlegel–Tieck translation of Shakespeare, 1825–33 (*The Merry Wives of Windsor, Measure for Measure, Comedy of Errors, Much Ado About Nothing, Love's Labour's Lost, Taming of the Shrew, All's Well that Ends Well, Henry VIII, Troilus and Cressida, Titus Andronicus, King Lear, Othello,* and *Antony and Cleopatra*). Baudissin also produced a modernized version of Hartmann's *Iwein* (q.v., 1845) and translated Molière (1865–7).

BAUER, LUDWIG AMANDUS (Orendelsall nr. Öhringen, Württemberg, 1803–46, Stuttgart), was a boyhood friend of the poet E. Mörike (q.v.), with whom he was at school in Urach. The two boys were the joint 'creators' of the imaginary island of Orplid (q.v.). Bauer, who became pastor in Ernsbach and later a teacher in a girls' school in Stuttgart, is the author of plays (including a verse trilogy on Alexander the Great and *Der heimliche Maluff,* 1828, which is connected with Orplid), and a novel, *Die Überschwänglichen* (1836). His *Schriften* appeared in 1847.

BAUER, WALTER (Merseburg, 1904–1976, Toronto), of working-class origin, became an elementary-school teacher, was taken prisoner by the British, returned to Germany after the war and emigrated to Canada in 1952, where, after many vicissitudes, he received a university appointment. His early poems were strongly political (*Kameraden zu euch sprech ich,* 1928; *Stimme aus dem Leunawerk,* 1929). Later, more conciliatory collections include *Gast auf Erden* (1943), *Dämmerung wird Tag* (1948), *Mein blaues Oktavheft* (1954), *Nachtwachen des Tellerwäschers* (1957—he was for a time a kitchen hand in Canada), and *Klopfzeichen* (1963). He is the author of several novels (*Ein Mann zog in die Stadt,* 1930; *Das Herz der Erde,* 1933; *Der Lichtstrahl,* 1937; *Besser zu zweit als allein,* 1950). A collection of prose and verse (1928–64), *Der Weg zählt, nicht die Herberge* appeared in 1964, and the volume *Lebenslauf* (1929–74) in 1976.

Bauer als Millionär, Der, see MÄDCHEN AUS DER FEENWELT, DAS.

BÄUERLE, ADOLF (Vienna, 1786–1859, Basel), was the author of more than seventy plays in the style of the Viennese popular theatre. Of middle-class origin, Bäuerle went on the stage. In 1806 he founded the *Theaterzeitung*, which appeared until 1860 and provided him with a livelihood. In 1813 he wrote his first play, *Kinder und Narren reden die Wahrheit*, for the Theater in der Leopoldstadt, in which his plays were to be acted for the next thirty years. He wrote three or four plays a year until 1827. From then until 1848 he restricted himself mainly to journalism. His opposition to the revolution of that year turned him from a popular figure into an object of hatred, and from then on his life was a hard struggle. With astonishing fertility he wrote in these years more than twenty novels dealing with Viennese life, all of which were avowed pot-boilers. His best work is in his plays (see LOKALPOSSE), in which details of Viennese life are combined with farcical episodes, all in Viennese dialect and seasoned with irresistible good humour. The most notable Lokalstücke are *Die Bürger in Wien* (1813, in which the comic character Staberl (q.v.) makes the first of his many appearances), *Staberls Hochzeit* (1814), *Der Fiaker als Marquis* (1816), *Die falsche Primadonna* (1818, owing something to Kotzebue's *Die deutschen Kleinstädter*, q.v.), and *Die schlimme Liesel* (1823). *Aline oder Wien in einem anderen Weltteile* (1822) and *Lindane* (1824) are Zauberstücke (see VOLKSSTÜCK). The popular Besserungsstück (q.v.) is most notably represented by *Wien, Paris, London und Konstantinopel* (1823).

BAUERNFELD, EDUARD VON (Vienna, 1802–90, Vienna), was the illegitimate son of a widow and bore his mother's maiden name, but his father married, not the widow, but her daughter, Bauernfeld's half-sister. The father, despite the illegitimacy, took a close interest in his education. Bauernfeld studied philosophy and law and in 1826 reluctantly entered the civil service, being much more inclined and gifted for the life of a man about town. He was a friend of the composer F. Schubert, of the artist M. von Schwind, and of his fellow civil servant Grillparzer (qq.v.); a gifted pianist himself, he frequented the musical soirées at which Schubert played and J. M. Vogl sang. Bauernfeld began with plays in the manner of Kotzebue (q.v.), but found his characteristic vein in the light comedy of manners, in which he was encouraged by J. Schreyvogel (q.v.), then in control of the Burgtheater in Vienna. *Leichtsinn aus Liebe* (q.v.) was successfully staged in 1831 and was quickly followed by *Das Liebes-Protokoll*, also played in 1831, *Helene* in 1833, *Die Bekenntnisse* in 1834, and *Bürgerlich und Romantisch* (qq.v.) in 1835. Bauernfeld was a rapid writer and wrote a number of other comedies which were produced in the 1830s and 1840s with less notable success. His 'dramatic fairytale' in verse, *Fortunat*, an early work refused by

the Burgtheater, made little impression when performed at the Theater in der Josefstadt in 1835. His principal achievements in the decade before the 1848 Revolution were the historical play *Ein deutscher Krieger*, performed in 1844, and the comedy *Großjährig* (q.v., 1846). Bauernfeld visited Paris and London in 1845, and was for a short time politically active as a Liberal, organizing in 1845 a petition for the abolition of the censorship of literature. In 1849 he resigned his civil service appointment and lived on his pension and the income earned by his plays. He continued to write indefatigably. The most noteworthy landmarks of his later career are the plays *Der kategorische Imperativ* (q.v.) produced in 1851, *Krisen* in 1852, *Aus der Gesellschaft* (q.v.) in 1867, *Moderne Jugend* in 1869, and *Die reiche Erbin* in 1876. An attempt with a historical verse play, *Landfrieden*, in 1869 was not successful. In his later years Bauernfeld was venerated as the symbol of an earlier Vienna and his seventieth and eightieth birthdays were accompanied by demonstrations of affection and by the conferment of honours. He wrote poetry (*Gedichte*, 1852) which was neat and unoriginal, and late in life he published a novel (*Die Freigelassenen*, 1875), which is interesting as a document of his age. The essays published in *Gesammelte Schriften* (12 vols., 1871–3) as *Aus Alt-und Neuwien* are a valuable reflection of Viennese intellectual life in the 19th c. This edition does not include all his plays. Mention should be made of Bauernfeld's political essays *Pia desideria eines österreichischen Schriftstellers* (1842) and *Schreiben eines Privilegierten in Österreich* (1847), liberal in message and moderate in tone.

Bauernkrieg, a peasants' revolt which began near Schaffhausen in 1524 and spread in 1525 over most of southern and central Germany. It was the most serious of six risings, of which the others took place in 1476 (see PFEIFER VON NIKLASHAUSEN), 1493, 1502, 1513 (see BUNDSCHUH), and 1514. Its causes were economic and social depression, coupled with the spiritual ferment of the Reformation (q.v.), and its aims, formulated in *Die Zwölf Artikel*, were also economic and social. They included free election of parish priests, abolition of tithes and of serfdom, extension of hunting and fishing rights to all, and the removal of all sovereigns except the Emperor. The principal leaders were the peasant Georg Metzler, the preacher Thomas Münzer (q.v.), and the knights Götz von Berlichingen and Florian Geyer (qq.v.). The initial successes of the peasants could not be maintained. Illdisciplined and badly led, they were defeated by trained forces at Böblingen (Württemberg), at Königshofen and Sulzdorf (Franconia), Schwer-weiler (Alsace), and Frankenhausen (Thuringia), all within the space of a few weeks. Luther (q.v.) at first sought to moderate the revolt (*Ermahnung zum Frieden*) and then condemned it root and branch (*Wider die räuberischen und mörderischen Rotten der Bauern*). Except at Weinsberg, the revolt was not marked by flagrant outrages. On its suppression the participants were punished with great ferocity. The failure of the revolt re-established the existing social order, increased the power of the princes, and turned large numbers of South Germans and Westphalians away from the Reformation.

The Bauernkrieg is portrayed by Goethe in *Götz von Berlichingen* (q.v., 1773) and by Gerhart Hauptmann in *Florian Geyer* (q.v., 1896).

Bauernspiegel, Der, oder *Lebensgeschichte des Jeremias Gotthelf, von ihm selbst beschrieben,* a novel written by Jeremias Gotthelf (q.v., Albert Bitzius) in 1836 and published in 1837. It is Gotthelf's first important work and reflects his concern as a pastor both with poverty and love of money among the rural population and with the administration of the meagre provision made by the state for paupers and orphans. The novel purports to be written by its principal character Jeremias Gotthelf, and the name was retained by the author as a pseudonym for his subsequent novels and stories.

Jeremias is the favoured grandson of a well-to-do farmer couple, who exploit their children, as was the custom, as unpaid farm servants. They arrange a match for the eldest son with a rich farmer's daughter and presently find themselves pushed on one side. Jeremias's father is so incensed at the attempt to hand him, with his brothers and sisters, over to his eldest brother as farm serfs, that he leaves and sets up as a tenant on a small-holding. But he is cheated by his landlord and is unable to make it pay. He is killed tree-felling. His estate is bankrupt, and Jeremias is auctioned as a pauper orphan. The book recounts his miseries, not usually caused by active maltreatment but by thoughtlessness and universal contempt for the pauper. He grows up a strong worker, but he can find no way of making himself independent. He and a maid, Anneli, fall in love, and the relationship is tenderly portrayed. They plan marriage, but the obstacles are insuperable. Anneli dies in childbirth and her child with her. Jeremias, now close to despair, takes to drink and brawling, and finds himself in trouble with the law. He flees to France and joins the Swiss Guard (it is in the time of Charles X), in which he pulls himself together. But the July Revolution breaks out, his regiment is disbanded, and he returns to Switzerland. Everyone looks askance at him, and when he falls ill he is poorly treated in hospital. He then inherits some money from a former comrade, and seeks to make himself respectable and useful. But no one will have him as schoolmaster or parish clerk, and he occupies himself as an amateur tutor to children of the village. At the close Jeremias is seriously ill, but the prospect of becoming parish clerk, if he recovers, has opened.

Der Bauernspiegel is a product of Gotthelf's social indignation, fed by first-hand experience. It is, as he puts it, a one-sided mirror which reflects the dark side of the peasant's life. In homely and robust language, he portrays Swiss country folk and their life, shows an unerring eye for concealed or unconscious motive, and has an infinite compassion for the suffering of the humble and the inarticulate.

Bauhaus, a 'comprehensive' art school, which was founded in Weimar in 1919 with official support (Staatliches Bauhaus) under the direction of W. Gropius (1883–1969). It stressed the interdependence of the plastic arts under the primacy of architecture, and the importance of craftsmanship. Gropius recruited a number of notable teachers, among them L. Feininger, W. Kandinsky, and P. Klee (qq.v.). In 1925 the Bauhaus moved to Dessau as the Hochschule für Gestaltung, though the familiar name remained in use. It was closed in 1932, reopened in Berlin in 1933, but was almost immediately closed for good by the National Socialists.

In 1937 the 'New Bauhaus' was opened in Chicago under the direction of L. Moholy-Nagy (1895–1946), one of the artist-teachers from Dessau.

BAUM, VICKI (Vienna, 1888–1960, Hollywood), studied music in Vienna, was an orchestral player in Darmstadt, and after marrying the conductor L. Lert entered the editorial office of the publishing firm Ullstein in 1926. She began in 1919 a series of light novels which combined dramatic events, erotic complications, a vivid and up-to-date contemporary social background, and a discreet dose of sentiment. Published under her maiden name, they include *Frühe Schatten* (1919), *Der Eingang zur Bühne* (1920), *Die Tänze der Ina Raffay* (1921), *Welt ohne Sünde* (sub-titled *Roman einer Minute* and an interesting essay in interior monologue, 1922), *Ulle, der Zwerg* (1924), *Ferne* (1926), *Hell in Frauensee* (1927), *Stud. chem. Helene Willfüer* (1929), and *Zwischenfall in Lohwinkel* (1930). Her outstanding international success was *Menschen im Hotel* (q.v., 1929), which was equally successful as the American film *Grand Hotel*. She went to America to supervise the filming of this work, remained there, and acquired American nationality. She continued until 1937 to write novels in German. From then until shortly before her death she wrote in English. These later novels were also published in German translation. She is also the author of Novellen (*Die andern Tage*, 1922) and several separate stories.

BAUMBACH, RUDOLF (Kranichfeld, 1840–1905, Meiningen), studied at various universities, became a private tutor, then a journalist, and finally librarian to the Duke of Meiningen. He wrote much unpretentious lyric poetry (*Lieder eines fahrenden Gesellen*, 1878; *Von der Landstraße*, 1881; *Spielmannslieder*, 1881; *Mein Frühjahr*, 1882; *Krug und Tintenfaß*, 1887) and two epics after the manner of J. V. von Scheffel, q.v. (*Horand und Hilde*, 1879, and *Frau Holde*, 1881). His song 'Keinen Tropfen im Becher mehr' is still sung by students. Baumbach is one of the poets against whom the term 'Butzenscheibenpoesie' (q.v.) was directed.

BÄUMER, GERTRUD (Hohenlimburg, Westphalia, 1873–1954, Bethel nr. Bielefeld), daughter of a Protestant school inspector, had her schooling at Halle and Magdeburg and became an elementary school teacher in 1892. From 1898 to 1904 she studied at Berlin University and thereafter devoted her life to women's causes. With Helene Lange, F. Naumann, and Theodor Heuß (qq.v.) as friends and collaborators, she was active in journalism, editing *Die Hilfe* (q.v., 1912–43) and *Die Frau* (1893–1944). From 1920 to 1933 she was a member of the Reichstag for the Democratic Party. With Helene Lange she edited the *Handbuch der Frauenbewegung* (1901–6), singling out the work of L. Otto-Peters (q.v.) in *Luise Otto-Peters* in 1939. She is the author of a study of Rilke (*Ich kreise um den Gott*, 1935) and of biographies of Dante (*Die Macht der Liebe*, 1942), Goethe's mother (*Frau Rat Goethe*, 1949), and R. Huch (q.v., *Ricarda Huch*, 1949). Her historical novels include *Adelheid, Mutter der Königsreiche* (1936–7), *Der Park* (1937), *Der Berg des Königs* (1938), and a three-volume novel on Otto III (q.v.), *Der Jüngling im Sternmantel, Größe und Tragik Ottos III* (1949), which is her most substantial work. *Im Licht der Erinnerung* (1953) is autobiographical.

BAUMGART, JOHANNES (Meißen, 1514–78, Magdeburg), a Lutheran pastor, is the author of a play, *Das Gericht Salomonis* (1561), in which the judgement of the men of the law is contrasted unfavourably with the wisdom of Solomon.

BAUMGARTEN, ALEXANDER GOTTLIEB (Berlin, 1714–62, Frankfurt/Oder), a disciple of C. Wolff (q.v.), appointed in 1740 professor at Frankfurt, was the first to use the term aesthetics to denote a sphere of philosophy (*Aesthetica acrodinatica*, 1750–8, unfinished). He is mentioned in Lessing's *Laokoon* (q.v.).

Bavaria, see BAYERN.

Bayerische Chronik, see FÜETRER, ULRICH.

Bayern (Bavaria), the largest state in the Federal Republic in area and the second largest in population, is predominantly Catholic. Bavaria was one of the early German duchies and Regensburg was its capital until 1255. The Wittelsbach (q.v.) family, though often divided, ruled from 1180 to 1918. The Duke of Bavaria became an Elector (Kurfürst) in 1623, in succession to the Elector of the Palatinate, Friedrich V (q.v.), who was deprived of his electoral privilege after defeat in the Thirty Years War (see DREISSIGJÄHRIGER KRIEG). The new Elector also received the Palatinate (see PFALZ, DIE), which continued under Bavarian rule until 1945. During the 18th c. Bavaria sought unsuccessfully to rival Austria as the leading state in Germany. In 1805 Napoleon conferred on the Elector the title of king (see MAXIMILIAN I). The electoral title lapsed in the following year with the extinction of the Holy Roman Empire. In the 19th c. Bavaria became a centre of the arts, especially painting and sculpture, which were greatly encouraged by King Ludwig I (q.v.). The Revolution (see REVOLUTIONEN 1848–9) provoked his abdication, and he was succeeded by Maximilian II (q.v.), who reigned in a more liberal style and encouraged literature and the arts. Munich (see MÜNCHEN) became the home of many men of letters, and a school of poets bears its name (see MÜNCHNER DICHTERKREIS). Maximilian was succeeded by Ludwig II (q.v.), who is well known for his patronage of R. Wagner (q.v.), his three new extravagant residences, his eccentricity, and his tragic death. He consented in 1871 to the inclusion of Bavaria in the new German Empire. After hard bargaining Bavaria retained certain privileges, including its own railways, post, diplomatic representation, and, with limitations, its own army. Ludwig's successor Otto was mentally incapable of ruling, though he lived until 1916. From 1886 to 1912 the Prince Regent Luitpold ruled shrewdly and unobtrusively. Otto remained nominally king until 1913, when Ludwig III succeeded him, only to flee abroad in 1918 without abdicating. The Freistaat Bayern was for a short time in 1918 a socialist republic under the Communist K. Eisner (q.v.), whose assassination in February 1919 led to the proclamation of a Soviet-type republic (Räterepublik). This was abolished by Prussian and Württemberg military intervention.

Bayreuth, town in Franconia included in Bavaria since 1810. Bayreuth, with Ansbach, was a small independent state until 1797 when it was incorporated in Prussia for nine years. It is best known for the Wagner festival now held annually in the Festspielhaus built in 1872-6 with the encouragement of King Ludwig II (q.v.) of Bavaria (q.v.). Richard Wagner is buried in the grounds of his house, the villa Wahnfried. The composer Franz Liszt (q.v.) also died at Bayreuth and is buried in the cemetery.

Bayrischer Erbfolgekrieg, the War of the Bavarian Succession, was precipitated by the extinction of the direct Wittelsbach line of Bavarian Electors in 1777. The succession passed to a collateral Wittelsbach, Karl Theodor, Elector of the Palatinate, who, having little interest in his new acquisition, agreed to cede Lower Bavaria to Austria (1778). Prussia and Saxony intervened to prevent this extension of Austrian territory and power, and an ineffective war broke out, in which the armies marched and countermarched without joining in any general engagement. Since most of the manœuvres were concerned with supply lines, the campaign was derided as the 'Potato War' (Kartoffelkrieg). By the Peace of Teschen 1779 Austria renounced its claim to all but a small area (Innviertel with the towns Braunau and Schärding); and the Prussian monarchy secured recognition of its hereditary rights in Ansbach and Bayreuth.

BEATUS RHENANUS, otherwise Beat Bild of Rheinau (Schlettstadt, 1485–1547, Strasburg), was a scholar of the New Learning of Erasmus (q.v.). By trade a printer, he lived in Basel, and in spite of his humanistic contacts remained loyal to the old religion. His historical work *Rerum Germanicorum libri tres* (1531) is an important step in the development of historical objectivity. He also wrote a commentary on the *Germania* of Tacitus and published editions of Velleius Paterculus, Tacitus (*Annales*), and Livy.

BEBEL, AUGUST (Cologne, 1840–1913, Passugg, Switzerland), a prominent German socialist leader, was a founder member of the Social Democratic Party in 1869 (see SPD). He entered the Reichstag in 1867, representing first a country constituency, then Dresden, and finally Hamburg; he was twice imprisoned. He wrote numerous political books of which the best known is *Die Frau und der Sozialismus* (1883).

BEBEL, HEINRICH (Ingstetten, Württemberg, 1472–1518, Tübingen), a peasant's son, became a professor at Tübingen in 1497. He was crowned poet (gekrönter Dichter, q.v.) by Maximilian I (q.v.) in 1501. Bebel was a skilled Latinist and the principal German exponent of the 'Facetie'

(see FAZETIE), a form in which he displayed wit, satirical edge, and elegant form. His collections bear the titles *Libri facetiarum iucundissimi* (1512) and *Novus liber facetiarum* (1514).

Bebenhausen, Cistercian monastery to the north of Tübingen, founded in 1185 and secularized in 1560. It is commemorated in *Bilder aus Bebenhausen,* 11 short poems in elegiacs, written in 1863 by E. Mörike (q.v.) on the occasion of a stay of some weeks in the former monastery.

BECHER, JOHANNES ROBERT (Munich, 1891–1958, Berlin), the son of a Bavarian judge, revolted in adolescence against his upper-middle-class environment, refused military service in 1914, and joined the Spartacists (see SPARTA-KUSBUND) in 1918 and the Communist Party in 1919. His early poetry is Expressionist and ecstatic in style (*Der Ringende,* 1911; *An Europa* and *Verbrüderung,* both 1916; *An Alle!,* 1919). He was a Communist deputy in the Reichstag and in 1925 was accused of treason on the ground of his poems *Der Leichnam auf dem Thron* (1925). The case, at first dropped, was renewed in 1927 after the publication of his anti-war novel *Levisite oder Der einzig gerechte Krieg* (1926); the proceedings, however, were again abandoned. In 1933 Becher emigrated, first to Austria, then to France, and finally, in 1935, to Russia. In 1945 he returned to Germany, settling in the East Sector of Berlin and becoming in 1954 Minister of Culture in the DDR. He was also president of the East German Akademie der Künste (1953–6). His autobiographical novel *Abschied* (1940) covers his early life up to the beginning of the 1914–18 War. In the 1930s, under Soviet influence, his style shifted from its earlier experimental vigour to socialist realism (see SOZIALISTISCHER REALISMUS), a development which became more pronounced after his return to Germany (*Neue deutsche Volkslieder,* 1950; *Deutsche Sonette 1952,* 1952). His *Schlacht um Moskau,* conceived in 1941, was subsequently retitled *Winterschlacht. Dramatische Dichtung* (1953). Some of his own poetry was influenced by the German baroque and classical tradition; he edited *Hölderlin. Dichtungen* (1952) and *Tränen des Vaterlandes. Deutsche Dichtung aus dem 16. und 17. Jahrhundert* (1954).

Der Aufstand im Menschen (partly unpublished prose) appeared in 1983 and *Gesammelte Werke,* ed. by the Johannes R. Becher Archiv der Deutschen Akademie der Künste in East Berlin (18 vols.) in 1966–81.

BECHSTEIN, LUDWIG (Weimar, 1801–60, Meiningen), a prolific poet and novelist, began life as an apothecary, and, following the publication in 1828 of *Sonettenkränze,* was assisted financially by the Duke of Meiningen, whose service he entered in 1831 as librarian. In addition to volumes of poetry, Bechstein wrote many historical novels (including *Die Weissagung der Libussa*, 1829; *Das tolle Jahr*, 1833; *Der Fürstentag*, 1834; *Der Dunkelgraf*, 1854), and collected fairy-tales, which he published in *Deutsches Märchenbuch* (1845), *Die Volkssagen, Märchen und Legenden des Kaiserstaates Österreich* (1848), and *Neues deutsches Märchenbuch* (1856).

BECK, HEINRICH (Gotha, 1760–1803, Mannheim), an actor and later a theatre director, was a minor dramatist whose plays include the Schauspiele *Verirrung ohne Laster* (1788), *Das Herz behält seine Rechte* (1788), *Lohn der Liebe* (1789), *Rettung für Rettung* (1801), and the comedies *Alles aus Eigennutz* (1793), *Die Quälgeister* (1802), *Das Chamäleon* (1803).

BECK, KARL ISIDOR (Baja, Hungary, 1817–79, Währing nr. Vienna), studied at Vienna and Leipzig universities, and lived a restless life in Vienna, Berlin, Budapest, Hamburg, and Switzerland. He was a supporter of the group Junges Deutschland (q.v.), but after 1849 became disillusioned with left-wing politics. He wrote poetry, both political and lyrical, collected as *Nächte, Gepanzerte Lieder* (1838), *Gedichte* (1844), *Monatsrosen* (1848), *Juniuslieder* (1853), and *Still und bewegt* (1870), and composed two verse epics, *Janke, der ungarische Roßhirt* (1841) and *Jadwiga* (1863). His work includes a tragedy, *Saul* (1841).

BECKER, CHRISTIANE (Crossen/Oder, 1778–97, Weimar), a gifted young actress of the Weimar theatre, was the daughter of an actor, Christian Neumann. She married the Weimar actor Heinrich Becker in 1793. As Christiane Neumann her first important part was Prince Arthur in Shakespeare's *King John*. By 1793 she was playing Minna von Barnhelm and Emilia Galotti. Her early death is lamented in Goethe's *Euphrosyne* (q.v.). Her monument, designed by H. H. Meyer (q.v.), was erected in the Weimar park in 1800.

BECKER, JUREK (Łódź, 1937–), spent almost his entire childhood in Jewish ghettos and concentration camps, which he and both his parents survived. The family settled in 1945 in East Berlin, where he learnt German, qualified for university entry, and studied philosophy. A free-lance writer since 1960, he gained prominence through his first novel, *Jakob der Lügner* (1969, DDR; 1970, BRD), and two more novels published in the DDR as well as in the West, *Irreführung der Behörden* (1973) and *Der Boxer* (1976). As one of the signatories to the

public protest letter concerning the expulsion of W. Biermann (q.v.) he was deprived of his SED membership, as a result of which he left the Schriftstellerverband in 1977, in which year his request for a two-year stay in the West on his condition of possible re-entry was granted.

Jakob der Lügner, which has had a wide circulation in the West (film version directed by F. Beyer), is unique among works dealing with the plight of the Jews in the Polish Ghetto during the final stages of the 1939–45 War. Jakob Heym, following an order by a German soldier on patrol, reports at the local military HQ for a supposed offence against curfew rules. Whilst waiting for a hearing, which results in his release, he listens to a radio report on the military situation, which he is not supposed to know. This gives him the idea of pretending ownership of a radio, the real possession of which carries with it the death penalty. The motive behind the deceit is to raise the spirits of his fellow sufferers by supplying them with 'authentic' news about the proximity of the Russian troops. He succeeds by his ingenious inventiveness. But Jakob cannot help those who are singled out for punishment, deportation, and death, and in a moment of physical and mental exhaustion he reveals the truth, the devastating effect of which results in the suicide of his close associate Kowalski. Jakob, feeling directly responsible for his death, is left in utter isolation. At this point the narrator relates the first of the novel's two alternative endings. Kowalski is not dead and Jakob, whilst pretending to own a radio, henceforth refuses to reveal any news and withdraws into isolation. His suffering is aggravated by the hostile reaction of a majority within the community. Again he gives up hope for the timely arrival of the Russian liberators and risks flight through the Ghetto's wire fencing. He is discovered and shot. In the second ending he is amongst those who find themselves in a train conveying Jews to a concentration camp; he himself has no illusion about its destination. The fictitious first-person narrator fulfils the function of the detached observer, who arranges the story and its possible endings (though in the first ending he joins those who are hostile towards Jakob). His own experience and sentiments dominate the formal frame.

The novel's powerful effect derives from the interplay of comedy and pathos, reminiscent of the Yiddish tradition and inherent in the emphasis on the fictitious nature of the story with its overriding pathos. It also promotes the reader's involvement in human reactions to a given situation. Stripped of its documentary character, the past is integrated into life itself. The will to live and to rebuild a future promising self-fulfilment also characterizes Aron Blank in *Der Boxer*. A survivor of the concentration camp who has lost his entire family, Aron persuades himself that he has found his own long-lost son in the boy Mark, but the relationship ends in failure and the presumed death of Mark in Israel. Aron is again left in utter loneliness. Works published only in the West include the novel *Schlaflose Tage* (1978). *Aller Welt Freund* appeared in 1982 (BRD; 1983, DDR). *Nach der ersten Zukunft* (1980) is a collection of twenty-five stories.

Following the publication of his first novel Becker was awarded the Heinrich-Mann-Prize and the Charles-Veillon-Prize in 1971.

BECKER, NIKOLAUS (Bonn, 1809–45, Hunshoven), a civil servant in the legal department in Bonn and, later, in Geilenkirchen, published a volume of poems (*Gedichte*) in 1841 which contained the famous patriotic Rhine song 'Sie sollen ihn nicht haben, den freien deutschen Rhein' (see FREIE DEUTSCHE RHEIN, DER).

BECKER, RUDOLF ZACHARIAS (Erfurt, 1751–1822, Gotha), a schoolmaster in Gotha, turned to literature and achieved a remarkable commercial success with his *Not- und Hilfsbüchlein für Bauersleute* (1788), which contributed much to the education of countless humble and inarticulate readers. It was followed by his *Didaktisches Not- und Hilfsbüchlein oder Lehrreiche Freuden- und Trauergeschichte des Dorfs Mildheim* (2 vols., 1797–8) and by his *Mildheimisches Liederbuch* (q.v., 1799). He edited the journal *Der allgemeine Reichsanzeiger* from 1791. Becker was an early champion of copyright (*Das Eigentumsrecht an Geisteswerken*, 1789). He visited the house of Schiller's future mother-in-law at Rudolstadt, and his character is praised in one of Schiller's letters. Public-spirited and patriotic, he was imprisoned for a time during the Napoleonic occupation. After Schiller's death he conducted an unsuccessful campaign for the erection of a monument to the poet.

BECKH, JOHANN JOSEPH, the dates of whose life and death are unknown, was a notary active in Strasburg in the middle of the 17th c. He was the author of religious poetry (*Geistliches Echo*, 1660, *Sichtbare Eitelkeit und unsichtbare Herrlichkeit*, 1671), and of a novel, *Die elbianische Florabella* (1677). His plays, written in prose, include *Theagenes und Chariklea*, *Der Schauplatz des Gewissens* (both 1666), *Die wiedergefundene Liarta*, and *Polinte* (both 1669). In his later years he is known to have lived in Eckernförde.

BECKMANN, MAX (Leipzig, 1884–1950, New York), an Expressionist painter, held a chair of art at the Städelschule in Frankfurt, from which he was dismissed in 1933. In 1937 he emigrated to Holland, and in 1947 to the U.S.A., teaching at St. Louis until his professorial appointment at the Brooklyn Museum in the year before his death. His graphic work includes a set of illustrations to Goethe's *Faust* (q.v., 1957). His diaries, *Tagebücher 1940–50*, and his letters, *Briefe im Kriege*, were both published in 1955.

Beckmesser, SIXTUS, a character in R. Wagner's *Die Meistersinger von Nürnberg* (q.v.); he is a clerk and also a Meistersinger. Beckmesser is the epitome of the narrow and pedantic critic.

BEER, JOHANN (St. Georgen, Austria, 1655–1700, Weißenfels), a Protestant by upbringing, emigrated with his parents from Austria to Regensburg in 1670. In 1676 he entered Leipzig University, but in the same year was accepted into the private choir of the Duke of Weißenfels at Halle. In addition to singing, he composed and wrote on music (*Musikalische Discurse*, published posthumously in 1719). He became the Duke's chief musician (Konzertmeister) in 1685, and court librarian as well in 1697. In 1680 he accompanied the court when it moved to Weißenfels.

A rich and ebullient character, Beer possessed also a considerable talent for story-telling, which he exercised in some twenty novels, all written before 1684, but published under pseudonyms between 1677 and 1704. They include *Der simplicianische Welt-Kucker* (1677–9), *Der Abentheuerliche und unerhörte Ritter Hopffensack* (1678), *Printz Adimantus und der Königlichen Princeßin Ormizella Liebes-Geschicht* (1678), *Jucundi Jucundissimi wunderliche Lebensbeschreibung* (1680), *Der berühmte Narren-Spital* (1681), *Der politische Feuermauer-Kehrer* (1682), *Der politische Bratenwender* (1682), *Zendorii à Zendoriis Teutsche Winternächte* (1682), *Die kurtzweiligen Sommer-Täge* (1683), and *Der verkehrte Staatsmann* (1700). Beer is a fluent narrator and has little concern for character or motive; he writes in a popular manner, which is sometimes vulgar and occasionally obscene.

Among Beer's outlandish pseudonyms were Wolfgang von Willenhag, Antonio Caminero, and Amando de Bratimero, and he also attributes novels to Simplicius (the character created by Grimmelshausen, q.v.) and to Expertus Robertus from Moscherosch's *Philander von Sittewalt* (see MOSCHEROSCH, J. M.). These attributions disguised his identity until the 20th c. His authorship was detected by R. Alewyn in 1932. His autobiography was discovered even more recently and was published in 1965 as *Johann Beer. Sein Leben von ihm selbst erzählt* by A. Schmiedecke with a preface by R. Alewyn.

BEER, MICHAEL (Berlin, 1800–33, Munich), brother of the composer Meyerbeer in 1823 wrote a one-act play, *Der Paria*, in support of Jewish emancipation. Goethe's advocacy secured its performance in Weimar in 1824. He also wrote the tragedies *Klytämnestra*, 1819, and *Struensee*, 1829; for the latter his brother provided incidental music.

BEER-HOFMANN, RICHARD (Vienna, 1866–1945, New York), a well-to-do Viennese of Jewish descent, was a friend of H. von Hofmannsthal and A. Schnitzler (qq.v.). He emigrated to the U.S.A. on the German invasion of Austria in 1938. He first published *Novellen* (1893), but he became better known as the author of the play *Der Graf von Charolais* (1904), based on *The Fatal Dowry* of 1632 by Massinger and Field. Towards the end of his life he concentrated on themes connected with Jewish history, particularly King David (*Jaákobs Traum*, 1918; *Der junge David*, 1933; *Vorspiel auf dem Theater zu König David*, 1936). A trilogy intended to end with David's death was never completed. In 1919 he published *Schlaflied für Mirjam*.
   Gesammelte Werke (1963, preface by M. Buber, q.v.) contains *Paula* (ed. O. Kallir), an autobiographical fragment. Correspondence with H. von Hofmannsthal (q.v.) appeared in 1972.

BEETHOVEN, LUDWIG VAN (Bonn, 1770–1827, Vienna), went to Vienna in 1792 and spent the remainder of his life there. Though never prosperous, he was able, as a musician of eminence, to frequent the houses of the nobility, including those of the Lobkowitz, Lichnowsky, and Brunswick families, and he was on particularly friendly terms with the young Archduke Rudolf, with whom the Sonata op. 81a (*Les Adieux*) is associated, and for whose consecration as archbishop of Olmütz the *Missa Solemnis* was designed.
   Beethoven wrote 9 symphonies, 7 concerti, 16 string quartets, and 32 piano sonatas, as well as other orchestral, chamber, and keyboard works. His only opera, *Fidelio* (q.v.), was performed in 1805. His incidental music for Goethe's *Egmont* (q.v.) was composed in 1810, and he also wrote an overture for the *Coriolan* of H. J. von Collin (q.v.). His best-known literary setting is the fourth movement of the Ninth Symphony, the text of which consists of the first five stanzas of Schiller's *An die Freude* (q.v.). The oratorio *Christus am Ölberge* (op. 85, 1802) uses a text by F. X. Huber (born 1760, date of death unknown).
   Beethoven wrote a large number of songs (Lieder); estimates vary from 67 to almost 100 according to the definition of 'Lied' (q.v.). He set

several poems by Goethe (*Mailied, Marmotte*, 'Kennst du das Land', *Neue Liebe, neues Leben*, the 'Flohlied' from *Faust* (q.v.), *Wonne der Wehmut, Sehnsucht, Mit einem gemalten Band*, 'Nur wer die Sehnsucht kennt', of which he made four settings, and the two songs from *Egmont*, q.v., 'Die Trommel gerühret' and 'Freudvoll und leidvoll'). Other poets set include A. Tiedge (two poems), C. F. Gellert (a group as *Sechs Lieder*), M. Claudius (*Urians Reise um die Welt*), G. A. Bürger (*Mollys Abschied, Das Blümchen Wunderhold, Seufzer eines Ungeliebten*), J. G. Herder (*Die laute Klage*), and F. von Matthisson (*Adelaide*, q.v., *Opferlied*, and *Andenken*, q.v.). The continuous song cycle *An die ferne Geliebte* (q.v.) sets six songs by Alois Jeitteles (q.v.).

Befehlerles, a short story in *Schwarzwälder Dorfgeschichten* (vol. 1, 1843) by B. Auerbach (q.v.). It consists of two anecdotes. In the first the farmer's son Matthes is wrongly kept in custody on a trivial charge. In the second the unjust magistrate of the first anecdote issues a decree that the traditional carrying of an axe by each married man shall cease. The villagers, led by Buchmaier, reject the decree. They are fined, but the magistrate is removed. The anecdotes are a denunciation of the arrogance of minor judicial officials.

'Befiehl du deine Wege', first line of the hymn *Christliches Wanderlied* by P. Gerhardt (q.v.).

Befreiungskriege, see NAPOLEONIC WARS.

Begegnung, a poem written by E. Mörike (q.v.) in 1828 or 1829. It has been set to music by Hugo Wolf (q.v.).

BEHEIM or **BEHAIM,** MICHAEL (Sülzbach nr. Weinsberg, 1416–*c.* 1472, Sülzbach), a wandering singer whose family originated in Bohemia (hence 'Beheim'). He began life as a weaver and was afterwards in the service of various great lords, including the Emperor Friedrich III, (q.v.) and the Count Palatine Friedrich I (1425–76). In later life he became mayor of his native town, and he there met his end by murder. He wrote a number of songs, which have affinities with Meistergesang (q.v.), and also two verse chronicles. The *Buch von den Wienern* describes the siege of the Hofburg in Vienna by the rebellious citizens in 1462, in the reign of Friedrich III. It is based on his own experiences and his fright is indicated by its heading, 'in grossen angsten'. In 1469 he wrote his second chronicle, *Das Leben Friedrichs I von der Pfalz*. Beheim was an undistinguished poet, whose pen secured him patronage.

BEHEIM-SCHWARZBACH, MARTIN (London, 1900–), was brought up in Hamburg, where he settled, and was called up in the closing stages of the 1914–18 War. His narrative and lyrical range extends from the humorous to the occult (abandoned in his later works). His versatile production includes Novellen (*Lorenz Schaarmanns unzulängliche Buße*, 1928; *Der kleine Moltke und die Rapierkunst*, 1929; *Das verschlossene Land*, 1932; *Der geölte Blitz*, 1933; *Die Todestrommel*, 1935; *Der Schwerttanz*, 1938) as well as the collections *Die Runen Gottes* (1927), *Gleichnisse* (1948), *Der magische Kreis* (1955), and *Das kleine Fabulatorium* (1959). He is the author of novels (*Die Michaelskinder*, 1930; *Die Herren der Erde*, 1931; *Der Gläubiger*, 1934; *Die Verstoßene*, 1938; *Der Unheilige*, 1948; *Die Insel Matupi*, 1955) and has written a biography of Novalis and Christian Morgenstern (qq.v.). He has continued to produce volumes of fiction and poetry (*Der Liebestrank*, 1975) into the 1980s.

BEHEMB, MARTIN (Lauban, Silesia, 1557–1622, Lauban), pastor in Lauban, wrote hymns (published in three collections 1606, 1608, and 1614) and also three plays for school performance, *Holofernes und Judith*, *Tobias*, and *Acolastus*, all printed in 1618. *Acolastus* treats the story of the Prodigal Son. The plays have a strong popular element with rural scenes and Silesian dialect. Behemb's name also appears as Behm, Böhme, and Bohemus.

Beherzigung, a humorous pseudo-philosophical poem written by Goethe about 1777 and first published in 1789. It has been set to music by Hugo Wolf (q.v.).

BEHM, MARTIN, see BEHEMB, MARTIN.

BEHRENS, BERTHA, wrote under the pseudonym W. Heimburg (q.v.).

BEHRISCH, ERNST WOLFGANG (Dresden, 1738–1809, Dresden), an eccentric friend of Goethe during the latter's residence and study at Leipzig University (1765–8). Behrisch, who was tutor to a young count, Graf Lindenau, is described by Goethe in *Dichtung und Wahrheit* (q.v.) as 'einer der wunderlichsten Käuze, die es auf der Welt geben kann'. He later became tutor in the family of the Prince of Anhalt-Dessau.

Bei Betrachtung von Schillers Schädel, see IM ERNSTEN BEINHAUS WAR'S.

Beichtformeln, designation for formulae of confession, of which a number survive from the Middle Ages. They were for the use of the priest, whether for public reading as *confessio publica*, or as a basis for the conduct of the *confessio privata*. The normal pattern is: (1) an introductory sentence; (2) the actual confession, which often includes a list of sins; and (3) a declaration of repentance. The oldest of these formulae are probably the *Lorscher* and *Würzburger Beichten*, written down at the end of the 9th c. Other early examples are the *Pfälzer* and *Reichenauer Beichten*, dating from the beginning of the 10th c., and the *Sächsische*, *Vorauer*, *Fuldaer*, and *Mainzer Beichten*, all written in the course of the 10th c.

Another group of formulae includes as well the confession of faith. Among these are examples from St. Gall, Wessobrunn, Benediktbeuren, and Bamberg.

Beiden Freunde, Die, a Novelle by Field-Marshal H. von Moltke (q.v.). It was written in 1827 when Moltke was a young officer, and was published in the same year under the pseudonym Helmuth in the magazine *Der Freimütige*. It appeared posthumously under Moltke's own name in *Gesammelte Schriften und Denkwürdigkeiten* (1891–3). It narrates a fictitious episode at the close of the Seven Years War. Two Prussian officers, Graf Warten and Baron Holt, successfully defend an old castle on the Elbe against an Austrian assault. Each is in love with an Ida von Eichenbach. Holt, convinced by Warten that the latter is the favoured one, intends to resign his claims, only to find himself clearly preferred. The confusion is dispelled when it is discovered that there are two girl cousins of the same name, and each officer is made happy with his Ida.

Bekennende Kirche (Confessing Church), a movement in the German Evangelical Church which was originally known as Bekenntniskirche (Confessional Church) to emphasize its adherence to the Confessions of the Reformation (q.v.). It was founded in 1934 by M. Niemöller (q.v.), who had already a few months previously created the Pfarrernotbund (Pastors' Emergency League). It aimed at maintaining Christian independence at a time when many elements in the Church complacently accepted National Socialism. Its principal theological representative was Karl Barth (q.v.). In spite of persecution and an attack in 1938 by the conforming Churches, the Bekennende Kirche survived in some measure as an opposition under the National Socialist regime. In 1948 it relinquished claims to Church government and became simply a movement for reform.

Bekenntnisse, Die, a three-act comedy by E. von Bauernfeld (q.v.), first performed at the Burgtheater, Vienna, in February 1834. It was published in 1837. Adolf von Zinnburg marries a girl, Julie Herrmann, who shortly after the marriage meets Assessor Bitter, with whom she has previously been in love. The old passion stirs, but Bitter is in pursuit of a rich widow. Adolf treats his wife with consideration, and confessions lead to a stable and harmonious relationship between the young wife and husband. Adolf for a time conceals Julie's identity from a rich uncle by disguising her as an officer. Bauernfeld acknowledged the help of Grillparzer (q.v.), notably in the third act of *Die Bekenntnisse*, which was one of his principal successes.

Bekenntnisse des Hochstaplers Felix Krull. Der Memoiren erster Teil, a picaresque comic novel by Th. Mann (q.v.) published in 1954. It was long in Mann's mind but remained unfinished at his death in 1955. The opening chapters were published in 1922 as an independent work: *Bekenntnisse des Hochstaplers Felix Krull. Buch der Kindheit.* Devoid of moral sense and showing in his boyhood astonishing gifts of simulation and dissimulation, Felix Krull is able to induce in himself the symptoms of illness and so evades conscription. He enters hotel service as a pageboy, is a willing partner in ecstatic and surprising erotic escapades in the hotel, steals jewellery, and leads a double life as the servant and the served. The most conspicuous episode in this phase of his career is his *affaire* with Mme Houpflé (Diane Philibert). He lives for a time on immoral earnings, and is later willingly persuaded to masquerade as the Marquis de Venosta, who pays handsomely for this substitution. He is on his way to Lisbon when he meets in the train the learned and voluble Professor Kuckuck, who introduces Felix (who goes by the name Armand) to his wife and daughter, who both succumb to the young man's charms. The unfinished book is in the first person and purports to be written from prison.

Though the whole sequence of adventures is treated with sovereign irony and with humour, the book affords a severely critical survey of the world in which Felix prospers so well merely through good looks, good acting, and alert quickwittedness. It is a continuation of the theme of the immorality of the artist in Mann's work, for in his own way Krull is an outstanding artist. (Film version by K. Hoffmann, 1957.)

Bekenntnisse einer schönen Seele, Bk. 6 of Goethe's novel *Wilhelm Meisters Lehrjahre* (q.v.). The original of the 'schöne Seele' in this interpolation was Susanna K. von Klettenberg (q.v.).

BEKKH, JOHANN JOSEPH, see BECKH, J. J.

'Bekränzt mit Laub den lieben vollen Becher', first line of *Rheinweinlied,* a popular

student song written by M. Claudius (q.v.) in 1775. The melody is by Johann André (1741–99).

Belagerung von Antwerpen durch den Prinzen von Parma in den Jahren 1584 und 1585, a short historical work written by F. Schiller in 1795 and published in the *Horen* (q.v.) in the same year. It is a pendant to Schiller's earlier *Geschichte des Abfalls der vereinigten Niederlande von der spanischen Regierung* (q.v.) and was included as an appendix (Beilage) in the second edition of the history (1801).

Belagerung von Mainz, Die, an autobiographical work by Goethe (q.v.), written in 1820 and published, together with the *Campagne in Frankreich* (q.v.), in 1822 as part of *Aus meinem Leben* (see DICHTUNG UND WAHRHEIT). It is based on the diary kept by Goethe during the siege of Mainz from his arrival on 26 May 1793 until 27 July. The besiegers were imperial troops and included a Prussian cuirassier regiment commanded by Duke Karl August (q.v.) of Saxe-Weimar, of whose suite Goethe was a member. The city was defended by French troops and pro-French revolutionaries. The surrender, on terms permitting the French troops to withdraw, took place on 22 June 1793.

Belle Alliance (Schlacht bei), German name, originating with Blücher (q.v.), for the Battle of Waterloo (see WATERLOO). The farm La Belle Alliance is on the road from Brussels via Waterloo to Charleroi, and was approximately at the centre of the French position. It was here that Wellington and Blücher met at 9.15 p.m. in the evening after the battle.

Belletristik, German equivalent of *belleslettres,* i.e. literature excluding historical, philosophical, and scientific writing. The word first appeared in German in the form *Belletrist,* 'writer of *belles-lettres*', about 1730. The form *Belliteratur* occurs in J. M. R. Lenz's *Anmerkungen übers Theater* (q.v.), 1774).

BELLOTTO, BERNARDO, also BELOTTO (Venice, 1720–80, Warsaw), was a nephew of the Italian architectural painter Antonio Canal (Canaletto). Bellotto himself adopted his uncle's painting name of Canaletto. In 1746–58 Bellotto was painter to the Saxon court in Dresden, which he revisited in the years 1761–6. Between 1758 and 1761 he spent two years in Vienna. His last years were passed in Warsaw. In each city he painted views of notable buildings which also include miniature figures and are of interest as a social record. His principal works are in the three cities he painted.

Belsazer, a ballad by H. Heine (q.v.) included in the *Buch der Lieder* (q.v.), 'Junge Leiden', No. X of the subdivision 'Romanzen'. It recounts in terse couplets the fate of Belshazzar as given in Daniel 5. The first edition used the spelling Belsatzar.

Belustigungen des Verstandes und Witzes (1741–5), a polemical periodical edited by J. J. Schwabe (q.v.) and intended to support J. C. Gottsched (q.v.) in his controversy with the Swiss critics J. J. Bodmer and J. J. Breitinger (qq.v.). The first of its eight volumes contained Gottsched's polemical satire, *Der Dichterkrieg* (1741).

Belvedere, see EUGEN, PRINZ.

BELZNER, EMIL (Bruchsal, 1901–79, Heidelberg), a journalist and minor novelist, is the author of a number of novels, including *Kolumbus vor der Landung* (1933, reissued as *Juannas großer Seemann*, 1936), *Ich bin der König!* (1940), and *Der Safranfresser* (1953). He also wrote verse epics (*Die Hörner des Potiphar*, 1924; *Iwan der Pelzhändler*, 1929). *Die Fahrt in die Revolution oder Jene Reise. Aide-Mémoire* (1969) was followed by the illustrated volume *Glück mit Fanny. Ein Katzenbuch* (1973).

BENDER, HANS (Mühlhausen, 1919–), had five years' war service, and was released from Russian captivity in 1949. His works include stories (*Die Hostie*, 1953; *Wölfe und Tauben*, 1957; *Mit dem Postschiff*, 1962), novels (*Eine Sache wie die Liebe*, 1954; *Wunschkost*, 1959), and poems (*Fremde soll vorüber sein*, 1951). Best known as the editor of *Akzente* (q.v.), Bender is a contributor to many leading periodicals. He is the editor of essays on poetry (*Was alles Platz hat in einem Gedicht*, in collaboration with M. Krüger, 1977) and of the anthology *In diesem Lande leben wir. Deutsche Gedichte der Gegenwart* (1978). The volume *Worte, Bilder, Menschen. Geschichten, Romane, Berichte* appeared in 1969, his short stories, *Die Wölfe kommen zurück*, in 1972, and the volume *Einer von ihnen. Aufzeichnungen einiger Tage* in 1979.

BENEDEK, LUDWIG AUGUST, RITTER VON (Ödenburg, Austria, 1804–81, Graz), Austrian general who distinguished himself at Solferino (1859) and was appointed Commander-in-Chief in Bohemia in 1866 against his own wishes, for he was unfamiliar with the conditions and terrain (see DEUTSCHER KRIEG). After the defeat of Sadowa (see KÖNIGGRÄTZ) Benedek was dismissed, and court-martial proceedings, stopped by the Emperor Franz Joseph

(q.v.), were instituted. Benedek loyally adhered to a promise not to attempt to rehabilitate himself. The belief that his conduct showed incompetence is now considered unjust.

Benediktbeurer Glaube und Beichte, see BEICHTFORMELN.

Benediktbeurer Spiele, two religious plays in medieval Latin preserved in the MS. of *Carmina Burana* (q.v.) discovered at Benediktbeuren, southern Bavaria. Both were written before the middle of the 13th c.

(1) *Benediktbeurer Weihnachtsspiel,* a Christmas play entitled *Ludus scenicus de nativitate Domini.* Among the scenes are the appearance of certain prophets, the Annunciation, the Nativity, the entry of the Magi, the Slaughter of the Innocents, and an unfinished episode concerned with Antichrist.

(2) *Benediktbeurer Osterspiel,* a passion play entitled *Ludus paschalis sive de passione Domini.* It consists chiefly of songs; the central figure beside Jesus is Mary Magdalene.

Benediktiner Regel, the code governing the rule of life of Benedictine monks; in Old High German literature the term is particularly applied to a copy of the *regula* in the library of St. Gall, which contains a partial interlinear translation. This version, which is a copy of the original, was made in Reichenau some time between 802 and 817. The translation into German of the *regula* belongs to Charlemagne's (see KARL I, DER GROSSE) campaign to raise the standard of monastic conduct as part of his educational and administrative policy.

BENEDIX, RODERICH (Leipzig, 1811–73, Leipzig), went on the stage at the age of 20, acting in plays and singing tenor parts in opera. In 1855 he became director of the Stadttheater at Frankfurt (Main), and from 1859 to 1861 was theatre director at Cologne and thereafter in Leipzig. He wrote more than 100 plays, well-constructed good-humoured comedies popular in almost all German theatres in the second half of the 19th c. Among the best known were *Das bemooste Haupt* (1841), *Doktor Wespe* (1843), *Das Gefängnis* (1859), *Die Dienstboten* (1865), *Die zärtlichen Verwandten* (1866), and *Aschenbrödel* (1868). He also wrote a novel with a theatrical background, *Bilder aus dem Schauspielerleben* (1847).

BENJAMIN, WALTER (Berlin, 1892–1940, Port Bou), an original and perceptive critic, his brilliant academic career was cut short by the rejection of his qualifying thesis (*Habilitationsschrift*) by Frankfurt University in 1925. It was published in 1928 (*Ursprung des deutschen Trauer-*

spiels) and, in revised form, in 1963. After the door to academic advancement had been closed to him Benjamin worked as a writer and critic. He became increasingly interested in Marxism, visiting Russia in 1926–7, and supported the writings of B. Brecht (q.v.). Vulnerable both as a Jew and a Marxist after 1933, Benjamin emigrated to Paris. In 1940, while attempting to reach Spain with the ultimate goal of the U.S.A., he found himself in danger of being betrayed to the Gestapo and committed suicide by taking poison. As a critic Benjamin abandoned the German tradition of Hegelianism and set out to interpret poetic work on its own terms. His principal publications are *Der Begriff der Kunstkritik in der deutschen Romantik* (1920), *Goethes Wahlverwandtschaften* (published in *Neue deutsche Beiträge*, 1924), and *Das Kunstwerk im Zeitalter seiner technischen Reproduzierbarkeit* (1936).

A selection of his work appeared as *Illuminationen* in 1961. The *Gesammelte Schriften*, planned in 6 vols., appeared in 1972 ff. and (12 vols.) in 1980.

BENN, Gottfried (Mansfeld, 1886–1956, Berlin), studied medicine, qualifying in 1912. His first volume of poems was partly medical in subject (*Morgue und andere Gedichte*, 1910). In 1914 he began to practise in Berlin, where he made many literary contacts and became a close friend of Else Lasker-Schüler (q.v.). He served as an army medical officer in Belgium and in 1917 became a consultant in Berlin. Benn followed his first collection of poems with others (*Söhne*, 1913; *Fleisch*, 1917; *Schutt*, 1924; *Spaltung*, 1925), and a comprehensive edition appeared in 1927 (*Gesammelte Gedichte*). In 1933 he associated himself with National Socialism but began to withdraw in the following year. In 1935 he entered the army medical service, in which he remained throughout the war, returning afterwards to private practice in East Berlin. From 1948, his poetic activity entered a new phase, in which linguistic experiment played a prominent part, and in this period he achieved a European reputation. *Fragmente* (1951), *Destillationen* (1953), and *Aprèslude* (1955) are volumes of poetry; the poems were again collected in 1956. Benn's work is primarily lyrical, but after the 1914–18 War he wrote several plays (*Etappe*, 1919; *Ithaka*, 1919; *Der Vermessungsdirigent*, 1919). His Novelle *Der Ptolemäer* has the postwar period (1945–6) as a background. He made notable contributions to criticism in *Goethe und die Naturwissenschaften* (1932) and *Probleme der Lyrik* (1951). He received the Büchner Prize in 1951.

Gesammelte Werke (4 vols.), ed. D. Wellershoff (q.v.), were published 1958–61, and *Gesammelte*

Werke in der Fassung der Erstdrucke, ed. B. Hillebrand, 1982 ff.

BENNIGSEN, Rudolf von (Lüneburg, 1824–1902, Bennigsen nr. Springe, Hanover), Hanoverian politician, who worked vainly for the neutrality of Hanover in 1866. After the incorporation of Hanover in Prussia (1867), Bennigsen joined the National Liberals, whom he led for thirty years. Until 1879 he was a supporter of Bismarck's policies, being prominent in the Kulturkampf (q.v.).

BENRATH, Henry, pseudonym of Albert Henry Rausch (Friedberg, Hesse, 1882–1949, Magreglio, Lake Como), a man of independent means and delicate health, who devoted himself to writing copious disciplined poetry in traditional form (*Der Traum der Treue*, 1907; *Die Urnen*, 1908; *Das Buch für Tristan*, 1909; *Nachklänge*, *Inschriften*, *Botschaften*, 1910; *Vigilien*, 1911; *Sonette*, 1912; *Kassiopeia*, 1919; *Atmende Ewigkeit*, 1928; *Stoa*, 1933; *Der Gong*, 1949; *Erinnerung an die Erde*, posth. 1953; *Liebe*, posth. 1955). He also wrote several novels, mostly historical but some containing contemporary life (*Pirol*, 1921; *Vorspiel und Fuge*, 1925; *Ball auf Schloß Kobolnow*, 1932; *Die Mutter der Weisheit*, 1933; *Die Kaiserin Galla Placidia*, 1937; *Die Kaiserin Theophano*, 1940; *Der Kaiser Otto III*, 1951; *Die Geschenke der Liebe*, 1952). This prolific, and almost forgotten, writer was also the author of a number of Novellen and stories.

BENZ, Richard (Reichenbach, Vogtland, 1884–1966), critic and *littérateur*, spent his life in Heidelberg. His principal interests have been Romanticism, folk literature, and music. His works include: *Märchendichtung der Romantiker* (1908), *Alte deutsche Legenden* (1910), *Die deutschen Volksbücher* (6 vols., 1911–24), an edition and translation of the *Legenda aurea des Jakobus de Voragine* (1917; see Jacobus de Voragine), *Volk und Kultur* (1919), *Die Stunde der deutschen Musik* (1923–7), *Die deutsche Romantik* (1937), *Stufen und Wandlungen* (1942), *Kultur des 18. Jahrhunderts* (1949–53).

BERCHTOLD, Leopold, Graf (Vienna, 1863–1942, Vienna), was Austro-Hungarian foreign minister from 1912 to 1915. His conduct of affairs between the assassination of the Archduke Franz Ferdinand (q.v.) on 28 June 1914 and the outbreak of war has been widely reprobated. In particular, the ultimatum to Serbia is seen as a cynical measure, since its rejection by the Serbs was expected, and Berchtold gambled on the ability of German assurances of support to prevent a general conflict. An elegant and accomplished man of the world, Berchtold was a statesman of limited horizons, who subsequently found an appropriate niche as Imperial Court Chamberlain.

Berengarius Turonensis, a work of ecclesiastical history published by G. E. Lessing (q.v.) in 1770. Based on a MS. discovered by him in the Wolfenbüttel library, it deals with a dispute on transsubstantiation in which the principal parties were Berengar of Tours (d. 1088) and Lanfranc of Bec and later of Canterbury (c. 1005–89). Lessing was drawn to write on Berengar because he saw him as a seeker after truth, persecuted by the orthodox.

BERG, ALBAN (Vienna, 1885–1935, Vienna), an Austrian composer who was a pupil of Arnold Schönberg (q.v.) and an exponent of musical Expressionism. Berg's principal work is the opera *Wozzeck*, which is based on the dramatic fragment *Woyzeck* (q.v.) by G. Büchner (q.v.). The opera consists of three acts (comprising 15 scenes) which are joined by symphonic interludes. Its first performance was on 14 December 1925 in Berlin. At his death Berg's unfinished works included a second opera, *Lulu*, based on the plays *Erdgeist* and *Die Büchse der Pandora* (qq.v.) by F. Wedekind (q.v.): it was first performed in Zurich in 1937. In *Der Wein* (1929) Berg set poems of Baudelaire in translations by Stefan George (q.v.).

BERG, LEO (Zempelburg, 1862–1908, Charlottenburg), a journalist in Berlin, was a founder member of Durch (q.v.) in 1886 and a founder of the Freie Bühne (q.v.) in 1889. He was an ardent supporter of the Naturalist movement (see NATURALISMUS), writing in 1892 an essay entitled *Der Naturalismus*. He showed himself an acute and acid critic in *Ibsen und das Germanentum in der modernen Literatur* (1887) and in *Der Übermensch in der modernen Literatur* (1897).

Berge in Flammen, see TRENKER, L.

BERGENGRUEN, WERNER (Riga, 1892–1962, Baden-Baden), by birth a member of the German colony in the Russian Baltic provinces, fought in the German army in the 1914–18 War and afterwards in the local campaign in Esthonia. He published his first novel (*Das Gesetz des Atum*) and two collections of Novellen (*Rosen am Galgenholz* and *Schimmelreuter hat mich gossen*) in 1923. A further collection (*Das Brauthemd*) appeared in 1925, followed by the novel *Das große Alkahest* in 1926 (as *Der Starost* in 1938). In 1927 Bergengruen settled in Berlin and in the same year published the novel *Das Kaiserreich in Trümmern* and a collection of eerie tales, *Das Buch Rodenstein*, which was enlarged

in 1930. In the next few years he wrote a succession of novels (*Herzog Karl der Kühne,* 1930; *Die Woche im Labyrinth,* 1930; and *Der goldne Griffel,* 1931) and volumes of Novellen (*Der Tolle Mönch,* 1930; *Die Ostergnade,* 1933; *Der Teufel im Winterpalais,* 1933). The Novelle *Die Feuerprobe* appeared separately in 1933. The novel *Der Großtyrann und das Gericht* (q.v., 1935) deals with the problem of authority and the temptations of power, and was the author's best achievement up to this point. In 1936 Bergengruen was converted to the Roman Catholic faith and in the same year he went to live in Munich, where he remained for six years. To these years belong two novels, *Im Himmel wie auf Erden* (1940) and *Das Hornunger Heimweh* (1942), two collections of stories (*Die Schnur um den Hals,* 1935 and *Der Tod von Reval,* 1939), and the separate Novellen *Die drei Falken* (1937), *Der spanische Rosenstock* (1942), and *Schatzgräbergeschichte* (1942). Bergengruen spent the later years of the war in Tyrol, moving in 1946 to Switzerland, where he lived in Zurich until 1958, and then to Baden-Baden, where he remained until his death.

To this period belongs the 'Rittmeister' trilogy (*Der letzte Rittmeister,* q.v., 1952, *Die Rittmeisterin,* 1954, and *Der dritte Kranz,* 1962). In the first, Bergengruen develops with graceful ease a form which presents a wealth of wide-ranging anecdotes in a unity provided by the character of the narrator, 'the last cavalry captain'. In its two sequels he traces round the plot arabesques of humour, fantasy, and irony. The Novellen of his later years include the series *Die Sultansrose* (1946), *Sternenstand* (1947), *Die Flamme im Säulenholz* (1953), *Bärengeschichten* (1959), and *Zorn, Zeit und Ewigkeit* (1960), as well as a number of separate Novellen, *Das Beichtsiegel* (1946), *Jungfräulichkeit* (1947), *Das Tempelchen* (1950), *Erlebnis auf einer Insel* (1952), *Nachricht vom Vogel Phönix* (1952), *Der Pfauenstrauch* (1952), *Die Sterntaler* (1953), *Das Netz* (1956), and *Der Herzog und der Bär* (1960). *Die Schwestern aus dem Mohrenland* appeared posthumously in 1963.

In his middle and later years Bergengruen was a prolific writer of lyric poetry, publishing *Die Rose von Jericho* (1934), *Der ewige Kaiser* (1937), *Die verborgene Frucht* (1938), *Dies Irae* (a cycle, 1945), *Lobgesang* (1946), *Der hohe Sommer* (1946), *Zauber- und Segenssprüche* (1947), *Dir zu gutem Jahrgeleit* (1949), *Die lombardische Elegie* (1951), and *Mit tausend Ranken* (1956). Towards the end of his life he himself collected all his poetry, except *Der ewige Kaiser*, in two volumes, *Die heile Welt* (1950) and *Figur und Schatten* (1958). A volume of previously unpublished poems appeared after his death (*Herbstlicher Aufbruch,* 1965). Mention should also be made of his travel writings, *Baedeker des Herzens* (1933; originally published as *Badekur des Herzens*, 1932), *Deutsche Reise* (1934), and *Römisches Erinnerungsbuch* (1949), of which the last two are partly autobiographical.

The basis of Bergengruen's work is religious, and his men and women, however extraordinary the events which befall them, move within a world which, despite its apparent confusion, is divinely ordered. His mastery of economical narration is especially evident in his Novellen, many of which are historical. The abundance of these stories and their terseness accord with his predilection for anecdote. In the later work the author often deliberately intrudes, and the ironic attitude becomes more pronounced. He is an adept in the use of traditional forms of structure and style.

Berge und Menschen, a novel by H. Federer (q.v.), published in 1911. Its central figure is Emil Manusch, an engineer who undertakes to build a railway up the Absomer (the Säntis in Appenzell is meant). The project encounters strong local resistance. Emil has recently married his cousin Sette, but the marriage does not work out well and the couple temporarily separate. While working on the Absomer Emil comes to believe that his 14-year-old assistant, Mang, is his son, conceived during a brief love-affair in Appenzell fifteen years before. When this proves to be true, the boy turns against him. A flood destroys the embankment built by Emil and causes the abandonment of the railway plan. In the face of a hostile population Mang rallies to his father, who confesses his paternity to Sette. The boy is adopted, and the marriage saved.

Berglinger, Joseph, a fictitious musician, the hero of a short narration, *Das merkwürdige musikalische Leben des Tonkünstlers Joseph Berglinger*, by W. H. Wackenroder (q.v.), published in the *Herzensergießungen eines kunstliebenden Klosterbruders* (1797), in which it is the final piece. Berglinger is thought to be in essence a projection of Wackenroder himself.

Bergroman, a novel by H. Broch (q.v.), begun in 1935; Broch worked at it intermittently until his death, intending to entitle it *Demeter*, but frequently referring to it as his Bergroman (novel of the mountains). His posthumous papers contained three drafts, which were arranged by F. Stössinger, and published as *Der Versucher* (1953). In 1968 the three versions were published together under the title *Bergroman. Die drei Originalfassungen* (4 vols., ed. F. Kress and M. A. Maier), the third draft having appeared separately in 1967, entitled *Demeter*.

It is a work of recollection, narrated by a country doctor serving an Alpine community,

Kuppron. In this rural environment, which also includes a community of miners, the mysterious stranger Marius Ratti exercises a disturbing and demoralizing influence; it manifests itself as a mass psychosis, culminating in the horrifying human sacrifice of the farmer's daughter Irmgard Wenter. The strength of tradition and humanity, concentrated in the rural matriarch, Mutter Gisson, is powerless to arrest the deterioration of human values.

Although the novel is primarily philosophical, parallels with the political trends of Broch's time are unmistakable. The setting in the everlasting hills enables Broch to prove himself one of the greatest literary portrayers of landscape.

Bergwerke zu Falun, Die, a story by E. T. A. Hoffmann (q.v.), written in 1818 and published in 1819 in vol. 1 of *Die Serapionsbrüder*. Its subject is an occurrence at Falun in Sweden, where the body of a miner was lost in 1670, discovered in 1719 in good preservation and recognized, it is said, by his former sweetheart. The story is also treated by H. von Hofmannsthal (see BERGWERK ZU FALUN, DAS).

Bergwerk zu Falun, Das, a prologue (Vorspiel) written in 1899 by H. von Hofmannsthal (q.v.), intended to accompany a play (*Das Bergwerk von Falun*, final title, 1933) based on a story by E. T. A. Hoffmann (see BERGWERKE ZU FALUN, DIE).

BERLICHINGEN, GOTTFRIED or GÖTZ VON (Jagsthausen, 1480–1562, Hornberg), a Franconian knight, was the original of Goethe's hero in *Götz von Berlichingen* (q.v.). Berlichingen lost his right hand in 1504 and its iron substitute earned him the appellation 'mit der eisernen Hand'. Berlichingen spent his life in private and public war, engaging in feuds with the city of Nürnberg and the Elector of Mainz, participating in 1519 in Ulrich von Württemberg's war against the Swabian League (see SCHWÄBISCHER BUND), and fighting for Karl V (q.v.) against the Turks (1542) and the French (1544). Under pressure he reluctantly joined the peasants in 1525 (see BAUERNKRIEG). He was twice outlawed and was imprisoned at Heilbronn (1519–22) and Augsburg (1528–30). He wrote an autobiography, first published nearly two centuries after his death (*Lebensbeschreibung Herrn Götzens von Berlichingen*, 1731), which was Goethe's source for his play.

Berlin, capital city of Prussia from 1709 and capital of the new German Empire (Deutsches Reich) from 1871 to 1945, since when it has occupied a special position. Berlin, which has on its arms a bear (a pun on the name), was founded in the 13th c. on the right bank of the Spree, the large island in the river being occupied by the rival township of Cölln. The two municipalities were combined in 1432, but this measure was countermanded in 1442 by the Elector Friedrich II (q.v.). The troubles which followed this episode are the subject of W. Alexis's novel *Der Roland von Berlin* (q.v.). At the end of the 15th c. the city became the Electoral Residence. The Great Elector, Kurfürst Friedrich Wilhelm (q.v.), and his successor (King Friedrich I, q.v.) extended the city westwards. An increase of population in the 18th c. was partly the result of the official encouragement of Huguenot immigrants from France. Th. Fontane (q.v.) was descended from such stock. In 1848 and in 1918 Berlin was the centre of revolution, and in 1920 of the Kapp-Putsch (q.v.). In 1933 the Reichstag building was burned (see REICHSTAG). From 1941 on Berlin was subjected to air attack, and in 1945 it was encircled and captured by the Russians. It was at first under Russian occupation, but the forces of the other allied powers entered their respective sectors in July 1945. The attempt by the Russians to incorporate the whole city into East Germany by blockade (24 June 1948–12 May 1949) failed thanks to the allies' provisioning of the city by air and to the Russian fear of atomic attack. The three Allied sectors are united as West Berlin and elect members of the Federal Diet (Bundestag) at Bonn, who have no vote. Communications between East and West Berlin became difficult after the unsuccessful rising in East Berlin on 17 June 1953, and virtually ceased on the building of the Berlin Wall by the Eastern authorities in 1961. The treaty of 1973 was intended to produce an easement.

The intellectual repute of Berlin was first established by the Prussian Academy of the Sciences (see AKADEMIEN) founded in 1700. In the 18th c. Lessing (q.v.) was active for a time in Berlin, but many found the intellectual climate at this time arid. In the 19th c. Berlin became the principal literary and cultural centre of Germany, and attracted numbers of men of letters. The city has provided the setting for many narrative works. They include H. von Kleist's *Michael Kohlhaas* (in part), W. Alexis's *Der Roland von Berlin, Ruhe ist die erste Bürgerpflicht*, and *Isegrimm*, Th. Fontane's *Vor dem Sturm*; works concerning modern Berlin include W. Raabe's *Chronik der Sperlingsgasse, Der Hungerpastor*, and *Die Akten des Vogelsangs*, eleven of the novels written by Th. Fontane, J. Stinde's *Die Familie Buchholz*, Kretzer's *Meister Timpe*, A. Döblin's *Berlin Alexanderplatz*, H. Mann's *Im Schlaraffenland* and *Der Kopf*, and the earlier novels of U. Johnson (qq.v.). Several plays by G. Hauptmann are also set in Berlin: *Der Biber-*

pelz, Michael Kramer, Der rote Hahn, Die Ratten, and *Peter Brauer*; likewise H. Sudermann's *Die Ehre* and *Sodoms Ende*, and E. von Wildenbruch's *Die Haubenlerche* (qq.v.).

Berlin Alexanderplatz, a novel by A. Döblin (q.v.), published in 1929. It is sub-titled *Die Geschichte vom Franz Biberkopf*, and was, both artistically and commercially, Döblin's outstanding success. Biberkopf serves four years for causing the death of his girl friend, and the book begins at the moment of his release from Tegel Prison. Though he wishes to go straight, he finds it difficult to pick up the strands of freedom. In a young widow he finds a steady friend, but when one of his associates deceives him with her, he begins to go to the bad again. He becomes involved with a criminal gang led by Reinhold, a sexual maniac. Reinhold, an utterly ruthless man, distrusts Franz, and throws him out of a moving car, injuring him so badly that he has an arm amputated. Biberkopf puts up with this, feeling that it serves him right, and once more he attempts to keep reasonably straight. He picks up Mieze, a girl who really loves him, and this new experience keeps him going. But Reinhold murders her and diverts suspicion to Franz, who is arrested. Mieze's murder stuns Franz; he goes out of his mind, and is confined for a time in the criminal mental institution at Buch. It now becomes clear to him that one act of his has precipitated a whole series of unavoidable catastrophes, the last of which is death, which he experiences as part of his demented state. He recovers, is a witness at the trial of Reinhold, who receives a ten-year sentence, and attains a normal life at last.

The story is told in a deliberately unliterary style full of ellipses and breaches of syntax, and of Berlin slang and dialect, with quotations of contemporary 'hits', well-known folk-songs, proverbs, placards, newspaper excerpts, and diverse montage. It is the most topographically precise novel of Berlin ever written, revolving round the Alexanderplatz in Berlin's East End, where Döblin for years had his surgery. It abounds in humour in spite of the brutality of the gangsters and the abattoirs. Each of the novel's nine books is prefaced by an oblique, ironic, and humorous summary. In its use of the ordinary man's speech, it might be said to be in the tradition of Luther. Moreover, Döblin repeatedly quotes Scripture—Adam and Eve, Job over and over again, Jeremiah, and Abraham's obedient readiness to sacrifice Isaac—and cites all these, not in ridicule, but to provide the ground bass for what is an unexpectedly religious work with a central motif of sacrifice and an apocalyptic climax. (Film versions by P. Jutzi, 1931, and R. W. Faßbinder, 1980.)

Berliner Abendblätter, a periodical edited and partly written by H. von Kleist (q.v.), which included patriotic material in its later numbers. It appeared from October 1810 to April 1811.

Berliner Ensemble, the company of actors founded by B. Brecht and Helene Weigel (qq.v.) in 1949 after the great performance of *Mutter Courage und ihre Kinder* (q.v.) at the Deutsches Theater in East Berlin. In 1954 the Ensemble acquired its own theatre, the Theater am Schiffbauerdamm. The Ensemble was supported by Brecht's old collaborators, above all his stage designer Caspar Neher (q.v.), Erich Engel, Elisabeth Hauptmann, and the composers Hanns Eisler and Paul Dessau (qq.v.), as well as by actors who had worked under Erwin Piscator (q.v.). The productions of Brecht's plays and of his adaptations of older works soon earned it an international reputation. Photographs of productions were assembled with comments in a *Modellbuch,* designed both as a historical record and for consultation by future producers. Brecht himself made good use of them. After his death in 1956 Helene Weigel continued in charge with M. Wekwerth as producer, and the Ensemble has been carried on since her death in 1971.

Berliner Kongress, the international congress called by Bismarck (q.v.) in 1878 to settle by negotiation the Russo-Turkish War and the tensions created by the Balkan situation. Bismarck coined for his role as intermediary and arbitrator the phrase 'ehrlicher Makler' (q.v.).

Berliner Priviligierte Zeitung, see Vossische Zeitung.

Berliner Volksblatt, title from 1884 to 1890 of the Social Democrat newspaper subsequently known as *Vorwärts* (q.v.).

Bern, archaic German form for Verona, chiefly occurring in connection with Dietrich von Bern (see Dietrichsage), i.e. Theodoric the Great (q.v.).

BERNARDON, stage name of J. J. F. von Kurz (q.v.), derived from a stage character created by him.

BERNAUER, Agnes (d. 1435, Straubing), was outstandingly beautiful and, according to most sources, the daughter of a barber surgeon of Augsburg, though it has also been suggested that she was his maid. During a tournament in 1428

Duke Albrecht (see Albrecht III) met and fell in love with her. He secretly married her probably in 1432, possibly in 1433, when he bought her an estate near Niedermenzig. While Albrecht's marriage to a commoner appears to have been welcomed by some of the populace, Albrecht's father, the reigning Duke Ernst (see Ernst I) of Bavaria/Munich, decided for dynastic reasons to have Agnes put to death. He used Albrecht's absence to have her arrested, convicted (presumably of witchcraft), and drowned in the Danube on 12 October 1435. Albrecht established for her a Foundation Mass in perpetuity. While accounts of her personality differ, chronicles of Bavarian history dispute the necessity for Agnes's death and speak of the cruel manner of it with great sympathy. In 1436, the year of Albrecht's remarriage and reconciliation with his father, Ernst erected by Agnes's grave in Straubing a chapel which contains her portrait. He also felt it prudent to establish a similar Foundation Mass. She was called 'Engel', the Swabian diminutive for Agnes, but she has become traditionally known, by virtue of her beauty, as the 'Engel von Augsburg'.

The tragedy of her love has made her the subject of numerous literary works, folk-ballads, plays, epic works, and operas, of which the following may be mentioned: *Die Bernauerin,* Volkslied, 15th c., written down in the 18th c., *Liebe zwischen Hertzog Ungenand und Agnes Bernin; Heldenbriefe* by C. H. von Hofmannswaldau (q.v., 1680); *Agnes Bernauerin. Ein vaterländisches Trauerspiel* by J. A. Graf von Törring (q.v., 1780); dramatic fragments *Engel von Augsburg* and *Agnes Bernauer* by O. Ludwig (q.v., 1840–64); *An Agnes Bernauerin. Gedicht* by Ludwig I of Bavaria (q.v., 1852); *Agnes Bernauer. Ein deutsches Trauerspiel* (q.v.) by F. Hebbel (q.v., 1855); *Herzog Albrecht. Dramatische Dichtung* by Melchior Meyr (q.v., 1862)—Hebbel's and Meyr's plays were both first performed in 1852, in Munich and Berlin respectively: *Agnes Bernauer. Volksschauspiel* by Arnold Ott (1889); *Agnes Bernauer. Vaterländisches Schauspiel* by Martin Greif (q.v., 1894); *Der Herzog und die Baderstochter* by Richard Billinger (q.v., 1934); *Die Bernauerin. Oper* by Carl Orff (q.v., 1946).

Berner Ton, see Eckenstrophe.

BERNGER VON HORHEIM, a Middle High German Minnesänger, was a nobleman from the region of Frankfurt, who is known to have been in Apulia with Heinrich VI (q.v.) in 1195–6. His poems, influenced by Friedrich von Hausen (q.v.), are courtly Minnelieder of great virtuosity.

BERNHARD, Thomas (Herleen, Holland, 1931–), was from a very early age brought up by his mother's parents, his grandfather being the writer Johannes Freumbichler, who died in 1949. Bernhard's father died in 1943 and his mother in 1950. He attended schools in Salzburg, where, in 1957, he completed his studies of music and drama, which he had begun in Vienna after prolonged stays at a sanatorium. In 1965 he settled in a farmhouse in Upper Austria (Ohlsdorf).

Bernhard emerged as a lyric poet with the volumes *Auf der Erde und in der Hölle* (1957), *In hora mortis* (1958), and *Unter dem Eisen des Mondes* (1958); the poem *Ave Vergil* appeared in 1981. He is obsessed with death and the stresses that bring the soul to the brink of insanity. An obvious influence discernible in his writing is Trakl (q.v.).

His narrative works pursue the same themes and break with traditional forms of fiction. Ingeborg Bachmann (q.v.) was among those who praised Bernhard's radical exposure of a disintegrating world and his uncomfortably cool, uncompromising style. They include the novels *Frost* (1963), *Verstörung* (1967), and *Das Kalkwerk* (1970), which centres on the figure of Konrad, the owner of a disused chalk works. He relates the story of his life, which has been one of suffering and failure. Years of an unhappy marriage end with the murder of his crippled wife and his arrest, and years of scientific research with his unsuccessful attempt to complete his manuscript. The stories include *Amras* (1964), the volume *Prosa* (1967), *Ungenach* (1968), *An der Baumgrenze, Ereignisse, Watten* (all 1969), and *Gehen* (1971). *Midland in Stilfs* (1971) contains, apart from the title-story, *Der Wetterfleck* and *Am Ortler. Nachricht aus Gomagoi.* The weather mark (Wetterfleck) symbolizes the individual in the masses, who is distinguishable only by an external mark of identity. The stories are also concerned with concrete reality, including the atrocities of the 1939–45 War. The volume *Der Kulterer. Eine Filmgeschichte* (1974) includes Bernhard's story *Der Kulterer.* Concerned with a prisoner about to be released, the story of Kulterer ends on a note of utter hopelessness, a mood conveyed by a close analogy to the departure of Lenz in Büchner's novella *Lenz* (q.v.).

The novel *Korrektur* (1975) again treats the theme of a man's struggle for a meaningful existence. Roithamer appears to succeed, but his sister, for whom he has built the 'Kegel' as a home, dies before its completion. He takes his own life. The novel incorporates aspects of the personality and work of L. Wittgenstein (q.v.). *Wittgensteins Neffe. Eine Freundschaft* (1982) is a tribute to Wittgenstein's deceased nephew Paul

and provides further insights into Bernhard's own personality. His five autobiographical works are *Die Ursache. Eine Andeutung* (1975), *Der Keller. Eine Entziehung* (1976), *Der Atem. Eine Entscheidung* (1978), *Die Kälte. Eine Isolation* (1981), and *Ein Kind* (1982). The stories *Ja* and *Der Stimmenimitator* appeared in 1978 and *Die Billigesser* in 1980; the novels *Beton* in 1982, *Der Untergeher* in 1983, and *Holzfällen. Eine Erregung* in 1984. First person narratives, these works on art and the artist, music and the theatre, writing and artistic performers, have an autobiographical stance. Obsessive striving for perfection is linked with tormenting frustration and encounters with death, themes symbolized by the titles. Thus *Ja* refers to an initially joking response to the suggestion of suicide, which is subsequently enacted; in *Beton* the concrete of the title refers to the burial place of a young couple in Palma, where they were spending a brief holiday. The husband dies from a fall from the hotel balcony on to the concrete below, his young widow takes her own life in order to be buried in the same concrete block. The narrator's horror and fear at this discovery demonstrates the point of the insertion of such episodes and chance encounters. In *Holzfällen*, tree-felling is associated with longing for the Alpine woods, but also with the 'felling' of reputations and the narrator's disgust at the artificial life of artistic Vienna. The novel's temporary confiscation in the year of its publication involved Bernhard in a dispute with the authorities; it became a bestseller. It was followed by the novel *Alte Meister* (1985).

Bernhard's play, *Ein Fest für Boris* (1970) was followed by *Der Ignorant und der Wahnsinnige* (1972) and *Die Macht der Gewohnheit* (1974); both plays had their premieres in Salzburg and subsequently appeared as *Die Salzburger Stücke* (1975). Others include *Die Jagdgesellschaft* (1974), *Der Präsident* (1975, on terrorism), *Die Berühmten* (1976; the 'famous' include M. Reinhardt (q.v.) and Toscanini), *Minetti. Ein Porträt des Künstlers als alter Mann* (1977, on Bernhard Minetti), *Immanuel Kant* (1978; a parrot echoes the philosopher, its owner is mentally deranged), *Der Weltverbesserer* (1979 in book form), *Vor dem Ruhestand. Eine Komödie von deutscher Seele* (1979, q.v.), *Am Ziel, Über allen Gipfeln ist Ruh. Ein deutscher Dichtertag um 1980* (both 1981), and *Der Schein trügt* (1983). The monologistic dialogue characterizing the plays is highly charged with irony, and the omission of punctuation promotes the rapid flow into inwardness. This type of theatre has become known as Theater der Neuen Subjektivität, of which Peter Handke and Botho Strauß (qq.v.) are also major representatives.

The volume *Die Erzählungen* appeared in

1979, and the plays, *Die Stücke 1969–81*, in 1983.

Bernhard's numerous awards include the Büchner Prize for 1970.

BERNHARDI, AUGUST FERDINAND (Berlin, 1769–1820, Berlin), a schoolmaster, who in 1808 became head of the Friedrichswerder Gymnasium, Berlin. In 1799 he married Sophie, sister of Ludwig Tieck (q.v.); the marriage was dissolved in 1804. Bernhardi contributed to the Romantics' periodical *Das Athenäum* (q.v.) and published *Die Bambocciaden* (1797–1800), which included Tieck's *Die verkehrte Welt*. His most notable achievement was the publication of his philological work *Sprachlehre* (1801–3).

BERNHARDI, FRIEDRICH VON (St. Petersburg, 1849–1930, Kunersdorf), Prussian general who held commands in the 1914–18 War. Bernhardi is best known for his aggressively nationalistic book, *Deutschland und der nächste Krieg* (1913), which aroused suspicion of German policy in Great Britain and France.

BERNHARD VON BREYDENBACH, an official of the diocese of Mainz who wrote an account of a journey to the Holy Land, which appeared in 1486 (*Reise ins Heilige Land*). Thanks to its literary qualities and to abundant illustration by woodcuts, it enjoyed great popularity.

BERNHARD VON SACHSEN-WEIMAR, HERZOG (Weimar, 1604–39, Neuenburg, Baden), a Protestant, had an adventurous career as a soldier of fortune and military leader, which proved detrimental to German interests in the crucial stages of the Thirty Years War (see DREISSIGJÄHRIGER KRIEG). After the death of Gustavus Adolphus (q.v.) at the battle of Lützen (q.v.) in 1631 Bernhard took over command of the troops. He became indispensable to Gustavus Adolphus's chancellor and successor, Axel Oxenstierna (q.v.), who gave him the bishoprics of Würzburg and Bamberg as Swedish fiefs. In return Bernhard took command of the combined forces of the League of Heilbronn in 1633. He was defeated at Nördlingen and lost his duchy of Franconia, but, although he disliked Richelieu, he concluded with him in 1635 the Treaty of Saint-Germain-en-Laye by which he thought he had secured Alsace. The fall of Breisach/Rhine in 1638, crowning the campaign, was in effect a triumph for France. On Bernhard's early death in 1639 his army passed into French service and the war on German soil into French hands.

BERNLEF, a Friesian bard of the 8th c., who was converted to Christianity *c.* A.D. 800 by St.

Liudger. The first recorded German poet, he is said to have been blind.

BERNSTEIN, AARON (Danzig, 1812–84, Berlin), was active as a political journalist, editing the *Volkszeitung* and campaigning for Jewish emancipation. He wrote stories of Jewish life (*Novellen und Lebensbilder*, 1838; *Vögele der Maggid*, 1860; *Mendel Gibbor*, 1860).

BERNSTEIN, ELSA, the real name of the pseudonymous dramatist E. Rosmer (q.v.).

Bernsteinhexe, Die, a play by H. Laube (q.v.), first performed at Hamburg in January 1844 and published in 1846. It is based on the novel *Maria Schweidler, die Bernsteinhexe* (q.v.) by J. W. Meinhold (q.v.). Marie, the daughter of the manse, rejects the advances of magistrate Wittich, whereupon she is accused by him of witchcraft, condemned to be burned, and, in the nick of time, saved and rehabilitated.

Bern und Freiburg, the earliest known historical folk-song, which tells in allegorical terms of the pact entered into in 1243 by the two hitherto hostile Swiss cities Berne and Fribourg.

BERNUS, ALEXANDER VON (Lindau, 1880–1965, Donaumünster nr. Donauwörth), was originally surnamed Grashey. He was adopted by an uncle and aunt on his mother's side, whose name he was given and whose considerable property he inherited. He devoted his life to poetry, generous hospitality, alchemy, and medieval medicine, varied occasionally with studies of theosophy and anthroposophy. A personality of character and distinction, he won numerous literary friends, including Wolfskehl, Däubler, Mombert, Edschmid, Bergengruen, Ricarda Huch, A. Paquet, F. Schnack, and S. Zweig (qq.v.), as well as the painters Kubin and Thylmann. He was especially interested in the transmigration of souls. He was a facile traditional, but not negligible, poet; for much of his verse has a bewitching musical quality. His collections of poetry include *Aus Rauch und Raum* (1903), *Leben, Traum und Tod* (1904), *Maria im Rosenhag* (1909), *Die gesammelten Gedichte* (1918), *Gold um Mitternacht* (1930, extended in 1948), and a final selection using an earlier title (*Leben, Traum und Tod*, 1962). He published only a portion of his output. He deserves credit for translations of many English poets, including Keats, Shelley, Blake, Morris, and Swinburne.

BERTHOLD VON HOLLE, a Middle High German epic poet, was a native of the district of Hildesheim; his patron was Duke Johann von

Braunschweig-Lüneburg, who reigned from 1252 to 1277. In spite of his Low German origin, Berthold wrote in the literary language of Thuringia. Three works by him are known. *Demantin* narrates a long succession of combats and tourneys, ending in the hero's marriage. *Crâne* tells the story of a king's son who runs away, undergoes various adventures, wins the hand of the emperor's daughter, revealing himself, since his father has died, as the rightful king of Hungary. Other adventures follow. Only slight fragments remain of the third work, *Darifant*. Berthold's poems abstain from fabulous episodes of oriental origin and make no mention of Artus (Arthur). They possess an element of realism and are sober and rather tedious accumulations of incident, representing a stage in the decline of the medieval epic.

BERTHOLD VON REGENSBURG (*c.* 1215–72, Regensburg), a Franciscan friar and the most powerful German preacher of the Middle Ages. Educated at Regensburg, probably under DAVID von Augsburg (q.v.), he set out *c.* 1250, accompanied by David, on a series of preaching journeys which took him all over the southern half of Germany and into Alsace, Switzerland, Silesia, Bohemia, and Austria. His reputation lived on for more than a century after his death, and during his lifetime his fame spread across the Channel, eliciting praise from Roger Bacon. The congregations listening to him are said to have been immense, though a contemporary giving the figure of 40,000 may well excite wonderment; the reformative consequences of his preaching are said to have been correspondingly remarkable. Berthold clearly possessed great rhetorical skill, and he must have been a preacher of genius with powers of empathy and a common touch, which effectively brought his words home to his hearers. It is difficult to determine the exact features of his style since the collections of sermons published under his name in 1862 and 1880 are in most cases free adaptations made after Berthold's death of the Latin translations which he himself had authorized. The sermons castigate the vices, especially that of avarice, and even in their corrupt Middle High German form reveal a dynamic eloquence and a sense of a personal address to the individual sinner. The fragments of a collection of sermons preached to nuns, known as the *St. Georgener Prediger* (so called because they were discovered at St. Georgen in the Black Forest), are also probably attributable to Berthold. The sermons were collected 1862–80, and reprinted in 1965.

BERTRAM, ERNST (Elberfeld, 1884–1957, Cologne), professor of German literature in Cologne from 1922 until 1946, when he was dis-

missed in the process of denazification, was a poet of some distinction who followed in the footsteps of Stefan George (q.v.). His principal poetic publications are *Gedichte* (1913), *Wartburg* (1933), *Griecheneiland* (1934), and *Die Fenster von Chartres* (1940). He wrote scholarly works on Nietzsche, Stifter, and H. von Kleist (qq.v.) among others. To Bertram's credit it should be said that he privately protested against the ceremonial burning of the works of Th. Mann and F. Gundolf (qq.v.) in 1933.

BERTUCH, FRIEDRICH JUSTIN (Weimar, 1747–1822, Weimar), was in the ducal service in Weimar but resigned in 1786. He wrote unsuccessful plays, and was chiefly active as a translator and popular publicist. Among the works he translated was *Don Quixote* (1775). He also issued a collection of Iberian works (*Magazin der spanischen und portugiesischen Literatur*, 1780–3) and founded the *Allgemeine Literaturzeitung* (1785). He published the first German fashion paper (*Journal des Luxus und der Moden*, 1786–9) and a series for children (*Bilderbuch für Kinder*, 1796–1834). In furtherance of these activities he established his own publishing firm in Weimar in 1791. He was on friendly terms with Goethe.

Bescheidenheit, Die, see FREIDANK.

BESSER, JOHANN VON (Frauenburg, Courland, 1654–1729, Dresden), the son of a country clergyman, spent the greater part of his life as a courtier. He entered Brandenburg service with the title Legationsrat in 1681, and was in London in 1684 on a diplomatic mission. In 1690 he became master of ceremonies to the Elector Friedrich III (see FRIEDRICH I, KÖNIG). Dismissed in 1713 by King Friedrich Wilhelm I (q.v.), he found employment four years later with Augustus the Strong, Elector of Saxony and King of Poland (see AUGUST II).

Besser was a skilful court poet who wrote deftly and efficiently the poems, dramatic scenas, and masquerades required for social occasions and court festivals. He also wrote ironical and witty pastoral poetry. His reputation remained high in the first half of the 18th c., and it was only after 1750 that it declined almost total obscurity.

Besserungsstück, a variety of the Viennese popular play (Volksstück, q.v.) in which the hero is cured of some evil or unfortunate propensity. Examples are *Die Musikanten am Hohen Markt* (1815) and *Der Berggeist* (1819), both by J. A. Gleich (q.v.). Raimund's *Das Mädchen aus der Feenwelt* (1826), *Der Alpenkönig und der Menschenfeind* (1828), and *Der Verschwender* (qq.v., 1834) are developments of the Besserungsstück.

So, too, is F. Grillparzer's *Der Traum ein Leben* (q.v., 1834), though this is not a Volksstück.

BESTE, KONRAD (Wendeburg, 1890–1958, Stadtoldendorf), North German humorist, wrote novels which rely for their effect on eccentric characters and the contrast between life and manners in a country town and the great city. Among these novels should be mentioned *Grummet* (1923), *Der Preisroman* (1927), *März* (1929), *Das heidnische Dorf* (1933), *Das vergnügliche Leben der Doktorin Löhnefink* (1934), *Die drei Esel der Doktorin Löhnefink* (1934), *Gesine und die Bostelmänner* (1936), *Das Land der Zwerge* (1939), *Boses absonderliche Brautfahrt* (1943), and *Löhnefinks leben noch* (1950). He also wrote comedies such as *Schleiflack* (1930) and *Seine Wenigkeit* (1936).

Besuch der alten Dame, Der, a play by F. Dürrenmatt (q.v.), published in 1956, and described as 'Eine tragische Komödie'. The scene is 'Güllen', a squalid, decaying township in central Europe. One of the richest women in the world, Claire Zachanassian, born and bred in the town, has announced her intention of re-visiting it. She had left with an illegitimate child and a bad reputation, and had been a prostitute in Hamburg until her good looks attracted a rich man. Since then she has grown richer and richer, no longer Klara Wäscher, but Claire etc. When the play begins she has acquired her seventh husband and she leaves with the ninth. Included in her luggage is an empty coffin. From her enormous wealth she proposes to pay Güllen and its inhabitants a milliard francs. Incredible wealth is at hand and the townspeople indulge in a spending spree before the money is paid out. But Claire, now a sexagenarian ruin, imposes one condition: the execution of Alfred Ill, now a small tradesman but formerly the young man who seduced Klara and denied the parentage of her child. This intended act of revenge, which she calls justice, is at first rejected by the notabilities of the town as well as by the townspeople. But greed and financial embarrassment caused by their newly incurred debts bring them gradually to see the necessity and the 'justice' of sacrificing Ill. Even Ill himself can make no serious effort to avert his doom. He is killed in a ritual sacrifice to the (financial and economic) welfare of the community. Claire departs with the corpse in the coffin, and resembles nothing more than a lifeless stone idol (a point made in a stage direction and emphasized by Dürrenmatt in a *Nachwort*). This grisly comedy is a parable on the vulnerability of ideals to money. When the mayor asserts that justice cannot be bought, Claire replies, 'Man kann alles kaufen.'

The opera *Der Besuch der alten Dame* by G. von

Einem (q.v.) is based on Dürrenmatt's play. It was performed at Glyndebourne in 1973 and 1974.

Besuch in der Kartause, an anecdotal poem, consisting of 141 lines of antique trimeters, written by E. Mörike (q.v.) in 1861. It is dedicated as an 'Epistel an Paul Heyse'. The narrator, speaking in the first person, records a visit to an inn, occupying a former Carthusian monastery, which he had known years earlier before its secularization. He meets the former conventual physician and learns from him of the life and death of the 'Pater Schaffner' and of his fear of the inscription on the clock 'una ex illis ultima'.

Besuch in Urach, a poem written by E. Mörike (q.v.) in 1827. In twelve stanzas of *ottava rima*, the poet, in sadness, recalls his lost childhood, but gathers new strength from the power of nature, expressed in a thunderstorm. Mörike was at school at Urach, near Reutlingen, from 1818 to 1822.

BETHGE, HANS (Dessau, 1876–1946, Kirchheim/Teck), German minor poet (*Die stillen Inseln*, 1898). He also wrote stories (*Der gelbe Kater*, 1902; *Satuila*, 1918; and *Ägyptische Reise*, 1926) and translated oriental poetry; his *Die chinesische Flöte* (1907) contains the poems that G. Mahler (q.v.) used as the text for *Das Lied von der Erde*.

BETHMANN-HOLLWEG, THEOBALD VON (nr. Eberswalde, 1856–1921, nr. Eberswalde), Chancellor of the German Empire from 1909 to 1917. After a civil service career, Bethmann-Hollweg became Prussian Minister of the Interior in 1905 and deputy chancellor in 1907. He succeeded Bülow on the latter's recommendation. Though he genuinely desired a relaxation of tension and a *rapprochement* with Great Britain, his intentions were sterilized by the consequences of Bülow's policy. In the events leading to the outbreak of war in 1914 Bethmann-Hollweg was a virtually helpless figure. His notorious reference to the guarantee of Belgian neutrality as 'a scrap of paper' ('einen Fetzen Papier') was an indiscretion widely publicized abroad. Gradually overshadowed by the politico-military partnership of Hindenburg and Ludendorff (qq.v.), he was unceremoniously dismissed in July 1917, some months after the failure of his peace overtures of December 1916.

Betschwester, Die, a comedy written by C.F. Gellert and published in vol. 2 of the *Bremer Beiträge* (q.v.) in 1745. It is described as *Ein Lustspiel in drei Aufzügen*. Simon and Ferdinand, two well-to-do gentlemen, have come from

Berlin to make arrangements for Simon's marriage to Christiane, daughter of Frau Richardin, a rich and miserly widow, the 'Betschwester' of the title, who makes a great display of piety by prayer-reading and hymn-singing, while conducting her affairs on strictly commercial and non-Christian principles. The contractual arrangements for the marriage run into difficulties and Simon turns away from the inexperienced Christiane to Lorchen, an attractive young woman living in the house. But Lorchen, overcoming her own desires, directs Simon back to Christiane, and Simon succeeds, with the help of handsome presents, in winning Frau Richardin's consent.

Bettelweib von Locarno, Das, a brief story by H. von Kleist (q.v.), published in the *Berliner Abendblätter* (q.v.) in 1810 and in vol. 2 of *Erzählungen* (1811). It relates the heartless treatment of an old beggar woman by a Marchese, in whose castle she has sought shelter. Unable to bear her presence, the Marchese orders the frail woman to move to a place behind the stove. She slips and dies as a result of her fall. The incident has disastrous consequences when, after the passage of years, the appearance of the dead woman's ghost drives the Marchese to commit suicide by setting fire to his castle. E. T. A. Hoffmann (q.v.) was among the first to recognize Kleist's ingenious handling of the horrific in the story.

Bettler, Der, one of the first Expressionist plays to be written. It is by R. J. Sorge (q.v.) and was published in 1912. Described as 'Eine dramatische Sendung' in five acts and written in a heightened form of prose with passages of ecstatic verse, it abjures reason and clarity, striving for mystic exaltation, as the central figures rise above all earthly ties. The 'beggar' of the title is the poet. There are no characters, but simply types indicated as 'Der Sohn', 'Der Vater', 'Der Dichter', etc. Though the play's idiom was new, its basic theme, the gap between generations, is perennial. It was first performed in 1917.

BETULEJUS, Xystus, see Birck, Sixtus.

BEUMELBURG, Werner (Trarbach/Moselle, 1899–1963, Würzburg), served in 1917–18, and was a ground staff officer in the German Air Force in the 1939–45 War. He wrote numerous nationalistic war books, notably *Douaumont* (1925), *Ypern* (1925), *Sperrfeuer um Deutschland* (1929), *Gruppe Bosemüller* (1930), and *Mont Royal* (1936). He was a senior member of the Nazi party, and was prominent in the National Socialist Deutsche Akademie der Dichtung from 1933 onwards.

BEYERLEIN, Franz Adam (Meißen, 1871–1949, Leipzig), wrote novels of military life, some of which (notably *Jena oder Sedan*, 1903, a best seller) are critical of militarism as represented by the German officer caste. Other titles include *Similde Hegewalt* (1904), *Stirb und werde* (1910), the trilogy *Friedrich der Große* (1922–4), *Der Brückenkopf* (1927), *Land will leben* (1933), *Der Ring des Lebens* (1938), and *Johanna Rosina* (1942). His play of military life *Zapfenstreich* (1903) also enjoyed outstanding success.

Bezauberte Rose, Die, a verse romance by E. Schulze (q.v.).

BGB, abbreviation for *Bürgerliches Gesetzbuch* (q.v.).

Biberpelz, Der, a four-act comedy by G. Hauptmann (q.v.), first performed in the Deutsches Theater, Berlin, in September 1893 and published in the same year. It is sub-titled 'Eine Diebskomödie', and is set in the outskirts of Berlin ('irgendwo um Berlin'). The action turns on two thefts, both organized by Mutter Wolffen, the central character, who is incorrigibly dishonest but quick-witted and good-natured. In the first two acts she arranges a theft of wood (and is the receiver of a roebuck taken by a poacher into the bargain), in the third and fourth acts Herr Krüger's fur coat is adroitly stolen. Both thefts come before the pompous and stupid magistrate, Amtsvorsteher von Wehrhahn, who has no inkling of Mutter Wolffen's real character and is chiefly occupied with sniffing out political agitators where there are in reality none. The play, the speech of which is deeply tinged with Berlin dialect, is a masterpiece of rich characterization and hilarious action. It has a sequel, *Der rote Hahn* (q.v.).

Bible, Translations of. The oldest translation of the Bible into any Germanic language is the Gothic rendering made c. A.D. 369 by Bishop Wulfila (q.v.), of which about half survives in the Codex argenteus (q.v.). Throughout the Middle Ages the Bible in use was the Latin Vulgate of St. Jerome (4th c.), and all medieval German translations, of which 43 complete MSS. exist, as well as many fragments, are versions of this model. The printing presses, invented by Johannes Gutenberg (q.v.), who himself designed a monumental Latin version of the Bible, known as the *Gutenberg-Bibel*, completed in 1455, produced, between 1461 and 1520, 18 Bibles in German. Of these 14 were in High German and 4 in Low German. All are based on an undistinguished 14th-c. translation of the Vulgate; only the Bible of 1475 makes significant improvements.

Martin Luther (q.v.), while at the Wartburg in 1521, began his painstaking translation from the original sources. His rendering of the N.T., based on the Greek, appeared in the autumn of 1522 as *Das Newe Testament Deutzsch*. It is sometimes called the *Septembertestament*. The O.T. followed in instalments, translated from the Hebrew, and his German Bible, including the Apocrypha, *Apokryphen*, was completed in 1534 (*Biblia/das ist/die gantze Heilige Schrifft Deudsch.* Mart. Luth., fac., 1935). The 'Ausgabe letzter Hand' (q.v.) appeared in 1545, adding after 'Deudsch' the words 'new zugericht'. This edition was reprinted in a delayed centennial issue in 1973 (2 vols.). Two factors contributed to the exceptional influence and cultural significance of Luther's version: first, Luther's return to the original sources, and second, his genius in creating a style which, in its simplicity and exactness, could be readily understood by all, while it preserved the authority of the subject (*Ein Sendbrief vom Dolmetschen*). The Lutheran Bible was a powerful support for the Reformation (q.v.). After its completion Luther continued to improve his translation, aided by Melanchthon (q.v.) and Aurogallus (otherwise Goldhahn). Revised versions appeared in 1892, 1912, and 1956–64.

Luther's Bible overshadows most later versions. The Bible of J. Piscator (1602–4) was not a serious rival. The *Perleburger* or *Berleburger Bibel* (1726–42) is a revision of the Lutheran Bible, which heightens the mystical aspects. Graf Zinzendorf (q.v.) produced a N.T. for Pietists in 1727. Several 18th-c. translations are written from a rationalistic standpoint, including the *Wertheimer Bibel* (1735) and a much satirized N.T. by K. F. Bahrdt, *Neueste Offenbarungen Gottes verdeutscht* (1773). Also worthy of mention is the O.T. of J. D. Michaelis, *Übersetzung des Alten Testaments mit Anmerkungen für Ungelehrte* (1769–83). The more important modern translations are those of H. Menge (1923–6, 'Menge-Bibel'), F. Pfäfflin (1939), L. Thimme (1936), and M. Buber (q.v., 1925–38).

Luther's Bible provoked Roman Catholic translations. These include Emser's N.T. (1523), which is heavily indebted to Luther, as is also Dietenberger's Bible of 1534, which had an extended life through a revision in 1662. In this way the influence of Luther's Bible extended into Catholic lands. The translation by Luther's opponent J. Eck (q.v., 1537) is inferior. The principal late translation of the Vulgate is by J. F. von Allioli (1830–2); it is still in use in the 20th c.

Biblia pauperum or *Armenbibel,* an illustrated manual used in the Middle Ages by the clergy for the exposition of the Bible. The illustrations preponderate over the text, which in some MSS. is in Latin, in others in German or in a mixture of the two. The parts of the Bible covered are the New Testament and some of the prophets. The oldest MS., dating from *c.* 1300, is at St. Florian in Austria.

Bibliothek der schönen Wissenschaften und freyen Künste, a critical review published from 1757 to 1765. Its first editors (1757–9) were F. Nicolai and M. Mendelssohn (qq.v.), who were succeeded in 1759 by C. F. Weiße (q.v.). The title was changed in 1765 to *Neue Bibliothek der schönen Wissenschaften und freyen Künste.* In this form it continued until 1806.

Bibliotheken, see LIBRARIES.

BICKEL, KONRAD, see CELTIS, KONRAD.

BIDERMANN, JAKOB (Ehingen nr. Ulm, 1578–1639, Rome), was educated at Augsburg and entered the Society of Jesus in 1594. After periods in Augsburg and Ingolstadt he became a professor at the Jesuit school in Munich, where he directed the dramatic activity from 1606 to 1614. From 1615 to 1622 he was at Dillingen University and was then posted to Rome. Bidermann was perhaps the most remarkable exponent of Jesuit drama (see JESUITENDRAMA). Of his plays (all originally in Latin) *Cenodoxus* (q.v., 1609) achieved a lasting success. Among other titles, he wrote *Belisar* (1607), *Macarius* (1613), *Philemon Martyr* (1618), *Josaphatus* (1619), and *Stertinius* (1620). His plays were not published till 1666.

BIEDERMAIER, GOTTLIEB, see EICHRODT, L., and SAUTER, S. F.

Biedermann, Der, a moralizing weekly published by J. C. Gottsched (q.v.) in 1727.

BIEDERMANN, WOLDEMAR, FREIHERR VON (Marienberg, 1817–1903, Dresden), a Saxon civil servant, published *Goethes Gespräche* in 10 vols., 1889–97.

Biedermann und die Brandstifter, a play by M. Frisch (q.v.), was first performed in the Zurich Schauspielhaus in March 1958, and published in the same year. It is sub-titled 'Ein Lehrstück ohne Lehre' in allusion to Brecht (q.v.), and is an adaptation of Frisch's radio play *Herr Biedermann und die Brandstifter* (1955). It achieved remarkable success. The play is a parable and it is 'ohne Lehre' because the 'Biedermänner' or Bürger, to whom it is addressed, are deaf to its message. Biedermann is not an individual but a type, a well-to-do middle-class manufacturer living in comfortable circumstances. His *bonhomie* is entirely superficial, his real character is brutal and cowardly. Because of his insincerity and lack of courage he allows into his house two suspicious lodgers, Schmitz (Sepp) and Eisenring (Willi). Both are pyromaniac terrorists, but Biedermann, who waxes indignant at the frequency of arson, is too much of a moral coward to want to see what dangerous guests he has admitted. Though purblind in a situation which requires moral courage, he has no compunction in driving his employee Knechtling to suicide, for Knechtling is a law-abiding Bürger, from whom Biedermann has nothing to fear. With Biedermann's knowledge and connivance Schmitz and Eisenring store petrol in the attic and in due course set the house ablaze, and Biedermann and his female counterpart, Frau Biedermann, perish in the flames. Simultaneously other fires break out all over the city, and the terrorist seizure of power is accomplished. The small part of the 'Dr. Phil.' is a satirical portrait of the intellectual who ideologically supports terror, only to quail before its reality. A chorus of Firemen, parodying Greek tragedy, provides a lugubrious commentary in verse.

Frisch later added a 'Nachspiel', which was first performed in 1958. Designed for 'deutsche Aufführungen', it shows Biedermann and his wife in Hell, unchanged and incorrigible. Eisenring, in a diabolical role, though simply styled 'Figur', and Schmitz, who plays Beelzebub, quench the fires of Hell and depart to resume their incendiarism on Earth. Though the significance of *Biedermann und die Brandstifter* is universal, it is aimed particularly at Swiss complacency.

Biedermeier, Das, a German style usually regarded as characteristic of the period 1815–48. Its use was originally pejorative, and the name derives from the fictitious naïve and unintentionally comic poet Gottlieb Biedermaier, lampooned in the Munich humorous weekly *Fliegende Blätter*. For details concerning these publications see EICHRODT, L., KUSSMAUL, A., and SAUTER, F. Biedermeier (the spelling with 'e' is now universal) is compounded of 'bieder', suggesting a somewhat contemptuous sense of 'worthy', and Meier, which (in various forms, including Meyer, Maier, and Mayer) is a common surname. In the second half of the 19th c. Biedermeier was proverbially used to express amused contempt of the parochial lives and unpolitical attitudes prevailing among the populace of the states and principalities of the German Confederation (see DEUTSCHER BUND).

At the beginning of the 20th c. a consciousness grew that this period had, at least in painting, in interior decoration, in furniture, and in architecture, a style which was sober, modest, unpretentious, and yet exhibited a prepossessing elegance. The spurious gilt of the *style Empire* gave way to homely, solid, impeccable craftsmanship, and a simple dignity replaced expansive pomp. Comfortable, though unluxurious, domestic interiors in subdued daylight or lit by oil lamps became favourite subjects of painters (e.g. F. Kersting, A. von Menzel, M. von Schwind, qq.v.). And an element of humour was extracted from the quainter semi-Dickensian aspects of this world by C. Spitzweg (q.v.).

The aptness of the term 'Biedermeier' to literature has, however, remained a subject of continuing controversy. The usual contention has been that the literature of the middle class before the industrial revolution and in the days of the repressive police state organized in the age of Metternich (q.v.) was one of withdrawal from politics and retreat into a private, domestic sphere; and that an attitude of personal quietism, inner security, and abstention from passion was simultaneously promoted. The 'quiet' writers to whom these 'Biedermeier' aspects have, in greater or lesser degree, been ascribed are poets such as A. von Droste-Hülshoff, J. Kerner, and E. Mörike, writers of fiction such as A. Stifter, B. Auerbach, and K. Immermann, and dramatists, especially Austrian, such as F. Grillparzer, E. Bauernfeld, and F. Raimund (qq.v.). Yet none of these writers could be fully comprehended under the designation 'Biedermeier' which has inherent in it a suggestion of mediocrity and 'averageness' which all the writers mentioned transcend. A more reasonable attitude suggests that 'Biedermeier' elements are discoverable in these authors, but the true 'Biedermeier' men of letters, if they exist, are to be found among writers of the second, or less than second, rank, such as K. J. P. Spitta, J. Rank, A. Weill (qq.v.), or even F. W. Hackländer (q.v.), who might be said to have invented a 'militärisches Biedermeier'. As a designation of period, 'Biedermeier' is only truly acceptable as an expression of manners, fashion, and domestic style in the years between 'Empire' splendour and 'new capitalist' affluent ornamentation and elaboration. Historically the idea is vitiated by the struggles of the Burschenschaft (q.v.) and the revolutionary agitation and opposition of men such as F. Lassalle, F. List, and G. Büchner (qq.v.), and, on the literary side, apart from Büchner, by the Jung-Deutschland group (see JUNGES DEUTSCHLAND). The term has been extended to include men of letters writing during the period of the foundation of the German Empire in 1871. A comprehensive revaluation of 'Biedermeier' has been undertaken by F. Sengle, *Biedermeierzeit. Deutsche Literatur im Spannungsfeld zwischen Restauration und Revolution 1815–1848*, 3 vols., 1971, 1972, 1980. (See also VORMÄRZ.)

BIENEK, HORST (Gleiwitz, 1930–), lived after the war in East Berlin, and from 1951 to 1955 in a labour camp in Siberia, and came to West Germany in 1956. He is the author of *Traumbuch eines Gefangenen* (1957), poems under the titles *was war was ist* (1966) and *Vorgefundene Gedichte* (1969), and a novel *Die Zelle* (1968). *Bakunin, eine Invention* (1970) is concerned with the last years in the life of Mikhail Bakunin (1814–76); the work was followed by the story *Der Verurteilte* (1972, with illustrations by C. H. Wegert) and the play *Im Untergrund. Nach Dostojewskij* (1972). Further collections of poetry include *Die Zeit danach* (1974) and *Gleiwitzer Kindheit. Gedichte aus zwanzig Jahren* (1976). Bienek wrote four novels on his native Gleiwitz that form a tetralogy: *Die erste Polka* (1975) takes place on 31 August 1939, the last day before the beginning of the war, *Septemberlicht* (1977) on 4 September 1939, *Zeit ohne Glocken* (1979) on Good Friday 1943, and *Erde und Feuer* (1982) at the end of the war in 1945.

Von Zeit und Erinnerung (1980) is a collection of stories, poetry, and essays, *Solschenizyn und andere* (1972) a collection of essays, and *Der Freitag der kleinen Freuden* (1981, with illustrations by B. and U. Schultze) a collection of stories. *Beschreibung einer Provinz. Aufzeichnungen, Materialien, Dokumente* appeared in 1983.

Biene Maja und ihre Abenteuer, Die, a tale for children (of all ages), by W. Bonsels (q.v.), published in 1912 and repeatedly reprinted over many years. Maja is a very special bee, whose curiosity leads her to explore the world of other insects and creeping creatures. Her investigations enable her to give the community advance warning of a hornet raid, and timely counter-measures are taken by the Queen Bee.

BIERBAUM, OTTO JULIUS (Grünberg, Silesia, 1865–1910, Dresden), studied at several universities (Zurich, Leipzig, Munich, Berlin) and, though he at one time contemplated a career in the consular service, devoted himself to literature, first in Munich and later in Berlin. From 1891 to 1894 he edited the *Moderner Musenalmanach*, and in 1893 *Die freie Bühne* (q.v.). He founded *Pan* (q.v.) in 1895, and collaborated in editing *Die Insel* (q.v.) in 1899–1902. In addition to this journalistic activity he wrote numerous deft, melodious, though often undistinguished

poems in the most varied styles—D. von Liliencron (q.v.), the German Anakreontik and Rokoko, and Minnesang (qq.v.) have all exercised a recognizable influence. The poems are incorporated in *Erlebte Gedichte* (1892), *Nemt, Frouwe, disen Kranz* (1894), *Der Irrgarten der Liebe* (1901), and *Maultrommel und Flöte* (1907). His fiction was of a more serious, though satirical, character. The Novelle *Die Schlangendame* (1896) has as its heroine a clergyman's daughter, Mathilde Holunder, who, 'going to the bad', lives with an idle student, keeps his nose to the grindstone, and eventually settles down in marriage with him. *Stilpe* (1897), sub-titled 'Ein Roman aus der Froschperspektive', tells the story of a gifted, yet undisciplined character who becomes a formidable journalistic critic, cannot resist dissipation and debauchery, and ends in suicide. *Prinz Kuckuck* (1907), sub-titled 'Leben, Taten, Meinungen und Höllenfahrt eines Wollüstlings', has as its central figure a would-be Nietzschean superman, who finds that he has not the strength and stamina to sustain the role, and so puts an end to his own life. Bierbaum also wrote the Singspiel *Die Hirten und der Schornsteinfeger* and a play, *Stella und Antonie* (1902), in which the poet J. C. Günther (under the transparent name Johann Christian) is attracted to two women. Other works by Bierbaum include the novels *Pankrazius Graunzer* (1895), *Das schöne Mädchen von Pao* (1899), and *Die Päpstin* (1909), and the comedy *Der Musenkrieg* (1907). *Gesammelte Werke in zehn Bänden*, ed. M. G. Conrad and H. Brandenburg, appeared in 7 vols. in 1912.

BIERMANN, WOLF (Hamburg, 1936–), the son of a Communist dockyard worker and resistance fighter who died in Auschwitz in 1943, settled in the DDR in 1953. He interrupted his studies of politics, philosophy, and mathematics at the Humboldt University for two years (1957–9) to work with the Berliner Ensemble. In the early 1960s he founded his own 'Berliner Arbeiter- und Studententheater' which met with official disapproval. A temporary ban on public performance in 1963 was followed by a permanent ban on performance and publication in the DDR in 1965. In 1976, whilst on a sponsored tour in the BRD, he was refused re-entry into the DDR and deprived of his DDR citizenship. The 'Biermann Affair', as it became known in the West, had serious consequences for prominent DDR authors, who put their signature to a public letter of protest that was directed against the principle underlying the expulsion. Most were subsequently deprived of membership of the SED.

Biermann returned to his native city. As a writer and singer he was influenced by Brecht,

Heine (qq.v.), F. Villon, and Béranger. In his music, for which he uses a variety of instruments, he is a follower of Dessau (q.v.) in that he uses song and accompaniment as a means of varying textual perspectives. An activist, he is bitterly critical of established socialism in both East and West Germany. All his publications appeared in the West, *Die Drahtharfe. Balladen, Gedichte, Lieder* in 1965, *Mit Marx- und Engelszungen. Gedichte, Balladen, Lieder* in 1968, *Der Dra-Dra. Die große Drachentöterschau in acht Akten mit Musik* (a show in which the dragon, representing capitalism, though slain, is succeeded by another exploiter) in 1970, and *Deutschland. Ein Wintermärchen, Für meine Genossen. Hetzlieder, Gedichte, Balladen*, and *Berichte aus dem sozialistischen Lager von Julij Daniel. Übersetzt von Wolf Biermann* in 1972; all titles are contained in the volume *Nachlaß 1. Noten, Schriften, Beispiele* (1977). *Preußischer Ikarus. Lieder/Balladen/ Gedichte/Prosa* appeared in 1978 and *Verdrehte Welt—das seh' ich gerne*, written between 1980 and 1982, in 1982.

BIERNATZKI, JOHANN CHRISTOPH (Elmshorn, 1795–1840, Friedrichstadt, Schleswig), a Protestant pastor on the Hallig (q.v.) Nordstrandischmoor, was appointed in 1825 to the pastorate of Friedrichstadt, where he spent the remainder of his life. He witnessed the great flood of 1825 on the North Sea coast and in the same year published a didactic religious poem *Der Glaube*, the proceeds of which he devoted to the victims of the disaster. He wrote stories, mostly connected with sea or coast. The most successful and interesting of these is *Die Hallig oder die Schiffbrüchigen auf dem Eilande in der Nordsee* (1836), the first known literary work to concern itself with these unusual islands and their inhabitants. *Der braune Knabe* appeared in 1840, and *Des letzten Matrosen Tagebuch* (1844) was published posthumously.

Bilder aus Bebenhausen, see BEBENHAUSEN.

Bilder aus der deutschen Vergangenheit, a historical work in 5 vols. by G. Freytag (q.v.), published 1859–67. It gives a popular account of German history by periods, with particular stress on social and economic aspects. The first volume, 'Aus dem Mittelalter', begins with Roman Germany and ends with the Hohenstaufens (q.v.). The second and third volumes (described by Freytag as 'Zweiter Band, erster [und] zweiter Teil') deal respectively with the later Middle Ages (up to the end of the 15th c.) and the 16th c. ('Das Jahrhundert der Reformation'). The fourth (in Freytag's numbering, third) volume concerns itself with the 17th c., 'Das Jahrhundert des großen Krieges', and the fifth

(officially fourth), headed 'Aus neuer Zeit', covers the 18th c. and the first half of the 19th up to the eve of the 1848 Revolution. The work is plainly and vigorously written and quotes, particularly in the more recent periods, at length from interesting personal documents. *Bilder aus der deutschen Vergangenheit* was immediately successful and first brought Freytag to the notice of Prussian royalty.

Bildungsroman, a novel in which the chief character, after a number of false starts or wrong choices, is led to follow the right path and to develop into a mature and well-balanced man. The form, which is more common in German literature than in English or French, was initiated by Wieland in his *Agathon* (q.v., 1765–6) and notable later examples are Goethe's *Wilhelm Meisters Lehrjahre* (q.v., 1795–6), Tieck's *Franz Sternbalds Wanderungen* (q.v., 1798), G. Keller's *Der grüne Heinrich* (q.v., 1854), G. Freytag's *Soll und Haben* (q.v., 1855), Stifter's *Der Nachsommer* (q.v., 1857), and W. Raabe's *Der Hungerpastor* (q.v., 1864). The Bildungsroman occurs more frequently in the 19th c. than in the 20th c., though H. Hesse's *Peter Camenzind* (1904) has been classified as a Bildungsroman, and Th. Mann's *Königliche Hoheit* (1909), *Der Zauberberg* (1924), and in particular his *Joseph und seine Brüder* (qq.v., 1933–42) might possibly be regarded as ironic instances of the form. Some critics differentiate between Bildungsroman, Erziehungsroman, and Entwicklungsroman, but these terms are barely distinguishable, and there is a perceptible tendency to adopt the last as the generic appellation in place of Bildungsroman. In so far as the element of self-realization is integrated in the author's presentation of society the Bildungsroman is synonymous with Zeitroman (q.v.).

Billard um halb zehn, a novel by H. Böll (q.v.), first published in 1959. The eightieth birthday of Geheimrat Fähmel on 6 September 1958 becomes the occasion for a family celebration with a difference. Three generations of architects reveal in a series of reminiscences, frequently in the form of interior dialogue, personal experiences determining their past conduct and shaping their sceptical attitude towards one another as well as towards society in general. In 1907, at the age of 29, Heinrich established his reputation by building the Abbey of St. Anton, which his son Robert, likewise at the age of 29, destroyed in the last phase of the 1939–45 War in protest against its architect. Joseph, now in his twenties, is engaged in the reconstruction of the Abbey. The end of the novel effects a measure of understanding between the generations through the old man's wisdom and ironical self-criticism: the spirit of the 'lamb' asserts itself against the false idol of the 'buffalo'. The full implications of the symbols consolidating the structure of this kaleidoscopic novel are illustrated by numerous characters and incidents exposing the false values and monuments of the 'Kaiserzeit' and the Hindenburg era, the tyranny of the Hitler period, and the courage of those who went into 'inner emigration' (see INNERE EMIGRATION). The title refers to Robert's daily game of billiards, which has a psychological and symbolical connection with the events and the characters.

BILLINGER, RICHARD (St. Marienkirchen nr. Schärding, 1893–1965, Linz, Austria), a country-bred Austrian, originally intended for the priesthood, studied philosophy at Innsbruck, Kiel, and Vienna universities and then settled as a writer, first in Munich, later by Lake Starnberg. Though for a time politically suspect (*Nachtwache* is said to have been partly written while he was under arrest in 1935), his novels and other works were sufficiently close to the soil (see HEIMATKUNST) to be published freely under the National Socialist regime. His more important works of fiction are a novel of childhood *Die Asche des Fegefeuers* (1931), the comic Austrian rural novel *Das Schutzengelhaus* (1934), and the tragic *Lehen aus Gottes Hand* (1935) and *Das verschenkte Leben* (1937). He is also the author of the stories *Ein Strauß Rosen* (1954). His numerous plays include *Das Perchtenspiel* (1928), *Rosse, Rauhnacht* (both 1931), *Lied vom Glück, Lob des Landes, Das Verlöbnis* (all 1933), *Stille Gäste* (1934), *Der Herzog und die Baderstochter* (1934), *Die Hexe von Passau* (1935), *Der Gigant* (1937), *Der Zentaur, Der Galgenvogel* (both 1948), *Traube in der Kelter* (1950), *Das nackte Leben* (1953), and *Menschen nennen es Schicksal* (three one-act plays, 1962). His principal volumes of verse are *Lob Gottes* (1923), *Sichel am Himmel* (1931), *Nachtwache* (1935), and *Holder Morgen* (1942). *Gesammelte Werke* (12 vols.), ed. H. Gerstinger, appeared in 1960.

BINDER, GEORG, see GNAPHEUS, WILHELM.

BINDING, RUDOLF GEORG (Basel, 1867–1938, Starnberg), despite his place of birth, was of German (Frankfurt) descent. His father was professor of criminal law at Basel, later in Freiburg, Strasburg, and Leipzig. Binding studied medicine and law without qualifying in either. His consuming passion was horses, of which he had a rare understanding. One of his most attractive later works is the *Reitvorschrift für eine Geliebte* (1926), which is as sensitive in its empathy as it is accurate in its equestrianism. Binding served as a cadet (see EINJÄHRIGER) in

the fashionable 14th Hussars and later as a reserve officer in the 18th Hussars. In his middle years he bred and dealt in horses and gained a reputation for a capacity to suit any purchaser with precisely the mount he required. His entry into literature was slow and tentative, beginning with the stories *Legenden der Zeit* (1909), which were followed in 1911 by the outstanding Novelle *Der Opfergang*, forming part of the collection *Geigen* (1911) which also contained *Die Waffenbrüder, Angelucia,* and *Die Vogelscheuche*. In 1914 at the outbreak of war, Binding, at 46, accepted the command of an independent squadron of dragoons attached as reconnaissance unit to a newly formed 'Jungdeutschlanddivision'. With this squadron he served in Flanders from October 1914 to August 1916. After four months as a staff officer in Galicia, he returned to the western front, where he served on a divisional staff until the closing weeks of the war.

His sober, balanced, humane diary entries and letter fragments, published as *Aus dem Kriege* (1925), make up his simplest yet most impressive work. The Novelle *Unsterblichkeit* (1922) has as its central figure a young Belgian noblewoman who falls instantly and irretrievably in love with an air ace, who is clearly based on Baron von Richthofen. *Wir fordern Reims zur Übergabe auf* is a short story from the phase of mobile warfare in September 1914. Autobiographical sketches appeared in 1927 as *Erlebtes Leben*. The collected war poems, stories, and recollections were published posthumously in 1939 as *Dies war das Maß*. Binding was never a member of the National Socialist Party and publicly dissociated himself from one of its actions; but his relationship to it was not unambiguous, for he saw it at times as an aspect of national revival. Binding's *Gesammeltes Werk* appeared in 3 vols. (2 vols. 1954 and 1 vol., *Briefe*, 1957).

BINZER, AUGUST, FREIHERR VON (Kiel, 1793–1868, Neiße), a journalist, was the author of the student song 'Stoßt an, Jena soll leben', and of 'Wir hatten gebauet ein stattliches Haus', which was sung at the dissolution of the Burschenschaft (q.v.) in 1819.

BIRCH-PFEIFFER, CHARLOTTE (Stuttgart, 1800–68, Berlin), noted actress and popular dramatist, made her début in Munich and played principal leads from 1818. She married C. A. Birch in 1825. For six years (1837–43) she was in charge of the Zurich theatre, returning to acting in Berlin in 1844. Her 74 plays are mainly adaptations of novels, executed with a keen eye for theatrical effect and predominantly sentimental in tone. They include *Das Pfefferrösel*

(1833), *Schloß Greifenstein* (1833), *Johannes Gutenberg* (1836), *Die Günstlinge* (1839), *Der Glöckner von Notre-Dame* (1839, after Victor Hugo), *Dorf und Stadt* (1847, based on Auerbach's *Die Frau Professorin*, q.v.), *Die Waise von Lowood* (1856, after *Jane Eyre*), and *Die Grille* (1856, after G. Sand's *La Petite Fadette*). (The dates are those of publication; performance was usually earlier.) Among other authors whose work she adapted for the stage were G. Döring, L. Storch, Lord Lytton, and A. Dumas *père*.

BIRCK, Sixtus or Sixt (Augsburg, 1501–54, Augsburg), a schoolmaster in Basel and Augsburg, became a Protestant and wrote plays with biblical subjects and a strong moral bias. They include *Ezechias*, *Zorobabel* (both 1530, but printed later), *Susanna* (1532), *Josef* (printed 1539), *Judith* (1532), and *Baal* (1535). He later remodelled his plays in Latin for school use. They appeared in 1547 as *Dramata sacra*. Birck latinized his name as Xystus Betulejus. *Sämtliche Dramen* (2 vols.) appeared 1969–76.

BIRKEN, Siegmund (von), also Betulius (Wildenstein nr. Eger (Cheb), 1626–81, Nürnberg), a minor poet, whose father was a Protestant pastor, was sent as a child to the Protestant town Nürnberg for protection against the perils of war. After studying at Jena he returned briefly to Nürnberg in 1645, where he became a member of the Hirten- und Blumenorden an der Pegnitz (q.v.), and then moved to Wolfenbüttel as tutor to the Brunswick princes (see Anton Ulrich, Herzog von Braunschweig). In Hamburg he met Johann Rist and Philipp Zesen (qq.v.). In 1648 he returned to Nürnberg and in 1662 became the head (Oberhirt) of the Blumenorden. Ennobled in 1654, he visited England and Holland between 1670 and 1672. Birken was a versatile baroque writer, whose often elaborate compositions include pastoral, historical, and religious poems, and festal works for great occasions in ode or dramatic form. Among his numerous publications are *Kriegs- und Friedensabbildungen* (1647), *Teutscher Kriegs Ab- und Friedens Einzug* (1650), *Die Friederfreuete Teutonia* (1652), *Geistliche Weihrauchkörner* (1652), *Passions-Andachten* (1653), *Androfilo* (1656), *Pegnesische Gesprächspiel-Gesellschaft* (1665), *Hochfürstlicher Brandenburgischer Ulysses* (1667), and *Der Chur- und Fürstlich Sächsische Heldensaal* (1677). Birken's treatise on poetics, *Teutsche Redebind- und Dichtkunst*, though written *c.* 1650, was not published until 1679 (reprinted 1973).

BISCHOF BONUS of Clermont, who died in 709, is the subject of a legend relating to the Virgin Mary written in the 12th c. in South Germany. It narrates, in 238 lines arranged in rhyming couplets, the apparition to him of the Virgin with saints and apostles, and the celestial gift of a magnificent vestment.

BISCHOFF, Friedrich (Neumarkt, Silesia, 1894–1976, Achern), radio administrator, producer, and scriptwriter in Breslau until 1946 and afterwards in Baden-Baden. He is the author of poetry of a reflective and mystical character (*Gottwanderer*, 1921; *Die Gezeiten*, 1925; *Schlesischer Psalter*, 1936; *Das Füllhorn*, 1940; *Der Fluß*, 1942) and of novels of similar temper (*Ohnegesicht*, 1922; *Die goldenen Schlösser*, 1935; *Der Wassermann*, 1937).

BISCHOFFWERDER, Johann Rudolf von (Cölleda, 1741–1803, Potsdam), at first a Prussian officer, became the favourite of King Friedrich Wilhelm II (q.v.); he was an initiate of Rosicrucianism (see Rosenkreuzer) and influenced the King in this direction. For a time he controlled policy, but on the accession of Friedrich Wilhelm III (q.v.) in 1798, he was dismissed, spending his last years as a country gentleman on his estate.

BISMARCK, Otto, Fürst von (Bismarck-Schönhausen) (Schönhausen, 1815–98, Friedrichsruh), of ancient landowning nobility with possessions in Pomerania (Kniephof) and Brandenburg (Schönhausen), was educated at the Friedrich-Wilhelm-Gymnasium in Berlin, at Göttingen University, where he studied law, and at Berlin. After a year's military service, he obtained posts in the administrative and judicial section of the civil service in Berlin and Aachen. On the death of his mother in 1839, he resigned in order to look after the family's Pomeranian estates. In their management he combined ability with application, and undertook additional responsibility as Deichhauptmann. In 1847 Bismarck began what was to be a long and happy marriage with Johanna von Puttkamer. At the approach of the 1848 Revolution (see Revolutionen 1848–9) Bismarck became a deputy of the Provincial Diet and, as a staunch royalist and Conservative, strongly supported the King in opposing constitutional demands made when the United Diet met in Berlin. In the following year he sharply criticized Friedrich Wilhelm IV (q.v.) for allowing himself to be intimidated by the populace. When in 1850 Prussia sought to lead a union of kingdoms (Prussia, Bavaria, Hanover, Saxony, and Württemberg), only to yield to Austrian pressure (see Olmützer Punktation), Bismarck showed mature political judgement in his appreciation of the situation. Capable of patient diplomacy in the struggle for Prussian hegemony, he was determined to realize this aim. Between 1850 and 1862 Bismarck was a member of the Prussian United Diet at Frankfurt and ambassador at St. Petersburg (1859) and at Paris (spring 1862). During this period his political career kept him at a distance from the centre of events, and he accepted the embassy at St. Petersburg with a sense of bitter frustration. This was the price he had to pay for his outspoken opposition to all who did not share his extreme Conservatism, for which he found a vehicle in the newly formed Conservative *Kreuzzeitung*. He had shown himself to be a forceful personality and was feared by numerous opponents. In 1862 King Wilhelm I (q.v.) needed precisely these qualities to save his crown. The critical difference of opinion between the King and his government arose out of the budget to provide for greater expenditure on the army, regarded as necessary by the King and his minister of war, von Roon (q.v.). Rather than accept defeat the King contemplated abdication; however, he decided, upon Roon's advice, to appoint Bismarck, whose deep-rooted loyalty to the Crown was known, to the office of chancellor and foreign secretary (September and October 1862 respectively). Bismarck overcame the crisis by blatantly infringing the constitution. His 'Lückentheorie' argued that the constitution made no provision for a situation in which there was no budget; the government was therefore obliged to levy the necessary taxes without a budget. Bismarck was well aware of the inadequacy of this 'theory', and therefore sought and won in 1867 the approval of the National Liberals for his initial unconstitutional conduct. Meanwhile he tried to silence public opinion, especially the bitter attacks of the newly formed Progressive Party (Fortschrittspartei, founded in 1861), by curbing the basic rights of the freedom of the press and of party political meetings.

In the following year Bismarck faced another crisis with Austria, and this continuing 'dualism' determined his policy during the first years of office. He abandoned the view that the problem of German unity could be resolved through diplomacy in a way acceptable to Prussia. He boycotted Austria's renewed attempt to settle the German constitution at a General Assembly of the German rulers in Frankfurt (1863) by not sending a Prussian representative. And he provoked Austria in the settlement of the issues involving the principalities of Schleswig and Holstein (see SCHLESWIG-HOLSTEINISCHE FRAGE), which resulted in a temporary settlement (Convention of Gastein, 1864). He secured French neutrality in the event of a war between Prussia and Austria. He cultivated good relations with Russia and made an alliance with Italy. But his attempt to counter Austria's move by summon-

ing a German Parliament at the Federal Diet at Frankfurt failed because of the attitude of the South German states. As Austria, encouraged by this, requested that the Frankfurt Parliament should decide the issue concerning Schleswig and Holstein, Bismarck made this breach of the Convention of Gastein an issue of war. The armed conflict was decided by the Prussian victory at Königgrätz (see DEUTSCHER KRIEG). The ending of the campaign at this early stage against the wishes of the King and the generals, and the subsequent Peace of Prague were not only a personal success, but a proof of great foresight. Bismarck already saw in Austria a future ally and wished to spare it undue humiliation. He had resolved the conflict, as he had predicted, with 'Eisen und Blut', but the solution was not an end in itself; it was the beginning of the second phase of the 'making of an empire'. Bismarck's next step was to form the North German Confederation (1867, see NORD-DEUTSCHER BUND). The whole of Germany with the exception of Austria was now virtually under Prussian sovereignty, and Bismarck was its chancellor (Bundeskanzler). Bismarck's handling of the crisis leading to the Franco-German War (see DEUTSCH-FRANZÖSISCHER KRIEG) and Prussian military success led directly to the foundation of the Second German Empire. On 18 January 1871 the Prussian king was proclaimed Deutscher Kaiser in the *Galerie des Glaces* at Versailles.

From now on Bismarck aimed at the maintenance of peace, and, in view of the annexation of Alsace and part of Lorraine, at the prevention of Germany's isolation in Europe. His policy of alliances (Bündnispolitik) served this end (see DREIKAISERBUND, ZWEIBUND, and RÜCKVERSICHERUNGSVERTRAG). At the Congress of Berlin in 1878 Bismarck acted as mediator between the powers in the settlement of the international crisis which had arisen out of the Russo-Turkish war (1877–8). In the intervening years he had, however, encountered serious difficulties in home affairs. He had to compromise and accept virtual defeat in the Kulturkampf (q.v.) and to face pressing social problems while resisting collaboration with the Social Democratic Party (see SPD). He also broke with the Liberals over the issue of fiscal policy. Bismarck used two attempts on the Emperor's life as a pretext for the dissolution and re-election of the Reichstag (October 1878). This enabled him to pass the anti-socialist law ('Gesetz gegen die gemeingefährlichen Bestrebungen der Sozialdemokratie'). Inflexibly adhering to an outworn reactionary feudal attitude, he refused to acknowledge the political implications of the industrial age, and sought to prevent democracy by social legislation. In 1881 he caused the Emperor to promise state assistance for the working class, a promise that was implemented in the Krankenversicherungsgesetz (1883), the Unfallsversicherung (1885), and the Invaliditäts- und Altersversicherung (1889). The death of Wilhelm I in 1888 removed the security of Bismarck's position and he was dismissed by Wilhelm II (q.v.) in 1890. He spent the remainder of his life in retirement on his estate at Friedrichsruh, where he worked on his memoirs.

Bismarck's autocratic statesmanship during his twenty-eight years of office, nineteen of which he served as the 'iron chancellor' of the Empire ('der eiserne Kanzler'), ended in bitter and angry retirement. He became a legendary and monumental figure, but his ruthless treatment of his opponents, his victimization of liberal, socialist, and progressive politicians, and his contemptuous attitude towards truly parliamentary legislative government had made him many enemies. He was so confident of the old Emperor's dependence on him that he did not hesitate to use the threat of resignation to force Wilhelm I into concurrence with his policy. He was the last German statesman to believe in the divine right of kings.

Bismarck's memoirs appeared as *Gedanken und Erinnerungen* (2 vols.) in 1898, to which was added in 1921 a third volume bearing the title originally intended by Bismarck for the whole work, *Erinnerung und Gedanke*. His collected works (*Friedrichsruher Ausgabe*), which include his speeches and letters, comprise 19 volumes (1924–35). Bismarck figures in numerous works of literature, poems, songs, fiction, and plays, among them *Bismarck*, an epic by G. Frenssen (q.v., 1914), the play *Bismarck* by F. Wedekind (q.v., 1916), and the biographical study *Bismarck* by Emil Ludwig (q.v., 1926). The social and political climate of the Bismarck era permeates the work of the writers of the day, including the representatives of Naturalism (see NATURALISMUS). Contemporary views of Bismarck are reflected in several novels of Th. Fontane and especially in *Irrungen Wirrungen* (q.v.).

BITEROLF, a 12th-c. medieval poet knight, a contemporary of HEINRICH von Veldeke (q.v.) and possibly at the Thuringian court of Landgraf HERMANN (q.v.). According to RUDOLF von Ems (q.v.), Biterolf wrote a verse epic, of which no trace remains, dealing with Alexander of Macedon.

Biterolf is the subject of a cycle of five poems by J. V. von Scheffel (q.v.), one of which, *Im Lager von Akkon*, has been set to music (first two stanzas only) by Hugo Wolf (q.v.).

Biterolf und Dietleib, a Middle High German epic poem written in the second half of the 13th

c. by a Styrian poet. Biterolf is the king of Toledo and Dietleib his son. The father sallies forth seeking adventures and the son later sets out to find his father. The central point of the poem is a combat at Worms in which Biterolf, Dietleib, and Dietrich von Bern fight Gunther, Gernot, and Siegfried and vanquish them. The author of *Biterolf und Dietleib* sees the heroic figures in terms of the chivalric ideal.

Bitterfelder Weg, term applied to the cultural programme of the DDR inaugurated at the first writers' conference at Bitterfeld on 23 April 1959. Arranged by the Mitteldeutscher Verlag (Halle) at a state-owned industrial concern, it directed writers on to the path of authentic literary production on workers, their problems, and responsibilities in industry and agriculture. Conceived as an implementation of socialist realism (see SOZIALISTISCHER REALISMUS), its objectives included the formation of circles by prospective writing workers. The programme was reviewed and reinforced in principle at a second Bitterfeld conference in 1964. Its participants included Christa Wolf and Erwin Strittmatter (qq.v.), whose works, *Der geteilte Himmel* and *Ole Bienkopp*, though exemplifying the new policy, had nevertheless incurred criticism.

BITZIUS, ALBERT, see GOTTHELF, JEREMIAS.

BLAICH, H. E., see OWLGLASS, DR.

BLANCKENBURG, CHRISTIAN FRIEDRICH VON (Moitzin nr. Kolberg, 1744–96, Leipzig), a nephew of the officer-poet Ewald von Kleist (q.v.), served in the Seven Years War (1761–3), resigning his commission in 1777. He then devoted himself to literary journalism in Leipzig. His *Versuch über den Roman*, published over the initial B. in 1774, while he was still serving in the army, is the first serious contribution in Germany to the aesthetics and history of the novel. He defines the field of the novel as 'Handlungen und Empfindungen des Menschen', drawing most of his examples, however, from drama. Wieland's *Agathon* (q.v.) is the novel which he most admires. Blanckenburg himself began, but did not complete, a novel—*Beyträge zur Geschichte teutschen Reichs und teutscher Sitte* (vol. 1, 1775).

Blankvers, an iambic pentameter without rhyme. It is almost always used in the plural (Blankverse). See JAMBEN.

Blasedow und seine Söhne, a three-volume novel by K. Gutzkow (q.v.), published in 1838. Written in conscious imitation of JEAN Paul (q.v.), it shows the failure of superficially understood theories of education. Blasedow launches each son into an unsuitable career. The name 'Blasedow' alludes to the well-known education-ist J. B. Basedow (q.v.).

Blasewitz, see GUSTEL VON BLASEWITZ.

BLASS, ERNST (Berlin, 1890–1939, Berlin), a bank clerk turned journalist, wrote Expression-istic poetry (*Die Straße komm ich entlang geweht,* 1912; *Die Gedichte von Trennung und Licht,* 1915; *Die Gedichte von Sommer und Tod,* 1918; *Der offene Strom,* 1921). Under the National Socialist regime he was denied permission to publish.

Blätter für die Kunst, Die, periodical founded in 1892 by Stefan George (q.v.) to publish the work of the esoteric group over which he pre-sided. Its aim was to cultivate 'die geistige Kunst auf Grund einer neuen Fühlweise und Mache'. Contributors included, apart from George him-self, H. von Hofmannsthal, M. Dauthendey, G. Vollmöller, F. Gundolf, E. Bertram, F. Wolters, and K. Wolfskehl (qq.v.). The periodical ap-peared until 1919, and there were in all 12 volumes, each of 160 pages.

Blaue Blume, the symbol of romantic longing. It first appears in 1802 in Novalis's *Heinrich von Ofterdingen* (q.v.).

Blaue Boll, Der, a play by Ernst Barlach (q.v.), published in 1926. The action is presented in seven loosely joined episodes (Bilder) and is set in a small town in North Germany. 'Guts-besitzer' Boll seeks to escape from his everyday material existence and find the 'other' Boll, who condemns the domination of the flesh. Grete Grünthal, known as 'the witch', is undergoing a similar crisis and exacts from Boll a promise to procure poison, with which she wants to put her three children to death, thus liberating their souls. Both, however, are ultimately reconciled with their lives. Grete's traumatic experience of hell at the tavern of Elias, the 'devil', becomes an integral part of Boll's process of rebirth ('Wer-den') in the church. Having kept watch over her throughout the night, he sends her to her children. But the Herr, also referred to as 'Herr-gott', who has joined him in the church, invites him to climb the church-tower once more to realize his earlier intention to kill himself by jumping down from it. Boll at first resists but overcomes his profound fear of death, which he has concealed beneath his excessive self-confidence. As he is about to ascend the tower, the Herr, however, intervenes, claiming that 'Boll has condemned Boll' and that the 'other' Boll has grown beyond such a 'primitive' method of self-realization as suicide. Boll, he

claims, is now ready to begin a new life accepting suffering and endeavour (Leiden und Kämpfen) as the true foundation of self-realization. Left to himself, Boll kneels in front of the statue of an apostle and, liberating himself from the fetters of compulsion, submits of his own free will to the challenge of 'der Herr', as if it were a divine revelation ('Boll muß? Muß? Also—will ich!'). The work is one of Barlach's most successful plays for stage presentation, showing his ability to blend realism and mysticism, grotesque and serene symbolical elements.

Blaue Reiter, Der, coterie of artists formed in Munich in 1911, of whom the most noted are F. Marc, P. Klee, A. Macke, and W. Kan-dinsky (qq.v.). The name derives from a pic-ture by Kandinsky, *Le Cavalier bleu,* which was reproduced on the cover of the group's Alma-nach, published in 1912. (See EXPRESSIONIS-MUS).

Blechschmiede, Die, a vast dramatic satirical extravaganza by A. Holz (q.v.), first published in 1902 and revised in 1917, 1921, and 1924. Largely in verse, it is written with the irrepress-ible verbal fantasy, luxuriance, and prolifera-tion of Holz's later manner. The object of its satire is contemporary literature and civiliza-tion, and the climax is the author's execution by his colleagues, from which he rises again to cock a final snook at them. The sub-title runs to 80 words (many of them long compounds of Holz's invention), of which the essential phrase is 'der umgekippte, umgewippte, umgeschwippte, um-gestippte, umgestürzte Wunderpapierkorb'. The list of characters covers fourteen closely printed pages. The headings of the five acts display a fantastic verbal buffoonery. The first is de-scribed as 'Der Kampf der Skalden, Barden, Minstrels, Lauten-, Lyrenschläger, Lurenbläser, Tubentuter, Dichter, Wagen, Helden, Rosse und Gesänge. Actus primus alias comotragicus. Allegro marciale risoluto quasi polifonia pom-posa bombastica. Tableau vivant isthmien', and the others, of which 'Moderne Walpurgisnacht' (II), 'Die Insel der Seligen' (III), 'Über die trauernde Harfe gebeugt, unter den hängenden Weiden an den weinenden, raunenden, rau-schenden, rollenden, grollenden Wässern Baby-lons' (IV), and 'Das Hochgericht' (V) are essential excerpts, match it in prolixity. In his verbal exuberance and fertility Holz anticipates some later poets. The beautifully produced work totals 515 large octavo pages.

Blechtrommel, Die, a novel by Günter Grass (q.v.) published in 1959. Set against the back-cloth of National Socialism, it is told in the first person by the central figure, Oskar Matzerath, tracing Oskar's history, beginning with his grandparents, and finishing at his thirtieth birthday (1954). Oskar is a dwarf, whose passion is his tin drum, which exercises some of the power of the Pied Piper's pipe, and he possesses a voice which is capable of breaking glass of all kinds at considerable range. The magic of Oskar's voice is matched by his ability to arrest his growth, but here, as elsewhere, the book moves on two planes, for the adult burgher world believes that his failure to develop is due to a fall. The grotesque figure of Oskar is accompanied by a grotesque series of happenings throughout his life, especially the eccentric deaths of those around him. So his mother dies of excessive fish-eating; his first putative father, Jan Bronski, is shot as a resistance fighter, though he wished ardently to avoid resistance; his second putative father, Alfred Matzerath, is suffocated by attempting to swallow his Nazi party badge; the greengrocer Greff commits ingenious suicide on the shop scales; and Oskar is finally condemned for a murder he has not committed and placed in a mental hospital. Oskar's detachment from the normal world enables him to comment upon it, and the book presents a dry and ironic review of the history of Oskar's times from the standpoint of Danzig, which was his home (and Grass's). Likewise Oskar, in his separateness, is exempted from all the inhibitions of the burgher, observing and reporting the sexual by-play, the lovemaking, and the natural functions which are usually screened from view.

Die Blechtrommel possesses unmistakable humanity and compassion. Grass shows himself to be a virtuoso of language, and deploys a prodigious vocabulary, often with minimal grammatical support. The book can be viewed as a revival of the picaresque novel (see SCHELMENROMAN). (Film version by V. Schlöndorff, 1979.)

BLEI, FRANZ (Vienna, 1871–1942, New York), emigrated to the U.S.A. in 1933. An essayist of great diversity and an accomplished ironist, his chief interests were the erotic and the religious. His collections include *Hippolyt* (1903), *Die galante Zeit* (1904), *Von amoureusen Frauen* (1906), *Die Puderquaste* (1909), *Der Geist des Rokoko* (1923), *Glanz und Elend berühmter Frauen* (1927), *Frauen und Männer der Renaissance* (1927), *Das Erotische* (1927), and *Männer und Masken* (1930). Early in his career he wrote three plays, *Die rechtschaffene Frau* (1893), *Thea* (1895), and *Die Sehnsucht* (1900), the last two, comedies. He is also the author of a tragedy (*Der dunkle Weg*, 1906) and the successful comedy *Logik des Herzens* (1915). Under the pseudonym Peregrinus Steinhövel he published the satire

Bestiarium literaticum (1920), which is better known by its later title *Das große Bestiarium der Literatur* (1924). A bibliophile, Blei issued in 1900–13 a beautifully produced edition of the works of J. M. R. Lenz (q.v., 5 vols.). *Vermischte Schriften* (6 vols.) appeared in 1911–12.

BLEIBTREU, CARL (Berlin, 1859–1928, Locarno), son of an artist, began writing as a belated Romantic with the epic *Gunnlaug Schlangenzunge* (1879), the novel *Traum aus dem Leben des Dichterlords* (Byron, 1880), and the plays *Feueranbeter* (1881) and *Lord Byrons letzte Liebe* (1881). He devised the form in which he was most successful—the novel with large-scale battle descriptions—in *Dies irae, Erinnerungen eines französischen Offiziers*. (His father, Georg Bleibtreu, 1828–92, was a notable battle painter.) Later works of this kind include *Friedrich der Große bei Kolin* (1888), *Napoleon bei Leipzig* (1884), *Schlachtenbilder* (1888), and *Cromwell bei Marston Moor* (1889). In the end he discarded the fictional trappings and wrote straightforward accounts of battles (*Der Kampf bei Mars-la-Tour*, 1897; *Aspern*, 1902; *Waterloo*, 1902; *Königgrätz*, 1903; and a number of others).

Bleibtreu was for a time prominent among the literary reformers of the 1880s, contributing to M. G. Conrad's *Die Gesellschaft* (q.v.), and writing a polemical essay *Revolution der Literatur* (q.v., 1886). A too prolific author, he wrote, as well as numerous novels several plays (mostly historical and military, e.g. *Wellington bei Talavera*, 1890), historical works, a history of English literature (*Geschichte der englischen Literatur*, 2 vols., 1888), and books on politics (*Staat und Christentum*, 1892) and on the authorship of Shakespeare's plays (*Der wahre Shakespeare*, 1906).

Blendung, Die, a novel by E. Canetti (q.v.) first published in 1935, when it failed to attract attention, and republished in extended form in 1963, since when it has been mentioned in the same breath as James Joyce's *Ulysses* and the novels of H. Broch (q.v.). The comparison with Broch is justified by the preoccupation with the relationship of the individual to the human mass around him, but the treatment of the theme differs radically. Canetti's central figure, the sinologist Kien, is an individualist of great eccentricity, whose world is his library, the symbol of concentrated intellectualism. Drawn into marriage with his uncomely housekeeper, Therese, he finds that the world outside intrudes upon him, and a series of grotesque Kafka-esque happenings and mishappenings ensues, ending in Kien's self-immolation by fire on an enormous pile of his books. It is known in English translation as *Auto da Fé.*

Blenheim, see Höchstädt.

BLIGGER VON STEINACH, name of a writer of Minnelieder and also of the author of one or more lost epic poems. It is uncertain whether the two are identical since the name Bligger was borne by all members of this family from Neckarsteinach near Heidelberg. Of the lyric poet, who was in the service of the Emperor Heinrich VI (q.v.) in Apulia and Sicily (1191–4), two poems are preserved. The epic poem of Bligger is praised for its formal qualities by Gottfried von Straßburg (q.v.), who, however, gives no hint of title or subject. It has sometimes been supposed that the word 'Umbehanc' (tapestry) used of this poem by Gottfried and repeated by Rudolf von Ems (q.v.), was its title; it is more likely that Gottfried was comparing the work to a tapestry.

BLOCH, Ernst (Ludwigshafen, 1885–1977, Tübingen), lived in exile from 1933 to 1948 and on his return became professor of philosophy at Leipzig University. In 1957 he left East Germany for the Federal Republic and became a professor at Tübingen University in 1961. He was the most distinguished heir to the dialectical processes developed in the 19th c. in a historical sequence from Hegel to Marx (qq.v.). By recognizing spiritual and cultural needs he enriched his 'philosophy of hope', which aimed at gradual social evolution and at the emancipation from the Marxist concept of human self-alienation. A thinker without illusions, he nevertheless sought to rouse society's latent potential, in a real, not abstract, Utopia. In 1967 Bloch was honoured with the Friedenspreis des deutschen Buchhandels (q.v.). His works include *Geist der Utopie* (1918), *Thomas Münzer als Theologe der Revolution* (1921), *Freiheit und Ordnung* (1946), *Das Prinzip Hoffnung* (1954–7), *Philosophische Grundfragen* (1961), *Verfremdungen* (2 vols., 1962–4), *Politische Messungen. Pestzeit, Vormärz* (1972), and *Experimentum Mundi—Frage, Kategorien des Herausbringens, Praxis* (1975).

Gesamtausgabe, 16 vols., appeared 1962–75.

Blocksberg, name given to various German mountains said to be frequented by witches, but chiefly to the Brocken (q.v.).

Blödigkeit, see Dichtermut.

BLOEM, Walter (Elberfeld, Wuppertal, 1868–1951, Lübeck), a lawyer, turned in 1904 to literature and from 1911 to 1914 was a theatrical producer. He served in both wars. His strongly nationalistic novels were widely read, but he was not whole-heartedly acknowledged

by the National Socialists. The best known of his novels are *Der krasse Fuchs* (1906), *Das eiserne Jahr* (1911), *Volk wider Volk* (1912), *Die Schmiede der Zukunft* (1913), *Das verlorene Vaterland* (1914), *Gottesferne* (1920), *Herrin* (1921), *Das Land unserer Liebe* (1924), *Teutonen* (1926), *Sohn seines Landes* (1928), *Wir werden ein Volk* (1929), *Frontsoldaten* (1930), *Heiliger Frühling* (1933), *Die große Liebe* (1935), *Der Volkstribun* (1936), and *Kämpfer überm Abgrund* (1944). He also wrote plays (*Caub*, 1897; *Heinrich von Plauen*, 1902; *Schnapphähne*, 1903; *Der neue Wille*, 1905; *Vergeltung*, 1910; *Dreiklang des Krieges*, 1918; *Helden von Gestern*, 1921). His son, Walter Julian Bloem (1898–1945), who used the pseudonym Kilian Koll, was also a writer of nationalistic novels. He served in the Waffen-SS and his presumed death in Berlin in 1945 has never been confirmed.

BLOMBERG, Barbara (*c.* 1527–97), the beautiful daughter of a citizen of Regensburg, was the mistress of the Emperor Karl V (q.v.) during the Diet of Regensburg in 1546 and bore him a son, who became Don John of Austria, the victor of Lepanto. Barbara Blomberg married in 1551 a man of her own class.

Blonde Eckbert, Der, a story (Märchen) by L. Tieck (q.v.), written in 1796 and published in 1797 in his collection *Volksmärchen*. Eckbert is the principal character of the frame (see Rahmen), which contains the story (Binnenerzählung) of Bertha, with whom he lives in childless marriage. Eckbert uses the occasion of the visit of an old friend, Philipp Walter, to persuade his wife to relate the strange tale of her childhood. Of poor country origin, Bertha ran away from home at the age of 7 to escape parental ill-treatment. She is saved from almost certain death in the wooded mountains by an old woman, who brings her up in her hut in a birchwood. At 14 Bertha is left by the old woman to look after the hut during her absence. Also in her care are a bird which lays diamonds, and a dog.

The longing for human company and the temptation of riches cause Bertha to disregard the old woman's instructions. She leaves hut and dog, and makes her way back into the world with the bird, which she kills for reminding her, in its song, of her disloyalty. She meets Eckbert and they marry. In ending her story Bertha finds to her surprise that she has forgotten the name of the dog; strangely, Walter is able to remind her that it was called Strohmian (Strohmi). This seemingly insignificant detail leads to Bertha's illness, of which she suddenly dies. Eckbert, also affected by the incident, loses control of his actions, shoots Walter in the woods, finds a new friend in Hugo von Wolfsberg, who assumes the features of Walter, and finally of the old woman. He is in a state of near madness when the old woman, whom he encounters near the birchwood, tells him that Bertha was his sister. Shattered by the news, he succumbs to a death of despair.

The element of atmosphere (Stimmung) and Romantic irony is enhanced by the bird's song expressing a longing for nature's peaceful solitude (Waldeinsamkeit); this song recurs with variations supporting the disciplined structure of the narrative. It is this feature, influenced by Goethe, which has led to the recognition of the tale as a Romantic Novelle (Novellenmärchen, see Novelle).

BLÜCHER, Gebhard Leberecht von (Rostock, 1742–1819, Krieblowitz/Silesia), Prussian field-marshal (Generalfeldmarschall), first served as an officer of Swedish cavalry. During the Seven Years War he transferred to the Prussian army, serving in a hussar regiment. In 1770 he left the army, incurring the displeasure of Friedrich II (q.v.) of Prussia. He returned to the army under Friedrich Wilhelm II (q.v.) in 1787, and distinguished himself in the French Revolutionary Wars in 1793–4 (see Revolutionskriege). He rose rapidly to general's rank and in 1806 fought at Auerstedt (q.v.) and, after the defeat, conducted a successful retreat to Lübeck, surrendering finally at Ratkau. Blücher held commands in Prussia during the period of subjugation, but his anti-French views led to his removal in 1811. With the outbreak of the War of Liberation in 1813 he was reinstated and was prominent at Großgörschen, Bautzen (see Napoleonic Wars), and, with Gneisenau (q.v.) as his chief of staff, at Katzbach, Möckern and, above all, Leipzig. Blücher commanded in the French campaign in 1814, and was at first successful but later suffered several reverses. He won a decisive victory at Laon, and shortly afterwards entered Paris. He was enthusiastically welcomed in England on a visit in 1814, receiving an honorary doctorate from Oxford University. When Napoleon returned from Elba, Blücher was placed in command of the Prussian army. At Ligny he was defeated and wounded, but was rescued by his adjutant von Nostitz. Two days later Blücher brought up his forces in time to outflank the right of the French position at Waterloo (see Waterloo, Battle of), meeting Wellington at La Belle Alliance in the moment of victory.

Blücher's outstanding qualities were his courage and energy (which gained him the nickname 'Marschall Vorwärts'), and also his loyalty, which led him, against contrary opinions, to come vigorously to the support of Wellington. He was proverbial for startling misspellings in his letters, which were first published

in 1876. He was created Fürst von Wahlstatt (near Katzbach in Silesia) in 1814. He is the subject of a cycle of five poems (*Blücher*) by F. Rückert (q.v.), in which reference is made to his visit to England and his passion for gambling. He figures in the fourth and fifth acts of C. D. Grabbe's *Napoleon oder Die hundert Tage* (q.v.). In the Prussian (later German) army the 5th (Pomeranian) Hussars bore his name.

BLUMAUER, Alois (Steyr, 1755–98, Vienna), a minor Austrian poet and a Jesuit until the Society of Jesus was banned in Austria in 1781, when he accepted the post of censor of books. In 1793 he became a bookseller. Blumauer wrote poetry in the elegant manner of Wieland and Bürger (qq.v.). He is best known for a frivolous and indecent parody of the *Aeneid* (*Virgils Aeneis oder Abenteuer des frommen Helden Aeneas*, 1783).

Blümchen Wunderhold, Das, a poem in twelve rhyming stanzas by G. A. Bürger (q.v.), referring to his second wife, Auguste Leonhart or Molly. It was first published in *Gedichte* (1789), and was set to music by Beethoven (q.v.).

Blumen der Tugend, Die, see Vintler, Hans.

Blumen-, Frucht- und Dornenstücke, oder Ehestand, Tod und Hochzeit des Armenadvokaten Fr. St. Siebenkäs im Reichsmarktflecken Kuhschappel, title of a novel by Jean Paul (q.v.) first published in 1796–7, commonly abbreviated to *Siebenkäs*. The poor man's lawyer Siebenkäs marries, and his bride Lenette believes him to be poor. In fact, he has a substantial legacy, which is in the hands of his guardian Blaise, but Blaise cheats him and the marriage comes into financial straits. Everard Rosa seeks unsuccessfully to seduce Lenette; Siebenkäs's jealousy is aroused and the young couple become estranged. Siebenkäs's friend Leibgeber hatches a plan by which Siebenkäs is to begin a new life by simulating illness and death. Leibgeber spirits Siebenkäs away and he becomes steward to Count Vaduz and falls in love with Natalie, a niece of Blaise. On revisiting Kuhschappel he finds that Lenette is dead and at her grave he encounters Natalie. They mourn together and their hearts are united. The book closes with a meditation on immortality headed *Rede des toten Christus vom Weltgebäude herab, daß kein Gott sei.*

BLUMENTHAL, Oskar (Berlin, 1853–1917, Berlin), a journalist who worked for the *Berliner Tageblatt* (1875–87) and became director of the new Lessing-Theater in 1888, a post which he held until 1897. He was the author of a number of highly successful but ephemeral comedies, including *Der Probepfeil* (1882), *Die große Glocke* (1885), *Die Fee Caprice* (1901), *Klingende Pfeile* (1904), *Der Schwur der Treue* (1905), *Das Glashaus* (1906), and *Der schlechte Ruf* (1910). Some of Blumenthal's box-office successes were written in collaboration with G. Kadelburg (1851–1925). The best known of these is *Im weißen Rößl* (1898); others include *Als ich wiederkam* (1902), *Das Theaterdorf* (1902), *Der blinde Passagier* (1902), *Großstadtluft* (1905), *Hans Huckebein* (1905), *Die Orientreise* (1905), *Der letzte Funke* (1907), and *Die Tür ins Freie* (1908). He collaborated with R. Lothar (q.v.) in the comedy *Die drei Grazien* (1910).

BLUNCK, HANS FRIEDRICH (Altona, 1888–1961, Hamburg), at one time a civil servant in Hamburg and later an administrative officer in Hamburg University (1925–8), took to authorship and made his name and fortune with numerous nationalistic and, later, National Socialistic novels and stories which place a strong stress on race and soil (see HEIMATKUNST). From 1933 to 1935 he was president of the Reichsschrifttumskammer. Though replaced by H. Johst (q.v.), he continued zealously to support the movement. Some of his better-known titles are *Nordmark* (1912), *Totentanz* (1916), *Peter Ohles Schatten* (1919), *Hein Hoyer* (1922), *Berend Fock* (1923), *Streit mit den Göttern* (1925), *Kampf der Gestirne* (1926), *Die Weibsmühle* (1927), *Die große Fahrt* (1935), *König Geiserich* (1936), *Wolter von Plettenberg* (1938), *Die Jägerin* (1940), and *Die Sardens und der Besessene* (1952). Blunck also wrote plays (*Die Frau im Tal*, 1920; *Die Lügenwette*, 1935; *Kampf um Neu-York*, 1940; *Das Londoner Frühstück*, 1955), and poems.

Blütenstaub, a collection of 114 prose 'fragments', many aphoristic in tone, contributed to *Das Athenäum* (q.v.) in 1798 by Novalis (q.v.). They are concerned with philosophy and psychology.

Blütezeit, this general term for an age of florescence has come in German medieval literature to refer to the optimal period of literary production from approximately 1180 to 1230, during which HARTMANN von Aue, WOLFRAM von Eschenbach, GOTTFRIED von Straßburg, and WALTHER von der Vogelweide (qq.v.) were active. See also HÖFISCHE DICHTUNG.

Blut- und Bodenliteratur, see HEIMATKUNST.

Blut und Eisen, phrase attributed to Bismarck (q.v.) in a speech in the Prussian House of Representatives on 30 September 1862. It is generally quoted as a terse summary of Bismarck's policy. Bismarck's actual words were, 'Nicht durch Reden und Majoritätsbeschlüsse werden die großen Fragen der Zeit entschieden —das ist der Fehler von 1848 und 1849 gewesen —sondern durch Eisen und Blut.'

BOBROWSKI, JOHANNES (Tilsit, 1917–65, Berlin), taken prisoner during the war on the eastern front, returned from Russia in 1949 and settled in East Berlin, where he worked as a publisher's reader. His melancholy and beautiful poems, largely inspired by the landscape of his native region near the Memel, a meeting-point and a focus of conflict between races, are laden with personal and historical recollection. His collections of poems are *Sarmatische Zeit* (1961), *Schattenland Ströme* (1962), and *Wetterzeichen* (posth., 1966). His two novels, *Levins Mühle* (1964) and *Litauische Klaviere* (posth., 1966), and his stories (*Boehlendorff, Mäusefest*, both 1965, and *Der Mahner, Im Windgesträuch*, posth., 1968 and 1970) are as regional in landscape, atmosphere, and outlook as the poems. His desire to contribute towards reconciliation and his Christian brand of socialism are fundamental features.

Bochum, industrial town in Nordrhein-Westfalen with an enterprising theatre which is noted for its productions of Shakespeare. The Ruhr-Universität, situated at Bochum, was founded in 1965.

BOCK, ALFRED (Gießen, 1859–1932, Gießen), is the author of novels and stories, mainly with a local (Gießen or Hessian) background. They include the Novellen *Aus einer kleinen Universitätsstadt* (1896) and *Hessenluft* (1907) and the novels *Der Flurschütz* (1901), *Kinder des Volkes* (1902), *Der Kuppelhof* (1905), *Die Oberwälder* (1912), *Grete Fillinger* (1918), *Der Schlund* (1918), *Der Elfenbeiner* (1922), and *Das fünfte Element* (1924).

BÖCKLIN, ARNOLD (Basel, 1827–1901, nr. Fiesole, Italy), Swiss painter, studied at Düsseldorf, Antwerp, Brussels, Geneva, and Paris, and spent the years 1850–7 in Rome. Though much admired in Germany in the last forty years of the 19th c., he spent only relatively short periods in Munich, being in Italy 1862–6, in Basel 1866–71, in Florence 1874–5, in Zurich 1885–92, and for the rest of his life in Fiesole. Böcklin was notable as a straightforward landscape painter and also as the creator of mythical or symbolical landscapes, of which the best known is *Die Toteninsel* (1882, existing in four versions). Classical nude groups such as *Triton und Nereide* (1875) and *Gefilde der Seligen* (1878) shocked his bourgeois admirers. He painted a self-portrait

(1872) in which a grinning skull and the upper part of a skeleton with violin and bow are partly hidden behind his head. Böcklin's painting was often literary in its use of myth, but his chief contact with literature was his close friendship with G. Keller (q.v.), hence his long stay in Zurich.

BODE, JOHANN JOACHIM (Brunswick, 1720–93, Weimar), a military bandsman who educated himself by reading, founded a printing and publishing business which issued *Der Hamburgische Correspondent* (1762–3), Lessing's *Hamburgische Dramaturgie* (q.v.), and Claudius's *Der Wandsbecker Bote* (q.v.). Lessing was for a time his partner, but the firm got into difficulties and Bode eventually went bankrupt. In 1778 he became agent to Countess Bernstorff in Weimar. Works translated by Bode include Sterne's *A Sentimental Journey* (*Yoricks Empfindsame Reise durch Frankreich und Italien*, 1768) and *Tristram Shandy* (*Tristram Shandys Leben und Meinungen*, 1774), Goldsmith's *The Vicar of Wakefield* (*Der Dorfprediger von Wakefield*, 1776), and Fielding's *Tom Jones* (*Geschichte des Thomas Jones*, 1786–8).

BODENSTEDT, FRIEDRICH VON (Peine, Hanover, 1819–92, Wiesbaden), worked in a counting-house, and then studied at Göttingen, Munich, and Berlin universities. He became a tutor in Moscow in 1840 and in 1844 wandered much further east, teaching at a school in Tiflis, where he learned some of the more obscure languages, including Georgian and Armenian, from a colleague named MIRZA Schaffy (q.v.). After his return to Germany in 1847 by way of Constantinople he was active as a journalist. He translated Russian authors, including Pushkin and Lermontov, and published a historical work on the Caucasus, *Die Völker des Kaukasus und ihre Freiheitskämpfe gegen die Russen* (1848), and the travel book *Tausend und ein Tag im Orient* (2 vols., 1849–50), which contained a collection of poetry entitled *Lieder des Mirza Schaffy*. These pseudo-oriental poems were at first believed to be translations, but Bodenstedt revealed in 1874 that they were his own work. In the meantime, in separate publication (1851), they enjoyed considerable success with the reading public. In 1854 Bodenstedt went to Munich as professor of Slavonic languages. He next diverted his interest to English studies, publishing *Shakespeares Zeitgenossen* (1858–60) and becoming in 1859 professor of Early English. Between 1862 and 1872 the translation of Shakespeare which bears his name was published; his numerous collaborators included O. Gildemeister, A. Wilbrandt, P. Heyse, and G. Herwegh (qq.v.). In 1867 he became director (Intendant) of the Court Theatre at Meiningen, receiving a patent

of nobility. He resigned this post in 1869, but continued to live in Meiningen until 1873. After a period in Berlin he moved to Wiesbaden in 1878, where he spent his last years. With the exception of *Die Lieder des Mirza Schaffy*, his poetry and plays evoked little response, and even a sequel to his great success, the poems *Aus dem Nachlasse Mirza Schaffys* (1874), failed to revive the old magic.

BODENSTEIN, ANDREAS, see KARLSTADT.

BODMAN, EMANUEL, FREIHERR VON UND ZU (Friedrichshafen, 1874–1946, Gottlieben nr. Constance), studied at Munich, Berlin, and Zurich universities. He was financially independent and devoted himself to consciously aesthetic literature in various forms. His poetry, which has a gentle lyrical quality, is contained in *Stufen* (1894), *Erde* (1896), *Neue Lieder* (1902), *Der Wanderer und der Weg* (1907), and *Mein Vaterland* (1914). *Jakob Schläpfle* (1901) and *Das hohe Seil* (1915) are collections of Novellen; the story *Erwachen* (1906) was published separately. Three tragedies, *Donatello* (1907), *Der Fremdling von Murten* (1907), and *Die heimliche Krone* (1909), were praised in their day for their classical idealism.

BODMER, JOHANN JAKOB (Greifensee, Switzerland, 1698–1783, Schönenberg nr. Zurich), travelled in Italy, became in 1720 an official (Staatsschreiber) in Zurich, and in 1725 a senior teacher (Professor) of history and politics at the Zurich grammar school, a post which he held until 1775. He became a city councillor in 1737. With his friend and colleague J. J. Breitinger (q.v.), Bodmer published a series of critical and aesthetic writings. In 1721 they founded a literary periodical entitled *Die Discourse der Mahlern*, which took its inspiration from Addison and continued until 1723. *Von dem Einfluß und dem Gebrauch der Einbildungskraft* (1727) treats imagination, not as something inspired, but as an agile faculty for the combination of perceptions. His interest in the English moralists drew his attention to Milton, who quickly aroused his enthusiasm. His translation of *Paradise Lost* into German (*Der Verlust des Paradieses*) was published in prose in 1732, in verse in 1742, and was further revised in 1754. In 1740 Bodmer published his *Critische Abhandlung von dem Wunderbaren in der Poesie*, which, coinciding in the same year with Breitinger's *Critische Dichtkunst*, provoked a quarrel with J. C. Gottsched (q.v.), which was further fanned by Bodmer's *Critische Betrachtungen über die Poetischen Gemählde der Dichter* (1741). From a distance of more than two centuries it is possible to discern considerable common ground of literary rationalism and common sense between the warring parties; but at the time Bodmer's and Breitinger's stand for a modest niche for the imagination seemed to mark a total opposition. Klopstock's religious epic *Der Messias* (q.v.), which began to appear in 1748, so captivated Bodmer that he invited the young poet to Zurich, only to find that the singer of the Messiah was more worldly than he had expected. Fired by Klopstock's example, however, he wrote a religious epic (*Noah*, 1750) which was quickly followed by *Jakob und Joseph* (1751), *Die Synd-Flut* (1751), and *Jakob und Rachel* (1752). In 1752 he invited C. M. Wieland (q.v.) to Zurich, but was once more, though rather less quickly, disappointed in his guest. He published in 1757 MS. C of the *Nibelungenlied* without acknowledgement to its true discoverer, J. H. Obereit (q.v.). In the 1770s Bodmer wrote mediocre tragedies on classical subjects (*Electra*, 1760; *Ulysses*, 1760; *Julius Caesar*, 1763), and a decade later tried his hand, equally unsuccessfully, at historical epic (*Conradin von Schwaben*, 1771). Bodmer's limited importance in the development of German literature ceased *c.* 1745. He was twice visited by Goethe, in 1775 and 1779. He is a minor character in G. Keller's Novelle *Der Landvogt von Greifensee* (q.v.).

A selection of Bodmer's works appeared as *Schriften* in 1938.

BODMERSHOF, IMMA VON (Graz, 1895–1982, Gföhl), *née* von Ehrenfels, studied philosophy and the history of art at Prague and Munich universities, and was acquainted with Rilke (q.v.) and members of George's circle (see GEORGE-KREIS). She married in 1925. Of her novels the best known is *Die Rosse des Urban Roitner* (1950). The others are *Der zweite Sommer* (1937), *Die Stadt in Flandern* (1939, reissued as *Das verlorene Meer*, 1952), *Sieben Handvoll Salz* (1958), and *Die Bartabnahme* (1967). Their main theme is man's discovery of his true self. She is also the author of Novellen (*Solange es Tag ist*, 1953).

BOEHRINGER, ROBERT, see GEORGE-KREIS.

BOHEMUS, MARTINUS, see BEHEMB, MARTIN.

BÖHLAU, HELENE (Weimar, 1859–1940, Widdersberg nr. Munich), a publisher's daughter, married in 1886 a Russian Jew, F. Arndt, who had become a convert to Islam and lived at Constantinople. After his death in 1910 she settled in Ingolstadt and later in Munich. She wrote many stories and novels in the Naturalist manner, of which the best known are the collections *Ratsmädelgeschichten* (q.v., 1888), *Altweimarische Geschichten* (1897), and the novels *Im*

frischen Wasser (1891), *Der Rangierbahnhof* (q.v., 1895), *Das Recht der Mutter* (1897), and *Halbtier* (q.v., 1899). *Isebies* (1911) is an autobiographical novel.

BÖHME, FRANZ MAGNUS (Willerstedt nr. Weimar, 1827–98, Dresden), a teacher of music at Frankfurt/Main and at Dresden, wrote on musical history (*Geschichte des Oratoriums*, 1887). He made several collections of folk-songs, some of which were in widespread family use: *Altdeutsches Liederbuch* (1877), *Erks deutscher Liederhort* (newly edited, 1893–4), and *Volkstümliche Lieder der Deutschen im 18. und 19. Jahrhundert* (1895). In 1886 he published *Geschichte des Tanzes in Deutschland*.

BÖHME, JAKOB (Seidenfeld nr. Görlitz, 1575–1624, Görlitz), came of peasant stock, was apprenticed to a shoemaker and in 1599 settled in Görlitz as a master. His unorthodox views, expressed in *Morgenröte im Aufgang* (1612), were assailed by the principal pastor of Görlitz, Gregor Richter, who elicited from him a promise to abstain from writing. Böhme, who in 1613 changed his trade to dealing in wool, complied for a time, devoting himself to study. In 1618 he began to write again, incurring in 1624 a new attack from Richter. He found support against his adversary, but died in the same year. A man of deep religious feeling, Böhme pursued his own idiosyncratic investigation of God's ways, in which pantheistic views are manifest and God is seen to include evil as well as good. He seeks to resolve the dualities, of which men are conscious, into a harmony. The soul should strive for a rebirth which will open the way to Grace. Böhme's style and thought are highly individual and not easily penetrable. All his works, except for two short tracts, were published after his death. *Morgenröte im Aufgang* (retitled *Aurora*) appeared in 1634, and an extended collection came out in 1682. In the 18th c. Böhme's work sank into obscurity, but interest in him revived in the Romantic movement. He appears as a character in the novel *Meister Joachim Pausewang* by E. G. Kolbenheyer (q.v.).

The 1730 edition of Böhme's complete writings (*Theosophia oder: Alle Göttlichen Schriften*, 4 vols.) was reprinted in facsimile arranged in 11 vols. (ed. W.-E. Peuckert, 1942–61).

BÖHME, MARTIN, see BEHEMB, MARTIN.

Böhmerschlacht, Die, title given to a short Middle High German poem written in the Rhineland by an unknown author *c.* 1300. The beginning and end are missing and there are four other lacunae. The subject of *Die Böhmerschlacht* is the battle of Dürnkrut (*Schlacht auf dem*

Marchfeld), in which Rudolf I (q.v.) of Habsburg defeated Ottokar of Bohemia in 1278. The poem is, however, not historical, treating the battle as a symbolical encounter of the two leaders in mortal but knightly single combat.

Die Böhmerschlacht is held by some authorities to be part of a poem composed of six separate episodes of which the others are *Die Schlacht bei Göllheim, Der Minnehof, Das Turnier, Die Ritterfahrt*, and *Der Ritterpreis* (qq.v.). All are of Rhenish origin and CILIES von Seyn (q.v.) has been proposed as their author. This attribution seems unlikely and the association of the poems in certain MSS. is probably fortuitous.

Böhmische Brüder or **Mährische Brüder,** known in England as the Moravian Brethren, were a sect which arose in Bohemia some years after the burning of J. Hus (q.v.). Their founder was Peter Chelczizky, who advocated the view that property is held in trust by the rich for the benefit of the poor. The sect obtained some official political recognition in 1457, and in 1467 the Brethren set themselves up as a Church of Bohemia independent of Rome. They eventually divided into a strict and a mild party, and under Lucas of Prague the latter gained the upper hand (1494). The Brethren gravitated in the 15th c. towards the Reformed Church. They suffered great persecution in the Thirty Years War, in which they were almost exterminated. In the 18th c. the sect was revived in Germany (see HERRNHUTER).

BOHSE, AUGUST (Halle, 1661–1730, Liegnitz), held various employments as private tutor, lawyer, and professor. He lived at different times in Hamburg, Dresden, Leipzig, Weißenfels, and Erfurt, and spent his last years as a professor at Liegnitz. Bohse, who used the pseudonym Talander, was a prolific writer of erotic novels, so-called 'galante Romane'. Apart from translations, which included Barclay's *Argenis*, he wrote 14 novels, in which the world is seen as a place of amorous intrigue and described in inflated language. Their titles are *Liebes-Cabinet der Damen* (1685), *Unglückliche Prinzessin Assinoe* (1687), *Alcestis aus Persien* (1689), *Amor am Hofe* (1689), *Die Eifersucht der Verliebten* (1689), *Der getreuen Bellamira wohlbelohnte Liebes-Probe* (1692), *Schauplatz der Unglückselig Verliebten* (1693), *Aurorens Königlicher Princessin aus Creta Staats- und Liebesgeschichte* (1695), *Der Liebe Irregarten* (1696), *Die Amazoninnen aus dem Kloster* (1696), *Die getreue Sklavin Doris* (1696), *Die Liebenswürdige Europäerin Constantine* (1698), *Ariadnens Königlicher Printzessin von Toledo Staats- und Liebesgeschichte* (1699), *Die verliebten Verwirrungen der Sicilianischen Höfe* (1725). These novels were enthusiastically received by Bohse's contemporaries.

BOIE, HEINRICH CHRISTIAN (Meldorf, 1744–1806, Meldorf), a minor poet who, after studying at Jena, became a private tutor at Göttingen, where he promoted the Göttinger Hainbund (q.v.). In 1770 he published in collaboration with F. W. Gotter (q.v.) the first *Musenalmanach* and in 1776 established a periodical, *Das deutsche Museum*. From 1781 until his death he was Danish commissioner (Landvogt) for Süderdithmarschen. He wrote poems, which have not yet been collected. Boie's sister Ernestine married J. H. Voß (q.v.) in 1777. His correspondence with Luise Mejer was published as *Ich war wohl klug, daß ich dich fand* (ed. I. Schreiber, 1961, repr. 1971).

BOIE, MARGARETE (Berlin, 1880–1946, Lüneburg), daughter of an officer, worked in a museum at Danzig and lived from 1919 to 1929 on the island of Sylt, from which period date her novels *Der Sylter Hahn* (1925), *Moiken Peter Ohm* (1926), and *Dammbau* (1930). She is also the author of Novellen (*Schwestern*, 1921; *Eine Wandlung*, 1932; *Die Tagfahrt der Preußen*, a collection, 1942; *Übers Jahr*, 1944).

BOISSERÉE, SULPIZ (Cologne, 1783–1854, Bonn), devoted his life to the study and revival of interest in German medieval art, and especially of Gothic architecture. With his younger brother Melchior (1786–1851) he built up a large and important collection of early German and Flemish paintings, which was acquired by King Ludwig I (q.v.) of Bavaria in 1827 and is now incorporated in the Alte Pinakothek in Munich. In 1810 he made the acquaintance of Goethe, a contact which ripened into friendship, and influenced Goethe to a more sympathetic view of Gothic art. From 1808 on, Boisserée studied Cologne Cathedral, and produced *Geschichte und Beschreibung des Doms von Köln* in 1823–32. With his brother he devoted himself to the campaign for the completion of the cathedral, the work for which was effectively begun in 1842 and brought to a successful conclusion in 1880. See KÖLNER DOM.

BÖLL, HEINRICH (Cologne, 1917–85, Bornheim-Merten nr. Cologne), the son of a joiner, was apprenticed to a bookseller in Bonn and completed one semester at Cologne University, studying Germanistic and Classical Philology, before being called up for Labour Service, and drafted into the army in 1939. Four times wounded, he was taken prisoner in 1945, but in the same year returned to settle in Cologne. He resumed his studies and family life, taking various jobs before devoting himself full-time to writing. Within a few years he became known as a formidable critic of authority who stood by his convictions and did not shun controversy. A Roman Catholic by upbringing, he directed his criticism against what he considered to be malpractices within the institution of the Church and against the encyclical *Humanae Vitae*, and in his tract *Brief an einen jungen Katholiken* (1958 and 1961) conveyed a frank expression of his views on the Church's approach to morality. In politics he was disillusioned by the decision of the Social Democrats (SPD) to join the Christian Democrats (CDU/CSU) in the so-called 'Grand Coalition' (Große Koalition) in 1965; he denounced the social malaise of the new Federal Republic, the spread of materialistic attitudes resulting from the 'economic miracle' (Wirtschaftswunder), and nuclear armament. He stood for Human Rights, and in 1974 was the host to the novelist and Nobel Prize winner A. Solzhenitsyn (b. 1918) immediately after the latter's eviction from Russia, which Böll himself had visited in 1962.

The absurdity and horror of war, as well as the problem of guilt, is the main concern of his early work which established his reputation as a writer of fiction. An exponent of the German brand of the genre, he began with short stories, *Die Botschaft, Kumpel mit dem langen Haar* (both 1947), *Der Zug war pünktlich* (1949), the collection of 25 short stories *Wanderer, kommst du nach Spa...* (1950), and *Die schwarzen Schafe* (1951), a story for which he received the prize of the Gruppe 47 (q.v.) in 1951. In the same year he published his first novel, *Wo warst du, Adam?*, which favours the episodic structure held together by a central figure, a technique that underlies his entire novelistic oeuvre; this form suited his professed aim to depict a variety of situations representing social and political forces that entrap, destroy, or challenge ordinary men and women, and families. The compassionate humanity and moral intent informing his realism is especially effective through his increasingly subtle use of different nuances of humour, irony, and satire.

His output after 1951 consisted mainly of novels and stories and a few radio plays. In the satire *Nicht nur zur Weihnachtszeit* (1952) an elderly lady is obsessed by the idea of celebrating Christmas in 1945 exactly as it was in the time of plenty before the war. The obsession persists beyond Christmas and through the year, with candles lit and carols sung on every night. Obsession, the result of deep-seated frustration and disillusionment, underlies also the title-story of the collection of satires *Doktor Murkes gesammeltes Schweigen und andere Satiren*

(1958). With *Und sagte kein einziges Wort* (q.v., 1953), Böll began a series of novels reflecting the conditions of the early post-war years. *Haus ohne Hüter* (q.v., 1954) deals with the difficulties of boys orphaned in the war. *Das Brot der frühen Jahre* (q.v., 1955) is a more positive work, in which materialistic values are rejected by a pair of young lovers. The stories *So ward Abend und Morgen*, containing among others *Die Postkarte*, also appeared in 1955, and *Unberechenbare Gäste* in 1956. These were written at various times between 1951 and 1956; *Die Waage der Baleks und andere Erzählungen* (1958) and *Der Bahnhof von Zimpren* (1959) unite stories from early collections with new ones. Böll's journeys to Ireland and his attractive *Irisches Tagebuch* (1957) were a kind of escape to a scene in which materialism and new-won affluence were less oppressive than at home. *Im Tal der donnernden Hufe* (1957) reflects the problems of puberty. Many German treatments of this subject end in suicide, but Böll implies that the crisis is surmounted. The title refers to an erotic fantasy invented by the two boys, Paul and Griff. The climax of Böll's work in the 1950s is the novel *Billard um halb zehn* (q.v., 1959), a reckoning between the new generation and the old. It was also the starting-point for a current of opinion critical of Böll's work. Two stories published in 1962 (*Als der Krieg ausbrach—Als der Krieg zu Ende war*) were originally broadcast as *Die Kaffeemühle meiner Großmutter*. They are reminders both of what 'the ordinary man' had gone through and of the diversity of 'ordinary men'. The novel *Ansichten eines Clowns* (q.v., 1963) marks a further stage in Böll's disapproval of the modern economic state and of that of the Church. The element of ironic masquerade, prominent in this novel, recurs in the story *Entfernung von der Truppe* (1964), which has a pronounced anti-militaristic message. A satirical appendix pokes fun at academic assessments and journalistic judgements. The same anti-militarism informs the novel *Ende einer Dienstfahrt* (1966). It is the story of a father and son, Gruhl by name, who, in an act of symbolical protest, deliberately set fire to a jeep of the Bundeswehr. They are given a light sentence and released. The mild ending reflects Böll's own aversion to violent action, but it is also meant to imply an uneasy conscience in the prosecuting authorities. His longest novel, *Gruppenbild mit Dame* (q.v., 1971) is a further exposure of a present he dislikes and a past he abhors, presented in an exceptionally complex manner. *Die verlorene Ehre der Katharina Blum oder Wie Gewalt entstehen und wohin sie führen kann* (1974) is a scathing indictment of journalistic and judicial malpractice, which is responsible for a young woman's 'lost honour', and in its approach bears some resemblance to Schiller's story *Der Verbrecher aus verlorener Ehre* (q.v.). Set against the background of terrorism involving the Baader-Meinhof group, it was a thinly disguised attack on *Bildzeitung*, one of the newspapers published by the group owned by Axel Springer. The novel *Fürsorgliche Belagerung* (1979) is similarly inspired by the years of terrorism and police protection in the 1970s. The story *Das Vermächtnis* (1982) centres on a soldier missing in Russia since 1943 who, it turns out, is dead, having been killed, not by the Russians, but by a German officer, once a schoolfriend, shortly before the German quarters were taken by the Russians. Schelling, who shot Schnecker, has survived and, with callous disregard of the past, still exploits life to the full. In contrast to him, the fictitious first-person narrator (Böll favours this technique), who breaks the news to the family, represents the conscience of that generation and its past that must not be forgotten. *Du fährst so oft nach Heidelberg* (1979) is a selection of short stories written between 1940 and 1979 (the title story is on the subject of Berufsverbot). The novel *Frauen vor Flußlandschaft. Roman in Dialogen und Selbstgesprächen* appeared after Böll's death in 1985.

Gesammelte Erzählungen (2 vols.) appeared in 1981, *Gedichte* (2 vols.) in 1972 and 1975 respectively, collections of radio plays, *Zum Tee bei Dr. Borsig. Acht Hörspiele* in 1964 and *Hörspiele* in 1980.

The *Frankfurter Vorlesungen* (1966) are based on lectures Böll delivered during his Gastdozentur für Poetik at Frankfurt University. Böll's polemical writings appeared in a number of collections, including *Erzählungen, Hörspiele, Aufsätze* (1961), *Hierzulande* (1963), *Aufsätze, Kritiken, Reden* (1967), *Neue politische und literarische Schriften* (1973), *Einmischung erwünscht. Schriften zur Zeit* (1977), *Gefahren von falschen Brüdern. Politische Schriften* (1980), *Spuren der Zeitgenossenschaft. Literarische Schriften* (1980), *Vermintes Gelände. Essayistische Schriften 1977–1981* (1982), and *Antikommunismus in Ost und West. Zwei Gespräche* (1982). *Über mich selbst* (1958), is a short autobiographical sketch that refers to Böll's native landscape and his forebears; *Was soll aus dem Jungen bloß werden? Oder: Irgendwas mit Büchern* (1981) covers Böll's grammar-school years, 1933–7. All his major work has appeared in translation.

Film versions of Böll's fiction include *Das Brot der frühen Jahre* by H. Vesely (1962), *Die verlorene Ehre der Katharina Blum* by M. von Trotta and V. Schlöndorff (1975), *Ansichten eines Clowns* by R. V. Jasný (1976), and *Gruppenbild mit Dame* by A. Petrović (1977).

Werke, 10 vols., ed. B. Balzer, were published

1977–80: *Romane und Erzählungen* (5 vols.), *Essayistische Schriften und Reden* (3 vols.), *Interviews* (1 vol.), and *Hörspiele, Theaterstücke und Drehbücher, Gedichte* (1 vol.). The collection *Ein- und Zusprüche. Schriften, Reden und Prosa 1981–1984* was published in 1984.

BÖLSCHE, WILHELM (Cologne, 1861–1939, Schreiberhau, Silesia), went to Berlin in 1887 after completing his studies in Bonn and Paris, and joined the group of Naturalist writers known as the Friedrichshagener Kreis (q.v.). He had no scientific training but an enthusiasm for Darwinism, positivism, and determinism, which he conceived as the basis for a new harmonious, non-religious, scientific world. His most notable contribution to the growing creed of Naturalism in art was the treatise *Die naturwissenschaftlichen Grundlagen der Poesie*, described by the author as 'Prolegomena einer realistischen Ästhetik' (1887). Bölsche had not the mental equipment for a profound work on this subject, and his treatise makes a number of general assertions, maintaining that literature should shun the abnormal and confine itself to 'normal' human beings, though he furnishes no criteria by which these can be identified. Bölsche's influence was exercised primarily through the impressive title of his tract and its general trend, rather than by cogent argument. He wrote two mediocre novels (*Der Zauber des Königs Arpus*, 1887, and *Die Mittagsgöttin*, 1891). Thirty years later he published two volumes of stories, *Der singende Baum* (1924) and *Von Drachen und Zauberkünsten* (1925).

BOLTZ, VALENTIN (b. Rufach, Alsace, ?–1560), a hospital chaplain in Basel, wrote religious plays (e.g. *St. Pauli Bekehrung*, 1546) and translated the comedies of Terence (1539).

BONAVENTURA, pseudonym used by F. W. J. Schelling (q.v.) for poems by him included in *Musenalmanach auf das Jahr 1802*. *Die Nachtwachen des Bonaventura* (q.v., 1805) have also been ascribed to Schelling, though this attribution is probably incorrect.

BONER, HIERONYMUS, recorded between 1527 and 1552 in Colmar, where he was mayor (Schultheiß), translated Thucydides (1533), Plutarch (1534), and Herodotus (1535) into German.

BONER, ULRICH, latinized as Bonerius, was a Dominican monk, recorded in Berne between 1324 and 1349, who translated from the Latin 100 fables for a patron in Berne, Johann von Ringgenberg. Boner entitled his collection, which he completed by 1350, *Der Edelstein*. His fables are simply told in unpretentious verse with emphasis on the moral. *Der Edelstein* was a popular book, existing in many MSS., and was among the first printed books (1461). The fables were reprinted in part by J. G. Scherz (1704–10), in full by J. J. Breitinger (q.v., 1757). Lessing, in *Über die sogenannten 'Fabeln aus den Zeiten der Minnesinger'* (1773–80), reasserted Boner's authorship against an error made by Gottsched (see ZUR GESCHICHTE UND LITERATUR. AUS DEN SCHÄTZEN DER HERZOGLICHEN BIBLIOTHEK ZU WOLFENBÜTTEL).

BONHOEFFER, DIETRICH (Breslau, 1906–45, Flossenburg Concentration Camp), was the son of a psychiatrist. He studied theology and became a distinguished scholar. In 1936 he was deprived of the right to hold an academic appointment. In the following year the Seminary of the Confessing Church (see BEKENNENDE KIRCHE), of which he had been director, was closed by the Gestapo. Bonhoeffer was also deprived of the right to preach and to publish, and in 1941 his works were proscribed. It would at various times have been possible for him to emigrate to the United States, but he chose to return from each visit abroad. In April 1943 he was arrested for his work for the Resistance (see RESISTANCE MOVEMENTS (2)). He was executed almost exactly two years later on 9 April 1945.

A Lutheran by choice and training, he was attracted by the dialectics of Karl Barth (q.v.), and evolved his own theory, which he first expressed in the curiously entitled *Nachfolge* (1937), exhibiting the notion of a dual justification, by faith (the Lutheran justification), and by obedience. The solitary meditations of his imprisonment are contained in letters smuggled out of Tegel prison. Going beyond Barth's positive existentialist theology, Bonhoeffer maintained that we now live in a godless world and yet are closer to God. In this sense he refers to the 'mündige Welt', a world that has come of age (e.g. in science and technology). While in prison Bonhoeffer expressed his thoughts in poetry.

BONIFACE, SAINT (Crediton, Devon, 672–754, Dokkum, Holland), often called 'Der Apostel der Deutschen'. St. Boniface, whose Saxon name was Wynfrith (Winfrid), studied at Exeter and Winchester. In 716 he was authorized by Pope Gregory II to go to Germany as a missionary to the Frisians. After a brief return to England he spent his life from 718 onwards in Germany, converting the heathens in the north, but also organizing the Church all over Germany. He was consecrated bishop in 722 and archbishop of Mainz in 739. He established a number of bishoprics, including Salzburg, Regensburg, Freising, Passau, and Würzburg, and founded the Abbey of Fulda. In 754, in the course of a further mission to the Frisians, he was murdered

with his companions at Dokkum. His remains were buried in Fulda. His work of ecclesiastical organization in Germany provided a foundation for the political and educational work of Charlemagne (see KARL I, DER GROSSE) a generation later.

Bonn, since 1949 provisional capital of the Federal Republic of Germany, belonged from 1273 to the Prince Archbishops of Cologne. After the Napoleonic Wars (q.v.) it passed with the Rhineland to Prussia. The city is a Roman foundation. The Emperor Karl IV (q.v.) was crowned German King (see DEUTSCHER KÖNIG) in the 12th-c. minster. Its most famous citizen is Beethoven, who was born there in 1770. The University of Bonn, originally an academy founded in 1777, was raised to university rank in 1796, but disintegrated during the troubled war years which followed. It was refounded by King Friedrich Wilhelm III (q.v.) of Prussia in 1818.

BONSELS, WALDEMAR (Ahrensburg, Holstein, 1881–1952, Ambach), an indefatigable globetrotter in his early years, visited every continent except Australasia. In 1919 he settled at Ambach on the Starnberger See, which remained his home for the rest of his life. Usually considered a neo-Romantic (see NEUROMANTIK), he began with erotic novels having subjects varying from the tender budding of first love (*Mare, die Jugend eines Mädchens*, 1907) to the conflict between passion and conscience (*Blut*, 1909, and *Der tiefste Traum*, 1911). His output was facile and prolific, and included the novels *Die Toten des ewigen Kriegs* (1911, title changed to *Wartalun* in 1920), *Indienfahrt* (1916), *Menschenwege* (1918), *Eros und die Evangelisten* (1921), *Narren und Helden* (1923, republished with the two preceding novels as a 3-vol. work under the new general title *Notizen eines Vagabunden* in 1925), *Marios Heimkehr* (1937), *Mario. Ein Leben im Walde* (1939), *Mortimer. Der Getriebene der dunklen Pflicht* (1946), and *Dositos* (1949), a novel dealing with Christ, which was republished in 1951 with the title *Das vergessene Licht*. He is the author of the stories *Ave vita, morituri te salutant* (1906), *Mario und die Tiere* (1927), *Mario und Gisela* (1930), *Die Reise um das Herz* (1938), and a volume of collected tales *Begegnungen* (1940). Bonsels had a deep love of nature, which is reflected in all his work, but most of all in his 'insect novel', really a kind of fairy-tale, *Die Biene Maja und ihre Abenteuer* (q.v., 1912), which is one of the outstanding publishing successes of the century. The best of his other narratives are probably the novel *Das Anjekind* (1913) and *Himmelsvolk* (1915), a tale for children. He also wrote an autobiography of his early years, *Tage der Kindheit* (1931), and the travel book *Der*

Reiter in der Wüste (1935). Always a mystic, Bonsels in his later years became increasingly involved in esoteric religious speculations.

BONSTETTEN, KARL VIKTOR VON (Berne, 1745–1832, Geneva), wrote on Swiss political and economic questions (*Briefe über ein schweizerisches Hirtenland*, 1781; *Über die Erziehung der Bernerschen Patrizier*, 1785; *Über Nationalbildung*, 1802). His *Schriften* appeared in 1792, his *Neue Schriften* in 1799–1801.

Bookesbeutel, Der, a dialect play written by H. Borkenstein (q.v.).

BOPPE, MEISTER, an itinerant Middle High German poet, a native of Switzerland, who is recorded between 1275 and 1287. Meister Boppe is a learned poet who parades his geographical knowledge and writes poems conveying zoological, petrological, or cosmological information. He is the author of one particularly abusive diatribe aimed at an unidentified adversary. A number of religious poems, including a cycle in which each poem begins *Ave Maria*, are wrongly attributed to him.

BORCHARDT, GEORG, pseudonym of G. Hermann (q.v.).

BORCHARDT, RUDOLF (Königsberg, 1877–1945, Trins am Brenner), was brought up in Berlin, where he studied archaeology and classics. The development of his artistic taste was influenced by visits to Italy and England, where the work of the Pre-Raphaelite painters especially interested him. He formed a close friendship with H. von Hofmannsthal and R. A. Schröder (qq.v.) and was influenced by Stefan George (q.v.). From 1903 until the 1914–18 War, in which he volunteered for military service, he lived in Italy, chiefly at Lucca. In 1922 he returned to Italy to live, was arrested by the Gestapo in 1944, then released, taking refuge at Trins, where he died. Borchardt believed passionately in the educative value of artistic experience, writing and lecturing in furtherance of his views with persuasive vigour and rhetorical emphasis. He is one of the few non-politicians of recent times to be an orator of note. His poetry, published in *Zehn Gedichte* (1896), *Jugendgedichte* (1913), *Die Schöpfung aus Liebe* (1923), and *Vermischte Gedichte* (1924), derives from classical and neo-Romantic inspiration and is animated by an intense consciousness of form. He was comparatively unsuccessful in the drama (*Päpstin Jutta*, 1920; *Die geliebte Kleinigkeit*, 1923; *Die Staufer*, 1936) and in narrative forms (*Das hoffnungslose Geschlecht*, 1929; *Die Begegnung mit dem Toten*, 1934; *Vereinigung durch den Feind*

hindurch, 1937). Much of his best work is to be found in his essays and speeches (*Rede über Hofmannsthal*, 1905; *Der Krieg und die deutsche Selbsteinkehr*, 1915; *Die Aufgaben der Zeit gegenüber der Literatur*, 1929). He was an outstanding anthologist (*Deutsche Denkreden*, 1925; *Ewiger Vorrat deutscher Poesie*, 1926; *Der Deutsche in der Landschaft*, 1927) and a translator of distinction (*Swinburne Deutsch*, 1919; *W. S. Landors Imaginäre Unterhaltungen*, 1923; *Dante Deutsch*, 1922–30). Borchardt's correspondence with H. von Hofmannsthal appeared in 1953. *Gesammelte Werke in Einzelbänden* (12 vols.), ed. M. L. Borchardt, appeared in 1955–79 (except for the last vol. containing translations).

BORCHERT, WOLFGANG (Hamburg, 1921–47, Basel), at first an assistant in a bookshop, became an actor. Called up in 1940, his military career was a succession of brushes with authority, which resulted in his serving two prison sentences. At the end of the war he returned to acting, and appeared in cabaret. His poems (*Laterne, Nacht und Sterne*) were published in 1946. His first success was *Hundeblume*, which appeared as the title story of a collection in 1947. His play *Draußen vor der Tür* (q.v., 1947), which had a world-wide success, was first performed on Hamburg Radio on 13 February 1947. The first stage performance took place, also at Hamburg, on 19 November 1947. Borchert was seriously ill the whole of that year, and he died on the day before the stage première of *Draußen vor der Tür*. Another volume of stories (*An diesem Dienstag*, 1947) appeared after his death; another posthumous collection was published in 1962 (*Die traurigen Geranien*). *Das Gesamtwerk* was published in one volume (with postscript by B. Meyer-Marwitz) in 1949.

BORCK or **BORCKE,** CASPAR WILHELM VON (Gersdorf, 1704–47, Berlin), Prussian diplomat, was for a time head of the Prussian Legation in London. In 1741 he published a translation in Alexandrine verse of Shakespeare's *Julius Caesar*, which, as the first rendering of Shakespeare (q.v.) into German, ranks as a historical landmark.

Bordesholmer Marienklage, title given to a quasi-dramatic poem in Middle Low German, expressing Mary's lamentations at the foot of the Cross. It has Latin stage directions and represents a form of monodic passion play. The author, who wrote the poem in 1475, gives his name as Reborch, a monk of Bordesholm Abbey in Holstein.

BORKENSTEIN, HINRICH (Hamburg, 1705–77, Hamburg), a Hamburg merchant, wrote comedies, of which *Der Bookesbeutel* (1742), in Hamburg dialect, is especially notable. It satirizes the haphazard ways of the older generation.

BORN, NICOLAS (Duisburg, 1937–79, Hamburg), participated in 1964–5 in the Berlin Colloquium, conducted by W. Höllerer (q.v.) and, after fifteen years in Essen, moved to Berlin and Dannenberg, Lower Saxony. In 1969 and 1970 he participated at the Writers Workshop of the University of Iowa (USA), and in 1975 was a visiting lecturer in contemporary literature at Essen University. Influenced by the style of the 'new realism' (see NEUER REALISMUS), he published after the novel *Der zweite Tag* (1965) three collections of poetry, *Marktlage* (1967), *Wo mir der Kopf steht* (1970), and *Das Auge des Entdeckers* (1972). They depict aspects of everyday life in a consumer society, the age of NATO, of nuclear armament, of television, and in the fourth floor flat of the lyrical I, whose presence adds to factual statements a subjective dimension that relates to an emotional void and ironic disillusionment with human relationships: 'Da steht nicht nur keine Linde' is the first line of *Vaterhaus*, and 'So wird der Schrecken ohne Ende langsam / normales Leben' opens *Entsorgt*, a poem from the collection *Keiner für sich, alle für niemand*, published with the earlier volumes in *Gedichte 1967–1978* (1978). The novel *Die erdabgewandte Seite der Geschichte* was published in 1976, and *Die Fälschung*, to which the figure of the war correspondent Laschen, in Beirut and Damur, is central, in 1979. Born is the author of the children's book *Oton und Iton* (1974) and of the radio plays *Schnee* (1966), *Innenleben* (1970), and *Fremdsprache* (1971).

BÖRNE, LUDWIG (Frankfurt/Main, 1786–1837, Paris), originally Löb Baruch, changed his name to Börne on conversion to Christianity in 1817 with the intention of qualifying for public service. He was employed in police administration in Frankfurt from 1811 to 1815, when he was dismissed on religious grounds. For the rest of his life he worked as a Radical journalist, editing *Die Waage* from 1818 until its suppression in 1821. He also published a short-lived journal entitled *Die Zeitschwingen*. For a short time in 1820 he was kept in custody, but the political charge against him was dismissed. After the July Revolution of 1830 (see JULIREVOLUTION) he moved to Paris, where he spent the rest of his life directing barbed and witty polemical writings against the German opponents of Radicalism. To this time belong *Briefe aus Paris* (1832–4) and *Menzel der Franzosenfresser* (1837, see MENZEL, W.). Börne expressly subordinated

literature to politics, but he had a warm corner in his heart for JEAN Paul (q.v.), whom he celebrated in *Denkrede auf Jean-Paul* (1825). Börne was not one of those named as a member of Jungdeutschland (see JUNGES DEUTSCHLAND) in 1835, but he sympathized with the views and aims of the group. His *Gesammelte Schriften* were published in 1868. *Sämtliche Schriften* (5 vols.), ed. I. and P. Rippmann, appeared in 1964–8. Of *Werke. Historisch-kritische Ausgabe in zwölf Bänden*, ed. L. Geiger et al., only 6 vols. appeared (1911–13).

BORNEMANN, JOHANN WILHELM JAKOB (Gardelegen, 1766–1851, Berlin), an official in the Prussian State Lottery and eventually its director, wrote poems in the Low German dialect of Prussia (*Plattdeutsche Gedichte*, 1810). He also wrote hunting and shooting anecdotes and poems (*Natur- und Jagdgemälde*, 1827, and *Humoristische Jagdgedichte*, 1855). His poem 'In grünbelaubter Heide' has virtually become a folk-song with the altered first line 'Im Wald und auf der Heide' (q.v.).

Bornhöved, a village near Segeberg, Schleswig-Holstein at which was fought in 1227 a battle between the forces of the Count of Holstein and a Danish army. The decisive victory of the Holsteiners removed the threat of a Danish annexation of Holstein and Lübeck.

Borte, Der, see DIETRICH VON DER GLEZZE.

Böse Adelheid, Die, an anonymous verse anecdote of the late 13th c. according to which Markhart, an Augsburg merchant, had a wife so contrary that she always did the opposite of what she was told. In the end, when told to keep away from the bank of the River Lech, she stepped closer, fell in, and was drowned.

Böse Frau, Die, an anonymous Middle High German poem of some 800 lines. It is in the form of a monologue spoken by the unfortunate husband of a vicious termagant. It contains many allusions to courtly literature, which in the context have satirical effect. The poem was written in the middle of the 13th c., and it is likely that the poet also wrote *Der Weinschwelg* (q.v.).

Böse Geist Lumpazivagabundus oder Das liederliche Kleeblatt, Der, a play by J. N. Nestroy (q.v.), written in 1833 and first performed at the Theater an der Wien, Vienna, in April 1833. It is an adaptation of his own comedy *Der Feenball* (1832). Described as 'Zauberposse mit Gesang in drei Aufzügen', *Lumpazivagabundus* was published in 1835. The spirit Lumpazivagabundus, the patron of the drunken and dissolute, appears only in the rather slender preliminary magic scene, in which Fortuna, arguing with Amorosa, maintains that money reforms people more effectively than love. The matter is put to the test: if two or three good-for-nothings, when given a fortune, make good use of it, then Fortuna triumphs. The trio of scamps consists of the joiner Leim, the tailor Zwirn, and the cobbler Knieriem. They put up in a hostel and all three dream the same lottery number. It proves to be the winner, and each finds himself possessed of a fortune. Knieriem and Zwirn spend all their money on drink and are soon in the condition in which they started. But Leim marries the girl he loves and turns over a new leaf. When Knieriem and Zwirn call on him he tries to change them, but they prove incorrigible. However, in a perfunctory magic scene at the end the two recalcitrants are after all transformed by love. Nestroy himself played Knieriem.

The source of *Der Feenball* and *Der böse Geist Lumpazivagabundus* was the title story in a collection of Novellen by K. Weisflog (q.v.), *Das große Loos*, 1827. Nestroy's vigorous adaptation of this story and his lifelike and ironic treatment of the scenes from real life immediately drew applause, and the play proved a phenomenal success, achieving its thousandth performance in the Nestroy cycle of 1881.

BOSSDORF, HERMANN (Wiesenburg, 1877–1921, Hamburg), worked as a Post Office official and wrote plays mostly in Low German (Plattdeutsch). They include *De Fährkrog* (1919, in High German in spite of its title), *Bahnmeester Dod* (1919), *Kramer Kray* (1920), *Dat Schattenspeel* (1920), and *De rode Ünnerrock* (1921).

BOTE, HERMANN (Brunswick, ?–1520, Brunswick), became the excise clerk of his native city. He is the author of a *Weltchronik* which begins with the Creation and goes as far as 1438. It was written not later than 1493. He also wrote a verse allegory dealing with the ranks and classes of society with the title *Boek van veleme rade* (*Buch von vielerlei Rädern*, c. 1490), a collection of proverbs and sayings in verse (*Koker*, i.e. *Der Köcher*), and a prose account of his own time with autobiographical material, including an account of his difficulties as excise clerk with the guilds (*Schichtbok*, 1510–13).

BOTE, KONRAD, a goldsmith in Brunswick, wrote a chronicle of Lower Saxony (*Sassenchronik*, 1492). It is sometimes referred to as the *Niedersächsische Bilderchronik*, on account of its numerous woodcuts.

Bötjer Basch, a Novelle by Th. Storm (q.v.), written in 1885–6. It was first published in the *Deutsche Rundschau* in 1886 under the title *Aus engen Wänden*. The action begins in the 1830s in the narrator's native city. The cooper Daniel Basch marries late in life, and his wife dies after she has given birth to a second child, which dies with her. Thus father and son grow up together until the young Fritz Basch emigrates to join the gold-diggers in California. After years with no news from Fritz, Daniel hears the rumour of his son's death in a fight, in which in reality he was only injured. When his sole companion, a tame singing bullfinch, has been stolen, Daniel attempts to drown himself. He is rescued but falls gravely ill. At this time Fritz returns and quickly establishes himself as a master cooper in his father's old business. Daniel recovers after Magdalena, the orphaned daughter of Fritz's schoolteacher, the Kollaborator, has returned the bird which, as she has discovered, had been taken by her brother. Daniel dies the day before Fritz and Magdalena marry.

BÖTTGER, ADOLF (Leipzig, 1815–70, Gohlis nr. Leipzig), who spent his life in his native town, wrote two once-popular verse romances, *Hyacinth und Liliade* (1849) and *Till Eulenspiegel* (1851).

BÖTTICHER, HANS, see RINGELNATZ, JOACHIM.

BÖTTIGER, KARL AUGUST (Reichenbach, 1760–1835, Dresden), was headmaster of the Weimar Gymnasium from 1791 to 1804. He contributed to *Die Horen* and *Die Propylaën* (qq.v.), and in 1797 became editor of *Der neue teutsche Merkur* (see TEUTSCHE MERKUR, DER). Regarded as a busybody by Goethe and Schiller, who called him 'Magister Ubique', he stood closest to Wieland (q.v.). His later years were spent as a headmaster and, from 1814, as museum director in Dresden.

BOY-ED, IDA (Bergedorf, Hamburg, 1852–1928, Travemünde), *née* Ida Ed, acquired a three-letter name by marriage and lived most of her life in Lübeck. She wrote a number of conventional but readable novels, of which the best-known are *Männer der Zeit* (3 vols., 1885), *Dornenkronen* (1886), *Ein königlicher Kaufmann* (1910), *Stille Helden* (1914), and *Erschlossene Pforten* (1918). Her collections of stories include *Getrübtes Glück* (1884), *Abgründe des Lebens* (1887), and *Geschichten aus der Hansestadt* (1909).

Božena, a novel by Marie von Ebner-Eschenbach (q.v.), published in 1876. It is a story of family intrigue, in which affection is subordinated to self-interest; it is spread over three generations, the wine-merchant Heißenstein and his second wife, his daughter Rosa by his first marriage, and her daughter Röschen. Božena is the Heißensteins' servant, who devotes herself to Rosa and, when the latter dies, fights for and eventually secures Röschen's inheritance.

BRACHMANN, LUISE (Rochlitz, 1777–1822, Halle), a talented young woman, who wrote poetry, some of which was published in Schiller's *Die Horen* (q.v.) and his *Musenalmanach*. Her later work, more prolific and more facile, includes stories and a romantic historical poem in 5 cantos, *Das Gottesurteil* (1818). She drowned herself in the River Saale because of an unhappy love-affair.

BRACHVOGEL, ALBERT EMIL (Breslau, 1824–78, Berlin), at first an engraver and sculptor, took a post as secretary of a theatre, was later employed in a telegraph office, and finally, from 1854 on, earned his living by authorship. His two outstanding successes were the tragedy *Narziß* (1857) and the novel *Friedemann Bach* (1858), based on the life of W. F., son of J. S. Bach (q.v.); this novel was reprinted many times and survived well into the 20th c. Brachvogel's novel *Das Rätsel von Hildburghausen* (1871) deals with the 'Dunkelgräfin' (q.v.). He is the author of a novel on C. F. D. Schubart (q.v.), *Schubart und seine Zeitgenossen* (1864, reissued 1926).

BRAHM, OTTO (Hamburg, 1856–1912, Berlin), a bank clerk, resigned in order to study at Berlin and Heidelberg universities. He became an influential critic in the late 1880s, writing on the theatre in the *Vossische Zeitung* (q.v.) and *Die Nation*. He was particularly active in furthering the Naturalist movement (see NATURALISMUS), and in 1889 directed the Freie Bühne (q.v.) in Berlin. From 1894 to 1904 he was in charge of the Deutsches Theater in Berlin, and then took over the Lessingtheater. His principal works of literary criticism were a monograph on H. von Kleist (q.v., 1884), and an unfinished book on Schiller (1888–92). His collected theatrical criticism was published posthumously (1913). His correspondence with G. Hauptmann (q.v.), *Otto Brahm – Gerhart Hauptmann. Briefwechsel 1889–1912*, ed. P. Sprengel, appeared in 1985.

BRAHMS, JOHANNES (Hamburg, 1833–97, Vienna), at first a piano virtuoso and later the outstanding classical composer of the second half of the 19th c. In addition to his instrumental music Brahms wrote many Lieder (see LIED) and choral works. The texts of *Ein deutsches Requiem* and *Vier ernste Gesänge* are from the Lutheran Bible, that of the *Schicksalslied* from Hölderlin (q.v.), and of the *Alto Rhapsody* from Goethe's

poem *Harzreise im Winter* (q.v.). His Lieder include settings of Goethe (*Trost in Tränen*), Hölty (*Mailied*), Eichendorff, Schenkendorf, Uhland, Rückert, Heine, Keller, Groth, Allmers, and Liliencron (qq.v.), as well as of many lesser poets.

BRÄKER, ULRICH (Näbis, Toggenburg, 1735–98, Wattwil, Switzerland), known as 'der arme Mann im Toggenburg', grew up in humble circumstances, was impressed into the army of Friedrich II (q.v.) of Prussia, deserted at Lobositz (see SIEBENJÄHRIGER KRIEG), and sought refuge in his native Switzerland, where he set up as a weaver in Wattwil. Lacking all formal education, he was nevertheless consumed with a thirst for knowledge, reading widely, and writing with simplicity and force. His enthusiasm for Shakespeare finds genuine expression in *Etwas über William Shakespeares Schauspiele*, written in 1777, first published in 1877 and again in 1942 (by W. Muschg). Bräker became known to his contemporaries through his autobiography, published in 1789 under the title *Lebensgeschichte und natürliche Ebentheuer des Armen Mannes im Tockenburg* (edited by H. H. Füßli, q.v.), a personal record of great sincerity and a valuable document of social history. A reprint, ed. W. Pfeiffer, appeared in 1965.

Bramarbas, signifying a boastful fellow, was first used in 1710 in a poem by J. B. Mencke (q.v.). It was popularized by its use as the title for a German translation of Holberg's *Jakob von Tyboe*, published in 1741 in Gottsched's *Deutsche Schaubühne*. The verb *bramarbasieren* was first used by Gellert (q.v.) in 1751.

Brandanlegende, the miraculous story of St. Brandan's (or Brendan's) voyage, is of Irish origin. It occurs in fragments of German verse legends of the 13th c. and 14th c., and in a chapbook (Volksbuch) of the 15th c. These were collected and published by S. Schröder as *Sanct Brandan* in 1871.

Brandenburg (1) *Mark* or 'March', formed the nucleus from which Prussia grew and was from 1815 to 1945 a Provinz of Prussia. In the DDR it was a Land until the Länder were abolished in 1952.

(2) City on the Havel in former Brandenburg.

BRANDES, GEORG (Copenhagen, 1842–1927, Copenhagen), whose real name was Cohen, was a Danish critic and propagandist for Naturalism (see NATURALISMUS). He was widely influential in Germany, particularly through *Hovedstrømninger i. de 19. Aarh. Literatur* (1872–94, translated as *Hauptströmungen in der Literatur des 19. Jahr-*

hunderts) and *Bjørnson og Ibsen* (1882). His interpretation of Goethe (*Wolfgang Goethe*) appeared in 1915, and *Briefwechsel mit A. Schnitzler*, ed. K. Bergel, in 1956 (see SCHNITZLER, A.).

BRANDES, JOHANN CHRISTIAN (Stettin, 1735–99, Berlin), became an actor in 1755. A spell of acting in Breslau (1762–3) coincided with Lessing's sojourn there, and brought him fresh stimulus. He had already written the comedies *Der Zweifler* (1760) and *Die Entführung* (1761). In 1766 he ventured on a tragedy, *Miss Fanny* (which he later entitled *Der Schiffbruch*); this was followed by *Olivie* (1775), *Die Medicäer* (1776), *Ottilie* (1780), *Hans von Zernow* (1785), and *Rahel die schöne Jüdin* (1789). Brandes also provided the libretto for a once-famous opera by the Bohemian composer G. Benda (1722–95), *Ariadne auf Naxos* (1774). His autobiography, *Meine Lebensgeschichte*, appeared in 1799 and 1800. His *Sämtliche dramatische Schriften* (8 vols.) appeared 1790 ff.

Brandopfer, Das, a story by A. Goes (q.v.), published in 1954. It purports to be told by a librarian. The background is the 1939–45 War. When Herr Walker, the butcher, is called up, his wife takes charge of the business, and the local Nazi Gauleiter instructs her to supply meat to the Jewish community. A sympathy springs up between her and her new clients, who in the course of time confide in her, relating their sufferings. As the war goes on, they disappear one by one, and Frau Walker, who knows their fate, suffers an acute sense of guilt, because she belongs to the community responsible for so much suffering. After an air raid she is found injured in her burnt-out house. The cause of the fire is not discovered, but it is suggested that her crisis of conscience induced her to sacrifice herself as a 'burnt offering', and that for this reason she had set fire to the house.

BRANDSTRÖM, ELSA (St. Petersburg, 1888–1948, Cambridge, Mass.), daughter of the Swedish Minister to Russia from 1908 onwards, devoted herself during the 1914–18 War to the welfare of German and Austro-Hungarian prisoners of war in Russia. Trained in nursing, she organized and participated in the nursing of the P.O.W. victims of a typhoid epidemic. She donated the proceeds of her book *Bland Krigsfångar i Ryssland och Sibirien* (1921, translated into German as *Unter Kriegsgefangenen in Rußland und Sibirien*) to the foundation of a sanatorium for former P.O.W.s in Russia and of an orphanage for the children of some of those who had died. Praise of her occurs in authors as remote from each other in their outlook as E. E. Dwinger

(*Die Armee hinter Stacheldraht*, q.v.) and H. von Doderer (*Der Grenzwald*, q.v.).

BRANT, SEBASTIAN (Strasburg, 1457–1521, Strasburg), who spent his childhood and later life in Strasburg, is known primarily for his long moral and satirical poem, *Das Narrenschiff* (q.v.). He studied at Basel University and lectured on law there from 1484 to 1500, when he returned to his native city with a municipal appointment. Brant's numerous Latin poems (*Varia carmina*, 1498) are of little importance. He translated into German the *Liber faceti* (1496) and *Dicta Catonis* (1498), and he published *Bescheidenheit* (1508) by Freidank (q.v.) in expanded and adapted form. Brant is a transitional figure, rooted in the Middle Ages, but already reflecting some aspects of the new outlook of the Renaissance. The edition *Narrenschiff* by F. Zarncke (1854) was reissued in 1961 and 1973.

BRAUN, FELIX (Vienna, 1885–1973, Klosterneuburg), a highly cultivated man and friend of H. von Hofmannsthal (q.v.), was a professor at Palermo from 1928 to 1937; he emigrated to England in 1939, where he remained until 1951, afterwards living in Vienna. His delicate poetry derives from Viennese neo-Romanticism and has a Roman Catholic background (*Gedichte*, 1909; *Das neue Leben*, 1913; *Hyazinth und Ismene, Das Haar der Berenike*, both 1919; *Das innere Leben*, 1925; *Viola d'amore*, 1953). He wrote verse dramas (*Tantalos*, 1917; *Kaiser Karl V*, 1936; *Rudolf der Stifter*, 1953). The novel *Agnes Altkirchner* (1927, renamed in 1959 *Herbst des Reiches*) has as its background the decline and fall of the Dual Monarchy. Other novels are *Die Taten des Herakles* (1921), *Der unsichtbare Gast* (1924), and *Der Stachel in der Seele* (1948). Braun's autobiography is entitled *Das Licht der Welt* (1949).

BRAUN, LILI (Halberstadt, 1865–1916, Berlin), a general's daughter (*née* von Kretschmann), married, as a widow, the socialist H. Braun (1854–1927), and became a member of the Social Democratic Party. She is known for one novel (*Liebesbriefe der Marquise*, 1912) and for her *Memoiren einer Sozialistin* (2 vols., 1909–11). She is also the author of a verse tragedy, *Mutter Marie* (1913).

BRAUN, VOLKER (Dresden, 1939–) worked as a printer and in industry before studying philosophy at Leipzig University (1960–4). Having worked for the Berliner Ensemble (1965–6), he became a producer at the Deutsches Theater, Berlin. He is the author of poetry, prose, and plays. Concerned with the self-realization of the individual within the general process of socialist evolution (which is central to one of his best stories, *Unvollendete Geschichte*, 1975) and critical of the lack of workers' involvement in management, Braun was obliged to modify his best plays, and even these have seen only few performances. *Kipper Paul Bauch* (1966) appeared in revised form as *Die Kipper. Schauspiel* in 1972. Introduced by a Brechtian prologue, the initially effective ideas of the lever operator Paul Bauch come to grief, though they are not entirely lost on the brigade. Bauch leaves his job and counts his losses with bitter-sweet humour. The carefully balanced plot exposes the dangers of political and industrial stagnation. Similarly the play *Hans Faust* (1968) appeared in revised form, as *Hinze und Kunze*, in 1973. The prologue makes the play's moral explicit and hints at the failure of the Faustian pact between Hinze and Kunze. A complication is introduced through Kunze's love for Marlies, the 'emancipated' woman, who appears as a principal figure in *Tinka*. *Schauspiel* (1973), a play that does not end with the will for a new beginning. Tinka, to whom love is less important than effectiveness at work, dominates Brenner, depriving him of his confidence at work, where she displays her intellectual superiority. Brenner kills her under extreme provocation. Another play, *Guevara*, was performed in Mannheim in 1977. A collection, *Im Querschnitt: Volker Braun. Gedichte. Prosa. Stücke. Aufsätze*, ed. H. J. Schubert with a preface by D. Schlenstedt, which includes a new play, *Großer Frieden*, appeared in 1978. *Training des aufrechten Gangs* (1979) is a collection of poetry and *Berichte von Hinze und Kunze* (1983) a collection of short prose. The three stories, styled 'reports', *Der Schlamm* (1959), *Der Hörsaal* (1964), and *Die Bretter* (1968), published in 1972 as *Das ungezwungne Leben Kasts*, are of interest for their autobiographical reflections on developments in the DDR and the Bitterfeld policy (see SOZIALISTISCHER REALISMUS).

Braun was awarded the Heine Prize (1971) and, in 1980, the Heinrich Mann Prize. His plays were published as *Stücke I* (1975), *Stücke II* (1981), and *Stücke III* (1983).

Braunschweig (Brunswick) (1) a former duchy of the German Confederation, the North German Confederation (1866–71), and of the German Empire (1871–1918), and afterwards a Land of the Weimar Republic. In 1946 it was included in the new Land Niedersachsen of the Federal Republic, except for a small eastern portion which is in the DDR. Brunswick was divided by inheritance into Braunschweig-Lüneburg (13th c.), and Braunschweig-Wolfenbüttel (14th c.–18th c.). From 1807 to 1813 it was part of the kingdom of Westphalia. When

the line of Braunschweig–Bevern died out in 1884, the Welf (q.v.) house of Hanover should have succeeded, but Bismarck (q.v.), who bore a strong resentment against it, dating from 1866 (see HANNOVER), prevented this and it was ruled by Prussian regents until 1913, when a reconciliation was achieved by the marriage of the rightful claimant to Princess Viktoria, the only daughter of the Emperor Wilhelm II (q.v.).

Two of the dukes have been notable writers, Heinrich Julius and Anton Ulrich (qq.v.). The Bibliotheca Augusta, founded in the 17th c. by Duke August, is housed at Wolfenbüttel, where G. E. Lessing (q.v.) was its librarian from 1770 until his death in 1781. Friedrich Wilhelm (1771–1815), who succeeded as duke in 1806, is 'Brunswick's fated chieftain' in Byron's *Childe Harold*.

(2) Historic city some 30 miles east of Hanover, which was the residence of Heinrich der Löwe (q.v.), who in 1166 erected the well-known bronze statue of a lion. It was the capital of the Duchy of Brunswick-Wolfenbüttel. The Collegium Carolinum, founded in 1745 by the enlightened Duke Karl I (q.v.), became in 1877 the Technische Hochschule, and in 1968 the Technische Universität Braunschweig. The city is the birthplace of J. H. Campe, and of Ricarda and Friedrich Huch (qq.v.); W. Raabe (q.v.) lived there from 1870 until his death in 1910.

Braunschweiger Reimchronik, an anonymous Middle Low German verse history of the Saxon princely house, up to Duke Albrecht I (1279). It is of moderate length (10,000 lines) and is written in rhyming couplets. Its tone is laudatory, and its aim exaltation of the dynasty.

BRAUTLACHT, ERICH (Rheinberg nr. Xanten/Lower Rhine, 1902–57, Kleve), a stipendiary magistrate and later judiciary civil servant (1953), wrote stories and novels in which feeling and humour are often combined. They frequently have the lawcourt background with which he was familiar, and are to some extent local in appeal. An imaginary small Rhenish town 'Pöppelswyck' provides the milieu for several of his books. His principal works are the novels *Einsaat* (1933), *Das Testament einer Liebe* (1936, republished in 1940 with the title *Das Vermächtnis einer Liebe*), *Magda und Michael* (1937), *Meister Schure* (1939), *Der Sohn* (1949), and *Das Beichtgeheimnis* (1956). His collections of stories are *Die Pöppelswycker* (1928), *Spiegel der Gerechtigkeit* (1942), and *Ignoto* (1947). In his last novel, *Versuchung in Indien. Warren Hastings*, published posthumously in 1958, he turned from his customary provincial settings to British colonial history.

Braut von Korinth, Die, a ballad written by Goethe in June 1797, and published in the *Musenalmanach* for 1798. He refers to it in letters as 'das vampyrische Gedicht', and indicates that it had undergone years of gestation before emerging complete in a few hours. Its eerie subject is the sexual union between a youth and the girl he loves, who proves to have died before their encounter. The macabre story is told in polished, reticent stanzas. A minor undertone implies hostility to the Christian negative attitude to gratification of the senses. The main lines of the story derive from a classical source.

Braut von Messina, Die, oder Die feindlichen Brüder, a tragedy written by Schiller in 1802-3, and first performed at Weimar on 19 March 1803. It was published in the same year, with the essay *Über den Gebrauch des Chors in der Tragödie* (q.v.) as preface. It is described on the title-page as *Ein Trauerspiel mit Chören. Die Braut von Messina* was intended as a re-creation of Greek tragedy and it restricts the number of characters, employs a chorus, and dispenses (in the original edition) with division into acts. The plot was invented by Schiller, though the motif of dream and interpretation is clearly adapted from the oracle in Sophocles' *Oedipus tyrannus*.

The action takes place in a medieval Messina of uncertain date. The ruling prince has died, leaving two sons whose mutual hostility has since defied the efforts of their mother Isabella to reconcile them. At last Isabella proves successful, and Don Manuel and Don Cesar are united. Each prince, however, is in love, and fate brings it about that each, unknown to the other, loves the same girl, Beatrice. Many years before, the father, responding to a dream of disaster, ordered the death of his infant daughter. Isabella saved her by sending her away and now hopes that the young woman will help to cement the new friendship of her sons. So Beatrice is both sister and object of desire to the two brothers. As Manuel, suspecting the truth, questions Beatrice, Cesar appears, and on an impulse stabs and kills his brother. When Cesar learns the true nature of his action he determines to avenge his brother and expiate his crime by taking his own life; he adheres to and executes this resolve in spite of the pitiful and urgent pleas of mother and sister. The chorus of attendant knights both comments on and participates in the action. The predominating speech of the principal characters is blank verse, though one important scene is written in Greek trimeters; the chorus makes use of lyrical measures. In the MS. sent to Hamburg, Schiller made the concession of attributing the choric passages to separate speakers, and divided the play into five acts.

Brautwahl, Die, a story by E. T. A. Hoffmann (q.v.), published in 1820 in vol. 3 of *Die Serapionsbrüder.*

BRAWE, JOACHIM WILHELM VON (Weißenfels, 1738-58, Dresden), was at Schulpforta (see FÜRSTENSCHULEN), and studied in Leipzig, where he met C. F. Gellert, E. von Kleist, and G. E. Lessing (qq.v.). He was a promising dramatist, who, in spite of his short life, left two tragedies, *Der Freigeist* (q.v.) and *Brutus.* The former was in prose, but the posthumously published *Brutus* (1768) was written in blank verse.

BRECHT, BERTOLT (EUGEN BERTHOLD FRIEDRICH) (Augsburg, 1898-1956, East Berlin), writer, poet, and 'Stückeschreiber', as he styled himself, had his schooling at Augsburg. In 1917 he matriculated at Munich University to read medicine and returned in the following year to Augsburg, where he worked as a medical orderly at a military hospital. He became a radical opponent of war and the nationalistic attitudes associated with it, which he saw as capitalism thinly disguised. In the decade up to his marriage with Helene Weigel (q.v.) in 1928, he established himself as a versatile writer of plays and lyric poetry and as a practical man of the theatre who emancipated himself resolutely from Expressionistic trends and experimented with new forms. *Baal* (q.v.) was written in Augsburg before he moved back to Munich. By 1922, when he was awarded the Kleist Prize, he had added *Trommeln in der Nacht* (q.v.) and *Im Dickicht der Städte. Leben Eduards des Zweiten von England* (after Christopher Marlowe) and *Mann ist Mann* followed his removal to Berlin to work for two years under Max Reinhardt (q.v.) at the Deutsches Theater. A new phase in his development began with his study of the 'political theatre' of Piscator (q.v.) and the dialectical materialism of Marx (q.v.). The climax of this decade occurs in his operas *Die Dreigroschenoper* and *Aufstieg und Fall der Stadt Mahagonny* (qq.v.), which he wrote in collaboration with the composer K. Weill (q.v.). In 1927 he published a cycle of poems, *Bertolt Brechts Hauspostille.* Its title suggests, like that of the *Kalendergeschichten* (q.v.) of 1949, a popular anthology in the moralizing tradition of J. P. Hebel (q.v.). Brecht's Lehrstücke (see LEHRSTÜCK), which result from his extensive studies of Communism, are more radical in their socialist aim. While all his subsequent plays are in this sense Lehrstücke, the short plays written between 1929 and 1930 differ from his later works in their elimination of the 'culinary' (kulinarisch) effect. They include *Das Badener Lehrstück vom Einverständnis, Der Ozeanflug, Die Maßnahme, Der Jasager und der Neinsager, Die Ausnahme und die Regel* (qq.v.). Der

Jasager is indebted to the Nō play. The best-known play of this time is, however, *Die heilige Johanna der Schlachthöfe* (q.v.).

Brecht was fifth on the black list of the NSDAP (q.v.) in 1923. When it came to power in 1933 he went into exile in Switzerland and then Denmark, where he stayed until 1939. He moved on to Finland, but the German occupation forced him to leave for Russia, though he did not settle there, moving to California, where he remained from 1941 to 1947 and worked with Charles Chaplin and Charles Laughton among others. He returned via Zurich to settle in East Berlin in 1949.

He devoted his remaining years to the Berliner Ensemble (q.v.), which he founded in 1949 in deliberate contrast to the Weimar theatre of Goethe and Schiller.

Brecht's longer prose works are considerable in bulk, e.g. *Dreigroschenroman*, 1934; *Die Geschäfte des Herrn Julius Caesar*, four of the intended six books of which were completed and published posthumously in 1957; the fragment *Der Tui-Roman* and the fragment *Me-ti/Buch der Wendungen*, 1935–9. His studies of stage techniques, philosophy, and literature of all kinds are contained in the intellectual symposium *Der Messingkauf* (q.v., 1937–51). In this lengthy tract the philosopher, the producer, and the actor, striving for the ultimate benefit of the worker, discuss for four nights running the function of the epic theatre (see EPISCHES THEATER), its origins, its experimental character, and possible shortcomings, only to stress again and again that what matters in the end is the 'point of view' (Standpunkt). Brecht cultivated a lapidary style to stimulate a 'thinking' as distinct from an 'educated' person.

Brecht wrote some 80 stories, and the *Geschichten vom Herrn Keuner*, written from 1930 onwards, the shortest of which is some four lines in length. The creation of Herr Keuner (Swabian/Bavarian dialect for *keiner*, nobody) demonstrates the dual and 'split' personality in his portrayal of character (in this case his own) which marks Brecht's dialectics and was second nature to him. The majority of his plays, both completed and projected, are based on models.

Brecht distinguishes sharply between feeling and the psychology of character. Aristotelian drama is didactic in terms of human ethics. Its adaptation by Lessing associated it with the middle class (see BÜRGERLICHES TRAUERSPIEL). For Christian ethics or fate Brecht substituted his Communist commitment. Feeling is subordinated to uncompromisingly revolutionary dialectics, psychology is simplified so as not to interfere with doctrines. For this purpose he favoured remote settings, inducing detachment, arousing curiosity, both of which stimulate

thought. After the Lehrstück phase Brecht revised his purely didactic view of the theatre. He wrote his best plays in the conviction that the public will learn more readily if it learns with pleasure, and as long as actor and stage designer (notably Caspar Neher and Teo Otto, qq.v.) promoted his aim he did not hesitate to modify his stage version.

The first collection of his works appeared under the title *Versuche*, which accords with his view that no work is complete without prolonged rehearsals complemented by performances, during which the author can study the reaction of the public. Indispensable for the epic theatre is the Verfremdungseffekt (q.v.) or V-Effekt, which Brecht discussed in many of his theoretical writings, the best summary of which is the *Kleines Organon für das Theater* (q.v., 1948). Nevertheless his greatest creations, *Mutter Courage und ihre Kinder* (1941), *Der gute Mensch von Sezuan* (1943), *Leben des Galilei* (1943), and *Der kaukasische Kreidekreis* (1948, qq.v.) have an independent life and can move any audience. Kattrin's death (*Mutter Courage*) is a conspicuous example of Brecht's power to appeal to the emotions, and so is the creation of his mother figures from Frau Carrar on; this gift became an acknowledged aspect of his technique.

Brecht's work for the theatre (which he viewed as the most effective means of communication) shows his ingenious and versatile use of language and poetic forms. He devised his own form of opera in protest against conventional opera (see EPISCHE OPER). Similarly he adapted the use of the Greek chorus. He used hexameters for *Das kommunistische Manifest*, and in a number of poems he employed the form of the sonnet in protest against its traditional tone (e.g. *Über Shakespeares Stück 'Hamlet'* and *Über Kleists Stück 'Der Prinz von Homburg'*). He used biblical language in a deliberately profane context. He imitated the language of the classics and of Büchner's *Woyzeck* (q.v.), as well as that of *Der Hofmeister* (q.v.) by J. M. R. Lenz, which he adapted for his own play *Der Hofmeister* (q.v., 1950). He used the language of the man in the street in Luther's manner, including the adoption of advertising slogans for self-education such as 'Wissen ist Macht' (*Turandot*, q.v.). The *Straßenszene* is the basic scene for his discussion of the epic theatre in *Der Messingkauf*. He collaborated with composers and himself set some of his songs to music. In his essay *Über reimlose Lyrik mit unregelmäßigen Rhythmen* (1939) he gave his reasons for changing from his earlier regular and rhyming verse to unrhymed and irregular verse, asserting that dialectics are more stimulating if they are not expressed in smooth language. But because the rhythm of language depends for its full effect on the speaker, Brecht prefers the term 'Gestus'.

Brecht's lyric poetry has affinities with that of H. Heine (q.v.) in its blend of irony and humanity. He uses the term 'Gebrauchslyrik' to draw attention both to the aesthetic and to the utilitarian function of his own poetry. His specific 'point of view' is no less idealistic than the classical brand of Idealism. In objecting to the classical concept of 'das ewig Menschliche' he wanted to demonstrate that change was both necessary and possible. The change he envisaged leads him at times to an over-simplification of the world he wishes to change, but his own sensitivity to the basic instincts and needs of human nature rises superior to the anti-Aristotelian theorist.

The following plays written during and since his exile represent his production in chronological order of writing: *Die Mutter* (1931), *Die Rundköpfe und die Spitzköpfe* (1931–4), *Die sieben Todsünden der Kleinbürger*, a libretto for ballet in 7 sections with music by K. Weill (1933), *Die Horatier und die Kuriatier*, another didactic play for schools in three episodes (1934), *Furcht und Elend des Dritten Reiches* (1935–8), *Die Gewehre der Frau Carrar* (1937), *Das Verhör des Lukullus* and the opera *Die Verurteilung des Lukullus* (1939), *Der gute Mensch von Sezuan* (1938–40), *Herr Puntila und sein Knecht Matti* (q.v., 1940), a highly effective example of Brecht's own conception of the Volksstück (q.v.), *Der aufhaltsame Aufstieg des Arturo Ui* (1941), *Die Gesichte der Simone Machard* (1941–3), *Schweyk im Zweiten Weltkrieg* (1943), *Die Tage der Commune* (1948–9), *Die Antigone des Sophokles* (1947), *Turandot oder der Kongreß der Weißwäscher* (1930–54), qq.v.

Brecht produced the *Herrnburger Bericht* with music by P. Dessau, which includes ten choral songs, for the World Festival of Democratic Youth (1950–1). Special mention should be made of the following collections of his lyric poetry: *Svendborger Gedichte* (1939), *Buckower Elegien* (1953), and *Die Kriegsfibel* (1955).

Versuche (15 vols.) appeared 1930–57, and *Vorläufige Gesamtausgabe in 33 Bänden* from 1953: *Stücke* (12 vols.), *Schriften zum Theater* (7 vols.), *Gedichte* (9 vols.), *Prosa* (5 vols.). In 1967 appeared *Gesammelte Werke. werkausgabe edition suhrkamp*, edited in collaboration with Elisabeth Hauptmann. *Arbeitsjournal 1938–1955* (2 vols.) appeared in 1973.

BREDEN, CHRISTIANE VON, see CHRISTEN, ADA

BREHM, ALFRED EDMUND (Renthendorf/Thuringia, 1829–84, Renthendorf), travelled in Africa and the wilder parts of Europe, pursuing zoological studies. In 1863 he was appointed

director of the Hamburg zoological garden. His principal work is his *Tierleben* (1864–9), a comprehensive treatise on the whole range of living creatures outside the microscopic field. A select Jubiläumsausgabe (8 vols., ed. C. W. Neumann) was published in 1928 under the title *Brehms Tierleben*. The readability of Brehm's vivid style secured it a wide circulation.

BREHM, BRUNO VON (Laibach, 1892–1974, Altaussee), son of a colonel, was commissioned in the Austrian artillery in 1913, wounded in 1914, taken prisoner by the Russians and returned in an exchange in 1916. He studied the history of art at Vienna University and from 1927 devoted himself to writing novels. He was, for a time, a National Socialist, welcoming the invasion and annexation of 1938, and from 1941 he served as an auxiliary staff officer (Ordonnanzoffizier). After the war he settled in Styria. For several years after 1927 Brehm wrote agreeable light Austrian novels such as *Der lachende Gott* (1928), *Susanne und Marie* (1929), and *Ein Graf spielt Theater* (1930). He is best known, however, for three politico-historical novels reflecting his own times and simultaneously chronicling and sternly analysing the decline and fall of the Habsburg monarchy and the German defeat in 1918: *Apis und Este* (1931), *Das war das Ende* (1932), and *Weder Kaiser noch König* (1933). They were published in 1951 as a trilogy with the title *Die Throne stürzen* (q.v.). After the 1939–45 War Brehm published the novels *Der Lügner* (1949), *Aus der Reitschul'!* (1951), *Dann müssen Frauen streiken* (1957), *Der Trommler* (1960), *Der böhmische Gefreite* (1960), and *Wehe den Besiegten allen* (1961). The last three were grouped as a trilogy, *Das zwölfjährige Reich* (1961) and refer to the Hitler period.

BREITBACH, JOSEPH (Koblenz, 1903–80, Munich), a businessman, published his first story, *Rot gegen Rot*, in 1928. In 1929 he settled in Paris and continued to write sporadically, usually in German, though the story *Le Liftier amoureux* (1948) is in French. He is the author of the novels *Die Wandlungen der Susanne Dasseldorf* (1932) and *Bericht um Bruno* (1962), a treatment of the problem of power. Breitbach's witty comedies *Fräulein Schmidt* (1932) and *Das Jubiläum* (1960) were successful on the stage. A selection of stories, *Die Rabenschlacht und andere Erzählungen*, appeared in 1973.

BREITINGER, JOHANN JAKOB (Zurich, 1701–76, Zurich), was appointed in 1731 professor of Hebrew and later of Greek at the Zurich grammar school, an appointment which he held for the rest of his long life. A close friend of J. J. Bodmer (q.v.) and his partner in critical enterprises, Breitinger has been somewhat overshadowed by his more mercurial and vocal colleague. He participated in the *Discourse der Mahlern* (q.v.) and in *Von dem Einfluß und Gebrauch der Einbildungskraft zur Ausbesserung des Geschmacks* (1727). It was the practice of the two friends for one to give support in a preface to a book written by the other, and Breitinger accordingly wrote the preface to Bodmer's *Critische Abhandlung von dem Wunderbaren* (1740). In the same year he published his most important work, his *Critische Dichtkunst worinnen die poetische Malerei in Absicht auf die Empfindung im Grunde untersuchet und mit Beispielen aus den berühmtesten Alten und Neuern erläutert wird* (with a preface by Bodmer). This two-volume treatise sets out the literary ideas current at the time—poetry is an imitation of nature and its aim is moral teaching; but Breitinger's rejection of rhyme and his insistence on the importance of the miraculous in poetry were felt by Gottsched (q.v.) to be a provocation. Breitinger's conception of 'das Wunderbare' is severely limited; he calls it 'ein vermummtes Wahrscheinliches'. An unostentatious but sound scholar, Breitinger produced editions of the Septuagint, of the fables of U. Boner (q.v.) and of Opitz's works, and was in charge of the Zurich revision of the Bible in 1772.

Bremer Beiträge, the usual designation of the weekly *Neue Beiträge zum Vergnügen des Verstandes und Witzes*, published at Bremen from 1745 to 1748. The editors were K. C. Gärtner, G. W. Rabener, and J. A. Cramer (qq.v.), who seceded from Gottsched's periodical *Belustigungen des Verstandes und Witzes* (q.v.) and founded the *Bremer Beiträge* in opposition to it. Among the contributors were J. A. Ebert, J. A. Schlegel, J. F. W. Zachariä, and, less frequently, Gellert, Giseke, E. von Kleist, and Ramler (qq.v.). The founders and chief contributors are sometimes referred to as 'Bremer Beiträger'. In their last year the *Beiträge* achieved their greatest fame by the publication of the first three cantos of Klopstock's *Der Messias* (q.v.).

BRENNGLAS, ADOLF, pseudonym of Adolf Glassbrenner (q.v.).

BRENTANO, BETTINA, see ARNIM, BETTINA VON.

BRENTANO, CLEMENS (Ehrenbreitstein, 1778–1842, Aschaffenburg), son of Maximiliane Brentano (*née* La Roche) and grandson of the novelist Sophie von La Roche (q.v.), was sent in his childhood from *pension* to *pension*, on a curious educational principle, which probably increased

his natural tendency to fight shy of systematic work. His mother's death in 1793 was a great blow. His father endeavoured to launch him in the prosperous family business, but abandoned the attempt in 1797, after the young Brentano had decorated business letters with caricatures. He was allowed to begin studies at Halle University and then switched to Jena (1798–1800), where he made contact with the older school of Romantics, including the Schlegel brothers, Fichte, and L. Tieck (qq.v.). He also made the acquaintance of the poetess Sophie Mereau (q.v.), whom he married in 1803.

His first publication was *Satiren und poetische Spiele von Maria. Erstes Bändchen. Gustav Wasa* (1800), a satire on the play *Gustav Wasa* by A. von Kotzebue (q.v.). This was followed in 1801–2 by *Godwi* (q.v.), which by its playfulness introduced a new structural variation to the Romantic novel and which includes much of his lyric poetry. Brentano stayed at Göttingen in 1801 and there met Achim von Arnim (see ARNIM, L. J. VON) with whom he formed a lifelong friendship. After his marriage he lived in Heidelberg in close contact with Arnim and J. J. von Görres (q.v.), and the three friends formed the nucleus of the Heidelberg group of Romantics (see ROMANTIK). Arnim and Brentano collected the folk-songs which they were to publish in *Des Knaben Wunderhorn* (q.v., 1805–8). Brentano also wrote a comedy, *Ponce de Leon* (q.v., 1804), and collaborated with Görres in *Wunderbare Geschichte von BOGS dem Uhrmacher* (1807). The curious appellation of the chief character is made up of the first and last letters of the authors' surnames. From 1804 to 1812 he worked fitfully at the collection of religious poems which was to appear after his death, still in uncompleted form, as *Romanzen vom Rosenkranz* (q.v., 1852). Sophie Mereau died in childbirth in October 1806, and Brentano made a rash second marriage with the 17-year-old Auguste Busmann in August 1807; this was dissolved in 1810. His comic pseudo-treatise *Der Philister vor, in und nach der Geschichte* was published in 1811. He was in Berlin with Arnim in 1810, visited a family estate (Bukovan) in Bohemia in that year and lived there from 1811 to 1813, a period which ended with a stay in Prague. At this time he wrote his diffuse drama *Die Gründung Prags* (1814). He spent 1813–14 in Vienna in contact with Adam Müller (q.v.), J. von Eichendorff (q.v.), whom he had known in Heidelberg, and the brothers A. W. and F. Schlegel. During another spell in Berlin he fell in love with Luise Hensel (q.v.), then employed as a governess, but was unsuccessful in persuading her to marry him. He returned at this time to the Catholic faith and, in his new-won piety, abandoned creative writing.

The stories, which appeared mostly in Brentano's middle life, date from his early years: *Aus der Chronika eines fahrenden Schülers* (q.v., 1818) dates from 1802–4, *Geschichte vom braven Kasperl und dem schönen Annerl* (q.v., 1817) was probably written not long before publication, and *Die mehreren Wehmüller und die ungarischen Nationalgesichter* (q.v., 1817) was most likely a product of the years 1811–14. The Märchen, chiefly written 1805–11, were published posthumously 1846–7, except for *Das Märchen von Gockel und Hinkel*, which appeared in 1838. In 1819 Brentano appointed himself companion and amanuensis to the former nun Katharina Emmerich, who lay ill in Dülmen, Westphalia, exhibiting from time to time the stigmata of the Crucifixion. Her account of her visions is contained in Brentano's publication *Das bittere Leiden unseres Herrn Jesu Christi. Nach den Betrachtungen der gottseligen Anna Katharina Emmerich* (1833). Two other books dealing with Katharina were taken from his posthumous papers (*Leben der heiligen Jungfrau Maria, nach den Betrachtungen der gottseligen Anna Katharina Emmerich*, 1852, and *Leben unseres Herrn und Heilandes Jesu Christi, nach den Geschichten der gottseligen Anna Katharina Emmerich*, 1856). After the death of Katharina in 1824, he lived with friends for varying periods in Koblenz, Paris, Frankfurt, Regensburg, and finally in Munich, but produced nothing more. All who knew Brentano spoke of him as a man of prodigious poetic fantasy, which he was unable and unwilling to control.

Sämtliche Werke, historisch-kritische Ausgabe, including correspondence (planned in 42 vols.), ed. J. Behrens, W. Frühwald, D. Lüders, appeared in 1974 ff, *Werke* (4 vols.), ed. F. Kemp, 1963–8, and *Briefe* (2 vols.), ed. F. Seebaß, 1951.

Brest-Litovsk, Treaty of, instituting a separate peace between the Central Powers and Russia, was signed on 3 March 1918, after a German ultimatum issued on 24 February. The negotiations for the treaty had begun in November 1917 and had been broken off in the following February. The very severe terms imposed on Russia included the loss of Courland, Livonia, Esthonia, Lithuania, Poland, Finland, the Ukraine, and parts of the Caucasus. By an additional treaty signed in August 1918, Russia agreed to pay an indemnity and to recognize the independence of Georgia. The treaty was effectively neutralized by the armistice terms at Compiègne (November 1918) and abrogated by the Treaty of Versailles in June 1919.

BRETZNER, CHRISTOPH FRIEDRICH (Leipzig, 1748–1807, Leipzig), a minor dramatist, whose plays include the comedy *Das Räuschchen* (1786),

the tragedy *Der Lüderliche* (1789), and *Belmont und Constanze*, which, in an adaptation by G. Stephanie (q.v.), served as the libretto for Mozart's opera *Die Entführung aus dem Serail* (1782). The songs 'Wer ein Liebchen hat gefunden' and 'Vivat Bacchus, Bacchus vivat!' are as Bretzner wrote them. He translated the Italian text of *Così fan tutte* (1790, see DAPONTE, L.) into German as *Weibertreu*. His plays were collected as *Schauspiele* (4 vols., 1820).

Brief des Lord Chandos, an essay in letter form, written by H. von Hofmannsthal (q.v.) in 1901 and published as *Ein Brief* in *Der Tag* in 1902, in book form in 1905. The fictitious writer is Philipp Lord Chandos, and the addressee is 'Francis Bacon, später Lord Verulam und Viscount St. Albans'. The letter is dated 22 August 1603. The details concerning Bacon are authentic, and so are the literary tastes of the fifth Baron Chandos, the presumed author of the essay collection *Horae subsecivae*. Lord Chandos writes his letter in response to an inquiry by Bacon about his two years of silence after five years of creative writing which he began at the age of 19. Now 26, Lord Chandos accounts for his silence, implicitly trusting in Bacon's sympathetic reception. He has come to denounce verbal splendour which fascinates for its own sake, and has made the relationship between language and existence his principal concern: 'Mein Fall ist, in Kürze, dieser: Es ist mir völlig die Fähigkeit abhanden gekommen, über irgend etwas zusammenhängend zu denken oder zu sprechen.' This statement is discussed and richly illustrated. Chandos creates an awareness of both physical and spiritual reality (Geist, Seele, and Körper), but his conviction that only thought can inspire him with a sense of peace prevails. Words distort meaning and the 'confusion of language' (Wirbel der Sprache) is destructive. Wit emerges as the only constructive means of communicating mature thought.

Briefe antiquarischen Inhalts, a series of controversial letters written by G. E. Lessing (q.v.) in 1768, several of which were first published in two Hamburg newspapers, and then collected with others in two parts (1768–9). There are 57 letters, all directed against Professor C. A. Klotz (q.v.) of Halle University. The quarrel arose out of a claim by Klotz to have convicted Lessing of errors in *Laokoon* (q.v.). Lessing, as usual, rounded on his adversary, trouncing him unmercifully, but the positive gains of these polemics were few. See also WIE DIE ALTEN DEN TOD GEBILDET.

Briefe, die ihn nicht erreichten, a novel in

the form of letters published anonymously in 1903, and actually written by Baroness Elisabeth von HEYKING (q.v.). It unfolds the love of a woman, caught in an unhappy marriage and touring the Far East with her brother. She falls in love with an acquaintance, who, going from Shanghai to Peking, finds himself cut off in the latter by the Boxer Rising. The anonymous woman, staying first in Canada, then in the U.S.A., next in Berlin, and finally again in New York, writes him a series of tender and somewhat melancholy letters, to which she receives no reply. Her husband dies in a private mental asylum, but her release brings no fulfilment, for the news of the relief of Peking is quickly followed by the intelligence of 'his' death (he, too is anonymous) in the last stages of the fighting. The fictitious letters cover a year, August 1899 to August 1900. In an epilogue her own death shortly afterwards is recorded. The book was a best seller for a quarter of a century.

Briefe, die neueste Litteratur betreffend, see LITERATURBRIEFE.

Briefe über die ästhetische Erziehung, see ÜBER DIE ÄSTHETISCHE ERZIEHUNG DES MENSCHEN IN EINER REIHE VON BRIEFEN.

Briefe über Merkwürdigkeiten der Literatur (1766–70), a series of periodic essays published, on the model of Lessing's *Literaturbriefe*, by H. W. von Gerstenberg (q.v.). They show a considerable breadth of interest, ranging over Edmund Spenser, Ariosto, Ossian, Old Norse poetry, Edward Young, Wieland, Rousseau, and Shakespeare. Gerstenberg was an enthusiastic admirer of Shakespeare, and the *Merkwürdigkeiten* contributed to his appreciation and also notably furthered the idea of the primacy of original genius. These letters are often known as the *Schleswigsche Literaturbriefe*.

Briefe über Don Carlos, a publication by Schiller which appeared in epistolary form in *Der teutsche Merkur* (q.v.) in 1788. The letters excuse and interpret ambiguities which had been found in *Don Carlos* (q.v.).

Briefroman, a genre derived in Germany from Samuel Richardson's *Pamela* and J.-J. Rousseau's *La Nouvelle Héloïse*. German epistolary novels fall into two categories. In the more frequent form all the letters are written by the same person, as in Goethe's *Die Leiden des jungen Werthers*, Hölderlin's *Hyperion*, and the anonymous *Briefe, die ihn nicht erreichten* (qq.v.). The second type, represented by J. T. Hermes's *Sophiens Reise von Memel nach Sachsen* and Sophie von La Roche's *Geschichte des Fräuleins von Sternheim* (qq.v.),

employ more than one letter-writer, though the letters of the principal character predominate. *Aristipp* by Wieland (q.v.), though also containing some dialogue, in the main conforms to the second type. The epistolary novel was most popular in the second half of the 18th c.

Brigitta, a Novelle by A. Stifter (q.v.) which was written in 1843 and published in the Taschenbuch *Gedenke mein!* in 1844. The revised version was published in vol. 4 of *Studien* (1847).

The Novelle is divided into 4 sections with the sub-titles 'Steppenwanderung', 'Steppenhaus', 'Steppenvergangenheit', and 'Steppengegenwart'. It is framed (see RAHMEN) by the narrator who, in section one, rides across the Hungarian plains to visit Major Stephan Murai on his remote estate, Uwar. In the second section the narrator observes the activities of the landowning community and the enlightened principles underlying Murai's agricultural developments on the fringe of the vast expanse of the Puszta. Murai's estate borders on the estate of Maroshely, which is run by Brigitta, a woman of outstanding qualities in character and management. The third section reveals the story of Brigitta from childhood to the time when she moved to Maroshely. Deprived of affection because of her ugly appearance, she found as a young woman a handsome suitor who was the first to discover the concealed beauty of her nature. They married and had a son, Gustav. But their happiness came to an abrupt end when Brigitta discovered that Murai (for he is the husband) had deceived her with a girl of striking beauty, Gabriele. Unable to understand and forgive a fleeting passion, she forsook Murai and withdrew again into her former isolation. The couple were divorced and Brigitta devoted her life to bringing up Gustav at Maroshely. Fifteen years afterwards Murai moved into his own country estate and a new and neighbourly relationship developed. In a dramatic scene in 'Steppengegenwart' Murai saves Gustav's life when he is attacked by a pack of wolves near the gallows on the edge of the Puszta. Gustav is not seriously injured, but over his sick-bed the parents confess to their old love: the years of separation had for both been a time of repentance, in which the 'gentle law of beauty' ('das sanfte Gesetz der Schönheit') had revealed itself to them. Soon after their reunion the narrator leaves for Germany.

The theme of beauty (see BUNTE STEINE) and marriage (Gattenliebe) achieves a climax of poetic realization in this Novelle, in which Stifter creates a new and distinct background in the Hungarian landscape and the development of a rural community.

BRINCKMAN, John (Rostock, 1814–70, Güstrow), studied at Rostock University and then went to America in 1839 as a political refugee. He returned in 1842 and spent the rest of his life as a tutor and schoolmaster. He wrote in Low German (Mecklenburger Platt), publishing *Aus dem Volk für das Volk*, a collection of stories, in 1854–5. It is in two volumes, individually entitled *Dat Brüden geit um* and *Kasper Ohm und ick*. *Vagel Griep* (1859) is a collection of dialect poetry. Brinckman also wrote the dialect novel *Uns Herrgott up Reisen* (1870). He is regarded as ranking with F. Reuter (q.v.) as a writer in Mecklenburg Low German.

BRION, Friederike (Niederrödern/Alsace, 1752–1813, Meißenheim nr. Lahr), daughter of the pastor in the Alsatian village of Sesenheim *c.* 30 miles north of Strasburg. Goethe (q.v.) met her in October 1770 and fell in love with her. Some of his freshest and most enchanting love poetry was written during the ten months that the love-affair lasted (see Willkommen und Abschied, Mit einem gemalten Band, Mailied). Goethe broke off the relationship on his return to Frankfurt in August 1771. He gives an account of the love-affair in *Dichtung und Wahrheit*, Bks. 10 and 11. It is also the subject of the operetta *Friederike* by F. Lehár (1870–1948). Goethe visited Friederike again while on his way to Switzerland in 1777. Though J. M. R. Lenz courted her in 1772, she remained unmarried.

BRITTING, Georg (Regensburg, 1891–1964, Munich), volunteered for war service in 1914, became an officer and was severely wounded. In 1920 he settled in Munich and devoted himself to writing. He began with a Novelle (*Der verlachte Hiob*, 1921) and later added further volumes of stories (*Die kleine Welt am Strom*, 1933; *Das treue Eheweib*, 1933; *Der bekränzte Weiher*, 1937; *Das gerettete Bild*, 1938; *Der Schneckenweg*, 1941). His only novel (*Lebenslauf eines dicken Mannes, der Hamlet hieß*, 1932) is a highly original variation on the plot of *Hamlet*. Britting's Hamlet is a man of indolence and gargantuan appetite, who avenges his father by provoking Claudius into such excessive gluttony that he dies of a stroke. Hamlet, who has a son by Ophelia, withdraws, in his disgust with life, into a monastery. Britting is chiefly known as an imaginative lyric poet, whose principal collections (*Gedichte*, 1931; *Der irdische Tag*, 1935; *Rabe, Roß und Hahn*, 1939; *Lob des Weines*, 1944; *Die Begegnung*, 1947; *Unter hohen Bäumen*, 1951) are characterized by vivid vocabulary and original, often conspicuous, rhyme.

BROCH, Hermann (Vienna, 1886–1951, New Haven, U.S.A.), of Jewish parentage, entered his father's textile concern in 1908 but when over 40 left it to study at Vienna University (mathematics, philosophy, psychology, 1928–31). He settled in the Salzkammergut (Altaussee) and then in Tyrol at Mosern near Seefeld and began to write novels. His first was the trilogy *Die Schlafwandler* (q.v., 1930–2); *Die unbekannte Größe* (q.v.) followed in 1933. On the National Socialist invasion of Austria, Broch was arrested, but influential friends, including James Joyce, obtained permission for him to emigrate. He settled in America and there wrote his best-known novel *Der Tod des Vergil* (q.v., 1945). *Die Schuldlosen* appeared in 1950, and his last novel, existing in three drafts, was published in 1953 as *Der Versucher* in a version which has given rise to controversy. It is vol. 4 in his *Gesammelte Werke* (10 vols., 1952–61). The author himself referred to this work as his *Bergroman* (q.v.). Another version, based on the third draft, was published as *Demeter* in 1967. Broch, a powerful and profound writer, was obsessed with the problems presented by death, and by mass psychology. He was also an outstanding essayist (*Werke*, vols. 6 and 7). His correspondence with Daniel Brody was published in 1971. *Kommentierte Werkausgabe*, ed. P. M. Lützeler, appeared in 1978–9 (vols. 1–12) and 1981 (vol. 13 in 3 vols., *Briefe* of the years 1913–51).

Brocken, the highest mountain in the Harz region. The Brocken was climbed by Goethe in winter in 1777, and is the scene of amusing passages in Heine's *Die Harzreise* (q.v., 1826). Under its alternative name 'Blocksberg' it was reputed to be the scene of a witches' sabbath each 'Walpurgisnacht', i.e. night of 30 April/1 May. The legend is incorporated in an important scene in Goethe's *Faust* (q.v.) Pt. I.

BROCKES, Barthold Hinrich (Hamburg, 1680–1747, Ritzebüttel), a prominent citizen of Hamburg with a gift for poetry, came of a patrician merchant family. He studied law at Halle (1700–2) and at the Reichskammergericht (q.v.) in Wetzlar. At 22 he went on the grand tour, visiting parts of Germany, Switzerland, Italy, France, Holland, and England, returning to Hamburg in 1704. For the next sixteen years he lived a life of leisure in Hamburg, devoting himself to poetry. He became a senator of Hamburg in 1720 and mayor of Ritzebüttel in 1735. Brockes founded in 1714 Die Teutschübende Gesellschaft which in 1716 changed its name to Die Patriotische Gesellschaft. Brockes's first poetic work is an oratorio in the highly ornamented and florid baroque manner (*Der für die Sünden der Welt gemarterte und sterbende Jesus*, 1712); it was set to music by G. R. Keiser, G. F. Handel, and G. P. Telemann. Seven numbers of this work were included in the St. John Passion of J. S. Bach (q.v.). Most of Brockes's poetry is contained in a series of 9 vols. (1721–48), each of which bears the title *Irdisches Vergnügen in Gott*. As the title indicates, the poems express pleasure at the divine pattern embodied in the sensual world. Brockes writes of sunshine and rain, snow and shadow, nightingale and cornflower, and many other details of nature, and though some of his poems are pedestrian or tritely moral, he often succeeded in expressing with precision things seen, heard, or felt, opening up to poetry a new world of exact perception. Some of the poems from the *Irdisches Vergnügen* were composed as arias by Händel. From 1724 to 1726 Brockes published a weekly, *Der Patriot*. He also translated Thomson's poem *The Seasons* (*Die Jahreszeiten*, 1745), which, in much altered form, eventually became the libretto of J. Haydn's oratorio *Die Jahreszeiten* (q.v.). His works (5 vols.), ed. J. J. Eschenburg, appeared in 1800; *Auszug der vornehmsten Gedichte aus dem Irdischen Vergnügen in Gott. Faksimiledruck nach der Ausgabe von 1738*, with postscript by D. Bode, in 1965.

BROD, MAX (Prague, 1884–1968, Tel Aviv), born and brought up an Austrian citizen, became a civil servant and was also dramatic and music critic of the Prague German language newspaper, *Prager Tagblatt*. His first book, *Der Weg des Verliebten*, a collection of lyric poems, appeared in 1907. He published more volumes of poetry in the next few years (*Tagebuch in Versen*, 1910; *Die Höhe des Gefühls*, 1910; *Das gelobte Land*, 1917; and *Das Buch der Liebe*, 1921), and a final volume (*Neue Gedichte*) appeared in 1949. His plays include *Abschied von der Jugend* (1912), *Eine Königin Esther* (1917), *Die Fälscher* (1920), and *Lord Byron kommt aus der Mode* (1929). Among his numerous novels, *Arnold Beer. Das Schicksal eines Juden* (1912) has a central figure suggestive of Brod's friend F. Kafka (q.v.); *Tycho Brahes Weg zu Gott* (1916) and *Rëubeni, Fürst der Juden* (1925), a historical novel in a Renaissance setting, were later combined with *Galilei in Gefangenschaft* (1948) to form the trilogy *Ein Kampf um die Wahrheit*. *Annerl* (1937) is an acute psychological novel. A lesser work, *Die Frau, nach der man sich sehnt* (1927), was his best-selling book. Brod became a Zionist in 1913, and many of his later works are concerned with Jewish problems. In 1939 he emigrated to Palestine, living in Tel Aviv as a theatre director.

He is perhaps best known as the biographer of Kafka (q.v., *Franz Kafka*, 1937), whose posthumous works he prepared for publication. It was he who, as Kafka's executor, took the decision, in disregard of the testator's wishes, to save the Kafka MSS. from destruction and to publish them. A biography of Heine appeared in 1934. Brod published his autobiography (*Streitbares Leben*) in 1960. He continued in old age to write novels, including *Der Meister* (1952), a study of Christ, *Mira* (1958), a story about H. von Hofmannsthal (q.v.), and *Die Rosenkoralle. Ein Prager Roman* (1961).

Brod und Wein, an elegy by F. Hölderlin (q.v.) written in 1800 and published in 1807. It begins in praise of night, laments the passing of Greece, and sees Christ as the successor to Dionysus. Bread and wine, the symbols of the ancient world, appear also as symbols of Christianity. The verse form is elegiac (see ELEGISCHES VERSMASS), and the poem comprises 80 distichs in 9 strophes.

BRÖGER, KARL (Nürnberg, 1886–1944, Nürnberg), a revolutionary and pacifist poet, who later went over to the National Socialists. He is best known for his partly autobiographical novel *Der Held im Schatten* (1919) and for his early antiwar poems, *Kamerad, als wir marschiert* (1916) and *Soldaten der Erde* (1918). *Sturz und Erhebung* is the title of a collected edition of his poetry, published in 1943; *Bekenntnis* (1954) is a selection.

BRONNEN, ARNOLT (Vienna, 1895–1959, Berlin), one of the more prominent left-wing experimental Expressionist dramatists, came to Berlin in the early 1920s. *Vatermord*, written in 1915 and published in 1920, virtually provoked a riot when it was performed in Berlin in 1922. It includes incest between mother and son, and ends with the son murdering his father on stage. It was in reference to his early work that the term Episches Theater (q.v.) was first used. *Vatermord* was followed by *Die Geburt der Jugend* (1922) and the crude comedy *Die Exzesse* (1923). *Katalaunische Schlacht* (1924), a play looking back to the war, is full of sensation and artifice. Steadily swinging to the right, Bronnen produced in 1925 a nationalistic play, *Rheinische Rebellen*. *Ostpolzug* (1926), with one character only (see MONODRAMA), treats Alexander the Great of Macedon. The play *Michael Kohlhaas* (based on Kleist's story of the same title, q.v.) appeared in 1929, and was revised in 1948. It was followed by *Sonnenberg* (1934). In 1937 Bronnen was dismissed from a post in the German film industry, and after the war turned to journalism in Linz, before working from 1955 as a theatre critic in East Berlin. His later plays are *Gloriana* (1941), *Die Kette Kolin* (1950), and *Die jüngste Nacht* (1952). His fiction includes the novel *O.S.* (1929), which deals with the clash between a 'free corps' (see FREIKORPS) and Poles irrupting into Silesia. In 1960 his *Tage mit Bertolt Brecht* appeared, and *Stücke* (1 vol.) in 1977.

BRONNER, FRANZ XAVER (Höchstädt, 1758–1850, Aarau), entered the Benedictine Order but ran away from his monastery in 1785 and lived for a time in Zurich. He returned to his Order at Augsburg, but again left it and became a professor at Aarau, and later at Kazan in Russia. He returned to Aarau in 1817, became a Protestant convert, and held a municipal appointment. He wrote *Fischergedichte und Erzählungen* (1787), *Neue Fischergedichte* (1794), and, in 1795, a somewhat premature autobiography.

Brot der frühen Jahre, Das, a story by H. Böll (q.v.), published in 1955. The maintenance electrician Walter Fendrich tells the story of his love for Hedwig Muller. It matures on a Monday in the spring of 1955 which, from morning till night, marks the time of action. His love effects a complete change of outlook in Fendrich's attitude to life, which has been shaped by the years of deprivation during his childhood. The lack of 'the bread of the early years' during the war and the immediate post-war period has left him with an anxiety complex over his material existence, which the relative prosperity and security of his more recent existence as an employee of a flourishing washing-machine company has not removed. This side of his life is reflected in his friendship with Ulla, the daughter of his employer, whom he abandons for the affection and stability that Hedwig inspires in him. (Film version by H. Vesely, 1962.)

Brücke, Die, a coterie of Expressionist painters in Dresden, 1905–13. Its best-known members were E. Heckel, E. L. Kirchner, E. Nolde, and K. Schmidt-Rottluff (qq.v.).

BRUCKNER, ANTON (Ansfelden, 1824–96, Vienna), Austrian composer of symphonies and church music, was appointed organist at the Abbey of St. Florian in 1849, and at Linz in 1856. In 1867 he became organist to the Court Chapel in Vienna, and four years later professor of composition at the Vienna Conservatoire. Bruckner's choral works are almost all settings of Latin liturgical texts (3 masses, *Te Deum*, motets).

BRUCKNER, FERDINAND, pseudonym of Theodor Tagger (Vienna, 1891–1958, Berlin), studied at Vienna University and the Sorbonne. In 1923 he founded the Renaissance-Theater in Berlin in his real name and directed it until 1928. His first publications were Expressionistic poems

(*Der Herr in den Nebeln*, 1917). These and the novels *Die Vollendung des Herzens* of the same year, and *Auf der Straße* (1920), appeared under his own name. His first partial success came with the plays published under his pseudonym, beginning with *Krankheit der Jugend* (1928, first performed in Hamburg and Vienna, 1926). The 'illness of youth' turns out to be the promiscuous life of a group of male and female medical students who view the future in utter disillusionment. *Die Verbrecher* (1928, first performed in Berlin and Vienna in that year) deals savagely with the incongruence of justice and law. It is technically an extraordinary play, with the stage in Act I subdivided into seven separate rooms, which are separately illuminated or obscured as the fragmented action requires; in the second act it is similarly divided into four court rooms, a common room, and a corridor. Other plays of this early period were *Te Deum* (1929) and *Die Kreatur* (1930). With *Elisabeth von England* (q.v., 1930) Bruckner achieved his only general success, interpreting history in the light of contemporary ideas and psychology and exploiting the theatrical techniques popularized by E. Piscator (q.v.). This was followed by *Timon. Tragödie vom überflüssigen Menschen in fünf Aufzügen* (1932). In 1933 Bruckner emigrated to Austria, and from 1936 until 1951 lived in the U.S.A. He spent his last years in Berlin. *Die Rassen* (1935), set in a German university in the spring of 1933, attacks National Socialist Germany. *Heroische Komödie* (1945) has as its principal characters Madame de Staël, Benjamin Constant, and Bernadotte. The later tragedies *Der Tod einer Puppe* and *Der Kampf mit dem Engel* (both 1956 and both in verse) made little impact. Among Bruckner's other plays are a version of *Die Marquise von O.* (1933, see KLEIST, H. VON), *Simon Bolivar* (1945), *Fahrten* (1949, retitled *Spreu im Wind*, 1952), *Pyrrhus und Andromache* (1951), *Früchte des Nichts* (1952), *Clarissa* (1956), and *Das irdene Wägelchen* (1957).

BRUDER KLAUS, see NIKOLAUS VON DER FLÜE.

'Brüder, lagert euch im Kreise', first line of a student drinking song first published in *Das akademische Lustwäldlein* in 1794.

'Brüderlein fein, Brüderlein fein', first line of a well-known song in *Das Mädchen aus der Feenwelt oder Der Bauer als Millionär* (q.v., 1826) by F. Raimund. It is sung by 'Jugend', who gently brings home to the 'Bauer' that all his riches cannot save him from the passage of time.

Bruder Rausch, an anonymous Low German poem of the late 15th c. It is a legend of the devil, who, in the shape of 'Bruder Rausch', completes the demoralization of an ill-conducted monastery. The original edition of 428 lines, published probably in 1488, was followed by a High German version in 1515. The story is analogous to a Schwank (q.v.).

Bruderzwist in Habsburg, Ein, see EIN BRUDERZWIST IN HABSBURG.

BRÜES, OTTO (Krefeld, 1897–1967), a journalist, who worked with the *Kölnische Zeitung* and later with the *Düsseldorfer Zeitung*. He is the author of entertaining light novels and stories, including among the former *Jupp Brand* (1927), *Der Walfisch im Rhein* (1931), *Die Wiederkehr* (1932), *Das Mädchen von Utrecht* (1933, a short novel of which *Die Affen des großen Friedrich*, 1939, is a sequel), *Der schlaue Herr Vaz* (1937), *Marie im neuen Land* (1938), and *Der Silberkelch* (1948). Among the stories are *Helden, Heilige, Narren und Musikanten* (1923), *Klas Pottbäcker* (1924), and *Die Sonate* (1941). He also wrote plays (*Der Prophet von Lochau*, 1923; *Der alte Wrangel*, 1925) and poems (*Gedichte*, 1926; *Die Brunnenstube*, 1948).

BRÜHL, FRIEDRICH ALOIS, GRAF VON (Dresden, 1739–93, Berlin), son of the Saxon minister Heinrich, Graf von Brühl (q.v.), held office in Poland during the reign of August III (q.v.). He was passionately devoted to drama, maintained a private theatre on his estate at Pförten in Lower Lusatia (Nieder-Lausitz), and wrote comedies in the French manner in which he himself acted. His collected plays were published as *Theatralische Belustigungen*, 1785–90.

BRÜHL, HEINRICH, GRAF VON (Weißensee, Thuringia, 1700–63, Dresden), became the all-powerful minister of August II ('der Starke') and August III (qq.v.) of Saxony. As a page he gained the confidence of his sovereign, and by his early thirties was in control of taxes and minister of the interior. By 1738 he was sole minister and in 1746 received the title Prime Minister (Premierminister). He was responsible for the disasters of Saxon foreign policy in the reign of August II, whose confidence in him remained unshaken. Brühl annexed the salaries of all the ministers whom he had displaced and also enriched himself liberally out of the public purse. He acquired a library of more than 60,000 volumes and a splendid collection of pictures; and he is said to have worn a new suit every day. After the death of August III he was dismissed and died a few weeks later. His fortune, at first sequestrated, was later released. His name survives in the Brühlsche Terrasse in Dresden.

BRÜLOW, KASPAR (nr. Pyritz, Pomerania, 1585–1627, Strasburg), a professor from 1609

and a laureate (gekrönter Dichter, q.v.), wrote the Latin plays *Andromeda* (1611), *Elias* (1613), *Chariklea* (1614), *Nebukadnezar* (1615), *Julius Caesar* (1616), and *Moyses* (1621).

BRUN, FRIEDERIKE (Grafentonna, Thuringia, 1765–1835, Copenhagen), German poetess, *née* Münter. After her marriage in 1783 to Konstantin Brun, a Dane, she travelled in Europe for some years, meeting Klopstock, Mme de Staël, and the poet Matthisson (qq.v.), whose work is said to have influenced her poetry. She returned to Copenhagen in 1810. Matthisson published her poems in 1795 and she issued further volumes in 1812 and 1820. She also published accounts of her travels. Her poem *Ich denke dein* (q.v.) so impressed Goethe in 1795 that he was stimulated to write a composition of his own with the same initial phrase and an identical rhythmic pattern. (See NÄHE DES GELIEBTEN.)

BRÜNING, HEINRICH (Münster, 1885–1970, Norwich, U.S.A.), was elected a member of the Reichstag in 1924 and in 1929 became leader of the Roman Catholic Centre Party (see ZENTRUM). In 1930, after the fall of the coalition government in consequence of the economic crisis, Brüning was appointed chancellor by President Hindenburg (q.v.). The budgetary measures he proposed were rejected by the Reichstag. His decision to dissolve the House in the hope of securing a safe majority resulted in a considerable increase in the number of National Socialist deputies. Brüning, who thereafter pursued a policy of economy and retrenchment by decree, was dismissed by Hindenburg in May 1932. In July 1933 the Zentrumspartei was dissolved and in 1934 Brüning took refuge in the U.S.A., becoming a professor at Harvard. From 1951 to 1954 he was professor of politics at Cologne University. He spent his last years in the U.S.A. His memoirs (*Memoiren, 1918–1934*) were published in 1970.

BRUNNER, THOMAS (b. Landshut, date unknown, d. 1571, Steyr, Austria), a schoolmaster at Steyr, wrote biblical plays for school performance. They include *Jakob und seine zwölf Söhne* (1566), *Tobias*, and *Isaak* (both 1569). Brunner used the pseudonym Pegaeus.

Brünner Programm, a programme of the Austrian Social Democratic Party, drawn up at its congress at Brünn (Brno) in 1899 under the leadership of Viktor Adler (q.v.). The programme envisaged the transformation of the autocratic dualistic state of Austria–Hungary into a democratic state, which would be divided into autonomous regions based on nationality.

BRUNNGRABER, RUDOLF (Vienna, 1901–60, Vienna), an early exponent of the technological novel dealing with specific materials (*Radium,* 1936; *Opiumkrieg,* 1939; *Zucker aus Kuba,* 1941; *Heroin,* 1951). In *Die Engel in Atlantis* (1938), which mingles the biblical and the futuristic, he rejects the conception of a perfect technological world and shows its annihilation in the Deluge, which only Noah, the man of God, survives.

BRUNO VON HORNBERG, a minor Middle High German poet of the mid-13th c., belonged to a family of the southern Black Forest in the region of Breisgau. Three Minnelieder by him and a Tagelied are preserved.

Brunswick, see BRAUNSCHWEIG.

BRUN VON SCHÖNEBECK, a patrician of Magdeburg, who lived in the second half of the 13th c., and appears to have played an active part in the life of his native city. He is best known as the author of a long paraphrase and interpretation of the Song of Songs (das Hohe Lied), conceived upon an original triple plan and plentifully interspersed with learned Latin quotations. It was written *c.* 1275. He also wrote an *Ave Maria,* a part of which survives, and possibly a life of Jesus. A book describing a festivity, which he helped to organize in his earlier years, is lost.

BRUNWART VON AUGHEIM, a minor poet of the second half of the 13th c. who is the author of four extant Minnelieder. Brunwart was a native of Auggen in Breisgau, south-west of Freiburg.

BRUSCH (BRUSCHIUS), CASPAR (Schlackenwald, Bohemia, 1518–57), was a humanistic disciple of Celtis (q.v.). He published collections of Latin poems, *Silvae* (1543) and *Poemata* (1554), and also geographical poetry in the manner of Celtis (*Elegia de Mulda flumine,* 1543). He was murdered while travelling between Windsheim and Rotenburg by noblemen who feared his pen.

BUBER, MARTIN (Vienna, 1878–1965, Jerusalem), Jewish philosopher who was educated at Vienna, Leipzig, Berlin, and Zurich universities. Though active in the Zionist movement, he later came to reject the exclusiveness of the Zionist outlook. His *Ekstatische Konfessionen* (1909) is an example of early Expressionist literature. Deprived of a professorship at Frankfurt University in 1933, he emigrated to Palestine in 1938 and until 1951 was professor of social philosophy at the Hebrew University in Jerusalem. Buber's

teaching and narrative work co-ordinates biblical interpretation with mystical Jewish Chassidism. One of his fundamental tenets was respect for the individual. He is the author of an outstanding translation of the Bible into German (*Die Schrift,* 15 vols., 1925–38, rev. 1954–62). The recipient of international honours, he was awarded the Friedenspreis des deutschen Buchhandels (q.v.) in 1953. *Werke* (3 vols.) were published 1962–3.

BUCER, MARTIN (Schlettstadt, 1491–1551, Cambridge), adherent of the Lutheran reform from 1518 on, mediated between Luther and Zwingli in their dispute concerning Communion. Bucer is believed to have written *Neu-Karst Hans* (though some have attributed it to U. von Hutten, q.v., who probably in any case had a hand in it). Invited to England by Cranmer, Bucer was elected to a chair at Cambridge in 1549. Bucer's *Opera omnia. Historisch-kritische Ausgabe* (planned in 30 vols. containing *Deutsche Schriften, Opera latina,* and correspondence), ed. R. Stupperich, appeared 1955 ff. (His name is also spelled Butzer.)

Buch der Abenteuer, see FÜETRER, ULRICH.

Buch der Bilder, Das, a volume of poetry by Rilke (q.v.), published in 1902 and consisting of two parts containing 47 poems. In 1906 a new edition converted the original volume into Bk. I, and added Bk. II with 42 new poems, many of them much longer than anything in I. The work represents a half-way stage between the vague emotionalism of Rilke's first phase and the precise and plastic style of his maturity. It is marked by an attempt to 'fix' poems structurally by the use of images which, besides existing for their own sake, also have an emotional significance. Not all the poems attempt this (e.g. *Die Stille, Herbsttag, Herbst*); but *Ritter, Pont du Carrousel,* and *Von den Fontänen* look forward to the later Rilke.

Buch der Könige alter und niuwer Ê, a Middle High German history of the ancient world, giving first the history of the Jews up to Esther and Judith (alte ê) and then the history of the Roman Empire (niuwe ê). The Jewish section is derived from the Bible and supplemented by the *Historia scholastica* of Petrus Comestor; the Roman portion turns the *Kaiserchronik* (q.v.) into prose. It is thought to have been written *c.* 1270–80 and is perhaps from the same hand as the *Schwabenspiegel* (q.v.).

Buch der Leidenschaft, by G. Hauptmann (q.v.), published in 1929 (2 vols.), purports to be a diary of unrevealed authorship. It records the diarist's struggle, extending over a period of ten years, to free himself from his marriage with 'Melitta'. In 1904 he enters a new marriage with 'Anya'. The year is that of Hauptmann's second marriage, and this diary is his most personal and candid book.

Buch der Lieder, H. Heine's best-known collection of verse, published in 1827. It brings together nearly all the poetry he had written up to that time. It is arranged in sections entitled *Junge Leiden 1817–1821* (subdivided into 'Traumbilder', 'Lieder', 'Romanzen', 'Sonette', and 'Lyrisches Intermezzo'), *Heimkehr 1823–1824, Aus der Harzreise 1824,* and *Die Nordsee 1825– 1826.* The vast majority of the poems treat of dreams and love, motifs which often occur in association, and they are frequently dominated by a self-conscious destructive irony. Except for the sonnets, 4-line stanzas of folk-song character predominate. The collection contains many of Heine's most famous poems, including *Die Grenadiere, Belsazer* (q.v.), the poems of Schumann's *Dichterliebe* (q.v.), which are taken from 'Lyrisches Intermezzo', 'Auf Flügeln des Gesanges', 'Ein Fichtenbaum steht einsam', 'Es fällt ein Stern herunter', 'Ich weiß nicht, was soll es bedeuten' (often referred to as *Die Loreley*), 'Wir saßen am Fischerhause', 'Das Meer erglänzte weit hinaus', 'Still ist die Nacht, es ruhen die Gassen' (*Der Doppelgänger*), 'Du bist wie eine Blume', 'Saphire sind die Augen dein', and 'Der Tod, das ist die kühle Nacht' (qq.v.). Many have been set to music, notably by F. Schubert and R. Schumann (qq.v.).

Buch der Rügen, the German translation of a Latin poem, *Sermones nulli parcentis,* which is a general directive on the manner and matter of sermons to be preached by Dominican monks. The original insists on sermons aimed directly at the individual, and contains a systematic denunciation of the sins and failings of all men, arranged hierarchically under two principal rubrics, clergy and laity. The *Sermones* were probably written *c.* 1230 and the translation was made in 1276–7 by a North Bavarian monk.

Buch der Zeit, Das, a volume of poetry written in 1884 by A. Holz (q.v.) and published in 1885. It is sub-titled 'Lieder eines Modernen'. It contains many poems in traditional rhyming forms, but also some free verse. Holz shows himself as a deft, alert, and terse satirist and ironist of literature, politics, and religion.

Büchelīn der Heiligen Margarēta, a Middle High German verse legend of St. Margaret,

written in the 14th c. It was a popular version of the legend.

BUCHHOLTZ, ANDREAS HEINRICH (Schöningen, Brunswick, 1607–71, Brunswick), was first a schoolmaster at Hamelin, and became in turn a Protestant pastor, a headmaster at Lemgo, and, from 1638 to 1647, a professor at Rinteln University. From 1647 until his death he was in Brunswick, latterly as Moderator and Inspector of Schools. Buchholtz translated Horace (*Odenbuch des Horatius, deutsch* and *Verteutschte Poetereikunst des Horatius,* both 1639) and Lucian (*Lucians satirische Geschichte, deutsch,* 1659). His religious poetry appeared in *Christliche Weihnachtsfreude* (1639), *Adventsgesang* (1640), and *Herzlicher Friedenswunsch* (1643). His most considerable works were his 'court' novels designed for the entertainment and instruction of the nobility and gentry. *Des christlichen teutschen Großfürsten Hercules und des böhmischen königlichen Fräuleins Valiska Wundergeschichte* appeared in 1659 and its sequel *Des christlichen königlichen Fürsten Herkuliskus und Herkuladesla auch ihrer hochfürstlichen Gesellschaft anmutige Wundergeschichte* in 1665. These long stories, which are intended to supersede the 'gallant' novel (see GALANTE DICHTUNG) after the pattern of *Amadis,* are plentifully furnished with edifying discourses providing information likely to be useful to persons of standing.

Buchholz, a family group created in a series of amusing stories and novels by J. Stinde (q.v.). Frau Wilhelmine Buchholz, a typical Berlin *bourgeoise,* is the central figure.

BÜCHLER, FRANZ (Strasburg, 1904–), is the author of the plays *August der Starke* (1938), *Herzog Bernhard* (1939), *Theseus* (1952), and the trilogy *Ree, Wind, Iris* (1961).

BÜCHMANN, GEORG (1822–84), a Berlin schoolmaster, compiled a well-known and longlasting dictionary of quotations, *Geflügelte Worte* (1864).

BUCHNER, AUGUST (Dresden, 1591–1661, Wittenberg), became professor of poetry and rhetoric at Wittenberg University in 1616. An admirer of Opitz (q.v.), he wrote both German and Latin poetry, publishing *Nachtmahl des Herrn* in 1628, *Weynacht-Gedanken* in 1638, and an opera libretto, *Orpheus* (1638), for which Heinrich Schütz (q.v.) composed the music. His Opitzian essay on poetics, *Anleitung zur deutschen Poeterey,* was published posthumously in 1663 (reprinted 1966).

BÜCHNER, GEORG (Goddelau nr. Darmstadt, 1813-37, Zurich), the son of a doctor, was at Darmstadt Gymnasium, and in 1831 commenced medical studies in Strasburg, which he continued from 1833 in Giessen, whence he fled back to Strasburg in March 1835, finally settling as a political refugee in Switzerland. His dissertation *Mémoire sur le système nerveux du barbeau* (1836) won him a doctorate from the University of Zurich and election as a corresponding member of the Société d'histoire naturelle in Strasburg; after a trial lecture, 'Über die Schädelnerven der Fische', at Zurich University in the autumn of 1836 he was appointed lecturer in comparative anatomy, and began what promised to become a distinguished career. He was in his twenty-fourth year when he died of typhoid. During his first stay in Strasburg Büchner became engaged to Minna (Wilhelmine) Jaeglé (q.v.), the daughter of a clergyman, J. J. Jaeglé, whom he hoped to marry as soon as he could set up house. Summoned to his sickbed, Minna arrived in Zurich two days before Büchner's death.

In Strasburg Büchner became keenly interested in the ideas and activities of movements against authoritarian government and political oppression, which he pursued with vigour upon his return to his native Hesse. He founded the Gesellschaft der Menschenrechte (q.v.) in March 1834, which was modelled on the Société des Droits de l'Homme et du Citoyen of 1830, and expressed his radical socialist ideas in the political pamphlet *Der Hessische Landbote* (q.v., July 1834). He aimed at this stage at a Hessian peasants' revolt, because he was convinced that only the use of force would effect social justice and remedy the distressing conditions of the lower classes. The mainspring of his courageous but dangerous political activities was his deep sympathy with social misery. In an age of economic crises and reluctant constitutional and fiscal reforms, the peasants in the upper part of the Grand Duchy of Hesse had reason to be particularly aggrieved at their lot. The July revolution in Paris had evoked in September 1830 a local revolt ('Blutbad von Södel'), which had quickly been suppressed by the army. Büchner associated with a poor ex-student of theology, August Becker, who later commented on Büchner: 'Die Grundlage seines Patriotismus war wirklich das reinste Mitleid und ein edler Sinn für alles Schöne und Große'. Through Becker, Büchner made contact with the teacher Friedrich Ludwig Weidig (q.v.). But his revolutionary activities came to an abrupt end with the arrest of his collaborators Karl Minnigerode, Becker, and Weidig after a treacherous denunciation. Büchner himself narrowly escaped. The arrest of the conspirators was not the only

cause of Büchner's disillusionment with his revolutionary activities; for the people themselves had failed, for fear of persecution, to respond to the co-ordinated mass movement he had envisaged.

Büchner wrote his first play, *Dantons Tod* (q.v.), early in 1835. In it is reflected his detailed study of the French Revolution seen in the light of his recent experience and disillusionment. During his second stay in Strasburg he followed up the traces of J. M. R. Lenz (q.v.), whose works and aesthetic views he greatly admired, to the point of making him the subject of a Novelle, *Lenz* (q.v., posth., 1839). Büchner translated Victor Hugo's *Lucrèce Borgia* and *Marie Tudor* (*Lukretia Borgia* and *Maria Tudor*), and began *Woyzeck* (q.v.), which, though continued in Zurich, remained, like *Lenz*, a fragment. His final intentions in these writings are uncertain.

Of all Büchner's works only *Dantons Tod* (1835) was published in his lifetime. The comedy *Leonce und Lena* (q.v.), written for a competition for the best German comedy, appeared in 1838. In 1850 Georg's younger brother Ludwig Büchner (q.v.) published *Nachgelassene Werke*, but *Woyzeck* was only discovered and published in 1879 by K. E. Franzos (q.v.). *Leonce und Lena* was first performed in 1895, *Dantons Tod* in 1902, and *Woyzeck* in 1913. Büchner did not fully come into his own until the 20th c., although K. Gutzkow, F. Hebbel, and G. Hauptmann (qq.v.) had perceived his unique qualities. Hauptmann introduced Büchner's work to the literary club Durch (q.v.) in 1887. Büchner expressed his views on aesthetics and his strong opposition to German classicism, especially to Schiller, in the central section of his Novelle *Lenz*. Classical idealism was to him a distortion of reality, a view related to the *Anmerkungen übers Theater* (q.v.) by J. M. R. Lenz. Büchner sought to give an impression of reality by the interaction of a variety of stylistic devices. His appeal to modern attitudes lies also in his questioning scrutiny of all forms of convention and positive beliefs (see also PUPPEN-SPIEL). His reported last words suggest that he had abandoned his early atheistic attitudes.

Büchner's posthumous papers were first published by his brother Ludwig Büchner in 1850. Of great influence was the edition of *Sämtliche Werke*, published by K. E. Franzos in 1879. *Sämtliche Werke*, including correspondence, ed. F. Bergemann, appeared in 1922 (reissued 1965) and the *historisch-kritische Ausgabe*, ed. W. R. Lehmann (2 vols.), 1967-71, second extended edn., planned as 4 vols., 1974 ff. The Büchner Prize for literature was instituted by the city of Darmstadt in 1923.

BÜCHNER, LUDWIG (pseudonym Karl Ludwig) (Darmstadt, 1824–99, Darmstadt), a brother of Georg Büchner (q.v.), studied medicine in Gießen and Strasburg. In 1852 he was appointed to the University of Tübingen but was dismissed after the publication of his book *Kraft und Stoff* in 1855 because of the materialistic views it expressed. His other publications include *Natur und Geist* (1857), *Physiologische Bilder* (2 vols., 1861–75), *Lassalle und die Arbeiter* (1862), *Sechs Vorlesungen über die Darwinsche Theorie* (1868; as *Die Darwinsche Theorie*, 1890), *Die Stellung des Menschen in der Natur* (1870; as *Der Mensch und seine Stellung in der Natur*, 1889), *Der Gottesbegriff und dessen Bedeutung in der Gegenwart* (1874; as *Gott und die Wissenschaft*, 1894), *Über religiöse und wissenschaftliche Weltanschauung* (1887), *Darwinismus und Sozialismus* (1894), *Meine Begegnung mit Lassalle* (1894), and *Am Sterbelager des Jahrhunderts* (1898). In 1850 he published his brother Georg's *Nachgelassene Schriften*.

BÜCHNER, LUISE (Darmstadt, 1821–77, Darmstadt), sister of Georg and Ludwig Büchner (qq.v.), actively supported the emancipation of women and, with Princess Alice, daughter of Queen Victoria and subsequent Grand Duchess of Hesse-Darmstadt, was founder of the Alice-Verein. She wrote lyric poetry (*Frauenherz*, 1864) and prose works, some of which deal with the position of women in society and at work (*Die Frau und ihr Beruf*, 1855; *Weibliche Betrachtungen*, 1869; *Praktische Versuche zur Lösung der Frauenfrage*, 1870; and *Über weibliche Berufsarten*, 1871).

Büchse der Pandora, Die, a play in three acts by F. Wedekind (q.v.), written between 1892 and 1901, and published in 1904 (revised 1906). It is a sequel to *Erdgeist* (q.v.). Lulu, rescued from gaol by her Lesbian friend Countess Geschwitz, becomes a prostitute, narrowly escapes being sold by a white slave trader to Cairo, and takes refuge in London, where she and the Countess are murdered by Jack the Ripper. (See also BERG, A.)

Büchsenbuch, Das, an illustrated treatise on fire-arms, fortifications, and war material, written by Augustinus Dachsberg of Munich in 1443.

Buch von der deutschen Poeterey, Das, a small manual of poetics written in 1624 by M. Opitz (q.v.). It was written at high speed, for Opitz, apologizing for possible errors, declares that he first set pen to paper five days before (VIII. Capitel). After justifying poetry in general, he defends German poetry and defines its forms: epic ('ein heroisch Getichte'), tragedy, comedy, satire, epigram, eclogue,

elegy, echo, hymn, *silva*, and lyric, i.e. poetry for musical composition. The most important section deals with metre and rhyme, recognizing the Alexandrine and the *vers communs* (rhyming pentameters) as the most suitable forms, insisting on accuracy of accentuation and purity of rhyme, and emphasizing that the basis of German prosody is stress.

Buch von guter Speise, Das, see WÜRZBURGER KOCHBUCH.

Buch von Ordnung und Gesundheit, Das, a late medieval treatise on medicine. It was dedicated by its unknown author to Count and Countess Rudolf von Hochberg and printed in 1472.

Buddenbrooks, Verfall einer Familie, the first novel by Th. Mann (q.v.), begun in 1897 and published in 2 vols. in 1901, winning Mann instant acclaim as a major writer. It is the epic of four generations of a family of corn factors in a North German city which is easily recognizable as Lübeck, Mann's native town. Originally planned as a shorter work on the life of the last descendant of the family, Hanno (Justus Johann Kaspar), who dies at the age of 15, the narrative grew into a 19th-c. chronicle, opening in 1835 and closing a few years after the death of Hanno's father, Thomas, in 1875. In 1868 the family celebrates the centenary of the firm of Johann Buddenbrook, which for two generations has prospered as one of the city's most wealthy and respected family businesses, the members of which have served the municipality with distinction. But the generations degenerate biologically, declining from the robust and jovial Johann to the intellectual and artistic attitudes unwittingly imported into the next generation by Thomas's choice of a wealthy Dutch wife, Gerda Arnoldsen. The culmination is Hanno's withdrawal from life before he succumbs to typhoid fever.

The new stately home of Johann senior represents by its decorative style and the wealth of family silver the material prosperity of the upper middle class, the patrician 'Bürger', who is married to the elegant Madame Antoinette. Relationships inside the family are, however, clouded by Johann's treatment of Gotthold, his eldest son by an earlier marriage, who, until his father's death, lives with his wife and three daughters in Hamburg, disgraced by his marriage, which is socially unacceptable to the family. The business passes into the hands of the second son, Consul Johann, the eldest of Antoinette's three children, who is married to the gracious and personable Elisabeth Kröger. The burden of the business soon falls on the

Consul's eldest son, Thomas, whose dedicated efforts to uphold the family's respectability and fortunes are impeded by the contrasting character of his brother Christian and by the misfortunes of his sister Antonie (Tony). Tony marries twice, experiencing after her separation from Bendix Grünlich (by whom she has a daughter, Erika) the bitter life of a divorcee. She hopes to regain respectability by making a second marriage with Herr Permaneder, who is, surprisingly, uninterested in money-making. The family's lawyer has, however, to effect a second divorce on the grounds of incompatibility, for which Permaneder's forthright Bavarian behaviour is held responsible. When Tony's daughter Erika Grünlich marries Weinschenk, the director of a fire insurance company, Tony grasps the chance of running the household and so 'marrying' for a third time. Although Erika gives birth to a daughter, Elisabeth, Tony's illusions are shattered by Weinschenk's imprisonment for embezzlement. In 1872 Elisabeth Buddenbrook dies of pneumonia. Soon after the funeral the family house in the Mengstrasse is sold at a disappointing price. It is bought and modernized by a rival in business, Consul Hagenström, who prospers, while Thomas loses his grip on life and watches the family fortunes dwindle. Apart from certain economies in the new and still grandiose household there are, however, no outward signs of change. Yet Thomas's wife Gerda devotes herself more intensely to music, and her exaggerated awareness that Thomas has no sensitivity for art causes her to cultivate the extraordinary musicality of her son. From the age of 7 Hanno enjoys lessons from the organist Pfühl, and becomes acquainted with Beethoven's *Fidelio* and the music of Wagner. From the age of 11 Hanno has to reconcile his artistic aspirations and frail health with the strict demands of school life at the Realgymnasium—his father's unavoidable contribution to the education of his son and heir in business. School life is made tolerable by Hanno's friendship with Kai, the son of the impoverished Count Mölln. Senator Thomas Buddenbrook succeeds in concealing his deepening sense of futility beneath an immaculate appearance, even when he suffers agony from a decayed tooth so that for the first time in his life he leaves a session of the Senate in the city hall before the conclusion of its business. On his way home from the family dentist he suffers a stroke, falls headlong into the slush of the gutter, and dies soon afterwards. His death allows his brother Christian at last to marry Aline Puvogel who attracts him by her vulgarity, and by whom he has a child. But he soon has to retire into a mental home. It is now revealed that in his will Senator Thomas Buddenbrook

has decreed the liquidation of the family business in anticipation of his son's failure to live up to the responsibility. Hanno dies and Gerda returns to her father's musical home in Amsterdam.

The novel is rich in episodes of shifting perspectives employing at times pathetic, at times grotesque, irony to enhance the sub-title. Decadence, disease, death, and the wish for death are absorbed by a documentary style appropriate to conditions before and after the foundation of the German Empire in 1871. The novel has a strongly autobiographical imprint and shows especially the influence of Schopenhauer (q.v.).

BUFF, LOTTE (Wetzlar, 1753–1828, Hanover), whose full names were Charlotte Sophie Henriette, was the eldest of the sixteen children of Heinrich Adam Buff, the official of the Teutonic Order (see DEUTSCHER ORDEN) at Wetzlar. Her mother died after the birth of the last child, and Lotte took her place in caring for the family. On 9 June 1772 Goethe, who had not previously met her, called for her on the way to a ball, and was entranced both by her charm and her motherly attention to the children. Only after he had fallen in love with her did he discover that she was engaged to J. G. C. Kestner (q.v.). After months of mixed pleasure and torment he broke away, remaining in contact with the engaged couple for another two years through letters. The first part of Goethe's novel, *Die Leiden des jungen Werthers* (q.v.), published in 1774, makes considerable use of material derived from this relationship. Lotte was married in 1773 to Kestner and the couple lived at Hanover. In 1816 Lotte, a widow since 1800, visited Weimar and met Goethe again, an episode which is the nucleus of Thomas Mann's novel *Lotte in Weimar* (q.v.).

BUGENHAGEN, JOHANN (Wollin, Pomerania, 1485–1558, Wittenberg), was from 1525 to 1539 a professor at Wittenberg University. He became a close collaborator of Luther (q.v.), assisting him in the translation of the Bible; noted for his administrative ability in the organization of the new Church (in Brunswick, Hamburg, Lübeck, Pomerania, Denmark, and Schleswig-Holstein, 1528–42), he was appointed to a high administrative post (Generalsuperintendent) in Saxony in 1539. He preached at Luther's wedding and at his funeral.

Bühnenaussprache, a standard pronunciation, free of dialectal peculiarities, used on the German stage. It was established in 1898 and its rules are contained in *Deutsche Bühnenaussprache— Hochsprache* (1898), by Theodor Siebs (1862–1941). See GERMAN LANGUAGE, HISTORY OF.

BULLINGER, HEINRICH (Bremgarten, Aargau, 1504–75, Zurich), Swiss religious reformer, succeeded Zwingli (q.v.) at the Großmünster as chief pastor of Zurich in 1531. He had a hand in the First Swiss Confession (1536) and wrote the Second (1566). In 1549 he reached an agreement with Calvin on the Lord's Supper (*Consensus Tigurinus*). Bullinger is the author of many theological works, and of plays on Roman and Swiss history, of which *Lucrezia* (full title *Schön spil von der geschicht der Edlen Römerin Lucretiae*, 1526), performed in Basel in 1533, is the most important.

BÜLOW, BERNHARD, FÜRST VON (Klein Flottbek nr. Hamburg, 1849–1929, Rome), was chancellor of the German Empire from 1900 to 1909. Bülow was created Graf in 1899, Fürst in 1905. He served as an officer in the war of 1870–1 and then entered the diplomatic service (1874), holding appointments in Athens, Paris, St. Petersburg, and Bucharest, and was ambassador in Rome (1893). In 1897 he became foreign secretary and three years later succeeded Hohenlohe (q.v.) as chancellor. Bülow's policy of holding aloof from other powers (notably Great Britain) in order to secure better terms of agreement was a failure which was marked by the formation of the Entente Cordiale in 1904, by German isolation in the Moroccan crisis of 1905, and by the Anglo-Russian agreement of 1907. His handling of the *Daily Telegraph* affair (see DAILY TELEGRAPH-AFFÄRE), which was partly due to his negligence, alienated the Emperor, and led eventually to his fall. Skilful as a negotiator, he lacked vision, and has passed into history as an 'Epigone' (q.v.), who squandered Bismarck's political heritage.

BULTHAUPT, HEINRICH (Bremen, 1849–1905, Bremen), was from 1879 chief librarian of the Bremen municipal library. He established his reputation as a forceful analyst of drama in relation to stage performance, resuscitating Dramaturgie (in allusion to Lessing's *Hamburgische Dramaturgie*, q.v.) as a discipline in its own right. His *Dramaturgie der Classiker* (vol. 1, Lessing, Goethe, Schiller, H. von Kleist, 1881; vol. 2, Shakespeare, 1882) was revised and extended in subsequent editions, as *Dramaturgie des Schauspiels* (vol. 3, Grillparzer, Hebbel, Ludwig, Gutzkow, Laube, 1890; vol. 4, Ibsen, Wildenbruch, Sudermann, G. Hauptmann, qq.v., 1901). He was an early exponent of Ibsen and an outspoken critic of early Naturalism. A *Dramaturgie der Oper* (2 vols.) appeared in 1887, and a biography of C. Loewe (q.v.) in 1898. He is the author of a number of stories and Novellen, of a volume of poetry (*Durch Frost und Gluten*, 1876), and of plays, of which *Die Arbeiter* (1877)

stands out for its social theme; *Die Malteser* (1883) is based on a plan by Schiller; *Imogen* (1885) and *Timon von Athen* (1892) are modelled on Shakespeare. *Literarische Vorträge*, the fruit of his lecture tours, ed. H. Kraeger, and his letters were published posthumously in 1912.

BULTMANN, RUDOLF KARL (Wiefelstede nr. Oldenburg, 1884–1976, Marbach), a Protestant theologian and New Testament scholar of Marburg University (1921–51), noted for his 'form criticism' (see FORMGESCHICHTE), which he elaborated under the influence of M. Heidegger and K. Barth (qq.v.). His *Offenbarung und Heilsgeschichte* (1941) contains the essay 'Neues Testament und Theologie', with which he first introduced his 'demythologization' (Entmythologisierung) of the New Testament. He differentiates between historical New Testament studies and acceptance of faith on a symbolical level in order to objectivize the belief of the spiritual Resurrection of Christ through the Word. *Die Geschichte der synoptischen Tradition* appeared in 1921. Later writings include *Theologie des Neuen Testaments* (2 vols., 1953), *Geschichte und Eschatologie* (1964), and *Das Verhältnis der urchristlichen Christusbotschaft zum historischen Jesus* (1961).

Bundesfürsten, the rulers of the various lands which made up the German Empire (1871–1918). They comprised the kings of Prussia, Bavaria, Saxony, and Württemberg, the grand dukes of Mecklenburg-Schwerin, Mecklenburg-Strelitz, Baden, Hesse, Oldenburg, and Sachsen-Weimar, the dukes of Brunswick, Sachsen-Meiningen, Sachsen-Altenburg, Sachsen-Koburg-Gotha, and Anhalt, and the princes (Fürsten) of Schwarzburg-Rudolstadt, Schwarzburg-Sondershausen, Waldeck, Reuß ältere Linie, Reuß jüngere Linie, Lippe, and Schaumburg-Lippe.

Bundeslied, see SIND WIR VEREINT ZUR GUTEN STUNDE.

Bundesrepublik Deutschland, BRD (Federal Republic of Germany) owes its origin to the divisions between the three western occupying powers on the one hand and the Soviet Union on the other, which culminated in the walk-out of the Russian representative from the Control Council (Kontrollrat) on 20 March 1948. On 1 July 1948 an assembly was convened to draw up a constitution. This 'Verfassungskonvent' sat in August and created the Parliamentary Council (Parlamentarischer Rat), which adopted the basic constitutional law (see GRUNDGESETZ) on 8 May 1949. It became effective on 23 May.

The head of state is the Federal President (Bundespräsident), elected for five years and eligible for a second term, but the political direction belongs, not to the President, but to a cabinet under the Federal Chancellor (Bundes-kanzler) representing a majority of the members of the elected single-chamber parliament (Bundestag). Election is by a complex system of direct and proportional votes designed to eliminate the small splinter groups which rendered the governments of the Weimar Republic (q.v.) unstable. The limitation of presidential powers is also intended to prevent rule by decree as it occurred in the Hindenburg (q.v.) era. The Federal Government controls foreign policy, defence (Bundeswehr, instituted 1956), railways, post and federal finance, but not police or education (including the universities), both of which pertain to the Länder.

The federal presidents have been Th. Heuß (q.v., 1949–59), H. Lübke (1959–69), G. Hei-nemann (1969–74), W. Scheel (1974–79), K. Carstens (1979–84), and R. von Weizsäcker (1984–   ). The chancellor from 1949 to 1963 was K. Adenauer (q.v.), 1963–6 L. Erhard, 1966–9 K. G. Kiesinger, 1969–74 W. Brandt, 1974–83 H. Schmidt, and 1983– H. Kohl. The Federal Republic joined NATO in 1954 and was one of the six original members of the European Economic Community (Europäische Wirtschaftsgemeinschaft), 1957. In 1973 a treaty of mutual recognition came into force between the Federal Republic and the German Democratic Republic (see DEUTSCHE DEMO-KRATISCHE REPUBLIK).

The Federal Republic is composed of West Berlin (see BERLIN) and 10 Länder (Baden-Württemberg, Bayern, Bremen, Hamburg, Hessen, Niedersachsen, Nordrhein-Westfalen, Rheinland-Pfalz, Saarland—since 1957—and Schleswig-Holstein, qq.v.). The capital (origin-ally 'provisional') is Bonn (q.v.).

Bundschuh denotes the crude shoe of the medieval peasant. It became a symbol for the peasant against the nobility in the rural unrest of the late 15th c. and was specifically applied as a name to a revolt in Alsace in 1493 and then to further revolts in Bruchsal in 1502 and in Breis-gau in 1513.

BUNSEN, ROBERT WILHELM (Göttingen, 1811–99, Heidelberg), the distinguished chemist, who occupied chairs successively at Marburg (1838), Breslau (1851), and Heidelberg (1852), retiring in 1889. With G. R. Kirchhoff (1824–87) he developed spectrum analysis. He invented many appliances, and is most widely known for the Bunsen cell (1841) and, above all, the Bunsen burner (1855).

Bunte Steine, a collection of six stories by A. Stifter (q.v.), published in 1853. Each story is named after a variety of stone or rock (of which Stifter was a keen collector): *Granit, Kalkstein, Turmalin, Bergkristall, Katzensilber, Bergmilch.* The volume was dedicated to his deceased friend Gustav Scheibert. Provoked by criticism that his work indulged in the treatment of 'small' and insignificant subject-matter (das Kleine), Stifter composed, in defence, a Vorrede to *Bunte Steine* containing the principles of his aesthetics. He bases the evaluation of 'great' and 'small' on the observation that seemingly 'great' occurrences are merely the *isolated* manifestation of a universal force ('Kraft') and are therefore less significant than the *general* manifestation of this force, which is inconspicuous, but constant and effective. Man should search for this 'gentle law' which applies to human existence as well as to nature: 'Wir wollen das sanfte Gesetz zu erblicken suchen, wodurch das menschliche Geschlecht geleitet wird.' This 'gentle law' expresses itself through man's urge to maintain a state of harmony and balance. It asserts itself through love in the small circle of the family and in the common striving for order and stability in society and states. It is, Stifter claims, the force underlying the evolution of mankind. The ethics of the 'gentle law' presuppose the prevalence of justice, mutual respect, and selflessness in man, the ultimate triumph of good over evil. There is an interesting complement to this theory in the descriptive essay *Die Sonnenfinsternis am 8. Juli 1842.*

BURCKHARD, MAX (Korneuburg, 1854–1912, Vienna), director of the Burgtheater, Vienna, from 1890 to 1898. A lawyer by pro-fession, he wrote a treatise on Austrian civil law in 1886, and became a civil servant. He raised the level of acting at the Burgtheater, and added modern drama (Ibsen, Hauptmann, Suder-mann, Schnitzler, qq.v.) to the repertoire. He was himself a dramatist, writing the comedies *Die Bürgermeisterwahl* (1898), *Die verflixten Frauenzimmer* (1902), and *Jene Asra* (1909). A genial and ironical personality, he is the original of Hofrat Dr. Winckler in Schnitzler's play *Pro-fessor Bernhardi* (q.v.).

BURCKHARDT, CARL JACOB (Basel, 1891–1974, Geneva), entered the Swiss diplomatic service but later took up an academic career, and in 1932 was invited to Geneva as professor of history. In 1937 he was entrusted with the difficult office of League of Nations high com-missioner in Danzig, an appointment which was terminated by the outbreak of war in 1939. From 1944 to 1948 he was president of the International Red Cross. As a young diplomat in Vienna in the 1920s he was a friend of H. von Hofmannsthal (q.v.). He is the author of *Erinnerungen an Hofmannsthal* (1948); the correspondence of the two men was published in 1956. His principal historical work is *Richelieu* (3 vols., 1935–66), which he completed at the age of 75. He published his *Werke* (6 vols.) in 1971. They include fiction, essays, and some of his vast correspondence stretching over more than six decades.

BURCKHARDT, JACOB (Basel, 1818–97, Basel), after studying theology for a time at Basel, transferred his interest to history and the history of art, which he studied at Berlin University under L. von Ranke (q.v.) and subsequently at Bonn. Visits to Italy focused his attention on the Renaissance period. He was appointed to a professorial chair at Zurich University in 1855, and three years later at Basel, where he remained for 35 years, retiring in 1893. Burckhardt's published work was concentrated into a comparatively short span of his professional career (1853–67), but his influence extended well into the 20th c. His principal works are *Die Zeit Konstantins des Großen* (1853), *Cicerone* (1855, an erudite guide to the art of Italy), *Die Kultur der Renaissance in Italien* (1860), which constitutes his outstanding achievement, and *Die Geschichte der Renaissance* (1867). Burckhardt strongly influenced the pessimism of the late 19th c. and early 20th c., seeing the modern world as undergoing a retrogression in cultural and spiritual values, and condemning, as a detached Swiss observer, the pursuit of power in the Bismarckian age. He gave expression to his views in numerous public lectures.

He left the MSS. of two works, which were published soon after his death: *Griechische Kulturgeschichte* (4 vols., 1898–1902) and *Weltgeschichtliche Betrachtungen* (1905). Burckhardt's *Gesammelte Werke* (10 vols.) appeared 1955–9.

BÜRGER, GOTTFRIED AUGUST (Molmerswende, Harz, 1747–94, Göttingen), a son of the manse, studied in Halle where he associated with the dissolute Professor C. A. Klotz (q.v.). In 1768 he migrated to Göttingen. In spite of his leanings to dissipation he was appointed magistrate at Altengleichen near Göttingen in 1772. He came into contact with the Göttinger Hainbund (q.v.) at this time, and in 1773 wrote his vivid popular ballad *Lenore* (q.v.), published in the *Göttinger Musenalmanach für 1774*. In 1774 he married Dorette Leonhart (q.v.), only to discover almost immediately that he preferred her sister, Auguste, who is the 'Molly' of his poems, of which the first collection (*Gedichte*) appeared in 1778. Bürger, who neglected his official duties

and got into financial difficulties, resigned in 1784 and became a lecturer at Göttingen. His wife died shortly afterwards and Bürger then married Auguste, who, however, died in childbirth in 1786. *Münchhausen* (see MÜNCHHAUSEN, K. F. H. VON), the grotesquely absurd fictitious adventures of an irrepressibly inventive liar, which was only in part Bürger's work (see RASPE, R. E.), appeared in the same year and was expanded in 1788. Bürger became a professor at Göttingen in 1789. His third bride was Elise Hahn (q.v.), but this marriage in 1790 was a catastrophe, and ended in divorce in 1792. His collected poems, published in 2 vols. in 1789, were savagely reviewed by Schiller in 1791 (*Über Bürgers Gedichte*). The double sensation of being discredited by the scandal of his third marriage and by Schiller's depreciation of his poems is thought to have contributed to his early death.

Bürger had a spontaneous and fluent talent which is apparent both in the originality of his ballads and in the successful exploitation of personal experience in his lyric poems. The judgement of his work has perhaps been too often coloured by the failures of his life. *Sämmtliche Schriften* (4 vols.), ed. K. Reinhard, were published 1796–1802 (reissued 1970). See also BLÜMCHEN WUNDERHOLD, DAS, and DES PFARRERS TOCHTER VON TAUBENHAIN.

BÜRGER, HUGO, pseudonym of H. Lubliner (q.v.).

Bürgergeneral, Der, a one-act comedy written by Goethe in 1793 and published in the same year. A good-for-nothing (Schnaps) masquerades as a general of the coming revolution in order to scare a rich peasant (Märten) into giving him a meal. Schnaps's pretensions are exposed. The play, which had a poor reception, was first performed in Weimar on 2 May 1793.

Bürgerliches Gesetzbuch (abbreviated BGB), the German code of civil law adopted in 1896.

Bürgerliches Trauerspiel, a term originating in the middle of the 18th c. to denote tragedy that is set in a 'bürgerlich', i.e. non-noble, environment and is written in prose. Lessing (q.v.) expressed both practical and theoretical dissent from Aristotle's assertion that only kings and princes are suitable subjects for tragedy. His *Miß Sara Sampson* (q.v., 1755), though called on the title-page 'Ein Trauerspiel', was revolutionary in German literature, both in its middle-class personae and its use of prose. It was influenced by G. Lillo's play *The London Merchant* (1731). *Emilia Galotti* (q.v., 1772) is Lessing's best example of the form, though in this tragedy he adheres to the simple term 'Trauerspiel'. In the *Hamburgische Dramaturgie* (q.v.) Lessing interpreted Aristotle's catharsis in terms which made 'Bürger' suitable subjects for audiences composed of 'Bürger'. That his 'Bürger' are well-to-do is immaterial, and he influenced, particularly through *Emilia Galotti*, those writers of Sturm und Drang (q.v.) who concentrated on the social problems of their own contemporary society. The term 'Bürgerliches Trauerspiel' is explicitly used by Schiller on the title-page of *Kabale und Liebe* (q.v., 1784), and on that of *Maria Magdalene* (q.v., 1844) by F. Hebbel, who reassessed the term, which in essence, if not in name, retains a place in modern social drama.

Bürgerlich und Romantisch, a four-act comedy by E. von Bauernfeld (q.v.), first performed at the Burgtheater, Vienna, in September 1835. It was first published in 1839 in the *Almanach für's Lustspiel*, ed. J. C. Zedlitz (q.v.), and then in *Gesammelte Schriften* (1871–3). The action takes place at a spa. One pair of lovers (Sittig and Cäcilie) fall out and are reunited; another pair (Baron von Ringelstern and Katharina) meet under misapprehensions, and are happily brought together. The noble pair count as the Romantics, the others as the representatives of the burgher's outlook, but all four declare their hostility to a narrow and petty (spießbürgerlich) outlook. This conventional and agreeable drawing-room comedy was one of Bauernfeld's greatest successes, surviving on the stage into the 20th c.

Bürger Schippel, a comedy by C. Sternheim (q.v.), written in 1912 and first performed in 1913 (Deutsches Theater, Berlin). Three amateur singers, all 'Bürger', the goldsmith Hicketier, the civil servant Krey ('fürstlicher Beamter'), and the owner of a printing firm, Wolke, have to replace the fourth member of their quartet after his sudden death before a contest in which they hope to win the 'Siegeskranz' for the third time running. Reluctantly they secure the co-operation of the workman Paul Schippel, whose outstanding voice brings them victory. Hicketier tries to marry his sister Thekla to Schippel after finding that she has had an affair with the ruling prince. Schippel rejects her out of pride and insults Krey who has won her hand instead. In a duel in defence of Thekla's honour, Schippel slightly wounds Krey, and is accepted by Hicketier as 'Bürger'. The comedy makes fun of 'Bürger' and workman alike, for while it exposes the snobbery of the 'Bürger', it portrays in Schippel one who hates the middle class and yet in the end naïvely delights in his 'promotion' ('Du bist Bürger, Paul').

Bürger von Calais, Die, a play by G. Kaiser (q.v.), described as 'Bühnenspiel in drei Akten', published in 1914 and first performed at the Neues Theater, Frankfurt, in January 1917. The work exists in three versions: (1) original MS., first published 1958; (2) the printed edition of 1914, MS. not extant; (3) a later as yet unpublished MS. The differences, though of interest to scholars, are not substantial. The source of the story is the well-known episode from the Hundred Years War, recounted by Froissart, according to which Edward III of England, besieging Calais in 1347, imposed as his terms for its surrender without sack and pillage the handing over of six citizens, each clad only in a shirt and with a rope round his neck. The action of Kaiser's play is broadly based on this incident, which afforded ample opportunity for the expression of his detestation of war and of man's inhumanity to man. He introduces, however, a variant of his own. The military governor who plans resistance to the uttermost is opposed by Eustache de Saint-Pierre (the name is invented by Kaiser), who urges the saving of life and offers himself as the first hostage. Six volunteers come forward, making the total one more than is necessary. The casting of lots is proposed. Eustache de Saint-Pierre does not appear when the six are gathered together, and murmurs of betrayal are heard among the crowd. In Eustache de Saint-Pierre's place his father appears, accompanying a bier. He removes the pall to show his dead son, who has taken his own life as an example of self-sacrifice to the rest. Written in an exalted style, but without the extremes of ecstasy or hysteria often found in Expressionist works, the play is generally acknowledged to be Kaiser's masterpiece.

Burggraf, the governor of a castle or fortified town in the Middle Ages, subsequently used as a title. The best-known Burggrafen were the Hohenzollerns, Burggrafen of Nürnberg, 1191–1427.

BURGGRAF, WALDFRIED, see FORSTER, F.

BURGKMAIR, HANS (Augsburg, 1473–1531, Augsburg), German artist, famous for his paintings and woodcuts. A friend of Dürer (q.v.), he studied in Strasburg and Venice, as well as Augsburg. He executed the woodcuts for *Weißkunig* (q.v.) and some of those in *Teuerdank* (q.v.). His greatest works are the altar-pieces in Munich and Augsburg.

Bürgschaft, Die, a ballad written by Schiller in 1798 and published in the *Musenalmanach auf das Jahr 1797*. It tells a story of fidelity triumphant. The friend of Möros (in later texts Damon), who has attempted to assassinate the tyrant Dionysius, stands hostage for the condemned youth while the latter visits his sister. As Möros returns he faces a concatenation of human and natural obstacles, but presses on and in the nick of time arrives to save his friend from execution in his stead. The tyrant is so impressed that he pardons Möros and offers friendship—

'Ich sei, gewährt mir die Bitte,
In eurem Bunde der Dritte.'

Burgtheater, or k.k. Theater nächst der Burg, and until 1918 also Hofburgtheater, the principal theatre of Vienna, and one of the outstanding theatres of the German world. The Burgtheater was founded in 1776 by the Emperor Joseph II (q.v.) as a national theatre. Its first home was the 'Ballhaus'; in 1888 it moved into a new and much larger theatre on the Franzensring. Severely damaged in 1944, it has been reconstructed. The Burgtheater, which achieved its great reputation in the middle of the 19th c., developed a stylized mode of diction (Burgtheaterstil). Among its more famous actors were H. Anschütz (q.v.), Christine Enghaus (q.v.), Sophie Schröder (q.v.), Charlotte Wolter, A. Sonnenthal, F. Mitterwurzer, J. Kainz, H. Thimig (q.v.), and Paula Wessely. The outstanding directors were J. Schreyvogel (1814–32), H. Laube (1849–67), F. Dingelstedt (1870–81), A. Wilbrandt (1881–7), M. Burckhard (1890–8, qq.v.), and H. Thimig (1912–17).

BURKHARDT, Georg, see Spalatin, Georg.

BURKHART VON HOHENFELS, a Middle High German Minnesänger, was a nobleman from the neighbourhood of Überlingen in Swabia, who is recorded between 1212 and 1242. He was in the service of the Emperor Friedrich II (q.v.), then of Heinrich VII (q.v.), and finally of the Bishop of Constance. Burkhart, 18 of whose songs survive, belongs to the generation following Walther von der Vogelweide (q.v.). Some of his poems lean towards allegory in presentation and mannerism in style, and frequently employ images drawn from hunting. His fresh and attractive dance songs are pastoral in character.

BURK MANGOLD, a squire in the service of Hugo von Montfort (q.v.), who composed the melodies for some of Hugo's poems.

BURMESTER, Heinrich (Niendorf/Lauenburg, 1839–89, Boitzenburg), wrote Low German (plattdeutsch) poetry: *Arm und Riek* (1872), *Schulmester Klein* (1873), *Landstimmen* (1881). Extreme poverty drove him to drown himself in the Elbe.

Burschenschaft, a term originally (*c.* 1790) applied to the student body at a university. From 1814 it was adopted for a student movement which grew out of the Wars of Liberation (see Napoleonic Wars). The Jena Burschenschaft was founded on 12 June 1815 and similar bodies followed at other universities; the Wartburgfest (q.v.) of 18 October 1817 was the symbol for the union of all these societies in one national body. The Burschenschaft was from the outset hostile to the reactionary policy pursued by many German heads of state, and desired the political unity of Germany. After the Wartburgfest the governments regarded it with disfavour, and the proclamation of an Allgemeine Deutsche Burschenschaft at Jena, followed by the assassination of A. von Kotzebue (q.v.) in 1819 by the Burschenschaft member K. L. Sand (q.v.), led to a sharp reaction embodied in the Karlsbad Decrees (September 1819, see Karlsbader Beschlüsse); as a result, the Burschenschaft was banned on 26 November 1819 and denounced as a 'demagogic movement'. The local Burschenschaft continued to meet clandestinely in many places, and the trend of the movement became more radical. After the Hambacher Fest (q.v., May 1832) an attack by students on the police headquarters at Frankfurt led to a wave of arrests all over Germany (which G. Büchner, q.v., escaped by flight) and to severe sentences. Among those so sentenced was F. Reuter (q.v.). Though students continued to be politically active in the forties, the Burschenschaften as such were quiescent, even though many of the politicians in the Frankfurt Parliament of 1848 were former Burschenschafter. In the second half of the century the Burschenschaft (officially united as Deutsche Burschenschaft in 1881) developed into a group of social clubs (see Studentische Verbindungen) of nationalistic and, latterly, anti-Semitic character. A rival and somewhat more liberal organization, the Allgemeiner Deutscher Burschenbund, also took a similar course and by 1924 both federations demanded proof of 'Aryan' descent. Under the National Socialist regime the two bodies were virtually fused in 1934 and dissolved in 1935. A new Burschenschaft was founded in 1950.

BÜRSTENBINDER, Elisabeth, wrote under the pseudonym E. Werner (q.v.).

BURTE, Hermann, pseudonym of Hermann Strübe (Maulburg, Baden, 1879–1960, Lörrach), a painter and the author of poems, stories, and plays. His novel *Wiltfeber, der ewige Deutsche* (1912) established his reputation as a writer, and the play *Katte* (1914) was his greatest stage success (see Katte, H. H. von). His other plays include *Der kranke König* (three one-act plays,

1907), *Herzog Utz* (1913), *Simson* (1917), *Der letzte Zeuge* (1921), *Apollon und Kassandra* (1926), *Krist vor Gericht* (1926), *Prometheus* (1932), *Warbeck* (1935). His stories include *Mit Rathenau am Oberrhein* (1925) and *Der besiegte Lurch* (1933). His lyric poetry comprises the following collections of sonnets and poems: *Patricia* (1910), *Flügelspielerin* (1915), *Madlee* (1923), *Ursula* (1930), *Anker am Rhein* (1937), *Die Seele des Maien* (1950).

Busant, Der, an anonymous Middle High German tale written early in the 14th c. Its theme is the love of the son of the King of England, who has been sent to Paris to study, for the daughter of the King of France. Since she is threatened with a diplomatic marriage, the couple elope. While they are in the forest a hawk (*Busant*) steals her ring. The prince pursues the hawk, loses his way, fails to find the princess again, and falls into madness. The story ends with their reunion after his recovery. The dialect of the poem is Alsatian.

BUSCH, WILHELM (Wiedensahl, Hanover 1832–1908, Mechtshausen, Harz), German artist and comic poet, was first trained as an engineer but at the age of 19 changed over to the study of art, which he pursued first at Düsseldorf, then at Antwerp. He contributed to weeklies in Munich from 1859 onward. In 1864 he returned to his birthplace, where he remained for thirty-five years, during which he published all the illustrated verse tales which made him famous. His speciality is pungent satire in humorous doggerel verse accompanied by large numbers of drastic and grotesque line drawings done with great verve and firmness. The first (and most famous) of these is *Max und Moritz* (1865), the two 'böse Buben'. It was followed by *Der Heilige Antonius von Padua* (1870), *Hans Huckebein* (1870) the raven, *Pater Filucius* (1872) an allegory of the Kulturkampf (q.v.), *Die fromme Helene* (1872), *Kritik des Herzens* (1874), *Abenteuer eines Junggesellen* (1875) with its sequels *Herr und Frau Knopp* (1876) and *Julchen* (1877), *Fipps der Affe* (1879), *Plisch und Plum* (1882), *Balduin Bählamm der verhinderte Dichter* (1883), and *Maler Klecksel* (1884). Much of Busch's satire is directed against the complacency of the older Germany, but in 1870 and the immediately following years there is a perceptible anti-Catholic element as well as a streak of Gallophobia. *Sämtliche Werke* (6 vols.), ed. O. Nöldeke, appeared in 1943 (3rd edn. 1955) and (2 vols.), ed. R. Hochhuth (q.v.), in 1959, and *Sämtliche Briefe. Kommentierte Ausgabe* (2 vols.), ed. F. Bohne, in 1968–9.

BUSSE, CARL (Lindenstadt in the former Prussian province Posen, 1872–1918, Berlin), was for a short time in his youth a journalist in Augsburg, then studied German literature at Berlin and Rostock universities, settling afterwards in Berlin. He published his poetry, which is deliberately popular in character but saved from banality by a finely tempered selective observation and sensitive response, in several collections (*Gedichte*, 1892; *Neue Gedichte*, 1896; *Vagabunden*, 1901; *Heilige Not*, 1910). He has often been compared with Liliencron, and his Impressionistic poem *Rote Husaren* has found its way into several anthologies. He was a successful regional novelist with *Jugendstürme* (1896), *Höhenfrost* (2 vols., 1897), *Jadwiga* (2 vols., 1899), *Die Referendarin* (2 vols., 1906), *Das Gymnasium zu Lengowo* (2 vols., 1907), and *Lena Küppers* (2 vols., 1910). He is also the author of several volumes of stories (*In junger Sonne*, 1892; *Stille Geschichten*, 1894; *Träume*, 1895; *Die Schüler von Polajewo*, 1901; *Im polnischen Wind*, 1906; *Flugbeute*, 1914; *Sturmvogel*, 1917). His narratives are mostly set in the eastern territory in which he grew up. Busse occasionally used the pseudonym Fritz Döhring.

BUSSE, HERMANN ERIS (Freiburg/Breisgau, 1891–1947, Freiburg), regional novelist and folklorist of Baden, is the author of the novels *Peter Brunnkant* (1927) and *Bauernadel* (1933), which is a trilogy composed of *Das schlafende Feuer, Markus und Sixta* (both 1929), and *Der letzte Bauer* (1930).

BUSTA, CHRISTINE (Vienna, 1915–), municipal librarian in Vienna, is the author of several collections of poetry which reflect sincere piety, deep compassion, and a feeling for nature (*Jahr um Jahr*, 1950; *Der Regenbaum*, 1951; *Lampe und Delphin*, 1955; *Die Scheune der Vögel*, 1958; *Die Sternenmühle*, 1959; *Unterwegs zu älteren Feuern*, 1965). Her language is traditional, showing indebtedness to the Bible as well as to the world of nature.

Büste des Kaisers, Die, a short tale by J. Roth (q.v.), first published in a French version as *Le Buste de l'Empereur* in 1934 and in 1964 in the original German. It is the touching and ironic story of Graf Morstin, a middle-aged eccentric, in whose country-house the Emperor Franz Joseph (q.v.) had once spent a week during summer manœuvres. This episode and Graf Morstin's loyalty are commemorated by a bust of the Emperor made by a local boy. The war begins and ends, and with it ends the Habsburg era. Graf Morstin, who is now resident in republican Poland, cannot understand the new world. He has the bust put in the cellar, and goes abroad to find his old world. In a cabaret in respectable Zurich he is shocked to see the memory of the Empire insulted. He returns home, sets up the bust, and puts on his old uniform as an Austrian captain of dragoons. None of the local peasants takes the slightest notice, but before long a government official demands that the bust be removed. Graf Morstin has it solemnly buried in the churchyard and goes abroad to end his days on the Riviera.

BÜTNER or **BÜTTNER,** WOLFGANG, pastor at Wolferstedt, Mansfeld, collected and published in 1572 a number of anecdotes concerning Claus Narr, a court jester and practical joker who lived in the late 15th c. and early 16th c.

BUTSCHKY, SAMUEL VON (1612–78), a publicist active in Breslau, was brought up as a Protestant and was converted to Roman Catholicism when in his forties. He is the author of a manual of correspondence, and of moral writings (*Pathmos*, 1677, and *Das wohlgebaute Rosen-Tal*, 1679); he also translated Seneca (1666).

Büttnerbauer, Der, a novel by W. von Polenz (q.v.), published in 1895. It tells, in Naturalistic style, of the ruin of a respected farmer, who through his father's intestacy finds himself in financial difficulties, and is driven to have recourse to money-lenders. His farm is sold up and he hangs himself. The rural sections of the book are more convincing than the urban scenes. The novel is in part an indictment of absentee landlords.

Butzenscheibenpoesie (or Butzenscheibenlyrik), derogatory designation given to a type of poetry of the late 19th c., which evinces a degenerate Romanticism. Its principal characteristics are its addiction to historical (primarily medieval) themes, its crude psychology, its nationalistic values, and its pseudo-medieval style. Its chief progenitor is J. V. von Scheffel (q.v.) and its other principal exponents are R. Baumbach and J. Wolff (qq.v.). The 'Butzenscheibe' is an old-fashioned target for archery and the term was coined by P. Heyse (q.v.) in 1884 in a verse prologue to an edition of poems by E. Geibel (q.v.).

BUTZER, MARTIN, see BUCER, MARTIN.

BUWENBURG, DER VON, a minor Middle High German poet, who probably lived in Swabia in the late 13th c. and early 14th c., is the author of six Minnelieder characterized by realistic images. Though he is thought to have been influenced by Steinmar, he evolved a style of his own.

C

Cabanis, a historical novel in 6 vols. by W. Alexis (q.v.), published in 1832. It is the first of his patriotic (vaterländisch) novels and is set in the time of Friedrich II (q.v.) of Prussia.

CAGLIOSTRO, ALESSANDRO CONTE DI, the assumed name of Giuseppe Balsamo (Palermo, 1743–95, San Leone nr. Urbino). Cagliostro conducted alchemical experiments and was a skilled and enterprising propagandist. He early established a scientific reputation by experiments and remarkable 'cures', and was invited to visit courts and great cities throughout Europe. Though expelled from St. Petersburg, he was successful in Warsaw and Strasburg, visited London, and established himself in Paris. Cagliostro was deeply implicated in the scandal of the queen's necklace in Paris, which discredited the innocent Marie Antoinette (q.v.); he was imprisoned in the Bastille and subsequently deported. Though he was less active in Germany than in some other European coun-

tries, Cagliostro's bold deceptions attracted much attention there, as is attested by Schiller's *Der Geisterseher* (1789, q.v.) and Goethe's *Der Groß-Cophta* (1791, q.v.), in which he appears as Conte Rostro. He was made the central figure in an operetta by Johann Strauß (q.v., *Cagliostro in Wien*, 1875).

CALAMINUS, GEORG (Silberberg/Silesia, 1545–95, Linz/Danube), German schoolmaster and humanist, wrote Latin plays for school performance, first at Strasburg and later at Linz in Austria. *Carmius sive Messias in praesepi* (1576) combines biblical and classical elements; *Helis* (1591) praises Senecan stoicism; and *Rudolphottocarus* (1594) portrays, from the Habsburg standpoint, the defeat of Ottokar of Bohemia by Rudolf I of Habsburg (qq.v.).

CALÉ, WALTER (Berlin, 1881–1904, Freiburg/ Breisgau), a student who wrote much poetry before committing suicide. He destroyed all his

MSS., but copies of some of them, which were in the possession of friends, were collected and published as *Nachgelassene Schriften* (1907).

Calvinism, a Protestant persuasion based on the views of J. Calvin (1509–64), which became influential during the Counter-Reformation (see GEGENREFORMATION). The rulers of the Palatinate and of Brandenburg turned Calvinist and these two states became places of refuge for the Huguenots (see HUGENOTTEN). Calvinism was the most puritanical and also the best organized of the Reformation movements, but it was not recognized until the Peace of Westphalia (see WESTFÄLISCHER FRIEDE).

Cambrai, Friede von, treaty agreed in 1529 between Margaret of Austria (see MARGARETE VON ÖSTERREICH), representing the Emperor Karl V (q.v.), and Louise of Savoy, mother of Francis I of France. By its terms France abandoned claims to Flanders, Artois, Milan, and Naples, and Karl gave up his right to Burgundy (which by the Treaty of Madrid in 1526 should have been surrendered to him). Because its principal signatories were both women the Peace of Cambrai is commonly called the 'Damenfriede' or 'Ladies' Peace'.

Cambridger Liederhandschrift (The Cambridge Songs), a collection of Latin songs made originally in the Rhineland and preserved in the Cambridge University Library. Of its 47 songs, two (*de Heinrico* and that usually called *Kleriker und Nonne*, qq.v.) are partly in Old High German. It also contains the Latin prayer *Carmen ad Deum* (q.v.). The MS. was written in the 11th c. and the songs date from the first half of that century and the late 10th c. *Die Cambridger Lieder, Carmina cantabrigiensa* were edited by K. Strecker, 1926.

CAMERARIUS, JOACHIM (Bamberg, 1500–74, Leipzig), a classicist and humanist, whose name is a latinized form of Kämmerer. In 1535 he was entrusted with the reorganization of Tübingen University, and in 1542 with that of Leipzig. He was a friend of Melanchthon (q.v.) and his biographer (1566). He is also the author of a life of the humanist E. Hessus (q.v., 1533). He is known for his Latin correspondence *Epistolae familiares* (1583). Camerarius edited many classical texts, of which his Plautus is the most noteworthy.

Campagne in Frankreich. 1792, an autobiographical work by J. W. Goethe (q.v.), written 1820–1, and published in 1822 as Part Five of *Aus meinem Leben* (see DICHTUNG UND WAHRHEIT). The present title dates from 1829 (*Ausgabe*

letzter Hand, q.v.). Goethe describes, in the form of a journal, the invasion of northern France by the imperial army, as he himself saw and experienced it. The journal begins at Mainz on 23 August, reaches the deepest point of invasion at Valmy on 19 September, and recounts the retreat which began on 29 September and ended at Trier on 23 October. It adds, after the conclusion of the campaign, the record of Goethe's visits in November to the Jacobis (see JACOBI, F. H. and J. G.) at Pempelfort, to Plessing at Duisburg, and to the Fürstin Galizin at Münster (q.v.). The account of the cannonade of Valmy includes Goethe's prophetic pronouncement, made, he states, to a group of comrades after the cessation of fire: 'Von hier und heute geht eine neue Epoche der Weltgeschichte aus, und ihr könnt sagen, ihr seid dabei gewesen.'

CAMPE, JOHANN HEINRICH (Deensen nr. Brunswick, 1746–1818, Brunswick), an educationist, studied theology, was employed as a tutor by the Humboldt family at Tegel (see HUMBOLDT, ALEXANDER VON and WILHELM VON), and then for a short time assisted J. B. Basedow (q.v.) at the Philanthropinum. Campe spent his life in the organization and administration of schools and in writing. He was sympathetically disposed to the French Revolution (q.v.) and visited Paris in 1789.

Campe wrote books for boys and girls as well as books on education. His *Sämtliche Kinder- und Jugendschriften*, the best known of which is an adaptation of *Robinson Crusoe* (*Robinson der Jüngere*, 2 vols., 1779–80), appeared in 37 vols., 1807 ff. The collection also included *Die Entdeckung von Amerika* (2 vols., 1780–1), another of his successes, and the didactic *Theophron. Ein Ratgeber für die Jugend* (1777). Campe edited an educational encyclopedia (*Allgemeine Revision des gesamten Schul- und Erziehungswesens*, 16 vols., 1785–92) and compiled a dictionary (*Wörterbuch der deutschen Sprache*, 5 vols., 1801–11). His poetic work, which is of minor importance, includes the satires *Das Testament* (1766) and *Satyren* (1768), and an epic poem *Der Candidat* (1769).

CANETTI, ELIAS (Rustchuk, Bulgaria, 1905–), descended from Spanish Jews, emigrated with his parents to Manchester, where he lost his father in the year of their arrival (1911). Having attended an English school, he moved with his mother to Lausanne, began to acquire the German language in preparation for his education in Vienna (from 1913), Zurich (from 1916), and Frankfurt (1921–4). In 1929 he completed his studies of chemistry at Vienna University with a doctorate. A decisive influence in Vienna, his 'literary home', was Karl Kraus (q.v.). He turned to writing, spent periods in Berlin, where he met H. Broch (q.v.), and Paris (with his mother, whose death in 1937 marked the end of a striking relationship). In 1938 he emigrated via Paris to London where he settled. His novel *Die Blendung* (q.v., 1936, *Auto da Fé*, transl. by C. V. Wedgwood, 1946), long virtually ignored, was republished in extended form in 1965. It shows an extreme individualist (and intellectual) at odds with the world around him, and this relationship of the individual to the mass and its relevance to fascism is the chief concern of Canetti's *œuvre* and the subject of a pervasive anthropological analysis in *Masse und Macht* (1960, 3rd edn. 1980, *Crowds and Power*, transl. by C. Stewart, 1962). He wrote three plays, *Hochzeit* (1932), *Komödie der Eitelkeit* (1950), and *Die Befristeten* (1964). Canetti's *Aufzeichnungen 1942–1948* and *Alle vergeudete Verehrung. Aufzeichnungen 1949–1960* appeared in 1965 and 1970 respectively, *Die Provinz des Menschen. Aufzeichnungen 1942–72* in 1973, and *Die Stimmen von Marrakesch. Aufzeichnungen nach einer Reise* in 1967. Canetti had initially begun his Aufzeichnungen as a therapeutic mental exercise during the decades of study for *Masse und Macht*, and he had chosen the term to distinguish his daily periods of personal contemplation from the conventional diary. The first two volumes of Canetti's autobiography, *Die gerettete Zunge. Geschichte einer Jugend* and *Die Fackel im Ohr. Lebensgeschichte 1921–1937* (the title alludes to *Die Fackel*, q.v.) appeared in 1977 and 1980, and the third volume, *Das Augenspiel. Lebensgeschichte 1931–1937*, in 1985. The first volume ends with his move from Zurich (enforced by his mother in 1921) to a tougher experience of reality in Germany (Frankfurt). *Der Ohrenzeuge—Fünfzig Charaktere* appeared in 1974. *Das Gewissen der Worte* (1975), a collection of essays from 1962 to 1974, partly overlaps with the earlier publication of essays and interviews, *Die gespaltene Zukunft* (1972); it includes essays on Karl Kraus and Kafka (qq.v.), *Karl Kraus. Schule des Widerstands*, *Der neue Karl Kraus* (largely concerned with the letters to Sidony Nádherný von Borutin), and *Der andere Prozeß. Kafkas Briefe an Felice* (1969, *Kafka's Other Trial*, transl. by C. Middleton, 1974). Written after the publication of the correspondence, the latter offers a dual perspective of power as represented by society (Felice) and by the man and artist in isolation.

The recipient of many European honours, Canetti was awarded the Büchner Prize in 1972, and in 1981 the Kafka Prize and the Nobel Prize for Literature.

CANITZ, FRIEDRICH RUDOLF, FREIHERR VON

(Berlin, 1654–99, Berlin), a distinguished Prussian diplomat, made the grand tour in his youth, visiting Italy and the countries of western Europe, including England. He served under the Great Elector (see FRIEDRICH WILHELM, DER GROSSE KURFÜRST), became a local administrator (Amtshauptmann) in Trebbin and Zossen near Berlin, and was later the Prussian envoy in the negotiations leading to the Treaty of Ryswijk (1697).

Canitz wrote poetry in his leisure hours and possessed a gift for satire, attacking the bombast fashionable in his day. He also wrote poems emphasizing the vanity of the world and the imminence of death, popular themes in the late 17th c. (*Bereitung zum Tode, Eitelkeit des Irdischen*). But his style was much more sober and direct than that of the baroque poets, and his poems, which were published posthumously (*Nebenstunden unterschiedener Gedichte*, 1700), had a distinct popularity and influence in the first half of the 18th c.; both J. U. König (1727) and J. J. Bodmer (1737, qq.v.) republished them.

Canossa, now a ruined castle in the northern Apennines to the south-west of Reggio nell'Emilia, was the scene in 1077 of the humiliation of the German King Heinrich IV (q.v.), who is said to have done penance barefoot in order to obtain release from the ban of excommunication imposed by Pope Gregory VII, then residing at the castle as guest of Mathilda of Tuscany. The details of the humiliation were probably exaggerated by papal propaganda. In 1872 Bismarck, in the course of the Kulturkampf (q.v.), uttered the phrase 'Nach Canossa gehen wir nicht', indicating Prussian refusal to capitulate to Roman Catholic demands.

Cantilena de conversione Sancti Pauli, authentic title of an Alsatian verse sermon urging repentance of sins. The surviving fragment of 56 lines, preserved in Colmar, makes no mention of St. Paul. It was probably written *c.* 1130.

Canut, a tragedy in Alexandrines by J. E. Schlegel (q.v.), published in 1746. Canute pardons his liegeman and brother-in-law Ulfo for rising against him. Ulfo, consumed by ambition, conspires a second time against his king, and Canute reluctantly orders his execution.

CANZ, WILHELMINE (Hornberg, Baden, 1815–1901, Großheppach), principal of an institution for the training of women teachers, published anonymously a novel, *Eritis sicut Deus* (1864), attacking the theological views of D. F. Strauß (q.v.).

CAPRIVI DE CAPRERA DE MONTECUCCOLI, GEORG LEO VON (Charlottenburg, 1831–99, Crossen), entered the Prussian army in 1849 and served in the wars of 1866 and 1870, attaining the rank of general (General der Infanterie). On the fall of Bismarck (q.v.) in 1890 Caprivi was appointed imperial chancellor by the Emperor Wilhelm II (q.v.) with the task of setting a new course in foreign affairs. An immediate consequence was the refusal to renew the Reinsurance Treaty with Russia (see RÜCKVERSICHERUNGS- VERTRAG). Under his administration the treaty was signed by which Britain and Germany exchanged Heligoland for Zanzibar. Caprivi incurred violent criticism for this step and also encountered opposition with an education measure for Prussia (1892), an increase in the strength of the army (1893), tariff reforms which affected the landed classes, and a revision of the criminal code. He was dismissed by the Emperor in 1894. Caprivi was a soldier and administrator, not a politician. After his fall he engaged in no polemics and wrote no memoirs.

Cardenio und Celinde oder Unglücklich Verliebte, a play written by Andreas Gryphius (q.v.) between 1647 and 1649 and published in 1657. Though designated Trauerspiel, it is not a tragedy, but a serious play or Schauspiel which ends in the reform of the principal characters. Cardenio, who has lost his love Olympia, plans to murder her husband, while Celinde, who nourishes an unrequited love for Cardenio, seeks to arouse in him a response by brewing for him a potion made from her former lover's heart. Cardenio is enticed by a double of Olympia, which changes into a skeleton, and he encounters Celinde in the act of robbing the corpse of its heart. Both are horrified, confess their sins, and amend their lives. The play, in Alexandrine verse, emphasizes by macabre examples the vanity of the flesh. It is also remarkable for the use in a serious play of characters from ordinary contemporary life.

CARL, KARL (Cracow, 1789–1854, Ischl), whose real name was Karl von Bernbrunn, was a prominent figure in the theatrical life of Vienna in the first half of the 19th c. After a successful directorship of the Isarthortheater in Munich (1822–6), he took over the Theater an der Wien in Vienna, and in 1838 bought the Theater in der Leopoldstadt, which he demolished and replaced by the Carl-Theater (1847). He continued as director of this theatre until his death. Carl, himself an actor and an author of farces, was particularly successful on the stage as Staberl (q.v.), a vulgarized version of a character of the same name in *Die Bürger in*

Wien by A. Bäuerle (q.v.). A brash and ruthless businessman, Carl built up a successful company which included J. N. Nestroy (q.v.), and he performed many of Nestroy's plays.

Carmen ad Deum, title given by W. Scherer (q.v.) to a Latin metrical prayer in praise of the Holy Trinity. A word-for-word German translation is preserved in a MS. in Munich. It is a copy of an original, probably from Reichenau, is in Bavarian dialect, and was made in the monastery of Tegernsee in Bavaria early in the 9th c., though the MS. is later by some fifty years or more. Five other copies of the prayer are extant. They have no German translation. One of these is in the Cambridger Liederhandschrift (q.v.) or *Cambridge Songs*, and it is from the superscription to this version that Scherer took the title *Carmen ad Deum*.

CARMEN SYLVA, pseudonym of Elisabeth, Queen of Romania (Schloß Monrepos, Neuwied, 1843–1916, Bucharest), was a daughter of Prince Hermann zu Neuwied-Wied, marrying in 1869 Karl von Hohenzollern, who in 1881 became, as Carol I, the first king of Romania. She wrote poetry, stories, novels, and plays, as well as translating Romanian poems into German. Her collections of poetry include *Sappho* (1880), *Stürme* (1882), *Jehovah* (1882), and among her novels are *Aus zwei Welten* (1884) and *Defizit* (1890). She was also a musician and a painter of merit. The dominating experience in her life was the loss of her only daughter.

Carmina Burana, a collection of medieval Latin songs, including some songs in German and mixed Latin and German, contained in a 13th-c. MS. discovered at the Abbey of Benediktbeuren in Bavaria, whence its title. The songs range from naïve piety to drunken ribaldry and are the work of strolling singers and clerics of the preceding two centuries. They were published in 1847. Some have been made the basis of 19th-c. student songs. Carl Orff (q.v.) used a selection as the basic text for his 'scenic cantata' *Carmina Burana* (1937).

Carolin, a former gold coin worth 6 Reichsthaler (see THALER), current chiefly in south Germany. It was roughly equivalent to the *Friedrich d'or* or *Louis d'or*.

Carolinum, the Collegium Carolinum in Brunswick founded by Duke Karl I (q.v.) von Braunschweig in 1745. In 1862 the Carolinum became the Polytechnic School, in 1877 a Technische Hochschule, and in 1968 the Technische Universität Braunschweig.

Carolus Stuardus, see ERMORDETE MAJESTÄT ODER CAROLUS STUARDUS, KÖNIG VON GROSS BRITANIEN.

CAROSSA, HANS (Tölz, Bavaria, 1878–1956, Rittsteig nr. Passau), studied medicine and practised in Passau and Munich. In 1914 he volunteered for the army and served as a medical officer (Bataillonsarzt). His first publication to make any impression was a volume of war reminiscence, unmilitaristic and humane, drawn from the Romanian campaign and entitled *Rumänisches Tagebuch* (1924, republished as *Tagebuch im Kriege*, 1935). *Doktor Bürgers Ende* (1913) did not attract attention until it reappeared in 1930 with the title *Die Schicksale Doktor Bürgers*. It is the most pessimistic of his writings and has been labelled 'Carossa's *Werther*'. Most of Carossa's work has autobiographical reference, and his *Eine Kindheit* (1922) and *Verwandlungen einer Jugend* (1928), published together in 1933, inhabit the borderland between recollection and fiction. Carossa's awareness of growth and sense of the integrity of the evolving personality in these and other works recall Goethe. In the short novel *Der Arzt Gion* (1932) the fictional element is more prominent, though the figure of the healing physician, whose personality as well as clinical knowledge benefit his patients, has an affinity with the author. The novel *Geheimnisse des reifen Lebens. Aus den Aufzeichnungen Angermanns* (1936) treats a complex series of personal relations in diary form. In 1941 Carossa, whose conciliatory personality did not easily form sharp political decisions, had forced upon him the presidency of the National Socialist Europäischer Schriftstellerverband. *Ungleiche Welten* (1951) deals frankly with this point in his career. A further autobiographical volume, *Aufzeichnungen aus Italien*, had appeared in 1947. In his later years Carossa, never a prolific writer, produced only a few stories, of which *Reise zu den elf Scharfrichtern* (1953) is the most notable. His harmonious and traditional early poetry (*Stella mystica*, 1907; *Gedichte*, 1910) aroused little attention and his best-known poem is the lengthy *Abendländische Elegie*, written during the war and secretly circulated, a poem of searching political clarity, in which hope for a human future persists through the dark vision of the present. It was published in 1945. Carossa's correspondence with H. von Hofmannsthal (q.v.) appeared in 1960. *Sämtliche Werke* (2 vols.) appeared in 1962 and *Briefe* (3 vols.), ed. E. Kampmann-Carossa, 1978–81.

Carsten Curator, a Novelle by Th. Storm (q.v.), first published in *Westermanns Monatshefte* in 1878. Carsten Carstens runs with his sister Brigitte a small family wool shop. He has won by popular acclaim the honorary title 'Curator' for his devoted services as legal adviser and custodian to bereaved families. At the age of 40 he marries a young orphan, Juliane, who is not of local parentage. She dies after giving birth to a son, Heinrich, who grows up as the image of his seductive and light-hearted mother, and causes Carsten's tragic ruin. In a vain bid to save Heinrich, Carsten's young foster-daughter Anna marries him. But soon after he has become the father of a son, Heinrich relapses into his old ways. On a wild night which brings storms and floods, Carsten refuses his son's desperate plea to save him from bankruptcy and hears no more of him. But during that night a man is drowned and only Carsten knows that it was his son. Personal responsibility weighs on Carsten's conscience when he allows Anna to marry his son without telling her the full truth about Heinrich's disreputable past. But the emphasis Storm places on heredity and a deterministic view of life draws attention to the conflicting forces of daemonic sensuality and dedicated spiritual awareness which are symbolized by the two wives, Juliane and Anna. The historical perspective opens with Napoleon's continental blockade which challenges the stability of the small North German shipping and trading community. The estate agent and 'Stadtunheilsträger' Jaspers adds an element of the grotesque.

CASANOVA, GIACOMO GIROLAMO, CHEVALIER DE SEINGALT (Venice, 1725–98, Château Dux/ Bohemia), an Italian publicist, intriguer, and adventurer who accumulated a vast experience of contemporary life, both political and amorous, during his extensive travels. His numerous contacts included Friedrich II (q.v.) of Prussia (1764), Voltaire (q.v.), and King Stanislaus of Poland. Of his many occupations, including secret missions, his position as director of the state lotteries in Paris after his escape from a Venetian prison (1756) was the most lucrative. He settled in 1785 as librarian to Count Waldstein in Bohemia, where he wrote his *Mémoires* (12 vols., Leipzig, 1826–38), which record his life up to 1774 and have become one of the most colourful commentaries of the social scene of his day.

Casanova is referred to in Grabbe's play *Scherz, Satire, Ironie und tiefere Bedeutung* (q.v., 1827), in which one of his books is used as a bait to trap the devil. The libretto of *Casanova* (1841), an opera by A. Lortzing (q.v.), is based on the French comedy *Casanova au Fort de Saint André* by Varin, Arago, and Desvergers. He is twice portrayed by A. Schnitzler, in the story *Casanovas Heimfahrt* (q.v., 1918) and the comedy *Die Schwestern oder Casanova in Spa* (q.v., 1919), and by R. Auernheimer in the comedy

Casanova in Wien (1924), by C. Sternheim (q.v., *Der Abenteurer*, 1924), and by H. Eulenberg (q.v.) in the play *Casanovas letztes Abenteuer* (1928). H. von Hofmannsthal's plays *Der Abenteurer und die Sängerin* (1899) and *Cristinas Heimreise* (1910, qq.v.) are based on episodes in Casanova's memoirs, though the adventurer appears under the names Baron Weidenstamm and Florindo in the respective works.

Casanovas Heimfahrt, a Novelle by A. Schnitzler (q.v.), which was published in 1918. In the collected works it was included with two earlier stories (*Frau Beate und ihr Sohn* and *Doktor Gräsler, Badearzt*, qq.v.) in the cycle *Die Alternden* (1924). The main action of the story concentrates on the forty-eight hours preceding the return of Casanova (q.v.) de Seingalt to his native Venice which he had left some twenty years previously in a daring escape from prison, and which is now prepared to receive him back on condition that he serves the city as a spy in its campaign against moral corruption. Invited to stay with Olivo on a country estate near Mantua, which Casanova's generosity has enabled him to acquire upon his marriage to Amalia, a former mistress, Casanova sees in their niece Marcolina, a girl whose beauty and outstanding intellect present an unrivalled challenge to him: this is heightened by his bitter resentment of approaching old age and by the loss of sexual attraction and vitality. He arranges a deal with Lorenzi, an unprincipled young officer, who is the secret lover of Marcolina and is also in desperate financial straits, enabling him to visit her at night in Lorenzi's guise. When Casanova leaves the house at dawn, Lorenzi confronts him and forces a duel, in which Casanova kills the younger man, whom he had recognized at first sight as an image of his own lost youth. Schnitzler uses this 'Doppelgänger' motif with bitter irony, for the apparent victory completes the defeat which Casanova has already experienced after Marcolina's disgust at the discovery of his identity. His ensuing flight to ignominious office in Venice adds to his humiliation. Although Schnitzler alludes to a few historical facts, the events of the story are freely invented.

CÄSARIUS VON HEISTERBACH (Cologne, *c.* 1180–*c.* 1240, Heisterbach nr. Königswinter), a learned monk of the Abbey of Heisterbach, compiled a catalogue of the archbishops of Cologne and wrote a life of Archbishop Engelbert, which is an important historical source. He also set down a number of legendary anecdotes, *VIII Libri miraculorum*, of which three books have survived. All the works of Cäsarius are written in Latin. His name has been conferred upon the central figure of a popular medieval and later legend concerning a kind of cloistered Rip van Winkle, usually referred to as 'Der Mönch von Heisterbach'.

Caspar Hauser oder Die Trägheit des Herzens, a novel by J. Wassermann (q.v.), published in 1908. Based on the historical K. Hauser (q.v.), it recounts the mysterious episode of Caspar's appearance in 1828 and his life in Nürnberg, and the rise of two opposing factions, one regarding him as an impostor, the other as a man grievously wronged. When the leader of the pro-Caspar side dies, the young man is murdered and the enigma of his life remains unsolved.

Casseler Gesprächsbüchlein and ***Casseler Glossen***, see VOCABULARIUS STI. GALLI.

CASTELLI, IGNAZ FRANZ (Vienna, 1781–1862, Vienna), an Austrian patriot and civil servant, was a conspicuous figure in the Viennese theatre in the first half of the 19th c., especially in the popular dialect theatre. In 1809 he attracted attention with his patriotic song 'Kriegslied für die österreichische Armee', which was banned by the French. From 1809 to 1813 he was a journalist, editing first *Der Sammler*, then the Viennese *Thalia*, before accepting various official administrative appointments. Castelli was an enthusiast for the theatre, playing as a young man in a theatre orchestra in order to be able to see the plays. He wrote close on 200 mainly dialect plays, the majority of them adaptations from the French. He wrote jointly with A. Jeitteles (q.v.) *Der Schicksalsstrumpf*, which was pseudonymously attributed to 'die Brüder Fatalis'; it is a parody of Müllner's *Die Schuld* (q.v.), a fate tragedy (see SCHICKSALSTRAGÖDIE). *Roderich und Kunigunde* pokes fun at fashionable plays of medieval chivalry.

Castelli's important collection of plays and play-bills is in the Nationalbibliothek in Vienna. *Sämtliche Werke* (16 vols.) appeared in 1843.

Catharina von Georgien oder Bewährte Beständigkeit, a tragedy in Alexandrine verse, written by Andreas Gryphius (q.v.) in 1647 and published in 1657. Its heroine is a Christian queen who falls into Persian hands. The Shah falls in love with her and proposes her conversion to the Muslim faith. Catharina spurns the suggestion, remaining steadfast in her Christian witness, and is cruelly tortured and finally burned at the stake without flinching in her fidelity to her religion. After her death, the Shah regrets his act, but is denounced by Catharina's ghost. The play includes a chorus.

Cécile, a novel by Th. Fontane (q.v.), published in instalments in 1886, and in book form in 1887. One of Fontane's social works, it is a triangular narrative with a Berlin setting. Its principal characters are von St. Arnaud, a retired colonel, his wife Cécile, formerly the mistress of a minor prince, and Herr von Gordon, a civil engineer and reserve officer. St. Arnaud neglects his wife, and Gordon associates with her. The affair culminates in Gordon's death in a duel at the hands of St. Arnaud, the latter's flight, and Cécile's suicide.

The novel is remarkable for its picture of a fringe of Berlin high society, and for the account of spa life at Thale in the Harz, where the first part of the novel is set. It is also notable for the compassionate picture of the fallen Cécile, whose life has been blighted in childhood by a seduction connived at by her mother.

CELADON or Seladon, pseudonym of Georg Greflinger (q.v.).

CELAN, PAUL, pseudonym of Paul Antschel (Czernovitz, Romania, 1920–70, Paris). Of Jewish parentage, Celan suffered during the occupation of Romania by the Germans in 1941–2, losing his parents in a concentration camp. He emigrated to Vienna in 1947, moving to Paris in 1948. Deeply affected by the catastrophes of the war and the horrors besetting the Jews, he wrote a number of powerful and highly original poems, of which *Todesfuge* and *Engführung* are the best known. The titles of his collections are *Der Sand aus den Urnen* (1948), *Mohn und Gedächtnis* (1952), *Von Schwelle zu Schwelle* (1955), *Sprachgitter* (1959), *Die Niemandsrose* (1963), *Atemwende* (1967), *Fadensonnen* (1968), *Lichtzwang* (1970), and *Schneepart: Letzte Gedichte* (1971). *Gesammelte Werke* (5 vols.), ed. Beda Allemann and S. Reichert, were published in 1983. Celan was awarded the Büchner Prize in 1960. He took his own life.

Celestina, see WIRSUNG, C.

CELTIS, KONRAD (also spelt Celtes), adopted name of Konrad Bickel (or Pickel) (Wipfeld nr. Würzburg, 1459–1508, Vienna), one of the most gifted of the German humanists. Brilliant, energetic, and restless, Celtis spent several years as a wandering scholar, studying at Cologne, Heidelberg, and Erfurt, visiting Italy, Poland, and Hungary. In his mature years he founded learned societies at his various places of call, such as the Sodalitas vistulana in Cracow, Sodalitas literaria Hungarorum, and the Sodalitas Rhenana. The most famous of these was the Sodalitas danubiana in Vienna, a forerunner of later academies. In 1494 Celtis was a

professor at Ingolstadt, and in 1496 in Vienna, where he enjoyed the favour of the Emperor Maximilian I (q.v.), who inaugurated for him a special faculty or 'Collegium'. He discovered and published in 1501 the plays of the 10th-c. nun Roswitha (q.v.) of Gandersheim and planned a survey of Germany, of which he wrote the section concerning Nürnberg (*De origine, situ et institutis Norimbergae libellus,* 1502). He wrote entirely in Latin, and his works include vivid erotic poems (*Amores,* 1502), which have often been interpreted as autobiographical. The *Peutingersche Tafel* (q.v.) was discovered by him and bequeathed to Konrad Peutinger (q.v.).

Cenodoxus, a Latin tragedy by Jakob Bidermann (q.v.), performed in 1602 and published in 1609. Its source is an episode in the legend of St. Bruno. The learned Dr. Cenodoxus disregards the divine commandments and, in a powerful final scene, is sentenced by Christ to eternal damnation. Translated into German in 1635 by Joachim Meichel, it is one of the few plays of its age to survive as a living work. In its own day it is said to have had a remarkable religious and moral influence.

CHAMBERLAIN, HOUSTON STEWART (Portsmouth, 1855–1927, Bayreuth), a diligent publicist, was born in England as the son of a general, and was educated in Switzerland. He wrote in praise of R. Wagner (1892, 1896, and 1900), married Wagner's daughter (1908), and became a Germanophile, an anti-Semite, a Pan-German, and an anglophobe. He wrote on Kant (q.v., 1905) and Goethe (1912) and published a controversial history (*Grundlagen des XIX. Jahrhunderts,* 2 vols., 1899).

CHAMISSO, ADELBERT VON (Boncourt, Champagne, 1781–1838, Berlin), born a French subject of noble descent, fled from the Revolution as a boy of 9 with his parents and brothers and spent the rest of his life in Prussia. The family château of Boncourt was razed, a fact referred to in Chamisso's poem, *Das Schloß Boncourt* (q.v.). Born Vicomte de Chamisso, he was baptized Louis Charles Adélaïde, and adopted the name Adelbert in Germany. Though this name is often given as Adalbert, signatures as Adelbert are extant. In 1796 he became a page to Queen Friederike Luise, consort of Friedrich Wilhelm II (q.v.), and two years later was commissioned as an ensign of foot in the Prussian army. When in 1801 his parents returned to France he adhered to his adopted career in spite of poverty and dissatisfaction. During his army years he began to write, at first using his mother tongue and later going over to German. He devoted his leisure to literary and philosophical studies. With Varnhagen von Ense, J. E. Hitzig (qq.v.), and others he took part in literary parties, which were dignified with the title 'Nordsternbund'. A product of this was a Musenalmanach, published in 1804, to which Chamisso contributed. In 1806 Chamisso's regiment was put on a war footing and stationed in Hamelin, where he was taken prisoner in November. Released on parole, he visited France, returning to Germany in 1807. In 1811–12 he was in Coppet at the house of Madame de Staël. Returning to Berlin, he devoted himself to scientific, especially botanical, studies. His story *Peter Schlemihls wundersame Geschichte* (q.v., 1814), which can be seen as a symbolical autobiography, achieved an immediate and lasting success. From 1815 to 1818 he was on a Russian-sponsored voyage round the world as expedition scientist (see REISE UM DIE WELT, 1834–5). He was appointed to the staff of the Botanical Gardens in Berlin in 1819. From 1833 until his death he was joint editor with G. Schwab (q.v.) of the *Deutscher Musenalmanach*.

Chamisso wrote much poetry of a mildly Romantic kind, but little of it is now read, apart from *Das Schloß Boncourt*, which occurs in anthologies, and the cycle *Frauen-Liebe und -Leben* (q.v.), which probably owes its survival to Schumann's sensitive setting. His ballads, which include *Die Weiber von Weinsberg* (q.v.) and the tersely dramatic, translated folk-ballad 'Es geht bei gedämpfter Trommel Klang' (q.v.), have lasted better. It is a curious fact that this foreign-born German poet, who could handle technically difficult forms like *terza rima* with ease, never felt completely at home with the German tongue. A *Gesamtausgabe* of Chamisso's works was edited by J. H. Hitzig (6 vols., 1836–9). A blank-verse play *Fortunati Glückssäckel und Wunschhütlein* was discovered in MS. and published in 1895. *Sämtliche Werke* (2 vols.), ed. J. Perfahl, were published in 1975.

Chandos-Brief, see BRIEF DES LORD CHANDOS.

Chaos, see GOETHE, OTTILIE VON.

CHARLES THE GREAT, see KARL I, DER GROSSE.

Charon, a literary periodical published 1904–14 and 1920–2, and also designation of the group responsible for it. Its editors were Otto zur Linde and R. Pannwitz (qq.v.). K. Röttger (1877–1942) and R. Paulsen (q.v.) were regular contributors. The literary views of *Charon* were akin to those of the later A. Holz (q.v.).

CHAUBER, THEOBALD, pseudonym used by B. Auerbach (q.v.) for his first work, a biography of Frederick the Great (see FRIEDRICH II, KÖNIG).

CHELIDONIUS [CHELIDONICUS], BENEDICTUS, an Austrian humanist of the late 15th c. and early 16th c., who was a monk at St. Aegidius's monastery at Nürnberg and later abbot of the Schottenstift in Vienna. He wrote a Latin drama, *Voluptatis cum virtute disceptatio* (performed in 1515), which takes the form of a trial at law between the two parties, lust and virtue. He also wrote Latin religious poems (*Carmina de vita Mariae*).

CHEMNITZ, MATTHÄUS FRIEDRICH (Barmstedt, Holstein, 1815–70, Altona), a lawyer in Schleswig, is the author of the patriotic song 'Schleswig-Holstein, meerumschlungen', first published in 1844 in the *Itzehoer Nachrichten* and publicly sung to a melody by C. G. Bellmann 1740–95 at a choral festival at Schleswig on 24 July 1844.

Cherubinische Wandersmann, Der, a volume of mystical epigrams by ANGELUS Silesius (q.v.), published in 1675; in earlier and shorter form it bore the title *Geistreiche Sinn- und Schlußreime* (1657).

CHÉZY, WILHELMINE CHRISTIANE VON (Berlin, 1783–1856, Geneva), usually known as Helmine von Chézy, was a granddaughter of the poetess Anna Luise Karschin (q.v.). Her maiden name was von Klencke. She had a turbulent marital career: she married Baron von Hastfer in 1799, was divorced in 1801, married M. de Chézy in Paris in 1805, and left him in 1810. She also had a love-affair with A. von Chamisso (q.v.). She wrote poetry (of which 'Ach, wie ist's möglich dann' survives) and provided the libretto for Weber's *Euryanthe* (1824). Her memoirs (*Unvergessenes*) were published posthumously in 1858.

CHIAVACCI, VINCENZ (Vienna, 1847–1916, Vienna), started his career on the railway in Hungary, but spent the latter part of his life (from 1887) as a Viennese journalist. He wrote Volksstücke in Viennese dialect (*Frau Sopherl vom Naschmarkt*, 1890; *Einer von der Burgmusik*, 1892; *Aus'm Herzen raus*, 1901). He collaborated with C. Karlweis (q.v.) in *Einer vom alten Schlag* (1886) and with F. von Schönthan (q.v.) in *Der letzte Kreuzer* (1893). He also wrote sketches of Viennese life (*Kleinbürger von Groß-Wien*, 1893; *Wiener Typen*, 1894; *Wiener vom alten Schlag*, 1895). Frau Sopherl was his most popular creation.

Chinesische Flöte, Die, a collection of poems by Hans Bethge (q.v.).

Chinesische Mauer, Die, a farce (Eine Farce) by M. Frisch (q.v.) published in 1947, revised

1955, 1965, and 1972. Its subject is Mans self-destruction. It has a short Vorspiel. The action takes place in China about 200 B.C. The Emperor Hwang Ti, having vanquished his enemies with the aid of his general Prince Wu Tsiang, builds the Great Wall to prevent future invasions. He proposes to marry his daughter Mee Lan to the prince, but she declines the match. A panorama of subsequent history is unrolled which includes not only historical persons such as Alexander the Great, Brutus, Pontius Pilate, Columbus, Philip of Spain, and Napoleon, but also the legendary figures of Romeo, Juliet, and Don Juan, and some anonymous persons of whom the most important is 'Der Heutige', the mediocre man of today, who is faint of heart and unable to retain the love which Mee Lan was at first willing to bestow on him. He is present when a dumb man, believed to be a dangerous rebel (Min Ko), is tortured because he will not speak, but has not the courage to make more than a half-hearted attempt to intervene, whereupon Mee Lan rejects him and disappears with Don Juan. 'Der Heutige' pulls himself together and makes an impassioned speech against the monstrous dangers of a technology which has produced the atom bomb. His hearers applaud, and the Emperor rewards his splendid oration with a gold chain, but no one pays any attention to his message. Revolution breaks out, and the Emperor and his court are swept away. The play contains verse parody and achieves farcical effects by the mingling of ancient Chinese and historical European characters.

CHLUMBERG, HANS, pseudonym of Hans Bardach, Edler von Chlumberg (Vienna, 1897–1930, Leipzig), an Austrian officer who wrote a symbolical pacifist play, *Wunder um Verdun* (1931), first performed at Leipzig in 1930. The play is set in the future (1939) and before the eyes of a German tourist who has fought at Verdun the dead of both sides rise from their graves in a plea for peace. By a tragic accident the author was killed at the dress rehearsal. He had previously written two plays, *Eines Tages* (1926) and *Das Blaue vom Himmel* (1929).

CHODOWIECKI, DANIEL NIKOLAUS (Danzig, 1726–1801, Berlin), German graphic artist, became Director of the Akademie der bildenden Künste in Berlin in 1797. Though also a painter in oils, Chodowiecki was particularly prolific as a book illustrator, and executed an enormous number of etchings and drawings, many of which, apart from their technical skill, are valuable as social documents. He illustrated works by Lessing, Bürger, Claudius, Goethe, and Schiller (qq.v.), among others. A parti-

cularly notable series, based on his own experience, is *Die Reise nach Danzig* (1773).

Choriambus, a metrical foot consisting of an amalgamation of a trochee (also termed *Choreus*) and an iamb. It occurs chiefly in Asclepiadic verse, as in the central feet of the following example from Hölderlin's *Dichtermut*:

> Sind denn / dir nĭcht vĕrwandt / ál-lĕ Lĕbén- / digen.

(See ASKLEPIADISCHE STROPHE.)

Chotusitz, Battle of, see SCHLESISCHE KRIEGE (1).

CHRÉTIEN DE TROYES, wrote in the second half of the 12th c. a number of Arthurian romances or *romans bretons*, which provided the sources for some important Middle High German works. These include Hartmann von Aue's *Erec* and *Iwein* (*Yvain*), Wolfram von Eschenbach's *Parzival* (*Perceval* or *Le Conte del Graal*), and the lost *Cliges* of Ulrich von Türheim (*Cligès*).

CHRIST, LENA (Glonn, Oberbayern, 1881–1920, Munich), an illegitimate child, wrote under her father's surname Christ (her baptismal name was Magdalena Bichler). Her married name was Benedix. She published an autobiographical work *Erinnerungen einer Überflüssigen* (1912) and two Bavarian regional novels of merit, *Mathias Bichler* (1914) and *Die Rumplhanni* (1916). Suffering from tuberculosis, she took her own life.

CHRISTEN, ADA, pseudonym for Christiane Frideriks (Vienna, 1844–1901, Vienna), became an actress while still a girl and experienced dire poverty after the death of her first husband, S. von Neupaur. She married a retired cavalry captain, A. von Breden, in 1873. Her poetry, inspired by the experience of her own poverty and by sympathy with the obscure and inarticulate, is contained in the collections *Lieder einer Verlorenen* (1868), *Aus der Asche* (1870), *Schatten* (1872), and *Aus der Tiefe* (1878). She wrote the novels *Ella* (1869) and *Jungfer Mutter* (1892), of which the latter was successfully adapted as a drama with the title *Wiener Leut'* in 1893. Her play *Faustina* (1871) was not a success. Late in life she wrote two comedies, *Hypnotisiert* (1898) and *Fräulein Pascha* (1899).

Christenheit oder Europa, Die, a historical essay written in 1799 by Novalis (q.v.) but not published until 1826, when it appeared in the fourth edition of Novalis's *Schriften*. The essay looks back to the Middle Ages as a golden age when Europe was united by a single religion. The division of the Church at the Reformation and the rationalism of the 18th c. are seen as elements of decline, and Novalis sees signs of the rise of a new unified religion: 'Nur Geduld, sie wird, sie muß kommen die heilige Zeit des ewigen Friedens, wo das neue Jerusalem die Hauptstadt der Welt sein wird.' Through circulation in manuscript the essay is thought to have influenced the historical and religious thinking of the early Romantics (see ROMANTIK).

Christian Wahnschaffe, a novel by J. Wassermann (q.v.), published in 1919 (2 vols.). It portrays the idle futile life of the wealthy nobility and the families of industrial magnates who associate with them. Christian Wahnschaffe is born into this useless world, and at first identifies himself with it. But he undergoes a conversion, rejects his family, and disappears with the intention of helping his fellow men. He does not reappear, but rumour connects him with rescue work at a mine disaster at Hamm, and with social work in the London and New York slums.

Christi Tageszeiten, a short anonymous Middle High German poem setting forth the seven canonical hours of prayer (Mette, Prime, Terze, Sexte, None, Vesper, Komplete) and linking them with the seven hours of the Passion. The poem was written in the 14th c.

Christlichsoziale Partei, name of two distinct political parties towards the end of the 19th c., the one in Germany, the other in Austria.

(1) In Germany the Christian Social Party was founded in 1878 as Christlichsoziale Arbeiterpartei by the court chaplain (Hofprediger) Adolf Stoecker (1835–1909) with a social programme of amelioration in conditions of work, combined with loyalty to Church and State. After a time it changed its title to Christlichsoziale Partei, its social element was reduced, its chauvinism increased, and anti-Semitism included in its programme. The party eventually disintegrated.

(2) In Austria the Christian Social Party was founded in 1887 and led by Karl Lueger (q.v.). It claimed to be the party of the 'little man' against the great corporations, financiers, and entrepreneurs, and included anti-Semitism in its platform from the outset. From 1897 it was the dominant party in the Vienna municipality. The party survived the 1914–18 War and achieved a majority and consequent control in 1920.

Christoph Marlow, a verse tragedy in four acts by E. von Wildenbruch (q.v.) published in 1885 and first performed at Hanover on 6 May

1884. The passionate Marlow (*sic*) seduces his benefactor's daughter, finds that in the theatre he must yield to the rising star of Shakespeare, and is killed by the fiancé of the girl he has seduced.

Christophoruslegende, two Middle High German poems of unknown authorship, both originating from Bavaria or Austria. The earlier was probably written in the second quarter of the 13th c. The other, which reveals popular elements, is of uncertain date, but most probably belongs to the 14th c.

Christus und die minnende Seele, a Middle High German verse dialogue between Christ and the soul written in the first half of the 15th c. The poem, of just over 2,000 lines, rejects the world and concludes with the mystical union of the soul with Christ. The two principal MSS. of the poem contain also *Die kreuztragende ·Minne* (q.v.).

Christus und die Samariterin, Old High German rhyming poem of 31 lines, describing Christ's meeting at the well with the woman of Samaria (John 4:6 ff.). It is believed to have been written by a monk of Reichenau (q.v.) at the beginning of the 10th c., possibly in 908. In its subject and outward form akin to Otfrid's *Evangelienbuch,* it is freer and more popular in style. The dialect shows Frankish and Alemannic traces. The MS. is in the Nationalbibliothek, Vienna.

Chronik der Sperlingsgasse, the first novel of W. Raabe (q.v.), written in 1854–5 and published in 1856. Its setting is Berlin in the first half of the century. The backbone of the novel is Johannes Wacholder, the narrator, an ageing bachelor, who has lived most of his life in the Sperlingsgasse and is its devoted, but unmethodical, historian. Wacholder sees life whole. He records the misery, want, and bitterness, and the many tragedies of the street, but he also sees humour and determination overcoming adversity. The continuous thread of happiness in the story is the love of his adopted daughter Elise and her orphaned cousin Gustav Berg. These two, whose lives began in a dark story of seduction and despair, prove themselves able to face the vicissitudes of life, making a serenely happy marriage. Though the novel is not free from sentimentality, Raabe shows an acute sympathy with human suffering, a strong sense of the values of courage and tenacity, and a lively awareness of the cathartic virtue of humour. The Berlin street in which Raabe lived at the time, the Spreegasse, has been renamed Sperlingsgasse after the book.

Chronik von Reichenau, a history of the Abbey of Reichenau in chronicle form, compiled between 1491 and 1508 by Gallus Oheim, chaplain of the abbey.

CHYTRAEUS, NATHAN (Menzingen, Pfalz, 1543–98, Bremen), professor of poetry at Rostock (1565) and later headmaster (Rektor) at Bremen, wrote Latin and German poetry (*Poematum libri septemdecem,* 1579, *Hundert Fabeln,* 1591, after Aesop), and a play (*Tragedia Abrahami Opfer,* 1595).

Cidli, name of Gedor's faithful and loving wife, who dies in Canto 15 of F. G. Klopstock's *Der Messias.* Klopstock used the name Cidli in his odes to refer to his bride Meta Moller (see KLOPSTOCK, META).

CILIES VON SEYN, who is the author of seven Middle High German *Sprüche* (see SPRUCH) preserved in the Jenaer Liederhandschrift (q.v.), lived in the Rhineland in the late 13th c. and early 14th c. Attempts to attribute to him the authorship of *Die Böhmerschlacht* (q.v.) and other associated poems have not won general acceptance.

Cissides und Paches, a short epic poem in three books written by Ewald von Kleist (q.v.) in 1758 and published in 1759. The poem, which is in blank verse, is set in Greece after the death of Alexander the Great. Cissides and Paches are Macedonian commanders and friends, who sacrifice their lives defending a frontier fortress against odds. The poem is a product of Kleist's patriotic enthusiasm in the Seven Years War.

CLAJUS, JOHANNES (Herzberg, Saxony, 1535–92, Bendeleben), a Saxon pastor, schoolmaster, and humanist of the school of Melanchthon (q.v.), was the author of Latin poetry of prosaic substance and elegant form (*Elegiae sacrae tres,* 1557; *Carminum sacrum libri tres,* 1568). Clajus produced the first grammar of the German language, *Grammatica Germanicae lingua* (1578, reprinted 1894). He also composed a Latin version of a portion of scripture (*Ecclesiastes Salomonis carmine redditus et enarratus,* 1583) and published a tract directed against alchemy (*Alcumistica,* 1586). His real name was Klaj, and he is sometimes referred to as Clajus or Klajus der Ältere to distinguish him from Johann Klaj (q.v.), also called Clajus or Klajus der Jüngere, who lived in the 17th c.

CLASS, HEINRICH (Alzey, 1868–1953, Jena), a lawyer in Mainz, was politically active in the Alldeutscher Verband (q.v.) and became its president in 1908. Under the pseudonym Ein-

hart he wrote a strongly nationalistic *Deutsche Geschichte* (1908), which went through many editions. For many years Claß exercised a baneful influence on German opinion by his extreme nationalism, his inveterate anti-Semitism, and, during the 1914–18 War, by his vigorous support for a policy of extensive annexation. Under the Weimar Republic (q.v.) he edited *Die deutsche Zeitung.*

Claudine von Villa Bella, ein Schauspiel mit Gesang, a Singspiel (q.v.) written by Goethe in 1774–5 and published in 1776. The sub-title was changed to 'Ein Singspiel' in 1788. It is a good-humoured romantic comedy in which the complication of the love-match between Claudine and Pedro hinges on the attraction of the vagabond Crugantino to Claudine. Crugantino turns out to be Pedro's lost brother, who had turned his back on society.

CLAUDIUS, HERMANN (Altona, 1878–1980, Grönwohld, nr. Trittau), a great-grandson of the 18th-c. poet M. Claudius (q.v.), was an elementary school teacher in Hamburg. Deafness caused him to retire in 1934, after which he lived in the near-by village of Hummelsbüttel. Claudius was primarily a dialect poet, dealing with aspects of the lives of working people and the moods of both the city and the rural scene. His principal collections are *Mank Muern* (1912), *Hörst du nicht den Eisenschritt* (patriotic, 1914), *Lieder der Unruh* (1920), *Heimkehr* (1925), *Daß dein Herz fest sei* (1935), *Zuhause* (1940), and *Das Wolkenbüchlein* (1948). He also wrote novels (*Das Silberschiff,* 1923; *Bertram von Mynden,* 1927) and stories, mostly connected with his own experiences. Among the collections of Novellen are *Wie ich den lieben Gott suchte* (1935) and *Mein Vetter Emil* (1938). In 1933 he was elected to the National Socialist Dichterakademie (see AKADEMIEN).

CLAUDIUS, MATTHIAS (Reinfeld, Holstein, 1740–1815, Hamburg), a pastor's son, attended the grammar school at Plön and then studied theology and law at Jena University. After a short spell as secretary to Graf Holstein in Copenhagen (1764–5), he worked on the *Andreßcomptoir-Nachrichten* (1768–70).

In 1771 he took over the editorship of the newspaper *Der Wandsbecker Bote,* with which he identified himself so successfully that he himself is often referred to as 'Der Wandsbecker Bote'. Claudius, who, under the pseudonym Asmus, was the chief contributor to the newspaper, managed to give it a popular style without cheapness or vulgarity, and a homely, friendly tone without seriously relaxing intellectual standards. The basis of his writing was religious, and he maintained a broad and tolerant judgement. *Der Wandsbecker Bote* ceased publication in 1776, but Claudius collected his prose and verse contributions and published them as *Asmus, omnia sua secum portans oder sämtliche Werke des Wandsbecker Botens* (1775), which, with new material, continued in seven further volumes up to 1812.

Herder (q.v.), a friend of Claudius, obtained for him in 1776 an administrative appointment in the government of Electoral Hesse at Darmstadt. But Claudius did not settle in public life and returned in 1777 to Wandsbeck. His marriage to a joiner's daughter in 1772 was uniformly happy. From 1785 he received a Danish pension. Among his friends at various stages of his life, apart from Herder, were J. H. Voß and Goethe (qq.v.).

Claudius, who is the father of German popular journalism, is also a poet with a delight in the minor happenings of life and a sincere simplicity in their expression. Three of his poems have become virtual folk-songs (*Abendlied, Der Tod und das Mädchen*, and *Rheinweinlied*). He is the originator of the designation 'Freund Hein' (q.v.) as a personification of death, and Asmus's works are accompanied by a line engraving of Freund Hein as a skeleton holding a scythe.

CLAUERT, HANS (d. 1566), a citizen of Trebbin, famous for his pranks and practical jokes and often called 'der märkische Eulenspiegel'. His exploits are recounted in *Hans Clawerts werckliche Historien* (1587), by Bartholomäus Krüger (q.v.). Clauert enjoyed the protection of the Elector Joachim II (q.v.).

CLAUREN, HEINRICH, pseudonym of Karl Gottlieb Samuel Heun (Dobrilugk, Saxony, 1771–1854, Berlin) and the name by which he is usually known. H. Clauren is an anagram of Carl Heun. Clauren was for most of his life a Prussian civil servant, and his abundant writing represents the product of his leisure hours. His novel *Mimili* (1816), which is overtly sentimental and has been held to be secretly titillating, was widely read and repeatedly reprinted. Clauren published many of his stories, with titles such as *Das Mädchen aus der Fliedermühle* and *Liesli und Elsi*, in his 'Taschenbuch' *Vergißmeinnicht* (annually 1818–34). He was one of the chief purveyors of fiction for the entertainment of the middle class in the first half of the century. He also wrote comedies. In 1826 W. Hauff (q.v.) published the parody *Der Mann im Monde* under the pseudonym H. Clauren, and was involved in a lawsuit, which Clauren won. Clauren's collected works (*Gesammelte Schriften*, 25 vols.) appeared in 1851.

CLAUSEWITZ, KARL VON (Burg nr. Magdeburg, 1780–1831, Breslau), Prussian general and military writer, entered the infantry in 1792 and almost immediately saw active service in the French campaigns of 1792–3. He came to the notice of Scharnhorst (q.v.) and after Jena, where he was taken prisoner, was active in the War Department. He emigrated to Russia and served in the War of 1812 and the War of Liberation (see NAPOLEONIC WARS). In 1814 he returned to the Prussian army and was present at Ligny and Wavre. After 1815 he held various high appointments, dying of cholera at Breslau.

Clausewitz wrote a number of military books, including accounts of the campaigns of 1813 and 1815. His most famous work, *Vom Kriege*, was left unfinished and published in the first three volumes of *Hinterlassene Werke* (10 vols., 1832–7). It sees war as a continuation of politics, the ultimate arbiter when all else has failed. He supported the conception of national war and insisted that wars must be conducted swiftly and ruthlessly in order to reach a clear decision in the minimum time. Clausewitz was not a warmonger; his doctrine is a logical deduction from existing conditions.

CLAUS NARR, court jester to the Elector of Saxony, enjoyed a reputation for his pranks and practical jokes, similar to that of Till Eulenspiegel (q.v.). His biographer was Wolfgang Bütner (q.v.).

Clavigo, a tragedy written by J. W. Goethe (q.v.) in 1774 and first published in the same year. Its source is in the *Quatrième mémoire à consulter pour Pierre Augustin de Beaumarchais*. The incidents took place in 1764, and the chief personage (Clavijo in real life) is historical. In the play Marie Beaumarchais has been courted, then abandoned, by Clavigo, and her brother comes to Madrid to put things right. He forces Clavigo into a written confession, but holds his hand in the hope of a reconciliation, which takes place. But Clavigo, persuaded by his worldly friend Carlos, again abandons Marie, who dies of the shock. Clavigo is killed by Beaumarchais. Goethe claims (in *Dichtung und Wahrheit*, q.v.) that he wrote the play in eight days at the request of a girl friend, Anna Sibylle Münch. *Clavigo* was performed at Mannheim in Goethe's presence in November 1779.

Clinschor, a character in Wolfram von Eschenbach's *Parzival* (q.v.). The name corresponds to the Klingsor of Novalis's *Heinrich von Ofterdingen*, R. Wagner's *Parsifal*, H. Hesse's *Klingsors letzter*

Sommer, and the Klinsor or Klingsor of the *Wartburgkrieg* (qq.v.).

CLOSENER, FRITSCHE (d. *c.* 1373, Strasburg), a canon of Strasburg, compiled, in the *Straßburger Chronik,* a prose chronicle of the Empire, devoting special attention to the history of his own city. It is modelled on the *Sächsische Weltchronik* (q.v.) of Eike von Repgau. Closener's *Straßburger Chronik* is a valuable source of information on the Flagellants.

Cluniac Reform, name given to a movement for monastic reform which originated in the Burgundian Abbey of Cluny, founded in 910, and spread through western Europe largely through the foundation of affiliated monasteries. The principles of the reform were strict asceticism and unconditional obedience, hence an accentuation of the hierarchical structure of Christianity. Underlying it were on the one side a new emphasis on a life of self-denial as the means of overcoming death and achieving salvation, on the other an emphasis on the Church as the supreme authority at the expense of the temporal power of the state. The Cluniac Reform, which was most successfully furthered by the abbots Odo (927–42) and Odilo (994–1048), did not at first affect Germany, though a parallel movement which concentrated on the ascetic way of life, ignoring the political implications, began to spread in German territory in the 10th c. from Gorze Abbey in Lorraine by way of Trier.

A direct influence of Cluny made its impact in Germany in the abbeys of Hirsau in Württemberg and St. Blasien in Baden. It spread rapidly through the Benedictine foundations in Germany, but whether it was the dominant factor underlying the literary remains of the period 1060–1170 is doubtful. Possibly under its impulse the use of German, rather than Latin, was revived in such a work as *Memento mori* (q.v.), forming a means of communicating to the laity the fear of death, the hope of salvation, and the necessity of mortifying the flesh to attain it. The impulse of the Cluniac Reform spent itself in the second half of the 12th c.

COCHLÄUS, JOHANN (Wendelstein, 1479–1552, Breslau), whose real name was Dobeneck, took holy orders, lived in various South German cities (Mainz, Frankfurt, Nürnberg, Eichstätt, Breslau) and vigorously defended the Roman Catholic faith against the Lutheran reformers. He wrote his polemics both in German (*Ein heimlich Gespräch von der Tragedia Johannis Hussen,* 1538) and in Latin (*Commentaria de vita et scriptis Lutheri,* 1549). Lessing (q.v.) essayed a rehabilitation of his tarnished reputation in *Rettungen,* 1754.

Codex argenteus, a 6th-c. Ostrogothic MS. preserved in the University of Uppsala. It is written in silver and gold letters on purplish-red parchment and originated in Northern Italy. It contains a translation into Gothic of the four Gospels by Bishop Wulfila (q.v.), representing a portion of Wulfila's Gothic Bible begun in 369. The MS. is no longer complete, consisting of only 187 out of 330 pages, and it has been in this state at least since the middle of the 17th c. Formerly in Werden Abbey, Westphalia, it passed into Habsburg hands, was seized by Swedes in Prague in 1648, and was for a time in the possession of Queen Christina. It was brought to Holland in 1655, and was presented to its present owners in 1662. It was first printed in 1665 and was reproduced in photographic facsimile in 1928.

Colberg, see KOLBERG.

COLIN, PHILIPP, see NEUE PARZEFAL, DER.

College Crampton, a comedy (Komödie) in five acts by G. Hauptmann (q.v.), first performed in the Deutsches Theater, Berlin, in January 1892 and published in the same year. It is a comedy of character built round one central figure, Crampton, a professor of painting at a school of art in a Silesian town. Gifted with much ability but no concentration, he has become an absent-minded alcoholic who no longer carries out his duties properly. He is dismissed and disappears, to the concern of his daughter Gertrud and of Max Strähler, the one pupil who believes in him; both fear that he has done away with himself. Ending his life has, however, not crossed his mind, he has simply gone to ground in a low pub and drowned his sorrows. From this he is rescued by Max Strähler, whose wealthy elder brother installs the incorrigible and irrepressible Crampton in a handsome new studio. Max completes the conventional comedy by becoming engaged to Gertrud.

COLLIN, HEINRICH JOSEF VON (Vienna, 1771–1811, Vienna), Austrian dramatist, whose classical verse plays, which had themes from either ancient or modern history, were regularly played in the Burgtheater at the beginning of the 19th c. They included *Regulus* (1801), his most popular play, *Polyxena* (1803), *Coriolan* (1804), *Balboa* (1805), *Bianca della Porta* (1807), *Mäon* (1808), and *Horatier und Curiatier* (1810). For *Coriolan* Beethoven composed a well-known overture. Collin's plays extolled patriotism in an age when Austria was repeatedly at war with

France, and in 1809 he published a volume of patriotic poems (*Lieder österreichischer Wehrmänner*). He was by profession a civil servant.

COLLIN, Matthäus von (Vienna, 1779–1824, Vienna), younger brother of Heinrich von Collin (q.v.), was professor of philosophy at Cracow (1808) and Vienna (1812). In 1815 he was appointed tutor to Napoleon's son, the Duke of Reichstadt (see Reichstadt, Herzog von). He wrote Austrian historical plays in Romantic tone, as well as poetry, but was more successful as a journalistic critic.

Colmarer Liederhandschrift, a 16th-c. collection of songs of the Meistersinger (see Meistergesang), which formerly belonged to the guild of shoemakers in Colmar and is now in Munich. It includes music as well as words.

CONRAD, Michael Georg (Gnodstadt, Franconia, 1846–1927, Munich), a farmer's son, studied in Switzerland, Italy, and France, and became a teacher at Geneva. In 1878 he met Émile Zola in Paris. He settled in Munich in 1882, founding and editing *Die Gesellschaft* (q.v., 1885–1901), the chief organ of Naturalism in its early phase. From 1893 to 1898 Conrad was a member of the Reichstag for the Demokratische Volkspartei. His creative works are less important than his enthusiasm and partisanship for the new movement. He published several volumes of stories (*Lutetias Töchter*, 1883; *Totentanz der Liebe*, 1885; *Fantasio*, 1889; *Erlösung*, 1891; *Raubzeug*, 1893) and six novels. The first three of these (*Was die Isar rauscht*, 2 vols., 1888; *Die klugen Jungfrauen*, 3 vols., 1889; *Die Beichte des Narren*, 1883) make up a trilogy ('Isar-Trilogie') of social criticism referring to Munich. *In purpurner Finsternis* (1895), which has been compared to *Erewhon*, betrays Nietzsche's influence, and the fifth novel, *Majestät* (1902), deals with Ludwig II (q.v.) of Bavaria. Conrad's last novel, *Der Herrgott am Grenzstein* (1904), is a revision of an earlier work, *Am Grenzstein* (1897). *Salve regina*, a volume of poems, appeared in 1899, and he also published the collections of critical essays, *Parisiana. Plaudereien über die neueste Literatur und Kunst der Franzosen* (1885), *Die Sozialdemokratie und die Moderne* (1891), and *Von Emile Zola bis Gerhart Hauptmann* (1902). He collaborated with L. Wilfried in two plays, the comedy *Die Emanzipierten* (1887) and the drama *Firma Goldberg* (1888).

CONRADI, Hermann (Jeßnitz, Anhalt, 1862–90, Würzburg), died of pneumonia soon after the end of his student career. He was a radical forerunner of Naturalism (see Naturalismus), and was in contact with the brothers Hart

(qq.v.). His publications include the stories *Brutalitäten* (1886) and the poems *Lieder eines Sünders*. He also wrote two novels, *Phrasen* (1887) and *Adam Mensch* (1889). The latter attracted a prosecution for obscenity. He died before the verdict, which was in his favour, was given. *Gesammelte Schriften* (3 vols.), ed. P. Ssymak and G. W. Peters, were published in 1911.

CONRADIN, see Konradin.

Consilium abeundi, decree of expulsion from a German university. It did not preclude the student from transferring to another university.

CONTESSA, Karl Wilhelm (Hirschberg, 1777–1825, Berlin), a native of Silesia of Italian descent, was a friend of E. T. A. Hoffmann and F. de la Motte Fouqué (qq.v.). In the *Serapionsbrüder* (q.v.) he figures as Sylvester. Contessa wrote a number of fairy-tales (*Kindermärchen*, 1817). His elder brother, Christian Jakob Contessa (1767–1825), was also active as a writer of plays and stories.

Contra caducum morbum, superscription of an Old High German prose spell against epilepsy preserved in a MS. in Paris. Another version, in Munich, is headed *pro cadente morbo*. On the same sheet of the Paris MS. is the spell *Ad equum errehet* (q.v.).

Contra malum malannum, superscription of an Old High German spell designed to cure ulcers, boils, etc.

Contra vermes, see Wurmsegen.

CONZ, Karl Philipp (Lorch, 1762–1827, Tübingen), minor poet and Professor of Classics in Tübingen. Conz, who became friendly with Schiller (q.v.) in Stuttgart in 1782 and visited him in Jena in 1792, wrote a tragedy (*Konradin von Schwaben*, 1782) and made a translation of the plays of Aeschylus, published in 1811.

CORDAY D'ARMONT, Charlotte (Champeaux, 1768–93, Paris), came from Normandy to Paris as an ardent revolutionary set upon avenging the fall of the Girondins which had led to the Jacobin Terror (see French Revolution). On 13 July 1793 she murdered J. P. Marat in his bath and was guillotined four days later. She is the subject of *Charlotte Corday* by C. M. Wieland (q.v., 1793), *Über Charlotte Corday* by Jean Paul (q.v., 1801), *Charlotte Corday* by Karl Frenzel (q.v., 1864), and she appears in a play by Peter Weiss (q.v.), *Die Verfolgung und Ermordung Jean Paul Marats, dargestellt durch die*

Schauspielgruppe des Hospizes zu Charenton unter Anleitung des Herrn de Sade (1964).

CORDUS, Euricius (Simshausen/Hesse, 1486–1535, Bremen), scholar and physician, became a professor of medicine at Marburg in 1527, moving to Bremen in 1534. A zealous supporter of Luther (q.v.) and a man of serious purpose, he distinguished himself in literature chiefly by his Latin epigrams, some of which were later translated by G. E. Lessing (q.v.). He satirized medical superstition in his *Liber de urinis* (pub. 1543) and also wrote a systematic treatise on botany (*Botanologicon*, 1534). His support of the Reformation was manifested in polemical writings directed against Eck and Emser (qq.v.). Cordus's real name was Heinrich Urban Solde.

CORINTH, Lovis (Tapian/East Prussia, 1858–1925, Zandvoort, Holland), was an early German Impressionist painter. He is best known for landscapes, but also painted a number of portraits, including among his sitters some literary figures (Eduard Graf Keyserling, P. Hille, G. Hauptmann, and G. Brandes). He also painted F. Ebert (q.v.), the first president of the Weimar Republic. He published books on art, *Das Erlernen der Malerei* (1908), *Legenden aus dem Künstlerleben* (1912), *Über deutsche Kunst* (1914). His collected writings were published in 1921, and an autobiography appeared posthumously in 1926.

Coriolan (Coriolanus), a tragedy written in 1804 by H. J. von Collin (q.v.), for which Beethoven (q.v.) wrote the overture in 1807.

CORNELIUS, Peter von (Düsseldorf, 1783–1867, Berlin), painter, spent some time in Rome (1811–19), where he joined the Nazarene school (see Nazarener). He sought to revive mural painting on the grand scale, working for the most part in Munich. Some of his large works were destroyed during the 1939–45 War, but those in the Ludwigskirche and Glyptothek survived. His name was held in high repute through the latter part of the 19th c. Early in his career (1816) he made a series of pen-and-ink drawings for Goethe's *Faust* (q.v.). In 1825 he was ennobled.

Cornet, common abbreviation for Rilke's *Die Weise vom Leben und Tod des Cornets Christoph Rilke* (q.v.).

CORRODI, Wilhelm August (Zurich, 1826–85, Zurich), an art master in Zurich, wrote the dialect idylls *De Herr Professor* (1858), *De Herr Vikari* (1858), and *De Herr Doktor* (1860). His dialect comedies include *De Ritchnecht* (1873), *De*

Maler (1875), *Eine Pfarrwahl* (1877), *D' Bademerfahrt* (1879), *Wie d' Warret wirkt* (1884), and *De Gast* (1885). His poems were published in 1853 (*Lieder*).

CORVINUS, Jakob, the pseudonym used by W. Raabe (q.v.) for the first publication of *Die Chronik der Sperlingsgasse* (q.v.) in 1857.

COTTA, Johann Friedrich (Stuttgart, 1764–1832, Stuttgart), printer and publisher, son of the founder of the firm, which he took over in 1787. Cotta made Schiller's acquaintance in Württemberg in 1794, undertook to publish *Die Horen* (q.v., 1795), and subsequently published Schiller's new productions as well as most of Goethe's. Throughout the greater part of the 19th c. the firm was the principal publisher of the German classics. Cotta was also known for his philanthropic activities which included the abolition of serfdom on his estates.

Counter-Reformation, see Gegenreformation.

Courasche, Landstörzerin, a minor figure in Grimmelshausen's novel *Der abenteuerliche Simplicissimus* (q.v.) and the principal character in his later novel, *Trutz Simplex: Oder Ausführliche und wunderseltzame Lebensbeschreibung der Ertzbetrügerin und Landstörtzerin Courasche* (q.v.).

Court Epic, see Höfisches Epos.

COURTHS-MAHLER, Hedwig (Nebra, Thuringia, 1867–1950, Rottach-Egern, Bavaria), *née* Mahler, a most prolific author of popular fiction. Between 1905 and the beginning of the Second World War barely a year passed, during which she did not publish new works (more than two hundred in all) that appealed to a broad public and, though not as much as in the 1920s and 1930s, still satisfy a need for the reading and viewing public.

Titles like *Die wilde Ursula* (1912, her first great success), *Der Scheingemahl* (1919), *Glückshunger* (1921), and *Des Herzens süße Not* (1932) mingle with *Hexengold* (1914), *Griseldis* (1916), *Die schöne Melusina* (1924), and *Aschenbrödel und Dollarprinz* (1928), and with *Es irrt der Mensch* (1910), *Das ist der Liebe Zaubermacht* (1924), and *Nur wer die Sehnsucht kennt* (1934, the first line of a song sung by Mignon in Goethe's novel *Wilhelm Meisters Lehrjahre*, q.v.). Not recognized as literature of aesthetic value, her work has been referred to as representing Un-Bildung, Un-Kunst, and Kitsch (q.v.); it is also relevant to the study of Trivialliteratur (q.v.). Many titles have appeared in translation.

CRAMER, Johann Andreas (Jöhstadt, Saxony, 1723–88, Kiel), studied in Leipzig and, as a young man, was one of the founders of the *Bremer Beiträge* (q.v.). He became a pastor in 1748 and achieved rapid preferment, becoming on Klopstock's recommendation court preacher (Hofprediger) in Copenhagen in 1754, and also professor of theology. He was implicated in Struensee's fall (1771, see STRUENSEE) and banished, but was recalled in 1772. In 1774 he settled in Kiel as professor of theology. In Copenhagen he edited *Der nordische Aufseher*, publishing in it many of his own productions, which incurred the critical condemnation of Lessing in the *Literaturbriefe* (q.v.). A collection of hymns (*Das Schleswig-Holsteinische Gesangbuch*, 1780) survived into the 19th c. His collected poems were published in 1782.

CRAMER, Karl Friedrich (Quedlinburg, 1752–1807, Paris), an original member of the Göttinger Hainbund (q.v.), was appointed professor ordinarius for Greek and Oriental languages at Kiel University in 1780. A supporter of the French Revolution, he was dismissed and banished in 1794, emigrating to Paris, where he became a bookseller, translator, and hack writer. He contributed poems and other minor writings to Claudius's *Der Wandsbecker Bote* (q.v.), and included among his translations versions of Chateaubriand, Racine, and Rousseau in German, and of Klopstock and Schiller in French.

CRAMER, Karl Gottlob (Prädelitz nr. Freyburg/Unstrut, 1758–1817, Meiningen), a forestry official, wrote numerous 'Gothic' novels (see RITTER- UND RÄUBERROMAN) of which the best known is *Hasper a Spada* (1792–3), set in the 13th c. He is also the author of a well-known song, 'Feinde ringsum', which first appeared in his novel *Hermann von Nordenschild* (1791).

CRANACH, Lukas (Kronach nr. Bamberg, 1472–1553, Weimar), painter and etcher, is known as 'der Ältere' because his sons Hans and Lukas were also artists. Cranach favoured religious subjects, including a Judith with the head of Holofernes, and was a noted portraitist. Among his sitters was Luther (q.v.), whose friend he was. Several portraits of Luther survive. Cranach der Ältere was outstanding in his treatment of woodcuts. In later life he painted nudes in a distinctive style.

CRANC, Claus, a Minorite friar in Thorn, recorded between 1323 and 1335, is the author of a verse translation of the prophetic books of the Bible, which he executed for Siegfried von Dahenfeld, marshal of the Teutonic Order from 1347 to 1359 (see DEUTSCHER ORDEN).

CREUZER, Friedrich (Marburg, 1771–1858, Heidelberg), a classical philologist, was professor at Marburg from 1802 to 1807 and at Heidelberg from 1807 to 1845. Creuzer published *Symbolik und Mythologie der alten Völker, besonders bei den Griechen* (1810–12) and in old age wrote an autobiography, *Aus dem Leben eines alten Professors* (1858). He fell in love with Karoline von Günderode (q.v.) and contemplated divorce, but became reconciled with his wife after she had tended him during a serious illness. This change of intention was followed by the suicide of Karoline.

Cristinas Heimreise, a comedy by H. von Hofmannsthal (q.v.), published in 1910. The plot, set in Venice, is drawn from the *Mémoires* of Casanova (q.v.). The young and handsome adventurer Florindo seduces Cristina, a beautiful country girl on a visit to Venice, and abandons her in the morning, leaving the sea captain Tomaso, a blunt, but kindly and reliable character, to escort her to her rural home.

Critische Abhandlungen von dem Wunderbaren, a critical essay by J. J. Bodmer (q.v.).

Critische Dichtkunst, worinnen die poetische Malerei in Absicht auf die Empfindung im Grunde untersuchet und mit Beispielen aus den berühmtesten Alten und Neuern erläutert wird, a treatise on aesthetics in two volumes by J. J. Breitinger (q.v.).

CRONEGK, Johann Friedrich von (Ansbach, 1731–57, Nürnberg), the son of a Prussian general, studied at Halle and Leipzig, and travelled for a year in Italy and France. He died of smallpox in the night between 31 December 1757 and 1 January 1758, which accounts for differences of opinion as to the year of his death. Cronegk wrote a tragedy, *Codrus*, which in 1758 won a competition instituted by F. Nicolai (q.v.), and left another, *Olint und Sophronia*, three-quarters finished. The latter was completed in 1764 by K. A. von Roschmann-Hörburg (q.v.). Both tragedies were published posthumously in 1760, are classical in form and written in Alexandrine verse. Cronegk also wrote melancholy poems and published a periodical, *Der Freund*, 1754–6.

CROTUS RUBEANUS or **RUBIANUS** (Dornheim/Thuringia, 1480–after 1539), was a member of the humanist group in Erfurt headed by MUTIANUS Rufus (q.v.), and his principal literary distinction is the authorship of the first part of the polemical Latin work, *Epistolae obscurorum virorum* (q.v.). He had early contacts with Luther (q.v.) and for a time supported him. In 1530 he was received back into the Roman Catholic Church, becoming a canon of Halle. He defended his action in a Latin *Apologia*. Luther derided Crotus as 'Dr. Kröte'. Crotus's real name was Johannes Jäger.

CRÜGER, JOHANN (Großbreesen/Guben, 1598–1662, Berlin), was organist at the Nikolaikirche, Berlin, for the forty years preceding his death. He composed the melodies for a number of well-known Protestant hymns, including some by P. Gerhardt (q.v.), who was pastor at the same church. They include 'Nun danket alle Gott', 'Jesus, meine Zuversicht', 'Jesus, meine Freude', 'Schmücke dich, o liebe Seele', 'Herzliebster Jesu', 'Gelobet seist du, Jesu Christ', and—by Gerhardt—'Befiehl du deine Wege', 'Fröhlich soll mein Herze springen', 'Ich steh' an deiner Krippe hier', and 'Geh aus, mein Herz, und suche Freud'.

CRUSIUS, JOHANNES PAUL (Strasburg, 1588–1629, Strasburg), a professor at Strasburg, wrote the Latin plays *Crösus* (1611) and *Heliodor* (1617).

CSOKOR, FRANZ THEODOR (Vienna, 1885–1969, Vienna), was an officer in the Austrian army 1915–18. He had previously been a *régisseur* in St. Petersburg, and after the war he returned to this occupation in Vienna, first at the Raimund-Theater and later at the Deutsches Volkstheater. In 1938 the German invasion of Austria drove him to Poland, Romania, and Yugoslavia. He was arrested on the island of Korčula, interned, and released in 1945. From 1946 he lived in Vienna. He was a prolific dramatist in the Expressionist manner (see EXPRESSIONISMUS). His first plays concerned humanity and war: *Die Sünde wider den Geist, Die rote Straße* (both 1918), and *Der Baum der Erkenntnis* (1919). *Gesellschaft der Menschenrechte* (1929) is a play about G. Büchner (q.v.) whom Csokor admired. He was inclined to group what were originally single plays into series, as in his *Europäische Trilogie*, which consists of plays written over seventeen years: *3. November 1918* (1936, a portrayal of the dissolution of the Austrian army), *Besetztes Gebiet* (1930, on the French occupation of the Ruhr), and *Der verlorene Sohn* (1947). The religious trilogy *Olymp und Golgatha* consists of *Kalypso* (1941), *Cäsars Witwe* (1953), and *Pilatus* (1949). He also wrote a so-called 'Dramatisches Diptychon', *Die Kaiser zwischen den Zeiten* (1965), dealing with Christianity under Diocletian and Constantine. Other plays are *Gewesene Menschen* (1932), *Gottes General* (1938), *Wenn sie zurückkommen* (1940), *Die Erweckung des Zosimir* (1960), and *Das*

Zeichen an der Wand (1962). Csokor also wrote comedies (*Die Weibermühle*, 1932, *Treibholz*, 1959). He is the author of poetry (*Der Dolch um die Wunde*, 1918, *Das schwarze Schiff*, 1944, and *Immer ist Anfang*, 1952). His only novel (about the Anabaptist rising in Münster 1534–5) appeared in 1933 as *Das Reich der Schwärmer* and was republished in 1955 as *Der Schlüssel zum Abgrund*. The collection *Ein paar Schaufeln Erde. Erzählungen aus 5 Jahrzehnten* appeared in 1965.

Cuius regio, eius religio, principle accepted and established in the Religious Peace of Augsburg (1555, see AUGSBURGER RELIGIONSFRIEDE). A compromise designed to cope with the existing situation, it laid down that the subjects of the states of the Holy Roman Empire must accept the religion of their respective rulers or emigrate.

CUSANUS, see NIKOLAUS VON CÜES.

CUSPINIAN, JOHANNES (Schweinfurt, 1443–1529, Vienna), an early humanist, was rector of Vienna University in 1500 and held office in the councils of the Emperor Maximilian I (q.v.). His history of Rome, *De Caesaribus atque Imperatoribus Romanis opus insigne*, was published posthumously in 1540 by Nikolaus Gerbel (q.v.). Cuspinian's real name was Hans Spießhaymer.

Custozza, close to Verona, was the scene of an Austrian victory over an Italian army on 24 June 1866. The victorious general was the Archduke Albrecht. The campaign was subsidiary to the war in Bohemia, in which the Austrians were defeated by the Prussians (see DEUTSCHER KRIEG). Consequently Custozza failed to save Venetia for Austria.

CZEPKO (VON REIGERSFELD), DANIEL (Koschwitz nr. Liegnitz, 1605–60, Wohlau), a religious poet, grew up in Schweidnitz, and studied law in Strasburg. After a period of practice at the imperial lawcourts at Speyer, he returned home and in 1636 married a rich wife. He lived on her estates until her death in 1658, and then took office in the administration of the Duchy of Brieg. Czepko, who was a Protestant, was a friend of Abraham von FRANCKENBERG (q.v.), with whose mystical religious views he sympathized. His work, which remained unpublished in his lifetime, consists of sonnets, pastoral poetry, including a 'pastoral epic' (*Corydon und Phyllis*), a play *Pierie*, and a series of distichs. In these he is a forerunner of ANGELUS Silesius (q.v.), who knew Czepko and read his epigrams in MS. Though much of Czepko's poetry is imitative of Opitz (q.v.), his tone has an invariable sincerity and personal simplicity. Most of his work had to wait for publication until 1930–2.

D

DACH, Simon (Memel, 1605–59, Königsberg) a lyric poet, became a schoolmaster in Königsberg in 1633, and professor of poetry at the university in 1639. By nature retiring, he led an uneventful life and mingled principally with the group of citizens who wrote poetry in their spare time and formed the so-called Königsberg school (see Königsberger Dichterschule). Shortly before his death, the Great Elector (see Friedrich Wilhelm, der Grosse Kurfürst) granted Dach a small estate.

Dach was gifted musically, as well as poetically. Much of his poetry is occasional, written to order for weddings or funerals, christenings or graduations, and provided him with a small income to supplement his meagre stipend. The most famous poem associated with his name, *Ännchen von Tharau,* is now regarded as probably the work of Dach's friend Heinrich Albert (q.v.), who certainly composed the tune. In addition to lyrical and occasional poetry, Dach wrote a play to celebrate the centenary of the university (*Prussiarcha,* 1644), which was performed by students. *Gedichte* (4 vols.), ed. W. Ziesener, were published 1936–8.

DACHSBERG, Augustinus, see Büchsenbuch, Das.

Dadaismus (Dada), the etymology of which has been alternatively attributed to inarticulate childish sounds and to the French 'dada' (hobby-horse), was an exaggerated form of revolutionary Expressionismus (q.v.), originating about 1916 in reaction against the 1914–18 War and continuing until about 1922. In February 1916 H. Ball (q.v.) founded the Cabaret Voltaire in Zurich. Its name was meant to be provocative and it was conceived as an international centre. Dada's principal German exponents were H. Arp, W. Mehring, K. Schwitters, R. Huelsenbeck, and G. Grosz (qq.v.). The use of colour by W. Kandinsky (q.v.) influenced the favoured Klanggedicht, sound-poems to which he himself contributed. After the war groups were formed in Germany, notably in Berlin and Cologne. The Dadaists were in principle against aesthetic systems and produced none, but they adopted the technique of simultanism and bruitism as an expression of revolt against bourgeois attitudes to art, including Expressionism as they saw it. The aim was a poetry 'Ohne-Sinn' (H. Arp) designed to create a new consciousness of (primitive) life with vague notions of a new beginning; their orien-

tation was psychoanalytical and political, though the radical Berlin Bolshevik Dadaists, including R. Huelsenbeck and W. Mehring, were not committed to Activist meliorism (J. C. Middleton).

Huelsenbeck is the author of *Erste Dada-Rede in Deutschland* (1918), *Dadaistisches Manifest, En avant Dada* and *Dada siegt,* and *Die dadaistische Bewegung,* and editor of the *Dada-Almanach* (all 1920) and of *Dada; eine literarische Dokumentation* (1964, reissued 1984). In 1957 he published *Mit Witz, Licht und Grütze. Auf den Spuren des Dadaismus.* H. Ball reflects on his association with Dada in *Die Flucht aus der Zeit* (1927). In 1957 Huelsenbeck, Arp, and Tristan Tzara published *Die Geburt des Dada* and in 1961 (Arp and Tzara) *Dada Gedichte. Dichtungen der Gründer.*

'Da droben auf jenem Berge', first line of a folk-song included in *Des Knaben Wunderhorn* (q.v.).

DAFFINGER, Moritz Michael (Vienna, 1790–1849, Vienna), a gifted Viennese miniaturist who executed numerous portraits of society figures. During the Congress of Vienna (see Wiener Kongress) he painted many political personalities including Metternich and Napoleon's son the Duke of Reichstadt. His large collection of water-colours of Austrian flowers is in the Akademie der bildenden Künste. A complex personality, he married in 1827 Marie von Smolenitz to whom Grillparzer (q.v.) was strongly attracted.

Dafnis, a pastiche of baroque poetry published by A. Holz (q.v.) in 1904. The work is described as 'Des berühmten Schäffers Dafnis sälbst verfärtigte, sämbtliche Freß-, Sauff- und Venus-Lieder benebst angehänckten Aufrichtigen und Reuemüthigen Buß-Thränen'.

DAHLMANN, Friedrich Christoph (Wismar, 1785–1860, Bonn), originally a classical philologist under F. A. Wolf (q.v.), became a historian and took up politics while professor of history at Kiel University (1813–25), participating in the national movement against Danish encroachment (see Schleswig-Holsteinische Frage). Appointed professor at Göttingen in 1825, he was the leading figure among the seven professors (see Göttinger Sieben) who protested against the abrogation of the constitution in 1837 and were dismissed. In 1842 Dahlmann accepted the chair of history and political science at Bonn. He was a member of the Frankfurt Parliament (see FRANKFURTER NATIONAL-VERSAMMLUNG and REVOLUTIONEN 1848–9). His liberal views were influenced by British models, and his writings reflect the nationalism generated during the Napoleonic period. In his view a united Germany should not include Austria (see KLEINDEUTSCH). This attitude underlies his *Quellenkunde der deutschen Geschichte* (1830). He also wrote a *Geschichte von Dänemark* (3 vols., 1840–3).

DAHN, FELIX (Hamburg, 1834–1912, Breslau), son of an actor, was brought up in Munich, studied law there and at Berlin University, and began an academic career in 1857. From 1863 to 1872 he was a professor at Würzburg University, was then called to Königsberg, and finally (1888–1910) to Breslau University. As a young man he published an epic poem (*Harald und Theano*, 1855) and a volume of poems (*Gedichte*, 1857). He continued to write poetry and also composed two tragedies (*König Roderich*, 1876; *Markgraf Ruediger von Bechelaren*, 1875) and a drama (*Deutsche Treue*, 1876), but made little impact until the publication of his first novel (*Ein Kampf um Rom*, 4 vols., 1876), which immediately popularized his name throughout Germany. It deals with the decline and fall of the Ostrogoth empire in Italy in the 6th c. A.D., and initiated a series of novels of early German history, including *Odhins Trost* (1880), *Kleine Romane aus der Völkerwanderung* (13 vols., 1892–1901), *Odhins Rache* (1891), *Julian der Abtrünnige* (3 vols., 1894), *Attila* (1895), and *Stilicho* (1901). Dahn's conscientiously detailed and highly coloured evocations of the Germanic past were overvalued at the end of the century and a sharp reaction followed, culminating in total neglect. But he was probably the most talented of the practitioners of the so-called Professorenroman (q.v.).

Daily Telegraph-Affäre, an episode which in 1908 endangered the position of the German emperor Wilhelm II (q.v.). The Emperor, at a time of stress in Anglo-German relations, gave an interview to Colonel Stuart Wortley, in which he spoke of Great Britain with well-intentioned but tactless and even reckless phrasing. Wortley secured the Emperor's consent to publication, and the German Foreign Office passed the document. Publication in the *Daily Telegraph* on 28 October 1908 provoked a storm of protest in Great Britain and equally vociferous criticism of the Emperor in Germany. The chancellor, Bülow (see BÜLOW, BERNHARD, FÜRST VON), who had neglected to check the document, failed to support the Emperor and abdication seemed possible. In the end Wilhelm had to give an undertaking to observe proper constitutional channels in future. Bülow's equivocal attitude contributed to his dismissal by the Emperor when opportunity offered two years later.

DALBERG, KARL THEODOR, REICHSFREIHERR VON (Herrnsheim, Worms, 1744–1817, Regensburg), scion of an influential family in the Rhineland and brother of Heribert von Dalberg (see DALBERG, W. H.), became in 1787 heir-designate to the electoral archbishopric of Mainz. He was at the time governor of Erfurt and, being a man of enlightened opinions and literary interests, entertained men of letters, including Schiller (q.v.) and his wife, to whom he held out hopes of patronage on his accession. The French Revolution, however, intervened. In 1800 Dalberg became bishop of Constance and at last, in 1802, Elector of Mainz; but his reign was short-lived, for the Holy Roman Empire was dissolved four years later. Karl Theodor then became the Primate of the Confederation of the Rhine (see RHEINBUND) instituted by Napoleon. The War of Liberation caused his dismissal in 1813, but he retained the archiepiscopal see of Regensburg.

DALBERG, WOLFGANG HERIBERT, REICHSFREIHERR VON (Herrnsheim, Worms, 1750–1806, Mannheim), was from 1778 to 1803 Director (Intendant) of the Mannheim National Theatre. He wrote a number of ephemeral plays and also made adaptations of Shakespeare, Southern, and Cumberland. Dalberg accepted Schiller's *Die Räuber* (q.v.) for first performance (1782), and later *Fiesco* and *Kabale und Liebe* (qq.v., both 1784). Though he appears in an unsympathetic light in Schiller's correspondence, he was a capable director.

Damenfriede, alternative designation for the Peace of Cambrai (see CAMBRAI, FRIEDE VON).

DAMON, poetic name adopted by S. G. Lange (q.v.) in the title *Thirsis und Damons freundschaftliche Lieder* (1745).

Dämonen, Die, a novel in three parts (2 vols.) by H. von Doderer (q.v.), published in 1956. Its title is derived from Dostoevsky's novel, variously styled in German *Die Dämonen* or *Die Teufel*, and in English *The Possessed* or *The Demons*. Doderer began the work in 1931, completed Part One in 1937, but did not resume writing until 1950. It is superscribed *Nach der Chronik des Sektionsrates Geyrenhoff*, who, in the chapter *Ouvertüre*, points out that passages of the book are also 'written' by Kajetan von Schlaggenberg and other characters. The first part, beginning

in 1926, deals with a group of people who have
lost their youth through the war. They consist
principally of Geyrenhoff, Schlaggenberg and his
sister, nicknamed Quapp, Imre von Gyurkiez,
and also René Stangeler and Rittmeister von
Eulenfeld, both characters in the *Strudlhofstiege*
(q.v.).

'Die Unsrigen', as Geyrenhoff calls them,
spend their time in introspective discussion in
the style of Dostoevsky. Contrasted with them
are the wives of well-to-do bankers and civil
servants, whose chief recreation is to gossip over
pastry and cream in the cafés of Vienna. Contrast of a different kind is provided by the
weaver Leonhard Kakabsa, who, by a chance
encounter, is led to apply himself to learning
Latin. As the book proceeds, the Dostoevskian
element diminishes, a host of diverse figures
move in a complex pattern, their lives now running parallel, now intersecting. They include the
sinister figure of the financier Levielle, the rich
and childless widow Friederike Ruthmayr, Frau
Selma Steuermann, the Hungarian diplomat
von Orkay, the honest banker Altschul, René
Stangeler's fiancée Grete Siebenschein, the
prostitutes Annie Gräven and Herta Plankl, the
latter's murderer Meiergeis, Jan Herzka, who
inherits Burg Neudegg, and many others, including Mary K., whose disastrous tram
accident is the climax of *Die Studlhofstiege*.

In spite of the initial prevalence of discussion,
Die Dämonen has a plot. René Stangeler, transformed by the reading of a 16th-c. MS. which he
finds at Burg Neudegg, discovers his real self,
resolves the eternal wranglings with Grete
Siebenschein, and marries her. Quapp proves to
be the illegitimate daughter of Rittmeister Ruthmayr, who in his last testament on a Galician
battlefield in 1914 made the amplest provision
for her, which has been withheld by the duplicity
of Levielle. Geyrenhoff comes to the rescue, her
fortune is restored, and she marries Geza von
Orkay. Geyrenhoff himself saves the fortune of
Friederike Ruthmayr from the impending
failure of the Bodencreditanstalt, and becomes
her second husband. The factory worker Leonhard, without losing his working-class identity,
becomes a scholar of note and librarian to the
Prince von Croix. He becomes devoted to Frau
Mary K., and his devotion is accepted.

The climax of the book is the (historical) 15
July 1927, on which day a strike and a demonstration by the electricity workers degenerate
into a conflict with the police and the storming
and burning of the Justizpalais. The worst
excesses are precipitated, not by the workers, but
by disreputable elements from the quarter of the
Prater. Doderer traces the movements of his
characters in relation to the riot throughout the
day, some evading the conflict entirely, some as
distant spectators watching the smoke from the
hills, others caught up in the tumult. The result
is a convergent method of presentation by which
the same events are seen from different angles.
Doderer writes of two realities, 'eine erste und
eine zweite Wirklichkeit', by which he seeks to
convey not only the 'layers' of personalities, but
also the obsessions by means of which a man
evades self-recognition.

Die Dämonen presents both a complex exercise
in analytical psychology and an intricate pattern
of facts, designed to 'fix' for ever a point in space
and time, Vienna 1927, 'Jenseits im Diesseits'.

Dämonisch, a word which Goethe defines in
the last book of *Dichtung und Wahrheit* (q.v.,
iv. 20), attaching it to an irrational phenomenon
('Wesen') of Nature (both inorganic and
organic) capable of determining man's destiny.
Detecting it in himself, he felt that it contra-
vened established laws of existence and moral
concepts. He experienced it as an irresistible
urge threatening his conscious will and conduct.
'Ich suchte mich vor diesem furchtbaren Wesen
zu retten, indem ich mich, nach meiner
Gewohnheit, hinter ein Bild flüchtete', he wrote
with reference to *Egmont* (q.v.). The 'daemonic'
manifests itself in Egmont's blind trust in the
rightness of his instincts. In concluding *Dichtung und Wahrheit* Goethe quoted once more
Egmont's own words in Act Two: 'Wie von
unsichtbaren Geistern gepeitscht, gehen die
Sonnenpferde der Zeit mit unsers Schicksals
leichtem Wagen durch; und uns bleibt nichts,
als mutig gefaßt die Zügel festzuhalten, und
bald rechts, bald links, vom Steine hier, vom
Sturze da, die Räder wegzulenken. Wohin es
geht, wer weiß es? Erinnert er sich doch kaum,
woher er kam.' The 'daemonic' embraces
Egmont's self-destructive impulsion. Other
creations from *Prometheus* to *Faust* (qq.v.) owe
their life-blood to this experience of the young
Goethe. However, Goethe also points out that
when it dominates a man it can manifest itself in
tyranny over others. He claims that he knew or
had heard of men who were invincible in human
terms, and to whom the words 'Nemo contra
deum nisi deus ipse' might well apply. Goethe
refrains from quoting examples. But he is
thought to have mentioned to Eckermann
Napoleon, Karl August (q.v.), Byron, Friedrich
II (q.v.) of Prussia, and Peter the Great of
Russia. The 'daemonic' is related to Goethe's
use of the word 'Dämon' in various works of his
later creative period (cf. the poem *Urworte.
Orphisch*, q.v.).

Daniel, a Middle High German biblical poem
written *c*. 1330 by an anonymous priest of the
Teutonic Order (see DEUTSCHER ORDEN). It

narrates the content of the Book of Daniel, adding allegorical interpretations intended to show how the book prefigures the life of Christ and the doctrines of Christianity. The author states that his poem was written at the request of the High Master of the Teutonic Order, Luder von Braunschweig.

DANIEL VON SOEST, whose real name was Gervin Haverland (*c.* 1490–still living in 1539), was a Minorite in charge of the priory at Soest. He was a gifted satirist, who defended the old faith and attacked the reformers in a Low German homely verse satire, *Eine gemeyne Bicht oder Bekennung der Predikanten to Soest* (1539), which, for all its indignation, is not lacking in humour and in vigour of description and presentation. His writings also include *Ketterspegel* (1533), *Dialogon,* and *Apologeticon* (both 1539).

Danklied, see NUN DANKET ALLE GOTT.

DANNECKER, JOHANN HEINRICH (Waldenburg nr. Stuttgart, 1758–1841, Stuttgart), a Württemberg sculptor, was the son of a groom in the ducal stables. Dannecker was educated in the ducal Militär-Akademie, where he was one of Schiller's companions. After study in Paris and Rome he was appointed professor of sculpture in the Karlsschule (q.v.). In 1794 he executed a bust of Schiller (q.v.), who was then on a visit to Stuttgart. It remains one of the most impressive portraits of the poet.

DANTON, GEORGES (Arcis-sur-Aube, 1759–94, Paris), a leading figure in the French Revolution (q.v.), educated as a lawyer, founded in 1790, with Camille Desmoulins and Marat, the radical Club des Cordeliers and took part in the storming of the Tuileries on 10 August 1792. As Minister of Justice he gained sudden prominence and was responsible (though historians are not unanimous as to what degree) for the September Massacres of 1792 (2–5 September). As a member of the National Convention he was the chief author of the Revolutionary Tribunal (March 1793). He made common cause with Robespierre against the Girondists and, in March 1794, against the Hébertists. Danton had for some time been suspect in the Jacobin Club as a moderate. He was arrested on 30 March 1794 and condemned by the Revolutionary Tribunal. He was executed six days later.

G. Büchner made free use of this background in his play *Dantons Tod* (q.v.). Danton is also an important figure in *Danton und Robespierre* by R. Hamerling (q.v.).

Dantons Tod, a play in four acts by G. Büchner (q.v.), which was written within five weeks in January and February 1835. In the same year it was published twice, in instalments by K. Gutzkow (q.v.) in the journal *Phönix* and complete (but with alterations to the text) by E. Duller (q.v.) under the title *Dantons Tod. Dramatische Bilder aus Frankreichs Schreckensherrschaft.* It was not performed until 1902 (Freie Volksbühne, Berlin). It has maintained its place on the 20th-c. stage and was performed in the Group Theatre in London in 1937 in an English version by Stephen Spender and Goronwy Rees.

Among Büchner's sources were the histories of the Revolution (see FRENCH REVOLUTION) by Thiers and Mignet and studies by Mercier and Riouffe; a German work, which Büchner consulted, by K. Friedrich, is conspicuous for drawing on Thiers's record of Danton's speeches at his trial, which include the words: 'le néant sera bientôt mon asyle'. The action is concentrated between 24 March and 5 April 1794.

Büchner introduces Danton (q.v.) as a man in isolation who has left the political scene. The chaos of the Revolution has led him to resign himself to an epicurean mode of life, and to the conviction that man is not the free agent of his actions. At his meeting with Robespierre, Danton urges his rival to put an end to further bloodshed. From Robespierre's monologues at the end of Act One it becomes clear that he shares Danton's fatalistic view of history although he is at the zenith of power and clings to his austere ideal of virtue. He is urged on by St. Just, who has no scruples and is ultimately responsible for the arrest of the Dantonists, among them Camille Desmoulins, Hérault-Séchelles, Lacroix, Philippeau, and Fabre d'Églantine. Büchner incorporates in the second act Danton's historic 'Ils n'oseront pas' ('sie werden's nicht wagen!'), when he has a final chance of flight. But the night before his arrest he discloses to his wife Julie the agony of conscience which haunts him for his involvement in the September Massacres. Danton's courageous denunciation of Robespierre and St. Just before the Revolutionary Tribunal is the culmination of the third act, but the people, easily roused and swayed, side with Robespierre. In the last act Julie poisons herself while Danton is in prison awaiting death. Danton himself shows at his execution an invincible humanity which emerges as the positive force of the play. Camille's young wife, Lucile, dominates the end of the play as she lingers by the guillotine, suddenly exclaiming 'Es lebe der König!' she makes sure that she will follow her love. She is promptly arrested 'in the name of the Republic'.

The play provided the text for the opera

Dantons Tod by G. von Einem (q.v.), first performed in 1947.

Danzig, one of the great Baltic ports situated at the mouth of the Vistula (Weichsel), and now, as Gdansk, included in Poland. In the 14th c. and 15th c. it was in the hands of the Teutonic Order (see DEUTSCHER ORDEN). In 1454 the relationship with the Order was terminated and Polish protection accepted. Danzig was then virtually a free city, possessing links with the Hanseatic League (see HANSE, DEUTSCHE), and it continued to enjoy this status after the first partition of Poland (see POLAND, PARTITIONS OF). In 1793, on the occasion of the second partition, it was incorporated in Prussia. During the Napoleonic domination (1807–14) it reverted to its independent position, but then again became Prussian. In 1920 it regained its status as Free City under the League of Nations. C. J. Burckhardt (q.v.) was its last high commissioner. Predominantly German in population, it was in 1939 a focus of National Socialist agitation, leading up to the 1939–45 War. Danzig is the principal scene of G. Grass's novels *Die Blechtrommel* and *Hundejahre* (qq.v.).

DAPONTE or **DA PONTE,** LORENZO (Ceneda, Italy, 1749–1838, New York), wrote the Italian libretti for Mozart's operas *Le nozze di Figaro*, *Don Giovanni*, and *Così fan tutte* (qq.v.). He spent the years from 1782 to 1791 in Vienna.

Darmstädter Kreis, a circle of society men and women with literary interests and sentimental outlook grouped around Landgräfin Karoline von Hessen at the Residenz, Darmstadt. The principal members of the group, which was active in the 1770s, were J. H. Merck, Herder's fiancée Caroline Flachsland (qq.v.), Henriette von Roussillon, Luise von Ziegler, and F. Leuchsenring (q.v.). In 1771 and 1772 Goethe, then living at home in Frankfurt, was a frequent visitor. The enthusiasm of the group was centred on Klopstock, Richardson, J.-J. Rousseau, Herder (qq.v.), and the writings of Goethe himself.

'Das Meer erglänzte weit hinaus', first line of an untitled poem by H. Heine (q.v.) included in the *Buch der Lieder* (q.v.) as No. XVI of 'Die Heimkehr'. It has been set by F. Schubert (q.v.) as No. 12 of *Schwanengesang* with the title 'Am Meer'.

'Das Moralische versteht sich immer von selbst', often-quoted remark from F. Th. Vischer's novel *Auch Einer* (q.v.) where it is constantly on the lips of the 'hero', A.E. It also occurs in the variant 'Das Höhere versteht sich ja immer von selbst'.

'Das Volk steht auf, der Sturm bricht los', first line of the poem *Männer und Buben* by Theodor Körner (q.v.). This vigorous patriotic poem of the War of Liberation was intended by Körner to be sung to the tune of 'Brüder mir ist alles gleich' (1708). It was subsequently set anew by Weber in 1814. The superscription to the poem indicates that it was written in bivouac in August 1813. It was posthumously published in the collection *Leyer und Schwerdt* (q.v., 1814).

DÄUBLER, THEODOR (Trieste, 1876–1934, St. Blasien), spent part of his youth in Venice and was bilingual in German and Italian. He served as a one-year officer cadet in the Austrian army (1898), and thereafter led a restless life in the great art centres of Europe (Paris, Venice, Florence, Rome, Naples, Vienna, Dresden, and Berlin) and was an early protagonist of Expressionist painting and sculpture (F. Marc and Barlach, qq.v.). His far-ranging journeys took him not only to Switzerland, Sicily, and Greece, but to North Africa and the Levant. In middle and later life he visited the countries of western Europe, including Great Britain, and was noted as a lecturer on art. He contracted tuberculosis, but died of a stroke at a sanatorium in the Black Forest. Däubler published nothing before 1910. Nevertheless he wrote much poetry, ranging from an epic, through lofty odes and visionary hymns, to simple nature poems. His principal poetic works are the epic *Das Nordlicht* (1910), *Oden und Gesänge* (1913), *Hesperien* (1915), *Der sternhelle Weg* (1915), *Hymne an Italien* (1916), *Die Treppe zum Nordlicht* (1920), *Päan und Dithyrambos* (1924), and *Attische Sonette* (1924). He also wrote autobiographical fragments (*Wir wollen nicht verweilen*, 1915), two novels (*L'Africana*, 1928; *Die Göttin mit der Fackel*, 1931), and a large number of essays, including art criticism. E. Barlach, a perceptive friend of the stockily built Däubler, made of him in 1929 a monumental wood-carving entitled 'Ruhender Däubler'. *Dichtungen und Schriften* (1 vol.), ed. F. Kemp, were published in 1956.

DAUMER, GEORG FRIEDRICH (Nürnberg, 1800–75, Würzburg), a schoolmaster in Nürnberg who gave up his post in order to devote himself to writing. His *Bettina* (1837) is a transmutation into poetic form of *Goethes Briefwechsel mit einem Kinde* by Bettina von Arnim (q.v.). *Die Glorie der heiligen Jungfrau* (1841) anticipates his conversion to Roman Catholicism in 1859. Daumer wrote pseudo-oriental poetry (*Orientalische Gedichte*, 1848) and translated Hafiz (*Liederblüten des Hafis*, 1846–52). Several of his poems, including some of the translations of Hafiz, have been set to music by Brahms (q.v.).

DAUN, LEOPOLD, GRAF VON (Vienna, 1705–66, Vienna), Austrian soldier who first distinguished himself in the Polish and Turkish wars. In the War of the Austrian Succession (see ÖSTER-REICHISCHER ERBFOLGEKRIEG) he was the defeated commander at Hohenfriedberg (1745). After the war he was in charge of the reorganization of the army and was promoted field-marshal in 1754. In the Seven Years War (see SIEBEN-JÄHRIGER KRIEG) he defeated Frederick the Great at Kolin (1757) and Hochkirch (1758) and forced the surrender of Finck's army at Maxen (1759). He became president of the War Council (Hofkriegsrat) in 1762. Daun was often reproached with undue caution, and some of his victories are in part attributable to the more aggressive spirit of his subordinate commanders, Laudon (q.v.) and Lacy.

DAUTHENDEY, MAX (MAXIMILIAN) (Würzburg, 1867–1918, Java), began a career as a painter and remained highly responsive to sensuous impressions, especially of colour. He led a restless life in Germany, Sweden, London, and Paris, extending his range later to the U.S.A., Mexico, Greece, Egypt, and the Far East. In 1914 he undertook a journey round the world, which included Arabia and New Guinea, but was interned in Java, where he died. An all-embracing Romantic and a solitary person, he published numerous volumes of verse, e.g. *Ultra-Violett* (1893, the title is said to indicate his own lonely obscurity), *Reliquien* (1899), *Die ewige Hochzeit* (1905), *Lusamgärtlein* (1909), and *Die geflügelte Erde* (1910). His Novellen comprise the collections *Lingam* (1909), *Die acht Gesichter am Biwasee* (1911), and *Geschichten aus den vier Winden* (1915); he also wrote a novel, *Raubmenschen* (1911), and a number of unsuccessful plays. Other prose works include *Der Geist meines Vaters* (1912) and *Gedankengut aus meinen Wanderjahren* (1913). Published posthumously were *Erlebnisse auf Java* (1924) and *Letzte Reise* (1925).

Davidsbündler, see SCHUMANN, ROBERT.

Davidstern, a six-pointed star commonly represented as two intersecting triangles and serving as a symbol of Jewry. The wearing of it, in the form of a yellow cloth star, was imposed as a humiliation on Jews by a National Socialist decree in 1941. In 1948 the Zionist flag with its blue Davidstern became the national flag of Israel.

DAVID VON AUGSBURG (*c.* 1210–72, Augsburg), a Franciscan monk in charge of the monastery school at Regensburg. He is believed to have given instruction to BERTHOLD von Regensburg (q.v.), whom he subsequently accompanied on the latter's preaching peregrinations. In later life David was at Augsburg. He wrote, mainly in Latin, tracts on the Christian virtues and way of life. A few German writings are attributed to him, of which *Die sieben Vorregeln der Tugend, Der Spiegel der Tugend*, and *Von der Erlösung des Menschengeschlechts* are probably authentic. David was a devout and gentle spirit, inclining to mysticism.

Decameron (*Il Decamerone*), written between 1348 and 1353 by Giovanni Boccaccio (1313–75), and first printed in 1521. The first German translation is that of Arigo (q.v., 1472). Some stories from the *Decameron* are told in the *Ehebüchlein* (1472) of ALBRECHT von Eyb (q.v.). Boccaccio uses the word *novella* for his stories, and encloses them in a framing narration which is provided by a group of Florentine men and women who tell 100 *novelle* in the space of ten days (*giornate*). This framing device was borrowed by Goethe for his *Unterhaltungen deutscher Ausgewanderten* (q.v., 1795), which is the starting-point for the German Novelle (q.v.).

In drama Lessing, in his *Nathan der Weise* (q.v., 1779), uses a story from the *Decameron* (I.3) in the parable of the ring, and F. Halm's verse play, *Griseldis* (q.v.), is a dramatization of *Decameron* X. 10. The translation made in 1827 by K. Witte (1800–83), and widely used in the 19th c., was superseded by the version of A. Wesselsky (1871–1939), published in 1909.

De captivitate babylonica ecclesiae praeludium, one of Luther's three important theological tracts of the year 1520. In it he denounces the papacy and attacks the doctrine of sacraments, of which he admits only two: Holy Communion and baptism. See LUTHER, MARTIN.

DEDEKIND, FRIEDRICH (Neustadt/Leine, 1525–98, Lüneburg), a Protestant pastor in Neustadt (1551) and Lüneburg (1575), is the author of *Grobianus* (q.v., the full title is *Grobianus sive de morum simplicitate libri II*, 1549), a famous satire ridiculing coarse and ill-mannered behaviour. Dedekind wrote in Latin and his book was freely translated into German rhyming verse by Kaspar Scheidt (q.v.) in 1551. In this form it enjoyed great popularity. Dedekind subsequently extended his satire to include Grobiana, the female counterpart of Grobianus (1552). His later works, *Der christliche Ritter* (1590) and *Papista conversus* (1596), are less successful.

Defenestration of Prague (Prager Fenstersturz), see MAJESTÄTSBRIEF.

De Heinrico, a political poem of 27 lines written in the 10th c. It is in mixed Latin and Old High German, the first half of each line being in Latin, the second in German, thus

Nunc almus assis filius . thero êuuigero thiernun.

It is a propagandist poem in praise of Duke Heinrich of Bavaria, recounting the gracious reception accorded him by the Emperor Otto. Which Duke Heinrich and which Emperor Otto are concerned has been much disputed. Modern scholarship has tended to reject the earlier view that the reconciliation of Otto I (q.v.) with his brother Heinrich is intended, and refers the poem to a meeting between Otto III (q.v.) and Heinrich der Zänker, his uncle. This view dates the poem, which is contained in the Cambridger Liederhandschrift (q.v.), between 996 and 1002.

DEHMEL, RICHARD (Wendisch-Hermsdorf, 1863–1920, Blankenese), studied science at Berlin and Leipzig, and then took to journalism, devoting himself after 1895 entirely to literature. A man who needed the stimulus of discussion and argument, he associated with the brothers H. and J. Hart, O. E. Hartleben, A. Holz, and Strindberg (qq.v.) in Berlin, and wrote revolutionary lyric poetry. He was also powerfully influenced by Nietzsche (q.v.), to whom his ecstatic rhetoric may be traced. He was a close friend of Liliencron (q.v.) from 1891 until the latter's death in 1909. In 1914 he volunteered for the army and served as a subaltern. He published several volumes of verse (*Erlösungen,* 1891; *Weib und Welt,* 1896; *Die Verwandlungen der Venus,* 1907; *Schöne wilde Welt,* 1913) and two collections, in which stories as well as poems were included (*Aber die Liebe,* 1893; *Lebensblätter,* 1895). The cyclical epic poem *Zwei Menschen* (1903) is often considered his masterpiece. He wrote several plays, including the Schauspiele *Der Mitmensch* (1895) and *Die Menschenfreunde* (1917) and the comedies *Michel Michael* (1911) and *Die Götterfamilie* (posthumous, 1921). Dehmel's collected works were published in his lifetime as *Gesammelte Werke,* 10 vols., 1906–9, followed by a shorter selection with the same title, 3 vols., 1913. His autobiography, *Mein Leben,* appeared posthumously in 1922. His letters (*Ausgewählte Briefe,* 2 vols., 1922–3) are of lasting interest.

De hoc quod spurihalz dicunt, superscription of an Old Low German spell intended to prevent or remedy lameness in a horse. It is written in prose and is presumed to belong to the 9th or 10th c.; it is contained in a MS. preserved in the Nationalbibliothek in Vienna. The word 'spurihalz' denotes lameness. See also TRIERER ZAUBERSPRÜCHE.

Deichgraf, in the coastal areas of the German states bordering the North Sea, title of an official charged with supervision of the maintenance of the dikes. The office was of great antiquity and was normally held by a farmer or notable citizen. Deichgrafen occur in the stories of Theodor Storm, notably in *Der Schimmelreiter* (q.v.).

DEINHARDSTEIN, JOHANN LUDWIG (Vienna, 1794–1859, Vienna), was a conspicuous figure in Viennese literary life during the Metternich era. He became professor of aesthetics at the Theresianum in 1827, and two years later began to edit the *Jahrbücher der Literatur,* continuing until 1849. From 1829 to 1848 he was also official censor of books. Artistic Director of the Burgtheater for nine years (1832–41), he was himself active as a dramatist, writing principally light comedies and Künstlerdramen (see KÜNSTLERROMAN). They include *Ehestandsqualen* (1820), *Hans Sachs* (1829), *Der Streitsüchtige* (1829), and *Garrick in Bristol* (1834). *Hans Sachs,* in an adaptation by Philipp Reger, was composed as an opera by Lortzing (q.v.). Deinhardstein's *Gesammelte dramatische Werke* were published 1848–57. He also wrote poetry, published in *Gedichte,* 1845.

Dekadenz, term applied to a trend of the 1890s. The 'decadents' are characterized by an air of world-weariness, moral negation, and self-absorption. A characteristic formulation is provided by H. von Hofmannsthal's line 'Frühgereift und zart und traurig' ('Einleitung' to Schnitzler's *Anatol,* q.v.). The 'decadents' were almost all young men, and most of them shed some of the symptoms of decadence as they matured. Among those to whom the term has been applied are P. Altenberg, H. Bahr, R. Beer-Hofmann, S. George, H. von Hofmannsthal, Hermann Graf Keyserling, Th. Mann, Rilke, R. von Schaukal, and Schnitzler (qq.v.). The *fin de siècle* mood was most evident in Vienna, and seven of those mentioned are Austrians.

De la littérature allemande, des défauts qu'on peut lui reprocher, quelles sont les causes, et par quels moyens on peut les corriger (1780), see FRIEDRICH II, DER GROSSE.

DELLE GRAZIE, MARIE EUGENIE, see GRAZIE, MARIE EUGENIE DELLE.

Demagogenverfolgung, a term applied to the political persecutions following the Karlsbad Decrees of 1819 (see KARLSBADER BESCHLÜSSE).

Dem aufgehenden Vollmonde, a poem written by Goethe on 25 August 1828, three days before his seventy-ninth birthday, and sent to Marianne von Willemer (q.v.).

Demeter, see BERGROMAN.

Demetrius, a tragedy planned and partly written by Schiller in 1804–5. Its completion was prevented by his death. The plot is drawn from Russian history in the time of Boris Godunov. Demetrius, a servant of unknown origin in an aristocratic Polish household, is revealed as Prince Demetrius (Dmitri), the heir to the Russian throne who, it was believed had been killed by Boris. Demetrius wins over the Polish Diet and undertakes with Polish support an invasion of Russia, in which Boris is killed. He learns meanwhile that he is not after all the true Demetrius, but an unwitting impostor. He refuses to turn from his course, however, and is murdered by nobles in revolt.

A scenario and many notes and sketches survive, together with sections of the play fully worked in verse. The most important of these are a first act in which Demetrius appears before the Polish Diet, and a scene for the second act, in which Marfa, the mother of the murdered prince Dmitri, learns of the supposed Demetrius, disbelieves his claim, but nevertheless determines to support him in order to be avenged upon Boris. There are also fragments of an earlier first act, in which Demetrius is identified.

Goethe (q.v.) thought of completing the play after Schiller's death. Hebbel (q.v.) attempted the task for the centenary of Schiller's birth, but abandoned the plan. He subsequently wrote his own *Demetrius,* a tragedy in verse which likewise remained a fragment, though he completed four acts before his death in 1863. His different treatment of the subject affects mainly the portrayal of Marfa and Demetrius's reaction to his discovery that he is an impostor, which impels him to renounce the crown. Schiller intended a tragedy of will and power, Hebbel one of conscience. Completed versions by F. G. Kühne and H. Laube (qq.v.) were published in 1859 and 1869 respectively.

Demian, a novel by H. Hesse (q.v.), described as 'Die Geschichte von Emil Sinclairs Jugend', and published in 1919. The book is a kind of Bildungsroman (q.v.), and Hesse draws freely on the experiences of his own childhood and youth. Emil Sinclair becomes acutely conscious of the duality of life, reflected in the tidy, sheltered, unexciting world of his home and in the dynamic, wicked, and cruel world outside, which impinges on him in many ways, but especially in the aggression of the bully Franz Kromer. He is rescued from his moral confusion by a new and older boy at the school, Max Demian, with

whom he later temporarily loses contact. He hears from Demian, who writes to him of the god Abraxas, in whom the divine and the diabolical are fused, and who represents the fulfilment of all the individual's impulses. Emil finds another worshipper of Abraxas in the young organist Pistorius, but is unable to accept Pistorius's negative attitude to sex. He again encounters Demian and the latter's mother, both of whom influence him profoundly.

Demian stresses the decline of European civilization and prophesies its impending catastrophic end in mystical and quasi-oriental terms, already implicit in the reference to the Greek Abraxas. He looks forward to a regeneration of the world. The disaster to civilization comes with the outbreak of war in 1914. Demian, who is an officer of the reserve, reports for duty, and Sinclair is called up. During his service the latter recognizes good impulses in men and deplores the fact that they seem to be evoked only in situations of emergency. But he also encounters a few individuals capable of detached heroism and devotion. Emil is wounded, and while in a field hospital has a visionary encounter with Demian, which implies the latter's death. He remains devoted to Demian's memory, and feels himself the inheritor of his friend's noble personality.

Dem Schmerz sein Recht, eleven poems by F. Hebbel (q.v.), arranged in a cycle in the final collection of his poems in 1857. The work includes poems from the cycle *Lebens-Momente,* arranged in 1836 and published separately in the earlier collections of 1842 and 1848. The brief 'Schlußgedicht' was written later to balance the temper of the cycle, which commemorates Hebbel's experience of isolation, despair, and longing for escape, which he began to express in poetry before he turned to drama. The tenth poem contains the line 'Alles Leben ist Raub', which suggests in the context that suffering, the consuming force of life, makes the gods indebted to him who transcends it and discovers in it a purpose.

DENAISIUS, PETRUS (Strasburg, 1560–1610, Heidelberg), a lawyer by profession, was employed on diplomatic missions by the Elector of the Palatinate. He is the author of a polemical poem attacking the Jesuits (*Jesuiterlatein,* 1607).

DENCK, HANS, see DENK, H.

DENIS, JOHANN NEPOMUK COSMAS MICHAEL (Schärding/Inn, 1729–1800, Vienna), entered the Society of Jesus in 1747 and was ordained priest in 1757. From 1759 he lived in Vienna as a schoolmaster, and in 1784 he was appointed chief librarian at the Imperial Court Library (Hofbibliothek). As pseudonym he used the anagram Sined. In his early years Denis wrote a Latin play for school performance. He translated Ossian into hexameters (*Gedichte Ossians*, 1768–9), and revised his translation after the appearance of Macpherson's second revised edition (*Lieder Ossians und Sineds*, 1784). As a 'bardic poet' (see BARDENDICHTUNG) he followed the lead of Klopstock (q.v., *Lieder Sineds des Barden*, 1772).

DENK, or DENCK, HANS (Habach, Oberfranken, *c.* 1495–1527, Basel), a schoolmaster in Nürnberg (Sebaldusschule), was expelled from the city for his Anabaptist beliefs. He settled in Basel and was conspicuous in religious controversy. With L. Hätzer (*c.* 1500–29) he made an excellent translation of the Hebrew prophets, sometimes referred to as the *Wormser Propheten* (1527).

Denk' es, o Seele!, a poem, consisting of two stanzas of irregular unrhymed verse, written by E. Mörike (q.v.) *c.* 1851. It is a reminder of the precariousness of life, and is aptly used as the conclusion to Mörike's Novelle *Mozart auf der Reise nach Prag* (q.v.). It has been set to music by Hugo Wolf (q.v.).

'Der Dichter steht auf einer höhern Warte/Als auf den Zinnen der Partei', see PARTEI, PARTEI, WER SOLLTE SIE NICHT NEHMEN.

'Der du von dem Himmel bist', first line of Goethe's poem *Wanderers Nachtlied* (q.v.).

DERFFLINGER, GEORG VON (Neuhofen, Austria, 1606–95, Gusow), was at first in Swedish service in the Thirty Years War (see DREISSIGJÄHRIGER KRIEG). In 1654 he transferred to the Brandenburg army and rose to field-marshal in 1670. The intervention of his cavalry decided the battle of Fehrbellin (q.v.) in 1675. His nickname 'Der alte Derfflinger' was earned by his continued active service into old age. At 72 he participated in the winter campaign in East Prussia, and, in his eighty-fourth year, was present at the siege of Bonn (1689). In his ballad *Der alte Derffling* (1846) Th. Fontane (q.v.) recalls that Derfflinger began as a tailor before becoming a soldier.

'Der Freiheit eine Gasse!', dying words attributed to Arnold Winkelried (q.v.).

'Der Gott, der Eisen wachsen ließ', first line of *Vaterlandslied*, a patriotic poem in six stanzas

by E. M. Arndt (q.v.), written in 1812. It is a call to battle on the eve of the War of Liberation (see NAPOLEONIC WARS).

Der Heiligen Leben, an anonymous Middle High German collection of lives of the saints arranged in the order of the calendar, which was written (in prose) about the beginning of the 15th c. It was a popular work and became one of the early printed books. The customary title derives from the first printed edition, made in 1471 at Augsburg.

Der Jäger Abschied, a poem by J. von Eichendorff (q.v.), the first line of which runs 'Wer hat dich, du schöner Wald . . .' It is a farewell song to the forest.

DERLETH, LUDWIG (Gerolzhofen, Lower Franconia, 1870–1948, San Pietro di Stabio, Switzerland), a classics teacher at a grammar school, became associated with Stefan George's circle (see GEORGE-KREIS), and devoted himself to writing. He lived at first in Munich (1895–1925), later in Rome, Basel, and finally, from 1935, in S. Pietro in Ticino, Switzerland. Derleth was a Roman Catholic of strong and militant outlook who conceived the establishment of a new lay order. His belief is expressed in *Der fränkische Koran* (1932), his life work, begun in 1892, and comprising 15,000 lines. He also wrote *Seraphinische Hochzeit* (1939) and *Der Tod des Thanatos* (1946).

'Der liebste Buhle, den ich han', first line of a well-known folk-song first printed as *Muskatellerlied* in Fischart's *Geschichtklitterung* (1575). The melody to which it is now sung was first recorded in 1603. See FISCHART, JOHANN.

Der maget krône, an anonymous Middle High German poem written probably in the 15th c., or possibly in the mid-14th c. The poem, preserved in an incomplete MS., recounts the life of the Virgin Mary and includes the legends of ten female saints, Barbara, Dorothy, Margaret, Ursula, Agnes, Lucy, Cecilia, Christiana, Anastasia, and Juliana.

'Der Mai ist gekommen', first line of an untitled poem by E. Geibel (q.v.), written in 1841 and often sung to a melody composed in 1843 by J.W. Lyra (1822–82).

Der minne fürgedanc, an anonymous Middle High German poem, written about the middle of the 13th c.; it gives an account of the virtues necessary to the lover and an ideal picture of the object of his *minne*.

Der Minne Gericht, see ELLENDE KNABE, DER.

Der Minne Lehre, an anonymous Middle High German poem of some 2,000 lines. Written in the second half of the 13th c. it was at one time wrongly attributed to HEINZELIN von Konstanz (q.v.). The first part of *Der Minne Lehre* is an allegory of love in which Venus and Cupid appear; the second narrates the course of a love-affair in an exchange of versified letters. It is preserved among the later additions to the Weingartner Liederhandschrift (q.v.).

Der Minne Spiegel, a Middle High German poem of 1,000 lines written in the 14th c. It is in the form of a dialogue between God and the soul, in which the latter expresses repentance, and God receives the soul into grace and mystical union.

'Der Mond ist aufgegangen', first line of the poem *Der Abend*, written in 1782 by M. Claudius (q.v.). In 7 six-line stanzas, it conjures up a night landscape of serene beauty which forms the background for childlike piety and submission to God's will.

Der neuen Liebe Buch, an anonymous Middle High German poem of 1,800 lines, written *c.* 1480. It is an allegory of conjugal love.

Der saelden hort, accepted designation of an untitled Middle High German poem written *c.* 1300 by a Swiss poet, probably resident in Basel. It is a life of Jesus and allocates disproportionate space to the stories of John the Baptist and Mary Magdalene.

Der Seele Rat, a Middle High German poem which takes the form of an allegorical presentation of the sacrament of penance. The opening is lost. The poet, who gives his name as Bruder HEINRICH von Burgus (q.v.) (Hainrich von Purgews), wrote the work between 1301 and 1320.

'Der Spiegel dieser treuen, braunen Augen', first line of the first of Mörike's so-called 'Peregrina-Lieder' (see PEREGRINA). Originally .entitled *Agnes die Nonne*, then (in *Maler Nolten*, q.v.) *Warnung*, it was finally published without a title in the Peregrina cycle.

Der Sünden Widerstreit, a Middle High German allegorical poem, which conceives the virtues as an army of spiritual knights, equivalent to a religious order of chivalry, such as the Teutonic Order (see DEUTSCHER ORDEN). The forces of virtue quickly overcome the hordes of vice. *Der Sünden Widerstreit* is not solely concerned with the allegorization of the moral conflict, but urges Christians to remember the infinite love of Christ. The unknown author wrote the poem, of approximately 3,500 lines, in the second half of the 13th c. He was possibly himself a member of the Teutonic Order.

'Der Tebel hohl mer' (Devil take me), favourite interjection of Schelmuffsky in Christian Reuter's novel *Schelmuffskys wahrhafftige, curiöse und sehr gefährliche Reisebeschreibung* (q.v., 1696).

'Der Tod, das ist die kühle Nacht', first line of an untitled poem by H. Heine (q.v.), included as No. LXXXVII in *Die Heimkehr* (q.v.).

Der Tugenden Kranz, a short Middle High German poem allegorically treating the seven principal virtues in the semblance of a wreath of flowers. The unknown author wrote the poem, of some 300 lines, in the 13th c.

'Der Wein erfreut des Menschen Herz', first line of a much-sung poem by K. F. Müchler (q.v.). It was first published in F. W. A. Schmidt's *Neuer Berliner Musenalmanach für 1797*. The line reproduces almost exactly v. 15 of Psalm 104 ('And wine that maketh glad the heart of man'), which in the Lutheran version runs 'und daß der Wein erfreue des Menschen Herz'.

Der Welt Lohn, a short verse narrative of 274 lines by KONRAD von Würzburg (q.v.). It recounts an episode which is attributed by Konrad to WIRNT von Grafenberg (q.v.). The knight, reading in his room, is visited by a beautiful woman who declares herself to be the mistress whom he has always served. She is Frau Welt (q.v.), and when she turns round he sees that her other aspect is nothing but loathsome decomposition, maggots and worms. The horrified knight undergoes a total change of outlook and sets out on a crusade. Konrad gives his little allegory its obvious moral: he who would save his soul must discard the world.

Der Wirtin Töchterlein, a short ballad in rhymed couplets written by L. Uhland (q.v.) in 1809 and published in 1813. Its first line runs 'Es zogen drei Bursche wohl über den Rhein'. The three young men contemplate the daughter as she lies dead in her coffin, and each expresses his love in different terms.

Des Entkrist Vasnacht, a satirical Fastnachtspiel (q.v.) written, probably in Switzerland, by an anonymous author in the 15th c. Antichrist offers the Emperor vast territories in return for his soul. The subject refers to the early, turbulent years of the reign of Karl IV (q.v.), *c.* 1350. The play is probably based on a lost work of the middle of the 14th c.

Des Feldpredigers Schmelzle Reise nach Flätz, a novel by JEAN Paul (q.v.) published in 1809. Written in the first person, it is a burlesque apologia by Army Chaplain Attila Schmelzle for his notorious cowardice, and is set out in the form of the record of a journey.

Des Knaben Wunderhorn, a collection of more than 700 German folk-songs made by L. J. von Arnim and C. Brentano (qq.v.) chiefly in the years 1804–7. The first volume was published in the autumn of 1805, though the title-page gives the year 1806; the second and third volumes followed in 1808. The title refers to the figure of a boy mounted on a horse and brandishing a horn, and this in turn illustrates *Das Wunderhorn,* the first poem in this anthology. It is a translation of an old French lay. The large collection covers a wide range, as the classified index, with its rubrics *Geistliche Lieder, Handwerkslieder, Historische Romanzen, Liebeslieder, Trinklieder,* and *Kriegslieder,* indicates. No attempt was made to preserve pure texts, and indeed the editors made frequent amendments in accordance with their own tastes; but the source, whether oral or printed, of many of the folk-songs is given. The *Wunderhorn* was criticized as an unscholarly publication, but it succeeded in its purpose of widely disseminating the extraordinary wealth of German folk-song, and has long been regarded as one of the most important and influential documents of the German Romantic movement.

Des Künstlers Erdewallen, a verse dialogue in two scenes, written by Goethe (q.v.), which shows the artist caught between the impulse to follow his own bent and the demands of daily life. It was written in 1773 or 1774 and published in the latter year. See also KÜNSTLERS APOTHEOSE.

Des Meeres und der Liebe Wellen, a blank-verse tragedy in five acts by F. Grillparzer (q.v.) on the ancient love story of Hero, priestess of Aphrodite, and Leander. Although Ovid and Virgil had treated it, Grillparzer used as his main source F. Passow's edition of the 'Epyll' of Hero and Leander by the Grammarian Musaeus (5th c. A.D.). The story was also known through a folk-song ('Es waren zwei Königskinder', q.v.) and a ballad by Schiller (*Hero und Leander*), and Grillparzer knew Christopher Marlowe's poem *Hero and Leander.* He elaborated the story

to suit his purpose, introducing supporting characters, of which the priest is the most significant, and reducing the time of action to three days. The project goes back to 1820, but it was not until 26 February 1829 that the play was completed. Its first performance on 5 April 1831 proved a failure, but H. Laube (q.v.) successfully revived it in revised form at the Burgtheater in 1851. It was published in 1839 (title-page gives 1840).

Hero's dedication to priesthood, entailing her renunciation of matrimony, opens the play and is treated with ominous ambiguity as the sight of Leander throws her into confusion during the ritual. The following night Leander returns from Abydos to the shores of Sestos, having swum across the Hellespont, and visits Hero in the priestess's tower. The next day Hero lives for her lover's promise to return at night, but the priest, suspecting a grave sin, exposes to the winds the lamp which is to guide the swimmer through the darkness to the tower; he thus aids the gods in their retribution. Hero dies of a broken heart after Leander's body, washed ashore the following morning, is removed, at the priest's command, for burial.

Grillparzer commented on the title of the play: 'Der etwas prätiös klingende Titel ... sollte im voraus auf die romantische oder vielmehr menschlich allgemeine Bedeutung der antiken Fabel hindeuten.'

Des Minnesangs Frühling, a collection of Minnelieder up to and including those of REINMAR der Alte (q.v.), but excluding those of WALTHER von der Vogelweide and later poets. There are some anonymous poems; the named poets are Der Kürnberger, MEINLOH von Sevelingen, the Burggrafen von Regensburg and Rietenburg, Spervogel, DIETMAR von Aist, FRIEDRICH von Hausen, HEINRICH von Veldeke, ULRICH von Gutenberg, RUDOLF von Fenis, ALBRECHT von Johannsdorf, HEINRICH von Rugge, BERNGER von Horheim, HARTWIG von Rute, BLIGGER von Steinach, der von Kolmas, HEINRICH von Morungen, ENGELHART von Adelnburg, REINMAR, and HARTMANN von Aue. *Des Minnesangs Frühling* was published in 1857 by M. Haupt (q.v.) from the papers of K. Lachmann (q.v.), and has been repeatedly republished and re-edited since, notably by F. Vogt (3rd edn., 1882) and Carl von Kraus (1940, and 30th edn., 1950). The names of both Lachmann and Haupt appear on the title-page as joint editors.

Des Pfarrers Tochter von Taubenhain, a ballad by G. A. Bürger (q.v.), first published in the *Göttinger Musenalmanach,* 1782. It is a story of the seduction of a pastor's daughter by a nobleman. She kills her illegitimate child and is put to death. Bürger adds a ghostly element. His plan to write a poem on this theme was finally crystallized by a case in his own jurisdiction as a magistrate. In considerably altered form the story is used by O. Ludwig (q.v.) in his tragedy *Die Pfarrose* (q.v.).

Des Quintus Fixlein Leben bis auf unsere Zeiten; in fünfzehn Zettelkästen, a short novel by JEAN Paul (q.v.), published in 1796. 'Zettelkästen' is here an eccentric designation for chapters. Fixlein is a village schoolmaster who hopes with the help of patronage to become a deputy headmaster (Konrektor). He duly receives the appointment but finds that the board has nominated him because it believes that he, like his predecessor in the office, will die in his thirty-second year. Meanwhile a pastorate, which he would much prefer, falls vacant; he is not the favoured candidate, but is appointed through an error in spelling, the clerk having written 'Fixlein' instead of 'Füchslein'. Fixlein marries Thiennette, a retainer of his deceased patroness in accordance with her testamentary provisions, but the prophecy of an early end (before 32) so weighs on him that he goes out of his mind. In this state he is persuaded that he is still a child and thereupon recovers. The manner of narration and the style are equally oblique, whimsical, and allusive.

Des Sängers Fluch, a ballad written by L. Uhland (q.v.) in 1814 and published in Uhland's *Gedichte* (1815). Its first line runs, 'Es stand in alten Zeiten ein Schloß, so hoch und hehr'. It tells of the brutal murder, by a king, of the younger of a pair of minstrels and of the curse pronounced by the surviving older minstrel. Its source is a Scots ballad published by Herder (q.v.) in his *Volkslieder*.

DESSAU, PAUL (Hamburg, 1894–1979, East Berlin), the composer who during his exile in the United States set poems and plays by Brecht to music, including *Mutter Courage und ihre Kinder, Der kaukasische Kreidekreis,* and *Die Verurteilung des Lukullus* (qq.v.), for which Stravinsky had declined a commission. He was readier than other composers working for Brecht to aim at what Brecht termed 'culinary' (kulinarisch) effects.

Des Teufels General, a three-act play by C. Zuckmayer (q.v.), first performed and published in 1946. Written in the U.S.A. in 1942, it bears a double dedication, the first (1942) 'Dem unbekannten Kämpfer', i.e. of the German Resistance, the second (1945) to W. Leuschner and

Count Helmuth James von Moltke (q.v.), who were executed in 1944 (see RESISTANCE MOVEMENTS, 2). The play is set in the early winter of 1941, and the last act is intentionally dated 6 December, the day before Pearl Harbor and five days before Germany declared war on the U.S.A. It successfully combines a study of 'the devil's general', General Harras, with the theme of resistance and the terrorism which it has to encounter. Harras is modelled on General E. Udet (1896–1941, suicide). In the first act (*Höllenmaschine*) Harras gives a party in a luxurious restaurant in Berlin which, unknown to host and guests, houses a concealed recording apparatus. A convincing impression of life at the higher levels of Nazi society in Berlin is evoked, including the rigid and the lax, the fanatics, the fellow-travellers, and the deluded honest supporters. Harras, an air ace of the 1914–18 War, now in charge of aircraft production, scathingly mocks the regime. For all his apparent gaiety he is in a critical situation. Crashes of aircraft owing to constructional faults have become frequent, he has been unable to trace the cause, and is under suspicion. His crisis is also moral. He fights for what he does not believe in. Flying has meant so much to him that he re-joined to build up the new German air force, though he knew how unscrupulous the new rulers of Germany were. Thus he has become 'the devil's general'. In the second act (*Galgenfrist oder die Hand*) the net closes round Harras, who is given ten days to solve the problem of possible sabotage. Harras, who in his heedless cynicism has set little value on his own life, falls in love with Diddo, a young actress, and recovers his will to live. News now arrives of the death of his closest friend, Colonel Eilers, whose aircraft has crashed through metal fatigue. The third act (*Verdammnis*) takes place at an airfield. Harras knows that he will be liquidated. A glimpse of organized resistance is given with the appearance of two factory workers under arrest. Eilers's widow denounces Harras for encouraging Eilers in what she now recognizes as an unjust cause. Finally Harras discovers that Oderbruch, the manager he has trusted utterly, is the chief figure in the sabotage of the aircraft. Harras does not betray Oderbruch, but takes up one of the batch of faulty aircraft and crashes to his death. Dr. Schmidt-Lausitz informs the Propaganda Ministry by telephone that all has gone well, and a state funeral is ordered. The play is held together by the brave, reckless, outspoken Harras who redeems himself in two actions, his interview with Hartmann which initiates the young man's disillusionment with the Nazi world, and his decision to save Oderbruch by sacrificing his own life. The play was one of the great successes of the immediate post-war stage. (Film version by H. Käutner, 1955.)

Des Teufels Netz, an anonymous Middle High German poem of some 13,000 lines. Its date is uncertain, but it seems likely that it was written at the time of the Council of Constance (1414–18). The dialect is that of the region of Lake Constance. *Des Teufels Netz* is a bitter and misanthropic satirical allegory. The vices of each station in society are exposed, and all are ensnared in the devil's net. The poem takes the form of a dialogue between the devil and a hermit.

Des Vetters Eckfenster, a prose dialogue by E. T. A. Hoffmann (q.v.), written in 1822 and published in the same year in *Der Zuschauer*. The cousin ('der Vetter'), an ailing author, sits at his window overlooking the market-place, and from his enforced seclusion imaginatively interprets to the first-person narrator the happenings and characters which he watches.

Detektivroman, see KRIMINALGESCHICHTE.

Detmold, capital town of Lippe (q.v.), one of the smallest German states to survive in the German Empire (1871–1945). It has often figured as the characteristic small provincial town, as in the humorous soldier's song

Lippe-Detmold, eine wunderschöne Stadt
Da drinnen ein Soldat.

Detmold was the birthplace and home of the dramatist C. D. Grabbe (q.v.). It is now in Land Nordrhein-Westfalen of the Federal Republic.

Dettingen, battle of (17 June 1743), see PRAGMATISCHE ARMEE and ÖSTERREICHISCHER ERBFOLGEKRIEG.

DEUTSCH, NIKLAS MANUEL, see MANUEL, NIKLAS.

Deutsche Akademie, see AKADEMIEN.

Deutsche Allgemeine Zeitung, see LEIPZIGER ALLGEMEINE ZEITUNG.

Deutsche Bibel, Die, an ode in five unrhymed stanzas written in 1784 by F. G. Klopstock (q.v.). It praises Luther (q.v.) as the creator of a language addressed to the common man.

Deutsche Bibliothek, see LIBRARIES.

Deutsche Bücherei, see LIBRARIES.

Deutsche Chronik, Die, a newspaper founded in Augsburg in 1774 by C. F. D. Schubart (q.v.) and written by him single-handed, first at Augsburg, then at Ulm, until his notorious imprisonment in Württemberg in 1777.

Deutsche Demokratische Republik, DDR (German Democratic Republic) was formed out of the Soviet zone of occupation, including the Soviet sector of Berlin, after differences between the Russians and the Allies (see BUNDESREPUBLIK DEUTSCHLAND). It was instituted on 7 October 1949. It is a democracy in the Soviet sense, and the resemblance to the Russian model has become closer since a revision of the constitution through the Second Constitution of 1968. Originally the DDR had a president (W. Pieck), but on his death in 1960 the office lapsed. The parliament (Volkskammer) is elected from a list of candidates arranged according to a party quota, the SED being the dominant party. Executive power is distributed among several organs of the Volkskammer, the most important of which are the Staatsrat and the Ministerrat. The formal head of state, who at the same time is Chairman of the Staatsrat, was W. Ulbricht (1893–1973) who was succeeded in 1971 by E. Honecker.

The DDR originally had five Länder, Brandenburg, Mecklenburg, Sachsen, Sachsen-Anhalt (mainly the former Prussian Provinz Sachsen), and Thüringen. In 1952 the Länder were abolished and their governments dissolved; and the state was reorganized in fifteen districts (Bezirke), each named after its administrative town: Cottbus, Dresden, Erfurt, Frankfurt/Oder, Gera, Halle, Karl-Marx-Stadt (Chemnitz), Leipzig, Magdeburg, Neubrandenburg, Potsdam, Rostock, Schwerin, Suhl, and East Berlin, which has also the status of a Bezirk. The object of this change was to diminish local influence to the advantage of centralized government.

East and West Germany have been separated from the beginning by the Iron Curtain (see EISERNER VORHANG). The principal events and features of the history of the DDR are its policy of collectivization, the workers' rising in East Berlin and elsewhere on 17 June 1953, the creation of the Nationale Volksarmee in 1955, accession to the Warsaw Pact in 1957, the building of the Berlin Wall (see BERLIN) in 1961, the introduction of universal military service in 1962, and the participation of East German troops in crushing the liberal regime in Czechoslovakia in 1968. A *détente* with the Federal Republic was initiated by a treaty (Grundlagenvertrag, DDR; Grundvertrag, BRD), signed in 1972.

Deutsche Gelehrtenrepublik, Die, a fantasy published by F. G. Klopstock (q.v.) in 1774. In full, the title runs *Die deutsche Gelehrtenrepublik, ihre Einrichtung, ihre Gesetze, Geschichte des letzten Landtags.* In the guise of an antiquarian, Klopstock treats the literary and intellectual life of contemporary Germany as a literary commonwealth, in which the highest virtue is originality and imitation a crime. More than half the work is devoted to the 'last Diet' (*Geschichte des letzten Landtages*), in which the morning debates of the twelve days of the Diet are summarized. Each *Morgen* (except the first, seventh, and twelfth) is followed by an *Abend* which discourses on linguistic, philological, or lexicographical problems, in which Klopstock is at times rash and arbitrary, but also original and stimulating; the *Abend* of the sixth *Morgen* is devoted to literary theory, in which his essential point is that literary rules must derive from experience. *Die deutsche Gelehrtenrepublik* was to have been continued, but only the first volume of 1774 was published.

Deutsche Hausvater, Der, a once famous play by O. H. von Gemmingen-Hornberg (q.v.).

Deutsche Museum, Das, a literary periodical published by H. C. Boie (q.v.). It appeared under his editorship from 1776 to 1791, though the title was changed in 1789 to *Das neue Deutsche Museum.*

Deutschen Kleinstädter, Die, a light comedy by A. von Kotzebue (q.v.), first performed in September 1802 at Mannheim and published in 1803. It is set in 'Krähwinkel' (q.v.), the archetype of a small provincial town in a petty German state. The Bürgermeister's daughter, Sabine Stahr, has fallen in love with Herr Olmers during a visit to the capital, and her love is returned. The Bürgermeister's household is ruled by his mother, Frau Unter-Steuer-Einnehmerin Stahr, who insists that Sabine shall marry a title and has chosen Herr Bau-, Berg-, und Weg-Inspectors-Substitut Sperling. Olmers comes to Krähwinkel with a warm letter of recommendation from Count Hochberg. Being, apparently, plain 'Herr', he makes little headway, and does not at first perceive the reason. After various farcical interludes he learns the situation from Sabine and produces his title, to which he has hitherto attached little importance. He is a 'Geheimder Commissionsrath', and this grandeur at once makes Olmers a most eligible suitor. In addition to satirizing the provincial passion for titles, Kotzebue mocks at the slow pace of the law in Krähwinkel. *Die deutschen Kleinstädter* was his most successful play and was performed as recently as 1962. He

exploited his success with two other 'Krähwinkel' plays, *Carolus Magnus* (1806) and *Des Esels Schatten oder Der Prozeß in Krähwinkel* (1810).

Deutschenspiegel, Der, also termed *Spiegel deutscher Leute,* is in part an adaptation and in part a translation (into Middle High German) of the Middle Low German *Sachsenspiegel* (q.v.). It is an attempt to formulate the body of law for the whole German people and was written down c. 1260 by a priest of Augsburg.

Deutscher Bund, the German Confederation unifying Germany in a loose confederation of states by the Federal Act (8 June 1815) of the Congress of Vienna (see WIENER KONGRESS), and by the Final Act (Wiener Schlussakte, 15 May 1820). Its thirty-nine members included foreign powers representing German states over which they were sovereigns (Denmark for Holstein and Lauenburg, England for Hanover, and the Netherlands for Luxemburg and Limburg), and four Free Cities (Hamburg, Lübeck, Bremen, and Frankfurt). Austria and Prussia were the two leading member-states. The appointed location for the Federal Diet (Bundestag) was Frankfurt. Metternich (q.v.) was the chief architect of the Confederation, and through his influence Austria assumed its presidency. The Federal Act, though making provision for the introduction of constitutional government in the various states, purposely aimed at a decentralization of national power, thus enabling Metternich to exercise considerable control (see KARLSBADER BESCHLÜSSE).

National and liberal aspirations for reform culminated in the 1848 revolutions (see REVOLUTIONEN 1848–9), and led to the formation of the Frankfurt Parliament (see FRANKFURTER NATIONALVERSAMMLUNG). Its failure to reach an agreement led to the revival of the Confederation, a triumph of Austrian policy under F. zu Schwarzenberg (q.v.) over Prussia (see OLMÜTZER PUNKTATION). When Austria ordered Federal execution (Bundesexekution, against Prussia as a result of Prussia's Schleswig-Holstein policy, which culminated in the War of 1866 (see DEUTSCHER KRIEG), Prussia withdrew its membership. In the Peace of Prague (23 August 1866) Austria recognized the dissolution of the Bund. (See also SCHLESWIG-HOLSTEINISCHE FRAGE.)

Deutscher Kaiser, a term used in the Middle Ages to denote 'der Deutsche König und Römische Kaiser' (see DEUTSCHER KÖNIG and DEUTSCHES REICH, ALTES). Strictly speaking, 'Deutscher Kaiser' was applicable only to those medieval German kings who were crowned emperor by the Pope at Rome. The title ceased with the dissolution of the Holy Roman Empire in 1806 (see HEILIGES RÖMISCHES REICH). In 1871 'Deutscher Kaiser' became the designation of the head of the new German empire, an office vested hereditarily in the king of Prussia. This title lapsed with the abdication of Wilhelm II (q.v.) in 1918.

Deutscher König, title in use since the 10th c. for the monarchical head of the German nation. From 962 the German King claimed the right to be crowned Emperor of the Roman Empire by the Pope (see DEUTSCHES REICH, ALTES and DEUTSCHER KAISER). When this ambition was achieved, the German King became Deutscher Kaiser. The Deutscher König was elected in Frankfurt and crowned in Aachen. The title was also borne by the son of an emperor elected as successor during his father's lifetime. In Latin documents the German King is referred to as *rex* or *rex Francorum* up to the late 11th c.; from the time of Heinrich IV (q.v.) he is *rex Romanorum*, and is occasionally styled Römischer König.

Deutscher Krieg, the war fought in 1866 between Prussia and Austria, in which states of the German Confederation also joined. Provoked by Bismarck (q.v.) through manipulation of the Schleswig-Holstein question (see SCHLESWIG-HOLSTEINISCHE FRAGE), it was in effect fought to decide hegemony in Germany. It lasted approximately six weeks, the critical point being the decisive defeat of the Austrians at Sadowa (see KÖNIGGRÄTZ) in Bohemia on 3 July. The smaller North German states, except Hanover and Hesse-Cassel, sided with Prussia, the southern and central states gave Austria lukewarm support.

In the Mainfeldzug the Prussians, after an initial Hanoverian success at Langensalza (27 June), defeated their enemies and occupied Frankfurt and Würzburg by 31 July. Though the King of Prussia and his generals wished to exploit the success against Austria, Bismarck, for reasons of future policy, insisted on a lenient settlement.

The consequences of the war were the exclusion of Austria from Germany, the foundation of the North German Confederation (see NORDDEUTSCHER BUND), and the annexation of Hanover, Hesse-Cassel, Nassau, the Free City of Frankfurt, and Schleswig-Holstein by Prussia. It represents a decisive step in Bismarck's policy of achieving a strong German state dominated by Prussia. Italy joined in the war in order to acquire Venice and the province of Venetia, in which it was successful, in spite of suffering military and naval defeat.

Deutscher Michel, see MICHEL, DEUTSCHER.

Deutscher Orden, known in English as the Teutonic Order, originated in Palestine as a religious order of knights analogous to the Knights Templar and the Knights of St. John. Founded by North Germans at first as a hospital at Acre in 1190 and sanctioned by the Pope in the following year, it assumed its definitive military form in 1199 under Pope Innocent III. The knights, known as *Deutschherren, Deutsche Ritter, Deutschordensritter,* or *Kreuzritter,* were Germans, and early in the 13th c. their crusading activity began to be transferred to eastern Europe, where it was exclusively concentrated after the fall of Acre in 1291. The aim of the Order became the conversion of Prussia and the Baltic lands, a task which was readily combined with the extension of power and the acquisition of territory. Under the High Master (Hochmeister) Hermann von Salza (1210–39) the Order began the conquest of Prussia (then Slavonic and heathen), which was completed in 1236. In the second half of the 13th c. it penetrated into Courland and Livonia, and by the end of the 14th c. had conquered Lithuania in a series of campaigns marked on both sides by exceptional ferocity. From 1309 its central seat was Marienburg, south-east of Danzig. The Order was followed by colonists who settled in new German cities, including Thorn, Elbing, Danzig, and Königsberg. From 1234 the Order owed allegiance only to the Pope, and ruled its territories as an efficient, autocratic aristocracy. It acquired commercial interests, which led to rivalry with the cities, and its continued military aggressiveness alienated neighbouring principalities.

The decline of the Order began with a defeat at Tannenberg (1410) in the course of war with Poland. The Order was, however, temporarily saved by the successful defence of Marienburg by HEINRICH von Plauen (q.v.), who became High Master. Heinrich attempted reforms and was promptly deposed in 1413. In the course of the Thirteen Years War (from 1454 to 1467) the Order lost West Prussia, though it retained East Prussia. Marienburg was surrendered to Poland, and the seat of the government transferred to Königsberg. The effort to maintain independence failed, and from 1525 the Order confined itself to administering its possessions in Germany, with Mergentheim as its headquarters. It was dissolved in 1809 by Napoleon, but was revived in 1834 by Franz I of Austria (see FRANZ II) as an order providing medical help in time of war.

The High Master, who was the head of the Order, was complemented by a German Master (Deutschmeister). From 1530, with the loss of the east European territories, the two titles were combined in one (Hoch- und Deutschmeister).

In the 13th c. the court of Marienburg was a well-known centre of knightly activity. Chaucer's knight, it will be recalled, had travelled in 'Pruce' and 'Lettwo', territories of the Order.

Deutsche Rundschau, a monthly periodical founded in 1874 by Julius Rodenberg (q.v.). It was suppressed in 1942, and revived in 1949 by R. Pechel (1882–1961), who had succeeded Rodenberg on his death in 1914 as chief editor. After the 1914–18 War the periodical underwent considerable policy changes. Its initial moderate national-liberal orientation became more nationalistic, but it actively opposed National Socialism.

Influenced by foreign periodicals, such as the *Quarterly Review*, its literary section published in its early years contemporary fiction, especially Novellen by G. Keller, C. F. Meyer, Th. Storm, and Th. Fontane (qq.v.). It gave prominence to critical and literary essays (among others by W. Dilthey and H. Grimm, qq.v.), and philosophical essays (notably by R. Eucken, q.v.). The Danish critic G. Brandes (q.v.) was the first to support Nietzsche (q.v.) in its pages. A great number of 20th-c. writers are represented in the *Deutsche Rundschau*.

Deutsche Sappho, a complimentary appellation bestowed on Anna Luise Karschin (q.v.) by J. W. L. Gleim (q.v.).

Deutsche Schaubühne, Die, a collection of plays, including translations from the French, published by J. C. Gottsched (q.v.) and his wife Luise Adelgunde, to which J. E. Schlegel (q.v.) and others contributed.

Deutsches Reich, official title of the united (federal) Germany proclaimed at Versailles on 18 January 1871. The title was retained by the Weimar Republic (q.v.), which followed in 1918, and by the National Socialist regime from 1933 to 1945. See also DEUTSCHES REICH, ALTES, and DRITTES REICH.

Deutsches Reich, Altes, the German Empire, reunited in A.D. 919 under Heinrich I (q.v.) after years of division since A.D. 843 (see KAROLINGER). It was ruled by an elected emperor, and consisted of duchies, counties, episcopal principalities, etc. owing feudal allegiance to the emperor. In practice the empire was frequently in a state of internal disturbance, in which the central ruler and the great nobles and prelates clashed. A considerable factor in the difficulties of the empire was the emperor's claim to sovereignty over Italy, which involved his frequent absence from Germany, during which the great nobles usurped his power, or quarrelled among themselves. The Emperor Friedrich II (q.v.), a strong ruler, spent most of his life in Italy, visiting Germany only twice to restore order. After his death in 1250 German imperial power both in Italy and at home waned.

Despite its electoral statutes, the empire became from time to time virtually hereditary. Thus Saxon emperors were elected from 919 to 1024, Franconians (Salier) from 1024 to 1125, Stauffer from 1138 to 1250, and Luxemburgs from 1347 to 1437. The first Habsburg was Rudolf I (1273, q.v.). From 1438 to 1806 the emperors were all Habsburgs.

Of the various names borne by the German Empire, *Romanum Imperium* dates from 1034, *Sacrum Imperium* from 1157, and *Sacrum Romanum Imperium* from 1254. From the 16th c. (unofficially already in the 15th c.) the official title of the German Empire was *Das Heilige Römische Reich Deutscher Nation* (in Latin documents *Sacrum Romanum Imperium Nationis Germanicae*, abbreviated S.R.I.N.G.). 'Roman' implied that the empire was a continuation of the Roman Empire, 'Holy' alluded to its character as a Christian empire, and the combination of the two suggested the connection with Italy and the Papacy.

Deutsches Wörterbuch, the monumental dictionary begun by Jakob Grimm in collaboration with his brother Wilhelm (qq.v.), and continued by other scholars, notably K. Weigand, R. Hildebrand, M. Heyne, and M. Lexer. Its sixteen volumes (32 bound in pairs) appeared from 1854 to 1960 (the instalments for the first volume began in 1852). Only the first four were completed at the time of Jakob Grimm's death. It is the most comprehensive work of modern German lexicography and gives illustrations of usage together with etymological information. Its standing is equivalent to that of the *Oxford English Dictionary*.

Deutsche Treue, a poem written by F. Schiller (q.v.) in 1800. It recounts an episode of fidelity between former enemies now reconciled, Ludwig IV and Friedrich der Schöne (qq.v.). The story is derived from *Geschichte der Deutschen* (1778) by M. I. Schmidt (1736–94). Written in distichs (elegiacs), the poem is included in the group 'Epigramme' in Schiller's collected poems. First publication was in the *Musenalmanach*, 1800.

Deutsch Franços, the hero of a comic work by J. C. Trömer (q.v.).

Deutsch-Französischer Krieg, the third and last of the three wars which marked successive stages in Bismarck's policy of German unification. The persistent antagonism between France and Prussia had become more acute after the Austrian defeat in 1866 (see DEUTSCHER KRIEG) and the consequent sudden increase in Prussian power and influence. In the immediately following years Prussia watched for an opportunity to isolate and humiliate France, and France sought to establish an overwhelming combination (with Austria and Italy) against Prussia.

Tension between the two countries was increased early in 1870 by the candidature of Prince Leopold of Hohenzollern-Sigmaringen for the vacant Spanish throne. When the prince withdrew, the French sought to press home a diplomatic advantage, but the incautious *démarche* of the French ambassador, Benedetti, led to a rebuff by the King of Prussia at Ems. By adroit abbreviation Bismarck (q.v.) turned the report of this encounter (see EMSER DEPESCHE) into a virtual challenge which provoked the outbreak of war.

The forces of the North German Confederation (see NORDDEUTSCHER BUND) were joined by the allied southern states, and Austria declined to intervene against Prussia. The better led and better organized German forces seized the initiative and pressed the French back in early engagements at Weißenburg (Wissembourg), Wörth, and Spichern without achieving the hoped-for annihilation. The forces of Bazaine withdrew into Metz, and a large army under Macmahon and Napoleon III, marching to its relief, was engaged, encircled, and destroyed on 1 September 1870 at Sedan (q.v.). Metz surrendered on 28 October. This was the final point of the first phase of the war, an orthodox campaign between regular, established forces.

There followed on the French side a national war, in which quickly raised and armed *levées-en-masse* sought unsuccessfully to defeat the less numerous but highly trained German forces. Meanwhile the Germans invested Paris, which was enclosed on 19 September and surrendered on 28 January 1871. The period of the siege provoked a sharp conflict over the bombardment of Paris between the political leader Bismarck, who wished for a quick decision, and the military commander Moltke (q.v.), who was for starving the city. A contemporary doggerel ran,

Lieber Moltke, gehst so stumm
Immer um den Brei herum.
Bester Moltke, sei nicht dumm,
Mach doch endlich bumm, bumm, bumm!

The peace which was signed at Frankfurt on 10 May 1871 annexed Alsace and much of Lorraine to Germany under the designation Elsaß-Lothringen (q.v.) and imposed a financial indemnity. The outstanding significance of the war in German history was the proclamation, in

fulfilment of Bismarck's plans, of the German Empire, which took place in the *Galerie des glaces* at Versailles on 18 January 1871, with King Wilhelm I (q.v.) of Prussia as German Emperor (Deutscher Kaiser, q.v.).

The victory was accompanied by a great upsurge of patriotic feeling, much of it brash and arrogant, and few significant works of literature arose from the war and its background. A handful of the multitude of patriotic poems has survived, but the outstanding work is probably to be found in the war sketches in prose and verse of Detlev von Liliencron (q.v.). Theodor Fontane (q.v.), who was a war correspondent, gives a vivid account of his falling into French hands and of his subsequent experiences in *Kriegsgefangen. Erlebtes 1870*.

Deutschherren, see DEUTSCHER ORDEN.

'Deutschland, Deutschland über alles', first line of the German national anthem 1922–45. Written by HOFFMANN von Fallersleben (q.v.) in 1841, it was frequently sung as a patriotic hymn before being made the official national anthem. It is known as the Deutschlandlied. The melody is that of the Austrian *Kaiserhymne*, composed by J. Haydn (q.v.) in 1797.

The national anthem of the Federal Republic (see BUNDESREPUBLIK) retains the tune, but the words are limited to the third stanza, 'Einigkeit und Recht und Freiheit'.

Deutschland. Ein Wintermärchen, a satirical epic by H. Heine (q.v.), written and published in 1844. It is written in four-line stanzas of rhyming verse and comprises 27 'chapters' (each designated a 'Kaput'). The poem arises partly out of the visit Heine paid to Germany at the end of 1843, but it does not reproduce his itinerary. Its basis is an imaginary journey from Paris via Aachen, Hagen, Bückeburg, and Hanover to Hamburg, and its substance is a cannonade of witty, yet serious, satire directed at German philistinism and political life. Its most famous passage deals with the idealization of the Middle Ages and the anachronism of monarchy under the image of the Emperor Barbarossa (Kaput XIV–XVII), otherwise Friedrich I (q.v.).

Deutschordensritter, see DEUTSCHER ORDEN.

Deutschstunde, a novel published in 1968 by S. Lenz (q.v.). It confines itself to the years 1943–5, except for its frame (see RAHMEN), which is set a few years later. The main events take place on the North Sea coast close to the Danish frontier. The story begins in the present, and from time to time reverts to it, the large interstices being filled with the narrator's recollections.

The narrator is Siggi Jepsen, a youth lodged in a detention centre (Institut für schwer erziehbare Jünglinge) situated on an island in the Elbe near Hamburg. It is only much later in the novel that we learn that his offence is the compulsive stealing of paintings.

The novel opens with Siggi confined to a cell with paper, pen, and ink, because he has not written a word on the subject set in class (*Deutschstunde*). This subject is 'Die Freuden der Pflicht' and it is one on which Siggi has, not too little, but too much to set down in a short space of time. In the end he writes his 'Strafarbeit'. Though intended only for an hour or two, it keeps him occupied for months, growing into an autobiography, which includes his experience and his boyhood environment for the years 1943–5. The story itself is partly a conflict between two men and partly a study of a deplorable family situation.

Siggi's father is the village policeman, who is on friendly terms with a well-known painter, M. L. Nansen, living near by. In 1943 the policeman is ordered to hand Nansen an order forbidding him to paint, and is also responsible for the observance of the ban. Although he has been Nansen's friend, the maintenance of the ban becomes an obsession ('eine Freude der Pflicht'). He spies, makes surprise visits, and tries to enlist Siggi's aid. Nansen goes on painting, and Siggi forms his irresistible habit of 'rescuing' paintings. Siggi's father, Jens Jepsen, alienates the boy by the rigidity of his attitude, as he alienates his other children.

After the capitulation the policeman is arrested, but three months later he is back on the beat, more obsessive than ever in his belief that Nansen is still under a ban. Meanwhile the grown-up children have left home for good; yet Constable Jepsen and his wife remain convinced that they have only done their duty, and that their children and their critical neighbours are perverse. That the essay 'The Pleasures of Duty' is set in the post-war period is intended to prove that the concept is not dead.

According to Lenz, the painter is modelled on E. Nolde, M. Beckmann, and E. L. Kirchner (qq.v.).

Deutsch Theologia, Eyn, see Theologia deutsch.

DEVRIENT, Eduard (Berlin, 1801–77, Karlsruhe), nephew of the actor Ludwig Devrient (1784–1832), went on the stage and became producer of the Dresden Court Theatre in 1844 and director of the Karlsruhe Court Theatre in 1852. His first, and best-known, dramatic work

is the libretto to Marschner's opera, *Hans Heiling* (q.v., 1833), but *Die Gunst des Augenblicks* (1835) and *Die Verirrungen* (1837) also had stage success. Eduard Devrient wrote a *Geschichte der deutschen Schauspielkunst* (1848–74).

Dialektdichtung, term applied to dialect works written after the establishment of the modern standard language, especially when the dialect is consciously used, either to maintain a threatened tradition or because of its freshness, vigour, or quaintness. Dialect literature (Viennese and a few other instances excepted) is primarily a feature of the 19th c. and the 20th c. The first outstanding instance is the volume *Alemannische Gedichte* by J. P. Hebel (q.v., 1803). In the middle of the 19th c. K. Groth and F. Reuter (qq.v.) wrote most of their works in their native Low German (Plattdeutsch) dialects, those of Holstein and Mecklenburg respectively. More than 500 exponents of High German dialect literature are recorded in works of reference, and a score or so for Low German. For High German, apart from Hebel, the following are among the more notable: L. Anzengruber, H. C. Artmann, I. F. Castelli, W. A. Corrodi, A. Frey, J. K. Grübel, G. Hauptmann, C. Karlweis, F. von Kobell, J. F. Kringsteiner, G. J. Kuhn, K. Malß, J. N. Nestroy, F. Raimund, P. Rosegger, J. V. Sailer, K. Schönherr, J. G. Seidl, F. Stelzhamer, L. Thoma, J. M. Usteri, and J. Weinheber (qq.v.). For Low German, in addition to Groth and Reuter, J. Brinckman, H. Burmester, H. H. Fehrs, R. Kinau, J. Stinde (qq.v.), and F. Stavenhagen deserve mention. Some of these dialect writers use a simplified form of the local language in order to make their works intelligible to a larger public, e.g. L. Anzengruber and G. Hauptmann, whose play *Die Weber* (q.v.) exists in modified dialect and also as *De Waber*, in full Silesian. A special form of dialect literature is found in the Viennese Volksstück (see VOLKSSTÜCK).

Diamant des Geisterkönigs, Der, a farce by F. Raimund (q.v.), written in 1824 and first performed at the Theater in der Leopoldstadt, Vienna, in December 1824. It was published posthumously in volume 1 of Raimund's *Sämtliche Werke* (1837). It is described as 'Zauberspiel in zwei Aufzügen'. Eduard is the son of a wealthy magician who forgot to bequeath the magic means of access to his treasures before he died. Eduard is helped by the King of Spirits, Longimanus, and in the second act has to find a girl who has never told a lie. He finds this paragon and all ends happily. The comedy is provided partly by the parodied aristocracy of Fairy-land, and partly by the Viennese characters Florian and Mariandel, who speak broad dialect. The play contains a number of musical numbers. The source of the story is the *Arabian Nights*.

DIBELIUS, MARTIN (Dresden, 1883–1947, Heidelberg), was from 1915 professor of New Testament theology at Heidelberg. His work *Die Formgeschichte des Evangeliums* (1919) greatly advanced the study of form criticism (see FORM-GESCHICHTE).

Dichter Firdusi, Der, a group of three poems by H. Heine (q.v.) included in the section *Historien* of the *Romanzero* (q.v.). The poet Firdusi receives his due reward only after his death.

Dichterkrieg, Der, a satire directed by J. C. Gottsched (q.v.) in 1741 against his adversaries in literary controversy, J. J. Bodmer and J. J. Breitinger (qq.v.).

Dichterliebe, a song cycle with piano accompaniment, composed by R. Schumann (q.v.) in 1840. It consists of sixteen poems by Heine (q.v.) drawn from the section *Lyrisches Intermezzo* of the *Buch der Lieder* (q.v.). They are: (1) 'Im wunderschönen Monat Mai'; (2) 'Aus meinen Tränen sprießen'; (3) 'Die Rose, die Lilie, die Taube, die Sonne'; (4) 'Wenn ich in deine Augen seh'; (5) 'Ich will meine Seele tauchen'; (6) 'Im Rhein, im heiligen Strome'; (7) 'Ich grolle nicht'; (8) 'Und wüßten's die Blumen'; (9) 'Das ist ein Flöten und Geigen'; (10) 'Hör ich das Liedchen klingen'; (11) 'Ein Jüngling liebt ein Mädchen'; (12) 'Am leuchtenden Sommermorgen'; (13) 'Ich hab im Traum geweinet'; (14) 'Allnächtlich im Traume'; (15) 'Aus alten Märchen winkt es'; (16) 'Die alten, bösen Lieder'.

Dichtermut, title of the first two versions of an Alcaic ode written in 1800–1 by F. Hölderlin (q.v.). The third and final version bears the title *Blödigkeit*. The ode expresses the poet's acceptance of suffering as a necessity of his mission.

Dichter und ihre Gesellen, a novel written in 1833 by J. von Eichendorff (q.v.) and published in 1834. Eichendorff eccentrically termed it a Novelle, although it extends to three Books.

Baron Fortunat visits Walter, his friend from university days at Heidelberg. Walter, still a young man but already a civil servant and steady burgher, takes Fortunat with him to visit the family of his (Walter's) future bride, Florentine. Near by is the castle of Graf Victor, a poet whom Fortunat admires, but the count is unfortunately absent. A gifted but unbalanced poet named Otto joins the friends. Meanwhile Fortunat feels an attraction to the engaged Florentine, and therefore leaves. He joins a company of itinerant actors, which includes a learned young man named Lothario. Lothario presently carries off a half-willing young countess, who in the end saves herself from infamy by setting her horse into a mountain stream, in which she drowns. The scene changes to Italy (where Fortunat falls in love with Fiammetta, the daughter of the Marchese, his host) and then back to Germany. Otto, who has wrecked his life by a rash marriage, dies, and Fortunat at last marries Fiammetta. Lothario, who proves to be Count Victor in disguise, takes holy orders in the Roman Catholic Church.

The title of the book is justified by the exhibition of three poets, Dryander, a member of the theatrical troupe, who is clever but insincere, Otto, the unbalanced Romantic, whose indiscipline destroys him, and Victor, the personality matured by experience, for whom religion comes first and poetry second. *Dichter und ihre Gesellen* contains poems by Eichendorff and some detached stanzas by him. They include *Die Nacht, Schöne Fremde,* 'Waldeinsamkeit! /Du grünes Revier', and *Sehnsucht* (qq.v.). In the novel they are all untitled.

Dichtung und Wahrheit, form in which the title of Goethe's autobiography *Aus meinem Leben. Dichtung und Wahrheit* is usually quoted. It deals with Goethe's early life up to his departure for Weimar in November 1775. It is divided into twenty numbered Books, which are grouped in four Parts, each of five Books. Part One, written 1810–11, was published in 1811; Part Two, written 1811–12, followed in 1812; Part Three, written 1812–14, appeared in 1814; Part Four is a later contribution, written 1824–31 and published posthumously in *Nachgelassene Werke,* vol. 8, 1833.

The contents of the four Books, briefly summarized, are as follows: Pt. I, Bk. 1: Birth, ancestry, parentage, early impressions (including the marionette theatre—Puppentheater—and the Lisbon earthquake of 1755), Frankfurt, and the paternal house. Bk. 2: Experiences of childhood, including further parental details, and the outbreak of the Seven Years War (see SIEBENJÄHRIGER KRIEG). Bk. 3: The requisitioning of part of the house for the French military commandant, Count Thoranc, the clash between the Count and Goethe's father, and an account of watching artists at work in the house. Bk. 4: Boyhood, with language study and theological comments, and an account of Goethe's earliest attempts at writing. Bk. 5: Account of the coronation of the Emperor Joseph II (q.v.) in Frankfurt, and Goethe's first love-affair (see GRETCHEN, FRANKFURTER).

Pt. II, Bk. 6: Youth, including early impressions of Leipzig University. Bk. 7: A review of German literature of the 18th c. up to the 1760s, Goethe's love-affair with Käthchen Schönkopf (q.v.), and the writing of *Die Laune des Verliebten* and *Die Mitschuldigen* (qq.v.). Bk. 8: Goethe's early artistic interests, his illness and return home, consequent tension in the family, and the influence of Fräulein von Klettenberg (q.v.). Bk. 9: Strasburg University, the experience of Gothic architecture, the passage of the Archduchess Maria Antoinette (the future Queen of France) through Strasburg. Bk. 10: Klopstock, Gleim (qq.v.), and Goethe's contact with Herder (q.v.), Goethe's initial interest in Götz von BERLICHINGEN and *Faust* (qq.v.), and his love for Friederike Brion (q.v.).

Pt. III, Bk. 11: Continuation of the relationship with Friederike and Goethe's return to Frankfurt. Bk. 12: The end of the love-affair, the months at Wetzlar, with Goethe's love for Charlotte Buff (q.v.). Bk. 13: Review of German drama in the 1770s, *Götz von Berlichingen*, *Die Leiden des jungen Werthers* (qq.v.). Bk. 14: Sturm und Drang, J. M. R. Lenz, H. L. Wagner, F. M. Klinger, Lavater, Merck, Basedow, and F. H. Jacobi's interest in Spinoza (qq.v.), and the plan for a play about Mahomet. Bk. 15: Preoccupation with the Wandering Jew (see EWIGER JUDE) and with *Prometheus* (q.v.), visit of Duke Karl August (q.v.) of Sachsen-Weimar, *Clavigo* (q.v.).

Pt. IV, Bk. 16: Influence of Spinoza, first meeting with Lili Schönemann (q.v.), Jung-Stilling (see JUNG, H.). Bk. 17: Engagement to Lili Schönemann. Bk. 18: Beginning of the Swiss journey with the brothers Stolberg (qq.v.), Karlsruhe, further meeting with Karl August, Zurich, walking in the Alps. Bk. 19: Alpine tour, including St. Gotthard, parting from Stolbergs, return to Frankfurt and Lili Schönemann, latent conflicts. Bk. 20: Work on *Egmont* (q.v.), end of engagement, invitation to Weimar, famous passage on 'das Dämonische' (q.v.), apparent failure of Weimar to proceed with invitation, Goethe's departure for Italy, arrival of the belated envoy from Weimar.

Goethe's ambiguous title indicates that his autobiography is a stylized formulation and presentation of his early life, not an exact record. Forty years separate even the latest of the events from the time of writing. Remoteness enabled the recollections to be presented as a rounded and consistent whole with the qualities of a work of literary art.

Goethe used the title *Aus meinem Leben* (omitting 'Dichtung und Wahrheit') for the first publication of the *Campagne in Frankreich* and *Die Belagerung von Mainz* (qq.v.), which he regarded as the Fifth Part (1822), and for his *Italienische Reise*, vols. 1 and 2, which appeared as Second

Section (Abteilung), First and Second Parts (1816–17). The Third Part of this section of *Aus meinem Leben* was published as vol. 29 of the *Ausgabe letzter Hand*, and described as *Zweyter Aufenthalt in Rom* (1829). Other autobiographical works by Goethe are the *Sankt-Rochus-Fest zu Bingen* (q.v., 1817) and the *Annalen* or *Tag- und Jahreshefte* (q.v., *Ausgabe letzter Hand*, vols. 31 and 32, 1830).

Dictionaries, see WÖRTERBÜCHER.

'Die Himmel rühmen des Ewigen Ehre', first line of a religious poem, *Die Ehre Gottes in der Natur*, written by C. F. Gellert (q.v.) in 1747. It was set to music by Beethoven in 1803.

'Die Liebe, sagt man, steht am Pfahl gebunden', first line of the fifth poem (a sonnet) in Mörike's so-called 'Peregrina-Lieder' (see PEREGRINA). It was first published in 1829 in the *Morgenblatt* (Stuttgart) under the title *Verzweifelte Liebe*, was included in revised form in *Maler Nolten* (q.v.) as *Und wieder*, and then published without title in the Peregrina cycle in *Gedichte*, 1838.

Die Musik kommt, a poem by D. von Liliencron (q.v.), published in *Gedichte* (1889). It is famous for its impressionistic, onomatopoeic language ('Klingling, bumbum und tschingdada', etc.).

Dies Buch gehört dem König, a book by Bettina von Arnim (q.v.), in which she develops her views on society, politics, and religion. It was published in 1843. The king of the title is Friedrich Wilhelm IV (q.v.) of Prussia, who gave her permission before the book was written to dedicate it to him.

The greater part of the book is a dialogue in which the chief participant is Goethe's mother (Frau Rat), to whom Bettina von Arnim attributes her own religious, political, and social views. Though not irreligious, she is hostile to the existing religions and has a keen appreciation of the misery of the oppressed and the arrogance of the aristocracy. It is the duty of the monarch, she considers, to put right the social injustice of the realm. An appendix headed 'Erfahrungen eines jungen Schweizers im Vogtlande' documents the misery of a Berlin slum outside the Hamburger Tor. In 1849 Bettina von Arnim published *Gespräch mit Dämonen*, a further plea for social reform, of which the central feature is a dialogue between the sleeping king and his good spirit (Dämon).

DIESEL, RUDOLF (Paris, 1858–1913, at sea), of German parentage, was sent to Augsburg when

12 years old. He invented the diesel engine in the 1890s and developed it through Krupp and Maschinenfabrik Augsburg–Nürnberg. He was accidentally drowned on a Channel crossing.

Dies Irae, used as the title of a medieval Latin hymn, the first lines of which are 'Dies irae, dies illa, / Solvet saeclum in favilla.'

It was formerly a part of the requiem mass, and in this context is used by Goethe in the scene *Dom* of *Faust I* and *Urfaust* (see FAUST) at the obsequies of Gretchen's mother.

Dietegen, see LEUTE VON SELDWYLA, DIE.

Diethelm von Buchenberg, principal character in B. Auerbach's Novelle *Die Geschichte des Diethelm von Buchenberg* (q.v.).

DIETMAR DER SETZER, a minor Middle High German poet of the 13th c., who is the author of four Sprüche (see SPRUCH). Though he is recorded with the prefix *her*, he was probably a commoner. He may be the *vagabundus dictus Seczere* mentioned in the *Chronicon Colmariense* c. 1240.

DIETMAR VON EIST, or **AIST,** is the name under which a number of Middle High German Minnelieder (37 strophes) appear in the MS. collections. It is the name of a nobleman of a family resident east of Linz, who is documented between 1139 and 1171. It is generally thought that this Dietmar lived too early to have written more than two of the stanzas attributed to the name. It is possible that the author of some of the poems may have been a later, unrecorded Dietmar. Some of them are clearly of much later date and must be considered as of unknown authorship. In consequence of this uncertainty the discussion of Dietmar's poetry is fraught with difficulty. The poems which are thought to be genuine embody the style of the early Austrian Minnelied, in which the courtly conception of *minne* is not yet fully developed.

Dietrichsage, the forms in which the story of the historic 5th–6th-c. Theodoric the Great (q.v.) remained current in Germanic lands in the Middle Ages. They represent a considerable distortion of history, and have become entwined with the story of the Nibelungen. Theodoric is usually called Dietrich von Bern (i.e. of Verona). Dietrich, in the legend, is an Amelung by descent (hence *Amelungensage*), his father is Dietmar (historically Theodemir). Dietrich's principal thane is Hildebrand, his opponent Ermenrich (only the *Hildebrandslied* mentions the true antagonist Odoacer, whom Theodoric put to death). Ermenrich drives Dietrich from his realm, and is successful in defeating him at the battle of Ravenna (Rabenschlacht). Dietrich returns to his country, after thirty years' exile, on the death of Ermenrich. The historical Ermenrich lived in the 4th c., and Attila (Etzel), with whom Dietrich is connected in the legend, was dead before Dietrich was born. The best-known works connected with this legend are the *Hildebrandslied* and the *Nibelungenlied* (qq.v.), but there are numerous others: *Ortnit, Wolfdietrich, Biterolf und Dietleib, Walther und Hildegund, Rosengarten, Dietrichs Flucht, Die Rabenschlacht, Alpharts Tod, Virginal, Sigenot,* the *Eckenlied, Goldemar, Laurin, Der Wunderer, Dietrich und Wenezlan,* and the *Jüngeres Hildebrandslied* (qq.v.). The story of Dietrich in its folk form is most fully told in the Norwegian *Thidreksaga* (c. 1250), which is based on German oral tradition, and is also concerned with the story of the Nibelungs.

Dietrichs Flucht, title usually given to a Middle High German epic poem, which its author entitled *Das Buch von Bern* because it treats the ancestry and early life of Dietrich von Bern, i.e. Verona. It is written in rhyming couplets and dates from the late 13th c. After cataloguing Dietrich's ancestors, it narrates his expulsion from Italy. He is received by Etzel (Attila), whose niece he marries. He returns to Italy and expels Ermenrich. The author gives his name as Heinrich der Vogler, and is believed to have been a wandering court poet of common birth. In the MSS. it accompanies the epic *Die Rabenschlacht* (q.v.). See DIETRICHSAGE.

Dietrich und Wenezlan, an anonymous Middle High German poem written towards the end of the 13th c., of which a fragment of 510 lines is preserved. It is designed as a laudatory poem for King Wenzel II of Bohemia, who reigned from 1285 to 1305, and narrates a fictitious combat between Wenezlan (Wenzel) and Dietrich von Bern (see DIETRICHSAGE). The extant fragment breaks off before the issue is decided.

DIETRICH VON APOLDA, see ELISABETH, HEILIGE and ELISABETHLEGENDE.

DIETRICH VON BERN, see DIETRICHSAGE and THEODORIC THE GREAT.

DIETRICH VON DER GLEZZE, a Silesian in the service of Wilhelm von Weidenau in the late 13th c., was probably a native of Glatz (Glezze). He is the author of *Der Borte,* an erotic verse tale, in which a marriage breaks down and is renewed. The wife of a knight gives herself to a lover for various presents, and particularly a girdle which ensures victory. The husband discovers her deceit and leaves her. Disguised as a man she follows him. He is ready to sleep with the supposed man, whereupon she discloses her identity and the couple are reconciled. The hint of homosexuality is socially noteworthy.

DIETTRICH, FRITZ (Dresden, 1902–64, Kassel), was a stretcher-bearer in the 1939–45 War and was a prisoner of war in Russia from 1945 to 1947. A Christian poet with leanings to classical, especially hymnic, forms, he published *Gedichte* (1930), *Stern überm Haus* (1932), *Der attische Bogen* (1934), *Mythische Landschaft* (1936), *Güter der Erde* (1940), *Hirtenflöte* (1940), *Aus wachsamem Herzen* (1948), *Sonette* (1948), and *Zug der Musen* (1948). His *Nukleare Ode* dates from 1956.

Die von Hohenstein, a long novel by F. Spielhagen (q.v.), published in 1864. It is set in 1848–9 in and around the fictitious town of Rheinstadt (see REVOLUTIONEN 1848–9). The Hohensteins are a numerous and powerful aristocratic family settled in the Rhineland as officers and officials after the acquisition of the province by Prussia in 1815. Arrogant, avaricious, and corrupt, old and young, menfolk and womenfolk, they are exhibited in the novel in their decline and fall. The one exception is Wolfgang von Hohenstein, and he no doubt owes his integrity to his mother, a commoner. Wolfgang, after deferring for a time to the Hohensteins, becomes a revolutionary and, when the revolution fails, lives in free Switzerland, marrying a commoner cousin. An important character is Dr. Münzer, said to be modelled on F. Lassalle (q.v.). He is a vehement revolutionary, who deserts his wife for the beautiful Antonie von Hohenstein, and is murdered by Colonel Guibert von Hohenstein. The one-sided, corrosive portrayal of the Hohensteins in the first half of the novel is impressive, but the latter part degenerates into sensationalism.

Die vor den Toren, a historical novel by C. Viebig (q.v.), published in 1910. It is set in Berlin in the 1870s and deals with the plight of those overtaken or displaced by its rapid southward expansion.

DILHERR, JOHANN MICHAEL (Themar in Thuringia, 1604–69, Nürnberg), pastor of St. Sebaldus's Church, Nürnberg, was a member of the Hirten- und Blumenorden an der Pegnitz (q.v.) and published several volumes of verse.

DILTHEY, WILHELM (Biebrich/Rhine, 1833–1911, Seis/Schlern), a philosopher, whose philosophical interests were mainly historical and literary. He occupied a succession of chairs of philosophy, beginning with Basel in 1866, then Kiel (1868), Breslau (1871), and finally Berlin in 1882. He was also a member of the Akademie der Wissenschaften in Berlin (see AKADEMIEN). Dilthey turned away from metaphysics and treated philosophical systems as historically determined. He closely related philosophical, historical, and literary studies to psychology, and established in each field a systematic typology. In *Der Aufbau der geschichtlichen Welt in den Geisteswissenschaften* (1910) he drew a firm distinction between *Geisteswissenschaften* and *Naturwissenschaften*, and in *Ideen über beschreibende und zergliedernde Psychologie* (1894) set out a method of approach. Particularly important for the study of literature were *Das Leben Schleiermachers* (1870), the essay *Die Einbildungskraft des Dichters* (1887), and *Das Erlebnis und die Dichtung* (1905), which comprises the five essays: *Gang der neueren europäischen Literatur, Gotthold Ephraim Lessing, Goethe und die dichterische Phantasie, Novalis,* and *Friedrich Hölderlin.* In the study of literature Dilthey's work was largely responsible for discrediting the historical methods of the 19th c., and for inaugurating the approach of Geistesgeschichte (q.v.), which dominated German scholarship in the first half of the 20th c. Through his psychological perception and typology he has had an even more lasting influence.

Diné zu Koblenz, see ZWISCHEN LAVATER UND BASEDOW.

DINGELSTEDT, FRANZ, FREIHERR VON (Halsdorf nr. Marburg, 1814–81, Vienna), born a commoner, studied at Marburg University and became a schoolmaster, first at Hanover, then at Kassel, and finally at Fulda (1834–41). While still a teacher he engaged in literary journalism, publishing the periodical *Die Wage* (in which his ironical *Spaziergänge eines Kasseler Poeten* appeared), and later contributing to A. Lewald's *Europa* (q.v.). A satirical novel, *Die neuen Argonauten* (1839), led ultimately to Dingelstedt's dismissal in 1841. He then became a full-time (and successful) journalist, acting as foreign correspondent in Paris, London, and Vienna for the *Allgemeine Zeitung* (q.v.) of Augsburg. In 1840 he published a second political novel, *Unter der Erde* (2 vols.), and, in 1841, a collection of rasping satirical poems (*Lieder eines kosmopolitischen Nachtwächters,* anon.). Up to this time he was a radical, having affinities with the Young German movement (see JUNGES DEUTSCHLAND); but he subsequently underwent a conversion, and, as a newly won supporter of the existing order, was appointed royal librarian in Stuttgart (1843). In 1851 he was appointed director of the Munich Court Theatre, where he put on a number of modern plays of merit. Dismissed in 1857, he was at once appointed to a similar post at Weimar, where he produced Hebbel's *Die Nibelungen* (q.v.) in 1861 and a cycle of Shakespearian histories in 1864. Dingelstedt, by this time a theatre manager with a great reputation, became director of the Vienna opera in 1867 and of the Burgtheater in 1870, a post which he retained until his death.

Dingelstedt is now remembered chiefly for his successful and sometimes courageous direction of the theatres under his control, often in the face of opposition, especially in earlier years; but he was also a prolific author of short stories, written primarily for entertainment. These appeared in the collections *Frauenspiegel* (1838), *Licht und Schatten in der Liebe* (1838), *Wanderbuch* (1839–40), *Heptameron* (1841), and *Sieben friedliche Erzählungen* (1844). *Die Amazone* (2 vols., 1868) was a novel of high society. Dingelstedt also had some theatrical success with a play *Das Haus der Barneveldt* (1850). His lyric poetry appeared in *Gedichte* (1845) and *Nacht und Morgen* (1851). His *Sämtliche Werke* (12 vols.) were published in 1877–8.

Dinggedicht, a term designating a poem focused upon an object, though the object frequently fulfils a symbolic function. Classical examples are Mörike's *Auf eine Lampe*, C. F. Meyer's *Der römische Brunnen*, and Rilke's *Das Karussell* (qq.v.).

Diogena, a parody in the form of a novel by Fanny Lewald (q.v.), published in 1847; its title-page falsely suggested the authorship of the writer it parodies, Ida, Gräfin von Hahn-Hahn (q.v.).

Diotima is the name of the priestess in Plato's *Symposium*. She is the subject of an essay by Friedrich Schlegel (q.v.) entitled *Über die Diotima* (1797). The name is, however, best known in German literature for its use by Hölderlin (q.v.) for the woman he loved, Susette Gontard (q.v.). Three poems, existing in differing versions, are headed 'Diotima', and Diotima is also the name of the woman loved by the hero in Hölderlin's novel *Hyperion* (q.v.).

In R. Musil's novel *Der Mann ohne Eigenschaften* (q.v.) Diotima is the name which Ulrich, in his own mind, gives to Frau Ermelinda Tuzzi. He does so in mental reference to Hölderlin's Diotima ('Also eine geistige Schönheit . . . eine zweite Diotima').

Diplomatische Revolution, the reversal of European alliances following the Peace of Aachen (1748, see ÖSTERREICHISCHER ERBFOLGE-

KRIEG) and preceding the Seven Years War (see SIEBENJÄHRIGER KRIEG). Conflicting interests between existing allies and potential enemies in Europe, America, and India led to a reappraisal of foreign relations. In 1748 Great Britain, Saxony, and Russia were allied with Austria, and France with Prussia. By 1756–7 Austria had become allied to France, Russia, and Saxony, while Prussia and Great Britain found common ground, not least by being the odd ones out.

The reversal of alliances was the idea of the youngest member of Maria Theresia's Cabinet, W. A. von Kaunitz (q.v.), whose prime concerns were Austrian prestige in Germany and the isolation of Prussia, which was to be forced to disgorge the former Austrian territories of Silesia (see SCHLESISCHE KRIEGE). He had the whole-hearted support of Maria Theresia (q.v.). The imminence of war between Great Britain and France in 1756 spurred on the powers to commit themselves after eight years of diplomacy. By the Treaty of Westminster (January 1756) Prussia undertook the protection of Hanover against a possible French attack, while Great Britain promised to safeguard Prussian territory in the event of a Russian invasion. Against this Austria succeeded, after some 200 years of rivalry between the Houses of Habsburg and Bourbon, in concluding the Treaty of Versailles (1 May 1756) with France and Russia. Because of the influence of the three women, the Tsarina Elizabeth of Russia, the Empress Maria Theresia, and Madame de Pompadour, mistress of Louis XV, this defensive alliance became known as the 'Alliance des trois cotillons'.

Fifteen days after the treaty had been signed Britain declared war on France, the least fortunate partner in an alliance which committed the French to support Austria against an invasion of Prussia, but not Austria to assist France against an attack by England. In the Second Treaty of Versailles (May 1757), completing the Diplomatic Revolution, the three powers consolidated their interests. By this time Friedrich II (q.v.) had already gone to war to fight for the survival of Prussia, which was threatened, apart from the main signatories, by Sweden and Saxony as well, and thus on all four frontiers.

Discourse der Mahlern, a literary weekly founded by J. J. Bodmer and J. J. Breitinger (qq.v.). It lasted from 1721 to 1723. The contributors signed their articles with the names of famous painters. A basic assumption was the fundamental identity of painting and poetry.

Disticha Catonis, also *Dicta Catonis*, title given to a Latin moral poem of the 3rd or 4th c. which had considerable popularity in Germany in the Middle Ages, especially as a school manual. It consists of a short preface (*epistola*); a group of rules for living, expressed in prose (*breves sententiae*); and four books of moral aphorisms in the form of hexametrical distichs. Sometimes referred to as *Dionysius Cato*, after a supposed author, the work is of unidentified authorship. Its general moral tone is stoical and utilitarian. An Old High German translation, made *c*. 1000 by Notker Labeo (q.v.), is lost. The oldest surviving German version of the *Disticha* is in Middle High German and dates from about the middle of the 13th c. It was made in Austria and is preserved in a MS. formerly at Zwettl Abbey in Lower Austria. An abbreviated version is also extant. The *Disticha* were translated in 1498 by Sebastian Brant (q.v.). Some MSS. contain a translation of a continuation of the *Disticha* known as *Novus Cato* or *Supplementum Catonis*, which is mainly concerned with rules of behaviour.

Distichon, a distich, two lines of verse, in which a classical hexameter is followed by a pentameter. It was used (both by the Ancients who devised it and by the Germans, who employed it chiefly in the 18th c. and 19th c.) singly or in pairs for epigrams, or in greater quantity in elegies; in this form distichs are known as elegiacs (see ELEGISCHES VERSMASS). The first German distichs, which were rhymed, were written in the 16th c. by Fischart and Claj (qq.v.). In the 18th c. unrhymed distichs were composed by Klopstock, and, abundantly, by Goethe and Schiller (qq.v.) in their classical period. The *Xenien* (q.v.) are written in this metre, as well as Schiller's *Der Spaziergang* and Goethe's *Römische Elegien* (qq.v.). Hölderlin also employs elegiacs, notably in *Brod und Wein* (q.v.). In the 19th c. they occur in the work of many poets, though nowhere so successfully as in the poetry of Mörike (e.g. *Häusliche Szene* or *Bilder aus Bebenhausen*, qq.v.). Recent poets have for the most part turned away from distichs, but Rudolf Alexander Schröder (q.v.) has written in this form with distinction.

Dithmarschen, a region of N. Germany, since 1474 legally a part of Holstein. It is the coastal strip by the North Sea running from the Eider estuary to the mouth of the Elbe and extending inland to include the towns of Heide, Wesselburen, Meldorf, and Marne. It is low-lying, largely agricultural land, and developed in the Middle Ages into what was virtually an independent republic dominated by farmers. Dithmarschen resisted attempts at incorporation, notably routing Duke Friedrich von Holstein at Hemmingstedt (q.v.) in 1500. Though

conquered in 1559, it retained its own administration, centred at Heide. In 1867, as a part of Schleswig-Holstein, it passed to Prussia. F. Hebbel and the lesser figure, A. Bartels (qq.v.), were both Dithmarscher, and G. Frenssen (q.v.) was born on the south-eastern edge of the region.

DÖBLIN, ALFRED (Stettin, 1878–1957, Emmendingen nr. Freiburg), grew up in Berlin and studied medicine (psychiatry and neurology) at Berlin and Freiburg universities, worked for a time in a mental hospital in Regensburg, and from 1911 practised in Berlin (Lichtenberg). From 1914 to 1918 he was a medical officer in the army. After the war he practised again in the slums of Berlin and supported the Social Democrats (he left the SPD in 1928). Endangered as both a Jew and a Socialist under the National Socialist regime, he took refuge in France in August 1933. In 1940 he escaped from France through the Iberian peninsula to the U.S.A. His experiences on this hazardous journey are described in *Schicksalsreise. Bericht und Bekenntnis* (1949), which also includes reflections on religion. In 1941 he converted to the Roman Catholic faith. In 1945 he returned to Germany as an American education officer in Baden. He lived for a time in Mainz, was in 1949 co-founder of the Akademie für Wissenschaften und Literatur and its vice-president, but was not at ease in West Germany and, a French national since 1936, settled in Paris in 1951. During his last years he suffered from illness and had lost the verve and spirit to which he owed his best-known work. He died in a nursing-home in Baden-Württemberg.

Döblin was an individualist acutely conscious of the vulnerability of the individual to collective pressures. He felt early drawn to Hölderlin, Schopenhauer, and Nietzsche (qq.v.), then showed an affinity to Jugendstil (q.v.) and *fin de siècle*, before becoming associated with Expressionism and *Der Sturm* (q.v.), in which he published his early stories, including *Die Ermordung einer Butterblume*, the title-story of a collection of 1913; this was followed by another collection of twelve stories, *Die Lohensteiner reisen nach Böhmen* (1917), a volume containing *Der Feldzeugmeister Cratz* and *Der Kaplan* (1926), and *Heitere Magie* (1948, *Märchen vom Materialismus* and *Reiseverkehr mit dem Jenseits*). But he was primarily a novelist and (with *Berlin Alexanderplatz*) an exponent of the modernist novel, which for him also implied radical rejection of Th. Mann's realism (*Buddenbrooks*, q.v.) with its reliance on the narrator's omniscience.

The novel *Die drei Sprünge des Wang-lun* (1915), set in 18th-c. China, describes an unsuccessful and tragic rebellion in which even the most peaceable are impelled to resist by force the tyranny of the state. In *Wallenstein* (2 vols., 1920) the strongest of men are overwhelmed by the events of history. *Berge Meere und Giganten* (1924) is a novel of a future age in which a proliferating technology threatens to subjugate the human race. In *Berlin Alexanderplatz* (q.v., 1929), Döblin's best-known novel, the depiction of the uphill struggle of a casualty of society to rehabilitate himself is matched by an uncompromising exposure of Berlin, expressed in a strikingly lively and flexible style. *Babylonische Wandrung oder Hochmut kommt vor dem Fall* (1934) is a grotesque novel with surrealist features. *Pardon wird nicht gegeben* (1935), the first novel written in exile in Paris, again shows the individual destroyed between opposed social forces, represented by the mother and her factory-owning family and her son Karl, whose true sympathies with the proletarians are suppressed until it is too late. The novel derives from autobiographical substance (the implied setting is Berlin) and differs from the first Berlin novel both in its structure and in its uncompromising treatment of fate and Nemesis to which the title alludes. The ambitious *Amazonas-Trilogie* (*Das Land ohne Tod, Der blaue Tiger, Der neue Urwald*, 1937–47) discovers the seeds of present ills in colonial and religious history. *November 1918. Eine deutsche Revolution* (a trilogy composed in its second version of the novels *Verratenes Volk, Heimkehr der Fronttruppen*, and *Karl und Rosa*, 1948–50) uses an abortive revolution (see WEIMAR REPUBLIC) to expose German society of the time and its attitudes. This large-scale work represents the climax of Döblin's political scepticism. His last novel, *Hamlet oder Die lange Nacht nimmt ein Ende*, begun in Hollywood and completed in Baden-Baden in 1946 (without then finding a publisher), comprises a main action, a psychoanalytical contribution to Heimkehrerliteratur (q.v.), and a cycle of stories. Edward Allison loses one leg in action and is brought home suffering from shock and neurosis. His stepfather, the writer Gordon Allison, arranges story-telling sessions with Edward, in the company of friends, in order to divert Edward's relentless search for those who are responsible for the war. The discovery of disturbing family relationships (his Hamlet experience) aggravates his disillusionment with the hypocritical society around him and ends with the disintegration of his home and the death of his mother and stepfather. After P. Huchel (q.v.) published sections of the work in *Sinn und Form* (q.v., 1954 and 1955), Döblin changed its original ending, according to which Edward, having given the proceeds of his inheritance to the poor, enters a monastery; in the final version he rejects this idea and decides to shake

off the ghost of the past and make a new beginning.

Reise in Polen (1925) is a stimulating *compte-rendu* of a visit to Poland. Döblin was a lively and perceptive essayist, convinced of his responsibility, as a writer, to be political without party-political commitment. He used the pseudonyms Linke Poot (*Der deutsche Maskenball*, 1921) and Hans Fiedeler (*Nürnberger Lehrprozeß*, 1946). Other collections include *Das Ich über der Natur* (1928), *Der Bau des epischen Werkes* (1929), *Wissen und Verändern* (1931), *Jüdische Erneuerung* (1933, essays and stories; *Der verlorene Sohn*; *Das Märchen von der Technik*), *Der historische Roman* (1936), *Sieger und Besiegte. Eine wahre Geschichte* (1946), *Der unsterbliche Mensch. Ein Religionsgespräch* (1946), *Die literarische Situation* (1947), *Unsere Sorge – der Mensch* (1948), and *Die Dichtung, ihre Natur und ihre Rolle* (1950).

Ausgewählte Werke in Einzelbänden, ed. W. Muschg (until 1965), H. Graber (until 1972), and A. W. Riley (from 1978), appeared 1960 ff., a Jubiläumssonderausgabe (7 vols.), ed. E. Pässler, in 1977. Two volumes of this edition, *Erzählungen aus fünf Jahrzehnten* (1979, also containing stories not previously published in book form) and *Autobiographische Schriften und letzte Aufzeichnungen* (1980), were reprinted as volumes of the Einzelbände, which include previously unpublished work. The volume *Schriften zur Politik und Gesellschaft. 1896–1951* appeared in 1972.

DODERER, HEIMITO VON (Weidlingau nr. Vienna, 1896–1966, Vienna), served as a reserve officer in an Austrian dragoon regiment, was captured by the Russians in 1916 and repatriated in 1920. His interest in Dostoevsky dates from these years. He studied history at Vienna University, obtaining a doctorate in 1925. Two years previously he had published his first book, the collection of poems *Gassen und Landschaft* (1923), and a short narrative, *Die Bresche*, followed a year later. A pronounced individualist, he was a member of no literary circle in Vienna, though he maintained a close friendship with A. P. Gütersloh (q.v.), whose sixtieth birthday he commemorated with the pseudonymous essay *Von der Unschuld des Indirekten* by 'René von Stangeler' (1947).

The novel *Das Geheimnis des Reichs* (1930) is based on his Siberian experiences. In 1931 he began the novel *Die Dämonen* (q.v.), which was to occupy him on and off for twenty-five years. The novel *Ein Mord, den jeder begeht* (q.v., 1938) unravels an unexplained murder in which the 'detective' proves to be the 'criminal'. This psychological thriller was followed in 1940 by the novel *Der Umweg*, set in the period following the Thirty Years War (see DREISSIGJÄHRIGER

KRIEG); its hero, Corporal Brandter, rescued from hanging, comes by a circuitous route to be hanged precisely through the circumstances which earlier had saved him from this fate. Neither of these novels, nor a third, *Die erleuchteten Fenster oder Die Menschwerdung des Amtsrates Julius Zihal* (q.v., 1950), attracted serious attention, though this last already shows mastery in the detailed portrayal of the Viennese scene.

Die Strudlhofstiege (q.v., 1951), forming a prelude to *Die Dämonen*, had an immediate impact, and Doderer was recognized almost overnight as an important novelist of the midcentury. The presentation of an age and locality in an immense and diverse web of intersecting and parallel lines characterizes also the 'chronicle' novel *Die Dämonen*, finally completed and published in 1956. *Die Strudlhofstiege* portrays pre-war (1910–11) and post-war (1923–5) Vienna; *Die Dämonen* presents in extraordinary completeness and complexity the immediately following years (1926–7). For this novel Doderer was awarded the Großer Österreichischer Staatspreis.

In 1959 he published a work on the novel, *Grundlagen und Funktion des Romans*. In this witty and penetrating three-part essay of some 50 pages he disposes of much theoretical lumber and many learned accretions, and asserts the novelist's task to be one, not of subjective self-indulgence ('fishing in his own well', as he puts it with reference to James Joyce) but, of the application of technique, using language both as creative material and as an instrument to achieve a true universality which excludes 'was alles man heute—*nicht* zu wissen braucht, um universal zu sein!'. This technique he develops by using material composed of many independent strands, which by their intertwining contribute to re-create a 'lost', because forgotten, reality.

Die Merowinger oder die totale Familie (1962) is an imaginative and coruscating fantasy, in which the Freiherr Childerich von Bartenbruch marries his own mother and grandmother, and so realizes his ideal of the total family, or 'la famille c'est moi!'. In this amusing and ambitious piece of persiflage Doderer may be said to mock, without discrediting, his own method and at the same time to direct his satire upon the age. In 1963 appeared *Die Wasserfälle von Slunj* (q.v.), the first novel of a tetralogy bearing the neutral collective title *Roman Nr. 7* (q.v.). The second volume, *Der Grenzwald* (q.v.), remained unfinished and was published posthumously in 1967. These two works constitute a monumental fragment, representing Doderer's attempt to achieve the 'Totaler Roman' or 'Universalroman' for the period 1880 to 1960. In 1966 appeared a volume of minor prose, *Unter schwarzen*

Sternen, which includes *Das letzte Abenteuer* (1953) and *Die Posaunen von Jericho* (1958). The collected stories, including some previously unpublished, appeared posthumously as *Die Erzählungen* (1973). The volume *Frühe Prosa*, containing *Die Bresche*, *Jutta Bamberger*, and *Das Geheimnis des Reichs*, ed. H. Fleisch-Brunningen, appeared in 1968, *Repertorium. Ein Begreifbuch von höheren und niederen Lebens-Sachen*, ed. D. Weber, in 1969, and *Die Wiederkehr des Drachen. Aufsätze, Traktate, Reden* in 1970. His diaries, *Tangenten. Tagebuch eines Schriftstellers 1940 bis 1950* (1964) were followed by *Commentarii 1951 bis 1956. Tagebücher aus dem Nachlaß* (ed. W. Schmidt-Dengler, 1976).

Doderer's fluent, precise, and often witty prose exhibits a mastery over words which permits fine gradations of irony, innuendo, and understatement; and he possessed the power to organize a stratified yet organically unified novel. He epitomized his own artistic approach in the words *facta loquuntur* ('deeds speak').

Doge und Dogaressa, a story by E. T. A. Hoffmann (q.v.), written in 1817, first published in the *Taschenbuch für Liebe und Freundschaft* (1818), and afterwards included in volume 2 of *Die Serapionsbrüder* (1819). The work took as its starting-point a picture by C. Kolbe, which Hoffmann saw at an exhibition. It portrays an aged doge of Venice with a young wife. Hoffmann found the story of Marino Falieri and his young wife Annunziata in a history of Venice. The newly elected 80-year-old Doge Falieri is plagued by jealousy after his marriage to the beautiful 19-year-old Annunziata; he alienates his supporters, conspires against the Signoria, and is executed. Annunziata, who loves a young man named Antonio, flees with him, but the boat is overwhelmed in a storm, and both are drowned. The story is enclosed by a framing narrative (see RAHMEN) relating to the painting.

DOHNA, ABRAHAM, BURGGRAF UND HERR ZU (Mohrungen, Prussia, 1579–1631, Schlobitten), a member of a great Prussian noble family studied at the universities of Rostock and Altdorf, acquired military experience in the Netherlands, and became an expert on fortification. He was Brandenburg envoy at the Diet of Regensburg (1613, see REICHSTAG). He left MS. diaries and a bitter verse satire in Alexandrines on the German nobility as he saw it at the Diet. These were published in 1896. He was a nephew of Fabian, Burggraf zu Dohna (q.v.).

DOHNA, FABIAN, BURGGRAF UND HERR ZU (Stuhm, Prussia, 1550–1621, Karwinden, Prussia), a member of a great Prussian noble family, was a soldier and an administrator in the service of the Count Palatine (Pfalzgraf) Johann Kasimir. He left a MS. autobiography, which was published in 1905. He was the uncle of Abraham, Burggraf zu Dohna (q.v.).

DÖHRING, FRITZ, occasional pseudonym of C. Busse (q.v.).

Doktor Faustus, a novel by Th. Mann (q.v.), written between 1943 and 1947, and published in 1947. It is Mann's judgement on the Germany of his lifetime. The sub-title runs 'Das Leben des deutschen Tonsetzers Adrian Leverkühn, erzählt von einem Freunde'. This friend, Serenus Zeitblom, a German 'Bürger', represents traditional values but lacks strength. In contrast, the dynamic Leverkühn is a man possessed. Two aspects of the German character are here embodied. Leverkühn, born in 1885, grows up at the time when Nietzsche (q.v.) was the idol of the young, and Mann establishes points of identity between him and the fictitious Leverkühn. But he also demonstrates Leverkühn's subjection to Wagner's overwhelming and seductive music. The genius of the German people is conceived as musical, and the composer Leverkühn continues the task of destruction begun by R. Wagner (q.v.). Zeitblom portrays his friend as intellectually ruthless, determined to penetrate that which should remain intact, and to explore paths which should remain barred. His mind is incapable of compassion. His life extends to 1940, embracing the growth of evil in Germany and its apparent triumph.

Leverkühn devotes himself entirely to a new kind of music. Only in parody do his compositions exhibit any link with musical tradition, and his life runs a destructive course parallel to that of his music. A decisive and symbolic incident occurs during his studies at Leipzig. He experiences a passionate attraction to a prostitute, whom he later calls the 'Hetaera Esmeralda'. He meets her in Preßburg, and, in defiance of her warnings, insists on spending a night with her. He contracts syphilis, for which he vainly seeks a cure. One of his doctors dies, another is arrested. He thereupon accepts his disease as the 'diabolical curse' taking charge of his life. He visits Italy, and during an attack of migraine the Devil appears to him and announces his ultimate enthralment at the expiry of the allotted span of twenty-four years. Leverkühn betakes himself to a Bavarian village, Pfeiffering, where he lives for nineteen years, writing most of his music in that time.

Three times this disciple of the Devil attempts a human relationship, and each ends in failure.

He falls in love and entrusts his wooing to a friend, the violinist Rudolf Schwerdtfeger. But Schwerdtfeger and Marie Godsau go off together, and Leverkühn loses at one stroke both friend and bride. When Nepomuk, the 5-year-old nephew to whom he is devoted, suffers an agonizing death from meningitis a few weeks after his arrival on a visit, Leverkühn becomes convinced that his curse afflicts those around him. His sanity begins to waver. He summons his acquaintances to hear his latest work, and informs them that he has a contract with the Devil and that the time has arrived at which the pact matures. He collapses, and the rest of his life is spent in imbecility, during which, like Nietzsche, he is tended by his mother.

Doktor Faustus is largely concerned with the nihilism which permeated European, and especially German, civilization in the 20th c., and Nietzsche and Wagner are exposed as its sponsors. The climax of Leverkühn's work of negation occurs in the composition and in the nature of his last work, the cantata *D. Fausti Weheklag.* To Zeitblom he exclaims 'es soll nicht sein', and goes on to explain that the 'es' of this cryptic prohibition is 'what is good and noble'. He further declares that he will cancel 'it', and 'it' is revealed to be the Ninth Symphony of Beethoven, the symbol of faith and optimism. This nihilistic cantata, brilliantly described by Mann, closes with an orchestral movement, in which, when all the other instruments have fallen silent, there is heard a prolonged diminuendo on G flat from a solo cello; to 'the listening soul' the recollection of this note 'steht als ein Licht in der Nacht'. This 'light in the night' ends Leverkühn's musical œuvre. It may express a hope of grace and give at least a hint that Mann's final attitude to German civilization is not one of total despair.

In 1949 Mann published *Die Entstehung des Doktor Faustus. Roman eines Romans.* It is not a novel but a discursive autobiographical essay covering the period January 1943 to the appearance of *Doktor Faustus* in the autumn of 1947. Discussions with Th. Adorno (q.v.) are reported, and the evil influence which Mann attributes to the work of Nietzsche and Wagner is confirmed. A film version by F. Seitz had its premiere in 1982.

Doktor Gräsler, Badearzt, a Novelle by A. Schnitzler (q.v.), published in 1917, and included in 1922 in *Die Alternden.* Dr. Gräsler, a bachelor in middle age, finds himself alone when his sister, who has kept house for him, takes her own life. Gräsler is virtually offered marriage by a most eligible and attractive young woman, but his irresolution estranges her. He picks up a girl of little education and thinks of marrying her, but she dies of scarlet fever. He finally marries a young and personable widow with a 7-year-old daughter. In spite of the 'happy ending' an autumnal melancholy pervades the story.

Dr. Katzenbergers Badereise, a novel by JEAN Paul (q.v.), published in 1809. It is a grotesquely humorous account of a visit to Bad Maulbronn by Dr. Katzenberger with his daughter Theoda. Theoda adores the poetry of Theudobach von Nieß, and when a handsome young army captain pretends to be the poet, she falls head over heels in love with him, to the discomfiture of the real poet. The novel, which contains a series of intentionally absurd incidents, ends with the marriage of Captain Theudobach and Theoda. The chapters are eccentrically designated *Summulae* (or 'Summeln') and the forty-second contains one of Jean Paul's famous evocations of nocturnal landscape.

Doktor Murkes gesammeltes Schweigen, a story by H. Böll (q.v.) which provides the title for a collection of 'satires' (Satiren) published in 1958. It pokes fun at the world of radio, in which words flow ceaselessly day after day. Dr. Murke, a programme editor, has developed such an intense dislike of words that he collects small strips of silent tape, which he splices and then 'plays' to himself at home. Böll weaves into this amusing nonsense a satire on a pompous literary figure who broadcasts regularly. In this story he is called Dr. Bur-Malottke. Having used the name 'Gott' in two of his recordings, he has second thoughts, and fears that anything so overtly religious may not accord with the current intellectual climate. He therefore proposes to substitute on each occurrence of 'Gott' the phrase 'jenes höhere Wesen, welches wir verehren'. Murke is given the task of complying with the great man's wishes and makes up for it by insisting on Bur-Malottke repeating the phrase in the nominative, genitive, etc. in order to annoy him.

DOLLFUSS, ENGELBERT (Texing, Lower Austria, 1892–1934, Vienna), an Austrian politician, became Federal Chancellor (Bundeskanzler) in 1932, when Austria was beset by Communists on the one side and National Socialists on the other. Leaning strongly on Italian Fascist support, Dollfuß set up a virtual dictatorship. A rising by the Social Democrats in February 1934 was suppressed by force. On 25 July he was assassinated in the Chancellery in an unsuccessful attempt by National Socialists to seize power.

He is caricatured by B. Brecht (q.v.) as Ignatius Dullfeet in *Der aufhaltsame Aufstieg des Arturo Ui* (q.v.).

DÖLLINGER, IGNAZ VON (pseudonyms Janus and Quirinus) (Bamberg, 1799–1890, Munich), a Roman Catholic professor of ecclesiastical history at Munich (1826–73), was excommunicated in 1871 for refusing to recognize the doctrine of Papal Infallibility (see KIRCHENKAMPF). At the Frankfurt Parliament (see FRANKFURTER NATIONALVERSAMMLUNG) he advocated the establishment of a German Church without state control, and linked with the papacy. He devoted himself in later life to the reunion of the Churches. In 1873 he succeeded J. von Liebig (q.v.) as president of the Bavarian Royal Academy of Sciences. His prolific writings include *Die Reformation* (3 vols., 1845–8), *Über die Wiedervereinigung der christlichen Kirchen* (1872), *Die Moralstreitigkeiten in der römisch-katholischen Kirche seit dem 16. Jahrhundert* (2 vols., 1889, with F. H. Reusch), and *Briefe und Erklärungen über die vatikanischen Dekrete* (1890).

DOMIN, HILDE (Cologne, 1912–), emigrated in 1934, and lived in England, S. Domingo, and the U.S.A. She is the author of the collections of poetry: *Nur eine Rose als Stütze* (1959), *Rückkehr der Schiffe* (1962), *Hier* (1964), *Höhlenbilder* (1968), and *Ich will dich* (1970). The story *Die andalusische Katze* (1971) was followed by an autobiographical volume, *Von der Natur nicht vorgesehen* (1974). In 1977 Hilde Domin edited *Gedichte* by Nelly Sachs (q.v.). *Abel steh auf. Gedichte, Prosa, Theorie* appeared in 1979, and the volume *Aber die Hoffnung. Autobiographisches. Aus und über Deutschland* in 1982. *Das zweite Paradies. Roman in Segmenten* (1968), her only novel, is based on her experience during her years in exile, her return to West Germany in 1954, and, finally, in 1960 when she settled in Heidelberg. She published a shortened version of the novel in 1980. She is also known as a translator (from four languages). In 1966 and, in extended form, in 1968 she published *Wozu Lyrik heute. Dichtung und Leser in der gesteuerten Gesellschaft*; a striking feature is her preoccupation with the function of poetry, which centres on the notion of the 'usefulness' of a poem to both author and reader, in that it creates a new awareness of different levels of experience. She is concerned with communication based on personal integrity.

Donaueschinger Passionsspiel, a passion play of the second half of the 15th c., preserved in a MS. at Donaueschingen.

Donaumonarchie, the Austro-Hungarian monarchy, 1867–1918.

Donauweibchen, Das, a Viennese Volksstück by K. F. Hensler (q.v.).

Don Carlos, Infant von Spanien, a verse tragedy in five acts by F. Schiller (q.v.), subtitled 'Ein dramatisches Gedicht'. He began it in 1783, published parts of it in 1785–6 in *Die Thalia* (q.v.), and completed it with difficulty for full publication in 1787. In its original form it was immensely long (6,282 lines), and in revisions in 1801, 1802, and 1805 Schiller reduced it to the still considerable length of 5,370 lines. It was first performed at Hamburg on 30 August 1787. Other theatres were unable to cope with the verse, and for them Schiller prepared a prose version, which survives as the *Rigaer Prosafassung*.

The plot is unusually intricate. Don Carlos, son of Philip II of Spain (q.v.), loves his stepmother Queen Elisabeth, who had originally been destined for him; he conceals this passion until his friend Posa arrives at court, to whom he confesses his legally incestuous longing. Posa secures for Carlos a meeting with the Queen, who seeks to divert Carlos's passions into political channels, persuading him to take up the cause, championed by Posa, of the persecuted Spanish Netherlands. King Philipp, however, refuses Carlos's request to be allowed to go to Brussels and appoints the reactionary Alba. Carlos is consoled on receiving a note of assignation from a lady, whom he wrongly believes to be the Queen. He keeps the appointment and discovers instead Princess Eboli. A scene of misunderstanding ends in wretchedness on both sides, and the Princess, determined to avenge herself on the Queen, denounces the supposed love-affair to Alba and Domingo, Philipp's confessor. These two sow suspicion in the King's mind. Philipp, distrusting all around him, interviews and appoints as his minister the apparently disinterested Posa, who pleads with great frankness for religious tolerance ('Geben Sie Gedankenfreiheit!'), and democratic rule. Posa seeks to divert suspicion from Carlos, but a series of mischances frustrates his intentions, and in the end he can only save Carlos by sacrificing himself. Hardly has Posa explained all to Carlos, than he is shot by an emissary of Philipp. Carlos in despair upbraids the King, his father, and then seeks to fulfil Posa's wishes by raising the standard of revolt in the Netherlands. As he takes his leave of Queen Elisabeth he is arrested by Philipp and handed over to the Inquisition.

Don Carlos, which contains the most vigorous and passionate blank verse ever written by Schiller, is confused in structure because of shifts in the author's intentions while he was writing the play. So obvious is the lack of unity that Schiller himself felt impelled to admit its faults and defend its psychology in his *Briefe über Don Carlos* (1788). Though Carlos is the hero of the title, Posa and Philipp have also been plausibly advanced as alternative principal figures. The events and characters of the play are only nominally historical. Schiller's main source was *Dom Carlos. Nouvelle historique* (1672) by the Abbé de Saint-Réal.

Don Giovanni, an opera by Mozart (q.v.) on the story of Don Juan. The full designation is *Il dissoluto punito, ossia il Don Giovanni*. The Italian libretto was provided by L. Daponte (q.v.), the music was composed in 1787, and the first performance took place at Prague on 29 November of that year. Mörike's Novelle *Mozart auf der Reise nach Prag* (q.v., 1856) is constructed around the first performance of the opera, and E. T. A. Hoffmann's Novelle *Don Juan* also has a stage production of *Don Giovanni* at its centre (see DON JUAN).

Don Juan, the hero of a 14th-c. Spanish legend, probably originating in Seville with a Don Juan Tenorio, an amorous adventurer. In this legend he kills in a duel the Count Ulloa who is defending the honour of his daughter, Donna Anna. He later invites the monument (a stone statue) on the Count's grave to supper. The statue appears, takes his hand, and sinks into the ground with him to the infernal regions. Don Juan's insatiable vitality and sensuality, and his condemnation, were first dramatized in a play attributed to Tirso de Molina (*El burlador de Sevilla y convidado de piedra*, a religious play, 1630). Tirso calls the count Don Gonzalo. The legend became known, with variations, in Italy and France (Molière, *Don Juan ou Le festin de Pierre*, first performed 1665), before L. Daponte (q.v.) wrote the libretto for Mozart's opera *Don Giovanni* (q.v.). Daponte's text is based on Giuseppe Gazzaniga's opera *Don Giovanni ossia Il Convitato di pietra* (first performed in Venice in 1787, the year of the first performance of Mozart's opera in Prague).

Among the many German treatments of the subject the following may be mentioned: E. T. A. Hoffmann's story *Don Juan* (q.v., 1813), C. D. Grabbe's tragedy *Don Juan und Faust* (q.v., 1829), *Faust und Don Juan* (3 vols., 1846) by G. Hesekiel (q.v.), N. Lenau's dramatic poem *Don Juan* (q.v., 1851); *Don Juans Ende* (1883) by P. Heyse (q.v.), *Don Juan* (1909) by Karl Sternheim (q.v.), *Don Juan* (1919) by Waldemar Bonsels (q.v.), *Don Juans Tod* by Georg Trakl (q.v., extant fragments are published in the collected edition of his works), *Don Juan oder die Liebe zur Geometrie* (1953) by Max Frisch (q.v.), and *Don Juan* (Tondichtung based on Lenau, 1889) by Richard Strauss (q.v.).

Don Juan, a Novelle by E. T. A. Hoffmann, first published in the *Allgemeine Musik-Zeitung* in

1813 and then in the first volume of his *Fantasiestücke in Callots Manier* in 1814. The story is sub-titled 'Eine fabelhafte Begebenheit, die sich mit einem reisenden Enthusiasten zugetragen'. A hotel guest hears music from next door, finds it is a theatre, and, discovering that the hotel has a private box, listens in seclusion to a remarkable performance of Mozart's *Don Giovanni*. He becomes aware of the presence of Donna Anna beside him as well as on the stage. He understands the opera as he has never done before, and afterwards sits up expounding it in a letter to a friend. At 2 a.m. he feels a peculiar shiver in the air, and learns the next day that at that precise moment the singer of the part of Donna Anna had died. The story combines realism, satire, and the uncanny, and also offers an original interpretation of the opera.

Don Juan. Ein dramatisches Gedicht, a poem by N. Lenau (q.v.), of 1,094 lines of rhyming iambic verse; written in 1844, it was published posthumously in 1851 by Anastasius Grün (q.v.) in Lenau's *Nachlaß*. Though it has its beginning and its end, a number of intervening scenes remained unwritten. Don Juan, whose pleasure it is to sample every variety of woman, promptly discarding each, in the end loses his impetus ('Es war ein schöner Sturm, der mich getrieben,/Er hat vertobt, und Stille ist geblieben') and allows himself to be killed by the son of a man he has murdered. He very properly makes testamentary provision for his numerous progeny.

Don Juan und Faust, a four-act tragedy by C. D. Grabbe (q.v.) in blank verse with some prose, written between 1823 and 1828. It was published in 1829 and in the same year had its first performance, in the Hoftheater in Detmold, with music by Albert Lortzing (q.v.). Notable treatments of Faust and Don Juan influenced Grabbe's conception of his play, viz. Goethe's *Faust* (q.v.) Pt. I, the *Faust* of E. A. F. Klingemann (q.v.), the opera *Faust* by L. Spohr (q.v.), Mozart's *Don Giovanni* (q.v.), and the *Don Juan* (1819–24) of Lord Byron.

Grabbe contrasts two themes (though he does not successfully co-ordinate them): the 'striving' of the German (and Northern) Faust and the ruthless sensuality of the Spanish (and Southern) Don Juan. Faust enters into a pact with the knight (Der Ritter) representing the Devil, the 'fallen angel', who also ironically reflects Faust's assertion of the 'superhuman' in man. Faust interferes with Don Juan's plan to seduce Donna Anna on her wedding day. With the help of the knight he transports Donna Anna to a magic castle on Mont Blanc after Don Juan has slain Don Oktavio, Donna Anna's bridegroom, and

her father, the governor Don Gusman. But as Donna Anna refuses to submit to Faust, even after he has killed his wife, Faust murders her too. He repents his rash deed and decides to surrender to the Devil before his time is up. He sees it as his last task to inform Don Juan in Rome (the principal location of the action) of Donna Anna's death. But Don Juan does not share his despair. He only increases Faust's awareness that he has destroyed the most worthy (and yet defiant) object of his aspirations from sheer want of comprehension. Faust gives himself up to the knight, resolved to fight him in hell in order to conquer immortality. Don Juan rejects such sentiments when, at the supper party with the statue of the governor, he refuses to repent for his salvation. He is still defiant as the knight drags him to join Faust since he too has pursued, in his own way, the same aim. Leporello, Don Juan's servant, has to follow his master amidst fire, thunder, and lightning. Both Faust and Don Juan actively seek self-realization. Faust, the idealist, fails tragically; Don Juan, the realist, succeeds by his undying defiance of any laws other than those of his nature. Herein lies his attraction for Donna Anna; she, however, sacrifices herself to virtue. In Don Oktavio, Grabbe satirizes the typical representative of the educated middle class and, in the governor, the man with an excessive sense of humour.

Donna Diana, title of a verse translation of Moreto's play *El desdén con el desdén* by J. Schreyvogel (q.v.), published in 1819 under the pseudonym Carl August West.

Doppeladler, the heraldic representation of two conjoined eagles, often erroneously construed as a two-headed eagle, was the central feature of the arms of the Holy Roman Empire (see DEUTSCHES REICH, ALTES) from 1401. After the extinction of the Empire in 1806, the double eagle passed to the Austrian Empire. The origin of the Doppeladler is the combination of the eagles representing the Western and the Eastern empires. In literature the word is frequently used to signify Habsburg or Austria, including the dual monarchy of Austria-Hungary.

Doppelgänger, Der, apocryphal title of Heine's poem beginning 'Still ist die Nacht, es ruhen die Gassen' (q.v.).

Doppelgänger, Ein, see EIN DOPPELGÄNGER.

Doppeltgänger, Die, a story by E. T. A. Hoffmann (q.v.), published in 1821 in the periodical *Feierstunden*. It was originally intended to be Hoffmann's contribution to a joint novel

(*Roman von Vieren*) in which A. von Chamisso, J. E. Hitzig, and K. W. Contessa (qq.v.) were also to participate. The plan came to nothing, probably because Chamisso left Berlin for his voyage round the world.

Dorfgeschichte, a story (novel, Novelle, or Erzählung) with village life as its subject. The earliest German example is H. Pestalozzi's *Lienhard und Gertrud* (q.v., 1781–7), closely followed by U. Bräker's *Der arme Mann in Toggenburg* (1789). The Dorfgeschichte, however, does not come into its own until the period of homely realism which followed the Romantic movement, when two types of Dorfgeschichte become distinguishable in the work of J. Gotthelf and B. Auerbach (qq.v.). A countryman himself, Gotthelf sought to establish communication with country folk. Auerbach, as a city dweller, was attracted by the idyllic aspects of rural life (cf. *Die Frau Professorin*). In subsequent years this distinction became less important. After Annette von Droste-Hülshoff's *Die Judenbuche* (q.v., 1844), O. Ludwig's *Die Heiteretei* (q.v., 1854), and G. Keller's *Romeo und Julia auf dem Dorfe* (q.v., 1856), the Dorfgeschichte tends to merge with Heimatkunst (q.v.). Among other writers of village stories in the second half of the 19th c. are H. Kurz, L. Anzengruber, P. Rosegger, Th. Storm, G. Frenssen, H. Voigt-Diederichs, J. Kneip, F. Nabl, J. F. Perkonig, J. M. Wehner, Lulu von Strauss und Torney, W. von Polenz, C. Viebig, L. Ganghofer, E. Zahn, H. Federer, L. Thoma, H. Hansjakob, H. Löns, W. Lobsien, and K. H. Waggerl (qq.v.).

DÖRFLER, Peter (Untergermaringen, 1878–1955, Munich), a Roman Catholic priest, was principal of an orphanage in Munich from 1915. He wrote novels of Catholic outlook and with realistic descriptions of peasant life, including *La Perniziosa* (1914, given the German title *Die Verderberin* in 1919), *Der Weltkrieg im schwäbischen Himmelreich* (1915), *Judith Finsterwälderin* (1916), *Neue Götter* (2 vols., 1920), *Der ungerechte Heller* (1922), *Die Papstfahrt durch Schwaben* (1923), *Die Schmach des Kreuzes* (1927, retitled *Heraklius* in 1950), *Apollonia-Trilogie* (1930–2), *Allgäu-Trilogie* (1934–6), *Auferstehung* (1938), *Die Wessobrunner* (1941), *Der Sohn des Malefizschenk* (1947), *Der Urmeier* (1948), and *Die Gesellen der Jungfer Michline* (1953). He also wrote biographies of Albertus Magnus (q.v.), St. Vincent de Paul, Neri, Niklaus von der Flüe (q.v.), and St. Ulrich.

Dorf- und Schloßgeschichten, a collection of five Novellen by Marie von Ebner-Eschenbach (q.v.), published in 1883. They are *Der Kreis-*

physikus, the story of the magnanimous act of a Jewish doctor in Galicia in 1848, *Jakob Szela*, *Krambambuli*, the tale of a faithful dog, *Die Resel*, and *Die Poesie des Unbewußten*, described as 'Novellchen in Korrespondenzkarten'. A further collection, *Neue Dorf- und Schloßgeschichten*, followed in 1886. It contains the three stories *Die Unverstandene auf dem Dorfe*, *Er laßt die Hand küssen*, and *Der gute Mond*. Of these the second, a story of aristocratic arbitrary, though unintended, cruelty, told within a frame (see RAHMEN) by the noble lady's grandson, is the best known.

Dorothe (in modern editions *Dorothee*), a three-volume historical novel by W. Alexis (q.v.), published in 1856. It is set in the time of the Great Elector, Friedrich Wilhelm (q.v.) of Prussia, and its principal figure, Dorothe, is his second consort (1636–89).

DORST, Tankred (Sonneberg, Thuringia, 1925–   ), called up in 1942, was taken prisoner and released in 1947. From 1952 he studied Germanistik, drama, and history in Munich, where he later lived. As a student he became involved in marionette theatre (with the student studio kleines spiel, for which he wrote four plays), recording this experience in *Das Geheimnis der Marionette* (1958) and *auf kleiner bühne* (1959). He achieved his early success with *Die Kurve* (1960), *Gesellschaft im Herbst* (1961), and his anti-war play *Große Schmährede an der Stadtmauer* (1961); *Die Mohrin*, a burlesque fairy-play, followed in 1964. His increasingly flexible dramatic technique aimed at reflective audience participation and employed all manner of well-tried distancing devices, including revue. His adaptation of plays was an aspect of this experimentation; they include *Der gestiefelte Kater oder Wie man das Spiel spielt* (1963, subsequently also published as *Der Kater oder Wie man das Spiel spielt*), *Rameaus Neffe* (1963, based on Diderot's *Le Neveu de Rameau*), *Der Richter von London* (1966, based on Thomas Dekker's Elizabethan comedy *Shoemaker's Holiday*). *Kleiner Mann, was nun?* (1972, based on a novel by H. Fallada, q.v.) was a revue notable for Dorst's collaboration with the director Peter Zadek, a collaboration which had begun with the television film version, entitled *Rotmord oder I was a German*, of *Toller* (1968), the first of a series of political plays. Set in the context of the Räterepublik (see BAYERN) and understandably viewed at the time of its first performance as a provocative play, Dorst's prime concern in his choice of Ernst Toller (q.v.) was 'to see the writer confronted with a concrete situation' (M. Jacobs). Karl Ludwig Sand and his murder of Kotzebue (qq.v.) is the subject of *Sand*.

Ein Szenarium (1971), and the aged Norwegian writer Knut Hamsun (1859-1950), who collaborated with the Nazis, that of *Eiszeit* (1973, film version by P. Zadek, 1975). The comedy *Auf dem Chimborazo* (1974), set 'about 1970', forms the last part of a family saga beginning in the 1920s. It was followed by *Dorothea Merz*, a work of fiction in which some 84 sections employing diverse narrative devices, including dialogue and scenario, present 'fragments from the life of a young woman in Germany 1925-34', *Klaras Mutter* (1978), and *Die Villa* (1980). The cycle is not chronologically arranged and its episodic presentation analyses (ultimately bitter) family relationships against the background of Germany's political and social developments leading up to and beyond 'Stunde Null'. These are also the subject of *Mosch* (1980). The terse scenario of *Die Reise nach Stettin* (1984, set in the latter half of the 1939-45 War) illustrates Dorst's technique of creating situations for his characters which appear to arise unexpectedly, take on unexpected dimensions, both tragic and comic, and result in self-scrutiny. The epic play *Merlin oder Das wüste Land* (1981) adapts the myths of the Middle Ages. In some of his works Dorst collaborated with Ursula Ehlers. Two volumes of collected plays appeared as *Stücke I* and *Stücke II* (postscript by G. Mensching) in 1978.

Drachenfels, peak of the Siebengebirge overlooking the Rhine near Königswinter. It is surmounted by a ruined castle, abandoned since the Thirty Years War (see DREISSIGJÄHRIGER KRIEG). Near by is the Drachenhöhle which, according to legend, was inhabited by the dragon slain by Siegfried (see NIBELUNGENLIED and NIBELUNGENSAGE, in which the character is known as Sigurd).

Dramaturgie, (1) an essay or treatise on the laws of drama. The classic example is Lessing's *Hamburgische Dramaturgie* (q.v.), a collection of essays and commentaries on dramatic theory and different types of drama; to this might be added the *Dramaturgie des Schauspiels* (4 vols.), by H. Bulthaupt (q.v.).

(2) The theory and practice of the work of the Dramaturg or dramatic adviser. He either performs the functions of a director or producer, or acts as a resident consultant.

DRANMOR, pseudonym of F. von Schmid (q.v.).

Draußen vor der Tür, a play by Wolfgang Borchert (q.v.), first performed on Hamburg Radio in February 1947, and on the stage in Hamburg in November 1947. It is described on the title-page as 'Ein Stück, das kein Theater spielen und kein Publikum sehen will'. It portrays the situation of a soldier, Beckmann, returning to a Germany in ruins, where he no longer has a home. It is the first play following the 1939-45 War to deal with this problem (see HEIMKEHRERLITERATUR).

Beckmann attempts suicide in the Elbe, but the (personified) river rejects him; he next meets the mysterious Other Man (der Andere), who also insists on life. He is briefly harboured by the Girl (ein Mädchen), and visits his former commander (ein Oberst), the well-fed Colonel, whom he bitterly reproaches for having sent him out in command of a patrol, of which eleven men were lost. His search for his parents ends with the discovery that they have committed suicide.

Beckmann's desperate situation remains unchanged, and the play ends with a question-mark: the impression of a nightmare persists. This is achieved by Borchert's Expressionistic technique involving dreams and characteristic, not individual, figures.

Drei, Die, a ballad by N. Lenau (q.v.). It has affinities with the Scots ballad *The Twa Corbies.* Its first line runs 'Drei Reiter nach verlorner Schlacht'.

Dreibund (Triple Alliance), an alliance between the German Empire, Austria-Hungary, and Italy, established in 1882. Its basis was the undertaking that, if two or more great powers attacked one of the parties, the other two parties to the Alliance would come to its assistance. It was additional to the secret Dual Alliance (see ZWEIBUND) between Germany and Austria-Hungary signed in 1879. Its original duration was for five years. It was renewed in 1887, 1891, 1896, 1902, 1907, and 1912. The Triple Alliance was brought to an end in 1914 by the refusal of Italy to enter into hostilities with the powers at war with Germany and Austria-Hungary.

Drei gerechten Kammacher, Die, a Novelle by G. Keller (q.v.), published in *Die Leute von Seldwyla,* Vol. 1 (1856). It is an essay in grotesque comedy (see GROTESKE, DAS). Each of three journeymen cherishes a hope of succeeding the master comb-maker by whom they are employed. Dietrich, the youngest, begins to court Züs Bünzlin, an unattractive spinster with a modest portion of 700 florins, and the other two become his rivals. The master declares that he must reduce his staff to one, and, since they cannot agree who should stay, he bids them all go outside the town walls and run home; the first to arrive will be re-engaged. When Züs learns this she announces that she will give her hand to the winner. She wishes to have one of the older men and therefore tries to hold Dietrich back, but he nevertheless wins and marries Züs, to be hen-pecked ever after. One of the unsuccessful runners hangs himself, and the other goes to the bad.

Dreigroschenoper, Die, an opera by B. Brecht (q.v.) with music by K. Weill (q.v.), written in 1928, published in 1929, and first produced in Berlin (Theater am Schiffbauerdamm) in August 1928. It is an adaptation of *The Beggar's Opera* by John Gay.

Brecht has more or less preserved Gay's plot, but has set the action against a background of Victorian London; he thus secured the opportunity to lay the blame for the existence of an underworld at the feet of bourgeois society. Most of the dialogue and the songs (19 in all) have been rewritten, revised, or extended, to make them topical for conditions in Germany. The new songs include 'Die Ballade vom angenehmen Leben der Hitlerstaaten' and the 'Schlußchoral'. In *Versuche*, 3, Brecht states: 'Die *Dreigroschenoper* ist ein Versuch im epischen Theater.' The application of his theories is elaborated in *Anmerkungen zur Dreigroschenoper. Winke für Schauspieler.* (Film version by G. W. Pabst, 1931.)

Drei Jünglinge im Feuerofen, Die, an anonymous narrative biblical poem in Middle High German, retelling the events of Daniel, chapter 3 (Shadrach, Meschach, and Abed-nego and the fiery furnace). The three youths are conceived as Christians, and the poem is analogous to a legend of martyrdom. In the Vorauer Handschrift (q.v.), in which the poem is preserved, it has, as a continuation, a poem on the story of Judith, which is, however, believed in reality to be a separate work (see JUDITH OR ÄLTERE JUDITH).

Dreikaiserbund, an alliance of the emperors of Germany, Austria-Hungary, and Russia, which resulted from their meeting in Berlin in 1873. It had been in preparation since the Franco-German War (see DEUTSCH-FRANZÖSISCHER KRIEG). In general terms, the emperors agreed to resist democratic reforms and to suppress revolutions, thus reflecting the spirit of the Holy Alliance (see HEILIGE ALLIANZ). There appears to have been no written agreement of this preliminary informal alliance. It was shaken by the Balkan crisis, in which Bismarck (q.v.) assumed the role of mediator between the powers at the Congress of Berlin (see BERLINER KONGRESS),

and in which Disraeli represented Great Britain. Alexander III of Russia felt himself deserted by the German emperor, and wrote to him to this effect in 1879, the year in which Bismarck concluded the Dual Alliance with Austria (see ZWEIBUND). At the same time Bismarck sought to improve relations with Russia.

In 1881 the Dreikaiserbund came into being officially; the three emperors undertook to remain neutral if one or more of the signatories found himself at war with a fourth party. In the special case of a war with Turkey, however, neutrality was obligatory only if agreement had previously been reached on the measures to prosecute a successful campaign. Germany and Austria also agreed to the Russian interpretation of the closure of the Dardanelles, according to which the warships of all nations were barred from passage.

The duration of the treaty was for three years and it was renewed in 1884. It expired in 1888, but in 1887 Bismarck had already forged another secret treaty with Russia to 'reinsure' peaceful relations. (See RÜCKVERSICHERUNGS-VERTRAG.)

Dreikönigsbündnis, the League of the kings of Prussia, Saxony, and Hanover, which was formed in May 1850 upon the initiative of Prussia. King Friedrich Wilhelm IV (q.v.) had rejected the imperial crown offered to him by the Frankfurt Parliament (see FRANKFURTER NATIONALVERSAMMLUNG), but hoped to achieve unity on a smaller scale by drawing Hesse and other smaller principalities into a Union with Prussia (Preussische Union). The author of this plan was the King's minister J. M. von Radowitz. By the time the Parliament met at Erfurt (March 1850), Saxony and Hanover had withdrawn from the League. Prussia's creation of a Union had little chance of survival and was finally dissolved at the settlement of Olmütz (see OLMÜTZER PUNKTATION).

Drei Lieder von der Jungfrau, see WERNHER, PRIESTER.

Drei Männer im Schnee, a novel by E. Kästner (q.v.), published in 1934. In a highly entertaining comedy of mistaken identities, Kästner upsets the social hierarchy, shows the hollowness of the standards of the Grand Hotel at Bruckbeuren, and emphasizes in his principal characters, the millionaire Tobler masquerading as a tramp, and the unemployed Dr. Hagedorn, mistaken for the millionaire, the true inner worth of both.

Drei Mönche zu Colmar, Die, a grotesque and macabre tale in verse (see SCHWANK) of the

13th c., in which three monks of Colmar use the confessional to entice a burgher's handsome wife by bribes. The husband, informed by his wife, arranges a trap, in which each monk perishes in boiling water. A drunken student is persuaded to throw the corpses into the Rhine. The unknown author gives his name as Nieman (Niemand). It is thought that this pseudonym may conceal a citizen of Colmar who intended the poem as an attack on the monks of the city.

Drei Novellen, a collection of three short works by Marie von Ebner-Eschenbach (q.v.), published in 1892 and included in Vol. 3 of *Gesammelte Werke.* They are *Oversberg. Aus dem Tagebuch des Volontärs Ferdinand Binder,* a narrative enclosed in a frame (see RAHMEN), *Der Nebenbuhler,* written in letter form (see BRIEFROMAN) and dated May–September 1875, and *Bettelbriefe,* using the dialogue form favoured by the early Naturalists (not the letter form). The Bettelbriefe are petitions which the young widowed Countess Beate receives from the poor. Experience teaches her that she must know the true plight of the petitioners when giving. She becomes dedicated to her social work and attains happiness through the help and love of a baron. All three stories expound social problems and human ethics, the example of Albrecht Oversberg giving the collection its substance. Regarded by most as a fool, he spends his life in the service of others, sacrificing his love for Lene, who becomes the wife of a rich Viennese businessman after Albrecht's impoverishment.

Drei Reiherfedern, Die, a dramatic poem ('Ein dramatisches Gedicht in fünf Akten') by H. Sudermann (q.v.), published in 1898. It is in rhyming verse. A symbolical mythological play, it shows how Prince Witte wins three egret feathers, the first of which, when burned, reveals the woman he loves; the second brings her to him; the third kills her. The theme is man's failure to recognize what he really desires, and his impulse to destroy what he loves.

Dreißigjähriger Krieg, a war waged in Germany from 1618 to 1648 which was the culmination of the Counter-Reformation (see GEGENREFORMATION). The prime objective of the Emperor Ferdinand II (q.v.), in conjunction with the Catholic League (see KATHOLISCHE LIGA), was the restoration of the Catholic faith in all dominions of the Empire, a task in which he failed. The war was not, however, a purely religious war, for political, dynastic, and national interests and attitudes also determined its course, and caused the involvement of most European powers, resulting in a decisive shift of the European power structure. The Empire ceased to exist as a political force and was henceforth identified with Austria. The territorial division of Germany was radically altered in the peace settlement, while German princes continued to rule absolutely in a particularist system consisting of some 200 independent states and sees. Of these Austria, Bavaria, Saxony, and Brandenburg (the future Prussia) were the most powerful states. The political link with Rome inherent in the medieval title of the 'Holy Roman Empire' (see DEUTSCHES REICH, ALTES) was at most a personal one of individual princes. Socially the protracted war developed into a serious disaster. The population of Germany was almost halved and the enormous task of economic reconstruction depended on the initiative of the princes. The devastation of the war had equally serious cultural consequences, which made themselves felt for at least a century.

The war developed in four phases:

1. 1618–25. Hostilities were first set in motion by the Bohemian revolt of 1618 (see MAJESTÄTSBRIEF), directed against Habsburg supremacy and achieving the deposition of Archduke Ferdinand (see FERDINAND II), hereditary King of Bohemia. Reverting to their traditional elective system, the Bohemians chose as their king the young Calvinist Friedrich V von der Pfalz, leader of the Protestant Union (see PROTESTANTISCHE UNION) and son-in-law of James I of England. He had the support of Charles Emmanuel of Savoy, who desired a lessening of Habsburg prestige. The Catholic League provided the main support for the Emperor Matthias (q.v.) and his heir Ferdinand. On 8 November 1620 the Bohemians under Christian of Anhalt were defeated by the League led by General Tilly (q.v.) in the battle of the White Mountain (see WEISSEN BERGE, SCHLACHT AM). Friedrich, henceforth derisively known as the 'Winter King', fled from Bohemia. The success of Ferdinand, emperor since 1619, alarmed other powers, and by 1625 England, Denmark, and part of North Germany had become allied against the Emperor, the Catholic League, and Spain. To meet the military challenge Ferdinand appointed Albrecht von Wallenstein (q.v.) general with an army of some 20,000 men.

2. 1625–30. During this period Wallenstein dominated the war. By 1628 he had overrun practically the whole of North Germany and the Danish possessions on the mainland. Only Stralsund (q.v.) successfully resisted his siege. On 12 May 1629 Christian IV of Denmark submitted to the Peace of Lübeck. But Ferdinand, encouraged by success, rejected moderation and issued the Edict of Restitution (Restitutionsedikt) according to which all ecclesiastical

estates secularized since the Peace of Augsburg (see AUGSBURGER RELIGIONSFRIEDE, 1555) were to be restored to the Church. But the Catholic League, headed by Maximilian I of Bavaria (see MAXIMILIAN I, KURFÜRST) and jealous of Wallenstein, urged Ferdinand at the Diet at Regensburg (1630) to dismiss him. Ferdinand gave in to this demand, although he stood no chance of applying the Edict without Wallenstein's aid.

3. 1630–4. Wallenstein's dismissal benefited the Swedish king, Gustavus Adolphus (q.v.), who took the lead in the Protestant cause with the backing of France and England. He was not welcome on German soil, but the sack of Magdeburg (q.v.) by Tilly in 1631 demanded decisive action. Georg Wilhelm (q.v.) of Brandenburg was forced into an alliance with Gustavus Adolphus, who was now subsidized by the United Provinces (Holland) as well as France. Hesse asked for Swedish help, and, under the threat of Tilly's advance into Saxony, Johann Georg (q.v.), the Saxon Elector, was forced to abandon his policy of neutrality. He, too, joined forces with Gustavus Adolphus. Tilly was defeated at Breitenfeld near Leipzig (17 September 1631). Gustavus Adolphus carried his campaign to defeat the forces of the League further south. In the battle at Rain (15 April 1632) Tilly was mortally wounded, and soon afterwards Munich was occupied by the Swedes. Simultaneously Johann Georg of Saxony entered Prague. Driven by desperation, Ferdinand recalled Wallenstein, who was defeated by the Swedish armies at Lützen (q.v., 16 November 1632), though Gustavus Adolphus was killed. His chancellor, Axel Oxenstierna (q.v.), became the new Swedish leader. Wallenstein thereupon opened peace negotiations with him and with Johann Georg. But the threat of his power led Spain and Maximilian to bring pressure on Ferdinand to dismiss Wallenstein a second time. He was condemned for treason, and on 25 February 1634 he was assassinated at Eger (now Cheb). Ferdinand recovered South Germany; in the Peace of Prague Saxony and the majority of the other Protestant powers came to an agreement to liberate Germany from foreign troops. Ferdinand at last abandoned the Edict of Restitution.

4. 1635–48. The efforts for peace, however, were frustrated by Cardinal Richelieu, whose intervention as an ally of Sweden gave the war new impetus. By becoming, in 1633, a party to the League of Heilbronn (see HEILBRONNER BUND) Richelieu had already entered German politics. When the Protestant forces suffered defeat at Nördlingen (q.v., 6 September 1634) the Emperor enjoyed a rare hour of triumph, but the defeated Germans were at the mercy of

Richelieu, who secured the assistance of the League of Heilbronn and BERNHARD von Sachsen-Weimar (q.v.) to fight against Spain. With the French declaration of war against Spain (19 May 1635) the religious war was finally over.

The last phase was on the one hand a war between two dynasties, the House of Bourbon against the House of Habsburg, in which France aimed at the acquisition of Alsace and Lorraine; and on the other a war between Sweden and the Empire, Sweden likewise aiming at territorial aggrandizement in the Baltic provinces.

After Bernhard von Sachsen-Weimar had captured Breisach in 1638 Richelieu was close to his aim, for, with Bernhard's death in the following year, his army passed entirely into French hands. Richelieu's policy was carried on successfully after his death in 1642 by Mazarin. Bavaria suffered the greatest devastation in the closing stages of the war, in which French and Swedish forces combined in their march towards the Danube to capture Donauwörth and Zusmarshausen (17 May 1648). In 1642 Friedrich Wilhelm (q.v.) of Brandenburg, the Great Elector, and in 1645 Johann Georg of Saxony, concluded a treaty of neutrality with Sweden. Negotiations for peace had been in preparation since 1642 and conducted at Münster and Osnabrück since 1644. It was due to the initiative of Queen Christina of Sweden and the French successes in Bavaria that the Peace of Westphalia (see WESTFÄLISCHER FRIEDE) was signed on 24 October 1648. It was not until the mid-fifties that the last foreign troops left.

Literature concerning the Thirty Years War includes the novel of a contemporary, Grimmelshausen's *Der abenteuerliche Simplicissimus* (q.v.), F. Schiller's study *Geschichte des Dreißigjährigen Kriegs*, and his trilogy *Wallenstein* (qq.v.), Grillparzer's play *Ein Bruderzwist in Habsburg* (q.v.), A. Stifter's Novelle *Der Hochwald* (q.v.), the novel *Der deutsche Krieg* by H. Laube (q.v.), C. F. Meyer's Novelle *Gustav Adolfs Page* (q.v.), the novel *Der große Krieg in Deutschland* by Ricarda Huch (q.v.), B. Brecht's play *Mutter Courage und ihre Kinder* (q.v.), and *Das Treffen in Telgte* by G. Grass (q.v.).

Dreizehnlinden, an epic by F. W. Weber (q.v.), published in 1878. It is his first and most popular work. It is in trochaic verse, written in four-line stanzas, recalling the similar verse of Heine's *Atta Troll* and *Deutschland. Ein Wintermärchen* (qq.v.). There are nearly 2,000 stanzas, arranged in 25 cantos. The background is the spread of Christianity in the 8th c. among the Saxons and their resistance to it. 'Dreizehnlinden' is the imaginary name of a monastery which Weber himself suggested might be identified with Corvey. In the last canto he insists on the claim of the spirit against the materialism of the age.

DREYER, MAX (Rostock, 1862–1946, Göhren, Rügen), a schoolmaster, turned to journalism in his middle twenties, and was on the staff of the *Tägliche Rundschau* in Berlin. In 1898 he resigned in order to devote himself solely to authorship. He was successful with Naturalistic plays (*Drei,* 1894; *Winterschlaf,* 1895; *Hans,* 1898; *Der Probekandidat,* 1899; *Die Siebzehnjährigen,* 1904) and also had a gift for neat comedy (*Eine,* 1896; *In Behandlung,* 1897; *Liebesträume,* 1898; and, most popular of all, *Das Tal des Lebens,* 1902). His poems were published in *Nah Huus* (1904). His novels and stories are mostly set in his home country. They include *Ohm Peter* (1908), *Der Siedler von Hohenmoor* (1922), *Das Gymnasium von St. Jürgen* (1925), and *Erdkraft* (1941). He was also the author of a volume of Novellen (*Strand,* 1910).

Dritte Buch über Achim, Das, a novel by U. Johnson (q.v.), published in 1961. Achim is modelled on the East German champion long-distance cyclist G. A. Schur, and the first two books 'about Achim' were actual books on Schur by K. Ullrich, a journalist. Johnson's novel tells of the attempt by the West German journalist Karsch to write the third book on Achim. Karsch fails because Achim insists on having himself presented as the conformist sporting hero of the DDR and as a lifelong supporter of Communism. Karsch, however, discovers that Achim was in the Hitlerjugend, was ready to denounce his father for listening to foreign broadcasts, and took part in the anti-Communist rising of 1953. Achim alternately recalls and denies or explains away, perceiving the world through a protective filter, which eliminates what he does not want to see, though his girl-friend, Karin, is unable to pardon Ulbricht (q.v., who throughout is disguised as *der Sachwalter*) for his betrayal of East Germany to the Soviet Union. Achim utters parrot-fashion all the clichés and slogans of the New Totalitarianism, and no real dialogue is possible between him and Karsch.

A postscript emphasizes that the book is concerned with the division of Germany, and the obstacles to understanding both between diplomats and between individuals. It is set entirely in East Germany, and is written and constructed with a detachment which does not exclude sympathy. Its structure is more immediately intelligible than that of *Mutmaßungen über Jakob* (q.v.) and the work displays a dry humour.

Dritte Reich, Das, a work of anti-democratic political theory by A. MOELLER van den Bruck (q.v.), published in 1923.

Drittes Reich, unofficial term for National Socialist Germany, much used by the Nazis themselves in the years after 1933, but banned after 1939. The official name of the state remained Deutsches Reich until 1945. The term 'the Third Reich' is widely used in Britain and America. The implied First and Second Empires are the 'Old German' or 'Holy Roman Empire' (see DEUTSCHES REICH, ALTES), and the Empire of the later Hohenzollerns (1871–1918); but 'Erstes Reich' and 'Zweites Reich' are not current expressions.

'Droben stehet die Kapelle', first line of the poem *Die Kapelle,* written by L. Uhland (q.v.) in September 1805 and published in 1806. In three stanzas, it reminds the listening shepherd, a symbol of youth, that death must come in the end.

DROLLINGER, KARL FRIEDRICH (Durlach, Baden, 1688–1742, Basel), wrote poetic fables and epigrams, as well as philosophical poems (*Lob der Gottheit, Über die Unsterblichkeit, Über die göttliche Fürsehung*). The only collected edition (*Gedichte*) appeared posthumously in 1743.

DRONKE, ERNST (Koblenz, 1822–?), a Radical journalist, wrote the poems *Armesünderstimmen* (1845), the stories *Polizeigeschichten* (1846) and *Die Maikönigin, ein Volksleben am Rhein* (1846). After the 1848 Revolution (see REVOLUTIONEN 1848–9) he is thought to have emigrated to America and his subsequent life is unknown.

DROSTE-HÜLSHOFF, ANNETTE, FREIIN VON (Hülshoff nr. Münster, Westphalia, 1797–1848, Meersburg/Lake Constance), who, with aristocratic prodigality, was christened Anna Elisabeth Franziska Adolfine Wilhelmina Luisa Maria, came of an old Westphalian Roman Catholic family, and was born in the castle which her ancestors had occupied for 400 years. A sense of tradition, both spiritual and aristocratic, never left her, and its presence makes her originality the more striking. She received a good education from private tutors, at the same time living the life of seclusion inevitable for a woman of her class in a country where great houses were scattered and communications poor. Though in delicate health, she spent much time out of doors, becoming a careful observer of landscape, birds, insects, and plants.

Her interest in poetry was early stimulated by

A. M. A. Sprickmann, once a friend of G. A. Bürger (qq.v.). In 1814 she began a two-act romantic verse tragedy (*Bertha*) which remained a fragment of some 2,000 lines. Other early writings were the epic poem *Walter*, completed in 1818, and *Ledwina*, an unfinished novel written in 1824, of which some fifty pages were posthumously published. In 1820 while on a visit to cousins, she fell simultaneously in love with two men, A. von Arnswaldt and H. Straube, but neither relationship lasted. Towards the end of that year she began to write for her grandmother a cycle of devotional poems reflecting the ecclesiastical calendar. *Das geistliche Jahr* (q.v.), published posthumously in 1851, at first proceeded rapidly then came to a halt and was resumed in 1839. The 72 poems of confessional character reflect not only piety but also a conscientious struggle with doubt. In 1826 Annette von Droste's father died, and she and her mother moved to a more modest, though still substantial, residence, the Rüschhaus, situated between Hülshoff and Münster.

After writing nothing for some years, between 1828 and 1838 she produced three epic poems, *Das Hospiz auf dem großen St. Bernhard*, *Des Arztes Vermächtnis*, and *Die Schlacht am Loener Bruch*. All three were published in her first collection of poems (*Gedichte*) in 1838. Annette's elder sister Jenny married Baron J. von Laßberg (q.v.) in 1834, and at their house at Eppishausen in Switzerland (Thurgau) Annette and her mother made a long stay in 1835–6. After her return to the Rüschhaus she met in 1837 Levin Schücking (q.v.), then 22, with whom she fell in love.

In 1841 Annette von Droste, whose health was giving cause for concern, went on a long visit to her sister, whose husband Laßberg had bought the castle of Meersburg in 1838. Schücking was there too, and these proved her happiest months, as well as her most productive, for she wrote at this time most of the landscape and nature poems (including the *Heidebilder*, q.v.) which are the most remarkable feature of the new volume (*Gedichte*) published in 1844. The Novelle *Die Judenbuche* (q.v.) was written between 1837 and 1841, and published in 1842. She continued to reside at Meersburg on account of her poor health, at the time believed to be caused by tuberculosis, though it is now thought that its origins may have been nervous; but in 1842, and again in 1845, she returned for a time to her Westphalian home, which she finally left in September 1846. The marriage of Schücking in 1843 to Luise von Gall is thought to have accelerated the decline in her health, and in the last two years of her life (1846–8) she wrote nothing more.

The unpublished poetry of the years before, containing some of her most moving poems, appeared posthumously in 1860 as *Letzte Gaben* (q.v.).

In 1840 Annette von Droste wrote a one-act comedy, *Perdu, oder: Dichter, Verleger und Blaustrümpfe*, in which the 'poet' is modelled on Freiligrath (q.v.), the critic (Rezensent) and 'part-time poet' on Schücking, and the 'bluestocking' (Blaustrumpf von Stande), Anna Freiin von Thielen, is the authoress herself, who withdraws her manuscript as soon as she senses that the publisher is only interested in business.

To call Annette von Droste the greatest German poetess, as is often done, is not to do justice to the strength and originality of her poetry. Many of her poems of reflection or observation show a bitter-sweet sense of transience, a keen and accurate vision, and a remarkable power in the choice and manipulation of words.

Her reputation as a writer of fiction depends on *Die Judenbuche*, an outstanding 19th-c. Novelle. She wrote brilliantly of her native Westphalia. According to her own comments, she did not complete her novel *Bei uns zu Lande auf dem Lande* because she was unable to prevent the characters resembling members of her own family. In 1845 *Bilder aus Westfalen* appeared anonymously under the title *Westfälische Schilderungen aus einer westfälischen Feder* in the *Historisch-Politische Blätter*. These sketches of Westphalia describe the differing mental and physical types found in the districts of Sauerland, Paderborn, and Münsterland. They are shrewd and bold, so much so that they could not appear under her name. But they are also written with the same empathy as *Die Judenbuche*, which resulted from her studies for this work; both works show the remarkable nuances of wit and irony which balance, but never obscure, the humanity of Annette von Droste's mature style.

Sämtliche Werke. historisch-kritische Ausgabe, 4 vols., ed. K. Schulte-Kemminghausen, appeared 1925–30 and *Historisch-kritische Ausgabe*, ed. W. Woesler from 1978.

DROSTE ZU VISCHERING, KLEMENS AUGUST, FREIHERR VON (nr. Münster, 1773–1845, Münster), was an intransigent Roman Catholic churchman who came into serious conflict with the Prussian state.

When Vikar of the cathedral chapter at Münster he forbade theologians of the diocese to attend the University of Bonn, and also adopted a rigid attitude in the question of mixed marriages. He became suffragan bishop of Münster in 1827 and was consecrated archbishop of Cologne in 1835. He immediately attacked Hermesianism (see HERMESIANISMUS) in the University of Bonn, and renewed his intransigent policy concerning marriages between Catholics and Protestants.

The clash with the Prussian state resulted in his suspension and imprisonment (1837–9). The conflict ended in 1842 when Archbishop von Droste gave up active participation in the affairs of the diocese. This period of tension is known as the Kölner Kirchenstreit.

'Du bist wie eine Blume', first line of an untitled poem by H. Heine (q.v.), included as No. XLVII in *Die Heimkehr*.

Ducus Horant, title given to a poem contained in a Hebraic MS. of 1382, which was discovered in Cairo and first published in 1957. Though the characters are Hebraic, the language is German. The poem combines elements from the stories of *Kudrun* and *König Rother* (qq.v.).

Duden, a series of dictionaries published by the Bibliographisches Institut, Mannheim. It originated with the volume *Vollständiges orthographisches Wörterbuch der deutschen Sprache* (1880), compiled by Konrad Duden (1829–1911), a schoolmaster; it became (and in its most up-to-date form remains) the standard authority for spelling.

This work is the foundation of *Der Große Duden*, which comprises ten volumes: (1) *Rechtschreibung*; (2) *Stilwörterbuch*; (3) *Bildwörterbuch*; (4) *Grammatik*; (5) *Fremdwörterbuch*; (6) *Aussprachewörterbuch*; (7) *Etymologie*; (8) *Vergleichendes Synonymwörterbuch*; (9) *Hauptschwierigkeiten der deutschen Sprache*; (10) *Bedeutungswörterbuch.*

A separate *Duden* is published in the DDR (Leipzig). The nearest rival of *Duden* is *Knaurs Rechtschreibung—Fremdwörter—Grammatik* (1973), which also provides interesting examples of changes in contemporary use of the German language. See also GERMAN LANGUAGE, HISTORY OF.

Duineser Elegien, a cycle of ten elegies in free hymnic verse by Rilke (q.v.), published in 1923. It is generally reckoned his highest achievement, apart perhaps from the *Sonette an Orpheus* (q.v.). The composition of the elegies extended over ten years, which were the most critical of Rilke's inner development. The first two elegies were written in 1912 at the castle of Duino on the Adriatic, from which Rilke took the title for the whole cycle. The third elegy was begun at Duino at the same time, and finished in Paris in 1913. Rilke wrote the fourth, and began the fifth, at Munich in 1915. There then followed a long pause until six more were written in an extraordinary outburst of creativity at Muzot,

Switzerland, between 7 and 15 February 1922. The last of these Rilke interpolated as the fifth, substituting it for the fragments of 1915. This outpouring formed part of an astonishing three-week period, in which Part I of the *Sonette an Orpheus* was written before the elegies, and Part II after.

The first six elegies lament the sorry plight of mortal man in his highest form as artist, and especially in Rilke, with his narcissism and his capacity for self-absorption. The seventh elegy, the first to be composed in the final burst, attempts a solution to the disharmony by transmuting the earth and its objects into art, something which outlasts the maker. Though Rilke did not expressly reject this impermanent extension of existence, grief and lamentation break out, even more despairingly than hitherto, in the eighth elegy. In the ninth Rilke achieves his final resolution of the mortal dilemma by suggesting that the earth and its objects, including man's artefacts, are raised again, insubstantially, in our souls: 'Erde, ist es nicht dies, was du willst: *unsichtbar*/in uns erstehn?' So Rilke's profound despondency is relieved by 'transforming external realities into invisible inward possessions in life as in death' (E. L. Stahl).

This highly personal form of perpetuation by 'inwardness' (to which Rilke in a letter written in French gave the words 'transmutation', 'resurrection', and 'transfiguration') enabled him to represent in the *Elegien* a long and arduous striving against the transience of the earth, ending in a personal victory. The last poem allegorically describes the journey into death in terms of mythical landscape. The verse, in which a wide range of imagery, including the well-known, non-theological angels who establish the mythical temper of the *Elegien* from the outset, is magnificently integrated, attains a consistent sublimity. Though they have been the subject of much speculative interpretation, the *Elegien* are rather to be understood as an architectonically arranged cycle of hymnic poetry, embodying and proclaiming an intensely felt personal process of experience, than as a metaphysical statement.

Dulciflorie, an incomplete erotic verse tale probably of the late 13th c. (though some scholars have dated it up to a century earlier). It recounts, with names taken from courtly poetry, the sexual surrender of a princess imprisoned in a tower, and has close affinities to *Der Sperber* (q.v.). The author is unknown.

DULLER, EDUARD (Vienna, 1809–53, Wiesbaden), a Radical, who left Vienna during the post-1815 period, and worked as a journalist in Bavaria and the Rhineland. His early tragedy

Meister Pilgram (1829) was performed in Vienna in 1827. He also wrote ballads, Novellen, and novels, but was best known as a popular Radical historian. It was Duller who, in 1835, first published a full (though inaccurate) version of G. Büchner's play *Dantons Tod* (q.v.).

Dummer August, the type of foolish clown known in English as an 'Auguste'.

DUNKELGRÄFIN, DIE, a figure of mystery, lived in seclusion at Schloß Elshausen near Hildburghausen (Thuringia) from 1810 until her death in 1837. It has been suggested that she was Princesse Thérèse-Charlotte, daughter of Louis XVI of France and Marie-Antoinette. She lived with a male companion, L. C. van der Valek, who became known as 'der Dunkelgraf'. Her story has been exploited in novels by L. Bechstein, A. E. Brachvogel, and K. Martens (qq.v.).

Dunkelmännerbriefe, German title for the *Epistolae obscurorum virorum* (q.v.).

Düppel, German name of the Danish village Dybbøl (Prussian from 1866 to 1920), the scene of fierce fighting in the Danish wars of 1848 and 1864 (see SCHLESWIG-HOLSTEINISCHE FRAGE). In 1848 Düppel witnessed a Danish victory; the storming of the Danish fortified position (Düppeler Schanzen) in 1864 was the outstanding feat of the war.

Durch, often in the form Der Verein Durch, name adopted for a group of *avant-garde* writers and thinkers in Berlin, who held regular club meetings in the late 1880s and energetically furthered the Naturalist movement (see NATURALISMUS). The originator was Dr. C. Küster and his closest associates were Leo Berg and Eugen Wolff (qq.v.). Durch was founded in 1886 and the name, devised by Küster, is the slogan 'Durch!', as much as to say, 'Cut through the web of convention!' Among the associates who attended the meetings were Arno Holz and Johannes Schlaf, J. H. Mackay, Wilhelm Bölsche, the brothers Julius and Heinrich Hart, and Gerhart Hauptmann, who joined in May 1887 (qq.v.). Durch provided a focal point for the predominantly realistic and left-wing movement in literature.

Durchwachte Nacht, a poem written towards the end of her life by Annette von Droste-Hülshoff (q.v.), published in *Letzte Gaben* (1860). In this poem of insomnia attention is concentrated on the noises and movements of the nocturnal world. Each of the first seven stanzas

notes in a refrain the striking of the clock from 10 to 4, the final one greets the sunrise.

DÜRER, ALBRECHT (Nürnberg, 1471–1528, Nürnberg), the son of a goldsmith, who settled in Nürnberg in 1455. Dürer studied painting in Nürnberg under Michael Wolgemut (q.v.) from 1486 to 1490, and then spent four years in various cities on the Rhine (Colmar, Basel, and Strasburg). He was married in Nürnberg in 1494, and visited Venice in 1495. From 1500 to 1506 he was again in Venice, visiting also other Italian cities. He was in the Netherlands in 1520–1. The greater part of his life was spent in his native city, painting, drawing, and engraving in what has since become known as the Dürerhaus.

Dürer painted some seventy works in oils, including the *Four Apostles* (Munich), the *Allerheiligenbild* (Vienna), and many portraits. He executed numerous woodcuts, including several biblical series (*Große* and *Kleine Passion, Marienleben*), and a considerable number of copper engravings and etchings, including portraits of his friend Pirckheimer, of Melanchthon, and of Erasmus (qq.v.), and the well-known plates *Ritter, Tod und Teufel* (q.v.) and *La Melancolia*. He also painted in water-colour and made a large number of drawings.

DÜRINC, DER, a minor Middle High German poet of the late 13th c. He was probably a strolling singer and, as his name suggests, a native of Thuringia. His Minnelieder imitate KONRAD von Würzburg (q.v.), whose elaborate style he exaggerates.

Dürnkrut, a village in the Marchfeld, east of Vienna, near which Ottokar II (q.v.) of Bohemia was defeated in 1278 by Rudolf I (q.v.) and slain. Dürnkrut is also the site of an earlier battle (1260), in which Ottokar defeated the Hungarians.

Dürnstein, a village on the left bank of the Danube in the Wachau, with a ruined castle in which King Richard I of England was held prisoner by Duke Leopold of Austria in 1193. The episode in which the minstrel Blondel is supposed to have discovered Richard is traditionally located at Dürnstein.

DÜRRENMATT, FRIEDRICH (Konolfingen, Canton Berne, 1921–), son of a pastor who settled in Berne in 1935. He pursued literary, philosophical, and scientific studies at the universities of Zurich and Berne. His study of the dramatic technique of B. Brecht (q.v.) was a decisive factor in his development as an outstanding dramatist of the post-war era. He is also an intriguing writer of fiction. From 1946 he settled in Basel, and in 1952 moved to Neuenburg (Neuchâtel).

Dürrenmatt's first play, *Es steht geschrieben*, performed at the Schauspielhaus, Zurich, in April 1947 and published before the year was out, is an unhistorical, tragicomic treatment of the Anabaptist rule of Münster from 1534 to 1536, and the subsequent retribution exacted by the Catholic bishop's soldiers on the fall of the city. The work is an original mixture of serious thought, violence, and grotesque comedy, in which Knipperdollinck, the man of simple-minded sincerity, and Bockelson, the selfish and ruthless agitator, both suffer death. As the troops are about to break into the city, Knipperdollinck and Bockelson execute a grotesque roof-top dance (an episode which provoked a disturbance at the first performance). As both men die, it is the passive victim Knippeldollinck who praises God and receives his grace. The disparate elements of the play foreshadow the character of much of Dürrenmatt's work. In 1967 a revised version of *Es steht geschrieben* appeared under the title *Die Wiedertäufer*. In 1948 *Der Blinde* was performed in Basel (published 1960). It is set a century later in the Thirty Years War (see DREISSIGJÄHRIGER KRIEG) and is an allegory of faith in the form of a parallel to Job. The blind Duke overcomes, through his credulous passivity, the machinations and outrages of his antagonist Negro da Ponte.

Dürrenmatt's highly successful play *Romulus der Große* was performed and published in 1949 (revised 1957). It presents the end of the Roman Empire in comic terms: intent on the ruin of the Empire, Romulus is more concerned with breeding poultry than with running the state. A passive martyr in the first version, he is not allowed to achieve martyrdom in the later, more comic, form which accentuates the political satire. Two minor narrative works (*Pilatus*, 1949, and *Der Nihilist*, 1950) preceded Dürrenmatt's first première outside Switzerland, the comedy *Die Ehe des Herrn Mississippi* (q.v., 1952, Munich). His next play, *Ein Engel kommt nach Babylon* (performed in Munich, 1953, published 1954), is a kind of prelude to the building of the Tower of Babel; it is his most poetic and harmonious play.

While Dürrenmatt was producing this succession of grotesque and paradoxical comedies— and abstaining from tragedy because in his view the world had reached a state in which the assumptions of tragedy appeared ridiculous—he was also busy writing a succession of prose works, 'potboilers', as he is reported to have called them. These narrative works comprise primarily three detective novels which have be-

come well known in English-speaking countries. *Der Richter und sein Henker* (q.v., 1952) begins the cycle, sharing its detective, Bärlach, with the second, *Der Verdacht* (q.v., 1953). Both deal with the fulfilment of justice, but with the special peculiarity, which is shared also by the succeeding story, *Das Versprechen* (q.v., 1958), that the detective suffers as greatly as the quarry he pursues. These works are marked by ingenious psychology and considerable power to evoke landscape.

Between the second and the third novel Dürrenmatt published a collection of short narratives, *Der Tunnel* (1952), and the 'Prosa-komödie' *Grieche sucht Griechin* (1955). The marriage advertisement 'Grieche sucht Griechin' leads to the incongruous coupling of the mild and simple Arnolph Archilochos and the sophisticated courtesan Chloë Saloniki, which for a time lifts Arnolph on to a social plane he has admired from below, only to dash him into revolutionary depths in which he seeks to assassinate the state president. Dürrenmatt gives the story alternative endings, 'Ende I' and 'das Ende für Leihbibliotheken'.

The story *Die Panne* (1956, originally a radio play, with a less radical ending, also version for television and stage performance) provides a ghostly court scene and another character, Alfredo Traps, who, like Knipperdollinck and later Alfred Ill in *Der Besuch der alten Dame*, passively accepts his guilt and, as a necessary consequence, his end. The title refers to a car accident with which the story opens, but also, with varied emphasis, to today's world, 'diese Welt der Pannen' (Erster Teil). Dürrenmatt published his early stories in *Die Stadt. Prosa I–IV* (1952), containing *Weihnacht, Der Folterknecht, Der Hund, Das Bild des Sisyphos, Der Theaterdirektor, Die Falle, Die Stadt, Der Tunnel*, and *Pilatus*.

In 1956 appeared his play *Der Besuch der alten Dame*, and in 1962 *Die Physiker* (qq.v.). Both works, which stand out clearly above the rest, deal with power, responsibility, and guilt; the first in the context of finance, the second in that of technology. The ordinary individual confronted with either of these forces of the modern world can, in Dürrenmatt's view, at best face up to it—'die Welt bestehen'. Between these two plays came *Frank der Fünfte. Oper einer Privatbank* (q.v., 1960), a work with some affinity to *Der Besuch der alten Dame*. *Herkules und der Stall des Augias*, adapted in 1962 from a radio play of 1954, twists the ancient myth to make a parable. Hercules does not shift the muck; Augias endures it and uses its fertilizing power to advantage. This play, which contains amusing comedy and brilliant satire, encountered fierce opposition. In *Der Meteor* (q.v., 1966) Dürrenmatt turned death into a subject for comedy. *Porträt eines Planeten* (1st version Düsseldorf, 1970; 2nd version Zurich, 1971), his bleakest play, presents archetypal characters (Adam, Eva, etc.). Placed on a bare stage, they conduct an arid, laconic dialogue, interrupted from time to time by addresses to the audience, explaining the end of our planet and its recommencement. The printed play is preceded by a paradoxical preface on the improbability of reality.

Among Dürrenmatt's radio plays are *Der Doppelgänger* (1946), *Der Prozeß um des Esels Schatten* (1958), *Nächtliches Gespräch mit einem verachteten Menschen* (1952), *Stranitzky und der Nationalheld* (1952), *Das Unternehmen der Wega* (1954), and *Die Abendstunde im Spätherbst* (1956).

The volume presenting Dürrenmatt's comedy *Der Mitmacher* (1976, first publication of the play 1973) is introduced as *Ein Komplex. Text der Komödie. Dramaturgie. Erfahrungen. Berichte. Erzählungen*. The latter items appear in the table of contents in two sections, *Nachwort* and *Nachwort zum Nachwort*, which together are some 200 pages longer than the play itself. They include comments on the play, dramaturgical suggestions, and an analysis of the philosophy underlying the notion of the 'conformer' and 'conforming' (Mitmachen). Everyone conforms, but there are two types of conformers: the positive conformer who acts with conviction (though his convictions can produce evil as well as good), and the negative conformer who acts out of self-interest and is a morally inferior and weak character. Set in an underground laboratory, accessible through a lift, the play is divided into two parts containing five monologues, spoken in turn by Doc, Ann, Bill, Boß, and Cop, Doc being the Conformer of the title. The one-syllable names of the figures indicate their purely functional capacity and the curt phrases of the dialogue the absence of genuine communication between them. The inventor of a necrodializator that dissolves corpses and hence makes a perfect murder possible, Doc works for the 'syndicate', a gang that supplies him with the corpses of its victims. A formerly prosperous biochemist, he is thus able to rebuild his existence, only to find that his conformism ends in even greater ruin. The volume as a whole adds further dimensions to Dürrenmatt's well-known attitudes and techniques.

Die Frist, a comedy, was published in 1977 and *Justiz*, a novel begun in 1957, in 1985.

Dürrenmatt is a skilful draughtsman, and his economical and firm line-drawings illustrate some of his works. He is also a keen critic and essayist, beginning with *Theaterprobleme* (1955), in which he promulgates his conception of comedy, the 'mutige Mensch', and the function of the grotesque. *Theater-Schriften und Reden I* appeared in 1966, *Theater-Schriften und Reden II* in 1972, the volume *Zusammenhänge. Essay über Israel. Eine Konzeption* in 1976. *Gesammelte Hörspiele* appeared in 1960, *Stoffe I–III* in 1981, and a comprehensive *Werkausgabe* (29 vols.) in 1980. Dürrenmatt has received prizes and honours in Switzerland, Austria, West Germany, the USA, Israel, and Great Britain.

'Du Schwert an meiner Linken', first line of the patriotic poem *Schwertlied* by Theodor Körner (q.v.), included in *Leyer und Schwerdt* (q.v.). Its popularity was partly due to the fact that the last stanza was composed early on 26 August 1813, the day on which Körner was mortally wounded. It was set to music by Weber (q.v.) in 1814.

DWINGER, EDWIN ERICH (Kiel, 1898–1981, Gund, Tegernsee), son of a German naval officer and his Russian wife, volunteered for service in 1914 and was wounded and taken prisoner on the eastern front in 1915 while serving as an officer-cadet (Fähnrich) with a dragoon regiment. After three years as a prisoner of war in harsh conditions (much of it was in Siberia) he joined the White Russian force of Admiral Kolchak. His books *Die Armee hinter Stacheldraht* (q.v., 1929) and *Zwischen Weiß und Rot* (q.v., 1930) deal with these phases of his career. He returned to Germany in poor health in 1921, and his account of the period after his return is contained in *Wir rufen Deutschland* (q.v., 1932). These books of recollection, with their combination of crass detail and nationalistic spirit, were widely read. They were followed by the stories *Zug durch Sibirien* (1934), *Und Gott schweigt?* (1936), and *Die eiserne Schar* (1941), and by the novels *Die letzten Reiter* (q.v., 1935), *Das namenlose Heer* (1935), and *Auf halbem Wege* (q.v., 1939). Dwinger, who became a horse breeder in Bavaria, was a war correspondent during the 1939–45 War with the rank of SS-Obersturmführer. After 1945 he was for a short time forbidden to publish. *Wenn die Dämme brechen* (1950) portrays in his characteristic journalistic manner the overrunning of East Prussia by the Russians in 1944–5.

E

Early New High German [ENHG], see GERMAN LANGUAGE, HISTORY OF.

East Prussia, see OSTPREUSSEN.

EBERHARD, PRIESTER, see GANDERSHEIMER REIMCHRONIK.

EBERHARD VON CERSNE, a canon of Minden who *c.* 1404 wrote a Minneallegorie (see MINNEREDE) entitled *Der Minne Regel.* It is a poem of nearly 5,000 lines, and consists of an introductory narration, a central portion containing 38 questions and answers on *minne* and its conduct, and a further narration, in which are set out another 31 questions on *minne*, framed this time in prose. The work is a free translation of the *Tractatus amoris* (*c.* 1185) of Andreas Capellanus. Eberhard also wrote twenty Minnelieder.

EBERHARD VON SAX, a priest who lived in Switzerland in the late 13th c. and early 14th c., is the author of a poem in praise of the Virgin Mary.

EBERHARD VON WAMPEN, a physician of repute in the island of Rügen, documented in 1330, is the author of the medical poem *Spiegel der Natur* (q.v.).

EBERLIN, JOHANN VON GÜNZBURG (near Ulm) (*c.* 1465–*c.* 1530, Wertheim), a Franciscan, was converted to Lutheranism in 1521. He advocated reform in eloquent and vigorous pamphlets (*Fünfzehn Bundesgenossen*, 1521–3). He was known also as an impressive and persuasive preacher.

EBERNAND VON ERFURT, author of a Middle High German poem which recounts, with legendary accretions, the lives of the Emperor Heinrich II and his consort Kunigunde (qq.v.). The Emperor was canonized in 1146 and Kunigunde in 1201; both are buried in Bamberg. Ebernand's poem (*c.* 1220) was topical, and, as he records, was written at the suggestion of a friend who was sacristan in Bamberg. The sources of the poem are two Latin lives of the saints.

EBERS, GEORG MORITZ (Berlin, 1837–98, Munich), born of well-to-do Jewish parents, studied Egyptology at Berlin university and was appointed to a professorial chair at Leipzig in 1870. He retired in 1884. He published a series of novels set in Egypt, Greece, and Rome, with authentic details derived from his researches. *Eine ägyptische Königstochter* (3 vols., 1864), which enjoyed a considerable success, came first and was followed by *Uarda* (3 vols., 1877), *Homo sum* (1878), *Die Schwestern* (1880), *Der Kaiser* (2 vols., 1882), *Serapis* (1885), *Die Nilbraut* (3 vols., 1886), *Per aspera* (2 vols., 1892), *Kleopatra* (1894), and *Arachne* (1897). Encouraged by the success of these novels on archaeological subjects (see PROFESSORENROMAN) Ebers ventured into the medieval field with the novels *Die Frau Burgemeisterin* (1882), *Ein Wort* (1882), *Die Gred* (2 vols., 1889), *Im Schmiedefeuer* (1894), and *Barbara Blomberg* (2 vols., 1896). He published an autobiography (*Die Geschichte meines Lebens*) in 1893. His collected works (*Gesammelte Werke*) appeared in 32 vols., 1893–7.

EBERT, FRIEDRICH (Heidelberg, 1871–1925, Berlin), was the first president of the Weimar Republic (q.v.). Originally a saddler, Ebert was a prominent Social Democrat (see SPD), becoming party secretary in 1905 and chairman in 1913. In November 1918 he succeeded Prince Max von Baden (q.v.) as chancellor; he was elected president of the new republic in February 1919. Supported by the moderate parties, he was the object of abuse by extreme right and left. He has been criticized for his readiness to rely on the army as a stabilizing force in the state.

EBERT, JOHANN ARNOLD (Hamburg, 1723–95, Brunswick), was educated at the Johanneum (q.v.), Hamburg, where he became friendly with J. B. Basedow and N. D. Giseke (qq.v.). In Leipzig, where he studied, he frequented literary circles which included among their members Gellert, Zachariä, and Cronegk (qq.v.). He was one of the contributors to the *Bremer Beiträge* (q.v.). Ebert became a schoolmaster at the Carolinum (q.v.), Brunswick, in 1748, and five years later received the title Professor. He was friendly with Klopstock (q.v.), who addressed the ode *An Ebert* (1748) to him. Ebert's principal poetic achievement is his translation of Edward Young's *Night Thoughts* (*Youngs Nachtgedanken*, 1751). His *Christliche Gedanken* were published in 1742, his *Episteln und vermischte Gedichte* (2 vols.) in 1789 and 1795.

EBERT, KARL EGON VON (Prague, 1801–82, Prague), spent most of his life as an administrator with the princely house of Fürstenberg writing poetry in his leisure hours. His poems (*Gedichte*) appeared in 1824, and were augmented in 1828 and 1845. He also published verse epics and romances, including *Vlasta* (1829), the idyll *Das Kloster* (1833), and *Eine Magyarenfrau* (1865). In his early life a passionate and reciprocated love for a princess ended in separation and renunciation.

EBNER, Christine (Nürnberg, 1277–1355, Engelthal), a Dominican nun and later prioress at Engelthal Abbey, who compiled, on the instructions of her superiors, a MS. recounting the lives and mystical visions of her fellow nuns. It is entitled *Büchlein von der Gnaden Überlast* and was written before 1346. Her own visions and life were recorded for her by another nun.

EBNER, Margareta (Donauwörth, 1291–1351, Medingen Abbey nr. Donauwörth), mystic and nun, who communicated her ecstatic visions as 'Offenbarungen' to Heinrich von Nördlingen (q.v.). Her spiritual and exalted correspondence with Heinrich over the years 1332–50 is preserved.

EBNER-ESCHENBACH, Marie, Freifrau von (Zdislavic, Moravia, 1830–1916, Vienna), *née* Countess Dubsky, was married to the officer, engineer, and professor Moritz von Ebner-Eschenbach in 1848. They lived at first in Moravia, then in Vienna. Her life was outwardly uneventful following the social round of her class, the winter in Vienna, the summer in the country. Early in her married life she conceived a passion for the theatre and the ambition to become a great dramatist. Her first play, *Maria Stuart in Schottland* (1860), was performed without success in Karlsruhe and severely criticized by O. Ludwig (q.v.). Other dramatic works, including *Marie Roland* (1860), followed and failed, though *Doktor Ritter* (1869, a play about the young Schiller) and *Die Veilchen* (1878, a one-act play) were performed in the Burgtheater and *Das Waldfräulein* (1873) was produced by H. Laube (q.v.) in the Vienna Stadttheater.

In the 1870s Baroness von Ebner-Eschenbach began to write fiction, and soon achieved the public recognition which was denied her in the theatre. A collection of stories (*Erzählungen*, 1875) was followed by the novel *Božena* (q.v., 1876) and by *Neue Erzählungen* (1881), which included *Die Freiherren von Gemperlein* (q.v.). Her first outstanding volume was *Dorf- und Schloßgeschichten* (q.v., 1883), which by its title suggests her characteristic broad social vision directed upon both the aristocracy to which she belonged

and the peasantry which she sympathetically observed. *Zwei Komtessen* (q.v.) followed in 1885, *Neue Dorf- und Schloßgeschichten* in 1886, and the four stories of *Miterlebtes* in 1889, whilst the novel *Das Gemeindekind* (q.v., 2 vols.), her best-known work, appeared in 1887. Equally powerful are the story *Lotti, die Uhrmacherin* (q.v., 1889), the aristocratic novel *Unsühnbar* (q.v., 1890) and the religious novel *Glaubenslos?* (q.v., 1893); among her later works the *Drei Novellen* (q.v., 1892) and the story *Rittmeister Brand* (q.v., 1896) particularly deserve mention. She continued to write into her eighties, producing a further novel (*Agave*, 1903) and several volumes of stories (*Alte Schule*, 1897; *Aus Spätherbsttagen*, 1901; *Altweibersonne*, 1909; *Genrebilder*, 1910; *Stille Welt*, 1915). In 1906 she published *Meine Kinderjahre* and in 1916 *Meine Erinnerungen an Grillparzer*. She was a novelist who combined acute realistic portrayal with reserved yet unflinching social criticism; at the same time she had compassion, which she extended to animals as well as to human beings. She had also at her disposal a dry humour and a caustic irony.

Marie von Ebner-Eschenbach's *Sämtliche Werke* (12 vols.) appeared in 1928 and *Gesammelte Werke* (9 vols.), ed. E. Gross, in 1961.

Ecbasis captivi, a poem written in Latin hexameters in the 10th c. by a monk of Toul, which was then within the borders of the Holy Roman Empire (see Deutsches Reich, Altes). It tells of a foolish calf which strays from its byre, falls into the clutches of the wolf, but is rescued by the herd. It has been supposed that the story refers to the flight of a young monk into the world and his recovery. It reveals the beast story as current coin in the monastic literature of the period.

Eccius dedolatus (1520), a drastic Latin satire directed against Johann Eck (q.v.), an opponent of Luther (q.v.). Eck undergoes a violent cure, being planed, shorn, and flayed. Its authorship, formerly attributed to Nikolaus Gerbel (q.v.), is uncertain. Some consider that it may be the work of Fabius Zonarius, a friend of Ulrich von Hutten (q.v.); more recently the view has been advanced that the author was a member of the circle of Willibald Pirkheimer (q.v.) in Nürnberg.

Echten Sedemunds, Die, a play in seven parts (Bilder) by Ernst Barlach (q.v.), published in 1920. The 'genuine' Sedemunds maintain their good name by despicable means. They are represented by Sedemund, his son Gerhard (der junge Sedemund), and Onkel Waldemar, Sedemund's brother. Gerhard is summoned home on the pretence that his father is critically ill, but really because the father wants to give his son, whose provocative and hostile views about society have brought the family name into disrepute, into the care of a mental home, in which Dr. Faßlich is already treating Grude, a married man, with apparent success. Gerhard discovers his father's true intentions from Grude, and together they pay Sedemund back in his own coin. Grude, on leave from the clinic to attend a funeral, tells Onkel Waldemar of Gerhard's sudden death. When Onkel Waldemar arrives with Sedemund at the hearse for the funeral, Gerhard confronts them and publicly exposes his father's responsibility for the death of his mother. Then, satisfied at having stirred his father's conscience, he decides to submit to Dr. Faßlich as a voluntary patient: Onkel Waldemar is to make his nephew's insanity responsible for the scandal, and thus restore the good name of the Sedemunds. In the constantly shifting action, in which a great number of characters are introduced, various social evils are exposed.

Barlach projects the action against the image of a lion. Grude secretly buys the skin of the circus lion Schesar, which has died that morning, from its owner Candido Franchi, and then spreads the news of its supposed escape. Panic ensues and rumour has it that the lion is hiding in the graveyard, where the dead worry and weep over the sins of the living. At the end Grude confesses to the deception by displaying the lion's skin: although the lion was dead, its soul lived in everyone. The image, which is associated also with the Crucifixion, is focused on man's conscience, his fear of God and truth. Grude prophesies a 'new age' (in the Expressionist manner) peopled with 'die echten Grudes'.

ECK, JOHANN (Egg, 1486–1543, Ingolstadt), whose real name was Maier or Mayer, was professor of theology at Ingolstadt and Luther's principal theological opponent. He was the adversary of Luther (q.v.) in the disputation at Leipzig in 1519, and the chief Catholic spokesman at the Diet at Augsburg (1530). He was satirized in the *Eccius dedolatus* (q.v.).

ECKART, DIETRICH (Neumarkt, Upper Palatinate, 1868–1923, Berchtesgaden), a reactionary anti-Semite, founded an anti-Communist society, and in 1919 made the acquaintance of Hitler (q.v.), whose views he influenced. From 1921 to 1923 he was editor of *Der Völkische Beobachter* (q.v.). He is said to have coined the successful slogan 'Deutschland erwache'. Before the 1914–18 War he translated Ibsen's *Peer Gynt* (1912) and also wrote plays.

ECKART, MEISTER, see ECKHART, MEISTER.

Eckenlied, a Middle High German verse tale of unknown authorship, of which various versions are wholly or partially extant. It belongs to the group of romances about Dietrich von Bern (see DIETRICHSAGE) and is Tyrolean in origin. It deals with a legendary giant. In its best-known Rhenish form (written in the second half of the 13th c.) the tale appears in courtly guise. Ecke is transformed from an uncouth giant into a powerfully built knight, who, seeking to prove himself, challenges Dietrich and is slain by him. Dietrich subsequently kills Ecke's brother Fasolt. The poem is written in a strophe of 12 lines, which has come to be known as the Eckenstrophe (q.v.) or Berner Ton. An early form of the tale occurs in the *Thidreksage* (see DIETRICHSAGE). An edition of the *Eckenlied*, by M. Wierschin, appeared in 1973 (*Altdeutsche Textbibliothek*, 78).

Eckenstrophe or Berner Ton, a stanza of 12 lines with the rhyme pattern aab aab cd cd ee. It is used in the Middle High German epic poems *Eckenlied, Goldemar, Sigenot,* and *Virginal* (qq.v.). It was possibly first used by ALBRECHT von Kemnaten (q.v.), the author of *Goldemar*.

ECKERMANN, JOHANN PETER (Winsen, 1792–1854, Weimar), was Goethe's companion and unpaid secretary in the last years of his life (1823–32). Eckermann fought as a volunteer in the War of Liberation (see NAPOLEONIC WARS), and afterwards studied at Göttingen. He made contact with Goethe in 1822 by sending him his *Beiträge zur Poesie mit besonderer Hinweisung auf Goethe* (published by Cotta in 1823). Eckermann actively assisted Goethe in the preparation of the definitive edition of his works, the *Ausgabe letzter Hand*. In 1837 he published the first two volumes of *Gespräche mit Goethe in den letzten Jahren seines Lebens*; the third volume followed in 1848. In spite of inaccuracies these conversations are a valuable source for Goethe's life and opinions in his last years. Eckermann was appointed ducal librarian in Weimar in 1836.

ECKHART or **ECKEHART VON HOCHHEIM,** MEISTER (Hochheim nr. Gotha, *c.* 1260–1327, Avignon), a great German mystic, was a Dominican monk who became prior at Erfurt in 1290, provincial prior for Saxony in 1303, and in 1307 vicar-general for Bohemia. Eckhart began by studying theology at Strasburg and Cologne, and continued at Paris (1300–2), where he received the licence to teach theology (Lesemeister), whence his normal style, 'Meister Eckhart'. His views, widely circulated in Latin and German writings and supported by sermons in German, were thought to be heretical and an inquiry was begun in 1326. Eckhard died in 1327 before the proceedings were complete. In 1329 twenty-eight of his statements of belief were condemned as heretical.

Eckhart's German works have for the most part not survived in authentic form. Only *Reden der Unterscheidung*, *Tischgespräche* (both early works, before 1298), and the *Buch der göttlichen Tröstung* are as he left them. His *Klosterpredigten* were mostly written down by hearers. The most important of the Latin works is the *Opus tripartitum*; it is, however, incomplete.

The basis of Eckhart's views is the *unio mystica* which takes place in the soul, a union between the innermost retreat of the soul and God, achieved in the act of recognition. In the secret recess of the soul is the spark ('Fünklein' or *scintilla*), which is ready to participate in the timeless existence of God. These views tend towards a unitarian and even pantheistic conception; Eckhart, however, insisted on maintaining a belief in and a place for the Trinity. And though the insistence on the quiet, secret places of the soul suggests a concentration on the *vita contemplativa*, he maintained the simultaneous importance of the *vita activa*. Eckhart's mysticism is logical rather than ecstatic, tracing its descent from Aquinas and neo-Platonism.

Die deutschen und lateinischen Werke, ed. J. Quint and J. Koch, began to appear in 1936. *Meister Eckharts Deutsche Predigten und Traktate*, ed. J. Quint, were published in 1955.

ECKSTEIN, ERNST (Gießen, 1845–1900, Dresden), a journalist with a classical education acquired at Giessen, Bonn, and Marburg universities, gained a reputation as a humorist with *Aus Sekunda und Prima* (1875). The volume contains the well-known 'Der Besuch im Karzer'. Eckstein is also the author of a series of novels of Roman life which had a considerable vogue at the end of the 19th c. (*Die Claudier*, 3 vols., 1881; *Prusias*, 3 vols., 1884; *Nero*, 3 vols., 1889).

Edda, two Old Icelandic works, which have a connection with Germanic mythology. The Prose Edda is believed to have been written 1220–30. It is a textbook of poetic expressions containing, in addition, a summary of Nordic mythology. The MSS. date from the 13th c. and the 14th c.

The Poetic Edda is a collection of Nordic songs written in the 13th c. The extant MS. of *c.* 1270 is not the original one. Among the songs are some from the Nibelungen legend (see NIBELUNGENSAGE); these introduce Atli (Etzel), Sigurd (Siegfried), and Gudrun (Kudrun). It is likely that they derive from Germanic sources. Some of the songs originate in the age of the Vikings.

Edelweiß, a novel by B. Auerbach (q.v.), published in 1861 and subsequently included in *Sämtliche Schwarzwälder Dorfgeschichten* (1884). It is the story of a village marriage between a clockmaker, Lenz, and an innkeeper's daughter, Annele. Lenz proves too gentle and Annele too rancorous to support adversity, which comes when Annele's father is declared bankrupt. At the climax of hatred and despair the lonely house is engulfed by an avalanche. In the face of imminent death Lenz and Annele slowly find their real selves and are reconciled.

The story is framed by a brief picture of the harmonious couple in later life when one of their grown-up children leaves home. The title alludes to a faded specimen of the alpine flower edelweiss (*Leontopodium alpinum*), preserved by Lenz's mother. In their night of terror under the snow Annele's hair has turned white and this is her 'Edelweiß'.

Edict of Restitution, see DREISSIGJÄHRIGER KRIEG.

Edolanz, a knight of the Round Table, is the subject of a Middle High German Arthurian romance in verse, which is lost except for fragments amounting to less than 400 lines. Edolanz goes to the assistance of Gawein and in some of his adventures has the help of a dwarf. The author, who wrote in the 13th c., is unknown.

EDSCHMID, KASIMIR, pseudonym of Eduard Schmid (Darmstadt, 1890–1966, Vulpera, Switzerland), by which he is invariably known. Edschmid began as an Expressionist and is seen as an eloquent theorist in *Über den Expressionismus in der Literatur und die neue Dichtung* (1919). His Expressionistic works include the Novellen *Sechs Mündungen* and *Das rasende Leben* (1916). Between the wars he travelled extensively, and wrote a number of novels for which the countries he visited provided the background. These novels include *Die achatnen Kugeln* (1920), *Die Engel mit dem Spleen* (1923), *Die geistigen Abenteuer des Hofrats Brüstlein* (1927, retitled *Pourtalès Abenteuer* in 1947), *Sport um Sagaly* (1928), *Lord Byron* (1929), *Glanz und Elend Südamerikas* (1931), *Das Südreich* (1933), *Der Liebesengel* (1937), *Das gute Recht* (1946), *Der Zauberfaden* (1949), *Der Marschall und die Gnade* (on Bolivar, 1954), *Drei Kronen für Rico* (1954), *Drei Häuser am Meer* (1958), *Whisky für Algerien?* (1964). He also wrote straightforward travel books, *Afrika—nackt und angezogen* (1930), *Zauber und Größe des Mittelmeers* (1932), and *Italien* (5 vols., 1935–48). A late essay on Expressionism appeared in 1961 (*Lebendiger Expressionismus*) and his diary (*Tagebuch 1958–60*) in 1960.

Eduard Allwills Papiere, a short work of fiction by F. H. Jacobi (q.v.). It is in the form of letters and was first published as a fragment in 1775 in the periodical *Iris* (q.v.) edited by the author's brother J. G. Jacobi (q.v.), and then, with additions, in Wieland's monthly *Der teutsche Merkur* (q.v.) in April, July, and December 1776. The author later revised and extended it, and this version was published in 1792 as *Allwills Briefsammlung.* Said to have been influenced by Goethe's *Leiden des jungen Werther's,* Jacobi's 'novel' presents, through their letters, two characters imbued with the introspective subjectivism of the age, Sylli Clerdon and Eduard Allwill. Sylli is beset with melancholy, Allwill, in Sturm und Drang fashion, questions values and asserts the free rights of the personality. There is no action and Jacobi found it necessary to explain in a preface who the characters are.

Eekenhof, a Novelle by Th. Storm (q.v.), written and published in the *Deutsche Rundschau* (q.v.) in 1879. Set in 17th-c. Schleswig-Holstein, the story traces the family history of the owners of an estate, which has fallen into decay. Herr Hennicke has acquired Eekenhof through marriage. His wife dies after giving birth to a son, Detlev. Through a second marriage, to Benedikte, he acquires another estate, on which he brings up the two sons of the second marriage. The only child he cherishes, however, is an illegitimate daughter, Hedwig. By his jealous hatred of Detlev, whom he deprives of his heritage, he destroys his hope of Hedwig's affection. The particular effect of this story derives from the mysterious power of love emanating from the portrait of Detlev's mother, which protects Detlev from the ominous intentions of Hennicke. Storm evokes atmosphere in this work to enhance the peculiarly legendary character of a dark age.

Effi Briest, a novel by Th. Fontane (q.v.), written mainly between 1891 and 1893. It first appeared in instalments in the periodical *Deutsche Rundschau* (q.v., October 1894–March 1895) and in book form in October 1895.

Effi is the 17-year-old daughter of a country gentleman and his wife, Herr und Frau von Briest. A marriage is arranged for Effi by her parents with a former suitor of the mother, Geert von Innstetten, once a hussar subaltern, now an ambitious, 38-year-old civil servant. Effi, flattered by the unexpected status of married woman, and quite liking Innstetten, acquiesces and, without any courtship, the betrothal takes place at once and the couple are married after two months. After a cultural honeymoon in Italy, which Effi finds rather boring, the couple arrive at Kessin (it has obvious resemblances to Swinemünde), where Innstetten is Landrat. Seven months later Annie, Effi's only child, is born. The house is dismal, and has the reputation of being haunted, something which Innstetten at best half-heartedly denies. Effi finds herself short of company, except for an eccentric and elderly 'platonic' admirer, the apothecary Gieshübler. Innstetten is wrapped up in his duties, the neighbouring nobility is stuffy, ill-educated, and narrow-minded, and the strange noises in the house get on her nerves. The situation improves when a new territorial district commander (Landwehrbezirkskommandeur) is appointed. He is Major von Crampas, a former cavalry officer in the same brigade as Innstetten, now aged 44, and with a reputation as a *coureur de femmes.* Crampas, who is married, sees that Effi is bored with Kessin and that Innstetten treats her as a child. She is captivated by his good looks and charm, and a brief liaison takes place while Innstetten is on an extended business visit to Berlin. He returns with the news of promotion to Ministerialrat which entails their move to Berlin.

Effi is immensely relieved at this escape from an *affaire* into which she has drifted from boredom and reaction against a neglectful husband. A new and quite different social life opens up, and six agreeable years pass. Since Effi has not again become pregnant she is sent to take the waters. While she is at Bad Ems there is a minor domestic crisis at home in Berlin. The 7-year-old daughter cuts her forehead, the maids force open Effi's drawer to find a bandage, and its contents are disarranged. Innstetten notices a bundle of letters. They prove to be love-letters from Crampas. He immediately sends a challenge and goes down to Kessin; in the ensuing duel Crampas is mortally wounded. Innstetten notifies Effi's parents, who undertake to support her but decline to receive her. She lives in Berlin with a faithful retainer, Roswitha, and shows signs of tuberculosis. She pines for her daughter, but when Innstetten reluctantly allows the 10-year-old Annie to see her mother, Effi finds that the girl has been indoctrinated against her. Her medical adviser persuades the parents to let her return home, and she spends a last winter and summer of subdued contentment, dying of tuberculosis in September.

Effi Briest is perhaps Fontane's best novel, developing effortlessly and dealing fairly and without comment with a human problem in which faults and virtues cannot be disentangled. The characterization and the dialogue are both on the highest level of his achievement. (Film version by R. W. Faßbinder, 1974).

EGENOLF VON STAUFENBERG, see PETER VON STAUFENBERG.

EGEN VON BAMBERG, MEISTER, a Middle High German poet active in the first half of the 14th c. (probably between 1320 and 1340), is the author of two Minnereden (see MINNEREDE) in verse, a *Klage der Minne* of c. 200 lines and *Das Herz* of 138 lines. Egen, who was admired and imitated by the author of the *Minneburg* (q.v.), has a highly ornamental style (geblümter Stil).

Egerer Fronleichnamsspiel, designation of a religious play of the 15th c., covering biblical history from the Creation to the Resurrection. Performed by schoolboys in the open-air at Corpus-Christi-tide, the full presentation occupied three days. It is known to have been performed annually, except for one or two years, from 1443 to 1517. The MS., formerly at Eger, is now at Nürnberg.

Eginhard, see EMMA UND EGINHARD.

EGMONT, also Egmond, prince of Gavre, whose full name was Count Lamoral Egmont (Hainaut, 1522–68, Brussels). He married Sabine of Bavaria, by whom he had eleven children. He served under Karl V (q.v.), who in 1546 made him a Knight of the Golden Fleece, and under his successor Philip II (q.v.), for whom he led the Spanish cavalry to victory at St. Quentin (1557) and Gravelines (1558). As governor of the provinces of Flanders and Artois he was implicated, together with William the Silent (q.v.) of Orange and Count Hoorn, in the opposition to Cardinal Granvella, the adviser to the regent Margaret of Parma.

In 1564, after the removal of the cardinal, Egmont, a loyal Catholic, failed to move Philip II to relax his regime in the Netherlands, which included persecution by the Spanish Inquisition. Still hoping for a reconciliation, he refused to join the subsequent open rebellion in the Provinces. On 9 September 1567 he was arrested after the arrival of Duke Alva in Brussels. On 4 June 1568 he was condemned to death for high treason by Alva's Council of Blood, and on the following day he was executed, together with Count Hoorn, in the market square of Brussels in front of the town hall. He is the subject of Goethe's tragedy *Egmont* (q.v.).

Egmont, a prose tragedy in five acts by J. W. Goethe (q.v.) written between 1775 and 1787, and mainly in the years 1775, 1778–9, 1782, and 1787. Goethe mentions in *Dichtung und Wahrheit* (q.v.) that he gave his father the impression in 1774 that the whole play was ready in his head.

None of his other plays, except *Faust*, was subject to so much revision and reconsideration. The final version, completed on 5 September 1787, reflects the decisive change from his early Sturm und Drang period to the classical conception which he formulated in Italy. The first performance in Weimar in 1791 was not a success. For this Goethe had entrusted Schiller with a stage adaptation designed for the actor A. W. Iffland (q.v.) at the Court Theatre. Schiller, who was particularly critical of the ending, drastically altered parts of the MS.; his stage version, which Goethe never brought himself to see, was used until productions, incorporating the music of Beethoven (q.v., overture, four entr'actes, and songs, 1810), revived the original text. The play was first published in 1788 in *Schriften*, Vol. 5 as 'Ein Trauerspiel'.

Goethe used as his sources *De bello Belgico decades duae* by the Jesuit Famiamus Strada (Mainz, 1651) and the *Eygentliche und vollkommene historische Beschreibung des Niederländischen Kriegs* by Emanuel van Meteran (Amsterdam, 1627). The historical event of the play took place against the background of the Counter-Reformation (see GEGENREFORMATION, DIE) in the Netherlands to which Count Egmont (q.v.) fell victim. Goethe has condensed the historical sequence of Egmont's arrest and execution into the last two acts of the play. Although Egmont is introduced amidst intensifying religious and political conflict, his personality is Goethe's own free creation. By his unique disposition, which Goethe, in his comments in *Dichtung und Wahrheit*, embraced in the Italian word *attrativa*, Egmont wins the affectionate acclamation of the people. He has endeared himself no less to Margarete von Parma, daughter of the late Emperor Karl V (q.v.), who acts as regent for the Netherlands until King Philip II (q.v.) of Spain causes her resignation and her replacement by Alba.

In his love for the burgher girl, Klärchen, Egmont seeks escape from reality. Klärchen herself has a strong and independent nature and remains adamant despite her mother's protestations that she should marry her patiently devoted suitor Brackenburg. Act Four opens with Alba's move to arrest Oranien (see WILLIAM THE SILENT) and Egmont. But Oranien, the political realist, foresees the event, and, to spare himself for active resistance at a more opportune time, flees, after vainly imploring Egmont to join him (Act Two). Egmont's total unawareness of the nature of his summons enables Alba to carry out the arrest. The news of Egmont's impending death reaches Klärchen, who, failing to rouse the intimidated people into revolt to free the prisoner, poisons herself. After the death sentence has been communicated to

Egmont in prison, Ferdinand, Alba's son, enters to make through his friendship and admiration a protest against his father's tyranny. The tragedy ends with Egmont's summons to his death, to which he goes with the conviction that his sacrifice will promote the just cause of his people.

The tragedy of the individual, resulting from the 'daemonic' (see DÄMONISCH), trusting, and unreflective impulsion in Egmont's character, represents the play's most cogent dramatic idea. It is linked, through Egmont's vision of Klärchen as 'Siegesgöttin', with the idea of political freedom based on freedom of speech and religious tolerance in a law-abiding society. The play thus closes with an attempted fusion of two distinct types, the political drama and the tragedy of the individual, which profits by Goethe's choice of an elevated rhythmic prose style.

Ehe des Herrn Mississippi, Die, a play by F. Dürrenmatt (q.v.), first performed in Munich in March 1952, and published in the same year. Its action takes place in a single luxurious room. Anastasia, a woman of impulse, sits at a table with four men; one of them is her husband, Mississippi, the man of rigid, inexorable justice, who has married her because he believes that marriage is the logical outcome of their mutual misdeeds. The other men are Diego, the opportunist, Saint-Claude, the mere legalist, and Graf Übelohe, the man of genuine feeling. Through the deliberately absurd dialogue the apparently Naturalistic stage setting becomes the background to an allegory, a cataclysm of destruction, which only Übelohe, the man of *caritas*, survives; the apparently solid room and its expensive contents disintegrate.

Ehen in Philippsburg, a novel by M. Walser (q.v.), published in 1957. It portrays a world of materialism, snobbery, and ruthless competition. Hans Beumann, who from humble beginnings has done well in his first job, is accepted into a representative group of the 'successful', not without experiencing a qualm of conscience and a pang of hatred for their empty world, in which adultery and corruption are the norm.

Ehre, Die, a play (Schauspiel) in four acts by H. Sudermann (q.v.). It was first performed in the Lessing-Theater, Berlin, in November 1889 and was published in 1890. It exploits a characteristic feature of German (here Berlin) city life, the addition to the rear of the apartments of the rich which face the street (Vorderhaus), of a tenement (Hinterhaus) inhabited by the poor (a somewhat similar juxtaposition occurs in Nestroy's *Zu ebener Erde und erster Stock*, q.v., 1835).

A member of the rich family (the Mühlingks) has seduced Alma Heinecke of the 'Hinterhaus'. Her well-educated brother Robert challenges the seducer, but his challenge is contemptuously declined because he is not considered of suitable social standing to participate in a duel. Robert, for his part, is in love with Lenore Mühlingk; his love is returned, but the Mühlingk family have quite other ideas for their daughter. In the end Robert and Lenore are united when Graf Trast, Robert's unconventional millionaire friend, makes the young man his heir. Once regarded as a masterpiece of Naturalism, it is more properly considered a satire, in which Graf Trast, the nobleman, makes devastating, quasi-Shavian comments on the aristocratic code of honour. The social theme is the relationship of honour to income. The play was an immense success when first performed.

Ehre Gottes in der Natur, Die, see DIE HIMMEL RÜHMEN DES EWIGEN EHRE.

Ehrenbreitstein, town and former fortress on the Rhine opposite Koblenz. The town was the home of the Laroche family (see LA ROCHE, SOPHIE AND MAXIMILIANE VON), whom Goethe visited there on his return from Wetzlar in September 1772.

Ehrenbrief, see PÜTERICH VON REICHERTSHAUSEN, JAKOB.

EHRENFRIED, GEORG, see GROSZ, GEORG.

EHRENSTEIN, ALBERT (Vienna, 1886–1950, New York), of Jewish descent, studied in the Philosophical Faculty of Vienna University, settled in Berlin in 1910, and in the next twenty years travelled extensively within Europe and on other continents. He began with stories (*Tubutsch*, 1911; *Der Selbstmord eines Katers*, 1912, retitled *Bericht aus einem Tollhaus* in 1919), but is known chiefly as an Expressionist poet, capable of lyrical ecstasy but more often angrily and contemptuously denouncing contemporary civilization. His volumes of poetry include *Die weiße Zeit* (1914), *Der Mensch schreit* (1916), *Die rote Zeit* (1917), *Die Nacht wird* (which also included Novellen, 1920), *Wien* (1921), *Briefe an Gott* (1922), *Herbst* (1923), and *Mein Lied* (1932). A collection of the earlier poems appeared as *Gedichte* (1920). He is the author of a novel, *Mörder aus Gerechtigkeit* (1931), and of translations from the Chinese.

EHRLER, HANS HEINRICH (Mergentheim, 1876–1951, Waldenbruch nr. Stuttgart), a Roman Catholic journalist, minor poet, and novelist. The background of his work is his native Swabia. He is the author of several volumes of verse (*Lieder an ein Mädchen*, 1912; *Frühlingslieder*, 1913; *Die Liebe leidet keinen Tod*, 1915; *Gedichte*, 1920; *Gesicht und Antlitz*, 1928; *Die Lichter schwinden im Licht*, 1932; *Unter dem Abendstern*, 1937); and his novels include *Briefe vom Land* (1911), *Die Reise ins Pfarrhaus* (1913), *Briefe aus meinem Kloster* (1922), *Wolfgang, das Jahr des Jünglings* (1925), *Die Reise in der Heimat* (1926), *Die Frist* (1930), *Die drei Begegnungen des Baumeisters Wilhelm* (1935), and *Charlotte* (1946).

Ehrlicher Makler ('honest broker'), a famous phrase coined by Bismarck (q.v.) to denote the role of Germany in the forthcoming Congress of Berlin (1878, see BERLINER KONGRESS). It was first used in a speech to the Reichstag on 19 February 1877.

EICH, GÜNTER (Lebus/Oder, 1907–72, Salzburg), while studying Chinese and law at the universities of Leipzig, Berlin, and at the Sorbonne, used the pseudonym Erich Günter when participating with M. Raschke and K. Mann (q.v.) in the *Anthologie jüngster Lyrik* (1927). In 1929 he published his first radio play, *Das Leben und Sterben des Sängers Caruso*, and in 1930 his first book of poems, *Gedichte*. He served in the war, and was taken prisoner by the Americans. After his release in 1946 he settled in Bavaria, marrying Ilse Aichinger (q.v.) in 1953. His last years were spent in Austria. He was awarded the Büchner Prize in 1959.

Except for the story *Katharina* (1934), Eich virtually limited himself to lyric poetry and radio plays. One of the most terse of poets, he expressed himself in verbal abstractions which are inaccessible to normal logical thought. His mature style sought to establish points of contact with reality which he saw as 'trigonometrische Punkte' in his poem *Der große Lübbe-See*.

The dry, matter-of-fact prisoner-of-war poem *Inventur* has often been cited as characteristic of the post-war generation's 'clean sweep' (see KAHLSCHLAG). An apostle of brevity, Eich wrote a number of poems consisting of Einzeiler (e.g. 'Katzenschatten, stille Feiung gegen das Glück'). His *Gedichte* were followed by *Abgelegene Gehöfte* (1948), *Untergrundbahn* (1949), *Botschaften des Regens* (1961), *Ausgewählte Gedichte* (1960, a selection from the earlier volumes), *Zu den Akten* (1964), and *Anlässe und Steingärten* (1966). His last two volumes, *Maulwürfe* (1968) and *Ein Tibeter in meinem Büro* (*Maulwürfe II*, 1970), have given rise to much controversy. The 'moles' characterize Eich as profoundly sceptical of literature, including his own, for 'Nachtigallen kann auf die Dauer nur ertragen, wer schwerhörig ist'.

Eich adapted the lyric poetry of his favourite poet Li Tai Pe and other oriental poetry, and was a pioneer of the modern German radio play. In 1966 he published a collection of radio plays, written between 1950 and 1964, under the title *Fünfzehn Hörspiele*. His principal successes in this form were *Das festliche Jahr* (1936), *Die Mädchen von Viterbo* (1952), *Zinngeschrei* (1955), *Die Brandung von Setúbal* (1957), *Allah hat hundert Namen* (1957). *Träume* (1963) and *Stimmen* (1958) were interim collections of *Hörspiele*. In the year of his death Eich edited a selection of his work in *Ein Lesebuch*.

Gesammelte Werke (4 vols.) appeared in 1973.

EICHENDORFF, JOSEPH, FREIHERR VON (Schloß Lubowitz nr. Ratibor, Silesia, 1788–1857, Neiße), son of a noble landed family, the fortunes of which were declining, spent his early years on the family estate, where he was educated by private tutors, and in leisure hours ranged freely through the forested Silesian countryside. The family tradition was Roman Catholic, and Eichendorff adhered unswervingly to his faith throughout his life. In 1801 he was sent with his elder brother Wilhelm, to whom he was deeply attached, to the grammar school (Gymnasium) at Breslau, where he remained until 1805. In the spring of that year he and his brother began to study law at Halle University, transferring in 1807 to Heidelberg. A reflection of Eichendorff's early life is to be found in an autobiographical fragment entitled *Der Adel und die Revolution*, and the impressions of his university years are mirrored in *Halle und Heidelberg*; both essays were published posthumously in 1866 in *Aus dem literarischen Nachlasse*.

In Heidelberg Eichendorff was deeply influenced by J. J. von Görres and by the Romantic poet O. H. von Loeben (qq.v.), as well as by the body of folk poetry in *Des Knaben Wunderhorn* (q.v.). The university years 1805–8 were followed by visits to Paris and Vienna, after which the two brothers returned home in 1809 to help manage the estate. Eichendorff had by this time written a number of poems, and had recently completed his first prose work, the story *Die Zauberei im Herbste* (q.v.). A visit to Berlin in the autumn of 1809 enabled him to meet L. J. (Achim) von Arnim, C. Brentano, H. von Kleist, and Adam Müller (qq.v.). In the following year he left Lubowitz, as it proved, for good, and took up studies in Vienna with the aim of qualifying as a civil servant, in which he succeeded in 1811. In Vienna he was in contact with Friedrich Schlegel and his wife, and with Philipp Veit (qq.v.), the painter, Frau Schlegel's son by her first marriage.

Eichendorff joined the volunteer forces on the outbreak of the War of Liberation (see NAPOLEONIC WARS) and was commissioned in October

1813. He was demobilized at the end of 1814 and immediately took up a minor civil service post in Berlin. He married Luise von Larisch in the early spring of 1815, but the return of Napoleon took him back to the forces and he did not return to civilian life until 1816. His first novel, *Ahnung und Gegenwart* (q.v.), appeared in 1815. From 1816 to 1819 he held a minor civil service appointment in Breslau, publishing the story *Das Marmorbild* (q.v.) in 1819. A posting to Danzig followed in a better-paid post as Regierungsrat (1821). He was transferred to Berlin in 1823 and then, with further promotion, to Königsberg in 1824.

His satire *Krieg den Philistern* appeared in 1824, and his best-known story, *Aus dem Leben eines Taugenichts* (q.v.), in 1826. His tragedy *Der letzte Held von Marienburg* (1830) was linked with the restoration of the castle by the Prussian government. He returned to Berlin in 1831, where he remained until his early retirement through ill-health in 1844. Minor narrative works appeared in these years, including *Viel Lärmen um Nichts* (q.v., 1832), *Das Schloß Dürande* (q.v., 1837), *Die Entführung* (1839), and *Die Glücksritter* (1841), as well as his comedy *Die Freier* (q.v., 1833); his most considerable work of this time was his second novel *Dichter und ihre Gesellen* (q.v., 1834). In the year of his retirement, his history of the efforts to restore Marienburg (*Die Wiederherstellung des Schlosses der deutschen Ordensritter in Marienburg*), written at the request of Th. von Schön (q.v.), had its official publication.

Eichendorff's remaining years were spent in Sedlnitz, Moravia (1845 and 1855–7), in Vienna (1846–7), in Köthen and Dresden (1848), in Berlin (1849–55), and finally (1855–7) in Neiße. His publications in these years were mainly concerned with the history of literature, which he approached from a religious standpoint in *Zur Geschichte der neueren romantischen Poesie in Deutschland* (1847), *Die geistliche Poesie in Deutschland* (1847), *Der deutsche Roman des achtzehnten Jahrhunderts in seinem Verhältnis zum Christentum* (1851), and *Geschichte der poetischen Literatur Deutschlands* (1857). *Libertas und ihre Freier* (Ein Märchen) is a satire on contemporary liberal and radical politics. Written in 1849, it was published posth. in 1866. Eichendorff's poems were liberally included in his narrative works, especially *Ahnung und Gegenwart*, the *Taugenichts*, and *Dichter und ihre Gesellen*. They were published in collected form (289 in all) in *Gedichte* (1837), and with additions in Vol. 1 of *Werke* (4 vols.) in 1841. The edition of *Sämtliche Werke. Historisch-kritische Ausgabe* by A. Sauer and W. Kosch was initiated in 1908, planned in 16 vols. (later 25); 6 vols. appeared in 1908–39. In 1950 publication was resumed, ed. H. Kunisch *et al.*

The two fundamental experiences underlying Eichendorff's work are the intimate association with landscape deriving from his early years and the religious faith which strengthened as his life progressed. His poetry has an apparent simplicity, which is belied by great subtlety of rhythm and mood. Much of it is poetry of joy, confidence, or resolution, but Eichendorff is constantly aware of dark forces, and the victory is not easily won.

The religious basis of his writings is even clearer in the narrative works, notably in *Ahnung und Gegenwart*, in which he is able to evoke the Romantic atmosphere of his vision of nature, and to assert at the same time religious criteria and stringent moral standards. The writings on the history of literature are original only in their religious emphasis, leaning heavily for their substance on the work of Gervinus (q.v.) and Vilmar (1800–68). The poetry, which is Eichendorff's greatest achievement, has outstanding singable qualities, and has been set in quantity by Robert Schumann and Hugo Wolf (qq.v.). Its magical power of evocation is best summed up in his poem *Wünschelrute*:

> Schläft ein Lied in allen Dingen,
> Die da träumen fort und fort,
> Und die Welt hebt an zu singen,
> Triffst du nur das Zauberwort.

Eiche und Angora, a play by M. Walser (q.v.), first performed in Berlin in 1962, later revised and performed in Zurich in 1963, and published in the same year. Sub-titled 'Eine deutsche Chronik', it is a succession of nine ironic, often farcical, scenes, beginning late in April 1945 and concluding seventeen years later. The old Nazis of 1945 are still there, but in new posts. The curious title refers to the oaks of the 'Eichkopf', a fictitious mountain where a last Nazi H.Q. is set up; 'Angora' alludes to the white fur pelts of the Angora rabbits which one of the characters breeds and sells to the townspeople for use as signs of surrender.

EICHRODT, Ludwig (Durlach, 1827–92, Lahr), went to school at Karlsruhe, and studied at Heidelberg and Freiburg universities before becoming an official in the Baden Ministry of Justice and later a judge (Oberamtsrichter) at Lahr. He wrote much humorous poetry, some of it in dialect. In 1855 he began to publish in *Fliegende Blätter* (Munich) some of the naïvely comic poems of S. F. Sauter (q.v.), and himself imitated and parodied Sauter's unintended humour. These poems (his own and Sauter's) were described as *Gedichte des schwäbischen Schulmeisters Gottlieb Biedermaier und seines Freundes Horatius Treuherz* (see BIEDERMEIER). They were collected and published in 1869 as *Biedermaiers Lebenslust*. His other collections of verse include *Gedichte in allerlei Humoren* (1853), *Lyrischer Kehraus* (2 vols., 1869), and *Reinschwäbische Gedichte in mittelbadischer Sprechweise* (1869). His *Gesammelte Dichtungen* were published in two volumes in 1890. His first book of poems appeared under the pseudonym Rudolf Rodt.

Eidechsenbund, a league founded in 1397 in opposition to the Knights of the Teutonic Order (see DEUTSCHER ORDEN), organized in 1411 an unsuccessful conspiracy against the Grand Master, Heinrich von Plauen (q.v.). It was later merged in the Prussian League.

Eiderdänen, adherents of a political party in Denmark in the 19th c., which sought to absorb Schleswig, separating it from Holstein. Its activities came to an end with the war of 1864, when Schleswig-Holstein was incorporated in Prussia. See SCHLESWIG-HOLSTEINISCHE FRAGE.

EIKE VON REPGAU (*fl.* Anhalt, 1209–33), a nobleman with legal experience. He compiled the *Sachsenspiegel* and was probably the author of the *Sächsische Weltchronik* (qq.v.).

EILHART VON OBERGE, 12th-c. poet from the neighbourhood of Brunswick, author of *Tristrant und Isalde* (q.v.). Nothing else is known with certainty of Eilhart, whose authorship is recorded in the poem.

Ein Bekenntnis, a Novelle (Rahmennovelle) by Th. Storm (q.v.) published in *Westermanns Monatshefte* in 1887. It deals with the problem of euthanasia in a contemporary setting. Franz Jebe, a noted gynaecologist, marries a young Swiss woman, Else Füßli, a great-niece of the painter Heinrich Füßli (q.v.). Their happy married life suffers a severe blow when Else is struck by an apparently incurable disease (cancer of the uterus). Unable to face the prolonged agony of physical pain, she asks her husband to give her a lethal injection. Some time after her death Franz reads in a medical journal of a possible cure for the disease by surgery. He succeeds in fact in achieving this with a woman patient, the Etatsrätin Roden. This success and the affection of Hilda, daughter of Frau Roden, offer him an opportunity to start a new life in marriage. But he rejects it on moral grounds. He also refuses to accept the law as an arbiter of his guilt. Instead he elects to expiate his guilt by spending the rest of his life working as a doctor for a missionary in remote East Africa. Franz relates the critical stages in

his life leading up to the murder of his wife (Binnenerzählung) to a friend of his youth, Hans, a lawyer. They meet one afternoon in June 1856 at Reichenhall, whence Franz departs for Africa the following morning. After thirty years Hans learns of his death during an epidemic. Although Storm leaves it to the reader to judge the doctor's actions, he appears to condemn the interference of an individual, in any circumstance, with another's life, viewing 'the flame of life' as sacred.

Ein Brief, see BRIEF DES LORD CHANDOS.

Ein Bruderzwist in Habsburg, a five-act tragedy in blank verse by F. Grillparzer (q.v.), planned in 1824, completed 1848–50, and published in 1873. It was performed in 1872 in Vienna in the Burgtheater and Stadttheater under the direction of Dingelstedt and Laube (qq.v.) respectively.

The play condenses historical events between 1581 and 1618, the years leading to the beginnings of the Thirty Years War (see DREISSIGJÄHRIGER KRIEG), advancing the 'Prager Fenstersturz' (see MAJESTÄTSBRIEF) from 1618 to 1612, the year of the death of the Emperor Rudolf II (q.v.). Rudolf's isolation and indecision in an age rent by political and religious strife retard but cannot halt the clash of arms. They encourage the personal ambition of his brother and heir, Matthias (q.v.), to succeed him. The motif of the hostile brothers merges with a powerful conspiracy against Rudolf, of which the bishop Melchior Klesel (see KHLESL) is the leader. Rudolf's brother Max and the Archduke Ferdinand II (q.v.), of Styria, Rudolf's nephew and an ardent Roman Catholic, agree that Matthias should assume the kingship of Hungary and the imperial rights. Only Archduke Leopold, another nephew of the Emperor, remains loyal. As Rudolf learns of the treacherous approach of Matthias's troops, the Bohemian Estates, exploiting the situation, secure his signature to the Letter of Majesty guaranteeing the right of free worship, which Rudolf foresees will have grievous consequences.

As a result of Leopold's unsuccessful attack on Prague in order to protect him, the Emperor Rudolf finds himself a prisoner in his residence. His personal tragedy assumes an ironic poignancy as he condemns his natural son Don Cäsar (already dying of a self-inflicted wound) for killing Lukrezia, the daughter of the commoner Prokop. Ferdinand and Max arrive too late at the Hradschin to reverse Rudolf's decision not to resist Matthias's bid for power. Justifying his abdication Rudolf insists on the indivisibility of Austria, Hungary, and Bohemia and on his ardent desire for peace, which is symbolized by the 'Friedensritterorden', with which he has invested Julius, the tolerant and loyal Duke of Brunswick (see HEINRICH JULIUS, HERZOG VON BRAUNSCHWEIG), a Protestant. In Vienna Ferdinand secures Klesel's surrender to the Papal Inquisition and the support of Wallenstein (q.v.) in the imminent war. Upon the news of Rudolf's death Matthias is hailed as emperor, but unlike Ferdinand he assumes his duties aware of guilt and disillusionment.

Grillparzer has been praised by historians for his portrayal of Rudolf II, who embodies features of the author's own personality. In particular, Grillparzer expresses in Rudolf his view of the representative function of the crown in a divinely appointed system of government during an age of crises common to both the 17th c. and the 19th c.

Ein Doppelgänger, a Novelle by Th. Storm (q.v.), published in *Deutsche Dichtung* in 1887. The narrator of the enclosing frame (see RAHMEN) stays with a forester and his wife Christine. He relates in the 'Binnenerzählung' the story of Christine's father, John Hansen, whose death during her childhood has left her with the puzzling image of a now violent, now gentle and loving, parent.

In his youth John Hansen, an ex-soldier condemned to idleness during a period of unemployment, once relieved his boredom by taking part in a robbery. For this he served six years in prison. On his release he soon becomes known as John Glückstadt after the town in which he was imprisoned. John vainly struggles against the public image of an ex-convict, his 'Doppelgänger'. A few years of happiness, which he finds in his marriage to Hanna, end in disaster when he unwittingly causes her death during a quarrel. Left to fend for his child, Christine, he meets his death when extreme poverty drives him to steal from a potato field. He trips at night into an unused well, around which the narration unfolds, and his body is never found.

In this tragic tale Storm expresses sharp social criticism, which is tempered in the closing part of the frame. Christine has found happiness in her marriage. Her husband, the son of her foster-parents, has treated her with consideration and kept her ignorant of her father's past. But she understands John Hansen and cherishes his memory.

Eine Handvoll Erde, a realistic novel by Clara Viebig (q.v.), published in 1915. Its central theme is the desire of the urbanized countrywoman to have some contact with the open air and the soil. The Reschke family lives in the

asphalt wilderness of Berlin. Mine, the wife, comes from the country, and the couple decide to rent a small weekend plot outside Berlin. The story takes a criminal turn when a double murder is committed near by, and the Reschke son is for a time suspected. In the end he is cleared, and a benevolent retired physician buys for Mine the freehold of her plot. Mine had previously appeared as a character in Clara Viebig's novel *Das tägliche Brot*.

EINEM, GOTTFRIED VON (Berne, 1918–), composer of the operas *Dantons Tod* (1947, based on *Dantons Tod*, q.v., by G. Büchner), and *Der Prozeß* (1953, based on Kafka's novel *Der Prozeß*, q.v.). *Der Besuch der alten Dame*, based on Dürrenmatt's play of this title (q.v.), was first performed in May 1971 at Vienna, and was given at Glyndebourne in 1973 and 1974.

Einen Jux will er sich machen, a four-act farce by J. N. Nestroy (q.v.), written in 1841–2 and first performed at the Theater an der Wien in March 1842. Published in 1844, it is a riotously funny piece in which Weinberl, a respectable clerk, is promoted to a partnership in his firm and feels that, before settling down for life, he must just for once kick over the traces. His escapade is slow to start, then gathers impetus and Weinberl is swirled along from one embarrassing and hilarious contretemps to another. At the end of all the absurd confusion and misunderstanding Weinberl proposes to and is accepted by an attractive widow, Frau von Fischer, and as there are two other minor strands of romance in the play, between Zangler and Madame Knorr, Sonders and Marie, it ends in the announcement of a triple wedding. Based on a one-act play by John Oxenford, *A Day Well Spent* (1835), Nestroy's play has itself been adapted both by Thornton Wilder (*The Merchant of Yonkers*, 1938, subsequently rewritten as *The Matchmaker*, 1955) and by Tom Stoppard (*On the Razzle*, 1981).

Eine Quarantäne im Irrenhause. Novelle aus den Papieren eines Mondsteiners, title of a narrative work by F. G. Kühne (q.v.), published in 1835. 'Mondstein' is a transparent disguise for Sonnenstein, a mental hospital in Saxony. The narrator, who is often identified with Kühne himself, is interned in the asylum because of radical agitation; he undergoes a searing experience through contact with the inmates, and emerges as a more balanced and tolerant individual. The story is a kind of Bildungsnovelle. The 'Mondsteiner' suffers a defeat, but is a better man for it. The work is more important for its criticism of the age than for its somewhat bizarre narrative.

Eine Stimme hebt an, a novel by G. Gaiser (q.v.), published in 1950. The soldier Oberstelehn returns from the war, not to his wife with whom he has lost any real contact, but to his native market town. He finds a situation of hunger, want, and suffering on the part of all except the farmers and the black marketeers, and becomes aware of a general decline in morals and a withering of compassion. He tries, not unsuccessfully, to help some of the best and neediest, and finally leaves to seek his wife, driven less by affection than by an impulse towards an ordered existence amidst disorder and corruption.

Ein Fastnachtspiel vom Pater Brey, dem falschen Propheten, a short satirical play written in Knittelverse (q.v.) by J. W. Goethe in Frankfurt in 1773, published in 1774. Pater Brey abandons his sanctity in order to make love to Leonora, whose betrothed Balandrino is absent. It is a drastic satire directed at Franz Leuchsenring (q.v.), who, during the absence of Herder (q.v.), had pressed his attentions on Caroline Flachsland (q.v.), Herder's fiancée.

Ein Fest auf Haderslevhuus, a Novelle by Th. Storm (q.v.), first published under the title *Noch ein Lembeck* in *Westermanns Monatshefte* in 1885, and in revised form in 1889. The narrator purports to relate the life of a medieval knight omitted from the chronicles of the 14th c. It is the story of Rolf, the eldest son of a knight of repute, Claus Lembeck of Dorning in Schleswig. Educated in Paris and Prague, and a lover of poetry, especially of Gottfried's *Tristan* (q.v.), Rolf marries Frau Wulfhild of Schauenburg, the widow of the Ritter Hans Pogwitsch, having received from his father the castle of Dorning, the ancient family seat of Claus Lembeck's late wife.

Rolf is unaware that his beautiful and possessive wife has murdered her first husband for his unfaithfulness in marriage. The new marriage also has a tragic end, for Rolf falls in love with Dagmar, the daughter of a neighbouring knight, Hans Ravenstrupp, who has lost his wife and all his other children in an epidemic in 1349. Brought up in solitude, Dagmar too grows into the world of poetry, in which Hartmann von Aue's *Der arme Heinrich* (q.v.) evokes her deepest response. Her elf-like nature realizes Rolf's dream of *minne* during his secret visits to her father's grounds, which he can reach by climbing a poplar tree. This summer of love comes to an end when Wulfhild causes the tree to be felled and Rolf is unable to see Dagmar. Dagmar dies of grief, but with her last breath reveals her lover's identity to her father. To revenge himself on the adulterer, he invites Rolf to his castle under the pretence that he is going to celebrate his daughter's wedding. Rolf arrives to find the 'bride' about to be buried: with her body in his arms he scales the castle wall, and plunges to his death at the spot where the poplar tree had stood.

'Ein' feste Burg ist unser Gott', the most famous hymn by Luther (q.v.), based on Psalm 46 ('God is our refuge and strength'), written in 1528, first printed 1529. It is familiar in English as 'A safe stronghold'. Luther almost certainly composed the tune to which it is sung.

'Ein Fichtenbaum steht einsam', first line of an untitled poem by H. Heine (q.v.) included in the *Buch der Lieder* (q.v.) as No. XXXIII of *Lyrisches Intermezzo.*

EINHARD (Maingau, *c.* 770–840, Seligenstadt), counsellor and later biographer of Charlemagne (see KARL I, DER GROSSE). Einhard was educated at Fulda, went to Charlemagne's court in 794 where he was trained by Alcuin (q.v.). He supervised building projects in Aachen (q.v.), and carried out a diplomatic mission to Rome in 806. He was afterwards in the service of Ludwig the Pious (see LUDWIG I, DER FROMME). He gave up office in 830, and founded an abbey at Michelstadt. Though a layman and married, he was a pluralist abbot, holding five abbeys. He is buried at Seligenstadt on the Main, an abbey which he founded in 828. His Latin biography of Charlemagne (*Vita Caroli Magni*) is modelled on Suetonius. He has often been falsely identified with the legend of *Emma und Eginhard* (q.v.).

EINHART, pseudonym under which H. Class (q.v.) published his *Deutsche Geschichte* (1908).

Einhorn, Das, a novel by M. Walser (q.v.), published in 1966. Its narrator and principal character is Anselm Christlein of the novel *Halbzeit* (q.v.). Anselm, married to Birga, signs a contract with a Swiss woman publisher, Melanie Sugg, to write a first-hand novel on love. Anselm's saga, which involves, among others, Frau Sugg and a Dutch-Indonesian 'dream girl' Orli, is documented by comic linguistic acrobatics, which mark successive coitions. The novel, in which the unicorn of the title assumes symbolic significance, is a *tour de force* of *grotesquerie*, a satire on contemporary life, and an inquiry into identity. In the end Anselm and Birga find themselves reunited. The novel is noteworthy for its immense verbal virtuosity and inventiveness.

Ein Hungerkünstler, a short story written by Franz Kafka in 1921–2, and published posthumously in 1924. Kafka corrected the proofs while on his death-bed. It is the story of a performer, whose feat is to starve for periods up to forty days, earning admiration and astonishment. But he admits in the end, as he dies, that this involved no skill; he had refrained from eating by compulsion, since he could not find food which he liked. It may be seen as a parable passing a negative judgement on art, but other interpretations have been suggested.

Ein Jäger aus Kurpfalz, German folk-song of *c.* 1750, said to refer to a hereditary forester of the Palatinate, F. W. Utsch, and attributed to M. Klein, a Carmelite monk. The melody is used allusively in the *Akademische Festouvertüre* by Brahms (q.v.).

Einjähriger, abbreviation for Einjährig-Freiwilliger, status in which it was possible for young men of education in Germany (mainly university students) to perform their military service. The Einjähriger, who could choose his arm of the service, and had to equip, clothe, and maintain himself at his own expense, was eligible for promotion to reserve officer. This system of one-year service was introduced in 1867 and lapsed in 1919. A similar system existed in the Austro-Hungarian Empire.

Ein Kampf um Rom, a four-volume historical novel by Felix Dahn (q.v.), published in 1876. Its subject is the decline of the Ostrogoth empire in Italy from its peak under Theodoric (q.v.) to its extinction under Teja (A.D. 526–52). The book was one of the outstanding best sellers of the late 19th c.

'Ein Männlein steht im Walde', first line of *Rätsel,* a poem for children, written in 1827 by Hoffmann von Fallersleben (q.v.). It is used, together with its folk-tune, by E. Humperdinck in the opera *Hänsel und Gretel* (q.v.). The 'Männlein' of the riddle is a toadstool.

Ein Mord, den jeder begeht, a novel by H. von Doderer (q.v.), published in 1938. A retrospective, semi-biographical, semi-detective novel, it begins with the childhood of Conrad Castiletz, who later becomes successful in the textile industry and marries a well-connected young woman, Marianne Veik. A few years before, Marianne's sister Louison had met her death in a train in circumstances suggesting murder for robbery. An impulse in Castiletz to clear up this episode repels and estranges his wife. In the end he discovers that he was himself involved.

Louison, travelling in the next compartment

to some young people, among whom was the young Castiletz, met her death by striking her head against a tunnel wall through fright at a foolish prank played by the young men next door. They left the train at the next station with no idea that a dead woman lay in the compartment next to theirs. The psychology is involved and penetrating, but connoisseurs of detective fiction will discover implausibilities. Unlike most of Doderer's work, the novel has no connection with Vienna, being mainly set in Stuttgart, Leipzig, Berlin, and along the railway from Stuttgart to Heilbronn.

Einsame Mann, Der, a novel by C. Viebig (q.v.), published in 1924. It deals with the devoted affection of a retired and ageing officer, Colonel Baron Rettberg, for a young boy, whom he virtually adopts. Hans Helmut, as he grows up, has an affair with a servant girl, Maria Kaspers, by whom he has a child. He drowns himself and Colonel Rettberg marries the girl to safeguard his protégé's son, accepting the fact that his sacrifice makes him *déclassé.*

Einsame Menschen, a play (Drama) in five acts by G. Hauptmann (q.v.), first performed by the Freie Bühne (q.v.) in Berlin in January 1891, and published in the same year. The scene is Friedrichshagen near Berlin. Johannes Vockerat, a young philosopher of apparent distinction, has an attractive wife, Käthe, who, though not unintelligent, cannot participate in Johannes's processes of thought. Their infant child is baptized with Johannes's reluctant consent while Frau Vockerat senior, is staying with them. Anna Mahr, unmarried, three years older than Käthe and also good-looking, calls. She is a student, a rare phenomenon in the 1890s, and is able to converse with Johannes as an intellectual equal. It soon becomes clear to all but Johannes that he is emotionally attracted to her. Anna Mahr makes an effort to end the relationship, but, after taking the decision to leave, is dissuaded by Johannes, who is consumed by self-pity.

Frau Vockerat in alarm summons her husband. Both are respectable, pious, and well-meaning. They call for an end of the friendship, but encounter hysterical outbursts from Johannes. Anna Mahr takes the final decision to leave, realizing that a continuation will break the marriage. After her departure Johannes rows out on the lake, the Müggelsee, and drowns himself.

Hauptmann endeavours to maintain a balance between two different points of view. Johannes and Anna are, of course, the 'lonely people' of the title.

Einsame Weg, Der, a play by A. Schnitzler (q.v.), published in 1904. It is a retrospective conversation piece in which two generations confront each other. It can scarcely be said to have a plot. The focus is the family of Professor Wegrat, whose wife dies early in the play. His remote and psychic daughter drowns herself after a brief love affair with Herr von Sala, who has a serious heart condition. Wegrat's sensitive yet balanced son is revealed as the son of another, the artist Julian Fichtner, but the ties of human relations, custom, and environment prove stronger than the blood-link, and Felix Wegrat turns from Fichtner to the 'father' in whose affection he has grown up.

The play is a dialogue under the shadow of death and its speech is heavy with an autumnal beauty foreshadowing an end. As Sala puts it, 'Sonderbar, in dieser klaren lauen Luft weht doch schon eine Ahnung von Winter und Schnee.'

Ein Sendbrief von Dolmetschen, see LUTHER, M.

EINSIEDEL, FRIEDRICH HILDEBRAND, FREI-HERR VON (nr. Altenburg, Thuringia, 1750–1828, Weimar), spent much of his life at the court of Duke Karl August (q.v.) at Weimar, first as a page and later as a lord-in-waiting (Kammerherr). After studying at Jena University, he became a member of the intimate circle round the Duke, to which Goethe and K. L. von Knebel (q.v.) belonged. He had theatrical interests, and translated and adapted works for court performance. He published his *Neueste vermischte Schriften* (2 vols., 1783–4) and later wrote a treatise on acting, *Grundlinien zu einer Theorie der Schauspielkunst* (1797).

Einsiedler, Der, a poem by J. von Eichendorff (q.v.). The first line, 'Komm, Trost der Welt, du stille Nacht', bears a close resemblance to 'Komm, Trost der Nacht, o Nachtigall', which is the opening line of the hermit's song in J. J. C. Grimmelshausen's *Der abenteuerliche Simplicissimus* (q.v., Bk. 1, Ch. 6). Eichendorff also wrote a short poem entitled *Einsiedler* with the first line 'Einsiedler will ich sein und einsam stehen'.

EINSTEIN, CARL (Neuwied, 1885–1940, by suicide while fleeing through southern France), is the author of a novel, *Bebuquin oder Die Dilettanten des Wunders* (1912, reprint 1963). He was a contributor to *Die Aktion* (q.v.) and *Die weißen Blätter*, and influenced the Expressionist style of writing. He was a friend of G. Benn (q.v.). An art historian, he was an expert on African art. He emigrated to Paris in 1928, and

later took part on the republican side in the Spanish Civil War. His *Werke* appeared from 1980 (vol. 1, 1908–18, ed. R.-P. Baacke, in 1980; vol. 2, 1919–28, ed. M. Schmid, in 1981; vol. 3, 1929–40, ed. M. Schmid and L. Meffre, in 1985).

Ein Stündlein wohl vor Tag, a poem written by E. Mörike (q.v.) in 1837. As the girl sleeps, the swallow before her open window sings of her lover's unfaithfulness. The words of the title are also the refrain of each of the three stanzas. The poem has been set to music by Hugo Wolf (q.v.).

Eins und Alles, a poem written by Goethe in 1821 and published in *Das Morgenblatt*, and in the same year in Goethe's own scientific periodical *Zur Naturwissenschaft überhaupt*. *Eins und Alles* sings of a universe constantly in flux and development, in which the individual aspires to fusion with the whole. See also VERMÄCHTNIS.

Ein treuer Diener seines Herrn, a five-act tragedy in verse by F. Grillparzer (q.v.), completed in 1826; it was originally commissioned to celebrate the coronation of the Empress Karoline Auguste, the fourth wife of the Emperor Franz II (q.v.) as queen of Hungary in 1825. The popular Hungarian national figure of Bancbanus, who served his country under the reign of Andreas II (1205–22), is central to the freely adapted plot, derived from J. A. Feßler's history of Hungary.

The King appoints Queen Gertrude and Bancban as co-regents, making them responsible for the maintenance of peace and order while he heads his army in defence of his claims on Galicia. Bancban pursues his duties with single-minded loyalty to the King even after the Queen's brother, Herzog Otto von Meran, causes the death of Bancban's young wife Erny, who stabs herself, seeing no other way of escaping from Otto's amorous advances. Bancban cannot quell a revolt, led by his brother Graf Simon, who, with Erny's brother, Graf Peter, pursues Otto, the Queen, and Bancban after their escape from the palace. Peter kills the Queen, who has deceived him about her identity in order to protect her brother, against whom the assault was aimed. Bancban continues his flight with Otto and succeeds in saving Bela, the King's son and heir. Upon the King's return all surrender to his judgement, which by its moderation humbly acknowledges the fallibility of man. Otto, the moral misfit, is sent into exile, and Bancban is honoured for his selfless loyalty; yet he relinquishes his offices to spend the evening of his life in the privacy of bereavement.

The title of the play, a secularized adaptation

of Matthew 25:21, stresses the moral message in its historical and human context. This was lost on the Emperor Franz, who interpreted the successful first performance in Vienna on 28 February 1828 as a lesson in revolt, but, though the Emperor (who attended three of the first four performances) sought to intervene, the play was given 15 times in just over a year. While the absolute principle guiding Bancban's conduct has proved to be the play's most controversial feature, the bold psychological motivation, steeped in subtle irony—directed by Grillparzer partly at himself—earned a special *Nachwort* by Heinrich Laube (q.v.) on the occasion of the publication of the first edition of Grillparzer's collected works.

'Ein Veilchen auf der Wiese stand', first line of an originally untitled poem written by J. W. Goethe in 1773 or 1774 and first published as part of *Erwin und Elmire* (q.v.) in *Iris* (q.v.) in 1775. In 1800 it appeared in Goethe's *Neue Schriften* with the title *Das Veilchen*. Set to music by Mozart (q.v.), it is the first Lied (q.v.).

Eipeldauer, or **Eipeldauer Briefe,** shortened title of a work of satirical journalism by J. Richter (q.v.). In full, *Briefe eines Eipeldauers an seinen Herrn Vetter in Kakran, über d'Wienstadt* (1785–1813), it was continued after 1813 by F. K. Gewey (1813–19) and A. Bäuerle (q.v.) until 1821.

EISENBART, JOHANN ANDREAS, also Eysenbart (Viechtach, 1661–1727, Hannoversch-Münden), has become the proverbial quack. His grotesque and crass methods of publicity were responsible for this reputation. Dr. Eisenbart was in reality one of the most skilful physicians of his day and was honoured by the king of Prussia for his talents and services. He is the subject of a novel (*Doktor Eisenbart*, 1929) by J. Winckler (q.v.).

EISENREICH, HERBERT (Linz/Danube, 1925–), a friend of H. von Doderer (q.v.), and author of the Erzählung *Einladung deutlich zu leben* (1952), the novel *Auch in ihrer Sünde* (1953), *Böse schöne Welt* (Novellen, 1957), *Der Urgroßvater* (1964), and *Sozusagen Liebesgeschichten* (Novellen, 1965). His radio play *Wovon wir leben und woran wir sterben* was broadcast in 1955. As an essayist he is sharply critical of the trends of the age (*Carnutum. Geist und Fleisch*, 1960, *Reaktionen*, 1964).

Eiserner Vorhang, the fire-proof curtain legally necessary in German as well as British theatres. Since 1946 the term has been applied to the frontier between West and East Germany, sealed by the DDR.

Eisernes Kreuz, Prussian military decoration, introduced by King Friedrich Wilhelm III (q.v.) at the outbreak of the War of Liberation (March 1813). It continued to be awarded until 1817. It was not revived for the wars of 1864 and 1866 but was again awarded in the Franco-Prussian War 1870–1 and in the 1914–18 and 1939–45 Wars. There were three types of award: Grand Cross, 1st Class and 2nd Class.

Eislauf, Der, an ode in 15 unrhymed strophes written by F. G. Klopstock (q.v.) in 1764, praising the pleasures and warning of the dangers of skating. Klopstock himself was an adept on the ice.

EISLER, HANNS (Leipzig, 1898–1962, Berlin), the composer son of the philosopher Rudolf Eisler (1873–1926), who compiled standard dictionaries of philosophical terminology. Hanns Eisler, who was a pupil of A. Schönberg (q.v.), went into exile in 1933, and taught from 1948 at the East Berlin Staatskonservatorium. As an exponent of the chanson he became a close collaborator of B. Brecht (q.v.) and other socialist writers (see ROTE SIGNALE) and producers. He composed the national anthem of the DDR.

EISNER, KURT (Berlin, 1867–1919, Munich), a political journalist of Jewish descent and socialist views, settled in Munich before the 1914–18 War, and went over in 1917 to opposition, and in 1918 was imprisoned for fomenting strikes. Released in October, he took a prominent part in the Munich revolution in November 1918, and became Minister President of the Bavarian socialist government. In February 1919 he was shot dead by a young Bavarian nobleman, Count Anton Arco auf Valley, an act which provoked a Communist revolution (see BAYERN).

EKHOF, KONRAD, also ECKHOF (Hamburg, 1720–78, Gotha), was one of the outstanding actors and impresarios of the mid-18th c. He acted with various troupes, including those of J. F. Schönemann (1704–82) and K. E. Ackermann, (1712–71), and was one of the principal actors of the short-lived Hamburg Nationaltheater (1767–9), of which Lessing was house critic. Afterwards he became a member of the Seyler troupe (see SEYLERSCHE TRUPPE), and was finally director of the theatre at Gotha. A master in the natural delivery of his lines, Ekhof included among his roles Mellefont, Tellheim, and Odoardo Galotti. He worked to raise the social condition of the actor by attempting to institute a provident fund to provide pensions (1778). In 1757 he founded in Schwerin a

'Theaterakademie zur Hebung der Schauspielkunst und des Schauspielerstandes'. Ekhof's own conduct helped to increase public esteem of the actor's profession.

EKKEHARD or **EKKEHART**, the name of four monks of St. Gall Abbey, Switzerland, who were active as scholars and writers in the 10th c. and 11th c.

Ekkehard I (d. 973, St. Gall), to whom the authorship of some Latin hymns is attributed, was long believed to have written the Latin epic *Waltharius* (q.v.). At the time of his death he was dean of the abbey. Ekkehard II (d. 990, Mainz), a nephew of Ekkehard I, was head of the abbey school, Latin tutor to Duchess Hedwig of Swabia (see HEDWIG, HERZOGIN VON SCHWABEN), and, at the time of his death, provost of Mainz Cathedral Chapter. He was for a time at the court of the Emperor Otto I (q.v.). He was the author of a *sequentia* (see SEQUENZ) and possibly also of the Latin poem *Hadewiga nobilissima femina*. Little is known of Ekkehard III, who was a cousin of Ekkehard II. The most important of the four is Ekkehard IV (*c*. 998–*c*. 1060), a pupil of Notker Labeo (q.v.), to whom he was related, who compiled the Latin chronicle of St. Gall (*Casus Sancti Galli*). Ekkehard IV was for a time in charge of the cathedral school at Mainz, returning to St. Gall in 1031.

The eponymous hero of J. V. von Scheffel's novel *Ekkehard* (q.v.) is a compound figure based upon the first two historical Ekkehards.

Ekkehard, a historical novel set in the 10th c., written by J. V. von Scheffel (q.v.) and published in 1857. Its hero, Ekkehard, is compounded of two historical figures, Ekkehard I and II of St. Gall Abbey (qq.v.); and its other principal historical character is the Duchess Hadwig of Swabia (see HEDWIG, HERZOGIN VON SCHWABEN).

The young monk Ekkehard is chosen to act as tutor in Latin to the widowed, but still young Duchess, and takes up residence in her castle of Hohentwiel. Imperceptibly a thread of love is spun between them, unperceived by the monk, whose emotional obtuseness gradually alienates Hadwig. An incursion by the Huns, in which Ekkehard distinguishes himself in battle, forms a central episode. Ekkehard at last recognizes his attraction to Hadwig, but she refuses to accept him. His attempt to embrace her in the Hohentwiel chapel is witnessed by monks, and he is cast into prison. He escapes and makes his way into the remote mountains of Appenzell, where he becomes a hermit, conquers his passion, and writes the Latin heroic epic *Waltharius* (q.v.). Eventually he returns to civilization and becomes the Emperor's chancellor. Chapter 24

consists in a German translation of *Waltharius*.

Perhaps the most widely read German novel of the 19th c., *Ekkehard* is a variant of the Bildungsroman (q.v.), for the monk, learning by experience and error, achieves a balanced maturity. A learned novel, with a formidable appendix of 279 notes containing references to sources, it is written in a consciously poetic style, deliberately archaic in tone.

ELBLIN VON ESELSBERG, a Middle High German poet, probably of the 14th c., of whom nothing is known except his authorship of two Minnereden (see MINNEREDE). One of these is untitled, the other is called *Das nackend pilde*. Both are didactic poems on the quality of true love.

Elbschwanorden or **Elbschwanenorden**, name of a society of poets founded in 1660 by J. Rist (q.v.) of Hamburg. It did not survive Rist's death in 1667.

Elegie, term used in German literature in two distinct senses: firstly for a poem written in 'elegiac verse', i.e. distichs composed of a hexameter followed by a pentameter; secondly for a poem with a subjective emotional content conveying melancholy, grief, or nostalgia. The double meaning arises in classical times; the first elegies are such in a purely formal sense, whereas Latin writers such as Catullus, Tibullus, Propertius, and Ovid compose elegies in which melancholy feelings are expressed.

Formal elegies appear in Germany in the Latin poetry of Lotichius Secundus (q.v.). Martin Opitz (q.v.), in his *Buch von der deutschen Poeterei* (1624), opts for the view that elegies are sad in tone: 'In den Elegien hat man erstlich nur trawrige sachen, nachmals auch buhlergeschäffte, klagen der verliebten, wündschung des todes, brieffe verlangen nach den abwesenden, erzelung seines eigenen Lebens unnd dergleichen geschrieben.' Mournful elegies written in strophic and, less often, in classical form abound in the middle of the 18th c. in the work of Klopstock, E. von Kleist, Hölty, Schiller, and, somewhat later, in that of Matthisson and Salis-Seewis (qq.v.).

The classicism of Goethe and Schiller revived the ancient elegy, conspicuously exemplified in Goethe's *Römische Elegien* and *Alexis und Dora* (qq.v.), and Schiller's *Nänie* and *Der Spaziergang* (qq.v.). The summit of the German classical elegy is achieved by Hölderlin in *Brod und Wein* (q.v.). Schiller in his essay *Über naive und sentimentalische Dichtung* defined the elegy. It belongs, according to him, to the sentimental category, and the elegiac poet, faced with the contrast of Nature and art, ideal and reality, dwells upon regret for what is past and unattainable, rather than on the defects of present reality. The classical elegy is continued in the 19th c. by Rückert and Platen, and above all by Mörike (qq.v.), who, however, was equally at home in the strophic modern form of elegy (e.g. *Besuch in Urach*, q.v.). The most remarkable modern elegies are the ten *Duineser Elegien* (q.v.) of Rilke.

Elegie, poem by Schiller, see SPAZIERGANG, DER.

Elegie auf dem Schlachtfelde bei Kunersdorf, see TIEDGE, C. A.

Elegisches Versmaß, metrical form borrowed in 18th- and 19th-c. German literature from the classics. It is composed exclusively of distichs, each of which consists of a classical hexameter, followed by a somewhat arbitrarily conceived dactylic pentameter. The original elegiac metre was based on quantity, which in the German version is replaced by stress. See ELEGIE.

Elektra, a tragedy in one act by H. von Hofmannsthal (q.v.), published in 1904. Hofmannsthal described it as 'frei nach Sophokles'. Elektra, one of Oedipus's two daughters, lives only for revenge on her mother the murderess, and Aegisth the murderer, which she believes will be carried out by Orest. Two strangers arrive and announce the death of Orest. Elektra now feels bound to do the deed herself, but her more cautious sister Chrysothemis holds back. The younger stranger reveals himself to be Orest and murders first Clytemnestra and then Aegisth.

The Greek play depends upon the dilemma of Orestes, who is bound to avenge his father, yet in doing so must commit matricide, one of the worst of crimes. Hofmannsthal's play conforms to Greek tragedy in the undivided long single act, but dispenses with a chorus. Written in blank verse, it invokes all the resources of modern psychology in portraying the transformation in Elektra, whose hatred has become obsessional to the point of madness.

In 1906 R. Strauss (q.v.) decided to use it as the libretto for an opera. In this form, which uses the original text, except for a few short cuts, *Elektra* was first performed on 25 January 1909. Strauss's choice of Hofmannsthal's work marks the beginning of the long collaboration between the two men.

Elementargeist, Der, a story by E. T. A. Hoffmann (q.v.), written and published in 1821 and included in *Letzte Erzählungen* (1825). The story is framed in a setting in eastern Belgium

a few months after Waterloo. Two Prussian officers meet, and one, Viktor, recounts his early history and especially the magic practices to which he was persuaded by the Irish emigrant Major O'Malley; he conjures up a beautiful woman, to whom Viktor almost—but not quite —loses his soul. Back in the present, Viktor is convinced that his hostess, the fat Belgian baroness, is his former spirit love. He tears himself away, but remains unmarried.

ELEONORE VON ÖSTERREICH, ERZHERZOGIN (Scotland, 1433–80, Innsbruck), a daughter of James I of Scotland, was married in 1449 to Siegmund, Archduke of Outer Austria (Vorderösterreich) and Tyrol. She translated a French prose romance into German under the title *Pontus und Sidonie* (q.v., 1456).

Elfenlied, a humorous poem written by E. Mörike (q.v.) probably in 1828. The elf hears the night-watchman call the hour ('Elfe!') and mistakes this for a summons. The poem, the first line of which runs 'Bei Nacht im Dorf der Wächter rief', was included in *Der letzte König von Orplid*, which in turn is incorporated in *Maler Nolten* (q.v.). It has been set to music by Hugo Wolf (q.v.).

Elfriede, a play (Schauspiel) in three acts by L. Anzengruber (q.v.), published in 1873. It was performed in the Carl-Theater, Vienna, on 24 April 1873, and in the Burgtheater on 29 April 1873. Elfriede comes to realize that her husband has treated her as a child in their marriage, but she represses her impulse to leave, and a reconciliation takes place on grounds of duty. The problem, though not its solution, is an interesting anticipation of *A Doll's House* (1879) by Ibsen (q.v.).

Elf Scharfrichter, Die, a cabaret founded and directed in Munich in 1901 by H. H. Ewers (q.v.) in collaboration with H. Gumppenberg. F. Wedekind (q.v.) was one of the performers.

El Greco malt den Großinquisitor, a Novelle by S. Andres (q.v.), published in 1936 and based on a portrait of Cardinal Niño de Guevara by El Greco. The picture, in a private collection in Winterthur, is interpreted in the story chiefly through dialogues between the Inquisitor, his doctor, Cazalla, and the painter.

ELHEN VON WOLFHAGEN, TILEMANN (Wolfhagen, Hesse, 1348–1420), a notary and town clerk of Limburg, compiled from 1378 onward the *Limburger Chronik* (q.v.).

ELISABETH, HEILIGE (Preßburg, 1207–31, Marburg), daughter of King Andrew II of Hungary, was betrothed at the age of 4 to Ludwig, son of Landgraf Hermann (q.v.) of Thuringia. Brought up at the lively Thuringian court, she held aloof from festivity, and devoted herself to religion and charitable works. She was married to Ludwig in 1221, but was widowed six years later by his death on a crusade. Persecuted by his successor, HEINRICH Raspe (q.v.), she was for a time almost destitute, but was then given asylum by her uncle, the Bishop of Bamberg, and finally by KONRAD von Marburg (q.v.), who, for the salvation of her soul, treated her with great severity. At Marburg she devoted herself to good works, especially the care of the sick. Four years after her death she was canonized by Pope Gregory IX.

The Gothic church of St. Elisabeth in Marburg was erected over her grave between 1235 and 1283. Her biography was written in Latin by the Dominican monk Dietrich von Apolda in Thuringia (*Vitae sanctae Elizabetae*, 1289). She is the principal character in Charles Kingsley's verse play *The Saint's Tragedy*, and, in much romanticized form, also appears in R. Wagner's *Tannhäuser* (q.v.). See also ELISABETHLEGENDE.

ELISABETH, KAISERIN VON ÖSTERREICH, KÖNIGIN VON UNGARN (Munich, 1837–98, Geneva), said in her girlhood to be the most beautiful princess in Europe, was married in 1854 to the Emperor Franz Joseph (q.v.). She was a granddaughter of King Maximilian I of Bavaria (q.v.). Impulsive and given to outdoor sports, especially riding, she found it difficult to accommodate herself to the rigid etiquette of the Viennese court. In the 1860s and 1870s she was a frequent visitor to England and Ireland, riding regularly to hounds. Her marriage was unhappy, and her life was marred by the suicide of her only son in 1889 (see RUDOLF, KRONPRINZ).

She made little attempt to exercise political influence, except in furtherance of the Ausgleich (q.v.) with Hungary in 1867. She was assassinated by an anarchist, Luigi Luccheni, at Geneva on 10 September 1898.

ELISABETH, QUEEN OF ROMANIA, see CARMEN SYLVA.

Elisabethlegende, an anonymous Middle High German poem, written c. 1300, which recounts the life of St. Elisabeth (see ELISABETH, HEILIGE). Her sufferings are interpreted in parallel with the life of Christ. The source of the poem is the Latin life of the saint by the Dominican monk Dietrich von Apolda in Thuringia. It has been thought by some that the unknown poet was

also the author of *Die Erlösung* (q.v.), but this opinion has little support.

ELISABETH CHRISTINE, KÖNIGIN VON PREUSSEN (Wolfenbüttel, 1715–97, Berlin), a daughter of Herzog Ferdinand Albrecht II von Braunschweig-Wolfenbüttel, was married to Crown Prince Friedrich (see FRIEDRICH II OF PRUSSIA) in 1733. She was a niece of Maria Theresia (q.v.) and of Friedrich Wilhelm I (q.v.) of Prussia, who aimed through this match at a closer link with Austria. Friedrich's invasion of Silesia in 1740 ruined this prospect (see SCHLESISCHE KRIEGE). Friedrich pitied her for having to marry him, and lived apart from her. She had no children.

Elisabeth von England, a play by F. Bruckner (q.v.), published in 1930 and described as 'Schauspiel in fünf Akten'. It is concerned on the personal level with the relationship between the Queen and the Earl of Essex, and on the political plane with the antagonism between England and Spain, symbolized by the Queen and Philip II (q.v.). The frequency with which the play was, for a time, performed is probably largely due to the use of the fashionable technique of dividing the stage for parallel representation of two separate environments.

ELISABETH VON NASSAU-SAARBRÜK-KEN (1397–1456, Saarbrücken), daughter of a duke of Lorraine, married the Count of Nassau-Saarbrücken, who died in 1429. She translated four French *chansons de geste* into German prose, completing the last in 1437. These narratives are the first German prose romances since the *Lanzelet* of c. 1225. *Loher und Maller,* written in French in 1405 by Elisabeth's mother, Margarethe von Lothringen, is the story of an illegitimate son of Charles the Great. *Huge Scheppel* (Hugo Capet) tells of the butcher's son who becomes king of France. *Herpin* has as its hero Herpin's son Löw. *Sibille* is a story of an unjustly slandered queen. The first three later achieved popularity as chapbooks (see VOLKSBUCH). *Huge Scheppel* (under the title *Hug Schapler*) was printed in 1500, *Loher und Maller* in 1513, and *Herpin* in 1514.

ELISABETH VON SCHÖNAU (c. 1129–64, Schönau nr. St. Goarshausen), a Benedictine nun, became prioress of her convent, and experienced visions, which she recorded in the *Liber visionum*. Her *Liber viarum Dei* is an admonition to follow the Lord's ways, and the *Liber Revelationum* gives a visionary account of the legend of St. Ursula. All are in Latin.

Elisa oder das Weib, wie es seyn sollte, a sentimental and moralizing novel by Wilhelmine Karoline von Wobeser (q.v.), published in 1795. It is the story of a girl compelled by her family to marry an unsympathetic and ill-behaved husband; in spite of maltreatment, she dutifully cherishes him and in the end redeems his character. In the decade following its publication it was perhaps the most widely read novel in Germany.

Elixiere des Teufels, Die, a novel by E. T. A. Hoffmann (q.v.), written in 1814–15 and published in 1815–16 (2 vols.). It bears the sub-title *Nachgelassene Papiere des Bruders Medardus, eines Kapuziners.* It is the life history, told in the first person, of the monk Medardus, who is descended from a sinful family and beset by evil powers.

Medardus, while still a young man, becomes a preacher of great power and repute, but he has a breakdown and loses his eloquence. He succumbs to the temptation to drink an elixir, which is a relic of St. Anthony's temptation by the Devil, and thereupon finds himself, almost unwittingly, committing crime after crime. He falls in love with a penitent, is responsible for the apparent death of Graf Viktorin (his half-brother, though Medardus does not know this), who falls over a precipice, and he consummates his desire with a married woman, Baroness Euphemie, whom he subsequently poisons. He murders her stepson Hermogen and assaults her stepdaughter Aurelie, who proves to be the penitent he already loves. After the murder of Hermogen a mysterious double appears.

Medardus now leads a worldly life at the Residency of the ruling prince, at whose court Aurelie appears and denounces him. He is arrested, but almost immediately released because his double, calling himself the monk Medardus, has confessed to the murder of Hermogen. The real Medardus wins Aurelie's love, but on the day of the wedding he confesses his guilt, stabs her (though, as we learn later, she recovers), and flees. He takes refuge in a monastery, where he finds papers which reveal his family history and in particular his descent from one who has tasted the Devil's elixir, as he himself has. Medardus returns to his own monastery, and submits to penance. He learns that Aurelie is about to take vows as a nun, and his passion flares up again, but she is stabbed to death during the ceremony in the convent chapel by Medardus's mysterious double, who is revealed to be his half-brother Viktorin, whom he had believed dead. Medardus writes his story and dies a penitent a year later, as a brief epilogue indicates. The novel is intimately connected with Hoffmann's passion for Julia Marc.

Ella Rose oder die Rechte des Herzens, a five-act play (Schauspiel) by K. Gutzkow (q.v.), first performed in 1856 at Dresden. It is set in England. Rose, Ella's husband, lives with a mistress; Ella goes on the stage, becomes a great actress, and attracts the love of the author Tailfourd. Rose sets her free, but she will have none of this and plans a dramatic suicide in the last scene of her farewell performance. But the actor-manager guesses her intention and saves her life by cancelling the performance. Ella is ostensibly reconciled with her husband. Gutzkow himself called this improbable play, in which the insignificant wife becomes a famous actress, a double tragedy, 'Beide [Ella and Tailfourd] sterben den Zivilisationstod unserer Institutionen'.

ELLENDE KNABE, DER, designation adopted by a poet of the 15th c., signifying his ill fortune in love. He is the author of *Der Minne Gericht* (1449) and *Die Liebe und der Pfenning,* two allegorical poems.

Ellernklipp, a story by Th. Fontane (q.v.), published in 1881 and sub-titled *Nach einem Harzer Tagebuch.* The orphan Hilde is brought up by a widower, the Heidereiter (or head gamekeeper) Baltzer Bocholt, with his son Martin. The two young people fall in love, but Bocholt, jealous of his son, pushes him over the edge of the cliff (Ellernklipp). Baltzer and Hilde are married, but some time later Baltzer perishes mysteriously at the scene of the still unsolved murder. Hilde devotes herself to works of charity and dies young.

Elpenor, a fragment of a play (Schauspiel) by Goethe, who worked at it from 1781 to 1783. The first act and part of the second, written in rhythmic prose, are extant. The fragment was published in a slightly revised form in verse in 1806. The course of the action is not clearly recognizable. Antiope has lost her husband, and apparently her son as well, through assassination. It has been thought that Elpenor, who is regarded as the son of her brother-in-law, would prove to be Antiope's son. Goethe's original intention was that *Elpenor* should be performed to celebrate the birth of an heir to Duke Karl August (q.v.), an event which took place in January 1783.

Elsaß-Lothringen, the German name between 1871 and 1918 for the French territory of Alsace (except Belfort) and a part of Lorraine including Metz annexed by Germany in 1871. The region was designated Reichsland and was ruled by a governor (Statthalter) nominated by the Emperor. At the time of the annexation German was the predominant language, but the sympathies of the inhabitants were in the main French. Attempts to conciliate the Alsatians and Lorrainers were largely frustrated by the behaviour of German functionaries, tourists, and the officers drawn from other parts of the German Empire (see ZABERN). In 1918 the French occupation of the territories was greeted as a liberation. By the Treaty of Versailles Alsace-Lorraine ceased to exist, the lands reverting to their previous French 'departmental' organization (see VERSAILLES, TREATY OF, 2).

Elsi, die seltsame Magd, a story by J. Gotthelf (q.v.), published in 1843. Set in the year 1796, it has as the background to its climax the French invasion of the Swiss cantons. Elsi is the orphan daughter of a spendthrift miller of good family. She seeks employment at a farm in a district where she and her family are not known. By her willingness and proficiency she wins the favour of the farmer's wife. Christen, the farmer's son, falls in love with her, and his mother approves his choice; but Elsi, though secretly returning his love, repulses him because she cannot bring herself to confess the disgrace of her father's bankruptcy.

War breaks out, and Christen leaves to join his battery, still without any response from Elsi. After his departure she breaks down and confesses the truth to his mother, who berates her for a humility which has turned into a kind of pride. The Landsturm (q.v.) of all able-bodied men is called out against the French, and Elsi goes with it, armed with a pitchfork. When others falter and retreat, she goes resolutely forward and reaches Christen's battery at the moment when it is taken in the rear by a squadron of hussars. Both Christen and Elsi fight desperately, and they sink down to die in the recognition of love fully reciprocated.

Among Gotthelf's shorter narratives only *Die schwarze Spinne* (q.v.) surpasses this study of the psychological barriers which can separate those who desire to be together.

Emblem, term used to designate a range of concrete symbols (such as the dove and olive-branch signifying peace) which occur throughout the ages, but most conspicuously in the 16th c. and 17th c. Emblem books contained a series of mottoes accompanied by a pictorial representation and a moral comment. Notable are *Emblematum liber* (1531, in Latin) by A. Alcaitus and the collections by Gabriel Rollenhagen (q.v., 1610–13). A monumental and profusely illustrated work on Emblematik by A. Schöne and A. Henkel (*Emblemata. Handbuch der Sinnbildkunst der 16. und 17. Jahrhunderte*) appeared in 1967.

Emilia Galotti, a tragedy written by G. E. Lessing (q.v.) between 1754 and 1772 (mainly in the latter year), first performed at Brunswick on 13 March 1772, and published in Lessing's *Trauerspiele* (1772). It is described as 'Ein Trauerspiel in fünf Aufzügen'. Its origin is the classical story of Virginia, and Lessing referred to it at an early stage (1758) as 'eine bürgerliche Virginia'.

Emilia Galotti, a beautiful girl of prosperous middle-class origins, is betrothed to Count Appiani. She is seen by the ruling prince, who, seized by desire for her, empowers his favourite, Marinelli, to stop the match by any available means. Marinelli, who has a grudge against Appiani, organizes a hold-up of the couple's coach; Appiani is killed and Emilia abducted under the pretence of rescue. Emilia learns from the prince's former mistress, Countess Orsina, the fate that awaits her and, rather than suffer it, persuades her father to kill her.

The play is Lessing's outstanding 'domestic tragedy' (see BÜRGERLICHES TRAUERSPIEL), possessing tension, vivid dialogue, and convincing characters, and had an immediate and lasting success. Though Lessing disliked the undisciplined writing of the Sturm und Drang, *Emilia Galotti*, by its social criticism, strongly influenced them. In Goethe's *Die Leiden des jungen Werthers* (q.v., 1774) a copy of *Emilia Galotti* lies open on Werther's desk when he takes his own life. This detail, which Goethe took from life (from Kestner's account of the death of K. W. Jerusalem, q.v.), is indicative of the significance of the play for Lessing's younger contemporaries. In Auerbach's novel *Auf der Höhe* (q.v., Bk. 3) a performance of *Emilia Galotti* plays a part in the action.

Emil und die Detektive, a children's book by Erich Kästner (q.v.), published in 1929. Emil, going on holiday, has his money stolen, while he dozes in the train. Instead of relying on adult help, he gets together a gang of children who follow up clues and succeed in catching the thief. The book was so successful that it has been widely translated.

Emma und Eginhard, a legend first noted in the 12th c. It concerns an apocryphal love-affair between Emma, daughter of Charlemagne (see KARL I, DER GROSSE), and Eginhard, who was formerly identified with the Emperor's biographer Einhard (q.v.). The tribulations of the lovers have a happy ending. See FOUQUÉ, F. H. K. DE LA MOTTE, and HOFFMANN VON HOFFMANNSWALDAU. F. H. Flayder (1596–1640) wrote a Latin play on this story (*Imma portatrix,* 1625).

EMMERAM, HEILIGER, is believed to have been an itinerant monk of the 8th c. who was specially active at Regensburg, where a noted monastery was dedicated to him.

EMMERICH, KATHARINA (d. 1824, Dülmen, Westphalia), a nun from a secularized convent at Dülmen, who was reported in 1813 to exhibit the stigmata every Friday. In 1818 C. Brentano (q.v.) went to Dülmen, and spent nearly five years of his life taking down the reports of her visions. He published these in 1833; two others left in manuscript at his death were published in 1852 and 1856.

Empfindsamkeit, used to denote the strain of sentiment in German literature of the middle and second half of the 18th c. Signs of it are perceptible even in earlier years. It runs parallel with the rationalism which is the most obvious feature of the Aufklärung (q.v.) and seems at first sight to be in opposition to it. Yet the sentimental trend can be discerned in the work of Lessing, the foremost exponent of the Aufklärung, in his play *Miß Sara Sampson* (q.v., 1755). The word Empfindsamkeit was first used with this connotation, at Lessing's suggestion, by J. J. Bode (q.v.) for the title of the latter's translation of Sterne's *Sentimental Journey* (*Yoricks empfindsame Reise*, 1768).

The seeds of this inwardly directed yet effusively expressed emotion are discernible in the early 18th c. among the Pietists (see PIETISMUS). Traces of the cult of emotion are also to be found in the poetic writings of A. von Haller (q.v.), in the principal novel of Schnabel (see INSEL FELSENBURG, DIE), in the poems of E. von Kleist, and the work and correspondence of Gellert (qq.v.). From the mid-century it is a conspicuous feature in *Der Messias* and the odes of Klopstock (qq.v.), the early writings of Wieland (q.v.), Goethe's *Werther* (see LEIDEN DES JUNGEN WERTHERS, DIE), the poets of the Göttinger Hainbund (q.v.), especially Hölty (q.v.), J. M. Miller (*Siegwart*, q.v.), and the brothers Stolberg (qq.v.). Sophie von la Roche (*Geschichte des Fräuleins von Sternheim*, q.v.) and Hermes (*Sophiens Reise nach Sachsen*, q.v.) provide further notable examples of the sentimental style. Goethe's short satirical play *Der Triumph der Empfindsamkeit* (q.v.) is symptomatic of a reaction, and by 1780 the heyday of Empfindsamkeit is past. Its late inheritors are minor figures such as F. von Matthisson (q.v.).

Emser Depesche, a telegram which in July 1870 contributed significantly to the outbreak of the Franco-Prussian War (see DEUTSCH-FRANZÖSISCHER KRIEG). Wilhelm I (q.v.) of Prussia was at the time staying in Bad Ems. The French ambassador sought an audience in order to present new demands, and this was refused. The news of this development was telegraphed to Berlin on 13 July by Geheimrat Abeken, and received by Bismarck (q.v.) while in consultation with the generals Moltke and Roon (qq.v.). By skilful cutting, Bismarck altered the tone of the message to convey the impression of a double affront by and to the French ambassador and published the telegram in this form. The consequence was a storm of indignation and hostility between the two countries, creating a climate favourable to war. Bismarck recounts the episode with complacency in *Gedanken und Erinnerungen*. The results of his editing of the telegram were aptly summarized by Moltke: 'So hat das einen andern Klang, vorher klang es wie Chamade, jetzt wie eine Fanfare in Antwort auf eine Herausforderung.'

EMSER, HIERONYMUS (Weidenstätten nr. Ulm, 1477 or 1478–1527, Dresden), studied with Reuchlin (q.v.), and at Erfurt was one of the teachers of Luther (q.v.). He produced a rival translation of the Bible (see BIBLE, TRANSLATIONS OF) in 1527, and engaged in anti-Lutheran polemics.

Endreim, denotes rhyme in the English sense, the consonance of the last one, two, or three syllables of adjacent or closely placed lines of verse. The prefix is used to distinguish it from alliteration (q.v.), in German called Stabreim. Assonance is termed Zwischenreim or Assonanz.

Enfant perdu, a poem by H. Heine (q.v.), included in the section *Lamentationen* of the *Romanzero* (q.v.). The twentieth and last poem in the subdivision *Lazarus*, it begins 'Verlorner Posten in dem Freiheitskriege'. Heine reviews, in a spirit of sovereign irony, his political activity.

ENGEL, GEORG (Greifswald, 1866–1931, Berlin), a graduate of Berlin University and later a dramatic critic in Berlin, wrote novels and plays in the Naturalistic manner characteristic of the late 19th c. and early 20th c. They include the comedies *Der Ausflug ins Sittliche* (1900) and *Der scharfe Junker* (1910), and the novels (many with a Pomeranian or Baltic background) *Zauberin Circe* (1894), *Die Last* (1898), *Hann Klüth, der Philosoph* (1905), *Der Reiter auf dem Regenbogen* (1908), *Claus Störtebecker* (2 vols., 1920, dealing with a pirate executed in 1402), and *Der Uhlenspiegel* (1927). He also published two collections of Novellen, *Das Hungerdorf* (1893) and *Die Leute von Moorluke* (1910).

ENGEL, JOHANN JAKOB (Parchim, Mecklenburg, 1741–1802, Parchim), studied at Rostock

University, intending to follow his father as a Lutheran pastor, but changed his mind and became a senior schoolmaster (Professor) at the Joachimsthaler Gymnasium in Berlin (1776–87). He was also tutor to the future King Friedrich Wilhelm III (q.v.), whose affection he retained to the end of his life. Before going to Berlin Engel had written a number of plays, mainly comedies after the manner of Lessing, including *Der dankbare Sohn* (1771), *Der Diamant* (1772), and *Der Edelknabe* (1774).

During his teaching career at the Gymnasium he published philosophical writings, notably on aesthetics. His *Ideen zu einer Mimik* (1785–6) led to appointment as chief director (Oberdirektor) of the Berlin Hoftheater (1787–94). After four years in retirement at Schwerin he returned to a sinecure in Berlin. His death at Parchim occurred on a visit to his mother.

The chief work of Engel's later years was the novel *Herr Lorenz Stark* (q.v., 1795–6), an acute study of a character clash within a family and its resolution. He also published *Der Philosoph für die Welt* (1775–1803), a popular four-volume collection of essays, edited and partly written by himself.

Engelhard, a verse romance of some 6,000 lines by KONRAD von Würzburg (q.v.). Known only through a late printed version of 1573, it is believed to be the earliest of Konrad's long romances and was probably written in the third quarter of the 13th c. The source is a medieval Latin poem. *Engelhard* combines the themes of loyal friendship and of the healing of a leper by innocent blood, which are found, separate or associated, in other medieval works (see ATHIS UND PROPHILIAS and DER ARME HEINRICH).

Engelhard and his friend Dietrich, who are as like each other as twins, visit the court of Denmark. Engelhard falls in love with the King's daughter Engeltrud, and the pair are observed at a nocturnal assignation by Prince Ritschier of England, who denounces them. In order to prove his innocence, Engelhard has to submit to ordeal by battle. He summons his double, Dietrich, who impersonates him and is victorious. Engelhard and Engeltrud are united, and she bears him children. Meanwhile Dietrich is afflicted by leprosy. He discloses that he can only be cured by the blood of innocent children. The surpassing friendship of Engelhard impels him, against Dietrich's wish, to sacrifice his own children, and Dietrich recovers. God intervenes with a miracle, restoring the children to life.

Konrad affirms at the outset his intention of exalting fidelity in a world which has ceased to value it. That fidelity should express itself in a fraud (Dietrich's impersonation of Engelhard) is a sign of the ethical limitations of Konrad and of the decline from the high standards of the preceding age.

ENGELKE, GERRIT (Hanover, 1890–1918, in a British military hospital in France), the son of poor parents, developed a passion for music and poetry, but was apprenticed to a house-painter. He began to write poetry, and in 1913 went to Blankenese in order to show his work to R. Dehmel (q.v.), who furthered his ambition by sending him to the Werkleute auf Haus Nyland (q.v.), where he was particularly befriended by J. Kneip (q.v.). His Expressionistic hymnic verse is compounded of a positive feeling for spiritual values and of a sense of waste and oppression in an environment of increasing industrialization and imminent war. After four years' war service he was mortally wounded a few days before the Armistice of 11 November 1918. *Das Gesamtwerk. Rhytmus des neuen Europa* (1 vol.), ed. H. Blome, was published in 1960.

ENGELS, FRIEDRICH (Barmen, 1820–1895, London), was the son of a factory owner and trained for his father's business. He did not become seriously involved with social problems until his close contact and collaboration with K. Marx (q.v.). From 1844 on he remained a loyal friend of Marx, whom he was able to help in London, while he worked in his father's business at Manchester. He was, however, not only a popularizer of socialism but a sociologist in his own right, as his research into the conditions of the British working class shows (*Die Lage der arbeitenden Klassen in England*, 1845). He became increasingly interested in the impact of materialism on philosophy (*Ludwig Feuerbach und der Ausgang der klassischen deutschen Philosophie*, 1886). After his death his study *Revolution und Konterrevolution in Deutschland* (1896) appeared under Marx's name. A *Marx-Engels Gesamtausgabe* was begun in 1926 and a new edition (under the auspices of the Institute for Marxism–Leninism of the German Democratic Republic) in 1975.

Engel und der Waldbruder, see LEGENDE VOM ENGEL UND VOM WALDBRUDER.

ENGHAUS, CHRISTINE (Brunswick, 1817–1910, Vienna), made her stage début at the age of 12 and was engaged at the Burgtheater in 1840, retiring in 1875. She married F. Hebbel (q.v.) in 1846, and created the roles of Klara (*Maria Magdalene*), Judith, Mariamne (*Herodes und Mariamne*), and Brunhild (*Die Nibelungen*, qq.v.).

Englische Fräulein, popular name for a Roman Catholic order for women formed in

Munich in 1630 by English emigrant ladies. Its official designation is Institutum Beatae Mariae Virginis. The order has always devoted itself to the education of girls.

Englische Komödianten, troupes of English actors who played in Germany in the late 16th and early 17th c. Their invasion of Germany began with Lord Leicester's troupe, which played in Dresden in 1586, returning to England in the following year. Other companies followed, and from 1592 some remained permanently in Germany, gradually recruiting native actors. This change in the national constitution of the groups initiated, *c.* 1604, the substitution of German for the original English. The name Englische Komödianten, however, persisted until *c.* 1650.

The companies sought the protection of princes, and obtained it at Kassel, Brunswick, Halle, and Dresden. They did not, however, spend their whole time at the courts, but went on tour, visiting at various times most of the important cities of Germany, playing on average for a fortnight in each place. Their repertoire included works of most of the Elizabethan and Jacobean dramatists, among them a number of plays by Shakespeare and others by Marlowe, Dekker, Kyd, Beaumont and Fletcher, Massinger, Heywood, Day, and Marston, all of which were performed in prose adaptations. As time passed they made increasing use of extemporization. The few German works eventually included were chiefly by Ayrer and Duke Heinrich Julius of Brunswick (qq.v.).

The success of the Englische Komödianten is attributed to their highly matured technique of declamation and gesture. They aimed at powerful and sensational effects, and supported their histrionics with the visual appeal of rich and varied costume and crassly realistic blood scenes. Their influence on 17th-c. German drama was considerable.

ENIKEL, see JANSEN ENIKEL.

ENKING, OTTOMAR (Kiel, 1867–1945, Dresden), a journalist in north-west Germany, and from 1944 writer and lecturer in Dresden, was the author of novels of village and small-town life. The best known is the first, *Familie P. C. Behm* (1903). Others are *Wie Truges seine Mutter suchte* (1908), *Kantor Liebe* (1910), *Momm Lebensknecht* (1911), *Matthias Tedebus* (1913), and *Claus Jesup* (1919).

Entartete Kunst, a National Socialist conception applied to virtually all non-representational painting, sculpture, and graphic art. An exhibition of 'degenerate art' was held in

Munich in 1937, and a number of public monuments and frescoes were destroyed or defaced. Some of the artists concerned emigrated; a ban was imposed on those who remained in Germany, but some continued their work in seclusion. A fictitious painter of 'entartete Kunst', who is subjected to 'Malverbot', is an important figure in the novel *Deutschstunde* (q.v.) by S. Lenz.

Entführung, Die, a Novelle by J. von Eichendorff (q.v.), written probably in 1836 and published in Brockhaus's *Urania* in 1839. The complex action is set in France in the early years of the reign of Louis XV. The central episode is the attempt by Diana, an attractive and masterful girl, to entrap Count Gaston into an elopement with her lady's maid; but Gaston outwits her, and she herself is carried off. In a mill where they halt she tries to kill herself by setting fire to the building, but Gaston rescues her. He gives his hand, however, to Leontine, a young countess. Leontine, who is in love with him, had earlier sought to protect him when she believed him to be of low degree and in mortal danger. The Novelle is studded with (untitled) poems and stanzas.

Entlaufene Hasenbraten, Der, see VRIOLSHEIMER, DER.

Entstehung des Doktor Faustus, Die, see DOKTOR FAUSTUS.

Entwicklungsroman, see BILDUNGSROMAN.

ENZENSBERGER, HANS MAGNUS (Kaufbeuren, 1929–), was enrolled at 15 in the Volkssturm (q.v.), studied philosophy and literature at Erlangen, Hamburg, Freiburg, and Paris universities (1949–55), and has since lived a restless life, visiting the U.S.A., Mexico, Russia, the Near East, Norway, and Italy. In the early 1970s he lived alternately in Norway and Berlin.

A drastic critic of present-day civilization, he has an asyntactical style of poetry, in which newly coined, modish, and foreign words are abundantly used (*verteidigung der wölfe*, 1957; *landessprache*, 1960; *blindenschrift*, 1964). He is also an outspoken essayist and critic (*Einzelheiten* 1962) and an idiosyncratic anthologist (*Museum der modernen Poesie*, 1960; *Allerleirauh*, 1961). Enzensberger introduced a select edition of his poetry, written between 1955 and 1970 and published in 1971, by declaring his intention of adjusting the reader to a 'political alphabetization' ('politische Alphabetisierung'). In 1972 he published a provocative novel *Der kurze Sommer der Anarchie*, which has the Spanish anarchist

Buenaventura Durruti, killed in 1936, as its hero. *Der Untergang der Titanic*, described as a comedy, appeared in 1978. Collections of poetry include *Mauseleum. Siebenunddreißig Balladen aus der Geschichte des Fortschritts* (1975), *Beschreibung eines Dickichts*, ed. with a postscript by K. Schuhmann (1979), *Die Furie des Verschwindens* (1980), and *Dreiunddreißig Gedichte* (1981). A collection of his poetry, *Die Gedichte*, was published in 1983. The volume *Politische Brosamen* (1982) comments on contemporary issues. Enzensberger is the translator and editor of a play by A. Suchowo-Kobylin, published as *Tarelkins Tod oder Der Vampir von St. Petersburg. Farce in drei Akten* (1981).

Enzensberger, whose numerous editions include *Der hessische Landbote* (q.v.), published in 1974, was awarded the Büchner Prize in 1963.

Christian Enzensberger, poet, critic, and Anglist, is a brother of Hans Magnus Enzensberger.

Enzyklopädie (also known in German as *Konversationslexikon*). The first notable encyclopedia completed in German is the translation of P. Bayle's *Dictionnaire historique et critique* (1695-7) by J. C. Gottsched (4 vols., 1741-4). Before this J. P. von Ludewig had begun to publish his *Großes vollständiges Universallexikon aller Wissenschaften* (1731), but the full total of 68 volumes was not completed until 1754. This encyclopedia is often called the *Zedlersches Lexikon* after the publisher.

Encyclopedias flourished in the 19th c. with the first *Conversationslexicon* of Brockhaus (6 vols., 1796-1808, purchased by the firm virtually complete in 1808, expanded to 10 vols. and renamed *Enzyklopädie*, 1819-20). Meyer's *Das große Conversations-Lexikon für gebildete Stände* was on a larger scale (44 vols., 1840-55). The first encyclopedia published by Herder in Freiburg appeared in 6 volumes (1853-7). The most ambitious project, *Die Allgemeine Enzyklopädie der Wissenschaften und Künste*, begun in 1818 by J. S. Ersch and J. G. Gruber, was abandoned, still incomplete, in 1890 after 167 volumes had been published.

The principal existing German encyclopedias are the *Brockhaus Enzyklopädie* (20 vols., 1966-74), *Meyers Enzyklopädie* (25 vols. planned, 9 published 1971-3), and *Der große Herder* (10 vols., 1952-62).

Epigone, a rare word in English for 'one of a later (and less distinguished) generation' (*Concise Oxford Dictionary*), is generally used in German to denote derivative writers in the shadow of greater predecessors. Its frequency is probably due to the use of the word as a title by K. L. Immermann for a widely read novel, *Die*

Epigonen (q.v., 1836). 'Epigone' is not confined to literature and the arts; it is often applied to the post-Bismarck politicians and diplomats of the Wilhelmine era (see WILHELMINISCHES ZEITALTER).

Epigonen, Die, a novel written between 1823 and 1835 by K. L. Immermann (q.v.) and published in three volumes in 1836. It is described as 'Familienmemoiren in neun Büchern'. The title signifies 'those born after', i.e. after a period of greatness and great men, and Immermann himself wrote in a letter that the subject of the novel was 'Segen und Unsegen des Nachgeborenseins'.

In Bk I (*Klugheit und Irrtum*) the German youth Hermann comes into contact with an impulsive young girl, Flämmchen, and with a duke and duchess. He fights a duel with a former guardian of Flämmchen, and is wounded. Bk. II is entitled *Das Schloß des Standesherrn*. Hermann has been cared for by the Duke and spends some time at his seat, falling in love with the Duchess and coming to believe (wrongly) that his love is returned. Among his new acquaintances at this time is a doctor who is to play a part in Bk. VIII.

In *Die Verlobung*, which occupies Bk. III, Hermann becomes engaged to Kornelie, whom he had known earlier as a young girl. He returns to the ducal castle (Bk. IV, *Das Karussell, der Adelsbrief*) and helps to arrange a medieval tournament, which proves a fiasco. He also finds a document which prevents the Duke's property from passing to Hermann's commercially astute uncle. Next he falls in with a group of Radical demagogues (Bk. V, *Die Demagogen*), whom he tries to argue out of their political beliefs. Ironically he is arrested as their leader. Through the help of a new friend, Medon, he is released (Bk. VI, *Medon und Johanna*). Medon's wife, Johanna, is a friend of the Duchess, and his brilliant personality charms Hermann. But in Bk. VII (*Byzantinische Händel*) Medon proves to be the head of a subversive movement and is arrested as a demagogue. Johanna disappears and Hermann, searching for her, discovers her with Flämmchen, now a rich widow. At a riotous party given by Flämmchen he goes to bed (so he believes) with Johanna, only to discover afterwards that she is his sister.

The eighth book (*Korrespondenz mit dem Arzte 1835*) takes the form of a correspondence between the narrator and the doctor of Bk. II. It reveals much of the relations of the characters and their fate. The Duke has discovered that he has no legal claim to his title or estates. Decently and without fuss he ends his life. Flämmchen has died in poverty and Hermann, oppressed by the thought of incest, has lapsed into melancholia.

Bk. IX (*Kornelie*) shows Hermann cured,

having discovered that his partner in the apparently incestuous cohabitation was Flämmchen, not Johanna. By a curious twist he accedes to the immense property of the former duke and marries Kornelie.

This rather convoluted novel, which is in part a 'Bildungsroman' (q.v.) owing something to Goethe's *Wilhelm Meisters Lehrjahre* (q.v.), is a document of the age, reflecting the decline of the aristocracy and the rise of Radicalism, which it rejects; it also deplores the new worship of money and the rise of industrialism. Its own implied norm is the middle way of integrity. To some extent it is a *roman à clef* (see SCHLÜSSEL-ROMAN); Medon is said to have had Karl Follen (q.v.) as a model, and Flämmchen may owe something to Bettina von Arnim (q.v.), while a judge appearing in Bk. VI is thought to derive from E. T. A. Hoffmann (q.v.).

Epik and Episch, normally applied in German usage not only to the epic poem (Epos) but also to narrative fiction.

Epilog zu Schillers 'Glocke', a memorial poem for Schiller written by Goethe. It is in the form of an epilogue and was spoken at the end of a public recitation of Schiller's *Lied von der Glocke* (q.v.) at Bad Lauchstädt on 10 August 1805. The poem was first published in the *Taschenbuch für Damen auf das Jahr 1806* and then in Goethe's *Werke*, 1808. In 1810 it was given in slightly extended form on a similar occasion, commemorating on 10 May the fifth anniversary of Schiller's death. A third, still further extended version was recited on the tenth anniversary, and it is this form, with the superscription *Wiederholt und erneut bei der Vorstellung am 10. Mai 1815*, which is usually printed in Goethe's works. It gives in 13 eight-line stanzas Goethe's tribute to the genius and personality of his friend and colleague and three times uses the well-known phrase, *Denn er war unser!*

Epiphanias, a humorous poem written by Goethe in 1781 and printed in 1811. It was devised for a festivity on 6 January 1781. The first, second, sixth, seventh, and eighth stanzas are sung by all three kings, and each sings one of the three intervening stanzas. The first king was sung by the actress Corona Schröter (q.v.). The poem has been set to music by Hugo Wolf (q.v.).

Epische Oper, a term introduced by B. Brecht (q.v.) to contrast his conception of opera with 'dramatic opera'. The principles of 'epic opera' correspond to those of 'epic theatre' (see EPISCHES THEATER), and stress the didactic function of music in providing an independent commentary on the text, instead of supporting

it as in traditional opera. He introduced the neologism 'Misuk' for Musik to underline his different attitude, which was, like that of his close collaborator, the composer Kurt Weill (q.v.), opposed to the Gesamtkunstwerk of R. Wagner (q.v.).

Episches Theater (Epic Theatre) is almost exclusively associated with B. Brecht (q.v.), though the term was used in the 1920s of the early plays of A. Bronnen (q.v.) and of A. Paquet (q.v.), whose play *Fahnen* (1924) was produced in Berlin by E. Piscator (q.v.) for his socialist theatre. Piscator's striking technical innovations for stage productions influenced Brecht, as well as the sardonic styles of the great cabaret artists K. Valentin and F. Wedekind (qq.v.).

The epic theatre of Brecht represents a clean break with established dramatic styles. It proceeds in the manner of a narrative or chronicle, avoiding an intertwining and concentrated construction. It aims at the spectator's detachment from the action, which is designed to instruct and, by reducing the emotional involvement of the spectator, to stimulate a critical scrutiny of reality. Though Brecht had in 1935 published an essay, *Über die Verwendung der Musik für eine epische Bühne*, and had repeatedly discussed his views (e.g. in *Der Messingkauf*, q.v.), it was not until 1949 that he published, as *Kleines Organon für das Theater* (q.v.), his principal theoretical work on epic theatre, In 1954 he expressed some hesitation about the suitability of the term Episches Theater. but he rejected the alternative Theater des wissenschaftlichen Zeitalters (*Nachträge zum Kleinen Organon*).

Aspects of epic theatre have been adopted, for different ends, by many dramatists, among them F. Dürrenmatt, T. Dorst, M. Frisch, P. Hacks, W. Hildesheimer, M. Walser, and P. Weiss (qq.v.).

Epistolae clarorum virorum, title used by J. Reuchlin (q.v.) in 1514 for a collection of letters of support written to him by eminent men during his controversy with J. Pfefferkorn (q.v.). See also EPISTOLAE OBSCURORUM VIRORUM.

Epistolae obscurorum virorum, a collection of Latin letters published in 1515 (expanded in 1516) and addressed to O. Gratius (q.v.), a professor of Cologne and a prominent Dominican supporter of J. Pfefferkorn (q.v.) in his attacks on J. Reuchlin (q.v.). The letters, intentionally couched in barbarous language, purported to be written by admirers of Gratius and Pfefferkorn, but were in fact a satire composed by Reuchlin's supporter CROTUS Rubeanus (q.v.) with the aim of making Reuchlin's persecutors ridiculous. A second part of the *Epistolae* (1517) was mainly the work of Ulrich von Hutten (q.v.). The title alludes to the *Epistolae clarorum virorum* (q.v., 1514), published by Reuchlin. *Dunkelmännerbriefe* is an alternative title.

Epos, normally bears the same meaning as the English noun 'epic', a long, elevated, narrative poem divided into books. In German it is occasionally extended to cover works of prose fiction.

Eraclius, see OTTE.

ERASMUS, DESIDERIUS (Rotterdam, 1466–1536, Basel), Dutch humanist, spent some time in France and England, visited Italy, and then settled in Switzerland. From 1521 to 1529 he lived in Basel, collaborating with the publisher Froben (q.v.), whose presses poured out a stream of humanistic works, including many by Erasmus himself. The Reformation (q.v.), which he came to regard as a mortal enemy of learning, though he had sympathized with its original aims, drove him to Freiburg, but he returned to Basel in 1535.

By his intense devotion to the New Learning, and by his capacity for human relationships, in both personal contact and correspondence, Erasmus acquired immense prestige in Europe, including Germany, as a champion of humanism; his repute is summed up in the phrase *homo pro se*, which occurs in the *Epistolae obscurorum virorum* (q.v.). Though Erasmus engaged in controversy with Luther (q.v.) on the question of free will (*De libero arbitrio*, 1524, and *Hyperaspistes*, 1527), he was a pioneer of the Reformation, as the then popular saying testifies, Erasmus laid the egg and Luther hatched it. His edition of the Greek New Testament with Latin translation (1516) was a basis for the Lutheran Bible. His tolerance and breadth of view were supplemented by wit and satire, evident in his best-known work, the wittily and learnedly urbane *Moriae encomium seu laus stultitiae* (1509, *Praise of Folly*). He was an indefatigable correspondent, and a large number of his elegant letters, including many addressed to notable personalities in Germany, survive.

Erbe am Rhein, Das, a trilogy of novels by R. Schickele (q.v.), comprising *Maria Capponi* (1925), *Blick auf die Vogesen* (1927), and *Der Wolf in der Hürde* (1931). The novels cover the 1914–18 War and the following decade. The central figure, Claus von Breuschheim, an Alsatian with a German-speaking father and a French mother, comes to reject both the French and the German solutions for his native province, hoping for international status in a unified Europe. This political thinking underlies the novels, which have complex plots.

Erbförster, Der, a five-act tragedy (Trauerspiel) by O. Ludwig (q.v.). After years of brooding and rewriting, Ludwig completed it in 1850 and it was first performed in Dresden in March 1850.

The scene is set alternately in the forester's lodge at Düsterwalde and a mansion at Waldenrode. Though it is not stated, the background is clearly Thuringian. A betrothal is about to take place between the forester's daughter Marie and his employer's son, Robert Stein. But Ulrich the forester and Stein senior fall out over a trifle and, in their obstinacy, become involved in a serious quarrel. A keeper is shot by a poacher, using a gun stolen from Ulrich's son Andres, who is in consequence suspected of the murder. Ulrich believes his son to be in danger, takes his gun, and fires at a human figure, which he believes to be Robert Stein, only to find that he has killed his daughter Marie. He takes his own life in the conviction that he is executing God's judgement. Much in *Der Erbförster* is melodramatic, and it leans heavily on coincidence; but Ulrich is a magnificently realized large-scale character, a personality of great forcefulness and complete integrity, but of limited imagination and intelligence; the successfully evoked forest background is a prominent feature.

Erbkaiserliche Partei, a predominantly Liberal party in the National Assembly (see FRANKFURTER NATIONALVERSAMMLUNG) of 1848–9, which supported the establishment of a hereditary Hohenzollern empire in Germany. It was responsible for the official offer of the imperial crown to Friedrich Wilhelm IV (q.v.), which the King declined. After the dissolution of the Assembly the party met at Gotha in 1849, and agreed to support the Prussian policy of unification without Austria.

Erdbeben in Chili, Das, a Novelle by H. von Kleist (q.v.) which was published in serial form in Cotta's *Morgenblatt für gebildete Stände* (q.v.) in 1807 with the title *Jeronimo und Josephe. Eine Scene aus dem Erdbeben von Chili, vom Jahre 1647*. The story was included in the first volume of *Erzählungen* (1810) under its present title. Kleist's source is unknown, but the earthquake to which he refers took place on 13 May 1647 in Santiago.

The action is confined to two days, and commences with the earthquake, which largely destroys the Chilean capital Santiago, on the day on which Donna Josephe, the daughter of Don Henrico Asteron, is to be executed, because, while in a Carmelite convent, she has given herself to her former tutor Jeronimo Rugera, by whom she has borne a son. The earthquake, resulting in the death of countless citizens, miraculously saves the lives of the lovers and their child. They are reunited outside the city, in the peaceful seclusion of a valley in which survivors have flocked together. Instead of using their freedom to flee the country, they decide to join a small party of noble citizens to return the following morning to the cathedral to attend a thanksgiving service. While they are listening to the priest praising God who, he claims, has passed judgement on the city's sinners, the shoemaker Meister Pedrillo, recognizing Josephe Asteron, creates a tumult among the panic-stricken congregation. Under the protection of one of their party, Don Fernando Ormez, the lovers succeed in leaving the cathedral. But as they step outside the church Jeronimo is killed by his own father. After Donna Constanze is murdered in mistake for Josephe, Meister Pedrillo corrects the error by killing Josephe, who has entrusted Donna Elvire's and her own babies to Don Fernando, Donna Elvire's husband. Don Fernando heroically defends the children, but the mob does not disperse until Meister Pedrillo, aiming a blow at Josephe's son, Philipp, kills Juan, Don Fernando's child. Don Fernando and Donna Elvire subsequently adopt Philipp as their own child. The survival of the 'little stranger' indicates the existence of a measure of true love and heroism in a world of apparent chaos.

Erde, a play by K. Schönherr (q.v.).

Erdgeist, a four-act tragedy in prose (Tragödie) by F. Wedekind (q.v.), published in 1895. It is preceded by a verse prologue in which an animal tamer appears and emphasizes the bestial aspects of the forthcoming performance. The principal character is Lulu, the completely amoral embodiment of female sexuality. Her first lover dies of a stroke when he finds her with another; her second cuts his throat, and she shoots the third dead. *Erdgeist* has a sequel in *Die Büchse der Pandora* (q.v.).

Erdgeist, the spirit which appears at Faust's behest in the first scene of *Faust* Pt. I (see FAUST). The Earth Spirit, before which Faust quails, represents the multiplicity and power of Nature. It is the same Erdgeist which Faust addresses in the scene *Wald und Höhle* as 'Erhabner Geist'.

Erec, a Middle High German poem of some 10,000 lines by HARTMANN von Aue (q.v.). It is an Arthurian romance, written *c.* 1180–5, and is a free translation of the *Erec* of Chrétien de Troyes. The action falls into two main sections. The first (ll. 1–2431) comprises a conventional Arthurian narrative. The knight Erec, riding out with the queen, avenges an insult and in doing so secures as a bride Enite, the beautiful daughter of a completely impoverished nobleman. He returns with her to Arthur's court, and the wedding is celebrated with great festivity.

The second part, which is conspicuously moral and didactic, begins with Erec's excessive devotion to a life of love with his wife which leads him to neglect his knightly duties. He falls into disrepute and, when he realizes the general disfavour, is angered with his wife. He sets out to encounter adventures, bidding Enite ride before him without warning him of any danger. She disobeys this injunction, setting his welfare above a literal conformity, and is each time upbraided by Erec. After various adventures Erec faints from wounds and appears to be dead. Enite is succoured by Count Oringles, who is inflamed by her beauty and, in the presence of the apparently dead Erec, brutally presses his attentions upon her. Erec, roused by her shrieks, slays Oringles, and the couple are reconciled. Unnecessarily pursuing further adventures, Erec is defeated and, looking back, perceives that only morally justified exploits are admissible. He undertakes such a virtuous adventure in challenging the knight Mabonagrin, who is bound by an oath to his lady to remain in the Garden of Love as long as he is undefeated. Erec vanquishes Mabonagrin, who is released from the garden to his own joy and, ultimately, that of his lady. Erec, restored to true knighthood, is received back into King Arthur's court and presently rules happily with Enite over his own land.

The story is educative, illustrating the element of virtue in true knightly conduct, commending self-control (*mâze*) and courteous behaviour, and condemning immoderate behaviour in love as in other activities. The episode of Mabonagrin contrasts unfavourably the courtly theory of love (Minnedienst) with the true conjugal love of Erec and Enite. The only (almost) complete MS. of *Erec* is contained in the Ambraser Handschrift (q.v.), from which two pages are missing.

Erfolg, Der, a novel by L. Feuchtwanger (q.v.), published in 1930. Sub-titled *Drei Jahre Geschichte einer Provinz,* it afterwards became the first volume of the tetralogy *Der Wartesaal* (*Zyklus aus dem Zeitgeschehen*). Its principal character is an art historian and museum official named Kruger, who is falsely accused of a crime, and dies in prison; but the book is a portrait of the times (1921–4) and is partly a Schlüsselroman (q.v.). The background includes the French occupation of the Ruhr, the time of inflation, and the Hitlerputsch. Hitler appears as 'Rupert Kutzner', Ludwig Thoma (q.v.) as 'Dr. Matthäi', B. Brecht (q.v.) as 'Kaspar Pröcke', and Ludendorff (q.v.) as 'General Vesemann'. The author himself is represented by the character Jacques Tüverlin.

Erfurter Judeneid, a formal oath for the use of Jews appearing before a court. It is preserved at Magdeburg in a MS. of the late 12th c. originating from Erfurt. A number of later versions exist. The oath is remarkable for its poetic language.

Ergo bibamus!, see HIER SIND WIR VERSAMMELT ZU LÖBLICHEM TUN.

Erinnerung, a poem written by E. Mörike (q.v.) in 1822, included in *Gedichte,* 1838, and revised in 1865. A poem of recollection of early love, it is dedicated to C. N. (Clara or Klara Neuffer, the poet's cousin). The first lines run 'Jenes war zum letzten Male,/Daß ich mit dir ging, o Klärchen!'

Erinnerungen von Ludolf Ursleu dem Jüngeren, a novel by Ricarda Huch (q.v.), published in 1893. The first of her narrative works, it employs an indirect technique of storytelling. The events are recalled by a first-person narrator, who is distinct from the author, and was not himself a participant in the action.

When Ludolf Ursleu sets down the story he has withdrawn from the world, and is a monk in Einsiedeln Abbey in Switzerland. He depicts the passions which accompanied the decline of his once prosperous and respectable family. At the centre of events are two of his cousins, Ezard and Galeide. Ezard marries Lucile, a Swiss governess, and then develops a passion for Galeide, which is returned. An epidemic of cholera breaks out and Lucile dies, leaving Ezard and Galeide free to marry. But Galeide falls in love with Lucile's brother Gaspard and takes her own life. Behind these passionate events is an equally strenuous commercial struggle which causes the suicides of Ludolf's father and uncle before it is finally resolved by Ezard. Both Gaspard and the narrator retire to a life of monastic renunciation.

Er ist's, a poem written by E. Mörike (q.v.) in 1829, and included in the novel *Maler Nolten* (q.v.). An expression of the coming of spring, it begins 'Frühling läßt sein blaues Band'.

'**Eritis sicut Deus, scientes bonum et malum',** Genesis 3:5, according to the Vulgate.

In *Faust I* by Goethe (see FAUST) these words are written in the student's album by Mephistopheles disguised as the learned Doctor Faust (Studierzimmer 2). The Authorized Version has, 'ye shall be as gods, knowing good and evil'.

Erklärung eines alten Holzschnittes, see HANS SACHSENS POETISCHE SENDUNG.

Erlebnis des Marschalls von Bassompierre, Das, a Novelle by H. von Hofmannsthal (q.v.), written and published in 1900. The story adapts, in part literally, an anecdote narrated by Karl in Goethe's *Unterhaltungen deutscher Ausgewanderten* (q.v.), which is itself largely a translation of a passage from the *Mémoires* of the Maréchal François de Bassompierre (q.v.). Hofmannsthal acknowledged both sources (Goethe and Bassompierre) in a footnote dated Vienna, 27 November 1900, when he published the second half of the Novelle in the Viennese weekly *Die Zeit,* adding: 'woselbst sich der Leser über das Verhältnis des Überlieferten zu meiner dichterischen Ausgestaltung des Stoffes orientieren kann.'

The story is an elaboration of the marshal's reminiscences of an amorous adventure with a woman of low station (Krämerin), who had irresistibly attracted him. Hofmannsthal's technique, using irony and symbolism, preserves the anecdotal frame (including the time of action between a Thursday and a Sunday night) and the central experience of the narrator, his encounter with love and death caused by the plague, to which the Krämerin falls victim, a fact which he makes more evident than does Goethe.

Erlebnisse eines Schuldenbauers, J. Gotthelf's last novel, written in 1852–3 and published in 1853. Hans Joggi and Anne Marei, a simple-minded peasant couple with little capital, are jockeyed by speculating agents into taking a farm which they cannot afford. They work hard but soon find themselves in difficulties, for money due to them is not paid, and so they cannot pay what they owe. After years of hard work, during which, too, their favourite child is drowned, they are sold up, in circumstances which all honest people of the neighbourhood reckon to be a scandal. In the end they are rescued by a crusty squire of the old order who engages Hans Joggi as his steward, and is delighted with his honesty and industry and that of his wife and children. The book reflects Gotthelf's indignation at the declining moral standards, particularly in financial matters, which he perceived in the post-1848 government.

Erlebte Rede (*style indirecte libre*), term introduced in 1921 by E. Lorck for a stylistic device, prominent in Impressionist and psychoanalytically oriented fiction. Related to the stream of consciousness techniques (e.g. Schnitzler's *Leutnant Gustl*, q.v., written 1900) which employs the first person technique deriving from monologue in drama, it is written in the third person and enables the writer to switch inconspicuously from third person narrative to passages revealing inner states of mind without involving the use of the first person or even the subjunctive of indirect speech (Kafka, ending of *Der Prozeß*, q.v.). Interesting variations displaying the narrator's consistent remote control can be found in Schnitzler's *Die Toten schweigen* (q.v., written in 1897); 'sagte sie für sich' and 'sie dachte' introduce soliloquies in direct speech, which then switches to erlebte Rede: 'Warum hatte sie den Kutscher weggeschickt? Was für ein Unsinn! ... Ja, was soll sie denn tun ... ?', before reverting to: 'Ich bin nicht allein mit ihm, fiel ihr ein.' Stream of consciousness technique (in German inadequately termed Bewußtseinsstrom), rigorously employed in *Leutnant Gustl*, dispenses altogether with the narrator's presence by making no concession to phrases like 'fiel ihr ein'.

Erleuchteten Fenster, Die, oder Die Menschwerdung des Amtsrats Julius Zihal, a humorous short novel by H. von Doderer (q.v.), published in 1950. It is an elaborate and wayward piece of irony. Zihal, a widower in the tax office, reaches retiring age and moves into a smaller flat. He discovers that none of the female inhabitants of the quarter seem to draw their blinds. Window observation in the dark becomes his new interest, and he buys a pair of binoculars. In true civil servant fashion he keeps a minute and orderly record of his observations. Impatient of the limited range of the binoculars, he buys a large telescope. But he breaks it as he stumbles in the dark on the way to answer a call at the door. These embarrassments precipitate a mild fever confining him to bed. A visit of the mature but attractive Fräulein Oplatek, to whom he gives his hand, sets the final seal on his recovery from his bizarre aberrations.

Erlkönig, a ballad written by Goethe in 1782, published in the same year as part of the Singspiel *Die Fischerin*, and separately in *Goethes Schriften* in 1789. It tells of a father's ride through the windy night bearing his sick son, who is rent from him by the sinister elfin king. The source is Danish and 'Erlkönig' is a corruption of *ellerkonge*. The poem has been powerfully set to music by both F. Schubert and C. Loewe (qq.v.).

Erlösung, Die, title given in 1858 by the Germanist K. Bartsch to an anonymous Middle High German poem recounting in 6,593 lines the history of man from the Creation to the Last Judgement. It was written in Hesse *c*. 1300. It begins with a prologue, which announces its intention of avoiding florid style because of the high matter of the poem. The poem portrays man after the Fall, incurring God's wrath, and narrates the plea of the Son of God for leave to redeem man. The prophets are passed in review, and the story of Jesus is told with special detail up to his Baptism; the Crucifixion, Resurrection, and Ascension are dealt with fully, and the poem concludes with the Day of Judgement. Some scholars have held that the author of *Die Erlösung* was also the poet of the *Elisabethlegende* (q.v.).

Ermächtigungsgesetz, a law enabling a government to override other existing laws. 'Enabling laws' were passed in 1914, 1919, 1920, 1921, and 1923 in times of crisis in Germany, but were regarded as temporary emergency procedures. The most famous Ermächtigungsgesetz is that passed on 24 March 1933 in the Reichstag on Hitler's direction (Gesetz zur Behebung der Not von Staat und Volk). It was used for the total and permanent nazification of Germany.

Ermittlung, Die, a so-called oratorio (Oratorium in elf Gesängen) by P. Weiss (q.v.), published and performed in 1965. It is concerned with the National Socialist policy of annihilation of the Jews which it regards from a Marxist standpoint as a consequence of capitalism. It takes the form of a documentary play based on the trial at Frankfurt (1963–5) of eighteen former members of the staff of Auschwitz concentration camp.

Ermordete Majestät oder Carolus Stuardus, König von Groß Britanien, a tragedy by A. Gryphius (q.v.), published in 1657 and revised in 1663. Its subject is the execution of Charles I in 1649, and the action is limited to an abortive plan of rescue. Gryphius began it in the same year, a proof of the sensation created by the regicide; it is also a rare instance of topical events treated in serious drama of the 17th c. Gryphius revised it in 1663 in the interests of historical accuracy and in the light of the Restoration. Written in Alexandrines, it is a static play covering only the period between Charles's condemnation and his execution, which takes place on the stage. The Christ-like king is conceived as a guiltless martyr.

ERNST I, Herzog von Bayern-München from 1373 to 1438 and father of Albrecht III (q.v.) 'der Fromme', whose life he saved in the battle of Alling in 1422, in which he won a decisive victory over Ludwig of Ingolstadt. He was responsible for the execution of Albrecht's wife, the commoner Agnes Bernauer (see BERNAUER, AGNES) in 1435. He ruled jointly with his brother Wilhelm, who died in 1435, at a time when Bavaria was divided into three parts, the other two, Ingolstadt and Landshut, being ruled by hostile relations, Ludwig der Bärtige and Heinrich respectively. Wilhelm's son Adolf, born in 1432, died in childhood (1437), leaving Albrecht sole heir of the Munich branch of the Wittelsbach dynasty (q.v.). Ernst was married to Elisabeth, Princess of Milan.

Ernst, Herzog von Schwaben, a historical tragedy in blank verse, written by L. Uhland (q.v.) in 1816–17, published in 1817, and first performed in Hamburg in May 1818. It is described as 'Trauerspiel in fünf Aufzügen' and is concerned with the historical character who has become a legend in the medieval poem *Herzog Ernst* (q.v.). The play is set in 1030 in the reign of the Emperor Konrad II (q.v.). Ernst has twice revolted against the Emperor, who is his stepfather. His mother, the Empress Gisela, pleads for his release, which the Emperor grants. But Ernst refuses to renounce his friend Werner von Kyburg, as the Emperor demands, and is outlawed. The imperial forces concentrate against Ernst and his friends, who are all killed. The ban is then lifted so that they receive proper burial. The conflict of loyalties and the faithfulness of friends are the play's central themes.

ERNST, PAUL (Elbingerode, Harz, 1866–1933, St. Georgen, Styria), son of a miner, was educated at Göttingen and Tübingen universities. He took part in Social Democratic politics in the 1890s, and associated with the Naturalist writers Holz and Schlaf (qq.v.). About the turn of the century he abandoned the political Left, and, after a visit to Italy, adopted the neo-Romantic style (see NEUROMANTIK). He soon turned to neo-Classicism (see NEUKLASSIZISMUS), of which he was the principal champion. His associates were W. von Scholz and S. Lublinski (qq.v.).

Ernst began his career as an author in his thirties with two one-act plays, *Lumpenbagasch* and *Im Chambre séparée*, both published in 1898. His tragedies, some of which were historical, include *Demetrios* (1905), *Das Gold* (1906), *Canossa* (1908), *Brunhild* (1909), and *Ninon de Lenclos* (1910). Neither these nor his neo-Classical plays, *Manfred und Beatrice* (1912) and *Preußengeist* (1914), were successful. His earlier novels, *Der schmale Weg zum Glück* (1904), in which he draws on his own experiences, and *Saat auf Hoffnung* (1916), are more engaging than

the later historical romances *Der Schatz im Morgenbrotsaal* (1926) and *Das Glück von Lautenthal* (1933). He is the author of many Novellen, often grouping them within a frame (see RAHMEN). They include the collections *Die Hochzeit* (1913), *Die Taufe* (1916), *Der Nobelpreis* (1919), and *Lustige Geschichten* (1930).

A critic and essayist, Ernst published *Der Weg zur Form* (1906), which is notable for the dogmatic presentation of his views, *Ein Credo* (2 vols., 1913), *Der Zusammenbruch des Idealismus* (1919), and *Der Zusammenbruch des Marxismus* (1919, revised as *Die Grundlagen der neuen Gesellschaft*, 1930). His once popular *Erdachte Gespräche* (1921) have affinities with his essays. In the 1920s he made a belated attempt to revive the verse epic with *Das Kaiserbuch* (3 vols., 1922–8) and *Der Heiland* (1930). He published two autobiographical works, *Jugenderinnerungen* (1930) and *Jünglingsjahre* (1931). He was nominated as one of the fourteen founder members of the National Socialist Dichter-Akademie, but he died before the first meeting.

ERNST AUGUST, KING OF HANOVER, DUKE OF CUMBERLAND AND OF BRUNSWICK-LÜNEBURG (Kew, 1771–1851, Hanover), was the fifth son of George III of Great Britain. He studied at Göttingen, and served in the Hanoverian forces in the Netherlands in 1793–4, losing an eye at Tournai. He then served in England, being promoted lieutenant-general in 1799. He was an extreme Tory, totally opposed to reform, and in 1810 an attempt was made to assassinate him. He commanded the Hanoverian forces in 1813–14. The unpopularity of his political views in England led to the parliamentary curtailment of his allowance, and he spent some years in Berlin, returning to England, however, under his brother George IV.

On the death of William IV in 1837, the accession of a queen in Great Britain dissolved the union with Hanover, and Ernst August succeeded to the throne as king of Hanover. He immediately embarked on a reactionary policy, revoking the constitutional law of 1833, and dismissing seven Göttingen professors who protested (see GÖTTINGER SIEBEN). He promulgated a new, illiberal constitution in 1840. In spite of great unpopularity he skilfully survived the revolutionary year of 1848.

Erschlagene Abel, Der, a prose 'idyll' by F. Müller (q.v.), first published anonymously in *Die Schreibtafel* (Mannheim) in 1775. It is linked with *Adams erstes Erwachen* (q.v.).

Erstes Liebeslied eines Mädchens, a poem written by E. Mörike (q.v.) in 1828, and sent as a wedding poem to his friend Friedrich Kaufmann. It has been set to music by Hugo Wolf (q.v.) who considered it one of his best works. The first line runs 'Was im Netze? Schau einmal.'

ERTL, EMIL (Vienna, 1860–1935, Graz), Librarian at the Technical University (Technische Hochschule) at Graz, wrote novels and stories of regional life, drawn both from his own and from historical times. His principal novels are the tetralogy *Ein Volk an der Arbeit*, consisting of: (1) *Die Leute vom blauen Guguckhaus* (1906); (2) *Freiheit, die ich meine* (1909); (3) *Auf der Wegwacht* (1911); (4) *Im Haus zum Seidenbaum* (1926); *Der Neuhäuselhof* (1913), *Der Anlaßstein* (1917), *Das Lattacherkind* (1929), and *Eingeschneit auf Korneliagrube* (1931). His shorter works, which are on the whole the more successful, comprise the collections *Opfer der Zeit* (1895), *Miß Grant* (1896), *Mistral* (1901), *Feuertaufe* (1905), *Gesprengte Ketten* (1909), *Nachdenkliches Bilderbuch* (2 vols., 1911–12), and *Teufelchen Kupido* (1925), as well as the separate Novellen *Die Perlenschnur* (1896), *Das Trauderl* (1918), *Der Berg der Läuterung*, and *Der Handschuh* (both 1922).

Erwählte, Der, a novel by Th. Mann (q.v.), published in 1951. It is a retelling and expansion of *Gregorius* (q.v.), Hartmann von Aue's medieval verse tale of unwitting incest and expiation. The story is told by the Irish monk Clemens, who, in Mann's words, embodies 'der Geist der Erzählung'.

Two noble children, Willigis and Sibylla, are left orphans. Being quite without guidance, they succumb to the temptations of the flesh. When they discover that Sibylla is with child they are filled with remorse. Willigis departs for the Holy Land, and is not seen again. Sibylla, acting on the advice of the counsellor who has arranged her lying-in, commits the child Grigorss (or Gregorius, both names are used) to God's care, and, having attached to him a document of identity, she sets him adrift in a boat. God protects him, he is found, cared for, and has the opportunity to lead a secluded life of repentance. But his knightly instincts are too strong, he sallies forth, and in due course saves the Lady Sibylla, whose land is beset by enemies. He is rewarded with her hand, and it is only after the birth of children that their mutual relationship is discovered. Doubly incestuous, they see themselves as the most depraved sinners of human kind. Gregorius seeks out a remote island where he lives without shelter or proper food. After seventeen years the Pope dies, and the Lamb of God appears in a dream to a wise Roman, Sextus Anicius Probus, instructing him how to

seek out the new pope. The search brings Probus and his companions to Gregorius, whose sins are now expiated. He is enthroned at Rome, and Sibylla, hearing of the virtuous new pope, without knowing his identity, comes to Rome to confess her sins. She, receives absolution, and a scene of recognition and thanksgiving follows. The story is told in chronicle style with deliberate archaic touches. In the brief epilogue Mann acknowledges his indebtedness 'in den Hauptzügen' to Hartmann.

ERWIN, F. TH., see KUGLER, FRANZ THEODOR.

Erwina, a journal published under the joint editorship of August and Adolf Stöber (qq.v.) in 1838–9.

Erwin und Elmire, ein Schauspiel mit Gesang, Singspiel (q.v.) written by Goethe in 1773–5 and published in J. G. Jacobi's periodical *Iris* in March 1775. It was first performed in Weimar on 1 March 1777. The sub-title was changed to *Ein Singspiel* in 1788. It is based on Oliver Goldsmith's *Edwin and Angelina*. Elmire regrets the coyness which has driven Erwin away. After a short separation the two lovers are united. The play contains the well-known song 'Ein Veilchen auf der Wiese stand' (q.v.). Othmar Schoeck (1886–1957) composed stage music for it and set the songs to music.

Erzählung is not only the process of story-telling but also the term used by some writers for a relatively short story which can hardly be called a 'Roman' and is straightforward in its narrative technique. As the term has never been closely defined, many writers have applied it to works to which the term Novelle (q.v.) could properly be applied.

ERZBERGER, MATTHIAS (Buttenhausen, Württemberg, 1875–1921, nr. Griesbach, Black Forest), a schoolmaster, entered politics, and in 1903 became a member of the Reichstag for the Centre Party (Zentrum, q.v.). In the years preceding the 1914–18 War Erzberger opposed the colonial policy of the German government; during the war he worked for the opening of negotiations for peace, and in 1917 was the proposer of a resolution to this effect which was passed by the Reichstag. In 1918 Erzberger, as leader of the German delegation to the Allied H.Q. at Compiègne, signed the Armistice. He was the target for abuse by German nationalists, among whom K. Helfferich was prominent. He was assassinated by two ex-officers.

Erzherzog, Erzherzogin (Archduke, Archduchess), titles borne respectively by the sons and daughters of the Habsburg emperors from 1453 to 1918.

ERZHERZOG KARL, see KARL, ERZHERZOG.

Erziehung des Menschengeschlechts, Die, a short philosophical tract by G. E. Lessing (q.v.). The first part (§§ 1–53) was published in Lessing's *Gegensätze* in 1777, the complete work appeared anonymously in 1780. In a preface (Vorbericht) and 100 numbered paragraphs Lessing sets out a philosophy of religious history, in which he traces a progress from polytheism to the monotheism of Judaism, from this to the ethic of Christianity, which in turn will give way to a world in which good is done for its own sake without the sanction of punishment or reward. In the final paragraphs Lessing tentatively advances a theory of metempsychosis.

Erziehungsroman, see BILDUNGSROMAN.

Esau und Jakob, see VON ESAU UND JAKOB.

Eselsfest, ecclesiastical custom, of medieval origin, in which an ass was led into church. Its chief associations were with the flight into Egypt and the entry into Jerusalem, hence its frequent connection with Palm Sunday. The custom died out at the beginning of the 19th c.

'Es fällt ein Stern herunter', first line of an untitled poem by H. Heine (q.v.) included in the *Buch der Lieder* (q.v.) as No. LIX of *Lyrisches Intermezzo.*

'Es fiel ein Reif in der Frühlingsnacht', first line of a poem by F. von Zuccalmaglio (q.v.), published in 1825 in the journal *Rheinische Flora* and republished in 1825 by Heine as the central section of the poem *Tragödie.* Heine asserted that it was a folk-song ('Dieses ist ein wirkliches Volkslied, welches ich am Rheine gehört'). The original folk-melody belonged to the folk-song 'Es fuhr ein Knecht'; settings have also been composed by R. Schumann and F. Mendelssohn (qq.v.).

'Es geht bei gedämpfter Trommel Klang', first line of the poem *Der Soldat* written in 1832 by A. von Chamisso (q.v.). It is spoken by a soldier in a firing party ordered to execute another, who is his closest friend. It is taken from a Danish poem by Hans Christian Andersen. The tune to which it is sung is by Silcher (q.v.).

'Es ist bestimmt in Gottes Rat', first line of the poem *Gottes Rat und Scheiden* by E. von

Feuchtersleben (q.v.), sung to a melody composed by F. Mendelssohn (q.v.) in 1839.

'Es ist ein Ros' entsprungen', first line of an ancient hymn, originally celebrating the Virgin Mary and later used as a Christmas hymn. Its well-known four-part setting is by Michael Prätorius (q.v.). It was first printed in 1599; its alternative form, 'Es ist ein Reis entsprungen', is first recorded in 1658.

'Es kann ja nicht immer so bleiben', first line of the poem *Ewiger Wechsel* (1802) by A. von Kotzebue (q.v.). The melody was composed by F. H. Himmel (1765–1814) in 1803.

Esra und Nehemia, title of a Middle High German biblical poem narrating the story of the Old Testament Books of Ezra and Nehemiah. The poem was written *c.* 1335 by a member of the Teutonic Order (see DEUTSCHER ORDEN). The warlike conditions of the story of Nehemiah offered a parallel to those facing the order.

'Es reden und träumen die Menschen viel', first line of Schiller's poem *Hoffnung,* first printed in *Die Horen* (q.v.) in 1797.

Essay, a term of French and English origin (Montaigne, 1580, Bacon, 1597), was first applied to German essays by Hermann Grimm (q.v., *Essays,* 4 vols., 1859–90), supplementing German terms such as 'Versuch', 'Entwurf', 'Fragment', 'Abhandlung', and 'Aufsatz', all of which had been employed since the mid-18th c. (e.g. by Winckelmann, Herder, Goethe, Schiller, F. Schlegel, Novalis, A. von Humboldt, qq.v.). 'Essayistik', the art of critical essay writing with its often frankly subjective bias, has been cultivated since Grimm by a great number of outstanding men of letters (e.g. P. Ernst, R. Kaßner, H. and Th. Mann, G. Lukács, R. A. Schröder, qq.v.).

ESSIG, HERMANN (Truchtelfingen, 1878–1918, Berlin-Lichterfelde), a pastor's son and a minor Expressionist dramatist with a leaning to social criticism conveyed by caricature. His principal plays are the tragedy *Mariä Heimsuchung* (1909), *Die Weiber von Weinsberg* (1909), *Die Glückskuh* (1910), *Der Frauenmut* (1912, all three comedies), *Ihr stilles Glück, Der Held vom Wald, Napoleons Aufstieg* (all 1912), *Überteufel* (a tragedy, also 1912), the comedy *Der Schweinepriester,* and *Des Kaisers Soldaten* (both 1915). He also wrote a novel, *Der Taifun* (posthumously published in 1919).

Esther, a Middle High German biblical poem written in the early 14th c. by a member of the Teutonic Order (see DEUTSCHER ORDEN).

Esto mihi, Quinquagesima, the Sunday before Lent. The designation is taken from the psalm for the day, 31, 'Sei mir ein starker Fels' (Luther). In the Vulgate the psalm is numbered 30 ('For thou art my rock', A.V.).

'Es war, als hätt' der Himmel', first line of Eichendorff's poem *Mondnacht* (q.v.).

'Es waren zwei Königskinder', first line of a folk-song deriving from the story of Hero and Leander. It was first printed in 1807, but is believed to have originated in the 16th c.

'Es weiß und rät es doch keiner', first line of the poem *Die Stille* by J. von Eichendorff (q.v.).

Etats Généraux, Die, a political ode in five unrhymed stanzas written in 1787 by F. G. Klopstock (q.v.) in anticipation of the convening of the French States General (see FRENCH REVOLUTION).

Ettal, Benedictine abbey in Alpine Bavaria near Oberammergau with a baroque church (St. Maria) erected 1710–26 and restored in its present form by J. Schmutzer after a fire in 1744. The Gothic church which it replaced was noteworthy in having been constructed in accordance with the description of the Grail temple in the medieval poem *Der jüngere Titurel* (q.v.). The abbey was founded in 1330, dissolved in 1803, and restored to the Order in 1900.

ETZEL, see ATTILA.

Etzel Andergast, a novel by J. Wassermann (q.v.), published in 1931. It is a sequel to *Der Fall Maurizius* (q.v.), in which Etzel is the son of the public prosecutor. In spite of the title, Etzel appears only in the second part of this later novel. Its principal figure is Joseph Kerkhoven. Engaged as physician to a wealthy man, Johann Irlen, who is suffering from an incurable disease, Kerkhoven undergoes a marked change of personality, which alienates him from his wife Marie. Years pass and Etzel Andergast, now a student, enters the couple's life and becomes Marie's lover. Kerkhoven, wishing to be reunited with his wife, discovers Etzel and turns him out. Etzel returns home to recover his mental equilibrium, and Kerkhoven's story is continued in *Joseph Kerkhovens dritte Existenz* (q.v.).

Etzels Hofhaltung, see WUNDERER, DER.

EUCKEN, RUDOLF (Aurich, 1846–1926, Jena), a German philosopher who was awarded the Nobel Prize for Literature in 1908. He wrote no works of literature. His philosophical treatises include *Die Einheit des Geisteslebens in Bewußtsein und Tat der Menschheit* (1888), *Der Kampf um einen geistigen Lebensinhalt* (1896), *Der Wahrheitsgehalt der Religion* (1901), and *Der Sinn und Wert des Lebens* (1908).

EUGEN, PRINZ VON SAVOYEN (Paris, 1663–1736, Vienna), joined the Austrian army in 1683 and immediately took part in the relief of Vienna from the Turkish siege and in the capture of Ofen (Buda). In 1687 he defeated the Turks at Mohács, and in a further campaign nine years later won a decisive victory at Zenta. In the War of the Spanish Succession he co-operated with Marlborough in the victory at Blenheim (Höchstädt) in 1704, carried out a successful campaign in North Italy (1705–6), and in 1708 and 1709 participated in the victories of Oudenarde and Malplaquet. He commanded in a further Turkish war from 1716 to 1718, gaining a victory at Peterwardein (q.v., 1716), and capturing Belgrade in 1719. The folk-song 'Prinz Eugen der edle Ritter' (q.v.) arose out of this exploit.

Prince Eugene became governor of the Austrian Netherlands in 1724 and proved himself a skilful and wise politician. Leibniz (q.v.) wrote his tract *La Monadologie* (1714) for his enlightenment. The Prince was a distinguished patron of the arts and poetry, and his handsome residence in Vienna, the Belvedere Palace, built by Lukas von Hildebrandt (q.v.) between 1713 and 1723, commemorates his taste. His town residence (Winterpalais des Prinzen Eugen), now housing the Ministry of Finance, was built by the same architect in collaboration with Fischer von Erlach (q.v.) at the same time as the Belvedere. Eugene's equestrian statue faces that of the Archduke Charles in the Heldenplatz of the Hofburg in Vienna.

Eugénie oder Die Bürgerzeit, a novel by H. Mann (q.v.), published in 1928. It is set in Lübeck not long after the War of 1870 (see DEUTSCH-FRANZÖSISCHER KRIEG) and has a well-to-do burgher, Jürgen West, and his French wife Gabriele as principal characters. Gabriele is unsettled and tempted to yield to the advances of the adventurer Pidohn, who also persuades Jürgen into reckless speculations. A playlet concerning the fallen Emperor Napoleon III and his consort Eugénie, whom Gabriele is said to resemble, is included in the action. Its 'author', Heines, is a portrait of the Lübeck poet E. Geibel (q.v.). Pidohn's risky operations result in his arrest. The West couple are unable to stave

off financial ruin, but Gabriele is saved from
elopement and they settle down, reconciled, to
live in relative contentment and obscurity.

EULENBERG, Herbert (Cologne, 1876–1949,
Kaiserswerth), abandoned a civil service career
to become a theatrical producer in Düsseldorf
(1906–9), and thereafter was a prolific, skilful
writer of second-rate plays with sensational
action directed against conventional society. His
best-known plays are *Anna Walewska* (1899),
Leidenschaft (1910, a military tragedy akin to
Hartleben's *Rosenmontag*, q.v.), *Ein halber Held*
(1903), *Der natürliche Vater* (1909), *Alles um Geld*
(1911), *Belinde* (1911), *Ernste Schwänke* (1913),
Zeitwende (1914), *Der Übergang* (1922), *Thomas
Münzer* (1932), and *Tilman Riemenschneider*
(1934). He is also the author of novels, stories,
and successful biographical sketches (*Schattenbilder*, 1910–11), as well as full biographies of
Heine (1947) and Freiligrath (1948). An edition of *Ausgewählte Werke* (5 vols.) appeared in
1925.

Eulenspiegel, see Till Eulenspiegel.

Euphorion, a character in Goethe's *Faust* (q.v.),
Pt. II, Act 3. He is the offspring of Faust and
Helena, and Goethe associated him with Byron
and Romanticism.

Euphrosyne, an elegy written by Goethe in the
winter of 1797–8, and first published in the
Musenalmanach für das Jahr 1799. It was composed in memory of the young Weimar actress
Christiane Becker (q.v.), who died at the age of
19 in October 1797. Though her death was not
unexpected, Goethe, who had helped to train
her talent, was deeply moved by it. The poem,
written in elegiac metre and, comprising 76
distichs, combines a lament for her death, the
news of which she brings under the name of
Euphrosyne, with an affirmation of art as the
true form of immortality—

Nur die Muse gewährt einiges Leben dem Tod.

EURINGER, Richard (Augsburg, 1891–1953,
Essen), an officer cadet (Fahnenjunker) in 1913,
served as a German army pilot in 1914–16 on
the western front, later with the Turks in Syria,
and finally as commander of the Flying School
at Lechfeld, Bavaria. He was unable to settle
down in the restless immediate post-war years,
and became an early member of the NSDAP
(q.v.). He wrote a number of novels, some based
on his war experiences, others reflecting his own
day: *Fliegerschule 4* (1929), *Vortrupp Pascha*
(1937), *Der Zug durch die Wüste* (1938), *Die
Arbeitslosen* (1930), and *Die Fürsten fallen* (1935).

Ludwigslegende (1934) deals with Ludwig II of
Bavaria (q.v.). Under National Socialism he
was municipal librarian in Essen and later
'Reichskultursenator'.

Europa, a periodical founded in Stuttgart in
1835 by A. Lewald (q.v.). G. Herwegh and F.
Dingelstedt (qq.v.) were contributors.

Europäisches Sklavenleben, a novel by
F. W. Hackländer (q.v.), published in 1854. It
deals with social questions, especially the economic dependence of women, in a structure recalling Gutzkow's novel *Die Ritter vom Geiste*
(q.v.).

Europamüden, Die, a novel by E. A. Willkomm (q.v.).

Eusebius, see Schumann, Robert.

Evangelienharmonie, a continuous account of
the life of Jesus on the basis of a combination of
the Gospels. The most important German
examples are the *Heliand* (q.v.) and Otfried's
Evangelienharmonie (see Otfried), both written in
the 9th c. Klopstock's *Der Messias* (q.v., 1748–
73) can also be regarded as a gospel harmony
with free additions; and a modern example is
Der Heiland (1930) by Paul Ernst (q.v.). All
these are in verse.

Evangelimann, Austrian term for a wandering
beggar who requited alms by reading extracts
from the Gospels. Such a figure is the principal
character in the opera *Der Evangelimann* (1895)
by Wilhelm Kienzl (1857–1941).

Evangelische Kirche, general term for the
Protestant Churches in Germany, whether
Lutheran, Reformed, or United. It was formerly
applied only to the United Church.

*Evchen Humbrecht oder Ihr Mütter,
merkt's euch!,* title of a stage adaptation by
H. L. Wagner (q.v.) of his tragedy *Die Kindermörderin* (q.v.).

Ewald Tragy, see Rilke, R. M.

EWERS, Hans Heinz (Düsseldorf, 1871–1943,
Berlin), founder and director of the Munich
cabaret Die elf Scharfrichter, was the author of
novels and stories, many of which deal in a
sensational manner with themes of perversion or
the occult. His novels include *Alraune* (1911),
Vampir (1921), *Reiter in deutscher Nacht* (1932),
and *Horst Wessel. Ein deutsches Schicksal* (1932).
Among the collections of stories are *Das Grauen*
(1908), *Die Besessenen* (1909), and *Nachtmahr*

(1922). Ewers was at first in favour with the Nazis, and was appointed to the new Dichter-Akademie in 1933. Later in that year he was denounced as a fabricator of 'Jewish' perverted literature, and his novel on Horst Wessel was withdrawn.

Ewige Braut, Die, refers to Katharina (Kathi) Fröhlich. See GRILLPARZER, F.

Ewiger Jude, the legendary figure of Ahasuerus, who, speaking roughly to Jesus, forbade him to rest by leaning against the house-wall, on the way to the Crucifixion, and was therefore condemned to wander restlessly about the world until the Second Coming. Hints of the story were current in the Middle Ages, but the legend first crystallized with the anonymous publication of a pamphlet, *Kurtze Beschreibung und Erzählung von einem Juden mit Namen Ahasverus* (Leyden, 1602), which reported his appearance in Hamburg in 1542. In the late 18th c. the legend came to be regarded as a symbol of the fate of the Jews. In 1774 Goethe wrote a fragment of vigorous narrative verse, facetiously headed *Des Ewigen Juden erster Fetzen*, and in 1783 C. D. Schubart (q.v.) composed a 'lyrische Rhapsodie' entitled *Der ewige Jude*, in which Ahasuerus is forgiven—'Gott zürnt nicht ewig'. In the 19th c. epic poems on the theme were written by J. Mosen (q.v., *Ahasver*, 1838) and by R. Hamerling (q.v., *Ahasuer in Rom*, 1866).

Ewiger Landfriede, a law passed at the Reichstag (Diet) at Worms on 7 August 1495, abolishing hitherto widely accepted rights of private vengeance (Fehde).

Exaudi, sixth Sunday after Easter (in the Anglican calendar the Sunday after Ascension). The name comes from the first words of the introit for that day ('Exaudi, Domine'—'Hear, O Lord', Psalm 27:7). In the numeration of the Vulgate it is Psalm 26.

Exil, a novel by L. Feuchtwanger (q.v.), published in Amsterdam in 1940. It is the third of the novels afterwards grouped together as *Der Wartesaal* (q.v.). It is concerned with the life of refugees, mainly Jewish, from National Socialist persecution who *c.* 1935 have found asylum in Paris. The National Socialist presence is felt through the German Embassy and the Party offices. The chief emigrant is a musician, Sepp Trautwein, who composes a 'Wartesaal-Symphonie', the waiting-room symbolizing the contingent nature of the refugees' lives.

Exilliteratur, a term for literature written by German authors while living in exile in countries in which they had sought political asylum

during the National Socialist regime. Its use is usually extended to 1949, the year of the foundation of the two new German states. The validity of a separate status for literature falling into this category has, however, been questioned. Notable examples of such writers are J. R. Becher, B. Brecht, H. Broch, F. T. Csokor, A. Döblin, L. Feuchtwanger, L. Frank, P. Gan, I. Goll, H. H. Jahnn, G. Kaiser, H. Mann, Th. Mann, J. Roth, E. Schaper, R. Schickele, A. Seghers, E. Toller, F. Werfel, C. Zuckmayer, A. Zweig, and S. Zweig (qq.v.). Since publication in Germany was not possible, publishers were chiefly found in Austria (until 1938), Holland (until 1940), Sweden, and Switzerland.

Existentialism in its various forms takes its origins from Kierkegaard (q.v.), whose works, subjected to violent attack in his lifetime, have in the 20th c. exercised a powerful influence which has been greatly increased by the effects of industrial development and of two world wars. The sense of threatened individuality, of dread (*Angst*), of solitude, and of tragedy springs from the conditions of life in this age, but a powerful formulation of them lay ready to hand in Kierkegaard's writings. It is only in France that there has arisen, around Sartre, a movement which can properly be called *existentialisme*. Germany has existential philosophers, some, like Jaspers (q.v.), willingly so called, others, like Heidegger (q.v.), rejecting the description. The preferred locution is *Existenzphilosophie*. Almost all writers of the 20th c. reflect existential problems. A symptom of the consciousness of the problem is the widespread use by critics of the neologism *existentiell*.

Expressionismus, a movement in art and literature in the 20th c. It first becomes prominent in painting and sculpture with the short-lived groups Der blaue Reiter and Die Brücke (qq.v.). Its principal exponents were the painters M. Beckmann, E. Heckel, E. L. Kirchner, O. Kokoschka, Paula Modersohn-Becker, E. Nolde, K. Schmidt-Rottluff, and W. Kandinsky (qq.v.). F. Marc (q.v.) is sometimes grouped with the Expressionists, though his mature style is closer to Cubism. E. Barlach (q.v.) was its most prominent sculptor. Its first exponent in music was A. Schönberg (q.v.), in film R. Wiene (1881–1938) with *Das Cabinett des Dr. Caligari* (1919). Its main journals were *Der Sturm* and *Die Aktion* (qq.v.), edited by H. Walden and F. Pfemfert (qq.v.) respectively. An extensive bibliography of journals (1910–25) and their contributors, *Index Expressionismus* (18 vols.), ed. P. Raabe, appeared in 1972.

Expressionism in literature manifests itself

about 1910, though signs of it are perceptible in certain plays of Wedekind (q.v.). The movement has some analogies to the Sturm und Drang (q.v.) and was accompanied by a renewed interest in the works of J. M. R. Lenz and G. Büchner (qq.v.). Expressionism reacted against Naturalism and the effeteness of neo-Romanticism (see NEUROMANTIK and DEKADENZ). Its aim is to show the truth within, not in psychological analysis, but by proclaiming man's aspirations, hopes, and fears. Its style, both in poetry and in drama, is exclamatory and elliptical (frequent omission of verbs and articles), and has been called 'Telegrammstil' (as distinct from the Naturalistic 'Sekundenstil'). Its tone is by turns ecstatic, minatory, satirical, grotesque, and also noted for the 'scream' ('Schreidrama'), in reaction to which G. Kaiser, in his tract *Vision und Figur* (1918) stressed the need for an articulate presentation of the 'vision of the New Man'. Expressionism developed the form of 'Stationendrama' (q.v.) and favoured characters that are indicated by a role, such as 'der Vater', 'der Ingenieur', etc. The Expressionists were radicals, concerned less with party politics than with a fundamental change of attitude, often expressed in terms of the regeneration of Man. In this respect they owe something to Nietzsche (q.v.). Among the variety of themes the conflict of generations is prominent and is at the centre of R. Sorge's *Der Bettler* (q.v.) of 1912, though O. Kokoschka's play on the conflict of the sexes, *Mörder, Hoffnung der Frauen* (1910) is regarded as the first Expressionist play.

The principal dramatists of Expressionism, many of whom are antagonistic to material progress and the rise of the industrial and technological society, are E. Barlach, B. Brecht (in his early work), R. Goering, W. Hasenclever, H. Johst, G. Kaiser, G. Sack, C. Sternheim, and F. von Unruh (qq.v.). Among the poets are G. Benn, J. R. Becher, G. Heym, Else Lasker-Schüler, E. Stadler, A. Stramm, F. Schnack, G. Trakl, and F. Werfel (qq.v.). Several of these later turned away from the Expressionistic style. Fiction commended itself less to Expressionist writers, but some novels of A. Döblin

and early stories by Kafka (qq.v.) have a kinship with the Expressionist outlook. As a movement Expressionism rose to its height in the 1914–18 War and died away in the early 1920s, but traces of its style persist.

Exsurge Domine, first words of the Bull, dated 15 June 1520, in which Pope Leo X threatened Luther (q.v.) with excommunication.

EYTH, MAX (Kirchheim, Teck, 1836–1906, Ulm), an engineer and inventor, worked for Fowler's of Leeds, steam plough manufacturers, from 1861 to 1882, travelling in Europe and America after a year in Leeds. In 1884 he founded the Deutsche Landwirtschaftsgesellschaft. He wrote the historical novels *Der Kampf um die Cheopspyramide* (2 vols., 1902) and *Der Schneider von Ulm* (2 vols., 1906), but is chiefly known for his autobiography *Hinter Pflug und Schraubstock* (2 vols., 1899).

Ezzolied, also called *Ezzos Gesang,* a choral hymn by Ezzo, a canon of Bamberg, probably composed in 1063 for the consecration of the reorganized monastery of St. Gangolph at Bamberg. An unusual feature in a medieval poem is that both the poet (Ezzo) and the composer of the music (Wille) are separately mentioned in an introductory strophe to one of the MSS.

The *Ezzolied* is a confident and deeply felt account of the history of the world from the Creation through the Fall of Man to the redemption wrought by Christ, which is foretold by the prophets and prefigured in various episodes of the Old Testament. Christ is present from the beginning as the Word which is in the beginning. The poem closes with a lyrical apostrophe of the Cross, which betokens the certainty of salvation. It was apparently sung on a pilgrimage to Palestine headed by Bishop Gunther of Bamberg 1064–5.

The hymn exists in two forms. The better text, in a MS. preserved at Strasburg, is incomplete. The complete version in the Vorauer Handschrift (q.v.) is inferior, the poem having been expanded without regard to the poetic qualities of the original.

F

Fabeln und Erzählungen, a collection of verse fables and tales published by C. F. Gellert (q.v.) in 1746. Additions were made in 1748. Some fables had been previously published in Schwabe's *Belustigungen des Verstandes und Witzes* (q.v., 1741–5). The fables are preceded by an essay, *Nachricht und Exempel von alten deutschen Fabeln,* in which Gellert reviews medieval and later German fables, mentioning those of Boner, as well as those of HUGO von Trimberg and Burkard Waldis (qq.v.). Gellert's own fables have a smooth, easy, conversational style and a clearly pointed moral.

Fabian. Die Geschichte eines Moralisten, a novel by E. Kästner (q.v.), published in 1931. It is a bitter yet entertaining satire on the life of a great city, here Berlin. Fabian, a well-meaning idealist, is overwhelmed by the profligacy, egoism, and Schadenfreude he encounters. He returns home where he, a non-swimmer, ironically perishes by jumping into the river to rescue someone, who, it turns out, is able to swim to the bank unaided.

FABRICIUS, JOHANNES (dates unknown, a native of the Duchy of Cleves), studied at Cologne, and in 1546 published a Latin epic on the Anabaptist troubles in Münster (q.v.) in 1546, drawing on first-hand knowledge (*Motus monasteriensis libri decem*).

Facetie, see FAZETIE.

Fackel, Die, a Viennese satirical journal published and largely written by Karl Kraus (q.v.). *Die Fackel* first appeared in 1899, coming out three times a month. Up to 1911 it included contributions by a great variety of authors, notably Altenberg, Bleibtreu, Strindberg (also 1912), Wedekind, Franz Werfel (qq.v.), and Oscar Wilde; its subsequent publication was irregular, and Kraus himself was sole author. It ceased publication in February 1936. In all, 922 numbers appeared. *Die Fackel* was well known, if not notorious, for its lively presentation, sharp wit, and remorseless satire. The complete run of *Die Fackel* was reprinted in 39 vols., ed. H. Fischer, 1968–73.

Fackelsonntag, see FUNKENSONNTAG.

FAESI, ROBERT (Zurich, 1883–1972, Zollikon, nr. Zurich), professor of German literature at Zurich University from 1922 to 1953, is the author of books on C. F. Meyer, Spitteler, G. Keller, and Th. Mann (qq.v.); he also wrote novels (*Die Stadt der Väter,* 1941; *Die Stadt der Freiheit,* 1944; *Die Stadt des Friedens,* 1952; *Alles Korn meinet Weizen,* 1961), poetry (collected as *Die Gedichte,* 1955), and plays.

Fähnlein der sieben Aufrechten, Das, a Novelle by G. Keller (q.v.), first published in *Auerbachs Volkskalender* in 1860 and contained in *Züricher Novellen* (q.v.). The 'Seven Upright Men' are radicals, who have continued their friendship after their political aims have been attained. Prominent among them are the master tailor Hediger and the master carpenter Frymann. Hediger's son Karl and Frymann's daughter Hermine are in love, but both fathers oppose the match because they believe that financial disparity (Frymann is the richer) will ruin their long-standing friendship. Frymann decides to marry Hermine to Ruckstuhl, a man of means, who is in reality a boor and a reckless speculator. Karl and Ruckstuhl are militiamen and have to spend a few weeks in barracks. At Hermine's suggestion Karl takes Ruckstuhl in hand. He and his comrades make Ruckstuhl so drunk that he insults an officer and is confined to barracks. As a result he cannot appear at a family meal with Frymann, who does not forgive this discourtesy. Yet Karl is still not regarded as an acceptable suitor for Hermine until he is given a chance to prove his worth at the national shooting festival of 1849 (Schützenfest). The 'Seven Upright Men' attend with their banner but without a speech which one of them is expected to deliver. To his dismay, Frymann is chosen by lot. Karl now offers himself as their standard-bearer and therefore as their spokesman. He acquits himself well and is much applauded. Frymann is still not fully reconciled, for he thinks that the gift of the gab is no criterion of character. However, Karl performs with equal distinction in the shooting match. Objections are now withdrawn, and Karl and Hermine are united.

Fahrende, fahrende Leute or fahrendes Volk, strolling jugglers, contortionists, etc., and, later, minstrels in the Middle Ages, who formed a distinct social group with minimal legal rights. They were also ecclesiastically under-privileged, being excluded from Communion. They provided an important element of popular entertainment. The minstrels (fahrende Sänger or Spielleute) were formerly thought to be the custodians of the tradition of the popular epic (see SPIELMANNSDICHTUNG), but this view has been largely abandoned.

FALK, ADALBERT (Metschkau, 1827–1900, Hamm), a lawyer by profession, was elected a member of the Prussian Parliament in 1858. In 1872 he was appointed Kultusminister, with responsibility for educational and religious matters. He was immediately involved, as Bismarck's agent, in the clash between the Prussian state and the Roman Catholic Church known as the Kulturkampf (q.v.). Falk introduced the May Laws and secured their passage, but when Bismarck (q.v.) abandoned the struggle he dropped Falk, attributing, with some ingratitude, the failure of the policy to this loyal supporter. Falk resigned in 1879, having achieved educational reforms in Prussia which can be set against the miscarriage of the anti-Catholic policy. These are the law concerning the inspection and supervision of schools and the introduction of Simultanschulen, i.e. schools to which all confessions were admitted.

FALK, JOHANNES DANIEL (Danzig, 1768–1826, Weimar), came to Weimar after completing his studies, and made close contact with Wieland (q.v.). From 1797 to 1806 he published *Das Taschenbuch für Freunde des Scherzes und der Satire*. He was honoured in 1806 with the title Legationsrat, and in 1813 he founded an orphanage at Weimar (das Johanneum). His satires (*Satiren*) appeared in 1800, followed by two plays (*Prometheus*, 1803; *Amphitryon*, 1804) and an autobiography (*Leben, wunderbare Reisen und Irrfahrten des Johannes von der Ostsee*, 1805). In 1832 he published recollections of Goethe (*Goethe, aus näherem persönlichem Umgange dargestellt*). He is now best known as the author of the first verse of the Christmas carol 'O du fröhliche, O du selige, gnadenbringende Weihnachtszeit!' (1816). He used on occasion the pseudonym Johannes von der Ostsee.

Falke, see NOVELLE.

FALKE, GUSTAV (Lübeck, 1853–1916, Groß-Borstel, Hamburg), was a bookseller and later a music teacher in Hamburg. His literary gifts were discovered by D. von Liliencron (q.v.). In 1903 he was granted a pension by the Hamburg Senate, and he devoted himself thereafter entirely to literature. He wrote smooth and simple verse of folk-song or Romantic character, some of it in dialect (*Mynheer der Tod und andere Gedichte*, 1892; *Tanz und Andacht*, 1893; *Zwischen zwei Nächten*, 1894; *Neue Fahrt*, 1897; *Mit dem Leben*, 1899; *Hohe Sommertage*, 1902; *En Handvull*

Appeln, 1906; *Frohe Fracht*, 1907). He is the author of novels and stories, principally of local interest (*Aus dem Durchschnitt*, 1892; *Landen und Stranden*, 2 vols., 1895; *Der Mann im Nebel*, 1899; *Die Kinder aus Ohlsens Gang*, 1908). A collection of Novellen, *Geelgösch*, as well as the separate story *Der Spanier*, appeared in 1910. *Die Stadt mit den goldnen Türmen* (1912) is his autobiography. His collected works, *Gesammelte Dichtungen*, were published in 5 vols. in 1912.

FALKE, KONRAD, pseudonym of Karl Frey (Aarau, 1880–1942, Eustis, Florida, U.S.A.), is the author of verse plays (*Francesca da Rimini*, 1904; *Die ewige Tragödie*, 1909; *Caesar Imperator*, 1911; *Astorre*, 1912), of novels (notably *Der Kinderkreuzzug*, 2 vols., 1924), and Novellen, as well as some poetry. He made a translation of Dante's *Divina Commedia* (*Die göttliche Komödie*, 1921). He was influenced by C. F. Meyer (q.v.). From 1910 he edited *Raschers Jahrbücher*, and from 1937, with Th. Mann (q.v.), the monthly *Maß und Wert*.

FALLADA, HANS, pseudonym of Rudolf Ditzen (Greifswald, 1893–1947, Berlin) who, after trying various careers, became a novelist. He began with *Der junge Goldeschall*, but his first success came with *Bauern, Bonzen und Bomben* (1931), a novel of social criticism, and he achieved world-wide success with his portrayal of 'Lämmchen', the young working-class wife of Pinneberg, the eponymous 'little man' in the novel *Kleiner Mann—was nun?* (1932, translated into many languages and filmed), which has as its background the social evils of the inflation of the 1920s, which are also the subject of his extensive novel *Wolf unter Wölfen* (1937). *Wer einmal aus dem Blechnapf frißt* (1934) deals with criminal psychology and the problem of recidivism. Some of his post-war novels have the Resistance Movement (q.v.) as their subject (*Jeder stirbt für sich allein*, 1947). *Damals bei uns daheim* (1941) and *Heute bei uns zu Haus* (1943) are autobiographical works. Numerous titles have repeatedly been reprinted since the 1950s; *Gesammelte Erzählungen* appeared in 1967.

Fall Maurizius, Der, a novel by J. Wassermann (q.v.), published in 1928. It is a story of wrongful condemnation for murder, and of its consequences. Leonhart Maurizius, an art historian, is found guilty of murdering his wife and condemned to death, but the sentence is commuted to life imprisonment. The public prosecutor in the case, Freiherr von Andergast, has a son Etzel, who, at the instigation of the condemned man's father, devotes himself to ascertaining the truth. He succeeds, after considerable research, in proving that Leonhart's sister-in-law, Anna Jahn, was the guilty person. Leonhart, who has served eighteen years, is pardoned. The prosecutor resigns and loses his reason. Leonhart can make nothing of his newly granted freedom and takes his own life. See also the sequel, *Etzel Andergast*.

Falschen Spieler, Die, a comedy (Lustspiel) by F. M. Klinger (q.v.). Herr von Stahl's prodigal son, known as Marquis Bellfontaine, is reclaimed from his life of dissipation by the love of Juliette, a distant cousin.

Falsche Woldemar, Der, a patriotic novel ('vaterländischer Roman') by W. Alexis (q.v.), published in 1842. Its subject is the appearance in 1347 of a pretender, claiming to be Woldemar (q.v.), Margrave of Brandenburg, who died in 1319. Alexis's Woldemar is a better man than any of the Bavarian Wittelsbachs who have become margraves in his place. After defeat in battle he resigns his claim to the margravate and retires to Dessau.

Falun, see BERGWERKE ZU FALUN, DIE.

Familiengemälde, term applied in the last thirty years of the 18th c. to realistic plays in prose dealing with social themes, ending happily, and set in middle-class families. J. M. R. Lenz (q.v.) applied it to his play *Die beiden Alten*, and A. W. Iffland (q.v.) was also partial to its use.

Familie Schroffenstein, Die, a blank-verse tragedy in five acts by H. von Kleist (q.v.). The original title was *Die Familie Ghonorez*, and this grew out of a scenario entitled *Die Familie Thierrez*. The setting was at first in Spain, but Kleist transposed it to Swabia for the anonymous publication of the play in 1803. Relentless family strife between Counts Rupert and Sylvester of Schroffenstein ends in the death of their children Ottokar and Agnes, who have fallen in love with each other. Distrust and hatred, instigated by Rupert, who blindly makes Sylvester responsible for the death of his son Peter, cause the fathers to kill the lovers, each aiming at the other's child while in fact killing his own, unrecognized because of disguise; for Ottokar has exchanged clothes with Agnes in an attempt to save her. The blind grandfather, Sylvius, discovers the true identity of the bodies. This, Kleist's first tragedy, owes much to Shakespeare, Lessing, and Schiller, though it establishes in many respects the pattern of Kleist's own conception of tragedy.

Familie Selicke, Die, a three-act play by A.

Holz and J. Schlaf (qq.v.), first performed in April 1890 by the Freie Bühne, Berlin, and published in the same year. It was reissued in *Neue Gleise* (q.v., 1892). Set in a Berlin tenement, it shows in Naturalistic speech the drab and cowed life of the Selicke family dominated by fear of the drunken father. The youngest child dies in the course of the play. Toni, the eldest, ardently wishes to escape from her oppressive and brutalizing environment, and has the chance in the offer of Wendt, their lodger, a young theology student; but the ties of her surroundings are too strong. The character of 'olle Kopelke' from *Die papierne Passion* (q.v.) reappears in this play, which, at the time of its production, was regarded as a landmark in the new drama.

Fanny, name by which Klopstock referred in his odes to Marie Sophie Schmidt (q.v.), with whom he was in love.

Fantasiestücke in Callots Manier, title used by E. T. A. Hoffmann (q.v.) for collections of his stories published in four volumes in 1814 and 1815. Their more important contents included *Ritter Gluck, Kreisleriana, Don Juan, Nachricht von den neuesten Schicksalen des Hundes Berganza, Der Magnetiseur*, and *Der goldne Topf* (qq.v.). In a short preface (*Jacques Callot*) Hoffmann expresses his admiration for the personality and work of the French graphic artist Callot (1592–1635).

Fasching, a Novelle by G. Hauptmann (q.v.), written and published in 1887. The story centres on the sailmaker Kielblock, who is so happily married to a young and robust wife that his inexhaustible gaiety becomes the envy of others. When he spends too much money he consoles himself with the thought of savings which his old mother accumulates in a green box, which one day will belong to him. At the annual village 'Fasching' he earns great admiration as a merry-maker by wearing a mask that makes him look like a dead man. But beneath his mask, his gaiety, and his greed, his soul is empty. The night after the ball he tries to cross the frozen lake at night, relying on the light of the moon and on his mother's lamp in his distant home. As both lights are extinguished, nature's forces direct him and his sledge, carrying his wife and child, towards the spot where the water is not frozen. The sledge vanishes with its occupants beneath the surface, and Kielblock experiences the true face of death before he sinks.

Fasching, see FASTNACHT.

Fastnacht or Faßnacht, Shrove Tuesday and, by extension, Shrovetide, traditionally celebrated by carousings, processions, masquerades, and, more recently, masked balls. Dramatic performances (see FASTNACHTSPIEL) were a feature of the period in the 16th c. and 17th c. The term 'Karneval' is applied in the Rhineland to Shrovetide, and 'Fasching' in Bavaria and Austria.

Fastnachtspiel, a play of popular character originally performed during Shrovetide. (See FASTNACHT.)

The Fastnachtspiel was characteristic of towns, and was performed by citizens. Its themes were primarily social, especially marriage (normally unhappy, with the woman at fault), and the misdeeds of soldiers or peasants; and its form is frequently that of a court hearing. Religious and political subjects occur, particularly after the Reformation. Some are simple comedies, dramatizations of Schwänke (see SCHWANK). Their typical medium is the Knittelverse (q.v.). They are frequently short, sometimes no more than 300 lines, but long examples exist, such as *Vom verlorenen Sohn* by Waldis (q.v.) and *Das Weinspiel* of H. R. Manuel (q.v.). Almost invariably their moral is clearly summed up in a final tag or speech. Most of the surviving Fastnachtspiele were written in Nürnberg but they were popular in most of Germany and Switzerland.

The origins of the Fastnachtspiel are to be found in the 15th c. in Shrovetide masquerades. The earliest-known writer of these plays is Hans Rosenplüt (q.v.), *c.* 1450. The heyday of the Fastnachtspiel is the 16th c. with Hans Folz and above all Hans Sachs (qq.v.); Niklaus and Hans Rudolf Manuel (qq.v.) were also prominent, *Der Ablaßkrämer* by N. Manuel being a particularly outstanding example. By the beginning of the 17th c. the plays appeared old-fashioned and lost their popularity. The last prolific writer was Jakob Ayrer (q.v.), who wrote thirty-six Fastnachtspiele. In the late 18th c. the form was revived as a literary (non-theatrical) drama by Goethe (see EIN FASTNACHTSPIEL VOM PATER BREY) and A. W. Schlegel (q.v.).

Fate Tragedy, see SCHICKSALSTRAGÖDIE.

Faust, a verse tragedy by Johann Wolfgang von Goethe (q.v.), published in two parts, *Der Tragödie erster Teil* (1808) and *Der Tragödie zweiter Teil* (1832).

The dedicatory ode (*Zueignung*, 1797) prefixed to Part One laments the passing of the years and the death of friends, and expresses Goethe's dedication to the work. The *Vorspiel auf dem Theater* presents the conflicting interests and arguments of the three theatrical powers, the Director, Poet, and Actor (*Lustige Person*). The business proper of *Faust* begins with the *Prolog im Himmel*. Here God, surrounded by his Archangels, admits the visit of Mephistopheles. Faust is mentioned, and Mephistopheles seeks and obtains permission to tempt him, while the Lord affirms confidence in Faust's steadfastness. Part One, which incorporates almost all the scenes in *Urfaust* (q.v., and see below), begins (in *Nacht*) with Faust's disgust with the unreality of accepted academic knowledge, his thirst for real, penetrating knowledge, and his summoning of the Earth Spirit (Erdgeist, q.v.). But the Erdgeist overwhelms him with its dynamism and stature. Faust is next contrasted with the pedant Wagner. Recollecting his failure in the presence of the Erdgeist, Faust is about to take poison, but is checked at the critical moment by the sound of Easter bells. *Vor dem Tor* presents a kaleidoscope of city life released into the spring landscape of Easter Day; Faust, seen in a state of increasing self-dissatisfaction, finds himself followed by a poodle, who in the next scene (*Studierzimmer* 1) proves to be Mephistopheles. A pact is suggested, but not yet drawn up. In *Studierzimmer* 2 Faust proposes the terms of a pact (variously referred to as 'Pakt', 'Wette', and 'Bündnis'). As long as he continues to strive, Mephisto shall serve him. If he ever subsides into satisfied indolence, Mephisto shall claim him ('Werd' ich beruhigt je mich auf ein Faulbett legen,/ So sei es gleich um mich getan!'). While Faust withdraws in order to prepare for the journey which the pair are to undertake, Mephisto dons Faust's gown, and gives cynical advice to a first-year student. Faust is thereupon rejuvenated for his new life by a witch's potion (*Hexenküche*), and begins his peregrination by watching Mephisto perform some hocus-pocus upon the drunken student clientele of *Auerbachs Keller*. Faust then encounters Margarete (Gretchen) and is seized by lustful desire, which is chastened by the simple integrity of her character. For a time he withdraws from her to the contemplation of nature in the wilderness (*Wald und Höhle*), but the pull of sex is stronger; he seduces her, persuading her at the same time to give her mother a sleeping potion, which proves fatal. Margarete, becoming pregnant, is distraught; her anguish is augmented when her brother, Valentin, is killed by Faust in a street affray. Mephisto transports Faust to the witches' sabbath (*Walpurgisnacht*), seeking to drown her memory of Gretchen in crude and earthy sensuality, but her image appears to Faust's eye and draws him back to her. He finds her in prison, condemned to death for infanticide, and seeks to effect her escape by Mephisto's magic, yet she, though the agony is upon her, refuses this evil opportunity, and maintains her integrity in death, invoking the comment of a voice from above 'Ist gerettet'. Faust is dragged from the scene by Mephisto.

The first part has been concerned with the so-called little world (kleine Welt), the world of private emotion. The scene of the second is the great world (große Welt), the field of public affairs, of politics, economics, and education represented by art. Act I sees Faust's recovery through the healing powers of nature from the shock of Gretchen's end (Anmutige Gegend), and presents the Emperor's court in satirical terms. The pleasure-loving Emperor demands money and more money, and the imperial coffers are empty. Faust, with Mephisto, solves his dilemma by the creation of paper currency. The Emperor, far from profiting by his new solvency, demands fresh extravagant entertainment, and Faust is persuaded to conjure up for him the shades of Paris and Helen. He succeeds in doing so, in spite of great difficulties and hazards; he summons the goddesses (see MÜTTER, DIE), but, when the figures appear, he is so taken with Helena's beauty that he attempts to seize her; the shades dissolve, an explosion shakes the hall, and Faust falls senseless to the ground.

Shattered for the second time in the play, Faust is borne back to his old study. While he lies insensible, Mephisto has a short encounter with the Student of the First Part, now grown into a brash, conceited bachelor of arts. Wagner, too, is present, a respected and competent scientist, though his vision still extends no further than the laboratory walls. He is engaged on making a synthetic man, and succeeds in creating a tiny creature (Homunculus), radiating light, and, since he has no true form, unable to exist outside the test-tube in which he is distilled. But Homunculus, though not embodied, has a penetrating intellect and wide knowledge; and he is able to indicate that Faust, if he is to recover, must be taken to the classical witches' sabbath (Klassische Walpurgisnacht, an invention of Goethe's). In the half-darkness of the moonlit southern night are assembled the grotesque, distorted, or horrifying figures of antiquity, griffons, sphinxes, sirens, lamias, etc., as well as the erotic nymphs. Faust begins a healing transition to the classical at the point at which it is nearest to Nordic myth. He meets Chiron, mounts him, and is carried off in search of Helena. The thread of Goethe's scientific interests is picked up in a cosmogonical argument between the believer in primal fire, Anaxagoras, and Thales the protagonist of a watery origin. The scene ends in the triumph of water, the pageant of Neptune, enshrining Galatea, to whom Homunculus is drawn so violently that his enveloping test-tube breaks, and he dissolves

in radiance upon the waters. It is a cryptic genetic poem of rich imagination, to the meaning of which Goethe has given no later extraneous hint.

Act III (Vor dem Palaste des Menelas zu Sparta) begins as a classical tragedy. Troy has fallen, Helen is returned to Menelas, who sends her with her handmaidens back to Sparta. They learn that it is his intention to execute them all. Phorkyas (Mephisto disguised as an old woman) offers them rescue. They are transported to a castle (Innerer Burghof) and welcomed by its lord, who is Faust. The advancing forces of Menelas are repulsed, Faust and Helena are united, and a son is born to them, named Euphorion. But Euphorion, in his irrepressible rashness, falls to his death, showing a momentary likeness to Byron as he does so. The catastrophe drives Helena and Faust apart.

The fourth act returns to the world of politics. The Emperor, ruling as badly as ever, is opposed by a rival emperor (Gegenkaiser). In this situation Faust comes to his rescue with magic, and demands for his reward a fief of land at present covered by sea.

When the fifth act begins, the aged Faust has satisfied a dream of activity and economic progress. He has reclaimed the land from the sea, peopled it, and given it prosperity. But his pleasure and pride are not complete. A freehold enclave held by an old couple, Philemon and Baucis (q.v.), disturbs the unity of his estate. He asks Mephisto to remove them, and the consequence is the burning-down of their house and their murder. Thereupon Faust is visited by Care (Sorge), who blinds him. He orders Mephisto to press on with work on the dam which excludes the sea, but cannot in his blindness perceive that the spades and shovels are in reality digging his grave. Delighted with the growth of his project, Faust, now one hundred years old, speaks a phrase of satisfaction,

Im Vorgefühl von solchem hohen Glück
Genieß' ich jetzt den höchsten Augenblick,

and falls back dead. Thereupon Mephisto steps in, claiming his own. Heavenly spirits, however, intervene, drive off Mephisto, and reclaim Faust. In the final mystical scene, Bergschluchten, Faust's soul is conveyed in a progress towards Heaven, amidst the intercessions of Gretchen (Una poenitentium), and the play closes with a mystic chorus.

The subject of Faust occupied Goethe's mind at intervals throughout almost the whole of his life. As a boy he enjoyed the puppet play (Puppenspiel von Dr. Faust) and read one of the later forms of the Faust Volksbuch. By 1773 he had begun to write a play on the theme, which was at first probably concentrated mainly on Faust's disgust with academic learning and on the shallowness of university life. It developed into the love tragedy in which Faust, with Mephisto's aid, gains the heart of a good and virtuous girl (Gretchen), and forsakes her, driving her to an infanticide for which she is condemned to death. In this form the play exists in a copy made in Weimar about 1776 by Luise von Göchhausen (q.v.); the copy was discovered among her papers in 1887, and published by Erich Schmidt (q.v.) as *Goethes Faust in ursprünglicher Gestalt*. The *Urfaust*, as this early version is generally known, contains Faust's opening monologue, the summoning of the Earth Spirit, Faust's interview with his *famulus* Wagner, Mephisto's mocking scene with the first-year student, the wine magic of Auerbach's cellar, and then the long series of short scenes which make up Gretchen's tragedy. Most of the play is in the irregular 16th-c. metre of Knittelverse (q.v.), but *Auerbachs Keller* and the final scene, *Kerker*, are in prose.

Goethe felt unable to advance the play further while in Weimar, but some years later he decided to incorporate it as a fragment in an edition of his works (*Schriften*, 1787–90). While in Italy and after his return he added the scenes *Hexenküche* and *Wald und Höhle*, turned the prose of *Auerbachs Keller* into verse, and cut off the last scenes so that the play breaks off at the end of *Dom*. He published it as *Faust, ein Fragment* in vol. 7 (1790) of *Schriften*.

It was at Schiller's instigation that Goethe began in 1797 to work again at *Faust*, and the idea soon arose of treating the story in two parts. In the next ten years he added the scene *Studierzimmer*, in which Faust enters into the pact with Mephisto, and the scenes *Vor dem Tor* and *Walpurgisnacht*, and published the work, together with the dedicatory ode, the prologue on the stage, and the prologue in Heaven, as Pt. 1 in 1808.

A section of the third act of Pt. 2 was written in 1800, and it is likely that certain other passages were written before Pt. 1 was completed. The main work on the second part fell in the years from 1826 on. Act III (Helena and Faust) was published in 1827 as *Helena. Klassischromantische Phantasmagorie. Zwischenspiel zu Faust*. Act I appeared in 1828, Act II (*Klassische Walpurgisnacht*) in 1830. In the same year Act V was completed, and in 1831 Act IV. The whole vast Second Part was published shortly after Goethe's death as vol. 1 (1832) of *Nachgelassene Werke*, supplemental to the *Ausgabe letzter Hand*.

Performances of parts of *Faust I* took place privately in Berlin in 1819, and publicly in Breslau in 1820. The first complete performance was at Brunswick on 19 January 1829 under

E. A. F. Klingemann (q.v.). The first performance in Weimar was given on 29 August 1829. *Faust II* was first given complete at Hamburg in 1854, though Act III had been performed at Weimar in the centennial celebrations of 1849. The first performance of the complete work (I and II) was given at Weimar in 1875.

The interpretations of *Faust* are innumerable, but it can at least be said that the play symbolically embraces human life, commenting with irony on human, social, and political phenomena, and insisting on endeavour, striving, and unremitting activity as the fundamental human virtue.

Faust is notable also for the range of verse forms in which it is written. *Faust I* contains, in addition to Knittelverse, blank verse, hymnic passages, and some strophic songs. *Faust II* includes *ottava rima*, *terza rima*, trimeters, and various rhyming measures, as well as the forms contained in Pt. 1.

Of English translations those of A. G. Latham (1902–5), Bayard Taylor (1908), L. MacNeice (1951), and Barker Fairley (1970) deserve mention.

FAUST, GEORG (probably Knittlingen nr. Bretten, *c.* 1480–1540 or 1541, Staufen, Breisgau), an obscure figure, half scholar, half quack, about whose exploits the myth of the Faust legend gathered. He visited various universities, was believed to have been at Heidelberg, and at Erfurt in 1513 or 1520, at Wittenberg in 1527, and at Ingolstadt in the following year. He seems to have made himself everywhere unpopular, and was more than once expelled. Faust's later life was spent in more favourable circumstances in the lower Rhenish region and Westphalia. The legends of his magic production of wine (later connected with Auerbachs Keller, q.v.), and of his evocation of Homeric figures became current during his lifetime.

Faustbuch, Spieß'sches is the best-known early literary version of the Faust legend, and was published by the Frankfurt printer Johann Spies or Spieß (q.v.), from whom the customary designation derives, in 1587. The usual form of the title is *Historia von D. Johann Fausten*, though in the original it is much lengthier, filling the title-page. Faust, in this popular Volksbuch, turns from theology to sorcery, makes a pact with the devil for a term of twenty-four years, lives richly and riotously, and at the expiration of the term is carried off to Hell. The university at which this Faust is active is Wittenberg. It is a book of strict moral intention, holding up Faust's intellectual curiosity and its consequences as a dreadful warning to others. The *Spieß'sches Faustbuch* is the source of Marlowe's *Doctor Faustus* (1589).

Faust. Der Tragödie dritter Theil in drei Acten. Treu im Geiste des zweiten Theils des göthe'schen Faust gedichtet von Deutobold Symbolizetti Alegoriowitsch Mystifizinsky, a parody of Goethe's *Faust* (q.v.). The absurd pseudonym conceals the authorship of F. Th. Vischer (q.v.). The principal characters in this mocking confusion of motifs, quotations, and misquotations from *Faust* are Faust himself, Valentin, Mephisto, Helena, and Lieschen (the girl who, in the original, gossips with Gretchen in the scene *Am Brunnen*). Vischer published the work in 1862 (revised 1886); an edition by H. Falkenheim appeared in 1921 (reprinted 1963 and 1969).

Faust, ein Fragment, Goethe's first published version of *Faust* (q.v.). It appeared in vol. 7 of *Goethes Schriften* (1790).

Faust. Ein Gedicht, a poetic work by N. Lenau (q.v.), first published in 1836 and then, in revised form, in 1840. Written in rhyming iambic verse (*c.* 3,400 lines), it is mainly in dramatic form, though there are passages of verse narrative and comment. Conceived under the shadow of Goethe's *Faust* (*Faust. II. Teil,* 1832), it is perceptibly derivative. There are no act divisions, but the scenes have titles, e.g. *Der Morgengang, Der Besuch, Die Verschreibung.* Faust enters into a pact with Mephistopheles, seduces a village girl (who later reappears as a worn-out beggar-woman), makes advances to a princess, whose portrait he is painting, and murders her betrothed. He repents, seeing himself at one with God, denounces Mephistopheles, and then stabs himself.

Fausts Höllenzwang, title of numerous handbooks of magic printed in the 16th c. and 17th c. They purport to contain magic spells devised by Faust.

Fausts Leben dramatisiert, an incomplete play dealing with Faust, which is the work of F. Müller (q.v.). It was published in 1778, and the author's name was given as 'Mahler Müller'. It is dedicated to Otto, Freiherr von Gemmingen-Hornberg (q.v.). Only Pt. I (Erster Teil) is given in the text. Except for a prologue of devils and a final encounter between Faust and devils (*Dunkler Wald. Kreuzweg*), the extant part of the play is set entirely at Ingolstadt. It shows Faust determined to gain or lose all, yet plagued by the vulgar irrelevancies of life, notably by his creditors. At the end he is about to enter into his pact with Mephistopheles. The scene *Marktplatz. Faust den Degen unterm Arm* satirizes C. Kaufmann (q.v.) under the nickname Gottesspürhund. At the end of his life Müller began to rewrite the play in verse. See also Ṣɪᴛᴜᴀᴛɪᴏɴ ᴀᴜѕ Fᴀᴜѕᴛѕ Lᴇʙᴇɴ.

Fausts Leben, Thaten und Höllenfahrt, a novel written by F. M. Klinger (q.v.) in 1791. Klinger attributes to Faust not only great knowledge but the invention of printing. Despite Faust's ingenuity, he is desperately poor, and his failure to secure financial reward drives him to use his magic arts. He gives himself to the devil, hoping to learn the reasons for the existence of evil and especially of human suffering. His guide, the devil Leviathan, takes him first to a German city (doubtless Frankfurt), then into families and convents. Faust begins by abandoning his own family, and everywhere seduces and forsakes women. He visits the France of Louis XI, the England of Richard III, and the Rome of the popes. Everywhere his hope of finding virtue is disappointed, and he finds instead treachery, villainy, and cruelty. But he fails to realize that he himself is guilty of great wickedness. The spirit of mankind appears and points to the path of virtue Faust has forsaken. It is too late, and Faust is conducted to eternal damnation along with Pope Alexander VI. A curious feature of the novel is the printing of substantial tracts of dialogue exactly as if in a play, including even occasional stage directions.

Fazetie, derived from Latin *facetia*, a neatly rounded, wittily or urbanely phrased anecdote which is often, though not necessarily, erotic in content. Fazetien appear in humanistic Latin literature in Germany (H. Bebel, q.v., *Facetiae,* 1508–12; N. Frischlin, q.v., *Facetiae, c.* 1600, which contains also some German, as do the *Facetiae Latinae et Germanicae,* 1486, by A. Tünger, q.v.). In the vernacular literature the Fazetie can hardly be said to exist, being replaced by the cruder and more homespun Schwank (q.v.).

FECHNER, Gᴜѕᴛᴀᴠ Tʜᴇᴏᴅᴏʀ (Groß Särchen/-Lausitz, 1801–87, Leipzig), attempted, as a psychologist and philosopher, to adjust idealistic philosophy to 19th-c. science; all branches of science should, according to him, be incorporated in metaphysics. He rejected the notion of an *a priori* concept of God, and sought to replace it by a pragmatic notion of God as we sense him in the world and within us.

Fechner's special contribution to psychology lay in his renewed investigation into the relationship of body and soul: according to his psychophysics, there is a parallel development in physical substance and the soul, there being no substance (including the planets, the stars, and the earth) without soul. Fechner's views on aesthetics, which he terms in his *Vorschule der Ästhetik* (2 vols., 1876, 2nd edn. 1898) 'Ästhetik von unten', are based on experiments with sensuous reactions. *Über das höchste Gut* (1848) deals with his views on ethics. Of his numerous publications *Elemente der Psychophysik* (2 vols., 1860) is the most important.

As a writer of humorous sketches etc. (*Stapelia mixta*, 1824; *Gedichte und Satiren*, 1841; *Rätselbüchlein*, 1850) he used the pseudonym Dr. Mises.

Fechter von Ravenna, Der, a blank-verse tragedy (Trauerspiel) in five acts, written in 1852–3 by F. Halm (q.v.), and first performed at the Burgtheater, Vienna, on 18 October 1854. It was published in 1856. The play is set in A.D. 41 at the end of Caligula's reign. Thusnelda, wife of the defeated Cheruscan leader, Arminius, is held captive in Rome. Her son Thumelicus has been trained as a gladiator (Fechter) at Ravenna. On the suggestion of his consort Cäsonia, Caligula plans to have Thumelicus killed in the arena in the presence of his mother. Thusnelda, however, kills her son and then herself. The play ends with a hint at a plot to put an end to Caligula's life and reign the next day. It is dedicated to the actress Julie Rettich (q.v.). *Der Fechter von Ravenna* proved a great stage success.

Halm submitted the play without indication of authorship, which was then claimed by a Bavarian schoolmaster, F. Bacherl (q.v.); a noisy controversy ensued which was not ended by a declaration of authorship issued by Halm on 27 March 1856.

FEDERER, HEINRICH (Brienz, 1866–1928, Zurich), was at the Benedictine school at Sarnen. After studying in Roman Catholic seminaries at Eichstätt, Freiburg, and St. Gall, he was ordained priest in 1893. After seven years of pastoral work, asthma compelled his retirement in 1900. He turned to journalism (remaining in holy orders), and edited the *Züricher Nachrichten*. A few years later he began to write fiction, which was widely read in the second and third decades of the 20th c.

His novels and stories are sharply divided by their subject-matter into two categories, realistic novels of Swiss and especially Alpine life of his own day, and historical stories set in Italy. The inspiration of the latter is commonly attributed to his reading of C. F. Meyer (q.v.). Both groups of works share deep psychological insight into a wide range of characters.

Federer began with an Alpine novel set in Appenzell, *Berge und Menschen* (q.v.), and *Lachweiler Geschichten* (both 1911), and followed these with another Alpine novel, *Pilatus* (q.v., 1912). The first of the Italian stories is *Sisto e Sesto* (q.v., 1913); others are *Das letzte Stündlein des Papstes* (1914), *Umbrische Geschichten* (1921), *Wander- und Wundergeschichten aus dem Süden*

(1924), and *Unter südlichen Sonnen und Menschen* (1926). The Swiss stories include *Jungfer Therese* (1913), the novel *Das Mätteliseppi* (1916), and *Regina Lob* (1925). For the story *Papst und Kaiser im Dorf* (1924) Federer received the Gottfried Keller prize. He wrote two autobiographical works, *Am Fenster* (1927) and *Aus jungen Tagen* (1928), and a volume of poems, *Ich lösche das Licht* (1930), appeared posthumously. A drama, *Thomas Becket*, was performed at Sarnen in 1898, but not published. Within somewhat restricted conventional forms Federer wrote with power and conviction.

Fehme, see FEMGERICHTE.

Fehrbellin, a small town in Brandenburg, which has given its name to a battle fought on 28 June 1675 between Brandenburg forces under the Great Elector (see FRIEDRICH WILHELM, GROSSER KURFÜRST) and a Swedish army under General Wrangel. Though the fighting was not decisive, the Swedish forces withdrew. Fehrbellin established Friedrich Wilhelm's fame, gained for him the title Great Elector, and initiated the rise of Prussia to the rank of a great power. The battle of Fehrbellin is the central point about which turns Heinrich von Kleist's play *Prinz Friedrich von Homburg* (q.v.).

FEHRS, JOHANN HINRICH (Mühlenbarbeck, Holstein, 1838–1916, Itzehoe), the son of a veterinary surgeon, was educated at Eckernförde, becoming a schoolmaster in 1862. He is the author of verse epics (*Krieg und Hütte*, 1872; *Eigene Wege*, 1873; *In der Wurfschaufel*, 1877), and published a volume of verse (*Zwischen Hecken und Halmen*, 1886). His main achievement is in his Low German (Plattdeutsch) dialect stories (*Lütj Hinnevk*, 1883; *Allerhand Slag Lüd*, 2 vols., 1887–91; *Ettgron*, 1901; *Ut Ilenbeck*, 1901) and the village novel *Maren* (q.v., 1907), the best of all his narratives.

Feind, Der, an unfinished story by E. T. A. Hoffmann (q.v.), published posthumously in the *Frauentaschenbuch* (1823). Hoffmann worked at it almost to his death. It is a story of Albrecht Dürer, (q.v.) who at the end of his life is confronted by an implacable enemy, named Solfaterra.

FEININGER, LYONEL (New York, 1871–1956, New York), an American who studied painting in Germany and France, living in Germany from 1887 to 1937. He was associated with the groups Die Brücke and Der Blaue Reiter (qq.v.), and from 1919 to 1933 taught at the Bauhaus (q.v.). He was influenced by Cubism.

Feldeinsamkeit, a poem by H. Allmers (q.v.). It has been set to music by Brahms (q.v.).

FELDER, Franz Michael (Schoppernau, Vorarlberg, 1839–69, Bregenz), self-taught and of peasant stock, wrote realistic narrative works, in which the social conditions of his home country were critically viewed, including the dominance over the peasantry of the Roman Catholic clergy. He published the story *Nümmamüllers und das Schwarzokaspale* (1863), a collection of Novellen, *Sonderlinge* (2 vols., 1867), and the novel *Reich und Arm* (1868). His work would probably never have been published but for the encouragement of Professor R. von Hildebrand (1824–92). His autobiography (*Aus meinem Leben*), of value both as a social document and as literature, was not published until 1904.

Feldherrnhügel, Der, a comedy by A. Roda Roda and C. Rößler (qq.v.).

Femgerichte, originally the local Westphalian courts of the Middle Ages, were developed into a secret organization over the whole of Germany, which reached its greatest influence in the first half of the 15th c. Proceedings and membership were secret (stilles Gericht), and death sentences were immediately carried out by hanging without possibility of appeal. In the latter part of the 15th c. the Femgerichte began to lose their prestige in the face of opposition from the territorial princes and the emperor. Though they forfeited all their important powers, they lingered into the 19th c. The best-known literary presentations of Femgerichte are in Goethe's *Götz von Berlichingen* and H. von Kleist's *Das Käthchen von Heilbronn* (qq.v.).

FERDINAND I, Kaiser (Alcalá de Henares, 1503–64, Vienna), a nephew of Maximilian I (q.v.) brought up in Spain, received, after Maximilian's death, the Austrian Habsburg territories (1521, see Wormser Vertrag). In 1526 he was elected king of Bohemia and Hungary. The early part of his reign was chiefly occupied in resisting Turkish pressure upon his lands, and he made efforts to reach a compromise with the Protestants (Treaty of Passau, 1552). Though he adhered to the Roman Catholic faith, he was ready to support ecclesiastical reform. He became emperor on the abdication of Karl V (q.v., 1556), but the eight years of his reign saw no significant change of policy.

FERDINAND II, Kaiser (Graz, 1578–1637, Vienna), son of Archduke Karl von Steiermark, and heir to the Emperor Matthias (q.v.), was elected hereditary king of Bohemia in 1617 and king of Hungary in 1618 in preparation for his succession to the imperial crown. An ardent Catholic, he recovered his Austrian possessions (Styria) for the Roman Church, and expelled the Protestants. The Bohemian Estates, dissatisfied with Habsburg rule (see Majestätsbrief) and fearing his intolerance, deposed him, and in 1619 elected Friedrich V (q.v.) of the Palatinate as their king. Friedrich's defeat at the battle of the White Mountain (see Weissen Berge, Schlacht am) restored Ferdinand's sovereignty.

Supported by the Catholic League (see Katholische Liga) and by the generals Tilly and Wallenstein (qq.v.), Ferdinand fought the Thirty Years War (see Dreissigjähriger Krieg) with uncompromising dedication to the Catholic cause. In 1629 he boldly attempted by means of the Edict of Restitution to put the clock back and restore all ecclesiastical estates which had been secularized since 1555. Two years before his death he had to abandon this reactionary policy in the Peace of Prague (1635). At his death he left the Empire to his son (see Ferdinand III), of whose fidelity he felt secure.

Ferdinand is, as Archduke, a character in Grillparzer's tragedy *Ein Bruderzwist in Habsburg* (q.v.), which deals with the reign of Ferdinand's cousin Rudolf II (q.v.).

FERDINAND III, Kaiser (Graz, 1608–57, Vienna), King of Hungary from 1625 and King of Bohemia from 1627, succeeded his father Ferdinand II (q.v.) as emperor in 1637. Though, like his father, he upheld the Roman Catholic cause, he was realistic in his assessment of the critical point in the Thirty Years War (see Dreissigjähriger Krieg), at which he was drawn into the political scene. He had already shown himself ready to support a policy of compromise without which the Peace of Prague (1635) would not have come into existence. Foreign intervention forced him to defend the Empire against France and Sweden, and he had to accept in the Peace of Westphalia (see Westfälischer Friede) severe territorial losses and the eclipse of the old imperial power. After the peace settlement he ably restored the administration of Austria, the seat of the Habsburg dynasty.

FERDINAND I, Kaiser von Österreich (Vienna, 1793–1875, Prague), eldest son of the Emperor Franz I (q.v.), came to the throne in 1835. Ferdinand was feeble-minded, and the business of state was carried on by a council of regency. In March 1848 he fled from Vienna to Tyrol, and thence to Olmütz, where he signed an act of abdication (2 December). He spent the rest of his life in Prague, living in the Hradschin and pursuing his hobbies. (See Revolutionen 1848–9.)

FERDINAND VON TIROL, Erzherzog (Linz, 1529–95, Innsbruck), second son of the Emperor Ferdinand I (q.v.), married a commoner, Philippine Welser (q.v.) in 1557 and, after his father's death in 1564 became governor of Tyrol. A zealous Roman Catholic, he did his best to extirpate Protestantism. Ferdinand was a connoisseur of art, and initiated the famous collection of *objets d'art* at his castle of Ambras or Amras near Innsbruck. He is the author of an ambitious play, *Speculum vitae humanae* (1584), in nine acts, the first and last of which frame the rest (see Rahmen). Each of the other seven portrays a work of mercy and the opposing deadly sin.

Fest auf Haderslevhuus, Ein, see Ein Fest auf Haderslevhuus.

Festschrift, a collection of essays by notable scholars designed as a tribute to a scholar of eminence on his retirement or on a birthday.

Festspiel in deutschen Reimen, a puppet-play written by G. Hauptmann in 1913 at the invitation of the student body at Breslau University, and intended as a contribution to the centennial celebrations for the battle of Leipzig (see Völkerschlacht). It met on the whole with disapprobation: the nationalists found it insufficiently heroic, the left wing too nationalistic, and both parties deprecated the use of puppets, which intentionally reduces the scale of the action (see also Puppenspiel). Hauptmann's 'German rhymes' are the irregular Knittelverse of Goethe's *Urfaust*. The play abandons realism, and presents the principal figures in succession, allowing them to justify their actions in speeches, which the assembled citizens applaud or condemn. It ends with a panegyric of peace.

FEUCHTERSLEBEN, Ernst, Freiherr von (Vienna, 1806–49, Vienna), son of an Austrian civil servant and nephew of a general, studied medicine from 1825 to 1833, interspersing his professional work with courses in philosophy and literature. He qualified in 1835, and in 1844 became a lecturer in 'ärztliche Seelenkunde', a forerunner of psychiatry. He published a number of contributions to medical knowledge and also popular philosophical writings (*Beiträge zur Literatur, Kunst- und Lebenstheorie*, 1837–41; *Zur Diätetik der Seele*, q.v., 1838).

His poems (*Gedichte*), published in 1843, are forgotten except for the well-known *Gottes Rat und Scheiden*, beginning 'Es ist bestimmt in Gottes Rat' (q.v.). Feuchtersleben belonged to the circles frequented by E. von Bauernfeld, F. Grillparzer, and M. von Schwind (qq.v.), for whom he wrote poems for drawings, published

in the *Almanach der Radierungen*, 1843. His collected works were edited by F. Hebbel (*Sämtliche Werke*, 7 vols., 1851–3; this edition excluded medical works).

FEUCHTWANGER, Lion (Munich, 1884–1958, Los Angeles), after studying at Munich and Berlin universities, became a dramatic critic. In July 1914 he was in Tunisia (then French), and was interned. In the post-war period he lived first in Berlin, then in Munich. Being a Jew he was proscribed in 1933, while visiting California. He lived in France and visited Russia. Vichy-France interned him, but he escaped to the U.S.A. Feuchtwanger occupied himself largely with historical novels and plays. His novels *Die häßliche Herzogin Margarete Maultasch* (q.v., 1923) and *Jud Süß* (q.v., 1925) enjoyed international success. Other novels include *Der tönerne Gott* (1910) and a sequence of four novels, subsequently turned into a tetralogy, *Der Wartesaal. Zyklus aus dem Zeitgeschehen* (q.v.): the separate parts are entitled *Erfolg* (1930), *Die Geschwister Oppenheim* (1933), *Exil* (1940), and *Simone* (qq.v., 1944). The Jewish historical novels *Der jüdische Krieg* (1932) and *Der Tag wird kommen* (1945), dealing with Josephus and the Jews in Roman times, are related to the situation of the Jews in Feuchtwanger's own time. Other novels were *Der falsche Nero* (1936), *Die Brüder Lautensack* (1944), *Waffen für Amerika* (2 vols., 1947–8, later entitled *Die Füchse im Weinberg*), *Goya* (1951), *Narrenweisheit oder Tod und Verklärung des J. J. Rousseau* (1952), *Spanische Ballade* (1955), and *Jefta und seine Töchter* (1957).

Feuchtwanger also wrote plays, beginning with short Old Testament dramas (*Kleine Dramen*, 2 vols., 1905–6, containing *Joel, König Saul, Das Weib des Urias* together with some non-biblical plays). English history (*Warren Hastings*) followed in 1916, and other plays were *Thomas Wendt* (1919), *Drei angelsächsische Stücke* (1927, (1) *Die Petroleuminseln*, (2) *Kalkutta, 4. Mai*, (3) *Wird Hill amnestiert?*), *Wahn oder der Teufel in Boston* (1948), and *Die Witwe Capet* (Marie Antoinette, 1956). In 1956 he collaborated with Brecht (q.v.) in a dramatized version of his novel *Simone* under the title *Die Gesichte der Simone Machard* (q.v.). In 1924 he made a German adaptation of Marlowe's *Edward II*.

Gesammelte Werke in Einzelbänden were published in 1959–84.

FEUERBACH, Anselm (Speyer, 1829–80, Venice), a painter, who was a nephew of L. Feuerbach (q.v.). He chose classical and Renaissance subjects for his principal works, for example *Iphigenie, Das Gastmahl des Plato*, and *Dante und die edlen Frauen von Ravenna*. After his death his stepmother Henriette Feuerbach

(1812–92) published *Das Vermächtnis* (1882); it contained an autobiographical fragment written in 1876 under the title *Aus meinem Leben*.

FEUERBACH, LUDWIG (Landshut, 1804–72, Rechenberg nr. Nürnberg), the son of a criminologist who became a judge, studied theology at Heidelberg, and philosophy under Hegel (q.v.) at Berlin. In 1828 he became a lecturer at Erlangen University, but resigned in 1832 because of his unorthodox views on religion. His earliest work, *Gedanken über Tod und Unsterblichkeit*, appeared anonymously in 1830. In his most influential work, *Das Wesen des Christentums* (1841), published under his own name in 1843, he maintained that the dogmas and beliefs of Christianity are figments of the human imagination, fulfilling a need inherent in human nature. He viewed theology as a branch of anthropology, and religion as 'the dream of the human spirit'. In this sense he conceded their evolutionary function: 'Das Bewußtsein Gottes ist das Selbstbewußtsein des Menschen, die Erkenntnis Gottes die Selbsterkenntnis des Menschen.' His conception of a trinity in man, a unity of reason, love, and will, underlies his phrase that Man 'ist, was er ist nur durch sie' which has been turned into a well-known pun, Man 'ist, was er ißt (eats)'. Feuerbach was most influential in the 1840s, and G. Keller (q.v.) was foremost among the writers susceptible to his views.

Feuerreiter, a figure of folk-lore in the form of a man with a red cap mounted on a lean horse, who is seen riding towards a building in which fire is about to break out. The ballad *Der Feuerreiter* (1824) by E. Mörike (q.v.) gives a vivid realization of this apparition.

Feuilleton, the part of a newspaper devoted to literary, artistic, and musical matters, etc. Formerly printed on the lower part of several pages (unter dem Strich), it is now usually printed on a separate page or pages.

FEYERABEND, SIGMUND (Heidelberg, 1528–90, Frankfurt/Main), a successful bookseller and publisher, whose 'best sellers' give an indication of the taste of the age. They include *Das Heldenbuch* (1560), a collection of crude moral tracts, *Theatrum Diabolorum* (1569), and *Das Buch der Liebe* (1578), which contains thirteen Volksbücher (see VOLKSBUCH).

Fibel, an ABC or first reading book. The word is a corruption of 'Bibel', for the Bible was used as the basic text in early examples.

FICHTE, JOHANN GOTTLIEB (Rammenau, Lausitz, 1762–1814, Berlin), studied theology at Jena and Leipzig universities, and in 1790 became an enthusiastic student of Kant's philosophy. He devised a system of his own, based on Kant's thought. He termed this development 'subjective idealism', and through it he exercised a considerable influence on the Romantic movement (see ROMANTIK). Fichte was an eloquent patriot, and his *Reden an die deutsche Nation* expressed his ardent nationalism. They were delivered at Berlin University 1807–8, and printed in 1808. Their influence on Fichte's contemporaries has sometimes been overrated. Even so, Fichte is associated with those who prepared for the Wars of Liberation (see NAPOLEONIC WARS). He died of typhoid contracted from his wife (a niece of Klopstock, q.v.), who was working as a volunteer nurse in a military hospital.

When Fichte began to interest himself in Kant (q.v.) he was earning a pittance as a private tutor. In order to bring himself to Kant's notice, he sent him the tract *Versuch einer Kritik aller Offenbarung*. It won the approval of Kant, who published it anonymously in 1792, and it was at first attributed to him. When its true authorship was revealed by Kant, Fichte's reputation was established. In 1794 he was appointed to a chair of philosophy at Jena University. There he met Goethe, Schiller, the Schlegel brothers, and W. von Humboldt (qq.v.). During the five years of his tenure he expounded his system in a series of treatises: *Über den Begriff der Wissenschaftslehre* (1794); *Grundlage der gesamten Wissenschaftslehre* (1794–5); *Grundriß des Eigentümlichen der Wissenschaftslehre* (1795); and *Grundlage des Naturrechts nach Prinzipien der Wissenschaftslehre* (1796). His ethics were set forth in *System der Sittenlehre nach Prinzipien der Wissenschaftslehre* (1798). Doubts about his theological orthodoxy were confirmed by *Über den Grund unseres Glaubens an eine göttliche Weltregierung* (1798), and he was deprived of his chair on grounds of atheism. After a few years in Berlin he became professor of philosophy at Erlangen University, resigned after one semester, and moved to Königsberg. In 1810 he was appointed rector of the new University of Berlin.

As Fichte developed his philosophy, he diverged in important particulars from his Kantian model. He rejected Kant's 'thing-in-itself' (*Ding-an-sich*), and saw existence solely in terms of the self. For him only the Ego exists 'in itself'. The world around it, comprehensively classified as the Non-Ego, is a creation of the Ego. Fichte preached moral virtues, especially patriotic ones. He seems to have been prepared to transfer the Ego (Ich) to the German nation, which would represent the supreme incarnation of the moral ideal. This transformation of the German people into a race morally superior to all others was an important source of German nationalism in the 19th c. and 20th c. Fichte is the subject of a study by R. Schneider (q.v., 1932).

Fichte also wrote specifically nationalistic works before he delivered the *Reden* of 1807–8; he published *Der geschlossene Handelsstaat* in 1800, *Die Grundzüge des gegenwärtigen Zeitalters* in 1804, and *Der Patriotismus und sein Gegenteil* in 1807. By 1805 a tendency towards mysticism had manifested itself in his thinking, which is perceptible in *Die Anweisung zum seligen Leben* (1806). The *Gesamtausgabe*, edited by R. Lauth and H. Jacob (26 vols.), began to appear in 1965.

Fidelio, Beethoven's only opera, first performed in Vienna on 20 November 1805. It originally bore the sub-title *Die eheliche Liebe*. The libretto by Josef Sonnleithner (q.v.) is an adaptation of *Léonore ou l'amour conjugal* by J. N. Bouilly, which had already served as text for an opera by P. Gaveaux performed in Paris in 1798. Fidelio is the assumed name of Leonora, who takes a post in a prison in the belief that her missing husband Florestan is imprisoned there. He is indeed held there by Pizarro, whose corruptness he had tried to expose. Pizarro, learning that an inspection of the prison is to take place, resolves to murder Florestan and takes Rocco, the chief gaoler, and Fidelio as assistants. A moving scene follows in the dungeon where Florestan is found chained and starving. A trumpet call warns Pizarro that the visiting minister is at the gates, but his attempt to dispose of Florestan forthwith is frustrated by Fidelio, who claps a pistol to his head and stands revealed as Leonora. All now ends in triumph and Pizarro is led away. When first played, *Fidelio* (then in three acts) was a failure. A revision compressing it into two acts achieved a triumphant success in 1814. An earlier revision had used the title *Leonora* (q.v.).

Fierabras, a Volksbuch printed in 1533. It is an adaptation of a French prose romance, and deals with the warlike feats of Fierabras, a giant who is the King of Spain's son, and with his eventual defeat by Oliver, the knight of Charlemagne (see KARL I, DER GROSSE).

Fiesco, see VERSCHWÖRUNG DES FIESCO ZU GENUA, DIE.

Figurentheater, see PUPPENSPIEL.

FINCKELTHAUS, GOTTFRIED (Lützen, 1610–47 or 1648, Leipzig), studied medicine at Leipzig and was a friend of Paul Fleming (q.v.). He travelled in France and Holland, and visited Brazil, finally settling in Leipzig as town clerk (Stadtschreiber). He wrote fresh, robust, and unpretentious lyric poetry which appeared in four collections: *Deutsche Gesänge* (c. 1640), *Dreißig teutsche Gesänge* (1642), *Deutsche Lieder* (1642), and *Lustige Lieder* (1647). Finckelthaus used the pseudonym Gregor Federfechter von Lützen.

FINCKH, LUDWIG (Reutlingen, 1876–1964, Gaienhofen nr. Radolfzell), a doctor, practised in Frankfurt and Aachen, and settled in 1805 near Radolfzell on Lake Constance. He was a close friend of H. Hesse (q.v.). A prolific writer of verse and fiction, he was among the authors approved by the National Socialist regime. Among his volumes of verse are *Frau du, du Süße* (1900), *Rosen* (1906), *Mutter Erde* (1917), and *Rosengarten* (1953). His novels include *Der Rosendoktor* (1906), *Die Reise nach Tripstrill* (1911), *Der Bodenseher* (1914), *Die Jakobsleiter* (1920, reissued 1943 as *Der Wolkenreiter*), *Der Vogel Rock* (1923), *Stern und Schicksal* (1931, dealing with the astronomer J. Kepler), *Ein starkes Leben* (1936), *Herzog und Vogt* (1940), *Das goldene Erbe* (1943).

Finckh published *Zaubervogel*, a collection of tales, in 1936, and the separate stories *Rapunzel* (1909), *Urlaub von Gott* (1930), and *Der Goldmacher* (1953). His autobiography is entitled *Himmel und Erde* (1961). *Das dichterische Werk* (7 vols.) appeared in 1926, and *Ausgewählte Werke* (2 vols.) 1956–66.

Findling, Der, a story by H. von Kleist (q.v.), published in 1811 in *Erzählungen* (vol. 2). The foundling, Nicolo, is an orphaned boy whom a wealthy Roman trader, Antonio Piachi, has adopted after his own son, Paolo, has fallen victim to the plague in Ragusa. Upon Nicolo's marriage Piachi transfers to him most of his property. But Nicolo outrageously deceives his foster-parents, Piachi and his wife Elvire, before evicting them from their own home; he wantonly desecrates Elvire's cherished memory of Colino, who, at the cost of his life, had rescued her in her youth in a fire.

As Nicolo acts within his legal rights, Piachi takes his own revenge; he murders Nicolo, and, accepting the consequences of his deed, submits to the authorities. He intends to continue his revenge in hell, where he is confident that he will find Nicolo again; he therefore refuses absolution, without which the execution cannot legally take place. A special papal decree enables the sentence of hanging to be carried out.

Findling, Der, a play in prose and mixed rhymed and unrhymed verse by E. Barlach (q.v.), described as Ein Spiel in drei Stücken, published in 1922. The theme of this mystery play centres on the birth of a saviour in a world of unbelief, corruption, guilt, and misery, in which mankind devours itself by eating its own flesh. A stone-breaker, working by a road peopled with starving refugees, kills the Red Emperor who comes to him for shelter, and with his body he feeds an unloved foundling (Vorspiel). Among others taking refuge with the stone-breaker are Mother Kummer's daughter Elise and the puppeteer's son Thomas, who recognize in each other kindred souls searching for redemption; they both turn away from the revolting greed of their elders. By slipping his hand inside his princess glove puppet Thomas animates the puppet, thus expressing his faith in a power within himself and Elise which will inspire them with new life (Mittelstück). After the stone-breaker has revealed to the people that he has killed the Red Emperor, who had prophesied the birth of a saviour, Elise finds the strength to touch the foundling, in whose terrifying appearance she sees the image of a desperate god, and the 'sorest wound of the world' ('die wehste Wunde der Welt'). Her action transforms the foundling into a beautiful child and, by obscuring sin and misery, illuminates eternity. Barlach created twenty woodcuts to illustrate the play.

FINK, GOTTFRIED WILHELM (Sulza/Ilm, 1783–1846, Leipzig), a writer on music, who became a lecturer at Leipzig in 1841 and compiled one of the best-known song books of the 19th c., *Musikalischer Hausschatz der Deutschen* (q.v., 1843). He also wrote poems (*Gedichte*, 1813), which he set to music himself. Later volumes include *Balladen und Romanzen* (1820) and *Familienunterhaltungen* (1825).

Finkenritter, Der, a Volksbuch printed in Strasburg c. 1560. It presents a series of impossible adventures sustained by the knight Polykarp von Kirrlarissa. Its attribution to Lorenz von Lauterbach, notary of Neustadt, is doubtful.

FINX, E., see FRANCISCI, E.

FISCHART, JOHANN (Strasburg, 1546–90, Forbach nr. Saarbrücken), a satirist of Lutheran persuasion, was educated in Strasburg at the school founded by Johannes Sturm (q.v.), and spent his later youth in the care of his cousin Kaspar Scheidt (q.v.) in Worms. He then studied in Italy, and travelled in France, England, and Holland. From 1571 to 1584 he was proof-reader for his brother-in-law, the printer B. Jobin. In 1580 he received an appointment at the Reichskammergericht in Speyer, and in the same year entered the service of a nobleman at Forbach, who installed him as magistrate there.

Fischart was an ardent champion of Protestantism who engaged in scathing attacks on Catholic personalities, including J. J. Rabe and J. Nas (qq.v.), and on Catholic institutions, such as mendicant orders and the Society of Jesus. To this phase of his work belong *Barfüßer Secten und Kuttenstreit* (1570), *St. Dominici Leben* (1571), *Der Bienenkorb* (1579), and *Jesuiterhütlein* (1580).

Many of his works are moral satires, with more or less strong didactic purpose, including a poetic version of *Till Eulenspiegel* (1572), *Aller Praktik Großmutter* (1572), an incursion against the prophecies of almanacs, and *Der Flöhhatz* (1573, rev. 1577), which castigates the vices and foibles of women. *Das Ehezüchtbüchlein* (1572) deals with the problems of marriage, and *Das Podagrammatische Trostbüchlein* (1577) offers humorous consolation to sufferers from gout. *Das Glückhafft Schiff von Zürich* (1577) is an account of a voyage to Strasburg by inhabitants of Zurich.

Fischart's best-known work is his translation of Rabelais's *Gargantua et Pantagruel* in which he gives his addiction to puns and extravagant verbal gymnastics the fullest scope. First published in 1575 with a comic title of enormous length, it has been known since the second edition of 1582 as *Geschichtklitterung*. It is no literal translation, but a free and expansive adaptation, in which Fischart's irrepressible verbal humour is the dominant characteristic. He shifts the centre of gravity of the work from Rabelais's humanism to his own position of moralist.

Werke. Eine Auswahl (3 vols.), ed. A. Hauffen, appeared in 1895 and was reissued in 1974.

FISCHER, JOHANN BERNHARD VON (Lübeck, 1685–1772, Hinterbergen nr. Riga), was brought up in Riga. After studying in Germany and travelling in western Europe (including a visit to England) he returned to Riga as a physician; for eight years (1734–42) he was medical adviser to the Empress Anna; he spent the remainder of his life at Hinterbergen. Fischer wrote an account of his life, *Der In Beruhigung und Friede wohnende Montan* (1745), in which title the capitals I, B, and F represent his initials. He also wrote verse idylls (*Empfindungen des Frühlings*, 1750; *Hirtenlieder und Gedichte*, 1753; *Daphnis an Silen*, 1754).

FISCHER, Johann Georg (Groß-Süßen, Württemberg, 1816–97, Stuttgart), a schoolmaster in various towns of Württemberg, but chiefly in Stuttgart (1845–85), was a friend of E. Mörike (q.v.), and was himself a prolific minor poet usually reckoned to be a member of the Swabian school. His collections are *Gedichte* (1838), *Dichtungen* (1841), *Gedichte* (1854), *Neue Gedichte* (1865), *Den deutschen Frauen* (1869), *Aus frischer Luft* (1872), *Neue Lieder* (1876), *Auf dem Heimweg* (1891), and *Mit 80 Jahren* (1896). He is the author of the plays *Saul* (1862, performed at Stuttgart in the same year), *Friedrich II von Hohenstaufen* (1863, performed in Weimar 1862), and *Florian Geyer* (1866).

FISCHER, Wilhelm (Csakathurn, Croatia, now Cakovec in Yugoslavia, 1846–1932, Graz), director of the Styrian Library (Steiermärkische Landesbibliothek) from 1901 until his death, and usually known as 'Fischer in Graz', or 'Fischer-Graz', was the author of poetic writings (the epic *Atlantis*, 1880; *Lieder und Romanzen*, 1884) and, later, of narrative work. His reputation, which is mainly local, rests on collections of Novellen (*Sommernachtserzählungen*, 1882; *Unter altem Himmel*, 1891; *Grazer Novellen*, 1898) and on the novels *Die Freude am Licht* (1902) and *Sonnenopfer* (1908).

Fischer, Der, a ballad written by Goethe in 1778 and first published in the following year. It sings of a nymph ('feuchtes Weib'), who entices the fisherman into the depths.

Fischerin, Die, a short Singspiel (q.v.) written by Goethe for open-air performance at Weimar, and published in 1782. It was first performed in the park at Tiefurt (q.v.) on the banks of the Ilm on the evening of 22 July 1782. Dortchen, who keeps house for her fisher father, is betrothed to the fisherman Niklas. Thinking that she is insufficiently appreciated, she hides, leaving her father and Niklas, when they return, to think that she has been drowned. They summon the neighbours to search, but Dortchen reappears and all is well. In a note added in 1807 Goethe indicated that the scene in which the agitated torches of the neighbours make a fascinating spectacle was the *raison d'être* of this slender piece. The ballad *Erlkönig* (q.v.) is sung by Dortchen as the opening number.

FISCHER VON ERLACH, Johann Bernhard (Graz, 1656–1723, Vienna), a leading architect of early baroque in Austria. Among his important works in Vienna (some of them completed after his death) are the Palais Trautson, the Finance Ministry (formerly Winterpalais of Prinz Eugen, q.v.), the Karlskirche, and the

Nationalbibliothek (formerly Hofbibliothek). He designed Schönbrunn Palace in 1695, but building was suspended in 1705 and resumed 1744–9 under other hands. In Salzburg he was responsible for the Dreifaltigkeitskirche.

FLACHSLAND, Caroline (Reichenweier, Alsace, 1750–1809, Weimar), lived with her sister and brother-in-law in Darmstadt, where, under the name Psyche, she was a prominent member of the Darmstädter Kreis (q.v.). In 1770 she became engaged to Herder (q.v.), whom she married in 1773.

FLACIUS ILLYRICUS (Albona, Istria, 1520–75, Frankfurt/Main), usual designation of Matthias Vlacich or Flacius, who studied under Luther (q.v.), became a professor at Wittenberg in 1544, and, after a short interval at Magdeburg, was appointed to a chair at Jena in 1557. He was dismissed in 1561, and was for brief periods in Regensburg, Antwerp, Frankfurt, and Strasburg. He was the most prominent and rigid Lutheran after Luther's death, and was involved in several doctrinal controversies. He planned and partly wrote a history of the Church (seen from the Lutheran point of view), which is known as the *Magdeburger Zenturien* (1559–74), from its division into centuries. His other works include *Catalogus testium veritatis* (1556) and *Calvis scripturae sacrae* (1567).

FLAISCHLEN, Cäsar (Stuttgart, 1864–1920, Gundelsheim/Neckar), at first a bookseller, studied at Berlin, Heidelberg, and Leipzig universities. In 1890 he settled in Berlin. His first book was a volume of poems, *Nachtschatten* (1884). This was followed in 1892 by poems in Swabian dialect, *Vom Haselnußroi*, and by various other volumes, *Vom Alltag und Sonne* (1898, recalling Walt Whitman in its rhythms and prose-like style), *Aus den Lehr- und Wanderjahren des Lebens* (1900), *Neujahrsbuch* (1907), *Zwischenklänge* (1909), to which were added in 1921 and 1924 two posthumous collections, *Mandolinchen, Leierkastenmann und Kuckuck* and *Von Derhoim ond Drauße*. Flaischlen was for a time caught up in the Naturalistic movement and planned a trilogy in this idiom. The first two plays were *Toni Stürmer* (1891), preoccupied with sex, and *Martin Lehnhardt* (1895), which deals with the problem of faith and doubt. The third play was not written, but came out instead as a novel, *Jost Seyfried* (2 vols., 1905). The eponymous hero had already appeared as a character in *Martin Lehnhardt*. Flaischlen's change of plan may have been connected with his shift of interest from Naturalism to *art nouveau* (see Jugendstil), for from 1895 to 1900 he was the editor of the periodical *Pan* (q.v.). *Jost*

Seyfried, which is said to contain autobiographical elements, deals with the role of the artist, in particular of the man of letters. Flaischlen, whose reputation was once considerable, remained something of a dilettante. *Gesammelte Dichtungen* (6 vols.) appeared in 1921.

FLAKE, OTTO (Metz, 1880–1963, Baden-Baden), after studying at Strasburg, devoted his whole life to literature, being especially prolific in the writing of novels, many of them in cyclical form. The scene of most of his work is either the western frontier territories of Germany, or France, and his outlook is oriented towards international, and particularly towards Franco-German, understanding. His earlier novels included Expressionistic experiments, and in *Die Stadt des Hirns* (1919) he attempted a new kind of philosophical novel. Its sequel *Nein und Ja* (1920) was concerned with theoretical and practical sexual promiscuity. Ruland, the ideal figure of the intelligent, independent, modern European, occurs in a whole cycle of novels, *Freitagskind* (1913), *Ruland* (1922), *Der gute Weg* (1924), *Villa USA* (1926), and *Freund aller Welt* (1928); the first of these was retitled *Eine Kindheit* (1928). They were later published as a cycle with the title *Romane um Ruland* (5 vols., 1926–8). The novel *Montijo oder Die Suche nach der Nation* was published in 1931. The 19th-c. novel *Hortense oder Die Rückkehr nach Baden-Baden* (1933) was the first of a series of novels with the social background of Baden-Baden, but it is not included in *Badische Chronik*, which is composed of the novels *Der junge Monthiver* (1934) and *Anselm und Verena* (1935), which many years later were reissued together as *Die Monthivermädchen* (1952). *Die Töchter Noras* appeared in 1934, and in 1948 under a new title, *Kamilla. Scherzo* dates from 1936. A third cycle was composed of *Fortunat* (2 vols., 1946) and *Ein Mann von Welt* (2 vols., 1947) which reflects the European intellectual and social outlook of the 19th c. in a figure linked with both Germany and France. Late novels were *Die Sanduhr* (1950) and *Schloß Ortenau* (1955). Flake was also an essayist (e.g. *Große Damen des Barock*, 1939). His autobiography was published in 1960 (*Es wird Abend*), and a statement of non-belief, *Der letzte Gott*, in 1961. Two early and atypical novels should be mentioned: *Schritt für Schritt* (1912) and *Horns Ring* (1916). *Werke* (5 vols.), ed. R. Hochhuth and P. Härtling (qq.v.), appeared 1973–6.

Flamingos, Die, poem by Rilke (q.v.), superscribed *Paris, Jardin des Plantes,* and included in *Neue Gedichte* (q.v.).

FLECK, KONRAD, author of the Middle High German poem *Floire und Blanscheflur* (q.v.),

assumed to have been written about 1220, and of a lost poem entitled *Clies*, doubtless based on the *Cligés* of Chrétien de Troyes, is known by two references to him made by RUDOLF von Ems (q.v.). Like Gottfried (q.v.), he is believed to have been an educated townsman, and to have lived in the region of the upper Rhine, possibly in Basel.

Flegeljahre, an unfinished novel by JEAN Paul (q.v.), published in 1804–5. It is sub-titled *Eine Biographie*. It is a comic novel dealing with the extraordinary conditions of inheritance imposed in a testament. Herr van der Kabel dies and leaves his house to whichever of the presumptive heirs can first shed a tear. But he leaves the rest of his enormous fortune to Gottwalt (Walt) Harnisch, provided that the latter repeats his benefactor's life in the following respects: (1) he must tune pianos for a day; (2) become for a time a gardener; and (3) a notary; (4) he must shoot a hare; (5) must correct 192 pages of proofs; (6) spend a week with each of the disappointed would-be heirs; and (7) become a pastor. The news reaches Walt just as he is about to become a notary. Near by but concealed in a tree is Walt's long-lost brother Vult now a distinguished flautist. As Walt sets out to begin the ordeal of qualifying for the legacy, Vult reveals himself, and the two brothers are united in blissful affection. Walt fulfils certain of the conditions clumsily and ineffectively, but all the time retains his endearing innocence and good nature. He falls in love with Wina, daughter of General Zablocki, and so does Vult, but Wina gives her love to Walt. Vult generously urges Walt's suit, and then strolls out of the story playing his flute, at which point the narrative breaks off. Full of irony, sudden starts and changes, and arabesques of humour and wit, *Flegeljahre* is also, with *Wuz* (q.v.), the most engaging, generous, and winning of Jean Paul's novels.

FLEISSER, MARIELUISE (Ingolstadt, 1901–74, Ingolstadt), was educated at a convent school and studied for a time at Munich University. She became acquainted with L. Feuchtwanger (q.v.), through whom she met B. Brecht (q.v.), then a young writer in Augsburg. Brecht's insistence on plain language had a decisive influence on her work. Her first play, originally called *Die Fußwaschung*, was performed, thanks to Brecht's intervention, at the Theater am Schiffbauerdamm in 1926 under the director's title *Fegefeuer in Ingolstadt*, which it has retained. The play's second version, written 1970–1, was published in 1972, a period that marks the promotion of the new Volksstück (q.v.). It is concerned with problems of adolescence in a small Catholic town. Her second play, *Pioniere* *in Ingolstadt*, described as a comedy, is a succession of short scenes in laconic dialogue. It is chiefly concerned with soldiers and sex, having as background to its action the presence of a battalion of bridge-building sappers in Ingolstadt. Revised within a year of its completion in 1928, its Berlin production of 1929 by Brecht, who transposed the action against Fleißer's wish from 1926 to 1910, caused a public scandal and Fleißer's rift with Brecht, though he promoted the production of a late Volksstück in Bavarian dialect, *Der starke Stamm* (1950). Fleißer revised her early play in 1968; performed in Munich in 1970, the third as well as the second version was published in 1972. Fleißer reflects on this episode, which caused her to turn to fiction, in *Avantgarde* (1963). The play *Tiefseefisch*, based on personal experience as a woman, begun 1929–30 and resumed in 1972, has remained unfinished. The tragedy *Karl Stuart*, on the preservation of inner freedom, was started in 1938 and published in 1946.

The collection of ten stories, *Ein Pfund Orangen* (1929), was followed by *Mehlreisende Frieda Geier* (1931, retitled *Eine Zierde für den Verein*, 1975), *Andorranische Abenteuer* (1932), and *Abenteuer aus dem englischen Garten* (1969). *Gesammelte Werke*, ed. G. Rühle, appeared in 3 vols., 1972.

FLEMING, PAUL (Hartenstein, Saxony, 1609–40, Hamburg), lyric poet, was at school in Leipzig and afterwards studied medicine there. In 1630 he met Opitz (q.v.), whose work he admired, and his own writing was encouraged by Adam Olearius (q.v.). Through Olearius he was invited to participate in a commercial expedition to Russia and Persia, sponsored by the Duke of Holstein-Gottorp, which in 1634 went as far as Moscow and then returned. Fleming spent a year in Reval waiting for the second stage of the expedition; he became engaged to Elsabe Niehus at this time. In 1636 he accompanied the expedition, which proved a commercial failure, to Ispahan, and returned in 1637. Elsabe Niehus had married another, and Fleming became engaged to her sister Anna. In 1639 he completed his medical studies at Leyden and moved to Hamburg, intending to practise there. He died shortly after his arrival.

In the poetic conventions he adopts, his use of antithesis and accumulation, and his insistence on transience, vanity, and a stoical resignation, Fleming is a man of his age, but many of his poems have also an immediacy of experience which was uncommon in the poetic writing of the 17th c. His earliest poems were versions of the psalms. Most of his poetry was published after his death (*Prodromus*, 1641, *Teutsche Poemata*, 1642). Shortly before his death he composed his own epitaph in the form of a sonnet. In his earlier years Fleming also wrote Latin poetry. His *Deutsche und Lateinische Gedichte* (3 vols.) were edited by J. H. Lappenberg, 1863–5; the *Deutsche Gedichte* (2 vols.) were reprinted in 1965.

FLEX, WALTER (Eisenach, 1887–1917, Ösel), studied at Erlangen and Strasburg, and became private tutor to descendants of Bismarck (q.v.). He volunteered in 1914, was commissioned, and was killed in the amphibious operation for the capture of the Baltic island of Ösel. Flex's once considerable popularity rested largely on his military career and death in battle; his most widely read work was the war book *Der Wanderer zwischen beiden Welten* (1916). He published two volumes of poems, *Im Wechsel* (1910) and *Vom großen Abendmahl* (1915), an historical tragedy in verse, *Klaus Bismarck* (1913), set in the 14th c. and dealing with an ancestor of Otto von Bismarck (q.v.), and some Novellen. His mystical, quasi-religious praise of war and of sacrificial death has diminished his popularity since 1945. *Gesammelte Werke* (2 vols.) appeared in 1925.

FLIEGEL, HELLMUTH, see HEYM, S.

Fliegende Holländer, Der, an opera by R. Wagner (q.v.), written in short lines of iambic verse, and composed between 1839 and 1841. It was first performed at Dresden on 2 January 1843. The source is an oral tradition of Dutch origin, given literary form by H. Heine (q.v., *Aus den Memoiren des Herrn von Schnabelewopski*). Wagner's thoughts are said to have been drawn to this sea subject by a storm experienced on a voyage to England in 1839; and the work is noteworthy for its storm music. The Dutchman, condemned to sail the seas for all eternity unless a pure woman sacrifices herself for him, makes one of his permitted septennial disembarkations in a Norwegian fjord. He is befriended by the Norwegian sea captain Daland, to whom he offers rich treasure in return for shelter. Discovering that Daland has a daughter, Senta, he asks for her hand, to which Daland consents. Senta, who is half promised to Erik, knows of the story, recognizes the Dutchman and gladly gives him her hand. Erik interposes and the Dutchman, assuming that Senta will throw him over, boards his ship, and casts off for eternity at sea. But Senta breaks free from family and friends, and, for the Dutchman's sake, hurls herself into the sea. The ship founders, and the forms of the Dutchman and Senta are perceived united and transfigured.

Floire und Blanscheflur, a Middle High German poem of some 8,000 lines, written probably c. 1220 by Konrad Fleck (q.v.). It tells the story of two faithful lovers. Floire, the son of the heathen king of Spain, falls in love with Blanscheflur, the daughter of a Christian captive. Floire's parents disapprove; he is sent away, and Blanscheflur sold into slavery and taken to Babylon. Floire's parents erect a handsome tomb for Blanscheflur, the description of which is the most celebrated element in the poem; and they convince Floire that she is dead. His despair is so intense that they reveal the truth. Floire leaves for Babylon and is reunited with Blanscheflur, but the pair are discovered and condemned to death. Their selfless devotion to each other makes so great an impression that they are pardoned and return to Spain. Meanwhile Floire's father has died, the son assumes the throne and the pair live happily, dying on the same day at the age of one hundred.

The source of the poem is French. It is narrated as a touching love story, inclining to the sentimental, but full of charm. It is known in two corrupt MSS. of the 15th c., and two fragments from the 13th c. In addition there is a prose version dating from the late 15th c. A modernized version was published in 1822 by Sophie Tieck (q.v.).

Florestan, see SCHUMANN, ROBERT.

Florian Geyer, a play by Gerhart Hauptmann (q.v.), first performed in January 1896 in the Deutsches Theater, Berlin, and published in that year. It is sub-titled *Die Tragödie des Bauernkrieges in fünf Akten mit einem Vorspiel.* A product of Hauptmann's first Naturalistic period, it traces the Peasants' War (see BAUERNKRIEG) from the siege of Würzburg to the final collapse of the peasant cause. In Act I the peasants lay siege to Würzburg. In Act II Geyer, an idealist and one of their leaders, learns of the disastrous failure of the assault. In the third act a worse catastrophe follows at Schweinfurt, and further ill tidings are received in the fourth. Finally Geyer, beaten and at the end of his tether, is assassinated in Wilhelm von Grumbach's castle. The author is clearly in sympathy with the insurgent peasants, but impartially displays the great variety of motives among the participants. It is a huge canvas with more than seventy successfully differentiated characters, and Geyer is not so much the hero as a device for focusing the historical presentation.

Flucht in die Finsternis, a Novelle by A. Schnitzler (q.v.), published in 1931. It is an almost clinical study of persecution mania, the story of a derangement which destroys the happiness of the hero Robert and his fiancée Paula, growing irresistibly until Robert, in fear,

murders his well-intentioned brother Otto; he runs out into the snowy night of the mountain landscape and there meets his death.

Flugschriften, 'broadsheets', anonymously published sheets conveying information, normally in a tendentious and frequently scurrilous tone. They were a powerful means of propaganda in the great controversies which followed the coming of the Reformation (q.v.). Their heyday is the 16th c. and 17th c. Though their tone is usually, in the most literal sense, 'vulgar', they are often the work of highly educated men. Many of them are illustrated with crude woodcuts.

Fluß ohne Ufer, a trilogy by H. H. Jahnn (q.v.), the separate novels of which are *Das Holzschiff* (1949), *Die Niederschrift des Gustav Anias Horn* (2 vols., 1949–50), and *Epilog* (1961).

Das Holzschiff, dealing with the mysterious voyage and freight of a full-rigged ship, the *Lais,* and its equally mysterious crew, is externally Naturalistic but in essence is tortuously and elusively symbolical. It has even been suggested that the attempt to penetrate the supposed concealed space above the keel, by which the timbers are opened and the ship sinks stern first, is a symbol of the self-destructiveness of Jahnn's ruthless introspective self-dissection. The previously unnoticed appearance of Venus as figurehead as the ship sinks is an evidently erotic, but unintegrated, symbol. The second part (*Die Niederschrift*) follows two of the survivors of the sinking ship (Horn and Tutein), united in blood brotherhood, through curious and often perverted adventures. One of these two, the murderer Tutein, dies and the survivor replaces him with Ajax von Uchri. The book ends with Horn's last testament with its expression of vain desire for 'Segen auf Wälder und Tiere', and a postscript explains that Horn is found with battered skull, his favourite mare and dog shot, Ajax having vanished. The third part, *Epilog,* remained a fragment.

Flußpiraten des Mississippi, Die, a three-volume novel of adventure by F. Gerstäcker (q.v.), published in 1848. Mississippi shipping in the neighbourhood of the town of Helena suffers heavy loss through the activities of a gang operating secretly from an island. Ships are sunk and murders committed. The authorities let sleeping dogs lie, but the Irishman O'Toole successfully probes the secret of the gang's whereabouts and activities. Dayton, the judge of Helena, proves to be identical with Kelly, the gang leader. In a spectacular scene on land and on the river the pirates are defeated and massacred by enraged backwoodsmen. In a preface Gerstäcker asserts that the story is based on fact.

FOCK, GORCH, pseudonym of Hans Kinau (Finkenwerder, 1880–1916, at sea), a deep-sea fisherman's son and brother of R. Kinau (q.v.), held a number of shore jobs connected with the sea. His pen-name 'Fock' (foresail) deliberately symbolizes his imaginative preoccupation with seafaring. He is the author of a number of separate stories and collections concerned with the sea, including *Schullengrieper und Tungenknieper* (1911), *Hein Godenwind, de Admiral von Moskitorien* (1912), *Fahrensleute* (1914), and *Hamborger Janmooten* (1914), which mingle High German narrative with Low German (Plattdeutsch) dialogue. His best-known book is the sea-fishing novel *Seefahrt ist not!* (q.v., 1913). Fock's work combines humour with a sense of the poetry of the sea and its desolation. He volunteered for the German navy in 1914, and met his end when the cruiser *Wiesbaden* was sunk at Jutland (31 May 1916). His diaries were published posthumously as *Sterne überm Meer* (1917) and *Ein Schiff! Ein Schiff! Ein Segel!* (1934), a title devised by its National Socialist publisher. *Sämtliche Werke* appeared in 5 vols., 1925.

FOLLEN, AUGUST or ADOLF (Gießen, 1794–1855, Berne), brother of K. Follen (q.v.), served in the War of Liberation, and was afterwards prominent in the Burschenschaft (q.v.). He was imprisoned from 1819 to 1821, and then lived in Switzerland, at first as a schoolmaster and later as a landowner. In 1819 he published a collection of political songs by various poets (*Freie Stimmen frischer Jugend*).

FOLLEN, KARL (Romrod nr. Alsfeld, 1796–1840, Lake Erie), a radical politician in the years after 1815, and brother of A. Follen (q.v.), fled from persecution into France, and then moved to Switzerland. He emigrated to America in 1824, and died in a fire on board a lake steamer. In 1818 his poem *Das große Lied* was distributed as a propagandist broadsheet.

FOLZ, HANS (Worms, *c.* 1450–not later than 1515, Nürnberg), a barber surgeon who settled in Nürnberg, became a prominent Meistersinger and extended the range of Meistergesang (q.v.) by insisting that the songs should not be confined solely to the traditionally accepted tunes. He himself devised 27 new 'Töne'. He wrote verse Sprüche of didactic or practical character, coarse and sometimes obscene Schwänke, and six Fastnachtspiele. One of his poems (*Von der collation Maximilians in Nürnberg zugericht*) tells of the visit of the Emperor Maximilian I (q.v.) to the city in 1491.

FONTANE, THEODOR (Neu-Ruppin, 1819–98, Berlin), came of a French family which had settled in Prussia in the 18th c. In 1827 the family moved to Swinemünde on the Baltic, but the boy spent 1833–6 in Berlin receiving a desultory schooling. In 1836 he was apprenticed to an apothecary (his father's profession), spent four years in Berlin, and was then an assistant successively in Burg, Dresden, and Leipzig. He did his military service in 1844 as a volunteer (Einjährig-Freiwilliger) in a guard regiment in Berlin, and during this time was able to make his first brief visit to England. He was also a member of the literary club Der Tunnel über der Spree (q.v.), in which some of his early poems were read. In 1849 he abandoned the practice of pharmacy and lived partly from his pen and partly from minor civil service posts. He married in 1850, and in the same year his first books of poetry (*Männer und Helden* and *Von der schönen Rosamunde*) were published. In 1852 he spent the summer in London as correspondent for German newspapers. From 1852 to 1855 he lived in straitened circumstances in Berlin, giving private tuition and again working in a government office. His account of his London experiences appeared in 1854 (*Ein Sommer in London*). In 1855 he returned to London as an official correspondent, and remained there until 1859, touring Scotland in 1858. On his return to Berlin he conceived the idea of writing a travel book on the landscape and historical associations of his own country, the outcome of which was the four volumes of *Wanderungen durch die Mark Brandenburg* (q.v., 1862–82), which were supplemented in 1889 by a fifth (*Fünf Schlösser*). In 1864, and again in 1866, he visited the scenes of war, and subsequently published *Der Schleswig-Holsteinische Krieg im Jahre 1864* (1865) and *Der Deutsche Krieg von 1866* (2 vols., 1869–70). After working for the *Kreuz-Zeitung* for some years, he removed in 1870 to the *Vossische Zeitung* (qq.v.) and in the same year entered France as a war correspondent. In October he was taken prisoner in Domremy, but was released in December. He incorporated his experiences in *Kriegsgefangen* (1871). In 1876 he was appointed secretary to the Berlin Academy of Arts, and almost immediately resigned, a step which imperilled his marriage. He then devoted himself to finishing a long historical novel (*Vor dem Sturm*, q.v., 4 vols., 1878).

After this late beginning (he was 58) he wrote fourteen more works of fiction in twenty years. *Grete Minde* (q.v., 1879), *Ellernklipp* (q.v., 1881), and *Schach von Wuthenow* (q.v., 1882, dated 1883) were all short works set in the past, as was the later *Unterm Birnbaum* (q.v., 1885). The first novel dealing with contemporary Berlin was *L'Adultera* (q.v., 1882). *Graf Petöfy* (1884) was an excursion into an Austro-Hungarian setting, but *Cécile* (q.v., 1887), *Irrungen, Wirrungen* (q.v., 1888), and *Stine* (q.v., 1890) established the Berlin series. *Quitt* (1890, dated 1891) was a venture into a part-American environment. Fontane's *Gesammelte Romane und Novellen* appeared in twelve volumes 1890–1 and contained all the fiction to this point, plus *Kriegsgefangen*. In 1891 *Unwiederbringlich* (q.v.), a novel of the Danish aristocracy, was published, followed by two outstanding novels of Berlin life, *Frau Jenny Treibel* (q.v., 1892, dated 1893) and *Effi Briest* (q.v., 1895). *Die Poggenpuhls* (q.v., 1896) followed the same pattern in a lower key, as did the unfinished *Mathilde Möhring* (1906; correct version, 1969). The discursive conversational Berlin novel *Der Stechlin* (q.v., 1898, dated 1899) provided a fitting epitaph. A number of the novels were first serialized in magazines; the dates given are those of appearance in book form.

In later years Fontane wrote two autobiographical works, *Meine Kinderjahre* (1894), which is one of his most attractive books, and *Von Zwanzig bis Dreißig* (1898). The poems (*Gedichte*), first published in 1851, were repeatedly expanded up to the fifth edition (1897). From 1870 to 1890 Fontane was dramatic critic for the *Vossische Zeitung*, and his notices include a favourable review of the first performance of G. Hauptmann's *Vor Sonnenaufgang* (q.v., 1889).

Long known only as the author of ballads, Fontane developed in old age into one of the most alert and subtle of German novelists. His numerous stories of Berlin life portray the social scene with delicate irony, and depict the characters with penetration and compassion. He is probably the most accomplished writer of dialogue among all the German novelists. His standing has risen steadily in the 20th c.

Sämtliche Werke (Nymphenburger Ausgabe), ed. E. Gross *et al.*, appeared in 24 vols., 1959–75; *Th. Fontane: Briefe*, ed. K. Schreinert and C. Jolles, were published in 4 vols., 1968–71.

Formgeschichte (Form Criticism) was introduced by the Old Testament scholar H. Gunkel (1862–1932), and applied to New Testament studies, notably by Dibelius and R. Bultmann (qq.v.). It arose out of a synthesis: (1) of literary criticism (originally applied to Old Testament stories and psalms), i.e. the analysis of textual structure; and (2) the study of sources (Quellenkritik), which examines the occurrence and function of the structural units in their original historical settings. Both types of criticism are applied to the narrators' (e.g. the evangelists') adaptations of established genres for a specific purpose.

FORSTER, FRIEDRICH (Munchengosserstadt nr. Altenburg, 1791–1868, Berlin), a regimental comrade of Theodor Körner (q.v.) in 1813, published in later life a biography of Körner (*Theodor Körners Leben*, 1862). Förster also wrote plays (*Das Hermannsfest*, 1815; *Der Sylvesterabend*, 1816; *Dankwarths Heimkehr*, 1826; *Gustav Adolf*, 1833), adapted Shakespeare (*Richard III*, *All's well that ends well*, *Julius Caesar*), and tried his hand at popular history (*Preußens Helden in Krieg und Frieden*, 1846–61). Employed first as a lecturer at the School of Artillery and later as curator of the Ethnographical Museum, he each time lost his post by publications which offended the authorities.

FORSTER, FRIEDRICH, pseudonym of Waldfried Burggraf (Bremen, 1895–1958, Bremen), an actor who became director of the Bavarian state theatres in Munich (1933–8). He is the author of plays, including *Sermon der alten Weiber* (1928), *Der Graue* (1931), the Volksstück *Robinson soll nicht sterben* (1933), *Alle gegen Einen, Einer für Alle* (1933), the comedy *Die Weiber von Redditz* (1935), *Ariela* (1930), *Gastspiel in Copenhagen* (1940), *Die Liebende* (1949). This last deals with Medea.

FORSTER, GEORG (Amberg, 1514–68, Nürnberg), studied medicine at Heidelberg and Wittenberg, and settled in 1544 at Nürnberg, where he practised as a physician. Between 1539 and 1556 he published five volumes of songs, which he had collected, together with the music for them. Love songs predominate, and the second and fifth volumes contain a number of folksongs. Some of the music was composed by Forster himself. Though the titles of the individual volumes vary (they appeared respectively in 1539 (1), 1540 (2), 1549 (3), 1556 (4 and 5)), they are usually known as *Frische teutsche Liedlein*.

FORSTER, GEORG or GEORGE (Nassenhuben nr. Danzig, 1754–94, Paris), of English descent (his ancestors had emigrated in the middle of the 17th c.), was a child prodigy, and was exploited in his childhood by his restless, speculating father, J. R. Forster (q.v.). In 1765 he was taken on a journey into Russia, and later lived in London. He accompanied his father on Captain Cook's second South Sea voyage (1772–5), and wrote an account of it in English, *A Voyage towards the South Pole and round the World* (2 vols., 1777). The German version appeared a year later as *Johann Reinhold Forsters Reise um die Welt* (2 vols., 1778–80). He was appointed to a professorship at Kassel in 1778, and in 1784 at Vilna in Poland. In the following year he married

Therese, daughter of C. G. Heyne (1719–1812), a famous professor at Göttingen University. He returned to Germany in 1788, and became librarian at Mainz. He was an enthusiastic adherent of the French Revolution, and advocated the incorporation of the Rhineland in France. In 1793 he was in Paris as a representative of the Mainz republicans. He was outlawed in Germany, and died in 1794 in penury. Therese Forster deserted her husband for L. F. Huber (q.v.), whom she married after Forster's death.

In addition to the great book of exploration, Forster wrote an account of the lower Rhine, *Ansichten vom Niederrhein* (1791–4). Based on a journey made with A. von Humboldt (q.v.) in 1790, it is one of the outstanding German travel books. Forster is the subject of a novel by Ina Seidel, *Das Labyrinth* (q.v., 1922). *Georg Forsters Werke*, planned in 18 vols. and edited under the auspices of the Akademie der Wissenschaften, DDR, contain 1958–82 *A voyage round the World, Reise um die Welt* (2 vols), *Streitschriften und Fragmente zur Weltreise, Kleine Schriften zu Kunst und Literatur, Schriften zu Philosophie und Zeitgeschichte, Ansichten vom Niederrhein, Rezensionen, Tagebücher*, and 5 vols. of *Briefe* (the 5th, comprising vol. 18 of the edition, consists of letters addressed to Forster).

FORSTER, JOHANN REINHOLD (Dirschau, 1729–98, Halle), a Protestant pastor, abandoned his cure, and in 1765 made an unsuccessful speculative journey into southern Russia. With his son Georg (see FORSTER, GEORG or GEORGE) he accompanied Captain Cook on his second South Sea voyage (1772–5). He was appointed to a professorship at Halle University in 1780. His writings, chiefly on exploration and geography, include *Geschichte der Entdeckungen und Schiffahrten im Norden* (1784).

FÖRSTER-NIETZSCHE, ELISABETH (nr. Lützen, 1846–1935, Weimar), was married to B. Förster and was the sister of F. Nietzsche (q.v.). She devoted herself to her brother during his illness and insanity, edited his papers, and wrote copiously on him, including a biography, *Das Leben Friedrich Nietzsches* (1895–7). She is no longer regarded as a reliable editress and scholar.

Fortunat, a five-act play in verse by E. von Bauernfeld (q.v.), first performed at the Theater in der Josephstadt, Vienna, in 1835. It is described as *Ein dramatisches Märchen*.

Fortunatus, a German Volksbuch (q.v.) first printed in 1509, recounting the hero's adventures. Fortunatus acquires from Fortuna an inexhaustible purse and steals a 'wishing hat', which confers invisibility and instantly transports its wearer to his desired destination. A sequel narrates the misfortunes of Fortunatus's sons.

The original Volksbuch was many times reprinted, and retained its popularity up to the end of the 18th c. In more sophisticated form it was treated dramatically by A. von Chamisso (q.v., *Fortunati Glückssäckel und Wunschhütlein*, 1806), Ludwig Tieck (q.v., in *Phantasus*, 1815), and Eduard von Bauernfeld (*Fortunat*, q.v., 1835). A poem in *ottava rima, Fortunat und seine Söhne*, was left unfinished by L. Uhland (q.v.). The source of the story, which is known in many European languages, including French, Italian, Dutch, and English (Dekker's *Old Fortunatus*, 1600), is uncertain. A French origin in the middle of the 15th c. has been suggested. The original Volksbuch was reprinted in 1973.

FOUQUÉ, FRIEDRICH HEINRICH KARL, FREIHERR DE LA MOTTE (Brandenburg, 1777–1843, Berlin), a grandson of a well-known general serving under Friedrich II (q.v.), entered the army in 1794 as a cornet in the Kürassier-Regiment Herzog von Weimar. At this time he saw active service. In 1803 he married Karoline von Briest (see FOUQUÉ, KAROLINE DE LA MOTTE) and resigned his commission, living on his wife's estate at Nennhausen, west of Berlin. In the same year three dramatic sketches by Fouqué, including *Der gehörnte Siegfried in der Schmiede*, were published by F. Schlegel (q.v.) in *Europa*; and A. W. Schlegel (q.v.) published Fouqué's *Dramatische Spiele* in the following year. Both these publications appeared under the pseudonym Pellegrin.

Fouqué's production of romantic novels, tales, and plays grew rapidly, and he became probably the most widely read of all Romantic authors. Such prolific writing resulted in loss of quality, and most of his works proved ephemeral. Fouqué himself collected what he believed to be his best works into twelve slim volumes, which he published as *Ausgewählte Werke. Ausgabe letzter Hand* in 1841, at a time when his rather obvious brand of Romanticism had lost its magic. The works which he singled out in this way are: the mythological trilogy *Der Held des Nordens* (q.v., 1810), consisting of *Sigurd der Schlangentöter, Sigurds Rache*, and *Aslauga*, the play *Eginhard und Emma* (1811), the Novelle *Undine* (q.v., 1811), which is indisputably his best work, the long novel *Der Zauberring* (q.v., 1813), the shorter work *Sintram* (1815), the stories *Die beiden Hauptleute* and *Der Geheimrath*, a few lesser narrative works, and a collection of poems. An edition of the poems had been published earlier in five volumes (*Gedichte*, 1816–27).

In 1813 Fouqué enlisted in the forces mobilized against Napoleon, and was soon a captain of horse in a regiment of volunteers. He served with distinction, participating in many engagements, and returned to civilian life in 1815. After his wife's death in 1831, he gave lectures in Halle, and was later invited to Berlin and granted a pension by King Friedrich Wilhelm IV (q.v.). Fouqué's combination of historical, mythological, and idyllic elements is usually only partially successful, but in *Undine* he created a work which has become a symbol for Romanticism.

FOUQUÉ, KAROLINE DE LA MOTTE (Nennhausen, 1773–1831, Nennhausen), *née* von Briest, was first married in 1791 to Herr von Rochow, from whom she was divorced in 1800. In 1803 she married the officer and author F. de la Motte FOUQUÉ (q.v.), with whom she lived on her estate at Nennhausen near Rathenow in Brandenburg. She was herself a prolific writer in the Romantic manner, chiefly of novels and Novellen. She used the pseudonym Serena.

Fragmente, (1) title given to scattered aphorisms, *aperçus,* and detached paragraphs concerned with criticism or aesthetics, written by Friedrich Schlegel (q.v.) in the years 1796–1801, and published chiefly in the *Lyzeum der schönen Künste* (1797) and *Das Athenäum* (q.v., 1798–1801). They are brilliant and illuminating improvisations which do not lend themselves readily to systematic classification. The most famous *Fragment* includes the definition 'Die romantische Poesie ist eine progressive Universalpoesie'.

(2) Common designation for Herder's *Über die neuere deutsche Litteratur* (q.v.).

Fraktur denotes the German black letter or 'Gothic' type founts which derive from medieval handwriting, whereas Antiqua (roman) denotes founts originating in the latinized script of the humanists. Fraktur became the accepted typeface for German books from the 16th c., but in the middle of the 18th c. Antiqua was adopted for many works of literary character; it was widely adopted in the early 20th c. In 1910 a society was formed to defend Fraktur (Bund für deutsche Schrift). Since 1945 Fraktur has yielded to Antiqua, and its use is now virtually restricted to ornamental functions (titles etc.).

FRANCISCI or **FINX,** ERASMUS (Lübeck, 1627–94, Nürnberg), a minor writer who continued the *Monatsgespräche* of Johann Rist (q.v.) after the author's death.

FRANCK, HANS (Wittenburg, Mecklenburg, 1879–1963, Frankenhorst), at first a school-master, worked in the theatre and became a prolific writer. His first play, a historical tragedy (*Der Herzog von Reichstadt,* 1910), was influenced by the work of F. Hebbel (q.v.). Much of his work consisted of novels, which are *biographies romanceés* of notable men and women: *Annette* (1937) deals with A. von Droste-Hülshoff (q.v.), *Marianne* (1953) with Marianne von Willemer (q.v.), and *Letzte Liebe* (1958) with Goethe and Ulrike von Levetzow (qq.v.). *Die vier großen B* (1955) consists of four Novellen on J. S. Bach, Beethoven, Brahms, and Bruckner (qq.v.).

FRANCK, JOHANN (Guben, 1618–77, Guben), a Protestant pastor, composed simply written, deeply felt hymns, which were published in *Hunderttönige Vaterunserharfe* (1646) and in two volumes of *Deutsche Gedichte* (1672–4). They include 'Schmücke dich, o liebe Seele' (1649), which Klopstock submitted to considerable adaptation in *Veränderte Lieder* attached to *Geistliche Lieder* (1758). Franck also wrote a German version of the *Te Deum,* composed in 1648 for the end of the Thirty Years War (see DREISSIGJÄHRIGER KRIEG).

FRANCK, also **FRANK,** SEBASTIAN (Donauwörth, 1499–1542, Basel), was at first a priest, then became a follower of Luther (q.v.) in Nürnberg, but later rejected Lutheranism because of its rigid adherence to dogma. His personal religious views, his tendency to mysticism, and his tolerance alienated both parties, and he was driven from Strasburg in 1533 and from Ulm in 1539. He spent the last three years of his life in Basel. Franck was a man of vision and courage in an age when the refusal to conform was accounted a crime. His religious views are expressed in *Paradoxa* (1534), *Die Guldin Arch* (1539), and *Das Verbütschiert Buch* (1539).

Franck was also a historian, notable for his independent judgement (*Chronika, Zeitbuch und Geschichtbibel,* 1531, and *Germaniae Chronikon,* 1538), and a geographer (*Weltbuch,* 1534). In his *Kriegbüchlein des Friedens* (1539) he rejected war. He made an important collection of proverbs (*Die deutschen Sprichwörter,* 1541), which succeeds in its avowed object of demonstrating the wealth of German proverbial speech.

FRANCKE, AUGUST HERMANN (Lübeck, 1663–1727, Halle), a pietistic theologian influenced by P. J. Spener (q.v.), had to face persecution for his beliefs before becoming a professor at Halle University in 1692. He appealed for deep personal piety and for the practical expression of devotion in works of charity and in service. He founded a number of charitable institutions, which became known as Franckesche Stiftungen.

Francke's published works include *Öffentliches Zeugnis vom Werk, Wort und Dienst Gottes* (1702), and *Segensvolle Fußtapfen des noch lebenden Gottes* (1709). His *Kurzer Unterricht, wie man die Heilige Schrift zu seiner wahren Erbauung lesen solle* is still included in some printings of the Lutheran Bible.

FRANCKENBERG, ABRAHAM VON (Ludwigsdorf, Silesia, 1593–1652, Ludwigsdorf), a mystical writer, was a Silesian nobleman who studied at Leipzig, Wittenberg, Jena, and Frankfurt/-Oder universities. In 1622 he came into contact with Jakob Böhme (q.v.), in whose mystical thought he became absorbed. Franckenberg travelled in Europe, and lived for a time in Danzig; he spent his last years (1649–52) in his ancestral home. His devotional works in verse and prose include *Andächtige Bet-Gesänglein* (1633), *Via veterum sapientium d.i. der Weg der alten Weisen*, and *Mir nach* (both posth. published in 1675). His admiration of Böhme is reflected in his biography *Bericht von dem Leben des Jakob Böhme* (1651).

Franco-Prussian War, see DEUTSCH-FRANZÖSISCHER KRIEG.

FRANÇOIS, LOUISE VON (Herzberg, Elster, 1817–93, Weißenfels), the daughter of a Prussian major of Huguenot descent, who died when she was an infant, lost her fortune through the mismanagement of her guardian—a disaster which also cost her her fiancé: she released him from his engagement when she discovered herself to be portionless. She remained in straitened circumstances, spending many years looking after her ailing mother. A woman of character, determination, and firm religious faith, she educated herself; when well into middle age, she began to write and publish fiction with the encouragement of G. Freytag (q.v.). Her first stories appeared anonymously in magazines. She began under her own name with two volumes of *Novellen* (1868), in which the best-known story is *Judith die Kluswirtin*. Her principal novel is the historical work *Die letzte Reckenburgerin* (q.v., 1871), and this was followed by *Frau Erdmuthens Zwillingssöhne* (2 vols., 1873), *Stufenjahre eines Glücklichen* (1877), and *Der Katzenjunker* (1879). Two volumes of *Erzählungen* appeared in 1871, and two further collections in 1874 (*Hellstädt*, 3 vols.) and in 1876 (*Natur und Gnade*, 3 vols.). Two stories appeared separately, *Phosphorus Hollunder* in 1881, and *Zu Füßen des Monarchen* in 1882. She also wrote a comedy, *Der Posten der Frau* (1882). A final volume of stories (*Das Jubiläum*) was published in 1886. Her works are predominantly historical, and she made careful studies of appropriate sources in preparation for them. They show perceptive characterization, a gift for realism, and some originality of construction. Louise von François was a friend and correspondent of both Marie von Ebner-Eschenbach and C. F. Meyer (qq.v.).

FRANK, BRUNO (Stuttgart, 1887–1945, Beverley Hills, California), studied at Tübingen, Munich, Strasburg, and Heidelberg universities, and travelled for some years, publishing 3 vols. of poetry (*Aus der goldenen Schale*, 1905; *Die Schatten der Dinge*, 1912; *Requiem*, 1913). He fought in the 1914–18 War, and in 1915 published his first novel, *Die Fürstin*. His inclination was towards historical fiction, and his best-known novel, *Trenck* (1926), sensationally exploits an adventurous figure of the reign of Friedrich II (q.v., who is the subject of his novel *Tage des Königs*, 1924, and see TRENCK, FRIEDRICH, FREIHERR VON DER). In 1933 Frank emigrated, staying successively in Austria, Switzerland, France, and England, and finally settling in the U.S.A. as friend and neighbour of Th. Mann (q.v.). Frank's *Cervantes* (1934) is a *biographie romancée*. Among other novels should be mentioned *Der Reisepaß* (1937) and *Die Tochter* (1943, a bitter attack on anti-Semitism in Poland). He also wrote successful plays, including the long-running comedy *Sturm im Wasserglas* (1930) and the historical play *Zwölf Tausend* (1927), dealing with the sale abroad of young men as soldiers to colonial powers, in which the dramatic intervention of Frederick the Great (in the form of a military envoy of high rank) ensures a happy end. The nucleus of this theme occurs as early as Schiller's *Kabale und Liebe* (q.v.).

FRANK, LEONHARD (Würzburg, 1882–1961, Munich), of working-class origin, took a number of jobs, such as cycle mechanic, chauffeur, house painter, and hospital porter. A talent for drawing and painting led to six years' study and practice of art in Munich (1904–10). He was in Berlin from 1910 and, being an ardent pacifist, took refuge in Switzerland in 1915. His first novel, *Die Räuberbande*, appeared in 1914. Most of his work is directed against the shortcomings of society, seen from a socialistic and humane standpoint. *Die Ursache* (1916), a Novelle, is an attack on capital punishment. Frank dramatized it in 1929. The collection of Novellen *Der Mensch ist gut* appeared in 1919. These stories reject existing society, and look forward to a humane fraternal future. In 1918 Frank returned to Germany. He lived in Munich and became a writer. In 1924 he published the novel *Der Bürger*, and in 1927 *Karl und Anna* (q.v., dramatized in 1929) and *Das Ochsenfurter Männer-*

quartett, which deals with the misery of the period of runaway inflation; it contains, however, an idyllic love story. In 1929 followed *Bruder und Schwester.*

In 1933 Frank emigrated, was in Switzerland, France (where he was for a time interned), Portugal, and, from 1940 to 1950, in the U.S.A. After the war he spent the rest of his life in Munich. The later works, in which the psychological interest increased at the expense of the political, include the novels *Traumgefährten* (1936), *Mathilde* (1948), *Die Jünger Jesu* (1949), and the story *Deutsche Novelle* (1954). His autobiography, published in 1952, is entitled *Links wo das Herz ist. Gesammelte Werke* (6 vols. and a supplementary volume), ed. G. Caspar, appeared in 1957–9.

Frank der Fünfte. Oper einer Privatbank, a comedy with musical numbers by F. Dürrenmatt (q.v.), first performed in 1960. Its style recalls Brecht's *Dreigroschenoper* (q.v.), though in a note to the revised version of 1964 (*Frank der Fünfte. Eine Komödie mit Musik von Paul Burkhard,* b. 1911) Dürrenmatt asserts an affinity with Shakespeare's *Richard III.* Frank V, in order to defend the merchant bank of which he is director by inheritance, callously commits crime after crime, including several murders, in all of which his wife Ottilie is an active collaborator. They are eventually outmanœuvred and exposed by Frank's own son, who locks his father up to die in the safe. The dramatic style includes deliberate absurdity, doggerel, and parody. The work is a parable with a serious message for democratic society.

Frankfurt am Main, Goethe's birthplace (see GOETHE-HAUS), is a city of Roman origin and was important in the Middle Ages. The German King (Friedrich I, q.v.) was crowned at Frankfurt in 1152, and from 1356 it was the appointed place for the election of the emperor. In 1375 it became a free imperial city (Freie Reichsstadt). In 1562 the first coronation of an emperor took place in Frankfurt, and these ceremonies were held there until the last emperor of the Holy Roman Empire was crowned in 1792. The ceremony of 1764 is graphically described by Goethe in *Dichtung und Wahrheit* (q.v., Bk. 5). Frankfurt was a free city in the German Confederation, and the Paulskirche was the seat of the National Assembly in 1848 (see FRANKFURTER NATIONALVERSAMMLUNG). In 1866 it was annexed by Prussia. The university was founded in 1914.

FRANKFURTER, PHILIPP, of whom little is known beyond his residence at Vienna between 1420 and 1490, is the author of a poem recounting in anecdotal form the pranks of the 'parson of Kalenberg' (the parish priest of Kahlenbergerdorf in Döbling on the northern outskirts of Vienna). *Der Pfaffe vom Kalenberg* is based on a real person transformed into a legend, and is set in the time of Otto the Merry of Austria (d. 1339). The episodes, which are frequently coarse and obscene, show the parson outwitting the peasants, and at loggerheads with his superiors, who are displayed as immoral and corrupt. The poem was probably written *c.* 1470, and was first printed in 1473.

FRANKFURTER GRETCHEN, see GRETCHEN, FRANKFURTER.

Frankfurter Nationalversammlung, the National Assembly or parliament, founded in 1848 and lasting until March 1849. The impact of the February Revolutions in France resulted in massive popular demands in Germany and the Austrian Empire for constitutional change (see REVOLUTIONEN 1848–9), and the Diet (Bundestag) of the German Confederation (see DEUTSCHER BUND) was driven to appoint a small committee of 17 to prepare for a revision of the constitution. By the end of March 1848 an independent preliminary parliament (Vorparlament) was established under the leadership of Heinrich von Gagern (q.v.). This appointed a committee of 50 members (Fünfzigerausschuß) whose call for the election of a National Assembly was accepted by the Diet, which thus ceased to function.

The Frankfurt Parliament opened its sessions on 18 May 1848 in St. Paul's Church (see PAULSKIRCHE) under the presidency of Heinrich von Gagern. Archduke Johann of Austria (see JOHANN, ERZHERZOG) was elected Reichsverweser. The Parliament consisted of representatives from the member states of the old Confederation, based on the ratio of 1:70,000. The 16 states which had a population of less than 70,000 were allocated one representative each. Independent male citizens over 24 were eligible. The aim of the Parliament was to create a united Germany on a parliamentary and constitutional basis. Representatives also included a minority of non-German members (e.g. Danes, Italians, Poles), and members were not necessarily in favour of national unity. The majority of representatives were educated men of liberal convictions, members of the academic and legal professions, a preponderance which earned it the nickname Professorenparlament.

Of the various committees, the constitutional committee (Verfassungsausschuß) was of central importance, but from the late autumn of

1848 the Assembly was divided between those who sought full unity (Großdeutsche) and those who wished to detach Austria from Germany (Kleindeutsche). In December 1848 von Gagern replaced the Austrian von Schmerling as head of the ministry, in order to resolve the deadlock by a compromise, in accordance with which there would be two federal states, a small one (Engerer Bund) headed by Prussia, and a large one (Weiterer Bund) including German Austria. But Schwarzenberg (q.v.), head of the Austrian government, succeeded in March 1849 in having the entire Habsburg Empire included, and the Assembly agreed to a modified constitution providing for a hereditary emperor, direct elections, an imperial ministry responsible to parliament, and an imperial diet (Reichstag) consisting of two houses, the Staatenhaus and the Volkshaus (Gesamtstaatsverfassung, 4 March 1849). On 28 March 1849 Friedrich Wilhelm IV (q.v.) of Prussia was elected emperor (receiving 290 votes with 248 abstentions), but he rejected the crown. Thereupon Austria and Prussia withdrew their representatives and other states followed their example, so that the Parliament shrank to a minority of the republican left. On finding themselves locked out of the Paulskirche, they moved to Stuttgart (30 May 1849), where the army intervened. There were fresh risings in Saxony (Dresden), the Palatinate (Pfalz), and Baden (resulting in the temporary institution of a republican government). The leaders were officers from Prussia's Polish territories. Many of them suffered imprisonment and execution when they were finally defeated at Rastatt by Prince Wilhelm of Prussia (see WILHELM I).

Since Friedrich Wilhelm's professed aim was to promote German unity, it is surprising that he withdrew at the precise moment when this goal had been achieved. The King's own explanation was that he could not accept a crown stained with the blood of revolutions, but this does not explain the hurried dissolution of the Parliament. Many reasons have been put forward for the failure of the Frankfurt Parliament. It has been held that it was doomed from the start because of its intellectual and ideological character. Its members tended to be men of lofty ideals and good will without experience in practical politics. Subsequent events revealed an important factor in the century-old rivalry between Austria and Prussia; Friedrich Wilhelm may have feared an armed conflict with Austria, which was indeed the ultimate outcome of the Austro-Prussian dualism (see DEUTSCHER KRIEG).

The reinstatement of the German Confederation was enforced by Austria's diplomatic victory over Prussia, the culmination of which was the treaty of Olmütz (1850, see OLMÜTZER PUNKTATION), at which Prussia's own plans to replace

the Parliament of all German states by a smaller parliament came to grief (see PREUSSISCHE UNION).

The writers Ludwig Uhland (q.v.) and Anton A. von Auersperg (see GRÜN, ANASTASIUS) were members of the Frankfurt Parliament. Hebbel (q.v.) was considered, but not elected. Noted members supporting von Gagern included Arndt, Jacob Grimm, Gervinus, and Dahlmann (qq.v.).

Frankfurter Spiele, designation of two medieval religious plays:

(1) *Frankfurter Dirigierrolle*, a 'producer's text' (Regiebuch) which gives, in Latin, instructions for the play, followed by the first line (in Latin or in German) of the following speech or song. It covers the life of Jesus from Baptism to Ascension and is provided with a prologue in which David, Solomon, and certain prophets appear. The book, which is entitled *Ordo sive Registrum*, was written c. 1350 by Baldemar von Peterwell, a canon of Frankfurt. The play, the progress of which it outlines, was intended to be performed in two parts on separate days. Its sources are the *St. Galler Passionsspiel* and *Die Erlösung* (qq.v.).

(2) *Frankfurter Passionsspiel von 1493*, written in more than 4,000 Middle High German lines, tells the story of Jesus from the summoning of the disciples to his interment. It was intended for production on three separate days.

Frankfurter Zeitung, a daily newspaper founded in 1856, and one of the principal organs of democratic opinion in Germany. After 1945 it continued to appear in Fraktur (q.v.), and this type-face was not abandoned until 1966.

Fränkisches Gebet, title given to a prayer of one sentence set down in Old High German in the monastery of St. EMMERAM (q.v.) in Regensburg between 789 and 821.

FRANZ I, KAISER (Nancy, 1708–65, Innsbruck), originally Franz Stephan, Herzog von Lothringen, and, from 1735, when he had to renounce Lorraine, Grand Duke of Tuscany, married in 1736 the Empress Maria Theresia (q.v.), who was then an Austrian archduchess. After her accession in 1740 Franz was her coregent. During the War of the Austrian Succession (see ÖSTERREICHISCHER ERBFOLGEKRIEG) he was elected emperor (1745) in succession to Karl VII (q.v.). As Maria Theresia, who cherished him as her husband and as the father of her children, preferred to conduct her policy on her own and with the help of her ministers, he had no political role. He used his financial acumen for the benefit of Maria Theresia's reforms, and for the family fortunes, and took a great interest in the cultural development of Vienna. He was a patron of the theatre, and the founder of the zoological and botanical gardens in Schönbrunn (see WIEN) and of the Natural History Museum.

FRANZ II (Florence, 1768–1835, Vienna), Deutscher Kaiser from 1792 until the end of the Holy Roman Empire (see DEUTSCHES REICH, ALTES) in 1806, and, as Franz I, Kaiser von Österreich from 1804 until his death. The son of Leopold II (q.v.), he came to the throne as the French Revolution approached its climax. Franz induced the states of the Empire to invade France in the summer of 1792 (see REVOLUTIONSKRIEGE). The expedition was a failure, and imperilled still further the lives of the French royal family (including Franz's aunt Queen Marie Antoinette), which it was intended to save. Though in 1795 Franz acquired 'Little Poland' with Lublin (see POLAND, PARTITIONS OF), he lost much territory to the French in Italy. He suffered repeated defeat in the Napoleonic Wars (q.v.), notably at Austerlitz in 1805, which was followed not only by the occupation of Vienna and the loss of territory, but also by the end of the Empire, and with it his traditional title. He had, however, made provision against this event by declaring the Austrian Crown Lands an empire (Kaiserreich) in 1804. He therefore reigned for two years as Franz II of the one empire, and as Franz I of the other. In 1809 he was again at war with France, but sustained defeat at Wagram. His attitude to popular support is indicated by his viewing the Tyrolean patriot Andreas Hofer (q.v.) as a disobedient subject. In 1810 he gave his daughter Marie-Louise to Napoleon in marriage, and in 1812 contributed a substantial force to the Grand Army for the invasion of Russia. After Napoleon's retreat he maintained a policy of caution, refraining from alliance with Russia and Prussia until August 1813.

In home policy he was a strict absolutist, seeing himself as God's delegate. He was one of the partners in the Holy Alliance (see HEILIGE ALLIANZ), and, guided by Metternich (q.v.), he exerted Austrian influence in the repression of liberal tendencies in Germany, notably in the Carlsbad Decrees (see KARLSBADER BESCHLÜSSE). He was married four times. He is the Franz of Haydn's Austrian national anthem, 'Gott erhalte Franz den Kaiser' (q.v.).

FRANZ FERDINAND, ERZHERZOG (Graz, 1863–1914, Sarajevo), also known as Franz Ferdinand von Österreich-Este, heir to the Austrian throne from 1896, and from 1898

responsible for Austro-Hungarian military affairs; he introduced reforms aimed at modernization, and became Inspector General in 1913. His uncle, the Emperor Franz Joseph (q.v.), who disliked him, kept him at a distance from other administrative functions. In politics Franz Ferdinand was opposed to Hungarian separatism, and cultivated the German alliance. He made a morganatic marriage with the Gräfin von Chotek (Herzogin von Hohenberg in 1909) of which the Emperor disapproved. Franz Ferdinand was assassinated, together with his wife, by a Serbian nationalist on 28 June 1914; the outrage led to the outbreak of the 1914–18 War five weeks later.

FRANZ JOSEPH, KAISER (Vienna, 1830–1916, Vienna), a grandson of Franz I of Austria, and nephew of Ferdinand I (qq.v.), became emperor of Austria in 1848 on the abdication of Ferdinand. After coming to the throne in the midst of a revolution (see REVOLUTIONEN 1848–9), Franz Joseph sought to re-establish absolute government, but in the late 1850s had to adjust to rapid changes. These were due to failures in foreign policy, notably the defeats suffered in Italy in 1859 and against Prussia in 1866 (see DEUTSCHER KRIEG), which resulted in the loss of the Italian provinces, in the exclusion of Austria from Germany (see AUSGLEICH, 1867), and the growth of separatist movements in Hungary and the Slav provinces. He had to adopt a moderate liberal constitutional government, and to reorganize the Austrian Empire. He achieved by his diligent, indefatigable devotion to duty considerable unsought popularity. He bore personal disasters with apparent fortitude: in 1867 his brother Maximilian (q.v.) was executed in Mexico, in 1889 his only son (see RUDOLF, KRONPRINZ) committed suicide; in 1898 his wife, the Empress Elisabeth (q.v.), was assassinated, as was in 1914 his heir, the Archduke Franz Ferdinand (q.v.).

FRANZOS, KARL EMIL (Czortkow, Galicia, 1848–1904, Berlin), the son of a doctor, studied law at Vienna and Graz universities, and then turned to journalism in Vienna. He travelled in south-eastern Europe and the Near East, exploiting some of his travels in *Halbasien, Kulturbilder aus Galizien, der Bukowina, Südrußland und Rumänien* (1878). He was primarily a writer of fiction, beginning with Novellen (*Die Juden von Barnow*, 1877) and then turning to novels (*Moschko von Parma*, 1880; *Ein Kampf ums Recht*, 1882). More Novellen appeared in 1886 (*Tragische Novellen*), and a series of novels, chiefly Galician in setting, followed (*Die Schatten*, 1888; *Judith Trachtenberg*, 1891; *Der*

Wahrheitssucher, 1893; and *Der Pojaz,* 1905). In 1879 Franzos edited the works of G. Büchner (q.v.), including *Woyzeck* (q.v.), which had not previously been published. From 1884 to 1886 he edited the Viennese *Neue Illustrierte Zeitung,* moving in 1886 to Berlin he founded the journal *Deutsche Dichtung,* which appeared successively in Stuttgart, Dresden, and Berlin.

Franz Sternbalds Wanderungen, a novel by L. Tieck (q.v.), written between 1795 and 1797, and published in 1798. Tieck planned it jointly with his friend W. H. Wackenroder (q.v.), but Wackenroder's long illness and his death in February 1798 prevented his participation after early discussions, and the novel is Tieck's own work. It is unfinished.

Franz Sternbald, a young painter and pupil of Dürer (q.v.), sets out from Nürnberg to seek professional improvement, first with colleagues in central Europe, notably Lucas van Leiden (van Leyden, 1494–1533), and then in Italy. He visits his humble parents, discovers that they are only foster-parents, but remains ignorant of his true identity. He cherishes a passion for a girl, once seen (Marie) and ever since adored, and he maintains a sentimental correspondence with his friend Sebastian in Nürnberg, who is thought to be a reflection of Wackenroder. Sternbald falls in with Rudolf Florestan, a young man of mobile, impulsive temperament, and travels with him. Under his influence the gentle, meditative Sternbald is transformed into a livelier and more sensual personality. After various erotic adventures in Germany and Italy Sternbald discovers his true love, Marie, at which point the fragment breaks off.

Tieck later indicated that Sternbald was to return to Nürnberg to devote himself to German art, and to infuse into it the warmth and vigour of his Italian experience. The book contains many poems and discussions of art and artists. An earlier draft (1795–6) grew out of the reading of Goethe's *Wilhelm Meisters Lehrjahre* (q.v.), and was published in 1836 as *Der junge Tischlermeister.* It has little resemblance to the novel.

FRAU AJA, nickname of Goethe's mother.

Frau Beate und ihr Sohn, a Novelle by A. Schnitzler (q.v.), published in 1913, and included in *Die Alternden* (1922). Frau Beate, widow of a great actor, has a son of 17, Hugo. While staying in the Austrian Alps with Hugo, she allows a young friend of his to become her lover. Hugo finds out, and while boating on the lake with him at dusk it becomes clear to her that he knows. In a close and passionate embrace

mother and son slip over the gunwale of the boat to their death.

Frau Berta Garlan, a Novelle by A. Schnitzler (q.v.), written in 1900, and published in 1901. Frau Berta, a young widow in a small Lower Austrian town, experiences a sexual reawakening, and renews acquaintance with a childhood friend, now a famous violin virtuoso. The night they spend together proves a somewhat squalid affair, and the added shock of the fate of another woman, Frau Rupius, restores Frau Berta to her accustomed domestic sphere; Frau Rupius, who tends her crippled husband, and seeks sexual compensation in Vienna, dies of the consequences of an illegal abortion.

Frauen-Liebe und -Leben, title of a cycle of nine poems written in 1830 by A. von Chamisso (q.v.), and first printed in 1831. The poems are set in the mouth of a young girl, and accompany courtship, marriage, the birth of her child, and the early death of her husband. The best-known settings are by Loewe (q.v., Nos. 1–7) and R. Schumann (q.v., Nos. 1–8). The tenderness of Schumann's music disguises the masculine egotism underlying the cycle.

Frauenlist, a short, erotic verse tale of the late 13th c. A wife takes as lover a student, whom the husband observes leaving her room, but she succeeds in persuading the latter that his senses have deluded him.

FRAUENLOB, see HEINRICH FRAUENLOB.

Frauentreue, an anonymous Middle High German verse tale of the 14th c. A knight in love with a citizen's wife receives in a tournament a dangerous wound, which the woman heals by removing the broken spear-point. He enters her room at night and is repulsed, his wound breaks open and he dies. At his burial the woman, who had loved him, discards her clothes and falls dead. The poem is related to *Der Schüler von Paris* and the Middle High German *Hero und Leander* (qq.v.).

Frauentrost, a Middle High German verse legend of the Virgin Mary, the author of which gives his name as Siegfried der Dorfer. Written in the 14th c., it tells of an unhappy wife who is dissuaded from suicide by a woman in grey, who proves to be the Virgin Mary. Afterwards the wife approaches her husband with a new humility, which so touches him that the two are reconciled. It comprises rather more than 600 lines.

Frauenturnier, Das, a short verse tale of the late 13th c., which tells how wives, in the absence of their husbands, arrange a tournament. The prize is won by a poor maid, whom the Duke of Limburg rewards with the hand of a rich man.

Frauenzucht, see SIBOTE.

Frau im Fenster, Die, a short verse play in one act by H. von Hofmannsthal (q.v.), published in the collection *Theater in Versen,* 1899. It has a Renaissance setting. Madonna Dianora, awaiting her lover, speaks a passionate monologue, pouring out her longing and desire. Instead of her lover, her husband, Messer Braccio, enters, and sees the rope-ladder intended for the lover. He ignores her eloquent and long-drawn-out reproaches and finally strangles her. The effect of the play derives in part from Braccio's impenetrable silences and undeviating will.

Frau Jenny Treibel, a short novel by Th. Fontane (q.v.). The fifth of his Berlin novels, it was published in 1892 in instalments, and in book form in 1893. It recounts in the shape of a witty, ironical comedy the short-lived betrothal of Leopold Treibel, the not too bright younger son of a well-to-do *bourgeois* family (the French word is several times used of the Treibels by Fontane), and Corinna, the highly intelligent daughter of an impecunious schoolmaster, Wilibald Schmidt. Frau Jenny Treibel is a good-looking, and, within limits, intelligent woman, who constantly insists on poetry and sentiment, but directs her conduct only to material advantage. In the past she has jilted Schmidt for the well-to-do Treibel, but maintains friendly relations with him. Corinna sets her cap at Leopold, deliberately rejecting an intelligent cousin in favour of Treibel's wealth, and she quickly draws Leopold into a proposal. Frau Jenny opposes, and, stirred by this assault of the impecunious, summons her hated daughter-in-law's pretty and empty, but well-portioned, sister as a counter-attraction. Leopold resists obstinately but dully, unable to shake off his mother's domination, and Corinna, realizing that she is pursuing a mistaken goal through a false sense of values, releases him and marries her sensible and agreeable cousin Marcell Wedderkop. The book contains highly amusing portraits of Frau Jenny and of Corinna's father, the genially egotistic schoolmaster Professor Wilibald Schmidt, and gives a sympathetic account of Schmidt's good-natured housekeeper, the widowed Frau Schmolke.

Fräulein Else, a Novelle by A. Schnitzler (q.v.), published in 1924. Else, on holiday with relatives, is pressed by her feckless and financially embarrassed father to approach a rich acquaintance, Herr von Dorsday, for help. Dorsday makes the condition that she should show herself naked in his room. Else, distraught and tormented, creates a scandal by appearing unclothed in the hotel lounge. Put to bed, she takes an overdose of Veronal and, it may be inferred, dies. The remarkable feature of *Fräulein Else* is the brilliant use of interior monologue, the whole action being conducted in the girl's articulate thoughts, in which Schnitzler's acute perception and his identification with Freudian psychology are evident.

Fräulein von Scuderi, Das, a Novelle by E. T. A. Hoffmann (q.v.), sub-titled *Erzählung aus dem Zeitalter Ludwigs des Vierzehnten.* After its completion in 1818, it was published in 1819 in the *Taschenbuch für Liebe und Freundschaft,* and in 1820 in the *Serapionsbrüder* (q.v., vol. 3). The chronicle of Nürnberg by J. C. Wagenseil (q.v.) drew Hoffmann's attention to the figure of the Paris goldsmith Cardillac, who was obsessed by the works he had created by his own hand.

Hoffmann demonstrates in Cardillac one of the darkest aspects of the Romantic artist, for the mania to recover his works drives the goldsmith to murder his clients, thus creating an atmosphere of panic among the Parisians. The murder of Cardillac himself in the manner of his own expert death-dealing stab emerges as the central event, which leads to the arrest of his apprentice Olivier as the suspected culprit.

The aged Madame de Scudéri, a respected woman of letters, who is in favour at court, is a key figure in the skilfully arranged sequence of events. Through her outstanding courage and virtue she becomes strangely and paradoxically involved in the destiny of Cardillac, and in the affairs of his daughter Madelon and Olivier, her lover. Olivier proves to be the son of a woman who, as a child, had been in the care of Madame de Scudéri. Louis XIV, trusting in Madame de Scudéri's unfailing sense of justice, clears Olivier's name, and allows him to settle with Madelon as his wife in Geneva.

The narrative technique, evidently inspired by H. von Kleist (q.v.), succeeds in evoking the suspense of a detective story, as well as portraying the wider background of a society corrupted by judicial anarchy. Hoffmann's personal interest focuses on the dual existence of Cardillac as a respected citizen (by day) and as an artist and criminal (by night). The story is a projection of its author's inmost experience of irreconcilable duality.

Fräulein von Scuderi, Das, a five-act play (Schauspiel) in blank verse completed by O.

Ludwig (q.v.) in 1848, and published posthumously in 1870. Avowedly based on E. T. A. Hoffmann's Novelle of the same title (q.v.), it faithfully follows its model.

Fräulein von Sternheim, Das, commonly used abbreviated title of *Geschichte des Fräuleins von Sternheim* (q.v.) by Sophie von La Roche (q.v.).

Frau ohne Schatten, Die, an opera in three acts composed by R. Strauss (q.v.) to a libretto provided by H. von Hofmannsthal (q.v.), who also wrote a story on this theme with the same title (1919). It was first performed on 10 October 1919 in Vienna.

The 'shadow' of the title is a symbol of fertility. The setting evokes an oriental fairy-land. The Empress, half spirit, half mortal, is warned by a supernatural messenger that, if she fails to cast a shadow, the Emperor will be turned to stone, and she will have to return to the spirit realm. She succeeds in buying a shadow from a poor woman (the wife of Barak, the dyer), but she returns it when she discovers that she has endangered the woman's life. Her refusal to acquire the shadow, together with her unswerving love for the Emperor, overcomes the spell, and she at last casts a shadow: 'Die Gattin blickt zum Gatten, / Ihr fällt ein irdischer Schatten / Um Hüfte, Haupt und Haar.'

Frau Professorin, Die, a Novelle by B. Auerbach (q.v.) published in 1846 in *Schwarzwälder Dorfgeschichten*. Reinhard, a high-spirited and boisterous artist, visits a Black Forest farm and falls in love with Lorle, the farmer's daughter, a girl of sterling character. Against his own real desires he obtains a well-paid post as an academy professor, in order to be able to marry Lorle. When they settle in the capital (Residenz) Lorle seems not to fit in, and he soon feels ashamed of her. Thrust back upon herself, Lorle helps the poor and sick. Reinhard becomes embittered over the loss of his freedom, and the precarious marriage is brought to an end by a drunken escapade on his part. Lorle returns to her village home and family, and devotes herself to good works. The story contains much social criticism.

In 1876 Auerbach published a sequel, *Des Lorles Reinhard*, in *Nach dreißig Jahren* (q.v.). A dramatization of *Die Frau Professorin* by Charlotte Birch-Pfeiffer (q.v.), entitled *Dorf und Stadt* (1847), had considerable success.

FRAU RAT, Goethe's mother. The designation refers to the title Rat acquired by Goethe's father in 1742.

Frau Regel Amrain und ihr Jüngster see LEUTE VON SELDWYLA, DIE.

Frau Sorge, a poem by H. Heine (q.v.), included in the section *Lamentationen* of the *Romanzero* (q.v.). No. 14 of the subdivision Lazarus; it begins 'In meines Glückes Sonnenglanz'. The poem contrasts the misery of the poet's sickbed with his carefree youth.

Frau Sorge, a novel published by H. Sudermann (q.v.) in 1887. It is set in his home country, the Lithuanian fringe of East Prussia. The narrative opens with the birth of Paul Meyerhöfer, the son of a shiftless father and a loving and industrious, but down-trodden, mother. Disregarded or despised by most of those around him, he grows up under the shadow of 'Frau Sorge', the personification of cares and worries, devoting himself to the welfare of his family. His rancorous father seeks to harm a rich family (Douglas) which has shown him nothing but benevolence, and the climax of the story occurs when Paul sets fire to their own family farm-house in order to prevent his father doing the same to the rich neighbour. He is successful, but is sentenced to two years' imprisonment, at the end of which he marries his childhood friend Elsbeth Douglas. Though the theme, with its acts of arson, is in places sensational, the environment of Sudermann's own childhood is well caught.

Frau Welt, an allegorical figure occurring frequently in medieval literature. Seen from the front she is beautiful; when she turns round, her back is a mass of decay, maggots, and noisome creatures. Frau Welt figures in the poem 'Frô Welt, ir sult dem wirte sagen' by WALTHER von der Vogelweide (q.v.), in Konrad von Würzburg's *Der Welt Lohn* (q.v.), and in poems of Der Guotaere and HEINRICH Frauenlob (qq.v.). She is the central figure in *Von der schönen verlorenen Frau* (q.v.), to which the title *Frau Welt* is sometimes given. The figure is also represented in medieval sculptures; a famous example is the statue on the south portal of Worms Cathedral.

FREDER [FREDERUS], JOHANNES (Köslin, 1510–62, Wismar), a Lutheran churchman, studied in Wittenberg, was a pastor in Hamburg, a professor in Greifswald (1549), and finally (1556) moderator (Superintendent) for Rügen and Wismar. He is the author of hymns in Low German.

Frei, a play written by A. Holz (q.v.) in collaboration with A. Jerschke, and published in 1905. Described as *Eine Männerkomödie*, it has no woman's part. It is set in Alsace-Lorraine, and deals with the attempt of a high-minded lawyer to expose a bogus bank, and the resistance he encounters.

Freiberg, town in Saxony, which is the home of the Bergakademie, founded in 1754 and reputed to be the oldest school of mining in the world. Novalis and F. K. von Baader (qq.v.) studied there.

Near Freiberg was fought on 29 October 1762 a battle in which Prussian forces under Prince Heinrich (q.v.) were victorious over the Austro-Imperial troops. This victory enhanced Prussian prestige at the negotiations for the Peace of Hubertusburg (see SIEBENJÄHRIGER KRIEG).

FREIBERG, HEINRICH VON, see HEINRICH VON FREIBERG.

FREIDANK, the author of a collection of Middle High German Sprüche entitled *Bescheidenheit.* Nothing is known with certainty of his life, but it may be inferred that he went on a crusade with the Emperor Friedrich II (q.v.) in 1228-9. It is believed that Freidank was his true name, though some have held it to be a symbolical pseudonym. He has sometimes, though uncertainly, been identified with Fridancus magister, whose death is recorded at Kaisheim near Donauwörth in 1233.

The title *Bescheidenheit* signifies wisdom, i.e. knowledge of the world derived from experience. The work is composed of a considerable number of sayings framed for the most part in couplets. They are arranged to read consecutively, though it is possible that this order is simply a grouping of detached sayings. Their main field is the conduct of man in the world, but the religious basis of Freidank's judgement is clearly emphasized. His wisdom is traditional, and originality is not attempted, though his aphorisms are crystallized in homely, pithy language. Certain Sprüche are severely critical to the point of satire, notably those relating to the crusade in the section *von Akers* (Acre). He is also sharp in his judgement of the lust for power and money, which he attributes to the papal court.

Freidank's *Bescheidenheit* was immensely popular and his Sprüche are often appropriated by others, with or without acknowledgement. He is mentioned by RUDOLF von Ems (q.v.), and HUGO von Trimberg (q.v.) repeatedly refers to him with respect and approval. The collection is thought to have been written between 1215 and 1230. The edition by H. E. Bezzenberger (1872) was reprinted in 1962.

Freie Bühne, Verein, a society in Berlin sponsoring by subscription private performances of realistic plays banned by the censor, or unlikely to be staged by a commercial theatre. The visit of the Théâtre libre (q.v.) of Paris to Berlin was a contributory factor to the foundation of the society, revealing new techniques of acting and production.

The Freie Bühne was inaugurated in 1889 by O. Brahm, M. Harden (qq.v.), P. Schlenther (1854-1916), and Th. Wolff (1868-1916), and gave its first performance, Ibsen's *Ghosts* (see IBSEN), on 29 September 1889. The second (20 October 1889) was devoted to G. Hauptmann's first play, *Vor Sonnenaufgang* (q.v.); this performance is usually reckoned to be a landmark in German Naturalism; it excited as much condemnation as praise. Brahm was the director from 1889 to 1893. Later productions of the Freie Bühne were *Die Familie Selicke* (q.v.) by A. Holz and J. Schlaf, Anzengruber's *Das vierte Gebot* (q.v.), Hauptmann's *Das Friedensfest,* and *Die Weber* (qq.v.), Tolstoy's *Die Macht der Finsternis* (*The Power of Darkness*), and plays by Bjørnson and Strindberg (q.v.). The aims of the society were also advanced by the periodical *Die Freie Bühne für modernes Leben,* founded in 1890 by O. Brahm, who was joined as editor by W. Bölsche in 1891; in 1893 J. Hart became editor, and was succeeded in the same year by O. J. Bierbaum (qq.v.); in 1894 O. Bie (1864-1938) took over, and continued the journal (see DIE NEUE RUNDSCHAU). The Verein Freie Bühne was dissolved in 1893.

Freie deutsche Rhein, Der, a patriotic poem by N. Becker (q.v.), written in 1840, and intended as a riposte to French territorial claims prominent at the time in the Paris press. The usual melody is by R. Schumann (q.v.), but it has also been set by several other composers. The first line runs 'Sie sollen ihn nicht haben, den freien deutschen Rhein'.

Freier, Die, a comedy by J. von Eichendorff (q.v.), published in 1833. Some scenes had appeared earlier in a magazine (1821-4). Graf Leonard is to marry Gräfin Adele, and he decides first to test his lady by appearing in disguise. The Countess learns of his intention and retaliates by changing clothes and roles with her maid. Leonard, however, wears an unexpected disguise, and so both parties are taken in. The disguised Leonard and the apparent maid fall in love; their affairs seem to be in a tangle, but identity is revealed and all is well. Additional comic elements are provided by subsidiary characters.

Freie Rhythmen, term applied to verse which is unmistakably rhythmical, yet is irregular in the length and numerical grouping of the lines.

'Free rhythms' (which have no rhyme) frequently claim their ancestry from Pindar. They are usually associated with a rapt, ecstatic, or rhapsodic tone. In German literature they appeared first in Klopstock (e.g. *Die Frühlingsfeier*, q.v.), and commended themselves to the exponents of the Sturm und Drang (q.v.) by their apparent freedom from rule or compulsion. Some of Goethe's finest early poems such as *Prometheus*, *Ganymed*, and *Harzreise im Winter* belong to this category. The greatest master of 'free rhythms' is Hölderlin (q.v.), with, for instance, *Schicksalslied*, 'Da ich ein Knabe war', *Der Rhein*, *Am Quell der Donau*, *Der Einzige*, *Patmos*, and *Hälfte des Lebens*; but many later poets, including Heine (*Nordseebilder*), Nietzsche, Rilke (*Duineser Elegien*), and Trakl (qq.v.) have written notable examples.

Freie Stadt, a city possessing its own government and jurisdiction. The majority of the early free cities were governed by bishops, whose tutelage they later succeeded in discarding. Mainz (1244), Regensburg (1245), Strasburg (1262), and Cologne (1288) are typical examples. Privileges of exemption, notably from imperial taxes, distinguished the free cities (also called free imperial cities—Freie Reichsstädte) from the imperial cities of which there were a great number (see REICHSSTADT).

By the constitution of the German Confederation (see DEUTSCHER BUND) Hamburg, Bremen, Lübeck, and Frankfurt became in 1815 component member states under the designation Freie Stadt. After the war of 1866 (see DEUTSCHER KRIEG) Frankfurt was annexed to Prussia, but Hamburg, Bremen, and Lübeck became free cities in the German Empire of 1871, with the title Freie und Hansestadt (Hamburg and Lübeck) and Freie Hansestadt (Bremen). This status persists for Hamburg and Bremen, but not for Lübeck.

Freigeboren, the last novel written by F. Spielhagen (q.v.), published in 1900. In it F. Lassalle (q.v.), on whom two earlier characters, Dr. Münzer (*Die von Hohenstein*, q.v.) and Leo Gutmann (*In Reih' und Glied*, q.v.), were based, appears under his own name.

Freigeist, Der, a comedy in five acts written by G. E. Lessing (q.v.) in 1749, and first published in 1755 in *Lessings Schriften* (vol. 5). It is a comedy in the French manner (Lessing acknowledges his indebtedness to de l'Isle), with two pairs of lovers at cross purposes. The intrigue is finally solved by an exchange of partners, but the real merit in the play lies in the portrayal of the two contrasting men, the tolerant and

balanced cleric, Theophan, and the irritable, biased, and intolerant free-thinker, Adrast. The comedy is an exemplification of Lessing's own spirit of tolerance, and also perhaps a tribute to his pastor father.

Freigeist, Der, a tragedy written by J. W. von Brawe (q.v.) in 1757, and published in 1758 by F. Nicolai (q.v.) in an appendix to the *Bibliothek der schönen Wissenschaften und freien Künste*. It is described as *Ein Trauerspiel in fünf Aufzügen*, and was first performed in 1767 in Berlin. Its theme is the corruption of a good but weak character, Clerdon, by a wicked atheist, Henley, who compasses Clerdon's fall out of jealousy; for Clerdon has won the love of Amalia Granville, whom Henley desires. Clerdon is persuaded to become a free-thinker, and, against all evidence, to believe that Amalia's brother is his greatest enemy. Finding that his father has died in want, Clerdon begins to repent, but Henley reasserts his influence, egging Clerdon on to murder Amalia's brother. No sooner is the deed done than Henley discloses to Clerdon his intention, now achieved, of destroying him. The overwrought Clerdon stabs Henley, who dies unrepentant, and then kills himself.

Freigeisterei der Leidenschaft. Als Laura vermählt war im Jahre 1782, a poem written by Schiller (q.v.) probably in 1784, and published in No. 2 of *Die Thalia* (q.v.), 1786. It reflects Schiller's passionate and hopeless attachment to Frau von Kalb (q.v.). Schiller republished it in severely abridged form with the title *Der Kampf* in his *Gedichte*, 1800. The superscription referring to Laura is a screen to disguise the identity of the addressee. It is not one of the *Laura-Oden* (q.v.).

Freiheit, Gleichheit, Brüderlichkeit, the German version of the slogan summarizing the aspirations of the French Revolution (q.v.) as expressed in the *Declaration of the Rights of Man*.

Freiheit in Krähwinkel, a satirical comedy by J. N. Nestroy (q.v.), written in 1848, first performed at the Carl-Theater, Vienna, on 1 July 1848, and published in 1849. It is described as a *Posse mit Gesang in zwei Abteilungen und drei Aufzügen*. The first section (Acts I and II) is entitled *Die Revolution*, the second (Act III) *Die Reaktion*.

The journalist Eberhard Ultra helps to bring the revolution in 1848 (see REVOLUTIONEN 1848–9) to the proverbially comic small town of Krähwinkel. A reactionary movement follows, but is defeated. The play has an extremely intricate plot in which Ultra appears in various disguises. Many elements in the plot are borrowed from the Viennese Volksstück *Die falsche Primadonna* (1818) by A. Bäuerle (q.v.). Nestroy's own contribution is the vigorous outspoken satire and mordant wit, which was able to find untrammelled expression after the collapse of the Metternich (q.v.) regime and the relaxation of the censorship.

The use of Krähwinkel in comic representations of a very small German (or Austrian) town derives from A. von Kotzebue's *Die deutschen Kleinstädter* (q.v.).

Freiherren von Gemperlein, Die, a humorous Novelle by Marie von Ebner-Eschenbach (q.v.), published in *Neue Erzählungen* (1881). Two eccentric bachelor brothers, who cannot agree, yet cannot bear to part, make various absurd, theoretical plans for entering into marriage. In the end both fall in love with the same woman, who turns out to be married, a possibility they had never contemplated. But the fiasco reconciles them, for each had gone to the lady, separately and secretly, to plead the other's cause.

Freikorps, volunteer units formed from demobilized soldiers, students, and adventurers in 1918 and 1919, which were active in repressing communist revolts in Germany and, principally, in the Baltic provinces of Prussia and of the former Russian Empire.

FREILIGRATH, FERDINAND (Detmold, 1810–76, Cannstadt nr. Stuttgart), left school at 16 and entered retail trade at Soest, transferring in 1831 to a bank at Amsterdam and obtaining in 1837 a post at Barmen. His *Gedichte* (1838), a varied collection containing the ballad *Prinz Eugen der edle Ritter* (q.v.) as well as exotic poems of the Orient and Africa, attracted considerable notice, and in 1839 he resigned his post in order to devote himself entirely to writing. He married in 1841 and settled in Darmstadt. In this year he was attacked for some lines in the poem *Aus Spanien* (1839) by G. Herwegh (q.v. and see PARTEI, PARTEI, WER SOLLTE SIE NICHT NEHMEN).

In 1842 Freiligrath was granted a pension by King Friedrich Wilhelm IV (q.v.) of Prussia, which his social conscience caused him to resign in 1844, the year of the publication of his political poems *Ein Glaubensbekenntnis* (containing a reprint of *Aus Spanien*). Exposed to political persecution, he emigrated to Belgium, moving afterwards to Switzerland, and from there to England. In the revolutionary year of 1848 (see REVOLUTIONEN 1848–9) he returned to Germany, settling in Düsseldorf, but a vehement poem on the victims of the fighting in Berlin

(*Die Toten an die Lebenden*) led to his arrest in August for sedition. He was tried in October and acquitted amidst a demonstration of popularity. After a collection of lyric poetry (*Zwischen den Garben*, 1849) two further volumes of political poetry appeared in 1849 and 1851 as *Neuere politische und soziale Gedichte*. The first of these included *Die Toten an die Lebenden*. Shortly before the publication of the second volume Freiligrath, rightly foreseeing further persecution, emigrated with his family to London, where he remained for seventeen years. He worked first in the counting-house of a Mr. Oxford and then in the London branch of a Swiss bank, in his spare time writing newspaper articles on German literature (in English). The bank failed in 1868, but a public subscription in Germany enabled Freiligrath to return and settle at Cannstadt close to Stuttgart. During the war of 1870–1 he wrote patriotic poetry, including the well-known *Die Trompete von Vionville* (q.v.), often called *Die Trompete von Gravelotte*. A final volume of lyric poetry appeared as *Neue Gedichte* in 1876. Freiligrath was a contributor to the English review *The Athenaeum*.

If Freiligrath lacked depth, he had a vivid imagination, a fertile gift for rhyme, and considerable technical facility (his *Gedichte* of 1838 contain successful essays in *terza rima*, Alexandrines, and trochaic verse). His poetry has a touch of declamatory brashness, but he was a successful translator and popularizer of French (e.g. V. Hugo, *Odes* and *Chants du crépuscule*) and more especially English, Scots, and Irish poetry by Herrick, Burns, Campbell, Scott, Moore, Felicia Hemans, Macaulay, and Tennyson. Among the more notable translations are 'O my Luve's like a red, red rose' ('Mein Lieb' ist eine rote Ros'), 'My heart's in the highlands' ('Mein Herz ist im Hochland'), *The Brook* ('Wo Rohrhuhn nistet, Reiher baut'), the lullaby from *The Princess* ('Süß und sacht, sachte weh'), *The Lady of Shalott, Ulysses,* and *Locksley Hall. Sämtliche Werke* (10 vols.), ed. L. Schröder, appeared in 1907 and *Werke in sechs Teilen*, ed. J. Schwering, in 1909 (reissued 1974).

Freischütz, Der, opera by C. M. Weber (q.v.), first performed in Berlin on 18 June 1821. The libretto is by F. Kind (q.v.), who based it on an episode from the *Gespensterbuch* of J. A. Apel (q.v.). The happy ending of the libretto replaces a tragic conclusion in the original story.

The opera begins with a shooting competition in which the hero, the young gamekeeper (Jäger) Max, fails dismally. He is deeply worried by his poor form because a more important shooting contest is to take place next day. At this he will be competing, in the presence of the reigning prince, for the hand of the girl he loves, Agathe, daughter of Kuno, the hereditary head gamekeeper (Erbjäger). Tempted by Kaspar, who has sold himself to Samiel, the 'black hunter', a supernatural being, Max seeks magic bullets (Freikugeln) which infallibly hit their mark. The casting of these bullets takes place at midnight in the Wolf's Glen (Wolfsschlucht), in a famous scene of Romantic eeriness and horror which owes its effectiveness almost entirely to Weber's music. On the following day Max fires at a white dove, which is suddenly revealed as Agathe, but the bullet is diverted by Samiel on to Kaspar, who, as he dies, is consigned to perdition. Agathe's miraculous escape is due to the presence of roses in her chaplet, which have been blessed by a pious hermit. Max is pardoned and all ends happily.

Freisinger Paternoster, title given to a translation and exposition of the Lord's Prayer in Old High German prose, made in the Monastery of Freising (Bavaria) and preserved in Munich. The text of each phrase is given first in Latin, then in German, and is then expounded in German. The translation represents an advance on that of the *St. Galler Paternoster* (q.v.). The MS. dates from the beginning of the 9th c.

FREISLER, ROLAND (Celle, 1893–1945, Berlin), German lawyer, was a Communist in 1920, joined the NSDAP (q.v.) in 1925, and was presiding judge of the Volksgerichtshof 1942–5 (see RESISTANCE MOVEMENTS 2). Notorious for his bullying demeanour, he conducted the trials of the unsuccessful conspirators and of numerous suspects after 20 July 1944. He died on 3 February during an American daylight air raid when the court building received a direct hit.

Freiwild, a play in three acts by A. Schnitzler (q.v.), first performed on 1 December 1896 in the Deutsches Theater, Berlin. In spite of its tragic ending it is called a Schauspiel. It is set in the summer resort of Baden by Vienna. In a brilliant first act the aggressive hussar subaltern Karinski provokes the civilian Paul Rönning by speaking insultingly of the young actress Anna Riedel, whom Rönning loves. Rönning strikes Karinski in the face. The code of honour prevailing among the military and the educated classes requires a duel between the men. The remaining two acts debate this duel, which Rönning refuses to accept. Karinski's brother officer Rohnstedt seeks to mediate, but is bound by his commission to support the duel. After Rönning has positively refused to comply with what he regards as a barbarous custom, he is

waylaid and shot dead by Karinski. The polemic against the duel is not wholly successful, since Rönning, when he is shot, is carrying a revolver in order to defend himself if he should encounter Karinski.

French Revolution, the French Revolution of 1789, the first of four major French revolutions in modern times, is the only one generally referred to as 'die französische Revolution'. It was in one sense the culmination of French enlightened political thought represented chiefly by Voltaire, Rousseau (qq.v.), and Mirabeau, which in Prussia (see FRIEDRICH II, DER GROSSE) and Austria (see JOSEPH II), but not in France, had in varying degrees introduced an era of enlightened despotism. French feudal tyranny and social injustice were not felt acutely until the reign of Louis XVI, when the state was feared to be heading towards bankruptcy. The King's attempts at reform through his ministers (Turgot and Calonne) failed, and as an extreme measure he summoned the States General (*États généraux*) for 1789 (5 May), an act which was anticipated by Klopstock's ode *Die Etats Generaux* (q.v.). The following month (17 June) the Third Estate, deciding upon independent action, styled itself the National Assembly (*Assemblée nationale*), and declared its right to reform the tax system and demanded a new constitution. On 9 July the Third Estate changed the National Assembly into the Constituent Assembly (*Assemblée constituante*), thus virtually abolishing the absolute monarchy. Encouraged by these events, the Parisians stormed the Bastille (14 July), the demolition of which became the symbol of a new age of freedom to many who watched events from Germany.

The Assembly voted the *Declaration of the Rights of Man* (*Déclaration des droits de l'homme*). Inspired by the American Declaration of Independence; it set out the principles of liberty, equality, and fraternity ('Freiheit, Gleichheit, Brüderlichkeit', a slogan which was soon common currency in Germany). The royal family was removed from Versailles to the Tuileries, and, attempting an escape soon after, was recaptured at Varennes, an episode which intensified Jacobin advocacy of violent action. The situation grew more acute as the emigrant French nobility sought to bring about armed intervention by the German states. Austria was particularly concerned since Marie-Antoinette was a daughter of the Empress Maria Theresia (q.v.) and a sister of the Emperor Leopold II (q.v.). On 20 April the Legislative Assembly (*Assemblée législative*), which in October 1791 had succeeded the Constituent Assembly, declared war on Austria and Prussia, thus opening the Revolutionary Wars (see REVOLUTIONS-KRIEGE).

The abolition of the French monarchy was an inevitable consequence, and the royal family, including the 15-year-old Madame Royale and the 8-year-old Dauphin, were imprisoned in the Temple (10 August 1792). Between 2 and 5 September 1792, 1,200 prisoners were massacred by a mob which had invaded the prisons. Danton (q.v.), as minister of justice, and the radical Marat were among those responsible for the September massacres (Septembermorde), which aroused a strong revulsion in Germany. On 21 September the National Convention (*Convention nationale*) replaced the Legislative Assembly, and proclaimed the Republic. Henceforth the Revolution turned into a struggle between the various rival factions, culminating in the Jacobin Terror (Schreckensherrschaft) under the Montagne (Bergpartei) led by Robespierre which lasted from July 1793 until July 1794. Louis XVI and Marie-Antoinette were executed (21 January and 16 October 1793) and the ailing Dauphin left to perish in prison. In ruthless persecutions countless victims were executed with or without trial. The Terror ended with the execution of Robespierre and his followers (27 July 1794). From October 1795 until December 1799 France was ruled by the Directory (*Directoire*), consisting of a legislative and an executive chamber with five elected directors, whose misgovernment provoked Napoleon Bonaparte's *coup d'état* leading to the establishment of the Consulate with Bonaparte as First Consul. The defeat of the absolutism of the *ancien régime* had thus been followed by the destruction of liberalism, and ultimately by Napoleonic dictatorship, of which Bonaparte's position as First Consul proved to be the first stage.

The French Revolution did not lead to revolutions in Germany and Austria as did the Revolution of 1848 (see REVOLUTIONEN 1848–9). Nevertheless there was for a time a strong republican movement in the Rhineland, particularly in the Electorate of Mainz, an episode which is reflected in Goethe's *Belagerung von Mainz*, and in Ina Seidel's novel *Das Labyrinth* (qq.v.). Politically its greatest impact was still to come, under new auspices, with Napoleon's European conquests (see NAPOLEONIC WARS).

The ideals of the French Revolution, represented at their best by the Girondins, found many followers in Germany, among both the older and the younger generations. Klopstock (like Schiller declared an honorary French citizen), Schelling, JEAN Paul, Hegel, Hölderlin, Herder, Wieland, Kant, Fichte, and Schiller (qq.v.) were among those who welcomed the Revolution during its early stages. Goethe maintained from the first an attitude of caution and scepticism, which is reflected in his play *Die natürliche Tochter* (q.v.). Like Schiller, many, though at first sympathetic, reacted decidedly against the Revolution after the King's execution. Some radicals, among them Georg Forster (q.v.), went to Paris to witness the emergence of the Republic on the spot. Iffland and Kotzebue (qq.v.) wrote burlesques on the Revolution, and a number of publications discussed the political implications. Friedrich von Gentz (q.v.) translated Burke's outspoken criticism of the Revolution of 1790, *Reflections on the Revolution in France* (*Betrachtungen über die französische Revolution*). In 1799 Novalis expressed in his essay *Die Christenheit oder Europa* (q.v.) an ideal of freedom on a universal religious basis which inspired the sentiments of the early Romantics (see ROMANTIK).

The French Revolution has frequently been the subject of works of literature, many of which concentrate on its leading figures. Among these are: Grabbe's play *Napoleon oder Die hundert Tage* (1831), Büchner's *Dantons Tod* (1837), *Mirabeau* (1850) by Ernst Raupach, *Graf Mirabeau* (1858) and *Robespierre* (1859) by Th. Mundt, *Die Göttin der Vernunft* (1871) by P. Heyse, *Danton und Robespierre* (1871) by R. Hamerling, *Der grüne Kakadu* (1899) by A. Schnitzler, *Joseph Fouché* (1929) and *Marie Antoinette* (1932) by Stefan Zweig, the Novelle *Die Letzte am Schaffott* (1931) by Gertrud von le Fort, and *Die Verfolgung und Ermordung Jean Paul Marats, dargestellt durch die Schauspielgruppe des Hospizes zu Charenton unter Anleitung des Herrn de Sade* (q.v., 1964) by Peter Weiss.

FRENSSEN, GUSTAV (Barlt, Dithmarschen, 1863–1945, Barlt), after theological studies at Tübingen, Berlin, and Kiel became pastor of Hennstedt (1890–2) and of Hemme (1892–1902), villages near his birthplace. Frenssen wrote of the people and the landscape of Dithmarschen, making his name as a representative of Heimatkunst (q.v.). His earliest novels, *Die Sandgräfin* (1896) and *Die drei Getreuen* (1898), attracted little attention, but *Jörn Uhl* (q.v., 1901) had a nation-wide success and remains his best-known work. The slow-moving, brooding figures of his novels are characteristic of the region and reflect Frenssen's own temperament. His religious belief became increasingly unorthodox, and in 1902 he resigned his pastorate in order to devote himself to literature and moved to Meldorf. *Hilligenlei* (1905, 'hillig' is Low German for 'heilig') caused controversy because of its unusual interpretation of the Gospels. In 1906 Frenssen moved to Blankenese near Hamburg, where he remained until 1912, when he returned to spend his later years in Barlt. His novels of Dithmarschen include *Peter Moors Fahrt nach Südwest* (1906), *Klaus Hinrich Baas* (1909), *Der Untergang der Anna Hollmann* (1911), *Die Brüder* (1917), *Der Pastor von Poggsee* (1921), *Lütte Witt* (1924), and *Meino der Prahler* (1933). *Otto Babendieck* (1926), though a novel, is in large part autobiographical. *Lebensbericht* (1940) is an avowed autobiography. In *Der Glaube der Nordmark* (1936) Frenssen finally rejected Christianity and adopted a pagan Germanic religion supported by Nordic mythology.

FRENZEL, KARL (Berlin, 1827–1914, Berlin), a schoolmaster who turned to journalism and wrote novels and Novellen, some of them historical (*Watteau*, 1864; *Charlotte Corday*, 1864; *La Pucelle*, 1871).

FRESSANT, HERMANN, town clerk of Ulm in the 14th c., is the author of a verse tale of moderate length in praise of marital constancy, *Der Hellerwertwitz*. It tells of a merchant with a wife and two mistresses. When he leaves for a journey the wife asks for 'a ha'porth of good sense' (einen Hellerwert Witz), and the mistresses demand rich presents. The merchant is advised to pretend on his return that he has lost his fortune. The mistresses reject him, but the wife makes him welcome and consoles him. So he discovers the meaning of the 'Hellerwert Witz'. Fressant was probably a native of Augsburg.

FREUD, SIGMUND (Freiburg/Moravia, 1856–1939, London), was brought up in Vienna, where he began his medical studies in 1873. His specialization in neurology received a decisive stimulus through his work with J. M. Charcot in Paris and H. Bernstein in Nancy in 1885–6. In his practice in Vienna he developed experimental therapy of mental disorders. In 1896 he introduced the term psycho-analysis to embrace his investigation of subconscious phenomena. It has remained associated with his name. He continued his work in the field of psychotherapy throughout his life despite a serious illness during his last sixteen years, and he applied his psychological findings to philosophy in a variety of contexts. His most important work extends over a period of more than thirty years, during which he found both convinced followers (among them for a time C. G. Jung and A. Adler, qq.v.) and severe critics. His first important contribution was his reassessment of dreams as revealing the subconscious activities of the mind, particularly as wish-fulfilment. He published *Die Traumdeutung* (*Interpretation of Dreams*), dated 1900, and five years later *Der Witz und seine Beziehung zum Unbewußten*, in the light of which a joke ceases to be a joke by the time

its underlying meaning has been apprehended. Not all types of joke are suitable for analysis, among them those with intellectual content, while tendentious jokes (i.e. obscene and hostile jokes) constitute the most revealing category. In the same year, 1905, Freud first published *Drei Abhandlungen zur Sexualtheorie*, which were followed by six revised editions spread over twenty years. They included his ideas on auto-erotism, narcissism, and object-choice, indicating three distinct stages in sexual development: the love of the self, the love of the image of the self in another person, and the sexual attraction to an external object. Where this is of an incestuous nature it can be referred to (as Freud first suggested in 1897) as an Oedipus complex since it can be detected in Oedipus (Freud also finds it in Hamlet). Its best-known form is that in which the child's love of his mother induces a hatred of the father. A later theory (published 1923) of the subconscious workings of the mind arises out of Freud's distinction between the *Ego*, the *Id* (*Das Ich und das Es*), and the *Superego*. These were designed to signify three different aspects of development, the *Id* denoting the primeval drives with which a child is born, the *Ego* the adjustment of the organism to reality, which gradually imposes upon the self a social consciousness, and with it a moral conscience corresponding to the concept of the *Superego*. Freud emphatically maintained that *libido*, the sexual drive, was the most important factor in human behaviour. He virtually anticipated a permissive society as the ideal, which he was led to formulate through investigations of the origins of anxiety and neurosis. He attributed these states to repression, and viewed the identification of symptoms as a preliminary to medical treatment. His wide erudition enabled him to illustrate his theories with literary examples, which stimulated further scientific discussion and also gave to writers of the medical profession, notably Schnitzler (q.v.), a new orientation. Among other writers influenced by him H. von Hofmannsthal, Musil, Sternheim, Hesse, Wassermann, Kafka, and Th. Mann (qq.v.) deserve particular mention. The influence of Schopenhauer's pessimism on Freud was a lasting one. Freud was a materialist and a determinist, and conceived civilization basically in terms of the antagonistic dualism inherent in the human species, the battle between Eros and Thanatos, the instinct or drive ('Trieb') of life and the instinct of death or destruction.

From 1912 Freud edited the journal *Imago* (q.v.). In 1933 his books were burnt in Berlin because of his Jewish descent. Among the honours bestowed upon him on the occasion of his eightieth birthday (1936) was the Fellowship of the Royal Society. In 1938 Freud left Vienna and, with the help of friends, went into exile in London. *Gesammelte Werke* (18 vols.), including his autobiography (vol. 14), appeared 1942–8.

Freund Hein. Eine Lebensgeschichte, a novel by E. Strauss (q.v.), published in 1902. It is the tragedy of a child. Heinrich Lindner's father, a respected lawyer, is determined that his son shall go one step higher than himself, and become a state prosecutor (Staatsanwalt). But the boy, who has great musical gifts, fails miserably at school. The irate father forbids Heinrich to practise his music, and the boy, seeing no other way out, shoots himself.

Freund Hein, a euphemism for death, invented by the poet M. Claudius (q.v.).

'Freut euch des Lebens, weil noch das Lämpchen glüht', first line of the poem *Gesellschaftslied*, written in 1793 by J. M. Usteri (q.v.). The tune was composed by H. G. Nägeli (1773–1836) in 1775.

FREY, ADOLF (Küttigen nr. Aarau, 1855–1920, Zurich), studied at Berne and Zurich, was a schoolmaster and from 1898 a professor at Zurich University. He wrote biographies of A. von Haller (q.v., 1879), J. G. von Salis-Seewis (q.v., 1889), his father J. Frey (q.v., 1897), C. F. Meyer (q.v., 1899), A. Böcklin (q.v., 1903), and R. Koller (*Der Tiermaler R. Koller,* 1906). He is also the author of novels (*Die Jungfer von Wattenwil,* 1912; *Bernhard Hirzel,* 2 vols., 1918), minor plays (*Erni Winkelried,* 1893; *Der Fürst der Hulden,* 1919), and much poetry (*Gedichte,* 1886; *Totentanz,* 1895; *Neue Gedichte,* 1913; *Stundenschläge,* 1920; *Aus versunkenen Gärten,* 1932). The poems in the collection *Duss und underm Rafe* (1891) are in Swiss dialect.

FREY, JAKOB (*c.* 1500–*c.* 1562), town clerk of Maursmünster in Alsace, is the author of biblical plays, including *Abraham und Isaak* (date unknown) and *Von dem armen Lazaro* (1533), and of a collection of Schwänke (see SCHWANK), *Die Gartengesellschaft* (1556), intended as light entertainment and spiced with obscenity.

FREY, JAKOB (Gontenschwyl, Aargau, 1824–75, Berne), a Swiss journalist and father of the poet and professor A. Frey (q.v.), is the author of a number of stories with Swiss settings. They include *Zwischen Jura und Alpen* (2 vols., 1858), *Die Waise von Holligen* (a novel, 1863), *Schweizer Bilder* (2 vols., 1864), *Neue Schweizer Bilder* (1877), *Der Alpenwald* (1885), and *Erzählungen aus der Schweiz* (1885), all of which, apart from the novel, are collections.

Freydal, intended as a counterpart to the *Teuerdank* and *Weißkunig* (qq.v.) sponsored by the Emperor Maximilian I (q.v.). It was to have dealt with courtly pleasures such as jousting, but it did not advance beyond the stage of planning.

FREYLINGHAUSEN, JOHANN ANASTASIUS (Gandersheim, 1670–1736, Halle), a Pietist, son-in-law of A. H. Francke (q.v.) and his close associate, wrote hymns and compiled a much-used collection, *Geistreiches Gesangbuch* (1704).

FREYTAG, GUSTAV (Kreuzburg, Silesia, 1816–95, Wiesbaden), son of a burgomaster, studied German literature at Breslau and Berlin universities (1835–8), and from 1839 to 1844 was a lecturer (Privatdozent) at Breslau. His first literary publication was a comedy, *Die Brautfahrt oder Kunz von Rosen* (1844); it was followed by the plays (Schauspiele) *Die Valentine* (1847) and *Graf Waldemar* (1850). In 1848 he joined Julian Schmidt (q.v.) in Leipzig as joint editor and owner of *Der Grenzbote,* a periodical of moderate Liberal leanings. With his comedy *Die Journalisten* (q.v., 1854) he achieved a spectacular success, and the social novel (see ZEITROMAN) *Soll und Haben* (q.v., 3 vols., 1855), praising the burgher in commerce as the solid kernel of the German state, was reprinted twenty-seven times in twenty-three years. A second novel, *Die verlorene Handschrift* (q.v., 1864), was less successful, but his popular survey of German history, *Bilder aus der deutschen Vergangenheit* (q.v., 5 vols., 1859–67), attracted much attention.

Freytag gave up his newspaper work in 1861, though he resumed it for three years in 1867. From that year till 1871 he was also a deputy in the Diet (Reichstag) of the North German Confederation (see NORDDEUTSCHER BUND). In August 1870, at the personal request of the Prussian Crown Prince (see FRIEDRICH III), he travelled with H.Q. to France, afterwards conceiving the idea of the patriotic, wide-flung, chronicle novel *Die Ahnen* (q.v., 6 vols., 1873–81). In 1851 he bought an estate at Siebleben, near Gotha, and in his later years he spent the summers there and, from 1879, the winters in Wiesbaden. Freytag also published an essay on drama (*Die Technik des Dramas,* 1863) and a volume of reminiscences (*Erinnerungen aus meinem Leben,* 1887). A robust and vigorous, if at times insensitive, writer, he was the characteristic German literary figure for a whole generation.

Fridericus Rex, a poem by W. Alexis (q.v.), written in 1820, and included in his novel *Cabanis* (1832). It was set to music by C. Loewe (q.v.) in 1837.

FRIED, ERICH (Vienna, 1921–), emigrated to London in 1938, and settled there permanently. He has made German translations of Dylan Thomas, T. S. Eliot, and Shakespeare. He is best known as a poet who expresses his preoccupation with contemporary problems—the persecutions of the Jews, the war in Viet Nam, scepticism in matters of religion—in terse, at times acid, language which aims at the intellectual response of the reader. His poetry has been characterized as 'denkende Dichtung', and Fried himself has referred to it in terms of 'Sprüche und Widersprüche'. He favours the adaptation of traditional forms and the use of aphorisms and epigrams. His poems, of which eight volumes appeared in the 1960s, include the collection *Gedichte* (1958), the cycles *Reich der Steine* (1963), *Warngedichte* (1964), *Überlegungen* (1965), *und Vietnam und* (1966), *Anfechtungen* (1967), *Die Freiheit den Mund aufzumachen* (1972), *Höre, Israel!* (1974), *So kam ich unter die Deutschen* (1977, the title is a quotation from *Hyperion*, q.v., by Hölderlin), *100 Gedichte ohne Vaterland* (1978), *Lebensschatten* (1981), and *Befreiung von der Flucht. Gedichte und Gegengedichte* (1983). He is also the author of a novel, *Ein Soldat und ein Mädchen* (1960).

Friedberger Christ und Antichrist, title given to fragments of a MS. dating from 1120–30, which originally formed part of a work dealing with the doctrine of salvation from the birth of Jesus to the Day of Judgement. It is in rhyming verse. The MS. was found at Friedberg (Hesse).

Friedberger Passionsspiel (Friedberg near Frankfurt), a late medieval passion play preserved in skeleton form (i.e. producer's directions or Regiebuch). The stage directions are given in Latin, each time followed by the first line (either in Latin or in German) of the following speech or song. It was written towards the end of the 15th c.

FRIEDELL, EGON (Vienna, 1878–1938, Vienna), studied at Berlin and at Heidelberg universities and was at various times cabaret director, dramatic critic, and actor. He is the author of the play *Judastragödie* (1920) and of parodies and Schwänke (with A. Polgar, q.v.), of essays and of stories, and of two immense and stimulating works on cultural history: *Kulturgeschichte der Neuzeit* (3 vols., 1927–32) and *Kulturgeschichte des Altertums* (2 vols., 1936–49). They reveal an astonishing verve, fertility of ideas, wit, and breadth of reading. Friedell, who was a Jew, committed suicide after the occupation of Austria in 1938.

His collected aphorisms were published in 1974 as *Egon Friedells Conversationslexikon* (ed. P. Haage).

Friedensfeier, a hymnic poem by F. Hölderlin (q.v.), celebrating peace in the cryptic prophetic language of his maturity. The occasion of the poem was the signing of the peace of Lunéville in 1801 (see NAPOLEONIC WARS). The poem was not known until the discovery of the MS. in 1954. Another untitled fragmentary version, beginning 'Versöhnender der du nimmergeglaubt', was already extant.

Friedensfest, Das, a play by G. Hauptmann (q.v.), first performed by the Freie Bühne in Berlin on 1 June 1890, and published in the same year. Sub-titled *Eine Familienkatastrophe in drei Akten*, it is set in the 1880s at Erkner near Berlin, where Hauptmann himself lived. The action is concentrated within a few hours. The reunion of a neurasthenic, disturbed family consisting of Dr. Scholz, his wife, and their two sons takes place on Christmas Eve and deteriorates into a succession of quarrels and neurotic outbursts. The younger son, Wilhelm, is the least disturbed, and is in love with Ida Buchner, a sensible and healthy girl, who hopes to rescue him from his debilitating environment; in this she enjoys her mother's support. The death of Dr. Scholz from a stroke terminates the family quarrels. The play ends in uncertainty over the future of Wilhelm and Ida. Can Wilhelm be saved from the Scholz neuroses?

Friedenspreis des deutschen Buchhandels was introduced in 1950 at the International Book Fair at Frankfurt in order to promote international understanding by the annual allocation of the Peace prize to an elected author of any race or nationality. Among the German-speaking recipients of the Peace Prize are A. Schweitzer, M. Buber, C. J. Burckhardt, H. Hesse, R. Schneider, K. Jaspers, Th. Heuß, P. Tillich, Nelly Sachs, and E. Bloch (qq.v.).

FRIEDENTHAL, RICHARD (Munich, 1896–1979, Kiel), worked for the publisher Knaur, editing the latter's *Konversationslexikon* (1931–2). He emigrated to England in 1938, where he settled. During the war he was on the staff of the BBC. He published poetry (*Tanz und Tod*, 1918; *Demeter*, 1924; *Brot und Salz*, 1943), single Novellen (*Der Fächer mit der goldenen Schnur*, 1924; *Der Heuschober*, 1925), collections (*Marie Rebschneider*, 1927; *Das Erbe des Columbus*, 1950), and novels (*Der Eroberer*, 1929, the conqueror is Cortez; *Die Welt in der Nußschale*, 1956). Friedenthal is best known in England for his biographies of Händel (1959), Leonardo da Vinci

(1960), Goethe (1960), and Luther (1967). All these have been translated.

FRIEDERICH, MATTHÄUS (d. 1559), a Lutheran pastor in Saxony, wrote a popular work, *Wider den Saufteufel* (1551), an example of the Teufelliteratur (q.v.) of the mid-16th c., in which vices are attributed to particular devils.

FRIEDLÄNDER, DER, refers to Wallenstein (q.v.), who was Duke of Friedland.

FRIEDRICH I, KAISER, nicknamed Rotbart or Barbarossa (*c.* 1122–90, in the River Calycadnus/ modern Geuksu, Cilicia), was the son of the Duke of Swabia of the powerful Hohenstaufen family (see STAUFEN). He was elected Deutscher König (q.v.) in 1152 on the death of his uncle, Konrad III (q.v.). Friedrich's policy had two principal aims: opposition to extension of the papal power in Germany, and the re-establishment of dominion in Italy. In 1155 he was crowned emperor at Rome, and, immediately returning to Germany, quickly established order, settling a quarrel over the Bavarian dukedom by confirming the claim of the Guelph Duke Heinrich (see HEINRICH DER LÖWE) but separating the eastern division of Bavaria (Ostmark) and establishing it as a separate dukedom (Österreich) for the Babenberg (q.v.) claimant. He later quarrelled with Heinrich, totally destroying his power in 1180.

His endeavours to assure imperial appointment of bishops in Germany were successful, but four campaigns in Italy between 1158 and 1176 achieved only partial success and ended in a catastrophic defeat at Legnano (1176). From 1159 to 1177 Friedrich was involved in a feud with Pope Alexander III. He secured, by negotiation with the Lombard League (1183) and by the marriage of his son to Princess Constance, heiress of Sicily, power which he had failed to achieve by force of arms. In 1189 he set out on a crusade, and was drowned either while crossing or while bathing in the River Calycadnus.

Though Friedrich's efforts in Italy were thwarted, he initiated a period of prosperity and good government in Germany. His personal popularity became legendary, and later resulted in the transference to him of the myth of the sovereign sitting inside the Kyffhäuser mountain in Thuringia ready to come to his country's aid in time of need. This legend was originally attached to the Emperor Friedrich II (q.v.). Friedrich Barbarossa appears as the sleeping emperor in a Volksbuch published in 1519 and in the ballad *Barbarossa* by F. Rückert (q.v., 1813), which gives the legend its classical form. Barbarossa also figures in the ballad *Schwäbische*

Kunde by Uhland (q.v., 1814) and is the subject of a tragedy by Grabbe, *Kaiser Friedrich Barbarossa* (q.v., 1829).

FRIEDRICH II, KAISER (Iesi nr. Ancona, 1194–1250, Fiorentino), was the son of the Emperor Heinrich VI (q.v.), and so came of the Staufen (q.v.) line. From his mother, before her marriage Princess Constance of Sicily, he inherited a claim to the Sicilian crown. He was elected Deutscher König (q.v.) at the age of 2 (1196); his father died in the following year, and his mother in 1198. He grew up in Italy, which was his true homeland, and in 1208 was declared of age by his guardian, Pope Innocent III. His claim to the imperial throne had lapsed in 1198 because of his tender age. In 1212, however, he appeared unexpectedly in Germany, was acclaimed on all sides, and staked anew his claim as king against Otto IV (q.v.), though he was obliged by the Golden Bull of Eger (1213) to make substantial concessions to the Papacy in Italy. By 1214 his position as German King was secure. In 1220 he had his 9-year-old son Heinrich crowned as German King, and thereupon returned to Italy, where he was crowned emperor in the same year.

Friedrich devoted himself to the ruthless and efficient reorganization of his Sicilian kingdom. He fell out with his former ally, Pope Gregory IX, and, while excommunicated, undertook a crusade in 1228, crowning himself king of Jerusalem in 1229. In 1234 his eldest son Heinrich VII (q.v.) revolted against his rule, but rapid and energetic intervention in Germany by Friedrich extinguished the revolt in 1235. In his later years he was for a second and final time excommunicated by Gregory (1239). An attempted reconciliation with Gregory's successor Pope Innocent IV failed. The Pope deposed him (1245) and encouraged the election of rival kings in Germany, but Friedrich maintained his supremacy until his death. He was followed by a younger son, Konrad IV (q.v.).

Friedrich's reign is remarkable for his ability to assert his rule while spending only two periods in Germany, from 1212 to 1220 and from 1235 to 1237. He maintained his position partly by the prestige of his Staufen descent and by his recognized energy and ability; he also pursued a policy of concessions to territorial princes and imperial cities, which were to make centralized rule more difficult in future. His real interest was the extension of his sway and the intensification of his administrative control in Italy.

A man of culture, who wrote poetry in Italian, Friedrich had an alert and open mind and was avidly interested in the scientific knowledge of his day. He was particularly attentive to natural history, and is the author of a notable treatise on falconry, *De arte venandi cum avibus*, which contains much accurate observation (printed in 1546, 1896, and 1942). At his court he showed religious tolerance and he employed Saracens in his army, but he saw to it that religious orthodoxy was maintained in his states. He supported the persecution of heretics in Germany in 1230 and 1232. He could also be ruthlessly cruel in the treatment of persons who evoked his disfavour. He is often regarded as a man born before his time, but it is easy to misread his medieval Italian character. Since his rule in Germany depended upon his personal standing, it is not surprising that after his death a period of disintegration supervened.

The legend, popularized soon after his death, that he was not really dead and would return reflects his prestige. It was afterwards Germanized, and by 1519 had been transferred to Barbarossa (see FRIEDRICH I, KAISER). Friedrich II is the king whom WALTHER von der Vogelweide (q.v.) mentions in a poem of gratitude for the grant of a fief: 'der edel künec, der milte künec'.

FRIEDRICH III (Innsbruck, 1415–93, Linz), German King and Emperor of the Holy Roman Empire (see DEUTSCHES REICH, ALTES) from 1440 to 1493, sought during his long reign to further the dynastic pretensions of the house of Habsburg. By arranging the marriage between his son Maximilian (see MAXIMILIAN I) and Marie of Burgundy (1477) he laid the foundation for the future greatness of Habsburg. Though ambitious, he was hesitant in decision, and slow to take action; his arrogant motto A.E.I.O.U. (*Austriae est imperare orbi universo*) was unsupported by his policy. He was the last emperor to be crowned in Rome, and he is also distinguished by having had in his service the noted Italian humanist Enea Silvio Piccolomini (q.v.). In 1486 he agreed to the election of his son Maximilian as German King, a step which ensured the Habsburg succession. His reign coincided with Turkish encroachments in southeast Europe, renewed French pressure on the western borders of the Empire, lawlessness at home, and continuing disquiet in the Church.

FRIEDRICH III, KAISER (Potsdam, 1831–88, Potsdam), German Emperor and King of Prussia, was the son of King Wilhelm I (q.v.). Friedrich married Victoria, the Princess Royal, daughter of Queen Victoria, in 1858, thirty years before his accession. In politics he was opposed to Bismarck (q.v.), especially to his handling of Schleswig-Holstein (see SCHLESWIG-HOLSTEINISCHE FRAGE) and to his censorship of the press, but he was careful to avoid indiscretion or embarrassment for his father. In the wars of 1866 and 1870 Friedrich held high commands. He was a strong supporter of the concept of a unified German Empire.

Unable as Crown Prince to find suitable expression for his Liberal opinions, he devoted himself to aesthetic projects, furthering the Berlin museums and supporting archaeological excavations at Olympia. When in 1888 he came to the throne he was suffering from cancer of the throat and died after a reign of ninety-nine days. The mistaken diagnosis of the Scots consultant Sir Morell Mackenzie led after the Kaiser's death to anti-British polemics in the German press. The Empress Victoria was known after his death as Kaiserin Friedrich (q.v.).

FRIEDRICH I, KÖNIG IN PREUSSEN (Königsberg, 1657–1713, Berlin), the only son of the Great Elector (see FRIEDRICH WILHELM, GROSSER KURFÜRST), succeeded his father as the Elector Friedrich III of Brandenburg in 1688. He left politics mainly to his ministers, concentrating his energies on the acquisition of a royal title, which he attained in 1701 by subservience to Austria. The territory which was held to entitle him to royal rank was the Duchy of Prussia (later East Prussia, see PREUSSEN and OSTPREUSSEN). Hence Brandenburg remained his Electorate (Kurfürstentum), and the prescribed new title was 'König in Preußen'. This form was dropped by his successors, who adopted the style 'König von Preußen'. In his reign were founded Halle University (1694), the Akademie der Künste (1696), and the Akademie der Wissenschaften (1701). Building of the Berlin Royal Palace, now destroyed, was begun in 1697.

FRIEDRICH II, DER GROSSE, KÖNIG VON PREUSSEN (Berlin, 1712–86, Potsdam), succeeded his father Friedrich Wilhelm I (q.v.) in June 1740. Friedrich Wilhelm had vainly tried to shape his heir in his own image. Failing that, he had, equally vainly, tried to force Friedrich to resign his rights of succession. The humiliations to which his father subjected him reached a crisis when Friedrich, at the age of 18, attempted to flee to Paris. The plot was discovered, and Friedrich was kept in solitary confinement at Küstrin for treason, and his friend, Lieutenant H. H. von Katte (q.v.), was executed in the courtyard below his window. But by the end of 1731 Friedrich was restored to his rank of colonel. He had learnt that life was bearable if he made his father believe that he was an obedient son, evidence of which he gave by inspecting troops, and by marrying, in 1733, Elisabeth Christine von Braunschweig-Wolfenbüttel, a niece of Maria Theresia (q.v.). There were no children of this marriage, and in later years Friedrich paid her a formal visit once a year on her birthday at her court. From 1736 to 1740 he resided at Rheinsberg, devoting his leisure to reading, writing, and music. He was an accomplished flautist, and more successful as a composer of music than as a writer of French verse. After his accession he maintained a private orchestra (Kapelle) which included the composers Quantz (q.v.) and C. P. E. Bach (see BACH, J. S.). In 1747 he suggested a theme for variations to J. S. Bach from which grew the *Musical Offering* (*Das musikalische Opfer*).

Within a few months of his accession he seized Silesia (see SCHLESISCHE KRIEGE), exploiting a constitutional crisis over the accession, in October 1740, of Maria Theresia (see PRAGMATISCHE SANKTION and ÖSTERREICHISCHER ERBFOLGEKRIEG). Friedrich thus realized longstanding Hohenzollern ambitions to extend Prussian territory and consolidate the prestige of the former Electorate, which had achieved the status of kingdom only in 1701. The whole of Friedrich's reign was determined by his early successful campaign which he owed to the efficient military and administrative system inherited from his father. He had provoked not only Austria, but the principal states of Germany and Europe as well. By 1745 he had successfully defended his new acquisition in the second of the Silesian wars, but in 1756 he entered the field again to foil the preparations of an Austrian coalition against him. His offensive developed into seven years of war (see SIEBENJÄHRIGER KRIEG), and earned Friedrich both hatred and admiration and, after an early victory at Roßbach (1757), the appellation 'the Great'. Although the war devastated the country and brought Friedrich to the verge of suicide, he emerged with increased prestige. The Seven Years War made history, not least by Prussia's endurance which was inspired by Friedrich himself, who was in the unique position of being his own commander-in-chief and head of government. It won him many sympathizers, among them Lessing and Goethe (qq.v.). After 1763 the prevention of further war and the recuperation of his country and army were his prime concern. He suggested the division of Poland (see POLAND, PARTITIONS OF) from which Russia, Austria, and Prussia were to benefit, and over which he hoped Russia and Austria would forget their differences. Friedrich's acquisition, in 1772, of the part of Poland renamed West Prussia (without Danzig and Thorn) was particularly valuable to him, as it linked East Prussia with Brandenburg. In 1778 he mobilized for the last time against Austria when Joseph II (q.v.) tried to acquire Bavaria upon the death of the Elector Maximilian (q.v., see BAYRISCHER ERBFOLGEKRIEG). Largely owing to Maria Theresia's own intervention, the crisis passed; but when the Prusso-Russian alliance was, in 1780, reversed into an Austro-Russian alliance, Friedrich found himself in isolation. To frustrate Joseph's plans to create a kingdom of Burgundy out of part of the Austrian Netherlands, Friedrich rallied the German princes, and in 1785 created the Fürstenbund (q.v.), which was joined by enough princes and bishops to act as an effective deterrent against Austria. Friedrich's sole concern was the consolidation of Prussian interests. Unlike his forebears he had no sense of loyalty to the authority of the Empire; he was contemptuous of the term 'Holy Roman Emperor' and, as a political realist, never aspired to this 'vain title'.

Friedrich's administration of the state followed the pattern established by his forebears. He was hard-working, efficient, and ruthless. He adhered to a strict caste-system by conviction. It was the weakness of his government that it depended on an experienced and knowledgeable ruler. Friedrich's heir, his nephew Friedrich Wilhelm II (q.v.), was ill equipped for this formidable task; nor did Friedrich provide responsible and independent ministers who could carry on his work. In his last years he regarded the future of Prussia with cynical resignation. Yet he left an army of some 200,000 men and a well-provided treasury, the result of his stringent mercantile policy and heavy taxation, especially of the middle class.

He disliked the German language and was contemptuous of German men of letters. In 1780 he published a treatise on the shortcomings of German literature: *De la littérature allemande, des défaults qu'on peut lui reprocher, quelles en sont les causes, et par quels moyens on peut les corriger*. A German translation, *Über die deutsche Litteratur*, appeared simultaneously and anonymously. The King's authorship was, however, an open secret. He blames the German language ('à demi barbare'), but believes that it is not beyond hope, if cultivated after the Ancient and French models. Canitz, C. F. Gellert, and S. Geßner (qq.v.) had made a beginning. Friedrich had no more than a random acquaintance with contemporary German literature (1740–80), which is also evident in his treatment of Lessing (q.v.). His mother Sophie Dorothea of Hanover and his favourite sister Wilhelmine (qq.v.) spoke French fluently, and French was from an early age Friedrich's favourite language. All his writings are in French. And while he practised religious tolerance out of a disbelief in all creeds, he was an admirer of French enlightenment, and while still at Rheinsberg had a profuse correspondence with Voltaire (q.v.). *Briefwechsel Friedrich des Großen mit Voltaire* was edited by Koser and G. Droysen (3 vols., 1908–11) and followed by *Nachträge zu dem Briefwechsel Friedrich des Großen mit Maupertuis und Voltaire* (1917). Mirabeau was one of his last visitors at Sanssouci (q.v.) before he died.

In Rheinsberg Friedrich pursued also historical studies and wrote his first political treatises on the principles of government and the critical analysis of European politics: *Considérations sur l'état présent du corps politique de l'Europe* was published anonymously at The Hague in 1738; in 1739 he wrote *L'Antimachiavel* (q.v.), a treatise on Machiavelli's *Il principe*, which Voltaire revised before it was published anonymously in 1740. His other writings include *Histoire de mon temps* (1746), and *Mémoires pour servir à l'histoire de la maison de Brandebourg* (1751); his *Histoire de la guerre de sept ans* was written after the Peace of Hubertusburg. The *Œuvres de Frédéric le Grand* (31 vols.) were edited by J. D. E. Preuß (1846–57). They did not include his two political testaments of 1752 and 1768. In 1771 he privately printed a short essay for Voltaire: *Essai sur les formes de gouvernement et sur les devoirs*. His works were published in German translation in 1912–14 (10 vols., with illustrations by A. Menzel, q.v.) and in 1916 (select edn., 2 vols.).

In *L'Antimachiavel* Friedrich first formulated the function of the ruler as 'le premier domestique de l'état' and his conviction of 'la raison d'état', a principle which he bore out in his service to Prussia and in his direction of politics. He developed an increasingly inaccessible personality. His affection concentrated on his sister Wilhelmine, who died in 1758, but he benefited from the outstanding ability of his brother Prince Heinrich (q.v.) in battle and in diplomacy. He practised self-criticism without vanity and did not seek popularity.

An abundance of literature has grown out of anecdotes of his life. Carlyle (*Frederick the Great*, 1858–65) praised him as a hero, and 19th-c. German historians of the Prussian school, Ranke, Treitschke (qq.v.), and Droysen, gave him fresh laurels in the name of nationalism. At the beginning of the 1914–18 War Th. Mann (q.v.) criticized the Prussian school (*Friedrich und die Große Koalition*, written in 1914). Of the many works on Friedrich the following may be mentioned (excluding those dealing with the Seven Years War): C. F. D. Schubart (q.v.), *Friedrich der Große*, a hymnic ode (1786); Willibald Alexis (q.v.), *Cabanis*, a novel (1832); Karl Gutzkow, *Zopf und Schwert*, a comedy (q.v., 1844); Theodor Fontane (q.v.), *Der alte Fritz*, poems (1851); Heinrich Laube (q.v.), *Prinz Friedrich*, a play (1854); Julius Mosen (q.v.), *Der*

Sohn des Fürsten, a tragedy (1858); Gustav Freytag, *Bilder aus der deutschen Vergangenheit* (q.v., 1859–67), a historical survey, Hermann Burte (q.v.), *Katte*, a play (1914); Paul Ernst (q.v.), *Preußengeist*, a play (1915), Walter von Molo (q.v.), *Fridericus*, a novel (1918); J. von der Goltz (q.v.), *Vater und Sohn*, a play (1921); Bruno Frank (q.v.), *Tage des Königs*, a novel (1924); Paul Alverdes (q.v.), *Die Flucht*, a story (1934); E. von Naso (q.v.), *Preußische Legende*, a story (1939); H. Mann, *Die traurige Geschichte von Friedrich dem Großen* (q.v., posth., 1960).

FRIEDRICH II, Kurfürst von Brandenburg (1413–71, Neustadt/Aisch), known as 'der Eiserne', acceded as Elector in 1440, and abdicated in 1470 in favour of his brother Albrecht Achilles. He devoted his attention to his own lands and came into conflict with Berlin and Cölln, an episode which is exploited by W. Alexis in his novel *Der Roland von Berlin* (q.v., 1840).

FRIEDRICH V VON DER PFALZ, Kurfürst (Amberg, 1596–1632, Mainz), son-in-law of James I of England, succeeded to the Palatine Electorate in 1610. He was the leader of the Protestant Union (see Protestantische Union). Hoping for assistance from the English king, the young Calvinist Elector rashly accepted the Bohemian crown in 1619 after the Bohemians had deposed Archduke Ferdinand (see Ferdinand II). Friedrich was defeated at the White Mountain (see Weissen Berge, Schlacht am) in November 1620 in a battle which virtually opened the Thirty Years War (see Dreissigjähriger Krieg). The Jesuits had predicted that Friedrich would only be a brief 'Winter King' (Winterkönig), and as such he became known in history. He fled to the Netherlands and remained a wretched, wandering outcast until he died of the plague at Bacharach. His electoral vote (Kur) was transferred to Maximilian I of Bavaria (q.v.). In the Peace of Westphalia (see Westfälischer Friede) Friedrich's son received a newly created eighth electoral vote, and was restored to the Rhenish Palatinate.

FRIEDRICH, Caspar David (Greifswald, 1774–1840, Dresden), German painter, studied art at Copenhagen, and in 1798 settled in Dresden. Friedrich painted chiefly landscapes and seascapes, with and without figures, architectural pictures, including a few of Dresden, and some religious subjects. Religious feeling and symbolism permeate his *œuvre*, of which the seascape with figures, *Die Lebensstufen*, is a characteristic example. He possessed considerable power to convey mood in landscape. Almost forgotten in the 19th c. and early 20th c., interest in his work increased considerably in the mid-20th c. He is hardly represented in Britain, but an exhibition of 112 of his pictures at the Tate Gallery in 1972 attracted much attention. F. G. Kersting (q.v.) was a friend of Friedrich.

FRIEDRICH, KAISERIN, see KAISERIN FRIEDRICH.

FRIEDRICH CHRISTIAN, HERZOG VON AUGUSTENBURG (Augustenburg, 1765–1814, Augustenburg), was Danish education minister under kings Frederick V and VI. In December 1791, with Graf von Schimmelmann (q.v.), he became a benefactor of Schiller (q.v.), sharing in a grant of a pension of 1,000 Thaler per annum for three years.

FRIEDRICH DER SCHÖNE (c. 1286–1330, Schloß Gutenstein), was in 1314 elected German King (see DEUTSCHER KÖNIG), in opposition to Ludwig IV (q.v.). From 1308 he was Friedrich III, Duke of Austria. He became known as Gegenkönig (i.e. a rival to the legitimate king) and never ruled. In 1322 he was defeated and imprisoned by his Bavarian rival Ludwig, who in 1325 came to an agreement with him confirming the dual kingship. Owing to the opposition of Friedrich's brother Leopold, this never took effect, and Friedrich spent the last years of his life in his native Austria.

FRIEDRICH DER STREITBARE (Altenburg, 1370–1428, Altenburg), became duke of Saxony, and led a life of vigorous campaigning in southern Germany. The Emperor Sigismund (q.v.) confirmed him in the dukedom, and in 1420 made him an Elector (Kurfürst). Leipzig university was founded during his reign.

FRIEDRICH DER WEISE (Torgau, 1463–1525, Lochau), Elector of Saxony from 1486 as Friedrich III, enjoyed a considerable respect among his contemporaries as a man of piety as well as wisdom. After the death of the Emperor Maximilian I (q.v.) he was offered the imperial crown, which he declined; his refusal led to the election of Karl V (q.v.). Friedrich is well known as a protector of Luther (q.v.), ensuring him a safe conduct for his appearance at the Diet at Worms (see WORMSER REICHSTAGE), and afterwards according him a refuge in the Wartburg. He was the founder of the University of Wittenberg (1502).

Friedrich d'or, gold coin introduced in 1750, also termed a Louis d'or or a Pistole, worth 5 Thaler (q.v.).

FRIEDRICH KÖDITZ VON SAALFELD, a monk and schoolmaster at Reinhardsbrunn,

wrote between 1315 and 1323 an account in Middle High German prose of the life of 'St. Ludwig' (see LUDWIG IV, LANDGRAF VON THÜRINGEN), who was the husband of Saint Elisabeth (see ELISABETH, HEILIGE). His story, based on the biography of Ludwig by Berthold, the Landgrave's chaplain, and the life of St. Elisabeth by Dietrich von Apolda (both in Latin), includes an account of the Wartburgkrieg (q.v.).

Friedrichshafen, a town of Württemberg, situated on the north shore of Lake Constance. Originally called Buchhorn, it was renamed in 1811 in honour of King Friedrich I of Württemberg. It is best known for its association with the building of experimental airships by Count Zeppelin (1838–1917). The town was an imperial city from 1275 to 1802, when it passed to Bavaria, being transferred to Württemberg in 1810.

Friedrichshagener Kreis, a small group of Naturalist writers and theorists, founded in Friedrichshagen in 1890. They included W. Bölsche and the brothers Hart (qq.v.).

Friedrichsruh, situated in the Sachsenwald (Duchy of Lauenburg/Schleswig-Holstein), is best known as the estate to which Bismarck (q.v.) retired after his dismissal in 1890. He was buried there, and the Mausoleum as well as the Bismarck Museum remain, though the Schloss has been rebuilt.

FRIEDRICH VON HAUSEN, a Middle High German Minnesänger, belonged to a well-known noble family from the region of Kreuznach south-west of Mainz. He was a member of the entourage of the Emperor Friedrich I (q.v.), and took part in two campaigns in Italy in 1186 and 1187. On the second of these his commander was Heinrich VI (q.v.), then German King. He died on Friedrich's crusade in 1190.

Hausen modelled his Minnelieder on Provençal and Old French poetry, and had for this reason an important historical formative role. In his songs, in which nature plays no part, *minne* is the courtly relationship of the lover in the service of his lady. Hausen's three crusade songs (Kreuzlieder), in which courtly *minne* and God are in conflict, and God in the end triumphant, are the summit of his achievement.

FRIEDRICH VON LEININGEN, a Middle High German Minnesänger, who is represented by only one poem, a farewell on departure for Apulia. The poem is thought possibly to refer to the crusade for which Friedrich II (q.v.) assembled forces in 1227. Leiningen belonged to a line of counts, and is mentioned in Alsace between 1214 and 1239.

Friedrich von Schwaben, a degenerate Middle High German verse romance of unknown authorship. A later hand has altered and made additions. Its motifs are principally drawn from fairy-tale. A noteworthy feature is the first occurrence in German literature of the name Wieland (Wayland the Smith), which is at one stage substituted for Friedrich. The poem, which is almost 5,000 lines in length, was written early in the 14th c., and certainly after 1314.

FRIEDRICH VON SONNENBURG, a minor Middle High German poet of the second half of the 13th c. Sonnenburg, a poet of marked individuality, who is designated as Meister, originated in Tyrol and was the author of a number of Sprüche (see SPRUCH) of moral or political content. Like some other itinerant poets he particularly praises the virtue of generosity (*milte*). In the political conflicts of the age he takes the side of the Pope.

FRIEDRICH WILHELM I, König von Preussen (Berlin, 1688–1740, Potsdam), was the son of the first king of Prussia, Friedrich I (q.v.). He came to the throne in 1713. His foreign policy, which aimed at the aggrandizement of Prussian territory, was relatively unsuccessful, and his robust personality expressed itself mainly in administration at home (he may be said to have created the Prussian civil servant) and in the expansion and organization of the army. He was excessively economical, and began his reign by dismissing a large number of royal servants and selling the handsome furniture of his palaces. He centralized the financial administration, and reduced imports, clearing off the debts with which his father had encumbered the state, and leaving a well-filled treasury at his death. Friedrich Wilhelm's passion was the army, the strength of which he gradually increased from 38,000 to 83,000. For so military a monarch he was remarkably unwarlike, fighting only minor wars in the early years of his reign. One of his principal amusements was the reviewing of his troops, and his great pride was his guard regiment of tall, upstanding grenadiers, the so-called 'lange Kerls'. No exceptionally tall man in his domains was safe from his recruiting officers. He conducted his court on military principles, and preserved a military discipline in his family. His short-sighted tyranny in his own house led to the serious and tragic clash with his son, the future FRIEDRICH II (q.v.), who was imprisoned and forced to witness the execution of his friend Katte. Friedrich Wilhelm's chief evening re-

laxation was his Tabaks-Kollegium, at which he and his favoured friends drank and smoked heavily, and indulged in crude pleasantries and horse-play. His military eccentricities and his Tabaks-Kollegium lent themselves to anecdotal treatment, and after his death he was often portrayed in a genial and homely light. Friedrich Wilhelm was a firm Protestant, yet in religious matters reasonably tolerant.

The best-known literary portrayal of him and his manners is found in K. Gutzkow's comedy *Zopf und Schwert* (q.v., 1844). J. Klepper's novel *Der Vater* (q.v., 1937) sees his conduct in a more favourable religious light. His military interests and achievements have earned for him the nickname Soldatenkönig.

FRIEDRICH WILHELM II, König von Preussen (Berlin, 1744–97, Berlin) was a nephew of Friedrich II (q.v.), and heir apparent from 1758, acceding in 1786. Though he was an enemy of the French Revolution, his eagerness to divide with Russia and the Austrian Habsburgs the remaining territory of Poland (see POLAND, PARTITIONS OF) rendered the campaign against France, begun in 1792, ineffectual. Friedrich Wilhelm was notorious for his mistresses, the best known of whom was Wilhelmine Ritz, created Gräfin Lichtenau (see LICHTENAU, GRÄFIN) in 1796. He was also interested in the mystical societies of the period, especially the Rosicrucians (see ROSENKREUZER). His military and political interests were slight, and he entrusted himself largely to his minister, J. R. von Bischoffwerder (q.v.).

FRIEDRICH WILHELM III, König von Preussen (Potsdam, 1770–1840, Berlin), son of Friedrich Wilhelm II (q.v.), came to the throne on the death of his unpopular father in 1797. His unassuming personality, his love match with Luise von Mecklenburg-Strelitz (see LUISE, KÖNIGIN), and his irreproachable family life secured him immediate popularity. Timid and irresolute, he was hesitant in carrying out reforms which he believed to be right. Thus between 1799 and 1805 the serfs on the royal domains were progressively freed, but the King feared to impose this measure on the landed nobility. In foreign affairs he was equally hesitant and preferred dilatory and cautious ministers. He did not join the Austro-Russian coalition against Napoleon in 1805 (see NAPOLEONIC WARS), but in autumn 1806, after Austria's defeat at Austerlitz, he plunged too late into war with France and was disastrously defeated at Jena. Virtually ignored at the discussion of peace at Tilsit in 1807, Friedrich Wilhelm had to accept the dismemberment of his state. He did, however, for a time employ activist reformers of strong and energetic character (see STEIN, SCHARNHORST, and GNEISENAU). The serfs throughout Prussia were liberated in 1807, and the out-of-date army was reformed and trained on modern principles. Friedrich Wilhelm feared his own people as much as the French, and resisted those who called for a *levée en masse*. He kept the peace with France in 1809 when Austria was again defeated, and condemned the action of General H. D. L. YORCK von Wartenburg (q.v.) in making a truce with the Russians on 30 December 1812. Only after the disasters to Napoleon's Grand Army did he take the decisive step of moving the government to Breslau. The popular proclamation *An mein Volk* (q.v.), though signed by the King, was not drafted by him. Influenced by the activists in Breslau, he made a declaration of war against France in February 1813. After 1815 it was expected that the King would promulgate a liberal constitution, but the reforming ministers (W. von Humboldt and K. A. von Hardenberg, qq.v.) failed to move him and the mistrustful King aligned himself with the Metternich (q.v.) system and the Holy Alliance (see HEILIGE ALLIANZ) for the remainder of his reign. He appears as a character in Fontane's novel *Schach von Wuthenow* (q.v.).

FRIEDRICH WILHELM IV, KÖNIG VON PREUSSEN (Berlin, 1795–1861, Sanssouci), the son of Friedrich Wilhelm III (q.v.), whom he succeeded in 1840. A man of humane outlook, he was not a clear political thinker and, though disposed to rule liberally, was not willing to accept a diminution of his powers. His thinking was in part shaped by the Romantic movement, which was dominant in his formative years (see ROMANTIK). He was a man of intelligence, piety, and taste, but, like his father, was hesitant and given to temporizing. His reign began auspiciously for the Liberals, with the termination of the persecution of so-called 'demagogues' (see DEMAGOGENVERFOLGUNG). Moreover he appointed the brothers J. and W. Grimm, Savigny, Schelling, and L. Tieck (qq.v.) to professorships at Berlin university, and in 1842 put an end to the ecclesiastical quarrel involving K. A. Droste zu Vischering (q.v.). In the matter of a constitution he hedged, and hoped to avoid radical changes. In the 1848 Revolution (see REVOLUTIONEN 1848–9) he was compelled to make an act of public mourning in the presence of the bodies of the revolutionaries who had been shot (19 March 1848). This act was bitterly resented by conservative forces in Prussia as a humiliation of the monarchy. Two days later Friedrich Wilhelm rode through the streets in a deliberate

but unsuccessful attempt to appear as a popular leader and take over the Revolution. He sought German unity, but had no wish to exclude Austria from a new German state. He rejected the imperial crown (Kaiserkrone) offered to him by the Frankfurt National Assembly (see FRANKFURTER NATIONALVERSAMMLUNG) on 3 April 1849, at first provisionally, and definitively on 28 April. Repelled by armed conflicts throughout, he gave way in the confrontation between Prussian and federal troops in Hesse. From 1857 onwards he was ailing, and his younger brother (see WILHELM I) acted as regent.

FRIEDRICH WILHELM, DER GROSSE KURFÜRST (Berlin, 1620–88, Potsdam), succeeded as Elector of Brandenburg in 1640. He had been brought up as a Calvinist, and in conformity with the policy of the Peace of Augsburg (see AUGSBURGER RELIGIONSFRIEDE), *cuius regio, eius religio*, soon to be confirmed in the Peace of Westphalia (see WESTFÄLISCHER FRIEDE) in 1648, made the Reformed Church the dominant church in Brandenburg. Friedrich Wilhelm was the first decisive character and skilled politician after a long line of insignificant rulers, and, for good or ill, set Brandenburg (later to expand to Prussia) on a course of opportunistic alliances for territorial gain—'Realpolitik' (q.v.), in fact, before the word was created. His acquisitions included Minden, Halberstadt, and Magdeburg. His great ambition to annex Vorpommern (the district round Stettin) failed, in spite of a brilliant victory over the Swedish army at Fehrbellin in 1675, through the opposition of Louis XIV. The Elector prudently encouraged the immigration of French Calvinists driven out by the revocation of the Edict of Nantes in 1685. He ruled despotically, but sought to relieve the burdens of the citizenry and to curb the independence of the landed nobility. He raised his military forces to a high level of efficiency. He is an important character in H. von Kleist's play *Prinz Friedrich von Homburg* (q.v.).

FRIKART, THÜRINGEN (Berne, 1429–1519, Berne), town clerk of Berne, vividly described in a prose chronicle a conflict in 1470 between the governing nobility and the common citizens of the city (Twingherrenstreit).

FRISCH, MAX (Zurich, 1911–), the son of an architect, qualified in the same profession in Zurich in 1941. He practised successfully for a time and maintained his contact with the architectural profession until 1954, when he made authorship his career. During the 1939–45 War he served as a part-time soldier in the Swiss artillery. A great traveller, Frisch went to the Balkans in 1933, and many times to West

Germany, France, and Italy. In 1948 he visited the countries of eastern Europe, and met B. Brecht (q.v.) in East Berlin, having already made his acquaintance in Zurich in the previous year. He was in the U.S.S.R. in 1966 and in 1968, the U.S.A. and Mexico in 1952 and 1956, and the U.S.A. in 1970, the Arab states in 1957, Israel in 1965, Japan in 1969, and China in 1977. In 1960 he made his home in Rome, but moved in 1965 to Ticino in his native Switzerland.

Frisch's first novel, *Jürg Reinhart* (1934), was the outcome of his Dalmatian journey; it was followed by *Antwort aus der Stille* (1937) and the novel *J'adore ce qui me brûle oder Die Schwierigen* (1943, new version entitled *Die Schwierigen oder J'adore ce qui me brûle*, 1957). *Bin oder die Reise nach Peking* (1945) relates the conversations of a young man with his *alter ego* (Bin) on an imaginary journey to Peking. Hardly more than part-time diversions, these early works made little impression, but the play *Nun singen sie wieder* (q.v., 1946, revised 1955 and 1972), a requiem for the dead and a warning to the living, attracted immediate attention, and was performed in Germany as well as in Switzerland. *Die chinesische Mauer* (q.v., 1947, revised 1955, 1965, and 1972), which, like some of his other plays, shows the influence of Brecht, is a protest in the form of a black comedy against nuclear war. *Santa Cruz*, a short play, also of 1947, deals with problems of identity, which were to become one of Frisch's chief preoccupations. In 1949 appeared the play *Als der Krieg zu Ende war*, a more conventionally realistic and melodramatic work, set in Berlin in the first days of the Russian occupation. The play *Graf Öderland* (q.v., 1951, revised 1957 and 1961) uses reality and dream to probe the latent potentialities of the individual. Frisch's next play, *Don Juan oder Die Liebe zur Geometrie* (1953, revised 1962), parodies the legend, suggesting that it was invented by Juan himself in order to allow him to escape from women into an existence devoted to his true love, geometry.

By the early 1950s Frisch was known internationally as a dramatist of originality. He turned next to fiction: the novel *Stiller* (q.v., 1954) treats a problem closely analogous to that of *Don Juan*. Stiller has disappeared for six years, and, when challenged, denies his identity, which in the end is proved irrefutably. *Homo faber* (q.v., 1957) demonstrates the inability of the scientist technologist to control events. It is a problem which Frisch, himself a technologist, was well equipped to understand. Abroad Frisch is best known for his two outstanding minatory plays, *Biedermann und die Brandstifter* (q.v., 1958) and *Andorra* (q.v., 1961), both harshly critical of the attitude 'it can't happen

here', and of the supineness of the bourgeoisie. Though both have a special application to Switzerland, their significance is universal. As *Biedermann* was too short for performance alone, Frisch wrote a one-act play to accompany it, *Die große Wut des Philipp Hotz* (1958). *Ausgewählte Prosa* and *Erzählungen des Anatol Ludwig Stiller* were both published in 1961. *Stücke* (2 vols.) appeared in 1962 (informative notes are appended): (vol. 1) *Santa Cruz. Eine Romanze*; *Nun singen sie wieder. Versuch eines Requiems*; *Die chinesische Mauer. Eine Farce*; *Als der Krieg zu Ende war. Schauspiel*; *Graf Öderland. Eine Moritat in zwölf Bildern*; (vol. 2) *Don Juan oder die Liebe zur Geometrie. Komödie in fünf Akten*; *Biedermann und die Brandstifter. Ein Lehrstück ohne Lehre*; *Die große Wut des Philipp Hotz. Ein Schwank*; *Andorra. Stück in zwölf Bildern*. The *Anhang* includes *Nachspiel zu 'Biedermann und die Brandstifter'*, which was omitted from editions from 1973.

Frisch's novel *Mein Name sei Gantenbein* (q.v., 1964) is perhaps the most complex of his fictional studies of identity and existence.

Biografie. Ein Spiel (1967, revised 1968) is a comedy which deals with life as it *might* have been. The enactment of unfulfilled potentialities concerns Herr Kürmann and his wife Antoinette. A Registrator supervises the experimental procedure ('wiederholen, probieren, ändern', *Anmerkungen*). *Triptychon. Drei szenische Bilder* (1978, working-title *Styx*), Frisch's most abstract play, views death as final, devoid of any form of permutation (demonstrated by the static central scene), and with no hope for any communication between the dead and the living (elaborated in the first and third scenes)— an attitude repeatedly expressed in his work and reminiscent of *Nun singen sie wieder*, though with a marked affinity to S. Beckett.

Based on an idea suggested by Brecht in the late 1940s, the satirical prose of *Wilhelm Tell für die Schule* (1971) demolishes the Swiss myth, not least the figure of Tell, in sharp contrast to Schiller's treatment of the national hero in his play *Wilhelm Tell* (q.v.); presented in the form of an annotated story with 74 notes, the text is implicitly directed at Frisch's own compatriots. The social criticism of *Dienstbüchlein* (1974) proceeds from Frisch's military experience. Frisch's later fiction shows increasingly a withdrawal from the contemporary world to the privacy of his personal life, which is central to his overtly autobiographical story *Montauk* (1975). A discomforting self-analysis presented through a technique of shifting perspectives, it consists of a series of diverse recollections. These are supported by captions, in English, some of which endorse the epigraph, a citation from Montaigne. *Der Traum des Apothekers von Locarno. Erzählungen aus dem Tagebuch 1966–1971*

(1978) was followed by *Der Mensch erscheint im Holozän* (1979), the story of Geiser, aged 73, who collects and recollects knowledge to test his memory, aware of the futility of such endeavours in his voluntary retreat, an isolated mountain village (the location is that of Frisch's own Italian home), where the rocks induce a sense of indifference to the passing of life. In *Blaubart* (1982) the central figure, Schaad, a medical doctor in his mid-fifties and known as Ritter Blaubart for having been divorced six times, is accused of the murder of one of his wives. Despite his subsequent acquittal he longs for an end to his existence, but even after a serious road accident he has to live on.

Frisch is a notable diarist. His early journal of his life as a soldier of a neutral state appeared in 1940 as *Blätter aus dem Brotsack*; *Tagebuch mit Marion* was published in 1947, and later reprinted as part of *Tagebuch 1946–49*, a diary written for publication, which appeared in 1950. This affords fascinating glimpses of Frisch's mind and his mode of creation, and the reader, even if he disregards Frisch's injunction in the preface to read it from cover to cover, will still gather a rich harvest. A second equally arresting diary, *Tagebuch 1966–71*, was published in 1972.

Gesammelte Werke (6 vols., as well as a paperback edition in 12 vols.), arranged in chronological order and edited by H. Mayer in collaboration with W. Schmitz, appeared in 1976 and the volume *Die Tagebücher* (1946–1949 and 1966–1971) in 1983.

Frisch was awarded the Büchner Prize in 1958 and has since received many German, Swiss, and American honours.

'Frisch auf zum fröhlichen Jagen', first line of the patriotic poem, *Kriegslied für die freiwilligen Jäger*, written in 1813 by Friedrich de la Motte-Fouqué (q.v.). It was intended to be sung to the tune of 'Auf, auf zum fröhlichen Jagen!' of 1724.

Frische teutsche Liedlein, an anthology of songs with music compiled by Georg Forster (q.v.) and published in 5 vols., 1539–56.

Frisch, fromm, froh, frei, also occurring in the variant 'Frisch, fromm, fröhlich, frei', was adopted by Turnvater Jahn (see JAHN, F. L.) as slogan for the German athletic movement (Turnerschaft). From it was derived in 1844 a badge consisting of four conjoined *F*s, in the form of an elaborate cross.

FRISCHLIN, NIKODEMUS (Balingen, 1547–90, Hohenurach), one of the last German humanists, led a turbulent life, which was a consequence of his stormy temperament. Twice he

quarrelled with his academic colleagues in Tübingen and had to leave. He journeyed through central Europe, visiting Laibach, Prague, Wittenberg, and Brunswick. Arrested and imprisoned in the castle at Urach (Hohenurach), he died while attempting to escape.

Frischlin wrote plays in Latin, including *Rebecca* (1576), *Susanna* (1578), and *Julius redivivus* (1584), a Latin comedy entitled *Phasma* (1592), dialogues and *facetiae* (see FAZETIE), and a Latin epic, *Hebraeis*. Some of his plays contain scenes or choruses in German. His only surviving German play is *Frau Wendelgard* (1579). Frischlin also translated Aristophanes into Latin (*Plutus, Knights, Clouds, Frogs*, and *Acharnians*).

FRITSCH, GERHARD (Vienna, 1924–69, Vienna), served in the last years of the 1939–45 War, studied German literature, and was a publisher's reader and a librarian before devoting himself to writing. He began with poetry, *Zwischen Kirkenes und Bari* (1952), *Lehm und Gestalt* (1954), and the large-scale poem *Dieses Dunkel heißt Nacht* (1955). In 1958 he published a further volume of verse, *Der Geisterkrug*. He wrote two novels, *Moos auf den Steinen* (1956) and *Fasching* (1967), both of which are concerned with the tenacious Austrian tradition, which the first regards with nostalgia and the second with anger. He also wrote several radio plays. He took his own life. *Gesammelte Gedichte* appeared in 1978.

FRITZ, WALTER HELMUT (Karlsruhe, 1929–), a reticent nature poet, is the author of the collections *Achtsam sein* (1956), *Bild und Zeichen* (1958), *Veränderte Jahre* (1963), and *Die Zuverlässigkeit der Unruhe* (1966). *Gesammelte Gedichte* appeared in 1979, *Wunschtraum, Alptraum. Gedichte und Prosagedichte 1978–1981* in 1981, *Gedichte. Werkzeuge der Freiheit* in 1983.

Fritzlarer Passionspiel, a medieval passion play which is preserved only in fragments amounting to *c.* 200 lines. Probably written *c.* 1460, it has links with the *Frankfurter Dirigierrolle* (see FRANKFURTER SPIELE) and the *Alsfelder Passionsspiel* (q.v.).

FRÖBEL, FRIEDRICH (Oberweißbach, Thuringia, 1782–1852, Marienthal, Thuringia), worked for two years (1808–10) with Pestalozzi (q.v.) at his educational institute at Yverdon, Switzerland, and arrived through his study of Romantic attitudes (see ROMANTIK) at new ideas in child education, which he put into practice after participating in the Wars of Liberation, founding his first pedagogical institute in Keilhau nr. Rudolstadt in 1817. Fröbel adapted the idea of the workings of a creative spirit in Nature to human development from the early stages of infancy. In 1837 he realized his aim to devote himself entirely to children of pre-school age, founding his Anstalt zur Pflege des schaffenden Tätigkeitstriebes in Blankenburg. His practical work was based on the methodical use of toys (Gaben), which he grouped and applied in stages in accordance with his assessment of a child's latent creativity. His six basic groups, starting with the soft ball, follow mathematical principles by the introduction of cubes, which, as building bricks, etc., stimulate constructive, purposeful, and satisfying activity. The Fröbel Kindergarten, established in modernized form in European countries and the U.S.A., derives from these principles. Fröbel introduced the term Kindergarten in 1840 and undertook the training of teachers.

In 1851 Fröbel's Kindergärten in Prussia were closed and not reopened until after his death. The charge of atheism which led to this is thought to have arisen from a confusion of identity with Julius Fröbel (1805–93), his nephew, who was prosecuted on a political charge during the revolutionary period, being condemned to death (but pardoned) in 1848 in Vienna.

Fröbel published *Die Menschenerziehung* in 1826 and *Plan zur Errichtung einer Armenerziehungsanstalt* in 1840, followed in 1844 by *Mutter- und Koselieder* which, with a number of short-lived periodicals, were designed to stimulate parental consciousness of a child's mental and emotional needs, which was a fundamental aspect of his whole approach. *Gesammelte pädagogische Schriften* (3 vols.), ed. W. Lange, appeared in 1862–74 (reissued 1966).

FROBEN, JOHANN (Hammelburg c. 1460–1527, Basel), a friend of Erasmus (q.v.) and well known as a printer and publisher, set up business in Basel in 1491. His firm published a Latin Bible, editions of Fathers of the Church, and the Greek New Testament prepared by Erasmus (1516). The firm was continued by his descendants until 1603.

Frohe Wandersmann, Der, a poem by J. von Eichendorff (q.v.), beginning 'Wem Gott will rechte Gunst erweisen'. It expresses simple faith, a love of nature, and the impulse to travel far afield on foot. It is contained (without title) in *Aus dem Leben eines Taugenichts* (q.v.).

FRÖHLICH, ABRAHAM EMANUEL (Brugg, Aargau, 1796–1865, Gebensdorf, Aargau), pastor and schoolmaster in Aarau and a friend of J. Gotthelf (q.v.), is the author of fables (*Fabeln* and *Hundert neue Fabeln*, both 1825), poems (*Schweizerlieder,* 1827; *Trostlieder,* 1851), stories (*Novellen, Figuren und Charakterbilder,*

1853), and an epic (*Johannes Calvin,* 1864). *Ausgewählte Werke* were published in 1884.

FRÖHLICH, KATHARINE (KATHI), see GRILLPARZER, F.

Fronleichnamsspiele, religious plays performed in the Middle Ages on Corpus Christi Day (instituted in 1264). They were first presented at halts of the procession at various points in the town, afterwards developing into continuous plays, often dealing with biblical stories from the standpoint of redemption, which the Holy Sacrament symbolizes. The principal Fronleichnamsspiele are the *Künzelsauer, Innsbrucker,* and *Egerer* (qq.v.).

FRONSPERGER, LEONHARD (Ulm, c. 1520–75, Ulm), wrote copiously on military matters, his expertise being recognized by the Emperor Ferdinand I (q.v.), who appointed him provost-marshal (Feldgerichtsschultheiß). His principal work is *Das Kriegsbuch* (1573).

Froschmeuseler, Der, a satirical epic by Georg Rollenhagen (q.v.).

Fruchtbringende Gesellschaft, Die, name of a society founded in 1617 by Prince LUDWIG von Anhalt-Köthen (q.v.) for the purification and promotion of the German language. It was modelled on the Florentine Accademia della Crusca (founded 1582), of which Prince Ludwig was a member. The society was founded on the suggestion of the Prince's chamberlain, Kaspar von TEUTLEBEN (q.v.), on 24 August 1617, when a number of Thuringian noblemen were gathered together for the funeral of the Prince's sister. Prince Ludwig became the first president of the society, which is sometimes called the Palmenorden from its emblem, a palm-tree. After the manner of the Florentine Accademia, the members, who were initially chiefly noblemen, adopted designations such as 'Der Nährende' (Prince Ludwig), 'Der Schmackhafte' (Duke Wilhelm of Saxony), 'Der Vielgekrönte' (Diederich von dem Werder), and 'Der Gekrönte' (Opitz). Among its members, in addition to those mentioned, were Duke Bernhard von Sachsen-Weimar, Georg Schottel, G. P. Harsdörffer, F. von Logau, and J. M. Moscherosch (qq.v.).

The Fruchtbringende Gesellschaft sought to encourage the use of German (in preference to French), and it made German versification its special province. Its actual achievements were inconsiderable, but its aristocratic prestige did something to maintain interest in German literature in an age in which French taste was widespread. The society ceased to be effective after the withdrawal of Prince Ludwig in 1650, and it was finally wound up in 1680. Friedrich Wilhelm (q.v.), the Great Elector of Brandenburg, was a member in its last phase.

Frühen Gräber, Die, a short ode in three unrhymed strophes written by F. G. Klopstock (q.v.) in 1764. It has been set to music by Gluck and F. Schubert (q.v.).

Frühling, Der, a poem published by E. von Kleist (q.v.) in 1749, and consisting of 400 lines of hexameters. Kleist revised it in 1756. Though inspired by James Thomson's *The Seasons,* it is an original poem. *Der Frühling* praises the harmony of nature, rural seclusion, and simplicity, casts a reproving glance at warring sovereigns, and regrets the poet's own separation from nature. It ends with the hope that he may find his last resting-place in the country. Kleist is said originally to have intended to write further poems on the other seasons to complete the work.

Frühlings Erwachen. Eine Kindertragödie, a play by F. Wedekind (q.v.), written 1890-1, and published in 1891. The first two acts deal tersely and realistically with puberty and the awakening of sexual feelings in Melchior Gabor, a sixth-former, and the 14-year-old Wendla Bergmann. Melchior's friend Moritz Stiefel commits suicide, and among his papers is found a document compromising Melchior. At a school inquiry (in which caricature replaces realism) Melchior is expelled and sent to a reformatory. Wendla dies from an abortion. In a symbolical last scene in the churchyard in which Wendla is buried, Moritz, representing suicide, and a disguised gentleman (vermummter Herr), representing life, argue for Melchior's soul. Melchior rejects Moritz's temptations to commit suicide and follows the anonymous man, who leads him back to life.

Frühlingsfeier, Die, one of the best-known odes of F. G. Klopstock (q.v.), written in 1759. It has 27 irregular, unrhymed strophes. Its sense of the presence and protection of God in face of the immensity of the universe is conveyed in a famous description of a thunderstorm, to which there is an allusion in Goethe's *Die Leiden des jungen Werthers* (q.v.) under the date 16 June 1772.

Frühneuhochdeutsch, Early New High German, see GERMAN LANGUAGE, HISTORY OF.

FRUNDSBERG, GEORG VON (Mindelheim nr. Memmingen, 1473-1528, Mindelheim), was a professional soldier of the Swabian League (see SCHWÄBISCHER BUND) who held commands

under the emperors Maximilian I and Karl V (qq.v.). He distinguished himself in 1525 at Pavia. In 1527 he raised a mercenary force to fight in Italy, but suffered a stroke when the soldiers mutinied near Bologna; he was carried back to Germany and died in his castle. He is sometimes called 'Vater der deutschen Landsknechte'. He appears in Hauff's *Lichtenstein* (q.v.).

Fuchs, a freshman at a German university. In some student corporations the Fuchs could not become a senior (Bursche) merely by keeping semesters, but had to pass a Burschexamen (largely a drinking test) and give proof of courage in a student duel. See STUDENTISCHE VERBINDUNGEN.

FUCHS, GÜNTER BRUNO (Berlin, 1928-77, Berlin), an artist by training, but also a man of many jobs, was a satirical poet. His published work includes *Der Morgen* (1954), *Nach der Haussuchung* (1957), *Pennergesang* (1965), *Blätter eines Hof-Poeten und andere Gedichte* (1967). *Brevier eines Degenschluckers* (1960) contains prose as well as verse. Posthumously published collections include *Die Ankunft des großen Unordentlichen in einer ordentlichen Zeit* (1978) and *Gesammelte Fibelgeschichten und letzte Gedichte* (1978).

'Fuchs, du hast die Gans gestohlen', first line of the children's song *Warnung,* written in 1824 by H. Anschütz (q.v.). It is sung to an older folk melody.

FÜETRER, ULRICH, who lived in the second half of the 15th c., surviving at least until 1502, originated in Landshut and was by trade a heraldic painter in the service of the Munich court. Between 1478 and 1481 he wrote a Bavarian chronicle in prose for Duke Albrecht IV. A friend of J. Püterich (q.v.), Füetrer shared the latter's enthusiasm for WOLFRAM von Eschenbach (q.v.), and his principal work, *Buch der Abenteuer,* is written in the Titurelstrophe (q.v.) as it appears in *Der jüngere Titurel,* then thought to be by Wolfram, and was written for Duke Albrecht about 1490. A poem of more than 400,000 lines, it strings together the substance of a great number of medieval romances, including *Merlin, Titurel, Parzival, Die Krone, Lohengrin, Wigalois,* and *Iwein.*

FUGGER, a prosperous merchant family in Augsburg, which became a great banking house in the later Middle Ages. The name is first mentioned in the 14th c., at the end of which the head of the family, Johannes (d. 1409), was a recognized patrician. His second son, Jacob

(d. 1469), greatly extended the activities of the house and founded its fortune. His three sons, who survived into the 16th c., made the family one of the richest in Europe. In 1514 Jakob (styled Jakob II, 1459–1525) founded the Fuggerei, a complex of almshouses for 106 families, which still exists in Augsburg. In 1535 the Fuggers received the privilege of minting coinage. The head of the house was made a count in 1530; from him sprang three noble lines, all of which survived into the 20th c. One branch was raised to the rank of prince (Fürst) in 1913.

FÜHMANN, FRANZ (Rokytnice, Czechoslovakia, 1922–84, Berlin), became a Communist while a prisoner of war in Russia (1945–9), and subsequently established himself as a prominent DDR writer whose work centred for many years on the evils of fascism and war. He is the author of an epic poem, *Die Fahrt nach Stalingrad* (1953), and of several volumes of poetry (*Die Nelke Nikos*, 1953; *Aber die Schöpfung soll dauern*, 1957; *Die Richtung der Märchen*, 1962). The collection *Das Judenauto; Kabelkran und Blauer Peter; Zweiundzwanzig Tage oder Die Hälfte des Lebens* (1979) presents a revised version of *Das Judenauto* of 1962. Consisting of 14 sections, it relates political and economic crises from 1929 to 7 October 1949, the day of the foundation of the DDR. Together with the reportage *Kabelkran* of 1961, it is described as a cycle in episodic form (Episodenzyklus) and conceived as a personal 'Prozeß der Selbstfindung' (*Nachwort*). The pithy and at times humorous prose of the *Zweiundzwanzig Tage* (1973) records in diary form experiences during a visit to Hungary. The entries include reflections on Fühmann's Nazi past, the war, and his change to Communism, which began in Russia (discussed more fully in some sections of *Das Judenauto*). Other stories include the Novelle *Kameraden* (1955), *Stürzende Schatten* (1959), *Die heute vierzig sind. Eine Filmerzählung, Spuk. Aus den Erzählungen des Polizeileutnants K.* (both 1961), the collection of Novellen *König Ödipus* (1966), *Der Jongleur im Kino oder Die Insel der Träume* (1970), *Barlach in Güstrow* (1973), *Bagatelle, rundum positiv* (1978), and *Der Geliebte der Morgenröte* (1978, in extended form 1979), a cycle of four stories adapted from Greek mythology. Written in stylized prose, they open with Aphrodite's adulterous love for Ares, and end with Hephaestus' dedication to his art. Other adaptations include *Das Nibelungenlied* (1971). He was also well known in the DDR for his books for children. A lengthy discourse on poems by G. Trakl (q.v.), initially conceived as a *Nachwort* to a select edition of Trakl's poetry, appeared under the title *Der Sturz des Engels: Erfahrungen mit Dichtung*

(BRD) and *Vor Feuerschlünden. Erfahrungen mit Georg Trakls Gedicht* (DDR, both 1982). The collection *Erzählungen 1955–1975* was published in 1977, *Essays. Gespräche. Aufsätze 1964–1981* (containing lectures and a tract on E. T. A. Hoffmann, q.v.) in 1983, and *Werke in Einzelbänden* 1977 ff. *Die Schatten* is a posthumous publication (1985).

Fuhrmann Henschel, a play by G. Hauptmann (q.v.), originally in dialect (1898, titlepage 1899), first performed in November 1899 in the Deutsches Theater, Berlin, and published in that year. The second version (*Neue, der Schriftsprache angenäherte Fassung*) retains an element of dialect. Termed a Schauspiel, it is more like a tragedy. It is set in the 1860s in a small Silesian spa. The wife of the carrier Henschel is dying; from her bed she watches Henschel, and detects an attraction to the robust young maid Hanne. Distrustful of Hanne, she exacts from Henschel a promise that he will not marry her. Frau Henschel's death is followed by that of her baby, and Hanne, exploiting the family tragedy, sets out to take her place. Henschel, in need of a woman to run the house, and susceptible to Hanne's attractions, decides to marry her, though only after Hanne threatens to leave. Proud as mistress of the carrier's house, but not satisfied with Henschel's attentions, Hanne takes a lover. Henschel succumbs to the stresses of his domestic life and to the voice of his sick conscience; his character deteriorates, and he alienates his friends. When he realizes the extent of his moral collapse, he takes his own life.

Fuhrmann Henschel resumes themes of Hauptmann's Naturalistic phase, achieving a powerful, claustrophobic action. The setting bears a close resemblance to the environment in which Hauptmann grew up, and the hotel lesseè, Herr Siebenhaar, has many points of resemblance to the author's father.

Fulda, city of the Rhön district, and seat of a Roman Catholic bishop, was one of the most important centres of learning in Germany in the early Middle Ages. Its focus was the great abbey founded by St. BONIFACE (q.v.) in 744. Among its abbots in the 9th c. was the theologian HRABANUS Maurus (q.v.). The present ecclesiastical buildings of Fulda are predominantly baroque, and are in part the work of J. Dientzenhofer (1663–1726).

FULDA, LUDWIG (Frankfurt/Main, 1862–1939, Berlin), studied German literature at Heidelberg, Berlin, and Leipzig universities, and began writing plays in his twenties. He lived mainly in Berlin, and in 1928 became president of the Abteilung für Dichtung of the Preussische Akademie der Künste. Fulda, who was a Jew, took his own life in 1939.

In his early years as a writer he was regarded as one of the great hopes of Naturalism (see NATURALISMUS), but his output consists in the main of fluent verse plays or neat drawing-room plays. His first play, a Künstlerdrama, was on the poet J. C. Günther (q.v., *Christian Günther*, 1882); he gained his initial success with a one-act verse comedy, *Die Aufrichtigen* (1883). Fulda was an extremely prolific author. Works which merit special mention include *Unter vier Augen* (1887), *Das Recht der Frau* (1888), *Das verlorene Paradies* (1892), *Die Sklavin* (1892), *Der Talisman* (which was nominated for the Schiller prize by the committee, but vetoed by the Emperor Wilhelm II, q.v., in 1903), *Die Kameraden* (1895), *Jugendfreunde* (1898), and *Maskerade* (1904). He was an industrious translator (Molière's works, 1897, Rostand's *Cyrano de Bergerac*, 1898, Shakespeare's *Sonnets*, 1913, Ibsen's *Peer Gynt*, 1915, and two volumes of Spanish comedies, 1926).

FÜLLEBORN, GEORG GUSTAV (Groß-Glogau, 1769–1803, Breslau), classics master (Professor für alte Sprachen) at St. Elizabeth's grammar school at Breslau, wrote a collection of Silesian fairy-tales, which was published in 1789 as volume 6 of *Volksmärchen der Deutschen* by J. K. A. Musäus (q.v.). He was also a parodist (in *Die Herren von Felsenau*, 1798) and a satirist (in *Schutzrede*, 1800).

FUNCKELIN, JAKOB (Constance, *c.* 1520–65, Biel or Bienne), a Swiss pastor, wrote biblical plays, including one on the theme of Lazarus (*Tragedi von dem armen Lazaro*, 1551), and a nativity play (*Geistliches Spiel von der Geburt Christi*, 1553). He was also the author of hymns, of which the best known is 'Ehre sei Gott im höchsten Thron'.

Funkensonntag is the first Sunday in Lent (in the Roman Catholic calendar *Invocavit*, q.v.), which is traditionally the occasion for bonfire-making. In the Rhineland it is known as Fackelsonntag.

Furcht und Elend des Dritten Reiches, a collection of 24 (originally 27) episodes depicting conditions in Germany from 1933 to 1938 by B. Brecht (q.v.). It is introduced by a prologue in verse, *Die deutsche Heerschau*. Each episode has a heading consisting of party political slogans (e.g. *Volksgemeinschaft, Winterhilfe, Die Volksbefragung*), ironically allusive compounds (e.g. *Das Kreidekreuz*), direct headings (e.g. *Die jüdische Frau*), or the biblical title, *Die Bergpredigt*, in an anti-religious context. Each episode

is introduced by a rhymed (aabccb) verse in 6 trochaic lines formulating the chaotic conditions described in the following dialogue. Each dialogue is in prose and illustrates situations as they were reported by eyewitnesses or newspapers. The scenes are not linked by any action, but are illustrations of the fear and suffering indicated in the political title of the work. The last episode is set against the background of Hitler's entry into Vienna on 13 March 1938 with the slogan 'Ein Reich, ein Volk, ein Führer'. The mother of a proletarian family reads a letter from a father, written on the evening before his execution, to his son, reminding him to remain loyal to the cause of the workers' movement. The original title, *Deutschland—ein Greuelmärchen*, recalls Heine's satirical poem *Deutschland. Ein Wintermärchen* (q.v.). The first Paris production in 1938 by a group of German emigrants consisted of a selection of scenes entitled 99%.

FÜRNBERG, LOUIS (Iglan, Moravia, 1909–57, Weimar), a member of the Communist Party, was a theatre director. After the occupation of Czechoslovakia he emigrated, eventually reaching Palestine. In 1946 he returned to Czechoslovakia, and became an official in the Czech embassy in the DDR. In 1954 he was made deputy director of the Nationale Forschungs- und Gedenkstätten, Weimar, founding the *Weimarer Beiträge* in 1955. His work is politically orientated, and much of it is deliberately simple. Poems predominate; published collections include *Lieder, Songs und Moritaten* (1936), *Hölle, Haß und Liebe* (1943), *Der Bruder Namenlos* (1947), *Wanderer in den Morgen* (1951), and *Das wunderbare Gesetz* (1956). Two Novellen deserve mention, *Mozart-Novelle* (1947), and *Die Begegnung in Weimar* (1952).

Gesammelte Werke (6 vols.), ed. L. Fürnberg and G. Wolf, appeared 1964–73.

Fürstenbuch, see JANSEN ENIKEL.

Fürstenbund, a league of princes set up in 1785 under Friedrich II (q.v.) of Prussia, in which first Saxony and Hanover and, later, many states of central and south Germany joined Prussia in an attempt to check the increase of Austrian power in central Europe. The Fürstenbund was especially aimed at preventing Joseph II (q.v.) from acquiring Bavaria in exchange for the Austrian Netherlands. The failure of Joseph's plans was due to foreign political developments, notably the negative attitude of France, and to the common front presented by the German princes of the Fürstenbund. After Friedrich's death in 1786 the Fürstenbund disintegrated. Popular hopes that it would be associated with reform were from the outset misconceived.

Fürstengruft, Die, a poem written by C. F. D. Schubart (q.v.) in 1779 or 1780, while he was incarcerated in the fortress of Hohenasperg (q.v.). In 26 four-line stanzas, it describes a vault in which the bodies of dead sovereigns are housed. The poem savagely denounces those who have misused their power, while praising the humane rulers. It was first published in the *Frankfurter Musenalmanach* (1781).

Fürstenschulen, three famous schools founded in the 16th c. by Duke MORITZ of Saxony (q.v., later Elector of Saxony). They were Schulpforta near Naumburg (1543), St. Afra at Meissen (1543), and St. Augustin at Grimma (1550). They were the successors of dissolved monasteries, and the name denoted the supreme authority of the prince, substituted for that of the Church. They were boarding schools with numerous, much sought after, scholarships. Klopstock (q.v.) was a pupil at Schulpforta, Lessing (q.v.) at St. Afra.

FUSELY, HENRY, see FÜSSLI, JOHANN HEINRICH.

FÜSSLI, HANS HEINRICH (Zurich, 1745–1832, Zurich), a politician and administrator in his native city, was also active as a publicist and journalist, editing the *Zürcher Zeitung*. His col-

lections include *Allgemeine Blumenlese der Deutschen* (1782–8) and *Helvetische Almanache* (1799 ff.). He is best known as the editor who made U. Bräker (q.v.) known to the public of his age by publishing in 1789 the *Lebensgeschichte und natürliche Ebentheuer des Armen Mannes im Tockenburg*.

FÜSSLI, JOHANN HEINRICH (Zurich, 1741–1825, Putney), commonly called Henry Fuseli or Fusely in England, where he spent most of his life, studied art, moving to Berlin in 1764 and in 1765 to London, where he worked as a book-illustrator. From 1770 to 1778 he studied in Italy, conceiving a boundless admiration for Michelangelo, whose monumental style he attempted to copy. In 1780 he settled permanently in England, painting pictures of extravagant fantasy and horror, as well as cycles of large works illustrating Milton, Shakespeare, Dante, Virgil, the *Nibelungenlied* (q.v.), and the Bible. Reynolds and Blake both influenced his work. He became an R.A. in 1790, professor at the Academy in 1799, and keeper in 1804. His best-known picture is *The Nightmare* (1782), said to be based on Reynolds's *Death of Dido*.

FUST, JOHANN (*c.* 1400–*c.* 1466), financed the printing-press of Gutenberg (q.v.), which he later took over. It was managed for him by his son-in-law Peter Schöffer (q.v.), who continued it after Fust's death. Fust is believed to have died of the plague in Paris while on a business visit.

G

Gabriel Schillings Flucht, a play (Drama) in five acts by G. Hauptmann (q.v.), first performed at Bad Lauchstädt in June 1912, and published the same year; it was written five years earlier. Gabriel Schilling, a not very gifted artist, endeavours unsuccessfully to escape from his wife and his mistress; he ends by drowning himself. The best feature of this play is Hauptmann's evocation of the atmosphere of the Baltic coast.

GAGERN, WILHELM HEINRICH, FREIHERR VON (Bayreuth, 1799–1880, Darmstadt), came into political prominence during the revolutionary period (see REVOLUTIONEN 1848–9), and made his name as the president and, later, minister of the Frankfurt Parliament (see FRANKFURTER NATIONALVERSAMMLUNG). He was a Liberal and

at the head of the ministry of Hesse-Darmstadt when he began to work for the unification of Germany. His proposals to solve the deadlock by the creation of a larger and a smaller federation (Weiterer and Engerer Bund) became known as the Gagernsches Programm. He ceased to support Prussian politics in the German sphere in the year Bismarck (q.v.) became chief minister. His endeavours to achieve a Großdeutsch (q.v.) solution of the German problem, in which Austria would play a major part, a shift from his original Frankfurt kleindeutsch (q.v.) attitude, remained ineffective.

GAISER, GERD (Oberriexingen, Württemberg, 1908–76, Reutlingen), son of a pastor, served throughout the 1939–45 War as a fighter pilot, and was taken prisoner on the Italian front in 1945. Having studied art before the war, he took up painting again before his return to Germany. In 1949 he became a grammar school teacher and in 1962 he was appointed an art lecturer at Reutlingen College of Education.

Gaiser is best known for three of his first four novels. *Eine Stimme hebt an* (q.v., 1950) is a variant of the novel of the returning soldier (see HEIMKEHRERLITERATUR), concentrating more on the general deterioration of life in Germany than on the situation of the soldier himself. *Die sterbende Jagd* (1953) draws on first-hand experience of the hitherto scarcely exploited subject of the war in the air (but see also EURINGER, R.); the action recounts the final gallant efforts against overwhelming odds of a fighter group along the North Sea coast of Germany and Scandinavia in 1945, and its destruction down to the last squadron. The materialistic and competitive world of the Wirtschaftswunder is the background to *Schluß-ball. Aus den schönen Tagen der Stadt Neu-Spuhl* (1958): thirty separate monologues, linked by the end-of-term ball of a sixth form, expose the rottenness of society. The preceding novel, *Das Schiff im Berg* (1955), attracted less attention. In *Einmal und oft* (1956) and in *Gib acht auf Domokosch* (1959) Gaiser included a number of effective, terse stories, which are often moving in their reticence (e.g. *Revanche* and *Der Mensch, den ich erlegt hatte* in the second of these collections). *Am Paß Nascondo* is the title story of a volume published in 1960. It is critical of the political divisions of the age; in 1965 followed a collection of short stories entitled *Gazelle, grün*, in 1966 a select edition of stories, *Der Mensch, den ich erlegt hatte*, in 1967 the collection *Vergeblicher Gang*, and in 1971 *Merkwürdiges Hammelessen*. His essays include *Moderne Kunst. Eine Einführung* (1958), *Klassiker der modernen Malerei* (1961), *Alte Meister der modernen Malerei* (1963), and *Umgang mit Kunst* (1974).

Galante Dichtung had its origin in the *poésie galante* of the French aristocratic *salons* of the mid to late 17th c., in which highly stylized eroticism found elegant expression. In Germany no corresponding poetic habitat existed, and the galante Lyrik of Germany is almost exclusively the imitation of French models at the turn of the century.

The great landmark of galante Dichtung is the collection *Herrn von Hoffmannswaldau und anderer Deutschen auserlesene und ungedruckte Gedichte* (7 vols.), compiled by B. Neukirch (q.v.) *et al.* between 1695 and 1727. Other anthologies were the anonymous *Des schlesischen Helicons*

auserlesene Gedichte (1699) and the three volumes of *Auserlesene und theils noch nie gedruckte Gedichte verschiedener berühmter und geschickter Männer* (1718–20) published by Menantes (see HUNOLD, C. F.). The Silesian poets were the chief purveyors of this stylized and suggestive erotic poetry, but traces of it may be found in poems by J. C. Günther and B. H. Brockes (qq.v.).

The galanter Roman is more openly erotic, more frivolous, and more explicit about the secret amorous exploits of its personages. Examples are Hunold's *Die liebenswürdige Adalie* (1702) and the novels of J. L. Rost (1688–1727). At times the galanter Roman approaches the pornographic, as in *Der im Irrgarten der Liebe herum taumelnde Cavalier* (1738) by J. G. Schnabel (q.v.), but a certain elegant lightness of touch is a redeeming feature. Faint echoes of this type of novel may be found with C. M. Wieland (q.v.). Parodies of it occur in the *Dafnis* of A. Holz (q.v.).

GALEN, CLEMENS AUGUST, CARDINAL GRAF VON, Bishop of Münster, see RESISTANCE MOVEMENTS (2).

Galgenlieder, a volume of nonsense verse by C. Morgenstern (q.v.), published in 1905, and reissued in 1932 together with his later poetry under the title *Alle Galgenlieder*.

GALLITZIN, FÜRSTIN, see MÜNSTER.

GAN, PETER, pseudonym of Richard Moering (Hamburg, 1894–1974), studied at Oxford before the 1914–18 War, in which he served as an officer in the German army. Afterwards he studied law at Marburg, Bonn, and Hamburg universities (1919–24). From 1927 to 1929 he lived in Paris. Until 1938, when he returned to Paris, he was a publisher's reader in Berlin. He was interned by Vichy-France, but reached Spain in 1942. He settled in Hamburg in 1958.

Gan writes fluent and often reflectively melancholy poetry in traditional forms. His collections of verse are entitled *Die Windrose* (1935), *Ausgewählte Gedichte* (1936), *Die Holunderflöte* (1949), *Preis der Dinge* (1956, a selection of previously published poems), *Schachmatt* (1956), *Die Neige* (1961), and *Das alte Spiel* (1965). He has translated English poetry, and one of his own poems begins with a German rendering of Herrick's line 'When I a verse shall make'. In 1935 he published a volume of essays, *Von Gott und der Welt*.

Gandersheimer Reimchronik, title given to a Middle Low German verse chronicle which is the work of a priest named Eberhard, belonging to the Abbey of Gandersheim in north-west

Germany. Eberhard is recorded between 1204 and 1207. The chronicle, which is relatively short (*c.* 2,000 lines), was written *c.* 1218. Its original purpose was to support the historic claims of the abbey in a dispute with the Bishop of Hildesheim; the latter part, however, gives an account of the Saxon emperors from Heinrich I (q.v.) to Heinrich II (919–1024), followed by a list of rulers down to Eberhard's own time.

GANGHOFER, LUDWIG ALBERT (Kaufbeuren, 1855–1920, Tegernsee), studied at Munich, Berlin, and Leipzig universities, and was a journalist and free-lance writer in Vienna and, from 1895, in Munich. He wrote a few rural Volksstücke with Bavarian settings and dialect (*Der Herrgottschnitzer von Ammergau,* 1880; *Der Prozeßhansl,* 1881; *Der Geigenmacher von Mittenwald,* 1884), but he is best known as the author of light novels and magazine stories with Bavarian mountain backgrounds; a number of these are historical. Of his prolific output the following titles deserve mention: the story *Der Jäger vom Fall* (1883) the novels *Edelweißkönig* (2 vols., 1886), *Die Martinsklause* (2 vols., 1894), *Schloß Hubertus* (1895), *Das Schweigen im Walde* (2 vols., 1899), and—perhaps his best work—*Der Dorfapostel* (1900). His autobiography *Lebenslauf eines Optimisten* appeared in 3 vols., 1909–11.

Gänsemännchen, Das, a novel by J. Wassermann (q.v.), published in 1915. The young composer Daniel Nothafft marries Gertrud Jordan, only to discover that he really loves her sister Lenore. Gertrud takes her own life, and later Lenore dies in childbirth. A happy marriage with the violinist Dorothea Döderlein is prevented by a malicious cousin, Philippine Schimmelweis. Nothafft breaks down, but recovers and accepts a life of renunciation. He devotes himself to the teaching of music, for which he becomes famous. The title alludes to the statue of the 'goose-man' (das Gänsemännchen) in Nürnberg, who carries a goose under each arm, by which Wassermann points to Daniel's relationship to the two sisters.

Gänslein, Das, an erotic verse anecdote of the late 13th c. An inexperienced monk, on first leaving his abbey, is told that the women he sees are geese. In all innocence he learns from them the pleasures of sex. On his return he praises geese so highly that the abbot looks into the matter; he learns in confession their true identity.

Ganymed, a poem written by Goethe probably in 1774 and first printed in 1789. Set in the irregular lines of his early hymnic manner, it conveys the poet's longing to be at one with the spring around him. *Ganymed* has been set to music by F. Schubert and by Hugo Wolf (qq.v.).

Garde, title of a number of élite regiments of the Prussian army in the 19th c. and 20th c. which eventually constituted an army corps composed of two infantry divisions and a cavalry division with supporting troops. All units bore the hyphenated prefix Garde-. The Guard Army Corps (Garde-Korps) was garrisoned in Berlin, Pots-dam, and Spandau, with headquarters in Berlin. After the unification of Germany in 1871 it remained Prussian, though some recruits were drawn from non-Prussian states. There were in addition certain local guard regiments in other component states of the Empire (Saxony and Hesse). The Prussian Guard originated early in the 18th c. under Friedrich I, and was notably developed under Friedrich Wilhelm I (qq.v.).

Gardes du Corps, Regiment der, the senior regiment of guard cavalry in the Prussian army, armed as cuirassiers. It was founded in 1740 and disbanded in 1918.

GART, THIEBOLT, of Schlettstadt, wrote in 1540 a powerful play treating the story of Joseph. *Josef* went through many editions and appears to have been a highly influential work.

Gartengesellschaft, Die, a collection of Schwänke (see SCHWANK) by Jakob Frey (q.v., 1556). M. Montanus (q.v.) published a second part, *Das Ander Theyl der Gartengesellschaft* (*c.* 1559).

Gartenlaube, Die, an illustrated weekly, intended for family consumption. Founded in 1853 by Ernst Keil (1816–78), it lasted until 1918, when it lost its identity in an amalgamation. W. Raabe, Th. Storm, L. Schücking, and E. Marlitt (qq.v.) were among its contributors.

GÄRTNER, KARL CHRISTIAN (Freiberg, Saxony, 1712–91, Brunswick), studied at Leipzig University, and came under the influence of Gottsched (q.v.). He was one of the first editors of the *Bremer Beiträge* (q.v.) which was set up in opposition to Gottsched by some of his former adherents; his pastoral play *Die geprüfte Treue* appeared in the first number. In 1746 Gärtner became a schoolmaster at the Carolinum (q.v.), Brunswick, where he remained for the rest of his life.

GARVE, CHRISTIAN (Breslau, 1742–98, Breslau) succeeded C. F. Gellert (q.v.) as professor of moral philosophy at Leipzig (1769), but soon resigned and returned to Breslau to devote himself to writing. His interest in English philosophy is exemplified in his translation of Adam Ferguson's *Institutes of Moral Philosophy* (*Grundsätze der Moralphilosophie*, 1772). He also translated Edmund Burke's *Philosophical Inquiry into the Origin of our Ideas on the Sublime and Beautiful* (*Über den Ursprung unserer Begriffe vom Erhabenen und Schönen*, 1773). In his early years Schiller (q.v.) was influenced by Ferguson through the medium of Garve's translation. Garve's principal philosophical interest was ethical and his method eclectic. His chief original works are *Über Gesellschaft und Einsamkeit* (1797–1800) and *Versuche über verschiedene Gegenstände aus der Moral, Litteratur und dem gesellschaftlichen Leben* (1792–1802).

Gas I and *Gas II*, two plays by G. Kaiser (q.v.), which, with *Die Koralle* (q.v.) as first part, make up the so-called *Gas-Trilogie*.

Gas — Erster Teil was first performed in the Schauspielhaus, Düsseldorf, on 28 November 1918, *Gas — Zweiter Teil* in the Neues Theater, Frankfurt, on 13 November 1920. They were published respectively in 1918 and 1920.

Gas I, which bears a motto drawn from *Die Koralle*, is in five acts. The Milliardärsohn from *Die Koralle* has taken over his inheritance, and now controls a gigantic gas manufactory, the symbol of technological power. He has devised a scheme of collective ownership and profit-sharing for those employed in this vast and efficient undertaking, but an irrational element intervenes. Though the formulae and calculations are correct and no mechanical defect exists, the gas explodes and shatters the works. Technology has extended beyond the control of man. The Miliardärsohn refuses for Man's own sake to rebuild the works and continue gas production, but the Engineer and the workers demand a resumption, a call which is backed in the end by the armed force of the state. The Milliardärsohn shoots himself, despairing of the regeneration of mankind, but his sister proclaims that she will give birth to the New Man.

Gas II, the final three-act play of the trilogy, looks into a mechanized future, in which men function as automata and the gas of the rebuilt works is produced for a war. In spite of the urgent appeals by a new figure, Der Milliardär-arbeiter, for brotherhood and peace, the workers continue to manufacture the gas, and, after being defeated by the enemy, use the Engineer's new poison-gas missile. The play, and the world, end in one cataclysm.

GAST, JOHANNES (?–1572, Basel), lived in Basel and is the author of a volume of *facetiae* written in Latin (*Convivales sermones*, 1509). See FAZETIE.

Gastfreund, Der, see GOLDENE VLIESS, DAS.

'Gaudeamus igitur', opening words of a German student song of great antiquity, existing in its present form since 1781 under the title *De brevitate vitae*. The tune is older (it was first recorded in 1717) and is used by Brahms (q.v.) in the *Akademische Fest-Ouvertüre*. It used to be the custom in German universities to stand for the singing of the song.

GAUDY, FRANZ, FREIHERR VON (Frankfurt / Oder, 1800–40, Berlin), son of a general of Scots descent, spent part of his boyhood in the company of the future Friedrich Wilhelm IV (q.v.), whose education was guided by Gaudy's father. From 1815 to 1818 he was at Schulpforta, and in 1818, at his father's wish, entered the First Footguards at Potsdam. He was commissioned in 1819, lived above his means, and was posted in 1821 to a line regiment, the 10th Grenadiers at Breslau. He continued to live recklessly, was a duellist, and on occasion a bully, to say the least (he cut off the ear of a dunning tailor with his sword); he incurred several spells of detention (Festungshaft), and in 1825 was posted to the 6th Grenadiers at Glogau as a disciplinary measure. While on Polish frontier duty in 1831 he suffered a severe attack of cholera but recovered, and while convalescing wrote a collection of whimsical essays, *Gedankensprünge eines der Cholera Entronnenen* (1831). In 1833 he resigned his commission in order to devote himself to literature in Berlin. Here he was in friendly contact with A. von Chamisso, F. Th. Kugler, and J. E. Hitzig, as well as F. de La Motte-FOUQUÉ and W. Alexis (qq.v.). His Napoleonic poems, *Kaiser-Lieder* (1835), made a favourable impression and were followed by the sketches *Aus dem Tagebuch eines wandernden Schneidergesellen* (1836). *Lieder und Romanzen* appeared in 1837, and *Venezianische Novellen*, a collection of seven stories, in 1838. He died of a stroke while preparing for a Mediterranean journey. *Poetische und prosaische Werke* (8 vols.), ed. A. Mueller, appeared 1853–4.

Gauriel von Muntabel, see KONRAD VON STOFFELN.

Gawein (or **Gawan**), nephew of King Arthur, and a knight of the Round Table, is an important figure in Hartmann von Aue's *Iwein* (q.v., referred to as Gawein), and Wolfram von Eschenbach's *Parzival* (q.v., referred to as Gawan).

Gebet, a poem of 9 lines by E. Mörike (q.v.), of which the last 5 appeared in *Maler Nolten* (q.v.). It was published complete in 1847. The poet resigns himself to God's will, while hoping to be spared both excessive joy and excessive suffering. It begins 'Herr, schicke was du willt'.

Gebete des Sigihart, title referring to two short old High German prayers, each of two lines of verse, written at the end of the Freisinger Handschrift (Codex Frisingensis) of Otfried's *Evangelienharmonie*. The copyist, Sigihart, has appended his signature beneath them. He completed the MS. (and added the prayers) between 903 and 905.

GEBHARD, TRUCHSESS VON WALDBURG (1547–1600, Strasburg), a member of a powerful noble family, became Elector and archbishop of Cologne in 1577. Wishing to marry his mistress Countess Agnes von Mansfeld, he declared in 1582 for Protestantism, but was evicted from his diocese in 1584 (see KÖLNISCHER KRIEG), and fled to the Netherlands.

GEBLER, TOBIAS PHILIPP, FREIHERR VON (Zeulenroda, Vogtland, 1726–86, Vienna), was a prominent civil servant in the chancellery (Hofkanzlei) in Vienna, where he settled in 1753. He wrote numerous plays, many in the manner of Lessing (q.v.), including *Der Minister* (1771).

Geblümter Stil, a style elaborately decorated with flowery metaphors, fanciful expressions, and artificial constructions, which was admired in later medieval poets. KONRAD von Würzburg, HEINRICH Frauenlob, and HERMANN von Sachsenheim (qq.v.) provide examples.

Geburt der Tragödie aus dem Geiste der Musik, Die, original title of a treatise on tragedy by Nietzsche (q.v.), published in 1872, and retitled *Die Geburt der Tragödie: oder Griechentum und Pessimismus* in 1886. It is Nietzsche's first major work, and was written in Basel between 1869 and 1871. In it he assails the view, current since the middle of the 18th c., that Greek civilization was predominantly serene. While accepting serenity as one aspect, which he calls Apolline, he emphasizes the dark mysterious forces which he associates with Dionysus, the representative of intoxication. The Apolline manifests itself primarily in sculpture; the Dionysiac expresses itself in music, and is the force inspiring Greek tragedy. Nietzsche sees a decline in tragedy, first brought about by Euripides, and subsequently by the Socratic rational principle which affected Greek as well as later civilizations. A rebirth of tragedy in irrational terms is, he believes, taking place in the work of Richard Wagner (q.v.). *Die Geburt der Tragödie* owes something of its plan and purpose to Nietzsche's friendship with Wagner, which was

shattered by personal and artistic disillusionment in 1876. The book proved from the outset highly controversial.

Gedächtnisfeier, a poem by H. Heine (q.v.), included in the section *Lamentationen* of the *Romanzero* (q.v.). No. 12 of the subdivision *Lazarus*, it begins 'Keine Messe wird man singen'. It is full of tender thought for the wife who will survive him.

Gedankenlyrik, term frequently used in writings on German literature for poetry which has as its theme an idea rather than an emotional experience, an impression, or a description. There is no equivalent English term, for 'philosophical poetry' suggests something too systematic and comprehensive, and 'metaphysical poetry' in English literature is limited to a specific period, but Wordsworth's *Ode to Duty* might serve as a typical example. German examples may be found in the sonnets of Gryphius, the epigrams of Angelus Silesius and Logau, and in many more recent poets: most obviously in Schiller, but also in some of Goethe's poems, in Hölderlin, Hebbel, George, and Rilke (qq.v.), and others.

In all good Gedankenlyrik image, eloquence, and intensity transfigure the idea. A more pedestrian kind of Gedankenlyrik is found in the early and mid-18th c., in, for example, Brockes and certain poems of Klopstock (qq.v.). The gradations are infinite, and no hard and fast lines can be drawn.

Gedanken über die Nachahmung der griechischen Werke in der Malerei und Bildhauerkunst, an epoch-making essay on Greek art by J. J. Winckelmann (q.v.), published in 1755. Though it was written before Winckelmann's departure for Rome, it already contains the ideas which inspire all his works. It praises Greek plastic art as the peak of human achievement and advocates the imitation of the antique. For Winckelmann the ancient Greeks have successfully combined natural beauty (sinnliche Schönheit) and ideal beauty (idealische Schönheit). In a famous passage, which was later to become the starting-point of Lessing's *Laokoon* (q.v.), Winckelmann asserts that the highest qualities of Greek art are 'a noble simplicity and a quiet grandeur': 'Das allgemein vorzügliche Kennzeichen der griechischen Meisterstücke ist endlich eine edle Einfalt und eine stille Größe, sowohl in der Stellung als im Ausdruck.' In these qualities the Laocoön group is superior to Virgil's shrieking Laocoön, a point over which Lessing took issue. Winckelmann sees Michelangelo as the one artist of later ages who has caught the antique spirit. The essay closes with a short discussion of Greek painting.

Gedanken zur Aufnahme des dänischen Theaters, an essay on drama by J. E. Schlegel (q.v.), published in 1747. In making suggestions as to how the theatre should be encouraged in Denmark, Schlegel shows a preference for English drama at a time when it was despised in Germany. Without neglecting didactic values, he emphasizes the importance of pleasure in the response to plays.

Gedichte eines Lebendigen, a collection of political poems of radical complexion, published by G. Herwegh (q.v.) in two volumes (1841–4). With the first volume Herwegh won fame over night. The title alludes polemically to the *Briefe eines Verstorbenen* of Fürst Pückler-Muskau (q.v.).

GEDRUT, see GELTAR.

Gefährliche Wette, Die, a short story, written in 1807 by Goethe and incorporated in *Wilhelm Meisters Wanderjahre* (q.v., Bk. III, Ch. 8). An impudent student wager is successful, but brings in its train the death of an innocent party, and, years later, a duel with permanent injury.

'Gefangener Mann, ein armer Mann', first line of the poem *Der Gefangene*, written in 1782 in prison by C. F. D. Schubart (q.v.).

Gefesselte Phantasie, Die, a play by F. Raimund (q.v.), written in 1826, and first performed at the Theater in der Leopoldstadt, Vienna, in January 1828. Described as *Original-Zauberspiel in zwei Aufzügen*, it was published posthumously in vol. 3 of *Sämtliche Werke* (1837). It is an allegory in praise of the poetic imagination.

The island of Flora is plagued by two malevolent fairy sisters, Vipria and Arrogantia. The oracle has pronounced that only when the Queen, Hermione, marries a worthy consort will the land be freed from its tormentors. Hermione, however, has sworn to marry a poet. Such a worthy poet-consort is available in Amphio, who is inspired by the allegorical character Phantasie. But the sisters, fearing their fall, entrap and enchain Phantasie, thus depriving Amphio of his poetic power. In the nick of time Phantasie is freed, Amphio sings his song, and is accepted by Hermione. Apollo intervenes to consign the wicked sisters to the underworld and all ends happily. Comic relief is provided by Nachtigall, a tavern musician from Vienna (speaking broad Viennese), who is put forward by the sisters as a rival to Amphio.

Gefesselte Strom, Der, a poem written probably in 1801 by F. Hölderlin (q.v.) in the form of an Alcaic ode. The image of the frozen river and its return to movement in spring may be seen as a symbol of the poet, and also of the hide-bound world which will some day quicken. It is an earlier version of *Der Abschied* (q.v.) and was first printed in 1826.

Geflügelte Worte, see BÜCHMANN, G.

Gefunden, a poem written by Goethe on 26 August 1813. It tells with magical simplicity of the flower which pleads not to be picked, and is then transplanted by the poet from wood to garden, where it continues to bloom. Goethe enclosed the poem in the letter he wrote to his wife, Christiane, on 28 August (his birthday); apart from its general symbolism, its reference to the beginning of their association twenty-five years before is apparent. It was first printed in Goethe's *Werke*, 1815.

Gegenreformation, Die (Counter-Reformation), developed as a consequence of the Reformation (q.v.), which spread from Germany to many European countries during the first half of the 16th c. It culminated a hundred years later in the Thirty Years War (see DREISSIG-JÄHRIGER KRIEG). The peace of Augsburg (see AUGSBURGER RELIGIONSFRIEDE) of 1555 generally held to mark the beginning of the Counter-Reformation. From the theological viewpoint, however, it began during the Renaissance as an internal Roman Catholic movement against detrimental developments in the 1520s, before the effects of the Protestant movement (see PROTESTANTISMUS) were fully evident through the Church of J. Calvin. Initially Luther (q.v.) had not aimed at a schism but at reforms within the body of the Church of Rome.

The reform of the Roman Catholic Church was not promoted as a uniform movement but by separate Orders which originated in the 1520s and 1530s, notably the Capuchins (1536), the Theatines, the Barnabites, and Ursulines of 1535 who combined charitable work with the promotion of spiritual regeneration and, by the end of the century, the education of girls. The powerful Society of Jesus (Jesuits) was founded by Ignatius Loyola in 1534, and in 1537 the Consilium de emendanda ecclesia began to investigate deficiencies within the Church. The progressive provocation of papal authority by the Protestant movement, of which many German princes took advantage in order to gain independence from Rome, resulted in a more radical assertion of constructive movements for reform within the Roman Catholic Church. This assertion increasingly took the form of repression and persecution by the Inquisition, which spread with secular support from Italy to all Catholic countries within the Holy Roman Empire (see DEUTSCHES REICH, ALTES). In 1559 the Index Librorum prohibitorum listed numerous books which were to be banned and burnt. During this time of change and schism, of scientific progress and ecumenical regeneration (especially after the opening of the Council of Trent in 1545), there were in the lower strata of the populace still sections practising witchcraft. By fighting against this and other misinterpretations and aberrations (including the doctrine of indulgences) the Roman Catholic Church hoped to create a general unifying ecumenical movement. This remained largely ineffective during the succeeding centuries, and has only assumed importance in the 20th c. by the creation of the World Council of Churches. From the point of view of the Roman Catholic Church the Counter-Reformation thus stretched over four centuries and ended with the recognition of the Protestant churches.

Viewed in its historical context, the second half of the 16th c. was increasingly dominated by unrest among the European countries, an unrest which was motivated by both religious intolerance and dynastic struggles for power on every conceivable level. That the Prager Fenstersturz (q.v.) should have gone down in history as the immediate cause of the bitter persecution and brutal armed conflict which seized a whole generation is a piece of grim irony. The middle of the 17th c. saw a new map of Germany (see WESTFÄLISCHER FRIEDE), the silent witness of untold suffering. Writers of the pre-war era include Th. Murner, J. Fischart, and J. Nas (qq.v.). Goethe's play Egmont, Schiller's Don Carlos, and Grillparzer's Ein Bruderzwist in Habsburg (qq.v.) are set against the background of this period. Galileo Galilei (1564–1642) had by the beginning of the war completed essential parts of the work which provoked his arrest and trial by the Inquisition. It is the subject of Brecht's play Leben des Galilei, G. von Le Fort's story Am Tor des Himmels, and the novel Galilei in Gefangenschaft by Max Brod (qq.v.).

Geharnschte Venus, Die, a collection of poetry, containing erotic poems, published in 1660 in Hamburg. Formerly thought to be the work of J. Schwieger (q.v.), it is now attributed to K. Stieler (q.v.).

Geheimnisse, Die, a projected religious epic poem in ottava rima, of which Goethe wrote the first 44 stanzas, plus one intended to come later. The stanzas of this esoteric work, which were written in 1783–4, introduce Bruder Markus, who arrives at a monastery, is admitted, and

discovers a mysterious brotherhood of twelve knights presided over by a superior named Humanus. In an essay (Die Geheimnisse. Fragment) published in Cotta's Morgenblatt für gebildete Stände in 1816, Goethe explained that the reader was to be led through an 'ideal Montserrat', and that each of the knights was to represent a religion, each with its attributes and special virtues. The poem Zueignung (q.v.), which Goethe later placed at the head of his collected poems, was originally intended as an introductory poem to Die Geheimnisse. The fragment of the epic was first published in Goethes Schriften, vol. 8, 1789.

Gehörnte Siegfried, Der, see DER HÜRNEN SEYFRID.

GEIBEL, EMANUEL (Lübeck, 1815–84, Lübeck), son of a pastor, was already a fertile poet in his schooldays. He studied at Bonn and Berlin universities, making the acquaintance of most of the literary celebrities of Berlin. In 1838 he accepted an invitation to become tutor to the family of the Russian minister in Athens. In 1839 he made a tour of the Aegean islands, returning to Lübeck in 1840.

Geibel had no inclination for any conventional employment—his chief wish was to write poetry —and in 1842 he had the good fortune to receive a pension from King Friedrich Wilhelm IV (q.v.) of Prussia. He lived for a time in St. Goar on the Rhine (1843), visited J. Kerner (q.v.) in Weinsberg in 1844, and then settled once more in Lübeck, though he soon resumed his travels in Germany. By this time he was the author of two volumes of poems (Gedichte, 1840; Zeitstimmen, 1841), a collection of translations (Volkslieder und Romanzen der Spanier, 1843), and a verse tragedy (König Roderich, 1844), which he afterwards discarded. He next published Zwölf Sonette (1846) and the Juniuslieder (1848). In 1852 he was appointed to an honorary professorship in Munich on the personal intervention of King Maximilian II (q.v.), and became a prominent member of the Munich School of poets (see MÜNCHNER DICHTERKREIS). Together with Paul Heyse (q.v.) he published in the same year the Spanisches Liederbuch, some poems of which were later set to music by Hugo Wolf (q.v.). He generally spent the summer in Lübeck. The remaining publications of his Munich years were the tragedies Brunhild (1858) and Sophonisbe (1868) and the translations Romanzero der Spanier und Portugiesen (1860), in which he co-operated with Count A. F. von Schack (q.v.), and Fünf Bücher französischer Lyrik (1862). A poetic encomium of King Wilhelm I (q.v.) of Prussia in 1868, when

Bavaro-Prussian relations were under strain, cost Geibel his professorship, but he was immediately compensated by the renewal of the Prussian pension. From then on he lived solely in Lübeck, writing patriotic and other poetry (*Heroldsrufe*, 1871; *Spätherbstblätter*, 1877).

Geibel's reputation as a poet of refinement and an apostle of Beauty stood high in the 19th c. It has declined since his lack of originality and his derivative, though elegant, diction have become apparent. The number of his poems in *The Oxford Book of German Verse* dropped from fourteen in 1911 to three in 1967. *Werke* (3 vols.), ed. W. Stammler, appeared in 1918–20.

GEILER VON KAISERSBERG, or KAYSERSBURG (Schaffhausen, 1445–1510, Strasburg), was an outstanding preacher in the years immediately preceding the Reformation (q.v.). He was brought up at Kaysersberg in Alsace and educated at the newly founded universities of Freiburg and Basel, becoming a doctor of theology at the latter in 1475; he took orders in 1470. In 1478 he was called to Strasburg, where he spent the remainder of his life as a priest of the Minster.

A man of great erudition and extensive reading, his sermons are marked by bold phrasing, homely turns of phrase, and striking, often unecclesiastical, illustration. The existing texts are frequently unreliable, since he spoke freely in German from a Latin draft and gave no heed to publication. Certainly genuine is the collection *Der Seelen Paradies von waren und volkomnen Tugenden*. Best known is his series of sermons in the form of a commentary on Sebastian Brant's popular satirical poem *Das Narrenschiff* (q.v.), *Navicula sive speculum fatuorum praestantissimi sacrarum literarum doctoris Joannis Geiler Keysersbergii* (1511). His life was written by his Alsatian compatriot BEATUS Rhenanus (q.v., *Vita Geileri*, 1510).

Geist der Zeit, Der, 1806, the first of the anti-French and anti-Napoleonic publications of E. M. Arndt (q.v.). In 1809 and 1813 respectively Arndt added a second and third volume. A fourth volume, published in 1818, was in the main a challenge to his opponents and detractors.

Geisterseher, Der, an uncompleted novel written by Schiller between 1786 and 1789. It was published in instalments in Schiller's periodical *Die Thalia* (1787–9), and as a book in 1789. It bears the sub-title *Eine Geschichte aus den Memoires des Grafen von O***. A young prince lives modestly in Venice with his companion Count O. A strange Armenian and a Sicilian adventurer seek to acquire an ascendancy over the prince, the Sicilian promising to summon up the prince's dead friend. He seems to do so, but his attempt is exposed as a fraud. The second book is in the form of letters. The prince's character has now changed. He falls in love with a beautiful Greek woman, who is involved in a conspiracy to convert him to Roman Catholicism. After her death he undergoes conversion, and at this point the story breaks off.

Der Geisterseher, of which Schiller soon wearied, proved to be one of the most popular of his works, chiefly because it treated in masterly style the then topical theme of experiments in the supernatural. It is mentioned with approval in E. T. A. Hoffmann's story *Das Majorat* (q.v.).

Geistesgeschichte, a term coined by F. Schlegel (q.v.) in 1808; it was only rarely applied to the study of literature before the late 19th c., when W. Dilthey (q.v.), in his publication *Einleitung in die Geisteswissenschaften* of 1883, initiated the reaction against the often pedestrian positivism of the school of W. Scherer (q.v.), reviving the term Geistesgeschichte. It attempts to exemplify the abstract Geist of a given epoch on the basis of a broad cultural conspectus involving religious, philosophical, social, as well as artistic, musical, and literary studies. J. Burckhardt and H. Wölfflin (qq.v.) were two of its most important sponsors; under their influence the genius of Dilthey applied the concept to literature with such success that the pursuit of Geistesgeschichte became supreme and remained virtually exclusive in the study of German literature during the first four decades of the 20th c. Among the most notable exponents were R. Unger, H. A. Korff, F. Strich, O. Walzel, and H. Cysarz. Medieval studies were slower to respond to the new approach, but it eventually became almost universal. A peculiarly idiosyncratic form of Geistesgeschichte was practised by F. Gundolf (q.v.) in, for example, *Shakespeare und der deutsche Geist* (1928).

The virtue of Geistesgeschichte in the treatment of literature is that it reveals and emphasizes continuity and development. Its drawback is that it concentrates primarily on common, shared, or characteristic factors, and tends to ignore (in theory, though not always in practice) the uniqueness of the individual writer and of his works. Hence it provides a technique which is more productive with authors of secondary rank than with poets, dramatists, and novelists of genius. It also encourages excessive abstraction.

Geistliche Jahr, Das, a collection of religious poetry begun by Annette von Droste-Hülshoff (q.v.) in 1820, and finished between 1837 and 1839. It was published in 1851 after her death. It takes the form of a cycle of 72 poems for the Sundays and feasts of the Christian year, the first poem being entitled *Zum neuen Jahr* and the last *Am letzten Tage des Jahres*. The 25 poems written in 1820 extend to Easter Monday. Each poem is headed by a reference to an appropriate passage of Scripture. Originally intended to give pleasure to Annette's grandmother, the work reflects in the later poems the struggle of faith and doubt.

Geistliche Lieder, a collection of 15 hymns written in 1799 by Novalis (q.v.). The first 7 were published in the *Musenalmanach* for 1802 and the whole collection appeared shortly after in vol. 2 of Novalis's *Schriften* (1802). All but one are in the strophic form of hymns for congregational use, and some (notably 'Was wär' ich ohne dich gewesen' (I), 'Wenn ich ihn nur habe' (V), and 'Wenn alle untreu werden' (VI)) are still in regular use. The exception (VII) to strophic form is a hymnic poem in which the sacrament of Holy Communion assumes pervasive erotic significance.

Geistliche Oden und Lieder, a collection of religious poems and hymns by C. F. Gellert (q.v.), published in 1757. With the *Fabeln und Erzählungen* (q.v.) it was Gellert's most popular work, and several of its hymns are still sung in German Protestant churches. Indeed, 'Jesus lebt, mit ihm auch ich' is in English use in the translation of F. E. Cox ('Jesus lives! no longer now/Can thy terrors, Death, appal us'). Six of Gellert's best-known poems from this collection were set to music by Beethoven (q.v.) as *Sechs Lieder Gellerts am Klavier zu singen*. They are *Bitten* ('Gott, deine Güte reicht so weit'), *Liebe des Nächsten* ('So jemand spricht: Ich liebe Gott'), *Vom Tode* ('Meine Lebenszeit verstreicht'), *Die Ehre Gottes aus der Natur* ('Die Himmel rühmen des Ewigen Ehre'), *Gottes Macht und Vorsehung* ('Gott ist mein Lied!'), and *Bußlied* ('An dir allein').

Geistliche Rat, Der, a Middle High German allegorical religious poem counselling nuns to use their ears and mouth, hands and knees, and especially their hearts, in the service of Christ. The short poem of 178 lines was probably written in the 14th c.

Geistliche Streit, Der, a Middle High German allegorical poem portraying the struggle of the soul, assisted by the seven cardinal virtues, against the seven deadly sins. The poem, which consists of 1,025 lines, was probably written by a nun *c.* 1320, and it is thought to have been intended for the edification of nuns. The dialect is Alsatian.

Geistliche Tageweise, designation of medieval songs based on the form of Minnelied known as the Tagelied. In the geistliche Tageweise the awakening is conceived symbolically as the awakening from sin. Some geistliche Tageweisen are new poems, others are Kontrafakturen, i.e. adaptations of profane songs to spiritual purposes (see KONTRAFAKTUR). The geistliche Tageweise, 'O starcker got . . .', by Count PETER von Arberg (q.v.), is mentioned under the year 1356 in the *Limburger Chronik* (q.v.).

Gekrönter Dichter, an honorary title, akin to Poet Laureate, conferred in Renaissance Italy and adopted in Germany in the 15th c. The first German gekrönter Dichter was K. Celtis (q.v.), who received the title from the Emperor Friedrich III (q.v.) in 1487. The principal qualification was expertise in Latin verse, and the title gave the right of teaching at a university. The Emperor Maximilian I (q.v.) crowned a number of poets himself and conferred the right to crown also on Celtis. Among well-known holders of the title were J. Locher, U. von Hutten, H. Glareanus, and, in later generations, M. Opitz and F. Rist (qq.v.). The proliferation of 'crowned poets' devalued the title, which, however, achieved prominence again when Gottsched (q.v.) acquired the right of coronation from Karl VI (q.v.).

Geld und Geist oder Die Versöhnung, a novel by J. Gotthelf (q.v.), published in three instalments. The first appeared in 1843 in vol. 2 of *Bilder und Sagen der Schweiz*, and the second and third in 1844 in vols. 4 and 5 of the same collection. Originally Gotthelf contemplated only the first part, but the second part (the second and third instalments are a continuous narration) grew out of it almost immediately. In Part One a rift occurs in the hitherto united and harmonious married life of Christen and Änneli, a farming couple. She finds him too slow and not sharp enough in money matters; he thinks she gives too readily and too generously to beggars. Smouldering resentments are set off by a trifle, a breach grows, and the whole family is affected. But Änneli goes to church at Whitsun, and, moved by the sermon, she takes the first step to a reconciliation to which Christen responds. Into the new harmony breaks the alarm of church bells warning of fire.

Part Two begins at the scene of the fire. Änneli's son Resli is there to help and so, too, is Anne Mareili, daughter of the Dorngrütbauer. On the way home Resli is knocked senseless by brawling youths, left on the ground and succoured by Mareili. Resli hopes to marry Mareili, but her miserly and curmudgeonly father is determined to dispose of her to an aged farmer, whose property he covets. In spite of rebuffs, Resli makes a serious attempt to win over the Bauer, driving over with four horses and a load of planks, an episode which is one of the finest passages in Gotthelf's works. But his gesture is of no avail. His mother Änneli goes to the help of a poor family suffering from dysentery, catches the disease, and dies. On her death-bed Anne Mareili, who has left her home, comes to her. Änneli dies, happy that Resli and Anne Mareili are at last united.

GELLERT, CHRISTIAN FÜRCHTEGOTT (Hainichen nr. Freiburg, 1715–69, Leipzig), was educated at St. Afra's school, Meißen (see FÜRSTENSCHULEN), 1729–34, and then at Leipzig University, 1734–8. He was a delicate child, and never enjoyed robust health. After various short appointments as a private tutor, he qualified to lecture in 1743. His first volume of poems (*Lieder*) appeared in 1743. In 1745 his comedy *Die Betschwester* (q.v.) was published in the *Bremer Beiträge* (q.v.). Its exposure of hypocrisy drew criticism upon him. A collection of comedies, including his first play and also *Das Los in der Lotterie, Die zärtlichen Schwestern, Das Orakel*, and *Die kranke Frau*, appeared as *Lustspiele* (2 vols., 1747). Gellert's unexciting and influential novel *Leben der schwedischen Gräfin von G . . .* (q.v.) was also published in two volumes in 1747–8. His popular fables were first printed in 1746 (*Fabeln und Erzählungen*, q.v.), and added to in 1748. A volume of model letters came in 1751 (*Briefe, nebst einer praktischen Abhandlung von dem guten Geschmacke in Briefen*), and in the same year Gellert, who had been lecturing since 1744, was elected to a professorship at Leipzig University. In his inaugural lecture *Pro commoedia commovente* (q.v.) he supported *comédie larmoyante* (*das weinerliche Lustspiel*). His *Lehrgedichte und Erzählungen* were published in 1754, and *Geistliche Oden und Lieder* (q.v.) in 1757. Gellert, who never married, was in failing health throughout his tenure of the chair of poetry, rhetoric, and ethics.

He is best known for his verse fables, written with charm and with the moral clearly pointed; his hymns (see GEISTLICHE ODEN UND LIEDER), often combining mellifluousness with an unexpected sense of grandeur, attracted Beethoven's attention, and still survive in church use. As a writer of comedy, Gellert inclined to the sentimental tone popularized by Nivelle de la Chaussée (1692–1754), and rarely succeeded in dispelling an air of insipidity. *Die schwedische Gräfin*, though a painstakingly flat production,

reveals with singular clarity the combination of reason and sentiment (in which the former controls the latter) which was characteristic of the 1740s.

Gellert was, however, if anything more important for his personal influence than for his published works. He lectured to large audiences on moral themes, and maintained a vast punctilious correspondence with strangers as well as friends, always ready, without distinction of rank, to advise those who wrote to him in moral perplexity. His wide-ranging popularity is symbolized by recorded gifts of firewood from a peasant and of a quiet horse from Prince Heinrich (q.v.) of Prussia. He was one of the few German authors approved by Friedrich II (q.v.), who termed him 'le plus raisonnable de tous les savants allemands', and received him in audience in order to listen to his fables in Leipzig in 1760. Gellert himself would probably not have objected to the view that he was even more a moralist than a man of letters. He preaches contentment with one's lot and the subjection of the passions to reason.

Gellert is a principal character in Heinrich Laube's comedy *Gottsched und Gellert* (q.v., 1847).

Sämmtliche Schriften. Neue rechtmäßige Ausgabe (10 vols.), ed. J. L. Klee, appeared in 1839 (rev. edn. 1867), and *C. F. Gellerts Briefwechsel*, ed. J. F. Reynolds, 1983 ff.

GELTAR, whom modern scholarship considers to be identical with Gedrut, was probably a minor Middle High German poet of the late 13th c., whose poems parody and satirize Minnedienst. Three songs are ascribed to this name, and one to Gedrut. It is possible that the name, which is given in the Heidelberg MS. (see HEIDELBERGER LIEDERHANDSCHRIFT, GROSSE), refers, not to an author, but to the owner of the earlier MS., from which the poems are copied.

Geltstag, Der, oder *Die Wirtschaft nach der neuen Mode*, a novel by J. Gotthelf (q.v.), published in 1845. The story begins with the funeral of the innkeeper Steffen, a good-for-nothing who has died in his sleep after one of his many drunken sprees. The widow, Eisi, with her numerous children, has to face being sold up. Gotthelf now looks backward and retraces the story of the innkeepers. He relates how they obtained one of the many new licences and began their innkeeping without knowledge or experience, and without any intention of working hard. A protracted battle developed between the two spouses in which Eisi proved the tougher and stronger.

The narrative returns to the present, and the auction of furniture and equipment of the inn occupies the greater part of the novel. Gotthelf uses several items (e.g. Steffen's gun, the chaise, Eisi's mirror) to reveal episodes and aspects of the couple's deplorable married life. At last Eisi and the children are left in the empty inn from which they will soon be evicted. But Eisi refuses to face reality, and believes she will be allowed to continue to run the inn. From this fool's paradise she is rescued by Steffen's godfather, who offers the family asylum on his farm. Eisi accepts, but shows no gratitude or willingness to work, and the children, ill-bred except for the two youngest, run riot. The godfather soon puts an end to this by distributing the misbehaved children to various farms. The two youngest are adopted by him and his wife. Eisi, glad to be rid of her children, marries again; Gotthelf discreetly hints that the new marriage is unlikely to be happier or more prosperous than the old. The novel was begun as a protest against the reckless liberality (as Gotthelf viewed it) with which inn licences were granted. It makes free use of Swiss dialect.

Gelübde, Das, a story by E. T. A. Hoffmann (q.v.), published in 1817 in vol. 2 of *Nachtstücke.* Set in Poland in the late 18th c., it narrates a remarkable event, and then gives its explanation. On the orders of Fürst Z. the burgomaster of L. receives a veiled woman to lodge in his house. Known as Cölestine, she proves to be pregnant, and a child is born. At the same time it is discovered that Cölestine wears beneath her veil a mask. A few weeks later an officer enters the house and carries off the child by force. Cölestine is taken away by her relatives, and shortly afterwards the burgomaster hears of the interment of a young woman at a neighbouring nunnery. It turns out that Cölestine was Countess Hermenegilda von C., that she was betrothed to Count Stanislaus, and was visited by his cousin Count Xaver, who fell in love with her. In a trance-like state she had the hallucination of going through the marriage ceremony with her betrothed, and afterwards—still in this state—gave herself to Xaver, whom she took to be her betrothed. By him she conceived a child. When the truth was brutally revealed to her, she became frenzied, but was restored to sanity by her father confessor, before whom she swore an oath never to reveal her face. It was Xaver who stole the child, which died immediately; years after he is discovered in a monastery in Italy. The story has a resemblance to H. von Kleist's *Die Marquise von O . . .* (q.v.), which is hardly fortuitous.

Gemeindekind, Das, a novel by Marie von Ebner-Eschenbach (q.v.), published in 1887. It is set in her Moravian homeland about 1860. A drunken rogue, Martin Holub, who tyrannizes his family, murders the parish priest. He is hanged, and his wife, who does not exculpate herself, is sentenced to ten years' hard labour. Milada, the daughter, is brought up by the lady in the great house, and is encouraged to undertake penance, which eventually undermines her health. Meanwhile Pavel, the thirteen-year-old son, falls on the parish (Gemeindekind) and is turned over to the most notorious family in the village, since no other will take a murderer's son. With bad examples constantly before him, Pavel goes to the bad, but a meeting with his sister brings the beginning of a change. With the help of the schoolmaster, and against the contempt and hostility of the village, he pulls round and becomes resolute, dependable, and honest. A few days after his sister dies, he is visited by his mother who has just been released from prison. It is meant as a fleeting visit, for the mother knows that her repute as a murderess will harm the son. But when she tells him that she is truly innocent he persuades her to stay. They will face the world together.

GEMMINGEN-HORNBERG, Otto Heinrich, Freiherr von (Heilbronn, 1755–1836, Heidelberg), became a civil servant and diplomat in the Palatinate and Baden, retiring in 1805. His play *Der deutsche Hausvater* (1780), which is indebted to Diderot's *Le Père de famille,* was one of the most popular stage works of the age. Wodmar, the aristocratic Hausvater, brings his erring sons and daughter into the path of virtue. This moral reform includes a marriage between his son Karl and a girl of lower class, as an atonement and amends for seduction. Gemmingen also adapted Shakespeare's *Richard II* (1782). He edited the journals *Der Weltmann* (1782–3) and *Magazin für Wissenschaft und Literatur* (1784–5).

GENAST, Eduard (Weimar, 1797–1866, Wiesbaden), son of the Weimar *régisseur* Anton Genast (1765–1831), was an actor and opera singer in Leipzig (1818–28) and Weimar (1829–64). Genast composed two operas, but he is best known for his memoirs, *Erinnerungen aus dem Tagebuch eines alten Schauspielers* (1862–6).

Genesene an die Hoffnung, Der, a poem written by E. Mörike (q.v.) in 1838. The sick man, recovered from his illness (Der Genesene), praises Hope, whom he had for a time forgotten.

GENGENBACH, Pamphilus (Basel, *c.* 1480–1525, Basel), a printer of Basel and adherent of the Reformation, wrote Fastnachtspiele (q.v.) of a serious nature (*Die Gouchmatt der Buhler,* 1510) as well as a morality play (*Das Spiel von den zehn Altern,* 1500).

Geniezeit, see STURM UND DRANG.

GENNEP, JASPAR VON (Cologne, before 1520–80, Cologne), a printer of Cologne, was active as a dramatist and controversialist in support of the Roman Catholic Church. His play *Homulus* (1539) is a version of the story of Everyman (see JEDERMANN).

Genoveva, a five-act tragedy in blank verse by F. Hebbel (q.v.), written in 1840–1, and published with a preface in 1843. It was first performed without Hebbel's knowledge on 13 May 1849 in a Czech translation in Prague. On 20 January 1854 the Vienna Burgtheater produced it under the title *Magellona.* Hebbel modified the work considerably for stage performance, and in 1851 added a *Nachspiel.* The play is based on the *Volksbuch von der Pfalzgräfin Genoveva* (1647, which derives from a French legend) and the plays *Golo und Genoveva* (*c.* 1780) by Maler Müller (see MÜLLER, F.) and *Leben und Tod der heiligen Genoveva* (1799) by L. Tieck (q.v.).

Hebbel sets his play in the Middle Ages, which he describes as the 'poetic' age. The play opens as Pfalzgraf Siegfried takes leave of his wife, Genoveva of Brabant, to join a crusade. He has chosen the young knight Golo to protect her during his absence. Golo witnesses the embrace of husband and wife, and becomes possessed with the urge to win Genoveva's love. When Genoveva rejects him, he imprisons her in the castle on a false charge of adultery with the loyal knight Drago, who is murdered on his orders. During her imprisonment Genoveva gives birth to Siegfried's son Schmerzenreich, and despite her sufferings remains faithful to her husband. Encouraged by the hag Margaretha, Golo now schemes her death. Genoveva is led into the forest, but, as her assassin is about to strike, he is attacked by a madman (der tolle Klaus). Thus unknown to Golo, Genoveva's life is providentially spared.

Siegfried returns from the crusade, and is persuaded into the belief that Genoveva has betrayed his love and borne another's child. Plagued by his conscience, Golo takes leave of Siegfried, ostensibly to set out on an expedition of knight-errantry. In fact he commands Caspar, whom he had involved in his plot against Genoveva, to blind him in the forest in which Genoveva rears Schmerzenreich, and to leave him to perish.

The character of Golo is, on Hebbel's admission, based on aspects of his own personality. In the *Nachspiel* Genoveva's sufferings are rewarded by Siegfried's discovery of wife and son, but joy is tempered by Genoveva's awareness of an early death.

Gensdarmes, Das Regiment, Prussian Guard cuirassier regiment raised as a single squadron in 1691, increased to a regiment in 1713, and disbanded in 1807 after the defeat at Jena. It figures in Fontane's novel *Schach von Wuthenow* (q.v., 1883), which bears the sub-title *Erzählung aus der Zeit des Regiments Gensdarmes.*

GENTZ, FRIEDRICH VON (Breslau, 1764–1832, Vienna), was successively in Prussian (1785–1802) and Austrian service, and is best known for his extensive and astute writings on the political scene of his age. A supporter of the French Revolution (q.v.) in its early stages, he turned against it and translated Burke's *Reflections on the French Revolution* of 1790 (*Betrachtungen über die französische Revolution,* 2 vols., 1793) and other political publications. After 1809 he became a close collaborator of Metternich (q.v.), whose policy he strongly supported during the settlements at the Congress of Vienna (see WIENER KONGRESS). He henceforth abandoned liberal and nationalist views, and became a strong proponent of German unification, looking to Austria to head a united Germany. His works include *Politische Paradoxien* (1799), *Fragmente aus der neuesten Geschichte des Politischen Gleichgewichts in Europa* (1804). In 1818 he founded the *Wiener Jahrbücher der Literatur.* Gentz's *Ausgewählte Schriften* were first published by W. Weick (5 vols., 1836–8).

GEORGE II, King of Great Britain and Ireland and, as Georg August, Kurfürst von Hannover (Herrenhausen, Hanover, 1683–1760, London), supported Austria in the War of the Austrian Succession (see ÖSTERREICHISCHER ERBFOLGEKRIEG) with the Pragmatic Army (see PRAGMATISCHE ARMEE), winning the battle of Dettingen in 1743. During the Seven Years War (see SIEBENJÄHRIGER KRIEG) he was an ally of Prussia. He founded Göttingen university (q.v.) and was a patron of G. F. Handel.

GEORGE, STEFAN (Rüdesheim, now part of Bingen, 1868–1933, Minusio nr. Locarno), the son of a prosperous wine merchant, spent his early childhood (1873–81) in Bingen, and was at the grammar school (Gymnasium) of Darmstadt from 1881 to 1886. He studied at Berlin University, but gave up after three semesters. He possessed ample means throughout his life, never engaged in any profession or trade, never married, and never established any sort of home. He travelled much, visiting most of the countries of central and western Europe, including Britain.

George determined as a quite young man to devote himself to poetry (his *œuvre* is almost exclusively lyrical) and to cultivate beauty for its own sake. The greatest formative influence inclining him to this decision was Mallarmé. George's conception of *l'art pour l'art* saw the beauty of poetry in the sensual, especially aural, presentation of a highly selective vocabulary in disciplined, deliberate organization. The themes of his poems are chiefly landscape, friendship, and art.

A strikingly private personality, possessed of great determination, and obsessed with a sense of lofty uniqueness, George was adverse to the usual forms of publication; his first volume of verse, *Hymnen* (18 poems, including the well-known *Der Infant*) was printed privately, and copies were presented to friends in 1890. In 1891 the volume *Pilgerfahrten* was printed and distributed in similar fashion; its twenty-two poems include 'Mühle, laß die arme still' and *Die Spange*. In 1892 *Algabal* (21 poems) appeared in like manner. In this year George, having discovered several kindred spirits, to whom he stood in the relationship of master to disciples (he was indeed usually addressed as Meister), began to publish his own poems, and some of those of his friends, in his own irregularly appearing periodical *Blätter für die Kunst* (q.v.), which continued until 1919. Consciously writing for an élite, he published in 1895 *Die Bücher der Hirten- und Preisgedichte der Sagen und Sänge und der hängenden Gärten*. This title embraces the twenty-five poems of *Das Buch der Hirten- und Preisgedichte* (including *Sporenwache*), and the thirty-one of *Das Buch der hängenden Gärten*. In 1897 appeared *Das Jahr der Seele*; its three sections are entitled *Nach der Lese*, *Waller im Schnee*, and *Sieg des Sommers*. The collection contains thirty-one poems (beginning with the untitled 'Komm in den totgesagten park und schau') and a number of verse dedications and miscellaneous poems. *Die Fibel*, a collection of his earlier verse, followed in 1901.

From about the turn of the century George saw himself as an educator, the leader in the renewal of a debased culture. The expression of this is to be found in *Der Teppich des Lebens* (1899), which is divided into the twenty-four poems of *Das Vorspiel*, and the twenty-four of *Der Teppich des Lebens* proper (including *Der Jünger*). To these forty-eight poems, which breathe a reticent didacticism, are appended the twenty-four poems of *Die Lieder von Traum und Tod*.

After 1900 George curtailed his foreign journeyings, and stayed mainly in Germany (Bingen, Heidelberg, Berlin, Marburg, Würzburg), occasionally visiting the Alps. He also spent some time in Munich, and here in 1903 he underwent a decisive experience in meeting Maximin, a 15-year-old boy, whom George found so handsome and perfectly formed that he saw in him an incarnation of the godhead. Maximin died in the following year. This en-counter, commonly referred to in writings on George as 'das Maximin-Erlebnis', directed George's thought to youth: *Der siebente Ring* (1907) is concerned with the new élite of youth which will effect the renewal of civilization. It is one of his largest collections, containing 184 poems, divided into the sections *Zeitgedichte*, *Gestalten*, *Gezeiten*, *Maximin*, *Traumdunkel*, and *Lieder*.

The tone of his poetry passes in *Der Stern des Bundes* (1914, divided into three books) to the prophetic, apocalyptic, and monumental, in such poems as 'Auf stiller stadt lag fern ein blutiger streif', 'Wer je die flamme umschritt', and 'Gottes pfad ist uns geweitet'. In his volume *Das neue Reich* (1928), George's inspiration, partly derived from Hölderlin, evokes the vision of a new Germany, which will be a realization of the ideal Hellas. Its opening section has no general heading but contains, among others, the poems *Goethes letzte Nacht in Italien*, *Hyperion I–III*, *Der Krieg*, *Einem jungen Führer im ersten Weltkrieg*, and *Der Brand des Tempels*. The second section is called *Sprüche* and the third *Das Lied*. The title *Das neue Reich*, unfortunate in its timing, was exploited after 1933 by the National Socialists, who sought to exhibit George as a prophet of the new state. After Hitler's appointment as chancellor, Stefan George was offered the presidency of the new Dichter-Akademie, which he declined. Soon afterwards he went to Switzerland, where he died in the following December.

From being a strictly private figure, George gradually and consciously moved into the public eye, but still addressed himself primarily to his own circle (see GEORGE-KREIS). The esoteric quality of his work and message was emphasized by the presentation of his books, which appeared on finest quality paper in a special typeface designed for him by M. Lechter (first used in 1897) with minimal punctuation devised by George himself, and, except for titles and the word opening a line, without capital initials to common nouns. Though much can be seen as affectation, there is no doubt of the seriousness and intensity of George's views, or of the well-wrought character and monumental beauty of much of his poetry.

George published only one prose work, *Tage und Taten* (1903). He translated Dante (a selection from the *Divina Commedia*), Shakespeare's Sonnets, and poetry by D. G. Rossetti, Swinburne, Verlaine, Mallarmé, Rimbaud, Baudelaire, Verhaeren, and other poets. His own uniform edition of his works (*Gesamt-Ausgabe*) comprises 18 volumes (1927–43, reissued 1964–9). George's correspondence with H. von Hofmannsthal (q.v.) appeared in 1938.

George-Kreis, the select circle of friends, or

rather disciples, who at one time or another surrounded S. George (q.v.), sharing his views and seconding his efforts to renew German civilization by the creation of disciplined poetic beauty. Included among them were E. Bertram, M. Dauthendey, L. Derleth, F. Gundolf, K. Wolfskehl, F. Wolters, and, for a relatively short time, H. von Hofmannsthal (qq.v.). Special mention must be made of Robert Boehringer (1884–1974), one of George's literary executors, author of *Mein Bild von Stefan George* (1951), and founder of the Stefan George Stiftung (1959).

Georgslegende, see REINBOT VON DÜRNE.

Georgslied, an Old High German hymn in praise of St. George, probably written *c*. 900. Its composition is perhaps connected with the transfer in 896 of the relics of the saint to the Church of St. George on Reichenau, founded in 888. The hymn narrates the saint's triple martyrdom and resurrection. The *Georgslied* is one of the most puzzling of Old High German documents because of its strangely affected spelling, which led the apparently despairing copyist Wisolf to abandon his task with the words 'nequeo Wisolf'. It is written on a spare page of the Heidelberg Otfried MS. (see OTFRIED).

GEORG WILHELM, KURFÜRST VON BRANDEN-BURG (Cölln/Spree, 1595–1640, Königsberg), was a son of the Elector Johann Sigismund (q.v.) His reign, beginning in 1619, covered the vital phases of the Thirty Years War (see DREISSIG-JÄHRIGER KRIEG), and brought Brandenburg to the brink of ruin, from which it was saved by his son (see FREIDRICH WILHELM, GROSSER KUR-FÜRST).

GERBEL, NIKOLAUS (Pforzheim, *c*. 1485–1560), a humanist who studied in Cologne, Vienna, and Bologna, sided strongly with Reuchlin and with Luther, and was a friend of Ulrich von Hutten (qq.v.). He was formerly believed to be the author of *Eccius dedolatus* (q.v.). In his later years he devoted himself to scholarship in Strasburg, publishing Cuspinian's history of Rome in Latin in 1540, and in German in 1542. He used the pseudonym Musophilus.

GERENGEL, SIMON, a parish priest at Aspang near Vienna in the 16th c., was converted to Lutheranism, and was imprisoned from 1531 to 1534 at Passau and Salzburg. He is the author of a play on John the Baptist (*Johannes der Täufer*, 1553) and of Meisterlieder.

GERHARDT, PAUL, also Paulus (Gräfenhainichen nr. Wittenberg, 1607–76, Lübben), a son of the mayor of Gräfenhainichen, became the most notable Protestant hymn writer. He was educated at Grimma (see FÜRSTENSCHULEN) and Wittenberg University, where he studied theology. For some years a private tutor in Berlin, he was appointed pastor at Mittenwalde near Berlin in 1651, and six years later went to St. Nicholas's Church (Nikolaikirche), one of the most important pastoral appointments in Berlin. He married in 1655 and his wife bore him four children, but she and all but one son died during his twelve years in Berlin.

To Gerhardt's personal sufferings were added tribulations in office. He was a staunch Lutheran but the Electors of Brandenburg had been Calvinistic since 1613. During Gerhardt's time a conflict arose, in the course of which the Electoral authorities ordered his dismissal, which was withdrawn, but Gerhardt, conscious that a principle was at stake, resigned in 1667 from St. Nicholas. For two years he was supported by devoted friends and parishioners, and in 1669 became pastor at Lübben, where he spent the remainder of his life.

Gerhardt's earliest hymns (18 in number) were published in 1647 in a collection (*Praxis pietatis melica*) made by J. Crüger (q.v.), the organist at St. Nicholas's Church, who provided the tunes. In the second edition of this hymnal (1653) Gerhardt's contribution rose to 63. At the same time 37 of his hymns were included in C. Runge's *D. M. Luthers und anderer vornehmem, geistreichen und gelehrten Männer geistliche Lieder und Psalmen* (1653). The largest collection to appear in Gerhardt's lifetime was *Pauli Gerhardi geistliche Andachten*, containing 120 hymns and published in 1667 by Crüger's successor as organist. Gerhardt's poetry, nearly all of which is intended to be sung, is devout yet unsentimental, sensitive yet robust, striking a balance between man's inner life and the visible world around him. Not only have a number of the hymns retained a place in anthologies of poetry, they also continue to be included in hymnals. Among the best known are *Abendlied* ('Nun ruhen alle Wälder'), *Morgensegen* ('Die güldne Sonne'), *Christliches Wanderlied* ('Befiehl du deine Wege'), *Sommerlied* ('Geh aus, mein Herz, und suche Freud''), and 'Ich bin ein Gast auf Erden'. Particularly widely known is 'O Haupt voll Blut und Wunden', the last of a set of seven Latin poems by St. Bernard of Clairvaux, which Gerhardt translated in 1656 with such success that they persist as German poems in their own right (*Sieben Lieder an die Gliedmaßen des Herrn Jesu*). The first two verses of 'O Haupt voll Blut und Wunden', as well as the first verse of 'Befiehl du deine Wege', are used as chorales in J. S. Bach's *St. Matthew Passion*. A single-volume *Gedichte und Schriften* was published in 1957. *Geistliche Andachten* was reprinted in 1974.

GERHARD VON MINDEN, a priest of Minden, is the author of a Low German translation of fables by Aesop (*c.* 1370).

GERHOH VON REICHERSBURG (Polling, Bavaria, 1093–1169, Reichersberg), one of the outstanding German theologians of his age, was educated at Augsburg, became a canon in the Augustinian Order, carried out various missions in Rome, and in 1132 was appointed provost (Probst) of the Abbey of Reichersberg. Gerhoh strongly advocated a strict and ascetic rule of life, and opposed clerical abuses and laxity. He wrote profusely in Latin; his principal works include *De investigatione Antichristi* (*c.* 1160), which contains an attack on theatrical performances in church, *Opusculum de edificio Dei*, and a commentary on the Psalms. These last two works are of unknown date.

Germania, a Latin monograph by Cornelius Tacitus (*c.* A.D. 56–*c.* 112/13), written in A.D. 98. Its full title is *De origine et situ Germanorum* (alternatively *Germania sive de situ moribus et populis Germaniae liber*). It comprises 46 numbered chapters (each no longer than a normal paragraph) and deals with the territory and the tribes (not necessarily all Germanic) east of the Rhine and north of the Danube. Chapter III has an interesting reference to their songs, chapter VI to their mode of warfare. A number of tribes are distinguished by name.

GERMANICUS, the name by which Gaius Julius Caesar Germanicus (Rome, 15 B.C.–A.D. 19, Antioch) is usually known. A successful and popular general, Germanicus avenged the defeat sustained by Varus in A.D. 9, defeating Arminius (q.v., Hermann) at Idistaviso on the Weser in A.D. 16, but was recalled for failing to exploit his success. On his return to Rome in A.D. 17 Arminius' wife Thusnelda was paraded as a captive. Germanicus was subsequently given command in the east and is believed to have died by poison.

Germanien, a hymnic poem written in 1801 by F. Hölderlin (q.v.), prophetically celebrating a new Germany which will be a reincarnation of Hellas.

Germanistik, a term current since the middle of the 19th c. for the study of and research into German literature and philology.

German Empire, see DEUTSCHES REICH.

German King, see DEUTSCHER KÖNIG.

German Language, History of. The oldest documents in German belong to the 8th c. of the Christian era. The history of the language before that time is a matter of reconstruction by inference, analogy, and comparative study. The ancestor of German, as of most European and many Asiatic tongues, is termed Indo-European, of which no trace remains. It was spoken for thousands of years, and was still current *c.* 2500 B.C. Authorities are divided as to whether the peoples who originally spoke it lived near the Baltic or the Caspian Sea. From Indo-European developed Primitive Germanic, a language which is also virtually undocumented. The most decisive change which differentiated it from Indo-European is the First Sound Shift (see SOUND SHIFTS), a process which was completed by 500 B.C. German is the successor to Primitive Germanic, from which it is distinguished, among other aspects, especially by the Second Sound Shift, which fixed the consonants of High German. High German early established its predominance, and persists as the normal spoken and literary language. (For Low German, which did not undergo this sound shift, see below.)

Documented High German is conventionally divided into the following four phases, the dates of which represent only a rough approximation: (1) Old High German (Althochdeutsch), *c.* 770–1050; (2) Middle High German (Mittelhochdeutsch), *c.* 1050–1350; (3) Early New High German (Frühmittelhochdeutsch), 1350–1650; and (4) New High German (Neuhochdeutsch). The initials OHG and MHG are frequently used in English for (1) and (2) respectively, and ENHG sometimes for (3). OHG and MHG are not unified languages, but groups of spoken dialects. The documentation of OHG was entirely in the hands of clerics at monastic centres of learning; for the laity, regardless of rank, was illiterate in the early Middle Ages. Even in MHG the work of monks predominates, since the ability to read and write, though no longer exclusive to ecclesiastics, was not a common accomplishment even at courts. The clerics, whether they were monks or administrators holding office at courts, usually preferred Latin to OHG, and this, as well as subsequent neglect, largely accounts for the scarcity of documents in this oldest form of recorded German.

Both OHG and MHG are too dissimilar from modern German to be accessible without special study. OHG is the richer in sound, preserving a variety of vowels in its inflections. Towards the end of the OHG period Notker Labeo (q.v.), a monk of St. Gall, developed a new fluency of expression in his translations of Latin authors. In MHG the inflectional sounds of OHG were reduced to a mute 'e', but the vowel sounds of the stems are closer to the old language than to modern German. The notation of modified vowels (Umlaut), which was irregular in OHG, is systematically observed in MHG, which also developed a flexible and easily manipulated syntax. MHG produced for the first time a virtually standardized literary language, which is the vehicle of Minnesang and the courtly and heroic epics. In common use MHG remained almost as dialectally variable as OHG.

Early New High German, which developed from the late 14th c. onwards, exhibits a more obvious kinship with the modern language. Its vowels approximate to later use, and simplification of inflections goes a stage further. A special feature of the ENHG period is the development of a form of official language to be used for purposes of administration (Kanzleistil or Kanzleisprache, i.e. Chancery style). The Kanzleisprachen of Prague, Meissen, Mainz, and Vienna are especially notable. At the end of the 15th c. new printing techniques became a powerful factor in the standardization of German. In the 16th c. the wide dissemination of the Lutheran Bible influenced, by its strong and direct prose, the language even of Catholic regions. In 1578 appeared the first German grammar, the *Grammatica Germaniae linguae* by J. Clajus (q.v.). In the 17th c. M. Opitz (q.v.) defended the merits of German and laid down rules for its literary employment; language societies, such as Die Fruchtbringende Gesellschaft (q.v.), sought to purify it. From 1687 onwards German very gradually made headway in the universities as the language of lecturing and of scholarship. A setback to German occurred, especially among the German aristocracy, through the political prestige of the French monarchy under Louis XIV, which encouraged the written and spoken use of French and French forms of literature.

New High German came completely into its own in the 18th c., largely through the achievement of talented men of letters of the rising middle class. Important landmarks in the development of this language are *Deutsche Sprachkunst* (1748) by J. C. Gottsched (q.v.), the lexicographical work of J. C. Adelung (q.v.) between 1770 and 1790, and the *Deutsches Wörterbuch* of J. and W. Grimm (qq.v.), begun in 1852 and completed (by other hands) in 1960. German orthography was standardized in 1880, and pronunciation was established in the form of 'Bühnendeutsch' by J. C. Siebs (1862–1941) in his *Deutsche Bühnenaussprache* (1898). Local dialects continue to be spoken and tinge the speech even of the educated, so that they indicate place of origin rather than social standing. In recent years literary German has gained greatly from a new and enterprising generation of writers. At the same time some of the special savour of German has been lost by the indiscriminate adoption in speech and in the press of American and English expressions.

Low German is distinguished from High German primarily by the fact that it has not undergone the Second Sound Shift, so that High German *Wasser* is Low German *Water*, HG *Teufel* is LG *Deubel*, etc. The area covered by Low German was the North German plain and the Netherlands. Its best-known ancient literary monument is *Der Heliand* (q.v.), and it achieved its greatest florescence at the time of the Hanseatic League (see HANSE) when it developed a standardized form for commercial use. Though in the west it prospers in the Dutch and Flemish languages (and, to a point, in English), in Germany it is a collection of dialects, known as Plattdeutsch, and is still spoken in country districts in the north. It differs from region to region; the best-known variations are Hamburger Platt, Holsteinisches Platt, and Mecklenburgisches Platt. All are in decline under the spread of High German as the language of the educated. Plattdeutsch has a literature, of which the best-known representatives are F. Reuter (Mecklenburg) and K. Groth (Holstein) in the 19th c., and R. Kinau (Hamburg) in the 20th c. (qq.v.).

GEROK, FRIEDRICH KARL (Vaihingen/Enz, 1815–90, Stuttgart), the son of a pastor, entered the Church in Württemberg and occupied various ecclesiastical posts in Stuttgart from 1849 onwards, becoming Oberhofprediger and Prälat in 1868. He was an industrious writer of religious poems and hymns, some of which are still sung. They appeared in the collections *Palmblätter* (1857), *Pfingstrosen* (1864), *Blumen und Sterne* (1868), *Eichenblätter* (1870), *Palmblätter. Neue Folge* (1878), *Der letzte Strauß* (1885), and *Unter dem Abendstern* (1886). In 1871 he published a volume of patriotic poems, *Deutsche Ostern*. He was granted a patent of nobility as Karl von Gerok in 1868.

GERSTÄCKER, FRIEDRICH (Hamburg, 1816–72, Brunswick), the son of an operatic tenor, went to the U.S.A. in 1837, where he lived a chequered life, plying many trades, including those of hunter in the backwoods, stoker, lumberjack, silversmith, and cook. After six years he returned to Germany and wrote his first and best-known novel of American life, *Die Regulatoren in Arkansas* (q.v., 3 vols., 1845). This was followed by *Die Flußpiraten des Mississippi* (q.v., 3 vols., 1848). Between 1849 and 1852 Gerstäcker made a grand tour of North and South America and Australasia; he revisited South America in 1860–1. During the intervals between these voyagings he lived in Saxony, publishing novels, including *Tahiti, Roman aus* *der Südsee* (1854), *Die beiden Sträflinge*, an Australian story (3 vols., 1856), *Gold*, sub-titled *Kalifornisches Lebensbild* (3 vols., 1858), and *Unter dem Äquator* (3 vols., 1860). His novels of adventure include also *Die Kolonie* (1864), *Unter den Penchuenchen* (3 vols., 1867), *Die Blauen und die Gelben*, set in Venezuela (1870), and *In Mexico* (1871).

By the early 1860s Gerstäcker had acquired a reputation as a traveller; in 1862 he accompanied the Duke of Sachsen-Coburg-Gotha to Egypt and Abyssinia. He made a final journey to North and Central America in 1867–8 before settling in Brunswick. His reputation rested chiefly on his keenly observed and lively descriptions of exotic lands and manners, but he also had some success with novels of German life, including *Der Wahnsinnige* (1853), *Der Kunstreiter* (1861), and *Die Franktireurs* (1871). His best books continued to be much read, well into the 20th c. His *Gesammelte Schriften* (1872–9) occupy 44 vols.

GERSTENBERG, HEINRICH WILHELM VON (Tondern, Schleswig, 1737–1823, Altona), after studying law at Jena, became in 1760 an officer in a Danish regiment of horse. He had already published two poetic collections, *Prosaische Gedichte* and *Tändeleyen* (both 1759), and he continued to write while serving, publishing in 1762 patriotic poems after the manner of J. W. L. Gleim's *Kriegslieder eines preußischen Grenadiers* (*Kriegslieder eines königlich dänischen Grenadiers*). In 1765 Gerstenberg accepted an administrative post in the war ministry in Copenhagen, where he was in contact with F. G. Klopstock (q.v.). His *Briefe über Merkwürdigkeiten der Literatur* (q.v.), often called the *Schleswigsche Literaturbriefe*, caused a stir when they appeared in 1766–70 by their emphasis on genius and their enthusiasm for Shakespeare. Gerstenberg's interest in primitive poetry is manifested in his *Gedicht eines Skalden* (1766), and later in the figure of the bard Ryno in his immense play (Melodrama) *Minona oder die Angelsachsen* (1785). A brief incursion into Greek mythology took place in 1767 with the play *Ariadne auf Naxos*. Gerstenberg's outstanding creative achievement is the agonizing Dantesque tragedy *Ugolino* (q.v., 1768).

From 1767 to 1771 he was a reviewer for the *Hamburgische Neue Zeitung*, in which he praises Lessing, Klopstock, Herder, and Winckelmann and criticizes Wieland (qq.v.) and French taste. These *Rezensionen* (published in collected form in 1904) show Gerstenberg working towards a new outlook which was to be that of the Sturm und Drang (q.v.). At this point he seems to have lost his impetus. He retired with the rank of captain in 1771, was Danish consul in Lübeck

from 1775 to 1783, spent two years at Eutin in contact with J. H. Voß (q.v.), and in 1786 moved to Altona, where he was lottery director from 1789 to 1812. German Classicism and Romanticism passed him by. His collected works (*Vermischte Schriften*, 3 vols., 1815–16) were published towards the end of his long life. An edition of *Ugolino* with a selection from the critical writings (ed. C. Siegrist) appeared 1966.

GERTRUD VON HACKEBORN (1241–98, nr. Eisleben), a nun of the Cistercian Abbey of Helfta near Eisleben who rose to be its abbess. She recorded, in the Latin *Liber specialis gratiae*, which was at one time attributed to GERTRUD von Helfta (q.v.). the mystical visions and ecstasies of MECHTHILD von Hackeborn (q.v.). The work was translated into German in the 14th c.

GERTRUD VON HELFTA (1256–1302, nr. Eisleben) spent her life as a nun of the Cistercian Abbey of Helfta near Eisleben. Her visions of Christ are recorded in Latin in *Legatus divinae pietatis*, and she also wrote a book entitled *Exercitia spiritualia septem*. Both were translated into German in the 14th c. Gertrud, who is also known as Gertrud die Große, was, with her sister MECHTHILD von Hackeborn (q.v.), an initiator of the adoration of the Sacred Heart.

GERVINUS, GEORG GOTTFRIED (Darmstadt, 1805–71, Heidelberg), was a pioneer in the study of the history of literature. Intended at first for a business career, he entered Gießen University in 1825, and migrated to Heidelberg in 1826. In 1830 he began to lecture at Heidelberg, and in 1835 published the first of 5 vols. of *Geschichte der poetischen Nationalliteratur der Deutschen*, which was completed in 1842. It presents for the first time a history of German literature in relation to the political and intellectual environment. Its title was subsequently changed to *Geschichte der deutschen Dichtung*. Gervinus was appointed to the chair of history and literature at Göttingen in 1835, but in 1837, in consequence of a courageous protest against the abrogation of the constitution by King Ernst August, he was dismissed with six of his colleagues (see GÖTTINGER SIEBEN). In 1844 he was appointed to a chair at Heidelberg, where he spent the remainder of his life. In 1848 he was elected a member of the National Assembly at Frankfurt (see FRANKFURTER NATIONALVERSAMMLUNG), but soon resigned.

Among his other works were *Grundzüge der Historik* (1837), *Shakespeare* (1849–52), and *Geschichte des 19. Jahrhunderts seit den Wiener Verträgen*, 8 vols. (1855–66). Gervinus held Liberal opinions of nationalistic tinge, and his scholarship exhibits a strong national bias. He did not approve of the policies by which Bismarck (q.v.) sought to achieve the unification of Germany. He died in the year in which the German Empire was proclaimed.

Gesamtkunstwerk, a term coined by R. Wagner (q.v.). It denotes a work of art to which music, poetry, mime, painting (in décor), etc. all contribute. It is clearly fashioned by Wagner to suit his own later works. Heinrich Mann, with characteristic paradox, affirmed in his autobiography (*Ein Zeitalter wird besichtigt*, 1945) that the true Gesamtkunstwerk is the 19th-c. novel.

Gesang der Geister über den Wassern, a poem written by Goethe on 9–11 October 1779 at Lauterbrunnen, Switzerland. It was sent to Charlotte von Stein on 14 October, and published in *Goethes Schriften* in 1789. The waterfall which inspired the poem is the Staubbach. The image of falling water and rising spray is extended to cover the course of the river down to the lake and provides a parable of the life of man. The title in the original version sent to Frau von Stein was *Gesang der lieblichen Geister in der Wüste*. Compositions, all for several voices, have been made by F. Schubert, C. Loewe, and F. Hiller (qq.v.).

Gesang im Feuerofen, Der, a play by C. Zuckmayer (q.v.), published in 1950. Its action juxtaposes two incidents reported in one issue of a Swiss newspaper (*Basler National-Zeitung*, 8 December 1948). The main theme is derived from the announcement of the execution of a former member of the French Resistance, whose treachery had caused his comrades to be burned alive. The secondary theme is the story of a school of whales stranded in Florida, which Zuckmayer uses as a portent of ultimate catastrophe.

Gesang Weylas, a short poem written by E. Mörike (q.v.) in 1831, beginning 'Du bist Orplid, mein Land'. Orplid (q.v.) is an imaginary land of fantasy created by the young Mörike, and Weyla is its goddess. The poem has been set to music by Hugo Wolf (q.v.).

Gesang zu zweien in der Nacht, a poem written by E. Mörike (q.v.), and included at first in an unpublished playlet, *Spillner*. In slightly extended form it was then published in *Der letzte König von Orplid* in *Maler Nolten* (q.v.), where it is divided between two characters. With the voices indicated by 'Sie' and 'Er', and with its present title, it appeared in *Gedichte* (1838).

Geschichte der Kunst des Altertums, a monumental and epoch-making work by J. J. Winckelmann (q.v.), published in 1764. It deals first with the origins of art, then with the art of the Egyptians, Phoenicians, Persians, and Etruscans. Winckelmann next treats Greek art, to which he devotes the greater part of the book, describing many works, especially of sculpture. A number of these are illustrated in line drawings. The erudite and plentifully annotated work establishes the Apolline view of Greek art as noble and serene, which Winckelmann had already adumbrated in his *Gedanken über die Nachahmung der griechischen Werke in der Malerei und in der Bildhauerkunst* (q.v., 1755). *Anmerkungen über die Geschichte der Kunst des Altertums* followed in 1766. G. E. Lessing's *Laokoon* (q.v.) is extensively concerned with points arising out of Winckelmann's great work, which Goethe took with him to Italy (in an Italian translation) as his guide to the remains of ancient art.

Geschichte der 1002. Nacht, Die, a short novel by J. Roth (q.v.), published in 1939. Set in Vienna of the late 19th c., and in a Galician garrison town, it successfully re-creates the atmosphere of both these environments. The title alludes to *The Arabian Nights*, known in German as *Tausendundeine Nacht* (q.v.). The '1002nd story' occurs during a visit of the Shah of Persia to Vienna. At a state ball, he is fascinated by the *décolleté* of the unveiled ladies, and, choosing a Gräfin von W., insists that she must spend the night with him. The embarrassing situation is solved by the dragoon Rittmeister Taittinger, on special duty at court. Mizzi Schinagel, who has borne him an illegitimate son, is the Gräfin's double, and she receives the Shah in the luxurious brothel she frequents. The following day he gives her, as a reward, pearls worth many thousands. Taittinger, however, is sent back to his regiment in Galicia. Mizzi sells the pearls and lives like a lady. The money soon dwindles, and her lover Lissauer invests what is left of it in a shop selling imitation Brussels lace. When this is discovered, Mizzi is sent to prison for 16 months.

The book traces the decline and fall of Rittmeister Taittinger, the typical heedless officer, who thoughtlessly allows the steward of his estate to swindle him, signs financial documents for a shifty cousin, becomes involved again with Mizzi Schinagel, gives money to a scandalmongering journalist, and finds himself at last in so compromising a situation that he is obliged to leave his regiment. Taittinger visits his estate in the Carpathians, but cannot endure it for more than a few weeks. Mizzi's son gets into trouble with the police, so that Taittinger (in the form of Graf ***) is again in the news. An attempt to re-enter the army as an infantry officer is shelved when the Shah comes on a second visit, and Mizzi, now the owner of a waxworks in the Prater, has bills posted recalling the episode of the Shah's last visit. Taittinger, having no further ideas for his future, shoots himself. The story is told with the ease of an ironical raconteur.

Geschichte des Abfalls der vereinigten Niederlande von der spanischen Regierung, an historical work written by F. Schiller in 1787–8, and published in 1788. It traces the beginnings of unrest in the Spanish Netherlands up to the departure of Margaret of Parma in 1567. Schiller intended to deal with the later phases of revolt up to 1581 in a further volume, which, however, was never written. The *Geschichte des Abfalls der vereinigten Niederlande* (as the work is generally called) achieved considerable success and was also instrumental in obtaining for Schiller appointment to a professorship of history at Jena University.

Geschichte des Agathon, a novel written by C. M. Wieland (q.v.) between 1761 and 1767, and published in 1766–7. It bears as motto the lines

Quid virtus et quid sapientia possit,
Utile proposuit nobis exemplum.

from Horace, *Epistolae*, Bk. 1, Ep. 2 ('He has furnished us with a useful example of what virtue and wisdom can achieve'). The novel is set in ancient Greece, a token of the rising interest in Greek antiquity among men of culture and breeding. Nevertheless, Wieland, in his preface (*Vorbericht*, 1766), indicates Agathon's identity with himself, and suggests that the events and some of the other characters are drawn from life. It has been suggested that Sophie Gutermann (see LA ROCHE, SOPHIE VON) was the original of Psyche, and Wieland's mentor Graf Stadion (1691–1768) that of Hippias.

Agathon, whose name symbolizes his innate goodness, is a young Greek of good parentage, and is reared in a temple. The priestess casts an eye upon him, but he is drawn to Psyche, a young devotee. Psyche is banished, and Agathon returns to his father at Athens. He is appointed to political office, but, after a time, is dismissed and banished. These preliminaries are revealed as the story progresses. The novel actually begins with the banished Agathon meeting Psyche again, only to be separated from her once more. Taken by pirates, he is sold at Smyrna to the sophist Hippias, who sustains a philosophy of hedonistic self-interest. Hippias seeks in vain to convert Agathon from idealism to self-interest,

but he brings about an acquaintance with Danae, a hetaera, whom Agathon comes to love; however, on learning her trade, he renounces her and leaves for Syracuse. After a further unsuccessful venture into politics he retires into private life. He goes to Tarento, where he admires the wise rule of Archytas; he discovers that Psyche is his sister, and encounters Danae, now ennobled and become altruistic through her love for Agathon. Agathon, abandoning vague idealism, contents himself with a limited sphere of useful activity, and adopts a philosophy which emphasizes the spirit without disregarding the senses.

The *Geschichte des Agathon*, told in smooth, lucid prose, is an early and competent example of the Bildungsroman (q.v.), in which the hero learns ultimate wisdom through errors. It enjoyed a considerable reputation in the 18th c. and was termed by Lessing 'der erste und einzige Roman für den denkenden Kopf von klassischem Geschmacke' (*Hamburgische Dramaturgie*, 69. Stück).

Geschichte des Diethelm von Buchenberg, a Novelle by B. Auerbach (q.v.), published in 1852 in *Schwarzwälder Dorfgeschichten* (q.v.). Diethelm, a former labourer who has married the farmer's widow, appears to make a success of his new life, and is widely respected. But he cannot restrain an impulse to speculate, and while all around flatter and defer to him, he knows that his financial position is seriously endangered. After much hesitation he resolves to carry out a plan to burn his house and claim the fire insurance money. His shepherd detects his preparations, but Diethelm binds him, and leaves him to perish in the burning house. For a time he battles successfully against conscience, but in the end he blurts out the truth, is tried, and is sentenced to life imprisonment. After three years he takes his own life in prison.

Geschichte des Fräuleins von Sternheim, a novel written by Sophie von LA ROCHE (q.v.), published for her, anonymously, by C. M. Wieland (q.v.) in 1771. The title is followed by the assertion that the work is written by a friend of the Fräulein: 'Von einer Freundin derselben aus Originalpapieren und anderen zuverlässigen Quellen gezogen'. Although the novel contains a few short narrative passages, it is mainly written in the fashionable letter form (see BRIEFROMAN). Most of the letters purport to be by Fräulein von Sternheim; others derive from a variety of characters.

Sophie von Sternheim is the daughter of parents of wealth and high character, who have made a love match which is well viewed by all their relatives except one, an aunt. Sophie's parents both die before she is of age. The aunt, who had resented the marriage of her parents, takes her into her house and presents her at court with the intention of making her the mistress of the ruling prince; she hopes through Sophie's influence to obtain a favourable judgement in an important lawsuit. Sophie's virtue renders her unsuitable for the plot, and in the end frustrates it. An anglophile, she falls in love with Lord Seymour, a visitor at court. When Seymour hesitates to return her love, Lord Derby, another visitor, and an experienced seducer, lays siege to her, feigning benevolence to the poor in order to impress her with his virtue. Sophie, in order to escape the prince, marries Lord Derby, only to find the ceremony a sham. Derby soon tires of her. She moves to the Low Countries, and, as Madame Leidens, devotes herself to good works. Derby meanwhile has married, and Sophie, meeting his mother-in-law at Spa, is invited to her house in England. Derby, fearing discovery, carries her off to the Highlands and seeks to bring about her death. She is saved by poor crofters, and eventually rediscovered at Hopton (which is perhaps Hopetoun House) by Lord Seymour. Derby dies, Sophie and Seymour marry and thenceforth live happily together.

Faded though the novel now seems, its insight into character and motivation, its advocacy of sentiment and of virtue based on education, made it a characteristic and popular work of the day.

Geschichte des Herrn William Lovell, a novel in letter form written by L. Tieck (q.v.), published in 3 vols. in 1795–6, and reprinted in 1813 in 2 vols. The hero is sent by his father on the grand tour in order to turn his mind from a love-affair. He soon succumbs to temptations, is converted to a ruthless hedonism, and seduces an innocent girl, who takes her own life. His increasing depravity is accelerated by the influence of a sinister figure, Andrea Cosimo. Lovell returns to England; he seduces a friend's wife, who also commits suicide; he next attempts to seduce his first love, now married. Pursued by her brother Wilmot, he flees to the Continent, but is overtaken and killed in a duel. Before he dies he realizes the extent of his moral turpitude.

Geschichten aus dem Wiener Wald, a Volksstück in three parts by Ödön von Horváth (q.v.). Published and first performed in Berlin in 1931, it describes the tragic victimization of Marianne, daughter of the Zauberkönig, owner of a toyshop selling 'magic' and toy soldiers (wounded ones are popular), and specializing in doll repairs. She has been promised to Oskar, a butcher in the same street (Vienna,

8th District). The Zauberkönig and his party enjoy a picnic outing to the Vienna Woods in celebration of the engagement. But Marianne, in love with the impecunious philanderer Alfred, breaks off the engagement when her romance by the Danube is discovered that night. Abandoned by her father, she lives with Alfred and bears a son. Hoping that she will earn some money, Alfred soon places the infant in the care of his mother, who lives with his grandmother at the foot of a ruin in the Wachau. Although she has served a prison sentence for a theft, allegedly committed while working in the cabaret 'Maxim'—the play's central scene (Act III, scene 2), in which she suffers the most devastating humiliation as a daughter and a woman—Marianne is allowed to return to her father. She owes this reconciliation to Valerie, proprietress of the tobacconist's shop next door, who herself forgives her disloyal Alfred. At long last, it seems, Marianne will be able to claim her child, to whom she is so devoted that she cannot even express regret for its birth in sin when pressed to do so by the priest. A happy party arrives in the Wachau, only to find that the infant has just died, having been exposed to cold air and fatal illness by the evil grandmother. Oskar, who has waited for the day when he will be able to claim Marianne, does so when she collapses into his arms.

The waltz (*G'schichten aus dem Wiener Wald*) by J. Strauß (q.v.) and the idyllic settings create true atmosphere, but they are also anti-illusionary devices; another is the frequent interruption of the dialogue (and music) by moments of silence (Stille) that are meant to encourage reflection on the part of a Kleinbürger society, which is as alienated from its innate humanity (for example, in its ever-present exploitation of women) as it is from its dialect. Written in High German with a touch of Viennese, the play's language is spiced with the pretentious, shallow, bourgeois Bildungsjargon of the period, heightens, like the schoolgirl's unmusical piano playing in the background, the play's ironic and grotesque effects. Nominated by his friend C. Zuckmayer (q.v.), Horváth was awarded the Kleist prize in the year of the play's publication. Arguably the most important 20th c. Volksstück, it was filmed by M. Schell in 1979 and translated into English by C. Hampton as *Tales from the Vienna Woods* (first performed at the National Theatre, London, in 1977).

Geschichte vom braven Kasperl und dem schönen Annerl, Die, a Novelle by C. Brentano (q.v.), published in 1817. Its frame (see RAHMEN) encloses two stories, the seduction of Annerl, based on folklore (the poem *Weltlich Recht*), and the suicide of the corporal Kasperl, based on a true incident. The major figures of the frame are the narrator (Schreiber) and Kasperl's grandmother, who represents honour before God, which is contrasted, through the stories, with the vanity of wordly honour. The closely wrought narrative, at the end of which Annerl, under sentence of death for infanticide, is executed because the Duke's pardon arrives too late, is renowned for the augmentation of tension reflecting the opposition between these concepts.

Geschichtklitterung, see Fischart, J.

Geschlecht der Maechler, Das, title of a trilogy of novels by H. Stehr (q.v.). They are *Nathanael Maechler* (1929), *Die Nachkommen* (1933), and *Damian oder Das große Schermesser* (published posth. in 1944). The novels trace the life of a Silesian family between 1848 and 1918.

Geschwister, Die, a short play (Ein Schauspiel) in one act written by J. W. Goethe in 1776, and first published in 1787. Wilhelm and Marianne are apparently brother and sister; in fact they are not blood relations, for Marianne is the daughter of the deceased Charlotte, once Wilhelm's love. He has adopted Marianne, who believes him to be her brother. A friend, Fabrice, proposes to Marianne; after some hesitation, she decides that she cannot leave her brother. Wilhelm, in delight, reveals their true relationship, and Marianne recognizes her love as conjugal. Fabrice accepts the situation. The play is thought to have some connection with Goethe's love for Frau von Stein (q.v.), whose attitude to him was at first sisterly. In the poem which constitutes Goethe's letter to Frau von Stein of 14 April 1776 occur the apposite lines:

> Ach du warst in abgelebten Zeiten
> Meine Schwester oder meine Frau.

Geschwister Oppenheim, Die, a novel by L. Feuchtwanger (q.v.), published in Amsterdam in 1933. It forms the second of four novels collected as a tetralogy, *Der Wartesaal. Zyklus aus dem Zeitgeschehen.* It is a story of the persecution of Jews by the National Socialist regime in January 1933 and the following months. Three well-situated middle-aged Jewish brothers, Gustav, Martin, and Edgar Oppenheim are driven to emigrate, and their assets are confiscated. Martin's son commits suicide. Gustav returns secretly, but is arrested and sent to a concentration camp. A friend secures his release, but he dies shortly afterwards in Czechoslovakia.

Geschwister von Neapel, Die, a novel by F. Werfel (q.v.), published in 1931. Don Pascarella, a widower of Naples, rules tyrannically over his six children, three boys and three girls, ranging in age from the early teens to the late twenties. His insistence on a narrow and rigid pattern of life is not intentionally cruel, but stems from the overpowering influence of his personality upon his family. The young people are cowed by his fierce outbreaks of rage and his withering sarcasm, and spellbound by his performance of the aria 'O monumento! Regia e bolgia dogale!' from Ponchielli's *La Gioconda,* an event which takes place on Sunday evenings, provided he is in humour.

The story, set in the 1920s, portrays the irruption of the outer world into this hermetically closed circle. Two of his children go to a ball without permission, and so enter into contacts outside the family. Pascarella himself is obliged to allow the world to change his routine when he is swindled by his partner and has to slave night and day to keep the creditors at bay. In order to help their father, the three sons decide to emigrate. He opposes their plan, but they refuse to be diverted and leave for Brazil, where Lauro dies, probably by his own hand; the other two brothers fail in their attempts to make money. Their youngest sister, Iride, dies after a lingering illness. Grazia, the second daughter, becomes engaged to an Englishman, who does all that he can to help Pascarella financially, but is unwelcome none the less. The eldest, Annunziata, enters a convent. The final blow is Pascarella's arrest by Fascisti on suspicion of conspiracy. He is released after six days with apologies, but the stain, to his mind, cannot be erased. When the family is gathered together, he attempts once more to sing 'O monumento!', but this time breaks down when a string of the dead Lauro's cello snaps. This final sound marks the end of an epoch: 'Das Zeitalter des Gesanges und Gesetzes ist nun zu Ende! Welches Zeitalter aber hat begonnen?'

Gesellschaft, Die, a periodical founded in Munich in 1885 by M. G. Conrad (q.v.). It was for a few years (*c.* 1885–90) the principal organ for the supporters of a radical renewal of literature on a realistic basis (see Naturalismus). It ceased publication in 1902.

Gesellschaft der Menschenrechte, a secret society founded by G. Büchner (q.v.) in March 1834 in Gießen, and, during Büchner's Easter vacation, in Darmstadt. Büchner attracted to its membership men of any station who shared his socialist, egalitarian, and republican ideas. He founded his society at the time when he was writing *Der Hessische Landbote* (q.v.) and preparing for active revolution. Büchner's model was the Strasburg Société des Droits de l'Homme et du Citoyen of 1830. The society was active for less than a year, and disintegrated after Büchner's flight to Strasburg.

Gesellschaftsroman, term used by German literary historians to classify novels in which the portrayal of a part or the whole of a society (usually contemporary) is at least as important as the individual characters in the narrative. Random examples are Gutzkow's *Die Ritter vom Geiste*, Auerbach's *Das Landhaus am Rhein*, Th. Mann's *Der Zauberberg* (qq.v.), and the novels of Th. Fontane and of Doderer (qq.v.). As with many such terms, its application is vague.

Gesetz zur Behebung der Not von Volk und Staat, see ERMÄCHTIGUNGSGESETZ.

Gesichte der Simone Machard, Die, a four-act play with four dream episodes by B. Brecht (q.v.). In prose with songs in rhymed verse, it was written in collaboration with Lion Feuchtwanger (q.v., who wrote a novel *Simone*) between 1941 and 1943, and published in 1956 in *Sinn und Form* (q.v.). It was first produced in Frankfurt on 8 March 1957. The action takes place in Saint-Martin, a small town on the main road from Paris to the South of France, between 14 and 22 June 1940. The play, a variation on the Joan of Arc theme, is set against the background of the German occupation of France. The young girl Simone sees in a dream her dead brother André. He appears as an angel, and calls on her to work for the resistance movement against foreign oppression and the Pétainistes of Vichy. In these dream scenes characters from the play assume historical identities, such as King Charles VII, for whom Joan (Simone) frees Orleans and Reims. Simone is in the end caught and sentenced to detention in the mental institution of Sainte-Ursule, but she has succeeded in promoting the resistance of the refugees, who follow her example by setting fire to the gymnasium. Simone sees the flames as she is driven to the nunnery.

Gesichte Philanders von Sittewald (in full *Wunderliche und Wahrhafftige Gesichte Phs. v. Sittewalt*), title of a satirical work by H. M. Moscherosch (q.v.), first published in 1641 or 1642. Its first part is an adaptation of *Los sueños* of Francisco Gómez de Quevedo y Villegas (1580–1645), its second part a free continuation.

GESNER, KONRAD (Zurich, 1516–65, Zurich), a polymath, who was a professor at Zurich University, compiled a catalogue of writers in the ancient tongues (*Bibliotheca universalis, seu*

catalogus omnium scriptorum locupletissimus in tribus linguis, Graeca, Latina et Hebraica exstantium, 1545–9). He edited classical texts and translated Greek works into Latin. Gesner was an important pioneer in scientific development, and is sometimes styled the German Pliny (der deutsche Plinius). His work *Historia animalium* (1551–8) is the most important zoological treatise of its time, and his botanical writings (not published in his lifetime, *Opera botanica*, 1753–9) contain evidence of first-hand observation. *De omni rerum fossilium genere, gemmis, lapidibus, metallis* (1555) has original illustrations of petrifacts and crystals. His *Mithridates* (1555) is a notable early example of the comparative study of languages.

GESSNER, SALOMON (Zurich, 1730–88, Zurich), a bookseller's son, was apprenticed to the bookseller Spener in Berlin. Giving up this employment, he lived for a time by painting and engraving, for which he had a considerable talent. In 1750 he settled in Zurich, continuing to live by painting, including painting on porcelain. He began to write idylls in poetic prose, beginning with *Daphnis* (1754). His *Idyllen* (1756) achieved a nation-wide success. In *Der Tod Abels* (1758) he attempted an epic in prose. Geßner illustrated his own works. From 1765 on he occupied various official positions, and in 1775 he inherited his father's business.

In his idylls, Geßner, who is indebted to Theocritus and Virgil, creates an idealized, orderly, almost horticultural state of nature, from which everything rough and craggy has been eliminated; his shepherds are similarly untouched by the ruder aspects of country life. His work embodies the city-dweller's longing for a nature which he does not know, and this explains its instant popularity. W. Raabe uses Geßner's *Idyllen*, the publication of which coincided with the outbreak of the Seven Years War (see SIEBENJÄHRIGER KRIEG), as a powerful contrast to rural realities in time of war in his novel *Hastenbeck* (q.v.). Geßner remained unspoilt by his success, acknowledging the limitations of his literary and artistic talents.

Gestapo, abbreviation for Geheime Staatspolizei, the secret political police organized with this title by H. Göring (q.v.) in Prussia in 1933 and subsequently in the whole of Germany. In 1936 they were made into a unified national force under H. Himmler (q.v.). Subject to no legal limitations, the Gestapo became the principal instrument of rule by terror in National Socialist Germany and in the occupied countries. Under Himmler, R. Heydrich was its most notorious leader.

Gesta Romanorum, a collection of anecdotes of widely varying character taken originally from Roman history narrated in Latin and assembled, possibly in England, in the 13th or 14th c. The Latin version was first printed in Germany at Cologne in 1472, and in 1489 a German translation was published at Augsburg.

Gestiefelte Kater, Der, a fairy-tale in dramatic form (*Ein Kindermärchen in drei Akten mit Zwischenspielen, einem Prologe und Epiloge*) by L. Tieck (q.v.), published in vol. 1 of his *Volksmärchen* in 1797 and also separately in the same year. It is a play within a play: within is a burlesqued version of Puss in Boots, and around this are grouped a stage audience, author, actors, and a stage carpenter, resulting in an agreeable mixture of boisterous humour and irony. Tieck claims that he almost finished it in one evening.

Gestrenge Herren, popular designation for the three days, 11–13 May, and sometimes extended to cover the 14th and 15th, a period which is traditionally supposed to coincide with a brief return of wintry weather. The Herren are Saints Mamertus, Pancras, and Servatius, whose feasts fall on 11, 12, and 13 May respectively.

Getreue Eckart und der Tannenhäuser, Der, a tale by L. Tieck (q.v.), based on two traditional stories, and first published in 1799. It is in two sections. The first tells the story of Eckart's fidelity. The Duke of Burgundy slays Eckart's sons; later Eckart finds the Duke exhausted in the forest, represses his impulse to vengeance, and carries his enemy to safety. In the second section Tannenhäuser reappears after spending years in the grotto of Venus. He recounts how he had loved a girl and murdered his rival in her love; he then entered the grotto, although the spiritual form of Eckart had sought to prevent him (this is the only link of this section with Eckhart). His interlocutor, Friedrich von Wolfsburg, explains that he is the successful rival, and that Tannenhäuser has not committed murder and suffers from a delusion. Tannenhäuser disappears, and Eckart discovers that his wife has been murdered by Tannenhäuser.

Gewehre der Frau Carrar, Die, a short play by B. Brecht (q.v.), written in 1937 in the first year of the Spanish Civil War, which provides the background. It was intended for performance by a German troupe in Paris. Acknowledgement is made to an idea in J. M. Synge's *Riders to the Sea.*

Teresa Carrar, who lives in an Andalusian fisherman's cottage, has lost her husband and fears for her sons Juan and José. For this reason she adopts a neutral attitude and hides her husband's rifles; she refuses to surrender them to her brother Pedro Jaquéras, who intends to join the fight against the rebel generals. José struggles with his mother and injures her foot. Juan's body is carried into the cottage on the sail of his boat. He has been shot while out fishing. Taking the rifles, Teresa now joins her brother and younger son to unite and fight for the common cause. The character of the Padre, who shares the suffering of the hungry and prays for them but does nothing to help them get their daily bread, is introduced by Brecht as a criticism of the Church's passive and appeasing attitude towards General Franco. The active resistance depicted at the end of the play indicates Brecht's support for a solution by force.

GEYER, FLORIAN (1490–1525, Rimpar), a 16th-c. nobleman and soldier, served in 1519 in the forces of the Swabian League which expelled Duke ULRICH von Württemberg (q.v.) from his territory. In 1525 he joined the peasant side in the Peasants' War (see BAUERNKRIEG), and was assassinated while on a diplomatic mission. He is the subject of G. Hauptmann's tragedy *Florian Geyer* (q.v.).

Ghasel, a form of stanza common in Arabic, Persian, and Turkish poetry; its chief characteristic is that it has a single rhyme, which is preserved throughout the stanza in alternate lines (aabacada, etc.); while the intervening lines are without rhyme. The ghasel is variable in the number of lines it comprises. Ghaselen were first written in German by F. Schlegel (q.v.). Frequently, as in the *Ghaselen* (1821) of A. von Platen (q.v.), the repetition of one word is substituted for normal rhyme. F. Rückert (q.v.) also wrote a collection of these poems with the title *Ghaselen* (1819). Lesser poets cultivating the ghasel were F. von Bodenstedt, E. Geibel, and G. Keller (qq.v.).

Gickelhahn or **Kickelhahn,** a mountain in the Thuringian forest near Ilmenau. It was here that Goethe wrote the poem 'Über allen Gipfeln' (q.v.), the second of his two poems entitled *Wanderers Nachtlied.*

GIESEBRECHT, LUDWIG (Mirow, Pomerania, 1792–1873, Jasenitz nr. Stettin), fought as a volunteer cavalryman in the War of Liberation (see NAPOLEONIC WARS) and then became a schoolmaster in Stettin. His poems, which include patriotic and North German dialect pieces, were published as *Gedichte* in 1836.

GILDEMEISTER, OTTO (Bremen, 1823–1902, Bremen), an administrator, was for twelve years Burgomaster of his native city. In his leisure time he was an active and competent translator. His achievements in this field include *Byrons sämtliche Werke* (6 vols., 1864–5), Shakespeare's Sonnets (*Sonette*, 1871), Ariosto's *Orlando furioso* (*Der rasende Roland*, 4 vols., 1882), and Dante (*Die göttliche Komödie*, 1888). Gildemeister also translated twelve plays of Shakespeare for the German version sponsored by F. von Bodenstedt (q.v.). Translations of four more Shakespearian plays (*Shakespeare-Dramen*, 1904) appeared posthumously.

GILM ZU ROSENEGG, HERMANN VON (Innsbruck, 1812–64, Linz), an Austrian civil servant, latterly in Linz, wrote poetry published in *Tiroler Schützenleben* (1863) and, posthumously, in *Gedichte* (1864–5). He is now only known for the poem beginning 'Stell auf den Tisch die duftenden Reseden' (*Allerseelen*), which still occurs in anthologies and is sung in a setting by R. Strauss (q.v.).

GINZKEY, FRANZ KARL VON (Pola, 1871–1963, Vienna), had a military career, serving from 1891 to 1897 as a regimental officer, and until 1914 at the Military Geographical Institute. He was in military employment in Vienna during the war, and after 1918 lived as an author in Salzburg, Vienna, and by the Attersee in the Salzkammergut. From the age of 30 onward he wrote graceful and unpretentious lyric poetry and numerous novels, some historical, some *biographies romancées*, others, such as *Jakobus und die Frauen* (1908), examples of Heimatkunst (q.v.). His later work shows a nostalgic attraction towards pre-1914 Austria–Hungary and its society.

GISEKE, NIKOLAUS DIETRICH (Nemes Cso, Hungary, 1724–65, Sondersheim), a contributor to the *Bremer Beiträge* (1741–5), was brought up in Hamburg after the death of his father, a Protestant pastor in Hungary. After a period as a private tutor in Brunswick when K. W. Jerusalem (q.v.) was his pupil, he received a pastorate in Trautenstein in the Harz. In 1754 he became court preacher in Quedlinburg, and in 1760 was appointed moderator (Superintendent) in Sondershausen. He wrote fluent poetry marked by tender feeling, and made a cult of emotional friendship. His poems were collected after his death by K. C. Gärtner (q.v.) under the title *Poetische Werke* (1767).

GISEKE, ROBERT (Marienburg, Prussia, 1827–90, Leubus), a great-grandson of N. D. Giseke

(q.v.), studied at Breslau, Halle, and Berlin universities, and was active as a journalist, first in Leipzig, then in Berlin. He suffered a mental breakdown in 1866 and died in an asylum. He is the author of novels critical of the age, of which the first, *Moderne Titanen. Kleine Leute aus großer Zeit* (1850), is the most notable. Other titles are *Pfarröschen* (2 vols., 1852), *Carrière* (1853), and *Kleine und große Welt* (3 vols., 1853). His plays include *Die beiden Cagliostro* (1858), *Moritz von Sachsen* (1860), and *Luzifer und die Demagogen* (1861).

GLAESER, ERNST (Butzbach, Hesse, 1902–62, Mainz), a journalist, emigrated in 1933, and returned in 1939. During the 1939–45 War he edited the soldiers' paper *Adler im Süden*. He remained, however, an object of suspicion in some NSDAP (q.v.) quarters. The author of realistic novels, he is chiefly known for his more serious first work, *Jahrgang 1902* (1928), a psychological novel dealing with the effect of the 1914–18 War on the adolescent.

GLAREANUS, HENRICUS (Mollis, Switzerland, 1488–1563, Freiburg Breisgau), a Swiss neohumanist, musicologist, and geographer, whose real name was Heinrich Loris (his Latinized name refers to his origin in Canton Glarus). A follower of Reuchlin and Erasmus (qq.v.), he lectured in Basel and became a professor at Freiburg University (1529). He published a treatise on tonality (*Dodekachordon*, 1547), and wrote a Virgilian Latin poem in praise of Switzerland (*Helvetiae descriptio*, 1514).

Glasperlenspiel, Das, a Utopian novel by H. Hesse (q.v.), published in two volumes in 1943 It bears the sub-title *Versuch einer Lebensbeschreibung des Magister Ludi Josef samt Knechts hinterlassenen Schriften*. It is set in the 23rd c. In the esoteric state of Kastalien, which devotes itself to the cultivation of letters and the arts, a ruling élite in the form of a quasi-monastic order develops the 'glass bead game'. From a simple pastime it has evolved into a sophisticated intellectual procedure, which symbolizes the pure spirit of order and harmony.

As a boy, Josef Knecht is selected for future office, and, after proving his intellectual and spiritual quality by his mastery of the Glasperlenspiel, becomes as, Josephus III, Ludi Magister of the Order. This preliminary information is conveyed in a long exordium entitled *Versuch einer allgemeinverständlichen Einführung in seine* (sc. des Glasperlenspiels) *Geschichte*. Before his election, Knecht is sent on a diplomatic mission to the Benedictine Abbey of Mariafels to persuade Pater Jacobus of the Abbey to act as intermediary in an approach by Kastalien to the Vatican. This he accomplishes successfully, but, after his return and election, doubts which Pater Jacobus has expressed about the Kastalian way of life persist. The very refinement and delicately poised equilibrium between the active and the contemplative life, of which the Glasperlenspiel is the highest expression, seem to him to be precarious. In Kastalien's rarefied civilization personal vanity and self-absorption constitute a perpetual threat.

Lacking full belief in what he represents, Knecht resigns and rejoins the outer world. Through a chance meeting with a former friend in the Alps he meets a boy, Tito, whom he sees as his successor in Kastalien. Knecht's end is shrouded in mystery; but it is thought that he plunges with the boy into the icy water of an Alpine lake and dies from the shock. From this moment Tito has a consciousness of a future mission. Appended to Knecht's story are his posthumous papers, consisting of thirteen poems and three prose works grouped as *Die drei Lebensläufe*, and separately called *Der Regenmacher*, *Der Beichtvater*, and *Indischer Lebenslauf*.

Though this visionary and mystical book with its echoes of oriental religion and philosophy is not Hesse's last work, it is often regarded as the quintessence of his outlook, and the culmination of a life which sought spirituality while rejecting the available formulations of it.

GLASSBRENNER, ADOLF (Berlin, 1810–76, Berlin), a Radical journalist, was chiefly active in Mecklenburg (1841–50) and Berlin (1858–76). He was a notable observer of Berlin life, and a talented satirist and humorist. His numerous writings include the series *Berlin, wie es ist—und trinkt* (1832–50), *Aus den Papieren eines Hingerichteten* (1834), *Leben und Treiben der feinen Welt* (1834), the series *Buntes Berlin* (1837–41), the comic epic *Neuer Reineke Fuchs* (1846), the 20 vols. of the *Komischer Volkskalender* (1846–67), the comedy *Kaspar der Mensch* (1850), and the stories *Humoristische Table d'hôte* (1859). He used the pseudonym Adolf Brennglas.

Glaubenslos?, a novel by Marie von Ebner-Eschenbach (q.v.), published in 1893. It is set in a remote rural parish in the Austrian mountains. The young curate Leo, who has arrived eager to educate and civilize, is discouraged by the obduracy of the peasants, of whom even the best are narrow-minded, apathetic, and superstitious, and the worst savagely cruel and ruthlessly selfish. His faith wavers, but he is restored to his task of dispelling ignorance and awakening humanity by his first success—the consent of Bauer Schloger's widow to the marriage of her daughter Vroni to Seppl; only a farmworker, he

has kept the farm going during Schloger's long illness. There is a particularly sympathetic portrait of the weary but kind old priest of the parish, Pfarrer Thalberg.

GLEICH, JOSEPH ALOIS (Vienna, 1772–1841, Vienna), from 1790 to 1830 an Austrian civil servant, was a prolific purveyor of popular reading matter in the form of robber and ghost stories. It is said that he wrote more than 300 of these. Gleich was also a dramatist of almost equal fertility; his output of plays exceeded 200. These stage works were designed for the popular Viennese theatres (see VOLKSSTÜCK) and cover a great range, including Lokalpossen mit Gesang, Zauberstücke mit Gesang, and Besserungsstücke (q.v.), as well as serious plays. Among his best-known titles are the Lokalposse *Die Musikanten am hohen Markt* (1815) and the Zauberstück *Ydor, der Wanderer aus dem Wasserreich* (1820). Gleich befriended F. Raimund (q.v.), and his daughter Luise Gleich became Raimund's wife. Popular favour deserted Gleich in the early 1830s, and he died in poverty. *Ausgewählte Werke*, ed. O. Rommel, 16 vols., appeared 1910 ff.

Gleichen, Die, a play (*Schauspiel in sechs Aufzügen*) by L. J. (Achim) von Arnim (q.v.), published in 1819. Its plot is an elaborate variation of the story of Count von Gleichen (q.v.). The three principal participants shrink from the traditional bigamous marriage and resolve to live in friendship. But the Count distrusts the women; his wife, the Countess, leaves him and marries Ritter Plesse, and the oriental woman, Amra, marries the Countess's sister. The plot is complicated by the existence of a feud between the houses of Neu-Gleichen and Alt-Gleichen. The first two and some of the later scenes are written in prose, but the greater part of the play is in verse.

Gleichen, Graf von, a legend first recorded in the 16th c., according to which, in the 13th c., a Graf von Gleichen returned home to his first wife with a Turkish woman, whom he had liberated from the Saracens and taken as his second wife. He lived henceforth in a double marriage. Among writers who have treated the legend are L. J. von Arnim, E. Hardt, W. Schmidtbonn, W. von Schütz, and F. J. H. von Soden (qq.v.).

GLEICHEN-RUSSWURM, EMILIE VON (Weimar, 1804–72, Greifenstein), Schiller's youngest daughter, married H. A. von Gleichen-Rußwurm (1803–87), in 1828. She published sections of Schiller's correspondence, notably the letters exchanged by her parents (*Briefwechsel von Schiller und Lotte, 1788–89*, 1856).

GLEIM, JOHANN WILHELM LUDWIG (Ermsleben, 1719–1803, Halberstadt), was a student at Halle, where he met Uz and Götz (qq.v.). The three friends developed a stylized and completely unsensual poetry in praise of love, wine, and song modelled ostensibly on Anacreon (see ANAKREONTIK). Gleim's Anacreontic poems were published as *Versuch in scherzhaften Liedern* in 1744. The purely formal and social quality of this poetry is confirmed in Gleim's own words to E. von Kleist (q.v.): 'Wir, die wir von Wein und Liebe gesungen, aber wenig getrunken und wenig geliebt haben.'

In 1745 Gleim saw something of war when, as secretary, he accompanied Prince Wilhelm von Brandenburg-Schwedt in the Second Silesian War (see SCHLESISCHE KRIEGE). After a short period in the service of Prince Leopold von Dessau, he became in 1747 a canon of Halberstadt, where he spent the remainder of his life. His poetic reputation was greatly enhanced by the publication, during the Seven Years War (see SIEBENJÄHRIGER KRIEG), of a volume of patriotic Prussian poetry, *Kriegs- und Siegeslieder der Preußen von einem preußischen Grenadier* (1758), later reissued under the title *Preußische Kriegslieder*. The original title implies that the poems were composed by a grenadier (who appears to have enjoyed a classical education). Though the *Kriegslieder* were enthusiastically received, even by so good a judge as Lessing, their inflated style and obviously derivative character soon became evident. Gleim's significance as poet ends with this collection; forty years later he was best known for his hospitality and generosity to young writers, earning the affectionate name Vater Gleim. *Sämtliche Werke* were edited by W. Körte (8 vols., 1811–41).

GLIERS, DER VON, a minor Middle High German poet of the late 13th c., is usually identified with Wilhelm von Gliers, a nobleman of a region of the Jura which is now French territory; he is known to have been still living in 1308. Der von Gliers is the author of three Leiche (see LEICH) and of an allegorical poem.

GLOGER, GEORG (Habelschwerdt, Silesia, 1603–31, Leipzig), a friend of both M. Opitz and P. Fleming (qq.v.), studied medicine in Leipzig. He wrote Latin and German poetry, but only his Latin poems were published by him (*Decas latino-germanicorum Epigrammatum*, 1631).

Paul Fleming: Lateinische und deutsche Gedichte,

ed. J. M. Lappenberg (3 vols., 1863–5, repr. 2 vols., 1965), also contains poems by Gloger.

Glossen, in medieval German literature are explanations or translations of Latin words or sentences or, less frequently, translations or adaptations of classical glossaries. Entries above the line are described as 'interlinear', those made to one side as 'marginal' or Randglossen; when included in the line they are known as Textglossen. They were important in the early history of the written language as a means of extending the field of expression and achieving precision. For philological research they provide important evidence on the state and evolution of the German language in the early Middle Ages and indicate the Latin texts in common use. Among the best known are the *Abrogans*, the *Vocabularius Sti. Galli*, the *Kasseler Glossen*, and the *St. Pauler Glossen* (qq.v.).

Glück auf!, the German miners' greeting. It occurs in the traditional miners' folk-song 'Glück auf! Glück auf! Der Steiger kommt'.

Glück im Winkel, Das, a play (Schauspiel in drei Akten) by H. Sudermann (q.v.), published in 1896. Elisabeth Wiedemann, who has married in order to live quietly, is tempted by Baron von Röcknitz into a grand passion. Determined to possess her, Röcknitz blackmails her with the threat of denunciation to her husband. Elisabeth, in desperation, decides to kill herself. Wiedemann intervenes, and reveals that he knows and understands all. His support gives her the courage to resist Röcknitz and content herself with her quiet corner (the 'Winkel' of the title).

Glückhafft Schiff von Zürich, Das, see FISCHART, J.

Glückliche Fahrt, see MEERESSTILLE.

Glücksritter, Die, a Novelle by J. von Eichendorff (q.v.), published in 1841 in *Rheinisches Jahrbuch für Poesie und Kunst*. It is an improbable story set at the end of the Thirty Years War (see DREISSIGJÄHRIGER KRIEG). Full of misunderstandings, mistaken identities, and disguises, it ends with the timely defeat of a gang of paid-off soldiers attempting a raid on the castle of Count Gerold, and with the happy reunion of two lovers, of whom the girl is the daughter of one of the soldiers. Its main interest resides in prolific comic incident and in the evocation of the romantic forest atmosphere, in which Eichendorff excels.

Glück von Edenhall, Das, a ballad written by L. Uhland (q.v.) in 1834 and published in the same year. The young Cumbrian lord of Eden Hall recklessly breaks the goblet which guarantees the good fortune of his house, and immediately his castle is stormed.

GMELIN, OTTO (Karlsruhe, 1886–1940, Bensberg nr. Cologne), volunteered for war service in 1914, and from 1917 to 1936 was a schoolmaster at Solingen. From his forties on he devoted himself to authorship and lost favour with the National Socialist regime. He wrote historical novels of strongly nationalistic tendency in which the Middle Ages were seen in the perspective of the present. His plays include *Das Angesicht des Kaisers* (1927), *Der Ruf zum Reich* (1936, later retitled *Die Krone im Süden*), and *Das neue Reich* (1939).

GNAPHEUS, WILHELM (The Hague, 1493–1568, Norden, Hanover), a Dutch Protestant who, because of religious persecution, emigrated to North Germany, and became rector of the grammar school at Elbing. He is the author of a Latin play, *Acolastus,* based on the story of the prodigal son; it was performed by his pupils in 1529, and translated into German in 1530 by G. Binder (d. 1545) of Zurich. Gnapheus taught at Königsberg University from 1544 to 1547, when, as a member of the Reformed Church (q.v.), he was evicted through the intolerance of the Lutherans. He became a tutor in the service of Countess Anna of East Frisia.

GNEISENAU, AUGUST WILHELM ANTON NEITHARDT, GRAF VON (Schildau nr. Torgau, 1760–1831, Posen), a Prussian general, was at first an officer of the Markgraf of Ansbach-Bayreuth, in which capacity he saw service with the British in America in 1782–3, entering the Prussian army in 1786. He served as an infantry officer in the Polish campaigns of 1793 and 1795, and in the French war of 1806 was present at Saalfeld and, as a staff officer, at Jena; in 1807 he successfully defended Kolberg. Until Stein's dismissal he was active in furthering the reorganization of the Prussian army. In 1809 Gneisenau visited London and St. Petersburg, returning to Berlin in 1811 as an advocate of resistance to Napoleon. In 1813, 1814, and 1815 he was Blücher's chief of staff, and was the inspirer of the strategic and tactical dispositions which led to Blücher's successes, including the Prussian intervention at Waterloo, though the decision to intervene was Blücher's. (See REVOLUTIONSKRIEGE and NAPOLEONIC WARS.)

In 1816 Gneisenau retired, and in 1818 he became Governor of Berlin. In 1831 he was appointed to the command of a corps destined to subdue a Polish rising, but died of cholera. His original name was Neithardt; in 1782 he assumed the name of the Austrian noble family of Gneisenau in the erroneous belief that he was a descendant. His title Graf dates from 1814. Gneisenau is the subject of a play, *Neidhardt von Gneisenau,* by W. Goetz (q.v., 1925).

Gnomische Dichtung, see SPRUCH.

GÖCHHAUSEN, LUISE VON (Eisenach, 1752–1807, Weimar), lady-in-waiting to the dowager Duchess Anna Amalia (q.v.) at the court of Weimar. Fräulein von Göchhausen was a member of the court circle when Goethe (q.v.) went to Weimar, and in 1776 made a copy of Goethe's *Faust* MS., which was discovered among the papers of a descendant in 1887 and published by Erich Schmidt (q.v.). It is the work now known as *Urfaust* (q.v.). She received the nicknames Gnomide from her mis-shapen figure and Thusnelda for her courage.

GOCLENIUS, RODOLPHUS (Corbach, 1547–1628, Marburg), a professor at Marburg, wrote Latin poetry and plays. His name is a latinization of Rudolf Göckel.

Godwi oder Das steinerne Bild der Mutter, a novel by C. Brentano (q.v.), published in 2 vols. in 1801, and described on the title-page as *Ein verwilderter Roman von Maria,* a pseudonym used by Brentano for his early works. It is a capricious Romantic Bildungsroman (q.v.), the first volume of which is in the form of letters written by Godwi himself, his half-brother Römer, the depressive Werdo Senne, the painter Antonio Firmenti, and the three women to whom Godwi is attracted, Molly Hodefield, Joduno von Eichenwehen, and Ottilie Senne. This volume is chiefly occupied with Godwi's emotional response to the three women. The second volume (Zweiter und letzter Teil) purports to be edited by friends of Maria, who has died (*Herausgegeben von den Freunden des Verstorbenen, mit Nachrichten von seinem Leben, seinen Arbeiten und seinem Tode*). It narrates obliquely and spasmodically an intricate story of family relations in which Godwi and Römer are involved, and spans two generations. Its final episode is a beautifully told story of the corruption and redemption of the young girl Violette. Appended to the story is an ironical obituary of Maria, i.e. of Brentano by himself.

Godwi is perhaps the most wayward of all Romantic novels, deliberately eccentric in structure, and repeatedly cut across by destructive irony. Though baffling and perplexing as a whole, it has passages of brilliant perception and narrative episodes of singular beauty. It is strewn with many of Brentano's finest early poems, including the ballad 'Zu Bacharach am Rheine' (first line), usually known as *Die Lore Lay* (see LORELEI).

GOEBBELS, PAUL JOSEPH (Rheydt, 1897–1945, Berlin, shot on his own instructions), began his adult life as a journalist and after joining the NSDAP (q.v.) took over in 1925 as party business manager for the Northern Rhineland and Westphalia (Gau Nordrhein-Westfalen). Soon after this he became editor of G. Strasser's *Nationalsozialistische Briefe*. In 1926 he was transferred to Berlin-Brandenburg; from 1927 to 1935 he edited the weekly *Der Angriff*. He entered the Reichstag as a member for the NSDAP in 1928. From 1929 he was the party's propaganda chief, and in 1933 he was appointed propaganda minister (Minister für Volksaufklärung und Propaganda). In this capacity he was closely associated wth Hitler (q.v.) during the war years. Goebbels died with his wife and six children in Hitler's bunker when defeat had become a certainty. His publications *Kampf um Berlin* (1932) and *Vom Kaiserhof zur Reichskanzlei* (1934) were given great prominence.

GOECKINGK, LEOPOLD FRIEDRICH GÜNTHER VON (Gröningen nr. Halberstadt, 1748–1828, Wartenberg, Silesia), descended from landed gentry, was at Halle University, and became a Prussian civil servant, working successively in Halberstadt, Ellrich in the Harz, Magdeburg, and Wernigerode. In Halberstadt he formed a friendship with J. W. L. Gleim (q.v.) and was in contact with J. G. Jacobi and W. Heinse (qq.v.). From 1793 until his retirement in 1806 he served in Berlin. His patent of nobility dates from 1789. Goeckingk wrote agreeable and elegant poetry of playful rococo character, but genuine emotion is apparent in *Gedichte zweier Liebenden* (1777). He was also a skilled epigrammatist, publishing 2 vols. of *Sinngedichte* in 1772 and a further volume in 1778. His collected poems (*Gedichte*) were published in 3 vols., 1780–2, his *Prosaische Schriften* in 1784. See also GÖTTINGER MUSENALMANACH.

GOEDEKE, KARL (Celle, 1814–87, Göttingen), studied at Göttingen, but did not graduate. He became a journalist and author, publishing *Novellen* (1840) and writing political poetry, which has been praised but not published. Goedeke, though not a professional academic—he received an honorary doctorate at Tübingen in 1862, and was appointed to a professorship at Göttingen university at the age of 59 (1873)—became one of the best-known historians of

literature and bibliographers of the 19th c. Among his publications were *Elf Bücher deutscher Dichtung. Von Sebastian Brant bis auf die Gegenwart* (2 vols., 1849), *Deutsche Dichtung im Mittelalter* (1852–4), an edition of Pamphilius Gengenbach (q.v., 1856), *Übersicht der Geschichte der deutschen Dichtung*, Vol. 1 (not completed), *Goethe und Schiller* (1865), *Bibliothek deutscher Dichter des 16. Jahrhunderts* (in collaboration with J. Tittmann, 1866 ff.), *Bibliothek deutscher Dichter des 17. Jahrhunderts* (1869 ff.), and *Goethes Leben und Schriften* (1874).

He is the author of a monumental bibliographical work, which is often referred to simply by the name Goedeke. Its full title is *Grundriß zur Geschichte der deutschen Literatur* (3 vols., 1859–81, revised 2nd edn. 13 vols., 1884–1953). Volume 4 appeared in a third edition (5 Abteilungen, 1910–60). Goedeke covered German literature only up to 1830. A *Neue Folge* (1955 ff.) covers the period 1830–80.

GOELI, HERR, a Swiss poet of the 13th c., who has been identified as Diethelm Goeli, a knight of Basel, where he is recorded between 1254 and 1276. His four extant poems derive from the village poetry of NEIDHART von Reuental (q.v.).

GOERDELER, CARL FRIEDRICH (Schneidemühl, 1884–1945, Berlin), pursued a career in municipal administration, becoming Oberbürgermeister of Leipzig in 1931. He continued in office under the National Socialist regime until 1937, when he resigned in protest against an act of anti-Semitism. During the war he was one of the principal figures of the resistance (see RESISTANCE MOVEMENTS, 2), preparing a programme of reform intended to become operative after Hitler's fall. Goerdeler would probably have become chancellor if the revolt had proved successful. After the abortive attempt to assassinate Hitler (20 July 1944), a price was put upon his head. He was eventually tried by a People's Court and executed.

GOERING, REINHARD (Schloß Bieberstein, 1887–1936, nr. Jena, by suicide), studied medicine, but suffered from tuberculosis and was at a sanatorium for most of the 1914–18 War. An Expressionist dramatist, he created the impressive and reticent naval tragedy *Seeschlacht* (1917) set in a gun-turret of a battleship at the battle of Jutland (31 May 1916). Later plays include *Scapa Flow* (1918), on the episode of the scuttling of the interned German warships, and *Die Südpolexpedition des Kapitän Scott* (1930), on Captain R. F. Scott's expedition in the *Terra Nova*.

GOES, ALBRECHT (Langenbeutingen, Württemberg, 1908–), a pastor's son, became himself a Lutheran pastor in Württemberg. He was an army chaplain during the 1939–45 War, and afterwards continued as a parish priest until 1953. Goes is the author of poems which were collected in *Gedichte 1930 bis 1950* (1950), and comprised originally five volumes (*Der Hirte*, 1934; *Heimat ist gut*, 1935; *Lob des Lebens*, 1936; *Der Nachbar*, 1940; *Die Herberge*, 1947), and of a number of volumes of essays, including *Die guten Gefährten* (1942), *Von Mensch zu Mensch* (1949), *Freude am Gedicht* (1952), and *Dichter und Gedicht* (1966). Goes is best known for two tales of the war years, *Das Brandopfer* (q.v., 1954) and *Unruhige Nacht* (1950), a story of a chaplain ordered in 1943 to give the final spiritual consolation to a condemned deserter. The chaplain is deeply conscious of guilt, for he regards himself as a part of the procedure of war. The collections *Tagwerk. Prosa und Verse* and *Lichtschatten Du. Gedichte aus fünfzig Jahren* appeared in 1976 and 1978, and the volume *Das Brandopfer*, which includes *Das Löffelchen* (1965), in 1980.

GOETHE, CHRISTIANE (Weimar, 1765–1816, Weimar), was married to J. W. von Goethe (q.v.) in 1806. As Christiane Vulpius, a good-looking, warm-hearted young woman working at an artificial flower factory in Weimar, she attracted the attention of Goethe after his return from Italy in 1788. The sensual poetry of the *Römische Elegien* (q.v., 1790) relates in part to her. Goethe took her into his house, and she lived with him as his mistress and bore him five children, only one of whom, August (see GOETHE, JULIUS AUGUST WALTHER VON), survived infancy. When Weimar was occupied by French troops after the battle of Jena on 14 October 1806, her resolute conduct and presence of mind saved Goethe from possible injury or death. Five days later Goethe regularized the union in a service of marriage in the court chapel at Weimar. She died on 6 June 1816. Her brother C. A. Vulpius (q.v.) wrote popular novels.

GOETHE, CORNELIA (Frankfurt/Main, 1750–77, Emmendingen, Baden), was the younger sister of J. W. Goethe (q.v.), and the chief companion of his childhood. She was married in 1773 to J. G. Schlosser (q.v.), magistrate in Emmendingen, Baden. The marriage was not happy. Her early death was a great shock for her brother. In 1777 Cornelia and her husband befriended J. M. R. Lenz (q.v.).

GOETHE, JOHANN WOLFGANG (Frankfurt/Main, 1749–1832, Weimar), who was elevated to the nobility as J. W. von Goethe in 1782, was born into a patrician household. On the mother's side the grandparents, Textor by name, were an old-established and influential family; the father, Johann Kaspar Goethe, was well-to-do, though not of ancient Frankfurt descent. The Goethes lived in a large and comfortable house in the Hirschgasse (see GOETHE-HAUS). Johann Caspar was a man of artistic interests, serious nature, and limited imaginative capacity. Katharina Elisabeth, Goethe's mother, known later as Frau Aja, was in contrast lively, perceptive, and full of fantasy. Their respective influences were humorously summed up by Goethe in the lines:

> Vom Vater hab ich die Statur,
> Des Lebens ernstes Führen;
> Vom Mütterchen die Frohnatur
> Und Lust zum Fabulieren.

Wolfgang Goethe was educated at home. In 1759 the family life was disturbed by French occupation and billeting, but the boy derived artistic stimulus and understanding from painters working for the French commander, who lived in the Goethes' house. In 1762 Goethe was sent to Leipzig University, where he neglected his studies, acquired fashionable manners, educated his taste in painting, fell in love (with Käthchen Schönkopf, see SCHÖNKOPF, A. K.), and learned to write elegant erotic poetry. The short pastoral play *Die Laune des Verliebten* (q.v.) also dates from this time. Goethe fell seriously ill in 1768 and had to return to Frankfurt, where for a time his life was despaired of. During this illness he was influenced by a devout friend of his mother's, Susanna von Klettenberg (q.v.), who is recalled in Bk. 6 of *Wilhelm Meisters Lehrjahre* (q.v.); he also made alchemical studies which were to have a bearing on *Faust* (q.v.). The play *Die Mitschuldigen* (q.v.), which in spirit belongs to the Leipzig years, was written down in this period of recuperation and reflection in Frankfurt.

In March 1770 Goethe arrived in Strasburg in order to complete his university studies in law. The eighteen months which he spent there represent a period of rapid development of his mind and unfolding of his talent.

J. G. Herder (q.v.), who spent the autumn in Strasburg, opened Goethe's eyes to new sources of poetry in folk-song, and this new valuation of simplicity and spontaneity coincided with an idyllic love affair with Friederike Brion (q.v.), daughter of the pastor of Sesenheim. *Mailied* and *Willkommen und Abschied* (qq.v.), two poems in the new manner, springing from the heart, are both related to his love for Friederike. In the autumn of 1771 Goethe, having completed his studies, returned to Frankfurt, breaking off his relationship with Friederike; he worked for a time in his father's legal practice. Influenced by Herder's appreciation of Shakespeare's genius (see ZUM SCHÄKESPEARS TAG) he wrote at speed a pseudo-Shakespearian tragedy, *Geschichte Gottfriedens von Berlichingen*. This, however, was so ill received by Herder that Goethe put it aside, revising it two years later, and publishing it in 1773 as *Götz von Berlichingen mit der eisernen Hand* (q.v.). In the spring of 1772 his father sent him to Wetzlar to gain experience of the Reichskammergericht (q.v.); he did little work, but fell desperately in love with Charlotte Buff (q.v.), who was betrothed to another. In September Goethe tore himself away and returned to Frankfurt, still tormented by his love. He resolved his anguish at last in 1774 in *Die Leiden des jungen Werthers* (q.v.), the sensationally successful novel of the hypersensitive outsider, for whom the world has no place.

In Frankfurt he became the centre of the group who formed the inner circle of the Sturm und Drang (q.v.), F. M. Klinger, J. M. R. Lenz, H. L. Wagner, and F. (Maler) Müller (qq.v.). He wrote stormy poetry in free rhythms (see FREIE RHYTHMEN), such as *Wanderers Sturmlied*, *Prometheus*, and *An Schwager Kronos* (qq.v.), drafted the scenes of a Faust play, now called *Urfaust* (q.v.), and composed brilliant and high-spirited satires, such as *Götter, Helden und Wieland* (in prose) and *Satyros* (in verse, qq.v.). He wrote the domestic tragedy (see BÜRGERLICHES TRAUERSPIEL) *Clavigo* (q.v.) in a week in 1774, and *Stella* (q.v.), the play of a man between two women, in 1775. In the summer of 1775 he began *Egmont* (q.v.), and all the time he wrote poems of notable originality and beauty, such as *Neue Liebe, neues Leben* and *Herbstgefühl* (qq.v.). Goethe also wrote about this time *Erwin und Elmira* and *Claudine von Villa Bella* (qq.v.), light plays with music (see SINGSPIEL), of which he was to produce three more in the years 1779–82 (see JERY UND BÄTELY and FISCHERIN, DIE).

In 1775 Goethe fell in love with Lili Schönemann (q.v.), a patrician's daughter, to whom he became engaged; her personality charmed him, her social environment jarred on his unconventionality. In the summer he unsuccessfully sought escape from his love for her in a Swiss tour with young and over-enthusiastic friends, the Counts C. and F. L. Stolberg (qq.v.). In September the engagement with Lili was broken off, and for a variety of reasons, including a certain impatience with his Sturm und Drang cronies, Goethe gladly accepted an invitation to visit the young Duke of Saxe-Weimar (see KARL AUGUST). Though the first months at Weimar were boisterous ones (the 18-year-old Duke expected from his guest co-operation in pranks

and wild-cat expeditions), Goethe rapidly developed—partly under the influence of the serene-tempered lady-in-waiting Frau von Stein (q.v.)—into a mature and balanced man with a gift for administration. In June 1776 he was appointed to the Duke's cabinet, and in the following ten years he took on many of the tasks of government, and was in effect the Duke's right-hand man. Goethe himself paid tribute to the calming influence which Frau von Stein exerted upon him in the difficulties and harassments of these busy years. It is often said that his poetry suffered under the distractions of governmental work, and he himself frequently deplored the calls upon his time; nevertheless his output was not inconsiderable, and its quality was of the highest. Poems such as *Harzreise im Winter, An den Mond, Gesang der Geister über den Wassern* (qq.v.), *Wanderers Nachtlied* (q.v., 'Der du von dem Himmel bist', and see ÜBER ALLEN GIPFELN), and some of the wonderful songs in *Wilhelm Meisters Lehrjahre* (q.v.) were written in these Weimar years. Numerous poems and songs were set to music by composers ranging from Mozart and F. Schubert (qq.v.) to Othmar Schoeck (1886–1957). In 1779 Goethe began the first version of *Iphigenie auf Tauris*; in the following year he started *Torquato Tasso* (qq.v., published in 1787 and 1790 respectively). In approximately eight years (from 1777) he completed the six substantial books of *Wilhelm Meisters Theatralische Sendung* (q.v., unpublished in Goethe's lifetime). His interest had swung away from the tempestuous energy of the Sturm und Drang and was moving towards the discipline, serenity, and balance of classical art.

The combination of a weariness with administrative tasks and a desire to study classicism at its source prompted Goethe to plan a temporary absence from Weimar. In September 1786 he set out for Italy, where he spent nearly two years, during which he experienced a sense not only of relaxation but of renewal. Italy released the springs of writing. *Iphigenie auf Tauris* was remodelled in verse, *Egmont* (1788) and *Torquato Tasso* (1790) completed. Two scenes of *Faust*, which had lain idle for ten years, were conceived and written, and the work came to provisional publication as *Faust. Ein Fragment* (1790). The Italian journey enabled Goethe above all to make a complete reorientation, to find, as it seemed to him, in harmony and balance the full and complete expression of his personality.

Goethe returned to Weimar in 1788, and at once found himself at a distance from his former friends, who did not comprehend the change in him. The estrangement, which extended to Frau von Stein, was augmented when Goethe took into his house as his mistress Christiane Vulpius (see GOETHE, CHRISTIANE), a handsome young woman of lower rank. Goethe sought and obtained release from his administrative responsibilities, and devoted himself to antique and scientific studies, and to the organization of his own collections of *objets d'art* and scientific specimens, which he substantially expanded as the years passed. In 1791 he nevertheless accepted appointment as director of the newly opened Weimar theatre, which he retained until 1817. The poetry of these years was almost exclusively classical, and principally in elegiac form. The elegy *Alexis und Dora* (q.v.) was written in 1789, and in 1795 the once notorious *Römische Elegien* (q.v.), neo-classical erotic poems as tender as they are sensual, appeared in Schiller's *Die Horen* (q.v.).

Goethe's first published testimonies of his scientific interests, *Versuch, die Metamorphose der Pflanzen zu erklären* and *Beiträge zur Optik*, appeared in 1790 and 1791 respectively. In 1790 he paid a second visit to Italy, going to Venice in connection with a visit of the Duchess Anna Amalia (q.v.), but the old magic of the south refused to rekindle on this semi-official occasion. To Goethe, in his dual absorption in the art of the ancient world and in science, the French Revolution (q.v.) seemed a deplorable irrelevance, but he found himself involved in its consequences when he was required to accompany Duke Karl August on the invasion of France by imperial troops in the autumn of 1792; he was present at the siege of Mainz in the following year. His autobiographical accounts of these events, *Campagne in Frankreich. 1792* and *Die Belagerung von Mainz* (qq.v.) were not written until 1820–1. Goethe's immediate response to affairs in France is incorporated in three comedies, which are among his weakest works: *Der Groß-Cophta* (1791), *Der Bürgergeneral* (1793), and the unfinished *Die Aufgeregten* (qq.v., 1791, published 1817).

An important new work of fiction, *Unterhaltungen deutscher Ausgewanderten* (q.v.) was published in 1795, and in 1795–6 there appeared the definitive revised version of *Wilhelm Meister* as the novel *Wilhelm Meisters Lehrjahre* (q.v.); concerned with the education of personality (see BILDUNGSROMAN), it was to fascinate the next generation (see ROMANTIK).

Schiller lived in Weimar from 1787 to 1789, when he moved to Jena, but Goethe's relationship with him remained cool until 1794, when overtures by Schiller penetrated Goethe's reserve, and inaugurated a lasting intellectual friendship; it was based on their common classicism and on their conviction of the central function of art in human affairs. Goethe's classical poetry had on the whole repelled the public, but his idyllically treated epic poem

Hermann und Dorothea (q.v., 1797) was favourably received. Both Goethe and Schiller felt a resistance in their contemporaries to their classical doctrines; in 1796 they carried the war into the hostile camp by publishing, in Schiller's *Musenalmanach*, the *Xenien* (q.v.), satirical epigrams in which they ridiculed the opposition. The *Musenalmanach* of the following year contained, as a positive sequel to the negative *Xenien*, ballads by both authors; Goethe's contribution included *Der Zauberlehrling* and *Die Braut von Korinth* (qq.v.). It was Schiller who, in 1797, succeeded in stimulating in Goethe a renewed interest in *Faust*, which was to preoccupy him intermittently for the next nine years, and again at intervals in later life (Part One appeared in 1808, Part Two in 1832). A classicistic journal for the arts, *Die Propyläen* (q.v.), which Goethe launched in 1798, did not succeed, and ceased publication after two years. The next years saw Goethe's literary classicism in its most uncompromising form with such works as the unfinished epic *Achilleis* (q.v.), *Die natürliche Tochter* (q.v., 1804), the first play of an uncompleted trilogy, and the Festspiel *Pandora* (q.v.). The death of Schiller in 1805 coincided with the end of this classical phase.

When in 1806 war broke out between France and Prussia, the decisive battle was fought at Jena (see NAPOLEONIC WARS), and French soldiers, occupying Weimar, broke into Goethe's house. Goethe believed that Christiane saved his life from these marauders, and a few days later he had their long-standing liaison legitimized in marriage. His new marital state did not inhibit a powerful attraction to Wilhelmine (Minchen) Herzlieb (q.v.) in the following year, which flowered for a few weeks before it was repressed. In 1809 appeared the subtle and problematical novel *Die Wahlverwandtschaften* (q.v.), treating the interrelations of two couples. It was in 1808 that Goethe's encounters with Napoleon took place at Erfurt and Weimar. He recognized a daemonic power more readily in Napoleon than in Beethoven (q.v.), whom he met without enthusiasm four years later. These years were largely spent in scientific work, and in writing his autobiography. His theory of light and colour, which ran counter to Newton's, was incorporated in *Zur Farbenlehre* (1810). The first three volumes of the autobiography *Dichtung und Wahrheit* (q.v., full title *Aus meinem Leben. Dichtung und Wahrheit*) appeared in 1811–14. The fourth and last volume was delayed until 1832 after Goethe's death. The whole autobiography covers the first twenty-six years of his life, ending with his removal to Weimar. His record of his Italian journey (*Italienische Reise*) appeared in 1816–17.

As Napoleon's grasp on Europe relaxed,

Goethe spent much time in Frankfurt, cultivating the friendship of a young married woman, Marianne von Willemer (q.v.), the most gifted and intelligent of the many women to whom he was attracted. His collection of pseudo-oriental poetry, *West-östlicher Divan* (q.v., 1819), is closely associated with her, and even contains several unacknowledged poems by her hand. The relaxed geniality of this volume is a symptom of Goethe's sense of renewal and serenity. In 1821 he took his farewell of Wilhelm Meister with the first publication of the desultory novel *Wilhelm Meisters Wanderjahre oder Die Entsagenden* (q.v.). The summers of 1821 to 1823 he spent in Marienbad, and here his last passion flamed up, the object of which was a 17-year-old girl, Ulrike von Levetzow (q.v.). The pain of renunciation in which this attraction ended is expressed in the three poems *Trilogie der Leidenschaft* (q.v.). Among his last works was *Novelle* (q.v., 1828), which had its genesis thirty years earlier. Goethe's old age was spent in increasing loneliness; Christiane had predeceased him in 1816, Frau von Stein died in 1827, Duke Karl August in 1828, and Goethe's son August in 1830. Goethe worked away at the last volume of *Dichtung und Wahrheit* and at *Faust II*, which he completed not long before he died on 22 March 1832.

The span of Goethe's eighty-two years covers a critical period in the development of the modern world. The Bastille was stormed when he was 39 and three years later his sense of historical awareness led him to say to his companions at Valmy 'Von hier und heute geht eine neue Epoche der Weltgeschichte aus und Ihr könnt sagen, Ihr seid dabei gewesen'. Rooted in the old world, he met the new with dispassionate understanding. He was a man of remarkable range. Fiery, energetic, and impatient in youth, he grew into a shrewd, resourceful, and tenacious administrator, and stylized himself in old age into a remote oracular figure of Olympian stature. Best known as a man of letters, he nevertheless also had a distinct talent for drawing, was interested in acting, and became a successful theatre director. His knowledge of antique art was comprehensive and profound. In science he concerned himself with biology, both in detail and in general evolutionary concepts, with optics, and with mineralogy. And his practical pursuits extended to mining, economics, architecture, horticulture, and landscape gardening. He is sometimes referred to as the last universal man. And he has the distinction of being perhaps the most fully documented creative artist.

For a man who is universally regarded as a writer of the first rank, Goethe's œuvre is surprisingly fragmented. Its diversity is remarkable, both in style (ranging from Sturm und Drang subjectivism to the conscious harmony of classicism) and in form, which includes lyric, epic, and ballad poetry, drama, novels, shorter tales, and autobiographical works. Many of these writings are imperfect, either fragments or works which seem to have a flaw of development, a fracture or rift. Perfection is only achieved in his lyric poetry and in a handful of works chiefly written in his classical period (*Iphigenie auf Tauris*, *Hermann und Dorothea*, and perhaps also the novel *Die Wahlverwandtschaften*). Fragmentariness is, however, in Goethe's work no defect; it is the essence of his literary genius. He himself said 'Alle meine Werke sind Bruchstücke einer großen Konfession'. It is the immediacy of Goethe's works which is their special characteristic. They reflect facets of an extraordinarily rich, multiple, Protean personality, in all its changing moods and varied experiences. Not surprisingly, Goethe is one of the most original and powerful German lyric poets, but the immense panorama of *Faust*, reflecting the developing vision of a lifetime, with its comedy and tragedy, pathos, wit, and satire, is a work of inexhaustible ambiguity and magical poetry.

Following the custom of his age, Goethe from time to time published collected editions of his works. These comprise *Schriften* (8 vols., 1787–90), *Neue Schriften* (7 vols., 1792–1800), *Werke* (13 vols., 1806–10, the first to be produced by J. F. Cotta, q.v.), and *Werke* (20 vols., 1815–19, expanded to 26 vols., 1820–2). The definitive edition begun in Goethe's lifetime is *Goethes Werke. Vollständige Ausgabe letzter Hand* (60 vols., 1827–42, see AUSGABE LETZTER HAND).

The comprehensive *Weimarer Ausgabe*, also called *Sophienausgabe* after its patroness, the Grand Duchess Sophie of Saxe-Weimar, appeared in 143 volumes (1887–1920). Other notable editions include the *Jubiläums-Ausgabe* (ed. E. von der Hellen, 41 vols., 1902–12), the *Propyläen-Ausgabe* (49 vols., 1909–32, chronological), the *Hamburger Ausgabe* (ed. E. Trunz, 14 vols. with Sachregister, 1948–64, widely used because of its critical apparatus), the *Ausgabe der Deutschen Akademie der Wissenschaften*, 1952 ff., and the *(Artemis) Gedenkausgabe der Werke, Briefe und Gespräche. 28. August 1949* (27 vols., incl. 3 suppl. vols.), ed. E. Beutler, 1948–71.

Selections of Goethe's letters include *Goethe-Briefe* (ed. P. Stein, 8 vols., 1924) and 4 vols. supplementary to the *Hamburger Ausgabe*.

See also ECKERMANN, J. P. and GOETHE-GESELLSCHAFT.

GOETHE, JULIUS AUGUST WALTHER VON (Weimar, 1789–1830, Rome), usually referred to as August von Goethe, was Goethe's only surviving child by Christiane Vulpius (see GOETHE, CHRISTIANE), whom the poet married in 1806 when August was 16. The boy was educated by Goethe himself and F. W. Riemer (q.v.). Though not without talent, he was inevitably overshadowed and dominated by his father. In 1813, when August joined the volunteers who were to march against Napoleon, Goethe, who employed him as a secretary, insisted upon his withdrawal, exposing the young man to derisive comment. In 1817 August married Ottilie von Pogwisch (see GOETHE, OTTILIE VON), even here not escaping from his father's shadow, for the bride was more attached to the father than to the son. The marriage was unhappy, and August consoled himself with other partners and with drink. In 1830, in order to escape from the unsatisfactory tenor of his life, he undertook a journey to Italy and died in Rome of scarlet fever. He was a contributor to his wife's periodical *Chaos*.

GOETHE, OTTILIE VON (Danzig, 1796–1872, Weimar), originally a baroness (Freiin) von Pogwisch, married Goethe's son August (see GOETHE, J. A. W. VON) in 1817. The marriage, which was not happy even at its beginning, was favoured by Goethe senior, whom Ottilie greatly admired. From 1829 to 1831 she edited a private weekly entitled *Chaos*, to which Goethe and his son, as well as she herself, contributed. Widowed in 1830, she removed after her father-in-law's death to Vienna.

Goethe-Gesellschaft, founded in 1885 for the furtherance of scholarship connected with Goethe. Its headquarters is in Weimar, and its membership and direction are not subject to division between East and West Germany. The *Schriften der Goethe-Gesellschaft* are published at intervals (1885 ff.).

The *Goethe-Jahrbuch*, edited by L. Geiger, was begun independently of the Society in 1880. From vol. 7 (1886) it was an official publication of the Goethe-Gesellschaft, and it continued under its original title until 1913 (34 vols.). It was succeeded by the *Jahrbuch der Goethe-Gesellschaft*, vols. 1–21 (1914–35). This was continued as *Goethe. Vierteljahresschrift der Goethe-Gesellschaft*, vols. 1–9 (1937–45). With vol. 10 it reverted in 1947 to annual publication as *Goethe. Neue Folge des Jahrbuchs der Goethe-Gesellschaft*.

Goethe-Haus, name given to two houses associated with J. W. Goethe (q.v.), one in Frankfurt and one in Weimar. In the Goethe-Haus in Frankfurt (Grosser Hirschgraben 23) the poet was born and had his home from 1749 to 1775. The house, originally two houses built *c.* 1590,

was acquired by Goethe's grandmother in 1733, and was rebuilt and extended upwards by his father in 1755. In 1795 it was sold by Goethe's mother, and in the 19th c. underwent various alterations. It was acquired in 1863 for the Freies Deutsches Hochstift, which restored the building to its condition in 1755 and opened it as a museum. The house was in great part destroyed in 1944, and has since been carefully rebuilt (so far as possible with use of the original materials) and fully restored.

The Goethe-Haus in Weimar, situated on the Frauenplan, was in Goethe's occupation from 1782 until his death fifty years later. Built in 1710, it was leased by Goethe in 1782. He was absent in Italy from 1786 to 1788, and on his return he occupied with Christiane Vulpius (see GOETHE, CHRISTIANE) a smaller house outside the Frauentor. In 1792 the house on the Frauenplan was bought for Goethe by Karl August, and, after alterations, he moved in finally in the same year. It remained in the family until 1885, when it was bequeathed to the state.

Goethe-Jahrbuch, see GOETHE-GESELLSCHAFT.

Goethes Briefwechsel mit einem Kinde, a pseudo-biographical work by Bettina von Arnim (q.v.), published in 1835. It is based primarily on her contacts with Goethe's mother in 1806, and with J. W. Goethe (q.v.) in 1807, 1810, and 1811. Bettina has often been attacked for her readiness to alter Goethe's letters or to invent new ones, and for her insistence on seeing herself as Goethe's Muse, appropriating to herself poems addressed to other women. But the work should properly be seen as an imaginative novel which brings Goethe back to life in essence if not in detail.

Goethezeit, expression used in histories of literature, usually to cover the period of Goethe's literary activity *c.* 1770–1830. It owed its vogue to H. A. Korff (*Geist der Goethezeit*, 4 vols., 1923–50). See also IRRATIONALISMUS.

GOETZ, CURT (Mainz, 1888–1960, nr. St. Gall, Switzerland), was an actor in Rostock, Nürnberg, and Berlin. He was a successful author of witty comedies with a leaning towards the grotesque; these include *Menagerie* (1920), *Ingeborg* and *Nachtbeleuchtung* (both 1921), *Der Lampenschirm* (1923), *Die tote Tante* (1924) on which the one-act comedy *Das Haus in Montevideo* (1953) is based, *Hokus Pokus* (1928), *Der Lügner und die Nonne* (1929), and *Dr. med. Hiob Prätorius* (1934). Goetz toured with several of his plays, acting the lead himself. In 1939 he emigrated to Hollywood, where several of his comedies were turned into films. He returned to Europe in 1945, living in Switzerland by Lake Thun. Here he wrote the very successful novel *Die Tote von Beverly Hills* (1951) and the one-act comedy *Miniaturen* (1958). He entitled his autobiography *Die Memoiren des Peterhans von Binningen* (1960); interrupted by his death, it was completed by his widow, the former actress Valérie von Martens, with two further volumes, *Die Verwandlung des Peterhans von Binningen* (1962) and *Wir wandern, wir wandern . . .* (1963). *Sämtliche Bühnenwerke* (1 vol.) appeared in 1963.

GOETZ, WOLFGANG (Leipzig, 1885–1955, Berlin), studied at Leipzig University, and from 1920 to 1929 was a civil servant in the film censorship office (Filmprüfstelle). He was esteemed by the National Socialist regime, primarily because of the pronounced nationalistic tendency of his dramatic work; from 1936 to 1940 he was president of the Gesellschaft für Theatergeschichte. After the 1939–45 War he was appointed editor of the *Berliner Hefte für geistiges Leben* (1946–9). His plays are mainly concerned with Prussian history; the most notable title is *Neidhardt von Gneisenau* (1925, see GNEISENAU, N. VON), which portrays Gneisenau as an irascible, repressed military genius who overcomes obstruction, engineers the Prussian victory at Waterloo, and finally becomes a true German (as opposed to Prussian) nationalist. The play *Der Ministerpräsident* (1936) lauds Bismarck. Other plays, written between the wars, include *Robert Emmett* (1928), *Kavaliere* (1930), and *Kampf ums Reich* (1939). Goetz's narrative work is mainly concerned with his own times, though *Der Mönch von Heisterbach* (1935) deals with a historical legend (see CÄSARIUS VON HEISTERBACH). The contemporary, often satirical, Novellen include *Die Reise ins Blaue* (1920), *Franz Hofdemel* (1932), *Der Herr Geheime Rat* (a collection, 1941), *Das Wiegenlied* (1946), and *Das Glück sitzt in der nächsten Ecke* (posth., 1958).

GOEZE, JOHANN MELCHIOR (Halberstadt, 1717–86, Hamburg), was Lessing's most formidable opponent in the religious controversy of 1778. Appointed chief pastor to St. Katharine's Church in Hamburg in 1755, Goeze had acquired a reputation for the energetic and vigorous defence of religious orthodoxy even before he came into conflict with Lessing. His name has been chiefly preserved in Lessing's virulent pamphlets *Anti-Goeze* (q.v., 1778). He also conducted a campaign against the theatre as an immoral institution. The irascible Goeze was nevertheless a serious scholar who wrote an important *Versuch einer Historie der gedruckten niedersächsischen Bibel* (1775).

Goeze is said to have been the model for the intolerant Superintendent Dr. Stauzius in F. Nicolai's novel, *Das Leben und die Meinungen des Herrn Magister Sebaldus Nothanker* (q.v.).

Goldemar, a Middle High German poem written by ALBRECHT von Kemnaten (q.v.).

Goldene Bulle, document promulgated in 1356 under the Emperor Karl IV (q.v.) by which the procedure for the election of the German King (see DEUTSCHER KÖNIG) was regulated. The number of electoral princes (Kurfürsten, q.v.) was fixed at seven. They were, in the order in which they recorded their votes: 1. Archbishop of Trier; 2. Archbishop of Cologne; 3. King of Bohemia; 4. Count Palatine (Pfalzgraf bei Rhein); 5. Duke of Saxony-Wittenberg; 6. Margrave of Brandenburg; 7. Archbishop of Mainz. The last named was convener. The electors received important commercial and political privileges and their jurisdiction was extended. At the same time, in order to avoid future disputes over inheritance, their territories were declared indivisible. A clause setting up an imperial council consisting of the emperor and the Electors, meeting annually, proved to be a dead letter. The Bull was designed to counter papal interference in the affairs of the Empire.

Goldene Horn von Gallehus, Das, which was discovered at Gallehus in Schleswig, bears an early Germanic runic inscription of the 5th c.:

Ek Hlewagastiz Holtijaz　　horna tawido
(Ich Hlewagast der Holting　machte das Horn).

See RUNEN.

Goldener Sonntag, term used with two totally different meanings. In the calendar of the Church it is applied to Trinity Sunday. In trade it denoted the Sunday before Christmas, when shops remained open.

Goldene Schmiede, Die, see KONRAD VON WÜRZBURG.

Goldene Spiegel, Der, oder Die Könige von Scheschian, political fantasy with a pseudo-oriental setting by C. M. Wieland (q.v.), first published in 1772. It bears the description *Eine wahre Geschichte aus dem Scheschianischen übersetzt* and is in two parts. The essence of the book is a series of episodes from the history of the kingdom of Scheschian illustrating features of government. The reign of Azor is that of a well-intentioned but indolent and pleasure-loving monarch, who leaves government to his unprincipled agents. Isfandiar reigns as an arbitrary tyrant. But Tifan governs in the interest of his people, observing laws which are secure against arbitrary change, and achieving the prosperity and happiness of the land. It is the 18th c. ideal of the enlightened despot, ruling in accordance with humanity and reason. Wieland sets his political parables in a frame (see RAHMEN); for they are told by the minister Danischmend to the Sultan Schach-Gebal in order to dispose His Serenity to sleep. The good impression made by this political novel secured Wieland's appointment in 1772 as tutor to the princes of Weimar. A sequel to *Der goldene Spiegel, Geschichte des weisen Danischmend und der drei Kalender*, appeared in 1775. It is marked by political scepticism.

Goldene Vließ, Das, (in modernized spelling often 'Vlies'), a verse tragedy in three parts (*Dramatisches Gedicht in drei Abteilungen*) by F. Grillparzer (q.v.). The trilogy consists of *Der Gastfreund. Trauerspiel in einem Aufzuge, Die Argonauten. Trauerspiel in vier Aufzügen, Medea. Trauerspiel in fünf Aufzügen*. Pt. One, the first two and most of the third act of Pt. Two were written by Grillparzer in 1818 before personal distress at the death of his mother interrupted the work for a year. Act Four and Pt. Three were written late in 1819 and completed on 27 January 1820. Blank verse predominates, but there are occasional patches of free verse. The first performance of the trilogy took place on 26 and 27 March 1821 in the Burgtheater, Vienna. In his drafts for a preface for its publication in 1822 Grillparzer alludes to the imperfections of the work and to the difficulties he encountered during its conception and states his refusal to defend the work in literary journals against adverse criticism. His extensive treatment and his psychological motivation of the legend of the child murderess Medea of Colchis and the Greek Jason come nearer to perfection than the other modern versions deriving from Euripides and from Seneca's *saevum monstrum*. Unknown to Grillparzer, Schiller had been attracted by the material and had stated that the story needed to be treated in its entirety in a cycle of plays.

Pt. One presents, in a one-act tragedy, the arrival of Phryxus, the Gastfreund, in Colchis, bearing the golden fleece. Tempted to gain possession of it, King Aietes ignores the rights of the guest which it is his duty, as host, to honour, and murders Phryxus when the Greek refuses to surrender the fleece. Phryxus dies, cursing the new unlawful owner of the fleece. Aietes has compelled his young daughter Medea, against her own better judgement, to assist him in the crime. The first part prepares for the action of the four-act tragedy, in which Jason and his followers, the Argonauts, claim the fleece. They succeed because Medea's love for Jason causes her to betray her father. As she leaves with Jason for Greece, the curse of vengeance is partly realized; Aietes dies, cursing his daughter, while her brother Absyrtus plunges to his death in the sea in order not to survive the disgrace.

Omitting the intervening years, the third part presents the events directly leading to Medea's murder of her two children. The breakdown of her marriage reaches its climax when Jason, having found asylum at King Kreon's court at Corinth, is outlawed for the murder of his uncle, King Pelias, which had caused his flight from Iolcus. Hoping that Jason will marry his daughter Kreusa, Kreon decrees under the pretence of her sole responsibility that only Medea should be banned. Granting her request to take one of the children with her into exile, he leaves the decision which of them is to go to the children themselves. When both refuse to follow their mother, Medea murders them and Kreusa as well. Although Jason has now to follow Medea into exile, Medea pursues her own separate path in order to restore the fleece to Phoebus's temple in Delphi from whence it came. The open ending expresses utter disillusionment with man's hopes and aspirations in this world ('der Erde Glück' and 'der Erde Ruhm'), while upholding Medea's moral stature.

Goldne Topf, Der, a story written by E. T. A. Hoffmann (q.v.) in Dresden in 1813, first published in 1814 in *Fantasiestücke in Callots Manier*, vol. 3, and described as *Ein Märchen aus der neuen Zeit*. It presents a battle royal of two worlds in which a realistic philistine sphere allies itself with demonic powers against a realm of poetry and fantasy; the battle ends in the victory of the poetic. The story comprises twelve chapters termed *Vigilien*. The student Anselmus, incurably clumsy in ordinary affairs, knocks over a market-woman's basket of apples and arouses the hostility of the owner, a sinister power of darkness referred to as the Äpfelweib. While smoking his pipe by the riverside (the scene is the Elbe near Dresden), Anselmus becomes aware of three attractive snakes in the tree over his head, with one of which, by name Serpentina, he falls in love. His prosaic burgher friends Konrektor Paulmann and Registrator Heerbrand represent the everyday world, in which Anselmus still has roots. Paulmann's daughter Veronika is determined to marry Anselmus when he has made a success of his career by becoming Hofrat. Veronika invokes the help of the Äpfelweib, and her rival Serpentina, who truly loves Anselmus, is aided by her own father Archivarius Lindthorst, apparently a man of this world, but in reality a prince of spirits and a salamander.

At one point, under the influence of a conbibation with the Konrektor and his friends and family, Anselmus wavers, and Lindthorst, as a punishment, seals him in a bottle. From this point of vantage he watches the battle between Salamander Lindthorst and the Äpfelweib, which ends in the latter's death. Anselmus is released from the bottle and wafted off to Atlantis with the lovely Serpentina; Veronika marries Heerbrand, who in the meantime has become Hofrat. The title refers to a magical golden vessel in Lindthorst's possession, a symbol of imagination, which is destined to be a dowry for his daughters. The story is as remarkable for its broad and infectious comedy as for its flights of fantasy.

GOLL, IVAN or YVAN, pseudonym of Isaac Lang (St. Dié, Alsace, 1891–1950, Paris), a Jew bilingual in French and German, and a German citizen. A pacifist, he avoided military service in 1915 by moving to Switzerland, where he associated with Romain Rolland in Geneva. From 1919 to 1939 he lived in Paris, then in New York until his return to France in 1947. Though he is the author of novels (*Lucifer vieillissant*, in French, 1934, among others) and plays (e.g. the grotesque *Methusalem oder Der ewige Bürger*, 1922), he is primarily a lyric poet, writing successfully in German and French, and attempting, inadequately, to write poems in English (*Fruit from Saturn*, 1946). His early poetry was Expressionistic (*Films*, 1914; *Der Panamakanal*, 1914; *Der neue Orpheus*, 1918). In association with James Joyce, Apollinaire, Le Breton, and Eluard, he turned to Surrealism, writing poetry containing highly original imagistic elements. His first acknowledged masterpieces are the *Poèmes d'amour* written with his wife Claire, of which the *Chansons malaises* (not published until 1967 in their German version, *Malaiische Liebeslieder*) form the most arresting section. *La Chanson de Jean sans Terre*, a series of ballads, appeared between 1936 and 1944. Some of Goll's most impressive German poems, including many of those which have appeared in anthologies, are in the collection *Traumkraut* (1951) in which biblical motifs and mineral, metallurgical, and chemical images are used to express the bitterness and disillusionment of Goll's life. The collection *Dichtungen, Lyrik, Prosa, Dramen*, ed. Claire Goll (1960) includes German translations of texts written in French. *Oeuvres* (2 vols.), ed. C. Goll and F. X. Jaujard, appeared 1968–70.

Golo und Genoveva, a play (*Ein Schauspiel in fünf Aufzügen*) by F. Müller (q.v.), first published in *Werke* (ed. L. Tieck, q.v.) in 1811. Two scenes were published separately in 1776, *Genoveva im Turm* (IV, 6) and *Die Pfalzgräfin Genoveva* (V, 11).

A large work, it is approximately twice as long as normal stage plays. In addition to three songs, it contains two short passages of verse. Its fragmented structure (52 scenes in five acts) is typical of the Sturm und Drang (q.v.). Genoveva, left at home by her husband Siegfried, becomes the victim of a diabolical intrigue between Golo, who desires her, and Mathilde, a Lady Macbeth-like character. Genoveva withstands Golo, and is sent into the forest to be murdered; but the would-be assassins relent and release her. She survives in the forest, and is reunited four years later with Siegfried. Golo dies by throwing himself upon the avenger's sword.

GOLTZ, BOGUMIL (Warsaw, 1801–70, Thorn), studied at Breslau University, and then became a farmer, but gave up farming for writing, making his home at Thorn in 1847. His works, mainly social, psychological, and ethnic studies and travel books, include *Buch der Kindheit* (1847), *Ein Jugendleben, biographisches Idyll aus Westpreußen* (3 vols., 1852), *Ein Kleinstädter in Ägypten* (1853), *Der Mensch und die Leute* (1858), *Zur Charakteristik und Naturgeschichte der Frauen* (1859), *Zur Physiognomie und Charakteristik des Volkes* (1859), *Die Deutschen, ethnographische Studien* (2 vols., 1860), *Die Bildung und das Gebildeten* (2 vols., 1864), and *Die Weltklugheit und die Lebensweisheit* (2 vols., 1869). He was a shrewd observer and ironist.

GOLTZ, JOACHIM, FREIHERR VON DER (Westerburg im Westerwald, 1892–1972), became a civil servant in 1914, but was called up in August as a reserve officer and saw three years' front-line service. After the war he lived in Baden, devoting himself to writing. He was for a time successful as a dramatist. *Vater und Sohn* (1921) presents the conflict between Friedrich Wilhelm I (q.v.) and his son, the future Friedrich II of Prussia (q.v.). Other plays include *Die Leuchtkugel* (1920), *Der Rattenfänger von Hameln* (1932), and the comedies *Das Meistermädchen* (1938) and *Peter Hunold* (1949). He published a war novel, *Der Baum von Cléry* (1934), which deals with the western front. His fiction includes *Der Wein ist wahr* (1928), *Von mancherlei Hölle und Seligkeit* (1936), *Die Ergriffenen* (1948). His poetry is published under the titles *Deutsche Sonette* (1916), *Ewig wiederkehrt die Freude* (1942), and *Mich hält so viel mit Liebesbanden* (1951).

GOMRINGER, EUGEN (Bolivia, 1925–), an exponent of concrete poetry (see KONKRETE

POESIE), expressed his views on this form of experimental poetry in *vom vers zur konstellation. zweck und form einer neuen dichtung* (1955), publishing *konstellationen* (1953), *33 konstellationen mit 6 konstellationen von Max Bill* (1960), and *das stundenbuch* (1965). Gomringer worked for a time at the Academy of Art (Hochschule für Gestaltung), Ulm, under Max Bill (1908–). *Josef Albers. Das Werk des Malers und Bauhausmeisters als Beitrag zur visuellen Gestaltung im 20. Jahrhundert* (with contributions by C. Diament de Sujo, W. Grohmann, N. Lynton, and M. Seuphor, 1968), is an assessment of the work of Josef Albers, known for his abstract paintings and Master of the Bauhaus (q.v.) from 1922 to 1933, the year of his emigration to the USA. Other publications include *Poesie als Mittel der Umweltgestaltung* (1969), the collections *worte sind schatten. die konstellationen 1951–1968* (1969) and *Konstellationen, Ideogramme, Stundenbuch* (1977), both introduced by H. Heißenbüttel (q.v.), *Texte 1970–1972* (1973), *Kein Fehler im System. Eine unsystematische Auswahl von Sätzen aus dem gleichnamigen, imaginären Computer-Lesebuch* (1978, with lithographs by G. Uecker). *Wie weiß ist wissen die weisen* (1975) is a text with illustrations by G. Uecker and *Distanzsignale* (1980) a text with illustrations by L. Gebhard. *Das Stundenbuch. The book of hours. Le livre d'heures. El libro de las horas. Timbok* (1980) is a collection of poetry with a postscript by W. Gössmann.

GONTARD, SUSETTE (Hamburg, 1769–1802, Frankfurt), *née* Borkenstein, wife of a Frankfurt banker, Jakob Friedrich Gontard, was loved by F. Hölderlin (q.v.) and returned his love. Hölderlin was employed as tutor to the eldest of her four children from December 1795 to September 1798. Her marriage was unhappy, and she did not long survive Hölderlin's enforced departure. Hölderlin idealized her in his poetry, and in his novel *Hyperion* (q.v.), as 'Diotima'. The touching letters written by Susette Gontard after the parting were published in 1921 as *Die Briefe der Diotima* (veröffentlicht von Frida Arnold, herausgegeben von Karl Viëtor).

GORGIAS, JOHANN (Kronstadt, Transylvania, 1640–84, Kronstadt), was a student at Wittenberg, and may have spent some years in Germany, for he became a member of Die Fruchtbringende Gesellschaft (q.v.). He spent his last years as a schoolmaster in his native Transylvania. Gorgias wrote satires directed against women (*Die buhlende Jungfer*, 1663, and *Jungfräuliche Erquickstunden*, 1665) and a misogynistic novel, *Der betrogene Frontalbo* (1670). His works were published under the pseudonyms Floridan,

Poliandin, and Veriphantor, of which the last was the most frequently used.

GÖRING, HERMANN (Rosenheim, 1893–1946, Nürnberg), was a fighter ace of the 1914–18 War and afterwards turned to civil aviation. He entered the NSDAP (q.v.) in 1922, and was wounded in the attempted Hitlerputsch (q.v.) of 1923. After 1928 he was Hitler's right-hand man, and after 1933 in the various offices he held—chief of secret police (see GESTAPO), Minister-President of Prussia, and Air Minister —he continued to prosecute Hitler's policy of unification of power and liquidation of opponents. At the outbreak of war in 1939 he was nominated by Hitler as his eventual successor, and was given the title Reichsmarschall des Großdeutschen Reiches. After the failure of the German Air Force to break British resistance Göring began to lose influence, and in 1945 was denounced by Hitler. He was captured in 1945 and condemned to death at Nürnberg, but took poison in his cell on the morning appointed for execution. He is a background figure in Zuckmayer's play *Des Teufels General* (q.v.).

GÖRRES, JOHANN JOSEPH VON (Koblenz, 1776–1848, Munich), a man of dynamic and choleric temperament who strongly supported the French Revolution while a student, and was an ardent republican, eager for the secession of the Rhineland. In 1799 he headed a delegation sent to Paris, but in 1800 he changed his views completely, becoming a fervent supporter of German nationalism. In 1806 he began to lecture on philosophy at Heidelberg University, and became a close friend of the young Romantics L. J. von Arnim and C. Brentano (qq.v.). With the latter he was joint author of the story *Des Uhrmachers BOGS wunderbare Geschichte* (1807). His essay *Die teutschen Volksbücher. Nähere Würdigung der schönen Historien, Wetter- und Arzneibüchlein, welche teils innerer Wert, teils Zufall, Jahrhunderte hindurch bis auf unsere Zeit erhalten hat* (1807) is, together with *Des Knaben Wunderhorn* (q.v.), one of the landmarks in the development of the German Romantic movement. His plan to complement this essay by reprinting a number of Volksbücher remained unfulfilled. From 1814 to 1816 Görres edited the newspaper *Der Rheinische Merkur* (q.v.), first supporting the struggle against Napoleonic domination, and later urging the moral regeneration and political reform of Germany by a return to a quasi-medieval society. In his later years he was a devout Roman Catholic. Görres wrote a *Mythengeschichte der asiatischen Welt* (1810), collected folk poetry and Meisterlieder *Altteutsche Volks- und Meisterlieder*, 1817; see

MEISTERGESANG), and published religious writings. His ennoblement dates from 1839. *Gesammelte Schriften* (9 vols., incl. 3 vols. correspondence), ed. M. Görres and F. Binder, appeared 1854–74 and the critical edition, ed. A. Dyroff *et al.*, 1926 ff.

GÖSCHEN, GEORG JOACHIM (Bremen, 1752–1828, Grimma), founded a book-selling and publishing business in Leipzig in 1785. Göschen was on friendly terms with Schiller and C. G. Körner (qq.v.), and published Schiller's *Don Carlos* (q.v.) in 1787. He was also the publisher of the first collected edition of the works of Goethe (q.v.), *Goethes Schriften*, 1787–90. His firm was acquired by Cotta (q.v.) in 1838, but became independent again in 1868. In 1919 it was merged with Walter de Gruyter & Co.

Gothaische Genealogische Taschenbücher, annual publications listing noble families, arranged in separate volumes according to rank and antiquity. The earliest is the *Hofkalender,* listing ruling and princely families, including those of other countries, which was first published in 1763. The volumes referring to counts, barons, and lower nobility are restricted to German families. They are commonly known in Britain as the *Almanach de Gotha.*

GOTTER, FRIEDRICH WILHELM (Gotha, 1746–97, Gotha), studied law in Göttingen, held for a short time an appointment as archivist at the court of Gotha, and then returned to Göttingen as private tutor to two young noblemen. He was briefly associated with H. C. Boie (q.v.) in the publication of the *Göttinger Musenalmanach* (q.v.). In 1772 Gotter met Goethe at Wetzlar. He spent the remainder of his later life as a civil servant in Gotha, occasionally visiting Weimar and actively furthering the theatre in Gotha. Gotter's taste, which was for French literature, ran counter to the trend of the age, expressed in Klopstock and the Sturm und Drang (qq.v.). He translated and adapted French plays for the German stage (Voltaire's *Mérope*, 1774, and La Harpe's *Mariane*, 1776), and wrote occasional poetry. His *Medea* (1775) is an example of the semi-musical form known as Melodrama (q.v.). His comedies of contemporary life (*Die Dorfgala,* 1772; *Der Ehescheue,* 1778; *Der Erbschleicher,* 1789) are slight but entertaining. Towards the end of his life he wrote an adaptation of *The Tempest,* which was praised by Goethe and published posthumously (1797) in Schiller's periodical *Die Horen* (q.v.).

Götterdämmerung, see RING DES NIBELUNGEN, DER.

Götter Griechenlands, Die, a poem written by Schiller (q.v.) in 1787 and first published in *Der teutsche Merkur* in March 1788. It consists of 25 eight-line stanzas. The poem is a lament for the lost beauty of the Hellenic world. The original version complemented this with sharp criticism of the Christian theology and iconography which have replaced it. In this form it was subjected to vigorous denunciation by Friedrich, Graf von Stolberg (q.v.). Some five years later Schiller began a revision, tempering the references to Christianity, and this version was published in vol. 1 of Schiller's *Gedichte* in 1800. Schiller did not, however, abandon the original but included it in the second volume of *Gedichte* in 1803.

'Gott erhalte Franz den Kaiser', first line of the former Austrian national anthem; Franz refers to the Emperor Franz II (q.v.) of the Holy Roman Empire (see DEUTSCHES REICH, ALTES), later Franz I of Austria. The words were written in 1797 by L. L. Haschka; the melody was specially composed by Josef Haydn (qq.v.). In 1854, under the Emperor Franz Joseph (q.v.), the line was altered to 'Gott erhalte, Gott beschütze' by J. G. Seidl (q.v.). See also DEUTSCHLAND, DEUTSCHLAND ÜBER ALLES.

Götter, Helden und Wieland, a short play (Eine Farce) by J. W. Goethe (q.v.), written in 1773 and published in 1774. In it Goethe lampoons Wieland (q.v.), whose *Alceste* appeared in 1773. Wieland, magically wafted from his bed, is confronted with the true figures of the Greek story of Admetus and Alcestis, together with their dramatist Euripides. Wieland is astonished at the Greek reality and especially confounded by the enormous proportions of Hercules, whom he had conceived 'Als wohlgestalter Mann mittlerer Größe'. Wieland took the satire well, and his unperturbed good humour paved the way for a friendly relationship when Goethe arrived in Weimar.

Gottesfreund vom Oberlande, Der große, a fictitious name to which, in all probability, R. Merswin (q.v.) attributed mystical writings of his own. The writings are in support of the mystic religious community of Gottesfreunde sponsored by Merswin in the 14th c.

GOTTFRIED VON NEIFEN, a Middle High German Minnesänger of the mid 13th c., was associated with Heinrich VII, son of the Emperor Friedrich II (qq.v.). Documentary references to him occur between 1234 and 1255. Neifen's poetry is for the most part confined within the accepted limits of the Minnesang; his distinguishing formal quality is a brilliant and virtuoso-like handling of rhyme. His songs are conceived in sensual terms, with references to sparkling eyes (*spilnde ougen*) and red lips (*rôter munt*), and happiness (*fröide*) is their basic tone; but they remain essentially conventional. A few poems written in reference to peasants or craftsmen, some of which (notably the *Büttner* poem) are obscene, are held by some authorities to be wrongly attributed to Neifen.

GOTTFRIED VON STRASSBURG, Middle High German poet, was the author of *Tristan* (q.v.). Little is known of Gottfried's life. He is presumed to have died *c.* 1210 and to have written his poem for one Dietrich, whose name is spelt out in an acrostic at the beginning. Unlike most of his poet contemporaries, he was not of knightly birth and is referred to as *meister*, not *herr*. Living in Strasburg at the linguistic meeting-point of two cultures, he handled French with fluency and correctness; and he was · also well acquainted with the classics. He was not a cleric, but he had probably enjoyed the educational opportunities of a monastery school. In *Tristan* he makes favourable mention of HEINRICH von Veldeke, HARTMANN von Aue, and BLIGGER von Steinach (qq.v.) among epic poets, and of Reinmar and WALTHER von der Vogelweide (qq.v.) among lyric writers, but is thought by some to express distaste for Wolfram von Eschenbach's *Parzival* (q.v.), though he abstains from mentioning Wolfram's name. Gottfried's epic poem is virtually his only work. The Minnelieder attributed to him in various MSS. are unauthentic; however, he is believed to have written two poetic Sprüche (see SPRUCH), which were formerly allocated to ULRICH von Lichtenstein (q.v.).

Gottgelobte Herz, Das, a historical novel by E. G. Kolbenheyer (q.v.), published in 1938. Its central figure is the 14th-c. nun, Margareta Ebner (q.v.). Kolbenheyer spells her first name Margarete.

GOTTHELF, JEREMIAS, pseudonym and now universally accepted designation of Albert Bitzius (Murten, 1797–1854, Lützelflüh nr. Berne), Swiss pastor and novelist. The son of a pastor, Gotthelf was at school in Berne, where he subsequently studied theology. After ordination in 1820 he was for a short time his father's curate at Utzenstorf, and then spent a year of further study at Göttingen University. He returned to assist his father again in 1822, but when the father died in 1824 he was, to his bitter disappointment, passed over for the living. He was next curate in Herzogbuchsee, where his

reforming temper brought him into conflict with the authorities, and in 1829 he was transferred to Berne. Two years later he was sent as curate to the remote village of Lützelflüh in the Emmental, becoming pastor there in 1832, and marrying in the following year. It was in Lützelflüh that he spent the remainder of his life and wrote his novels.

Gotthelf found his vocation as a novelist in 1836, when he wrote with great rapidity the powerful and gloomy didactic story of Swiss peasant life, *Der Bauernspiegel* (q.v., 1837). For all his subsequent novels he used as his pseudonym the name of Jeremias Gotthelf, the principal character of this story, who purported also to be its narrator. From this time on, while still active as pastor of Lützelflüh, Gotthelf wrote and published a series of Swiss rural novels marked by strong social commitment, a vivid eye for character, detailed psychological penetration, and a concise, relaxed, and yet forceful style. *Leiden und Freuden eines Schulmeisters* (q.v., 2 vols., 1838–9) grew out of his preoccupation with education. *Wie Uli der Knecht glücklich wird* (q.v., 1841), a novel of back-slidings and renewed resolve, ends in achievement, and this optimistic tone persists in its belated sequel *Uli der Pächter* (1849).

Meanwhile Gotthelf had written the two great novels *Wie Anne Bäbi Jowäger haushaltet* (q.v., 2 vols., 1843–4) and *Geld und Geist* (q.v., 3 vols., 1843–4). To these he later added *Der Geltstag* (q.v., 1845), *Käthi die Großmutter* (q.v., 2 vols., 1847), *Die Käserei in der Vehfreude* (q.v., 1850), *Zeitgeist und Berner Geist* (q.v., 2 vols., 1853), and *Erlebnisse eines Schuldenbauers* (q.v., 1853). Gotthelf was equally successful with shorter stories and Novellen, publishing the collections *Bilder und Sagen aus der Schweiz* (6 vols., 1842–6), *Hans Joggeli der Erbvetter* (1848), and *Erzählungen und Bilder aus dem Volksleben der Schweiz* (5 vols., 1850–5). Among the large number of impressive Novellen the best known are *Die schwarze Spinne* (q.v., 1842 in *Bilder und Sagen aus der Schweiz*) and *Elsi, die seltsame Magd* (q.v., 1843, later included in *Erzählungen und Bilder*).

Gotthelf's burning interest in the welfare of the peasantry prompted him also to undisguised political and social writing, of which the most notable example is the long essay *Die Armennot* (1840). He continued to write uninterruptedly until his death in 1854, brought about by dropsy. Although well known in his own day in Switzerland, his writings did not at first make an impact on the wider German public, though a collected edition, *Gesammelte Schriften* (24 vols.), was published in Berlin, 1855–8.

Gotthelf's reputation has risen rapidly in the 20th c. His avowed didactic purpose may seem at first homespun and his range of environment

parochial, but his infallible sense for character, his instinctive grasp of emotion and motivation, his natural gift for story-telling, and his racy, dialect-touched style make him one of the most readable of German novelists; one who, seemingly without effort, creates a world in all its particulars.
R. Hunziker and H. Bloesch *et al.* edited *Gesamtausgabe* (24 vols., 1911–32, supplementary vols. 1922 ff.); a select edition, *Jeremias Gotthelfs Werke* (20 vols.), ed. W. Muschg, appeared in 1948–53.

Göttingen University, was founded in 1737 as the university for Hanover by the Elector Georg August, who was also King George II of the United Kingdom. From him it derives the title *Universitas Georgia Augusta*. It quickly attained a reputation, especially in the faculty of law and in the study of history. Among noted professors in the 18th c. were A. von Haller and C. G. Heyne (appointed on foundation), J. D. Michaëlis, J. H. Gesner, and G. A. Bürger (q.v.), and in the 19th c. the brothers Grimm (qq.v.). In 1772 some students of Göttingen founded the Göttinger Hainbund (q.v.), which played a part in the new florescence of German literature. Heine was an alumnus of the university, but, as *Die Harzreise* (q.v.) indicates, he did not find it much to his taste. In 1837 it was in the political forefront through the dismissal by King Ernst August (q.v.) of seven protesting professors (see GÖTTINGER SIEBEN).

Göttinger Hainbund (also referred to as Göttinger Hain and Göttinger Dichterbund), name given by J. H. Voß (q.v.), one of the members, to a league of students and their friends interested in poetry. They were as passionately attached to the new, apparently spontaneous, emotional poetry of Klopstock (q.v.) as they were repelled by the elegant artificial verse of Wieland (q.v.). The Bund began to germinate in 1771 around H. C. Boie (q.v.), editor of the *Musenalmanach* (see GÖTTINGER MUSENALMANACH). The first to send contributions were the students L. P. Hahn, G. A. Bürger, L. H. C. Hölty, and J. M. Miller (qq.v.). To these were added in 1772 K. F. Cramer (q.v.), Voß, and the Counts C. and F. L. von Stolberg (qq.v.). Boie, who was rather older than the majority, dropped into the background without open secession. The Bund was created spontaneously on a moonlit woodland walk on 12 September 1772, when those present (Voß, Miller and his cousin G. D. Miller, Hahn, Hölty, and J. T. L. Wehrs), inspired by the poetry of Klopstock and the beauty of the night, joined hands and danced round an oak-

tree, vowing eternal friendship. From this forest background and from Klopstock's ode *Der Hügel und der Hain* the group, which elected Voß as leader, took its name. The young men met and read their poems of Nature, friendship, and love, and on Klopstock's birthday, 2 July 1773, they held a ceremony exalting Klopstock and consigning the works of Wieland to the flames. The brothers Stolberg, already contributors to the *Almanach*, formally joined late in 1772, Count C. A. von Haugwitz in 1773, and J. A. Leisewitz (qq.v.) in 1774. Bürger, though associated with the group, was never a member. The Hainbund could, by its nature, not last long, since university students are transient, and it faded rapidly in the mid-1770s. But, either at the time or later, Voß, Hölty, and the Stolbergs (as well as Bürger) made their mark in German poetry, as did J. M. Miller in the novel, and Leisewitz in drama. The episode, coincident with the Sturm und Drang (q.v.), remains an important, if brief, symptom of the reaction against rationalism, artificiality, and respectable convention.

Göttinger Musenalmanach, an annual collection of poetry published at Göttingen, and edited from 1770 to 1775 by H. C. Boie (q.v.). It appeared under the title *Musenalmanach für das Jahr 1770*, etc., *Göttinger* being used to distinguish it from other similarly titled publications. The first volume, of which F. W. Gotter (q.v.) was joint editor, included chiefly Anacreontic poetry and poems by K. W. Ramler (q.v.). The second, however, went over to Klopstock in its selection, and from 1772 it published mainly poems by the group of poets known as the Göttinger Hainbund (q.v.). It ceased to be influential after 1775, when J. H. Voß (q.v.) took over the editorship—to be succeeded later in the year by L. F. G. von Goeckingk (q.v.). Voss meanwhile continued, so that there were two parallel *Musenalmanache*. G. A. Bürger (q.v.) was the editor from 1779 to 1794, and Carl von Reinhart (1769–1840) from 1794 to 1804, except for 1803, when it was edited by Sophie Mereau (q.v.). It ceased publication with the issue for 1804. The *Göttinger Musenalmanach* was an important vehicle for the propagation of the poetry and outlook of the Hainbund. See also BALLADENALMANACH.

Göttinger Sieben, seven professors of Göttingen University who protested against the abrogation of the constitution of Hanover by King Ernst August (q.v.) in 1837. They were W. Albrecht (1800–76), F. C. Dahlmann (q.v.), G. H. A. Ewald (1803–75), G. G. Gervinus, Jakob and W. Grimm (qq.v.), and W. Weber

(1804–91). Their dismissal caused widespread indignation in Germany.

Göttingisches Magazin für Wissenschaft und Literatur, periodical founded in 1780 by G. C. Lichtenberg and G. Forster (qq.v.). It continued until 1785.

Göttliche, Das, a poem by Goethe, which opens with the lines:
> Edel sei der Mensch,
> Hilfreich und gut!

Written c.1783, it was published by F. H. Jacobi (q.v.) in *Über die Lehren des Spinoza* in 1785, and in 1789 by Goethe in *Goethes Schriften*. Composed in irregular, short-lined, unrhymed strophes, it accepts the limitations of humanity and affirms man's noble potentialities.

GOTTSCHALL, RUDOLF VON (Breslau, 1823–1909, Leipzig), the son of an officer, was a student successively at Königsberg, Breslau, and Berlin universities, and then took to writing, at first as a political opponent of the existing order. In later years he became a generally esteemed literary figure. His political poems appeared as *Lieder der Gegenwart* in 1842. He wrote a number of plays, including *Ulrich von Hutten* (1843), *Robespierre* (1845), *Lambertine von Méricourt* (1850), and the comedy *Pitt und Fox* (1854). His complete plays filled twelve volumes (*Dramatische Werke*, 1865–80). In later life he was also a prolific writer of novels, among which *Im Banne des schwarzen Adlers* (3 vols., 1876), *Welke Blätter* (3 vols., 1877), *Das goldene Kalb* (3 vols., 1880), *Die Erbschaft des Blutes* (3 vols., 1882), *Verkümmerte Existenzen* (2 vols., 1892), and *Moderne Streber* (2 vols., 1896) deserve mention. Gottschall also made a name for himself among his contemporaries by his writings on recent literature (*Die deutsche Literatur der ersten Hälfte des 19. Jahrhunderts*, 1855, which was ultimately expanded into *Die deutsche Nationalliteratur des 19. Jahrhunderts*, 4 vols., 1901–2). He was granted a patent of nobility by the German Emperor in 1877.

GOTTSCHED, JOHANN CHRISTOPH (Judittenkirchen nr. Königsberg, 1700–66, Leipzig), son of a Protestant pastor, began to study theology at Königsberg University when still under 15, later changing his course to philosophy and literature. In 1723 he qualified to give lectures at the university. Because of his height and strong build (Goethe described him as 'der große, breite, riesenhafte Mann') Gottsched was in danger of compulsory enrolment in King Friedrich Wilhelm I's pet regiment of tall grenadiers, and he therefore slipped out of

Prussia to Saxony, establishing himself at Leipzig. Up to this point, his writings had consisted mainly of occasional poems. In Leipzig he became tutor to the sons of Johann Burkhard Mencke (q.v.), president of the Deutschübende Poetische Gesellschaft in which Gottsched became 'Senior' in 1726. He reorganized the society in the following year, changing its title to Die Deutsche Gesellschaft, and using it to serve his own reformative literary purposes. He was appointed a supernumerary professor at Leipzig University in 1730 and Professor of Logic and Metaphysics in 1734.

Gottsched, who was a man of strong character, energy, and determination, conceived the idea of a linguistic reform which should establish a single German educated tongue, based on Saxon usage, and of a literary renewal which was to assimilate German poetry and drama to the admired French model. Nearly all the works he published in the 1730s were designed to support this policy, for which he could claim Opitz (q.v.) as an antecedent. They include the literary periodical *Beyträge zur kritischen Historie der deutschen Sprache, Poesie und Beredsamkeit* (1732–44) and especially his *Versuch einer critischen Dichtkunst vor die Deutschen* (q.v., 1730). His efforts to reform the theatre were seconded by an alliance with the troupe of Friederike Neuber 1697–1760, which performed his tragedy *Der sterbende Cato* (q.v., 1732), and by the translations and original writings of his wife, Luise Adelgunde Gottsched (q.v.). In conjunction with Frau Neuber's troupe a ceremony was held in 1737 banning the clown (see HANSWURST) from the stage. Gottsched set about establishing a repertoire of modern German plays after the French classical model, publishing from 1740 to 1745 the six volumes of *Die deutsche Schaubühne* (Pts. 4–6 reprinted 1972). He also wrote a compendium of philosophy, *Erste Gründe der gesamten Weltweisheit* (2 vols., 1734), based on the teachings of C. Wolff (q.v.).

Gottsched's considerable reputation began to wane about 1740. His obstinate character involved him in sterile and sometimes ridiculous disputes, and his dictatorial attitude provoked a growing opposition. His authority was flouted by the Swiss professors J. J. Bodmer and J. J. Breitinger (qq.v.), whom he attacked as Merbod and Greibertin in his satire *Der Dichterkrieg* (1741); and in 1741 Frau Neuber turned against him and burlesqued him on the stage. In his last years Gottsched was a lonely figure, whose remarkable combination of dignity and indignity is portrayed by Goethe in *Dichtung und Wahrheit*, Bk. 7. His services to German literature have often been ridiculed along the lines of Lessing's savage attack in the 17th *Literaturbrief* (1759; see LITERATURBRIEFE).

His Francophile ideal of literature was hardly suited to German conditions, and his inflexibility was a serious hindrance, but he took German literature seriously and induced others to do the same. His philological efforts (*Grundlegung einer deutschen Sprachkunst*, 1748) have had a better reception, and he also compiled a valuable bibliographical work, *Nötiger Vorrat zur Geschichte der deutschen dramatischen Dichtkunst* (1757–65). Gottsched was a pioneer with moralizing weeklies (see MORALISCHE WOCHENSCHRIFTEN), publishing *Die vernünftigen Tadlerinnen* (1725–6) and *Der Biedermann* (1727–9). He was, in addition, the editor of the literary periodical *Das Neueste aus der anmutigen Gelehrsamkeit* (1751–62). His translations include Bayle's *Dictionnaire*, plays of Racine, and the *Théodicée* of Leibniz. Gottsched is a leading character in H. Laube's comedy *Gottsched und Gellert* (q.v.).

Ausgewählte Werke (14 vols.) were edited by J. and B. Birke, 1968 ff.

GOTTSCHED, LUISE ADELGUNDE VICTORIA, *née* Kulmus (Danzig, 1713–62, Leipzig), often known in 18th-c. style as 'Die Gottschedin', married J. C. Gottsched (q.v.) in 1735. She collaborated with him before as well as after marriage, notably in the periodicals *Beyträge zur kritischen Historie der deutschen Sprache, Poesie und Beredsamkeit* (1732–44) and *Das Neueste aus der anmutigen Gelehrsamkeit* (1751–62). She identified herself completely with Gottsched's literary policy and supported him by writing comedies in the French manner (*Die Pietisterey im Fischbeinrocke*, 1736; *Das Testament*, 1745). She was also active as a translator of Addison (*Spectator* as *Der Zuschauer*, 1739–43), Pope (*Rape of the Lock* as *Der Lockenraub*, 1744), and Molière.

Gottsched und Gellert, a five-act comedy by H. Laube (q.v.), first performed at Dresden in October 1845, published in 1847. Laube describes it as a comedy of character (Charakterlustspiel). The play is set in Leipzig towards the end of the Seven Years War (see SIEBENJÄHRIGER KRIEG). Both Gottsched and Gellert (qq.v.), find themselves in difficulties with the occupying Prussian army, and are extricated by the arrival of Prince Heinrich (q.v.) of Prussia, who is particularly well disposed towards Gellert. The essential character of the piece is the contrast between the unassuming Gellert and the arrogant Gottsched.

Gottsucher, Der, a novel by P. Rosegger (q.v.), published in 1883. Set in an unspecified past which could be the 16th or the 17th c., it tells the story of the parish of Trawies, in which the intolerant, avaricious, and worldly parish priest is murdered. The village is excommunicated, and the subsequent moral and material decline of this community without religion leads to further murder, until Wahnfred vom Gestade introduces a religion of fire; the consequential total destruction of the village by fire brings expiation.

Gott und die Bajadere, Der, a ballad with the sub-title *Indische Legende* written by J. W. Goethe (q.v.) in June 1797, and published in the *Musenalmanach* for 1798. It is the story of an Indian prostitute who is visited by the god Siva, unrecognized in human form, and is seized with true love for him. The human shape is found lifeless in the morning, and the Bajadere immolates herself upon its pyre. She ascends to heaven in the arms of the god.

Göttweiger Trojanerkrieg, Der, title by which a mediocre Middle High German verse romance dealing with the lives of Paris and Hector is known. It runs to more than 25,000 lines and is indebted to WOLFRAM von Eschenbach (q.v.), with whom the author in certain passages identifies himself. The MS. was found in Göttweig Abbey near Krems in Lower Austria, but the work appears to have been written by a Swiss, probably c. 1300.

GÖTZ, JOHANN NIKOLAUS (Worms, 1721–81, Winterburg nr. Kreuznach), while a student at Halle was a friend of J. W. L. Gleim and J. P. Uz (qq.v.). He became a private tutor in Emden and in Forbach, Lorraine, and from 1748 to 1751 served as a chaplain in the French army. His later life was spent first as pastor and then as moderator (Superintendent) in the Palatinate towns Hornbach, Meisenheim, and Winterburg. He wrote Anacreontic poems (see ANAKREONTIKER) which he published as *Versuch eines Wormsers in Gedichten* (1745); he also translated the *Anacreontea* or *Pseudo-Anacreon* (*Die Oden Anakreons in reimlosen Versen*, 1746).

Götz von Berlichingen mit der eisernen Hand. Ein Schauspiel, the play which first established Goethe's reputation on its publication in 1773. Its principal source was the *Lebensbeschreibung des Herrn Götz von Berlichingen* (see BERLICHINGEN, GOTTFRIED VON), first published in 1731, long after its author's death in 1562. Goethe originally wrote the play in Frankfurt, in the autumn of 1771 after his return from Strasburg. Largely because of adverse criticism by Herder, he deferred publication and rewrote it in the early months of 1773, changing the title from *Geschichte Gottfriedens von Berlichingen mit der eisernen Hand dramatisiert* to its present form. The earlier version was not published until 1832, when it appeared in the *Ausgabe letzter Hand*. *Götz von Berlichingen* is constructed as a chronicle play with a multiplicity of scenes (56), a kaleidoscopic juxtaposition which was intended as an imitation of Shakespeare.

Götz, a robust man of action, is engaged in a feud with the Bishop of Bamberg, and in this opposition he is seen as the representative of freedom and natural right, whereas the Bishop stands for privilege, corruption, and retrogression. Götz captures the Bishop's knight Weislingen, who pledges peace with Götz, to whose sister Maria he is then affianced. But Weislingen, pressed by the Bishop and lured by the bewitching temptress Adelheid von Walldorf, breaks his word to Götz and to Maria, and reopens hostilities against his friend and former captor. Besieged by an overwhelming force, Götz surrenders on terms which are immediately breached by the victors. Götz is arraigned at Heilbronn, but is rescued with the help of his ally Franz von Sickingen. The Peasants' War (see BAUERNKRIEG) meanwhile breaks out, and Götz accepts under duress a command in the peasant forces. He is captured and condemned to death, but is saved by Maria, who persuades Weislingen to quash the death sentence. Weislingen perishes, poisoned by his mistress Adelheid, and Götz dies in prison with the words 'Freiheit! Freiheit!' on his lips.

The play was first performed in Berlin in 1774, and although Goethe had conceived it as a play for reading, it was an immediate popular success, becoming one of the first landmarks of the Geniezeit or Sturm und Drang (q.v.). *Götz von Berlichingen* was adapted for the English stage in 1965 by John Arden under the title *Iron Hand*.

GOUE or **GOUÉ**, AUGUST SIEGFRIED VON (Hildesheim, 1742–89, Steinfurt), was at the time of Goethe's stay in Wetzlar at the Reichskammergericht (q.v.). He was later in the service of the counts of Bentheim, but was dismissed because of dissolute conduct. He is believed to be the model for Satyros in Goethe's play (see SATYROS). Goue wrote poems and plays, including *Masuren oder Der junge Werther* (1775).

Gouverneur, Der, a historical novel by E. Schaper (q.v.), published in 1954. It bears the alternative title *oder Der glückselige Schuldner*. It is set in Reval at a time when it is likely to be invested by the Russians after the defeat of Charles XII of Sweden at Poltava in 1709. Patkul, the Swedish governor, weighed down by the military hopelessness of the defence of Reval, is determined to avoid useless bloodshed by capitulating on reasonable terms. Despite failing health, he summons all his energies to save a Swedish captain of horse, Rittmeister Cronstedt, who stands wrongly accused of a brutal murder. Cronstedt was present when his two companions in the escape from Poltava murdered seven members of a family of eight. The only survivor is the young Baroness Maria Barbara, whom Cronstedt protects from the other two. He escorts her to Reval, and the two fall deeply in love. The elders of the city and the populace demand Cronstedt's head. But Patkul, who accepts Cronstedt's word of honour that on the whole journey he has not committed murder, stands out against general pressure and succeeds in procuring Cronstedt's repatriation to Sweden, sending Maria Barbara with him. By this humane action, achieved while battling vainly against worsening illness, Patkul is redeemed from his many past sins. He dies of heart failure as the ship carrying the lovers to safety reaches the open sea.

GRABBE, CHRISTIAN DIETRICH (Detmold, 1801–36, Detmold), only son of the head warder of the local prison, attended the Detmold Gymnasium, and was at an early age a prolific reader, gifted, imaginative, ambitious, but increasingly self-conscious about his humble origin. From 1820 to 1822 he studied law at Leipzig University, but dissipation began to undermine his health. He spent a year in Berlin, completing his first dramatic works: *Herzog Theodor von Gothland*; *Scherz, Satire, Ironie und tiefere Bedeutung*; *Marius und Sulla* (qq.v.); and *Nannette und Maria*. Grabbe, who was too self-centred and quick to jealousy to make friends easily, also had a hatred of Jews, which caused him to throw away a chance of friendship with H. Heine (q.v.). He vainly tried to become an actor, first in Dresden through the help of Tieck and subsequently in Hanover, Brunswick, and Bremen. After his return to Detmold in 1824, he passed his law examinations, and from 1826 to 1834 he worked in his native town in the army legal branch (Militärauditeur).

In 1827 he published through G. F. Kettembeil, a friend of his Leipzig days, his *Dramatische Dichtungen* (2 vols.) in which he included his work to date and a critical essay *Über die Shakespearo-Manie* (q.v.). By 1829 he had completed *Don Juan und Faust* (q.v., 1829), *Kaiser Friedrich Barbarossa* (q.v., 1829) and *Kaiser Heinrich der Sechste* (q.v., 1830, the only two of a projected cycle of Hohenstaufen plays), *Napoleon oder Die hundert Tage* (q.v., 1831), and the play (Märchenlustspiel) *Aschenbrödel* (1835), which Kettembeil refused to publish after its completion. In 1831

Grabbe became engaged to Henriette Meyer, who after a few months jilted him. In March 1833 he married Luise (after her marriage Lucie) Clostermeier, a woman ten years his senior, from whom he soon became estranged. He expressed his bitter disillusionment with life in *Hannibal* (q.v.), perhaps his finest tragedy, written during the last months in Detmold.

As a dramatist he had at least one exhilarating experience during these years: on 29 March 1829 *Don Juan und Faust* was successfully performed with music by Albert Lortzing (q.v.), who also played Don Juan. Grabbe's move to Frankfurt in 1834, without his wife, failed to improve his prospects for a more settled and less isolated existence. His friendship with his publisher Kettembeil came to a sudden end, and with it his purpose of coming to Frankfurt. Eduard Duller, one of his biographers, was virtually his only companion. To relieve his loneliness and also to find a publisher for *Hannibal*, Grabbe wrote to W. Menzel and Immermann (qq.v.) for help. Immermann had already visited him in Detmold in 1831, and invited him now to join him in Düsseldorf, where Grabbe arrived in December 1834. Through Immermann, *Hannibal* and the revised *Aschenbrödel* found a publisher. But although Grabbe supported Immermann's theatre by favourable criticism and a pamphlet *Das Theater zu Düsseldorf*, Immermann did not offer Grabbe the obvious service of producing one of his plays. For a short time Grabbe found in the young composer Norbert Burgmüller (1810–36) a new friend and fellow-sufferer. After he had written him a libretto, *Cid*, for an opera, Burgmüller's sudden death ended his stay in Düsseldorf. Grabbe returned to Darmstadt. Here he completed his last play, *Die Hermannsschlacht* (q.v., 1838). He still had many plans, among them projected plays on Christ, Alexander the Great, Frederick the Great (see FRIEDRICH II OF PRUSSIA), and Shakespeare, whom, in his own way, he admired and emulated.

Grabbe's egocentric, exhibitionistic, cynical, and yet naïve character, and his dissipated life are attributable to a possibly inherited psychopathic condition and to his humble parentage and spoilt upbringing. His odd appearance, slight physique, large head, and noble forehead have been described by contemporaries as representing the contradictions inherent in the man.

Grabbe's contribution to German drama has been appreciated more fully since realistic drama became predominant. A forerunner of the epic theatre (see EPISCHES THEATER), Grabbe could penetrate beneath the surface of reality by brilliant application of the grotesque, ruthlessly exposing hypocrisy. His historical plays are his greatest achievement, and are less remote from the modern stage than they were from the stage of his own day (several of his plays, including his first, were produced in the early 1970s). Grabbe's cult of the great man anticipates Nietzsche (q.v.).

Sämtliche Werke, including letters, were published by E. Grisebach (4 vols., 1902–3), and by S. Wukadinović (6 vols., 1912); *Werke und Briefe. Historisch-kritische Gesamtausgabe* (6 vols.), ed. A. Bergmann, appeared in 1960–73.

Graf Essex, a five-act tragedy by H. Laube (q.v.), first performed at the Burgtheater, Vienna, in February 1856, and published in that year. It is chiefly in blank verse with some humorous scenes in prose. Remarkably competent, though uninspired, it treats the well-known story of the rebellion of the Earl of Essex, and of the opportunity to save him by the ring entrusted to Lady Nottingham.

Laube's Essex is not the victim of the rancour of a spurned woman; he refuses to send the ring to Elizabeth because it is the pledge of an impure love. The Nottinghams and Cecil intrigue against Essex, who is absent in Ireland and has secretly married Lady Rutland. Elizabeth suspects Essex and turns against him, a process completed by his arrogant conduct to her in audience. His rebellion is quelled, and his life declared forfeit. Lady Nottingham, beset by remorse, seeks to persuade Elizabeth to pardon Essex. The Queen will do so if he sends to her the ring she once gave him. He refuses, preferring execution.

GRAFF, ANTON (Winterthur, 1736–1813, Dresden), a successful 18th-c. portrait painter, was principally active in Dresden, where he held a teaching appointment at the Akademie. More than 1,500 portraits by him are known, and his sitters included many of the great names of his age. His well-known portrait of Schiller (q.v.) was begun in 1786 but not completed until 1791.

Graf Öderland, a play (*Eine Moritat in zwölf Bildern*) by M. Frisch (q.v., and see MORITAT), first published in 1951 (revised in 1957 and 1961). All versions have been performed.

The state prosecutor, suffering from overwork, refuses to be persuaded by his wife to go to bed. The following morning he is due to prosecute in a murder case, in which he feels some sympathy for the murderer. When his wife has left his room, a maid enters, believing he has rung for her. In conversation she tells him the legend, still alive in the countryside, of the great liberator, Graf Öderland, who will one day return. In the rest of the play, which appears to be a dream, the prosecutor is welcomed by the country folk in the role of Graf Öderland. He leads them against the authorities, murdering, burning, and pillaging, and takes over the government. Suddenly he finds himself again in his study, and must clearly have been dreaming. But was it a dream? The maid draws attention to his muddy boots which belong to the dream. And at once the dream returns. Has he dreamed that he has been dreaming? Frisch offers no answer to the prosecutor's dual identity. Is he the respectable lawyer, or the murderous agitator?

Graf Rudolf, a 12th-c. Middle High German narrative poem, of which only fragments of a MS. amounting to *c.* 1,300 lines, exist. The authorship is unknown. The time of composition is believed to lie between 1170 and 1185. The exact course of the story is uncertain, but from the fragments it appears that the Pope summons Christendom to a crusade, and Rudolf is among the crusaders. He quarrels with the King of Jerusalem, and is next found with the Saracens. The Sultan resists a summons to hand Rudolf over to the Christian king. Rudolf, who loves the Sultan's daughter, is presumably (a gap occurs at this point) separated from her. The heathen princess is later found, baptized, in Constantinople, and Rudolf is discovered wounded and in captivity. He escapes, joins the princess, and together they secretly leave Constantinople.

The later development of the story can only be conjectured. *Graf Rudolf* has a dual emphasis, on the military and moral virtues of the knight, and on love (Minne). It is noteworthy in recognizing knightly virtues in the Saracen, so exhibiting a tolerance which implies a kind of freemasonry of chivalry. The work was edited by P. F. Ganz (1964).

Graf von Ratzeburg, Der, an unfinished play in stylized prose by E. Barlach (q.v.). As the revised MS. is presumed lost, an earlier MS. of 1927 was published posthumously in 1951. The play, which comprises ten acts (Bilder) and includes grotesque elements, is a symbolical drama of martyrdom in which biblical, legendary, and historical figures and motifs are combined to indicate the timelessness of experience. In Graf Heinrich von Ratzeburg, Barlach portrays the 'seeker's' quest for spiritual recognition: 'Es geht nicht ums Gelten, es geht ums Sein!'

Heinrich leaves his position as ruler to go on a pilgrimage to the Holy Land, but falls into captivity and is enslaved. At Smyrna he witnesses a martyrdom, that of the widow Chansa, who is a victim of Marut, Statthalter Satans auf Erden. Heinrich is freed from his chains by

Offerus (Christoffer), and listens to Hilarion's counsel of Christian asceticism which the hermit himself practises.

Heinrich's illegitimate son Wolf is accused of robbery and murder, and Heinrich follows him into captivity, which brings him back to his castle in Ratzeburg, which his brother Jos has seized. Abandoned by his other son, Rolf, Bishop of Ratzeburg, as well as by Jos, Heinrich follows Wolf on his way to execution in Mölln. Christoffer frees Wolf, whose mockery Heinrich patiently bears. Christoffer's journeys and services as Knecht are portrayed as a parallel to Heinrich's Wegsuche, and culminate in his baptism by the child he carries across the river. Heinrich rejects Christoffer's proposal to escape and follow him on his journeys. He accepts death, content that his search has reached a goal: 'Ich habe keinen Gott, aber Gott hat mich'. He has, in spite of appearances, remained his own master by his humility which inspires him with a vision of the glory of eternity.

GRASS, GÜNTER (Danzig-Langfuhr, then Free State of Danzig, 1927–), of part Polish (more precisely, Keshubian) descent, was educated in Danzig, the seizure of which by the National Socialist regime occurred when he was 11. In 1944–5 he served in the German army, then worked as a labourer, and in 1947 was apprenticed to a stonemason. From 1949 to 1956 he studied art at Düsseldorf, then spent some years in Paris. At this time he appeared on the literary scene, first with surrealist poems illustrated by himself (*Die Vorzüge der Windhühner,* 1956), then with grotesque plays (*Hochwasser,* 1956, revised 1963; *Onkel-Onkel, Noch zehn Minuten bis Buffalo, Die bösen Köche,* all 1957), and, as climax, with the large-scale and highly original novel, *Die Blechtrommel* (q.v., 1959), which was an immediate commercial success and was translated into many languages (into English as *The Tin Drum*). A volume of poems with his own drawings, *Gleisdreieck,* followed in 1960, and the play *Die bösen Köche* was performed in 1961. *Katz und Maus* (1961, later subtitled *Danziger Trilogie 2* to indicate its relationship to *Die Blechtrommel* and *Hundejahre*), a Novelle, chooses as its central figure a character with a striking, if less sensational, physical abnormality than Oskar Matzerath's in *Die Blechtrommel*; it is an enormous Adam's apple, but it similarly enables him to exploit his peculiarity for his own ends in a series of events both baffling and comic. *Hundejahre* (q.v., 1963), which was almost as successful as Grass's first novel, contains, like *Die Blechtrommel,* innumerable recollections from the author's early and adolescent years. The trilogy is held together by the theme of collective guilt

(Vergangenheitsbewältigung) that underlies all three works. The play *Die Plebejer proben den Aufstand* (performed 1965, published 1966) is, theoretically, concerned with the East Berlin revolt of 17 June 1953, but is in reality directed against B. Brecht (q.v.), called in the play 'der Chef', who is so concerned with the spell of his own dramatic activity that the Plebejer do not actively make a revolution; they merely rehearse it. But Grass also hints at the inability of Germans to combine theory and practice. Grass was awarded the Büchner Prize in 1965.

Grass is a committed Socialist (he refers to himself as a revisionist), and his election speeches appeared in print in 1965 and a further volume of illustrated poems, *Ausgefragt*, in 1967. A third novel, *örtlich betäubt* (q.v.), set in the dentist's chair, met with a mixed reception in 1969. It includes the episode involving Philipp Scherbaum and Veronika Lewand drawn from his play *Davor* (first performed in February 1969). The complete plays, including *Davor*, were also published in 1969.

Grass's fourth novel, *Aus dem Tagebuch einer Schnecke* (1972), is a complex production, based on his election activities in 1969 and told in answer to his children's questions. Into this basic scheme are woven recollections of Danzig, notably the horror of the extermination of the Jews; a study of melancholy, *Melancolia I*, induced by Dürer's *La Melancolia*, and connected with the celebrations of the 500th anniversary of Dürer's birth (1971, see Dürer, A.); and in addition the recurrent appearance of the snail, both as a creature and a metaphor (for West Germany's painfully slow progress towards democracy); for Grass suggests that he himself, Willi Brandt (Chancellor of the Bundesrepublik, q.v., 1969–74), and the SPD (q.v.) have some affinity with the snail, a creature which is generally regarded more amiably in Germany than in Britain and figures in many children's books. The lengthy novel *Der Butt* appeared in 1977, and *Das Treffen in Telgte* in 1979, the fictitious 'meeting' being concerned with a discussion among a group of 17th-c. writers on a crisis of language analogous to a similar one leading to the formation of the Gruppe 47 (q.v.). The translation of this story by R. Manheim (*The Meeting at Telgte*, 1981) is an annotated edition with a postscript by L. W. Forster. The text of *Kopfgeburten oder Die Deutschen sterben aus* (1980, dedicated to N. Born, q.v.) deals overtly with a variety of mainly global contemporary problems.

Gesammelte Gedichte appeared in 1971, *Mariazuehren* in 1973, *Liebe geprüft* in 1974, *wie ich mich sehe* in 1980, and *Ach Butt dein Märchen geht böse aus*, a volume with Grass's own etchings, in 1983. *Über das Selbstverständliche. Reden, Aufsätze,* *offene Briefe*, is a collection of 1968. *Zeichnungen und Texte 1954–1977*, vol. 1, ed. A. Dreher, appeared in 1982 and a volume of essays, *Aufsätze zur Literatur 1957–79* in 1980. *Widerstand lernen. Politische Gegenreden 1980–1983* (1984, preface by O. Lafontaine) contains political speeches for the peace movement. Grass has usually collaborated closely with the translators of his works.

GRATIUS, Ortwinus (Holtwick, Westphalia, c. 1480–1542, Cologne), whose name is Latinized from Ortwin de Graes, was educated in Deventer, Netherlands, and became professor at Cologne. In the conflict with Reuchlin (q.v.) he sided with Pfefferkorn (q.v.), and in the ensuing polemics he was the addressee of the *Epistolae obscurorum virorum* (q.v.).

GRAZIE, Marie Eugenie delle (Weißkirchen, Hungary, 1864–1931, Vienna), was educated in Vienna from 1872 and, after being trained as a teacher, devoted herself to writing. She began with verse epics in the elevated style of the late 19th c. (*Hermann*, 1884; *Robespierre*, 1894) and with lyric poetry which varies from the rhetorical to the reflective (*Gedichte*, 1882). She was then attracted by the Naturalistic Movement (see Naturalismus) and took to writing plays, in which for a time she passionately espoused radical causes (*Schlagende Wetter*, 1900; *Der Schatten*, 1901; *Narren der Liebe*, a comedy, 1904; *Ver sacrum*, 1906). She also wrote a novel at this stage, *Vor dem Sturm* (1910). In 1912 she returned to the Roman Catholic faith, and her subsequent narrative works bear the imprint of this change. They are the novels *O Jugend* (1917), *Der Liebe und des Ruhmes Kränze* (1920), *Unsichtbare Straße* (1927), and the Novelle *Die weißen Schmetterlinge von Clairvaux* (1925).

GREFF, Joachim (Zwickau, c. 1500–?), a schoolmaster in Halle, Magdeburg, and Wittenberg, wrote biblical plays, mainly on Old Testament subjects (*Spiel vom Patriarchen Jakob*, 1534; *Judith*, 1536; *Mundus*, 1537; *Lazarus*, 1545; *Zacheus*, 1547). Praised by Luther and Melanchthon (qq.v.), they were performed, unlike most early 16th-c. plays, by burghers, not by schoolboys.

GREFLINGER, Georg (Regensburg, c. 1620–c. 1677, Hamburg), was orphaned early, and led a restless life as student and soldier, eventually settling in Hamburg in 1647 as a notary. He was a member of the Elbschwanenorden (q.v.) in which he went by the name Seladon (or Celadon), which he also used as a pseudonym. Greflinger wrote lyric poetry (particularly love poems), epigrams, plays, and historical works. His chief collections of poetry are *Seladons* *beständige Liebe* (1644), *Deutscher Epigrammatum erstes Hundert* (1645), *Lieder über die jährlichen Evangelien* (1648), *Seladons Weltliche Lieder* (1651), and *Celadonische Musa* (a collection of odes and epigrams, 1663). The play *Ferrando und Dorinde* (1644) is probably the libretto for an opera. Greflinger also translated Pierre Corneille's *Le Cid* as *Die sinnreiche Tragicomödia genant Cid*, and Mira de Amescua's *El palacio confuso* as *Verwirrter Hof, oder König Carl* (1652). His *Diarium Britannicum* (1652) is a chronicle of English affairs through the Civil War, and *Der Deutschen dreißigjähriger Krieg* (1657) also recounts events of his own time (see DREISSIGJÄHRIGER KRIEG).

Gregorius, a Middle High German poem of *c.* 4,000 lines by HARTMANN von Aue (q.v.). *Gregorius* is a religious legend, and Hartmann is frequently supposed to have been drawn to the subject by a crisis caused by the death of his lord, to whom he was deeply attached. It was probably written between 1187 and 1189, though some authorities set it as late as 1195. Hartmann's source was an anonymous 12th-c. French poem, *Vie du pape Grégoire*, which does not refer, however, to any of the historical popes named Gregory.

The poem begins with a prefatory passage in which Hartmann confesses his own sinfulness, exemplified in his profane writings (*Erec* and the *Büchlein* are presumably intended). Gregorius, who is termed the good sinner ('vom guoten sündaere'), is born of an incestuous union of brother and sister, models of courtly conduct, who are enmeshed by the Devil's wiles. The father, overcome with remorse, sets out for the Holy Land, dying on the way. Gregorius's mother launches the child in a boat, entrusting him to God's care. He is preserved and brought up under the guardianship of an abbot, who in his deep kindness and goodness is one of the most attractive characters in the poem. Blood tells, and Gregorius insists on becoming a knight. He succours an oppressed lady, deprived of her dominions, and is rewarded with her hand. She proves, however, to be his mother, so that Gregorius and she are for the second time involved in incest. Gregorius leaves her, and spends seventeen years with fettered feet on a rock set in a lake in an almost uninhabited region expiating his sins. Then the papal throne falls vacant and God communicates to two 'wise Romans' that a hermit (who proves to be Gregorius) is His appointed choice. Gregorius is discovered, brought to Rome and crowned. His mother, ignorant of the identity of the new Pope, comes to Rome seeking absolution. Gregorius recognizes her; both live their lives in piety and virtue, winning a forgiveness of sins which includes also the dead father. The poem is vividly written with much realistic detail and a tone of deep sincerity.

Gregorius was translated into Latin not long after its completion, under the title *Gesta Gregorii peccatoris ad penitenciam conversi et ad papatum promoti*. This was the work of an abbot, Arnold von Lübeck, who attempted to reproduce the German metre. There exists also a free translation into Latin hexameters, which is preserved in a 14th-c. MS. at Munich. As *Gregorius auf dem Steine*, written in German prose, the legend became a Volksbuch (q.v.), first printed in 1471, and several times reprinted. Th. Mann's novel *Der Erwählte* (q.v.) is also a retelling of this story.

GREGOROVIUS, FERDINAND (Neidenburg, East Prussia, 1821–91, Munich), studied history at Königsberg University and developed into an outstanding art historian. From 1852 to 1874 he lived in Rome, afterwards moving to Munich, where he was a friend of Graf A. F. von Schack (q.v.). Gregorovius's principal works are *Geschichte der Stadt Rom im Mittelalter* (8 vols., 1859–72), *Lucrezia Borgia* (2 vols., 1874), and *Geschichte der Stadt Athen im Mittelalter* (2 vols., 1889). In literature Gregorovius was no more than a dilettante. He wrote the tragedy *Der Tod des Tiberius* (1851) and a short epic, *Pompeji Euphorion* (1858), and translated the songs of Giovanni Meli (1856). A volume of poems was published after his death by Graf Schack (1891). *Lucrezia Borgia* was used by C. F. Meyer as the source for his Novelle *Angela Borgia* (q.v.).

Greibertin, anagram of J. J. Breitinger (q.v.), used by J. C. Gottsched (q.v.) in his satire *Der Dichterkrieg*.

GREIF, MARTIN (Speyer, 1839–1911, Munich), was at first a Bavarian artillery officer (1857–67), and then settled in Munich, travelling extensively in Europe and devoting himself to writing. Greif's real name was Hermann Frey, but in 1882 his pseudonym became his legal name by the German equivalent of deed poll.

Greif wrote a number of plays deriving from the classicism of Schiller (q.v.). They include *Corfiz Ulfelde* (1873), *Nero* (1877), *Marino Falieri* (1879), *Francesca da Rimini* (1892), and *Agnes Bernauer* (1894), all of which are tragedies, as well as the Schauspiele *Hans Sachs* (1866), *Prinz Eugen* (1880), *Heinrich der Löwe* (1887), *Ludwig der Bayer* (1891), and *General Yorck* (1899). His only success with these on the stage was with the Bavarian patriotic play *Ludwig der Bayer*. His best work was in the field of nature poetry, incorporated in his *Gedichte* (1868, augmented in later editions up to 1886).

GREIFFENBERG, CATHARINA REGINA VON (Seyßenegg nr. Amstetten, Austria, 1633–94, Nürnberg), a distinguished religious poetess, was, before her marriage to Hans Rudolf von Greiffenberg, a baroness (Freiin) von Seyssenegg. A Protestant in Catholic Austria at the time of the Counter-Reformation, she more than once took refuge at Nürnberg. She fought a long losing battle for her property rights in Austria, finally giving up and settling in Nürnberg in 1679. Her writing was encouraged by S. von Birken and by P. von Zesen (qq.v.), but her religious poetry is independent and individual. She wrote principally sonnets, which are both sensitive and closely wrought, reflecting a mind and character of distinction (*Geistliche Sonette*, 1662). She also wrote a book of devotional meditations in which prose and poems, especially sonnets, are intermingled (*Der Allerheiligst-und Allerheilsamsten Leidens und Sterbens Jesu Christi Zwölf andächtige Betrachtungen*, 1672).

A selection (*Gedichte*) was edited by H. Gersch (1964).

GREINZ, RUDOLF (Pradl nr. Innsbruck, 1866–1942, Innsbruck), studied at Graz and Innsbruck universities, and in 1889 became an author. He lived at Merano (then Austrian), Innsbruck, and at Munich (1911–33), finally returning to a village near Innsbruck. His work belongs almost exclusively to the world of Heimatkunst (q.v.), but several of his novels and collections of stories reached a wider public than that of his native Tyrol. His principal Tyrolean novels are *Das Haus Michael Senn* (1909), *Allerseelen* (1911), *Die Stadt am Inn* (1917), *Der Garten Gottes* (1919), and *Mysterien der Sebaldusnacht* (1926). Even more widely read were the volumes of stories *Im Herrgottswinkel* (1905), *Bergbauern* (1906), *Aus'm heiligen Land'l* (1908), and *Das fröhliche Dorf* (1932). Early in his career he produced in collaboration with J. A. Kapferer two collections of dialect poetry (*Tiroler Schnadehupfeln*, 2 vols., 1889–90, and *Tiroler Volkslieder*, 2 vols., 1889–93). When writing for the journal *Jugend* in Munich he used the pseudonym Tuifelermaler Kassian Kluibenschädel.

Grenzen der Menschheit, a poem written by J. W. Goethe (q.v.), probably in 1781, and first published in 1789 in *Goethes Schriften*. Written in the short unrhymed lines of Goethe's hymnic style, it expresses an acceptance of the limitations of human life.

Grenzwald, Der, an unfinished novel (Fragment) by H. von Doderer (q.v.), published posthumously in 1967. The fragment bears also the title *Roman Nr. 7* (q.v.) *Zweiter Teil*, and is

accompanied by extracts from Doderer's diaries, and a four-page note by D. Weber. Doderer clearly regarded this as a new and important advance in his work, an emancipation from the quasi-autobiographical elements of his earlier works. The Grenzwald, the thinning wood, of which the original is at the edge of the garden of a now demolished villa in the Haltertal near Hütteldorf (Wiener Wald) is meant to be a symbol of his own emergence into clarity.

The novel begins in 1919 with the return of a prisoner of war to Vienna, a marginal figure from *Die Wasserfälle von Slunj* (q.v.); it then switches, in retrogressive chronology, to pre-war days depicting Dr. Halfon and his putative father, Ephraim, and his mysterious mother, who was thirty-five years younger than her husband and died at the age of twenty-three. The book then moves to the middle of the war, to the capture of large numbers of officers in the Brussilov offensive (1916), and so to a large P.O.W. camp for officers at Krasnoyarsk in Siberia. Here among many others are the medical officer Dr. Halfon, Baron Ernst von Rottenstein (briefly mentioned in *Die Wasserfälle von Slunj*), and Heinrich Zienhammer, the real father (though neither knows it) of Halfon. The almost idyllic camp life is little disturbed by the March and October Revolutions, but the coming of the White Russian forces, including the Czech Legion, changes the picture. Two companies of Russian conscripts mutiny and are massacred, and, in the Austrian camp, nine Hungarian officers are accused of plotting a *coup*, condemned without evidence, and shot. Elsa Brandström (q.v.) makes unavailing efforts to save them. Zienhammer ·is called upon to identify the Hungarians and is widely believed to have denounced them. The book breaks off with the prospect of his return with others via the Far East. From the jottings of Doderer it is clear that Zienhammer, alarmed by the prosecution of officers responsible for the deaths of comrades in Russia, shoots Ernst von Rottenstein in the Grenzwald, because he believes him to be a witness of his apparent complicity in the shooting of the Hungarians. Section 7 contains a notable tribute to Elsa Brandström. The work has greater tension and fewer arabesques than the earlier novels.

Gretchen, the principal female character in Goethe's *Urfaust* (q.v.). The name is often used also of the same character in *Faust I*, though there she is known in full as Margarete except in the scenes *Gretchens Stube*, *Gretchen am Spinnrade*, *Am Brunnen*, *Zwinger*, *Nacht. Straße vor Gretchens Tür*, *Dom*, *Walpurgisnacht*, and here and there in *Kerker*.

GRETCHEN, FRANKFURTER, customary designation of a young woman named Gretchen, with whom Goethe fell in love at 14 (1764). She associated with rather questionable friends and had to leave Frankfurt as a result of police inquiries into the activities of her circle, but her conduct towards Goethe seems to have been irreproachable. The story is told by Goethe in *Dichtung und Wahrheit* (q.v.), Bk. 5. Nothing is known of it from any other source.

Grete Minde, a story by Th. Fontane (q.v.), published in 1879. It bears on the title-page the words *Nach einer märkischen Chronik,* and is in fact based on two such chronicles dating from the 17th c. Grete Minde, daughter of a patrician of Tangermünde by his second marriage, is brought up after his death by the wife of her older half-brother, and, when this woman bears a child, is treated as a servant in the house. She runs away with a boy friend, but he dies three years later, and she is left destitute with her child. She returns to Tangermünde ready to abase herself before her half-brother, who rejects her and refuses her her inheritance. When the municipality supports his action, she, in half-crazed revenge, sets fire to the town. The historical fire of Tangermünde took place in 1617. (Film version by H. Genée, 1977.)

GRETSER or **GRETSCHER, JAKOB** (Markdorf, Swabia, 1562–1625, Ingolstadt), entered the Society of Jesus in 1578, and was a professor at Ingolstadt from 1588 to 1616. He was one of the outstanding exponents of Jesuit drama (see JESUITENDRAMA), and his plays (in Latin) include *Comoedia de Timone* (1584), *Comoedia de Lazaro resuscitato* (1584), *Dialogus de Nicolao episcopo* (1586), *Comoedia de Nicolao Unterwaldio* (1586 dealing with NIKOLAUS von der Flüe, q.v.), *Comoedia de Itha Doggia* (1587), and *Dialogus de Udone Archiepiscopo* (1587, a story of the damnation of a wicked prelate, which is his principal work).

GRIECHEN-MÜLLER, nickname of Wilhelm Müller (q.v.).

GRIEPENKERL, ROBERT (Hofwyl, Switzerland, 1810–68, Brunswick), studied at Berlin University and became a professor at the Brunswick Carolinum (q.v.). A man of musical interests, he published in 1843 *Die Oper der Gegenwart.* He wrote several plays, among which those dealing with subjects taken from the French Revolution (q.v., *Maximilian Robespierre,* 1851, and *Die Girondisten,* 1852) deserve mention. He also wrote a play on the exiled Napoleon (*Auf St. Helena,* 1862) and a volume of *Novellen* (1868).

GRIES, JOHANN DIEDERICH (Hamburg, 1775–1842, Hamburg), studied from 1795 to 1801 in Jena, where he met most of the great literary personalities of Weimar and Jena; he became particularly friendly with the group of Romantic writers living there at the time. He wrote poems, but is especially noteworthy for his translations from Italian and Spanish. *La Gerusalemme liberata* appeared as *Tassos Befreytes Jerusalem* (1800–3), *Rolando furioso* as *Ariostos Rasender Roland* (1804–8), and a seven-volume translation of Calderón (*Calderons Schauspiele*) was published between 1815 and 1842.

GRIESE, FRIEDRICH (Lehsten, Mecklenburg, 1890–1975), of peasant stock, became a schoolmaster in Mecklenburg. He is the author of regional novels (see HEIMATKUNST), which emphasize the power of the elements and the recalcitrance of the soil. He leaned towards the ideas of Blut und Boden promoted by the National Socialist regime. His rather melancholy novels include *Winter* (1927), *Der ewige Acker* (1930), *Das Dorf der Mädchen* (1932), and *Das letzte Gesicht* (1934). He also wrote a historical story set in Napoleonic times, *Die Wagenburg* (1935).

Grillenvertreiber, Der, see SCHILDBÜRGER, DIE.

GRILLPARZER, FRANZ (Vienna, 1791–1872, Vienna), the son of a lawyer, studied law at Vienna University, graduating with distinction in 1811. His father's death in 1809 imposed family responsibilities, which he faithfully performed. He is thought to have derived his literary gift from his mother's side, the Sonnleithner family (see SONNLEITHNER, J.). For a time he held an unpaid post in the Court (now National) Library (see LIBRARIES). In 1814 he began a career in the civil service, becoming greatly indebted to the sympathetic support of Count Stadion (q.v.) who, as finance minister, promoted him in 1823 to a post (Ministerialkonzipist) which made Grillparzer virtually his personal assistant and enabled him to look after Grillparzer's interest more effectively. This happy relationship, emphatically acknowledged by Grillparzer in his autobiography, ended with Stadion's death in the following year. In 1832 Grillparzer became director of the Hofkammerarchiv, but his heart was never in his work. He would have preferred a library appointment, which he failed to obtain in 1834 at the University Library, and in 1844 at the Court Library, when F. Halm (q.v.) was the successful candidate. Grillparzer resigned in 1856. In spite of some patriotic verse and poetic compliments to the imperial family, Grillparzer's relationship with the court was always uneasy. A poem written in Italy, *Die Ruinen des Campo vaccino in Rom* (q.v., 1819), was judged by the Emperor Franz (see FRANZ II) to be an offence against the Catholic Church. Grillparzer, however, owed the production of his first historical play to the intervention of Franz's consort, the Empress Caroline Augusta. He never abandoned his devotion to the House of Habsburg or his belief in the divine sanction of monarchy. He expressed it in the year of crisis (see REVOLUTIONEN 1848–9) by composing a fervent patriotic poem, *Feldmarschall Radetzky*, but he nevertheless held liberal views. In 1861 he became a member of the new Herrenhaus (Upper House), and voted in 1868 against the Concordat with Rome.

The tragedies in his family and the complex story of his relations with women intensified Grillparzer's hypochondria and prolonged periods of depression. In 1819 his mother, to whom he was devoted, committed suicide. His brother Adolf drowned himself at the age of 17, and his brother Karl gave himself up to the police for a murder he had not committed. From his youth, Grillparzer was torn between fulfilment in love and an all-absorbing devotion to his creative impulses. His first love was for Lotte Pichler in 1817. For some years he loved Charlotte von Paumgartten, who had married his cousin, a high-ranking civil servant, in 1818. In 1822 he discovered that he had been the object of the passionate love of Marie von Piquot (1798–1822). The enigmatic beauty and youth of Marie von Smolenitz kindled another passion which persisted after her marriage to M. M. Daffinger (q.v.). His relationship with Katharina (Kathi) Fröhlich (1800–79), one of a quartet of musical sisters, whom he had first met in the winter of 1820–1, resulted in an engagement. But he was too certain of her devotion, of the likeness of their temperaments, and of his own vocation, to bring himself to the point of marriage. From 1826 he lived, except for the years 1830–49, in a house in the Spiegelgasse in which the Fröhlich sisters had their home. Kathi, the ewige Braut, as she is often called, cared for him until his death and was his literary executor.

Grillparzer owed his first success on the Viennese stage to J. Schreyvogel (q.v.), who was in charge of the Burgtheater; persuaded by his friend J. L. Deinhardstein (q.v.), he had published a sample of Calderón's *La vida es sueño* in German trochaic verse. Schreyvogel, who read it, recognized his unusual talent, and gave him encouragement. The one-act comedy *Die Schreibfeder* was completed in 1808–9, and a similar play, *Wer ist schuldig?*, in 1811. In 1810 *Blanka von Kastilien*, a tragedy influenced by

Schiller's *Don Carlos* (q.v.), was rejected for performance, but *Die Ahnfrau* (q.v., 1817) was an immediate fashionable success. He was resolved, however, not to write another fate-tragedy (see SCHICKSALSTRAGÖDIE). The influence on his next play was Goethe's *Torquato Tasso* (q.v.). Notwithstanding his personal identification with Tasso, Grillparzer succeeded in giving *Sappho* (q.v., 1819) his own, distinct signature. The play earned him the status of Theaterdichter of the Burgtheater. He wrote two more classical tragedies, the trilogy *Das goldene Vlieβ* (q.v., performed in 1821 and published in 1822), and *Des Meeres und der Liebe Wellen* (q.v., completed in 1829 and published in 1840), a treatment of the tale of Hero and Leander. The dedication to the trilogy, *Zueignung an Desdemona*, refers to Charlotte von Paumgartten, who died in 1827. Both plays proved too demanding for the Viennese public; in both, Grillparzer demonstrated his reaction against Weimar classicism. He detected in himself a kinship with Schopenhauer (q.v.) long before the latter's influence upon men of letters became clearly discernible. Grillparzer's presentation of the destructive dualism within man was a firmly established feature of his portrayal of character. It is balanced by variations on the Greek concept of fate, and of gods who represent an absolute ideal of justice tempered by neither love nor hatred. This, his own observation, explains his critical response to Goethe's *Iphigenie auf Tauris* (q.v.). His idea of writing a second part to Goethe's *Faust I*, first contemplated *c.* 1811, materialized only in a brief scene (1813, *Waldgegend*). The ideas went into other plays as diverse as *Sappho* and *Der Traum ein Leben* (q.v., completed in 1831–2 and published in 1839).

Grillparzer turned to patriotic drama in *König Ottokars Glück und Ende* (q.v., written in 1823, performed and published in 1825). The shift of emphasis from hubris towards human fallibility, and from ruler to servant, is a distinguishing feature of the tragedy *Ein treuer Diener seines Herrn* (q.v., completed in 1826 and published in 1830), which shows the influence of Lope de Vega (1562–1635); Grillparzer sought to reconcile the conflicting forces of body and mind. His fragment *Hannibal und Scipio* (published in *Album* under the title *Hannibal* in 1838, performed in 1869) was in part provoked by his dislike of Grabbe's tragedy *Hannibal* (q.v.). In the late 1840s and the early 1850s he completed *Ein Bruderzwist in Habsburg*, *Libussa*, and *Die Jüdin von Toledo* (qq.v.), but withheld them from publication. Among other historical projects *Esther* stands out. The fragment was published in *Dichteralbum* (ed. E. Kuh, q.v.) in 1862 and performed in 1868.

Like most Viennese, Grillparzer had a great interest in the Viennese Volksstück (q.v.) as well as the Singspiel. Seelenmusik was his own description of Mozart's opera *Figaros Hochzeit* (see Mozart). The idea of writing for opera had begun with *Sappho*, and in 1823 he wrote *Melusina* (q.v.) as a libretto for Beethoven (q.v.). He was not discouraged when it was not set to music; the contact with the admired composer sufficed, and his sense of affinity with Beethoven's individuality grew almost to the point of self-identification. He last met Beethoven in 1826, more than twenty years after their first acquaintance. *Der Traum ein Leben* was Grillparzer's closest approach to the Volksstück. His comedy *Weh dem, der lügt!* (q.v., 1838) failed signally at its first performance on 6 March 1838. For Grillparzer, who, though proof against malicious attacks in the press, was sensitive to the judgement of the theatre-going public, this failure was decisive, and he turned his back on the theatre. Thirteen years later H. Laube (q.v.) reintroduced Grillparzer's plays to the Burgtheater.

Grillparzer wrote two Novellen, *Das Kloster bei Sendomir* (q.v., published in *Aglaja* in 1828) and *Der arme Spielmann* (q.v., published in *Iris* in 1848). The element of irony which this work contains was already evident in 1808 in a fragment called *Das Narrennest*.

Grillparzer's poetry covers a wide range. He had a gift for graceful occasional poetry and for the formulation of epigrams. *Abschied von Gastein* (1818) and *Entsagung* (written in Paris in 1836) open a selection of the 1840s. The cycle *Tristia ex Ponto* (q.v.), containing 17 poems, was published in *Vesta* in 1835. In *Die Schwestern* (1836) Grillparzer laments the deterioration of German literary style, the two sisters being 'Prosa' and 'Poesie'.

Grillparzer wrote extensively on his travels to Italy (1819), and France and England (1836), and to Greece and Turkey (1843). His diary complements his essays on aesthetics. His *Selbstbiographie*, written in 1853–4, covers his life up to 1836. His poem *An die Sammlung* (1833) expresses the need for prolonged periods of complete seclusion communing with the creations of his imagination. He had contacts with many Viennese writers, musicians, and artists, among them Bauernfeld, F. Schubert (qq.v.), who set to music Grillparzer's *Berthas Lied in der Nacht* (1819), Raimund, the singer J. N. Vogl, and M. von Schwind (qq.v.). Grillparzer's tour of Germany in 1826 culminated in three meetings with Goethe in Weimar, at the end of September and on 1 and 2 October. The last of these was terminated abruptly as a result of Grillparzer's inborn shyness. Grillparzer was a founder member of the Austrian Akademie der Wissenschaften in 1847, and from 1856 bore the honorary title

Hofrat in recognition of his services. Five years after the separation of Austria from Germany, the Empress Augusta of the new German Empire honoured him on the occasion of his eightieth birthday.

In 1872 H. Laube and J. Weilen edited Grillparzer's collected works; *Sämtliche Werke, historisch-kritische Ausgabe*, edited by A. Sauer and R. Backmann, 42 vols., was published between 1909 and 1948. *Werke* (4 vols.), ed. P. Frank and K. Pörnbacher, appeared 1960–5. Grillparzer is the subject of a Novelle, *Grillparzers letzter Ausgang* (in *Unsterbliche*, 1919), and of the play *Gewitter im Vormärz* (1943), both by R. Hohlbaum (q.v.), and of a novel by F. Schreyvogl (q.v.), *Sein Leben ein Traum* (1937, the new title of *Grillparzer*, a novel published in 1935).

GRIMM, Hans (Wiesbaden, 1875–1959, Lippoldsberg/Weser), worked in a London office in 1896, went to Port Elizabeth, Cape Colony, in 1897, and set up in business in East London in 1901. He returned to Germany in 1910, and from 1911 devoted himself to writing, simultaneously studying politics, first at Munich University, then at the Colonial Institute in Hamburg. His first published writings were the collections of Novellen *Südafrikanische Novellen* (1913) and *Der Gang durch den Sand* (1916), the second of which contains, among other tales, some stories concerning the South African invasion of German South West Africa in 1914. From 1914 to 1917 Grimm served in the German artillery in Europe, but was detailed in 1917 to prepare a document alleging French atrocities in the conquest of the German colony of Togoland. The outcome of this was *Der Ölsucher von Duala* (1918), described as *Ein afrikanisches Tagebuch bearbeitet von Hans Grimm*. The book, which is strongly nationalistic and anti-French, is not a true diary since Grimm uses fictitious names and concedes that the first part and the conclusion are also fictitious, but he avers (in a preface to the second edition, 1931) that the book is based on authentic documents. Grimm is best known for *Volk ohne Raum* (2 vols., 1926). This novel, set in southern Africa, is spread over the years 1887 to 1925; it advocates a vigorous German imperialist policy. Its title became one of the principal slogans of the NSDAP (q.v.); the novel itself is diffuse and lacking in power. Thereafter Grimm made himself the leading propagandist for the restoration of the former German colonies and especially of South West Africa, which he visited in 1927–8. *Das deutsche Südwester-Buch* appeared in 1929, and the seven Novellen grouped as *Lüderitzland* in 1934. Grimm's best work is in his first two published volumes, the tersely written and compact Novellen of the South African veldt.

His late works are autobiographical (*Rückblick*, 1950; *Leben in Erwartung*, 1952; *Suchen und Hoffen*, 1960).

GRIMM, HERMAN, also HERMANN (Kassel, 1828–1901, Berlin), eldest son of W. Grimm (q.v.), devoted himself to the history of literature and art, and was, from 1873, Professor für neuere Kunstgeschichte (history of modern art) at Berlin University. His works include *Das Leben Michelangelos* (2 vols., 1860–3), and *Goethe* (2 vols., 1877), which initiated a revival of interest in Goethe in the late 19th c. In his earlier years Grimm wrote the plays *Armin* (1851) and *Demetrius* (1854), a collection of *Novellen* (1856), and a novel, *Unüberwindliche Mächte* (1867). He was particularly notable as an essayist: *Essays* (4 vols.) appeared 1859–90.

GRIMM, JACOB (Hanau, 1785–1863, Berlin), after studying law at Marburg University from 1802 to 1805, entered the Hessian civil service at Kassel. In 1808 he became Librarian to Jérome, King of Westphalia, in whose realm Kassel was included. He was in the Hessian delegation at the Congress of Vienna (see WIENER KONGRESS), and on his return resumed employment as a librarian. Jacob Grimm was always in the closest collaboration with his brother Wilhelm (q.v.); in 1830 both were appointed to chairs at the Hanoverian university of Göttingen. By this time the two brothers, of whom Jacob was the leading and more original spirit, had written their two famous works of folk-lore in the spirit of the Romantic movement (see ROMANTIK), the *Kinder- und Hausmärchen* (q.v., 2 vols., 1812–14) and *Deutsche Sagen* (2 vols., 1816–18). They had also laid the foundations of their Germanistic studies in the periodical *Altdeutsche Wälder* (1813–16), while Jacob had earlier published *Über den altdeutschen Meistergesang* (181·). Jacob alone was responsible for the systematic philology of the historical *Deutsche Grammatik* (1819–37), and his *Deutsche Mythologie* (1835) made an important contribution to the Germanic theory of folk-lore. In 1837 the two brothers were among the seven dissident professors (see GÖTTINGER SIEBEN) dismissed by Ernst August (q.v.), who succeeded William IV of Great Britain as king of Hanover.

Grimm returned with his brother to Cassel until in 1841 both brothers once more received a simultaneous invitation, this time to Berlin, to become members of the Prussian Akademie der Wissenschaften. Jacob Grimm's achievement in his Berlin years was his *Geschichte der deutschen Sprache* (2 vols., 1848). His most ambitious project, the great dictionary (*Deutsches Wörterbuch*), the German equivalent of the *Oxford English Dictionary*, began to appear in

1852, but got no further than four volumes in his lifetime. This monumental enterprise, in which Wilhelm Grimm assisted his brother until his death in 1859, was continued by later scholars, and was completed in 1960 after more than a century of lexicographical and etymological research. A complete edition of the writings of J. and W. Grimm in 62 vols. began to appear in 1974.

GRIMM, WILHELM (Hanau, 1786–1859, Berlin), was the younger brother of Jacob Grimm (q.v.) and his closest collaborator. He studied with his brother at Marburg university, was in library administration with him at Kassel, and was elected simultaneously with Jacob to a chair at Göttingen; both brothers were among the professors dismissed in 1837 for their political stand (see GÖTTINGER SIEBEN). Together the two brothers were summoned to Berlin as members of the Prussian Akademie der Wissenschaften in 1841, and indeed the main outlines of their lives differ only in their dates of birth and death. If Jacob was the scholar of genius, Wilhelm was the more poetically gifted and had the greater hand in the story-telling of the fairy-tales (*Kinder- und Hausmärchen*, q.v., 2 vols., 1812–14). He also collaborated in *Deutsche Sagen* (2 vols., 1816–18), in the periodical *Altdeutsche Wälder* (1813–16) and, until his death, in the monumental *Deutsches Wörterbuch* (1852–1960). He wrote *Die deutsche Heldensage* (1829) and published a number of medieval works, including *Graf Rudolf* (1828), Freidank's *Bescheidenheit* (1834), the *Rolandslied* (1838), Konrad von Würzburg's *Goldene Schmiede* (1836), and *Silvester* (1841, qq.v.). A complete edition of the writings of J. and W. Grimm (62 vols.) began to appear in 1974.

GRIMMELSHAUSEN, JOHANN (HANS) JAKOB CHRISTOFFEL VON (Gelnhausen, Hesse, 1622–76, Renchen, Baden), came of a noble family which had sunk below its original station. As a boy he was carried off by Hessian and Croat soldiers in 1635, spending the next fourteen years first as a boy soldier, then as a musketeer, and finally, thanks to his quick and alert mind, as regimental clerk. His service with various regiments in the Thirty Years War (see DREISSIGJÄHRIGER KRIEG) was not confined to one side. In 1649 he married, and became steward at Gaisbach to his former commanding officer, Colonel von Schaffenburg. In this period of relaxation after the hard and restless years of soldiering, Grimmelshausen began to write. In 1662 he became castellan at Ullenburg Castle near Gaisbach, and is believed to have made good use of its substantial library. In 1665 he took an inn (Zum Silbernen Stern) in Oberkirch, but gave this up two years later to become magistrate of Renchen in Baden, where he spent the remainder of his life. In age he resumed the 'von' of nobility.

Grimmelshausen's novels were written in the last ten years of his life. He began with translations before writing *Der satirische Pilgram* (1666), followed in the same year by *Der keusche Joseph*. His masterpiece, *Der abenteuerliche Simplicissimus Teutsch* (q.v.) appeared in 1669, the first edition being dated 1668. It should be noted that the spelling of titles varies throughout, at times owing to the different spellings of editions published in Grimmelshausen's lifetime, or to uncertainties about their authenticity, or simply to modernizations in subsequent editions. The second edition, *Neueingerichteter und vielverbesserter abentheurlicher Simplicissimus* (1669), appeared virtually simultaneously with the *Continuatio des abentheurlichen Simplicissimi oder Schluß desselben*, a further volume resulting from the immediate success of the work, and bringing it to a total of 6 vols.

The next cycle of works, known as *Simplicianische Schriften* (which include the *Continuatio*), followed in swift succession: *Trutz Simplex: Oder Ausführliche und wunderseltzame Lebensbeschreibung der Ertzbetrügerin und Landstörtzerin Courasche* (q.v., 1669), the origin of Mother Courage in Brecht's play *Mutter Courage und ihre Kinder* (q.v.), *Der seltzame Springinsfeld* (q.v., 1670), *Das wunderbarliche Vogel-Nest* (q.v., 1672), in which Springinsfeld again appears. The popular moralizing works *Der erste Bärenhäuter* and *Die Gaukeltasche*, and *Die zweiköpfige Ratio Status* appeared in 1670, as did also the almanac *Ewig währender Calender* containing autobiographical material. The almanac *Wundergeschichten Kalender* came out in three successive years (1670–2), *Der Teutsche Michel*, *Der Stolze Melcher*, and *Proximus und Lympida*, Grimmelshausen's last novel, in 1672.

A further title adds to the range of Grimmelshausen's immense output: the courtly novel *Dietwalts und Amelinden anmuthige Lieb- und Leidsbeschreibung* (*Dietwalt und Amelinde*) of 1670, which extends the scope of his literary models; none influenced him more than the picaresque Abenteuerroman and Schelmenroman (qq.v.). *Simplicissimus* nevertheless stands by itself as the most remarkable single literary work in German of the 17th c.

Grimmelshausen cautiously availed himself of pseudonyms which are anagrams of his name; they include German Schleifheim von Sulsfort, Samuel Griefnson von Hirschfeld, Philarchus Grosses von Trommenheim, and Michael Rechulin von Sehmsdorf. His authorship was not established until the 19th c.

The first editions of Grimmelshausen's col-

lected works appeared in 1683–4, 1685–99, and 1713; the 19th-c. annotated edition by H. Kurz stands out (4 vols., 1863–4); *Gesammelte Werke in Einzelausgaben*, ed. R. Tarot, 1966 ff., includes *Der Abenteuerliche Simplicissimus Teutsch und Continuatio des abentheurlichen Simplicissimi, Lebensbeschreibung der Ertzbetrügerin und Landstörtzerin Courasche, Der Seltzame Springinsfeld*, and *Das wunderbarliche Vogel-Nest*.

Grimmsches Wörterbuch, see WÖRTERBÜCHER.

GRISEBACH, EDUARD (Göttingen, 1845–1906, Charlottenburg), was a German diplomat, retiring in 1889. He wrote the verse epics *Der neue Tanhäuser* (1869) and *Der Tanhäuser in Rom* (1875), and is the author of a biography of Schopenhauer (q.v., 1897).

Griselda, the heroine of the last story (*Giornata decima*, Novella X) in the *Decameron* (q.v.), which was retold in Latin (*Griseldis*) by Petrarch in 1373. Griselda, married to Marchese Gualtiere di Saluzzo, is subjected by him to a succession of humiliating and painful tasks, all of which she performs humbly and obediently, thereby gaining his love and esteem. Among German versions of the story are those by E. Groß, H. Steinhöwel (qq.v., both in the 15th c.), F. Halm's play *Griseldis* (q.v.), the play *Griselda* by G. Hauptmann (q.v.), and the ballad *Graf Walter* by G. A. Bürger (q.v., 1789).

Griseldis, a blank-verse play (dramatisches Gedicht in fünf Akten by F. Halm q.v.), written in 1833–4, first performed at the Burgtheater, Vienna, on 30 December 1835, and published in 1837. It adapts Boccaccio's *novella* of Griselda (q.v.). Halm has set the play at the court of King Arthur and given it a new psychological twist; for, when Griseldis, after patient and steadfast endurance of the trials imposed upon her, discovers that her misfortunes were fabricated, she turns from her husband Sir Percival and leaves him for ever. The play enjoyed an immense success, which was partly due to the performance of Julie Rettich (q.v.), who took over the part of Griseldis after the first performance.

GROB, JOHANNES (Enzenschwyl, 1643–97, Herisau), Swiss satirist, who served in his youth in the household troops of the Elector of Saxony, and in 1664 had a linen business in Switzerland (Lichtensteig and Enzenschwyl); he took part in public affairs but withdrew because of the prevailing religious stringency; in 1675 he moved to Herisau. Grob wrote conventional odes, but his best work is in his numerous satirical epigrams. He published *Dichterische Versuchsgabe* (written in 1666) in 1678 and *Der Treugemeinte Eidgenössische Erwecker*, a political tract, in 1689; his *Poetisches Spazierwäldlein* appeared posthumously in 1700.

Grobe Hemd, Das, a satirical comedy (Volksstück) by C. Karlweis (q.v.), first performed in the Deutsches Volkstheater, Vienna, in February 1897, and published in 1901. Max Schöllhofer returns home full of socialistic ideas expressed in pompous clichés and adopts a superior tone to his father, a rich and good-humoured burgher. His sister Franzi is engaged to Rudolf, son of Wendelin, a civil servant. The Wendelins welcome this engagement because of Schöllhofer's wealth. Schöllhofer, tired of being harangued by his son, simulates poverty, selling his apartment and furniture. Max finds the reality of poverty different from what he had imagined and soon abandons his lofty views. Franzi, however, adapts herself to the new situation, although Rudolf Wendelin, on his mother's orders, gives her up. Rudolf, however, pulls himself together, makes it up with Franzi, and defies his mother. Schöllhofer then reveals his ruse, and a more sensible Max is united with Wendelin's ward Christine.

Grobianus, a satirical Latin poem by F. Dedekind (q.v.). Published in 1549 and subtitled *sive de morum simplicitate, libri duo*, it ridicules vulgar, coarse, and selfish behaviour in social intercourse, especially at table, ironically praising the advantages of boorishness. In 1552 Dedekind added a third book which includes Grobianus's feminine counterpart, Grobiana. The popularity of the work is attested by twenty editions in the 16th c., followed by others in the 17th c. and a last edition in 1704. Even more successful was the German translation by K. Scheidt, q.v. (*Grobianus, Von groben sitten und unhöflichen geberden*), a poem of 5,000 lines of four-stress couplets, published in 1551. Apart from its intrinsic interest as satirical poetry, *Grobianus* supplies valuable material for the history of manners.

GROGGER, PAULA (Öblarn, Styria, 1892–   ), a Roman Catholic village teacher, wrote novels and tales set in the environment of her native province as well as religious legends and simple poetry. Her first work, the historical novel *Das Grimmingtor* (1926), made a deep impression. Subsequent works include the tales *Das Gleichnis von der Weberin* (1929), *Der Lobenstock* (1935), *Unser Herr Pfarrer* (1946), *Die Mutter*, and *Die Reise nach Salzburg* (both 1958). Among the legends are *Die Sternsinger* (1927), *Die Räuberlegende* (1929), *Das Röcklein des Jesukindes* (1932),

and *Der Antichrist und Unsere liebe Frau* (1949).

Her poems appeared as *Das Bauernjahr* (1947) and *Gedichte* (1954). The collection *Aus meinem Paradiesgarten* was published in 1962 and *Sieben Legenden* in 1977. Recognized as a major author of Heimatdichtung, she is the recipient of a number of Austrian awards.

Groschen, small coin current until the adoption of the Mark (q.v.) 1871–3. 16 Groschen made 1 Gulden (q.v.), 24 Groschen 1 Reichsthaler (see THALER). 'Groschen' is still used colloquially to denote the 10 Pfennig piece.

GROSS, ERHART (Nürnberg, *c.* 1400–50), a Carthusian monk, is best known for his *Grisardis* (*c.* 1436), a translation of the Latin novella *Griseldis*, which is itself an adaptation of the last novella in the *Decameron* (q.v., *Giornata decima*, Novella X), in which the lady is styled Griselda (q.v.). Groß also wrote devotional books, including *Das Nonnenwerk* (1436) and *Das Laiendoctrinal* (1443). *Das Witwenbuch* (1446), a work of solace for a widow, Margarete Mendel of Nürnberg, is in dialogue form.

Groß-Cophta, Der, a comedy (Lustspiel) in five acts written by Goethe in 1791, performed in Weimar on 17 December, and published in the same year. It treats in disguise the famous episode of the Queen's necklace in France, which involved Marie Antoinette, the Cardinal de Rohan, and Cagliostro (q.v.). The Groß-Cophta is Conte Rostro (Cagliostro), de Rohan is represented by the Domherr, and the Comtesse de la Motte is the Marquise. The web of intrigue, controlled by Conte Rostro, is finally torn apart and the parties are arrested. Though described as a comedy, it is one of the more heavy-footed of Goethe's works.

Großdeutsch, term applied in the 19th c. to supporters of a united Germany that would include the German-speaking territories of the Habsburg Empire. The parties favouring the exclusion of Austria and Prussian pre-eminence were termed kleindeutsch.

GROSSE, JULIUS WALDEMAR (Erfurt, 1828–1902, Torbole, Lake Garda), studied painting in Munich and then turned to poetry, becoming one of the poets of the Munich School (see MÜNCHNER DICHTERKREIS). From 1855 to 1867 he was a literary journalist, chiefly in Munich, but was in Leipzig for a short time in 1861. In 1870 he became secretary of the German Schiller Foundation (Schillerstiftung) in Weimar. Grosse was a prolific author of poems, both lyrical and epic, verse plays, and stories and novels. In the Munich manner he laid stress in poetry on formal beauty and elegance. His first work was the tragedy *Cola di Rienzi* (1851), and among his numerous other plays only the tragedy *Tiberius* (1876) deserves mention. His poetry appeared in 2 vols., both entitled *Gedichte* (1857 and 1882). His principal epics were *Gundel vom Königssee* (1864), *Der Wasunger Not* (1872), and *Das Volkramslied* (1889). His prose fiction included *Novellen* (3 vols., 1862–4) and the novels *Der getreue Eckart* (2 vols., 1885), *Das Bürgerweib von Weimar* (2 vols., 1887), and *Der Spion* (2 vols., 1887). His autobiography, *Ursachen und Wirkungen*, appeared in 1896.

Große Netz, Das, a novel by H. Kasack (q.v.), published in 1952. In its recollections of National Socialism, its Kafka-esque frustrations, and its technique of revealing an eerie city by following the impressions of a stranger, it recalls the earlier *Die Stadt hinter dem Strom* (q.v.). This town, from which roads radiate only to lead back into it again, is tyrannically ruled by the 'I.F.E.'. This, it is later disclosed, is a gigantic film corporation using the town to make a film demonstrating the decline of Europe, in order to prevent further decline. Nevertheless, the I.F.E. has no compunction in exploding a bomb which kills 30,000 people but adds realism to the film. Fantasy and satire are extravagantly mixed in the novel, which aims at ironical criticism.

Großjährig, a two-act comedy by E. von Bauernfeld (q.v.), first performed at the Burgtheater, Vienna, in November 1846, and published in *Gesammelte Schriften* (1871–3). It takes the form of a light, almost flippant, drawing-room comedy, but beneath the surface is a political satire directed against the vacuity of the Metternich (q.v.) regime (represented by Herr Blase) on the one hand, and the pretentious talk of the Liberal opposition (Herr Schmerl) on the other. Hermann, representing the German nation, is declared of age and emancipates himself from Blase, his guardian. Blase's plan to retain control of Hermann by marrying him to his niece Auguste collapses. But Hermann after all determines of his own free will to marry her.

GROSSMANN, GUSTAV FRIEDRICH WILHELM (Berlin, 1746–96, Hanover), was for a short time in Prussian diplomatic service, became an actor, and later a theatre director, acting and directing in Gotha, Hanover, and Bremen. He wrote many plays, including a number of semi-realistic Rührstücke (q.v.) in the then fashionable sentimental manner. His best-known play, *Nicht mehr als sechs Schüsseln!*, was performed in 1780. Its curious title alludes to the basic conflict of the work, in which an aristocratic wife

takes offence at her burgher husband's refusal to serve more than six courses at a meal for her relatives. The quarrel is resolved by the conversion of the wife to her husband's standpoint.

In 1784 Schiller attended a performance of *Kabale und Liebe* (q.v.) given at Frankfurt by Großman's troupe.

Großmütiger Rechtsgelehrter oder Sterbender Ämilius Paulus Papinianus, a tragedy written by Andreas Gryphius (q.v.), published in 1659. The hero, a distinguished lawyer and highly regarded personage at the imperial court in the 3rd c. A.D., is pressed to give legal justification to the murder of Geta by his stepbrother, the Emperor Bassianus Caracalla. Papinianus refuses to compromise with justice, and he and his son are executed. It is a tragedy of innocent suffering and stoical endurance, heightened by vivid acts of horror on the stage; at the same time Papinianus' resistance unto death is a triumph for right.

Großtyrann und das Gericht, Der, a novel published by W. Bergengruen (q.v.) in 1935. It bears the motto from the Lord's Prayer 'Ne nos inducas in tentationem'. The book is a parable of the temptations and corruptions of power, and its moral is clearly stated in a Präambel. The story, set in an Italian Renaissance city state, turns upon the murder of a monk of dubious character, Fra Agostino. The despotic ruler (the Großtyrann) bids his police chief Nespoli discover the murderer in three days on pain of death. Nespoli produces as supposed murderess the corpse of a half-witted girl who has drowned herself. Next Nespoli's mistress attempts to divert suspicion from her lover by forging a confession in the name of her newly dead husband. The mist of deceit and suspicion spreads, demoralizing the city. After Sperone the dyer has confessed to the murder in order to put an end to the confusion, the Großtyrann summons all parties and admits that he himself killed Agostino, who was a traitor. Though the criticism of dictatorial government is mild, it is nevertheless noteworthy that the book achieved publication under the National Socialist regime. It may be seen as an example of innere Emigration (q.v.).

GROSZ, GEORG, assumed name of Georg Ehrenfried (Berlin, 1893–1959, Berlin), a brilliant cartoonist who satirized the military, the industrialists, and the bourgeoisie. His cartoons more than once brought him into conflict with the law. He was also a book illustrator. He lived in America 1932–59 and died shortly after his return to Germany. His autobiography *A little Yes and a big No* appeared in

1946, and in German as *Ein kleines Ja und ein großes Nein* in 1955.

Groteske, Das, a term that derives from the Italian *grotta* (grotto) and was adopted by art historians to define ancient ornamental art before it became a literary term. As a stylistic device it is particularly favoured by writers cultivating irony as an expression of pessimism. Thus the simple technique of employing 'like a distorting mirror of existence' the grotesque and monstrous to evoke horror and to appeal to a crude sense of humour was turned into a means of obtaining a detached view of reality. The grotesque can be used to expose irreconcilable and absurd dichotomies or, didactically, to support reformative aims. It can function in drama, narrative works, and lyric poetry. J. M. R. Lenz, JEAN Paul, L. A. von Arnim, E. T. A. Hoffmann, C. D. Grabbe, G. Büchner, G. Keller, Th. Storm, Th. Mann, F. Kafka, G. Kaiser, C. Sternheim, E. Barlach, J. Ringelnatz, C. Morgenstern, F. Dürrenmatt, and G. Grass (qq.v.) provide in their works a wide range of examples. W. Kayser's tract *Das Groteske* (1957) has proved seminal for a variety of more elaborate definitions.

GROTH, KLAUS JOHANN (Heide, Holstein, 1819–99, Kiel), a countryman by birth and upbringing, from the district of Dithmarschen, became in his teens factotum to the parish clerk, and was then trained at Tondern as a teacher. From 1841 to 1847 he taught in a girls' school in Heide; he spent his leisure hours in the study of the dialect of the region, working so hard that in 1847 he had a breakdown. From then until 1853 he lived at the home of a friend on the Baltic island of Fehmarn, writing his best-known work, the collection of poems in Low German (Holsteiner Platt), which was published in 1852 as *Quickborn*. From 1853 to 1855 he worked at Kiel with K. Müllenhoff (1818–84) on the orthography of Holstein dialect, and in the following years visited Bonn, Leipzig, Dresden, and Weimar. He married in 1858 and became a professor at Kiel University in 1866. He visited England in 1872–3.

Groth's other publications did not equal *Quickborn* (which was expanded in 1853, 1854, and especially in 1871) either in popularity or importance. They include *Hundert Blätter* (1854, in High German), *Vertelln* (2 vols. of stories, 1855–9), the verse idyll *Rotgetermeister Lamp un sin Dochder* (1862), and a further set of stories (*Ut min Jungsparadies*, 1876). He also published treatises on dialect (*Briefe über Hochdeutsch und Plattdeutsch*, 1858; *Über Mundarten und mundartige Dichtung*, 1876). He was awarded the Schiller prize in 1890. Groth's importance lies in his

combination of dialect poet, scholar, and propagandist. The seriousness with which he approached the matter is reflected in his rejection of Reuter's *Läuschen un Rimels* (q.v.) as frivolous. His poetry is simple in structure, warm in tone, and inclined to sentimentality.

Gesammelte Werke appeared in 1893 (4 vols.) and *Sämtliche Werke*, 1952–65 (8 vols.).

GRÜBEL, JOHANN KONRAD (Nürnberg, 1736–1809, Nürnberg), a master tinsmith of Nürnberg, wrote poetry in Nürnberg dialect, which was collected as *Gedichte in Nürnberger Mundart* (1798–1803).

GRUMBKOW, FRIEDRICH WILHELM VON (Berlin, 1678–1739, Berlin), was minister of King Friedrich Wilhelm I (q.v.) from 1713. He was responsible, under his master, for the reorganization and augmentation of the army. In foreign policy Grumbkow sought an understanding with Austria. He appears as a character in Gutzkow's historical comedy *Zopf und Schwert* (q.v., 1844).

GRÜN, ANASTASIUS, pseudonym of Anton Alexander, Graf von Auersperg (Laibach, Austria, now Ljubljana in Yugoslavia, 1806–76, Graz). It is the name by which he is generally known. He was at school in Vienna, and studied law at Graz and Vienna universities. In 1830 he published a volume of love poetry, *Blätter der Liebe*, which bore his true name. The first work to appear under his pseudonym was *Der letzte Ritter* (1830), a cycle of poems written in the Nibelungenstrophe (q.v.), making up a kind of epic on the career of the Emperor Maximilian I (q.v.), the 'last knight' of the title.

The political and satirical collection, *Spaziergänge eines Wiener Poeten* (q.v., 1831), written in the same metre, is aimed at the repressiveness of the Metternich (q.v.) regime. Auersperg, whose father had died in 1818, took over his estates in 1831. For the rest of his life he lived primarily at his castle of Thurn in Carniola (Krain), developing and cultivating his vineyards and forests, and embellishing his park. He married a lady of his own rank in 1839. His pseudonym, which was necessary as a protection against the secret police, was used again for *Schutt* (1835), a cycle of long romances in four-line stanzas rhyming in couplets, which looked forward to a new Austria rising from the rubble of the early 19th c. The identity of Auersperg and Grün being eventually discovered, he had some trouble with the censorship and abandoned political poetry. His lyrical poems (*Gedichte*) were published in 1837. He travelled abroad, visiting England in winter (November 1837) and hastily retiring from the fogs.

After some years of silence Grün published a

comic epic in the Nibelungenstrophe, *Nibelungen im Frack* (1842), which has an eccentric Duke of Merseburg of the early 18th c. as its central figure. In 1848 he was a member of the National Assembly (see FRANKFURTER NATIONALVERSAMMLUNG), from which, however, he resigned. Another comic epic, *Der Pfaff von Kahlenberg*, retells the amusing exploits of the 14th-c. Pfaffe von Kalenberg (see FRANKFURTER, P.). It is in Knittelverse (q.v.). The death of N. Lenau (q.v.) in 1850 ended a friendship which had lasted more than twenty years.

In his later years Auersperg (Grün still retained this name) was prominent in Austrian politics, being a member of the Reichsrat in 1860, a member of the House of Lords (Herrenhaus) in 1861, and president of the Reichsratsdelegation in 1868. He published only one more poetic work, the poems *In der Veranda* (1876). He translated Slovenian folk-poetry of his home region (*Volkslieder aus Krain*, 1850) and English ballads (*Robin Hood*, 1864).

GRÜN, MAX VON DER (Bayreuth, 1926–), a member of the Gruppe 61 (q.v.), who from 1951 to 1964 worked as a miner in the Ruhr district before living solely by his pen. He began with the novels *Männer in zweifacher Nacht* (1962), *Irrlicht und Feuer* (1963), *Zwei Briefe an Pospischiel* (1968), and *Stellenweise Glatteis* (1973), and established his reputation as the most representative writer of West German workers' literature. Other works include the children's book *Vorstadtkrokodile* (1976 BRD and DDR), the novel *Flächenbrand* (1979), and the story *Späte Liebe* (1982). *Klassengespräche. Aufsätze, Reden, Kommentare* was published in 1981. Grün edited with Fritz Hüser, and in collaboration with Wolfgang Promies, *Aus der Welt der Arbeit. Almanach der Gruppe 61 und ihrer Gäste* (1966).

Gründerzeit, the years after 1870, in which, partly as a result of industrial development and partly through the considerable sums obtained as reparations from the French, numbers of companies were floated in Germany, many of which failed, inflicting widespread and severe financial losses. The Gründerzeit occurred in Austria as well as in the German Empire, and was marked by a severe slump in 1873.

GRÜNDGENS, GUSTAV (Düsseldorf, 1899–1963, Manila, while on tour), a leading producer and actor from the 1920s, began in the Hamburg Kammerspiele in 1923, and, as theatre director in Düsseldorf (from 1947) and at the Hamburg Deutsches Schauspielhaus (1955–63), contributed to the establishment of these cities as cultural centres in the Federal Republic. From

1928 until the end of the 1939–45 War he was in Berlin, first at the Deutsches Theater, and from 1934 at the Staatliches Schauspielhaus. Although he is best known for his production of Goethe's *Faust* (q.v.) and established his reputation in Berlin with the part of Mephistopheles (1932), he promoted Shakespearian and classical plays and also modern works ranging from Offenbach to Brecht (*Die heilige Johanna der Schlachthöfe*, q.v., first posthumous production with C. Neher, q.v., 1959). Other productions include plays by T. S. Eliot, Cocteau, Sartre, Mozart's *Die Zauberflöte* (q.v., 1938 in Berlin, 1948 in Vienna), and Verdi's *Don Carlos* (1958 in Salzburg).

Grundgesetz, the legal instrument which embodies the constitution of the Federal Republic of Germany (see BUNDESREPUBLIK DEUTSCHLAND). It was passed on 8 May 1949 by the Parliamentary Council (Parlamentarischer Rat) set up by the three western occupying powers, and came into force on 23 May 1949.

Grüne Heinrich, Der, a novel by G. Keller (q.v.), which exists in two distinct versions with the same title. The second version is the one familiar to most readers. Both versions consist of 4 vols.; they appeared respectively in 1854–5 and 1879–80. To a large extent *Der grüne Heinrich* is an autobiographical novel, but since it was conceived when Keller was 28, and written in his early thirties (1850–3 in Berlin), Heinrich's middle and later life is fictitious, and remains so largely in the second version. In his preface to the original edition, Keller compares the work to a long letter written at widely separated intervals. This image justifies its shapelessness, and also its changes of direction.

The first version of 1854–5 begins in the third person with Heinrich Lee's departure from his Swiss home to study painting. 'Green' because he always wears green, he is the only son of a devoted mother, whose love he is not fully able to appreciate. On his journey he is offered a lift in a coach by a friendly count, who is taking leave of his sister and daughter; to both Heinrich is immediately attracted. In Bk. 1, Chap. 4, on his arrival in the capital city (Keller had Munich in mind), Heinrich takes out a manuscript which is set before the reader and continues until Bk. 3, Chap. 3. Entitled *Eine Jugendgeschichte*, it is written in the first person and provides the retrospective information necessary for understanding Heinrich's character. Heinrich's account of his childhood is followed, in 17th-c. language, by the story of the 7-year-old supposed witch Emerentia (das kleine Meretlein), who is driven to her death in the name of religion. Recollections of school include an episode in which Heinrich brings about the punishment of four boys by lying, an act of which he feels deeply ashamed in later years. The first stirring of sex occurs when he is allowed to take the part of one of the monkeys (Meerkätzchen) for a company playing *Faust* (q.v.) in a one-night stand. He falls asleep in the theatre and on waking beats a thundering tattoo on the tympani, until the actress who plays Gretchen retrieves him and allows him to spend the night at the foot of her bed.

At senior school he becomes marginally involved in an act of mass misbehaviour against a teacher, is treated as the main culprit, and expelled. To his perplexed mother he declares his intention of becoming an artist. While staying with relatives in the country he finds himself attracted simultaneously to two women, the young Anna, and Judith, a handsome widow of 30; with them he leads a double life until his departure for home. There he finds the study of art with his teacher Habersaat unprofitable and works on his own. He pays a further visit to the country, but the atmosphere has changed; Anna is away at finishing school, and he and Judith avoid each other. When the rural community celebrates Shrove Tuesday with an open-air performance of Schiller's *Wilhelm Tell* (q.v.), he takes the part of Rudenz and has some happy moments with Anna, who plays Berta, but afterwards Judith persuades him to spend the evening alone with her.

Bk. 3 begins with Heinrich's decision next morning to return to his vocation as a painter with a competent instructor, Römer; but, owing to the mental derangement from which Römer suffers, the relationship ends in a deplorable breach. Anna is ill, and Heinrich spends much time at her sick-bed; but once more he leads a double life with both women, and on one occasion gazes spellbound on Judith's moonlight bathe. Before he is called up for military service Anna dies, and he breaks with Judith. While he is on parade he detects her as a passenger on one of two wagons which pass, laden with emigrants and their belongings. With this episode *Eine Jugendgeschichte* ends, and the narrative reverts to Munich.

Heinrich works at his painting and makes two artist friends, the Dutchman Lys and the Scandinavian Erikson. All three participate in a pageant representing Nürnberg in the late Middle Ages. Lys, a rich man, has attached himself to an ethereal girl named Agnes; Erikson courts a wealthy and beautiful widow whom Lys, deserting Agnes, attempts to win. Heinrich challenges Lys and seriously wounds him in a duel.

In Book 4 Erikson and Lys go their separate ways, and Heinrich finds his artistic powers flagging. He begins desultory studies at Munich University, and adopts a philosophical pessim-

ism. Having become destitute after four years, he decides to return home on foot. In a state of exhaustion he meets the Count (of Bk. 1, Chap. 3) with the young girl to whom he had been attracted. Now grown to womanhood, she proves to be the Count's foster-child. The Count, who has bought Heinrich's pictures, insists on Heinrich's return to Munich to prove himself as an artist. Heinrich receives two substantial legacies, one from Lys, who has died of his wound, the other from the dealer, who had bought Heinrich's pictures. Once more he decides to go home, but now as a wealthy man. He arrives for his mother's funeral, and learns of the disappointments and hardships of her last years.

The second version of *Der grüne Heinrich* reproduces much of the earlier version in a more logical order. Abandoning the frame story (see RAHMEN), Keller begins with *Eine Jugendgeschichte* in slightly modified form. Anti-clerical passages are toned down, and Judith's moonlight bathe is omitted; two new chapters are inserted (Bk. 3, Chaps. 8 and 9): *Das Pergamentlein*, in which the orphan commissioners reluctantly hand over Heinrich's small inheritance, and *Der Schädel*, in which a skull serving Heinrich for study is made the pretext for a short Novelle about Albertus Zwiehans, whose skull it is believed to be.

At the end of the Shrove Tuesday festivities Lys retires from the duel after a few minutes' fencing. In Bk. 4, Chap. 4, an entirely new and brilliantly executed episode, *Das Flötenwunder*, leads Heinrich to realize that he can temporarily stave off destitution by selling his possessions. The buyer is the curiosity-shop owner, Schmalhöfer. As in the first version, Heinrich sets out for home and is admitted, hungry and wet, to the Grafenschloß. He falls in love with the Count's foster-daughter, Dortchen Schönfund, but cannot utter the words of love which might have transformed his life. In this version Heinrich finds his mother on her death-bed, still able to see him but past speech. He takes a post in local government and resigns himself to a restricted, if useful, life. Judith returns from America as devoted to him as ever. The lovers meet, but no marriage takes place, and after some years Judith dies of an illness which she contracts while nursing sick children.

Der grüne Heinrich is often cited as a classical example of the Bildungsroman (q.v.), but this is true only of the edition of 1879–80. Neither version has a satisfactory conclusion. The first finishes in romantic fantasy; the second has a conciliatory but rather drab ending, suggested by Th. Storm (q.v.). Keller was evidently unable to invent an ending which would be both adequate and consistent.

GRÜNEWALD, MATHIAS (*c.* 1470–1528,

Halle), the traditional name of a painter whose real name is believed to have been Nithardt or Neithardt. He was active at Isenheim nr. Colmar *c.* 1514 and in Frankfurt *c.* 1526. Grünewald's principal work, an elaborate altarpiece of many panels painted for the monastery at Isenheim (Isenheimer Altar), is now in the Colmar museum. He is the subject of an opera *Mathis der Maler* (1938) by Hindemith (q.v.), who wrote his own libretto.

Gruppe 47, a loose association of authors founded in September 1947 by H. W. Richter and A. Andersch (qq.v.), after the journal *Der Ruf* (q.v.), in which both collaborated, had been forbidden by the censorship of the American Military Government. Richter and Andersch at first considered a satirical periodical with the title *Der Skorpion.* The conference at which this plan was discussed and rejected became the first of the annual meetings of the Gruppe 47, held in autumn, and intended to be a forum for reading and criticism. The word member can only be used in a loose sense. Among those associated with the Gruppe 47 from the beginning, in addition to Richter and Andersch, were W. Kolbenhoff, W. Schnurre, and W. Weyrauch (qq.v.). Others who attended regularly or occasionally include Ilse Aichinger, Ingeborg Bachmann, H. Böll, G. Eich, H. M. Enzensberger, G. Grass, W. Höllerer, W. Jens, U. Johnson, S. Lenz, M. Walser, and P. Schallück (qq.v.). Apart from readings and discussions the only function of the Group was to award the prize of the Gruppe 47. The Group had no political or social programme, but encouraged criticism of political or social conditions, and was generally opposed to the values and standards of the Federal Republic and to the type of socialism practised in the DDR. The last full conference of the Group was held in 1967.

Gruppe 61, a short-lived association of working-class writers centred at Dortmund in the industrial Ruhr region. Their aim was to encourage writing about work by the men who did the work. Among the best-known members of the Group (Arbeitskreis) are Max von der Grün, Josef Reding (qq.v.), and Wolfgang Körner (*Versetzung*, 1964). Günter Wallraff (b. 1942) is the best-known representative of reportage, a radical form of workers' literature that is not contained in the original programme. His attempt to promote it led to the formation of a Werkkreis (1970). *Aus der Welt der Arbeit. Almanach der Gruppe 61 und ihrer Gäste* was edited by M. von der Grün and the Group's father, Fritz Hüser (1908–79), director of the city library, who in 1958 had created the Archiv für Arbeiterdichtung und soziale Literatur (since 1974 Institut für deutsche und ausländische Arbeiterliteratur of the city of Dortmund).

Gruppenbild mit Dame, a novel by H. Böll (q.v.), published in 1971. It tells the life story of a war widow, Leni Pfeiffer, *née* Gruytens, born in 1922, who has lived through the Nazi period, the war, the post-war shortages, and the years of affluence. During this time she undergoes a progressive social decline. The portrait of this rather inarticulate and not very intelligent lady (Dame) is only one aspect of the novel, whose principal concern is a reckoning with the past and a criticism of the new world of affluence. The structure of the novel is based on what Böll terms his 'Facetten-Technik' by which the book purports to have been put together by an anonymous narrator, who repeatedly refers to himself in book-review jargon as Verf. (i.e. Verfasser). This narrator composes his book from shreds of evidence (facets) provided by third parties (Zeugen). Through these fragments we learn of Leni's parents, of her brief marriage, of her wartime love-affair with a Russian pioneer officer, Boris Lvovíc Koltowski, of her illegitimate son Lev, and of celestial visions on the switched-off TV screen. The personnel of the Gruppenbild multiplies in the course of these revelations. Böll's political commitment emerges less obviously than in some of his other works. The book is dedicated to three of its characters, Leni, Lev, and Boris.

GRYNAEUS, SIMON (1725–99), who lived in Basel, translated Shakespeare's *Romeo and Juliet* into German blank verse in 1758. This and Wieland's tragedy *Lady Johanna Gray* (q.v.), of the same year, are the first German plays written in this metre.

GRYPHIUS, ANDREAS (Glogau, 1616–64, Glogau), poet and dramatist, was orphaned early. He was educated at Glogau and Fraustadt. From 1634 to 1636 he was a private tutor in Danzig and then accepted a similar appointment under Georg von Schönborn in Freistadt. After Schönborn's death in 1637 Gryphius continued to educate his late employer's sons, and accompanied them on a journey to Danzig and subsequently to Holland, where he himself pursued studies at the University of Leyden. The influence of the Dutch dramatist J. van Vondel is perceptible in his work.

In 1644 Gryphius began a grand tour in the course of which he spent a year and a half in Paris, also visiting Florence, Rome, Ferrara, and Venice, and staying for a year in Strasburg. His first publications appeared during these years abroad and included Latin poetry (notably the religious epic *Olivetum*, 1646). His German poetry is especially rich in sonnets; in addition to an early volume of *Sonnete* (1637), he published *Sonn- und Feiertags Sonette* (1639), *Oden* and *Epigrammata* (1643), *Oden* (with other works, 1650), and *Kirchhofsgedanken*, *Oden* (1655), and *Sonette* (also with other works in *Deutscher Gedichte erster Teil*, 1657). His poems are predominantly sombre, embodying Christian reflections on the vanity and brevity of human life and earthly values. Gryphius returned to Fraustadt in Silesia in 1647, where he married in 1649. Determined to remain in Silesia, he declined offers of chairs in Frankfurt/Oder, Uppsala, and Heidelberg and became in 1650 secretary to the estates (Syndikus) in the principality of Glogau. Though he continued to write lyric poetry, Gryphius in his later years was primarily a playwright; none of his tragedies, however, was performed in his lifetime. *Leo Armenius* (q.v., written in 1646) was included in his *Teutsche Reim-Gedichte* (1650), and *Catharina von Georgien* (q.v., written in 1647), *Ermordete Majestät oder Carolus Stuardus, König von Groß Britanien* (q.v., written 1649), *Beständige Mutter oder Die Heilige Felicitas* (a translation from a work by N. Canisius), and *Cardenio und Celinde oder Unglücklich Verliebte* (q.v., written 1647–9) appeared with the *Kirchhofsgedanken* (1657, reprinted with additions in 1663). *Cardenio und Celinde* is unlike any other of Gryphius's plays. To these tragedies he added *Großmütiger Rechtsgelehrter oder Sterbender Ämilius Paulus Papinianus* (q.v., 1659). His tragedies portray with horrific detail the stoically borne sufferings of great or noble men and women, who are held up for our admiration.

The comedies, marked by an extraordinary linguistic fertility, are entertaining satires. Gryphius is best known as the author of *Absurda Comica oder Herr Peter Squentz* (q.v., 1657 or 1658), commonly referred to as *Peter Squentz*, and *Horribilicribrifax Teutsch oder Wählende Liebhaber* (q.v., 1663). Other titles include *Das verliebte Gespenst* and *Die geliebte Dornrose* (both 1661).

Gesamtausgabe der deutschsprachigen Werke (7 vols.), ed. M. Szyrocki and H. Powell, appeared 1963–8.

GRYPHIUS, CHRISTIAN (Glogau, 1649–1706, Breslau), son of A. Gryphius (q.v.), studied in Jena and Strasburg and later settled in Breslau, where in 1686 he became head (Rektor) of the grammar school. His poems were collected in 1698 as *Poetische Wälder*.

GUARINONI, HIPPOLYT (Trento, 1571–1654, Hall, Tyrol), physician to the Emperor Ferdinand II (q.v.) in Prague and later to a convent at Hall, is the author of *Greuel der Verwüstung menschlichen Geschlechtes* (1610), a work of mixed form containing poems, stories, anecdotes, etc., exposing human folly, vice, and depravity.

Gudrun, see KUDRUN.

GUGGENHEIM, KURT (Zurich, 1896–), a Swiss bookseller turned novelist, is the author of the novels *Entfesselung* (1935), *Riedland* (1938), *Wilder Urlaub* (1941), *Die heimliche Reise* (1945), *Wir waren unser vier* (1949), *Alles in allem* (a tetralogy, 1952–5), *Der Friede des Herzens* (1956). He is also a playwright. *Einmal nur. Tagebuchblätter* (2 vols., 1925–50 and 1951–70) appeared in 1981–2.

Gulden, a silver coin of lower value than the Thaler (q.v.). 16 Groschen made 1 Gulden. In the 19th c. the Gulden was replaced in Germany by the Mark and in Austria by the Krone (qq.v.).

GUMPPENBERG, HANNS, FREIHERR VON (Landshut, 1866–1928, Munich), a Bavarian journalist who settled in Munich in 1898, wrote a number of tragedies, including *Der Messias* (1891), *Alles und Nichts* (1894), *Die Verdammten* (1901), and *Überdramen* (a trilogy, 1902); he distinguished himself notably in parody with *Deutsche Lyrik von Gestern* (1891) and *Das teutsche Dichterroß in allen Gangarten geritten* (1901). He collaborated with H. H. Ewers (q.v.) in founding the Munich cabaret Die elf Scharfrichter (q.v.).

GUNDACKER VON JUDENBURG, a Middle High German poet, who is the author of the poem called *Christi Hort*, written c. 1300. Gundacker was a native of Styria, and, though he possessed some learning, was probably a layman. His poem, which consists of some 5,000 lines, tells the story of mankind from the theological standpoint of salvation, but substitutes for the Last Judgement the legend of Pontius Pilate's life and end. The structural oddities of *Christi Hort* have led some scholars to suppose that more than one author is involved.

GUNDELFINGER, MATTHIAS, probably a priest of Constance, was the author in 1494 of a passion play entitled *Die Grablegung Christi*.

GÜNDERODE, KAROLINE VON (Karlsruhe, 1780–1806, Winkel, Rhineland), wrote, under the pseudonym Tian, romantic poems which were published in 1804 and 1805 (*Gedichte und Phantasien* and *Poetische Fragmente*) and intimately corresponded with Bettina von Arnim (q.v.). Deeply in love with Professor Friedrich Creuzer (q.v.), she took her own life when Creuzer decided against dissolving his marriage.

GUNDLING, Jakob Paul (Hersbruck, 1673–1731, Potsdam), was appointed press adviser and historiographer to King Friedrich Wilhelm I (q.v.) of Prussia. A man of appreciable learning and great vanity, Gundling was included in the King's smoking and drinking circle (Tabakskollegium), where he came to play the part of a court buffoon and was often the butt of practical jokes. His appointment in 1718 to succeed Leibniz (q.v.) as president of the Prussian Akademie der Wissenschaften is thought by some to have been a gesture of contempt for men of learning on the part of the King. His elevation to the peerage in 1724 as Freiherr von Gundling may be regarded as another of the King's practical jokes. It was Gundling whom the theologians of Halle used to provoke Friedrich Wilhelm in 1723 to dismiss C. Wolff (q.v.) from his chair at Halle. He evidently did not always relish his position at court, for he made one unsuccessful attempt to flee. A heavy drinker, he was buried in a wine cask at Bornstedt. An account of his grotesque obsequies is given by Theodor Fontane in *Wanderungen durch die Mark Brandenburg* (q.v., vol. 3).

GUNDOLF, Friedrich, the pseudonym of Friedrich Gundelfinger (Darmstadt, 1880–1931, Heidelberg). Gundolf was a disciple of Stefan George (q.v.), collaborating in *Die Blätter für die Kunst* (q.v.), in which his two volumes, *Fortunat* (1903) and *Zwiegespräche* (1905), appeared. *Shakespeare und der deutsche Geist* (see Shakespeare) was published in 1911, followed by his monumental monograph *Goethe* in 1916 and by *Stefan George* in 1920. In that year he was appointed to a chair at Heidelberg, which he held until his death. *Kleist* appeared in 1922, and *Cäsar* in 1924. For some years after the 1914–18 War he enjoyed an almost pontifical authority. He is also noteworthy as a translator of Shakespeare (1908–14).

GÜNTHER, Agnes (Stuttgart, 1863–1911, Marburg), *née* Breuning, married Professor R. Günther, a theologian. Her long, sentimental novel, *Die Heilige und ihr Narr* (q.v., 2 vols.), was published posthumously in 1913 and became a best seller, being reprinted more than a hundred times. A shorter narrative work, *Von der Hexe, die eine Heilige war*, was also a posthumous publication (1913).

GÜNTHER, Johann Christian (Striegau, Silesia, 1695–1723, Jena), was the son of a doctor in Striegau who fell out with his son during the latter's schooldays and steadfastly refused to receive him back into favour. His first and principal love-affair, with Leonore Jachmann, was the cause of the first quarrel with his father. Günther, who evinced an affectionate but all too mobile temperament, early showed poetic ability. He studied medicine at Frankfurt/Oder, Wittenberg, and Leipzig (1715–19), but neglected his work, and is said to have acquired an addiction to drink, though this is probably an exaggeration. His candidature for the post of Saxon court poet, backed by J. B. Mencke (q.v.), was rejected, allegedly because he was drunk at the interview. During his university years at Leipzig he was captivated by the charms of another Leonore (die Leipziger Leonore). In 1719 he moved to Breslau, restoring his affections to his original love, Leonore Jachmann. But his health failed after years of irregular and dissolute living and privation, and in 1720 Günther freed Leonore from her engagement. He then attempted unsuccessfully to run a medical practice at Lauban, and finally sought escape from his poverty and distress in a marriage with a pastor's daughter, a plan which failed because the girl's father insisted that Günther should be reconciled with his father, who, however, remained obdurate. In 1722 Günther removed to Jena, where an opportunity to resume his medical studies arose—too late, for his health was completely undermined.

At school Günther wrote a tragedy (*Die von Theodosio bereute Eifersucht*, 1715), but his mature works consist of lyric and occasional poetry. The latter served at times as a means of livelihood. His finest occasional poems are patriotic, such as the ode *Auf den zwischen Ihro Kayserl. Majestät und der Pforte An. 1718 geschloßnen Frieden*, an extended essay in florid baroque style. But his most remarkable and original verse is the intensely personal love poetry, especially that addressed to Leonore Jachmann. Günther expresses himself in these poems with an uninhibited directness new to German literature, so that they vibrate with an almost tactile warmth of life, virtually identifying his poetry with his private experience. Günther has usually, and probably unjustly, been summed up in a famous sentence written by Goethe in *Dichtung und Wahrheit* (Bk. 7): 'Er wußte sich nicht zu zähmen, und so zerrann ihm sein Leben wie sein Dichten.'

Günther's poetry was not collected until after his death (*Sammlung von deutschen und lateinischen Gedichten*, 1724–35). *Sämtliche Werke*, ed. W. Krämer (6 vols.) appeared 1930–7 and were reprinted in 1964.

GÜNTHER VON DEM FORSTE, a minor Middle High German poet of Thuringian origin. He probably lived in the second half of the 13th c. and is thought by some to have been a nobleman. His poems include a Tagelied (q.v.) of exceptional length, the style of which exhibits popular as well as courtly elements.

GUOTAERE, DER, a Middle High German poet of the late 13th c., who is the author of a number of Sprüche (see SPRUCH). He is best known for a connected series of five Sprüche dealing with Frau Welt, and deriving from Konrad von Würzburg's *Der Welt Lohn* (q.v.). He furnishes his didactic poetry with homely, proverbial illustrations. Der Guotaere was probably a native of Bavaria or Austria.

GURK, PAUL (Frankfurt/Oder, 1880–1953, Berlin), was a clerk and later an administrator in the municipality of Berlin. He gave up his post in 1924 to devote himself to writing. An individualist, aligned with no school, his output consists chiefly of plays and novels. He wrote more than forty plays, mostly historical, few of which were performed, while a number of them have never been printed. The most notable are probably *Thomas Münzer* (1922) and *Wallenstein und Ferdinand II* (1927). Of his novels *Berlin* (1934), a bitter book, sub-titled *Ein Buch vom Sterben der Seele,* is the best known, but *Meister Eckehart* (1925), *Palang* (1930), *Judas* (1931), *Tresoreinbruch* (1934), and *Tuzub* (1937) should also be mentioned. His late trilogy of novels, *Wendezeiten* (1940–1), appeared under the pseudonym Franz Grau. His last novel was *Der Kaiser von Amerika* (1949).

Gustav Adolfs Page, a Novelle by C. F. Meyer (q.v.), published in the *Deutsche Rundschau* in 1882 under the title *Page Leubelfing.* Meyer presents in five episodes the life and death of the page of Gustavus Adolphus (q.v.), in reality a girl, Auguste Leubelfing, who will be buried and remembered as August Leubelfing, a page who was mortally wounded in the battle of Lützen, in which the Swedish king died. The wealthy Nürnberg businessman Leubelfing has rashly offered his son August as page to the Swedish king. When Gustavus Adolphus in due course sends for the boy to replace the third page he has already lost in battle, neither the father, who has put his son's life at stake without the boy's knowledge, nor young August appreciates the privilege. But August's girl cousin, whose father has been killed in a duel, proudly puts on her father's uniform and follows the Swedish king as August Leubelfing. The true August adopts the name Laubfinger, leaves Nürnberg on his father's business, and by chance witnesses his cousin's death, at which her true sex is discovered. His name, however, remains with her, and he is left only with his alias, Laubfinger. The

page embodies the love and devotion the Swedish king inspires; for 'he' unhesitatingly stakes all to defend the king against the evil intentions of the Duke of Lauenburg.

GUSTAVUS II ADOLPHUS, King of Sweden (Stockholm, 1594–1632, Lützen), succeeded Charles IX in 1611 and in 1620 married Marie Eleonore of Brandenburg. He had proved his skill as a general when he landed on German soil in 1630 for his decisive intervention in the Thirty Years War (see DREISSIGJÄHRIGER KRIEG), which was determined by the advance of Wallenstein (q.v.) to the Baltic coast. His quick successes in battle made him the protagonist of the Protestant cause. In his last battle, at Lützen (q.v.) in 1632, in which he was killed, his army gained a victory as Bernhard von Sachsen-Weimar (q.v.), took command. The Elector Johann Georg (q.v.) of Saxony, in whose country he died, was the most sceptical of the Protestant rulers about his interference in German affairs. On Gustavus Adolphus's death even his enemies expressed admiration for his sense of dedication, his leadership and personal bravery, while those who saw him as a liberator idealized his personality. His dedication to the Protestant cause is undoubted, but he also came as a conqueror, taking possession of Pomerania and planning a *Corpus Evangelicorum* (Nürnberg, 1632) consisting of four 'Circles', the Upper and Lower Rhine, Swabia, and Franconia, which was Swedish-occupied territory. This confederation under the presidency of the Swedish Crown did not materialize, but the Swedish chancellor Axel Oxenstierna (q.v.) continued the struggle for supremacy in the name of Protestantism with varied fortune until the end of the war.

Gustavus Adolphus is prominent in C. F. Meyer's Novelle *Gustav Adolfs Page* (q.v.) and is a background figure in A. Stifter's Novelle *Der Hochwald* (q.v.) and Brecht's play *Mutter Courage und ihre Kinder* (q.v.).

Gustel von Blasewitz, the name by which in Schiller's *Wallensteins Lager* the Erster Jäger knows the Marketenderin. Schiller is said here to have used the name of an innkeeper's daughter in Blasewitz near Dresden, who was well known to him and his friends in Dresden. See WALLEN-STEIN by F. Schiller.

Gute Frau, Die, usual title of an anonymous Middle High German poem, written probably in the second quarter of the 13th c. Its source is a French poem of which the author heard the substance from the French chaplain of a margrave, who has been identified, though uncertainly, with Heinrich V of Baden, who died in 1242. The heroine, 'die gute Frau' or 'la bone dame', is a count's daughter who marries a knight of low degree. They decide, as the result of a religious experience, to abandon wealth and live as beggars. They become separated from each other and from their children, and the wife in time contracts fresh marriages, first with the Count of Blois and then with the King of France; but both marriages are preserved by God from consummation. Finally the original family, husband, wife, and children, are re-united in wealth and prosperity.

Gute Gerhard, Der, a Middle High German poem of almost 7,000 lines written *c.* 1220 by RUDOLF von Ems (q.v.). Its hero is not a noble-man, but a rich merchant, 'der Gute Gerhard' of Cologne. Gerhard has earned his name by the goodness of his heart and the charitable use he has made of his great wealth, notably in journeying to redeem Christian captives from the infidel. Among these is a princess of Norway betrothed to the missing King William of Eng-land. When, after two years, there is still no sign of William, Gerhard's son is betrothed to the princess. But on the eve of the wedding William appears and Gerhard respectfully waives his son's claim.

The appearance of a merchant commoner as central figure hints at a shift in the social struc-ture, but by Gerhard's ready submission Rudolf makes it clear that the fundamental hierarchy of the state is still valid. The poem provides an example for imitation, as appears from the frame in which it is set: the Emperor Otto I (q.v.), warned by an angel against overweening conceit and told of Gerhard as an example of goodness, hears the merchant's story from his own lips. The source, possibly a Latin work, is unknown.

Gute Kamerad, Der, a short poem written by L. Uhland (q.v.) in 1809, and published in 1811. It is a soldier's lament for his comrade killed at his side in battle. Its first line runs 'Ich hatt' einen Kameraden', and it is sung to a folk-melody.

Gute Mensch von Sezuan, Der, a play (Parabelstück) by B. Brecht (q.v.), which was written in 1939–41 and first performed in 1943 at Zurich. The parable aims to demonstrate the moral that to be 'good' man must be cruel as well. Three gods visit the Chinese city of Sezuan, capital of the province of Szechwan, in search of a 'good' human being to restore their blemished image in a world rent with poverty and evil. They find what they seek in Shen Te, a prostitute with a bad reputation but a kind heart. She offers them shelter and is rewarded by the gods with money, which enables her

to secure a living. She rents a tobacco shop, which she considers as the gift of the gods, dedicating its proceeds to good works. But she does not succeed in keeping the balance between goodness (she is referred to as the Engel der Vorstädte) and holding her own in business without the help of her 'cousin', who is her hard and calculating *alter ego* appearing in the dis-guise of Shui Ta. Shui Ta exploits labour to build up a flourishing tobacco factory (he becomes known as Tabakkönig). At the end he appears in the courts, charged with the murder of Shen Te who is expecting the child of Sun, a pilot whom she loves although he has ruthlessly exploited her. When Shui Ta finds that the gods themselves are his judges, he tears off his mask to reveal his true identity as Shen Te. Shen Te confesses that she has failed in her desperate attempt to live up to the gods' image of 'der gute Mensch'. The gods, however, tired and disillu-sioned by their travels on earth, praise her for the sake of their own morale. Yet they are unable to offer a solution to her dilemma other than allowing her to assume the identity of the 'cousin' whenever her existence is threatened. The actor speaking the epilogue urges the public to find a truly 'good ending'.

The play consists of ten scenes and a number of interludes (Zwischenspiele) in which the water-seller Wang, who also figures in the action, communicates with the gods and plays the role of commentator. Among the apposite lyrics accompanying the action are *Lied des Wasser-verkäufers im Regen* (Wang), *Das Lied von der Wehrlosigkeit der Götter und Guten* (Shen Te/Shui Ta), *Das Lied vom Sankt Nimmerleinstag* (Sun), and *Lied vom achten Elefanten* (Arbeiter).

GUTENBERG, JOHANNES (Mainz, *c.* 1398–1468, Mainz), whose name derives, not from his father Friele Gensfleisch, but from his father's residence, the Haus zum Gutenberg in Mainz, invented the printing-press using movable metal type which revolutionized the process of print-ing. Gutenberg is thought to have begun printing in 1436 during a period spent in Strasburg (1434–44). A loan from J. Fust (q.v.) enabled him to build his press, but after some seven years Fust proceeded against Gutenberg to recover the money. By 1458 Gutenberg was bankrupt, and the business was continued by Fust and his son-in-law P. Schöffer (q.v.). From 1465 Gutenberg enjoyed the patronage of the Archbishop of Mainz. The present University of Mainz (q.v.) is named after him.

Gutenberg's most famous production was his Latin Bible (*Gutenberg-Bibel*), completed in 1455 (see BIBLE, TRANSLATIONS OF). Of its 180 copies 30 were printed on vellum. A facsimile edition appeared 1913–23.

GUTERMANN, Sophie von, see La Roche, Sophie von.

'Guter Mond, du gehst so stille', first line of a poem entitled *Mondlied*. Its author is unknown, but it is believed to have been written c. 1780–1800. The tune dates from 1795.

GÜTERSLOH, Albert Paris, pseudonym of Albert Conrad Kiehtreiber (Vienna, 1887–1973, Baden, nr. Vienna), a gifted painter, worked under Max Reinhardt (q.v.) as a stage designer. In the 1914–18 War he was in the official Press H.Q. at Vienna. After the war he continued his artistic activities, in 1929 became a professor at the Kunstgewerbeschule, and, after being forced out of office in 1938, from 1945 at the Wiener Akademie der Bildenden Künste, whose rector he was in the mid-1950s. He was also active as a literary Expressionist, contributing to *Die Aktion* (q.v.); he wrote the novels *Die tanzende Törin* (1910, revised in 1913), an Expressionist work, reissued with a postscript by W. Rasch in 1973, *Innozenz oder der Sinn und Fluch der Unschuld*, and *Der Lügner unter Bürgern* (both 1922 and together receiving the Fontane Prize for 1923). His principal novel, the work of many years, is *Sonne und Mond. Ein historischer Roman aus der Gegenwart* (1962, reissued with an essay by H. Heißenbüttel, q.v., 2nd edn., 1984). It was followed by *Die Fabel von der Freundschaft. Ein sokratischer Roman* (1969; on Faust and Mephistopheles). As a writer he is noted for his eccentric, convoluted style, often described as baroque. He was a close friend of H. von Doderer (q.v.). Collected editions of his work include *Zur Situation der modernen Kunst* (1963) and *Zwischen den Zeiten. Texte und Miniaturen* (1967). A volume of poetry, *Treppe ohne Haus oder Seele ohne Leib. Späte Gedichte*, was published posthumously in 1974 and *Beispiele. Schriften zur Kunst. Bilder. Werkverzeichnis*, with an introduction by H. Hutter, in 1977.

Gute Zeit, Die, a ten-act play in stylized prose by E. Barlach (q.v.), published in 1929. The play, opening with grotesque parodies of an age of disbelief, is a symbolical drama of redemption (Erlösungsdrama). Atlas, the self-styled 'absolute king' of a southern island, promises his disciples through his 'Absolute Insurance' scheme (Absolute Versicherung or AV) the 'gute Zeit' available to people of wealth and no responsibility. However, Countess Celestine, an expectant mother who has come to the island to convalesce, cannot evade a sense of guilt and responsibility aroused by her unborn child, the Erbprinz. Syros, the old white-bearded 'royal'

father of the mountains, who likewise seeks the 'gute Zeit', has tried to attain it by killing his issue, except for Idaos and Kastro who are left behind in the 'böse Zeit' of toil and worry. When Kastro's son Vaphio is to be crucified for a murder, Kastro goes to Syros demanding that he should die in Vaphio's place since he, as father, bore the guilt of creating life in the 'böse Zeit' ('die Schuld an dem Sein im hündischen Leben'). Syros admits his guilt, but breaks his promise to die for Vaphio. Celestine alone has the vision and the will for redemption; for time does not exist apart from man, but through man, and time is good and evil as man is good and evil. In atoning for Vaphio, Celestine accepts the cross which extends towards heaven 'wo die Herrlichkeit der guten Zeit Gestalt bekommen soll'.

GUTTENBRUNNER, Michael (Althofen, Austria, 1919–), was severely wounded in the 1939–45 War. A passionately committed poet, he is the author of verse which commemorates suffering and condemns the political and literary perpetrators of man's 'crucifixion' (his second collection of verse, following *Schwarze Ruten*, 1947, is significantly called *Opferholz*, 1954). Later volumes are *Ungereimte Gedichte* (1959), *Die lange Zeit* (1965), *Der Abstieg* (1975), and *Gesang der Schiffe* (1980). *Spuren und Überbleibsel* (1947) is autobiographical. He has something of the profound melancholy of Trakl and also of the stern linguistic stringency of Karl Kraus (qq.vv.). He is critical alike of the rich style of J. Weinheber and of the extreme experimentalism of Heißenbüttel (qq.vv.). Under the pseudonym Straßburg he contributed for a time to the *Surrealistische Publikationen* edited by M. Hölzer.

GUTZKOW, Karl Ferdinand (Berlin, 1811–78, Sachsenhausen nr. Frankfurt/Main), was the son of a groom employed to school horses in the service of a Prussian royal prince. He was educated at the Friedrichwerdersches Gymnasium in Berlin and then studied theology and philosophy at Berlin University. The outbreak of the July Revolution in Paris (see Juli-revolution) provoked him to break off his studies in 1830 in order to devote himself to politics, in which he was at this time a patriotic radical of the Burschenschaft (q.v.) school. He became a political journalist in 1831, assisting W. Menzel (q.v.) on the *Literaturblatt* in Stuttgart. At this time he wrote a novel (*Maha Guru*, 1833) satirizing, in the guise of an oriental story, the Christian religion as practised in Germany.

In 1833 Gutzkow left Menzel and briefly resumed his studies, first at Heidelberg University, then in Berlin. After an Italian journey with H. Laube (q.v.), his increasingly radical views on social matters led to a breach with Menzel, but Gutzkow found a home for his rapidly executed writings in the *Morgenblatt*, Cotta's Stuttgart journal. Here he published a series of articles, later collected and published as *Öffentliche Charaktere*, and also the Novelle *Der Sadduzäer von Amsterdam* (q.v., 1834). In 1835 his novel of the emancipated woman, *Wally die Zweiflerin* (q.v.), achieved notoriety and drew a savage attack from his old chief Menzel. The attempt to use literature as a lever to shift rooted prejudices and to set up a new social and political order was brought to an abrupt halt in December 1835 when a federal decree instructed member states to act against the new subversive writers, including Gutzkow (see JUNGES DEUTSCHLAND). Gutzkow was summoned, on the evidence of *Wally die Zweiflerin*, for blasphemy and for bringing the Christian religion into contempt. He was acquitted of blasphemy, but convicted on the second charge and sentenced to a month's imprisonment (1836).

Gutzkow married in the summer of 1836, and soon after became editor of the *Frankfurter Börsenzeitung*. In 1837 he founded the periodical *Der Telegraph für Deutschland*, which was taken over in 1838 by Hoffmann und Campe in Hamburg, who retained Gutzkow as editor. The novels *Seraphine* and *Blasedow und seine Söhne* appeared in 1838. The former is partially autobiographical; *Blasedow* satirizes the educational views of J. B. Basedow (q.v.). At this point Gutzkow turned away from the novel and devoted himself to writing plays which dealt, in historical disguise, with political and social ideas of the day, and not infrequently with his own emotional life. He had already tried his hand at a play (Tragikomödie), *Nero* (1835), which was in mixed verse and prose, and also had written a fragment of a verse play (*Marino Falieri*). Three scenes of *Hamlet in Wittenberg* bring Hamlet and Ophelia into contact with Faust. Gutzkow's career as a dramatist, however, takes its real departure with the prose tragedy *Richard Savage* (q.v., 1839), which was followed by some seventeen other plays mostly performed in the years 1840–56. Of these, the most important are *Werner oder Herz und Welt* (Schauspiel, q.v., 1840), the comedies *Zopf und Schwert* (q.v., 1844) and *Das Urbild des Tartüffe* (q.v., 1845), the tragedies *Patkul* (1842), *Uriel Acosta* (q.v., 1846), *Pugatscheff* (1847), and *Jürgen Wullenweber* (1848, see WULLENWEVER, J.), and *Ella Rose* (Drama, q.v., 1856). The dates are those of performance. Gutzkow's *Dramatische Werke* were published 1842–57 (9 vols.). Mention should also be made of the occasional play *Der Königsleutnant* composed for the Goethe Centenary in 1849.

Gutzkow grew tired of editing *Der Telegraph* and moved to Frankfurt in 1842 after making a visit to Paris. In Hamburg he had begun a liaison with a Frau Therese von Bacheracht (see STRUWE, Th. VON), which produced a triangular situation that taxed the nerves of all three participants. In 1846 he accepted an appointment as Dramaturg at the Court Theatre, Dresden, where he remained until 1848. He was in Berlin at the time of the Revolution, but took no active part in the events of the day. Here his wife died, and in 1849 he made a second marriage (not with Frau von Bacheracht).

At this point Gutzkow's dramatic production began to slacken and he turned again to writing novels. *Die Ritter vom Geiste* (q.v.), a ninevolume work referred to as a *Zeitgemälde* and dealing mainly with contemporary social, political, and philosophical ideas, appeared in 1850–1. Gutzkow regarded it as a new type of novel, der Roman des Nebeneinander, a crosssection of intellectual life at one point of time. From 1852 to 1862 he edited the *Unterhaltungen am häuslichen Herd*, in which some of his own shorter writings appeared. A second large-scale novel, *Der Zauberer von Rom* (9 vols., 1859–61), dealt with the Roman Catholic Church and particularly with Ultramontanism.

From 1861 to 1864 Gutzkow was secretary of the Schiller Foundation in Weimar, where his pathological nervousness led to difficulties and friction. His state deteriorated into persecution mania and in 1865 he made an attempt at suicide in Friedberg, Hesse. After his recovery he continued to write novels, publishing *Hohenschwangau* (5 vols., 1867–9), *Die Söhne Pestalozzis* (3 vols., 1870, see HAUSER, KASPAR), and *Fritz Ellrodt* (2 vols., 1872). He became increasingly restless and frequently changed his place of residence, living in turn in Vevey, Hanau, Berlin (1869–74), Heidelberg, and Sachsenhausen (Frankfurt). He completed his last novel (*Die neuen Serapionsbrüder*, 3 vols., 1877) in 1875. He died of suffocation by smoke, and it is supposed that he overturned a lamp while under self-administered sedation.

The first collection of his works, *Gesammelte Werke* (13 vols., 1845–52), was followed in his last years by *Gesammelte Werke* (12 vols.), 1873–6. *Ausgewählte Werke* were edited by H. H. Houben (12 vols., 1908). *Werke in 15 Teilen* (7 vols.), ed. R. Gensel, appeared in 1912 (reissued 1974).

Though much of his work was outwardly historical, Gutzkow's main interest lay with topical ideas and attitudes, to which he was often unable to give convincing expression in terms of human character; but he handled conventional dramatic forms with competence. His tendency to hurried writing was aggravated by the need to live by his pen, and much of his work represents an uneasy compromise between the man of letters and the journalist. But his immense novel *Die Ritter vom Geiste* is a remarkable achievement which has been consistently underrated.

Gyges und sein Ring, a five-act tragedy in blank verse by F. Hebbel (q.v.), first planned under the title *Rhodope,* completed in 1854, published in 1856, and first performed, long after Hebbel's death, at the Vienna Burgtheater. Hebbel failed to have the play performed in Dresden, Paris, and Vienna, where he met with a rejection from H. Laube (q.v.). Hebbel used as his main source an entry on Gyges in *Pierers Universallexikon,* which refers to the tale of Gyges and Kandaules in Herodotus. Hebbel describes the action as prehistorical and mythological, and confines it to a short space of time, and to three principal characters.

The young Greek Gyges is staying at the court of Kandaules, King of Lydia, whose admiration he wins by his successes in the games and contests. Before participating, he presents to the King a ring that confers invisibility on the wearer. Kandaules, in his pride, which Gyges's success has challenged, cannot refrain from impressing the handsome Greek with the unequalled beauty of his wife, Rhodope, his most precious possession, and the only one which he cannot display for all the world to see; for Rhodope insists on maintaining the oriental seclusion to which she has been accustomed

from childhood. Kandaules succeeds in bringing the reluctant Gyges, rendered invisible by the ring, into the royal bedchamber. The sight of her beauty evokes from him a sigh, which betrays his presence to Rhodope, whose modesty is so deeply wounded that she insists that Kandaules challenge Gyges to a duel which must end in the death of one of the two contestants. Kandaules is not only at a disadvantage in age but he is also oppressed by the burden of guilt, and it is he who dies. As Rhodope's honour can only be fully restored if the man who has seen her marries her, she insists that Gyges, the successor to Kandaules's royal status, go through the ceremony of marriage in the temple of Hestia. In an act of atonement, Gyges has already caused the ring to be buried with Kandaules, in order to dispel the temptation of evil which no man can withstand. He consents to the ceremony of marriage, but, as an added gesture of expiation, is intent on leaving Rhodope and abandoning the crown despite the acclaim of the people, led by Thoas, who see in him an ideal ruler. Rhodope, however, submits to her own moral law, and kills herself as soon as the ceremony is completed, in order to accomplish the total purification of her soul. Hebbel has woven into the play another element. Kandaules, as well as Gyges, is a progressive, who glimpses the future, but is unable to persuade the people, and least of all his own queen, to take a single step to meet it. The play represents in form and content Hebbel's most overt response to Goethe's classicism.

H

HABE, Hans, pseudonym of Hans Bekessy (Budapest, 1911–77, Locarno), a journalist in Vienna, served in 1939 in the French army and in 1940 escaped to the U.S.A., returning in 1945 in American service to be editor of the *Neue Zeitung* in Munich. He is the author of successful topical novels, including *Eine Welt bricht zusammen* (1937), *Zu spät?* (1940), *Der Weg ins Dunkel* (1948), *Off Limits* (1955), *Im Namen des Teufels* (1956), *Die rote Sichel* (1959), *Ilona* (1960), *Die Tarnowska* (1960), *Die Mission* (1965), *Christoph und sein Vater* (1966), *Das Netz* (1969), *Palazzo* (1975), and *Staub im September* (1979). His essays appeared in four volumes in 1976.

Habsburg, family name of the emperors of the Holy Roman Empire (see DEUTCHES REICH, ALTES) from 1438 to 1806, and emperors of Austria from 1804 to 1918. The family is of Swiss origin, and its earliest seat is the Habsburg, a castle not far from Aarau in Canton Aargau. The Habsburgs were already a powerful family in the 10th c. and possessed extensive lands. Count Rudolf was elected German King (see RUDOLF I) in 1273, and acquired the duchies of Austria and Styria through his victory over Ottokar II of Bohemia (q.v.) in 1278. Rudolf's grandson failed to secure election, and the control of the Empire passed to the Luxemburg family. In the 14th c. revolts deprived the

Habsburgs of their Swiss territories. From Albrecht II (q.v.) until the end of the Holy Roman Empire the imperial dignity was virtually hereditary in the Habsburg dynasty. By astute political marriages the family became the most powerful in Europe.

In the 18th c. the marriage of Maria Theresia (q.v.) and the Duke of Lorraine inaugurated the line of Habsburg-Lothringen, the first representative of which was her son, the Emperor Joseph II (q.v.). The last Habsburg was the Emperor Karl (q.v.) who abdicated in 1918. See also KARL V, FRANZ JOSEPH, FRANZ FERDINAND.

HACKLÄNDER, FRIEDRICH WILHELM, RITTER VON (Burtscheid nr. Aachen, 1816–77, Villa Leoni am Starnberger See), the son of a teacher, was apprenticed to a merchant in Elberfeld in 1830, volunteered for service in the Prussian artillery, and then returned to business. He began to write, capitalizing his soldiering experiences (*Bilder aus dem Soldatenleben im Frieden,* 1841, later abbreviated to *Das Soldatenleben im Frieden,* and *Wachtstubenabenteuer,* 1845), and in 1842 gave up his career in order to visit the East as companion to a Württemberg nobleman. From 1843 he was employed at the Württemberg court as secretary to the Crown Prince. During this period he published *Der Pilgerzug nach Mekka* (1847) and *Humoristische Erzählungen* (1847). His employment was terminated in 1849, and thereupon he became a war correspondent in North Italy. There followed *Bilder aus dem Soldatenleben im Kriege* (2 vols., 1849–50) and the novels *Handel und Wandel* (2 vols., 1850) and *Eugen Stillfried* (3 vols., 1852). Hackländer founded the magazines *Hausblätter* (1855) and *Über Land und Meer* (1858). He took part in the war in North Italy in 1859, receiving in 1861 the Austrian title 'Ritter von'. From 1859 to 1864 he was in charge of the office for the royal buildings and gardens in Stuttgart. Thereafter he lived as a private gentleman and man of letters in a villa by the Starnberger See in Bavaria. Among his later novels were *Europäisches Sklavenleben* (4 vols., 1854), *Der Augenblick des Glücks* (1857), *Fürst und Kavalier* (1865), *Künstlerroman* (5 vols., 1866), *Der letzte Bombardier* (4 vols., 1870), *Der Sturmvogel* (4 vols., 1874), and *Nullen* (3 vols., 1878). *Der letzte Bombardier,* though a humorous novel of peace-time service, contains some serious criticism of aristocratic officers. His collections of stories include *Namenlose Geschichten* (3 vols., 1851) and *Neue Geschichten* (2 vols., 1867). He also wrote comedies and an autobiography, *Roman meines Lebens* (2 vols., 1878). Hackländer was a fluent and facile author; his collected works (*Werke,* 1860–73) run to 60 volumes. Though he has fallen into complete neglect, some of his books,

notably the early sketches of army life, still make lively and entertaining reading.

HACKS, PETER (Breslau, 1928–   ), studied at Munich and settled in East Berlin in 1955. In the 1950s a follower of Brecht (q.v.), he propagated Marxist ideas in plays with historical subjects and in adaptations of literary works by others (e.g. H. L. Wagner's *Die Kindermörderin,* q.v., and John Gay's *Polly*). His works include the plays (Stücke) *Eröffnung des indischen Zeitalters* (1954, on Columbus), *Das Volksbuch vom Herzog Ernst oder Der Held und sein Gefolge* (1953, see HERZOG ERNST), *Die Schlacht bei Lobowitz* (1956), *Der Müller von Sanssouci* (1958), *Die Sorge und die Macht* (1958, revised after being banned, 1962), and *Der Schuhu und die fliegende Prinzessin* (1966, a fairy-tale, dramatized in collaboration with Uta Birnbaum). In the three-act comedy in blank verse *Amphitryon* (first performed in the Deutsches Theater, Göttingen, in 1968, published in 1970) Hacks presents in Jupiter and Alkmene his image of the two sexes, perfect in body and mind, and in Sosias the useless man of wisdom who does not insist upon his human identity. The comedy is consciously contrasted with traditional myth. In a note on the play Hacks briefly assesses previous treatments of the subject (though he omits Giraudoux). Kleist's *Amphitryon* (q.v.) provided the most obvious challenge. *Ausgewählte Dramen* (1972) contains *Columbus, Die Schlacht bei Lobowitz, Moritz Tassow, Amphitryon,* and *Omphale. Ausgewählte Dramen 2* (1976) contains *Herzog Ernst, Die Sorgen und die Macht, Margarete in Aix, Prexaspes,* and *Ein Gespräch im Hause Stein über den abwesenden Herrn von Goethe. Ausgewählte Dramen 3* (1981) contains *Der Müller von Sanssouci, Adam und Eva, Die Fische, Senecas Tod,* and *Musen.* The libretto *Die Vögel,* based on Aristophanes, is central to the volume *Oper. Inhalt: Geschichte meiner Oper. Omphale. Die Vögel. Versuch über das Libretto* (1975 DDR and 1976 BRD).

As an author of stories and verse for children, Hacks began with *Das Windloch. Geschichten von Henriette und Onkel Titus* (1956). Other titles include *Der Flohmarkt. Gedichte für Kinder* (1965), *Der Schuhu oder Die fliegende Prinzessin* (1966), *Kathrinchen ging spazieren* (1973), *Der Mann mit dem schwärzlichen Hintern* (1980), and *Juls Ratte oder selber lernen macht schlau* (1981).

A volume *Lieder. Briefe. Gedichte* appeared in 1974, and a collection of essays written between 1961 and 1977, *Die Maßgaben der Kunst,* in 1977.

HADAMAR VON LABER (Bavarian Palatinate, c. 1300–c. 1360) was of knightly birth and served under the Emperor Ludwig IV (q.v.). He is the author of a poem *Die Jagd* (1355–40), which treats the pursuit of love allegorically under the image of the chase. It is written in Titurelstrophen (see TITURELSTROPHE), a form borrowed from WOLFRAM von Eschenbach (q.v.). *Die Jagd* enjoyed great popularity and was widely imitated.

HADLAUB or **HADLOUB**, JOHANS, a Swiss Minnesänger of citizen stock, lived in Zurich, where he is recorded as having bought a house in 1302. His year of birth is unknown, as is also that of his death, but he was no longer alive in 1340. Hadlaub is the author of 54 extant poems, one of which refers to the Manesse collection of Minnelieder preserved in the Manessische Liederhandschrift, a designation which is now replaced by the Große Heidelberger Lieder-handschrift (see HEIDELBERGER LIEDERHAND-SCHRIFT, GROSSE).

Hadlaub's derivative and rather flat poetry, which is indebted to WALTHER von der Vogel-weide, NEIDHART von Reuental, and Steinmar (qq.v.), is not without ingenious invention. In addition to Minnelieder his range includes the Tagelied (q.v., four examples), and the Leich (q.v.), as well as harvest and autumn songs. He is the subject of G. Keller's Novelle *Hadlaub* (q.v.).

Hadlaub, a Novelle by G. Keller (q.v.), included in *Züricher Novellen* (q.v.). It is con-cerned with the love of the medieval poet Had-laub (q.v.) for Fides, which leads to their marriage, and also with the making of the great MS. collection of medieval poetry, formerly known as the Manessische Handschrift (see HEIDELBERGER LIEDERHANDSCHRIFT, GROSSE). This was once attributed to Hadlaub. See also NARR AUF MANEGG, DER.

HADWIGER, VIKTOR (Prague, 1878–1911, Prague), wrote stories which look forward, though without distinctive originality, to the coming movement of Expressionismus (q.v.). His works include volumes of poems (*Gedichte*, 1900; *Ich bin*, 1903), stories (*Blanche. Des Affen Jugo Liebe und Hochzeit* and *Der Empfangstag*, both 1911), and the novel *Abraham Abt* (1912). He was a contributor to the periodical *Die Aktion* (q.v.).

HAECKEL, ERNST (Potsdam, 1834–1919, Jena), was professor of zoology at Jena from 1862 to 1909 and devoted himself to the advance-ment of Darwin's views. He formulated the so-called 'biogenetic law' (biogenetisches Grund-gesetz), according to which each living creature passes in its embryonic stage through all the phases through which the species has evolved. His zoological writings include *Natürliche Schöpfungsgeschichte* (1868), *Anthropogenie* (1874),

and *Der Kampf um den Entwicklungsgedanken* (1905). In a philosophical work intended for laymen as well as specialists (*Die Welträtsel*, 1899) Haeckel rejected dualism (including Christianity) and propounded a monistic philo-sophy, which is virtually pantheistic. At a late stage in his academic career Haeckel published 2 vols. describing his extensive zoological travels, *Kunstformen der Natur* (1904) and *Wanderbilder* (1905). These are illustrated by his own paint-ings and drawings.

HAFNER, PHILIPP (Vienna, 1735–64, Vienna), Austrian lawyer turned dramatist, is reckoned to be the founder of the Viennese Lokalposse (see VOLKSSTÜCK). His numerous plays, all written within a few years before his premature death from consumption, are early examples of the Singspiel (q.v.), and have local settings, dialect speech, and a number of songs, though some of these were inserted in later adaptations. Most of his plays were performed while still in MS. and there is much confusion about dates of writing, performance, and publication, which was mostly posthumous. *Megära, die förchterliche Hexe* was so successful that Hafner wrote a sequel (Zweiter Teil) in 1765. In the same year as *Megära, Die bürgerliche Dame oder Die bezähmten Ausschweifungen eines zügellosen Eheweibes* and *Etwas zum Lachen im Fasching* also played to full houses, and *Evakathel und Schnudi* followed in 1765. *Die reisenden Komödianten* and *Der Furchtsame* date from *c.* 1762 and 1763 respectively.

Hafner substituted good-humoured farce for the obscenities of much earlier extemporized Viennese comedy; his plays, in one guise or another, were still in the Viennese repertoire in the first half of the 19th c.

HAGEDORN, FRIEDRICH VON (Hamburg, 1708–54, Hamburg), after studying at Jena, received a diplomatic appointment and spent two years (1729–31) in London as secretary to the Danish Minister. A man of means, Hagedorn devoted his years in Hamburg from 1731 on-wards to an active social life and the cultivation of poetry. He was called in his day 'the German Horace', but was influenced by English poetry as well as classical models. He wrote, with a sure, deft touch, elegant, unpretentious poetry which was collected in *Versuch einiger Gedichte* (1729), *Versuch in poetischen Fabeln und Erzählungen* (1738), *Oden und Lieder* (1742–52), and *Moralische Gedichte* (1750). Hagedorn's poetry has a distinct rococo charm and also exhibits some feeling for nature. (See ANAKREONTIKER and ROKOKO.)

Sämtliche poetische Werke were published in 3 vols. (1757), and a 5-vol. edition appeared in 1800. For many years he was remembered only for the verse fable *Johann der Seifensieder* (q.v.), printed in anthologies, but his poetry was revived in *Gedichte*, ed. A. Anger, and published in 1968.

HAGELSTANGE, RUDOLF (Nordhausen, Harz, 1912–84, Hanau), at first a journalist, was called up in 1940, taken prisoner in 1945, and released shortly afterwards. After the war he lived for several years at Unteruhldingen on Lake Constance, moving in 1968 to Erbach in the Odenwald (Hesse). He is the author of the novels *Spielball der Götter* (1959), *Altherrensommer* (1969), and *Der General und das Kind* (1974), and of the stories *Ich bin die Mutter Cornelias* (1939), *Balthasar* (1951), *Wo bleibst du, Trost . . .* (1958), *Der Krak in Prag* (1969). He is best known as a religious poet, writing in modern idiom. From *Venezianisches Credo* (1945) on, through the *Meersburger Elegie* (1950) to the *Ballade vom verschütteten Leben* (q.v., 1952) and *Zwischen Stern und Staub* (1953), the terse power of his work matches its sincerity. His collected poems were published in 1961 as *Lied der Jahre* (in extended form 1964), as *Der Krak in Prag* in 1969, and as *Gast der Elemente. Zyklen und Nachdichtungen 1944–72* in 1972. *Die Puppen in der Puppe; eine Rußlandreise* appeared in 1963. *Menschen und Gesichter* (1982) is a volume of reminiscences.

Hagen, a character in the story of the Nibelungs, appears in the earliest form as Högni, son of Gjuki, brother of Gunnar, and murderer of Sigurd. In somewhat analogous form, Wagner makes him in *Der Ring des Nibelungen* (q.v.) a son of Alberich and the warrior of Gunther, as well as the slayer of Siegfried. In the *Nibelungenlied* (q.v.) Hagen is the staunch and ruthless adherent of the Burgundian royal family, who kills Siegfried, and in the end faces death unflinchingly. He is referred to as Hagen von Tronie or Tronege, which is possibly identifiable with Tronia, now Kirchheim, in the Palatinate.

A Hagen appears as a king in one of the three generations whose story appears in *Kudrun* (q.v.). Hagen also occurs in the *Walthersage*, which survives in the medieval Latin epic *Waltharius* (q.v.).

Hagen figures as a character in the many treatments of the *Nibelungenlied*, notably in Hebbel's trilogy *Die Nibelungen* (q.v.).

HAGEN, ERNST AUGUST (Königsberg, 1797–1880, Königsberg), became in 1825 a professor of the history of art at Königsberg University. His romantic poem *Olfried und Lisena* was published in 1820, and his shorter poems (*Gedichte*) in 1822. His principal work is *Norica* (1827), 2 vols. of stories about Nürnberg. They purport to be based on an unspecified old MS. Hagen's

Künstlergeschichten (4 vols., 1833–40) were also widely read in his day. In 1863 he published a biography of the poet Schenkendorf (q.v., *Max von Schenkendorfs Leben*).

HAGEN, GODEFRIT, a citizen of Cologne, who in 1270 wrote in the regional dialect a *Chronik der Stadt Köln* (*Boech van der stede van Coelne*). In rhyming couplets amounting to some 6,000 lines it traces the history of the city from its Christian conversion to the writer's own time, with special concentration on the years 1250–70, and offers an interesting insight into the political life and intrigues of a medieval city.

HAGER, GEORG (Nürnberg, *c.* 1560–*c.* 1640, Nürnberg), a shoemaker, who was probably a pupil of Hans Sachs (q.v.). As a young man he spent some years in Breslau and was active in the school of Meistergesang (q.v.) there. He was a devoted adherent of the art and maintained the tradition of the earlier Meistersinger well into the 17th c., writing more than 1,000 songs, over 600 of which survive in his own hand. They are printed in *Georg Hager, a Meistersinger of Nürnberg*, ed. C. Bell (4 vols., 1947, Berkeley, U.S.A.).

Hagestolz, Der, a Novelle by A. Stifter (q.v.) written in 1843 and first published in shortened form in the periodical *Iris* in 1844. Its full text was published in vol. 5 of *Studien* in 1850. *Der Hagestolz* is an encounter between youth and age. Viktor, an orphan brought up in a loving and sheltered environment, is summoned to spend some weeks with his uncle, whom he has never seen, before taking up an official post. At the uncle's bidding he makes the journey on foot and is then rowed to the remote island where the uncle lives. His reception is bleak, and the atmosphere of the uncle's house is stern and suspicious. Gradually the freshness and integrity of the young man impinge upon the uncle, whose harsh personality undergoes a partial thaw. The two learn mutual respect and a certain affection, but the uncle can no longer change his personality and secluded life. Through his integrity and generosity Viktor is able to abandon a cramping administrative career and to live independently. He marries his foster-sister, Hanna.

The most remarkable features of the story are the powerful portrayal of the stern, rugged uncle, and the evocative and sympathetic description of the anonymous but obviously Upper Austrian Alpine lake landscape.

Hagestolzen, Die, a play (Ein Lustspiel in fünf Aufzügen) written by A. W. Iffland (q.v.). It was first performed in 1793, and in March 1796 was played at the Weimar Court Theatre with Iffland himself as guest artist. Hofrat Reinhold, a wealthy bachelor who has hitherto been persuaded by his domineering and avaricious sister not to marry, contemplates marriage at 40. The extravagant young woman to whom he is attracted injures his *amour propre* and he decides against marriage. At this point he discovers that his sister and his manservant are in league against him and that they have been privately running an extortionate money-lending business. In great perturbation he goes to stay with his tenant farmer. Here, in idyllic rural peace, he is so charmed with his tenant's unspoilt and good-hearted young sister-in-law Margrethe that he marries her and dismisses his manservant. Reinhold's sister departs in dudgeon. In the fifth act Margrethe sings the song 'Was frag ich viel nach Geld und Gut' by J. M. Miller (q.v.).

HAHN, ELISE (Stuttgart, 1769–1833, Frankfurt/Main), without knowing G. A. Bürger (q.v.) proposed to him by means of an original poem which was published for her by a Stuttgart newspaper in 1789 and sent to Bürger at Göttingen. Bürger married her in 1790. The marriage was a disastrous failure and was dissolved in 1792 on the ground of her adultery. She wrote plays (including *Adelheid, Gräfin von Teck*) and stories, most of which were published in 1799. Her later life was spent mainly in Leipzig and Dresden. She is often termed 'das Schwabenmädchen' or 'Schwabenmädle', and used the latter form herself.

HAHN, JOHANN FRIEDRICH (Gießen, 1753–79, Zweibrücken), studied at Göttingen University, where he became one of the founder members of the Göttinger Hainbund (q.v.). His poems were not collected until a century after his death (*Gedichte und Briefe*, 1880).

HAHN, KARL FRIEDRICH, GRAF VON (Remplin, Mecklenburg, 1782–1857, Altona), father of the novelist Ida Gräfin von Hahn-Hahn (q.v.), served in the Wars of Liberation, 1813–15 (see NAPOLEONIC WARS) and then devoted himself with boundless enthusiasm to the stage, consuming most of his fortune in theatrical ventures and earning the nickname 'der Theatergraf'.

HAHN-HAHN, IDA, GRÄFIN VON (Tressow, Mecklenburg, 1805–80, Mainz), the daughter of Count Karl Friedrich von Hahn (q.v.), acquired her hyphenated dual name by marrying a cousin, Count Adolf von Hahn in 1826. The marriage was dissolved in 1829, and the Countess travelled for some years, visiting the Near East in 1843. She published several volumes of poetry (*Gedichte*, 1835; *Neue Gedichte*, 1836; *Venezianische Nächte*, 1836; *Lieder und Gedichte*, 1837). These were followed by a series of social novels focused exclusively on the aristocratic society of her time, and dealing especially with passionate, emotionally dissatisfied women: *Aus der Gesellschaft* (1838), *Der Rechte* (1839), *Gräfin Faustine, Ulrich* (both 1841), *Sigismund Forster* (1843), and *Cecil* (1844). In 1844 she changed the title of the first to *Ilda Schönholm* (the name of its heroine) and made *Aus der Gesellschaft* the title of the whole series, in which some later works were also included. Mention should also be made of the novels *Zwei Frauen* (1846), *Gräfin Clelia Conti* (1846), and *Sibylle* (2 vols., 1846). After the 1848 Revolution (see REVOLUTIONEN 1848–9) she amended her critical view of society. In 1850 she adopted the Roman Catholic faith and in 1854 founded a convent at Mainz, in which she lived, though without taking conventual vows, until her death. She justified her conversion in *Von Babylon nach Jerusalem* (1851), and her later novels (*Maria Regina*, 2 vols., 1860; *Die Geschichte eines armen Fräuleins*, 2 vols., 1869; *Nirwana*, 2 vols., 1875; *Der breite Weg und die enge Straße*, 1877; *Wahl und Führung*, 1878) show a strong Roman Catholic influence. She also wrote a religious song-cycle, *Unserer lieben Frau* (1871). Her collected works (*Gesammelte Werke*) were published in 21 parts (Teile) in 1851.

Haimonskinder, Die, a legend of French origin in which the four sons of Count Haimon of Dordogne fight against Charlemagne (see KARL I, DER GROSSE); they are Adelhart, Ritsart, Witsart, and Rainalt. The story began as a *chanson de geste*, was translated into Flemish, and, in the 15th c., from Flemish into Middle High German under the title *Reinolt* (or *Reinalt*) *von Montalban*. In the 16th c. the *Haimonskinder* was translated three times from the Flemish and published as a printed book, first in 1531, then in 1535; in 1604 it appeared as the well-known Volksbuch *Von den vier Haimonskindern*. The author of this version was Paul van der AELST (q.v.). It was included by L. Tieck (q.v.) in *Volksmärchen herausgegeben von Peter Leberecht* (1797); it was retold in the 19th c. by O. Marbach (1838), K. Simrock (q.v., 1845), and G. Schwab (q.v., 1859).

Hainbund, see GÖTTINGER HAINBUND.

Hakenkreuz (Swastika), notorious as the National Socialist symbol, is a sign of great antiquity, which was formerly best known for associations with Indian Buddhism. In 1910 it was adopted by Guido List as an aryan symbol, and so became associated with anti-Semitic groups, hence its adoption by the National Socialist Party (see NSDAP).

HALBE, Max (Guettland nr. Danzig, 1865–1944, Neuötting, Bavaria), the son of a country gentleman, studied from 1883 to 1888 at Heidelberg, Munich, and Berlin universities, turned to writing, and in 1895 settled permanently in Munich. A follower of the Naturalistic movement (see NATURALISMUS), he was active both as a dramatist and as a writer of fiction. His early plays, *Ein Emporkömmling* (1889), a tragedy, *Freie Liebe* (1890), and *Eisgang* (1892), made little impression, but *Jugend* (q.v., 1893) was an immediate and immense success, which he was never able to repeat.

The most notable of his later plays was *Mutter Erde* (q.v., 1897). He also wrote the Renaissance tragedy *Der Eroberer* (1899) and *Die Heimatlosen* (1899), *Das tausendjährige Reich* (1900), *Haus Rosenhagen* (1901), *Der Strom* (1904), and *Das wahre Gesicht* (1907, also set in the Renaissance), *Der Ring des Gauklers* (1911), *Freiheit* (1913), *Schloß Zeitvorbei* (1917), *Die Traumgesichte des Adam Thor* (1929), and the historical play *Heinrich von Plauen* (1933). The early comedy *Der Amerikafahrer* (1894) was written in verse; his other comedies were *Lebenswende* (1896), *Walpurgistag* (1903), *Die Insel der Seligen* (1906), *Blaue Berge* (1909), and the grotesque *Kikeriki* (1921). *Hortense Ruland* (1917) was a late Naturalistic tragedy and *Schloß Zeitvorbei* (1917) a dramatic legend.

Some of Halbe's best work is in his fiction, which began with *Frau Meseck* (1897), the story of an aged peasant woman. *Die Tat des Dietrich Stobäus* (q.v., 1911) is set in the country round Danzig and on the Baltic coast. The novel *Io* (1917) alludes in its title to Io's erotic experience as imagined in Correggio's picture in Vienna. Halbe's late works include the novels *Generalkonsul Stenzel und sein gefährliches Ich* (1931) and *Die Elixiere des Glücks* (1936) and the plays *Erntefest* (1936) and *Kaiser Friedrich II* (1940), a return to historical drama. *Sämtliche Werke* (14 vols.) appeared in 1945–50; the autobiographical volume *Jahrhundertwende. Geschichte meines Lebens 1893–1914* (1935) was published in 1976 with the subtitle *Erinnerungen an eine Epoche.*

Halbe Birne, Die (*Diu halbe bir*), an obscene anecdote in verse written in the 13th c. It purports to be a work of KONRAD von Würzburg (q.v.), but the attribution has been widely doubted.

Halberstädter, sobriquet for Bismarck (q.v.), who held a commission in the 7th (Magdeburg) Cuirassiers, a regiment permanently garrisoned in Halberstadt.

Halbtier, a novel published by Helene Böhlau (q.v.) in 1899. It is a propagandist feminist novel, protesting against the sexual exploitation of the female by the male. The discarded Isolde Frey in the end shoots her lover and, apparently, herself.

Halbzeit, a novel by M. Walser (q.v.), published in 1960. Taking as its central figure the commercial representative Anselm Christlein, who in order to do his job must continually metamorphose himself into a different person to match his client, Walser analyses contemporary society with acid humour, and exposes the depreciation and imprecision of language, extending his criticism of society by implication to its critics also.

Hälfte des Lebens, one of the most beautiful and poignant of the shorter poems of Hölderlin (q.v.). Consisting only of two 7-line stanzas, it was written not later than 1803 and published in 1805.

Halle, city on the River Saale in the German Democratic Republic (see DEUTSCHE DEMOKRATISCHE REPUBLIK), and also, since 1952, the name given to a district (Bezirk). In the Middle Ages Halle belonged to the archbishops of Magdeburg. In 1680 it passed to Brandenburg-Prussia, and in 1916 to the newly constituted Prussian Province of Saxony. From 1949 to 1952 it was in the Land Sachsen Anhalt of the DDR. It is the birthplace of G. F. Händel (anglicized Handel).

Halle's famous university was founded in 1694, and from its early years attracted outstanding scholars, including Thomasius, A. H. Francke, and C. Wolff (qq.v.). J. von Eichendorff (q.v.) gives a nostalgic account of student life in the first decade of the 19th c. in *Halle und Heidelberg*. In 1817 Wittenberg University was united with Halle University, and the full title is now Martin-Luther-Universität Halle-Wittenberg. Halle University was the first university in Germany (and probably in the world) to confer a medical degree on a woman (1754).

HALLER, Albrecht von (Berne, 1708–77, Berne), was primarily a physician but was also in his early years a poet of note. Haller studied medicine at Tübingen and Leyden. In 1728 he visited the Alps, subsequently writing his didactic and satirical poem *Die Alpen* (q.v.), which was the most important item in *Versuch schweizerischer Gedichte* (1732). This journey also laid the foundation of his botanical interests. Having practised as a physician in Berne since 1729, he was appointed in 1736 to the new Hanoverian university of Göttingen as professor of anatomy, botany, and surgery. This promotion coincided with the death of his young wife, whom he commemorated in the *Trauerode, beim Absterben seiner geliebten Mariane*. Between his late twenties and his sixties Haller confined himself to scientific publications, written for the most part in Latin and dealing with anatomy, physiology, and botany. In 1753 he resigned on grounds of ill health and returned to Berne, where he took a minor post in the city administration, finally retiring in 1773. Haller was regarded in 18th-c. Germany as the foremost medical authority. He founded the Sozietät der Wissenschaften at Göttingen and was its president until his death. Towards the end of his working life he wrote three novels of a politico-didactic character, *Usong* (1771), *Alfred* (1773), and *Fabius und Cato* (1774).

Halle und Jerusalem, a play by L. J. von Arnim (q.v.), published in 1811. It is a free adaptation and extension of the play *Cardenio und Celinde* by A. Gryphius (q.v.).

Hallig, name applied to nine North Sea islands off the Frisian coast of Germany, separated from the mainland by flooding and erosion; the most disastrous floods occurred in 1362, 1532, 1570, and 1634. The Halligen are Hooge, Süderoog, Norderoog, Südfall, Nordstrandischmoor, Oland, Gröde, Habel, and Langeneß. A tenth Hallig, Jordsand, was ceded to Denmark in 1919. The Hamburger Hallig is no longer an island, being connected with the mainland by a causeway. Some of the islands are permanently inhabited by farmers. They figure in the work of Th. Storm (q.v.), but the writers who have most fully evoked the peculiar life and culture of their inhabitants are J. C. Biernatzki, W. Lobsien, and F. Zacchi (qq.v.).

HALLMANN, JOHANN CHRISTIAN (Breslau, *c.* 1640–?1704, Breslau?), trained as a lawyer but took to the stage, writing plays for performance by his own itinerant company. At some point he became a Roman Catholic convert, but whether this was the cause of his abandoning a stable career is unknown. Some phases of his life are completely obscure. Hallmann wrote a number of tragedies, some of which are preserved only in their titles or in scenarios. They have sensational plots and contain scenes of torture or martyrdom presented with detail on the stage. *Verführter Fürst Oder Entseelter Theodoricus* (1666) deals with a conspiracy against the emperor (see THEODORIC THE GREAT); *Mariamne* (1670) treats Herod's murder of his wife, a theme derived from Josephus and treated in the 19th c. by F. Hebbel (q.v.); *Sophia* (1671), set at the court of Hadrian, portrays the martyrdom of the empress.

Hallmann also wrote plays (both tragic and allegorical) for special festive occasions; they included *Sinnreiche Liebe Oder Der Glückseelige Adonis und Die Vergnügte Rosibella* (1673). In later life his style approached opera, as exemplified in *Die sterbende Unschuld Oder Die Durchlauchtigste Catharina Königin von England*, the date of which is uncertain. Its heroine is Catherine Howard, fifth queen of Henry VIII. It has been generally accepted that Hallmann died in poverty in Breslau in 1704, but this is not certain, and it is possible that he died in Vienna in 1716. An edition of *Mariamne* (ed. G. Spellerberg) was published in 1973. *Sämtliche Werke* (6 vols.), ed. G. Spellerberg, began to appear in 1975.

HALM, FRIEDRICH (Cracow, 1806–71, Vienna), pseudonym and customary designation of Eligius Franz Joseph, Freiherr von Münch-Bellinghausen. The family moved to Vienna in 1811, and young Münch was educated at the monastery school at Melk, the Schottengymnasium in Vienna (1816–22), where N. Lenau and E. Bauernfeld (qq.v.) were schoolfellows, and Vienna University. In 1826 he married a young noblewoman, Sophie von Schloißnigg, and in the same year he received a probationary civil service appointment, rising to senior rank in 1840.

Münch first used the pseudonym Friedrich Halm in 1834, and his first stage success, *Griseldis* (q.v., 1837), took place at the Burgtheater, Vienna, under this name in 1835. The performance of the actress Julie Rettich (q.v.) in the title role was the beginning of a lifelong friendship between her and Halm. His plays, which were regularly accepted by the Burgtheater, were mostly designed to provide her with a suitable and important part. The tragedy *Der Adept* (1838) was performed in 1836, the one-act play *Camoens* (1838) in 1837, *Imelda Lambertazzi* (1842) in 1838, and *Ein mildes Urteil* (1857) in 1840. *König und Bauer* (performed 1841) is an adaptation of a play by Lope de Vega (*El villano en su rincón*). A particular success was *Der Sohn der Wildnis* (1843), produced in 1842; the tragedies *Sampiero* (1857) and *Donna Maria de Molina* and the comedy *Verbot und Befehl* (1857), performed respectively in 1844, 1847, and 1848, made little mark. In 1844, to the bitter disillusionment of Grillparzer (q.v.), who was a candidate, Halm was appointed custodian of the Court Library (Hofbibliothek, now Nationalbibliothek) in Vienna. In 1854 the anonymous verse tragedy *Der Fechter von Ravenna* (q.v.) was performed in Vienna and elsewhere. Authorship was claimed in 1855 by a Bavarian schoolmaster named Bacherl (q.v.), and a *cause célèbre* developed, which was not immediately ended by Halm's declaration in 1856 that he was the author. The play remained a considerable stage success. The classical *Iphigenie in Delphi* (1864) was not well received in the theatre in 1856. Halm's last stage works were *Wildfeuer* ('dramatisches Gedicht', 1864) and the Indian tragedy *Begum Somru* (1872), in which Warren Hastings is one of the characters. Both were produced in the 1860s.

Halm, to whom honours came easily, was appointed in 1867 Intendant of the two Court theatres (Oper and Burgtheater), and immediately clashed with the Burgtheater director, H. Laube (q.v.), by arrogating to himself choice of play and engagement and casting of actors. Laube resigned, and, realizing that the task of directing the theatres was beyond him, Halm himself abandoned his post in 1870. He published 2 vols. of poetry (*Gedichte*, 1850, and *Neue Gedichte*, 1864). Of his stories, the best known is *Die Marzipanliese* (q.v., 1856).

A characteristic figure of the mid-century in Vienna, Halm embodied the advantages of privilege and financial security. His poetic writing is a self-conscious tribute to an ideal of beauty, which excluded all that was crass or harsh, and enveloped its serenity or melancholy in uniformly mellifluous verse. Though in his own day he was rated among the great, his reputation had paled by the end of the century and has not since been revived.

Halm's complete writings (*Werke*, 12 vols.) appeared between 1856 and 1872.

HAMANN, JOHANN GEORG (Wendisch-Ossig, Oberlausitz, 1697–1733, Hamburg), a cousin of J. G. Hamann (1730–88, q.v.), became a private tutor and was active as a publicist, contributing to Gottsched's periodical *Die vernünftigen Tadlerinnen*. He is best known for his sequel (2 vols., 1724) to *Die Asiatische Banise* by Heinrich Anselm von ZIEGLER und Kliphausen (q.v.).

HAMANN, JOHANN GEORG (Königsberg, 1730–88, Münster), the son of a surgeon (a profession of low standing in former times), studied theology at Königsberg University and then held various posts as private tutor. In 1755 he turned to commerce, and in 1757 went to London on behalf of his principal. A spell of dissolute living at this time culminated in a religious experience of conversion and repentance. His anguished and self-tormenting mood is preserved in *Gedanken über meinen Lebenslauf* (1758). Returning to Königsberg in 1759, he spent eighteen years in various minor posts. A long liaison with a woman, who bore him several children, estranged his fellow citizens. Towards the end of his life he was given asylum by the Princess Gallitzin in Münster (q.v.).

In a rationalistic age Hamann spoke up decisively for intuition. His *Sokratische Denk-würdigkeiten* (1759) put the case, in cryptic and sibylline language, for the genius's capacity for knowing by instinct. 'Was ersetzt,' asks Hamann, 'bei Homer die Unwissenheit der Kunstregeln, die ein Aristoteles nach ihm erdacht, und was bei einem Shakespeare die Unwissenheit oder Übertretung jener kritischen Gesetze? Das Genie, ist die einmütige Antwort. Sokrates hatte also freilich gut unwissend sein; er hatte einen Genius, auf dessen Wissenschaft er sich verlassen konnte.' In this emphasis on the rightness of genius Hamann proved himself a precursor of the Sturm und Drang (q.v.).

Die Kreuzzüge eines Philologen (1762), the centre of which is *Aesthetica in nuce*, anticipates Herder in proclaiming that poetry is the original state of language ('Poesie ist die Muttersprache des menschlichen Geschlechts'), and that images are its central treasure ('In Bildern besteht der ganze Schatz menschlicher Erkenntnis und Glückseligkeit'). In later years Hamann entered into controversy with Kant (q.v.) in *Metacritik* (written 1784, published 1800) and *Golgatha und Scheblimini* (1784).

Though Hamann had a firm command of language and could express himself lucidly, he deliberately chose a rhapsodic, exclamatory, and provocative style. He was known to his contemporaries as Der Magus im Norden.

Schriften (9 vols.) appeared 1821–43 and *Sämtliche Werke. Historisch-kritische Ausgabe* (6 vols.), ed. J. Nadler, 1949–57.

Hambacher Fest, designation of a revolutionary assembly held at Hambach near Neustadt in the Bavarian Palatinate west of Speyer from 27 to 30 May 1832. The aims proclaimed by the speakers (notably J. G. A. Wirth and J. Siebenpfeiffer) were the setting-up of a republic and the unification of Germany. Sympathy was also expressed with the victims of the Polish revolt of 1830. The Hambacher Fest provoked federal decrees severely limiting the press and suspending the right of assembly.

Hamburg, Freie und Hansestadt, the largest city in West Germany, and a constituent Land of the Federal Republic (see BUNDESREPUBLIK DEUTSCHLAND). Its territory, which is chiefly urban with commuters' suburbs, lies mainly on the right bank of the Elbe, though in 1937 Harburg, on the left bank and formerly Hanoverian, was included in Hamburg. At the same time Altona and Wandsbek, both hitherto in Schleswig-Holstein, were added to Hamburg.

The legislative body of the city of the Freie und Hansestadt Hamburg is the Bürgerschaft. The Land government is the Senat. Hamburg is not only the largest German port but a great manufacturing and commercial centre with important cultural institutions, notably the Staatsoper, Deutsches Schauspielhaus, Thalia-Theater, an important art gallery (Kunsthalle), museums (especially the Museum für hamburgische Geschichte), and the famous zoological gardens (Tierpark) founded by C. Hagenbeck, and old botanical gardens and new exhibition centre (known by the dialect name Planten un Blomen) near the city's central feature, the Alster (q.v.), a small river bulging into a large lake. Hamburg University was founded in 1919.

The origins of the city go back to the 9th c., but it did not become a town until the 12th c. It was a prominent member of the Hanseatic League (see HANSE, DEUTSCHE), and in 1510 was confirmed as a Free Hanseatic city. Hamburg suffered two destructive disasters, the Great Fire of 1842 and massive bombing (mainly by night) from 24 July to 3 August 1943.

By tradition Hamburg was republican and patrician. It played an important part in the development of German literature in the 18th c. Brockes and Hagedorn (qq.v.) were Hamburg patricians: Klopstock and Gerstenberg (qq.v.) lived in Altona (originally a separate town built 'all zu nah', all too near to Hamburg, so it is said). Lessing (q.v.) accepted the post of house critic at the National-Theater on the Gänsemarkt in 1767, and remained as a publisher until 1770 Here he met both H. S. Reimarus and Pastor J. M. Goeze (qq.v.). Heine (q.v.) worked for two years in his uncle's bank at Hamburg, and was, allegedly, attracted by his girl cousins. Hebbel (q.v.) was invited there by Amalie Schoppe (q.v.). D. von Liliencron (q.v.) lived the last twenty years of his life in the suburbs (Ottensen and Rahlstedt, 1889–1909). G. Falke and R. Dehmel (qq.v.) spent an appreciable span of their lives in Hamburg, and E. Barlach (q.v.) studied (1888–91) and worked in the city (1897–9). H. H. Jahnn and W. Borchert were both born in Hamburg, and spent part of their lives there. Jahnn died in Hamburg, and Borchert is buried there. Hamburg is the birthplace of a number of writers, including Count C. Stolberg, F. Gerstäcker, F. Dahn, O. Ernst, G. Fock and his brother R. Kinau, and H. E. Nossack (qq.v.).

Hamburger Jüngstes Gericht, a fragment of a poem narrating the appearance of Jesus as judge at the Last Day. The poet, whose outlook derives from the sense of sin and imminent damnation associated with the Cluniac Reform (q.v.), is a powerful rhetorician, who delivers a swingeing sermon in verse. The fragment dates from 1140–50.

Hamburgische Dramaturgie, a critical work by G. E. Lessing (q.v.), written and published as a periodical in 1767–8, and in 1769 in book form. The *Dramaturgie* was intended by Lessing to incorporate criticism of current productions at the Hamburg National Theatre, to which he was attached as house critic. His comments were not welcomed by the actors, and he devoted more and more of his periodical, which eventually totalled 104 numbers, to theoretical discussion of the drama, in which two related matters are conspicuous: a destructive criticism of Voltaire's plays and an interpretation of the passages in Aristotle's *Poetics* relating to tragedy. Voltaire, viewed as typical of the French conception of Aristotelian drama in the 18th c., serves for Lessing as a springboard by which he launches himself into his argument. In the most important numbers (73–83), which include a discussion of Shakespeare's *Richard III*, Lessing interprets catharsis as purification (Reinigung) and conceives pity as that which is felt for the tragic hero and fear as that which is felt by the spectator, lest he should find himself in a like position. It is a view which is best realized when the fictitious character closely resembles the onlooker; Lessing provides, in fact, a defence of domestic tragedy (see BÜRGERLICHES TRAUERSPIEL), and calls Aristotle as a corroborative witness. The last four numbers are in the nature of a personal confession and apologia, and contain Lessing's well-known disclaimer 'Ich bin weder Schauspieler noch Dichter', which is, however, considerably qualified in the ensuing paragraph.

HAMERLING, ROBERT, pseudonymous name for Rupert Hammerling (Kirchberg am Walde, Lower Austria, 1830–89, Graz), a weaver's son who was educated partly at Zwettl Abbey, and partly in Vienna. In 1854 he became a classical master at Graz Gymnasium, but was transferred in the following year to Trieste, where he remained until illness compelled his retirement in 1866. He spent his later years as an invalid in Graz, where he befriended the young P. Rosegger (q.v.).

Hamerling's early publications were poetic. *Ein Sangesgruß vom Strand der Adria* (1857) was a collection of poetry accompanied by extracts from his poem *Venus im Exil* (1858), a pseudophilosophical work comprising five cantos in *ottava rima*. *Ein Schwanenlied der Romantik* (1860), in the form of 55 poems set in Nibelungenstrophen (see NIBELUNGENSTROPHE), looks idealistically to a future in which true civilization, freedom, and right (Gesittung, Freiheit, Recht) will prevail. His patriotism expressed itself in a short epic, *Germanenzug* (1863), and this was followed by his best-known work *Ahasuer in Rom* (Dichtung in sechs Gesängen, 1866). This blank-verse epic recounts the Wandering Jew's appearance in Nero's Rome and it depicts the burning of Rome and Christians torn asunder in the arena by ravenous beasts. Ahasuerus, whom Hamerling identifies with Cain, is made to represent humanity (Hamerling calls him 'den ewigen Menschen'). A collection of his lyric poetry appeared in 1868 as *Sinnen und Minnen*, the title of a much smaller edition of 1859. *Aspasia*, a three-volume novel of ancient Greek life, was published in 1876. In his last years Hamerling wrote two more verse epics, *Amor und Psyche* (1882) and *Homunkulus* (1888), of which the latter is a satire on the modern world. He had also tried his hand at tragedy (*Danton und Robespierre*, 1871) and had translated Leopardi (*Leopardis Gedichte*, 1865).

Hamerling's glowing diction and lofty idealism won him a following in his own age, which has since faded away. His autobiography *Stationen meiner Lebenspilgerschaft* appeared in 1889, and *Sämtliche Werke* (16 vols., ed. J. Böck-Gnadenau) in 1911.

HAMMER-PURGSTALL, Joseph, Freiherr von (Graz, 1774–1856, Vienna), a noted orientalist, served at first as an Austrian diplomat in the Near East. Posted back to Vienna in 1807, he made it his business to disseminate knowledge of the Levant. His many books, almost all now forgotten, were hastily written treatises on oriental history, institutions, manners, and literature. In 1812 he published a translation of the *Divan* of Hafiz, which provided Goethe with the impulse to write the poems of the *West-östlicher Divan* (q.v.).

HAMMERSTEIN-EQUORD, Hans August, Freiherr von (Schloß Sitzenthal, 1881–1947, nr. Micheldorf, Upper Austria), descended from a Hanoverian noble family and, on his mother's side, from the poet Graf F. L. zu Stolberg (q.v.), was a student in Marburg and Vienna, entered the Austrian civil service in 1908, and served as an officer in the 1914–18 War, returning to the civil service in the reduced Austria of 1919. He became chief magistrate (Bezirkshauptmann) in Braunau/Inn and, after various other promotions, was made Minister of Justice. He was removed on the German invasion in 1938, but not seriously molested until July 1944 when he was sent to Mauthausen Concentration Camp, from which the Americans freed him in 1945.

Hammerstein (he dropped the name Equord for his books) was a sensitive writer and a devout Roman Catholic. His poetry comprises the collections *Zwischen Traum und Tagen* (1919), *Das Tagebuch der Natur* (1920), and *Der Wanderer im Abend* (1936). His fiction includes the historical

novels *Februar* (1916), *Ritter, Tod und Teufel* (1921), *Mangold von Eberstein* (1922), *Wald* (1937, but originally published as a shorter tale in 1923), *Die finnischen Reiter* (1933), and *Die gelbe Mauer* (1936).

Hammer und Amboß, a five-volume novel by F. Spielhagen (q.v.), published in 1869. It is in two parts. Told in the first person by the hero Georg Hartwig, it narrates his life from schooldays, including his participation in smuggling and his consequent imprisonment, in the course of which he rescues the governor during a convicts' mutiny. On release he begins as a workman, makes his way, marries the boss's daughter, takes over the firm, and introduces profit-sharing.

Spielhagen makes use of sensational incidents and exaggerated characters, which tend to weaken the novel. The title alludes to Goethe's second *Kophtisches Lied*, which ends with the line 'Hammer oder Amboß sein'; Spielhagen's amendment to 'Hammer *und* Amboß' is explained in Chap. 27 of Pt. I.

HANDEL-MAZZETTI, Enrica, Freiin von (Vienna, 1871–1955, Linz/Danube), the daughter of a Roman Catholic father (an Austrian officer) and a Protestant mother of Hungarian aristocracy, was brought up a Roman Catholic. At the age of 15 she spent a year in the convent of the Englische Fräulein at St. Pölten. She remained a devout Catholic, maintaining a humane and tolerant standpoint. Although she is the author of several plays (*Nicht umsonst*, 1892, *Pegasus im Joch*, 1895, *Geistige Werdejahre*, 1911), she is chiefly known for her historical novels which are set in the 17th c. and 18th c. and deal with the problem of divided confessions.

Meinhard Helmpergers denkwürdiges Jahr (1900) tells of an honourable atheistic father tortured to death by Lutheran intolerance, and of his son who becomes a Roman Catholic through the simple piety and kindness of a monk. In *Jesse und Maria* (1906) Jesse Helfried, a Protestant, is denounced by Maria, the wife of his friend, and tried for heresy. He commits an act of violence when browbeaten during his trial and is condemned to death by decapitation. Jesse is the father of a baby whose mother cannot feed him. Maria, moved with remorse and compassion, takes charge of the infant. In *Die arme Margaret* (1910) a Protestant widow maintains herself in spite of persecution, and, at the climax, kills a Roman Catholic who tries to seduce her. These themes are resumed in two later trilogies, *Stephana Schwertner* (1912–14) and *Frau Maria*

(1929–31); the former is set in the 17th c., and the latter in the 18th c. and deals with Maria Aurora von Königsmarck (q.v.).

The Napoleonic Wars (q.v.) and their culmination in the Wars of Liberation form the background to *Der deutsche Held* (1920), a preliminary to a trilogy on K. L. Sand (q.v.), the murderer of A. von Kotzebue (q.v.): *Das Rosenwunder* (1925), *Die deutsche Passion* (1926), and *Das Blutzeugnis* (1927). Less successful is a novel on contemporary life, *Ritas Vermächtnis* (1922). A volume of poetry, *Deutsches Recht und andere Gedichte*, appeared in 1908. A select edition of her works (ed. K. Vansca, 1959) is entitled *Ein groß Ding ist die Liebe*.

HANDKE, PETER (Griffen/Carinthia, Austria, 1942–   ), studied law at Graz University and lived for short periods in West Germany and Paris until, in 1979, soon after his return from his third visit to the U.S.A., he moved to Salzburg. His early writing is indebted to his contact with a small circle of poets and artists in Graz and the Wiener Kreis (q.v.); his writing for the theatre, to which he largely owes the wide publicity which he sought and received during his twenties, owes much to Beckett: it is decidedly anti-Brechtian.

Handke began to write almost simultaneously with prose pieces, novels, poetry, and plays (Sprechstücke). His object-orientated style aims at self-awareness in political, social, and psychographic contexts and at alienation from literary conventions, of form and speech. His aesthetics search for 'heile Natur', which he associates with Goethe (in *Das Gewicht der Welt*), and which in his own work is manifest in the recurring motif of the child and its inherent sense of purpose and authority. It supports his consistent technique of confrontation, ultimately, within the 'inner world'.

His publications began with the novels *Die Hornissen* (1966, revised 1968 and 1978) and *Der Hausierer* (1967), which derives from the detective novel; the prose of *Begrüßung des Aufsichtsrats* (1967, several times revised); the poetry of *Die Innenwelt der Außenwelt der Innenwelt* (1969); and plays, *Publikumsbeschimpfung und andere Sprechstücke* (1966): the volume includes *Weissagung* and *Selbstbezichtigung*, but it is the radical anti-theatre of *Publikumsbeschimpfung* that achieved remarkable success and notoriety. The detective story *Die Angst des Tormanns beim Elfmeter* (q.v., 1970) followed in the wake of *Kaspar* (q.v., 1968), a much performed play based on Kaspar Hauser (q.v.), *Das Mündel will Vormund sein* (1969), and *Quodlibet* (1970), a preamble to *Der Ritt über den Bodensee* (1971); this last title derives from the ballad (a genre Handke rejects) 'Der Reiter und der Bodensee'

(1826) by G. Schwab (q.v.), in which a man, having unwittingly crossed the frozen lake on horseback, dies of shock when hearing of his feat. An anarchic piece, in which the theatre again ceases to be a theatre and the actors appear not as actors but as themselves, it attaches the illustrious names of five actors of a previous generation (among them Emil Jannings and Elisabeth Bergner) to the figures' anonymous function. It met with little response, and it marked a point from which Handke himself felt he could not proceed. *Wind und Meer* (1970) contains four radio plays, and *Chronik der laufenden Ereignisse* (1971) is a film script. Collected plays, *Stücke I* and *Stücke II*, were published in 1972 and 1973; *Prosa, Gedichte, Theaterstücke, Hörspiel, Aufsätze* in 1969, and *Ich bin ein Bewohner des Elfenbeinturms*, tracts on writers who have influenced Handke's own development as well as on other literary topics, in 1972. The ivory tower of the title is a symbol of Handke's insistence on the autonomy of literature, which in 1966 was the cause of his argument with Gruppe 47 (q.v.) at its meeting in Princeton. *Der Rand der Wörter. Erzählungen, Gedichte, Stücke* appeared in 1975, *Das Ende des Flanierens* in 1977 and, in extended form with essays, speeches, and reviews, in 1980; *Das Gewicht der Welt. Ein Journal (November 1975–März 1977)* of 1977 records 'Momente der Sprachlebendigkeit' (Preface) that describe Handke's mental orientation rather than events, in this way distancing itself from the conventional journal.

The stories *Wunschloses Unglück* and *Der kurze Brief zum langen Abschied* (both 1972) proceed from personal experience, the former from the suicide of Handke's mother in 1971, and the latter from estrangement in marriage, presented in the form of a detective story; in reading Keller's *Der grüne Heinrich*, the pursued narrator relates to the Bildungsroman (q.v.) which, through Goethe's *Wilhelm Meister*, underlies the conception of the figure of Wilhelm in *Falsche Bewegung* (1975). *Die Stunde der wahren Empfindung* (1975) and *Die linkshändige Frau* (1976), in which prose, reminiscent of a film script, relates the perceptions of a woman searching for independence after her husband has left her and her young child, was followed by *Langsame Heimkehr* (1979); presented in long syntactical sequences and complex verbal structures, its principal figure, Sorger, devotes himself to the scientific laws inherent in nature in order to find a 'personal law' for his own existence. The novel is designed to have a thematic link with *Die Lehre der Sainte-Victoire* (1980), *Kindergeschichte*, and *Über die Dörfer. Dramatisches Gedicht* (both 1981); performed in 1982 in Salzburg, it relates to *Das Salzburger große Welttheater* (q.v., 1922) by Hofmannsthal. *Die Geschichte des Bleistifts* (1982), another 'journal' (though not described as such), dated 1976–80, was followed by *Der Chinese des Schmerzes* (1983), a story focusing on the motif of the threshold (Schwelle) to vistas of inner experience. Handke represents a trend styled Neue Subjektivität.

He was awarded the Büchner Prize in 1973.

HANDL, JAKOB (Reifnitz, 1550–91, Prague), Bohemian composer of masses and motets, who was active in Olmütz and in Prague. He used the name Gallus, a Latinization of his surname.

Handschuh, Der, a ballad written by Schiller in 1797 and published in the *Musenalmanach auf das Jahr 1798*. It narrates in irregular rhyming stanzas an anecdote from *Essais historiques sur Paris* by Saintfoix (1754). At the court of Francis I of France a lady tests her lover by throwing her glove into a lions' cage and requesting him to retrieve it. He does so, and promptly throws the glove in her face. This climax was too ungentlemanly for Frau von Stein (q.v.), and in deference to her Schiller amended the glove-throwing into a deep bow. He later reverted, however, to his original text.

Handvoll Erde, Eine, see EINE HANDVOLL ERDE.

Hanneles Himmelfahrt, a two-act play (Traumdichtung) by G. Hauptmann (q.v.), first performed in the Königliches Schauspielhaus in November 1893, and published in the same year. Set in a sordid workhouse, the play opens with Naturalistic dialogue between the miserable and quarrelsome occupants. Hannele, deprived by death of her mother's protection, and brutally battered by a drunkard father, is brought into the workhouse hall to die. She is tended by the nun Sister Martha. Hannele dreams, first of her dead mother, then of angels; after a brief awakening the dream returns, but this time with the Angel of Death, and finally with a Christ-figure appearing as Der Fremde, whose speech is in blank verse. The play is brought to a close in the real world as the doctor pronounces her dead. Hauptmann's admirers were shocked by its deviation from the principle of stringent Naturalism and accused him of sentimentality.

Hanne Nüte un de lütte Pudel, a verse epic in Lower German dialect (Mecklenburger Platt) by F. Reuter (q.v.), published in 1860. Partly idyllic in character, it brings human beings (Hanne Nüte and his love Fike) into close contact with the bird world. In the later stages of the narrative the birds indicate the true perpetrator of a murder of which Hanne is accused.

Hannibal, a tragedy by C. D. Grabbe (q.v.), written in 1834 and published in 1835. It was first performed in Munich in 1918. The play was begun in blank verse but was turned into prose upon the advice of K. Immermann (q.v.), to whom Grabbe dedicated the work in gratitude for his encouragement. It is divided into five acts or sections with headings, and marks the climax of Grabbe's technique of loose scenic development. The action, a free adaptation of historical facts, covers the failure of Hannibal's Italian campaigns after his great victory at Cannae, and his return to Africa to defend Carthage against the Roman troops. At Zama he meets Scipio the Younger, leader of the enemy forces, to negotiate a truce, but Scipio proudly rejects Hannibal's terms. In the ensuing battle Hannibal is defeated, finds refuge with King Prusias of Bithynia, and poisons himself when he discovers that his host has betrayed him to the Romans.

Unlike the historical general, Grabbe's hero is a man of admirable qualities, who, by virtue of personality rather than nationality, commands the complete loyalty of his troops. He fails because the Carthaginians, headed by the Triumvirs (Dreimänner) Melkir, Hanno, and Gisgon, deny him the country's support. Gisgon's belated change of heart achieves nothing but an honourable death on the battlefield. Since Carthage is doomed, Alitta, of the noble house of Barkas, Hannibal's loyal kinsman, sets her house alight and heroically immolates herself, an example followed by the Carthaginian women, so that the Romans find only ashes on the spot where the great commercial city had flourished. Hannibal's tragedy reveals the dependence of a man, however great, on environment, circumstance, and chance, represented by the short-sighted and profit-seeking mercantile society. It reveals also the superiority of the national and military state, represented by Roman imperialism, which triumphs because it is dedicated to a greater cause; this, it appears, is more important than brilliant leadership. *Hannibal* is Grabbe's most personal play, reflecting his own bitterness and frustrations.

Hannover (1) **Kurhannover** (the Electorate of Hanover), previously a principality, attained its electoral status (see KURFÜRSTEN) in 1692, and was legally styled *Kurbraunschweig*, though George I of Great Britain was always called the Elector of Hanover before his accession, and the state continued to be known as *Kurhannover* or *Kurfürstentum Hannover* until the dissolution of the Holy Roman Empire (see DEUTCHES REICH, ALTES) in 1806.

Occupied during the Napoleonic Wars (q.v.) by Prussia and France alternately, Hanover was made a kingdom in 1814, and George III, George IV, and William IV of Great Britain were kings of Hanover. On Queen Victoria's accession, Hanover passed to George III's fifth son, the Duke of Cumberland, Ernst August (q.v.). His successor took the Austrian side in the war of 1866 (see DEUTSCHER KRIEG), and was deprived of his throne; his kingdom was annexed to Prussia as the Provinz Hannover. A strong movement for independence from Prussia persisted into the 20th c., and was represented by the Deutsch-Hannoversche Partei, but at a referendum in 1924 the majority voted for the retention of provincial Prussian status. After the 1939–45 War Prussia was dissolved, and Hanover became the capital city of Lower Saxony (Land Niedersachsen) in the Federal Republic (see BUNDESREPUBLIK DEUTSCHLAND).

(2) The picturesque old centre of the city of Hanover survived until the 1939–45 War, in which the 17th-c. Leibnizhaus (in which Leibniz lived and Iffland, qq.v., was born) and the palace of Herrenhausen were destroyed. The former College of Technology (Technische Hochschule, originating in 1831) was converted into the Technical University of Hanover in 1968. The city was the birthplace of J. A. Leisewitz, A. W. and F. Schlegel, and F. Wedekind (qq.v.).

Hanns Frei, a five-act comedy (Lustspiel) in rhyming verse, written in 1843 by O. Ludwig (q.v.) and published posthumously in 1891. The play is set in 16th-c. Nürnberg. Two fathers plan a match for their reluctant children, Albert and Engeltraut. A kinsman, Hanns Frei, undertakes to overcome the difficulty by ostensibly choosing new and disagreeable spouses for the young couple. Albert and Engeltraut discover that they love each other and all ends happily, though not before they have played a trick on Hanns Frei.

Hans Dampf in allen Gassen, a humorous expression formerly used of a busybody. The name occurs in a poem sometimes attributed to C. F. D. Schubart (q.v.), *Hans Dampf der Dichter Lobesam* (1775). *Hans Dampf in allen Gassen* is also the title of a story written in 1814 by H. Zschokke (q.v.).

Hanse, Deutsche, a trading organization of North Sea and Baltic German towns, which had its beginnings in the 12th c., and effectively came to an end in the 17th c. The first Hanseatic organizations were the offices set up in foreign countries (in England, 1157, and somewhat later in Norway) by groups of German traders from Cologne and Bremen. Separate associations, based on Lübeck (newly founded in 1158), grew up in the late 12th c.; one such was the Association of Gothland Merchants (Gemeinschaft der Gothlandfahrer), which cultivated and protected trade in the Baltic and achieved economic penetration into Russia. These associations grew rapidly in the 13th c. under the leadership of Lübeck, and more and more German cities, in the interior as well as near the coast, joined the Hanseatic League which, though having no formal constitution and revenues, succeeded in exercising considerable economic and political power. In all, and at different times, 164 cities belonged to the League, but the most prominent members apart from Lübeck were Hamburg, Lüneburg, Bremen, Cologne, Soest, Brunswick (see BRAUNSCHWEIG), Stralsund, Dortmund, Stendal, Bergen (Norway), Danzig, Reval, Riga, and Wisby. The last five, though outside the boundaries of the Empire, were predominantly German cities. In western Europe the most important transit port of the Hansa was Bruges. By the 14th c. the Hanseatic League dominated European maritime trade, and its power was such that it was able to undertake a successful war against Denmark in 1367–8, and to exact stringent peace terms in the Treaty of Stralsund (1370). The Hanse also again maintained its privileges by force in the 15th c. against Denmark (1435) and England (1474). In the course of the 15th c. the diverging interests of individual members, and the rise of the Dutch as maritime traders, began to threaten the Hanseatic maritime hegemony; in the 16th c. Lübeck, under the revolutionary government of Jürgen Wullenwever (q.v.), undertook a campaign against Denmark which ended in disaster (1536). The League, having lost its monopoly, could no longer offer its members attractive privileges and gradually diminished in extent and influence. The last diet (Hansetag), held in 1669, was attended only by a rump consisting of six cities, Bremen, Brunswick, Cologne, Danzig, Hamburg, and Lübeck. The League was never officially dissolved, and its traditions were maintained by the three municipalities still known as Hanseatic Cities (Hansestädte), Hamburg, Bremen, and Lübeck.

Hänsel und Gretel, an opera (Märchenspiel) on a fairy-tale by E. Humperdinck (q.v.), first performed in 1893. The libretto is by Humperdinck's sister Adelheid Wette. Based on *Hänsel und Gretel,* the traditional fairy-tale of innocent children lost in a forest and enticed to the gingerbread house (Knusperhaus) of the witch, it ends with the release of Hänsel and Gretel and the death of the witch. In the form in which the story is used in the opera it derives from the collection of Lugwig Bechstein (q.v.). *Hänsel und Gretel* is the most popular opera in Germany for children.

Hans Heiling, a legendary mountain spirit who loves a mortal girl and marries her. The marriage fails, and Hans Heiling transforms the girl into a rock. The story is Bohemian in origin, and is connected with the hills around Karlsbad (Karlovy Vary) where there is a Hans-Heiling-Felsen. The story is used by Th. Körner (q.v.) in the narrative (Erzählung) *Hans Heilings Felsen*; Körner retains the motif of jealousy, but eliminates the marriage. His Heiling acts in spite against a girl who has rejected him. Körner also wrote a poem dealing with this rock, entitled *Hans Heilings Felsen* (in *Erinnerungen an Karlsbad*, 1811).

The story is the subject of an opera, *Hans Heiling* (1833), by H. Marschner (1795–1861). The libretto, by E. Devrient, abandons the petrification and ends happily in a renunciation by Heiling and the union of a pair of human lovers.

HANSJAKOB, HEINRICH (pseudonym Hans am See) (Haslach, Baden, 1837–1916, Haslach), was ordained a Roman Catholic priest in 1863. He worked as a schoolmaster in Waldshut from 1865 to 1869, when he was dismissed and twice briefly imprisoned by the secular authority at a time of conflict with the Church. He was a parish priest in Hagnau on Lake Constance (1869–84), and from 1884 to 1913 a member of the cathedral clergy in Freiburg.

Hansjakob's early work was autobiographical: *Auf der Festung* (1870), *Im Gefängnisse* (1874), *Aus meiner Jugendzeit* (1880). His fiction depicts Black Forest rural life and is collected under such titles as *Wilde Kirschen* (1888), *Dürre Blätter* (2 vols., 1889–90), *Schneeballen* (3 vols., 1892–4), *Bauernblut* (1896), *Waldleute* (1897), and *Erz-bauern* (1898). *Ausgewählte Schriften* (8 vols.) appeared 1895–6.

Hans Sachsens poetische Sendung, customary abbreviated title of a poem by Goethe, the full superscription of which runs *Erklärung eines alten Holzschnittes vorstellend Hans Sachsens poetische Sendung*. The poem was written in 1776, and published in the same year in *Der teutsche Merkur*. It expresses Goethe's admiration for the freshness and directness of Hans Sachs (q.v.), who, in the fictitious woodcut, is associated with allegorical figures, Ehrbarkeit, Historia, and Der Narr, and is visited by the Muse. The metre used is a free Goethean version of Sachs's accustomed Knittelverse (q.v.).

Hans und Heinz Kirch, a Novelle by Th. Storm (q.v.), written in 1881–2 and published in 1882.

Hans Kirch, in full Hans Adam Kirch, has worked his way up in the hierarchy of the seafaring community of a coastal town, recognizably Storm's native Husum. The proud owner of his own ship and his own export and import business, he is possessed by the ambition that his son should extend the business and become one of the city's aldermen.

Heinz, after confirmation, joins his father's ship as a deck hand, and at 17 sails in a large ship for the China Sea. The night before his departure he has a rendezvous with his childhood sweetheart, Wiebke, and does not return until late at night. Away from home, he duly writes to his father, who, however, on hearing that Heinz has spent his last evening with the socially unacceptable Wiebke, replies in anger. Two years pass and Hans receives an unstamped letter from Heinz, which he refuses to accept. After fifteen years, during which Heinz's mother dies of grief for the lost son, Hans learns that Heinz is in Hamburg. He goes and brings his son home, but only Heinz's sister Lina offers him a warm welcome. She has meanwhile married Christian Martens, a wealthy young man, according to her father's standards, who is running the Kirch business with success. Heinz, prematurely aged and disfigured by smallpox, is acutely aware that he is not wanted. Rumour throws doubt upon his true identity; it is said that he is in reality Hasselfritz, a poor orphan, who has disappeared. A confrontation between father and son reveals that Heinz, when in Santiago had received back the unopened letter in which he had desperately pleaded for a reconciliation. Nevertheless, Hans Adam's humanity remains subordinated to his monomanic obsession to have a son cast in his own mould; using the doubts about Heinz's identity as a pretext, he allows Heinz the legal benefit of the doubt by placing an envelope with money, carefully calculated, in his room. Heinz takes a small sum, leaving the remainder, and quits the house.

Hans hears Heinz leave by night, but does not stop him; 'his' Heinz left him finally seventeen years ago. Soon after he sees Heinz drown in a dream vision. He suffers a stroke and ends his days a feeble-minded and broken man, whose sole comfort is Wiebke. But he remains fully conscious that in eternity he will have to account for having twice abandoned his own son.

The Novelle centres on the tragic conflict between father and son, but it also critically reviews a narrow-minded, gossiping community.

It ranks, with *Carsten Curator* (q.v.), as Storm's most concentrated achievement in the tragedy of the Kleinbürger.

HANS VOM NIEDERRHEIN, Bruder, a monk who, at the end of the 14th or beginning of the 15th c. (after 1391), composed a cycle of seven songs in praise of the Virgin Mary, which can also be conceived as sections of one poem.

HANS VON BÜHEL (*c.* 1360–after 1429 and before 1444), came from the district of Rastatt and was in the service of at least two noblemen, the Electoral Archbishop of Cologne and the Margrave of Hochberg. He is the author of two long, undistinguished poems, *Die Königstochter von Frankreich* and *Diokletians Leben*. *Die Königstochter*, finished in 1401, is a version of the legend of the slandered princess found in *Mai und Beaflor* (q.v.) and in the legends of Genoveva and Crescentia (see also Schondoch). Bühel completed *Diokletians Leben* in 1412 at the court of the Archbishop. It is in the form of a narration (see Rahmen), enclosing a sequence of stories which are a version of those in the *Historia septem sapientum* (*Story of the Seven Wise Men*, *Die sieben weisen Meister*). Diokletian, who is educated by the seven sages, declines the advances of his stepmother, the Empress, and is himself falsely accused by her of seduction, and condemned. The seven wise men defer the day of execution by telling in turn, each day, a story. Finally the treachery of the Empress is exposed and Diokletian vindicated.

Hanswurst, term applied in the 18th c. to the principal comic character or clown in plays. The character lasted longest in Austria, where it was created by J. A. Stranitzky (Wienerischer Hanswurst) and played by the comedians, G. Prehauser and J. J. F. von Kurz (qq.v.), all of whom were masters of comic improvisation and 'gagging'.

The name Hanswurst first occurs in 1519 in a Low German (Rostock) version of Sebastian Brant's *Narrenschiff* (q.v.) in the form Hans Worst. It was used pejoratively by Luther (q.v.) in a tract of 1530, and became current for a stage character (also known as Pickelhering) about the middle of the 16th c. In the 1730s J. C. Gottsched (q.v.) waged a campaign against the extemporized comedy of which Hanswurst was the centre, on the grounds of its vulgarity and obscenity. In 1737 he organized an official ceremony in Leipzig banning the character. A fragment of a comedy by Goethe bears the title *Hanswursts Hochzeit oder der Lauf der Welt* (1773–5).

HAPPEL, Eberhard Werner (Kirchhain, Hesse, 1647–90, Hamburg), had to interrupt his studies at Marburg, when his pastor father was dismissed from his cure. He led a precarious life, chiefly in Hamburg, as a novelist and a provider of occasional writings to order. In a later age he would probably have been a journalist. His novels, designed for the exotic taste of the age, combine a mass of supposed geographical and historical facts with current anecdotes and occurrences, all connected by an extravagant story. They include *Der Asiatische Onogambo* (1673), *Der Insulanische Mandovell* (1682), which purports to give descriptions and histories of all the islands there are, as well as telling a story, *Der Italienische Spinelli* (4 vols., 1685–6), *Der Spanische Quintana* (4 vols., 1686–7), and *Der Academische Roman* (1690), which satirizes all imaginable types of student.

HARDEKOPF, Ferdinand (pseudonym Stefan Wronski) (Varel, Oldenburg, 1876–1954, Burghölzli, Switzerland), a shorthand writer employed in the Reichstag, moved in 1916 to Switzerland, returned briefly to Berlin (1921), from where he went to France and then to Switzerland. An exponent of early Expressionism, he was a contributor to *Die Aktion* and close collaborator with F. Pfemfert (qq.v.), and published poetry and prose (*Lesestücke*, 1916), which appeared with *Die Aeternisten. Erste Proklamation des Aeternismus* in *Aktionsbücher der Aeternisten*. During the years in France and Switzerland he published translations of works by French writers, including A. Gide, A. Malraux, Balzac, and Maupassant. *Gesammelte Dichtung* was published posthumously in 1963.

HARDEN, Maximilian, name adopted in 1876 by Felix Ernst Witkowski (Berlin, 1861–1927, Montana, Switzerland). Of Jewish extraction, Harden also used the pseudonyms Kent, Apostata, Proteus, and Kunz von der Rosen. He began as an actor, and in 1889 co-operated in founding the Freie Bühne (q.v.) in Berlin. From then until 1914 he was a controversial and much read critic and publicist; he was an early champion of Ibsen, Strindberg (qq.v.), Maeterlinck, and Dostoevsky. He became notorious for his criticism of Wilhelm II (q.v.) and his circle of advisers, though his vicious attack upon Fürst Philipp zu Eulenburg was almost certainly a fabrication instigated by F. von Holstein (q.v.). From 1892 to 1922 he edited the weekly *Die Zukunft*. During the 1914–18 War he was at first patriotic and in favour of annexation but later became a pacifist; after 1918 he bitterly opposed the Weimar Republic (q.v.).

Harden was a complex, paradoxical character,

subject to violent and unstable loves and hates. He was savagely, and not entirely unjustifiably, attacked by Karl Kraus (q.v.). His works include *Berlin als Theaterhauptstadt* (1888), *Essays* (2 vols., 1892), *Literatur und Theater* (1896), *Kampfgenosse Sudermann* (1903), *Köpfe* (4 vols. of essays, 1910–24), and *Krieg und Friede* (2 vols., 1918).

HARDENBERG, FRIEDRICH VON, see NOVALIS.

HARDENBERG, KARL AUGUST, FÜRST VON (Essenrode, 1750–1822, Genoa), a Hanoverian by birth, was in Hanoverian and later in Brunswick service. In 1791 he transferred to Prussia, and was at first administrator of Ansbach-Bayreuth. After his appointment as peripatetic envoy he was Prussian negotiator for the Peace of Basel (1795, see REVOLUTIONSKRIEGE). In 1804 he succeeded his friend Haugwitz (q.v.) as Prussian foreign minister for a few months, until Haugwitz was reinstated. After the Treaty of Vienna (Schönbrunn, 1805, see NAPOLEONIC WARS) Hardenberg was dismissed at Napoleon's request.

In 1810 Hardenberg was appointed chancellor, and immediately pursued a policy of social reform, along lines similar to those of Stein (q.v. with whom he at first co-operated. Notable steps were the implementation of Stein's edict for the emancipation of serfs, and the emancipation of the Jews in 1812. Hardenberg persuaded Friedrich Wilhelm III (q.v.) to break with the French after the capitulation of Yorck (q.v.) at Tauroggen, and conducted the negotiations with the other powers for the Grand Alliance of 1813. In 1814, after the Peace of Paris, he was created Fürst. He was Prussian representative at the Congress of Vienna, and has often been regarded as a tool in the hands of Metternich (q.v.), but this view has been disputed. Though Hardenberg was a liberal and enlightened man, he was unable to persuade Friedrich Wilhelm III to adopt a generous domestic policy after 1815. He died shortly after the conclusion of the Congress of Verona. His *Denkwürdigkeiten*, referring to 1805–7, were published by L. von Ranke (q.v.) in 1877.

HARDT, ERNST, pseudonym regularly used by Ernst Stöckhardt (Graudenz, 1876–1947, Ichenhausen nr. Augsburg), who travelled much in Europe in his youth, was a journalist at Dresden, and devoted himself to authorship in Berlin. From 1919 to 1924 he directed the German National Theatre at Weimar; in 1925 he moved to a similar post at Cologne, where he switched in 1926 to broadcasting. He was dismissed in 1933.

Hardt's early publications were Novellen,

beginning with the collections *Priester des Todes* (1898) and *Bunt ist das Leben* (1902). His best work in this form is the separately published Novelle *An den Toren des Lebens* (1904). His work has links with Jugendstil (q.v.), and much of his poetry (*Aus den Tagen des Knaben*, 1904) echoes Stefan George (q.v.). His Ibsenite first play *Tote Zeit* (1898) was followed by a more successful family drama *Der Kampf ums Rosenrote* (1903). From this point he diverged into neo-Romantic writing with forced and affected language, beginning with the one-act play *Ninon de Leuchs* (1905), achieving a temporary sensation with *Tantris der Narr* (1907), and declining visibly with *Gudrun* (1911), *Schirin und Gertraude* (1913, parodying the legend of the Graf von Gleichen, q.v.), and *König Salomo* (1915). *Gesammelte Erzählungen* appeared in 1919. His last work, the story *Don Hjalmar*, appeared in 1947.

HARINGER, JAKOB (Dresden, 1898–1948, Zürich), lived for the most part near Salzburg until the German invasion of Austria in 1938, when he emigrated to Switzerland. An unstable character, he lived in poverty with some support from friends. His poetic and dramatic works are uneven and contrasting in style, reflecting his own disunited temperament; violence alternates with simplicity. He published the poetic collections *Hain des Vergessens* (1919), *Die Kammer* (1921), *Heimweh* (1925), *Abschied* (1930), and *Das Fenster* (1946). *Die Dichtungen* appeared in 1925. Two posthumous collections of verse are *Der Orgelspieler* (1955) and *Lieder eines Lumpen* (1962). His stories include *Das Räubermärchen* (1925), *Weihnacht im Armenhaus* (1925), and *Der Reisende oder die Träne* (1932).

HARRING, HARRO (Ibenhof nr. Husum, 1798–1870, London), a stormy petrel of 19th-c. demagogy, took part in the Greek War of Liberation in 1821, travelled restlessly in Europe, and in 1828 was for a few months house dramatist in the Theater an der Wien, Vienna. Later in that year he obtained a commission in a Russian guard regiment stationed in Warsaw, but returned to Germany when the July Revolution broke out in 1830.

Over the next twenty-five years he was repeatedly expelled as an agitator from various German states, from Switzerland, from Norway, and from Denmark. His points of rest were the U.S.A., South America (Rio), and London, where he was a member of the European Democratic Central Committee. He also succeeded in getting himself expelled twice from Heligoland, then a British colony. He died of self-administered poison.

Harring was a prolific writer, chiefly of

political poetry, novels, and satire. His publications included the poems *Blüten der Jugendfahrt* (1821) and *Republikanische Gedichte* (1848), historical plays (among them *Moses zu Tanis*, 1859), and stories (*Zypressenlaub*, 1825; *Erzählungen*, 1826). Noteworthy is the account of his European experiences published in 1828, *Rhon gar Jarr, Fahrten eines Friesen in Dänemark, Deutschland, Ungarn, Holland, Frankreich, Griechenland und der Schweiz*. He astutely obtained his entry to the exclusively noble Russian Imperial Guard by producing his father's commission as Deichgraf (dyke warden) in his native Schleswig.

HARSDÖRFFER, Georg Philipp (Nürnberg, 1607–58, Nürnberg), studied at Altdorf University, travelled through Europe, visiting Switzerland, France, Holland, England, and Italy. From 1634 he was in the administration of his native town. He became a member of Die Fruchtbringende Gesellschaft (q.v.) in 1642, taking the name Der Spielende, and in 1644 he founded with Johann Klaj (q.v.) a society of poets, Der löbliche Hirten- und Blumenorden an der Pegnitz (q.v.).

Harsdörffer's earliest publication was *Frauenzimmer-Gesprechspiele* (after the first volume simply *Gesprechspiele*, 1641–9), a succession of intellectual amusements (parlour games, in a sense), expounded in dialogue for high society. The series includes an opera, *Seelewig* (1644), one of the first written in German; S. G. Staden (q.v.) composed the music. In his poetry, which shows sensitiveness to sound-values, Harsdörffer is most addicted to the pastoral, but he also wrote hymns. His treatise on poetry, *Poetischer Trichter, Die Teutsche Dicht- oder Reimkunst ohne Behuf der lateinischen Sprache in VI Stunden einzugieß*en (1647–53, reissued 1975), by its flippant sub-title, damaged Harsdörffer's reputation for later generations, albeit unjustly; for, although he provides precepts for verse writing, he knows and states that poets are not made.

HART, Heinrich (Wesel, 1855–1906, Tecklenburg), was at school in Münster and studied at Halle, Munich, and Münster universities. From schooldays onwards he was in close association with his younger brother Julius (q.v.). He settled in Berlin in 1877, whither his brother followed him, and the two men devoted themselves to critical journalism. They began with *Deutsche Monatsblätter* (1878–9), edited the *Deutscher Literaturkalender* (1879–82), and then produced in *Kritische Waffengänge* (q.v., 1882–4) a work of urgent propaganda for a new realism in literature.

Both brothers belonged to the literary club Durch (q.v.). Their fame rests upon their pioneering work for the new Naturalism (see NATURALISMUS) of the late 1880s. In his later life Heinrich Hart was best known as a dramatic critic. His poetic production was unimportant, and scarcely conformed to his strongly held critical views. His poems (*Weltpfingsten*) appeared in 1872, a tragedy (*Sedan*) in 1882; an ambitious epic, planned for 24 vols. (*Das Lied der Menschheit*), got no further than the third (1888–96). *Gesammelte Werke* (4 vols.) appeared 1907–8.

HART, Julius (Münster, 1859–1930, Berlin), was closely associated with his elder brother Heinrich (q.v.). After schooling at Münster he studied law at Berlin University for a short time, and then became a dramatic critic in Bremen. He is best known for his participation in *Kritische Waffengänge* (q.v., 1882–4). He was a member of the club Durch (q.v.). His plays include the tragedies *Don Juan Tenorio* (1881), *Der Rächer* (1884), *Die Schauspielerin* (1884), and *Sumpf* (1886). He also published several volumes of verse (*Sansara*, 1879; *Homo sum*, 1890; *Triumph des Lebens*, 1898). Editions of his fiction (*Prosa*) appeared in 1899–1902 (2 vols.) and in 1905. In his later theoretical writings he turned away from his early attitude in *Der neue Gott* (1899) and *Revolution der Ästhetik* (1909).

HARTLAUB, Felix (Bremen, 1913–45, Berlin), a student at Heidelberg and Berlin from 1934 to 1939, was called up in 1939 and from 1941 onwards was employed in archival and historical work in a department of the OKW (q.v.). Called to active service in the defence of Berlin against the Russians, he is presumed to have been killed. A story was published posthumously (*Parthenope*, 1951), but Hartlaub is chiefly known for his notes on the situation in Berlin, published as *Im Sperrkreis, Aufzeichnungen* in 1955. In the same year his *Gesamtwerk* was published. Its contents include *Tagebuch aus dem Kriege* and sketches for three substantial narrative works, *Schlitters, Kinderkreuzzug*, and *Das Abenteuer des Tobias*, as well as a reprint of *Parthenope*.

HARTLEBEN, Otto Erich (Clausthal, 1864–1905, Salò, Lake Garda), was a law student at Berlin and Leipzig universities. He entered the civil service in 1889, but resigned in the following year in order to devote himself to writing. The immense success of his first tragedy *Rosenmontag* enabled him to spend much time in Italy by Lake Garda. In his last years tensions developed in his married life and at the time of his early death he was contemplating divorce. In his writing Hartleben was anti-bourgeois, but he had no socialist leanings; his attitude was more akin to the irony and mockery of O. J. Bierbaum (q.v.) and his anti-philistinism made him many enemies.

Hartleben's earliest works were deft and amusing erotic comedies. *Angele* (1891) exposes what he sees as the fickleness of women and their fundamental selfishness in the pursuit of love. In *Die Erziehung zur Ehe* (1893) a wealthy mother makes a mistake in choosing an uncle from the country to introduce her son to life in town. The uncle interprets the task in his own way in a series of hilarious episodes. *Hanna Jagert* (1893) has as its central figure a working-class girl of Social Democrat opinions, who takes a feminist standpoint, and then succeeds by means of a pregnancy in becoming a baroness. The comedy *Die sittliche Förderung* followed in 1897. Three years later Hartleben took all the theatres of Germany by storm with the tragedy *Rosenmontag* (q.v., 1900), the success of which is attributable to its military setting and sensational action.

Hartleben is the author of a number of light, fluent stories, including *Die Geschichte vom abgerissenen Kopf* (1893) and other collections, among them *Vom gastfreien Pastor* (1895), *Der römische Maler* (1898), and *Liebe kleine Mama* (1904), which is also the title of the second story in the volume, a Novelle in short letters. Hartleben also wrote mellifluous and often charming verse, contained in the volumes *Meine Verse* (1895) and *Von reifen Früchten* (1902). He collected his poetry in 1902, repeating the title *Meine Verse*. His *Tagebuch* was published posthumously in 1906, *Aphorismen* in 1920, and letters (*Briefe an eine Freundin*) in 1910. *Ausgewählte Werke* appeared in 1909 (3 vols.).

HARTLIEB, DR. JOHANN (b. Neuburg, Danube, 14..?, d. between 1468 and 1474), was the personal physician, from 1440 on, of dukes Albrecht III (q.v.) and Sigismund of Bavaria, and also Bavarian court poet. He translated extensively from Latin works (*Alexanderbuch, c.* 1440 from *Liber de preliis*; *Tractatus amoris* of Andreas Capellanus; and *Dialogus miraculorum* of Cäsarius von Heisterbach, q.v.). He also translated medical works, including *Das Buch Trottula über die Krankheiten der Weiber*, which is a version of *De secretis mulierum* by Albertus Magnus, and attacked necromancy in the *Buch aller verbotenen Kunst* (1456).

HÄRTLING, PETER (Chemnitz, now Karl-Marx-Stadt, 1933–), was brought up in Württemberg and began as a journalist in Stuttgart. From 1967 to 1974 he worked for the publishing house S. Fischer (from 1968 as a director) before settling as a writer in Walldorf (Hesse). He is a poet of fantasy and humour, and has published the collections *Poeme und*

Songs (1953), *Jamins Stationen* (1955), *Unter den Brunnen* (1958), *Spielgeist-Spiegelgeist* (1962), and *Neue Gedichte* (1972), and the volume *Anreden. Gedichte aus den Jahren 1972–1977* (1977).

Härtling's first novel, *Im Schein des Kometen* (1959), a work of Heimkehrerliteratur (q.v.), was followed by three novels that thematically make up a trilogy. The first, *Niembsch oder Der Stillstand* (1964) centres on Nikolaus Lenau (q.v.), whose real name, chosen because of its unfamiliarity, was Nicolaus Franz Niembsch, Edler von Strehlenau, and whose letters to Sophie von Löwenthal (q.v.) Härtling edited in 1968 (*Briefe an Sophie von Löwenthal 1834–45*). In the second, *Janek. Porträt einer Erinnerung* (1966), the central figure, the ballad-singer Janek Bialas, is based on the Viennese actor and comedian Max Pallenberg (1877–1934); it is set against the background of the German invasion of Czechoslovakia. The third, *Das Familienfest oder Das Ende einer Geschichte* (1969), moves from the mid-19th c. (Part One) to a family reunion in 1967 (Part Two). All derive from Härtling's scepticism towards history and recollection in relation to truth and identity. This 'Widerstand zur Geschichte', as he termed it, led him to reject historically documented fiction; he projects in its place a mental process of retrospection and anticipation that effects a fusion of different layers of time. In Part One of *Das Familienfest* the fictitious historian Georg Lauterbach attempts on his return to his native Nürtingen in 1857 to establish the cause of a fire from which he, at the time still a youth, had a near escape. He fails because of inconsistent evidence from differently orientated observers. He gradually arrives at a new awareness of the manipulating power of social and ideological forces whose victim he becomes when he is about to oppose them. The theme is varied in Part Two through the failure of the physics teacher Brenner to write a factually reliable family chronicle. The play *Gilles. Ein Kostümstück aus der Revolution* (1970) was followed by the story *Ein Abend eine Nacht ein Morgen* (1971), and two novels set in the war and post-war period, *Eine Frau* (1974) and *Hubert oder Die Rückkehr nach Casablanca* (1978). The novel *Hölderlin* (1976) centres on Hölderlin and *Die dreifache Maria* (1982) on E. Mörike (qq.v.). The novel *Das Windrad* appeared in 1983, *Felix Guttmann* in 1985.

Zwettl. Nachprüfung einer Erinnerung (1973) and *Nachgetragene Liebe* (1980) are autobiographical. (In 1945–6 Härtling had been in Zwettl, Lower Austria, with his father, whom he lost at that time, and his mother, who died shortly afterwards.) Härtling is the author of several volumes of novels, stories, and poetry for children. A selection of his numerous essays, *Meine*

Lektüre. Literatur als Widerstand (1981, ed. K. Siblewski) contains a tract on his views on children's literature, *Das Ende der Geschichte*, and other previously published essays that are of particular relevance to his novels. *Der spanische Soldat oder Finden und Erfinden* (1984) contains Härtling's lectures from his Gastdozentur at Frankfurt University during the winter semester 1983–4. Härtling is editor of the work of Otto Flake (q.v.).

HARTMANN, ANDREAS, an ecclesiastical administrator active in Dresden in the late 16th c., is the author of a play, *Martin Luther* (1600). It was intended as the first part of a trilogy, and takes the story of Luther (q.v.) as far as the Wartburg.

HARTMANN, DER ARME, a religious poet of the 12th c., author of *Rede vom Glauben* and of a lost poem on the Day of Judgement. He himself originated his designation, describing himself as 'ih arme Hartman'. He is believed to have been a nobleman, who turned from the world and its work under the influence of the new ascetic outlook of his time, and, entering a monastery, adopted a life of repentance and self-mortification. He was a Rhinelander or a Thuringian.

Die Rede vom Glauben is a poem of nearly 4,000 lines which is formally an exposition of the Nicene Creed. It is, however, more than a dogmatic work; it seeks, in the tone and rhetorical style of a sermon, to persuade men to leave wealth and family, and to seek salvation in repentance and seclusion. The MS., which is in Strasburg, is an Alsatian copy. One sheet with approximately 400 lines is lost. The poem was written *c.* 1140–50.

HARTMANN, EDUARD VON (Berlin, 1842–1906, Großlichterfelde, Berlin), was obliged to abandon his career as an officer in the Guard Artillery (Garde-Artillerie) because of a knee injury. From 1865 onwards he devoted himself to philosophical studies, of which the first published result was the successful *Die Philosophie des Unbewußten* (1868). Later editions were expanded to 3 vols. In this book there are links with Schelling and foreshadowings of Freud (qq.v.).

Hartmann concerned himself with many aspects of philosophy, and he is most important for his work on epistemology. He may be reckoned a founder of Critical Realism (Kritischer Realismus), which in part accepts Kant's theory of knowledge (Erkenntnistheorie, see KANT) but denies that what is commonly called reality is unknowable. Hartmann's principal later works are *Die Phänomenologie des sittlichen Bewußtseins* (1879), *Das religiöse Bewußtsein der Menschheit* (1881), *Die Religion des Geistes* (1882), *Das Grundproblem der Erkenntnistheorie* (1889),

Kategorienlehre (3 vols., 1896), *Geschichte der Metaphysik* (2 vols., 1899–1900), *Die moderne Psychologie* (1901), and *Das Problem des Lebens* (1906).

Ausgewählte Werke (13 vols.) appeared 1885–1901; *System der Philosophie im Grundriß* (8 vols.) was published posthumously 1906–9.

HARTMANN, MORITZ (Duschnik, Bohemia, 1821–72, Oberdöbling, Vienna), studied at Prague University, became a private tutor and was politically active on the Radical side. He was elected a member of the Frankfurt Parliament (see FRANKFURTER NATIONALVERSAMMLUNG), took part in the rising in Baden, and, after its suppression, fled to Switzerland. His political opinions moderated and he returned to Germany, working in Stuttgart as a journalist; in 1868 he received an appointment on the staff of the *Neue Freie Presse* (q.v.) in Vienna.

Hartmann's political poetry includes *Kelch und Schwert* (1845), *Neuere Gedichte* (1846), and the satire *Reimchronik des Pfaffen Mauritius* (i.e. Moritz, 1849). In middle life he wrote the idyll *Adam und Eva* (1851), and later published several volumes of stories (*Erzählungen eines Unsteten*, 1858; *Novellen*, 1863; *Nach der Natur*, 1866).

HARTMANN VON AUE or **OUWE** (b. 1160–70, d. after 1210), Middle High German epic and lyric poet, is one of the three great names of the Blütezeit (q.v.) in the field of narrative poetry, the others being WOLFRAM von Eschenbach and GOTTFRIED von Straßburg (qq.v.) Hartmann was born in Swabia, but to which place the Aue refers is uncertain; the modern view inclines to Eglisau on the Rhine in Canton Zürich. A *ministeriale* in the service of a great lord, he was a man of education, acquainted with the Classics, and reading French with fluency; presumably he received his early education in a monastery, perhaps Reichenau.

Hartmann's early works consist of Minnelieder, the Arthurian epic *Erec* (q.v.), and a verse tract on the theory of love (*minne*) known as *Das Büchlein*. The death of Hartmann's lord may possibly have contributed to an emotional crisis, in the course of which he vowed to participate in a crusade. It has been held by some that an expression of this crisis is to be found in *Gregorius* (q.v.), a legend of atonement. Opinions differ as to the date of the crusade in which he may have taken part, but it is probable that it was in 1189–90, though some suggest 1197. In this experience Hartmann is believed to have recovered his peace of mind, and the outward sign of this is perhaps the narrative poem *Der arme Heinrich* (q.v.), his best known and, in the opinion of many, his finest work. About 1200 he composed a second Arthurian epic, *Iwein* (q.v.), his last work. He appears to have still been alive in 1210, according to a reference in Gottfried von Straßburg's *Tristan* (q.v.)

Hartmann's Minnelieder regard the service of the inaccessible adored as morally educative. Among them, however, is one poem in which the poet proposes to abandon the service of his unyielding mistress and to seek love among humbler folk; it is regarded as the earliest example of niedere Minne. Hartmann also wrote three crusaders' songs (Kreuzlieder), reflecting the crusade as a means of reconciling the conflict between God and the world.

Hartmann, though not notably original, combined a sense of balanced and harmonious form with deep sincerity, genuine piety, and great humanity. His poetry was highly esteemed by his contemporaries, especially GOTTFRIED von Straßburg, HEINRICH von dem Türlin, who lamented his death in *Diu Crône*, and RUDOLF von Ems (qq.v.).

HARTWIG VON RUTE or RAUTE, a Middle High German Minnesänger of whose life virtually nothing is known. Hartwig appears to have been a native of Salzburg or Upper Bavaria; he was apparently a member of the minor nobility, living in the second half of the 12th c. Of his poems only slight fragments are extant. In contrast to the accepted tone of Minnesang, they seem to reveal traces of passion and sensuality.

Harzburger Front, a public demonstration of unity by the Nationalist Party, the Stahlhelm (q.v.), and the NSDAP (q.v.) against the government of Brüning (q.v.) held on 11 October 1931 at Bad Harzburg. The respective leaders, Hugenberg, Seldte, and Hitler (qq.v.), were present. The Harzburger Front collapsed in 1932, when Hugenberg's party supported Hindenburg (q.v.) instead of Hitler for the presidency.

Harzreise, Die, a whimsical account of a journey in the Harz mountains by H. Heine (q.v.), based on a walking tour made in 1824, and published in Vol. 1 (1826) of *Reisebilder*. It is partly idyllic and partly satirical, and the principal objects of its satire are Göttingen University and the self-satisfied German citizen. It contains eight poems of folk-song character; the best known are the cycle of three poems beginning 'Auf dem Berge steht die Hütte', 'Tannenbaum mit grünen Fingern', and 'Still versteckt der Mond sich draußen', and the separate poem 'Ich bin die Prinzessin Ilse'.

Harzreise im Winter, a poem written by Goethe in December 1777, while on a visit to the Harz mountains. It reflects his own confident sense of mission, and portrays in its central section the dejection of the character who cannot accommodate himself to the world, which originated in Goethe's visit to a young man named Plessing, led to despair, it was alleged, by the reading of *Werther* (see LEIDEN DES JUNGEN WERTHERS, DIE).

The central section of the poem forms the text of the *Alto Rhapsody* by Brahms (q.v.).

HAS, KUNZ, a citizen of Nürnberg, who in the later years of the 15th c. wrote several poems, including a eulogy of his native city and lamentations over the moral corruption of the inhabitants.

HASCHKA, LORENZ LEOPOLD (Vienna, 1749–1827, Vienna), a Jesuit until the suppression of the Order in 1773, became university librarian in 1797, and was also professor of aesthetics at the Theresianum. A patriotic poet, he is best known as the author of the Austrian national anthem. See GOTT ERHALTE FRANZ DEN KAISER.

HASENCLEVER, WALTER (Aachen, 1890–1940, Les Milles, Aix-en-Provence), studied at Oxford, Lausanne, and Leipzig universities. He volunteered in 1914 but almost at once became a pacifist. After the 1914–18 War he worked as a newspaper correspondent and went to the U.S.A. and (1924–30) to Paris. In 1933 he was deprived of German citizenship, and spent restless years of exile in England, Italy, Yugoslavia, and France. It was here that he was interned at the beginning of the 1939–45 War. He was released, but interned for a second time at the commencement of the German offensive in 1940; he took an overdose of a drug on the collapse of French resistance.

The first stirrings of Expressionism (see EXPRESSIONISMUS) caught Hasenclever's imagination, and in 1913 he published a volume of Expressionist poetry, *Der Jüngling*. His views were influenced by K. Pinthus and F. Werfel (qq.v.), and his play *Der Sohn* (q.v., 1914) is one of the earliest important dramatic works of the Expressionist movement; it is a protest of the young generation against the old. *Antigone* (1917, in free verse) shapes the Greek legend into a passionate anti-war play; pacifism is the theme of another play written during the war, *Der Retter* (1919).

Towards the end of the war Hasenclever entered a phase of mysticism which culminated in the publication of a translated selection from Swedenborg's works (*Himmel, Hölle, Geisterwelt*, 1925). The first product of this experience was the compressed, laconic play *Die Menschen* (1918) aiming at a shattering effect ('zu erschüttern'); the last was the play *Mord* (1926).

Hasenclever turned to comedy with *Ein besserer Herr* (1926) and *Ehen werden im Himmel geschlossen* (1928), a title which the producer substituted for Hasenclever's own choice (*Doppelspiel*). Hasenclever mocks at Fascism in the comedy *Napoleon greift ein* (1929), in which Napoleon begins and ends as a wax figure in the Musée Grévin in Paris. *Christoph Columbus* (unpublished, performed in London in 1932), was written in collaboration with K. Tucholsky (q.v.), Hasenclever's friend; it is a satire, in which Columbus is cheated of his rights by manipulators at home.

Hasenclever's later works were neglected. The comedy *Münchhausen*, written in 1934, was performed in 1948, and published in 1963; in the same year appeared his novel *Die Rechtlosen*, which was written in France at the time of Hasenclever's internment (1939–40), the theme of the work. The unpublished *Ehekomödie* was performed in London in 1935 as *What should a husband do?* A selection from his work, *Gedichte, Dramen, Prosa*, edited by K. Pinthus, appeared in 1963; it includes *Der Sohn* and other titles mentioned among his first publications. A further MS., the novel *Irrtum und Leidenschaft*, was published in 1969.

Häslein, Das, an erotic verse tale of the late 13th c. It turns on the contrast between sexual innocence and knowledge. The victim, a naïve girl of noble birth, is tricked into yielding herself to a knight. Later, when the knight is about to marry another, he finds his bride too experienced, renounces her, and takes the innocent as his wife. The rabbit (*Häslein*) of the title is the object for which the girl gives her *minne*.

Hasper a Spada, a novel by K. G. Cramer (q.v.).

HASSENPFLUG, HANS DANIEL (Hanau, 1794–1862, Marburg), was chief minister to the Elector Wilhelm II (1821–47) of Hesse-Cassel, whom he served from 1832 to 1837, and to his son, the Elector Friedrich Wilhelm I (ruled first as co-regent, 1831–66), whom he served from 1850 to 1855. Hassenpflug's name was a byword for repression in 19th-c. Germany.

Häßliche Herzogin Margarete Maultasch, Die, an historical novel by L. Feuchtwanger (q.v.), published in 1923. Set in Tyrol in the 14th c., it tells a complex story of intrigue, jealousy, deception, and murder among the great, including the great ones of the Empire in contact with the 'Ugly Duchess', and culminates in her excommunication, and the subsequent sufferings of her people from fire, flood, and the plague. The sensational events are accompanied by a portrait in grotesque caricature of the Duchess herself, who, at the end of the book, has degenerated into a hideous glutton. It is by no means certain that the historical duchess was called 'Maultasch' because of personal ill-looks (see MARGARETE VON TIROL). Persecution of the Jews is an element in the story.

Hastenbeck, a narrative (Eine Erzählung) by W. Raabe (q.v.), written in 1895–8, and published in 1899. It is Raabe's last completed work, and bears as a motto the words of Freiherr vom Stein (q.v.): 'Ich habe nur *ein* Vaterland, das heißt *Deutschland*.'

Set in the second year of the Seven Years War (see SIEBENJÄHRIGER KRIEG), it tells of Pold (Leopold) Wille, a young painter in the service of the Duke of Brunswick, who is conscripted into the Hanoverian army, and deserts after the Convention of Kloster Zeven following the defeat at Hastenbeck. Ill, exhausted, and a fugitive, he finds shelter in the village of the girl he loves; she is Hannchen, a foundling brought up in the parsonage of Pastor Holtnicker and his wife. An old widow, known as die Wackerhahnsche, takes charge of the couple. Pold is befriended by Captain Uttenberger, a Swiss in French service billeted as a sick man in the parsonage. With his aid Pold escapes a French provost patrol, and makes his way with his sweetheart and die Wackerhahnsche through forest country to the Duke of Brunswick's temporary refuge in Blankenburg. On the way die Wackerhahnsche persuades Pastor Störenfreden of Derenthal, who was to have received Hannchen from her foster-parents as his wife, to marry the couple; she also wins the Duke's favour for the couple, thus ensuring their safety.

This slender story attracts an immense and vivid historical background. Raabe uses the characters to evoke the intense suffering and misery brought by the war. Simultaneously he uses literary motifs, especially the pastoral poetry of Salomon Geßner (q.v.), to suggest the literary renaissance, and the power of the enlightened spirit, which was to mark the age following the Seven Years War. In this sense the story is a patriotic historical novel, aptly matched by Raabe's choice of motto, and conceived in his humane manner with no trace of chauvinism.

HATZFELD, ADOLF VON (Olpe, Westphalia, 1892–1957, Bad Godesberg), an officer cadet (Fahnenjunker) at Potsdam, lost his sight in an unsuccessful attempt at suicide. In spite of his blindness he studied at several universities and obtained a doctorate. He settled at Bad Godesberg in 1925. A deeply religious writer, his best-known novel is *Franziskus* (1918), which has the saint as its central figure. He wrote two other novels, *Die Lemminge* (1923) and *Das glückhafte Schiff* (1931), some stories (*Der Flug nach Moskau*, 1942, and the volume *Zwischenfälle*, 1952), and published several collections of religious verse (*Gedichte*, 1916; *An Gott*, 1919; *Ländlicher Sommer*, 1926; *Gedichte des Landes*, 1936). His poems were published complete as *Melodie des Herzens* (1951).

HÄTZLERIN, CLARA, a nun of Augsburg who was a professional copyist. She wrote out in 1471 to the order of Jörg Roggenburg a collection of poems and songs (*Liederbuch der Klara Hätzlerin*) in which HEINRICH der Teichner, Suchenwirt, Suchensinn, Muscatblüt, Rosenblüt, HERMANN von Sachsenheim, and OSWALD von Wolkenstein (qq.v.), as well as a number of others, are represented. The MS. is in Prague.

Haubenlerche, Die, a play (Schauspiel) in three acts by E. von Wildenbruch (q.v.), first performed in the Deutsches Theater, Berlin, in September 1891. It is Wildenbruch's first attempt to keep up with the new Naturalist movement (see NATURALISMUS), and has as its theme love and seduction set against an up-to-date factory background. The crested lark of the title is Lene Schmalenbach, a pretty factory girl, who is pursued by the owner's dissolute brother. The owner himself, August Langenthal, intervenes and proposes marriage to Lene, who, though loving Paul Ilefeld, a worker in the factory, accepts because her sick mother will then be able to have medical treatment. In the end August perceives Lene's love for his workman and withdraws, so allowing her to be united with Ilefeld.

Noteworthy is the character of the philanthropic factory owner.

HAUCH VON KÖLN, see STEPHANUSLEGENDE.

HAUFF, WILHELM (Stuttgart, 1802–27, Stuttgart), son of a Württemberg civil servant, who died when the boy was 6, was at school at Tübingen and Blaubeuren, and then studied theology at the Tübinger Stift and Tübingen University. In 1824, as a young clergyman, he became private tutor in a Stuttgart family, and in 1826 undertook a journey to Paris and then to Bremen and Hamburg; in 1827 he was appointed editor of Cotta's daily *Morgenblatt für gebildete Stände* (q.v.). He married early in 1827 and died the following winter, having caught a chill while attending a funeral.

In his short life Hauff was involved in a literary controversy when, in the novel *Der Mann im Monde* (1826), he used the pseudonym H. Clauren (q.v.), whose style he successfully and satirically imitated. He lost a consequent lawsuit, but gained a moral victory by his *Controvers-Predigt über den Mann im Monde* (1827).

Hauff is the author of the historical novel *Lichtenstein* (q.v., 1826), and of a satirical novel, *Mitteilungen aus den Memoiren des Satan* (1826–7). He became well known for his fairy-tales and stories. His *Märchenalmanach* of 1826 included *Das Wirtshaus im Spessart* (q.v.) and *Der Scheik von Alexandrien und seine Sklaven* (see HENZE, H. W.) appeared in the *Märchen-Almanach* of 1827. In the same year he published the stories *Phantasien im Bremer Ratskeller* (q.v.). *Jud Süß* (q.v.) is one of the stories collected posthumously in *Novellen* (1828). Hauff's small output of poems includes the rewritten folk-song 'Morgenrot,/Leuchtest mir zum frühen Tod', and 'Steh ich in finstrer Mitternacht', entitled *Soldatenliebe*.

Hauff possessed an inventive mind and a gift for story-telling, but he died before he could establish his potential originality.

Sämtliche Werke (3 vols.), ed. S. von Steinsdorff and U. Schweikert, appeared in 1970.

HAUG, BALTHASAR (Stammheim nr. Calw, 1731–92, Stuttgart), a Württemberg clergyman, who became in 1773 a professor at the Stuttgart Gymnasium, transferring later to the ducal Akademie. In 1774 he founded a journal which is best known under the title it bore from 1775 to 1781, *Schwäbisches Magazin von gelehrten Sachen*. Some of Schiller's early poems were published in it. Haug prophetically described Schiller as *os magna sonaturum* ('a mouth which will sound forth great things'). Haug wrote both prose and poetry, though without achieving distinction in either. The poet Friedrich Haug (q.v.) was his son.

HAUG, FRIEDRICH (Niederstolzingen, Württemberg, 1761–1829, Stuttgart), in full Johann Christoph Friedrich, son of the Stuttgart professor Balthasar Haug (q.v.), was conscripted into the ducal Akademie, in which he was a pupil of unusual distinction and a friend of Schiller. In 1763 he was given an important confidential post in the higher administration of the Duchy (later kingdom), and in 1816 he became royal librarian. He was a prolific lyric poet and a writer of distinctive epigrams. His *Sinngedichte* appeared in 1791, *Epigramme und vermischte Gedichte* in 1805, and a selection (*Gedichte*) in 1827.

HAUGWITZ, AUGUST ADOLF VON (Übigau, Oberlausitz, 1654–1706, Übigau), governor of the province of Oberlausitz, was the author of a tragedy on Mary Queen of Scots (*Maria*

Stuarda, 1683) and of a Turkish play entitled *Soliman* (1684). His poems were published in *Prodromus Poeticus* (1684).

HAUGWITZ, CHRISTIAN AUGUST, GRAF VON (Peucke nr. Öls, 1752–1831, Venice), a native of Silesia, entered the Prussian service in 1791, and was employed as minister to Vienna; in the same year he became cabinet minister in charge of foreign affairs.

Haugwitz was responsible for the Peace of Basel (1795, see REVOLUTIONSKRIEGE) in which Prussia withdrew from the wars against France. In 1805 he signed the Peace of Vienna (Schönbrunn, see NAPOLEONIC WARS), by which Prussia exchanged outlying territories (Ansbach, Kleve, Neuchâtel, in German Neuenburg) for Hanover. An almost immediate consequence of this was the isolation of Prussia. The Prussian defeat in 1806 ended Haugwitz's political career, and his name became an object of vilification for the Prussian patriots in the years after the Treaty of Tilsit (1807). Haugwitz nevertheless was not, as was thought, a supporter of Napoleon; he lacked strength of character, and allowed himself to be drawn into implementing policies of which he disapproved.

In 1775 Haugwitz, as a young man, accompanied Goethe and the counts C. and F. L. Stolberg (qq.v.) on a journey to Switzerland.

HAUPT, MORITZ (Zittau, 1808–74, Berlin), one of the founders of Middle High German studies in the 19th c., was elected a professor at Leipzig in 1841 and dismissed in 1850 for participation in the Revolution of 1848. In 1853 he was appointed to Berlin. His publications, apart from *Des Minnesangs Frühling* (q.v., 1857), are in the main concerned with HARTMANN von Aue and NEIDHART von Reuental (qq.v.). In 1841 he founded the *Zeitschrift für deutsches Altertum*, and collaborated with Hoffmann von Fallersleben (q.v.) in *Altdeutsche Blätter* (1836–40).

Haupt was also a classical philologist, publishing an edition of Aeschylus in 1852 and editions of Horace (1851), Catullus, Tibullus, and Propertius (together in 1853), and Virgil (1858). He is said to have been Freytag's model for Professor Werner in *Die verlorene Handschrift* (q.v.).

HAUPTMANN, CARL (Obersalzbrunn, 1858–1921, Schreiberhau, Silesia), the less well-known elder brother of G. Hauptmann (q.v.), and son of a Silesian hotelier, studied science at Jena and Zürich universities. After two years (1889–91) in Berlin, he settled for the rest of his life in the small mountain resort of Schreiberhau

in the Riesengebirge. He began with Naturalistic dialect plays set in his native Silesia (*Marianne*, 1894; *Waldleute*, 1896; *Ephraimsbreite*, 1900), but the comments and poems of *Aus meinem Tagebuch* (1900) revealed an anti-rationalistic and mystical approach. The stories *Aus Hütten am Hange* (1902) have a predominantly tragic tone, but the novel *Mathilde* (1902), recounting a woman's life story, shows character triumphing over environment. The novel *Einhart der Lächler* (2 vols., 1907) surveys sympathetically the total career of an artist. *Ismael Friedemann* (1913) deals with the situation of the half-Jew. Of the later dramatic works *Die Bergschmiede* (1902) had the greatest success.

HAUPTMANN, GERHART (Obersalzbrunn, Silesia, 1862–1946, Agnetendorf, Silesia), son of a hotelier and younger brother of C. Hauptmann (q.v.), took to farming on leaving school, but abandoned his training to attend the Breslau Academy of Art with the intention of becoming a sculptor (1880–2). He spent the winter semester 1882–3 at Jena University, where he was influenced by the lectures of E. Haeckel (q.v.). In 1883 he visited Italy with the hope of settling in Rome as a sculptor, but returned home by the summer of 1884. After a short spell at the Dresden Academy, this time working at graphic art, he studied history for a year at Berlin, and took lessons in acting. A turning-point in his life was his marriage in 1885 to Marie Thienemann, whose wealth made him independent. They settled at Erkner about 15 miles from Berlin, where Hauptmann began to write.

Hauptmann's first production was an unoriginal epic in *ottava rima*, *Promethidenlos* (1885). About this time he made contact with other writers, notably with the Naturalists of the newly founded group Durch and the Friedrichshagener Kreis (qq.v.), which together comprised most of the advanced writers of the day. The tale *Fasching* appeared in 1887 and his outstanding narrative work *Bahnwärter Thiel* in 1888 (qq.v.). In the autumn of 1889 his first play, *Vor Sonnenaufgang* (q.v.), was privately produced by the Freie Bühne (q.v.), and this performance made him famous overnight. In that year he moved to Charlottenburg, and two years later also bought a house at Schreiberhau in Silesia. He moved from Schreiberhau to Haus Wiesenstein, in Agnetendorf, which remained his Silesian residence.

Hauptmann quickly showed his ability to handle the new Naturalistic idiom in a succession of plays, *Das Friedensfest* (q.v., 1890), *Einsame Menschen* (q.v., 1891), the unexpected comedy *College Crampton* (q.v., 1892), the novel social play in which a group of weavers takes the place of an individual character as the hero, *Die Weber* (q.v., 1892), and the Berlin dialect comedy *Der Biberpelz* (q.v., 1893). The next play, *Hanneles Himmelfahrt* (q.v., 1893), startled his admirers, for it set Naturalistic and visionary elements side by side. *Florian Geyer* (q.v., 1896) reassured them, but was not entirely successful in mastering the problem of the Naturalistic treatment of history. With *Die versunkene Glocke* (q.v., 1896) Hauptmann moved towards neo-Romanticism, writing a verse parable on the artist's lot (see KÜNSTLER-DRAMA). Since 1894 he had been involved in a marital crisis, which was not to be solved until 1904, when his first marriage was dissolved, and he married Margarete Marschalk.

Hauptmann's personal difficulties did not stem the flow of his writing. One of his most impressive tragedies, *Fuhrmann Henschel* (q.v., 1899), reverted to the Naturalist vein; as did also *Michael Kramer* (q.v., 1900), which resumes the problem of the artist. At this time he began to adapt older literature. *Schluck und Jau* (q.v., 1900) reflects his interest in Shakespeare (traits of *The Taming of the Shrew* and even *As You Like It* are recognizable), and *Der arme Heinrich* (1902) is based on Hartmann von Aue's verse tale *Der arme Heinrich* (q.v.). *Elga* (1905) is a dramatization of Grillparzer's story *Das Kloster bei Sendomir* (q.v.). Meanwhile he had returned to the characters of *Der Biberpelz* in *Der rote Hahn* (q.v., 1901), but with a less assured touch. He again fully vindicated his capacity to write serious and compassionate Naturalist drama in the rural play of sexual attraction, *Rose Bernd* (q.v., 1903). In the strange visionary fantasy of *Und Pippa tanzt* (q.v., 1906) he mingled myth, earthiness, and sentimentality into a concoction which puzzled its audience. The comedy *Die Jungfern vom Bischofsberg* (1907) was found amusing, but showed signs that his dramatic powers were falling away.

In the spring of 1907 Hauptmann paid a three-month visit to Greece, which inspired *Griechischer Frühling* (1908), a fresh and enthusiastic response to the landscape of antiquity. With *Kaiser Karls Geisel* (1908), Hauptmann, using the gossip of a 16th-c. Italian (Sebastiano Erizzo), wrote an inflated blank-verse play representing the elderly Charlemagne's supposed irresistible infatuation for a young girl. *Griselda* (1909), a half-length play in seven scenes, is a feeble treatment of a traditional story of loyalty and devotion (see GRISELDA). Hauptmann's long and ambiguous novel on the theme of religious delusion, *Der Narr in Christo Emanuel Quint* (q.v.), appeared in 1910. This was his first narrative work since the short *Der Apostel* (q.v.) of 1890. Another novel appeared in 1912, *Atlantis*, a story of the wreck of a liner, to which is incon-

gruously attached a double love-story. In between these second-rate narrative works, Hauptmann made a surprisingly successful return to Naturalism with the socially critical tragicomedy of *Die Ratten* (q.v., 1911). The play *Gabriel Schillings Flucht* (q.v., 1912), also in the Naturalist manner, reverts to the overworked treatment of the exceptional individual, an artist, entangled between the women in his life. *Peter Brauer*, written in 1911, but not published till 1921, is a routine tragicomedy in prose, faintly recalling *College Crampton* and certainly more comic than tragic.

In 1912 Hauptmann was awarded the Nobel Prize for Literature. *Das Festspiel in deutschen Reimen* (q.v., 1913), specially commissioned for the centenary celebrations of Napoleon's expulsion from Germany in 1813, surprised its sponsors by its levity in presenting the stirring times of the Wars of Liberation (see NAPOLEONIC WARS) as a puppet-play, and by its unassertive but firm negation of war. *Der Bogen des Odysseus* (1914), a blank-verse classical play in five acts said to have been conceived in 1907, brings little that is new to the theme except an unfavourable reinterpretation of Penelope. The bloody legend presented in the play (dramatische Dichtung) *Winterballade* (1917), derived from the Swedish authoress Selma Lagerlöf, shows Hauptmann in a phase of fascination with savagery. In striking contrast is the sun-drenched pagan atmosphere of the Novelle *Der Ketzer von Soana* (q.v., 1918).

Der weiße Heiland and *Indipohdi* (both 1920) embody Hauptmann's response to the 1914–18 War. The former play (Phantasie) treats the story of Montezuma in trochaic verse; *Indipohdi* (Dramatisches Gedicht) places Prospero in a Renaissance Mexican setting, and shows him confronting the measureless evil of the world. To the end of his life Hauptmann remained irrepressibly productive, but the flashes of inspiration became rarer, the repetition of earlier motifs and the resuscitation of worn subject-matter more noticeable. Two exercises in epic poetry came between the wars, *Anna*, a rural idyll (ein ländliches Liebesgedicht) in Goethean hexameters (1921), and the lengthy comedy of human life, set in the same measure, *Des großen Kampffliegers, Landfahrers, Gauklers und Magiers Till Eulenspiegel Abenteuer, Streiche, Gaukeleien, Gesichte und Träume* (1928). The novels *Der Phantom* (1923) and *Die Insel der großen Mutter* (q.v., 1924) miscarry, the Nordic mythical flights of the verse tragedy *Veland* (1925) show him at his worst. The tragedy *Dorothea Angermann* (1926) returned to Naturalism and the representation of intense sexuality. *Das Buch der Leidenschaft* (q.v., 1929) adapts personal records of a period of divided love (1894–1904). *Vor Sonnenuntergang* (Schauspiel, q.v., 1932), treats an old man's love without bringing the theme to life. *Hamlet in Wittenberg* (1935), an attempted prelude to *Hamlet*, draws what little vitality it has from Shakespeare, and is written in brittle verse. *Im Wirbel der Berufung* (1936) and *Das Abenteuer meiner Jugend* (2 vols., 1937) are autobiographical, the latter ending with a reading of his first play in 1889.

Mention should also be made of the drama *Das Hirtenlied* (published in part in 1921, completed in 1935), the novel *Wanda* (1928), the short plays *Die schwarze Maske* and *Hexenritt*, grouped as *Spuk* (1929), the Novellen *Die Spitzhacke* (1931), *Der Schuß im Park* (1941), and *Mignon* (posth., 1947), and of the plays *Die goldene Harfe* (1933), *Das Abenteuer der Kathedrale* (1939), *Die Finsternisse* (1947), and the fragment *Herbert Engelmann*, which C. Zuckmayer (q.v.) completed (1952). The chief work of Hauptmann's last years was the *Atriden-Tetralogie* (q.v., 1941–8), his response to the 1939–45 War, seen in the light of fate and Nemesis.

Hauptmann had many of the gifts of the great dramatist, but he had not in sufficient measure the intellectual power, the depth of vision, or the integrity of spirit which could develop those gifts to their full advantage. His prose dramatic work particularly is impressive in power and scope. Hauptmann died in his home in Agnetendorf, but was buried on the Baltic island of Hiddensee near Rügen.

In 1932 and 1935 appeared a *Gesamtausgabe*, *Das dramatische Werk* (6 vols.) and *Das epische Werk* (6 vols.) to mark Hauptmann's 70th birthday; the *Centenar-Ausgabe*, ed. H. E. Hass, was published in 11 vols., 1962–74, *Diarium 1917–1933* in 1980, *Notiz-Kalender 1881–1891* in 1982, and *Tagebuch 1892–1894* in 1985 (all ed. by M. Maschatzke). Correspondence with O. Brahm (q.v.), *Otto Brahm – Gerhart Hauptmann. Briefwechsel 1889–1912*, ed. P. Sprengel, appeared in 1985.

Hauptmann von Köpenick, Der, a comedy by C. Zuckmayer (q.v.), published in 1931. Its central figure, Wilhelm Voigt, finds it impossible on his release from prison to establish himself in civil life without a pass; he conceives the idea of raiding the town hall of Köpenick near Berlin and stealing an identity card. From a pawnshop he obtains a cap, a guards captain's greatcoat, and a sword, commandeers a detachment of guardsmen, and seizes the town hall, only—crowning irony—to discover that Köpenick is not an important enough town to deal in passes. The play ends with the amusement of the Emperor off-stage and Voigt's pardon. (Film version by H. Käutner, 1956.)

This satire on the prestige of the uniform, which was immensely successful in the years 1931–3, had as its basis an episode in 1906, when a Wilhelm Voigt impersonated a Guards captain at Köpenick, and escaped with 4,000 Marks.

Haupt- und Staatsaktion, term originating *c.* 1700, applied to the popular (and often partly extemporized) plays performed in barns and market-places by strolling troupes of actors. Concerned with the sudden fall of kings and princes, with dark conspiracies and executions, they were corrupt and debased versions of literary plays of the 17th c., either Shakespearian or German. From 1730 onwards the term was used chiefly in contempt, e.g. by J. C. Gottsched (q.v.), by G. E. Lessing (q.v.) in the variation Staats- und Heldenaktion, and by Goethe in *Faust* (q.v.), Pt. I: 'Und höchstens eine Haupt- und Staatsaktion' (l. 583).

HAUSENSTEIN, WILHELM (Homberg, Black Forest, 1882–1957, Tutzing), was a talented essayist and travel writer, who in 1907 became a militant Social Democrat. In 1919 he adopted a Christian standpoint and was later converted to Roman Catholicism. He was the first ambassador of the Federal Republic to France (1950–5), laying the foundations of reconciliation. His best work is his autobiography (mainly concerned with his early life) *Lux perpetua* (1947).

HAUSER, KASPAR (?, 1812–33, Ansbach), a foundling who became legendary. Discovered in 1828 in Nürnberg, Hauser carried with him an unsigned letter addressed to a cavalry officer (Rittmeister), telling of the year of his birth and his upbringing in secrecy and isolation; it also stated that the boy wanted to be 'what his father had been', a mounted soldier (Reiter). Strange in appearance and awkward in movement, Hauser would initially only eat bread and drink water. After acquiring intelligible speech, he maintained before the courts that he had lived in a dark room for as long as he could think, not knowing who had looked after him. Anselm, Ritter von Feuerbach (1775–1833), who first took him into his care, later wrote a case history entitled *Kaspar Hauser. Beispiel eines Verbrechens am Seelenleben des Menschen* (1832, reprinted 1983). He placed Hauser into the care of G. F. Daumer (1800–75), then a teacher (Professor) at the local grammar school, but in October 1829 he was found with an injury, recovered, and was successively placed into the homes of two local dignitaries. In 1831 Lord Stanhope (Philip Henry, Earl of Stanhope), who subsequently wrote *Materialien zur Geschichte Kaspar Hausers* (1835), adopted him, entrusting an Ansbach teacher with his care and education. From December 1832 he worked as a clerk for the local court of appeal, and on 13 December 1833 suffered a stab wound, from which he died three days later. Although Hauser himself said that he had been attacked by a stranger, it has been suggested that the wound was self-inflicted, though not with suicidal intent. It has also been maintained that he was the son and heir of Grand Duke Karl of Baden (d. 1812). Contributions to the Kaspar-Hauser-Forschung have been published since the 1870s.

The enigma of Hauser's origin, childhood, and early youth, and the known facts have aroused sustained interest in him on the purely human, psychological, and social, as well as political, level. He is the subject of the novels *Die Söhne Pestalozzis* (3 vols., 1870) by K. Gutzkow (q.v.) and *Caspar Hauser oder Die Trägheit des Herzens* (q.v., 1908) by J. Wassermann. The chant *Gaspar Hauser* by Verlaine (contained in *Sagesse*, 1881) was translated by R. Dehmel and S. George (qq.v.). G. Trakl (q.v.) wrote a poem, *Kaspar Hauser Lied* (contained in the cycle *Sebastian im Traum*), P. Handke a play, *Kaspar* (q.v., 1967), and Dieter Forte (b. 1935) a play, *Kaspar Hausers Tod* (1979). In all, some twenty plays and three films deal with the 'Child of Europe'. In 1974 Werner Herzog directed the film *Jeder für sich und Gott gegen alle* (English title, *The Enigma of Kaspar Hauser*). An illustrated and documented volume on Hauser, *Kaspar Hauser. Das Kind von Europa*, by Johannes Mayer and Peter Twardowsky appeared in 1984.

HAUSHOFER, ALBRECHT (Munich, 1903–45, shot by the Gestapo in Berlin-Moabit prison), professor of political geography at Berlin University, participated in the resistance movement which culminated in 20 July 1944 (see RESISTANCEMOVEMENTS,2),andwasarrestedand executed. His Roman plays, *Scipio* (1934), *Sulla* (1938), and *Augustus* (1939), symbolically criticize his times. His *Moabiter Sonette*, written in prison, and published posthumously in 1946, are among the most powerful and poetically moving documents of the resistance movement.

Häusliche Szene, a poem in elegiacs (51 distichs), written by E. Mörike (q.v.) in 1852. It is a 'curtain dialogue' headed *Schlafzimmer. Präzeptor Ziborius und seine junge Frau. Das Licht ist gelöscht.* A mild quarrel develops because the wife has used some special vinegar, the distilling of which is the husband's hobby. In the end harmony is restored.

HAUSMANN, MANFRED (Kassel, 1898–), served in both the 1914–18 and the 1939–45 Wars. After completing his studies at Göttingen and Munich universities he was for a time a

lecturer at Heidelberg University, a dramatic producer, a business trainee, and then a journalist (Feuilletonredakteur) with the *Weserzeitung*; from 1929 to 1939 he lived as a writer at Worpswede. After the 1939–45 War he was, until 1952, editor of the newspaper *Weserkurier*. A Protestant, he sympathized in his later years with the views of Karl Barth (q.v.).

Hausmann is a prolific author, and in his twenties and thirties wrote novels of romantic vagabond life tinged with sadness, and sometimes rent by sudden tragic and horrifying events. The best known of these novels are the two Lampioon books (*Lampioon küßt Mädchen und kleine Birken*, 1928, and *Salut gen Himmel*, 1929), and the popular story of the adventures of a group of young people who take an unauthorized holiday in a yacht, *Abel mit der Mundharmonika* (1931). There followed the novels *Abschied von der Jugend* (1937) and, years after the war, *Liebende leben von der Vergebung* (1953). Hausmann has also published several volumes of verse, beginning with *Jahreszeiten* (1924) and including *Jahre des Lebens* (1938), *Alte Musik* (1941), *Füreinander* (1946), *Die Gedichte* (1949), and *Irrsal der Liebe* (1960). He has also written plays (chiefly after the war, including *Der dunkle Reigen*, 1951, and *Die Zauberin von Buxtehude*, 1959), and a number of stories, which were collected in *Der Überfall* (1956). *Gesammelte Werke* (12 vols.) appeared in 1983; vols. 1–6 contain the novels, vols. 7–10 the stories, and vols. 11–12 the poetry from 1922–82.

Hausmeier were holders of the office of supervisor of the royal household in the Frankish empire in the 7th c. and 8th c. The Hausmeier in the three kingdoms of Austria, Neustria (see AUSTRIEN and NEUSTRIEN), and Burgundy became virtual rulers. The office was abolished in 751. The word derives from the Latin *majordomus*.

Haus ohne Hüter, a novel by H. Böll (q.v.) which was first published in 1954. It describes with the aid of a variety of characters, drawn from different walks of life, social and human problems caused by the 1939–45 War. But it is the 'house without guardian' which is central to this novel. Two boys, Heinrich Brielach and Martin Bach, have never known their fathers who have been killed in the war. Although Heinrich's mother is poor while Martin's mother and grandmother benefit from an old family enterprise, a jam factory, their experience is very similar, for neither mother has recovered sufficient emotional stability to give her child a stable home life.

Heinrich's mother shares her home with a

succession of 'uncles' whom he learns not to trust. At the end she leaves Leo, a bus conductor, for a better room at the house of her employer, a married baker, and yet a happy home is not in sight. Martin has in Albert a fatherly friend who, for his sake, would marry his mother, Nella, but she refuses to accept more than friendship from him, and, as the widow of Rai, a young poet, lives in a world of cinematic unreality. In order to break the spell of the past, Albert moves with Nella and Martin from their city dwellings to his mother's country inn. On the day of their move Nella and her formidable mother-in-law succeed in achieving a measure of revenge against Gläseler, the ex-officer who was responsible for Rai's death. Frequent changes of perspective from the boys to their elders enhance the psychological and social interest of the novel and support its slender action.

HAUSRATH, ADOLF (Karlsruhe, 1837–1909, Heidelberg), a professor at Heidelberg from 1870, wrote a number of novels based on archaeological study, and known as Professorenromane; the best-known titles are *Antinous* (1880) and *Jutta* (1884). His works were published under the pseudonyms George Taylor or (less frequently) Konrad Mähly. *Pater Maternus* (1898) was the only novel to bear his real name. His two-volume life of Luther (q.v., *Luthers Leben*) appeared in 1906.

HAYDN, JOSEPH (Rohrau/Leitha, 1732–1809, Vienna), a choirboy at St. Stephen's Cathedral, Vienna, until his voice broke, spent most of his life in the service of the Hungarian princely family of Esterhazy as director of music at the two palaces of Eisenstadt and Esterhaz. He was pensioned in 1790, visited London 1791–2, and was there again 1794–5. On his first visit to England he received the honorary degree of Doctor of Music from Oxford University. His Symphony No. 92, though not written specially for the occasion, was played at Oxford on the day of the conferment, and has since been known as the 'Oxford' Symphony. Haydn's last twelve symphonies (Nos. 93–104) were commissioned by the London impresario Salomon, and are sometimes called the 'Salomon' Symphonies.

In addition to orchestral and chamber music Haydn composed a number of Latin masses, including six of exceptional merit composed for the name day of his patroness Princess Maria Hermenegildis Esterhazy in the years 1796–1802. One of these, the D minor, was named the 'Nelson' Mass after it had been performed in Nelson's presence in 1800. Haydn's principal non-liturgical works are the oratorios *Die Schöpfung* (1798), based on *Paradise Lost*, and *Die*

Jahreszeiten (1801), based on James Thomson's *The Seasons*. The German libretto for each of these works was prepared by Baron G. van Swieten (q.v.).

In connection with his visits to Britain Haydn set some Welsh and Scottish folk-songs to music. He also composed the tune of the Austrian national anthem, 'Gott erhalte Franz den Kaiser' (q.v.).

HAYM, RUDOLF (Grünberg, 1821–1901, St. Anton, Arlberg), was at first a schoolmaster, and in 1848 was elected to the Frankfurt National Assembly (see FRANKFURTER NATIONALVERSAMMLUNG). In 1860 he became professor of philosophy and of the history of literature at Halle University. Haym wrote a massive book on the Frankfurt Parliament (*Die deutsche Nationalversammlung*, 3 Pts., 1848–50), but his reputation rests chiefly on his writings on German literature, especially *Die romantische Schule* (1870). He also wrote on Wilhelm von Humboldt, Hegel, Schopenhauer, and Herder (qq.v.). *Herder* (2 vols., 1880–5), one of his last works, was reissued in 1958. A selection of his writings on philosophy and literature, ed. E. Howald (*Zur deutschen Philosophie und Literatur*), appeared 1963.

HAYNECCIUS, MARTIN (Borna, Saxony, 1544–1611, Grimma), the Rector of the well-known school at Grimma, wrote plays for school performance, including *Hans Pfriem* (1582), a satirical comedy which was several times reprinted, and *Almansor, der Kinder Schulspiegel*, an encouragement of industry and castigation of indolence at school. *Almansor* appeared in Latin in 1578, and in German in 1582.

HEBBEL, FRIEDRICH (Wesselburen, Dithmarschen, 1813–63, Vienna), a mason's son, lived in poverty in his native market town until he was 22. His father, who had tried in vain to force the boy to manual labour, died in 1827. Hebbel's mother was devoted to her sons, but could make no provision for them. For eight years Hebbel carried out duties as clerk (Kirchspielschreiber) to J. J. Mohr, the parish mayor and magistrate (Kirchspielvogt), with whom he lodged. Though highly sensitive about his humble station, he benefited from his employer's library. In 1835, through the influence of Amalie Schoppe (q.v.), he went to Hamburg in order to prepare for university study.

In 1836 Hebbel sampled university life at Heidelberg, attending lectures on law, history, and philosophy. Six months later he moved to Munich where he found greater opportunities to pursue his intellectual interests. In 1839 poverty forced him to return on foot to Hamburg. Meanwhile he had heard of his mother's death, and of that of Emil Rousseau, a young lawyer, who had been his only close friend in Heidelberg. From his early Hamburg days until the age of 30 Hebbel remained dependent on the generosity of an unmarried woman, Elise Lensing, who shared her modest earnings as a seamstress with selfless devotion. Upon his return she nursed him through a serious illness caused by his privations. She bore him two sons who died in infancy.

The decade of Hebbel's close relationship with Elise, whom he refused to marry, were years of intense inner conflict. He believed in his vocation as a writer, to which he was prepared to sacrifice all. He first proved his aptitude as a dramatist in *Judith* (q.v., 1841). This prose play was followed by the verse *Genoveva* (q.v., 1843). His first comedy, *Der Diamant* (1847, completed in 1841), derives from established models, principally Kleist's *Der zerbrochene Krug* (q.v.). The play's bitter temper reflects Hebbel's natural inclination towards tragedy. The distinction between tragedy and comedy was to him a matter not of idea, but of form. In 1842 he published a collection of poems (*Gedichte*). His writings secured him a grant from Christian VIII, King of Denmark and sovereign of his native province, which he used to satisfy his urge for travel. He studied for a few months in Copenhagen, where he was in contact with the Danish sculptor Thorwaldsen (1768–1844). Having applied unsuccessfully for a chair of aesthetics at Kiel University, he turned his back on the north and went to Paris for a year (1843–4). He met Heine (q.v.), but he was lonely in the French capital, crippled by poverty and personal problems. He completed the domestic tragedy *Maria Magdalene* (q.v., 1844), and on the advice of a loyal friend, Felix Bamberg, wrote the *Vorwort zur Maria Magdalena betreffend das Verhältnis der dramatischen Kunst zur Zeit und verwandte Punkte* (1844, see MARIA MAGDALENE). The *Vorwort* and an earlier tract on theory, *Mein Wort über das Drama!* (1843), written in response to provocative criticism by J. L. Heiberg, earned him a doctorate from Erlangen University (1844). A further essay, *Über den Stil des Dramas*, was written in 1847. A despondent Hebbel resumed his travels and journeyed to Italy. In Naples he met H. Hettner (q.v.). The southern environment and climate afforded a respite, but the Danish grant ran out and was not renewed. The decision not to return to Elise and to Hamburg was virtually taken when Hebbel set out for Vienna, where, in fact, he settled for the remainder of his life.

Hebbel's love for the gifted actress Christine Enghaus (q.v.) and their subsequent marriage (1846) introduced stability and a lively partnership into his life, though H. Laube (q.v.) for years obstructed their collaboration with the Burgtheater. In Vienna Hebbel completed a one-act tragedy in blank verse, *Ein Trauerspiel in Sizilien* (1847), which contains a strong element of social criticism and is an attempt at tragicomedy, and a three-act tragedy in prose, *Julia* (1851), a successor to *Maria Magdalene*. The following years were highly productive. *Herodes und Mariamne* (q.v., 1849) was nearing completion during the street fighting of 1848 (see REVOLUTIONEN 1848–9). The three-act Märchen-Lustspiel in blank verse, *Der Rubin* (1851), begun in 1849, is based on Hebbel's own Märchen *Der Rubin* (published by Th. Mundt, q.v., in 1843), which he had first planned in 1836. He dedicated the play to F. G. Kühne (q.v.). From his early conception of this work derives his motto 'Wirf weg, damit du nicht verlierst!', suggesting that the retention of one's inner worth depends on the abandonment of one's most precious possession. Many variations on this seemingly paradoxical principle can be traced throughout his work, including his treatment of the theme of the Nibelungen treasure. Hebbel's Märchen-Lustspiel is set in Baghdad. The youth Assad, by throwing away a precious ruby, breaks the evil spell which has turned Princess Fatime into a unique precious stone. He thus unwittingly wins her hand and the Kalif's crown. The play was not appreciated at its first performance on 21 November 1849 at the Burgtheater (q.v.).

Hebbel turned to a religious drama, *Moloch*, and then wrote his two-act play in flowing blank verse, *Michel Angelo*, which he dedicated to Robert Schumann (q.v.). The play is a kind of Künstlerdrama, and owes its inspiration to Kühne, who had impressed Hebbel by his understanding of the artist who pursues his creative work regardless of public indifference, a view which Hebbel endorses in the play. It is an introduction to the first of his two 'deutsche Trauerspiele', the five-act tragedy in prose *Agnes Bernauer* (q.v., 1852). It has as its theme the necessity for sacrifice on the part of the ruler and, more strikingly, for the sacrifice of an innocent individual. This theme is linked with the idea that perfect beauty constitutes tragic guilt, a conception peculiar to Hebbel. *Gyges und sein Ring* (q.v., 1856), a five-act tragedy in blank verse, originates from Hebbel's desire to accommodate political and ethical ideas in classical form. Hebbel's last completed tragedy and his second German tragedy is the dramatization of the medieval *Nibelungenlied* (q.v.), *Die Nibelungen. Ein deutsches Trauerspiel in drei Abteilungen* (q.v., 1862). To him the Nibelungen epic represented Germany's only national literary heritage of universal appeal. The evolution of mythologies and of history, and the exposure of the workings of the human mind are the forces directing the 'watchmaker' who claimed, after he had been assured of the success of his *Nibelungen*, that he had merely restored an old clock.

Though Hebbel had by now abandoned his *Moloch*, he still contemplated a *Christus*, a project which he envisaged as a combination of drama and oratorio. His interest in the relationship between music and drama is reflected in *Ein Steinwurf oder Opfer um Opfer. Ein musikalisches Drama* (posth., 1883), which he wrote for the Russian composer Anton Rubinstein (1830–94), who paid for it but did not set it to music. Conscious of the power of his vision, but hindered by the unequal flow of creativity, Hebbel concentrated more and more, as death drew near, on *Demetrius* (q.v.). He had originally planned to complete Schiller's fragment in time for the centenary of Schiller's birth. The work reached the fifth act and, although unfinished, assimilated ideas which show a pervasive grasp of present and future political issues and their dependence on the intricacies of human ethics and spirituality.

Hebbel referred to his brief novel *Schnock. Ein niederländisches Gemälde* (1850, written in 1837) as 'ein kleines Büchlein'. It is his longest narrative work, and owes its inspiration to early reading of JEAN PAUL (q.v.), and to the cholera epidemic of the thirties, which shocked him into his grim brand of humour. In 1855 Hebbel published seven *Novellen*, notable for their density of form: *Matteo, Herr Haidvogel und seine Familie, Anna, Pauls merkwürdige Nacht, Die Kuh, Der Schneider Nepomuk Schlägel auf der Freudenjagd*, and *Eine Nacht im Jägerhause*. Hebbel's only epic, *Mutter und Kind. Ein Gedicht in sieben Gesängen* (q.v., 1859) is written in the flexible hexameters of Goethe's *Hermann und Dorothea* (q.v.), and earned him a prize instituted to commemorate Goethe's work. Hebbel published *Neue Gedichte* in 1848, and in 1857 *Gedichte von Friedrich Hebbel*, his final collected and revised edition. His poetry includes the cycles *Dem Schmerz sein Recht* (q.v.), *Des Dichters Testament*, and *Sonette*, as well as epigrams. His extensive writings on literature, art, and travel also include commentaries on the events of 1848. They reveal his preoccupation with human existence in isolation, with ethical and metaphysical philosophy, mythology, and religion. *Aufzeichnungen aus meinem Leben* were written between 1846 and 1854, *Selbstbiographie für Saint René Taillandier* and *Selbstbiographie für F. A. Brockhaus* (modelled on the former) in the fifties. Hebbel's valuable *Tagebücher* were begun on 23 March 1835 and continued into the year of his death. The sub-title to the *Erstes Tagebuch* is indicative: 'Reflexionen über Welt, Leben und Bücher, hauptsächlich aber über mich selbst, nach Art eines Tagebuchs'. Emil Kuh (q.v.) was

Hebbel's first biographer and editor (*Friedrich Hebbels sämmtliche Werke*, 12 vols., 1865–7). *Sämtliche Werke*, edited by R. M. Werner, 1901 ff., consist of three parts: *Werke*, 12 vols., *Tagebücher*, 4 vols., and *Briefe*, 8 vols.; *Werke*, 5 vols., edited by G. Fricke, W. Keller, and K. Pörnbacher, appeared in 1963–7, and *Neue Hebbel-Briefe*, ed. A. Meetz, in 1963.

HEBEL, JOHANN PETER (Basel, 1760–1826, Schwetzingen), born in humble circumstances, lost his father in infancy, and his mother when he was 13. Their savings and the generous help of friends enabled him to escape the worst consequences of being orphaned. Schooling, first at Hausen, then at Karlsruhe, prepared him for theological study at Erlangen (1778–80), and he took holy orders in the Protestant Church in Baden. Hebel became a private tutor in Hertingen for a short time, and was then appointed a teacher (Präzeptoratsvikarius) at Lörrach just inside the Baden frontier with Basel (1783–91). In 1791 he was appointed a teacher at the Gymnasium in Karlsruhe, rising to the rank of professor in 1798. Ten years later he became head of the school, and thereby also *ex officio* a member of the ecclesiastical authority for Baden. In 1819 he was appointed Prälat, the highest office in the Evangelical Church of the land. He died in Schwetzingen while making an official visit to north Baden.

Hebel had begun to write poetry in about 1800, and his first collection of 32 poems in south Swabian dialect appeared in 1803 as *Alemannische Gedichte* (q.v.). They achieved a modest success, being reprinted three times up to 1820. In 1803 he began to contribute to the official Calendar for the Lutheran Church in Baden, and in 1807 he became its editor and principal author. In 1808 its title was changed to *Der Rheinländische Hausfreund*. In this form Hebel edited it from 1808 to 1815, and again in 1819. His contributions to it in the form of pithy anecdotes in prose were collected and published by Cotta in Stuttgart in 1811 as *Schatzkästlein des Rheinischen Hausfreundes* (q.v.).

Hebel possessed a gift for simple, direct writing and a sincere and unaffected mind. His popular poetry and prose, whether in dialect or in educated German, owe nothing to movements, but speak to the common man in his own tongue.

A *Gesamtausgabe*, ed. W. Zentner (3 vols.), appeared 1968–72.

HECKEL, ERICH (Döbeln, Saxony, 1883–1970), an Expressionist painter and one of the founders of the Brücke (q.v.) in 1905. Under the National Socialists his work was regarded as 'degenerate' (see ENTARTETE KUNST).

HEDRICH, FRANZ (Podskal nr. Prague, 1823?–95, Edinburgh), a political Radical, and a member of the Frankfurt Parliament (see FRANKFURTER NATIONALVERSAMMLUNG). He collaborated in some measure with Alfred von Meißner (q.v.), whose novels Hedrich later claimed to have written. After Meißner's death he wrote a tract entitled *A. Meißner–F. Hedrich* (1890). The full truth is not known with certainty, but it is generally thought that at the very least Hedrich overstated his claims.

HEDWIG, HEILIGE (1174–1243, Trebnitz, Breslau), Duchess of Lower Silesia (Niederschlesien), became a Cistercian nun and was canonized in 1267. She is the patron saint of Silesia.

HEDWIG or **HADWIG,** HERZOGIN VON SCHWABEN (d. 994), was the wife of Duke Burchard II; after his death in 973 she resided in Hohentwiel Castle. A woman of intelligence and education, she devoted herself to classical studies. She is one of the characters in Scheffel's *Ekkehard* (q.v.).

HEER, FRIEDRICH (Vienna, 1916–83, Vienna), pseudonym Hermann Gohde, studied philosophy at Vienna University where he obtained an appointment in 1949, and in 1962 a professorship, having joined the Burgtheater in 1961 as principal dramatic adviser (Chefdramaturg). After the 1939–45 War and participation in the Austrian Resistance he contributed to the monthly *Wort und Wahrheit* (1946) and to the weekly *Die Österreichische Furche* (1948), and became co-editor of *Neues Forum*. He is the author of a number of works on cultural history, beginning with *Die Stunde des Christen* (1947), which was followed by *Aufgang des Abendlandes* (2 vols., 1949), *Das Experiment Europa* (1952), *Die Tragödie des Heiligen Reiches* (1952, and see DEUTCHES REICH, ALTES), *Europäische Geistesgeschichte, Grundlagen der europäischen Demokratie der Neuzeit* (both 1953), *Koexistenz, Zusammenarbeit, Widerstand* (1956), *Die dritte Existenz, Land im Strom der Zeit* (both 1959), *Europa—Mutter der Revolutionen* (1964), *Das Heilige Römische Reich, Gottes erste Liebe* (both 1967), *Der Glaube des Adolf Hitler* (1968), and *Scheitern in Wien* (1974).

HEER, JAKOB CHRISTOPH (Töß nr. Winterthur, Switzerland, 1859–1925, Rüschlikon), was first a teacher and then a journalist in Zürich. For three years (1899–1902) he was editor of *Die Gartenlaube* (q.v.) at Stuttgart. Heer wrote novels with an Alpine background, varying from Wallis in *An heiligen Wassern* (1898) to Grisons (Graubünden) in *Der König der Bernina* (1900). Skilful description and a regional atmosphere were usually woven into a sentimental love-story. This tendency to novels of light entertainment became more marked with the later works, *Felix Notvest* (1901), *Joggeli* (1902), *Der Wetterwart* (1905), *Laubgewind* (1908), and *Der lange Balthasar*. Of these *Joggeli*, which contains autobiographical material, is probably the best. *Tobias Heider* (1923) is virtually an autobiography.

Romane und Novellen (10 vols.) appeared in 1927, *Gedichte* in 1913, and *Erinnerungen* in 1930.

HEERMANN, JOHANNES (Raudten, Silesia, 1585–1647, Lissa), a Protestant author of hymns, became in 1612 pastor in Köben, but resigned his cure in 1638 because of ill health. His religious poetry, quickened by a deep personal sense of relationship to Jesus, is contained in several volumes; *Devoti musica cordis* (1630) and *Sonntags- und Fest-Evangelia durchs gantze Jahr* (1636) deserve special mention.

Geistliche Lieder, ed. Philipp Wackernagel, appeared in 1856, and a selection entitled *Johannes Heermanns frohe Botschaft,* ed. R. A. Schröder (q.v.), in 1936.

HEGEL, GEORG WILHELM FRIEDRICH (Stuttgart, 1770–1831, Berlin), philosopher, was educated at the Tübinger Stift (q.v., 1788–93), where Hölderlin and Schelling (qq.v.) were among his friends. In 1801 he wrote the tract *Differenz des Fichteschen und Schellingschen Systems der Philosophie,* and settled in Jena as Privatdozent to prepare his own system designed to supersede the theories of relativity (Relativitätstheorien) of Herder and Kant (qq.v.), and to set the course for later idealistic philosophy.

Hegel aimed at an almost Aristotelian comprehensiveness. He set out in three parts his system of logic based initially on the substantiation of the 'idea', which he describes as *Logos,* and which in turn develops and manifests itself in *Nature* before it reaches the third stage embracing *Ethics* (concepts of morals as well as the legal administration of justice). These three, *Sitte, Recht,* and *Staat,* are compressed in a single term, culture. This third stage (culture) in turn forms the basis for a parallel development of God's self-realization in man, and of man's increasing consciousness of a dialectical evolution which determines Hegel's approach to history. *Die Phänomenologie des Geistes* was ready in proof form for his pupils in Jena in 1806, the year in which he had to leave to escape the Napoleonic occupation (see NAPOLEONIC WARS). Over the next ten years Hegel was for a short time successively editor of the *Bamberger Zeitung,* and headmaster (Rektor) at Nürnberg Gymnasium. In 1817 he became professor of philosophy at Heidelberg University, and in 1818 was appointed to the corresponding chair at Berlin

University, vacated by Fichte (q.v.). He died of cholera two years after becoming university rector.

Hegel opens his speculative philosophy of history by defining three categories of historical development. (1) Earlier recorded history (e.g. Herodotus); (2) the period in which history is recorded in a 'reflective' style taking account of the attitude of the historian to his material, and constituting pragmatic history; (3) philosophical history; this he conceives in terms of intellectual contemplation searching for a pattern of historical evolution, which should yield *a priori* concepts appropriate to logical thought as he understood it ('Europa ist schlechthin das Ende der Weltgeschichte, Asien der Anfang'). Hegel sought to devise a concept of logic which is not based on empirical cause and effect, but on a dialectical process adapted from that which Fichte had already introduced. He sees historical evolution in terms of a triadic pattern for which he first establishes a 'thesis' which provokes an 'antithesis', out of which emerges a 'synthesis'. This 'synthesis' again forms a new 'thesis', but on a higher level than the preceding 'thesis'. Each stage contains in reflective form a 'thesis', since it grows out of an existing situation by both 'preserving' (Hegel uses the term 'aufgehoben') and reacting against it. No stage in this dialectical progression is repetitive.

In his *Philosophie der Geschichte* Hegel classifies the three geographical parts of the world (Africa, Asia, and Europe), and abstracts from them the following phases: the Oriental, the Greek, the Roman, and the Germanic phase, or world. The Germanic world he traces from its beginnings to the Middle Ages, a period which forms the first stage in its development. The second stage generates a feudal monarchical system, while the third opens with the Reformation (q.v.). It closes, in Hegel's perspective, with the age of Enlightenment (see AUFKLÄRUNG) and the ensuing revolutionary period.

Oriental history is dominated by despotism, Greek history by democracy, and European history by aristocracy culminating in constitutional monarchy ('Der König ist der Punkt auf dem i in Verfassungsfragen'). Criticisms of Hegel's conception of monarchy are directed against his favourable appraisal of the Prussian state. In the development of the various forms of government, which are co-ordinated with the history of art and religion, Hegel finds many variations, but tends to subordinate historical facts to his dialectics.

This applies also to Hegel's metaphysics. His pantheistic approach is contained in his vague concept of a Weltgeist objectivizing and manifesting itself in history. Hegel distinguishes in his philosophy of the spirit, the Geist transcend-

ing nature, between the subjective, objective, and absolute spirit. The objective spirit, as opposed to the limited subjective spirit, represents the ethics of communities, from the small unit of the family to that of the state, and establishes the laws containing the highest forms of ethics. Above and beyond this, the absolute spirit permeates the three spheres of art, religion, and philosophy. While the subjective and objective spheres of the spirit generate the forces of history, the absolute spirit induces, through its conciliatory and harmonic properties, a sense of purity and perfection. In this Hegel sees the goal of aesthetics in art, as elaborated in *Vorlesungen über die Aesthetik*.

Historically Hegel emerged after Kant, Fichte, and Schelling as the last Romantic idealist, whose views led antithetically to the dialectics of Marx and of L. Feuerbach (qq.v.).

The principal works appearing during Hegel's lifetime, and containing titles already mentioned, are *Wissenschaft der Logik* (1812–16) and *Die Encyclopaedie der philosophischen Wissenschaften im Grundrisse* (1817, and, extended, 1827). A third edition followed in 1830. Hegel's lectures, among them *Philosophie der Weltgeschichte* and *Geschichte der Philosophie*, were edited after his death by E. Gans, and, in extended form, by Hegel's son Karl Hegel. The first complete edition of his works, which had in part to be pieced together from the lecture notes of his students and Hegel's own notes, appeared in 19 vols. from 1832 to 1840. This edition formed the basis for Hegel's influence throughout the 19th c. as well as the 20th c., making itself felt at various levels of German, and especially Prussian, cultural development in secondary school education, as well as Geistesgeschichte. It dominated German approaches to the history of literature, notably with W. Dilthey (q.v.), who wrote on Hegel's early work (1905), F. Gundolf (q.v.), H. A. Korff, and F. Strich. Hegel's life of Christ (a fragment) was not published until 1906, but his Religionsphilosophie has influenced German theology as well as aspects of English theology.

The 200th anniversary of Hegel's birth was marked by a publication of his works (20 vols., 1970). A *Neue kritische Gesamtausgabe* of approximately 40 vols. is being edited by F. Nicolin, the director of the Hegel-Archiv at Bochum University.

HEGNER, JOHANN ULRICH (Winterthur, 1759–1840, Winterthur), an administrator and judge in Switzerland (Zürich and Winterthur), devoted his later years to literature, writing novels with Swiss settings (*Die Molkenkur*, 1812; *Salys Revolutionstage*, 1814; *Suschens Hochzeit*, 1819).

HEIBERG, HERMANN (Schleswig, 1840–1910, Schleswig), a bookseller, and later a bank director, is the author of several novels, including *Ausgetobt* (1883), *Apotheker Heinrich* (q.v., 1885), *Esthers Ehe* (1886), and *Dunst aus der Tiefe* (1890). In the 1880s he was esteemed as a realist who managed to avoid the sordid.

Heidebilder, a series of nineteen poems published by Annette von Droste-Hülshoff (q.v.) in *Gedichte* (1844). Though they refer to the heaths of her Westphalian homeland, they were written at Meersburg, Lake Constance, in 1841–2. They are remarkable as much for their grasp of the tone of the landscape as for their observation of detail. The group includes *Die Lerche, Die Jagd, Der Weiher, Die Mergelgrube, Das Hirtenfeuer, Der Heidemann, Das Haus in der Heide,* and *Der Knabe im Moor.*

HEIDEGGER, GOTTHARD (Stein am Rhein, 1666–1711, Zürich), a Protestant pastor in St. Margrethen (1688), Rorbas (1697), and Zürich (1703), is the author of a tract entitled *Mythoscopia Romantica: oder Discours von den sogenannten Romans* (1698); it severely criticizes the extravagant novels of his age.

HEIDEGGER, MARTIN (Meßkirch, Baden, 1889–1976, Freiburg), a philosopher, whose esoteric terminology makes his thought virtually inaccessible to the layman, became a lecturer (Privatdozent) at Freiburg University in 1915, and was appointed professor of philosophy at Marburg in 1923, and at Freiburg in 1928. In 1933, under the National Socialist regime, he was appointed Rector in place of a dismissed colleague. In 1945 he himself was dismissed, but was later reinstated in the chair at Freiburg University.

Heidegger, who was influenced by Kierkegaard (q.v.) and is preoccupied with existence, declared that his philosophy differed radically from Existentialism as understood by Sartre and others. His basic problem, as stated by himself, is the 'Sinn von Sein', a phrase loaded with ambiguities. His thought was first developed in *Sein und Zeit* (1927). His linguistic speculation over what is meant by 'ist' leads to the view that 'Dasein' (existence) is threatened by 'Sorge' (anxiety), behind which is concealed the temporality of existence (Zeitlichkeit). Heidegger, as he unfolds his linguistic structures, develops a peculiar vocabulary in which philosophy is turned into 'Seinsdenken', language into 'Haus des Seins', Man into 'Hirte des Seins', and the business of philosophy, it appears, is to listen to the silence of existence (Sein). Affirmations are made, such as 'das Nichts nichtet' (his terminology includes 'Nich-

tung') and 'die Angst ängstet' and the 'Kehre des Denkens' leads to the 'Denken der Kehre'. Heidegger's thought amounts to a disavowal of science and technology, which are associated with nihilism, and an affirmation of a mystical self-liberation.

Heidegger's considerable influence extends to theology (see BULTMANN, R. K.), and he contributed to literary criticism in *Erläuterungen zu Hölderlins Dichtung* (1944) and *Hebel—Der Hausfreund* (1957) in a less opaque style. His works include *Was ist Metaphysik?* (1929, inaugural lecture at Freiburg), *Vom Wesen des Grundes* (1929), *Vom Wesen der Wahrheit* (1943, a lecture), *Was heißt Denken?* (1954), *Der Satz vom Grund* (1957), *Die Frage nach dem Ding* and *Die Technik und die Kehre* (both 1962). His collected essays are entitled *Wegmarken* (1967). A complete edition of his works (planned in *c.* 70 vols.) appeared 1975 ff.

Heidelberg was the residence of the counts of the Palatinate from the 13th c. to 1720. Its Renaissance palace (Heidelberger Schloss) was destroyed by the French in 1689 and 1693. In 1803 Heidelberg was incorporated in Baden. From the time of the Romantic movement onwards, its picturesque situation on the Neckar and its student life have repeatedly been the subject of poetic praise, of which the best-known example is Scheffel's 'Alt-Heidelberg, du feine' (q.v.). Heidelberg was the centre of the activities of the Romantic writers J. J. von Arnim, C. Brentano, J. J. von Görres (qq.v., and see ROMANTIK). J. von Eichendorff (q.v.) recorded his impressions of his stay in Heidelberg (1807–8) in *Halle und Heidelberg*, and F. Hölderlin (q.v.) gave the title *Heidelberg* to one of his odes. See also HEIDELBERGER FASS and HEIDELBERG UNIVERSITY.

Heidelberg, an ode written in 1798–1800 by F. Hölderlin (q.v.) in praise of the beauty of Heidelberg. It is in Asclepiadic verse and was printed in 1801. (See ASKLEPIADISCHE STROPHE.)

Heidelberger Faß, a huge barrel with a capacity of approximately 5,000 gallons, preserved in the Castle of Heidelberg. It is referred to in Heine's poem 'Die alten bösen Lieder' (*Lyrisches Intermezzo* No. LXV in *Buch der Lieder,* q.v., 1827).

Heidelberger Katechismus, a catechism for the Reformed Church (see REFORMIERTE KIRCHE) devised in 1563 by two theologians of Heidelberg, Zacharias Ursinus (1534–83) and Caspar Olevianus (1536–87), at the instigation of the Elector of the Palatinate, Friedrich III, der Fromme (1515–76), a Protestant, who had become a Calvinist in 1561. His successor, the Elector Ludwig VI, was a Lutheran, and soon after Friedrich's death Ursinus and Olevianus had to leave Heidelberg.

Heidelberger Liederhandschrift, Große, formerly known as the Manessische Liederhandschrift, is a MS. of large format, containing poems by 140 poets, with 138 full-page illustrations giving imaginary portraits of the authors. It is believed to have been planned and commissioned by Rüdeger Manesse, a patrician of Zurich and a patron of the arts, together with his son Johannes, who died in 1297. The work was executed after their death, between 1310 and 1330. It passed into the hands of the Fugger family (see FUGGER), by whom it was sold in 1571 to the Elector of the Palatinate, Friedrich III, der Fromme (1515–76), who placed it in his library at Heidelberg. It was sold again in 1657 to Louis XIV of France, and was long in the Bibliothèque nationale in Paris, for which reason it has also been known as the Pariser Liederhandschrift. In 1888 an exchange secured its return to Heidelberg, where it is preserved in the University Library. In consequence of its Swiss origin, Swiss and south-west German poets are strongly represented. It contains no record of the music to which the songs were sung.

The manuscript, the purpose of which was the preservation of the lyric poetry of the 13th c., remains one of the most important repositories of Minnesang. Its beautiful and copious illustrations make it also artistically the most valuable of the medieval collections of songs. It is referred to in G. Keller's Novelle *Hadlaub* (q.v.).

Heidelberger Liederhandschrift, Kleine, an important collection of Middle High German poems written down in Strasburg during the 13th c. It contains rather fewer poems than the Große Heidelberger Liederhandschrift (see HEIDELBERGER LIEDERHANDSCHRIFT, GROSSE), and has no illustrations. It is the oldest of the great manuscript collections. See also JENAER LIEDERHANDSCHRIFT.

Heidelberger Passionsspiel, a medieval passion play preserved in a MS. at Heidelberg. The date of the MS., which is incomplete, is 1514. The work, of some 6,000 lines, is based on the *Frankfurter Dirigierrolle* (see FRANKFURTER SPIELE). It contains a number of scenes from the Old Testament which prefigure the episodes from the New.

Heidelberg University, one of the oldest of German universities, founded in 1386 by the Elector of the Palatinate Ruprecht I (1309–90); except for Prague and Cracow, which are no longer German-speaking universities, only Vienna (1365) is an older foundation.

Heidelberg University received a new impulse in the late 15th c. from the Elector Philipp, who made it a centre of humanism at which were welcomed among others R. Agricola, Celtis, Reuchlin, and Wimpfeling (qq.v.). In 1556 the university became Protestant, and was associated with the *Heidelberger Katechismus* (q.v.) of 1563. The world-famous library (Biblioteca Palatina), built up from the earliest years, contains priceless treasures, including the Große Heidelberger Liederhandschrift (see HEIDELBERGER LIEDERHANDSCHRIFT, GROSSE).

During the Napoleonic period (see NAPOLEONIC WARS) Heidelberg passed to Baden, and in 1803 its university was refounded by Margrave Karl Friedrich, hence its present double title Ruprecht-Karl-Universität Heidelberg. Its popularity is reflected in the presence of members of the Heidelberger Romantik (see ROMANTIK), including L. J. von Arnim, C. Brentano, Görres, and Eichendorff (qq.v.). Among its scientists were R. W. Bunsen (1811–99) and H. L. F. Helmholtz (1823–94). In more recent times F. Gundolf and K. Jaspers (qq.v.) have held chairs at the university.

In the 19th c. Heidelberg enjoyed a romantic, legendary image, fostered by Scheffel's 'Alt-Heidelberg' in *Der Trompeter von Säkkingen* (q.v., 1854), and Eichendorff's nostalgic *Halle und Heidelberg* (posth. 1866).

HEIDENBERG, JOHANNES VON, see TRITHEMIUS, JOHANNES VON.

Heidenröslein, one of the best known of Goethe's early poems, written in Strasburg in 1771, and published in 1789. The erotic symbolism of the boy who plucks the flower, in spite of its reluctance, is obvious, and the poem is one of those originating from the time of Goethe's love for Friederike Brion (q.v.). The fresh simplicity of its style was entirely new at the time, and Herder (q.v.) quoted from it in 1773 (in *Von deutscher Art und Kunst,* q.v.) in terms that led to the false assumption that it was a folksong. Though there are settings by R. Schumann and Brahms, that by F. Schubert (qq.v.) is the best known.

Heidepeters Gabriel, a novel by P. Rosegger (q.v.), published in 1882. Gabriel is the child of a poor farmer who suffers persecution from his neighbours and from the great landowner Graf Frohn; he develops a poetic talent, and goes out into the world. Returning to his mountain homeland he meets Anna Mildau, whom he marries and lives with happily for two years until she dies of heart failure. The story has some affinity with Rosegger's own life up to the death of his first wife in 1875, but a preface added in 1912 discourages an autobiographical interpretation.

Heidin, Die, an anonymous Middle High German verse tale of erotic character, written probably about the middle of the 13th c. It tells of a knight who seeks the beautiful wife of an oriental potentate, is at first repulsed, but later welcomed. The lady proposes that he should have a half-share in her. He chooses the upper half, and, having made the correct choice, attains by a ruse enjoyment of the whole woman. For he orders her mouth, which 'belongs' to him, to address the husband in such a way that the latter is provoked to beat her, whereupon the 'whole woman' gives herself to the knight. The episode represents a remarkable intrusion of the broadly comic into the dignified knightly world.

Heilbronner Bund, the League of Heilbronn, April 1633, was formed and directed by the Swedish chancellor Oxenstierna (q.v.), with Bernhard von Sachsen-Weimar (q.v.) as military leader, in order to unite the rulers of Swabia, Franconia, and the upper and lower Rhine with Sweden against the Emperor Ferdinand II (q.v., and see DREISSIGJÄHRIGER KRIEG). The League broke up after the battle of Nördlingen in 1634.

'Heil dir im Siegerkranz', first line of the former Prussian national anthem, sung to the tune of 'God save the King'. The words were written by Pastor Heinrich Harries in 1790 for the Danish king, and were adapted for Prussia by B. G. Schumacher in 1793.

Heilige, Der, Novelle by C. F. Meyer (q.v.), published in 1879 in the *Deutsche Rundschau*. Expounding his source, *Histoire de la Conquête de l'Angleterre par les Normands* by J. N. A. Thierry (1825), Meyer presents the saint of the title, Thomas à Becket, through the eyes of a fictitious narrator (see RAHMEN), a Zurich crossbowmaker (Hans der Armbruster), in the year of Becket's canonization. Hans relates his tale, the climax of an adventurous life, to a Chorherr of a Swiss monastery, where he has been offered accommodation for the night. Meyer skilfully places Hans in the complex pattern of events as a close observer and confidant of Becket and of Henry II, to whom the Swiss craftsman has commended himself by the invention of a new crossbow.

The account of Becket's services to Henry as chancellor and as educator of his sons stresses the cultured man of the world, who places his rare gifts in dealing with affairs entirely at the King's disposal. The close bonds with Henry arising from the King's dependence on his chancellor lead to Becket's election as archbishop of Canterbury, and to a radical change in his relationship with Henry. The meeting of both men in France and the failure of an effort at reconciliation dominate the latter part of the story. When the King's outburst of anger subsides, he orders Hans to follow the four knights, who have already set out for England on their murderous errand, to ensure that no harm comes to Becket, who has resumed his clerical duties. Hans fails to persuade Archbishop Becket to follow him to safety, and is present when Becket, defying the King's interference, is murdered at vespers in the sanctuary in Canterbury Cathedral (in reality the assassination took place in the north transept). The King's remorseful return, submission to scourging, and death in agony conclude the historical record, into which Hans's personal experiences, culminating in marriage in his native country, are woven.

Meyer's portrayal of Becket's character includes an invented episode in the past. Becket's (fictitious) daughter Gnade embodies the 'grace' of unblemished beauty, which King Henry has destroyed by seduction, despite the precautions taken by Becket to keep her isolated from the court and from worldly life. Her death, for which the King is responsible, precedes his fateful decision to have Becket appointed archbishop. All subsequent events are thus poised between the King's initially repressed sense of personal guilt towards Becket and the open wound of Becket's grief, his urge for revenge, and his inability to bestow 'grace' upon the King in the name of Christ.

Heilige Allianz, Holy Alliance, conceived by the Tsar Alexander I, and agreed on 26 September 1815 at Paris by the three powers Russia, Austria, and Prussia through the signature of the three monarchs. After its publication in 1816 almost all continental sovereigns acceded to it. The document contained no stipulations but was a declaration of intent, affirming the will for peace, the love of justice, and the sanctity of religion. The Holy Alliance was held in detestation by Liberal opinion in Europe.

Heilige Cäcilie oder Die Gewalt der Musik, Die, a story (Legende) by H. von Kleist (q.v.), first published in a short version in 1810 in the *Abendblätter*, and in full in Kleist's *Erzählungen*, vol. 2, in 1811. Against a background of the Counter-Reformation (see GEGENREFORMATION) and of violent iconoclasm, the story relates the abortive attempt of four brothers to destroy, with the help of a mob, a convent near Aachen. The miracle which saves the convent is ascribed to the intervention of Saint Cecilia, who, mysteriously replacing the sick nun Antonia, conducts the old Italian mass chosen for the celebration of Corpus Christi Day.

The ethereal quality of the *Gloria* effects the instantaneous conversion of the four brothers, who are seized with religious frenzy. They spend the rest of their lives in a mental asylum singing the *Gloria* at each midnight with the tormented voices of lost sinners. They are otherwise docile and die in peace.

Heilige Experiment, Das, a five-act play (Schauspiel) by F. Hochwälder (q.v.), written in 1942, first performed in the Städtebund-theater Biel, Solothurn (Switzerland), in March 1943, and published in 1953. The action, based on the expulsion of the Jesuits from Paraguay in the 18th c., is invented, and purports to have taken place on 16 July 1767 in the Jesuit College in Buenos Aires. The 'Holy Experiment' refers to the Jesuit state embracing various Indian settlements, which represent the realization of God's Kingdom on Earth. This state is run on Christian principles and has, by its increasing economic prosperity, developed, in the eyes of Spain and Portugal, into a rival secular power. It has also aroused the displeasure of the Church. Its head, the S. J. Provincial Alfonso Fernandez, is ordered by the Spanish king, represented by Don Pedro de Miura, and by the S. J. General, represented in secret by Lorenzo Querini, to dissolve the settlements, which implies the return of some 150,000 Indians into slavery. The Provincial complies with the order, shaken by the realization that the Indians only honour Christ for shelter and food. He is mortally wounded in a short-lived revolt instigated by the Jesuit military commander Ladislaus Oros, who refuses to abandon the Jesuit ideal to change this world. Before he dies, the Provincial absolves Oros, who has been arrested and condemned to death by the Spaniards, and recovers his own faith in his ideal of a Christian state, which remains embodied in the spirit of St. Francis Xavier.

Structurally and in the presentation of arguments the play follows the traditional idealistic tragedy and 'Ideendrama'. It was performed in London in 1955–6 under the title *The Strong are Lonely*.

Heilige Johanna der Schlachthöfe, Die, a play in 11 scenes in mixed prose and free verse written by B. Brecht (q.v.) in 1929–30, published in 1932, and first produced in Hamburg (Deutsches Schauspielhaus) on 30 April 1959. A radio version was broadcast from Berlin on 11

April 1932. The play is based on *Happy End* by Dorothy Lane, an arrangement of which by Elisabeth Hauptmann Brecht produced in 1929, also contributing six songs.

The scene is Chicago in the 1920s, and the slump is engineered by the capitalists, represented by the meat-king Pierpont Mauler. A group of members of the Salvation Army conduct a religious mission in the factories, which is a failure. Mauler, on the other hand, is moved by the idealistic sense of mission of one of the girls, Johanna Dark, and expresses respect for her sympathy with the workers. He hopes to help them through his bold business speculations; these, however, fail. The factories close, and unemployment causes great suffering among the workmen, which Johanna decides to share with them. The leaders of the workers' union urge a general strike, and entrust Johanna with a letter proclaiming it. Her humanity revolts against the use of force and she does not deliver the letter, on which the strike depends, thus betraying her 'mission'. The desperate proletarians treat her with callous antipathy, and through prolonged exposure to the bitter Illinois winter she dies of pneumonia. As she dies she affirms her Marxist political faith, and confirms her final abandonment of her original mission with the Salvation Army. Her words, however, become inaudible as she is 'canonized' by Mauler and the other managers of the meat trade, Paulus Snyder, Graham, Slift, and the butchers.

Whole stretches of the play are deliberate parody of Goethe's and Schiller's classical idealism. Brecht states that the play should illustrate 'die heutige Entwicklungsstufe des faustischen Menschen'. This programmatic comment prepares for the note of parody, for Brecht's 'fist' (Faust) is not Goethe's character, but the fist of revolution as well as the savage fist which the unemployed workers show in their struggle for survival. Johanna becomes the counterpart not only of Schiller's Joan in *Die Jungfrau von Orleans* (q.v.), but also of Goethe's Iphigenie (see IPHIGENIE AUF TAURIS) with her faith in the basic goodness of men. In the final scene the chorus of the butchers parodies the chorus of the angels in the finale of *Faust* (q.v.) Pt. 2, and Johanna's death, including the covering of her body with the flag, is an equally obtrusive parody on the death of Schiller's Joan. The play was written during the period of, and in the same vein as, Brecht's Lehrstücke (see LEHRSTÜCK), and the overt parody is an aspect of his technique of Verfremdung (see VERFREMDUNGS-EFFEKT).

Heiligenhof, Der, an individualistic religious novel by H. Stehr (q.v.), published in 2 vols. in

1918. At its centre is the farmer Sintlinger, whose soul is awakened by the ethereal quality of his blind daughter Helene. A shock gives her sight, she falls in love with Peter Brindeisener (a character from an earlier novel), is jilted by him, and drowns herself. Sintlinger is rescued from despair by Faber, who is held by many to be an agitator but is in reality a deeply religious man.

Heiligenstädter Testament, title given to a moving document written by Beethoven (q.v.) in October 1802 at the village of Heiligenstadt, now a suburb of Vienna. It marks the utter despondency induced by the loss of his hearing, and possibly also by disappointment in love. It is uncertain whether his despair foresaw a fatal illness or whether he contemplated suicide.

Heiliges Römisches Reich Deutscher Nation, see DEUTSCHES REICH, ALTES. Its designation became the subject of satire which is crisply contained in the song in Goethe's *Faust* (q.v., Pt. 1, *Auerbachs Keller*): 'Das liebe heilge römsche Reich,/Wie hälts nur noch zusammen?'.

Heilige und ihr Narr, Die, a novel by Agnes Günther (q.v.), published in two volumes in 1913. It tells of the love of an orphan princess for a Count Harro Thorstein, whom she marries despite his lower station. She endures a long illness and dies young. This princess (who is constantly referred to as Seelchen) can perhaps be likened to Dickens's Little Nell transferred to a princely environment. The long and rather sentimental book achieved phenomenal success.

Heimat, a drama (Schauspiel in vier Akten) by H. Sudermann (q.v.), published and first performed in 1893. Magda, who has had a career as the great Italian *prima donna* dall'Orto, has consented to sing in a concert in a German provincial town. In reality she is the estranged daughter of one of the inhabitants, the retired Lieutenant-Colonel von Schwartze, and her coming to the town is motivated by a secret desire for reconciliation. A somewhat strained reunion takes place, which is at once endangered by Magda's confession that she has an illegitimate child by Herr von Keller, one of the local worthies. Schwartze insists on a marriage. Magda at first agrees, but, when she finds that Keller will not recognize her child because to do so would imperil his career, she retracts her consent and quarrels violently with her father, who tries to compel her at pistol point. When Magda hints that she has had other lovers, Schwartze, who has recently suffered a stroke, has a further seizure and dies. The sensational, but powerfully written and brilliantly con-

structed, play was immensely successful, and the role of Magda was taken by the great actresses of the 1890s, including Eleonora Duse and Sarah Bernhard. Outside Germany it goes by the title *Magda*.

Heimat, a collection of eight stories by C. Viebig (q.v.), published in 1914. Later editions of the book contain an autobiographical sketch. The homeland of the title is the region of the Eifel near Trier, where the authoress spent some of her early years.

Heimat, Die, a poem by F. Hölderlin (q.v.), written in 1800 in the form of an Alcaic ode. It is a song of praise for home, of mourning for lost love, and of the poet's acceptance of suffering. A much shorter version dates from 1798.

Heimatkunst, term adopted at the end of the 19th c. in a conscious anti-urban and anti-cosmopolitan reaction by certain German writers, notably A. Bartels, F. Lienhard (qq.v.), T. Kröger (1844–1918), and H. Sohnrey (1859–1948). They encouraged authors to write rural fiction on the region in which they lived and knew intimately, hoping to produce a truly German literature. Lienhard, an Alsatian, was less committed to the nationalistic trend. Bartels openly combined it with anti-Semitism.

The Heimatkünstler regarded as their ancestors such eminent writers as J. Gotthelf, Stifter, J. P. Hebel, P. Rosegger, Th. Storm, and F. Reuter (qq.v.), but minor writers, such as P. A. de Lagarde (q.v.), and J. Langbehn, exercised great influence. Two periodicals were launched in support of Heimatkunst: *Der Bote für deutsche Literatur* (1897, from 1900 to 1904, when publication ceased, entitled *Deutsche Heimat*) and *Der Türmer* (1898–1943). Apart from its nationalistic bias and excesses, the movement encouraged some good writing, including the Eifel and West Prussian novels of C. Viebig, the Saxon novels of W. von Polenz, and the sketches from the Lüneburg Heath of H. Löns (qq.v.). Other lesser exponents were the Holsteiner H. Voigt-Diederichs (q.v.), the Hamburger O. Ernst, the Bavarian L. Ganghofer (q.v.), the Hessian W. Holzamer, the Thuringian G. Schroer, the Black Forest priest H. Hansjakob (q.v.), the Lower Saxons L. von Strauss und Torney (q.v.), and F. Stavenhagen, and the Friesians W. Lobsien, F. Zacchi, and W. Jensen (qq.v.). Much trivial literature was also thus fostered.

The unattractive nationalistic aspects of Heimatkunst appealed to the National Socialist regime, which turned it into a Blut- und Boden-literatur to serve its racial policy. The work of the Holsteiner G. Frenssen, aged 70 in 1933, was exploited to this end; its principal writers were H. F. Blunck, F. Griese, H. Stehr, and W. Schäfer (qq.v.).

Switzerland and Tyrol had by their isolation a tendency towards regional literature, which was encouraged by Heimatkunst. J. C. Heer, H. Federer, and E. Zahn represent Switzerland, and, in some of his plays, K. Schönherr Tyrol; with them the nationalistic note is less insistent or absent. For Austria, mention should be made of R. H. Bartsch, J. F. Perkonig, and E. Ertl (qq.v.), though Austrian rural literature had passed its peak with P. Rosegger as well as L. Anzengruber (qq.v.).

HEIMBURG, W., pseudonym of Bertha Behrens (Thale, 1850–1912, Kotzschenbroda), who wrote numerous light novels published serially in the periodical *Die Gartenlaube* (q.v.). She completed the last, unfinished, novel of E. Marlitt (q.v.), *Das Eulenhaus*. Her *Gesammelte Romane und Novellen* (10 vols.) were published 1890–3.

Heimgarten, Der, a monthly magazine published from 1876 to 1935. Its editor until 1910 was P. Rosegger (q.v.).

Heimkehr, Die, a collection of lyric poems by H. Heine (q.v.), published in the *Buch der Lieder* (q.v., 1827), where it follows *Lyrisches Intermezzo*. It consists of 88 songs, mainly on love, and 5 longer poems, of which the best known is *Die Wallfahrt nach Kevlaar*. The songs include many well-known poems, some of which have been set to music by F. Schubert or R. Schumann (qq.v.). Heine gives 1823–4 as the date of composition.

Heimkehrerliteratur, term adopted shortly after the 1939–45 War for fiction and plays concerning the demobilized soldier and the problems which face him on his return to a world in disturbance and confusion. Typical examples are W. Borchert's *Draußen vor der Tür* (q.v.), G. Gaiser's *Eine Stimme hebt an* (q.v.), and H. Böll's *Wo warst du, Adam?* (q.v.). Though elements of this situation frequently recur in later works (e.g. G. Grass's *Hundejahre*, q.v.), the term is usually limited to the early post-war years. L. Frank's *Karl und Anna* (q.v., 1927) is one of the few forerunners of this type of literature after the 1914–18 War.

Heimliche Bote, Der, a short Middle High German didactic poem, giving advice on the proper choice in love and, to a lesser extent, on conduct. It is in two parts, the first addressed to a woman, the second to men; and it seems likely that they were not originally parts of the same poem. The woman is counselled to disregard

outward advantages in a man, and to esteem courtliness. In the second part men are advised to gain the approbation of the world by virtue and good education. The not particularly skilful poem was written some time after the middle of the 12th c. It is also known by the title *Ratschläge für Liebende*.

Heimweh, title of two poems by J. von Eichendorff (q.v.). The better-known poem, beginning 'Wer in die Fremde will wandern', is contained in *Aus dem Leben eines Taugenichts* (q.v.). The other, the first line of which runs 'Du weißts, dort in den Bäumen', is a poem of brotherly love with the sub-title *An meinen Bruder*.

Heimweh, two poems by E. Mörike (q.v.), written respectively in 1828 and *c.* 1830–2. The earlier poem expresses the lover's increasing sadness as he goes further and further away from his sweetheart's home. This poem, the first line of which runs 'Anders wird die Welt mit jedem Schritt', has been set to music by Hugo Wolf (q.v.). The first line of the later poem runs 'Zu den altgewohnten Orten'.

HEINE, HEINRICH (Düsseldorf, 1797–1856, Paris), originally Harry and of Jewish family, was educated at the Lyzeum of Düsseldorf (1807–14), and then sent to learn business at Frankfurt. In 1816 he went to Hamburg to work in the office of his banker uncle, Salomon Heine. In 1818 the uncle launched him into business on his own account (Harry Heine u. Co.), a venture which ended in bankruptcy (1819).

The uncle, convinced that Heine's gifts were not for commerce, agreed to finance him at a university, and in 1819 Heine matriculated at Bonn and a year later migrated to Göttingen, studying law. He nevertheless attended literary lectures, particularly those of A. W. Schlegel (q.v.) at Bonn. In January 1821, because of his involvement in an intended duel, he was rusticated for six months. Instead of biding his time and returning to Göttingen, he moved on to study for two years (1821–3) in Berlin, where he frequented the *salon* of Rahel Varnhagen von Ense (q.v.), the rendezvous of the most distinguished intellectual figures, including Alexander von Humboldt, Bettina von Arnim, L. Ranke, A. von Chamisso, and F. de la Motte Fouqué (qq.v.).

In Heine's early 20s some of his poems had appeared in magazines, and a collection was published in December 1821 (*Gedichte*, 1822). *Tragödien, nebst einem lyrischen Intermezzo* appeared in 1823, containing *Almansor* and *William Ratcliff* (qq.v.), separated by the *Intermezzo*. At this time Heine became an ardent supporter of the Verein für Kultur und Wissenschaft der

Juden. In 1824 he visisted North Germany (Lüneburg, Cuxhaven, Heligoland, Hamburg), and then went on a walking tour in the Harz, which formed the basis for his satirical-idyllic prose book *Die Harzreise* (q.v., in vol. 1 of *Reisebilder*, q.v., 4 vols., 1826–31). He completed his studies in Göttingen in 1825, taking his doctorate, and at the same time was baptized, a step intended to smooth his career in the civil service, the law, or the academic profession, three spheres of employment which he essayed unsuccessfully or failed to enter. After a further visit to the north (including this time Norderney, 1825) he made an English tour in 1827, and on his return was for a short time a journalist in Munich. The *Buch der Lieder* (q.v., 1827) laid the foundation of his reputation as a poet, though the common belief that it reflects a supposed love for his cousins Amalie and Therese Heine rests on tenuous evidence. Disappointed in his hope of a professorship in Munich in 1828, he passed a few months in Italy, spent the spring of 1829 in Berlin, and then moved to Hamburg where he remained for two years, interposing visits to the North Sea coast and to Heligoland.

The July Revolution roused visions of a new world, and in May 1831 Heine made Paris his home for the rest of his life. He lived by his pen, writing at first especially for the *Allgemeine Zeitung* of Augsburg, and for various French journals. In 1836 he was granted a small pension by the French government. During his first decade in Paris Heine's writing was mainly concerned with politics, social questions, and the history of literature. *Französische Zustände* (first published in the *Allgemeine Zeitung*) appeared in collected form in 1833 and *Die Romantische Schule* (an expansion of the earlier *Geschichte der neueren schönen Literatur in Deutschland*, 1833) in 1836. Larger collections of his journalistic work were published in 4 vols. as *Der Salon* (1834–40), including *Französische Maler, Aus den Memoiren des Herrn von Schnabelewopski, Florentinische Nächte, Elementargeister, Der Rabbi von Bacherach*, and, in vol. 2, *Zur Geschichte der Religion und Philosophie in Deutschland*. In 1844 a further volume of partly previously published poems appeared as *Neue Gedichte* (q.v.).

From 1834 Heine lived with Eugénie Mathilde Mirat, a Frenchwoman without education, whom he married in 1841. In December 1835 his works were officially branded as subversive in a federal declaration, which was principally directed against Young Germany (see JUNGES DEUTSCHLAND). In 1843 and 1844 he made his last visits to Germany, and in the same years he published his two masterpieces of verse satire on German affairs, *Atta Troll* (q.v., 1843, revised 1847) and *Deutschland. Ein Wintermärchen* (q.v., 1844, separately and in *Neue*

Gedichte). In December 1843 Heine first met Karl Marx (q.v.).

Symptoms of illness manifested themselves in the mid-thirties, and in 1848 he was found to be suffering from a spinal tuberculosis (*tabes dorsalis*) of syphilitic origin. He rapidly became paralysed, and spent the last eight years of his life bedridden; he himself called his sick-bed 'die Matratzengruft'. In these last years he wrote some of his finest poetry, contained in the *Romanzero* (q.v., 1851) and *Gedichte 1853 und 1854* (in *Vermischte Schriften* of 1854), poems in which he faces the moral and physical suffering of his slow death with courageous irony. *Lutetia. Berichte über Politik, Kunst und Volksleben* also appeared in *Vermischte Schriften*. The last months of his life were lightened by the spiritual love he felt for Elise Krinitz, who appears in his poetry as 'die Mouche' (q.v.).

Though Heine's output of critical and expository prose was considerable, only the early *Die Harzreise* has attracted lasting attention. The prose of his middle period, together with such poetic works as *Atta Troll* and *Deutschland. Ein Wintermärchen*, reflects a complex conflict between revolution and tradition which remained unresolved. It is Heine's poetry, and primarily the lyric part of it, which has ensured him a European reputation. Heine has long and deservedly been valued for the *Buch der Lieder*, with its poetry of dreams, and the Romantic gesture of self-conscious and studied melancholy. Gradually the worth of later poems has become more apparent, with their terse and direct formulations of real suffering (in *Romanzero* and *Gedichte 1853 und 1854*, which were later augmented and posthumously entitled *Letzte Gedichte*, q.v., in *Letzte Gedichte und Gedanken*, 1869).

The first edition of *Heines Werke* (22 vols.) was published in 1861–6. The *Säkularausgabe. Werke. Briefwechsel. Lebenszeugnisse* (planned in *c.* 45 vols.) began to appear in 1970. *Sämtliche Werke* (*Historisch-kritische Gesamtausgabe*, 16 vols.) appeared as the Düsseldorfer Ausgabe, ed. M. Windfuhr *et al.*, 1973 ff., *Sämtliche Schriften*, 12 vols., ed. K. Briegleb *et al.*, in 1981.

HEINRICH I, DER VOGLER, DEUTSCHER KÖNIG (876–936, Memleben), became duke of Saxony by inheritance in 912, and in 919 was elected Deutscher König (q.v.). Largely by negotiation he extended the authority of the empire eastwards into Schleswig, Holstein, and Brandenburg. In 923 he bought off the Hungarians, and ten years later, having used the respite to rearm, defeated them on the Unstrut. Heinrich, the father of Otto I, der Große (q.v.), was never crowned emperor, though at the time of his death he was contemplating a journey to Rome. He was buried in Quedlinburg.

A legend of the 12th c. recounts that the news of his election was brought to him while he was occupied with a bird decoy, and this episode is the subject of a well-known ballad by J. N. Vogl (q.v.), *Heinrich der Vogler* (see HERR HEINRICH SITZT AM VOGELHERD), which was set to music by C. Loewe (q.v.). He is also the subject of plays by A. Gryphius (*Heinrich der Vogler*, *c.* 1660) and J. Mosen (*Heinrich der Finkler*, 1835), and of poems by Klopstock and J. P. Conz (qq.v.).

HEINRICH II, DER HEILIGE (Abbach, Bavaria, 973–1024, Göttingen), became duke of Bavaria on his father's death in 995, and was elected Deutscher König (q.v.) in 1002. His coronation as emperor in Rome followed twelve years later. His reign was largely spent in conflict with rival claimants to the throne, foreign powers on his borders, and rebellious nobles at home. He was a friend of Odilo, Abbot of Cluny, and furthered ecclesiastical and monastic reform within his borders. He is buried with his consort Kunigunde (q.v.) in Bamberg Cathedral. Their statues on the cathedral portal date from 1230; the funeral monument by Tilman Riemenschneider (q.v.) was completed in 1513. Heinrich was canonized in 1146 by Pope Eugenius III.

Heinrich is the subject of *Das Nest der Könige* in G. Freytag's novel *Die Ahnen* (q.v., vol. 2, 1872).

HEINRICH III, KAISER (1017–56, Bodfeld, Harz), the son of the Emperor Konrad II (q.v.) and the Empress Gisela, was elected Deutscher König (q.v.) in 1026, and crowned in 1028. He succeeded his father in 1039. Heinrich's reign marks the highest point of medieval imperial power. He established his rule fully in Germany, subdued or won over his turbulent eastern neighbours, and overcame the resistance of Burgundy. In Italy he successfully exercised his right to elect the pope. Heinrich III, a man of piety, conscious of his role as the anointed of the Lord, and a strong administrator, lent his support to the movement for ecclesiastical and monastic reform.

HEINRICH IV, KAISER (1050–1106, Liège), son of Heinrich III (q.v.), was elected Deutscher König (q.v.) two years before his father's death in 1056. His minority was a period of internecine strife. In 1065 he was declared of age, though he did not actively assume government until 1069. A serious revolt of the Saxons in 1073 threatened his power, but in 1075 he was able temporarily to destroy the insurgent forces. In the same year Pope Gregory VII opened the Investiture Contest (see INVESTITURSTREIT) by forbidding lay appointment of bishops. Seeing this as a threat to the imperial power, Heinrich secured the denunciation of the Pope, who responded with excommunication. Heinrich's power was in jeopardy in Germany, where many princes were in opposition, and the Saxons again in arms against him.

In 1077 Heinrich endeavoured to save himself from deposition by submitting to Pope Gregory VII at Canossa (q.v.), the well-known occasion on which he is said to have done penance barefoot in the snow, details which are probably exaggerated. His humiliation did not resolve his difficulties, for Rudolf von Schwaben was elected as rival king. In the ensuing conflicts, Heinrich was again excommunicated by Pope Gregory VII, and he in turn declared the Pope deposed. Rudolf was killed in action in 1080, and in the following year Heinrich invaded Italy, captured Rome, though not until 1084, and secured his coronation. Gregory died in 1085, but his successor continued the struggle. At home, too, Heinrich faced opposition, and from 1093 to 1097 was practically powerless. Although he secured the election of his son, the future Heinrich V (q.v.), as Deutscher König in 1098, young Heinrich joined the German princes against his father, who was captured in 1105 and forced to abdicate. Having escaped and resumed the struggle, he died in the following year. He was buried in Speyer Cathedral.

Heinrich IV's greatest misfortune was the death of his father when he was still an infant; the prolonged regency weakened control. In the matter of investiture he nevertheless maintained the imperial rights to the end.

German dramatists since the 18th c. have interpreted Heinrich IV as a nationalist fighting a foreign yoke. The first play on him was written by Bodmer (q.v.), *Kaiser Heinrich IV* (1768); others include *Leben und Tod Kaiser Heinrichs IV* (1784) by Julius Graf von Soden (q.v.), *Kaiser Heinrich IV* (1844) by Rückert (q.v.), *Hildebrand* and *Heinrichs Tod* (1863–7) by F. von Saar (q.v.), and *Das neue Gebot* (1886) and *Heinrich und Heinrichs Geschlecht* (1896) by E. von Wildenbruch (q.v.). Pope Gregory VII and Heinrich IV are central to the play *Gregor und Heinrich* (1934) by E. G. Kolbenheyer (q.v.).

HEINRICH V, KAISER (1081–1125, Utrecht), son of the Emperor Heinrich IV (q.v.), was elected Deutscher König (q.v.) in 1098. In 1104 he revolted against his father, whose abdication he compelled. Heinrich V resumed the struggle with the Papacy which had dominated his father's reign over the issue of investiture (see INVESTITURSTREIT). An attempt at a compromise in 1111 failed through the resistance of the German bishops, at whose expense it would have been achieved. The Investiture Contest was not settled until 1122 (see WORMSER KONKORDAT, DAS). At home Heinrich V lost ground in the struggle for power with the German princes with whom he had made common cause against his father. He was buried, like his father, in Speyer Cathedral.

HEINRICH VI, KAISER (Nijmegen, 1165–97, Messina), son of the Emperor Friedrich I (q.v., Barbarossa), was elected Deutscher König (q.v.) in 1169. He acquired by marriage rights of succession in Sicily. In 1189 Heinrich assumed government of the Empire on the sudden death of his father, who had set out on a crusade. Almost immediately the Sicilian throne became vacant, but was seized by the pretender Tancred. Breaking off hostilities with the Duke of Saxony, Heinrich der Löwe (q.v.), Heinrich hurried to Italy, was crowned emperor in 1191, but failed to subdue Naples. Following unrest in Germany, he returned home, and was for a time in serious straits. From these he was extricated by the ransom of Richard I of England, who had been taken prisoner while returning from the crusades and had pressed for a compromise settlement during his captivity. In 1194 Heinrich returned to Italy, and successfully subdued Naples and Sicily.

Heinrich planned to extend his power over the whole of Europe, and to make the imperial crown hereditary in his house. He did not achieve his goal, and died after contracting a chill while hunting. In 1196 his son, the future Friedrich II (q.v.), was crowned Deutscher König, Heinrich having thus prepared for his succession.

Heinrich VI is believed to be the Minnesänger referred to in the MSS. as Kaiser Heinrich (q.v.). He is the subject of C. D. Grabbe's tragedy *Kaiser Heinrich der Sechste* (1830).

HEINRICH VII (c. 1275–1313, Buonconvento nr. Siena), Count of Luxemburg, was elected Deutscher König (q.v.) in 1308. His election was a move against the plans of Philip the Fair of France, who urged the candidature of his brother Charles de Valois. Heinrich VII, a ruler with only limited private resources, gained considerable backing from Bohemia by marrying his son John to the Bohemian princess Elizabeth. His main policy was aimed at the restoration of imperial power in Italy. Though crowned in 1312 at Rome, he was unable to make headway against the perpetual strife and intrigue of the Italian scene. His desire had been to bring peace, but he left Italy a prey to worse dissensions. He died of malaria, and is buried at Pisa.

HEINRICH (VII) (Sicily, 1211–42, Martirano), son of the Emperor Friedrich II (q.v.), was elected Deutscher König (q.v.) in 1220 while still a boy, and in 1234 rebelled against his father. His revolt collapsed, he surrendered, and was imprisoned at San Felice in Apulia, and later at Martirano in Calabria. He is believed to have committed suicide.

HEINRICH, PRINZ VON PREUSSEN (Berlin, 1726–1802, Rheinsberg), a brother of Friedrich II (q.v.), der Grosse, and the thirteenth child of Friedrich Wilhelm I (q.v.), achieved recognition as an outstanding military commander in the Seven Years War (see SIEBENJÄHRIGER KRIEG). For his bravery at Hohenfriedberg (q.v.) Friedrich gave him his Rheinsberg residence (q.v.) for life. He promoted him at the age of 19 to the rank of major-general, and at 30 to the rank of lieutenant-general. Prince Heinrich, in winning the Prussian victory at Freiberg/Saxony (q.v., 1762), furthered peace and prestige for Prussia in the Treaty of Hubertusburg (see HUBERTUS-BURGER FRIEDE). Friedrich built for him a palace in Unter den Linden, which later accommodated the University of Berlin. On his visit to St. Petersburg in 1770 he won the friendship of Catherine the Great, and was Friedrich's mediator for the Partition of Poland (see POLAND, PARTITIONS OF) in 1772. He shared Friedrich's admiration of French culture and the philosophy of enlightenment, and visited Paris in 1784, meeting Goethe on his way at Weimar. He was tutor to Friedrich's heir, Friedrich Wilhelm II (q.v.), and Friedrich meant him to act as regent in case of his early death. But Prince Heinrich was a man of great personal vanity as well as of ability, and friction between him and Friedrich persisted throughout his life. He frequently disagreed with his brother in matters of politics. During the War of the Bavarian Succession (see BAYRISCHER ERBFOLGEKRIEG) he asked to be relieved of his command. He criticized the King in public, and never forgave him for disgracing their brother August Wilhelm in 1757. He is recorded as having made a present of a quiet horse to C. F. Gellert (q.v.), and he is often mentioned in Th. Fontane's *Wanderungen durch die Mark Brandenburg* and *Vor dem Sturm* (qq.v.).

HEINRICH CLUSENER, see VOM ARMEN SCHÜLER.

HEINRICH DER GLÎCHEZAERE was the author of the Middle High German beast epic, *Reinhart Fuchs*. Only the name Heinrich refers to him with certainty; the appellation Glîchezaere (*Gleisner*, 'dissembler', which might seem more apt for the fox) may be a misunderstanding on the part of the adapting scribe responsible for the only MS. preserved almost intact, a clumsy version made in the late 13th c. The original exists only in imperfect fragments preserved at Kassel. The title *Reinhart Fuchs* is appropriate to the subject, but its original title was *Isengrînes nôt*, in allusion to *Der Nibelunge Nôt*. The source is French.

Reinhart Fuchs narrates the adventures of the fox Reinhart and the wolf Isengrin. Reinhart tempts his companion into rash undertakings, in which the wolf comes off badly, while the fox adroitly evades the dangers. The lion, the king of the animal realm, is persuaded by a number of wronged animals to arraign Reinhart. It happens that the king is unwell, an ant having lodged unrecognized in his ear. Reinhart extricates himself from his precarious situation, by 'curing' the king, for which the skins of Isengrin, the bear, and the cat are necessary, and his enemies and rivals are all in various ways discomfited. Finally Reinhart poisons the king, completing the image of a state in which force and cunning are the means to power, and of these the greater is cunning. (See also REINKE DE VOS.)

HEINRICH DER LÖWE (1129–95, Brunswick), Duke of Saxony 1142–80, son of Heinrich der Stolze (q.v.), reconquered Bavaria, which his father had lost, and held it as duke from 1156 to 1180. He quarrelled with the Emperor Friedrich I (q.v.) and was outlawed in 1179, and deprived of his possessions in 1180. He was in exile from 1182 to 1185, living with Henry II of England in Normandy. In 1189 and 1190 he regained some of his possessions. He is buried in Brunswick Cathedral, which he built 1173–95 (see BRAUNSCHWEIG). His symbol, the bronze lion (1166), stands in the Burgplatz in front of the Burg Dankwarderode, which he erected *c.* 1175. He was the patron of the Pfaffe Konrad who wrote the *Rolandslied* (q.v.) for him and his consort *c.* 1170. Heinrich der Löwe is the subject of plays by E. Klingemann and M. Greif (qq.v.) and of the novel *Kaiser und Herzog* by W. Beumelburg (q.v.); he is an important figure in the play *Kaiser Friedrich Barbarossa* by C. D. Grabbe (q.v.).

HEINRICH DER STOLZE (Heinrich X) (d. 1139, Quedlinburg), Duke of Bavaria and, from 1137 to 1139, Duke of Saxony, supported the Emperor Lothar III (q.v.) in his conflict with the Staufer (see HOHENSTAUFEN). Heinrich was probably the patron for whom the *Kaiserchronik* (q.v.) was undertaken, though he died before its completion.

HEINRICH DER TEICHNER (d. Vienna, *c.* 1377), an Austrian, who probably lived in Vienna, was a commoner and layman praised by his contemporary Peter Suchenwirt (q.v.) for

his wholesome and godly life. His poems, in the form of Reimreden (see REIMREDE) are moral and descriptive, concerned with aspects of the citizen's life or with religion. Sober in style, he was also prolific; his extant poetry comprises more than 700 Reden, totalling more than 70,000 lines.

HEINRICH DER VOGLER, see HEINRICH I.

HEINRICH FRAUENLOB, MEISTER (d. 1318, Mainz), was regarded in his own day as a poet of exceptional quality. He was born probably *c.* 1250–60 in the district of Meissen, and is sometimes referred to as Heinrich von Meißen, or even (confusingly, as there is another author so named) as Der Meißner. His title Frauenlob has been variously attributed to his authorship of a *leich* (q.v.) to the Virgin Mary, to his general praise of women, and to the support he gave, in a literary controversy with Regenboge (q.v.), to the designation Frau in preference to Weib.

Frauenlob seems to have been a youthful prodigy, and, in maturity, a poet sought after and welcomed by the great, in contrast to the majority of itinerant commoner poets, who mostly querulously protest at their intended patrons' lack of generosity. Frauenlob visited Bavaria, Tyrol, and Carinthia, was present at the court of Bohemia in 1286, attended splendid festivities at Rostock in 1311, and settled soon after in Mainz. He was buried in Mainz Cathedral, and was reputed to have been borne to his grave by ladies, a legend which is thought to have its origin in a depiction on his gravestone of a coffin carried by allegorical female figures. The tradition that he founded a Meistersänger school in Mainz is regarded as doubtful.

More than 450 poems are attributed to Frauenlob, and, though some are spurious, his output was certainly large. Apart from three *leiche* and a handful of Minnelieder his poems comprise Sprüche (see SPRUCH) on religious, political, social, and literary matters. His intentionally obscure and alembicated style owes something to WOLFRAM von Eschenbach, and also to KONRAD von Würzburg (qq.v.) whom he commemorates in a poem of quite exceptional obliquity. Frauenlob is a master of the mannerist style, which was to become fashionable in the 14th c. and this up-to-dateness is probably a reason for his success. In the controversy with Regenboge he rates himself higher than the revered masters of medieval poetry, REINMAR der Alte, WALTHER von der Vogelweide (qq.v.), and Wolfram von Eschenbach. This high self-estimation combined with the obscurity of his style was to give rise in the 19th c. to doubts of his sanity.

HEINRICH HETZBOLD VON WEISSEN-SEE, a minor Middle High German poet of Thuringia, who is recorded in the period 1312–45. His eight poems are Minnelieder in homely dress.

HEINRICH JULIUS, Herzog von Braunschweig (Hessen, Brunswick, 1564–1613, Prague) was, in spite of being a Protestant, installed as bishop of Halberstadt in 1578. He succeeded to the dukedom in 1589. His reign was largely spent in quarrels with the municipality of Brunswick. He was a friend of the Emperor Rudolf II (q.v.), at whose court he died. In 1592 a troupe of English actors under Thomas Sackville played at Brunswick, and fired Heinrich Julius with enthusiasm for the drama. In 1593 and 1594 he wrote ten plays, all in prose, exploiting the English style of realistic acting: four tragedies, two tragicomedies, and four comedies. All are straightforward didactic plays, in which vice is punished or exposed. In *Der Fleischawer* the cheating butcher is executed; the illicit lovers in *Buler und Bulerin* perish; while in *Die Ehebrecherin* husband and wife both die; *Von einem ungeratenen Sohn* is the story of Nero, replete with horrors. The tragicomedies are *Susanna*, which is based on Frischlin's Latin play (see FRISCHLIN, N.), and *Von einem Wirte oder Gastgeber.* Of the comedies, *Von einem Wirte* (not the play just mentioned) exposes cunning and credulity; *Vincentius Ladislaus* pillories the braggart; *Von einem Edelmann* contrasts the effete and privileged with the poor and godly; and *Von einem Weibe* ridicules a foolish cuckold. The tragedies abound in crass and bloody effects, and all the plays have broadly realistic scenes.

Heinrich Julius figures in Grillparzer's play *Ein Bruderzwist in Habsburg* (q.v.) as Julius, Herzog von Braunschweig.

HEINRICH RASPE (d. 1247), Landgrave of Thuringia from 1227 to 1239, and again from 1241 until his death. In a turbulent life Heinrich, after his brother's death, evicted his sister-in-law, Landgräfin Elisabeth (see ELISABETH, HEILIGE) with her children; he, in turn, was deposed by his nephew Hermann in 1239, but repossessed himself of the state when Hermann died two years later. He was prominent in imperial politics, acting as deputy for Konrad IV, son of Friedrich II (qq.v.) and accepting nomination as an anti-king to Konrad, whom he defeated in 1246. The papal support which gave him his brief spell as German King (see DEUTSCHER KÖNIG) earned him the derisive appellation Pfaffenkönig.

Heinrichs Litanei, title given to a Middle High German litany in poetic form, the author of which gives his name as Heinrich. He is believed to have been an Austrian cleric, and to have written the litany between 1160 and 1170. The poem, which is imbued with a deep sense of sinfulness, testifies to the growing devotion to the Virgin Mary towards the end of the 12th c. It is preserved in two MSS., one in Graz, the other in Strasburg-Molsheim. The name of Heinrich as author occurs only in the Graz MS.

Heinrich Stillings Leben, the autobiography of J. H. Jung (q.v.).

Heinrich und Heinrichs Geschlecht, a historical trilogy by E. von Wildenbruch (q.v.), described as *Tragödie in zwei Abenden* for performance on two evenings; the first comprises the prelude (Vorspiel) *Kind Heinrich,* and the play *König Heinrich*; the second is devoted to the five-act tragedy *Kaiser Heinrich.* The trilogy was published in 1896, and first performed on 22 January and 1 December 1896 at the Deutsches Theater, Berlin. The three parts cover the life of Heinrich IV (q.v., 1050–1106) from childhood to burial, giving the trilogy the character of a chronicle.

In the prelude Heinrich is separated from his mother and entrusted to Anno of Cologne. *König Heinrich* portrays the quarrel with Pope Gregory VII, and Heinrich's submission at Canossa. In *Kaiser Heinrich,* Heinrich shows himself to be a supporter of the common people against the oppression of the nobles; he dies in Act III, and the last two acts are devoted to his burial and his vindication.

HEINRICH VON BERGEN, see THEOLOGIA DEUTSCH.

HEINRICH VON BERINGEN, a canon of Augsburg recorded between 1282 and 1320, is the author of a chess book (Schachbuch) in some 10,000 lines of Middle High German verse, which is a free translation of the *Solatium ludi scaccorum* of Jacopo Dacciesole (see SCHACH-BÜCHER). The poem was probably completed between 1290 and 1300. Four other shorter poems by Heinrich are extant.

HEINRICH VON BRESLAU (PRESSELA), HERZOG, who ruled from 1270 to 1290, is the author of two attractive Minnelieder, one of which is in the form of a complaint to a number of powers (beginning with May and ending with Venus) about the coyness of his mistress. In the end, however, the poet takes his lady's side.

HEINRICH VON BURGUS, BRUDER, a Franciscan friar of Burgus, Vintschgau, South Tyrol (present-day Val Venosta in Italy). Bruder

Heinrich wrote the poem *Der Seele Rat* (q.v.) at the beginning of the 14th c.

HEINRICH VON DEM TÜRLÎN, a Middle High German poet of the 13th c., was a commoner who lived and wrote in Carinthia, Austria, possibly at St. Veit. He is the author of *Die Krone (Diu Crône)*, and probably also of *Der Mantel*, a poem of which only some 900 lines remain. It was formerly thought to be the beginning of a poem about Lancelot, but this is now regarded as doubtful. *Die Krone*, written *c.* 1230, is a huge Arthurian romance, 30,000 lines long, packed with fantastic adventures, the narration of which is Heinrich's chief interest. At first Arthur (Artus) is the hero, but in the last two-thirds the focus is on Gawein. Heinrich decks out his miraculous episodes with realistic detail. An interesting feature of one adventure is its transfer from the perpetual spring of Arthurian romance to the snow-covered mountains of the Carinthian winter.

HEINRICH VON DER MURE or MUORE, a minor Middle High German poet of the 13th c., also known as Heinrich von der Mauer; he was probably a native of Styria. His four extant poems, one of which records his entry into holy orders, suggest an individual poetic personality.

HEINRICH VON FREIBERG, a Middle High German poet, was a commoner, whose home was at Freiberg near Meißen. In the late 13th c. he was in the service of noblemen at the Bohemian court of Wenzel II (q.v.). About 1290 Heinrich completed Gottfried von Strassburg's *Tristan* (q.v.). His stylistically competent continuation gives the story a religious significance, setting the love of God above the human love of Tristan and Isolde. To Heinrich are also commonly attributed two poems of moderate length, a verse legend, *Die Legende vom Kreuzesholz*, and the rhymed accounts of a journey in Paris, in which a Bohemian knight distinguishes himself (*Die Ritterfahrt des Johann von Michelsberg*); some scholars, however, consider that the two poems are by different authors.

HEINRICH VON HESLER, a Middle High German poet connected with the Teutonic Order (see DEUTSCHER ORDEN). He was a knight originating probably in the district of Naumburg (a rival view suggests Gelsenkirchen), and is thought by some to be identical with a Henricus de Hesler of the Teutonic Order, who is documented in Zschillen nr. Rochlitz in 1341–2. Heinrich is the author of three poems, the first of which, *Die Erlösung*, only survives in fragments which deal with Lucifer's fall and the Fall

of Man. *Die Apocalypse* is a poem paraphrasing the Book of Revelation, and the *Nicodemusevangelium* recounts the Passion according to the synoptic gospels, together with the legend of Joseph of Arimathea, the descent into Hell, the story of Saint Veronica, and the destruction of Jerusalem, all of which is based on the apocryphal gospel of Nicodemus. The poem has an introduction, which is concerned with the problem of free will and predestination. Heinrich, who gave his attention to metrical problems and wrote careful and shapely verse, was doubtless influenced by the poetry of GOTTFRIED von Straßburg (q.v.).

Heinrich von Kempten, the title given to a short verse narration by KONRAD von Würzburg (q.v.). It has also been known as *Otte mit dem Barte*, and in some MSS. is designated *Von Keiser otten*. It was written in Strasburg probably *c.* 1270 for Berthold von Tiersberg of the Cathedral Chapter.

Heinrich von Kempten incurs the anger of the Emperor Otto, who swears by his beard to be avenged upon him. Heinrich seizes the Emperor by the beard and secures his life at the point of the sword. Some years later he saves the Emperor's life, rescuing him from the hand of robbers, and receives, in return, forgiveness for his early offence, and also rich reward. The figure of Kaiser Otte is a fusion of the historical emperors Otto I and II (qq.v.)'.

HEINRICH VON KROLEWIZ, a Saxon priest, who gives his home as the district of Meißen. Heinrich wrote an exposition of the Lord's Prayer in verse. The poem, strictly didactic in aim, consists of nearly 5,000 lines, and took him three years to write, from 1252 to 1255. It contains digressions referring to precious stones and animals, drawing for the latter on the *Physiologus* (q.v.).

HEINRICH VON LAUFENBERG (*c.* 1390–1460, Strasburg) was first a priest in Freiburg/Breisgau and then a monk of the Order of St. John at Strasburg. He is the translator of a large number of hymns from Latin originals, including such well-known examples as *Ave maris stella*, *Salve regina*, *Stabat mater*, *Puer natus*, *Surrexit Christus*, *Veni redemptor*, *Veni sancte spiritus*, and *Pange lingua*. He also translated a long theological poem, *Speculum humanae salvationis* (*Spiegel des menschlichen Heils*, 1437), and an *Opus figurarum* (*Buch der Figuren*, 1441), dealing symbolically with Old Testament history, as well as a treatise on medicaments, *Regimen sanitatis*. A collection of sermons is lost.

HEINRICH VON MEISSEN, Markgraf (1218–88), son of Dietrich von Meißen, is the author of six pleasing, but unoriginal Minnelieder suggesting the manner of Reinmar der Alte (q.v.).

HEINRICH VON MELK, 12th-c. author of two Early Middle High German poems, *Von des tôdes gehugde* (Erinnerung an den Tod) and *Das Priesterleben*. In the 14th-c. manuscript, which contains these two poems, 'Haïnrîchen dînen armen chnecht' is named as the author of the first. The second is attributed to him on convincing stylistic grounds. Heinrich's association with Melk (q.v., Lower Austria) is based on circumstantial evidence. He appears to have been a nobleman who abandoned the world and sought salvation in a life of seclusion and mortification as a lay brother in Melk Abbey. The spirit of his work is the denial of the world, the fear of damnation, and the affirmation of the ascetic life first propagated by a movement for monastic reform known as the Cluniac Reform (q.v.). The personal note is sombre and almost fanatical.

Von des tôdes gehugde (the words occur in the text—the poem has no title in the MS.) is an eloquent and powerful sermon in verse comprising roughly 1,000 lines. It is divided into two parts, first *von dem gemaïnen lebene*, which ruthlessly castigates the sins of the nobility and clergy with a sidelong glance at the lower orders; and secondly, *von dem tôde*, which is the reminder of death. In this section Heinrich portrays with merciless insistence the face of death with all the macabre details of decomposition. It contains two famous episodes, in one of which a widow is shown the body of her dead husband, whilst in the other a damned father addresses his son.

Das Priesterleben, which is somewhat shorter, castigates the vices of the clergy, notably gluttony, simony, and unchastity; but asserts the validity of the mass, notwithstanding the sinfulness of the celebrant. Both poems were probably written in the decade 1150–60.

HEINRICH VON MORUNGEN, a Middle High German Minnesänger from the neighbourhood of Sangerhausen near Merseburg, where there are ruins of the family castle, served as a *ministeriale* with Dietrich von Meißen (d. 1221). He left his property to St. Thomas's Abbey in Leipzig, where he died in 1222. His 23 extant poems are exclusively concerned with the unfulfilled devotion to the remote mistress, which is termed hohe Minne, but show intensity and depth within the narrow limits of the form,

exalting the daemonic power of love. His poetry is visually conceived with images of light and fire, as may be seen in his powerful and passionate erotic Tagelied (q.v.). His command of form shows indebtedness both to Provençal and to Latin poetry.

His name passed into legend, becoming identified with the hero of a 15th-c. ballad, *Vom edlen Möringer*.

HEINRICH VON MÜGELN (Mügeln nr. Pirna, Saxony, *c.* 1320–72) was a commoner and a learned poet in the service of various princes in Prague, Vienna, Styria, and perhaps also in Hungary. His most substantial work is *Der Maide Kranz*, an allegorical poem dealing with the twelve liberal arts (expanded from the traditional *septem artes liberales*) and with the twelve virtues. It was written before 1358, and its source is the *Anticlaudianus* of Alanus ab Insulis (*c.* 1120–1203) and the poem *Von Gottes Zukunft* by Heinrich von Neustadt (q.v.), which derives from the *Anticlaudianus*. His fulsome Meisterlieder and verse fables enjoyed high repute, and the later Meistersinger included him in the Zwölf alte Meister (q.v.). He also wrote a Marienlied entitled *Der Dom*, and a poem on Virgil the Sorcerer. His works also include a Hungarian chronicle in Latin verse (1352–3, which was translated into German prose, possibly by Heinrich himself) and translations of a work by Valerius Maximus (*Factorem et dictorum memorabilium libri novem*), as well as a commentary on the Psalms by Nicola de Lyra (1270–1340).

HEINRICH VON NEUSTADT, author of two widely differing Middle High German poems, originated in Wiener Neustadt and practised as a physician in Vienna, where he is recorded in 1312. He lived in the Graben, a street that is still in existence.

His verse romance *Apollonius von Tyrland* is believed to be the first of his extant works. It is a long poem of more than 20,000 lines based on the Latin *Historia Apollonii regis Tyrii*, which in turn is a translation of a lost Greek original, and is also the source of Shakespeare's *Pericles, Prince of Tyre*. Heinrich's *Apollonius* begins with the incestuous relationship of the wicked King Antiochus and his daughter. Antiochus sets the daughter's suitors a riddle, and puts to death those who fail to solve it. The eponymous hero succeeds in the task, but flees Antiochus's fury. In Pentapolis he marries a princess, Lucina, who gives birth to a daughter and afterwards falls into a coma which is mistaken for death; she is placed in a coffin and buried at sea. Apollonius leaves his daughter at Tarsus. These episodes, which in the main conform to the original, are followed by a succession of tenuously connected adventures, partly of chivalric and partly of oriental origin, which occupy more than three-quarters of the romance. Meanwhile Lucina's coffin is washed ashore at Ephesus, and she is resuscitated; Apollonius is reunited with his wife and daughter, converted to Christianity, and appointed emperor of Rome. The poem with its wealth of barely connected material was probably intended for the entertainment of the rising middle class.

Either simultaneously or later Heinrich wrote a religious poem, *Von Gottes Zukunft* (*Concerning the Coming of God*), which begins with a translation of the *Anticlaudianus* of Alanus ab Insulis, a 12th-c. French theologian. This section recounts, in allegorical fashion, Nature's endeavour, with the help of the virtues, and ultimately of God, to create a perfect man. The work continues through the remainder of Bk. I and Bks. II and III, first with an account of the life and passion, of Jesus, in which the Virgin Mary is the central figure, and then with a vision of the Day of Judgement. In one MS. a section of some 600 lines is interpolated, giving a translation of the medieval Latin *Visio Philiberti*. It is uncertain whether this was conceived as an integral part or was a separately written poem, later incorporated in the larger work. *Von Gottes Zukunft* is a work of devotion, written, in Heinrich's phrase, 'in rehter andâht' (in richtiger Andacht).

HEINRICH VON NÖRDLINGEN, a 14th-c. priest, who translated *Das fließende Licht der Gottheit* by Mechthild von Magdeburg (q.v.) from Middle Low German into Middle High German. Heinrich was driven from Nördlingen during the religious troubles under Ludwig the Bavarian (see LUDWIG IV), and in 1339 took refuge in Basel. He was one of a group of mystics who styled themselves Gottesfreunde, and corresponded on spiritual matters with the nun Margareta Ebner (q.v.) of Medingen Abbey near Donauwörth.

HEINRICH VON OFTERDINGEN appears in the Middle High German poem *Der Wartburgkrieg* (q.v.) as the unsuccessful challenger. Though later ages assumed Heinrich to be a historical personage, no evidence attests his existence, and it is probable that he is fictitious. His mysterious figure attracted the Romantics, and he appears as the hero of Novalis's fragmentary novel *Heinrich von Ofterdingen* (q.v.), in E. T. A. Hoffmann's *Der Kampf der Sänger*, and as the fictitious author of poems by J. V. von Scheffel (q.v.) published in *Frau Aventiure*.

Heinrich von Ofterdingen, an unfinished novel by Novalis (q.v.), published posthumously

in 1802; it consists of Bk. 1, entitled *Die Erwartung*, and the opening of Bk. 2, *Die Erfüllung*. Its eponymous hero is the legendary medieval poet who plays a leading part in the poem *Der Wartburgkrieg* (q.v.).

Novalis's Heinrich is an ingenuous youth of great poetic potentiality, who leaves Eisenach and goes to Augsburg with his mother. Before the departure he dreams a wonderful dream in which a blue flower is the central point of attraction ('eine hohe lichtblaue Blume'). This blue flower, the symbol of Heinrich's longing and his love, was destined to become the widely recognized symbol of all Romantic longing. At Augsburg Heinrich meets Mathilde, whom he loves; he associates her with the blue flower of his dream, and they are betrothed. The first book concludes with discussions on poetry between Heinrich and the poet Klingsohr. The second book resumes the story at a point at which Mathilde is dead.

Novalis began *Heinrich von Ofterdingen* in reaction against the worldly tenor of Goethe's *Wilhelm Meisters Lehrjahre* (q.v.), and his pages contain an astute analysis and justification of the poet from the Romantic standpoint.

HEINRICH VON PLAUEN (*c.* 1370–1429, Lochstedt), a commander of the Teutonic Order (see DEUTSCHER ORDEN), successfully defended Marienburg after the defeat of the Order at Tannenberg (1410), and, being appointed High Master, made satisfactory terms of peace with Poland in 1411. His efforts to reform the Order resulted in his deposition in 1413, and he was imprisoned for ten years. His story is the subject of plays by A. von Kotzebue (*Heinrich Reuß von Plauen*, 1805), Eichendorff (*Der letzte Held von Marienburg*, 1830), and Max Halbe (*Heinrich von Plauen*, 1933), and also of a three-volume novel by Ernst Wichert (*Heinrich von Plauen*, 1881).

HEINRICH VON RUGGE, a Middle High German Minnesänger and a nobleman; he was *ministeriale* to the counts of Tübingen. He is known to have been alive between 1175 and 1191, and is believed to have died on a crusade *c.* 1191. The authenticity of many of the poems attributed to him has been disputed, but a handful of poems are regarded as unquestionably his own. Rugge appears in his songs as a man of robust and pious character with a leaning towards didacticism. He is also the author of a longer poem (a *leich*, q.v.) mentioning the death of Barbarossa (see FRIEDRICH I, KAISER), and extolling, on religious grounds, participation in the crusades.

HEINRICH VON SAX, a minor Middle High German Minnesänger from Vorarlberg, wrote five extant songs in traditional praise of courtly *minne*, one of which is a *leich* (q.v.). He is believed to have lived in the second half of the 13th c.

HEINRICH VON STRETELINGEN, a minor Middle High German poet of the second half of the 13th c. A Swiss nobleman from the district of Lake Thun, he is the author of three Minne-lieder, which recall motifs of GOTTFRIED von Neifen and WALTHER von der Vogelweide (qq.v.).

HEINRICH VON VELDEKE, 12th-c. Middle High German poet, whose biography is scantily known from an epilogue attached, possibly by another hand, to his *Eneit*. Heinrich is believed to have been born between 1140 and 1150, and to have died before 1210. He was a Rhenish knight originating from the region of Maastricht, and later in attendance at the Thuringian court of Landgraf Hermann (q.v.).

Veldeke's first substantial work, preserved in Low German form, was a religious poem of some 6,000 lines, recounting the legend of St. Serva-tius, the patron of Maastricht, who, according to the poem, converted Attila temporarily to Christianity. The anonymous poem *Moriz von Craûn* (q.v.) mentions another poem by Veldeke, referred to as *Salomo und die Minne*, which is otherwise unknown.

The *Eneit*, an adaptation of the anonymous French *Roman d'Enéas*, is Veldeke's principal work. He began it at Maastricht *c.* 1170, and the unfinished MS. was lent to a countess of Cleves, from whose court it disappeared in 1174. It was handed back to him nine years later at the Thuringian court, whereupon he resumed work on it, completing it in 1189. In contrast to his poem on St. Servatius the language is a form of High German. It is a long poem, recounting the substance of the *Aeneid* in smooth, level verse; its highlights are the passion of Dido, Aeneas's journey to the Underworld, and the love of Lavinia. The action displays two widely differing instances of the power of *minne*; in the one it destroys Dido, in the other it overwhelms Lavinia's filial dutifulness. Veldeke's Trojan and Latin heroes are medieval knights, a sign perhaps less of *naïveté* than of freedom from ecclesiastical prejudice. The work accorded with the classical tastes of the Thuringian court, and its formal qualities, its metrical facility and its purity of rhyme, ensured it considerable influence, which is attested not only by the number of extant MSS., but by the tributes of contemporaries. GOTTFRIED von Straßburg and WOLFRAM von Eschenbach (qq.v.) both praised Veldeke, and RUDOLF von Ems (q.v.) wrote 'Von Veldeke der wîse man, der rehte rîme alrerst began' ('who wrote the first proper verse').

Veldeke has also left a handful of lyric poems (Minnelieder), which, if less distinguished for formal qualities, have a touch of originality and an attractive freshness. Nature appears in some of the poems to take precedence over *minne* in the poet's pleasure; joy predominates in them, and love is usually fulfilled, or capable of fulfil-ment. These poems show traces of Veldeke's native dialect, which has encouraged the assumption (not necessarily true) that they are mostly products of his youth.

HEINSE, JOHANN JAKOB WILHELM (Lange-wiesen nr. Weimar, 1746–1803, Aschaffenburg), usually known as Wilhelm Heinse, studied law at Jena University (1766), but came under the influence of F. J. Riedel (q.v.) and migrated in his wake to Erfurt University in 1768, where he applied himself to journalism. In 1771 he became secretary to a nobleman, for whom he translated the *Satyricon* of Petronius. In 1772 he became a private tutor in the house of Herr von Massow in Halberstadt, a position he owed to the recom-mendation of J. W. L. Gleim (q.v.). He fell in love with the mother of his pupil, and regarded her in later years as the awakener of his genius. In 1774 he joined J. G. Jacobi (q.v.) at Pempel-fort, collaborating on Jacobi's periodical *Iris* (q.v.). It was here that he first met Goethe. In 1780 Gleim and Jacobi sponsored a visit of Heinse to Italy, where he remained for three years, meeting Maler Müller (see MÜLLER, F.) and responding especially to Renaissance painting.

Heinse, who had become a Roman Catholic, was appointed reader to the Electoral Arch-bishop of Mainz in 1786. His best-known work, the sensual Utopian novel *Ardinghello* (q.v.), was published in 1789. He was appointed electoral librarian in 1787, and in 1794 removed the contents of the library to Aschaffenburg in order to protect them from destruction or dispersal during the troubles of the French Revolution (q.v.). His second and partly autobiographical novel, *Hildegard von Hohenthal* (1795-6), has a central character based on Frau von Massow. A third novel, *Anastasia und das Schachspiel*, followed in 1803. He translated Tasso's *La Gerusalemme liberata* (*Das befreite Jerusalem*, 1781) and Ariosto's *Orlando furioso* (*Roland der Wütende*, 1782-3). Both versions are in prose. In 1796 Heinse met F. Hölderlin in Westphalia; he was the dedicatee of Hölderlin's poem *Brod und Wein* (q.v.).

Heinse belonged to the generation of Sturm und Drang (q.v.), which he resembled in his enthusiasm for genius, his unconventional men-tal energy, and his dynamic emotional tempera-ment. His passion for Italian art and his pronounced sensual interest led him away from introspection towards a standpoint which had links both with Goethe's classicism and with Romanticism (see ROMANTIK). *Sämmtliche Werke* (10 vols.), ed. K. Schüddekopf (last two vols. A. Leitzmann), appeared 1903–25 and correspondence with Gleim, *Briefwechsel zwischen Gleim und Heinse* (2 vols.), ed. K. Schüddekopf, 1894–5.

HEINZELIN VON KONSTANZ was steward to Count Albrecht von Hohenberg, a canon of Constance, and later bishop of Würzburg. He wrote, probably in the late 13th or early 14th c., two poems in the form of disputations.

In *Von dem Ritter und von dem Pfaffen*, a poem of *c.* 400 lines, two women argue about the merits of a knight and a priest as lovers. *Von den zwein Sanct Johansen*, of similar length, shows two nuns disagreeing on the respective worthiness of St. John the Baptist and St. John the Evangelist. Each saint appears to his advocate in a dream and convinces her of the folly of the argument.

Heinzelin von Konstanz was formerly believed to be the author of *Der Minne Lehre* (q.v.).

HEISELER, BERNT VON (Braunenburg/Inn, Bavaria, 1907–), devoted himself to a literary career, writing poems, plays, and fiction reflecting a Protestant and nationally self-conscious, though unaggressive, background. His earliest publication was a volume of poems (*Wanderndes Hoffen*, 1935); it was followed by *De profundis* (1947), *Spiegel im dunkeln Wort* (1950), and *Gedichte* (1957). He is the author of the novels *Die gute Welt* (1938) and *Versöhnung* (1953) and of a number of stories. His plays fall chiefly into two groups, classical and historical; his classical plays include *Cäsar* (1941), *Semiramis* (after Calderón, 1948), and *Philoktet* (after Sophocles, 1948), and his historical plays *Schill* (1937), *Hohenstaufentrilogie* (1948), and *Die Malteser* (1957), a reconstruction of Schiller's long-considered, but never executed, project .

Heiseler is the author of several literary biographies (*Stefan George*, 1934; *Kleist*, 1939; *Schiller*, 1959) for the general reader. His father was the minor dramatist Henry von Heiseler (1875–1928), whose biography Bernt von Heiseler wrote in 1932.

HEISSENBÜTTEL, HELMUT (Rüstringen nr. Wilhelmshafen, 1921–), served in the war and was wounded in 1941. After the war he worked for a publisher and in broadcasting. His anti-grammatical concrete poetry (see KONKRETE POESIE), working by isolation of words or new arrangements of them in which their associations are released, appeared, together with prose, in six numbered volumes, *Textbuch 1*,

Textbuch 2, etc. They were published as follows: 1. 1960; 2. 1961; 3. 1962; 4. 1964; 5. 1965; 6. 1967 and appeared as *Das Textbuch* in 1970.

The volume *Über Literatur* (1966) was followed by *Briefwechsel über Literatur* (1969, written in collaboration with H. Vormweg), *Was ist das Konkrete an einem Gedicht?* (1969), and *Zur Tradition der Moderne. Aufsätze und Anmerkungen* (1972). Heißenbüttel's theories expounded in these essays underlie his radio plays, *2 oder 3 Porträte, Was sollen wir überhaupt senden?* (both 1970), and *Marlowes Ende* (1971). *Projekt Nr. 1. D'Alemberts Ende*, a 'novel' (1970, republished 1981), is the first of a new cycle consisting of 'projects', *Das Durchhauen des Kohlhaupts. Dreizehn Lehrgedichte* (1974) the second. The third is divided into three parts, *Eichendorffs Untergang und andere Märchen. Projekt 3/1* (1978), *Wenn Adolf Hitler den Krieg nicht gewonnen hätte. Historische Novellen und wahre Begebenheiten. Projekt 3/2* (1979) and *Das Ende der Alternative. Einfache Geschichten. Projekt 3/3* (1980). The designations referring to accepted literary genres are deliberately chosen in order to distance the reader from conventions, whilst their effectiveness presupposes familiarity with major figures and works since the age of enlightenment. Through the combination of quotations, many of which are authentic, Heißenbüttel's own ingenious verbal and syntactical permutations in different contextual perspectives invite the reader to probe their possible implications. The collection *Gelegenheitsgedichte und Klappentexte* (1973) was followed by the collections *Ödipuskomplex made in Germany. Gelegenheitsgedichte, Totentage, Landschaften 1965–1980* (1981), and *Von fliegenden Fröschen, libidinösen Epen, vaterländischen Romanen, Sprechblasen und Ohrwürmern. 13 Essays* (1982). *Der fliegende Frosch und das unverhoffte Krokodil. Wilhelm Busch als Dichter* appeared in 1976, *Mümmelmann oder Die Hasendämmerung* in 1978, and *Die goldene Kuppel des Comes Argobast oder Lichtenberg in Hamburg* in 1980. (Heißenbüttel had published an essay on G. C. Lichtenberg, q.v., in 1974: *Georg Christoph Lichtenberg, der erste Autor des 20. Jahrhunderts?* in *Aufklärung über Lichtenberg. Mit Beiträgen von Helmut Heißenbüttel u.a.*)

He was awarded the Büchner Prize in 1969.

Heiteretei, Die, a Novelle by O. Ludwig (q.v.), first published in instalments in *Die Kölnische Zeitung* in 1855–6, and in book form in 1857 under the title *Die Heiteretei und ihr Widerspiel*, together with the story *Aus dem Regen in die Traufe* (q.v.). The volume was intended as the first of a series entitled *Thüringische Naturen*. The work is set in a small market town in Ludwig's native Thuringia.

Heiteretei is a nickname for Annedorle and depicts her striking and cheerful appearance, which is matched by a robust and upright character. A poor orphan girl, Annedorle is house-proud, and as yet unattached. On her way home from market she meets Holders-Fritz, locally renowned as the toughest and strongest young man. An indecisive trial of strength takes place between them, and she gives him such a piece of her mind that he is shamed into silence. Though it is not apparent to either of them, they begin to fall in love from this moment. Fritz turns over a new leaf and lives sensibly, but his attempts to approach Annedorle are misinterpreted as a prelude to revenge. This misconception is nourished by a group of self-appointed female guardians of propriety, the town's 'große Weiber'. By their presence in her cottage they work Annedorle into so nervous a state that on her way home one dark night she panics, and runs Fritz into the river with her barrow. She realizes her folly, dismisses the women, and puts an end to their boycott. They have been aided in their hostility by continuous rain which gradually causes her cottage to collapse. Overcoming the dilemma, Fritz masters himself, and woos and wins Annedorle.

Unexpectedly, the story takes a new turn depicted in the second part of the title (*Widerspiel*). Annedorle regrets the loss of her independence, and wilfully misinterprets Fritz's thoughtfulness and generosity. The engagement is threatened, but Fritz, holding up a mirror to Annedorle, as she had once done to him, shames her into self-knowledge and cures her misguided pride. The story thus portrays a double process of education (Bildung) with acute psychological perception as well as brilliant caricatures. At the same time the immense strength of Holders-Fritz and of Annedorle and the latter's invincible courage add a heroic touch, and with it a mythical element.

Heizer, Der, see VERSCHOLLENE, DER.

Held des Nordens, Der, a trilogy by F. de la Motte-FOUQUÉ (q.v.), published in 1810. The three plays, each 'Ein Heldenspiel' in verse, are entitled *Sigurd der Schlangentöter, Sigurds Rache,* and *Aslauga.* Fouqué based his plays on the *Edda* (q.v.), which he read in the original Icelandic.

In *Sigurd der Schlangentöter,* Sigurd slays the dragon Faffner, acquires his treasure, and learns to understand the speech of the birds. He enters the fire-girt castle of Hindarfiall, frees Brynhildur from her enforced sleep, and espouses her. Sent forth on further exploits he comes to King Giuke's castle by the Rhine. Queen Grimhildur, seeking to obtain the treasure, gives Sigurd a potion that makes him forget Brynhildur; he is betrothed to Gudruna, the daughter of Giuke and Grimhildur, and swears comradeship with her brothers Gunnar, Högne, and Guttorm. Guttorm, moved by covetousness and Brynhildur's desire for vengeance, murders Sigurd, and Brynhildur stabs herself and throws herself on Sigurd's funeral pyre. In *Sigurds Rache,* Gudruna is brought by one of Grimhildur's potions to forget Sigurd and marries King Atli. Accused of adultery, she clears herself by surviving the ordeal by boiling water. Atli entices her brothers to visit her, in the hope of obtaining the treasure. Högne and Gunnar perish, and Gunnar, before dying, reveals that the treasure is lost for ever in the Rhine. With the help of Högne's son Niflung, Gudruna slays Atli.

Aslauga deals with the daughter of Sigurd and Brynhildur. After her parents' death, Aslauga is protected by King Heimer, who hides her in a zither. He takes shelter in a hut, but is murdered by the peasant occupants, who bring her up as their daughter; in the end Aslauga marries Ragnar, King of Denmark. Each of the three plays is preceded by a verse dedication to J. G. Fichte (q.v.).

Heldenepos denotes a type of Middle High German epic poem derived from remote Germanic historical traditions and emphasizing heroism and grandeur in tragic situations. The outstanding example of the heroic epic is the *Nibelungenlied* (q.v.), which is unique in its power and stature; almost on the same level is *Kudrun* (q.v.). The authorship of these poems is unknown; they are believed to be the work of commoners following poetry as a calling and patronized at courts.

The heroic epic flourished at the same time as the court epic (see HÖFISCHES EPOS), and similarly continued in gradually degenerating form in subsequent times. The commonest subject is some version of the story of the legendary figure of Dietrich von Bern (derived from the historical Theodoric of Verona, see DIETRICHSAGE). Examples of this are *Alpharts Tod, Dietrichs Flucht, Das Eckenlied, Laurin, Ortnit, Die Rabenschlacht, Sigenot, Virginal,* and *Wolfdietrich* (qq.v.).

The anonymous epic *Biterolf und Dietleib* (q.v.) occupies a special position, sharing the characteristics of both types of epic.

Helgoland (Heligoland), situated in the North Sea some 30 miles from the German mainland, belonged in the 14th c. to Schleswig, and passed to Denmark in the 18th c. During the Napoleonic Wars (q.v.) it was occupied by the British (1807) and was used as a base for breaching the Continental Blockade. In 1814 the Danes, under the Peace of Kiel, ceded the island to Great Britain, which in turn ceded it to Germany in 1890 in return for Zanzibar. The deal was regarded in Germany as an interference with German colonial policy. In 1919 the fortifications were dismantled in accordance with the Treaty of Versailles, but it was refortified after 1936. It has long been an important research station for the study of bird migration.

H. Heine (q.v.) visited Heligoland in 1829 and 1830.

Heliand, title given by the first editor, J. A. Schmeller (1830), to a religious epic dealing with the life of Jesus and written in Old Saxon during the reign of Ludwig the Pious (814–40, see LUDWIG I). The poem contains 5,983 lines of alliterative verse (Stabreim). The end is missing. The author is unknown, but the Latin preface, printed in 1562 by Matthias Flacius Illyricus, speaks of him as a man already skilled in poetry. He is believed to have been a monk or at least to have been educated in a monastic school, probably Fulda. The preface also states that the work was commissioned by Ludwig the Pious in order to spread knowledge of Jesus among the people. The primary sources of the epic are the gospel harmony of *Tatian* (q.v.) and the Latin commentary on St. Matthew by Hrabanus Maurus (q.v.). The style derives from the Anglo-Saxon biblical epic. Christ appears as a Germanic king, the disciples as his noble warriors. Nevertheless the essence of the work is truly Christian, and, unlike heathen epic, it speaks to the heart.

There are two principal MSS.: one, in Munich, was formerly in Bamberg; the other, originally in the library of Sir Robert Cotton (1571–1631), was transferred in 1753 to the British Museum.

HELL, THEODOR, pseudonym of Karl Gottlieb Theodor Winkler (Waldenburg, 1775–1856, Dresden), a publicist, who began as an archivist in Dresden and was later secretary and then deputy director of the Court Theatre. His own works, both narrative and dramatic, were largely adaptations and translations from the French and included E. Scribe's *Un Verre d'eau* (*Ein Glas Wasser*). He translated Camoens (*Die Lusiaden von Camoëns, deutsch,* 1807, with F. A. Kuhn, 1774–1844) and Byron's *Mazeppa* (1820). In 1821 he published *Lyratöne,* a collection of his own poetry.

Hell was particularly active in the publication of annuals (Taschenbücher), including *Penelope* (1811–48) and *Dramatisches Vergißmeinnicht* (1824–49). From 1817 to 1843 he was editor of the Dresden *Abendzeitung.* In these publications appeared chiefly writings by authors such as W. Alexis, Luise Brachmann, H. Clauren, E. C. Houwald, and Graf Loeben (qq.v.).

Heller, small coin, equivalent to the Pfennig, at first of silver but from the 18th c. of copper, whence the expression 'roter Heller'. The Heller (or Häller) took its name from Schwäbisch Hall, where the coin was minted until 1494.

HELMBOLD, LUDWIG (Mühlhausen, 1532–98, Mühlhausen), was crowned poet (see GEKRÖNTER DICHTER) by the Emperor Maximilian II (q.v.) in 1566. He wrote Latin poetry, and hymns in German, including 'Ich weiß, daß mein Erlöser lebt'.

HELWIG, see MAERE VOM HEILIGEN KREUZ.

HELWIG, JOHANN (Nürnberg, 1609–74, Regensburg), a physician, was a member of the Hirten- und Blumenorden an der Pegnitz (q.v.). He published poetry and a translation of Boethius's *De consolatione philosophiae* (1660).

Hemmingstedt, a village in Dithmarschen (q.v.); on 17 February 1500 it was the scene of a battle in which an army of Dithmarschen peasants defeated the Danish army of Duke Friedrich of Schleswig-Holstein. The hard fighting and the victory of the peasants are the subject of a number of poems and novels. F. Hebbel wrote *Der Sieg bei Hemmingstedt* (1832), and contemplated a play on the theme. His patroness Amalie Schoppe (qq.v.) is the author of a novel with the same title (1840). In 1851 Th. Fontane (q.v.) wrote a ballad, *Der Tag von Hemmingstedt,* and in 1898 A. Bartels (q.v.) published a novel, *Die Dithmarscher.*

HENCKELL, KARL FRIEDRICH (Hanover, 1864–1929, Lindau), was prominent in the left-wing literary movement *c.* 1890, writing proletarian poetry, some of which was forbidden in Germany as inflammatory. He evaded the ban by setting up as a publisher in Zurich. His style now seems uncomfortably rhetorical. His collections of poems include *Umsonst* (1884), *Poetisches Skizzenbuch* (1885), *Strophen* (1887), *Amselruf* (1888), *Diorama* (1889), *Gründeutschland* (1890), *Trutznachtigall* (1891), *Aus meinem Liederbuch* (1892), *Zwischenspiel* (1894), *Gipfel und Gründe* (1894), *Gedichte* (1898), *Neues Leben* (1900), *Schwingungen* (1906), *Im Weitergehn* (1911), and *Weltmusik* (1918).

Henker, Der, a novel by E. Schaper. See SIE MÄHTEN GEWAPPNET DIE SAATEN.

HENRICI, Christian Friedrich (pseudonym Picander) (Stolpen, 1700–64, Leipzig), wrote comic poetry (*Ernst- Schertzhaffte und Satyrische Gedichte*, 1727–37), but is now best known as the author of texts for choral works by J. S. Bach (q.v.), notably the St. Matthew Passion and a number of cantatas. Henrici wrote much occasional poetry, and some crude comedies of social satire: *Der Säuffer* (1725), *Die Weiberprobe oder die Untreue der Ehe-Frauen* (1725), and *Academischer Schlendrian* (1726). He was one of the Leipzig writers whom J. C. Gottsched opposed. A volume of poems was published after his death, *Sammlung vermischter Gedichte* (1768).

Henri Quatre-Romane, two consecutive historical novels by H. Mann (q.v.), *Die Jugend des Königs Henri Quatre* (1935) and *Die Vollendung des Königs Henri Quatre* (1938), which have no authorized collective title but make up in effect one great novel. Both were written during Mann's sojourn in France after his flight from Germany in 1933, and the choice of hero, 'le bon roi Henri IV', and the relaxed tone reflect the congeniality of Mann's French environment. The Renaissance always had an attraction for him, and in this late novel he gratified it to the full. He made serious historical studies but did not hesitate to depart from known fact where his design demanded it. Among his inventions are Henri's interviews with Montaigne and Montaigne's conversation with Michelangelo. Moreover, Mann plainly hints that these novels are in part a moral commentary on the events of his own day by using words with a contemporary National Socialist ring such as *Arbeitsdienst*, *Wehrpflicht*, and *Gauleiter*, and by implicitly equating the *Ligue* headed by the Guise family with Hitler's SA (q.v.).

The two novels cover Henri's lifetime (1553–1610), from childhood to his assassination by Ravaillac. There is no organic separation between the two books. *Die Jugend des Königs Henri Quatre* follows his childhood and youth to the eve of kingship in 1589. It shows the humiliations suffered by his mother at the hands of Catherine de Medici (Katharina von Medici) his own compulsory conversion to Roman Catholicism in Paris at the age of 12, and his reversion to Protestantism on his return to the south-west two years later. The world in which he lives is pervaded by savage and merciless intrigue, and his own mother is poisoned, apparently by Catherine. Henri, whose sexual diversions are almost as conspicuous as his good-will and balanced tolerance, marries Margot, daughter of Catherine. Though he genuinely loves her, he is at all times unable to resist the attraction of a pretty woman.

Mann portrays the feeble reigns of Charles IX and Henri III, and gives a brilliant and graphic description of the horrors of the massacre of St. Bartholomew's Day, 1572. Henri's fortunes veer from captivity in Paris (1572–6) to the prospect of succession to the throne, which becomes a fact when Henri III, who has brought about the murder of the hostile Duc de Guise and his brother, the Cardinal of Lorraine, is himself assassinated. The first novel closes with Henri's victories at Dieppe and Arques, but with no certainty that the opposition of the *Ligue* is destroyed.

In *Die Vollendung des Königs Henri Quatre* Henri is converted to Roman Catholicism in order to ensure the throne for himself, but he desires the crown, not from ambition, but for the tolerance, unity, and goodwill which he can offer the people of France. He is crowned at Chartres and, with the assistance of Sully and Mornay, rules wisely. The climax of his reign is the great decree of tolerance, the Edict of Nantes (1598), which officially terminates sectarian conflict. Included in this novel are further love affairs, of which the most notable is with Gabrielle d'Estrées. Numerous attempts at assassination culminate in Henri's murder by Ravaillac, a tragedy which provokes a great demonstration of popular affection and grief.

The two novels are arranged, not in numbered chapters, but in sections and sub-sections. In *Die Jugend des Königs Henri Quatre* each of the nine sections is concluded by a *moralité* in French, summarizing the situation, and two of the nine are addressed to Henri himself. *Die Vollendung des Henri Quatre* has no such reminders, but its end is immediately followed by a speech; Henri makes it while experiencing a vision in which he stands erect upon a cloud. It is addressed to Mann's own fellow-men during the National Socialist regime, is spoken in French, and bears the heading *Allocution d'Henri quatrième, Roi de France et de Navarre, du haut d'un nuage qui le démasque pendant l'espace d'un éclair, puis se referme sur lui*. This speech, couched in sober rhetoric, is a *moralité* for the whole work.

The combined *Henri Quatre* novels are generally held to be Heinrich Mann's masterpiece, more positive and balanced than anything he wrote before or after, displaying his hatred of evil, but revealing an appreciation of kindness and a stress on tolerance which are not elsewhere united in his work. In the figure of the gentle, civilized Montaigne and in his (fictitious) relationship with Henri, this pervasive harmony is especially emphasized. The active moralist prevails over the negative satirist.

HENSEL, Luise (Linum, Brandenburg, 1798–1876, Paderborn), daughter of a Protestant pastor, became a governess in Berlin; in 1817 she was ardently courted by C. Brentano (q.v.), whom she rejected, while successfully encouraging him to return to the Roman Catholic Church, in which he had been brought up. In 1818 she herself adopted the Catholic faith. She spent her life in the service of others, as companion (in Münster and Düsseldorf), as governess in various houses, as nurse to the sick in Koblenz, as a teacher in Boppard and Aachen, and finally as a governess in Cologne and Wiedenbrück. In 1874 she withdrew to a convent. She wrote religious poetry, published in *Gedichte* (1858) and *Lieder* (1869). Her brother Wilhelm Hensel (1794–1861) was a painter and book illustrator.

HENSLER, KARL FRIEDRICH (Vaihingen/Enz, 1759–1825, Vienna), a young man of good family, was sent to Vienna to train for the diplomatic service, but became instead a playwright. His large and varied output included Soldatenstücke (q.v.), Ritterstücke, Räuberstücke (see RITTER- UND RÄUBERROMAN), and Viennese Lokalpossen (see LOKALPOSSE). His most successful play was *Das Donauweibchen. Ein romantisch-komisches Volksmärchen mit Gesang* (2 Pts., 1798). Among other successes were *Das Petermännchen* (1794), *Alles in Uniform für unsern König* (1794), *Der alte Überall und Nirgends* (1794), *Die zwölf schlafenden Jungfrauen* (1797), and *Rinaldo Rinaldini* (1799) based on a popular novel by C. A. Vulpius (q.v.). He provided many of his plays with a sequel.

Hensler was an early exponent of the Viennese Volksstück (see VOLKSSTÜCK) in its best period. He was at one time director of the Theater in der Leopoldstadt in Vienna, and was later at the Theater an der Wien.

HENZ, RUDOLF (Göpfritz, Lower Austria, 1897–), a devout Roman Catholic, was an officer in the 1914–18 War, and afterwards studied literature and music at Vienna University. He worked as a journalist, and was also director of adult education for the Volksbund der Katholiken Österreichs. From 1931 until his dismissal in 1939 by the National Socialists he was programme director of the Austrian Broadcasting Service and became president of the Austrian Kulturvereinigung. He worked as editor (and co-editor) of the literary periodicals *Wort in der Zeit* (1955–65) and *Literatur und Kritik* (from 1966), and of *Dichtung und Gegenwart* (from 1950). His Austrian awards include the Staatspreis für Literatur (1953). During the war he occupied himself with stained glass and the restoration of church windows. In 1945 he was reinstated in his post in broadcasting. His whole work is an affirmation of his religious faith.

Henz's first poems, *Lieder eines Heimkehrers* (1920), were followed by the collections *Unter Brüdern und Bäumen* (1929), *Döblinger Hymnen* (1935), *Österreichische Trilogie* (1950), and *Lobgesang auf unsere Zeit* (1956). A selection (*Wort in der Zeit*) was published in 1945 and (*Neue Gedichte*) 1972, followed by the volume *Die Gedichte* (1984). *Der Turm der Welt* (1951) is a huge religious epic in *terza rima* (see TERZINEN). His novels include *Die Gaukler* (1932), *Dennoch Mensch* (1935), *Begegnung im September, Die Hundsmühle* (both 1939), *Der Kurier des Kaisers* (1941, rev. edn. 1982), *Der große Sturm* (1943), *Ein Bauer greift an die Sterne* (1943, republished as *Peter Anich, der Sternsucher*, 1947), *Das Land der singenden Hügel* (1954), *Die Nachzügler* (1961), *Der Kartonismus. Ein satirischer Roman* (1965), *Unternehmen Leonardo* (1973), and *Wohin mit den Scherben?* (1979). He is also the author of historical and religious plays (*Kaiser Joseph II.*, 1937; *Die Erlösung*, a passion play, 1949; *Die große Entscheidung*, 1952; *Tollhaus Welt* (1970) is a collection. His autobiography is entitled *Fügung und Widerstand* (1963, ext. edn. 1981).

HENZE, HANS WERNER (Gütersloh, 1926–), composer of orchestral music, including five symphonies, and of a number of operas, of which the most notable are: *Boulevard Solitude* (1951), *Das Ende einer Welt* (1953, libretto by W. Hildesheimer, q.v.), *König Hirsch* (1956, revised 1963), *Der Prinz von Homburg* (1960, libretto by I. Bachmann, after H. von Kleist, qq.v.), *Elegie für Liebende* (1961, libretto by W. H. Auden and C. Kallmann), and *Der junge Lord* (1965, libretto by I. Bachmann, after *Der Scheik von Alexandrien und seine Sklaven* by W. Hauff, q.v.). *Die Bassariden* (1966, libretto by W. H. Auden and C. Kallmann after the *Bacchae* of Euripides) had a successful production in London (as *The Bassarids*) in 1974.

After this work, which some hold to be his best, Henze espoused Marxist views. The most striking work, for which E. Schnabel (q.v.) provided the text, is *Das Floß der Medusa* (1968), based on the following occurrence, which is imaginatively recorded by the painter Géricault in *Le Radeau de la Méduse* (1819, Louvre). In July 1816 the French frigate *Méduse* was lost off the West African coast. While the officers and some of the crew escaped in boats, 149 persons were left on a raft. Fifteen survivors, five of whom were dying, were eventually rescued by the frigate *Argus*. The opera *We come to the River* (libretto by Edward Bond), German title: *Wir erreichen den Fluß*, was first performed in 1976 (London and Berlin). Other compositions include a Fantasy for string sextet for a film (*Der junge Törless*, 1966) based on Musil's *Die Verwirrungen des Zöglings Törleß* (q.v.). *Musik und Poli-*

tik. Schriften und Gespräche 1955–1975, ed. with an introduction by J. Brockmeier, appeared in 1976, and *Die englische Katze. Ein Arbeitstagebuch 1978–1982* in 1983.

HENZI, SAMUEL (nr. Berne, 1701–49, Berne), Swiss reformer and conspirator, was exiled from Berne in 1744 because of his support of a petition for reform of the cantonal government. Pardoned in 1748, he was appointed to a librarian's post. In the following year he took part in an abortive conspiracy against the patrician oligarchy, was tried and executed. Lessing (q.v.) began in the year of Henzi's execution a verse tragedy dealing with the conspiracy. It remained unfinished, and was published as a fragment entitled *Samuel Henzi. Ein Trauerspiel* in *Schriften* (1753, vol. 2, *Briefe*, 22 and 23).

HERÄUS, KARL GUSTAV (Stockholm, 1671–c. 1725, Mürzzuschlag, Styria), an antiquarian and numismatist, was in the service first of the court of Schwarzburg-Sondershausen, then in that of Vienna. He invented a new verse form by arranging distichs so that hexameter rhymed with hexameter, pentameter with pentameter, i.e. ababcdcd (*Versuch einer neuen deutschen Versart*). His *Gedichte und lateinische Inschriften* appeared first in 1721.

HERBERGER, VALERIUS (Fraustadt, Poland, 1562–1627, Fraustadt), a schoolmaster and later pastor in Fraustadt, is the author of Protestant hymns, one of which ('Valet will ich dir sagen, du arge falsche Welt') has remained in use.

HERBORT VON FRITZLAR, author of a long Middle High German poem, *Das Lied von Troja*, written *c.* 1190 or, less probably, 1210, for the Thuringian court of Landgraf Hermann (q.v.). Nothing is known of Herbort's life except that he was a cleric belonging to the college of Fritzlar. His poem is a translation and abridgement (18,000 lines) of an immense French original by Benoît de Sainte More. It shows little understanding of courtly ideals, embodying gruesome details of battles, and crass abuse uttered by lords and ladies. Its style is pedestrian and clumsy, but touches of dry irony reveal an independent and even original personality.

Herbstgefühl, a poem written by Goethe in 1775, and first published in the same year in Jacobi's *Iris* (q.v.) with the title *Im Herbst 1775*. The present title was adopted in 1789. It is remarkable for the rich maturity and concentration of its style.

Herbstgefühl, a lyric poem by N. Lenau (q.v.), written in 1831, and beginning 'Mürrisch braust der Eichenwald'.

Herbstklage, a lyric poem by N. Lenau (q.v.), written probably in 1831, and beginning 'Holder Lenz, du bist dahin'.

HERDER, JOHANN GOTTFRIED (Mohrungen, East Prussia, 1744–1803, Weimar), the son of a parish clerk, grew up in humble circumstances, but received a grammar school education, enabling him to study theology at Königsberg University where Kant (q.v.) taught. There Herder became friendly with J. G. Hamann (q.v.). In 1764 he was appointed as teacher at the cathedral school at Riga, becoming also a popular preacher in the cathedral. In 1767 he published anonymously *Über die neuere deutsche Literatur*, a brilliant review of the current literary situation in Germany, interspersed with original and fertile ideas. This work is often referred to as Herder's *Fragmente* because its three parts are so described. *Kritische Wälder* (q.v., 3 vols., 1769) includes discussions on J. J. Winckelmann (q.v.) and on G. E. Lessing's *Laokoon* (q.v.), and foreshadows Herder's concept of the totality of the human personality.

In 1769 Herder, an enthusiast for the ideas of J. J. Rousseau, resigned his appointments and set out for France, travelling by ship to Nantes. His views and feelings on this journey are recorded in the *Journal meiner Reise im Jahr 1769* (q.v.), a diary first published posthumously in 1846. While in France he was offered an appointment as travelling tutor to a prince of Holstein-Eutin. On the way to his new post he met Lessing and M. Claudius (q.v.). He soon set out with his charge on what was intended as a three-year tour. In Darmstadt he met Caroline Flachsland (q.v.), whom he married in 1773, and in Strasburg in Sept. 1770 he encountered the young Goethe. Herder stayed there to receive medical treatment for his eyes, and the contract with the prince was dissolved. In 1771 he became court preacher to the petty court of Schaumburg-Lippe at Bückeburg. In the year following the appearance of his philological tract *Über den Ursprung der Sprache* (q.v., 1772) Herder published a notable series of essays, *Von deutscher Art und Kunst* (q.v.), to which he and Goethe were the principal contributors. His early historical essay *Auch eine Philosophie der Geschichte zur Bildung der Menschheit* (q.v.) and the theological work *Älteste Urkunde des Menschengeschlechts* appeared in 1774.

In 1776 Herder, whose restless temperament made him easily dissatisfied with any situation, was glad to accept an invitation to Weimar, where Goethe had secured him a post as Moderator (Generalsuperintendent). For the next twelve years Herder lived in harmony with his environment, and entered a new phase of productivity. It began with the publication of a collection of folk-songs, *Volkslieder* (2 vols., 1778–9), which is better known under the title *Stimmen der Völker in Liedern* (q.v., 1807). This collection marks the climax and the end of Herder's association with the Sturm und Drang (q.v.), to which he had made two major contributions: firstly, the idea of perception by the total personality (replacing the belief in the primacy of reason); and secondly, the high estimation of those poets and literary works which are closest to nature (notably the Bible, Homer, Shakespeare, Ossian, q.v., and folk-song). *Vom Geist der ebräischen Poesie* (q.v., 1782–3), arising out of *Briefe, das Studium der Theologie betreffend*, applies aesthetics to theology. The great achievement of Herder's first decade in Weimar is his most systematic and complete historical work, *Ideen zur Philosophie der Geschichte der Menschheit* (q.v., 1784–91), which combined his cyclical, organic conception (the birth, growth, and death of civilizations and their manifestations) with an idea of progress, which he was one of the first not to perceive in terms of a linear development.

Impatient to move, Herder travelled to Italy in 1788, but abandoned the tour within a year, preparing his *Briefe zur Beförderung der Humanität* (1793–7), ten collections originating from his response to the ideals of the French Revolution (q.v.). In the following decade his publications were mainly theological, and included *Von der Auferstehung als Glauben, Geschichte und Lehre* (1794) and *Sammlung christlicher Schriften* (1794–8). His later years in Weimar were marked by increasing dissatisfaction and disharmony with Duke Karl August (q.v.) and Goethe. He also embarked on a campaign to discredit the philosophy of Kant (q.v.), publishing *Verstand und Erfahrung. Eine Metakritik zur Kritik der reinen Vernunft* (2 vols.) in 1799, and *Die Metakritik der Urteilskraft*, known as *Kalligone*, in 1800, both of which encountered severe criticism. His selection of his later writings resuming the themes of his early works, *Adrastea* (6 vols., 1801–4), made his supporters in Weimar alert to the weaker aspects of his intellect and personality. It included the cycle *Der Cid*, his translation (largely based on a French version) of the Spanish poem *El Cid*. His renderings of foreign folk-songs and of the Greek Anthology were also in Herder's own eyes far more noteworthy than the original poetry and the few plays he wrote.

By the turn of the century Herder's influence had declined, but his contribution to German literary and historical thought remains outstanding. His stature depends less on comprehensiveness than on fertility of ideas and suggestive originality. He was noted for an uneven temper emanating from his hypersensitive personality; but he also possessed great charm and a persuasive eloquence.

Sämtliche Werke (45 vols.) appeared 1805–20, the *Historisch-kritische Ausgabe* (33 vols.), ed. B. Suphan, 1877–1913, *Herders Briefwechsel mit Nicolai* (1 vol.) in 1975, and *Aus dem Nachlaß* (1 vol.) in 1976.

HERGÊR, see SPERVOGEL.

HERIBERT VON SALURN (Salurn, Tyrol, 1637–1700, Merano), ecclesiastical name of Anton Mayr, a friar who gained a reputation for sermons in popular style with plentiful illustration by familiar proverbs and turns of speech.

HERLOSSOHN, KARL (Prague, 1804–49, Leipzig), left Bohemia for Saxony and wrote sketches (*Weihnachtsbilder*, 1846), and historical novels, including *Der Ungar* (3 vols., 1832), *Die Hussiten* (4 vols., 1843), and *Die Mörder Wallensteins* (3 vols., 1847). He also translated Serbian folk-songs into German.

HERMAN (also spelt **HERMANN**), NIKOLAUS (Altdorf nr. Nürnberg, c. 1480–1561, Joachimsthal, Bohemia), schoolmaster and cantor in Joachimsthal, is the author of two collections of hymns with tunes which he composed himself, *Sonntagsevangelia* (1560) and *Historien von der Sindflut* (1562); they were intended for household use in replacement of profane songs, of which they are *contrafacta* or adaptations. He also translated *De libero arbitrio* by Erasmus (q.v.) into German (1526).

HERMANN, LANDGRAF VON THÜRINGEN (d. 1217, Gotha), who ruled from 1190, played a dubious part in the conflict for the imperial crown between Philipp of Swabia and Otto IV (qq.v.). His court was the resort of the Minnesänger and poets of the age, including for a time WOLFRAM von Eschenbach and WALTHER von der Vogelweide (qq.v.). Walther, in a well-known Spruch (q.v.), criticizes the noisy restless life of the court and the Landgrave's extravagance. Hermann figures as a prominent character in R. Wagner's opera *Tannhäuser und der Sängerkrieg auf der Wartburg* (q.v.).

HERMANN, GEORG (Berlin, 1871–1943, Auschwitz), was the author of novels of Jewish life in old Berlin. Best known are *Jettchen Gebert* (1906) and *Henriette Jakoby* (1908).

HERMANN DER CHERUSKER, see ARMINIUS.

HERMANN DER DAMEN (or Hermann Damen), a minor Middle High German poet of the late 13th c. Hermann was a nobleman of Brandenburg whose home was in the basin of the River Dahme to the south of present-day Berlin. He wrote religious and social Sprüche (see SPRUCH) of didactic tone.

Hermannsschlacht, usual appellation since the middle of the 18th c. for the battle in which Varus and his legions were defeated in A.D. 9 by Arminius (q.v.).

Hermannsschlacht, Die, a five-act play (Ein Drama) in blank verse by H. von Kleist (q.v.), written in 1808–9, and published by L. Tieck (q.v.) in 1821. It was written under the impact of the Napoleonic occupation (see NAPOLEONIC WARS) to serve the patriotic necessity of the moment. In this sense Kleist made it a gift to the German people, asking for no more than its immediate stage production. His call went unheeded; in the 1860s a stage version of the play was produced through the influence of H. von Treitschke (q.v.).

The plot revolves round the historic battle in the Teutoburger Wald (q.v.), in which Hermann der Cherusker (see ARMINIUS), the Germanic chieftain, combined forces with various German tribes, and defeated Varus, the Roman commander in Germany; as a result the Roman occupation forces had to withdraw to the Rhine. The play ends with the death of Varus at the hands of Fust, the chieftain of the Cimbri and a former ally of Varus. Fust celebrates liberation and fraternity with Hermann, and Marbod, chieftain of the Suevi and king, passes the crown to Hermann, hailing him king of Germania.

The theme of liberation achieved principally by intrigue and deceit is by itself unattractive. Kleist's treatment of it, in its expression of frenzied hatred and violence, is perhaps the most extreme example among literary works dealing with Arminius. The analogy with the disunited Germanic tribes is directed at Austria and Prussia who failed, to Kleist's exasperation and that of other patriots, to make common cause against Napoleon. It is remarkable that Kleist has depicted a chivalrous enemy (Ventidius excepted), so reflecting his admiration for French civilization.

Hermannschlacht, Die, a play by C. D. Grabbe (q.v.), completed after prolonged revision in July 1836, some three months before he died, and published in 1838. The play is divided into eight sections (*Eingang*; *Erster Tag*; *Erste Nacht*; *Zweiter Tag*; *Zweite Nacht*; *Dritter Tag*; *Dritte Nacht*; *Schluß*), and describes the central action, the battle in the Teutoburger Wald (q.v., A.D. 9), in a series of short episodes. Grabbe knew Kleist's *Die Hermannsschlacht* (q.v.), but to him the fascination of the subject lay in its association with his native landscape, the scene of the battle and the home of those who fought it.

Hermann follows Varus with the intention of betraying him as soon as he feels sure that his countrymen have suffered enough injustice and contempt from their Roman overlords to be prepared to fight for their liberation. There are various pitched battles in drenching rain in the dense woods, into which Hermann lures the enemy, and the losses on both sides are considerable. On the third night Hermann is victorious, and Varus kills himself to escape surrender. The battle costs Rome three legions, and in the last section the dying Emperor Augustus predicts to Tiberius the decline of Rome, the emergence of the Germanic realm as a rival power, and the advent of a new religion through Christ. Apart from Fürst Hermann and his wife Thusnelda, Grabbe portrays a variety of characters which emerge less as Tacitean figures from a distant past than as Grabbe's own fellow-countrymen.

Hermann und Dorothea, an epic poem in hexameters, written by Goethe in 1796–7, and published in 1797 as the *Taschenbuch für 1798*. Each of its nine books bears the name of one of the Muses, as well as a sub-title: I. *Kalliope. Schicksal und Anteil*; II. *Terpsichore. Hermann*; III. *Thalia. Die Bürger*, IV. *Euterpe. Mutter und Sohn*; V. *Polyhymnia. Der Weltbürger*; VI. *Klio. Das Zeitalter*; VII. *Erato. Dorothea*; VIII. *Melpomene. Hermann und Dorothea*; IX. *Urania. Aussicht*. The sources of the story are *Vollkommene Emigrationsgeschichte von denen aus dem Erzbistum Salzburg vertriebenen Lutheranern* (1734) by G. G. G. Göcking and the anonymous *Das liebtätige Gera gegen die Saltzburgischen Emigranten* (1732), but Goethe has transferred the scene from Salzburg to the Rhineland at the time of the French Revolutionary troubles of his own day (see REVOLUTIONSKRIEGE). Refugees are passing close to the village, and Hermann, son of the host of the Goldener Löwe inn, is sent to bring gifts of food and clothing to those in want. He sees a girl, Dorothea, with whom he instantly falls in love. On his return his father declares that he will not hear of a penniless daughter-in-law. The downcast Hermann is encouraged by his mother to try again, and at last wins his father over. With the parson and the apothecary, friends of his father, he sets out to fetch Dorothea; his two companions hear nothing about her that is not favourable, but Hermann flinches from declaring his love, and brings her back under the pretext that she is to be engaged as a maid. Misunderstandings arise, but they are cleared up and the handsome couple, well matched in character as in appearance, are betrothed.

One of Goethe's most popular works in the

19th c., *Hermann und Dorothea* later lost favour for a time on the grounds of incompatibility of form and content. Yet the piquant contrast between classical hexameter and modern ambience is one of its assets. For all its stylization and its ritual Homeric epithets, it is a poem of rich individual characterization and great tenderness, set in an idyllic landscape.

HERMANN VON FRITZLAR, author of a Middle High German collection of legends of the saints known as *Das Heiligenleben*. It is in prose, and was probably written about the middle of the 14th c.

HERMANN VON SACHSENHEIM (1363 or 1365–1458), a Swabian knight, originated in the region of the River Enz to the west of Stuttgart. Hermann, who held office under the Duke of Württemberg, wrote a number of long poems, some of which may be classed as epics. Their chronology is uncertain, but most are believed to have been written in old age, some when he was close on 90 years old. Hermann was a learned poet, and liked to parade his knowledge. *Die Mörin, c.* 6,000 lines, was written in 1453 for the Palatine Count Friedrich der Siegreiche and his sister Mechtilt. The poet on a summer walk in the woods is seized and taken captive by Eckhart, whisked through the air to Venus's island, and there tried for inconstancy. The suit is prosecuted by Brinhilt (the blackamoor woman of the title), and the poet is defended by Eckhart. Eventually Venus relents and he is released. The poem, which employs traditional motifs (see TANNHÄUSER, and DER GETREUE ECKART), appears to be a somewhat inconsistent allegory on the blameworthiness of sexual infidelity. *Des Spiegels Abenteuer* repeats the message of *Die Mörin*. A knight protests his constancy, but falls passionately in love with a female image seen in a magic mirror. His wife, on the other hand, remains faithful in deed as well as word. *Das Schleiertüchlein* is a sad love-story. A young knight takes with him on his journey to the Holy Land a kerchief soaked in his true love's blood, and on his return he finds her dead. *Von der Grasmetzen* tells of the indecent proposals of an old man to a country girl, and of her rebuffs and mockery. It is a parody of the cult of *minne*. Hermann constructs an elaborate religious allegory in *Der goldene Tempel* (1455), which derives from *Die goldene Schmiede* of KONRAD von Würzburg (q.v.), and even more from the anonymous *Minneburg* (q.v.). In *Jesus der Arzt* the wound of sin is healed by Christ.

HERMANN VON SALZBURG, a 14th-c. monk of Salzburg, to whom are ascribed a large number of hymns (geistliche Lieder), translated from the Latin, it is said, on behalf of Archbishop Pilgrim of Passau, who died in 1396. He is also known as Johann von Salzburg and as Der Mönch von Salzburg.

Hermeneutik, see INTERPRETATION.

HERMES, JOHANN TIMOTHEUS (Petznick nr. Stargard, 1738–1821, Breslau), a Protestant pastor, taught in the 'noble school' (Ritterakademie) at Brandenburg, served for a time as chaplain in a regiment of dragoons, became court preacher of Anhalt-Köthen, and then held a cure, with the title provost (Probst), in Breslau. Hermes was an enthusiastic admirer of the novels of Samuel Richardson and himself wrote novels based on Richardson's work. The *Geschichte der Miss Fanny Wilkes* (1766) closely follows *Sir Charles Grandison*. He achieved great success with his second novel, *Sophiens Reise von Memel nach Sachsen* (q.v., 1769–73).

Hermesianismus, a rational trend in Roman Catholic theology promoted by Georg Hermes (1775–1831), who became professor of dogmatics at the universities of Münster, in 1807, and Bonn, in 1820. Hermes was particularly influential in Bonn, but his views were eventually repressed by the fanatical Archbishop von Droste of Cologne (see DROSTE ZU VISCHERING, K. A.).

Hermetisch, an adjective first applied to modern German poetry *c.* 1900. It is perhaps best translated as 'cryptic'. German hermetic poetry has included, according to those using the term, work written by S. George, G. Eich, P. Celan, J. Bobrowski, and K. Krolow (qq.v.).

HERMLIN, STEPHAN, pseudonym of Rudolf Leder (Chemnitz, now Karl-Marx-Stadt, 1915–), a Communist poet, emigrated in 1936 and now lives in East Berlin. He first used his pseudonym in *Zwölf Balladen von den großen Städten* (1945) and *Wir verstummen nicht* (1945, in collaboration with L. Aichenrand and J. Mihaly). Other collections include *Zweiundzwanzig Balladen* (1947) and four stories, including the title story, *Die Zeit der Gemeinsamkeit* (1950). The *Mansfelder Oratorium* (1950, set to music by E. H. Meyer) was followed by the story *Die Zeit der Einsamkeit* (1951), the collection of poetry *Der Flug der Taube* (1952), *Dichtungen* (1956), and *Nachdichtungen* (1957; Hermlin is a noted translator of poetry from four languages). *Der Tod des Dichters* (1958 and included in *Gedichte und Prosa* of 1965), a eulogy on J. R. Becher (q.v.), marks the end of a creative phase which was curbed by the apparent unsuitability of his poetic (at times surrealist) style for a broad public. The publication of *Die Kommandeuse*, on the Berlin revolt of 17 June 1953, had already provoked a crisis that caused him to resign from official offices with which he had been honoured during his first years in the DDR, notably his appointment as Secretary to the Academy of Arts. Several editions of his prose and poetry appeared in the 1960s, including *Erzählungen* (1966), a collection which includes *Die Kommandeuse* and *Der Leutnant Yorck von Wartenburg* (1945); the central figure of the story faces death as a result of his involvement in the revolt of 20 July 1944 (see YORCK VON WARTENBURG, PETER, GRAF VON). *Scardanelli* (1970), a radio play, centres on Hölderlin's reaction against a bourgeois society and includes, apart from the Gontards, the figures of Goethe, Bettina von Arnim, and F. W. Waiblinger (qq.v.). *Lektüre 1960–1971* (1973) contains essays on a great variety of writers, critics, and one on Mozart (q.v.), *Mozarts Briefe*; it was followed by *Deutsches Lesebuch. Von Luther bis Liebknecht* (1976), *Gesammelte Gedichte* (1979), and the autobiographical *Abendlicht* (1979), which conveys through the controlled fusion of different layers of style a pervasive sense of human dignity against the political background of the Weimar and National Socialist period and its victims; it also contains allusions to poets who have influenced Hermlin's own creative writing and choice of poetic form, including Pindar, Keats, Shelley, and Hölderlin. The collection of stories, *Lebensfrist. Gesammelte Erzählungen und Aufsätze. Reportagen. Reden. Interviews*, ed. U. Hahn, appeared in 1980 and *Äußerungen 1944–1982*, ed. U. Dietzel, in 1983.

Herodes und Mariamne, a five-act tragedy in blank verse by F. Hebbel (q.v.), completed in 1848, published in 1850, and first performed in April 1849 in the Burgtheater, Vienna. Hebbel used as his source Josephus' *The Jewish War* and *The Jewish Antiquities*. At the height of his power, yet threatened by conspiracies fomented by Mariamne's mother Alexandra, Herodes has caused Mariamne's brother, the High Priest Aristobolus, to be drowned, ostensibly by accident. At the risk of losing his power and his life, he sets out for Alexandria to account to Antonius for Aristobolus' death. He orders Joseph, his viceroy, to kill Mariamne should he not return. When he returns; he finds that Joseph has failed to keep his order secret, and he therefore has him executed. Mariamne, deeply wounded by Herodes' order, for she would have taken her own life in the event of his death, vainly hopes that Herodes will sense her dedication. His suspicion that he has forfeited her love through her knowledge of the true circumstances of her brother's death, and

his jealousy of Antonius' designs on Mariamne, blind him to Mariamne's wish not to lose his trust in her power of love and forgiveness.

As Herodes is called away for a second time, he passes his order to Soemus, governor of Galilei, a repetition already contained in Hebbel's sources, but one which he put to effective dramatic use. Mariamne is not spared the knowledge of this second order, and, in her despair, threatens to kill herself. Alexandra, in the hope of winning Mariamne's support against Herodes, intervenes, but her attitude kindles in Mariamne the idea of revenging herself, not from political motives or on account of her brother's death, but because Herodes' conduct is a violation of her basic sense of human worth and dignity. She organizes a festive ball, intending Herodes on his return to find her rejoicing at his supposed death. Her ruse succeeds. Herodes arrives, has her tried and condemned to death. The Roman officer Titus is the only person to whom Mariamne confides her true motives on condition that he keeps them secret. But Titus feels free to reveal the truth to Herodes after her death. This coincides with the arrival of the Three Kings heralding the birth of the Son of God. Shattered at his self-destructive deed, Herodes revenges this challenge to his power by ordering the Massacre of the Innocents.

Heroldsdichtung or **Wappendichtung,** a form of poetry written chiefly in the 14th c. and 15th c., but originating in the 13th c. with Konrad von Würzburg (especially in *Das Turnier von Nantes*, q.v.). It consists in descriptions of coats of arms, etc., either alone or—more often—incorporated in eulogies of living or dead persons. The best-known heraldic poets are P. Suchenwirt, H. Rosenplüt, and WIEGAND von Marburg (qq.v.).

Hero und Leander, the legend of the Greek lovers separated by the Hellespont, in which Leander meets his death while swimming across to Hero, first occurs in German literature in a short Middle High German verse tale written in the 14th c. by an unknown author. It bears the title *Hero und Leander*. This is also the title of one of Schiller's longer ballads, published in 1802, and the story is the basis of the folk-song 'Es waren zwei Königskinder' (q.v.). The most important treatment of the subject in German is Grillparzer's tragedy *Des Meeres und der Liebe Wellen* (q.v.).

HERRAD VON LANDSBERG, abbess of St. Odile in Alsace from 1167 to 1195, is the author of the *Hortus deliciarum*, a Latin work which combines passages from the Bible with the most varied information. The beautifully illuminated

MS. was destroyed by fire during the bombardment of Strasburg in 1870.

HERRAND VON WILDONIE, a 13th-c. nobleman, is the author of four short Middle High German narrative poems. Herrand, who married a daughter of ULRICH von Lichtenstein (q.v.), was a nobleman of Styria and was involved in the Austrian troubles which followed the extinction of the Babenberg line in the middle of the century. His poems are thought to have been written between 1257 and 1275. *Die getriuwe kone* (*Die treue Gattin*) is the story of a wife whose husband loses an eye and feels unworthy to appear again in her presence. The wife, in her great love for him, re-establishes equality by removing her own eye. The gruesome episode is intended as an example of profound and true love. *Der verkêrte Wirt* (*Der betrogene Gatte*) tells of a wife who successfully and repeatedly deceives her husband. *Von dem blôzen Keiser* (*Vom nackten Kaiser*) is a political parable. The Emperor Gorneus, deprived of his clothes, is seen as a mere man, and subjected to unaccustomed humiliation. An angel assumes his imperial place and discharges the duties which he has neglected. Gorneus is presently restored to his throne and henceforth rules justly. The most attractive of the four tales is *Von der Katzen*, in which a tom-cat deserts his consort in search of a powerful wife who will be more worthy of him. He proposes to the sun, who refers him to the mist as able to overcome his rays. The mist passes him to the wind, the wind to the resistant wall, the wall to the mouse which weakens his foundations. The mouse, however, sends him to the cat of whom she stands in fear. And so he completes the circle, and, after mild reproof, is received back into favour.

Each poem is given a moral application and in each Herrand names himself as author. All four stories are found in other versions.

'Herr Bacchus ist ein braver Mann', first line of a poem by G. A. Bürger (q.v.), published in the *Göttinger Musenalmanach* in 1771; it has long been a popular student song. The melody is by J. A. P. Schulz (q.v.).

Herr Etatsrat, Der, a Novelle by Th. Storm (q.v.), published in *Westermanns Monatshefte* in 1881. The Etatsrat Sternow has two children, a son, Archimedes, and a daughter, Sophie (Phia), at whose birth his wife has died. Both children grow up lonely and loveless, and both die young. Archimedes, a student of mathematics, dies of an illness resulting from excessive drink and work. The delicate and unhappy Phia dies in childbirth together with her child, having been seduced by her father's assistant Käfer. Contrasting with the tender love between brother and sister is the heartless figure of the father, who is ultimately responsible for his children's tragedies. His grotesque portrayal, inspired by E. T. A. Hoffmann, enlivens the cold inhumanity of his character which, unlike other comparable characters in Storm's work, shows no redeeming features. Storm uses the theme of seduction for his forthright criticism of the self-righteousness of society.

'Herr Heinrich sitzt am Vogelherd', first line of the ballad *Heinrich der Vogler* by J. N. Vogl (q.v.). It narrates the bringing to Heinrich, Duke of Saxony, of the news of his election as Deutscher König in 919 (see HEINRICH I). The ballad has been set to music by Carl Loewe (q.v.).

Herr Lorenz Stark, a short novel by J. J. Engel (q.v.), first published in Schiller's *Die Horen* (q.v.) in 1795–6. It appeared in book form in 1801, and is aptly described as *Ein Charaktergemälde*. The eponymous hero is an elderly merchant, who, while successful in business, contrives to remain generous to the needy. At the same time he has the faults of a strong character, especially a marked self-righteousness. He is at loggerheads with his son, whom he believes to be a wastrel. The novel deals with the reconciliation of son and father, brought about by Herr Stark's daughter and son-in-law, who reveal the son to be as generous as the father, and as secretive in his generosity. The son makes a love-marriage with a virtuous young widow, and Herr Lorenz retires, making way for the next generation. The novel, which was widely read, shows a fine perception of motive and reaction.

HERRMANN-NEISSE, MAX (Neiße, 1886–1941, London), was in reality Max Herrmann, to which name he added the name of his birthplace. A frustrated and clamorous Expressionist, he contributed to *Die Aktion* (q.v.) and published several volumes of rhapsodic verse. His best work was written in exile after 1933, and published posthumously as *Letzte Gedichte* (1941). A selection of his verse bears the title *Heimatfern* (1946).

Herrnhuter or **Herrnhuter Gemeine,** known in Britain and the USA as the Moravian Church, an association of pietistic Christians, the origin of which was a community of religious refugees settled by Count Nikolaus Ludwig von Zinzendorf (q.v.) on his estate at Herrnhut in Saxony in 1722. The movement attracted great attention in the 18th c., and no less a figure than Lessing (q.v.) was moved to write as its apologist in the fragment *Gedanken über die Herrnhuter* (written 1750, published 1784).

Herr Peter Squentz, see ABSURDA COMICA.

Herr Puntila und sein Knecht Matti, a play (Volksstück) by B. Brecht (q.v.), written in 1940, first performed in June 1948, and published in 1959, though a Finnish version appeared in 1946. An unusually amusing comedy, it is based on stories and on a projected play by the Finnish writer Hella Wuolijoki (*Versuche*, vol. 10, *Vorspruch*). The farmer Puntila alternates between sobriety and inspired intoxication; during the former state he has all the drawbacks, from the Marxist viewpoint, of the orthodox landowner, and during the latter state he behaves like a human being among other human beings. Matti, Puntila's chauffeur, in the end abandons his unpredictable employer, singing a song ending ' 's wird Zeit, daß deine Knechte dir den Rücken kehren./Den guten Herrn, den finden sie geschwind./Wenn sie erst ihre eignen Herren sind'. The *Puntilalied* was set to music by P. Dessau (q.v.).

HERTZ, WILHELM (Stuttgart, 1835–1902, Munich), studied philology at Tübingen University, was for a short time an officer in the Württemberg army, began an academic career in 1861, and was appointed professor of German literature at Munich in 1869, where he associated with E. Geibel, P. Heyse, and H. Lingg (qq.v.) of the Munich School of poets (see MÜNCHNER DICHTERKREIS). He wrote lyric (*Gedichte*, 1859) and epic poetry (*Lanzelot und Ginevra*, 1860; *Hugdietrichs Brautfahrt*, 1863), and translated from French medieval literature (*Das Rolandslied*, 1861; *Marie de France*, 1862; *Aucassin et Nicolette*, 1865). He also made a modern version of the *Tristan* (q.v.) of Gottfried von Straßburg.

HERWEGH, GEORG (Stuttgart, 1817–75, Baden-Baden), was in 1835 a pupil at the Tübinger Stift (q.v.), which he left in the following year to take up journalism, writing for the periodical *Europa* (see LEWALD, A.). In 1839 he was in danger of being court-martialled for insubordination during his military service and deserted to Switzerland. In Zurich he published the revolutionary poetry of *Gedichte eines Lebendigen* (1841), which promptly established his reputation as the prophet of a new age. This collection contained the poem *Die Partei* directed against Freiligrath (q.v.) (see PARTEI, PARTEI, WER SOLLTE SIE NICHT NEHMEN). A second volume of *Gedichte eines Lebendigen* followed in 1843, as well as the poems *Einundzwanzig Bogen aus der*

Schweiz. On a tour of Germany in 1842 he was lionized and received in audience by King Friedrich Wilhelm IV (q.v.) of Prussia, whom he afterwards offended by a tactless letter. Expelled from Prussia, he found himself unwelcome in Switzerland, and moved in 1844 to Paris. In 1848 he led an invasion of Baden by a revolutionary column, an enterprise which was quickly disposed of by regular troops (see REVOLUTIONEN 1848–9). The opinion that Herwegh's personal conduct in this episode left much to be desired was not confined to his detractors; and in consequence his quickly won reputation collapsed. After his flight from Baden, Herwegh lived in Switzerland until 1866, publishing little, though in 1863 he composed for Lassalle's *Allgemeiner Deutscher Arbeiterverein* the song 'Mann der Arbeit, aufgewacht' (see LASSALLE, F.). He spent his last years in a suburb of Baden-Baden. The poems of his later years were published posthumously (*Neue Gedichte*, 1877). As a young man he translated the works of Lamartine (12 vols., 1839–40); he contributed seven plays, including *Coriolanus*, *King Lear*, and *Troilus and Cressida* to the Shakespeare translation (1866–72) edited by F. von Bodenstedt (q.v.).

Herwegh possessed a gift for vigorous denunciatory poetry, simple in form and unsubtle in thought. His opinions were straightforwardly socialistic and anti-clerical. *Werke* (3 vols.), ed. H. Tardel, appeared in 1909.

Herzemaere, Das, a short verse romance (often termed a 'verse Novelle') by KONRAD von Würzburg (q.v.). It is believed to be an early work. The story is that of a married woman and a knight, who are deeply in love. In order to allay her husband's suspicions, the woman sends her lover on a crusade; but he dies of heartbreak after giving instructions that his heart is to be sent to her. The husband intercepts it, has it cooked and set before his wife; when she has consumed it, he reveals the truth. The horrified wife declares that after this meal she will eat no further food, and, like her lover, she dies of a broken heart. Konrad adds a moral conclusion, proclaiming his story as an example of true love.

Herzensergießungen eines kunstliebenden Klosterbruders, a collection of essays on art and music written by W. H. Wackenroder and L. Tieck (qq.v.) which appeared in 1797; the title was suggested by the publisher Reichardt. Only four of the essays are wholly or partly by Tieck (*Sehnsucht nach Italien, Ein Brief des Malers Antonio, Brief eines jungen deutschen Malers*, and *Die Bildnisse der Maler*). Wackenroder's 14 contributions are more important, as well as more numerous. They include two essays on Raphael, one on Francia, one on Leonardo, one on Dürer, one on Piero di Cosimo, and one on Michelangelo, as well as a musical essay, *Das Leben des Tonkünstlers Joseph Berglinger*. Wackenroder, convinced that intellectual analysis destroys appreciation, is deliberately unsystematic in his criticism. His attitude is one of reverence before the sacred mystery of artistic creation. The essay on Joseph Berglinger shows some parallels with his own life.

HERZL, THEODOR (Budapest, 1860–1904, Edlach, Lower Austria), was the founder of modern Zionism. A prominent journalist with the *Neue Freie Presse* in Vienna, Herzl came to Jewish political consciousness during the Dreyfus affair in France. He formulated the idea of a Jewish state in *Der Judenstaat* (1896), and worked tirelessly for its realization in Palestine. His ideals are embodied in the political novel *Altneuland* (1902).

HERZLIEB, WILHELMINE, also MINNA, MINCHEN (Züllichau, 1789–1865, Görlitz), adopted daughter of the Jena publisher Friedrich Frommann. In 1807 Goethe was for a time passionately attracted to her. Features of Minchen Herzlieb's personality have been incorporated in the figure of Ottilie in *Die Wahlverwandtschaften* (q.v.). In 1821 she made an unhappy marriage with Professor Karl Walch of Jena.

HERZMANOVSKY-ORLANDO, FRITZ, RITTER VON (Vienna, 1877–1954, Merano), author whose collected works have been edited posthumously by F. Torberg (q.v.) as *Gesammelte Werke*, 1957–63. They consist of 4 vols., of which only the novel contained in the first, *Der Gaulschreck im Rosennetz*, was published during the author's lifetime (1928). The other three are *Maskenspiel der Genien, Lustspiele und Ballette*, and *Cavaliere Huscher und andere Erzählungen*. Steeped in the Austrian tradition, he has become recognized as its outstanding satirist. As a graphic artist, he was a friend of Alfred Kubin (1877–1959); their correspondence was published in 1977 and as *Der Briefwechsel mit Alfred Kubin 1903 bis 1952*, ed. M. Klein (1983) as vol. 7 of *Sämtliche Werke* (10 vols.), ed. W. Methlagl, W. Schmidt-Dengler *et al.*, 1983 ff.

HERZOG, RUDOLF (Barmen, 1869–1943, Rheinbreitach nr. Unkel), worked first in the dyeing industry, then studied at Berlin University, and became a journalist at Darmstadt and later in Berlin. His considerable financial success as a historical novelist with nationalistic leanings enabled him in 1908 to settle in the castle at Rheinbreitach as a prosperous author, whose works were eagerly awaited. His principal novels were *Der Graf von Gleichen* (1901), *Die vom Niederrhein* (1903), *Das Lebenslied* (1904), *Die Wiskottens* (1905), *Der Abenteurer* (1907), *Hanseaten* (1909, one of his most successful), *Die Burgkinder* (1911), *Das große Heimweh* (1914), *Die Stoltenkamps und ihre Frauen* (1917), *Wieland der Schmied* (1924), *Das Fähnlein der Versprengten* (1926), *Wilde Jugend* (1929), and *Über das Meer Verwehte* (1934). These, for the most part, entertaining books make no serious literary pretensions. *Gesammelte Werke* (18 vols.) appeared 1921–5.

Herzog Ernst, a medieval epic poem, the subject of which appears later in various other forms. It was composed towards the end of the 12th c. by a cleric living possibly in Bamberg. It tells of a quarrel between an emperor and his stepson, which ends in reconciliation. The poem falls into two parts. In the first, Duke Ernst's mother marries the Emperor Otto, and a harmonious relationship ensues, which is disturbed by an intriguer (Pfalzgraf Heinrich) who poisons the Emperor's mind against his stepson. The Emperor turns against Ernst, who, with his faithful follower Wetzel, avenges himself by assassinating Heinrich. Ernst is outlawed, resists for a time, until, finally overcome, he sets out with his followers on a crusade. Ernst's journey, which occupies the second part, takes him to legendary regions and adventures in the Orient. He encounters a people with cranes' bills, and rescues an Indian princess, whom they have made captive. His ships are wrecked on a magnetic cliff, though he and his companions escape by an ingenious ruse. He secures a magic stone (Waise, q.v.), which later becomes an ornament of the imperial crown. He serves the king of the Arimaspi, and defeats various strange peoples. Eventually he returns and is reconciled with his stepfather, the Emperor.

The first part of *Herzog Ernst* fuses two historical occurrences, the quarrel of Duke Liudolf with his father Otto I (q.v.) in 953–4, and the revolts of Duke Ernst I of Swabia against his stepfather, King Konrad II (q.v.) in 1026 and 1027, which ended with Ernst's death in 1030. The journeys of the second part have late classical and oriental origins, reflecting new horizons opened by the crusades. Behind all the interest of story-telling is a political element, favourable to Bavaria, but firmly rooted in a sense of unity of the Empire (see DEUTSCHES REICH, ALTES) and harmony between its princes.

The great popularity of the story of *Herzog Ernst* is attested by its recurrence over centuries in new adaptations. The original exists only in two fragments, but a complete MS. written in 1441 and preserved at Nürnberg, though an adaptation, is not far removed from the original. There are two other versions of less importance, one of which may be by ULRICH von Etzenbach (q.v.). *Herzog Ernst* also exists in two Latin translations of the 13th c., one in verse by Odo, a priest of Magdeburg, and one in prose. This prose version was in turn translated into German prose, and became in this form a popular Volksbuch. Interest in the story was revived again in the Romantic period, and it is the subject of a verse tragedy, *Ernst, Herzog von Schwaben*, by Ludwig Uhland (q.v., 1817). *Das Volksbuch vom Herzog Ernst oder Der Held und sein Gefolge* is a play by P. Hacks (q.v., 1953). Based on the Volksbuch of 1493, it serves his aim of debunking misguided nationalistic and fascist notions of heroism during two World Wars.

Herzog Friedrich von der Normandie, a lost Middle High German epic poem, of which a Swedish translation, made in 1308, is preserved. The original was written *c.* 1250 by an unknown poet, who was probably commissioned by Duke Otto of Brunswick. The poem tells how the hero woos and elopes with a princess, whose father opposes the match. It ends in reconciliation. A magic element is provided by a ring which makes its wearer invisible. This ring, which serves in the abduction of the princess, comes to Herzog Friedrich as a gift from a dwarf king whom he has succoured.

Herzog Theodor von Gothland, a five-act tragedy in blank verse by C. D. Grabbe (q.v.), written between 1818 and 1822, and published in 1827 in *Dramatische Dichtungen*. This enormous first play (5,577 lines) is clearly influenced by *Othello*. The action presents a violent protest against a supposedly evil world. Berdoa, the Negro commander of the Finns, seeks in the destruction of the Swedes an outlet for his urge to avenge himself against the 'Europeans', at whose hands in his native Africa he had suffered humiliation and ill-treatment. He singles out Theodor, Herzog von Gothland, as the principal object of his revenge, because of Theodor's innate goodness. By brutal deceit he succeeds in convincing Gothland that his brother Friedrich, chancellor to Olaf, King of Sweden, was responsible for the sudden death of their brother Manfred. Gothland kills Friedrich, and, on finding out the deceit, rejects remorse and turns against God and the world. Outlawed in his country and fleeing his pursuing father, the aged duke, Gothland becomes king of the Finns and Swedes, only to be betrayed again by Berdoa, who turns Gothland's son Gustav against his father before murdering the youth. Gothland avenges himself on Berdoa before being killed by Graf Arboga. King Olaf is restored to power, and the aged and despairing duke witnesses the extinction of the house of Gothland.

Grabbe gives the action many unexpected turns in order to portray the treachery and brutality which drive the characters, who are both participants and victims, to the verge of insanity. Heine (q.v.) was among the first to recognize Grabbe's originality in the play, which was not accepted for stage performance. Tieck (q.v.) was disquieted by the self-destructive streak which he detected in the dramatic speech. The play figured in the repertory of the German theatre in the early 1970s.

Herz von Douglas, Das, a well-known ballad by M. von Strachwitz (q.v.).

HESEKIEL, GEORG LUDWIG (Halle, 1819–74, Berlin), journalist and poet, became editor of the *Kreuzzeitung* (q.v.). A robust Prussian patriot, he wrote large numbers of patriotic poems, published as *Gedichte eines Royalisten* (1841), *Neue Preußenlieder aus dem Dänenkriege* (1864), *Neue Gedichte* (1866), and *Gegen die Franzosen* (1870), thus saluting each war with a new paean. He also wrote historical novels which appeared in serial form: *Das liebe Dorel* (1851), *Von Jena nach Königsberg* (1860), and *Unter dem Eisenzahn* (1864). He published a life of Bismarck (q.v.) only a few years after the latter took office as chancellor: *Das Buch vom Grafen Bismarck* (1868). Th. Fontane (q.v.) knew him well and wrote of him in *Von Zwanzig bis Dreißig*.

HESELLOHER, HANS, (*fl.* 1450–83), was a village magistrate and wrote seven poems satirizing the peasantry after the manner of NEIDHART von Reuental (q.v.). Some of them are incomplete.

Hesperus oder 45 Hundsposttage, eine Biographie, title of a novel by JEAN Paul (q.v.), written 1792–4, and published in 1795. The eccentric sub-title refers to the chapters, which are designated *Hundsposttage*, and are supposed to have been brought to the author's friend by a Pomeranian dog. Written in Jean Paul's characteristic whimsical style, the book has a complex and absurd plot. Lord Horion has taken the ruling prince's sons, and seeks to turn them into good rulers. The chief of these is Flamin, a youth of promise, who is brought up as the supposed son of pastor Eymann. The pastor's son Viktor is adopted by Lord Horion. Viktor loves Flamin's half-sister Klotilde, but when he discovers his true identity he believes that class barriers will prevent their marriage. Complications arise: Flamin is wrongly accused of killing

Klotilde's father in a duel, and intrigues are fomented by the courtier Matz. In the end the true identities are revealed, the difficulties solved, and Viktor and Klotilde united.

The action is really secondary. Jean Paul concerns himself successfully with the expression of emotion, and writes passages of landscape description of great sensitivity and beauty.

HESS, RUDOLF (Alexandria, 1894–), a military pilot in the 1914–18 War, joined the NSDAP (q.v.) in 1920, and participated in 1923 in Hitler's attempted *coup d'état* in Munich. Like Hitler he served for this a term of imprisonment at Landsberg. In 1933 Hitler appointed Hess as his deputy in the NSDAP. A member of Hitler's cabinet, Hess flew to Scotland on 10 May 1941 apparently with the intention of initiating peace negotiations. He was interned, tried in 1945–6 at Nürnberg, and condemned to life imprisonment at Spandau.

HESSE, HERMANN (Calw/Swabia, 1877–1962, Montagnola/Tessin), spent his early childhood (1881–6) in Basel and up to the age of 14 possessed Swiss citizenship, which he resumed in 1923. His father, who had been a missionary in India, moved to Calw to collaborate with his father-in-law, the Orientalist Dr. Gundert, in the Calwer Verlagsverein. Hesse was influenced throughout his life by the Pietist tradition of his parents and the family's scholarly oriental background. As his parents wished him to study theology, he acquired Württemberg citizenship to obtain admission to Maulbronn (q.v.) seminary, which he left in 1892 after a stay of only six months. After a brief spell in a bookshop in Esslingen he became an apprentice mechanic in Calw, but in 1895 entered a publishing firm in Tübingen. From 1899 to 1904 he worked in the book trade in Basel before deciding to devote himself entirely to writing.

In 1899 Hesse published his first volume of poetry under the title *Romantische Lieder*, and it is the Romantic style which characterizes his early work. In the same year appeared *Eine Stunde hinter Mitternacht*, nine vignettes, which found favour with Rilke (q.v.). Two years later appeared *Hinterlassene Schriften und Gedichte von Hermann Lauscher. Herausgegeben von Hermann Hesse* (revised 1907 as *Hermann Lauscher* and containing *Tagebuch 1900*). In 1904 Hesse published two monographs, *Boccaccio* and *Franz von Assisi*. *Peter Camenzind* (q.v., also 1904), a novel whose artist hero seeks to become a writer, brought him much-needed success. His novel *Unterm Rad* (q.v., 1906) reflects the recurring theme of the inner conflicts of boyhood. The novel *Gertrud* (Künstlerroman, q.v., 1910) treats

the tribulations of a physically handicapped composer. *Knulp. Drei Geschichten aus dem Leben Knulps* (q.v., 1915), begun in these years, marks the culmination of Hesse's Romantic phase.

During the early years of his marriage to Maria Bernoulli, Hesse enjoyed family life in the secluded village of Gaienhofen on Lake Constance as well as the friendship of many artists and writers in the district, notably L. Finckh (q.v.). From 1901 he travelled extensively in Italy, and his interest in art was stimulated by the art historian H. Wölfflin (q.v.). The culmination of his travels and lecture tours was his journey to India in 1911 with the painter H. Sturzenegger. *Aus Indien. Aufzeichnungen einer indischen Reise* (1913) was a preliminary to the more substantial post-war writings. In 1912 Hesse moved to Berne, where he remained until 1919, when he decided to leave his family. *Roßhalde* (1914 in book form) reflects the disintegration of his marriage to Maria, who suffered from a progressive mental illness.

During the 1914–18 War Hesse worked for the Deutsche Gefangenenfürsorge Bern. He devoted himself to editorial and library work, establishing the Bücherei für deutsche Kriegsgefangene. From the outset of the war Hesse adopted firm pacifist views which he expressed in *O Freunde, nicht diese Töne!*, an article published in the *Neue Züricher Zeitung*. After the war this appeal for the preservation of humanity and its cultural heritage was followed by a plea for spiritual regeneration in *Zarathustras Wiederkehr. Ein Wort an die deutsche Jugend* (1919).

Personal problems and the stress of war intensified Hesse's study of psycho-analysis and he himself underwent prolonged treatment by a disciple of C. G. Jung (q.v.). The collection of essays *Blick ins Chaos* (1920) reflects this phase, during which he used the pseudonym Emil Sinclair (after Isaak von Sinclair, a friend of Hölderlin, q.v.), notably for his novel *Demian. Die Geschichte von Emil Sinclairs Jugend* (q.v. 1919). His psycho-analytical preoccupations are evident in a number of his fairy stories contained in *Märchen* (1919, e.g. *Eine Traumfolge*, *Iris*), and in *Piktors Verwandlungen* (1925, published with Hesse's own illustrations in 1954), as well as in the strongly autobiographical work *Klingsors letzter Sommer* (q.v., 1920), which was published together with *Kinderseele* and *Klein und Wagner* (q.v.). In *Siddhartha. Eine indische Dichtung* (q.v., 1922) Hesse gives poetic expression to Indian philosophy. His novel *Der Steppenwolf* (q.v., 1927) met with considerable public response, which was renewed in the early 1970s. The novel *Narziß und Goldmund* (q.v., 1930) is his last principal work of the 1920s. Because of his preoccupation with man's dual nature Hesse's fiction tends to concentrate on two contrasting characters who are irresistibly drawn together, and yet feel impelled to express their individuality to the full.

Hesse's stories reappeared in the 1930s with new titles (notably *Weg nach Innen*, q.v., 1931), and during this period he wrote only a number of tales. The story *Morgenlandfahrt* (1932) appeared at a time when he was preparing his Utopian novel of the year 2400, which represents the quintessence of his vision and outlook, *Das Glasperlenspiel. Versuch einer Lebensbeschreibung des Magister Ludi Josef Knecht samt Knechts hinterlassenen Schriften. Herausgegeben von Hermann Hesse* (q.v.), published in 1943, the year in which Hesse's name was put on the black list of authors in Germany.

From 1919 Hesse lived in Montagnola. His third, and lasting, marriage to Ninon Dolbin took place in 1931. His publications following the 1939–45 War include reminders that his individualism was neither excessive nor aloof. They express a humanitarian attitude which repeats his identification with Goethe's Weltbürgertum, his respect for Christian piety, and his faith in the spirituality of all mankind, for which he coined the term Weltglaube. His pacifism was based on these convictions and is reflected in the periodical *Vivos Voco*, which he edited from 1919 to 1921, in *Dank an Goethe* (1946, written in 1932 to oblige his friend Romain Rolland), in *Der Europäer* (1946, five essays written between 1918 and 1945), and in *Krieg und Frieden. Betrachtungen zu Krieg und Politik seit dem Jahr 1914* (1946), in which his *Brief nach Deutschland* forms the epilogue. In 1961 Hesse resumed this open letter form in his *Brief an Peter Weiss* (see WEISS, P.). Hesse received many honours, both Swiss and German, culminating in the award of the Nobel Prize in 1946. In the same year he accepted the Goethe-Preis, because Frankfurt, with which it is associated, represented to him a humane tradition with which he could identify himself despite the bitter memory of political denunciations to which he had been subjected in Germany since 1914. In 1950 Hesse, who had visited Raabe (q.v.) in 1909 in Brunswick, was awarded the Wilhelm-Raabe-Preis. In 1955 he was admitted to the Friedensklasse des Ordens POUR le mérite (q.v.). Fiction written from 1944 to 1950 appeared as *Späte Prosa* (1951), and was followed by *Frühe Prosa* (written between 1899 and 1907) in 1960, the year in which he published *Letzte Gedichte*. An extensive selection of *Briefe* appeared in 1959. *Gesammelte Dichtungen* (6 vols., 1952) appeared in an enlarged edition (7 vols.) as *Gesammelte Schriften* in 1957. A select *Werkausgabe in zwölf Bänden*, edited by V. Michels, was published in 1970. Ten poems by Hesse were set to music by Othmar Schoeck (1886–1957).

HESSE, MAX RENÉ (Wittlich, 1885–1952, Buenos Aires), emigrated to Argentina as a doctor, and was also a big-game hunter (1910–27). He is the author of two novels concerned with a doctor in South America, *Morath schlägt sich durch* (1933) and *Morath verwirklicht einen Traum* (1935). The former deals with a marriage under strain, whilst in the latter the doctor finds his mission as a public health officer in undeveloped territory. Hesse also wrote a large-scale family trilogy, *Dietrich Kattenburg*, in a pre-1914 setting. It consists of the novels *Dietrich und der Herr der Welt* (1937), *Jugend ohne Stern* (1943), and *Überreife Zeit* (1950). Most of his novels were written while living in Germany, 1927–43. In 1943 he went to Spain, and in 1944 returned to Argentina.

HESSE, MEISTER, a scribe and lawyer of Strasburg in the 13th c., was regarded as an authority on writing. He is mentioned in Bk. II of *Willehalm von Orlens* by RUDOLF von Ems (q.v.), who consulted him. Meister Hesse appears to have edited Gottfried von Straßburg's *Tristan* (q.v.)

Hessen (Hesse), a constituent state (Land) of the Federal Republic of Germany. Frankfurt is the largest city, but the capital is Wiesbaden, and the administrative centres of its two regions (Regierungsbezirke) are Kassel and Darmstadt. Hesse emerged from the Middle Ages as an agglomeration of spiritual and temporal principalities. In the 16th c. it was nominally a landgravate, but was divided into the northern Hessen-Kassel and the southern Hessen-Darmstadt. Between these was the important free imperial city (Freie Reichsstadt) of Frankfurt. At the beginning of the 19th c. (1803) the northern (Hessen-Kassel) ruler was elevated to the rank of an Elector, a short-lived glory, for the Holy Roman Empire was dissolved in 1806. The Elector of Hesse, however, stuck to this title until his deposition and the annexation of his country by Prussia in 1866. G. Büchner (q.v.) was a sharp critic of the country's despotic rule (see also HESSISCHE LANDBOTE, DER). Nassau, Frankfurt, and the Landgravate of Hessen-Homburg (1622–1866, the home of H. von Kleist's *Prinz Friedrich von Homburg*, q.v.) were similarly annexed, and the four territories combined to make the Prussian province of Hessen-Nassau. The southern state Hessen-Darmstadt escaped the fate of Hessen-Kassel through its situation south of the Main, and so retained its identity and semi-independence as the Grand Duchy of Hesse (Großfürstentum Hessen). Its ruler abdicated in 1918, but it survived as a Land from 1919 to 1945. The fragments were reunited to make in 1949 the Land Hessen of the Federal Republic (see BUNDESREPUBLIK).

Hessische Landbote, Der, a subversive political pamphlet by G. Büchner (q.v.) which bears the motto: 'Friede den Hütten! Krieg den Palästen!' echoing a French revolutionary slogan. Büchner completed the MS. in March 1834, and it was ready for distribution among the peasants to whom it was addressed in July 1834. It was the fruit of Büchner's association with F. L. Weidig (q.v.), and was clandestinely printed. Büchner's manifesto concludes with a reminder to the 700,000 peasants that they pay the Grand Duchy 'six million' for a 'handful of people', on whose whim their life and property depended. The 'sword' of the people should assert itself against the oppressors: 'Deutschland ist jetzt ein Leichenfeld, bald wird es ein Paradies sein.' The *Landbote* failed to stimulate active resistance. Intimidated by persecution, most peasants handed their copies to the police. The conspirators were denounced by one of their own ranks, who turned out to be a paid government informer, Konrad Kuhl.

Hessisches Weichnachtsspiel, a Middle High German nativity play written in Friedberg (Hesse) about the middle of the 15th c. It begins with the Annunciation and includes the visit of the shepherds but not that of the Three Kings. It is homely in tone with much robust humour.

HESSUS, (HELIUS) EOBANUS (Halgehausen, 1488–1540, Marburg), whose real name was Koch, was one of the most prolific and admired of the Latin poets of the New Learning, and the most gifted of the Erfurt group of humanists headed by Mutianus Rufus (q.v.). He translated the Psalms and the *Iliad* (1540) into Latin verse. Notoriously given to drink, he led a restless life, of which the principal stations were Erfurt, Nürnberg (1526–33), and Marburg. In the religious quarrels of the age he sided with Luther (q.v.). His first name is often dropped, hence he is best known as Eobanus Hessus.

HETTNER, HERMANN (Nieder-Leisersdorf, 1821–82, Dresden), was a lecturer at Heidelberg, where he became acquainted with G. Keller (q.v.). He became a professor at Jena in 1851, and from 1855 was a museum director in Dresden. His main concern in literature was with the 18th c., and his monumental *Literaturgeschichte des achtzehnten Jahrhunderts* (1856–70) continued in use long after his death; but his treatise *Das moderne Drama* (q.v., 1852) is a stimulating consideration of contemporary dramatic trends.

HETZBOLT VON WEISSENSEE, see HEINRICH HETZBOLD VON WEISSENSEE.

HEUN, KARL GOTTLIEB SAMUEL, see CLAUREN, H.

HEUSS, THEODOR (Brackenheim, 1884–1963, Stuttgart), first president of the Federal Republic (see BUNDESREPUBLIK). He was a friend of F. Naumann (q.v.), and edited the periodical *Die Hilfe* (q.v.), from 1905 to 1912. In 1909 he published an anthology of contemporary Swabian poetry, *Sieben Schwaben*. In politics he was a Liberal (Fortschrittliche Volkspartei, 1910–18). In 1918 he joined the Deutsche Demokratische Partei, and was twice elected to the Reichstag (1924–8 and 1930–3). During the National Socialist regime he was banned from writing, but continued journalistic activity under the pseudonym Brackenheim. He was Minister for Education in Württemberg-Baden (U.S. Zone, see WÜRTTEMBERG) in 1945–6, joining the Freie Demokratische Partei in 1946. In 1947 he was appointed professor of modern history and politics at the Technische Hochschule, Stuttgart. He was a member of the Landtag, and its representative on the Parliamentary Council (Parlamentarischer Rat) instituted to plan the Federal Republic. In September 1949 he was elected president of the Republic. He was re-elected for a second term in 1954, retiring in 1959. As President he paid a visit to Queen Elizabeth II in 1958. In addition to political tracts, his publications include agreeable essays on topography, poetry, and the fine arts contained in *Von Ort zu Ort* (1959), *Lust der Augen* (1960), and *Vor der Bücherwand* (1961). His memoirs appeared as *Erinnerungen 1905–33* (2 vols.) in 1964.

HEUSSGEN, JOHANNES, see OEKOLAMPADIUS, JOHANNES.

Heutelia, a satirical work in prose published anonymously in 1658, and now attributed to F. Veiras (1577–1672).

Hexameter, the classical, dactylic hexameter, which was effectively introduced into German literature (with the substitution of stresses for quantities) by Klopstock with the first three cantos of *Der Messias* (q.v.) in 1748. Hexameters were also employed by E. von Kleist (q.v.) in *Der Frühling* (q.v., 1756), and with notable success by Goethe in *Hermann und Dorothea* (q.v., 1797) and *Achilleis* (q.v., 1799) and by J. H. Voß in his *Idyllen*, of which *Luise* (q.v., 1795) is the best known. Hölderlin excelled in the form, above all in *Der Archipelagus* (q.v., 1800). In the 19th c. Hebbel wrote his *Mutter und Kind* (q.v.)

and Mörike his *Märchen von sichern Mann* and *Idylle vom Bodensee* (qq.v.) in this metre; in the 20th c. G. Hauptmann adopted it in *Anna* (1921) and *Till Eulenspiegel* (1928). All these examples are epic, but Mörike also successfully used hexameters in lyric poems, e.g. *An Klara* and *Im Weinberg*.

The hexameter combined with a pentameter to form a distich (see DISTICHON) is much used in German literature from the time of Goethe and Schiller, especially for epigram and elegy. See ELEGISCHES VERSMASS.

HEY, WILHELM (Leina, 1790–1854, Ichtershausen), a Lutheran clergyman, writer of fables, which were popular in the mid-19th c. (*Fünfzig Fabeln für Kinder*, 1833; *Noch fünfzig Fabeln*, 1837).

HEYDEN, FRIEDRICH AUGUST VON (Nerfken nr. Heilsberg, East Prussia, 1789–1851, Breslau), fought as a volunteer in the Wars of Liberation (see NAPOLEONIC WARS), and then became a civil servant (Oberregierungsrat) in Breslau. He was a prolific spare-time writer of plays, novels, and poems. His most popular work was the poem *Das Wort der Frau* (1843). His collected plays (*Theater*), which include the historical play *Konradin*, were published in 1842.

HEYKING, ELISABETH, FREIFRAU VON (Karlsruhe, 1861–1925, Berlin), *née* Gräfin von Flemming, a grand-daughter of L. J. von Arnim (q.v.), was the anonymous authoress of a best seller, a novel in the form of letters (see BRIEFROMAN), *Briefe, die ihn nicht erreichten* (q.v., 1903). Her other novels, written under her own name, did not achieve the same success; they include *Der Tag Anderer* (1905), *Ille mihi* (2 vols., 1912), *Tschun* (1914), which has a setting in modern China, *Das vollkommene Glück* (1920), and *Tagebücher aus vier Weltteilen* (1926). Her husband, Baron von Heyking, was a diplomat, and at one time in China, hence her knowledge of foreign countries.

HEYM, GEORG (Hirschberg, Silesia, 1887–1912, Berlin), a young man of good family, studied law at Würzburg, Berlin, and Jena universities, and became a civil servant. He was accidentally drowned while skating on the Havel in January 1912. Beset by a sense of *malaise* at the political and social situation, and possessed of considerable poetic gifts, Heym wrote a number of visionary and apocalyptic poems which are among the best works produced by the early Expressionists. The poems *Der ewige Tag* were published in 1911, *Umbra vitae* in 1912. *Dichtungen und Schriften. Gesamtausgabe in 4 Bänden*

(vol. 1 *Lyrik*, vol. 2 *Prosa und Dramen*, vol. 3 *Tagebücher, Träume und Briefe*, vol. 4 *Dokumente zu seinem Leben*), ed. K. L. Schneider, appeared 1960–8.

HEYM, STEFAN, pseudonym of Hellmuth Fliegel (Chemnitz, 1913–   ), who fled abroad in 1937, was in the USA, joined the American army, and later settled in East Berlin. A signatory of the letter of protest against the expulsion of W. Biermann (q.v.) in 1976, he was deprived of his SED membership in 1979. He is a social and historical novelist with orthodox Eastern political views. Most of his works first appeared in English. *Hostages* (1942) came out in Germany as *Der Fall Glasenapp* (1958, DDR; 1965, BRD); *The Crusaders* (1948) was published as *Kreuzfahrer von heute* (DDR, 1950) and as *Der bittere Lorbeer* (BRD, 1950) and *The Eyes of Reason* (1951) as *Die Augen der Vernunft* (DDR, 1955). *Goldsborough* appeared in both languages (1953, DDR; 1954, USA), as did *The Lenz Papers* (UK, 1964) though under separate German titles: *Die Papiere des Andreas Lenz* (2 vols., DDR, 1963, revised edn. 1972) and *Lenz oder Die Freiheit. Ein Roman um Deutschland* (BRD, 1965). Heym found it increasingly difficult to adjust to official policies and since 1965 his works have appeared first, and in some cases only, in West Germany. During the early 1980s his novels were republished as Werkausgaben; *Der bittere Lorbeer, Goldsborough, Lenz oder die Freiheit, Lassalle. Ein biographischer Roman* (1969 in English, as *Uncertain Friend*, and in German, revised, 1974) appeared in 1981, *Der Fall Glasenapp* and *Die Augen der Vernunft* in 1982, and *5 Tage im Juni* (1974; *Five Days in June*, 1977) in 1983.

Hostages, turned into *Der Fall Glasenapp* and set at the time of the German occupation of Czechoslovakia, presents as a major theme the torture and killing of hostages who are held responsible for the suicide of a German officer. *Goldsborough*, on a miners' strike in the USA (Pennsylvania) in the late 1940s, was published in 1953, the year Heym moved to the DDR. *Lassalle* (1974), a psychological study depicting the last year in the life of F. Lassalle (q.v.) and the struggle for power in the context of the formation of the Allgemeiner Deutscher Arbeiterverein (see SPD), includes in its appendices excerpts from correspondence between Marx and Engels (dated 1861–9) on Lassalle. *Lenz oder Die Freiheit*, the first novel he wrote in the DDR, centres on the abortive 1849 revolt in Baden; its central figures, Andreas Lenz and the peasant Christoffel, are fictitious. *Der König David Bericht* (1972, German title of the *King-David-Report*, 1973), like *Ahasver* (1981), makes skilful satirical use of biblical background and language and allusively interspersed contemporary terms of reference. The author and historian Ethan, in writing the chronicle of King David at King Solomon's behest, defies the king's instructions by not concealing corruption. The king displays his 'wisdom' by 'silencing' the author in not publishing his work, instead of pronouncing the expected verdict of death. The theme of the writer and his awareness of censorship recurs in *Collin* (1979; English version, 1980), in which heart failure silences the author. Heym's major novel *5 Tage im Juni* (original title *A Day marked X*; *Der Tag X*), begun after the workers' rising on 17 June 1953, several times revised, and rejected by publishers, appeared some twenty years later in the BRD (1974), only to meet with strong criticism concerning the involvement of West German agents. The central figure, Witte, Chairman of his factory's union organization, experiences a conflict between his loyalty to the Party and to his union, but during the revolt makes every effort to restrain the workers. Compelled to relinquish his post, he agrees to spend a year at the Parteischule, the outcome of which is deliberately left open. Through the introduction of diverse figures whose past unfolds in the novel's analytical structure, Heym constructs a panorama of contemporary conditions and studies of individual behaviour. The novel *Schwarzenberg* appeared in 1984.

Collections of stories include *Shadows and Lights* (1963), German version *Schatten und Licht. Geschichten aus einem geteilten Land* (1960) and *Die richtige Einstellung und andere Erzählungen* (1977), which contains the German version of *The Queen against Defoe* (1974), entitled *Die Schmähschrift oder Königin gegen Defoe* (1970). The volumes *Auskunft. Neue Prosa aus der DDR* and *Auskunft 2. Neueste Prosa aus der DDR* appeared in 1974 and 1978, and *Wege und Umwege. Streitbare Schriften aus fünf Jahrzehnten*, ed. P. Mallwitz, in 1980. It should be noted that Heym continued to write the first versions of his fiction in English after leaving the USA, allegedly because of the greater succinctness of that language.

HEYMEL, ALFRED WALTER VON (Dresden, 1878–1914, Berlin), a man of means, and a connoisseur of art, was a co-founder with O. J. Bierbaum and R. A. Schröder (qq.v.) of the monthly *Die Insel* (q.v.). He published several volumes of poetry, among them *In der Frühe* (1898) and *Zeiten* (1907). He was ennobled in 1907.

HEYNE, CHRISTIAN LEBERECHT (Leuben nr. Meißen, 1751–1821, Hirschberg), pseudonym

Anton Wall, is the author of a number of plays and stories which enjoyed a passing success in the late 18th c. The plays include *Der Arrestant* (1780), *Karoline* (1780), and *Der Herr im Hause* (1783).

HEYNECCIUS, MARTIN, see HAYNECCIUS, MARTIN.

HEYNICKE, KURT (Liegnitz, 1891–), a bank clerk, turned to theatre direction in 1924 at Düsseldorf. He was one of the early writers of Expressionist poetry (see EXPRESSIONISMUS), contributing to the periodical *Der Sturm* (q.v.) and publishing the collections *Rings fallen die Sterne* (1917), *Gottes Geigen* (1918), *Das namenlose Angesicht* (1920, for which he was awarded the Kleist Prize), and *Das Leben sagt Ja* (1930). He is the author of plays (e.g. *Kampf um Preußen*, 1926), and of popular novels (Unterhaltungsromane), including *Fortuna zieht in die Welt* (1929), *Herz, wo liegst du im Quartier?* (1938), *Rosen blühen auch im Herbst* (1942), and *Der Hellseher* (1951). He is a prominent writer of radio scripts.

HEYSE, PAUL (Berlin, 1830–1914, Berlin), son of a notable philologist, was educated at the Friedrich Wilhelm-Gymnasium, Berlin, and at Berlin and Bonn universities. In 1851 he made an extended study tour of Italy, returned to Berlin, and in 1854 was invited to Munich and granted a pension by King Maximilian II (q.v.). With E. Geibel (q.v.), his older personal friend, he was the leading figure in the Munich group of poets (see MÜNCHNER DICHTERKREIS). When Geibel's pension was revoked in 1868, Heyse renounced his, though he continued to live in Munich in the summer, migrating to Gardone on Lake Garda each winter.

Heyse published his first story, *Jungbrunnen*, in 1849, and devoted his life to the writing of verse tragedies, Novellen, and poems. The tragedies begin with *Francesca da Rimini* (1850) and continue with *Meleager* (1854), *Die Sabinerinnen* (1859), *Elfride* (1877), *Graf Königsmarck* (1877), and *Alkibiades* (1880). His plays include the Schauspiele *Ludwig der Bayer* (1862), *Elisabeth Charlotte* (1864), *Hadrian* (1865), *Maria Maroni* (1865), *Hans Lange* (1866), *Colberg* (1868), *Die Göttin der Vernunft* (1870), and *Die Weiber von Schorndorf* (1881). In 1884 he was awarded the Schiller Prize.

Heyse wrote well over 100 Novellen, which are little known, with the possible exception of *L'Arrabbiata* (1855 with others, 1858 separately). His principal collections are *Novellen* (1855) *Neue Novellen* (1858), *Neue Novellen* (1862), *Meraner Novellen* (1864), *Fünf neue Novellen* (1866), *Moralische Novellen* (1869), *Das Ding an sich*

(1879), *Troubadour-Novellen* (1882), *Unvergeßbare Worte* (1883), *Himmlische und irdische Liebe* (1886), and *Novellen vom Gardasee* (1902). With H. Kurz (q.v.), and later with other help, he published a huge collection of German Novellen (*Deutscher Novellenschatz*, 24 vols., 1870–6). He is the author of seven novels: *Kinder der Welt* (3 vols., 1873), *Der Roman der Stiftsdame* (1887), *Merlin* (3 vols., 1892), *Über allen Gipfeln* (1895), *Crone Stäudlin* (1905), *Gegen den Strom* (1907), and *Die Geburt der Venus* (1909).

Only in the field of poetry has Heyse's work survived, chiefly through his great gift for translation and the remarkable compositions of Hugo Wolf (q.v.): *Spanisches Liederbuch* (with Geibel, 1852) and *Italienisches Liederbuch* (1860). A fluent and easy writer, he devoted himself to an ideal of beauty detached from everyday reality, and his 19th-c. public endorsed his views. The coming of Naturalism (see NATURALISMUS) and the changing standards of the early 20th c. exposed Heyse's writings as unreal and affected. In 1910 he was awarded the Nobel Prize for Literature and ennobled. *Gesammelte Werke* (38 vols.) appeared 1871–1914 and *Gesammelte Werke* (15 vols.), ed. E. Petzet, in 1924.

HGB, abbreviation for *Handelsgesetzbuch*, the German code of commercial and company law.

Hiddensee, a small island close to Rügen in the Baltic which is famous for the discovery of important Viking remains from the 10th c. The apparently misleading name has nothing to do with 'See' but is a corruption of the Scandinavian word for 'island' (Hiddensö or ø). G. Hauptmann (q.v.) was buried on the island in 1946.

'Hier sind wir versammelt zu löblichem Tun', first line of Goethe's student drinking song *Ergo bibamus!*, written in 1810. The words 'ergo bibamus!' occur as a refrain at the end of each of the four stanzas. The poem first appeared in *Gesänge der Liedertafel* (1811). The tune was composed by Traugott (Maximilian) Eberwein (1775–1831) in 1813.

Hildebrandslied, a fragment of Old High German alliterative heroic poetry. It narrates the encounter between Hildebrand and his son Hadubrand. Hildebrand, a voluntary exile with his king, Dietrich, returns in his lord's train. As the chosen champion, he faces a young man whom he recognizes as his own son. His conciliatory overtures are rejected as treacherous cunning, and battle becomes inevitable. The end of the poem, omitted in the MS. apparently for lack of space, was certainly tragic, Hadubrand falling to Hildebrand's sword. The conflict is largely portrayed through the dialogue of the two champions. The poem represents the ethical outlook of a warrior caste, for which military honour and unconditional loyalty are the highest values. Fate dominates the soldier's life, and he is its active executant.

The *Hildebrandslied* is the sole surviving fragment of a German heroic lay sung by a minstrel before an exalted audience. It possesses an economical and effective structure and splendid and powerful verse. Once regarded as an expression of Germanic paganism, it is now seen by some to belong to an era in which the warrior, while retaining his ancient virtues, had accepted Christianity.

The story of Hildebrand is regarded as being of Langobardic origin (see LANGOBARDS), and is supposed to have originated in the 7th c., combining a widely disseminated legend (cf. Matthew Arnold's *Sohrab and Rustum*) with a known personality of the recent past (Dietrich von Bern, see DIETRICHSAGE). The *Hildebrandslied* is basically Bavarian but has been adapted for a northern audience whose language was Low German. Even so the text is corrupt, and the attribution of some of the lines uncertain. The MS. represents a copy made c. 810 by two monks, probably in Fulda, of this adaptation. The language is philologically problematical, and has provoked various divergent hypotheses. The poem was preserved by a fortunate chance, having been copied on to two blank pages in a theological codex. Formerly in the possession of the Landesbibliothek, Kassel, the MS. disappeared in 1946, though the second of the two sheets was recovered in Los Angeles in 1950, and the first in Philadelphia in 1972. Both have been returned to Kassel, whence they had been removed for safety in 1939.

Hildebrandston, a stanza form which is a variant of the Nibelungenstrophe (q.v.). Its distinction is that the rhyme occurs not only at the end of each long line (Langzeile), but also before the caesura (Zäsurreim); often the half-lines are written separately, so turning the four-line stanza into one of eight lines, as in the following example from *Der hürnen Seyfrid* (q.v.),

> Es saß im Niderlande
> Ein Künig so wol bekandt,
> Mit grosser macht und gewalte
> Sigmund was er genant,
> Der het bey seyner frawen
> Ein sun, der hieß Seyfrid,
> Des wesen werdt jr hören
> Alhie in disem Lied.

The stanza takes its name from its use in the *Jüngeres Hildebrandslied* (q.v.).

HILDEBRANDT, Johann Lukas von (Genoa, 1668–1745, Vienna), distinguished baroque architect mainly active in Austria. His principal work in Vienna is the Belvedere, built for Prinz Eugen (q.v.). He was also responsible for Göttweig Abbey in Lower Austria, for Pommersfelden Palace near Bamberg, and for parts of the Archbishop's Palace (Residenz) in Würzburg.

HILDEGARD VON BINGEN, Heilige (Bermersheim, 1098–1179, Bingen), a mystic and a copious religious writer, who was a nun and, from 1136, prioress of a convent (later elevated to an abbey) at Rupertsberg near Bingen. Hildegard recorded her visions in *Scivias*, sc. *Sci vias*, 'Know the ways (of the Lord)', and wrote also *Liber vitae meritorum, Liber divinorum operum*, and some Sequenzen (see Sequenz). Skilled in medicine, she is the author of a medical work, *Hildegardis causae et curae*. All her writings are in Latin. Hildegard enjoyed a considerable reputation in her lifetime, and was canonized after her death.

HILDESHEIMER, Wolfgang (Hamburg, 1916–), was educated in England, lived in Palestine from 1933 to 1936, and from 1936 to 1939 studied art in London; he was for many years uncertain whether to give priority to his painting or his writing. He decided on the latter in 1950, though he continued to paint. In the 1939–45 War he was a British intelligence officer in Palestine. He returned to London in 1946, and was appointed an interpreter at the Nürnberg War Crimes Trials (1946–9). He settled in Switzerland and belonged to the Gruppe 47 (q.v.).

Hildesheimer's prose began with *Lieblose Legenden* (1952, revised and extended 1962), followed by the novel *Paradies der falschen Vögel* (1953) and the story *Ich trage eine Eule nach Athen* (1956). In his next works his satirical and even grotesque mode of writing shows a predilection for the absurd and also a more radical reaction against conventional forms of fiction. The pervasive reflective style of *Tynset* (1965), a monologistic prose text, proceeds from verbal play on sound and objects that convey the narrator's confinement within an alienated reality. The remote place of Tynset containing the narrator is associated with the sound of a gong (Tyn-) and its cessation (-set) in time and space and functions as a leitmotif (W. Rath). It has a sequel in *Masante* (1973), a text in aphoristic form and set in Meona, on the fringe of the desert, to which the narrator has moved from his Masante home. The surrounding void is an essential setting for an exhaustive range of reflections. *Zeiten in Cornwall* appeared in 1971

and, after prolonged preparation, *Mozart* in 1977. Not intending it as a biography, Hildesheimer proceeds from the thesis that documents on Mozart's life and state of mind obscure rather than reveal the truth. The work is based on his study of Mozart's work, correspondence, and biographers. The conventional view of Mozart is demolished and replaced by a decidedly more ordinary man of his time. In deliberate contrast, *Marbot. Eine Biographie* (1981), is based on fictitious documentation and centres on the young Sir Andrew Marbot, to whom art offers a refuge from life. His incestuous relationship with his mother and his increasing awareness that he lacks creative genius account for his suicide in 1830. The definability of art, represented by painting, and other wide-ranging themes are presented in different styles including parody. *Die Exerzitien mit Pabst Johannes. Vergebliche Aufzeichnungen* (1979) consists of fragments, including one of an abandoned novel on Hamlet. Hildesheimer's preoccupation with James Joyce is reflected in *Interpretationen. J. Joyce. G. Büchner* (1969). Of greater relevance for his plays and radio plays is Beckett; *Wer war Mozart? Becketts 'Spiel', Über das absurde Theater* appeared in 1966. Early plays include *Der Drachenthron* (1955), published in revised form as *Die Eroberung der Prinzessin Turandot* (1961), *Spiele, in denen es dunkel wird* (1958), a volume which also contains *Pastorale oder Die Zeit für Kakao, Landschaft mit Figuren*, and *Die Uhren*. The volume *Theaterstücke* (1976) contains *Die Verspätung, Nachtstück, Pastorale*, as well as the tract on the Theatre of the Absurd. *Mary Stuart. Eine historische Szene* appeared in 1971.

Hildesheimer wrote the libretto for the opera *Das Ende einer Welt* (1953) composed by H. W. Henze (q.v.). Some of his plays are also radio plays; the volume *Hörspiele* (1976) contains *Das Opfer Helena, Herrn Walsers Raben, Unter der Erde*, and *Monolog*; other radio plays include *Hauskauf* (1974) and *Biosphärenklänge* (1977), dealing with the Holocaust. The collection *Ende der Fiktionen. Reden aus 25 Jahren* appeared in 1984.

Hildesheimer was awarded the Büchner Prize in 1966.

HILDIKO, see Ildiko.

Hilfe, Die, a political and literary weekly founded in 1895 by F. Naumann (q.v.), it ceased publication in 1943. Th. Heuß (q.v.), later President of the Federal Republic, was its editor from 1905 to 1912, and was succeeded by Gertrud Bäumer (q.v.).

HILLE, Peter (Erwitzen nr. Höxter, 1854–1904, Großlichterfelde, Berlin), a gifted and arch-bohemian writer (in the French sense), who lost or mislaid many of his own writings. He published three novels, *Die Sozialisten* (1886), *Semiramis* (1902), and *Cleopatra* (1905), and a tragedy on Petrarch (*Des Platonikers Sohn*, 1896). His poems were published posthumously as *Blätter vom 50jährigen Baum* (1905) and *Aus dem Heiligtum der Schönheit* (1909). Years later appeared *Das Mysterium Jesu* (1921).

HILLER, FERDINAND (Frankfurt/Main, 1811–85, Cologne), conductor and composer, set Goethe's *Gesang der Geister über den Wassern* (q.v.) to music. He is the author of a book on Beethoven (q.v.), *L. von Beethoven* (1871), and on Goethe, *Goethes musikalisches Leben* (1883).

HILLER, JOHANN ADAM (Wendisch-Ossitz nr. Görlitz, 1728–1804, Leipzig), composer and impresario, who in 1768 took over the Leipzig concerts later known as Gewandhauskonzerte. He moved to Berlin in 1785, but returned to Leipzig to take charge of the music at St. Thomas's Church (Thomaskantor), a post once held by J. S. Bach (q.v.). He composed the music for many Singspiele (see SINGSPIEL), including the remarkably successful *Der Teufel ist los* (q.v., 1752 and 1766), *Lottchen am Hofe* (1767), *Die Jagd*, and *Der Dorfbarbier* (both 1770).

Hiller published a series of biographies of musicians entitled *Lebensbeschreibungen berühmter Musikgelehrten und Tonkünstler* (1784), and edited the first musical periodical *Wöchentliche Nachrichten und Anmerkungen, die Musik betreffend* (1766–70).

HILTBOLT VON SCHWANGAU (modern Hohenschwangau in Bavaria), a name to which certain Middle High German lyrics, included in *Des Minnesangs Frühling* (q.v.), are attributed. As all the knights of Schwangau were called Hiltbolt, his identity is uncertain, and some have thought that the authorship is divisible between two similarly named members of the family, the more so as features of some of the poems have led to widely differing dating. Here and there the influences of REINMAR der Alte and FRIEDRICH von Hausen, and even WALTHER von der Vogelweide (qq.v.), are traceable. The poems are among the less important Minnelieder.

Himelriche, see VOM HIMMELREICH.

Himmel und Hölle, a powerful description in prose of Heaven and Hell, written probably in Bamberg in the second half of the 11th c. under the influence of the Cluniac Reform (q.v.) emanating from Hirsau. The description of Heaven derives from Revelation, and shows the greater variety both in substance and in syntax. Hell is portrayed in one enormous sentence with endless predicates of infernal torment. The language is a closely wrought, glittering, rhythmic prose, though some early editions wrongly print it as free verse.

HIMMLER, HEINRICH (Munich, 1900–45, Lüneburg), trained at an agricultural college, took to politics in 1923, participating in Hitler's attempted *coup d'état* in Munich. In 1929 he was placed at the head of the SS (q.v.). After 1933 he was prefect of police in Munich and then head of the Gestapo, in which capacity he evinced a merciless efficiency. The concentration camps, the extermination of the Jews and the terrorization of the populations of occupied countries came within his scope. In 1944 Himmler was given military command, which in 1945 included two army groups, but his talents were not soldierly. In the final stages he attempted unsuccessfully to negotiate with the West. He was arrested while seeking to escape in disguise, and took poison.

Himmlische Jerusalem, Das, an Early Middle High German poem based on Revelation 21 : 2–26, giving a description of 'the holy city, new Jerusalem', and especially the walls and the precious stones with which they are garnished. These are given an allegorical interpretation. The poem, which is contained in the Vorauer Handschrift, comprises 473 lines, and was written by an Austrian cleric c. 1140.

HINDEMITH, PAUL (Hanau, 1895–1963, Frankfurt/Main), one of the leading German composers in the 1920s and 1930s, emigrated in 1938, returning to Germany in 1951. He formulated the conception of Gebrauchsmusik ('music for use', a twin to Gebrauchslyrik), which had connections with the didactic and revolutionary aims of K. Weill and B. Brecht (qq.v.). He later turned away from this movement.

Hindemith's best-known work is the opera *Mathis der Maler* (1934–5, concerning the painter Mathias Grünewald, q.v., with a libretto by the composer). Other works which should be mentioned are his ballet *Nobilissima visione* (1938) and his *Requiem* to words by Walt Whitman (1946).

HINDENBURG, PAUL VON BENECKENDORF UND VON (Posen, 1847–1934, Neudeck, West-Prussia), served first in the Prussian Guard (3. Garde-Regt. zu Fuss) in which he took part in the wars of 1866 and 1870–1 (see DEUTSCH-FRANZÖSISCHER KRIEG). His career as a staff officer began in 1878, but in 1893 he returned to regimental duty as O.C. Infantry Regiment No. 91 (Oldenburg).

Three years later he received a high staff appointment, and in 1900 became G.O.C. 28. Division in Karlsruhe. In 1904 he was given command of the IV Army Corps with H.Q. at Magdeburg, and in 1905 he was promoted General of the Infantry. In 1908 his conduct of manœuvres was criticized by the Emperor, and in 1911 he retired at his own request. Recalled in 1914 and sent in August to take over the 8th Army in the East from General von Prittwitz, Hindenburg, with Major-General Ludendorff (q.v.) as his chief of staff, immediately won decisive victories at Tannenberg (q.v.) and the Masurian Lakes (see MASUREN). In the following two years he commanded in the East with consistent if less spectacular success. In August 1916 he replaced Falkenhayn as commander-in-chief, taking with him Ludendorff as his chief of staff. In the last two years of the war he became politically influential, pressing the policy of annexationist expansion and unrestricted submarine warfare. His final offensives in the West in the spring of 1918 all failed after initial successes, and after the defeats and withdrawals of the summer he pressed for an armistice and the Emperor's abdication, though he himself evaded direct responsibility for it. After peace was signed he retired to Hanover. His reappearance in 1925 as a candidate for the presidency caused widespread surprise in view of his age (77) and his abstention from politics for the six years preceding. Hindenburg, who was re-elected in 1932, faithfully discharged his constitutional duties, though his personal inclinations lay well to the right, leaning to a restoration of the monarchy. He at first supported, then (possibly for personal reasons) dropped Brüning (q.v.), and felt most at ease with the ex-Ulan officer von Papen as chancellor. In January 1933 he reluctantly agreed to the appointment of Hitler (q.v.) as chancellor. The control of events slipped from his hands, but he survived until August 1934.

Hiob, a Middle High German verse paraphrase of the Book of Job, completed in 1338 by an anonymous poet who was probably a member of the Teutonic Order (see DEUTSCHER ORDEN). It is thought by some that TILO von KULM (q.v.) might be the author. The narrative is interspersed with passages of interpretation.

Hiob, Roman eines einfachen Mannes, a novel by J. Roth (q.v.), published in 1930. It is the story of a Jewish family with three sons which emigrates from Russian Galicia to the U.S.A., leaving behind one son who is too sickly to travel and another who is serving in the army. As Mendel Singer (the Job of the title) and his wife are about to send for their ailing son, the 1914–18 War breaks out, and their two soldier sons are killed, one in the Russian army, the other with the U.S. forces. These tragic events culminate in the death of the mother, which leaves the father in despair. But the unexpected happens. The invalid son is cured and joins his father in the U.S.A. A gifted musician, he has become a conductor of distinction, and this late blessing brings Mendel Singer back to his faith in the God of his forefathers.

HIPPEL, THEODOR GOTTLIEB VON (Gerdauen, East Prussia, 1741–96, Königsberg), the son of a headmaster, began to study theology at Königsberg University in 1756. After two years' intermission (1760–2), first as companion to a Russian officer and then as a private tutor, he returned to the university to study law, qualifying in 1765. From these early years date some comedies, of which the one-act *Der Mann nach der Uhr* (1760) was praised by Lessing (q.v.), and also a collection of poems (1757). In 1772 Hippel entered the administration of Königsberg, being appointed mayor (Erster Bürgermeister) and chief of police in 1780. He received various honours, including the title Stadtpräsident in 1786, and amassed a fortune. In 1791 he received a patent of nobility, which was intended as a restoration, for the family was supposed to be of noble descent. Hippel, an eccentric who combined religious sentimentality with sharp common sense, is best known for his novel *Lebensläufe nach aufsteigender Linie* (q.v., 4 vols. 1778–81), a whimsical disguised autobiography. His equally ironical second novel *Kreuz- und Querzüge des Ritters A. bis Z.* (1793–4) was less successful. A kind of Bildungsroman (q.v.), it reflects the enthusiasm of the time for exclusive societies devoted to enlightenment and good works, but it is also directed against freemasonry, to which Hippel was strongly opposed. In both of these works of fiction he was influenced by Sterne. Although Hippel was a confirmed bachelor, he is the author of *Über die Ehe* (1774), an essay on marriage, and in a later essay, *Über die bürgerliche Verbesserung der Weiber* (1792), he discussed the position of women, advocating a measure of equality. *Sämtliche Werke* (14 vols. 1827–39) were reprinted in 1978.

Hirsau, see CLUNIAC REFORM.

HIRSCHFELD, GEORG (Berlin, 1873–1942, Munich), studied at Munich University, after which he returned to Berlin and engaged in writing. In the 1890s he was regarded as potentially a writer of the first rank. His great success was the play *Die Mütter* (q.v., 1896), but apart from the one-act *Zu Hause* (also 1896), his dramatic work made no lasting impact. It includes *Agnes Jordan* (1897), *Der Weg zum Licht* (1902), *Nebeneinander* (1904), *Das zweite Leben* (1910), *Überwinder* (1913), and *Das hohe Ziel* (1920), as well as the comedies *Pauline* (1899), *Der junge Goldner* (1901), and *Mieze und Marie* (1907). He was the author of Novellen, *Dämon Kleist* (a collection, 1895) and *Der Bergsee* (1896), and of several novels, *Der Kampf der weißen und der roten Rose* (1912), *Die Belowsche Ecke* (1914) *Der Mann im Morgendämmer* (1925), and *Die Frau mit den hundert Masken* (1931). In 1925 Hirschfeld published *Otto Brahm, Briefe und Erinnerungen* (see BRAHM, O.).

Hirten- und Blumenorden an der Pegnitz, Löblicher, name of a society of poets founded at Nürnberg in 1644 by G. P. Harsdörffer and Johann Klaj (qq.v.), who took the pastoral names Strephon and Clajus respectively. The Blumenorden, whose spiritual ancestor was the tradition of the Guild of the Meistersinger (see MEISTERGESANG) in Nürnberg, cultivated poetry as an elegant social art. Apart from its two founders, the principal members were S. von Birken, J. M. Dilherr, J. Helwig, J. Rist, G. Schottel, and S. G. Staden (qq.v.). The principles to be followed in poetry were set down by Harsdörffer in his treatise on poetry *Poetischer Trichter* (1647–53), and later by M. D. Omeis (q.v.), who became head of the Order in 1697.

HIRZELIN (b. *c.* 1270), a strolling singer from the region of Lake Constance, wrote a poem on the battle of Göllheim (q.v., 1298), of which only a fragment survives. Hirzelin takes the Austrian side in contrast to the (unknown) author of the better-known poem *Die Schlacht bei Göllheim.*

Historia von D. Johann Fausten, see FAUSTBUCH, SPIESS'SCHES.

Historien der Alden É, a clumsily written Middle High German poem summarizing the story of the Old Testament. It was composed *c.* 1340 by a member of the Teutonic Order (see DEUTSCHER ORDEN), whose source was the Latin *Historia scholastica.*

Historie von dem Ritter Beringer, Die an anonymous comic poem written towards the end of the 15th c. It recounts the exposure of a boastful and cowardly knight.

HITLER, ADOLF (Braunau, Austria, 1889–1945, Berlin), the self-styled Führer of Germany from 1933 to 1945, was the son of an Austrian customs official, and grew up in Linz. He hankered after

a career as an artist, but was refused admission to the Vienna Art Academy. He began to take an interest in politics about 1909, instructing himself by indiscriminate reading. In 1913 he moved to Munich, and in August 1914 he volunteered for service in the Bavarian army. He served throughout the war, was twice wounded, and received both classes of Iron Cross. In September 1919 he entered the Deutsche Arbeiterpartei in Munich, and quickly proved himself an able and persuasive speaker, denouncing the Revolution of November 1918 and the Treaty of Versailles (see VERSAILLES, TREATY OF). In 1921 he became chairman of the party, which had changed its name in 1920 to Nationalsozialistische Deutsche Arbeiterpartei (see NSDAP). Hitler rapidly influenced middle-class and military elements in Bavaria, and mounted a *coup d'état* on 9 November 1923 (see HITLERPUTSCH), which failed totally and for a time lost the party its conservative support. Hitler was sentenced to five years' imprisonment, and was released within a year from Landsberg Fortress.

In Landsberg Hitler wrote the first volume of *Mein Kampf* (q.v., 2 vols., 1925–6). The NSDAP was revived in 1925. After his release Hitler sought to obtain power by constitutional means. His programme was a nationalistic one of German revival and expansion, coupled with virulent anti-Semitism. He established himself firmly as party leader, and set in motion the myth of the Führer. The economic crisis of 1929 gave him renewed opportunity to gain members among the unemployed, and allies among the right-wing parties (see HARZBURGER FRONT). With the parliamentary crisis of 1930 the prestige of Hitler and his party gained momentum, and NSDAP representation in the Reichstag increased. In 1932, after two years of demagogic oratory and street violence, Hitler unsuccessfully sought election as president. Although the party also suffered an electoral setback in that year, the impasse into which the Republic had drifted facilitated negotiations for Hitler's inclusion in a new government. On 30 January 1933 President Hindenburg (q.v.) acquiesced in Hitler's appointment as chancellor (Reichskanzler) in a cabinet consisting largely of conservatives.

By vigorous terrorism and astute political moves Hitler disposed of his conservative allies and made the NSDAP the instrument of rule in Germany. In addition to the persecution of former political opponents and of the Jews, he eliminated by planned assassination the leaders of the SA (q.v.). On Hindenburg's death he assumed the presidential powers, becoming head of the armed forces (see REICHSWEHR). He defied the military provisions of the Treaty of Versailles by the introduction of conscription in 1935 and the occupation of the Rhineland in 1936. In 1936 he also made an alliance with Mussolini's Italy. Up to a certain point Hitler displayed a truer view of the realities of power politics than his advisers and generals, gauging accurately the slow or timid reactions of foreign powers. Taking one step at a time, and on each occasion declaring that it was his last, he secured Austria and the borderlands of Czechoslovakia (Sudetenland) in 1938, and Czechoslovakia itself and Memel in 1939, and he turned the settlement with Chamberlain (Sept. 1938) to his advantage.

Hitler was unable to annex Poland without a European war (see WELTKRIEGE, II). He took an active part in military planning, and his early campaigns (Blitzkriege) were strikingly successful, strengthening his megalomania and delusions of infallibility. In 1941 he assumed direct command of the armed forces, but from the autumn of 1942 his manic inflexibility provoked and exacerbated a series of military disasters, including the catastrophe of Stalingrad. On 20 July 1944 an attempt to assassinate him failed (see RESISTANCE MOVEMENTS, 2). He continued to attempt to hold all conquests, wasting his military assets in so doing. With Germany invaded from east and west, and Berlin partly in Soviet hands, Hitler went through a marriage ceremony with Eva Braun, his mistress for twelve years, and committed suicide with her on 30 April 1945. A MS. by Hitler dealing with foreign policy and written in 1928 was published in 1961, *Hitlers zweites Buch*.

Hitlerputsch, the abortive revolt attempted by Hitler (q.v.) in Munich in 1923 at a time when the climate of opinion was hostile to the constitutional government after the French occupation of the Ruhr and the crisis of inflation. On 8 November Hitler, with a unit of storm-troopers (SA, q.v.), kidnapped the heads of the Bavarian government and declared a revolution. The ministers Kahr, Lossow, and Seißer, however, escaped. A second attempt on 9 November took the shape of a march on the War Ministry headed by Hitler and General Ludendorff (q.v.). In the Residenzstraße near the Odeonsplatz the police opened fire and dispersed the rebels. Hitler was arrested two days later, tried and sentenced to five years' imprisonment, of which he served nine months.

HITZIG, JULIUS EDUARD (Berlin, 1780–1849, Berlin), a civil servant, became a bookseller and publisher in 1808 without abandoning his state employment, in which he rose to high office, retiring in 1835. Hitzig's sympathies were with Romantic literature, and he was a friend of Z. Werner and of E. T. A. Hoffmann (qq.v.), whose

posthumous papers he published with a biography (*Aus E. Th. A. Hoffmanns Leben und Nachlaß*, 1839). He also performed a similar service for A. von Chamisso (q.v., *Leben und Briefe von A. von Chamisso*, 1839). The family name was originally Itzig. Hitzig was the founder of the Mittwochsgesellschaft (q.v.).

Hobellied, a song included in F. Raimund's play *Der Verschwender* (q.v.). It is sung by Valentin to a melody by Konradin Kreutzer (1780–1849). The first line runs 'Da streiten sich die Leut' herum'.

Hochdeutsch, commonly used for the standard literary and spoken language of the educated in Germany, Austria, and, with reservations, in Switzerland.

HOCHHUTH, ROLF (Eschwege nr. Kassel, 1931–), made his name as a writer with his first two plays, both linked with the 1939–45 War and both passionately concerned with man's inhumanity to man: *Der Stellvertreter* (q.v., 1963) and *Soldaten* (q.v., 1967). Hochhuth's special technique is to base a plausible case (for each play represents a *plaidoyer*) on an abundant and carefully selected stock of documents, and to use as the antagonist a prominent character of recent history (Pope Pius XII and Sir Winston Churchill) against whom he directs sweeping accusations. The protagonist is a poetically conceived representative of humanity. The dramatic construction shows a lively imagination, and the speech is often powerful. The plays have been the object of violent controversy. *Der Stellvertreter* offended countless Roman Catholics. *Soldaten* was the subject of a successful action for libel brought in the English High Court by General Sikorski's pilot. Other litigation followed. Hochhuth's third tragedy, *Guerillas* (1970), did not arouse the same intensity of public discussion. It marks, however, his turn to topical political debate, which he continued in his comedies. His first comedy, *Die Hebamme*, was performed in May 1972. A second, *Lysistrate*, performed in an abridged version as *Lysistrate und die NATO*, had an unsuccessful première at Essen in 1974. *Zwischenspiel in Baden-Baden*, a story, appeared in 1974, *Die Berliner Antigone*, a volume of prose and verse with an introduction by Nino Erné, in 1975, and *Tod eines Jägers*, a monodrama, in 1976. Based on documentary material, it deals with E. Hemingway and covers the hours preceding his suicide. The novel *Eine Liebe in Deutschland* (1978) was followed by two plays, *Juristen. Drei Akte für sieben Spieler* (1979) and *Ärztinnen*, in five acts (1980). *Räuber-Rede. Drei deutsche Vorwürfe. Schiller, Lessing, Geschwister Scholl* and a collection of essays, dialogues, and sketches, *Spitze des Eisbergs. Ein Reader*, ed. D. Simon, appeared in 1982. Hochhuth is editor of the works of Wilhelm Busch and Otto Flake (qq.v.).

Hochkirch, battle of (14 October 1758), in which the Austrian commander Daun (q.v.) defeated Friedrich II (q.v.) in the Seven Years War (see SIEBENJÄHRIGER KRIEG). Hochkirch lies in Upper Lusatia (Oberlausitz) to the west of Görlitz.

Höchstädt, a town on the Danube near Dillingen, which has given its name to two battles fought in the War of the Spanish Succession: on 20 September 1703 and on 13 August 1704. In the former the Elector of Bavaria defeated the imperial troops, and in the latter, known in England also as Blenheim, the tables were turned, the combined forces of Marlborough and Prince Eugene (see EUGEN, PRINZ VON SAVOYEN) routing the French and Bavarians.

Hochwald, Der, a Novelle written by A. Stifter (q.v.) in 1841, and published in 1842. It is set in the time of the Thirty Years War (see DREISSIGJÄHRIGER KRIEG). Two sisters, Clarissa and Johanna, daughters of a nobleman, are sent into refuge on the shore of a remote and barely accessible lake in the Bohemian Forest, because the passage of the Swedish armies is imminent. For weeks they live in a specially built wooden house, protected by Gregor, a faithful friend of their father, looking each day from a vantage point at their castle home, visible through a telescope. Their solitude is temporarily interrupted by Ronald, a Swedish prince in disguise, who renews and declares his love for Clarissa, by whom it is returned. Soon after his departure the castle is one day invisible, hidden by clouds and smoke, and the following day clearly seen to be roofless and burnt out. The two girls return to the ruins, in which they are furnished with improvised shelter. They discover the circumstances of the tragic death of their father and brother, as well as of Ronald.

Each chapter heading includes the word Wald: *Waldburg, Waldwanderung, Waldhaus, Waldsee, Waldwiese, Waldfels,* and *Waldruine.* The great merit of the work is the all-pervading presence of the forest conveyed in noble and sensitive prose. The final heading depicts the forest into which Gregor withdraws when he feels the end of his days and the ruin in which the two sisters remain in loyalty to their family and to Ronald. The opening section shows how time has merged the castle ruins with nature. There are obvious parallels with Goethe's *Novelle* (q.v.).

HOCHWÄLDER, FRITZ (Vienna, 1911–), was an artisan and later a trade union official. He emigrated in 1938 to Zurich, where he has remained. He is the author of a number of plays, mainly traditional in form but related to contemporary problems of conscience viewed from a religious standpoint. They include *Jehr* (1933), *Esther* (1940), *Meier Helmbrecht* (1941), *Das heilige Experiment* (q.v., 1947), which enjoyed European acclaim, *Der Flüchtling* (1948, revised 1955), *Virginia* (1951), *Donadieu* (q.v., 1953), which has also met with success abroad, *Der öffentliche Ankläger* (1954), *Die Herberge* (1956), *Donnerstag* (a mystery play, 1959), and *Der Befehl* (1967). Hochwälder's comedies, which are influenced by the tradition of the Viennese Volksstück (see VOLKSSTÜCK), include *Liebe in Florenz* (1936), *Die verschleierte Frau, Hôtel du Commerce* (both 1946), *Der Unschuldige* (1949, revised 1956), and *Der Himbeerpflücker* (1964). A collection of his plays appeared in 2 vols. (1959 and 1964), and a single volume followed in 1968. In *Über mein Theater,* a lecture delivered in 1956 and published in 1959, Hochwälder also discusses the Viennese tradition.

Hochzeit, Die, Early Middle High German poem, allegorically interpreting marriage in theological terms, relating it to the Song of Songs. It contains a detailed and robust description of a contemporary wedding, and the view has been advanced that an older theological poem has been adapted by the country priest who wrote *Vom Rechte* (q.v.). It was probably written in its present form after 1150, and is contained in the Milstätter Handschrift (q.v.).

Hochzeit der Feinde, Die, a novel by S. Andres (q.v.), published in 1947. Its theme is Franco-German hostility, exemplified in two families from opposite sides of the frontier, and its transcendence by love and common sense.

Hochzeit der Sobeide, a play (dramatisches Gedicht) by H. von Hofmannsthal (q.v.), published in his *Theater in Versen* in 1899. Sobeide, who loves Ganem, is to be married against her will to a rich merchant (Reicher Kaufmann). She is brought to his house, and there she confesses to her intended husband that she loves another. He magnanimously frees her to go to her lover Ganem, but she finds him with Gülistane. Broken and contrite, she returns to the merchant's house and hurls herself from the top of a tower. As she dies in the merchant's arms she recognizes his compassion and his love, which she herself, too late, returns.

Hochzeit des Mönchs, Die, a Rahmennovelle (see RAHMEN) by C. F. Meyer (q.v.), published in the *Deutsche Rundschau* (q.v.) in 1883–4. The tale of the monk Astorre is told by Dante to the Cangrande Scaliger and his Veronese court, which includes his consort, his young mistress, and the fool. Dante, a fugitive from Florence, joins the party to keep warm by the fire on a wintry day in November. He reveals as the key to the 13th-c. action an inscription on a grave: 'Hic jacet monachus Astorre cum uxore Antiope. Sepeliebat Azzolinus.' He weaves his tale as he goes along, welcoming the comments of his listeners, whom he studies and absorbs into the action, reflecting himself in Astorre and in the expelled Florentine goldsmith, who sells his booth on a bridge over the River Brenta in Padua. Ezzelin is at the time, which falls in the reign of the Emperor Friedrich II (q.v.), the head of the state.

The action is concentrated within some two weeks. Having lost a beloved wife during the plague, Umberto has been urged by his father, the head of the wealthy Vicedomini family, to marry Diana, daughter of the equally powerful Pizzaguerra. But the wedding barge on the Brenta sinks, and only Diana is rescued by the monk Astorre, Umberto's brother. Vicedomini does not survive the shock, but before he dies he extracts from Astorre the promise to marry Diana to save the Vicedomini name from extinction. A dedicated Franciscan friar of good repute thus becomes a traitor not only to his habit, but also to himself, for he has the inclination neither to abandon his Order nor to marry Diana. As he buys the wedding ring on the Brenta bridge, chance has it that of the two rings which he fancies, one drops to the ground and finds its way on to the finger of Antiope Canossa, at whom Astorre gazes spellbound. At the betrothal Diana receives the larger ring, which Astorre has kept, and places a ring of her own on Astorre's finger. Having thus completed the ceremony of mutual dedication, she promises fidelity, though she warns Astorre of her quick temper. But Olympia, Antiope's mother, protests, and claims Astorre's hand for her daughter, who, she asserts, was the first to receive a ring from him. In her fury at this insult Diana aims a blow at Olympia, which strikes Antiope, who has flung herself forward to protect her mother. To restore Antiope's honour, Germano, Diana's brother, sets out, with Astorre as escort, to ask for Antiope's hand in marriage. Antiope, however, refuses. Instead she is married during the night to Astorre. Both have thus surrendered to a secretly cherished love. The following day Ezzelin hears the two parties in order to pass judgement. He reconciles Astorre and Diana's father, and orders the wedding to be celebrated at the customary masked ball. But Diana insists that

Antiope humbles herself before her, and surrenders her ring. Antiope reluctantly complies, and discovers Diana at the ball dressed up as the goddess of her name. When she removes the ring from her hand, Diana kills her with one of her arrows. Astorre thereupon kills Germano with the arrow, and, in the act of killing, falls on Germano's sword. Ezzelin arrives in time to close the eyes of the dead.

HOCK, THEOBALD (? Limbach, Palatinate, 1573–after 1618), a man of classical education, was condemned to death in Bohemia as a result of a dispute over ecclesiastical rights, but was saved by the revolt in Prague which began the Thirty Years War (see DREISSIGJÄHRIGER KRIEG). He seems to have become an officer, though his later career is unknown. He wrote 92 poems which present a kind of conspectus of morals and suggest a transition to baroque style (*Schönes Blumenfeldt*, 1601).

HODDIS, JAKOB VAN, pseudonym of Hans Davidsohn (Berlin, 1887–1942, a victim of persecution, place unknown), was an early Expressionist poet, who from 1912 on was mentally ill. His poem *Weltende* was published in 1911.

HOFER, ANDREAS (St. Leonhard, Tyrol, 1767–1810, Mantua), during the Napoleonic Wars (q.v.) fought repeatedly and successfully against the French and their allies. Hofer was an innkeeper's son in the Passeiertal, and inherited the inn Am Sande, from which derives his nickname of Sandwirt. In the earlier French wars (1796–1805) he fought first as a rifleman (Schütze) and later as a captain of rifles. In 1808 he was invited to Vienna by the Archduke to discuss the plan of a Tyrolean rising, which took place with startling success in April 1809. The pro-French Bavarian garrisons were driven out after the two battles of Berg Isel at the end of May. In July, however, Austria, defeated at Wagram, signed the armistice of Znaim, and undertook to withdraw from Tyrol. On the entry of a French army under General Lefebvre, Hofer called for a new rising, which was at first successful. All Austrian help being eliminated by the Treaty of Schönbrunn (October 1809), Hofer's prospects faded. His forces were overwhelmed, and Hofer himself went into hiding, but was betrayed, carried off to Italy, and shot under martial law on 20 February 1810. In 1823 his remains were reburied in the Hofkirche in Innsbruck. His resistance and fate have attracted writers, notably K. L. Immermann (q.v.) with *Trauerspiel in Tirol* (1828, revised and retitled *Andreas Hofer*, 1834), B. Auerbach (q.v.) with the tragedy *Andre Hofer* (1850), F. Kranewitter (q.v.) with a play of the same title (1902), and K. Schönherr (q.v.) with *Der Judas von Tirol* (1927). Hofer is the subject of the song 'Zu Mantua in Banden' (*Andreas Hofer*, 1831) by Julius Mosen (q.v.), which became a well-known patriotic song (to an old melody), first recorded in 1844.

HOFFMANN, ERNST THEODOR AMADEUS (Königsberg, 1776–1822, Berlin), whose third baptismal name was Wilhelm, adopted the name Amadeus as a token of his reverence for Mozart (q.v.). The child of a broken home, Hoffmann was brought up by relatives. At school his principal friend was Th. G. Hippel (1775–1843, nephew of the writer of the same name, q.v.). Hoffmann's father and some of his mother's family had been law officers of the Prussian Crown, and, although he displayed considerable artistic gifts, he was destined for the same profession, studying law at Königsberg University (1792–5), and obtained appointment at Groß-Glogau in 1796. He made rapid progress, serving from 1798 in Berlin, and from 1800 in Posen. He devoted his spare time to drawing, painting (he was successful with portraits), and extensive reading; he composed music, chiefly ecclesiastical works and incidental music to plays. But he also drew caricatures, some of which made fun of his seniors, and rashly allowed them to circulate. As a result he received in 1802 a disciplinary posting (which did not affect his promotion) to the dull Polish town of Plozk. He married Michaeline (Mischa) Rorer and also discovered his ability to write. In 1804 he was transferred to Warsaw, where he resumed his artistic pursuits in a more congenial environment. During these rather isolated years in Poland he found stimulating friends in Julius Hitzig and Zacharias Werner (qq.v.).

After the defeat of Prussia by the French in 1806, the Prussian administration was dismissed, and Hoffmann found himself virtually without means until he was appointed orchestral conductor to the theatre at Bamberg. For the next few years he sought to live for his music alone. The theatre soon fell on evil days, and Hoffmann left it for a time in 1809, supporting himself and his wife by giving music lessons. Among his pupils was Julia Marc (q.v.), a girl with a beautiful voice, to whom Hoffmann became passionately attached. In 1812 she entered a short-lived marriage with a Hamburger, who was an alcoholic. Hoffmann's love for her recurs in the form of an idealization of music in many of his writings. In 1813 he obtained a post as musical director with a company that performed at Dresden and Leipzig.

By now Hoffmann was an accomplished writer. In musical reviews published 1809–14 he proved himself not only an acute as well as passionate admirer of Mozart, but also showed sympathy for baroque music, and notably for the work of J. S. Bach (q.v.), at that time subject to neglect. Furthermore, he was an informed and enthusiastic supporter of his contemporary Beethoven (q.v.). Not surprisingly, his early fiction is closely linked with music (*Ritter Gluck*, q.v., 1809; *Kreislers musikalische Leiden*, 1810; *Ombra adorata*, 1812; *Don Juan*, q.v., 1813). The last of these includes a noteworthy contribution to the interpretation of Mozart's *Don Giovanni* (q.v.). In 1814 the first volumes of *Fantasiestücke in Callots Manier* (q.v.), a collection of stories, including *Ritter Gluck* and *Don Juan*, appeared, and this was quickly followed by two more volumes, in which, among others, the new stories *Der goldne Topf* and *Abenteuer der Sylvester-Nacht* (qq.v.) appeared. Hoffmann's first novel, *Die Elixiere des Teufels* (q.v.), written in Leipzig and Berlin, was also published in 1815–16.

After his disappointing theatrical experiences in Bamberg and Leipzig, Hoffmann abandoned his intention to depend solely on his pen and his music, and sought reappointment in the Prussian civil service. He retained, however, his desire to compose, and completed the opera *Undine*, an adaptation of F. de la Motte-Fouqué's story, for which Fouqué (q.v.) himself supplied the libretto. The opera was eventually performed in the Berlin Schauspielhaus in August 1816, and achieved some fourteen performances before the theatre was destroyed by fire. Hoffmann was duly reappointed a Crown law officer, at first without pay, but in 1816 he was restored to his former seniority. He achieved a reputation for punctiliousness and humanity in the discharge of his duties, but undermined his health by heavy drinking, a habit acquired during his years in Poland. His writings were as much in demand as ever. *Nachtstücke* (q.v., 2 vols., including *Ignaz Denner*, *Der Sandmann*, *Das Sanctus*, and *Das Majorat*, qq.v.) came out in 1816–17, and in 1819–21 followed *Die Serapionsbrüder* (q.v., 4 vols., including *Der Artushof*; *Nußknacker und Mausekönig*; *Rat Krespel*; *Doge und Dogaressa*; *Meister Martin, der Küfner, und seine Gesellen*; *Das Fräulein von Scuderi*; *Der unheimliche Gast*; and *Der Baron von B.*, qq.v.). Simultaneously, volumes 1 and 2 of Hoffmann's unfinished second novel, *Lebensansichten des Kater Murr* (q.v.), appeared (vol. 1 in 1819, vol. 2 in 1821), as well as the three fantasies *Klein Zaches genannt Zinnober* (q.v., 1819), *Prinzessin Brambilla* (1820), and *Meister Floh* (1822).

Hoffmann died after an illness of some months. This last phase was embittered by an inquiry into his work involving charges that he did not pursue agitators with sufficient vigour. The dialogue *Des Vetters Eckfenster* (q.v., 1822) appeared shortly before his death; *Meister Johannes Wacht* and the unfinished *Der Feind* (qq.v.) were both dictated by the sick poet, and published posthumously in 1823. A projected work of autobiographical character, *Schnellpfeffers Flitterwochen vor der Hochzeit*, was never written. For further information on individual works see DIE AUTOMATE (1814), PRINZESSIN BLANDINA (1815), DER KAMPF DER SÄNGER (1818), DIE BERGWERKE ZU FALUN (1819), DIE DOPPELTGÄNGER (1821), and DER ELEMENTARGEIST (1821 and 1825).

It is a remarkable feature of Hoffmann's career that he did not regard himself as a writer until he was in his thirties and then, in the fourteen years before his death at the age of 46, wrote two novels and some fifty, mostly substantial, stories. He possessed extraordinary inventiveness, a remarkably fertile imagination, a vision at once childlike and sophisticated, and a gift for vivid evocation of scenes, often in realistic terms. His stories range from the pure fantasy of the fairy-tales *Der goldne Topf*, *Nußknacker und Mausekönig*, and *Klein Zaches genannt Zinnober* (which nevertheless have an element of parable) to dark, sinister, frightening tales, such as *Das Majorat*, *Ignaz Denner*, and *Das Fräulein von Scuderi*. In between these extremes are neatly contrived narrations such as *Doge and Dogaressa* and *Meister Martin, der Küfner, und seine Gesellen*. Of the novels, *Die Elixiere des Teufels* shows horror stretched to the point of insanity, while the *Lebensansichten des Kater Murr* maintains a subtle balance between the dark irrational elements and a palpable reality. Hoffmann is the central figure in the opera *Les Contes d'Hoffmann* (première 1881) by Offenbach.

Gesammelte Schriften (15 vols.) appeared 1827–39. 20th-c. editions include the incomplete *Historisch-kritische Gesamt-Ausgabe* (vols. 1–4 and 6–10, 1908–28), ed. C. G. von Maaßen, *Kritische Gesamtausgabe der musikalischen Werke* (3 vols., incomplete, 1922), ed. G. Becking, *Dichtungen und Schriften sowie Briefe und Tagebücher. Gesamtausgabe* (15 vols., 1924), ed. W. Harich, *Briefe*, 4 vols. (incl. diaries), ed. F. Schnapp, 1967–71, and *Werke in fünfzehn Teilen*, ed. G. Ellinger (1912, 2nd edn. 1927).

HOFFMANN, HANS (Stettin, 1848–1909, Weimar) a schoolmaster in Stettin and Stolp in Pomerania, resigned to become a journalist, and then lived as a successful author in various towns in Germany and Austria (Freiburg, Bolzano, Potsdam). From 1902 until his death he

was secretary to the Schiller Foundation in Weimar. Hoffmann's popularity has now faded, and his name has disappeared from many works of reference. He was a prolific author of Novellen of which the best known were *Unter blauem Himmel* (1881), *Der Hexenprediger* (1883), a story of witch-hunting in the 17th c., and *Im Lande der Phäaken* (1884). Among his historical novels *Der eiserne Rittmeister* (3 vols., 1890) and *Wider den Kurfürsten* (1894) may be mentioned. The cycle *Das Gymnasium zu Stolpenburg* (1891), originating in his own experiences in a small and strait-laced provincial town (Stolp), is probably his best work.

HOFFMANN VON FALLERSLEBEN, the name by which August Heinrich Hoffmann (Fallersleben, 1798–1874, Corvey) is usually known. In 1823 Hoffmann was appointed university librarian at Breslau, where he became a professor of German language and literature in 1830. In 1842 he was dismissed, paradoxically, because of the political significance of his *Unpolitische Lieder* (1840). In 1860 he became librarian at Schloss Corvey, where he remained until his death. Hoffmann was a notable collector and publisher of folk-songs (see VOLKSLIED), and also first published *Merigarto* (q.v., 1834) and other medieval works. Though his poetry was largely derivative, echoing the Romantics, some of his homely or vigorous poems enjoyed a remarkable success, notably 'Deutschland, Deutschland über alles' (q.v.) and 'Zwischen Frankreich und dem Böhmer Wald'.

HOFFMANN VON HOFFMANNSWALDAU or HOFMANNSWALDAU, CHRISTIAN (Breslau, 1616–79, Breslau), son of a nobleman and Kaiserlicher Kammerrat, travelled in western Europe, including England, and returned home in 1641. He became a man of mark in Breslau, was a city councillor (Ratsherr) in 1646, and received a title (Kaiserlicher Rat) in 1657. He became president of the council in 1677. Hoffmannswaldau was a man of remarkable technical poetic gifts, which he developed partly in emulation of M. Opitz (q.v.). He is an acknowledged master of the Alexandrine, handling also sonnets and poetic epistles (e.g. *Eginhard und Emma*, see EMMA UND EGINHARD) with ease. His range of stanza forms extends beyond that of most of his contemporaries. Hoffmannswaldau declined involvement in his themes, writing with equal expertise and deftness, but without personal feeling, reflections on morality or verses of erotic gallantry. Some of his poems are composed almost solely of accumulations (Häufung). Hoffmannswaldau's technical conception of his art is an assertion of his sense of craftsmanship. His poems were published in *Deutsche Übersetzungen Und Getichte* (1679, reprint ed. F. Heiduk 1984 f.), and, together with those of other poets, in collections. Among these anthologies were the seven volumes of *Herrn von Hoffmannswaldau und andrer Deutschen auserlesener und Bißher ungedruckter Gedichte erster theil* etc., edited by Benjamin Neukirch (q.v.) *et al.*, 1695–1727. A reprint, edited by A. G. de Capua *et al.*, appeared from 1961. Some of the attributions to Hoffmannswaldau in the original edition are false.

Höfische Dichtung, courtly literature, falls into two categories in German literature:

(1) the lyric (see MINNESANG and SPRUCH) and epic (see HÖFISCHES EPOS) of the period 1150–1250, the so-called Blütezeit of Middle High German literature, in which the poems were written by the knights or their more literate retainers, not for reading, but for performance, especially in the long winter evenings, in song, chant, or declamation in the courts of the princes, to whom the knights paid homage or from whom they sought favour. They were designed to entertain, but often also to express the high ideal of the knightly life.

(2) The literature of the absolutist courts of the 17th c., in which a man of lower rank would hold appointment as court poet (Hofpoet). Among the better-known court poets are F. R. von Canitz, B. Neukirch, J. von Besser, and J. U. König (qq.v.). Court poetry is a literature which sets out to entertain, and often to flatter, the prince and his intimates.

Höfisches Epos, term used to denote one type of Middle High German poem, often of considerable length, read or more frequently declaimed or intoned at the more sophisticated courts in Germany in the 12th c. and 13th c. The courtly epic sought not only to entertain, but also to embody the chivalric virtues (*mâze*, maintaining the golden mean or a harmonious self-control, *milte*, generosity, and *minne*, respectful love) which this esoteric society esteemed. The poems were chiefly derived, in free translation or adaptation, from French sources, notably several epics of Chrétien de Troyes and one by Thomas de Bretagne. Their subjects come from legends of Brittany, especially the Arthurian tradition.

The history of the German court epic begins with Eilhart von Oberge's *Tristrant und Isalde* (q.v.) and the *Eneit* of HEINRICH von Veldeke (q.v.) *c.* 1170. It quickly reaches its peak in the work of three great poets, HARTMANN von Aue (q.v.) with *Erec* and *Iwein*, WOLFRAM von Eschenbach (q.v.) with *Parzival*, and GOTTFRIED von Straßburg (q.v.) with *Tristan*, all three of whom were writing *c.* 1200. There are numerous other contemporary and later writers of court epics, including ULRICH von Zazikhofen (*Lanzelet*), HEINRICH von dem Türlin (*Die Krone* and *Der Mantel*), WIRNT von Grafenberg (*Wigalois*), der Stricker (*Daniel vom blühenden Tal*), K. Fleck (*Floire und Blanscheflur*), and der Pleier (*Garel vom blühenden Tal, Meleranz,* and *Tandareis und Flordibel*, qq.v.). Mention should also be made of RUDOLF von Ems (q.v., *Alexander*) and of KONRAD von Würzburg's three late courtly epics, *Engelhard, Partonopier und Meliur,* and *Der Trojanerkrieg* (qq.v.), which mark a decline in the form, though courtly epics continued to be copied and circulated in the 14th c. and even the 15th c. The earliest works of this kind were written by noblemen, but, as time passed, more and more were composed by commoners whose calling was poetry.

HOFMANNSTHAL, HUGO VON (Vienna, 1874–1929, Rodaun), of part Austrian Jewish, part Italian extraction, was brought up in Vienna in well-to-do circumstances. At the Akademisches Gymnasium he acquired an exceptionally thorough knowledge of the French and Greek languages and literatures, and was accomplished in Italian. A youthful prodigy, he published his first poetry at the age of 16 under the pseudonym Loris, though he used the pseudonym Theophil Morren for *Gestern* (1890), a one-act play in rich voluptuous verse with a Renaissance setting. At Vienna University (1892–4) he studied aesthetics, but his principal subject was law, in which he took the first Staatsexamen. In 1895 he served in the 6th Dragoons in Göding (Moravia), and was transferred with the rank of Wachtmeister to the 8th Ulanen, and promoted to the reserve of officers before he resumed his studies at Vienna University, where he took up Romance literature and in 1899 gained a doctorate for his thesis *Über den Sprachgebrauch bei den Dichtern der Plejade*. In 1901 he completed his Habilitationsschrift *Studie über die Entwicklung des Dichters Viktor Hugo*. But he decided against an academic career, and later (1905) also resigned his commission, devoting himself entirely to writing. He married the daughter of the General Secretary of the Anglo-Austrian Bank, Gerty Schlesinger. The couple made their home in Rodaun near Vienna. The eldest son of the marriage committed suicide two days before Hofmannsthal's death. From early adolescence Hofmannsthal combined an introspective nature with a deep sense of commitment to the preservation of European culture. During the 1914–18 War he served for a year as a reserve officer before obtaining an appointment in the War

Office. He accompanied missions to Switzerland and Scandinavia.

As a young man Hofmannsthal made the acquaintance of men of letters and artists in the literary *salon* of Josephine von Wertheimstein. He regularly frequented the Café Griensteidl with L. von Andrian, A. Schnitzler, H. Bahr, R. Beer-Hofmann (qq.v.). For a short time he came into close contact with Stefan George (q.v.), but George's authoritarian stance caused Hofmannsthal to break away from his circle. Up to 1904 some of his writings were published in George's *Blätter für die Kunst* (q.v.).

Hofmannsthal's early work is characterized by a luxuriant aestheticism and a *fin de siècle* melancholy. It includes *Der Tod des Tizian* (1892, a fragment), *Der Tor und der Tod* (q.v., 1893), *Der Kaiser und die Hexe, Der weiße Fächer, Die Frau im Fenster* (qq.v.), *Das kleine Welttheater*, all 1897, and the *Vorspiel* to *Das Bergwerk zu Falun* (q.v., 1899). *Die Hochzeit der Sobeide* and *Der Abenteurer und die Sängerin* (qq.v.) both appeared in 1899. By this time Hofmannsthal had begun to react against the magniloquence of his lyrical vein. This reaction is expressed in the *Brief des Lord Chandos* (q.v., written in 1901, and published in 1902). Despite the historical perspectives contained in the *Brief*, it is generally regarded as an autobiographical document as well as part of his narrative fiction, in which he explored new modes of poetic expression in objective form. *Das Märchen der 672. Nacht* (published in *Die Zeit* in 1895) was followed at the turn of the century by *Reitergeschichte* (q.v., 1899) and *Das Erlebnis des Marschalls von Bassompierre* (q.v., 1900). *Andreas oder Die Vereinigten* (written in 1912–13, but not published until 1930) was planned as a novel (Entwicklungsroman, see BILDUNGSROMAN) in 1907. A fragment from this year bears the title *Venezianisches Reisetagebuch des Herrn von N.* Hofmannsthal's narrative work reflects what he variously expressed as a 'Sprachkrise', 'Lebenskrise', and 'seelische Krise'. He used prose in the Venetian comedy *Cristinas Heimreise* (q.v., 1910). But he also explored a new path expressing subconscious motivation in the disciplined verse of *Elektra* (q.v., 1904), *Ödipus und die Sphinx* (1906, *König Ödipus* (1907), and Otway's *Venice Preserved* presented as *Das gerettete Venedig* (1905). *Kleine Dramen*, a collection of his earlier plays, appeared in 1907.

In 1900 R. Strauss (q.v.) rejected Hofmannsthal's scenario for a ballet (*Der Triumph der Zeit*), but in 1906 he chose *Elektra* as the basis for the libretto of an opera (*Elektra*, 1909), thus opening a period of close collaboration which lasted until Hofmannsthal's death. The resulting operas are *Der Rosenkavalier* (q.v., 1911), *Ariadne auf Naxos* (q.v., 1912), *Die Frau ohne Schatten* (q.v., 1919),

Die Ägyptische Helena (q.v., 1928), and *Arabella* (q.v., 1933). They also co-operated in the ballet *Josephslegende* (1912).

Hofmannsthal became closely associated with the Salzburger Festspiele (q.v.), for which he wrote *Jedermann* (q.v., 1911) and *Das Salzburger große Welttheater* (q.v., 1922). The 1914–18 War destroyed Hofmannsthal's Austro-Hungarian world, and the reflection of its qualities and the echo of its fall are perceptible in all his post-war writings, in which his witty, nostalgic Viennese comedies *Der Schwierige* (q.v., 1921) and *Der Unbestechliche* (performed 1923, revised version posth. 1956) stand out. *Der Turm* (q.v., 1925), his last tragedy, is the outcome of a protracted confrontation with the problems of power, violence, and humanity. His posthumously published works include not only the last opera written with Strauss, but also, in 1933, the completed *Das Bergwerk von Falun* (the final title) and a revision of *Der Turm* (1956).

As an essayist Hofmannsthal made a notable contribution with *Der Dichter und diese Zeit* (1906) and with *Über die Pantomime* (1911), which sums up his approach to ballet, dance, rhythm, and mime. The post-war essays are devoted to his endeavour to conserve and restore the values of a threatened civilization. They include *Beethoven* (1920), *Blick auf den geistigen Zustand Europas* (1921), *Griechenland* (1922), *Vermächtnis der Antike* (1926), and *Schrifttum als geistiger Raum der Nation* (1927).

Most of Hofmannsthal's poems were written early, and published separately and anonymously. A selection, *Ausgewählte Gedichte*, appeared in 1903, the *Gesammelte Gedichte* in 1907. Among the best known of his poems are *Vorfrühling, Ballade des äußeren Lebens,* 'Manche freilich . .', and *Terzinen über Vergänglichkeit.*

Hofmannsthal's *Aufzeichnungen* (1959) include the important document *Ad me ipsum*, which provides the word 'Präexistenz' for his early phase of devotion to the magic of words, out of which the way to 'Existenz' is found through introversion and self-sacrifice, motifs which recur, often in mystical form, in his later work, though the division of his work into distinct phases is an over-simplification.

Hofmannsthal was a notable anthologist. He was general editor of *Die österreichische Bibliothek* (26 vols., 1915–17), to which he himself contributed the Grillparzer (q.v.) volume. He also edited *Deutsche Erzähler* (1921), *Deutsches Lesebuch* (2 Pts., 1922–3), *Deutsche Epigramme* (1923), and *Wert und Ehre deutscher Sprache* (1927).

Letters from his vast correspondence were published posthumously in *Briefe 1890 bis 1909* (2 vols., 1935–7). Other editions of letters (with date of first publication) include his correspondence with R. Strauss (1926, revised 1964),

Wildgans (q.v., 1971), George (q.v., 1938), E. von Bodenhausen (1953), R. Borchardt (q.v., 1953), C. J. Burckhardt (q.v., 1956), Carossa (q.v., 1960), Schnitzler (q.v., 1964), Helene von Nostitz (1965), E. K. von Bebenburg (1966), L. von Andrian-Werburg (1968), Graf Kessler (1968), W. Haas (1969), J. Redlich (1971), R. Beer-Hofmann (q.v., 1972), S. and H. Fischer, O. Bie, and M. Heimann (1973), Rilke (q.v., 1978), and M. Mell (q.v., 1982). *Gesammelte Werke in Einzelausgaben* (15 vols.), ed. H. Steiner, appeared 1945–59 and *Sämtliche Werke. Kritische Ausgabe* (*c.*38 vols.) from 1975.

Hofmeister, Der, an adaptation by B. Brecht (q.v.) of J. M. R. Lenz's comedy *Der Hofmeister oder die Vorteile der Privaterziehung* (q.v.), written for the Berliner Ensemble (q.v.), and produced in April 1950. The last couplet of the rhymed prologue, spoken by the tutor Läuffer, explains the purpose of the play: 'Wills euch verraten, was ich lehre:/Das ABC der Teutschen Misere!'. The epilogue is spoken by the actor representing Läuffer. He appeals to future generations of teachers to liberate themselves from the tutor's 'Knechtseligkeit'. Lenz's intentions were not contrary to Brecht's views, but in order to present the message in his own way, Brecht has entirely rewritten the dialogue, while retaining the names of the characters. Elaborate notes on the rehearsals over a period of nine weeks are attached, and record the reactions of the joint producers C. Neher (q.v.), Monk, and Besson, and other participants and onlookers. Brecht also wrote a sonnet *Der Hofmeister* in a collection of *Sozialkritische Sonette*.

Hofmeister, Der, oder die Vorteile der Privaterziehung, a play (Eine Komödie) in five acts, containing 34 scenes, published anonymously by J. M. R. Lenz (q.v.) in 1774, and performed in November 1778 in Berlin. At first Goethe was widely thought to be the author.

Major von Berg's daughter, Gustchen, who really loves, and is loved by, her cousin Fritz von Berg, is seduced by the family tutor Läuffer. She runs away, bears a child, and tries—unsuccessfully—to drown herself. Läuffer, who has taken refuge with Wenzeslaus, a poor village schoolmaster, hears that Gustchen is dead, and, in remorse, castrates himself. Gustchen is found by her father, all is forgiven, and she is united with Fritz. In a minor intrigue an attempted seduction at a university is frustrated, and another happy couple, Pätus and Jungfer Rehaar, are united.

The strong didactic trend (the sexual dangers to which tutors and their charges are exposed) is underlined in the last words of the play, spoken by Fritz to Gustchen's baby: 'Wenig-stens, mein süßes Kind! werd' ich dich nie durch Hofmeister erziehen lassen'. Lenz's technique is that of short snippets of reality. See also Brecht's adaptation HOFMEISTER, DER.

HOHBERG, WOLFGANG HELMHARD, FREIHERR VON (Ober-Thumritz, Austria, 1612–88, Regensburg), an Austrian nobleman, served from 1632 in the Emperor's forces in the Thirty Years War (see DREISSIGJÄHRIGER KRIEG), and afterwards managed his own estates. A Protestant, he emigrated to Regensburg in 1664. Most of his works were written after his admission in 1652 to Die Fruchtbringende Gesellschaft (q.v.). He began with pastoral poems (*Hirtenlieder*, 1661). His *Georgica* (1682) show an interest in rural wild life unusual in his day. He wrote a three-volume epic (*Der habsburgische Ottobert*, 1664), which is said to be the only German epic poem to be completed in the 17th c., and he also translated a number of psalms (*Lust- und Arzneygarten des königlichen Propheten Davids*, 1675).

Hohenasperg, a castle near Stuttgart, built in 1530, and later used as a prison. In the 18th c. Jew Süss (see SÜSS-OPPENHEIMER J.) and C. F. D. Schubart (q.v.), and in the 19th c. B. Auerbach, F. List (qq.v.), and K. Hase were incarcerated there.

HOHENBURG, MARKGRAF VON, a name to which six Middle High German Minnelieder are attributed in the Große Heidelberger Liederhandschrift (see HEIDELBERGER LIEDERHANDSCHRIFT, GROSSE); as they appear in other collections under other names, his authorship is doubtful. He has been identified, though uncertainly, with Graf Diepolt von Vohburg in Bavaria, a nobleman who was in Southern Italy in the suite of the Emperor Heinrich VI, returned to Germany in 1212 with Friedrich II (qq.v.), and died in 1226.

Hohenfriedberg, battle of (4 June 1745), ended with a decisive victory of Friedrich II (q.v., see SCHLESISCHE KRIEGE, 2). The Prussian king, it is said, composed a military march, *Hohenfriedberger Marsch*, to mark the occasion.

Hohenfriedberg (also Hohenfriedeberg) lies in Silesia south of Liegnitz.

HOHENHEIM, FRANZISKA, REICHSGRÄFIN VON (nr. Aalen, 1748–1811, Kirchheim), the daughter of a baron von Bernardin, was married young to Baron von Leutrum. Duke Karl Eugen (q.v.) of Württemberg was attracted to her, and, after dissolution of her marriage in 1770, installed her in 1772 at Ludwigsburg as his acknowledged mistress. In 1774 he persuaded the Emperor Joseph II (q.v.) to create her a countess, and after the death of his wife in 1785 he married her. Franziska von Hohenheim is chiefly remarkable for the good use she made of her power over the Duke, encouraging his interest in education and moderating the violence of his temper.

HOHENLOHE-SCHILLINGSFÜRST, CHLODWIG, FÜRST ZU (Rotenburg/Fulda, 1819–1901, Ragaz, Switzerland), a Bavarian nobleman of ancient lineage and liberal opinions, became prime minister and foreign minister of Bavaria in 1866, working for closer association with Prussia. He resigned in 1869, but after the foundation of the German Empire (1871) became ambassador in Paris in 1874. In 1885 he was appointed governor of Alsace-Lorraine, and nine years later chancellor (Reichskanzler) and Prussian prime minister. He resigned in 1900.

Hohenstaufen, German dynastic house from which the medieval German emperors and kings were drawn for more than a century. They are Konrad III, Friedrich I, Heinrich VI, Philipp von Schwaben, and Friedrich II (qq.v.). Their reigns covered the years 1138–1250, except for the period 1198–1218 (see OTTO IV). They are commonly spoken of as Staufer.

Hohenstaufendramen, a cycle of Hohenstaufen plays planned by C. D. Grabbe (q.v.). Only two Hohenstaufen plays were written: *Kaiser Friedrich Barbarossa* and *Kaiser Heinrich der Sechste* (qq.v.).

Hohentwiel, a conspicuous isolated mountain in the district of Hegau near the western end of Lake Constance. It is surmounted by a ruined castle. In the 10th c. the Hohentwiel was the residence of the Duchess Hedwig of Swabia (see HEDWIG, HERZOGIN VON SCHWABEN) and it plays a conspicuous part in J. V. von Scheffel's novel *Ekkehard* (q.v.). In the 18th c. it was a Württemberg prison, in which for a time Schiller's godfather Colonel Rieger was incarcerated.

Hohenzollern (1) formerly the smallest Prussian province enclosed within the boundaries of the former states Baden and Württemberg. It was created in 1849 by the union of the principalities Hohenzollern-Hechingen and Hohenzollern-Sigmaringen. Though technically a Provinz, it was administered as the Regierungsbezirk Sigmaringen of the Rhine Province. Since 1952 it has been an integral part of the Land Baden-Württemberg (see WÜRTTEMBERG). Hohen-

zollern contained the Castle (Burg) of Hohenzollern, the ancestral home of the Prussian rulers of the German Empire. Perched on the Zollernberg, it was built in the 11th c., destroyed in 1423, rebuilt in 1454, and almost razed in the Thirty Years War (see DREISSIGJÄHRIGER KRIEG). Under Friedrich Wilhelm IV (q.v.) rebuilding was begun in 1850 to restore it to its appearance in the 14th c. The task was completed in 1867.

(2) The family name of the ruling Prussian-Brandenburg royal family from 1415 to 1918. They originated in the region of Hohenzollern, becoming Burggrafen of Nürnberg in 1191, and electors (Kurfürsten) of Brandenburg in 1415. Friedrich III acquired the dignity of kingship in 1701 as King Friedrich I (q.v.). King Wilhelm I (q.v.) became German Emperor in 1871. The last ruling Hohenzollern, Wilhelm II (q.v.), was obliged to relinquish the throne in 1918.

HOHLBAUM, ROBERT (Jägerndorf, Austria, 1886–1955, Graz), studied at Graz and Vienna universities, and served as an Austrian officer in the 1914–18 War. A supporter of the pan-German policy of the National Socialist regime from 1937, he held librarianships in Duisburg and Weimar before returning to Austria (Henndorf and Graz). He became a prolific author of fiction and poetry early in life, publishing most of his work in the 1920s and 1930s. He wrote on strongly nationalistic historical subjects and figures, as well as on musicians and writers. His novels include the trilogies *Frühlingssturm* (1924–6) and *Volk und Mann* (1931–5). His collection of Novellen, *Unsterbliche* (1919), includes *Grillparzers letzter Ausgang; Die Herrgotts-Symphonie* (1925) is a Novelle on Bruckner (q.v.), on whom he wrote the novel *Tedeum* (1950). Early writings on Goethe were followed by his novel *Sonnenspektrum* (1951).

HOLBEIN, HANS, DER JÜNGERE (Augsburg, 1497–1543, London), famous both as a draughtsman and as a portrait-painter, was also a great fresco-painter but all the examples of his work in this field have perished. He spent most of his life away from Germany. In 1515 he went to Basel where he met Erasmus (q.v.), for whose *Laus stultitiae* he made a series of drawings for woodcuts. He interrupted his stay in Basel for an extended visit to Italy (1516–17), and then returned to live there until 1524. During this period he painted several portraits of Erasmus (1523) and of other inhabitants, notably Burgomaster Meyer (1516). He also made the series of drawings *Der Totentanz* and *Das Todesalphabet* (1524). Religious paintings by Holbein from this period are preserved in Basel and Freiburg. After a visit to France in 1524 he spent eighteen months in England (1526–8) painting portraits, and then returned to Basel. The religious turmoil of the Reformation (q.v.) is believed to have driven him abroad again, and in 1532 he settled in England permanently. He became court painter to Henry VIII in 1536. Though many of his portraits remain, a considerable number, including that of Sir Thomas More and family, have been destroyed or lost. His father, Hans Holbein der Ältere (c. 1454–1524), was a noted painter of religious works.

HÖLDERLIN, Friedrich, in full Johann Christian Friedrich (Lauffen/Neckar, 1770–1843, Tübingen), invariably referred to as Friedrich Hölderlin, was the son of an estates bailiff, who died when the boy was two. In 1774 the mother married Burgomaster Gock of Nürtingen, who died five years later. Hölderlin's childhood was happy, and his relations with his mother, sister, and half-brother were harmonious. In 1784 he was sent to a boarding-school at Denkendorf, and transferred in 1786 to a similar establishment (Klosterschule) at Maulbronn (q.v.), where he began to write poetry. In 1788, while at Maulbronn, he became engaged to Luise Nast, but broke off the engagement in April 1789. At 18 he went to the theological seminary at Tübingen (see Tübinger Stift), studying for entry into the Evangelical Church. Hegel and Schelling (qq.v.) were his contemporaries and friends at Tübingen. Hölderlin became disinclined for the Church, but for the sake of his family continued his studies. His poetry at this time is full of a political idealism fostered by the French Revolution (q.v.).

In 1793 Hölderlin completed his course at the seminary and became, through Schiller's mediation, private tutor to the son of Frau von Kalb (q.v.) at Waltershausen near Gotha. He had begun to write a novel with a Greek setting, Hyperion (q.v., 1797–9). In 1794–5 he was for a few months in Jena, where he made contact with Schiller. In December 1795 he took up a new post as tutor in the house of a Frankfurt banker, J. F. Gontard. Here he fell in love with his employer's young wife Susette (see Gontard, Susette), who returned his affection. She became for him an embodiment of the Hellenic ideal, symbolized by the name Diotima (q.v.), by which he refers to her in his poems and in Hyperion. In 1798 he left after a scene with Gontard, and spent the next two years at Homburg with a devoted friend, Isaak von Sinclair (1775–1815). He did not see Susette Gontard after 1799. It was during the years 1796–9 that Hölderlin's characteristic style of poetry developed; the change is perceptible in the drafts made in 1797–9 for a tragedy on

Empedocles. In 1800 he returned home for a time, becoming a tutor once more at Hauptwil near St. Gall in 1801, and again at Bordeaux in 1802. In the same year he returned home to Nürtingen in a seriously disturbed state of mind. He recovered, made a visit to Regensburg with Sinclair, and was appointed librarian at Homburg. But the mental illness recurred, and after a period at an institution in Tübingen he was entrusted to the care of a local master carpenter, named Zimmer. With him Hölderlin spent the rest of his life (1807–43).

During his lifetime Hölderlin published, apart from Hyperion (which includes his best-known poem, the Schicksalslied), two volumes of translations of Sophocles (1804). Some of his poems appeared in periodicals, and the first collected edition (Gedichte) was published in 1826. Hölderlin was a man of intellectual passion, and his early humanitarian ideals were increasingly dominated by an aching longing for Greece, with which he successfully fused his love for Diotima. Conscious of his poetic powers, he believed that he had a mission to regenerate Germany in an age in which all that ancient Greece had stood for seemed lost, and he found himself tragically unable to fulfil this task.

Hölderlin's early poetry leans on Schiller's ideals, and on Schiller's strophic manner. His poems in classical metres, written mainly between 1796 and 1801, include a number of epigrams, among them Ehmals und Jetzt and Lebenslauf, the Diotima poems, including Der Abschied (q.v.), and the odes Mein Eigentum, Heidelberg, Der Neckar, Die Heimat, Dichtermut, and Der gefesselte Strom (qq.v.). Towards the end of this period he wrote the three great elegies Der Archipelagus, Menons Klagen um Diotima, and Brod und Wein (qq.v.). In the last years of his sanity, after the loss of Diotima, he turned to hymnic verse, writing poems of haunting beauty in free rhythms; many of these exist only as unfinished drafts. They include Am Quell der Donau, Germanien, Der Rhein, Friedensfeier, and Patmos (qq.v.). In some of the later poems Hölderlin seeks to reconcile Christianity (which he never completely abandoned) with his beloved Hellas. Hyperion and the dramatic fragments of Der Tod des Empedokles (q.v.) are concerned with the Greek ideal, the mission of the poet, and the deafness of the world around him.

An edition of Hölderlin's works appeared as Sämtliche Werke in 1846. His works, letters, and documents are included in the Große Stuttgarter Ausgabe, ed. F. Beißner and A. Beck (planned in 8 vols., of which the last is a concordance, 1946 ff.). Part I of Wörterbuch zu Friedrich Hölderlin, ed. H.-M. Danhauser et al. appeared in 1983. Sämtliche Werke. Frankfurter Ausgabe, ed. D. E. Sattler, appeared 1976 ff. Notable translations were furnished by J. B. Leishman in *Selected Poems* (1944) and M. Hamburger in a bilingual edition, *Poems and Fragments* (1966).

HOLLAENDER, FELIX (Leobschütz, 1867–1931, Berlin), worked at the Deutsches Theater, Berlin, under M. Reinhardt (q.v.) from 1908 to 1913, and was in charge of the Grosses Schauspielhaus, Berlin, from 1920 until his death. He wrote Naturalistic novels of socialistic trend (*Jesus und Judas*, 1891; *Magdalena Dornis*, 1892; *Das letzte Glück*, 1895; *Sturmwind*, 1895; *Erlösung*, 1899). His most ambitious novel was *Der Weg des Thomas Truck* (2 vols., 1902). His later novels aimed primarily at sophisticated entertainment (e.g. *Der Tänzer*, 1918; *Der Demütige und die Sängerin*, 1925).

HÖLLERER, WALTER (Sulzbach-Rosenberg, Upper Palatinate, 1922–), taught at Frankfurt University before being appointed to the chair for Germanistik at the Technische Universität, Berlin (1959). From 1954 to 1967 editor of *Akzente* (q.v.), and since 1961 editor (with N. Miller) of *Sprache im technischen Zeitalter*, he has directed the Literarisches Colloquium Berlin since its inception in 1963. *Autoren im Haus. Zwanzig Jahre Literarisches Colloquium* was published in 1982. The author of a novel (*Die Elefantenuhr*, 1973) and of a comedy (*Alle Vögel alle*, 1978), but mainly an exponent of experimental lyric poetry, he published his own collection, *Gedichte 1942–82*, in 1982. Well-known for his promotion of young authors, he edited *Transit. Lyrikbuch der Jahrhundertmitte* (1956) and *Spiele in einem Akt. 35 exemplarische Stücke* (1961) as well as other publications.

HOLLONIUS, LUDWIG, a pastor in Pölitz, Pomerania, in the early years of the 17th c., whose dates of birth and death are not known, is the author of two comedies, *Freimut d. i. Vom Verlorenen Sohn* (1603) and *Somnium vitae humanae* (1605). The latter is a successful treatment in short rhyming couplets of the theme familiar to English readers in the story of Christopher Sly in *The Taming of the Shrew*. Jan der Ebriack finds himself duke for a day. The play, which uses North German dialect (Plattdeutsch) for the characters of low life, is conceived as a moral warning against the vanity of this world.

HOLMSEN, BJARNE P., pseudonym used by Arno Holz and Johannes Schlaf for their joint work *Papa Hamlet* (q.v., 1889). The choice of name reflects the prestige which Scandinavian authors enjoyed at the time, as a consequence of the success of Ibsen (q.v.) and Bjørnson.

HOLSTEIN, FRIEDRICH VON (Schwedt, 1837–1909, Berlin), a German Foreign Service official, usually seen as an *éminence grise* (Graue Eminenz) behind the scenes in the Foreign Office of the Empire. From 1861 Holstein was a diplomat attached to Bismarck (q.v.), who used him as a kind of spy on the German Embassy in Paris, and he was both humiliated and embittered when Bismarck, contrary to a previous undertaking, forced him to appear in open court as a witness against Count Harry von Arnim (q.v.). As a result Holstein incurred widespread social ostracism, and avenged himself by collecting and indexing private and potentially damaging information on public personalities. From 1876 until 1906 Counsellor (Vortragender Rat) in the German Foreign Office, he intrigued against his chief over the Reinsurance Treaty (see RÜCKVERSICHERUNGSVERTRAG), passing secret information to the Austrian Foreign Office. After Bismarck's fall (1890) he was virtually Foreign Secretary, since the official holders of the office, the Imperial Chancellors Caprivi (q.v., 1890–4) and Hohenlohe-Schillingsfürst (q.v., 1894–1900), were inexperienced in diplomacy, and he continued to exercise considerable power under B. von Bülow (q.v.).

Holstein bears the moral responsibility for the policy of aloofness towards Great Britain, and for the German blustering over Morocco in 1905, which collapsed at the Algeciras Conference. He made a feint of resignation in 1906, which, to his chagrin, was accepted. He took his revenge by using his private collection of compromising information in prompting M. Harden (q.v.) to accuse Prince Philipp zu Eulenburg (q.v.), a personal friend of the Emperor Wilhelm II (q.v.), of homosexuality, an aspersion which was almost certainly false.

HOLTEI, KARL VON (Breslau, 1798–1880, Breslau), a prolific and popular writer of works for entertainment, was at school in Breslau, and in 1815 served as a volunteer rifleman in the last phase of the Wars of Liberation (see NAPOLEONIC WARS). He then studied law at Breslau University until he went on the stage in 1819. Unsuccessful as an actor, he became a theatrical administrator in Breslau and subsequently in Berlin. His first wife, the actress Luise Rogée, died in 1825, and he married another actress, Julie Holzbecher, in 1829. He left Berlin in 1828, was next at the Darmstadt theatre, then in Berlin, and in 1837 became theatre director in Riga. His second wife died in 1838.

Holtei began to write plays in his early Breslau days, and was especially successful with Liederspiele, light-hearted comedies with inserted songs modelled on the French vaudeville. The best-known examples are *Die Wiener in Berlin* (1824), *Die Berliner in Wien* (1825), *Die deutsche Sängerin in Paris* (1826), *Der alte Feldherr* (1829), and *Lenore* (1829), in which his second wife had a notable success as the eponymous heroine. *Lenore* also contains the song 'Schier dreißig Jahre bist du alt', known as the *Mantellied* because it refers to an old coat. Holtei's collected plays were published in 1867 as *Theater* (6 vols.).

Holtei left Riga shortly after the death of his second wife, and became an itinerant recitalist, giving readings from Shakespeare in many German cities. In 1847 he settled in Graz, and began to write novels, some of which have, perhaps generously, been compared to those of Dickens. *Die Vagabunden* (4 vols., 1852) and *Der letzte Komödiant* (3 vols., 1863) are essays dealing with artists (see KÜNSTLERROMAN). Of Holtei's other novels, *Christian Lammfell* (5 vols., 1853) and *Ein Schneider* (3 vols., 1854) deserve mention. His *Erzählende Schriften* (41 vols.) were published 1861–6. He also wrote an autobiography, *Vierzig Jahre* (8 vols., 1843–50). Holtei spent his last years (1864–80) in Breslau.

HOLTHUSEN, HANS EGON (Rendsburg, 1913–), served throughout the 1939–45 War, then joined a Bavarian anti-Hitler liberation group (see also RESISTANCE MOVEMENTS, 2) in the final overthrow of the National Socialist regime. Since the war he has been an author and influential critic. His works include essays on Rilke (q.v.), *Der späte Rilke* (1949) and *R. M. Rilke* (1958), and *Der unbehauste Mensch* (1951), *Ja und Nein* (1954), *Das Schöne und das Wahre* (1958), *Kritisches Verstehen* (1961), and *Plädoyer für den Einzelnen* (1967). A lyric poet of the Christian tradition with a consciousness of the existentialist's standpoint, he is the author of volumes of verse, which are perceptibly influenced by Rilke, T. S. Eliot, and Auden. They are *Klage um den Bruder* (sonnets, 1947), *Hier in der Zeit* (1949), and *Labyrinthische Jahre* (1952). The poem *Tabula rasa*, which has attracted particular attention, is included in *Hier in der Zeit*, though written in 1945. In 1968 Holthusen became president of the Bavarian Akademie der schönen Künste.

HÖLTY, LUDWIG CHRISTOPH HEINRICH (nr. Hanover, 1748–76, Hanover), German poet, who, as a student of theology at Göttingen University, was an original member of the Göttinger Hainbund (q.v.). Hölty's uneventful life was ended when he was 27 by tuberculosis. Of amiable and gentle character, he wrote a number of poems, some of them in classical metres, in which the perpetual nostalgic accentuation of May and springtime is linked with the awareness of impending death. Their bitter sweetness has been caught in musical settings, especially by Brahms (q.v.) in *Die Mainacht*. Only separate poems appeared in Hölty's lifetime. His *Gedichte* were published in 1783 by his friends F. L. Stolberg and J. H. Voß (qq.v.). *Sämtliche Werke*, including letters, ed. W. Michael (2 vols.) appeared as *Kritische Ausgabe*, 1914–18, and *Werke und Briefe*, ed. U. Berger, 1966.

Holy Roman Empire, see DEUTSCHES REICH, ALTES.

HOLZ, ARNO (Rastenburg, East Prussia, 1863–1929, Berlin), son of a pharmacist, had a good schooling in Berlin, but did not go to a university. He started, not very successfully, as a journalist, but became one of the foremost supporters of the Naturalistic movement (see NATURALISMUS). A member of the club Durch (q.v.), he also edited the journal *Freie Bühne* (see FREIE BÜHNE, VEREIN). His early collections of poetry, *Klinginsherz* (1883) and *Deutsche Weisen* (in collaboration with O. Jerschke, q.v., 1884), gained him some recognition, but his first real success came with the poetry of *Das Buch der Zeit* (q.v., 1885). In 1888 began his friendship and collaboration with J. Schlaf (q.v.). In the following year they published their joint work *Papa Hamlet* (q.v.), described as *Drei Skizzen*, under the pseudonym Bjarne P. Holmsen, convinced that Norwegian authorship would attract instant attention. The true authors were soon discovered and they subsequently published under their own names. *Die Familie Selicke* (q.v., 1890), another landmark of the Naturalistic movement, was performed by the Freie Bühne, and the collection *Neue Gleise* (q.v., 1892) contained new prose sketches (in Section One, entitled *Die papierne Passion*), as well as their other two works. In the same year a bitter quarrel separated Holz and Schlaf for life.

In 1890 Holz wrote a theoretical essay supporting the movement which he had helped to set in motion (*Die Kunst. Ihr Wesen und ihre Gesetze*, 2 vols., 1890–2), but his next work, the comedy *Sozialaristokraten* (q.v., 1896), shows signs of a swing away from the Naturalistic manner. In the following years Holz plunged into poetic experiments with words and rhythms, first in *Dafnis* (q.v., 1904), but even more in the extravagant and typographically elaborate *Phantasus* (q.v., 1898). Holz's later works include the comedies *Traumulus* (q.v., 1904) and *Frei!* (1907, both in collaboration with O. Jerschke) and the tragedies *Sonnenfinsternis* (q.v., 1908) and *Ignorabimus* (q.v., 1913). An intolerant, irascible personality, Holz quarrelled publicly with Schlaf, as well as with the Berlin literary historian R. M. Meyer (1860–1914).

Holz published his own works as *Werke* (10 vols.) in 1924–5, and as a *Monumentalausgabe* (12 vols.) in 1926. *Werke* (7 vols.), ed. W. Emrich and Anita Holz, appeared 1961–4, and a selection of correspondence, *Briefe. Eine Auswahl*, ed. Anita Holz and M. Wagner, 1949.

HOMBURG, ERNST CHRISTOPH (Mihla nr. Eisenach, 1605–81, Naumburg), a minor poet, studied in Wittenberg, visited Holland, and in 1642 settled in Naumburg, where he practised law. His poetry shows Dutch influence. He published *Schimpff- und Ernsthaffte Clio* in 1638 under the pseudonym Erasmus Chrysophilus Homburgensis. He also wrote hymns and translated the Odes of Horace. As a member of Die Fruchtbringende Gesellschaft (q.v.), he was known as Der Keusche.

Homo Faber, a novel by M. Frisch (q.v.), published in 1957. The title virtually subsumes the book. Walter Faber (it really is his surname) represents the species *homo faber*, an ironic taxonomic invention by Frisch to denote the modern technologist. The book is divided into two stations (*Stationen*). Faber, an engineer, is due to go from New York to Caracas by air. A halt is made at Houston, Texas, during which Faber, irritated by his German neighbour in the aircraft, tries to miss his flight. But he misinterprets the departure announcements and finds himself once more aboard in the same company. As a result of engine failure the plane makes an emergency landing in the Mexican desert. During the four days before their rescue Faber becomes friendly with Herbert, the German he had at first detested, and finds that he is the brother of Joachim, once Faber's closest friend, that Joachim married Hanna, with whom Faber had lived, and that the couple are now divorced. Faber decides to go with Herbert to the Guatemalan tobacco plantation which Joachim is now managing. They arrive to find that Joachim has hanged himself, probably out of loneliness and depression in the appalling climate. Faber, the man who organizes his life on tidy administrative lines, and cannot bear other people interfering or impinging upon it, has now already made serious departures from his schedule. He returns to New York, meets his woman friend Ivy, but curtails his stay with her as it threatens to become too intimate, and chooses the sea-crossing to Europe because it means an earlier departure. On the liner he meets a girl, Elisabeth; a sympathy springs up between him and Sabeth, as he calls her, and although he has always evaded marriage, Faber proposes to her. She refuses, but he calls on her in Paris, and they

go south in his car. At Avignon she comes to his room, and they continue the journey as lovers. In the Campagna Faber discovers, in casual conversation, that Hanna, now living in Athens, is Sabeth's mother. He already knows that Hanna was pregnant before her marriage and had a daughter by him. They cross to Greece, and near Athens Sabeth is bitten by a snake, and in her fright has a bad fall. She is taken to hospital, and in consequence Faber and Hanna meet again. The serum cures Sabeth's snake bite, but she dies suddenly of an undiagnosed minor fracture of the skull, a result of her fall. This fulfilment of fate, a case of Oedipus in reverse, completes the first station.

In the second station, a mere 50 pages, Faber completes the curse by dying of abdominal cancer. One act of 'chance' after another has involved the completely disengaged and aloof technologist in an archetypal human situation. *Homo faber* is no advance on *homo sapiens*.

HOPFEN, HANS (Munich, 1835–1904, Groß-Lichterfelde, Berlin), a lawyer who turned to writing in the self-consciously artistic manner of the Munich school (see MÜNCHNER DICHTERKREIS). He was a prolific author of novels and Novellen. Among his better-known titles are the sensational novels *Verdorben zu Paris* (1867) and *Arge Sitten* (1869), the success of which is said to have induced him to produce too many stories to the same formula. His best-known volumes of Novellen are *Bayrische Dorfgeschichten* (1878) and *Die Geschichten des Majors* (1879). He also wrote stories of student life, of which *Die fünfzig Semmeln des Studiosus Taillefer* (1891) is an example.

Horacker, a short novel written by W. Raabe (q.v.) in the second half of 1875, and published in the following year. Horacker, a youth who has escaped from a reformatory, haunts the woods, and rumour falsely attributes all manner of outrageous crimes to him. On the afternoon of the story he accidentally encounters two schoolmasters and is brought by them, starving and exhausted, to the parsonage of the village of Gansewinckel. There, too, is Lottchen Achterhang, an orphan like Horacker, who is deeply attached to him and has made her way on foot, having heard that he is at large and in trouble. The inflated bubble of Horacker's alleged crimes bursts, and the story ends with the couple asleep in the parsonage with their troubles assuaged. The quality of the book lies not in this simple plot, but in the idyllic atmosphere and in the skilful convergent structure. The whole action takes place in the few hours of a blazing July afternoon and cloudless evening. The dozen or so characters are brought to the Gansewinckel parsonage, including the ripe and genial eccentrics, Konrektor Eckerbusch and his wife Ida, Zeichenlehrer Windwebel, and Staatsanwalt Wedekind. At the centre are the rich characters of parson Winckler and his wife Billa. A dash of vinegar, provided by the pedantic and conceited young schoolmaster Dr. Neubauer, tempers the sweetness.

Horatier und die Kuriatier, Die, a play (Schulstück) in mixed prose and verse by B. Brecht (q.v.), written in 1934, and consisting of three short episodes. The *Vorspruch* published in *Versuche* (vol. 14) describes it as 'ein Lehrstück über Dialektik für Kinder'.

Horen, Die, a monthly periodical founded by Schiller in 1795 and edited by him. Its name (after the *horae*) is symptomatic of his classical orientation at the time. Its aim was aesthetic, the cultivation and extension of literary and artistic taste under the category of truth. In Schiller's own words 'Man widmet sie der *schönen* Welt zum Unterricht und zur Bildung, und der *gelehrten* zu einer freien Forschung der Wahrheit und zu einem fruchtbaren Umtausch der Ideen.' Political and religious subjects were excluded.

Die Horen was published by Schiller's new-won friend in Stuttgart, Cotta (q.v.). Goethe, W. and A. von Humboldt, Herder, and A. W. Schlegel (qq.v.) participated. Outstanding contributions were Schiller's *Über die ästhetische Erziehung des Menschen in einer Reihe von Briefen, Über naive und sentimentalische Dichtung* and *Die Belagerung von Antwerpen,* and Goethe's *Römische Elegien, Unterhaltungen deutscher Ausgewanderten,* and *Benvenuto Cellini.* The journal, which achieved only a modest circulation, ceased publication with the third volume in 1797.

Höret die Stimme, a biblical novel by F. Werfel (q.v.), published in 1937. Its subject is the life and prophecies of Jeremiah (Jirmijah) and the failure of the Israelites to take heed of them. His warnings are ignored first by Josiah (Josijah), who rashly makes war on the Egyptians and dies after defeat at Megiddo. Jeremiah prophesies the death of King Jehoiakim (Jojakim), and with difficulty escapes with his life. Faced with the armies of Babylon, Jehoiakim takes his own life, and is succeeded by Zedekiah (Zidkijah). Zedekiah is initially successful against the Babylonians and continues to disregard Jeremiah's warnings. The Israelites are overwhelmed, Zedekiah blinded, and the Babylonian captivity begins. Jeremiah has a vision in which the meaning of the defeats becomes clear in the ultimate life of the spirit.

The novel is set in a frame (see RAHMEN) in Werfel's time, in the course of which the non-practising Jew Jeeves has a vision which brings home to him the link between himself and his ancestors and the God-given destiny of the Jews. In the version published in 1956 as *Jeremias,* the frame is omitted on the instructions of Werfel's widow, who maintained that Werfel himself had rejected it. The book is thus converted into an historical novel, not linked to the present-day, a change which has been the subject of controversy.

HORMAYR, JOSEPH, FREIHERR VON (Innsbruck, 1782–1848, Munich), an Austrian civil servant, took part in 1809 in the rising against the French in his native Tyrol (see NAPOLEONIC WARS). Employed from 1803 in the national archives of Vienna, he was a fertile historian, publishing numerous works on Austrian history, including *Österreichischer Plutarch oder Leben und Bildnisse aller Regenten des österreichischen Kaiserstaats* (20 vols., 1807–14), *Allgemeine Geschichte der neuesten Zeit vom Tode Friedrichs des Großen bis zum zweiten Pariser Frieden* (3 vols., 1817–19), and *Wien, seine Geschichte und Denkwürdigkeiten* (5 vols., 1823–4). From 1828 he was in Bavarian employment, his last appointment being that of director of the Bavarian state archives (1846).

HORN, FRANZ CHRISTOPH (Brunswick, 1781–1837, Brunswick), a schoolmaster in Berlin and Bremen, and finally (1810) a private tutor in Berlin, is the author of numerous light novels, including *Guiskardo der Dichter* (1801), *Viktors Wallfahrten* (1802), *Henrico* (1804–5), *Der Traum der Liebe* (1806), *Otto* (1810), and *Bertha* (1819). He also wrote poems (*Gedichte,* 1820) and stories (*Leben und Liebe,* 1817; *Novellen,* 1819–20), and was a popular writer on the history of literature (*Geschichte und Kritik der deutschen Poesie und Beredsamkeit,* 1805; *Die schöne Literatur Deutschlands während des 18ten Jahrhunderts,* 1812–13; *Die Poesie und Beredsamkeit der Deutschen von Luthers Zeit bis zur Gegenwart,* 1822–9). Horn is ridiculed by Heine in *Atta Troll* (q.v., Kaput XVIII).

HORN, HEINRICH MORITZ (Chemnitz, 1814–74, Zittau), wrote a number of novels, but was chiefly known for his late Romantic verse narratives *Die Pilgerfahrt der Rose* (1852) and *Die Lilie vom See* (1853).

Hornberger Schießen, a proverbial expression used to denote futile efforts. It goes back to an episode in 1519, when, in an encounter between the inhabitants of Hornberg Gutach (Baden) and Villingen, the former were said to have fired over a hundred rounds without inflicting any loss on their adversaries.

HORNBURG VON ROTENBURG, Lupold, a native of Rothenburg/Tauber, who lived in the 14th c. and is believed to have been a layman, wrote five poems in the form of Reimreden (see Reimrede). These are: (1) *Die Landpredigt,* dilating on the various plagues and natural catastrophes visited on Central Europe in 1347 and 1348 as a punishment for the misrule of the princes; (2) *Die Rede von des Reiches Klage,* which is virtually identical with a poem by Otto Baldemann von Karlstadt; (3) *Der Zunge Streit,* a polemical poem directed at the Emperor Karl IV (q.v.) on account of his support for 'the false Woldemar' (see Woldemar, Markgraf von Brandenburg); (4) a eulogy (Ehrenrede) of Baron Konrad von Schlüsselberg. These four poems date from 1347–8. (5) *Von allen Singern,* not later than 1355, mentions briefly twelve Minnesänger and Spruchdichter (see Spruch) of the 13th c.

Horn von Wanza, Das, a short novel written 1879–80 by W. Raabe (q.v.) and published in 1881. It is a Rahmenerzählung (see Rahmen), and the frame is at least as significant as the narrative it contains. The story takes place in 1869, the year before the Franco-Prussian War and the proclamation of the German Empire. The family Grünhage, consisting of the father (a not very prosperous doctor), four daughters, and a son, lives in the Lüneburger Heide. The son decides to visit an aunt by marriage who lives at Wanza, a fictitious town in the Harz, and simultaneously to look up an old university acquaintance who is the Bürgermeister. There are scarcely any events in this part of the story, but a group of colourful characters is revealed, all, except Bürgermeister Dorsten, reaching back to the early part of the century. They are Aunt Sophie (Frau Rittmeisterin), the 80-year-old and blind Thekla Overhaus, and the night-watchman Marten. In the inner part of the story these three narrate their intertwining lives in Wanza. The Rittmeisterin, married off perforce to a wild, foul-mouthed ex-officer, survives her misery and subdues her husband. Thekla Overhaus loses her fiancé at the battle of Leipzig and keeps his memory green, and Marten devotes his life to the service of both of them. The characters touch German history at many points, always from a personal angle, and the night-watchman's horn (das Horn von Wanza), now no longer used, is a symbol of the old Germany yielding to the new. The sole event in that part of the story which frames the historical recollections is the planning of the golden jubilee of the arrival of Frau Rittmeisterin in Wanza in 1819. To the celebration young Grünhage's family are invited, and with their coming the story ends.

Horribilicribrifax Teutsch oder Wählende Liebhaber, a comedy (Schertz-Spiel), written by A. Gryphius (q.v.), probably *c.* 1648–50, and published in 1663. The plot, though unimportant, is complex, since it involves the varying fortunes of no fewer than seven pairs of lovers. The principal features of the comedy are firstly the caricature of the two vainglorious but cowardly soldiers (*milites gloriosi*), Horribilicribrifax and Daradiridatumtarides, and of the pedant Sempronius, and secondly the intricate and dazzling pyrotechnics of Gryphius's comic style.

Hörspiel, strictly a play written for broadcasting by radio, though the term is often extended to include adaptations for radio of plays originally written for stage performance or of works of fiction.

Horst-Wessel-Lied, see Wessel, Horst.

Hortus Deliciarum, see Herrad von Landsberg.

HORVÁTH, Ödön von (Fiume, now Rijeka, 1901–38, Paris), the son of a Hungarian diplomat, was brought up in Belgrade, Budapest, Munich, Preßburg, and Vienna before studying drama, philosophy, and Germanistik in post-war Munich. He then lived as an author at Murnau and Berlin, where he gained recognition, emigrating in 1933, as an anti-National Socialist, to Austria, though until 1936 he had a permit enabling him to return to Berlin. With his plays banned, his main interest was the study of Nazism. On the German annexation of Austria he emigrated to Zurich, but in late May 1938 visited Paris. On 1 June he was killed by a branch that fell from a tree beneath which he had sought shelter during a thunderstorm on the Champs Élysées. He was a successful writer of satirical, often political, comedies and Volksstücke (see Volksstück). Among these are *Geschichten aus dem Wiener Wald* (q.v., 1931), *Italienische Nacht* (1931), *Kasimir und Karoline* (1932), *Hin und Her* (1934), *Die Unbekannte aus der Seine* (posth. 1949), and *Figaro läßt sich scheiden* (1937). Other plays included *Revolte auf Côte 3018* (1927, retitled *Die Bergbahn,* 1929), *Sladek der schwarze Reichswehrmann* (1929), *Glaube Liebe Hoffnung* (1933), and *Der jüngste Tag* (first performed 1937). He is also the author of three novels, *Der ewige Spießer* (1930), *Jugend ohne Gott,* and *Ein Kind unserer Zeit* (both 1938). These last two works are concerned with dictatorship and individual responsibility, and were reissued together as *Zeitalter der Fische* (2 vols., 1953). Interest in Horváth, whose astringent satire is tempered by humour, revived in the 1960s.

Gesammelte Werke (4 vols.), ed. W. U. Huder *et*

al., appeared 1970–1 and *Gesammelte Werke* (8 vols.), ed. D. Hildebrandt and T. Krischke, in 1972.

Hose, Die, a comedy (bürgerliches Lustspiel) by C. Sternheim (q.v.). Written in 1909–10 (published 1911) as the first part of the *Maske-Tetralogie* (q.v.), it deals with a Bürger family symbolically named Maske, typifying the German middle class in the decades preceding the 1914–18 War. By conforming to its standards, by wearing the 'mask' of Bürgerlichkeit, while at the same time leading a life of self-indulgence, Sternheim's Maske characters achieve social standing and economic success.

When Theobald Maske's wife Luise loses her drawers in public, Theobald is convinced that she has jeopardized his position as a civil servant. In this he proves wrong, but the two men who happen to have noticed the incident call at his flat to rent a room. They are Scarron, an ineffective poet, and Mandelstam, a hairdresser; both, however, abandon their original intention of seducing Luise. Theobald himself makes an easy conquest of Gertrud Deuter, an unmarried neighbour resigned to watching the illicit love-affairs of others. To his unsuspecting wife he announces that they will now be able to start a family, for the income from the lodgers has secured the necessary money, and the invention of the press-stud brings the consolation that Luise's drawers will never again threaten his existence. Luise herself, who had hoped for an adventure, takes her place once more in the kitchen.

The play (with the other Maske dramas) belongs to the cycle of social comedies *Aus dem bürgerlichen Heldenleben.* Theobald's 'heroism' consists in the unashamed egotism with which he exploits his environment without sacrificing his trivial individuality. He has cultivated the ability to turn fetters into freedom and to reject unrealistic ambitions. The comedy was first performed in February 1911 (Berlin Kammerspiele) with the title *Der Riese,* a concession to the Berlin Polizeipräsident who had four days previously forbidden public performance of *Die Hose* on moral grounds.

Hosen des Herrn von Bredow, Die, a historical novel (Vaterländischer Roman, 2 vols.) by W. Alexis (q.v.), published 1846–8. It is set in Brandenburg in the early 16th c. when Joachim I (q.v.) was Elector. Götz von Bredow, lord of Hohenziatz, a staunch and rough-and-ready character, possesses, as a kind of heirloom, an ancient pair of leather breeches, from which he will not be parted for so much as one day. When he returns home from a carouse and begins to sleep off the effects, his wife Brigitte takes possession of this garment in order to wash it. It is stolen by a pedlar, who is in turn robbed and bound to a tree by a predatory noble, Wilkin von Lindenberg. Götz's adopted son, Hans Jürgen, sets free the pedlar, first retrieving the leather breeches. News of the robbery incenses the Elector, and Lindenberg persuades Götz von Bredow to take the blame for it; but the real offender is identified by the pedlar, whereupon he is condemned by the Elector and executed.

This judicial severity provokes the nobility to a conspiracy, which Götz von Bredow is minded to join. But the resourceful Frau Brigitte conceals his indispensable leather breeches, and so prevents his participation. Hans Jürgen discovers the movements of the conspirators and warns the Elector. In consequence the plot fails and the conspirators are executed.

Hosenrolle, a male part in a play or opera which is designed to be played by a female. An outstanding example in German literature is the part of Octavian in *Der Rosenkavalier* (q.v.).

HOUWALD, ERNST CHRISTOPH, FREIHERR VON (Straupitz, Saxony, 1778–1845, Lübben), a Saxon country gentleman, held, from 1821 onwards, elective administrative office in Lower Lusatia (Niederlausitz), devoting his spare time to writing. He began with two volumes of stories, *Romantische Akkorde* (1817) and *Erzählungen* (1819), and then took to writing plays. Starting with one-act plays (*Die Spielkameraden, Die Freistatt,* both 1819), he wrote a parody of a fate tragedy, *Seinem Schicksal kann niemand entgehen* (1819), before turning seriously to this genre (see SCHICKSALSTRAGÖDIE). His first two examples were *Die Heimkehr* and *Das Bild* (both 1821); *Der Leuchtturm* (q.v., also 1821) proved by far the most successful. Houwald followed this up with *Der Fürst und der Bürger* (1822), *Die Feinde* (1825), and *Die Seeräuber* (1830), but his reputation as one of the chief exponents of fate tragedy rests on his first plays. All his works depended on sensationalism, and soon lost their public.

HOWARD, CATHERINE, fifth queen of King Henry VIII, was the subject of an opera libretto (*Die sterbende Unschuld oder die Durchlauchtigste Catharina Königin von England*) written by J. C. Hallmann (q.v.), probably *c.* 1673.

HOYERS, ANNA OWENA (Koldenbüttel, Schleswig, 1584–1655, Sweden), daughter of an astronomer and wife of a gentleman farmer at Hoyerswort, emigrated to Sweden because, through her sympathies with the Schwenckfeldianer (see SCHWENCKFELD, KASPAR VON), she encountered local hostility after her husband's death. She wrote *Geistliche und weltliche Poemata* (1650), some of which are in Low German (Plattdeutsch).

HRABANUS MAURUS (Mainz, c. 776–856, Winkel nr. Rüdesheim), Abbot of Fulda from 822 to 842, became Archbishop of Mainz in 847. A pupil of Alcuin, he helped to develop and spread classical learning in German lands. His works, of which the handbook *De clericorum institutione* was the best known, are written exclusively in Latin. He is believed to be the author of the hymn 'Veni creator spiritus' (q.v.).

HROTSVITH, see ROSWITHA VON GANDERS-HEIM.

HUBER, KURT (Chur, Switzerland, 1893–1943, Munich), became in 1926 professor of philosophy and sound psychology at Munich University. He was an authority on the German Volkslied. From 1942 until his arrest he was the moving spirit behind the student movement Weiße Rose (see RESISTANCE MOVEMENTS, 2). He was condemned to death by a Volksgerichtshof, and executed on 13 July 1943. A number of his works were published posthumously, among them *Musik-Ästhetik* (1954) and *Grundbegriffe der Seelenkunde* (1955).

HUBER, LUDWIG FERDINAND (Paris, 1764–1804, Ulm), was brought up in Leipzig, and, as a friend of C. G. Körner (q.v.), was one of the four young people who wrote to Schiller in 1784 and welcomed him in Leipzig in 1785. Huber, who was in the Saxon legation at Mainz in 1787, entered into a liaison with Therese Forster (see HUBER, THERESE), whom he married in 1794, after resigning his post. He was active as a writer, chiefly in the political field, and as a translator; in his last years he edited Cotta's *Allgemeine Zeitung* (q.v.).

HUBER, THERESE (Göttingen, 1764–1829, Augsburg), a daughter of a distinguished professor of Göttingen, C. G. Heyne. In 1784 she married the explorer and travel writer Georg Forster (q.v.), whom she abandoned for L. F. Huber (q.v.) in Mainz. After Forster's death in 1794 Therese Forster married Huber. She wrote profusely in her middle and later years, and during Huber's lifetime her writings appeared under her husband's name. Her best-known works are the novels *Emilie von Barmont* (1794), *Die Familie Seeldorf* (1795), and *Die Ehelosen* (1829). She was also active as a journalist, editing Cotta's *Morgenblatt für gebildete Stände* (q.v.) from 1816 to 1823.

Hubertusburger Friede (the Peace of Hubertusburg), signed on 15 February 1763, concluded the Seven Years War (see SIEBENJÄHRIGER KRIEG). By its provisions Austria renounced all claims to Silesia and Glatz, and Prussia restored Saxony to independence under its ruler, the Elector August III (q.v.), who died in that year and was succeeded by his son Friedrich Christian.

Hubertusorden, the Order of St. Hubert, founded in 1444, was until 1919 the premier Bavarian order of chivalry.

HÜBNER, TOBIAS (Dessau, 1577–1636, Dessau), tutor to the princes in Dessau and, as Der Nutzbare, a member of Die Fruchtbringende Gesellschaft (q.v.), claimed to have been the first to write Alexandrines (see ALEXANDRINER) in German, but he seems to have been anticipated by Lobwasser and Melissus (qq.v.). He translated into German the unfinished epic *Les Semaines* of Guillaume du Bartas (1544–90) as *Die ander Woche* (1619) and *Herrn zu Bartas Erste Woche* (1640).

HUCH, FRIEDRICH (Brunswick, 1873–1913, Munich), after completing his university studies at Munich, Berlin, and Paris, was for a short time a private tutor, and then lived in Munich as an independent author. A sensitive and careful writer, interested in the psychology of childhood and youth, he wrote novels of a serious, yet popular character. They include the story of a village boy, *Peter Michel* (1901), *Geschwister* (1903) and its sequel *Wandlungen* (1905), and *Mao* (1907). His most successful work is a complicated web of love-affairs under the title *Pitt und Fox, die Liebeswege der Brüder Sintrup* (1909)—the names are eccentric first names, the novel is set in the Germany of Huch's day and has nothing to do with the English politicians. Huch also wrote stories (*Träume*, 1904; *Erzählungen*, published posthumously in 1914). He was a cousin of Ricarda Huch (q.v.).

Gesammelte Werke (4 vols.) appeared in 1925.

HUCH, RICARDA (Brunswick, 1864–1947, Schönberg/Taunus), came of an educated family, studied at Zurich University (German universities did not then admit women), and obtained a D.Phil. in 1891. After a period at the Zurich Zentralbibliothek (1891–7), she became a teacher in Zurich and later in Bremen. By this time she had published 2 vols. of poems (*Gedichte*, 1891 and 1894), a play (*Evoe*, 1892), her first novel, *Erinnerungen von Ludolf Ursleu dem Jüngeren* (q.v., 1893), the story *Der Mondreigen von Schlaraffis* (1896), and a collection of tales,

HUFELAND

Erzählungen (3 vols., 1897). In 1898 she married an Italian dentist in Vienna with whom she took up residence in Trieste (then Austrian). During the marriage, which was dissolved in 1906, she produced a further volume of stories, and made a study of the Romantic movement (see ROMANTIK), which resulted in *Blütezeit der Romantik* (1899) and *Ausbreitung der Romantik* (1902, published in 1908 as *Die Romantik*, 2 vols.). Though not intended to be handbooks of literary history, these works show great empathy and skill in imaginative reconstruction. While in Switzerland she had come to admire Keller (q.v.), on whom she published a study, *Gottfried Keller* (1904), and Gotthelf (q.v.), to whom she paid tribute in a lecture, *Jeremias Gotthelfs Weltanschauung* (1917).

Her immediate experience of the poverty and squalor of Trieste produced the most powerful of her early works of fiction, *Aus der Triumphgasse* (1902), described as Skizzen. The novel *Vita somnium breve* (1902, retitled *Michael Unger* in 1913) was less successful. In 1905 she published a collection of stories, *Seifenblasen*, one of which, *Lebenslauf des heiligen Wonnebald Pück*, appeared as a separate publication in 1913. During the Trieste period she moved from purely imaginative fiction and works on literature to the exploration of history. Her Garibaldi studies resulted in *Die Geschichten von Garibaldi* (2 vols., 1906–7). In 1908 she published *Risorgimento*, a short treatise, and in 1910 appeared her last work to present historical biography in the form of a novel, *Das Leben des Grafen Federigo Confalieri*. She had by now moved to Brunswick and married a lawyer cousin. The marriage was dissolved in 1910. For the next ten to fifteen years she devoted herself extensively to historical studies, which, combined in her subsequent works with a disciplined imagination, enabled her to present a plausible vision of past ages. *Der große Krieg in Deutschland* (3 vols., 1912–14) was renamed after the 1914–18 War *Der dreißigjährige Krieg* (1937). *Deutsche Geschichte* (3 vols., 1934–49) consists of: (1) *Römisches Reich deutscher Nation*; (2) *Das Zeitalter der Glaubensspaltung*; (3) *Untergang des römischen Reiches deutscher Nation*. These major works were supported by other publications, notably *Wallenstein* (1915), *Michael Bakunin und die Anarchie* (1923), *Freiherr vom Stein* (1925), and studies of German cities in the past, *Im Alten Reich* (2 vols., 1927–9).

The third phase of Ricarda Huch's development, which grew out of both her own meditation and her historical interests, is her affirmation of Christianity, to which in her early life she had been indifferent or hostile. *Luthers Glaube* (1916), *Der Sinn der Heiligen Schrift* (1919), *Entpersönlichung* (1921), and *Urphänomene* (1946)

reflect her religious attitudes. She presented a resolute front to the menace of National Socialism, and her letters to the president of the 'reformed' Akademie der Künste (see AKADEMIEN), written in March/April 1933, are models of calm, determined courage. She and her family had trouble with the authorities, but they escaped imprisonment. Thomas Mann (q.v.) described her as 'Deutschlands erste Frau'. *Gesammelte Werke* (11 vols.), ed. W. Emrich, appeared 1966–74.

HUCHEL, PETER (Berlin-Lichterfelde, 1903–), early devoted himself to writing; he served in the war from 1940 until 1945, when he was appointed director of the East Berlin radio. From 1948 to 1962 he was editor of *Sinn und Form* (q.v.), the East German literary periodical. His inadequate orthodoxy led to his dismissal in 1962. He is a melancholy, elegiac poet, fundamentally religious, and deeply imbued with the image and spirit of the landscape of the Mark Brandenburg, in which he spent his childhood and youth. His poems, published separately over many years, are available in *Gedichte* and *Chausseen Chausseen* (1963), *Gezählte Tage* (1972), and *Die neunte Stunde* (1977). *Gesammelte Werke* (2 vols.) were published in 1984.

HUDEMANN, LUDWIG FRIEDRICH (Friedrichstadt/Eider, 1703–70, Henstedt, Dithmarschen), who practised law in Hamburg, was a prolific poet. His epics include *Luzifer* (1765) and *Der auferstandene Messias* (1767); among his tragedies are *Der Brudermord Kains* (1765) and *Der Tod des Johannes des Täufers* (1770); and he translated some of the tragedies of Racine and P. Corneille. He also wrote lyric poetry.

HUELSENBECK, RICHARD (Frankenau, Hesse, 1892–1974, Minusio, Tessin), a founder of Dada and principal author of its literature (see DADAISMUS), spread the movement from Zurich to Berlin, and emigrated to the U.S.A. in 1936 to practise medicine and psychiatry. In 1970 he returned and settled in Switzerland. His works include fiction (*Azteken oder die Knallbude, Verwandlungen*, both Novellen, 1918; *Doctor Billig am Ende*, 1921, *China frißt Menschen*, 1930, *Der Traum vom großen Glück*, 1933, all novels), and poems (*Schalaben, Schalomai, Schalamezomai*, 1916; *Phantastische Gebete*, 1916; *Die New Yorker Kantaten*, 1952, and *Die Antwort der Tiefe*, 1954). In 1959 he published *Sexualität und Persönlichkeit*.

HUFELAND, CHRISTOPH WILHELM FRIEDRICH (Langensalza, 1762–1836, Berlin), son of the Weimar court physician, studied medicine and then himself practised in Weimar, where he attended Duke Karl August, Goethe, Schiller, Wieland, and Herder (qq.v.). The Duke nominated him in 1792 to a chair of medicine at Jena. In 1798 he moved to Berlin, and in 1809 became a professor at the newly founded university. He practised in the highest circles and attended Queen Luise (q.v.) on her flight to Memel after the Prussian defeat in 1806.

Hufeland addressed his writings to a lay public. *Makrobiotik, oder die Kunst sein Leben zu verlängern* (1796) attracted Karl August's favour even before its publication. It was followed by an almost equally popular work, *Guter Rat an Mütter über die wichtigsten Punkte der physischen Erziehung der Kinder in den ersten Jahren* (1799).

HÜGEL, FRIEDRICH, FREIHERR VON (Florence, 1852–1926, London), first studied law, then turned his attention to theology. Hügel, a Roman Catholic of broad outlook, lived in England and wrote in English on theological subjects.

HUGENBERG, ALFRED (Hanover, 1864–1951, Kükenbruch), a co-founder of the Pan-German League (see ALLDEUTSCHER VERBAND), was a director of various large concerns, and chairman of Krupp. He founded the Hugenberg-Konzern, which gained control over leading, mainly nationalistic, newspapers, and the film industry (Ufa). From 1928 leader of the Nationalist Party (Deutschnationale Volkspartei), he campaigned bitterly against the Weimar Republic, and in October 1931 achieved an alliance (see HARZBURGER FRONT) of Nationalists, Stahlhelm, and NSDAP (qq.v.) against the minority government of Brüning (q.v.). In 1932 the Harzburger Front collapsed through inability to agree about the presidential election. Hugenberg nevertheless was a member of Hitler's government in January 1933, but was dropped in June of that year. In 1927 he published *Streiflichter aus Vergangenheit und Gegenwart*.

Hugenotten, the French Huguenots who, as adherents of Calvinism (q.v.), suffered persecution in their native country, especially after the revocation of the Edict of Nantes in 1685. A number of German writers are of Huguenot origin, e.g. A. von Chamisso, F. de la Motte FOUQUÉ, and Th. Fontane (qq.v.). The large number of distinguished army officers with French names derive from the same source.

HUGGENBERGER, ALFRED (Bewangen nr. Winterthur, 1867–1960, Dießenhofen), a Swiss farmer, wrote humorous and sentimental novels and stories set in the rural Swiss background he knew. Of his novels *Die Bauern von Steig* (1913), *Die Frauen von Siebenacker* (1925), *Der wunderliche Berg Höchst und sein Anhang* (1932), and *Die*

Schicksalswiese should be mentioned. The most notable collections of tales are *Bauernbrot* (1941) and *Liebe auf dem Lande* (1943). He also published several volumes of verse and a book of reminiscence, *Die Brunnen der Heimat* (1927).

HUGO VON LANGENSTEIN, a Swabian nobleman who took holy orders, was the author of a legend of St. Martina, a saint new to the German calendar. Hugo, whose family owned land in Hegau and the island of Mainau in Lake Constance, entered the Teutonic Order (see DEUTSCHER ORDEN) in 1272, and is last mentioned as a priest in Freiburg (Breisgau) in 1298. *Die Heilige Martina*, an immense poem of 33,000 lines, recounts the saint's sufferings, steadfast faith, and martyrdom. Its length derives partly from prolixity of style and partly from the application to the saint's legend of the method of allegorical interpretation. The poem is dated 1293. Hugo seems to have been influenced by KONRAD von Würzburg (q.v.), whom he knew personally. He has also been credited with the authorship of the *Mainauer Naturlehre* (q.v.).

HUGO VON MONTFORT (1357–1423), a great nobleman and Middle High German poet, was a member of a powerful family of Vorarlberg. He was prominent in the political life of Austria under Duke Leopold III (reigned 1365–86), took part in campaigns in 1377 and 1382, and was governor of Aargau and Thurgau in Switzerland in 1388 and of Styria in 1415. He was married three times, each time happily. His artless songs, which record his domestic affections and reflect his sincere piety, are attractive in their touch of feeling, though technically they mark a stage in the decline of Minnesang.

HUGO VON TRIMBERG, a Middle High German didactic poet, and a native of the Würzburg region, who is believed to have been born c. 1230 and to have died c. 1313. For some forty years he was a schoolmaster in the abbey school of St. Gangolf in Teuerstadt, a suburb of Bamberg. He wrote industriously and claims to have composed seven German and 'four and a half' Latin works. Three of the latter survive, a list of saints and saints' days (*Laurea sanctorum*), a collection of moral examples (*Solsequium*), and a verse history of literature (*Registrum multorum auctorum*). From his German output the huge moral poem *Der Renner* (q.v.), a work of his last years, alone survives.

Hug Schapler, a Volksbuch printed in 1500. See ELISABETH VON NASSAU-SAARBRÜCKEN.

HUG VON WERBENWAG, a minor Middle High German poet of the middle of the 13th c., is recorded in Swabia between 1258 and 1279. Five of his Minnelieder are preserved, one of which treats his love for his lady as a cause at law, which he will take from instance to instance to achieve satisfaction.

Hühnchen, Leberecht, the principal figure in a series of popular stories by H. Seidel (q.v.). Hühnchen lives his life barely above the level of poverty, but maintains an irrepressible good humour and optimism.

Huldigung der Künste, a brief allegorical dramatic work in verse (Ein lyrisches Spiel) by Schiller, composed at short notice, at the request of Goethe, to mark the arrival in Weimar of the Russian bride of the heir apparent to the Duchy of Weimar, who until her marriage was the Grand Duchess Maria Paulovna. It was performed on 12 November 1804. Figures representing the arts of the theatre, architecture, sculpture, painting, music, dance, acting, and poetry, agree to co-operate: 'Denn aus der Kräfte schön vereintem Streben/Erhebt sich, wirkend, erst das wahre Leben.'

Humanismus, the aspect of the Renaissance which sought, by renewal and extension of knowledge of the Ancients, and by an unprejudiced examination of their works, to exalt and ennoble man. Its impetus came from Italy, and it found a more fruitful soil in the countries of western Europe than in Germany. The history of humanism in Germany begins with JOHANN von Neumarkt (q.v.) *c.* 1400 at the court of the Emperor Karl IV (q.v.) at Prague, and the contemporary JOHANNES von Tepl (q.v.), also a Bohemian, is one of the earliest to show humanistic influences. The most brilliant figure in the early phase of German humanism appeared some forty years later, and was not a German, but the versatile, learned, and dynamic Enea Silvio PICCOLOMINI (q.v.), who was for a few years a prominent figure at the court of the Emperor Friedrich III (q.v.) at Vienna. His exclusive use of Latin in his numerous learned works set the pattern for German humanism; for the German language, in contrast to French and English, seemed at that time to be recalcitrant to the spirit of the classical tongues.

The German humanists were masters of the principal classical verse forms, as well as of Latin oratorical and expository prose. Apart from Prague and Vienna, small groups arose in flourishing cities such as Nürnberg, Strasburg, Augsburg, and Heidelberg, and notably in the universities of Erfurt, Tübingen, and Ingolstadt, all of which were visited by C. Celtis (q.v.), the most elegant poet and most energetic propagandist of German humanism. Its heyday is in

the last decades of the 15th c. and the first decade of the 16th c. Apart from Celtis, the principal German humanists were R. Agricola, J. Aventinus, H. Bebel, S. Brant (who also wrote in German), GEILER von Kaisersberg, H. Glareanus, E. Hessus, P. Melanchthon, K. Peutinger, W. Pirkheimer, Regiomontanus, J. Reuchlin, CROTUS Rubeanus, MUTIANUS Rufus, H. Schedel, Ulrich von HUTTEN (who used German in his late writings), J. Vadianus, and J. Wimpfeling (qq.v.), who made German translations from Latin. Erasmus (q.v.), the greatest of all humanists, was not a German, though he spent some time in Basel and at Freiburg.

HUMBOLDT, ALEXANDER, FREIHERR VON (Berlin, 1769–1859, Berlin), traveller, geographer, mineralogist, and botanist, was a younger brother of W. von Humboldt (q.v.). After a childhood spent mainly in Berlin he studied at Frankfurt/Oder and Göttingen universities (1787–90), and finally at the Academy of Mining at Freiberg (1791–2). In 1791 he made a tour of France, the Low Countries, and England in the company of Georg Forster (q.v.). Humboldt entered the Prussian Department of Mines in 1792 and received rapid promotion to higher grades (1794 and 1795). For some years he had planned an ambitious voyage of exploration to Central America, and in 1796 he resigned his appointment in order to give time to completing his preparations. In December 1794 he had made the acquaintance of Goethe and Schiller in Jena, which he visited again from March to May 1797. In 1795 he published in Schiller's periodical *Die Horen* (q.v.) a curious allegorical story, *Die Lebenskraft oder Der rhodische Genius.* The influence of Weimar classicism remained with him throughout his life. In 1798 he went to Paris and there met the French botanist Aimé Bonplan (1773–1858), who was to become his companion on the expedition.

Humboldt and Bonplan left Paris in December 1798 and journeyed slowly across Spain, studying the geography of the country. In May 1799 they set sail from Corunna, arriving at Cumanà, Venezuela, on 16 July. Their journey, which took five years, covered Mexico, Cuba, and northern South America (the Orinoco basin of Venezuela, Colombia, Ecuador, and Peru). They also paid a short visit to the United States. After his return in 1804 Humboldt went briefly to Berlin, and then settled until 1827 in Paris, working at his notes and specimens. His immense account of the expedition, written in French, appeared in 35 large volumes (*Voyage aux régions équinoxiales du Nouveau Continent,* 1808–27). *Ideen zu einer Geographie der Pflanzen, nebst einem Naturgemälde* appeared in 1807, and his most widely known and most readable work,

Ansichten der Natur (2 vols.), in 1808. His *Versuch über den politischen Zustand des Königreichs Neu-Spanien* (i.e. Mexico) was published 1809–14.

In 1814 Humboldt accompanied Friedrich Wilhelm III (q.v.) to London, and in 1822 to Verona. He received a Prussian royal pension for many years, and from 1827 was required to reside at Berlin. In 1829 he took part in a scientific expedition to southern Russia and Siberia, which he recorded in *Asie centrale. Recherches sur les chaînes de montagnes et la climatologie comparée* (3 vols., 1843–4). During the July Monarchy (1830–48) he visited Paris several times on diplomatic missions. From 1834 onwards he devoted most of his time to his monumental work *Der Kosmos. Entwurf einer physischen Weltbeschreibung* (5 vols., 1845–62), which attempts to harmonize knowledge of the physical environment with the classical ideal of humanity championed by his brother and the Weimar friends of his youth.

A complete bibliography, compiled by J. C. Löwenberg, was published as vol. 2 of *Alexander von Humboldt. Eine wissenschaftliche Biographie* (3 vols., 1872) by K. C. Bruhns. *Gesammelte Werke* (12 vols.) appeared in 1853, and *Mexico-Atlas*, ed. H. Beck and W. Bonacker, in 1969.

HUMBOLDT, WILHELM, FREIHERR VON (Potsdam, 1767–1835, Schloß Tegel, Berlin), was a writer, philologist, and a prominent representative of the 18th- and early 19th-c. humanistic school of thought; he introduced far-reaching school and university reforms into Prussia as part of the general reforms preceding and following the Wars of Liberation (see NAPOLEONIC WARS).

Humboldt studied at Frankfurt/Oder (where Kant, q.v., taught) and Göttingen universities, and in 1789 went to Erfurt (where he met Karoline von Dacheröden, his future wife), to Weimar, and to Jena, forming friendships with K. Th. von Dalberg, the brothers A. W. and F. Schlegel, F. H. Jacobi, Goethe, and Schiller (qq.v.). He collaborated in the *Propyläen* and *Die Horen* (qq.v.), and in 1792 wrote his *Ideen zu einem Versuch, die Grenzen der Wirksamkeit des Staates zu bestimmen.*

For a time (1790–1) Humboldt worked at the Berlin Kammergericht, earning the title Legationsrat. He left Prussian service to devote himself to learning and travel, spending some time in Paris, Spain (1797–9), and Rome (1801–8), where he acted as Minister in the Prussian Legation, while pursuing extensive studies in ancient history. In 1809 he was appointed to the Prussian Ministry of the Interior (on the recommendation of Freiherr vom Stein, q.v.), and worked for the Prussian Academy (see AKADEMIEN). It was during his short period as director of culture and education (Kultur- und

Unterrichtswesen) that he put a number of his ideas on education into practice, the most conspicuous results being the foundation of Berlin University (Friedrich-Wilhelm, since 1945 Humboldt-Universität), and the reform of secondary education. Between 1810 and 1815 he was Prussian ambassador at Vienna, representing Prussia at the Congress (see WIENER KONGRESS). He expressed his views on the unification of Germany frankly in *Über die Behandlung der Angelegenheiten des Deutschen Bundes durch Preußen* (1816). In 1817 he was appointed ambassador in London, and he completed his public service as Minister für ständische und kommunale Angelegenheiten, retiring in December 1819 in protest against the repressive Carlsbad Decrees (see KARLSBADER BESCHLÜSSE). He spent the remainder of his life at Schloß Tegel devoting himself to private study and writing.

As an educationalist Humboldt was influenced, apart from Kant, by F. A. Wolf, Fichte, Schelling, and Schleiermacher (qq.v.), and he himself possessed a strong sense of individualism. He aimed at the development of a cultured personality, and sought a radical division between the education of the individual based on an idealized classical (Greek) model and vocational training (Fachschulausbildung). He envisaged education as a dialectical process in which the pupil (individuality) learns ancient culture (universality) to live for a humanist ideal (totality). His humanistic Gymnasium (humanistisches Gymnasium or Gelehrtenschule) was to serve this end, and Latin and Greek were prominent in the curriculum. Humboldt envisaged educational institutions as working communities entailing the exchange of ideas between professors and students. Matriculation (Abitur, q.v.), qualifying for university entrance was revised, and students were free to migrate to different universities as part of their academic freedom (akademische Freiheit). Universities were to be autonomous, educating a cultural, philosophically orientated society. Accordingly, Religionsphilosophie, Rechtsphilosophie, and Naturphilosophie replaced the existing methods of study in theology, law, and the natural sciences. The state appointed university teachers, who thereby gained in status and income.

The reforms which Humboldt inaugurated had far-reaching social and political consequences. His ideal that the educated personality realized the highest ethical potential in man was conceived as a bulwark against materialism.

Humboldt's anthropological research was based on the study of languages as an expression of human character and the mentality and cultural standard of different peoples throughout the ages (*Über die Kawisprache auf der Insel Jawa*, 4 vols., 1836–40). In his *Ideen* of 1792 Humboldt sought to limit the authority of the state over the individual. They were included in the first (incomplete) edition of his works (*Gesammelte Werke*, 1841–52, vol. 7), and aim at the realization of human dignity on the basis of individual freedom. F. Lassalle (q.v.) ridiculed Humboldt's conception of the state as a 'Nachtwächterstaat'. Humboldt's idealism made little allowance for man's moral and intellectual limitations. His broad views were nevertheless a factor in the development of 19th-c. liberalism.

Humboldt's correspondence with Schiller, published as *Briefwechsel zwischen Schiller und Wilhelm von Humboldt* (1830), is preceded by a highly perceptive and appreciative essay, *Über Schiller und den Gang seiner Geistesentwicklung*. *Gesammelte Schriften*, authorized by the Prussian Academy, ed. A. Leitzmann and B. Gebhardt (17 vols.), appeared 1903–36, reissued 1968. Leitzmann also edited various other writings, among them Humboldt's correspondence with A. W. Schlegel (1908), Schiller (1935); *Wilhelm und Karoline von Humboldt in ihren Briefen* (7 vols), ed. Anna von Sydow, appeared 1906–16. Wilhelm was the brother of Alexander von Humboldt (q.v.). *Werke* (5 vols.), ed. A. Flitner and K. Giel, were published in 1960 ff.

HUMPERDINCK, ENGELBERT (Siegburg, 1845–1921, Neu-Strelitz), composer, became in 1881 assistant to R. Wagner (q.v.), and based his style on Wagner's. He composed *Hänsel und Gretel* (q.v., 1893), and achieved with it a considerable success; of his other works, only the tragic opera *Die Königskinder* (1910) attracted more than passing attention. Humperdinck composed incidental music for several plays of Shakespeare (*The Merchant of Venice*, *Twelfth Night*, *A Winter's Tale*, *The Tempest*).

Hund des Generals, Der, a story and a play (Schauspiel) by H. Kipphardt (q.v.). The story, written in 1957, was published in *Die Ganovenfresse* (1964). It is based on an actual war incident. The play, which is adapted from the story and is better known, was first performed in Munich in 1962, and published in 1963. It takes the form of a belated war-crimes investigation, and is presented with minimal scenery and theatrical trappings, and with numerous flashbacks, film shots, and tapes re-enacting the events at Demodowo after the failure of Hitler's autumn and winter offensive in south-east Russia (1943).

Before the investigation commission is General Rampf, who is accused of having avenged the accidental shooting of his dog (an Alsatian) by sending the culprit, Pfeiffer, and some sixty other soldiers to almost certain and unnecessary death. Pfeiffer survives. Rampf turns out to be a double war criminal, since he had denounced anti-Hitler officers connected with the attempted revolt of 20 July 1944. Nevertheless, the state prosecutors accept the plea of orders from a superior officer and fealty to the oath of allegiance to Hitler. Only Professor Schweigeis (in whom we may glimpse the author) asserts the voice of conscience and the need for justice.

Hundejahre, a novel by G. Grass (q.v.), published in 1963. It is in three sections. In the first, which is divided into 33 'early shifts' (Frühschichten), a West German factory owner, Herr Brauxel (whose uncertain identity is suggested by his alternative spellings Brauksel and Brauchsel), writes of the childhood in Danzig of Eddi Amsel (which we learn much later is Brauksel's original name) and of Walter Matern. Amsel is of partially Jewish descent, Matern is the son of a flour miller with psychic and prophetic gifts. An early antagonism between the two boys gives way to friendship, and Matern seconds and protects Amsel, as the latter develops an astonishing fecundity of imagination and technical adroitness in the construction of scarecrows.

The second section, entitled *Liebesbriefe*, views the story from another angle. It consists of a large number of letters purporting to be written by the cabinet-maker's son Harry Liebenau to his cousin Tulla (Ursula) Pokriefke. Its background is chiefly the years of National Socialism, including the period of the war. Matern, first Communist, then SA man, leads a brutal assault on Eddi Amsel, after which the latter disappears, and is next heard of as the impresario Herr Haseloff and later as 'Goldmäulchen'. Tulla herself is an outrageous, dynamic, and totally amoral *enfant terrible*, who repeatedly instigates mischief and contrives that the blame shall fall on others. An important figure in this section is the black Alsatian dog Harras, whose personality and behaviour are described with impressive observation and empathy. It is to Harras's kennel that Tulla retires for a week after she has lost her brother by drowning. Harras has sired a police dog named Prinz, who becomes Hitler's favourite dog. Harras is finally poisoned for 'political' reasons by Matern, who has been expelled from the Nazi party.

In the third section, composed of 103 'Materniads' (Materniaden), Matern, with the dog Prinz, who has abandoned the Führer in Berlin in 1945, and now goes by the name of Pluto, travels about Western Germany, using Cologne as a base, on a private campaign of 'denazification', performing on the way prodigies of erotic prowess. However, in a radio discussion conducted by Harry Liebenau Matern's own Nazi past is brought to light (the chapter is a brilliant satire on the rage for public discussion, characterizing German intellectual life in the 1960s). In disgust Matern leaves for East Germany, but in Berlin he falls in with Amsel-Haseloff-Goldmäulchen-Brauchsel, and is taken by him to the Harz, where he is shown over Brauxel's subterranean scarecrow factory, a world of horrifying caricature, which he denounces as 'Hell', to which Brauksel's retort is 'Der Orkus ist oben!'

The novel, written with extraordinary verve, gives a brilliant conspectus of the Germany of Grass's lifetime, and sparkles with irrepressible, often grotesque and obscene humour. Events and dialogue are strung together with linguistic virtuosity, an encyclopedic range of vocabulary and allusion (Grass interpolates a parody of the philosophical jargon of Heidegger, q.v., in the first section) and a bold disregard of grammar, in which anacoluthon is raised to a prime instrument of communication.

Hungerpastor, Der, the first of W. Raabe's longer novels, written in 1862–3, and published in 1864. Raabe gives it a motto from Sophocles: 'Nicht mitzuhassen, mitzulieben bin ich da' (*Antigone*, l. 523). *Der Hungerpastor* tells of the birth, childhood, youth, and manhood of Hans Unwirrsch, the poor cobbler's son, whose manifold hungers are eventually satisfied in ministering as a pastor to the material and spiritual hunger of a remote village on the Baltic coast.

Hans is born late into his parents' hitherto childless marriage; his father dies not long after his birth, and he is brought up by his mother, who works long hours as a washerwoman, and by an aunt, Base Schlotterbeck, aided by lengthy, and often superfluous, advice from his guardian, Oheim Niklas Grünebaum. Hans's father and the consumptive teacher of the pauper school, Lehrer Silberlöffel, both incorporate the hunger for knowledge in Hans's early childhood. Hans forms a friendship with Moses Freudenstein, the son of a Jewish pawnbroker, and the two boys progress through the classical school and enter the university. Though the friendship persists for some time, it becomes apparent that their outlooks are divergent. Hans is a dreamer, and a kind and dutiful son; Moses, a sharp-eyed materialist. Soon after graduating Hans learns that his mother is dying. On the journey home he makes the passing acquaintance of Lt. Rudolf Götz and his orphaned niece, Franziska, who are later to play an important part in his story, and he hears sinister references to Moses's conduct.

After burying his mother, Hans, now a pastor but without a living, becomes a private tutor. He is successful and happy with the first family, that of a country squire, but has to leave when a rich relative takes a dislike to him. The second post, with an industrialist, is tolerable but comes to an end when, at a time of bread riots, Hans shows strong sympathies with the workers. His third post is obtained for him by Lt. Götz, who has seen an advertisement inserted by Hans, and recommends him to his civil servant brother, Theodor Götz, in Berlin, in the naïve expectation that Hans can act as a protector to Franziska, now living in the family of Geheimer Rat Götz. Here all is outwardly decorous, but the house is ruled by the proud and tyrannical wife; the Geheimer Rat is a nonentity at home, and the daughter, Cleophea, is in open opposition to her mother. Soon Moses Freudenstein appears in Berlin, now known as Dr. Théophile Stein and as a popular lecturer, especially to female audiences. He insinuates himself with Hans's reluctant assistance into the house of the Geheimer Rat, captivates the hostess, and presently elopes with Cleophea. In the shock of this event Hans and Franziska Götz discover their love for each other. After the Stein–Freudenstein catastrophe, Hans is dismissed, and loses contact with Franziska when he returns to Neustadt to bury Base Schlotterbeck and Oheim Grünebaum. On his return he traces Lt. Götz at the house of a genial and eccentric squire, Colonel Bullau, at Grunzenow on the Baltic, and sets out with their encouragement to find and fetch Franziska. This Hans achieves. He is betrothed to Franziska, abandons his early dreams of far-reaching influence and activity, contenting himself with the prospect of succeeding Pastor Tillenius in the parish of Grunzenow. On the day of his wedding, a burning ship is sighted. The passengers and crew are rescued, and among the former is a disillusioned and repentant Cleophea, who dies in Franziska's care a few months later.

Der Hungerpastor was Raabe's first real success. He himself termed it 'nicht nur *ein* deutches Volksbuch, sondern *das* deutsche Volksbuch'. Hans is seen as an epitome of the German soul in his fundamental goodness, dreamy unpracticality, and aspiration for the heights. The novel suffers, however, from the sharp black-and-white delineation, especially in the unrelieved evil of the Jew, Moses Freudenstein, and in the bitter caricature of proud gentility (Frau Götz). All the warmth and humanity are devoted to Hans and his friends, among whom is a rich gallery of eccentrics, in particular Base Schlotterbeck with her gift of second sight, Oheim Grünebaum with his convoluted epistolary and oratorical style, Lt. Götz, and Colonel Bullau.

Hunnenschlacht, the battle fought in 451 near Châlons-sur-Marne on the Campi Catalaunici

(Katalaunische Felder), in which the Huns of Attila (q.v.) were defeated by an allied army of Romans and Goths. The defeat put an end to the westward advance of the Huns.

HUNNIUS, Ägidius (Winnenden, Württemberg, 1555–1603, Wittenberg), a professor at Wittenberg, wrote biblical plays in Latin, including *Josephus* (1584) and *Ruth* (1586).

HUNOLD, Christian Friedrich (Wandersleben, 1680–1721, Halle/Saale), pseudonym Menantes, settled in Hamburg in 1700 after studying at Jena University. His novel *Der satyrische Roman* (1706) provoked a scandal which led to his flight, after which he established himself as a lecturer at Halle. His poems (*Edle Bemühungen*) were published in 1702 and in 1706 (*Theatralische, Galante und Geistliche Gedichte*). He is the author of four erotic novels: *Die verliebte und galante Welt* was published in 1700, and *Der Europäischen Höfe Liebes- und Helden-Geschichte* in 1705; *Der satyrische Roman*, mentioned above, recklessly exposed scandals in Hamburg society, and *Die Liebens-Würdige Adalie* (1702) gives a free version of the love story of Eléonore Desmier d'Olbreuse, mistress of the Duke of Lüneburg-Celle, who in 1676 became the duchess. Hunold tells the story with disguised names, and partly follows *L'Illustre Parisienne* (1679) by Jean de Préchac.

Hürnen Seyfrid, Der, an Early New High German poem, probably written in the late 15th c., but preserved only in printed editions of the 16th c. and 17th c. The earliest of these dates from 1527. It is in 179 six-line stanzas of a type known as the Hildebrandston (q.v.). Its hero, Seyfrid, is the Siegfried of the Nibelungen story (see Nibelungensage). It begins with a short account of Seyfrid's youth, in which he slays dragons, smears himself with their melted skin, and so acquires a horny (hürnen) skin, except for one vulnerable, and later fatal, spot between the shoulders. The story continues with the abduction of Krimhilt by a dragon, which Seyfrid, some three years later, seeks out, and slays with the help of a treacherous giant and a benevolent dwarf. He acquires the Nibelungen treasure, which he throws into the Rhine. The poem closes with a brief glimpse into the future, revealing Seyfrid's death and Krimhilt's revenge.

The two parts are in reality distinct and partly contradictory, the 'poet' having crudely combined two medieval sources dating from the 12th c. and 13th c. The story of *Der hürnen Seyfrid* also appears in a play by Hans Sachs (q.v., *Der hürnen Seufrid. Tragoedie in sieben Acten*, 1557). It re-emerges in the 18th c. as a Volksbuch (q.v.), *Von dem gehörnten Siegfried* (1726).

HUS or **HUSS,** Johannes (Husinetz, 1369–1415, Constance), a Czech priest, taught at the University of Prague, at which he was Rector in 1402–3, furthering Czech national consciousness. Hus adhered to Wycliffe's teachings, strenuously opposing the increasing worldliness of the clergy. In 1402 he became a parish priest in Prague, enjoying high protection and earning extensive support. In 1408 he came into conflict with his ecclesiastical superiors, and in 1411 he was excommunicated. His tract *De ecclesia* insists on the truths of the Christian faith, but condemns the existing hierarchical structure of the Church. When the Council of Constance was summoned in 1414 the restoration of ecclesiastical order and authority in Bohemia was one of the tasks before it. Hus was invited to the Council and given a safe-conduct by the Emperor Sigismund (q.v.). He was nevertheless arrested, arraigned, condemned, and executed by burning.

Hus's followers, the Hussites, thereupon demanded reforms, and maintained their demands against the Emperor by force of arms. The Hussite Wars (Hussitenkriege) lasted from 1420 to 1433, and the papal and imperial parties were eventually obliged to compromise. Sigismund's claim to the crown of Bohemia was finally acknowledged in 1436.

Husum, North Sea coastal town in Schleswig, made famous by Theodor Storm's poem *Die Stadt*, in which he addresses it as 'Du graue Stadt am Meer'. Storm (q.v.) was born there, and spent his early life in Husum until 1852; he lived there again from 1864 to 1880.

HUTTEN, Ulrich von (Burg Steckelberg nr. Fulda, 1488–1523, Ufenau nr. Zurich), was the most dynamic personality among the German humanists. Born of impoverished noble parents, he was at first intended for a monastic life. At 17 he ran away and wandered for some years from university to university, visiting among others Cologne, Erfurt, Greifswald, Wittenberg, Pavia, and Bologna. Frequently destitute, he served for a time as an ordinary soldier. In 1514 he was in favour at the court of Mainz, and seemed likely to devote himself to learning. In 1516, however, his cousin was murdered by Duke Ulrich of Württemberg (q.v.), and Hutten embarked on a feud against this powerful prince. He directed against Ulrich a Latin dialogue (*Phalarismus*, 1516), which developed into a general denunciation of tyrants. His pugnacity soon extended to other fields, and he plunged into the controversy between the Cologne

Dominicans and Reuchlin (q.v.), siding with the latter and taking a considerable hand in the second part of the *Epistolae virorum obscurorum* (q.v.).

Hutten was for a time protected by the Emperor Maximilian I (q.v.), and was crowned as a poet by him at Augsburg. In 1519 he participated in the war which evicted Duke Ulrich from his dominions. His quickly stirred aggressive resentment of wrongs led him to attack the Papacy in the Latin dialogue *Vadiscus* (1520). His lively nationalism found expression in *Arminius* (1524). He applied the dialogue also to his personal desires (*Fortuna*, 1519) and misfortunes, especially his sickness (*Febris*, 1519). From 1517 onwards, Hutten translated some of his dialogues into the vernacular and wrote new ones in fluent, vigorous German (*Gesprächsbüchlein*, 1521). Constantly entangled in conflicts with powerful forces, he became involved in the defeat of Sickingen (q.v.), and eventually took refuge with Zwingli (q.v.) in Switzerland, where he died of syphilis which had afflicted him for several years.

Hutten supported the Reformation (q.v.), but his services to it were not as great as was formerly supposed. A man of frail constitution but tremendous energy, he was by temperament ready to defend the weak and oppose the strong; his aggressiveness and combativeness went hand in hand with recklessness and lack of judgement. His mottoes 'Alea jacta est' and 'Ich hab's gewagt' are characteristic.

A fierce and feared satirist, Hutten is notable for his development of the dialogue form borrowed from Lucian and, in his last years, for his racy German style. A select edition of his works, *Huttens deutsche Schriften* (2 vols.), ed. H. Mettke, appeared 1972–4 and in 1963 a reprint of the critical edition by E. Böcking, *Opera quae reperiri potuerunt omnia* (5 vols., 1859–62).

Huttens letzte Tage, an epic poem by C. F. Meyer (q.v.), first published in 1871. It is written in iambic pentameters arranged in rhyming couplets, and consists in its final version of 71 poems, divided into 8 sections: *Die Ufenau, Das Buch der Vergangenheit, Einsamkeit, Huttens Gast, Menschen, Das Todesurteil, Dämonen,* and *Das Sterben.* Meyer referred to this epic as his first work, though he had already published two collections of poems. It is the product of his long-standing interest in Ulrich von Hutten (see HUTTEN, ULRICH VON), for which the biography *Ulrich von Hutten* (2 vols., 1858) by D. F. Strauß (q.v.) served as his principal source.

Hutten's varied experience is condensed in concentrated episodic and symbolic form in his last illness. On the small island of Ufenau in

Lake Constance, Hutten finds a solitary refuge to end his days. At the opening of this cycle of poems he steps on to the shore. In the last poem he briskly calls for the last boat of which he knows that the ferryman is death. A sick man on arrival, Hutten settles in his abode in a monastery on Ufenau by symbolically abandoning his sword, which he hangs on the wall together with a print by Dürer (q.v.), *Ritter, Tod und Teufel.* The knight devotes himself henceforth to his pen until he delivers himself up to death. His last direct contacts with the world outside are visits from Paracelsus (q.v., his doctor) and Ignatius von Loyola, while places and characters of the past, both fact and fiction, pass as memories and visions through his mind and elicit his response. Among them are the authors of the *Epistolae obscurorum virorum,* Herzog Ulrich von Württemberg, Bayard (Bajard), Pope Leo X, Faust, the Kurfürst von Mainz, Karl V, Zwingli, Ariosto, Luther, Erasmus, the Bilderstürmer, 'Der arme Heinrich', Holbein, and scenes derived from Meyer's two visits to Italy.

Meyer, in his meticulousness, made some amendments to each of the 15 editions which appeared in his lifetime. He achieved the suppression of subjective elements and the creation of an authentic historical atmosphere, which he linked with unobtrusive hints at the problems facing his own age.

HUYSMAN, ROELOF, see AGRICOLA, RUDOLF.

Hymne, in German use, is a song of praise in rapt or enthusiastic tone, closely related to and often barely distinguishable from the Ode. The term Hymne (used in the 16th c. in the form Hymnus for songs of praise written in Latin) was first employed for poems in German by M. Opitz (q.v.). An exalted hymnic style was developed in the 18th c. by Klopstock (q.v.). Many of the early poems of Goethe can be regarded as Hymnen, notably *Wanderers Sturmlied* and *An Schwager Kronos.* The poetry of Hölderlin (q.v.) is the climax of the German Hymne, and examples extend from his early work to poems in hexameters, such as *Der Archipelagus,* and in free rhythms in *An den Rhein* and *Patmos.* Maler Müller (see MÜLLER, F.) invented the Hymne in prose, and Novalis's *Hymnen an die Nacht* (q.v.) employ both verse and prose. Hymne is not used in German for the hymn which is sung in church by a congregation. This is the Kirchenlied (q.v.).

Hymnen an die Nacht, a work composed of 6 hymns (see HYMNE) written in 1799 by Novalis (q.v.). They constitute a homogeneous work and are Novalis's poetic response to the emotional crisis of 1797 provoked by the deaths of his fiancée Sophie von Kühn and his brother Erasmus. Night is here the symbol for death, in which the spiritual union with his lost bride will take place. The first three hymns are written in ecstatic prose, the fourth ends with a poetic coda, and the fifth contains two poems. The sixth and last is a strophic hymn in rhyming verse, entitled *Sehnsucht nach dem Tode*.

Hyperion, oder Der Eremit aus Griechenland, Hölderlin's only novel, the first volume of which was published in 1797, the second in 1799; fragments appeared in Schiller's periodical *Die neue Thalia* in 1794. It bears a Latin motto, *Non coerceri maximo, contineri minimo, divinum est.* The second part, written after Hölderlin's parting from Susette Gontard (q.v.), is a vicarious reckoning with this traumatic experience.

Hyperion, a Greek youth of Hölderlin's day, possessed of an exalted idealism and of a quivering sensibility, yearns to free his country from Turkish occupation. In Alabanda, another Greek, he makes a friend after his own heart, but after a time the paths of the two men separate. Hyperion falls in love with the Greek girl Diotima (Hölderlin's name for Susette Gontard), who represents for him the epitome of beauty in the Greek world. Alabanda summons Hyperion to participate in a revolt against the Turks. Though this is attended by success, the cruelty and the selfish opportunism of the rebels disgust Hyperion, who seeks death in a sea battle, first communicating his intention to Diotima. He survives, is reconciled to life, and seeks to be reunited with Diotima, only to find that she has died of a broken heart. He seeks consolation in communion with Nature.

A famous passage near the end denounces the Germans of Hölderlin's day, and the book closes with a prose hymn to Hyperion's and Diotima's love. *Hyperion* contains the well-known poem *Schicksalslied* (q.v.). The novel, in which Hyperion's exalted love of Greece and his ecstatic love for Diotima are fused, ends in a renunciation of earthly happiness. It is written in the form of letters (see BRIEFROMAN), nearly all of which are addressed by Hyperion to his friend Bellarmin or to Diotima.

Hyperions Schicksalslied, see SCHICKSALSLIED.

I

Ibrahim Bassa (1653) and *Ibrahim Sultan* (1673), the first and last tragedies written by D. C. von Lohenstein (q.v.).

IBSEN, HENRIK (Skien, 1828–1906, Oslo), the Norwegian dramatist, spent the most influential period of his life in Germany, living in Munich and Dresden from 1864 to 1891. The plays written before his voluntary exile from Norway were almost all historical. Soon after his arrival in Germany he wrote two verse dramas of epic proportions, *Brand* (1866) and *Peer Gynt* (1867). In 1877 he published his first play dealing with contemporary society, *Samfundets støtter* (*Stützen der Gesellschaft, Pillars of Society*), which was performed in Germany and Austria in 1878 without making any deep impression. There followed a series of bitter satirical comedies and harsh tragedies expressed in a strikingly realistic idiom. They comprise *Et dukkehjem* (1879, *Nora* or *Ein Puppenheim* in German, *A Doll's House*), *Gengangere* (1881, *Gespenster, Ghosts*), *En folksfiende* (1882, *Ein Volksfeind, An Enemy of the People*), *Vildanden* (1884, *Die Wildente, The Wild Duck*), *Rosmersholm* (1886), *Fruen fra havet* (1888, *Die Frau vom Meere, The Lady from the Sea*), and *Hedda Gabler* (1890). These were translated and performed in Germany, but it was not until 1887 that Ibsen made his full impact on the German theatre and public as the foremost pioneer of Naturalism (see NATURALISMUS). His plays became the subject of exaggerated propaganda on the one hand, and of savage abuse on the other.

A Doll's House, An Enemy of the People, and *Pillars of Society* were the most frequently performed. *Ghosts* was chosen as a controversial work for the first performance by the Freie Bühne (q.v.) in 1889. Though Ibsen's social plays were a reflection of peculiarly Norwegian conditions, they came to be regarded as characteristic portrayals of the European *bourgeoisie*. His later symbolical plays, from *Bygmester Solness* (1896, *Baumeister Solneß, The Master Builder*) to *Når vi døde vågner* (1899, *Wenn wir Toten aufwachen, When we Dead awaken*), were for some time regarded in Germany as evidence of failing powers.

'Ich bete an die Macht der Liebe', first line of the hymn *Abendsegen* by Gerhard Tersteegen (q.v.), written *c.* 1750.

'Ich denk' an euch, ihr himmlisch schönen Tage', first line of the poem *Sehnsucht* (1802) by S. A. Mahlmann (q.v.).

'Ich denke dein', opening words of Goethe's poem *Nähe des Geliebten* (q.v.), and of the poem *Ich denke dein* by Friederike Brun (q.v.).

'Ich fahr dahin, wann es muß sein', first line of the anonymous 15th-c. poem *Ritters Abschied*.

'Ich hab's gewagt', motto of Ulrich von Hutten (q.v.).

'Ich komme vom Gebirge her', first line of the poem *Der Wanderer* by G. P. Schmidt (q.v.), set to music by F. Schubert (q.v.).

'Ich weiß nicht, was soll es bedeuten', first line of Heine's poem *Die Loreley* (q.v.).

'Ich will einst bei Ja und Nein', first line of *Zechergelübde*, a student song written in 1777 by G. A. Bürger (q.v.), the tune of which was composed in 1784 by J. A. P. Schulz (1747–1800). It is based on the medieval drinking song *Mihi est propositum in taberna mori* (q.v.).

ICKELSAMER, VALENTIN, a schoolmaster living in the early part of the 15th c., wrote schoolbooks, including *Rechte Weis auf kürzist lesen zu lernen* (1527), and in 1534 published the first German grammar.

Ideale, Die, a poem written by Schiller in 1795, and published in 1796. The original version was slightly cut by Schiller, and it now consists of 11 eight-line stanzas. It is a deeply felt lament for the losses imposed on the human mind by the passage of time.

Idealismus, term used in various senses in German, including the familiar one of an elevated awareness of moral or social responsibility. It principally denotes the tradition of German philosophy, as developed especially by Kant, Fichte, and Hegel (qq.v.), in which the ideal, as opposed to the material, is conceived as the true reality. The word is further used to describe an important phase in German literature in the years *c.* 1790–1805, when Schiller was both influenced by Kant and in alliance with Goethe. German philosophical idealism, though it has lost many adherents and much of its momentum, has persisted and has influenced some later writers.

Ideal und das Leben, Das, a philosophical poem by Schiller, first published in 1795 in No. 9 of *Die Horen*, q.v., with the title *Das Reich der Schatten*. Schiller changed this in 1800 to *Das Reich der Formen*, and adopted the present title in 1804.

Ideen zur Philosophie der Geschichte der Menschheit, a historical work by J. G. Herder (q.v.), published in 4 vols., 1784–91. It is divided into four parts (Teile), each of which is subdivided into five books (Bücher). It is a survey of history in the most extended sense, beginning with cosmology (Bk. 1), passing on to geology and geography, considering biology, including human biology, and asserting that man is still at an early stage of evolution (Bks. 2–5). In Bks. 6–10 Herder considers the social organization of various regions, deals with environmental factors, considers language, and examines the mythology of the Creation and early man. Bks. 11–15 are a survey of ancient history. And the last five books treat the history of the western peoples and the rise of Christianity.

Herder sees history as organic, i.e. developing and declining on the analogy of a living creature, but he also emphasizes, in distinction from his early views, human moral progress. Herder's comprehensive view of history was one of the most original features of this work.

Idylle, a literary form, usually in verse, but occasionally in prose, originating in Greek literature. It presents rural situations of joy and contentment with a tendency towards idealization.

The idyll flourished in Germany chiefly in the 18th c. E. von Kleist (q.v.) was the first conspicuous writer of idylls, but S. Geßner (q.v.), using a rhythmic prose, popularized the form. He was followed by Maler Müller (see MÜLLER, F.) with two idylls in dialogue intermingling prose and verse, *Die Schafschur* and *Das Nußkernen*. These semi-dramatic works, however, had no progeny. The revival of classicism brought a renewal of the idyll written in classical elegiacs, examples of which are J. H. Voß's *Luise* and *Der siebzigste Geburtstag* (qq.v.), and Goethe's *Hermann und Dorothea* and *Alexis und Dora* (qq.v.). Schiller's *Über naive und sentimentalische Dichtung* (q.v.) describes the idyll as the 'sentimental' poet's imagined realization of his ideal. Among the 19th-c. poets Mörike wrote two notable idylls, *Der alte Turmhahn* and *Idylle vom Bodensee*, and Hebbel *Mutter und Kind* (qq.v.).

Idyllen, a volume of pastoral idylls written in poetic prose by S. Geßner (q.v.), and published in 1756. The book was illustrated by Geßner himself.

Idylle vom Bodensee, an idyllic poem in seven cantos of classical hexameters (*c.* 1,500 lines) written by E. Mörike (q.v.) 1845–6, and published in 1846. The basis of the poem is the story of the boatman Tone, who, jilted by his sweetheart Gertrud, finds a better choice in the shepherdess Margarete. The youths of the village play a prank on Gertrud and her bridegroom, the rich but dull-witted son of the miller of Barnau, by unloading their household goods in the middle of a wood. This story is framed (see RAHMEN) in the account of a trick, which Martin, the fisherman of the village, plays on the village tailor, leading him to try to make off with and sell a non-existent church bell.

The most attractive feature of the poem is the evocation of the fertile landscape around Lake Constance.

IFFLAND, AUGUST WILHELM (Hanover, 1759–1814, Berlin), ran away from home in order to go on the stage. He had the good fortune to be engaged at Gotha by the well-known actor K. Ekhof (q.v.), to whom he was to owe his training. In 1779 he was offered an engagement at the Mannheim National Theatre under W. H. von Dalberg (q.v.); he rapidly became the principal actor, and during the troubles of the French Revolution was temporary director. In 1796 Iffland, resenting unjust strictures by Dalberg, accepted a call to Berlin as director of the Royal Theatre, remaining there for the rest of his life.

Iffland was an outstanding character actor; he created the part of Franz Moor in Schiller's *Die Räuber* in 1782 and later played the hero in Goethe's *Egmont,* and Octavio in Schiller's *Wallenstein* (qq.v.). He encouraged a realistic style of acting, and, as director, favoured a lavish *mise en scène.* He wrote an account of his early professional life (up to 1796) in *Über meine theatralische Laufbahn* (1798).

Iffland is the author of 65 plays, competent examples of craftsmanship, intended for the theatres in which he acted and directed. The most successful of them were *Das Verbrechen aus Ehrsucht* (1784), *Die Jäger* (1785), and *Die Hagestolzen* (qq.v., 1793). They are among the best examples of the sentimental moralizing play (see RÜHRSTÜCKE), which was the popular theatrical fare of the middle classes. Iffland, whose acting powers were admired by Schiller, invented the title *Kabale und Liebe* (q.v.) for the play which Schiller had intended to call *Luise Millerin.*

Ignaz Denner, a horror story by E. T. A. Hoffmann (q.v.), written in 1814 and published in 1816 in vol. 1 of *Nachtstücke.* Hoffmann at first intended to entitle it *Der Revierjäger.* The scene is set in Germany near Fulda. Andres, a gamekeeper of Count Vach, marries an Italian orphan girl and brings her back to Germany. They are miserably poor, and the wife is in danger of death after the birth of her first child, when Ignaz Denner, a supposed merchant, effects a cure, so earning Andres's gratitude. After two years Denner reveals himself for what he is, a robber chief, and demands Andres's help, under threat of killing the latter's wife and child. Andres helps reluctantly, though his only act is to rescue the wounded Denner. He learns that his wife, who has borne him a second child, has inherited a sum of money, and goes to Frankfurt to fetch it. He returns to find his younger child murdered and his wife in despair. He also learns that robbers have stormed Count Vach's castle and killed the Count. All is the work of Denner. He is arrested, and brazenly inculpates Andres, who confesses to the deed under torture. Andres is about to be executed when evidence is given that he was in Frankfurt at the time of the crime.

What has long been suspected now becomes clear: Denner is in league with the Devil. Andres is released, but his wife dies. Denner escapes, persuades Andres to help him by revealing that he is his father-in-law, and then abuses Andres's kindness by attempting to murder the elder son, whereupon Andres shoots him. The diabolical past of Denner and his father, whose real name was Trabacchio, is revealed.

Ignorabimus, an enormous tragedy of 454 pages by A. Holz (q.v.), published in 1913. Concerned with the limits of knowledge, whether sought scientifically or otherwise, it demonstrates the extraordinary verbal fecundity of Holz's later manner.

'Ihr Kinderlein kommet, o kommet doch all', first line of a Christmas carol written by C. von Schmid (q.v.) *c.* 1850.

ILDIKO, or **HILDIKO,** wife of Attila (q.v.), who died by her side in the night following the wedding banquet. Though his death is believed to have been caused naturally, a legend arose that Ildiko had murdered him. It has been thought that she is the original of Kriemhild (q.v.).

Illuminatenorden, a society of freemason-like character, founded in 1776 by A. Weishaupt (q.v.) of Ingolstadt. Strongly anti-Jesuit, it was suppressed in 1784 by the Elector Karl Theodor of Bavaria (q.v.) at the instance of the Society of Jesus. Among its members at various times were Duke Karl August of Saxe-Weimar, J. G.

Herder, J. H. Pestalozzi, and A. von Knigge (qq.v.).

Ilmenau, a poem written by Goethe in 1783, and published in 1815. Ilmenau, some 25 miles south-west of Weimar, was the scene of hunting expeditions of Duke Karl August (q.v.), on which Goethe accompanied him, and the date beneath the title, 3 September 1783, is the Duke's twenty-sixth birthday. The poem, written partly in couplets and partly in alternately rhyming groups of four lines, combines several elements: a sense of the healing power of nature, a friendly caricature of court figures (Knebel and Seckendorff, qq.v.), a retrospect of Goethe's life, a tribute to the Duke, and a solemn word on a ruler's responsibilities. The unifying factor in this conglomerate poem is the stable and steady gaze of the poet, which makes each point of comment a part of his own experience.

Imago, a novel by C. Spitteler (q.v.), published in 1906. It is the story of a sexual obsession. Viktor, long pursued by the image he constructs of the woman he vainly loves, at last sublimates his obsession in devotion to art. The novel has autobiographical elements, and in its exploratory penetration of the mind suggests psychoanalytical procedures.

Imago. Zeitschrift für Anwendung der Psychoanalyse auf die Geisteswissenschaften, a journal edited by S. Freud (q.v.) from 1912. It derives its title from *Imago* (q.v.) by Spitteler and contained tracts on art and literature.

Im alten Eisen, a short novel (Eine Erzählung) by W. Raabe (q.v.), written in 1884–6, and published in 1887. Raabe gives it the Latin motto *Similia similibus.* Set in Berlin, the story occupies four days, from the death of the penniless Erdwine Wermuth on Sunday to her pauper funeral on Wednesday. When Erdwine dies, her two children are left untended through the caution and cowardice of neighbours, and for three days the 13-year-old boy Wolf and his little sister Paula keep watch, unvisited, over the corpse. Wolf has as his sole possession his grandfather's sword. When he perforce pawns it for a handful of coffin-nails, the first step is, by a strange chance, taken. The sword is recognized by two acquaintances from Erdwine's days in Lübeck. They are the former theatre directress, Frau Wendeline Cruse, now a general dealer, to whom Wolf brings the sword, and the rolling stone Peter Uhusen. They recruit a third acquaintance to trace the children, Hofrat Dr. Albin Brokenkorb, a spoilt darling of fortune and of the ladies, whose nerves are not up to the

exhausting task. Eventually the children are found with the help of a fourth character, referred to as the light-of-love Rotkäppchen, who knew Erdwine. Henceforth Wolf and Paula are cared for, though only after harrowing experiences culminating in the hastily arranged interment.

Repeated references to Märchen reinforce the atmosphere of this modern fairy-tale, which shows Berlin grim, grimy, ugly, and unfriendly. Only the three eccentrics Frau Wendeline, Uhusen (who is constantly referred to as der Schmied von Jüterbog), and the mercurial Rotkäppchen measure up to the test; the neighbours, respectable and less respectable, and the sophisticated man of culture, Hofrat Brokenkorb, fail in humanity.

'Im ernsten Beinhaus war's', first line of an untitled poem written by Goethe in September 1826, and published in the same year. It is sometimes printed with the title *Bei Betrachtung von Schillers Schädel,* but this, though apt, does not come from Goethe, and moreover makes overt reference to Schiller which Goethe avoids. The occasion was the clearance of the cemetery in which Schiller's remains were interred. Goethe secured the skull and kept it for a time in his house. The poem is written in *terza rima.*

Im Frühling, a poem in four irregular stanzas by E. Mörike (q.v.), written in 1828, and included in the novel *Maler Nolten* (q.v.). Its first line runs 'Hier lieg' ich auf dem Frühlingshügel'.

IMHOFF, Amalia, Freiin von (Weimar, 1776–1831, Berlin), was brought up at Mörlach near Nürnberg and at Erlangen. In 1790 she went to Weimar with her widowed mother and came into contact with Goethe and Schiller. Her early poetic attempts elicited praise, and in 1797 Schiller published in *Die Horen* (q.v.) her poem in six cantos, *Abdallah und Balsora.* Her principal work is the idyll *Die Schwestern von Lesbos* (q.v., 1800). In 1803 she married a Swedish officer, K. G. Helvig, and moved to Stockholm. She returned to Germany in 1810 and lived in Heidelberg. Her husband transferred in 1815 to the Prussian army and she joined him in Berlin. Her other works include a series of idylls *Die Tageszeiten* (1812), the poem *Die Schwestern auf Corcyra* (1812), and a novel *Helene von Tournon* (1826).

'Im Krug zum grünen Kranze', first line of *Brüderschaft,* a drinking song written by Wilhelm Müller (q.v.) in 1821. The melody is a folk-tune adopted from the ballad 'Ich stand auf hohem Berge'.

Immensee, a Novelle by Th. Storm (q.v.), written in 1849, published in *Biernatzkis Volksbücher* in 1850, and reissued in revised form in 1851. One of Storm's early lyrical Novellen, it first established his reputation as a writer of fiction. The story is in the form of an Erinnerungsnovelle, being seen in recollection by one of the participants, Reinhard Werner, whose solitary old age frames the narrative (see RAHMEN).

The childhood companionship of Reinhard and Elisabeth comes to an end when Reinhard goes away to university. Their mutual love grows during Reinhard's absence, and Elisabeth at first rejects the advances of a new suitor, Reinhard's friend Erich. In the end she yields to her mother's persuasion and marries Erich, who offers them both financial security at his estate Immensee (Bee's Lake). This background to the marriage is underlined by the song sung by Elisabeth, 'Meine Mutter hat's gewollt', when Reinhard, who has himself remained unmarried, visits Immensee some years later. The symbol of unattainable love is a water-lily on the lake, to which Reinhard unsuccessfully tries to swim at night. Leaving his hosts at dawn, he confronts Elisabeth for the last time in a parting which both recognize as final.

The story ends in sadness; none of the principal characters has achieved happiness, and none can really be blamed for its loss. Storm develops the theme of love, frustration, and resignation in eight episodes aiming at atmosphere (Stimmung) rather than story-telling, in pursuance of his technique of suggestive art (andeutende Kunst). The lyricism is enhanced by the use at climaxes of songs in the style of folk-poetry, and by the choice of symbols.

IMMERMANN, KARL LEBERECHT (Magdeburg, 1796–1840, Düsseldorf), offspring of a family with a Prussian civil service tradition, began to study law at Halle University in 1813. The dissolution of the university on French orders interrupted his course, and he became a volunteer rifleman, but was prevented by illness from taking part in the campaign of 1813 (see NAPOLEONIC WARS). On the renewal of hostilities in 1815 Immermann volunteered again, served at Waterloo, and was in the force which entered Paris. He was commissioned, but was demobilized in December, and returned to his studies at the reinstituted university at Halle. Here he made a name for himself by opposing the student corporation (see BURSCHENSCHAFT) Teutonia. In 1818 he was appointed to the Prussian civil service and posted first to Oschersleben, and in 1819, after brief service in Magdeburg, was sent to Münster. Here a close friendship began between Immermann and

Countess Elisa von Lützow, wife of General Adolf von Lützow (q.v.). The marriage was dissolved in 1825, and the Countess, declining to remarry, lived with Immermann until his marriage in 1839 to Marianne Niemeyer. Meanwhile, Immermann was posted, at his own request, to Magdeburg, where he served as a judge. Three years later, in 1827, he became Landgerichtsrat in Düsseldorf, where he remained for the rest of his short life. At the time of his death he was at work on his memoirs, *Memorabilien* (1840–3), planned in three parts, which remained unfinished.

Immermann wrote several, largely derivative, plays, including the comedies *Die Prinzen von Syrakus* (1821) and *Das Auge der Liebe* (1824), the tragedy *König Periander und sein Haus* (1823), and an adaptation of Gryphius's *Cardenio und Celinde* (q.v.) with the same title (1826). His most successful dramatic work, *Das Trauerspiel in Tyrol* (1828), was later revised and retitled *Andreas Hofer* (1834). Finding himself mocked in Platen's comedy *Der romantische Ödipus* (q.v.), he replied with *Der im Irrgarten der Metrik umhertaumelnde Kavalier* (1829), a title which alludes to a novel by J. G. Schnabel (q.v.). The verse satire *Tulifäntchen* and a verse drama (Eine Mythe) *Merlin* followed in 1832. Immermann's interest in the theatre led him to found a theatrical society in Düsseldorf, and from 1835 to 1837 he acted as director to the city theatre, for a time supporting C. D. Grabbe (q.v.).

In 1822 Immermann completed his first novel, *Die Papierfenster des Eremiten,* but he wrote his principal works in the 1830s: *Die Epigonen* (q.v., 3 vols., 1836) and *Münchhausen* (q.v., 4 vols., 1838–9), the latter containing the village story which was published separately in 1863 as *Der Oberhof,* and is probably the best, and certainly the best known, of all his writings.

Despite obvious talent, Immermann did not rise above literary dilettantism until a few years before his death. His numerous verse plays and epics lean heavily upon earlier models, notably Shakespeare and Romanticism, but in his late novels, especially in *Münchhausen,* he sought to exploit simultaneously fantasy and realism, with irony as the link between these two elements; it seems that his real gift lay with the early style of poetic realism (see POETISCHER REALISMUS) rather than with the past generation of Romantics.

Schriften (14 vols.) appeared 1835–43, *Sämtliche Werke,* ed. R. Boxberger (20 pts. in 8 vols.), 1883, *Historisch-kritische Ausgabe,* ed. M. Koch (4 vols.), 1887–8, select *Werke,* ed. H. Maync (5 vols.), 1906, and (6 pts. in 3 vols.), ed. W. Deetjen, 1908–11 (reissued in 1923).

Impressionismus, a familiar and precise term

in reference to painting, is also a useful, though less exact, expression when applied to literature. It is primarily used to signify passages or short works, whether in prose or verse, which evoke atmosphere (Stimmung). Impressionism in this sense can be found in a number of works in the 19th c. and 20th c., especially in the period 1880–1930. It can be validly applied to poems by Liliencron, Dehmel, Rilke, and Hofmannsthal (qq.v.) among others, and to passages from the narrative work of many writers, including Th. Storm, Fontane, Th. Mann, and Schnitzler (qq.v.).

Im Schlaraffenland, a novel by H. Mann (q.v.) published in 1900, and sub-titled *Ein Roman unter feinen Leuten.* It describes the attempt, at first successful but ending in failure, of a young provincial student to enter the world of letters and *nouveau riche* society (the Cockaigne or *pays de cocagne* of the title). It is set in Berlin about 1893. Andreas Zumsee finds that literary and intellectual abilities, with which he is not particularly well furnished, matter less than the considerable sexual attraction he exerts, especially on mature women. The world he enters is virtually ruled by the Jewish financier Türkheim.

Andreas captivates Frau Adelheid Türkheim, who makes his fortune, and sets him up in a luxury flat as her lover. Meanwhile Türkheim spends vast sums on keeping as his mistress a vulgar 17-year-old product of the gutter. This 'Achnes' Matzke, renamed on the rise of her fortune Bien-aimée, conducts a barely secret liaison with Andreas. The discovery of this provides a humiliating experience for both Türkheims, who are, however, powerful enough to force the young couple into marriage with financial provision, and to exclude them from their society.

At the centre of the work is a performance of *Rache,* a vicious parody of G. Hauptmann's *Die Weber* (q.v.), which is enthusiastically applauded by a bourgeois audience. *Im Schlaraffenland* is a savagely satirical novel, which makes liberal use of caricature and the grotesque.

Im Schloß, a Novelle by Th. Storm (q.v.), written in 1861, and published in 1862 in *Die Gartenlaube* (q.v.). An Erinnerungsnovelle, it consists of five parts dealing with the last descendants of the aristocratic owners of a castle. Anna is the author of the central section of the Novelle, *Die beschriebenen Blätter,* in which she relates the story of her life leading to her separation from her husband and her return to the solitary Schloß. But the tragic events of her life, the death of her brother in childhood, the death of her own child, and her social isolation

in the castle because of her alleged unfaithfulness to her husband, are in the end compensated by the prospects of a happy future. The sudden death of her husband enables her to marry Arnold, her brother's former tutor and a noted scholar, to whom her true affection has always belonged. Anna's first marriage was arranged by her father, who was intent on maintaining the family's superior social status. Storm makes Anna's consent to this marriage her sole guilt. As long as her husband lives she remains faithful to him, despite rumours accusing her of disloyalty; but she emerges from this first marriage emancipated from social prejudice.

Storm uses Arnold and the figure of Anna's uncle Christoph to express his most personal views on a changing social structure, in which both the aristocracy and the institution of the church are seen as historical forces of the past which are now being superseded by a humane, free-thinking, and educated middle class. Christoph is a philosopher and a naturalist, and through his influence Anna abandons the faith of her childhood and understands nature's cruelty which man can mitigate through love, an emotion born of man's fear of loneliness. This and the awareness of the transience of life, to which the family portraits of past times also bear witness, form the ideological background to her maturing love for Arnold, who is a descendant of the humbly born youth whose portrait inspired her fantasy as a girl, and which forms the story's principal leitmotiv.

'Im tiefen Keller sitz' ich hier', first line of a well-known song written in 1786 by K. F. Müchler (q.v.), and published in *Der Kritikaster und der Trinker* (1802) with a melody by the singer Ludwig Fischer (1745–1825) to suit his own renowned bass, for which Mozart (q.v.) had some twenty years previously created Osmin (*Die Entführung aus dem Serail*).

'Im Wald und auf der Heide', current first line of the poem *Jägerlied* by J. W. J. Bornemann (q.v.), published in 1816 in *Forst- und Jagdarchiv für Preußen.* The original first line ran 'In grünbelaubter Heide'. The composer of the tune (1827) is not known with certainty, but may have been one F. L. Gehricke.

Im Westen nichts Neues, a war novel by E. M. Remarque (q.v.).

In der Fremde, title of two poems by J. von Eichendorff (q.v.). The earlier one begins 'Ich hör die Bächlein rauschen' and was written between 1810 and 1812, a time when Eichendorff was separated from his fiancée Luise von Larisch; it was included in the first volume of

Werke (1841) in which it was attached to a cycle of six poems entitled *Der verliebte Reisende,* all originating from these years.

The first line of the later poem runs 'Aus der Heimat hinter den Blitzen rot'; it is contained in the satirical Novelle *Viel Lärmen um Nichts* (q.v., 1832). R. Schumann (q.v.) set both poems as part of his *Liederkreis* (q.v.).

In der Frühe, a poem written by E. Mörike (q.v.) in 1828. It is a poem of insomnia, beginning 'Kein Schlaf noch kühlt das Auge mir'.

In der Strafkolonie, a short story written by F. Kafka (q.v.) in 1914, and published in 1919. A traveller (Der Reisende), visiting a penal colony, is shown the extraordinarily ingenious machine used for protracted and harrowing execution. It is the invention of a former commandant. The officer who shows off the machine deliberately places himself upon it and is killed by it. The story has been variously interpreted as existentially determined torment, as a symbol of the emptiness of modern humanitarianism, and as a denunciation of Christianity; other interpretations have also been offered.

'In des Waldes düstern Gründen', first line of a robbers' song in the novel *Rinaldo Rinaldini* (1798) by C. A. Vulpius (q.v.). It is often misquoted in the form 'In des Waldes tiefsten Gründen'.

In Dingsda, prose sketches of lyrical character by J. Schlaf (q.v.), published in 1892. They present an idyllic, though at times faintly ironic, picture of an unspecified country town ('Dingsda'), which some writers have identified with Schlaf's native Querfurt.

'In dulci jubilo', a German Christmas carol dating from the 14th c. It is a macaronic poem with alternating Latin and German lines: 'In dulci jubilo/singet und sit vro'. It is played as an interlude by the shepherds in the Nativity in the biblical play *Von dem Anfang und Ende der Welt* (1579) by Bartholomäus Krüger (q.v.). It is still sung in a modernized version ('In dulci jubilo/ Nun singet und seid froh').

Infant, Der, a poem by S. George (q.v.) included in *Hymnen.*

Infanterie greift an, a book of war recollections by E. Rommel (q.v.), published in 1937 and intended as an aid in training for battle. Sub-titled *Erlebnis und Erfahrung,* it covers engagements in Belgium and France (1914–15), Romania (1916–17), and Italy (1917). Each section is followed by *Betrachtungen,* which point out the military lessons to be drawn from the action. In spite of a dry style, the episodes are often vividly evoked.

INGOLD, MEISTER, a Dominican friar of Strasburg, wrote *c.* 1432 *Das goldene Spiel,* a prose allegory after the manner of the chess allegories (see SCHACHBÜCHER). *Das goldene Spiel* includes not only chess as an allegory of pride, but other pastimes: a board game symbolizes gluttony, dice betoken covetousness, cards unchastity, dancing sloth, skittles anger, and lute-playing envy and hatred. A partial source is the *Solatium ludi scaccorum* of Jacopo Dacciesole, also known as Jacobus de Cessolis, a Dominican monk of the 13th c., whose monastery was near Alessandria in Lombardy.

Ingolstadt, a Bavarian city which was a ducal residence from 1392 to 1505. At its university, founded in 1472, many well-known humanists studied, including Celtis, Reuchlin (qq.v.), and, it is said, the historical Dr. Faust (see FAUST, GEORG). The university, which from 1549 was dominated by the Jesuits, was transferred to Landshut in 1800, and thence to Munich in 1826. In the Thirty Years War (see DREISSIGJÄHRIGER KRIEG) Ingolstadt survived a siege by Gustavus Adolphus (1632).

Innere Emigration, term used for the state of mental reservation which those dissenting from National Socialism were obliged to impose upon themselves if they were unwilling to incur draconian penalties by expressing their disagreement. The expectation that the end of the National Socialist regime in 1945 would reveal a considerable number of important MSS. for publication as a result of 'inward emigration' proved to be mistaken. Another, more active, form of innere Emigration was the publication of works which, though outwardly unpolitical, were capable of political interpretation. Bergengruen's *Der Großtyrann und das Gericht* (q.v., 1935) and E. Jünger's *Auf den Marmorklippen* (q.v., 1939) are examples.

Innerer Monolog, *monologue intérieur,* interior monologue or stream of consciousness. Its earliest well-known application in German literature is in Schnitzler's *Leutnant Gustl* (q.v., 1901). (See also ERLEBTE REDE.)

Innocens, a Novelle by F. von Saar (q.v.), published in 1865 and included in 1877 in *Novellen aus Österreich* (q.v.). The story of Father Innocens, who as a young priest is briefly attracted to a young girl and overcomes this temptation at the graveside of another young woman, is framed in a setting in Prague drawn from Saar's early years as an Austrian officer.

Innsbrucker, coins of the value of 6 Kreuzer, minted in Tyrol and circulating also in other parts of Germany. They were first minted in 1482.

Innsbrucker Fronleichnamsspiel, a religious play of simple character, consisting of detached speeches, and intended to be performed on Corpus Christi Day. It dates from the early 14th c. and comes from Thuringia. It is preserved in a MS. of 1391 at Innsbruck.

'Innsbruck, ich muß dich lassen', first line of a 15th-c. folk-song. The song was set to music by H. Isaak in 1502, and later converted into a hymn with the first line 'O Welt, ich muß dich lassen' (see KONTRAFAKTUR). The melody was later adopted for the hymn 'Nun ruhen alle Wälder' by P. Gerhardt (q.v., 1649).

Innviertel, the border district of Upper Austria (Oberösterreich) stretching on the right bank of the Inn from Braunau to Schärding. Originally Bavarian, it passed to Austria in 1799 at the conclusion of the War of the Bavarian Succession (see BAYRISCHER ERBFOLGEKRIEG).

In Reih und Glied, a novel in five volumes by F. Spielhagen (q.v.), published in 1867. Leo Gutmann, after participating in an unsuccessful revolt, flees to America. When, some years later, he returns with his socialist principles unchanged, he is able to persuade a prince to put into operation his plans for social amelioration and reform. But his schemes fail in consequence of working-class resistance. Leo, who appears to be modelled on F. Lassalle (q.v.), is simultaneously involved in intricate love-affairs, and is killed in a duel arising out of one of these.

I.N.R.I., a religious novel by P. Rosegger (q.v.), published in 1905, and sub-titled *Frohe Botschaft eines armen Sünders.* The initials stand for *Iesus Nazarenus Rex Iudaeorum* (the words over the Cross of Jesus). The novel has an encircling narration (see RAHMEN). Konrad Ferleitner is condemned to death for attempted murder. He suffers torments at the thought of his coming execution, which are not lessened when a petition for mercy lengthens the suspense. He asks a visiting monk for a copy of the New Testament, and when the request is refused, has the idea of writing the story of Jesus as he remembers it. His own version of the New Testament is the central portion of the book. Soon after its completion he dies, pardoned by God just before the news is brought that no pardon is to be granted by man.

Insel, Die, a monthly periodical for bibliophiles founded and edited by O. J. Bierbaum, A. W. Heymel, and R. A. Schröder (qq.v.) in 1899. It ceased to appear in 1902.

Insel der großen Mutter oder das Wunder von Ile des Dames, Die, a novel by G. Hauptmann (q.v.), published in 1924. It tells of a company of shipwrecked ladies stranded with one boy on an uninhabited island. A female state is established, and, since in time births of boys become too numerous, the boys are relegated to a distant part of the island. In the end the male state overcomes the female, and male supremacy is re-established.

Insel Felsenburg, Die, accepted title of a novel by J. G. Schnabel (q.v.), published in four parts in 1731, 1732, 1736, and 1743 respectively. It appeared under the pseudonym Gisander with the descriptive title reading in full: *Wunderliche Fata einiger See-Fahrer, absonderlich Alberti Julii, eines gebohrnen Sachsens, Welcher in seinem 18den Jahre zu Schiffe gegangen, durch Schiff-Bruch selb 4te an eine grausame Klippe geworffen worden, nach deren Übersteigung das schönste Land entdeckt, sich daselbst mit seiner Gefährtin verheyrathet, aus solcher Ehe eine Familie von mehr als 300. Seelen erzeuget, das Land vortrefflich angebauet, durch besondere Zufälle erstaunens-würdige Schätze gesammlet, seine in Teutschland ausgekundschafften Freunde glücklich gemacht, am Ende des 1728sten Jahres, als in seinem Hunderten Jahre, annoch frisch und gesund gelebt, und vermuthlich noch zu dato lebt, entworffen von dessen Bruders-Sohnes-Sohnes-Sohne, Mons. Eberhard Julio, Curieusen Lesern aber zum vermuthlichen Gemüths-Vergnügen ausgefertiget, auch par-Commission dem Drucke übergeben von Gisandern.* According to this title, which gives a succinct account of Albert Julius's success story on the island, the author is Eberhard Julius, and Gisander merely the editor, a piece of early 18th-c. realistic bluff. The narrative is set in a frame (see RAHMEN).

On the occasion of a visit by a descendant, Albertus Julius tells the story of his life. After considerable sufferings in youth in Europe, he sets out on a voyage, in the course of which the ship is wrecked in the South Atlantic. Albertus and three others, two men and a woman, survive. The two men are eliminated, one killing the other, and the murderer spiking himself accidentally on Albertus's sword. So Albertus and the woman, Concordia, marry and have numerous progeny, and Albertus rules harmoniously over a flourishing patriarchal state. The later volumes partly repeat the first and add new anecdotes. The story combines exciting adventures, erotic experiences, and the spectacle of a godfearing ideal state as a refuge from corrupt Europe. Popular at the time, it was forgotten in the course of the 18th c. and was then rediscovered by L. Tieck (q.v.), who published a version under the now current title *Die Insel Felsenburg* (1828).

Insomnis cura parentum, a German work by H. M. Moscherosch (q.v.), first published in 1643. It contains in 32 short chapters his advice to his children on their conduct in the world, reinforced by general moral reflections.

In Stahlgewittern, a war book by E. Jünger (q.v.), published in 1920. Though described as *Ein Kriegstagebuch* and narrated in the first person, it is not a true diary but a vivid later recollection of events with only occasional references to dates. It is an account of Jünger's experiences from 1914 to 1918, for a short time as a private, and then as a company officer. It provides interesting material on the psychology of war, particularly in the chapter on the German March offensive in 1918, in which a horrifying brutality (which Jünger neither extenuates nor rejects) emerges, giving one of the few convincing pictures of hand-to-hand fighting under conditions of extreme exertion and ceaseless, overwhelming din.

Interim, term applied to three provisional settlements of the religious conflict under Karl V (q.v.), which was precipitated by the Reformation (q.v.). The Regensburg Interim (1541) was not adopted and remained a dead letter. The Augsburger Interim (1548) made some concessions to the Reformation, and was enforced by the Emperor's authority in South Germany, though against opposition; it was rejected by the Northern princes. A third attempt (Leipziger Interim, 1548) met with no greater success.

Interimstaler were coins minted at Magdeburg in 1549 with an inscription deriding the Augsburg Interim: 'Packe di Satan, du Intrim' (see INTERIM).

Interpretation, term used for the study and exposition of a work of literature by itself and on its own terms. This approach, familiar in Britain and America since the 1920s (I. A. Richards, F. R. Leavis, and the New Criticism), was established in German studies in German-speaking countries by, among others, E. Staiger (*Grundbegriffe der Poetik*, 1946, and *Die Kunst der Interpretation*, 1955) and W. Kayser (*Das sprachliche Kunstwerk*, 1948). In its extreme form, in which all extraneous factors, biographical or historical, are rigorously excluded, it is known as werkimmanente Interpretation. The Graecism Hermeneutik, introduced by W. Dilthey (q.v.), is in use as an alternative term. (See also GEISTESGESCHICHTE and REZEPTIONSÄSTHETIK.)

Interregnum, term applied in German history to the period between the death of Konrad IV (q.v.), the last Staufen, in 1254 and the election of Rudolf I (q.v.), the first Habsburg, in 1273. It was marked by the election of rival kings from ruling houses abroad (Richard of Cornwall and Alphonso of Castile in 1257), who failed to establish themselves in Germany. The absence of any central authority was exploited by the territorial princes to augment their powers and domains, and the period was one of widespread anarchy.

The election of Rudolf of Habsburg in 1273 after his destruction of the power of Ottokar II (qq.v.) of Bohemia, did not at once rectify the disorder, but began the process of restoration. The Interregnum is the severest phase of a much longer crisis, styled by Schiller 'die kaiserlose Zeit', which lasted from the deposition of Friedrich II (q.v.) in 1245 to the coronation of Karl IV (q.v.) in 1355.

Investiturstreit, the conflict between the Pope and the temporal rulers in western Europe concerning the right of appointing bishops and abbots. Until 1075 this right was traditionally exercised by the sovereign, but in that year Pope Gregory VII forbade investiture by laymen. The ensuing struggle for power lasted nearly fifty years in Germany and involved fierce struggles, particularly in the reigns of the emperors Heinrich IV and Heinrich V (qq.v.). The Investiture Contest was ended in 1122 by the Concordat of Worms (see WORMSER KONKORDAT), a compromise by which the real power of appointment was vested in the papacy, while the emperor retained feudal rights over the temporal estates of the spiritual lords.

Invocavit, first Sunday in Lent, so called from the introit for the day, Psalm 91:15 (Lutheran and A.V.) or 90:15 (Vulgate). The verse in the Vulgate begins *Clamabit*, but *Invocavit*, occurring in the Old Latin Psalter, was already established and persisted in common usage.

Iphigenie auf Tauris, a verse play (Schauspiel) by Goethe, published in 1787 in *Goethes Schriften*, vol. 3. Goethe began it in Weimar in 1779, writing at that time in prose, and finished it in the space of about ten weeks. In this version (published in 1854) it was performed on 6 April 1779 at Ettersburg by an amateur company with Corona Schröter (q.v.) as Iphigenie, and Goethe as Orest. The reshaping of the play in blank verse was accomplished in Italy in 1786–7. In this, its final form, it was performed in Vienna on 7 January 1800, and at Weimar on 15 May 1802 (in an adaptation prepared by Schiller).

A play with little outward action, it presents Iphigenie prominently in each of the five acts in alternate confrontation with King Thoas of the barbaric land of the Tauri (Crimea), who is supported by Arkas, and Orest, her brother, and his friend Pylades. Iphigenie has been spirited away from the intended sacrifice in Aulis, the land of her fathers, and serves as priestess in Diana's temple in the kingdom of Thoas, her widowed host. Iphigenie has come to serve Diana with reluctance because of her greater love for her homeland to which she yearns to return. She has persistently evaded proposals of marriage from Thoas. These are repeated (Act I), first through Arkas, and then by Thoas himself. Impatient at Iphigenie's unsympathetic response, Thoas reinstates the old barbarian custom of the ritual sacrifice of strangers, which had lapsed over the years through Iphigenie's gentle influence on the King and the inhabitants. The strangers who have landed on the island are Orest and Pylades. Their identity is revealed to Iphigenie indirectly by Pylades (Act II), and fully at the climactic meeting between Iphigenie and Orest (Act III), which reveals the tragic happenings following her rescue from the sacrificial altar. Orest, who has murdered his mother because she had murdered his father, has learned from the Delphic oracle that he can expiate his crime by bringing 'the sister' back from the land of the Tauri. Orest has interpreted this as meaning the statue of Diana. Brother and sister discover their identity and face the prospect that he, the stranger, is to be sacrificially killed by Iphigenie, Diana's priestess; Orest undergoes a seizure in the course of which he has a cathartic experience ridding him of his despair and sense of persecution. He and Pylades resolve to escape with Iphigenie. Iphigenie herself experiences a conflict between her desire to flee with her brother and her sense of obligation towards Thoas, whose esteem she has won by her truthfulness (Act IV). The conflict is reinforced by the respective arguments of Arkas and Pylades, which bring Iphigenie to the verge of despair expressed in the song of the Parcae (Parzenlied) concluding the act. Orest and Pylades are caught while preparing for flight, and Iphigenie and Orest are both at the King's mercy (Act V). Iphigenie relieves her conscience by making a full confession, and Orest is inspired to refer the true meaning of the oracle to Iphigenie as 'the sister', and not to Diana's statue. Thoas overcomes his urge for vengeance, and magnanimously bids farewell to the Greeks, forgiving Orest's intended desecration of his temple, and still cherishing Iphigenie.

Among the elements in Goethe's inner life which contributed to *Iphigenie* are his longstanding yearning for the Mediterranean south, his love for Charlotte von Stein (q.v.), and his (temporary) sense of the power of moral integrity over barbarism. It is the first important work manifesting his classicism. Among Goethe's own comments are his description of the play in a letter to Schiller as 'ganz verteufelt human' (19 January 1802), and the lines written in 1827 on the flyleaf of a copy which end 'Alle menschliche Gebrechen/Sühnet reine Menschlichkeit'.

Irdisches Vergnügen in Gott, a series of 9 vols. of collected poetry written by B. H. Brockes (q.v.), and published between 1721 and 1748.

Iris, (1) a literary quarterly published by J. G. Jacobi (q.v.) in Düsseldorf, and intended chiefly for female readers. It ran from 1774 to 1777. Gleim, Goethe, Heinse, and J. M. R. Lenz (qq.v.) were among the contributors. Goethe's *Willkommen und Abschied* and *Maifest* (*Mailied*) were first printed in the periodical. The title was revived by Jacobi 1803–13.

(2) A journal (Taschenbuch) edited by Graf Johann Majláth, and published from 1839 to 1849 by G. Heckenast in Pest (Hungary).

Irrationalismus, a term used by many German critics and historians of literature. Introduced in the early 20th c., it acquired wide circulation in German studies through the works of H. A. Korff, who established the antithesis Rationalismus/Irrationalismus in *Der Geist der Goethezeit* (1923–50). It has the terminological drawback of stating only what it is not, i.e. Rationalismus. Irrationalismus is mainly useful as a blanket term covering many manifestations of the human spirit, including emotion, atmosphere (Stimmung), mystery, magic, awe, reverence, revealed religion, pantheistic yearnings, the numinous, enthusiasm, exaltation, the uncanny, the horrifying, and hallucination.

The term has been chiefly used in connection with the Sturm und Drang (q.v.), the Romantic movement (see ROMANTIK), and Expressionism (see EXPRESSIONISMUS).

IRREGANG, MEISTER, author of a short Middle High German poem, which presents, in apparently autobiographical form, the way of life of a travelling entertainer (fahrender Mann). The poet's name, given by him in the poem, is clearly adopted to match his life.

Irrungen, Wirrungen, a novel by Th. Fontane (q.v.), written between 1884 and 1887, and published in 1888. It is the third in Fontane's series of social novels set in Berlin. A short narrative of *c.* 150 pages, it falls into two parts (without formal division). It begins with a liaison between a girl of humble family, Lene Nimptsch, and Baron Botho von Rienäcker, an officer in the 'Kaiserkürassiere', a fictitious designation which conceals the identity of the Garde-Kürassier-Regiment of the Prussian Guard. The love of both is genuine, and they know in their hearts that their relationship cannot withstand the inevitable social pressures. It soon becomes apparent that Botho is being urged to make a 'good' marriage in order to restore the depleted family fortunes. A twenty-four-hour excursion to the lonely hostelry Hankels Ablage proves for Botho and Lene to be their last occasion of full happiness. He accepts his social duty, and she recognizes the overriding claims of his family. Botho and his pretty and vivacious bride settle down to the social round of Berlin.

The second phase of the story is separated from the first by the space of two and a half years. Botho's marriage, outwardly happy, does not satisfy him, and Lene is often in his thoughts. He burns her letters, and sternly resolves for duty. He is visited, on Lene's suggestion, by a Herr Gideon Francke, a religious sectarian, and an honest man who wishes to marry Lene. Botho satisfies Francke, the marriage takes place, and the novel ends on a note of resignation and duty and also of courage. In a frequently quoted conversation with a brother officer, Botho recognizes both the necessity of conformity and, by implication, the inextinguishable tie of his former love. Fontane treats his characters and their relationships sensitively, even tenderly, yet without sentimentality. And he convincingly and good-humouredly etches in, largely through dialogue, two contrasting social backgrounds of Berlin in the late 19th c.

Isabella von Egypten, Kaiser Karl des Fünften erste Jugendliebe, a Novelle by Achim von Arnim (see ARNIM, L. J. VON), published in *Vier Novellen* (1812). Isabella, an orphaned gipsy princess living in poverty in Flanders, meets by chance the Erzherzog Karl (the future Karl V, q.v.), and falls in love with him. She invokes magic to assist her, digging up from the foot of the gallows a mandrake, 'Cornelius Nepos', who provides her with the wealth necessary to approach Karl. She sees in her union with him an opportunity of redemption of the gipsy nation through their future offspring. Karl and Isabella meet and she arouses in him a love equal to hers, but their union is at first thwarted by a 'golem', a magic soulless double of Isabella. This rival is, however, disposed of, and the love of Karl and Isabella is consummated. She flees, and in Bohemia bears Karl's son, to whom she gives the name Lrak (an anagram of Karl).

Isegrim, name for the wolf in traditional animal stories, notably the group concerning Reynard the Fox (see REINKE DE VOS). Originally the name of a Germanic hero, the beast stories have led to its application to disagreeable, rasping, rapacious characters.

Isegrimm, a three-volume historical novel by W. Alexis (q.v.), published in 1854. A sequel to *Ruhe ist die erste Bürgerpflicht* (q.v.), it has the regeneration of Prussia after the defeat of Jena and the ensuing Wars of Liberation (see NAPOLEONIC WARS) as its background. It is notable for its description of landscape.

ISIDOR (*c.* 560–636), Bishop of Seville. The Old High German version of his *De fide catholica contra Judaeos* is the oldest surviving genuine translation of a substantial work into German (*Der althochdeutsche Isidor*). The German version exists in two partial texts, the more important one in Paris, the other in Vienna (see MONSEER FRAGMENTE). The translator's identity and the place at which he worked have not been established, though the latter was formerly believed to be Murbach in Alsace. The dialect of the translation appears to be an otherwise unrecorded form of Franconian, having close affinities to Rhenish Franconian (Rheinfränkisch).

ISIDORUS ORIENTALIS, pseudonym of OTTO Heinrich, Graf von LOEBEN (q.v.).

ISRAEL, SAMUEL (born in Strasburg in the second half of the 16th c., d. 1633), a Protestant pastor, wrote biblical plays including one on Pyramus and Thisbe (*Sehr lustige neue Tragödia von der großen unaussprechlichen Liebe zweier Menschen Pyrami und Thysbes,* 1604), and *Susanna* (1607).

ITTNER, JOSEPH ALBRECHT VON (Bingen/Rhine, 1754–1825, Constance), studied at Mainz and Göttingen universities and became an administrator, being at one time at the Reichskammergericht in Wetzlar, later chancellor of the Knights of Malta at Heitersheim near Freiburg, and then in Constance and Frankfurt. He published stories and essays in J. G. Jacobi's periodical *Iris* (q.v.), including *Der Prälat, Schiffskapitän Ali, Lob der Böcke, Über die Beschränkung der Eßfreiheit,* and *Hero und Leander am Bodensee.* His works were collected in 4 vols. after his death (*Schriften,* 1827).

Itzehoe (pronounced to rhyme with 'froh'), a small town in Schleswig-Holstein, which has become known in literature through a real person, the novelist Müller von Itzehoe (see MÜLLER, J. G.), and a fictitious character in

Schiller's *Wallenstein* (q.v.), 'der lange Peter von Itzehoe', otherwise Erster Jäger (*Wallensteins Lager*, l. 127).

Iwein, a Middle High German Arthurian epic poem of some 8,000 lines, written by HARTMANN von Aue (q.v.). It is probably Hartmann's last work, and was written *c.* 1200, being completed by 1203. Its source is *Le Chevalier au Lion* by Chrétien de Troyes. The knight Iwein hears of a dangerous adventure experienced by Sir Kalogreant at a magic well. He goes there, simply for the sake of adventure, is successful in defeating the knight defending the well, and pursues him to his castle. The knight dies, but Iwein is trapped in the castle and in great danger, from which he is rescued by Lunete, a lady-in-waiting, who gives him a ring making him invisible. The unseen Iwein is seized with love for the dead knight's widow Laudine, and with Lunete's help quickly wins her hand.

King Arthur and his knights come to Iwein's castle, and Sir Gawain reminds Iwein that he should not remain with his wife in unknightly sloth. Accordingly Iwein obtains Laudine's leave on condition that he returns a year later. But Iwein forgets the day of his return, and Lunete appears at Arthur's court, upbraids Iwein, and demands the return of Laudine's ring. Iwein is rendered insane by the shock and lives as a wild man in the woods. From this state he is restored with the aid of a magic salve administered by a passing lady. He then embarks on a series of adventures, the first of which is the rescue of his benefactress. He rescues a lion from a dragon and the lion is henceforth his faithful companion. Other adventures include the rescue of Lunete, imprisoned and threatened with execution. Eventually Iwein is drawn by his longing for Laudine to revisit the well, and he and his wife are reconciled.

Iwein, formally the most perfect of Hartmann's works, was one of the most popular medieval epics, and is preserved in twenty-five MSS., of which fifteen are complete.

J

JACOBI, FRIEDRICH HEINRICH (Düsseldorf, 1743–1819, Munich), studied philosophy, took over his father's business in 1764, and in 1772 became a civil administrator. In the 1770s he met Wieland, Hamann, Herder, and W. Heinse (qq.v.), and formed a friendship with Goethe. His first novel, *Eduard Allwills Papiere* (q.v., 1775–6), had its beginnings in his admiration for *Die Leiden des jungen Werthers* (q.v.), being also written in letter form (see BRIEFROMAN). Jacobi's second novel, *Woldemar* (1779, revised 1794 and 1796), shows a sensitive and emotional character attracted simultaneously by two women. The stress of the French Revolution along the German western border drove Jacobi to leave the Rhineland in 1794 for Eutin, which he left again in 1804 for Munich. Here he was entrusted with the reorganization of the Bavarian Academy of which he became the president. In his later years Jacobi, who had a deeply religious nature, drifted away from Goethe.

In 1780 Jacobi visited Lessing (q.v.), and the conversations he then held are included in *Über die Lehre des Spinoza* (1785). The suggestion that Lessing before his death inclined to pantheism and Spinozism caused a sensation, and greatly offended M. Mendelssohn (q.v.).

Werke (6 vols., vols. 1–3 ed. by Jacobi himself, vols. 4, 1–2 by F. Köppen, and 4, 3–6 by F. Roth) appeared 1812–25 (reissued 1968) and *Briefe,* ed. M. Brüggen and S. Sudhof, 1981ff.

JACOBI, JOHANN GEORG (Düsseldorf, 1740–1814, Freiburg/Breisgau), studied at several German universities, becoming a professor at Halle in 1766. A friend of J. W. L. Gleim (q.v.), he was made in 1768 a lay canon of Halberstadt. In 1784 he was elected to a chair at Freiburg University, where he remained for the rest of his life. His poetry, collected in *Poetische Versuche* (1764), *Abschied an den Amor* (1769), *Die Winterreise* (1769), and *Die Sommerreise* (1770), is mixed with rhythmic prose and consists in the main of pseudo-Anacreontic trivialities. From 1774 to 1777 he edited *Iris* (q.v.), a periodical in which a number of poems by Goethe, with whom Jacobi was acquainted, were first published. Late in life he resumed periodical publication with *Taschenbücher* (1795), the title of which was changed in 1803 to *Iris*. This second *Iris* appeared until 1813.

Jacobowsky und der Oberst, a play (Komödie einer Tragödie) by F. Werfel (q.v.), published in 1944. Set in the France of 1940, it follows the flight before the oncoming German troops of the Polish Jew Jacobowsky and the

Polish cavalry colonel Stjerbinsky, an anti-Semite. Ill-assorted companions in the same car, they make an absurd diversion to Brittany to visit the colonel's lady friend Marianne. Amidst the tragedies around them, and their own apparently tragic situation, comedy is extracted from incongruity and absurdity. The hopelessness of their position is suddenly relieved by the appearance of a R.N. officer, who rescues them.

JACOBUS DE VORAGINE or VARAGINE, see LEGENDA AUREA.

JACQUES, NORBERT (Luxemburg, 1880–1954, Coblence), lived most of his life at Lindau, but spent his last years at Hamburg. The author of many sensational novels, he created the character of Dr. Mabuse, internationally known through the cinema (*Dr. Mabuse, der Spieler,* 1921).

JAEGLÉ or JAEGLE, MINNA (Strasburg, 1810–80, Strasburg?), in full Wilhelmine, was the fiancée of G. Büchner (q.v.), after whose death in 1837 she remained single. The year of Büchner's death coincided with that of her father Johann Jakob (b. 1771), pastor of the church of St. Wilhelm in Strasburg, with whom Büchner had lodged when he resumed his medical studies after taking refuge in Strasburg. During Büchner's stay with him Jaeglé wrote poems in praise of the Polish patriots on the occasion of their reception in Strasburg (*Abschied und Willkomm an die Jahre 1831 und 1832*). He was already the author of poetry (*Cypressenhain. Ein Vorrat christlicher Trostgedichte,* 1830). Minna kept house for her widowed father.

Jagd der Minne, Die, an anonymous Middle High German poem of nearly 500 lines dating from the second half of the 14th c. It is an allegory of love based on *Die Jagd* by HADAMAR von Laber (q.v.).

Jagd von Württemberg, Die, an anonymous Middle High German poem of some 600 lines. It is also known by the title *Der Württemberger,* and two other versions are styled respectively *Des von Wirtemberg Buch* (436 lines) and *Der Ritter mit den Seelen* (713 lines). A knight participating in a hunting party given by the Count of Württemberg loses his way and falls in with a procession of knights and ladies. A lady riding alone warns him that they are all dead persons who must expiate a sinful love relationship, and he receives visible proof that they are tormented with purgatorial fire. The knight, who knows the lady's still living partner, informs her, and the two set out for the Holy Land in order to shorten the lady's period of torment. The poem was written in the 13th c.

JAGEMANN, CAROLINE (Weimar, 1777–1848, Dresden), a noted actress and singer, was trained in Mannheim under Iffland (q.v.). In 1797 she was engaged for the Weimar Court Theatre under Goethe's direction, and quickly became the leading actress. She also became, in an equally short time, the mistress of Duke Karl August (q.v.), who granted her the style Frau von Heygendorff. Goethe praised her abilities in generous terms ('Sie war auf den Brettern wie geboren und gleich in allem sicher und gewandt und fertig, wie die Ente auf dem Wasser'), but her assertive personality brought her into conflict with him. In 1808 her intrigues induced Goethe to offer his resignation, and in 1817 she succeeded, through her influence with the Duke, in forcing Goethe's departure from the theatre. She remained the Duke's mistress for thirty years, leaving Weimar after his death in 1828. Her greatest strength was in opera, but she created the roles of Thekla in *Wallenstein* and Elisabeth in *Maria Stuart* (qq.v.).

Jäger, Der, a folk-song-like poem written by E. Mörike (q.v.) in 1828 and included in the novel *Maler Nolten* (q.v.). It begins 'Drei Tage Regen fort und fort'.

Jäger, Die, a play (*Ein ländliches Sittengemälde in fünf Aufzügen*) by A. W. Iffland (q.v.), published and first performed in 1785. The honest forester (Oberförster) Warberger has a son, Anton, who loves his virtuous cousin Friederike. But his mother plans a wealthy marriage with the daughter of the prefect (Amtmann). Anton and Friederike are united, but Anton is arrested on a charge of unlawfully wounding the Amtmann's servant. The evidence against him is shown to be false, and all ends happily for the lovers. The background to this simple action is on the one hand the integrity of the rural group (Oberförster and family, pastor and Schulze, i.e. magistrate), and on the other the corruption, venality, and arrogance of the urban governing group.

JÄGER, JOHANNES, see CROTUS RUBEANUS.

Jäger aus Kurpfalz, Ein, see EIN JÄGER AUS KURPFALZ.

Jägerlied, a short poem in two four-line stanzas written by E. Mörike (q.v.) in 1837. All that the hunter sees in nature reminds him of his beloved. The poem begins 'Zierlich ist des Vogels Tritt im Schnee'.

Jägers Abendlied, a poem written by Goethe late in 1775 or early in 1776. It was printed with this title in *Goethes Schriften* in 1789, but the original version, published in *Der teutsche Merkur* in 1776, bore the superscription *Jägers Nachtlied*. The restless love it expresses refers either to Lili Schönemann or, more probably, to Charlotte von Stein (qq.v.). The poem has been set to music by F. Schubert (q.v.).

JAHN, Friedrich Ludwig (Lanz, 1778–1852, Freyburg/Unstrut), after abandoning his university studies, began as an enterprising philologist concerned with word usage (*Bereicherung des hochdeutschen Sprachschatzes*, 1806), and in 1810 took up teaching appointments in Berlin. It was here that he became a pioneer in physical training, earning the widely known sobriquet Turnvater. A strong exponent of nationalism, he became a victim of political persecution during the restoration. Allegedly involved in the formation of the Burschenschaften (q.v.) and in subversive activities, he was detained in the fortresses of Spandau and Küstrin from 1819 to 1825; his sports grounds, first founded in 1811, were closed. In 1848–9 he became a member of the Frankfurt Parliament (see FRANKFURTER NATIONALVERSAMMLUNG), commemorated in his *Schwanenrede* (1848).

Jahn's writings are on the subject of the German national character and on physical training, and were published as *Gesammelte Werke* (3 vols.), ed. C. Euler, 1883–7.

JAHNN, Hans Henny (Stellingen, Hamburg, 1894–1959, Hamburg), as a young pacifist took refuge in Norway from 1915 to 1918. After the 1914–18 War he became an expert on organs and their construction, and up to 1933 was official adviser on organ matters to the Hamburg municipality. In this year his writings were banned by the National Socialist regime, and Jahnn emigrated, first to Switzerland, then to Bornholm, his Danish refuge, where he farmed, breeding horses, and carrying out genetic research. A German citizen, he was expropriated by the Danish authorities in 1945; he returned to Hamburg, where he set up as a publisher and resumed his writing. In 1950 he was elected president of the Freie Akademie der Künste.

Deeply influenced by S. Freud (q.v.) as well as Expressionism (see EXPRESSIONISMUS), Jahnn wrote a series of plays in which violent and perverted sensuality is in conflict with a desire to achieve a harmony of balanced impulses. These include *Pastor Ephraim Magnus* (1919), *Die Krönung Richards III.* (1921), *Der Arzt, sein Weib, sein Sohn* (1922), *Der gestohlene Gott* (1924), *Medea* (1926), *Straßenecke* (1931), and *Neuer Lübecker Totentanz* (1931, revised 1954). After the 1939–45 War followed the plays *Thomas Chatterton* (1955) and *Die Trümmer des Gewissens* (published posthumously and produced under the title *Der staubige Regenbogen* in 1961).

As a novelist Jahnn began with *Perrudja* (2 vols., 1929), but his best-known work is the trilogy of novels *Fluß ohne Ufer* (q.v.) consisting of *Das Holzschiff* (1949), *Die Niederschrift des Gustav Anias Horn* (2 vols., 1949–50), and *Epilog* (unfinished and posthumously published, 1961). His other novel is *Die Nacht aus Blei* (1956). Jahnn was a writer torn between elemental impulses, expressed in brutality and terror, but also possessed by a yearning for harmony in nature, manifesting itself in kindness to man and beast. His standing as a writer has been hotly disputed by rival schools of criticism. *Werke und Tagebücher in sieben Bänden*, ed. Th. Freeman and Th. Scheuffelen, appeared in 1974.

Jahr der Seele, Das, a volume of poetry by S. George (q.v.), published in 1897.

'Jahre kommen, Jahre schwinden', first line of a once well-known poem by K. von Reinhard (1769–1840), first published in the *Göttinger Musenalmanach für 1794.*

Jahrestage. Aus dem Leben der Gesine Cresspahl, a four-volume novel by U. Johnson (q.v.), the first three published 1970–3, vol. 4 in 1983. The novel covers a year, from August to December 1967 (vol. 1.), from December 1967 to April 1968 (vol. 2), from April to June 1968 (vol. 3), and from June to 20 August 1968 (vol. 4). The central character, Gesine Cresspahl, appears briefly in *Mutmaßungen über Jakob* (q.v.).

Gesine is an unmarried mother who, with her daughter Marie, born in 1957, lives in New York, working for an American firm. The novel uses a technique of montage that includes notes, jottings, newspaper cuttings, and recollections. Dialogue, soliloquy, descriptions of persons and scenes alternate with extracts from the *New York Times*, and in a series of flashbacks a chronicle of the times through which Gesine and Lisbet, her mother, and her father, a joiner, have lived, is unfolded.

Four main themes run through the work: firstly, the relationship of Gesine and her daughter to each other and to their American environment, which is foreign to Gesine, but home for her young daughter; secondly, the condition of the world with the war in Vietnam; thirdly, the racial problem of the U.S.A., and the urban disintegration of New York. The fourth, and perhaps most important, theme is the history of Germany seen through the life of Cresspahl, Gesine's father, who married the local squire's daughter in Mecklenburg, and set up a business in England. Lisbet moved back to Germany when her child (Gesine) was due to be born in 1932, and Cresspahl followed her shortly afterwards. The years during the National Socialist regime bring the suicide of the pious Lisbet after the outrages against Jews in Kristallnacht (1938, q.v.), and Cresspahl's years of widowhood. The British take Mecklenburg but withdraw, and Soviet occupation ensues. Cresspahl, appointed Bürgermeister of Jerichow by the British, is arrested and maltreated by the Russians. At the end of vol. 3 Gesine contemplates a journey to Czechoslovakia during the Czech 'thaw', and it is with the commencement of this journey and after the death of Dr Erichson has deprived her of the prospect of happiness and security, that vol. 4 ends. The 'Last and Final' section of the fourth volume, dated 20 August 1968, records her stop in Copenhagen, where she and Marie meet Dr Kliefoth, Gesine's old teacher in English and deportment. An octogenarian representing her Mecklenburg homeland, he has attended her father's funeral as a friend; Gesine now hands him her story and the three stroll for a while along the water's edge. The reunion of three generations affirms a natural order that contrasts with the volume's sharp focus on the disturbing political scene, a poignant aspect of which is the fate of children. On the final page, Johnson states the time during which he wrote the tetralogy: 29. Januar 1968, New York, NY–17. April 1983, Sheerness, Kent.

The work is clearly planned with the utmost care to produce a coherent point–counterpoint. It is written with discipline and precision, and gives a feeling of balance and fairness. Although only the modern U.S.A. and the years from 1936 to 1959 were experienced at first hand by Johnson, his re-creation of Germany in the earlier years is fully convincing. The restrained style is frequently ironic, but also has passages of atmospheric description. *Jahrestage* combines originality of structure with a high level of craftsmanship.

Jahreszeiten, Die, an oratorio composed by J. Haydn (q.v.), and first performed in Vienna in 1801. The libretto by G. van Swieten (q.v.) took its starting-point from James Thomson's poem *The Seasons* (1730), but the adaptation is so free that the original is no longer recognizable.

Jahrgang 1902, a novel of youth in wartime and at war by E. Glaeser (q.v.), published in 1928.

Jahrmarktsfest zu Plundersweilern, Das, a comic dramatic scene written by Goethe in 1773, and published in 1774. It has no plot, but presents a quick changing dialogue in Knittel-verse (q.v.) among those attending the market. Goethe gives it the sub-title *Ein Schönbartsspiel* (see SCHÖNBARTLAUFEN).

Jakob der Letzte, a novel by P. Rosegger (q.v.), published in 1888. Jakob Steinreuter lives on the ancestral farm in the parish of Altenmoos, which is gradually bought up by a wealthy man who turns it into a sporting estate. The farmers sell out one after another, but Steinreuter stands firm. His wife dies, his elder son disappears, the younger is killed in war, his daughter and her husband are ruined, but he stands his ground and refuses to give way to persistent minor persecution. In the end he hears from his elder son, who now has a prosperous farm in Oregon. But Steinreuter stays in Altenmoos; having shot the gamekeeper, he drowns himself.

JAKOBS, KARL HEINZ (Kiauken, East Prussia, 1929–　), studied at the Leipzig Johannes R. Becher Institute for Literature and worked for a time as a bricklayer. His publications include the collection of poetry *Guten Morgen, Vaterlandsverräter* (1959), and the novels *Beschreibung eines Sommers* (1961), the subjective element in which was acclaimed by Christa Wolf (q.v.), *Eine Pyramide für mich* (1971), *Die Interviewer* (1973), *Tanja, Taschka und so weiter* (1975) a documentary travel novel, *Wüste kehr wieder* (1976), and *Fata Morgana* (1977). A volume on theory, *Heimatländische Kolportagen. Ein Buch Publizistik* was published in 1975. One of the signatories of the public letter of protest concerning W. Biermann (q.v.), he was deprived of his SED membership in 1977. His novel *Wilhelmsburg* (1979), covering the first decades of the DDR, highlights critically the phenomenon of careerism. It was his first work to be published only in the BRD. In the year of its publication, 1979, Jakobs was deprived of his membership of the Schriftstellerverband and settled in Bochum (BRD). In the novel *Die Frau im Strom* (1982), Liesbeth Koslowski, an unmarried pregnant woman, has been callously drowned in the river Elbe by a woman friend, Edith Schilder, whose motive is jealousy. Both she and Gustav Sasse, head of the borough surveyor's office, who has fathered the child, are convicted of murder. Sasse dies in prison, but Schilder, a grandmother by the time of her release, settles happily in her family circle. A social and psychological novel introducing a great variety of figures, it is set in the 1950s. *Das endlose Jahr* was published in 1983. Jakobs is the editor of the volume *Das große Lesebuch vom Frieden* (1983).

Jakobsbrüder, Die, see KUNZ KISTENER.

Jakob von Gunten. Ein Tagebuch, a novel by R. Walser (q.v.), published in 1909. Jakob, the youngest member of an ancient family, attends a small Swiss boarding-school, the Institut Benjamenta. His aim is to rid himself from the ties of his once illustrious, and still prosperous, mercantile family tradition, and to learn to live in humble service and self-discovery. His experiences at school turn out to be very different. The headmaster, Benjamenta, and his sister Lisa, the teacher, are in their different ways strange and bewildering personalities. Jakob is fascinated by the reserved Lisa who kindles hitherto unknown regions of his mind. Her death causes the dissolution of the school. The pupils disperse, but Jakob follows Benjamenta into a new way of life which resembles neither that which he set out to abandon nor the one for which he had searched: 'Ich gehe mit Herrn Benjamenta in die Wüste.' The fascination of the domineering Benjamenta prevails.

Impressionism, surrealism, and realism intertwine in this work, which was highly thought of by Kafka (q.v.), who detected affinities between himself and attitudes expressed by Walser.

Jamben, iambic verse, and especially the iambic pentameters known as blank verse. See BLANKVERS.

JANDL, ERNST (Vienna, 1925–), became a soldier in 1943, was taken prisoner, returned to Vienna in 1946, and studied Germanistik and English at Vienna University. He then turned to teaching in a Viennese grammar school. Primarily known as a pioneer of concrete poetry (see KONKRETE POESIE), he was associated with the Wiener Gruppe (q.v.) and in 1973 became a co-founder of the Grazer Autorenversammlung. The constructive social (and political) aim of his poetry is made evident through the choice of a (familiar) word, the poem's object, and its title (an early poem, *die zeit vergeht*, is a splendid example). Jandl's feel for the right nuance of humour, wit, and irony informs his experimental use of sound, speech, and space. Collections of poetry include *Hosi-Anna, Laut und Luise* (both 1966), *sprechblasen. gedichte* (1968), *der künstliche baum* (1970), *übung mit buben*, designed for children (1973), *dingfest*, containing poems from the 1950s and 1960s and a postscript by H. Mayer (1973), *der versteckte hirte* (1975), *die bearbeitung der mütze*, written during May 1975 and January 1978 (1978), *der gelbe hund*, written during June and August 1979 (1980), and *selbstporträt des schachspielers als trinkende uhr* (1983). Jandl has increasingly aimed in his use of language to achieve objectification and to eli-

minate sound effects in favour of speech (Sprechgedicht) and reading. Considerably varied in form, his poems tend to concentrate on ordinary objects or types ('man' and 'woman'), and to convey disillusionment and isolation in the context of social attitudes.

The collection of plays *Ernst Jandl, für alle* (1974) includes *szenen aus dem wirklichen leben* (1966), *parisätes stück*, and *der raum* (both 1970). The volume *Ernst Jandl, Sprechblasen* (1979) contains another early play, *gestern: ein spiel* (1964); the one-act play *die humanisten*, performed in 1976, was broadcast as a radio play in 1977.

A number of radio plays and a television play (*Traube*, 1971) were written in collaboration with Friederike Mayröcker (q.v.), notably *Fünf Mann Menschen* (first broadcast in 1968). The collection *Fünf Mann Menschen. Hörspiele* (1971) contains, apart from the title play, *die auswanderer* (1957), *Der Uhrensklave* (1969), and *Das Röcheln der Mona Lisa* (1970). A new experiment is *Aus der Fremde. Sprechoper in 7 Szenen* (1980). *Alle freut, was alle freut. Ein Märchen in 28 Gedichten*, based on drawings by Walter Trier, appeared in 1975, *die schöne kunst des schreibens* in 1976, and *Augenspiel. Gedichte*, ed. J. Schreck, in 1981. *Das Öffnen und Schließen des Mundes, Frankfurter Poetik-Vorlesungen* appeared in 1984. *Gesammelte Werke. Gedichte. Stücke. Prosa.* (3 vols.), ed. K. Siblewski, appeared in 1985.

Jandl was awarded the Trakl Prize in 1974 and the Büchner Prize in 1984.

JANSEN ENIKEL, a citizen of Vienna in the 13th c., wrote two chronicles in his later years (c. 1280), the *Weltchronik* and the *Fürstenbuch*. Both are written in Middle High German rhyming couplets. The *Weltchronik* covers, with little political sense and an avid appetite for anecdote, the whole of history from the Creation to the death of the Emperor Friedrich II (q.v.). The *Fürstenbuch*, which is unfinished, is a chronicle of Austria and Styria, from the foundation of Vienna onwards. The name Jansen Enikel (modern *Enkel*) means grandson of Jans or Johannes.

Jasager und Der Neinsager, Der, two short plays (see LEHRSTÜCK) by B. Brecht (q.v.), written in 1929 (*Der Jasager*, first version) and 1930 in the form of plays with music (Schulopern). They are adaptations of *Taniko*, from *The Nō Plays of Japan* (1921), translations by Arthur Waley. The music is by Kurt Weill (q.v.), and both plays, consisting of mixed prose and free verse, are intended to be sung together with orchestra and six soloists: the tutor (Der Lehrer), the boy (Der Knabe), his mother (Die Mutter), and three students (Die 3 Studenten).

The commentary is provided by a chorus (Der grosse Chor).

The two plays, consisting of ten scenes, have identical parts with contrasting conclusions. In *Der Jasager* the boy crosses the mountains with the other characters in order to fetch medicine for his sick mother from the city. He does not accomplish the feat and falls ill on the way. As it is impossible for the students to carry him along the perilous path to the town, he is, according to custom, to be thrown to his death in the valley to save him a slow and lonely death in the mountains. Custom also prescribes that he should first be asked, and that he should say yes. The boy conforms to the custom.

In *Der Neinsager* the boy refuses to conform to the custom on the grounds that new situations require new actions ('customs'). The students agree to carry him back to his mother, and in this they succeed. It is no disgrace to defy traditional laws if it is reasonable to do so in a given situation.

JASPERS, KARL (Oldenburg, 1883–1969, Basel), was a professor of psychiatry at Heidelberg University from 1916, turned to philosophy, and held the chair of philosophy from 1921 until 1937 when he was dismissed for political reasons. In 1948 he accepted a chair at Basel. Jaspers is noted for his conception of Existentialism, which he views in terms of transcendence and communication. Reality reveals itself when man's existence enters a final phase of consciousness in which time and eternity coincide. This he calls the Grenzsituation (border-situation) in which the experience of guilt, suffering, and a sense of failure (Scheitern) merges with that of death. Existence is analogous to time, and eternity to transcendence, which reveals itself in ciphers (symbols). Man can only transcend (transzendieren) the barriers of communication in the spheres of temporality and spirituality by means of reason. Thought should manifest itself in action for the sake of both self-revelation and self-discipline. Jaspers was a resigned empiricist and tolerant towards men of different views and institutions. He was nevertheless outspoken in his criticism of totalitarian systems of government and of the development of nuclear arms.

In psychology his principal work is *Allgemeine Psychopathologie* (1913). The major philosophical works are *Philosophie* (1932), *Von der Wahrheit* (1947), *Der philosophische Glaube* (1948), *Die großen Philosophen* (1957, incomplete). Among influential general writings are *Die Schuldfrage* (1946), *Vom europäischen Geist* (1946), *Vernunft und Widervernunft in unserer Zeit* (1950), *Rechenschaft und Ausblick* (1951), *Die Atombombe und die Zukunft des Menschen* (1957), *Freiheit und Friede* (1958), *Freiheit und Wiedervereinigung* (1960), *Lebensfragen der deutschen Politik* (1963), *Hoffnung und Sorge* (1965), *Wohin treibt die Bundespolitik?* (1966), *Provokationen* (1968).

It is in *Von der Wahrheit* that Jaspers expresses his views on tragedy (*Über das Tragische*).

JEAN PAUL (Wunsiedel, Fichtelgebirge, 1763–1825, Bayreuth), pseudonym of Johann Paul Friedrich Richter and the name by which he is universally known. The son of a schoolmaster in humble circumstances, who later became a Lutheran pastor, Jean Paul spent his childhood in Joditz and in Schwarzenbach. He was educated at the grammar school at Hof from 1779 to 1781, when he became a student at Leipzig University. His financial straits were such that he felt obliged in 1784 to abandon his studies. By this time he had published his first (unsuccessful) work, *Grönländische Prozesse* (1783–4). All his early years were overshadowed by poverty and misfortune, which included the suicide of his brother in 1790; a crisis of scepticism is reflected in *Rede des toten Christus* (q.v., first drafted in 1789). Jean Paul spent the years 1786 to 1790 as a private tutor, and from 1790 to 1794 as a schoolmaster. In these years he wrote *Auswahl aus des Teufels Papieren* (1789) and *Die unsichtbare Loge*, in which the story *Leben des vergnügten Schulmeisterleins Maria Wuz* (q.v., both 1793) is contained. These works began his rise to fame and affluence, which the publication of *Hesperus* (q.v., 1795) confirmed.

A celebrity almost overnight, Jean Paul was taken up by various notabilities, particularly Herder (q.v.), and by patrons such as Frau von Kalb (q.v.), who was the first to invite him to Weimar. His eccentric and discursive novels, full of humour, sentiment, and irony, were among the most widely read books, especially in the first two decades of the 19th c. After various intermediate stations at Hildburghausen, Berlin, Meiningen, and Koburg, he married in 1801 and settled in 1804 in Bayreuth, where he spent the rest of his life.

Blumen-, Frucht- und Dornenstücke (q.v., commonly called *Siebenkäs*) appeared in 1796–7, *Des Quintus Fixlein Leben* (q.v.) in 1796, *Titan* (q.v.) in 1800–3, and his greatest (though unfinished) work *Flegeljahre* (q.v.) in 1804–5. In 1808 he was granted a pension by Prince Karl Theodor von DALBERG (q.v.), and later received support from the Bavarian government. The sequence of Jean Paul's fashionable novels closes with *Dr. Katzenbergers Badereise* and *Des Feldpredigers Schmelzle Reise nach Flätz* (qq.v., both 1809). His later stories (*Leben Fibels des Verfassers der Bienrodischen Fibel*, 1812, and *Der Komet oder Nikolaus Marggraf*, 1820–2) were less successful. Mention should also be made of the short narrative *Das Kampanertal oder über die Unsterblichkeit*, of the prose idyll *Der Jubelsenior* (both 1797), and of the satirical *Palingenesien* (1798), which began as a rewriting of *Auswahl aus des Teufels Papieren* and revives the characters Siebenkäs and Leibgeber. Jean Paul's desultory ars poetica, *Vorschule der Ästhetik* appeared in 1804. In it he opposes both 'poetic nihilists' (Goethe, Schiller, and Romantics such as Novalis, qq.v.) and 'poetic materialists' (such as Brockes or Gellert, qq.v.). The true poet maintains a middle way between these two extremes, 'clothing Nature in ideal infinity' ('begrenzte Natur mit der Unendlichkeit der Idee umgeben'). *Levana*, published in 1807, is a treatise on education, the aim of which is the elevation of the human soul above the limitations of its age ('Erhebung über den Zeitgeist').

These theoretical works are wayward and discursive like the novels. The qualities of variability and discontinuity, which commended all Jean Paul's works to a whole generation (the magic still worked for Carlyle and Hebbel, q.v.), afterwards became reasons for his decline. The sentiment, the humour, the irony, and the verbal arabesques, which once delighted, in the long run seemed too deeply steeped in self-indulgence. Nevertheless, *Flegeljahre*, *Wuz*, and *Quintus Fixlein* have by their deep humanity escaped the oblivion into which much of his work has fallen.

Sämtliche Werke. Historisch-kritische Ausgabe, ed. E. Berend, appeared from 1927 and *Werke*, ed. G. Lohmann (until 1970) and N. Miller (from 1974 as *Sämtliche Werke*) from 1959.

Jedermann, a play by H. von Hofmannsthal (q.v.), published in 1911, and first performed at the Salzburger Festspiele (q.v.) in August 1920. It is based on the traditional religious parable of the rich man who is unexpectedly faced with death, and, for the first time, gives thought to the salvation of his soul. This is believed to be of Dutch origin, though early English and German versions are also known (see GENNEP, J. VON, MACROPEDIUS, G., NAOGEORGUS, T., SACHS, H., STEPHANI, C., and STRICKER, J.). In his preface Hofmannsthal regrets the neglect of these plays, and offers his own version in order that this dramatic parable may again be heard.

Jedermann eats, drinks, and is merry; he has no ear for his mother's warnings, no compassion for his debtor's distress. In the middle of a banquet with his paramour (Buhlschaft) and his fellow winebibbers, his usual sense of well-being deserts him, and Death appears, to warn him that his end is near. In vain he pleads for time. He begs the friends of his prosperity to accompany him on his dark journey, but without avail. He bids his servants carry his treasure chest on the journey; but they flee. As he seeks to console himself at least with the possession of the chest, the lid opens, and Mammon emerges and mocks him. The only companion willing to go with him is Good Works (Werke who, like Glaube, is a female allegorical figure), but his good works have been so few that she is too frail and feeble to endure the journey. Faith (Glaube) then appears, and in her presence Jedermann undergoes a conversion. His repentance restores Good Works to vigour and beauty. An attempt by the Devil to carry Jedermann off to Hell is thwarted by an angel, and Everyman descends with a contrite and believing heart into his grave.

Jehuda ben Halevy, a group of four poems (the last of which is unfinished) included by Heine in the last section, *Hebräische Melodien*, of the *Romanzero* (q.v.). In the person of the medieval Hebrew poet Juda Ha-Levi, Heine celebrates the great poetry of the Psalms.

JEITTELES, ALOIS (Brünn, Czech Brno, 1794–1858, Brünn), after studying medicine at Prague, Brünn, and Vienna, spent his life as a physician at Brünn. His comedies included a parody of fate tragedy (see SCHICKSALS-TRAGÖDIE) written jointly with I. F. Castelli (q.v., *Der Schicksalsstrumpf*, pseudonymously attributed to 'die Brüder Fatalis'). He is now only remembered as the author of the six poems grouped as *An die ferne Geliebte* (q.v.), which were set as a cycle by Beethoven in 1816.

JELLAČIĆ, JOSEF, GRAF, in full Jellačić von Bužim (Peterwardein, 1801–59, Agram), an Austrian general, was prominent in the suppression of the revolts in Vienna and Hungary during the Revolution of 1848 (see REVOLUTIONEN 1848–9). After the restoration of imperial power he was appointed governor of Croatia and Slavonia. In 1815 he published a volume of verse, *Gedichte*. He was created count in 1855.

Jena, Battle of, was fought on the plateau immediately north-west of the town on 14 October 1806. The Prussian army under General Hohenlohe (Fürst F. L. zu Hohenlohe-Ingelfingen) was outmanœuvred and outfought by the numerically superior French forces under Napoleon, and finally dissolved into flight. Simultaneously a second Prussian army was routed a few miles to the north at Auerstedt (q.v.). The consequence of the double defeat was the overrunning of Prussia by the French armies and its subjugation and dismemberment in the Treaty of Tilsit in 1807 (see NAPOLEONIC WARS). Of the two battles it is Jena which has become the proverbial symbol of Prussian defeat.

Jenaer Liederhandschrift, a MS. written in the middle of the 14th c. It contains primarily poems by central and North German poets, and includes many Sprüche (see SPRUCH). It has no illustrations, but is especially notable for the inclusion of the music to the songs. At one time in Wittenberg, it was brought to Jena in 1548.

JENATSCH, GEORG (Jürg) (Engadin, 1596–1639, Chur, murdered), a Swiss patriot of Canton Graubünden (Grisons), became a pastor like his father. In the Thirty Years War (see DREISSIGJÄHRIGER KRIEG), in which the cantons were involved, he was responsible for the murder of Pompejus Planta, head of the Roman Catholic patriots, who were allied with Spain. Jenatsch sought to secure the independence of his canton by serving the French under the Duc de Rohan, who liberated Graubünden from Spanish and Austrian occupation. In 1635 Jenatsch became a Catholic in the hope of enlisting Spanish support, and two years later he conspired against Rohan, forcing him to leave the canton. His name is known in literature mainly through C. F. Meyer's novel *Jürg Jenatsch* (q.v.).

Jena University, founded in 1558 by Johann Friedrich der Großmütige, Elector of Saxony (1503–54). The discrepancy between the dates of the Elector's death and the foundation is explained by a delay, occasioned by the refusal of the Emperor Karl V (q.v.) to grant a charter, which meant that the actual establishment took place under Karl's successor, the Emperor Ferdinand I (q.v.). Jena was a Lutheran university, hence the imperial reluctance. It was noted for its school of medicine, but was also notorious in the 17th c. and early 18th c. for the lawlessness and licentious conduct of its students. A caricature of the Jena student of this phase is to be found in *Der Renommist* (q.v., 1744) by J. F. W. Zachariä (q.v.). The hard drinking and roistering repelled Klopstock (q.v.), who spent less than a year there (1745-6). In the late 18th c. Jena became one of the principal centres disseminating the philosophy of Kant (q.v.). Schiller was a professor at Jena from 1789 until his death, though he ceased to lecture in 1791. In 1792 it was the scene of student riots and a threatened emigration to Erfurt, which was avoided by Goethe's conciliatory intervention. J. G. Fichte (q.v.) held the chair of philosophy 1794-9. In the last years of the century he was joined by the early Romantics A. W. and F. Schlegel (qq.v.) who were the centre of the Jena group (see ROMANTIK).

JENS, WALTER (Hamburg, 1923–), had his schooling at the Johanneum (q.v.) in Hamburg,

and studied at Hamburg and Freiburg universities. He became professor of classical literature and rhetoric at Tübingen University in 1956. Jens combines critical and scholarly work (notably on Euripides) with creative writing. He is the author of the novels *Nein. Die Welt der Angeklagten* (1950) and *Vergessene Gesichter* (1952), and of stories, including *Das weiße Taschentuch* (1947), *Der Blinde* (1951), and *Das Testament des Odysseus* (1957), as well as of plays for radio and television. *Der Fall Judas* and the libretto *Der Ausbruch* appeared in 1975, *Republikanische Reden* in 1976, and *Zur Antike* in 1978. *Aischylos: Die Orestie. Agamemnon/Die Cheophoren/Die Eumeniden* (1979) is a free rendering from the Greek original. Jens was one of the Gruppe 47 (q.v.), and his early writing was influenced by Kafka (q.v.).

Well known as the writer of critical essays (*Statt einer Literaturgeschichte*, 1957, and *Literatur und Politik*, 1963), Jens published his own translation of St. Matthew's Gospel, entitled *Am Anfang der Stall—am Ende der Galgen* (1972), and in *Der barmherzige Samariter* (1973) he collected a number of modern interpretations of the parable of the Good Samaritan. The volume *Ort der Handlung ist Deutschland. Reden in erinnerungsfeindlicher Zeit* was published in 1981.

Jenseits von Gut und Böse, a philosophical work by Friedrich Nietzsche (q.v.), written in 1885, and published in 1886. It bears the subtitle *Vorspiel einer Philosophie der Zukunft*, and consists of 296 numbered paragraphs, arranged in nine main sections (Hauptstücke) with the following headings: I. *Von den Vorurteilen der Philosophen*; II. *Der freie Geist*; III. *Das religiöse Wesen*; IV. *Sprüche und Zwischenspiele*; V. *Zur Naturgeschichte der Moral*; VI. *Wir Gelehrten*; VII. *Unsere Tugenden*; VIII. *Völker und Vaterländer*; IX. *Was ist vornehm?* The book, which followed *Also sprach Zarathustra* (q.v.), is written in an aphoristic and ironical style.

Jenseits von Gut und Böse derides the levelling tendencies of the modern world, and attacks existing morality which bolsters the weak and restricts the strong. To 'Herdenmoral' Nietzsche opposes a 'Herrenmoral' asserting a scale of human beings, 'eine Rangordnung zwischen Mensch und Mensch'. He has no good words for the new nationalistic Germany and castigates anti-Semitism. The book is a preparation for his intended but uncompleted masterpiece, *Der Wille zur Macht* (q.v.). It concludes with an epilogue (Nachgesang), a poem entitled *Aus hohen Bergen*, written in 1884, which expresses a longing for friendship; the last two stanzas, written later, convey deep disillusionment.

JENSEN, WILHELM (Heiligenhafen, Holstein,

1837–1911, Munich), studied medicine and history at a number of universities, but mainly at Kiel, where he took a doctorate in history. Of Frisian stock, Jensen is also the author of works of Heimatkunst (q.v.), and of poetry, songs, and sketches (a collection, *Ausgewählte Gedichte*, ed. Th. von Sosnosky, appeared in 1912). His tragedies (*Dido*, 1870, and *Juanna von Kastilien*, 1872) attracted little attention. Jensen edited the *Schwäbische Volkszeitung* in Stuttgart (1868) and the *Norddeutsche Zeitung* in Flensburg (1869–72) before taking up writing as a career, settling in 1888 in Munich.

Apart from *Nirvana. Drei Bücher aus der Geschichte Frankreichs* (1877), Jensen's main works deal with German history; they include *Um den Kaiserstuhl. Roman aus dem dreißigjährigen Krieg* (1878), *Vom römischen Reich deutscher Nation* (3 vols., 1882), *Aus den Tagen der Hansa* (3 Novellen, 1885), *Am Ausgang des Reichs* (1886), *Der Hohenstaufer Ausgang* (1896), and *In majorem Dei gloriam* (1905).

Jensen's native region appears in the novels *Flut und Ebbe* (1877), *Luv und lee*, and *Aus See und Sand* (both 1897), which are remarkable for their nature description. Jensen wrote an appreciation of W. Raabe (q.v.) in *Wilhelm Raabe* (1901).

Jeremias, see HÖRET DIE STIMME.

Jerominkinder, Die, a novel by E. Wiechert (q.v.), published in two volumes, 1945–7. The first was written in 1940–1, the second in 1946. The novel is a chronicle of the Masurian village of Sowirog, concentrating chiefly on the Jeromin family headed by the pious grandfather Michael, a centenarian, and on the period 1900–39. The most prominent characters are the charcoal-burner Jakob Jeromin and his five sons and two daughters, whose lives are the main substance of the book.

Jakob, the embodiment of simple goodness and faith, is called up in 1914 and dies on the eastern front. Rittmeister von Bulk, a retired captain of Uhlans, outwardly hard, but generous and just in his actions, steps in as squire of the village, and enables Jons Jeromin, the idealist youngest son, to study medicine in order to devote his skill to the welfare of the villagers. The patriarchal life of the village persists undisturbed by private misfortune until 1933, when the National Socialist regime begins to cast its shadow, causing acts of violence, of which the most serious are the murder of von Bulk and the suicide of the Jewish doctor, Jens's friend Lawrenz. Wiechert commented that the novel needed a third volume to cover the 1939–45 War, which he felt powerless to write.

JERSCHKE, OSKAR (Lähn, Silesia, 1861–1928, Berlin), a lawyer by profession who, though not a writer of importance, collaborated with A. Holz (q.v.) in *Traumulus* and *Frei!*

JERUSALEM, KARL WILHELM (Wolfenbüttel, 1747–72, Wetzlar), son of Lessing's friend J. F. J. Jerusalem (1709–89), was at Wetzlar in the diplomatic service of the Duchy of Brunswick. His relations with his principal, Graf Bassenheim, developed unfavourably, and his personal situation was aggravated by a hopeless love for a married woman. Lapsing into a state of depression, Jerusalem shot himself on the night of 29/30 October 1772. Goethe, who had known him in Wetzlar, was much affected by this tragedy, which was to influence *Die Leiden des jungen Werthers* (q.v.), Bk. II. The house in which Jerusalem committed suicide still stands, facing the Schillerplatz. A selection of Jerusalem's writings was published in 1776 by Lessing (q.v.) under the title *Philosophische Aufsätze*.

Jery und Bätely, a Singspiel (q.v.) written by Goethe in 1779, and published in 1790. Bätely is a Swiss smallholder's daughter, who refuses to marry Jery, a worthy young man who loves her. The returned soldier Thomas undertakes to put matters right, but makes such a nuisance of himself to Bätely and her father that they call in Jery to drive him away. Though he is not able to cope with Thomas, Bätely is now willing to marry Jery, and all are reconciled. The play was performed in Weimar in June 1804.

Jesuitendrama, a collective term for Latin plays written by Jesuit fathers for performance in their schools in the late 16th c. and in the 17th c. Performances took place in the hall of the schools each September. Since many of the audience could not understand the words, a synopsis was commonly provided, and the plays themselves tended, for the same reason, to rely on visual effects and on ancillary music. The purpose of the plays was religious and missionary, and they served as one of the instruments of the Counter-Reformation (see GEGENREFORMATION). Their subjects were mainly drawn from the Old Testament and from the legends of the saints. The earliest beginnings of Jesuit drama are seen c. 1565, and the heyday is the early 17th c., though its manifestations continue into the 18th c. The principal exponents are J. Gretser, J. Bidermann, N. Avancini (qq.v.), and J. B. Adolf (1657–1708). Jesuit drama flourished most conspicuously in the south, especially in Vienna and Munich.

Jesuitengesetz, see KULTURKAMPF.

'Jesus, meine Zuversicht', opening words of an anonymous hymn, first recorded in 1644 and at one time attributed to the Electress Luise Henriette of Brandenburg (1627–67). The tune is by J. Crüger (q.v.). It is usually sung in the Protestant burial service.

'Jetzt gang i ans Brünnle', first line of a Swabian folk-song published by F. Silcher (q.v.) in 1825. Another version appears in *Des Knaben Wunderhorn* (q.v.).

JOACHIM I (1484–1535, Stendal), in full Joachim Nestor, was Elector of Brandenburg from 1499 until his death. He founded the University of Frankfurt/Oder (1506), added the county of Ruppin to his dominions in 1516, and established a claim to the inheritance of Pomerania in 1529. He was an opponent of the Reformation (q.v.). Joachim is a prominent figure in two novels by W. Alexis (q.v.), *Die Hosen des Herrn von Bredow* and *Der Wärwolf* (qq.v.).

JOACHIM II (1505–71, Köpenick), in full Joachim Hektor, succeeded his father Joachim I (q.v.) in 1535, and in 1539 went over to the Reformation (q.v.). A man of culture and taste, he was also an extravagant builder, and his activities severely burdened the finances of Brandenburg. The establishment of Berlin as the permanent electoral residence (and thus as capital) dates from his reign.

JOACHIM FRIEDRICH (1546–1608), was Elector of Brandenburg from 1598 until his death. He is best known as the founder in 1607 of the Joachimsthalsches Gymnasium (q.v.). By a treaty signed in his reign (Geraischer Hausvertrag, 1599) the various marches and possessions of the Brandenburg electors were indissolubly linked, so that they could not subsequently be separated by inheritance.

JOACHIMSTHALER, silver coins first minted in 1515 from silver mined in Joachimsthal, Bohemia. From Joachimsthaler is derived the word Thaler (q.v.), later Taler, for more than three centuries the commonest German silver coin.

Joachimsthalsches Gymnasium, a classical grammar school founded in 1607 by the Elector Joachim Friedrich (q.v.) of Brandenburg in the town of Joachimsthal north-east of Berlin. In 1650 the school was transferred to Berlin, and in 1912 was removed to Templin, some 50 miles to the north.

Jobsiade, Die, see KORTUM, K. A.

JOBST VON MÄHREN (d. 1411), Margrave of Moravia from 1375, was a nephew of Karl IV (q.v.), and a cousin of the German and Bohemian King Wenzel. He acquired by purchase the Margravate of Brandenburg as well as Lusatia, Luxemburg, and part of Hungary. On the death of King Ruprecht in 1410 Jobst was elected German King by one group of Electors, while the remainder elected his cousin Sigismund (q.v.). The deadlock was resolved by Jobst's death in the following year.

JÖCHER, CHRISTIAN GOTTLIEB (Leipzig, 1694–1758, Leipzig), was successively a professor (1730) and University Librarian (1742) at Leipzig. In 1750 he published the *Allgemeines Gelehrten-Lexikon* in 4 vols., which, with its 76,000 articles, was one of the principal lexicographical works in Germany in the 18th c.

Johann der Seifensieder, a verse fable by F. von Hagedorn (q.v.), published in 1738 in *Versuch in poetischen Fabeln und Erzählungen.* Johann, 'der muntere Seifensieder', sings at his work, disturbs a rich man, and is persuaded to silence by a gift of money. But he pines, and soon returns the gift in order to resume his singing and his life of cheerful poverty. Hagedorn's source was La Fontaine's *Le Savetier et le Financier.*

Johannes, a tragedy (Tragödie in fünf Akten mit einem Vorspiel) by H. Sudermann (q.v.), published in 1898. It depicts the career of John the Baptist, ending with his murder and the promise of the coming of the new King of the Jews.

Johannes Baptista serves as title for three Middle High German poems from the first half of the 12th c., all dealing with John the Baptist, and all preserved only in fragments. They are: a poem by Priester Adelbrecht, who wrote in Carinthia *c.* 1120–30; the *Baumgartenberger Johannes Baptista* from the Austrian abbey of Baumgartenberg on the Danube below Linz; and slight relics of a poem preserved in Klagenfurt. There is also a poem on this subject by Frau AVA (q.v.).

Johannes der Schreiber, the fictitious narrator in C. M. Brentano's story *Aus der Chronika eines fahrenden Schülers* (q.v.). He derives from the *Limburger Chronik* (q.v.).

JOHANNES VON DER OSTSEE, pseudonym of J. D. Falk (q.v.).

453

JOHANNES VON FRANKENSTEIN, see KREUZIGER, DER.

JOHANNES VON FREIBERG, gives his name as the author of the 13th-c. verse Schwank, *Das Rädlein* (q.v.). He is not otherwise known.

JOHANNES VON INDERSDORF, father confessor to Duke Albrecht III (q.v.) of Bavaria and his duchess, prepared for them a prose tract (*Fürstenlehren*) setting forth the manner by which a prince might rule virtuously. It is largely a translation of Roger Bacon's *Aristoteles Secretum secretorum*.

JOHANNES VON POSILGE (Posilge, Prussia, *c.* 1340–1405), parish priest of Deutsch-Eylau, and later assistant to the Bishop of Riesenburg, wrote a prose chronicle of Prussia from 1360 which, after his death, was continued by another pen up to 1420.

JOHANNES VON RINGGENBERG, a knight of Thurgau in Switzerland, who lived in the 13th-c. and is the author of Minnelieder and Sprüche, of which a small number have survived.

JOHANNES VON TEPL (Tepl, now Tepla, Czechoslovakia, *c.* 1360–1414?, Prague), also called Johann von Saaz, now known to be the author of *Der Ackermann aus Böhmen* (q.v.), was for twenty-eight years town clerk of Saaz in Bohemia. In 1411 he became a notary in Prague. He also used the name Johannes Henslini de Sitbor, after his father Henslini de Sitbor.

JOHANNES VON WINTERTHUR (*c.* 1300–after 1348), a Franciscan friar, known to have been at Basel, Schaffhausen, Villingen, Lindau, and Zurich, wrote in Latin a Swiss chronicle covering the period 1250–1348.

Johanneum, Das, famous classical grammar school for boys, founded at Hamburg in 1529 and still in existence.

JOHANN GEORG I, KURFÜRST VON SACHSEN (1585–1656), succeeded the Elector Christian I in 1611. As a conservative and a patriot he was against individual German liberties, and supported the Emperor for the sake of national unity. As a devout Lutheran he opposed the policy of the Counter-Reformation, but was also unsympathetic to the Calvinists. Holding these principles, he found himself during the Thirty Years War (see DREISSIGJÄHRIGER KRIEG) in a dilemma. In the event he was forced to distribute his loyalties. In 1620 he refused to support the Elector Palatine as king of Bohemia (see FRIEDRICH V VON DER PFALZ) against the deposed Habsburg king (see FERDINAND II). He did not abandon his policy of neutrality until his country was in danger of being overrun by imperial and Swedish troops alike. Reluctantly he joined forces with Gustavus Adolphus (q.v.), who was killed on Saxon soil in 1632 (see LÜTZEN). After the battle of Nördlingen (q.v.) had restored Habsburg prestige in the face of the Swedish invaders, he came to terms with Ferdinand II at the Peace of Prague, hoping that it might be followed by a general peace. Only after the truce of Kötzschenbroda (1645) did Swedish troops finally leave Saxony.

At a time when France and Spain already influenced German fashion and food, Johann Georg remained aloof, devoting himself to beer (he was nicknamed 'Bierjörge'), hunting, and music. He was a patron of the composer Heinrich Schütz (q.v.).

Johanniterorden, the order of the Knights of St. John, which since its foundation in the 11th c. spread throughout Christendom, had in Germany a particularly close association with the march (Mark) Brandenburg, and later with Prussia and its ruling house, the Hohenzollern (q.v.). It was reorganized after the 1939–45 War with its seat in Bonn. In the German wars of the 19th c. members of the Order (Johanniter) served as stretcher-bearers and nurses, thus fulfilling its original function, the care of the sick. In the 16th c. the Order moved under the protection of Karl V (q.v.) from Rhodes, which was captured by the Turks, to Malta. During the reign of Philipp II (q.v.) Johann La Valette was Grand Master of the Order (1557–68). The siege of Malta by the Turks, which reached its climax in 1565, forms the background of Schiller's fragment *Die Malteser*; and in *Don Carlos* (q.v.) Marquis Posa is said to have distinguished himself in the Order's defence of Malta.

JOHANNSEN or **JOHANNSSEN,** MICHAEL (Bergedorf nr. Hamburg, 1615–79, Altengamme nr. Hamburg), pastor in Altengamme, published in 1662 a volume of Christmas poems (*Sulamitische Christ- und Freudenküsse*).

JOHANN SIGISMUND, KURFÜRST VON BRANDENBURG (1572–1619), ruled from 1608. By marriage he acquired Prussia (see OSTPREUSSEN) and a hereditary claim on the duchies of Cleves, Jülich, Berg, and Ravenstein, extending Brandenburg-Prussian interest to the west and to both sides of the Rhine. To strengthen his claim he went over from Lutheranism to Calvinism.

JOHANN VON BRABANT, HERZOG (d. 1294), is the author of seven preserved Minnelieder, including a Pastourelle (q.v.). Herzog Johann appears as one of the participants in Konrad von Würzburg's *Turnier von Nantes* (q.v.), and actually met his death as the result of a wound received at a tournament.

JOHANN VON MORSHEIM (d. 1516) was of knightly birth and spent his life in the service of the Count Palatine. He is recorded as governor of Germersheim in 1487, and in 1500 as controller of the household to Count Ludwig. He wrote in 1497 a bitter verse satire on the corruption of courts with the title *Spiegel des Regiments*, to which he adds 'inn der Fürsten Höfe, da Fraw Vntrewe gewaltig ist'. It leans heavily upon earlier authors, notably the *Schachzabelbuch* of KONRAD von Ammenhausen (q.v.), and *Das Narrenschiff* by Sebastian Brant (q.v.). It appeared as a printed book in 1515.

JOHANN VON NEUMARKT (Hohenmaut, *c.* 1310–80, Leitomischl), a Silesian cleric and translator, became chancellor to the Emperor Karl IV (q.v.) at Prague. He spent two years in Italy, and corresponded with Petrarch and Cola di Rienzo. He sought, with some success, to create a new German style for official communications, based on medieval Latin (see KANZLEISTIL).

JOHANN VON SAAZ, see JOHANNES VON TEPL.

JOHANN VON SALZBURG, see HERMANN VON SALZBURG.

JOHANN VON SOEST (Unna, 1448–1506, Frankfurt/Main) is the translator of a huge verse romance, *Margarete von Limburg*, work on which he completed *c.* 1476. He wrote an autobiographical poem from which the details of his life are obtained. The son of a mason, he was brought up in a choristers' school by Duke John of Cleves, wandered for some time in the Low Countries, and was then choirmaster to the Elector of the Palatinate at Heidelberg. Later he went to Italy and graduated as a physician at Pavia. He spent his last years at Frankfurt. He wrote other poems, including three religious works and a poem in honour of Frankfurt.

JOHANN VON WÜRZBURG, a scribe or clerk living in Esslingen, completed in 1314 the Middle High German verse romance *Wilhelm von Österreich*. The poem, of some 19,000 lines, narrates a story of Johann's invention, though he draws liberally upon established motifs. The subject is the love of the Austrian prince Wilhelm and the heathen princess Aglye, which is subjected to many trials. It ends with the lovers' reunion. Johann expresses admiration for WOLFRAM von Eschenbach and GOTTFRIED von Straßburg (qq.v.). His style is successfully modelled on the 'geblümte Rede' of KONRAD von Würzburg (q.v.). A prose version of *Wilhelm von Österreich* was printed in 1481 (*Ein schön vnd gantz kurtzweilige Historien von Hertzog Wilhelm aus Österreich vnd eins Königs Tochter aus Zisia, Agley genandt*).

Johan ûz dem Virgiere, a Middle High German verse epic of some 3,000 lines written by an unknown hand (probably from Rhenish Hesse) in the second half of the 14th c. It is an adaptation of a Flemish poem. It tells of a handsome child discovered by the Emperor Sigemunt in his garden. Brought up as a knight, he distinguishes himself in battle and proves to be the son of the Count of Artois. In the end he receives the hand of the Emperor's daughter and himself becomes emperor.

John Riew', a Novelle by Th. Storm (q.v.), published in the *Deutsche Rundschau* in 1885. John Riewe, a retired sea-captain, tells the story of his friend Rick Geyers who, when drunk, fell to his death from a Hamburg canal bridge. John brings up his daughter Anna, who is seduced after getting drunk at a ball. Soon after giving birth to a boy, Rick, she commits suicide by drowning at the spot where her father had met his death. John continues to look after Anna's mother, Riekchen, and after her death successfully brings up Rick to take up the career of his father and foster-father, and teaches him to resist drink. The pervasive feature of this Rahmennovelle (see RAHMEN) is the humane and unprejudiced treatment of alcoholism.

JOHNSON, MATTHIAS, whose biography is unknown, lived in the 17th c. and wrote an idyllic novel of young married love entitled *Damon und Lisille* (2 vols., 1663–72).

JOHNSON, UWE (Kammin, Pomerania, 1934–84, Sheerness, Kent, UK), spent his youth in Mecklenburg, was a student at Rostock and Leipzig universities, but broke off his studies without graduating. Publication of his first novel, *Ingrid Babendererde*, was forbidden by the East German authorities unless alterations were made and it was not published until 1985. Knowing that his next novel would also be unacceptable in East Germany, Johnson emigrated in 1959 to West Berlin. He spent the years 1966–8 in New York. His first published novel, the experimental *Mutmaßungen über Jakob* (q.v., 1959), is deeply concerned with the division of Germany, focusing largely on the contradictions of the East German state (see DEUTSCHE DEMOKRATISCHE REPUBLIK) and probing problems of communication and identity. The same themes are common to *Das dritte Buch über Achim* (q.v., 1961) and *Zwei Ansichten* (q.v., 1965).

Johnson's experimentalism is renewed in *Jahrestage. Aus dem Leben der Gesine Cresspahl* (q.v., vols. 1–3, 1970–3), a monumental sequel to *Mutmaßungen über Jakob*, which transfers the scene to the U.S.A. and retrospectively surveys Germany from the 1920s to the 1960s. *Eine Reise nach Klagenfurt* (1974) was followed by *Skizze eines Verunglückten* (1981) which is set in the U.S.A. In 1980 Johnson published *Jahrestage. Zwei Kapitel aus der letzten Lieferung* (in *Text + Kritik*, Heft 65/66) and explained in *Begleitumstände. Frankfurter Vorlesungen* (1980) why the book was to remain unfinished. He nevertheless completed the *Jahrestage* tetralogy with the appearance of *Jahrestage 4. Aus dem Leben der Gesine Cresspahl* (20 June to 20 August 1968) in 1983. *Begleitumstände* presents a record of Johnson's work as a visiting professor of Aesthetics (Gastdozent für Poetik) at Frankfurt University in 1979. Apart from views expressed on professional issues, it proffers interesting comments on his approach to his own works. Johnson's style is remarkable for its free syntax, its meticulous precision, its discreet irony, and its quality of control. He is also the author of *Karsch und andere Prosa* (1964). *Berliner Sachen* (1975) is a collection of essays. Johnson was awarded the Büchner Prize in 1971.

JOHST, HANNS (Seershausen, nr. Riesa, 1890–1978, Ruhpolding), served in the 1914–18 War, and then settled as an author in Bavaria. He produced several dramas in the surge of Expressionism (*Der junge Mensch*, 1916; *Stroh*, 1916; *Der Einsame*, with Grabbe, q.v., as hero, 1917; *Wechsler und Händler*, a comedy, 1923). He became a National Socialist and his nationalistic play *Schlageter* (1933) was widely acclaimed.

JOKOSTRA, PETER (Dresden, 1912–), escaped from East Berlin and settled in Munich. He has published several volumes of poetry of symbolic nature (*An der besonnten Mauer*, 1958; *Magische Straße*, 1960; *Hinab zu den Sternen*, 1961; *Die gewendete Haut*, 1967), and a novel entitled *Herzinfarkt* (1961).

JOLANDE VON VIANDEN, GRÄFIN (*c.* 1230–83, Marienthal, Luxemburg), daughter of a great noble house, insisted, in spite of the opposition of her powerful family, on entering an obscure nunnery at Marienthal. She resided there from 1248 until her death, becoming prioress in 1258, and enjoying a wide repute as a woman of holiness. Jolande is the subject of a verse life, in the manner of a saint's legend, though she was never canonized. Its author is a Bruder Hermann, belonging, like Jolande, to the Dominican Order, who wrote it not long after her death, possibly with her ultimate canonization in view.

JONAS, JUSTUS (Nordhausen, 1493–1555, Eisfeld), studied at Erfurt and Wittenberg. He was present as a supporter of Luther (q.v.) at the Diet of Worms, the religious discussions at Marburg, and the Diet of Augsburg. He became moderator at Halle in 1541 and at Eisfeld in 1553. In between he was court preacher at Coburg (1551–3). He wrote a number of Lutheran hymns. His correspondence was published in 1884–5.

JORDAN, WILHELM (Insterburg, 1818–1904, Frankfurt/Main), was a student at Königsberg University, took to writing poetry (*Glocke und Kanone*, 1841; *Irdische Phantasien*, 1842; *Schaum*, 1846), got into trouble with the Saxon censorship in the repressive years before 1848, and found asylum in Bremen. During the Revolution he was a prominent member of the Frankfurt Parliament, sitting at first on the left and later with the Erbkaiserliche Partei (q.v.), and becoming a secretary in the naval department. With the collapse of the democratic movement Jordan retired into private life in Frankfurt. His work (Mysterium) *Demiurgos* (1852–4), partly in epic and partly in dramatic form, reflects his philosophical response to the political events he had lived through.

Jordan devoted many years to a refashioning of the Nibelungen epic, adhering to alliterative verse, but abandoning the mythical in favour of an interpretation by character, and inventing matter to fill the gaps in the original. *Die Nibelunge* appeared in two parts, *Sigfridsage* (1867) and *Hildebrants Heimkehr* (1874), and for a time was very popular; he travelled widely in Germany, and even abroad, giving readings from the work. His family novel *Die Sebalds* (2 vols., 1885) is a plea for religious toleration. He also wrote a second, less successful, novel, and some facile verse comedies (*Die Liebesleugner*, 1855; *Tausch enttäuscht*, 1856; *Durchs Ohr*, 1870; *Sein Zwillingsbruder*, 1883).

Jörn Uhl, a novel by G. Frenssen (q.v.), published in 1901, and set in Dithmarschen, Frenssen's home country. Jörn (Low German for Georg) is the one sound and reliable member of a farming family. He has a widowed, shiftless father who sets a poor example to his brothers, and a sister who develops into a flighty young woman and leaves the farm. Jörn abandons his wish to study and keeps the farm going by his own efforts. During his absence for military service and his recall to the colours for the Franco-Prussian War (see DEUTSCH-FRANZÖSISCHER KRIEG) the situation on the farm deteriorates. Jörn's father suffers a stroke, and the creditors stipulate that the idle brothers shall be excluded from the farm. Jörn tackles the work doggedly, but luck is against him. His wife dies in childbirth, the wheat crop fails, and the farm buildings are struck by lightning and burned down. On his father's death Jörn gives up the farm, and, with a friend from his service days, establishes a cement works. He does valuable work, reinforcing canals and dikes, and at the end of his career can look back on a life of labour that is not without achievement.

JOSEPH I, KAISER (Vienna, 1678–1711, Vienna), the elder son of Leopold I (q.v.), became German King (see DEUTSCHER KÖNIG) in 1690, and succeeded his father as emperor in 1705. His short reign was occupied by the War of the Spanish Succession and troubles with Hungarian rebels. Thanks to Prince Eugene (see EUGEN, PRINZ), he was successful in the war, and his own energy and astuteness contributed to the reduction of the Hungarian revolt.

JOSEPH II, KAISER (Vienna, 1741–90, Vienna), was the eldest son of the Emperor Franz I and Maria Theresia (qq.v.). He was elected German King (see DEUTSCHER KÖNIG) in 1764, and became emperor on his father's death in 1765. In his mother's lifetime he had only a partial say in Austrian affairs. The many differences, primarily over foreign policy, between mother and son turned upon the traditional sympathies of the former and the reforming zeal of the latter. A particular subject of dispute was the first partition of Poland, which Joseph supported and Maria Theresia opposed (see POLAND, PARTITIONS OF). After her death in 1780 Joseph had a free hand in foreign politics, in which he sought principally to augment Austrian influence in Germany, thereby coming into conflict with Friedrich II (q.v.) of Prussia. In particular his efforts to add Bavaria to the Austrian territories were twice frustrated by the Prussian king, in 1778 (see BAYRISCHER ERBFOLGEKRIEG) and in 1785 (see FÜRSTENBUND).

At home Joseph carried out numerous reforms, liberating the serfs, mitigating the severity of the penal code, extending primary education, establishing religious toleration, and reducing the power of the Church. If he was truly enlightened, he was also truly despotic, and ruthlessly enforced the policies which appealed to him. In particular his rash and autocratic introduction of administrative changes in Bohemia and his attempt to make German the official language in Hungary laid up trouble for the future. Although an ardent supporter of toleration, Joseph was, in fact, an intolerant ruler who was unable to understand how any sensible man could oppose his own eminently reasonable views. Two notable benefits conferred by Joseph on Vienna were the founding of the Burgtheater (q.v.) in 1776, and the opening of the imperial park, the Prater, to the populace of Vienna.

Joseph im Schnee, a Novelle by B. Auerbach (q.v.), published in 1860, and subsequently included in *Sämtliche Schwarzwälder Dorfgeschichten* (1884). It is a Christmas story, in which the illegitimate boy Joseph wanders away from his unmarried mother Martina in the hope of finding his father, Adam Röttmann, a physical Hercules, who is nevertheless under the thumb of a violent and aggressive mother. Joseph is rescued from the snow, and his dangerous escapade reconciles both parents and their families, except for the unforgiving Röttmännin. Adam and Martina are married and all ends happily.

Josephinismus, term for the policy of introducing liberal reforms and diminishing the influence of the Roman Catholic Church by secularization, adopted by the Emperor Joseph II (q.v.).

Joseph Kerkhovens dritte Existenz, a novel by J. Wassermann (q.v.), published in 1934. It takes up again the story of the physician Joseph Kerkhoven, an important character in *Etzel Andergast* (q.v.). In the first book Kerkhoven and Marie, after a separation intended for socially useful work, are reunited. The second book is devoted to Kerkhoven's successful efforts to overcome by psychiatry the difficulties of another married couple, Alexander and Bettina Herzog. In Book III, Kerkhoven has a breakdown and believes himself beset by evil powers. He attains spiritual harmony in religious belief.

'Joseph, lieber Joseph mein', a hymn of 14th-c. folk-song origin with the title *Mariä Wiegenlied.* The tune is first recorded in 1544.

Joseph und seine Brüder, a biblical tetralogy of novels by Th. Mann (q.v.). It consists of four novels in three volumes. Vol. 1 contains both *Die Geschichten Jaakobs* and *Der junge Joseph*; vol. 2 comprises *Joseph in Ägypten,* and vol. 3 *Joseph der Ernährer. Die Geschichten Jaakobs* and *Der junge Joseph,* both written before Mann's emigration in 1933, appeared in 1933 and 1934 respectively. *Joseph in Ägypten* was written during his voluntary exile in Switzerland, and published in 1936. *Joseph der Ernährer,* written in the U.S.A., followed in 1942.

This work of *c.* 2,000 pages is an expansion and interpretation of Genesis 12–50, of which the most important chapters for Mann's purpose are 37–50. An immense corpus of historical, ethnic, and geographical knowledge is embodied in the work, particularly in *Joseph in Ägypten.* The novel is concerned with the dawn of religion and the emergence of God from myth. The first novel recounts retrospectively Jaakob's past history. Joseph, the central figure of the novel, corresponds in his early stages to the self-absorbed artist, clearly aware of his separation from others, and so providing one more example of a theme which is essential to Mann's work. But Joseph undergoes two purgatorial experiences, first when he is cast into the pit and sold into Egypt, the second (at the transition from the third to the fourth novel) when he is judged by Peteprê (Potiphar, for Mann uses both authentic and traditional nomenclature) after Potiphar's wife (Mut-em-enet) has abandoned her attempts to seduce Joseph and has turned against him. A psychological mutation, characteristic of Mann, replaces the familiar figure of the lustful wife evoked in Genesis 39 by a sex-starved woman whose husband is a eunuch. Joseph learns from his dual experience, and matures into the man of wisdom, judgement, and practical ability who proves his stewardship, first in Peteprê's house, then in Egypt, and finally as the economic saviour of his own tribe, emerging as a complete, balanced, and influential personality.

Joseph und seine Brüder is a Bildungsroman (q.v.) on the grand scale, and ends in expansive serenity. Th. Mann treats the myth, the biblical personages, and God himself with sympathy and respect, but with sovereign condescension. Mann himself referred to the work as a 'comedy', which has been interpreted (dubiously, since Mann was neither Christian nor Jew) as a Dantesque *Divina Commedia.* It is clear that the writing of it afforded Mann a special pleasure and relaxation. Owing to political circumstances the complete work was published in Sweden.

Journalisten, Die, a four-act comedy (Lustspiel) by G. Freytag (q.v.), published in 1854. It was first performed at the Hoftheater, Karlsruhe, in 1852, and rapidly established itself in the repertoire of all German theatres, drawing audiences for over thirty years. It is a good-humoured comedy of intrigue with settings in drawing-room and newspaper office, and contains a noteworthy scene showing journalists at work.

The journalist and Liberal candidate Professor Oldendorf loves Ida, the daughter of Colonel Berg, who disapproves of Oldendorf's political activities. Moreover the colonel has leanings to politics himself and writes unsigned articles which are attacked in Oldendorf's paper, the *Union.* The colonel is nominated in opposition to Oldendorf, but is defeated in the election. A kind and rich friend, Adelheid von Runeck, points out to the colonel how the friends who encouraged him to stand had played him false. She buys the *Union;* Oldendorf, no longer a journalist, is now acceptable to the colonel, and is reunited with Ida; and Adelheid becomes engaged to Konrad Bolz, an eccentric and irrepressible character, who has been a colleague of Oldendorf on the newspaper, and has played a genial part in the tangle of human relationships.

Journal meiner Reise im Jahr 1769, an autobiographical work by J. G. Herder (q.v.), published posthumously in 1846. Though called a journal, it is a continuous essay and not in diary form. It was written during the journey Herder made in 1769, leaving Riga by ship on 23 May, and disembarking in July at Nantes. He remained in Nantes for three months, then moved to Paris, leaving again for Germany early in 1770. The *Journal* records certain biographical facts, notably the storms Herder endured at sea, but it is in the main devoted to his states of mind and his revolutionary thinking. An egocentric work, it is also a mine of fertile ideas on language, art, learning, and education.

Jüdel, Das, a Middle High German poem of rather less than 500 lines, narrating a legend of the Virgin Mary. A Jewish boy, who has come to adore the Blessed Virgin Mary, is thrust by his kin into a heated oven; through the miraculous intervention of the Virgin he remains unharmed, and devotes himself thereafter to the conversion of the Jews. The poem was probably written in the first half of the 13th c.

Juden, Die, a one-act play written by G. E. Lessing (q.v.) in 1749, and first published in 1754 in *Lessings Schriften* (vol. 4). It aims at combating anti-Jewish prejudice. A baron who is set upon by robbers and rescued by a stranger believes his assailants to be Jews. They prove, however, to be the baron's Christian servants, and it is the resolute rescuer who is revealed as a Jew. This plea for tolerance, in dramatic form, probably reflects Lessing's regard for his friend, the young Jew M. Mendelssohn (q.v.).

Judenbuche, Die, a Novelle by Annette von Droste-Hülshoff (q.v.), first published in instalments in the *Morgenblatt für gebildete Leser,* April to May 1842. The title was adopted, without Droste's permission, by the editor, H. Hauff, brother of W. Hauff (q.v.). Droste-Hülshoff herself referred to the story as *Eine Kriminalgeschichte, Friedrich Mergel,* and intended it as a sketch for *Bilder aus Westphalen,* sub-titled *Ein Sittengemälde aus dem gebirgigten Westphalen.* The story was first published in book form in 1851 by L. Schücking (q.v.), in *Letzte Gaben. Nachgelassene Blätter von Annette Freiin von Droste-Hülshoff.* The principal source is an article published by an uncle of the authoress, Freiherr A. von Haxthausen, in the *Wünschelruthe. Ein Zeitblatt* in 1818 under the title *Geschichte eines Algierer Sklaven.* It records the murder of a Jew by a farm-hand, Hermann Johannes Winkelmann (Winkelhannes), his escape, his adventures and sufferings as a slave in Algiers, his return to his native Ovenhausen in the bishopric of Corvey, and his subsequent suicide.

Friedrich Mergel, the son of Margreth, is born in 1738 in the village of B. His father kills himself in a drinking bout in a forest, the Brederholz, when the boy is 9. When he is 12 his uncle Simon takes charge of him, and employs him in dishonest business deals. Since Mergel died without absolution, the superstitious villagers believe that his ghost haunts the forest, and Friedrich finds himself ostracized. His only companion is Johannes Niemand, so called because no one admits to being his father. Johannes figures in the whole story as Friedrich's double. Through Simon, Friedrich becomes involved with a gang of timber thieves, known as the Blaukittel. Though the squire, Herr von S., has hitherto turned a blind eye to wood-stealing, the Blaukittel, who work with speed at night, break all the written and unwritten laws of the district by their ruthless felling of forest trees.

One night in July 1756, the forester Brandes, who is on the track of the Blaukittel, finds Friedrich at his normal occupation minding the cows at the forest edge. After an exchange of words during which the suspicious forester insults Friedrich and his mother, Friedrich gives him false information. In the morning Brandes's body is discovered. Although a blood-stained axe is found, its owner is not traced. Conscience causes Friedrich to seek the comfort of confession. As he is about to steal out of his uncle's house, Simon detects him, and claims that he

who bears witness against another is unworthy of Holy Communion. Friedrich succumbs to the argument because he suspects that his uncle is the owner of the axe.

In the summer of 1760 Friedrich attends a wedding. He is accused by another youth, Wilm Hülsmeyer, of not having paid for his silver watch, and the Jew Aaron demands payment. Friedrich, thus publicly disgraced, leaves the party. The following day Aaron is found dead beneath a large beech-tree in the Brederholz. Friedrich is suspected of the murder, but he disappears. The Jewish community buys the beech-tree, on which they inscribe letters in Hebrew. They are confident that the murder will be avenged where it was committed.

Over twenty-seven years later, on Christmas Eve 1787, Johannes Niemand, who disappeared with Friedrich, returns to the village. He is unrecognizable as a result of the ill treatment he has endured as a slave in Turkey. The aged Herr von S. provides him with shelter. From him Johannes learns that Friedrich is no longer suspected of the murder. The following September, at the equinox, Johannes disappears, and a few days later the son of the murdered forester Brandes discovers him hanging from a branch of the beech-tree which the Jews had bought. A scar reveals that the dead man is Friedrich Mergel. By returning under a false name, he had hoped, vainly, for Christian burial. The end of the Novelle gives the Hebrew inscription in German. The humanity of this outstanding crime story is stressed in a prefatory poem.

Juden von Zirndorf, Die, a novel by J. Wassermann (q.v.), published in 1897. The first part is set in the 17th c., the second in the 19th. The village of Zionsdorf (later Zirndorf) is founded by Jews seeking a Messiah, who proves to be false. Two hundred years later one of their descendants appears to work healing miracles, but is convinced by an older man that the true wisdom is of this world and religion a delusion. The two parts are chiefly held together by the basic humanistic attitude.

Jüdin von Toledo, Die, a five-act tragedy in blank verse, with an opening dialogue in Spanish-type trochees, by F. Grillparzer (q.v.). The play was first planned in 1824 but not completed until after 1851; it was published, and first performed in Prague, in 1872. Although based on *Las paces de los Reyes y Judía de Toledo* by Lope de Vega, the plot is motivated differently. Alfons der Edle, King of Castile, falls in love with Rahel, daughter of the Jew Isaak, who sets her cap at him and succeeds in arresting his attention. The King neglects his country's defence against the Moors. The Queen, Eleonore, daughter of Henry II of England, and the grandees, headed by Count Manrique, secretly resolve Rahel's death. The King discovers the plot too late to prevent the murder. As he confronts the conspirators to pronounce them guilty, he accepts Manrique's charge implicating him in guilt as well. He appoints the Queen as regent for his young son, and moves off with his men against the Moors. Humility and repentance dominate the end of the play as Rahel's sister Esther overcomes grief and bitterness to accept, on behalf of the victims, her share of guilt.

It is likely that Grillparzer had the episode of Lola Montez (q.v.) and King Ludwig I (q.v.) of Bavaria in mind when finally working at the play. Through his passion the noble-minded young king is thrust into an inner conflict from which he emerges matured but disillusioned. The designation of the play as a historical tragedy (Historisches Trauerspiel) does not appear in the MS.

Judith, a Middle High German biblical poem dated 1254 in the MS. but possibly written in 1304. It is probably the work of a brother of the Teutonic Order (see DEUTSCHER ORDEN), to which the story of Judith's heroic deed against the military might of the heathen was particularly appropriate and even topical. The MS. was found at Mergentheim, headquarters of the Order from 1526.

Judith or **Ältere Judith,** an anonymous Middle High German poem recounting Judith's heroic assassination of Holofernes. In strophic form, written in the early 12th c., it is regarded as one of the earliest ballads. It is preserved in the Vorauer Handschrift (q.v.), where it follows immediately on *Die drei Jünglinge im Feuerofen* (q.v.), the whole appearing to form one poem. It is believed, however, that the two were originally quite distinct. The end of *Judith* is missing.

Judith, Jüngere, title given to an anonymous Middle High German epic poem dealing with the story of Judith and Holofernes. The title indicates that it is the later of two poems on this theme (see JUDITH OR ÄLTERE JUDITH). Written in Austria *c.* 1140, it is a work of limited poetic quality.

Judith, a tragedy in five acts by F. Hebbel (q.v.), written in 1839–40, and published in 1841. The play is based on the apocryphal Book of Judith: Judith, the widow of Manasses, murders Holofernes in his camp outside the besieged city, thus saving Bethulia from heathen conquest. But,

unlike the biblical figure, Hebbel's Judith, who has, by God's mysterious design, remained a virgin (an intended variation on Joan of Arc, the virgin), is irresistibly attracted by Holofernes. She murders him in an act of personal revenge for his humiliation of and contempt for her. When she realizes that she has betrayed her divine mission she asks the priests for no other reward than that they should kill her, should she request it: she feels she must die, should she have conceived a son by Holofernes, the outcome depending on God's mercy.

Judith's tragic experience is based on spiritual as well as psychological conflicts. The prophecy of the dumb Daniel urging the people not to open the city-gates to the enemy is an important factor supporting the divine sanction of Judith's mission. Hebbel's bold and involved treatment of the apocryphal story led Nestroy to write a parody of it (*Judith und Holofernes*, q.v.).

Judith und Holofernes, Travestie mit Gesang in einem Aufzug, a play by J. N. Nestroy (q.v.), written and published in 1849, and first performed in March of that year in the Carl-Theater, Vienna. The music is by M. Hebenstreit. It is a parody of Hebbel's first tragedy *Judith* (q.v.), with satirical allusions to contemporary conditions in Vienna. The parody turns on Nestroy's invention that Judith is impersonated by her brother Joab, who stages a mock-murder before Holofernes is led away, a prisoner, deflated and ridiculed.

Jud Süß, a Novelle by W. Hauff (q.v.), published in *Novellen* (1828). It recounts the fall and execution, after the death of the reigning duke, of the Jew Süß (see SÜSS-OPPENHEIMER, J.), the corrupt finance minister of Württemberg, in 1737–8. Süß's overthrow is the result of a conspiracy, which is interwoven with a love-affair between Gustav Lanbeck, son of one of the conspirators, and Lea, Süß's sister. Lea, it is implied, drowns herself, and young Lanbeck remains unmarried.

Jud Süß, an historical novel by L. Feuchtwanger (q.v.), published in 1925. Süß (see SÜSS-OPPENHEIMER, J.) is fiscal aide to the extravagant Duke Karl Alexander of Württemberg, and satisfies the latter's unceasing financial demands by burdensome taxation. When the Duke desires Süß's daughter Naëmi as his mistress, Süß turns secretly against him and betrays his political plans. The Duke dies, and long-restrained popular hatred of Süß results in his trial and execution for peculation and abuse of powers.

Jugend, a Naturalistic play by M. Halbe (q.v.), published in 1893, and performed with immense success in the same year. It tells of two young lovers Hans Hartwig and Anna. Anna is shot dead by her jealous stepbrother Amandus. The play is set in the West Prussian countryside in which Halbe grew up.

Jugend, a mainly satirical periodical which appeared 1896–1940. It was founded by the publisher Georg Hirth (1851–1916) and described as *Münchner illustrierte Wochenschrift für Kunst und Kultur.* It was an organ for young writers and artists and gave its name to the German style of *art nouveau* (see JUGENDSTIL). Like its contemporary weekly *Simplicissimus* (q.v.) its historical interest is confined to the first fifteen to twenty years of its existence. Many German and Austrian writers and art critics were among its contributors, but other European authors, including d'Annunzio, Bjørnson, Gorki, Kipling, Maeterlinck, Strindberg (q.v.), and Zola, were also represented. 'Unsere Zeit ist nicht alt, nicht müde!', wrote Hirth in 1899 in response to the *fin de siècle* mood. Illustrations by young artists were a special feature. A characteristic title-page was designed by Barlach (q.v.).

Jugend des Königs Henri Quatre, Die, see HENRI QUATRE-ROMANE.

Jugenderinnerungen eines alten Mannes, a well-known book of memoirs by W. von Kügelgen (q.v.).

Jugendstil, designation for a stylistic trend in the arts *c.* 1895–1905 (cf. *art nouveau*). In Germany it was propagated by the periodical *Jugend* (q.v., which gave it its name) primarily in reaction against the style of the preceding decades, the Gründerzeit (q.v.). It encouraged functional, linear ornamentation modelled on the Pre-Raphaelite movement in England and on Japanese art. In literature the term has been used in connection with, among others, Dehmel, Flaischlen, S. George, Hofmannsthal, and Rilke (qq.v.), especially in their early poetry. See also SEZESSIONEN.

Julirevolution, the revolution in Paris in the days of 27–9 July 1830, which encouraged political unrest in Germany and the production of politically orientated literature, especially by writers referred to as Junges Deutschland (q.v.).

Juliusturm, a tower of the fortress of Spandau in which was stored between 1871 and 1914 the German financial reserve in the event of war. The money was originally derived from French reparation payments, but was later augmented from home sources.

Julius von Tarent, a tragedy by J. A. Leisewitz (q.v.), written in 1775, and published in 1776. It was submitted by Leisewitz's friend J. H. Voß (q.v.) as an entry for a prize offered by the impresario F. L. Schröder (q.v.) of Hamburg. The successful play was Klinger's *Die Zwillinge* (q.v.), and Leisewitz, after what he regarded as his failure, published no more plays. *Julius von Tarent* was, however, performed by Schröder, and it achieved considerable popularity both on the stage and in print. The play presents the tragic conflict of two brothers. Julius, the elder, is an introverted, sentimental dreamer; Guido is a man of action, swift and impetuous, and angrily jealous of Julius. Both love the same woman, Blanca, and Julius is killed by his brother while attempting to carry her off from a convent, where she is lodged for her safety. Guido, conceding his guilt and accepting his fate, is put to death by his father, the Prince of Taranto, who thereupon abdicates and withdraws from the world.

The play is loosely based on the story of Cosmo I (1519–74) of Florence and his sons. Leisewitz transfers it to the end of the 15th c. See STURM UND DRANG.

JUNG, CARL GUSTAV (Kesswyl, Switzerland, 1875–1961, Zurich), Swiss psychiatrist, and leading figure in the Zurich school of analytical psychology, studied medicine at Basel University. Having specialized in psychiatry, he was appointed in 1900 to a post at Burghölzli Hospital, Zurich. He acquired a large private practice, and resigned from the hospital in 1909. In 1903 Jung was impressed by the dream psychology in *Traumdeutung* by S. Freud (q.v.), whom he visited in Vienna in 1907. Jung remained under Freud's influence until 1912, when he published *Wandlungen und Symbole der Libido* in which he expressed strongly independent views, which resulted in a permanent breach.

Jung was unable to accept Freud's view of sex as the sole determinant of action, and greatly widened the meaning of *libido.* He also attached less importance to repression as a factor in neurosis. He developed a psychology of the 'collective-unconscious' originating in the most primitive ages. The patterns of its life and thought, termed 'archetypes', survive in modern man and make up a substantial element in the unconscious of each individual. The personal elements of the unconscious in men develop by a process which he terms 'individuation'. Jung's horizon includes religious, mystical, mythical, and occult phenomena, and his views have been rejected by some specialists as unscientific. His influence is nevertheless considerable. Like Freud, he was eminently successful in the treatment of neuroses, and acquired a world-wide reputation. The familar words 'extravert' and 'introvert' were introduced by Jung.

In addition to the work mentioned above (the title of which was changed in 1952 to *Symbole der Wandlung*), Jung's principal writings include *Psychologische Typen* (1921), *Die Beziehungen zwischen dem Ich und dem Unbewußten* (1928), *Psychologie und Religion* (1939), *Psychologie und Alchemie* (1944), *Die Psychologie der Übertragung* (1946), *Über psychische Energetik und das Wesen der Träume* (1948), *Symbolik des Geistes* (1948), *Aion. Untersuchungen zur Symbolgeschichte* (1951), *Antwort auf Hiob* (1952), *Synchronizität als ein Prinzip akausaler Zusammenhänge* (1952), *Von den Wurzeln des Bewußtseins* (1954), *Welt der Psyche* (1954), *Versuch einer Darstellung der psychoanalytischen Theorie* (1954), *Mysterium conjunctionis* (with M. L. von Franz, 3 vols., 1955–7), and *Ein moderner Mythus* (1958).

Jung's collected works (*Gesammelte Werke*, 16 vols.) began to appear in 1958. The autobiography, written in his last years, 1957–61, with the assistance of Aniela Jaffé and edited by her (*Posthume Autobiographie*), was published in 1962. *Septem sermones ad murtuos* by 'Philemon' (a mythical figure created by Jung) were privately printed in 1916.

JUNG, JOHANN HEINRICH, commonly called Jung-Stilling (Grund nr. Hilchenbach, Siegerland, 1740–1817, Karlsruhe), grew up in humble circumstances and in a deeply pietistic environment. He worked as a schoolmaster and on the land, and at 29 began to study medicine. In 1772 he started to practise at Elberfeld, making a name for his skill in operating for cataract. In 1778 he embarked on an academic career as a professor of economics, first at Kaiserslautern, then at Heidelberg (1784), Marburg (1787), and again at Heidelberg (1803). His last years (1806–17) were spent in Karlsruhe, where he received a pension from the Margrave of Baden. Jung-Stilling wrote a progressive autobiography, of which the first and best-known volume, *Heinrich Stillings Jugend*, appeared in 1777. *Heinrich Stillings Jugend-Jahre* and *Heinrich Stillings Wanderschaft* followed in 1778. Then came *Heinrich Stillings häusliches Leben* in 1789, and *Heinrich Stillings Lehr-Jahre* in 1804. The five volumes were published together as *Heinrich Stillings Leben* in 1806. A sixth volume, *Heinrich Stillings Alter*, was published in 1817. This autobiography is marked by simple directness and by 'Stilling's' conviction that his life is visibly directed at all points by the hand of God. *Das Heimweh*, a four-volume novel (1794–6, reissued

1949), champions simple faith against the impiety of the 18th c.

Jung-Stilling's writings include some poetry, the novels *Die Geschichte des Herrn von Morgenthau* (1779), *Die Geschichte Florentins von Fahlendorn* (3 vols., 1781–3, vols. 1 and 2 reissued 1948), and publications on veterinary, medical, and economic subjects, notably his *Lehrbuch der Vieharzneykunde* (2 vols., 1785–7), *Methode den grauen Staar auszuziehen und zu heilen* (1791), and *System der Staatswirthschaft* (1792). Jung-Stilling, who was a friend of Goethe at Strasburg in 1770–1, is portrayed in *Dichtung und Wahrheit* (q.v., Bk. 16).

Sämtliche Schriften (14 vols.), ed. J. N. Grollmann, appeared 1835–8, *Briefe an seine Freunde*, ed. A. Vömel, 1905, and a new select edition of correspondence entitled *Wenn die Seele geadelt ist. Aus dem Briefwechsel Jung-Stillings*, 1967.

Junge Gelehrte, Der, a satirical comedy (Lustspiel) in three acts written by G. E. Lessing (q.v.) between 1745 and 1747, and first published in Lessing's *Schriften* (vol. 4, 1754). It was first performed at Leipzig in 1748 by the company of Frau Neuber (1697-1760).

Junge Leiden, a collection of poems which forms the first section of Heine's *Buch der Lieder* (q.v., 1827). *Junge Leiden* is divided into 'Traumbilder', 'Lieder', 'Romanzen', and 'Sonette'. The date of composition is given as 1817–21.

JÜNGER, ERNST (Heidelberg, 1895–), son of a pharmacist, spent his youth in Hanover, and at 18 enlisted in the French Foreign Legion, from which he was brought home. In 1914 he volunteered for service, was enrolled in the 73rd Hanoverian Fusiliers, reached the western front in December, and served as an officer throughout the war with intermissions to recover from his wounds. He seven times sustained a double wound. He received the highest Prussian award for bravery, the order Pour le mérite (q.v.). After the war he served as an officer in the Reichswehr (q.v.) until 1923, undertook a short period of study, and then devoted himself to authorship until 1939, when he returned to the army as a captain. He was dismissed in 1944. After the war he settled in Ravensburg.

Jünger was able to adapt himself to war as an element, and the years 1914–18 were probably the principal experience of his life. His first book, *In Stahlgewittern* (q.v., 1920), anticipates by several years the spate of war recollections, and implicitly elevates the soldier's life, isolated from normal humanity, into a mystical experience. It was followed by the essays *Der Kampf als inneres*

Erlebnis (1922), *Das Wäldchen* (1925), and *Feuer und Blut* (1925), which is a retelling of a crucial episode of *In Stahlgewittern*. The story *Afrikanische Spiele* appeared in 1936, and the allegorical and anti-totalitarian novel *Auf den Marmorklippen* (q.v.) in 1939.

After the 1939–45 War Jünger published the novels *Heliopolis* (1949) and *Gläserne Bienen* (1957). Many of his works are introspective recollections or broodings in essay form, among them *Das abenteuerliche Herz* (1929), *Strahlungen* (1949), *Der Waldgang* (1951), *Der Gordische Knoten* (1953), and *Jahre der Okkupation* (1958). In 1974 appeared *Zahlen und Götter. Philemon und Baucis. Zwei Essays*. The novel *Die Zwille* appeared in 1973, *Eumeswil* in 1977, and a late novel, *Eine gefährliche Begegnung*, in 1985. Several volumes of diaries appeared in the 1950s and 1960s. Diaries from the years 1965–80, *Siebzig verweht* (2 vols.) were published 1980–1 and *Sämtliche Werke* (18 vols.) 1978–83.

JÜNGER, FRIEDRICH GEORG (Hanover, 1898–1977, Überlingen), younger brother of E. Jünger (q.v.), served in the 1914–18 War, in which he was wounded, and afterwards as an officer in the Reichswehr. After studying law and practising for a short time he turned to writing. He is the author of much harmonious, disciplined, traditional poetry (*Gedichte*, 1934; *Der Krieg*, 1936; *Der Taurus*, 1937; *Der Missouri*, 1940; *Der Westwind*, 1946; *Die Perlenschnur*, 1947). His best-known poem is the political elegy *Der Mohn*. Many of his reflective writings are directed against technological civilization (*Die Perfektion der Technik*, 1946; *Maschine und Eigentum*, 1949). In his sixtieth year Jünger published his autobiography, *Spiegel der Jahre* (1958), and ten years later another volume of poetry, *Es pocht an der Tür* (1968). *Gesammelte Erzählungen* appeared in 1967.

Jüngeres Hildebrandslied, title given to a Middle High German poem which is a version of the story of Hildebrand and Hadubrand (see HILDEBRANDSLIED). It has a happy ending. 'Maister Hiltebrant' rides home after thirty-two years, and on the way does battle unwittingly with his son Alebrant, whom he overcomes. There is a recognition scene, following which Alebrant escorts his father to his home, where a reunion takes place with Hiltebrant's wife Ute. The poem, in rhyming verse, is of ballad-like character. It exists in two forms of different length, both written in the 13th c., and enjoyed a considerable popularity well into the 17th c. (See also DIETRICHSAGE.)

Jüngere Titurel, Der, see TITUREL.

Junges Deutschland, term first used in print by L. Wienbarg (q.v.) in *Ästhetische Feldzüge* (1834). It denotes a trend towards liberal ideas in politics, religion, and morality at a time of repression caused primarily by the policy of Metternich (q.v.). The impulse for Junges Deutschland (Young Germany) derived from the July Revolution in France in 1830. The name Junges Deutschland became known, and was wrongly thought to indicate a compact group in consequence of allegations such as the savage attack in 1835 by W. Menzel (q.v.) on Gutzkow's novel *Wally, die Zweiflerin* (q.v.). The attention of the political authorities having been aroused by this denunciation, the Federal Diet (Bundestag) issued on 10 December 1835 the warning: 'Sämtliche deutsche Regierungen übernehmen die Verpflichtung, gegen die Verfasser, Verleger, Drucker und Verbreiter der Schriften aus der unter der Bezeichnung "das junge Deutschland" oder "die junge Literatur" bekannten literarischen Schule, zu welcher namentlich Heinr. Heine, Karl Gutzkow, Heinr. Laube, Ludolf Wienbarg und Theodor Mundt [qq.v.] gehören, die Straf- und Polizeigesetze ihres Landes, sowie die gegen den Mißbrauch der Presse bestehenden Vorschriften, nach ihrer vollen Strenge in Anwendung zu bringen . . .'

This document encouraged the belief in a 'Young German School', whereas it would be more proper to speak of a Young German trend. Heine was at this time a self-styled exile in Paris, and most of the Young Germans (among whom were also L. Börne, F. G. Kühne, and E. A. Willkomm, qq.v.), far from conspiring together, were mutually unsympathetic, and sometimes actually hostile to each other. Their works nevertheless followed a common line of thought, turning away from German Idealism (see IDEALISMUS) and Romanticism (see ROMANTIK), to political reform, religious toleration, and emancipation from accepted sexual morality. The bolder spirits emphasized that action, not theory, was required; but action, when it came in 1848 (see REVOLUTIONEN 1848–9), led to disillusionment and to the decline of Junges Deutschland.

Important works associated with Junges Deutschland are, apart from those mentioned, Gutzkow's *Maha Guru* (1833), *Das Urbild des Tartuffe* (q.v., 1845), and *Uriel Acosta* (q.v., 1846), the cycle of novels *Das junge Europa* (1833–7) by Laube, *Französische Zustände* (1833), *Der Salon* (1834–40), and *Deutschland. Ein Wintermärchen* (q.v., 1844) by Heine, *Madonna* (1835) by Mundt, *Briefe aus Paris* (1832–4) by Börne, Kühne's *Eine Quarantäne im Irrenhaus* (q.v., 1835), and *Die Europamüden* (1838) by Willkomm.

The quality of Young German writing is uneven; but it marks the beginning of a phase in which German literature seeks to come closer to the conditions of contemporary life. The most remarkable works connected with it appeared when it was virtually a spent force. They are Gutzkow's novels *Die Ritter vom Geiste* (q.v., 1850–1) and *Der Zauberer von Rom* (q.v., 1859–61).

In the European context Junges Deutschland was not an isolated trend, and owes some of its impetus to La jeune France, Giovina Italia, and the Swiss Das junge Europa.

Jungfrauen, Spiel von den zehn, see LUDUS DE DECEM VIRGINIBUS.

Jungfrau von Orleans, Die, a verse play on Joan of Arc by Schiller, sub-titled *Eine romantische Tragödie.* It was written in 1800–1, first performed at Leipzig on 18 September 1801, and published in the same year. It is composed of a prelude (termed *Prolog*) and five acts. In the *Prolog* Johanna (Joan) is reproached by her narrowly religious father Thibaut for her refusal to marry. She breaks her silence to declare in inspired tones her mission to free France. In the play proper the King's cause is in a desperate state, when hope is revived by the confident arrival of Johanna. The English, in alliance with the Burgundians, are defeated by her, and she soon succeeds in uniting the dissident Burgundians with the French forces. Johanna ruthlessly slays the young Montgomery in battle, but later, when faced with the English commander Lionel, she is unable to force home her advantage because she is seized with feelings of love. Since God had forbidden her to experience earthly passion, she is overwhelmed with a sense of guilt. At the coronation of the Dauphin at Reims, Johanna is accused by her father of witchcraft. Seeing this as her penance for disobeying God's decree, she makes no answer and success returns to the English armies. Johanna falls into their hands, but in the critical engagement she is freed by a miracle, turns the tide of battle, and falls mortally wounded. She dies surrounded by testimonies of affection and respect, while in the words of the stage direction, 'a rosy hue suffuses the sky'.

The play is written mainly in blank verse, though Johanna speaks in the *Prolog* in eight-line stanzas, and once meditates in *ottava rima*. The scene with Montgomery employs the trimeters of Greek tragedy. *Die Jungfrau von Orleans* was a play to which Schiller himself was particularly attached. It makes greater use of pageantry and music than any other of his works.

Jüngling, Der, a Middle High German poem of *c.* 1,200 lines, written in the latter part of the 13th c. by KONRAD von Haslau (q.v.). It is a handbook of conduct for young noblemen, encouraging decent behaviour and manners, and condemning vice. It ignores *minne*, chivalric sports, and training in arms. Its language has conspicuous Austrian features.

JUNGMANN, JOSEPH ANDREAS (Sand in Taufers, Tyrol, 1889–1975), an eminent Roman Catholic liturgical scholar at Innsbruck University, whose work was influential at the Second Vatican Council. *Die Stellung Christi im liturgischen Gebet* (1929) was the first of his writings to appear, and *Liturgie der christlichen Frühzeit* (1967) the last.

JUNG-STILLING, HEINRICH, see JUNG, JOHANN HEINRICH.

Jung Volker, and *Jung Volkers Lied,* two songs by E. Mörike (q.v.) included in *Maler Nolten* (q.v.), where they are attributed, in an inserted story, to the robber leader 'Marmetin, gennent Jung Volker'.

Junker und der getreue Heinrich, Der, a Middle High German verse tale of some 2,000 lines written by an unnamed author in the 14th c. The young knight drives generosity (*milte*) to excess and is continually in financial difficulties, out of which he is extricated by his faithful servant Heinrich. Eventually his embarrassments are removed by his marriage to a princess of Cyprus, whose hand he wins at a tournament. The story also includes a supernatural element in the form of a magic stone through which the knight can transform himself into a bird, in which shape he wins the princess's love before he gains her hand by victory in the joust.

Jürg Jenatsch, a narrative (Bündnergeschichte) by C. F. Meyer (q.v.), published in 1876. Meyer's principal source was *Georg Jenatsch, Graubündens Pfarrer und Held während des 30jährigen Krieges* (1860) by B. Reber (see also JENATSCH, GEORG). The story is divided into three parts (Erstes Buch: *Die Reise des Herrn Waser*; Zweites Buch: *Lukretia*; Drittes Buch: *Der gute Herzog*). Herr Waser, future Amtsbürgermeister of Zurich, discovers on his travels a plot against his old schoolfriend Jenatsch and the Protestant community. He warns Jenatsch in his manse, but too late. In a massacre (1620) which claims many victims, Lucia, Jenatsch's wife, is shot by her own brother. Jenatsch escapes with her body, which he lays to rest. The loss of

Lucia is the decisive crisis in his life. In revenge he and his followers murder Pompejus Planta, head of the Catholic community. An ancient tradition of vendetta dictates that the closest relative of the deceased must avenge him. This is Lucretia, who falls in love with Jenatsch and serves his patriotic cause. In contrast to the gentle Lucia, she has a strong and passionate nature, and her sense of family allegiance bids her kill the man she loves. This conflict of divided loyalty is central to the end of the work, when Jenatsch returns to celebrate his triumph in freeing his canton, and to claim Lucretia as his bride.

Jenatsch achieves this goal by a policy of ruthless betrayal. After enjoying the full confidence of the Huguenot Duc de Rohan on the latter's mission to liberate Grisons from Austrian occupation, he betrays him when the 'good duke', as he is called, fails to secure from Richelieu the independence of the canton. Once a fanatical opponent of Rome, he betrayed his faith by becoming a Catholic in order to enlist Spanish assistance (1635). Jenatsch thus returns as a victor to be feared and distrusted. This situation is exploited by his enemies. Encouraged by Spain, Rudolf Planta plots his assassination at the masked ball arranged to celebrate his return. While Jenatsch is surrounded by the conspirators, the old family servant Lukas emerges from the crowd with the axe which had slain Pompejus. He kills Rudolf, but Lucretia wrests the axe from him and gives Jenatsch the *coup de grâce*, thus saving him from being killed by his enemies. Jenatsch recognizes this with gratitude as he falls.

Meyer's hero meets his death as a necessary, and even welcome, release from the daemonic forces which had held him in their grip. Meyer disclaimed the designation 'novel' for this work, which from its episodic composition might be seen as a very long Novelle, but it is as a historical novel that it is generally regarded.

JUSTINGER, KONRAD (d. *c.* 1425), who was town clerk of Berne in the early 15th c., wrote a chronicle of the city from its foundation in 1152 to his own day (1420).

Jutta, Frau, the legendary woman who is said to have become pope in the 9th c. See SPIEL VON FRAU JUTTEN.

K

Kabale und Liebe, a tragedy (Ein bürgerliches Trauerspiel in fünf Aufzügen) written by Schiller in 1782–3, published in 1784, and performed in the same year on 13 April at Frankfurt/Main and on 15 April, in Schiller's presence, at Mannheim. Schiller's original title was *Luise Millerin*, and the present title was suggested by A. W. Iffland (q.v.) early in 1784. The action deals with two unhappy lovers separated by class, and the setting is a despotic state in Schiller's own day.

Ferdinand von Walter, the son of the all-powerful minister (Präsident), and Luise Miller, a musician's daughter, are in love, but the Präsident plans to marry his son to the Prince's mistress, Lady Milford. Ferdinand resists, and his love-affair and intended marriage to a commoner become public. A diabolical plot by the Präsident and his helper Wurm provides Ferdinand with false evidence (written under compulsion by Luise herself) of her infidelity. He poisons her. As she dies, she discloses the machinations of the Präsident and Wurm to Ferdinand, who commits suicide to die with her. The Präsident and Wurm, overwhelmed by the catastrophe, submit to justice. Lady Milford, disgusted at the corruption of the court, and frustrated at her efforts to achieve virtue, leaves the principality. *Kabale und Liebe* is Schiller's most realistic play with a dialogue which in places is startlingly lifelike. It is also one of his most tautly constructed pieces, and has some affinity (not coincidental) with Lessing's *Emilia Galotti* (q.v., and see BÜRGERLICHES TRAUERSPIEL).

KAFKA, FRANZ (Prague, 1883–1924, Klosterneuburg nr. Vienna), was the son of a prosperous Jewish businessman, who insisted on his studying law. Kafka completed his studies in 1906, and two years later took up an appointment at Prague in a workers' accident insurance company. In 1910 he began to keep a diary in which his inner life is relentlessly analysed. From 1912 to 1917 he maintained a characteristically equivocal relationship with a young woman from Berlin, Felicie (Felice) Bauer, to whom he was twice briefly engaged. In 1917 he was found to be suffering from tuberculosis. He resigned his appointment shortly afterwards, and in 1919 stayed in various sanatoria. In 1920 he met Milena Jesenska-Pollak, with whom he later corresponded. In 1923 he met Dora Dymant and lived with her for a time in Berlin. His progressive illness drove him home to Prague before he entered a sanatorium near Vienna. Kafka spent the last stage of his illness in a nursing home at Kierling, where Dora Dymant helped to nurse him to the end.

Kafka published few works in his lifetime and left a testimentary direction that his unpublished writings should be destroyed, which was disregarded by his friend and executor Max Brod (q.v.). Brod prepared Kafka's two major works for publication within a year after his death. *Der Prozeß* (q.v.) appeared in 1925, and *Das Schloß* (q.v.) in 1926. A fragment of a novel, *Der Verschollene* (q.v.), was published in 1927 entitled *Amerika*. The select edition of Kafka's shorter fiction, *Beim Bau der chinesischen Mauer* (the title of a story) of 1931 is the last of the early posthumous publications and precedes the major publications of his works and correspondence of the mid 1930s. The revised and extended editions of the 1950s established Kafka's prominent position.

Kafka himself published *Betrachtung, Der Heizer* (both 1913), *Das Urteil* (q.v., 1916), *Die Verwandlung* (q.v., 1915), the collection *Ein Landarzt* and *In der Strafkolonie* (both 1919); and *Ein Hungerkünstler*, printed separately in 1922, also appeared as the title story of a collection of 4 stories, which appeared in the year of his death. The story *Ein Landarzt* reflects the impotence of physicians and the omnipotence of disease, a subject treated with irony in *Ein Bericht für eine Akademie*. The works published before 1919 were written before the 1914–18 War. *In der Strafkolonie* was written shortly after the outbreak of the war, and *Ein Landarzt* during the war. *Josefine, die Sängerin oder Das Volk der Mäuse*, and *Ein Hungerkünstler* contained in the collection of the latter title, are Kafka's last stories. They deal with the plight of the Jews, the community into which Kafka was born, and with that of the artist, the solitary man, and his emotional and spiritual needs. They are terse examples of Kafka's highly symbolical and oblique style of writing which makes his work subject to widely divergent interpretations. The most complex single composition of his quasi-Freudian self-analysis is his *Brief an den Vater* (written in 1919), which his father never received.

Kafka's *Briefe an Felice* appeared in 1967, *Briefe an Milena* in 1952, and *Briefe an Ottla* in 1974. *Gesammelte Schriften* (6 vols.), ed. M. Brod and H. Politzer, appeared 1935–7, the last volume containing correspondence and Kafka's diary. The collection was extended by M. Brod and appeared 1946 ff. (New York) parallel with *Gesammelte Werke* (unnumbered, 1950 ff., Frankfurt). The critical edition, *Schriften, Tagebücher, Briefe* (4 vols.), ed. J. Born, G. Neumann, M. Pasley, and J. Schillemeit, appeared 1982 ff.

Kahlenberg, a hill to the north-west of Vienna with a splendid and famous view across the city. It appears frequently in Viennese novels and plays. Until the late 18th c. it was known as the Sauberg. The nearby village of Kahlenbergerdorf is the scene of the legendary pranks of a 14th-c. parish priest in the poem *Der Pfaffe vom Kalenberg* by P. Frankfurter (q.v.).

Kahlschlag, term in forestry denoting the complete felling of trees and clearance of undergrowth, was first used as a literary metaphor by W. Weyrauch (q.v.) in an essay introducing his anthology of stories *Tausend Gramm* (1949). It quickly became the slogan for a complete break with literary tradition in matter and manner, being especially fashionable in the early and mid 1950s. The poem *Inventur* by G. Eich (q.v.) has frequently been quoted as Kahlschlag in action.

KAISER, FRIEDRICH (Biberach, 1814–74, Vienna), an unusually prolific dramatist who, although born in Württemberg, was the son of an Austrian officer. Kaiser began a career as a civil servant, but resigned his post in 1838. By then some of his plays had already been produced in the Leopoldstädter Theater (later Carl-Theater). In 1840 Kaiser signed a contract with the director, Karl Carl (q.v.), and provided the theatre with a number of effective comedies and farces in the tradition of the Viennese Volksstück (q.v.), including *Der Schneider als Naturdichter* (1843) and *Die Schule des Armen* (1847). Exploited by Carl and other directors, he derived little financial benefit from his numerous stage successes, but he influenced Anzengruber (q.v.), who admired his work. In his later years he treated social problems, including prostitution (*Ein verrufenes Haus*, 1872) and also composed a patriotic play dealing with General G. E. von Laudon (q.v.). He wrote some 150 plays in all, of which more than 70 have never been printed.

Kaiser is also the author of the historical novels *Ein Pfaffenleben* (2 vols., 1872), on ABRAHAM a Santa Clara (q.v., the subject also of a play in the same year), and *Unter dem alten Fritz und Kaiser Joseph* (2 vols., 1877), on Friedrich II of Prussia and the Emperor Joseph II (qq.v.). In 1871 appeared the memoirs (Bunte Bilder) *Unter fünfzehn Theaterdirektoren* and in 1914 *Ausgewählte Werke*, ed. O. Rommel. His description of the revolutions (see REVOLUTIONEN 1848–9), *1848. Ein Wiener Volksdichter erlebt die Revolution*, a selection from posthumous papers, ed. F. Hadamowsky, appeared in 1948.

KAISER, GEORG (Magdeburg, 1878–1945, Ascona), a bookseller's assistant in Magdeburg, worked his passage to South America in a freighter in 1898, and tried to find congenial work in Buenos Aires. In 1901 he returned with malaria, taking years to recover and being supported for a time by his family. In 1908 he married a well-to-do woman, and until the early 1920s enjoyed a comfortable existence. When money ran short, he sold his wife's estate, lived in a rented villa, pawned its furniture, and was charged with theft. His defence in court, expressed in a notorious speech stressing the artist's privilege of immunity from the common cares of life, injured his cause and revealed the extent of his self-absorption. It perhaps provides the key to his anti-bürgerlich attitude, to his desire for another kind of man, and to his praise of the woman who is ready to sacrifice her life for a man. Kaiser was convicted and sent to prison for six months.

Kaiser began to write about 1904. His first play, *Die jüdische Witwe* (not published until 1911), is a Freudian interpretation of Judith's killing of Holofernes. *König Hahnrei* (1913) parodies Wagner's *Tristan und Isolde* (q.v.). Subsequently Kaiser made his name as an Expressionist writer whose work contains much that is noble, humane, and compassionate. His programmatic tract *Vision und Figur* (1918) is a résumé of his involvement (see EXPRESSIONISMUS). *Die Bürger von Calais* (q.v., 1914) is generally regarded as his best work, though its quality was only perceived when it was first performed in 1917. *Von morgens bis mitternachts* (q.v., 1916), a characteristic example of the Stationendrama (q.v.), was performed in the same year, as was also *Die Koralle* (q.v., 1917). *Die Koralle* was to develop into a trilogy (not so called by Kaiser) by the addition of the two plays entitled *Gas*, and generally referred to as *Gas I* and *Gas II* (qq.v., 1918 and 1920). Kaiser wrote with excessive facility, completing in all some 70 plays, not all of which have been published.

In 1918 appeared three plays, *Rektor Kleist* (an immature work written in 1905), the comedy *Der Zentaur* (retitled *Konstantin Strobel* in 1920), and *Das Frauenopfer*, which has the purifying power of sacrificed love as its theme, illustrated in a historical example drawn from France in Napoleonic times. *Der Brand im Opernhaus* (1919) has an analogous story of self-sacrifice. Further variations on this theme are contained in *Gilles und Jeanne* (1923, on Gilles de Rais and Joan of Arc) and *Oktobertag* (1928). *Hölle Weg Erde* (1919) continues the theme of the New Man,

and sees him for once victorious, converting his fellows, who have lived as automata. *Der gerettete Alkibiades* (1920) presents a conflict between intellect and life, introducing Socrates as well as Alcibiades. *Kanzlist Krehler* (1922) repeats the formula of *Von morgens bis mitternachts*. Kaiser wrote *Noli me tangere* (also 1922) while in prison. *Nebeneinander* (1923) shows the hero ending his own life in the ecstatic awareness that he has done all he could to help another despairing human being. *Kolportage* (1924) reverts to cynical comedy, which was followed up in *Die Papiermühle* and *Der Präsident* (both 1927), and in *Zwei Krawatten* (1930). In *Gats* (1925) Kaiser presents the image of a Utopia which is to be realized by compulsory sterilization of a proportion of the population; the originator of this idea is murdered. Other exemplifications of Kaiser's aspirations are contained in *Zweimal Oliver* (1926) and *Die Lederköpfe* (1929), which has a repellent plot involving self-mutilation. Kaiser wrote the librettos for *Der Protagonist* (1926) and *Der Zar läßt sich photographieren* (1927) by K. Weill (q.v.), who later composed music for Kaiser's *Silbersee* (1933).

In 1933 the first performance of *Silbersee* was the occasion for a disturbance created by National Socialists. Kaiser was expelled from the Prussian Academy of Arts and forbidden to write. He remained in Germany until 1938, when he emigrated to Switzerland, where he spent the remainder of his life. His family remained in Germany. In exile he was befriended by the writer C. von Arx (q.v.). The works of the period of 'inward emigration' (see INNERE EMIGRATION) are not important, except for *Rosamunde Floris* (written in 1936, published 1940), another play of passion and purification. Kaiser is the author of two novels, *Es ist genug* (1932), on the theme of incest, and *Villa Aurea* (1940), set in Russia and Sicily. In 1940 he also published a notable pacifist play, *Der Soldat Tanaka*; Tanaka, a Japanese soldier, turns against war and is executed. Other plays, written at this time and later, were published posthumously. *Klawitter* (1949, written 1939–40) and *Der englische Sender* (1947, written 1941) endeavour to treat the National Socialist regime as a subject for satirical comedy. The theme of dictatorship is continued in the tragicomedy *Napoleon in New Orleans* (1950). In *Das Floß der Medusa* (1942) thirteen evacuee children are in a lifeboat in the Atlantic after the ship taking them to Canada in 1940 has been torpedoed; one is sacrificed, one refuses to be rescued, and eleven are saved. The title refers to the episode of 1816, which is the subject of Géricault's well-known picture *Le Radeau de la Méduse* (Louvre). Part of the play was published in the year of Kaiser's death, and the complete play in 1963. In his three Greek plays, *Pygmalion, Zweimal Amphitryon,* and *Bellerophon,* Kaiser reaffirms his self-confidence and his belief in regeneration. They were published in 1948. *Werke* (6 vols.), ed. W. Huder, appeared in 1970–2 and *Briefe,* ed. G. M. Valk, in 1980.

Kaiserchronik, an immense Early Middle High German poem of more than 17,000 lines written in Regensburg, probably between 1135 and 1150. It is likely that it was undertaken under the patronage of Duke Heinrich der Stolze (q.v.) of Saxony. The view has been advanced that it is the co-operative work of a group of priests. The earlier attribution to the Pfaffe Konrad (see KONRAD, PFAFFE) has been abandoned.

The work is a huge verse chronicle, covering the Roman kings and emperors, and the modern 'Roman' emperors from Charles the Great (see KARL I, DER GROSSE) to Konrad III (q.v.). It breaks off with the summons of Bernard of Clairvaux to the Second Crusade in 1147. The standpoint is clerical, and the rulers are judged according to their piety and their moral qualities. Among the more recent monarchs the chronicler approves those who were in harmony with the Papacy, and condemns those in opposition. Heinrich IV (q.v.) is particularly severely treated. The ideal is the Christian king. In spite of the abundance of warlike incident, the accent is on justice rather than heroism.

The work is rich in legendary incidents and fantastic episodes, and includes some fictitious emperors. The noble society portrayed is modelled on that of the chronicler's own day. It is the knightly world immediately preceding the courtly era.

The *Kaiserchronik* ranks as the first German historical work. Its popularity is attested by the existence of several MSS., some of which are adaptations. The most important is the Vorauer Handschrift (q.v.), in which this is the most substantial item.

Kaiser Friedrich Barbarossa, a five-act tragedy in mixed verse and prose by C. D. Grabbe (q.v.), published in 1829, and designed as the first of a cycle of six to eight Hohenstaufen plays, of which only this and *Kaiser Heinrich der Sechste* (q.v.) were completed. Its main source is *Geschichte der Hohenstaufen und ihrer Zeit* (1823–5) by F. L. G. von Raumer (1781–1873). It was not performed until the end of the 19th c. (in Schwerin, Stuttgart, and Berlin). The principal characters are based on the Emperor Friedrich I (q.v.) Barbarossa and Heinrich der Löwe (q.v.), the Welf (q.v.) Duke of Saxony.

Friedrich faces a crisis in his Italian wars. The Lombards prepare for the destruction of Milan, and, when Friedrich is about to challenge them, Heinrich deserts him; Friedrich is defeated by the Lombards at Legnano, the place of the historical battle of 1176. Friedrich makes his peace with Pope Alexander III and with the Lombards in Venice, in order to return to Germany to meet Heinrich in a (fictitious) battle by the River Weser. Heinrich is defeated and banished, and leaves for the shores of England, the home of his wife Mathildis. Having thus consolidated his power in Germany, Friedrich prepares for his last crusade. Since his son, Prince Heinrich, has meanwhile married Konstanze, heiress to Naples, the Waiblinger dynasty emerges more powerful than ever. Its triumph and fall at the death of Friedrich's son and successor, the Emperor Heinrich VI (q.v.), are the subject of Grabbe's second Hohenstaufen play.

The chief interest lies in the relationship between Friedrich and Heinrich. Both are men of greatness, whose rivalry for power cannot erase mutual feelings of true friendship and human sympathy. Grabbe sees the principal cause of Heinrich's defection in the compulsion by which he is driven to follow his own sense of destiny.

KAISER HEINRICH, author of three Minnelieder, is generally identified with Heinrich VI (q.v.). The poems, which are among the earlier Minnelieder, belong to his youth.

Kaiser Heinrich der Sechste, a five-act tragedy in mixed verse and prose by C. D. Grabbe (q.v.). The second of his two Hohenstaufen plays, it continues the action of the first, *Kaiser Friedrich Barbarossa* (q.v.), and was completed a few months later, in 1829, and published in 1830. It depicts the reign of the Emperor Heinrich VI (q.v.) from his accession to the death of his father, the Emperor Friedrich I (q.v.), to his own sudden death, through heart failure, at the height of his power. The middle act, with the death of the aged Heinrich der Löwe (q.v.), Duke of Saxony, in the presence of Heinrich VI at Brunswick, sees the consolidation of the Emperor's power in Germany and the reconciliation of Welf (q.v.) and Waiblinger. At the end of the play Heinrich stands on Mount Etna, visualizing in his insatiable drive for power the realization of his plan to make the imperial crown hereditary, to abolish the Papacy, and to conquer the whole of Italy and Africa. When death suddenly comes to him, he foresees the disintegration of his lands and uses his last breath to curse his existence.

KAISERIN FRIEDRICH, title of the Empress Victoria (London, 1840–1901, Cronberg,

Taunus) after the death of her husband, the Emperor Friedrich III (q.v.).

Kaiserjäger, four Austrian regiments of Tyrolean Rifles, the first of which was raised in 1816. They were highly regarded in the Austrian and Austro-Hungarian empires, and were trained in mountain warfare. They were disbanded at the end of the 1914–18 War.

Kaiser Octavianus, a Volksbuch the full title of which is *Die schöne und kurzweilige Historie von dem Kaiser Oktavian, seinem Weib und zweien Söhnen, wie sie in das Elend verschickt und wunderbarlich in Frankreich bei dem frommen König Dagoberto wiederum zusammengekommen sind.* It is a translation from the French by one Wilhelm Salzmann, and was printed in 1535.

Felicitas, wife of the Emperor, is cast out on the false suspicion of infidelity. Her two baby sons are carried off in a wood by animals, and grow up separated. Both become knights and fight against the Saracen, as does their father. One son rescues his brother and father from heathen captivity; there is a general recognition, and all ends happily. The story was dramatized by L. Tieck under the title *Kaiser Oktavianus* (q.v.).

Kaiser Oktavianus, a play (Lustspiel) by L. Tieck (q.v.), published in 1804, and based on the Volksbuch *Kaiser Octavianus* (q.v.). It consists of two parts (Erster Teil, divided into scenes, and Zweiter Teil in five acts), and is preceded by a prologue, *Prolog. Der Aufzug der Romanze,* a romantic allegory which has no connection with the action; it epitomizes the Romantic spirit and ends with the famous lines spoken in chorus: 'Mondbeglänzte Zaubernacht,/Die den Sinn gefangen hält,/Wundervolle Märchenwelt,/Steig auf in der alten Pracht!' This unwieldy play, designed to revive the tradition of the Volksbuch, is noteworthy for its use of unaccustomed verse forms, including sonnets, *ottava rima* (q.v.), and the trochaic verse of Spanish drama.

In Part One the Emperor Oktavianus is led by an intrigue to banish his wife Felizitas and their twin children Leo and Florens. The children are separated from the mother, but Leo is miraculously restored to her, while Florens is adopted by a burgher of Paris. Oktavianus meanwhile discovers the intrigue and regrets his action. Part Two portrays the battle waged in France by Christendom against the advancing Muslims. Florens falls in love with the Sultan's daughter, liberates France, but is captured. He is rescued by Leo, the Sultan turns Christian, and the happy ending follows the source.

Kaiserrecht, see SCHWABENSPIEGEL.

Kaiserreich-Trilogie, not an authorized title but the customary designation for a group of three satirical novels by H. Mann (q.v.) focused on the German Empire of Wilhelm II (q.v.). They are *Der Untertan* (q.v., part published early 1914, complete 1918), *Die Armen* (q.v., 1917), and *Der Kopf* (q.v., 1925).

Kaiser und der Abt, Der, a comic ballad by G. A. Bürger (q.v.), first published in the *Göttinger Musenalmanach,* 1785. It is an adaptation of *King John and the Abbot of Canterbury* in Percy's *Reliques of Ancient English Poetry.*

Kaiser und die Hexe, Der, a one-act play in verse by H. von Hofmannsthal (q.v.), written in 1897, and published in 1900. For seven years the Emperor of China has been under the spell of a witch, and can liberate himself only by abstaining from contact with her for seven days. The action takes place as the term approaches. The witch flaunts all her attractions, but the Emperor, fortified by the compassion which the suffering of others arouses in him, sustains the ordeal.

The play is written in the trochaic metre of Grillparzer's *Der Traum ein Leben* (q.v.), and ends with the Emperor's prayer of thanksgiving, in which he includes himself among those who have strayed from the way but found the path of redemption: 'Die dem Teufel sich entwanden/ Und den Weg nach Hause fanden'.

Kajütenbuch, Das, a two-volume novel by C. Sealsfield (q.v.), published in 1841. It is in the form of a frame story with inset Novellen (see RAHMEN), one of which, *Die Prärie am Jacinto* (q.v.), is Sealsfield's best-known work.

KALB, CHARLOTTE VON (Waltershausen, 1761–1843, Berlin), originally a Fräulein Marschalk von Ostheim, was married in 1783 to Heinrich von Kalb, a German officer in French service, garrisoned at Landau. Separated from her husband (whom she in any case did not love) by French military custom, she made Schiller's acquaintance in Mannheim in May 1784, and a close friendship developed between them. In December of that year she rendered Schiller an appreciable service by bringing him to the notice of Duke Karl August (q.v.) of Weimar, then on a visit to Darmstadt. The intense and passionate relationship with Schiller was broken by his removal in April 1785 to Leipzig. They met again in 1787 in Weimar, but were on a more distant footing. In 1796 Frau von Kalb was for a time passionately attracted to J. P.

Richter (see JEAN PAUL). A woman of intelligence, but of neurotic temperament, she was unable at any point to stabilize her life, which was subjected to many stresses, including the loss of her fortune and, in age, the loss of her sight. She is the author of a high-pitched and highly-strung novel, *Cornelia*, and of reminiscences (*Erinnerungen*) which are in places barely intelligible. Both were published posthumously, in 1851 and 1879 respectively.

Frau von Kalb appears to have been in her youth a beautiful woman. Schiller's poem *Freigeisterei der Leidenschaft* refers to their relationship, and she is the original of Linda in Jean Paul's *Titan*. She also appears in the novel *In Reih und Glied* (1867) by F. Spielhagen (q.v.).

KALCKREUTH, WOLF, GRAF VON, see RILKE, R. M.

KALDENBACH, CHRISTOPH (Schwiebus, Silesia, 1613–98, Tübingen), a schoolmaster at Königsberg (1646), then a professor at Tübingen (1656), was one of the group of poets around H. Albert, S. Dach, and R. Roberthin (qq.v., and see KÖNIGSBERGER DICHTERSCHULE). He wrote lyric, religious, and occasional poetry in both German and Latin, and also two plays (*Herkules am Wege der Tugend und Wollust*, 1635, and *Der Babylonische Ofen oder Tragödie von den drei jüdischen Fürsten in dem glühenden Ofen zu Babel*, 1645).

Kalendergeschichten, an anthology of prose and verse by B. Brecht (q.v.), published in 1949, and including the short stories *Der Augsburger Kreidekreis* (q.v.), *Die zwei Söhne*, which is set in the 1939–45 War, *Das Experiment*, dealing with Francis Bacon and a pupil (the original title was *Der Stalljunge*), *Der Mantel des Ketzers* (the coat is that of Giordano Bruno who was condemned by the Inquisition in 1600 and burnt at the stake as a heretic), *Cäsar und sein Legionär* (on the subject of the dictator and related to Brecht's unfinished novel *Die Geschäfte des Herrn Julius Caesar*), *Der verwundete Sokrates*, which is an adaptation of the play *Der gerettete Alkibiades* (1920) by G. Kaiser (q.v.), *Die unwürdige Greisin*, and some *Geschichten vom Herrn Keuner*. The verse contributions include *Gleichnis des Buddha vom brennenden Haus*, *Kinderkreuzzug 1939*, for which Benjamin Britten wrote a cantata (*Children's Crusade*, first performed in May 1969), and the superbly balanced *Legende von der Entstehung des Buches Taoteking auf dem Wege des Laotse in die Emigration* (1939).

KALFF, PETER, see REDENTINER OSTERSPIEL.

KALISCH, DAVID (Breslau, 1820–72, Berlin), invented the Berlin Lokalstück (q.v.) of which

Hunderttausend Taler, Berlin bei Nacht, Ein gebildeter Hausknecht, and *Berlin, wie es weint und lacht* are examples; they are contained in his collection *Lustige Werke* (5 vols., 1870–1), which followed an earlier collection of his plays, *Berliner Volksbühne* (4 vols., 1864). Kalisch was the founder of the humorous and satirical Berlin weekly *Kladderadatsch* (1848).

Kalliasbriefe, title usually given to an extended essay on aesthetics sent by Schiller to his friend C. G. Körner (q.v.) in four letters dated 8, 18, 23, and 28 February 1793. In an earlier letter (21 December 1792) Schiller had mentioned his intention of writing a dialogue entitled *Kallias, oder über die Schönheit*. Though this plan was never carried out, the February letters give the gist of the projected treatise. Schiller asserts that objectivity ('reine Objektivität') is the characteristic of great art.

Kalte Licht, Das, a play by C. Zuckmayer (q.v.), published in 1955 and widely performed at that time. Its theme is the disclosure to a foreign power of secret information concerning atomic weapons. Zuckmayer himself stated that the trial of Dr. Fuchs, a German refugee scientist who acquired British nationality, and passed secret information on atomic research to Russia, inspired the play. Zuckmayer's central figure Kristof Wolters corresponds closely to Fuchs, and it is difficult to see the play as more than a re-enactment of events reported in the press. Zuckmayer, however, concentrates on the relationship between Wolters and his sympathetic interrogator Northon, to whom he is impelled to confess his treasonable actions, obtaining a measure of relief in so doing. The impact of the play, which includes a conventional love-affair, was chiefly due to its topical subject-matter.

Kammermusik, a volume of poems published by J. Weinheber (q.v.) in 1939 in which he attempts to fathom the music of words, giving many of the poems titles such as *Erste Geige*, *Sinfonia domestica, Schlanke Flöte* . . , *Orgel* (dedicated to A. Bruckner, q.v.), *Capriccio, Notturno, Adagio*, and *Dissonanz*. The title of the volume is also that of its second poem (sub-title *Eine Variation*), in which four stanzas, each attributed to one of the instruments of a string quartet, dilate on human destiny.

Kammersänger, Der, a play (Drei Szenen) by F. Wedekind (q.v.), written in 1897, and published in 1899. The ruthlessly determined opera singer Gerardo resists the distraught appeals made to him by a love-struck girl, a frustrated composer, and his own mistress, Helene. Helene shoots herself, but this does not prevent Gerardo from catching his train in order to sing Tristan in Brussels. Though not acknowledged by Wedekind to be a conversation piece (Konversationsstück), it consists of three long conversations, and its only action is Helene's suicide at the end.

Kampf, Der, see FREIGEISTEREI DER LEIDENSCHAFT.

Kampf der Sänger, Der, a story by E. T. A. Hoffmann (q.v.), first published in 1818 in *Urania* and then included in vol. 2 of *Die Serapionsbrüder* (q.v.). The title is followed by the words 'Einer alten Chronik nacherzählt'. It is a retelling of the story of *Der Wartburgkrieg* (q.v.) as recounted by J. C. Wagenseil (q.v.) in *Von der Meister-Singer Holdseligen Kunst* (1697), and lays particular stress on the daemonic art of Heinrich von Ofterdingen.

Kampf um Rom, Ein, see EIN KAMPF UM ROM.

Kampl oder das Mädchen mit Millionen und die Nähterin, a farce (Posse mit Gesang) in four acts by J. N. Nestroy (q.v.), written in 1852, first performed at the Carl-Theater, Vienna, in March 1852, and published in the same year. It continues the trend of the social play, begun by Nestroy with *Der Unbedeutende* (q.v.), and has as its themes the contrast of wealth and modest means, and the confrontation of aristocratic corruption and burgher probity.

The improbable plot presents two girls who, without knowing it, are sisters. Pauline, the heiress, is besieged and extravagantly flattered by suitors. Nettchen, brought up as the adopted daughter of an honest locksmith, is loved by Ludwig, a young man who is in reality the son of an arrogant baroness, Sidonia von Auenheim. Consent to the marriage is refused by Sidonia. Pauline meanwhile discovers the true worth of flattery when she goes incognito and in poor clothing to a ball. Most of the men ignore her; only Wilhelm, Nettchen's supposed cousin, befriends her, and soon comes to love her for herself. Eventually the pride of Sidonia is subdued, the two girls are united with their honest lovers, and Nettchen is discovered to be Pauline's sister and co-heiress. Dr. Kampl, the eponymous hero, moves through the play as its presiding genius, sniffing out Nettchen's true situation, befriending the modest and true, and thwarting and provoking the pretentious and the dowry-hunters. The source of *Kampl* was the novel *La Duchesse* by Eugène Sue, but Nestroy's play has

become an original work rather than an adaptation.

KANDINSKY, WASSELY (Moscow, 1866–1944, Neuilly-sur-Seine), studied law and economics at Munich University and travelled in France and Italy before devoting himself to art in Munich (see BLAUE REITER, DER) and, after a spell in his native Russia (1914–21), in Weimar and Dessau (see BAUHAUS). A close friend of Schönberg and Klee (qq.v.) and a contributor to *Der Sturm* (q.v.), he is noted for his lucid writing on aesthetics. He defined his conception of abstract art (gegenstandslose Kunst) in an influential tract relating to Expressionist literature, *Über das Geistige in der Kunst* (1912, 8th edn. 1965). The postulate of the title that art should reflect an experience of the soul presupposes a progression from outer reality towards inwardness in which the word is instrumental. Thus Kandinsky proceeds from the notion that a word, used once, merely singles out an object, whereas its repetitive use causes the object's form and colour to disintegrate and produce echoes that are released from the poet's innate spirit; the title of the English translation is *The Art of Spiritual Harmony*. Other writings include an autobiography (1913), *Punkt und Linie zur Fläche* (1926), and *Essay über Kunst und Künstler 1912–43,* ed. M. Bill (1955).

Kannegießer, an obsolete word for a person who assiduously talks rubbish about politics. The word has a literary derivation, originating with Holberg's Danish comedy *Den politiske Kandestøber* (1722, *Der politische Kannegießer*).

KANT, HERMANN (Hamburg, 1926–), a prominent DDR author and since 1978 President of the Schriftstellerverband, trained as an electrician and was conscripted in the army towards the end of the war, after which he spent four years as a POW in Warsaw, where he was active as a committee member and teacher of the 're-education school' (Antifa-Schule). After preparing for matriculation at the Workers' and Peasants' Faculty (ABF, Arbeiter-und-Bauern-Fakultät) at Greifswald University he studied at the Humboldt-University in East Berlin (1952–6) and worked for a brief spell as an editor and journalist before becoming a free-lance writer. After publishing a volume of stories, *Ein bißchen Südsee* (1962), he established himself as a major writer with his first novel *Die Aula* (1965). A largely autobiographical work in episodic form, it uses as a frame an invitation received by the central figure, Robert Iswall, from Meibaum, head of the ABF, to make a speech in the assembly hall (Aula) on the occasion of its closure. Iswall re-examines his own development since his apprenticeship as an electrician, including his studies at the ABF, instituted in 1949 in order to promote workers' education and eventual integration into leading positions. His reminiscences and discussions with former associates, Filter, Trullesand, and Riek, who now runs the pub 'Zum toten Rennen' in Hamburg, are characterized by an ironic, colloquial style and deal satirically with a great variety of topics. The central moral problem of the novel concerns Iswall's manipulation of his comrade Trullesand, a rival in love, into marriage and temporary exile. Iswall, when studying reports in the archives of the ABF, comes across an assessment of his own examination essay stating that his work met the requirements of socialist realism. This reminder of the early phase of socialist realism (see SOZIALISTISCHER REALISMUS), from which Iswall has emancipated himself, explains his ironic reaction to the assessment.

In *Das Impressum* (1972) David Groth, nominated for a ministerial post, reflects on his working life from his first job as messenger in a newspaper publishing house in the 1950s. The various episodes survey the development of the DDR and include references to international crises since the 1960s. The ambivalence of Kant's critical stance is epitomized by Groth's disinclination for promotion into a post that removes him from his job as editor of the *Neue Berliner Rundschau*, though he finally accepts out of a sense of duty. *Eine Übertretung* (1975), a volume of stories, was followed by the autobiographical novel *Der Aufenthalt* (1977), which deals mainly with Kant's experience as a prisoner of war.

The volume *Zu den Unterlagen. Publizistik 1957–1980* (1981) includes speeches addressed to the Schriftstellerverband. Kant has been awarded a number of major DDR prizes.

KANT, IMMANUEL (Königsberg, 1724–1804, Königsberg), though one of the most difficult German philosophers to read, possessed a strong desire to address himself not only to academics, but also to the common sense inherent in everyone. Kant studied mathematics and physics, and his first treatise deals with Newton. In the year of its completion, 1755, he began his academic career as Privatdozent at Königsberg University. In philosophy he accepted for many years the thought of Leibniz (q.v.). In 1770 he completed his dissertation *De Mundi Sensibilis et Intelligibilis Forma et Principiis*, and became professor of logic and philosophy. His rational disposition has become proverbial through the precise regularity with which he organized his bachelor life under the

care of his servant Lampe. His contribution to moral philosophy, epitomized by his 'transcendental idealism', matured late in life. Despite his complex mode of expression and his at times inconsistent use of terminology, his basic strategy of thought is direct and balanced.

The most readable statement of Kant's philosophy is the *Prolegomena zu einer jeden künftigen Metaphysik, die als Wissenschaft wird auftreten können* (1783), which he wrote in defence of criticisms of the first of his three *Critiques* (*Kritik der reinen Vernunft*, 1781); known as the Garve-Federsche review, it appeared in the *Göttinger gelehrten Anzeigen*. This opens with the statement that the 856 pages of Kant's *Critique* constantly exercised the reader's mind without yielding proportionate information. Few will take exception to the verdict of Schopenhauer (q.v.) that the *Prolegomena* is the most beautiful and comprehensible of Kant's major writings. With his second *Critique* (*Kritik der praktischen Vernunft*, 1788) Kant placed an introduction before the main work, the *Grundlegung zur Metaphysik der Sitten* (1785), which matches the *Prolegomena* in scope and intention. The third *Critique* (*Kritik der Urteilskraft*, 1790) was, according to Kant's own introductory remarks, to be the end of his task: 'Hiermit endige ich also mein ganzes kritisches Geschäft'. He intended to devote the remainder of his life to other problems which he dealt with in *Die Religion innerhalb der Grenzen der bloßen Vernunft* (1793), *Zum ewigen Frieden. Ein philosophischer Entwurf* (1795), and *Die Metaphysik der Sitten in zwei Teilen* (1797).

In the *Kritik der reinen Vernunft*, which opens his 'critical period', Kant conceives 'pure' reason as knowledge which exists as such (*a priori*), as opposed to knowledge acquired by experience (*a posteriori*). Of the kind of knowledge which the metaphysical approach (precluding sense perception) probes, the concepts of God, freedom, and immortality represent the ultimate goal. This implies the basic query, avoided by rationalistic philosophy, whether metaphysics are acceptable as a science in the same way as pure mathematics and pure natural science, which Kant negates with the apology that he had to 'abolish knowledge to make room for faith'. The principal constructive argument arising from Kant's investigation into the nature of 'pure reason' is that his strict limitation of the scope of human knowledge and transcendental cognition points to an inscrutable source of all phenomena existing beyond the consciousness of time, space, and motion. This he calls the 'thing-in-itself' (*das Ding-an-sich*), which cannot be known, but by which creative and intellectual perception can be apprehended. He bases this philosophical term on the word 'thing' to stress its relation to objects, 'things'.

In his *Kritik der praktischen Vernunft* Kant elaborates what he has already defined in the *Grundlegung zur Metaphysik der Sitten* as the 'categorical imperative' (kategorischer Imperativ). This postulate aims at the co-ordination of human intellect and conscience, and is formulated with variants, but basically in this form: 'handle nur nach derjenigen Maxime, durch die du zugleich wollen kannst, daß sie ein allgemeines Gesetz werde'. The 'categorical imperative' represents a climax of Kant's reaction against the 'Glückseligkeitslehre' of the Aufklärung (q.v.). He stresses that moral action may be accompanied by pleasure, but should on no account be determined by it. Moral conduct should be guided only by the criterion of duty, in which he saw the sole foundation of human freedom. This conception of freedom rests upon the submission of the will to the sublime moral law, which Kant elaborates in his late work *Religion innerhalb der Grenzen der bloßen Vernunft* (1794). He draws the distinction between 'religion' and 'faith' by defining 'Theodizee' (q.v.) as 'religion', and the various confessions and sectarian denominations (his own background was pietistic) as 'faith'. He bases his definition on the subjective cognition that man's moral consciousness, the world as an ethical entity, and the immortality of the soul rest in God, who rewards human virtue with bliss ('Glückwürdigkeit'). Through this approach he maintains a point of contact with the ethical aspect of Christian doctrine.

In the final *Critique* (*Kritik der Urteilskraft*) Kant defines *Urteilskraft* as a concept linking the intellect (*Verstand*) with reason (*Vernunft*) by examining *a priori* principles in relation to the subjectivity of judgement. The first part is devoted to taste (*Geschmack*) in aesthetics, which affords 'pleasure without interest' ('Wohlgefallen ohne Interesse'), but cannot be gauged by any logical process of thought. In beauty Kant sees the symbol of virtue, and he adheres to the principle that art should be an adjunct to ethics, and that the encouragement of morality should be an aesthetic criterion. He rejects the portrayal of ugly and offensive objects in art. In this section he also comments on the concept of *Genie*, maintaining that genius is endowed by nature to present 'ideas' with exemplary originality. He amplifies this approach in his *Beobachtungen über das Gefühl des Schönen und Erhabenen*. On all important topics Kant consistently reiterates the link between ethics and religion. *Über das Mißlingen aller philosophischen Versuche in der Theodicee* singles out the example of Job whom, he believes, most theological or ecclesiastical authorities would have condemned. Yet God rewarded Job for basing faith on morality and not morality on faith.

In the political sphere Kant favoured the ideas underlying the French Revolution (q.v.). In *Von dem Verhältnis der Theorie zur Praxis im Staatsrecht* he argues with Hobbes and reveals himself as a republican, for a republic represents to him the only form of government which preserves the freedom of the individual. He disapproves of 'paternal' government (*imperium paternale*), including the 'benevolent' monarch, because he sees in the dependence of the people on a ruler a violation of human dignity. He advocates equality of opportunity and prerogative by ability, thus rejecting prerogative by birth and right by inheritance. His view of the electoral system as an instrument of judicial government shows the influence of Rousseau's *Du Contrat social*. In *In weltbürgerlicher Absicht* he distinguishes between 'civilization' and 'morality', commenting that while the modern age (as seen by him towards the end of the 18th c.) is in effect over-civilized in its social codes of behaviour, it is far from being morally educated. In his tract *Zum ewigen Frieden* Kant repeats his ideal of a Weltbürgertum reviving the ancient ethics of hospitality and introducing the idea of general disarmament. As long as nations depend for a peaceful coexistence merely on the principle of a balance of power, there can, in his view, be no prospect of lasting peace.

Kant's attempt to link reason and belief led to the speculative idealism of Romantic philosophy (Fichte, Schelling, qq.v.). There exists hardly any philosophical system which is not in one way or another indebted to Kant. Many men of letters responded to his writings. His influence upon Schiller is particularly important. Goethe alludes in a number of his writings (prose and epigrammatic) to his critical reading of Kant (e.g. *Wilhelm Meister*, q.v.). A crisis in the life of H. von Kleist (q.v.), precipitated by his study of the first part of the *Critique*, is known as his 'Kantkrise'. Bauernfeld finds ingenious variations on Kant's moral postulate in his comedy *Der kategorische Imperativ* (q.v.).

Kant's *œuvre*, including posthumous papers and correspondence, has been published by the Prussian Academy as *Kants gesammelte Schriften* (23 vols.), 1900–55.

KANTZOW, THOMAS (Stralsund, *c.* 1505–42, Stettin), secretary to the dukes of Pomerania, wrote a chronicle of the land (*Pommersche Chronik*). At first written in Low German, it was turned into High German by its author before 1538.

Kanzleischrift, an ornamental form of handwriting used in documents and especially in their headings and superscriptions. Since the middle of the 19th c. its use has been restricted to ceremonial documents.

Kanzleistil, an elaborate legal style of writing German, with complicated constructions, abundant borrowings from French and Latin, and frequent archaisms, used in official documents and letters by the courts and offices of the German states into the 19th c. See GERMAN LANGUAGE, HISTORY OF.

KANZLER, DER, a Middle High German poet of the 14th c., was a schoolmaster of Offenburg in Baden, who is documented in 1312 and 1323. His Minnelieder are usually compared unfavourably with the poetry of KONRAD von Würzburg (q.v.), but taken by themselves they have a distinct charm in spite of obvious deliberation. He is more important as an author of critical or didactic Sprüche (see SPRUCH). Der Kanzler, who takes an exalted view of the mission of the true poet, was a learned man who included cosmology among his themes.

Kapital, Das, see MARX, K.

Kaplied, a poem written by C. F. D. Schubart (q.v.) in 1787. Its 12 five-line stanzas are spoken by a soldier in a battalion recruited for Dutch service at the Cape, and are a farewell to his native land. As Schubart's own sovereign, Karl Eugen (q.v.) of Württemberg, and other German princes used such recruitments for foreign service to raise funds for themselves, Schubart's poem implies a topical accusation. The *Kaplied* and *Die Fürstengruft* (q.v.) were Schubart's most widely read poems.

KAPP, WOLFGANG (New York, 1858–1922, Leipzig), served in the East Prussian administration, and from 1906 held the appointment of Generallandschaftsdirektor. His political sympathies were Pan-German (see ALLDEUTSCHER VERBAND), and during the 1914–18 War he was bitterly opposed to any peace involving conciliation or concession. In 1917 he was, with Tirpitz (q.v.), co-founder of the Deutsche Vaterlandspartei (dissolved Dec. 1918). In March 1920 he participated in the so-called Kapp-Putsch (q.v.), and was proclaimed chancellor. After the rapid collapse of the revolt he fled to Sweden. In 1922 he surrendered to justice, but died a natural death during the early stages of the legal proceedings taken against him.

Kapp-Putsch, the name given to an unsuccessful attempt by right-wing elements to overthrow the legitimate government of the Weimar Republic (q.v.). The dissidents were inflamed by a defeat of the right in the Reichstag on 9 March 1920 and encouraged by the light sentence passed three days later on the would-be murderer of Erzberger (q.v.). On 13 March troops, includ-ing notably the Marine-Brigade Ehrhardt, occupied Berlin under General von Lüttwitz, and proclaimed an obscure right-wing politician, Wolfgang Kapp (q.v.), as chancellor. The legitimate government fled, but the passive resistance of the civil service and a general strike of the trade unions discouraged the rebels, who abandoned Berlin after four days, whereupon the insurrection collapsed.

Kapuzinergruft, Die, a short novel by J. Roth (q.v.), published in 1938. Its central figure and first-person narrator is Trotta, a connection of the line of Trottas in *Radetzkymarsch* (q.v.). An account of the disintegration of society, it begins shortly before the 1914–18 War, in the early days of which Trotta is taken prisoner by the Russians. The story continues into the period after the war, which has left Austria a shadow of its former self and in which Trotta can discover no reasonable part to play. The death of his mother, a symbol of the old order, in the middle of the brief civil war in 1934, foreshadows the end of his world. The climax is reached in a café, in which the former *jeunesse dorée* is gathered. A single NS storm trooper enters. It is 11 March 1938, the eve of Hitler's seizure of Austria. Trotta makes a symbolical midnight visit to the tomb of Franz Joseph (q.v.) in the Kapuzinergruft, the burial vault of the Habsburgs, before taking his own life.

KARL I, DER GROSSE (742–814, Aachen), equally well known in German and, as Charlemagne, in French history, was the founder of the Carolingian Empire, and ranks as Charles I of France. Karl was the son of King Pippin of the Franks. On Pippin's death he became king of the northern portion of the Frankish realm (768), and on the death of his brother Carloman (Karlmann) in 771 he seized the remainder. In 773 he intervened in Italy to protect the papal territories against the Langobards, whom he defeated and subdued, becoming King of the Langobards in 774. When the eastern frontier of the Frankish kingdom was threatened by the Saxons, Karl made successful war upon them in 772 and 775. Three years later they revolted, and in 780 were partially subdued. There followed a programme of forcible conversion to Christianity, in which refusal of baptism meant death. The formula of baptismal promises used for these occasions is preserved (see TAUFGELÖBNISSE). Renewed Saxon revolts were ferociously put down by measures that included a massacre of hostages at Verden in 782. By 787 the areas south of the Elbe were conquered, and by 805 the conquest was complete. Bavaria, which had broken away from the Kingdom of the Franks, was reconquered in 788. Meanwhile

Karl intervened in Spain against the Moors, and, after varied fortunes, including the defeat of Roncesvalles, conquered the country as far as the Ebro (803). Thus in some thirty years Karl created by conquest a vast Western Empire to rival the Byzantine Empire.

In 800 Karl was crowned by the Pope, and proclaimed Emperor of the Romans in 801. His anger over this ceremony, recorded by Einhard (q.v.), is believed to have been directed at the Pope's implied assumption of power. Karl was buried in Aachen Cathedral.

Karl was an energetic and successful administrator, who ruled his empire by creating marches (Marken) ruled by dukes or counts (Markgrafen) as his delegates, and he maintained control over them by royal emissaries (*missi dominici*). Churchmen, as the literate social group, provided him with an administrative class. He actively furthered learning and education, employing Alcuin (q.v.) as his chief educational administrator. The active Latin culture of the age is often referred to as the Carolingian Renaissance. Karl's vigorous, alert, and tireless character dominated and upheld his empire, which did not long survive his death.

Karl's biography was written by his friend Einhard in *Vita Caroli Magni*. In the Middle Ages Karl became a legendary figure, an ideal Christian king; his exploits, notably the Spanish campaign, became a subject for poetry (see ROLANDSLIED), and a vast cyclical collection of poems dealing with them is preserved in a 14th-c. MS. (see KARLMEINET).

KARL II, DER KAHLE (Charles the Bald) (Frankfurt, 823–77, Avrieux), son of Ludwig I (q.v.), is better known as Charles II of France. After his father's death he quarrelled with his brothers Ludwig II, der Deutsche, and Karlmann (qq.v.) and by the Treaty of Verdun (843) received the West Frankish kingdom. After Ludwig's death in 875 he gained the imperial title as Kaiser Karl II.

KARL III, DER DICKE (839–88, Neidingen/ Danube), a son of Ludwig II, der Deutsche (q.v.), was crowned emperor in 881. He acquired the East Frankish (i.e. German) kingdom in 882 and the West Frankish kingdom in 885. He abdicated in 887. He is Charles III of France.

KARL IV, KAISER (Prague, 1316–78, Prague), Roman emperor and German King of the house of Luxemburg, was brought up in France. His father, King John of Bohemia, employed him as his representative in Italy and in 1334 allocated to him the margravate of Moravia. In 1346 he

was elected German King, in opposition to King Ludwig the Bavarian (see LUDWIG IV). Karl's rival died in 1347 and, since other claimants failed to secure support, he was crowned at Aachen in 1349. Karl maintained a peaceful policy of alliances, and adroitly made use of his many children to secure by marriage extensive future inheritances. Abandoning effective claim to rule in Italy, he contented himself with achieving coronation as emperor in Rome in 1355. In Germany he acquired by purchase or marriage the Upper Palatinate (Oberpfalz), Lusatia (Lausitz), and Schweidnitz, and he purchased Brandenburg from its weak Wittelsbach rulers. In his later years, in spite of many obstacles, he secured the election of his son Wenzel (q.v.) as German King, and hence as his own successor, and he attempted to gain for his son Sigismund (q.v.), the Elector of Brandenburg, the right of succession to the thrones of Poland and Hungary. One of the most important acts of his reign was the promulgation of the Golden Bull (see GOLDENE BULLE) in 1356, by which the procedures of election of the German King were systematized. For a time he supported the claim to the margravate of Brandenburg made by the impostor, 'der falsche Woldemar' (see WOLDEMAR).

A patron of learning, Karl founded the German university of Prague in 1348, and employed the humanist JOHANN von Neumarkt (q.v.) at his court.

KARL V, KAISER (Ghent, 1500–58, San Geronimo de Yuste, Spain), grandson of the Emperor Maximilian I (q.v.), and son of Philip of Burgundy and a Spanish mother (Joana), acceded to the Burgundian estates (Netherlands and Franche Comté) in 1506, and was declared of age in 1515. Through the death of his maternal grandfather he came to the throne of Spain in 1518, though only as joint monarch with his mother. By the death of Maximilian in 1519, he succeeded to the Habsburg possessions, and in 1520 was crowned emperor at Aachen. Finding himself within these few years arrived at a position of extraordinary power, he devoted his energies to the attempt to neutralize French expansion, and to the assertion of temporal and spiritual unity in imperial central Europe in face of the Reformation (q.v.), which broke out at the beginning of his reign. By the decisions of the Diet of Worms in 1521 (see WORMSER REICHSTAGE) he sought to break the force and spirit of the new movement, but the rivalry with France immediately drew him away to Italy, leaving Germany without a resident ruler for nine years, during which the Reformation established itself. Karl V took Milan, defeated and captured Francis I at Pavia in 1525, subdued Pope

Clement V, and sacked Rome in 1527. In 1530 he returned to Germany, presiding over the Diet of Augsburg (see AUGSBURG), at which he vainly hoped to initiate the suppression of Protestantism. Turkish pressure drew him eastward in 1529 and 1532, and he was obliged to make concessions to the Lutheran party in the Religious Peace of Nürnberg (1532, see NÜRNBERGER RELIGIONSFRIEDE).

The next years were taken up with further successful struggles against France, but in 1546 Karl began a campaign against the Protestant forces of the Schmalkaldic League (see SCHMALKALDISCHER KRIEG), which culminated in his decisive victory in 1547 at Mühlberg, in which the Elector Johann Friedrich of Saxony was taken prisoner. A well-known equestrian portrait by Titian shows Karl V as a knight in this battle. At the Diet of Augsburg (1547) Karl again sought unsuccessfully to stem Protestant influence. The Interim (q.v.) of Augsburg, however, made no headway towards unification. In the 1550s the tide turned against Karl, who faced a rising of the northern princes (the Fürstenrevolution of 1552), in the course of which his former lieutenant, Moritz von Sachsen (q.v.), reduced him to military impotence. Karl, who had ceded Verdun, Toul, and Metz to France, without gaining the support he had hoped, wearied of a conflict to which he had fruitlessly devoted his whole life.

After the Religious Peace of Augsburg (1555, see AUGSBURGER RELIGIONSFRIEDE), by which the spiritual division of Germany was officially recognized, Karl abdicated in favour of Philip II, as to the throne of Spain, and of Ferdinand I (q.v.) in respect of the imperial crown, withdrawing to a monastery in Estremadura. His undoubted political talent and force of character were worn down by the complexity of the situations he faced, and the tide of human thought and belief which he sought to arrest.

KARL VI, KAISER, and KÖNIG KARL III VON UNGARN (Vienna, 1685–1740, Vienna), was the second son of the Emperor Leopold I (q.v.). In 1703 he became king of Spain (as Carlos III), and had to fight the Spanish War of Succession against a rival French claimant, Philip V. On the death of the Emperor Joseph I (q.v.) in 1711, he succeeded to the imperial title and the Habsburg dominions. In the Peace of Rastatt (March 1714) Philip won the Spanish crown, while Karl received Milan, Sardinia (which he exchanged in 1720 for Sicily), Naples, and the Spanish Netherlands. Prince Eugene (see EUGEN, PRINZ) fought for Karl against Turkey (1716–18). From 1713 Karl tried to win approval for the Pragmatic Sanction (see PRAGMATISCHE SANKTION), by which he sought to secure the succession of his daughter Maria Theresia (q.v.). In the last decade of his reign he squandered his resources in the War of the Polish Succession (see AUGUST III) and in a campaign against Turkey, leaving Maria Theresia a financially exhausted heritage.

KARL VII, KAISER (Brussels, 1697–1745, Munich), was Kurfürst von Bayern from 1726, and contested the Pragmatic Sanction (see PRAGMATISCHE SANKTION) designed by the Emperor Karl VI to secure the succession of his daughter Maria Theresia (qq.v.). Karl, the Elector, based his rights for the imperial title on the descent of his wife, who was a daughter of the Emperor Joseph I (q.v.). In 1741 he invaded Austria, thus opening the Austrian War of Succession (see ÖSTERREICHISCHER ERBFOLGEKRIEG) backed by France, and was elected Emperor as Karl VII in January 1742. Although he retained the title until his death, it did not help Bavarian prestige. The occupation of his country by Maria Theresia's troops forced him to give up his residence in Munich until 1744, and the election of Maria Theresia's husband Francis of Lorraine (see FRANZ I) in the year of his death satisfied the traditional Habsburg claim for the highest imperial office. K. Th. von Heigel published Karl's diary of the War of Succession in 1883.

KARL I, HERZOG VON BRAUNSCHWEIG (Brunswick, 1713–80, Brunswick), came to the ducal throne in 1735. He removed the residence from Wolfenbüttel to Brunswick, and lived on a scale which overburdened the resources of his duchy. He was greatly interested in art, literature, and intellectual matters generally, founded the Collegium CAROLINUM (q.v.) in 1745, and in 1770 appointed Lessing (q.v.) to be his librarian. Duke Karl makes a brief appearance as a character at the end of W. Raabe's novel *Hastenbeck* (q.v.).

KARL, ERZHERZOG (Florence, 1771–1847, Vienna), Austrian field-marshal and third son of the Emperor Leopold II (q.v.), was in 1793–4 governor of the Austrian Netherlands. In command of the Austrian forces in 1796, he drove the French forces back from central Germany to the Rhine, and was also successful in his campaign in 1797 against Jourdan and Masséna. He became president of the War Council in 1801. Appointed Austrian Commander-in-Chief in 1805, he reorganized the army. He was opposed to the reopening of hostilities in 1809. After several setbacks he gained a victory at Aspern (q.v.), Napoleon's first defeat, but was himself finally defeated at Wagram (q.v.). After the armistice of Znaim he resigned the command. Though cold-shouldered by the emperor, Karl came to be regarded as a national hero. His equestrian statue faces that of Prince Eugene (see EUGEN, PRINZ) in the Heldenplatz of the Hofburg in Vienna.

Karl was the author of various works on the theory, practice, and history of war: *Grundsätze der Strategie erläutert durch die Darstellung des Feldzugs von 1796 in Deutschland* (3 vols., 1814) and *Geschichte des Feldzugs von 1799 in Deutschland und der Schweiz* (2 vols., 1819).

KARL AUGUST, HERZOG VON SACHSEN-WEIMAR (Weimar, 1757–1828, Graditz nr. Torgau), succeeded to the ducal throne in infancy (1758), and passed his boyhood under the guardianship and regency of his mother, the Duchess Anna Amalia (q.v.). In 1772 C. M. Wieland (q.v.) was appointed his tutor. Karl August took over the government in 1775, marrying Princess Luise of Hesse-Darmstadt in the same year. He invited Goethe to Weimar in 1775, at first as a boon companion, and then gradually entrusted more and more of the administration to him. The Duke, with Goethe as his adviser, ruled on enlightened principles, encouraging industry, including textiles and glass, and mining. He improved the educational facilities of the state, and inaugurated a botanical garden and an art museum. He rebuilt the Court Theatre (1791) and established a permanent company with Goethe as director. His great energy, and his interest in military affairs, which his little state could not satisfy, led to tensions and to absences in Prussia; his liaison with the principal actress, Caroline Jagemann (q.v.), endangered both the theatre and his relationship with Goethe. Nevertheless, the mutual regard of the two men persisted in spite of occasional estrangement. Weimar during Karl August's reign was the literary focus of Germany, attracting Wieland, Herder, and Schiller as well as Goethe, and becoming an indispensable place of call for the grand tour. In 1815 Karl August's title was elevated to Grand Duke (Großherzog). He was one of the first German rulers to grant a constitution (1816).

KARL EUGEN, HERZOG VON WÜRTTEMBERG (Brussels, 1728–94, Hohenheim), came to the throne in 1737 and took over active government in 1744. He gave himself up to extravagance and dissipation, undertaking a building programme intended to rival Versailles, and indulging in sexual promiscuity which made him notorious. Ministers retained his favour as long as they raised the vast sums he needed. A long conflict with the Estates of Württemberg ended with a compromise in 1775, and Karl Eugen emerged in a new role as an enlightened despot. Under the influence of his mistress, Franziska von Hohenheim (q.v.), he founded a military academy (see KARLSSCHULE) which became one of his principal interests. Among its alumni was Schiller. In 1777 Karl Eugen again became notorious, this time for his vindictive arrest and imprisonment of the poet and musician C. F. D. Schubart (q.v.).

KARL FRIEDRICH, MARKGRAF VON BADEN, later GROSSHERZOG (Karlsruhe, 1728–1811, Karlsruhe), commonly referred to as Markgraf Friedrich von Baden, succeeded as Margrave at the age of 10 in 1738; with the reorganization of Germany in 1803, his title was changed to Elector (Kurfürst), and in 1806 to Grand Duke. He was one of the most remarkable and successful examples of the enlightened despot. His economic measures were based on the doctrines of the French Physiocrats, and his interest in economic matters expressed itself in an anonymous tract in French of which he was the author (*Abrégé des principes de l'économie politique*, 1772). He abolished serfdom in Baden in 1783. Friedrich was also a liberal patron of the arts and letters, inviting F. G. Klopstock (q.v.) to Karlsruhe, and appointing Jung-Stilling (see Stilling, H.) to a professorship at Heidelberg University, and later pensioning him in Karlsruhe; his idea of a German Academy for men of scholarship and science was taken up by J. G. Herder (q.v.).

KARLMANN (*c.* 829–80), eldest son of Ludwig II (see LUDWIG II, DER DEUTSCHE), was from 856 ruler of the Eastern Franks, and in 876 acquired Bavaria and Bohemia. In 877 he was crowned King of Italy, but abdicated in 879 in favour of his brother Karl III (q.v.).

Karlmeinet, title given since 1858 to an anonymous medieval poem of some 35,000 lines, recounting the life of Charlemagne (see KARL I, DER GROSSE). The title denotes the young Karl and is properly applicable only to the first part. *Karlmeinet* is a compilation of four existing poems with two connecting episodes which may also be adaptations of older works. Its compilation is probably the work of a single writer, active early in the 14th c. in the lower Rhine region. Aachen has been suggested as its place of origin, but without proof; its dialect is that of the Cologne district. Its first section (*Karlmeinet* proper) is the story of the young Karl, who, persecuted by disloyal regents, flees to Spain, fights under the heathen Galafer, and becomes a knight. After reconquering his kingdom he returns to Spain and elopes with Galafer's daughter Galîe, whom he marries. The source is a French poem in a Flemish translation. There follows the poem *Morant und Galîe* (q.v.), which in turn is succeeded by an account of Karl's wars derived from a Netherlandish translation of the *Speculum historiale* of Vincentius Bellovacensis (Vincent of Beauvais). The fourth narration is the story *Karl und Elegast*, according to which Karl is enjoined by God to embark on robbery, and thereby discovers a treasonable conspiracy as well as the fidelity of his outlawed knight Elegast. The episode is of Low Country origin. Next follows the story of Roland, derived from various versions, especially that of Der Stricker (q.v.). And finally Karl's end is portrayed after the account in the *Saeculum historiale*.

Karlsbader Beschlüsse, decrees drawn up by Metternich (q.v.) at a meeting in Karlsbad with representatives of eight other German states, and ratified by the Frankfurt Diet (see DEUTSCHER BUND) in September 1819. Provoked by the assassination of A. von Kotzebue (q.v.) in March of that year by K. L. Sand (q.v.), they were designed to suppress national and liberal movements, notably in the universities (see BURSCHENSCHAFTEN). In effect they were enforced by a stringent police system and became an instrument for ruthless persecution; they were largely responsible for the minor revolutions in 1830 (see JULIREVOLUTION) and for the radical revolutions in 1848 (see REVOLUTIONEN 1848-9).

The decrees included control of all educational institutions, the prohibition of political meetings, and the censorship of the press. Among the nine sovereigns at the meeting only the King of Württemberg protested strongly against them.

Karlsschule, the usual designation of Die Hohe Karlsschule, the combined school and university at Stuttgart from 1781 to 1794. The school originated in an orphanage founded in 1770 by Duke Karl Eugen (q.v.), and was reorganized in 1771 as a school for the training of officers and administrators, with the title Militärische Pflanzschule. It was situated at Schloß Solitude (see SOLITUDE). In 1773 its title was changed to Herzogliche Militär-Akademie. In 1775 it was transferred to Stuttgart and was raised in 1781 to the rank of a university as Hohe Karlsschule. It was dissolved under Karl Eugen's successor, Duke Ludwig Eugen, in 1794. Among its alumni were Schiller, the zoologist Cuvier, the sculptor Dannecker (q.v.), and the composer Zumsteeg (q.v.). Schiller, having left early in 1781, was a pupil of the Akademie but never, strictly speaking, a Karlsschüler.

Karlsschüler, Die, a play by H. Laube (q.v.), first performed at Dresden in November 1846, and published in 1847. It was extremely popular, owing its success to its subject; for it is a sentimentalized account of the young Schiller's conflict with Duke Karl Eugen (q.v.), and of his escape from Württemberg on 16 September 1782. Schiller is in love with Laura, a ward of the Duke. The Duke disapproves and interviews Schiller, who delivers a swingeing denunciation of tyrants. Schiller escapes, and the Duke is told in a letter (from W. H. von Dalberg, q.v.) how great a genius the young man is. Much of the play is written in a hollow rhetorical prose.

KARLSTADT (Karlstadt, *c.* 1480–1541, Basel) was so called after his native town. His real name was Andreas Bodenstein. Karlstadt was an academic colleague of Luther (q.v.) at Wittenberg, who at once sided with Luther in his disputes with the Papacy. In 1519 he supported Luther at the disputation in Leipzig. Though he was not politically active, his theological views were more radical than Luther's, and led in 1522 to image-breaking riots in Wittenberg, in which Luther intervened. Karlstadt's puritanical extremism alienated him from Luther, and he inclined towards Swiss religious views, spending his later life as professor at Basel University.

KARL THEODOR, KURFÜRST VON PFALZ-BAYERN (Sulzbach, 1724–99, Munich), son of the Count Palatine Johann Christian Joseph, became the ruling count on his father's death in 1733. In 1742 he acquired the Electoral Palatinate by inheritance, and in 1777 he succeeded Maximilian III Joseph as Elector of Bavaria. In 1778 he agreed to cede Lower Bavaria to the Habsburg Empire in return for the Austrian Netherlands, but was prevented by Prussia and Saxony, who began the War of the Bavarian Succession (see BAYRISCHER ERBFOLGEKRIEG). At the Peace of Teschen (1779) Karl Theodor abandoned the proposed exchange and ceded the Innviertel (q.v.) to Austria. He encouraged the arts in Mannheim, completing the Schloß in 1760; music and the theatre flourished in Mannheim during his rule. In 1784, at the instance of the Jesuits, he suppressed the Illuminatenorden (q.v.).

Karl und Anna, a short novel by L. Frank (q.v.), published in 1927. It is an early example of the novel of the soldier returning from captivity and of his problems (see HEIMKEHRER-LITERATUR). Richard, Anna's husband, works together with Karl in a P.O.W. camp in Siberia, and talks endlessly of every detail, essential and inessential, of his and Anna's life together. Karl finds himself in love with this unseen image. He returns earlier than Richard, whom he resembles, and goes to Anna, pretending to be her husband. Though she is not deceived, she accepts him. When Richard comes home, Anna, now expecting a child by Karl, declares her inability to return to her true husband, and the two, Karl and Anna, leave the house together. The story was successfully dramatized by Frank in 1929.

Karl und Elegast, title given to an episode in the early 14th-c. *Karlmeinet* (q.v.), which is also found independently in the Low Country poem *Carel ende Elegast.*

KARLWEIS, C., pseudonym of Karl Weiß (Vienna, 1850–1901, Vienna), a railway inspector who wrote a series of broad comedies in Viennese dialect (see VOLKSSTÜCK), which enjoyed considerable success on the popular stage. They include *Cousine Melanie* (1879), *Einer vom alten Schlag* (1886, with A. Chiavacci), *Aus der Vorstadt* (1893, with H. Bahr, q.v.), *Der kleine Mann* (q.v., 1894), *Goldene Herzen* (1895), *Das liebe Ich* (1895), and *Das grobe Hemd* (q.v., 1901), which was Karlweis's outstanding success. He is the author of novels (*Wiener Kinder*, 1887; *Ein Sohn seiner Zeit*, 1892; *Reichwerden*, 1893) and of Novellen (*Geschichten aus Stadt und Dorf*, 1889; *Adieu Papa*, 1898).

Karolinger, the dynasty of German emperors and kings which began with Charlemagne (see KARL I, DER GROSSE) and received its name from him. The line divided into three with the Treaty of Verdun between the sons of Charlemagne (843). The German line of Ludwig the German (see LUDWIG II, DER DEUTSCHE) reigned in Germany until 911, when it ended with the death of Ludwig the Child (893–911). The Lotharingian branch, begun by Lothar I, elder brother of Ludwig the German, reigned in Lorraine and Italy until 875, and the youngest line, that of Charles the Bald (Karl II, der Kahle, son of Ludwig I, der Fromme, qq.v., ruled 840–77), ruled in France until 987. The Carolingian Age (Karolingerzeit) is reckoned from 751, the year of the election of Pippin the Short (der Kleine, le Bref, died in 768) as German King, to 911.

Karolinische Schenkung, a promise made in 774 by Charlemagne (see KARL I, DER GROSSE) to place large tracts of Northern and central Italy under papal rule. It was made in renewal of a promise given twenty years earlier by his father Pippin (see KAROLINGER), but, like the earlier undertaking (Pippinische Schenkung), it was never fully carried out.

KARSCHIN, ANNA LUISE (Tierschtiegel nr. Crossen/Oder, 1722–91, Berlin), *née* Durbach or Dürbach, was twice married, in 1738 and 1749. The name of her first husband was Hirsekorn, of the second Karsch. Both marriages ended in separation. 'Die Karschin', as she was called, had little formal education, but read voraciously. She began to write poetry, and some of her poems were published during the Seven Years War. She was taken up by a Baron von Kottwitz and sent to Berlin, where she was helped by K. W. Ramler, G. E. Lessing, and M. Mendelssohn (qq.v.), who regarded her as a natural, untutored poetess (Naturdichterin). Her poems were published in 1764 by J. W. L. Gleim (q.v.), with whom she corresponded for many years. A further collection was published as *Neue Gedichte* (1772). Friedrich II, der Große (q.v.), received her in audience in 1763, and Friedrich Wilhelm II (q.v.) built her a house in 1789. 'Die Karschin', an open and affectionate personality, had considerable facility in expressing her emotions in conventional terms, and wrote much occasional poetry to order; the appellation 'deutsche Sappho', given her by Gleim, has done a disservice to a small but genuine talent.

Karsthans denotes, in the broadsheets and polemical pamphlets of the Reformation (q.v.) period, the peasant labouring with the hoe (Karst). Karsthans first appears in the dialogue *Karsthans und Kegelhans* (1521) by Vadian (q.v.). A further dialogue, *Neu-Karst Hans*, which has been attributed to Ulrich von HUTTEN and to M. Butzer (qq.v.), followed in the same year. Karsthans became a symbolical figure in the social and political unrest leading to the peasants' revolt (see BAUERNKRIEG) of 1524–5.

Kartätschenprinz, Der, nickname of Prince Wilhelm, later Wilhelm I (q.v.), German Emperor and King of Prussia. It alludes to his part in the military suppression of the 1848 Revolution in Berlin (see REVOLUTIONEN 1848–9).

Kartoffelkrieg, see BAYRISCHER ERBFOLGEKRIEG.

Karussell, Das, poem by Rilke (q.v.), marked *Jardin du Luxembourg,* and included in *Neue Gedichte* (q.v.).

KASACK, HERMANN (Potsdam, 1896–1966, Stuttgart), after working with the publishers Kiepenheuer transferred to the S. Fischer-Verlag, devoting himself from 1927 to authorship and broadcasting. In 1933 his activities were restricted by the National Socialist regime. From 1941 to 1949 Kasack was a reader with the Suhrkamp Verlag before returning to authorship. He was notably helpful in furthering the work of young writers.

Kasack began as an Expressionist, publishing exclamatory poems in *Der Mensch* (1918) and *Die Insel* (1920), but adopted a more sober lyrical tone in *Das Echo* (1933), *Das ewige Dasein* (1943), *Aus dem chinesischen Bilderbuch* (1955), *Antwort und Frage* (1961), and *Wasserzeichen* (1964). His narrative work has some resemblance to that of Kafka (q.v.), and both Surrealism and Existentialism have played a part in it. In 1947 his novel *Die Stadt hinter dem Strom* (q.v.) attracted attention by its combination of reality and ghostliness, and its sense of desolation. It was followed by *Das große Netz* (q.v., 1952). Kasack's shorter fiction includes *Der Weberstuhl* (1949), to which *Das Birkenwäldchen* was added on reprinting in 1958, and *Fälschungen* (1953). Kasack is also the author of plays, including one on van Gogh (*Vincent van Gogh*, 1934), and essays (*Mosaiksteine*, 1956).

KASCHNITZ, MARIE LUISE, pen-name of Freifrau Marie Luise von Kaschnitz-Weinberg, *née* von Holzing-Berstett (Karlsruhe, 1901–74, Rome), the widow of a baron who held a professorship in archaeology, and the daughter of a Baden nobleman. The position of her husband, which entailed years of residence in Rome (1924–32, 1955–8), made her familiar with the classical world and its setting. Between the wars she wrote two novels, both concerned with the love problems of young women and related in part to her own life: *Liebe beginnt* (1933) and *Elissa* (1937). After the 1939–45 War she emerged as a lyric poet who combined Christian faith with an unflinching vision of the contemporary world. She adopted traditional forms, including the sonnet, but developed a verse style of her own. Her first volume, *Gedichte*, was followed by *Totentanz und Gedichte zur Zeit* (both 1947), and *Zukunftsmusik* (1950), *Ewige Stadt* (1952), *Neue Gedichte* (1957), *Dein Schweigen—meine Stimme* (a moving collection inspired by her husband's death, 1962), the volume *Ein Wort weiter* (1965), which sounds a note of hope, and the selection from the years 1928–65, *Überallnie* (1965).

Marie Luise Kaschnitz is also the author of short narrative works, notably the collection *Lange Schatten* (1960), short stories of great variety but with a predominantly lyrical tone. *Wohin denn ich* (1963), though fiction, has autobiographical elements. *Beschreibung eines Dorfes*, also autobiographical, is an experiment in the technique of prose composition, and was followed by *Ferngespräche* (both 1966), and *Vogel Rock: unheimliche Geschichten* (1969). *Griechische Mythen* (1943) and *Menschen und Dinge* (1945) are collections of contemplative essays, and

Hörspiele (1962) contains her radio plays. *Haus der Kindheit* (1956), *Tage, Tage, Jahre* (1968), and *Orte. Aufzeichnungen* (1973) are autobiographical works. Marie Luise Kaschnitz spent the last fourteen years of her life in Frankfurt (she died during a visit to her daughter in Rome) and in 1960 was a visiting professor of Aesthetics (Gastdozent für Poetik) at Frankfurt University. Her publications in the 1970s include radio plays, *Gespräche im All* (1971), a collection of partly previously published prose and poetry, *Nicht nur von hier und heute* (1971), and *Eisbären* (1972). *Gesang vom Menschenleben* (1974) is her last volume of poetry. *Zwischen Immer und Nie. Gestalten und Themen der Dichtung* (1971) is a collection of essays on a variety of plays and poems, including *Hälfte des Lebens* by Hölderlin and *Die Wildente* by Ibsen (qq.v.), and on Ibsen and Tennessee Williams, as well as *Schwierigkeiten, heute die Wahrheit zu schreiben*. Mainly concerned with 'truth in art' (künstlerische Wahrheit), the essay insists on the writer's critical commitment to historical and sociological experience as well as on his own individual integrity. Even in irrational poetry both must be recognizable. The recipient of many honours, she was awarded the Büchner Prize in 1955 and the Orden pour le Mérite in 1967.

Gesammelte Werke (7 vols.), ed. C. Büttrich and N. Miller, appeared 1981ff.

Käserei in der Vehfreude, Die, a novel by J. Gotthelf (q.v.), written in 1849, and published in 1850. Sub-titled *Eine Geschichte aus der Schweiz*, it is concerned less with individuals than with a whole Swiss village known as 'Die Vehfreude'. The village elders decline to finance a new school, deciding instead to build a co-operative cheese factory, an innovation then being widely introduced in Switzerland. The first year of the Käserei is traced, with much instructive detail, from the planning stage, through the building, the appointment of the operator (Senn), the buying of suitable cows, the making of the cheeses, the sale of the cheeses with all its attendant difficulties, their delivery, and the problem of feeding the large number of cattle in the ensuing winter. The book is diversified by a wide range of characters, careful farmers, improvident farmers, the well-to-do magistrate (Ammann), the village mischief-maker and crook Eglihannes, the wild young men led by Felix, the stupid farmer Peterli, and his pathological termagant of a wife, Eisi.

As the novel proceeds, a thread of romance is spun. Felix, the son of the Ammann, is attracted to Änneli, the modest sister-in-law of Sepp, the Nägelibodenbauer, not one of the richest but one of the soundest farmers in the village. Though she is below Felix socially and financially, he persuades his parents to consent to the marriage. Among many lively and vigorous scenes, one of the best occurs when Felix, leading a procession of waggons returning from market, tries with his four-horse team to overtake Eglihannes's vehicle at a spanking trot; he collides with it in the dusk, knocking down Änneli, who is making her way home on foot, and happily escapes serious injury. Also remarkable are the scenes presenting the periodical meetings of the co-operative, in which Gotthelf relentlessly, yet amusingly, examines the competing interests at loggerheads in such deliberations.

Kaspar, a play by P. Handke (q.v.), published in 1967. A grotesque piece, and Handke's major Sprechstück, it is based on K. Hauser (q.v.) and consists of two parts and 65 numbered scenes. Kaspar, resembling Frankenstein's monster (or King Kong) in appearance and a puppet (Marionette) in movement, is subjected to language exercises by anonymous speakers (Einsager). Together with silent Kaspar figures speaking through poses they represent society, a member of which he feels himself at the end of the first part. In the second, his reflections on his achievement reveal his autonomous self.

The technique deliberately attacks established dramatic concepts. The verbal game makes non–sense of the clichés of established society, and the 'hero' (HELD) is manipulated by disembodied voices speaking through a microphone to the beat of rock music. Because of its cacophonous effect, Handke suggests 'Sprechfolterung' as an alternative title. The stage accessories are basic living-room objects, but the chairs are arranged in a disorder that indicates Kaspar's disorientated state of mind and, as objects of his self, are rearranged accordingly. Speech evolves in direct relation to objects, syntax, and grammar; an adaptation of the historic phrase 'Ich möcht ein solcher werden wie einmal ein andrer gewesen ist' forms the play's leitmotif.

Kasperltheater, a form of Viennese popular comedy originated by the actor J. J. Laroche (q.v.). Laroche created Kasperl or Käsperle, the apparently stupid, but in reality sly and mischievous, servant in 1781 at the Leopoldstädter Theater, which for a time acquired the nickname Kasperletheater. The figure survived into the 19th c., and was eventually taken over by puppet shows.

Kassandra, a story by C. Wolf (q.v.). Published in 1983, it draws on the legend of the Greek prophetess, whose warnings were left unheeded, and consists of the recollections of Priam's daughter when facing death, a perspective that determines the narrative stance against the background of the protracted Trojan War and its preliminaries. The plight of other women as well as her own inspires Cassandra to probe its cause. She 'sees' through experience, the most harrowing of which is Achilles' sadistic murder of Troilus; other male representatives include Anchises, whom she admires as a 'free man' with balanced judgement, and the man closest to her, Aeneas, who has fathered her twin children and offered her help to begin a new life, away from the ruins of Troy. Despite fear, the story's pervasive leitmotif, she is unrepentant, for she cannot associate with one who will be proclaimed hero. She also stands by her refusal, anticipating her death, to keep the betrayal of Paris secret; her protest is directed against the misuse of Polyxena by a male war council. She feels akin to Penthesilea, who fought all men in a spirit of hopelessness. Agamemnon's sacrifice of Iphigenia elicits the query whether victimization can be turned into meaningful sacrifice. Her acceptance of death, in her own mind a free decision, poses questions for posterity. Stylistic devices support the actuality of the story's arguments, which are more fully explained in Wolf's *Poetik-Vorlesungen* of 1983.

Kasseler Glossen, see VOCABULARIUS STI. GALLI.

KASSNER, RUDOLF (Großpawlowitz, Moravia, 1873–1959, Sierre, Switzerland), a cultural philosopher who propagated his eclectic views on human life and art, beginning with neo-Romanticism at the end of the 19th c., and gravitating in later years to Christianity. Among his numerous books are *Die Mystik, die Künstler und das Leben; über englische Dichter und Maler im neunzehnten Jahrhundert* (1900), *Die Moral der Musik* (1904), *Von den Elementen der menschlichen Größe* (1911) and *Die Chimäre* (1914), *Zahl und Gericht* (1919), *Die Grundlagen der Physiognomik* (1922), *Von der Einbildungskraft* (1936), *Der Gottmensch* (1938), *Die Geburt Christi* (1951), and *Das inwendige Reich: Versuch einer Physiognomik der Ideen* (1953). 'The major preoccupation of Kassner's physiognomics is nothing other than a synthesis of the Christian and the pagan' (E. C. Mason). His autobiography is given in the *Buch der Erinnerung* (1938), *Die zweite Fahrt* (1946), and *Umgang der Jahre* (1949). In earlier years he published a volume of short stories, *Der Tod und die Maske* (1902). Kassner's life-long friendship with Rilke (q.v.) is reflected in *Rilke. Gesammelte Erinnerungen 1926–56* (1976). *Sämtliche Werke,* ed. E. Zinn, appeared in 1969 ff.

KÄSTNER, ERHART (Augsburg, 1904–74, Staufen, Breisgau), a librarian in Dresden from 1927 to 1936, was secretary to G. Hauptmann (q.v.) 1936–8. During the 1939–45 War he was in the army, serving in Greece and Crete. Taken prisoner in the closing stages, he was held in Egypt 1945–7. From 1950 to 1968 he was director of the Herzog August-Bibliothek at Wolfenbüttel (a post Lessing, q.v., held nearly two centuries before). He is the author of some attractive travel books, *Griechenland* (1942, republished in 1953 as *Ölberge, Weinberge*), *Kreta* (1946), *Die Stundentrommel vom heiligen Berg Athos* (1956), and *Die Lerchenschule. Geschrieben auf der Insel Delos* (1964). He recorded his experience as a prisoner of war in *Zeltbuch von Tumilad* (1949).

KÄSTNER, ERICH (Dresden, 1899–1974, Munich), served in the army 1917–18, studied Germanistik (q.v.) at Berlin, Rostock, and Leipzig universities, and launched into authorship in 1927. An individualist with left-wing sympathies, intolerant of humbug, and possessed of a sharp and critical eye, he began with two volumes of ironical and humorous poems (*Herz auf Taille*, 1927; *Lärm in Spiegel*, 1929). In *Emil und die Detektive* (q.v., 1929) he created an international boys' and girls' book, which moved inside the children's own world. Poems, novels, and children's books followed in succession: the volume of poetry *Ein Mann gibt Auskunft* (1930), the boy's book *Pünktchen und Anton*, and the satirical novel *Fabian* (q.v., both 1931). Kästner maintained this rhythm of production in Zurich, where he settled after 1933, in which year he was banned from writing by the National Socialist regime; his books were publicly burnt. Before the outbreak of the 1939–45 War Kästner nevertheless returned to Germany, living in Dresden, and after the war settled in Munich, preferring the Bundesrepublik, though maintaining his contacts with writers of the left, among them Brecht (q.v.).

Kästner's poetry includes the volumes *Herz auf Taille* (1928), *Gesang zwischen den Stühlen* (1932), *Lyrische Hausapotheke* (1936), *Der tägliche Kram* (chansons and prose, 1948), *Die kleine Freiheit* (1952), and *Die 13 Monate* (1955). His children's books and childhood recollections continued to make a wide appeal: *Das fliegende Klassenzimmer* (1933), *Das doppelte Lottchen* (1949), *Die Konferenz der Tiere* (1949), and *Als ich ein kleiner Junge war* (1957). Further novels include *Drei Männer im Schnee* (q.v., 1934), *Die verschwundene Miniatur* (1936), and *Der kleine Grenzverkehr* (1938), a comic narrative of confused identity. All his best-known works have been filmed. Kästner's satirical comedy *Die Schule der Diktatoren* appeared in 1957, the year

in which he was awarded the Büchner Prize. In 1960 appeared *Notabene 45*, a diary and a preliminary to an extensive work on Hitler's Germany which did not materialize. But in this year he was awarded the Hans Andersen Prize, an international prize for children's books. *Gesammelte Schriften* (7 vols.) appeared in 1959 and *Gesammelte Schriften für Erwachsen* (8 vols.) in 1969; correspondence, *Mein liebes gutes Muttchen Du! Briefe aus dreißig Jahren*, appeared in 1981.

Katalaunische Felder, see HUNNENSCHLACHT.

Kategorische Imperativ, Der, a comedy in three acts and a prelude (Vorspiel) by E. von Bauernfeld (q.v.), first performed in the Burgtheater, Vienna, in March 1850. The prelude begins in the spring of 1815 at the time of the Congress of Vienna (see WIENER KONGRESS), and serves to mock the Viennese political police. The play has a simple plot ending in a double marriage. Countess Flora, a young and coquettish, though virtuous, widow, indignantly repulses the amorous advances of the supposed Don Juan, Colonel Wildenberg. But Napoleon returns to France, and war breaks out, administering a psychological shock to both parties. When Wildenberg returns wounded, Countess Flora accepts him. Her cousin Elise marries the German Burschenschafter Lothar, who seeks to live according to Kant's Categorical Imperative (see KANT, I.). The play was quickly seen to have symbolical significance, Lothar with his sense of German, not Austrian or Prussian, nationality representing the patriotic 'Bürger' of the coming age. The important incidental figure of the genial Baron, who believes that money can move mountains and control men's minds ('Nun, Geld regiert den Geist—das versteht sich von selber'), is a caricature of the banker Salomon Rothschild. The line 'Deutsch müssen wir endlich werden, da wir's immer noch nicht sind' became a 'geflügeltes Wort'.

Kategorischer Imperativ, see KANT, IMMANUEL.

Kater Murr, frequently used abbreviated form of the title of E. T. A. Hoffmann's novel *Lebensansichten des Kater Murr nebst fragmentarischer Biographie des Kapellmeisters Johannes Kreisler* (q.v.).

Käthchen von Heilbronn, Das, oder die Feuerprobe, a five-act play in blank verse (ein großes historisches Ritterschauspiel) by H. von Kleist (q.v.), published in 1810 after its first performance in Vienna (Theater an der Wien), the first two acts having appeared separately, with variants, in *Phöbus* (q.v.) in 1808. The only German performance during Kleist's lifetime took place in Bamberg after both the Dresden Hoftheater and the Berlin Schauspielhaus had rejected the MS. Kleist made free use of *Graf Walter* by G. A. Bürger (q.v.), who adapted *Childe Waters* from Percy's *Reliques*. The action is set in Swabia. Käthchen, the daughter of a Heilbronn armourer, Theobald Friedeborn, acquires through a visionary dream the unshakable conviction that she will marry a count, Friedrich Wetter, Graf vom Strahl. She follows him against his will, though she cannot prevent his engagement to Kunigunde von Thurneck. But Kunigunde's jilted suitor, the Rheingraf vom Stein, attacks the castle of Thurneck, setting it on fire. At Kunigunde's behest Käthchen enters the burning building to rescue a portrait, though Strahl protests. Käthchen succeeds, and miraculously emerges from the ordeal unharmed. This episode (Act Three) explains the sub-title, *die Feuerprobe*. But Strahl does not admit his true love for Käthchen until the Emperor reveals that she is his illegitimate daughter, and Theobald her foster-father. Kleist adds Romantic motifs to the traditional Ritterschauspiel.

Käthi die Großmutter, oder Der wahre Weg durch jede Not, a novel by J. Gotthelf (q.v.), written in 1846, and published in 1847. It is described as 'Eine Erzählung für das Volk'. The events take place in 1845 and 1846. Käthi is a grandmother of 70. Her children have died, except for a daughter, who has married into a remote district, and a son, Johannes, a widower who works as a farm labourer a few miles away. Käthi, a hard-working woman of great goodness of heart and humanity, devotes her meagre resources to bringing up Johannes's 5-year-old son, Johannesli. Johannes does little to help, spending what money he has on drink. Käthi's unpretentious goodness and sincerity earn her friends, and she is especially helped by her landlord's wife. Nevertheless, owing to a flood, and then to potato disease, she has the greatest difficulty in making ends meet. When Johannes is injured in a brawl and returns home, poor Käthi tries to feed and look after him as well. He comes to understand her goodness and repents of his past behaviour, but is unable to make full amends, because his injury hampers him. His opportunity, however, comes when the River Emme again overflows its banks; he saves Käthi by carrying her through the waters on his back. Mother and son are finally extricated from their poverty when Johannes marries Bäbeli, a girl with good temper and a handsome portion, whom he had first met on the day of the brawl.

Katholische Liga, Die, was formed in July 1609 in Munich by Maximilian I (q.v.) of Bavaria in answer to the provocative formation of the Protestant Union (see PROTESTANTISCHE UNION). Its members included the Catholic bishops of south Germany (Würzburg, Augsburg, Constance, Passau, and Regensburg) and, by 1619, most Roman Catholic Estates. It had the backing of the Pope and of Spain, and was a valuable instrument of the Counter-Reformation (see GEGENREFORMATION) and the Habsburg dynasty. The League took a prominent part in the Thirty Years War (see DREISSIGJÄHRIGER KRIEG), in which its forces were commanded by Count Tilly (q.v.). It was dissolved through the Peace of Prague in 1635.

KATTE, HANS HERMANN VON (Berlin, 1704–30, Küstrin), a son of the Prussian Field-Marshal H. H. von Katte, a lieutenant, and a close friend of Friedrich II (q.v.), when the latter was still Crown Prince. He was an accomplice to Friedrich's plan to flee the country in despair over humiliations inflicted upon him by his father (see FRIEDRICH WILHELM I). When the plan came to the King's notice, Katte was ordered to be tried by court martial and condemned to be beheaded in the courtyard of Küstrin where Friedrich was imprisoned. Friedrich was ordered to witness the execution from the window, to which Katte bowed as a final token of his loyalty. An essay on Katte is included in vol. 2 of Th. Fontane's *Wanderungen durch die Mark Brandenburg (Die Katte-Tragödie)* (q.v., 1870). He is also the subject of a drama (*Katte*) written in 1914 by Hermann Burte (q.v.).

Katzensteg, Der, an historical novel by H. Sudermann (q.v.), published in 1889. Set in East Prussia between the abdication of Napoleon in April 1814 and the battle of Ligny in June 1815 (see NAPOLEONIC WARS), it is the story of a son's attempt to overcome the disrepute into which his tyrannical father has brought the family name. Baron von Schranden dies execrated by his tenants, because in 1807 he had connived at a Prussian defeat by showing French troops the way across the brook by the footbridge (der Katzensteg). Boleslav, his son, has sought to expiate this act by serving, under the false name of Baumgart, in the War of Liberation. The peasants refuse to bury squire Schranden, and Boleslav, helped by the maid Regine, does it with his own hands. He despises Regine as his father's creature but, faced with the brutality of the peasantry, he comes to recognize her sterling qualities, though the full realization only comes at the moment when Regine is shot by her father. He goes off to the new war and falls at Ligny. In 1917 Sudermann adapted *Der Katzen-*

steg as a play (deutsches Volksstück in fünf Akten und einem Vorspiel).

KAUFMANN, CHRISTOPH (Winterthur, 1753–95, Gnadenfeld, Silesia), an enthusiastic follower of J. C. Lavater (q.v.), was trained as a physician. He became an extravagant propagandist for the religious aspect of the Sturm und Drang (q.v.), and after receiving initial support was widely derided. Goethe mocked him as 'Gottes Spürhund', and F. M. Klinger (q.v.) made him the absurd hero of *Plimplamplasko, der hohe Geist* (1780). He has also frequently been termed a 'Kraftapostel'. In this century an attempt has been made to rehabilitate Kaufmann's reputation (W. Milch, 1932). Kaufmann coined the term 'Sturm und Drang' as a better title for F. M. Klinger's *Der Wirrwar* (1776), and thereby unintentionally provided the designation by which the Geniezeit has since generally been known.

KAUFRINGER, HEINRICH, a Bavarian commoner poet born at Landsberg probably in the second half of the 14th c. He wrote some thirty narrative poems of various kinds, including crudely comic anecdotes, moral stories, and religious parables. He also versified a few religious prose works by others, notably a sermon by BERTHOLD von Regensburg (q.v.). His style is in part derived from KONRAD von Würzburg and Der Teichner (qq.v.).

Kaukasische Kreidekreis, Der, a play by B. Brecht (q.v.) with music by P. Dessau (q.v.). It was written in 1943–4 after the story *Der Augsburger Kreidekreis* (q.v.), and was first performed in Nourse Little Theatre, Northfield (Minnesota) in 1948 (German première at the Theater am Schiffbauerdamm, Berlin, in 1954), and published in 1954. The play incorporates an adaptation of the 13th-c. Chinese parable figuring in *The Chalk Circle* by Li Hsing Tao, which in 1925 was adapted by Klabund (q.v.); the parable is also prefigured in Solomon's judgement (1 Kings 3:16–28). The play consists of six episodes, the first introducing the parable (the play within a play) which in turn falls into two parts: the Grusche action (episodes two to four), and the story of the judge, Azdak (episode five). The last episode brings the two together.

The play opens in post-war Russia. Two villages dispute the ownership of a valley after its liberation from German occupation. The claims of the Galinsk commune, to whom the valley previously belonged, are challenged by the Rosa Luxemburg (q.v.) commune, which has designed an irrigation system enabling it to turn the valley into fertile land for orchards and vineyards. It is agreed that the latter claim is the stronger one. To illustrate the transfer of the valley to the more productive and useful commune, the Sänger Arkadi Tscheidse and a group of actors stage a play depicting the age of tyranny in a distant past.

On an Easter Day a group of minor princes revolt against their rulers. Prince Arsen Kazbeki orders the execution of the powerful governor of Grusinien. His wife Natella flees, deserting her small child Michel. But her maid Grusche saves him from persecution and looks after him for two years, when the ruler of Grusinien returns to power, and Natella reappears to claim her child. As Grusche refuses to hand him over, the judge Azdak has him placed in the middle of a chalk circle from which the two women are to claim him. Natella wrests her son from Grusche, but the judge gives Grusche possession of the child: in refusing to harm the boy she has proved herself the better mother. Azdak is satisfied of her ability to make the child a useful member of the community. Her greatest sacrifice for the child was her marriage to the farmer Jussup. Azdak divorces her, enabling her to marry her love, the soldier Simon Chachava, who has returned from the war. The fifth episode reverts to the time at which the Grusche action opens, to show how Azdak, the village clerk, becomes the judge of the poor, and how he defends their rights (including his own) by the unwritten laws of his social conscience and by his wit. At the end of the play he vanishes into anonymity. While Brecht develops Grusche (against his original intentions) from a simpleminded maid into an ideal mother, Azdak becomes increasingly a 'realist': he is content not to sacrifice his life in an act of senseless heroism for his socialist principles. In his *Materialien zum Kaukasischen Kreidekreis* (1966) Brecht comments that the inner action is not a true parable (although it contained a 'certain kind of wisdom'), thus justifying the absence of a final episode (Nachspiel) to balance the outer action of the first episode (originally the Vorspiel).

KAUNITZ, WENZEL ANTON, GRAF VON (Vienna, 1711–94, Vienna), was from 1753 until 1792 state chancellor of Austria and the chief architect of home and foreign policy. After an education that included travel to France, the Low Countries, and Italy, he entered the civil administration in 1735. He first made his mark as Austrian representative at the peace negotiations at Aachen in 1748 (see ÖSTERREICHISCHER ERBFOLGEKRIEG), when his ability earned the regard of Maria Theresia (q.v.), despite the modest results of his efforts. In the following years he sought an alliance with France, preparatory to an attempt to regain Silesia from Prussia, and was eventually successful in winning

French support in 1755 (see DIPLOMATISCHE REVOLUTION). The Seven Years War (see SIEBENJÄHRIGER KRIEG), however, ended with the coalition in ruins, and the Peace of Hubertusburg marked the failure of Kaunitz's foreign policy. He nevertheless retained the confidence of the Empress, and in his later years succeeded in adding important new territories to the Austrian dominions, including large tracts of Poland in 1772 (see POLAND, PARTITIONS OF) and the Innviertel in 1779 (see BAYRISCHER ERBFOLGEKRIEG). In his conduct of foreign affairs in the last decade of Maria Theresia's reign, he strengthened the ambitions of her co-regent, the Emperor Joseph II (q.v.), against the wishes of the Empress, seeking to maintain Austrian prestige against Prussian rivalry.

Kaunitz was created Prince (Fürst) von Kaunitz-Rittberg in 1764. He was a man of considerable eccentricity (he could not abide fresh air and took exercise only indoors), but he did not allow his vanity or his foibles to prejudice his service to the state.

'Keinen Tropfen im Becher mehr', first line of a student drinking song entitled *Der Lindenwirt*, written in 1876 by R. Baumbach (q.v.), and sung to a tune by Franz Abt (1819–85).

'Kein Feuer, keine Kohle', first line of a folksong first recorded c. 1790. It bears the title *Heimliche Liebe*.

Kein Hüsung, a short epic poem by F. Reuter (q.v.), published in 1857. It contains 13 cantos with rhyming verse and short lines written in the Low German dialect of Mecklenburg (Mecklenburger Platt). It is set in Mecklenburg at a time when serfdom (abolished in 1820) was still in force. Jehann and Mariken cannot marry until they have a domicile, and the squire, having a score to settle against Mariken's father, will do nothing to help. Jehann, provoked by the squire, kills him with a pitchfork and flees. Mariken is left with her child and presently takes her own life. A few years later Jehann returns and claims his son. In its strong sense of social injustice the poem has affinities with J. H. Voß's *Die Leibeigenen* (q.v.).

'Kein schönrer Tod ist auf der Welt', first line of a song extolling death in battle. It is an adaptation, made c. 1820, of the stanza beginning 'Kein seeligr Tod ist in der Welt' from J. Vogels's *Ungrische Schlacht* (1625). The tune is by Silcher (q.v.). Its popular appeal has promoted sentimental nationalism.

KELLER, GOTTFRIED (Zurich, 1819–90, Zurich), was the son of a turner, who had married a doctor's daughter. His father died in 1824, and his mother maintained the family by letting apartments in her substantial house. Keller was expelled from school in 1834, an act which he resented as an injustice. Its echoes are heard in *Der grüne Heinrich* (q.v.). He resolved to become a painter, and was sent for a time to live among his numerous relations around Glattfelden, his father's birthplace. On his return he had instruction from two teachers in Zurich, Peter Steiger and Rudolf Meyer. In 1840 he realized his modest assets and established himself in Munich, at that time the artistic metropolis of Germany. His efforts to develop as a painter were frustrated by poverty and, though he did not realize this at the time, by insufficiency of talent. He returned home in 1842.

In the 1840s Keller's literary talent developed; he wrote many poems, some of faded romantic character but mostly political or reflective. He was at this stage a keen Radical, admiring the poetry of G. Herwegh and A. Follen (qq.v.). He was also active in journalism, notably as an art critic. In 1846 a volume of his poems appeared as *Gedichte,* and a further volume followed in 1851 (*Neuere Gedichte*). In 1848 Keller received a grant from the cantonal government to enable him to study at a university, and he spent three semesters at Heidelberg, from the autumn of 1848 to the spring of 1850. This period was decisive in his development. The lectures of L. Feuerbach (q.v.) undermined his belief in Christianity, and a friendship with H. Hettner (q.v.), then a young lecturer, directed his attention to Goethe and to the drama. In April 1850 he travelled to Berlin, where he was to remain for five years. He did not like the city, and his literary contacts were mostly fleeting (they included Scherenberg and the members of Der Tunnel über der Spree, including Th. Fontane, qq.v.), but he received some help from VARNHAGEN von Ense (q.v.).

Keller's intention had been to become a dramatist. He did not complete a single play, but he discovered his capacity for fiction. Not only did he write the massive novel *Der grüne Heinrich*, which appeared in its first version in 1854–5, he also began his most popular work, the collection of Novellen *Die Leute von Seldwyla* (q.v., 1856). This contains five stories (the remainder did not appear until 1873–4). Late in 1855 Keller returned to Zurich, where his Novellen had established his reputation as a writer. He did not leave Zurich again. He maintained a correspondence with several men of letters, especially Th. Storm and P. Heyse (qq.v.).

In 1861 Keller was appointed clerk to the canton of Zurich (Erster Staatsschreiber), and for fifteen years he meticulously discharged the duties of his office. His writings in these busy years were sparse, consisting essentially of the charming *Sieben Legenden* (q.v., 1872) and the re-issue of *Die Leute von Seldwyla* with the additional volumes of 1874. He resigned in 1876 in order to devote himself to writing, publishing the 2 vols. of *Züricher Novellen* (q.v., 1878–9), the revised version of *Der grüne Heinrich* (1879–80), *Das Sinngedicht* (q.v., 1881–2), and his second and last novel, *Martin Salander* (q.v., 1886). His collected poems appeared as *Gesammelte Gedichte* (1883).

Keller did not marry, though in 1866 he became engaged to a young woman of 22, who drowned herself in the same year, apparently in consequence of doubts about Keller's character provoked by slanders published in newspapers which were politically hostile to him. That he was susceptible to the attractions of women is clear from his works. He proposed (by letter) to two young women, Luise Rieter in 1847 and Johanna Kapp in 1849. Both declined, and he discovered that Johanna and Feuerbach, a married man, were in love. In Berlin he was attracted to Betty Tendering, the unmarried sister of the wife of Duncker, his publisher. He was not able to bring himself to declare his love for her. She is generally regarded as the original of Dortchen Schönfund in *Der grüne Heinrich*. Keller's mother, born in 1787, lived with her son in Zurich; after her death his sister, who died in 1888, kept house for him.

If Keller's philosophical ideas often seem austere, his personality, as his works suggest, was lively and humorous, and possessed a tenderness which he often disguised. His range encompasses profound passion and grotesque comedy, political preaching and sensitive legends, providing a conspectus of human nature. He is one of the foremost representatives of Poetic Realism (see POETISCHER REALISMUS). *Historisch-kritische Ausgabe* (22 vols.), ed. J. Fränkel and C. Helling, appeared 1926–49, and *Gesammelte Briefe* (4 vols.), ed. C. Helbing, 1950–4.

KELLER, PAUL (Arnsdorf, Silesia, 1873–1932, Breslau), a primary schoolteacher, took a post in Breslau in 1898, and resigned in 1908 in order to devote himself to writing. In 1912 he founded a periodical, *Die Bergstadt*. A Catholic and a local Silesian writer, he reached a wide public with well-told, often atmospheric, somewhat sentimental novels, usually with a Silesian background. Among them are *Waldwinter* (1902), *Die Heimat* (1903), *Der Sohn der Hagar* (1907), *Die alte Krone* (1909), *Die Insel der Einsamen* (1913), *Ferien vom Ich* (1915), *Hubertus* (1918), and *Drei Brüder suchen das Glück* (1929). Mention should also be made of the volumes of stories *In*

deiner Kammer (1903), *Die fünf Waldstädte* (1910), and *Stille Straßen* (1912).

KELLERMANN, BERNHARD (Fürth, 1879– 1951, nr. Potsdam), a journalist and foreign correspondent, wrote five novels between 1904 and 1913. *Yester und Li* (1904), sub-titled *Geschichte einer Sehnsucht*, records a poet's unexpressed and unfulfilled love. *Ingeborg* (1906), set in Russia, is also a story of frustrated love, and so, too, is *Der Tor* (1909), whose central figure, Vikar Grau, is attracted simultaneously to two women, loses one by death, and is rejected by the other. *Das Meer* (1910), set in Brittany, is a story of love and seduction against the background of the ocean. *Der Tunnel* (1913), Kellermann's one international best-seller, is a sensational technological work with elements of crude satire. During the 1914–18 War Kellermann was a war correspondent, writing, in addition to his newspaper columns, two longer accounts, *Der Krieg im Westen* (1915) and *Der Krieg im Argonner Wald* (1916), which are emotional and chauvinistic. *Der neunte November* (1920, the title refers to November 1918) and *Die Brüder Schellenberg* (1925, about a materialistic society) are satirical novels.

Kellermann was banned by the National Socialist regime, and after the 1939–45 War established himself in East Berlin, founding the League for the Democratic Renewal of Germany (Kulturbund zur demokratischen Erneuerung Deutschlands). *Aufsätze, Briefe, Reden 1945–51* appeared in 1952. Kellermann's novels include *Die Stadt Anatol* (1932), *Das blaue Band* (1938), *Georg Wendtlandts Umkehr* (1941), and *Totentanz* (1948). His only play deals with an episode in the Reformation (q.v.): *Die Wiedertäufer von Münster* (1925). His selected works, *Ausgewählte Werke in Einzelausgaben*, ed. E. Kellermann and U. Dietzel (until 1973), appeared 1958 ff.

KERNER, JUSTINUS (Ludwigsburg, 1786–1862, Weinsberg), in full Justinus Andreas Christian, was the son of a Württemberg official, who died when the boy was 13. The ensuing financial difficulties led to Justinus being apprenticed in 1801 to a cabinet-maker. Professor K. P. Conz (q.v.), a minor poet himself, became acquainted with Kerner's early poetic efforts, and enabled him to abandon craftsmanship for university study. In 1804 Kerner matriculated at Tübingen, deciding to study medicine. Throughout his life he was a rich and eccentric personality and became the centre of a group of friends, including Ludwig Uhland (a distant kinsman), Karl Mayer, Gustav Schwab, and VARNHAGEN von Ense (qq.v.). During his years of medical study he was instructed to observe Hölderlin (q.v.), then living mentally deranged in Tübingen. After qualifying in 1808, he made a German tour with the object of visiting his elder brother, a physician in Hamburg. In Berlin he met, among the prominent names of Romanticism, F. de la Motte Fouqué and A. von Chamisso (qq.v.), and in Vienna he became acquainted with F. Schlegel and his wife Dorothea (qq.v.). On his return to Württemberg in 1810 he began to practise in Wildbad, moving in 1812 to Welzheim. In 1813 he married Friederike Ehmann. The beginning of their courtship in 1807 makes a romantic anecdote. Observing, on an excursion, the sad demeanour of a young woman, Kerner is said to have addressed to her Goethe's stanza from *Trost in Tränen*: 'Wie kommts, daß du so traurig bist,/Da alles froh erscheint?/Man sieht dirs an den Augen an,/ Gewiß du hast geweint?'; she rejoined with the second stanza: 'Und hab ich einsam auch geweint,/So ists mein eigner Schmerz,/Und Tränen fließen gar so süß,/Erleichtern mir das Herz.' Their engagement, it is said, was the consequence of this poetic dialogue.

Various poems by Kerner were published in magazines, but his first book was *Reiseschatten* (1811), one of the more eccentric and subjective Romantic novels, which includes not only a number of incidental poems but also a play entitled *Nachspiel der zweiten Schattenreihe oder Der Totengräber von Feldberg*. In 1819 Kerner settled in Weinsberg as district physician, and he remained there for the rest of his life, pursuing occult as well as medical interests, devoting himself to spiritualism and studying closely the psychic case of one Friederike Hauffe, about whom he wrote a book, *Die Seherin von Prevorst* (1829). His original and expansive personality attracted many visitors from all ranks of society, and he repeatedly enlarged the house he had built in 1822 (Kernerhaus) to accommodate them. D. F. Strauß and Mörike (qq.v.) were frequent callers. He published his recollections in *Bilderbuch aus meiner Knabenzeit* (1849). His poetry, which is of a predominantly sombre tone, shows the influence of folk-song. Most of it belongs to his early years, and was collected in *Gedichte* (1826). Later collections were *Letzter Blumenstrauß* (1852) and *Winterblüte* (1859). A curious posthumous publication was *Kleksographien* (1890), a collection of grotesque drawings based on the folding of paper containing blots of ink. Among his best-known poems are 'Poesie ist tiefes Schmerzen', *Der Wanderer in der Sägemühle*, *Wer machte dich so krank?*, and *Wanderlied* opening with the line 'Wohlauf! noch getrunken'. *Sämtliche poetische Werke in vier Bänden*, ed. J. Gaismaier, appeared in 1905.

KERR, ALFRED, name adopted by Alfred Kempner (Breslau, 1867–1948, Hamburg). It is not a normal pseudonym, for the surname is used by his descendants. Kerr was one of the most influential dramatic critics of the years 1895–1920, writing on works which appealed to him with rare insight, and also with elegance, wit, and irony. He was one of the most forceful supporters of Naturalism, and of freedom in the theatre. His criticisms appeared in *Der Tag, Die neue Rundschau* (q.v.) and *Das Berliner Tageblatt*. From 1912 to 1914 he edited the periodical *Pan* (q.v.). In 1933 he emigrated, settling in London.

In 1904 Kerr published *Schauspielkunst*, and in 1917 a selection of his criticism, *Die Welt im Drama* (5 vols.). A much travelled man, he was the author of *O Spanien* (1924), *Yankeeland* (1925), *Es sei wie es wolle, es war doch so schön!* (1928), and *New York und London* (1923). His poetry includes the collections *Die Harfe* (1917), *Caprichos* (1926), *Melodien* (1938), and the posthumous edition *Gedichte* (1955). The twelve poems of *Der Krämerspiegel* (1921) were set to music by R. Strauss (q.v.). He expressed his views on the contemporary stage in *Was wird aus Deutschlands Theater?* (1932) and on National Socialism in *Die Diktatur des Hausknechts* (1934). *Ich kam nach England* was published in 1979 and *Mit Schleuder und Harfe. Theaterkritiken aus drei Jahrzehnten* in 1982.

KERSTING, FRIEDRICH GEORG (Güstrow, 1785–1847, Meißen), a painter, mainly of interiors, some of which anticipated the Biedermeier (q.v.) style. Kersting volunteered for war service in 1813, and was later in charge of the painting department of the porcelain factory at Meißen. He was a friend of the painter C. D. Friedrich (q.v.).

KESTEN, HERMANN (Nürnberg, 1900–), was editorial manager of the Kiepenheuer-Verlag, emigrated in 1933 to Holland, directing the Emigrantenverlag Allert de Lange in Amsterdam, and moved in 1940 to the U.S.A. After the 1939–45 War he returned to Europe, eventually settling in Rome. He is the author of novels which satirize the age, and was a bitter opponent of Hitler and of religious intolerance. The novel *Die Zwillinge von Nürnberg* (1947) is a critical conspectus of his times. Kesten has grouped four of his novels, published over a long period, into a tetralogy entitled *Bücher der Liebe. Vier Romane* (1960); they are: (1) *Josef sucht die Freiheit* (1927), which tells of the disillusionment of an idealistic schoolboy through the sexual promiscuity of his mother and sisters; (2) *Glückliche Menschen* (1931), in which the ruthless build their happiness on the despair of others; (3) *Die Kinder von Gernika* (1939), which has at its centre the German air raid on Guernica in the Spanish

Civil War; and (4) *Die fremden Götter* (1949), which presents an irreconcilable clash between an orthodox Jewish father and his daughter, who, owing to war conditions, is brought up as a Roman Catholic and refuses to abandon her faith. Kesten also wrote a trilogy of historical novels set in Spain: *Ferdinand und Isabella* (1936, retitled *Sieg der Dämonen* in 1953), *König Philipp II* (1938, changed to *Ich der König* in 1950), and *Um die Krone. Der Mohr von Kastilien* (1952). *Ein Mann von sechzig Jahren* (1972) deals with a marriage in which there is a difference of twenty-eight years between the spouses, and tells the story (with flashbacks to the 1939–45 War and the persecution of the Jews) through separate monologues by the four people involved in the erotic tangle.

In addition to novels Kesten wrote some plays, Novellen, biographies (*Casanova*, 1952; *Copernicus*, 1953), essays (*Der Geist der Unruhe*, 1959), and recollections (*Meine Freunde, die Poeten*, 1953; *Lauter Literaten*, 1963). In 1964 he edited correspondence of writers opposing the National Socialist regime: *Deutsche Literatur im Exil. Briefe europäischer Autoren 1933–49* (see also EXIL-LITERATUR). *Gesammelte Werke* appeared 1969 ff. He was awarded the Büchner Prize in 1974.

KESTNER, JOHANN GEORG CHRISTIAN (Hanover, 1741–1800, Hanover), while attached to the Reichskammergericht (q.v.) at Wetzlar became engaged to Lotte Buff (q.v.), whom he married in 1773. Kestner served in the administration of Hanover. Goethe, while in Wetzlar in 1772, was friendly with Kestner and in love with Lotte Buff. The friendship withstood the strain, but a partial estrangement followed the publication of *Die Leiden des jungen Werthers* (q.v.).

Ketzer von Soana, Der, a story by G. Hauptmann (q.v.), in which Nature and civilization are sharply opposed. It was published in 1918. The story purports to have been handed in manuscript form to the narrator, and is left unfinished. It concerns a family living wild in a forest in the southern Alps, reputedly brother and sister with their offspring. It recounts the attempt of a young priest to bring the couple into the fold of the faithful. Instead the priest succumbs to the temptations of the flesh. The document breaks off, and the narrator describes an encounter in the forest with a young woman with a child. The reader is left to interpret the close, but it is apparent that Nature has prevailed over civilization.

KEYSERLING, EDUARD, GRAF VON (Schloß Paddern, Kurland, 1855–1918, Munich), a descendant of an ancient noble family originating in Westphalia, grew up in his native Latvian countryside, studied at Dorpat University, and took to writing. He lived in Vienna, in Italy, and, from 1899, in Munich. In 1907 he went blind. He is the author of novels and Novellen, many of which are set in his homeland and in the aristocratic circles he knew. Keyserling is primarily a delicate analyst of love portrayed in circumstances in which the emotions are hemmed in by convention and tradition.

Notable works are the novels *Beate und Mareile* (1903), *Dumala* (1907), *Wellen* (1908), *Abendliche Häuser* (1914), *Am Südhang* (1916), *Fürstinnen* (1917), *Feiertagskinder* (1919), and the collections of Novellen *Schwüle Tage* (1906), *Bunte Herzen* (1908), and *Im stillen Winkel* (1918). At the turn of the century he successfully tried his hand at drama (*Frühlingsopfer*, 1899; *Dummer Hans*, 1901). Keyserling was a first cousin once removed of the younger philosopher Graf Hermann Keyserling (q.v.).

Gesammelte Erzählungen (4 vols.), ed. E. Heilborn, appeared 1922, and *Die gesammelten Werke*, ed. Keyserling-Archiv, 1956 ff. (6 vols.; 2 vols. 1956–8).

KEYSERLING, HERMANN, GRAF (Könno, 1880–1946, Innsbruck), a German-speaking nobleman of the Russian Baltic provinces, studied science, and in 1911–12 made a world tour which provided the basis for his first important work, *Reisetagebuch eines Philosophen* (2 vols., 1919), which persuasively directed European eyes to the values of oriental cultures. Having lost his family fortune in the Russian Revolution, he settled in Darmstadt, and in 1919 married a granddaughter of Bismarck (q.v.). His provocative books, combining cultural history and a personal and anti-rational philosophy, were widely read in Europe and America between the wars (*Europas Zukunft*, *Was uns nottut*, both 1919; *Politik, Wirtschaft, Weisheit*, *Philosophie als Kunst*, and *Schöpferische Erkenntnis*, all 1922; *Die neuentstehende Welt*, *Menschen als Sinnbilder*, both 1926; *Wiedergeburt*, 1927; *Amerika*, 1930; *Das Buch vom persönlichen Leben*, 1936). In the bombing of Darmstadt Keyserling lost his property for the second time. It included his valuable and extensive library. He maintained in Darmstadt the Schule der Weisheit (also destroyed), and edited the annual *Der Leuchter* (1921–2).

Keyserling himself summed up the difference between his philosophy and the systems of his contemporaries in these words: 'sie [i.e. meine Philosophie] geht von der lebendigen Seele im Unterschied vom abstrakten Menschen aus' (*Menschen als Sinnbilder*). He was a first cousin once removed of Graf Eduard von Keyserling (q.v.).

The Keyserling-Gesellschaft was founded

when the philosopher was forty (1920), and re-founded after the 1939–45 War. *Gesammelte Werke*, ed. Keyserling-Archiv, appeared 1958 ff.

KHLESL (also **KLESL** and **KLESEL**), MELCHIOR (Vienna, 1553–1630, Vienna), the son of a Protestant, became a Roman Catholic priest and in 1598 was consecrated bishop of Vienna. From 1599 on he was chancellor to the Archduke Matthias (q.v.). His policy of allowing recognition of Protestantism became particularly influential after Matthias became emperor in 1612. In order to deal more firmly with the Bohemian revolt following his coronation as hereditary king of Bohemia, Ferdinand II (q.v.) had Khlesl arrested in 1618 and brought to the castle of Ambras. After his acquittal (1623) in Rome, he returned in 1627 to Vienna. As Klesel he appears as a character in Grillparzer's tragedy *Ein Bruderzwist in Habsburg* (q.v.).

KHUEN or **KUEN**, JOHANN or JOHANNES (Moosach nr. Munich, 1606–75, Munich), a Roman Catholic parish priest, composed religious poems and hymns for the use of his flock, using language with a dialect flavour. They are contained in *Epithalamium marianum* (1636), *Die geistlich Turteltaub* (1639), and many other volumes, of which the last, *Charismata meliora* (1674), appeared in the year before his death.

Kickelhahn, see GICKELHAHN.

KIERKEGAARD, SØREN AABY (Copenhagen, 1813–55, Copenhagen), trained for the Lutheran ministry, but developed into a 'religious thinker' (his own term) who was severely critical of organized Christianity. His philosophical education was based on Hegel (q.v.), whose system still dominated German and Danish universities, though Hegel himself had died by the time Kierkegaard began his studies. In his late twenties he criticized Hegel's dialectic as an abstraction, and asserted that the 'synthesis' had no reality, since in life either the 'thesis' or the 'antithesis' must prevail ('Either/Or'). Kierkegaard rejected collective thinking, and insisted on the importance of the individual. With this belief, which concentrates on the individual 'existing before God', in immediate relation to God, Kierkegaard combined a profound melancholy verging on despair. The trend of his thought was in favour of a personal religion more intense than Luther's, and against the Church in its collectiveness. Kierkegaard's preoccupation with the self and existence, and the accompanying dread (Angest, German 'Angst') and suffering, made him the father of modern Existentialism (q.v.). Though little attention

was paid to his work until the late 19th c., his influence in the 20th c. is immense. Th. Mann and Kafka (qq.v.) are indebted to Kierkegaard, who has also, either directly or indirectly, affected almost every Existentialist, pessimistic, or nihilistic German writer since 1920.

Kierkegaard's most influential writings are *Enten/Eller* (1843, *Either/Or*), *Frygt og Baeven* (1848, *Fear and Trembling*), *Begrebet Angest* (1848, *The Concept of Dread*), and *Sygdommen til Døden* (1849, *Sickness unto Death*). His influence on theology is best seen in K. Barth (q.v.), that on philosophy in M. Heidegger (q.v.). Though his last work was a violent onslaught on the Church (*Øjeblikket*, 1855, *The Moment*), Kierkegaard's religious writings *Christelige Taler* (1850, *Christian Discourses*) and *Indøvelse i Christendom* (1850, *Training in Christianity*), which have attracted less attention, are as profound and penetrating as his philosophical writings.

KINAU, HANS, see FOCK, GORCH.

KINAU, RUDOLF (Hamburg, 1887–1975, Finkenwerder, nr. Hamburg), a fisherman in his early years, was from 1916 to 1936 clerk in the Hamburg fish market, giving up this post to devote himself entirely to writing humorous Low German (Plattdeutsch) stories and sketches. Many of these first appeared in newspapers. They have been collected as *Sternkiekers* (1917), *Blinkfeuer* (1918), *Thees Bott dat Woterküken* (1919), *Lanterne* (1920), *Strandgoot* (1921), *Hinnik Seehund* (1923), *Dörte Jessen* (1925), *Muscheln* (1927), *Schreben Schrift* (1929), *Frische Fracht* (1931), *Mien bunte Tüller* (1948), and *Sunnschien un gooden Wind!* (1953). He also wrote in High German (*Kamerad und Kameradin*, 1939 and *Ein fröhlich Herz*, 1941). In 1957 he published a short autobiography, *Mit eegen Oogen*. He was popular as a reader of his own works. Hans Kinau (see FOCK, GORCH) was his brother.

KIND, FRIEDRICH (Leipzig, 1768–1843, Dresden), a lawyer practising in Dresden, devoted himself, from 1817, to periodical journalism, editing the Dresden *Abendzeitung* for a time with Theodor Hell (q.v.), and later the monthly *Die Muse*. He is best known as the author of the libretto of Weber's opera *Der Freischütz* (q.v., 1821).

Kinderbuch, Das, a book on the ailments of children written by Bartholomäus Metlinger of Augsburg and printed in 1457. Its two editions were respectively entitled *Wie die kint in Gesuntheit und in krankheiten gehalten werden Sollen* and *Ein regiment der jungen kinder*. It is the first German work on paediatrics.

KINDERMANN, Balthasar (Zittau, 1629–1706, Magdeburg), a Protestant clergyman, who was also active as a schoolmaster, wrote moral satires directed against brutality in schools (*Schoristenteufel*, 1661) and against women (*Die bösen Sieben*, 1661). He is also the author of a *Lobgesang auf das Zerbster Bier* (1658) and a *Lehrbuch der Dichtkunst* (1664). He used the pseudonym Kurandor.

Kindermörderin, Die, a tragedy (Trauerspiel) by H. L. Wagner (q.v.). Published anonymously in 1776, it is in six acts, Wagner's protest against the conventional 'Aristotelean' pattern of five. The action is set in Strasburg. The first act (which the 19th c. held to be too crass to be performed) shows Evchen Humbrecht and her mother inveigled by the officer von Gröningseck into a house of bad repute. Von Gröningseck drugs Frau Humbrecht with a sleeping potion and seduces Evchen. He later repents and promises to marry Evchen, but she believes herself abandoned and kills her child. The play closes as Gröningseck sets out for Versailles in the hope of obtaining a royal pardon for Evchen. Wagner later modified his play so that it could be produced in the theatre. He suppressed the first act and spared the life of the child. This version, entitled *Evchen Humbrecht oder Ihr Mütter, merkt's Euch!*, was performed in Frankfurt in 1779.

The play is characteristic of the Sturm und Drang (q.v.). Though the theme of infanticide is influenced by the Gretchen tragedy in Goethe's *Faust* (q.v.), *Die Kindermörderin* contains a strong element of social criticism chiefly directed against petty class prejudice within the middle class. This element influenced Hebbel's *Maria Magdalene* (q.v.).

Kinderseele, a story by H. Hesse (q.v.), published in 1920, and included in the collection *Weg nach Innen* (q.v., 1931). A first-person narrative, it deals with the psychology of a boy who envies other boys able to cope with life. He dislikes school and authority, having a complex relationship with a father figure, represented by his own father and by God. As in similar fiction by Hesse (e.g. *Unterm Rad*, q.v.), the emotional approach to the themes of guilt and fear is the basis of this portrait of a child's mind, which is endowed with unusual sensitivity and imagination. The boy accepts punishment for minor misconduct which is followed by a complete reconciliation with his father; but his father remains associated with the resentment the boy feels at an inexplicable compulsion to sin.

Kinder- und Hausmärchen, the collection of German stories known in English-speaking countries as *Grimms' Fairy Tales*. Under the influence of the Heidelberg Romantics, Jakob and Wilhelm Grimm (qq.v.) collected more than 200 stories and gave them their literary form, a task in which Wilhelm was particularly successful. The *Kinder- und Hausmärchen* were published in two volumes in 1812 and 1814. A third volume containing notes was added in 1822. Among the better-known stories are *Von dem Fischer un syner Fru, Die Bremer Stadtmusikanten, Rapunzel, Hänsel und Gretel, Aschenputtel* (a version of *Cinderella*), *Von dem Machandelboom, Hans im Glück, Rotkäppchen, Die beiden Künigskinner, Sneewitchen, Rumpelstilzchen,* and *Die Gänsemagd.* Modern editions use High German titles such as *Vom Fischer und seiner Frau, Aschenbüttel, Vom Machandelbaum, Schneewittchen,* etc.

Kinder von Finkenrode, Die, a novel by W. Raabe (q.v.), written 1857–8, and published in 1859. The narrator, a young journalist in the 'great city' (Berlin), learns that his uncle has died in the small provincial town of Finkenrode, and that he is the heir. He returns to Finkenrode, his home town, full of sentimental longings to relive the idyll of his early years. But it proves impossible to revive it, and he has to be extricated from his illusions by the 'realist' editor of his newspaper, Weitenweber. The character Bösenberg reappears as a minor figure in *Alte Nester* (q.v.).

KINKEL, Gottfried (Oberkassel nr. Bonn, 1815–82, Zurich), studied at Bonn University, and lectured for a time in ecclesiastical history in Berlin University. After a grand tour of Europe he became a schoolmaster in Bonn in 1839, and a Protestant curate (Hilfsprediger) in Cologne in 1841. He turned away from the Christian faith, and moved to the left in politics. In 1845 he became professor of cultural history at Bonn University, and edited the *Bonner Zeitung.* From 1840 to 1846 he was, together with Johanna Matthieux (*née* Mockel, 1810–58), whom he married in 1843, prominent in the Rhenish literary group the Maikäferbund (q.v.). During these years he published his first works, the play *König Lothar von Lotharingien* (1842), *Gedichte* (1843), the verse epic *Otto der Schütz* (1846), and *Margret* (1847), a village story.

In the Revolution of 1848 (see REVOLUTIONEN 1848–9) Kinkel was a Republican, took part in the rising in Baden, was wounded, taken prisoner, and condemned to life imprisonment. In 1850 he escaped from Spandau and fled to England, where he remained and taught, undertaking in the early 1850s a lecture tour to the U.S.A. In 1866 he was appointed to a professorship in Zurich University. His later works include the tragedy *Nimrod* (1857) and a further collection of poems, *Gedichte* (1868); an autobiography, *Selbstbiographie 1838–48*, ed. R. Sander, appeared in 1931.

KIPPHARDT, HEINAR (Heidersdorf, Silesia, 1922–82, Munich), qualified in medicine at Düsseldorf University, and from 1951 to 1959 was dramatic adviser (Dramaturg) at the Deutsches Theater in East Berlin. In 1959 he came to West Germany, settling at Munich. He is the author of the satirical comedies *Shakespeare dringend gesucht* (1954, performed 1953) and *Die Stühle des Herrn Szmil* (1958, performed 1961). *Der Aufstieg des Alois Piontek* (1956, performed 1960) is described as a tragicomic farce. *Der Hund des Generals* (q.v.) first appeared as a story, written in 1957 and published in *Die Ganovenfresse* in 1964. It is a bitter satire on war.

Kipphardt later became known for the documentary plays *In der Sache J. Robert Oppenheimer* (televised 1964), dealing with the injustice of McCarthyist hysteria in the U.S.A. in the 1950s, and *Joel Brand. Die Geschichte eines Geschäfts* (1965), which is concerned with the proposal to barter Hungarian Jews for lorries, to which attention was drawn by the trial of K. A. Eichmann in Israel. Kipphardt also wrote the comedy *Die Nacht, in der der Chef geschlachtet wurde* (1967); he adapted *Die Soldaten* (q.v.) by J. M. R. Lenz for the modern stage. His primary concern is with injustice and conscience. The novel *März* (1976) takes its title from the central character, Alexander März. Written in the form of a (fictitious) montage, it argues with modern psychiatry and ends with the suicide of März. It was followed up by the März poems and by the play *März. Ein Künstlerleben* (1980). The poems *Angelsbrucker Notizen* and the three stories of *Der Mann des Tages* (including *Der Deserteur*) appeared in 1977, two volumes of *Theaterstücke* in 1978 and 1981, the play *Bruder Eichmann*, begun at the time of publication of *Joel Brand*, in 1983, and *Traumprotokoll* (dated 1978–81) in 1981.

Kirchenkampf, term used to denote the resistance movement (see BEKENNENDE KIRCHE) within the German Evangelical Church against interference by the state under the National Socialist regime. The term is also used in the general sense of interdenominational conflicts, and conflicts between the state and the Christian Church.

Kirchenlied, a hymn to be sung in church. In the Middle Ages hymns were chiefly in Latin, but it is widely held that German translations of Latin hymns, although not a part of the liturgy itself, were interpolated, particularly on special occasions. Examples have been quoted at Aachen (11th c.), Seckau (1345), and Breslau (1417), and in 1482 the vernacular hymn *Christ ist erstanden* was prescribed by the bishop as an introduction to the Office for Easter Day. These forerunners of the later Kirchenlied were mostly translations of *Sequentiae* (see SEQUENZ, MARIENSEQUENZ VON ST. LAMBRECHT, MARIENSEQUENZ VON MURI, and MELKER MARIENLIED).

The ecclesiastical hymn (in German always Kirchenlied or geistliches Lied, not Hymne, q.v.) is primarily a product of the Reformation (q.v.) and the continuing Protestant tradition of worship in the vernacular. The principal German hymn writers, in roughly chronological order, are: 16th c.: Luther, P. Speratus, M. Weiße, P. Nicolai, B. Ringwaldt, N. Selnecker; 17th c.: M. Rinkhart, P. Fleming, J. Heermann, P. Gerhardt (perhaps the most outstanding of all), S. Dach, A. Gryphius, ANGELUS Silesius, G. Arnold, E. Neumeister, J. Neander, S. Tersteegen; 18th c.: C. F. Gellert, M. Claudius, Novalis; 19th c.: E. M. Arndt, M. von Schenkendorf (qq.v.). Many of these writers include translations from the Latin among their hymns. In Lutheran use the term Choral is virtually synonymous with Kirchenlied. In England it is best known through the often elaborately harmonized hymn verses which are found in many of the cantatas etc. of J. S. Bach (q.v.) and his contemporaries. A number of German hymns have passed into English hymnals.

KIRCHHOFF or **KIRCHHOF,** HANS WILHELM (Kassel, *c.* 1525–*c.* 1603, Spangenberg, SE. of Kassel), studied at Marburg and served as a mercenary. In 1584 he became governor (Burggraf) of Spangenberg. Kirchhoff wrote a large and varied collection of Schwänke (see SCHWANK) entitled *Wendunmuth*. It appeared in seven volumes between 1563 and 1603. In addition to the social and sexual episodes characteristic of this form, it contains historical anecdotes.

KIRCHNER, ERNST LUDWIG (Aschaffenburg, 1880–1938, Frauenkirch nr. Davos), a sculptor, draughtsman, and painter and co-founder of the Brücke (q.v.). His early Expressionism later gave way to abstract conceptions. Under the National Socialist regime his work was regarded as 'degenerate' (see ENTARTETE KUNST) and was impounded. He took his own life.

KIRSCH, SARAH (Limlingerode, 1935–), after a short period as a factory worker studied biology at Halle University (according to her own account influenced by A. Stifter, q.v., in her choice of subject) and at the Johannes R. Becher Institute for Literature in Leipzig. She became a free-lance writer in 1968. At one time a committee member of the Schriftsteller-verband and member of the SED, she was a signatory to the public letter of protest concerning W. Biermann (q.v.) and in 1977 moved from East Berlin to West Berlin. A prominent author of lyric poetry, she has developed an individual style, favours the use of the first person, and is concerned with the role of women in modern (notably DDR) society. She avoids, with exceptions, direct political involvement, though references to political crises, notably war (Vietnam), and critical views concerning the DDR, are clearly discernible. An example is the title poem of her collection *Landaufenthalt* (1967, re-edited 1977), which also exemplifies her suggestive syntax and line division. *Zaubersprüche* (1973) contains a number of poems that are transparently related to the contemporary political scene. The poem *Der Droste würde ich gern Wasser reichen* acknowledges the influence of A. von Droste-Hülshoff (q.v.) by depicting aspects of identification with her 'sister' of a bygone age. The collection *Rückenwind* appeared in 1976 and contains in the section *Wiepersdorf* a poem addressed to Bettina von Arnim (q.v.), which expresses a sense of personal affinity, but is also demonstratively political.

Musik auf dem Wasser (1977) is the last collection of previously published poems to appear in the DDR. All the other collections had appeared first in the DDR and later in the BRD. *Katzenkopfpflaster* (1978) corresponds to the DDR volume of 1977, *Drachensteigen* (1979) contains forty new poems written in the DDR, BRD, and Italy. The poem *Der Rest des Fadens*, the last lyric written in the DDR, expresses in seven terse lines the loyalty of an ideological idealist. The collection *Erdreich* appeared in 1982, *Katzenleben* in 1984, *Landwege. Eine Auswahl 1980–1985* in 1985, and texts, *La Pagerie*, in 1980. Kirsch was for a number of years married to the author Rainer Kirsch (b. 1934).

KIRSCHNER, ALOYSIA, see SCHUBIN, OSSIP.

Kitsch, a term originally applied to ephemeral and trashy works, especially sentimental novels and novelettes, and their illustrations and graphic equivalents, and to poetry of like character.

Among writers of minor standing, E. Marlitt, H. Courths-Mahler (qq.v.), L. Ganghofer, and even Karl May (qq.v.) are often regarded as purveyors of Kitsch in fiction, and K. Groth (in his High German poems) and H. Hesse (qq.v.) in poetry. Some patriotic writing falls into the same category. There is a tendency to extend the scope of the term to work of some real

literary merit which, in the critic's view, in some respect falls short of particular standards. K. H. Deschner's *Kitsch, Konvention und Kunst* (1957) is largely an attack on H. Hesse (q.v.) as a writer of Kitsch, and W. Killy's illustrated anthology entitled *Deutscher Kitsch* (1962) includes such names as Rilke and Weinheber, as well as those of R. G. Binding, H. Sudermann, and E. von Wildenbruch (qq.v.). The term is in some danger of degenerating into a personal expression of literary antipathy. See TRIVIAL-LITERATUR.

K.K. or **k.k.**, abbreviations for kaiserlich-königlich applied to all Austrian institutions and officials up to the Ausgleich (q.v.) of 1867. From 1867 to 1918 it was used only for purely Austrian (i.e. non-Hungarian) authorities and officials. K. u. k. (kaiserlich und königlich) was devised in 1867 and applied to the few extremely important institutions that were common to the Empire of Austria and the Kingdom of Hungary, of which the most obvious were the Ministry of Foreign Affairs and the regular army. The Austrian Landwehr (second-line troops) were purely Austrian and therefore remained k.k. Its Hungarian equivalent was the Königlich ungarisches Honvéd. K.k. provided R. Musil with an ironical name for Austria-Hungary, 'Kakanien' (see MANN OHNE EIGENSCHAFTEN, DER).

KLABUND, pseudonym of Alfred Henschke (Crossen/Oder, 1890–1928, Davos), who suffered from pulmonary tuberculosis from his seventeenth year, and made frequent stays in sanatoria in Switzerland. He was a close friend of G. Benn (q.v.). He published a number of novels on historical figures (*Moreau*, 1916; *Mohammed*, 1917; *Franziskus*, 1921; *Borgia*, 1928; and *Rasputin*, 1929) and wrote ecstatic Expressionistic poetry (*Morgenrot! Klabund! Die Tage dämmern!*, 1912), but his principal novel is about a vagabond named Bracke (*Bracke*, 1918), who has been compared to Till Eulenspiegel (q.v.), though he possesses an altruistic trait which is foreign to Till. The novel is set in the 16th c. Klabund is best known for his adaptations from the Chinese, especially of the play *Der Kreidekreis* (1925), which Brecht used when writing *Der kaukasische Kreidekreis* (q.v.). Klabund's *Gesammelte Werke* (1930) amount to 6 vols., and a selection, *Der himmlische Vagant*, ed. M. Kesting, appeared in 1968, *Gedichte und Prosa*, ed. W. E. Richartz, in 1978. Benn wrote *Totenrede für Klabund* (1928); and *Briefe an einen Freund*, ed. E. Heinrich, was published in 1963.

Kladderadatsch, a humorous and satirical

weekly, published in Berlin, and founded in 1848 by D. Kalisch (q.v.).

Klage, Die, see NIBELUNGENLIED.

Klage um eine edle Herzogin, an anonymous Middle High German poem in the form of an allegorical lament for Beatrix, wife of Duke Heinrich of Carinthia. The poem was written c. 1331, in which year the Duchess died. The author, a Swabian knight, also wrote *Das Kloster der Minne* (q.v.).

KLAJ, JOHANN (Meißen, 1616–56, Kitzingen/ Main), a minor poet, studied in Wittenberg and became a private tutor in Nürnberg (1644), where he became acquainted with G. P. Harsdörffer (q.v.) with whom he founded the Löblicher HIRTEN- und Blumenorden an der Pegnitz (q.v.). In 1650 he became pastor in Kitzingen. Klaj (who used the pastoral pseudonym Clajus) wrote pastoral poetry, including the volume *Pegnesisches Schäfergedicht* (1644) in which he collaborated with Harsdörffer. In some of his poetry rich sound is achieved by internal rhyme. His most original works are his oratorios, in which declamation and choruses alternate, and a wide variety of metres is used, often to distinguish different characters. These include *Die Aufferstehung Jesu Christi* (1644), *Die Höllen- und Himmelfahrt Jesu Christi* (1644), *Der leidende Christus* (1645), *Herodes der Kindermörder* (1645), and *Freudengedichte der seligmachenden Geburt Jesu Christi* (1650). Klaj also wrote religious poetry, published in *Andachts Lieder* (1646).

Klassik, Deutsche, the period of the classical interests and endeavours of Goethe and Schiller and their sympathizers, including in a special sense Hölderlin (q.v.). The classicism of Winckelmann (q.v.) provided a basis. Elevated classical idealism covers roughly the years 1780–1810, but is at its peak between 1786 (Goethe's departure for Italy) and 1805 (Schiller's death). Among the most notable poetic products of this period are Goethe's *Iphigenie auf Tauris, Torquato Tasso, Hermann und Dorothea, Alexis und Dora, Die natürliche Tochter,* and Act III of *Faust,* Pt. II (qq.v.), Schiller's *Wallenstein, Maria Stuart,* and *Die Braut von Messina,* together with many of his poems, especially *Der Spaziergang* and *Nänie* (qq.v.), the *Ion* of A. W. Schlegel (q.v.), and numerous poems of Hölderlin (q.v.), together with his *Hyperion* and *Der Tod des Empedokles* (qq.v.). Though homely and thoroughly German in content, the idylls of J. H. Voß (q.v.) are written in classical verse and conceived as a modern counterpart to the antique idyll. Voß also produced one of the most widely read

translations of Homer (*Odyssee,* 1781, *Homers Werke,* 1793).

Deutsche Klassik was not a widespread movement in German literature. Many writers declined to fall in with Goethe and Schiller, who, though they certainly enjoyed a high reputation, had not the commanding position which historical perspective now appears to give them. Moreover, the stinging epigrams which Goethe and Schiller published jointly as *Xenien* (q.v., 1796) angered enemies of their classicism and created new opponents.

Klassizismus, the imitation or cult of the literature, art, and thought of classical antiquity. The Middle Ages concerned themselves with classical formal conceptions only in a Latin context; the vernacular writers were attracted by classical subjects, especially the *Aeneid* and the story of Alexander the conqueror (see HEINRICH VON VELDEKE and ALEXANDERLIED). The first, and perhaps most thoroughgoing, movement of classicism in Germany is a part of European florescence of Latin literature in Humanism (see HUMANISMUS). The 17th c. and early 18th c. in Germany adopted classicism through the medium of French classicism. The classical phase of Goethe and Schiller (see KLASSIK, DEUTSCHE) is a landmark in German classicism, and the work of Hölderlin (q.v.) is one of its greatest individual achievements. Later examples of classical poetry are chiefly isolated experiments, in which Mörike, J. Weinheber, and R. A. Schröder (qq.v.) were notably successful. A remarkable feature of German classicism is the success with which the hexameters and distichs of the Ancients, and even Greek trimeters, all based originally on quantity, have been adapted to the accentuated system of German verse. That German classicism has continued well into the 20th c. is shown by G. Hauptmann's *Atriden-Tetralogie* (q.v.), the four plays of which were published under separate titles between 1941 and 1948.

KLAUS, BRUDER, see NIKOLAUS VON DER FLÜE.

Klaus Bur, a Fastnachtspiel (q.v.) written by M. Bado, a disciple of Erasmus, in 1523. Klaus, a shrewd north German peasant, argues successfully with his learned superiors, and converts the parish priest to Lutheranism.

KLEE, PAUL (nr. Berne, 1879–1940, Locarno), was in touch with Expressionist artists in Munich, and developed an independent, virtually Surrealistic, style which has been termed Bildsprache. In 1921–30 he was a professor at the Bauhaus (q.v.), and then until 1933 at Düsseldorf before emigrating to Berne on the establishment of the National Socialist regime. In 1937 his work was declared 'degenerate' (see ENTARTETE KUNST), and a large number of his works were impounded. His early publications include *Pädagogisches Skizzenbuch* (1925) and *Wissenschaftliche Experimente im Bereiche der Kunst* (1928); posthumous publications include *Über die moderne Kunst* (1945), *Das bildnerische Denken* (ed. J. Spiller, 1956), and *Tagebücher 1898–1918* (ed. F. Klee, 1957).

Kleider machen Leute, a Novelle by G. Keller (q.v.), published in 1874 as the first story in the second volume of *Die Leute von Seldwyla* (q.v.). The journeyman tailor Wenzel Strapinski loses his job in Seldwyla, and sets out on foot for Goldach, dressed above his station in a fur-trimmed cloak and cap. He is given a lift by a coachman, who drives him to the inn Zur Wage where his appearance causes him to be received as a man of quality. The respect with which he is treated increases when the mischievous coachman ennobles him as 'Wenzel, Graf Strapinski'. The coach drives off, Wenzel remains, lacking the courage to clear up the misunderstanding. The notabilities of Goldach pay court to him, and he falls in love with Nettchen, the daughter of the magistrate (Amtsrat), who is an heiress in her own right. She returns his love, but the betrothal party, arranged at an inn, is rudely disturbed by a malicious crowd from Seldwyla, summoned by Melcher Boehni, a rival suitor, who unmasks Strapinski's passive deception. Wenzel leaves in despair, and walks away through the snow till he drops from fatigue. After the first shock, his intended bride proves her sterling character. She takes a coach, finds him near death in the snow, and brings him to a farmhouse, where he is revived. Then she calls him to account. He confesses, and makes it clear that he had no wish to deceive but did not know how to extricate himself from the web of misunderstandings. Moreover, once he had met Nettchen, the will to escape had left him. She recognizes his fundamental integrity and forgives him. They are married, and he sets up business as a tailor in Seldwyla. He and Nettchen prosper and raise a large family. They do not forget the malice of the Seldwyla people, and finally move with their valuable custom to Goldach.

KLEIN, JOHANN AUGUST (Koblenz, 1778–1831, Koblenz), was the author of books on travel in the Rhineland, of which *Rheinreise von Mainz bis Cöln* (1828) was adapted by Baedeker (q.v.) in 1832.

KLEIN, JULIUS LEOPOLD (Miskolcz, Hungary, 1810–76, Berlin), a dramatic critic, and the author of a history of drama (*Geschichte des Dramas*, 13 vols., 1865–76) which he was not able to complete. He also wrote plays (collected in *Dramatische Werke*, 7 vols., 1871–2). The individual titles were *Maria von Medici* (1841) and *Luines* (1842, two historical tragedies linked together), *Zenobia* (1847), *Die Herzogin* (1848), *Kavalier und Arbeiter* (1850, a contemporary social tragedy), *Ein Schützling* (1850), *Moreto* (1859), *Maria* (1860), *Alceste, König Albrecht, Strafford, Voltaire* (all 1862), *Heliodora* (1867), and *Richelieu* (1871).

Kleindeutsch, term used in the 19th c. to denote the policy of unifying Germany to the exclusion of Austria. (See GROSSDEUTSCH.)

Kleine Lucidarius, Der see SEIFRIET HELBLING.

Kleine Mann, Der, a satirical comedy by C. Karlweis (q.v.), first performed in the Raimundtheater in Vienna in 1894, and published in the same year. Its action involves an election, the climax of which occurs after the poll, when Rohrbeck, the successful candidate, insults the electors, only to learn that the poll has been declared null and void and that a new election must be held.

Kleines Organon für das Theater, was written by B. Brecht (q.v.) to elaborate his theory of epic theatre (see EPISCHES THEATER), and published after its completion in 1948 in a special number of *Sinn und Form* (q.v.) in 1949. It has links with his other tracts on theory, notably *Über eine nichtaristotelische Dramatik*, and is followed by *Nachträge des Kleinen Organon* and a brief *Verteidigung des Kleinen Organon*, the '*Katzgraben*'-*Notate* (on the comedy *Katzgraben*, 1954, by E. Strittmatter, q.v., which is set in a rural community in the DDR), the *Stanislawski-Studien*, and *Die Dialektik auf dem Theater*.

The *Kleines Organon* consists of a preface and 77 sections in which Brecht examines Aristotle's *Poetics* and opposes to it his own views on aesthetics appropriate to the new scientific age which, in his own career as a dramatist, had entered the atomic era, making it all the more essential for the public to become fully aware of the changing world: 'Das Theater muß sich in der Wirklichkeit engagieren, um wirkungsvolle Abbilder der Wirklichkeit herstellen zu können . . .' (§23). But the reproduction of reality is only effective if the spectator sees reality on the stage as both familiar and unfamiliar ('Verfremdung des Vertrauten', §44, see VERFREMDUNGSEFFEKT), a paradox intended to provoke critical detachment and reassessment. In Brecht's view the empathy associated with Aristotle impedes rational thought. Brecht aimed at a synthesis of reason and feeling which allows for a degree of involvement on the part of both the actor and the spectator, while preventing them from identifying themselves with the characters portrayed. The actor must break with tradition and cultivate a style of acting which, through his detachment from his role, enhances the contradictions in the character. The 'split' personality thus portrayed is intended to provide a constant impediment of illusion. The mime (Gestus) of the actor and the projected story (Fabel) should be developed with maximum versatility to further the ideology of a new social order. The types of characters depicted serve to express the political dialectics of the plays.

Many of the theoretical pronouncements are not borne out by Brecht's plays, which, despite his one-sided criticism of Aristotle, banish neither illusion nor emotion from the theatre. On the subject of politics in the theatre Brecht emphasizes in his introduction to the 'Katzgraben'-Notate that the public should not only learn, but enjoy learning: 'Es ist nicht genug verlangt, wenn man vom Theater nur Erkenntnisse . . . verlangt. Unser Theater muß die Lust am Erkennen erregen, den Spaß [Brecht's emphasis] an der Veränderung der Wirklichkeit organisieren.' This statement resumes the concluding paragraph of the Kleines Organon. The Nachträge repeat once more Brecht's contention that 'subjective' (i.e. traditional) acting is 'asocial' (i.e. bürgerlich), so confirming his tendency, as a theorist, to over-simplification.

Kleine Stadt, Die, a novel by H. Mann (q.v.), published in 1909. Its setting is a small Italian city (Palestrina, which Mann knew, has been suggested), and its action takes place about 1870 at the time of the dissolution of the Papal State. It is Mann's first expressly political work. The city is dominated by a conservative faction led by the priest Don Taddeo. The progressives, headed by the lawyer Belotti, strive for a democracy. Belotti invites an opera troupe, which the Taddeo faction, because of the presumed immorality of the singers, condemns. The women singers attract the menfolk, and the tenors and baritones run after the good-looking daughters. The opera, invented by Mann and the subject of one of his Novellen (Die arme Tonietta), recalls Leoncavallo's I Pagliacci. In moral, yet criminal, indignation, Don Taddeo sets fire to the inn where the troupe is staying, and then bravely rescues one of the actresses, in whom he is (improperly) interested. The crisis is solved by Taddeo's confession from the pulpit, and an agreement to run the city democratically is achieved. The work, which is on the whole goodhumoured, ends in a sudden jolt of tragedy when one of the tenors, about to elope with a

local girl, is found murdered. The novel is in five long chapters, a form purposely devised to simulate the five acts of a dramatic work.

Klein-Paris, 18th-c. designation for Leipzig, referring to its then elegant and fashionable francophile tone. It occurs in Goethe's Faust (q.v.), Part I in the scene Auerbachs Keller: 'Es ist ein klein Paris, und bildet seine Leute'.

Klein und Wagner, a story by H. Hesse (q.v.), published in 1920, and included in the collection Weg nach Innen (1931). It was Hesse's first work written in Tessin after he had abandoned his family and left Berne. It consists of five episodes dealing with the last phase of Friedrich Klein's life, in which he seeks to reconcile his inner conflicts. His departure for a new environment and an independent existence is presented as a dreamlike flight. In the train taking him to Italy he remembers W., a teacher, who had murdered his family and then tried to kill himself. The case had aroused publicity, and Klein had joined his colleagues in condemning the murderer. He now becomes fully conscious of his sympathy for the criminal. On recollecting his name, Wagner, Klein associates him in his mind with the composer Richard Wagner (q.v.), whom Klein had formerly severely criticized. At his hotel Klein is fascinated by the dancer Teresina, but spends a night with the simple wife of an innkeeper. At the casino Teresina teaches him to gamble. He loses his money but wins her love. Yet only in death, which he seeks in the lake, can he find in a nameless god the sense of regeneration for which he had searched. He becomes reconciled with his life both as Klein and as Wagner, and overcomes the daemonic aspect of Goethe's Egmont (see DÄMONISCHE, DAS) with which he had felt an affinity.

Klein Zaches genannt Zinnober, a story (Ein Märchen) written in Berlin in 1818 by E. T. A. Hoffmann (q.v.), and published in 1819. Klein Zaches is a hideous dwarf born to a peasant woman. The fine lady Fräulein von Rosenschön (in reality Fairy Rosabelverde) passes by and confers on Zaches a magic gift, by which he receives the credit for whatever gracious or pleasing thing is said or done in his presence. He is adopted by a pastor, comes to the university as Herr Zinnober, makes his way at the prince's court and rises to power and affluence, all at the expense of the meritorious, whose merit he steals. He is about to be betrothed to Candida, loved by the student Balthasar, when the magician Prosper Alpanus intervenes, and breaks the magic spell, whereupon Zinnober-

Zaches, exposed in his misshapen form and character, is killed, and the way is made free for Balthasar to marry Candida.

KLEIST, EWALD CHRISTIAN VON (Zeblin nr. Köslin, Pomerania, 1715–59, Frankfurt/Oder), a Pomeranian nobleman, was at school in Danzig and studied at Königsberg University. He entered the Danish army in 1736, resigning his commission in 1740. In the same year he was commissioned in the Prussian army and posted to Potsdam. Kleist, who did not enjoy garrison life, made literary friends, including J. W. L. Gleim, K. W. Ramler, and F. Nicolai (qq.v.). He saw active service in 1744 and 1745, and was promoted captain in 1749. His early poetry of the 1740s is pessimistic in tone. In 1749 his poem *Der Frühling* (q.v.), the starting-point of which was James Thomson's *The Seasons*, achieved a considerable success. On a recruiting mission to Switzerland Kleist met Bodmer and Breitinger and the idyll-writers Hirzel and Geßner (q.v.). By the beginning of the Seven Years War (see SIEBENJÄHRIGER KRIEG) he was a major. Kleist was stationed in Leipzig in 1758 and there became a close friend of Lessing (q.v.). During the war years he wrote patriotic poetry, including an *Ode an die preußische Armee* (1757) and the short epic *Cissides und Paches* (q.v., 1759). Kleist died of wounds received at the battle of Kunersdorf (q.v.). He is believed to have been the model for Tellheim in Lessing's *Minna von Barnhelm* (q.v., 1767). *Sämtliche Werke* (ed. J. Stenzel) appeared in 1 vol., 1971.

KLEIST, HEINRICH VON (Frankfurt/Oder, 1777–1811, Wannsee nr. Berlin), came of an ancient military family, and was a distant relative of Ewald von Kleist (q.v.). He entered the First Foot Guards at Potsdam, and served as an ensign in the campaign of 1793 (see REVOLUTIONSKRIEGE). His father had died in 1788, and in 1793 he lost his mother. His distaste for garrison life led him to resign his commission in 1799, despite his forthcoming betrothal to Wilhelmine von Zenge, a general's daughter. He was left with only a modest inheritance, and reluctantly began to prepare himself for entry into the civil service by studying at Frankfurt/Oder University. He studied science and mathematics, and cultivated an interest in music and philosophy, which led to an acquaintance with the work of Kant and Fichte (qq.v.). His tracts *Aufsatz, den sicheren Weg des Glücks zu finden* and *Über die allmähliche Verfertigung der Gedanken beim Reden* (unpublished at the time) fall into this period. The optimism of the Aufklärung (q.v.) was fast coming to an end, and as early as 1799 Kleist expressed in a letter to his half-sister

Ulrike his intuitive apprehension that man cannot shape his own destiny. This passage is one of the earliest indications that he was to become the first radical exponent of the unpredictability of human destiny in modern German literature.

At the turn of the century, Kleist was as yet unaware of his potential as a writer. In 1801 he suffered a serious crisis in his development, usually referred to as the 'Kant crisis' (Kantkrise). The reading of Kant's first *Critique* convinced him of the futility of his aim of acquiring knowledge by solely intellectual methods. In a despairing letter to his fiancée (22 March 1801) he declared that his 'only and highest goal' had vanished. In order to recover his equilibrium he left Prussia for Paris, accompanied by Ulrike. A period of restless travel followed. He found a certain stabilizing influence in the writings of Rousseau, which appealed to his longing for closer union with Nature.

In the decade leading to his suicide Kleist made repeated efforts to come to terms with his own destiny and intensely complex individuality. In 1802 he went to Switzerland, where he met H. D. Zschokke (q.v.) and completed his first tragedy, *Die Familie Schroffenstein* (q.v.), which showed him in revolt against predominant influences in aesthetics. He also planned his comedy *Der zerbrochne Krug* (q.v.) and a tragedy on the subject of Robert Guiskard. During this time his engagement to Wilhelmine ended as a result of the uncertainties about his future. After recovering from illness in Berne he continued his restless travels in the autumn of 1802, spent the beginning of 1803 in the home of Wieland (q.v.), the father of his friend Ludwig, moved in the spring to Leipzig, then to Dresden, where he began work on *Amphitryon* (q.v.); in July he travelled from Leipzig to Switzerland, visited Italy, and in October arrived via Switzerland in Paris. Despite caring friends he now suffered a crisis that caused him to burn his MS on Guiskard, of which he later rewrote a portion: the fragment *Robert Guiskard. Herzog der Normänner* (q.v.) expresses defiance in the face of overpowering fate and an implicit admiration for Napoleon, whose army for the intended invasion of England he twice attempted to join. His motives for this extraordinary detour to the coast are not fully known, but when he turned up again in Paris after his second unsuccessful attempt he was urged to return to Germany. However, he broke off his journey in Mainz (then French) and did not arrive in Berlin until June 1804. After months of uncertainty he was received into the civil service and in May 1805 was posted to Königsberg. In August 1806 he was granted prolonged sick-leave, which he devoted to writing, but the Prussian defeat at Jena in October of that year once more radically changed his life. Perhaps in an attempt to participate in Prussia's liberation, he went at the beginning of 1807 to occupied Berlin, was arrested on suspicion of being a spy, and for the next six months found himself imprisoned, finally in Fort de Joux in the Jura and in Châlons-sur-Marne. During his solitary confinement he laid the foundations for his tragedy *Penthesilea* (q.v.).

After his release from the French fortress Kleist turned to Dresden (August 1807) in pursuit of literary contacts and work. His Ritterschauspiel *Das Käthchen von Heilbronn* (q.v.), the first version of which he read to L. Tieck (q.v.) in the summer of 1808, was written during this period, which initially brought him the happiness of recognition, though his repeated attempts to win Goethe's favour had in the spring of that year, after an unsuccessful performance of an unauthorized version of his first comedy in Weimar, ended in bitter antagonism. He was by now in dire financial straits. He begn to collaborate with Adam Müller (q.v.), with whom he founded the journal *Phöbus* (q.v.), but the project survived only a few months, failing to fulfil any of Kleist's hopes. He next moved to Austria, where the conflict between the Austrian and French forces was imminent. *Die Herrmannsschlacht* (q.v.) was intended to promote Austrian national aspirations for the liberation of Germany. Kleist's single-minded engagement in the national cause, intensified by the defeat of the Austrians at Wagram, is expressed in patriotic poems, in his project *Germania* (1809), and in his activities in Berlin, to which he returned in 1810. He completed his last play, *Prinz Friedrich von Homburg* (q.v.), at a time when, as editor of the *Berliner Abendblätter*, he hoped to realize the frustrated hopes of his earlier journalistic enterprises. The diverse articles which he wrote for this paper include his essay *Über das Marionettentheater* (q.v.). By March 1811 he had once more to admit defeat and financial disaster.

Some of Kleist's remarkable Novellen had already appeared in journals, though for the most part in fragmented form. They were collected in 2 vols. in the autumn of 1810 and in June 1811 as *Erzählungen*: (1) *Michael Kohlhaas, Die Marquise von O . . .*, and *Das Erdbeben in Chili*, and (2) *Die Verlobung in St. Domingo, Das Bettelweib von Locarno, Der Findling, Die heilige Cäcilie*, and *Der Zweikampf* (qq.v.).

For most of his life, Kleist's sister Ulrike gave her brother loyal support and companionship, and his cousin Marie von Kleist had used her influence at court to help him. For each Kleist

left a final letter. That to Ulrike was written on the shore of the Wannsee, where Kleist shot Henriette Vogel at her request in order to cut short her suffering from an incurable disease (probably cancer), and then killed himself. This letter to Ulrike is dated 'Am Morgen meines Todes' (21 November 1811). It is a message of reconciliation and contains the pronouncement 'daß mir auf Erden nicht zu helfen war'.

Kleist's literary work, his essays and letters, bear witness to his struggle to come to terms with the political and personal depression dominating the years of his manhood. He attempted to secure a livelihood without debasing his poetic gifts. His major works received no recognition during his lifetime. Ten years after his death L. Tieck (q.v.) undertook an edition of *Hinterlassene Schriften*, in which were published *Prinz Friedrich von Homburg, Die Hermannsschlacht*, and *Robert Guiskard* for the first time (1821). Kleist's preoccupation with ethics links him with German classicism; but he decisively rejected the stoicism which Winckelmann (q.v.) had considered to be a fundamental feature. He was also greatly influenced by Romanticism, its philosophy, and its interest in the subconscious workings of the human mind (see SCHUBERT, G. H.). But he resists classification under either literary school.

Sämtliche Werke (7 vols., including correspondence), ed. H. Sembdner, appeared in 1961 (6th revised edn. 1977) and *Werke und Briefe* (4 vols.), ed. S. Streller *et al.*, in 1978.

Klene Wegekörter, De, a Low German collection of comic tales (see SCHWANK), drawn from Latin and High German sources, which was published in 1592.

KLEPPER, JOCHEN (Beuthen, 1903–42, Berlin), a pastor's son, studied theology, and became a journalist and novelist. His sincere Christianity is an essential component of his work. His first, light-hearted, novel, *Der Kahn der fröhlichen Leute* (1933), was followed by *Der Vater* (q.v., 1937), a deeply serious novel on the father of Friedrich II of Prussia, Friedrich Wilhelm I (qq.v.). It is an attempt at a Christian interpretation of the unsympathetic figure of the king. Klepper left also a short fragment of a novel on the wife of Luther (q.v.); the single chapter has been published posthumously as *Die Flucht der Katharina von Bora* (1951). Two collections of poetry appeared during his lifetime, *Du bist als Stern uns aufgegangen* (1937) and *Kyrie* (1938), as well as the essay *Der christliche Roman* (1940).

In 1942 Klepper and his Jewish wife and stepdaughter, faced with the certainty of removal to a concentration camp, took their own lives. His detailed diaries extending over the period from the beginning of the National Socialist regime to the family's final personal decision are among the most moving and tragic documents of that time; they were published as *Unter dem Schatten deiner Flügel* (1956) and *Überwindung* (1958). Some essays, stories, and poems from his posthumous papers appeared in 1960, a Novelle, *Das Ende*, in 1962, and correspondence, *Briefwechsel 1915 bis 1942*, ed. E. G. Riemenschneider, in 1973.

Kleriker und Nonne (also called *Liebesantrag*), a poem of ten stanzas written partly in Latin and partly in Old High German. It has been suggested that its subject is a proposal of love to a nun, and that it ends with the victory of virtue, but the state of the text precludes any certain conclusion. It is contained in the Cambridger Liederhandschrift (q.v.) and must have been written in the late 10th c. or early 11th c. The poem has been crossed out in the MS., no doubt because of its subject, perhaps also because of more specific erotic features, so that much of it is illegible. It is regarded as important evidence for the development of the form of Minnesang (q.v.).

KLESEL, Melchior, see Khlesl, M.

KLETTENBERG, Susanna Katharina von (Frankfurt/Main, 1723–74, Frankfurt), a lady of devout life and deeply pious outlook, was a friend of Goethe's mother, and made a deep impression on the 19-year-old Goethe during his illness in 1768. She is the original of the 'schöne Seele' of the sixth book of *Wilhelm Meisters Lehrjahre* (q.v.) which bears the title *Bekenntnisse einer schönen Seele*. She wrote a few hymns which were published anonymously.

KLIMT, Gustav (Vienna, 1862–1918, Vienna), the principal Austrian painter of *art nouveau* (Jugendstil), and the founder of the Viennese Sezession (q.v., 1898–1903).

KLINGEMANN, Ernst August Friedrich (Brunswick, 1777–1831, Brunswick), was from 1814 a director of a minor theatre in Brunswick, and from 1826 director of the Braunschweiger Hofbühne (though he also worked for a few years as a professor at the Collegium Carolinum, see Carolinum). He was the author of novels and plays as well as of essays on dramatic production and the function of the theatre. These interests are reflected in his principal work

Kunst und Natur. Blätter aus meinem Reisetagebuch (3 vols., 1819–29). His early works include a pseudo-medieval novel, *Wildgraf Eckard von der Wölpe* (1795), and the tragedies *Die Maske* (1797) and *Heinrich von Wolfenschießen* (1806). His collection of historical plays (*Theater*, 3 vols., 1808–20) includes *Martin Luther* (1809) and *Faust* (1815), a tragedy which was one of the influences on Grabbe's *Don Juan und Faust* (q.v.). Klingemann produced the first stage performance of Goethe's *Faust I* on 19 January 1829 in Brunswick. One of his last works is the tragedy *Ahasver* (1827).

KLINGER, Friedrich Maximilian (Frankfurt/Main, 1752–1831, Dorpat, now Yuriev, Russia), the son of an army N.C.O., was early orphaned and brought up in poverty. He was befriended by Goethe (who was two years older) and, partly with Goethe's assistance, was sent to Giessen University, where he studied law from 1774 to 1776. He was one of the small group of friends around Goethe who formed the inner circle of Sturm und Drang (q.v.). During his student years he began to write plays, completing six in two years (*Otto*, q.v., 1775; *Das leidende Weib*, 1775; *Die neue Arria*, 1776; *Die Zwillinge*, 1776; *Simsone Grisaldo*, 1776; and *Sturm und Drang*, qq.v., 1776). These treat, with greater or less extravagance of language and technique, extreme situations, dynamic figures, or tragic sufferers. In 1776 Klinger met Christoph Kaufmann (q.v.), who was responsible for changing the title of Klinger's play *Der Wirrwarr* to *Sturm und Drang*, so creating the name by which the movement has since been known. *Die Zwillinge* was Klinger's successful entry for the Hamburg competition organized by F. L. Schröder (q.v.). In 1776 Klinger visited Goethe in Weimar, where he was coolly received. He spent two years with the Seylersche Truppe of actors, writing for them more conventional plays. After the tragedy *Stilpo und seine Kinder* (q.v., written in 1777 and published in 1780) Klinger turned away from the Sturm und Drang to comedies such as *Die falschen Spieler* (1782) recalling those of Holberg (q.v.), and historical tragedies such as *Konradin* (1784). In 1778 he was commissioned in the Austrian army and took part in the brief War of the Bavarian Succession (see Bayrischer Erbfolgekrieg). Demobilized afterwards, he wrote the satire *Plimplamplasko* (q.v.) while waiting to enter Russian service as a subaltern of marines (1780). He attracted the attention of Tsar Paul and served in his personal suite, and later he stood well with Alexander I. He became commandant of the cadet school at St. Petersburg and attained the rank of lieutenant-general. He was also for some years the head of

Dorpat (Yuriev) University. In his early years in Russia Klinger wrote a number of philosophical novels, including *Fausts Leben, Thaten und Höllenfahrt* (q.v., 1791), *Geschichte Giafars des Barmeciden* (1792–4), *Der Faust der Morgenländer* (1797), *Geschichte eines Teutschen der neuesten Zeit* (1798), and *Der Weltmann und Der Dichter* (1798). He renewed contact with Goethe by letter in 1814. Goethe's *Dichtung und Wahrheit* (q.v.) contains in Bk. 14 a sympathetic sketch of Klinger's character. Klinger's development shows some similarity to that of Goethe, passing from youthful excess and extravagance to a measured maturity, and achieving success in practical life. *Werke* (12 vols.) appeared 1809–16 and *Werke. Historisch-kritische Ausgabe*, ed. S. L. Gilman, U. Profitlich *et al.*, 1975 ff.

Klingsors letzter Sommer, a story by H. Hesse (q.v.), written in 1919 in Italy (which provides its setting), published in 1920, and included in the collection *Weg nach Innen* (q.v., 1931). The former writer Hermann has become a painter and adopted the name of Klingsor, the name of the evil magician (Clinschor) in Wolfram von Eschenbach's *Parzifal* (q.v.), which was also used by Novalis in *Heinrich von Ofterdingen* (q.v.). At the age of 42 (at which Hesse himself had taken up painting), Klingsor has turned his back on the past and experiences in the course of a single summer the fullness of life, the magic of colour, and the intoxication of wine. The work consists of episodes, showing him in the company of Louis, a vagrant painter and radical individualist, among whose pictures Klingsor admired the merry-go-round (Karussel). In a section headed *Die Musik des Untergangs* he watches the merry-go-round at a village fair and experiences the chaos and joy of existence. He meets an Armenian astrologer, who confirms that Klingsor's premonition of impending doom at the height of summer (it is July, the month of his birth) is written in the stars. As a writer Klingsor has adopted the name of his favourite Chinese poet Thu Fu. He includes Thu Fu's poem on death and regeneration in the section *Abend im August*. On this evening a countrywoman, his 'Madonna', yields to a last, fleeting experience of love. The final section is entitled *Das Selbstbildnis*. Klingsor's self-portrait expresses an incoherent state of ecstatic fantasy. It is designed as the 'confession' (Konfession) of the man and artist.

KLOPSTOCK, FRIEDRICH GOTTLOB (Quedlinburg, 1724–1803, Ottensen nr. Hamburg), was educated (1739–45) at Schulpforta (see FÜRSTENSCHULEN), where he received a sound classical training and was influenced by Pietistic ways of thought (see PIETISMUS). Here he became acquainted with Milton's *Paradise Lost* and resolved to write a great religious epic, an intention which he expressed in his passing-out speech. He studied briefly at Jena (1745–6) and then migrated to Halle (1746–8). In 1748 the first three cantos of his religious epic *Der Messias* (q.v.) were published in the *Bremer Beiträge* (q.v.). The poem, of which the twentieth and last canto did not appear until 1773, is written in classical hexameters; its combination of stylistic originality and religious subject-matter quickly forged for Klopstock a nation-wide reputation as a poetic genius of extraordinary promise. From 1748 to 1750 he was a private tutor to his cousins in Langensalza, and during this time he fell in love with a cousin of another branch of the family, Marie Sophie Schmidt, whom he idealized in odes as Fanny. Fanny did not return his passion. The odes, which he began to write in 1747 and continued to compose all his life, constitute his most original and persisting achievement. Klopstock's reputation as a seraphic singer of the Messiah won him an invitation to visit J. J. Bodmer (q.v.) in Zurich, but the acquaintance soon led to disillusionment, probably on both sides. Certainly Klopstock proved more normal than Bodmer had supposed, and was not only addicted to manly sports, but liked the company of young women. His ode *Der Zürcher See* (1750) commemorates, in lofty idealization, a boating picnic in mixed company.

In 1751 Klopstock was invited to Copenhagen by King Frederick V at the instigation of Count Bernstorff and received a pension which continued throughout his life. In 1754 he married Meta Moller (see KLOPSTOCK, META); their happy marriage was terminated in 1758 by her death in childbirth. In his middle years Klopstock wrote his first play, *Der Tod Adams* (1757); it was followed some years later by *Salomo* (1769) and *David* (1772). An attempt at hymnwriting (*Geistliche Lieder*, 1758, second volume 1769) produced religious poems, but no singable hymns for liturgical use. While in Copenhagen Klopstock expressed his growing interest in the German past in the first of his patriotic historical plays, to which he applied the term Bardiet (q.v.), *Hermanns Schlacht* (1769), to which he later added *Hermann und die Fürsten* (1784) and *Hermanns Tod* (1787).

In 1770 the fall of his patron Count Bernstorff and the latter's replacement by the free-thinker Count J. F. Struensee (1731–72) led Klopstock to leave Denmark and settle in Hamburg. Soon afterwards the first collected publication of his odes (*Oden*, 1771) renewed his influence upon the younger generation, especially the circle around the young Goethe in Strasburg and Frankfurt, and even more upon the group of young poets and enthusiasts in Göttingen (see

GÖTTINGER HAINBUND). In 1774 Klopstock visited Karlsruhe (taking in Göttingen on his way) at the invitation of Margrave Karl Friedrich von Baden (q.v.). This journey, on which he was much lionized, marked the zenith of his influence. His *Die deutsche Gelehrtenrepublik* (q.v., 1774) is an account of a fictitious ideal state in which poets and thinkers come into their own. Klopstock also wrote a tract on spelling reform (*Über die deutsche Rechtschreibung*, 1778), as well as short works on language, many of which contain stimulating and fruitful ideas (e.g. *Grammatische Gespräche*, 1794). His later odes are mostly concerned with political themes, including the French Revolution, which Klopstock at first supported with enthusiasm, only to turn away as the Terror developed (see FRENCH REVOLUTION).

In his own day Klopstock's reputation rested first upon *Der Messias*, and then on the odes. His remarkable lyric gifts were not matched with the sustained plastic imagination necessary for a long epic poem, and so the focus has gradually shifted to his lyric poetry in which he is a bold and original manipulator of language with a remarkable, though sometimes exaggerated, power to fix in monumental terms even the fleeting and evanescent. He has been variously regarded as the originator of a new age in poetry, as a late culmination of baroque, and as a rather boring Historical Figure. Nevertheless, the best of the odes, including long poems such as *Der Zürcher See* and *Die Frühlingsfeier* (1759), and shorter ones like *Die frühen Gräber* (1764) and *Die Sommernacht* (1766), can hold their own in almost any company.

Gesammelte Werke (4 vols.), ed. F. Muncker, appeared 1887. *Werke und Briefe, Historisch-kritische Ausgabe* (*c.* 36 vols.), ed. H. Gronemeyer, E. Höpker-Herberg, K. Hurlebusch, and R.-M. Hurlebusch, began publication in 1974 (*Hamburger Ausgabe*).

KLOPSTOCK, META (Hamburg, 1728–58, Ottensen nr. Hamburg), was before her marriage to Klopstock in 1754 Meta (Margareta) Moller or Möller. After her death in childbirth Klopstock published her writings as *Hinterlassene Schriften von Margareta Klopstock* (1759). Largely derived from her husband's style and views, they include two religious odes, a prose tragedy (*Der Tod Abels*), and a satirical essay on fashion. In his odes Klopstock refers to her as Cidli.

In 1791 Klopstock took as his second wife his first wife's niece, Johanna Elisabeth von Winthem (1747–1821) *née* Dimpfels, who had been widowed two years previously, having married Johann Martin von Winthem in 1765.

Kloster bei Sendomir, Das, a Novelle by F. Grillparzer (q.v.), published in *Aglaja* (q.v.) for 1828 with the sub-title *Nach einer als wahr überlieferten Begebenheit*. The story of Walter of Aquitania in the Grimms' *Altdeutsche Wälder* (vol. 1; see GRIMM, J., and GRIMM, W.) provided the source. The narration is set in a frame (see RAHMEN) in which two knights arrive at the abbey of Sendomir at night seeking accommodation. A monk tells them the story of the founder of the thirty-year-old monastery. It becomes apparent that this monk is the founder; he is Count Starschensky, who has endowed the monastery in atonement for murdering his wife Elga. Preserving his anonymity, Starschensky tells his story in the third person. The daughter of the Starost von Laschek, Elga, had married Starschensky, who had been a generous benefactor to her impoverished family. Some two years after the birth of a daughter, Starschensky discovered that his wife had deceived him with a distant relation of the Lascheks, Oginsky, who was in fact the father of the child. Starschensky succeeds in transporting Oginsky as his prisoner to the watch-tower of his estate, where he confronts Elga with the truth and challenges Oginsky to a duel. But Oginsky escapes, and Starschensky revenges himself on his unfaithful wife. He confesses the murder to his king, who grants him permission to dispose of his fiefs so as to benefit the monastery. There Starschensky spends the rest of his days. Grillparzer represents Starschensky's state as psychopathic.

G. Hauptmann (q.v.) based a play, *Elga*, on this Novelle.

Kloster der Minne, Das, an anonymous Middle High German poem of some 1,800 lines. It depicts an assemblage of knights and ladies, forming an order after the manner of a religious order, and living a courtly life according to strict rules. The poem was probably written *c.* 1350 by a Swabian knight, who was also the author of a *Klage um eine edle Herzogin* (q.v.).

KLOTZ, CHRISTIAN ADOLF (Bischofswerda, 1738–71, Halle), became a professor at Göttingen at 24, transferring in 1765 to Halle. An extremely active scholar, Klotz edited the *Acta literaria* from 1764 and the *Deutsche Bibliothek der schönen Wissenschaften* from 1767 until his early death. He wrote a number of Latin satires (*Mores eruditorum, Genius saeculi*, both 1760, and *Antiburmannus*, 1762), which were translated into German after his death. His special professional interests were archaeological, and he published energetically in this field. His works include *Über das Studium des Altertums* (1766), *Beitrag zur Geschichte des Geschmacks und der Kunst aus Münzen* (1767), and *Über den Nutzen und Gebrauch der alten geschnittenen Steine und ihrer Abdrücke* (1768), through which he became involved in a fierce controversy with Lessing, who directed against him the *Briefe antiquarischen Inhalts* (q.v., 1768–9). Klotz, a man of dissolute habits, was likewise the subject of attack by Herder in *Kritische Wälder* (q.v., 1769). The blame for the corruption of the morals of G. A. Bürger (q.v.), who was a student at Halle (1764–7), has been laid at Klotz's door.

KLUGE, KURT (Leipzig, 1886–1940, Liège), a professor of bronze sculpture in Berlin, developed a talent for amusing, whimsical fiction displayed in the novels *Der Glockengießer Christoph Mahr* (1934), *Der Herr Kortüm* (1938, his most successful work), and *Die Zaubergeige* (1940). He died while on a visit to the western front.

KLUIBENSCHÄDEL, TUIFELERMALER KASSIAN, journalistic pseudonym of R. Greinz (q.v.).

KNAUST, HEINRICH (Hamburg, c. 1522–77, Erfurt), a schoolmaster, became an official of the Berlin Consistory and then reverted to teaching. He wrote plays for school use, including *Agapetus* (1562) and *Dido* (1566), and was the author of devotional lyrics, of which the best known is 'O Welt, ich muß dich lassen' (1555), a Kontrafaktur (q.v.) of the 15th-c. secular song 'Innsbruck, ich muß dich lassen'. Knaust's name also appears in latinized forms, Knaustinus, Knustius, or Chnustius.

KNEBEL, KARL LUDWIG VON (Wallerstein, 1744–1834, Jena), served for nine years as a Prussian officer, and in 1774 was appointed tutor to Prince Constantine of Weimar, the younger brother of Prince, later Duke, Karl August (q.v.). He accompanied the princes in 1775 on a tour which included a visit to Frankfurt and acquaintance with Goethe. In 1793 he retired with a pension and the rank of major. Knebel was a dilettante poet and a translator of some merit. In the 1770s and 1780s he was a member of the small intimate circle of the ducal house, which also included Goethe.

KNEIP, JAKOB (Morshausen, Hunsrück, 1881–1958, Mechernich, Eifel), was brought up in a Roman Catholic country region which is reflected in his novels. The most notable is *Hampit der Jäger* (1927), a pleasant, humorous story set in the Hunsrück. *Porta Nigra* (1932, the title refers to Trier, q.v.) is based on his own experiences; it developed into a trilogy of novels of semi-autobiographical character through the addition of *Feuer vom Himmel* (1938) and *Der Apostel* (1955). The title *Porta Nigra* was extended to cover the whole trilogy. With J. Winckler and H. Lersch (qq.v.) Kneip founded in 1905 the

Werkleute auf Haus Nyland (q.v.). He lost his life in a railway accident.

KNIGGE, ADOLF FRANZ FRIEDRICH, FREIHERR VON (Bredenbeck nr. Hanover, 1752–96, Bremen), served at the court and in the administration of Hesse-Cassel and then, for a time, attended to his own estates. A rationalist and Freemason, he was from 1780 to 1784 a member of the Illuminatenorden (q.v.), seceding just before the Order was proscribed. In 1791 he took up an administrative post as Oberhauptmann in Bremen. Knigge is the author of several, mostly satirical, novels, including *Der Roman meines Lebens, in Briefen* (1781–7), *Geschichte Peter Clausens* (1783), *Die Verirrungen des Philosophen oder Geschichte Ludwigs von Seelberg* (1787), *Geschichte des armen Herrn von Mildenberg* (1789), *Das Zauberschloß oder Geschichte des Grafen Tunger* (1790), *Benjamin Noldmanns Geschichte der Aufklärung in Abyssinien* (1791), *Des seligen Etatsrats Schaafskopfs hinterlassene Papiere* (1792), and *Geschichte des Amtsrats Gutmann* (1794). He also wrote three novels on travel modelled on Sterne, of which the best known is *Die Reise nach Braunschweig* (q.v., 1792). Knigge's plays are mostly adaptations from the French. He is now chiefly remembered as the author of his didactic work *Über den Umgang mit Menschen* (q.v., 1788). The phrase 'nach Knigge' is still a proverbial reference in queries on manners and social conduct.

KNIPPERDOLLINK, BERNHARD (?–1536, Münster), one of the leaders of the anabaptist regime in Münster (see WIEDERTÄUFER), was executed after the capture of the city. He is a character in the play *Es steht geschrieben* by F. Dürrenmatt (q.v.).

Knittelverse, a verse form in which four stresses occur with an irregular number of unstressed syllables, varying from four to eleven, so that the full line can contain from eight to fifteen syllables. The lines usually occur as rhyming pairs. Knittelverse were first used in the 15th c. and were the commonest form of verse in the 16th c., after which they passed out of use. They were revived in the late 18th c., notably by Goethe in *Faust, Satyros,* and *Hans Sachsens Poetische Sendung* (qq.v.), and were employed by Schiller as the metre for *Wallensteins Lager* (see WALLENSTEIN).

KNOBELSDORFF, GEORG WENZESLAUS VON (Gut Kuckädel nr. Crossen, 1699–1753, Berlin), built the palaces of Rheinsberg, Charlottenburg, and Sanssouci (q.v., 1745–7), which was, however, largely designed by his patron and friend from his early military days, Friedrich II (q.v.) of Prussia.

KNOOP, Wera Ouckama (1902–21), a young dancer, the daughter of a minor novelist, Gerhard Ouckama Knoop (1861–1913). The news of her death at the age of 19 influenced Rilke (q.v.) in the writing of his *Sonette an Orpheus* (q.v.). Though he had only a slight acquaintance with her, she became identified in his mind with Eurydice, and the cycle is dedicated as a memorial (*Grab-Mal*) to her.

KNORR VON ROSENROTH, Christian (Alt-Raudten, 1636–89, Sulzbach), a pastor's son, was at school at Fraustadt and Stettin, and then studied at Leipzig and Wittenberg universities. From 1663 to 1665 he travelled abroad, visiting England, France, and Holland. In 1666 he entered the service of Count Christian August of Sulzbach. Knorr was addicted to alchemistic and cabbalistic studies, which appealed also to his employer. He was a Christian of Pietistic leanings (see Pietismus), and wrote religious poetry which is collected in *Neuer Helikon mit seinen neun Musen d.i. geistliche Sittenlieder* (1684); it includes the well-known poem *Morgen-Andacht*, beginning 'Morgen-Glantz der Ewigkeit'.

Knorr translated an important medical work, the *Ortus Medicinae* (1648, posth.) by the Flemish physician J. B. van Helmont (1579–1644) as *Aufgang der Artzney-Kunst* (2 vols., 1683). A reprint of this translation appeared in 1973.

Knulp. Drei Geschichten aus dem Leben Knulps, a short novel in three parts by H. Hesse (q.v.), published in 1915. The first part, *Vorfrühling*, appeared separately in 1908 in the *Neue Rundschau* (q.v.). The second part is entitled *Erinnerungen an Knulp*, and the third *Das Ende*. The first story is set in the 1890s. Knulp, who has led a vagrant life since his early youth, resumes his travels on his recovery from an illness. It is February, and bad weather induces him to stay for a while with his artisan friend Emil Rothfuß, who is married but childless. The wife, who is in no haste to bear her husband children, is attracted to the vagrant, but Knulp prefers to lavish his affection on a lonely girl, Barbara, a servant next door. Resuming his travels, Knulp visits a tailor friend. Now a weary and poor family man with five children, the tailor welcomes Knulp's cheerful presence and envies his single state. But Knulp's seemingly carefree life turns out to be not without guilt and sorrow. He has a 2-year-old son, whose mother

has died in childbirth. The boy has been adopted and will never know his true father.

The second part dwells on Knulp's sense of loneliness and fear which underlies his sensitivity to beauty. Knulp explains to a friend, who acts as narrator, that he cannot enjoy love and beauty without being aware of its transitoriness. The third part shows Knulp during his last illness. He is irresistibly drawn to return to his native Gerbersau, which he had left after an unhappy love affair. He discovers that the girl who had been the cause of his desertion of society has died, but his sense of homelessness and disillusionment persists. He rejects the advice of a doctor, a former school-fellow, that he should enter hospital, preferring to meet his death in the snow-covered country outside the town. He hears God's voice assuring him of his participation in his life, and finds peace in the awareness that God shares his sense of frustration and waste—a theme connected with Hesse's conception of the artist. Like Eichendorff's Taugenichts (see Aus dem Leben eines Taugenichts), who kindles happiness in others, Knulp was born to give, but Hesse is more concerned with the darker recesses in the character of the good-fornothing.

KOBELL, Franz, Ritter von (Munich, 1803–82, Munich), became in 1826 a professor of mineralogy in Munich and is now remembered for his poetry in the Bavarian and Palatine dialects. He published *Gedichte in oberbayrischer Mundart* (1839–44), *Gedichte in hochdeutscher und pfälzischer Mundart* (1843), *Gedichte in pfälzischer Mundart* (1844), *Schnadahüpfln und Sprüchln* (1845). He was also the author of *Gedichte* (1852, in High German), an extended poem *Urzeit der Erde* (1856), the stories *Pfälzische Geschichte* (1866), *G'schpiel, Volksstücke und Gedichte in oberbayrischer Mundart* (1868), and *Schnadahüpfln und G'schichtle* (1872).

Kobell was a hunting companion of King Maximilian I (q.v.), and his interest in shooting led him to publish in 1859 *Wildanger. Skizzen aus dem Gebiete der Jagd und ihrer Geschichte*. His scientific publications included *Die Galvanographie* (1842) and *Geschichte der Mineralogie 1650–1860* (1864).

KOBER, Tobias (b. Görlitz, date of birth and death unknown), graduated at Helmstedt as a doctor of medicine in 1595, and was later in Hungary and Löwenberg, Silesia. He is the author of Latin plays, including tragedies (*Troja*, 1593; *Palinurus*, 1593; *Anchises exul*, 1594), a satirical comedy, *Hospitia* (1594), and a historical play dealing with the Turkish war of 1529 (*Mars sive Zedlicîus*, 1607). This work, which has as its hero Christoph Zedlitz, an officer who remains a steadfast Christian in Turkish captivity, also appeared in Latin (*Idea militis vere Christiani*). Zedlitz was a real person.

KOEPPEN, WOLFGANG (Greifswald, 1906–), grew up in East Prussia, was a student at Hamburg, Greifswald, Berlin, and Würzburg universities, and then led a restless life, now as journalist, now as dramatic adviser (Dramaturg), now as actor. As a novelist he began in 1934 with *Eine unglückliche Liebe*, in which the two worlds of disorder and anarchy on the one hand, and a stable middle-class existence on the other, are at loggerheads with each other, though they come eventually to an uneasy truce. Friedrich, the son of a respectable family, falls in love with Sibylle, an actress. He follows her and tries to enter her world, then thinks of bringing her into his. In the end they remain together, yet acknowledge an invisible barrier between them. A similar problem is the subject of the novel *Die Mauer schwankt* (1935 and 1939 in revised form under the title *Die Pflicht*), in which order is faced by disorder, not only through external forces, but within the central character himself, the architect Johannes von Süde.

Koeppen is best known for his three strongly negative, satirical novels of criticism of contemporary life, *Tauben im Gras* (q.v., 1951), *Das Treibhaus* (q.v., 1953), and *Der Tod in Rom* (q.v., 1954). Since this period of intensive fiction writing, he has published books on travel, *Nach Rußland und anderswohin* (1958), *Amerikafahrt* (1959), *Reisen nach Frankreich* (1961), and *New York* (1965). *Die ernsten Griechen*, appeared in 1962, prose, *Romantisches Café*, in 1972, *Jugend* (autobiographical) in 1976, and essays, *Die elenden Skribenten*, in 1981. He received the Büchner Prize in 1962.

KOKOSCHKA, OSKAR (Pöchlarn/Danube, Austria, 1886–1980, Villeneuve), one of the greatest painters of the first half of the 20th c., is best known for his dynamic Expressionist pictures of figures and cities. Late in life he turned to mythological themes. From 1920 to 1928 he held a professorship in the Dresden Academy, but thereafter devoted himself to painting and travel, which were intimately linked. He emigrated from Prague to England in 1938, moving to Villeneuve, Switzerland, in 1954. For many years he held a summer school in Salzburg.

In 1907 Kokoschka wrote the playlet *Mörder, Hoffnung der Frauen* (furnished with illustrations). Its treatment of sex was radically novel, textually and in its stage artistry. Pub-

lished in 1910 in the first number of *Der Sturm* (q.v.), it came to be regarded as the first Expressionist play. Its 1921 edition was set to music by Hindemith (q.v.). The short play *Sphinx und Strohmann* (written 1907, published 1913) was turned into three acts, re-titled *Hiob*, and, in 1917, produced by Kokoschka himself. Another play on sex, written in 1911, was published in 1913 as *Der brennende Dornbusch*. War experience led to the conception of his only full-length play, *Orpheus und Eurydike* (1919). *Die träumenden Knaben* (1907) and *Der gefesselte Kolumbus* (1920) are notable cycles of drawings that are accompanied by a verse text. In 1956 appeared *Spur im Treibsand*, a volume of stories, in 1971 his autobiography, *Mein Leben*, *Oskar Kokoschkas Schriften 1907–55*, ed. H. M. Wingler in 1956, *Das schriftliche Werk*, 4 vols., in 1973–6.

KOLB, ANNETTE (Munich, 1870–1967, Munich), daughter of a German architect employed by the Munich Botanical Gardens and of a French pianist, was the author of a handful of novels, many essays, and of the biographies *Mozart* (1937) and *Schubert* (1941). The novels, written in middle life, move in the higher strata of European society, mainly in Southern Germany, though *Das Exemplar* (1913) is set in England. Of her other novels, *Die Last* (1918), *Daphne Herbst* (1928), and *Die Schaukel* (1934), the last has special interest for its autobiographical elements.

Having both French and German sympathies, Annette Kolb strove for the reconciliation of the two nations. She spent the 1914–18 War in Switzerland, returned to Germany after it, but emigrated to Paris in 1933. In 1940 she took refuge in the U.S.A., coming back to Germany in 1945 and living by turns in Paris, Badenweiler, and Munich, where she died at the age of 97. Her intelligence, wit, and generosity of heart manifest themselves in her essays, which were collected in the following volumes: *Sieben Studien* (1906), *Wege. und Umwege* (1914), *Briefe einer Deutsch-Französin* (1916), *Zarastro* (1921), *Die kleine Fanfare* (1930), and *Beschwerdebuch* (1932). The *Versuch über Briand* appeared separately in 1929. The study *König Ludwig II und Richard Wagner* (1947) is a fragment of biography. Her recollections appeared as *Memento* (1960) and *1907–1964.—Zeitbilder* (1964); a posthumous edition of novels was published in 1968. In 1955 Annette Kolb was honoured with the Frankfurt Goethe-Preis, and in 1966 with the Pour le mérite für Kunst und Wissenschaft. She figures in *Meine Freunde, die Poeten* by H. Kesten (q.v.).

KOLBENHEYER, ERWIN GUIDO (Budapest, 1878–1962, Munich), was a student at Vienna

University and hoped for an academic career; he gave up this ambition after the success of his earliest works, and devoted himself entirely to authorship. In 1919 he moved to Tübingen, and in 1932 to Munich. In 1926 he was elected to the Section for Literature of the Prussian Academy (see AKADEMIEN). His strong National Socialist leanings led in May 1933 to his nomination as one of the fourteen founder members of a newly constituted Dichterakademie. Kolbenheyer was among its most conspicuous members, and in 1933 undertook lecturing abroad on the 'Führerprinzip' of the National Socialist regime. After the 1939–45 War, as a result of these activities, he was forbidden to publish for five years. In fact, his only post-war work was autobiographical (*Sebastian Karst über sein Leben und seine Zeit*, 3 vols., 1957–8).

Kolbenheyer held philosophical and mystical ideas which, on supposed biological grounds, devalued the individual, insisting on the primacy of race. He possessed a distinct talent for historical evocation, and, in spite of intellectual perverseness, was an intelligent and fluent narrator. His principal novels, which concentrate chiefly on thinkers of the past, are *Amor Dei* (1908, Spinoza, q.v.), *Meister Joachim Pausewang* (1910, in which, though Pausewang is fictitious, Jakob Böhme, q.v., is prominent), a trilogy on Paracelsus (q.v.), *Die Kindheit des Paracelsus* (1917), *Das Gestirn des Paracelsus* (1922), *Das dritte Reich des Paracelsus* (1926), and *Das gottgelobte Herz* (1938, a reversion to the 14th c., dealing with the mystic and nun, Margareta Ebner, q.v.). In all these novels, except the first, Kolbenheyer adopted an archaic style of writing. His contemporary fiction includes the novel *Das Lächeln der Penaten* (1927) and the three stories of *Ahalibama* (1913). His first play, *Giordano Bruno* (1903), was followed by a second version entitled *Heroische Leidenschaften*, and by the plays *Die Brücke, Jagt ihn—ein Mensch* (all 1929), and *Gregor und Heinrich* (1934). The tetralogy *Götter und Menschen* (1944) is his last dramatic work. His verse is contained in *Lyrisches Brevier* (1929) and *Vox humana* (1940).

Kolbenheyer gave to his philosophy the curious name 'Bauhüttenphilosophie' (perhaps best, though freely, translated as 'corporative philosophy') and expounded it in the tracts *Die Bauhütte* (1925) and *Bauhüttenphilosophie* (1942). *Gesammelte Werke* (8 vols.) appeared 1939–41, and *Gesamtausgabe der Werke letzter Hand* (14 vols., 1957 ff.) is being completed posthumously.

KOLBENHOFF, WALTER, pseudonym of Walter Hoffmann (Berlin, 1908–), was of proletarian origin. After a nomadic early life he

took refuge in Denmark in 1933, was called up in 1942, and taken prisoner by the Americans in 1944. He returned to Germany in 1946, and was prominent in the Gruppe 47 (q.v.) in its early days.

Kolbenhoff's first novel, *Untermenschen* (1933), is a book of social and anti-National Socialist protest. *Von unserem Fleisch und Blut* (1947), written in a P.O.W. camp, exposes the fixed ideas of some of his more fanatical fellow prisoners in a story about a member of an imaginary pro-National Socialist underground movement just after the war (see WERWOLF). *Heimkehr in die Fremde* (1948) was followed in 1960 by *Die Kopfjäger*; described as a detective novel (Kriminalroman), the latter attacks the prosperous materialistic Germany of the 1950s. *Das Wochenende: Ein Report* (1970) is a work of fiction in the documentary manner.

Kolberg or **Colberg,** Prussian Baltic coastal town, situated in Pomerania. It is famed for its successful resistance under Gneisenau and Nettelbeck (qq.v.) to a French siege in 1807, at a time when, under the shock of the disaster of Jena, the majority of Prussian fortified places surrendered without firing a shot. (See NAPOLEONIC WARS.)

Kolin, town in Czechoslovakia near which a battle was fought on 18 June 1757 (see SIEBENJÄHRIGER KRIEG). The Prussians under Friedrich II (q.v.) were defeated by the Austrians commanded by Daun (q.v.). The battle is referred to in the poem *Wer weiß wo?* by D. von Liliencron (q.v.).

KOLL, KILIAN, pseudonym of Walter Julian Bloem (see BLOEM, W.).

KOLLWITZ, KÄTHE (Königsberg, 1867–1945, Dresden), *née* Schmidt, married a doctor, and spent most of her life in Berlin, where she was a member and, from 1919, a professor of the Academy. Under the National Socialist regime her work was regarded as 'degenerate' (see ENTARTETE KUNST). Her remarkable drawings and etchings are based largely on sketches of poverty, grief, and anxiety in the slums of Berlin, emotively portrayed. Two woodcuts express her political sympathies: *Zum Gedächtnis für Karl Liebknecht* (1919) and *Proletariat* (1925). War, based on her experience in 1914–18, is a recurrent theme in her work. *Tagebuchblätter und Briefe*, ed. H. Kollwitz, appeared in 1949.

E. Barlach (q.v.) was a great admirer of her art. His anti-war war memorial, *Der schwebende Engel* (Güstrow, DDR), was designed to bear her features.

KOLMAR, Gertrud, pseudonym of Gertrud Chodziesner (Berlin, 1894–1943) and the name by which she is known. A Jewess of well-to-do family, she became a teacher and taught for a time in Dijon before returning to Berlin. She was arrested in 1943, and perished in one of the extermination camps (Vernichtungslager). She was a poetess of considerable power, able to 'exploit a wide range of forms and drawing constantly on images from nature to express a sense of solitude. Critics have compared her verse to that of A. von Droste-Hülshoff (q.v.), but her idiom is modern. Two vols. were published in her lifetime, *Preußische Wappen* (1934) and *Die Frau und die Tiere* (1938). *Das lyrische Werk* appeared in 1955 (enlarged edition, 1960) and the story *Eine Mutter* in 1965.

Kolmarer Liederhandschrift, see Colmarer Liederhandschrift.

Köln (Cologne), a city of Roman origin on the left bank of the Rhine, derives its name from the Roman Colonia Claudia Ara Agrippinensis. The archiepiscopal see of Cologne was founded under Charlemagne (see Karl I, der Grosse) in 785, and throughout the Middle Ages the archbishops were imperial princes and later electors (see Kurfürsten) of the Holy Roman Empire (see Deutsches Reich, Altes). The city itself achieved independence and became a free imperial city in 1288. It lost its privileges during occupation by the French in 1794, and in 1803 became a part of France. The Congress of Vienna (see Wiener Kongress) allocated it with the Rhineland to Prussia. In the 19th c. it attracted attention through the Kölner Kirchenstreit involving the reactionary archbishop Droste zu Vischering (q.v.) and through the movement to complete the cathedral (see Kölner Dom). In the 1939–45 War the city was severely damaged. The University of Cologne was founded in 1388, dissolved in 1797, and refounded in 1919. In the Middle Ages Albertus Magnus (q.v.), Duns Scotus, and Meister Eckhart (q.v.) were all active for lengthy periods in Cologne and in the 17th c. F. von Spee (q.v.) spent much of his life there.

Kölner Dom, the cathedral church of Cologne, a Gothic building begun in 1248 on the site of a 10th-c. church. The choir was completed in 1322, and work on the west front and towers went on throughout the 14th c., and intermittently in the 15th c., ceasing in 1520. The fragmentary cathedral, in two parts separated by the gap of the non-existent nave, was the characteristic feature of the view of the city from the 16th c. to the early 19th c. The Romantic movement rescued Gothic architecture from earlier neglect and instigated the idea of completing Cologne Cathedral. The chief supporter of this project was Sulpiz Boisserée (q.v.), who tirelessly propagated it in his writings and in personal conference. In 1814 he won the lasting support of the Prussian Crown Prince, later Friedrich Wilhelm IV (q.v.). In 1816, when Cologne had become Prussian territory, King Friedrich Wilhelm III (q.v.) was persuaded to commission K. F. Schinkel (1781–1841) to inspect the cathedral, and in 1823 the overdue work of preservation of the existing fabric was begun. In 1842, not long after Friedrich Wilhelm IV's accession, the task of completion was begun. The project developed gradually into a symbol of the new Germany, and the association with patriotism was more effective than aesthetic interest in speeding the flow of funds. The cathedral was completed in 1880.

Kölnischer Krieg, name given to a war occasioned by the conversion of the Electoral Archbishop Gebhard I of Cologne to Protestantism in 1582. He was overwhelmed in 1584 by combined Roman Catholic forces and fled. His defeat was the decisive factor in the maintenance of Catholicism in the Rhineland and Westphalia. (See Gebhard, Truchsess von Waldburg.)

Kölnische Zeitung, prominent German newspaper, published in Cologne. It first appeared under this title in 1798, but is descended from the *Postzeitung,* founded in 1651, which was located in Cologne from 1762.

Kolportageroman, the sensational or sentimental cheap novels, formerly hawked around the countryside by itinerant book-vendors (Kolporteurs).

KOLROSS, Johannes (Basel, c. 1500–58, Basel), a Protestant pastor, is the author of hymns and of a morality play, *Von fünferlei Betrachtnussen* (1532), in which the consequences of virtue and vice are demonstrated in the fate of two young men, one of whom finally goes to Heaven, the other to Hell. The play has choruses in Sapphic strophes. Kolross was also the author of a handbook of orthography (*Enchiridion,* 1529).

KOLUP, Tile, a pretender who alleged in Cologne in 1284 that he was the Emperor Friedrich II (q.v.), who had died in 1250. Kolup, after initial successes, was disbelieved and regarded as insane, but he established himself at Neuß and set up court there. He later moved to Wetzlar, where he was arrested and burned at the stake.

Kommersbuch, a song book for students' ceremonial drinking sessions (Kommers). The first appeared in 1781 at Halle, and the title *Kommersbuch* dates from 1810. The usually accepted standard *Kommersbuch,* many times reprinted, is that of 1858.

'Komm, Trost der Nacht, o Nachtigall', first line of the religious song sung by the hermit in Grimmelshausen's novel *Der abenteuerliche Simplicissimus* (q.v.). It occurs in Bk. 1, ch. 7.

'Komm, Trost der Welt, du stille Nacht!', first line of Eichendorff's poem *Der Einsiedler* (q.v.).

Kommunistische Partei Deutschlands, see KPD.

Kommunistisches Manifest, the fundamental socialist document drawn up by Karl Marx and Friedrich Engels (qq.v.) in London in 1847–8 and issued in 1848. The Manifesto contains the well-known summons to class war, 'Proletarier aller Länder, vereinigt Euch!'

Komödie, strictly a comedy, but used also, mainly in the 18th c., of any theatrical work, of the theatre itself (e.g. 'in die Komödie gehen'), or metaphorically.

KOMPERT, LEOPOLD (Münchengrätz, Bohemia, 1822–86, Vienna), son of a poor Jewish family, was at Prague University and was afterwards by turns a private tutor and a journalist. He wrote a number of ghetto stories (*Aus dem Ghetto,* 1848; *Böhmische Juden,* 1851; *Neue Geschichten aus dem Ghetto,* 1860; *Geschichten einer Gasse,* 1865) and three novels (*Am Pflug,* 1855; *Zwischen Ruinen,* 1875; *Franzi und Heini,* 1881). His complete works (*Sämtliche Werke,* 10 vols.) were published posthumously in 1906.

Komtesse, used to denote the daughter of a count (Graf) in her childhood, teens, and early marriageable years, though she is actually a Gräfin. Examples of the use of Komtesse in literary titles are *Zwei Komtessen* (q.v., 1885) by M. von Ebner-Eschenbach and *Komtesse Mizzi* (1907) by Schnitzler (q.v.).

KONEMANN, PFAFFE, a priest of Dingelstädt in Thuringia, is the author of a Middle Low German allegorical poem, *Sante Marien wortegarte* (*St. Mariens Wurzgarten*), completed in 1304, which deals with the story of man's fall and salvation from the Fall of Lucifer to the Day of Judgement. He is probably also the author of a Middle Low German didactic poem, *Der*

Kaland, setting out the statutes of a pious fraternity which include an early example of burial insurance.

KÖNIG, JOHANN ULRICH (Eßlingen, 1688–1744, Dresden), Saxon court poet 1729–44, visited the Netherlands and Hamburg as companion to a young nobleman. In Hamburg, where he remained from 1710 to 1716, he made the acquaintance of B. H. Brockes (q.v.), with whom he collaborated. Later he was Saxon court jester in Dresden and became friendly with J. Besser (q.v.), whom he succeeded as court poet to August II (q.v., der Starke) in 1729. Tradition has it that he made the rival candidate, J. C. Günther (q.v.), drunk before the interview. König's earlier poems, which include opera libretti, were collected in *Theatralische, Geistliche, Vermischte und Galante Gedichte* (1713). A further volume (*Gedichte,* 1745) appeared after his death. In 1731 he published the first canto (which remained the only one) of a grandiose epic poem dealing with some large-scale military manœuvres ordered by Augustus (*August im Lager*).

Königgrätz (Sadowa), battle of, was fought on 3 July 1866 (see DEUTSCHER KRIEG). The Prussian armies under the command of King Wilhelm I, with Moltke (qq.v.) as his chief of staff, outflanked the Austrian forces under Benedek (q.v.), which retreated in disorder.

König in Thule, Der, a ballad by Goethe in *Faust* (q.v.), Pt. I, where it is sung by Gretchen in the scene *Abend. Ein kleines reinliches Zimmer.* It was written in 1774 and is included in the *Urfaust.* It was first published in 1782, with the title *Der König von Thule.*

Königliche Hoheit, a novel by Th. Mann (q.v.), published in 1909. It is the story of a German prince, Klaus Heinrich, younger brother of the heir to a fictitious grand-ducal throne. Klaus Heinrich is born with a malformed left arm, and an important part of his education is devoted to the concealment of this handicap. When the Grand Duke dies, his successor, Albrecht II, lives in seclusion, delegating representational duties to Klaus Heinrich. Klaus Heinrich meets an American multimillionaire, Samuel N. Spoelmann, and his attractive and intelligent daughter, Imma. Klaus Heinrich is fascinated by her astute independence, but a match would be a *mésalliance.* Fortunately, the fiscal advantage to the miniature state of having a resident millionaire is so considerable that economics takes precedence over lineage, and the marriage of the lovers concludes this modern fairy-tale.

König Ottokars Glück und Ende, a blank-verse tragedy in five acts by F. Grillparzer (q.v.), written within four weeks in 1823. It is his first mature historical play; for it he used sources (including the medieval *Österreichische Reimchronik,* see OTTOKAR VON STEIERMARK) from the age in which Rudolf I (q.v.) founded the Habsburg dynasty, thus putting an end to the powerful rule of Ottokar II (q.v.) of Bohemia. It was not until 1825 that the play was passed for production at the Burgtheater by the Austrian censorship, and even then only after the intervention of the Empress. But, largely owing to the strong objections of the Czechs to Grillparzer's portrayal of Ottokar, it failed to win for Grillparzer the acclaim as a national dramatist for which he had hoped, thus proving a bitter blow to his patriotism and to his plan to proceed with a variety of similar dramatic projects.

The plot shows Ottokar at the height of his power and happiness after his divorce from Margarete of Austria, through whom he had acquired Austria and Styria, and his marriage to Kunigunde of Massovia, granddaughter of King Bela of Hungary. He is hailed as the future emperor before the news of Rudolf's election as emperor provokes him to defiance (Grillparzer uses the imperial title, aimed at but not attained by the historical Rudolf I, symbolically). After defeat and betrayal within his own ranks Ottokar submits to Rudolf in the seclusion of the imperial tent. As he kneels to receive the fiefs of Bohemia and Moravia, Zawisch von Rosenberg, who has deceived him with his unfaithful young wife, opens the tent to effect a public exposure of Ottokar's homage. Unable to bear this humiliation, Ottokar engages in renewed aggression and is killed by the young Seyfried Merenberg (whose father is the victim of Ottokar's impetuous tyranny), acting against Rudolf's express order not to harm the King of Bohemia except in self-defence. Rudolf honours the slain king, who at the time of his death had overcome the pride which had deprived him of loyalty, affection, and good fortune.

Rudolf I is portrayed as the ideal ruler, just and yet magnanimous, humble and yet dignified (perhaps a tribute to the enlightened despotism of Joseph II, q.v.). Some aspects of the complex character of Ottokar were inspired by Grillparzer's study of Napoleon. Ottokar's death is preceded by that of his first and loyal wife Margarete, at whose bier the repentant king pays homage.

König Rother, a Middle High German epic poem recounting a double (i.e. repeated) wooing. Rother, whose capital is Bari in Apulia, decides to wed, in order to ensure an heir to his kingdom. He is advised to woo the daughter of the Emperor Konstantin in Constantinople, who opposes any marriage of his daughter and executes the suitors. Rother is well received by the princess and, outwitting Konstantin, carries her off to his own land. Konstantin, however, sends an emissary, who entices her on board ship and carries her back to Constantinople. Rother undertakes a second wooing in disguise, is captured, identified, and condemned to death. He is rescued in the nick of time and sails home with his bride.

In its fresh and vivid pleasure in the treatment of a secular story the poem represents a new phase in the development of medieval German poetry. The view of it as a minstrel's epic (see SPIELMANNSDICHTUNG) has, however, been abandoned, though the term is still often used for convenience. The author is held to have been a Bavarian priest who wrote the work *c.* 1150 for a noble public. It is a strikingly successful and entertaining adventure novel in verse.

Königsberger Dichterschule, a group of poets active in Königsberg *c.* 1625–*c.* 1650. The central personalities were the composer-poet Heinrich Albert and Robert Roberthin (qq.v.). The best-known poets were Simon Dach (q.v.), Valentin Thilo der Jüngere (1607–62), and Christoph Kaldenbach (q.v.). Other members were Andreas Adersbach (1610–60), Michael Behm (1612–50), Jonas Koschwitz (1614–64), Albert Lingemann (1603–53), and Christoph Wilkau (1598–1641). They forgathered in summer in Albert's garden in a hut, known as the 'musikalische Kürbishütte' because Albert there carved lines of verse in the skin of pumpkins. In 1638 M. Opitz (q.v.) was a guest of the circle, and later J. P. Titz (q.v.) of Danzig, a disciple of Opitz, was in touch with the group, whose members frequently wrote poems to order for special occasions.

Königsberger Jagdallegorie, an anonymous Middle High German allegorical poem of the 14th c. dealing with the pursuit of love in terms of a hunt. The woman (das Wild) is seen as unworthy game and the poem is satirical. It is thought to derive from *Die Jagd* of HADAMAR von Laber (q.v.). Written in Alemannic dialect, it takes its name from the former location of the MS. in the library of Königsberg University.

Königshofen, Chronik von, see TWINGER VON KÖNIGSHOFEN, JAKOB.

Königsleutnant, Der, a four-act comedy by K. Gutzkow (q.v.), performed in the Frankfurt Stadttheater on 27 August 1849, the eve of the centenary of Goethe's birth, for which it was written. It was published in 1852. The Königsleutnant is Count Thoranc, the French military governor of Frankfurt (1759–63) during the Seven Years War (see SIEBENJÄHRIGER KRIEG), who resided for a time in the Goethe family house. The slender *pièce d'occasion* freely adapts and amplifies events recorded by Goethe in *Dichtung und Wahrheit* (q.v., Bk. 3).

KÖNIGSMARCK, MARIA AURORA, GRÄFIN VON (Stade nr. Hamburg, 1662–1728, Quedlinburg), a descendant of the Swedish line of an old noble family of the Old March (Altmark) of Brandenburg, was a woman of exceptional beauty and intelligence. She attracted the attention of August II (q.v., der Starke), who made her his mistress in 1694. Their illegitimate son, born in 1696, became the French commander, the Maréchal de Saxe (see SAXE, MAURICE DE). Aurora's liaison with August was of short duration, and in 1700 she became lady provost (Pröbstin) of the Protestant convent for noblewomen at Quedlinburg, where she is buried. She is the subject of several novels, including *Frau Maria* (1929–31) by Enrica von Handel-Mazetti (q.v.).

KÖNIG VOM ODENWALD, DER, style adopted by a poet of the 14th c. who lived and wrote in the archiepiscopal state of Würzburg. The title probably signifies that he was the head of a group or guild of poets and minstrels. His twelve poems in the loose form of Reimrede (q.v.), are mostly bald accounts of material things (e.g. in *Der gense lob*, the usefulness of geese for their flesh and feathers). A tendency to parody is perceptible, and it is sometimes difficult to determine whether or not a poem or passage is serious. Two of the Reden are fables with animals as characters.

Koninc Ermenrikes Dot, a Middle Low German poem in 24 eight-line stanzas (see HILDEBRANDSTON), preserved in a broadsheet printed in Lübeck in 1560. The text is corrupt. It recounts a victory by Dietrich over Ermanarich, in which the latter is slain (see DIETRICHSAGE).

Konkrete Poesie, a form of *avant-garde* writing in which words are used as objects without reference to grammatical position or function, and in some cases regardless of their meaning, associations, or allusiveness. For the poet experimenting with this form, words—or even detached letters—are raw material to be arranged in a pattern. Since these patterns are frequently visual, they can be appreciated at a glance without being read, and some are deliberately unreadable and represent an extreme form of 'visuelle Poesie'. Concrete poetry at its best is a 'three-dimensional—semantic, phonetic, visual—exploration' (M. Butler, with special reference to E. Jandl, q.v.). Initial capitals are not used. Readings of sound and speech poems (Sprechgedichte) are an aspect of this type of experimental poetry which is published in print as well as on records.

The immediate forerunners of German concrete poetry are considered to be H. Arp and K. Schwitters (qq.v.), who were both involved in Dada (see DADAISMUS); both were painters and inclined to the visual. The best-known names among the German Konkretisten, apart from Jandl, are F. Achleitner, H. C. Artmann, E. Gomringer, H. Heißenbüttel, F. Mon, and G. Rühm (qq.v.): K. Bayer styled Rühm and Achleitner 'concrete poets' in the context of the Wiener Gruppe (q.v.). The anthology *konkrete poesie. deutschsprachige autoren*, ed. E. Gomringer (1972), contains work by all of these and by twelve other such poets. Much ingenuity has been devoted to concrete writing, especially in typography. One text by Rühm forms a square composed of the letter u (152 times) with a single d at its centre (it may be noted that 'du' is intimate German for 'you'). This ingenious form of literature gives ample scope for playfulness, some poets aiming at a purely aesthetic effect, though most are designed to convey social (and political) criticism. Affinities between concrete poetry and the language games of L. Wittgenstein (q.v.) point to the influence of his approach to reality.

KONRAD I, KÖNIG (d. 918), was elected German King in 911 in circumstances that remain obscure. He was related to the Carolingian line only by marriage, and his election therefore marks the end of the dynasty. His short reign was occupied by ceaseless and often fruitless struggles to maintain his power, and he is said, in the end, to have proposed Heinrich von Sachsen (see HEINRICH I), his principal enemy, as his successor. Konrad was buried at Fulda.

KONRAD II, KAISER (c. 990–1039, Utrecht), was elected German King at Mainz in 1024 and crowned emperor in Rome in 1017. Konrad, the first of the Salic emperors (see SALIER), succeeded Heinrich II (q.v.), and profited by Heinrich's prestige. In 1026 and in 1027 he suppressed revolts by his stepson Ernst von Schwaben, who was later outlawed and killed in 1030.

Konrad established his power in Italy, and in 1032 secured the lines of communication between Germany and Italy by the inheritance of the crown of Burgundy, in accordance with a treaty entered into by Heinrich II. His claim was contested by Odo of Champagne, whose attempts to wrest Burgundy from the Emperor by force were defeated. Konrad restored order on the eastern frontiers of Germany, and pushed forward into southern Italy. From 1034 it became customary to apply the term *Romanum Imperium* to the whole of the empire, and no longer to Italy alone (see DEUTSCHES REICH, ALTES). Konrad is buried in Speyer Cathedral, the building of which was begun in 1030 under his protection.

KONRAD III, KÖNIG (1093 or 1094–1152, Bamberg), the first Hohenstaufen (q.v.) ruler, was elected German King in 1138 after the death of the Emperor Lothar III (q.v.) in December 1137, having been elected anti-king (Gegenkönig) by a faction of the territorial princes in 1127. The early years of his reign were occupied in the attempted suppression of a series of revolts by dissident princes, notably Heinrich der Stolze and his son Heinrich der Löwe (qq.v.). In 1147 Konrad was persuaded to take part in the Second Crusade, which was a tragic failure. He himself was wounded, and he returned in 1150 to find the realm in disorder. His reign proved to be an interlude of struggle and failure between the successful reigns of Lothar and Friedrich I (q.v.). In a restless age, which coincided with stormy events for the papacy, Konrad was not able to achieve coronation at Rome, and therefore never bore the imperial title. He died while preparing an expedition to Rome.

KONRAD IV (Andria, Apulia, 1228–54, Lavello), younger son of the Emperor Friedrich II (q.v.), was elected German King in 1237 at the age of 9. It fell to him to sustain the imperial cause against the anti-kings (Gegenkönige) HEINRICH Raspe and WILHELM von Holland (qq.v.), against whom he fought from 1246 to 1251. After his father's death in 1250, Konrad set out for Italy to defend his possessions there against the papacy. In this he succeeded, but his death in 1254 precipitated a new crisis. His son, only 2 years old at Konrad's death, bore the same name and has gone down in history and literature as Konradin (q.v.).

KONRAD, DER JUNGE, KÖNIG, see KON-RADIN.

KONRAD, PFAFFE, a priest of Regensburg probably in the service of Duke Heinrich der

Löwe (q.v.), in the second half of the 12th c., and author of the *Rolandslied* (q.v.). He is no longer regarded as the author of the *Kaiserchronik* (q.v.).

KONRADIN (Wolfstein, Bavaria, 1252–68, Naples), name derived from the Italian diminutive Conradino, by which the boy Duke Konrad of Swabia is usually known. Konradin was the son of the German King Konrad IV (q.v.), and in 1267 sought to assert his right to Sicily, which Pope Clement IV had granted to Charles d'Anjou in 1265. Konradin crossed with a small force from Germany into Italy, where he was enthusiastically received. In August 1268, however, he was defeated at Tagliacozzo by the forces of Anjou. He escaped to Rome and thence to the coast, where he was taken prisoner and handed over to his enemy, who had him tried for treason and executed at Naples (he is buried there in the church of Santa Maria del Carmine.)

Konrad was barely 16½ years old when he died, and at his death the house of Hohenstaufen (q.v.) became extinct. Contemporary records speak of his handsome appearance, and it is not surprising that the tragic fate of so young and good-looking a prince has tempted many writers to treat the story in dramatic or narrative form. Something like a hundred such works are known, though none has secured general recognition, and many have remained fragments. The more notable completed works are: K. C. Beyer, *Commedia von der Histori Herzog Conrads Schwaben*, 1585; J. J. Bodmer, *Konradin von Schwaben* (verse narrative), 1771; C. P. Conz, *Konradin von Schwaben* (tragedy), 1782; F. M. Klinger, *Konradin* (tragedy), 1784; G. Schwab, *Konradin in Deutschland* (poem), 1828; E. Raupach, *Die Hohenstaufen* (tragedy), 1837; M. Greif, *Konradin, der letzte Hohenstaufe* (tragedy), 1889; Konrad Weiß, *Konradin von Hohenstaufen* (drama), 1938. Among the authors of unfinished works are J. A. Leisewitz, Theodor Körner, J. von Eichendorff, A. von Platen, L. Uhland, and Gerhart Hauptmann (qq.v.).

Konradin was noted among his contemporaries for his command of Latin. It seems likely that he was also gifted in the handling of poetic language in the vernacular; for in the Große Heidelberger Liederhandschrift (see HEIDELBERGER LIEDERHANDSCHRIFT, GROSSE), two poems are attributed to him under the style *Kunig Chuonrat der junge*, and an illustration depicts him mounted on a grey horse and flying a gyr-falcon. The poems are in the manner of GOTTFRIED von Neifen (q.v.), and were probably written in Germany in 1267.

KONRAD VON AMMENHAUSEN, from

Canton Thurgau, was a monk at Stein am Rhein (between Constance and Schaffhausen). He is the author of *Das Schachzabelbuch*, which is a free and expanding translation, in nearly 20,000 lines of Middle High German verse, of the *Solatium ludi scaccorum* of Jacopo Dacciesole (see SCHACHBÜCHER). Konrad's diffuse and pedestrian work, which was completed in 1337, became one of the most popular books of the late Middle Ages, probably because it contains social criticism based on experience and observation.

KONRAD VON FUSSESBRUNNEN, an Austrian nobleman, was the lay author of a biblical epic of some 3,000 lines, written *c.* 1200. The work, which is known as *Die Kindheit Jesu*, is in rhyming couplets, and the style suggests the influence of the early works of HARTMANN von Aue (q.v.). *Die Kindheit Jesu* begins with Jesus's parentage, and narrates with naïve charm the Annunciation and the Virgin Birth, the Flight into Egypt, and the return to Nazareth. It is plentifully supplied with extra-biblical details by its source, the apocryphal gospel of the childhood of Jesus. Konrad himself acknowledges 'Meister Heinrich', the presumed author of a life of the Virgin, as his forerunner.

KONRAD VON HASLAU is the author of the Middle High German poem *Der Jüngling* (q.v.). His name is known only through a reference in *Seifriet Helblinc* (q.v.) to 'von Haslou meister Kuonrât'. He is presumed to have lived in the 13th c. and was apparently an Austrian commoner who used his talents at various courts.

KONRAD VON HEIMESFURT, author of two Middle High German religious poems in rhyming couplets, *Die Himmelfahrt Mariä (Von unser vrouwen hinvart)* (*c.* 1225) and *Die Auferstehung (diu urstende)* (*c.* 1230), lived in the first half of the 13th c. Konrad was a cleric, and Heimesfurt is thought to be the modern Hainsfarth in Bavaria. *Die Himmelfahrt Mariä* narrates the death of the Virgin in the presence of the Apostles, her awakening by Jesus, her assumption, and the giving of her girdle to Thomas. The source, given by Konrad, is a Greek work by Bishop Melito of Sardes in Asia Minor; Konrad read it in a Latin version, *De transitu Mariae virginis*. The poem is a notable document of the cult of the Blessed Virgin. *Die Auferstehung* is an account of the resurrection, using apocryphal material, notably the miracle of the flags in the presence of Pilate, the miraculous delivery of Joseph of Arimathea from prison, and Christ's descent into Hell, the source of which is the late Gospel of Nicodemus.

KONRAD VON HELMSDORF, see SPECULUM HUMANAE SALVATIONIS.

KONRAD VON KILCHBERG, GRAF, a minor Middle High German poet of the 13th c., was a Swabian recorded between 1255 and 1268. Four Minnelieder and a dance song survive, in which Konrad's familiarity with REINMAR der Alte and GOTTFRIED von Neifen (qq.v.) is perceptible.

KONRAD VON LANDECK, a minor Middle High German poet, of the second half of the 13th c., was cup-bearer (Schenk) of St. Gall. One of his 22 poems purports to be written before the walls of Vienna, when he was with the besieging army in 1276. Another, which laments the hard winter weather in France and yearns for spring by Lake Constance, is remarkable for its profusion of topographical names. Konrad's smooth and agreeable verses are substantially indebted to GOTTFRIED von Neifen (q.v.).

KONRAD VON MARBURG (Marburg, *c.* 1180–1233, nr. Marburg), who probably belonged to the Dominican Order, was in 1227 appointed by the Pope to reform the clergy and to pursue heresy, to which duties he attended with a zeal which reached the point of terrorism. He was the confessor of Saint Elisabeth (see ELISABETH, HEILIGE), whom for her soul's good he subjected to painful and humiliating discipline. His excesses provoked concerted action against him, and he was waylaid and murdered by a group of noblemen. He is an important character in Charles Kingsley's verse play, *The Saint's Tragedy*.

KONRAD VON MEGENBERG (nr. Schweinfurt, *c.* 1309–74, Regensburg), a canon of Regensburg, is known for his scientific writings in German and for his political writings in Latin. Konrad studied in Paris and gave lectures there. In 1342 he moved to Vienna and shortly after to Regensburg, where he spent the remainder of his life. His *Buch der Natur* (1349–50) is a survey of the scientific knowledge of his time based on the Latin *Liber de natura rerum* of Thomas Cantipratensis (*c.* 1240). In *Die deutsche Sphaera* he provided a treatise on astronomy and physics (the astronomy is homocentrically conceived as astrology), using as his source the *Sphaera mundi* of Johannes de Sacro Bosco, the latinized name of John Holywood (d. 1244 or 1256).

In the conflict between the Emperor Ludwig IV (q.v., der Bayer) and the papacy at Avignon, Konrad intervened powerfully on the Emperor's side with the Latin tract *Planctus ecclesiae in Germaniam* (1338).

KONRAD VON STOFFELN is the presumed author of the Middle High German verse romance, *Gauriel von Muntabel*. He was possibly a native of Hohenstoffeln in Hegau north-west of Lake Constance. The date of the poem is uncertain, but it was probably written in the mid-13th c.

Gauriel loves a fairy. By revealing his love he loses her and also his good looks. Both are regained after a series of combats in which he defeats three Arthurian knights. Somewhat unnecessarily Gauriel is accompanied by a goat, and is known as 'der Ritter mit dem Bock'. This and other features are derived from Hartmann's *Iwein* ('der Ritter mit dem Löwen'); Konrad is also indebted to *Lanzelet*, *Wigalois*, and Der Pleier (qq.v.).

KONRAD VON WÜRZBURG (Würzburg, *c.* 1225–87, Basel), a Middle High German poet of great versatility and formal mastery, was a commoner by birth and spent his life as a professional poet, writing for wealthy patrons in Strasburg and then in Basel, which became his second home. The clergy, the city patricians, and the newly risen merchant class commissioned or encouraged his poetry, and he mentions several of these patrons by name.

The bulk of Konrad's poetry is considerable and he was active in almost all the forms current in his day. His 23 Minnelieder, in which nature plays a conspicuous part, comprise summer and winter songs and Tagelieder (see TAGELIED), and are distinguished chiefly for their smooth rhythms and the virtuosity of their rhyme schemes. His Sprüche (see SPRUCH) are largely conventional, dealing mainly with moral and religious matters; politics seem to have had little interest for him. His outstanding lyrical poems are a religious Leich and a Minneleich (see LEICH), which are remarkable for the rich elaboration of their style. *Die Klage der Kunst*, a poem on a larger scale with 32 eight-line stanzas, is an allegory, in the form of a court of law, in which false generosity, which rewards others than the true artists, is condemned. In its stylistic intricacy it is a striking example of the style which Konrad terms 'geblümte Rede' ('flowered speech').

Konrad's best work is probably to be found in his shorter verse romances, which range in length from less than 300 to more than 1,300 lines. They comprise *Das Herzemaere*, *Der Welt Lohn*, *Heinrich von Kempten*, and *Der Schwanritter* (qq.v.); the last-named is a version of the story of Lohengrin. Similar in style and scope is a verse account of an imaginary tournament,

which includes much heraldic description of remarkable virtuosity (*Das Turnier von Nantes*, q.v.). Konrad also wrote three verse legends of saints: *Silvester*, an early work written for Leuthold von Röteln, later bishop of Basel, which recounts a legend ot Pope Silvester; *Der heilige Alexius*, written in Basel; and *Pantaleon*, the story of a Roman physician and Christian martyr, the latest of the three, and stylistically the most mature. *Die goldene Schmiede*, an exalted hymn to the Virgin Mary, in which the abundant stylistic detail suggests the art of the goldsmith, was the most widely read of Konrad's religious poems. It is believed to have been commissioned by Konrad von Lichtenberg, Bishop of Strasburg.

Of three long verse romances, *Engelhard*, *Partonopier und Meliur*, and *Der Trojanerkrieg* (qq.v.), the last was interrupted, probably by Konrad's death in 1287, and was finished by another hand. The expansiveness of these poems no doubt suited the taste of the age, but the stylistic skill appears an inadequate compensation for the *longueurs* resulting from the poet's failure to maintain the thread and movement of his story.

Konrad was one of the most influential poets of an age which valued the skill of the master craftsman rather than originality. He is a virtuoso of rhyme and a master of mellifluous verse, of the 'geblümte Rede' which he also terms, in *Die goldene Schmiede* 'der süezen rede bluot' (der süßen Rede Blüte).

Konstanzer Konzil, Das, a great ecclesiastical council called by Pope John XXIII at the request of the Emperor Sigismund (q.v.) in 1414. It was intended to end the Great Schism, to reform the Church, and to deal with the alleged heresy of John Hus (q.v.) and his Bohemian followers. The Council in fact saw John XXIII and the two other rival popes succeeded by Martin V, but it made little progress in the matter of reforms before its dispersal in 1418. In dealing with Hus it was merciless. In spite of a safe-conduct, he was condemned and burned in 1415.

Kontrafaktur, term for the adaptation of secular songs to religious purposes by alteration or substitution, or (more rarely) for the reverse process. Recognizable Kontrafakturen occur as early as the 13th c., but they are particularly frequent in the 15th c. The process continues with diminishing frequency into the 17th c. Well-known examples of Kontrafaktur are Luther's 'Vom Himmel hoch' (q.v.) and 'O Welt, ich muß dich lassen', by H. Knaust (q.v.). The term is also applied to the substitution of a new text for an old one, to be sung to the tune of the rejected words.

Konversationslexikon, see ENZYKLOPÄDIE.

Kopf, Der, a satirical novel by H. Mann (q.v.), published in 1925. It is concerned with the ruling powers of Wilhelmine Germany, covers the period 1891–1914, and forms the third part of the trilogy known as the *Kaiserreich-Trilogie.* It is a *roman à clef*; the Emperor Wilhelm II (q.v.) appears by name, Lanna is the fictitious name for Bülow, Fischer for Tirpitz, Tolleben for von Bethmann-Hollweg, and Gubitz for F. von Holstein (qq.v.). Two important characters, Mangolf and Terra, are entirely fictitious. Mangolf, the so-called realist, and Terra, the devious idealist, both seek power. The former, to attain his ends, is willing to surrender his integrity by subservience to Lanna (Bülow). After the (fictitious) death of Tolleben (Bethmann-Hollweg) at the outbreak of the 1914–18 War, Mangolf becomes his successor. He recognizes that the German defeat at the Marne means that the war is lost, but he cannot stop it because the true power is held by the generals and admirals. Both Mangolf and Terra (whose personal affairs also occupy a substantial part of the novel) commit suicide.

The incongruity of the factual and the fictional may be one reason for the comparative neglect of this novel.

KOPISCH, AUGUST (Breslau, 1799–1853, Berlin), began as a painter, studying in Dresden, though handicapped by an accident to his right hand. In 1823 he went to Italy, staying for five years, and discovering, while on a swim, the Blue Grotto of Capri. In Italy he met A. von Platen (q.v.), and was presented to the future king of Prussia, Friedrich Wilhelm IV (q.v.). In 1828 he returned to Germany, living in Berlin and then in Potsdam. Friedrich Wilhelm IV commissioned *Die Schlösser und Gärten zu Potsdam,* which was published posthumously in 1854. Kopisch translated Dante's *Divina commedia* (1837) and wrote humorous fairy and elfin poetry (*Der Nock,* set by C. Loewe (q.v.), is a good example), and agreeable drinking songs, including 'Als Noah aus dem Kasten war'.

Koralle, Die, a play (Schauspiel) by G. Kaiser (q.v.) first performed at the Kammerspiele Munich, in October 1917 and published in the same year. It is often reckoned as the first part of a trilogy (*Gas-Trilogie*), though this title was not authorized by Kaiser. The connection between the three plays is nevertheless clear. The five acts are written in Expressionist prose. The principal figure is the Milliardär. He has begun at the bottom, climbing from poverty to fabulous wealth, and eases his conscience by distributing money and gifts to those in need. At the end of Act One he learns that his son, rather than live idly on his father's wealth, has chosen to sign on as a stoker in a collier. This news shatters the Milliardär. Although his son returns, there is no reconciliation, and the Milliardär seeks to escape from his own existence. He has a secretary (Der Sekretär), who is his double, and only distinguishable from his employer by a piece of coral (Koralle) attached to his watch-chain. The Milliardär murders him and assumes his identity, attaching the fragment of coral to his own watch-chain. He is mistaken for the secretary and charged with murdering his employer, the 'Milliardär', i.e. himself. He is tried and condemned, and goes to his death with firm step, content that he has successfully shaken off his own self. See also GAS I and GAS II.

KOREFF, DAVID FERDINAND (Breslau, 1783–1851, Paris), studied in Berlin, practised as a physician in Paris 1804–11, then returned to Berlin, where he frequented circles of Romantic *littérateurs* such as A. von Chamisso, K. A. VARNHAGEN von Ense, and E. T. A. Hoffmann (qq.v.), who included him as Vinzenz in the introductory narrative (see RAHMEN) to *Die Serapionsbrüder* (q.v.). In 1816 he became professor of medicine at Berlin University. Koreff translated Tibullus (*Tibulls und der Sulpicia Elegien,* 1810) and wrote two opera libretti (*Don Tacagno,* 1819; *Aucassin und Nicolette,* 1822). His poems (*Lyrische Gedichte*) appeared in 1815.

KÖRNER, CHRISTIAN GOTTFRIED (Leipzig, 1756–1831, Berlin), friend of Schiller, studied at Göttingen and was for a short time a lecturer (Dozent) in law at Leipzig University. In 1783 he received a minor administrative appointment (Oberkonsistorialrat) in Dresden. In 1875 Körner, his friend Huber, and their fiancées, Minna and Dora Stock, invited Schiller to stay with them at Leipzig, thus extricating him from an emotional and financial impasse. The Körners moved to Dresden, accompanied by Schiller, who remained with them until 1787. The life-long friendship is commemorated in their correspondence (*Schillers Briefwechsel mit Körner,* 1847), and Körner remained a valued critic and commentator. In 1815 Körner entered the Prussian civil service in Berlin. He contributed occasionally to Schiller's periodicals, and edited Schiller's works, writing the first biography (1812–15). His only son, the poet Theodor Körner (q.v.), was killed in 1813 in the Wars of Liberation.

KÖRNER, Theodor, in full Karl Theodor (Dresden, 1791–1813, Gadebusch, Mecklenburg), only son of Schiller's friend C. G. Körner (q.v.), was sent down from Leipzig University because of a duel and went in 1811 to Vienna, where he began to write plays. One of these, the tragedy *Zriny*, was performed at the Theater an der Wien and shortly afterwards he was appointed house dramatist (Hoftheaterdichter) to the Burgtheater (1813). In March 1813 he joined Lützow's Free Corps, was commissioned in April, and in August was mortally wounded in a skirmish near Gadebusch in Mecklenburg.

Körner was a precocious and prolific writer, who left five tragedies, including *Toni* (q.v.), an adaptation of the story *Die Verlobung in St. Domingo* (q.v.) by H. von Kleist, and five comedies, of which the best known is the one-act *Der Nachtwächter* (1812). He also wrote a one-act anecdote based on his war experiences (*Josef Heyderich*), and four stories. Little attention was paid to his early poetry (*Knospen*, 1810), but his patriotic poems, collected by his father and published posthumously as *Leyer und Schwerdt* (1814), were read all over Germany. They are the best work of a facile author, whose reputation was enhanced for more than a century by the patriotic manner of his death in the Wars of Liberation (Befreiungskriege, see Napoleonic Wars). *Werke* (3 vols.), ed. A. Stern, appeared in 1890 and *Werke* (2 vols.), ed. H. Zimmer in 1916.

KORNFELD, Paul (Prague, 1889–1942, Lodz, Poland), worked in the theatre in Frankfurt, Berlin, and Darmstadt as dramatic adviser (Dramaturg), and wrote ecstatic Expressionist plays. More clearly than many Expressionist writers, Kornfeld reveals the essential religious striving which animates the condemnation of society and the groping, searching quality of the exclamatory language. This is not contradicted by the sensational aspect of his plots, for his plays *Die Verführung* (1916) and *Himmel und Hölle* (1919) operate on two planes. He is also responsible for the important essay *Der beseelte und der psychologische Mensch* (1918), which explains the fundamental primacy of spirituality in his Expressionism. After the war he turned to comedies (*Der ewige Traum*, 1922; *Palme oder Der Gekränkte*, 1924; *Kilian oder Die gelbe Rose*, 1926). *Jud Süß* (1930, see Süss-Oppenheimer, J.) is his last Berlin play. His only novel, *Blanche oder Das Atelier im Garten* (1957) was written in Prague before his arrest in 1941. He died in a concentration camp. *Paul Kornfeld. Revolution mit Flötenmusik und andere kritische Prosa* appeared in 1977.

KORTUM, Karl Arnold (Mülheim an der Ruhr, 1745–1824, Bochum), who became a physician in his native town, was an eccentric with an interest in the occult. In 1782 he published a work on secret codes, *Entzifferungskunst deutscher Zifferschriften*. His principal work, however, is a comic epic with the intentionally cumbersome title *Leben, Meynungen und Thaten von Hieronimus Jobs, dem Kandidaten, und wie Er sich weiland viel Ruhm erwarb, auch endlich als Nachtwächter zu Sulzburg starb*, which appeared in 1784. It narrates, in jolting Knittelverse (q.v.), grouped in four-line stanzas, Jobs's childhood, university days as a student of theology, and his subsequent decline and fall. The illustrations, woodcuts in the manner of the 16th c., were provided by the author. In 1799 it was provided with a continuation in which Jobs, resuscitated, becomes a model pastor. This expanded version is entitled *Die Jobsiade*. Kortum was also the author of a popular medical work (*Skizze einer Zeit- und Literaturgeschichte von Arzneikunde*, 1809).

KOSEGARTEN, Ludwig Theobul (Grevesmühlen, Mecklenburg, 1758–1818, Greifswald), a pastor's son, took orders after studying theology at Greifswald University. He was baptized Ludwig Gotthard, but substituted Theobul for his second name. After some years in various parts as private tutor he was placed in charge of a school at Wolgast (1785), and was later (1792) a senior pastor at Altenkirchen on Rügen. In 1808 he was called to a chair of theology at Greifswald. In his early years Kosegarten wrote poems (*Gedichte*, 1776), and composed a derivative tragedy in the manner of Sturm und Drang (q.v.). This work, *Darmond und Allwina* (1777), was dedicated to J. A. Leisewitz (q.v., 'dem Verfasser des Julius von Tarent zugeeignet'). He is best known for his two short sentimental epics in hexameters, *Jucunde. Eine ländliche Dichtung in fünf Eklogen* (1803) and *Die Inselfahrt oder Aloysius und Agnes. Ländliche Dichtung in sechs Eklogen* (1805). Kosegarten was a rapid writer, claiming to have written *Jucunde* in five days and *Die Inselfahrt* in six. In 1804 he published 2 vols. of *Legenden*.

KOSSUTH, Lajos (Ludwig von) (Monck/Zemplén, 1802–94, Turin), was the leader of the Hungarian movement for national independence during the revolutionary wars (see Revolutionen 1848–9). A lawyer, he had served as a member of the Hungarian Diet (Landtag) before conviction for treason in 1837. After his release (1840) he resumed his political activities and had great popular support as a champion of freedom. The lack of co-ordination between him and Görgei, the commander-in-chief of the Hungarian forces, obstructed the national cause before the Russian military intervention. Having been for a brief period Reichsverweser (Imperial Vicar), he had to flee from Hungary upon its defeat. He never returned, but his image as a national hero remained.

KOTZEBUE, AUGUST VON (Weimar, 1761–1819, Mannheim), the son of a Weimar civil servant, studied law, and soon after qualifying went to Russia as secretary to the head of the General-Gouvernement St. Petersburg. The preposition of nobility (von) was conferred upon him in 1785, and he rose to high office in the province of Esthonia. In 1797 he became a theatre director in Vienna, and in 1799 was for a time in Weimar. He returned to Russia in 1800, fell from grace, and spent a few months as a convict in Siberia. He was then restored to favour as director of the St. Petersburg theatre. When his patron the Tsar Paul was assassinated, he tried his hand in Weimar, where he was unable to get on with Goethe, and then settled in Berlin. On the French occupation in 1806 (see NAPOLEONIC WARS), he fled to Russia and carried on a journalistic campaign against Napoleon in the periodicals *Die Biene* and *Die Grille*. From 1813 he was again in Russian service; in 1817 he was appointed to the Russian foreign service and sent to Germany as political informant to the Tsar Alexander I. His political activities were suspect to Liberal Germans, and in 1819 he was stabbed to death by the German student Karl Ludwig Sand (q.v.).

Kotzebue was a gifted but conscienceless writer, who was more than once convicted of shameless plagiarism. His first play, the tragedy *Menschenhaß und Reue* (q.v., 1789), written for private theatricals in Reval, was for a time his most famous, though the comedies *Die beiden Klingsberg* (1801) and *Die deutschen Kleinstädter* (q.v., 1803, ed. H. Schumacher, 1964) in the end surpassed it in popularity, and retained its place in the repertoire into the 20th c. His play *Das Kind der Liebe* (1790) appears as the subject of private theatricals, entitled *Lovers' Vows*, in Jane Austen's *Mansfield Park*. Other, once well-known, titles are *Die Indianer in England* (1790, with a naïve and, at the time, proverbial female character named Gurli), *Die Spanier in Peru, Die Unglücklichen* (both 1797), *Die Hussiten vor Naumburg* (1803, parodied by S. A. Mahlmann, q.v.), *Heinrich Reuß von Plauen* (1805, see HEINRICH VON PLAUEN), and *Rudolf von Habsburg und Ottokar* (1815). Kotzebue is also the author of novels, *Ich, eine Geschichte in Fragmenten* (1781), *Die Geschichte meines Vaters* (1785), and *Die gefährliche Wette* (1790). An attack on J. G. Zimmermann (q.v.) in dramatic form, written by Kotzebue and maliciously attributed to A. F. F. von Knigge (q.v., *Doktor Bahrdt mit der eisernen Stirn*, 1790), brought him for a time into general disrepute. Kotzebue's plays (40 vols.) appeared 1840–1 and *Schriften* (selected, 45 vols.), 1842–3.

KOTZEBUE, OTTO VON (Reval, 1787–1846, Reval), was a son of the dramatist August von Kotzebue (q.v.). A Russian naval officer, he three times commanded a ship on a cruise of exploration round the world (1803–6, 1815–18, and 1823–6), discovering a large number of islands in the Pacific. He published two books recording his voyages, *Entdeckungsreise in die Südsee und nach der Beringstrasse zur Erforschung einer nördlichen Durchfahrt in den Jahren 1815–18* (original in Russian, 1821–3; in German translation, 1821) and *Neue Reise um die Welt in den Jahren 1823–26* (2 vols., 1830). On Kotzebue's second voyage the poet A. von Chamisso (q.v.) sailed in the ship as a scientist to the expedition.

Kotzemaere, Das, a short Middle High German verse tale written *c.* 1300. The 'kotze' of the title is a rough blanket. An old man hands over his property to his son, who maltreats him, keeping him short of food and without protection against the cold. The son's little son succours his grandfather, and begs a blanket as a covering. The son hands it over grudgingly, first cutting it in two. The boy, however, begs for the other half, and on being pressed for his reason, declares that he wants it so that he can give it to his father when he is old and unprotected. The boy's words bring home to his father the wickedness of his conduct. The grandfather is now well treated and the poem thus points the moral. Five versions of the story are known.

KPD, abbreviation for Kommunistische Partei Deutschlands, founded in 1919 (see SPARTAKUSBUND) and dissolved in 1933 by the National Socialist regime. It continued to work underground. After the 1939–45 War it became the ruling party in the DDR (see DEUTSCHE DEMOKRATISCHE REPUBLIK), and was banned in the Federal Republic (see BUNDESREPUBLIK DEUTSCHLAND).

KRAFFT-EBING, RICHARD, FREIHERR VON (Mannheim, 1840–1902, Vienna), was professor of psychiatry in Vienna, and published works on forensic psychiatry and sexual perversion. His most widely known work is *Psychopathia sexualis* (1886). Krafft-Ebing was the inventor of the term masochism, which he derived from the name of Leopold von SACHER-Masoch (q.v.), in whose personality and works he detected the symptoms of male sexual pleasure in submitting to female physical or mental domination.

Kraftgenie, term used in the Sturm und Drang (q.v.) for mental originality coupled with dynamic energy and scant regard for convention. Goethe and the members of his circles in

Strasburg and Frankfurt regarded themselves as Kraftgenies, and Götz (see GÖTZ VON BERLICHINGEN MIT DER EISERNEN HAND), Faust (in *Urfaust*, see FAUST), and the eponymous hero of Klinger's *Simsone Grisaldo* (q.v.) are literary examples.

KRAFT VON TOGGENBURG, GRAF, a minor Middle High German poet of Switzerland, probably of the late 13th c., is the author of seven Minnelieder, which attractively combine conventional imagery of nature with the symbolism of the 'roter Mund' derived from GOTTFRIED von Neifen (q.v.).

Krähwinkel, fictitious small town, which is the setting of Kotzebue's *Die deutschen Kleinstädter* (q.v., 1803). It is associated with narrow-minded and self-important provincial manners. See BÄUERLE, A., and FREIHEIT IN KRÄHWINKEL by Nestroy.

KRALIK, RICHARD, RITTER VON MEYRSWALDEN (Eleonorenhain, at that time Austrian Bohemia, 1852–1934, Vienna), a minor Austrian poet and dramatist, published the volume of verse *Büchlein der Unweisheit* (1884). His plays included 3 vols. of *Osterfestspiele* (1895) and *Das Veilchenfest zu Wien* (1905), *Donaugold* (1905), *Revolution* (1908), and *Der heilige Gral* (1912). Kralik, who sought to reconcile Catholicism and Germanic myth, wrote *Das deutsche Götter- und Heldenbuch* (6 vols., 1900–4) and the historical works *Österreichische Geschichte* (1913) and *Allgemeine Geschichte der neuesten Zeit* (6 vols., 1914–23). A work of cultural history is *Geschichte der Kaiserstadt Wien und ihrer Kultur* (1912).

Krambambuli, Der, a students' drinking song written in 1745 by Christof Friedrich Wedekind (pseudonym Creszentius Koromandel, 1709–77). Krambambuli is a name applied to certain liqueurs. A well-known story by M. von Ebner-Eschenbach (see DORF- UND SCHLOSSGESCHICHTEN) is entitled *Krambambuli* after a dog who is named after a brand of cherry brandy.

KRANEWITTER, FRANZ (Nassereith, Tyrol, 1860–1938, Nassereith), a Tyrolean writer who achieved no fame outside his native province. He wrote many plays, of which the Volksstück *Um Haus und Hof* (1898) and the local historical play on Andreas Hofer (q.v.), *Andre Hofer* (1902), are characteristic.

Kraniche des Ibykus, Die, a ballad written by Schiller in 1797, and published in the *Musenalmanach* for 1798. It recounts the murder of the poet Ibykus in ancient Greece, while on his way to the games at Corinth. The only witness of the crime is a flight of migrating cranes, whom the dying man invokes. Ibykus is universally mourned, but the crime is for a time unsolved. At a performance of a tragedy, in which the idea of Nemesis is prominent, a flight of cranes passes over the theatre, and the murderers, shocked by the apparent return of the witnesses of the murder, unwittingly disclose their guilt. The ballad, written in 23 eight-line stanzas, is one of the most poetic and successful of Schiller's achievements in this form.

Kranz der Engel, Der, see SCHWEISSTUCH DER VERONIKA, DAS.

KRAUS, GEORG MELCHIOR (Frankfurt/Main, 1737–1806, Weimar), a painter, studied under J. H. Tischbein the Elder (1772–89) in Kassel, and in 1774 came to Weimar, where he was appointed a drawing master and in 1779 principal of the art school. He was in close touch with Goethe, was with him on his tour in the Harz in 1784, and also accompanied the Weimar contingent on the campaign in northern France in 1792.

Kraus painted or sketched a number of landscapes in the neighbourhood of Weimar, especially in the park. His best-known works, apart from a self-portrait, are an oil portrait of Goethe (1778), a water-colour of the performance of *Die Fischerin* (q.v.) at Tiefurt, and a water-colour of the evening circle of the Duchess Anna Amalia (see ANNA AMALIA, HERZOGIN).

KRAUS, KARL (Gitschin, now Jičin in Czechoslovakia, 1874–1936, Vienna), came with his Jewish parents as a 3-year-old to Vienna, where he spent the whole of his life except for brief visits to other cities. He broke off his studies at Vienna University when he discovered his ability to earn his living by writing. He began his journalistic work in 1892 with a review of Hauptmann's *Die Weber* (q.v.) for the *Wiener Literatur-Zeitung*. In the 1890s he contributed to a number of German as well as Viennese journals and published two pamphlets, *Die demolirte Litteratur* (1897, repr. 1972, first in serial form in the *Wiener Rundschau*, 1896–7) and the anti-Zionist *Eine Krone für Zion* (1898). The former marks his breach with his associates of the 'Jung-Wien' circle, the latter a disillusion that led to his renouncing the Jewish faith in 1899 and joining the Roman Catholic Church, from which he seceded in 1923.

A pronounced individualist, Kraus founded in 1899 his own periodical, *Die Fackel* (q.v.); it continued to appear, though latterly at irregular intervals, until his death. A man of great intellectual integrity and equal mental agility, Kraus proved himself a formidable, often dreaded and hated, satirist. For years he attacked the Viennese press, because he regarded its imprecision of language and its literary pretensions as corrupting cultural influences. The *Neue Freie Presse* was his principal target, partly because it was the most 'literary' of the Viennese newspapers. Kraus identified personal and political integrity with integrity of expression, and carried on a relentless single-handed campaign against slovenly, pretentious, or deceptive language. He combined a concern for social reform with a rigorous cultural conservatism, but refused to involve himself in party politics. He attacked individuals as well as parties, groups, and ideas; particular targets were H. Bahr, A. Kerr, and M. Harden (qq.v.). One of his causes was penal reform (*Sittlichkeit und Kriminalität*, 1908, and *Die chinesische Mauer*, 1910). His reaction to the 1914–18 War was repulsion, rising to violent hostility expressed in the huge, apocalyptic tragedy *Die letzten Tage der Menschheit* (q.v., final version 1926). Collected anti-war essays appeared in book form as *Weltgericht* (2 vols., 1919). In the 1920s he published a succession of satirical plays, including *Literatur* (1921), a riposte to F. Werfel's *Spiegelmensch* (q.v.), and *Die Unüberwindlichen* (1928). Kraus made himself unpopular with many former associates through his support of Dollfuß (q.v.), whom he regarded as a bulwark against National Socialism. The disaster which 1933 represented evoked no commensurate public utterance from Kraus, who, however, wrote a long prophetic denunciation intended as an issue of *Die Fackel* in 1933 but held it back from publication in book form, partly out of fear of consequences for his friends in Germany. It appeared as *Die dritte Walpurgisnacht* in 1952. Among other publications mention should be made of the collections of aphorisms, *Sprüche und Widersprüche* (1909), *Pro domo et mundo* (1912), and *Nachts* (1918), and of the tracts *Maximilian Harden. Eine Erledigung* (1907) and *Nestroy und die Nachwelt* (1912). His poems appeared as *Worte in Versen* (9 vols., 1916–30). No writer has so vigorously and tenaciously held that language is a moral criterion, and that its abuse or misuse proves moral corruption. His essays on language appeared in 1937 as *Die Sprache* (ed. P. Berger).

In his last years Kraus adapted some plays of Shakespeare as *Shakespeares Dramen. Für Hörer und Leser bearbeitet* (2 vols., 1934–5, ed. by H. Fischer as supplement to *Werke*, 2 vols., 1970), and in 1933 published the Sonnets as *Shakespeares Sonette. Nachdichtung von Karl Kraus* (repr. 1964). He condemned the renderings by S. George (q.v.) in *Sakrileg an George oder Sühne an Shakespeare?* (*Die Fackel*, Dec. 1932). One of the greatest satirists, Kraus was devoted to what he conceived to be the highest spiritual values. Devastating in his criticism of the contemporary theatre, including the Burgtheater and the Salzburg Festival, he advanced as an alternate form of performance the notion of the 'Theater der Dichtung' that underlies his extensive activity as a reader of plays and lecturer. These public performances served to promote both his own works and views and the works of favourites such as Nestroy (q.v.) and Offenbach (1819–80). His polemics against Heine (q.v.) have remained controversial.

Die Fackel appeared in 39 vols., 1968–73, and *Werke* (14 vols.), ed. H. Fischer, 1954–67, *Frühe Schriften* (2 vols.), ed. J. J. Braakenburg, in 1979, and *Briefe an Sidonie Nádherný von Borutin* (2 vols.), ed. H. Fischer *et al.*, in 1974.

Kreisauer Kreis, a group of dissident Germans formed on the initiative of Graf Helmut James von Moltke (q.v.), from whose estate (acquired in 1866 by Field-Marshal Graf Helmuth von Moltke, q.v.) it received its name. Adam von Trott zu Solz (1909–44) and Father A. Delp, S.J. (1907–45) were among the members of the circle, which was Christian in basis and more concerned with planning a new liberalized Germany than with conspiracy to liquidate the National Socialist hierarchy. Its members were arrested and executed during the radical persecution of the resistance (see RESISTANCE MOVEMENTS, 2).

Kreisleriana, generic designation for a number of humorous essays dealing with music and a fictitious musician (Johannes Kreisler), written by E. T. A. Hoffmann (q.v.). They appeared at various dates between 1812 and 1820, but the principal group was published under the title *Kreisleriana* in *Phantasiestücke in Callots Manier* (1814). These comprised *Des Kapellmeisters Johannes Kreislers musikalische Leiden, Über den hohen Wert der Musik, Ombra adorata, Höchst zerstreute Gedanken, Der vollkommene Maschinist,* and *Beethovens Instrumentalmusik.* Other *Kreisleriana* were *Lichte Stunden eines wahnsinnigen Musikers, Der Freund, Der Dichter und der Komponist, Kreislers musikalisch-poetischer Klub, Kreisler und Baron Wallborn, Kreislers Lehrbrief, Nachricht von einem gebildeten jungen Mann,* and *Der Musikfeind.* The central figure, the half-crazed, passionately musical Kreisler, is an ironical caricature of Hoffmann himself, and some of the stories are autobiographical in essence (e.g. *Kreislers musikalische Leiden* and *Ombra adorata,* which is concerned with his passion for Julia Marc). They also contain brilliant *aperçus* of musical criticism and a far-sighted perception of the worth of then neglected composers such as J. S. Bach (q.v.) and Corelli, as well as enthusiastic understanding for Mozart and for Beethoven (qq.v.), Hoffmann's contemporary.

R. Schumann (q.v.) used the title *Kreisleriana*, in conscious allusion to Hoffmann, for a collection of piano pieces written in 1838 (Op. 16).

KRETSCHMANN, Karl Friedrich (Zittau, 1738–1809, Zittau), studied law at Wittenberg and practised in his native town. Following in Klopstock's footsteps, he wrote a bardic poem, *Gesang Ringulphs des Barden als Varus geschlagen war* (1768), to which he added in 1771 *Die Klage Ringulphs des Barden*. Kretschmann also wrote pastoral poetry, epigrams, and comedies. His collected works were published in 1799.

KRETZER, Max (Posen, 1854–1941, Berlin), grew up in poverty, took a job at 13, and suffered an accident at work, after which he turned to literature and journalism. Writing at first hand about the squalor of the slums and the miseries of the workman's life, Kretzer seemed likely to develop into the true novelist of the working class. *Die beiden Genossen* (1880), *Die Betrogenen* (2 vols., 1881), and *Die Verkommenen* (2 vols., 1883) were early Naturalistic fiction preceding his best work, *Meister Timpe* (q.v., 1888), a novel of economic change and social decline. *Der Millionenbauer* (2 vols., 1891) was a successful novel at the time, and appeared in the same year as a play (Volksstück), but it stands apart from the socialistic and religious themes to which he devoted himself. He aroused considerable controversy with *Das Gesicht Christi* (1897), one of a number of works to which he drew attention by means of conspicuous titles (e.g. *Die Bergpredigt*, 2 vols., 1889, *Stehe auf und wandle*, 1913).

Kretzer's output, which included Novellen, some sketches, and a few plays, was prolific, and economic pressure may well have driven him to write too much. He did not repeat his initial successes. Titles of the early 20th c. include *Der Holzhändler* (1900), *Treibende Kräfte* (1903), *Familiensklaven* (1904), *Der Mann ohne Gewissen* (1905), *Söhne ihrer Väter* (1908), *Reue* (1910), and *In Frack und Arbeitsbluse* (1911). Among his collections of Novellen were *Der Baßgeiger—Das verhexte Buch* (1894) and *Ausgewählte. Novellen* (1912). In 1927 he published a novel on his native city (*Posen*). *Der Rückfall des Doktor Horatius* (1935) was his last novel. In 1938 his confessions appeared as *Ohne Gott kein Leben*. Berlin, which forms the background of so much of his work and experience (including *Berliner Geschichten*, 1916), is central to his memoirs and last publication, *Berliner Erinnerungen* (1939).

KREUDER, Ernst (Zeitz, 1903–72, Darmstadt), entered journalism in 1926, and in 1932 joined the staff of *Simplizissimus* in Munich. From 1934 to 1940 he was a free-lance writer; he served as an A.A. gunner in the war. He is the author of imaginative narratives with a strong, sometimes surrealist, element of fantasy, fairytales with elements of modern life, spiced with humour and irony and rejecting a drab realism. His principal works are the novels *Die Unauffindbaren* (1948, which treats a theme recalling Kafka's *Das Schloß*), *Agimos oder die Weltgehilfen* (1959), and *Hörensagen* (1969), and the stories *Die Gesellschaft vom Dachboden* (1946), *Herein ohne anzuklopfen* (1954), *Spur unterm Wasser* (1963), and the collection *Tunnel zu vermieten. Kurzgeschichten, Grotesken, Glossen, Erzählungen* (1966), which contains previously unpublished work and the title-story of the collection *Luigi und der grüne Seesack und andere Erzählungen* (1980). He received the Büchner Prize in 1953.

Kreuzelschreiber, Die, a three-act comedy (Bauernkomödie mit Gesang) by L. Anzengruber (q.v.), first performed at the Theater an der Wien, Vienna, in October 1872, and published in the same year under the pseudonym L. Gruber. Written in a generalized Austrian dialect, the play is set at the time of the promulgation of the doctrine of Papal Infallibility (1870). An influential farmer persuades the men of the village to sign a declaration supporting a dissenting theologian. The parish priest imposes as penance a pilgrimage to Rome and instructs the wives to go on 'sex strike' until their husbands return. The village philosopher, Steinklopferhans, resolves the situation by urging each man to take a girl friend with him, whereupon the wives' resistance collapses.

Kreuzer, a small coin formerly current in southern Germany and Austria. The name derives from a double cross in the original design. Kreuzer were in use from 1270 to 1892. 15 Kreuzer were equivalent to 1 Groschen (q.v.).

Kreuzesholz, Legende vom, see Heinrich von Freiberg.

Kreuziger, Der (Crüzigère), a long Middle High German biblical poem by Johannes von Frankenstein, a knight of St. John from Silesia, who was sent to Vienna and wrote the 11,000 or so lines of his poem in the house of his Order there. Its date is uncertain, for, though Johannes indicates 1300 as the year, his Latin source is thought by some scholars to be the work of a theologian born in 1340 (Matthew of Cracow).

Der Kreuziger gives a deeply felt account of the Passion and provides it with a commentary.

Kreuznach, a prose rhapsody (described as a Hymne, q.v.), written by F. Müller (q.v.) and published in 1778 in *Die Schreibtafel* at Mannheim. Kreuznach, Müller's birthplace, is first depicted as the home of his mother and sisters and then shown in 1279 as the scene of a heroic act by Michael Mort in defence of his lord.

Kreuztragende Minne, Die, a 15th-c. Middle High German poem of 72 lines in dialogue form, the subject of which is the rejection of the world and the assumption of the burden of Christ. The poem occurs in two of the MSS. of *Christus und die minnende Seele* (q.v.).

Kreuzzeitung, usual designation of the *Neue Preußische Zeitung*. The paper bore as emblem a reproduction of the Iron Cross (see EISERNES KREUZ), hence its popular name. Founded in 1848, it was the organ of the extreme Prussian conservatives, and figures frequently in literature as a symbol of right-wing views. From 1860 to 1870 Th. Fontane (q.v.) was on the editorial staff, though he was not in sympathy with its politics.

During the first three years of its existence Bismarck (q.v.) was closely associated with the paper, and in 1932 it became the organ of the Stahlhelm (q.v.). It ceased publication in 1939. In 1911 it changed its title to *Neue Preußische Kreuz-Zeitung*, and from 1932 appeared as *Kreuzzeitung*.

Kreuzzeitungspartei, term applied to the right-wing of the Prussian conservatives. It is derived from their newspaper, the *Kreuzzeitung* (q.v.).

Krieg, a novel by L. Renn (q.v.), published in 1928. It tells in direct language the experiences of a front-line infantry soldier from mobilization in 1914 to the return march in November 1918. The sections of the novel are entitled *Vormarsch, Stellungskrieg* (which has the sub-sections *Der Stellungskrieg vor Chailly, Die Sommeschlacht, Verwundet, Die Aisne-Champagneschlacht 1917, Der Stellungskrieg 1917/18*, and *Die Märzoffensive 1918*), and *Zusammenbruch*. It is told in the first person by Corporal Renn without self-pity or pose. It draws abundantly on the author's experience, but the work, though without plot, is a novel, not an autobiography; the author, for instance, was not an N.C.O. in a line regiment but a regular officer in the Saxon Guards.

Krieg und Friede, a poem by D. von Liliencron (q.v.) beginning 'Ich stand an eines Gartens

Rand/Und schaute in ein herrlich Land.' The poet revisits the scene of a battle fought in 1866 (see DEUTSCHER KRIEG), and contrasts the din and shock of arms, which he vividly recollects, with the idyllic spring landscape stretching out before him.

Kriemhild, Chriemhild, in the German version of the Nibelungensage (q.v.) is the wife of Siegfried and wreaks vengeance for his death on Hagen and her brothers. In the *Nibelungenlied* (q.v.) she and her opponent Hagen are the principal characters. In the first part of the epic she is the tender, deeply loving wife; Siegfried's death transforms her into a vengeful fury who shrinks from nothing in exacting full retribution.

It has been suggested that the name derives from Ildiko (q.v.) or Hildiko, the girl who was (probably wrongly) thought to have killed Attila, plus the appropriate attribute 'grim', i.e. Grimhild. In the Nordic version of the Nibelungen legend Kriemhild is replaced by Gudrun.

Kriminalgeschichte, originally a somewhat loose term, has acquired the meaning of a specialized form in which a mysterious crime is investigated and explained, and has been further narrowed down to the detective novel (Detektivroman). Much German reading matter of this kind was until recently provided by translations of American, English, and French works; since the early 1970s the rororo thriller series has promoted the work of native writers.

The vivid and often morbid interest in violent crime was catered for in literature for the three centuries after the invention of printing by broadsheets (see FLUGSCHRIFTEN). A more literate public was supplied with true criminal stories, notably in the *Causes célèbres et intéressantes* (20 vols., 1734 ff.) by F. Guyot de Pitaval (1673–1743), which was translated into German and acquired a colloquial title, *Der Pitaval* (9 vols., 1747–68). A French continuation appeared in German in 1792–5 (4 vols.), and both series excited Schiller's interest. A similar collection of German cases was published by J. E. Hitzig (q.v.) and W. Häring (see ALEXIS, W.) in 1842–65 as *Der Neue Pitaval* (6 vols.).

Of original German crime stories Schiller's *Der Verbrecher aus verlorener Ehre* (q.v., 1786) is particularly noteworthy as anticipating the later trend towards explaining violent crime in terms of the criminal's environment and history. Notable early works dealing with the unravelling of crime include Schiller's unfinished novel *Der Geisterseher* (1787–9), E. T. A. Hoffmann's *Das Fräulein von Scuderi* (q.v., 1819), A. von Droste-Hülshoff's *Die Judenbuche* (q.v., 1842), and Th. Fontane's *Unterm Birnbaum* (q.v., 1885). The first German detective story is almost certainly the virtually forgotten *Der Kaliber* (1829) by A. Müllner (q.v.). Among works of this kind of the last 90 years are W. Raabe's *Stopfkuchen* (q.v., 1891), *Der Fall Maurizius* (1928) by Wassermann (q.v.), E. Kästner's *Emil und die Detektive* (q.v., 1929), Bergengruen's *Der Großtyrann und das Gericht* (q.v., 1935), Doderers *Ein Mord, den jeder begeht* (q.v., 1938), and Dürrenmatt's *Der Richter und sein Henker* (q.v., 1952), *Der Verdacht* (q.v., 1957), *Das Versprechen* (1958), and *Justiz* (1985), *Der Hausierer* (1967) by Peter Handke (q.v.), and *Die Schattengrenze* (1969) by D. Wellershoff (q.v.). Successful authors of the BRD writing in a specifically German context include Irene Rodrian (b. 1937), Hansjörg Martin (b. 1920), and –ky (pseudonym of Horst Bosetzky, b. 1938). Fritz Erpenbeck (1897–1975) has significantly promoted the genre in the DDR.

KRINGSTEINER, JOSEPH FERDINAND (Vienna, 1775–1810, Vienna), an Austrian civil servant, wrote in his spare time plays in Viennese dialect for the Viennese popular stage. He is said to have written more than 25 of these, but only 14, mostly Lokalstücke (q.v.), survive. The best known is *Der Zwirnhändler aus Oberösterreich* (1807), in which a *nouveau riche* loses his fortune through his infatuation for a scheming woman. *Werthers Leiden* (1806) parodies *Die Leiden des jungen Werthers* (q.v.) by Goethe. Kringsteiner died of consumption.

KRINITZ, ELISE, see MOUCHE.

Kristallnacht, also known as the Reichskristallnacht, the night of 9/10 November 1938, when demonstrations by National Socialists, ordered by Goebbels (q.v.), burned down synagogues and destroyed thousands of Jewish shops and homes all over Germany. 91 Jews lost their lives. These organized riots were presented as a reprisal for the shooting on 7 November of Herr vom Rath of the German Embassy in Paris. He was killed by mistake, for the 17-year-old Jewish murderer had intended to assassinate the ambassador and mistook Rath for him. The Kristallnacht (the name alludes to the smashing of plate-glass shop windows) aroused considerable indignation abroad and some disquiet in Germany (cf. U. Johnson's *Jahrestage*, q.v.).

KRISTAN VON HAMLE, a Middle High German Minnesänger, to whom half a dozen derivative songs are attributed. He was a Thuringian, but nothing is known of his life. The indebtedness of his poems to HEINRICH von Morungen, WALTHER von der Vogelweide, and GOTTFRIED von Neifen (qq.v.) suggests that he lived in the middle of the 13th c.

KRISTAN VON LUPPIN or **LUPIN,** a minor Middle High German poet, who was a native of Thuringia and is recorded between 1292 and 1312. His seven Minnelieder derive from GOTTFRIED von Neifen and HEINRICH von Morungen (qq.v.) and are marked by bold hyperbole.

Kritische Dichtkunst, abbreviated title of J. C. Gottsched's *Versuch einer critischen Dichkunst vor die Deutschen* (q.v.). It is also the form in which the *Critische Dichtkunst* (1740) of J. J. Breitinger (q.v.) is sometimes given.

Kritische Waffengänge, a series of six critical brochures written and published by the brothers H. and J. Hart (qq.v., 1882–4). They denounced the literature of the 1870s for its remoteness from life and paved the way for Naturalism (see NATURALISMUS).

Kritische Wälder, a series of critical essays by J. G. Herder (q.v.); the first three 'Wäldchen' appeared in 1769, the fourth, though written at the same time, was published in 1846, long after Herder's death. The *Kritische Wälder* bear the sub-title *Betrachtungen, die Wissenschaft und Kunst des Schönen betreffend, nach Maßgabe neuerer Schriften.* The opening pages contain a famous comparison between Winckelmann and Lessing (qq.v.). The *Kritische Wälder,* while acknowledging the value of Lessing's *Laokoon* (q.v.), seek to expand Lessing's system into a more flexible and dynamic range of views, in which poetry is concerned with the concerted operation of the senses of sight, hearing, and touch.

KROETZ, FRANZ XAVER (Munich, 1946–), reacted at an early age against his middle-class and Roman Catholic upbringing, trained as an actor in Munich and at the Vienna Max-Reinhardt-Seminar, and in the mid-1960s resumed his studies and qualified as an actor. Apart from short theatre engagements in Munich (Büchner-Theater) he earned his living in a variety of temporary jobs before establishing himself as a playwright. His career was promoted by a number of grants and prizes. During the 1970s he was a member of the West German Communist Party (DKP), which he left in 1980. To some extent influenced by Horváth, Marieluise Fleißer, and Brecht (qq.v.), he has become a leading exponent of the 'critical' Volksstück (q.v.). A prominent feature of his 'realism' is the use of 'intervals' (Schweigen), dialect, simple syntax to indicate inarticulate modes of communication, and the absence of stage decorum. *Gesammelte Stücke* (1975) contains thirteen plays published between 1970 and 1973.

Heimarbeit typifies his early plays with a rural Bavarian setting. A succession of short episodes, with intervals that are longer than the duration of the spoken dialogue, it demonstrates a crisis in a family of three. Martha, a woman in her mid-thirties, expects a child not fathered by her husband Willy. A primitive attempt at abortion accounts for the abnormality with which the child is born. Martha leaves the family, and her husband, unwilling to adopt and care for the baby, drowns it in its bath. Martha returns; life can go on, since the legal consequences of the infanticide are likely to be slight in view of mitigating circumstances. The play's dialectics impart a positive view of life, evoke pity for the family's dilemma, and avoid tragic effects. *Oberösterreich* deals with the same theme involving the contemplated abortion of an unwanted child. Set in a provincial town that has conditioned a materialistically minded young couple, its social criticism is more direct. The 'modernization' of Hebbel's domestic tragedy *Maria Magdalene* (q.v.), which in 1977 was followed by an adaptation of *Agnes Bernauer* (q.v.), shows that the manipulation of literary models exerts an inhibiting effect despite sympathy with Hebbel's social criticism. This Kroetz conveys in his *Maria Magdalena. Komödie in drei Akten* through the bitter comedy around Marie's (unsuccessful?) suicide attempt. *Münchner Kindl*, written for a political demonstration, takes the form of a documentary discrediting Bavarian social services and planning policies. The three-act play *Dolomitenstadt Lienz*, described as *Posse mit Gesang*, deals with three men who share a prison cell whilst awaiting trial. A card game assists the slow-motion 'intervals' that are determined by the typography of the pages, each representing a time unit the length of which remains constant throughout—a technique reminiscent of S. Beckett. The sparse dialogue serves the social perspective and the songs evoke compassion, but also contribute to the critical stance that is accentuated in the final pages, during which a game of chess is played in silence; despair has been overcome, but the rational game imposes reflection.

Other plays in the collection are *Wildwechsel* (film version by R. W. Faßbinder, 1972), *Hartnäckig, Männersuche, Lieber Fritz, Stellerhof, Geisterbahn, Wunschkonzert*, and *Michis Blut*. Kroetz's 'positive' realism also forms the basis of the later plays that aim at further thematic and technical variations. A collection containing *Mensch Meier, Der stramme Max*, and *Wer durchs Laub geht* ... appeared in 1979, and another edition containing *Nicht Fisch nicht Fleisch, Verfassungsfeinde*, and *Jumbo-Track* in 1981.

Furcht und Hoffnung der BRD. Das Stück, das Material, das Tagebuch (1984) consists of three parts subtitled *Ein Stück in 15 Szenen aus dem deutschen Alltag* (in collaboration with Alexandra Weinert-Purucker), *Skizzen, Notate, Varianten*, and *Das Tagebuch. 28.8.82–4.9.83*. Kroetz describes this collection of scenic compositions with commentary as an experiment in Minutendramaturgie aiming at 'quick' communication.

Selected texts, radio plays, essays, and interviews are contained in *Weitere Aussichten ... Ein Lesebuch* (1976), ed. T. Thieringer *et al.*, and in *Weitere Aussichten ... Neue Texte* (1976, DDR), ed. W. Schuch *et al. Chiemgauer Geschichten. Bayrische Menschen erzählen ...* , a documentary with a preface, appeared in 1977, the novel *Der Mondscheinknecht* in 1981, and *Der Mondscheinknecht. Fortsetzung* in 1983.

Krokodil, an informal club formed in the 1850s by poets of the Munich school (see MÜNCHNER DICHTERKREIS).

KROLOW, KARL (Hanover, 1915–), educated at Göttingen and Breslau universities, devoted himself from 1942 to the writing of poetry. His early verse, in the collections *Hochgelobtes gutes Leben* (1943), *Gedichte* (1948), *Heimsuchung* (1948), and *Auf Erden* (1949), was influenced by the nature poetry of W. Lehmann and O. Loerke (qq.v.). French influence (notably that of Apollinaire) followed, and is clearly to be seen in *Die Barke* (1948), containing translations of modern French verse. Krolow's later verse is surrealistic, coherent, though not with the logic of prose, 'open' ('offen' is his own word), in contrast to the rounded, enclosed poetry of G. Benn (q.v.), and also 'transparent'. His later volumes of verse are: *Die Zeichen der Welt* (1952), *Wind und Zeit* (1954), *Tage und Nächte* (1956), *Fremde Körper* (1959), *Unsichtbare Hände* (1962), *Landschaften für mich* (1966), and *Alltägliche Gedichte* (1968). A volume of *Gesammelte Gedichte* appeared in 1965. His views on poetry are expressed in six lectures given in 1960–1 at Frankfurt University, published as *Aspekte zeitgenössischer deutscher Lyrik* (1961). Further collections of his own poetry include *Nichts weiter als Leben* (1970), *Zeitvergehen* (1972), *Gesammelte Gedichte 2* (from the years 1965–74, 1975), *Der Einfachheit halber* (1977), *Gedichte* (1980), a selection with postscript by Gabriele Wohmann (q.v.), *Sterblich* (1980), *Herbstsonett mit Hegel* (1981), *Zwischen Null und Unendlichkeit* (1982), and *Schönen Dank und vorüber* (from the years 1981–3, 1984). He has also published a broadly based anthology of French verse translated by himself, entitled *Nachdichtungen aus fünf Jahrhunderten französischer Lyrik* (1948). Other notable translations are *Apollinaires Bestiarium* (1959), *Spanische Gedichte des 20. Jahrhunderts* (1962), *Guillaume Apollinaire: 'Bestiarium oder das Gefolge des Orpheus'* (1978), and *Samuel Beckett: Flötentöne* (1982, in collaboration with E. Tophoven). His prose includes *Poetisches Tagebuch* (1966), *Minuten-Aufzeichnungen* (1968), the story *Das andere Leben* (1979), and *Im Gehen* (1981). The volume *Selbstdeutungen. Interpretationen. Aufsätze* appeared in 1973. *Karl Krolow. Ein Lesebuch*, ed. W. H. Fritz, appeared in 1975. Krolow has edited several volumes of poetry, including *Literarischer März. Lyrik unserer Zeit* (2 vols., 1979 and 1981), *Deutsche Gedichte* (2 vols.), and *Poesie der Welt. Deutschland* (both 1982).

The recipient of many honours, Krolow was awarded the Büchner Prize in 1956.

Krone, gold coin which succeeded the Gulden (q.v.) in Austrian currency in 1892. It was replaced in 1924 by the Schilling.

Kronenwächter, Die, an unfinished novel by Achim von Arnim (see ARNIM, L. J. VON), published in 1817. It bears the sub-title *Bertholds erstes und zweites Leben*. The incomplete MS. of a second volume was found in Arnim's posthumous papers and was published in 1854. The novel is set in the time of Maximilian I (q.v.) at the beginning of the 16th c. The Crown Guards (Kronenwächter) are a secret society who watch over the descendants of the Hohenstaufen emperors in readiness for the time when the dynasty can regain its former greatness. Berthold, one of these descendants, is left when a baby by the Guards with the watchman of Waiblingen in Württemberg and is brought up by him. He becomes a clerk, but the instincts of his unrecognized Staufen (see HOHENSTAUFEN) descent lead him to acquire the Hohenstaufen castle, though he can think of nothing better to do with it than turn it into a textile factory. He makes his pile and becomes burgomaster of Waiblingen, but his force is spent and his 'first life' ('erstes Leben' of the sub-title) is coming to an end when Dr. Faust takes him in hand and miraculously gives him new strength.

His 'second life' sees him full of energy and taking part in political life, meeting the Emperor Maximilian and assisting Luther (q.v.), but he dies without having furthered the Hohenstaufen cause.

In the second (uncompleted) volume Berthold's half-brother Anton was to take over as representative of the Staufen house. In spite of military successes he comes to doubt the value of his dynastic mission, and devotes himself to art, studying with Dürer and Cranach (qq.v.).

An interesting feature of the story is the treatment by which Dr. Faust (i.e. Georg Faust, q.v.) prepares Berthold for his second life, for it includes a blood transfusion.

KRÜDENER, BARBARA JULIANE, FREIIN VON (Riga, 1764–1824, Karazubazar), *née* von Vietinghoff, was married in 1782 and divorced in 1796. In 1801 she was at Coppet with Madame de Staël. Under the influence of Pietists, including H. Jung-Stilling (see JUNG, J. H.), she developed a mystical religious attitude. She is said, through her influence on Tsar Alexander I, to have contributed to the formation of the Holy Alliance (see HEILIGE ALLIANZ). Her two-volume novel *Valérie* (1803), written in French, was translated into German in 1804. Maria Meyer (q.v.), Mörike's 'Peregrina', is said to have been influenced by Frau von Krüdener's circle.

KRÜGER, BARTHOLOMÄUS (Sperenberg, Brandenburg, 1540–after 1597, Trebbin), town clerk and organist of Trebbin, wrote two plays, one of which is a religious drama (*Eine schöne vnd lustige neve Action Von dem Anfang vnd Ende der Welt, c.* 1579). It covers biblical history in selected episodes from the Fall of the Angels and the Fall of Man, and ends with the Day of Judgement. Strongly pro-Lutheran (see LUTHER), it consigns the papists to damnation. The other play is a social drama, depicting the wrongful condemnation by a village court of a mercenary soldier (*Ein Newes Weltliches Spiel, Wie die Pewrischen Richter, einen Landsknecht unschuldig hinrichten lassen, c.* 1579). Krüger is also the author of a collection of stories recounting the pranks of Hans Clauert (q.v.), a Brandenburg 'Eulenspiegel' (*Hans Clawerts Werckliche Historien*, 1587). These anecdotes concerning a real person were collected by Krüger from oral accounts and retold by him in simple style.

KRÜGER, HERMANN ANDERS (Dorpat, 1871–1945, Neudietendorf, Thuringia), a devout Protestant, was a professor in Hanover, then a librarian in Gotha (1925). He wrote two realistic novels of school life in a pietistic community, *Gottfried Kämpfer* (2 vols., 1904–5) and *Kaspar Krumbholz* (2 vols., 1909–10).

Krügerdepesche, also called Krügertelegramm, a telegram of congratulation sent by Emperor Wilhelm II (q.v.) in 1896 to President Krüger of the South African Republic after the defeat of the Jameson Raid. Originating with the Emperor, but drafted by the German Foreign Office and sent with the knowledge of Prince Hohenlohe (q.v.), the Chancellor, the telegram aroused a storm of protest in Britain and seriously damaged Anglo-German relations.

Krüginger, JOHANN, a 15th-c. Saxon schoolmaster, is the author of two plays, *Comödia von dem reichen Mann und armen Lazaro* (1543) and *Herodes und Johannes* (1545).

Krümpersystem, popular term for an aspect of the military policy of Scharnhorst (q.v.) in the years 1808–12. Short-term service for recruits enabled substantial numbers of reservists to be trained while the limitation of the Prussian army to 43,000 men (imposed by the Treaty of Tilsit in 1807, see NAPOLEONIC WARS) was strictly observed. The word *Krümper* means a cripple or deformed person, and its use for these recruits may have begun as an expression of popular humour.

KUCHIMEISTER, CHRISTIAN, see ST. GALLER CHRONIK.

'Kuckuck, Kuckuck ruft aus dem Wald', first line of the poem *Frühlingsbotschaft* written in 1835 by HOFFMANN von Fallersleben (q.v.). It is sung to the tune of an Austrian folk-song.

Kudrun, a Middle High German heroic epic, second in stature only to the *Nibelungenlied* (q.v.). The poem, which was written by an Austrian poet probably between 1230 and 1240 in stanzas resembling those of the *Nibelungenlied*, is known only through a later MS. in the *Ambraser Heldenbuch* (see AMBRASER HANDSCHRIFT) prepared between 1502 and 1515; and the unreliable text has led to much speculation. *Kudrun*, which is in thirty-two *Aventiuren*, tells the story of three generations. Avv. 1–4 tell of the birth to Siegebant and Uote of Hagen, who is carried off by a griffon, which he later kills; he finds three ladies, also victims of the creature, and eventually returns home with them, marrying one, Hilde by name. In the second stage (Avv. 5–8) Hagen and Hilde, reigning in Ireland, have a daughter, also called Hilde, whose suitors Hagen slays—a familiar motif in medieval literature. King Hetel, wishing to woo Hilde, sends a party led by Wate the warrior, and Horant, the minstrel. Horant's music wins Hilde and she flees with Hetel's men, pursued by her father, who is, however, reconciled to her marriage with Hetel. These two stages are a foundation for the main story of Kudrun in the remaining twenty-four *Aventiuren*. Kudrun, the daughter of Hetel and Hilde the younger, is wooed by Hartmut and Herwig, whose suit is refused. Herwig, however, is later accepted, and the betrothal takes place. Hartmut kidnaps Kudrun and carries her home to Ormanîe

(Normandy). Her relatives pursue, but are defeated in a battle on the Wülpensand. Kudrun steadfastly refuses to marry Hartmut and for thirteen years suffers ill-treatment at the hands of his mother Gerlint. At last the rescuing army appears, Kudrun is released, Gerlint slain, and Hartmut taken prisoner. They return home to great festivities, Kudrun and Herwig are married, and Hartmut is pardoned and married to Hildeburg, Kudrun's companion in captivity.

The poem, which has not the heroic stature of the *Nibelungenlied*, being gentler and more conciliatory, is remarkable for the figure of the heroine, who resists with a constancy which is a form of passive heroism, and also for the portrayal of the ever-present sea. This environment derives from the Scandinavian and Frisian origins of the story. *Kudrun* contains two celebrated episodes, firstly the persuasive singing of Horant in Av. 6 (*wie suoze Horant sanc*), which is a rapturous exaltation of music, and secondly the moment at which Kudrun, washing for Gerlint on the shore, barefoot in the March snow, learns of her impending rescue and casts the clothes into the sea.

KUFFSTEIN or **KUFSTEIN,** HANS LUDWIG, FREIHERR VON (1587–1657, Linz/Danube), is the author of pastoral romances, including a translation of Montemayor's *Diana* (1619, in extended form in 1646 by G. P. Harsdörffer, q.v.). *Gefängnüß der Lieb* (1624, ed. G. Hoffmeister, 1973) is an adaptation of Diego de San Pedro's *La Cárcel de amor*.

KÜGELGEN, WILHELM VON (St. Petersburg, 1802–67, Bernburg), the son of a well-known painter, G. von Kügelgen (1772–1820), studied art in Dresden, went to Russia in 1827, and in 1834 became official painter to the Duke of Anhalt-Bernburg. His memoirs, *Jugenderinnerungen eines alten Mannes*, were published posthumously in 1870. Written with urbanity and discreet humour, they are a valuable document of Kügelgen's times.

KÜGELIN, KONRAD, an Augustinian monk of Waldsee, Württemberg, was the author, in the 15th c., of *Das Leben der guten Beth*, a nun (Elsbeth Achler, d. 1420, known as Bona). It is written in prose, and was first published in 1881.

KUGLER, FRANZ THEODOR (Stettin, 1808–58, Berlin), studied both literature and architecture, and became in 1835 a professor at the Akademie der Künste in Berlin. He was a friend of J. E. Hitzig, J. von Eichendorff, and especially of R. Reinick (qq.v.), with whom he published a *Liederbuch für deutsche Künstler* in 1833. He was prominent in the literary club Der Tunnel über der Spree (q.v.), of which the young Th. Fontane (q.v.) was a member. He published a *Skizzenbuch* (1830), *Legenden* (1831), *Gedichte* (1840) and, under the pseudonym F. Th. Erwin, a historical story *Der letzte Wendenfürst* (1837).

Kugler was a systematic and scholarly art historian, publishing a *Handbuch der Geschichte der Malerei* (1837), a *Handbuch der Kunstgeschichte* (1841–2), and an unfinished *Geschichte der Baukunst* (1855–9), which his friend and former pupil, Jakob Burckhardt (q.v.), completed. He is now best known for his popular biography of Frederick the Great (*Geschichte Friedrichs des Großen*, 1842) with copious pen-and-ink illustrations by Adolph Menzel (q.v.), which has been reprinted many times. He wrote tragedies which were published in 1850 in *Belletristische Schriften*, and is also the author of the song 'An der Saale hellem Strande'. (q.v.). He was much given to entertaining artists and men of letters in Berlin; P. Heyse (q.v.) was his son-in-law.

KUH, EMIL (Vienna, 1828–76, Merano), journalist and teacher of literature, was prominent in Viennese literary circles. He was for years (1843–59) a close friend of F. Hebbel (q.v.), whose life he wrote (*Hebbels Leben*, 2 vols., posth. 1877). He also wrote poems (*Gedichte*, 1858) and stories (*Drei Erzählungen*, 1857).

KUH, EPHRAIM (Breslau, 1731–90, Breslau), a Jewish poet, whose poems were posthumously published (*Hinterlassene Gedichte*, 2 vols., 1792). He is the central figure of the novel *Dichter und Kaufmann* by B. Auerbach (q.v.).

KUHLMANN, QUIRINUS (Breslau, 1651–89, Moscow), a poet and religious fanatic influenced by J. Böhme (q.v.), studied in Jena and Leyden, from which he was expelled because of his revolutionary religious ideas. After visiting England he went to Constantinople in order to convert the Sultan (1678), returned to Holland, and finally journeyed to Moscow to proclaim his idiosyncratic Kingdom of God ('Kühlmonarchie'); he was denounced as an enemy of religion and the state, condemned, and burned at the stake. He married three times in his short life.

Kuhlmann wrote, prophesied, and acted with a manic obsessiveness. His propagandist religious poetry is couched in extravagant and frenetic language, though it adheres for the most part to the orthodox forms of the 17th c., the Alexandrine couplet, the sonnet, and the hymn-like strophe. His volumes of verse bear paradoxical or extraordinary titles; they include *Unsterbliche Sterblichkeit oder Hundert spiel-ersinnliche vierzeilige Grabsschriften* (1668), *Himmlische Li(e)-beskitsse* (1671), *Gechicht-Herold* (1673), *Neube-geisterter Böhme begreiffend Hundertfünftzig Weissagungen* (1674), *Schleudersteine wider den Goliath aller Geschlechter* (1680), and, best known of all, *Der Kühlpsalter* (q.v., 3 vols., 1684–6), a title which plays upon his name.

Kühlpsalter, Der, a work of ecstatic, hyperbolical religious poetry in the form of eight books of 150 'psalms', published by Q. Kuhlmann (q.v.) in 3 vols. (1684–6). The first 15 psalms had been previously published as *Funffzehn Gesänge* (1677). The title *Kühlpsalter* alludes to the Kingdom of God proclaimed by Kuhlmann as 'Kühlmonarchie'; in it Jesuel, the son of Jesus, is to reign. Kuhlmann sometimes wrote his own name with 'ü'. A selection from the work (Psalms 1–15, 73–93), ed. H. L. Arnold, appeared in 1973. *Der Kühlpsalter* was reprinted in 1962 ff. and (ed. R. L. Beare, 2 vols.) in 1972.

KUHN, GOTTLIEB JAKOB (Berne, 1775–1849, Berne), a Swiss clergyman, was curate in Sigriswil, then pastor in Rüderswil (1812), and finally in Burgdorf (1824). He is the author of many dialect poems of folk-song character, a number of which were collected in *Volkslieder und Gedichte* (1806). A fuller collection was published posthumously in 1879. Kuhn also edited with J. R. Wyß (father of J. D. Wyß, q.v.) the almanach *Die Alpenrosen* (1810).

KÜHN, SOPHIE VON (1782–97), was betrothed to the poet Friedrich von Hardenberg (see NOVALIS) in 1795, and died of tuberculosis two years later.

KÜHNE, FERDINAND GUSTAV (Magdeburg, 1806–88, Dresden),. was educated at the Joachimsthalsches Gymnasium (q.v.), Berlin (where Theodor Mundt, q.v., was a younger contemporary), and at Berlin University. From 1835 to 1859 he was active in Leipzig as a journalist with the *Zeitung für die elegante Welt* (q.v., 1835–42) and *Europa* (1846–59), of which he was owner as well as editor. He spent the last decades of his life in Dresden.

Kühne is now chiefly known for the title of his novel *Eine Quarantäne im Irrenhause* (q.v., 1835), but he was also the author of a number of Novellen (*Novellen*, 1831, *Klosternovellen*, 1838), and of the novels *Die Rebellen von Irland* (3 vols., 1840) and *Die Freimaurer* (3 vols., 1855). His poems (*Gedichte*) appeared in 1831, and in 1859 he published a completion of Schiller's *Demetrius* (q.v.). Kühne was an accomplished essayist (*Weibliche und männliche Charaktere*, 1838; *Portraits und Silhouetten*, 1843; *Deutsche Männer und Frauen*, 1851). His autobiography, *Mein Tagebuch in bewegter Zeit* (1863), is of historical value.

Though he was not named in the federal denunciation of Young Germany (December 1835, see JUNGES DEUTSCHLAND), Kühne's views and writings show affinity with this movement.

KULMANN, JELISAWETA BORISOVNA (St. Petersburg, 1808–25, St. Petersburg), daughter of a Russian- and German-speaking officer, was a child prodigy, who learned the principal European languages in childhood and translated into German parts of the works of Milton, Metastasio, Alfieri, and Camoens. She translated *Anacreon* into several languages and wrote original poetry in German and Russian. Her works (*Sämtliche Dichtungen*) were published posthumously in 1835.

Kulturkampf, the conflict in the German Empire of 1871 between the state and the Roman Catholic Church. Its name is derived from a political slogan coined by R. Virchow (q.v.): 'ein Kampf für die Kultur'. The Kulturkampf lasted effectively from 1872 to 1878 and was fought principally in Prussia, though Baden and Hesse had parallel Kulturkämpfe on a smaller scale.

Three factors precipitated the struggle. The new Germany included large Roman Catholic populations in the southern states, and in consequence a new party, the Zentrum (q.v.), opposed in the Reichstag the policy and personality of Bismarck (q.v.). Secondly, the Prussian state enjoyed a traditional authority over the Church, which in Prussia was predominantly Protestant. Thirdly, the promulgation of the Papal Dogma of Infallibility in 1870 created widespread anti-papal feeling, both in Germany and elsewhere, leading even to schism in the Roman Catholic Church.

The first step in the struggle was the abolition of the Roman Catholic office of the Prussian Ministry of Religion (Kultusministerium). In March 1872 Roman Catholic schools in Prussia were placed under government control; in July the Jesuitengesetz banned the Jesuits; and in May 1873 A. Falk (q.v.), the Prussian Kultusminister, secured the passing of the May Laws (Maigesetze) limiting the disciplinary powers of the Roman Catholic episcopate in respect of laymen, restricting clerical appointments to German nationals, regulating the training of ordinands (including Roman Catholics), and instituting for them a state qualifying examination. The Roman Catholic bishops refused obedience, and a campaign of persecution began, in the course of which two archbishops, several bishops, and many priests were suspended, fined, or imprisoned. In 1875 a law ordered the dissolution of the monasteries in Prussia, and civil marriage was introduced (in Prussia in 1874, and in the Empire in 1876).

Although Bismarck had promised a government victory ('Nach Canossa gehen wir nicht' nicht'), he miscalculated the strength of Roman Catholic resistance and became aware that the unity of the new German state was endangered. From 1878 he began a progressive withdrawal, in which he was helped by the conciliatory attitude of Pope Leo XIII, who in that year succeeded Pius IX. Most of the anti-Catholic measures were unostentatiously repealed. By 1887 the archbishoprics and bishoprics were filled, the monastic orders readmitted, and their property restored; with the disappearance of the punitive laws the Pope proclaimed the Kulturkampf to be at an end. The ban on Jesuits remained complete, however, until 1904, and in part until 1917. Civil marriage was also continued.

G. M. Hopkins's poem *The Wreck of the Deutschland* is an elegy for five nuns, 'exiles from the Falk laws', drowned in December 1875.

Kunersdorf, battle of (12 August 1759), in which the combined Russian and Austrian forces under Laudon (q.v.) inflicted a heavy defeat upon Friedrich II (q.v.) of Prussia (see SIEBENJÄHRIGER KRIEG). Kunersdorf lies close to Frankfurt/Oder. Among the casualties was the officer and poet E. von Kleist (q.v.), who died of his wounds a few days later. The poem *Elegie auf dem Schlachtfelde bei Kunersdorf* is by C. A. Tiedge (q.v.).

KUNERT, GÜNTER (Berlin, 1929–), had to abandon grammar school education during the National Socialist regime and studied at the Berlin Kunsthochschule after the war, but according to his own statement had decided at the age of seventeen to devote himself to writing. He is noted for the firm stand he takes against the misuse of language.

Influenced by Brecht (q.v.), his songs and notion of Gestus, he has also expressed a sense of affinity to N. Lenau and Heine (qq.v.). Deeply committed to the memory of the victims of persecution and the uncertainties of the future, he deals mainly with contemporary concerns expressed in exquisitely pointed parabolic modes and a density of motifs deriving from fairy-tales, myth, and literary models. A master of the short aphoristic form and of irony, he evolves a subtle technique of integrating nature into his dialectics. Collections of his lyrics include *Wegschilder und Mauerinschriften* (1950), *Unter diesem Himmel* (1955), *Tagwerke* (1961), *Das kreuzbrave Liederbuch* (1961), *Erinnerungen an einen Planeten* (1963), *Der ungebetene Gast* (1965, in extended form as *Verkündigung des*

Wetters, 1966), *Unschuld der Natur* (1966), *Notizen in Kreide*, *Warnung vor Spiegeln* (both 1970), *Offener Ausgang* (1972), *Im weiteren Fortgang* (1974), and *Unterwegs nach Utopia* (1977). Prose includes the volumes *Tagträume in Berlin und andernorts*. *Kleine Prosa*. *Erzählungen*. *Aufsätze* (1964), *Kramen in Fächern, Die Beerdigung findet in aller Stille statt* (both 1968), *Betonformen*. *Ortsangaben* (1969), *Gast aus England* (1973). *Im Namen der Hüte* (1967), the only novel, deals with the fight for survival of the Jew Henry and displays the range of Kunert's sensitively serious, ironic, and grotesque mode of writing. The hats of the title, adapted from a fairy-tale motif, enable the wearer to find out facts about their owners. *Der andere Planet*. *Ansichten von Amerika* (1974), observations on travels in the U.S.A. in a social and political context, and on German emigrants, includes an autobiographical postscript. The volume exemplifies Kunert's approach to his extensive travels which is marked by an astute assessment of different locations and its people. It was followed by *Ein englisches Tagebuch*, the poetry of *Bomarzo*, *Camera obscura* (all 1978), and *Abtötungsverfahren* (1980). *Verspätete Monologe* (1981) is a collection of prose and glossaries and *Diesseits des Erinnerns* (1982) a collection of essays. *Die Schreie der Fledermäuse*, ed. D. E. Zimmer (1979), contains lyrics, prose, and essays.

Kunert has been living in the West since 1979.

Künftige Geliebte, Die, an ode in 49 elegiac distichs written by F. G. Klopstock (q.v.) in 1748. The motif of the unknown future beloved is taken up again in the ode *An Ebert* (q.v.).

KÜNG, HANS (Sursee, Canton Lucerne, 1929–), a Roman Catholic, appointed professor at Tübingen University in 1960. He served the ecumenical movement and the Second Vatican Council (1962–5). Among his works are *Rechtfertigung*. *Die Lehre Karl Barths und eine katholische Besinnung* (1957, see BARTH, K.), *Konzil und Wiedervereinigung* (1960), *Wahrhaftigkeit* (1968), *Christ sein* (1974), and *Um nichts als die Wahrheit* (1978).

KUNIGUNDE, HEILIGE (d. 1033, nr. Kassel), consort of the Emperor Heinrich II (q.v.), founded the convent of Kaufungen, in which she died. She was buried beside her husband in Bamberg Cathedral, where they are commemorated by statues (1230) and a monument by Tilman Riemenschneider (q.v., 1513). There is also a baroque statue of her on the Untere Brücke in Bamberg, and another in the abbey church of Rott/Inn, Bavaria. She was canonized in 1200 by Pope Innocent III.

Künstler, Die, a philosophical poem of 481 lines written in 1788–9 by Schiller. The original version, which is not preserved, was less than 200 lines in length, and was expanded partly as a result of discussions with Wieland (q.v.). The poem was published in *Der teutsche Merkur* (q.v.) for March 1789, and is written in iambic lines of varying length, with an irregular rhyme pattern. It powerfully affirms belief in moral progress and in the role art has to play in it.

Künstlerroman, Künstlernovelle, and Künstlerdrama, designations of novels, Novellen, and plays which have an artist (in any branch of creative art) as the central character. The artist is often a historical figure, as in the novels *Schillers Heimatjahre* by H. Kurz, *Friedemann Bach* by A. E. Brachvogel, *Verdi* by F. Werfel, *Der Schiller-Roman* by W. von Molo, and Mörike's Novelle *Mozart auf der Reise nach Prag*. Fictitious artists are more common, and examples of these are found in Goethe's *Wilhelm Meisters Theatralische Sendung*, Tieck's *Franz Sternbalds Wanderungen*, Grillparzer's *Der arme Spielmann*, Mörike's *Maler Nolten*, G. Keller's *Der grüne Heinrich*, J. Wassermann's *Das Gänsemännchen*, and Th. Mann's *Tonio Kröger* and *Dr. Faustus*.

Characteristic examples of historical or fictitious Künstlerdramen are: Goethe's *Torquato Tasso*, Grillparzer's *Sappho*, Hebbel's *Michelangelo*, and G. Hauptmann's *Die versunkene Glocke*, *Michael Kramer*, and *Gabriel Schillings Flucht*. None of the works mentioned is earlier than 1788, and the rise of the genre, which is a conspicuous feature of German literature, coincides with the high estimation of the artist and man of genius since the late 18th c. The examples given could be multiplied many times.

Künstlers Apotheose, a short dramatic verse sketch written by Goethe in 1788 and published in the following year. It forms a sequel to *Des Künstlers Erdewallen* (q.v.). The painter, who in life has starved, achieves artistic immortality after his death with the picture of Venus Urania, which in his lifetime was ignored. The poem is written in the irregular Knittelverse (q.v.) of Goethe's Sturm und Drang (q.v.) years.

Künstlers Erdewallen, see DES KÜNSTLERS ERDEWALLEN.

Kunst und Altertum, see ÜBER KUNST UND ALTERTUM.

Kunst und die Revolution, Die, title of a tract by R. Wagner (q.v.), published in 1849. Written with the Revolution of 1848 (see

REVOLUTIONEN 1848–9) fresh in Wagner's mind, it views art as in need of renewal along with the renewal of man, and proposes that the theatre should be opened wide with free entry for all.

Kunstwart, Der, a literary and artistic fortnightly (initially with the sub-title *Rundschau über alle Gebiete des Schönen*), founded in 1887 in Munich by Ferdinand Avenarius (q.v.). It continued until 1931; in 1932 it was replaced by the *Deutsche Zeitschrift*. Largely conservative and bürgerlich, it was successful with the educated middle class.

Kunstwerk der Zukunft, Das, a long essay by R. Wagner (q.v.), published in 1850. It suggests that the art of the future will be found in the theatre in the collaboration of all branches of art, music, architecture, painting, acting, singing ('Das wahre Drama ist nur denkbar als aus dem *gemeinsamen Drange aller Künste* zur unmittelbaren Mitteilung an eine *gemeinsame Öffentlichkeit* hervorgehend').

KUNZE, REINER (Oelsnitz/Erzgebirge, 1933–), the son of a labourer and a mother who helped to make ends meet as an outworker, Kunze's education was financed by the state. He studied journalism at the Leipzig Institut für Publizistik (1951–5), subsequently incorporated into Leipzig University, where he began an academic career that he abandoned on political grounds. After a spell as an industrial worker, interrupted by ill-health, he was engaged by the Schriftstellerverband, travelled frequently to Czechoslovakia, and married a Czech dentist, with whom he settled in Greiz (DDR) as a free-lance writer in 1962. He established an international reputation with his renderings of Czech lyrics. A member of the SED since the age of sixteen, he left it in 1968. Mounting official criticism culminated in his expulsion from the Schriftstellerverband in 1976, when he sought permission to leave the DDR. In 1977 he settled in the West. All his work published up to 1979 was written in the East.

Kunze's humanitarian attitude shows a special concern for the young, and many aspects of conditions in the DDR are the target of uncompromising criticism. Familiar things, a soldier's helmet, a letter-box, a flower, or an animal, figure in many lyrics with pointed paradox and irony. Influenced by Brecht, apart from Heine (qq.v.), his most obvious model, his preoccupation with A. Camus helped him to adopt the view that the absurd, once recognized, must be overcome. It is discernible in a great variety of his lyrics and texts. The collections *Vögel über dem Tau* (1959) and *Aber die Nachtigall jubelt. Heitere Texte* (1962) contain only a small

number of works with which Kunze still identifies. They are contained in *Brief mit blauem Siegel* (1973, DDR). *Sensible Wege. Achtundvierzig Gedichte und ein Zyklus* (1969), containing poems written in the 1960s, is the first collection published only in the West. The edition was followed by *Der Löwe Leopold. Fast Märchen, fast Geschichten. Kinderbuch* (1970, extended edn. 1974). The lion of the title story enjoys making the circus audience happy with his skills, though he refuses to perform under the threat of his trainer's whip. *Zimmerlautstärke,* a collection of lyrics, appeared in 1972 and *Die wunderbaren Jahre* in 1976 (film version 1980; written and directed by Kunze himself, it involves the suicide of Stephan, a seventeen-year-old schoolboy). The title, a quotation taken from Truman Capote, alludes bitterly to the author's experience of life in the DDR. The volume's short texts are presented in terse, pointed prose; the final sections depict the years 1968 and 1975. At the time a bestseller, the book was translated into some eleven languages. A volume of verse, *Auf eigene Hoffnung,* appeared in 1981. Mention should be made of *Reiner Kunze. Materialen und Dokumente,* ed. J. P. Wallmann (1977). The recipient of a Czech prize and of a number of awards in the West, Kunze received the Büchner Prize in 1977.

Künzelsauer Fronleichnamsspiel, a religious play, covering biblical history from the standpoint of the redemption of man and intended to be performed at the Feast of Corpus Christi and on following days. It was written in 1479, perhaps at Künzelsau or, at any rate, in the district of Hohenlohe (Württemberg), in which Künzelsau is situated.

KUNZ KISTENER, a poet of Strasburg, who composed about the middle of the 14th c. the Middle High German poem *Die Jakobsbrüder.* The action of this work of some 1,400 lines takes place in two stages. Two young men are on their way to Compostela. One dies and his companion carries the body to the shrine, where it is restored to life by St. James. Later the faithful friend is smitten with leprosy; his grateful companion sacrifices his own son, and sprinkles the blood upon the friend. The leprosy is cured, and the saint restores the child to life. The story has, in its second stage, a distinct resemblance to Konrad von Würzburg's *Engelhard* (q.v.), and Kunz is indebted to Konrad also in his style.

Kürbishütte, see KÖNIGSBERGER DICHTERSCHULE.

Kurfürsten, the princes of the Holy Roman Empire (see DEUTSCHES REICH, ALTES), entitled to elect the German King (see DEUTSCHER

KÖNIG), who was normally later crowned emperor. Originally all the ruling princes elected the German King, but from the middle of the 13th c. this right was hereditarily exercised by a limited group of seven: the archbishops of Mainz, Trier, and Cologne, the Count Palatine (Pfalzgraf bei Rhein), the king of Bohemia, the duke of Saxony, and the margrave of Brandenburg. The electoral rights, which were a matter of dispute in the 13th c., were finally established under the Emperor Karl IV (q.v.) in the Golden Bull of 1356 (see GOLDENE BULLE). In 1400 Bohemia ceased to exercise its electoral rights, but resumed them three centuries later in 1708. A change in the constitution of the group of seven Electors occurred during the Thirty Years War (see DREISSIGJÄHRIGER KRIEG), when the Pfalzgraf was outlawed, and his electoral rights transferred to Bavaria in 1623. At the Peace of Westphalia in 1648, the ruler of the Palatinate was restored to his former rights as an eighth elector. In 1692 a ninth electorship was established for the ruler of Braunschweig-Lüneburg (usually termed the Elector of Hanover). In the last years of the Empire a series of rapid changes occurred in consequence of the fluid political situation of the Revolutionary and Napoleonic wars (see REVOLUTIONSKRIEGE and NAPOLEONIC WARS). In 1803 Cologne, Trier, and Mainz dropped out; Regensburg and Salzburg succeeded as spiritual electorates, and three new temporal electors were instituted, the rulers of Württemberg, Baden, and Hesse-Cassel. In 1805 Salzburg gave way to Würzburg, but in the following year the end of the Empire terminated the electoral function, though the ruler of Hesse-Cassel continued to use the now meaningless title until his deposition in 1866.

Kurmark, Brandenburg except for the Neumark, i.e. the Altmark, the Mittelmark, the Uckermark, Prignitz, Beeskow, and Storkow. The term Kurmark (Electoral March) dates from the confirmation of the hereditary electoral office of the Brandenburg princes (see KURFÜRSTEN) in the Golden Bull (1356) and remained in use until the dissolution of the Holy Roman Empire in 1806 (see DEUTSCHES REICH, ALTES).

KÜRNBERGER, DER, or DER VON KÜRENBERG, was possibly the earliest Middle High German lyric poet, writing c. 1150–75. He was an Austrian noble, perhaps from the vicinity of Linz. Fourteen stanzas, some of which group themselves into poems, are reproduced in Des Minnesangs Frühling (q.v.). The Kürnberger's songs were probably written before the full formulation of the ideal of courtly love (minne) and have as their subject a more direct and less stylized relationship. Some of them are in dialogue form. The best known is the longer poem on a falcon, beginning 'Ich zôch mir einen valken mêre danne ein jâr', which has been subjected to varying interpretations. It seems probable that both stanzas are spoken by a woman.

Kürnberger's poetry, with that grouped under the name DIETMAR von Eist (q.v.), suggests the existence of an indigenous lyric poetry before the impact of Provençal influence.

KÜRNBERGER, FERDINAND (Vienna, 1821–79, Munich), a Liberal of proletarian origin, was involved in the 1848 Revolution (during which he had to flee abroad) and in the rising in Dresden in 1849, for which he had to serve a term of imprisonment. He then lived in various German cities, returning to Austria in 1864. From 1864 to 1867 he was secretary to the German Schiller Foundation. He wrote among other plays a Catilina (1854); his novels include Der Amerika-Müde (1855), dealing with N. Lenau (q.v.), Der Haustyrann (1876), and Das Schloß der Frevel (posth., 1904). He also published Novellen (3 vols., 1861–2) and essays (Siegelringe, 1874, and Literarische Herzenssachen, 1877). Gesammelte Werke (4 vols.) appeared in 1911.

KURZ, HERMANN (Reutlingen, 1813–73, Tübingen), who in his teens lost both his parents, attended theological college (Tübinger Stift, q.v.) from 1831 to 1835, afterwards becoming a curate at Ehningen. In 1836 he gave up his curacy to live by his pen. One of his college friends was L. Uhland (q.v.), and later he was in close contact also with E. Mörike and G. Schwab (qq.v.). Kurz had little financial success with his writings and was glad in 1863 to accept the position of deputy librarian at Tübingen University, which he occupied until his death.

Kurz published poems (Gedichte, 1836) and was a prolific translator, rendering into German English poetry including Byron, Cervantes, and the Orlando furioso of Ariosto; he also made a modern version of the Tristan (q.v.) of Gottfried von Straßburg. His novels include Schillers Heimatjahre (1843), the central figure of which is not Schiller, but the fictitious Heinrich Roller whose name is taken from Die Räuber (q.v.), and Der Sonnenwirt (1862). Of his shorter stories, Unter dem Tannenbaum (1856), also known by the title Der Weihnachtsfund, achieved popularity. From 1871 he collaborated with P. Heyse (q.v.) in publishing vols. 1–21 of the Deutscher Novellenschatz. A hitherto unpublished novel, Lisardo, was printed in 1919. Kurz's collected works (Gesammelte Werke) were published in 10 vols. 1874–5, and his complete works (Sämtliche

Werke) in 12 vols. in 1904. The novelist Isolde Kurz (q.v.) was his daughter.

KURZ, Isolde (Stuttgart, 1853–1944, Tübingen), the daughter of Hermann Kurz (q.v.), lived for many years (1877–1913) after her father's death with her family among the German artistic circle in Florence. From 1913 to 1943 she lived in Munich. She belonged to no school and had mystic and mythological inclinations, expressed in the verse works *Die Kinder der Lilith* (1908) and *Leuke* (1926). She published two volumes of poems, *Gedichte* (1889) and *Neue Gedichte* (1905); a selection of these was published as *Auslese* in 1933. Her best works are probably the stories with Italian settings (*Florentiner Novellen*, 1890; *Italienische Erzählungen*, 1895). Of the novels, *Nächte von Fondi* (1922), *Der Caliban* (1925), and above all the largely autobiographical *Vanadis* (1931, pronounced with the accent on the first syllable) deserve mention. *Die Pilgerfahrt nach dem Unerreichlichen* (1938) is autobiographical.

KURZ, Johann Josef Felix von (Vienna, 1717–83, Vienna), was a noted actor in the theatre of extemporization popular in the 18th c. in Vienna (see Volksstück). Kurz played the stock part of Hanswurst (q.v.) in a new adaptation named Bernardon, and this became his stage name. He was also a prolific author of plays, of which some 300 titles are known. They include a Faust (*Das lastervolle Leben und erschröckliche Ende des weltberühmten und jedermänniglich bekannten Erzzauberers Doctoris Joannis Faust, Professoris Doctoris Wittenbergensis*). Kurz

received his patent of nobility in Poland towards the end of his life.

Kurzgeschichte, a term introduced into German to translate the English 'short story'. A corresponding narrative form has arisen in German literature. Kurzgeschichten are terse and pointed, as well as short. Examples can be found in the work of many recent or contemporary writers of fiction, among them Bergengruen, Böll, Ingeborg Bachmann, Dürrenmatt, S. Lenz, and Hans Grimm (qq.v.).

KUSSMAUL, Adolf (Graben nr. Karlsruhe, 1822–1902, Heidelberg), was a professor of medicine successively at Erlangen and Freiburg universities. He published a pseudonymous volume of verse, *Poetische Jugendsünden des Dr. Oribarius* (1893), and two autobiographical works *Jugenderinnerungen eines alten Arztes* (1899) and, posthumously, *Aus meiner Dozentenzeit* (1903). He is also well known for having directed the attention of L. Eichrodt (q.v.) to the naïve poems of the Swabian village poet S. F. Sauter (q.v.) and having taken some part in their publication as 'Biedermaier's' poems.

Kyffhäuser, a mountain ridge in Thuringia, north of Weimar. It is the legendary site of the cavern in which Barbarossa (see Friedrich I, Kaiser) waits to come to the assistance of the German nation in time of need. In 1896 a colossal monument, known as the Kyffhäuserdenkmal, was erected on the Kyffhäuser commemorating both the Emperor Wilhelm I (q.v.) and Barbarossa. The Kyffhäuser became a symbol of German nationalism and of pan-German sentiment.

L

Labyrinth, Das, a novel by Ina Seidel (q.v.), published in 1922. It is a *vie romancée*, based on the life of Georg(e) Forster (q.v.). It tells of his precocity as a child and its exploitation by his father, of his travels in Russia, England, and in the South Seas, of his love for Therese Huber and eventual marriage, and finally of his lonely end in Paris. It evinces deep human sympathy for one lost in the labyrinth of life.

LACHMANN, Karl (Brunswick, 1793–1851, Berlin), one of the pioneers of the study of Germanic philology. Lachmann studied at Leipzig and Göttingen and took part in the War

of Liberation in 1813 as a volunteer. He was appointed professor at Königsberg in 1818 and at Berlin in 1825; in 1830 he became a member of the Prussian Akademie der Wissenschaften. Trained as a theologian and classical philologist, he applied classical methods to the then new study of medieval texts.

Lachmann's enormous industry and thoroughness produced a succession of editions of important and, at the time, little-known or unknown authors and works, including Walther von der Vogelweide, Wolfram von Eschenbach, *Iwein*, and *Gregorius* (qq.v.), as well as publishing important treatises touching on

Old High German metrics (*Über althochdeutsche Betonung und Verskunst*, 1831), the *Hildebrandslied* (*Über das Hildebrandslied*, 1833), and *Parzival* (*Über den Eingang des Parzival*, 1835). His theory that the *Nibelungenlied* is an amalgamation of songs, propounded in *Über die ursprüngliche Gestalt des Gedichts von der Nibelungen Noth*, 1816, soon lost support. Lachmann began the edition of medieval poetry completed after his death by M. Haupt (q.v.) and published in 1857 as *Des Minnesangs Frühling* (q.v.), and he found time to issue an important edition of Lessing's works (1838–40). In his youth he wrote poetry. In 1820 he published the first complete German translation of Shakespeare's Sonnets and in 1829 a translation of *Macbeth*.

Lady Johanna Gray oder Der Triumph der Religion, a tragedy in blank verse by C. M. Wieland (q.v.), first published in 1758. The alternative title was first included in the edition of 1762. The tragedy, in five acts, deals with the brief reign of Lady Jane Grey, beginning at the death of Edward VI. Lady Johanna accepts the crown with reluctance, and after the overthrow of Northumberland she faces death with stoical fortitude. The only emotional climax occurs in the fourth act, when she is offered her life if she will embrace Catholicism. Her relatives at first press her to do so, but she spurns life on such a condition. The play is based on the tragedy *Lady Jane Grey* (1715) by Nicholas Rowe (1674–1718).

LAFONTAINE, August (Brunswick, 1758–1831, Halle/Saale), of French descent, became an army chaplain and was later a canon of the Magdeburg chapter. An extremely prolific author, Lafontaine wrote between 1791 and 1820 a quantity of moralizing novels in which vice is described as well as cured. Some of his sentimental stories are presented in the form of letters. He was one of the most widely read German authors at the turn of the 18th c., and his novels are said to have been the favourite reading of King Friedrich Wilhelm III (q.v.) of Prussia.

LAGARDE, Paul Anton de (Berlin, 1827–91, Göttingen), whose real name was Bötticher, studied theology, became a schoolmaster in Berlin, and in 1869 was elected professor of oriental languages at Göttingen University. He was one of the prime movers in the revival of anti-Semitism in the later years of the 19th c., demanding a national Christianity purged of Jewish and, in particular, Pauline elements. He is also typical of a substantial number of German academics of the time, who distorted their researches for nationalistic ends. Among his polemical works are *Arica* (1851), *Semitica* (1878–9), and *Deutsche Schriften* (1886). Lagarde was also active as a lyric poet, publishing *Gedichte* (1885) and *Am Strande* (1887). He assumed the name de Lagarde from his maternal grandfather.

Lalenbuch [Lalebuch], Das, a Volksbuch printed in 1597. It is a collection of Schwänke (see SCHWANK) centred on the inhabitants of the fictitious town of Lalenburg, and exists in an alternative form as *Die Schildbürger* (1598). A later version (1603) is entitled *Der Grillenvertreiber*.

LAMPE, Friedo (Bremen, 1899–1945, Berlin), a librarian in Hamburg, then in Stettin, finally became a publisher's reader in Berlin. He was killed on the day of the capitulation of Berlin. His story *Am Rande der Nacht* (1934) evokes the sensations and impressions of a September evening in Bremen with its charm and tenderness, its squalor and its lust, held together by the thread of the melodies of Bach, which Herr Berg plays upon his flute. The rats on the bank and the swans on the river serve as contrasting symbols and explain the title given to the 1950 edition, *Ratten und Schwäne* (which includes two other small pieces, *Laterna magica* and *Nach hundert Jahren*). Lampe also wrote the short novel *Septembergewitter*, which is the principal work in the posthumous collection *Von Tür zu Tür* 1936. An edition of his works, *Das Gesamtwerk*, appeared in 1955.

LAMPEL, Peter Martin, real first names Joachim Friedrich Martin (Schönborn, Silesia, 1894–1965, Hamburg), volunteered for war service in 1914, and later became an air pilot. He was active in a Freikorps (q.v.) in 1919, and was for a short time a police superintendent (Oberleutnant). After periods of study of both academic subjects and art, he held various jobs, including that of gymnastic and sports teacher. He was successful in the late 1920s with plays dealing with young people (*Giftgas über Berlin*, 1929; *Revolte im Erziehungshaus*, 1929; *Alarm im Arbeitslager*, 1932). His novels *Verratene Jungen* (1929) and *Jörg Christoph, ein Fähnrich* (1935) underline his interest in youth. He emigrated in 1933. After his return to Germany he wrote the novel *Die Geschichte von Billy the Kid* (1952) and the play *Drei Söhne* (1959).

LAMPRECHT, Pfaffe, a priest living in the first half of the 12th c., originating probably in Trier and later active in Cologne, who is believed to be the author of two Middle High German works, an epic poem on the subject of

Alexander the Great (see ALEXANDERLIED), and a shorter poem on Tobias, preserved only in fragments. Lamprecht is the first German writer known to have written both religious and profane poetry.

LAMPRECHT VON REGENSBURG, a citizen of Regensburg in the first half of the 13th c. who was influenced by the pulpit oratory of BERTHOLD von Regensburg (q.v.), and became a Franciscan monk. Before his entry into the Order he wrote *c.* 1240 *Sanct Francisken Leben,* a rendering in German verse of the life of the saint (*Vita S. Francisci, c.* 1230) by Thomas of Celano. Later, as a Franciscan friar, he composed *c.* 1250 *Die Tochter Syon,* a poetic allegory of the mystical union of the soul with God. The soul is the Daughter of Zion, who is brought to Jesus by Faith and Hope, by Wisdom and Charity, by Love, and finally by Mercy. The source of the poem is a Latin prose work, *Filia Syon.* A much shorter version in Swiss dialect, written at the end of the 13th c. or beginning of the 14th c. and known as the *alemannische Tochter Syon,* once attributed to Der Mönch von Heilsbronn (q.v.), is of unknown authorship.

Land, as a political term, denotes the separate components of the Weimar Republic (q.v.) 1919–33, the Federal Republic (see BUNDESREPUBLIK DEUTSCHLAND) since 1949, and the German Democratic Republic (see DEUTSCHE DEMOKRATISCHE REPUBLIK) 1949–52.

Landeskirchen, were introduced at the Diet of Speyer (1529, see SPEYER, REICHSTAG ZU), and from 1555 (see AUGSBURGER RELIGIONSFRIEDE) became evangelical state churches. The constitutional and legal position of the Landeskirchen, which varied in different states, has been reorganized since their foundation, especially during the 19th c. and after 1945. Since 1969 the Confederation of the Evangelical Church (Bund der evangelischen Kirche) in the DDR (see DEUTSCHE DEMOKRATISCHE REPUBLIK) has been independent of the Evangelical Church in Germany.

Landesvater, a ceremony of student 'Korps' (see STUDENTISCHE VERBINDUNGEN), in which the song 'Alles schweige' (q.v.) is sung, while the caps of those present are transfixed with a duelling sword. The name derives from an older song used for this ceremony, 'Landesvater, Schutz und Rater'. Each 'Korps' was originally drawn from a particular German state, and the ceremony was a declaration of loyalty to state and prince (Landesvater). It is now a declaration of loyalty to the 'Korps'. The custom dates from the 17th c.

Landfriede, a legal measure promulgated in the Middle Ages to keep internal peace and check feuds and lawless violence. A Landfriede either proscribed feuds or insisted upon formal declaration of a feud; it protected important buildings and chattels, such as churches, mills, and agricultural implements, and also a number of categories of persons, particularly the clergy, women, and certain other people in the exercise of their calling. Landfrieden, which were proclaimed for a term of years, began with the edict of 1103. The Landfriede of Mainz (Mainzer Reichslandfriede) of 1235 is notable as being the first imperial law drafted in German instead of Latin. The Ewiger Landfriede (q.v.) proclaimed in 1495 by the Emperor Maximilian I (q.v.) forbade private feud and made further laws of Landfriede unnecessary.

Landgraf Ludwig von Thüringen, a Middle High German epic poem written by an unnamed priest for Duke Bolko I of Schweidnitz-Jauer, probably *c.* 1300. A clumsily written poem of some 8,000 lines, it narrates the heroic actions of Ludwig III of Thuringia (d. 1190), elder brother of the well-known patron of literature Landgrave Hermann (q.v.), on a crusade in the Holy Land. The basis of the poem, which relies, at least in part, on Latin sources, is historical, though it confuses events and persons. The poet was influenced by ULRICH von Etzenbach (q.v.), and his portrayal of the Saracens as chivalrous warriors recalls WOLFRAM von Eschenbach (q.v.).

Landhaus am Rhein, Das, a five-volume novel by B. Auerbach (q.v.), published in 1869. It is divided into 15 books.

Herr Sonnenkamp, a German who has made a vast fortune in the U.S.A., has built himself a magnificent house overlooking the Rhine (the indications in the book lead us to imagine it a few miles upstream from Koblenz). Here he lives in princely luxury with his neurotic wife and his young son Roland. A daughter, Manna (Hermanna), is in a pension attached to a convent. Sonnenkamp seeks a tutor for his son, and a nobleman of the region, Otto von Prancken, recommends a former regimental comrade, Captain Erich Dournay, son of a deceased professor. Roland quickly takes to Erich, who is a man of elevated Liberal principles, as well as great strength of character. Prancken hopes to marry Manna and is soon treated by Sonnenkamp as his prospective son-in-law. Erich's position is strengthened by a friendship with a neighbour of Sonnenkamp, Graf Clodwig Wolfsgarten, whose beautiful young wife (Bella) seems outwardly content and considerate, but is inwardly embittered and ruthless. Sonnenkamp decides to seek ennoblement, and his supporters set in motion a train of intrigue and bribery. It has in the meantime become apparent that there are sinister secrets in his past. Indeed Manna's announced intention of taking the veil is due to a suspicion of her father's iniquity, for which she proposes to atone. The progress of the patent of nobility is slow, but at last all is in order. At the formal audience with the Prince of the land, Sonnenkamp is dramatically revealed to have been a slave trader and to have committed murders, and the question of ennoblement lapses. He returns embittered and frustrated to his home, where his daughter has turned from Prancken and engaged herself to Erich Dournay.

Sonnenkamp quickly rallies from his disgrace at court, for he is a man of immense power and dynamism. The American Civil War is about to break out and he decides to join the cause of the Southern States. He takes with him all the money he has earned by slaving, leaving the rest, and he persuades the freshly widowed Bella, Gräfin Wolfsgarten, to accompany him. Erich and Manna are married, and they, together with Roland, who has imbibed Liberal principles from Erich, also set out for America, where the two men intend to join the anti-slavery forces of the North. An epilogue of letters (Bk. 15) reveals the deaths of Sonnenkamp and Gräfin Bella and reports the assassination of Abraham Lincoln.

The novel is peopled by a vast number of characters, who are persuasively presented and astutely motivated. They symbolize, and frequently debate, the urgent social problems of the day, particularly the structure of society, the present and future roles of religion, especially the Roman Catholic Church, and the burning issue of the day, Negro slavery in the U.S.A. Though the plot may seem artificial, it is sustained by many unexpected details.

Landplagen, Die, a poem published by J. M. R. Lenz (q.v.) in 1769. Described as *Ein Gedicht in sechs Büchern,* it is written in hexameters recalling those of Klopstock. The poem deals with six catastrophes which can befall the countryman: (1) Der Krieg; (2) Die Hungersnot; (3) Die Pest; (4) Die Feuersnot; (5) Die Wassersnot; and (6) Das Erdbeben. The subject gives scope for the accumulation of horrors, and the poem ends with the approach of a seventh 'Landplage', the Day of Judgement.

Landshuter Erbfolgekrieg, a war of the succession of Bayern-Landshut, fought from 1503 to 1505 between the Munich line and Rupert of the Palatinate. The succession finally passed to the Duke of Bayern-München, who was supported by the Emperor Maximilian I (q.v.).

Landsturm, a uniformed reserve set up in the German Empire in time of war consisting of former reservists aged 40 to 44, and of all citizens between 18 and 44 who had not been called up for normal service. This Landsturm was established by a law passed in 1888 and ceased in 1919. The term had been used earlier (notably in the 17th c. and in Prussia 1808–14) for any call-up of the male population in time of national emergency.

In Austria-Hungary the Landsturm was equivalent to the German Landwehr (q.v.).

Landvogt von Greifensee, Der, a Novelle by G. Keller (q.v.). It is the third in the group of three stories which opens the *Züricher Novellen* (q.v., 1878) and has a narrative frame (see RAHMEN) enclosing a cycle of short tales.

The Landvogt (or governor) of Greifensee in 1783 is Colonel Salomon Landolt, a bachelor of 42. On 13 June as he rides off from a parade, in which his battalion has performed with credit, he encounters a coach containing a young married woman to whom he had once unsuccessfully proposed marriage. He has a nickname for each of several young women he has courted, and this one is 'Der Distelfink'. The meeting suggests to him the idea of issuing an invitation to the five ladies whose hand he has sought. He persuades Frau Marianne, his housekeeper, to participate in the plan. Her life story is first told, followed by sketches of the five ladies, including in each case the episode of the Landvogt's unsuccessful courtship. Frau Marianne, a woman of courage and character, eloped with a student, married him, and, when he turned soldier, followed him as a *cantinière*. She bore him nine children, all of whom died, then, finding that his love for her had also died, sent him packing, and became housekeeper to the Landvogt, whom she has loyally served.

The Landvogt explains his curious plan to Frau Marianne by giving her an account of his relationship to the five ladies, which he accomplishes in a series of miniature Novellen. Salome Alt ('der Distelfink') was attracted to him, but he described his own character and his ancestry in such disparaging terms that she turned him down and married another. She was followed by Figura Leu, whom he calls 'Hanswurstel'; a mercurial character, she nevertheless has the resolution, in spite of her love for Landolt, to refuse him because she has given her dying mother a promise never to marry. The principal feature of her story is a party, at which the writers J. J. Bodmer and Salomon Geßner (qq.v.), both of Zurich, are present. The Landvogt's third refusal came from Wendelgard Gimmel, whom he calls 'der Kapitän' after her father, a hard-drinking officer. It almost comes to a match, but Figura Leu intervenes because she distrusts Wendelgard's character. Wendelgard marries Figura's brother Martin, conquers a penchant for gambling, and becomes a good wife to him. Landolt is content and does not resent Figura's interference. The fourth girl, Barbara Thumeysen, who is 'die Grasmücke' for Landolt, takes fright at her suitor's eccentricity. The fifth, Aglaja or 'die Amsel', was not in love with Landolt and only wished to have his good offices in persuading her family to allow her to marry a young pastor. The Landvogt co-operated, and the marriage took place. The pastor died some years later, having proved to be avaricious, ambitious, and worldly.

All five ladies are invited to Schloß Greifensee on 31 May. The Landvogt arranges first for their presence in his magistrate's court, so that they shall have a chance to see how understandingly he adjudicates in matrimonial cases. Afterwards he invites them to choose him a wife, either Frau Marianne or her pretty young attendant. They unanimously choose the young girl, who is immediately shown to be a boy disguised in female clothing. The Landvogt, having enjoyed his gentle revenge, resigns himself to bachelordom. He takes an active part in the fighting in Switzerland during the Napoleonic Wars (q.v.), and dies at Andelfingen in 1818, aged 76.

Landwehr, army formations composed in Germany of reservists and in Austria-Hungary of supplementary troops. The first organized Landwehr in German lands was established in Austria in 1809 by the Archduke Karl (q.v.). A Prussian Landwehr was set up in 1813.

The Austrian Landwehr was a small regular force composed of normal recruits drawn from the German-speaking lands. Its Hungarian equivalent was the Honvéd. The Landwehr bore the title k. k. (q.v., kaiserlichköniglich) in subtle distinction from the k. u. k. (kaiserlich und königlich) of the regular army of the Dual Monarchy.

The Prussian Landwehr was regulated from 1815 to 1867 by the Landwehrordnung, and was composed of two categories of reservists, aged 25 to 32 (1. Aufgebot) and 33 to 39 (2. Aufgebot). This system, with modifications in the larger formations of troops, was adopted by the German Empire. The Landwehr was abolished under the terms of the Treaty of Versailles.

LANG, Isaac, see GOLL, I.

LANGBEIN, August Friedrich (Radeberg nr. Dresden, 1757–1835, Berlin), lived in Dresden and, after 1800, in Berlin. A purveyor of light literature in short story and anecdotal form

(*Schwänke*, 1792; *Feierabende*, 1793), he also wrote poems, plays, and several novels.

LANGE, Helene (Oldenburg, 1848–1930, Berlin), a prominent educationist and feminist, urged in a memorandum of 1887 that the education of girls should be taken out of the hands of men and entrusted to women. Her ideas were expressed in the periodical *Die Frau*, founded in 1893, and in the *Handbuch der Frauenbewegung* (edited jointly with Gertrud Bäumer, q.v., 5 vols., 1901–6). A selection of her propagandist work appeared in *Kampfzeiten* (2 vols., 1928).

LANGE, Samuel Gotthold (Halle, 1711–81, Laublingen nr. Halle), minor poet and translator, studied theology at Halle and was afterwards pastor at Laublingen. Together with his friend J. I. Pyra (q.v.) he founded in Halle, while still a student, the Gesellschaft zur Förderung der deutschen Sprache, Poesie und Beredsamkeit (1733). He sought to counter Gottsched's francophile literary policy by advocating the adoption in German of classical metres. In 1745 Pyra's and Lange's poems were published by Bodmer (q.v.) without Lange's consent (Pyra had died the year before) as *Thirsis und Damons freundschaftliche Lieder*, Damon being Lange. Lange's *Freundschaftliche Briefe* were published in 1746 and *Horazische Oden* in 1747. His translation of the *Carmina* and *Ars poetica* of Horace (*Des Qu. Horatius Flaccus Oden fünf Bücher und von der Dichtkunst ein Buch poetisch übersetzt*, 1752) was philologically inaccurate and presented a Horace who was too close to the pattern of the Lutheran pastor.

It was a misfortune for Lange that so formidable a critic as the young Lessing (q.v.) set about his translation; for, in consequence of Lessing's campaign, Lange is now chiefly known as the butt in Lessing's title *Vademecum für den Herrn S. G. Lange*, 1754. The best of Lange's rather thin poetry was written under the stimulus of his more gifted friend Pyra.

Langensalza, a town in the former Prussian province of Saxony (Provinz Sachsen), near which in the war of 1866 (see Deutscher Krieg) the Hanoverian forces first won a victory (27 June) over the Prussians, but were then encircled by superior forces; on 29 June they capitulated.

LANGER, Anton (Vienna, 1824–79, Vienna), wrote Viennese Volksstücke (see Volksstück), especially farces (*Wiener Volksbühne*, 4 vols., 1859–64), and humorous and sentimental Viennese stories and novels. He was a master of Viennese dialect.

LANGEWIESCHE, MARIANNE (Ebenhausen nr. Starnberg, Bavaria, 1908–), whose married name is Kuhbier (also spelt Coubier), became a journalist and achieved a reputation with popular historical novels, of which *Königin der Meere* (1940), dealing with the foundation of Venice, was the greatest success. Her other novels are *Die Ballade der Judith van Loo* (1938), *Die Allerheiligen-Bucht* (1942), *Die Bürger von Calais* (1949), and *Der Ölzweig* (1952), which treats of Noah and the Flood. She is also the author of the historical Novellen *Castell Bô* (1948) and *Der Garten des Vergessens* (1953).

LANGGÄSSER, ELISABETH (Alzey, 1899–1950, Rheinzabern), a Roman Catholic elementary school teacher, began to write in her twenties. Her volume of poems, *Der Wendekreis des Lammes* (1924), was followed in 1932 by the short novel *Proserpina* and the group of tales *Triptychon des Teufels*. The poems *Die Tierkreisgedichte* appeared in 1935, and the novel *Der Gang durch das Ried* in 1936. In this year she was banned from writing, since she was of partially Jewish descent. She resumed after the war with the striking novel *Das unauslöschliche Siegel* (q.v., 1946), which was followed some years later by her final novel, *Märkische Argonautenfahrt* (q.v., 1950). Other later publications were the stories *Das Labyrinth* (1949) and some works of poetry, in which particular use is made of flowers as significant symbols: *Der Laubmann und die Rose* (1947), *Kölnische Elegie* (1948), and *Metamorphosen* (1949). She was also the author of essays relating to religion (*Geist in den Sinnen behaust*, posth. 1951; *Das Christliche der christlichen Dichtung*, 1961). Her disciplined work is preoccupied with the problem of evil and lays stress on Grace. Correspondence, ... *Vergänglichkeit. Briefe 1926–1950*, ed. W. Hoffmann, appeared in 1954 and *Gesammelte Werke* (5 vols.) in 1959–64.

LANGMANN, ADELHEID (d. 1376, Engelthal), a Dominican nun of Engelthal Abbey who wrote down her visions. They coincide in part with the visions of Christine Ebner (q.v.).

LANGMANN, PHILIPP (Brünn or Brno, 1862– 1931, Vienna), originally a workman, became a manager but abandoned industry for writing and journalism, at first in his native town and from 1900 in Vienna. In a style of closely detailed realism he wrote stories which were published in the collections *Arbeiterleben* (1893), *Realistische Erzählungen* (1895), *Ein junger Mann* (1895), *Verflogene Rufe* (1899), *Erlebnisse eines Wanderers* (1911). He is the author of Naturalistic plays, including *Bartel Turaser* (1897), his best-known drama which deals with a strike, the comedy *Die vier Gewinner* (1898), and the dramas

Unser Tebaldo (1898), *Gertrud Antleß, Korporal Stöhr* (both 1900), *Die Herzmarke* (1901), and *Gerwins Liebestod* (1903). A novel entitled *Leben und Musik* appeared in 1904.

Langobards, a Germanic people who, in the 5th c. and 6th c. A.D., invaded and settled in Austria and Hungary, and then passed into Italy, where they established themselves in the north. Their kingdom was annexed by the Emperor Otto I (q.v.) in 962. It is from the Langobards that the name Lombardy is derived.

Lanzelet, see ULRICH VON ZAZIKHOFEN.

Lanzelot, a prose work recounting at immense length the career of Lancelot, the quest for the Grail, and the death of Arthur. The earliest MS. has been dated *c.* 1225, but belongs more probably to the late 13th c. *Lanzelot* is remarkable as a prose work of substance written at a time when verse otherwise prevailed unchallenged. It is possibly the translation of a lost Flemish original.

Lanzenfest, the Feast of the Holy Lance, centred upon the instruments of the Passion of Our Lord, the spear or lance and the nails. Restricted to Germany, it was first celebrated in 1353. It is now only observed in certain localities. The Feast was celebrated on the Friday after the Easter Octave, i.e. on the second Friday after Easter Day.

Laokoon, oder Über die Grenzen der Malerei und Poesie, a treatise on aesthetics begun by G. E. Lessing (q.v.) in 1763, written principally in 1765, and published in 1766. It is described as Erster Teil and was intended to be the first of three volumes, the second and third of which, however, were never written. Lessing hurried . the publication of the work in the illusory hope that it would secure his appointment as Royal Librarian in Berlin. The starting-point of the book and the reason for the title is a passage from *Gedanken über die Nachahmung der griechischen Werke in der Malerei und Bildhauerkunst* (1755) by J. J. Winckelmann (q.v.). Winckelmann had contrasted the stoical bearing in the statue of Laocoön with the loud cries which Virgil causes him to utter, and had interpreted the difference as a superior serenity in Greek art. Lessing dissented, asserting that the differing conditions of plastic and literary art imposed different means of expression. In the critical passage in Chapter XVI he contends that objects with their visible attributes are the concern of painting (which presents coexisting things, i.e. Gegenstände) and that actions (Handlungen) are the affair of poetry, which follows a consecutive pattern. Lessing accordingly rejects in a well-known passage the descriptive poetry of Haller's *Die Alpen* (q.v.). The volume discusses other relevant examples in learned and scholarly fashion.

The sculpture, which represents Laocoön and his two sons caught in the coils of the serpent, was excavated at Rome in 1506 and is in the Vatican Museum.

LAPPE, KARL GOTTLIEB (Wusterhausen, Pomerania, 1773–1843, Stralsund), a schoolmaster of Stralsund, wrote lyric poetry (*Gedichte*, 1801; *Blüten des Alters*, 1841). His poem beginning 'Nord oder Süd, wenn nur im warmen Busen', now forgotten, was popular in the 19th c.

LAROCHE, JOHANN JOSEPH (Preßburg or Bratislava, 1745–1806, Vienna), a talented actor, was from 1781 until his death the principal comic actor in the Theater in der Leopoldstadt (see LEOPOLDSTÄDTER THEATER). He was best known in the Viennese role of Kasperle (see KASPERLTHEATER).

LA ROCHE, MAXIMILIANE VON (Frankfurt/Main, 1756–93, Frankfurt/Main), younger daughter of Sophie von La Roche (q.v.), was married in 1774 to Pietro Brentano and was the mother of Clemens and Bettina Brentano (see BRENTANO, C. and ARNIM, BETTINA VON). Goethe, visiting the household on his way back from Wetzlar in 1772, was attracted briefly to Maximiliane.

LA ROCHE, SOPHIE VON (Kaufbeuren, 1731–1807, Offenbach/Main), *née* Sophie Gutermann, was brought up in Augsburg, and as a girl paid frequent visits to Biberach, where her cousin C. M. Wieland (q.v.) fell in love with her. In 1754 she was married to Georg von La Roche, an illegitimate son of Count Stadion (1691–1768), whose family estates La Roche supervised until 1771. When in 1766 Frau von La Roche's daughters were sent to boarding school at Strasburg, the mother extricated herself from a crisis of desolation by writing a novel, the *Geschichte des Fräuleins von Sternheim* (q.v.), which was published by Wieland in 1771. In that year La Roche received an administrative appointment from the Electoral Archbishop of Trier, for which the family moved to Ehrenbreitstein opposite Koblenz. Here Goethe visited them in 1772 on his return from Wetzlar and was attracted by the younger daughter Maximiliane (q.v.). La Roche retired in 1780 and died in 1786, and Frau von La Roche spent her old age at Speyer and Offenbach, with visits to her

daughters. She wrote several other novels, but none equalled the success of *Das Fräulein von Sternheim*. Their titles include *Rosaliens Briefe an ihre Freundin*, a collection of letters, educative in purpose (1780–1), *Geschichte von Miss Lony* (1789), *Rosalie von Cleberg auf dem Lande* (1791), and *Fanny und Julia* (1802).

L'ARRONGE, ADOLF (Hamburg, 1838–1908, Kreuzlingen nr. Constance), a gifted musician, was a conductor in Cologne and (1866–9) in Berlin. From 1874 to 1878 he was director of the Lobe-Theater in Breslau. In 1881 he acquired the Friedrich-Wilhelmstädtisches Theater in Berlin, which he directed with great success until 1894. His opera *Das Gespenst* was produced under his own direction in 1860, and in later life he was an industrious and skilful purveyor of light comedies. Many were written in collaboration with G. von Moser (q.v.) Among his own works were the Volksstücke (see VOLKSSTÜCK) *Mein Leopold* (1873), *Alltagsleben* (1873), and the comedies *Wohltätige Frauen* (1879) and *Der Weg zum Herzen* (1884). His *Dramatische Werke* appeared in 8 vols. 1879–86 and in 4 vols. in 1908. The family name was originally Aaron; L'Arronge was adopted by the dramatist's father.

LASKER-SCHÜLER, ELSE (Elberfeld, 1869–1945, Jerusalem), was a Jewess. Her maiden name was Schüler, to which she added the name of her first husband. She frequented the *avant-garde* literary circle of Berlin in the 1900s, which included P. Hille, Th. Däubler, G. Trakl, G. Benn, F. Werfel, R. Schickele, and the painter F. Marc (qq.v.). She emigrated in 1933, stayed in various countries, but in 1937 settled in Palestine, where she died in poverty. Her life had for years been eccentric and unpredictable, and she tended to live in a world of her own imagining. Her poetry, which is symbolical and sometimes playful, is at the same time religious and concerned with the decline of the modern world. She published the volumes of verse: *Styx* (1902), *Der siebente Tag* (1905), *Hebräische Balladen* (1913), *Die gesammelten Gedichte* (1917), *Die Kuppel* (1920), *Theben* (1923), and *Mein blaues Klavier* (1943). Her first play, *Die Wupper*, written in 'one night' in 1909 and using her native dialect ('Wupperdhalerplatt'), was published in its final form in 1919, the year of its performance at the Deutsches Theater, Berlin. Described by A. Kerr (q.v.) as 'phantasto-naturalistisch', it is one of the most interesting modernist plays and has been revived since its Cologne production by Hans Bauer in 1958 (Heinz Herald directed the 1919 production and Jürgen Fehling the Berlin Staatstheater production of 1927). *Konzert*, a volume of prose and verse, and her second play, *Arthur Aronymus und seine Väter*, relating to her father's childhood, was published in 1932, the year in which she was awarded the Kleist Prize. A third play, *Ichundich*, was written *c.* 1943 and published in 1980. Her fiction includes the novel *Mein Herz* (1912). *Dichtungen und Dokumente*, ed. E. Ginsberg, appeared in 1951; *Gesammelte Werke* (3 vols., vols. 1 and 2 ed. by F. Kemp, vol. 3, *Verse und Prosa aus dem Nachlaß*, ed. by W. Kraft) in 1959–62; *Sämtliche Gedichte*, ed. F. Kemp, in 1966; 2 vols. of correspondence, ed. M. Kupper (*Lieber gestreifter Tiger* and *Wo ist unser buntes Theben*) in 1969, and *Briefe an Karl Kraus*, ed. A. Gehlhoff-Claes, in 1959. A bilingual volume of select poetry, translated by R. P. Newton, was published in 1982.

LASSALLE, FERDINAND (Breslau, 1825–64, Geneva), whose name until 1846 was Lassal, was the son of a Jewish businessman, studied at Breslau and Berlin universities, and in 1863 founded the socialist Allgemeiner Deutscher Arbeiterverein (see SPD). He was the author of its programme, written at the request of German socialist groups for their congress at Leipzig in that year.

Although Lassalle contributed for a short time to the *Neue Rheinische Zeitung* edited by Marx (q.v.), the influence of Louis Blanc, whom he first met in Paris in 1845, remained a lasting one. His socialist programme (Arbeiterprogramm) was particularly concerned with the workers' share in productivity (drawn up in his 'Ehernes Lohngesetz'). At the time of his death in a duel he had not yet publicly expressed his opposition to the principle of monarchy. In the late 1850s and during the 1860s he supported Prussian policy for the unification of Germany without Austria.

Lassalle's principal works are *Die Philosophie des Herakleitos* (2 vols., 1858) and *Das System der erworbenen Rechte* (2 vols., 1861), both based on Hegel (q.v.); the former work represents the culmination of his scholarship. Lassalle originally contemplated an academic career. He is the author of a play on the 15th-c. rebel F. von Sickingen (q.v.), *Franz von Sickingen* (1859).

LASSBERG, JOSEPH, FREIHERR VON (Donaueschingen, 1770–1855, Meersburg), spent a large part of his life as a civil servant, settling in 1838 in the castle of Meersburg, where he was host to his sister-in-law, the poetess Annette von Droste-Hülshoff (q.v.), in her last years. Laßberg was an antiquary and a collector of MSS., acquiring as his outstanding treasure MS. C of the *Nibelungenlied* (q.v.). He published a number of medieval poems, including the *Nibelungen* text, in the 4 vols. of the *Liedersaal* (1820–5, repr. 1969). The most important items of his library are now at Donaueschingen.

Laubacher Barlaam, a Middle High German poem dealing with the story of Barlaam and Josaphat, which has also been treated by RUDOLF von Ems (q.v.). The *Laubacher Barlaam*, so called because the MS. was discovered at Laubach in Hesse, is the work of Bishop Otto of Freising, Bavaria, who died in 1220. The poem recounts in some 16,000 lines the story of Josaphat, son of an oriental prince, who turns from the transient wealth of the world to the religious teachings of the monk Barlaam, converts his father to the true faith, and then joins Barlaam as an anchorite. The poem, which contains much dialogue of didactic character, is closely modelled on a Latin version of the legend.

LAUBE, HEINRICH (Sprottau, Silesia, 1806–84, Vienna), in full Heinrich Rudolf Constanz, grew up in straitened circumstances in a remote country town and was sent to grammar school at Glogau and Schweidnitz, and in 1826 to Halle University, where he was supposed to study theology (the poor man's faculty); in fact he spent most of his time away from lectures, becoming an expert student duellist. In 1827 he migrated to Breslau, and inclined more and more towards literary studies. In 1829 Laube rashly founded a literary journal (*Aurora*), which ceased publication after a few months, leaving him with substantial debts that took years to pay off. In order to earn his living he became in 1830 private tutor in the house of Dr. Rupricht at Kottwitz, and in 1831 moved to a similar post with a family named Nimptsch at Jäschkowitz near Breslau.

While in this employment Laube wrote his first book, the political essays *Das neue Jahrhundert* (2 vols., 1833). He left Jäschkowitz in 1832 and moved to Leipzig to seek employment with the publishers Brockhaus; from 1833 to 1834 he was editor of the *Zeitung für die elegante Welt* (q.v.), a journal which, despite its title, was now radical in tendency. At this time he began a political and social novel, *Das junge Europa*, which appeared in 3 vols. entitled *Die Poeten* (1833), *Die Krieger* (1837), and *Die Bürger* (1837). In the same year (1833) he undertook with a new acquaintance, K. Gutzkow (q.v.), a journey to Italy which provided the stimulus for his 6 vols. of *Reisenovellen* (1834–7). Laube's journalism now took a more radical turn, becoming one of the conspicuous manifestations of the Young German movement (see JUNGES DEUTSCHLAND). In the repressive climate of those days it was not surprising that he was expelled from Leipzig. Ignoring advice to go abroad, Laube visited Berlin, was arrested in July 1834, and detained in prison for nine months while his alleged subversive activities were investigated. After his release he lived in Naumburg, writing *Moderne Charakteristiken* (1835), which reveal a more cautious and restrained trend. The persecution of writers associated with Junges Deutschland at the end of 1835 led to Laube receiving in December 1836, a month after his marriage, a sentence of seven years' detention (Festungshaft), six of which referred to his membership of the Burschenschaft in university days. An appeal to the king for mercy resulted in the reduction of the term to one and a half years, which he was allowed to serve, together with his wife, in comfortable circumstances on the estate of Prince Pückler-Muskau (q.v.). During this 'imprisonment' he wrote a history of German literature (*Geschichte der deutschen Literatur*, 4 vols., 1839–40). The shooting expeditions which he secretly made at Muskau had a literary result in *Jagdbrevier* (1841), a collection of poems with an afterword entitled *Die Gemse*, to which a glossary of hunting jargon (Jagdsprache) is appended.

Laube was released in January 1839, and from May until February 1840 the couple travelled abroad, visiting Holland, Belgium, Paris, and Algiers. In 1840 Laube published *Französische Lustschlösser* (3 vols.), further evidence of his swing away from the dangerous theme of politics. In the following years he was active as a journalist and dramatic critic, first in Leipzig, then (1845) in Vienna. He wrote stories and a novel (*Gräfin Chateaubriant*, 3 vols., 1843), but devoted himself especially to drama. His first play was *Monaldeschi* (q.v., performed 1841, published 1845); this historical tragedy was followed by the historical comedy *Rokoko* (q.v., performed 1842, published 1846). Other plays include the tragedy *Die Bernsteinhexe* (q.v., performed 1844, published 1846), the Danish tragedy *Struensee* (q.v., performed 1845, published 1847), the comedy *Gottsched und Gellert* (q.v., performed 1845, published 1847), *Die Karlsschüler* (q.v., performed 1846, published 1847), dealing with Schiller, and a play about the young Friedrich of Prussia (see FRIEDRICH II, DER GROSSE), *Prinz Friedrich* (performed 1848, published 1854).

In less than a decade Laube became one of the best-known playwrights in Germany. In 1848 he was elected to the Frankfurt Parliament (see FRANKFURTER NATIONALVERSAMMLUNG), but resigned early in the following year. His appointment in 1849 as Director of the Burgtheater in Vienna came as a surprise in view of his political

past. He held this post until 1867, when he resigned on the appointment of Friedrich Halm (q.v.) as his superior.

After a brief period as theatre director in Leipzig (1869–71), Laube returned to Vienna to found the Stadttheater, which he directed, with one short interruption, until 1880. During his theatrical years Laube wrote further plays, achieving his greatest success with *Graf Essex* (q.v., 1856). It was followed by *Montrose* (1859), *Der Statthalter von Bengalen* (1866), *Böse Zungen* (1868), and a completion of Schiller's *Demetrius* (1869). At the same time he resumed his novel-writing, publishing notably a fictional evocation of the Thirty Years War (see DREISSIGJÄHRIGER KRIEG), *Der deutsche Krieg* (9 vols., 1865–6), and *Die Böhminger* (1880).

Laube published three important works of theatrical recollections, *Das Burgtheater* (1868), *Das Norddeutsche Theater* (1872), and *Das Wiener Stadttheater* (1875). As a theatre director in Vienna he set and maintained high standards, and from 1849 to 1880 was the dominant figure in the theatrical life of the city. In his early years as director he suppressed, however, the work of Hebbel, while the revival of that of Grillparzer (qq.v.) is entirely due to his initiative. He is the author of a biography of the latter, *Franz Grillparzers Lebensgeschichte* (1884). He has the credit of having introduced Ibsen (q.v.) to German audiences (*Pillars of Society*, in German *Stützen der Gesellschaft*, 1878).

An energetic and rapid writer, Laube was appreciated most in his own day. His earlier political work now appears hasty, and in his later creative work he has little to say. His documentary writings provide a fund of valuable information. *Gesammelte Schriften* (16 vols.) appeared 1875–82, and *Gesammelte Werke* (50 vols. in 20), ed. H. H. Houben, 1908–9.

Lauchstädt, Bad, a spa near Weimar with a small theatre which was acquired by the Weimar Court Theatre in 1791 and used for regular summer seasons. Closed in the 19th c., the theatre was reopened in 1908. It was at Lauchstädt that occurred in 1803 the incident recorded by Schiller, in which a violent thunderstorm provided a perfectly timed cue for the lines in *Die Braut von Messina* (q.v.) beginning 'Wenn die Wolken getürmt den Himmel schwärzen,/Wenn dumpftosend der Donner hallt . . .'

LAUDON, GIDEON ERNST, FREIHERR VON (Totzen, Livonia, 1717–90, Neutitschein (Novejiçin), Moravia), was one of the most distinguished Austrian commanders under Maria Theresia (q.v.). He was of Scottish descent (the name is also spelt Loudon), and served in Russia before entering the Austrian army in 1743. He was a regimental officer in the Second Silesian War (see SCHLESISCHE KRIEGE), and rose to general's rank in the Seven Years War (see SIEBENJÄHRIGER KRIEG), in which he fought successfully at Prague, Kolin, Domstadtl, and Hochkirch, and especially at Kunersdorf (q.v.). He was also responsible for the storming of Glatz and the taking of Schweidnitz. His last important success in the field was the capture of Belgrade in 1789. Laudon was a bold commander, sometimes to the point of rashness, and was in frequent conflict with the cautious Daun (q.v.). Laudon is the subject of a patriotic Volksstück by F. Kaiser (q.v.), *General Laudon* (1874).

Laufen, Der, a Novelle by E. Strauß (q.v.), published in the collection *Hans und Grete* in 1909. The rapids (Laufen) of the Rhine pass Laufenburg on the Swiss border. They are the scene of a triangular love relationship which ends in the drowning of the engaged couple Siddy and Albiez; the survivor relates the incident some twenty years later. The Laufen in spate is a symbol of the surging torrent of the emotions.

LAUFF, JOSEPH VON (Cologne, 1855–1933, Bad Kirchen/Moselle), was a regular officer in the Prussian army from 1877 to 1898, when he became producer to the Court Theatre at Wiesbaden. He found favour with Wilhelm II (q.v.) for his vacuous historical plays on the Hohenzollerns, and received from him the 'von' of nobility in 1913. He was more successful with regional (Rhineland) novels, such as *Regina coeli* (2 vols., 1894), *Die Tucher von Köln* (1909), *Die Brinkschulter* (1913), *O du mein Niederrhein* (1930), and *Sinter Klaas* (1931).

Laune des Verliebten, Die, a one-act pastoral play (Schäferspiel), written by Goethe in Leipzig in 1767, and performed in Weimar in May 1779. It was first published in 1806 in vol. 4 of *Goethes Werke* (1806–10). Written in Alexandrine verse, it depicts a pastoral lover (Eridon) cured of jealousy of his shepherdess (Amine) by being tempted into kissing another (Egle).

Laura-Oden, name by which a group of six early poems by Schiller (q.v.) is known. They comprise *Phantasie an Laura, Laura am Klavier, Die seligen Augenblicke, Vorwurf, Das Geheimnis der Reminiszenz,* and *Melancholie.* The *Laura-Oden* are remarkable for their combination of sensuality and self-analysis. All were published anonymously in the *Anthologie auf das Jahr 1782,* though *Die seligen Augenblicke* had already appeared under the title *Die Entzückung an Laura* in Stäudlin's *Schwäbischer Musenalmanach auf das Jahr 1782.*

LAUREMBERG, JOHANN (Rostock, 1590–1658, Sorø, Denmark), satirical poet, was a professor of poetry in Rostock from 1618 to 1623 and thereafter a professor of mathematics at Sorø in Denmark. Lauremberg wrote Latin poems, but his principal work (*Veer Schertz Gedichte,* 1652) was written in his native Low German. These four satirically humorous poems are entitled *Van der Minschen jtzigem Wandel und Maneeren, Van Alamodischer Kleder-Dracht, Van vormengder Sprake und Titeln,* and *Van Poesie und Rymgedichten.* They appeared under the pseudonym Hans Willmsen L. Rost, of which the last two elements denote 'Lauremberg Rostochiensis'. Lauremberg, a supporter of good sense, opposed foreign extravagance in manners, clothing, and language, expressing himself in straightforward, often downright, Low German.

LAURENTIUS VON SCHNÜFFIS or SCHNIFIS (Schnifis nr. Feldkirch, Vorarlberg, 1633–1702, Constance), originally Johannes Martin, was at first an actor, then studied theology in Innsbruck and in 1665 entered the Capuchin Order, residing at Zug in Switzerland. He composed in 1682 a cycle of allegorical pastoral poems in strophic form, presenting the conversion of the soul, represented by Clorinda, to Christ, symbolized by Dafnis. To this work he gave the title *Mirantisches Flötlein,* in which the first six letters are an anagram of his real surname. The cycle is an original attempt to adapt the pattern of the pagan pastoral to devotional ends.

Laurin, an anonymous Middle High German heroic epic, set in Tyrol and existing in several versions written probably in the second half of the 13th c. The best is known as *Laurin A.*

Dietrich von Bern (see DIETRICHSAGE) learns from his henchman Witege of the physical prowess of the dwarf Laurin (*laur* = cunning), who fiercely defends his rose garden against all intrusion. Dietrich, accompanied by Witege and Dietleib, sets forth to prove his superiority. Witege is defeated by the diminutive Laurin, whose strength is derived from a magic girdle, which Dietrich succeeds in cutting, whereupon Laurin is easily overpowered. Laurin reveals, however, that he has married Dietleib's sister Künolt and invites all to a feast in his palace. He dopes the wine and, when his guests sink into a torpor, imprisons them. They are rescued by Künolt with the aid of a magic ring, and Laurin is carried off to Bern (Verona), where he is baptized and becomes a thane of Dietrich. Rose Garden (Rosengarten) is a place-name in Tyrol, which alludes to the sunset glow (Alpenglühen).

The poem was edited by G. Holz in 1897.

Läuschen un Rimels, 2 vols. of dialect verse by F. Reuter (q.v.), published in 1853 and 1858. They are described on the title-page as *Plattdeutsche Gedichte heiteren Inhalts in mecklenburgisch-vorpommerscher Mundart.* Vol. 1 contains 68 poems and vol. 2 69. They are mostly simple anecdotes of rural life, comic episodes or accounts of shrewd, humorous retorts, and are put in homely dialect, catching the tone of rural Mecklenburg.

Laus podagrae, see PIRKHEIMER, W.

Lautverschiebung, see SOUND SHIFTS.

LAVANT, CHRISTINE (Groß-Edling, Carinthia, Austria, 1915–73, Wolfsberg), partially blind and deaf from birth, she is best known for her lyric poetry, in which the combination of echoes of her Roman Catholic background and crass expressions of suffering and a sense of hopelessness, and the creative use of metaphors and of seemingly conventional rhymed verse are distinguishing features. Her collections include *Die unvollendete Liebe* (1949), *Die Bettlerschale* (1956), *Spindel im Mond* (1959), *Sonnenvogel* (1960), *Der Pfauenschrei* (1962), and *Hälfte des Herzens* (1966). Her stories grow out of her rural background and compassion for deprivation (she herself was the daughter of a miner). They include *Das Kind* (1948), *Das Krüglein* (1949), *Maria Katharina* (1950), *Baruscha* (1952), *Die Rosenkugel* (1956), *Der Lumpensammler* (1961), *Das Ringelspiel* (1963), and *Nele* (1969). A select edition (including the title-story), *Wirf ab den Lehm,* by W. Schmied, appeared in 1961 and *Kunst wie meine ist nur verstümmeltes Leben. Nachgelassene und verstreut veröffentlichte Gedichte—Prosa—Briefe* posthumously in 1978.

LAVATER, JOHANN CASPAR (Zurich, 1741–1801, Zurich), entered the Church and immediately engaged in a campaign against corruption in high places in Canton Zurich, which ended with the successful prosecution of the influential Landrat Grebel in 1762. Having thereby understandably made enemies, he travelled for a year in North Germany, making contact with several writers of the day, including F. G. Klopstock, J. W. L. Gleim, M. Mendelssohn, and K. W. Ramler (qq.v.). On his return he began to write, publishing *Gereimte Psalmen* (1768) and *Zwey Hundert Christliche Lieder* (1771) and editing a moralizing weekly, *Der Erinnerer.* In 1769 he was appointed deacon (Diakonus) to the Zurich orphanage and soon acquired a reputation as an eloquent and original preacher. In 1768 he began to express his emotional Christianity in *Aussichten in die Ewigkeit,* which was followed by the pietistic *Geheimes Tagebuch*

von einem Beobachter seiner selbst (1771). His best-known work is the *Physiognomische Fragmente zur Beförderung der Menschenkenntnis und Menschenliebe,* which appeared in 4 vols., 1775–8. The work was abundantly illustrated by engravings. The *Physiognomische Fragmente* sets out to interpret the links between the face and the soul, but the work is rashly executed without an adequate basis of investigation. In this it corresponds to the author's impulsive and enthusiastic character.

In 1778 Lavater was appointed pastor to St. Peter's church in Zurich. His friendship with Goethe, begun on their Rhine journey of 1774 (see ZWISCHEN LAVATER UND BASEDOW), was terminated by Goethe in 1786. Lavater extended his Christian physiognomical studies to anthropology in *Pontius Pilatus oder der Mensch in allen Gestalten* (1782–5). In 1800, when Zurich was taken by the French, he was shot while carrying out spiritual ministrations, and died of the consequences many months later. Goethe ridiculed Lavater in the *Xenien* and in *Faust,* Part One, as the crane (Kranich) in the *Walpurgisnachtstraum,* but he passed a more considered judgement in Bk. 14 of *Dichtung und Wahrheit* (qq.v.).

Lazarus, heading of a group of powerful poems in the section *Lamentationen* of Heine's *Romanzero* (q.v.).

Leben der schwedischen Gräfin von G . . ., a novel published anonymously by C. F. Gellert (q.v.). It appeared in two parts in 1747 and 1748 and was reprinted three times in Gellert's lifetime. The Swedish countess is the narrator of her own life. In Pt. 1 she is orphaned and brought up by a cousin in Livonia (astride former Latvia and Estonia). At 16 she marries a Swedish nobleman, Count G . . ., and lives happily with him. In the Polish wars he is reported killed, and the Countess, in order to escape the attentions of a Swedish prince, flees abroad to Holland with the Count's friend R., whom she marries. Count G . . ., however, reappears, and R. immediately renounces his claim to the Countess. The three friends continue, in spite of this potential emotional tangle, to live in friendship and harmony.

In Pt. 2 the Count recounts his experiences in Russian captivity, singling out an uncivilized girl and a Jew for their humane conduct. When Count G . . . dies, the marriage with R. is resumed. These examples of controlled emotion are contrasted with a secondary story of two young people who succumb to passion and end miserably.

Leben des Galilei, a play (Schauspiel) in 15 episodes by B. Brecht (q.v.) with music by Hanns

Eisler (q.v.), which was written in exile in Denmark in 1938–9 after Otto Hahn and his team of scientists had ('according to newspaper reports', as stated in *Versuche* vol. 14, 1955) succeeded in splitting the atom. The play was performed in Zürich in 1943 by German actors in exile. Brecht was at this time in the U.S.A., where he wrote, in collaboration with Charles Laughton, a second version in English which was performed in Beverly Hills (California) in 1947 with Laughton in the title role, and in the following year in New York. The third and final version is based on the English text and the rehearsals of the play by the Berliner Ensemble (q.v.) for performance at the Theater am Schiffbauerdamm. The main changes in these two versions are the omission of the final scene and the use of the *Vorspruch* of sc. 15 as an epilogue. While preserving the best facets of Brecht's epic technique, the play is written after the manner of Shakespearian historical drama. It opens when Galilei is 46 and closes in 1637, when his former pupil Andrea smuggles the *Discorsi* out of Italy. Galilei, by now almost blind, remains until his death (which took place in 1642) under house arrest, but has the satisfaction that he has deceived the Inquisition over his book, since Andrea carries a copy, made secretly while the original was being confiscated page by page as the book was written.

In the first episode Galilei tries to prove to the boy Andrea the Copernican theory which replaced the Ptolemaic system, the only one acceptable to the Church as being compatible with the Scriptures. Because of his poverty he uses information about a new Dutch telescope, obtained from the wealthy Ludovico who hopes to marry Virginia, Galilei's daughter, to make an instrument, and offers it as his own invention to the Republic of Venice for the money he sorely needs. His ruse succeeds. After his return to Padua (episode 3) Galilei finds with the aid of the telescope proof of the validity of the Copernican system (10 January 1610). He fails to convince the Grand Duke of Florence of his discoveries, but is successful in Rome, where the astronomer Clavius of the Collegium Romanum of the Vatican confirms his findings.

In the central episode (7) the Inquisition places the Copernican system on the Index (5 March 1616), and Cardinal Bellarmin impresses upon Galilei the need for silence. Eight years later the mathematician Cardinal Barberini succeeds as Pope Urban VIII. Galilei resumes his research, and his theories become the theme of a popular carnival in Florence and elsewhere celebrating Galilei as image-breaker. The following year (1633) Galilei is brought to Rome to face the Inquisition, an episode which culminates in his recantation of his theories (sc. 13).

The motivation of the recantation is a prominent feature of the play, and the apparently simple arguments become increasingly complex, although the ideological approach with its rational dialectics never shifts its ground. There remains a strong trace of pessimism, which is explained by the dropping of the atom bomb on Hiroshima. Galilei's retort to Andrea's accusation of cowardice, that only an unhappy country is in need of heroes ('Unglücklich das Land, das Helden nötig hat'), has an air of resignation which enhances his stature as he emerges from the trial, no hero, but one who bears the signs of brutal torture in his barely recognizable physical appearance.

The integration of dialectics into the portrayal òf character gives this play its special balance. Galilei is a hedonist, but Brecht so presents all ethical concepts that they appear, through the deliberately calculated situations which prompt the argument, to yield only relative judgements. Whether Galilei is a fool or a 'hero' when he risks his life during the plague (episode 5) is the most striking query in connection with his conduct at the trial, and yet the obsession of the scientist in his search for truth for its own sake is more compelling than Brecht's dialectical preoccupations. The play stresses the virtue of reason only to expose its subordination to opportunism on two levels, that of the populace, representing society, and the Church, representing secular authority. Both the strength and the limitations of the play, its characterization of human nature and its ideological commitment, derive from this.

Leben des vergnügten Schulmeisterleins Maria Wuz in Auenthal, a whimsical story written in 1791 by JEAN Paul (q.v.), and first published with *Die unsichtbare Loge* in 1793. It is described as 'Eine Art Idylle'. Wuz is an old-fashioned and eccentric country schoolmaster who, in spite of poverty and rural remoteness, has lived a serene and contented life, thanks to his cheerful sunny temperament. Jean Paul recounts how Wuz, too poor to buy books, writes his own versions using the titles of new books published by others. There is an amusing yet touching account of his courtship and marriage, and a moving description of his death. Wuz, in all his endearing eccentricities, is a symbol of humanity's desire for happiness. Later in life Jean Paul sometimes used the spelling Wutz.

Leben ein Traum, Das, title of J. Schreyvogel's translation of Calderón's play *La vida es sueño*, published in 1816.

Leben Jesu, kritisch bearbeitet, Das, a critical, analytical biography of Jesus by D. F.

Strauß (q.v.), published in 2 vols. 1835–6. It quickly became a focus of theological and religious controversy, and was one of the most quoted and discussed works of the mid-19th c.

Lebensansichten des Kater Murr nebst fragmentarischer Biographie des Kapellmeisters Johannes Kreisler in zufälligen Makulaturblättern, grotesque title of an unfinished novel by E. T. A. Hoffmann (q.v.), who names himself on the title-page as 'editor' (Herausgeber). The first volume was written and published in 1819, the second in 1821. The projected third and final volume remained unwritten. The eccentric counterpoint of the novel is based on an intentionally absurd fiction. The literate tom-cat Murr has written his autobiography, using as paper for his MS. the back of galley-proofs of the story of the musician Kreisler. The printer has taken the whole as one work, so that a dozen or so pages of Murr alternate with sections on Kreisler. The uncompleted biography of the latter, which is full of sombre events and appears to be heading for tragedy, is thus repeatedly interrupted by the oblique and humorous satire implicit in the self-satisfied philistinism of the cat, whose most characteristic utterance is: 'Gibt es einen behaglicheren Zustand, als wenn man mit sich selbst ganz zufrieden ist?'

The story of Kreisler is extremely complex and remains obscure since the final phase, in which the mysterious relationships between the characters would have been explained, was never written. Kreisler, the eccentric composer of genius who had already been the centre of *Kreisleriana* (q.v.), comes to a petty court in which the prince, Irenäus, though he no longer has a land to rule over, continues to behave as if he had. On the fringe of this court is a friend of Kreisler, Meister Abraham, an organ builder and a conjuror, who is also a man of wisdom and integrity. More intimately connected with the court are the Rätin Benzon and her daughter Julia. Frau Benzon, a former mistress of Prince Irenäus, knows and suspects Kreisler because she fears that he may, by attracting Julia, interfere with plans for the aggrandizement of her family. Julia, childlike and musically gifted, is friend and companion to Princess Hedwiga. Both Julia and the Princess are drawn to Kreisler.

A sinister figure, Prince Hektor from Naples, comes to the court as suitor to the Princess. He seeks to seduce Julia, but is frustrated by Kreisler. In the night the Prince's aide-de-camp shoots at Kreisler and wounds him slightly; Kreisler, however, mortally wounds the would-be assassin. Kreisler flees and takes refuge in a monastery, where he is temporarily happy composing and performing church music. This idyll is disturbed by an ascetic monk, Bruder Cyprian, who proposes to banish music from the monastery. In an altercation between Kreisler and Cyprian the latter discovers that Kreisler has evidence of a murder committed by Cyprian in a state of jealousy provoked by his brother Prince Hektor. Kreisler receives news that Prince Hektor, who had fled, has returned to the court and also that the Rätin Benzon, now Gräfin von Eschenau, proposes to marry Julia to the imbecile Prince Ignaz. The story breaks off at this point, and a postscript announces the death of Kater Murr. If the novel had been concluded it is likely that close family relationships would have been revealed between many of the characters.

In the figure of Kreisler Hoffmann incorporates his own passionate devotion to music, and in Kreisler's relationship to Julia he revives the passionate attachment to Julia Marc (q.v.) which he experienced in his years at Bamberg. In spite of the broad humour of the feline counterpoint and the ironical eccentricity of much of Kreisler's behaviour, the novel is a serious commentary on the tragedy of human life and on the elements which, in Hoffmann's view, redeem it—music and art.

Lebensläufe nach aufsteigender Linie, a novel by Th. G. Hippel (q.v.) published in 4 vols. 1778–81. It was intended to cover three generations (hence the title). Only the life of the third generation (i.e. of Hippel himself) was completed. Though it is basically autobiographical, Hippel has disguised his identity and introduced fictitious episodes. He deals with the childhood of his hero (Alexander), his first love-affair, which ends with the death of the beloved, his university studies, his life as a tutor in a Baltic nobleman's house, his exploits in war, the restoration of the family nobility, and his marriage. (Hippel himself never married.) The story is told in a wayward manner derived from Sterne. It is of sociological interest in its portrayal of the now extinct Baltic German community.

Leben und die Meinungen des Herrn Magister Sebaldus Nothanker, Das, a novel in 3 vols. published by F. Nicolai (q.v.) 1773–6. The baptismal name of its hero and that of his wife are taken from the 'prose poem' *Wilhelmine* (q.v., 1764) by M. A. von Thümmel, and the novel begins immediately after the wedding which is the climax and end of Thümmel's work.

Sebaldus is a country parson of great sincerity, who is brought before Superintendent Stauzius, arraigned for unorthodoxy, and dismissed. His eldest son has gone for a soldier, his wife Wilhelmine dies partly from the shock of Sebaldus's unjust fate, and the younger daughter dies of smallpox. Sebaldus, now unfrocked, goes with his elder daughter Mariane to Leipzig, where he corrects proofs for the kindly bookseller Hieronymus. Sebaldus helps when Stauzius's son is in danger of being punished as a deserter, but finds that Stauzius's promises of help are mere words. Meanwhile Mariane becomes a governess, falls in love with and is loved by one of the gentry (Säugling), and is persecuted by a private tutor. Sebaldus is robbed and made destitute, makes a living copying music, tries to emigrate and is shipwrecked. Appointed reader to Säugling's father, he wins a large prize in the state lottery, whereupon Säugling père is willing to bless his son's marriage to Mariane. Mariane's one-time would-be seducer, Rambold, turns out to be Sebaldus's long-lost son and therefore her brother.

The novel implicitly praises the truly charitable Christian and castigates the harsh intolerance of the self-righteously orthodox. Nicolai is said to have taken himself as model for the bookseller Hieronymus and to have based Dr. Stauzius on Pastor J. M. Goeze (q.v.) of Hamburg.

LEBERECHT, PETER, pseudonym used by L. Tieck (q.v.) in publishing his *Volksmärchen* (1797). Leberecht is the eponymous hero of an earlier novel by Tieck, *Peter Leberecht, eine Geschichte ohne Abenteuerlichkeiten* (q.v.).

Leberecht Hühnchen, a volume of stories by H. Seidel (q.v.).

Lechfeld, a plain to the south-east of Augsburg. It was the scene in 955 of a decisive victory won by the Emperor Otto I (q.v.) over the invading Hungarians (Schlacht auf dem Lechfelde), putting an end to their incursions which for half a century had brought slaughter and pillage.

LE FORT, GERTRUD, FREIIN VON (Minden, 1876–1971, Oberstdorf, Bavaria), the daughter of a Prussian colonel of Swiss Protestant descent, was educated privately and at Hildesheim, and later, as a mature student, read history at Heidelberg, Marburg, and Berlin universities. The climax of her academic work was the publication in 1925 of the posthumous *Glaubenslehre* of the Protestant theologian E. Troeltsch (1865–1923), whom she had revered as scholar and friend for many years. Soon after this she became a convert to the Roman Catholic Church, and, although she had earlier published a few stories and poems in magazines, and a volume of poetry, *Hymnen an die Kirche* (1924), this moment of decision marks the real beginning of her career as an authoress.

For some twenty years (1918–39) Gertrud von le Fort lived in Baierbrunn, Bavaria, before making Oberstdorf in the Bavarian Alps her permanent home, except for a sojourn of three years in Switzerland (1946–9). From 1925 she paid frequent visits to Italy, especially to Rome.

Gertrud von le Fort's œuvre is primarily narrative. The *Hymnen an die Kirche* were followed by one more collection of rhapsodic verse, *Hymnen an Deutschland* (1932), and her various poems were collected as *Gedichte* (1949, reissued with additions 1953). The literary fruits of her conversion were contained in the novel *Das Schweißtuch der Veronika* (q.v., 1928), to which a sequel, *Der Kranz der Engel*, was added in 1946. In 1958 the title of the earlier volume was changed to *Der römische Brunnen* and the original title was used to embrace both works. The novel *Der Papst aus dem Ghetto* (1930) deals with the bitter and bloody conflicts between Christendom and Jewry in the early Middle Ages. In *Die Letzte am Schafott* (q.v., 1931) the French Revolution provides the setting for a finely wrought Novelle of martyrdom; the background of the novel *Die magdeburgische Hochzeit* (q.v., 1938) is the Thirty Years War. The outstanding work of her old age is the Novelle *Am Tor des Himmels* (q.v., 1954), in which Galileo is the principal figure (Brecht's *Leben des Galilei*, q.v., was not known to her when she wrote the work). Her numerous other short narrative works include the Novellen *Die Frau des Pilatus* (1955), *Der Turm der Beständigkeit* (1957), and *Die letzte Begegnung* (1959), the stories (Erzählungen) *Die Opferflamme* (1938), *Die Abberufung der Jungfrau von Barby* (1940), *Das Gericht des Meeres* (1943), *Die Consolata* (1947), *Die Tochter Farinatas*, *Plus ultra* (both 1950), *Die Verfemte* and *Die Unschuldigen*, which were published together as *Gelöschte Kerzen* (1953), and *Das fremde Kind* (1961). *Das Reich des Kindes* (1933), *Die Vöglein von Theres* (1937), *Die Tochter Jephtas* (1964), and *Das Schweigen* (1966) are Legenden. Her essays appeared in several volumes entitled *Die ewige Frau* (1934, enlarged edition 1960), *Aufzeichnungen und Erinnerungen* (1951, enlarged 1956), *Die Frau und die Technik* (1959), and *Woran ich glaube* (1968). *Mein Elternhaus* (1941) and *Hälfte des Lebens* (1965) are autobiographical.

Gertrud von le Fort had the benefit of a Huguenot family tradition, which after her conversion combined with her new confession to develop in her a profound sensitivity to problems of faith and conscience. Her sense of permanence manifests itself not only in her subject-matter, but also in the firm structure of her work and the disciplined traditionalism of her style. The religious basis of her œuvre is unmistakable yet unobtrusive.

Legenda aurea, a collection of legends of the saints, written in Latin c. 1270 by Jacobus de Voragine or Varagine (1230–c. 1298), Archbishop of Genoa. It circulated widely and had a considerable influence on the writing of legends in Germany.

Legendar, term for collections of legends especially those arranged according to the calendar. Legendare were particularly numerous in the 14th c. See DER HEILIGEN LEBEN, LEGENDA AUREA, DAS PASSIONAL, and DAS VÄTERBUCH.

Legende, in German literature, term for narrations of the lives of saints, including the Virgin Mary, and martyrs, or of episodes from their lives. Legend in the English secular use (e.g. Arthurian legend) is always rendered in German by Sage, never by Legende.

Legenden were originally read in church (hence the name), and large collections in Latin were made in the Middle Ages, including notably the *Legenda aurea* (q.v., c. 1270) and *Das Passional* (c. 1300). The largest collection is the Bollandist *Acta sanctorum*, begun in 1643 by the Dutch Jesuit J. Bolland (1596–1665) and continued to the present day by members of the Society of Jesus (67 vols. by 1967). Among legends in the German vernacular *Gregorius* and *Der arme Heinrich* (qq.v.) by Hartmann von Aue are particularly notable. Medieval collections in German are not of high quality, but *Das Buch der Märtyrer* (c. 1300) and the 14th-c. *Heiligenleben* of HERMANN von Fritzlar (q.v.) deserve mention. A collection of a different kind is the *Sieben Legenden* (1872) of G. Keller (q.v.) which secularizes and gently parodies some legends by L. T. Kosegarten (q.v.). H. von Kleist wrote the Legende *Die heilige Cäcilie* (q.v.). Among modern writers of Legenden Gertrud von LE FORT (q.v.) is conspicuous.

Legende vom Engel und vom Waldbruder, an anonymous Middle High German verse legend, some 500 lines in length, probably written early in the 14th c. Its story, of oriental origin and already known in Latin versions, tells of a hermit visited by an angel who conducts himself in apparently unangelic fashion, burning the hermit's hut and committing dreadful crimes. At last the angel reveals himself and explains his conduct as a lesson to the hermit, whose desire to see Christ was *superbia*. The whole transpires to be a dream, and the hermit dies three years later in sanctity.

Legnano in northern Italy was the scene in 1176 of a defeat suffered by the Emperor Friedrich I (q.v., Barbarossa) at the hands of the troops of Milan and other Lombard cities.

LEHMANN, WILHELM (Puerto Cabello, Venezuela, 1882–1968, Eckernförde), a schoolmaster, spent the last twenty-four years of his career until his retirement in 1947 in Eckernförde, Schleswig-Holstein. He served in the 1914–18 War and spent a short time as a prisoner of war in England. His early works were novels and stories. The first was *Der bedrängte Seraph*, published in a periodical in 1915 and in book form in 1924. The novels *Der Bilderstürmer* (1917), *Die Schmetterlingspuppe* (1918), *Weingott* (1921), and the stories *Der Sturz auf die Erde* (1923) and *Die Hochzeit der Aufrührer* (1934) followed. Lehmann also wrote in the mid-twenties a semi-autobiographical novel, *Der Überläufer*, which remained unpublished until 1962. He is, however, known primarily as a lyric poet, the more remarkably since his first volume of verse (*Antwort des Schweigens*, 1935) did not appear until he was over 50. This was followed by the collections *Der grüne Gott* (1942), *Entzückter Staub* (1946), *Noch nicht genug* (1950), *Abschiedslust* (1962), and *Sichtbare Zeit* (1967). Lehmann combined a sensitive and subtle feeling for nature ('the green god' of his early poems) with a full awareness of the destructive powers of the age in which he lived, and his poetry is sober and precise, as well as memorable. A late novel, *Ruhm des Daseins*, appeared in 1953. He was a friend of O. Loerke (q.v.). His translations of Rudyard Kipling appeared as *Kleine Geschichten aus den Bergen* (1925), *In Schwarz und Weiß* (1926), and *Drei Soldaten* (1934). His other publications include essays on poetry. *Sämtliche Werke* (3 vols.) appeared in 1962 and *Gesammelte Werke* (8 vols.), ed. A. Weigel Lehmann, H. D. Schäfer, and B. Zeller, 1982 ff.

Lehnhold, Der, a Novelle by B. Auerbach (q.v.), published in 1853 in *Schwarzwälder Dorfgeschichten* (q.v.). The 'Lehnhold' (owner by inheritance of a large undivided farm) is the Furchenbauer, who has two sons, Alban and the younger Vinzenz whom he had struck with the haft of a horsewhip so that the boy lost an eye. The reproaches of conscience impel him to swear to give the estate (which he will not divide) to Vinzenz, against all law and rural tradition. A feud develops between Vinzenz and Alban, and in an encounter in the mountainous forest Alban grapples with Vinzenz after the latter has set his dog on him. They fall from a ledge, Vinzenz is killed, and soon after Alban dies of a brain fever. The new heir is the steward Dominik, who has married the Furchenbauer's daughter Ameili. Dominik promises the dying Alban that he will in time divide the estate justly among all of his children.

Lehrlinge zu Sais, Die, fragments of an unfinished novel by Novalis (q.v.), written in 1800, and first published posthumously in *Novalis Schriften* (1802). The first fragment is headed *Der Lehrling*, the second, which contains a short fairy story (*Hyazinth und Rosenblütchen*), is supercribed *Die Natur*.

Lehrstück. The form derives from traditional literary didactic works (notably those intended for school performance) and is associated with the early stages of the epic theatre (see EPISCHES THEATER) of B. Brecht (q.v.). The didactic element is political and grew out of Brecht's studies of Marx (q.v.). Those plays which Brecht specifically referred to as Lehrstücke (he generally preferred the collective term Lehrtheater) are: *Der Ozeanflug, Das Badener Lehrstück vom Einverständnis, Der Jasager und Der Neinsager, Die Maßnahme*, and *Die Ausnahme und die Regel* (qq.v.). In order to emphasize the general applicability of his message Brecht favours in his Lehrstücke anonymous characters in the Expressionist tradition (see EXPRESSIONISMUS).

Lehrtheater, see EPISCHES THEATER and LEHRSTÜCK.

Leibeigenen, Die, a poem by J. H. Voß (q.v.), published in *Idyllen* in 1801. Written in classical hexameters, it is in the form of a dialogue between two oppressed serfs on a great North German estate, and illustrates Voß's detestation of tyranny. *Die Erleichterten* forms a sequel to it, and is followed by *Die Freigelassenen*. A fuller poetic treatment of the theme of serfdom is found in F. Reuter's *Kein Hüsung* (q.v.).

LEIBNIZ, GOTTFRIED WILHELM VON (Leipzig, 1646–1716, Hanover), philosopher, mathematician, and polymath, entered the service of the Electoral Archbishop of Mainz. In 1672 he visited Paris in order to persuade Louis XIV to campaign against the Turks in Egypt and so to divert him from plans of conquest in west Germany. After visiting London in 1673 and 1676, Leibniz was appointed librarian in Hanover to the Duke of Brunswick-Lüneburg, a position which he occupied for the remainder of his life.

Leibniz's many activities included diplomatic missions and the foundation in 1700 of the Sozietät der Wissenschaften (later Preußische Akademie der Wissenschaften, see AKADEMIEN), of which he was the first president. He invented the infinitesimal calculus independently of Newton and almost simultaneously. He participated in plans for reuniting the religious denominations of Western Christendom. His publications, which were in Latin or French,

refer chiefly to mathematics, to history, and (in anonymous or pseudonymous tracts) to politics. Leibniz's principal published philosophical work is the *Essais de Théodicée sur la Bonté de Dieu, la liberté de l'homme et l'origine du mal* (1710). In this treatise he outlined an optimistic philosophy which explained evil in the world as necessary. The argument is briefly as follows. God alone is perfection. The world, God's creation, not being God, cannot be perfect. God in his goodness could not make any world but the least imperfect, so that this world is the best of the possible worlds ('le meilleur des mondes possibles'), a conclusion which half a century later was mocked by Voltaire in *Candide* (1759).

Leibniz also advanced a theory on the composition of the universe, set out in his *Monadologie* for the benefit of Prince Eugene (see EUGEN, PRINZ), and published in German in 1720. The world is made up of monads (Monaden), and these simple entities group themselves into more complex monads to make up all that is animate and inanimate. This theory has been seen as an imaginative anticipation of later physics. Faced with the problem of spirit and matter, free will and deterministic (or mechanical) causation, Leibniz offers as solution the concept of a pre-established harmony (prästabilierte Harmonie), which he illustrates by the example of two clocks which keep perfect time. Their simultaneity can be accounted for by one of three assumptions: (1) they are connected mechanically; (2) someone is concealed in one clock moving the hands to keep time with the other; (3) both clocks have been made by so skilful a clockmaker that they perpetually keep the same time. The third solution (with God as clockmaker) is the right one. Leibniz, who had not only one of the greatest but also one of the most inquisitive minds, was never able in his ceaseless inquiries on the most diverse matters to take the time to set forth a coherent system. There exist the two treatises and a number of disconnected essays, a vast quantity of letters on scientific subjects, numerous unpublished papers and jottings, and it is virtually impossible, because of inherent contradictions, to arrive at a co-ordinated systematic conspectus. For this reason Leibniz's views have met with more divergent interpretation than those of most philosophers.

In spite of his intellectual stature, Leibniz had little direct influence in Germany. Christian Wolff (q.v.) of Halle University expounded systematically ideas which he derived from Leibniz, but he himself admitted that there was much in Leibniz's thought that he could not understand.

The works of Leibniz are published by the Deutsche Akademie der Wissenschaften (until 1945 Preußische Akademie) as *Sämtliche*

Schriften und Briefe (*c.* 40 vols., 1923 ff.). Ten vols. had appeared by 1979.

Leich, a complex Middle High German poetic form, derives from the Latin *sequentia*. It was originally choral, consisting of an opening section sung by all the voices, which was succeeded by passages sung alternately by the two sections of the divided choir; finally the choir united again to sing a concluding passage. The intermediary passages for divided choir were not limited in number, but each was sung twice, half-choir B repeating what half-choir A had sung. Occasionally a passage for undivided choir was inserted in the middle. The *leich* could be religious or secular; the secular examples were either dance songs (*tanzleich*) or love songs (*minneleich*). One or more examples are found in the work of many Middle High German poets (including WALTHER von der Vogelweide). Those whose preserved *leiche* are the most numerous are Tannhäuser (6), RUDOLF von Rotenburg (5), and ULRICH von Winterstetten (5), who developed the form into new complexities. More than one example is extant in the work of KONRAD von Würzburg, Gliers, Hadlaub, and Frauenlob (qq.v.).

Leichtsinn aus Liebe, a four-act comedy by E. von Bauernfeld (q.v.), first performed at the Burgtheater, Vienna, in January 1831. It was first published in *Lustspiele* (1833). It is light entertainment, in which two well-matched couples are, for a time, kept apart by unreal difficulties made by the men. Apart from its easy dialogue its chief feature is the acute analysis of the masochistic jealousy of one of the lovers, Dr. Heinrich Frank.

Leiden des jungen Werthers, Die, a novel in two parts (Erstes Buch and Zweites Buch) written by Goethe within three months (March until May) in 1774, and published in the same year. Some prefer the title *Die Leiden des jungen Werther*, which Goethe himself chose for the last edition of 1824. It was substantially revised for a new edition in 1787, and it is now usually read in this later form. Apart from the 'editor's' introductory note eliciting the reader's admiration and sympathy for Werther's individuality, and emphasizing the cathartic effect which his fate is meant to produce, the first part of the novel is told mainly in letters (see BRIEFROMAN) dated from May until September 1771; the second part, in the latter half of which a fictitious narrator takes up the story, is dated from October 1771 to December 1772. The letters are written by Werther, and are addressed to his friend Wilhelm, who figures indirectly in the letters as a man of contrasting character.

At the opening of Bk. One Werther, a highly strung young man of great sensitiveness, is sent on a journey by his mother to deal with a legacy. The district is new to him, and he is carried away by the idyllic beauty of the rural scene in spring. After a few weeks he meets Lotte, a young woman, the eldest of a family of motherless young children, to whom she herself has become a mother. He accompanies Lotte to a ball and falls deeply in love with her. She is, however, promised to another, named Albert, and Werther, though he apprehends the situation, is unable to separate himself from Lotte. For a time a curious triangular state persists, in which each endeavours to be fair to the others. At the beginning of September Werther at last tears himself away and leaves Lotte and Albert.

In Bk. Two, which begins a month later, Werther is on the staff of a legation, ill at ease and restless. In March he sustains what he feels is an unmerited snub and hands in his resignation. In May, a year after his first arrival, he is back in Lotte's town. His passion soon dominates his life, and the awareness of being in an impasse from which there is no escape awakens destructive impulses. Distraught to the point of utter despair he takes a last leave of Lotte and shoots himself with Albert's pistol. It is the tragedy of a man who is not only disappointed in love, but also feels himself alienated from the world around him.

Although the novel is not autobiography, Goethe's experiences in Wetzlar in the spring and summer of 1772 are vividly recalled. Lotte Buff, with whom Goethe fell in love, and J. G. C. Kestner (qq.v.), her fiancé, are reflected in Lotte and Albert, and the details of the death of Werther follow very closely the suicide of K. W. Jerusalem (q.v.) at Wetzlar in October 1772.

By its intense sensibility and astonishing lifelikeness *Werther* conquered the reading public, not only of Germany, but of Europe. Probably the best-known work of the Sturm und Drang (q.v.), it made Goethe famous overnight and generated a wave of imitation. Young men of means found it *de rigueur* to wear the Werther dress of blue coat and yellow breeches, and Napoleon is said to have read the work seven times. In some quarters it was attacked as deleterious to morals, and it was parodied by F. Nicolai (q.v.) in *Die Freuden des jungen Werthers* (1775) and by J. F. Kringsteiner (q.v.) in *Werthers Leiden* (1806). Some suicides were attributed to it, notably that of Christel von Laßberg, who in January 1777 drowned herself in the Ilm near Goethe's house with a copy of

Werther in her pocket. The melancholia of V. L. Plessing, with whom the poem *Harzreise im Winter* (q.v.) is linked, was also attributed to his reading this powerful novel. (See also PLENZ-DORF, U.)

Leidende Weib, Das, a tragedy (Trauerspiel) in five acts by F. M. Klinger (q.v.), written probably in 1774-5 and published anonymously in 1775. It is a chaotically constructed work said to have been written in four days. Klinger excluded it from the canon of his works, and L. Tieck (q.v.) attributed it in 1828 to J. M. R. Lenz (q.v.). In a welter of 31 detached scenes Klinger presents a world of corrupt intrigue in which the central point is the passionate, but adulterous, love of two noble characters, the Gesandtin and Brant, to which both in an irresistible moment have succumbed. It ends in catastrophe. *Das leidende Weib* owes much to *Der Hofmeister* (q.v.) by J. M. R. Lenz.

Leiden eines Knaben, Das, a story (Rahmennovelle, see RAHMEN) by C. F. Meyer which was published in 1883 in *Schorers Familienblatt* under the title *Julian Boufflers. Das Leiden eines Kindes.* The tale is based on a reminiscence recorded in the *Mémoires* of Saint-Simon (edn. of 1711, vol. 5, ch. 31). Julian is the son of Marshal Boufflers's first wife. He is not mentally gifted, and although he turns out to be handsome, an excellent fencer, and endowed with artistic potentialities, he is quite unsuited for the monastic Jesuit education on which his father insists. Julian is 13 when he becomes fully aware of his humiliating limitations, which he nevertheless bears with good grace, trying desperately to keep up with a curriculum for boys well below his own age. As long as the Jesuits believe that he is a natural son of Louis XIV they take kindly to his deficiencies. But their treatment changes radically when they learn his true identity, and he becomes the weapon for revenge on the Marshal, his father. Their ill-treatment reaches its climax when Père Tellier brutally chastises him, although he is innocent of the offence for which he is being punished. Four days later he is dead. He has been taken home and the doctor, Fagon, reveals to the father the sufferings of the boy. To ease his death, the father makes Julian believe, in his delirium, that he is fighting in battle as had been his wish. Julian dies gladly with the words 'Vive le roi!' on his lips.

The frame gives this tale its point. The narrator is Fagon, the king's doctor, who comes to see Louis as he relaxes before dinner with Madame de Maintenon. Fagon has just heard that the 'wolf' Le Tellier is going to be the king's personal confessor. He reveals Le Tellier's brutal conduct, but the king, listening with condescension, shows no sign of changing his decision over the appointment of his confessor, who has been directly responsible for the death of the boy; indirectly the person responsible is the king himself, and the intolerance, favouritism, and intrigues of his rule, of which the innocent Julian is ignorant in his sincere devotion.

Leiden und Freuden eines Schulmeisters, a two-volume novel by J. Gotthelf (q.v.) published in 1838-9. It is told in the first person by the village schoolmaster Peter Käser. Its starting-point is the impoverished Käser's confident hope that his salary will be increased under a new law (1836); his hopes are dashed because he does not measure up to the requirements of the School Commissioners. It ends with the news that the higher salaries are to be paid to *all* schoolmasters (1837), and so Käser's poverty is at an end. Gotthelf himself was passionately engaged in this piece of educational legislation, and his powerful sense of the injustice done to many schoolmasters is the germ of the novel.

Within the frame given by the law of 1836 and the new interpretation of 1837 Käser tells his life story. It begins with his family life in which want creates bitterness, and poor Peter, at first a spoilt child, becomes a virtual slave to his father without prospects of any kind. Out of this he is rescued by the suggestion of a schoolmaster that Peter too should become a schoolmaster; and he does so, without any training except a little practical teaching with his benefactor. After spells as a helper, he obtains a schoolmaster's post, but, young and inexperienced as he is, he gets into trouble for going about with the lads of the village, and runs up debts. He is unintentionally involved in a scandal with a designing young woman and finds he has made his position impossible. Fortunately he obtains another school, at Gytiwyl, and here he sets about his work conscientiously. He marries the village cobbler's daughter (Mädeli), who in her kindness, good sense, and intuitive moral feeling is the perfect partner for him. But as their family grows, so their finances become more and more straitened until at last the rise in salary lifts them out of their cul-de-sac of penury.

The book is overtly didactic and is plentifully strewn with the sage observations of Käser, who conceals Gotthelf; his deep insight into the human heart is shown most clearly in the characters of Peter and Mädeli. A whole world of Swiss peasants is evoked, tough and robust, but narrow and selfish. The novel contrives to be a tract on the inadequate training of schoolmasters, and at the same time a living and moving work.

LEIP, HANS (Hamburg, 1893–1983, Fruthuilen, Switzerland), a versatile writer, grew up in the atmosphere of the Hamburg docks, where his father worked, going to sea in trawlers during his holidays. He served in the 1914–18 War in the Guards (Garde-Füsilier-Regt.) and in 1915 wrote the sentimental poem *Lilli Marleen* (q.v.), which achieved fame 25 years later. He earned his living for a time by teaching, but was chiefly active as a journalist, working especially, both as humorous writer and as draughtsman, for the weekly *Simplicissimus* (q.v.). He illustrated many of his own books. His unpretentious novels of sea and harbour life are full of vigour and adventure; they include *Godekes Knecht* (1925), *Miß Lind und der Matrose* (1928), *Die Blondjäger* (1929), *Die Lady und der Admiral* (1933), *Fähre 7* (1937), and *Das Muschelhorn* (1940). He touched other themes as well, inflation in *Der Pfuhl* (1923), and the return home of the soldier in *Tinser* (1926). Some of his novels are primarily boys' books (*Der Nigger auf Scharhörn*, 1927; *Jan Himp und die kleine Brise*, 1934; and the post-war novel *Drachenkalb singe*, 1949). Other later novels are *Die Sonnenflöte* (1952) and *Des Kaisers Reeder* (1956).

Leip also wrote much poetry ranging from catchy popular songs to intricate and sophisticated verse (*Die Nächtezettel der Sinsebal*, 1927; *Die Hafenorgel*, 1947; *Eulenspiegel*, a book of his drawings with his own verse text, 1942; *Kadenzen*, 1942; *Der Mitternachtsreigen*, 1947; *Pentamen*, 1963; and *Garten überm Meer*, 1968). His few plays are not important, but he wrote an interesting book on sailing, *Segelanweisung für eine Freundin* (1932). Leip did not support the National Socialists, though he does not appear to have been seriously molested after 1933. *Das Lied vom Schütt*, the aftermath of an air raid, is a clear expression of his disapproval of the National Socialist regime. The autobiographical volume *Das Tanzrad oder Die Lust und Mühe eines Daseins* was published in 1979.

Leipzig, the regional centre of the Bezirk Leipzig in the DDR (see DEUTSCHE DEMOKRATISCHE REPUBLIK). It is an important industrial town and was formerly the centre of the German book trade, including publishing.

With Dresden one of the two great cities of Saxony (see SACHSEN), Leipzig received its first charter in 1170. The famous Leipzig Fair (Leipziger Messe) for general trade and books, and later including industrial and technical products, was first authorized in 1497 by the Emperor Maximilian I (q.v.). In 1519 Leipzig was the scene of the formal disputation between Luther, Karlstadt, and J. Eck (qq.v.) in the presence of the Elector Friedrich (q.v., der Weise). Leipzig's cultural florescence was most notable in the first half of the 18th c., when J. S. Bach (q.v.) was Cantor at St. Thomas's Church (1723–50), and Gottsched and Gellert were the conspicuous names in the leading university of Germany (founded in 1409), at which Lessing and Goethe (qq.v.), as well as many future men of note, were students. The Gewandhaus concerts and orchestra have retained their reputation down to the present day.

Leipzig is also known for the Battle of Leipzig in October 1813 (see VÖLKERSCHLACHT) in which Napoleon was decisively defeated by the armies of Austria, Prussia, and Russia (see NAPOLEONIC WARS).

Leipziger Allgemeine Zeitung, a newspaper founded in 1837 by the firm of Brockhaus. Because of its pronounced Liberal views it was forbidden in Bavaria in 1842 and in Prussia in 1843, whereupon it changed its title to *Deutsche Allgemeine Zeitung*, continuing until 1879, when it ceased publication.

Leis, term deriving from the *Kyrie Eleison* from the Ordinary of the Mass. 'Kyrrieleison' occurs in a non-liturgical context in the *Ludwigslied* (q.v., 881). The word *leis*, originally probably only denoting the response, has been extended to cover verses constituting prayers of intercession, in which each stanza is concluded by the refrain *Kyrie eleison*. A notable early example is the OHG *Petruslied* (q.v., 9th-c.), in which each of the three short strophes is concluded by the *Christe* as well as the *Kyrie*. It was probably processional, and the vernacular portions were no doubt sung by a priest, the people contributing the refrain. The intercession of this poem has its origin in the Litany. *Leise* of this kind were sung at festivals and on other special occasions, and some authorities consider that they were, from time to time, intercalated in the liturgy. A later example of a *leis* is the vernacular Easter hymn *Krist ist erstanden* (12th-c.). It is maintained by some scholars that the *leis* is the seed from which later germinated the first beginnings of vernacular hymnody.

'Leise flehen meine Lieder/Durch die Nacht zu dir', opening lines of a poem written by L. Rellstab (q.v.) in 1827, set to music by F. Schubert (q.v.) and included in Schubert's posthumous volume of songs, *Schwanengesang* (1828).

LEISEWITZ, JOHANN ANTON (Hanover, 1752–1806, Brunswick), the author of one of the best-known plays of the Sturm und Drang (q.v.), *Julius von Tarent*, was the son of a wine-merchant of Celle. After his father's early death he grew up in Hanover, and in 1770 matriculated at

Göttingen University. In 1774 he became a member of the Göttinger Hainbund (q.v.), returning a few months later to Hanover, where he set up in practice as a lawyer, though with scant interest in the law. In 1775 he wrote *Julius von Tarent* (q.v.) as an entry for a competition, and was deeply disappointed and inhibited by its failure to win the prize. Two impressive dramatic prose sketches, *Die Pfändung*, which is an essay in social realism, and *Der Besuch um Mitternacht*, were published in the *Göttinger Musenalmanach für 1775*.

In 1777 Leisewitz became informally engaged to the 15-year-old Sophie Seyler, daughter of the deceased actor Abel Seyler (see SEYLERSCHE TRUPPE) and adopted child of a relation of Leisewitz. Being poorly off, Leisewitz sought stable employment and became a secretary in the administration of the Duchy of Brunswick. His marriage in 1781 proved an exceptionally happy union, clouded however by delicate health, hypochondria, and financial worries. The generosity of the Duke of Brunswick in 1800 removed the pecuniary cares, and Leisewitz devoted his remaining years to a successful reform of the care of the poor in the duchy, concerning whom he published a pro memoria, *Über die bei Einrichtung öffentlicher Armenanstalten zu befolgenden Grundsätze überhaupt und die Einrichtung der Armenanstalt in Braunschweig insbesondere* (1802–3). Leisewitz was incurably diffident and pessimistic, and on his death-bed he ordered the destruction of all his writings, including a history of the Thirty Years War and two plays, one of which (surprisingly in so melancholy a man) was a comedy.

Leisewitz's surviving writings, including some letters, were collected in one volume in 1838 as *Sämtliche Schriften*. They were edited under the same title by A. Sauer in 1883. The letters to his fiancée were edited by H. Mack (*Briefe an seine Braut*, 1906) and his diaries by H. Mack and J. Lochner (*Tagebücher*, 2 vols., 1916–20).

'Leise zieht durch mein Gemüt liebliches Geläute', first line of an eight-line untitled poem by Heine included in the section *Neuer Frühling* of *Neue Gedichte* (q.v., 1844). It was written in 1820 and has been set to music by Mendelssohn (q.v.).

LEITGEB, JOSEF (Bischofshofen, Austria, 1897–1952, Innsbruck), served in both wars, in 1914–18 with the Kaiserjäger (q.v.). Between the two wars he was a teacher in Tyrol, and after 1945 a school inspector there. His sensitive poetry was published in *Musik der Landschaft* (1935), *Läuterungen* (1938), *Vita somnium breve* (1943), and *Lebenszeichen* (1953). His novels, *Kinderlegende* (1934), *Christian und Brigitte* (1936), and, last and

best, perhaps because it is based largely on his own early life, *Das unversehrte Jahr* (1948), are concerned with the lives and problems of young people. *Sämtliche Gedichte* were published in 1953 and *Abschied und fernes Bild. Erzählungen und Essays* in 1959.

Leitmotiv, term coined by the musicologist H. P. von Wolzogen (q.v.) for basic themes with specific associations as used by R. Wagner (q.v.) in his music dramas (Musikdramen). They provide a concentrated and economical method of cross reference and dramatic comparison. Wagner's own word for what is now universally called the Leitmotiv is Grundthema. The word has been extended, by analogy, to literature, and is used to denote the recurrence of set phrases or sentences as exemplified, for instance, in the work of Thomas Mann (q.v.), who was influenced by Wagner's technique. In English usage leitmotiv is normally synonymous with motif.

LEMM (LEMNIUS), SIMON (St. Maria, Münstertal, Switzerland, 1511–50, Chur), a Swiss studying in Wittenberg, incurred the enmity of Luther (q.v.) by praising the Archbishop of Mainz in his Latin *Epigrammaton libri duo* (1538). Lemm reacted to his eviction from Wittenberg with the scurrilous satire (also in Latin) *Monachopornomachia* (1539). He spent the rest of his life in Chur. He wrote erotic poems (*Amores*), a Virgilian epic dealing with the Swiss war of 1499 (*Libri IX de bello suevico*), and a translation (into Latin) of the *Odyssey*.

G. E. Lessing (q.v.) devoted Letters 1–8 of his *Kritische Briefe* of 1753 to a defence of Lemm against Lutheran denigration.

LENAU, NIKOLAUS, is the pseudonym invariably used by Nicolaus Franz Niembsch, Edler von Strehlenau (Csatàd, Hungary, 1802–50, Oberdöbling nr. Vienna), who was the child of an unhappy marriage between a dissipated Austrian cavalry officer and a Hungarian girl of good family. His father died when he was 5, and in his youth he was the subject of an emotional tug-of-war between his impoverished mother and his paternal grandparents. He developed into an impulsive, depressive, deeply disturbed personality. He studied at various universities, switching from faculty to faculty, and though he remained longest with medicine, he never qualified. In Vienna he had friendly contact with Bauernfeld and Grillparzer (qq.v.), and was especially close to Graf Auersperg (see GRÜN, ANASTASIUS). After receiving a substantial inheritance in 1830, he moved in 1831 to Stuttgart where he had regular contact with Uhland, Schwab, and K. Mayer and was especially befriended by J. Kerner (qq.v.). For a time he was deeply attached to Lotte Gmelin, a niece of Schwab. Under the impulse of a romantic idealization he emigrated in 1832 to the United States, but was quickly disillusioned and returned in 1833 to Stuttgart.

In 1834 Lenau fell in love with Baroness Sophie von Löwenthal (q.v.) in Vienna, to whom he remained attached for the rest of his life. He conducted a dual correspondence with her, 'official' literary letters on the one hand, passionate private notes on the other. Her letters have been destroyed; Lenau's were published in 1906 and 1968 (ed. P. Härtling, q.v.). He became increasingly subject to moods of despair from which neither the persisting link with Sophie von Löwenthal nor a betrothal to the actress Caroline Unger could distract him. A second engagement to Marie Behrends, the daughter of a Frankfurt patrician, preceded a complete mental breakdown in 1844. Lenau did not recover his sanity and eventually became almost totally paralysed. He was taken to the asylum at Winnental in Württemberg and was later transferred to Döbling (near Vienna), where he died.

Lenau's first published work was a collection of poems (*Gedichte*, 1832), and these were supplemented in 1838 by *Neuere Gedichte*. In 1844 he published a further volume entitled *Gedichte*, which contained new poems as well as reprinting those which had appeared in the earlier collections. Lenau's *Faust. Ein Gedicht* (q.v.), of some 3,000 lines, is partly indebted to Goethe's work and was published in 1836; *Savonarola* (q.v.), a religious epic in the form of a cycle of verse romances, followed in 1837; and a second epic, *Die Albigenser* (q.v.), in the form of a collection of longer poems described as 'Freie Dichtungen', was published in 1842. Lenau's last work of importance, the epic *Don Juan* (q.v.), was published posthumously in 1851 by his friend Anastasius Grün, together with other shorter poems (*Nachlaß*). Of the longer works this is the best, but none of them has a vigour of imagination proportionate to its length.

Lenau's reputation rests chiefly on the lyric poetry, with its haunting rhythms, its pantheistic vision of nature, and its range of feeling from sadness to despair. Lenau was a violinist of merit, and was one of the first to appreciate the Beethoven of the Ninth Symphony and the late quartets. Under the name of Dr. Moorfeld he is the central figure in the novel *Der Amerika-Müde* (1855) by F. Kürnberger (q.v.). *Sämtliche Werke und Briefe*, (6 vols.), ed. E. Castle, appeared 1910–23 and (based on this edition) *Sämtliche Werke und Briefe* (2 vols.) in 1970. *Briefwechsel. Unveröffentlichtes und Unbekanntes*, ed. J. Buchowiecki, appeared in 1969.

Lennacker, a work of fiction by Ina Seidel (q.v.), described as *Das Buch einer Heimkehr*, and published in 1938. Hans Jacob Lennacker, demobilized in 1918, visits his only surviving relative, who presides over a home for aged daughters of Protestant pastors. He falls ill, and during the twelve nights of his delirium the history of his family and, simultaneously, that of the Lutheran Church are presented.

Since the Reformation (q.v.) all the Lennackers have been Lutheran pastors, except Hans Jacob's father. When Hans Jacob recovers, he is pressed by his relative and her friends to return to the clerical tradition of the Lennackers; he decides to serve God and mankind by becoming a doctor and commences the study of medicine.

In the use of a frame (see RAHMEN) which encloses connected stories, *Lennacker* has some resemblance to G. Keller's cyclical Novellen, *Das Sinngedicht* and *Der Landvogt von Greifensee* (qq.v.).

Lenore, title of a well-known ballad by G. A. Bürger (q.v.), published in the *Göttinger Musenalmanach für 1774*. Based on the Scottish ballad *Sweet William's Ghost*, it tells of Lenore's immoderate and blasphemous grief at the death of her lover in the Seven Years War (see SIEBENJÄHRIGER KRIEG), of Wilhelm's appearance on horseback beneath her window, and of her night-ride, pillion-wise, till the lover reveals himself as a grisly skeleton and Lenore sinks down dead. The ballad is notable for its infectious rhythms, its simple direct language, and its bold use of onomatopoeic forms such as 'trapp, trapp, trapp' and 'hurre, hurre, hopp, hopp, hopp'.

Le nozze di Figaro, opera in four acts by W. A. Mozart (q.v.), with an Italian libretto by L. Daponte (q.v.), based on Beaumarchais's *Le Mariage de Figaro* (1785). It was first performed at Vienna on 1 May 1786. Figaro, Count Almaviva's valet, is about to marry Susanna, the Countess's personal maid. But the Count, who has tired of his wife, has an eye on Susanna. The engaged couple and the Countess scheme to frustrate the Count's amorous intentions, and in the course of their manœuvres the Count accuses the Countess of infidelity. His attempted amours are eventually exposed, and the comedy ends in reconciliation and happiness. An important incidental character is the adolescent Cherubino, who finds all the women irresistible. When performed in German, the opera is known as *Figaros Hochzeit*.

LENSING, ELISE, see HEBBEL, FRIEDRICH.

Lenz, a Novelle by G. Büchner (q.v.), written in 1835–6 and published by K. Gutzkow (q.v.) in

1839 in the *Telegraph für Deutschland* under the title *Lenz. Eine Reliquie von Georg Büchner*. This publication formed the basis of the text in *Nachgelassene Schriften* (1850, ed. by his brother Ludwig Büchner, q.v.), under the title *Lenz. Ein Novellenfragment*. The story is based on, and is partly a literal adaptation of, a report by J. F. Oberlin (q.v.) on J. M. R. Lenz's stay from 20 January until 10 February 1778 (Büchner has 8 February) at his rectory in Waldersbach (Waldbach in Büchner's Novelle) at the foot of the Vosges between Strasburg and Colmar. It came to Büchner's notice through his friend August Stöber (q.v., and see STÖBER, D. E.). Lenz (q.v.) was at that time 26 and already in an advanced stage of the mental disease from which he died fourteen years later. He had been sent to Oberlin by C. Kaufmann (q.v.), with whom he had spent the previous year in Switzerland and who also figures in the Novelle, at the end of which Lenz is sent by Oberlin to Strasburg. Here Lenz had met Goethe in 1771 and, after Goethe's departure for Frankfurt, had fallen in love with Friederike Brion (q.v.). The unhappy course of this love is alluded to in the Novelle. Oberlin himself appears as the kind and well-loved Protestant pastor, and the story portrays the rare combination of spiritual devotion and practical initiative by which he improved the lot of the villagers.

Büchner portrays Lenz's schizophrenia by elaborating his search for God and a sign of his grace on the one hand, and on the other a sense of voidness, the result of unrelieved suffering which thrusts him into atheism. A sermon preached by Lenz in the village church forms the climax of the beneficial influence of Oberlin in the first part of the Novelle; Lenz's vain attempt to raise a dead child to life at Fouday is central to the second part, which also records his attempts at suicide before he is sent in a state of mental numbness to Strasburg.

Although Ludwig Büchner (and August Stöber) called this Novelle a fragment, it does not make a fragmentary impression and can be viewed as a consistent whole.

LENZ, JOHANN or HANS, a native of Rottweil in Swabia and from 1494 to 1496 a schoolmaster in Freiburg (Fribourg, Switzerland), witnessed the Swabian War (see SCHWABENKRIEG), of which, at the turn of the 15th c. and 16th c., he wrote a rhymed account. The historical events of 1488–1500 are enclosed within a narrative frame (see RAHMEN) involving the author and a fictitious hermit. An extant 16th-c. copy was published (ed. H. von Diesbach) in 1849.

LENZ, JAKOB MICHAEL REINHOLD (Sesswegen, Livland, 1751–92, Moscow), the son of a pastor who rose to be a Generalsuperintendent, spent his school-days in Dorpat, and studied theology at Dorpat and Königsberg universities. While a student he published his first work, *Die Landplagen* (q.v., 1769), a poem in Klopstockian hexameters dealing with catastrophes in rural life. In 1771 he took up a post as tutor and companion to two young barons von Kleist and settled with them in Strasburg, where they took commissions in the French army.

In Strasburg, Lenz became friendly with Goethe, and absorbed the ideas and attitudes of the Sturm und Drang (q.v.). The free translations *Lustspiele nach dem Plautus für das deutsche Theater* and Lenz's first original play, the eccentric didactic comedy *Der Hofmeister* (q.v.), as well as his theoretical manifesto *Anmerkungen übers Theater* (q.v.) appeared in 1774. Lenz had a tendency to fall in love with women connected with other men. At Strasburg he unsuccessfully courted Friederike Brion (q.v.), whom Goethe had forsaken in 1771. He fell in love with Cleophe Fibich, who had been the fiancée of one of the Kleist brothers, and in 1775 he believed himself to be deeply in love with Henriette von Waldner, an aristocratic lady; she was engaged, and scarcely knew of Lenz's existence. In the same year Lenz was recommended to Goethe's married sister, Cornelia Schlosser, at Emmendingen. He was irresistibly attracted to Goethe's orbit, following him in 1776 to Weimar, where his tactless and provocative eccentricity led to his expulsion.

Lenz had by now passed the zenith of his creativity as a writer. His mental powers flagged and he showed symptoms of derangement, from which he was not to recover. Upon leaving Weimar he went to stay with C. Kaufmann (q.v.) in Switzerland, where he also visited Lavater (q.v.). But in January 1778 Kaufmann put Lenz in the care of Oberlin (q.v.), a pastor in Waldersbach in the Steintal. After less than three weeks Oberlin realized that he could not help him and sent him to Strasburg. For a short period he was looked after by Schlosser in Emmendingen while arrangements were being made to send him to Riga, where his family reluctantly received him back. Three years later Lenz went to Russia, first to St. Petersburg and then to Moscow, where he was found dead in the street on a May night, eleven years after leaving his Baltic homeland.

Lenz's best work, the play *Die Soldaten* (q.v.), which reflects some of his experiences in Strasburg, was published anonymously in 1776. His dramatic works also include the comedies *Die Freunde machen den Philosophen* and *Der neue Menoza* (both 1776), *Der Engländer* (1776), a play connected with his love for Henriette von Waldner, and *Die sizilianische Vesper. Ein historisches Gemälde*

(1782, in *Liefländisches Magazin der Lektüre*). The literary satire *Pandämonium Germanikum* (q.v.), written in 1775 in dramatic form, was published in 1819. Lenz also wrote some short narrative works, including *Der Waldbruder* (q.v., published by Schiller in 1797), which is also linked with Fräulein von Waldner, *Zerbin* (1776), and *Der Landprediger* (1777), which is inspired by his contact with Oberlin. Numerous dramatic fragments were published long after his death. Among them may be mentioned *Henriette von Waldeck oder Die Laube* (written 1776, published 1884, and again concerning his feelings for the remote noblewoman) and *Catharina von Siena. Ein Künstler-Schauspiel*, originally described as 'Ein religiöses Schauspiel' (1884). The MS. of *Freundschaft geht über die Natur oder Die Algierer*, a play long believed to be lost, was discovered in Hamburg in 1971. *Die moralische Bekehrung eines Poeten*, an imaginative account of Lenz's relationship with Cornelia Schlosser (see GOETHE, CORNELIA), was published in 1889.

In *Dichtung und Wahrheit* (q.v., Bk. 14) Goethe dwells on Lenz's instability; but G. Büchner's Novelle *Lenz* (q.v., written 1835–6) gives a more sympathetic insight into Lenz's character and aims, particularly his rejection of classical conceptions, his search for realism, and his advocacy of Shakespeare as a model for an epically conceived theatre. The Naturalists rescued Lenz at the end of the 19th c. from the neglect into which his work had fallen, and it can be seen that he was an important forerunner, not only of Naturalism, but of the Expressionistic and epic theatres of the 20th c.

Lenz's works were first collected by L. Tieck (q.v.) in *Gesammelte Schriften* (3 vols., 1828). *Gesammelte Schriften* were edited by F. Blei (5 vols., 1909–13), and *Gesammelte Werke* by R. Daunicht (4 vols., 1967 ff.). The letters were edited by K. Freye and W. Stammler (*Briefe von und an J. M. R. Lenz*, 2 vols., 1918, repr. 1969). Lenz also wrote poems, mostly published in magazines. They are included in the collected works and published separately in *Gedichte* (ed. H. Haug, 1968).

LENZ, SIEGFRIED (Lyck, East Prussia, 1926–), served in the German navy towards the end of the 1939–45 War, and under difficult post-war conditions pursued philosophical, German, and English studies at Hamburg University before turning, against his original intention to enter the teaching profession, to journalism and authorship. A member of the Gruppe 47 (q.v.), he was on the staff of the newspaper *Die Welt* 1950–1. He made his home in Hamburg and the Danish island Als. His first novel, *Es waren Habichte in der Luft* (1950), deals with an escaped prisoner of war on the run in Finland. *Duell mit*

dem Schatten (1953) raises Lenz's constant preoccupation, the problem of guilt and 'heroism'; it recounts how, when a former German colonel and his daughter revisit North Africa, the daughter gradually becomes aware of a war crime he has committed, so that an unbridgeable gulf opens between father and daughter. *Der Mann im Strom* (1957), a novel with the background of Hamburg harbour and the life of deepsea divers, is concerned with personal, rather than collective, guilt.

Jäger des Spotts (1958) is a collection of 14 shorter tales with backgrounds as various as Kenya in the time of the Mau Mau, international athletics, the Arctic, and Sardinia. The title of the collection is that of the ninth story, which concerns an unlucky and unsuccessful hunter of polar bears. The novel *Brot und Spiele* (1959) also deals with athletics, and portrays the long-distance runner in his last race; it employs a retrospective technique to reveal the problems besetting the successful sporting personality in the modern world. Both in its symbolic allusiveness and in its subject it has close affinities with the story *Der Läufer* in *Jäger des Spotts*. *Das Feuerschiff* (1960), the principal work in a collection with this title, is a dramatic story of a lightship taken over by gangsters, who hold up the crew and in the end kill the captain. The kernel of the story is the tension between the generations, for the captain's son only becomes reconciled with his father when the latter is mortally wounded. The story raises questions of guilt, duty, and humanity. It was reissued separately in 1964. *Der Spielverderber* (the title of a story and that of a collection published in 1965) first appeared in a magazine as *Der Asoziale* in 1962. Its principal figure is an *enfant terrible* who infallibly remembers the unpleasant things people wish to forget, so that he becomes an unwelcome embodiment of conscience. The novel *Stadtgespräch* (1963) is set in occupied Norway and deals with the Norwegian resistance, the occupying troops, and the ordinary inhabitants. In the unavoidable involvement of the common man in the responsibility for harrowing decisions, the story has analogies with *Das Feuerschiff*. An interesting psychological aspect is the shift of attitude in the inhabitants as the tensions of war recede, so that the hero of the war years comes to be regarded as guilty of the death of the 44 hostages shot because he did not surrender.

Lehmanns Erzählungen oder So schön war mein Markt. Aus den Bekenntnissen eines Schwarzhändlers (1964) is set in the Germany of 1945–6, the heyday of the Black Market, and illustrates with satire and humour the proverb that 'pitch defiles'. The fullest development of Lenz's searching retrospective outlook, both in content and form, is the substantial novel *Deutschstunde* (q.v., 1968, also adapted as a television play). Another side of Lenz is seen in the two collections of stories *So zärtlich war Suleyken* (1955) and *So war das mit dem Zirkus* (1971), in which he tells with sympathy and enjoyable humour a series of anecdotes and tales connected with the fictitiously named village Suleyken of his native Masuria. Lenz's outstanding dramatic work is the play *Zeit der Schuldlosen* (q.v., 1961), which was followed by the comedy *Das Gesicht* (1963). Other plays include the television play *Inspektor Tondi* (shown in 1952), and the radio plays *Das schönste Fest der Welt* (1953), *Die Muschel öffnet sich langsam* (1956), and *Haussuchung* (1963). Since the 1960s Lenz has become a well-known political journalist and broadcaster. *Beziehungen*, a volume of personal and critical essays, appeared in 1970, and *Das Vorbild* (after *Deutschstunde* his major work) in 1973. The novel, set in Hamburg and interspersed with a variety of stories and characters, sets such diverse individuals as the retired schoolmaster Valentin Pundt, the young teacher Janpeter Heller, and the unfeminine female with pedagogic pretensions, Rita Süssfeldt, the task of composing a model educational reader, the 'Vorbild' of the title.

With the novel *Heimatmuseum* (1978) Lenz returns to his homeland. A fictitious first-person narrator, Zygmunt Rogolla, has set fire to the regional museum that is in his care and its contents have been consumed by the flames. Zygmunt's justification of the arson establishes the novel's historical and ideological function. Retrospective glances at the historical background are a reminder of power politics and racial prejudice that discredited the term Heimat and the Heimatroman (q.v.) during the National Socialist regime. Of equal relevance is Zygmunt's consciousness of the value of tradition and culture, now irrevocably destroyed. Reminiscences reconstruct a small world that fulfils a basic human need. The homeland lives in the heart, but the greater importance of sacrifice, acceptance of guilt, and adjustment to change has been demonstrated. The novel *Der Verlust* appeared in 1981, the 'loss' of the title alluding to the loss of the power of speech suffered by the principal figure, Ulrich Martens. The terse story *Ein Kriegsende* (1984) deals with a significant moral dilemma involving crew and captain of a mine-sweeper after Germany's partial surrender in 1945 and ends with the court martial and execution of two crew. It was followed by the novel *Exerzierplatz* (1985). The volume *Gespräche mit Manès Sperber und Leszek Kolakowski* was published in 1980, *Gesammelte Erzählungen* in 1970, *Die frühen Romane* in 1976, *Ein Haus aus lauter Liebe* in 1977, and *Drei Stücke* in 1980.

A master in the short story, with the occasional Kafka-esque touch, Lenz contrives to combine atmosphere with narrative suspense and psychological insight with tantalizing irony. He has published collections of tales under new titles. *Stimmungen der See* (1962) includes, apart from two previously unpublished stories, a short autobiographical sketch. *Einstein überquert die Elbe bei Hamburg* (1975) contains 13 stories written between 1966 (the title-story dates from 1969) and 1975.

Leo Armenius oder Fürstenmord, a tragedy in Alexandrine verse written by Andreas Gryphius (q.v.) in 1646 and first printed in 1650. Its subject is the assassination of Leo Armenius, Emperor of Byzantium. Leo has, by means of a palace revolution, made himself emperor. He discovers a plot to overthrow him and arrests its leader, Michael Balbus, whom he intends to put to death. But he defers the execution till the Feast of Christmas is over, whereupon the conspirators take heart and murder Leo. The play is formally constructed, with long tirades, passages of stichomythia, and a chorus.

LEON, JOHANN (d. 1597, Wölfis nr. Ohrdruf, Thuringia), a Lutheran pastor, wrote hymns, books of devotion (*Trostbüchlein*, 1589), and a Christmas play (*Erfurter Weihnachtsspiel*, 1553, revised 1566).

Leonce und Lena, a comedy written by G. Büchner (q.v.) between February and July 1836 as an entry for a competition for the best German comedy, sponsored by the publisher Cotta (q.v.); it was too late to qualify. Büchner revised the play in the autumn and winter of 1836–7. Gutzkow (q.v.) published it after Büchner's death in the *Telegraph für Deutschland* in May 1838, but edited the first act in fragmentary form. Ludwig Büchner (q.v.) published the complete work (possibly with alterations) in the *Nachgelassene Schriften* of 1850. Its first performance in May 1895 was organized by M. Halbe (q.v.) and produced by E. von Wolzogen (q.v.) in Munich. Büchner drew from a variety of sources, among which *Ponce de Leon* by C. Brentano (q.v.) and *Fantasio* by Alfred de Musset are most conspicuous, while the motto for the first act acknowledges the inspiration of Shakespeare's *As you like it* ('O! that I were a fool').

The plot shows us the eponymous hero and heroine, a prince and a princess, in protest against an enforced political marriage before they have even seen one another. Both escape from their respective petty courts in the imaginary states of Popo and Pipi, meet by chance at night and fall in love, unaware of each other's identity. After vainly inviting Lena, his ideal of nature's beauty ('flower'), to share in a Liebestod and seeking death by drowning, Leonce accepts the bet of Valerio the court fool that the prince will marry his still unidentified love at his father's (King Peter's) court. Wearing mask upon mask Valerio presents Leonce and Lena, who likewise appear in disguise and consent to stand in for the missing bride and bridegroom, so that the marriage ceremony can go ahead by proxy. When they remove their masks after the wedding they find in fact that they have married their own choice. Leonce will henceforth initiate his fairyland rule with Lena as his queen and Valerio as his minister.

The play is a satire on several levels, making fun of literature, philosophy (in King Peter), the petty states of the German Confederation, and the corruption of the ruling classes; and its ironic motto (Alfieri: E la fama? Gozzi: E la fame?) touches on Büchner's familiar theme of poverty and starvation. The lightness of touch fits this deliberate incursion into the absurd.

LEONHARD, RUDOLF (Lissa, 1889–1953, Berlin), was a volunteer at the beginning of the 1914–18 War, a Spartacist (see SPARTAKUSBUND) at its end. He remained a communist, removed to France in 1927, was interned in 1939, then imprisoned, but escaped and went underground, returning to Paris in 1944. During his internment he wrote *Geiseln* (1947, reissued 1952), said to be the first German play on the French Resistance Movement. In 1950 he returned to East Berlin. His early work, which was Expressionistic, included collections of poems (*Der Weg durch den Wald*, 1913; *Über den Schlachten*, 1914; *Katilinarische Pilgerschaft*, 1919; *Spartakussonette*, 1922; *Das Chaos*, 1923) and plays (*Segel am Horizont*, 1925; *Tragödie von heute*, 1927; *Traum*, 1933). His later work includes the collection *Deutsche Gedichte* (1947), *Der Tod des Don Quijote* (2 vols., 1938, in extended form 1951), *El Hel. Wolf Wolff* (1939), and radio plays. His numerous essays include *Deutsche Arbeiter, ihr seid die Hoffnung* (1938, in collaboration with H. Mann, L. Feuchtwanger, qq.v., and Gustav Regler, 1898–1963). *Ausgewählte Werke in Einzelausgaben* (4 vols.), ed. M. Scheer, appeared 1961–80.

LEONHART, DORETTE (1756–84) and AUGUSTE (1758–86), were sisters, who became the first and second wives of G. A. Bürger (q.v.). Dorette, who was married to Bürger on 22 November 1774, died of consumption on 30 July 1784. Bürger had fallen in love with the younger sister in 1775 and she had borne him a son in 1782. On 17 June 1785 Bürger married Auguste, who died in childbirth on 9 January 1786. She was called Molly by Bürger, and as Molly (q.v.) she appears in his poems.

Leonora, title at one time contemplated by Beethoven for his opera *Fidelio* (q.v.) and actually used in 1806 for a revised version (*Leonora oder der Triumph der ehelichen Liebe*). The title still survives in the three rejected overtures, *Leonora Nr. 1, 2, 3.*

LEONORE, name of two young women whom the poet J. C. Günther (q.v.) loved. Many of his poems are addressed to Leonore Jachmann ('die Schweidnitzer Leonore') and a few to the 'Leipziger Leonore'.

LEOPOLD I (Vienna, 1640–1705, Vienna), Emperor of the Holy Roman Empire, was intended for the Church, but became the Habsburg heir on the death of his elder brother in 1654. In 1658 he succeeded his father as emperor. Though a man of peace, he spent the greater part of his life at war in successful efforts to maintain the state against powerful and aggressive neighbours, the France of Louis XIV, and the Turks. The three French wars took up in all twenty years of his reign, and he left the last of them, the War of the Spanish Succession, unfinished, dying soon after Blenheim (see HÖCHSTÄDT). Leopold was engaged in intermittent wars with the Turks from 1663 to 1683, in which year Vienna successfully withstood a famous siege. Further Turkish wars (in which Prince Eugene, see EUGEN, PRINZ, distinguished himself) occurred in 1687 and 1696.

During Leopold's reign, though the Empire grew weaker, Austria rose to the rank of a European power; and the dynasty consolidated its position in Hungary at the Diet of Preßburg in 1687, which confirmed the hereditary Habsburg right to the Hungarian crown. Like many others of his family, Leopold was conspicuously intolerant in religious matters, and vigorously persecuted Protestants, notably in Hungary.

LEOPOLD II (Vienna, 1747–92, Vienna), Emperor of the Holy Roman Empire from 1790 to 1792, was the third son of Franz I and Maria Theresia (qq.v.). He became grand duke of Tuscany in 1765, carrying out a series of successful reforms and materially raising the duchy's standard of prosperity. On the death of his brother Joseph II (q.v.) he succeeded to the imperial crown and immediately took conciliatory measures to allay the unrest caused by Joseph's rapid and intransigent reforms. Leopold sought to avoid entanglement in the affairs of royalist and revolutionary France, proving himself an astute politician. He died suddenly after a reign of almost exactly two years.

LEOPOLD, Fürst von Anhalt-Dessau, 'der alte Dessauer' (Dessau, 1676–1747, Dessau), the first of several Prussian commanders whose names became household words in the 18th c. 'Der alte Dessauer' served with distinction in the War of the Spanish Succession, notably at Blenheim (see Höchstädt), Cassano, and Turin, and was made a Prussian field-marshal in 1712. In 1715 he fought against the Swedes, capturing Rügen and Stralsund. His brilliant victory at Kesselsdorf in 1745 decided the Second Silesian War (see Schlesische Kriege). He introduced marching in step into the Prussian army and substituted iron for wooden ramrods. He made a love match outside his class (with an apothecary's daughter), and was successful in winning recognition for his wife and succession rights for the sons of the marriage. One of the best-known German military marches bears his name. The poem *Der alte Dessauer* (1847) by Th. Fontane (q.v.) mentions beside his military prowess his notorious phobia for learning and letters. H. Mann in his posthumous fragment *Die traurige Geschichte von Friedrich dem Großen* (q.v.) gives a hostile portrait of him.

Leopoldstädter Theater, one of the best-known Viennese theatres in the late 18th c. and early 19th c., was founded in 1781 by K. von Marinelli (q.v.). Officially known as the Theater in der Leopoldstadt (now II. Bezirk), it was the home of the popular Viennese Volksstücke (see Volksstück) of such writers as Bäuerle, Gleich, Meisl, and F. Raimund (qq.v.). It was demolished and replaced in 1847 by the Carl-Theater owned by K. Carl (q.v.).

LEOPOLD VON TIROL, Erzherzog (Graz, 1586–1633, Graz), was a son of the Archduke Karl von Steiermark, a cousin of the Emperor Rudolf II (q.v.), and brother of the Emperor Ferdinand II (q.v.) who, in 1619, gave him the Tyrol. In 1610–11 he vainly opposed Archduke Matthias (see Matthias, Kaiser) in Bohemia with the support of his army (known as the 'Passauer', probably because he had been bishop of Passau since 1605). In 1625 he was secularized by the Pope so that he could marry, but his line died out after one generation. He appears as a character in Grillparzer's tragedy *Ein Bruderzwist in Habsburg* (q.v.).

LEPEL, Bernhard von (Meppen, 1818–85, Prenzlau), of a military family, entered the Prussian army and was stationed in Berlin, serving as an officer in the regiment of the Guards, in which his friend and fellow-member of the club Der Tunnel über der Spree (q.v.), Th. Fontane (q.v.), was a one-year volunteer in 1844–5. In 1848 Lepel took part with his regi-ment in the Danish War. He retired to live on his private estate, but was recalled to the colours as a major in 1866. He published a volume of anti-Catholic poetry, *Lieder aus Rom* (1846), a book of humorous verse, *Die Zauberin Kirke, Heitere Reime* (1850), and a further volume of poems, *Gedichte* (1866).

Lerche, Die, a poem by Annette von Droste-Hülshoff (q.v.) published in *Heidebilder* (of which it is the first) in her volume of *Gedichte* (1844).

LERNET-HOLENIA, Alexander (Vienna, 1897–1976, Vienna), an officer in the Austrian cavalry in the 1914–18 War and an extensive traveller, began his career as an author with several successful plays, mainly comedies, including *Demetrius* (1926), *Österreichische Komödie* (1927), *Ollapotrida* (1927), *Erotik* (1927), *Parforce* (1928), *Die nächtliche Hochzeit. Haupt-und Staatsaktion* (1929), and *Die Frau des Potiphar* (1934). His first volume of poetry, *Pastorale*, appeared in 1921, and was followed by *Das Geheimnis Sankt Michaels* (1927), *Die goldene Horde* (1935), *Die Trophae* (1946, vol. 1 *Gedichte*, vol. 2 *Szenen*), and *Das Feuer* (1949). His fiction of the 1930s includes detective and adventure novels, *Die Abenteuer eines jungen Herrn in Polen* (1931), *Ich war Jack Mortimer* (1933, a mixture of the erotic and the criminal), *Die Auferstehung des Maltravers* (1936), the story of an aristocratic swindler, and *Strahlenheim* (1938). In other works he draws on his war experiences; *Die Standarte* (1934 in a magazine, as *Das Leben für Maria Isabella* in 1966) is set in Serbia in 1918 and portrays the collapse of the Austro-Hungarian Empire, depicting mutiny and counter-action in an Austrian dragoon regiment; among his short stories *Der Baron Bagge* (1936, reissued in 1978 with a postscript by H. Spiel), with a background from the 1914–18 War, stands out.

Lernet-Holenia also served in the 1939–45 War, for a time with an army film unit. *Mars im Widder* (1941 in a magazine, then banned, 1947 in book form) is set in the beginning of the war; *Beide Sizilien* (1942) is concerned with violent crime. In 1955 appeared the novel *Das Finanzamt. Aufzeichnungen eines Geschädigten*, containing elegant satire. A facile and entertaining writer, he looks back to the old Austria in later works such as the biography of Eugen, Prinz von Savoyen (q.v.) *Prinz Eugen* and the stories included in *Mayerling* (both 1960). *Die Geheimnisse des Hauses Österreich. Roman einer Dynastie* appeared in 1971. *Konservatives Theater* (1973) is a collection of plays. He was the recipient of a number of Austrian awards (from 1945).

LERSCH, Heinrich (München-Gladbach, 1889–1936, Remagen), as a youth, joined the association Werkleute auf Haus Nyland (q.v., 1905). A boilermaker by trade, he eulogized the working man's life, was an anti-Marxist, a Roman Catholic, and after 1933 a supporter of the NSDAP (q.v.). Ill health made him retire from manual work in 1925. He wrote poetry (*Wir Volk*, 1924; *Stern und Amboß*, 1927) and a number of works of fiction, of which the novel *Hammerschläge* (1930) and the volume of tales *Im Pulsschlag der Maschinen* (1934) deserve mention. Posthumous publications include *Ausgewählte Werke* (2 vols.), ed. J. Klein, 1965–6.

Lerse, a character in Goethe's play *Götz von Berlichingen mit der eisernen Hand* (q.v.). One of Götz's men, he is the embodiment of generous loyalty. He bears the name of one of Goethe's Strasburg friends, Franz Christian Lerse (1749–1800), who is characterized in *Dichtung und Wahrheit* (q.v., Bk. 9).

LESSING, GOTTHOLD EPHRAIM (Kamenz, Saxony, 1729–81, Brunswick), the son of a Protestant pastor, was educated at St. Afra's School, Meißen (see FÜRSTENSCHULEN), from 1741 to 1746, and passed then to Leipzig University. He at first studied theology, but his interests were in letters and the theatre, and in 1748 he transferred to medicine. He soon left Leipzig to escape arrest for debt, spending a short time at Wittenberg University.

While at Leipzig and, for a brief period, at Berlin, Lessing collaborated with his cousin C. Mylius (q.v.). On leaving Wittenberg he returned to Berlin, earning his living by journalism. He edited *Beyträge zur Historie und Aufnahme des Theaters* and contributed to *Das Neueste aus dem Reiche des Witzes*. His first comedy, *Der junge Gelehrte* (1748), was performed while he was still a student. In 1749 he wrote *Der Freigeist* and *Die Juden* (qq.v.), two competent and original comedies, and began an interesting experiment in tragedy on a contemporary subject, *Samuel Henzi* (q.v.), which remained unfinished. He made lasting friendships with M. Mendelssohn and F. Nicolai (qq.v.), and fell out with Voltaire. He published *Die Theatralische Bibliothek* (q.v., 1754) and his *Schriften* (6 vols., 1753–5), which contained his poems, *Briefe* (often termed *Briefe 1753* or *Kritische Briefe*, which include a defence of S. Lemm, q.v.), the *Rettungen* of Horace, Cardanus, the anonymous Ineptus Religiosus, and Cochläus (1754; see RETTUNGEN), and the plays written up to 1755, including *Miß Sara Sampson* (q.v., 1755), his domestic tragedy (see BÜRGERLICHES TRAUERSPIEL). With his *Vademecum* (q.v., 1754), directed against S. G. Lange (q.v.), he first proved himself to be a formidable literary controversialist.

Grown weary of Berlin, Lessing set out to travel in 1756 as a young man's companion, but the outbreak of the Seven Years War (see SIEBENJÄHRIGER KRIEG) curtailed the journey, and he spent some time in Leipzig, where he formed a close friendship with the poet Ewald von Kleist (q.v.), then a Prussian major. In 1758 Lessing returned to Berlin, and in 1759 edited and partly wrote a new critical journal, the *Literaturbriefe* (q.v.). In the same year he published his fables with a treatise on the fable (*Abhandlungen über die Fabel*) and the quasi-classical tragedy *Philotas* (q.v.). The latter part of the war and the post-war period (1760–5) he spent in Breslau as secretary to General Tauentzien (1710–91).

In 1766, hoping for appointment as royal librarian in Berlin, Lessing, to support his claim, published his epoch-making essay on poetry and the plastic arts, *Laokoon* (q.v., 1766), which earned him the nickname 'Scheide-Kunst'. But he was disappointed in his hope of appointment. His studies of classical art had as an aftermath the *Briefe antiquarischen Inhalts* (1768–9), directed against C. A. Klotz (q.v.), and the essay *Wie die Alten den Tod gebildet* (1769). His comedy of the post-war period, *Minna von Barnhelm* (q.v.), was published in 1767; at the same time he became engaged in critical writing on the theatre in a new capacity as house critic to the National Theatre at Hamburg. The outcome of this activity was the *Hamburgische Dramaturgie* (q.v., 1767–8).

In 1770 Lessing accepted the post of librarian to the Duke of Brunswick at Wolfenbüttel. There he wrote his tragedy *Emilia Galotti* (q.v., 1772) and engaged in scholarly work based on material in the library. In 1775–6 he accompanied the Duke's son on an Italian tour, and towards the end of 1776 he married, but his wife, a widow, Eva König, died in childbirth at the beginning of 1778.

In 1777 Lessing began to publish fragments from a book by the free-thinker H. S. Reimarus (q.v.); these provoked a storm of criticism, to which Lessing responded with the polemical essays *Über den Beweis des Geistes und der Kraft*, *Eine Duplik*, and *Eine Parabel* (all 1778). The climax of this theological argument was the bitter controversy with Pastor J. M. Goeze (q.v.) of Hamburg, to whose animadversions Lessing replied with the eleven essays of *Anti-Goeze* (q.v., 1778). The Duke intervened, silencing Lessing, who after a time constructively closed the wrangle with his drama of tolerance and intolerance, *Nathan der Weise* (q.v., 1779). In his closing years Lessing wrote a dialogue on freemasonry (*Ernst und Falk*, 1778) and his moral testament, *Die Erziehung des Menschengeschlechts* (q.v., 1780).

Lessing possessed a formidable intellect and great versatility. He made his mark on the

German theatre with his four great plays, writing the first domestic tragedy and the first truly modern comedy, as well as introducing blank verse to German drama. His self-deprecatory comments have tended to obscure his real creative ability. As a critic Lessing may be said to have lifted German criticism from a provincial to a European level. He exercised a lasting influence on aesthetics with his *Laokoon*, and he gave to Aristotle's views a modern interpretation. In his philosophical and theological writings he pleaded powerfully for balance and tolerance. His rigorous mind had immense destructive power, but negation was always followed by an impulse to construct.

Sämtliche Werke, historisch-kritische Ausgabe, ed. K. Lachmann (q.v.) and F. Muncker, appeared 1886–1924 (23 vols., reissued 1968) and includes the letters. *Lessings Werke*, ed. J. Petersen and W. von Olshausen (18 vols., 1925–35), contains additional writings by Lessing. *Gesammelte Werke* (10 vols.), ed. P. Rilla, appeared 1954–8 and *Werke. Vollständige Ausgabe in fünfundzwanzig Teilen*, ed. H. G. Göpfert *et al.*, 1970 ff.

LESSING, KARL GOTTHELF (Kamenz, Saxony, 1740–1812, Breslau), younger brother of G. E. Lessing (q.v.), became director of the mint at Breslau. He wrote plays, including an adaptation of H. L. Wagner's *Die Kindermörderin* (q.v.), but is chiefly known for his biography of his famous brother, *G. E. Lessings Leben nebst seinem noch übrigen literarischen Nachlasse*, published in 3 vols., 1793–5.

Letter of Majesty, see MAJESTÄTSBRIEF.

Letzte am Schafott, Die, a Novelle by Gertrud von LE FORT (q.v.), published in 1931. It takes the form of a letter, written in Paris in October 1794 by an eyewitness of the reign of terror (see FRENCH REVOLUTION), von (*sic*) Villeroi. The addressee, a lady, is a French *emigrée*, and the theme of the correspondence is the nature of heroism in relation to faith. The central figure of the narrative is Blanche de la Force. An orphaned girl, she becomes a novice in the Carmelite convent of Compiègne, but abandons it shortly after her admission as a nun in the name Blanche de Jésus au jardin de l'Agonie. Blanche witnesses the execution of sixteen nuns after the closure of the convent by the National Assembly. The nuns encounter their ordeal singing the *Veni creator spiritus* (q.v.). As the voice of the last nun, that of Marie de l'Incarnation, becomes silent, Blanche carries on with the song of the martyrs on the Place de la Révolution, and is killed by the mob. Blanche, born at a time when the old order showed the first signs of doom, represents the fear which releases chaos. Her singing is not portrayed as an act of heroism, but as a miracle which dispels her weakness. A little wax figure of the infant Jesus, le petit Roi de Gloire, functions as a symbol of divine grace. It is associated with the destiny of the Dauphin, the convent, and Blanche.

Letzte Gaben, a collection of 64 poems written by Annette von Droste-Hülshoff (q.v.), mostly in 1844 and 1845, and published by Levin Schücking (q.v.) in 1860. Six other poems were later added. The collection contains, among others, *Durchwachte Nacht, Mondesaufgang, Im Grase*, and *Silvesternacht*.

Letzte Gedichte, a collection of poems by H. Heine (q.v.), most of which appeared as *Gedichte 1853 und 1854*. It forms part of *Letzte Gedichte und Gedanken*, published posthumously in 1869. Some poems written in 1855 reflect Heine's harrowing suffering and ironic courage during the later stages of his illness and include the finest poems he ever wrote, especially *Morphine*, 'Ich seh im Stundenglase schon', 'Laß mich mit glühnden Zangen kneipen', 'Es kommt der Tod', and *Der Scheidende*.

Letzte Held von Marienburg, Der, a tragedy by J. von Eichendorff (q.v.), published in 1830. It is written in the main in blank verse, but some scenes in the first three acts are in prose. Its subject is the unsuccessful struggle of HEINRICH von Plauen (q.v.), High Master of the Teutonic Order (see DEUTSCHER ORDEN), against disaffection and corruption in the Order in the early 15th c. Eichendorff was drawn to the subject through being associated as a civil servant with Theodor von Schön (q.v.), who was responsible for the rebuilding and restoration of the Order's Castle at Marienburg (q.v.).

Letzte König von Orplid, Der, a dramatic fantasy (Ein phantasmagorisches Zwischenspiel) by E. Mörike (q.v.), included in the novel *Maler Nolten* (q.v.). See ORPLID.

Letzten Reiter, Die, a strongly nationalistic novel by E. E. Dwinger (q.v.), published in 1935. It deals with the gallant, but ultimately vain efforts, of a mounted 'free corps' (see FREIKORPS) to secure Courland (Latvia) against Bolshevist forces.

Letzten Tage der Menschheit, Die, a gigantic Expressionistic drama by Karl Kraus (q.v.), written in wartime between 1915 and 1917. The first version (Akt-Ausgabe) appeared in 1918–19 in *Die Fackel* (q.v., Sonderhefte), the second version (Buchausgabe) in 1922, and the final version in 1926. It is in five acts with a prelude (*Vorspiel*), and an epilogue called *Die letzte Nacht*. In all, including the epilogue, there are 220 scenes. Kraus points out in a preface that performance would take something like ten evenings, and claims that the work is destined for a Martian theatre, an allusion to the apocalyptic end, which is the extermination of mankind by forces from Mars.

Die letzten Tage der Menschheit is a kaleidoscopic presentation of the 1914–18 War, as seen from a Viennese observation point. Kraus's bitter and ruthless satire ferociously passes in review the ruling house, the army commanders, the politicians, the press, the profiteers, the brutal regimental officers, the gullible newspaper readers, and the fickle crowd. The focus of his satire is the universal corruption of language, by which thought and feeling are falsified and smothered and the truth perverted by the press and the politicians. Through the whole play wanders the figure of the faultfinder (der Nörgler), easily identifiable with Kraus, who makes his searing comments on the bestiality and stupidity paraded. The treatment of the closing stages is visionary, and *Die letzte Nacht* takes on a deliberate resemblance to the last act of *Faust* (q.v.) Pt. II.

Letzten Zehn vom vierten Regiment, Die, a ballad linked with the contemporary Polish revolt, written by J. Mosen (q.v.) in 1831. It is quoted in Th. Fontane's Novelle *Unterm Birnbaum* (q.v.).

Letzte Reckenburgerin, Die, an historical novel by Louise von François (q.v.), published in 1871. Set in the 1790s and the first quarter of the 19th c., the narrative uses a complex retrospective technique by which it unfolds a mystery of identity, that of the soldier August Müller. The introduction is set in Belgium in 1817. In it the soldier Müller tells his *cantinière* wife of his obscure childhood. From his account the wife concludes that he is the son of Hardine (Eberhardine) von Reckenburg, and names her baby daughter Hardine. A few years later the wife dies, and Müller sets out for Saxony with his daughter in order to find his presumed mother, Fräulein Hardine von Reckenburg. He reaches the castle, but is turned away, falls ill, and dies in 1823.

Hardine von Reckenburg, last of her line, who looks after Müller's young daughter Hardine, is the narrator of the main section of the novel. She knows Müller's identity, but is not his mother, who turns out to be Dorothee, a companion of lower rank with whom she was brought up. A woman lacking stability, Dorothee

was seduced by a young prince connected with the Reckenburg family, who was killed at Valmy in 1792. She gave birth to a child, but took no interest in it, and Fräulein Hardine made arrangements for its education. When Müller arrived at Reckenburg, she recognized that he was the child. She travelled to find Dorothee, who had married a Doctor Faber although she did not love him. She found Dorothee, from whom she wished to obtain permission to reveal the facts to Müller, on her death-bed; she also found that Faber had never been told the truth.

Eberhardine brings up Müller's daughter Hardine, not out of love, but out of a sense of humanitarian duty. Silently and uncomplainingly she endures the social ostracism of all those who believe that Müller was her illegitimate son. The girl Hardine thaws her guardian's heart in the end, marries Nordheim, the steward, and finds herself heiress to the Reckenburg fortune. This somewhat austerely written novel by a noblewoman has an unobtrusive bias in favour of the burgher class.

Letzte Ritter, Der, see MAXIMILIAN I, KAISER.

Letzte Rittmeister, Der, a volume of 24 short stories by W. Bergengruen (q.v.), published in 1952. Some had already appeared in previous publications. They are anecdotes and tales with a predominantly cavalry background, and are enclosed and linked by a frame (see RAHMEN); its special feature is the relationship which the narrator establishes with the endearing personality of the Rittmeister, the 'last cavalry captain' of the title, an impoverished Russian émigré officer who has found refuge in Ticino, Switzerland. Bergengruen introduces the narrator to launch the book, and he recurs from time to time in *intermezzi*, visiting at the end the peaceful grave of the now deceased Rittmeister. The immediate sequel, *Die Rittmeisterin* (1954), is a more ambitious work but lacks the qualities of this successful volume. A third work linked with the Rittmeister, *Der dritte Kranz*, appeared in 1962.

LEUCHSENRING, FRANZ MICHAEL (Langenkandel, Alsace, 1746–1827, Paris), was for a time in the employment of the court of Hesse-Darmstadt as a tutor, and was later in a similar capacity in the service of Friedrich II (q.v.) of Prussia. In 1792 the French Revolution (q.v.) attracted him to Paris, where he spent the remainder of his life in obscurity.

Leuchsenring was a sentimentalist and an ingratiating ladies' man. His conduct and character led Goethe, who met him in Darmstadt, to lampoon him in *Ein Fastnachtspiel vom Pater Brey*, *Der Triumph der Empfindsamkeit*, and *Das Jahrmarktsfest zu Plundersweilen* (qq.v.). Leuchsenring is also the original of Prediger Frank, an episodic character in L. J. von Arnim's *Armut, Reichtum, Schuld und Buße der Gräfin Dolores* (q.v.).

Leuchtturm, Der, a two-act tragedy (Trauerspiel) written in trochaic verse by E. C. von Houwald (q.v.) in 1819 and published in 1821 (see SCHICKSALSTRAGÖDIE). Dorothea, the lighthouse keeper's daughter, in listening to the protestations of her lover Walter, forgets to watch the light. The keeper, Kaspar, happens to have a mad uncle, who extinguishes it, and a ship is in consequence wrecked. Walter rescues one survivor, whom he believes to be his father, though it presently appears that he is Graf Holm, the seducer of Walter's mother, who has been drowned in the wreck. The mad uncle of Dorothea, who turns out to be the real father, drowns himself and Graf Holm proposes to do the same, but is dissuaded and all ends happily. The vogue of the play was short-lived.

Leute aus dem Walde, Die, a novel sub-titled *Ihre Sterne, Wege und Schicksale*, written in 1861–2 by W. Raabe (q.v.) and published in 1863 (3 vols.). The people from the forest, the Winzelwald, are drawn from two groups: the poorest of down-trodden commoners and the squirearchy of Poppenhof.

The orphan Robert Wolf from the Eulenbruch in the Winzelwald comes to the great city, where he is befriended and adopted by the kindly police official (Polizeischreiber) Fiebiger. Robert at first loves Eva Dornbluth, also from the Winzelwald, but she is claimed as bride by Friedrich Wolf, Robert's long-lost brother, returned from America. Friedrich and Eva marry and go to the United States. Robert falls in love with the rich banker's daughter Helene Wienand, but the sly and corrupt Leon von Poppen, son of the squire of Poppenhof, ensnares her father into giving him her hand. News comes from America that Friedrich is dead, and Robert goes out to help Eva, whom he finds dying. He discovers gold and becomes a rich man. Leon von Poppen has meanwhile died and Robert returns in time to marry Helene and purchase the Poppenhof, thus symbolically reversing the old social order.

Leute von Seldwyla, Die, a collection of Novellen by G. Keller (q.v.), published in 2 vols. in 1856. It contained five stories and a preamble explaining that Seldwyla is a fictitious small town in Switzerland, and that the eccentric behaviour of the inhabitants is illustrated in the stories.

In 1873–4 Keller republished *Die Leute von*

Seldwyla unchanged except that the 2 vols. were now described as *Erster Band*. The title, however, now covered also a *Zweiter Band* (again in 2 vols.), adding five more stories and an introduction indicating that the inhabitants of Seldwyla have not basically changed.

The contents are as follows: the original work opens with *Pankraz der Schmoller*. The sulky boy Pankraz runs away from home and returns years later cured of this blemish in his character. He recounts his years of foreign service in the tropics, the climax of which is an encounter with a lion which banishes for ever all sulkiness. The second story is the classic *Romeo und Julia auf dem Dorfe* (q.v.) and the third *Frau Regel Amrain und ihr Jüngster*, which is basically a sermon on civic duty. Frau Regel, a widow, by her discipline and public spirit brings her youngest son up to be a conscientious citizen. Keller's model for Frau Regel is believed to have been his mother. This is the only story in the collection previously published (in the Zürich *Volkszeitung*, September–October 1855). *Die drei gerechten Kammacher* (q.v.) is the fourth story, and the fifth is *Spiegel das Kätzchen*, a humorous fairy-tale about a cat who talks himself out of an uncomfortable bargain with a wizard.

The first three of the five stories of the continuation were written in the 1850s, the last two in the 1860s. *Kleider machen Leute* (q.v.) is followed by *Der Schmied seines Glückes*, the sad tale of John Kabys, who tries to make an easy fortune by inheritance, but goes too far when, in order to make sure of the bequest, he makes love to the young wife of his benefactor. He begets a child, whom the wife declares to be legitimate, so John ceases to be the heir, and returns home poorer in prospects, but improved in character. The third story is *Die mißbrauchten Liebesbriefe* (q.v.). The next, *Dietegen*, is a tale of the late 15th c.: Küngolt, a young girl condemned to death for alleged witchcraft, is rescued in the nick of time by young Dietegen, who invokes a law by which she may be saved if she is married on the scaffold. His act of mercy and love is a requital of a similar act by Küngolt, who, when still a child, had saved the life of the boy Dietegen. *Das verlorene Lachen* completes the collection; a harmonious marriage breaks down for a time, but is in the end restored to its initial unity. The partners are Jukundus Meyenthal and Justine, *née* Glor. Keller devotes the story largely to discussions on social, political, and anti-religious topics.

Leuthen, Battle of (5 December 1757), in which Friedrich II (q.v.) of Prussia defeated the Franco-Imperial army (see SIEBENJÄHRIGER KRIEG). Leuthen lies to the west of Breslau.

LEUTHOLD, HEINRICH (Wetzikon nr. Zurich, 1827–79, Burghölzli nr. Zurich), of humble rural origin, studied at Zurich and Basel universities and, after a period of travel, settled in 1854 in Munich where he became an associate of the Krokodil (q.v.) group and a friend of E. Geibel (q.v.). For some years he was a journalist, and he was also a poet of some distinction, though he had not sufficient originality to transcend the narrow limits of the formal Munich school of poets (see MÜNCHNER DICHTERKREIS). Leuthold collaborated with Geibel in *Fünf Bücher französischer Lyrik* (1862) and wrote the verse epics *Penthesilea* (1868) and *Die Schlacht bei Sempach* (1870). From 1877 on he was insane. His poems (*Gedichte*, 1879) appeared in the year of his death.

LEUTHOLD VON SEVEN, a minor Middle High German Minnesänger, whose few surviving poems show a facility which is underlined in an ironical comment on his indefatigable fertility by REINMAR der Fiedler (q.v.). Seven's identity is obscure, but he is believed to have been a Swabian.

Leutnant Gustl, a Novelle written by A. Schnitzler (q.v.) in 1900, and published in book form in 1901. Gustl, an infantry subaltern garrisoned in Vienna, attends a concert. At the cloak-room on the way out he has an altercation with a civilian, who returns insult for insult, and prevents Gustl from drawing his sword. According to the rules, Gustl is now dishonoured and has only the alternatives of suicide or resignation from the army. He spends the night wandering about resolving to shoot himself. In the early morning, on his way to his room to put an end to his life, he learns that the civilian, who was known to him, has died during the night of a stroke. Now no one will ever know, and Gustl relaxes as life reasserts its claims.

Leutnant Gustl is a satire on the official code of military honour and its publication caused a scandal in army and other conventional circles, in consequence of which Schnitzler had to resign his commission as a medical officer in the reserve. The story is an early example of interior monologue, the whole sequence of events, as well as the exposition of Gustl's character, being conveyed by his own articulate thoughts, the words he speaks and the words he hears. Schnitzler's only other essay in this form is *Fräulein Else* (q.v., 1924). (See INNERER MONOLOG.)

LEVETZOW, ULRIKE VON (Leipzig, 1804–99, Triblitz, Bohemia), eldest of three sisters, who with their widowed mother spent the summers of 1821, 1822, and 1823 at Marienbad. There

Ulrike met Goethe, who also spent part of each of these three summers at Marienbad and became deeply attached to her. In 1823, then aged 74, Goethe made an indirect proposal to the 19-year-old girl, which was evasively answered. This last intense passion of his life inspired the poem *Elegie* (also referred to as *Die Marienbader Elegie*) which constitutes part of the *Trilogie der Leidenschaft* (q.v.). Ulrike von Levetzow remained unmarried.

LEVIN, RAHEL, see VARNHAGEN VON ENSE, RAHEL.

LEWALD, AUGUST (Königsberg, 1792–1871, Munich), a man of the theatre, was first an actor and then a stage manager in Hamburg. In 1834 he settled in Stuttgart as a journalist, founding and editing (1835–46) the periodical *Europa*; in 1849 he became a theatrical producer. His principal literary works comprise *Novellen* (1831–3), the six-volume *Aquarelle aus dem Leben* (1836–40), and the long, autobiographical novel *Theaterroman* (5 vols., 1841). He was a cousin of Fanny Lewald (q.v.).

LEWALD, FANNY (Königsberg, 1811–89, Dresden), born a Jewess, adopted Lutheran Christianity in 1828 in order to marry a young theologian, who, however, died before the wedding. She began to write at the age of 30, dealing particularly with social problems and marriage: Her first novel, *Clementine* (1842), was followed by *Jenny* (1843) and *Eine Lebensfrage* (1845). In 1845, while in Italy, she met Adolf Stahr (q.v.) and lived with him, marrying him after the dissolution of his marriage in 1854. Her novel *Diogena* (1847), published as 'Roman von Iduna Gräfin H. H.' (i.e. Ida, Gräfin von Hahn-Hahn, q.v.) parodied the Countess's *Gräfin Faustine*. *Prinz Louis Ferdinand* (1849) was a historical novel with Rahel von Varnhagen (q.v.) as its central figure; *Von Geschlecht zu Geschlecht* (1863–5) and *Die Familie Darner* (3 vols., 1887) both deal realistically with the successive generations of a family.

Fanny Lewald was also the authoress of a number of travel books (*Römisches Tagebuch*, 1845–6; *Italienisches Bilderbuch*, 1847; *Reisetagebuch aus England und Schottland*, 1852) and of an autobiography *Meine Lebensgeschichte* (1861–2). She was a cousin of August Lewald (q.v.).

Leyer und Schwerdt, title given to a collection of patriotic poems written in 1813 by Theodor Körner and published posthumously by his father C. G. Körner (qq.v.) in 1814. Many of them have an impetuous rhythm and an exalted enthusiasm. Among the better known are *Die Eichen*, 'Wo ist des Sängers Vaterland?', *Lied der schwarzen Jäger, Bundeslied vor der Schlacht, Gebet während der Schlacht, Lützows wilde Jagd*, 'Das Volk steht auf, der Sturm bricht los', and *Schwertlied*.

Libraries. In the Middle Ages the monasteries possessed in their collections of MSS. the only real libraries. Those of Fulda, Reichenau, and St. Gall were especially notable. Princely and royal libraries were instituted from the 16th c. on, partly by the secularization of monasteries. Among these new libraries were the Hofbibliothek in Vienna, founded in 1526, and the electoral collections in Dresden (1556) and Munich (1558), all of which have within the last century become national or state libraries.

The Bibliotheca Palatina, founded in Heidelberg in 1555 and augmented by the Fuggers (see FUGGER) in 1584, was the greatest German library of the 16th c., but its later history is complex, the most important MSS. and incunabula being surrendered to the Pope in 1633. Many of these were seized by the French in Rome in 1797 and brought to Paris. A number were returned to Heidelberg in 1814–16. The greater part of the Bibliotheca Palatina is, however, still in the Vatican Library. The Bibliotheca Augusta, founded by the Duke of Brunswick and housed at Wolfenbüttel (1604), had Leibniz and G. E. Lessing (qq.v.) among its librarians. The first properly organized university library was that of the new University of Göttingen (1737) in the Electorate of Hanover.

The modern German universal library is the Deutsche Bücherei, founded in Leipzig in 1912. After the division of Germany in 1945, a parallel institution was founded in West Germany, the Deutsche Bibliothek (1946). All the German Länder, universities, and larger cities have substantial libraries available for public use, and many specialized libraries have also been instituted, of which the best known are the Goethe-Nationalmuseum in Weimar, the Freies Deutsches Hochstift Goethe-Museum in Frankfurt, the Schiller Nationalmuseum in Marbach including the Deutches Literaturarchiv, and the Goethe- und Schiller-Archiv in Weimar.

Libussa, a five-act tragedy in blank verse by F. Grillparzer (q.v.). One of Grillparzer's early projects, the play was completed in 1848 and published in 1872. It was performed at the Burgtheater, Vienna, in 1874. The tragedy differs from Grillparzer's other historical plays in expressing a philosophical view of the evolution of history and civilization against the background of a mythical past, the foundation of Prague. Libussa and her two sisters, Kascha and

Tetka, daughters of Fürst Krokus of Bohemia, are the last descendants of a ruling house endowed with prophetic powers. Libussa is chosen by lot, on her father's death, to succeed him, but her matriarchal rule, relying on an instinctive sense of right and goodwill towards others, fails. Libussa's marriage to Primislaus, a commoner, meets with the approval of the people. Primislaus, a born ruler, is prepared to enforce justice, and he approaches the needs of the state rationally. In order to compete with the growing prosperity of an expanding civilization and to safeguard the future of Bohemia, he decides to build a town on the banks of the Moldau, a threshold (i.e. 'Praga') leading to trade and communication, to fame and good fortune. Libussa recognizes the necessity of the people's integration into the evolutionary process, but she foresees the evils accompanying the progressive emancipation of man. These include alienation from the pure spirit pervading nature, of which the three sisters, in virtue of their divine origin, are representative. She nevertheless believes in human goodness and prophesies the ultimate return of God into the soul of mankind. Her own death and the ensuing homelessness of Kascha and Tetka symbolize the tragic eclipse of an age of primal innocence, but her love impels her to bless the new city of Prague. The relationship between Libussa and Primislaus, an integral part of the basic design, dominates the first four acts, highlighting Grillparzer's ethos of individualism and his mastery of psychological realism.

LICHTENAU, WILHELMINE RITZ, GRÄFIN VON (Dessau, 1753–1820, Berlin), mistress of King Friedrich Wilhelm II (q.v.) of Prussia, was made a countess in 1796. After the King's death she was imprisoned for a time.

LICHTENBERG, GEORG CHRISTOPH (Oberramstadt, 1742–99, Göttingen), a pastor's son, was crippled by an accident in childhood. He studied mathematics and science at Göttingen University. After visiting England, where he attracted the attention of George III, he was appointed to an assistant professorship in physics at Göttingen in 1770. Lichtenberg visited England again in 1774–5, acquiring an interest in English life and art which was to find expression later in a comprehensive interpretation of Hogarth's engravings (*Ausführliche Erklärung der Hogarthschen Kupferstiche*, 14 instalments, 1794–1835, of which Lichtenberg was responsible for the first five, 1794–9). In 1775 he was appointed Professor Ordinarius at Göttingen, and he spent the remainder of his life there, though the pattern of his private affairs departed from the

Göttingen norm; for he took Maria Dorothea Stechard, a 13-year-old girl, into his house as his mistress in 1777, and, after her death in 1782, replaced her with Margarete Kellner, whom he married in 1789.

Lichtenberg wrote no major creative works. His first important writing is a brilliant satire on Lavater's *Physiognomische Fragmente* (see LAVATER, J. C.), appearing under the title *Über Physiognomik, wider die Physiognomen* (1778). The famous parody *Fragment von Schwänzen* (1783) is also directed against Lavater, as is the earlier satire *Von Konrad Photorin* (1773). From 1778 Lichtenberg edited the *Göttinger Taschenkalender* and from 1780 the *Göttingisches Magazin*, in which Georg Forster (q.v.) collaborated. His aphorisms, which are his finest achievement, are scattered throughout these and other publications. They were first collected in the posthumous edition *Vermischte Schriften* (9 vols. 1800–5). They are equally notable for the dispassionate clarity with which they analyse and ironically spotlight foibles, muddle, and obscurantism, and for the simple, relaxed directness of their form. Lichtenberg's irony is directed against hypocrisy, false sentiment, pretentiousness, and emotional inflation, in whatever intellectual camp these are to be found. He is one of the sharpest intellects and finest prose writers of the 18th c.; that he has been somewhat neglected is due to the fact that he wrote only in minor forms.

Lichtenberg's *Gesammelte Werke* (2 vols.), ed. W. Grenzmann, appeared in 1949, *Schriften und Briefe* (4 vols.), ed. W. Promies, 1967 ff., *Aphorismen, Briefe, Satiren*, ed. H. Nette, in 1948, and an extensive edition of correspondence, *Briefwechsel*, ed. U. Joost and A. Schöne, in 1983 ff.

Lichtenstein, an historical novel by W. Hauff (q.v.), published in 1826. Written in conscious emulation of the Waverley Novels, it is described as *Eine romantische Sage aus der württembergischen Geschichte*. It is set in the 16th c. in the time of Duke Ulrich (q.v.), and follows Scott's pattern in using fictitious principal characters with historical figures in the background. The political events narrated are the eviction of Ulrich in 1519 from his duchy by the Swabian League, his victorious return, and subsequent re-expulsion.

In the foreground is the romantic love of the hero, Georg von Sturmfeder, for Marie von Lichtenstein. Georg begins on the side of the League, but goes over to the Duke, whom he supports through thick and thin. In the end he saves the Duke's life by impersonating him and is taken prisoner. But he is well treated, marries Marie, and lives in Lichtenstein castle. Among the historical characters who appear prominently or fleetingly are Georg von Frundsberg (Frondsberg), TRUCHSESS von Waldburg, Franz von Sickingen, and Ulrich von Hutten (qq.v.).

LICHTENSTEIN, ALFRED (Berlin, 1889–1914, Reims, killed in action), an early Expressionist writer, whose speciality was the grotesque. His poem *Die Dämmerung* was published in a journal in 1911. His collected works (*Gedichte und Geschichten*) were published posthumously in 2 vols. in 1919.

LICHTWER, MAGNUS GOTTFRIED (Wurzen nr. Leipzig, 1719–83, Halberstadt), studied at Wittenberg University and lectured there for two years (1747–9). From 1752 he held various ecclesiastical and legal appointments at Halberstadt. His verse fables, published as *Vier Bücher Äsopischer Fabeln* (1748), were widely read and retained their popularity into the 19th c. He is also the author of a didactic poem written in Alexandrines, *Das Recht der Vernunft* (1758).

LIDA, the name by which Goethe refers in his poetry to his close friend in Weimar, Charlotte von Stein (q.v.).

'Liebchen ade! scheiden tut weh!', first line of the folk-song *Abschied*. It is also found in the form 'Schätzchen ade!' The second and third stanzas are an early 19th-c. addition by Ottmar Schönhuth (q.v.).

Liebe in der Fremde, a poem by J. von Eichendorff (q.v.), the first line of which runs 'Über die beglänzten Gipfel'. It occurs (untitled) in the story *Das Marmorbild* (q.v., 1819).

Liebelei, a play (Schauspiel in drei Akten) by A. Schnitzler (q.v.), written in 1894, published in 1895, and first performed in the Burgtheater, Vienna, on 9 October 1895. Two young men-about-town, Fritz and Theodor, out to amuse themselves, have made the acquaintance of two girls from the humbler suburbs. For one of them, Christine, this means real love and devotion to Fritz, who up to a point responds to her sincere attachment. But he is already entangled in a relationship with a married woman. A light-hearted party of the four young people is interrupted by the aggrieved husband of Fritz's existing liaison, and a duel is arranged. Fritz takes leave of Christine without disclosing his dangerous situation. Between the second and third acts the duel has taken place, and Christine, learning that the man she has loved has died for another woman and is already buried, rushes out. We are left to assume that she takes her life.

LIEBERMANN, MAX (Berlin, 1847–1935, Berlin), a leading exponent of Impressionism whose working life, except for a spell in Munich (1878–84), was spent in Berlin where, in 1893, he founded the Berlin Sezession (see SEZESSIONEN). He is a particularly noted for his Dutch scenes (including *Judengasse in Amsterdam*, 1905). His early work is Naturalistic and his late style (from 1914) is represented in paintings of his garden (in Wannsee). His self-portrait dates from 1901. Liebermann's writings include *Degas* (1898) and *Die Phantasie in der Malerei* (1916); *Gesammelte Schriften* were published in 1922.

Liebesfrühling, a collection of more than 400 love poems by F. Rückert (q.v.), published in 1844. They are linked with his courtship of Luise Wiethaus, whom he married in 1821. The poems are grouped in five books, each of which is called a 'posy' (Strauß), headed respectively *Erwacht, Geschieden, Gemieden, Wiedergewonnen*, and *Verbunden*.

Liebesgeschichten und Heiratssachen, a three-act farce (Posse mit Gesang) by J. N. Nestroy (q.v.), written in 1843 and first performed at the Theater an der Wien in March 1843. Based on an obscure English original (by J. Poole), it was published in 1891. Three courtships run parallel through the play, each of which is complicated by difference of class or wealth.

Alfred, the son of Marchese Vincelli, conceals his identity and enters the house of Herr Fett, a wealthy retired butcher, because he loves Ulrike Holm, a relative of Fett. He is turned down by Fett as ineligible. Fett's daughter Fanny loves Bucher. When Bucher was well off and Fett poor, the latter had approved, but their situations are now reversed, and Fett will not hear of a marriage. The third affair involves Nebel, an enterprising valet, who passes himself off as a nobleman and wins the hand of Fett's sister-in-law, the old maid Lucia Distl. After a series of amusing confusions Alfred and Ulrike, and Bucher and Fanny are allowed to marry. Nebel is exposed and makes off unabashed.

Liebes-Protokoll, Das, a three-act comedy by E. von Bauernfeld (q.v.), first performed at the Burgtheater, Vienna, in August 1831. It was first published in *Lustspiele* (1847). It is an agreeable drawing-room piece, in which two pairs of lovers, Baron Fels and Rosalie, and Ritter von Bergheim and Adelaide, disentangle themselves from unwilling engagements with the wrong partners and are happily coupled in the way they desire. An important comic character is Adelaide's father, Herr Müller, a Jewish millionaire banker, who is ennobled in the last act.

LIEBIG, JUSTUS VON (Darmstadt, 1803–73, Munich), a distinguished German chemist, had a brilliant career at Bonn and Erlangen universities and was appointed a professor at Gießen University at the age of 21. He remained there until 1852, establishing the most successful school of chemical research in Europe at that time. In 1852 he became a professor at Munich. During his fruitful years at Gießen he discovered chloroform and chloral, and made important advances in the study of alimentary chemistry and metabolism. His patent of nobility was conferred in 1845 in recognition of his scientific distinction. Liebig's laboratory, which is preserved as a museum, survived the destruction of Gießen in 1945. The University of Gießen now bears his name (Justus-Liebig-Universität).

LIEBKNECHT, KARL (Leipzig, 1871–1919, Berlin), son of the veteran socialist Wilhelm Liebknecht (q.v.), qualified as a lawyer; he devoted himself to politics, and in 1912 became a Social Democrat (see SPD) deputy of the Reichstag. Imprisoned for treason (i.e. publicly opposing the 1914–18 War), he founded the Spartakus League (see SPARTAKUSBUND) on his release and sought unsuccessfully to establish a German Soviet Communist Republic. In an attempt to overthrow the new regime in January 1919 (Spartakusaufstand) he was arrested and shot, allegedly while attempting to escape. His numerous writings are propagandist. He and Rosa Luxemburg (q.v.) are figures in the novel *Karl und Rosa* (1950) by Alfred Döblin (q.v.).

LIEBKNECHT, WILHELM (Gießen, 1826–1900, Charlottenburg), an early German socialist, took part in the 1848 Revolution (see REVOLUTIONEN 1848–9), fled to Switzerland, from which he was expelled in 1850, and spent some time in London in close contact with Karl Marx (q.v.). In 1867 he became a Social Democrat (see SPD) deputy of the Diet of the North German Confederation (see NORDDEUTSCHER BUND), and in 1874 won a seat in the Reichstag. He was once imprisoned (1872–4) and several times expelled from various German states.

Liebknecht took a prominent part in the editing of *Vorwärts* (q.v.), the Social Democrat newspaper (1890–1900). In addition to propagandist works he wrote a *Geschichte der französischen Revolution* (1890). Karl Liebknecht (q.v.) was his son.

Liechtenstein, a remote Alpine principality (Fürstentum) with only 21,000 inhabitants in the upper Rhine valley between Switzerland and Austria. It is a possession of the Liechtenstein family. It was given its present status by the Emperor Karl VI (q.v.) in 1719. After the dissolution of the Holy Roman Empire it was included in the French-sponsored Confederation of the Rhine (1806–14, see RHEINBUND), and in 1815 became one of the member states of the German Confederation (see DEUTSCHER BUND). With the collapse of the Confederation after the War of 1866 (see DEUTSCHER KRIEG), Liechtenstein entered into tariff and other relationships with Austria-Hungary. When the Dual Monarchy disintegrated in 1918, the little state, long wisely ruled (1858–1929) by Johann II (b. 1840), prudently put itself under the wing of Switzerland, remaining nominally independent, but entering into monetary, postal, and economic union with the Swiss, who also represent its interests abroad.

Many of the princes of Liechtenstein held high office under the Habsburgs, and the family owned a baroque palace in the Herrengasse, Vienna, and a summer palace (Sommerpalais, 1701–12) in the Rossau, IX. Bezirk. This long housed what was probably the finest private collection of pictures in the world (Liechtensteinsche Gemäldesammlung). Though some of the pictures have been sold, the bulk of the collection is now in Vaduz, capital of Liechtenstein.

The medieval poet ULRICH von Liechtenstein (q.v.) was not connected with this region but belonged to a collateral branch of the family in Styria which became extinct in 1619.

Lied, (1) normally a poem intended to be sung or suitable in its form and content for singing. The Latin *Carmina* are medieval strophic songs, but the MHG word *liet* meant only a single verse or strophe, although the plural *diu liet* is commonly used for a group of verses.

Lieder are now usually understood as poems of two or more verses, generally of identical form. The commonest type uses the four-line verse with either alternating rhymes or rhyming of the second and fourth lines only. Lieder have been written by most German poets, and from the 16th c. onwards there are many folk-songs (Volkslieder) whose original authors are unknown. In the course of time these have often undergone considerable adaptation. Specialized types of Lied include the hymn (see KIRCHENLIED), soldiers' songs (Soldatenlieder), the closely related marching songs (Marschlieder), student songs (Studentenlieder), Wanderlieder, love songs (Liebeslieder), drinking songs (Trinklieder), etc. The most prolific period for song writing was the Romantic Age (see ROMANTIK), in which Novalis, C. Brentano, Eichendorff, Schenkendorf, Rückert, Kerner, Uhland, and Heine (qq.v.) are notably conspicuous. Good

song-writing is, however, already found in the 17th c. with F. von Spee, Tscherning, Fleming, Rist, Albert, Dach, Hofmann von Hofmannswaldau, and Abschatz (qq.v.). In the pre-Romantic period Goethe was a most accomplished song writer, as were also M. Claudius, G. A. Bürger, Hölty, and Schiller (*An die Freude*) (qq.v.). After the Romantics the most prominent names are Mörike, A. von Droste-Hülshoff, Th. Storm, and D. von Liliencron (qq.v.). An enormous amount of inferior poetry, some of it political, has also appeared in song form. For Meisterlied see MEISTERGESANG.

(2) A special form is the musical Lied of the 19th c. and early 20th c. Mozart wrote the first examples, and Beethoven, in addition to many individual songs, composed the first song cycle (Liederzyklus) in *An die ferne Geliebte* (q.v.), but the real creator of the musical Lied was F. Schubert (q.v.), who used melody, modulation, and significant accompaniment to draw out the full meaning of poems in a remarkably expressive way. Schubert wrote more than 600 Lieder, including two song cycles (*Die schöne Müllerin* and *Die Winterreise*, qq.v.), and tended in his later work to diverge from strophic musical repetition. R. Schumann and Brahms (qq.v.) developed the forms which Schubert had originated, but H. Wolf (q.v.) broke away almost completely from strophic form, shaping each Lied as an entity of words and music, voice and accompaniment. The last master of this form is R. Strauss (q.v.).

(3) The word Lied has also not infrequently been used to denote a large-scale work both in poetry (e.g. Schiller's *Lied von der Glocke*, q.v.) and in vocal music (e.g. Mahler's *Lied von der Erde*, q.v.). The ballads of C. Loewe (q.v.) are also sometimes styled Lieder.

Liederkreis, Op. 39, a cycle of 12 poems by Eichendorff (q.v.) set to music with piano accompaniment by R. Schumann (q.v.) in 1840. They are *In der Fremde, Intermezzo, Waldgespräch* (q.v.), *Die Stille* (q.v.), *Mondnacht* (q.v.), *Schöne Fremde* (q.v.), *Auf einer Burg, In der Fremde, Wehmut, Zwielicht* (q.v.), *Im Walde*, and *Frühlingsnacht*. The two poems entitled *In der Fremde* (q.v.) begin respectively with the first lines 'Aus der Heimat hinter den Blitzen rot' and 'Ich hör die Bächlein rauschen im Walde her und hin'.

Liederliche Kleeblatt, Das, alternative title of Nestroy's comedy *Der böse Geist Lumpazivagabundus* (q.v.). It refers to the trio of disreputable journeymen, Knieriem, Zwirn, and Leim.

Lied vom Winde, a poem written by E. Mörike (q.v.) in 1828. Included in the novel *Maler*

Nolten (q.v.), it has been set to music by Hugo Wolf (q.v.).

Lied von Bernadette, Das, a novel by F. Werfel (q.v.), published in 1941 and largely based on documents. It was an immediate world-wide success, except in those countries dominated by the National Socialists, where it was banned. The work was undertaken in fulfilment of a vow, which Werfel made at Lourdes in 1940 while uncertain whether he and his wife would be able to make good their escape from France into safety in Spain.

The story of Bernadette's visions of the Lady begins in 1858, when she is 14. The Lady, in answer to Bernadette's unspoken question, declares herself to be the Immaculate Conception, words which have for Bernadette no meaning. Bernadette, who uncovers a hitherto unknown spring of water, is beset by disbelievers, persecuted by the civil authorities, and treated with cautious scepticism by the Church. She never, however, diverges in any point from her original declarations. She is received into a convent at Nevers, where she dies in 1879. On her death-bed the vision reappears. The book closes with her canonization at Rome in 1933.

Lied von der Erde, Das, a work for two solo voices and orchestra composed by Gustav Mahler (q.v.) between 1907 and 1911, and first performed in 1911 after Mahler's death. It is based on six poems drawn from *Die chinesische Flöte* by Hans Bethge (q.v.).

Lied von der Glocke, Das, a poem completed by Schiller in 1799, and first published in the *Musenalmanach* for 1800. It is constructed with a frame (see RAHMEN), the basis of which is the forging of the bell to which the bell-founder provides the commentary. Between its various stages are interpolated passages dealing with the course of human life, which the completed bell will accompany, including childhood, courtship, marriage, early death, and catastrophe by fire and revolution, so that a kind of counterpoint ensues between the two elements of the poem. The underlying ethic is humanitarian, and is symbolized by the name given to the bell, Concordia.

LIENHARD, FRIEDRICH (Rothbach, Alsace, 1865–1929, Weimar), a private tutor and journalist, was from 1920 to 1929 editor of the literary periodical *Der Türmer* (1898–1943). After Naturalistic beginnings (e.g. the drama *Weltrevolution,* 1889) he turned to a combination of neo-Romanticism and regional literature, which he defended in the essay *Die Vorherrschaft*

Berlins (1900). He wrote some romantic historical plays, notably his *Wartburg-Trilogie* (1903–6), composed of *Heinrich von Ofterdingen, Die heilige Elisabeth,* and *Luther auf der Wartburg.* His novels *Oberlin. Roman aus der Revolutionszeit im Elsaß* (1910), and *Der Spielmann* (1913), in which he, an Alsatian, is critical of German nationalism, deserve mention.

Lienhard und Gertrud. Ein Buch für das Volk, a novel by J. H. Pestalozzi (q.v.), published in 1781. Openly didactic in intention, it shows the household of Gertrud endangered by the intemperance of her husband Lienhard, a skilled mason who, liking his drink too much, has got into the debt of the innkeeper and estate steward Hummel. Hummel's rule has been oppressive to all and Gertrud approaches Arner, the land owner, unmasking the steward; he is punished and old wrongs righted. Written (partly in dialogue) with the utmost simplicity and directness, the book achieved a considerable success. Even before it was published, Pestalozzi conceived the idea of a second part, which appeared in 1783. In 1785 he added a third, and in 1787 a fourth. These three additions, though they retain the names of Gertrud and Arner, are primarily a statement of Pestalozzi's educational and social views, developing the patriarchal relationship and organization which is to abolish want. In the final version the original story has become Pt. 1 of a didactic work which progressively loses shape, though it retains sociological interest.

Lila, a Singspiel (q.v.) written by Goethe in 1776 and published in *Goethes Schriften,* 1790. Lila suffers from pathological melancholia provoked by false news of the death of her husband. She is cured by the performance by her friends of a kind of play, which enters into her fantasies instead of contradicting them. *Lila* was performed at Weimar on 30 January 1777, the birthday of the Duchess Luise, and it has been suggested that Goethe sought in the play to alleviate the Duchess's melancholy and enclosed state of mind.

Lilie, Die, a Middle High German mystical flower allegory, written by an unknown priest of Cologne in the second half of the 13th c. It is based on a Latin model and is written in rhythmic prose tending to verse. (See REDE VON DEN FÜNFZEHN GRADEN.)

LILIENCRON, DETLEV VON, the name under which Friedrich, Freiherr von Liliencron (Kiel, 1844–1909, Rahlstedt nr. Hamburg) published his writings. The son of a customs inspector of baronial descent, Liliencron had his schooling in

Kiel and, briefly, in Erfurt. He entered the Prussian army in 1863, training in Mainz with the 37th (Westphalian) Fusiliers, in which he was commissioned in the same year. He was by temperament a keen soldier and proved himself both brave and humane on active service. His regiment was kept on garrison duty in the Polish provinces of Prussia during the war of 1864, but Liliencron served actively in the war of 1866 as a company officer, seeing mounted action when detached as a liaison officer, and was wounded at Nachod in Bohemia. In the war of 1870 he served with the 81st (1st Hessian) Infantry Regiment and saw fighting at Metz and, during the winter of 1870–1, in northern France. He was wounded at St. Remy in October, but returned to duty after a short absence. Several of his poems incorporate, openly or under a thin disguise, experiences from his two campaigns (see DEUTSCHER KRIEG and DEUTSCH-FRANZÖSISCHER KRIEG).

Liliencron, with no private means, was not able to make ends meet as an officer, and his insolvency obliged him in 1875, while serving with the 54th (7th Pomeranian) Infantry Regiment, to resign his commission. He never fully recovered from this enforced abandonment of a career to which he was devoted. He emigrated to America, but failed to make money or to settle. He returned to Germany in 1877, and became in 1878 a probationary civil servant. In 1882 he was appointed governor (Landesvogt) of the island Pellworm (a minor post) and in 1884 was moved to the parish of Kellinghusen in Holstein. His inability to meet his debts led him to resign this post also (1887). He next lived at Ottensen near Hamburg, attempting, with little success, to make a living by writing. Official recognition with a royal pension came in 1901. He spent his last years in Rahlstedt, close to Hamburg. R. Dehmel and G. Falke (qq.v.) were among his friends. He was married three times.

Liliencron's first published work, *Adjutantenritte* (q.v.), a collection of war poetry and prose, appeared in 1883. He next turned to drama, writing five historical plays in verse (*Knut der Herr*, 1885; *Die Rantzow und die Pogwisch*, 1886; *Der Trifels und Palermo*, 1886; *Pokahontas*, not published until 1904; and *Die Merowinger*, 1888). *Arbeit adelt* (1887) is a short play dealing with a German officer who emigrates to America, and *Sturmflut* (published posth.) treats the disastrous coastal flood of 1821. None of these derivative plays enjoyed success. Liliencron wrote to live, and his output consisted largely of short works, stories, and Novellen, of which the following are the principal collections: *Eine Sommerschlacht* (1886), *Unter flatternden Fahnen* (1888), *Der Mäcen* (1889), *Krieg und*

Frieden (1891), *Kriegsnovellen* (1895), *Könige und Bauern* (1900), *Roggen und Weizen* (1900), *Aus Marsch und Geest* (1901), *Letzte Ernte* (1909). The distinctions are not always clear, and Liliencron himself classified the title-story of *Der Mäcen* (there are two others in the collection) as a novel. His novels *Breide Hummelsbüttel* (1887) and *Mit dem linken Ellenbogen* (1899) mingle realistic elements with sensational and melodramatic features. *Leben und Lüge* (1908, sub-titled *Biographischer Roman*), draws heavily upon the experiences of his own life. His *Poggfred, Kunterbuntes Epos in neunundzwanzig Kantussen* (1908, originally 12 cantos, 1896) is a shapeless panorama of his own world, sometimes in *ottava rima*, sometimes in *terza rima*. It perhaps descends from *Childe Harold*, and is impregnated with humour and irony.

Liliencron's greatest achievement is in his shorter poetry, ballads and especially lyrics. Apart from *Adjutantenritte*, they appeared in *Gedichte* (1889), *Der Haidegänger* (1890), *Neue Gedichte* (1893; reissued as *Nebel und Sonne*, 1900), *Bunte Beute* (1903), and *Gute Nacht* (1909). The collected poems were published in 3 vols. in *Gesammelte Gedichte* (1897–1900). This poetry is uneven, slipping easily into the commonplace, especially when reflection is attempted; but at his best Liliencron has a vigorous freshness of expression, which, disregarding convention, can boldly capture a fleeting moment, riveting the attention on vividly observed or glimpsed detail. The energy and uninhibited forcefulness of his language commended him to the Naturalists. Particularly notable poems are *Die Musik kommt*, *Krieg und Friede*, *Viererzug*, and *Zwei Meilen Trab* (qq.v.). Liliencron's *Gesammelte Werke* (10 vols., including letters) were published by R. Dehmel, 1910–12.

Lilis Park, a humorous poem by Goethe, satirizing the environment of his fiancée, Lili Schönemann (q.v.), in terms of a menagerie, in which he himself is the bear. It was written in 1775, and first published in 1789. See also NEUE LIEBE, NEUES LEBEN and AN BELINDEN.

Lilli Marleen, a sentimental poem, written by H. Leip (q.v.) in 1915 and, with a tune by N. Schultze (1911–), revived a generation later, when it became the German soldier's favourite song in the 1939–45 War. It was recorded and sung repeatedly on radio by Lale Andersen (1913–72). Its fame even spread to other armies. The National Socialist authorities, who reprobated its unsoldierly qualities, were helpless against its irresistible popularity. Goebbels, however, forbade Lale Andersen to sing it herself after she had boxed the ears of one of his assistants. Lale Andersen's real name was Bunterberg, and after marriage Wilke. Her autobiography, *Der Himmel hat viele Farben*, appeared in the year of her death.

Limburger Chronik, common title of *Fasti Limpurgenses*, a chronicle of the city of Limburg from 1336 to 1398 compiled by the town clerk Tilemann ELHEN von Wolfhagen (q.v.). It is based on first-hand knowledge, and contains much material of value to social historians. It also quotes a number of poems (Minnelieder and folk-songs) either complete or in part. The *Limburger Chronik* was continued after Elhen's death by other hands, and in 1617 was printed as *Fasti Limpurgenses*. A perusal of this chronicle led C. Brentano to write *Aus der Chronika eines fahrenden Schülers* (q.v., 1818).

LINCK, HIERONYMUS, of Nürnberg, is the author of Meisterlieder and of a play, *Ritter Julianus* (1564), portraying a parricide.

LINDAU, PAUL (Magdeburg, 1839–1919, Berlin), son of a Lutheran pastor, was a journalist who acquired a reputation for daring and frivolous wit. He later founded and edited *Die Gegenwart* (1872–81) and was also editor of *Nord und Süd* (q.v.). In 1895 he became director of the Meiningen theatre, afterwards successively directing the Berliner Theater (1899) and, briefly, the Deutsches Theater, Berlin (1904). Lindau is best known as a purveyor of entertaining lightweight comedies and garish dramas for the German stage (*Theater*, 4 vols., 1873–81), but he also wrote a number of social novels (*Der Zug nach Westen*, 2 vols., 1886; *Arme Mädchen*, 2 vols., 1888; *Die Gehilfin*, 1894; *Die blaue Laterne*, 2 vols., 1907). His collected fiction appeared as *Illustrierte Romane und Novellen* (10 vols., 1909–12). An autobiography, *Nur Erinnerungen*, was published in 1916.

LINDE, OTTO ZUR (Essen, 1873–1938, Berlin), belonged to the circle contributing to *Charon* (q.v.) and wrote mythical poetry influenced by Nietzsche (*Gedichte, Märchen, Skizzen*, 1901; *Die Kugel, eine Philosophie in Versen*, 1909).

LINDEMAYR, MAURUS (Neukirchen, Upper Austria, 1723–83, Neukirchen), was a pioneer of dialect literature in Austria. His dialect comedies were not published until 1875, when they appeared with the title *Maurus Lindemayrs sämtliche Dichtungen in obderennsischer Volksmundart* (Ob der Enns is the old name for Oberösterreich).

LINDENER, MICHAEL (Leipzig, c. 1520–62, Friedberg), a strolling scholar, who was at various times a schoolmaster in Nürnberg, Ulm, and Augsburg, was executed for murder. He is the author of two collections of crude and often obscene Schwänke (see SCHWANK), *Katzipori* and *Rastbüchlein* (both 1558).

LINGG, HERMANN (Lindau, 1820–1905, Munich), became a medical officer in the Bavarian army in 1846, resigned his commission in 1851 and settled in Munich, where he devoted himself to writing poetry. He was a friend of E. Geibel (q.v.), and was one of the Munich school of poets (see MÜNCHNER DICHTERKREIS). In 1854 he was given a pension by King Maximilian II (q.v.). His first volume of poems, *Gedichte*, appeared in 1854, and he later published five other volumes (*Vaterländische Balladen und Gesänge*, 1868; *Zeitgedichte*, 1870; *Schlußsteine*, 1878; *Jahresringe*, 1889; *Schlußrhythmen*, 1901). Lingg also wrote two epics (*Die Völkerwanderung*, 1865–8 and *Dunkle Gewalten*, 1872), some verse plays, of which *Catilina* (1864) and *Der Doge Candiano* (1873) may be mentioned, and stories (*Byzantinische Novellen*, 1881). He had a predilection for historical subjects, which he handled with vigour. His autobiography (*Meine Lebensreise*) appeared in 1899.

Linzer Antichrist, an Early Middle High German eschatological poem in three sections with Latin headings: I. *De anticristo*; II. *De signis XV dierum ante diem judicii*; III. *De adventu christi ad judicium*. It is pervaded by the spirit of scholasticism. The original was probably Swiss, written c. 1170. The MS. is preserved in Linz (Austria).

Lippe, or **Lippe-Detmold,** one of the tiniest states of Germany until 1945; it is now a part of Niedersachsen. Before 1919 it was a principality (Fürstentum), and from 1919 to 1945 a Land. Its capital was Detmold (q.v.).

LISCOW, CHRISTIAN LUDWIG (Wittenburg, Mecklenburg, 1701–60, Eilenburg, Saxony), a lively and reckless satirist, studied at Rostock, Jena, and Halle, and in 1736 entered the service of the exiled Duke Karl Leopold of Mecklenburg, by whom, however, he was soon dismissed. In 1740 he was secretary to Graf Danckelmann, a post which also soon came to an end through intrigue. A third appointment, in 1749, as secretary to Graf Brühl (q.v.) in Dresden, was ended by Liscow's imprisonment for rash comment on his employer's policy. On his release in 1750 he joined his wife in Saxony.

Liscow's satirical writings belong to the earlier part of his life. They include academic satires from his student days and *Von der Vortrefflichkeit und Notwendigkeit der elenden Scribenten* (1734). These were collected in *Sammlung satyrischer und ernsthafter Schriften* (1739). Liscow, who has been compared with Lessing (q.v.), has a biting wit but little of Lessing's honesty of purpose. Liscow's *Sämtliche satirische Schriften* (3 vols., ed. K. Müchler) were published in 1806. A selection was published by A. Holder in 1901.

LISSAUER, ERNST (Berlin, 1882–1937, Vienna), a Jewish writer of patriotic German poetry, is remembered for his *Haßgesang gegen England*, composed during the 1914–18 War, a poem which he afterwards regretted. He was also the author of *Luther und Thomas Münzer* (1929), a play dealing with the conflict between these two men (qq.v.) on the eve of the Peasants' War (see BAUERNKRIEG).

LIST, FRIEDRICH (Reutlingen, 1789–1846, Kufstein), an economist who, though self-taught, became one of the greatest authorities of the 19th c. on political economy (Nationalökonomie).

A man of strong liberal and democratic convictions, he was an early advocate of a protective customs system within the German Confederation. This led in 1820 to his dismissal from a chair at Tübingen University, to which he had been appointed only three years previously. He was still less fortunate in his native Württemberg, where he was twice sentenced to a period of detention (Festungshaft) and secured his release from the fortress Hohenasperg only by the undertaking to emigrate to America. He stayed there for eight years (1824–32), devoting himself to farming. He continued to concern himself with economics and returned to Leipzig as North American consul. He was denied an appropriate position by the Saxon and other governments, but his advocacy of a German railway system led to the development of the Saxon railway (the Leipzig–Dresden line); similarly, his views on a protectionist policy influenced the Customs Union (see ZOLLVEREIN). In 1834 List became a co-founder of the first *Staatslexikon*, and in 1843, after a spell in Paris (from 1837), he founded the *Zollvereinsblatt*. His writings included the proposal for an alliance between Great Britain and Germany .In 1841 appeared his principal work, *Das nationale System der politischen Ökonomie* (unfinished). List took his own life. *Schriften* including speeches and correspondence (12 vols.), ed. E. von Beckerath, appeared 1927–36 under the auspices of the Friedrich-List-Gesellschaft founded in 1925, and refounded in 1954.

List is the subject of the novel *Ein Deutscher ohne Deutschland* (1931) and the play *Friedrich List* (1932) by Walter von Molo (q.v.).

LISZT, FRANZ (Raiding, formerly in Hungary, now in Burgenland, Austria, 1811–86, Bayreuth), perhaps the greatest piano virtuoso of the 19th c., and a notable composer.

Liszt (the true form of the name is said to be List) gave recitals in all countries of Europe. He lived in Paris 1823–35, and for some years in Italy with his mistress, the Comtesse d'Agoult, who bore him three daughters, of whom Cosima became R. Wagner's wife. In 1842 he moved to Weimar as director of music to the Court. In 1861 Liszt moved to Rome and took minor orders, and so becoming the Abbé Liszt of society in the 1870s and 1880s. He gave up recitals, devoting himself to composition and teaching.

Liszt's compositions are mostly literary or pictorial in origin, and the majority are linked with Italy or Switzerland (e.g. *Symphonie zu Dantes göttlicher Komödie*, 1855, *Années de Pèlerinage* for solo piano). He originated the symphonic poem, composing, among other examples, *Les Préludes* (1848), *Tasso, lamento e trionfo* (1849), *Orpheus* (1854), *Bergsymphonie* (1854, known outside Germany as *Ce qu'on entend sur la montagne*), *Prometheus* (1855), *Mazeppa* (1856), and *Die Hunnenschlacht* (1857). Liszt's *Faust-Symphonie* (1855) has three movements representing Faust, Gretchen, and Mephistopheles, and the whole is concluded by a choral setting of the last eight lines of Goethe's *Faust*, Pt. II. His best-known choral work is the *Graner Festmesse* (1855). He furthered modern music, notably that of his son-in-law R. Wagner (q.v.). He took no fees for teaching, being content to foster promising talent. His writings (in French, subsequently translated into German) include notable works on Berlioz, Chopin, and gipsy music.

Gesammelte Schriften (6 vols., ed. La Mara and L. Ramann), appeared 1880–3, and *Gesamtausgabe* of his musical works (34 vols., ed. F. Busoni *et al.*), 1907–36. Both are incomplete editions; they were reissued 1966. *Briefe* (8 vols., ed. La Mara) appeared 1893–1905.

Litauische Geschichten, a collection of stories by E. Wichert (q.v.), published in 1881.

Litauische Geschichten, a collection of four stories by H. Sudermann (q.v.), published in 1917. They are reckoned by some to be his best work, and are for the most part ironical narratives set in the countryside in which Sudermann grew up.

Die Reise nach Tilsit tells of a husband's plan to murder his wife, from which he recoils at the last moment, losing his life in saving hers. *Miks Bumbullis* is a tale of a murder at last discovered. *Jons und Erdine* recounts the endeavour of two smallholders to give their daughters a better life, but the girls show scarcely more gratitude than do Goneril and Regan. *Die Magd* deals with seduction; the young farmer's maid of the title tries to drown herself, but is rescued and sees a happier future in marriage with a man she cares for.

Literarischer Sansculottismus, an essay by Goethe published in No. 5 of Schiller's periodical *Die Horen* (q.v.) in 1795. In it he attacks presumptuous, ill-mannered, and unjustified criticism, to which he finds many of his German contemporaries too readily addicted.

Literaturbriefe, commonly used abbreviated title for *Briefe die neueste Litteratur betreffend,* a literary weekly which appeared from 1759 to 1765. The first editor was G. E. Lessing (q.v.), who contributed 55 letters between the beginning of 1759 and the autumn of 1760. He was assisted principally by M. Mendelssohn and rather less by F. Nicolai (qq.v.). The identity of the editor was successfully concealed, and contributions were signed only with initials or letter groups. Lessing's letters are recognized by one of the following ciphers, A., E., G., L., O., or Fll. After Lessing's withdrawal in 1760, Nicolai acted as editor; Thomas Abbt (q.v.) became a contributor and was joined at the end by Herder (q.v.).

The form of these vivid and pungent commentaries on contemporary literature was determined by the fiction that they were intended to keep an officer on active service informed of the progress of literature. The letters written by Lessing are noteworthy for their severe criticism of mediocrity and their stringent insistence on competence and common sense. Notable letters concern Wieland (7–13 and 63, 64) and Klopstock (19, 51, and 111), whom Lessing praises with discretion. The most famous letter of all is the seventeenth, in which Lessing annihilatingly criticizes Gottsched, rejects French taste in the drama, and praises Shakespeare.

Livländische Reimchronik, title used to designate a Middle High German chronicle of the military feats of the Teutonic Order (see DEUTSCHER ORDEN) in Livonia. It comprises nearly 13,000 lines and was written between 1291 and 1297, probably by a knight of the Order.

Lob der Torheit, German title of the satire *Moriae encomium* (also *Laus stultitiae*) of Erasmus (q.v.).

'Lobe den Herren, den mächtigen König der Ehren', first line of a Protestant hymn written by J. Neander (q.v.) in 1679. The tune is older.

Lobgesang auf den Heiligen Gallus, a hymn in praise of St. Gall, originally written *c.* 880 in Old High German rhyming verse by Ratbert, a prominent monk of St. Gall Abbey. The German original is lost, but the poem is known through a Latin translation made in the first half of the 11th c. by Ekkehard (q.v.).

Löblicher Hirten- und Blumenorden an der Pegnitz, see HIRTEN- UND BLUMENORDEN AN DER PEGNITZ, LÖBLICHER.

Lobositz in Bohemia, battle of (1 October 1756), an indecisive battle opening the Seven Years War (see SIEBENJÄHRIGER KRIEG). Lobositz lies on the Elbe some sixty miles north of Prague. *Die Schlacht bei Lobositz* is the title of a comedy (1956) by P. Hacks (q.v.).

Lob Salomons, an Early Middle High German poem included in the Vorauer Handschrift (q.v.). It recounts how God gave to Solomon wisdom, power, and wealth, and how Solomon was visited by the Queen of Sheba. The story is given an allegorical Christian interpretation. An interesting feature of the work is an interpolated middle section based on Jewish legend.

LOBSIEN, WILHELM (Foldingbro, Denmark, formerly in Prussia, 1872–1947, Niebüll), a schoolmaster, first in Hoyer (since 1919 Danish), then in Kiel, was a prolific writer of poems, stories, and novels, almost invariably associated with the North Sea coast of Schleswig, especially the great expanse of tidal flats known as 'das Wattenmeer'. He is the only German writer to have made the unique off-shore world of the Halligen (see HALLIG) the subject of several works.

Lobsien began with volumes of verse, *Strandblumen* (1894), *Ich liebe dich* (1902), *Dünung* (1904), and *Blau blüht ein Blümelein* (1905). He is the author of collections of Novellen, *Wellen und Winde* (1908), *Friesenblut* (1925), and *Wind und Woge* (1947), and also of collections of short narratives (*Uthörn,* 12 stories, 1940; *Ekke Nekkepen,* fairy-tales, 1918; *Sterne überm Meer,* Christmas tales, 1938; and *Koog und Kogge,* 21 short stories, posth., 1950). Separately published stories include *Hinterm Seedeich* (1907), *Trutz, blanker Hans* (1912), *Heilige Not* (1914), *Letzte Fahrt* (1923), *Karsten Deichfahrer* (1925), *Klaus Störtebeker* (1926) and *Jürgen Wullenweber* (1930, both historical), *Gesa Früddens Weg* (1932, one of his best for atmosphere and character), and *Halligleute* (1935).

Lobsien's best novel is *Der Halligpastor* (1914). Others which should be mentioned are *Pidder Lyng* (1910), *Wattenstürme* (1910), *Ebba Enevolds*

Liebe (1919), *Landunter* (1921), *Der Pilger im Nebel* (1922), *Der Heimkehrer* (1941), and *Segnende Erde* (1942).

As a practitioner of Heimatkunst (q.v.), Lobsien follows closely in the wake of G. Frenssen (q.v.); though less well known than the latter, he had a considerable reading public in Schleswig-Holstein.

LOBWASSER, AMBROSIUS (Schneeberg, 1515–85, Königsberg), professor of law in Königsberg (1563), travelled in his earlier years, and in France became acquainted with Huguenot psalm-singing. He translated the Psalms and published them with French melodies in 1573 as *Der Psalter des Königlichen Propheten Davids*, a version which continued in use well into the 18th c.

LOCHER, JAKOB (Ehingen, 1471–1528, Ingolstadt), a humanist, wandered from university to university in search of knowledge, studying under Sebastian Brant (q.v.) at Basel, and under Celtis (q.v.) at Ingolstadt, afterwards visiting Italy. He taught at Freiburg and Ingolstadt, where he became a professor in 1506. He strongly supported the New Learning against the adherents of scholasticism, and became involved in bitter and scurrilous controversy with Wimpfeling (q.v.).

Locher is the author of two patriotically Turcophobe classical plays with chorus. His free Latin translation of Sebastian Brant's *Das Narrenschiff* (q.v.), *Stultifera navis* (1497), had a success even greater than that of the original.

LOEBEN, OTTO HEINRICH, GRAF VON (Dresden, 1786–1825, Dresden), the son of a minister in the Saxon Electoral government, was a student at Heidelberg, where he became a friend of J. von Eichendorff, L. J. von Arnim, C. Brentano, and J. J. von Görres (qq.v.), with whose Romantic ideas he sympathized. He wrote at this time a novel, *Guido* (1808), under the pseudonym Isidorus Orientalis, and also poems (*Blätter aus dem Reisebüchlein eines andächtigen Pilgers*, 1808; *Gedichte*, 1810). He served as an officer in the 1814 campaign of the Wars of Liberation (see NAPOLEONIC WARS), afterwards living in Dresden. His later works include the novel *Arkadien* (1811–12), a volume of poems (*Der Schwan*, 1816), and *Ritterehre und Minnedienst, alte romantische Geschichten* (1819).

LOEN (also **LOÖN**), JOHANN MICHAEL, FREIHERR VON (Frankfurt/Main, 1694–1776, Frankfurt), studied letters and law at Marburg and Halle universities, travelled, and then spent some years at home. In 1752 he became governor of Lingen and Tecklenburg. His writings include

the political novel *Der redliche Mann am Hofe oder die Begebenheiten des Grafen von Rivera* (1740), in which the misgovernment of a despotic state is rectified by a virtuous minister, and a two-volume work entitled *Die einzig wahre Religion* (1750), which expresses a plea for tolerance.

Loen is mentioned in *Dichtung und Wahrheit* (q.v., Bks. 1 and 2) by Goethe (his great-nephew), who erroneously states that he was not born in Frankfurt and quotes the titles of his works in slightly different form. Loen is sometimes regarded as the original of the Oheim, whose mansion Wilhelm visits in *Wilhelm Meisters Wanderjahre* (q.v.).

LOERKE, OSKAR (Jungen, West Prussia, now Poland, 1884–1941, Berlin), came of a farming family, was at school at Graudenz and a student at Berlin University. He became drama reader and adviser to a publisher (from 1917 with S. Fischer-Verlag), and was from 1927 to 1933 secretary to the literary section of the Prussian Academy of Arts (see AKADEMIEN).

Loerke had a deep feeling for nature, which characterizes, usually in a melancholy mode, most of his poetry. It is collected in the volumes *Wanderschaft* (1911), *Gedichte* (1916, reissued 1929 as *Pansmusik*), *Die heimliche Stadt* (1921), *Der längste Tag* (1926), *Atem der Erde* (1930), and *Der Silberdistelwald* (1934). Latterly he became more conscious of the political threat of evil masquerading as good, and, though he retained his post as a silent dissenter (but not his secretaryship of the Academy), he was deeply depressed by what he termed in his diaries 'das Unheil'. Loerke both influenced and encouraged poets, notably W. Lehmann and E. Langgässer (qq.v.).

H. Kasack (q.v.) edited a number of Loerke's writings, including *Tagebücher 1903–1939* (1955). *Literarische Aufsätze aus der Neuen Rundschau, 1909–1941*, ed. R. Tgahrt, appeared in 1967, and *Gedichte und Prosa*, ed. P. Suhrkamp, in 1958.

LOEWE, CARL (Löbejün nr. Halle, 1796–1869, Kiel), composer, spent the greater part of his life (1820–66) in Stettin, where he taught music and was director of the city's musical institutions. Loewe was primarily a composer of vocal music, and his output includes some oratorios, but he is chiefly remembered for his arresting ballads for solo voice and piano accompaniment. He composed more than 400 songs, mostly ballads, and many poets are represented in his œuvre. The following are among his best-known productions: *Prinz Eugen* (Freiligrath, q.v.), *Der Nöck* (Kopisch, q.v.), *Heinrich der Vogler* (Vogl, q.v.), *Der Wirtin Töchterlein*, *Des Goldschmieds Töchterlein*, and *Graf Eberstein* (Uhland, q.v.), *Die nächtliche Heerschau* (Zedlitz, q.v.), *Archibald Douglas* (Fontane, q.v.), *Erlkönig* and *Der Fischer* (Goethe, q.v.), *Süßes Begräbnis* (Rückert, q.v.), *Die Uhr* (J. G. Seidl, q.v.), *Der Graf von Habsburg* (Schiller, q.v.), *Fridericus Rex* (traditional), *Edward* (folk-song translated by Herder, q.v.), and *Tom der Reimer* (Scottish ballad).

LOEWENSTERN, MATTHÄUS APELLES VON (Neustadt, Silesia, 1594–1648, Breslau), was the author of hymns, including the well-known 'Nun preiset alle Gottes Barmherzigkeit'. He was in the service of the Emperor Ferdinand II (q.v.).

LÖFFELHOLZ, FRANZ, see MON, F.

LOGAU, FRIEDRICH, FREIHERR VON (Brockuth, Silesia, 1604–55, Liegnitz), an epigrammatist of noble descent, was orphaned early. He was educated at the grammar school in Brieg, and was a page to the Duchess of Brieg. From 1644 to 1653 he was in the ducal administration, finally moving to Liegnitz. Logau is not known to have written anything other than epigrams.

The first collection appeared in 1638 as *Zwei Hundert Teutscher Reimensprüche*, and a much larger edition, which included more polished versions of items from the earlier book, followed in 1654 as *Salomon von Golaw Deutscher Sinn-Getichte Drey Tausend*. An epigram by Logau is used by G. Keller as the starting-point of *Das Sinngedicht* (q.v.). Logau's epigrams of social and political satire, embodying a comprehensive criticism of his time, are mostly set in one or two couplets of Alexandrines. His style is unusually unostentatious and direct for the age in which he lived. After his death Logau's work was forgotten until it was revived by G. E. Lessing and K. W. Ramler (qq.v.) in *Friedrichs von Logau Sinngedichte* (1759). Logau's edition of 1654 was reissued as *Sinngedichte*, ed. E.-P. Wieckenberg, in 1984.

Lohengrin, a Middle High German epic poem of unknown authorship, written probably *c.* 1280. At one point, however, an acrostic gives the name Nouhusius, which has been interpreted as Neuhaus or Neuhäuser, a poet not otherwise recorded. Neuhaus is believed to have been a *ministeriale* in the service of the Duke of Bavaria. *Lohengrin* consists of 767 ten-line stanzas. Of these the first 67 are borrowed, in part from the *Wartburgkrieg* (q.v.) and in part from an earlier Lohengrin poem.

The story is that of the hard-pressed Duchess of Brabant, whose rights are maintained by the Knight of the Swan, whom she then marries.

When, however, she later transgresses an injunction not to ask his name, Lohengrin departs. The story is plentifully interspersed with descriptions of chivalric and courtly life, battles, banquets, tourneys, and hunts, and in the end devotes itself to a history of the reigns of Heinrich I and Heinrich II (qq.v.). The author appears to have been more interested in politics and the social life of the nobility than in the potentially moving fates of his principal figures.

The source of *Lohengrin* is believed to have been a lost poem dealing with a character called Lorengel, of which a 16th-c. adaptation in the form of a Meisterlied (see MEISTERGESANG) exists. A substantial allusion to Lohengrin (as Loherangrin) occurs in the closing stages of Wolfram von Eschenbach's *Parzival* (q.v.).

Lohengrin, a romantic opera (Romantische Oper) by R. Wagner (q.v.), written between 1845 and 1848. The text is in rhyming verse. It was first performed in Weimar on 28 August (Goethe's birthday) 1850. The production was superintended by F. Liszt (q.v.) during Wagner's absence in exile. Wagner adapted the medieval legend of *Lohengrin* (q.v.).

Elsa of Brabant is accused of murder and of unchaste living by Telramund in the presence of König Heinrich der Vogler (q.v.). The mysterious knight of the Swan appears to defend her cause. But first he bids her never ask his name. The knight defeats Telramund, and he and Elsa are united. Telramund and his spouse, Ortrud, seek to provoke Elsa into asking the fatal question, but she resists their efforts. In the end her own unrest and curiosity, however, prove too powerful, and she demands the knight's name. He reveals himself as Lohengrin, and the swan as Elsa's brother, supposed murdered. He sadly departs, and his loss kills Elsa.

LOHENSTEIN, DANIEL CASPER VON (Nimptsch, Silesia, 1635–83, Breslau), dramatist and novelist, had as his surname Casper, adopting Lohenstein when he was ennobled in 1670, but the tradition of referring to him as Lohenstein is firmly entrenched. The son of a tax-collector, he was educated in Breslau, Leipzig, and Tübingen, travelled abroad, then began practice as a lawyer in Breslau, and later was a senior officer (Syndikus) of the city. In 1657 he made a wealthy marriage.

Lohenstein was a busy administrator, whose writing was a part-time activity. Of his six tragedies, two have oriental themes and the remainder draw on ancient history. *Ibrahim Bassa,* believed to have been written in 1650, when he was only 15, deals with the downfall of the hero, whose death takes place unnecessarily, since his master Soliman repents of the order for execution only to find that it has been carried out. In his *Cleopatra* (1656, published 1661), the heroine betrays Antonius, then finds herself betrayed; she kills herself to escape humiliating captivity. *Agrippina* and *Epicharis*, both published in 1665, revolve around Nero in a tangle of conspiracy and passion, at the end of which both heroines die with stoical steadfastness. *Epicharis*, in particular, has scenes of torture displayed upon the stage. *Sophonisbe* (1666, published 1680), set in Numidia in Roman times, also portrays a heroic woman, who meets her end with noble equanimity. Lohenstein's last play, *Ibrahim Sultan* (1673), is more perfunctory.

Towards the end of his life Lohenstein wrote a substantial novel in 2 vols., which was published in 1689 after his death (*Großmütiger Feldherr Arminius oder Hermann als ein tapferer Beschirmer der deutschen Freiheit nebst seiner durchlauchtigsten Gemahlin Thusnelda in einer sinnreichen Staats- und Liebes- und Heldengeschichte dem Vaterlande zu Liebe dem deutschen Adel aber zu Ehren und rühmlichen Nachfolge vorgestellet*). It is a frankly political and patriotic novel in which the primitive Germans are favourably contrasted with the corrupt, civilized Romans. Replete with learning, it is the work of an industrious polymath, and it has been suggested that Lohenstein intended it as an encyclopedic repository of knowledge. Others have seen it as an allegory of the political situation of his own day. Lohenstein's style displays the ornateness of baroque in an exceptionally high degree.

The plays (3 vols.) were published as *Türkische Trauerspiele*, *Römische Trauerspiele*, and *Afrikanische Trauerspiele*, ed. K. G. Just, 1953–7.

Lokalposse, designation of a type of comedy popular in Vienna in the late 18th c. and early 19th c. and known in full as 'Wiener Lokalposse mit Gesang'. The Lokalpossen deal farcically with local themes, are liberally spiced with local and topical allusions, and contain songs. The principal authors were J. A. Gleich, K. Meisl, A. Bäuerle, and J. Nestroy (qq.v.). See also VOLKSSTÜCK.

Lokalstück, term used to denote a play intended for a popular audience, set in the urban or rural environment of the audience, and employing dialect, either in authentic or in modified form. The Lokalstück first flourished in Vienna at the end of the 18th c., usually in the form Lokalposse (q.v.). Most examples are comedies, but towards the end of the 19th c. serious Lokalstücke such as Anzengruber's tragedy *Das vierte Gebot* (q.v.) were written. Other cities for which Lokalstücke have been written include Berlin, Hamburg, Frankfurt/

Main, and Munich. Rural Lokalstücke are most frequent in Tyrol and Upper Austria.

LÖNS, HERMANN (Kulm, West Prussia, 1866–1914, nr. Reims, killed in action), is generally known as the poet of the Lüneburg Heath (see LÜNEBURGER HEIDE), though he was born in a Prussian province, which is now part of Poland. He spent part of his boyhood in Westphalia, and was from 1893 to 1907 a journalist in Hanover, then for a short time in Bückeburg. In spite of his age of almost 48 he volunteered at the outbreak of war and fell in the first weeks. He was a great walker, an acute and practised observer of nature and an ardent shooter of game. He wrote many sketches and stories of heath, forest, and wild life, including *Mein grünes Buch* (1901), *Was da kreucht und fleugt*, and *Aus Wald und Heide* (both 1909). As a novelist he belonged to the trend of Heimatkunst (q.v.), writing of a specific region with a nationalistic bias.

Löns's principal novel, *Der Wehrwolf* (1910), is a historical story full of violence and atrocities. It is set in the Thirty Years War (see DREISSIGJÄHRIGER KRIEG) and recounts the efforts of Harm Wulf (Der Wehrwolf) of Oedringen in the Lüneburger Heide and of his fellow heath farmers and labourers to protect and, where necessary, avenge their families and property against marauding soldiery. Other novels by Löns are *Der letzte Hansbur* (1909), *Dahinten in der Heide* (1910), *Das zweite Gesicht* (1911), and *Die Häuser von Olenhof* (posth., 1917). His lyric poetry (*Mein goldenes Buch*, 1901; *Der kleine Rosengarten*, 1911) does not rise above mediocrity, but has sometimes a folk-song-like character, and some of his poems are sung as Volkslieder.

Sämtliche Werke (8 vols.), ed. F. Castelle, appeared 1924 and *Nachgelassene Schriften* (2 vols.), ed. W. Deimann, 1928.

Lorelei, the name of a picturesque rocky promontory on the right bank of the Rhine below St. Goarshausen. The older form of the name is Lurlei or Lurleiberg. Since the appearance in 1802 of Brentano's poem *Die Lore Lay* it has signified a beautiful woman seated on the cliff, who, siren-like, by her entrancing song lures boatmen to shipwreck and death on the rocks below. Through Heine's poem *Die Loreley* (q.v., 1823), this Romantic creation became a figure of folk-lore. In Eichendorff's poem *Waldgespräch* (q.v., 1815), she appears as 'die Hexe Lorelei', a woman of seductive beauty who brings death, equivalent to Keats's *La belle dame sans merci*. In medieval legend the rocky cliff concealed the treasure of the Nibelungs.

Loreley, Die, an early poem by H. Heine (q.v.), written in 1823. It is contained in the section *Die Heimkehr* of *Das Buch der Lieder* (q.v., 1827), where it is untitled. Set to music by F. Silcher (q.v.), it has virtually become a folksong. For the legend it embodies, see LORELEI.

Lorengel, see LOHENGRIN.

LORIS, early pseudonym of Hugo von Hofmannsthal (q.v.).

LORM, HIERONYMUS, pseudonym of Heinrich Landesmann (Nikolsburg, now Mikulor, Moravia, 1821–1902, Brünn, now Brno), who, though deaf from adolescence and virtually blind, contrived to work as a political journalist in Vienna (which his radical opinions obliged him later to leave), Dresden, and Brünn. He wrote a number of novels, including *Abdul* (1852) and *Die schöne Wienerin* (1886), as well as Novellen.

Lorscher Bienensegen, a magic spell, enjoining bees to return home and to refrain from flying away. The dual function of this spell has led to the view that it is a combination of two older spells. The tone is remarkably gentle, and the relationship to the bees almost intimate. The spell was written down in the 10th c. in a MS. which is now in the Vatican, but originally came from Lorsch Abbey. (See also ZAUBERSPRÜCHE.)

LORTZING, ALBERT (Berlin, 1801–51, Berlin), was the son of two opera singers and followed their profession. He also composed light operas in the manner of the Singspiel (q.v.), among them *Zar und Zimmermann* (1837), *Der Wildschütz* (1842), and *Der Waffenschmied* (1846). He composed one serious opera, *Undine* (1845), based on Fouqué's *Undine* (q.v.).

LOTHAR I, KAISER (795–855, Prüm Abbey), eldest son of Ludwig I (q.v.), was involved in hostilities with his father from 829 in consequence of a proposed change of inheritance. After his father's death in 840 he came into conflict with his brothers, was defeated in battle at Fontenoy (841), and consented to a division of the Empire (see VERDUN, VERTRAG VON).

LOTHAR II (?–869, Piacenza), second son of the Emperor Lothar I (q.v.), succeeded in 855 as king of the northern portion of his father's domains, which acquired from him the name Lotharingia (Lorraine).

LOTHAR III, KAISER (?–1137, Breitenwang, Außerfern, Tyrol), known as Lothar von Sachsen or von Supplinburg, was elected German King in 1125 on the death of Heinrich V (q.v.), in preference to Duke Friedrich von Schwaben (1090–1147), Heinrich's nephew. Lothar was in

conflict with the Staufen faction (see HOHENSTAUFEN) led by Friedrich, which he subdued successfully by 1135. He pursued the quarrel with the Pope over the investiture of bishops, and and he adopted a policy of strengthening the eastern frontiers against the Slavs. He died on the way back from Rome. His body was brought home and buried at Königslutter Abbey, his own foundation, not far from Brunswick.

LOTHAR, RUDOLF, true name Spitzer (Budapest, 1865–1943, Budapest), was on the editorial staff of the *Neue Freie Presse* (q.v.) in Vienna, and was later with the *Lokal-Anzeiger* in Berlin. He was a prolific and successful provider of light fare for the theatre. His comedies include *Satan, Tantaliden* (both 1886), *Zauberlehrling* (1888), *Frauenlob* (1895), *Königsidyll* (1896), *Glück in der Liebe, Die Königin von Zypern* (both 1903), *Das Fräulein in Schwarz* (1907), *Die Hofloge* (1918), *Casanovas Sohn* (1920), *Der Werwolf* (1921), *Die Frau mit der Maske* (1922), *Die schwarze Messe* (1923), *Der sprechende Schuh* (1924), *Erlebnis, Der gute Europäer* (both 1927), *Das Märchen vom Auto* (1930), *Ist das denn so wichtig?*, *Besuch aus dem Jenseits*, and *Der Papagei* (all 1931).

Lothar collaborated with others, including O. Blumenthal (q.v.), F. Gottwald (b. 1896), and P. G. Wodehouse. He also tried his hand at serious plays (*Herzdame*, 1902; *Die Rosentempler*, 1905) and was well known as a librettist, providing in addition to numerous operetta books,the text for d'Albert's *Tiefland* (1904). Two novels of Berlin are *Kurfürstendamm* and *Der Herr von Berlin* (both 1910). In 1899 he published a history of the Burgtheater (q.v., *Wiener Burgtheater*).

LOTICHIUS, PETRUS, commonly called Lotichius Secundus (Niederzell, Hesse, 1528–60), was the author of Latin elegies, *carmina*, and eclogues, in which elegant treatment of the classical tongue is matched by genuine emotion and true feeling for nature. Like many humanists of his age he led a restless wandering life, which included service as a soldier at Mühlberg (q.v., 1547); he finally became a professor at Heidelberg University.

Lotte in Weimar, a novel by Th. Mann (q.v.), published in 1939. Its slender factual basis is the visit to Weimar in 1816 of the widowed Lotte Kestner (see BUFF, LOTTE), whom Goethe had known, loved, and last met at Wetzlar in 1772 (see GOETHE and LEIDEN DES JUNGEN WERTHERS, DIE). Other characters figuring in the action, which consists primarily in a dinner-party given by Goethe for Lotte and her daughter, are Adele Schopenhauer, Goethe's secretary F. W. Riemer, and his son August von Goethe (qq.v.).

The narrative is a *tour de force* in its creation of

characters, the analysis of their innermost feelings, and the verbatim report of their conversations, which reflect, of course, Thomas Mann's view of Goethe and his circle. The work is constructed with immense ingenuity and presents the transience of once overpowering emotion in a variety of modulations, chiefly through Lotte. Though more than forty years have passed, she finds it difficult to accept that the recollection of that summer in Wetzlar evokes no response from the 66-year-old Goethe.

Lotti, die Uhrmacherin, a short novel (Erzählung) by Marie von Ebner-Eschenbach (q.v.), published in 1889. Lotti, the skilled daughter of a master watchmaker, Johannes Fessler, has inherited from him a valuable collection of watches. Her former fiancé, a writer, finds himself in financial straits, and Lotti, though no longer caring for him, sells the collection in order to give him anonymous help. She marries the companion of her childhood, her foster-brother Gottfried.

Louis d'or, see FRIEDRICH D'OR.

LOUIS FERDINAND, PRINZ VON PREUSSEN (Friedrichsfelde, Berlin, 1772–1806, Saalfeld, killed in battle), a nephew of Friedrich II (q.v.) of Prussia, was one of the most colourful figures in Berlin at the turn of the 19th c. An advocate of reforms, he was critical of the royal family and the Prussian nobility alike, and enjoyed his contacts with the Romantics, whose literary circles he visited (see VARNHAGEN VON ENSE, RAHEL). He was a gifted pianist, a minor composer, and a patron of painters; and he was as much attracted to women as to the arts.

Louis Ferdinand distinguished himself in the campaign of 1792 (see REVOLUTIONSKRIEGE), but anticipated the national disaster that lay ahead. He died in hand-to-hand fighting in 1806, a few days before Jena (see NAPOLEONIC WARS). His *Kriegstagebuch 1806*, ed. E. Berner, appeared 1905 (in Hohenzollern-Jahrbuch 9), and *Musikalische Werke*, ed. H. Kretzschmar, 1910. The dramatist Ernst von Wildenbruch (q.v.) was a son of one of Prince Louis Ferdinand's illegitimate sons.

Prince Louis Ferdinand appears as a character in Th. Fontane's novel *Schach von Wuthenow* (q.v.) and is the subject of a play, *Louis Ferdinand, Prinz von Preußen*, by F. von Unruh (q.v.).

LÖWENTHAL, SOPHIE, FREIFRAU VON (Vienna, 1810–89, Vienna), *née* Sophie Kleyle, married Max, Freiherr von Löwenthal (1799–1872), a highly placed Austrian civil servant. From 1834 until his death she was the object of the love of the poet Lenau (q.v.), which she

returned. Lenau's letters to her have been published, but he destroyed in 1844 the letters she had written to him. She is the authoress of a novel, *Mesalliiert*, published posthumously in 1906.

Lübeck, city situated close to the Baltic with a harbour and a seaside resort (Travemünde). Lübeck, which lies close to the East German frontier, is in the Land Schleswig-Holstein of the Federal Republic (see BUNDESREPUBLIK DEUTSCHLAND). Founded by the Duke of Holstein in 1143 and renewed on a larger scale by Heinrich der Löwe (q.v.) in 1159, it was annexed in 1181 by the Emperor Friedrich I (q.v.). A threat from the Danes was decisively repelled at Bornhöved (q.v., 1227). Lübeck, a Reichsstadt (q.v.), became a great centre of trade, and was a prominent member of the Hanseatic League (see HANSE, DEUTSCHE). In the 16th c. it was involved by J. Wullenwever (q.v.) in an unsuccessful and costly war. After the collapse of the Holy Roman Empire (see DEUTSCHES REICH, ALTES) in 1806, Lübeck retained its independence, becoming a Free Hanseatic City (Freie und Hansestadt), and so remained until 1937, when the National Socialist regime deprived it of its separate status.

A bishopric of Lübeck was created in 1160, but from the 13th c. onwards the bishop resided in Eutin. Formerly studded with medieval, 16th-c., and 17th-c. buildings, Lübeck was heavily damaged in a British air raid in 1942, but the exteriors of churches, public buildings, and conspicuous houses have been restored. The massive twin-spired gate, the Holstentor, was built in 1477. The city possesses a famous grammar school (Gymnasium), Das Katharineum.

E. Geibel and G. Falke were both born and bred in Lübeck, as were the Mann brothers, Heinrich and Thomas (qq.v.). All four attended the Katharineum, and so, for a short time (1835–7), did Th. Storm (q.v.). The house (Mengstr. 4) in which the Mann family lived from 1842 to 1891 was destroyed except for the façade, which now conceals a new building. It is popularly known as the Buddenbrookhaus because it was the setting for Th. Mann's novel *Buddenbrooks* (q.v.). His *Tonio Kröger* (q.v.) is partly set in the town, as are H. Mann's novels *Professor Unrat* (q.v., also entitled *Der blaue Engel*) and *Eugénie* (q.v.). A eulogy of the city is contained in Th. Mann's speech *Lübeck als geistige Lebensform*, published in *Die Forderung des Tages* (1930).

LUBLINER, HUGO (Breslau, 1846–1911, Berlin), came to Berlin from Silesia as a boy in 1858, and in the 1870s, 1880s, and 1890s was a popular dramatist. His facile production includes the plays *Der Frauenadvokat* (1873), *Der Name* (1888), *Der kommende Tag* (1891), *Das fünfte Rad* (1898), and *Der blaue Montag* (1902). The most successful of all was *Der Jourfix* (1882). His novels (*Die Gläubiger des Glücks*, 1886; *Die Frau von neunzehn Jahren*, 1887) aroused less attention. He at first used the pseudonym Hugo Bürger.

LUBLINSKI, Samuel (Johannisburg, East Prussia, 1868–1910, Weimar), began with theoretical and critical writings and with essays (*Neu-Deutschland*, 1900) and also wrote historical tragedies in a Naturalistic manner (*Hannibal*, 1902; *Elisabeth und Essex*, 1903). Dissatisfied with Naturalism, he became one of the spokesmen for the short-lived trend of neo-Classicism (see NEUKLASSIZISMUS), supporting it in *Die Bilanz der Moderne* (1904) and *Der Ausgang der Moderne* (1909). To this phase belong his tragedies *Gunther und Brunhild* (1908) and *Kaiser und Kanzler* (1910).

Lubowitz, estate and country-house near Ratibor in Silesia, now under Polish administration. Schloß Lubowitz was the birthplace of the poet J. von Eichendorff (q.v.).

Lucidarius, also known as *Der große Lucidarius* to distinguish it from *Der kleine Lucidarius* (see SEIFRIED HELBLING), was a textbook of medieval knowledge written in prose and widely used in schools. Its method is a reversal of that of the catechism, the pupil puts his questions and the master gives the answers. It is divided into three books, the first of which deals with material things (cosmography, geography, physiology), the second and third are theological, concerned with doctrine, liturgy, and eschatology. It was written c. 1190, apparently at the instigation of Duke Heinrich der Löwe (q.v.), who wished it to be called *Aurea gemma*. MSS. were made throughout the later Middle Ages and the work appeared as a printed book in the 16th c.

Lucinde, a novel written by F. Schlegel (q.v.), published in 1799. It is described as 'Ein Roman. Erster Teil', but no second part followed. It has virtually no plot and is concerned mainly with the hero's erotic feelings for Lucinde. Though it caused a scandal, it is hardly concerned with physical detail. In so far as it has a development, it is the education of Julius to true love with Lucinde, after earlier experiences of false love. It is partly in the form of letters (see BRIEFROMAN) and includes a passage of dialogue. The prose, particularly towards the close, is rhythmic and at times ecstatic. The autobiographical element, as well as Schlegel's deliberate departure from the model of Goethe's Bildungsroman (q.v.), are noteworthy features.

LUDAEMILIA, Gräfin von Schwarzburg-Rudolstadt (1640–72, Rudolstadt), a member of one of the great noble families, was the authoress of some Protestant hymns. Countess Aemilia von Schwarzburg-Rudolstadt (q.v.) was her sister-in-law.

LUDENDORFF, Erich (nr. Posen, 1865–1937, Tutzing, Munich), Prussian general famous as the chief of staff of Hindenburg (q.v.) under whom he served in the east (1914–16) and at German General Headquarters. An efficient and ruthless organizer, Ludendorff was largely responsible for the temporary military successes of the spring of 1918. With his superior, he became increasingly influential politically in 1917 and 1918, but showed little understanding of political forces and problems. His right-wing extremism manifested itself after the war in sympathy with the Kapp-Putsch (q.v.) and in participation in Hitler's revolt in Munich in 1923 (see HITLERPUTSCH). For this he was tried and acquitted.

Ludendorff's second wife Mathilde (1877–1966) was an active publicist, attacking freemasonry which she believed to be responsible for the deaths of Mozart and Schiller (qq.v.). She was a prominent anti-Semite.

Ludendorff's reminiscences appeared as *Meine Kriegserinnerungen* (1919). He also published a number of tendentious works.

LUDER, Peter (dates unknown), an early humanist, led a restless and stormy life. He was in Heidelberg in 1431, visited Greece, studied in Italy, and returned to Heidelberg in 1456, whence he was expelled in 1460; thereafter he appeared in Erfurt, Leipzig, Basel, and Padua. Luder castigated the barbarism of medieval Latin, advocating a return to the classics.

LUDER VON BRAUNSCHWEIG, see MAKKABÄER.

Ludlamshöhle, a Viennese club of artists and writers which in the 1820s met at various inns for the free exchange of views. The name is derived from a play by Oehlenschläger (q.v.). Its members included J. C. von Zedlitz and M. M. Daffinger (qq.v.). Grillparzer (q.v.) was admitted in 1826 shortly before the inn was raided upon the instigation of a minor police official, as a result of which the club was disbanded. The political persecution of its members, which included searches of their private residences and the indiscriminate confiscation of MSS. and papers, resulting in a number of convictions, was eventually exposed as judicial misconduct. The incident illustrates repressive conditions in Vienna during the early phase of the Metternich (q.v.) regime.

Ludus de Antichristo, a medieval Latin drama written in the second half of the 12th c., probably between 1160 and 1190. The MS. derives from Tegernsee Abbey (hence the alternative title *Tegernseer Antichrist*), and it is possible that the play was written there. In the first act the Emperor, with his vassal kings, defeats the infidel. In the second Antichrist, supported by Hypocrisy and Heresy, wins over the Christian peoples, but is at length struck down by God's lightning flash, whereupon Ecclesia triumphs. The author was an admirer of Friedrich I (q.v.), with whom his emperor is probably identifiable. The source is a legend formulated in the 10th c. by Adso von Toul (see ANTICHRIST).

Ludus de decem virginibus (*Spiel von den zehn Jungfrauen*), a Middle High German religious play concerned with the parable of the five wise and five foolish virgins (Matthew 25:1–13). It was probably written by a Dominican monk of Eisenach towards the end of the 13th c. The main weight of the play falls on the lament of the foolish virgins to whom Jesus denies admission, saying 'Ich enweiz nicht, wie ir sit' (Ich weiß nicht, wer ihr seid). The perils and finality of damnation are conveyed with considerable poetic power. At a performance at Eisenach in 1322 Landgrave Friedrich of Thuringia is said to have been so affected that he had a stroke of which he died two years later. Though the text is in Middle High German, the brief directions are in Latin.

LUDWIG I, KAISER, DER FROMME (778–840, nr. Mainz), the third son of Charlemagne (see KARL I, DER GROSSE), was proclaimed emperor by his father in 813 in order to secure the succession. He took over the government of the empire on his father's death, and spent the greater part of his reign attempting, with indifferent success, to quell revolts; these included those of his own sons Lothar (see LOTHAR I, KAISER), Pippin (*c.* 797–838), and Ludwig (see LUDWIG II, DER DEUTSCHE).

In 817 Ludwig I divided the empire among these three sons, but, having married again and begotten a fourth son by the Empress Judith, he overthrew the existing arrangement in 829 and allocated a large territory to Karl at the expense of his three half-brothers. He was defeated in 830 and obliged to reverse the arrangements; but when his sons quarrelled, he was again able to assert himself, until betrayed at the Field of Lies (833, see LÜGENFELD) and imprisoned in a monastery. Renewed dissensions between the brothers led to a compromise with Lothar, and hostilities with the others were still in progress when Ludwig died. He was buried at Metz.

Since he reigned over the undivided Frankish Empire of Charlemagne, he is also reckoned as King Louis I of France. He is styled 'the Pious' because of his liberality to the Church and his efforts to banish immorality from his court.

LUDWIG II, KAISER (*c.* 822–75), son of Lothar I (q.v.), became emperor on his father's death in 855. His reign was spent in wars in Italy where he fought successfully against the Saracens in the south. Ludwig had no male issue.

LUDWIG III, KAISER, also known as Ludwig der Blinde and Louis de Provence (d. 928), was crowned emperor in Rome in 901, and captured at Verona in 905 by Berengar von Friaul or Friuli, who had him blinded. He was a great-great-grandson of Ludwig I (q.v.).

LUDWIG IV, KAISER, DER BAYER (1287–1347, Fürstenfeld nr. Munich), inherited Bavaria-Ingolstadt in 1310. In 1314 he was elected German King (see DEUTSCHER KÖNIG), though simultaneously Friedrich der Schöne (q.v.) was elected in opposition (Gegenkönig). Ludwig defeated Friedrich at Mühldorf in 1322, but was not able to reap the benefits of his victory because of papal opposition.

In 1323 Ludwig became involved in a long quarrel with the Pope, first with John XXII and, after the latter's death in 1334, with Benedict XII. He was excommunicated in 1324, retorting by accusing the Pope of heresy. He intervened in Italy in 1327, relieving Milan which was at that time besieged by papal forces, and was crowned in Rome in 1328, though not by the Pope. French support for the papacy and French rivalry for the imperial throne frustrated Ludwig's repeated efforts to achieve a reconciliation. At the Electoral Diet at Rhens (Rhenser Kurfürstentag) Ludwig proclaimed that a duly elected German King had no need of papal approbation and published this principle in an imperial law (*Licet juris*), promulgated at a Diet in Frankfurt.

Ludwig attempted to offset French pressure by an alliance in 1337 with Edward III of England, whom he appointed Imperial Vicar (Reichsverweser) for the western parts of Germany. The alliance lapsed in 1341, when Ludwig believed he saw a possibility of reconciliation with the papacy. The attempt, made in 1343, failed, and Pope Clement VI (elected 1342) instigated the election in 1346 of a rival German King (Ludwig's second Gegenkönig) in the person of Karl IV (q.v.). Before the opposition between the two rivals could be settled by arms, Ludwig died of a stroke while bear-hunting.

Ludwig augmented the possessions of the

house of Wittelsbach (q.v.), of which he was the head, by the acquisition of Brandenburg, Tyrol (see MARGARETE VON TIROL), Holland, Zealand, and Hainault, and the rapacity with which he pursued his policy of family aggrandizement brought him many enemies.

Ludwig IV is the subject of a poem by Schiller (*Deutsche Treue*, q.v., 1800) and of plays by L. Uhland (q.v., 1819), Paul Heyse (q.v., 1859), and Martin Greif (q.v., 1891).

LUDWIG II, DER DEUTSCHE, KÖNIG DER OST-FRANKEN (*c.* 804–76, Frankfurt/Main), the third son of the Emperor Ludwig I (q.v.), received Bavaria as his portion when Ludwig I divided the empire in 817.

Ludwig began to rule at 22 in 826, and in 829 he revolted with his brothers against his father because the latter had made a new division of the empire. He continued hostilities against his father intermittently until the latter's death. Thereupon strife developed between the brothers, which was resolved in the Treaty of Verdun (843, see VERDUN, VERTRAG VON). After the death of his brother Lothar in 855 (see LOTHAR I, KAISER) a new division was determined in the Treaty of Mersen (870). Ludwig was buried in Lorsch near Bensheim. The style 'the German' (Ludwig der Deutsche) derives from his portion at the division of 843, which was the land east of the Rhine and the Aare, therefore entirely German and German-speaking territory.

LUDWIG III (Louis III of France) (*c.* 863–82, Saint-Denis), King of the West Franks, son of Ludwig der Stammler (Louis II) and grandson of Charles the Bald (see KAROLINGER). He reigned from 879 to 882 and in 881 gained a victory over the invading Normans at Saucourt in Picardy, which is celebrated in the *Ludwigslied* (q.v.).

LUDWIG I, KÖNIG VON BAYERN (Strasburg, 1786–1868, Nice), fought on the Napoleonic side during the rule of his father (Maximilian I, q.v.). He succeeded his father as king in 1825, ruling in the manner of an enlightened despot of the 18th c. and restored Munich's reputation as a centre of the arts. His own literary success, however, rested on his status rather than on intrinsic merit. The first collection of his poems (1829) was marred by an (for the German language) inappropriately liberal use of the participial style, which Heine, his sharp critic, parodied in *Atta Troll* (q.v.). His infatuation for Lola Montez (q.v.) led to great unpopularity. He lost his throne in 1848 (see REVOLUTIONEN 1848–9).

LUDWIG II, KÖNIG VON BAYERN (Nymphenburg, 1845–86, Starnberger See), King of Bavaria 1864–86, son of Maximilian II and grandson of Ludwig I (qq.v.), at first maintained an . anti-Prussian policy, engaging Bavarian forces on the Austrian side in the War of 1866 (see DEUTSCHER KRIEG). In 1870 Ludwig supported Prussia against France and, as hostilities drew to a close, agreed under pressure and persuasion from Bismarck (q.v.) to a German Empire under the Prussian king (see DEUTSCH-FRANZÖSISCHER KRIEG).

A man of strikingly handsome appearance, Ludwig was appreciative of the arts to a high degree, and pathologically sensitive to the point of withdrawing himself from human contacts. Ludwig encouraged R. Wagner (q.v.), inviting him to Munich in 1864, and supporting the scheme for a Wagner opera house in Bayreuth. His predilection for opera performance at which he was the only spectator is well known. Ludwig indulged in a mania for building, constructing in remote situations three fantastic palaces which are monuments of extravagance, perhaps of megalomania, and certainly of a poetic, if distorted, imagination (Herrenchiemsee, Linderhof, Neuschwanstein, q.v.). The enormous cost of these buildings, coupled with increasing pathological eccentricity, led to his deposition on 10 June 1886. On his transfer from Neuschwanstein, where he was arrested, to Berg, he threw himself overboard while crossing the Starnberger See, pulling, according to one account, the escorting psychiatrist with him. Both men were drowned. The exact circumstances are not clear.

LUDWIG, LANDGRAF VON THÜRINGEN (*c.* 1200–27, Otranto), Ludwig IV of his house, was known as Ludwig der Heilige, though he was never officially canonized. Ludwig was the son of Landgrave Hermann and the husband of St. Elisabeth (see ELISABETH, HEILIGE). His death took place on the way to the crusades. A life of Ludwig, with legendary accretions, was written at Reinhardsbrunn, where he was buried (see FRIEDRICH KÖDITZ VON SAALFELD).

LUDWIG, EMIL, real name COHN (Breslau, now Vroclav, 1881–1948, Ascona), worked as a journalist and wrote numerous *biographies romancées*, in which much carefully gathered source material was fused and presented in a manner more appropriate to fiction. In the 1920s and 1930s he had a considerable vogue. His works include *Bismarck* (1926), *Richard Dehmel* (1913), *Wagner oder die Entzauberten* (1913), *Goethe* (1920), *Wilhelm der Zweite* (1925), *Juli 14* (1929), *Michelangelo* (1930), *Schliemann* (1932), *Roosevelt* (1938), and *Napoleon* (1939).

He became a Swiss citizen in 1932 and settled in the U.S.A. in 1940. *Gesammelte Werke* (5 vols.) appeared 1945–6.

LUDWIG, Otto (Eisfeld, 1813–65, Dresden), was the son of the town clerk of Eisfeld, from whom he inherited a delicate constitution. When he was 7 his father was accused of peculation, and, though his innocence was proved, he encountered further trials when his house was burned down, probably by arson. Ludwig's father died in 1822, and the boy, who possessed musical talent, was apprenticed to his merchant uncle; after his mother's death in 1831 he went to the Lyzeum at Saalfeld until ill health obliged him to abandon formal study. From 1834 to 1838 he occupied himself with the private study of music, and in 1837 and 1838 aroused attention with a musical play (Singspiel), *Die Geschwister*, and an opera, *Die Köhlerin*. A grant for three years from the Duke of Hildburghausen enabled him to study under F. Mendelssohn (q.v.) in Leipzig, but illness and adverse criticism from Mendelssohn led him to give up his training and he returned to Eisfeld. At this time he began to occupy himself with plans for plays, and on his return to Leipzig in 1842 he was already in his own mind a man of letters rather than a musician.

Ludwig was in Leipzig and Dresden from 1842 to 1844, during which time his first publication, the story *Die Emanzipation der Domestiken*, appeared in the *Zeitung für die elegante Welt* (1844). After a stay of three years in Meissen he settled in Dresden, where he married in 1852; he lived a secluded life with a devoted wife.

Most of Ludwig's literary production falls in the thirteen years 1843–56. The prelude to a play about Friedrich II (q.v.) of Prussia was published by H. Laube (q.v.) in 1844 with the title *Die Torgauer Heide* (q.v.). The two plays which were performed after completion and made his name as a playwright, *Der Erbförster* and *Die Makkabäer* (qq.v.), were published in vols. 1 and 2 of *Dramatische Werke* (1853). The story *Die Heiteretei* and its sequel *Aus dem Regen in die Traufe* (qq.v.) appeared together in book form in 1857, and *Zwischen Himmel und Erde* (q.v.) in 1856. Ludwig left a large number of unpublished plays of which he completed *Das Fräulein von Scuderi* (based on E. T. A. Hoffmann's story *Das Fräulein von Scuderi*, q.v.), *Die Pfarrose, Hanns Frei*, and *Die Rechte des Herzens* (qq.v.); others remained fragments despite years of effort devoted to carefully selected subjects. Both *Der Engel von Augsburg* and *Agnes Bernauerin* are contrasting studies of Agnes Bernauer (q.v.), on whom he also wrote two poems which are contained in the former; other plays are *Genoveva, Die Makkabäerin*, and

Tiberius Gracchus. All were published posthumously.

Ludwig's introspective nature, perhaps intensified by increasing ill health, led him to spend his later years in endless, though acute and perceptive, analysis of Shakespeare's technique, by which he hoped to discover the recipe of dramatic perfection. His *Shakespeare-Studien* were published by his friend G. M. Heydrich (1820–85, himself a minor playwright) in 1871. In his *Dramaturgische Aphorismen* and in his comments on his own approach to writing (e.g. *Mein Verfahren beim poetischen Schaffen*) Ludwig criticizes his contemporary F. Hebbel (q.v.), whom he sought to rival in his two major completed plays as well as in his dramatizations of Agnes Bernauer and Genoveva. Ludwig enjoyed the continued encouragement of Laube, who produced *Die Makkabäer* and *Der Erbförster* in Vienna, the latter in preference to Hebbel's *Maria Magdalene* (q.v.) to which it is indebted. Both Ludwig's choice of projects and his critical writings reveal an excessive degree of dependence on literary models, which inhibited his creative impulse, and he came fully into his own only in his narrative work. *Zwischen Himmel und Erde*, by which he is best known, is a remarkable examination of the border territory between psychological normality and pathology, and in *Die Heiteretei* he improves upon the type of village story written by his friend B. Auerbach (q.v.).

Ludwig is a representative of Poetic Realism (see POETISCHER REALISMUS), and his *Dramatische Aphorismen* contain a succinct definition of his own approach to stylized realism ('künstlerischer Realismus'), aiming at a balance between stylized idealism and Naturalistic realism. Despite the verse of *Die Makkabäer* and his success in the prose of his plays, he needed disciplined epic description to achieve his aim of combining atmosphere and character.

Ludwig's *Gesammelte Werke* (4 vols.) of 1870 contained, as previously unpublished material, *Das Fräulein von Scuderi, Der Engel von Augsburg*, and *Tiberius Gracchus*. The *Gesammelte Schriften* (6 vols.) of 1891 added *Die Pfarrose* and *Hanns Frei*, his only comedy in rhymed verse, as well as *Die wahrhaftige Geschichte von den drei Wünschen, Aus einem alten Schulmeisterleben*, and *Maria*, stories written in the early 1840s. Of the 18 vols. planned for his *Sämtliche Werke*, ed. P. Merker, six appeared 1912–22; a definitive edition, sponsored by the Deutsche Akademie der Wissenschaften, began to appear in 1961 (3 vols. of *Agnes-Bernauer-Dichtungen* by 1973). *Briefe*, ed. K. Vogtherr, appeared in 1935.

***Ludwig der Bayer**,* a historical play (Schauspiel in fünf Aufzügen) in blank verse, written by L. Uhland (q.v.) in 1818 and published in 1819. Uhland intended it as an entry for a competition organized by the Court Theatre in Munich for a play on a Bavarian historical subject. It was not successful in gaining the prize, and was first performed at Munich in 1826.

The action begins in 1314, when Ludwig der Bayer (see LUDWIG IV, KAISER), elected German King, faces a rival claimant, Friedrich der Schöne (q.v.) of Austria. Friedrich is defeated, captured, and released by Ludwig on condition that his adherents give no further trouble. But Friedrich cannot control them, and, being thus unable to keep his word, he surrenders voluntarily to Ludwig. A reconciliation follows. It is a play in which trustworthiness and fidelity, magnanimity and clemency are the moving principles.

Ludwigslied, a poem in Old High German rhyming verse, written to celebrate the victory of Ludwig (Louis) III (q.v.), King of the West Franks, over the Normans at Saucourt in 881. Religious in conception, it sees the Norman invasion as a punishment sent by God, and Ludwig as God's chosen instrument to end the visitation. Ludwig is praised as much for his piety as for his courage. The poem of 59 lines is contained in a MS. at Valenciennes.

LUDWIG VON ANHALT-KÖTHEN, FÜRST (Dessau, 1579–1650, Köthen), patron and cofounder of Die Fruchtbringende Gesellschaft (q.v.). His name in this society was Der Nährende. His interests were philological and prosodic.

LUEGER, KARL (Vienna, 1844–1910, Vienna), an Austrian politician, who from 1897 until his death was mayor of Vienna. Lueger was at first associated with the German National Party and its agitator, G. von Schönerer, with whose anti-Semitism he sympathized. He then joined the Christian Social Party, founded in 1887, and soon established himself as its leader. He maintained the platform of anti-Semitism, declared his party to be the party of the 'little man', and directed his attack against corruption and fraud. He was first elected mayor (Bürgermeister) of Vienna in 1895, but imperial sanction was repeatedly refused until 1897.

Lueger carried out an energetic programme of modernization in the city, laying out parks, building hospitals, and bringing the tram service, gas, and electricity under municipal ownership.

Lügenfeld, name given to a meeting near Colmar in 833 between the Emperor Ludwig I (q.v.) and his rebellious sons, Lothar, Pippin, and Ludwig, at which Ludwig I was deserted by his army and temporarily banished to a monastery. The name is said to derive from the treachery and deceit exercised against Ludwig.

Luise Millerin, original title before publication and performance of Schiller's *Kabale und Liebe* (q.v.).

LUISE, KÖNIGIN (Hanover, 1776–1810, Schloß Hohenzieritz nr. Neustrelitz, Mecklenburg), consort of King Friedrich Wilhelm III (q.v.) of Prussia. A princess of Mecklenburg, she was married to Friedrich Wilhelm in 1793. Her unpretentious way of life and kindness of heart won her wide popularity. After the Prussian defeat at Jena she suffered hardship in the flight to East Prussia. In 1807 she sought an audience of Napoleon in order to plead for lenient terms of peace, but was brusquely received because she had earlier influenced Prussian policy in a pro-Russian sense. Soon after the royal family returned to Berlin she fell ill and died in a few months.

Luise was the mother of King Friedrich Wilhelm IV and of King Wilhelm I (qq.v.), who became the first German emperor. Her attractive character, the hardships she suffered, and her early death have made of her a national figure, commemorated by women's charities and organizations, and by numerous historical pictures and works of literature, mostly of a nationalistic character. Among these are two novels by Walter von Molo (q.v.), *Luise* (1919) and *Luise im Osten* (1937), and a Novelle by Eckart von Naso (q.v.), *Die Begegnung* (1936).

LUISE, HERZOGIN (Darmstadt, 1757–1830, Weimar), a princess of Hesse-Darmstadt, was married to Duke Karl August (q.v.) of Saxe-Weimar in 1775. Of a sensitive and withdrawn temperament, she was not well matched with her robust extravert husband. She won the praise of Napoleon for her resolute and courageous conduct after the battle of Jena in 1806 (see NAPOLEONIC WARS).

Luise. Ein ländliches Gedicht, an idyllic poem written by J. H. Voß (q.v.) and published in 1795. It is 'in drei Idyllen'. The first idyll is *Das Fest im Walde*, the second *Der Besuch*; and the third, which is subdivided into two cantos (Gesänge), is entitled *Die Vermählung*.

The poem has no real action. The first section describes a water picnic, the second a visit by Luise's fiancé, the third portrays the preparations for the wedding and the impetuous celebration of the marriage by Luise's father, the pastor of Grünau, the evening before the appointed day. It ends with the newly married pair going to the bridal chamber. The idyll (see

IDYLLE) is written in classical hexameters and deliberately employs repeated stereotyped 'Homeric' epithets (e.g. 'der würdige Pfarrer von Grünau', 'die verständige Mutter'). Its picture of a patriarchal rural community is idealized into complete harmony. Voß twice refers to himself in the poem as 'unser Gast von Eutin' and 'der Eutinische Freund', and once to his wife, 'der freundlichen Ernestine'. Grünau has been identified as Malente in Holstein.

LUKÁCS, GEORG or GYÖRGY (Budapest, 1885–1971, Budapest), the most influential Marxist literary critic of the 20th c., was active in the revolutionary government in Hungary in 1919, and prominent in the 1920s as a Communist intellectual. He emigrated to Moscow in 1933, returning to Hungary in 1944,became a member of the Hungarian parliament, and was appointed to the chair of aesthetics and cultural philosophy at Budapest University. He supported the less rigorous socialist policy of I. Nagy in 1956, was appointed minister of education (Volksbildung), suffered temporary deportation, and on his release returned to Budapest, resuming work as a private scholar.

Under the influence of Hegel (q.v.), his main concern was the reassessment of changing forms of realism in literature, notably in the novel, in his view the most comprehensive genre for the study of the social and sociological conditions of a writer's age; these form the premise of his ethics and aesthetics which sought to demonstrate the continuity and 'Totalität' of the realist tradition. He first aroused interest with the collections of essays *Die Seele und die Formen* (1911), *Die Theorie des Romans. Ein geschichtsphilosophischer Versuch über die Formen der großen Epik* (1920, reissued with a preface in 1963), *Geschichte und Klassenbewußtsein. Studien über marxistische Dialektik* (1923), and *Lenin. Studie über den Zusammenhang seiner Gedanken* (1924). From time to time Lukács was forced to modify or 'recant' his views, which he subsequently reasserted, insisting on his own interpretation of socialist realism (see SOZIALISTISCHER REALISMUS), for example in his essay on Gorki (1936). His essay '*Größe und Verfall' des Expressionismus* (1934 in *Internationale Literatur*, Heft 1) led to the so-called Expressionismusdebatte, which turned on a reassessment of Marxist interpretations of realism and formed the background of his essay *Es geht um den Realismus* (1938 in *Das Wort*, Heft 6; see also WALDEN, H.). Lukács's methodology was particularly irreconcilable with the theories of Brecht, but he also provoked the criticism of E. Bloch and H. Eisler (qq.v.) among others. *Existentialismus oder Marxismus?* (1951) is the fruit of his debate with Sartre. His studies of

Goethe (*Goethe und seine Zeit*, 1947), Keller (*Gottfried Keller*, 1940), Th. Mann (*Thomas Mann*, 1949) (qq.v.), but also of Balzac (*Balzac und der französische Realismus*, 1952), Pushkin, Dostoevsky, and Tolstoy are central to his work as a literary critic. An eminent essayist, Lukács edited the first 15 vols. of *Werke* (1962 ff.) himself; they include the 2 vols. of *Ästhetik Teil I: Die Eigenart des Ästhetischen* (1963, vol. 11 and 12), *Deutsche Literatur in zwei Jahrhunderten* (1964, vol. 7), *Probleme des Realismus I. Essays über Realismus* (1971, vol. 4), and *Probleme des Realismus II. Der russische Realismus in der Weltliteratur* (1964, vol. 5). *Frühe Schriften zur Ästhetik I. Heidelberger Philosophie der Kunst 1912–1914* (vol. 16) and *Frühe Schriften zur Ästhetik II. Heidelberger Ästhetik 1916–1918* (vol. 17), both ed. by G. Márkus and F. Benseler, appeared in 1974, and *Zur Ontologie des gesellschaftlichen Seins* (vol. 13), ed. F. Benseler, in 1984; *Briefwechsel 1902–1917*, ed. E. Karádi and E. Fekete, appeared in 1982. From the mid-1950s his work ceased to be published in the DDR.

Lulu-Stücke denotes F. Wedekind's plays *Erdgeist* and *Die Büchse der Pandora* (1895 and 1904, qq.v.), both of which have Lulu as the principal character. *Lulu* is also the title of an unfinished opera based on these plays by Alban Berg (q.v.).

Lumpazivagabundus, see DER BÖSE GEIST LUMPAZIVAGABUNDUS.

LUND, ZACHARIAS (Nübel, Schleswig, 1608–67, Copenhagen), a minor poet, who was for a time in Leipzig and later lived in Denmark, wrote Latin poems (*Poematum juvenilium libri IV*, 1635) and German student songs (*Allerhand artige deutsche Gedichte*, 1636).

Lüneburger Heide, a region of low moorland in North Germany lying between the Aller and the Elbe and extending from Celle in the south almost to Hamburg in the north, and from Ülzen in the east to Bremen in the west. Roughly at its centre lies the town of Soltau. The characteristics of the region are tracts of heather, scattered birch wood, upright or contorted junipers, and a wealth of prehistoric burial places. It is grazed by a breed of sheep known as Heidschnucken. The outstanding poet of the Lüneburger Heide is Hermann Löns (q.v.). The area of true heath is now reduced by agriculture and afforestation, but a substantial nature reserve has been established. Formerly in Hanover, it has been since the reorganization in 1946 a part of Lower Saxony (Niedersachsen).

Lustspiel, light or amusing stage play. The term should not be used in the extended sense of Dante's poem *La divina commedia* or Galsworthy's cycle of novels *A Modern Comedy*; for such works Komödie is employed.

Lustspiele nach dem Plautus fürs deutsche Theater (1774), title of a collection of five plays by Plautus freely adapted in German by J. M. R. Lenz (q.v.). They are: 1. *Das Väterchen (Asinaria)*; 2. *Die Aussteuer (Aulularia)*; 3. *Die Entführungen (Miles gloriosus)*; 4. *Die Buhlschwester (Truculentus)*; 5. *Die Türkensklavin (Cistellaria)*.

LUTHER, MARTIN (Eisleben, 1483–1546, Eisleben), the dominating personality of the Reformation (q.v.), was the son of a peasant who had abandoned agriculture for mining and won a modest competence. The boy received his schooling at Mansfeld, Magdeburg, and Eisenach. From 1501 to 1505 he studied successfully at Erfurt University and then entered an Augustinian monastery. His first years of monastic life were consumed in a struggle for the clarification of inner religious conflicts. He was ordained in 1507 and in 1508 taught philosophy at the then recently founded University of Wittenberg (see WITTENBERG UNIVERSITY). In 1509–10 he lectured at Erfurt and developed his knowledge of Hebrew and Greek. In 1512 he received the degree of Doctor of Theology at Wittenberg, became professor of biblical exegesis there, and delivered between 1513 and 1518 a series of lectures which attracted wide attention and drew students from other parts of Germany.

The beginning of Luther's career as a reformer came almost accidentally when, on 31 October 1517, he nailed to the door of the castle church at Wittenberg his 95 theses (Thesen, see THESES, 95), directed against the sale of indulgences (see ABLASSKRAM). The theses rapidly circulated both in the original Latin and in German translation, and soon adversely affected this source of revenue for the Church. Invited to a disputation at Leipzig with Johann Eck (q.v.), Luther asserted that the papacy was a historical, not a divine, institution, and denied the authority of rulings by the oecumenical councils. On his return to Wittenberg he plunged into a ferment of theological writing, producing his three outstanding tracts *An den christlichen Adel deutscher Nation, De captivitate Babylonica (Von der Babylonischen Gefangenschaft der Kirche)*, and *Von der Freiheit eines Christenmenschen* (qq.v., 1520). Through the support of his sovereign, the Elector Friedrich der Weise (q.v.) of Saxony, Luther was able to escape a summons to Rome, but the investigations of Cardinal Cajetan at Augsburg were followed by the Papal Bull

Exsurge Domine (1520), demanding recantation. Luther's decisive answer to this was the public burning of the Bull at Wittenberg on 10 December 1520. A Bull of Excommunication was immediately issued (January 1521).

By this time Luther had a large following in Germany, and this popular support aided the Elector in his desire to defer the operation of the Bull until Luther had an opportunity to defend himself before the imperial court. Accordingly Luther was summoned to appear before the Diet at Worms in April 1521 (see WORMSER REICHSTAGE). His courageous and unflinching defence (narrated in a famous passage of Ranke, q.v.) was unavailing and he was outlawed by the Edict of Worms (May 1521). Luther meanwhile was spirited away by his patron's orders and given asylum under the assumed name of Junker Jörg in the Wartburg (q.v.). Here he spent ten months largely devoted to the translation of the New Testament which appeared in 1522 (see BIBLE, TRANSLATIONS OF). (It was during this period that Luther's legendary confrontation with the devil was alleged to have taken place.) In March 1522 he emerged from hiding and hurried at the risk of his life to Wittenberg, in order to check the excesses of the reforming party led by Karlstadt (q.v., *Predigten vom Sonntag Invocavit bis Reminiscere*).

Luther's opposition to violent insurrection and his insistence on temporal authority became even clearer with the outbreak of the Peasants' War (see BAUERNKRIEG) in 1525 (*Ermahnung zum Frieden* and *Wider die räuberischen und mörderischen Rotten der Bauern*). In that year he married a fugitive nun, Katharina von Bora (1499–1552), an act intended as a symbolical denial of the principle of clerical celibacy; it also initiated a lifelong happy union. In 1526 appeared Luther's order of service for the new Church (*Deutsche Messe*). In the same year he broke with Erasmus (q.v.), whose defence of free will in *De libero arbitrio* he rejected with *De servo arbitrio*. From 1526 to 1529 he was active in the reorganization of the Church in Saxony. As an outlaw he was necessarily absent from the Diet at Augsburg (1530), at which a compromise between Protestant and Roman Catholic standpoints was unsuccessfully attempted. Luther's protracted and devoted work on the translation of the Bible came to a provisional end with the publication of the complete work in 1534, but he continued to improve his rendering right up to his death (Ausgabe letzter Hand appeared in 1545). He was buried in the castle church (Schloßkirche) at Wittenberg.

The number and bulk of Luther's works, partly in Latin but principally in German, is immense. The Weimar edition, begun in 1883, exceeds 100 vols.; the greater part of it belongs to doctrinal history and theological polemics. In addition to the works cited, mention should be made of the sober and moving account of the martyrdom of Heinrich von Zütphen (*Heinrichs von Zütphen Märtyrertod im Jahr 1524*), and above all of the masterly tract on translation *Ein Sendbrief von Dolmetschen* (*Ein sendbrieff D. M. Luthers. Von Dolmetzschenn und Fürbit der heiligenn*, 1530). Further, Luther was a poet of power who created, partly by translation from Latin or adaptation, a number of moving hymns (*Geystliche gesangk-Buchleyn*, and *Enchiridion geystlicher gesenge*, both 1524, the latter expanded repeatedly and last reprinted in Luther's lifetime in 1545); some have found their way into English hymnals. Among the most famous are 'Ein' feste Burg ist unser Gott' and the tender Christmas hymn 'Vom Himmel hoch' (qq.v.). A great mass of observations on the human soul and human affairs is preserved in the *Tischreden*, collected by his friends and published in 1566. He also translated into German prose a number of fables of Aesop (*Etliche Fabeln aus Esopo*, 1530).

Luther's greatest literary monument is his translation of the Bible. Though it is now recognized that he owed more to his predecessors than was formerly thought, his ability to render the original into truly German phrasing, direct, memorable, and apt, is unsurpassed. He applied to his work the principle he enunciated in the *Sendbrief*: 'den man mus nicht die buchstaben inn der lateinischen sprachen fragen, wie man sol Deutsch reden, wie diese esel thun, sondern man mus die mutter jhm hause, die kinder auff der gassen, den gemeinen man auff dem marckt drumb fragen, und den selbigen auff das maul sehen wie sie reden, und darnach dolmetzschen so verstehen sie es den und mercken, das man Deutsch mit jn redet.'

To these qualities of homeliness, urgency, and power (achieved by extreme conscientiousness as well as good sense) Luther adds a highly developed sense of rhythm, which is especially conspicuous in his rendering of the Psalms. The Lutheran Bible has become, in consequence, a book of immense influence, and reflections of its prose are detectable in many later authors. His idiomatic prose style marks the beginning of a new age in the history of the German language (see GERMAN LANGUAGE, HISTORY OF).

The essential features of Luther's doctrine are the affirmation of justification by faith, the rejection of justification by works (in *Von der Freiheit eines Christenmenschen*), and the assertion of direct communication with God without priestly mediation. The immense and rapid success of his teaching over large tracts of Europe was made possible not only by the abuses of the existing Church, but also by the widespread spiritual malaise accumulating through plague,

famine, and wars in the preceding century. This is evident in the early tendency of the Reformation to be associated with revolutionary and social unrest. Luther himself denounced the subversive movements, insisting on the duty of the human being to fulfil his obligations within the existing social framework. He thus strengthened the hands of the ruling princes, who gained by the attack on the papacy, and at the same time prepared the swift change of Lutheranism from a dynamic spiritual to a rigid conservative force.

To see Luther as a nationalist is a distortion. He attacked the papacy for spiritual, not political, reasons, and his support of the existing political structure discouraged any trend towards German unity. By concentrating the minds of his contemporaries exclusively on theological and ecclesiastical problems he virtually destroyed in Germany the new humanism whose representative was Erasmus.

So dominating a figure was bound to become a subject of later literary works, among which may be mentioned: Z. Werner's play *Martin Luther oder Die Weihe der Kraft* (q.v., 1807), H. von Kleist's Novelle *Michael Kohlhaas* (q.v., 1810, in which Luther makes an important appearance), poems by Uhland (q.v., *Die Ulme von Hirsau*) and by C. F. Meyer (q.v., *Lutherlied*), and novels by L. Schücking (q.v., *Luther in Rom*, 1870) and W. von Molo (q.v., *Mensch Luther*, 1928). Among contemporary works concerned with Luther were *Die wittembergisch Nachtigall* (q.v., 1523) by Hans Sachs, and the hostile *Von dem Großen Lutherischen Narren* by Th. Murner (q.v., 1522) and *Monachopornomachia* by S. Lemm (q.v., 1539).

Werke. *Kritische Gesamtausgabe*, planned in four sections, appeared 1883 ff.; *Ausgewählte Werke* (8 vols.), ed. H. H. Borcherdt, 1914–25; and *Ausgewählte Schriften* (6 vols.), ed. K. Bornkamm and G. Ebeling, in 1982.

Lutherische Strebkatz, Die, a verse satire by an anonymous Lutheran author, published in 1524. Its irony is directed against the Pope and papal adherents such as Th. Murner and H. Emser (qq.v.), who appear as animals. It was formerly attributed to Lazarus Spengler (q.v.). Probably by the same author is *Der Sieg der Wahrheit*, which was also formerly believed to be by Spengler.

Lützen, a small town near Leipzig, was the location of a battle between the forces of Gustavus Adolphus and Wallenstein (qq.v.) on 16 November 1632 (see DREISSIGJÄHRIGER KRIEG). Gustavus Adolphus forced the battle, knowing that Wallenstein was vulnerable owing to the detachment of his general Pappenheim (q.v.) with some 10,000 men to Halle. At the opening of the battle the Swedes, some 16,000 strong, outnumbered Wallenstein's army, which was already defeated when Pappenheim, hastily summoned, checked the Swedish advance but sustained a mortal wound. Piccolomini (q.v.) then took his place. By that time Gustavus Adolphus had been killed and Duke Bernhard of Saxe-Weimar (see BERNHARD VON SACHSEN-WEIMAR) had taken over the command. Wallenstein set Lützen on fire in order to use it as a smokescreen and, at nightfall, withdrew his forces towards Halle.

LÜTZOW, ADOLF, FREIHERR VON (Berlin, 1782–1834, Berlin), served as a regular officer in the Prussian army until 1808, participated in the private war of Major von Schill (q.v.) in 1809, and was recalled to the army in 1811. In 1813 he was authorized to recruit a corps of volunteers, which was known as the Lützowsches Freikorps or, from its black uniform, Die schwarze Schar. Among well-known volunteers were Theodor Körner, Jahn, and Eichendorff (qq.v.). In June 1813 the corps sustained a serious reverse. It was then reorganized and in September distinguished itself at Göhrde. Lützow himself was twice wounded, was captured at Ligny and released at Waterloo. By then, however, the volunteer unit had been disbanded, and its cadre used to form two regular regiments. The corps is celebrated in Th. Körner's poem *Lützows wilde Jagd* (1813), set to music by C. M. von Weber (qq.v., 1814), which has ensured its continuing popularity.

LÜTZOW, THERESE, FREIFRAU VON, see STRUWE, THERESE VON.

Luxemburg, Grand Duchy (Großherzogtum), adjoining Belgium, France, and West Germany. Originally a county, its count became Deutscher König (q.v.) in 1308, and in 1312 the Emperor Heinrich VII (q.v.). His grandson, the Emperor Karl IV (q.v.), raised it to the rank of a duchy. Luxemburg was sold to Burgundy by the Emperor Wenzel (q.v.), and in 1477 became with Burgundy an Austrian Habsburg (q.v.) possession by marriage. In 1555 the duchy passed to the Spanish Habsburgs. Louis XIV of France annexed it in 1684, but in 1697 it returned to Austrian ownership until the Revolutionary and Napoleonic Wars (q.v.), when it again became French (1794–1815). It was created a grand duchy at the Congress of Vienna (see WIENER KONGRESS), and until 1890 was linked with the Netherlands by personal union.

The neutrality of Luxemburg (and of Belgium) was decreed and guaranteed by the great powers in the Treaty of London in 1867. In 1914

German troops invaded Luxemburg on their way to France with the consent of the Grand Duchess Marie-Adelheid. On the German defeat in 1918 she was obliged to abdicate in favour of her sister Charlotte. When Luxemburg's neutrality was infringed again in 1940, the Grand Duchess with her ministers took refuge abroad until 1945. In 1948 Luxemburg became one of the three states of the Benelux union, and the condition of neutrality was abrogated at approximately the same time. In 1961 the Grand Duchess Charlotte abdicated in favour of her son Jean. The official language of Luxemburg is French, but French and German are taught on equal terms in schools.

LUXEMBURG, HEINRICH GRAF VON, see HEINRICH VII.

LUXEMBURG, ROSA (Zamosc, Russia, 1870–1919, Berlin), a Marxist politician, studied in Zurich; she settled in Germany, where she became prominent in the radical wing of the Social Democratic Party (see SPD). She was imprisoned in Russia after the revolution of 1905 and in Germany during the 1914–18 War. With Karl Liebknecht (q.v.) she formed the Spartacus League (see SPARTAKUSBUND), which after the armistice sought a Communist revolution. She joined the Communist Party in 1918. In January 1919 she was arrested, ill-treated, and shot. She published a number of books propagating social theory, of which the most important is *Die Akkumulation des Kapitals* (1913). She is a central figure in the novel *Karl und Rosa* by Alfred Döblin (q.v., 1950).

Luxemburger. The Emperors and German Kings of the house of Luxemburg comprise the following: Kaiser Heinrich VII (1308–13), Kaiser Karl IV (1346–78), König Wenzel (1378–1400), and Kaiser Sigismund (1410–37), qq.v.

Lyrisches Intermezzo, a collection of lyric poems by H. Heine (q.v.), first published in *Tragödien, nebst einem lyrischen Intermezzo* (1823), where, placed between the two tragedies, it is truly an intermezzo. Heine gives the date of composition as 1822–3. *Lyrisches Intermezzo* was afterwards included in the *Buch der Lieder* (q.v., 1827). It contains 66 poems, many of which are among Heine's finest lyrical productions. Among them are the sixteen songs included in R. Schumann's *Dichterliebe* (q.v.). They are almost exclusively concerned with the theme of unhappy love.

M

MAASS, Joachim (Hamburg, 1901–1972, New York), journalist and author in Hamburg and Altona until 1933, emigrated to the U.S.A. and settled in New York, though after the 1939–45 War he returned temporarily to Germany. He was primarily a novelist whose psychology suggests Dostoevsky and whose irony recalls Thomas Mann (q.v.). His principal novels are *Bohème ohne Mimi* (1930), *Der Widersacher* (1932), *Die unwiederbringliche Zeit* (1935), *Ein Testament* (1939), *Das magische Jahr* (1945), and (usually reckoned his masterpiece) *Der Fall Gouffé* (1952), based on a notorious French murder case of the 19th c. In 1957 he published a biography of H. von Kleist (q.v.) entitled *Kleist, die Fackel Preußens.*

Maastrichter Osterspiel, a medieval religious play, probably of the 14th c. Its action, which is unfinished, runs from the Creation to the Garden of Gethsemane.

Mabuse, Dr., see Jacques, N.

Maccaronische Dichtung, normally a form of light or comic poetry in which foreign (usually Latin) words or phrases are introduced into vernacular sentences and vernacular words tricked out with Latin inflections, making intentionally barbarous neologisms. The origination of this type of verse, which spread to all civilized countries of Europe, is attributed to the Italian humanist Tifi degli Odasi (d. 1488), and it was the humanists' preoccupation with Latin which fostered the rise of the genre. The best-known German examples (all anonymous) are *Die Flohiade* (1593), *Cortum Carmen de Rotrockis atque Blaurockis* (1600), *Delineatio Lustitudinis studenticae* (1627), and *Gaudium studenticum ex autographo* (1693). The potentialities of this rather crude humour were limited and it was rarely attempted after 1700. Two exceptional attempts in serious macaronic verse, both attributed to one anonymous author, are the nuptial poems *Rhapsodia versu heroico macaronico ad Brautsuppam* and its similarly titled sequel *Rhapsodia andra*, etc., written in the early 18th c. Although the expression only dates from the late 15th c., it is now sometimes used as a convenient designation for medieval poetry in alternate lines of Latin and German.

MACKAY, John Henry (Greenock, 1864–1933, Berlin-Charlottenburg), was brought to Germany in infancy and was educated there, studying at Kiel, Leipzig, and Berlin universities. After some years of travel he settled in Charlottenburg in the 1890s. As a left-wing thinker with anarchistic leanings Mackay was in disfavour with the authorities. He is the author of the novels *Die Anarchisten* (1891), *Der Schwimmer* (1901), *Der Sybarit* (1903), and *Der Freiheitssucher* (1920, a sequel to his first novel). He also published a volume of poetry (*Sturm*, 1887) and collections of stories (*Die Menschen der Ehe*, 1892; *Staatsanwalt Sierlin*, 1927).

Mackay wrote a biography (1898) of Max Stirner (q.v.), and was associated with the club Durch (q.v.). *Gesammelte Werke* (8 vols.) appeared in 1911, a further volume in 1928.

MACKE, August (Meschede, 1887–1914, killed in action, Champagne), a painter who trained at Düsseldorf and Paris, influenced F. Marc (q.v.) in the direction of Expressionism. He was, with Marc and Kandinsky (qq.v.), a member of the Blauer Reiter (q.v.) group. Macke's own Expressionist style remained less abstract than that of his two colleagues. He wrote *Im Kampf um die Kunst* (1911), *Der blaue Reiter* (1912), *Kunst und Künstler, 12* (1914).

MACROPEDIUS, Georgius, pseudonym for Joris von Langenfeldt (Gemert, Holland, 1475–1558, 's Hertogenbosch), who belongs primarily to the literary history of the Netherlands, is the author of a number of Latin plays which are important also for German humanism. Among them are *Comediae sacrae*, for example, *Asotus* (1537), dealing with the prodigal son, *Lazarus* (1541), and *Hecastus*, which treats the theme of Everyman. His comedies (*fabulae ludicrae*) include *Aluta* (1535), directed against the peasant, and *Andrisca* (1537), the taming of a shrew.

Mädchen aus der Feenwelt, Das, better known by its sub-title *Der Bauer als Millionär*, a play by F. Raimund (q.v.), written in 1825–6, and first performed at the Theater in der Leopoldstadt, Vienna, in November 1826. It was published posthumously in vol. 2 of Raimund's *Sämtliche Werke* (1837), and is described as *Romantisches Original-Zaubermärchen mit Gesang in drei Aufzügen.*

Lottchen, the daughter of a fairy, wants to marry Karl, a fisherman. But Wurzel, her adoptive father, a farmer who has found a fortune and so become a millionaire, refuses to allow the match. With fairy aid, Wurzel's palace vanishes, and his wealth is dispersed. But now hostile spirits tempt Karl, offering him immense wealth. Faced with the choice of Lottchen or wealth, he opts for Lottchen and all ends happily. The play demands lavish spectacle in the magic scenes and is spiced with broad comedy in Viennese dialect. It contains musical interludes, of which the best known are the duet 'Brüderlein fein' and the aria known as *Der Aschenmann* ('So mancher steigt herum').

Mädchen aus der Fremde, Das, a poem by Schiller, first published in the *Musenalmanach* for 1797. The maiden (Das Mädchen) is a symbol for poetry.

Mädl aus der Vorstadt, Das, oder Ehrlich währt am längsten, a three-act farce (Posse) by J. N. Nestroy (q.v.), written in 1841 and first performed at the Theater an der Wien, Vienna, in November 1841. It was published in 1845. It has an implausible plot in which an attractive young widow (Frau von Erbsenstein) escapes an unsuitable second marriage and gives her hand instead to a middle-aged man of integrity and worth (Schnoferl). The rejected suitor is happily united with Thekla, the 'girl from the suburb' (Mädl aus der Vorstadt), and Schnoferl unmasks the dishonesty of Frau von Erbsenstein's uncle Kauz, who has allowed Thekla's innocent father to bear the blame for a theft. The play is an adaptation to a Viennese environment of a French piece, *La Jolie Fille du faubourg* (1840) by Paul de Kock and C.-V. Varin.

Maere vom Heiligen Kreuz, a Middle High German poem of roughly a thousand lines, the author of which gives his name as Helwig ('von Waldirstet Helwic'). He continues the history of the Holy Cross beyond the Crucifixion. Helwig appears to have been a Thuringian, and wrote his poem *c.* 1350.

Magda, title by which H. Sudermann's play *Heimat* (q.v.) is known outside Germany. It refers to the heroine, Magda Schwartze.

MAGDALENA SIBYLLE, HERZOGIN VON WÜRTTEMBERG (Darmstadt, 1652–1712, Stuttgart), though of Hessian birth, spent her childhood in Sweden, and was married to Duke Wilhelm Ludwig von Württemberg. She wrote two books of devotion, *Kreuzpresse, das ist das mit Jesu gekreuzigte Herz* (1690) and *Andachtsopfer* (1706).

Magdalis, name by which the poet J. C. Günther (q.v.) refers to his love Leonore Jach-

mann in his poems. On occasion he addresses her as Lenchen.

Magdeburg, German city, formerly in the Prussian Province of Saxony (Provinz Sachsen) and since 1949 in the German Democratic Republic (see DEUTSCHE DEMOKRATISCHE REPUBLIK). The city's coat of arms dates from the early 16th c. and represents a maiden (Magd) with a virginal wreath. Magdeburg accepted the Reformation (q.v.) in 1524. The most famous event in its history is the capture and sack of the city by Tilly's troops in the Thirty Years War (20 May 1631, see DREISSIGJÄHRIGER KRIEG). The disaster was augmented by a fire, which destroyed the greater part of the city; something like 30,000 of the inhabitants died by fire or sword. This episode of horror has long ranked as one of the great catastrophes of German history, and literary allusions to it are frequent. In 20th-c. literature two treatments stand out: that of Brecht in *Mutter Courage und ihre Kinder* and that of Gertrud von LE FORT in *Die Magdeburgische Hochzeit* (qq.v.).

Magdeburger Schöppenchronik, a chronicle of Magdeburg and the Saxon lands compiled between 1380 and 1516. It is the work of several hands; the additions made for the years 1422 to 1516 are sparser than the earlier portions.

Magdeburgische Hochzeit, Die, a historical novel by Gertrud von LE FORT (q.v.), published in 1938. Its theme is the siege and sack of Magdeburg (q.v.) in 1631 (see DREISSIGJÄHRIGER KRIEG). The dilemma facing both parties, the imperial marshal Tilly (q.v.) and the city council of Magdeburg, is expounded with sympathy and impartiality.

Magdeburg has been Lutheran for a century; the Edict of Restitution (1629) reimposes Roman Catholicism. Should the city yield and risk religious persecution, or should it resist its sovereign and rely on the support of Sweden? This dilemma forms the background; it is October 1630. Tilly disapproves of the Edict, but fails to secure its withdrawal. On both military and humane grounds he seeks to achieve a pact, but cannot arrive at an agreement. His generals are eager to storm the city. The indecisive council is overruled by the mob and by the stern single-minded Colonel Falkenberg, who represents the Swedes. An attempt at compromise by the young councillor Willigis Ahlemann results in his expulsion from the city, driving him into the camp of the besiegers.

The siege and storm are presented in allusion to the name Magde-burg (and the maiden in the city's coat of arms) as a wooing and marriage; this interpretation is repeated on an individual level, for at the time of his expulsion Willigis is about to marry a Magdeburg girl, Erdmuth Plögen. Erdmuth, the foolish maiden, throws herself at the head of the Swedish governor, believing that he will marry her.

The story is divided into the traditional phases of a citizen's wedding: *Der Jungfrauenabend, Der Ehrentanz,* and *Das Brautgemach.* In the 'bridal chamber' (*Das Brautgemach*) containing the third section, Willigis is sent by Tilly as a *parlementaire* to the city council bearing an ultimatum. The council hesitates, cowed by Falkenberg, and the time-limit passes. Tilly still hopes that the first phases of the storm will force the council to capitulate, but his aim is frustrated by the precipitate action of his general of horse, Pappenheim (q.v.). A number of survivors take refuge in the cathedral, where Tilly stations a guard for their protection. Among them is Erdmuth Plögen, carried there by Willigis after being ravished by a Croatian soldier. Willigis and Erdmuth are married on the spot, at Willigis's request, by Pastor Bake. Bake's wife and four children are also among the refugees, and the wife bears his fifth child in that fearful night. The following day Tilly reluctantly agrees to the celebration of a *Te Deum* in the cathedral. Bake, entering his beloved cathedral for the last time, is offended by the Roman liturgy. But when the Nicene Creed (common to Lutheran and Roman) is sung he realizes that the faith is one and indivisible, and he joins in at the words: 'Confiteor unum baptisma in remissionem peccatorum'.

This charitable and compassionate, and yet gripping novel avoids any excess of sentiment largely by the extensive use of reported speech (oratio obliqua).

Magelone, see SCHÖNE MAGELONE, DÍE.

Magezoge, Der, a Middle High German poem of unknown authorship. It comprises some 400 lines and purports to be a mirror of virtue and a tutor to youth: 'Ich heize ein spiegel der tugende/unde ein magezoge der jugende.'

It gives in clumsy verse practical advice for the young nobleman. The author, who wrote about the middle of the 13th c., is presumed to have been an Austrian.

Magischer Idealismus, term employed by Novalis (q.v., *Neue Fragmentensammlung,* 1798) to express his conception of Romantic idealism seeking to transcend the idealism of Kant and of Fichte (qq.v.). The concept 'magic' links up with Novalis's preoccupation with 'Magie' as expressed, for example, in the statement: 'Alle

Erfahrung ist *Magie*—nur magisch erklärbar' (*Fragmente und Studien,* §410).

Magischer Realismus, term used by G. Saiko (q.v.) to denote his manner of writing.

Magnetiseur, Der, a story by E. T. A. Hoffmann (q.v.), written in 1813 and published in 1814 in vol. 2 of *Fantasiestücke in Callots Manier.* It is described as *Eine Familienbegebenheit.* Alban, gifted with hypnotic powers, stays with a baron's family, subdues the daughter Marie to his will, and causes her death. A later version of the same theme is *Der unheimliche Gast* (q.v.), in which the young woman (Angelika) escapes from the power of her sinister lover.

Magus im Norden, Der, name by which J. G. Hamann (1730–88, q.v.) was known to his contemporaries.

MAHLER, GUSTAV (Kalischt, Bohemia, 1860–1911, Vienna), Austrian composer, was conductor at the opera houses of Prague (1885), Leipzig' (1887), Budapest (1891), Hamburg (1897), and Vienna (1907). Mahler's music, almost exclusively orchestral with or without voices, contains many settings of German texts. These include the last scene of Goethe's *Faust* (q.v.) Pt. II (8th Symphony), poems from *Des Knaben Wunderhorn* (q.v., 2nd, 3rd, and 4th Symphonies, and *Lieder aus des Knaben Wunderhorn*), Rückert, q.v. (*Kindertotenlieder*), Klopstock, q.v. (2nd Symphony), Nietzsche, q.v. (3rd Symphony), and translations from the Chinese by Hans Bethge, q.v. (*Das Lied von der Erde*). *Das klagende Lied* and the *Lieder eines fahrenden Gesellen* have texts written by Mahler himself; and the first movement of the 8th Symphony is a setting of the medieval Latin hymn *Veni creator spiritus* by HRABANUS Maurus (q.v.).

Mahler's widow married the writer F. Werfel (q.v.) in 1929. *Briefe,* ed. Alma Mahler, appeared 1924, and *Gesamtausgabe,* ed. E. Ratz, 1960 ff.

MAHLMANN, SIEGFRIED AUGUST (Leipzig, 1771–1826, Leipzig), was a private tutor until 1799 and thereafter a bookseller and editor of journals (from 1806 to 1816 of the *Zeitung für die elegante Welt,* q.v.). He wrote stories and a novel (*Albano der Lautenspieler,* 1802), as well as a parody of A. von Kotzebue's *Die Hussiten vor Naumburg* (*Herodes vor Bethlehem oder Der triumphierende Viertelsmeister. Ein Schau-, Trauer- und Tränenspiel in drei Aufzügen. Als Pendant zu den vielbeweinten Hussiten vor Naumburg,* 1803). His poems (*Gedichte,* 1825) were much read in the 19th c. Particularly popular were *Sehnsucht* ('Ich denk' an euch, ihr himmlisch schönen Tage',

1802) and *Weinlied* (1808), which contains the line 'Mein Lebenslauf ist Lieb' und Lust'.

Mahomets-Gesang, a hymnic poem written by Goethe in 1772 or 1773. It was first printed in the *Göttinger Musenalmanach* for 1774. The growth of genius is symbolized in the course of the stream growing irresistibly into a great river. The poem was intended to form part of an unwritten drama on Muhammad.

Mährische Brüder, see Böhmische Brüder.

MAIER, Jakob (Mannheim, 1739–84, Mannheim), an administrator in Mannheim, wrote the historical plays *Sturm von Boxberg* (1778) and *Fust von Stromberg* (1784).

Maigesetze, see Kulturkampf.

Maigraf, a male equivalent to the English Queen of the May. The custom of choosing a Maigraf and carrying him in procession was widespread in the Hanseatic cities of North Germany.

Maikäferbund, a Rhenish literary group (1840–6) which wore at its weekly sessions the cockchafer (may-bug) as its emblem. It was founded by G. Kinkel; K. Simrock and J. Burckhardt (qq.v.) were among its members. The only extant edition of its proceedings, the last (7 July 1846), is in the library of Bonn University.

Mailied, a poem written by Goethe at Strasburg in 1771 and first printed in 1775 in Jacobi's *Iris* (q.v.). It is an ecstatic outpouring in which love and nature coalesce. The present title dates from 1789; it replaces the original heading *Maifest*.

Mainauer Naturlehre, a Middle High German treatise which is concerned with geography, astronomy, and medicine and noteworthy for its assertion that the earth is spherical. It was written in the 13th c., possibly *c.* 1300. It has been attributed by some to Hugo von Langenstein (q.v.), which would imply a later date.

Mainlinie, a division between North and South Germany made by the course of the River Main. From 1867 to 1871 the Mainlinie formed the actual frontier between the North German Confederation (see Norddeutscher Bund) and the southern German states. Generally speaking, Prussian influence predominated to the north of the Main, whereas until 1870 the southern states inclined towards Austria.

Mainz, one of the oldest cities in Germany, grew up around the Roman fortress of Maguntiacum. It was from early times the seat of a bishop, and became under Boniface an archbishopric (747); its holders, one of whom was Hrabanus Maurus (q.v.), were the primates of the German Church. By the 12th c. Mainz was the centre of an important temporal state ruled by the Archbishop, who was hereditary chancellor of the empire and later an Elector. It was the scene in June 1184 of a brilliant courtly festival (Mainzer Pfingstfest) under the Emperor Friedrich I (q.v.). In the 15th c. Mainz had a special significance as the centre of printing, the inventor of which, Gutenberg (q.v.), was born in Mainz *c.* 1398 and spent much of his life there. During the French Revolution (q.v. and see Revolutionskriege) Mainz was for a short time (1792–3) the centre of a separatist movement under Georg (or George) Forster (q.v.). The University of Mainz was founded in 1477 and dissolved in 1798. It was refounded in 1946.

Mai und Beaflor, an anonymous Middle High German narrative poem of nearly 10,000 lines, written in the second half of the 13th c. It is a story of slandered innocence. Beaflor, escaping from the sexual pursuit of her father, marries Count Meie. During his absence she is accused by her mother-in-law of infidelity and undergoes much suffering with unchanging patience and humility before she is reunited with her husband.

Majestätsbrief (Letter of Majesty), a document issued on 9 July 1609 by the Emperor Rudolf II (q.v.) to the Bohemian estates. It guaranteed the Bohemians the free exercise of religious beliefs (Gewissensfreiheit) without social discrimination and gave the estates the right to erect churches and schools. 'Defensors' were to administer the interests of the Protestant communities. The concessions failed to secure for Rudolf the loyalty of the Bohemians (he was deposed in 1611), but, for the Protestant estates, they were a temporary triumph against the measures of the Counter-Reformation (see Gegenreformation). The Letter of Majesty and other concessions made by Rudolf's brother Matthias (q.v.) to Moravia and Hungary succeeded only in shelving the conflict which resulted in the outbreak of the Thirty Years War (see Dreissigjähriger Krieg) in 1618, when a renewed dispute over the interpretation of the terms of the Letter of Majesty led to a revolt in Bohemia which was initially a purely internal conflict. An incident in the Hradschin in Prague (23 May 1618) in which two governors for Matthias (Jaroslav Martinitz and William Slavata) were thrown out of the window by angry Protestant representatives, headed by

Count von Thurn, sparked off the armed conflict and is known as the Prager Fenstersturz. The officials survived their rough treatment, falling into the moat, though others say that they landed on a manure heap. The *coup d'état* was intended as a first step towards the dethronement of Ferdinand (see FERDINAND II), whose election as hereditary king the Bohemians by now bitterly regretted. After the defeat of the Bohemian 'Winter King' (see FRIEDRICH V VON DER PFALZ) in 1620, Ferdinand (since 1619 emperor) demonstrated his future policy in Bohemia by literally tearing up the Letter of Majesty.

MAJOR, JOHANNES (Joachimstal, Bohemia, 1533–1600, Wittenberg), a humanist writer of Latin poetry, was also a bitter theological controversialist, especially hostile to FLACIUS Illyricus (q.v.) in *Synodus avium* (1557).

Majorat, Das, a story by E. T. A. Hoffmann (q.v.) published in 1817 in vol. 2 of *Nachtstücke.* It is a ghost story with a partial explanation. The narrator, Theodor, visits, with his lawyer great-uncle V., the great house of a baron on the coast of Courland and experiences in the first two nights ghostly moanings and scratchings. They are exorcized by V., and Theodor then spends happy days playing the piano to the baron's young wife, to whom he becomes greatly attracted. He recounts to her his ghostly experiences, whereupon she falls ill. Theodor leaves with great-uncle V. and does not return. Some months later V. reveals a secret: when the entailed estate came to the present owner's father, he had reason to suspect the castellan Daniel of conspiring against him. Daniel had then murdered his new master by pushing him off a ledge of a part of the building which had collapsed. V. discovers this and Daniel dies while sleep-walking at the scene of his crime. The moanings and scratchings heard by Theodor are produced by Daniel's ghost.

Makkabäer, a Middle High German biblical poem written in the 14th c. by a knight of the Teutonic Order (see DEUTSCHER ORDEN), possibly Luder von Braunschweig (d. 1335), later High Master of the Order. The heroic story of the Maccabees (contained in the Vulgate and in the Protestant Apocrypha) was particularly appropriate to the conditions and mode of life of the Teutonic Knights.

Makkabäer, Die, a five-act tragedy (Trauerspiel) in blank verse completed by O. Ludwig (q.v.) after a third and final revision in 1852, and first produced in Vienna by H. Laube (q.v.) towards the close of the same year. It was published in 1854. Its subject is the Maccabean

revolt against Antiochus Eupator, and the play is set between 167 and 161 B.C. Its political theme is the struggle of the Jews under the leadership of Judas Maccabaeus ('Judah' in Ludwig's play) which, after many vicissitudes, culminates in victory. With it is interwoven the tragedy of Judah's mother Lea, who puts her favourite son before Judah, and in the end loses this child as well as two others. The second version was entitled *Die Mutter der Makkabäer.*

Makkaronische Dichtung, see MACCARONISCHE DICHTUNG.

MALER MÜLLER, see MÜLLER, FRIEDRICH.

Maler Nolten, a two-volume novel by E. Mörike (q.v.). Written between 1828 and 1832 (mainly in 1830) and published in 1832, it is one of the many novels of the early 19th c. which are indebted to Goethe's *Wilhelm Meisters Lehrjahre* (q.v.). *Maler Nolten* is a confused novel in which the principal characters are depressives and hypochondriacs, who succumb to the burdens of life.

Theobald Nolten himself is an artist who has forsaken his rustic love Agnes for Countess Konstanze. His actor friend Larkens endeavours to bring Nolten and Agnes together again and is at first successful; but Larkens and Agnes put an end to their lives, and Nolten dies because he lacks strength to face life. A mysterious vagrant, Elisabeth, who intervenes in Nolten's life, also perishes. The first book contains the verse play *Der letzte König von Orplid* (see ORPLID), which Larkens and Nolten perform at court. They are suspected of seditious political symbolism and are imprisoned for a time.

The novel contains a number of Mörike's best-known poems, including part of *Der Feuerreiter, Gesang zu zweien in der Nacht, Elfenlied* (these three in *Der letzte König von Orplid*), *Das verlassene Mägdlein, Er ist's, Im Frühling, Der Jäger, Agnes, Jung Volker,* four *Peregrina* poems, a stanza from *Gebet,* the sonnet *An die Geliebte,* and *Lied vom Winde* (qq.v.). The titles of these poems are omitted in the novel.

The first published version of the work, also referred to as the *Urnolten,* was termed Novelle (*Maler Nolten, Novelle in zwei Teilen von Eduard Mörike. Mit einer Musikbeilage*). Throughout his life, and especially in the last two decades, Mörike, who was dissatisfied with his novel, worked at a revision, which was approaching completion at his death. It was published in 1877 by J. Klaiber, who disregarded some of Mörike's amendments and made his own arbitrary alterations.

MALSS, KARL (Frankfurt/Main, 1792–1848, Frankfurt), served as a volunteer in the Wars of Liberation, becoming a captain of horse. He afterwards studied at Gießen and qualified as an architect. In 1827 he was appointed director of the Frankfurt theatre. He himself wrote plays (Lokalstücke, q.v.) in Frankfurt dialect. Titles include *Die Entführung oder Der alte Bürgerkapitän* (1820), *Das Stelldichein* (1832), *Herr Hampelmann oder die Landpartie nach Königstein* (1833), *Herr Hampelmann im Eilwagen* (1833), *Herr Hampelmann sucht ein Logis* (1834), and *Die Jungfern Köchinnen* (1835). They were collected in *Volkstheater in Frankfurter Mundart* (1849).

Malteser, Der, designation of Marquis Posa in Schiller's *Don Carlos* (q.v.); it refers to his brave conduct in the siege of Malta in 1565 (see JOHANNITERORDEN).

Malteser, Die, a play projected by Schiller. Although Schiller worked at it intermittently throughout the 1790s until after the turn of the century, it remained unwritten. The action was to centre on the siege of Malta by the Turks in 1565, and it was to portray the pure spirit ('reinen Geist') of the Knights of St. John (see JOHANNITERORDEN).

MALTITZ, GOTTHILF AUGUST, FREIHERR VON (Königsberg, 1794–1837, Dresden), served as a volunteer in the Wars of Liberation and was afterwards a forestry supervisor (Oberförster). He began to write plays (*Schwur und Rache*, 1826; *Hans Kohlhaas*, 1827; *Der alte Student*, 1828) and was in trouble with the censorship, the instructions of which he disregarded in the performance of his third play. He had in consequence to give up his post, and left Prussia. After a spell in Hamburg and a visit to Paris at the time of the July Revolution, he settled in Dresden in 1831.

Manche freilich . . ., heading of a poem by H. von Hofmannsthal (q.v.), the first line of which runs 'Manche freilich müssen drunten sterben'.

Manessische Liederhandschrift, see HEIDELBERGER LIEDERHANDSCHRIFT, GROSSE.

MANGOLD, BURK, see BURK MANGOLD.

MANN, GOLO, real name Gottfried (Munich, 1909–), historian, writer, and publicist and third child of Th. Mann (q.v.), qualified for an academic career under K. Jaspers and F. Gundolf (qq.v.) and taught from 1929 at Heidelberg University. In 1933 he emigrated to France, and 1937–40 edited the periodical

Maß und Wert in Zurich. In 1940 he was interned in France, but escaped to the U.S.A. and became professor of history at Olivet College (1942–3) and Claremont Men's College (California, 1947–57). From 1960 to 1964 he was professor of political science at Stuttgart University (Technische Hochschule).

Golo Mann is the author of a major work on 19th- and 20th-c. German history (*Deutsche Geschichte des 19. und 20. Jahrhunderts,* 1959). In his historical writings he has made particular contact with literature, and has stated this approach in newspaper publications and as an essayist (*Geschichtsschreibung als Literatur*). In 1964 he published a study of the Emperor Wilhelm II (q.v.), *Wilhelm II*, and in 1971 his extensive work on Wallenstein (q.v.), *Wallenstein. Sein Leben*.

Golo Mann's interest in the function of politics in literature emerges in the early monograph on F. von Gentz (q.v., *Friedrich von Gentz*, 1947), and he engaged in polemics in his study of the plays of Brecht (q.v.), *B. B.—Maß oder Mythos? Ein kritischer Beitrag über die Schaustücke Bertolt Brechts* (1958). He received the Büchner Prize in 1968, and in 1971 he published a study of G. Büchner (q.v.), *Georg Büchner und die Revolution*. In 1968 appeared *Preußen. Porträt einer politischen Kultur*. His views on Heinrich Mann (q.v.) aroused controversy. *Mein Vater Thomas Mann* appeared in 1970. Subsequent publications include *Geschichte als Ort der Freiheit* (1974), *Zeiten und Figuren* (1979), *Was ist Demokratie* (1980), and *Nachtphantasien* (1982). Golo Mann is the editor of and contributor to a ten-volume history of the world (*Propyläen-Weltgeschichte,* 1960–5).

MANN, HEINRICH (Lübeck, 1871–1950, Los Angeles), in full Luiz Heinrich Mann, was the eldest son of a well-to-do corn factor of Lübeck and the elder brother of Th. Mann (q.v.). He had his schooling at the Katharineum in Lübeck, and in 1869 was sent to Dresden to train for the book trade. In the following year he joined S. Fischer-Verlag, then newly founded in Berlin. The death of his father in 1891 closed the family home, and Heinrich Mann, after a short stay in Munich with his mother, moved to Italy, settling in Florence, spending some time in Palestrina, and visiting sanatoria in Switzerland and northern Italy.

Mann's first novel, *In einer Familie*, set in Dresden, appeared in 1894; he revised and republished it in 1924, and considered *Im Schlaraffenland* (q.v.) as his first work. His travels in southern Europe are reflected in the settings of many of his novels, but in this one he wrote a satire upon high life in Berlin.

Mann's next important publication was an

exotic trilogy, *Die Göttinnen oder Die drei Romane der Herzogin von Assy* (1903). Powerfully influenced by Nietzsche (q.v.) and d'Annunzio, the three novels are devoted to the dynamic duchess's vain endeavour to justify her existence, first in politics in an imaginary Dalmatian state, then in art, and finally, in the sultry third book, in love. The novels of the Duchess Violante d'Assy are entitled *Diana*, *Minerva*, and *Venus*. The hectic and caricatured erotic element is also dominant in Mann's next novel, *Die Jagd nach Liebe* (also 1903), set in Munich and Berlin; it is partly a *roman à clef*, since two of the principal characters are modelled on the Munich connoisseur and dilettante A. W. Heymel and the poet R. A. Schröder (qq.v.).

The now well established tendency of Mann towards caricature and sarcasm found its full expression in his novel *Professor Unrat oder Das Ende eines Tyrannen* (q.v., 1905), and to this his native Lübeck is central; it became years later by far the best known of his works, for in 1930 it was converted into a film under the title *Der blaue Engel* with Emil Jannings and Marlene Dietrich in the principal roles. On republication in 1947 the book bore, with Mann's reluctant consent, the film title, *Der blaue Engel*. The next novel, *Zwischen den Rassen* (1907), is concerned, not with anti-Semitism, on which Mann also wrote, but with the contrast between Nordic and Latin, of which he, as the son of a Lübeck father and a Brazilian mother, was especially conscious. Lola, the actress heroine, fails in her career, but is saved from a sensual surrender to the possessive and aggressive Italian Pardi by the determined intervention of the more congenial Arnold Acton. The novel *Die kleine Stadt* (q.v., 1909) appeared in the year in which Mann went with his brother Thomas to Italy; the work turns from private life to democratic politics in a small Italian town. In 1914 Mann married the Czech actress Maria (Mimi) Kanova. The marriage lasted until 1930. Another liaison, with Nelly Kroeger, was legalized in 1939; she took her own life in 1944. With the suicide of his actress sister in 1910 this was the second suicide in Mann's immediate family; a third was that of his nephew Klaus (in 1949, see MANN, K.).

In 1915 Heinrich Mann responded to public approval of the war expressed by his brother Thomas with an essay, *Emile Zola* (published in the periodical *Die Weißen Blätter*), in which he, a pacifist, attacked misguided nationalism. The relationship between the brothers never fully recovered from this public rift. In the years preceding the 1914–18 War Mann had begun work on three satirical novels which analyse and deride the Germany of Wilhelm II (q.v.). Jointly they are referred to as *Die Kaiserreich-Trilogie*, though the three components, *Der*

Untertan (1918), *Die Armen* (1917), and *Der Kopf* (qq.v., 1925), were published as separate works. *Der Untertan*, instalments of which were published 1911–13, was delayed by the war. The relentlessly savage and negative tone of the trilogy gives way to a more conciliatory note in the succeeding novels *Mutter Marie* (1927), *Die große Sache* (1930), and *Ein ernstes Leben* (1930), which, however, aroused little response since critics held, rightly or wrongly, that they were written without full conviction. A novel of more persuasive character appeared between the first and second of these works, *Eugénie oder Die Bürgerzeit* (q.v., 1928).

Mann, seeing clearly the dangers into which Germany was declining, was increasingly active with political speeches and essays. In 1931 he became president of the literary section of the Prussian Academy. In 1933 he was promptly dismissed from this office and deprived of his German citizenship. He was granted Czech nationality in 1936. Mann took refuge in France. French intellectual life was so congenial to him that he hardly felt himself in exile. In this temporary contentment he wrote his most conciliatory work, the double historical novel on the life of King Henry IV of France, *Die Jugend des Königs Henri Quatre* (1935) and *Die Vollendung des Königs Henri Quatre* (1938, see HENRI QUATRE-ROMANE), which is generally held to be the summit of his achievement.

In 1940 Mann escaped through Spain and Portugal to the U.S.A., where he settled in Los Angeles, close to his brother. He did not feel at ease in his new environment, and this may be in part responsible for a perceptible recession in his later novels. *Lidice* (1943) treated an atrocity committed in 1942 on orders from Berlin (in Czechoslovakia) in a tone which seemed (and not only to the Czechs) improper for such a tragedy. *Der Atem* (1949) and *Empfang bei der Welt* (posth., 1956) are structurally weak and unnecessarily multilingual. The unfinished historical novel on Friedrich II (q.v.) of Prussia, *Die traurige Geschichte von Friedrich dem Großen* (q.v., posth., 1960), which is written in the form of a dialogue, suggests a negative counterpart to the warmth and sympathy of the novels on Henry IV. After the 1939–45 War, Mann hoped for a recall to Germany, but until the end of the 1940s only feelers without precise proposals came from East Germany. In 1949 he accepted appointment as president of the newly founded Academy of Arts of the German Democratic Republic. Mann died while preparing for the move.

Mann was an impatient writer, and his insufficient attention to precision and neatness of style, coupled with his addiction to an irony so harsh that it becomes sarcasm, accounts, at least in part, for his failure to secure instant recognition. These characteristics are also evident in his numerous Novellen, written in bursts of activity, of which *Pippo Spano* and *Abdankung* (both 1905), *Die Branzilla*, *Das Herz*, and *Die arme Tonietta* (all 1908), and *Kobes* (1925), a caricature of the industrialist Stinnes, deserve mention; they are likewise manifest in his plays, two of which are Novellen in dialogue and were subsequently performed as one-act plays (*Der Tyrann*, 1908, and *Die Unschuldige*, 1908). Of his full-length plays *Schauspielerin* (1911), *Die große Liebe* (1912), and *Madame Legros* (1913) are the most noteworthy. Among his numerous, often polemical, essays the collections *Geist und Tat* (1931) and *Der Haß* (1933) should be mentioned. Reminiscences and observations are included in *Ein Zeitalter wird besichtigt* (1946).

Heinrich Mann was subject to powerful and conflicting emotions. Anger is linked with compassion, though it contrives to be more conspicuous, and both are part of his remarkable love–hate relationship with his native Germany. In many essays he expresses his democratic views, which inclined towards an idealistic communism, but he was repelled by the totalitarian aspects developed under official Communism. The high inner tension of his personality appeared to relax in Italy and France, and of this a passion for Puccini's operas is an unexpected symptom. The one work in which warmth of feeling, compassion, and good humour are fully revealed is the double novel written in France and dealing with a French theme, the life of Henry of Navarre.

Gesammelte Werke (12 vols.) appeared 1958–74, and *Gesammelte Werke* (planned in 25 vols.), ed. S. Anger, 1965 ff. A select edition of essays, ed. H. M. Enzensberger (q.v.), appeared 1968 as *Politische Essays*, and in the same year appeared *Thomas Mann und Heinrich Mann. Briefwechsel*, ed. U. Dietzel and H. Wysling.

MANN, KLAUS (Munich, 1906–49, by suicide, Cannes), in full Heinrich Klaus Mann, a son of Th. Mann (q.v.), worked as a dramatic critic and advisor (Dramaturg) in Berlin, where he collaborated for a time with Pamela Wedekind (daughter of F. Wedekind, q.v.), G. Gründgens (q.v.), and his sister Erika Mann (1905–69). After his emigration he collaborated for two years with A. Huxley, A. Gide, and Heinrich Mann (q.v.) in *Die Sammlung*, a periodical for emigrants. In 1936 he went to the U.S.A. where he continued to work as a journalist, which involved his return to Europe (1938) as a correspondent for the Spanish Civil War. His tract *Gottfried Benn, die Geschichte einer Verirrung* (1937 in *Das Wort*, Heft 9) opened the so-called

Expressionismusdebatte (see LUKÁCS, G., and WALDEN, H.).

His creative writing, begun in the second half of the 1920s, ceased after the beginning of the 1939–45 War. His more important works are the novels *Symphonie pathétique* (1935), which has Tchaikovsky as its subject, *Mephisto. Roman einer Karriere* (1936), and *Der Vulkan. Roman unter Emigranten* (1939). *Kind dieser Zeit* (1932) is an autobiography. He is the author of a study on Gide, written in English (1943) and published in German entitled *André Gide. Die Geschichte eines Europäers* (1948).

Klaus Mann was overshadowed by his father and disorientated by the experiences of his time, on which he commented in *The turning point. Thirty-five years in this century* (1942, posthumous publication in German, *Der Wendepunkt*, 1952). This turning-point came when Klaus and Erika Mann urged their parents to remain in Switzerland, where they were on holiday (February 1933). It was the beginning of emigration.

Two vols. of essays, *Prüfungen. Schriften zur Literatur* and *Heute und morgen. Schriften zur Zeit* (from 1922–49), ed. M. Gregor-Dellin, appeared 1968–9 and two vols. of correspondence, *Briefe und Antworten*, ed. M. Gregor-Dellin, in 1975, *Die Erzählungen* in 1976, *Woher wir kommen und wohin wir müssen. Frühe und nachgelassene Schriften* in 1980, *Werke in Einzelausgaben* 1963 ff.

MANN, THOMAS (Lübeck, 1875–1955, Kilchberg nr. Zurich), was the son of a prosperous corn factor and patrician of Lübeck, who was a member of the Lübeck Senate. Mann's mother, Julia da Silva-Bruhns, was of South American descent, part Portuguese and part Creole. The vivid contrast between the two parental lines provides a motif which runs through most of Mann's work. His elder brother was the novelist Heinrich Mann (q.v.). After the death of their father in 1891 the family moved to the more congenial atmosphere of Munich, where Thomas Mann stayed until 1933 with only the interruption of short spells in Italy, for a time with his brother Heinrich, in the late 1890s. In 1905 he married Katja Pringsheim, the daughter of a professor of mathematics at Munich University who was also a well-known authority on R. Wagner (q.v.). The couple had six children.

Thomas Mann's career as a writer began in the 1890s. He entered a bank in 1894, resigned to go to Italy, and on his return to Munich abstained from any formal study; an appointment on the editorial staff of *Simplicissimus* (q.v., 1899–1900) was abandoned with a sense of relief. At this time he was known as the author of short stories which appeared as *Der kleine Herr Friedemann* (the title of one of the stories) in 1898. In 1901 the two vols. of the epic family chronicle *Buddenbrooks* (q.v.) established his name as a writer and secured the family existence; its subject is the decline of a Lübeck patrician family through four generations. Mann's poised style and keen irony, basic features of his work, are clearly evident. In 1903 followed two of his principal Novellen, *Tonio Kröger* and *Tristan* (qq.v.). Both set burgher and artist in contrast, revealing the insensitiveness of the former and the decadence of the latter; both show the influence of Schopenhauer (q.v.) and Wagner on Mann's early phase of writing. In 1906 he made an unsuccessful excursion into drama with *Fiorenza* (performed 1907), and in 1909 published his second novel, *Königliche Hoheit* (q.v.). 'His Royal Highness' is Klaus Heinrich (the name of Mann's oldest son, see MANN, K.), son of the fictitious Grand Duke Albrecht III and his consort Dorothea. Klaus Heinrich marries a commoner whom he loves, and the wealth of her American father becomes an asset to the state. The writing of *Wälsungenblut* (1921) falls into this period. In 1912 appeared Mann's third great Novelle, *Der Tod in Venedig* (q.v.), in which the incompatibility of a respectable life with artistic talent is treated in a manner differing from that of the earlier works. A collection entitled *Tonio Kröger* and the Novelle *Das Wunderkind* appeared in the year of the outbreak of the 1914–18 War. Mann's support of the war led to a radical change in his relationship with his brother Heinrich, with whom he had in the 1890s contemplated joint authorship in *Buddenbrooks*. Heinrich, a strong pacifist, scorned his brother in thin disguise in *Emile Zola* (1915) and, despite a partial reconciliation in 1922 and common soil in the U.S.A. in the 1940s, personal and political differences kept them apart. The immediate outcome of this sad chapter was the lengthy tract *Betrachtungen eines Unpolitischen* (1918). The transition into the 1920s was made with the agreeable story *Herr und Hund* (1919). The *Bekenntnisse des Hochstaplers Felix Krull. Buch der Kindheit* (1922) was never completed, but appeared greatly extended as Mann's last novel in 1954 (*Bekenntnisse des Hochstaplers Felix Krull*, q.v.). It is a revival of the Schelmenroman (q.v.) or picaresque novel.

In 1924 appeared the 2 vols. of Mann's second great novel, *Der Zauberberg* (q.v.); it presents an assessment of the intellectual state of the period preceding the 1914–18 War and acutely diagnoses its ills. *Unordnung und frühes Leid* (q.v., 1926) portrays a cross-section of family life. In the 1920s Mann supported the Weimar Republic on his many lecture tours in Germany and abroad, and in 1929 was awarded the Nobel Prize for Literature.

With *Mario und der Zauberer* (q.v., 1930) Mann first came to grips with fascism and its abuse of power. He was on holiday in Switzerland when he was warned about the political climate in Munich by his own children (Erika and Klaus). From February 1933 he stayed in Switzerland, and in 1936 publicly dissociated himself from the National Socialist regime in an open letter to Eduard Korrodi (1885–1955), the Swiss publicist who was on the editorial staff of the *Neue Zürcher Zeitung*. In 1938 Mann went as a visiting professor to Princeton before settling in Pacific Palisades in California (1941–52), where he was in close touch with other distinguished German emigrant writers and artists. Deprived of his German citizenship, he was granted Czech citizenship in 1936, and in 1944 he became a citizen of the U.S.A. He did not return to live in Germany, but spent the remainder of his life in Switzerland.

In the 1930s Mann wrote the biblical tetralogy *Joseph und seine Brüder* (q.v., 1933–42). This work afforded him precious relaxation in years of stress; he regarded it as an elevated comedy. Another serene and penetrating work of reconstruction appeared in the year of the outbreak of the 1939–45 War: *Lotte in Weimar* (q.v.) is based on a visit paid by Lotte Kestner (see BUFF, LOTTE) to Goethe's Weimar in 1816. A very different aspect of Mann found expression in *Doktor Faustus* (q.v., 1947), in which the composer Leverkühn epitomizes the degeneration of Germany in Mann's lifetime. Mann wrote, as an appendix to this novel, *Die Entstehung des Doktor Faustus* (1949, see DOKTOR FAUSTUS). *Der Erwählte* (q.v., 1951), which attracted less attention than most of Mann's novels, is a retelling of the legend recounted in Hartmann von Aue's *Gregorius* (q.v.). Mann's last completed work is the Novelle *Die Betrogene* (1953). It was written while he was fashioning Krull, the figure of the artist, in new, frankly immoral form, as the accomplished swindler and impostor.

After the 1914–18 War, Mann became a prolific essayist, writing on political and cultural themes, and using the essay as an instrument of education. His published collections include *Rede und Antwort* (1922), which contains notable essays on Th. Fontane and G. Keller (qq.v.), *Von deutscher Republik* (1923), *Bemühungen* (1925), *Pariser Rechenschaft* (1926), *Die Forderung des Tages* (1930), a renewed appeal in support of the Weimar Republic (q.v.), *Leiden und Größe der Meister* (1935), *Freud und die Zukunft* (1936), *Achtung Europa!*, *Dieser Friede* (both 1938), *Das Problem der Freiheit* (1939), *Neue Studien* (1948), and *Goethe und die Demokratie* (1949). Though Mann wrote no full-length studies of authors or artists, his essays include many stimulating and penetrating excursions into criticism: *Goethe und Tolstoj* (1923), *Freud, Goethe, Wagner* (1937), *Schopenhauer* (1938), *Michelangelo in seinen Dichtungen* (1950), and *Versuch über Schiller* (1955). A collection of previously published essays appeared in 1945 as *Adel des Geistes*. Mann also delivered notable addresses, including *Lübeck als geistige Lebensform* (1926), *Goethe als Repräsentant des bürgerlichen Zeitalters* (1932), *Deutschland und die Deutschen* (1947), *Ansprache im Goethejahr* (1949), and *Meine Zeit* (1950). *Prosa 1951–1955* appeared posthumously (1956).

Mann's decision to settle after the war in Switzerland was influenced by the attitude of sections of the German public as well as writers of the 'inward emigration' (see INNERE EMIGRATION). His prestige outside Germany, especially in the U.S.A., was remarkable, and to the Anglo-Saxon world he represented what he felt himself to be at a time of bitter controversy, a good German. Mann possessed immense creative and intellectual power, and a faculty for assimilating knowledge and injecting life into it, which is perhaps most clearly demonstrated in *Joseph und seine Brüder* and *Lotte in Weimar*. His vision, especially after 1918, embraced the temper and the problems of the Europe of his day. His style is intentionally mannered, yet lucid, and as an analyst he shows penetrating acuteness. He established as the dominant feature of his work a pervasive irony, which he sometimes directs upon himself but applies universally to his characters, whom he appears to survey from an eminence. He himself considered his many nuances of parody, which culminated in his portrayal of Krull, to be his peculiarly constructive contribution to the German literary tradition since Goethe and Schiller.

Gesammelte Werke were published in 1922–5 (10 vols.), and again 1929 ff. The *Gesamt-Ausgabe*, published in Stockholm (12 vols.), appeared 1938–56, and *Gesammelte Werke* (12 vols.) 1960–74. Three vols. of *Briefe*, ed. Erika Mann, appeared 1961–5, and *Thomas Mann und Heinrich Mann. Briefwechsel 1900–1949*, ed. U. Dietzel and H. Wysling, 1968; *Meine ungeschriebenen Memoiren* by Katja Mann, 1974, and *Briefe und Tagebücher*, 9 vols., 1976–82. *Gesammelte Werke in Einzelausgaben* (Frankfurter Ausgabe) appeared in 1980 ff. and *Tagebücher* in 1977 ff., both ed. by P. de Mendelssohn *et al.*

'Mann der Arbeit, aufgewacht', first line of a song written in 1863 by G. Herwegh (q.v.) for the Allgemeiner Deutscher Arbeiterverein, which was headed by F. Lassalle (q.v.).

MÄNNLING, JOHANN CHRISTOPH (Wabnitz, Silesia, 1658–1723, Stargard, Pomerania), a Protestant pastor in Kreuzburg (1688–1700),

then in Stargard, published an *ars poetica* (*Europäischer Parnassus*, 1685), which he expanded in a second edition (*Der europäische Helikon oder Musenberg*, 1704). An admirer of Lohenstein (q.v.), he published two books of extracts from his works (*Arminius enucleatus*, 1708, and *Lohensteinius sententiosus*, 1710). Männling wrote poetry published in *Poetischer Blumengarten oder Teutsche Gedichte* (1717). His hymns include 'Mein Jesus, der ist tot' and 'Gottlob, es ist nunmehr zu Ende'.

Mann ohne Eigenschaften, Der, an unfinished novel by R. Musil (q.v.). Begun in 1924, the First Book, of 123 chapters, was published in 1930. The general structure was indicated at the end of this volume. The *Erstes Buch*, composed of the *Erster Teil: Eine Art Einleitung* and the *Zweiter Teil: Seines Gleichen geschieht*, were to be followed by the *Zweites Buch* comprising *Dritter Teil: Die Verbrecher* (for which Musil had at one time in mind the alternative title *Das tausendjährige Reich*) and *Vierter Teil: Eine Art Ende*. The intended structure was symmetrical, the first and fourth parts being relatively short, and the third and fourth inordinately long. A second volume appeared in 1933, containing only the first 38 chapters of Bk. II, Third Pt. Musil was still working at it when he died, and a further volume, described as *Dritter Band*, was published in 1943 by his widow, leaving much MS. material unpublished. The entire incomplete novel was published in 1952 by A. Frisé, whose arrangement of the difficult additional material encountered criticism. The novel in this form comprises about 1,600 pages, of which the original *Erstes Buch* of 1930 occupies 680 pages. The *Zweites Buch*, so far as it goes, has 128 chapters, of which 49 are newly added or interpolated by Frisé in a hypothetical sequence.

This immense torso has an action, which changes direction at the beginning of Bk. II. The Introduction unveils the figure of Ulrich, 'the man without qualities', explains his detached and independent situation, contrasts him with his friends the married couple Walter and Clarisse, and expresses his interest and somewhat detached sympathy for the condemned sex-murderer, Moosbrugger. In the second part, Ulrich finds himself secretary to the 'Parallelaktion', which is to prepare a jubilee celebration for the seventieth anniversary of the accession of the Emperor Franz Joseph (q.v.) in 1918 eclipsing the celebration, planned in the German Empire, for the thirtieth year of the accession of Wilhelm II (q.v.). The irony underlying these two operations, begun in 1913, can escape no reader, since both nations were involved in war in the following year, Franz Joseph died in 1916 aged 86, and Wilhelm II was deposed in the year intended for celebration. The 'Parallelaktion' in particular allows Musil to set forth an elaborate, highly intellectualized portrait of the Austria of his day, with a reflection of the German Empire in the character of the Prussian industrialist Arnheim (based on W. Rathenau, q.v.). The shallowness of the glittering Austrian society is revealed, the foibles and futilities of human beings exposed in the many subsidiary figures, especially Walter, Clarisse, Bonadea, Diotima, Graf Leinsdorf, and Direktor Leo Fischl; the sombre forces below the surface are realized in Moosbrugger.

In the Second Book, Ulrich, after his father's death, turns to his married sister Agathe, whom he had hitherto disregarded. They become absorbed in each other in an incestuous love, which is an attempt to escape from the world and may be interpreted as a symbol of return to the womb. Ulrich is himself clearly related to Musil, and his innate repulsion of qualities (Eigenschaften) is an encapsulation against surrounding reality.

The novel is brilliantly written, with wit, urbanity, and searching analysis, but Ulrich regards all around him in a curiously dispassionate way. Excessively egocentric, Musil worked unceasingly in the endeavour to produce a masterpiece of outstanding quality. The work may be seen as an enormous sequence of essays, among which is a well-known ironical, yet sympathetic, lament for the old Austria (Kakanien, ch. 9, the name being derived from the abbreviation k.k., q.v.).

Mann von funfzig Jahren, Der, the longest of the stories inserted by Goethe into his novel *Wilhelm Meisters Wanderjahre* (q.v.). Written partly in 1807 and partly in 1828, it occupies chapters 3, 4, and 5 of Bk. II. The 'man of fifty' is a retired major, alert in mind and body, who finds that his prospective daughter-in-law, Hilarie, has, by imperceptible degrees, fallen in love with him. He visits his son Flavio and is consoled to find that Flavio is in love with a young widow. But the widow is drawn to the major, who reciprocates her love; and Flavio, after the initial shock, begins to love Hilarie, who returns his affection. All seems set for a happy ending, but the sensitive Hilarie feels that the episode of her love for the father has made it impossible for her to marry the son. Later in the novel (ch. 14, Bk. III) we learn that Hilarie has adjusted herself and has married Flavio, while the major has married the widow. Both couples live a harmonious married life.

Mantellied, a song occurring in the musical comedy (Liederspiel) *Lenore* (1829) by K. von Holtei (q.v.). Its first line is 'Schier dreißig Jahre bist du alt', words which (like the designation 'Mantellied') refer to an old coat.

MANUEL, Hans Rudolf (Erlach, 1525–71, Morsee), sixth son of Niklas Manuel (q.v.), was Landvogt (governor) of Morsee in Switzerland. He was author of a notable Fastnachtspiel (q.v.), *Ein holdsäligs Faßnachtspil, darin der edel wyn von der Truncknen rott beklagt, vonn Räblüthen gschirmbt un von Richtern ledig gesproche wirt,* generally known as *Das Weinspiel* (1548). Of unusual length (it has more than 4,000 lines), it shows, in robust comic dialogue, wine absolved of vice, and its accusers punished for abusing by their intemperance one of God's gifts.

MANUEL, Niklas or Niklaus (Berne, 1484–1530, Berne), also known as Niklas Manuel Deutsch, propagandist for the Reformation (q.v.) and a member of the council governing Berne from 1512 on. In his earlier years he was mainly occupied as a painter, executing altar-pieces, murals (including a dance of death since destroyed), drawings, and woodcuts. In 1522 he took part in a campaign in Italy. For the next six years he was active as a publicist on behalf of the Reformation. His principal works are propagandist plays. *Vom Papst und seiner Priesterschaft* (*Vom pabst und siner priesterschaft*), also known as *Die Totenfresser* attacks the sale of indulgences (see ABLASSKRAM) and the saying of mass for the dead. Together with *Vom Papst und Christi Gegensatz* (*Unterscheid zwischen de Papst, vnd Christu Jesum*) it was performed in 1523. The criticism of indulgences was renewed in *Der Ablaßkrämer* (1525, ed. P. Zinsli, 1960). Manuel wrote nothing after 1528, devoting himself in his last years to his governmental duties.

Marat/Sade, see VERFOLGUNG UND ERMORDUNG JEAN PAUL MARATS DARGESTELLT DURCH DIE SCHAUSPIELGRUPPE DES HOSPIZES ZU CHARENTON UNTER ANLEITUNG DES HERRN DE SADE, DIE.

Marbach, a small town in Württemberg, situated on the River Neckar to the north of Stuttgart, is the birthplace of Friedrich Schiller. The house in which he was born was bought in 1859 by the Marbacher Schiller-Verein, restored, and opened to the public. Schiller left the town at the age of two. Marbach also houses the Schiller-Nationalmuseum, opened in 1903, with Schiller archives, library, portraits, mementoes, and furniture, together with material relating to other Swabian poets. This also contains the Deutsches Literaturarchiv.

Marburg, German university town on the Lahn, now in the Land Hessen of the Federal Republic (see BUNDESREPUBLIK DEUTSCHLAND), and formerly the residence of the Landgraves of Hesse. It was at Marburg that Saint Elisabeth (see ELISABETH, HEILIGE) took refuge after her expulsion from the Wartburg, and the existing St. Elisabeth's church was erected over her grave between 1235 and 1283. The city became Protestant at the Reformation (q.v.) and was the scene of a famous religious discussion between Luther and Zwingli (qq.v.) in 1529, known as the Marburger Religionsgespräch (see MARBURGER ARTIKEL). The university, which was founded in 1527 by Landgraf Philipp von Hessen, der Großmütige (q.v.), was the first Protestant university foundation in Germany.

Marburger Artikel, a document of 15 articles issued after the conversations between Luther and Zwingli (qq.v.) at Marburg (q.v.) in 1529 (Marburger Religionsgespräch), defining the points of agreement and disagreement between the Lutheran and Reformed parties in the Protestant Church. Agreement was reached on all but transubstantiation (No. 15). Apart from Luther and Zwingli, the signatories were M. Bucer, Oekolampedius, Melanchthon (qq.v.), and K. Hedio. See REFORMATION.

MARC, FRANZ (Munich, 1880–1916, killed in action nr. Verdun), a painter, was trained in Munich, visited Italy in 1902, Greece in 1906, and Paris in 1907 and 1912. From 1904 his residence was in Bavaria. With W. Kandinsky and A. Macke (qq.v.) he founded the group Blauer Reiter (q.v.) in 1911. He developed an Expressionist technique, painting principally animals. Among his best-known works are *Blaue Pferde, Der Tiger,* and *Rehe.*

MARC, JULIA (*c.* 1795–after 1822), daughter of a widow living in Bamberg, was the possessor of a beautiful voice. She was taught singing by E. T. A. Hoffmann (q.v.), who fell deeply in love with her. There is no evidence that she returned his love. In 1812 she was married to an elderly, well-to-do Hamburger named Groepel, who was an alcoholic. The marriage was dissolved after a few years. Julia Marc was the model for several female characters in Hoffmann's stories, most notably for Julia in *Lebensansichten des Kater Murr* (q.v.), and for Cäcilie in *Nachricht von den neuesten Schicksalen des Hundes Berganza* (q.v.).

Märchen, a fairy-tale in the wider sense, containing a supernatural element, but not necessarily introducing fairies. When such stories derive from popular oral tradition they are termed Volksmärchen. Many of these are of oriental origin. Its basic structural features include a simple plot of mythic or moral import, with archetypal figures, set in an unspecified time and place. Märchen occur among early printed books, and interest in them, after lapsing in the 17th c., revived in the 18th c. and 19th c., notably in the Romantic period (see ROMANTIK). J. K. A. Musäus (q.v.) published *Volksmärchen der Deutschen* (1782–6). The outstanding German collection is *Kinder- und Hausmärchen* (q.v., 1812–14) of the brothers J. and W. Grimm.

Märchen invented or substantially altered by writers since the late 18th c. are styled Kunstmärchen. Examples have been written by Goethe (*Das Märchen,* q.v.), L. Tieck, C. Brentano, F. de la Motte FOUQUÉ, E. T. A. Hoffmann, and W. Hauff (qq.v.), as well as by many others. Mörike's *Märchen vom sichern Mann* (q.v.) is in verse.

Märchen, Das, a fairy-tale (Kunstmärchen) by Goethe, published as the closing story of *Unterhaltungen deutscher Ausgewanderten* (q.v.) in 1795. Set outside the framed cycle of stories, it moves into the world of fantasy, taking in oriental and Romance motifs in symbolical and, at times, allegorical manner; in this respect it stands apart from Romantic fairy-tales as well as from Goethe's own later examples.

Das Märchen is one of the most cryptic of Goethe's narratives. Introducing a multiplicity of figures, creatures, objects, and happenings, Goethe constructs with disciplined artistry an enigmatic pattern of situations, evolving around the theme of doom and regeneration. The following are basic motifs which intertwine. The river, central to the tale, needs a bridge which may be crossed in either direction; the green serpent provides it by sacrificing itself and dissolving into a pile of precious stones. The realm needs a king and a queen, represented by a youth and the beautiful lily; but the promised land appears to be out of reach, since the chosen couple are beset with sorrow and death. However, in the temple, deep beneath the ground, dwell three kings in statues, one of gold, the second of silver, and the third of bronze; the first represents wisdom (Weisheit) and endows the youth with the idea of striving for the highest aims; the silver statue represents appearance (Schein) and tells him to tend his flock; the third statue represents force (Gewalt) and gives the youth a sword. Through his union with the lily the youth acquires love, which is defined as a creative power, higher than that of ruling ('Die Liebe herrscht nicht, aber sie bildet, und das ist mehr'). A final threat is overcome when the giant turns into a statue, whose shadow indicates the passing of time not in figures, but in hieroglyphics. The temple becomes henceforth a temple of pilgrimage for the happily united people.

Märchen vom sichern Mann, a humorous verse fairy-tale written by E. Mörike (q.v.) in 1838. In 292 classical hexameters, it tells the story of a Black Forest giant who is persuaded by Lolegrin, son of the goddess Weyla (see ORPLID), to write a book on the Creation. He steals byre doors to make the leaves of his book, and when it is written takes it to the Underworld. Mocked by the Devil, he retaliates by pulling the Devil's tail out and uses it as a bookmark.

MARCHWITZA, HANS (Scharley, Silesia, 1890–1965, Potsdam), was a miner in the Ruhr and a member of the KPD (q.v.). His trilogy of novels *Die Kumiaks* (1934), *Die Heimkehr der Kumiaks* (1952), and *Die Kumiaks und ihre Kinder* (1959) is widely read in the German Democratic Republic. *Werke in Einzelausgaben* (9 vols.) appeared in 1957–61 and *Gedichte* in 1965.

MARCO POLO (1254–1324), a Venetian, travelled in the Far East between 1271 and 1295, and dictated his *Travels* in 1298–9, while in Genoese captivity. A German version, *Das puch des edeln Ritters vn Landfahrers Marcho polo* (1477) was among the earliest German printed books.

Maren, a novel of village life by J. H. Fehrs (q.v.), published in 1907. It is written in the Low German (Plattdeutsch) dialect of Holstein. It is the story of a competent and energetic farmer's sister who makes herself indispensable on a rich neighbour's farm, marries the ineffective farmer, runs the farm with great success, but in the end regrets her marriage. She dies in childbirth.

Märe vom Feldbauer, a short Middle High German comic poem (see SCHWANK) which tells how the author was tricked of his money by the ingenious fabrications of a miner. The poem was written *c.* 1300. Its author is not known.

Margarete von Limburg, an immense Middle High German poem (of more than 23,000 lines), translated from the Flemish in the 15th c. by JOHANN von Soest (q.v.). Its hero is Heinrich von Limburg and its setting the crusade of Friedrich II (q.v.) from 1227 to 1230. Heinrich's sister Margarete is taken by pirates and sold into slavery, and love springs up between her and the son of her master the Count of Athens. Heinrich meanwhile performs remarkable feats

and sustains many adventures. The romance closes with the double marriage of Heinrich and Margarete to the partners of their choice. The poem is compounded of episodes from many medieval poems. Sometimes known by the alternative title *Die Kinder von Limburg*, it was completed in 1476 or soon after.

MARGARETE VON ÖSTERREICH (Brussels, 1480–1530, Malines), daughter of Maximilian, who became emperor in 1493 (see MAXIMILIAN I, KAISER), was betrothed at the age of 2 to the future Charles VIII of France, and was brought up at the French court. The betrothal was annulled by the French side for political reasons, and Margaret was married in 1496 to Don Juan, heir to the Spanish throne. After his death in 1497 she was married to Duke Philibert of Savoy and was again widowed in 1504. In 1507 she was appointed governor of the Netherlands by her father Maximilian, an office which she continued to discharge with diplomatic and administrative skill under her nephew Karl V (q.v.). Her best-known diplomatic action was the Peace of Cambrai (1529), known in German as 'der Damenfriede' because the representative of France was also a woman, Louise of Savoy, mother of Francis I. Margaret was a woman of intellectual distinction, interested in literature and the arts. Her writings were published posthumously as *Couronne Margarétique* (1529).

MARGARETE VON TIROL, HERZOGIN (Vienna, 1318–69, Vienna), in 1335 inherited Tyrol from her father, Duke Heinrich of Carinthia. Married at the age of 12 to Johann Heinrich von Luxemburg, she expelled him from Tyrol in 1341, and in 1342 bigamously married Ludwig, Margrave of Brandenburg, son of Ludwig IV (q.v.). In consequence she was excommunicated until 1359. Her son Meinhard having died, she ceded Tyrol in 1363 to Duke Rudolf IV of Austria (1339–65). Her nickname 'Maultasch' is attributed by some sources to the ugly shape of her mouth, by others, however, to one of her properties, Schloß Maultasch in Tyrol. L. Feuchtwanger's sensational novel about her (see HÄSSLICHE HERZOGIN, DIE) opts for and exaggerates the former interpretation.

MARIA, pseudonym used by C. Brentano (q.v.) for his first publications, *Satiren und poetische Spiele* and *Godwi* (q.v.).

Maria Magdalene, a three-act tragedy by F. Hebbel (q.v.); it is also known under the title *Maria Magdalena*, which Hebbel, alluding to the biblical figure, intended for the play. The variant is due to a printing error in the first edition of 1844. The work is Hebbel's only important play with a contemporary background. He aimed at a new form of domestic tragedy, as stated in his *Vorwort zur Maria Magdalena*, written after its completion. His criticism in this preface was aimed mainly at Schiller's *Kabale und Liebe* (q.v.) and he advocated the creation of a social drama confined to a single class, the Kleinbürgertum.

The place of the action is a small town ('eine mittlere Stadt'). Klara, the daughter of Meister Anton, a carpenter, is pregnant by her fiancée Leonhard, a clerk of calculating, mean character, who has seduced Klara out of jealousy of her love for the Sekretär Friedrich, who, so she believes, has abandoned her. Her pregnancy makes her marriage to Leonhard, which her well-meaning mother encourages as a socially suitable match, urgent. But the family is suddenly struck by disaster as Klara's brother Karl is arrested on suspicion of theft. Therese, the mother, dies of shock, and Anton extracts from Klara an oath that she will not dishonour his name. He reinforces this with his own oath that he will kill himself should she fail him. Karl proves to be innocent. But Leonhard has made the arrest a pretext for breaking off the engagement, as he has found out that Anton is no longer in a position to provide the expected dowry for his daughter. In order to save her father, Klara decides to sacrifice herself and dies by jumping into a well. Meanwhile the Sekretär, aware of Klara's true feelings, has killed Leonhard in a duel. He himself is mortally wounded. Karl decides to leave home in protest against a domineering father. Anton is thus left alone in a world in which he had sought to uphold a concept of integrity, unaware of his self-righteousness. The disastrous collapse of this world is indicated in his final words: 'Ich verstehe die Welt nicht mehr'.

The play embodies Hebbel's severe social criticism; yet the title points to a moral and religious content designed to create an awareness of the power of love and to present an absolute concept of ethics lifting the tragedy above the level of an ordinary 'bürgerliches Trauerspiel'. The play was rejected by the Berlin Hoftheater because of Hebbel's treatment of Klara's pregnancy. It was first performed on 13 March 1846 in Königsberg.

Mariamne, a tragedy by J. C. Hallmann (q.v.).

Marianne, a Novelle by F. von Saar (q.v.), published in 1869 and included in 1877 in *Novellen aus Österreich* (q.v.). In the form of letters from a young artist in Vienna, it tells of the unfulfilled and barely expressed love between him and a young married woman. The story lasts over a summer from May to August and ends with her death by a stroke after a last dance with him at a wedding party at Grinzing.

Maria Schweidler, die Bernsteinhexe, a novel by J. W. Meinhold (q.v.), published in 1838 in the guise of a 17th-c. chronicle, of which he claimed that he was merely the editor. His mystification of his readers was so successful that he later had difficulty in establishing his authorship. The book is an amplification of his earlier story *Die Pfarrerstochter zu Koserow* (1826). The virtuous Maria Schweidler, a modern Susannah, repulses the indecent advances of an elder of the village, is accused of witchcraft, tortured, and condemned to be burned at the stake, from which she is rescued in the nick of time by her noble lover, who exposes her persecutors.

The novel was adapted as a play by H. Laube (q.v. and see BERNSTEINHEXE, DIE) and enjoyed a considerable success when translated into English in 1844 by Lady Duff-Gordon (*née* Lucie Austin) as *Mary Schweidler, the Amber Witch*.

Maria Stuart, a five-act tragedy (Ein Trauerspiel) in blank verse by Schiller, written 1799–1800 and published in 1801. It was first performed at Weimar on 14 June 1800. The tragedy portrays the end of Mary Queen of Scots, but, although Schiller made careful studies, historical accuracy was not his aim.

Maria is first shown imprisoned at Fotheringhay, deprived of luxury and denied privacy. Her gaoler's nephew, Mortimer, offers her assistance, and through him she seeks contact with Lord Leicester. At the court of Elisabeth the execution of the capital sentence against Maria is debated, and it is apparent that the English queen desires Maria's death; she has the strong support of Lord Burleigh, but is unwilling to take responsibility for Maria's execution. In an effort at conciliation a meeting is arranged by Lord Shrewsbury between the two queens. Its result is disastrous. In a scene, which is a bravura piece for the actress playing Maria, Elisabeth insults her enemy, whereupon Maria vehemently denounces her. An attempt to rescue Maria, to which Mortimer is a party, fails through a premature attempt on Elisabeth's life by one of the conspirators. Maria, discovering that Mortimer's motives are sexual, is filled with horror and remorse.

Elisabeth signs the death warrant and manœuvres her secretary Davison into a position in which the responsibility for its execution falls on him. Maria goes to her death with dignity in a deeply moving scene, protesting her innocence of the plots against Elisabeth, but accepting her death as a penance for her earlier complicity in the murder of Darnley. She receives the viaticum according to the Roman rite from Melvil.

Elisabeth loses all whom she had involved. Mortimer, whom she had rashly trusted, kills himself on his arrest. She banishes Lord Burleigh, orders the arrest of Davison; Lord Shrewsbury, who has throughout questioned the justifiability of the sentence, resigns his position. She has no hope of regaining the support of Lord Leicester, who is exposed by both queens for his duplicity; broken by witnessing Maria's death, he takes ship for France.

MARIA THERESIA, Kaiserin, the Austrian Empress Maria Theresa (Vienna, 1717–80, Vienna) was the eldest daughter of the Emperor Karl VI (q.v.) and as such an archduchess. In 1736 she was married to Duke Francis of Lorraine (see Franz I), with whom she lived in happy marriage while maintaining her sovereign authority intact. When she succeeded her father in 1740 under the provisions of the Pragmatic Sanction (see Pragmatische Sanktion), her position was immediately contested by the young Friedrich II (q.v.) of Prussia, who promised to support her imperial election, but only on the condition that she ceded Lower Silesia. She promptly refused, and he invaded Silesia (see Schlesische Kriege). In 1741, when a rival claimant, the Elector Karl of Bavaria (see Karl VII, Kaiser), invaded Austria, she became involved in a general war known as the War of the Austrian Succession (see Österreichischer Erbfolgekrieg). After Karl was elected emperor in January 1742, to the great satisfaction of France, George II of England, as Elector of Hanover, went to her aid with the Pragmatic army (see Pragmatische Armee). Maria Theresia's own loyal Hungarians repulsed the Bavarians and occupied Bavaria. On the death of Karl VII in 1745, Maria Theresia's husband was elected emperor, but the Peace of Aachen between France, Austria, and England was not concluded until 1748.

This turbulent opening of her reign injured Maria Theresia's sense of justice and strengthened her patriotism. In the following years she strongly supported the policy of Kaunitz (q.v.), which was to seek the recovery of Silesia (see Diplomatische Revolution). But seven years of war (1756–63, see Siebenjähriger Krieg) ended, in spite of many Austrian successes, in the *status quo ante bellum*. After the sudden death of Franz I in 1765, her eldest son acceded as the Emperor Joseph II (q.v.), and he and his mother ruled in uneasy partnership until her death in 1780.

Maria Theresia's greatest achievements lay in the field of home policy. She reformed and centralized the Austrian administration, and laid the foundation for a system of elementary schools. This she accomplished partly by reducing the independence of the Roman Catholic Church in order to make it more amenable to state interests. At the same time she took heed of the susceptibilities of nationalities, and of attachment to tradition. She combined tenacity and strength of purpose with an unmistakably feminine temperament. Her contented family life and the rectitude of her conduct won for her respect and affection at all levels of the population. The 6-year-old Mozart (q.v.) enchanted her with his playing in Schönbrunn in 1762, and was rewarded with the court dress which he wears in a well-known portrait. He played to her again in 1768.

MARIE-ANTOINETTE, originally Maria Antoinette (Vienna, 1755–93, Paris), Archduchess of Austria and youngest daughter of the Empress Maria Theresia (q.v.), was married to the Dauphin (later Louis XVI) in 1770. Her reception in Strasburg on her way to her marriage is described by Goethe in *Dichtung und Wahrheit* (q.v., Bk. 9). She became queen of France in 1774. Her political ineptness, her occasional frivolity, and the scandal involving Cardinal Rohan and the famous diamond necklace (in which she was innocent) made her eventually extremely unpopular in France, where she was suspected of political bias, and generally referred to as 'l'Autrichienne'. The unsuccessful imperial invasion of France in 1792 was partly motivated by the dynastic connection with Austria, which was at the time ruled by Leopold II (q.v.).

Nine months after the execution of Louis XVI, Marie-Antoinette was tried on 14 October 1793 and guillotined two days later. She is the subject of a novel by Luise Mühlbach (q.v., *Marie Antoinette und ihr Sohn*, 6 vols., 1867), and of a biography by Stefan Zweig (q.v., *Marie Antoinette*, 1932). There are frequent references to her in literary works dealing with the French Revolution (q.v., and see Revolutionary Wars).

MARIE LOUISE, originally **MARIA LOUISE** (Vienna, 1791–1847, Vienna), Empress of the French, was the eldest daughter of the Emperor Franz II (q.v.). In 1810, possibly through the agency of Metternich (q.v.), she was married to Napoleon, who had recently divorced the Empress Josephine. A proxy wedding took place in the Augustinerkirche in Vienna in March, followed in April by civil and religious ceremonies in Paris. In 1811 she bore a son, the King of Rome and later Duke of Reichstadt (q.v.). In 1813 and 1814 she was regent of France during Napoleon's absence at war (see NAPOLEONIC WARS). On Napoleon's fall she fled to Austria and refused to rejoin her husband on Elba. The Congress of Vienna (see WIENER KONGRESS) granted her the duchies of Parma, Piacenza, and Guastalla, where she reigned from 1816 until her death. On Napoleon's death she married morganatically Count Neipperg, with whom she had lived for some years, and, after his death in 1834, Count Bombelles. In the early years of her reign she proved an enlightened ruler and introduced many social and judicial reforms.

Marienbader Elegie, see TRILOGIE DER LEIDENSCHAFT.

Marienburg, a town in Poland in the former Prussian province of West Prussia, is dominated by the castle of the Teutonic Order (see DEUTSCHER ORDEN). Begun before 1276, the castle became in 1309 the residence of the High Master. In 1410, under HEINRICH von Plauen (q.v.), it successfully resisted a Polish siege, but passed into Polish hands in 1460. In the first partition of Poland in 1772 (see POLAND, PARTITIONS OF) Marienburg came into the possession of Prussia. The castle was used as a granary; in 1803 it was threatened with demolition, which was averted by the efforts of the 20-year-old poet, Max von Schenkendorf (q.v.). In the 19th c. it underwent restoration between 1817 and 1830 (under the leadership of Theodor von Schön, q.v.), and again between 1882 and 1914. An account of the earlier restoration is given by J. von Eichendorff (q.v.) in *Die Wiederherstellung des Schlosses der deutschen Ordensritter in Marienburg* (1844).

Mariendichtung, an important element in medieval religious poetry, first appears in Latin poems, of which in particular the hymn *Ave praeclara maris stella* (9th c.) and the antiphons *Salva regina* (11th c.) were held in great affection in Germany and were frequently translated. It is usual to classify Marian poetry into hymns of praise and adoration, lives of the Virgin, and legends, but the distinction between the last two is not easy to draw. Most of the poetry is moreover either translation or adaptation of Latin sources. Only four prominent names are connected with this literature: WALTHER von der Vogelweide (one *leich*, q.v.), KONRAD von Würzburg (*Die Goldene Schmiede*, q.v.), REINMAR von Zweter, and HEINRICH Frauenlob (qq.v.). For other authors of Marian poetry or for anonymous titles see entries on the following: (1) for hymns and prayers: HEINRICHS LITANEI; MELKER MARIENLIED; MARIENSEQUENZEN VON

MURI and VON ST. LAMBRECHT; ARNSTEINER MARIENGEBET; ST. TRUDPERTER HOHES LIED; RHEINISCHES MARIENLOB; HEINRICH VON LAUFENBERG; MARIENKLAGE; (2) lives: WERNHER, PRIESTER; PHILIPP DER KARTHÄUSER, BRUDER; KONRAD VON HEIMESFURT; KONRAD VON FUSSESBRUNNEN; WALTHER VON RHEINAU; and (3) legends: PASSIONAL; THOMAS VON KANDELBERG; FRAUENTROST.

Mariengruß, denotes a medieval poem based on the *Ave Maria* or *Englischer Gruß* (angelic greeting). The best-known example is a Bavarian poem written in ornate style in the 14th c.

Marienklage, a poem, in monologue form, expressing the lamentations of the Virgin Mary as she watches the sufferings of Jesus on the Cross. *Marienklagen,* written in the 13th c., 14th c., and 15th c., were sometimes performed in church and represent an intermediate stage between narrative and passion play. The principal German examples are *Unser Vrouwen Klage,* the *Wolfenbüttler Marienklage,* and the *Bordesholmer Marienklage* (qq.v.).

Mariensequenz von Muri, a Sequenz (q.v.) in Early Middle High German. Its startingpoint is the first strophe of the Latin Sequentia or hymn *Ave praeclara maris stella,* but it is an independent poem expressing a fervent prayer to the Virgin. It was probably written in Switzerland towards the end of the 12th c. The MS. came from Muri Abbey in Canton Aargau.

Mariensequenz von St. Lambrecht (or *von Seckau*), a Sequenz (q.v.) in Early Middle High German intended for the feast of the Assumption. It is a version of the Latin sequentia *Ave praeclara maris stella.* The end is missing. It was probably composed in St. Lambrecht Abbey, Styria (Austria), about the middle of the 12th c.

MARINELLI, KARL VON (Vienna, 1744–1803, Vienna), a nobleman by birth, became an itinerant actor. In 1781 he founded the Leopoldstädter Theater (q.v.), which became the home of the Viennese Volksstück (see VOLKSSTÜCK).

Mario und der Zauberer, a Novelle by Th. Mann (q.v.), published in 1930. It has an important narrative frame (see RAHMEN). The narrator and his family, on holiday at Torre di Venere, are required to leave the resort because they have allowed a child of 6 to run naked on the beach, which is treated as an offence against public morality. The family spends the last evening watching a hypnotist's performance. Cipolla, the hypnotist (Zauberer), deprives people of their wills and makes them do his.

bidding. He humiliates Mario, a local young man, who recovers from the hypnosis and shoots his tormentor dead. The story, written in Mann's most accomplished and relaxed style, is a symbolical representation of Italian fascism, of which Cipolla is seen as characteristic in his contempt for human dignity.

Marius und Sulla, an unfinished tragedy in five acts in mixed verse and prose by C. D. Grabbe (q.v.). Grabbe wrote two extant versions of this, his first realistic historical drama, begun in 1823 and abandoned in 1827. The second version was published in *Dramatische Dichtungen* (1827). Plutarch's biographies of Marius, Sulla, and Sertorius, and Appian's *Roman history* (Bks. 13–17) were the principal sources for Grabbe's portrayal, by means of two great men, of a critical epoch in the history of Rome, in which he saw an apt parallel to the French Revolution (q.v.) and the period of Restoration.

The aged and banished Marius makes a vain effort to recover his stature as a political force. He revenges himself by taking Rome, but in the hour of triumph learns of the approach of the younger rival Sulla. As he is about to lead his troops against the enemy, he dies with a vision of his past glory, of which the massacre of his loyal 'Marianer' by the troops of Sertorius and Cinna cannot deprive him. Sulla emerges victorious over all his enemies, including King Mithridates and the young Marius, who has taken his father's place. According to Grabbe's synopsis of the fifth act, Sulla is celebrated as the all-powerful dictator who, having taken measures to save Rome from prolonged anarchy, relinquishes his position and retires with his wife, Metella, to the privacy of his country home.

Mark, a coin first recorded in the 16th c. in Lübeck and Lüneburg. In 1873 the Mark replaced the Thaler as the standard silver coin in the German Empire. Later designations are the Rentenmark, introduced in 1923 after the inflationary collapse of the Mark, the Reichsmark (RM), a renaming of the Rentenmark, and the Deutsche Mark (DM, DM-Ost in the Russian Zone of Occupation), which replaced the Reichsmark in the currency reform (Währungsreform) of 1948. The Federal Republic retained the DM, the DDR changed the designation to Mark der Deutschen Notenbank (MDN) in 1964.

Märkische Argonautenfahrt, a novel by E. Langgässer (q.v.), published in 1950. Set in the period immediately after the 1939–45 War, it traces the journey of refugees from Berlin (who in their passions and lusts represent the works of the Devil) to the refuge of the convent of Anastasiendorf (a symbol of the Resurrection, *Anastasis*). Their journey is to be interpreted as a pilgrimage in search of Grace. The novel is curiously constructed, with virtually unrelated additional sections including episodes of past childhood and scenes of Russian piety.

Marktsänger, see BÄNKELSÄNGER.

MARLITT, EUGENIE, pseudonym of Eugenie John (Arnstadt, 1825–87, Arnstadt), who was one of the most successful writers of serial novels in the periodical *Die Gartenlaube* (q.v.). She was trained as a singer but had to abandon this career in consequence of a defect of hearing. For many years she was companion to a princess of Schwarzburg-Sondershausen. In 1863 she returned to Arnstadt and began an active career as a writer of light novels. These include *Goldelse* (1867), *Das Geheimnis der alten Mamsell* (2 vols., 1868), *Die Reichsgräfin Gisela* (2 vols., 1869), *Das Haideprinzeßchen* (2 vols., 1872), *Die zweite Frau* (2 vols., 1874), *Im Hause des Kommerzienrats* (1877), *Im Schillingshof* (1879), *Amtmanns Magd* (1880), *Die Frau mit den Karfunkelsteinen* (2 vols., 1885), and *Das Eulenhaus*, which was completed by W. Heimburg (q.v., 1888). She also wrote Novellen (*Thüringer Erzählungen*, 1869). Though unoriginal and stereotyped in character portrayal, she was a skilful narrator, who possessed also a talent for the description of landscape. Her collected works (10 vols.) appeared 1886–90.

Marmorbild, Das, a story by J. von Eichendorff (q.v.), published in 1819 in the *Frauentaschenbuch* edited by F. de la Motte FOUQUÉ (q.v.). Florio travels to Lucca where he is attracted to Bianca, a charming young woman. He is enticed from her by a beautiful Venus-like figure, an apparition, which seems from time to time to turn into a statue. Florio tears himself away from this seductress, whereupon she turns definitively to stone and her palace crumbles into ruin. He returns to Bianca, thus, in the symbolism of the story, rejecting pagan beauty and pagan morals for the Christian ideal.

MARNER, DER, the author of Middle High German Sprüche (see SPRUCH) and Minnelieder, written in the second half of the 13th c. Probably a Swabian, he was a man of education who spent his life as an itinerant singer and met his end, when old and blind, by murder, a fact recorded by RUMELANT von Sachsen (q.v.). Der Marner, who acknowledged WALTHER von der Vogelweide (q.v.) as his master, was, like other learned poets, easily involved in controversy. He wrote also Latin poems, five of which survive.

MARQUARD VOM STEIN, see RITTER VOM THURN, DER.

Marquise von O . . ., Die, a Novelle by H. von Kleist (q.v.), published in *Phöbus* (q.v.) in 1808, and in *Erzählungen* (vol. 1) in 1810. The heroine of the title, Julietta, Marquise von O . . ., is a widow, mother of two children, and daughter of the Obrist von G . . . She finds herself, through the discovery of an inexplicable pregnancy, in a situation of embarrassment and distress; for she experiences not only public scandal and expulsion from her father's house, but also the protestations of her own innocent heart against a seemingly impossible condition. Her unusual situation causes her to take the equally unusual step of appealing publicly, through a newspaper, to the father of her expected child to confess to his identity, adding that she feels obliged to marry him. To her horror she finds the man who joyfully presents himself to be Graf F . . ., an officer in the Russian army. He had been of late the object of her gratitude and admiration, for she owed her life to his courageous conduct. It was he who rescued her from rape at the brutal hands of Russian soldiers who had stormed the fortress, of which her father was commandant during an invasion of northern Italy. Reconciled with her parents, but not with the Count, she weds him in fulfilment of her promise. However, a year after the birth of her son she comes to terms with the fallibility of her heroic rescuer, whose image had been shattered by the revelations following her advertisement. Graf F . . . had taken advantage of a swoon which left her unconscious in his arms at her father's fortress. Recognizing his devotion and repentance to be true and constant, and resigning her ideal of perfection, Julietta celebrates in a second wedding the fulfilment of earthly happiness in true love and forgiveness.

The ingenious narrative technique, opening with the newspaper advertisement, the application of paradox and symbolism (with its central image of the swan) are outstanding features of this Novelle. (Film version by E. Rohmer, 1976.)

Marquis von Keith, Der, a play (Schauspiel) in five acts by F. Wedekind (q.v.), written in 1900 and published in 1901. Keith, a bogus nobleman, is a swindler who, by fascination and intellect, persuades others to entrust him with large sums for a fraudulent theatrical project. When he is exposed, his thoughts turn to suicide, but as soon as a way out is offered by which his

disappearance from Munich is required and the demand accompanied by a modest payment, he puts down the revolver, pockets the cheque, shrugs his shoulders, and goes, recognizing with nonchalance that life has its downs as well as ups. The world in which Keith seeks to exploit morality proves as bogus and immoral as he.

MARSCHALL VORWÄRTS, nickname of Blücher (q.v.), first given him by the Russians in the campaign of 1813 (see NAPOLEONIC WARS); it refers to his aggressive energy.

MARTENS, KURT (Leipzig, 1870–1945, Dresden), was for a short time a civil servant, resigning in 1896 in order to devote himself to writing. From 1899 to 1927 he lived in Munich and thereafter in Dresden. Martens took his own life after the air raid on Dresden in February 1945. He wrote mainly novels and Novellen, and among the novels are *Roman aus der Décadence* (1898), *Die Vollendung* (1902), *Kreislauf der Liebe* (1906), *Jan Friedrich* (1916), *Gabriele Bach* (1935), *Die junge Cosima* (1937), and *Graf Benyowsky verachtet den Tod* (1944). *Verzicht und Vollendung* (1941) treats the story of the 'Dunkelgräfin' (q.v.). Martens's volume of Novellen is entitled *Katastrophen* (1904). He also wrote an autobiography (*Schonungslose Lebenschronik,* 2 vols., 1921 and 1924).

MARTENS, VALÉRIE VON, see GOETZ, C.

MARTINA, HEILIGE, see HUGO VON LANGENSTEIN.

Martini or *Martinstag,* 11 November, treated as the end of the agricultural year, was the date for the annual engagement of farm servants. The rural dish for the feast of St. Martin is the 'Martinsgans'.

Martin Luther, oder Die Weihe der Kraft, a five-act tragedy in verse by Z. Werner (q.v.), published in 1807 with a prologue. The action covers Luther's burning of the papal bull of excommunication, which is followed by the Diet at Worms, where Luther defends his creed before the Emperor Karl V (qq.v.), who saves him from the stake, granting him safe conduct to return to Wittenberg but at the same time passing sentence of outlawry on him as a punishment for his heresy. From here the action proceeds to Luther's removal to the Wartburg, whence he returns to Wittenberg to condemn the activities of the iconoclasts. The imperial message that the ban has been lifted restores his position as a public figure in the forefront of the conflict. Through the love of Katharina von Bora he fully recognizes God's blessing. Katharina is the only nun who remains in the Augustinian convent at its enforced closure.

Werner explains in the prologue the 'blessing of strength' as consisting of art, faith, and purity, and it is revealed to Luther during critical periods of his life, culminating in Katharina's holy love ('heilge Minne'). Katharina von Bora condemns the burning of the papal bull, but when she sees Luther she recognizes in him her image of love and spirituality, to which she sacrifices the love of the knight Franz von Wildeneck. Her alienation from her life as a nun is reflected in the increasing loneliness of her 9-year-old foster-daughter, Therèse, who symbolizes pure faith through her care of the garden of hyacinths in which she dies. In a parallel arrangement Luther's 15-year-old Famulus Theobald with his flute symbolizes art, on which Luther draws for strength in his faith. Theobald gives his life for Luther when Franz von Wildeneck attacks the latter in a fit of passionate jealousy.

Werner's deft realism in the presentation of the popular image of Luther and in his colourful portrayal of the scenes in Worms is an achievement that was recognized at the production of the play by Iffland (q.v.) in Berlin in 1806. To Werner the development of mysticism, obtrusive though it may be, justified the designation of the play as a tragedy.

In 1814 Werner, after his conversion to Roman Catholicism, published a sequel, *Die Weihe der Unkraft* (q.v.). The sensation caused by the original performances in Berlin is depicted by Th. Fontane in *Schach von Wuthenow* (q.v.).

Martin Salander, a novel by G. Keller (q.v.), serialized in the *Deutsche Rundschau,* Jan.–Sept., 1886, and published in book form in the same year, thirty-two years after his only other novel, *Der grüne Heinrich* (q.v.). It is a conspectus of the social and political world of the Swiss middle class, especially in Canton Zurich, as Keller saw it in the 1870s and early 1880s.

Martin Salander, a teacher in Münsterburg (i.e. Zurich) had once stood surety for his friend Louis Wohlwend, who then failed to meet his obligations. Salander paid the sum with most of his assets and, leaving wife and children in Switzerland, emigrated to Brazil to make his fortune. He succeeds and, at the beginning of the novel, returns to Münsterburg, only to find that the bank in which he has deposited his money has failed. Once again it is Wohlwend who causes the disaster, for, unknown to Salander, the bank was under his former friend's control. Wohlwend emigrates to Hungary, and Salander for the second time packs his bags for South America. Within three years he is back

in Münsterburg with a new fortune. He settles this time and takes an active interest in social and political affairs. He is disappointed in the behaviour of his marriageable daughters who, against their father's wishes, insist on marrying twins from a family of lower social standing. Before long the twins are convicted of embezzlement and sentenced to long terms of imprisonment, whereupon their wives divorce them. Contacts with Wohlwend, who is repaying his debts to Salander by instalments, are renewed. The hitherto unimpeachable Salander now succumbs to an infatuation for Wohlwend's sister-in-law, Myrrha Glawicz. From this he is extricated by the return from Brazil of his industrious and upright son Arnold.

Keller portrays a society which is beginning to disintegrate, and puts forward Arnold as the representative of commercial and civic virtues which must be restored.

MARTIN VON COCHEM (Cochem/Mosel 1634–1712, Waghäusel nr. Bruchsal), a Capuchin friar, rose to high ecclesiastical office in the dioceses of Mainz and Trier. In his middle years he visited Austria and Bohemia. Martin was a prolific writer of devotional and ecclesiastical works. He began with a *Kinderlehrbüchlein* (1666), and wrote a life of Jesus (*Das Große Leben Jesu,* 1681) and a *Historie- und Exempel-Buch* in four volumes (1696–9), which is intended as a kind of repertoire for the composers of sermons. The 2 vols. of his *Kirchenhistorie* (1693) present a balanced view, and the book is notable as one of the first serious Catholic historical works since the Reformation (q.v.) to abstain from polemics.

Martin's exposition of the mass (*Meßerklärung*) was published in Latin in 1697, and in German in 1702. He also wrote an encyclopedic work in four volumes covering the lives of the saints (*Neue Legend der Heiligen,* 1708). Martin wrote in a deliberately straightforward style (in his own words 'klar und fließend'), because he wished to be understood by the common man.

Märtyrerbuch, Das, title given to a Middle High German collection of verse legends of martyred saints, arranged according to the calendar. It was written in Swabia *c.* 1320–30 by a priest, little skilled in poetry, for a Countess of Rosenberg.

MARX, KARL HEINRICH (Trier, 1818–1883, London), the son of a Jewish lawyer converted to Protestantism, studied philosophy and law at Bonn and Berlin universities. At Berlin he came under the influence of Hegel (q.v.). His radical political views debarred him from an academic career, and he took to journalism in Cologne, only to find his newspaper, the *Rheinische Zeitung*, suppressed by the censor. He left Germany for Paris, where he met H. Heine, L. Börne (qq.v.), and Arnold Ruge (1803–80), with whom he edited the *Deutsch-Französische Jahrbücher*, in which he introduced his major polemics against Hegel, *Zur Kritik der Hegelschen Rechtsphilosophie* (1844). From his stay in Paris dates his friendship with F. Engels (q.v.). His reactions to the representative of the Hegelian Left, Bruno Bauer (1809–82), are reflected in *Die Heilige Familie*. In 1845 he wrote (again jointly with Engels) the eleven *Thesen über Feuerbach*, which criticize L. Feuerbach (q.v.) for totally failing to establish a link between materialism and history, and *Die deutsche Ideologie*, which prepares the way for Marx's conception of historical materialism. His tract *Misère de la philosophie. Réponse à la philosophie de la misère de M. Proudhon* was composed in Brussels (1847, published in German in 1885 as *Das Elend der Philosophie*) after the Prussian government had effectively pressed for his expulsion from France. The tract links him with Pierre-Joseph Proudhon (1809–65), whose views on the function of economic forces in social revolution Marx considered to be insufficiently radical. French socialism represents the third major influence upon Marx's work.

Marx and Engels joined the Communist League in Brussels, and together they drafted the Manifesto (see KOMMUNISTISCHES MANIFEST) at a conference held in London in 1847. In the year of revolution (see REVOLUTIONEN 1848–9) Marx promoted his socialist and republican ideas for the last time on German soil by editing the *Neue Rheinische Zeitung* in Cologne. For the remaining thirty years of his life he settled in London, from where he kept in touch with international affairs and expressed his views in journalistic and polemical writings. He disagreed with the Darwinist Karl Vogt (1817–95) and with Ferdinand Lassalle (q.v.). From 1864 he headed the General Council of the International Working Men's Association (Internationale Arbeiterassoziation), which dissolved in 1876 in New York. In 1875 he criticized the relatively moderate programme of the German Socialist Party (Sozialdemokratische Partei Deutschlands, see SPD).

The systematic reappraisal of materialistic socialism, at which Marx aimed, progressed in fragmented form. In 1859 he published *Zur Kritik der politischen Ökonomie*, which contains the seed of his principal work, *Das Kapital*. Marx himself only published the first volume (1867), but Engels, using Marx's notes and drafts, completed vols. 2 (1885) and 3 (1895). In his theory of production, based on a dialectical triad, his only concession to Hegel, Marx distinguishes between the condition of production (Produktionsverhältnisse) and the efficiency of production (Produktionskräfte). He traces their historical development in Hegelian fashion and speculates on their function in an age of rapid industrialization, in which the 'material' threatened to become more important than Man. He sees the beginnings of the expansion of productivity in the slavery of antiquity, and in feudalism he sees the precursor of capitalism. In each phase the individual worker has been exploited, the difference being only one of degree and manner of exploitation. Marx sees in the surplus value (Mehrwert) the crucial margin by which the class owning the material and controlling the process of production exploits proletarians. Marx's appeal for a classless society in which all shared in the process of production on an equal basis determines his rudimentary theory of revolution.

Marx took a firm stand against the charge that his ideology represented an 'abstract utopia'. On the occasion of the 150th anniversary of his birth, Ernst Bloch (q.v.) reviewed it in terms of a 'concrete utopia'. The chief advocate of Marxism in German literature in the first half of the 20th c. is Brecht and in literary criticism and aesthetics G. Lukács (qq.v.). Marx's own sustained interest in literature and aesthetics (re-examined by S. S. Prawer in *Marx and World Literature*, 1976) resulted in his appraisal of 'realistic' (as distinct from 'naturalistic') literary production, and in all manner of imaginative writing he insisted on the congruence of form and convincing substance. *Werke, Schriften, Briefe* (7 vols.) ed. H. J. Lieber, P. Furth, and B. Kautsky, appeared 1960–9. A complete edition of the works of Marx and Engels (some 100 vols.) began to appear under the auspices of the Institute for Marxism–Leninism of the German Democratic Republic in 1975.

Marzipanliese, Die, a Novelle by F. Halm (q.v.), published in *Unterhaltungen am häuslichen Herd* (edited by K. Gutzkow, q.v.) in 1856. It is a crime story with a moral. A young Austrian, because of his engaging manner, is offered employment as book-keeper by Horváth, a rich Hungarian farmer. Under the name Ferencz he commends himself to the household by his efficiency and his ingratiating manners, especially with the women, and chiefly with Horváth's daughter Czenczi, but his past remains obscure.

A chance visitor recounts a mysterious murder in Bruck an der Mur. A young man had insinuated himself into the confidence of a money-lending widow, and drafted for her a will in his favour. This widow (the Marzipanliese, so called because she tells her money-borrowing clients that their compassionate stories are just marzipan to her) is found murdered. Her will leaves all to charity, and the young man shortly afterwards disappears. By this time it is obvious to the reader that Horváth's book-keeper is the murderer. Horváth, however, sees only that Ferencz has captivated Czenczi, and he therefore dismisses him. Ferencz determines to run away with her, and gets her to lock him in a cellar until they can make their escape. She, however, falls ill of a delirious fever, and, by the time she recovers, Ferencz is dead. Inquiries make it clear that he was the murderer of Marzipanliese. Horváth dies, and Czenczi enters a convent to which she gives her entire property.

Maschinenstürmer, Die, a verse play in five acts by E. Toller (q.v.), published in 1922. It deals with the Luddites and is set in London and Nottingham in 1812–15. The inspiration of the play is derived from Lord Byron's maiden speech in the House of Lords (27 Feb. 1812), of which Toller gives his version in the Prolog. The riots fail and the leader Jimmy Cobbett is killed.

Maske-Tetralogie, designation for four social plays by C. Sternheim (q.v.) which are contained in the cycle *Aus dem bürgerlichen Heldenleben* and present episodes in the fortunes of a bourgeois family, whose name is Maske (meaning 'mask'): *Die Hose* (q.v., 1911), *Der Snob* (1914), *1913* (1915), and *Das Fossil* (1925).

Masse-Mensch, an Expressionist play by E. Toller (q.v.), published in 1921. A note on the flyleaf indicates that it was written 'im Oktober 1919, im ersten Jahr der deutschen Revolution'. This note was added by Toller in Niederschönenfeld prison. The play is dedicated to 'den Proletariern'. It is divided into seven scenes, of which the second, fourth, and sixth are described as dream visions (Traumbilder). Of the other four, only in the first is reality suggested ('angedeutet'), as the preparations for a revolution in wartime are conveyed in free verse. The third, fifth, and seventh scenes are shown, as the title-page indicates, 'in visionärer Traumferne'.

The leader of the revolution is Sonja Irene L., the only named character, and she combines the desire for a revolution by the mass ('Masse') with respect and love for individuality ('Mensch'). When the mass-revolution breaks out it indulges in mass executions and outrages, which are then avenged in like manner by the state in the name of order. Die Frau (Sonja) loses control to 'Der Namenlose', the faceless image of mass violence. The revolution fails, and Sonja is condemned and shot. She states her

creed in the fifth 'Bild'. 'Masse soll Volk in Liebe sein./Masse soll Gemeinschaft sein./Gemeinschaft ist nicht Rache.' The play, moved by a profound pacifism, deplores violence, whether in war, revolution, or retribution.

MASSMANN, HANS FERDINAND (Berlin, 1797– 1874, Muskau, Silesia), served in the Wars of Liberation and was one of the demonstrators at the Wartburgfest (q.v.) in 1817. A friend of Jahn (q.v.), he was, with Jahn, the founder of gymnastic instruction in Germany. From 1827 to 1843 he was gymnastic instructor at the Cadet School in Munich and simultaneously (from 1829) professor of Germanic philology at the university. From 1843 he combined similar, at first sight incompatible, appointments in Berlin. He published poems (*Lieder für Knaben und Mädchen,* 1832; *Deutsche Gedichte,* 1834), a history of the literature of his own youth (*Das vergangene Jahrzehnt der deutschen Literatur,* 1827), and books on gymnastics (*Altes und Neues vom Turnen,* 1849). He is an object of Heine's satire in *Atta Troll* (q.v.), Kaput IV.

As a Germanist, Maßmann is principally noteworthy for his edition of the *Kaiserchronik* (q.v.), *Der keiser und die kunige buoch, oder die sogennante Kaiserchronik* (3 Pts., 1849–54, re-issued in 1963); other works which were reissued in the 1960s include *Die deutsche Abschwörungs-, Glaubens-, Beicht- und Betformeln vom 8.–12. Jahrhundert* of 1839 (1968), *Die Literatur der Todtentänze* of 1840–50 (1963), *Gedrängtes althochdeutsches Wörterbuch* of 1846 (1963), and *Eike von Repgow, Das Zeitbuch* of 1857 (1969).

Maßnahme, Die, a play (Lehrstück, q.v.) by B. Brecht (q.v.) of the years 1929–30 with music by Hanns Eisler (q.v.) designed 'ein bestimmtes eingreifendes Verhalten einzuüben' (*Versuche,* vol. 4). The text was revised in 1938. The play consists of a prologue and 8 scenes, the prologue serving to introduce the experience of the four Moscow communist agitators (Agitatoren) who report to the Kontrollchor that they have killed a young comrade (Der junge Genosse) and disposed of his body in a chalk-pit. The Kontrollchor approves of the execution and the play proceeds to re-enact in a Chinese setting the events leading to the 'death sentence'. They are accompanied by a number of 'songs of praise' by the Kontrollchor (e.g. *Lob der illegalen Arbeit, Lob der Partei*). The didactic element prevails throughout in comments provoking discussion as to the rights and wrongs of the agitators' conduct. The agitators cannot fulfil their task of spreading the communist message unless they observe strict anonymity. The guilt of the young comrade, which he readily accepts, rests on his inability to subordinate his compassion for the maltreated people to the rational behaviour which his mission requires. The crisis occurs when he pulls off his mask, exposing himself and the four agitators to prosecution. The agitators decide on his execution, not to save their own skin, but in proper subordination to their political task. A young man's humanity has stood in the way of the ultimate good, which can only be achieved by radical measures which 'die Maßnahme' attempts to illustrate.

Masuren, a district of the former Prussian province of East Prussia (see OSTPREUSSEN), largely covered with forest and with innumerable lakes. During the 1914–18 War it was the scene of two German victories. The Battle of the Masurian Lakes (Schlacht an den masurischen Seen), fought between 5 and 15 September 1914, was a sequel to the Battle of Tannenberg (q.v.), ending in the defeat of the Russian army of Rennenkampf. In the Winter Masurian Battle (Winterschlacht in Masuren, 4–22 February 1915), strong Russian forces were defeated and a partial breakthrough made into Russian Poland. The German commander of both battles was Hindenburg, with Ludendorff as chief of staff (qq.v.).

Masuren is now a part of Poland. Its landscape provides the background to some of the works of H. Sudermann and S. Lenz (qq.v.), both Masurians.

MATHESIUS, JOHANNES (Rochlitz, Bohemia, 1504–65, Joachimsthal), a friend of Luther (q.v.) in his early years, remained a lifelong disciple. From 1542 until his death he was pastor in Joachimsthal. He collected and published Luther's *Tischreden*, and is the author of the first biography of Luther (*Historien Von des Ehrwirdigen in Gott Seligen thewren Manns Gottes, Doctoris Martini Luthers, anfang, lehr, leben und sterben*, 1566). A volume of his own sermons appeared in 1562 as *Sarepta oder Handpostille*. He is also the author of hymns.

MATTHESON, JOHANN (Hamburg, 1681–1764, Hamburg), a musician, produced the first German imitation of *The Spectator* (see MORALISCHE WOCHENSCHRIFTEN) in his weekly periodical *Der Vernünftler* (1713–14), and later translated the novels of Samuel Richardson into German. His rendering of *Pamela* (1740) appeared in 1742; *Clarissa* (1747) followed in 1748, and the *History of Sir Charles Grandison* (1753) in 1754.

MATTHIAS, KAISER (Vienna, 1557–1619, Vienna), the son of the Emperor Maximilian II (q.v.) and emperor from 1612, was a supporter of the Counter-Reformation (see GEGENREFOR-

MATION), and in 1595 his measures to suppress Protestantism led to a revolt in Upper Austria. In 1606 he relaxed this attitude in order to win sovereignty over Austria and Hungary. In 1608 the Emperor Rudolf II (q.v.), his brother, conferred on him, under pressure, the government of both countries, which were granted the free exercise of their religion. This move of Matthias against his brother had the approval of other senior members of the Habsburg dynasty, who were alarmed at the Emperor's incapacity. In 1611 Rudolf was obliged to pass the Bohemian kingship to Matthias. Upon Rudolf's death in 1612 Matthias was elected emperor. Matthias depended for his policy upon the counsel of Khlesl (q.v.). Having no issue, he took steps to assure the succession of his cousin Ferdinand (see FERDINAND II). The strife between the two brothers Rudolf and Matthias is the subject of Grillparzer's tragedy *Ein Bruderzwist in Habsburg* (q.v.).

MATTHISSON, FRIEDRICH VON (Hohendodeleben nr. Magdeburg, 1761–1831, Wörlitz, Anhalt) studied at Halle University and afterwards taught at the Philanthropinum (1781–4), the school founded by J. B. Basedow (q.v.) at Dessau. He became a tutor and travelled in Germany, meeting Klopstock, J. H. Voß, and M. Claudius (qq.v.). Duke Friedrich of Württemberg became his patron, granting him in 1809 a patent of nobility and appointing him theatre director in 1811, and librarian as well in 1812. Matthisson retired in 1828. He published his first collection of poems in 1781 as *Lieder*, changing the title for a further edition in 1787 to *Gedichte*.

Matthisson's elegant, decorous, and lightly sentimental poetry, often in classical metres, was widely read in the late 18th c. and early 19th c. Schiller reviewed it favourably in *Über Matthissons Gedichte* (published in the *Allgemeine Literatur-Zeitung*, 1794), though in his letters he shows little enthusiasm for Matthisson or his poetry. Beethoven (q.v.) set to music the poems *Adelaide*, *Opferlied*, and *Andenken*.

Gesammelte Schriften (8 vols.) appeared 1825–9, posthumous writings, ed. F. R. Schoch (*Literarischer Nachlaß*, 4 vols.), 1832, and a critical edition of *Gedichte* (2 vols.), ed. G. Bölsing, 1912–13.

Maulbronn, Kloster, a well-preserved medieval abbey situated in Württemberg to the north-west of Stuttgart. Originally a Cistercian community, it was made a Protestant school in 1558, and converted to a Protestant seminary in 1806. Hölderlin, Georg Herwegh, and Hermann Hesse (qq.v.) were all pupils at Maulbronn, and

Hesse's story *Unterm Rad* (q.v.) relates to his experiences there.

MAULTASCH, MARGARETE, see MARGARETE VON TIROL, HERZOGIN.

MAUPERTUIS, PIERRE-LOUIS MOREAU DE (St. Malo, 1698–1759, Basel), a French scientist, was a member of the Académie des Sciences (Paris) and of the Royal Society (London) before being appointed president of the Prussian Académie des Sciences et Belles-Lettres (Berlin; see AKADEMIEN) by Friedrich II (q.v.) of Prussia in 1740. In a learned dispute between Maupertuis and the mathematician Samuel König, Voltaire, who was at the Prussian Court (1750–3), took König's side, ridiculing his countryman in the *Diatribe du Docteur Akakia*, which he published without Friedrich's consent.

Maupertuis resigned his post in 1757. Correspondence between the King and Maupertuis is contained in *Briefwechsel zwischen Friedrich dem Großen, Grumbkow und Maupertuis, 1731–59*, edited by Koser (1898), and *Nachträge zu dem Briefwechsel Friedrich des Großen mit Maupertuis und Voltaire*, edited by Koser and G. Droysen (1917).

MAURER, GEORG (Sächsisch-Regen, Romania 1907–71, Potsdam), a socialist poet, is the author of hymnic verse in the manner of Hölderlin, and of sonnets. His publications include *Ewige Stimmen* (1936), *Gesänge der Zeit*, including sonnets (1948), *Zweiundvierzig Sonette* (1953), *Die Elemente* (1955), *Poetische Reise* (1959), *Gedichte* (1962), a selection, and *Variationen* (1965). *Essay I* appeared in 1968 and *Essay II*, ed. G. Wolf, in 1973.

Mäuseturm, a 13th-c. tower built on a rocky islet in the Rhine near Bingen as a customs station. It is popularly associated with Bishop Hatto II of Mainz, who lived in the 10th c. and is supposed to have been devoured there by mice in retribution for his refusal to share his stocks of grain in time of famine. The story is told in *Deutsche Sagen* by Jakob and Wilhelm Grimm (qq.v.).

MAUTHNER, FRITZ (Hořitz, Bohemia, 1849–1923, Meersburg), was for many years a journalist in Berlin, retiring in 1911 to Meersburg on Lake Constance. He is the author of a number of novels, the best known of which is the trilogy of social criticism *Berlin W* (1886–90). His other novels include *Der neue Ahasver* (2 vols., 1882) and *Die böhmische Handschrift* (1897). He had a gift for parody, which he exploited in *Nach berühmten Mustern* (collected 1898), and showed his satirical talent in *Schmock* (1883). The term Sprachkritik entered philosophical

currency through his *Beiträge zu einer Kritik der Sprache* (3 vols., 1901–3, reprint 1967); it is used with direct and critical reference to Mauthner by Wittgenstein (q.v.) in his *Tractatus logico-philosophicus*.

MAXIMILIAN I, KAISER (Wiener Neustadt, 1459–1519, Wels), was the son of the Emperor Friedrich III (q.v.) and his consort, a Portuguese princess. By his marriage to Marie of Burgundy in 1477 he united the extensive Burgundian territories (including the Netherlands), with the Habsburg dominions. He was elected German King (see DEUTSCHER KÖNIG) in 1486, and in 1493 became Emperor of the Holy Roman Empire (see DEUTSCHES REICH, ALTES). He conducted numerous campaigns in the Netherlands, France, Italy, and Switzerland, the majority of which ended in defeat.

Although Maximilian's political career was largely unsuccessful because of his instability, recklessness, and inordinate self-esteem, he enjoyed considerable popularity and after his death became a legendary figure. An attractive personality and a gifted speaker, he was regarded as a symbol for German national aspirations, though his aims were in reality dynastic. Known as 'der letzte Ritter', he was skilled in knightly accomplishments, and well versed in hunting, in architecture, and in horticulture. He was also a patron of the New Learning, encouraging scholars. He was himself an author, writing essays on his favourite arts (*Die Baumeisterei, Die Gärtnerei, Geheimes Jagdbuch*) and contributing to two quasi-autobiographical works, *Teuerdank* and *Der Weißkunig* (qq.v.). A third work, *Freydal*, remained a plan.

Although he was buried in Wiener Neustadt, his ornate funeral monument (1584) is in the Hofkirche in Innsbruck. Contemporary likenesses exist by Dürer (q.v.) and Lucas von Leyden (1494–1533), and a posthumous portrait was painted by Rubens. He is the subject of a cycle of romances, *Der letzte Ritter* (1830), by Anastasius Grün (q.v.).

MAXIMILIAN II, KAISER (Vienna, 1527–76, Regensburg), son of Ferdinand I (q.v.), was brought up in Spain, and from 1548 to 1550 was the deputy there of his uncle, the Emperor Karl V (q.v.). Despite his Catholic upbringing, Maximilian maintained cordial relations with Protestant princes, and was believed by some to be a secret sympathizer with the Protestant cause. It is now thought that he may rather have been a Christian humanist of the Erasmian tradition with little interest in the dogmas of either faction. In 1562 he was elected German King, and in 1564 he succeeded his father as emperor. At a Diet at Augsburg in 1566 he maintained a neutral position in religious matters, opposing concessions to the Protestants, yet declining to support action against them. He engaged in war against the Turks, who threatened his eastern frontiers, but after indecisive fighting made a compromise treaty in 1568. In 1573 he was elected king of Poland, but was unable to establish himself on the throne in the face of opposition from the estates of the Empire.

MAXIMILIAN I, König von Bayern (Zweibrücken, 1756–1825, Nymphenburg, Munich), became in 1799 Elector of the Palatinate and Bavaria (Kurfürst von Pfalz-Bayern). In 1805 Napoleon rewarded him for his support by making Bavaria a kingdom (see Napoleonic Wars). Maximilian abandoned the French cause in 1813 just in time to save his crown. After 1815 he patronized the arts and introduced reforms, including the granting of a constitution in 1818.

MAXIMILIAN II, König von Bayern (Munich, 1811–64, Munich), son of Ludwig I (q.v.), succeeded to the throne in 1848 on his father's abdication. He devoted himself principally to encouraging the arts and sciences in Munich. He was succeeded by his son Ludwig II (q.v.).

MAXIMILIAN I, Herzog von Bayern, Kurfürst (Munich, 1573–1651, Ingolstadt), a Wittelsbach (q.v.), succeeded his father Duke Wilhelm V in 1597, and emerged as an influential statesman before and during the Thirty Years War (see Dreissigjähriger Krieg). Through the formation of the Catholic League (see Katholische Liga) he became a powerful figure in the defence of the Catholic and imperial cause. He supported the Edict of Restitution, which Ferdinand II (q.v.) issued in 1629, but urged the Emperor to dismiss Wallenstein (q.v.), whom he regarded as a rival threatening his dynastic interests. The Peace of Westphalia (see Westfälischer Friede) confirmed his status as Elector, an imperial reward for defeating Friedrich V (q.v.) of the Palatinate as king of Bohemia (see Weissen Berge, Schlacht am). He was ably served by general Tilly (q.v.). An efficient and austere administrator, he had, even by the standards of his time, an exaggerated sense of responsibility for the morality of his subjects. He forbade his peasants to dance, and he introduced the death penalty for adulterers.

Maximilian was a protector of the arts (a collector of Dürer, q.v.) and built the Munich Residenz. He gave the Bibliotheca Palatina of Heidelberg (see Libraries) to the Pope.

MAXIMILIAN II, Emanuel, Kurfürst von Bayern (Munich, 1662–1726, Munich), was a son of the Elector Ferdinand Maria (ruled from 1651), whom he succeeded in 1679. During a term as Statthalter of the Netherlands, an appointment made by Leopold I (q.v.) in 1691, ambition to annex this province led him to fight in the Spanish War of Succession on the side of France against Austria. Defeats in the battles of Donauwörth (July 1704) and Höchstädt (August 1704), known in England as Blenheim, forced him to flee from Bavaria. By the Peace of Rastatt (1714) he was restored to his possessions, and the imperial ban was lifted.

MAXIMILIAN, Erzherzog (Vienna, 1832–67, Mexico), was a younger brother of the Emperor Franz Joseph (q.v.). A man of intelligence and energy, he rose to high rank in the Austrian navy, and in 1857 became governor of Lombardy and Venetia, retiring from this office at the beginning of the war of 1859. In 1863 he was offered the throne of Mexico with the title Emperor, which, in spite of initial reluctance, he accepted. But the republican party proved stronger than the new imperialist faction, the supporting French troops were withdrawn, and Maximilian was taken prisoner, tried, and executed by a firing squad.

Maximin-Erlebnis, see George, S.

MAX VON BADEN, Prinz (Baden-Baden, 1867–1929, Constance), heir to the Grand-ducal throne of Baden, was appointed chancellor (Reichskanzler) in October 1918 by Kaiser Wilhelm II (q.v.), in order to negotiate an armistice with the Allies, and to prepare for parliamentary concessions. On 9 November Prince Max announced the abdication of the Emperor, who had not in fact been able to bring himself to take the final decision. At the same time he resigned the chancellorship in favour of F. Ebert (q.v.).

MAY, Karl (Ernsttal, Saxony, 1842–1912, Radebeul nr. Dresden), grew up in poverty. Through his weaver-father's determination to improve the boy's lot he became a schoolmaster. He was almost immediately convicted of the theft of a watch, which, he claimed, was lent to him, and he was consequently dismissed from his employment. A psychological crisis marked by delinquencies followed, and he served prison sentences amounting in all to seven years. After writing a few sentimental village stories, he fell into the hands of an unscrupulous publisher, for whom he wrote (anonymously) a large number of trashy novelettes (see

KOLPORTAGEROMAN). He then turned his attention to stories of American Indians, after the manner of Fenimore Cooper. In the last quarter of the 19th c. he was perhaps the most popular author of boys' books in Germany. He described these novels as Reiseschilderungen, and used the first-person narrative in order to give the impression of actual experience. His best-known characters were the Indian Winnetou and the white man Shurehand.

Among numerous titles, some of the best known are *Im fernen Westen* (1880), *Helden des Westens* (1890), *Winnetou* (3 vols., 1893–1910), and *Old Shurehand* (1894). May is the author of similar novels set in the Near East and South America, including *Im Lande des Mahdi* (1895) and *Das Vermächtnis des Inka* (1895). Having amassed a fortune, he wrote late in life for his own pleasure the symbolical novel *Ardistan und Dschinnistan* (1909). In an imperialistic age he took up a pacifist standpoint, which he defended in polemical writings. His autobiography *Mein Leben und Streben* (1910) was reissued posthumously entitled *Ich* (1917). His popularity extended well into the 20th c.

Gesammelte Reiseerzählungen (33 vols.) appeared 1892–1910, and two editions of *Gesammelte Werke* appeared concurrently, 1913 ff. and 1950 ff.

MAYER, KARL FRIEDRICH HARTMANN (Neckarbischofsheim, 1786–1870, Tübingen), a friend of L. Uhland (q.v.) from his student days at Tübingen, spent most of his life as a law officer in Ulm and Eßlingen (1818–24), in Waiblingen (1824–43), and finally in Tübingen (1843–57). Mayer wrote agreeable nature poetry, publishing *Lieder* in 1833 and *Gedichte* (an expansion of the *Lieder*) in 1840. He edited Lenau's letters (*Nikolaus Lenaus Briefe*, 1831) and published a memoir of Uhland, (*Ludwig Uhland, seine Freunde und Zeitgenossen*, 1867). G. Schwab and J. Kerner (qq.v.) were among his friends.

Mayerling, in Lower Austria near Baden by Vienna, is the site of a shooting lodge in which the Habsburg Crown Prince Rudolf (q.v.) met his end on 30 January 1889. The lodge was demolished by order of the Emperor Franz Joseph (q.v.) and replaced by a Carmelite convent.

MAYFART, JOHANN MATTHÄUS (Jena, 1590–1642, Erfurt), schoolmaster at Coburg, became a university professor, and simultaneously pastor, at Erfurt. Mayfart wrote hymns, but he is more remarkable for his humanitarian attitude in a superstitious age, exemplified in his opposition to witch-hunting and witch-trials (*Traktat über die Hexen*, 1636). He also opposed the traditional tyranny of senior students over juniors at the university.

MAYRHOFER, JOHANN (Steyr, Austria, 1787–1836, Vienna, by suicide), at first a cleric at St. Florian's Abbey, Linz, studied law at Vienna University, becoming a friend of the composer F. Schubert (q.v.), who set many of his poems. Mayrhofer published *Gedichte* (1824) and *Erinnerungen an Franz Schubert* (1829). A posthumous collection of poems appeared in 1843, and M. M. Rabenlechner produced an edition (*Gedichte*) in 1938.

MAYRÖCKER, FRIEDERIKE (Vienna, 1924–), an *avant-garde* poet, has published poetry and prose, using montage as her principal technique (*Larifari*, 1956; *Tod durch Musen. Poetische Texte*, 1966; *Minimonsters Traumlexikon, Texte in Prosa*, 1968). Collections of poetry include *Blaue Erleuchtungen. Erste Gedichte* (1972), *Ausgewählte Gedichte 1944–1978* (1979), *Schwarze Romanzen* (1981), and *Gute Nacht, guten Morgen. Gedichte 1978–1981* (1982), and prose *Das Licht in der Landschaft* (1975), *Fast ein Frühling des Markus M.* (1976), *rot ist unten* (1977), *Heiligenanstalt* (1978), a composition on composers, *Die Abschiede* (1980), *Magische Blätter* (1983), and *Reise durch die Nacht* (1984), in which a return journey from France forms the framework for daunting inner experience, rich in metaphor and stylistic experiment. She has also written radio plays, collaborating with E. Jandl (q.v.), notably in *Fünf Mann Menschen* (1968).

Mâze, Die, title given to a Middle High German poem of some 200 lines, giving advice to men and to women on the value and application of *mâze*, interpreted as 'moderation' (*Mäßigkeit*) It is chiefly concerned with the relations of the sexes. The poem has been dated as early as 1170 (on stylistic grounds) and as late as the beginning of the 14th c. Since it contains an allusion to a Spruch by WALTHER von der Vogelweide (q.v.), the correct date is probably late in the 13th or early in the 14th c. The title of the rather clumsy and prosaic poem derives from the first line, which refers to *mâze* as 'muoter aller tugende' (Mutter aller Tugenden).

MECHOW, KARL BENNO VON (Bonn, 1897–1960, Emmendingen), of a military family, volunteered in 1914 and was commissioned in an Uhlan regiment, fighting on the eastern front. After 1918 he farmed, first in Brandenburg, then in Bavaria. His books draw upon his feeling for nature, including agriculture, and on his war experiences. Among his novels are: *Das ländliche Jahr* (1930), the story of a conscientious but too sensitive bailiff, who for a year looks after an estate, is dismissed through intrigue, and finds his happiness in a smallholding of his own; *Das Abenteuer* (1930), a tragic story which closely follows an escape from Mechow's own experience as an Uhlan officer on the Russian front; and *Vorsommer* (1933), a kind of Bildungsroman (q.v.), in which Praetorius, a disoriented soldier-farmer returned from the 1914–18 War, finds his bearings through the help of a true-hearted, hard-working girl, Ursula. The book, telling only of the beginnings of their summer (Vorsommer), implies that a full summer will follow.

MECHTHILD VON HACKEBORN (1241–99, nr. Eisleben) was, with her sister GERTRUD von Helfta (q.v.), a nun at the Cistercian Abbey of Helfta near Eisleben. Her mystical and ecstatic visions of Christ and of the Sacred Heart are recorded in Latin in the *Liber specialis gratiae*, which was formerly attributed to her sister Gertrud but is now thought to be probably by the Abbess GERTRUD von Hackeborn (q.v.).

MECHTHILD VON MAGDEBURG (*c.* 1212–82, nr. Eisleben), who became a beguine *c.* 1235 and a Cistercian nun at Helfta near Eisleben *c.* 1270, experienced mystical visions, which she recorded together with her thoughts and aspirations under the title *Das fließende Licht der Gottheit.* The Low German original has not survived, and her visions of Christ and the Holy Trinity are now known only in a High German version made by HEINRICH von Nördlingen (q.v.) *c.* 1344.

MECKEL, CHRISTOPH (Berlin, 1935–), an art student who developed into a Surrealist poet (*Hotel für Schlafwandler*, 1958; *Nebelhörner*, 1959; *Tarnkappe*, 1961; *Wildnisse*, 1962; *Bei Lebzeiten zu singen*, 1967). *Land der Umbramauten* (1961) is a prose fantasy. Other publications in the DDR and BRD include *Werkauswahl. Lyrik, Prosa, Hörspiel* (1971), a volume of stories with drawings, *Kranich*, the novel *Bockshorn* (both 1973), a select edition of poetry, *Ausgewählte Gedichte 1955–1978* (1979), and a volume of stories and poetry, *Die Sachen der Liebe* (1980). The volume *Nachricht für Baratynski* appeared in 1981 and the story *Der wahre Muftoni*, with drawings by the author, in 1982. The collection *Die Gestalt am Ende des Grundstücks. Erzählungen*, a DDR publication, appeared in 1975 and *Suchbild. Über meinen Vater* in 1980; *Ein roter Faden. Gesammelte Erzählungen* in 1983 and *Gesammelte Werke* 1977 ff.

MEDING, OSKAR, see SAMAROW, GREGOR.

Meeresstille, a short poem written by Goethe probably in 1795, though some have dated it as early as 1787. It is coupled with the poem *Glückliche Fahrt*, and the two poems refer respectively to a ship becalmed and to one sailing before a favourable wind. Both poems were first published in the *Musenalmanach für das Jahr 1796.*

Meersburg, Swabian town on the northern shore of Lake Constance with a medieval castle, which was bought in 1838 by the connoisseur and bibliophile Joseph von Laßberg (q.v.). Laßberg's sister-in-law, Annette von Droste-Hülshoff (q.v.), spent some of her later years (from 1841 on) at the castle and died there in 1848. Her rooms are preserved with some of her furniture. She was instrumental in obtaining for the writer L. Schücking (q.v.) an appointment as librarian to Herr von Laßberg. Formerly in Baden, Meersburg is now in the Land Baden-Württemberg.

MEGERLE, JOHANN ULRICH, see ABRAHAM A SANTA CLARA.

Mehreren Wehmüller und die ungarischen Nationalgesichter, Die, a whimsical story (Novelle) by C. Brentano (q.v.), written probably between 1811 and 1814, and published in the periodical *Der Gesellschafter* in 1817. A wildly improbable story full of exuberant humour, it tells of a travelling painter in Hungary, who paints 39 'national' faces and lets each client choose the one who seems to resemble him most.

Wehmüller seeks to join his wife who is in an area beset by plague and isolated by a sanitary cordon. He learns to his alarm that a double of himself has been observed. A night at an inn gives an opportunity for the insertion of some tales. Wehmüller, attempting to cross the cordon, encounters yet another double (Doppelgänger). However, all is cleared up, for this second double turns out to be his wife, who is seeking to join him, while the first is a rival painter attempting to steal a march on him. The two painters join in partnership.

The story reflects Brentano's fertile imagination and mercurial intellect. It was reissued in 1833 in a joint edition with a story by Eichendorff (q.v.) entitled *Viel Lärmen um Nichts. Von Joseph Freiherrn von Eichendorff und: Die mehreren Wehmüller und ungarische Nationalgesichter von Clemens Brentano. Zwei Novellen.*

MEHRING, WALTER (Berlin, 1896–1981, Zurich), studied history of art at Berlin and Munich universities and served in the 1914–18 War (from 1916). From 1915 he was a contributor to *Der Sturm* (q.v.), and in the 1920s to a number of other journals including *Weltbühne* (see C. VON OSSIETZKY). A co-founder of the Berlin circle of Dada (see DADAISMUS), he founded the Politisches Cabaret (1920) and contributed to the cabaret Schall und Rauch (see M. REINHARDT). His satirical poetry is contained in *Gedichte, Lieder und Chansons des W. M.* (1929). In 1933 he emigrated, was interned in France in 1939, but in 1940 escaped to the U.S.A. After the war he moved to Switzerland. His fiction includes the satirical novel *Müller. Die Chronik einer deutschen Sippe von Tacitus bis Hitler* (1935) and *Die Nacht der Tyrannen* (1938). *Die verlorene Bibliothek. Autobiographie einer Kultur* appeared in 1964. Written in New York, it first appeared in English in 1951. *Werke in Einzelausgaben* (10 vols.), ed. C. Buchwald, appeared 1978–83; they include the two volumes *Chronik der Lustbarkeiten. Gedichte, Lieder und Chansons 1918–1933* and *Staatenlos im Nirgendwo. Die Gedichte, Lieder und Chansons 1933–1974* (1981).

Meier Helmbrecht, a didactic and satirical Middle High German poem, written probably between 1237 and 1290 by Wernher der Gartenaere, who names himself in the last line. He is not otherwise known and is thought to have been either a strolling poet (Fahrender) or, less probably, a cleric. The poem has 1,934 lines and is in rhyming couplets.

Helmbrecht is a farmer's son, who apes courtly dress and is determined to become a knight who robs and oppresses his neighbours. With nine companions he commits a series of violent and atrocious crimes. He returns home and, like the Prodigal Son, is made welcome. But his object is not to repent, but to persuade his sister Gotelind to become the bride of one of the robbers. She flees with him, but at the wedding feast the members of the gang are surprised and taken by the officers of the law. The nine are hanged, and Helmbrecht is blinded and mutilated. He returns home again, and this time is sternly rejected by his father. A year later he is recognized by peasants as their former oppressor, and hanged. The forceful tale is followed by a moral application, revealing Helmbrecht's fate as a consequence of his desertion of the 'estate of the plough' and of his failure to observe his father's sound advice. The poem is a document reflecting changing forces and a shifting balance in society, in which the poet sides with the threatened feudal order.

Neither of the existing MSS. is the original. In the older one (see AMBRASER HANDSCHRIFT) the location is the Innviertel (q.v.). The former Berlin MS. (now in Tübingen) sites the events

in Upper Austria between Kremsmünster and Wels. Neither location is necessarily that of the original.

Meierin mit der Geiß, Die, a short verse tale of the late 13th c., in which a farmer's wife sends her husband off by means of a ruse, so that her lover, a knight, can take his place in bed.

Meine Göttin, a poem written by Goethe on 15 September 1780 and sent at once to Charlotte von Stein (q.v.). It was first published in 1789 in *Goethes Schriften*. Written in hymnic form, the poem is a panegyric of the poetic imagination.

Meineidbauer, Der, a three-act play (Volksstück mit Gesang) by L. Anzengruber (q.v.), first performed in December 1871 at the Theater an der Wien, Vienna, and published in 1872. It is in a generalized Austrian dialect. Matthias Ferner, with a serious perjury on his conscience, brings his son Franz up to be a priest, so that he may one day confess his crime in safety. But Franz has other ideas, resolving to become a farmer and marry Vroni, whom he loves. In a fearful quarrel Matthias attempts to murder Franz, believes (erroneously) that he has succeeded, and dies of a stroke. Franz and Vroni face the future together in confidence and hope.

Mein Eigentum, an ode in Alcaic verse by F. Hölderlin (q.v.) written probably in 1799 and printed in 1846. Its theme is the poet's dependence on his creativity, his sole possession.

MEINHARD, JOHANN NIKOLAUS (Erlangen, 1727–67, Berlin), who as a private tutor and companion made two extensive European tours (1756 and 1763), translated into German the *Elements of Criticism* of the Scottish philosopher Henry Home (Lord Kames, 1696–1782) under the title *Grundsätze der Kritik* (1763–6). His *Versuche über den Charakter und die Werke der besten italienischen Dichter* (1763–4) earned Lessing's praise in the *Literaturbriefe* (q.v., Pt. 23, No. 332).

MEINHOLD, JOHANNES WILHELM (Netzelkow, Insel Usedom, 1797–1851, Berlin-Charlottenburg), a pastor's son, studied theology at Greifswald and became a private tutor, a schoolmaster, and finally a pastor, first at Krummin (1828–40) and then at Rehwinkel near Stargard (1840–50). He published a volume of poems (*Gedichte*) in 1835, but is chiefly known for his novels, *Maria Schweidler, die Bernsteinhexe* (q.v., 1838) and *Sidonie von Bork, die Klosterhexe* (1847). *Maria Schweidler* purported to be a 17th-c. chronicle, and Meinhold had later some difficulty in proving his authorship. Meinhold, who enjoyed the patronage of King Friedrich Wilhelm IV (q.v.) of Prussia, leaned in later years towards Roman Catholicism and resigned his living in 1850. His collected works (*Gesammelte Schriften*, 9 vols.) appeared 1846–58.

Meininger, a troupe of actors recruited and maintained by Georg II, Herzog von Sachsen-Meiningen (1826–1914), from 1874. Their high standards of acting and production greatly influenced the German drama and theatre on the eve of the Naturalistic movement (see NATURALISMUS), for which their insistence on exact historical detail was preparatory.

Mein Kampf, Adolf Hitler's only completed book, the first volume of which, written in 1924 in Landsberg prison, was published in 1925. The second volume, written in 1925, was published in 1926. An unabridged edition in one volume appeared in 1930. *Mein Kampf* is less an autobiography than an emotional and subjective political treatise. The first volume, entitled *Die Abrechnung,* gives Hitler's views on the world in which he grew up, on the 1914–18 War, and on the collapse of Germany in 1918. It expresses especially his anti-international and anti-Semitic opinions, and his obsessive conception of German living space (*Lebensraum*). The second volume, entitled *Die nationalsozialistische Bewegung,* is mainly concerned with the political struggle and the future National Socialist state. It contains, either developed or in embryo, all the terroristic aspects of Hitler's policies when he achieved power, but little attention was paid to it at the time. After 1933 this work, which has justifiably been called the Bible of National Socialism, was distributed in enormous quantities; up to 1940 the total of copies reached 9,000,000.

'Mein Lebenslauf ist Lieb' und Lust', much-quoted line occurring in the poem *Weinlied* (1808) by S. A. Mahlmann (q.v.).

MEINLOH VON SEVELINGEN, one of the earliest Middle High German Minnesänger, was a knight, whose home was Söflingen, near Ulm. His poems, presumed to have been written in the late 12th c., reflect two conceptions of love, the attraction which can achieve consummation, and the formal conception of courtly *minne* as service to a remote noble lady. It has been suggested that Meinloh's original disposition to sing of mutual love has had superimposed upon it, perhaps through his western situation, the new courtly attitude of French origin.

Mein Name sei Gantenbein, the third novel written by M. Frisch (q.v.), published in 1964. It is the most baffling of the three, yet probably the most widely read. Underlying the novel is Frisch's distrust of words and especially metaphors. This novel is intended as an illustration of the difficulty of communication by means of words and simultaneously as a demonstration of Frisch's virtuosity in their employment. It has no plot, but many incidents. The first-person narrator conceals three separate identities: Theo Gantenbein himself, Felix Enderlin, and Frantisek Svoboda. All three are related to one person, Lila. Gantenbein is her second husband, Enderlin her lover, Svoboda her first husband. The book consists of episodes in the lives of these, mostly related in the present tense, in words which the author intends us to perceive as an arbitrary convention, while the speaker uses them because the actual episode is incommunicable by words. It is frequently spoken of as a treatment of the problem of identity, but more important is Frisch's sense of existence in terms of transience, what he calls our perpetual awareness, not of time, but of what he calls 'Vergängnis', and the banal concluding words 'ich lebe gern' express an unpretentious existentialism.

Mein Wort über das Drama, see HEBBEL, F.

MEISL, KARL (Laibach, now Ljubljana, 1775–1853, Vienna), quarrelled with his parents and joined up as a private. He achieved rapid promotion, became a quartermaster, and in 1820 was transferred as a well-placed civil servant to the Admiralty (Austria had then an Adriatic coast and fleet). He retired in 1844. In 1801 he began to write popular plays (see VOLKSSTÜCK) and, after Raimund and Nestroy, was with Bäuerle and J. A. Gleich (qq.v.) one of the principal purveyors of such pieces. His last work had its première in 1845, by which time he had already outlived his reputation. Like other writers of Viennese Volkstücke, he had to write too much and too fast in order to meet the insatiable demands of the Viennese theatre-goers. He is reputed to have written more than 200 comedies, of which less than 40 were printed. Of the titles that survive several, including *Die Frau Ahndel* (1817), *Der lustige Fritz oder Schlaf, Traum und Besserung* (1818), *Das Gespenst auf der Bastei* (1819), *Das Gespenst im Prater* (1821), and *Die Geschichte eines echten Schals in Wien* (1820), have been reprinted in this century. Other titles which should be remembered are *Ein Tag in Wien* (1812), *Die Damenhüte im Theater* (1818), *Überall ist's gut, aber zu Hause am besten* (1823), and the remarkable conspectus of the centuries, *1723, 1823, 1923* (1823). He had a talent for parody: *Die Frau Ahndel* guys Grillparzer's *Die Ahnfrau*, q.v., and *Die travestierte Zauberflöte* (1818) takes off Mozart's opera. When the Theater in der Josefstadt was rededicated after rebuilding in 1822, Beethoven's overture *Zur Weihe des Hauses* was followed by a play by Meisl with the same title.

MEISSNER, DER, was a Middle High German poet, who wrote learned Sprüche (see SPRUCH), which include an attack on Der Marner (q.v.) for his want of learning. Der Meißner, whose poems belong to the late 13th c., may possibly be a name covering the work of more than one poet. The MSS. refer both to 'der alte' and 'der junge Meißner'; in neither case is the identity clear.

MEISSNER, ALFRED VON (Teplitz, 1822–85, Bregenz), of mixed Austrian, Jewish, and Scottish parentage and a nephew of A. G. Meißner (q.v.), studied medicine at Prague University and subsequently participated in the radical movements leading to the Revolutions of 1848–9 (q.v.), travelling much and joining in the revolution in Prague in 1848. In 1869 he settled in Bregenz where he committed suicide in the year after he was ennobled by King Ludwig II (q.v.) of Bavaria. Meißner first published a volume of poems (*Gedichte*, 1845), and followed this with a historical epic *Ziska* (1846), dealing with the Hussites in Bohemia. His plays, *Das Weib des Urias* (1850), *Reginald Armstrong oder die Macht des Geldes* (1853), and *Der Prätendent von York* (1857), aroused little interest. Meißner published a number of novels which in greater or less degree were collaborations with his friend F. Hedrich (q.v.). The two friends later fell out, Hedrich claiming chief credit for the novels published under Meißner's name. It is generally assumed that Hedrich exaggerated any claims he may have had, and it has been suggested that Meißner's suicide is attributable to Hedrich's blackmail. The novels concerned included *Der Freiherr von Hostiwin* (2 vols., 1855, revised and retitled *Sansara* in 1858), *Zur Ehre Gottes* (2 vols., 1860), a novel directed against the Jesuits, *Neuer Adel* (2 vols., 1861), *Schwarzgelb* (8 vols., 1862–4), and *Die Kinder Roms* (6 vols., 1870). Meißner's *Gesammelte Schriften* (18 vols.) appeared in 1871–3.

MEISSNER, AUGUST GOTTLIEB (Bautzen, 1753–1803, Fulda), became a professor in Prague in 1785, and in 1805 director of the Lyceum (university college) at Fulda. He wrote plays and stories and was active as a translator. His work has a pronounced moralizing trend.

The novelist Alfred von Meißner (q.v.) was his nephew.

MEISTER, ERNST (Hagen-Haspe, Westphalia, 1911–79, Hagen), an office worker and spare-time poet, whose delicate and sensitive verse appeared in *Ausstellung* (1932), *Unterm schwarzen Schafspelz* (1953), *Dem Spiegelkabinett gegenüber* (1955), *... und Ararat* (1956), and *Flut und Stein* (1962). A collection was published in 1964 as *Gedichte 1932–64*. Further collections of poetry include *Es kam die Nachricht* (1970), *Sage vom Ganzen den Satz* (1972), *Schatten*, with lithographs by the author (1973), *Im Zeitspalt* (1976), *Wandloser Raum* (1979), and *Ausgewählte Gedichte 1932–1976* (1977), a volume edited by B. Allemann that was extended and edited with a postscript in 1979, *Ausgewählte Gedichte 1932–79*. Thirteen radio plays by Meister were broadcast between 1963 and 1975. He was awarded the Büchner Prize posthumously in 1979.

MEISTER ECKHART, see ECKHART, MEISTER.

Meister Floh. Ein Märchen in sieben Abenteuern zweier Freunde, a story by E. T. A. Hoffmann (q.v.), written in Berlin in 1821–2 and published in 1822. At the end of many extravagant adventures two pairs of lovers are united. George Pepusch and Princess Gamaheh die in the act of union (a Liebestod), while Peregrinus Tyss and Röschen Lämmerhirt live on happily. The supernatural repeatedly intrudes into common life. The character Meister Floh is the principal flea in a flea circus, and at the same time a benevolent personality in the lives of the lovers; Hoffmann invents a magic microscope which enables Peregrinus to read people's thoughts.

Meistergesang, or **Meistersang,** a form of poetry set to music and sung solo and unaccompanied by members of guilds of Meistersinger. The music descended from Gregorian chant and other sources, and the poems adhered to the triple structure of Minnesang (q.v.), but laid down that a song (Bar) must consist of three strophes. Meistergesang was both spiritual and secular. The religious poetry was sung at the Singschule held in church; the worldly songs accompanied the ensuing celebration in the tavern (Zechsingen). The lay themes were anecdotal (analogous to the Schwank, q.v.), historical, legendary, or even classical (e.g. Ovid).

Meistergesang originated possibly in lay participation in ecclesiastical music, and developed in the course of the 15th c., but its origins are not fully clear. A school of Meistersinger appears first to have arisen in Mainz, followed by Worms and Strasburg. From Mainz Meistergesang spread eastward in the 16th c. through cities of Swabia (Nördlingen, Ulm, Rothenburg/Tauber, Augsburg), and of Franconia (Nürnberg). Its furthest extension was to Silesia (Görlitz, Breslau) and Upper Austria. Further schools arose in Freiburg/Breisgau, Colmar, Frankfurt, and Eßlingen.

The Meistersinger were Bürger and for the most part respected master craftsmen. The Singschule, which conformed to strict rules, is perceptibly descended from the medieval disputation, even to the sitting posture of the singer. One or more 'Merker' judged the singer. The location in a consecrated building reinforced the gravity of the proceedings. Strict rules (Tabulatur) governed the composition of the song, and at first only tunes (Töne) of twelve medieval singers might be used. Hans Folz (q.v.), however, breached this tradition at Nürnberg (after failing to do so in Worms) and throughout the 16th c. the composition of a new 'Ton', as well as poem,'was required of a candidate for the title Meister. The tunes were distinguished by fanciful names, e.g. 'spitzige Trunkschuh-Weise', 'schreckliche Donnerweise', etc. Apart from this the Meistergesang underwent virtually no development. The strictly esoteric nature of the guilds (which forbade public performance or the printing of Meisterlieder) and the extreme rigidity of the Tabulatur condemned the form to sterility. The best-known Meistersinger, apart from Folz, were Hans Sachs (q.v.), Lienhard Nunnenbeck and Benedikt von Watt, both in Nürnberg in the 16th c., Onophrius Schwarzenbach (*fl. c.* 1550) and Johannes Spreng (q.v.) in Augsburg, and Adam Puschman and Georg Hager (qq.v.) in Breslau. The enclosed nature of the proceedings prevented these masters from influencing in any important degree the development of German poetry. Their heyday was the first half of the 16th c., but they continued into the 19th c.; the date of the dissolution of the last school in Memmingen (Bavaria) has been given by various authorities as 1852, 1875, and 1880. This chronological uncertainty underlines the obscurity into which the activity had lapsed.

In *Die Meistersinger* (q.v.) R. Wagner gives a partly amusing and partly endearing re-creation of the Nürnberg school, which, however, inflates its contemporary prestige.

Meister Johannes Wacht, a story by E. T. A. Hoffmann (q.v.), written in 1822 and first published posthumously in 1823 in the *Frauentaschenbuch*. It appeared subsequently in *Letzte*

Erzählungen (1825). Johannes Wacht, a master carpenter possessing exceptional skill, integrity, and strength of character, loses his 18-year-old son in a fire, and the master's wife dies from the shock. Wacht overcomes his grief and continues his work. He still has two daughters, Rettel and Nanni, and he takes into his house Sebastian, the son of another carpenter who has died. A second son of weak physique, Jonathan, is intent on studying law, to Wacht's disgust, for he hates all lawyers as men who distort justice for gain. Sebastian goes away and nothing is heard of him, meanwhile Jonathan and Nanni, to Wacht's chagrin, fall in love. Wacht is at first deaf to all their pleas, but at last is won over when he discovers how generously Jonathan, who has been the victim of an attack by Sebastian, treats his stepbrother. The focus of the story is the robust, engaging, kindly character of Wacht with its streak of unreasoning prejudice.

Meisterlied, see MEISTERGESANG.

MEISTERLIN, SIGMUND or SIGISMUND (*c.* 1425–90), a Benedictine monk of Augsburg, who is also recorded in Pavia, Würzburg, and Nürnberg. He is the author of a chronicle of Augsburg (*Chronographie Augustensium* or *Augsburger Chronik* 1457), which was printed in 1522, and of a chronicle of Nürnberg written in Latin in 1485 and printed in German (*Nürnberger Chronik*) in 1488. He also wrote a Latin life of St. Sebaldus, patron saint of Nürnberg (*Vita Sancti Sebaldi, c.* 1480).

Meister Martin, der Küfner, und seine Gesellen, a story by E. T. A. Hoffmann (q.v.), written in 1817–18, first published in the *Taschenbuch zum geselligen Vergnügen* (1818) and afterwards included in vol. 2 of *Die Serapionsbrüder* (1819). The story is set in Nürnberg in the late 16th c.

Meister Martin, a master-cooper, is elected president of the coopers' guild and immediately shows that he knows his own worth. He has a beautiful daughter named Rosa, whom he plans to marry only to a cooper, because he interprets in this way the words of his grandmother on her deathbed. His whim brings him journeymen who are not really coopers. Rosa soon has three suitors, Friedrich, Reinhold, and Konrad. Konrad will not tolerate the master's roughness, returns blow for blow, and leaves the house. Reinhold, in reality a painter, gives up the suit. Friedrich, who has concealed his true trade as a goldsmith and is Rosa's real choice, is driven away by Meister Martin. In the end the master discovers that he has misinterpreted the prophecy and gives Rosa to Friedrich. Konrad appears at the wedding with a wife, and proves to be a nobleman, Konrad von Spangenberg; Reinhold also returns, and all ends happily.

Meister Oelze, a play in three acts by J. Schlaf (q.v.), published in 1892. It is written in Berlin dialect. It shows the murderer Oelze (his crime has not been discovered) beset by his half-sister Pauline, who is convinced of his guilt. Pauline, reduced to poverty by Oelze's crime, besieges the consumptive murderer with insinuations, and plays upon his fears of the supernatural, but Oelze dies undefeated and unconfessing. The play was widely abused when performed, though Nietzschean critics of the day applauded Oelze's self-reliance and resistance to religious pressures.

Meister Reuaus, an anonymous satirical Middle High German poem of some 660 lines, written early in the 15th c. The author, an Austrian, directed his poem against the seven deadly sins. Reuaus is a devil in the guise of a quack, and the sins are the medicines he retails. The work is influenced by *Der Renner* (q.v.) by Hugo von Trimberg.

Meistersang, see MEISTERGESANG.

MEISTER SEPP VON EPPISHUSEN, a pseudonym used by Freiherr Joseph von Laßberg (q.v.).

Meistersinger, see MEISTERGESANG.

Meistersinger von Nürnberg, Die, an opera by R. Wagner (q.v.), begun in 1849, written and composed at Triebschen (1866–68), and first performed at Munich on 21 June 1868. It is the only comedy in the canon of Wagner's mature works. Set in 16th-c. Nürnberg, it has the cobbler-poet-composer Hans Sachs (q.v.) as its central figure; the Guild of Meistersinger, to which Sachs belongs, plays an important part.

Veit Pogner, goldsmith and Meistersinger, has promised his daughter Eva to the musician successful in the festival of Midsummer. The handsome but impoverished nobleman Walther von Stolzing falls in love with Eva and she with him. He consents to give the Meistersinger a sample of his musical skill, but is turned down because he infringes the rules, on the observation of which his rival for Eva's hand, the pedant Beckmesser, insists. The lovers are at a loss and think of elopement, which Sachs, in his good

sense, frustrates. But Sachs also derides Beckmesser's singing at night in a Nürnberg street in a scene which culminates in a riot, though it closes in moonlight serenity. The next morning Sachs, whose plan is now complete, leaves Walther's proposed song on the table and, as he expects, Beckmesser, calling for his new shoes, steals it in the belief that it is by Sachs. At the great prize singing festival Beckmesser tries to sing Walther's song, fails to make head or tail of it, and is laughed off the rostrum. When Walther sings it with understanding and warmth, all the listeners are carried away in spite of its neglect of the rules, and after initial reluctance on the part of the Meisteringer, the prize, Eva's hand, is awarded to him. A rich comedy in its own right, the opera also symbolizes Wagner's musical originality, its reception by the critics, and its justification.

Meister Timpe, a novel by M. Kretzer (q.v.), published in 1888. Timpe, a master-turner who employs eight journeymen, finds himself in competition with men with modern massproducing methods, and goes to the wall. His downfall is accelerated by the treachery of his son, who goes over to the bourgeois factoryowner Urban and even steals Timpe's designs. The ruined Timpe goes to a Social Democrat meeting, at which he makes an inflammatory speech. Threatened with eviction by Urban, who owns his house, Timpe barricades himself in, and dies as entry is forced. At the same time the first train of the new Berlin Stadtbahn steams past, a symbol of the new industrial age which has no room for Timpe.

Meister von Palmyra, Der, a play in blank verse (dramatisches Gedicht) by A. Wilbrandt (q.v.), published in 1889. An attempt to present philosophical ideas in classical form with a minimum of action, it has as its dominating figure Pausanias, the master of Palmyra, a symbolic and consolatory being who brings death, and with it peace, to the painter Apelles. The play enjoyed considerable success, being regarded as a blow struck for beauty at a moment in which modern literary art, in the form of Naturalism, was believed to be turning to ugliness.

MELANCHTHON, PHILIPP (Bretten, Baden, 1497–1560, Wittenberg), was appointed professor of Greek in Wittenberg in 1518 and soon became a firm friend and supporter of Luther (q.v.). His *Loci communes rerum theologicarum* (1521) is a formulation of the theology of the Reformation (q.v.). After Luther's death in 1546 Melanchthon became the leader of the

Lutheran party, but his flexibility, and perhaps weakness, exposed him to fierce and intolerant attacks from both sides, which saddened and embittered his last years. His theological repute has undergone considerable fluctuations.

Melanchthon was a Greek scholar of outstanding merit and was devoted to the cause of education and learning. He earned the title Praeceptor Germaniae and may be regarded as the creator of the German classical grammar school (Gymnasium). His adopted name of Melanchthon was intended as a Greek equivalent of his true surname, Schwarzert, interpreted as 'schwarze Erde' ('black earth'). *Opera quae supersunt omnia* (28 vols.), ed. K. G. Bretschneider and H. E. Bindseil, appeared in Latin 1834–60 (reissued 1963) and *Werke in Auswahl*, ed. R. Stupperich *et al.*, 1951 ff.

MELISSUS, PAULUS, the pseudonym of Paul Schede (Mellrichstadt, Franconia, 1539–1602, Heidelberg), who in the course of a nomadic life visited England in 1585–6, after which he settled in Heidelberg as librarian to the Elector of the Palatinate. He wrote religious poems in Latin, which are concerned more with personal feeling than with formal elegance. He translated the French versions of 50 psalms into German, observing the French rules of rhyme and prosody, so taking the first step towards the long supremacy in German poetry of French metrical procedures (*Die Psalmen Davids in teutische gesangreymen nach Frantzösischer unt sylben art mit sönderlichem Fleise gebracht*, 1572). He was also interested in the reform of German spelling, and was the first to use *terza rima* in German verse.

Melk, Abbey of, founded in 985 as a college of secular canons and converted into a Benedictine abbey in 1089 during the period of Cluniac Reform (q.v.). The present buildings overlooking the Danube constitute one of the most splendid baroque monuments in Europe and were constructed between 1702 and 1738 according to the plans of Jakob Prandtauer (1660–1726), who supervised the work until his death. In the 12th c. HEINRICH von Melk (q.v.) is believed to have been a lay brother at the Abbey and Frau Ava (q.v.) a recluse under its protection.

Melker Marienlied, an Early Middle High German hymn in praise of the Virgin Mary. It consists of 14 verses, each followed by the refrain *Sancta Maria*, and is one of the earliest German poems in stanza form. It is of Austrian origin and has been dated between 1140 and 1180. The MS. is in Melk Abbey.

MELL, MAX (Marburg, Austria, 1882–1971, Upper Austria), a student of Vienna University, served in the Austrian artillery in the 1914–18 War. Previously he had published Novellen (*Die Grazien des Traumes*, 1906; *Jägerhaussage*, 1910) and poems (*Das bekräntze Jahr*, 1911). The collapse of the Austro-Hungarian Empire and the consequent threat to its cultural traditions was a decisive experience for Mell, who devoted himself to an assertion of religious continuity in a series of plays in archaic verse, dealing with Christian themes, and especially with Redemption. They include *Das Wiener Kripperl von 1919* (1921), *Das Apostelspiel* (1923), *Das Schutzengelspiel* (1923), *Ein altes deutsches Weihnachtsspiel* (1924), and *Das Nachfolge-Christi-Spiel* (1927). He turned to classical themes and Nordic myth in the tragedy *Die Sieben gegen Theben* (1932) and the play *Der Nibelunge Not* (1951). He also wrote a play on Joan of Arc, *Jeanne d'Arc* (1957), and one on Paracelsus (q.v.) entitled *Paracelsus und der Lorbeer* (1964).

Mell's works of short fiction since the 1914–18 War include *Morgenwege* (1924) and *Das Donauweibchen* (1938); *Verheißungen* and *Gabe und Dank* appeared in 1943 and 1949 respectively. *Gesammelte Werke* (4 vols.) were published in 1962 and correspondence with H. von Hofmannsthal (q.v.) in 1982.

Melodrama in German literature signifies a passage or scene in which either a speaking voice or a musically declamatory voice is accompanied by music. Examples of the one kind are the finale of Goethe's *Egmont* (q.v.) and Johanna's monologue in Act IV of Schiller's *Die Jungfrau von Orleans* (q.v.); of the other the accompanied recitative in Beethoven's *Fidelio* or Weber's *Der Freischütz* (qq.v.).

Monodrama (q.v.) is sometimes included under the heading Melodrama, as is also the recitation of a poem against a background of music, e.g. the composition by M. von Schillings (q.v.) of the *Hexenlied* by Wildenbruch (q.v.); in this sense the original performance of Walton's *Façade* would be accounted a Melodrama. The word does not include in German the crass and sensational plays known in English as melodramas.

Melusina, a libretto (Romantische Oper in drei Aufzügen) written by F. Grillparzer (q.v.) in 1823 to be set by Beethoven (q.v.), who did not pursue the project. It was first performed with music by the German composer Konradin Kreutzer (1780–1849) in February 1833 in Berlin, and in April 1835 in Vienna; it was published in 1833. Written in the tradition of the Viennese Zauberstück (see VOLKSSTÜCK), it is a variant of the *Schöne Melusine* (q.v.), from which it also adopts the name of the principal male character, Raimund.

The fairy Melusina and Raimund unite in love, Melusina against the warnings of her two fairy sisters, and Raimund in defiance of his human companions and of his pledge to Bertha, his betrothed. On condition that Raimund always keeps Melusina's magic ring and renounces all activity associated with temporality, Melusina promises him the delights of heaven and eternity. Raimund fails her, causing her doom, but in dying achieves their reunion, and heaven receives them both. Unlike the other characters, Melusina and Raimund perceive the spirit linking their realms, but Raimund must die to make Melusina's highest worth apparent.

The allegorical presentation of the action, based on the conflict between life and art, contains motifs treated in *Sappho* (q.v.) and other works by Grillparzer, including *Libussa* (q.v.).

Melusine, see SCHÖNE MELUSINE and NEUE MELUSINE, DIE.

Memento mori, title given by the first editor to an Early Middle High German poem probably written in Hirsau *c.* 1070. It addresses the laity in the style of a sermon, insistently warning of the inevitability of death and the peril of damnation. It rejects the temptations of the world and points out the social levelling wrought by death. Its spirit is that of the asceticism associated with Hirsau Abbey, the first German monastery to be touched by the Cluniac Reform movement (q.v.) in Germany. The stanzas are of 8 lines and the lines rhyme, often impurely, in pairs. A cryptic last line seems to hint at authorship (daz machot all ein Noker), possibly 'Nogger' (Swiss Notker), a monk of Hirsau, who later became Abbot of Zwiefalten. The older view that 'Noker' is Notker Labeo (q.v.) of St. Gall is certainly incorrect.

MENCKE, JOHANN BURKHARD (Leipzig, 1674–1732, Leipzig) the son of a professor of Leipzig University, was himself elected to a professorship in 1699. Mencke, who befriended the poet J. C. Günther and later gave his protection to Gottsched (qq.v.), was himself a minor poet, publishing under the pseudonym Philander von der Linde *Galante Gedichte* (1705), which were followed by *Scherzhaffte Gedichte* (1706), *Ernsthaffte Gedichte* (1706), and *Vermischte Gedichte* (1710). He is best known for his *De charlateneria eruditorum* (1713 and 1715), two orations deriding the affectations and pretensions of scholars. Mencke continued the learned journal *Acta eruditorum* (q.v.) founded at Leipzig in 1682 by his father Otto Mencke (1644–1707).

MENDEL, GREGOR JOHANN (Heinzendorf/Austria, 1822–84, Brünn, now Brno), laid the foundation for a scientific approach to human genetics by assessing hereditary factors with the aid of mathematical formulae. An Augustinian monk (1843), he studied science at Vienna. By his experimental research he established theories which formed the basis for progressive investigations and are known as the *Mendelsche Gesetze* (Mendel's Laws, 1865). Mendelism has since the turn of the century been increasingly applied in the medical and sociological field. In 1860 Mendel became abbot of his monastery at Brünn.

MENDELSSOHN, MOSES (Dessau, 1729–86, Berlin), at first a tutor and later secretary to a silk manufacturer in Berlin, was a friend of G. E. Lessing and F. Nicolai (qq.v.). His writings are primarily philosophical, and derive mainly from Leibniz (q.v.), Locke, and Shaftesbury. His principal works are *Brief über die Empfindungen* (1755), *Abhandlung über die Evidenz in den metaphysischen Wissenschaften* (1764), and *Phädon oder über die Unsterblichkeit der Seele* (1767), which his contemporaries regarded as his most important contribution. In *Über die Hauptgrundsätze der schönen Künste und Wissenschaften* (1757) and *Betrachtungen über das Erhabene und Naive in den schönen Wissenschaften* (1758) he grappled with aesthetic problems, and in the latter he anticipated in some degree Schiller's views. In the purely literary field Mendelssohn contributed to Nicolai's *Bibliothek der schönen Wissenschaften und der freien Künste* and to Lessing's *Literaturbriefe* (q.v.). Lessing cited his character to justify the courageous and gentlemanly Jew (Der Reisende) in *Die Juden* (q.v.), and probably had him in mind in the composition of *Nathan der Weise* (q.v.). Himself an outstanding representative of the enlightened Jew of high character and humane rationalism, he strove in his later years for Jewish emancipation (*Jerusalem oder über religiöse Macht und Judentum,* 1783). His rejection of the assertion, made in 1775 by F. H. Jacobi (q.v.), that Lessing inclined to Spinoza's philosophy (*Über die Lehre des Spinoza*) is conveyed in *Moses Mendelssohn an die Freunde Lessings* (1786).

Gesammelte Schriften (7 vols.) appeared 1843–5; *Schriften zur Philosophie, Ästhetik und Apologetik,* ed. M. Brasch (2 vols., 1881), *Gesammelte Schriften. Jubiläumsausgabe* (planned in 16 vols.), ed. I. Elbogen and A. Altmann, 1929 ff. (6 parts, ed. I. Elbogen, were reissued in 1968).

MENDELSSOHN-BARTHOLDY, FELIX (Hamburg, 1809–47, Leipzig), a grandson of

Moses Mendelssohn (q.v.), was a child prodigy as pianist and a mature composer at an early age. In his short life he composed five symphonies, three concertos, a number of concert overtures, works of chamber music, and piano pieces. His first oratorio, *St. Paul* (1836), was very successful in England, and *Elijah* (1846, biblical libretto adapted by J. Schubring, translated by W. Bartholomew) had English audiences primarily in mind. He visited England ten times between 1829 and 1847, and the overture *Fingal's Cave* (*Hebriden-Ouvertüre*) was inspired by his Scottish tour in 1829. His third symphony in A minor is described as the *Scottish Symphony*. Mendelssohn wrote no opera, but he composed songs, some of which set poems of Heine (q.v.). He met Goethe, on whom he made a favourable impression, in 1817, 1821, 1825, and 1830; and his overture *Meeresstille* refers to Goethe's poem of this title (q.v.). The overture *Die schöne Melusine* is based on the well-known legend (q.v.). Mendelssohn retrieved J. S. Bach's *St. Matthew Passion* from long neglect, and gave the first performances of a forgotten MS. symphony by Schubert (q.v.), now known as 'the great C Major'.

Though of Jewish descent, Mendelssohn was a Lutheran. The National Socialist regime attempted to erase his memory, but the rectitude and generosity of his character, as well as the excellence of his compositions, have been recognized before and since.

Menonit, Der, a verse tragedy (Trauerspiel) in four acts by E. von Wildenbruch (q.v.), first performed at Frankfurt/Main in November 1881, and published in 1882. The unusual spelling 'Menonit' instead of 'Mennonit' is Wildenbruch's. It is a patriotic play set near Danzig in 1809 at the time of the French occupation. The background is a settlement of Mennonites, a pacifist Protestant sect. The austere elder Waldemar betroths his daughter Maria to the equally rigid Mathias. Maria, however, loves the more human Reinhold, who protects her from the insults of a French officer while her fiancé Mathias stands by unmoved. Reinhold plans to flee with Maria and to join the patriotic conspiracy of Schill (q.v.), but the attempt is frustrated. Maria dies (of what is not clear) and Reinhold, after he has shot Mathias, is seized by the French and at the end awaits execution.

Menons Klagen um Diotima, an elegy by F. Hölderlin (q.v.) in nine sections, comprising in all 65 distichs of elegiac verse (see ELEGISCHES VERSMASS). Written in 1799–1800 and published in 1802, it is a lament for lost love, and an expression of desolation, which is finally resolved in the determination to live to fulfil the mission of a poet (see also DIOTIMA).

Menschenhaß und Reue, a play (Ein Schauspiel in fünf Aufzügen) by A. von Kotzebue (q.v.), published in 1789. It is the most famous of Kotzebue's serious works. At a country mansion belonging to Count von Wintersee, an obscure Madame Müller lives in retirement and quietly performs good works. In the same district there is staying a mysterious and misanthropic stranger, who, paradoxically, is also actively charitable. The Countess and her brother, Major von der Horst, come to stay at the mansion. Horst is in love with Madame Müller, who rejects him. The Countess, pressing her brother's suit, learns from Madame Müller that she is in reality a Baroness Meinau, who has been seduced by a friend and has left her husband. Meanwhile Horst seeks out the stranger who has gallantly rescued the Count from a fall into the river. To his astonishment he recognizes in him his old friend Baron Meinau. Horst, abandoning his suit, seeks to reconcile the married couple. Though both desire it in their hearts, they obstinately resist, Meinau on the principle of honour, his wife insisting on her own perpetual repentance. A ruse by which their own children surprise them throws them into each other's arms.

The play was performed in the Weimar Court Theatre on the occasion of a visit by A. W. Iffland (q.v.) as guest artist in 1798.

Menschen im Hotel, a novel (Kolportageroman) by Vicki Baum (q.v.), published in 1929. Set in Berlin, in the luxury Grand Hôtel, it portrays vividly the kaleidoscopic life within it. In a quasi-cinematographic technique various characters are shown both separately and entwined in groups. The ageing ballet star Grusinskaja spends a night with the crook Baron Gaigern. The seedy clerk with heart trouble, Herr Kringelein, feels the urge for once to stay in a grand hotel and, after various humiliations, finds in the secretary Flämmchen a sweet-tempered companion to share his bed. General-direktor Preysing, the businessman, is less successful with Flämmchen and runs into serious trouble when, erroneously believing himself to be in danger, he shoots Baron Gaigern. These and other figures pass and re-pass; the good at heart (or some of them) are rewarded, the mean come to grief.

The novel was an enormous success, not only in Germany, but also in England and the U.S.A., as *Grand Hotel*, and under this title it was turned into an even more successful American film.

Menschliches, Allzumenschliches, a polemical work by F. Nietzsche (q.v.), published in 1878–80. Sub-titled *Ein Buch für freie Geister,* it is in 2 vols., the first divided into a *Vorrede,* 9 *Hauptstücke,* and a *Nachspiel,* the second containing the two essays *Vermischte Meinungen und Sprüche* and *Der Wanderer und sein Schatten.* Perhaps his most passionately written book and a personal apologia, it marks his breach with R. Wagner (q.v.).

Mensur, term used for formal student duelling conducted in private by various student associations (see STUDENTISCHE VERBINDUNGEN). It is synonymous with Pauken. The term Mensur originally referred only to the line drawn on the floor between the combatants, and the general use of Mensur is derived from the expression 'auf die Mensur gehen'. The duellists (Paukanten) are protected over most of the body and wear safety spectacles. The weapons are sharp-edged swords with a basket hilt. The point is not used. The duellists fight a prescribed number of rounds, each delivering a slashing blow in turn, which the adversary endeavours to parry. The only wound normally sustained is a cut on the cheek or side of the head. The resultant scar was formerly regarded as evidence of social standing.

MENZEL, ADOLPH VON (Breslau, 1815–1905, Berlin), painter, draughtsman, and etcher, first made his name by his numerous line illustrations of Kugler's *Geschichte Friedrichs des Großen* (see KUGLER, F. Th.). He was prolific in his drawings, constantly sketching from life and executing immense tasks of historical graphic work, e.g. 600 lithographs of Prussian uniforms (*Friedrichs des Großen Armee in ihrer Uniformierung,* 3 vols., 1842–57). As a painter he is an untaught forerunner of Impressionism, rendering commonplace subjects with a great sense of atmosphere (*Balkonzimmer, Des Malers Schwester, Abendgesellschaft*), and public events or functions both in interior scenes (*Théâtre Gymnase, Im Opernhaus, Ballsouper*) and in the open air (*Aufbahrung der Märzgefallenen, Abreise Wilhelms I. zur Armee* 1870) They are sometimes marked by deft touches of humour. He also painted in 1847 a remarkable picture, *Die Berliner-Potsdamer Eisenbahn,* which is well in advance of his time. Menzel illustrated H. von Kleist's *Der zerbrochne Krug* (q.v.) in 1877.

MENZEL, WOLFGANG (Waldenburg, Silesia, 1798–1873, Stuttgart), a prominent founder member of the Burschenschaft (q.v.), spent the years 1820–4 in exile in Switzerland, afterwards settling in Stuttgart where he was active as a journalist of pronounced nationalistic and antiradical opinions. In the *Literaturblatt* (see MORGENBLATT FÜR GEBILDETE STÄNDE) he developed a fierce opposition to the left wing represented by H. Heine, L. Börne (qq.v.), and writers associated with Junges Deutschland (q.v.), including his own collaborator K. Gutzkow (q.v.): the authorities responsible for the federal decree of 1835 denouncing these writers were aware of Menzel's attitude. Börne's *Menzel der Franzosenfresser* (1837) was the most conspicuous of the ensuing counterattacks. Menzel made a reputation for himself as a savage and unscrupulous critic, attacking Goethe as well as the writers associated with Junges Deutschland in *Die deutsche Literatur* (1827, extended 1836), and as a well-informed but tendentious historian of literature in *Deutsche Literatur von der ältesten bis auf die neueste Zeit* (3 vols., 1858–9). He was the author of a historical novel, *Furore* (1851), and of popular historical works with a nationalistic bias.

Mephistopheles, a character in Goethe's *Faust* (q.v.) and *Urfaust,* the diabolical tempter who seeks Faust's undoing.

MERBOD, anagram of J. J. Bodmer (q.v.), used by J. C. Gottsched (q.v.) in his satire *Der Dichterkrieg* (1741).

MERCATOR, GERHARD (Rupelmonde, Flanders, 1512–94, Duisburg), a latinization of Kremer, the surname of the German geographer who invented the cartographic projection which bears this name. It was first used on a map of the world published in 1569. Mercator, who began as a mathematician, engraved his first map in 1537. In 1552 he settled at Duisburg, publishing an important map of Europe in 1554, and preparing the atlas which is his principal work, *Atlas, sive cosmographicae meditationes de fabrica mundi et fabricati figura,* published at Duisburg in the year after his death.

MERCK, JOHANN HEINRICH (Darmstadt, 1741– 91, Darmstadt), was appointed in 1767 to an administrative post in Darmstadt, where he was one of the literary circle known as the Darmstädter Kreis (q.v.). He became acquainted with Goethe in 1771, introduced him into the Darmstadt circle, and encouraged him to publish *Götz von Berlichingen* (q.v.). He was a severe critic of much of Goethe's early work. Goethe met him repeatedly between 1777 and 1783 at Weimar, where he stood well with the court. As a poet (*Lindor,* 1781) he was not successful, but his journalistic writings on art were collected and published in 1840 by A. Stahr, *Ausgewählte Schriften zur schönen Literatur und Kunst* (reissued 1965). *Fabeln und Erzählungen,* ed. H. Bräuning-Oktavio, appeared 1962, and select *Werke und Briefe* (2 vols.), ed. A. Henkel and H. Kraft, 1968. Merck took his own life after getting into insoluble financial and other difficulties.

MEREAU, Sophie (Altenburg, 1770–1806, Heidelberg), *née* Schubert or Schubart, is usually known by this, the name of her first husband F. E. K. Mereau. She married Mereau in 1793 and lived with him in Jena, where he was a professor. In her twenties she began to write novels, including *Das Blütenalter der Empfindung* (1794) and *Amanda und Eduard* (1803), a novel in the form of letters (see Briefroman), which evoked favourable comment from Schiller and was published in part in his periodical *Die Horen* (q.v., 1797). Her poems were published in two volumes as *Gedichte* (1800–2). She edited the *Göttinger Musenalmanach* (q.v.) for 1803. In 1801 she was divorced and two years later married Clemens Brentano (q.v.). She died in childbirth. She was a competent translator, particularly of English and Italian works. A periodical, *Kalathiskos*, which she edited 1801–2 was reprinted, ed. P. Schmidt, in 1969.

Mergelgrube, Die, a poem by Annette von Droste-Hülshoff (q.v.) included in *Heidebilder*, published in *Gedichte* (1844). It reflects her interest in fossils.

MERIAN, Matthäus der Ältere (Basel, 1593–1650, Schwalbach), a copperplate engraver, as a journeyman visited Strasbourg, Nancy, and Paris and in 1620 returned to Basel. In 1624 he took over his father-in-law's publishing and engraving house in Frankfurt. His great work is a collection of views of European cities (*Topographien*, i.e. *Topographia Helvetica*, *T. Austriaca*, etc., 16 vols., 1642–88), which was completed after his death by his son, Matthäus Merian der Jüngere (1621–87). The text was furnished by M. Zeiller (1589–1661). Merian also began the publication of the historical *Theatrum Europaeum* (1635–1738).

Merigarto, title given by the first editor, H. Hoffmann von Fallersleben (q.v.), in 1834, to remains, preserved in Prague, of an Early Middle High German poem dealing with geography. The extant fragments of this cosmography treat of Iceland and of some of the waters of the world. In the main it rests upon Latin sources, but the section on Iceland is derived from oral communication from a cleric who had visited the island. The information is haphazard and largely fantastic, the style inept. It was written *c.* 1090 in Utrecht by a cleric who

had sought refuge there from an episcopal dispute, probably in Würzburg.

Merker (1) Word occurring frequently (in its Middle High German form) in medieval poetry, denoting persons seeking to frustrate the union of a pair of lovers or to detect or betray them.

(2) In Meistergesang (q.v.) the word was applied to the judge or judges who, using fixed criteria, estimated the value of a song in performance. Beckmesser, satirically exploited in *Die Meistersinger von Nürnberg* (q.v.) by R. Wagner to castigate his critics, is better known than any instance or reference in the period of Meistergesang; Sachs, in the same opera, performs a parody of the Merker's function (Act II).

Merowinger, the first house of Frankish kings, ruling from about 430 to 751. In the last century of their rule the real power was in the hands of their ministers (see Hausmeier). The last Merovingian king was Childerich III, who was deposed in 751.

Merseburger Zaubersprüche, two magic spells entered in the 10th c. on a blank page of a missal preserved in the Capitular Library at Merseburg and first published by J. Grimm (q.v.) in 1842. They are the only spells which clearly reveal pre-Christian origins. The first spell is designed to secure the release of a prisoner of war, the second to heal a lame horse. They have a dual structure: first, a successful similar past occurrence is narrated, whereupon a magic imperative formula is uttered. The narrations in these two spells are heathen, concerning in (1) *idisi*, probably a kind of Valkyrie, and in (2) Germanic gods, Phol and Wotan. The spells are composed in alliterative verse. (See also Zaubersprüche.)

MERSWIN, Ruland (Strasburg, 1307–82, Strasburg), a patrician of Strasburg, who endowed the house of the Order of St. John in Strasburg and furthered the cause of the mystics who were described as Gottesfreunde. Merswin is held by some to be responsible for launching the fiction of the mystic 'der große Gottesfreund vom Oberlande' (q.v.), one of the successful deceptions of literary history. Merswin himself wrote mystic tracts and fictitious lives based on the writings of J. Tauler and H. Seuse (qq.v.).

MESMER, Franz Anton (Iznang, 1734–1815, Meersburg), qualified at Vienna University as a physician. There is some doubt as to whether his first baptismal name was Franz or Friedrich. Mesmer, believing that all physical phenomena are interconnected, evolved the theory that living creatures influence each other by an omnipresent tenuous substance which he termed 'animal magnetism' ('tierischer Magnetismus'). He was principally active in Paris, where he arrived in 1778, attracting great attention by the cures he effected and by the scene-setting in which they took place (semi-darkness, soft music, mysterious gestures, etc.).

Mesmer was attacked by the medical faculties as a charlatan, but he appears not to have been a conscious impostor. 'Animal magnetism' came to be known as Mesmerism, and by it Mesmer frequently induced hypnotic states in his patients. Mesmerism is employed as a literary motif by some Romantic writers (see ROMANTIK).

Messias, Der, a religious epic begun by F. G. Klopstock (q.v.) in youth and finished shortly before he reached the age of 50. It was consciously intended to rival Milton's *Paradise Lost.* Written in hexameters, it is composed in 20 cantos (Gesänge). The first three of these were published in 1748 in the *Bremer Beiträge* (q.v.), and these, together with Cantos 4 and 5, formed the first volume, published in 1751. Vols. 1 and 2 (Cantos 1–5 and 6–10) were issued together in 1755, vol. 3 (Cantos 11–15) followed in 1768, and vol. 4 (Cantos 16–20) in 1773.

The first ten cantos portray the Passion from the Mount of Olives to the death of Christ on the Cross ('Und er neigte sein Haupt und starb'). The second half is devoted to redemption, the Resurrection, and the triumph of Christ ('Indem betrat die Höhe des Thrones/Jesus Christus und setzete sich zu der Rechte des Vaters'). Klopstock adds an epilogue in the form of an ode of dedication and satisfaction at a task completed (*An den Erlöser*). The story of redemption unrolled in *Der Messias* (its first line is 'Sing, unsterbliche Seele, der sündigen Menschen Erlösung') is seen as a battle between the forces of Heaven led by God and supported by his angels and the powers of Hell commanded by Satan. The infernal assault begins in Canto 2, reaches a climax in Canto 10, and is finally and utterly defeated in Canto 16. Klopstock introduces a moving figure in the repentant fallen angel Abbadona (Canto 2).

The first three cantos of *Der Messias* were received with widespread enthusiasm, of which Bodmer's invitation to Klopstock to join him in Zürich was one manifestation. Goethe records in *Dichtung und Wahrheit* (q.v., Bk. 2) how Rat Schneider, no great friend of books, read the first ten cantos of *Der Messias* regularly in Holy Week. The generation of the 1770s turned away from the work and K. P. Moritz records in *Anton Reiser* (q.v., 1785–90) the terrible boredom ('entsetzliche Langeweile') it aroused. Later judgements have been a shade less adverse, but

in general it is clear that Klopstock's gifts were lyrical rather than epic, and that his imagination was emotional rather than visual. *Der Messias* (2 vols.), ed. E. Hopker-Herberg, appeared in 1974 in the Hamburger Klopstock-Ausgabe.

Messingkauf, Der, an extended, but unfinished, dialogue by B. Brecht (q.v.), written between 1937 and 1951, and published in vol. 5 of *Schriften zum Theater* (7 vols., 1963–4). It discusses epic theatre (see EPISCHES THEATER) and extends over four nights, during each of which a specific aspect is reviewed. In the epilogue the Dramaturg, Schauspieler, and Philosoph are joined by the Arbeiter, who speaks the words: 'Ändert die Welt, sie braucht es!' The title alludes to a story of a man who buys brass musical instruments solely for their metal.

Meteor, Der, a two-act comedy (Komödie) by F. Dürrenmatt (q.v.), first performed in the Zurich Schauspielhaus in January 1966 and published in the same year. Its remarkable subject, a kind of dance of death, is presented in terms of farce, driving the audience to irresistible laughter. Yet, farcical though it is, it has a bitter sting. The great writer Wolfgang Schwitter, a Nobel Prize winner, has died in a nursing home, and risen again—this we learn after the curtain goes up, for the play opens with Schwitter in pyjamas, a fur coat, and carrying two large candles and suitcases. He has come to his old studio of forty years ago to die again, away from the paraphernalia of medicine, the press, and state visits. The whole play takes place in this room. At the end of the first act he dies his second death. The second act begins with a parody of a funeral oration. Shortly afterwards Schwitter rises again from the dead, and the act ends with him waiting for his final end. In the meantime this twice dead and always dying man has dominated the scene, causing directly or indirectly four deaths, an arrest for murder, and the destruction of the reputation of the famous consultant Schlatter, who has twice scientifically established his death. Dürrenmatt makes it clear that Schwitter really has died each time and has been the object of two miracles. The play ends with a triumphant hymn from the Salvation Army, who see their faith demonstrated in him.

METLINGER, BARTHOLOMÄUS, see KINDERBUCH, DAS.

METTERNICH, WENZEL LOTHAR, FÜRST VON (Koblenz, 1773–1859, Vienna), Austrian statesman, studied in Strasburg, Frankfurt, and Mainz, where he acquainted himself with the political principle of a 'just equilibrium'; this and his horror of the French Revolution (his tutor Simon died as a Jacobin terrorist), resulting in his distrust of any kind of participation of the people in government, formed the basis of his policies from 1809 to 1848. After appointments at the embassies at Dresden and Berlin, he became in 1806 Austrian ambassador in Paris. His knowledge of the French scene and his French contacts led him to work for an improvement in Austro-French relations after the Emperor Franz I (q.v.) appointed him minister for foreign affairs in 1809, the year of Austria's defeat by Napoleon (see NAPOLEONIC WARS). He arranged Napoleon's marriage to the Austrian Princess Marie Louise, but failed to exert any moderating influence upon Napoleon's expansionist ambitions.

During the European settlement and the German reorganization at the Congress at Vienna (see WIENER KONGRESS) he succeeded in realizing a long-standing ideal: the German Confederation (see DEUTSCHER BUND) prevented the reunification of Germany on a nationalistic basis and secured the autonomy of the sovereign princes over whom he exerted considerable influence. His policy was reactionary in that he suppressed liberal aspirations by persecution and press censorship, creating the police state, though he was not directly responsible for some of its excesses; he believed to the end that he served the interest of the people by safeguarding an absolute and monarchical system of government.

Metternich convinced Friedrich Wilhelm III (q.v.) of Prussia of the rightness of his policy of preventing the spread of revolution in Europe, when he issued the Carlsbad Decrees (see KARLSBADER BESCHLÜSSE) in 1819. His policy of European alliances ('balance of power') helped substantially to secure the prolonged period of peace following the Napoleonic era. Metternich's influence began to wane after 1825. He was dismissed and had to flee from Austria during the Revolution of 1848 (see REVOLUTIONEN 1848–9). In 1851 he was allowed to return to Vienna, having stayed in the meantime in London, Brighton, and Johannisthal, where Bismarck (q.v.) called on him; he did not return to politics, though he entertained friendly relations with the young Emperor Franz Joseph (q.v.). The first of his three marriages was to a granddaughter of Wenzel Anton, Graf von Kaunitz (q.v.). Nine volumes of *Aus Metternichs nachgelassenen Papieren* were edited by R. M. and A. von Klinkowström (1880–9).

MEUSEL, ANDREAS, see MUSCULUS, ANDREAS.

MEYER, CONRAD FERDINAND (Zurich, 1825–98, Kilchberg nr. Zurich), a Swiss, bilingual in French and German, wrote most of his *œuvre* between 1870 and 1887, choosing German, after some hesitation, in preference to French. Of patrician descent, he lost his father, who had a brief but distinguished career in the Zurich administration and as a teacher, in 1840. As long as Meyer remained in the care of his puritanical and over-anxious mother, his development and his artistic inclinations were severely checked. In 1843 his mother sent him for a year to be cared for by Louis Vuillemin, a family friend in Lausanne, who helped him to recover his self-confidence. He made a half-hearted attempt to study law at Zurich University. A simultaneous attempt at training as a painter failed through lack of ability, but Meyer continued to write poetry. During the next few years his susceptibility to a psychopathic condition became evident. It was checked for a time by energetic exercise. He became a keen mountain walker, swimmer, and fencer. In 1852 his mental stability broke down and he had to enter a mental asylum. After seven months' treatment he returned home to begin a new phase of life dominated by intensive study of literature and history. He turned his back on the Romanticism of his youth, and devoted himself to the study of Renaissance art and the cultural and historical works of J. Burckhardt, L. von Ranke, and Th. Mommsen (qq.v.), which were of great importance to his development. His mother's suicide in 1856 left him and his sister Betsy with considerable means, which enabled them to travel abroad and so consolidate his studies.

Meyer's visits to Paris and the Louvre, to Munich, and above all to Italy (1858), the world of classical and Renaissance painting, sculpture, and architecture, determined his predilection for a style which aimed at concreteness of presentation and at the absorption of the visual arts into literature. This aspect, more than any other feature of his work, marks the place he was to take among the writers of Poetic Realism (see POETISCHER REALISMUS). It proved particularly congenial to his vulnerable nature, which left him hypersensitive to the exposure of personal sentiments. His style thus became also a means of 'masking' experience by adopting the disguise of historical objects and figures. Detachment in literature was an intellectual and a psychological need for him, and this also accounts for the late release of his creative powers. Meyer's studies in Lausanne (ending in 1860) are his last attempt to return and complete the studies begun at Zurich. His sister understood that he needed proof that his work was fit for print, and she succeeded in getting his first collection of poetry published anonymously as *Zwanzig Balladen von einem Schweizer* (1864). Some ten years of close collaboration with his sister in Zurich, and another visit to Italy (1872), established him in his career as a writer. In 1875 he married Luise Ziegler and settled in Kilchberg. He was in contact with many men of letters and artists, but his close friends François and Eliza Wille afforded him the staunchest support. In 1887, however, he began to show serious signs of a renewal of the mental illness of his youth, from which he did not recover. His literary production ceased at this point.

Meyer's first publication under his own name (he added his father's Christian name Ferdinand to his own to avoid confusion with Conrad Meyer, another Swiss writer) was a collection of poetry, *Romanzen und Bilder* (2 sections, *Stimmung* and *Erzählung*, with 46 titles). In 1882 Meyer's poetry was published in the single volume *Gedichte*. The poems are compact, restrained, well proportioned, and perfectly balanced. Many of them can be said to anticipate the 'Dinggedicht' (q.v.) of the 20th c. They are inspired by things seen, such as landscapes, architecture, statues, and paintings. In *Eingelegte Ruder* the poet observes the drops falling from his oars as he pauses while rowing on a lake; *Lethe* is derived from a painting; *Der römische Brunnen* indicates its model, and, in its combination of precision, limpidity, and clarity, is possibly his finest poem. The high quality of Meyer's poetic achievement was not recognized in his lifetime.

In Meyer's own eyes, the epic *Huttens letzte Tage* (q.v., 1871) was his 'first' work. It was followed by his idyll *Engelberg* (1872), which he had first conceived some ten years previously. The use of rhyming verse couplets with four stresses, stretching over some 1,800 lines, makes his portrayal of what he called the medieval psyche a *tour de force*. Meyer's prose narrative works which followed are select rather than profuse, and are shaped with the same stringent consciousness of form. Having abandoned initial leanings towards drama, he adopted the Novelle, and for his most ambitious projects favoured the frame technique (see RAHMEN). His highly successful novel, *Jürg Jenatsch* (1876 under the title *Georg Jenatsch*), which followed the publication of *Das Amulett* (1873), prepares for the style of his subsequent narrative works, *Der Schuß von der Kanzel* (1878), *Der Heilige* (1879), *Plautus im Nonnenkloster* (1881), *Gustav Adolfs Page* (1882), *Das Leiden eines Knaben* (1883), *Die Hochzeit des Mönchs* (1883–4), *Die Richterin* (1885), *Die Versuchung des Pescara* (1887), and *Angela Borgia* (1891), qq.v. All these works are set in the Middle Ages, the Renaissance, the

Reformation, or the Counter-Reformation, Meyer's favourite periods. He never confines himself to a narrow Swiss patriotism, but prefers to demonstrate the secular and religious forces shaping history in relation to what he himself termed the human psyche. He presents his characters as enigmatic figures, torn by conflicting emotions and attitudes, whose actions determine, at times almost accidentally, the course of history. *Der Heilige* (referring to Thomas à Becket) displays with characteristic mastery Meyer's tendency to deliberate ambiguity. He chooses well-known figures from history, or places fictitious characters at the centre of formidable historical events. Yet, however remote these characters and times may be, they reflect the scepticism of Meyer's own age, of which he, a characteristic Swiss, is a keen and critical observer. His astute intellect and his innate despondency combine to give to his fiction the imprint of a predominantly objective, ironic, and tragic portrayal of existence.

Sämtliche Werke (4 vols.) first appeared in 1926, and *Sämtliche Werke, historisch-kritische Ausgabe* (15 vols.), ed. H. Zeller and A. Zäch, 1958 ff.

MEYER, HANS HEINRICH (Stäfa, Switzerland, 1760–1832, Weimar), Swiss painter and pupil of J. H. Füßli (q.v.), became a close friend of Goethe, whom he met in Rome. Goethe later secured an appointment for him at the Weimar School of Art. In 1807 Meyer became its director, a post he held until his death. From 1792 until his marriage he lived in Goethe's house. Meyer developed into an art historian and theorist, whom Goethe consulted as an authority. He collaborated with Goethe in the periodical *Die Propyläen* (1798–1800), to which he contributed *Raphaels Werke* (1799), *Masaccio* (1800), and *Mantua im Jahre 1795* (1800). He had previously published *Beiträge zur Geschichte der neueren bildenen Kunst* in Schiller's periodical *Die Horen* (q.v., 1795). Another essay by him on a closely related subject (*Entwurf einer Geschichte der neueren bildenden Kunst* in Schiller's periodical Goethe's *Winckelmann und sein Jahrhundert* (q.v., 1805).

MEYER, JOHANN HEINRICH OTTO (Wilster, Holstein, 1829–1904, Kiel), school-teacher, journalist, and finally head of an institution for subnormal children, published some High German poetry (*Lyrische Gedichte*, 1856; *Dithmarscher Gedichte*, 1858–9) and then turned with more success to Low German dialect (Dithmarscher Platt) in the epic *Gröndunnersdag bi Eckernför* (1873) and the plays *Opn Amtsgericht* (1880), *Uns' ole Modersprak* (1880), *To Termin* (1890), and *En lütt Waisenkind* (1892).

MEYER, MARIA, a young woman whom E. Mörike (q.v.) met in 1823 and with whom he fell deeply in love. His break with her some months later was an act of violence to his own feelings. She is the Peregrina (q.v.) of the cycle of poems bearing that title. Maria Meyer, when Mörike met her, was 25, and is said to have been exceptionally beautiful. She was of Swiss birth and belonged to a well-to-do family, but had left her parents and joined the Pietistic sect of Frau von Krüdener (see KRÜDENER, B. J. VON). Her parents refused to take her back, and she led a nomadic life for some years. In the spring of 1823 she was found half insensible near Ludwigsburg and was befriended by citizens moved by her plight, her good looks, and her refined manners. She won the heart of the 19-year-old Mörike during his vacation at Ludwigsburg, but then disappeared. At the beginning of 1824 she was reported in Heidelberg and in July she appeared in Tübingen, but Mörike refused her pleas for a meeting. Soon afterwards she disappeared again and her further life is not known. An epileptic, she seems to have had a disturbed and hypersensitive personality, given to dreams and visions.

MEYER-FÖRSTER, WILHELM (Hanover, 1862–1934, Berlin), is well known as the author of the great box-office success, *Alt-Heidelberg* (1903), the dramatization of his own sentimental novel *Karl Heinrich* (1899). He went blind in 1904.

MEYER VON KNONAU, JOHANN LUDWIG (Weiningen nr. Zurich, 1705–85, Weiningen), a patrician of Zurich, published in 1744 *Ein halbes Hundert neuer Fabeln* with a preface by J. J. Bodmer (q.v.). Meyer was an early observer of animal characteristics.

MEYR, MELCHIOR (Ehringen nr. Nördlingen, 1810–71, Munich), a farmer's son, was at school at Nördlingen and studied at Munich University, where he was influenced by Schelling (q.v.). In 1840 he moved to Berlin, and in 1852 returned to Munich. A prolific and versatile writer, he was particularly successful with stories set in Ries, his home district; they appeared as *Erzählungen aus dem Ries* (1856) and *Neue Erzählungen aus dem Ries* (1859). Forerunners of Heimatkunst (q.v.), their regional peculiarities are enhanced by the integration of dialect expressions. Meyr's rural poem *Wilhelm und Rosine* (1835) was followed by *Gedichte* in 1857. He is the author of plays on historical subjects, including *Franz von Sickingen* (1851), *Herzog Albrecht. Dramatische Dichtung*, which deals with Agnes Bernauer (q.v.), and *Karl der Kühne* (both 1862), and of novels, including *Vier Deutsche* (3

vols., 1861), *Ewige Liebe* (2 vols., 1864), and *Duell und Ehre* (2 vols., 1870). The last of his numerous writings on philosophical and religious subjects appeared in the year of his death in the form of forty letters (*Die Religion und ihre jetzt gebotene Fortbildung*, 1871). A posthumous collection, ed. Count Bothmer and M. Carrière, appeared in 1874 as *Gedanken über Kunst, Religion und Philosophie*.

MEYRINK, GUSTAV (Vienna, 1868–1932, Starnberg), the illegitimate son of a Swabian baron and a Bavarian actress, Marie Meyer, from whom he adapted his name, was a convert to Buddhism and a devotee of the occult and the grotesque. He is best known for his novels, especially the eerie story of the Prague ghetto, *Der Golem* (1915). His other novels include *Das grüne Gesicht* (1916) and *Der weiße Dominikaner* (1921).

MEYSENBUG, MALVIDA, FREIIN VON (Kassel, 1816–1903, Rome), younger sister of a reactionary politician of the Grand Duchy of Baden, was a revolutionary in 1848 (see REVOLUTIONEN 1848–9), and was banned from Berlin in 1852 because of her contacts with revolutionary figures. She emigrated to London, took employment as a governess and also acted as a newspaper correspondent. In 1862 she moved to Italy. She became an admirer of R. Wagner (q.v.) and from 1872 was a frequent visitor to his Bayreuth home. She was also a friend of Nietzsche (q.v.), spending the winter of 1876–7 with him at Sorrento. In 1877 she settled in Rome. She was on friendly terms with many prominent figures, including Garibaldi, Mazzini, Liszt (q.v.), and Romain Rolland. She published memoirs (*Memoiren einer Idealistin*, 3 vols., 1875) and a later volume of autobiography, *Der Lebensabend einer Idealistin* (1898); her correspondence with Nietzsche was published posthumously (1905). She is the author of one novel, *Phädra* (3 vols., 1885).

M.G.K., a war book by F. Seldte (q.v.), published in 1929, in which he narrates, in thinly disguised form, his recollected experiences of 1914 from mobilization, through the battles of Mons, Le Cateau, the Marne, the Aisne, and Arras, to the end of the year. The initials M.G.K. stand for 'Maschinengewehrkompanie'.

MHG, Middle High German, see GERMAN LANGUAGE, HISTORY OF.

MICHAEL, FRIEDRICH (Ilmenau, 1892–), worked for the Insel-Verlag (1945–60). He wrote novels and comedies, of which *Der blaue*

Strohhut (1942) achieved exceptional success on the stage and as a film. His poetry was collected in *Blume im All* (1940). His writings as a theatre critic and historian include *Die Anfänge der Theaterkritik* (1918) and *Deutsches Theater* (1923); the latter appeared in revised form in 1969 as *Geschichte des deutschen Theaters*. In 1952 appeared *Weltliteratur* and *Dank ans Theater*, in 1965 *Von der Gelassenheit*, and in 1967 *Gastliches Haus*.

MICHAEL, WILHELM, pseudonym of W. M. Schneider (q.v.).

Michael Kohlhaas. Aus einer alten Chronik, a Novelle by H. von Kleist (q.v.), published in fragmented form in the periodical *Phöbus* (No. 6) in 1808, and in full in the first volume of *Erzählungen* (1810). The chronicle providing the source of Kleist's story records the quarrel of Kohlhase, a cattle dealer of Cölln (nr. Berlin), with a Junker von Zaschwitz, in the course of which Kohlhase, failing to obtain satisfaction through a lawsuit begun in 1532, took to robbery and murder; he was executed in 1540.

Kleist sets the action in Saxony and Brandenburg. He introduces the horse-dealer Michael Kohlhaas of Kohlhaasenbrück as 'one of the most virtuous and most terrifying men of his time', and one whose sense of justice (Rechtgefühl) caused him to become a robber and murderer. The grievance which prompts Kohlhaas to go to law arises out of an incident on his way to Leipzig, which takes him through the estates of Junker Wenzel von Tronka. When crossing the border he is unexpectedly required to produce a permit. Promising to return with the required document Kohlhaas leaves two of his horses under the care of his groom Herse in the Junker's stables. In Dresden he learns that no permit was necessary. He obtains, however, a document which should satisfy the Junker's request, but finds upon his return to the Tronkenburg that his groom has been sent away and that his horses are in a state of dire neglect. Having satisfied himself that the Junker is responsible for the ill-treatment of his horses as well as of poor Herse, Kohlhaas commences, through the proper channels of the law, his vain attempts to obtain justice by having his horses restored to their former strength and value before their transfer from the Junker's stables to his own. Kohlhaas abandons his legal battle only after he has exhausted all possibilities, the last of which, a petition to his sovereign in Brandenburg, inadvertently costs him the life of his wife Lisbeth.

Obsessed by the urge to see justice done, Kohlhaas now arms a band of followers, storms the Tronkenburg in the name of every good Christian, and, failing to capture Wenzel, takes Wittenberg, where he suspects the Junker to be in hiding. At Mühlberg he defeats an army under the command of Friedrich of Meißen, but suffers a personal loss in the death of the loyal Herse. The Elector mobilizes two thousand men to protect Leipzig from fire and murder. When Dresden, too, is threatened, Martin Luther intervenes by publicly condemning Kohlhaas in the name of God and justice. A meeting between both men procures for Kohlhaas not the desired absolution, but the promise of a fresh hearing and an amnesty from the Elector of Saxony on condition that Kohlhaas disbands his army.

This he does, but in Dresden, where he settles with his children, he finds himself virtually under house-arrest. His desperate position is demonstrated by the appearance of his two horses in the market-place; they are in the hands of a knacker and therefore by custom dishonoured. A letter in which Kohlhaas resumes his contact with Nagelschmidt, a former associate, in an effort to flee the country, is intercepted and swiftly made the pretext for a brief trial, in which he is condemned to be tortured, quartered and his body afterwards burned.

In the last part of the story the Elector of Brandenburg, suspicious of Saxon justice, intervenes on behalf of his subject, effects Kohlhaas's extradition, and obtains an imperial verdict on the issues involved: Kohlhaas is to receive full satisfaction in his case against Junker Wenzel von Tronka before dying by the sword for crimes committed during his unlawful invasion of Saxon territory. In the hour of his execution Kohlhaas receives his horses, restored to honour and fitness, and the assurance of the Elector of Brandenburg that his every complaint against Junker Wenzel, who is to serve a two-year sentence, has been met.

Kohlhaas submits to his death, reconciled with God and the world, but he revenges himself on the Elector of Saxony for having violated the conditions of the amnesty by swallowing a capsule which the Elector has spared no effort to obtain; it contains a paper scrap on which a mysterious gipsy woman has predicted his future. This episode recedes into the background as the magnanimous Elector of Brandenburg confers knighthoods on Kohlhaas's two sons.

In spite of structural weaknessses, *Michael Kohlhaas* is Kleist's most significant prose work, by which he has earned recognition as the innovator of a specifically German form of Novelle. (Film version by V. Schlöndorff, 1969.)

Michael Kramer, a play in four acts by G. Hauptmann (q.v.), first performed in November 1900 in Berlin, and published in the same year.

Kramer is an artist whose greatest ambition is to paint a huge picture of Christ with a crown of thorns in order to convey his idea of the synonymity of art and religion. He sees in Beethoven (q.v.) the symbol of this fusion. Because he knows in his heart of hearts that he lacks the genius of a Beethoven, he stakes his hopes on Arnold his son, Arnold, however, who has grown into a young man with a misshapen figure, squanders his great gifts on caricatures expressing his contempt of the world. His resentment against his domineering father leads in the central scene to a breakdown of their relationship, while his figure makes him feel that he is a laughing-stock in the eyes of women and of society. His sense of inferiority turns into aggressiveness, and, when he is mocked by one of the regulars at Bänsch's inn, he draws a revolver (which is wrenched from him), flees to escape arrest, and drowns himself. His father now pronounces a funeral oration for his son, whom he has underrated during his lifetime and now idolizes with excessive pathos. Through Kramer's daughter Michaline and his former pupil Lachmann, Hauptmann brings variations into the play's central theme, the artist in society.

Michel, Deutscher, proverbial figure of easy-going good-nature and simple-mindedness, representing the German populace. He is first mentioned in 1541 in the collection of proverbs *Die deutschen Sprichwörter* by Sebastian Franck (q.v.), appears in a favourable light in *Wunderbarliche und Wahrhafftige Gesichte Philanders von Sittewald* (1640–3) by H. M. Moscherosch (q.v.), is the eponymous hero of *Der Teutsche Michel* (1672) by Grimmelshausen (q.v.), and in the 19th c. and 20th c. has been a favourite figure with cartoonists and political satirists. It is believed that this figure of fun derives, in some devious and unexpected way, from the Archangel Michael. During and after the 1914–18 War Michel was re-endowed with heroic attributes.

A gallant officer of horse, Michael Obentraut (1574–1625), who served in the Thirty Years War (see DREISSIGJÄHRIGER KRIEG), also came to be known for a time as 'Der deutsche Michel'.

Michelangelo, see HEBBEL, F.

MICYLLUS, JACOBUS (Strasburg, 1503–58, Frankfurt/Main), the pseudonym of Jakob Moltzer, a humanist and the author of facile Latin elegies and didactic poems, including one on astrology. He translated Livy in 1532 and Tacitus in 1535.

MIEGEL, AGNES (Königsberg, Prussia, now Kaliningrad, 1879–1964, Bad Salzuflen), was educated in Weimar and Paris and at Clifton High School for Girls, Bristol, and became a journalist in Königsberg. Before 1926 her literary production was exclusively poetic (*Gedichte,* 1901; *Balladen und Lieder,* 1907, reissued in 1939 as *Frühe Gedichte; Gedichte und Spiele,* 1930). She showed an inclination for the ballad and the simple strophic form of the Volkslied. Her collected poems (*Gesammelte Gedichte*) were published in 1927; later volumes of verse were *Herbstgesang* (1932), *Kirchen im Ordensland* (1933), *Ostland* (1940), and *Flüchtlingsgedichte* (1949). Beginning with *Geschichten aus·Alt-Preußen* (1926), she wrote a number of stories, some in groups (*Gang in die Dämmerung,* 1924; *Noras Schicksal,* 1936; *Katrinchen kommt nach Hause,* 1937; *Im Ostwind,* 1940; *Wunderliches Weben,* 1940; *Die Blume der Götter,* 1949; *Der Federball,* 1951; *Truso,* 1958; *Heimkehr,* 1962), some published singly (*Die schöne Malone,* 1926, and *Das Bernsteinherz,* 1937). *Kinderland* (1930), *Der Vater* (1932), *Unter hellem Himmel* (1936), and *Die Meinen* (2 vols., 1951) are books of reminiscence. She was principally a regional and religious writer. *Gesammelte Werke,* 6 vols., appeared 1952–5.

Mignon, an enigmatic character in Goethe's novel *Wilhelm Meisters Lehrjahre* (q.v.); she proves to be the child of the Harfner by his sister Sperata. She sings the songs 'Kennst du das Land', 'Heiß mich nicht reden', 'Nur wer die Sehnsucht kennt', and 'So laßt mich scheinen, bis ich werde'.

'Mihi est propositum in taberna mori', first line of a Latin song included in *Estuans intrinsecus ira vehementi* by the Archipoeta (q.v.).

Mildheimisches Liederbuch, a collection of 518 songs by various authors published in 1799 by R. Z. Becker (q.v.). It was in widespread use as a parlour song book in the first half of the 19th c.

MILLER, JOHANN MARTIN (Ulm, 1750–1814, Ulm), poet and novelist, was a pastor's son, who followed his father's vocation, first studying theology at Göttingen. Here he met kindred spirits with whom he founded the Göttinger Hainbund (q.v.) on 12 September 1772. He composed a number of largely derivative poems of folk-song character including the well-known 'Was frag' ich viel nach Geld und Gut', and devoted himself vigorously to the writing of sentimental novels. These include *Beytrag zur Geschichte der Zärtlichkeit, Briefwechsel dreier akademischer Freunde,* and his best seller *Siegwart,* *eine Klostergeschichte* (q.v.), all three published in 1776.

Miller taught at Ulm Gymnasium, and in 1780 took a country parish at Jungingen, returning to Ulm three years later as one of the minster preachers. In these years he wrote two further novels, *Geschichte Karls von Burgheim und Emiliens von Rosenau* (1778) and *Geschichte Gottfried Walthers* (1786). He was also active as a journalist, publishing a moral weekly *Beobachtungen zur Aufklärung des Verstandes und Besserung des Herzens* from 1779 to 1782. In 1797 he became professor at the Gymnasium, and in 1810 dean of Ulm. His collected poems were published in 1783. He was a noted preacher and his sermons were published in *Predigten fürs Landvolk* (1776–84) and *Predigten über verschiedene Texte und Evangelien* (1790). Miller had a gift for observation and for realistic presentation, but he was primarily a moralist, and too often a facile writer.

Milstätter Blutsegen, title given to a Middle High German spell of the 12th c., designed to stanch the flow of blood.

Milstätter Genesis und Exodus, see MILLSTÄTTER HANDSCHRIFT, WIENER GENESIS, and WIENER EXODUS.

Milstätter Handschrift, important medieval MS. from Milstatt Abbey, now preserved in Klagenfurt, Austria. It contains Genesis and Exodus (see WIENER GENESIS and WIENER EXODUS), which are separated by the rhymed *Physiologus* (q.v.), *Vom Rechte, Die Hochzeit,* and *Milstätter Sündenklage* (qq.v.), as well as one or two lesser works. It was written in the second half of the 12th c.

Milstätter Sündenklage, a confession of sins in Early Middle High German rhyming verse. It conforms to the threefold structure of the formula of confession and, in spite of its use of the first person, is a general, rather than individual, catalogue of sins. Underlying it is the profound sense of sin and peril of damnation emphasized through the movement of reform often associated with Cluny (see CLUNIAC REFORM). It forms part of the Milstätter Handschrift (q.v.), comprises nearly 900 lines, and was probably written *c.* 1130.

Minna von Barnhelm oder Das Soldatenglück, a comedy (Ein Lustspiel in fünf Aufzügen) by G. E. Lessing (q.v.), published in 1767 and first performed in Hamburg on 30 September of that year. Although written mainly in 1766, Lessing states that it was accomplished in 1763 ('verfertigt im Jahre 1763')

in order to underline its connection with the Seven Years War (see SIEBENJÄHRIGER KRIEG).

The action centres on a pair of lovers who have lost contact since their betrothal in Saxony during the war; they are Major von Tellheim, an officer in Prussian service, and a Saxon lady, Minna von Barnhelm. Tellheim finds himself suspected of misappropriation of funds, and deems his honour so impugned as to render him ineligible as a partner for Minna, whom he deeply loves. Minna, hearing nothing from him, comes to Berlin to seek him out. She chances upon him at once, but finds him adamant for a separation because of his supposed dishonour. A complex intrigue ensues, in the course of which Minna makes use of Tellheim's ring, which he has pledged with the innkeeper. By simulating poverty she neatly dislodges Tellheim from his moral pinnacle, but in excessive self-confidence nearly overdoes matters. The knot is cut on the arrival of her wealthy and benevolent uncle and all ends happily. The pair of noble lovers is paralleled by the lady's maid, Franziska, and a sergeant, Werner; the other characters include a rude but loyal batman, Just, a comic grasping innkeeper, and a rascally but absurd French adventurer, Riccaut.

Minna von Barnhelm is remarkable for its topicality, its lifelike dialogue, and its indestructible good humour. Goethe, in *Dichtung und Wahrheit* (q.v.), calls it 'die erste aus dem bedeutenden Leben gegriffene Theaterproduktion von spezifisch temporären Gehalt', and refers to its political symbolism, suggesting a union of opposed Germans (Prussian with Saxon), though Tellheim, who has fought for Friedrich II (q.v.) of Prussia, is not himself a Prussian.

Minneallegorie, see MINNEREDE.

Minneburg, Die, an anonymous Middle High German allegorical poem of some 5,000 lines. It was probably written in Eastern Franconia, in the episcopal state of Würzburg, *c.* 1340. The somewhat confused allegory gives a spiritual view of love, condemns infidelity, and includes a battle between the virtues and the vices. The more lyrical portions of the poem are written in an excessively 'flowery style' (see GEBLÜMTER STIL). The author cites Meister EGEN von Bamberg (q.v.) as his stylistic model. A prose version is also extant.

Minnehof, Der, a short Middle High German poem which is an allegory of love in the guise of court proceedings. The poem is one of a group of six MSS. occurring together in MSS. and sometimes thought to constitute a single work (see BÖHMERSCHLACHT, DIE). The unknown author was a Rhinelander, writing *c.* 1300.

Minne im Garten, Die, an anonymous Middle High German poem of nearly 400 lines, written in the 14th or 15th c. It is an allegory of love.

Minnerede denotes a didactic poem on a theme of love, which is treated in narrative style without, as in Minneallegorie, the use of persons to represent abstractions. Examples of Minnerede are the poems of ELBLIN von Eselsberg (q.v.).

Minnesang, in the strict sense the formal love poetry of the chivalric age, attaining its peak from *c.* 1180 to 1220, and continuing in slow decline to the 14th c. A late degenerate descendant is Meistergesang (q.v.).

The term Minnesang is sometimes loosely extended to cover all lyric poetry of the age, including the Spruch and the *leich* (qq.v.). The primary sources of true Minnesang are the Provençal poetry of the troubadours and the Old French poetry of the *trouvères*. Minnesang was invariably true song, but evidence of its musical aspect is unfortunately scanty.

Minnesang is based on a given and only slightly variable situation. The knight adores his lady (who is of higher degree) without expectation of gratification, singing her praises and expressing his longing and his gratitude for any token of favour. It has also been commonly held that the lady is a married woman, but evidence for this is regarded by some scholars as lacking. The situation is a parallel (and it is thought by many to be consciously so) to the feudal relationship of a noble to his liege lord. The lady may at the same time be seen as a civilizing and educative influence on her knight. This tenuous relationship is portrayed as secret and endangered by watchers, spies, and envious rivals. The convention is most rigidly observed in the heyday of Minnesang, but even then two important variations occur. The first, the Kreuzlied, is a lament at separation when a knight leaves for or is on a crusade. The second, a borrowing from the Provençal *alba*, is the dawn song (Tagelied, q.v.), in which the lovers actually meet privily and lament the coming of dawn, which compels them to part.

The form of Minnesang in the earliest examples is relatively simple and uniform, but under Provençal influence a complex verse structure was evolved, which was both binary and triadic. The *liet* (the term for a single strophe) is composed of a first part consisting of two formally identical groups of lines, linked together by rhyme, and a final section or coda with different rhymes. Later the Meistergesang called the quatrains Stollen, and the two combined formed the Aufgesang, while the coda was termed the Abgesang. Many writers have found it convenient to apply this later terminology to

Minnesang itself. The strictness of this form demanded of the poet considerable subtlety and fine workmanship if variety of expression was to be attained, and it is not surprising that the flowering of this stringent art lasted only a few decades.

With the passage of time, the obligatory formalism relaxed to some extent and subjects not originally approved (e.g., songs of ordinary love and satire) became acceptable. Other poets developed excessive elaboration of form for the sake of mere ingenuity.

For the principal poets (Minnesänger) see separate entries on: KÜRNBERGER, DIETMAR VON EIST, FRIEDRICH VON HAUSEN, RUDOLF VON FENIS, HEINRICH VON MORUNGEN, REINMAR DER ALTE, WALTHER VON DER VOGELWEIDE, WOLFRAM VON ESCHENBACH, NEIDHART VON REUENTAL, ULRICH VON LIECHTENSTEIN, GOTTFRIED VON NEIFEN, ULRICH VON WINTERSTETTEN, REINMAR VON ZWETER, KONRAD VON WÜRZBURG, HEINRICH FRAUENLOB, HADLAUB, and OSWALD VON WOLKENSTEIN. Entries are also included on lesser figures to whom strophes of Minnesang have been attributed.

Minnesangs Frühling, see DES MINNESANGS FRÜHLING.

Minneturnier, Das, title given to an anonymous Middle High German poem of the 15th c. It is a threadbare allegory of love in about 1,300 lines. The modern title does not correspond to the content.

MIRZA SCHAFFY, a Georgian schoolmaster in Tiflis under whom F. Bodenstedt (q.v.) studied languages, and whose name Bodenstedt used for his own pseudo-oriental poems, *Die Lieder des Mirza Schaffy* (first separate publication 1851). 'Mirza' is a description, meaning 'man of letters', not a name.

MISES, DR., see FECHNER, G. Th.

Missa sine nomine, a novel by E. Wiechert (q.v.), published in 1950. It is a reckoning with the National Socialist period. Three middle-aged to elderly brothers, Amadeus, Erasmus, and Ägidius, Barons von Liljekrona, are reunited at the end of the 1939–45 War in a hut close to their ancestral castle, now requisitioned. All three need to come to terms with themselves.

Ägidius has the simplest problem. His interest is in farming; he marries a farmer's widow and sets to work to plough and sow. Erasmus, a retired general, has held aloof from the National Socialist regime, but his conscience troubles him, because he once failed to succour refugees in need. He devotes himself to the refugees in the castle and thus attains inner harmony. The central figure is Amadeus, who has endured the sufferings of a concentration camp, from which he has returned in a state of mental numbness. He eventually overcomes this through the actions of his enemies. He had been denounced by an estate forester, at whose trial for collaboration with the National Socialist regime Amadeus gives evidence on the forester's behalf. Further, Amadeus helps to track down a predatory Nazi in hiding, known as der Dunkle, 'the dark man', who has murdered lonely inhabitants to obtain food. Amadeus is present at his capture. Der Dunkle is tried and executed, but his mistress, the forester's daughter, who is still a Nazi, decides to avenge him. She lures Amadeus into an ambush and stays to watch him die. The wounded Amadeus tells her that by this act she is murdering her unborn child; aroused by compassion, she goes for help. Through this experience Amadeus emerges from his mental torpor with a conviction that evil can be converted into good. He also realizes that he himself, when ambushed and firing back, had instinctively shot wide. The forester's daughter becomes mentally deranged and believes Amadeus to be the father of her child. Amadeus devotes his life to maintaining this illusion; in so doing he achieves spiritual peace. The narrative is written in a markedly biblical style.

Mißbrauchten Liebesbriefe, Die, a Novelle by G. Keller (q.v.), published in 1856 in the *Deutsche Reichszeitung* and included in *Die Leute von Seldwyla* (q.v., vol. 2) in 1874. Viggi Störteler, a businessman, has literary aspirations. He makes frequent journeys in connection with his business, and insists that Gritli, his attractive young wife, write him elaborate high-flown letters, which he intends subsequently to weave into a novel. Gritli, who is at her wit's end, enlists the help of a teacher, Wilhelm (Lehrer Wilhelm); she passes Viggi's letters on to him and receives warm replies, for Wilhelm, a shy young man, secretly admires her. Gritli copies out Wilhelm's letters and sends the copies to Viggi, who is delighted; but, discovering the deception on his return, he maltreats Gritli, locks her in the cellar, and finally turns her out of the house. Divorce proceedings follow, and Gritli successfully claims the return of her dowry, while Wilhelm, dismissed from his teaching post because of his involvement in the divorce, seeks solitude and takes up viticulture. He is sought out by Gritli, who overcomes his shyness. They marry and prosper. Viggi makes a second marriage with Kätter Ambach, a bluestocking, who spends his money and makes his life a misery.

Keller uses the tale to satirize some pretentious, but ungifted, literary contemporaries.

Miß Sara Sampson, a tragedy (Ein Trauerspiel in fünf Aufzügen) in prose written by G. E. Lessing (q.v.) in Potsdam in 1755 and published and performed (Frankfurt/Oder, 10 July) in the same year. It is the first German domestic tragedy of importance (see BÜRGERLICHES TRAUERSPIEL).

The plot concerns the consequences of an elopement. The virtuous Sara, daughter of Sir William Sampson, elopes with the weak, yet well-meaning, Mellefont. Remaining immaculate, she is still beset by remorse. Her father arrives with the best intentions of forgiveness, but before he and Sara meet, she is poisoned by Mellefont's discarded mistress, Marwood. Though the dialogue now seems sentimental, it passed for reality in an age which was devoted to the novels of Richardson, for these, even more than the acknowledged source, Lillo's *The London Merchant*, seem to have given it its tone. It was immediately and immensely successful.

Mit einem gemalten Band, superscription of a poem by Goethe, sent in 1771 with a present of a decorated ribbon to Friederike Brion (q.v.). It was first published in Jacobi's *Iris* (q.v.) in 1775.

'Mit Mann und Roß und Wagen', first line of an untitled popular poem written during the Wars of Liberation (see NAPOLEONIC WARS) by an otherwise unknown author, E. F. August.

Mitschuldigen, Die, a verse comedy in three acts written by Goethe in Frankfurt in 1769. Composed in Alexandrine verse, it exists in three slightly differing versions, of which only the third was published in Goethe's lifetime (*Goethes Schriften*, vol. 2, 1787). It was first performed at Weimar by amateurs in 1777 and by professional actors in 1805.

The scene of this rather cynical little comedy is an inn; there are four characters: the innkeeper, his daughter, her good-for-nothing husband, and a guest attracted to the daughter. Each commits a misdemeanour. The innkeeper breaks into the guest's room, the husband steals the guest's money, and the wife and the guest keep an assignation. All are discovered, each is compromised in the eyes of the others, and they agree to call it a day.

Mittelfränkische Reimbibel (formerly known as the *Mittelfränkisches Legendar*) is a retelling in rhyming verse of the Bible, including the Apocrypha, together with a number of legends of the saints. Fragments only survive, totalling less than 1,500 lines. Its purpose was to present the doctrine of salvation in the broadest narrative terms. It was probably written *c.* 1120.

Mittelhochdeutsch, Middle High German, often abbreviated in English to MHG. See GERMAN LANGUAGE, HISTORY OF.

MITTERER, ERIKA (Vienna, 1906–), whose married name is Petrowsky, conducted a correspondence in verse with Rilke (q.v.), which was published in 1950 as *Briefwechsel in Gedichten mit R. M. Rilke.* Earlier she had published two volumes of melodious traditional verse, *Dank des Lebens* (1930) and *Gesang der Wandernden* (1935). Her novel *Der Fürst der Welt* (1940) is set in the first years of the 16th c. It portrays the spread of evil in individuals and in the Church, which is corrupted by the unjust methods and evil will of the Inquisition. By its act two sisters, the one sinning, the other innocent, are alike destroyed. Her other novels include *Wir sind allein* (1945), *Die nackte Wahrheit* (1951), *Kleine Damengröße* (1953), and *Wasser des Lebens* (1953).

Mittwochsgesellschaft, a Berlin literary circle which had its regular meetings on Wednesdays and was founded in 1824 by J. E. Hitzig (q.v.). Its members included at various times W. Alexis, A. von Chamisso, J. von Eichendorff, and K. von Holtei (qq.v.). Among its critics were K. Gutzkow and M. G. Saphir (qq.v.), founder of the Berliner Sonntagsverein, renamed Der Tunnel über der Spree (q.v.).

MOCERUS, ANTONIUS (Hildesheim, 1535–1607), a German humanist, whose real name was Anton Mocker, wrote Latin didactic poems, including *Bellum scholasticum* (1564) and *Psychomachia* (1594).

Modellbuch, see MUTTER COURAGE UND IHRE KINDER.

Moderne Drama, Das, a treatise by H. Hettner (q.v.), published in 1852. It is divided into three sections: (1) *Die historische Tragödie*; (2) *Das bürgerliche Drama*; (3) *Die Komödie*. The first attacks those dramatists who imitate Shakespeare's 'Histories' under the impression that epic breadth is the essence of tragedy. In Hettner's view historical tragedy must have roots in the age in which it is written. In the second section Hettner expresses approval of social drama, but condemns the work of existing practitioners, with the exception of Hebbel (q.v.). The third section is partly concerned with R. Wagner (q.v.): while believing that Wagner (whose most recent work when Hettner wrote was *Lohengrin*, q.v.) overstates his case, Hettner

puts forward the view that comedy should be serene, rather than witty or ingenious, and that it should make considerable use of music.

MODERSOHN-BECKER, PAULA, married name of P. Becker (Dresden, 1876–1907, Worpswede), a powerful Expressionist painter, whose works, depicting peasant women and children and also still life, are primarily represented in the Paula-Modersohn-Haus in Bremen. Her pictures convey deep compassion. She lived and worked in the colony of painters in the remote fenland village of Worpswede, and was a friend of Rilke (q.v.) and his wife, the sculptress Klara Westhoff. Her husband Otto Becker (1865–1943) was also an artist. Her *Briefe und Tagebuchblätter* were published in 1917.

Modus, term applied in early medieval times to tunes, and extended to signify the text, in the form of *sequentiae* (see SEQUENZ), to which the tunes were sung. Four such Latin poems survive with the titles *Modus qui et Carelmanninc, Modus florum, Modus Liebinc,* and *Modus Ottinc.* The *Modus qui et Carelmanninc,* also called *Karlmannsweise,* dates from *c.* 1020 and is a hymn in praise of Christ. The *Modus florum* is a comic poem in which a Swabian wins a king's daughter through the magnitude of the hunting lies he tells. Its author terms it *mendosam cantilenam.* It belongs to the early 11th c. The comic poem *Modus Liebinc* exists also in a Middle High German version as *Das Schneekind* (q.v.). The Latin poem was probably written in the 11th c. The *Modus Ottinc* is a poem in praise of Otto the Great (see OTTO I) and his line, in which Otto III (q.v.) provides the climax. It was written in the latter's lifetime, towards the end of the 10th c.

MOELLER VAN DEN BRUCK, ARTHUR (Solingen, 1876–1925, Berlin, by suicide), a man of means with a deep interest in civilization, lived for years in France and Italy, and in 1914 volunteered for military service. After the 1914–18 War he was active as a political and cultural publicist in Berlin. Primarily a writer on art and civilization, he was also an amateur anthropologist.

Moeller's first works are on literature and include *Die moderne Literatur in Gruppen und Einzeldarstellungen* (1900–3), *Das Variété, eine Kulturdramaturgie* (1901); his eight-volume *Die Deutschen, unsere Menschengeschichte* (1904–10) was supplemented by *Die Zeitgenossen* (1905). *Die italienische Schönheit* (1913) is copiously illustrated with photographs of buildings and paintings. Moeller's best-known political work is entitled *Das Dritte Reich* (1923), and expresses anti-Liberal and anti-parliamentary views, by which he asserts what he termed a revolutionary conservatism; the title made it inevitable that he was regarded as one of the fathers of the National Socialist Party (see NSDAP).

Moisasurs Zauberfluch, a play (Zauberspiel in zwei Aufzügen) by F. Raimund (q.v.), written in 1827, first performed at the Theater an der Wien, Vienna, in September 1827, and published posthumously in *Sämtliche Werke* (1837, vol. 2). The play is set in a make-believe India. Moisasur is a cruel and wicked god whose worship is abolished. In revenge he curses the young empress, turning her into an old hag shedding diamond tears. She is released from the curse by tears of pure joy when she learns that her husband is willing to sacrifice himself for her.

This Zauberspiel (see VOLKSSTÜCK), which represents a determined attempt by Raimund to turn the magic play into something serious and allegorical, was less successful than its comic predecessors.

MÖLLHAUSEN, BALDUIN (Bonn, 1825–1905, Berlin), emigrated in 1850 to America, accompanied Duke Paul of Württemberg on a tour of the Rockies, lived for a time in the real Wild West, and returned to Germany in 1854. He was appointed royal librarian in Potsdam and then revisited the far west of the U.S.A. in 1857–8.

Möllhausen is the author of many novels of American life, including *Der Halbindianer* (1861), *Der Majordomo* (1863), *Das Mormonenmädchen* (6 vols., 1864, republished as *Die Kinder des Sträflings*, 1876), *Der Meerkönig* (1866), *Der Hochlandpfeifer* (6 vols., 1868), *Der Schatz von Quivira* (1880), *Haus Montague* (3 vols., 1891), and *Der Vaquero* (1898). Accounts of his travels are contained in *Reise in den Felsengebirgen Nordamerikas* (1861). His works were published posthumously as *Illustrierte Romane, Reisen und Abenteuer* (30 vols.) in 1906–8.

Molly, name which G. A. Bürger (q.v.) gave to his sister-in-law Auguste Leonhart (q.v.), his second wife. Many of his poems are addressed to or refer to Molly. Among them are *Molly's Abschied, Molly's Wert, Das Blümchen Wunderhold,* and *Das Mädel, das ich meine.*

MOLO, WALTER, REICHSRITTER VON (Sternberg, Moravia, 1880–1958, Murnau), studied at the Viennese college of technology and became an engineer. From 1904 to 1913 he worked in the patents office in Vienna, then moved to Berlin and devoted himself to writing. He had by this time already published four novels. From 1928 to 1930 he was President of the Sektion Dichtkunst of the Prussian Academy (see AKADEMIEN). He was not well viewed by the National Socialist regime and was at times in difficulties.

Molo's speciality was the *biographie romancée* of a prominent historical or literary figure, for which he made careful studies, though his method tended towards over-dramatization. His works include *Der Schiller-Roman* (4 vols., 1912–16) and the three national novels *Fridericus Rex* (1918), *Luise* (1919, dealing with the consort of Friedrich Wilhelm III, q.v.), and *Das Volk wacht auf* (1921, the War of Liberation), which were united in 1922 as a trilogy entitled *Ein Volk wacht auf.* Further noteworthy works of this kind are *Martin Luther* (1928), *Ein Deutscher ohne Deutschland. Der Friedrich List-Roman* (1931, followed by a play on F. List, q.v., *Friedrich List,* 1832), and *Eugenio von Savoy* (1936, on Prince Eugen, q.v.). His work on H. von Kleist (q.v.), *Geschichte einer Seele* (1938), was published in 1958 under a new title, *Ein Stern fiel in den Staub. So wunderbar ist das Leben* (1957) and *Wo ich Frieden fand . . .* (posth. 1959) are autobiographical.

MOLTKE, HELMUTH, GRAF VON (Parchim, 1800–91, Berlin), Prussian field-marshal, was chiefly responsible for translating the foreign policy of Bismarck (q.v.) into military success in the years 1866–71. Moltke began his career in the Danish army, transferring to the Prussian army in 1822. He served in the 8th Grenadiers, a crack regiment. His intellectual qualities were soon recognized in high quarters, though promotion was slow in the peacetime army of the twenties and thirties. As a captain he was given leave in 1835 to serve in Turkey, returning in 1839. He married a 16-year-old English girl (Mary Burt) in 1840, when a major on the staff. She died in 1868 after a happy, though childless marriage. In 1858 he became Chief of the General Staff. When war broke out with Denmark in 1864, Moltke was at first retained in Berlin, but his eventual arrival at the front turned an unsatisfactory campaign into a decisive victory. As the Prussian king's chief of staff in 1866 he was responsible for the overwhelming victory over the Austrians at Sadowa (see KÖNIGGRÄTZ and DEUTSCHER KRIEG). At this point he first came into conflict with Bismarck, desiring to exploit his military success, which Bismarck, for political reasons, forbade. In 1870 he achieved the encirclement of the French field armies at Sedan (q.v.) and deep penetration into France (see DEUTSCH-FRANZÖSISCHER KRIEG). He again came into conflict with Bismarck, opposing the latter's wish for the early bombardment of Paris. After the war he reorganized and expanded the General Staff into a formidable machine for the preparation of war.

In his later years Moltke was regarded as a

national hero; an exceptionally unostentatious mode of life and remarkable taciturnity created the legend of 'der große Schweiger'. His military successes were due to his insistence on concentric attack (his slogan was 'getrennt marschieren und vereint schlagen') and to swift adaptation to changing circumstances.

Though miserly with speech, Moltke was a noted writer on travel and military matters. His *Briefe über Zustände und Begebenheiten in der Türkei aus den Jahren 1835–39*, published in 1841, provide a valuable and reliable picture of the moribund Ottoman Empire. *Geschichte des Deutsch-Französischen Kriegs* appeared in 1891, and he also published *Briefe aus Rußland* (1877) and a *Wanderbuch* (1879). His only purely literary work is a Novelle, *Die beiden Freunde* (q.v.), written and published pseudonymously in 1827. He was an excellent letter-writer and his letters appeared as *Briefe an seine Braut und Frau* (1893) and *Briefe* (1922). His *Gesammelte Schriften und Denkwürdigkeiten* were published posthumously (8 vols., 1891–3).

MOLTKE, HELMUTH VON (Gersdorff, Mecklenburg-Schwerin, 1848–1916, Berlin), 'der jüngere Moltke', nephew of Field-Marshal Graf Helmuth von Moltke (q.v.), entered the army in 1870, commanded the 1st Guards Division in 1902, and became Chief of General Staff in 1906. In the summer campaign of 1914 (see WELTKRIEGE, I) Moltke failed to control the field armies operating at great distances from his headquarters at Coblence and was removed from his command in mid-September. He was subsequently criticized for altering the Schlieffen-Plan (q.v.) before the war by taking troops from the right wing to reinforce the left, and for detaching two army corps for use against the Russians while the campaign in the west was still undecided.

Erinnerungen, Briefe, Dokumente 1877–1916, ed. Eliza von Moltke, were published in 1922.

MOLTKE, HELMUT JAMES, GRAF VON (Kreisau, Silesia, 1907–45, Berlin-Plötzensee, by execution), a great-nephew of Field-Marshal Graf Helmuth von Moltke (q.v.), formed a group of dissident Germans after the establishment of the National Socialist regime in 1933 (see KREISAUER KREIS). Suspected of collaboration with the resistance (see RESISTANCE MOVEMENTS, 2), he was arrested in January 1944 and executed in January 1945.

Moltke's *Letzte Briefe aus dem Gefängnis Tegel* were published in 1953.

MOLTZER, JAKOB, see MICYLLUS, JACOBUS.

MOMBERT, ALFRED (Karlsruhe, 1872–1942, Winterthur, Switzerland), was a lawyer who turned to writing. He was influenced by Nietzsche (q.v.) and devoted his efforts to writing rhythmic works of mythological character. He was also interested in the occult and in the doctrine of the transmigration of souls. As a Jew he was expelled from the Sektion Dichtkunst of the Prussian Academy (see AKADEMIEN) in 1933, declined to emigrate, and was sent in 1940 to a concentration camp. When seriously ill he was rescued by a friend, H. Reinhart, in 1941 and allowed into Switzerland. Among his numerous poetic works are *Der Glühende* (1896), *Die Schöpfung* (1897), *Der Denker* (1901), *Die Blüte des Chaos* (1905), and *Der Held der Erde* (1919). He also wrote a dramatic trilogy *Aeon*, composed of *Aeon der Weltgesuchte* (1907), *Aeon zwischen den Frauen* (1910), and *Aeon von Syrakus* (1911).

Mombert's collected works, *Dichtungen* (3 vols.), ed. E. Herberg, appeared in 1963. Collections of correspondence include *Briefe an R. und Ida Dehmel*, ed. H. Wolffheim (1956) and *Briefe 1893–1942*, ed. B. J. Morse (1961).

MOMMSEN, THEODOR (Garding in Schleswig, 1817–1903, Charlottenburg), studied law for a time with his brother Tycho and Th. Storm (q.v.), and specialized in Roman law, on which he became a widely acknowledged authority. His student friendship with Storm resulted in a collection of legends and the *Liederbuch dreier Freunde* (1843). In the year of the Revolutions, 1848 (see REVOLUTIONEN 1848–9), he became professor of Roman law at Leipzig. He also taught at Zurich and Breslau before he settled in Berlin.

Mommsen was as closely involved with his scholarship as with the critical political developments of his day. During the Revolutionary period he sympathized with the left and was praised by K. Gutzkow (q.v.) as well as the young Gustav Freytag (q.v.) for his lively, anthropological approach to history, most obvious in his famous and massive *Römische Geschichte* (3 vols., 1854–6). This is a work complete in itself, from the beginnings to Julius Caesar. A 'fifth' volume, published in 1885, was an account of the Roman provinces. (There was no fourth volume.) An illustrated abridged edition of the three-volume work (over 1,000 pages in one volume) was a best seller in 1932.

Mommsen also wrote on Roman law (*Römisches Staatsrecht*, 2 vols., 1871–5, expanded to 3 vols., 1887–8). In this field he was influenced by Savigny (q.v.), and in history by Edward Gibbon and B. G. Niebuhr (1776–1831). By his own example Mommsen proved his maxim that history is not merely for the scholar but should promote the political consciousness of all. In the *Römische Geschichte*, which was translated into nine languages, he purposely modernized Roman terminology (e.g., *Bürgermeister* for *consul*). He devoted much time to politics and was for several years a member of the Prussian Landtag and of the Reichstag, distinguishing himself as a keen critic of Bismarck (q.v.) as well as of the nationalistic and anti-Semitic historian Treitschke (q.v.). He was the first German to receive the Nobel prize for Literature (1902).

Gesammelte Schriften (8 vols.) appeared in 1905–10 (reissued 1966), and published correspondence includes *Briefwechsel mit Th. Storm*, ed. H.-E. Teitge (1966).

MON, FRANZ, pseudonym of Franz Löffelholz (Frankfurt/Main, 1926–   ), published *avant-garde* poetry (*artikulationen*, 1959; *protokoll an der kette*, 1960; *verläufe*, 1962, *herzzero*, 1968) and is an exponent of concrete poetry (see KONKRETE POESIE). The radio play *das gras wächst wies wächst* appeared in 1969, *Texte über Texte* in 1970, an extended edition of *Lesebuch* (1968) in 1972, and *maus in mehrl* in 1976. The volume *Fallen stellen. Texte aus mehr als elf Jahren* appeared in 1981.

Monaldeschi, a tragedy by H. Laube (q.v.) written in 1840, published in 1845, and first performed at Stuttgart in November 1841. Its five acts are preceded by a prologue. The principal character is the Monaldeschi who became the favourite of Queen Christina of Sweden and was executed on her orders at Fontainebleau in 1657. In Laube's play Monaldeschi, who disapproves of the Queen's abdication, is murdered by his rival Santini.

Mönch Felix, an anonymous Middle High German verse legend, written, it is presumed, in the 13th c. in Thuringia by a Cistercian monk. Felix hears the song of a bird announcing to him the joys of Heaven. After an hour he returns, but the monastery is changed. His hour of celestial joy has corresponded to a hundred years on earth.

MÖNCH VON HEILSBRONN, DER, a member of the Cistercian Order, who belonged to a monastery at Heilsbronn between Ansbach and Nürnberg, and is the author of a religious poem, *Von den sieben Graden*, and a prose work, *Von den sechs Namen des Fronleichnams*. *Von den sieben Graden* treats of prayer, defining seven stages, each of which approaches closer to God. *Von den sechs Namen des Fronleichnams* expounds the names by which the host is known. *Eucharistia, Donum, Cibus, Communio, Sacrificium, Sacramentum*. The two works were written early in the 14th c.

'Mondbeglänzte Zaubernacht', famous line from *Prolog. Der Aufzug der Romanze*, the Prologue to L. Tieck's play (Lustspiel) *Kaiser Oktavianus* (q.v.).

Mondesaufgang, a poem written towards the end of her life by Annette von Droste-Hülshoff (q.v.) and published posthumously in *Letzte Gaben* (1860). It is a poem of moonrise over Lake Constance.

Mondnacht, a poem by J. von Eichendorff (q.v.), the first line of which runs 'Es war, als hätt' der Himmel'. The setting of it by R. Schumann (q.v.) is one of his best-known songs.

Monodrama, in principle a drama in which only one speaking character (as distinct from a chorus) appears, as in early Greek drama. In German literature it is usually taken to refer to a play, or a part of a play, in which a single voice speaks against a background of music. Its starting-point was J.-J. Rousseau's *scène lyrique Pygmalion* (performed 1772–3 in Weimar); and *Ariadne auf Naxos* (1774) by J. C. Brandes (q.v.) is usually quoted as the first German example. Other instances are the end of Goethe's *Egmont* (q.v.) and Johanna's extended monologue in Act IV of Schiller's *Die Jungfrau von Orleans* (q.v.). The use of monologue in the unfinished play *Catharina von Siena* by J. M. R. Lenz (q.v.) is also related to monodrama. A 20th-c. example is *Ostpolzug* by A. Bronnen (q.v., 1925). See also MELODRAMA.

Monseer Fragmente or **Monsee-Wiener Fragmente,** fragments of a theological MS. in Old High German written early in the 9th c. The MS. originated in Monsee Abbey in Bavaria, not far from Salzburg, and the separate sheets were later used for binding. Two are in Hanover and thirty-nine were found in the former Hofbibliothek (now Nationalbibliothek) in Vienna. The fragments contain passages translated from: (1) Isidor's *De fide catholica contra Judaeos*; (2) St. Matthew's Gospel (according to the Vulgate); (3) an anonymous tract, *De vocatione gentium*; (4) Sermon LXXVI of St. Augustine; and (5) an unidentified work. In each case the Latin original is on the back of the sheet. The *Fragmente* were first published in 1834 by Stephan Endlicher (1804–49) and HOFFMANN von Fallersleben (q.v.) with the title *Fragmenta theotisca*. The translations are believed to have been undertaken as a consequence of Charlemagne's endeavour to further Christian learning (see KARL I).

MONTANUS, MARTIN (b. Strasburg, c. 1537), the author of collections of frequently crude and obscene anecdotes (see SCHWANK) under the titles *Der Wegkürtzer* (1557) and *Das Ander Theyl der Gartengesellschaft* (*c.* 1559; the first part was by Jakob Frey, q.v.).

MONTEZ, LOLA (Limerick, 1818–61, New York), a dancer of mixed Scottish and Creole extraction, became the mistress of Ludwig I of Bavaria (q.v.) in 1846, and was created Gräfin von Landsfeld. She conducted herself extravagantly and interfered in politics. In 1848 she fled abroad. Her purported memoirs (*Memoiren der Lola Montez,* 1849) are a forgery. (See also DIE JÜDIN VON TOLEDO.)

Monumenta Germaniae Historica (also *MGH*), a collection of medieval documents which were assembled by the concerted efforts of notable historians and form the principal source of our knowledge of German medieval history. The idea of this documentation originated with the Freiherr vom Stein (q.v.), who to this end founded in 1819 the Gesellschaft für Deutschlands ältere Geschichtskunde. The enterprise was planned in detail by G. H. Pertz, who directed its extensive publications until 1873.

The Society was reorganized in 1937 into the Reichsinstitut für ältere deutsche Geschichtskunde. A further reorganization took place in 1946. Since 1959 it has borne the title Monumenta Germaniae Historica. Deutsches Institut für Erforschung des Mittelalters.

Moralische Wochenschriften, periodicals published in Germany through the greater part of the 18th c., modelled more or less closely on *The Spectator* and *The Tatler.* They met the need of the rising middle class, supplying it not only with unexceptionable entertainment, but also with moral and educational guidance. German moral weeklies were, on the whole, more occupied with education and with matters of taste and aesthetic principles than their English models; on the other hand they were almost entirely unpolitical, which is scarcely surprising in a land consisting largely of small states ruled by petty despots.

Hamburg, which had close commercial links with England, was the nursery of the moralische Wochenschrift. *Der Vernünftler* (1713–14), edited by J. Mattheson (q.v.), adapted essays from *The Spectator* and *The Tatler* to local conditions in Hamburg. Most of these German periodicals were short-lived, and Mattheson was editor for only two years having produced 100 issues. *Die lustige Fama* appeared in Hamburg in 1718, but lasted only for a few months, and in 1719 an already inaccurate French translation was used as the basis for a German version of *The*

Spectator. Two decades later this received better treatment as *Der Zuschauer* in a rendering by Luise A. Gottsched (q.v., 1739–43). The Swiss scholars Bodmer and Breitinger (qq.v.) were stimulated by Addison's work to attempt an imitation, *Discourse der Mahlern* (1721–3). J. C. Gottsched also published a weekly, the title of which, *Die vernünftigen Tadlerinnen* (1725–6), derives from a misunderstanding of *The Tatler.* These two weeklies, one produced in Zürich, the other in Leipzig, were the first to circulate outside their own localities. Gottsched's second weekly, *Der Biedermann* (1727–9), was less successful, but the fashion was set, and numerous periodicals of this kind came into existence all over Germany. Their number runs into hundreds. Among the more notable were *Der nordische Aufseher* (1758–61) by J. A. Cramer (q.v.), to which Klopstock (q.v.) was a contributor, and *Der Fremde* (1745–6) by J. E. Schlegel (q.v.).

Moralische Wochenschriften declined in popularity and number towards the end of the 18th c.

Morant und Galie, an anonymous Middle High German epic. In original form only fragments amounting to some 500 lines survive in a MS. believed to date from the 13th c. The complete poem occurs, however, in the 14th-c. MS. of *Karlmeinet* (q.v.). It was written in the lower Rhineland and is based on a French original which has, however, not been identified. It deals with an episode from the legendary life of Charlemagne (see KARL I). The knight Morant is falsely accused of illicit relations with Karl's consort Galie; eventually the slanderers are confounded and punished.

Moravian Brothers, see BÖHMISCHE BRÜDER and HERRNHUTER.

Mord, den jeder begeht, Ein, see EIN MORD, DEN JEDER BEGEHT.

Morgenblatt für gebildete Stände, a literary daily founded in Stuttgart in 1807 by J. F. Cotta (q.v.). It continued until 1865. Its first editor was F. C. Weißer. Therese Huber (q.v.) edited it from 1816 to 1823, and in 1827 Wilhelm Hauff (q.v.) and his brother Hermann (1800–65) became joint editors. Wilhelm died in the same year, and Hermann continued as editor until 1863. In 1820 a literary supplement (*Das Literaturblatt*) was added, which was edited first by Adolf Müllner and then by Wolfgang Menzel (qq.v.). From 1839 it bore the title *Morgenblatt für gebildete Leser.*

'Morgen kommt der Weihnachtsmann', first line of a poem entitled *Weihnachtsbescherung* written in 1835 by HOFFMANN von Fallersleben (q.v.).

Morgenlied, a short poem written by L. Uhland (q.v.) in 1811 and published in 1813 as No. 6 of nine *Wanderlieder*. Its first line runs 'Noch ahnt man kaum der Sonne Licht'.

'Morgen muß ich fort von hier', first line of the folk-song usually entitled *Lebewohl*. The poem originated in the 17th c. and was first printed in 1690. It was included in *Des Knaben Wunderhorn* (q.v.) by Arnim and Brentano. The tune to which it is nowadays sung was composed in 1827 by Friedrich Silcher (q.v.)

'Morgenrot,/Leuchtest mir zum frühen Tod', first two lines of the poem *Reiters Morgengesang* by Wilhelm Hauff (q.v.). It was written in 1824 and is (as Hauff acknowledges) an adaptation of a Swabian folk-song, which, however, was a popularized version of a poem by J. C. Günther (q.v.). Hauff's poem, too, has become popular, and is usually known as *Reiters Morgenlied*; in the singing version the first word, 'Morgenrot', is repeated.

MORGENSTERN, CHRISTIAN (Munich, 1871–1914, Merano, South Tyrol), son of an artist, spent his adolescent years in Breslau and was a student of Breslau University. He visited Norway, Switzerland, and Italy and acquired a command of Norwegian and Swedish, which enabled him later to translate Ibsen (q.v.), Bjørnson, Strindberg (q.v.), and Hamsun. He worked for a time as a journalist in Berlin, though pulmonary tuberculosis was diagnosed in 1893, and he battled with the disease for twenty years, the last of which were spent at Merano, then Austrian. Much of his lyric poetry (which constitutes almost his entire *œuvre*) was fundamentally religious, though not Christian. For a time he was fascinated by Nietzsche (q.v.), turned to Buddhism and finally to the theosophy of R. Steiner (q.v.). This serious poetry, which has on the whole been neglected, is contained in the volumes *Ich und die Welt* (1898), *Und aber ründet sich ein Kranz* (1902), *Melancholie* (1906), *Einkehr* (1910), *Ich und Du* (1911), and *Wir fanden einen Pfad* (1914).

The common image of Morgenstern is as a writer of nonsense poetry, not unlike Edward Lear's, which revolves round the absurd characters Palmström and Herr von Korf, displaying a genuine humour, yet often touching the grotesque or burlesque to reveal a sharp edge of satire. These well-known poems are contained in *Galgenlieder* (1905), *Palmström* (1910), and in the posthumous volumes *Palma Kunkel* (1916), *Der Gingganz* (1919), and *Die Schallmühle* (1928). A general collection under the title *Alle Galgenlieder* was published in 1932. An early volume of humorous verse not connected with Palmström appeared in 1895 (*In Phantas Schloß*), and a volume of aphorisms and diary notes, together with an autobiographical sketch, was published posthumously as *Stufen* (1918).

Sämtliche Dichtungen in drei Abteilungen (17 vols. with 1 vol. index), ed. H. O. Proskauer, appeared 1971–80.

MORGNER, IRMTRAUD (Chemnitz, 1933–), studied Germanistik at Leipzig University and worked for two years on the editorial staff of *Neues Deutschland*. In 1958 she settled as a free-lance writer in Berlin. Her fiction began with the story *Das Signal steht auf Fahrt* (1959), the novel *Ein Haus am Rand der Stadt* (1962), and *Hochzeit in Konstantinopel* (1968), the first work to be published in the BRD (1969) and one in which the element of fantasy becomes prominent. It was followed by *Gauklerlegende. Eine Spielfraungeschichte* (1970, DDR; 1971, BRD) and *Die wundersamen Reisen Gustavs des Weltfahrers. Lügenhafter Roman mit Kommentaren* (1972, DDR; 1973, BRD), a short novel presenting Gustav's adventures as recorded by himself and containing an introduction and postscript by his granddaughter, Beda H., dated January 1972. The technique links two generations, the materially deprived Gustav, a socialist, who followed the Münchhausen (q.v.) and Sinbad tradition, and his educated granddaughter, who edits his story as 'das große Leben eines kleinen Mannes' (introduction to the first of the seven travels).

Morgner's major work treats of history, myth, fantasy, and feminism on an epic scale. Intended as the first part of a trilogy, the novel *Leben und Abenteuer der Trobadora Beatriz nach Zeugnissen ihrer Spielfrau Laura. Roman in dreizehn Büchern und sieben Intermezzos* (1974) was followed in 1983 by a similarly extensive second part, *Amanda. Ein Hexenroman*, consisting of 139 chapters, a Vorspiel, and a Nachspiel. Proceeding from JEAN Paul and E. T. A. Hoffmann (qq.v.), the fantastic is used to highlight by contrast the function of women as it is and as it should be in an envisaged future society. The argument derives from the dichotomy underlying the obligatory integration of women in the DDR into the male work force and their conventional role in the private sphere; it can be removed only through a fundamental change of attitude in patriarchal systems, which have created the threat of nuclear war.

The dialectics are illustrated through the

experience of the troubadour Beatriz de Dia and Laura, unlike her Petrarchan namesake a tramdriver. Beatriz enters the contemporary world after an 830-year slumber into which she withdrew in escape from male-dominated medieval society. At first in France, where she lands in 1968, she meets politically and personally with a good measure of disillusionment, which hastens her decision to move to the DDR. In 1973, at the end of the first part, she dies following a fall and seven years later re-emerges as a siren and as the narrator of the second novel, in which Amanda represents the fantastic 'half' of Laura (hexische Hälfte von Laura Amanda Salman). The epilogue preparing for the third part begins on New Year's Eve 1980. The work's prolific montage of diverse quotations and adapted motifs is an essential source of humour, irony, and the grotesque.

MORHOF, DANIEL GEORG (Wismar, 1639–91, Lübeck), was appointed professor of poetry and rhetoric at Rostock at the age of twenty-one and, on the foundation of Kiel University in 1665, became its first professor of poetry and, later, Rector. He visited England in 1660 and again in 1670. Morhof wrote undistinguished poetry in Latin and in German (*Epigrammatum et jocorum centuria*, 1659; *Teutsche Gedichte*, 1682); his chief and considerable merits were as a scholar. He is the author of a remarkable early history of German literature which includes the first German mention of Shakespeare, *Unterricht von der deutschen Sprache und Poesie, deren Ursprung, Fortgang und Lehrsätzen* (1682, reissued 1969). After 1673 he turned his attention also to historiography and completed two volumes of a history of learning, *Polyhistor sive de notitia auctorum et rerum commentarii* (1688–92).

MÖRIKE, EDUARD FRIEDRICH (Ludwigsburg, 1804–75, Stuttgart), invariably known as Eduard Mörike, was the son of a physician who suffered a paralytic stroke in 1815 and died in 1817. The boy was sent to school in Stuttgart and then in 1818 to Urach, where he remained until 1822. In these three years falls an attachment to his cousin Klara Neuffer (q.v.). Following the accustomed path for a future Württemberg pastor, he was next admitted to the Tübinger Stift (q.v.), from which he emerged as a clergyman in 1826. Among his fellow-students his particular friends included L. Bauer, F. Th. Vischer, and D. F. Strauß (qq.v.). In 1823 he became deeply attached for a time to the mysterious vagrant Maria Meyer (q.v.), who appears in his poetry as Peregrina (q.v.).

Mörike spent the next eight years as a curate, serving in nine different parishes, of which the more important were Möhringen (1827),

Plattenhardt (1829), Owen (twice), and Ochsenwang (1832). He soon discovered that by temperament he was ill suited to clerical life and, as early as 1827, sought and obtained leave of absence to seek other employment. No sooner had he obtained a post in periodical journalism than he regretted the step and returned to clerical duties. Mörike's first published poems appeared in 1828 in the Stuttgart *Morgenblatt für gebildete Stände*, and his first substantial publication was the romantic novel *Maler Nolten* (q.v., 1832), through which were scattered some thirty more poems. In 1829 he became engaged to Luise Rau (q.v.), a pastor's daughter, but after a four-year betrothal with little prospect of a benefice the engagement was dissolved. In the following year (1834), however, the long-awaited cure of souls was given to him at Cleversulzbach near Weinsberg, and here Mörike, by now a confirmed valetudinarian, lived with his mother and youngest sister, devoting as little time as possible to his pastoral cares. His first lyrical collection, *Gedichte*, which comprised 143 poems, was published in 1838. This collection was expanded in new editions in 1847, 1856, and 1867, reaching in the last a total of 226 poems. His mother died in 1841, and two years later Mörike, with whose inactivity the parishioners had expressed some dissatisfaction, was pensioned at his own request. The pension was minimal, and, after staying a few months with a clerical friend, he lived with his sister in very modest circumstances, first at Schwäbisch-Hall, then (1844) at Bad Mergentheim, where he met his future wife, Margarete von Speeth. They were married in 1851 and lived first at Stuttgart where Mörike had been appointed to a part-time post at a young ladies' seminary, teaching German literature for two hours a week. The marriage began inauspiciously, for Mörike's sister was unwilling to relinquish her long-standing position as housekeeper and chief companion to him. After long years of bickering (*perturbatio domestica* was Mörike's recurring note in his diary) the couple separated in 1873, though Margarete returned when Mörike was dying.

Of Mörike's other works the *Idylle vom Bodensee* appeared in 1846, the fairy-tale *Das Stuttgarter Hutzelmännlein* in 1853, and the Novelle *Mozart auf der Reise nach Prag* (q.v.) in 1855. In the 1850s Mörike's increasing poetic reputation led to a friendship and correspondence with Theodor Storm (q.v.), and he was also on close terms with the painter Moritz von Schwind (q.v.), who illustrated some of his works. From 1852 he began to receive tokens of respect and admiration in the form of honorary doctorates, an honorary professorship, decorations, and a modest pension from the Schillerstiftung. He resigned his part-time teaching appointment in 1866, though the small stipend was continued.

Mörike's ineffectualness in the practical affairs of life was probably related to the exceptional sensitiveness of his mind. The subject-matter of his poetry is highly personal, yet fully accessible, and he exhibits a remarkable power of entering sympathetically into the minds of others. His poetry has a truth of feeling and a subtlety of perception which few poets have equalled; his craftsmanship, exemplified in forms and metres varying from folk-song-like stanzas to classical hexameters and elegiacs, is superb; and he possesses a highly flexible and adaptable sense of rhythm. Though his calibre was not fully realized in his lifetime, the 20th c. has recognized in the inadequate village pastor a poet of the first rank with a special niche, shared by no other, in which the popular and homely encounter the subtle and refined. Mörike had pronounced musical sensitivity and his poems lend themselves readily to musical treatment. The perceptive settings by Hugo Wolf (q.v.) of 53 poems (*Mörike-Lieder*, 1888) did much to broaden the appeal of Mörike's poetry.

Werke and *Briefe*, ed. H. Göpfert, appeared in 1954 and *Sämtliche Werke. Historisch-kritische Gesamtausgabe* (15 vols.), ed. H.-H. Krummacher, H. Meyer, B. Zeller in 1967 ff. Editions of Mörike's correspondence with M. von Schwind, Th. Storm, H. Kurz (q.v.), and F. Th. Vischer (q.v.), Margarethe von Speeth, and Luise Rau were published between 1918 and 1926. The first edition of *Briefe Mörikes* (2 vols.), ed. K. Fischer, was published 1903–4, and *Unveröffentlichte Briefe*, ed. F. Seebaß, in 1941 (revised edn. 1945).

Mörin, Die, see HERMANN VON SACHSENHEIM.

Moritat. Moritaten were sensational stories of murders, executions etc., sung in ballad form by Bänkelsänger (q.v.) in the 18th c. and 19th c. The word is derived from 'Mordtat'.

Morituri, group title of three one-act plays by H. Sudermann (q.v.), published in 1896. They are *Teja, Fritzchen*, and *Das Ewig-Männliche*. The first is a symbolical prose tragedy, in which the besieged Teja, King of the Goths, learns from his bride the sweetness of life at the moment at which death becomes certain.

Das Ewig-Männliche is a rather feeble satirical verse comedy. *Fritzchen* (Drama in einem Akt), however, is possibly Sudermann's finest dramatic achievement. Fritzchen, the adored son of a retired major and his ailing wife, comes to say good-bye before leaving for a duel which he

knows must have a fatal outcome for him. He conceals the matter as long as he can, but the major finds out and father and son leave on their last journey together.

MORITZ, KARL PHILIPP (Hamelin, 1756–93, Berlin), had a hard and unhappy childhood and was apprenticed to a hat-maker in Brunswick. He tried his hand at acting before becoming a schoolmaster under J. B. Basedow (q.v.) at the Philanthropinum in Dessau, from where he moved to the military orphanage in Potsdam, and, in 1780, to a post (Konrektor) at the Gymnasium am Grauen Kloster in Berlin, where he accepted four years later another appointment (Professor) at the Köllnisches Gymnasium. He became engaged in journalistic editorial work, and in 1781 published a play (Schauspiel) entitled *Blunt, oder der Gast*, which has been termed the first German fate tragedy (see SCHICKSALSTRAGÖDIE).

In 1785 the first volume of Moritz'a autobiographical novel *Anton Reiser* (q.v.) appeared, the second and third in 1786, the fourth and last following in 1790. This, his principal work of fiction, is both the key to his tormented personality and a cultural document of the age. In his other novels self-analysis and psychological portrayal of character recede in favour of a symbolical style: *Andreas Hartknopf* (1786), to which he wrote a continuation, *Andreas Hartknopfs Predigerjahre* (1790), and *Die neue Cecilia*, which remained unfinished (fragment published posthumously 1794). In 1786 Moritz gave up his teaching post and went to Italy in order to acquire the breadth of knowledge which his hard youth had denied him. There he met Goethe, who became his friend.

Moritz's essay *Versuch einer deutschen Prosodie* (1786), which asserted the principle of stress against that of quantity, was valued highly by Goethe, and Moritz thereafter devoted himself chiefly to writings on aesthetics. These include the important *Über die bildende Nachahmung des Schönen*, published in 1788, *Vorlesungen über den Stil, Götterlehre oder Mythologische Dichtungen der Alten* (on Greek mythology, both 1791), and *Über ein Gemälde von Goethe* (1792), which is an examination of the style of *Die Leiden des jungen Werthers* (q.v.). His extensive studies on psychology are contained in the periodical *Magazin für Erfahrungsseelenkunde* (10 vols., 1783–93), the first of its kind to appear in Germany. He wrote about his travels to England (in 1782) in *Reisen eines Deutschen in England* (1783), and to Italy in *Reisen eines Deutschen in Italien* (3 vols., 1792–3). From 1789 until his death from tuberculosis Moritz held a professorship at the Academy of Arts in Berlin, where he is believed to have died, though some authorities have stated that his death took place during a visit to Dresden. A number of his works, including his novels, have been reissued in the 20th c.; *Schriften zur Ästhetik und Poetik. Kritische Ausgabe*, ed. H. J. Schrimpf, appeared in 1962 and *Werke* (3 vols.), ed. H. Günther, in 1981.

MORITZ VON HESSEN, LANDGRAF, known as Moritz der Gelehrte (Kassel, 1572–1627, Eschwege), reigned in Hesse from 1592, introduced Calvinism in 1605, and in 1627 was forced to abdicate after forfeiting to the Emperor Ferdinand II (q.v.) a substantial part of his territory, the result of his policy during the Thirty Years War (see DREISSIGJÄHRIGER KRIEG).

Moritz, who founded the Collegium Mauritianum in 1599, was a man of intellectual gifts and made contributions in various fields of knowledge, including philogy, mathematics, and botany. He was an excellent linguist and made appointments to promote the teaching of Italian in his land. He was also a composer of liturgical music. He had a passionate interest in the theatre and maintained an English company at Kassel for which he built a permanent theatre, the Ottoneum. He wrote plays (some in German, some in Latin), including biblical plays, adaptations of Terence, a drama of martyrdom, *Sophronia*, and a Latin comedy *Praemium pietatis*. In some cases he contented himself with sketching out the play, leaving execution to another hand.

MORITZ VON SACHSEN (Freiberg, 1521–53, Sievershausen), was Elector of Saxony from 1547 to 1553. Moritz was a son of Duke Heinrich der Fromme of Saxony (reigned 1539–41), at a time when the Saxon lands were divided into an Albertine Duchy and an Ernestine Electorate. He succeeded his Albertine father as duke in 1541, became estranged from the Elector (of the Ernestine branch), and, being a man of considerable ambition, aspired to possession of the Ernestine lands as well as his own.

Moritz sided with the Emperor Karl V (q.v.) at the time of the War of Schmalkalden (see SCHMALKALDISCHER KRIEG), and after the capture and deposition of the Elector Johann Friedrich (reigned 1532–47), Moritz succeeded him. He did not obtain from the Emperor all that he had been promised and by 1550 he was making preparations for hostilities against Karl V, whom he attacked by surprise in 1552. Moritz successfully concluded the Treaty of Passau, but was unable fully to maintain his advantage. In 1553 he was mortally wounded at the battle of Sievershausen, in which his forces defeated his former ally the Margrave of Brandenburg.

He has been regarded as a champion of Protestantism and as the instigator of Germany's fragmentation. He was certainly an extremely ambitious man, and was perhaps no more than an able opportunist. He was the founder of the three Saxon 'Fürstenschulen' (q.v.). The marshal who served Louis XV as Maurice de Saxe (see SAXE, MAURICE DE, MARÉCHAL) is also known in German as Moritz von Sachsen.

Moriz von Craûn, title given to an anonymous Middle High German narrative poem of nearly 2,000 lines. It is based on a lost French poem and was written in the Rhineland (Alsace or Palatinate). It has been attributed to the decades 1180–90 and 1210–20; recent opinion inclines to the earlier date. The poem has an introductory section outlining the history of knighthood, beginning with Greece, passing through Troy and Rome, arriving at Charlemagne (see KARL I) and his paladins. The court of King Arthur is not mentioned.

The story proper is a scandalous episode said to have taken place in France. Its title figure, Maurice de Craon in Anjou, is historically authenticated; he died in 1196. Moriz courts his neighbour's wife, the Countess de Beaumont, and is promised due reward on condition that he organizes a splendid tournament. He does so, and is led to a handsome chamber where the lady is to join him. Overcome with fatigue he falls asleep, the lady is incensed and orders his dismissal. Moriz, who has overheard the Countess's angry words, follows her to her room. Here the Count is wakened, mistakes Moriz for the ghost of an adversary whom he has slain, and swoons. Moriz climbs into the Countess's bed and takes his 'due reward', after which he upbraids the lady for her earlier conduct and rides away. In the closing scene the Countess sadly gazes into the landscape, regretting the loss of her lover. Strange as it may seem, the poem is moral in its intention, reproaching the lady for her failure to reward Moriz as the laws of love (*staete minne*) demand. The final episode of the lady's remorse seems, however, to belong to a less artificial and more real world, and the poem has been interpreted as an ironical commentary on the ideal of courtly love.

MORRÉ, KARL (Klagenfurt, 1832–97, Graz), a civil servant in Styria, first in Graz then in Bruck an der Mur, wrote popular plays (see VOLKSSTÜCK). They include *Die Familie Schnock* (1881), *Die Frau Rätin* (1884), *'s Nullerl* (1885), and *Der Glückselige* (1886). *'s Nullerl* was an outstanding success and was widely performed even outside Austria. Morré, who was a Liberal, sat in the Austrian lower house from 1891.

MOSCHEROSCH, Johann Michael (Willstädt nr. Strasburg, 1601–69, Worms), a moralist, was educated at Strasburg. He travelled in France and Switzerland, then worked as a private tutor, and in 1630 became magistrate (Amtmann) at Kriechingen near Metz. Moscherosch suffered the hardships and hazards of war, fleeing to Strasburg in 1636 (and losing his second wife on the journey), and sustaining financial loss and damage to his property. His first notable work began as a translation of Quevedo's *Sueños*, under the title *Wunderliche und Warhafftige Gesichte Philanders von Sittewalt* (1640–3). The work was continued in a second book which is the original work of Moscherosch. The whole is a satire on the manners and morals of the time, which particularly assails the imitation of all that is foreign in both language and life. *Insomnis cura parentum* (1643, written in German in spite of the Latin title) is a sincere and homely tract of moral counsel based on the family and directed ostensibly to his children, yet applicable to all. Towards the end of the Thirty Years War (see DREISSIGJÄHRIGER KRIEG) he was given an appointment in the service of the occupying Swedish armies and spent his later years in the service of various German princes, including the Elector of Mainz and a Landgravine of Kassel. From 1645 he was a member of the Fruchtbringende Gesellschaft (q.v.). A didactic poem by Moscherosch, *Patientia*, unpublished in his lifetime, first appeared in 1897. He died in Worms while on a journey.

A selection, *Wunderliche und Warhafftige Gesichte Philanders von Sittewald*, ed. F. Bobertag (1884), was reissued in 1964.

MOSEN, Julius (Marieney, Vogtland, 1803–67, Oldenburg), set up in practice as a lawyer in 1834 in Dresden and was appointed director of the Court Theatre at Oldenburg in 1844. A steadily advancing paralysis first made him a cripple and finally left him completely bed-ridden. He published two epics early in his career, the *Lied vom Ritter Wahn* (1831) and a version of the legend of the Wandering Jew, *Ahasver* (1838). Both are written in *terza rima*. His two-volume novel based on recent political events, *Der Kongreß von Verona*, which some have held to be his best work, appeared in 1842; his *Novellen* (1837) and his gently ironic tales of Dresden life, *Bilder im Moose* (2 vols., 1846), were also held in repute for a time. He owed his theatrical appointment to his verse plays, the first of which, *Heinrich der Finkler* (1836, on Heinrich I, q.v., der Vogler), was successfully revived in open-air performance in 1917. He published the tragedy *Cola Rienzi* in 1837 and a collection of plays (*Theater*) in 1842, which included *Kaiser Otto III*,

Die Bräute von Florenz, and *Wendelin und Helene*. Two late verse tragedies were *Herzog Bernhard von Weimar* (1855) and *Der Sohn des Fürsten* (1858) which dealt with Friedrich II (q.v.) of Prussia. Mosen is now virtually known only by four poems, *Die letzten Zehn vom vierten Regiment* (linked with the Polish revolt of 1831), *Andreas Hofer* (beginning 'Zu Mantua in Banden', first published 1831 and widely sung, see HOFER, A.), *Der Nußbaum* (set to music by R. Schumann, q.v.), and *Der Trompeter an der Katzbach*. The poems appeared in 1836 as *Gedichte*.

Sämtliche Werke (8 vols.) were published posthumously in 1863.

MOSENTHAL, Salomon, Ritter von (Kassel, 1821–77, Vienna), after studying in Karlsruhe, came to Vienna as a private tutor in 1843, obtained a civil service appointment in 1850, and rose to high office as Regierungsrat and custodian of the library of the Ministry of Education. Mosenthal was a skilful, though sentimental and conventional, dramatist, and is best known for his highly successful plays (Volksschauspiele, see VOLKSSTÜCK) *Deborah* (1849), which had a long run in London, and *Der Sonnwendhof* (q.v., 1857). He is also the author of a play based on the life of G. A. Bürger (q.v.), *Ein deutsches Dichterleben* (1850), and of some opera libretti, of which the best known is *Die lustigen Weiber von Windsor* (1849), composed by O. Nicolai (q.v.).

MOSER, Friedrich Karl (Stuttgart, 1723–98, Ludwigsburg), ennobled as Freiherr von Moser in 1767, was the son of J. J. Moser (q.v.). Moser spent his life in political and administrative service, notably in Hesse-Homburg and Hesse-Darmstadt. His *Der Herr und der Diener* (1759) criticized the misgovernment of the petty German states, and in the Heldengedicht in prose, *Daniel in der Löwengrube* (1763), he wrote a political parable directed against absolutism (see also NOVELLE by Goethe). Moser took the imperial side against Friedrich II (q.v.) of Prussia in *Vom deutschen Nationalgeist* (1765) and in *Patriotische Briefe* (1767). His *Über die Regierung der geistlichen Staaten von Deutschland* and *Geschichte der päpstlichen Nuntien in Deutschland* (both 1787) are critical of the spiritual principalities, coming out in favour of secularization.

MOSER, Gustav von (Spandau, 1825–1903, Görlitz), served as an officer in a Prussian Jäger battalion (1843–56), and in middle and later life was the author of numerous successful light comedies and farces. Many of these were written in collaboration with others, notably A. L'Arronge and F. von Schönthan (qq.v.). The best-known titles are *Der Veilchenfresser* (1874), *Der*

Bibliothekar (1878), and *Krieg im Frieden* (1881), which was written jointly with Schönthan.

MOSER, JOHANN JAKOB (Stuttgart, 1701–85, Stuttgart), a professor at Tübingen University at the age of 19, had a long and honourable career as an academic and as an administrator in various German states. He left Tübingen for Vienna, and was later in Stuttgart (1726) and Prussia (1736). From 1739 to 1747 he retired into private life at Ebersdorf. After a short period in Hesse-Homburg (1747–9) and another in Hanau (1749–51), he returned to Stuttgart, where his courageous opposition to the extortions of Duke Karl Eugen (q.v.) resulted in his imprisonment in Hohentwiel castle (1759–64).

Moser was distinguished for his Christian charity, his rectitude, and his extraordinary industry. His writings were mainly political. The *Grundriß der heutigen Staatsverfassung* appeared in 1731. While at Ebersdorf he wrote most of the 52 volumes of his *Deutsches Staatsrecht* (1736–53), to which he later added *Neues deutsches Staatsrecht* and *Zusätze*, amounting to a further 24 volumes. He is also the author of *Versuch des neuesten europäischen Völkerrechts* (1777–80). He is said to have originated the name 'Magus im Norden' for J. G. Hamann (q.v.). He was the father of F. K. Moser (q.v.).

MÖSER, JUSTUS (Osnabrück, 1720–94, Osnabrück), was a successful lawyer in administrative employment in the district of Osnabrück, who eventually (1783) reached the rank of Justizrat. A man of vigorous patriotism with a gift for persuasive popular expression, Möser wrote a patriotic tragedy (*Arminius*, 1749), but is especially known as the author of *Patriotische Phantasien* (1774–8), a series of newspaper essays published in collected form under this title by his daughter Jenny von Voigts (q.v.). From his *Osnabrückische Geschichte* (2 vols., 1768), a part of the preface was abstracted by J. G. Herder (q.v.) and published in the manifesto *Von deutscher Art und Kunst* (q.v., 1773). In his historical writing Möser gave weight to legal and economic factors. Dissatisfied with the essay *De la littérature allemande* by Friedrich II (q.v.) of Prussia, Möser published a rejoinder, *Über die deutsche Sprache und Literatur* (1781).

Twelve volumes of *Sämtliche Werke, historisch-kritische Ausgabe* appeared 1944–68.

MOSHEIM, JOHANN LORENZ VON (Lübeck, 1694–1755, Göttingen), became a professor at Helmstedt University in 1723 and at Göttingen in 1747. A noted preacher, he was also a theologian and an ecclesiastical historian. His principal work is the Latin *Institutiones historiae ecclesiasticae* (1726, rev. 1755), which appeared

in German translation (by J. R. Schlegel and J. Fraas) in 1770–88 (7 vols.). His *Versuch einer unparteyischen und gründlichen Ketzergeschichte* (2 vols.) was published in 1748–50.

Mouche, affectionate name given by H. Heine (q.v.) to Elise Krinitz (1830–97). Married early and abandoned by her husband, 'die Mouche' called on Heine in July 1855, bringing greetings from an admirer of the poet in Vienna, and she continued to visit him until his death. One of his most moving poems is superscribed *Für die Mouche*. Late in life, under the pseudonym Camille Selden, she published her recollections of Heine (*Les Derniers Jours de Henri Heine*, 1883).

MOZART, WOLFGANG AMADEUS (Salzburg, 1756–91, Vienna), Austrian composer and musical prodigy, toured Europe with his father and sister when he was seven, visiting, among other cities, London and Paris. In 1762 and 1768 he played to Maria Theresia (q.v.) at Schönbrunn. He composed numerous symphonies, concertos, and divertimenti, as well as much chamber and keyboard music. The catalogue of his works (*Köchel-Verzeichnis*, by L. Köchel), contains 626 K numbers, as well as a number of interpolations marked by an added 'A'. Of Mozart's liturgical music, all of which is to Latin texts, the best-known work is the unfinished Requiem Mass.

Mozart composed a number of songs with German words, including poems by Uz, J. C. Günther, Canitz, J. M. Miller, Hermes, C. F. Weiße, Goethe, Blumauer, Hagedorn, J. G. Jacobi, and Hölty (qq.v.). Most of his operas have Italian libretti, *Il rè pastore* (1775) by Metastasio, *Idomeneo, Rè di Creta* (1781) by Varesco; the libretti for *Le nozze di Figaro* (q.v., 1786), *Don Giovanni* (q.v., 1787), and *Così fan tutte* (1790) are by L. Daponte (q.v.). The libretto of the last Italian opera, *La clemenza di Tito* (1791), again goes back to Metastasio.

Mozart's two German operas are of Singspiel (q.v.) type: *Die Entführung aus dem Serail* (1782, text by G. Stephanie after C. F. Bretzner, qq.v.) and *Die Zauberflöte* (q.v., 1791), text by E. Schikaneder (q.v.).

Mozart auf der Reise nach Prag, a Novelle by E. Mörike (q.v.) which was published in Cotta's *Morgenblatt für gebildete Leser* and in book form in 1855 in time for the centenary of the birth of Mozart (q.v.). The incidents which befall Mozart on the journey to Prague for a performance of *Don Giovanni* (q.v.) are freely invented. Travelling by coach with his wife Konstanze, Mozart halts in a village and, while resting in the park of Count von Schinzberg, absent-mindedly picks an orange and cuts it in two. The angry gardener reports this untrust-worthy-looking stranger whose name, he understands, is 'Moser', to his master. As soon as the identity of Mozart is established he and his wife are invited to share in the festivities at the castle celebrating the engagement of Eugenie, a niece of the count, to a nobleman. The following morning the count presents Mozart with a coach, in which he resumes his journey to Prague.

The sophistication and playfulness of rococo, now nearing its end (it is 1787, two years before the French Revolution), form the essential background to Mörike's portrayal, penetrating with a fluid change of mood into the sphere of creative inspiration. Mozart's abundant fantasy and gaiety are focused in Susanna's aria in *Le nozze di Figaro* (q.v.), which Eugenie sings, and in the rustic dance of Zerlina's 'Giovinette, che fatte all' amore'; this Mozart had composed that day in the garden by the orange-tree which forms the central symbol of the Novelle. But in the composition of Don Giovanni's final supper Mozart apprehends supra-natural daemonic forces, by which Mörike fuses the artistic and historical elements of his story. His short poem 'Denk' es, o Seele!' (q.v.) concludes this Novelle.

MÜCHLER, KARL FRIEDRICH (Stargard, Pomerania, 1763–1857, Berlin), a civil servant in Prussia, and later in Saxony, wrote abundantly in his spare time. His innumerable publications consist chiefly of novels and stories, sometimes in large collections such as the five-volume *Kleine Frauenzimmerbibliothek* (1782–6). He also wrote poems, two of which, 'Der Wein erfreut des Menschen Herz' (1797) and 'Im tiefen Keller sitz' ich hier' (q.v., 1802) are well-known items in almost all German song-books. *Die Weihe der Unkraft von Z. Werner, nebst einer Antwort von einem Deutschen* (1814) is a response to Werner's recantation of his play on Luther, *Die Weihe der Unkraft* (1814, see MARTIN LUTHER, ODER DIE WEIHE DER KRAFT).

MÜGGE, THEODOR (Berlin, 1806–61, Berlin), intended to become an artillery officer, but abandoned this career in order to participate in Bolivar's struggle for freedom in South America. The campaign had ended successfully by the time Mügge reached London, and he returned to Germany, where he became active as a Liberal journalist, helping to found the *Nationalzeitung* in 1848. He wrote stories and several novels, of which the most notable were *Die Vendéerin* (1837), *Toussaint* (1840), *Der Vogt von Sylt* (2 vols., 1834–5), *Afraja* (1854), *Erich Randal* (1856), and *Der Prophet* (1860). His favourite theme is the tragic predicament of the revolutionary.

Mügge's *Gesammelte Novellen* (15 vols.) were published between 1836 and 1845. He is the author of books on travel, *Streifzüge in Schleswig-Holstein* (2 vols., 1846) and *Bilder aus Norwegen* (1856). His collected novels (*Romane*, 18 vols.) were published 1857–62, followed by a *Gesamtausgabe* (33 vols., 1862–7).

MÜHLBACH, LUISE, pseudonym of Klara Müller (Neubrandenburg, 1814–73, Berlin), who married in 1839 Theodor Mundt (q.v.), the publicist and academic associated with Junges Deutschland (q.v.). She was a well-known Berlin hostess in the middle of the century. An exceptionally prolific author of facile historical novels, her œuvre was (literally) extremely voluminous, several of her works appearing in ten or more volumes, and one, *Deutschland im Sturm und Drang*, in seventeen.

Mühlberg, near Merseburg, has given its name to the battle fought on the Lochauer Heide on 24 April 1547 between the forces of the Emperor Karl V (q.v.) and those of the Elector Johann Friedrich of Saxony. It ended in the defeat of the Saxons and the capture of the Elector. The battle put an end to the War of Schmalkalden (see SCHMALKALDISCHER KRIEG) and, simultaneously, to the League of Schmalkalden. The well-known equestrian portrait of Karl V by Titian purports to represent him at the battle of Mühlberg.

MÜHLPFORT, HEINRICH (Breslau, 1639–81, Breslau), registrar (Registrator) in his native city, wrote Latin love poems and German occasional poems modelled on those of Opitz (q.v.). They were published after his death (*Poemata*, 1686; *Teutsche Gedichte*, 1686).

MÜLLER, ADAM HEINRICH (Berlin 1779–1829, Vienna), ennobled (1826) as Ritter von Nittersdorf and known as Adam Müller, was the son of a Prussian civil servant, and was for a time a private tutor in various aristocratic families. His philosophical essay *Die Lehre vom Gegensatze* (1804) is said to have influenced Hegel (q.v.). From 1805, when he became a Roman Catholic while on a visit to Vienna, he lived in Dresden, collaborating in 1808 with H. von Kleist (q.v.) in publishing the periodical *Phöbus* (q.v.). In 1809 he published the political treatise *Elemente der Staatskunst*. From then until 1811 he was again in Prussia, engaged in politics and actively opposing the reforms initiated by Freiherr vom Stein and Hardenberg (qq.v.). In 1811 he returned to Vienna, becoming an Austrian civil servant and accompanying Metternich (q.v.) to Paris in 1815.

Müller held strong conservative and authoritarian views, in economics as well as politics, and advocated them in various publications, including *Theorie der Staatshaushaltung* (2 vols., 1812) and *Versuch einer neuen Theorie des Geldes mit besonderer Rücksicht auf Großbritannien* (1816). He strongly upheld the principle of nationality, believing the various states to exist by divine sanction. In this religious cast of political thought (demonstrated also in *Nothwendigkeit einer theologischen Grundlage der gesammten Staatswissenschaft*, 1819), he shows his affinity with Romanticism (see ROMANTIK). He was a friend of F. von Schlegel (q.v.), who shared his attitudes. Müller held various Austrian diplomatic posts up to 1827, after which he was employed in Vienna.

Kritische, ästhetische und philosophische Schriften (2 vols.), ed. W. Schroeder and W. Siebert, were published in 1967.

MÜLLER, FRIEDRICH (Kreuznach, 1749–1825, Rome), who usually called himself Maler Müller, developed a double talent, for painting and literature. The son of an innkeeper, who died early, he attracted attention by his drawings and was apprenticed in 1767 to the court painter at Zweibrücken, passing into the service of the Duke of Pfalz-Zweibrücken. He fell into disfavour, probably because he got an unmarried girl into trouble, left Zweibrücken and lived in Mannheim from 1774 to 1778, where he enjoyed the protection of the Elector Karl Theodor and the patronage of Goethe (qq.v.).

In Mannheim Müller's poetic talents awakened, and within a short time he wrote a number of works of diverse character. These include biblical idylls (*Der erschlagene Abel*, 1775, and *Adams erstes Erwachen und erste seelige Nächte*, 1778, q.v.), classical idylls (*Der Satyr Mopsus, Bacchidon und Milon*, and *Der Faun*, all 1775), and, above all, his Palatine idylls (Pfälzische Idyllen) *Die Schaaf-Schur* (q.v., 1775) and *Das Nußkernen* (q.v., 1811), which infuse a robust realism into an idealizing form. These works are in prose, which varies from the ecstatic to the rhythmic. The patriotic prose poem *Kreuznach* (1778) was described as a hymn (see HYMNE). His diffuse play *Golo und Genoveva* (q.v.) was begun at this time, and two scenes were published in 1776, but it was not finished until 1781 and not published until 1811. Müller's interest in Faust was first apparent in *Situation aus Fausts Leben* (1776), and his *Fausts Leben dramatisiert* (q.v.) was published in 1778. *Niobe. Ein lyrisches (musikalisches) Drama* appeared in the same year. Müller also wrote a number of poems in these years, of which *Soldatenabschied* ('Heute scheid' ich, heute wandr' ich') and *Die Zeugen* ('Du grün bewachs'nes Thal') are the best known.

In 1778, with the financial assistance of the Elector and of Duke Karl August (q.v.) of Saxe-Weimar and of Goethe, Müller went to Rome to advance his skill in painting. He never returned. He was converted to Roman Catholicism in 1781 (it has alternatively been suggested that the conversion took place secretly before his departure), and the alienation produced at home by this change of front was augmented by the receipt of boldly executed paintings lacking in harmony and finish. Müller made his living in Rome, partly by his painting, but mainly by acting as a guide to visitors. He also wrote art criticism, and one of his essays was published in 1797 in Schiller's *Die Horen* (q.v.).

Müller's works were published by L. Tieck (q.v.) and others as *Schriften* (3 vols., 1811). A selection entitled *Maler Müllers Werke*, ed. M. Oeser, appeared 1916–18 (2 vols.; a popular edition, Volksausgabe, 2 vols. in one, 1918), and the complete *Idyllen* (3 vols.), ed. O. Heuer, 1914. A two-volume edition, *Dichtungen*, by H. Hettner (q.v., 1868) was reissued in 1968. Müller's artistic *œuvre* includes religious and classical pictures and vigorous engravings of animals and landscapes.

MÜLLER, HEINER (Eppendorf, Saxony, 1929–   ), after years in administration worked as a journalist and editor, and from 1958 to 1959 at the Maxim-Gorki-Theater in Berlin. His main achievement lies in the field of drama. He began with a series of plays demonstrating the socialist Brigadestück. Indebted to Brecht's Lehrstück (q.v.), it aimed at stimulating constructive critical debate as the theatre's contribution to the DDR during its formative years. *Der Lohndrücker* (1957, in book form 1958), written in collaboration with his wife Inge (1923–66), was based on the achievement of the DDR's first workers' hero, the bricklayer Hans Garbe, the brigade worker Balke of the play, which is set in 1948–9 at the time of reconstruction. Balke's repair of a red-hot chimney on which the uninterrupted flow of production in a brick factory depends is sabotaged by fellow workers, who believe that Balke's determined effort damages their interests (of which the issue of wage claims is singled out in the play's title). Balke is subjected to humiliations, but resumes work with the worker who has beaten him up. The play's dialectics, however, raise wider issues of policy that remain open, including that of the motivation of Balke's heroism. The different treatment of the Garbe story (on which Brecht also collected material for a play in the early 1950s) and other sensitive aspects resulted in the play's suppression. *Die Korrektur. Ein Bericht vom Aufbau des Kombinats 'Schwarze Pumpe'*, written as a radio play in collaboration with Inge Müller (1957, published 1958), had to undergo revision (second version 1959), and the comedy *Die Umsiedlerin oder Das Leben auf dem Lande* (1961), based on a story by Anna Seghers (q.v.), was immediately suppressed and revised as *Die Bauern* (1964). *Der Bau* (1965, written 1963–4 for the Deutsches Theater) is based on the novel *Spur der Steine* (1964) by Erik Neutsch (b. 1931). First performed by the Berlin Volks-bühne in 1980, it is concerned with the work of a brigade in the context of DDR industrial policy in the early 1960s. An unborn child, disowned by his father for the sake of his position in the party, emerges in the final episode headed 'Schnee' as the symbol of a true socialist state of the future.

Müller's political adaptations of Greek plays include *Philoktet* (1965, written 1958–64), *Herakles 5* (1966), *Ödipus Tyrann* (1967, based on the translation of Sophocles' play by Hölderlin, q.v.), and *Prometheus* (1968). *Philoktet, Der Horatier* (1973, written in 1968) and *Mauser* (1976 BRD and USA, first performed in Texas 1975, written in 1970) demonstrate Müller's radical rejection of Brecht's Lehrstück, notably *Die Maßnahme* (q.v.), as being outmoded. The Horatian kills his enemy, thus securing Rome's victory, but he also kills his sister who mourns her betrothed in the victim. The people crown the Horatian as their victor and execute him as a murderer. *Mauser* adapts a theme from M. Sholokhov's novel *And Quiet Flows the Don* (German title: *Der stille Don*) and introduces a chorus and two figures designated A and B. A is about to be liquidated after having been relieved of his appointment as executioner, which involved the killing of his predecessor B for having felt compassion for a condemned peasant. The chorus and A, who protests against his impending death, debate its necessity. It is left open whether A dies convinced that he is an enemy of the revolution, but the play's quasi-Greek form implicitly queries the revolution's absolute judgement. *Zement* (1974), written for the Berlin Ensemble and based on a novel by F. Gladkov, was first performed on the 55th anniversary of the Russian Revolution which provides its setting and wide-ranging issues. Of these the depersonalization of the individual and the radical emancipation of women culminate in the episode entitled 'Medeakommentar'; it brings to a climax the motif of the returning soldier (Heimkehrer) Gleb Tschumalow, whose emancipated wife Dascha Tschumalowa denies her husband and herself the right of possession in love and sex, the family unit having already disintegrated through the death of their pre-revolutionary child. As in *Der Bau*, the cement of the title

symbolizes the construction of a new society by an intelligentsia that aims at the hardening of man. *Drachenoper* (1970, a libretto for the opera *Lanzelot* by P. Dessau, q.v.) and *Weiberkomödie* (1971, based on *Die Weiberbrigade*, a radio play by Inge Müller) were followed by *Macbeth* (1972), *Germania Tod in Berlin* (1977, completed 1971), *Die Schlacht. Szenen aus Deutschland* (1975), and *Leben Gundlings Friedrich von Preußen Lessings Schlaf Traum Schrei. Ein Greuelmärchen* (1977). Müller's increasing withdrawal from readily accessible communication culminates in the manipulation of Shakespeare's *Hamlet*, a motif favoured by contemporary West German writers, always subjective, but rarely more abstruse than in the short composition *Hamletmaschine* (1977) with its extreme pessimism. *Der Auftrag. Erinnerung an eine Revolution* (1979) is based on the story *Das Licht auf dem Galgen* by Anna Seghers. *Quartett* was published in 1981, *Verkommenes Ufer Medeamaterial Landschaft mit Argonauten* and *Wladimir Majakowski Tragödie* (in *Theater heute*, with an interview with K. Völker: *Ein Stück Protoplasma. Heiner Müller über Majakowski*) in 1983. *Geschichten aus der Produktion* (2 vols., 1974) includes prose and lyrics. Müller's numerous comments and discussions on his own plays and on his adaptations and translations, including his views on Brecht, have appeared in a variety of publications in East and West. Müller has become the most radical and effective representative of experimental theatre and its reliance on audience participation. He was awarded the Büchner Prize in 1985.

MÜLLER, JOHANN, see REGIOMONTANUS.

MÜLLER, JOHANNES VON (Schaffhausen, 1752–1809, Kassel), Swiss historian, who began as a schoolmaster, published his first historical work, *Die Geschichten der Schweizer*, in 1780; this later developed into the five-volume *Geschichten schweizerischer Eidgenossenschaft* (1786–1808) which takes the history of the Swiss to 1489. In 1807 he edited J. G. Herder's *Stimmen der Völker in Liedern* (q.v.).

Müller became a much sought-after scholar and held appointments successively in Kassel, Mainz, Vienna, Berlin, and again in Kassel. His title of nobility dates from 1791. His other historical writings include *24 Bücher allgemeiner Geschichte* (3 vols., 1811) and works on the Papacy and the League of Princes (see FÜRSTEN-BUND), *Darstellung des Fürstenbundes*, published anonymously in 1787. Among his admirers were Goethe and Schiller; the latter used Müller's Swiss history as one of his sources for *Wilhelm Tell* (q.v.).

Müller's works were edited by his brother J. G. Müller as *Sämtliche Werke* (27 vols.), 1810–

19, and (40 vols.), 1931–5. *Schriften in Auswahl*, ed. E. Bonjour, appeared in 1953, and correspondence, *Briefwechsel mit Herder*, ed. K. E. Hoffmann, in 1952.

MÜLLER, JOHANN GOTTWERT (Hamburg, 1743–1828, Itzehoe), usually known as Müller von Itzehoe, was a bookseller, first in Hamburg then in Itzehoe. Towards the end of his life he received a pension from King Frederick VI of Denmark. Müller is the author of a number of novels, of which the best known is the satire of Pomeranian squirearchy, *Siegfried von Lindenberg. Ein komischer Roman* (1779). Müller also translated stories from French and Spanish. He was a contributor to F. Nicolai's *Allgemeine Deutsche Bibliothek* (q.v.). His *Komische Romane aus den Papieren des braunen Mannes und des Verfassers des Siegfried von Lindenberg* (8 vols.) was published 1784–91.

MÜLLER, KLARA, see MÜHLBACH, LUISE.

MÜLLER, WENZEL (Tyrnau, Moravia, 1767–1835, Baden nr. Vienna), music director of the Theater in der Leopoldstadt in Vienna from 1786, was a prolific composer of Singspiele, the forerunners of operettas (see SINGSPIEL). He composed the music for Raimund's *Der Alpenkönig und der Menschenfeind* (q.v., 1828), including the famous sextet 'O leb denn wohl, du stilles Haus'.

MÜLLER, WILHELM (Dessau, 1794–1827, Dessau), son of a master tailor, began to study at Berlin University and then volunteered for the army in 1813, serving in a regular battalion (Garde-Jäger). He returned to his classical and Germanistic studies in 1814 and made many friends among the Romantic writers, including Arnim, Brentano, and Fouqué (qq.v.). In 1817 he visited Italy as companion to a nobleman and remained there until 1819. On his return to Dessau he became a schoolmaster and also held the post of ducal librarian. He married in 1821.

Müller is best known as a fluent and sensitive, but perhaps too facile, lyric poet. His early poems appeared in periodicals, and his first collection was *Siebenundsiebzig Gedichte aus den hinterlassenen Papieren eines reisenden Waldhornisten* (1821), which contains the cycle *Die schöne Müllerin* (q.v.). His *Lieder der Griechen* (1821–4), prompted by the Greek rising against the Turks, earned him the nickname 'Griechen-Müller'. The cycle of poems *Die Winterreise* (q.v.) first appeared in *Urania* in 1823 and was reprinted in *Zweites Bändchen der Gedichte aus den hinterlassenen Papieren eines reisenden Waldhornisten* (1824). Müller was also a scholar and a good linguist,

translating Marlowe's *Dr. Faustus* in 1818 and modern Greek folk-songs (*Neugriechische Volkslieder*, 1825). The *Homerische Vorschule* (1824) is an introduction to the study of Homer. Müller, whose poetry has had little attention for its own sake, is chiefly known through Schubert's settings of *Die schöne Müllerin* (1824) and *Die Winterreise* (1827, see SCHUBERT, F.). *Gedichte. Vollständige kritische Ausgabe*, ed. J. T. Hatfield, appeared in 1906.

MÜLLER, WOLFGANG (Königswinter, 1816–73, Neuenahr), generally known as Müller von Königswinter, was a physician in the Rhineland. In 1848 he was a member of the Frankfurt Parliament (see FRANKFURTER NATIONALVERSAMMLUNG), and in 1853 gave up his practice and turned to writing. He wrote several long narrative verse poems of a kind then popular with the reading public. They include *Rheinfahrt* (1846), *Die Maikönigin* (1852, a romance of village life which was his most popular work), *Prinz Minnewin* (1854), *Der Rattenfänger von St. Goar* (1856), and *Johann von Werth* (1858). He is the author of the poem beginning 'Mein Herz ist am Rhein'. A selection of his poems is contained in *Dichtungen eines rheinischen Poeten* (6 vols., 1871–6). His plays, which include the comedy *Der Einsiedler von Sanssouci* and the tragedy *Die Rose von Jericho* (both 1865), appeared as *Dramatische Werke* (6 vols.) in 1872.

Müller von Sanssouci, see ARNOLDSCHER PROZESS. Sanssouci refers to Friedrich II (q.v.).

MÜLLNER, ADOLF (Langendorf nr. Weißenfels, 1774–1829, Weißenfels), in full Amadeus Gottfried Adolf, was a nephew of G. A. Bürger (q.v.). He was educated at Schulpforta and studied law in Leipzig (1796–8). From 1798 to 1815 he was in legal practice in Weißenfels. He was an enthusiastic actor and in 1810 started a theatre at Weißenfels for amateur performances. He began his own writing with a novel (*Der Incest*, 1799) and then wrote, for private theatricals, a number of comedies (published with other plays in *Spiele für die Bühne*, 1815), one of which, *Die Verlobten*, was successful in professional performance in Vienna.

In 1812 he wrote a fate tragedy (see SCHICKSALSTRAGÖDIE) in avowed imitation of Werner's *Der vierundzwanzigste Februar* (q.v.). This horror play, *Der neunundzwanzigste Februar* (q.v.), intended to out-Werner Werner, aroused widespread repugnance, and Müllner remodelled it as *Der Wahn* (1818). Müllner's principal success was the fate tragedy *Die Schuld* (q.v.), written in 1812 and published in 1816. It was performed throughout Germany, and Müllner sought further successes with the tragedies *König*

Yngurd (1817) and *Die Albaneserin* (1820), noteworthy as extreme examples of a temporary aberration in public taste.

In his later years Müllner was active as a journalist, editing *Das Literaturblatt*, the literary supplement of Cotta's *Morgenblatt* (see MORGENBLATT FÜR GEBILDETE STÄNDE) from 1820 to 1825, and afterwards taking over the editorship of the *Mitternachtszeitung* (1826–9). In the year of his death he published the first German detective story, *Der Kaliber* (1829, see KRIMINALGESCHICHTE). Müllner's plays were published in *Dramatische Werke* in 1828 in 8 vols.; 4 vols. containing his poems and critical writings were added in 1830.

Mummenschanz, a parade of masked characters (or masquerade) performed on Shrove Tuesday. The best-known Mummenschanz in German literature is that in *Faust* (q.v.) Pt. II, Act I, in the scene *Weitläufiger Saal*.

MÜNCH, ANNA SIBYLLE (b. 1758), daughter of a patrician of Frankfurt, was long believed to have been Goethe's partner in a party game which consisted in arranging mock marriages by drawing lots. No name is given by Goethe in the passage in *Dichtung und Wahrheit* (q.v.) in which the game is mentioned, and it is now thought that the young woman concerned was Anna Sibylle's elder sister Susanne Magdalene (b. 1753). Goethe indicates that his partner in this game would have been welcome to his parents as a daughter-in-law, and he also credits her in the same work with stimulating him to write *Clavigo* (q.v.).

München (Munich), the capital of Bavaria, and the most important city of southern Germany, was founded in 1158. It has architectural monuments, dating (apart from the Gothic Frauenkirche) mainly from the 17th c. and 18th c. (Cuvilliés-Theater, Asamkirche, Theatinerkirche, Schloß Nymphenburg), and, thanks to a succession of Bavarian rulers interested in the arts, has outstanding collections of pictures (Alte Pinakothek etc.) and an active artistic and literary life (see MÜNCHNER DICHTERKREIS). The artists' quarter is Schwabing on the northern edge of the inner city.

The writers J. R. Becher, L. Feuchtwanger, and C. Morgenstern (qq.v.) were born in Munich. F. W. Schelling (q.v.) was a professor at the university for fourteen years, and J. J. Görres (q.v.) for twenty-two. The following lived for extensive periods in the city: C. M. Brentano, E. Geibel, W. H. Riehl (43 years), P. Heyse (60), Eduard Graf Keyserling, H. Bahr, F. Wedekind (28), H. Mann (27), Th.

Mann (40), and J. Ponten (20, qq.v.). R. Wagner (q.v.) was in Munich in 1864–5 as the protégé of Ludwig II (q.v.); F. von Dingelstedt (q.v.) directed the theatre from 1851 to 1857. Rilke (q.v.) spent the years 1914–18 in Munich, and G. Keller and B. Brecht were students there. In 1923 Munich was the scene of the Hitler-putsch (q.v.) and of the 'settlement' with Chamberlain, which strengthened Hitler's hand.

The university (Ludwig-Maximilians-Universität), established in Munich in 1826, is descended from Ingolstadt University (founded 1472), which was removed to Landshut in 1802 for twenty-four years.

Münchhausen, a novel written by K. L. Immermann (q.v.) in 1838–9 and published in 4 volumes also in 1838–9. Described as *Eine Geschichte in Arabesken*, it is divided into four parts, each of two books, and is preceded by an ironic motto from Horace's *De arte poetica*: 'Non fumum ex fulgore, sed ex fumo dare lucem/ Cogitat, ut speciosa dehinc miracula promat,/ Antiphatem Scyllamque et cum Cyclope Charybdim. (He does not seek to produce smoke from brightness, but brightness from smoke, so that splendid wonders are shown forth: Antiphates and Scylla and Charybdis with the Cyclope.)

An intentionally eccentric novel, it begins with chapters 11–15 and continues with 1–10. It is woven out of two elements, the comic, and even crazy, story of Münchhausen with its satire on contemporary trends and attitudes, and a solid story of Westphalian rural life (later published separately as *Der Oberhof*, 1863). The two strands alternate, running parallel. The incorrigible and indefatigable liar Münchhausen establishes himself in the castle of Schnick-Schnack-Schnurr, although he is not always welcomed by the eccentric owners, promising, when his host tires of him, to promote a company for the solidification of air. The extravaganza is pushed to the point at which three satirical figures of Hegelian philosophers appear in the novel as well as Immermann himself.

The story of the Oberhof contrasts with Münchhausen's lunacy in its four-square robustness. The farm (Oberhof), its inhabitants, and their customs, and especially a rural wedding, are described with relish. Within this framework the hunter (Jäger) Oswald and the foundling Lisbeth successfully maintain their love against the initial opposition of the village magistrate (Oberschulze) and their own misunderstandings.

MÜNCHHAUSEN, BÖRRIES, FREIHERR VON (Hildesheim, 1874–1945, Windischleuba nr. Altenburg, by suicide), a descendant of the noble family which produced the notorious inventor of tall stories (see MÜNCHHAUSEN, K. F. H. VON), studied at several universities (Heidelberg, Berlin, Munich, Göttingen) and then lived and travelled as an independent gentleman. In the 1914–18 War he served as a captain in the Saxon Horse Guards. After the war he was a well-to-do landowner. From 1897 to 1923 he edited a *Göttinger Musenalmanach* and was particularly active in writing historical, legendary, and contemporary ballads in rhythmical, memorable verse, embodying chivalrous and romantic values.

Münchhausen's *Balladen* appeared in 1901, *Die Balladen und ritterlichen Lieder* in 1908, *Das Herz in Harnisch* in 1911, and *Das Balladenbuch* in 1924. In *Meisterballaden* (1923) he set out his theory of the ballad. His *œuvre* was published posthumously in *Das dichterische Werk* (2 vols., 1950–3) and *Aus letzter Hand* (1959).

MÜNCHHAUSEN, KARL FRIEDRICH HIERONYMUS, FREIHERR VON (Bodenwerder, 1720–97, Bodenwerder), a member of an ancient North German noble family, served as an officer, taking part in two campaigns against the Turks, and travelled much in his early years. Settling on his ancestral estate, he became a mighty hunter and a great raconteur of extraordinary, improbable, or impossible stories, with which he regaled his guests. So well known did he become for his rodomontades that, as with Till Eulenspiegel (q.v.), he has developed into a figure of popular myth, and is sometimes referred to as 'der Lügenbaron'. He certainly told many tall stories, but some of those which are attributed to him were current long before his time. Seventeen of his (alleged) stories were published in the *Vademecum für lustige Leute* between 1781 and 1783, and in 1785 an expanded English version appeared, published at Oxford as *Baron Münchhausen's Narrative of his Marvellous Travels and Campaigns in Russia.* The compiler was R. E. Raspe (q.v.). This version, in its second edition, was translated and expanded by G. A. Bürger (q.v.) as *Wunderbare Reisen zu Wasser und zu Lande, Feldzüge und lustige Abenteuer des Freyherrn von Münchhausen* and first published (in London) in 1786. Bürger's version is the popular one and has contributed most to Münchhausen's fame as the inventor of absurdly impossible adventures.

Münchhausen's principal later reappearances in literature are with K. Immermann and W. Hasenclever (qq.v.). The poet Börries von Münchhausen (q.v.) was a descendant of the family.

Münchner Dichterkreis or **Dichtergruppe,** a group of poets, many of them from other parts of Germany, who wrote in Munich in the years *c.* 1850–64 with the active encouragement of King Maximilian II (q.v.) of Bavaria. They included E. Geibel, P. Heyse, M. Greif, H. Lingg, Graf von Schack, F. Dahn, and F. Bodenstedt and such minor figures as J. Große, W. Hertz, and H. Leuthold (qq.v.). They held meetings to discuss new work in an official Symposion under royal patronage, and in an unofficial club, the Krokodil. Their writings, largely lyric and epic poetry, are marked by conscious devotion to a narrowly conceived formal beauty, which assumes that reality is ugly and therefore unaesthetic.

Mundartdichtung, see DIALEKTDICHTUNG.

MUNDT, KLARA, see MÜHLBACH, LUISE.

MUNDT, THEODOR (Potsdam, 1808–61, Berlin), was educated at the Joachimsthalsches Gymnasium (q.v.) in Berlin and at Berlin University, and after graduating devoted himself to journalism, editing *Die Blätter für literarische Unterhaltung* (1832) in Leipzig and *Literarischer Zodiacus* (1835). A short politico-philosophical novel (*Moderne Lebenswirren,* 1834) satirized Germany as 'Kleinweltwinkel', and his feministic and sensual novel *Madonna* (1835) earned him inclusion in the federal denunciation of Junges Deutschland (q.v.) in December 1835.

From 1835 Mundt sought to qualify as a university teacher, encountering for years determined political opposition. He was finally admitted in 1842. Meanwhile he had continued his journalism with the periodicals *Dioskuren für Kunst und Wissenschaft* (1836–7), *Der Freihafen* (1838–44), and *Der Pilot* (1840–3), and had written a memoir of Charlotte Stieglitz (q.v., *Charlotte Stieglitz. Ein Denkmal,* 1836). His *Die Kunst der deutschen Prosa* urges that German written prose should discard its intricacies and base itself on speech. Mundt became a lecturer at Berlin in 1842, was appointed professor at Breslau in 1848 and at Berlin in 1850. He wrote abundantly on contemporary affairs and literature (*Charaktere und Situationen,* 1837; *Spaziergänge und Weltfahrten,* 1838–9; *Geschichte der Literatur der Gegenwart,* 1842; *Geschichte der Gesellschaft,* 1844; *Ästhetik,* 1845; *Geschichte der deutschen Stände,* 1854). *Ästhetik* was republished in 1966 and *Die Kunst der deutschen Prosa* in 1969 (both ed. H. Düvel). Mundt's later novels (*Thomsa Münzer,* 1841; *Mendoza,* 1846–7; *Die Matadore,* 1850; *Graf Mirabeau,* 1858) are of slight importance. In 1839 he married Klara Müller, who is known as a novelist under the pseudonym Luise Mühlbach (q.v.).

Munich, see MÜNCHEN.

Münster, city of the Land Nordrhein-Westfalen and the capital of the former Prussian province of Westphalia (1815–1945). In the 16th c. it was held for over a year (1534–5) by Anabaptists (Wiedertäufer), who proclaimed the Millennium (Tausendjähriges Reich) and maintained a reign of terror until they themselves were put down by terror. In 1648 the Peace of Westphalia (see WESTFÄLISCHER FRIEDE) was signed in the town hall. Münster again became prominent in the late 18th c., when it was noted for a Catholic circle centred on the Princess (Fürstin) Gallitzin (1748–1806, *née* von Schmettau). Goethe was greatly impressed by her and characterizes her in the first pages of the final section (*Münster, November 1792*) of the *Campagne in Frankreich* (q.v.). The university (Westfälische Wilhelms-Universität) was founded in 1773.

MÜNSTER, SEBASTIAN (Ingelheim, *c.* 1489–1552, Basel), originally a Franciscan friar, became a Protestant and taught theology and Hebrew at Heidelberg and, from 1529, at Basel. In 1534–5 he edited the first complete version of the Hebrew Bible (2 vols.) to appear by a German scholar. He was notable as a geographer producing in 1544 the *Cosmographia Universalis*, profusely illustrated with woodcuts and repeatedly reprinted, especially in the 16th c. and 17th c.

MÜNZER or **MÜNTZER,** THOMAS (Stolberg im Harz, 1490–1525, Mühlhausen, Thuringia), a social agitator and reformer, went over to the Reformation, but diverged from the religious reformers on the social application of their doctrine. He was pastor in various localities (Zwickau, Prague, Allstedt, Mühlhausen between 1520 and 1525), but his Christian communism, which gained him a considerable popular following, brought him into sharp opposition to the governing powers. In 1525 he led a local revolt in Mühlhausen, which spread throughout Thuringia as part of the Peasants' War (see BAUERNKRIEG). After defeat at Frankenhausen he went into hiding, was captured, and executed, having, under torture, recanted his opinions. In his polemics with Münzer and notably in his letter *Brief an die Fürsten zu Sachsen von dem aufrührerischen Geist,* Luther (q.v., 1524) condemns Münzer's revolutionary spirit. Münzer, who called himself in his appeals to the oppressed sections of the populace (mainly the peasants) 'Thomas Münzer, ein Knecht Gottes wider die Gottlosen', figures in a number of literary works, including the novel *Thomas Münzer* by Th. Mundt (q.v., 1841) and the plays *Luther und Thomas Münzer* by E. Lissauer (q.v., 1929) and *Thomas Münzer* by H. Eulenberg (q.v., 1932).

Thomas Münzer. Schriften und Briefe. Kritische Gesamtausgabe, ed. G. Franz, appeared in 1968, and *Politische Schriften,* ed. C. Hinrichs, in 1953.

Murbacher Hymnen, a collection of 26 Latin hymns in the manner of St. Ambrose, usually styled 'ambrosianische Hymnen', with word-for-word interlinear translation. The MS., which is in the Bodleian Library, is a copy of an original made in Reichenau (q.v.) and was probably in the ownership of the monastery of Murbach. It is presumed to have been written between 802 and 817.

MURNER, THOMAS (Oberehnheim, Alsace, 1475–1537, Oberehnheim), a learned Franciscan friar, satirized in his writings the follies of the age and, in later years, attacked Luther and the Reformation (qq.v.).

Murner grew up in Strasburg, studied theology in Freiburg, Cologne, Paris, Rostock, and Cracow, and later (1518) law in Basel. Between 1512 and 1515 he wrote a series of satirical rhyming poems, the first of which, *Die Narrenbeschwörung* (1512), is consciously modelled on Brant's *Das Narrenschiff* (q.v.). *Die Schelmenzunft* (1512) is a rogue's gallery and is appropriately bitter in tone. *Die Gauchmatt* (1515, published 1519) exposes the follies of lovers disporting themselves on the 'fools' meadow' (*Matt*'); it is particularly harsh on women, who are represented as perpetually scheming to subjugate foolish men. *Die Mühle von Schwindesheim* (1515) renews the satire of *Die Schelmenzunft*.

The division of the Church and the success of Luther drove Murner, who, though critical of abuses, was a faithful adherent of orthodoxy, to the bitter satire *Von dem großen Lutherischen Narren* (1522), which attacked its object, after the manner of the age, with grotesque and scurrilous exaggeration. Murner's Lutheran opponents applied to him in retaliation the sobriquet Murrnarr.

Murner wrote fluent and racy verse, employing homely images and familiar proverbs. He was a powerful and popular preacher, and his satirical style owes much to his pulpit manner. His works were plentifully illustrated with woodcuts based on his own sketches. His Latin theological works and his translation of Virgil (1515) have been overshadowed by his vernacular verse.

Murner's *Deutsche Schriften* (9 vols.), ed. F. Schultz, appeared in 1918–31, and *Satirische Feldzüge wider die Reformation,* ed. A. Berger, in 1932.

Murr, Kater, a literate tom-cat who is an important character in E. T. A. Hoffmann's novel *Lebensansichten des Kater Murr* (q.v.).

Musarion oder Die Philosophie der Grazien, a poem in three books by C. M. Wieland (q.v.), published in 1768. Set in a fictional Arcadia, it shows how Musarion, the charming representative of grace, tames and converts to a balanced sensualism the resentful, temporarily cynical Phanias. Simultaneously she shows up the hollow pretensions of Cleanth, a stoic, and Theophron, a Pythagorean philosopher. *Musarion* represents, according to Wieland, the poet's own view of life—'Ihre Philosophie ist diejenige, nach welcher ich lebe'. The poem, in three books of 1,441 lines in all, is written in accomplished, freely handled rhyming verse.

MUSÄUS, JOHANN KARL AUGUST (Jena, 1735–87, Weimar), studied theology at Jena, but spent his life in teaching. He became tutor to the Weimar court pages in 1763 and a senior teacher (Professor) at the grammar school (Gymnasium) in 1769. An amateur actor and a man of sociable temperament, Musäus was held in high regard in Weimar. His first book was a parody of Richardson's *History of Sir Charles Grandison*, *Grandison der Zweite* (1760–2), which is better known by the title of the second edition, *Der deutsche Grandison* (1781–2). The novel satirizes Richardson and his imitators, and, in the later version, also turns on the generation of Sturm und Drang (q.v.). *Physiognomische Reisen* (1778–9) uses the manner of Sterne to mock Lavater (q.v.).

Musäus is the author of fairy-tales, partly derived from printed and partly from oral sources and published as *Volksmärchen der Deutschen* (8 vols., 1782–6). They were reprinted, ed. P. Zaunert, in 1912 and 1965. A selection, *Märchen und Sagen* (2 vols.), appeared in 1972.

MUSCULUS, ANDREAS, real name Meusel (Schneeberg, 1514–81, Frankfurt/Oder), from 1542 taught theology at Frankfurt/Oder University, having first completed his studies at Wittenberg University. A staunch Lutheran, he was from 1556 also moderator (Generalsuperintendent). His tracts denouncing vice include *Der Hosenteufel* and *Der Eheteufel* (both 1556). He collaborated with S. Feyerabend (q.v.) in the *Theatrum Diabolorum* (1569).

Musensohn, Der, a poem written by Goethe in 1799, and published in *Neue Schriften* in 1800. It was set to music by F. Schubert (q.v.).

Musikalischer Hausschatz der Deutschen, an anthology of songs published in 1843 by G. W. Fink (q.v.). It was one of the most widely used song books in the 19th c. It contains many folk-songs as well as works of known authorship.

MUSIL, ROBERT, EDLER VON (Klagenfurt, 1880–1942, Geneva), who in his writing omitted the title gained by his father in 1917, was the son of an engineer, who eventually became a professor of his subject. Musil was sent to a cadet school, later discovered his mathematical and engineering abilities, and transferred to engineering studies at Brünn (Brno). For a short time (1902–3) a demonstrator at the Technical College (Technische Hochschule), Stuttgart, he studied mathematics, philosophy, and psychology at Berlin University, obtaining a doctorate in 1908, though he refrained from the pursuit of an academic career. In 1911 he took up a post as librarian at the Technical College in Vienna, which he abandoned in favour of an appointment to the editorial staff of *Die neue Rundschau* (q.v.) in Berlin. At the outbreak of the 1914–18 War he was called up for the Austrian army, serving chiefly on the Italian front, being decorated, and reaching the rank of captain. Evidence of Musil's response to the war is contained in the article *Europäertum, Krieg, Deutschtum* (1914); steeped in the traditions of imperial Austria, he emerged from the end of the war with an even greater consciousness of its heritage, in which he remained rooted throughout his life. From 1918 to 1922 he was a civil servant at the War Office, but lost this post as a result of the financial crisis, which also cost him his private income. In 1924 his parents died. For the remainder of his life Musil subsisted on his writings, and on journalistic work, which included contributions to the *Prager Presse*. From 1931 he lived in Berlin, moved in 1933 to Vienna, and in 1938 to Switzerland, staying in Zurich before settling in Geneva.

While still a student, Musil published the short novel *Die Verwirrungen des Zöglings Törleß* (q.v., 1906), and in 1911 *Vereinigungen*. During the 1920s he made his name with shorter works of fiction and two plays, *Die Schwärmer* (1920) and *Vinzenz oder die Freundin bedeutender Männer* (1921), which were performed in Berlin and Vienna. In 1924 he published a group of three stories, *Grigia*, *Die Portugiesin*, and *Tonka*, under the collective title *Drei Frauen*, and in 1927 a further ingenious and enigmatic tale, *Die Amsel* (q.v.). Apart from his shorter writings, however, his real preoccupation was with the novel that became his life's work, *Der Mann ohne Eigenschaften* (q.v.); it is also the work through which interest in him was revived after the 1939–45 War. The first volume of this work (never completed, but constantly revised and replanned) appeared in 1930, and part of the second volume in 1933, the remainder being published posthumously in 1943. An extended version was edited by A. Frisé in 1952 (revised 1965). In it imperial Austria of 1913 found a brilliant analyst.

Musil's principal characters, including those of his plays, are obsessed by being different from others who typify the society to which they belong, and experience reality as a labyrinth of irrationality, for which Musil develops a complex symbolical style. At his best a writer of extraordinary discipline and intellectual capacity, he suffered from the awareness that he could not achieve in his novel the comprehensive vision for which he strove. He entitled a collection of 1936, during which time his health had already been affected, *Nachlaß zu Lebzeiten*.

Gesammelte Werke in Einzelausgaben (3 vols.), ed. A. Frisé, appeared in 1952–7 and contained his novel (vol. 1), *Tagebücher, Aphorismen, Essays und Reden* (vol. 2), and *Prosa, Dramen, Späte Briefe* (vol. 3); *Theater. Kritisches und Theoretisches*, ed. M.-L. Roth, appeared in 1965. A. Frisé edited *Tagebücher* (2 vols., 1976), *Briefe* (2 vols., 1981), and *Gesammelte Werke* (9 vols., 1978).

MUSKATBLÜT or MUSKATPLÜT, lived in the late 14th c. and early 15th c., and probably came from east Franconia, the north-eastern region of modern Bavaria. He visited Mainz and was at Constance at the time of the Council (1414–18, see KONSTANZER KONZIL).

Muskatblüt is the author of more than a hundred poems, written for various courts between 1410 and 1438. Many of his songs are religious, and include mystical poems in praise of the Virgin Mary. His love-songs owe something to Minnesang (q.v.) and contain, like some of the Marienlieder, conventional passages of nature poetry. His most interesting poems are his Rügelieder in which he criticizes contemporary life and morals, castigating all classes of society and directing his satire chiefly against lust and the love of money.

Muskatblüt's *Lieder* were published by E. von Groote in 1853.

MUSOPHILUS, see GERBEL, N.

Muspilli (destruction of the world by fire), title given in 1832 by its first editor, J. A. Schmeller, to a fragment of 103 lines of an eschatological poem written in alliterative verse (see ALLITERATION). The extant text comprises the middle section; the beginning and end are lost. The poet first describes the struggle between angels and devils for a departed soul, and the bliss or torment that awaits it, passes on to the Day of Judgement and the end of the world, describing briefly the battle between Elias and Satan, and then portrays God's appearance at his judgement seat. The rhetorical tone is partly that of pulpit oratory.

The text is corrupt and *Muspilli* provides a wide field for philological dispute; even the

title-word, which occurs in line 57, is the subject of inconclusive argument. The poem is believed to represent a late and degenerate development in alliterative verse and to have been written in the first half of the 9th c. It may be an adaptation of a lost Anglo-Saxon poem. The MS., which is in Munich, originated in the monastery of St. Emmeram in Regensburg and was written in Bavarian dialect towards the end of the 9th c. It is thought likely that the original was written in Fulda, but proof is lacking.

Mußpreußen, name given humorously or in bitterness to those inhabitants of German states which after the War of 1866 (see DEUTSCHER KRIEG) were annexed to Prussia. The states so embodied were Hanover, Electoral Hesse (Kurhessen), Nassau, and Frankfurt.

MUTH, KONRAD, see MUTIANUS RUFUS.

MUTIANUS RUFUS (Homberg, Hesse, 1471–1526, Gotha), latinized his name from Konrad Muth; the 'Rufus' refers to the colour of his hair. Mutianus was a prominent humanist in Erfurt. He probably collaborated in, and certainly sympathized with, the *Epistolae obscurorum virorum* (q.v.). He gave Luther partial support, though he was apprehensive of the effects of a complete rift in the Church. From 1503 he was a canon of Gotha.

Mutmaßungen über Jakob, a novel by U. Johnson (q.v.), published in 1959. Its crucial episode, hinted at at the beginning and expressly stated near the end, is the death of Jakob Abs, a young German railway signal-foreman, who is run down by a train while crossing the tracks in misty weather. The underlying theme of the book is the coexistence of the two disparate Germanies, neither of which can understand the other, and it is significant that the important junction at which Jakob meets his death carries the through traffic between East and West. The fog in which he dies is a symbol of the difficulty of communication between the two Germanies, and corresponds to the deliberately obscure structure of the novel. This very original work is written in a series of multiple perspectives, unspecified conversations, internal monologues, and anonymous speculations (*Mutmaßungen*), in which the reader is only occasionally told who is speaking and is himself obliged to infer and speculate. The simple action is also geographically linked by Jakob's love for Gesine Cresspahl, who has gone to West Germany. She pays a risky and fleeting visit to her home in 'Jerichow' in Mecklenburg, and Jakob goes surreptitiously to visit her in Frankfurt, a visit which has important consequences in the later novel

Jahrestage (q.v.). His death occurs on his return. A sub-plot involves Jakob's friend, the university lecturer Dr. Jonas Blach, who attends a meeting of mild political dissent, is dismissed, and, at the end of the book, arrested.

Jakob emerges as a likeable, industrious, quiet, and highly efficient railway technician, a convinced socialist, accepting what he has been taught about politics at school, yet bewildered and dissatisfied by the way in which it is applied by the East German government. At the same time his principles debar him from remaining in the West with Gesine and accepting its conditions of life. The ruthlessness of the East German regime (as well as its subservience to Soviet Russia) is hinted at, and the reader is left uncertain whether Jakob's death was an accident, a political liquidation, or a suicide. The occurrences are fixed in 1956 by occasional mention of the Hungarian rising and the Franco-British Suez operation.

Mutter, Die, a play in 14 episodes in prose and verse by B. Brecht (q.v.) described as *Leben der Revolutionärin Pelagea Wlassowa aus Twer* and based on M. Gorky's novel *Mother* (1907). It was written (in collaboration with G. Weisenborn, q.v.) in 1931 and first performed on 17 January 1932, the anniversary of the death of 'the great revolutionary Rosa Luxemburg' (*Versuche*, vol. 7) in Berlin (Theater am Schiffbauerdamm). (See LUXEMBURG, R.)

The worker Pavel Vlassow and other collaborators in .the revolutionary movement win his mother's support, which becomes stronger after Pavel's arrest for his part in the demonstrations of 1 May 1905, his exile in Siberia, and his death on his flight towards the Finnish frontier. Rejecting the consolation of religion, Pelagea controls her grief at her son's death, for the cause to which she is committed can only be promoted by rational conduct; it is better for her son to have died than to lead a bad life with religion, which in her (and her party's) view has done nothing to relieve the lot of the poor. Her sufferings nevertheless cause her to fall ill. When she hears of the Tsar's mobilization and the outbreak of the war, she, too, decides to mobilize: the sick mother decides to lead the sick party and is braver than the workers although she is beaten up by the police. In 1917 the 60-year-old 'mother of the revolution' heads with the red flag the anti-war demonstrations of the strikers and naval deserters, and appeals in the epilogue for an all-out fight for victory.

Mütter, Die, a play by G. Hirschfeld (q.v.), published in 1896. It deals with a weakling who wavers between his mother and the unmarried mother of his future child.

Mütter, Die, goddesses in Goethe's *Faust* (q.v.), whom Faust must visit in order to summon the spirits of Helen and Paris for the entertainment of the Emperor (Pt. II, Act I, *Finstere Galerie*). These goddesses, whom Goethe invented, are referred to by Mephisto: 'Göttinnen, ungekannt/ Euch Sterblichen, von uns ungern genannt.'

Mutter Courage und ihre Kinder, a play by B. Brecht (q.v.), written in exile in Denmark 1938–9. The revised version was first produced in the Schauspielhaus, Zurich, in April 1941. Its first performance in Germany was in the Deutsches Theater, Berlin, in January 1949. The play consists of twelve scenes, each of which is preceded by chapter headings (Spruchbänder) giving a summary of the plot (Fabel). Brecht freely adapted the principal figure, Courasche, from the novel *Trutz Simplex* (q.v.) by J. C. Grimmelshausen.

The play covers some twelve years (1624–36) of the Thirty Years War (see DREISSIGJÄHRIGER KRIEG), depicting at various locations in Sweden, Poland, and Germany (among them Magdeburg, Ingolstadt, the Fichtelgebirge, and Halle), Mutter Courage's struggle for existence. A *cantinière*, she follows the troops engaged in fighting for the Protestant cause through victory and defeat, trailing the covered wagon (Planwagen) which houses her goods and is the family home. She has three children of different paternity, Eilif, Schweizerkas, and the dumb Kattrin, all of whom are, at different times and for different reasons, killed. Eilif is executed during the short-lived truce after the battle of Lützen for his brutality against peasants, which had brought him credit during the fighting. Schweizerkas dies for his honesty and loyalty: he refuses to surrender the regimental cash-box to the enemy, whose side his mother has prudently and temporarily taken. Kattrin dies when she asserts her compassion: she is shot on the roof of a farmhouse outside the city of Halle as she successfully beats the alarm, warning the inhabitants for the sake of their children of the approaching imperial forces. At the end of the play Mutter Courage alone pulls the wagon, which at the opening was drawn by her sons, determined to follow her regiment, to restore her business, and to find Eilif, of whose execution she is ignorant.

A number of characters, denoted by their rank and occupation rather than name, support the action, which is designed to expose the horrors of war as experienced by the ordinary people, soldiers, and peasants. Mutter Courage depends on the war for her living; her business fluctuates with the ups and downs of the fortunes of her regiment, but peace threatens to ruin it. Brecht leaves it to his audience to comprehend the futility. Several songs, among them *Das Lied vom Weib und dem Soldaten, Das Lied vom Fraternisieren, Das Lied von der Großen Kapitulation,* and *Das Lied von Salomon* support the didactic intentions of the play, and 'Eia popeia', ·Mutter Courage's cradle song by her daughter's body, marks its most moving climax. Brecht's comments on the play include his instructions for production contained in the *Couragemodell* of 1949 (1958) and this statement: 'Die Courage—dies sei gesagt, der theatralischen Darstellung zu helfen—erkennt zusammen mit ... nahezu jedermann das rein merkantile Wesen des Kriegs: das ist gerade, was sie anzieht ... Dem Stückeschreiber obliegt es nicht, die Courage am Ende sehend zu machen—, ... ihm kommt es darauf an, daß der Zuschauer sieht.' (*Anmerkungen zu 'Mutter Courage und ihre Kinder'.* See also EPISCHES THEATER.)

Mutter Erde, a play by M. Halbe (q.v.), published in 1897. Paul Warkentin, who has made a marriage which has gone stale, revisits his rural home district and meets again his former love Antoinette. The two see no possible future, since Stella, Paul's wife, will not give way to Antoinette, and the latter and Paul take their own lives together.

Mutter und Kind. Ein Gedicht in sieben Gesängen, an epic idyll (see IDYLLE) by F. Hebbel (q.v.), published in 1859. In 1857, when the work was still in MS., it received the Tiedge prize (see TIEDGE, C. A.) established by the Tiedge-Stiftung for a worthy successor of Goethe's *Hermann und Dorothea* (q.v.). Not unlike the latter, it has a contemporary background

and serves a didactic purpose. Its original setting, Gmunden near Vienna, was transposed to a less specific place of action in the Harz mountains before its publication.

A well-to-do merchant and his wife decide to adopt a child to end the misery of their childless marriage. Their maid, Magdalena, loves Christian, a man of her own class, but poverty prevents them from marrying. The merchant enables the couple to marry on condition that he and his wife adopt their firstborn. Magdalena, having given birth to a son, cannot face the parting and runs away with the baby. Christian finds and understands her. Their common feeling of parental love and duty proves stronger than the contract with their benefactors, and they go into hiding. In the end the merchant and his wife face the situation with magnanimity and indeed repentance, and succeed in finding and helping the fugitives.

MYLIUS, CHRISTLOB (Reichenbach, 1722–54, London), a cousin of G. E. Lessing (q.v.), was with the latter in Leipzig (1746–8) and in Berlin (1748–53). Mylius was active as a journalist, a reviewer, and a writer of comedies (*Die Ärzte,* 1745; *Der Unerträgliche,* 1746; *Der Kuß,* 1748; and *Die Schäferinsel,* 1749). After his premature death, Lessing published a selection of his works (*Vermischte Schriften,* 1754).

MYNSINGER, HEINRICH, see VON DEN FALKEN, PFERDEN UND HUNDEN.

MYSTIFIZINSKY, DEUTOBOLD SYMBOLIZETTI ALLEGOROWITSCH, a pseudonym of F. Th. Vischer (q.v.).

N

NABL, FRANZ (Lautschin, Bohemia, now in Czechoslovakia, 1883–1974, Graz), was a journalist before becoming an author writing fiction with Austrian settings in realistic style and with a marked feeling for nature. His best-known novel is *Ödhof* (2 vols., 1911–14) which followed *Hans Jackels erstes Liebesjahr* (1908); *Das Grab des Lebendigen* (1917) was republished in 1936 as *Die Ortliebschen Frauen. Die Galgenfrist* (1921) and *Ein Mann von Gestern* (1935) are novels, and his later shorter fiction includes *Johannes Krantz* (1948, enlarged edn. 1958) and *Das Rasenstück* (1953). Nabl wrote two plays, *Trieschübel* (1925) and *Schichtwechsel* (1929), and autobiographical

works, among them *Steirische Lebenswanderung* (1938) and *Die zweite Heimat* (1963). Select *Werke* (4 vols.) appeared in 1966.

'Nach Canossa gehen wir nicht', see CANOSSA.

Nach dem Großen Kriege, a story in letter form (see BRIEFROMAN) written by W. Raabe (q.v.) in 1860 and published in 1861. It is described as 'Eine Geschichte in zwölf Briefen', and has as its background the Napoleonic Wars (q.v.) and their aftermath. The twelve letters purport to be written in 1816 and 1817 by

Fritz Wolkenjäger to his friend Sever. Fritz has a new appointment as schoolmaster and tells of his attraction to and eventually of his courtship of Annie, a foundling picked up on the battle-field of Talavera and cared for by Wolfgang Bart, a soldier of the German Legion. Annie lost her memory on the day of Talavera. With Fritz she hears the dark past of the house of Rhoda, especially the action of a Herr von Rhoda, who arrested and helped to condemn one of Schill's hussars (see SCHILL, F.). Her memory returns and she realizes that she is Anna von Rhoda. Herr von Rhoda, who proves to be her father, is found unconscious near the scene of the arrest and dies. Anna and Fritz marry with hopes of a brighter future after the sombre events of past years.

Nach dreißig Jahren, a group of three stories by B. Auerbach (q.v.), published in 1876 and later included in vols. 9 and 10 of *Sämtliche Schwarzwälder Dorfgeschichten* (1884). Each deals with the later life of characters from an earlier story. *Des Lorles Reinhard,* a sequel to *Die Frau Professorin* (q.v.), records the death of the ageing Reinhard on the threshold of a second marriage. In *Der Tolpatsch aus Amerika,* Tolpatsch's son, Aloys the younger, returns from America to his father's village and marries Marannele, the daughter of the Marannele who let down his father (see TOLPATSCH, DER). *Das Nest an der Bahn* links up with the story *Sträflinge* (q.v.).

Nachfolgestaaten, the succession states which, by the treaties of St. Germain and Trianon, were established in 1919 on former Austro-Hungarian territory (Austria, Hungary, Czecho-slovakia) or took over large areas of it (Poland, Rumania, Yugoslavia). Italy, which acquired South Tyrol, was nevertheless not included among the succession states, which were held responsible for proportionate amounts of the Austro-Hungarian national debt.

'Nach Frankreich zogen zwei Grenadier', first line of H. Heine's poem *Die Grenadiere,* written *c.* 1820 and included as No. VI of *Junge Leiden* in *Das Buch der Lieder* (q.v., 1827). The well-known setting is by Robert Schumann (q.v.).

Nachricht von den neuesten Schicksalen des Hundes Berganza, a dialogue written by E. T. A. Hoffmann (q.v.) in 1813 and published in 1814 in vol. 2 of *Fantasiestücke in Callots Manier.* The title alludes to a dialogue between two dogs ('El coloquio de los perros'), one of whom is named Berganza, in one of Cervantes's *Novelas ejemplares* (1613). Hoffmann read Cervantes's dialogue in a German translation,

Lehrreiche Erzählungen (1800–1) by D. W. Soltau (1745–1827). Hoffmann's dialogue takes place, not between two dogs, but between the narrator and Berganza. It is mainly concerned with the contrast between dilettantism and devotion to art, but it also gives a thinly dis-guised account of the behaviour of Madame Marc in marrying her daughter Julia (see MARC, J., Cäcilie in the dialogue) to a wealthy, uncivilized drunkard.

Nachsommer, Der, a novel by A. Stifter (q.v.), published in 1857. The hero Heinrich Drendorf is the narrator. He traces his development from a happy childhood in a balanced family environ-ment, through a serious life devoted to studies, chiefly botanical, geological, and meteorological, to a mature serenity achieved in marriage with Natalie, the woman he loves. It is a Bildungs-roman (q.v.), in which Heinrich's education proceeds steadily and without setbacks. Linked with the story of Heinrich is that of another and less obviously happy pair of lovers.

On one of his journeys Heinrich is entertained at a country-house (das Rosenhaus) inhabited by a solitary nobleman, Freiherr von Risach, who, having loved but missed fulfilment, spent his prime working for the state, and now in his retirement devotes himself to the loving care and management of the Rosenhaus and the attached estate. In age he discovers his youthful love Mathilde, who is Natalie's mother, and learns that his early love was returned; though the couple do not marry they maintain a harmonious friendship, experiencing the Indian summer to which the title *Der Nachsommer* alludes. It is a long and uneventful novel which steadily pursues its course in serene and tranquil beauty.

Nacht, Die, a poem by J. von Eichendorff (q.v.), the first line of which runs 'Wie schön, hier zu verträumen'. It is a magical evocation of the forest night which the poet dreams through. It occurs (without title) in Eichendorff's novel, *Dichter und ihre Gesellen* (q.v., 1834). He also wrote a cycle of four poems entitled *Nacht.*

Nachtbüchlein, a collection of Schwänke (see SCHWANK) by Valentin Schumann (q.v.).

Nachtfeier der Venus, a verse translation by G. A. Bürger (q.v.) of the anonymous Latin poem *Pervigilium Veneris* (*The Vigil of Venus*). It was first published in *Der teutsche Merkur,* 1773.

NACHTIGAL, KONRAD, a Meistersinger (see MEISTERGESANG) of the 16th c., is mentioned by

Hans Sachs (q.v.). A baker by trade, he was a notable musician, whose tunes were used by other Meistersinger. As Kunz Nachtigal he appears in Richard Wagner's *Die Meistersinger von Nürnberg* (q.v.).

Nachtigall, Die, a short verse tale of the late 13th c. of mildly erotic character. Written with some charm, it deals with two young people who sleep together and are then united in marriage by their parents.

Nachtigallen, Die, a poem by J. von Eichendorff (q.v.), the first line of which runs 'Möcht wissen, was sie schlagen'. It occurs (untitled) in the story *Die Glücksritter* (q.v., 1841). There is also a poem by Eichendorff entitled *Nachtigall* ('Nach den schönen Frühlingstagen').

Nächtliche Heerschau, Die, a poem by J. C. von Zedlitz (q.v.), first published in the *Taschenbuch für Damen,* 1829, and included in his collection *Gedichte* (1832). A product of the Napoleonic legend, it begins 'Nachts um die zwölfte Stunde'.

Nachtlied, a poem by J. von Eichendorff (q.v.), the first line of which runs 'Vergangen ist der lichte Tag'. It is a religious song, in which poet and nightingale together praise God. It occurs (untitled) in the novel *Ahnung und Gegenwart* (q.v., 1815).

Nachtstücke, a collection of stories by E. T. A. Hoffmann (q.v.), published in 2 volumes in 1816–17. Volume 1 contains *Der Sandmann* (q.v.), *Die Jesuiterkirche in G., Das Sanctus* (q.v.), and *Ignaz Denner* (q.v.); vol. 2: *Das öde Haus, Das Majorat* (q.v.), *Das Gelübde,* and *Das steinerne Herz.*

Nachtwachen. Von Bonaventura, an anonymous work, published in 1804 (date on titlepage 1805). Its authorship, attributed to F. G. Wetzel (q.v.) and since the 1970s to E. A. Klingemann (1777–1831), is still uncertain. An ingenious product of Romanticism, it has also been attributed to F. W. J. Schelling, E. T. A. Hoffmann, and C. Brentano (qq.v.). It exhibits in detached scenes, observed by the watchman during 16 nights (Nachtwachen), the corruption and hypocrisy of life, but also its values, though it emerges as a nihilistic work.

Nähe des Geliebten, a poem written by Goethe in 1795, and first printed in the *Musenalmanach für das Jahr 1796.* Goethe composed it after reading the poem *Ich denke dein* by Friederike Brun (q.v.). He retained the opening words

and the alternation of long and short lines. The poem has been set by Beethoven (q.v.).

Naiv, term used by Schiller as a category for the characterization of poets and poetry. See ÜBER NAIVE UND SENTIMENTALISCHE DICHTUNG.

NAKATENUS, WILHELM (München-Gladbach, 1617–82, Aachen), a Roman Catholic priest, a professor at Münster University, and a popular preacher, was the author of a book of devotional poems, *Himmlisch Palmgärtlein* (1660), which continued to be reprinted up to the end of the 19th c. Nakatenus published the poems of Friedrich SPEE von Langenfels (q.v.) as *Trutznachtigall* in 1649, after the poet's death.

Nänie, a poem in elegiac metre by Schiller, first published in *Gedichte,* 1800. It is a dignified and moving lament for the passing of beauty.

NAOGEORGUS, THOMAS, real name Kirchmair (nr. Straubing, c. 1508–63, Wiesloch), a Protestant cleric, held cures in Saxony (Salza and Kahla) and, later, in Bavaria (Kaufbeuren and Kempten). He is the author of Latin plays attacking the papacy and the higher Catholic clergy. His early works *Pammachius* (1538) and the comedy *Mercator* (1540) are the most successful. *Pammachius* is left incomplete, because the religious conflict which it embodies was still undecided. *Mercator* is the story of Everyman (see JEDERMANN). Among Naogeorgus's later polemical plays are *Incendia* (1541), *Hamanus* (1543), *Hieremias* (1551), and *Judas* (1552). Naogeorgus also translated the tragedies of Sophocles into Latin. *Werke* (9 vols.) began to appear in 1975.

Napoleonic Wars, the wars which Napoleon conducted in Germany, began in 1805, four years after the termination of the Revolutionary Wars and after Napoleon I, proclaimed emperor in 1804, had consolidated his position at the Reichsdeputationshauptschluß (see REVOLUTIONSKRIEGE). From 1803 Napoleon tried to implicate Prussia in the war between France and England; in 1805 he offered Hanover to Friedrich Wilhelm III (q.v.), who refused to accept it. During the war of the Third Coalition (England, Austria, Russia, and Sweden) against France Napoleon directed his troops through Prussian Ansbach after defeating the Austrians at Ulm on 17 October 1805. Following this violation of Prussian territory, Friedrich Wilhelm prepared for action. He expressed to Russia his intention of supporting the Coalition (Treaty of Potsdam, 3 November 1805), and agreed to accept British subsidies in return for Prussian military assistance and to present his terms to Napoleon; these included French recognition of the independence of Germany as well as French concessions in Italy in favour of Austria. But Napoleon's victory over the Austrians at Austerlitz (2 December 1805) enabled him to impose his own terms on Austria in the Treaty of Pressburg, and on Prussia in the Treaty of Schönbrunn.

In the Treaty of Preßburg Napoleon forced Austria to cede Venetia to Italy, to cede Tyrol and Vorarlberg as well as other temporal and ecclesiastical principalities to the newly created kingdom of Bavaria, and to hand over some of its German territories to Württemberg, which he raised to a kingdom, and to Baden.

In his treatment of Prussia in the Treaty of Schönbrunn Napoleon was contemptuous, forcing Prussia to accept Hanover, and to close its northern ports to British ships and cargoes, thus attempting to force Prussia into war against Great Britain. Prussia had also to cede Ansbach to Bavaria.

Napoleon continued his reorganization of Germany by the creation of the Confederation of the Rhine (Treaty of Paris, 17 July 1806; see RHEINBUND), which provided him with backing among the western and southern German states. While the Confederation supported Napoleon by its subordination to French foreign policy and military command, Napoleon dissolved the Imperial Diet at Regensburg (1 August 1806). By formally renouncing the title Holy Roman Emperor (having previously assumed the title Emperor Franz I of Austria) Franz II (q.v.) completed the dissolution of the old Empire (6 August 1806; see DEUTSCHES REICH, ALTES).

To boost its position, Prussia hoped for the creation of a confederation of the northern states with the Hohenzollern king as emperor. Instead, Napoleon used Prussian-occupied Hanover as a bait in his peace negotiations with Great Britain. This breach of the Treaty of Schönbrunn caused Friedrich Wilhelm to mobilize. When Napoleon provoked Prussia still further by ordering the execution of the Nürnberg bookseller Palm (q.v.) for having sold an anonymous pamphlet, *Deutschland in seiner tiefsten Erniedrigung*, Friedrich Wilhelm declared war on France (1 October 1806). Within three weeks Napoleon disposed of Prussia's military forces. The decisive battles were fought on 14 October 1806 at Jena and Auerstedt (qq.v.). Prussian fortresses surrendered one after another, among them Erfurt, Halle, Spandau, Prenzberg, Stettin, Lübeck, and Magdeburg. Only Kolberg (q.v.) held out until the end of hostilities (1807). On 25 October 1806 Napoleon entered Berlin and decreed the Continental

Blockade (Kontinentalsperre) against England. Friedrich Wilhelm and Queen Luise (q.v.) fled to East Prussia.

After defeat at Friedland (14 January 1807) Russia withdrew from the war, and Tsar Alexander met Napoleon on a raft on the Niemen to fix the terms for the Treaty of Tilsit, which Prussia had to accept on 9 July 1807. Prussia ceded its territory west of the Elbe to the newly created kingdom of Westphalia, in which Napoleon installed his brother Jérome as king. Prussian territories in Poland (see POLAND, PARTITIONS OF) were turned into the Grand Duchy of Warsaw, which was given to the Duke of Saxony. Danzig (q.v.) was declared a Free City and remained under French occupation. At Alexander's persuasion Napoleon refrained from annihilating Prussia, but he reduced it to about half its size.

By 1808 Europe was ruled by the French emperor and his family. In the autumn of that year Napoleon held a congress at Erfurt, making sure of Alexander's support, staging splendid festivities, and receiving homage and admiration from many Germans, including Goethe and Wieland (qq.v.). From here he moved to liberated Spain. This first sign of revolt against Napoleon, resulting in the deposition of his brother Joseph as king of Spain, was followed by a peasant revolt in Tyrol (January 1809) and by another Austrian declaration of war (9 April 1809), obliging Napoleon to return from Spain. In Germany French troops suppressed the revolts of Baron von Dörnbert (1768–1850) in Hesse, of the Prussian major Ferdinand von Schill (q.v.) in Stralsund, and the young Duke Friedrich Wilhelm von Braunschweig (see BRAUNSCHWEIG) which followed the national appeal of Archduke Karl (q.v.) of Austria.

On 13 May 1809 Napoleon moved into Vienna, but Austrian resistance continued, and in the two-day battle of Aspern (q.v., 20–1 May) Napoleon suffered his first defeat. Owing to the lack of co-ordination between the Austrian commanders, he fully recovered his position and, by winning the battle of Wagram (6 July 1809), forced the Emperor Franz to accept the armistice of Znaim (12 July 1809). In Tyrol Andreas Hofer (q.v.) maintained his position for a few weeks after the Austrians had accepted the Treaty of Vienna (signed at Schönbrunn, 14 October 1809).

The severity of the terms imposed upon Austria by the Treaty of Vienna is comparable with that of the Treaty of Tilsit imposed upon Prussia. Russia and the Grand Duchy shared Galicia, and the Austrian possessions on the Adriatic were annexed by France as the Illyrian Provinces; Tyrol, Vorarlberg, and

Salzburg were among Austrian possessions passing into the hands of Bavaria.

Napoleon, having divorced his wife Josephine, married the Austrian Archduchess Marie Louise (q.v., 1 April 1810), a step which was followed by a peace lasting for two years. But, far from consolidating his power, tensions arose from the effects of the Continental Blockade and from an increasingly strained relationship between Napoleon and the Russian tsar, which in turn imposed a new dilemma on Prussia, already threatened by Napoleon with the loss of Silesia if it did not pay its war indemnities. Without offering any concessions, Napoleon concluded in 1812 (24 February) a treaty with Prussia, obliging it to support his campaign against Russia, and stipulating free military passage for his armies and the provision of 40,000 troops. In a treaty concluded with Austria, Napoleon secured 30,000 troops. His Russian campaign lasted six months and involved 600,000 troops of which 200,000 were German. By December 1812 he was on his way back to Paris, his occupation of the deserted and burning city of Moscow having ended in a crippling retreat. The crossing of the Beresina (November 1812) alone imposed immense losses and suffering upon the Grand Army. For Prussia these months marked the turning-point towards a recovery which culminated in the Wars of Liberation (Befreiungskriege).

The first initiative for the Wars of Liberation came from Freiherr vom Stein and General Yorck (qq.v.), who in the Convention of Tauroggen (30 December 1812) promised the Tsar the neutrality of his troops. Friedrich Wilhelm III, while apologizing to Napoleon, yielded to pressure at home and concluded at Breslau the Treaty of Kalisch (28 February 1813), by which Russia undertook to assist Prussia in the recovery of its position before the Treaty of Tilsit in return for its Polish territories.

The Wars of Liberation developed into a European war on a scale hitherto unknown. On 17 March 1813 Prussia declared war on France. Extensive reforms, inaugurated since 1807 and permeating all spheres of life, military, civil, economic, and intellectual, preceded this declaration of war. Among those associated with the changes in Prussia were Scharnhorst, Gneisenau, Clausewitz, K. A. von Hardenberg (after Stein), Fichte, Schleiermacher, W. von Humboldt, E. M. Arndt, A. von Lützow, and F. L. Jahn (qq.v.). Appeals to rouse all Germans for the war against Napoleon failed. The Confederation of the Rhine supported Napoleon's army, along with contingents from Holland, Belgium, Switzerland, Italy, and Poland.

The first battle, at Großgörschen (2 May 1813), was won by Napoleon, who recaptured Dresden (14 May) and won another battle at Bautzen (21 May). The allies withdrew towards Silesia, while Napoleon, trying to strengthen his forces, agreed in the truce of Pläswitz to a seven weeks' armistice. During this period Metternich (q.v.), alarmed at the popular response to the Prussian king's appeal (see AN MEIN VOLK) and at the national wave of liberalism as well as at the influence of Russia, attempted a settlement by means of diplomacy. The new Austrian approach aimed at adjusting the European equilibrium, in which France would still be a significant force. By the Treaty of Reichenbach (27 June 1813) Austria agreed to join the alliance against France if Napoleon refused to withdraw to the left bank of the Rhine and restore the Illyrian Provinces to Austria. But at the Congress of Prague (summer 1813) Napoleon refused a settlement, and on 11 August Franz I declared war against France. Bavaria, too, joined the alliance (8 October), now consisting of Russia, Prussia, Austria, Great Britain, and Sweden. Schwarzenberg, Blücher (qq.v.), and the Swedish Crown Prince Bernadotte were the chief allied commanders.

The second phase of the war was marked by Napoleon's failure to advance upon Berlin in spite of successes, including one at Dresden (26–7 August 1813), which was his last victory on German soil. Allied victories were achieved by Blücher at Katzbach (27 August), by General von Kleist (1762–1823) at Nollendorf (29–30 August), and by General F. W. von Bülow (1755–1816) at Dennewitz (6 September). Napoleon's troops stationed near Hamburg tried to join him, but were held back in a number of battles, in one of which, fought at Gadebusch (26 August), Th. Körner (q.v.) was mortally wounded.

The decisive battle of the Wars of Liberation was fought at Leipzig; it became known as the Völkerschlacht (q.v., the Battle of the Nations), and terminated the Napoleonic empire in Germany.

There followed a period of negotiations by the Confederation of the Rhine. The Treaty of Töplitz (September 1813) guaranteed the independence of its member states, except Saxony, whose king had been taken prisoner. The Primate of the Confederation, K. Th. von Dalberg (q.v.), fled to Switzerland, and Jérome left Westphalia. The allies, although refraining from immediate pursuit, renewed their campaign after vainly offering Napoleon peace on terms broadly based on France's position in 1792. In the winter France was invaded, although negotiations with Napoleon were resumed (at Chatillon, February 1814). The decisive advance was accomplished in Blücher's victories at La Rothière (1 February) and Laon (9 March 1814). Meanwhile Wellington invaded the south of France. Paris surrendered on 30 March 1814, and Tsar Alexander and the Prussian king Friedrich Wilhelm entered the city the following day. Napoleon, deposed by the French Senate (1 April), signed an instrument of abdication at Fontainebleau on 6 April 1814, and went into exile at Elba. Louis XVIII was installed as king of France.

The First Treaty of Paris (30 May 1814) terminated the war. Alsace, the Palatinate (up to Landau), and the Saar district (including Saarlautern) remained French, and Austrian possessions in Italy and Bavaria were restored, but the future reorganization of Germany was referred to a congress to be held at Vienna (see WIENER KONGRESS).

On 6 March 1815 the Congress of Vienna was disrupted by the news of Napoleon's return from Elba. On 20 March he installed himself in Paris and prepared for a new campaign. To check it, the allies advanced through the Netherlands and the middle and upper Rhineland. On 16 June Napoleon repulsed Blücher at Ligny, and Wellington suffered a setback at Quatre-Bras. Napoleon himself now advanced on the road to Brussels, encountering at Waterloo (q.v., 18 June 1815) Wellington's, and later also Blücher's, troops for the decisive battle which ended in complete victory for the allies and determined pursuit of the routed French.

On 7 July 1815 Paris was occupied by the Prussians under Blücher after Napoleon had already abdicated for the second time (22 June 1815). In the Second Treaty of Paris (20 November 1815) France was reduced to its position of 1790. Austria received Landau for Bavaria, and Prussia the Saar district. France was obliged to return many of the confiscated works of art. Napoleon, having surrendered to the British, was sent as a prisoner to St. Helena. On the day the treaty was signed, the Great Powers formed the Quadruple Alliance. The Holy Alliance (see HEILIGE ALLIANZ), too, was the product of this final phase of the Napoleonic Wars.

Napoleon oder Die hundert Tage, a play by C. D. Grabbe (q.v.), written in 1830, published in 1831, and first performed in Frankfurt/Main in 1895. Divided into five acts, it is essentially an epic drama reflecting the period of the French Revolution and Restoration. The brilliantly conceived first act revives the memory of history through the eyes of the people, old Bonapartists, survivors of the Russian campaign (Vitry and Chassecœur),

returned émigrés, the Duke and Duchess of Angoulême, and Louis XVIII, and before it closes Napoleon decides to leave the shores of Elba for his renewed bid for power. The flight of the King and Napoleon's return to the empty throne in the Tuileries, the revival of the Paris mob (headed by Jouve, a fictitious figure), and the frustration of the plans of the Liberals Fouché and Carnot are among the events preceding Napoleon's victory at Ligny. The last act (complete in itself and often referred to as a Schlachtendrama) portrays Napoleon's defeat at Waterloo (q.v.), which ends his hundred-day rule.

Grabbe made use of a variety of sources as well as of his experience of the contemporary scene, culminating in the July revolution of 1830 (see JULIREVOLUTION), to air his views on the subject of revolution and the reactionary regimes of his day. The kaleidoscopic presentation, by juxtaposition rather than integration, frequently achieves masterly characterization; it is serious, satirical, and grotesque by turns, and conveys a mood of all-pervading disillusionment. Such glimpses as the play offers of Napoleon himself reveal him as an individual possessed by a sense of personal achievement and destiny rather than patriotism. Grabbe portrays Napoleon as the child of the Revolution, and above all as a chosen tool of the 'universal spirit' (Weltgeist, a concept derived from Hegel, q.v.), which deserts him. On the other hand, pointed critical comments prevent Napoleon from becoming the subject of hero worship. Blücher represents true patriotism and popular leadership. His final comments, closing the play, inspire little confidence in a future worthy of those who fought the great battle; they are an echo of Napoleon's conviction that the forthcoming period of peace will be ineffective and insignificant, with many petty states in the hands of as many petty tyrants, all aiming at the oblivion of the political consciousness of the people.

Narr auf Manegg, Der, a Novelle by G. Keller included in Züricher Novellen (q.v., 1878). It is the second story in the collection, and is a sequel to Hadlaub (q.v.). It records the preservation of the so-called Manessische Handschrift (see HEIDELBERGER LIEDERHANDSCHRIFT, GROSSE) from destruction by fire.

Narrenschiff, Das, a long satirical poem by Sebastian Brant (q.v.), first published in 1494. It is composed of more than 7,000 lines, arranged in a preface (Vorred) and 115 sections. The general idea, indicated by the title and outlined in the preface, is that of a ship manned by all imaginable representatives of folly. The image of the ship is, however, not maintained in the satirical sections that follow. Brant takes folly in a wide and moral sense to cover all human failings, and his book is a castigation of vice and wickedness as well as of imprudence and eccentricity. Among the evils he pillories are familiar sins and vices such as adultery, gluttony, and blasphemy, and lesser faults such as love of litigation and taking the cares of the world upon one's shoulder. In the latter part of the poem the satire is broadened into an attack upon the present state of Church and Empire. Brant criticizes the princes and comes out in support of the Emperor Maximilian I (q.v.). The general standpoint of the work is one of robust common sense, and the morality implies utilitarian standards.

The verse, with a standard line of four iambic feet with eight or nine syllables, has a brisk emphatic rhythm, and the formulations are crisp and often memorable. The technique of the sections resembles that of a sermon in the driving home of points, not by development, but by repetition in different terms. The accompanying woodcuts enhanced the book's effect.

Das Narrenschiff was one of the most popular works of its age. Six authorized editions (twelve in all) had appeared by 1521 when Brant died, and it continued to be published throughout the 16th c. Even more successful was a Latin translation, *Stultifera navis*, by Jakob Locher (q.v.), which formed the basis of English (1507 f.), French (1497 f.), and Dutch (1500) translations. The preference for the Latin version reflects the extent to which the neo-humanism of the age had penetrated.

Narr in Christo Emanuel Quint, Der, a novel by G. Hauptmann (q.v.) published in 1910. It is a sympathetically told account of a man, poor, despised, rejected, who has visions and is for a time believed by some to be Christ in his Second Coming. Maltreated at home and rebuked by a magistrate and by his father, Quint goes away and is inspired to preach a sermon in the market-place. He takes refuge with obscure dissenters, whose father he appears to cure, and is baptized by another, Nathanael Schwarz. Quint is followed by disciples, is stoned and reviled, is tempted in the wilderness; he himself comes to believe that Christ is present in him. Taken up by the police, he is returned to his home and again maltreated. A noblewoman (das Gurauer Fräulein) takes an interest in him and gives him asylum. He is wrongly believed to have led astray the gardener's daughter, Ruth Heidebrandt, and is dismissed by the Lady of Gurau. As he wanders on, he acquires the firm conviction that he is Christ. Ruth is murdered

and Quint is held for a time in custody as a suspect, but he is shown to be guiltless and is released. His disciples have now deserted him and he wanders about.

At the beginning of this last phase of Quint's life Hauptmann takes the stance of a narrator who does not know how Quint met his end, merely suggesting that he may be identical with a man who announced himself at many houses as Christ and ultimately perished in a snowstorm in the Alps.

Narziß und Goldmund, a novel by H. Hesse (q.v.), published in 1930. It is set in the Middle Ages. Narziß, a gifted pupil in a monastic school at Mariabrunn, is conscious of his vocation for the monastic life. He becomes the mentor of Goldmund, a younger pupil, whose unsuitability for a celibate life soon becomes clear to Narziß. Goldmund, who is the central figure of the novel, leads a life rich in emotional and artistic experiences. Unobserved, he leaves the monastery for a vagrant existence, which is punctuated by a series of love-affairs. He becomes a sculptor, and his masterpiece (a figure of St. John) bears the features of his friend Narziß. He survives an epidemic of the plague, which, however, carries off Lene, whom he deeply loves. He consoles himself with Agnes, the mistress of a great nobleman. This relationship is the apogee of Goldmund's life, but it is discovered, and he is condemned to death. The priest who comes to shrive him before execution proves to be Narziß, who at once successfully exerts his influence to obtain a pardon for Goldmund. Together they return to the monastery, where Goldmund resumes his carving. Once more he leaves, but a final experience of love leads to renunciation, and he returns to the monastery to die. At last he appreciates the depth of Narziß's love for him, which he returns.

The book expresses an ultimate dissatisfaction with the temporal, which it probes at length, and a corresponding longing for something transcending mortality. Goldmund seeks it in a sublimation of human relationships, both erotic and filial (with a maternal orientation), and in art. He dies in peace, with the maternal image predominant in his mind.

NAS, JOHANNES (Eltmann nr. Bamberg, 1534–90, Innsbruck), a Lutheran journeyman tailor, was converted to the Roman Catholic faith and entered the Franciscan Order as a lay brother. In 1557 he was ordained; in 1572 he became a cathedral cleric at Innsbruck, and in 1580 was consecrated bishop of Brixen (Bressanone) in South Tyrol). Nas engaged in anti-Reformation theological controversy, and his best-known work is the *Sex Centuriae* (1565–9, in German notwithstanding the Latin title) attacking respectively Rauscher, FLACIUS Illyricus (q.v.), Jakob Andreä, Coelestin, Cyriakus Spangenberg (q.v.), and Osiander (q.v.). Nas was also a noted preacher. He is attacked by J. Fischart (q.v.) in *Der Barfüßer Secten und Kuttenstreit* (1570) and in *Von Sanct Dominici des Prediger münchs und Sanct Francisci Barfüßer artlichem Leben* (1571), to which Fischart gave the alternative title *Nasenspiegel*.

NASO, ECKART VON (Darmstadt, 1888–1976, Frankfurt/Main), a general's son, studied law. He served in the 1914–18 War and was wounded. He devoted himself to the theatre as producer (Regisseur) and adviser (Dramaturg) in the Berlin Staatliches Schauspielhaus 1918–45. From 1954 to 1957 he was at the Staatstheater in Stuttgart. Naso wrote few plays (*Die Insel*, 1918; *Die Frau im Garten*, 1921), preferring narrative fiction. He is best known for the nationalistic historical novels *Seydlitz* (1932) and *Moltke* (1937), and for the Novelle *Die Begegnung* (1936), which deals with the attempt by Queen Luise (q.v.) of Prussia to intercede with Napoleon at Tilsit. Other titles include the novels *Menschen unter Glas* (1930), *Scharffenberg* (1935), *Der Halbgott* (1949, with Alcibiades as principal figure), *Die große Liebende* (1950, of which the heroine is Ninon de l'Enclos, the French courtesan, 1620–1705), and *Eine charmante Frau* (1950). The Novellen *Preußische Legende* (1939) and *Der Rittmeister* (1942) should also be mentioned. Naso is the author of the biographies *Moltke* (1937) and *Caroline Schlegel* (1969). *Ich liebe das Leben* (1954) and *Glückes genug* (1963) are autobiographical.

Nathan der Weise, a verse play (Ein dramatisches Gedicht in fünf Aufzügen) written by G. E. Lessing (q.v.) in 1778, and published in 1779. It was first performed in Berlin on 14 April 1783. The time is that of the Crusades and the place Jerusalem. A young Templar (Tempelherr), captured and unexpectedly released by Saladin, rescues Recha, a Jew's adopted daughter, and after much hesitation seeks her hand. But Nathan, the adoptive father, temporizes. The Templar, in anger which is inflamed by the well-meaning Christian servant Daja, who believes that Nathan has seduced his adoptive daughter from Christianity to Judaism, comes near to betraying Nathan to the intolerant Christian patriarch of Jerusalem. Meanwhile Saladin, in need of money, summons Nathan and sets out to sound his quality, asking the Jew which of the three great religions is the true one. Nathan answers with a parable of three identical rings (adapted from Boccaccio, *Decameron*, I.3),

and his wise tolerance gains him Saladin as a friend. As a result of their meeting it is discovered that the Templar and Recha are brother and sister, children of Saladin's deceased brother and his Christian wife.

The play is a noble plea and argument for religious tolerance. For Lessing it was a counterpart on a higher plane to the embittered controversy with Pastor Goeze (q.v.), in which he had been silenced by ducal censorship. *Nathan der Weise*, noteworthy for its pioneering of blank verse in German drama, is a masterpiece of comedy, wit, and noble serenity.

NATHUSIUS, MARIE (Magdeburg, 1817–67, Neinstedt), *née* Scheele, a pastor's daughter and philanthropist, was the author of a large number of edifying stories for the young. Among her popular works were *Das Tagebuch eines armen Fräuleins* (c. 1853), *Die Geschichten von Christfried und Julchen*, and *Elisabeth* (both 1858), her last novel. Her songs include the well-known 'Alle Vögel sind schon da', welcoming spring. *Gesammelte Werke* (15 vols.) appeared in 1858–68.

Nationalkomitee Freies Deutschland, an anti-National-Socialist body set up in Russia in 1943, composed of German Communist refugees. E. Weinert, W. Ulbricht, J. R. Becher (qq.v.), W. Pieck (1876–1960), and A. Ackermann (1905–) were prominent members. The Committee had a military counterpart, the Bund Deutscher Offiziere. This, though opposed to National Socialism, was not Communist. Its most prominent members were General W. von Seydlitz (1888–1976) and Field-Marshal F. Paulus (1890–1957). Both organizations were dissolved in 1945.

Naturalismus denotes a form of exact and detailed realism which developed in Europe in the late 19th c. Its special characteristic is its claim to be scientifically based, and it is in part the outcome of developments in 19th-c. science, especially in the fields of heredity and evolution. Naturalism began in France in the fiction of the brothers Goncourt and of Zola c. 1860–70, in Russia with Tolstoy, and in Scandinavia with Ibsen (q.v.) and Bjørnson. In Germany theoretical tracts and periodicals advocating Naturalism appeared from about 1885. German Naturalism reached a climax in the early 1890s and thereafter gradually declined, though its influence remained appreciable for many years.

The first signs of the movement in Germany are seen in Berlin with the publication of *Kritische Waffengänge* (q.v., 1882–4) by the brothers H. and J. Hart (qq.v.), and in Munich with the periodical *Die Gesellschaft* (q.v.) edited by M. G. Conrad (q.v.). Further symptoms of its progress are two theoretical tracts, *Revolution der Literatur* (1886) by K. Bleibtreu (q.v.), and *Die naturwissenschaftlichen Grundlagen der Poesie* (1887) by W. Bölsche (q.v.). *Das Buch der Zeit* (1885) by A. Holz (q.v.) contains the beginnings of proletarian poetry more in its choice of subjects than in its manner. With *Die Kunst. Ihr Wesen und ihre Gesetze* (1890–2), Holz produced the principal theoretical tract, which includes the lapidary formulation 'Kunst = Natur – x', amplified at the end of the essay in the following: 'Die Kunst hat die Tendenz, wieder die Natur zu sein. Sie wird sie nach Maßgabe ihrer jedweiligen Reproduktionsbedingungen und deren Handhabung.' This sentence was simplified by Holz in 1899: 'Die Kunst hat die Tendenz, die Natur zu sein; sie wird sie nach Maßgabe ihrer Mittel und deren Handhabung.' G. Hauptmann's *Bahnwärter Thiel* (1888, q.v.) achieved Naturalistic representation, though poetic elements are also perceptible. The social plays of Ibsen were a powerful influence on the German *avant-garde* in the 1880s, especially *Ghosts*, because it treated a subject (syphilis) which was taboo, and laid great emphasis on heredity.

Naturalism received a new impetus from the visit to Berlin of the Théâtre libre (q.v.) in 1887, and its increased momentum is seen in the founding of the Verein Freie Bühne (see FREIE BÜHNE, VEREIN) and its first two productions in 1889, single performances of *Ghosts* (*Gespenster*, September) and G. Hauptmann's *Vor Sonnenaufgang* (q.v., October). The title of the last work was in itself a slogan. *Papa Hamlet* (q.v.) by Holz and J. Schlaf (q.v.), published in 1889 under the pseudonym Bjarne P. Holmsen, developed a technique of minute realism midway between narrative and drama. This innovation became known as 'Sekundenstil'. By 1892 Naturalism was accepted in the commercial theatre, and, having become respectable, began within a few years to lose its revolutionary impetus. *Die Familie Selicke* (q.v., 1890) by Holz and Schlaf and *Die Weber* (q.v., 1892) by G. Hauptmann are notable works from the heyday of the movement. Commercialization set in early, especially in the theatre, with the skilfully tailored plays of H. Sudermann (q.v.), *Die Ehre* (q.v., 1890) and *Heimat* (q.v., 1893). The principal novelist of thoroughgoing Naturalism was M. Kretzer (q.v.), though most novels in the first half of the 20th c. show some influence of the trend. The limitations of Naturalism soon became apparent, and as early as 1891 H. Bahr (q.v.) published his essay *Die Überwindung des Naturalismus*. Naturalism may be considered to be the dominating influence only up to about 1898.

Other writers who are notable for Natural-istic writing in at least a phase of their career are F. Adamus, Helene Böhlau, M. Halbe, O. E. Hartleben, G. Hirschfeld, W. von Polenz, and Clara Viebig (qq.v.).

Natureingang, term used in connection with Minnesang (q.v.) to denote an introductory passage using images drawn from nature. Such images are conventional and depend only remotely on observation, being limited to such concepts as *bluomen, walt, vogelin*, and so on. The Natureingang, which was extensively employed, is first found in early poems such as the following, ascribed in the manuscripts to DIETMAR von Eist (q.v.): 'Sô wol dir sumerwunne!/daz vogelsanc ist geswunden,/alse ist der linden ir loup./(Wohl dir Sommerwonne!/Der Vogelsang ist verschwunden,/sowie der Linden Laub.)'

Natürliche Tochter, Die, a tragedy (Trauerspiel) written by Goethe in 1801–3, published in 1803, and first performed in Weimar on 2 April and in Berlin on 12 July of that year. It was intended to be the first part of a trilogy, the second and third parts of which were never written. The plan probably dates from 1799. The play is set in France in the closing years of the *ancien régime*. Eugenie, the illegitimate daughter of the Duke—she is the natural daughter of the play's title—is drawn from obscurity and presented, a paragon of beauty and education, to the King. A splendid prospect seems to open before her, but a plot against her is ripening. She is abducted and faced with the threat of exile to tropical America or marriage to a commoner, the Gerichtsrat, which will destroy her hopes of a brilliant life at court. After vainly seeking to evade the dilemma, she opts for marriage with the Gerichtsrat, but exacts a promise that he will remain remote from her unless she summons him.

The play embodies the corruption and violence of the old society and sounds a repeated warning of coming revolution. Written in blank verse, it has stylized characters denoted by rank, and its action is conducted statically in long set speeches. The source of this notable example of Goethe's classicism is the *Mémoires historiques de Stephanie-Louise de Bourbon Conti. Écrites par elle-même* (1798).

NAUBERT, BENEDIKTE (Leipzig, 1756–1819, Leipzig), *née* Hebenstreit, was twice married, first to L. Holderieder and then to J. G. Naubert. Her home during both marriages was in Naumburg. She wrote many historical novels (see RITTER- UND RÄUBERROMAN) and published a collection of fairy-tales, *Neue Volksmärchen der Deutschen*, in five volumes (1789–93).

NAUMANN, FRIEDRICH (Störmthal nr. Leipzig, 1860–1919, Travemünde nr. Lübeck), was active as a political journalist who sought to combine ideas of social reform with nationalism. His open and persuasive personality and transparent integrity won him the steadfast loyalty of a small group of adherents, some of whom, including Th. Heuß (q.v.), assisted him with his journal Die Hilfe (q.v.) and with the short-lived daily Die Zeit (1896–7), which later reappeared as a weekly (1901–3). Against his enlightened social aspirations must be set his emphasis on national power, expressed in Demokratie und Kaisertum (1900). He joined the Liberal Freisinnige Vereinigung in 1903 and worked for the unification of the various Liberal factions. His efforts culminated in the formation of the Fortschrittliche Volkspartei (1910). In Mitteleuropa (1915) he proposed an economic union of central Europe and pressed in 1917 for a constitutional monarchy (Der Kaiser im Volksstaat). He was a member of the Reichstag from 1907 to 1918 (except for a brief period 1912–13). In November 1918 he founded the Deutsche Demokratischc Partei, of which he became the leader, and he was active in the Weimarer Nationalversammlung, the body which established the Weimar Republic (q.v.). He died too early to influence the policies of the new state.

Werke (6 vols.), ed. under the auspices of the Friedrich-Naumann-Stiftung, appeared in 1964.

Nausikaa, title given to a number of fragments of dialogue in blank verse, written by Goethe for an intended play on the Greek legend of Nausicaa and Odysseus (*Odyssey,* vi. 15–315 and viii. 457–68). The play was first mentioned by Goethe in October 1786 with the title *Ulysses auf Phäa,* and the fragments were written in Sicily early in 1787. They were published in 1827 in vol. 4 of the *Ausgabe letzter Hand* with the title *Nausikaa. Ein Trauerspiel.*

Nazarener, a term applied, originally in derision, to the Lukasbund, a group of Romantic artists founded in Vienna in 1809 by J. F. Overbeck (1789–1869) and F. Pforr (1788–1812). Among thc members were also P. Cornelius (q.v.), W. Schadow (1788–1862), J. Schnorr von Carolsfeld (1794–1872), and Philipp Veit (q.v.) and his brother Joseph. The object of the Nazarener was a renewal of religious painting, based on Raphael and Perugino. The League (Bund) exercised some influence on the English Pre-Raphaelites.

Nazi Party, see NSDAP.

NEANDER, JOACHIM (Bremen, 1650–80, Bremen), was appointed headmaster of the Düsseldorf grammar school in 1674 and pastor of the Reformed Church in Bremen in 1679. He wrote hymns which were collected in A und O. Glaub- und Liebes-Übung (1680, enlarged edn. 1707). Perhaps the best-known hymn, for which he also composed the music, is 'Lobe den Herren, den mächtigen König der Ehren'. The Neanderthal, famous since 1856 for the discovery in the Neanderhöhle of the skull of Neanderthal Man, is named after him because, so it is said, it was a favourite haunt of his during his time at Düsseldorf.

NEBEL, GERHARD (Dessau, 1903–74, Hohenlohe), was a teacher and wrote travel books (*An der Mosel,* 1948; *Phäakische Inseln,* 1954; *An den Säulen des Herkules,* 1957). He was also an amateur cultural philosopher concerned with the regeneration of man and published several volumes in support of his ideas. His original standpoint was classical and aesthetic (*Feuer und Wasser,* 1939; *Griechischer Ursprung,* 1948), but his later writings are based on Protestant Christianity. They include *Weltangst und Götterzorn* (1951), an interpretation of Greek tragedy; *Das Ereignis des Schönen* (1953); *Die Not der Götter* (1957); *Homer* (1959); and *Die Geburt der Philosophie* (1967).

Neckar, Der, a river rising in the southern Black Forest near Schwenningen, flowing past Rottweil, Tübingen, Heilbronn, and Heidelberg, and entering the Rhine by Mannheim. Next to the Rhine, it is perhaps the German river most celebrated in song (see BALD GRAS' ICH AM NECKAR, BALD GRAS' ICH AM RHEIN), partly because of its association with the ancient university of Heidelberg. Hölderlin wrote an ode entitled Der Neckar (q.v.).

Neckar, Der, an Alcaic ode written by F. Hölderlin (q.v.) in 1800 and printed in 1801. It maintains his attachment to the landscape of the River Neckar in his native land side by side with his longing for the landscape of Greece.

NEHER, CASPAR (Augsburg, 1897–1962, Vienna), the stage designer whose name is closely linked with the theatre of B. Brecht (q.v. and see BERLINER ENSEMBLE) and the development of the Raumbühne (q.v.). He aimed at productions which discouraged illusion, and shared with Brecht a predilection for austere colour (the famous background in subdued shades of grey) and bright stage lighting. His collaboration with Brecht, however, went well beyond designing. Brecht's own tribute to him is one of unqualified enthusiasm; Neher's decorations were steeped in the spirit of the plays and induced the actors to live up to their artistry. Neher's reputation does not rest only on his collaboration with Brecht, which began with the *Dreigroschenoper* (q.v.) in 1928: he directed many important productions at the Deutsches Theater, the Vienna Staatsoper, and the Salzburger Festspiele, and designed settings to operas by Mozart (q.v.). He wrote an opera libretto, *Die Bürgschaft* (1932), which K. Weill (q.v.) set to music.

Neidharte, designation of songs celebrating the pranks of NEIDHART Fuchs (q.v.).

Neidhart Fuchs, a legendary figure, based on the 13th-c. poet NEIDHART von Reuental (q.v.). The addition of Fuchs (fox) denotes his cunning. Legend placed Neidhart Fuchs and his pranks at the Viennese court of Otto der Fröhliche (the Merry), who died in 1339.

Neidhart Fuchs, title given to an anonymous collection of comic anecdotes (see SCHWANK) in verse centring on the cunning with which NEIDHART Fuchs (q.v.) outwits the peasants. It runs to nearly 4,000 lines, was compiled in the late 14th c. or early 15th c., and was printed at the end of the 15th c., and twice in the 16th c. (1537 and 1566).

NEIDHART VON REUENTAL (c. 1185–c. 1240), also Neithart, a Middle High German poet of remarkable originality. The name Reuental is sometimes symbolically interpreted by Neidhart himself, but it has also been supposed, though without documentary evidence, that his home actually was the village of Reuental in Bavaria, not far from Tegernsee. He was from the lower ranks of the nobility and participated in a crusade either in 1217–19 or in 1228–9. Not long after the latter date he lost the favour of his overlord in Bavaria, but found a new patron in Friedrich der Streitbare (the Quarrelsome) of Austria (reigned 1230–46).

Neidhart revolutionized lyric poetry, abandoning the strict Minnesang (q.v.), and creating a form which is characterized by a sharp social dissonance. The peasant appears in Neidhart's poems in a form varying from realism to caricature. Nearly all the poems are dance songs, and belong to two distinct groups, summer songs and winter songs. The summer songs have an introductory passage of conventionalized nature description, and then pass to the environment of the peasant girl determined to go out to dance on the green, frequently in opposition to her mother's wishes. Crass domestic scenes occur, in which mother and

daughter come to blows. Neidhart himself (der von Reuental) is mentioned, often as the object of the girl's desire. The winter songs portray more drastic scenes within doors, which have been compared with the genre scenes of Flemish painters. Here, too, Neidhart plays a part, but he is worsted by the aggressive rustics. A powerful element of social satire pervades these poems, which reflect the decline in the standing of the knight and the rising pretensions of the prosperous peasants. A recurrent symbol is the theft by a brash lout from the peasant girl Friderun of a mirror given to her by Neidhart. In four poems he openly laments the decay of his world. Neidhart's style is remarkable for its vigorous rhythms, its realistic touches and genre sketches, and the disharmony which is its fundamental tone. He is anything but a spiritual poet; even his crusade songs are mainly complaints of hardship and wishes for the comfort of home.

Neidhart's poetry appealed to its contemporaries because it caught the spirit of the age. It became even more popular in the 14th c. and 15th c., when numerous imitations were attributed to Neidhart. He became a legendary figure, appearing in chapbooks (see VOLKSBUCH) and Fastnachtspiele (q.v.). *Die Lieder Neidharts*, ed. E. Wiessner, appeared in 1963.

Neinsager, Der, see DER JASAGER UND DER NEINSAGER.

Neptunismus denotes the theory that all rocks are of marine origin. The principal German Neptunist was A. G. Werner (1750–1817). A dispute between a Neptunist and his vulcanist opponent, supporting the volcanic origin of rocks, in *Faust* (q.v., Part II, Act II) is resolved in favour of the former, indicating Goethe's Neptunist sympathies.

NESTROY, JOHANN NEPOMUK (Vienna, 1801–62, Graz), son of a Viennese lawyer, began to study law, but in 1822 decided on a theatrical career. The possessor of a well-trained bass voice of high quality, he was immediately engaged by the Viennese Court Opera and made his debut in the important role of Sarastro in Mozart's *Die Zauberflöte* (q.v., 24 August 1822). In the years 1823–31 he worked, first as a singer but increasingly as a comic actor, in Amsterdam, Brünn (Brno), where his engagement was cancelled by the police in 1826 because his frequent extemporization was contrary to the strict censorship regulations, and Graz.

He began to write plays for the Graz theatre, his principal success at this time being *Dreißig Jahre aus dem Leben eines Lumpen* (1828). The only earlier plays to survive are a one-act comic dialogue, *Der Zettelträger Papp* (1827), and a verse drama, *Prinz Friedrich* (written *c.* 1828, performed as *Rudolph, Prinz von Korsika* in 1841). Nestroy had made an early marriage (1823) which proved a failure, and his wife left him in 1827. In Graz he met the actress Marie Weiler and lived with her in an unofficial union for the remainder of his life. In 1831 he returned to Vienna as a star in the company of Karl Carl (q.v.) with which he was to remain for 29 years, first at the Theater an der Wien, from 1839 also at the Theater in der Leopoldstadt (rebuilt in 1847 as the Carl-Theater). Tall and angular, he formed a celebrated comic partnership with the corpulent Wenzel Scholz (1787–1857); his reputation as an actor was such that he was sought after for guest appearances in many German cities, and his agreeable and kindly personality off stage increased his popularity.

Nestroy was also one of the principal purveyors of plays for Carl. He drew his plots from a variety of sources, including contemporary French and English novels (*Die Anverwandten*, 1848, is based on Dickens's *Martin Chuzzlewit*) and popular comedies and melodramas of the Paris and London stage; into these plots he injected sustained wordplay, interspersed with solo scenes consisting of satirical monologues and songs, mainly for the central figures that he himself acted. Between 1827 and his death he wrote more than 80 plays, and from 1832 his output was intended exclusively for Carl and his theatre. Between 1832 and 1834 he burlesqued the tradition of the Viennese magic play (Zauberstück, see VOLKSSTÜCK), achieving a resounding success with *Der böse Geist Lumpazivagabundus oder Das liederliche Kleeblatt* (q.v., 1833) for which he provided a sequel in the farce *Die Familien Zwirn, Knieriem und Leim* (1835). He also produced a drastic parody of the then fashionable opera of Meyerbeer, *Robert le diable*, under the dialect title *Robert der Teuxel* (1833). After 1834 Nestroy virtually abandoned the magic element and concentrated on the rival form of Viennese Volksstück, the Lokalposse (q.v.). From 1835 to 1845 he wrote some 30 plays, many ephemeral, but some of outstanding quality which are still performed. Among his successes in this period was *Zu ebener Erde und erster Stock* (q.v., 1835), which boldly divides the stage into two, in order to exhibit social contrast. Critical rejection of a local satire, *Eine Wohnung ist zu vermieten* (1837), was followed by a spell of experimentation, with the stylized comedy of *Das Haus der Temperamente* (1837) using a stage divided both horizontally and vertically, until he found his way to a settled and distinctive pattern of satirical farce. The next six years brought a series of brilliant successes, ushered in by *Glück, Mißbrauch und Rückkehr* (1838) and *Die verhäng-*

nisvolle Faschingsnacht (1839, a parody of a play by K. von Holtei, q.v.); these were followed in 1840 by *Der Färber und sein Zwillingsbruder* and by *Der Talisman*, which is generally recognized as his masterpiece, the 'classic Posse' (F. H. Mautner), and then by other popular successes including *Das Mädl aus der Vorstadt* (q.v., 1841), *Einen Jux will er sich machen* (q.v., 1842), *Liebesgeschichten und Heiratssachen* (q.v. 1843), and *Der Zerrissene* (q.v., 1844).

Nestroy's dramatic production took a new turn with *Der Unbedeutende* (q.v., 1846), a comedy in which social questions take on a new importance. It proved to be the most impressive success of his career and was followed by other similar plays, notably *Der Schützling* (1847) and *Der alte Mann mit der jungen Frau* (q.v., written in 1848 or 1849, but withheld by Nestroy for political reasons). In the revolutionary year of 1848 (see REVOLUTIONEN 1848–9), when censorship was briefly lifted, Nestroy satirized the outbreak of the revolution in *Freiheit in Krähwinkel* (q.v.), and in 1849 produced a brilliant parody of Hebbel's *Judith* (q.v., *Judith und Holofernes*, q.v.). His rate of production now became slower, and his last notable full-length play was performed in 1852 (*Kampl oder Das Mädchen mit Millionen und die Näiherin*, q.v.) In 1854 Karl Carl died and Nestroy agreed to become director of the Carl-Theater. He resigned in 1860 and retired to Graz, but made guest appearances in Vienna in 1862 in his last two successful one-act satires, *Frühere Verhältnisse* and *Häuptling Abendwind* (the latter based on Offenbach's *Vent du soir*).

Even in his heyday, Nestroy was repeatedly criticized by conservative reviewers (including the combative Viennese journalist M. G. Saphir, q.v.), who measured his work against the often sentimental 'humour' of the old genial Viennese Volksstück and particularly against the work of Raimund (q.v.). But the old forms had outlived their natural vitality and could not survive in the new age initiated in 1830 by the July Revolution and established by the turmoil of 1848. Nestroy's caustic satirical and parodistic wit established a new and more modern tone in the Volksstück and exercised a dominant influence on other popular playwrights such as Friedrich Kaiser (q.v.); but his basic attitude is humane and positive, not the nihilistic cynicism that was formerly attributed to him.

Nestroy's funeral in Vienna was the occasion of a great public tribute of affection and esteem, but his work almost at once fell into a phase of neglect, which was ended by a sensationally successful Nestroy cycle in the Carl-Theater in 1881. His literary reputation rose with the publication in 1890–1 of a collected edition of his works (*Gesammelte Werke*, 12 vols.) by V. Chiavacci and L. Ganghofer (qq.v.), and was established by the energetic advocacy of K. Kraus (q.v.), who adjudged him the wittiest satirist in the German language and who in 1920 and 1925 respectively published adaptations of two of his plays, *Die beiden Nachtwandler* (1836) and *Der konfuse Zauberer* (1832). While much admired by contemporaries such as Grillparzer, Hamerling, and Kierkegaard (qq.v.), Nestroy is, like G. Büchner (q.v.), a dramatist whose stature has been recognized increasingly in the 20th c.; his work has been of influence on modern dramatists including F. Dürrenmatt and Ö. von Horváth (qq.v.) and has been championed by prominent critics including F. Torberg and H. Weigel (qq.v.).

Sämtliche Werke, historisch-kritische Ausgabe, ed. F. Brukner and O. Rommel (15 vols.), appeared in 1924–30 and *Komödien* (3 vols.), ed. F. H. Mautner, in 1970. A thoroughly revised *Historisch-kritische Ausgabe*, ed. J. Hein and J. Hüttner *et al.* (*c.* 33 vols.) began to appear in 1977. It includes correspondence (*Briefe*, 1 vol., 1977), ed. W. Obermaier.

NETTELBECK, JOACHIM (Kolberg, 1738–1824, Kolberg), after a career at sea, set up in Kolberg as a brandy-distiller and brewer. In 1807 he prevented the surrender of the fortress to the French and besought King Friedrich Wilhelm III (q.v.) for help. The King replaced the commandant with Gneisenau (q.v.), whose adjutant (Bürgeradjutant) Nettelbeck became. As a result of these measures Kolberg (q.v.) successfully defied the besiegers until the end of hostilities. Nettelbeck received as a reward permission to wear Prussian admiral's uniform, a somewhat striking distinction in a country which at that time was virtually without a navy. He was also awarded a pension in 1817. The autobiographical *Joachim Nettelbeck, Bürger zu Colberg. Eine Lebensbeschreibung* (3 vols.), ed. J. Haken, which was published 1821–3, appeared in revised form, ed. K. Burow, under the title *Nettelbeck. Seefahrer, Sklavenhändler und Patriot* in 1969. Some historians consider that Nettelbeck's role in the siege has been over-emphasized.

NEUBECK, VALERIUS WILHELM (Arnstadt, 1765–1850, Altwasser), a physician in Liegnitz, is the author of a didactic poem, *Der Gesundbrunnen* (1794), and of other poetry in classical measures.

Neue Arria, Die, a play (Schauspiel) written by F. M. Klinger (q.v.) in 1775, and published anonymously in 1776. It was acknowledged by Klinger in 1786, when it appeared in *F. M. Klingers Theater*. Two dynamic, more than life-size lovers, Julio and Solina, conspire against Duke Galbino, a dissolute tyrant. The conspiracy is discovered and Julio is arrested. Solina joins him in prison, and the two decide to end their lives. The tragic action and its theme are characteristic of the Sturm und Drang (q.v.).

Neue Beiträge zum Vergnügen des Verstandes und Witzes (1745–8), full title of a periodical better known as *Bremer Beiträge* (q.v.).

Neue Bibliothek der schönen Wissenschaften und freyen Künste, see BIBLIOTHEK DER SCHÖNEN WISSENSCHAFTEN UND FREYEN KÜNSTE.

Neue Deutsche Museum, Das, see DEUTSCHE MUSEUM, DAS.

Neue Dorf- und Schloßgeschichten, see DORF- UND SCHLOSSGESCHICHTEN.

Neue Freie Presse, leading Viennese newspaper founded in 1864 and locally referred to as *Die Presse*. It was the butt of much satire by Karl Kraus (q.v.). It closed down in 1939, but in 1946 a new paper began to appear under the title *Die Presse* (from 1948 as a daily).

Neue Gedichte, a collection of poems by H. Heine (q.v.), published in 1844. The volume, in addition to the 166 *Neue Gedichte* proper, contains also *Zeitgedichte* and *Deutschland. Ein Wintermärchen* (q.v.), though the latter was omitted from subsequent editions. The *Neue Gedichte* are subdivided into *Neuer Frühling*, *Verschiedene*, and *Romanzen*. The section *Verschiedene* contains three short poems grouped as *Tragödie*, of which the central one ('Es fiel ein Reif in der Frühlingsnacht', q.v., by Zuccalmaglio) is given by Heine as 'ein wirkliches Volkslied' ('a real folk-song').

Neue Gedichte, a collection of poems by Rilke (q.v.), published in two successive volumes (*Neue Gedichte*, 1907; and *Der neuen Gedichte anderer Teil*, 1908). Most were written in the years 1906–8 during Rilke's residence in Paris. They represent the apex of his development in the single short poem (for his later masterpieces are the cycles *Duineser Elegien* and *Die Sonette an Orpheus*, qq.v.). The first volume contains 83 poems, the second 107. Their general quality is so high, even by the stringent standards of the mature Rilke, that the mention of individual poems seems invidious.

In the first volume *Der Panther*, *Das Karussell*, and *Orpheus. Eurydike. Hermes* are among the most widely known; but many others can be quoted to demonstrate both excellence and range. They include the cathedral group, *L'Ange du méridien*

(*Chartres*) to *Das Kapital*, *In einem fremden Park*, *Abschied*, *Letzter Abend*, *Alkestis*, and the numerous non-religious poems on themes drawn from Christianity, such as *Der Ölbaumgarten*, *Pietà*, and *Sankt Sebastian*. The second volume, in which the poems tend to be on a slightly larger scale, includes several classical motifs (e.g. *Archaïscher Torso Apolls*, *Leda*, *Delphine*, *Klage um Antinous*), a number of Old Testament poems, the well-known *Papageienpark* and *Die Flamingos*, and some poetry on Venetian subjects. Throughout, visual and pictorial objects (including paintings, sculptures, and cathedrals) are prominent, not only as marks of Rilke's admiration for their intrinsic beauty, but as a means of expressing his inner self in tangible things (see DINGGEDICHT). This kind of poetry represents Rilke's solution to the problem of escaping from excessive subjectivity, a problem which had beset him since he encountered the poetry of S. George (q.v.) and the sculpture of Rodin.

Neue Gleise, a collection (Gemeinsames) of both new and previously published work by A. Holz and J. Schlaf (qq.v.) which appeared in 1892. It contains three sections, each introduced by a preface (Vorwort): I. *Die papierne Passion*; the title is that of the first of the four prose sketches presented in this section; the others are entitled *Krumme Windgasse 20*, *Die kleine Emmi*, and *Ein Abschied*. Section II, *Papa Hamlet*, contains the three sketches already published in 1889, *Papa Hamlet* (q.v.), *Der erste Schultag*, and *Ein Tod*. Section III, entitled *Die Familie Selicke* (q.v., 1890), is remarkable for its preface consisting of a large collection of reviewers' comments, *pro* and *contra*, printed in the original French, Dutch, or Danish, as well as German.

Neue Liebe, neues Leben, a poem written by Goethe in 1775, and published in Jacobi's *Iris* (q.v.) in the same year. It springs from his relationship with Lili Schönemann (q.v.) and reflects the ambiguous nature of his love, in which attraction is balanced by the desire to escape. See also AN BELINDEN and LILIS PARK.

Neue Melusine, Die, a story inserted in Goethe's novel *Wilhelm Meisters Wanderjahre* (q.v.), where it occupies Bk. III, Ch. 6. It is a variant of the legend (see SCHÖNE MELUSINE) and probably dates from 1770 or 1771, but was dictated by Goethe in 1807, and first published separately in the *Taschenbuch für Damen* (Pt. 1 in 1817, Pt. 2 in 1819).

A barber meets a beautiful and wealthy woman, and the pair fall in love. She poses certain conditions which he repeatedly infringes, and in the end she leaves him and returns to her own kingdom, for she is in reality a minute dwarf. In order to stay with her, he accepts dwarf form, but soon tires of it and returns to normal human stature, without wealth or bride.

Neuenburg, French Neuchâtel, a town and canton in French-speaking Switzerland, which in 1707 elected as its sovereign King Friedrich I (q.v.) of Prussia. Except for a brief period of French rule (1806–13), Neuchâtel continued to be ruled by the Prussian kings (who wisely interfered little) for exactly a century and a half. Tension grew after the 1848 Revolution (see REVOLUTIONEN 1848–9), and in 1856 a war between Switzerland and Prussia over Neuchâtel seemed imminent. In the following year, however, Friedrich Wilhelm IV (q.v.) renounced his rights in the territory while retaining the title Fürst von Neuenburg.

Neue Parzefal, Der, title given to an immense interpolation in a MS. of Wolfram von Eschenbach's *Parzival* (q.v.), which is at Donaueschingen. It runs to more than 36,000 lines, and is inserted between Bks. XIV and XV in Wolfram's work. It is a crude translation of a French original, and the authors of the German version were two impoverished goldsmiths of Strasburg, Philipp Colin and Claus Wisse, who completed the work between 1331 and 1336. Since neither understood French, it was explained to them by one Samson Pine. Wisse, Colin tells us, wrote the beginning; Colin, it may be assumed, wrote the rest (the greater part). His patron was an Alsatian nobleman, Ulrich von Rappoltstein, and his motive was frankly financial. The enormous conglomeration of adventures is crudely rendered into German verse, contrasting conspicuously with the work into which it is inserted.

Neue Preußische Zeitung, see KREUZZEITUNG.

Neue Reich, Das, a volume of poetry by S. George (q.v.), published in 1928.

Neuere politische und soziale Gedichte, a collection of political poems by F. Freiligrath (q.v.) published in two volumes (Hefte) in 1849 and 1851.

Neuer Kurs, denotes the policy, under the Emperor Wilhelm II (q.v.), of General Caprivi, who succeeded Bismarck (qq.v.) as chancellor. The 'new course' abandoned the attempt to maintain a balance between Austria-Hungary and Russia, opting for alliance with the former. The expression alludes to a statement, made by Wilhelm II on his accession, that the 'old course' would be maintained: 'Der Kurs bleibt der alte, und nun Volldampf voraus!'

Neuer Realismus, also Kölner Schule des neuen Realismus, term for a group of young writers who, in the early 1960s, agreed on a literary programme designed to express an instinctive, biologically informed self-awareness that derives from the anthropological philosophy of Arnold Gehlen (b. 1904). Proceeding from the *nouveau roman* (notably by A. Robbe-Grillet, N. Sarraute, and C. Simon), it involves alienation from modes of behaviour imposed by convention and society in the technological age. *Ein Tag in der Stadt. Sechs Autoren variieren ein Thema,* is an anthology, published in 1962. The six authors were Rolf Dieter Brinkmann, U. Chr. Fischer, Ludwig Harig, Günter Herburger, Robert Wolfgang Schnell, and Günter Seuren. The School's originator was D. Wellershoff (q.v.). Other authors linked with it included N. Born (q.v.), Günter Steffen, Renate Rasp, Paul Pörtner, and Sigrid Brunk. Though most of these authors became better known through their later work, the School was seminal to their development.

Neue Rundschau, Die, a German monthly, published in Berlin and concerning itself with letters, the arts, and the sciences. It dates from 1894 and bore until 1903 the title *Neue deutsche Rundschau,* which in turn had grown out of the journal *Freie Bühne* (see FREIE BÜHNE, VEREIN).

Neue Sachlichkeit, a term brought into use *c.* 1925 in reaction against the ecstatic, exclamatory tone of the Expressionists (see EXPRESSIONISMUS), and applied to cover a wide variety of sober, earthy, or realistic writing. Among authors who have been included in this category are F. Bruckner, H. Carossa, A. Döblin, Marie Luise Fleißer, O. von Horváth, E. Kästner, C. Zuckmayer, and A. Zweig (qq.v.). The term signified no close-knit group and was little more than a tentative, convenient description.

Neueste aus dem Reiche des Witzes, Das, a monthly supplement written by G. E. Lessing for the *Vossische Zeitung* (q.v.) in 1751. Its contents are a mixed bag, but the June number contained an important assertion that literary rules arise out of experience and are modified by each new genius.

Neue Thalia, Die, see THALIA, DIE.

NEUFFER, CHRISTIAN LUDWIG (Stuttgart, 1769–1839, Ulm), was educated at the Tübinger Stift, where he became a friend of Hölderlin

(q.v.); he had his first pastoral appointment at Stuttgart and was transferred to Wertheim in 1803, and in 1819 to Ulm. He is the author of a poem in nine cantos, *Die Herbstfeier, ein Sittengemälde* (1801), and of the idyll *Der Tag auf dem Lande* (1802). The latter, in ten cantos, was published anonymously and was for a time attributed to J. H. Voß (q.v.). Neuffer's poems appeared in 1816 (*Auserlesene lyrische Gedichte*) and his poetic works in 1827–8 (*Poetische Schriften,* 3 vols.). He made a verse translation of Virgil's *Aeneid,* which was published in 1815. An early fragmentary ode by Hölderlin (*An Neuffer*) is addressed to him.

NEUFFER, KLARA OR CLARA, a cousin of the poet E. Mörike (q.v.) and his first love, was the daughter of the pastor of Benningen (later of Bernhausen, near Stuttgart). She was the same age as Mörike, to whom her engagement in 1823 to a curate named Schmid came as a great shock. Mörike's poem *Erinnerungen* ('Jenes war zum letzten Male') is addressed to her.

NEUHAUS or **NEUHÄUSER,** the supposed author of *Lohengrin* (q.v.).

NEUKIRCH, BENJAMIN (Reinicke, Silesia, 1665–1729, Ansbach), was a lawyer in Breslau (1687–91); he later abandoned the law and attempted to live by his pen and by giving popular lectures. In 1703 he was appointed a professor at the Ritterakademie in Berlin, and fifteen years later became tutor to the heir apparent of Ansbach-Bayreuth. Neukirch's original works (*Galante Briefe und Gedichte,* 1695; *Satyren und poetische Briefe,* 1732) are of slight importance. His most notable publication was a collection of verse by HOFFMANN von Hoffmannswaldau (q.v.) and other primarily Silesian poets. Neukirch also translated Fénelon's *Télémaque* into German Alexandrines (1727 ff.).

Neuklassizismus, term for a short-lived reaction against Naturalism (see NATURALISMUS), manifesting itself in the first decade of the 20th c. It is more evident in theoretical works based on German classicism of a century before (see KLASSIK, DEUTSCHE), than in creative literature. Its principal supporters were P. Ernst (q.v., *Der Weg zur Form,* 1906), W. von Scholz (q.v., *Gedanken zum Drama,* 1904), and S. Lublinski (q.v., *Die Bilanz der Moderne,* 1904, *Der Ausgang der Moderne,* 1909).

NEUMANN, ALFRED (Lautenburg, West Prussia, now Poland, 1895–1952, Lugano), was active as a dramatic adviser (Dramaturg) in Munich and Fiesole. In 1941 he emigrated to the U.S.A. and became an American citizen, returning to Europe in 1949 and settling at Lugano. He is the author of a number of historical novels concerned with the problems of power and dictatorship and ruthlessly analysing the motives of action.

Neumann's outstanding works are the story *Der Patriot* (q.v., 1925), the novels *Der Teufel* (q.v., 1926) and *Die Rebellen* (1927, the Italian revolt against Austrian hegemony in the north), and the trilogy of novels about Napoleon III (*Neuer Caesar*, 1934; *Kaiserreich*, 1936; *Die Volksfreunde*, 1941). (The last of these was retitled *Das Kind von Paris* in 1952.) He also wrote the stories *Lehrer Taußig* (1924) and *Viele heißen Kain* (1950), and the novels *Die Brüder* (1924), *Guerra* (1929), *Der Held* (1930), *Narrenspiegel* (1932), *Die Goldquelle* (1938), *Es waren ihrer sechs* (1945), and *Der Pakt* (1949). The story *König Haber* (1926) reappeared as *Dr. Danieli* in 1932. He wrote poetry of his own and translated that of A. de Musset, A. Lamartine, and A. de Vigny.

NEUMANN, BALTHASAR (Eger, 1687–1753, Würzburg), one of the most prominent architects of German baroque, served as an officer in the Turkish wars of the early 18th c., attaining the rank of colonel. In 1719 he was appointed architect for the diocese of Würzburg and designed and executed numerous buildings, as well as making important additions to others. His activities were not confined to Würzburg. His masterpieces are the Residence in Würzburg, Schloß Werneck (1731–7), the staircase in Schloß Brühl near Bonn (1743–8), the pilgrimage church of Vierzehnheiligen near Bamberg (1745–72), and the abbey church of Neresheim near Nördlingen (1745–92). Neumann lectured on architecture at the University of Würzburg. He is the subject of a novel, *Balthasar Neumann*, by E. Ortner (q.v.).

NEUMANN, CHRISTIANE, see BECKER, CHRISTIANE.

NEUMANN, ROBERT (Vienna, 1897–1975, Monaco), studied medicine, chemistry, and Germanistik, and after a business career in which he lost all his money, went for a time to sea as a deck-hand. In 1934 he emigrated and settled in England at Cranbrook (recalled in *Mein altes Haus in Kent*, 1957) and after the war moved to Locarno. A master of parody, he published the volumes *Mit fremden Federn* (1927, ext. 1955) and *Unter falscher Flagge* (1932), issued as *Die Parodien* (1962). His fiction, some of which appeared first in English, centres on the 1920s, fascism, and the sufferings of Jews; it includes the novels *Die Sintflut* (1929, the deluge

of the title is the inflation of currency) and *Die Macht* (1932, reissued as *Macht*, 1964), *Sir Basil Zaharoff, der König der Waffen* (1934), *Struensee* (1935, reissued as *Der Favorit der Königin*, 1953), *An den Wassern von Babylon* (1945), *Die Puppen von Pohansk* (1952, English title *Insurrection of Pohansk*), and *Kinder von Wien* (1948, in English in 1946), a strikingly original and moving story of a group of children, survivors of war and persecution.

His stories include the satirical *Hochstaplernovelle* (1930, reissued as *Die Insel der Circe*, 1952) and *Karrieren* (1931); both are contained in the collection *Karrieren* (1966), a volume of *Gesammelte Werke in Einzelausgaben* (1959 ff.). Other volumes include the novels *Die dunkle Seite des Mondes* (1959), *Olympia* (1961), *Der Tatbestand oder Der gute Glaube der Deutschen* (1965); the parodies *Dämon Weib oder Die Selbstverzauberung durch Literatur* (1969), *Nie wieder Politik oder Von der Idiotie der Schriftsteller* (1969), and *Vorsicht Bücher* (1969); the autobiographical works *Ein leichtes Leben. Bericht über mich selbst und Zeitgenossen* (1963), *Vielleicht das Heitere. Tagebuch aus einem andern Jahr* (1968), and *Oktoberreise mit einer Geliebten. Ein altmodischer Roman* (1970).

Neumark, a territory acquired by the margraves of Brandenburg in 1250 and long known as 'das Land jenseit der Oder'. It stretches from the River Oder on either side of Küstrin northeastwards, but does not reach the Baltic. The designation Neumark (New March) first occurs towards the end of the 14th c. From 1402 to 1455 it was leased to the Teutonic Order (see DEUTSCHER ORDEN). In the 19th c. and 20th c. it formed part of the Prussian Provinz Brandenburg. After the 1939–45 War it passed to Poland.

NEUMARK, GEORG (Langensalza, 1621–81, Weimar), a minor baroque poet and novelist, was first a private tutor at Kiel, then studied at Königsberg, where he met Simon Dach (q.v.). In 1651 he became librarian and archivist in Weimar, was elected to Die Fruchtbringende Gesellschaft (q.v.) in 1653 and became its secretary in 1656. He was also a member of the Hirten- und Blumenorden an der Pegnitz (q.v.). He published several volumes of poetry (including *Poetisch und Musikalisch Lustwäldchen*, 1652; *Fortgepflanzter Musikalisch-Poetischer Lustwald*, 1657; and *Lieder*, 1675), but he is now almost exclusively known for the religious poem *Trostlied* ('Wer nur den lieben Gott läßt walten'), written in Kiel in 1640. Neumark also wrote novels (*Betrübt-verliebter doch endlich hocherfreuter Hirt Filamon wegen seiner edlen Schäfernymphen Belliflora*, 1640, and *Keuscher Liebesspiegel*, 1649), and in 1766 published the collection *Poetisch-historischer Lustgarten.*

NEUMEISTER, ERDMANN (Üchteritz nr. Weißenfels, 1671–1756, Hamburg), was pastor at Bibra from 1679 to 1704 and moderator (Superintendent) at Sorau from 1706 to 1715, after which he moved to Hamburg as principal pastor. He published several volumes of hymns between 1717 and 1729, as well as sermons and devotional works (*Der Zugang zum Gnadenstuhl Christi*, 1705). He was the author of a Latin survey of German literature (*Specimen diss. hist. crit. de poëtis Germanicis*, 1695). He used at times the pseudonym Adam Martini.

Neunundzwanzigste Februar, Der, a play by Adolf Müllner (q.v.), written in 1812, and published in 1815 in Müllner's *Spiele für die Bühne.* It was first performed by amateurs in Weißenfels where Müllner lived. It consists of one act comprising 935 lines of trochaic verse. The play is heavily indebted to Z. Werner's *Der vierundzwanzigste Februar* (q.v.), a much better work. Both are examples of the Schicksalstragödie (q.v.).

The scene opens on a married couple, Walter and Sophie; it is the 29th of February, a day of ill omen for them. On this day Walter's father and, a few years later, his daughter had died. An uncle arrives, who is likely to leave them his property. Instead of consolation, the uncle's visit brings the terrible revelation that Walter and Sophie are brother and sister. Walter kills his son, born in incest, and then prepares to surrender to justice. Once more the 29th of February has proved a fatal day for the family.

Neuromantik, a once fashionable term used to denote a vague category of writers opposed to Naturalism (see NATURALISMUS). Among the large number of names associated with neo-Romanticism, the following are most justifiably linked with it: the early H. von Hofmannsthal, K. Vollmoeller, and R. Beer-Hofmann, and, on occasion, G. Hauptmann (qq.v.).

Neuruppiner Bilderbogen, brightly coloured crude drawings of political events, crimes, etc., with brief texts, printed at Neuruppin in Brandenburg (north-west of Berlin). The pictures, naïve representations with no pretence at topographical accuracy, circulated widely throughout Germany in the 19th c. and early 20th c. The principal producers were the printing firms of Gustav Kühn and Öhmigke und Riemschneider.

Neuschwanstein, perhaps the most famous of the extravagant building projects of King Ludwig II (q.v.) of Bavaria. Built between 1869

and 1886, principally by E. Riedel (1813–85), it is a medieval-style castle in the pinnacled fairyland manner, perched on a mountainside not far from Füssen. It was in Neuschwanstein that the declaration of his deposition was communicated to Ludwig.

NEUSER, ADAM (d. 1576), a German Unitarian, emigrated to Turkey, where he adopted the Muslim faith. In a contribution to *Zur Geschichte und Literatur* (*Von Adam Neusern einige authentische Nachrichten*, 1774) G. E. Lessing (q.v.) explains this as a flight from Christian persecution.

Neustrien, coinciding roughly with present France, less its north-eastern, and eastern regions, formed one part of the Merovingian Frankish Empire (see MEROWINGER).

Nibelungen, Die. Ein deutsches Trauerspiel in drei Abteilungen, a dramatic trilogy by F. Hebbel (q.v.), was planned in 1855 as a dramatization of the medieval *Nibelungenlied* (q.v.) in two parts with ten acts. The final arrangement in three parts consists of *Der gehörnte Siegfried. Vorspiel in einem Akt*, a one-act prelude, and the two five-act parts *Siegfrieds Tod* and *Kriemhilds Rache*; all are in blank verse. The writing of the work took Hebbel seven years. H. Laube (q.v.) declined to produce the Siegfried tragedy at the Burgtheater (1857). The first two parts were produced in Weimar by F. Dingelstedt (q.v.) in January and May 1861, Hebbel's wife Christine taking alternately the part of Brunhild and of Kriemhild. In February 1863 Laube produced the first two parts in Vienna. Dingelstedt arranged the performance of the complete trilogy at the Burgtheater in 1871. The work was published in 1862. Hebbel was familiar with the medieval tale from his youth, and knew versions of the subject by F. de la Motte-Fouqué (*Der Held des Nordens*, q.v., 1810), E. Raupach (q.v., *Der Nibelungenhort*, 1834), and E. Geibel (q.v., *Brunhild*, 1858). He intended to cover the legend in its entirety and to adjust it to the 'reale Bühne' (*Vorwort*). In his view the author of this 'Nationalepos', as he referred to the *Nibelungenlied*, struck a balance between mythology and character. Hebbel made it his task to turn the dramatic epic into an epic drama.

The *Vorspiel* leads to Siegfried's and King Gunther's departure for Isenland. *Siegfrieds Tod* opens in Isenland at the Queen's court. Against the advice of Frigga, Brunhild accepts Gunther's challenge. She follows him to Worms, ignorant of the deceit by which Siegfried and Gunther vanquish her, but she is aware of her love for Siegfried, which Hebbel makes the main motivation for the Siegfried tragedy. When Brunhild does not submit to Gunther's love-making, Siegfried allows himself to be persuaded to aid the King once more, concealed by his cloak of darkness (Tarnkappe). Kriemhild, in an assertion of pride, reveals the deceit to Brunhild. Hebbel ascribes the motivation of Hagen Tronje's murder of Siegfried during the bear hunt not only to Hagen's personal jealousy of Siegfried's superhuman powers, but also to his sympathy for Brunhild's outraged humanity. In ostentatiously taking Siegfried's sword as he lies in state in the chapel, Hagen ignites Kriemhild's resolve for revenge.

Kriemhilds Rache opens at Gunther's court at Worms. Rüdiger has arrived to escort Kriemhild on her way to be married to King Etzel in Heunenland. Kriemhild accepts the proposal as an opportunity to avenge Siegfried's death, aiming at the destruction of Hagen, regardless of the slaughter of all who are innocent and dear to her. She uses her last strength to recover Siegfried's sword, Balmung, with which she kills Hagen. She herself is slain by Hildebrandt. The senseless fury of Kriemhild's passion is set against the indestructible will of Hagen, who keeps to himself the secret of the disappearance of the Nibelungen gold (Hort). Hebbel places the emphasis of the finale on Etzel, whose humanity is shattered by the unfathomable self-destructive urge within man, which confounds all conceivable standards of justice. He hands his crown to Dietrich von Bern, who accepts 'Im Namen dessen, der am Kreuz erblich!' (the last line of the trilogy). Hebbel's treatment of the subject differs substantially from Wagner's *Der Ring des Nibelungen* (q.v.).

Nibelungenhort, Der, a tragedy by E. Raupach (q.v.).

Nibelungenlied, commonly used title of a Middle High German heroic epic poem, formerly also known as *Der Nibelunge Nôt*, after its last half-line, 'diz ist der Nibelunge nôt'. It was written in Austria, almost certainly in the decade 1200–10; its author is unknown, but he is believed to have stood in some relationship to Bishop Wolfger of Passau (1194–1204). The poem is written in more than 2,300 rhyming four-line stanzas, each line being subdivided and each subdivision having three stresses, except for the last line, which is longer (see NIBELUNGENSTROPHE).

Though presented as a continuous poem, the story is told in two distinct phases; the climax of the first is the death of Siegfried; the second recounts Kriemhild's revenge upon her Burgundian brothers. Its opening is set in Worms at the Burgundian court of three royal brothers, Gunther, Gernôt, and Gîselher, and it begins

with the dream of Kriemhild, in which her noble falcon is destroyed by two eagles, a symbolical presage of the tragedy of the first part. To the Burgundian court comes the young prince Siegfried, who distinguishes himself by his strength and his knightly accomplishments. By his extraordinary prowess he defeats a threatened invasion by the Saxons to the north-east. He sees Kriemhild on his return from the campaign and desires to marry her. Gunther wishes to marry Brunhild the Queen of Isenstein (Iceland), who possesses fabulous strength, and will only give her hand to one who can defeat her in an athletic contest. Siegfried undertakes to achieve this for Gunther on condition that he may marry Kriemhild. Siegfried is able to vanquish Brunhild partly by his own strength, but partly also by magic means; for he is the owner by conquest of the Nibelung treasure, the sword Balmung, and the cloak of darkness (Tarnkappe), which he has wrested from the dwarf Alberich; and furthermore he has acquired an invulnerable skin, by bathing in the blood of a dragon he has killed. Invisibly supporting Gunther with the aid of the Tarnkappe, he defeats Brunhild in the games, whereupon all return to Worms for the weddings and accompanying festivities. The marriage of Siegfried and Kriemhild is a perfect match, but all is not well with Gunther. On the bridal night Brunhild wrestles with him and binds him, and Siegfried's assistance is sought once more. Hidden by the Tarnkappe he subdues Brunhild, who is henceforth content, believing herself to have been overcome by Gunther. Ten years pass in placidity and contentment. On a visit to Worms, however, Brunhild and Kriemhild quarrel about precedence, and in the heat of the moment Kriemhild discloses the deceit practised on Brunhild with Siegfried's help. Brunhild demands revenge, Gunther demurs, but Hagen, Gunther's nobleman-at-arms, presses for Siegfried's death. Hagen tricks Kriemhild into disclosing Siegfried's one vulnerable spot, and in the course of a hunt, arranged for the purpose, Siegfried is stabbed as he stoops to drink from a spring. Hagen, whose guilt is revealed to Kriemhild, further injures her by making off with the Nibelungen treasure, which he throws into the Rhine. Thirteen years of mourning follow for Kriemhild, and this period forms the division between the first and second parts of the story.

Etzel (Attila) has lost his wife, desires to remarry, and hears of Kriemhild's beauty. His messenger Ruedeger prevails upon her to accept Etzel's proposal; she does so because of the opportunity of vengeance on Hagen. She journeys to Vienna to marry Etzel and they reside at Etzelnburg (Gran, Hungarian Esztergom). After another period of thirteen years

Kriemhild invites the Burgundians (in this second part of the poem they are often referred to as Nibelungs) to Etzel's court. Hagen divines Kriemhild's purpose and advises against the journey, but nevertheless agrees to accompany them. On the way there various prophecies indicate that none will return. At Etzel's court the King receives them courteously, but Kriemhild provokes various lords to attack the Burgundian Nibelung party and their men. These attacks are at first unsuccessful, but, after Hagen has decapitated Kriemhild's young son, Etzel's full forces are gradually drawn into the combat. Bloedelin, Iring, and Ruedeger of Etzel's court are all slain, and eventually the Nibelung kings Gernôt and Gîselher and the warriors Volker and Dancwart all fall. Dietrich finally captures and binds first Hagen, then Gunther. Kriemhild has Gunther decapitated, confronts Hagen with the head and demands the Nibelung treasure. On his refusal to disclose where it is hidden she cuts off his head with Siegfried's sword. Dietrich's lord Hildebrand, enraged at this atrocity, cuts Kriemhild to pieces, and the story has an end.

To the *Nibelungenlied* is attached a brief inferior sequel called *Die Klage*, dealing with the mourning for the heroes and for Kriemhild, the coronation of Gunther's son, and the departure of Dietrich and Hildebrand. It is written in the rhyming couplets of courtly epic.

The events and characters of the *Nibelungenlied* arise from remote legend, but the poem absorbs them almost completely into the contemporary courtly world. Siegfried and Kriemhild, Gunther and his brothers, Dietrich, Hildebrand, Ruedeger, and even the heathen Etzel, exhibit good breeding (*zuht*) and self-restraint (*mâze*). The life of this social stratum is punctuated by great festivals (*hôchgezîte*) distinguished by lavish hospitality, knightly contests, and splendid liberality (*milte*). Where the ancient mythological elements proved indigestible (e.g. Siegfried and the Nibelungs or Brunhild's Icelandic fastness) they are treated cursorily. The courtly world here portrayed is closer to reality than the fantasies of Arthurian epic, and is set in recognizable geographical locations. Siegfried's father holds court at Xanten; Siegfried is slain in the Spessart; Ruedeger's castle is at Bechelaren (Groß-Pöchlarn). The eastward journeys pass through Passau, Enns, Melk, and Tulln; in addition Worms, Vienna, and Gran are important locations.

The *Nibelungenlied* is intensely tragic; it ends with the violent death of almost all the characters, and every festival leads to suffering and tragedy. This sense of inevitable doom and the decline of all happiness to grief is a motif running through the whole poem from Kriemhild's 'wie liebe mit leide ze jungest lônen kan' to the penultimate stanza 'mit leide was verendet des küniges hôchgezît'. Though the world portrayed conforms without scepticism to Christian usage, its ultimate values are heroic: loyalty, courage, and steadfastness under the blows of inescapable fate. This is the clue to the predominance and importance of Hagen, a figure in many respects repellent to a modern sensibility. Ruedeger and Dietrich represent a more humane conception. A remarkable feature of the poem is the sensitivity of the poet to emotional values, so that the note of faithful love and unassuaged grief is manifested in Kriemhild at the very moment at which she strikes down Hagen.

The *Nibelungenlied* is one of the most impressive, and certainly the most powerful, of the German epics of the Middle Ages. It was clearly popular, being preserved in whole or part in 34 MSS. or fragments of MS. The three principal MSS. are A (superscribed 'Der Nibelunge nôt', discovered at Hohenems and now at Munich), B (similarly entitled, also discovered at Hohenems and now at St. Gall), and C ('Der Nibelunge liet', Donaueschingen), to which should perhaps be added MS. d (Ambras), MS. D, and MS. I. Of the more important ones, A, B, and C belong to the 13th c., D and I to the 14th c. The order and rank of these MSS. have been the subject of fierce contention, but since the publication in 1963 of H. Brackert's investigations the attempt to claim one of them (usually B) as the original or a copy of the original has been abandoned. The title *Ältere Nôt*, occurring in treatises on the *Nibelungenlied*, refers to a completely hypothetical epic of *c.* 1160, alleged to resemble in its contents the Norwegian Thidrek saga (see DIETRICH-SAGE).

The popularity of the *Nibelungenlied* lasted into the 16th c. Then, by a remarkable revolution of taste, it lapsed into oblivion. It was rediscovered by J. H. Obereit (q.v.) and published by J. J. Bodmer (q.v., *Chriemhilden Rache und die Klage*, 1757). See also NIBELUNGENSAGE.

The 15th edition of the *Nibelungenlied*, ed. K. Bartsch (original title *Der Nibelunge Not*, 3 vols., 1870–80), appeared in 1959. Text and parallel modern translation with *Nachwort* and *Anhang* were published as *Das Nibelungenlied. Mittelhochdeutscher Text und Übertragung*, ed. H. Brackert, 2 vols., 1970–1.

Nibelungensage, a Germanic legend, was particularly well developed in Old Norse literature and in the German *Nibelungenlied* (q.v.). The centre of all the legends is the treasure (Hort) of the Nibelungs. In the Nordic form, contained in the Eddas (see EDDA), Wolsungen saga, Thidrek saga (see DIETRICH-

SAGE), and Faroëse songs, the three gods Odin, Hoenir, and Loki seize the dwarf Andvari and force him to yield up his treasure of gold. The treasure is guarded by a dragon, who is slain by Sigurd, the son of the King of the Franks, and he thereupon acquires the gold. Sigurd finds the Valkyrie Brynhild, put to sleep by Odin, wakes her, and they swear fidelity to each other. Sigurd later associates with King Gjuki on the Rhine and his sons Gunnar, Högni, and Gutthorm. Their sister Gudrun gives Sigurd a potion engendering oblivion. He forgets Brynhild and marries Gudrun. By magic means he enables Gunnar to overcome and marry Brynhild. Brynhild and Gudrun quarrel, and Sigurd is killed by Gutthorm. The brothers acquire the Hort. Gudrun is then married to Atli (see ATTILA), who seeks to wrest the gold from her brothers. When they refuse to yield it, he has them killed. He, in turn, is killed by Gudrun. This Nordic version is the oldest recorded form of the story, but its origin is probably Frankish.

The earlier part is largely mythological; from the point at which Sigurd and Gudrun are married it has a basic resemblance to the German *Nibelungenlied*. The legend has a historical foundation in the time of the migration of peoples, and particularly in two episodes: the destruction of the Burgundian kingdom at Worms in 437 by the Huns under Attila, and the death of Attila in bed with a Germanic girl, Ildico (q.v.), in 453. Attila's death was probably due to haemorrhage, possibly cerebral, but was popularly turned into an act of assassination by the girl. A further element is the legend of Dietrich von Bern. The combination and distortion of these tales, with the fusion of mythological elements, have produced the story, which has been treated in many adaptations in German literature. The principal Nibelungen works are the anonymous *Nibelungenlied* and *Der hürnen Seyfrid* (qq.v.), *Der Nibelungen Hort* by E. Raupach (q.v.), *Nibelungen im Frack* by A. Grün (q.v.), Hebbel's *Die Nibelungen* and Wagner's *Der Ring des Nibelungen* (qq.v.), *Nibelunge* by W. Jordan (q.v.), and *Der Nibelunge Not* by M. Mell (q.v.), a 20th-c. treatment.

Nibelungenstrophe, the stanza form of the *Nibelungenlied* (q.v.). It is composed of four long lines (Langzeilen), which rhyme in pairs (aabb); each 'Langzeile' is divided into two short sections (Kurzzeilen), of which the first has a 'feminine', the second a 'masculine' ending. The last 'Kurzzeile' is usually extended, as in the following example:

Der was der selbe valke, / den si in ir troume sach,

den ir beschiet ir muoter. / wi sêre si daz rach an ir næhsten mâgen, / di in sluogen sint! durch sîn eines sterben / starb vil maneger muoter kint.(19)

(Der war derselbe Falke, den sie in ihrem Traume sah,
Den ihre Mutter erklärte. Wie sehr rächte sie das
An ihren nächsten Verwandten, die ihn nachher erschlugen!
Durch das Sterben dieses Einen, starb so mancher Mutter Kind)

A closely similar stanza form occurs in the poetry of Der Kürnberger (q.v.), written earlier than the *Nibelungenlied*. The Nibelungenstrophe is used in slightly varied form in other Middle High German epics, such as *Kudrun*, and *Die Rabenschlacht* (qq.v.).

Nibelungentreue, expression coined by the German Chancellor Prince Bernhard von Bülow (q.v.) in a speech to the Reichstag on 29 March 1909. Bülow used it to underline the loyalty of the allies, Germany and Austria-Hungary, to each other. The pretentious phrase caught on in the newspapers and was derided by the Austrian satirist Karl Kraus (q.v.).

Nibelungias, a hypothetical Latin poem on the story of the Nibelungs, supposed to have been written in the 10th c. An analogy for such a poem is to be found in *Waltharius* (q.v.). Its existence has been postulated on the basis of certain passages in *Die Klage*. No trace has ever been found and the theory of its existence must be regarded as unproved. (See NIBELUNGEN-LIED.)

Nicht mehr als sechs Schüsseln!, a once popular play by G. F. W. Großmann (q.v.).

Nicht zu weit, a short, fragmentary narrative incorporated by Goethe in his novel *Wilhelm Meisters Wanderjahre* (q.v.), where it occupies Bk. II, ch. 10. It is the story of a marriage which, through the vanity of the wife, is in danger of catastrophe.

NICOLAI, FRIEDRICH (Berlin, 1733–1811, Berlin), a bookseller's son who followed his father's trade, educated himself by wide reading. In the 1750s he became a friend and collaborator of G. E. Lessing and M. Mendelssohn (qq.v.), editing the *Bibliothek der schönen Wissenschaften und der freien Künste* (1757) and participating in the *Literaturbriefe* (q.v., 1759).

Nicolai's first independent publication was a brisk survey of contemporary German literature

(*Briefe über den itzigen Zustand der schönen Wissenschaften in Deutschland*, 1755). From 1765 he edited the *Allgemeine Deutsche Bibliothek* (q.v.). A firm adherent of rationalistic Enlightenment (see AUFKLÄRUNG), Nicolai gradually fell behind the times, as his friend Lessing and the younger generation advanced to new standpoints in the late 1760s. His three-volume novel *Das Leben und die Meinungen des Herrn Magister Sebaldus Nothanker* (q.v., 1773–6) is primarily concerned with pillorying the religious intolerance of the excessively orthodox. Fascinated yet repelled by Goethe's novel *Die Leiden des jungen Werthers* (q.v.), Nicolai sought to neutralize it by means of a parody, *Die Freuden des jungen Werthers. Leiden und Freuden Werthers des Mannes* (1775, reissued, ed. C. Grützmacher, 1972). By similar satire he endeavoured to ridicule Herder's cult of folk-song in *Eyn feyner kleyner Almanach vol schönerr echterr liblicherr Volckslieder, lustigerr Reyen unndt kleglicher Mordgeschichte* (1777–8, the absurd archaic spelling is part of the mockery). His literal and pedagogic mind discovered so much of interest on a German tour in 1781 that he wrote an account of it in twelve volumes (*Beschreibung einer Reise durch Deutschland und die Schweiz im Jahre 1781*, 1783–96). His scorn for the idealistic philosophy of Kant (q.v.) found expression in the novel *Geschichte eines dicken Mannes* (1794) and (with a slant towards Fichte, q.v.) in *Leben und Meinungen Sempronius Gundiberts, eines deutschen Philosophen* (1798). And his resentment of the new Romanticism (ROMANTIK) appears in *Vertraute Briefe von Adelheid B. an ihre Freundin Julie S.* (1799). Nicolai's resolute opposition to change and the ineffectiveness of his parodies made him a popular butt in the world of letters at the end of the century. Goethe and Schiller mocked him in the *Xenien* and he appears in a ridiculous light as the Proktophantasmist in the *Walpurgisnacht* of *Faust* (q.v.), Pt. I.

NICOLAI, Otto (Königsberg, 1810–49, Berlin), a pupil of K. F. Zelter (q.v.), became a composer and conductor in Vienna (1837) and Berlin (1847). He is remembered as the composer of *Die lustigen Weiber von Windsor*, an opera after Shakespeare for which S. Mosenthal (q.v.) wrote the libretto.

NICOLAI, PHILIPP (Megeringhausen, Waldeck, 1556–1608, Hamburg), was a Protestant pastor, whose succession of livings in mixed Protestant and Catholic districts brought him a turbulent life, until he became in 1601 *pastor primarius* at St. Katharine's (Katharinenkirche), Hamburg. Nicolai is the author of deeply felt hymns, including 'Wie schön leuchtet der Morgenstern' and 'Wachet auf, ruft uns die Stimme', both

contained in *Freudenspiegel des ewigen Lebens* (1599). He was active as a polemical writer on theological questions. He was also the author of poetic oddities, including a long Latin poem in hexameters, in which every word begins with C, *Certamen corvorum cohabitum columbis*, written in 1573, to which he added in the following year another, in which P is the exclusive initial letter, *Pacis pietatisque periclitatio*.

NIEBELSCHÜTZ, WOLF VON (Berlin, 1913–60, Düsseldorf), originally an art critic in Magdeburg, then Essen, served throughout the 1939-45 War, and afterwards settled as an author near Düsseldorf. His principal work is *Der blaue Kammerherr* (2 vols., 1949), a novel impregnated with baroque fantasy. Baroque inspiration underlies his other works, of which the most notable are the novel *Die Kinder der Finsternis* (1959), the story *Verschneite Tiefen* (1940), the Festspiel *Eulenspiegel in Mölln* (1950), and the volumes of verse *Preis der Gnaden* (1939) and *Die Musik macht Gott allein* (1942). A posthumous collection, *Gedichte und Dramen*, appeared in 1962. His essays include *Goethe in dieser Zeit* and *Jacob Burkhardt* (see BURKHARDT, J., both 1946). Posthumous papers, ed. I. von Niebelschütz, appeared in 1961 under the title *Freies Spiel des Geistes. Reden und Essays*.

NIEBERGALL, ERNST ELIAS (Darmstadt, 1815–43, Darmstadt), pseudonym E. Streff, a schoolmaster of intemperate habits, was the author of fiction and of plays, including the local plays (see LOKALSTÜCK) in Darmstadt dialect *Des Burschen Heimkehr oder der tolle Hund* (1837) and *Datterich* (1841, ed. V. Klotz 1963). As a student Niebergall was a friend of G. Büchner (q.v.). The first posthumous publication of *Dramatische Werke*, ed. G. Fuchs, appeared in 1894, and *Gesammelte Erzählungen*, ed. F. Harkes, in 1896.

Niederösterreich (Lower Austria), one of the federal Länder of the Austrian Republic, occupying the Danube basin from just below Linz almost to Preßburg (Bratislava). Under the Habsburgs (see HABSBURG) it was an archduchy with the alternative title Österreich unter der Enns. Its chief city was Vienna (see WIEN), which, however, became a separate Land in 1920.

Niedersachsen (Lower Saxony), a traditional geographical name for the region formerly divided politically into Hanover, Oldenburg, Brunswick (see BRAUNSCHWEIG), and Lippe, which in the 13th c. belonged to the Duchy of Saxony. In 1945 the region was reunited and is now Land Niedersachsen of the Federal Republic (see BUNDESREPUBLIK DEUTSCHLAND). Bremen and Hamburg (q.v.) are independent Länder.

NIEMBSCH, NICOLAUS FRANZ, EDLER VON STREHLENAU, see LENAU, N.

NIEMÖLLER, MARTIN (Lippstadt, 1892–1984, Wiesbaden), a naval officer in the battleship *Thüringen* in the first year of the 1914–18 War, transferred to the submarine service, in which he had a distinguished career, becoming a U-Boat captain and receiving the order Pour le mérite (q.v.). Disoriented for a time after the war, he became convinced of his vocation to preach the Christian message. After studying theology at Münster he was ordained in 1924.

Niemöller was director of the Innere Mission (see PROTESTANTISMUS) until 1930, and in 1933 founded the Pfarrernotbund, which in 1934 grew into the Bekennende Kirche (q.v.). Though pensioned in 1934, he continued to oppose the regime and was arrested in 1937 and detained in concentration camps until 1945. After his release he held important offices in the German Evangelical Church, but became a controversial figure through his rejection of confessional Lutheranism and his emphatic pacifism, aimed at an active engagement of the Church for the prevention of a nuclear war and for closer co-operation with the DDR (see DEUTSCHE DEMOKRATISCHE REPUBLIK) and Eastern Europe. Niemöller published in 1954, jointly with W. Lüthi and others, *Frieden. Der Christ im Kampf gegen die Angst und den Gewaltgeist der Zeit. Reden* (4 vols.) appeared in 1957 ff. The early years of Niemöller's career are recounted in *Vom U-Boot zur Kanzel* (1934).

NIETZSCHE, FRIEDRICH (Röcken nr. Lützen, 1844–1900, Weimar), a pastor's son, showed remarkable gifts at an early age. He was at Bonn and Leipzig universities and was elected at 25 (1869) to a chair of classical philosophy at Basel University. In the Franco-Prussian War (see DEUTSCH-FRANZÖSISCHER KRIEG) he was a volunteer in the medical service. In 1879 he resigned his chair on the ground of ill-health.

During his tenure of the professorship Nietzsche wrote *Die Geburt der Tragödie aus dem Geiste der Musik* (q.v., 1872), in which, in addition to underlining the Dionysiac element in Greek civilization and its expression in tragedy, he praised the work of R. Wagner (q.v.), whose friend he had been since 1868. *Die Philosophie im tragischen Zeitalter der Griechen* (1873) extended his revolutionary view of classical philology by diminishing the emphasis on Plato and Socrates in favour of earlier Greek thought. His preoccupation with cultural values is most clearly seen at this time in the four long essays (termed *Stücke*) of *Unzeitgemäße Betrachtungen*; the first of these to appear was *David Strauß, der Bekenner und der Schriftsteller* (1873); the second, *Vom Nutzen und Nachteil der Historie* (1874), made the cleavage between Nietzsche's attitude and that of contemporary historians unmistakably clear; the third and fourth were devoted to two figures of whom Nietzsche then approved, *Schopenhauer als Erzieher* (1875) and *Richard Wagner in Bayreuth* (1876). But the adoration for Wagner waned from 1876 on. A complete breach came with the publication of volume 1 of *Menschliches, Allzumenschliches* (q.v.) in 1878–80. This had an epilogue ten years later in Nietzsche's *Der Fall Wagner* (1888).

Between his resignation in 1879 and 1888 Nietzsche was a sick man suffering from violent headaches and eye trouble. In this period he made frequent changes of residence from Sils Maria in Switzerland, to Sorrento, thence to Genoa, Turin, and Nice, but his symptoms continued unabated. He also suffered from personal crises. The breach with Wagner remained unhealed; difficulties arose with his sister Elisabeth (see FÖRSTER-NIETZSCHE, E.) and his mother; and in 1882 he fell in love with Lou Andreas-Salomé (q.v.), proposed to her by proxy and was rejected. In the clear periods of his mind, however, he pursued the process of relentless disillusionment and demasking of false idealisms with *Vermischte Meinungen und Sprüche* and *Der Wanderer und sein Schatten* (1880) in a second volume of *Menschliches, Allzumenschliches*. The 1880s began with an attack on morality in *Morgenröte. Gedanken über die moralischen Vorurteile* (1881), a work of 575 numbered paragraphs divided into five books, the underlying idea of which is the condemnation of mere obedience to tradition—'Sittlichkeit ist Gehorsam gegen die Sitten'. The assault on accepted morality, and especially Christian morality, is continued in *Jenseits von Gut und Böse* (q.v., 1886) and *Zur Genealogie der Moral* (1887), which, written at speed in three weeks, savagely underlines some of the points of its predecessor. Nietzsche attempted something more positive in *Die fröhliche Wissenschaft* (1882, revised version 1886), which sets out his idea of the true man, free of prejudice, a European in his thought, rather than a German. The book, though cast in the familiar form with 383 aphoristic paragraphs in five books, betrays the lyrical and subjective element, both in the groups of poems with which it begins and ends, and in the heading given to the fifth book, *Wir Furchtlosen*. The third book ends with a Nietzschean catechism (§§ 268–73). The subjective tone becomes much more marked in the rhapsodical utterances of Nietzsche's best-known work, *Also sprach Zara-*

thustra (q.v.), of which Pts. I–III were published in 1883–4 and Pt. IV privately in 1885, though the complete work did not appear until 1892. It is in *Also sprach Zarathustra* that the 'superman' (Übermensch) appears. Related to this work are the rhapsodical poems in free verse, often impressive, but of uneven quality, published as *Dionysos-Dithyramben* (1888). In the same year appeared *Der Antichrist*, his bitterest attack on Christianity with its 'Sklavenmoral', to which he opposed his own 'Herrenmoral'.

As early as 1883 Nietzsche conceived the idea of a systematic master-treatise on his philosophical views, which were otherwise scattered throughout diverse writings. To this conception he gave in 1885 the title *Der Wille zur Macht. Versuch einer Welt-Auslegung.* He never completed this work, for which he made abundant notes and aphoristic formulation, but it was published after his death, edited and arranged by his sister and P. Gast, under the title *Der Wille zur Macht. Versuch einer Umwertung aller Werte* (q.v.) in 1906. Its presentation has been the subject of much controversy, and a scholarly edition of what Nietzsche actually wrote did not appear until 1960 under the title *Aus dem Nachlaß der Achziger Jahre.* In addition to *Der Fall Wagner* of 1888, Nietzsche worked in this year on *Götzendämmerung, oder: Wie man mit dem Hammer philosophiert* (1889, English title *Twilight of the Idols*), which is concerned with his admiration for the highly developed instinct of the Greeks to be strong, to have, in his own term, the 'Will to Power'. *Ecce homo*, not published until 1908, is a work of self-justification, written in a tone of overconfident assertion. The last ten and a half years of Nietzsche's life were spent in increasing physical paralysis and mental illness, in which he was devotedly cared for by his mother and sister in Naumburg and, after 1897, in Weimar.

Nietzsche's searing and destructive criticism of the culture of his age is a remarkable individual achievement, and he may be said to have foreseen with accuracy many of the developments of the 20th c. Though some still see him as a systematic philosopher, he is more generally viewed as a cultural critic of outstanding vision, integrity, and ruthlessness. Nietzsche's personality was undoubtedly deeply split, and the consequent contradictions and inconsistencies have laid his work exceptionally open to misinterpretation and exploitation. The gentleness and pure love of truth which were in him were overcompensated by an insistence on brutality and an arrogant dogmaticism. In reality no one was less like the 'blonde beast' ('blonde Bestie') or the 'superman', two of the clichés most frequently cited from his work. The cleavage in the personality is probably one reason why the more poetic works seem less satisfactory as time

gives more perspective. But as a clear-eyed critic Nietzsche deserves the highest reputation.

The phrases 'Herrenmoral' and 'Sklavenmoral', the idea of the 'blonde beast', and the conception of the 'will to power' were easily adapted and prostituted to nationalistic ends in Germany, and this occurred in the Wilhelmine era before 1918, and, more brazenly, in the National Socialist period.

Nietzsche's *Werke, kritische Gesamtausgabe,* ed. G. Colli and M. Montinari, planned in 30 vols., began appearing in 1967 and correspondence (*Briefwechsel*) (20 vols.) in 1975 ff.

NIGRINUS, GEORG, real name Schwarz (Battenberg, Hesse, 1530–1602, Alsfeld), a school rector and pastor in Gießen and later a moderator (Superintendent) in Alsfeld, was polemically active on the side of Fischart (q.v.) against Johannes Nas (q.v.). His principal pamphlet is *Von Bruder Joh. Nasen Esel und seinem rechten Titel* (1570), in which Nas figures as a donkey.

NIKOLAUS VON CÜES, also known as Nikolaus von Cusa, and Cusanus (Cües/Mosel, 1401–64, Todi, Umbria), was a noted churchman of German origin (his family name was Krebs). After studying at Heidelberg and Padua, he was a priest at St. Florian in Koblenz. In 1432 he was present at the Council of Basel, and was papal legate at the Diets of Mainz, Nürnberg, and Frankfurt. A cardinal in 1448, he was consecrated bishop of Brixen (Bressanone) in south Tyrol in 1450. Nicholas, who was much concerned with the reform and unity of the Church, was accessible to humanistic learning, proposing a reform of the calendar (*De reparatione Calendarii,* 1436) and anticipating Copernicus's discovery of the rotation of the earth. He produced the first map of central Europe, which was engraved in Eichstätt in 1491, twenty-seven years after his death.

NIKOLAUS VON DER FLÜE (1417–87), a Swiss hermit, who lived in Melchtal, was declared Blessed in 1669 and canonized in 1947. His intervention, preventing a Swiss civil war in 1481, has made him a national hero, popularly known as Bruder Klaus. His biography has been written by H. Salat, and he is the subject of a play by C. von Arx (qq.v.).

NIKOLAUS VON JEROSCHIN, a priest of the Teutonic Order (see DEUTSCHER ORDEN), wrote a life of St. Adalbert and a verse chronicle of the Order from its beginnings to 1330. The life of St. Adalbert, Bishop of Prague, who was killed in 991 in an attempt to convert the heathen in Prussia, is a translation of a Latin

Vita. Die Kronike von Pruzinlant was written between 1330 and 1340 and adapts a Latin chronicle by another member of the Order, Peter von Dusberg (probably Duisburg), written in 1326 (*Chronica terre Prussie*). It is a substantial work of nearly 28,000 lines, which, though ill-constructed and often repetitive, reveals poetic and metrical skill.

NIKOLAUS VON LANDAU, a Cistercian monk of Otterberg Abbey near Kaiserslautern, completed in 1341 a collection of sermons (part of which is lost). The methodology is scholastic, but the sermons contain mystical elements.

NIKOLAUS VON LÖWEN, Nicholas of Louvain (1339–1402), secretary to Heinrich Blankhart, a bookseller in Strasburg, and possibly later to Ruland Merswin (q.v.), may have originated the myth of the mystic 'der Große GOTTESFREUND vom Oberlande' (q.v.).

Nimmersatte Liebe, a poem written by E. Mörike (q.v.) in 1828.

NISSEL, FRANZ (Vienna, 1831–93, Gleichenberg), son of an actor, spent his life as a mostly unsuccessful dramatist. His highest point was reached in the verse tragedy *Agnes von Meran* (1877), which was awarded the Schiller prize in 1878 and for a short time held a place in the repertoire of the Burgtheater. *Ein Wohltäter* (1854) and *Die Zauberin am Stein* (1864) also made brief appearances in the Burgtheater. A selection of his numerous comedies and pseudo-Schillerian tragedies appeared in *Ausgewählte dramatische Werke* (1892).

NITHART, an early German (or Frankish) historian, grandson of Charlemagne (see KARL I, DER GROSSE), wrote in Latin a chronicle of the wars between his cousins, the sons of Ludwig I, der Fromme (q.v.), under the title *Historiarum libri quatuor*. He was killed in these wars, probably in 844.

NIVARDUS OF GHENT, a Flemish cleric, who wrote *c.* 1150 *Ysengrimus*, the beast story of the fox and the wolf, using the Germanic names which later occur in the German beast stories (see HEINRICH DER GLÎCHEZAERE and REINKE DE VOS).

Nobel Prize for Literature, conferred anually in October. German recipients (including Swiss who write in German) are: Th. Mommsen (1902), R. Eucken (1908), P. Heyse (1910), G. Hauptmann (1912), C. Spitteler (1919), Th. Mann (1929), H. Hesse (1946), Nelly Sachs (1966), H. Böll (1972), and E. Canetti (1981), qq.v.

NOBILING, KARL EDUARD (Kolno, 1848–78, Berlin), a minor civil servant, made an attempt to shoot the Emperor Wilhelm I (q.v.) on 2 June 1878. Nobiling, whose motives were anarchistic, tried to commit suicide and died of his wound three months later. This outrage, occurring only a few weeks after an earlier attempt by one Hödel on the Emperor's life, was used by Bismarck (q.v.) as a pretext for rapid anti-socialist legislation, though Nobiling's links with the Social Democratic Party (see SPD) were tenuous.

Noch ein Lembeck, original title of Th. Storm's Novelle *Ein Fest auf Haderslevhuus* (q.v.).

'Noch ist Polen nicht verloren', first line of a poem published in *Polenlieder* (1831) by E. Ortlepp (q.v.).

NOLDE, EMIL (Südtondern, 1867–1956, Seebüll, Schleswig-Holstein), painter and draughtsman, had his first training in Flensburg, was afterwards in Munich and Paris, then settled in Berlin. He was a member of Die Brücke (q.v.), but was not a true Expressionist. In 1913–14 he visited the Far East. Nolde spent the rest of his life between Seebüll and Berlin, painting highly individual works with brilliant, even glaring, colour contrasts. Many of his later works are religious. In 1933 he was put on the list of degenerate artists (see ENTARTETE KUNST) and forbidden to paint. He is an outstanding master of the woodcut and his etchings are also of high quality. His real name was Hansen, and he is one of the three painters mentioned by S. Lenz (q.v., the others are Beckmann and Kirchner, qq.v.), as models for the character Max Nansen in the novel *Deutschstunde* (q.v.).

NONNE, JOHANN HEINRICH CHRISTOPH (Lippstadt, 1785–1853, Schwelm), became pastor in Schwelm in 1815. He is the author of a popular patriotic poem, 'Flamme empor!'.

Norddeutscher Bund (North German Confederation), a federal association founded in 1866 after the defeat of Austria by Prussia (see DEUTSCHER KRIEG). It was designed by Bismarck (q.v.), possibly as a half-way house to a unified German state without Austria. A constitution was adopted which included a North German Reichstag and named the king of Prussia as its federal president. The constitution was so drawn up that it could easily be modified at a future date.

Prussia was overwhelmingly powerful in the Confederation, since it had increased its existing majority of the population by the annexation of those North German states which had taken the Austrian side (Hanover, Electoral Hesse, Nassau, and the Free City of Frankfurt), as well as the duchies of Schleswig and Holstein, about which the war was fought (see SCHLESWIG-HOLSTEINISCHE FRAGE). The original members of the North German Confederation, apart from Prussia, were Oldenburg, Bremen, Hamburg, Lübeck, Mecklenburg-Schwerin, Mecklenburg-Strelitz, Lippe, Schaumburg-Lippe, Braunschweig, Anhalt, Sachsen-Weimar, Sachsen-Altenburg, Sachsen-Coburg-Gotha, Schwarzburg-Sondershausen, Schwarzburg-Rudolstadt, Waldeck, and Reuß jüngere Linie, to which were added certain opponents of Prussia, the kingdom of Saxony, the territory of Grand Ducal Hesse (Hessen-Darmstadt) north of the Main, Sachsen-Meiningen, and Reuß ältere Linie. The Norddeutscher Bund ceased to exist with the proclamation of the German Empire (Deutsches Reich) in January 1871.

Nordische Aufseher, Der, see CRAMER, J. A.

Nord-Ostee-Kanal (formerly Kaiser-Wilhelm-Kanal), commonly known as the Kiel Canal, connects the North Sea with the Baltic, running from Brunsbüttelkoog on the Elbe to Holtenau near Kiel. It was constructed between 1887 and 1895 to serve both naval and mercantile purposes. The foundation stone was laid by the Emperor Wilhelm I and the canal was opened by Wilhelm II (qq.v.), hence the name Kaiser-Wilhelm-Kanal, by which it was known officially until 1918. It was widened 1909–15.

Nordsee, Die, a collection of poems by H. Heine (q.v.), forming the last section of the *Buch der Lieder* (q.v., 1827); it is arranged in two cycles of twelve and ten poems. The date for the composition of these poems, in which seascapes and love intermingle, is given as 1825–6. *Die Nordsee I* first appeared in vol. 1 of *Reisebilder* (q.v., 1826), *Die Nordsee II* was first published in vol. 2 (1827) and transferred in 1831 to vol. 1. The prose essay *Die Nordsee III*, written on Norderney, was included in *Reisebilder*, vol. 2.

Nordsternbund, a group name adopted for a short time *c.* 1804 by a number of Romantic writers then living in Berlin; they included A. W. Schlegel, Chamisso, F. de la Motte Fouqué, VARNHAGEN von Ense, J. E. Hitzig, and D. F. Koreff (qq.v.). They were associated with *Der grüne Almanach* (1804–6).

Nord und Süd, a political monthly founded in 1877 by Paul Lindau (q.v.). It ceased to appear in 1930.

NOSKE, GUSTAV (Brandenburg, 1868–1946, Hanover), a working man, became a political journalist in Chemnitz, and was from 1906 to 1918 a member of the Reichstag for the Social Democratic Party (see SPD). After the split in 1916 he belonged to its right wing (Mehrheitssozialisten). In the provisional council of representatives (Rat der Volksbeauftragten) during the winter of 1918–19 he acted as minister for defence affairs. In January 1919, with the co-operation of General von Lüttwitz, he suppressed the Spartacist revolt (see SPARTAKUSBUND); he was responsible, as first war minister (February 1919), for founding the new army (see REICHSWEHR). The Kapp-Putsch (q.v.) in 1920 cost him his office. In the same year he became Oberpräsident of the Prussian province of Hanover. During the National Socialist regime he was deprived of office and twice interned, in 1939 and in 1944. His writings include *Von Kiel bis Kapp* (1920) and *Erlebtes aus Aufstieg und Niedergang einer Demokratie* (1947).

NOSSACK, HANS ERICH (Hamburg, 1901–77, Hamburg), the son of a coffee merchant, studied law and philosophy and from 1956 lived entirely by his writings, begun through the encouragement of H. Kasack (q.v.). In 1933 he was banned as an author, and in 1943 he lost all his early manuscripts (including a play on Lenin and one dealing with the circle of G. Büchner and *Der Hessische Landbote*, qq.v.) in the severe bombing of Hamburg from 24 July to 3 August. The harrowing experience of those days is faithfully reproduced in *Der Untergang*, published in 1948 with nine other episodes dealing with his experience of the 1939–45 War and its immediate aftermath under the title *Interview mit dem Tode*; in 1950 this collection was renamed *Dorothea*. It was preceded by another prose work, entitled *Nekyia. Bericht eines Überlebenden* (1947), a web of dream, fantasy, and recollection passing through the mind of a survivor of an appalling catastrophe.

Nossack's work is permeated by the horror of the world, as he has come to see it, and the groping for a means of living in and with it. His novels include *Spätestens im November* (q.v., 1955), *Spirale. Roman einer schlaflosen Nacht* (1956), written in the form of visions of the unforeseeable, *Der jüngere Bruder* (1958), and *Nach dem letzten Aufstand* (1961). In spite of nihilistic appearances Nossack (who came under the influence of Camus) is fundamentally in search of hope, of which a ray is suggested in the novel *Der Fall d'Arthez* (1968), which was followed by *Dem unbekannten Sieger* (1969) and *Die gestohlene Melodie* (1972). He is represented in *PEN. Neue Texte deutscher Autoren*

(1971) with the story *Um es kurz zu machen*, a macabre piece of black humour. *Begegnung im Vorraum* (1958) is the title-story of a collection published in 1963; *Das Mal und andere Erzählungen* followed in 1975. Nossack was vice-president of the Akademie für Wissenschaften und Literatur in Mainz; *Das Verhältnis der Literatur zu Recht und Gerechtigkeit* (1967) is an address to the Academy. *Rede am Grabe* (1960) contains his funeral oration for H. H. Jahnn (q.v.) who, with E. Barlach (q.v.), had influenced his work. Nossack received the Büchner Prize in 1961.

NOSTRADAMUS, latinized form of the name of Michel de Notredame (St. Remy, 1503–66, Salon), an astrologer who achieved a European reputation. He is mentioned in *Faust* (q.v.), Part I: 'Und dies geheimnisvolle Buch, / Von Nostradamus' eigner Hand.'

Nothanker, Sebaldus, see LEBEN UND DIE MEINUNGEN DES HERRN MAGISTER SEBALDUS NOTHANKER, DAS.

Nötiger Vorrat zur Geschichte der deutschen dramatischen Dichtkunst (1757–65), a bibliographical work published by J. C. Gottsched (q.v.).

NOTKER BALBULUS (Notker der Stammler) (Elgg, Switzerland, *c.* 840–912, St. Gall), a learned monk of St. Gall Abbey, wrote *Gesta Karoli Magni* and originated the Sequenz (q.v.) in Germany, leaving some forty examples with their music. He wrote exclusively in Latin.

NOTKER LABEO (*c.* 950–1022, St. Gall), also called Notker Theutonicus because of his devotion to the German language, was in charge of the school at St. Gall Abbey. He came of a family that provided several notable monks, including Ekkehard I (q.v.), who was Notker's uncle. He appears to have been a man of attractive personal qualities. Notker's field of activity was limited to the monastic school, and his influence on wider circles is no longer believed to have been considerable. A careful scholar devoted to classical interests, he subordinated these to theological ends out of a sense of religious duty. Of his works, all designed for school use, his translations of Boethius and Martianus Capella, of Aristotle's *Hermeneutics*, and of the Psalms survive. Versions of *Disticha Catonis*, Virgil's *Bucolics*, Terence's *Andria*, and Gregory the Great's commentary on Job are lost.

Notker is especially remarkable for the command of language which enabled him to translate with a previously unequalled accuracy and flexibility. He combined this capacity for expression with a devotion to his mother tongue, which made him, linguistically, the outstanding German writer of his century.

NOVALIS, the pseudonym and universally accepted designation of Friedrich, Freiherr von Hardenberg (Oberwiedstedt nr. Mansfeld, 1772–1801, Weißenfels), was formerly accented on the second syllable, but it has been established that the poet himself placed the stress on the first syllable. The young Hardenberg grew up in a strictly pietistic household and was educated by private tutors until, at 18, he was sent for a year to the grammar school at Eisleben. In 1790 he entered Jena University, where he made the acquaintance of Schiller, and was one of those students who helped to watch at Schiller's bedside during the latter's critical illness in 1791. From Jena he migrated to Leipzig University, where he studied with his brother and met the young Friedrich Schlegel (q.v.). He completed his university career at Wittenberg in 1792, and in 1793 was sent to learn the technique of administration at Tennstedt. In 1794–5 he devoted his spare time to philosophical studies, especially of Fichte and Kant (qq.v. and see MAGISCHER IDEALISMUS).

On a visit to Grüningen in November 1794 Novalis met the 12-year-old Sophie von Kühn, with whom he immediately fell deeply in love. They were betrothed four months later, and in the same year Sophie developed pulmonary tuberculosis. During her illness Novalis was working as an administrative assistant in the salt-mine offices at Weißenfels and in the stress of these months, which was augmented by the illness and death of his brother, he underwent a profound religious experience. The death of Sophie in March 1797 led to a crisis, a reckoning with death, which finds expression in the *Hymnen an die Nacht* (q.v., written 1799, published in *Das Athenäum*, q.v., 1800). In the same year (1797), he entered the school of mining at Freiberg, Saxony, at the same time continuing his philosophical reading and contributing *Blütenstaub* (q.v.) to *Das Athenäum* in 1798. Towards the end of his stay he became engaged to Julie von Charpentier. He returned to mine management at Weißenfels in 1799, renewing contact with Friedrich Schlegel and making visits to Jena, where both Schlegels (see SCHLEGEL, A. W. VON) and Schelling (q.v.) were living. At this time he became a close friend of Ludwig Tieck (q.v.). In the summer of 1800 symptoms of tuberculosis appeared and Novalis died after an illness of seven months.

In his last years (1799–1801) Novalis not only completed his *Hymnen an die Nacht* but also wrote the hymns of *Geistliche Lieder* (q.v.), the historical essay *Die Christenheit oder Europa* (q.v.), and the fragmentary novels *Die Lehrlinge zu Sais* and *Heinrich von Ofterdingen* (qq.v.), in the latter of which he created the Romantic symbol of the blue flower (die blaue Blume). Novalis, both by temperament and by creative gifts, was the truest poet of the first Romantic School (see ROMANTIK), though the philosophical notes and fragments in his posthumous papers are greater in bulk than his creative writings.

Schriften (2 vols.), ed. F. Schlegel and L. Tieck, were published in 1802; *Schriften, historisch-krititische Ausgabe* (4 vols.), ed. P. Kluckhohn and R. Samuel, in 1929 and, revised, 1960–75; 3rd rev. edn. 1977 ff.

Novara, scene of a battle won by the Austrians under Radetzky (q.v.) on 23 March 1849 over Italian forces. See REVOLUTIONEN 1848–9.

Novelle became prominent in German literature both as a term and as a genre of fiction at the end of the 18th c. It was introduced by Goethe with *Unterhaltungen deutscher Ausgewanderten* (q.v.) in 1795, which is modelled on the *novella* as it occurs in Boccaccio's *Decameron* (q.v.), from which it adapted the cyclic frame (see RAHMEN). Prominent features of the Novelle are the concentrated presentation of an action which arouses suspense and contains an element of surprise leading to an unexpected ending. This element of surprise is the 'new' (*novella*) feature of the narrative, exposing seemingly inexplicable aspects of reality which are inaccessible to reason.

The Romantics favoured supernatural motifs to convey irrational experience (Novellenmärchen), the first example being Tieck's *Der blonde Eckbert* (q.v., 1797). The various Novellen inserted in Goethe's novels substantiated the association of the genre with society and its moral values. This is likewise a feature of the narrative work of Heinrich von Kleist (q.v.), the first important exponent of the German Novelle, who is noted for the metaphysical content of his stories. Kleist, like many writers who are generally referred to as the authors of Novellen, published his stories as Erzählungen. A variety of definitions of the genre emerged from the turn of the 19th c., at first based on non-German models, especially Boccaccio (e.g. Friedrich and August Wilhelm Schlegel, qq.v.), but, in view of the proliferation of the genre during the first two decades of the 19th c., on German models as well. Mindful of this as well as of the need for a title for one of his own creations, Goethe pronounced a simple solution (see NOVELLE by Goethe), which by its forceful wording has become an indispensable reference for the definition of the genre, the 'sich ereignete unerhörte Begebenheit' (1827). A. W. Schlegel

had regarded turning-points in the narrative as a formal requirement; L. Tieck (1829) stressed the need for a point (Wendepunkt) at which the story took an unexpected, decisive turn, a theory which has been compared with the Aristotelian peripety, so suggesting a parallel with drama, which was emphasized by Th. Storm (q.v., 1881). P. Heyse (q.v.) in his introduction to the *Deutscher Novellenschatz* (1871–6) claimed that the Novelle should have a definite silhouette contained in a brief summary distinguishing it from all others, quoting the falcon image deriving from the *Decameron* (V.3) as an exemplary feature of the symbolical allusiveness of the narrative (Falkentheorie).

The development of the German Novelle as a major genre of fiction is above all due to the high level achieved by writers associated with Poetic Realism (see POETISCHER REALISMUS). Prominent authors writing in the German language in the late 19th c. and in the early 20th c. have applied the term Novelle to works whose diversity accounts for continued attempts at detailed definition.

Novelle, a Novelle by Goethe (q.v.), published in 1828. First conceived as an epic, *Die Jagd,* in 1797, it took Goethe almost thirty years to find for the projected material a suitable form, that of the Novelle, of which he could by now claim that it was a genre harbouring 'gar vieles wunderliche Zeug'. As early as 1795 Goethe himself had introduced this genre into German literature with the publication, in Schiller's *Die Horen* (q.v.), of the *Unterhaltungen deutscher Ausgewanderten* (q.v.) which he had modelled on Boccaccio's *Decameron.* He called the story, for want of a thematically characterizing title (*Die Jagd* was no longer apt), simply *Novelle*: 'Wissen Sie was, wir wollen es die *Novelle* nennen; denn was ist eine Novelle anders als eine sich ereignete unerhörte Begebenheit' (to Eckermann, 29 January 1827). The *Novelle* was published in vol. 15 of the *Ausgabe letzter Hand. Daniel in der Löwengrube* (by F. K. Moser, q.v., 1763) and the pietist environment in Frankfurt (Susanna von Klettenberg, q.v.), as well as the tale of Prince Achmed and the fairy Pari Banu from the *Arabian Nights,* are considered to have influenced the conception of the story.

Moving between the new and the old castle of a Fürst, it relates the adventure of a small ducal party which, on riding towards the ruined old castle in the wooded hills, finds itself confronted by a tiger. The animal escaped from its cage when fire broke out in the menagerie by the market-place in the small town in the valley below. In order to protect the Fürstin, Honorio, a courtier, promptly shoots the tiger, hoping, too, thereby to win the favour of the lady. Its death is bitterly lamented by the family owning the beautiful and harmless animal. As a lion has also found its way into the wood, its owner pleads with the Fürst not to let it suffer the fate of the tiger; and to the amazement of the party, the owner's little boy safely entices the lion by the gentle melody of his flute and song into the old courtyard as if into an arena, whence it can be returned to its cage. The incident is heightened by a hint of legend (Androcles, St. Jerome) as the child removes a thorn from the lion's paw and dresses it with his silk neckerchief.

The story is an outstanding example of the stylized prose of Goethe's last creative period, which sets out to harmonize opposites at all levels by showing the interaction of art and nature as it is inherent in his conception of polarities. The finale demonstrates, in close analogy to the prophet Daniel, the triumph of the naïve and trusting child over the seemingly ferocious force of nature represented by the lion ('Tyrann der Wälder' and 'Despot des Tierreiches').

Novellen aus Österreich, a collection of five Novellen (*Innocens, Marianne, Die Steinklopfer,* qq.v., *Die Geigerin,* and *Das Haus Reichegg*), published by F. von Saar (q.v.) in 1877. Saar used the same title for an expanded collection published in two volumes in 1897. This included the above five Novellen and nine others. Of these *Vae victis!* (q.v.), *Der 'Exzellenzherr',* and *Tambi* had been published in *Drei neue Novellen* (1883), *Leutnant Burda, Seligmann Hirsch,* and *Die Troglodytin* in *Schicksale* (1889), and *Ginevra* and *Geschichte eines Wiener Kindes* in *Frauenschicksale* (1892). The fourteenth Novelle, *Schloß Kostenitz,* had appeared separately in 1893.

Saar's later collections *Herbstreigen* (1896), *Nachklänge* (1899), *Camera obscura* (1901), and *Tragik des Lebens* (1906) have often been included by later editors in *Novellen aus Österreich,* but this arrangement was not sanctioned by the author.

Novemberrevolution, the revolution in Germany which began with the mutinies of the High Seas Fleet in Kiel and Wilhelmshaven on 28 October 1918 and ended with the formation of a legitimate republican government on 13 February 1919, after the election of F. Ebert (q.v.) as provisional President. The revolution was provoked firstly by the unsuccessful course of the war, then by the reluctance of the imperial and Prussian governments to make appreciable social concessions, by the irresolution of the High Command, and finally by the unwillingness of the Emperor Wilhelm II (q.v.) to accept the necessity of abdication. The occasion of the naval mutiny was a fleet order to sail, presumably against England. The sailors occupied the

northern cities, Hamburg, Bremen, and Lübeck, and the mutiny turned to revolution.

On 7 November a republic was declared in Bavaria, and on the 8th in Brunswick. On the 9th the German Republic was proclaimed by the moderate socialist P. Scheidemann (1865–1939) in Berlin, and on the 10th Wilhelm II left G.H.Q. at Spa and took refuge in Holland. In the last weeks of the year a test of strength took place between the extreme left, in which the Spartacus League (see SPARTAKUSBUND) was prominent, and the majority socialists led by Ebert, Noske (q.v.), and Scheidemann. The aim of the Communist Party (see KPD), into which the Spartakusbund converted itself in December 1918, was to prevent a general election and to introduce a 'dictatorship of the proletariat' on the Russian model. Street fighting culminated in the murder of K. Liebknecht and Rosa Luxemburg (qq.v.) on 15 January 1919, and elections duly took place on 19 January. The National Assembly (Nationalversammlung) met in Weimar on 6 February, and a provisional constitution was adopted on 10 February. The appointment of a government on 13 February may be regarded as the conclusion of the Revolution.

NSDAP, abbreviation for Nationalsozialistische Arbeiterpartei. Its origin was in a small extremist party, located chiefly in Bavaria (Deutsche Arbeiterpartei) and led by Anton Drexler (1884–1942). Hitler (q.v.) was made a member in 1919 and soon became its dominating personality. Its platform was nationalistic and anti-Semitic, and in its first years it gathered members at only a moderate rate. Its full title, signified by NSDAP, dates from February 1920. In November 1923 the Munich branch of the party, headed by Hitler, attempted an armed revolt, which was easily put down (see HITLER-PUTSCH). Hitler was tried and condemned to five years' imprisonment, and the NSDAP (at that time flourishing only in Bavaria) was banned. Hitler was released late in 1924 and reconstituted the party in February 1925. Its aim was now the achievement of power without infringement of the constitution. After four years of slow expansion, the NSDAP suddenly found a favourable climate for growth in the economic crisis of 1929 and the following years. In 1928 it had obtained 12 seats in the Reichstag, but at the elections of September 1929 the number rose to 107. From this point on, the party was a political force of national significance.

In October 1931 a demonstrative assembly at Bad Harzburg affirmed an alliance (see HARZBURGER FRONT) between the ultra-conservative parties (Deutschnationale and

Stahlhelm, q.v.) and the NSDAP. A period of stress followed, marked by Hitler's unsuccessful candidature for the presidency against Hindenburg, by street rowdyism and violence, and by a ban on the SA (q.v.). Elections held in July 1932 resulted in the return of 230 NSDAP deputies, making the party the strongest in the Reichstag. Pressure mounted for National Socialist representation and for the inclusion of the leaders in the cabinet. In November, at a further general election, the NSDAP representation dropped to 196. The feeling that the tide had turned helped to make the conservative leaders more ready to tolerate Hitler, but negotiations for the chancellorship failed at that time, only to be renewed in January 1933, when a cabinet composed mainly of conservatives was formed with Hitler as chancellor (30 January).

Hitler soon outmanœuvred his conservative colleagues, establishing a completely National Socialist government, and on 14 July 1933 the NSDAP was declared the only political party. This situation continued until the collapse of Germany in 1945. In 1934, however, the party had undergone a crisis, arising out of dissatisfaction in the leadership of the SA. This movement was scotched by Hitler in the murderous purge of 30 June 1934 (see Röhm, E.). A feature designed to impress both Germany and the world was the annual Party Rally (Reichsparteitag) held at Nürnberg, with mass parades of disciplined formations and skilful handling of propaganda techniques.

Numerus clausus, the limitation of numbers at universities.

'Nun danket alle Gott', hymn written by Martin Rinkart (q.v.) and published in 1636. It is sung in English-speaking countries to the words 'Now thank we all our God'. The tune is by J. Crüger (q.v.) and dates from 1649. It is also called *Das Danklied*.

'Nun preiset alle Gottes Barmherzigkeit', first line of a hymn by M. A. von Loewenstern (q.v.).

'Nun ruhen alle Wälder', first line of the hymn *Abendlied* by Paul Gerhardt (q.v.).

Nun singen sie wieder. Versuch eines Requiems, a play by M. Frisch (q.v.), written and performed in Zurich and Munich in 1945, and published in 1946. It consists of seven scenes (Bilder), four of which depict the land of the living (Part One), while the last three are set in a monastery symbolizing the land of the dead (Part Two). The first part shows a variety of characters caught up in the 1939–45 War and

in war crimes. When the action opens, Karl, upon the command of Herbert, has shot twenty-one hostages, among them women and children, whom the Orthodox priest buries before he is shot by Herbert, since Karl refuses this time to obey and deserts. He arrives home, confronts his father, the teacher (Oberlehrer), with their common guilt, and hangs himself at the sound of the sirens. His wife Maria and his child become victims of the air raid. The Oberlehrer is subsequently shot by his star pupil Herbert at the edge of the 'pit without hope', from whence he goes to join the dead (Scene Six).

Members of the enemy air force likewise represent different types in reaction to the war. Some of them find themselves in the land of the dead where they are offered bread and wine by the priest (Pt. Two). The dead watch the living but cannot communicate with them. Jenny, the widow of the army captain (Hauptmann), encourages her son to grow up like his father, ignorant of the dead father's warning to live a new, purposeful life and do away with ambition and ostentation. The song of the dead hostages is heard whenever an evil action occurs. The play is thus not only a commemoration (Requiem); it conveys a moral, which the survivors in the play do not sense because they do not realize that those who have gone through death have changed in spirit. The death of the victims of the war will have been in vain, Frisch implies, unless the audience accepts their message.

Nürnberg (Nuremberg), the second largest city of Bavaria, into which it was incorporated in 1806 on the dissolution of the Holy Roman Empire (see Deutsches Reich, Altes), having been a Free Imperial City (Freie Reichsstadt) since 1472. Founded *c.* 1000, it was ruled from 1105 to 1472 by Burggrafen, who from 1191 were Zollerns, ancestors of the Hohenzollern (q.v.) dynasty of Prussia. The city, being rich, acquired its independence by purchase. It was an important centre of trade in the Middle Ages and had outstanding local craftsmen, among them V. Stoß (*c.* 1445–1533), A. Krafft (*c.* 1460–*c.* 1508), and P. Vischer (*c.* 1460–1529), some of whose work can still be seen in the city churches, St. Sebaldus and St. Lorenz.

Nürnberg was the birthplace and home of Albrecht Dürer (q.v.), and is well known for its guild of Meistersinger (see Meistergesang), who are idealistically presented in Wagner's *Die Meistersinger von Nürnberg* (q.v.). Hans Sachs and H. Folz (qq.v.) are the best-known members of the guild. Nürnberg was also the centre of 16th-c. humanism (see Humanismus), represented by W. Pirkheimer and C. Celtis (qq.v.).

In 1835 the first German steam railway was

opened between Nürnberg and Fürth 4 miles away (Ludwigs-Eisenbahn, named after Ludwig I of Bavaria, q.v.). Nürnberg's maze of medieval streets made it a favourite city of the Romantics, beginning with Wackenroder and L. Tieck (qq.v.). In the 20th c. the National Socialist regime exploited the city's prestige by its annual rallies (Reichsparteitage). The war-crime trials following the 1939–45 War were held in Nürnberg, which lost most of its old houses through bombing in the later stages of the war. The principal larger buildings, including churches and the city walls, have either survived or been restored.

NÜRNBERGER, WOLDEMAR, pseudonym M. Solitaire (Sorau, Lausitz, 1818–69, Landsberg/Warthe), a physician at Landsberg in the Neumark, west of Küstrin, wrote the poem *Josephus Faust* (1842) and a number of stories, which include *Dunkler Wald und gelbe Düne* (1855), *Erzählungen bei Nacht* (1858), *Erzählungen bei Licht* (1860), and *Erzählungen bei Mondenschein* (1865). All his works appeared under his pseudonym.

Nürnberger Chronik, see MEISTERLIN, SIGMUND.

Nürnberger Religionsfriede, a religious truce between the Catholic and Protestant factions entered into by Karl V (q.v.) and the estates of the Empire in 1532; it was intended to last until the next ecumenical council and closes the second phase of the Reformation (q.v.).

Nürnbergischer Dichterkreis, see HIRTEN- UND BLUMENORDEN AN DER PEGNITZ.

Nürtingen, a small town in Württemberg north-east of Tübingen. It was the home of F. Hölderlin (q.v.) from 1774 to 1784 and was often visited by him later, as his mother continued to reside there. Hölderlin's stepfather, who died in 1779, was burgomaster of Nürtingen. The mother of E. Mörike (q.v.) moved to Nürtingen in 1825 and the poet often visited her, and in 1870–1 he himself settled there.

Nußbaum, Der, a poem by J. Mosen (q.v.), which has been beautifully set to music by R. Schumann (q.v.).

Nußbraune Mädchen, Das, a story woven into the plot of Goethe's *Wilhelm Meisters Wanderjahre* (q.v., Bk. I, ch. 11). The girl, by name Nachodine, is the daughter of a tenant farmer who has been evicted by Lenardo's uncle. The girl appealed to Lenardo for help, and the latter feels in his conscience that he did not do enough for her. Later in the novel Nachodine, now called Susanne and referred to by Goethe as 'die Schöne-Gute', is discovered by Wilhelm Meister in a weavers' settlement. She is drawn into the circle of Makarie and the Society of the Tower (Gesellschaft vom Turm), and the prospect of marriage with Lenardo is opened.

Nußkernen, Das. Eine pfälzische Idylle, an idyll (see IDYLLE) by F. Müller (q.v.), written in prose with inserted songs *c.* 1781, and published in *Werke* (ed. L. Tieck, q.v.) in 1811. A group of villagers in the Palatinate has gathered to shell nuts (Das Nußkernen). While the characters are busy shelling and talking, one of them, a young man who is the only newcomer, is shown to be the long-lost son of a villager in the group. The young man falls in love with one of the girls present and later becomes engaged to her. A contrasting theme, frequently treated by the Sturm und Drang (q.v.), is interpolated: the moving story of a girl executed for infanticide, which is related by the schoolmaster.

Nußknacker und Mausekönig, a fairy-tale (Kunstmärchen) by E. T. A. Hoffmann (q.v.), published in 1819 in vol. 1 of Hoffmann's *Die Serapionsbrüder* (q.v.). It is a fantasy in which a child (Marie) falls in love with a nutcracker she has received as a Christmas present. In the night the nutcracker comes to life and fights a battle with the seven-headed mouse king. The nutcracker turns out to be a young man bound under a spell. Marie releases him and they float away and are married. The story alternates between a vividly realized everyday world and a realm of fantastic imaginings. Obergerichtsrat Drosselmeier leads a double life, moving equally easily in each of these two worlds. The story forms the basis of Tchaikovsky's *Casse-Noisette* ballet and suite.

Nyland, see WERKLEUTE AUF HAUS NYLAND.

O

Oberaltaicher Sammlung, a collection of 63 Middle High German sermons made in the 13th c. and 14th c. It originated in Oberaltaich Abbey, not far from Straubing in eastern Bavaria.

Oberammergauer Passionsspiel, the most celebrated of German traditional plays presenting the life and passion of Jesus, was first performed in 1634 in pursuance of a vow made by the inhabitants of Oberammergau when plague threatened in 1633. Since 1680 it has been performed at ten-year intervals in the last year of the decade (1680, 1690, etc.). The original play was devised on the basis of various earlier plays, including the *Augsburger Passionsspiel* and a passion play (1566) by Sebastian Wild (q.v.) of Augsburg. The oldest existing text is that of 1662; modifications were made to it in 1680, 1750, 1780, and 1811. The Festspielhaus in its present form dates from 1930.

OBEREIT, JAKOB HERMANN (Lindau?, 1725–98, Jena?), grew up in Lindau, where he afterwards practised medicine. In 1776 he moved to Winterthur, and in later life he ranged restlessly from place to place. He was a pietistic Christian of mystical bent, but left no writings of importance. He, and not J. J. Bodmer (q.v.), is the true discoverer of MS. C of the *Nibelungenlied* (q.v.).

OBERKOFLER, JOSEPH GEORG (St. Johann/Ahrn, South Tyrol, 1889–1962, Innsbruck), brought up in Brixen (now Bressanone and Italian), fought in the Austrian army in the 1914–18 War and was afterwards a journalist at Bolzano, finally moving to Innsbruck. He wrote verse, much of it hymnic (*Stimmen aus der Wüste*, 1918; *Gebein aller Dinge*, 1921; *Triumph der Heimat*, 1925); and is the author of several regional novels (*Sebastian und Leidlieb*, 1926; *Das Stierhorn*, 1938; *Das rauhe Gesetz*, 1938; *Der Bannwald*, 1939; *Die Flachsbraut*, 1942). Later volumes of verse are *Nie stirbt das Land* (1937) and *Verklärter Tag* (1950). *Wie eine Mutter ging* (1960) is autobiographical.

Oberlande, Der große Gottesfreund vom, see GOTTESFREUND VOM OBERLANDE, DER GROSSE.

OBERLIN, JOHANN FRIEDRICH (Strasburg, 1740–1826, Waldersbach), was from 1767 until his death Protestant pastor in Waldersbach in the Steintal at the foot of the Vosges between Strasburg and Colmar. Oberlin made it his task to improve the lot of the villages in his cure materially, culturally, and spiritually. He introduced cotton weaving, founded nursery centres (Kinderbewahranstalten, the first of their kind), for which he trained women teachers, and offered asylum to French refugees during the revolutionary years. His *Lebensgeschichte* and *Gesammelte Schriften* were published in four volumes in 1843 (ed. W. Burckhardt, Hilpert, and D. E. Stöber, q.v.). Several notable men of letters and learning were counted among his friends (J. C. Lavater, J. H. Jung, qq.v.). G. H. von Schubert (q.v.), who in 1835 published *Züge aus dem Leben von Oberlin*, published the third edition of *Die Symbolik des Traumes Mit einem Anhange aus dem Nachlasse eines Visionärs des J. F. Oberlin, gewesenen Pfarrers im Steinthale* (1840; in accordance with Schubert's wish the Anhang was not included in the fourth edition of 1862).

Oberlin's life and work have been the subject of various works of literature (among them the novel *Oberlin*, 1910, by F. Lienhard, q.v.) especially since the revival of interest in J. M. R. Lenz and G. Büchner (qq.v.). In his Novelle *Lenz* (q.v.) Büchner follows Oberlin's record of Lenz's stay with him early in 1778. This record was first published by A. Stöber (q.v., son of D. E. Stöber, the co-editor of Oberlin's works and author of a biography of Oberlin, 1831) in 1839 in the journal *Erwina*. The father of Büchner's fiancée, J. J. Jaeglé (see JAEGLÉ, M.), delivered the funeral sermon on Oberlin at Fouday, where he was buried.

Oberon, a verse romance (Ein Gedicht in zwölf Gesängen) written by C. M. Wieland (q.v.) in 1778–80 and published in 1780. Written in *ottava rima*, it comprises *c.* 7,000 lines. It combines the story of the quarrel between Oberon and Titania with the story of Hüon de Bordeaux.

Hüon has the misfortune to kill Charlemagne's son in self-defence. The Emperor, after threatening him with death, imposes banishment unless Hüon succeeds in plucking the beard and pulling four teeth of the Sultan of Baghdad, and running away with the Sultan's daughter. In Syria Hüon finds an honest German, Scherasmin, who becomes his squire; and he receives powerful protection from Oberon, who gives him a magic horn and goblet. Hüon encounters the Sultan's daughter Rezia, and the two fall in love and elope. Oberon relieves Hüon of the distasteful task of dentistry and beard-plucking, and the loving pair set out on their return with Scheras-min and Rezia's maid Fatme. Oberon makes a condition of chastity, which they infringe. From this moment perils beset Hüon and Amanda (as Rezia is now called). They are thrown into the sea, stranded on a desert island; they lose the child which has been born to them, and Amanda is carried off by pirates to Tunis. The other characters are brought to the same spot by Oberon and Titania, but Hüon, disguised as a gardener, is unable to reach Amanda, who is in the Sultan's palace. In his efforts to find her he meets the Sultana, who pursues him after the manner of Potiphar's wife. He remains steadfast, is accused by her of sexual assault, and con-demned to be burned. Amanda repulses the Sultan and is to suffer the same fate. Oberon intervenes; the four characters are transported to France, and Hüon, after triumphing in a tournament, is restored to favour.

It is a poem of light touch, ironical yet affectionate, celebrating constancy, and is primarily intended to delight, an end which it achieves in a civilized and elegant manner. Its orientalism, its magic, and fairy-tale character, though not taken seriously by Wieland, suggest the coming of Romanticism. The poem begins with the lines 'Noch einmal sattelt mir den Hippogryphen, ihr Musen, / Zum Ritt ins alte romantische Land!' But it is also viewed as the last climax of literary rococo (see ROKOKO).

Oberösterreich (Upper Austria), a consti-tuent Land of the Austrian Federal Republic. Under the Habsburgs (see HABSBURG) it was an archduchy with the alternative title Kronland Österreich ob der Enns. It includes the Innviertel (q.v.) round Schärding, ceded by Bavaria in 1779, and the larger part of the Salzkammergut. Its capital is Linz/Donau.

Ode an die preußische Armee, a patriotic ode written by Ewald von Kleist (q.v.) in 1757, praising the army and its commander-in-chief, Frederick the Great (see FRIEDRICH II OF PRUSSIA).

Oder-Neiße-Linie, the frontier between Ger-many and Poland agreed by the Western Allies and Russia at Teheran in 1943. It was intended to compensate Poland for the cession to Russia of the territory east of the 'Curzon Line', by drawing the Polish frontier further to the west. The German inhabitants of the areas concerned, viz. Pomerania, the New March (Neumark), Lebus, Lusatia (Lausitz), and Silesia, fled or were expelled. The transfer to Polish administra-tion began before the German surrender in 1945. The line, which is that of the River Oder and a tributary, the Neiße, was ambiguous, since there is an eastern and a western Neiße. The Russians and Poles chose the western line and adhered to it in spite of Allied protests at the Potsdam Conference in 1945. The German Democratic Republic (see DEUTSCHE DEMO-KRATISCHE REPUBLIK) officially recognized the frontier in 1950. In 1971 it was accepted in an agreement between Poland and the Federal Republic.

Odfeld, Das, a short novel (Eine Erzählung) by Wilhelm Raabe (q.v.), written in 1886–7, and published in 1889. It is a historical narrative, resembling *Hastenbeck* (q.v.) in having the Seven Years War (see SIEBENJÄHRIGER KRIEG) as its background. The action covers little more than twenty-four hours and deals with the adventures and sufferings of a group of people who find themselves entangled in the military operations of 5 November 1761.

The French are withdrawing westward, and the allies, Brunswickers, Hanoverians, and British, engage them on the Odfeld and attempt to encircle them. But the French elude the pincers and continue their retreat. The futile battle nevertheless leaves plenty of corpses and dying and wounded men scattered on the ground. Both sides burn houses, and plunder, bully, and batter the inhabitants. The little group of civilians caught up in this cumbersome and cruel military machine are a couple of young farm servants in their twenties, the village Amtmann's niece, Mamsell Selinde Fegebanck, and her admirer, young Thedel von Münch-hausen, and the old, superannuated, apparently useless schoolmaster, Magister Noah Buchius. Thedel, not much more than a boy, was Buchius's 'dearest and naughtiest' pupil, we are repeatedly told. In the course of the day's fighting the two girls and the two mature men are drenched and scratched and knocked about, but they survive. Young Thedel meets his death on horseback late in the day, guiding two squadrons of Elliott's dragoons. In the end Magister Buchius, whose courage, tenacity, and kindliness have sustained the others, returns to the old monastery which the school had formerly inhabited, where he still has his lodging to find as its occupant a crow (or is it a raven? Raabe's taxonomy is uncertain), which he had rescued the day before. The crow flies wild and insists on leaving, and the Magister reluctantly opens the window, allowing Wotan's bird to return to the Odfeld (Odins or Wodansfeld). This yielding by Buchius is the symbol of his acquiescence in the stream of history, with its inexplicable suffering and sacrifice from pre-historic times to his own.

'O du fröhliche, o du selige, gnaden-bringende Weihnachtszeit', first line of a Christmas carol, the first verse of which is by J. D. Falk (q.v.). The authorship of the remainder is obscure. In its present form it dates from 1816. In its original state (*Alldreifeiertags-lied*) its three verses referred respectively to the feasts of Christmas, Easter, and Whitsun. Each was later expanded, so that each feast now has a three-verse hymn.

OEHLENSCHLÄGER, ADAM GOTTLOB (Vesterbro, Copenhagen, 1779–1850, Copenhagen), Danish writer of partial German descent, was introduced to the German Romantic movement (see ROMANTIK) by H. Steffens (q.v.), whom he met in Copenhagen in 1802. He translated some works by Tieck (q.v.) into Danish, but later adopted a critical attitude towards the Romantics, which he explained to F. Hebbel (q.v.), who was greatly stimulated by contact with him while staying in Copenhagen in the early 1840s. On his visits to Germany, Oehlenschläger met Goethe, Schelling, and Hegel (qq.v.). He wrote in German as well as in Danish. His tragedies were influenced by Goethe and Schiller, *Correggio* (1808) being written in German. Most of his works written in Danish appeared in German as well: *Gesammelte Schriften* (21 vols.) in 1829–39, poetry, *Meine Gedichte*, in 1817 and in 1844, and *Neue dramatische Dichtungen* in 1850. In the same year appeared *Meine Lebenserinnerungen* (4 vols.).

OEKOLAMPADIUS, JOHANNES, real name Huszgen or Husschin (Weinsberg, Swabia, 1482–1531, Basel), an early supporter of the Reformation, was a pupil of Reuchlin (q.v.) in Basel and for a time assisted Erasmus (q.v.). Originally a Roman Catholic priest, he went over to the Reformation, being closest to Zwingli (q.v.) among the leaders. His letters were posthumously published in 1536.

Briefe und Akten zum Leben Oekolampadius (2 vols.) and *Das theologische Lebenswerk J. Oekolampadius*, ed. E. Staehelin, appeared in 1927–34 and 1939 respectively.

Ofen, the German name for Pest, the Hungarian city opposite Buda. The form Ofen-Pest was formerly in use.

Offiziere, a play (Drama) by F. von Unruh (q.v.), published in 1911. It deals in ejaculatory Expressionist language with the opposition between the exalted military idealism of youth (Ernst von Schichting and many of his comrades) and the cold formalism of duty, represented by Colonel von Kracht. The first two acts are set in the officers' mess in Germany. Act I passes in trivial futility, Act II offers the chance to volunteer for active service in South West Africa. Act III is set on board the troop ship, and Act IV brings hunger, thirst, and battle; it ends with Ernst's death and the valedictory volleys fired over his grave.

Ogier, an anonymous Middle High German epic written in the 15th c. It tells first of Ogier's childhood and of his feats against the Saracens, and then recounts a number of adventures, including a war against Charlemagne (see KARL I, DER GROSSE).

'O Haupt voll Blut und Wunden', first line of a hymn by Paul Gerhardt (q.v.).

OHG, Old High German, see GERMAN LANGUAGE, HISTORY OF.

OHL, abbreviation for Oberste Heeresleitung, the German supreme command in the 1914–18 War, mostly used in reference to Hindenburg and Ludendorff (qq.v.).

OKOPENKO, ANDREAS (Kosiče, formerly Kaschau, Czechoslovakia, 1930–), an Austrian writer of radical views, is the author of collections of poetry *Grüner November* (1957), *Seltsame Tage* (1963), and *Warum sind die Latrinen so traurig? Spleengesänge* (1969). In his experimental novel *Lexikon einer sentimentalen Reise zum Exporteurtreffen in Druden* (1970) he constructs alphabetical units for the reader to arrange in whatever order he pleases, a kind of do-it-yourself novel kit. *Der Akazienfresser* (1973) is sub-titled *Parodien, Hommagen, Wellenritte*. In it Okopenko proposes, among other matters, a new punctuation mark to signify boredom. His collected poetry, *Gesammelte Lyrik*, was published in 1980.

OKW, abbreviation for Oberkommando der Wehrmacht, the German supreme command in the 1939–45 War.

OLEARIUS, ADAM, real name Oelschläger (Aschersleben, c. 1599–1671, Gottorp, Holstein), was referred to by his contemporaries as 'der Holsteinische Plinius' and 'der Gottorfische Ulysses' (because of his domicile in Gottorp). After studying in Leipzig, Olearius became a schoolmaster. In 1633 he was appointed to participate in a commercial expedition to Russia and Persia, sponsored by the Duke of Holstein-Gottorp, securing also an invitation for his younger friend, the poet Paul Fleming (q.v.). After his return in 1637 he settled in

Gottorp, becoming court librarian. His account of the Persian journey, *Offt begehrte Beschreibung der Newen Orientalischen Rejse* (1647) is often regarded as the first scientific travel book in German. It was expanded as *Vermehrte Newe Beschreibung der Muscowitischen und Persischen Reyse* (1656, reissued 1971). He published a volume of translations of Persian poetry, *Persianisches Rosental* (1654, reissued 1970), in which he also broke new ground.

Olint und Sophronia, a tragedy (Trauerspiel) written in Alexandrine verse by J. F. von Cronegk (q.v.) and left unfinished at his death. Cronegk completed the first three acts and part of the fourth, and the fragment was printed in his posthumous *Schriften* (ed. J. P. Uz, q.v., 1760). A completion by Baron K. A. von Roschmann-Hörburg (q.v.) was published in 1764. The source of the tragedy is *La Gerusalemme liberata* of Torquato Tasso, Canto secondo (*La pudica Sofronia e Olindo ardito*). The play is set in Jerusalem at the time of the crusades. Olint, an agent of Aladin (sc. Saladin), is a secret Christian and loves the Christian Sophronia, who, desiring martyrdom, confesses to a crime she has not committed; Olint, who has really committed it, confesses too. Each vies in eagerness for martyrdom, and the conclusion by Roschmann includes their deaths. A harsh review of the play by G. E. Lessing appears in *Die Hamburgische Dramaturgie* (q.v., 1. *Stück*).

Olle Kamellen (old tales), title given by F. Reuter (q.v.) to his dialect narrative works. They consist of seven volumes: 1. *Woans ick tau 'ne Fru kam* and *Ut de Franzosentid*, 1859; 2. *Ut mine Festungstid*, 1862; 3–5. *Ut mine Stromtid*, 1862–4; 6. *Dörchläuchting*, 1866; 7. *De mecklenburgischen Montecchi un Capuletti oder De Reis' na Konstantinopel*, 1868.

Olmützer Punktation, a settlement between Austria and Prussia arranged at a meeting of the Austrian minister F. zu Schwarzenberg (q.v.) and the Prussian minister O. von Manteuffel (1805–82) at Olmütz in November 1850. Prussia, under Friedrich Wilhelm IV (q.v.), had to agree to the restoration of the German Confederation (see DEUTSCHER BUND), which had ceased to function upon the opening of the Frankfurt Parliament (see FRANKFURTER NATIONALVERSAMMLUNG), and not to interfere with the restoration of the Elector of Hesse by the Federal Diet (Bundestag). The Punktation marks the height of Austrian prestige in Germany and a humiliating diplomatic defeat for Prussia, which was forced to dissolve its Union (see DREIKÖNIGSBÜNDNIS) in order to

avoid an armed conflict with Austria. Prussia had already suffered a setback through the Vierkönigsbündnis (q.v.) which had been created under Schwarzenberg's auspices. Prussia did not fully recover her prestige in Germany until the War of 1866 (see DEUTSCHER KRIEG).

OMEIS, MAGNUS DANIEL (Nürnberg, 1646–1708, Altdorf), who was in Protestant orders, was at first a private tutor in Vienna, then moved to Altdorf University (q.v.) in 1674, occupying the Chair of Ethics from 1677, and of Poetry from 1699. He became a member of the Hirten- und Blumenorden an der Pegnitz (q.v.) at Nürnberg in 1667, and in 1697 rose to be its president (Oberhirte). He is the author of Latin treatises on the ancient Teutons and of an essay in poetics, *Gründliche Anleitung zur Teutschen accuraten Reim- und Dicht-Kunst, durch richtige Lehr-Art, deutliche Reguln und reine Exempel vorgestellet* (2 Pts., 1704). He also published a volume of religious verse, *Geistliche Gedicht- und Lieder-Blumen* (1706).

OMPTEDA, GEORG, FREIHERR VON (Hanover, 1863–1931, Munich), pseudonym Georg Egestorff, served from 1883 to 1892 as an officer in the 18th (1st Saxon) Hussars at Großenhain, retiring with the rank of captain. He spent some time in travel before settling in Merano (then Austrian), and later in Munich. In his narrative writings, the first of which appeared under his pseudonym, he adopted a realistic style influenced by Maupassant's stories, which he translated (*Maupassants Gesammelte Werke*, 20 vols., 1898–1903).

Ompteda published one Novelle (*Freilichtbilder*, 1891) during his service, and two novels (*Sünde*, 1892, and *Drohnen*, 1893) shortly after leaving it. His most important work is a trilogy of novels portraying the decline of the aristocratic world which he knew at first hand, *Deutscher Adel um 1900*; it is composed of *Sylvester von Geyer* (2 vols., 1897), *Eysen* (2 vols., 1900), and *Cäcilie von Sarryn* (2 vols., 1901). The cycle was revised and reissued in three volumes in 1923. Among his numerous other novels, which include light and entertaining fiction, are *Es ist Zeit. Tiroler Aufstand 1809* (1921), *Ernst III* (1925), *Die kleine Zinne* (1931), and *Die schöne Gräfin Cosel* (1932); *Sonntagskind* (1928), his autobiography, reveals a sunny temperament, much favoured by his early environment.

Oper und Drama, a treatise in three volumes by R. Wagner (q.v.), published in 1852. It consists of an introduction and three sections, I. *Die Oper und das Wesen der Musik*, II. *Das Schauspiel und das Wesen der dramatischen Dichtkunst,*

III. *Dichtung und Tonkunst im Drama der Zukunft.*
The work argues passionately Wagner's view
that the words and music must combine in an
intimate union to produce the drama of the
future, a view which is plentifully supported by
analogies to the consummation of human love.

Opfergang, a war book written by Fritz von
Unruh (q.v.) while on active service in the
prolonged assault on Verdun (see VERDUN,
SCHLACHT UM) in 1916; its publication was
forbidden by the censor, and it did not appear
until 1919 after the fall of the monarchy. A
relatively short work (204 pp.) based on first-
hand experience, it deals with the period of
suspense before the offensive and with the
horrifying and heroic details of attack and
counter-attack. It is hectically written, generat-
ing a continuous and exhausting tension.
Portrayed at company level, with frequent side
glances at general and staff, it indulges in the
early stages in reflections on the futility of war.
Towards the end it stresses the gulf between the
soldiers on the ground who fight and the general
and staff, perched literally and symbolically in a
tower, issuing paper orders and spoken commands which cannot be carried out, although
the attempts to obey them involve enormous loss
of life. The feverish style is partly Expressionist
(see EXPRESSIONISMUS).

OPITZ, MARTIN (Bunzlau, Silesia, 1597–
1639, Danzig), a son of middle-class parents,
received his schooling at Bunzlau and Breslau.
He then studied at Beuthen, Frankfurt/Oder,
and Heidelberg, taking refuge from the disturbances of war in Holland, where he became
a friend of Heinsius, in 1620. He was in Denmark
in 1621, and in 1622 accepted a teaching
appointment in remote Transylvania, which he
held for a year.

In 1617 Opitz wrote *Aristarchus sive de con-
temptu linguae Teutonicae,* an essay in Latin
maintaining the suitability of the German
language for poetic use. His Transylvanian year
is commemorated in the idyllic poem *Zlatna
oder von der Ruhe des Gemütes* (1623). On his return
he entered the service of the Duke of Liegnitz
and almost immediately (1624) published three
important works. First came his paraphrases of
the Psalms, followed by a collection of poetry by
himself and others, *Teutsche Poemata;* finally *Das
Buch von der deutschen Poeterey* (q.v., ed. R. Alewyn,
1963) provided a theoretical gloss to his poems.
These, which include sonnets, are characterized
by neat, elegant versification and pure rhymes.
In 1625, in which year Opitz was crowned
laureate (see GEKRÖNTER DICHTER) by the
Emperor Ferdinand II (q.v.), he translated
Seneca's *Troades* into Alexandrines (*Troerinnen*).

From 1626 to 1632 Opitz, a Protestant, was
secretary to Count Dohne, a Roman Catholic
who was charged with establishing the Counter-
Reformation (see GEGENREFORMATION) in
Breslau. At Dohne's request he translated *De
veritate religionis christianae* (1622) by H. Grotius
(H. de Groot, 1583–1645), *Von der Wahrheit der
christlichen Religion* (1631). In 1626 Opitz translated Barclay's novel *Argenis* (see BARCLAY, J.),
his major contribution to the development of
the German novel, and rendered German versions of Jeremiah and the Song of Songs. In 1627
he was ennobled by the Emperor (Martin Opitz
von Boberfeld). In the same year he translated
Rinuccini's *Daphne* for performance in Dresden
to music by H. Schütz (q.v.). In 1629 Opitz
published a translation of Sir Philip Sidney's
Arcadia, and in 1630 made his own contribution
to pastoral poetry with *Die Schäfferey von der
Nimpfen Hercinie.* His *Trostgedichte in Wider-
wertigkeit des Krieges* appeared in 1633. After two
more years (1633–5) in the service of the Duke
of Liegnitz, Opitz moved to Danzig as court
historiographer. There he translated Sophocles'
Antigone and published the medieval *Annolied*
(q.v.). He is also the author of a geographical
didactic poem, *Vesuvius* (1633). By his precise,
orderly mind, his facility for quick, clear
expression, and the aptness of his outlook to his
age, Opitz acquired an immense reputation
which lasted into the 18th c. Little of his work
was original, but he had drive and energy in his
self-imposed role of aesthetic educator. He died
of the plague.

Gesammelte Werke (5 vols.), ed. G. Schulz-
Behrend, appeared in 1968 ff.

OPPELN-BRONIKOWSKI, FRIEDRICH VON
(Kassel, 1873–1936, Berlin), was commissioned
in 1892 in the 14th Hussars but had to resign
in 1896 as the result of a riding accident. He
turned to writing, publishing sketches (*Aus dem
Sattel geplaudert,* 1898), two novels (*Fesseln und
Schranken,* 1905, later re-titled *Der Rebell,* and
Schlüssel und Schwert, 1929), and a Novelle
(*Zwischen Lachen und Weinen,* 1912). Some of these
drew on his own experiences. He also wrote
popular historical books, mostly of patriotic
tendency (*Abenteurer am preußischen Hof 1700–
1800,* 1927; *Liebesgeschichten am preußischen Hof,*
1928; *Der Baumeister des preußischen Staates,* 1934;
Der große König als erster Diener seines Staates,
1934; *Der alte Dessauer,* 1936). He was an industrious and competent translator, turning, among
others, the works of Stendhal (1921–4) and of
Maeterlinck (1924–9) into German.

Orbis pictus, customary title of an illustrated
encyclopedic reader first published by the
Czech J. A. Comenius (1592–1670) at Nürn-

berg in 1654, and widely used for educational purposes. Its full title is *Orbis sensualium pictus, hoc est omnium fundamentalium in mundo rerum et in vita actionum pictura et nomenclatura.*

Ordnung und Gesundheit, Das Buch von, see BUCH VON ORDNUNG UND GESUNDHEIT, DAS.

Orendel, a late medieval poem contained in a 15th-c. MS. (destroyed in the bombardment of Strasburg, 1870) and also known in an early printed form (1512). There is a prose printed version of the same date. The story, in the form of an oriental wooing, is concerned with Jesus's seamless coat, preserved in Trier and first exhibited in 1512.

Orendel, son of a king of Trier, sets out to woo Queen Bride of Jerusalem. He encounters various adventures, finds the sacred coat, wins his lady, and frees the Holy Sepulchre. He returns with his queen and deposits the coat in Trier. They set out again for the Holy Land and free the Sepulchre once more from the heathen. Finally, they and their companions enter on a conventual life. The verse is crude and unskilful, the tone popular. The Nordic names of the characters and their associations have given rise to the belief that the origin of *Orendel* is a lost early heroic work, first Christianized and finally debased.

H. E. Berger's edition of 1888 was reprinted in 1974.

ORFF, CARL (Munich, 1895–1982, Munich), German composer and teacher of composition. He derives inspiration in part from Renaissance and medieval music. His works include the operas *Der Mond* (1939) and *Die Bernauerin* (1946), for both of which he wrote the libretto. He is best known for his 'trionfi' *Carmina Burana* (1937) and *Catulli Carmina* (1943). For the Sophoclean *Antigone* (1949) and *Oedipus der Tyrann* (1959) he used the translations of F. Hölderlin (q.v.). His characteristic style, which admits of theatrical or concert presentation, is retained in *De temporum fine comoedia* (1973).

ORLAMÜNDE, AGNES, GRÄFIN VON, according to legend, murdered her two children after her husband's death in 1293, in order to facilitate her marriage with Burggraf Albert of Nürnberg, a Hohenzollern. She was said to haunt the Hohenzollern castles as 'weiße Frau' ('woman in white'). Her story is treated by Christian Graf Stolberg (q.v.) in a ballad cycle *Die weiße Frau* (1814) and there are allusions to her in the work of Th. Fontane, notably in *Vor dem Sturm* (q.v.).

Orpheus. Eurydike. Hermes, a poem by Rilke included in *Neue Gedichte* (q.v.), I.

Orplid, the name of a fantastic island created in imagination by E. Mörike and his friend L. A. Bauer (qq.v.) when they were boys at school at Urach. Its topography and history (which the two boys gradually elaborated) are outlined by the actor Larkens in the first part of Mörike's novel *Maler Nolten* (q.v.). It was conceived to be in the Pacific between New Zealand and South America and its principal city also bore the name Orplid.

The island was protected by the goddess Weyla. In addition to heroic human inhabitants there was a population of elves, cobolds, and fairies. In time the island became too civilized and incurred the wrath of the gods; its people (except for one survivor) died out, and its buildings, except for the town and castle Orplid, fell into ruins. The dramatic fantasy (Ein phantasmagorisches Zwischenspiel) entitled *Der letzte König von Orplid,* which is included in *Maler Nolten,* portrays the landing on the island, centuries later, of the survivors of a shipwreck and the death of the last king of ancient Orplid, himself centuries old.

ORTLEPP, ERNST (Droyssig, 1800–64, Almrich nr. Naumburg), a pastor's son, became politically suspect and led a restless and unhappy life. His works include the tragedy *Der Cid* (1828), the novels *Cölestine* (1833) and *Friedemann Bach* (1836), and much political poetry, the principal collections of which were *Polenlieder* (1831) and *Lieder eines politischen Tagwächters* (1843). The *Polenlieder* contain the poem 'Noch ist Polen nicht verloren'. He was also active as a translator of Byron and Shakespeare. He published *Gesammelte Werke* in 1845. He died by drowning in a ditch.

örtlich betäubt, a novel by G. Grass (q.v.), published in 1969. Set in the time of the so-called Grand Coalition (Große Koalition, CDU/SPD, 1966–9), it is a work of dialogue and fantasy in which the one unmistakable reality is the dentist's chair, in which the local anaesthetic of the title is injected. Even the apparently infallible dentist, a symbolical technocrat, who proposes to solve all the world's problems by a world health service, proves in the end to have a vulnerable spot; the remaining characters, the narrator Studienrat Starusch, his colleague Irmgard Seifert, their pupils Philipp Scherbaum and Veronika Lewand, all reveal themselves as protesters in varying degrees, and all demonstrate their inadequacy in this role. The novel involves political protest, firstly against former National Socialists holding office (Kiesinger) or tolerated by society (the absurd Field-Marshal Krings) and secondly against the war in Viet Nam. Philipp Scherbaum (repeating exactly the situation in Grass's play *Davor*, 1969) occupies almost the whole of the latter part of the novel with his proposal to burn his dog on the Kurfürstendamm as a protest which the Berliners will take note of, since they are, he claims, more attached to dogs than to human beings. Grass's idiosyncratic combination of humour, irony, and the grotesque prevails throughout. The television screen of his up-to-the-minute dentist enables the author to interweave current events in a process of montage. The setting is Berlin, though recollection stretches westwards to the Voreifel (Andernach, Mayen, and the Laacher See) and eastwards to the Vistula and the author's native Danzig. *örtlich betäubt* ends with the words 'Nichts hält vor. Immer neue Schmerzen', but the local anaesthetic and the analgesic Arantil cope with them.

ORTNER, EUGEN (Glaishammer nr. Nürnberg, 1890–1947, Traunstein), a school teacher and journalist, devoted himself to authorship from 1928. His plays include *Meier Helmbrecht* (1928) and *Jud Süß* (1933), and among his cultural novels is one dealing with the great baroque architect, *Balthasar Neumann* (1937, see NEUMANN, B.); a two-volume novel has the Augsburg merchant family as its subject, *Geschichte der Fugger* (1939–40, see FUGGER). The novel *G. Fr. Händel* appeared in 1942, and one on the poet Günther (q.v.), *J. Ch. Günther*, in 1948.

Ortnit, a Middle High German epic preserved in the Ambraser Handschrift (q.v.) to which *Wolfdietrich* (q.v.), in the same manuscript, forms a sequel. The name Ortnit derives from Russian legend. King Ortnit of Lamparten (Lombardy) sets out to woo a heathen princess, whose father puts all suitors to death. With the help of the dwarf Alberich, who possesses magic powers, he wins his bride and returns to his own land. But the heathen king in revenge sends him two dragon's eggs which hatch. Ortnit intends to slay the dragons, but s caught sleeping and killed and eaten by one of them. The dragon in turn is slain by Wolfdietrich.

This poem and Wolfdietrich were both written shortly before 1250 by the same poet, a skilled story-teller chiefly interested in action and fabulous incident.

OSIANDER, ANDREAS (Gunzenhausen nr. Nürnberg, 1498–1552, Königsberg), was a Lutheran pastor and theologian in Nürnberg from 1522 and in Königsberg from 1549. He became involved in a long controversy with his fellow Lutherans on the theological doctrine of justification (the Osiandrischer Streit), and was

the butt of Johannes Nas (q.v.) in the latter's *Centuria*, No. 6. He expressed his convictions in a number of works, including *Von dem einigen Mittler Jesu Christo* (1551). His views were formally rejected by Melanchthon (q.v.) and his adherents.

Osnabrück, Westphalian city in the Land Niedersachsen of the Federal Republic (see BUNDESREPUBLIK DEUTSCHLAND). The bishopric of Osnabrück was founded by 783, and the city is first mentioned in 1078. With its surrounding territory it became in the 13th c. an episcopal principality. During the Thirty Years War (see DREISSIGJÄHRIGER KRIEG) it was, with Münster, the scene of negotiations which resulted in the Peace of Westphalia in 1648. From this time the principality of Osnabrück was subject to the peculiar condition that its rulers should be chosen in accordance with a rota of three categories, a Roman Catholic bishop, a Protestant bishop, and a prince of the house of Brunswick-Lüneburg. The principality was incorporated into Hanover in 1803 by decree of the Reichsdeputationshauptschluß (see REVOLUTIONSKRIEGE). It next formed part of Jérome's Kingdom of Westphalia in 1807, reverting to Hanover in 1815. In 1867 it became, with Hanover, Prussian (Provinz Hannover). It was heavily damaged in the 1939–45 War, but the principal ancient buildings, including the cathedral, have been restored. Since 1971 it has had a university.

OSSIAN, more properly Oisin, a legendary Gaelic warrior, to whom (as Ossian) James Macpherson (1736–96) attributed 'poetic translations' which were in reality of his own composition. They consist of *Fragments of Ancient Poetry collected in the Highlands of Scotland* (1760), *Fingal* (1762), and *Temora* (1763). The authenticity of these volumes of Celtic legend in rhythmic prose, though impugned by Dr. Johnson, was for a time widely accepted, and a complete German translation by J. N. C. M. Denis (q.v.) appeared in 1768–9. The writers of the Sturm und Drang (q.v.), notably Herder and Goethe, received the Ossianic poems with enthusiasm. Herder included in *Von deutscher Art und Kunst* (q.v., 1773) his own rhapsodic essay entitled *Auszug aus einem Briefwechsel über Ossian und die Lieder alter Völker*, and Goethe incorporated passages translated into German by himself in the second part of *Die Leiden des jungen Werthers* (q.v., 1774). Translations continued to appear until well into the 19th c. (1847) and a belated straggler was published in 1924.

OSSIETZKY, CARL VON (Hamburg, 1889–1938, Berlin), became a pacifist after the 1914–18 War. He expressed his views as a journalist, first with *Die Berliner Volks-Zeitung* (1920–4), then with the periodical *Das Tagebuch* (1924–6), and finally was editor of *Die Weltbühne* (1927–33, reprinted 1978). He was arrested in 1933 and sent to a concentration camp in 1934. In 1935 he was awarded the Nobel Prize for Peace, which he was not allowed to receive in person. Released in that year (but subjected to police supervision), he died in a Berlin hospital of the effects of his imprisonment. Two volumes of *Schriften*, ed. B. Frei and H. Leonard, appeared in 1966 and *Rechenschaft. Publizistik aus den Jahren 1913–1933*, ed. B. Frei, in 1972.

Ostelbien, the German territories to the east of the River Elbe, viz. Mecklenburg and the Prussian provinces of Pomerania, Brandenburg, Silesia, and East Prussia. Some writers have extended it to Prussia and Posen. Since the dissolution of Prussia in 1945 the name has fallen into disuse.

Österreich, originally the name of a margravate with its centre at Vienna, which then became a duchy (1156) and was ruled by the Babenberg family. This house in time acquired Styria and what is now Oberösterreich. In the 13th c. Austria passed to Ottokar II of Bohemia (q.v.). On his defeat by the German King Rudolf I (q.v.) of Habsburg in 1278, it came into the possession of the latter's family, which continued to rule it until 1918. From 1273 to 1308 and from 1438 to 1806 the Habsburg dukes of Austria were also emperors of the Holy Roman Empire (see DEUTSCHES REICH, ALTES). The Reformation (q.v.) made only temporary headway in Austria, which in the 17th c. became the strongest element in the Counter-Reformation (see GEGENREFORMATION). In the Thirty Years War (see DREISSIGJÄHRIGER KRIEG) it was the strongest German power, only held in check by Swedish military force. By then Austria was a multi-racial state including, as well as Austro-Germans, the Czechs of Bohemia, the Hungarians, and the Croats.

In the 18th c. Austria became engaged under the Empress Maria Theresia (q.v.) in a struggle for hegemony with the newly emergent power, Prussia. The Napoleonic Wars (q.v.) brought military set-backs and the dissolution of the Holy Roman Empire. The Emperor Franz II (q.v.) had meanwhile (1804) assumed the title of emperor of Austria as Franz I. The Austrian monarchy was able to reassert a dominating position at the Congress of Vienna (see WIENER KONGRESS) in 1815, and until 1848 it was the principal power in central Europe. After the war of 1866 (see DEUTSCHER KRIEG), Bismarck (q.v.) secured the exclusion of Austria from Germany (dissolution of the German Con-

federation). This set-back provoked trouble with the Hungarians, who demanded additional rights, and from 1867 to 1918 the country was designated the Austro-Hungarian Monarchy (Österreichisch-Ungarische Monarchie) or, more commonly, Austria-Hungary (Österreich-Ungarn).

After the removal of the Habsburgs in 1918 the Empire split into the so-called successor-states (Czechoslovakia, Hungary, Poland, Yugoslavia, and Romania), leaving only the former nucleus of Austria (plus Salzburg, acquired 1805, and Tyrol, less South Tyrol), but reunion with Germany (Anschluß) was forbidden by the Treaty of St. Germain-en-Laye (see also VERSAILLES, TREATY OF). The province of Burgenland was, however, transferred from Hungary in 1921. The Federal Republic of Austria existed from 1919 to 1938, coming under increasing pressure from National Socialist forces within and without. In 1938 the country was occupied by Germany of which it became a part with the designation Ostmark. On 13 April 1945 Vienna was taken by Soviet forces, and in May other Allied forces entered Austria and four zones of occupation were established. The Federal Republic was reconstituted with K. Renner (q.v.) as first president, but wrangles between the occupying powers delayed final recognition. In 1955 an Austrian declaration of neutrality satisfied the Russians, and the forces of the four powers were withdrawn.

Österreichische Chronik, a history of Austria in the form of a prose chronicle, written *c.* 1394–8 for Duke Albrecht III of Austria. Its authorship, formerly ascribed to one Gregor Hagen, is not known.

Österreichische Reimchronik, see OTTOKAR VON STEIERMARK.

Österreichischer Erbfolgekrieg, the War of the Austrian Succession, ensued after Maria Theresia (q.v.) succeeded her father Karl VI (q.v.) on the Habsburg throne in accordance with the terms of the Pragmatic Sanction (see PRAGMATISCHE SANKTION). Several sovereigns withdrew their assent to the Sanction, and Karl Albert of Bavaria (see KARL VII) and Friedrich August III of Saxony (see AUGUST III) renewed their claims to the imperial throne since they were married to the daughters of the Emperor Joseph I (q.v.). None exploited the crisis more bluntly than Friedrich II (q.v.), who succeeded to the Prussian throne in 1740, the year of Maria Theresia's accession. Resurrecting the super-annuated claims of his forebears to Silesia, he informed the Viennese court that he would support the Pragmatic Sanction if Austria yielded Lower Silesia to Prussia. Maria Theresia rejected the King's proposal in spite of the superior military power which Friedrich Wilhelm I (q.v.) had bequeathed to his son. Friedrich was quick to employ it (16 December 1740), and the first Silesian War (see SCHLE-SISCHE KRIEGE) gave him Silesia.

The rape of Silesia, however, became the cause of a wider armed conflict involving the European powers in the War of the Austrian Succession as well as, in 1756, the Seven Years War (see SIEBENJÄHRIGER KRIEG). For France, aiming at the fall of the House of Habsburg, invaded Austria with a combined French and Bavarian army. Deserted by all, Maria Theresia, as queen of Hungary, rallied the Hungarian Diet to her cause. Hungarian troops occupied Munich, ironically on the same day on which Karl of Bavaria was crowned as the Emperor Karl VII, having been elected on 24 January 1742. Maria Theresia was fortunate in that England, under Walpole's successor Carteret, decided actively to support the Pragmatic Sanction by the formation of the Pragmatic Army (see PRAGMATISCHE ARMEE), commanded by George II of England, who was also Elector of Hanover, which defeated the French at Dettingen (June 1743). This success encouraged Maria Theresia to persist in her cause; but it did not entirely resolve her problem. With France now officially at war with Great Britain, a French alliance was not in sight. But two events furthered her aims. Early in 1745 Great Britain, the Netherlands, August III (as Elector of Saxony and King of Poland), and Austria concluded the Treaty of Warsaw in which the Pragmatic Sanction was recognized. Moreover, the death of Karl of Bavaria removed the rival emperor, and his young son Maximilian II (q.v.) formally renounced all rights to the imperial title in the Treaty of Füssen (April 1745). In September 1745 Maria Theresia's husband Franz of Lorraine/Tuscany was elected emperor as Franz I (q.v.). In the spring of that year Maria Theresia had tried once more, with the help of Saxony, to recover Silesia from Prussia, but without success.

In the Treaty of Dresden (December 1745) Austria and Saxony concluded peace with Prussia, in which Friedrich II recognized Franz I as emperor. There remained the task of reconciling Great Britain, France, and Austria, which was accomplished by George II in the Peace of Aachen (or Aix-la-Chapelle) of October 1748. The treaty put an end to the War of the Austrian Succession. In effect, however, it was a truce rather than a peace as none of the signatories was truly satisfied with its terms, which included evacuation by France of the Austrian Netherlands and Austrian acquiescence

in the loss of Silesia and Glatz. Nor was Maria Theresia happy about the conciliatory attitude of George II towards Prussia. The years between the Peace of Aachen and the beginning of the Seven Years War were spent in active diplomacy effecting a reversal of alliances, which is known as the Diplomatic Revolution (see DIPLO-MATISCHE REVOLUTION).

Österreich, wie es ist, see AUSTRIA AS IT IS.

Osterspiel von Muri, an Easter play written in Middle High German in the Aargau, Switzerland, early in the 13th c. Its focal points are the encounter of the risen Christ with the women, and the narration of the Resurrection. The play, which has survived only in fragments, is written in a courtly style and metrically correct verse. It is the earliest drama in the German language.

Ostfränkisches Reich, designation of the eastern part of the Carolingian Empire (see KAROLINGER) which was allocated to Ludwig the German (see LUDWIG II, DER DEUTSCHE) when the empire was divided into three by the Treaty of Verdun in 843. It lasted effectively until Ludwig's death in 876 and finally lost its separate existence at the death of Ludwig IV, das Kind, in 911 (b. 893, acceded 900). It included Germany between the Rhine and the Elbe, Swabia, Bavaria, and most of modern Austria.

Ostgoten or Ostrogoths, a tribe of Goths which was subjugated by the Huns in the 4th c. After the collapse of the Hunnish kingdom they settled in Pannonia and percolated into north Italy, becoming a valuable source of recruitment for the later Roman armies. Theodoric the Great (q.v.), who vanquished Odoacer, was their greatest king, with dominions extending over Italy, Sicily, the eastern Alps, and the northern Balkans. Theodoric died in 526, and in the late 6th c. the Ostrogoths were finally defeated.

Ostmark, early medieval designation of the eastern marches of Germany, subsequently called Österreich (q.v.). The name was revived by the National Socialist regime in 1938.

Ostpreußen (East Prussia), a former Prussian province (Provinz), is now partly Russian and partly Polish territory. The north-eastern area around Königsberg (now Kaliningrad) has been incorporated into Russia, while Masuria (see MASUREN) and the western region are Polish.

In the Middle Ages the originally Slavonic region was conquered by the knights of the Teutonic Order (see DEUTSCHER ORDEN), and in 1525 became the Duchy of Prussia. In 1618 it passed after a long period of confusion and misrule into the hereditary possession of the Elector of Brandenburg, Johann Sigismund (q.v.). It was at that time horseshoe-shaped, the central portion (Ermland), including a stretch of the Baltic coast, belonging to Poland; the duchy was also separated from Brandenburg by a broad belt of Polish territory. The Duchy of Prussia did not take kindly to Hohenzollern rule, but in 1662 disorders and risings were finally quelled by the Great Elector (see FRIEDRICH WILHELM, DER GROSSE KURFÜRST). Friedrich I (q.v.) extended the name to his other dominions, including Brandenburg and Pomerania, and the duchy was renamed East Prussia.

The first Partition of Poland (see POLAND, PARTITIONS OF) in 1772 gave to Prussia the territory linking East Prussia with Pomerania and Brandenburg. The third Partition (1795) added to East Prussia vast Polish territories, including Ostrolenka and Bialystok, designating them New East Prussia (Neuostpreussen). By the Treaty of Tilsit (see NAPOLEONIC WARS) this was permanently lost by East Prussia. From 1815 to 1945 East Prussia was a Prussian province. In 1919 it was separated from the rest of Prussia by a strip of Polish territory reaching the coast at Gdynia, known as the Polish Corridor, and by the Free State of Danzig (q.v.), established under the Treaty of Versailles (see VERSAILLES, TREATY OF).

East Prussia has repeatedly suffered from the ravages of war, notably in the Seven Years War (see SIEBENJÄHRIGER KRIEG), in 1914 and 1915, when it was the scene of the battle of Tannenberg (q.v.) and of the two Masurian battles, and finally in 1945 (see WELTKRIEGE I AND II). In the course of the 1939-45 War some 600,000 inhabitants lost their lives by enemy action or in attempts to escape across the ice when East Prussia was cut off in 1945. After the end of the war the greater part of the German population was expelled.

Ostrogoths, see OSTGOTEN.

OSWALD, ST., see ST. OSWALD.

OSWALD VON WOLKENSTEIN (Tyrol, 1377–1445, Hauenstein nr. Bolzano), a restless and dynamic figure, left 125 Middle High German songs. Oswald, who was of knightly birth, ran away from home at the age of 10, and for fourteen years earned his keep in various menial posts, including those of stable-boy, boatman, and cook, and also as minstrel. He wandered through almost the whole of Europe, northwards to Prussia and Sweden, westwards to England, southwards to Italy, and, as he himself indicates, to the Near and Middle East, including Turkey, Persia, and the Holy Land.

After the death of his father in 1401, Oswald returned home to take possession of the family castle of Hauenstein, but became involved in a dispute over ownership which was not decided in his favour until 1427. A man of strong sensual impulses, he was long powerfully attracted by Sabine, the daughter of the rival claimant to his estate, Martin Jäger, an attraction which he continued to feel even after his marriage in 1417 to Margarete von Schwangau, to whom he was greatly attached. He was employed by the Emperor Sigismund (q.v.) at the Council of Constance (see KONSTANZER KONZIL) and on a mission to Spain and Portugal. In 1421 he was trapped with the aid of Sabine Jäger and held prisoner for some time. In his later years his turbulent nature expressed itself in various local feuds.

Oswald's lyrical production is varied and includes Tagelieder (see TAGELIED), Minnelieder, dance songs, a Fehdelied ('feud song'), Reiselieder, and religious songs. His poems are frequently put in dialogue form, and their style is often boldly original and even brash. Many of the songs refer to his own life and experience and the phrase 'lyrische Selbstbiographie' has been coined for them. A development in originality and power of expression is traceable in the chronological sequence of his poems.

'O Täler weit, o Höhen', first line of Eichendorff's poem *Abschied* (q.v.).

'O Tannenbaum, o Tannenbaum, wie treu sind deine Blätter', first line of an adaptation from folk-songs made in 1819 by J. A. Zarnack (1777–1827). In this form it is a song of forlorn love; its use as a carol dates from a further adaptation made in 1824 by E. Anschütz, a schoolmaster in Leipzig. The tune, a traditional one already in use for student songs, is identical with that to which, in England, *The Red Flag* is sung.

OTFRIED or **OTFRID,** a monk of the Benedictine Abbey of Weißenburg (Wissembourg in Alsace), wrote in the 9th c. his *Evangelienbuch*, finishing it between 863 and 871. Nothing is known of Otfried beyond the name of his teacher (Salomon, Bishop of Constance) and the names of two friends, monks in St. Gall. His poem is an Evangelienharmonie combining material from the synoptic gospels. It is in five

books and is preceded by five prefaces in verse
which indicate his aim, viz. to produce a
religious work to supplant profane epic. It is
thought that he envisaged the reading of the
work to noble audiences as a substitute for the
recitation of heroic lays. Otfried's verse becomes
noticeably more fluent as the work progresses.
It is the first known substantial German work
in rhyme. Otfried's aim was to illustrate the
theological significance of Jesus's life, and his
poem is interspersed with passages of dogmatic
exposition, which he usually headed *moraliter,
spiritualiter,* or *mystice.* He was not a poet in the
modern sense, but a Latin scholar diligently and
faithfully performing a task in the service of the
Church.

The principal MS., in the Nationalbibliothek,
Vienna, is known to have been corrected by
Otfried himself. Other manuscripts are in
Heidelberg and in Munich (a Bavarian version
executed in Freising); fragments of a fourth
exist in other libraries. The first modern editor
gave the poem the title *Krist* (E. G. Graff, 1831),
but the designation *Otfrieds Evangelienbuch* has
been current since 1856. Otfried prefaced the
work with the superscription *Incipit liber
Evangeliorum Domini gratia theotisce conscriptus.*

ÔTLOH (*c.* 1000–*c.* 1070), a monk of the
Benedictine Abbey of St. Emmeram at Regensburg, was in charge of the abbey school from
1032 to 1062. At various times he was also at
Hersfeld and Hirschau, and towards the end of
his life in Fulda and Amorbach. Ôtloh's life
coincides with new reforming trends in the
Church, tending to a rejection of worldly values
and classical literature, and to a strict clerical-
ism. He was a prolific writer in Latin, and his
works included legends of the saints and an
account of his own inner life (*Libellus de suis
tentationibus*). He himself made a German prose
translation of a Latin prayer which he had
composed (*Ôtlohs Gebet*). It represents the first
German prayer of personal character.

Ottava rima, see STANZEN.

OTTE, author of the Middle High German
poem *Eraclius,* written *c.* 1210 probably for the
Thuringian court of Landgraf Hermann. The
poem (of some 5,000 lines) is based on a French
original. It falls into three parts. The first, of
oriental origin, recounts Eraclius' childhood
and his acquisition of the magic gift of distinguishing the true from the false in precious
stones, horses, and women. The second takes
place while Eraclius is away at the wars with the
Emperor Focas, whom he serves; Athenais, the
Emperor's young consort, is unfaithful, and her
infidelity is detected by Eraclius, who also

achieves a reconciliation. The third part is the
legend of Eraclius, who regains from Cosdras,
King of the Persians, the stolen Holy Cross.

Otte mit dem Barte, see HEINRICH VON
KEMPTEN.

OTTENHALER, PAUL, a humanist living in
Tyrol, studied at Ingolstadt, and is recorded in
Innsbruck between 1554 and 1571. He is the
author of a satirical poem, *Der Schmarotzer Trost*
(1569).

Ottenton, term given to a group of six political
Sprüche (see SPRUCH) by WALTHER von der
Vogelweide (q.v.), supporting Otto IV (q.v.)
in his conflict with the papacy. Three are
addressed to the Emperor and three directed
against the Pope.

OTTO I, KAISER OTTO DER GROSSE (912–73,
Memleben), succeeded his father Heinrich I
(q.v.) in 936 as German King (see DEUTSCHER
KÖNIG), and immediately faced dangerous
revolts by sections of the nobility, including his
own brothers. Within a few years Otto established his power, relying upon the bishops to
provide an effective support against an ambitious
aristocracy. Otto overcame a second crisis in
954; and in 955, by his victory on the Lechfeld,
he permanently freed the empire from marauding Hungarian armies. In his later years he
strengthened his hand in Italy, obtaining in
962, the year in which he was crowned emperor
in Rome, the right of veto in the papal election.
Otto further secured control of the Langobard
kingdom in North Italy and had his son Otto II
(q.v.) crowned emperor in 967. He is buried in
Magdeburg. The predicate 'the Great' was used
already in his lifetime.

OTTO II, KAISER (955–983, Rome), son of
Otto the Great (see OTTO I) was crowned
German King (961) and emperor (967) in his
father's lifetime. He was married in 972 to the
Byzantine princess Theophano. In the early
years of his short reign (973–83) he faced
serious revolts at home (973–8) and an invasion
from France (978). In both conflicts he was
eventually victorious. In the years 980–3 he was
in Italy advancing into the south against the
Saracens, at first successfully, but then suffering
defeat, which his death from illness shortly
afterwards made final. He is buried in St.
Peter's, Rome.

OTTO III, KAISER (980–1002, Paterno nr.
Viterbo), was only three and a half when his
father (see OTTO II) died, but was already
designated as his successor. He became the

centre of a struggle for power, in which Duke Heinrich of Bavaria (deposed by Otto's father) was eventually worsted by Otto's mother, Theophano, who ruled as regent. Declared of age in 996 and crowned emperor in the same year, Otto devoted his short reign primarily to the establishment of his power in Italy and especially to securing the Emperor's right to nominate the Pope. He is buried in Aachen. He is the subject of a play by Jakob Ayrer (q.v., *Tragedia von Keiser Otten dem dritten*, 1618) and of a ballad (*Klaglied Kaiser Otto des Dritten*, 1833) by A. von Platen (q.v.).

OTTO IV, KAISER (Argenton, France?, *c.* 1174–1218, Harzburg), was the second son of Heinrich der Löwe (q.v.), Duke of Saxony and Brunswick, and was brought up at the court of Richard I of England. On the death of the Emperor Heinrich VI (q.v.) in 1197 Philipp (q.v.) of Swabia was elected German King (see DEUTSCHER KÖNIG), but a number of dissident princes persuaded Otto to accept election as a rival king. Otto's election took place first at Cologne and, more legally, at Aachen. Otto, partly through papal support purchased by territorial concessions in Italy, at first gained the upper hand, but then suffered serious reverses. In 1208 the assassination of Philipp transformed his situation, and he was again elected king and crowned emperor at Rome in 1209. He quarrelled, however, with the Pope, who excommunicated him, and many princes at home rose against him. In 1211 he was deposed in favour of Friedrich II (q.v.), and in 1214 his power was finally destroyed by a total defeat at the hands of the French at Bouvines, after which he took refuge in Thuringia. Otto is buried in Brunswick.

WALTHER von der Vogelweide (q.v.) was for a time in his retinue, supporting his cause in political poems (see SPRUCH).

OTTO IV, MARKGRAF VON BRANDENBURG (1266–1309), a poet as well as a territorial prince, became Elector of Brandenburg in 1281. In 1280 he was wounded in the head by an arrow, which remained embedded for a year, hence his sobriquet Otto mit dem Pfeil. Otto bore the reputation of being an accomplished knight, skilled in arms and chivalrous in conduct. His seven Minnelieder are simple and unpretentious.

Otto, a tragedy (Trauerspiel) by F. M. Klinger (q.v.), written in 1774 and published in 1775. Klinger's first work, it is an imitation of Goethe's *Götz von Berlichingen* (q.v.) and likewise uses motifs from Shakespeare. Its five acts are divided into numerous short scenes. The incoherent plot concerns an attempt by the wicked

Bishop Adelbert to overthrow Duke Friedrich by poisoning the latter's mind against his son Karl, and so causing a civil war, from which Adelbert will profit in territory and power. The Otto of the title is one of the Duke's knights whom Adelbert alienates from his lord by false report. Duke Friedrich is killed by Adelbert's poison, Otto kills himself when he discovers Adelbert's treachery. Karl survives and undertakes to avenge his father. In a minor plot Adelbert wantonly causes the death of his honest counsellor von. Hungen at the hands of the Inquisition.

OTTO, TEO (Remscheid, 1904–68, Frankfurt), was the youngest of the stage designers who, after the 1914–18 War, developed the anti-illusionary theatre (see RAUMBÜHNE). From 1927 to 1933, when he emigrated to Zurich, he worked at the Staatstheater in Berlin. A collaborator of B. Brecht (q.v.), he worked for noted producers, designing after the 1939–45 War the production of Goethe's *Faust* in Hamburg by G. Gründgens (q.v.). Otto worked with increasing emphasis on abstract and concrete art, which a comparison of his designs for *Mutter Courage und ihre Kinder* (q.v.) from 1941 and 1963 demonstrates. *Meine Szene* appeared in 1965 with a preface by F. Dürrenmatt (q.v.).

OTTO DER GROSSE, see OTTO I.

OTTO DER RASPE, probably a canon of Brixen (Bressanone in South Tyrol), in the 14th c., is the author of a work portraying a suit brought by the Devil against Christ. It is based on the *Processus Belial* (see PROZESS BELIAL).

OTTOKAR II, KÖNIG VON BÖHMEN, or Otakar (*c.* 1230–78, nr. Dürnkrut, Marchfeld), was a son of King Wenceslas I and grandson of Ottokar I. Ottokar came to the throne in 1253 and was able during the Interregnum (q.v.) to acquire the duchies of Austria, Styria, Krain, and Carinthia. In his attempt to found a huge east European state he warred against the heathen Prussians and Lithuanians and founded Königsberg (q.v., which owes its name to him). His reign was marked by good administration and the economic growth of his territories, especially of Bohemia. In the end Ottokar overreached himself, seeking to become German King (see DEUTSCHER KÖNIG) and being passed over in 1273 in favour of Rudolf of Habsburg (see RUDOLF I). In 1274 he ignored an order to restore the lands seized since 1254, and in 1276 he was defeated and forced to make humiliating submission to Rudolf. He is said to have knelt in his regalia to Rudolf who sat in a workman-like leather jerkin. In 1278 Ottokar rose again and was defeated on the Marchfeld near Dürnkrut. He was waylaid and killed while fleeing from the battlefield.

Ottokar's conflict with Rudolf has been portrayed in *Rudolf von Habsburg und Ottokar* by A. von Kotzebue (q.v., 1815) and, more notably, by F. Grillparzer in *König Ottokars Glück und Ende* (q.v., 1825).

OTTOKAR VON STEIERMARK (*c.* 1260–*c.* 1320), an Austrian knight, was the author of the *Österreichische Reimchronik*, a huge work of *c.* 100,000 lines, written in Middle High German rhyming couplets, tracing Austrian history from the death of the Emperor Friedrich II (q.v., 1250) to 1309, at which point the poem ceases abruptly. The work has an ideal basis, for Ottokar sees the hand of God, particularly in a retributive sense, in the events of history. The historical standpoint is that of the knightly class to which he belongs. The most famous section of the *Österreichische Reimchronik* deals with the fall of Acre in 1291. Ottokar also wrote a chronicle of the emperors which is lost, and planned a history of the popes. He mentions, as his teacher, an otherwise unknown minstrel, meister Kuonrât von Rôtenbere, and developed his style from courtly models.

OTTO MIT DEM PFEIL, see OTTO IV, MARKGRAF VON BRANDENBURG.

Ottonen, generic designation for the house of German kings and emperors reigning from 936 to 1002. They comprise the three early Ottos, I (the Great), II, and III (qq.v.). See also OTTONISCHE RENAISSANCE.

Ottoneum, Das, the first permanent theatre in Germany, built in 1604–5 at Kassel by Landgrave Moritz (q.v.) of Hesse. Its name derived from Moritz's heir, Prince Otto.

Ottonische Renaissance, term used to denote the florescence of art under the Saxon kings and emperors Heinrich I, Otto I, Otto II, Otto III, and Heinrich II (qq.v.), who reigned from 919 to 1024. It is applied principally to architecture, the illumination of MSS., fresco painting, and the art of the goldsmith and the ivory carver. Among notable surviving works are the abbey church of Gernrode (961), the crypt of the abbey of Quedlinburg (*c.* 970), and the Cologne crucifix known as the Gerokreuz (*c.* 970).

The term was formerly applied also to the Latin literature which was cultivated to the exclusion of German in this age Its use, however, has declined, especially since *Waltharius manu fortis* (q.v.), once thought to be the principal

literary work of the Ottonian Renaissance, has been ascribed to a writer of the 9th c.

OTTO-PETERS, LUISE (Meißen, 1819–95, Leipzig) *née* Otto, pseudonym Otto Stern, supported the radical, political, and social ideas of the repressed opposition, and after 1848 became an ardent feminist, founding a women's journal (1848–50) and publishing polemical writings. In 1858 she married the left-wing politician August Peters, who died in 1864. In the following year she founded, with three other women, the Allgemeiner deutscher Frauenverein and edited its journal, *Neue Bahnen*. In the years before the 1848 Revolution (see REVOLUTIONEN 1848–9) she wrote several novels of left-wing social tendency (*Ludwig der Kellner*, 1843; *Die Freunde*, 1845; *Schloß und Fabrik*, 1846; *Römisch und deutsch*, 1847) and also published political poetry (*Lieder eines deutschen Mädchens*, 1847). In later years her writings were principally feminist articles and tracts. Among 20th-c. writers recognizing her work in this field were Helene Lange and Gertrud Bäumer (qq.v.).

OTTO VON BOTENLAUBEN, GRAF, a Franconian nobleman, whose castle of Botenlauben was situated near Kissingen, enjoyed a reputation in his lifetime as a Minnesänger. He belonged to the powerful family of the counts of Henneberg and was in the suite of the Emperor Heinrich VI (q.v.). He took part in a crusade in 1197 and remained in the Holy Land for some twenty years. He spent his last years in the monastery of Frauenrode. He died *c.* 1244 and is commemorated, together with his wife, by a handsome monument (still extant).

Botenlauben's poems include songs of a love which, in contrast to *hohe minne*, can be fulfilled, Tagelieder (see TAGELIED), crusaders' songs, which are not in the least religious, and a *leich* (q.v.).

OTTO VON FREISING, BISHOP, see LAUBACHER BARLAAM.

OTTO VON WITTELSBACH, PFALZGRAF (?–1209, nr. Regensburg), was led by a personal grievance to assassinate Philipp von Schwaben (q.v.) at Bamberg in 1208. Outlawed by Otto IV (q.v.), he was killed in the following year. He is the subject of a tragedy by J. M. von Babo (q.v.).

OTTO ZUM TURNE, a minor Middle High German poet of the early 14th c., resided at Lucerne, where he is recorded between 1312 and 1331. He is the author of five Minnelieder and a *leich* (q.v.), which are stylistically indebted to GOTTFRIED von Neifen (q.v.). The tone of the poems inclines to sentimental preciosity.

Oversberg. Aus dem Tagebuch des Volontärs Ferdinand Binder, the first of three short works contained in *Drei Novellen* (q.v.), published by Marie von Ebner-Eschenbach (q.v.) in 1892.

OWLGLASS, DR., pseudonym of Hans Erich Blaich (Leutkirch, 1873–1945, Munich), who was latterly a specialist for pulmonary diseases, but from 1912 to 1924 and from 1933 to 1935 was editor of the Munich weekly *Simplicissimus* (q.v.). His comic poetry includes the collections *Der saure Apfel* (1904), *Gottes Blasebalg* (1910), and *Käuze* (1917).

OXENSTIERNA, AXEL, COUNT (Län Uppsala, 1583–1654, Stockholm), who studied at the German universities of Rostock, Wittenberg, and Jena, became in 1612 chancellor to King Gustavus Adolphus (q.v.). After Swedish intervention in Germany (see DREISSIGJÄHRIGER KRIEG) he worked for the realization of the King's German policy. Unlike Gustavus Adolphus (d. 1632) he was not a great military leader, but he was a determined and gifted diplomat. Through the formation of the League of Heilbronn (see HEILBRONNER BUND), without the Elector Johann Georg I (q.v.) of Saxony, he became the leader of German Protestantism. But partly through military defeat at Nördlingen and partly through the divergent interests of his French ally Richelieu, his influence was checked. He approached the negotiations leading to the Peace of Westphalia for which the Swedish Queen Christina pressed, primarily from the standpoint of his country's prestige and of war reparation. At first regent for the Queen, he became, after her majority, her chancellor. In 1645 he was created count. He is a minor figure in Brecht's play *Mutter Courage und ihre Kinder* (q.v.).

Ozeanflug, Der, a play (Ein Radiolehrstück für Knaben und Mädchen, see LEHRSTÜCK) in 17 short sections of irregular unrhymed verse for solo voice and choir by B. Brecht (q.v.), first published in the magazine *Uhu* in 1929 under the title *Der Lindberghflug,* which Brecht abandoned on the occasion of the play's production by the Süddeutscher Rundfunk in 1949 in order to express his disapprobation of Lindbergh's attitude to the National Socialist regime. Brecht saw the Lehrstück as an experiment using 'Dichtung für Übungsstücke' (*Versuche,* vol. 1). The musical accompaniment was composed by K. Weill and Hindemith (qq.v.). The theme is the first flight across the Atlantic by Captain Charles Lindbergh (then aged 25) on 20–1 May 1927. Lindbergh is the only character; Brecht uses him and his machine to praise a new technical age which conquers the elements ('das Primitive'), erasing the belief in God.

P

PAALZOW, HENRIETTE (Berlin, 1788–1847, Berlin), *née* Wach, married a Major Paalzow in 1816. The marriage was dissolved in 1821. She wrote historical novels, which her contemporaries compared with those of Sir Walter Scott. Her principal novels (each of 3 vols.) are *Godwie-Castle* (1838), *St. Roche* (1839), *Thomas Thurnau* (1843), and *Jakob van der Rees* (1844).

Paläophron und Neoterpe, a short occasional play (Festspiel) of allegorical character, written by Goethe for performance on the birthday of the dowager Duchess Anna Amalia (q.v.) of Weimar, 31 October 1800. Written in a few days, it was published in the *Neujahrs-Taschenbuch von Weimar auf das Jahr 1801.*

Palatinate, see PFALZ, DIE.

PALM, JOHANN PHILIPP (Schorndorf, 1768–1806, Braunau), a Nürnberg bookseller, who published during the Napoleonic occupation (see NAPOLEONIC WARS) an anonymous pamphlet entitled *Deutschland in seiner tiefen Erniedrigung.* Napoleon ordered his arrest and execution in 1806. This act caused deep indignation, and Palm was admired for the courage displayed during his brief trial when he refused to reveal the name of the author of the pamphlet. He became the subject of a number of biographies, plays, and works of fiction, mainly in the 20th c.

Palmström, a volume of nonsense poetry by C. Morgenstern (q.v.), published in 1910. Palmström is a character who recurs throughout the poems.

Pan, a literary periodical founded in 1895 in Berlin by O. J. Bierbaum (q.v.) and others and from Heft 3 (1895) edited by Cäsar Flaischlen (q.v.). It ceased to appear in 1900. A second periodical entitled *Pan* appeared from 1910 to

1914; it was run by A. Kerr (q.v.) and notably promoted avant-garde literature. (Kerr was its editor from 1912.)

Pandämonium Germanikum. Eine Skizze, a boisterous satire in dramatic form written by J. M. R. Lenz (q.v.) in 1775 and published posthumously in 1819. It is in two acts. Rationalists and men of the Aufklärung (q.v.), including Wieland, are abused, while Lessing, Herder, Shakespeare, and Klopstock (qq.v.) are praised.

Pandora, a verse play written by Goethe in 1807 in response to a request by L. von Seckendorff (1775–1809) and a Dr. Stoll for a contribution to an annual of verse (Musenalmanach) entitled *Pandora.* Part of Goethe's *Pandora* was published in the first two instalments of the annual (1808); it was published complete in a special number described as *Pandora von Goethe. Ein Taschenbuch für das Jahr 1810.*

Pandora uses the myth of Pandora's descent to earth bringing ambiguous gifts, and combines it with the myth of the two Titans Prometheus and Epimetheus. Epimetheus mourns the loss of Pandora and yearns for her return. She has taken with her one daughter, Elpore, who represents hope. A second daughter, Epimeleia, remains with him, and she is loved by Phileros, son of Prometheus. Suspecting her of infidelity, Phileros pursues and wounds her. Epimetheus and Prometheus intervene. Both lovers come close to perishing, she by fire and he by water. As dawn gives way to day, both are rescued and united.

The allegory is concerned with man's loss of powers and resignation in age, counterbalanced by confidence based on the succession of generations. *Pandora* was originally to include a continuation in which Pandora herself would return, but this remained unwritten, and the proposed title *Pandorens Wiederkunft* was therefore modified. The work is a product of Goethe's classical period and is written chiefly in iambic trimeters, with interpolated classical lyrical measures. The classical mood is also emphasized by the scenic description 'im großen Stil nach Poussinischer Weise'.

Pan-German League, see ALLDEUTSCHER VERBAND.

Pankraz der Schmoller, see LEUTE VON SELDWYLA, DIE.

PANNWITZ, RUDOLF (Crossen/Oder, 1881–1969, Astano, Ticino), except for a short period as a private tutor after completing his university studies, spent his life in the study of

philosophy, arts, letters, and civilization in general. His philosophical works are chiefly concerned with civilization ('culture' in the German sense) and include such titles as *Die Krisis der europäischen Kultur* (1917), *Nietzsche und die Verwandlung des Menschen* (1940), *Der Nihilismus und die werdende Welt* (1951), and *Beiträge zu einer europäischen Kultur* (1954). He is not an originator, being chiefly indebted to Nietzsche and S. George (qq.v.). His visionary poetry also bears the stamp of Nietzsche's rhapsodic manner and George's selectivity. Among his poetic writings are the epics *Prometheus* (1902) and *König Laurin* (1956), the poems *Urblick* (1926) and *Hymnen aus Widars Wiederkehr* (1927), and the dramatic works *Dionysische Tragödien* (1913) and *Die Erlöserinnen* (1922). His interest in mythology is expressed in his verse *Mythen* (1919–21). With O. zur Linde (q.v.) he was a co-founder of the periodical *Charon* (q.v.) and its associated group.

An autobiography, *Nach siebzig Jahren* (1951), is included in *Über den Denker Rudolf Pannwitz* by U. Rukser (1970).

Panther, Der, a poem by Rilke (q.v.), superscribed *Im Jardin des Plantes, Paris* and included in *Neue Gedichte* (q.v.).

Panthersprung and **S. M. S. Panther,** see AGADIR INCIDENT.

PAOLI, BETTY, pseudonym of Barbara Elisabeth Glück (Vienna, 1815–94, Baden by Vienna), a once well-known Viennese poetess of Jewish descent, was from 1843 to 1848 companion to a Princess Schwarzenberg. Collections of her poems appeared as *Gedichte* in 1841, as *Nach dem Gewitter* in 1843, and as *Neue Gedichte* in 1850, from which year on she devoted herself entirely to authorship. Her publications include *Neueste Gedichte* (1870) and fiction, *Die Welt und mein Auge* (3 vols., 1844). Shortly after the death of Grillparzer (q.v.) she published the biography *Grillparzer und seine Werke* (1875).

Papageienpark, a poem by Rilke (q.v.), superscribed *Paris* and included in *Neue Gedichte* (q.v.).

Papa Hamlet, a collection of three stories published in 1889 with the sub-title *Drei Skizzen,* and alleged to be translated from the Norwegian of one Bjarne P. Holmsen. In fact they were German originals written jointly by A. Holz and J. Schlaf (qq.v.). Their true authorship was acknowledged when they were reprinted in 1892 in *Neue Gleise* (q.v.).

The stories are *Papa Hamlet, Der erste Schultag,* and *Ein Tod.* All three are experiments, in which dialogue and soliloquy are used to turn the sketch into a hybrid between narrative and drama. *Papa Hamlet* presents the death of an ailing infant in poverty-stricken circumstances, and the eponymous principal speaker is an unsuccessful actor who had once played Hamlet. This story employs a minutely detailed realism which acquired the term Sekundenstil (see NATURALISMUS). *Der erste Schultag* portrays in a similar manner the anxieties and perplexities of Jonathan's first day at school. *Ein Tod* has as its subject the death by sepsis of a student wounded in a duel.

Papierne Passion, Die, a sketch written by A. Holz and J. Schlaf (qq.v.) and published in *Neue Gleise* (q.v., 1892). It portrays a half-hour at Christmas time in a Berlin tenement, using a semi-dramatic technique of dialogue set in normal type and simple stage-direction-like narration in small type. The title derives from the skill of old Kopelke, one of the characters, in cutting the symbols of the Passion out of a sheet of paper.

Papinianus, see GROSSMÜTIGER RECHTSGELEHRTER.

PAPPENHEIM, GOTTFRIED HEINRICH, GRAF ZU (Pappenheim/Altmühl, 1594–1632, Leipzig), of an ancient noble family formerly occupying the hereditary marshalcy (Reichserbmarschallamt), was a Roman Catholic convert who served King Sigismund of Poland before he became, under the Catholic League (see KATHOLISCHE LIGA), a participant in the Thirty Years War (see DREISSIGJÄHRIGER KRIEG and WEISSEN BERGE, SCHLACHT AM). In the service of the Emperor Ferdinand II (q.v.) he commanded a regiment of cuirassiers (Pappenheimer Kürassiere), fighting in Lombardy and supporting Tilly (q.v.) in the siege of Magdeburg (q.v., 1631). Impatient with the slow pace of this ageing general, he provoked the battle of Breitenfeld which gave the Swedish King Gustavus Adolphus (q.v.) his first decisive victory over the imperial army. At the battle of Lützen (q.v.) he arrived in time to avert the worst consequences of Wallenstein's defeat, before being mortally wounded.

Pappenheimer, Die, a cavalry song (Reiterlied) by Julius Wolff (q.v., 1889) referring to the 17th-c. regiment of horse known as 'die Pappenheimer' after their commander Count G. H. zu Pappenheim (q.v.).

Soldiers from this regiment (commanded by Max Piccolomini) play a part in Schiller's trilogy *Wallenstein* (q.v.).

PAQUET, ALFONS (Wiesbaden, 1881–1944, Frankfurt/Main), began in business and then turned to journalism. From 1903 to 1908 he was a student successively at the universities of Heidelberg, Munich, and Jena. He was widely travelled, visiting Siberia and Asia Minor as well as European countries. Paquet, a Quaker, was strongly pacifist. His poetic works, *Lieder und Gesänge* (1902), *Held Namenlos* (1911), and *Erwähnung Gottes*, are in free rhythm recalling Walt Whitman. He also wrote Expressionist plays (*Fahnen*, 1923; *Markolph*, 1924; *Sturmflut*, 1926) and the novel *Kamerad Flemming* (2 vols., 1911). Of the plays, *Fahnen*, which deals with the trial of anarchists at Chicago in 1889, was produced by E. Piscator (q.v.) at his 'Proletarisches Theater' (Zentraltheater, Berlin) in 1922. *Gesammelte Werke* (3 vols.), ed. H. M. Elster, appeared in 1970.

PARACELSUS, THEOPHRASTUS (Einsiedeln, Switzerland, 1493–1541, Salzburg), physician, original thinker, and legendary figure, whose true name is variously given as Theophrast von Hohenheim (in Württemberg) and Th. Bombast von Hohenheim. In the latinized form the names Philippus and/or Aureolus sometimes precede the Theophrastus.

Paracelsus was the son of a Württemberg doctor who moved to Villach, Carinthia, in 1502. He studied medicine, possibly at Tübingen, and certainly at Ferrara. In 1527 he became professor of medicine at Basel, where he was in contact with Erasmus and the printer Froben (qq.v.). Paracelsus was the first to deliver a course of lectures at a German-speaking university in German instead of the traditional Latin, an innovation which aroused opposition among his colleagues. He was soon at odds with the medical faculty and left Basel hastily to escape prosecution. He then led a restless, wandering life, being recorded at Colmar, St. Gall, Amberg, Innsbruck, Merano, Ulm, and Augsburg, as well as turning up in Moravia and Hungary. In 1541 he was offered a permanent post in Salzburg, but died there after a few months.

Paracelsus revolted against traditional medicine, based on Galen and Avicenna, and his unconventionality led to his being regarded by his numerous enemies as a quack and charlatan. His unpopularity was augmented by his flamboyant and egregious personality. He was, however, clearly a physician with a gift for healing and, in addition, with a perception that the human body possesses wonderful recuperative powers, which the physician must help. He had an instinctive awareness of biological chemistry. He perceived the human body as a 'microcosm' corresponding to the 'macrocosm' of Nature and held that remedies for all diseases could for this reason be found in Nature if the relations between 'microcosm' and 'macrocosm' were fully understood. *Die große Wundarznei* (1537), written, contrary to the medical practice of the age, in German, was the only one of his works published in his lifetime. His tense, complex, and obscure style corresponds to his striving to express truth.

The philosophical views of Paracelsus reveal him as a man linked both with the Middle Ages and the New Humanism. His opinions, though unorthodox, are firmly theistic; at the same time modern neo-Platonist views, derived from Pico della Mirandola and others, contributed to form his thought. A complex system of relationships enables God to be recognized in Nature, and Nature to be known in God.

The strange personality of Paracelsus, with its blend of arrogance and compassion, overconfidence and devotion, as well as his unusual and ambiguous philosophy, have attracted subsequent writers. Goethe studied him in his early years; Schnitzler (q.v.) devoted a verse play (*Paracelsus*, 1899) to him; and E. G. Kolbenheyer (q.v.) made him the central figure of his sizeable trilogy of novels (*Paracelsus-Trilogie*, 1917–26).

The first edition of his works was published as *Werke*, ed. J. Huser (1589–91) and reissued (11 vols.) 1968 ff. *Sämtliche Werke* (14 vols.), ed. K. Sudhoff, appeared in 1922–33.

Pariser Tagzeiten, designation of an anonymous Middle High German poem of more than 4,000 lines dealing with the canonical hours of prayer and written in the 14th c. The beginning of the MS. is missing. (See also CHRISTI TAGESZEITEN.)

PARRICIDA, JOHANN, name given to Johann von Schwaben (1290–1313, Pisa), the nephew of the Emperor Rudolf I (q.v.) of Habsburg and son of Duke Rudolf of Habsburg. Impelled by a sense of grievance at the refusal of what he believed was his inheritance, Johann murdered his uncle Albrecht I (q.v.), Deutscher König, in May 1308, and the name Parricida refers to this deed. He is the original of the character Johannes Parricida, whom Schiller introduces into the fifth act of *Wilhelm Tell* (q.v.) in order to demonstrate by contrast the purity of motive prompting Tell's assassination of Geßler.

Parsifal, a 'Bühnenweihfestspiel' by R. Wagner (q.v.), first conceived in 1845. Wagner completed the text in 1877 and the musical composition continued until 1882. *Parsifal* was first performed in the Festspielhaus, Bayreuth, on 26

July 1882. The primary source is Wolfram von Eschenbach's 13th-c. epic poem *Parzival* (q.v.).

The centre of the action is the Grail Castle (Gralsburg), which housed the Holy Grail, here understood as the vessel from which Christ drank at the Last Supper. The castle is garrisoned by the Grail Knights, whose head, Titurel, has abdicated in favour of his son Amfortas. Amfortas has succumbed to sensual love, and has been wounded with the Holy Spear in the hands of the magician Klingsor, a former Grail Knight, now a renegade. The wound will not heal until a 'perfect and innocent fool' touches Amfortas with the Holy Spear. Parsifal arrives by chance at the Grail Castle and Gurnemanz, the aged knight, sees in the guileless boy a possibility of healing for Amfortas. But Parsifal watches Amfortas in silent compassion without asking the question which could lead to the end of the older man's suffering. Klingsor, aided by the ambivalent Kundry, who serves the magician yet desires salvation, tempts Parsifal with sensual delights, failing, however, to seduce him. Kundry curses Parsifal, obliging him to wander for years. He returns, a man with experience of the world, yet still pure of heart, and bearing the Holy Spear which he has wrested from Klingsor. He heals Amfortas's wound with the Spear and takes over the kingship of the Grail. Kundry attains salvation as she dies.

Until 1913 *Parsifal* could only be given at Bayreuth. Its performance was treated as a religious occasion and, by convention, the audience abstained from applause.

'Partei, Partei, wer sollte sie nicht nehmen', first line of the poem *Die Partei* by G. Herwegh (q.v.), written in 1841 as a rejoinder to the loftier view of F. Freiligrath (q.v.) expressed in the lines 'Der Dichter steht auf einer höhern Warte,/Als auf den Zinnen der Partei'. The lines are taken from the poem *Aus Spanien*, first published in 1839 and later included in *Ein Glaubensbekenntnis* (1844).

Partonopier und Meliur, a Middle High German verse romance of more than 21,000 lines by KONRAD von Würzburg (q.v.). Its date is uncertain, but it is believed to be the second of Konrad's longer verse narratives and was written in Basel in the second half of the 13th c. In the introduction Konrad names his patrons, Peter der Schaler and Arnold der Fuchs. The original is the French poem *Partonopeus de Blois* by Denis Piramus; and Konrad, who had not sufficient French, acknowledges the help of Heinrich Marschant in interpreting the original.

The poem is a romance of adventure and magic. The young Count Partonopier of Blois loses his way when hunting, arrives at the seashore and is wafted in a magic ship to a deserted castle. Unseen hands serve his meal, and at night Princess Meliur, who has inveigled him thither by magic means, shares his bed. She imposes the condition that he shall not behold her for three and a half years. After a year he secures leave to visit his parents. Religious scruples about Meliur's magic are roused, and on his return he illuminates his beautiful companion with a magic lamp. He is thereupon banished from her, but after long adventures gains her hand at a tournament. It is noteworthy that Meliur is not the supernatural creature that the legend implies, but a normal mortal skilled in necromancy.

Konrad's method of narration is to follow the pattern of his original and to expand the episodes, so that his poem is almost double the length of his model. Its chief merit is its stylistic virtuosity.

Parzival, a Middle High German epic poem by WOLFRAM von Eschenbach (q.v.). Its source is the unfinished *Li contes del graal* by Chrétien de Troyes, but Wolfram has completed the story and freely adapted Chrétien's work to his purpose. It runs to nearly 25,000 lines and is believed to have been begun *c.* 1200 and to have been finished *c.* 1210. It is written in an idiosyncratic and often obscure style. Though it has an Arthurian element, its central point is not Arthur's court. It tells the story of Parzival from his birth until he becomes the Grail King.

Parzival's father, Gahmuret, undertaking knightly adventures in the East, marries the Moorish queen Belakane, by whom he has a son, Feirefiz. He deserts her and later marries Herzeloyde of Waleis, who bears him Parzival. Meanwhile he returns to the Orient and is killed. Herzeloyde, grieving over the death of her husband, brings the boy up in ignorance of the world in the seclusion of the forest. But the sight of passing knights attracts him to the great world and he insists on leaving her to become a knight. She dresses him in clothes befitting a simpleton, hoping thus to repel the world from him, but as he leaves she dies of heartbreak unnoticed by Parzival. His ignorance leads him to brutal action in robbing a lady (Jeschute) of her ring and, later, killing a knight (Ither) with an unknightly weapon. He takes the knight's armour, but the simpleton's clothing beneath it still symbolizes his ignorance. Meanwhile he has learned his name from his unrecognized kinswoman Sigune (who is faithfully mourning her dead husband) and has visited Arthur's court, though he is still too raw to be accepted as a knight of the Round Table. He next visits the castle of Gurnemanz, who instructs him in courtly behaviour. He saves a lady (Condwiramurs) from her oppressor and marries her.

This rescue marks his achievement of knightly status. Parzival next comes to the Castle of the Grail (Gralsburg at Munsalvaesche), and unwittingly misses the opportunity to become Grail King, by following too literally Gurnemanz's counsel to refrain from questioning; for he omits to put to the sick Grail King Anfortas the humane inquiry about the reason for his suffering which would release Anfortas and cause Parzival to succeed him. The next day the castle is almost deserted, and, as he leaves, an insulting reproach is shouted after him. He encounters Sigune again, who explains that Anfortas is Herzeloyde's brother and curses him for his failure to relieve his uncle's suffering. He returns to Arthur's court and becomes a knight of the Round Table. But Cundrie, the messenger of the Grail, enters and pronounces the curse of the Grail upon him. He leaves dishonoured and turns against God, who, he feels, has misled him. The story then turns for a time to the Arthurian circle and traces the knightly adventures of Gawan. Parzival is next led to the hermit Trevrizent, who explains to him the misfortunes of Anfortas and the Grail community and brings him back to God and absolves him of his sins. Trevrizent subsequently proves to be the brother of Herzeloyde and Anfortas and thus Parzival's uncle. At this point Parzival is ready to revisit Munsalvaesche, but, before he does so, the story resumes the adventures of Gawan. These include the release of noble ladies imprisoned in Schastel Marveile by Clinschor, a decadent knight turned magician. Finally Parzival returns to the Gralsburg, puts the question which releases Anfortas, and becomes Grail King. In the later stages of his quest he is assisted by his half-brother, the heathen Feirefiz, who is later baptized and with his consort establishes Christianity in India. At the end the story is briefly told of Parzival's son Lohenrangrin, the Knight of the Swan, who marries a lady of Brabant on condition of concealing his identity, and sadly leaves when years after she asks his name.

The complex story has been subjected to various interpretations, but all agree upon the ethical seriousness which animates Wolfram. Parzival, emerging from complete simplicity and ignorance reaches, by experience and instruction, the level of the courtly world, but is not content with its extrovert joys and superficial conventions. His isolation from God is explained by his misconception of divinity. He sees God anthropomorphically as a feudal lord who has failed to help him. Through the hermit (seconded by two preceding minor episodes) he is brought to a true conception of religion and so enabled to reach the maximum ethical development. Wolfram's knightly world here reaches its highest level of religious consciousness. It is not, however, an ecclesiastical conception, and it is noteworthy that Wolfram abandons Chrétien's form of the Grail, a vessel associated with the Host, and substitutes for it a stone of miraculous properties, *lapsit exillis*, as he terms it. The popularity of *Parzifal* is reflected in the existence of more than seventy MSS., of which fifteen are more or less complete. See also NEUE PARZEFAL, DER and PARSIFAL.

Passion, Die, a novel by C. Viebig (q.v.), published in 1925. It is the life story of a sweet-natured, physically handicapped, illegitimate child who, after her mother's death, goes into service, battles against ill health, and, after an unsuccessful attempt to gas herself, dies of a heart attack.

Passional, Das, a huge Middle High German collection of verse legends amounting to 100,000 lines. It consists of three sections: (1) a life of Christ, which is focused on the Virgin Mary and contains a number of legends referring to her; (2) the lives of the apostles and evangelists together with those of Saint John the Baptist and Saint Mary Magdalene; (3) seventy-five legendary lives of saints. The principal source of the poem, which was completed *c.* 1300, is the *Legenda aurea* (q.v.). The author was a cleric connected with the Teutonic Order (see DEUTSCHER ORDEN) and was probably also the author of *Das Väterbuch* (q.v.).

Pastourelle, pastoral song in dialogue form. Of Provençal origin, it occurs occasionally in Middle High German poetry.

Pater Brey, see EIN FASTNACHTSPIEL VOM PATER BREY.

Paternoster or **Auslegung des Vaterunsers,** designation of an Early Middle High German interpretation of the Lord's Prayer in terms of the mystic number seven. It is in verse and consists of twenty stanzas, each of twelve lines. The principal MS. is at Innsbruck. See also HEINRICH VON KROLEWIZ.

Patmos, a hymnic poem written in 1801–3 by F. Hölderlin (q.v.). In irregular unrhymed verse the poet conjures up images of Christianity and Hellas to form a complex and numinous pattern. The exaltation, power, and sublimity of the language express awe, apprehension, and hope in a poem which is not reducible to any prose formula. Complicated by the existence of more than one version, it has become the subject for interpretative controversy.

Patricius, fragment of an Early Middle High German poem dealing with the miracles of St. Patrick, written probably *c.* 1150.

Patriot, Der, a historical story by A. Neumann (q.v.), published in 1925. It is set in St. Petersburg in 1801 and deals with a conspiracy led by Count Pahlen to remove, whether by abdication or assassination, the cruel and tyrannical Tsar Paul I. The conspiracy culminates in the Tsar's murder, whereupon Pahlsen proves the purity of his motives by suicide.

Patriotische Phantasien, title of 4 vols. of essays by Justus Möser (q.v.), published from 1774 to 1778 by his daughter, Jenny von Voigts (q.v.). They had been originally published in the *Wöchentliche Osnabrückische Intelligenzblätter*.

Pauken, see MENSUR.

PAULI, JOHANNES (Pfeddersheim, Alsace, *c.* 1450–*c.* 1533, Thann, Alsace), an Alsatian monk, who compiled a collection of Schwänke (see SCHWANK) under the title *Schimpf und Ernst* (1522). The 232 anecdotes, varying in length from a few lines to a couple of pages, are all classified either as 'schimpff' (i.e. 'Scherz') or 'ernst'. Pauli's aim was strictly moral and practical: 'Schimpff und ernst findestu in diesem Bůch, kurzweilig, und auch das ein iechlich mensch im selben davon exempel und leren nemen mag, und ist im nützlich und gůt.'

Nevertheless the book, with its racy colloquial style and terse narration, achieved considerable popularity in the 16th c. and 17th c. as a work of entertainment. Pauli also translated into German in 1520 the Latin sermons of GEILER von Kaisersberg (q.v.) on *Das Narrenschiff* (q.v.).

Schimpf und Ernst, ed. H. Oesterley, was published in 1866; a reprint appeared in 1967.

PAULSEN, RUDOLF (Berlin, 1883–1966, Berlin), the son of a professor of philosophy, was associated with O. zur Linde and R. Pannwitz (qq.v.) in the group Charon and with its periodical *Charon* (q.v.). A lyric poet of modest achievement, he published volumes of verse, including *Gespräche des Lebens, Lieder aus Licht und Liebe* (both 1911), *Im Schnee der Zeit* (1922), *Die kosmische Fibel* (1924), *Die hohe heilige Verwandlung* (1925), and *Das festliche Wort* (1935). He sought to reconcile dionysian and Christian thought. A select edition of his poetry appeared under the title *Musik des Alls und Lied der Erde* in 1954; it was followed by *Träume des Tritonen* (1955) and *Werte bewahrt im Wort* (1960). In 1936 he published his autobiography, *Mein Leben*.

Paulskirche, a neo-classical Lutheran church in Frankfurt/Main, built between 1789 and 1833. Circular in plan, it was used in 1848 for the sessions of the Frankfurt Parliament (see FRANKFURTER NATIONALVERSAMMLUNG). Hence the name Paulskirche is often used to denote the Assembly itself. The church, damaged in 1944, was restored in 1948.

PAUMGARTTEN, CHARLOTTE VON, see GRILLPARZER, F.

Peasants' War, see BAUERNKRIEG.

PEGAEUS, pseudonym of Thomas Brunner (q.v.).

Pegasus im Joche, an allegorical poem by Schiller on the subject of poetry. It was first published in the *Musenalmanach* in 1796 with the title *Pegasus in der Dienstbarkeit.*

Pegnesischer Blumenorden, see HIRTEN-UND BLUMENORDEN AN DER PEGNITZ.

Pegnitz, name of the river on which the city of Nürnberg is situated, often used in allusion to the city.

Pennalismus, a custom in German universities in the 16th c. and 17th c. by which the freshmen (Pennälen) were required to perform menial duties for seniors. The system, which was seriously abused, was abolished in the middle of the 17th c.

Penthesilea, a one-act tragedy in more than 3,000 lines of blank verse by H. von Kleist (q.v.) written in 1807, and published in *Phöbus* (q.v.) in 1808. In its 24 scenes Kleist reversed the post-Homeric legend of the slaying of the Amazon Penthesilea by Achilles during the Trojan War.

The action opens with Penthesilea in pursuit of Achilles as she leads the Amazons against the Greeks. According to Amazon law the women warriors are bound to make war in order to take male prisoners, who at the subsequent festival of roses (Rosenfest) will provide for the continuity of the state. In singling out Achilles, Penthesilea breaks the special law forbidding the Amazons to choose their individual opponents. She fails to win Achilles, but the campaign is otherwise successful. Ignoring the warnings of the High Priestess, Penthesilea sets out in quest of Achilles. She suffers defeat and loses consciousness, and he follows her into the Amazon camp. As Penthesilea regains consciousness she cherishes the delusion that she has gained Achilles in fulfilment of a prophecy made by her mother. Achilles, who loves her and wishes to carry her off, undeceives her. When the situation becomes clear to him, he sends her a challenge with the intention of surrendering to her, and goes forth unarmed. Penthesilea mistakes his action for scorn, and in a fury of mad despairing rage sets her hounds on him and joins them in rending his body. When she becomes aware of what she has done she defies state and god, casts away her sword, and through the power of her will undergoes a death of repentance, love, and hope, which looks forward to her reunion with Achilles in the Elysian realm.

Goethe rejected Kleist's radical presentation of tragedy after Kleist had sent him the MS. in the hope that it might be performed under Goethe's auspices in Weimar. The play embodies, in its action and its free adaptation of classical form, a powerful denial of the classical ideals of the Weimar stage. It was not performed until 1876, and even then in an adaptation in three acts. Othmar Schoeck (1886–1957) adapted it for an opera (1st performance 1927, Dresden; revised version 1928, Zurich).

PENZOLDT, ERNST (Erlangen, 1892–1955, Munich), son of a professor at Erlangen university, studied art at the Weimar and Kassel academies, becoming a sculptor in Schwabing, the artists' quarter of Munich, but is better known for his whimsical and sensitive writings. He wrote a certain amount of poetry (*Der Gefährte,* 1922; *Zwölf Gedichte,* 1937; *Fünfzehn Gedichte,* 1954), but was mainly an author of fiction, some of which he dramatized himself. His novels are *Der Zwerg* (1927, re-titled *Die Leute aus der Mohrenapotheke* in 1938), *Der arme Chatterton* (1928, a *biographie romancée* of the poet), *Die Powenzbande* (q.v., 1930), *Kleiner Erdenwurm* (1934), and *Der Kartoffelroman* (1948). Shorter narrative works are *Etienne und Luise* (1929, dramatized 1930), *Die Portugalesische Schlacht* (1930, a collection of stories, the first of which was turned into a play in 1931), *Idolino* (1935), *Korporal Mombour* (1941), *Zugänge* (1947), *Bitternis* (1951, a collection), and *Squirrel* (1954, dramatized 1955). He also wrote a tragedy, *Karl Ludwig Sand* (1931, see SAND, K. L.), and two comedies, which were not adaptations of stories, *So war der Herr Brummell* (1933) and *Die verlorenen Schuhe* (1946). Penzoldt, who wrote with charm and wit, was also an essayist (*Der dankbare Patient,* 1937; *Episteln,* 1942; *Tröstung,* 1946; *Causerien,* 1949). His reminiscences are entitled *Das Nadelöhr* (1948). *Gesammelte Schriften* (4 vols.) appeared in 1949–62.

Peregrina, a poetic designation applied by E. Mörike (q.v.) to Maria Meyer (q.v.). Mörike used the name as a collective title for the five poems connected with her: (1) 'Der Spiegel dieser treuen, braunen Augen'; (2) 'Aufgeschmückt ist der Freudensaal'; (3) 'Ein Irrsal kam in die Mondscheingärten'; (4) 'Warum, Geliebte, denk' ich dein'; (5) 'Die Liebe, sagt man, steht am Pfahl gebunden'. (1) and (5) (qq.v.) especially are profoundly moving expressions of the emotional disturbance induced by his love for her.

PERFALL, ANTON, FREIHERR VON (Landsberg/Lech, 1853–1912, Schliersee, Bavaria), was a prolific author of Bavarian rural fiction, including the novels *Dämon Ruhm* (2 vols., 1889), *Gift und Gegengift* (1890), and *Das verlorene Paradies* (1896) and the stories *Aus meinem Jägerleben* (1906), *Jagd- und Berggeschichten* (1909), and *Meine letzten Waidmannsfreuden* (1914). K. T. von Perfall (q.v.) was his brother.

PERFALL, KARL, FREIHERR VON (Munich, 1824–1907, Munich), a gifted musician, became director of the Royal Conservatoire and Intendant of the Hoftheater in Munich in 1867. His operas include *Sakuntala*. He wrote books of theatrical history (*Fünfundzwanzig Jahre Münchner Hoftheater-Geschichte*, 1892; *Die Entwicklung des modernen Theaters*, 1899). His nephews A. and K. T. von Perfall (qq.v.) were novelists.

PERFALL, KARL THEODOR, FREIHERR VON (Landsberg/Lech, 1851–1924, Cologne), entered the Bavarian civil service, but turned to journalism, becoming editor of the *Kölnische Zeitung*. He wrote numerous light novels. He was a brother of A. von Perfall (q.v.).

PERINET, JOACHIM (Vienna, 1763–1816, Vienna), an actor in the popular theatres of Vienna, particularly the Theater in der Leopoldstadt, wrote numerous Singspiele (see SINGSPIEL), mostly adaptations of popular works of the previous generation, especially those of Philipp Hafner (q.v.). They include *Das neue Sonntagskind* (1793, Hafner's *Der Furchtsame*), *Die Schwestern von Prag* (1794, originally *Der von dreyen Schwiegersöhnen geplagte Odoardo*), and *Die Hexe Megära* (1804, an adaptation of Hafner's *Mägera, die förchterliche Hexe*). Perinet is the author of the well-known song 'Wer niemals einen Rausch gehabt'. It occurs in *Das neue Sonntagskind*. Perinet's *Der travestierte Telemach* and its sequel *Antiope und Telemach*, examples of burlesqued mythology on the stage, were performed in 1805.

PERKONIG, JOSEF FRIEDRICH (Ferlach, Carinthia, 1890–1959, Klagenfurt), was a primary school teacher and from 1922 a professor at the college of education in Klagenfurt. He wrote novels and stories set in the Carinthian highlands, and was chiefly read in his native province. His novels include *Bergsegen* (1928, retitled *Auf dem Berge leben* in 1934), *Mensch wie du und ich* (1932, revised 1954), *Nikolaus Tschinderle*,

Räuberhauptmann (1936), *Patrioten* (1950), and *Ev und Christopher* (1952). *Dorf am Acker* (1926) and *Ein Laib Brot, ein Krug Milch* (posth., 1960) are collections of stories. Selected *Werke* appeared in 1965 ff.

PESTALOZZI, JOHANN HEINRICH (Zurich, 1746–1827, Brugg, Switzerland), real name Pestalutz, was brought up as an orphan. The great interest of his life was the provision and reform of education, especially for the poor. At first a farmer, he established a school for orphaned and deprived children at his farmhouse (Gut Neuhof), which lasted until 1779. Thereupon he set about furthering his ideas by writing, beginning with *Die Abendstunde eines Einsiedlers* (1780) and followed by his best work *Lienhard und Gertrud. Ein Buch für das Volk* (q.v., 1781), a short village novel, which he subsequently extended with didactic material in three additional parts (1783–7).

Pestalozzi sought financial support from various German sources, but with insufficient success, and from 1790 onwards changed his residence several times. He was at one time in Leipzig, ran a silk factory in Zurich, and was active as a political publicist of patriotic Swiss persuasion. From 1799 to 1804 he was again in charge of a school, this time at Burgdorf. His principal educational treatise, *Wie Gertrud ihre Kinder lehrt* (1801), appeared at this time in the form of letters linking up with his first novel. In 1804 he was invited to direct a pedagogical institute from 1805. His sense of dedication is reflected in *Selbstschilderung* (1802).

Pestalozzi sought to shape education so that the child became increasingly aware of his relationship to the world around him; his elementary method (Elementarmethode) relied on a mother's capacity to develop the physical and mental potential of her child. Pestalozzi's theoretical writings enlarging on this attitude include *Über die Idee der Elementarbildung* (1809), *An die Unschuld, den Ernst und den Edelmut meines Vaterlandes* (1815), and *Über die Naturgemäßheit in der Erziehung* (1826). He influenced the 19th-c. educationalist F. Fröbel and the writings of J. Gotthelf (qq.v.). Pestalozzi's Neuhof was reinstituted as a Swiss foundation, and a number of educational institutions commemorate his devoted work as a pioneer in the care of deprived children, including the international Pestalozzi villages. A novel by Wilhelm Schäfer (q.v.), of which Pestalozzi is the subject, is entitled *Der Lebenstag eines Menschenfreundes* (1915). *Pestalozzi*, a play by A. Steffen (q.v.), appeared in 1939.

Sämtliche Werke, ed. L. W. Seyffarth, appeared in 1869–73 and 1899–1902 (12 vols.), and of the critical edition (ed. A. Buchenau, E. Spranger, and H. Stettbacher) 21 vols. appeared in 1927–64. Correspondence, *Briefe*, published under the auspices of the Pestalozzianum, Zurich, amounts to 13 vols., 1946–71.

Peter Camenzind, the first novel of H. Hesse (q.v.), published in the *Neue deutsche Rundschau* in 1903, and in 1904 in book form. Peter grows up in the isolated Swiss village Nimikon before he is sent to a grammar school and, after the death of his mother, to Zurich University. His student life brings out his divided personality: he longs for friendship, and finds it in his fellow student Richard, but he reacts against society. He turns to writing, and twice goes to Italy, where he roams, searching for inner peace, in the footsteps of St. Francis. He is attracted especially by the natural beauty of Italy, not by its cultural tradition. The accidental death of Richard by drowning is a shattering blow to Peter. His love for Erminia Aglietti, a painter who represents his ideal of southern beauty, ends in rejection. He settles in Basel and takes to journalism. An unfulfilled love for Elisabeth, a Swiss girl who influences him both before and after her marriage to another, saves him from declining into alcoholism. He finds a purpose in life in caring for Boppi, a congenital cripple whose capacity for bearing suffering teaches Peter the *ars moriendi*. After Boppi's death Peter returns to his native village, accepts its way of life, and looks after his sick and aged father. The novel may be regarded as an Entwicklungsroman (see BILDUNGSROMAN). An autobiographical element and the influence of G. Keller's *Der grüne Heinrich* (q.v.) are both prominent. The great and lasting success of the work lies chiefly in Hesse's sensitive depiction of nature which he sees as 'the language of God'. The predominantly lyrical qualities of the novel are epitomized in the poem *Wolkenlied*. In 1951 Hesse referred to his Camenzind as a youth who undergoes, on a smaller scale, the 'halb tapfere, halb sentimentale Revolte Rousseaus' through which he becomes a writer.

Peter Leberecht, eine Geschichte ohne Abenteuerlichkeiten, a short novel published by L. Tieck (q.v.) in 1795. It is the autobiography of the hero, dealing in the trivial and simple events of life, as the sub-title implies, but including nevertheless a sensational event in Leberecht's unwitting approach to incest, through betrothal to a girl who proves to be his sister. Tieck used Peter Leberecht as a pseudonym for his *Volksmärchen* (1797).

Peter Lewen or *Leu,* see WIDMANN, ACHILLES JASON.

PETERS, LUISE, see OTTO-PETERS, LUISE.

Peter Schlemihls wundersame Geschichte, a story by Adelbert von Chamisso (q.v.), published in 1814. The preface indicates that it was written in Kunersdorf (1813).

Young Peter Schlemihl approaches a wealthy merchant with a letter of introduction. Bidden to follow the merchant and his friends, he notes a man in inconspicuous grey. Presently the man in grey takes Schlemihl on one side and buys his shadow, giving in payment the inexhaustible purse of Fortunatus. But the shadowless Schlemihl cannot enjoy his wealth; he is reviled and persecuted and loses the woman he loves. The man in grey reappears and offers to return the shadow if Schlemihl will yield him his soul. Schlemihl, after an inner struggle, rejects the offer and frees himself from the sinister man in grey by throwing away the magic purse. His shadowless misery continues until he happens to discover a pair of seven-league boots with the help of which he devotes himself to scientific inquiry and research and so compensates himself, by useful endeavour, for his loss. The story, which achieved instant success, offers opportunity for diverse interpretations, both on the political and the personal plane.

Peter Squentz, abbreviated title of *Absurda Comica oder Herr Peter Squentz* (q.v.), a comedy by Andreas Gryphius.

PETER VON ARBERG, GRAF, a Swiss nobleman of the 14th c. to whom are ascribed 'geistliche Tageweisen' and also profane Tagelieder (see TAGELIED) preserved in the Colmarer Liederhandschrift (q.v.). The poem 'O starcker got . . .' is mentioned under 1356 in the *Limburger Chronik* (q.v.).

Peter von Staufenberg, a Middle High German tale in verse written by Egenolf von Staufenberg *c.* 1310. The story is related to the legend of the Schöne Melusine (q.v.). Peter, who is a knight of exemplary conduct, meets a beautiful woman who professes supernatural powers. They live together on the condition that, should he marry another woman, he will die three days after the wedding. Peter in the end marries a feudal heiress and accordingly dies. An adaptation by Bernhard Schmidt, with a preface and prologue by Johann Fischart (q.v.), was printed in 1588.

Peterwardein, former Austrian town which has belonged to Yugoslavia (as Petrovaradin) since 1918. At the battle of Peterwardein, fought on 5 August 1716, Prince Eugene (see EUGEN, PRINZ) decisively defeated a Turkish army.

Petrarkismus, a style of love poetry deriving from the Italian poet and humanist Francesco Petrarch (1304–74). His cycle of *Canzoniere* (posth. 1470), consisting mainly of sonnets, was especially influential in European poetry. Petrarch was first introduced into German by Opitz (q.v.) who translated two of the sonnets. Distinctive poetic devices are the use of antithesis, the canon of motifs (e.g. golden hair) and images derived from nature, classical eroticism and mythology, conceits, and forms of rhetoric including oxymoron. P. Fleming (q.v.) is the most creative of 17th c. exponents of the Petrarchan style, echoes of which have been traced in the 18th c. (Günther, Lenz, Goethe, qq.v.), in Galante Dichtung (q.v.) and anacreontic poetry (see ANAKREONTIKER), and, in inverted form, in Heine (q.v.). Modern reappraisals began in the early 1930s (H. Pyritz) and were reassessed by L. W. Forster (*The Icy Fire. Five Studies in European Petrarchism*, 1969). (See SONETT.)

Petruslied, an Old High German prayer of intercession in Bavarian dialect. It is designed to be sung in procession and the third, sixth, and last of its nine lines are the prayer 'Kirie eleyson, Christe eleyson', which was sung as a refrain by the people. The *Petruslied* has been dated before Otfried's *Evangelienbuch* (863–71, see OTFRIED), in which case it would be the oldest German rhyming poem; but opinions remain divided, and many scholars consider it to be later than Otfried's poem.

PETZOLD, ALFONS (Vienna, 1882–1923, Kitzbühel), of working-class origin, held various jobs, including those of waiter and builder's labourer. He wrote poetry which was both socialistic and religious (*Trotz alle dem*, 1910; *Der Ewige und die Stunde*, 1912; *Der heilige Ring*, 1914; *Der stählerne Schrei*, 1916; *Der Dornbusch*, 1919). A selection appeared in 1922 as *Gesang von Morgen bis Mittag*. His fiction includes the novels *Erde* (1913), *Der feurige Weg* (1918), and *Das rauhe Leben* (1920), which is largely autobiographical. The selection *Pfad aus der Dämmerung. Gedichte und Erinnerungen* appeared in 1943.

PEUCKER, NIKOLAUS (*c.* 1623–74, Cölln/Spree), a municipal officer of Cölln near Berlin, was the author of simple homely poems and songs, later collected as *Wohlklingende lustige Pauke* (1702). He enjoyed the favour of the Great Elector (see FRIEDRICH WILHELM, DER GROSSE KURFÜRST).

PEUTINGER, KONRAD (Augsburg, 1465–1547, Augsburg), a neo-humanist of Augsburg, who was town clerk from 1493. Educated in Italy, he was a noted lawyer and enjoyed the confidence of the Emperor Maximilian I (q.v.). In a life devoted to legal duties he found time to occupy himself with classical studies (*Sermones mirandis Germaniae antiquitatibus*, 1506). The *Peutingersche Tafel* (q.v.) of which he was the possessor, received its name from him. His correspondence, *Peutingers Briefwechsel*, ed. E. König, was published in 1923.

Peutingersche Tafel (Tabula peutingeriana), a Roman atlas of 11 sheets dating from the 4th c. A.D. The twelfth sheet of the original (Iberia and Britain) is missing. In 1507 the atlas was given by Celtis (q.v.) to Konrad Peutinger (q.v.), who intended to publish it but died before he could do so. Since 1738 it has been in the possession of the Court (now National) Library at Vienna (see LIBRARIES). A French edition appeared in 1869–76, and a German one, ed. K. Miller (*Die Weltkarte des Castorius, genannt die Peutingersche Tafel*), in 1888.

Pfaffe Amis, Der, a verse satire by Der Stricker (q.v.).

Pfaffe vom Kalenberg, Der, see FRANKFURTER, PHILIPP and GRÜN, ANASTASIUS.

Pfalz, Die, the Palatinate or County Palatine, was also known as Kurpfalz, since the Pfalzgraf was an elector or Kurfürst (q.v.). In 1947 it became a part of the Land Rheinpfalz of the Federal Republic (see BUNDESREPUBLIK DEUTSCHLAND). It lies to the west of the Rhine and consists of a stretch of the flat Rheingau and the hilly regions of the Pfälzer Bergland and the Pfälzer Wald. At its centre is the town of Kaiserslautern. Worms and Speyer are its best-known cities.

From 1214 the Palatinate was ruled by the house of Wittelsbach (q.v.), one branch of which also provided the dukes (later electors and kings) of Bavaria (see BAYERN). In 1542 the Palatinate became Protestant. After the failure of Friedrich V (q.v.) of the Palatinate to establish himself as king of Bohemia in 1619, territory and electoral rights were given in 1623 to Duke Maximilian I (q.v.), Elector of Bavaria. In 1648 the son of Friedrich V, Karl Ludwig (reigned until 1680) recovered the Lower Palatinate (Niederpfalz), but the Upper Palatinate (Oberpfalz) remained Bavarian. The complete union of Bavaria and the Palatinate was achieved in 1816. In English history and literature the Palatinate is best known through Prince Rupert, a Cavalier and an Englishman by adoption (see RUPRECHT, PRINZ).

Pfälzische Idylle, sub-title of two idylls by Maler Müller (see MÜLLER, F.), *Die Schaaf-Schur* (1775) and *Das Nußkernen* (1811, qq.v.). It is sometimes used in the plural as a collective designation for them.

Pfarrer von Kirchfeld, Der, a four-act play (Volksstück mit Gesang) by L. Anzengruber (q.v.), first performed at the Theater an der Wien, Vienna, in November 1870 and published in 1871 under the pseudonym L. Gruber. It is in Austrian dialect. Its background is the conflict between Liberalism in the Roman Catholic Church in Austria and the intransigence of Rome. The Liberal priest Father Hell converts the sceptic village philosopher Wurzelsepp, but is reproved by his bishop for his unorthodoxy and sent to Rome. A tender relationship between Hell and Anna Birkmeier, which has grown unperceived, ends in renunciation by both, in a sense of duty and propriety. The immensely successful play profited (accidentally) by its timing, for it coincided with the revocation by the Austrian Government of the Concordat with Rome.

PFARRER ZU DEM HECHT, DER, wrote *Das mitteldeutsche Schachbuch,* a Middle High German verse translation (in some 6,000 lines) of the *Solatium ludi scaccorum* of Jacopo Dacciesole (see SCHACHBÜCHER). The Pfarrer's version, which adheres fairly closely to the original, was completed in 1355.

Pfarrose, Die, a five-act tragedy by O. Ludwig (q.v.), written in 1846–7 and published posthumously in 1891. Ludwig adopted the unusual spelling 'Pfarrrose'. The plot is based on the ballad *Des Pfarrers Tochter von Taubenhain* (q.v.) by G. A. Bürger, and the resultant play (much altered) may be termed a domestic tragedy (see BÜRGERLICHES TRAUERSPIEL) in a rural setting.

Rose, the pastor's daughter, loves the nobleman Friedrich von Falkenstein, who returns her affection. But Falkenstein is under pressure to marry an aristocratic heiress. An incautious letter by Rose is used by Falkenstein's friends to discredit her and she is dismissed from his mansion and rejected by her parents. She goes into an Ophelia-like madness, and when Falkenstein eventually seeks her out, she dies in his arms. The play abounds in melodramatic situations.

PFAU, LUDWIG (Heilbronn, 1821–94, Stuttgart), in full Karl Ludwig, was a revolutionary in 1848 (see REVOLUTIONEN 1848–9), took refuge in Switzerland and, later, in Paris. He returned to Germany in 1865. He wrote poetry (*Gedichte,* 1847; *Deutsche Sonette auf das Jahr 1850,* 1849),

but was primarily a journalist. His collected criticism was published as *Kunst und Kritik* (4 vols., 1888). While in France he translated Breton folk-songs and some of the novels which appeared under the name Erckmann-Chatrian.

PFEFFEL, GOTTLIEB KONRAD (Colmar, 1736–1809, Colmar), studied law at Halle University before becoming a Protestant educational administrator in Colmar, despite blindness incurred in his early 20s. In his fables (*Fabeln,* 1783) he appears as a sympathizer with the ideals of the coming French Revolution. A selection of his works, *Skorpion und Hirtenknabe. Fabeln, Epigramme, poetische Erzählungen,* ed. R. J. Unbescheid, was published in 1970.

PFEFFERKORN, JOHANNES (Nürnberg, *c.* 1469–*c.* 1522), was the opponent of Reuchlin (q.v.) in a famous controversy. Pfefferkorn, Jewish by birth, was baptized in 1505, becoming zealous for the conversion of the Jews and intolerant of persisting Jewry. He expressed his views in a series of tracts (*Der Judenspiegel,* 1507; *Die Judenbeichte,* 1508; *Der Judenfeind,* 1509), going so far as to advocate the destruction of all Jewish literature except the Old Testament. The Emperor Maximilian I (q.v.) consulted Reuchlin, who declared himself against Pfefferkorn's anti-literary zeal. Pfefferkorn, with the support of the Dominicans of Cologne, attacked Reuchlin in *Der Handspiegel* (1511) and a polemic developed, the most notable contribution to which is the *Epistolae obscurorum virorum* (q.v.) or *Dunkelmännerbriefe* by Reuchlin's humanist friends.

Pfeiferstube, Die, a story (Erzählung) by P. Alverdes (q.v.), published in 1929. It is a moving tale of comradeship between three German soldiers in hospital with severe throat wounds, who, after initial hostility, admit to their friendship a similarly wounded English prisoner of war. The title refers to the whistling noises made by the tubes inserted into their throats.

PFEIFER VON NIKLASHAUSEN, DER, the name by which Hans Böhm (d. 1476) was known. Böhm, a shepherd, publicly demanded in 1476 the execution of clerics and the abolition of property, and denounced the Pope and the Emperor. The favourable reception by the peasantry of his revolutionary doctrine led to his arrest and execution at Würzburg. His was the first of a series of revolts which culminated in the Peasants' Revolt in 1524–5 (see BAUERNKRIEG).

PFEMFERT, FRANZ (Lötzen, East Prussia, 1879–1954, Mexico City), a left-wing political journalist, became editor of *Der Demokrat* in

1910 and founded the journal *Die Aktion* (q.v.) in 1911. He also edited anthologies of Expressionist poetry, *Aktions-Lyrik* (7 vols., 1916–22, published in his own Verlag Die Aktion, reprinted in 1973), as well as *Aktionsbücher der Aeternisten* (10 vols., 1916–21, reprinted in 1973), with contributions by F. Hardekopf and C. Einstein (qq.v.), *Politische Aktions-Bibliothek* (14 vols., 1916–30), and *Der rote Hahn* (60 vols., 1917–25), with contributions by Y. Goll, F. Mehring, Sternheim, J. R. Becher (qq.v.), and V. Hugo; vol. 14–15, *Bis August 1914* (1918), containing essays, was reprinted in 1973. A selection, *Die Revolution G.M.B.H. Agitation und politische Satire in der 'Aktion'*, ed. K. Hickethier, W. H. Pott, and K. Zerges, appeared in 1973, and *Ich setze diese Zeitschrift wider diese Welt. Sozialpolitische und literaturkritische Aufsätze*, ed. W. Haug, in 1985. A Communist opposed to social democracy, he later opposed Stalinism. He emigrated to Czechoslovakia in 1933, to France in 1936, and to Mexico in 1941. His recollections, intended for publication, are presumed lost.

Pfennig, German coin introduced in Carolingian times (see KAROLINGER) and frequently known in the early Middle Ages as a *denar* (from the Latin *denarius*). At first of silver, it was degraded to a copper coin in the 16th c. The regulation of the Pfennig at 100 to the Mark (q.v.) dates from 1871.

PFINTZING, MELCHIOR (Nürnberg, 1481–1535, Mainz), served the Emperor Maximilian I (q.v.) as chaplain and private secretary. Following Maximilian's instructions, he composed the verse romance *Teuerdank* (q.v.).

PFISTER, ALBRECHT, one of the first printers in the 15th c., died not long before 1466. His presses in Bamberg produced the first editions of *Der Ackermann aus Böhmen* (q.v.), Boner's *Edelstein* (see BONER, ULRICH), and the *Biblia pauperum* (q.v.).

Pfisters Mühle, Ein Sommerferienheft, a short novel written by W. Raabe (q.v.) in 1883–4, and published in 1884. As the sub-title indicates, it purports to be written down by Ebert Pfister while on holiday with his young wife at the family mill, which is about to be demolished. Ebert tells of the mill's prosperity in the days when the miller not only ground his flour but served drinks in his garden to numerous citizens and students. But the mill stream became polluted, the guests, repelled by the smell, departed, and Ebert's father had to face the decline of his salubrious old mill in the rising age of industrial

development. Though he discovered the source of contamination in the sugar refinery of Krickerode upstream and won a lawsuit against it, the whole experience brought a realization of the inevitable retreat of his own world in the new technical age, so that he lost heart and died.

Raabe's sympathies are clearly with the values that are fading, but his book, which uses Pfisters Mühle and Krickerode as symbols for the old world and the new, gives the modern development, in the person of Dr. Asche, the Pfisters' progressive friend, a fair hearing.

PFITZNER, HANS (Moscow, 1869–1949, Salzburg), a composer of orchestral and vocal music. His principal operas are *Der arme Heinrich* (1895), after HARTMANN von Aue (q.v.); *Palestrina* (1917), the second act of which presents a session of the Council of Trent (1545–63, see GEGENREFORMATION); and *Das Herz* (1931), for which Pfitzner himself wrote the libretto with the assistance of a pupil, H. Mahner-Mons. Pfitzner's cantatas include *Von deutscher Seele* (1921), which sets a number of poems by Eichendorff (q.v.), and *Urworte. Orphisch* (unfinished, performed 1952), a setting of Goethe's cycle of poems (see URWORTE. ORPHISCH). *Gesammelte Schriften* (3 vols.) appeared in 1927–9, *Über musikalische Inspiration* in 1940, and a posthumous selection, ed. W. Abendroth (*Reden, Schriften, Briefe*), in 1955.

PFIZER, GUSTAV (Stuttgart, 1807–90, Stuttgart), son of a law officer of the Württemberg Crown, studied theology at Tübingen University and then taught at the Tübinger Stift, admired Schiller and was associated with the Swabian Romantics. He published in 1831 a first volume of poems (*Gedichte*) and a second (*Gedichte. Neue Sammlung*) in 1835. In 1834 he turned to journalism, but took up a senior teaching post at the Stuttgart Obergymnasium in 1847, retiring in 1872. He translated Byron (1835–9) and Bulwer-Lytton (jointly with F. Notter, 1838–43) and is the author of *Martin Luthers Leben* (1836) and *Uhland und Rückert* (1837).

PFORR, ANTONIUS VON (Breisach, ?–1483), a cleric of Rottenburg/Neckar, translated for the Palatine Countess Mechtild the Indian *Panchatantra* as *Buch der Beispiele der alten Weisen*. His source was not the original, but a Latin version. It was published, ed. W. L. Holland, in 1860, facsimile 1925.

Pfründner, Die, a Novelle by F. von Saar (q.v.), published in the *Neue Freie Presse* in 1906 and then in *Tragik des Lebens*. It is a pessimistic story of the destruction of the last faint hopes of two inmates of a workhouse.

PFUNDT, Georg, pseudonym Georg Pondo, a sexton at Cölln (now part of Berlin), was the author of plays, many of which are lost. Extant are *Isaaks Heirat* (1590), *Historia Walthers und Griseldas* (1590, see Griselda), the *Knabenspiegel* (1596), an elaboration of the *Speculum puerorum* of G. Wickram (q.v.), *König Salomo* (1601), and *Susanna* (1605, after Heinrich Julius (q.v.), Duke of Brunswick).

Phantasien im Bremer Ratskeller, ein Herbstgeschenk für Freunde des Weins, a fantasy by W. Hauff (q.v.), published in 1827. Its genesis was a visit to Bremen in 1826. The narrator spends the night by himself drinking famous wines in the cellar, is visited by the Twelve Apostles, by the statue of Roland (q.v.), and other unexpected figures, is whisked away over the roofs by the Apostles in their bacchanalian dance, falls to the ground, and awakes to find himself on the floor beside his chair.

Phantasien über die Kunst, für Freunde der Kunst, a collection of enthusiastic essays on art, partly written by L. Tieck (q.v.) and partly drawn from the posthumous papers of his friend W. H. Wackenroder (q.v.), published in 1799. Tieck gave almost the same title (*Phantasien über die Kunst, von einem kunstliebenden Klosterbruder*) to a reprint of Wackenroder's essays in this collection and in *Herzensergießungen eines kunstliebenden Klosterbruders* (1797), which he published as a memorial to his friend in 1814.

Phantasus, title given by L. Tieck (q.v.) to a compendium of his shorter prose works published in 1812–17. Its 3 vols. contained *Der blonde Eckbert* (q.v.), *Der getreue Eckart und der Tannenhäuser* (q.v.), *Der Runenberg* (q.v.), *Liebeszauber*, *Die schöne Magelone* (q.v.), *Die Elfen*, *Der Pokal*, *Leben und Tod des kleinen Rotkäppchens*, *Ritter Blaubart* (q.v.), *Der gestiefelte Kater* (q.v.), *Die verkehrte Welt*, *Leben und Taten des kleinen Thomas, genannt Däumchen*, and *Fortunat* (see Fortunatus).

Phantasus, a gigantic poetic work in the form of a series of lengthy poems by A. Holz (q.v.), first published in 2 vols. in 1898. In 1916 it appeared transformed in an outsize format with carefully arranged extravagant typography and, in final form (1,345 pages), was published in 3 vols. in 1925. Holz shows himself to be a juggler with words and sounds, of scintillating brilliance and inexhaustible fertility. The work, which purports to give the author's view of life, has been variously regarded as a masterpiece of poetic imagination and as an interminable act of self-indulgence. The volumes of 1898 were illustrated in Jugendstil (q.v.).

PHILANDER VON SITTEWALD, pseudonym used by J. M. Moscherosch (q.v.).

Philanthropinum, a school founded at Dessau in 1774 by J. B. Basedow (q.v.).

Philemon and Baucis, a contented old couple who appear as characters in Goethe's *Faust* (q.v.), Pt. II (Act V). The names are derived from Ovid, *Metamorphoses*, viii.

PHILIP II (Valladolid, 1527–98, El Escorial), son of the Emperor Karl V and brother of the Emperor Ferdinand I (qq.v.), was king of Portugal and, after his father's abdication in 1556, king of Spain. He became a leading figure in the Counter-Reformation (see Gegenreformation). His suppression of the Reformed Church in the Netherlands, over which he had sovereignty, involved extensive persecution, including the execution of Counts Egmont (q.v.) and Hoorn.

Philip is a character in Schiller's *Don Carlos* (q.v.) and in the play *Elisabeth von England* by F. Bruckner (q.v.), and a background figure in Goethe's *Egmont* (q.v.). H. Kesten (q.v.) wrote the novels *Ferdinand und Isabella* (1936, republished in 1953 under the title *Sieg der Dämonen*) and *König Philipp II* (1938, republished in 1950 as *Ich der König*). R. Schneider (q.v.) published a biography of Philip II in 1931.

PHILIPP DER KARTHÄUSER, Bruder, was a Carthusian monk, originating in Middle Franconia, who, while at Seitz Abbey in Styria, wrote a versified life of the Virgin Mary (*Marienleben*). The poem, of more than 10,000 lines, was probably written at the beginning of the 14th c. Its source was the *Vita Beatae Mariae Virginis et salvatoris rhythmica* (q.v.). It achieved considerable popularity.

PHILIPPI, Felix (Berlin, 1851–1921, Berlin), a journalist in Munich and (from 1891) in Berlin, was a prolific writer of ephemeral plays and, in later years, of Novellen and novels. The former appeared in a collection entitled *Pariser Schattenbilder* (2 vols., 1906–7). *Alt-Berlin* (2 vols.) followed in 1912–14. His last novels are *Das Schwalbennest* (1919) and *Liebesfrühling* (1920). In 1884 he published *Die Münchener Oper und das Münchener Schauspiel*.

PHILIPPI, Fritz (Wiesbaden, 1869–1932, Wiesbaden), a clergyman who in 1927 became dean and Landeskirchenrat in Wiesbaden, worked during an early stage of his career as a prison chaplain. Much of his writing (plays and works of fiction) emanates from his religious convictions and his zeal for prison reform. Titles include the novels *Adam Notmann* (1906), *Vom Pfarrer Matthias Hirsekorn und seinen Leuten* (1924), and *Pfarrer Hirsekorns Zuchthausbrüder* (1925). *Im Netz* (1912), an early collection of Novellen, was followed by *Aus dem Westerwald* in 1927, the title recalling his *Westerwälder Volkserzählungen* of 1906. His plays include *Pfarrer Hellmund* (1913), *Belial*, and *Mose* (both 1924). *Aus der andern Wirklichkeit* (1926) is a collection of sermons.

PHILIPP VON HESSEN, LANDGRAF (Marburg, 1504–67, Kassel), also known as Philipp I and Philipp der Großmütige, acceded to the Landgravate of Hesse in 1509. He adopted the new ideas of the Reformation (q.v.) on the persuasion of Melanchthon (q.v.) in 1524 and became, with the Elector Johann Friedrich of Saxony, the principal defender of the new religion. Philipp was the leader of the Schmalkaldischer Bund (see SCHMALKALDISCHER KRIEG), of which he was co-founder with the Elector of Saxony, and in 1534 secured the return of the banished Protestant Duke Ulrich of Württemberg (q.v.) to his dukedom. In 1541 the threat of prosecution for bigamy induced him for a time to forsake the Protestant cause. In the Schmalkaldic War Philipp was defeated by the Emperor Karl V (q.v.) at Mühlberg (1547), was taken prisoner, and was not released until 1552. In 1527 Philipp founded the University of Marburg, the first Protestant university in Germany.

PHILIPP VON SCHWABEN (*c.* 1180–1208, Bamberg) was the youngest son of the Emperor Friedrich I and brother of Heinrich VI (qq.v.). After Heinrich's death he was elected German King (see DEUTSCHER KÖNIG) in 1198 by a majority of the princes, but a North-west German minority elected an anti-king (Gegenkönig), Otto IV (q.v.), who was supported by the Pope. Philipp nevertheless was able to establish his power and was crowned at Aachen in 1205. In 1208 he was assassinated by Otto von Wittelsbach (q.v.). WALTHER von der Vogelweide (q.v.) was in Philipp's suite in 1198 and for some years after, and some of his political poems (see SPRUCH) support Philipp's cause.

Philotas, a short tragedy (Ein Trauerspiel) written by G. E. Lessing (q.v.) in 1759 and published in the same year. Philotas, a king's son captured in battle, learns that his captor's son has had the same misfortune. Thereupon Philotas kills himself, and so puts his own father in a position of political advantage. This exaltation of a patriotism which willingly embraces death, but does not disdain a measure of deceit, is presented in starkly simplified terms, with no act-divisions and with a minimum of action. Lessing clearly intended it to be a modern

equivalent to a Greek tragedy. The rather cold and sober play was recast in verse by Gleim (q.v.) and published in this form in 1760.

Phöbus, Ein Journal für die Kunst, a monthly periodical, edited jointly by H. von Kleist and Adam Müller (qq.v.) in 1808 in Dresden. The editors failed to secure contributions by Goethe and other prominent men of letters and remained themselves the principal contributors. *Michael Kohlhaas, Penthesilea, Robert Guiskard*, and parts of *Der zerbrochne Krug* (qq.v.) were first published in *Phöbus*. The project, which ran to twelve issues, ended in financial disaster after disagreement between the editors.

Physiker, Die, a comedy (Komödie) in two acts by F. Dürrenmatt (q.v.), performed in Zurich and published in 1962. It is a grim and grotesque play dealing with sanity and insanity, the impossibility of demarcation, and the reversibility of diagnosis. In the first act the murder of a nurse by one of the three inmates of an expensive psychiatric nursing home has just taken place. The murderous patient claims to be Einstein. An identical murder has taken place recently, committed by a patient claiming to be Newton. The third patient, Möbius, a physicist by profession, claims that King Solomon regularly appears to him. And he, the third patient, murders a third nurse. In the second act all three reveal themselves as sane men masquerading as madmen. Einstein and Newton are agents of rival states interested in Möbius's research. Möbius has feigned madness in order to protect his results which, if applied, would destroy the world. He has just burned all his work when it is revealed that the lady psychiatrist, Dr. Mathilde von Zahnd, who owns the home, has had all their conversations bugged and has photocopied Möbius's work. She has a gigantic organization, with the aid of which Möbius's formulae will subjugate the whole world. The three pretended madmen are now faced by a madwoman. Or is she pretending too? One thing is clear to them: each has committed a murder and can only avoid the consequences by maintaining the pretence of madness. The resemblance of the end to that of Pirandello's *Enrico IV* is striking.

Physiognomische Fragmente, a once famous work by J. C. Lavater (q.v.).

Physiologus, a group of descriptions of animals and birds, fabulous either in themselves or in the qualities attributed to them. It originated in Alexandria in the 2nd c. A.D. The appearance or behaviour of the animals was used to illustrate symbolically articles of the Christian faith. In its

Latin form it was a popular work in the Middle Ages, serving as a textbook in education.

Three medieval German versions of the work exist: (1) *Älterer Physiologus*, a translation which includes only 12 out of 29 animals. The rendering is free, tending towards compression and simplification. The MS., in Vienna, is an incomplete copy of the original, which is believed to have been made in Hirsau *c.* 1070. (2) *Jüngerer Physiologus*, a complete version adhering closely to the Latin text *Dicta Chrysostomi de naturis bestiarum* (usually referred to as Pseudo-Chrysostomus). It was written in Bavaria *c.* 1130. The MS. is that of the *Wiener Genesis* and *Wiener Exodus* (qq.v.), between which the *Physiologus* is placed. (3) *Milstätter Physiologus*, (a clumsy rhyming version of the *Älterer Physiologus* made between 1130 and 1150, and set between the MSS. of the *Milstätter Genesis und Exodus* (q.v.). The conversion into verse was presumably done to render it suitable for the instruction and entertainment of the laity.

PICANDER, pseudonym of C. F. Henrici (q.v.).

PICCOLOMINI, ENEA SILVIO (Corsignano, 1405–64, Ancona), was elected pope in 1458, taking the name Pius II. Piccolomini, a highly accomplished humanist, was present at the Council of Basel (see BASELER KONZIL) and became secretary to the last anti-pope (Felix V) in 1440. In 1442 he entered the chancellery of the Emperor Friedrich III (q.v.) and was his envoy in Rome in 1445. He became bishop of Trieste in 1447 and a cardinal in 1456. As pope he maintained the position of the papacy by the same skilful diplomacy that he had displayed in imperial service.

Piccolomini exerted an influence on the growing trend of humanism in Germany, and his *Historia de duobus amantibus* (1444) was translated into German by Niklas von Wyle (q.v.).

PICCOLOMINI, JOSEPH SILVIO MAX (d. 1645), a nephew of and heir to Octavio Piccolomini (q.v.) and a colonel of horse, was killed in the battle of Jankau against the Swedes. As Max Piccolomini he is a character in Schiller's trilogy *Wallenstein* (q.v.). Schiller makes him Octavio's son.

PICCOLOMINI, OCTAVIO (PICCOLOMINI-PIERI), Prince, Duke of Amalfi (Florence, 1599–1656, Vienna), gained experience in the Spanish and Tuscan armies and from 1627 served under Wallenstein (q.v.) in the Thirty Years War (see DREISSIGJÄHRIGER KRIEG). He rose to be one of Wallenstein's most trusted generals, distinguish-

ing himself in the battle of Lützen (q.v.) in 1632. Piccolomini headed the conspiracy which led to Wallenstein's assassination in 1634. Having supplied the Emperor Ferdinand II (q.v.) with details of Wallenstein's secret plans and negotiations with the enemy, he was entrusted with the arrest of Wallenstein and was rewarded with a share of his estates. He continued to serve the Habsburg cause against France and Sweden and in 1648 was promoted field-marshal. In 1650 he was created a prince of the Empire (Reichsfürst).

Octavio Piccolomini is a character in Schiller's trilogy *Wallenstein* (q.v.).

Piccolomini, Die, see WALLENSTEIN.

PICHLER, ADOLF, RITTER VON RAUTENKAR (Erl nr. Kufstein, 1819–1900, Innsbruck), a schoolmaster and later professor of geology at Innsbruck, was ennobled for military service as a volunteer officer in 1877. Pichler was an ardent Tyrolean and at the same time an Austrian nationalist, participating as a captain in the fighting against Italy (see REVOLUTIONEN 1848–9) and publishing his experience in *Aus dem Wälschtirolischen Kriege* (1849). He wrote blank-verse epics (*Marksteine*, 1874; *Der Hexenmeister*, 1871; *Fra Serafico*, 1879; *Zaggler-Franz* 1889) in which the setting is usually Tyrol. He also published several collections of lyric poetry (*Frühlieder aus Tirol*, 1846; *Lieder der Liebe*, 1852; *Gedichte*, 1853), and a number of plays, none of which achieved success. *Gesammelte Werke* (17 vols.) appeared in 1904–8.

PICHLER, KAROLINE (Vienna, 1769–1843, Vienna), *née* von Greiner, was married in 1796 to Andreas Pichler, a civil servant. She was a well-known literary hostess in Vienna, with leanings towards the Romantic movement. A prolific writer herself, she had her greatest success with the novel *Agathokles* (1808) and the series of Austrian historical novels *Die Belagerung Wiens* (1824), *Die Schweden in Prag* (1827), and *Die Wiedereroberung von Ofen* (1829). She wrote numerous other novels and stories and also some plays, including *Rudolf von Habsburg* (in *Neue dramatische Dichtungen*, 1818). Her complete works (*Sämtliche Werke*, 1828–44) comprise 60 volumes. Her memoirs (*Denkwürdigkeiten aus meinem Leben*, 4 vols., ed. F. Wolf, 1844) reflect the intellectual life of the day. A selection, *Auswahl aus dem Werk*, ed. K. Adel, was published in 1970.

Pickelhering, a comic character in German drama in the 17th c. and early 18th c. He originated with the strolling troupes of English players in Germany at the beginning of the 17th c., and was exploited in German comedy especially by Christian Weise (q.v.). His role was to comment on the action, extemporizing gags.

Pietismus, a religious trend manifesting itself in Germany in the late 16th c. and early 17th c. (c. 1680–1740) among Protestants dissatisfied with the rigid and formal orthodoxy of the Lutheran Church of the time. The Pietists maintained a personal devotion to God and encouraged emotion in their religious life, and in so doing they fostered a tendency to introspection. J. Arndt (q.v.), who lived in the early 17th c., is often quoted as a forerunner, but, in spite of similarity of outlook, there is no direct connection between him and what was later recognized as Pietism. The Pietists tended to form closely knit groups, of which the best known are those headed by G. Arnold (q.v.) in Quedlinburg, Arnold's friend P. J. Spener (q.v.) in Strasburg and Frankfurt, A. H. Francke (q.v.) in Halle, and, most prominent of all, the fraternity fostered by Count Zinzendorf (see ZINZENDORF UND POTTENDORF, N.L., GRAF VON) on his estate at Herrnhut in Silesia, which acquired the name Herrnhuter (q.v.). The Pietists sought to live quietly in devout and heart-searching meditation and were also active in unostentatious acts of charity. Though their importance in church history declined after c. 1740, individuals of pietistic outlook made contributions to literature well after that date, notably Klopstock (q.v.), Jung-Stilling (see JUNG, J. H.), and J. C. Lavater (q.v.), while Goethe was for a time (1768–9) deeply influenced by the Pietist Susanna von Klettenberg (q.v.), whom he commemorated as late as 1795 in Bk. VI. *Bekenntnisse einer schönen Seele* of *Wilhelm Meisters Lehrjahre* (q.v.). Fontane in *Vor dem Sturm* (q.v.), which is set in 1812, gives a credible portrait of a pietistic lady (Tante Schorlemmer) attached to a country gentleman's household. In the second half of the 18th c., Pietism, with its stress on emotion, tended to encourage the expression of deep feeling, which not infrequently degenerated into sentimentality. Though the antithesis of Rationalism, Pietism was in agreement with the ideals of tolerance and humanitarianism. Indeed, Lessing intended (1750) a defence of the Pietists in his unfinished essay *Gedanken über die Herrnhuter*. The Pietists were themselves for the most part only interested in literature as a means of edification, but some of their number, notably Arnold, Spener, G. Tersteegen (q.v.), J. Neander (q.v.), and Zinzendorf wrote hymns of merit.

Pietisterey im Fischbeinrocke, Die, a comedy by Luise A. V. Gottsched (q.v.).

PIETSCH, JOHANN VALENTIN (Königsberg, 1690–1733, Königsberg), a physician who had also studied poetry, wrote a panegyric on the victory of Prince Eugene (see EUGEN, PRINZ) at Peterwardein (*Über den ungarischen Feldzug des Prinzen Eugen,* 1717) which made so great an impression that he was elected professor of poetry at Königsberg University. He was influenced by B. H. Brockes, whom he had met; J. C. Gottsched (qq.v.) was his pupil. *Gesammelte Poetische Schrifften* were published in 1725, and *Gebundene Schriften,* ed. J. G. Bock, in 1740.

Pilatus, an incomplete Middle High German poem, written between 1170 and 1180, recounting the medieval legend of Pontius Pilate. In style it is a forerunner of courtly poetry.

Pilatus, a novel (Eine Erzählung aus den Bergen) by H. Federer (q.v.), published in 1912. The Pilatus of the title is the well-known mountain which rises to the south-west of the Lake of Lucerne.

The central figure of the story is Marx Omlis, son of a farmer on the lower slopes. He is a youth capable of both great generosity and savage aggression. His alcoholic father neglects the farm and dies when it is sold up, but Marx retains a small-holding much higher up the mountain. He is partly responsible, through his taunts, for the death of a friend, Florin Lauscher, who falls in an attempt to pluck an edelweiss flower. Marx marries Agnesli Dannig against the wishes of her mother, whom he regards as his greatest enemy. Though he loves Agnes, she is unhappy in the remote cottage and frightened by his daring climbing in pursuit of game. He builds a dam to protect the cottage in case of flood, but torrential rains cause such a disaster that, though most of the water is diverted and causes a catastrophe in many farms lower down, Marx's dam, too, is breached and his cottage destroyed; his wife gives birth prematurely and dies.

Virtually outlawed by the community, who blame him for the disaster, Marx moves to Grindelwald and becomes one of the best guides in the Berner Oberland. But his unstable temper remains his enemy. He despises the tourists and in the end only climbs willingly with two well-to-do young men from Ulm (Lucian and Emil Brunner). When he finds that they put their careers before their summer climbing, he turns away in disgust and returns to his cottage on Pilatus, which his now repentant and reconciled mother-in-law has had restored. A final blow awaits him: the orphan commission (Waisenamt) will not allow his helper Balzli to be in such supposedly dangerous company. Marx leaves on a wild walk and climb, and plunges to his death trying unsuccessfully to rescue a wild mountain kid. The work is especially notable for its evocation of the desolation, sterility, and grandeur of the high Alpine scene.

Pilgernde Törin, Die, a story inserted into Ch. 5, Bk. I of Goethe's *Wilhelm Meisters Wanderjahre* (q.v.). It is a translation, made by Goethe, of a French story, *La Folle en pèlerinage,* published (in French) in 1789 by H. A. O. Reichard in *Cahiers de lecture.*

Pilgerschaft des träumenden Mönchs, designation of an allegory in the form of a dream. The monk's dream symbolizes the journey through life with its encounters with the vices. There are four versions: (1) *Berleburger Handschrift* in verse (nearly 14,000 lines), early 15th c.; (2) *Kölner Fassung* in verse (of similar length) by Peter van Meroede, *c.* 1430; (3) and (4) in prose, *Hamburger Handschrift* and *Darmstädter Handschrift,* 15th c. The source is a French poem *Le Pélerinage de la vie humaine* by Guillaume de Deguileville.

PINTHUS, KURT (Erfurt, 1886–1975, Marbach), Berlin critic, emigrated to the U.S.A. in 1933. An authority on the theatre, he held an appointment at Columbia University, New York, from 1947 to 1961. His *Menschheitsdämmerung* (1920) is a well-known anthology of Expressionistic verse which has been repeatedly reissued.

PIONTEK, HEINZ (Kreuzburg, Silesia, 1925–), is chiefly known as a lyric poet with a dry, clear, visual style, fixing moments in nature or figures in landscape, or responding to pictures. His verse has been collected in *Die Rauchfahne* (1953), *Wassermarken* (1957), *Mit einer Kranichfeder* (1962), *Randerscheinungen* (1965), *Klartext* (1966), *Tot oder lebendig* (1971), and *Wie sich Musik durchschlug* (1978). His novels include *Die mittleren Jahre* (1967), *Dichterleben* (1976), and *Juttas Neffe* (1979), and the stories and texts *Vor Augen* (1955), *Kastanien aus dem Feuer* (1963), *Außenaufnahmen* (1968), *Liebeserklärungen in Prosa* (1969), *Die Zeit der anderen Auslegung* (1976), and *Dunkelkammerspiel* (1978). He is also the author of radio plays and the editor of an anthology of religious verse, *Aus meines Herzens Grunde. Evangelische Lyrik aus vier Jahrzehnten* (1959). *Männer, die Gedichte machen–Zur Lyrik heute* (1970) is an international anthology with comments and essays. The sketches *Helle Tage anderswo. Reisebilder* appeared in 1973, collected stories, *Wintertage: Sommernächte* in 1977, *Handwerk des Lesens* in 1979, and *Die Zeit einer Frau* (postscript by R. Malkowski) in 1984; *Werke* (6 vols.) in 1982 f. Piontek was awarded the Büchner Prize in 1976.

Pippinsche Schenkung, see KAROLINISCHE SCHENKUNG.

PIRKHEIMER, WILLIBALD, also **PIRCKHEIMER** (Eichstätt, 1470–1530, Nürnberg), was a prominent personality among German humanists. Educated in Italy at Padua and Pavia, he was a man of wealth and a city councillor of Nürnberg. He was a connoisseur of art, a bibliophile, and a close friend of Dürer (q.v.), who executed and engraved a portrait of him which is still extant. He served as an officer in the Swiss war of 1499. Pirkheimer's writings, all in Latin, include historical works and translations from Greek, and also an ironical laudation of gout (*Apologia seu podagrae laus,* 1510), commonly called *Laus podagrae.* Like many educated men of his age he welcomed the Reformation (q.v.), only to react later against its extremism. A select edition by M. Goldast of 1610, *Opera politica, historica, philologica et epistolica,* was reprinted in 1969.

PISCATOR, ERWIN (Ulm, 1893–1966, Starnberg), producer and theatre director, sought after the 1914–18 War to create a Proletarisches Theater (1919–22, in hired halls, 1923 in the Zentraltheater Berlin) in order to promote his strong pacifist and communist convictions. He became the principal producer of the Berlin Volkstheater, where he revolutionized traditional stage techniques for the strictly political purpose of keeping the abortive revolution of 1918 alive. His innovations included the use of projectors, filmstrips, placards, loudspeakers, and Laufbänder running parallel or criss-cross over the stage, so that actors could give a realistic impression of marching without actually moving from the spot, as well as divisions of the stage into sections. This spatial exploitation of the stage was extended vertically by sinking floors. In a famous production of Schiller's *Die Räuber* (q.v.) in 1926 he turned Spiegelberg into a revolutionary in the mask of Lenin. This was perhaps the most provocative performance of his anti-Aristotelian epic and documentary theatre from which Brecht (q.v.) learnt much. Both Brecht and K. Kraus (q.v.) commented critically on this venture. Brecht records his indebtedness to Piscator in *Der Messingkauf* (q.v., 3. Nacht).

Piscator spent 1931–6 in Russia, stayed some time in Paris, and then went to New York until his return to the German Federal Republic in 1951, where he directed productions in a number of cities until 1962, when he became director of the Berlin Freie Volksbühne. He established contact with a new generation of playwrights, who wrote under the impact of the 1939–45 War, directing the first production of R. Hochhuth's *Der Stellvertreter* (q.v., 1963), *In der Sache J. Robert Oppenheimer* by H. Kipphardt (q.v., 1964), and *Die Ermittlung* (q.v., 1965) by P. Weiss.

In 1929 Piscator published *Das politische Theater* (reprinted in 1964); *Schriften zum Theater* appeared posthumously in 1968.

PITAVAL, FRANÇOIS GAYOT DE (Lyons, 1673–1743, Lyons), a French legal writer, compiled 20 vols. of *Causes célèbres et intéressantes* (1734 ff.), which were translated into German between 1747 and 1768. Schiller (q.v.) wrote an introduction to a new edition of this translation (1742–5). The name Pitaval has since been used in the titles of several similar German collections viz. *Der neue Pitaval* edited by J. E. Hitzig and W. Alexis (qq.v., 30 vols., 1842–62), augmented in a second edition to 36 vols. (1857–72), *Der neue Pitaval neue Serie*, 24 vols. (1866–90), *Der Pitaval der Gegenwart* compiled by R. Frank and others (1909–13), and *Der sächsische Pitaval* (3 vols., 1861–2).

PLATEN, AUGUST, GRAF VON (Ansbach, 1796–1835, Syracuse, Sicily), in full Karl August Georg Maximilian, Graf von Platen-Hallermünde, spent his early years, in spite of his high-sounding name and ancient lineage, in modest burgher surroundings, for his father possessed little fortune and earned his living as a forestry official. In 1806 the boy was given a free place at the newly founded Cadet School (Kadettenkorps) in Munich, an environment for which his sensitiveness made him ill suited. In 1810 he was transferred to the Pages' School (Pagerie), in which young noblemen were educated while they took turns in pages' duties at court.

In 1813, to the surprise of his family, Platen decided to enter the army. He was commissioned in March 1814 in the 1st Bavarian Infantry Regiment, which was then serving in France. Platen, however, remained at the depot in Munich. The regiment was again mobilized in 1815 and marched into France, but arrived long after the Hundred Days (see NAPOLEONIC WARS) were over. Platen, dreamy, eccentric, and increasingly isolated by a homosexual disposition, was unhappy in garrison life and in 1818 obtained indefinite leave of absence. During his years as an officer he had devoted much time to study and to poetry, as well as to travel. He now began university studies in law, with the ultimate aim of a civil service post, matriculating at Würzburg, where his stay ended with a catastrophic love for a male friend Eduard Schmidtlein (Adrast). Platen was next in Erlangen, where he remained until 1826, except for journeys through Germany and, in 1824, to Italy.

His first volume of poems (*Ghaselen*) was published in 1821, and was followed by *Lyrische Blätter* (1821), 2 vols. of *Schauspiele* (including *Der gläserne Pantoffel*, 1824, and *Der Schatz des Rhampsinit*, 1828), and *Sonette aus Venedig* (1825). After an unpleasant experience in 1825 when he was imprisoned for overstaying his leave, Platen went to Italy, where he spent the remainder of his life in voluntary exile. During his years of study and his sojourn in Italy he was partly maintained by royal grants, but remained a poor man. He lived first in Naples, briefly visited Munich in 1833 and 1834, returned to Florence, and finally fled to Sicily in fear of cholera. There he fell ill, believing himself to have cholera, though the medical evidence contradicts this. He died of an undefined gastric fever and is said to have exacerbated this by secretly taking supposed remedies for cholera in massive doses.

Platen's years in Italy were expected by his patron to produce a flowering of his genius, but most of his later projects, including plays and epics, remained unfinished. He completed two satirical plays, *Die verhängnisvolle Gabel* (1826, parodying the Schicksalstragödie, q.v.) and *Der romantische Ödipus* (1829), directed at Immermann (q.v.), and a brief historical verse play in three acts, *Die Liga von Cambrai* (1833). *Die Abassiden*, an epic in nine cantos, appeared in 1835. Platen also wrote a historical work, *Geschichten des Königreichs Neapel von 1414–43* (1833). A meticulous craftsman with a fine ear, he devoted much time to a revision of his early poems (*Gedichte*, 1828). His poetry ranges over many forms, including oriental and Latin measures, but Platen, with all his desire for poetry and his formal gifts, seems to have had little to say.

A historisch-kritische Gesamtausgabe (12 vols.), ed. M. Koch and E. Petzet, appeared in 1910 (reprint 1969).

PLANCK, MAX (Kiel, 1858–1947, Göttingen), the prominent physicist, contributed to the advancement of modern science notably through his radiation law, known as the Plancksche WARS) were over. Platen, dreamy, eccentric, and quantum theory, and also through his strong support of the theories of Albert Einstein. For 25 years he was (in succession to H. L. F. von Helmholtz) one of the four permanent secretaries of the Prussian Academy (see AKADEMIEN). In 1918 he received the Nobel Prize for Physics. His presidency of the Kaiser-Wilhelm-Gesellschaft zur Förderung der Wissenschaften was commemorated after the 1939–45 War (during which his son Erwin, b. 1893, was executed for his participation in the 20 July 1944 plot), when the society was renamed the

Max-Planck-Gesellschaft zur Förderung der Wissenschaften. He left his mark as a man of exceptional culture and integrity.

PLATTER, Thomas (Grächen, Wallis, 1499–1582, Basel), was for more than thirty years headmaster of Basel grammar school. He began life as an Alpine goatherd, then journeyed through Germany on foot, and was for a time a printer for Calvin. Known in later life as a humanist, he is the author of an autobiography (*Selbstbiographie*) which, after its publication in 1840, came to be recognized as an important cultural document of the 16th c.

Platter's elder son Felix (1536–1614) was professor of medicine at Basel, serving city and science throughout his life, and his younger son Thomas (Thomas Platter der Jüngere, 1574–1628) was a distinguished and much-travelled physician. Felix's autobiography was published jointly with that of his father (1840) and was used by G. Freytag in *Bilder aus der deutschen Vergangenheit* (q.v.). Correspondence between Platter and Felix, *Briefe an seinen Sohn Felix*, ed. A. Burckhardt, was published in 1890. Thomas Platter's autobiography, ed. W. Muschg, was republished in 1944.

Plautus im Nonnenkloster, a story (Rahmennovelle, see Rahmen) by C. F. Meyer (q.v.), published in the *Deutsche Rundschau* (q.v.) in 1881 under the title *Das Brigittchen von Trogen*. At a social gathering in Florence the host, Cosmus Medici, invites the humanist Poggio Bracciolini to relate one of his *facezie inedite*. Poggio chooses a reminiscence which he entitles the 'discovery of Plautus' (*Fund des Plautus*). It is designed to justify his acquisition of a work by Plautus which he has discovered in the library of a convent.

Poggio detects that an ancient custom of the convent, by which novices have at their initiation to carry an exceedingly heavy cross to demonstrate the operation of a divine miracle, is based on deceit, for the abbess provides the novices with a copy of the old relic, which is light enough to be carried by any young woman without effort. A new novice, Gertrude, is to be inaugurated; she would sooner enter marriage than the convent but for a promise given to the Virgin Mary as a child. With the help of Poggio's hints she discovers the deceit, breaks the false cross, and carries the actual relic until she breaks down 'under its weight. Relieved at this 'sign' that the Virgin Mary does not insist on her service, she gives her hand in marriage to her lover Anselino. Poggio leaves the convent with the treasured Plautus in his pocket as the bribe for his promise to the abbess not to disclose the blasphemous deceit. The working of conscience

in Gertrude is the central motif giving weight to his tale.

Plebejer proben den Aufstand, Die, a play by G. Grass (q.v.), published in 1966.

PLEIER, DER, a Middle High German poet, author of three Arthurian verse romances. He was probably of minor nobility, originating from the neighbourhood of Schärding on the Inn, to the south of Passau. His poems were probably written between 1260 and 1280.

The earliest and best of the three romances is *Garel vom blühenden Tal,* the title of which is modelled on *Daniel vom blühenden Tal* by Der Stricker (q.v.). *Garel* seems to be a retort to Der Stricker's poem, reaffirming the qualities of Arthurian poetry in contrast to the unchivalric aspects of *Daniel.* Garel rides out on various adventures, proves victorious not by cunning, but by his valour, and the work ends with a great feast and rejoicing at Arthur's court. Incidents from the poem are portrayed in a series of frescoes, painted c. 1400, in Castle Runkelstein near Bolzano.

Meleranz (reprinted 1974), which is based on Albrecht von Scharfenberg's *Seifrid,* and *Tandareis und Flordibel* are similar sequences of adventures which end happily in celebrations at the court of Arthur. Der Pleier is indebted to various models, but the influence of Hartmann von Aue (q.v.) is especially notable.

PLENZDORF, Ulrich (Berlin–Kreuzberg, 1934–), studied philosophy at the Leipzig Franz Mehring Institute and, after working in the theatre and completing a year's military service, resumed studies at the Babelsberg Filmhochschule (1959–63). He then began work as producer and scriptwriter for the DEFA in Berlin. He is best known for *Die neuen Leiden des jungen W.* First conceived as a film script, it appeared in book form and as a stage play in 1972 (film version 1976). Skilful use is made of quotations from Goethe's epistolary novel *Die Leiden des jungen Werthers* (q.v.) to underline Edgar's emotional subjectivity. Edgar Wibeau, aged seventeen, is depicted in his relationship to society and to Charlie, a woman already engaged. The story opens after Edgar's death, by accident, on 24 December. The reconstruction of his life is partly based on tapes, which he has recorded for his friend Willy. He leaves his mother and his apprenticeship in the VEB factory in Mittenberg, and runs away to Berlin. The idea of constructing a hydraulic spraygun of his own invention is central to his urge for self-realization and may be seen as his individual contribution to industrial progress. He appears to have succeeded, but his equipment for testing his spraygun has the wrong voltage, as he is well aware. He nevertheless proceeds with the test and is electrocuted. There are indications that he might have intended to take his own life, an issue that is left open, partly because the portrayal of suicide is not favoured in the DDR, though the depiction of individualism had at the time of writing become a feature of DDR literature. The presentation of Edgar's problems is calculated to encourage a flexible assessment of one who faces isolation at a difficult stage of adolescence. *Die Legende von Paul & Paula. Filmerzählung* (1974) appeared as *Legende vom Glück ohne Ende* in 1979. Paul and Paula's belated fulfilment in love is short-lived. Paula dies after giving birth to a third child, the only one fathered by Paul, whose own life is finally ruined by an accident that leaves him paralysed from the waist down. The poignancy of Paula's tragedy derives from her knowledge that on medical grounds she is not likely to survive if she risks another pregnancy. Paul's accident bears some resemblance to Edgar's. Although social problems are raised, notably marriage, divorce, and careerism (Paul opts out of a promising position), the motivation of the plot lacks the control and conviction of the more widely discussed first work. In using teenage jargon and stylized colloquialism, Plenzdorf draws primarily on his experience as a scriptwriter. *Gutenachtgeschichte* was published in 1983.

PLIEVIER, THEODOR (Berlin, 1892–1955, Avegno, Switzerland), a worker's son, ran away from home at 17 and led a vagrant existence in central Europe until the 1914–18 War, in which he served in the German navy. He participated in the naval mutiny, and during the Weimar Republic took a variety of jobs (mostly manual) at home and abroad. As an avowed communist he fled abroad in 1933, was in Czechoslovakia, then Sweden and, finally, Russia. He returned to Berlin in 1945 and became a journalist and publisher. In 1947 he moved to West Germany, and was later in Ticino.

Plievier, who at first wrote under the thin pseudonym Plivier, was primarily a war writer. The early novels, *Des Kaisers Kulis*, *Zwölf Mann und ein Kapitän* (both 1930), *Der Kaiser ging, die Generäle blieben* (1932), *Der 10. November 1918* (1935), and *Das große Abenteuer* (1936), attracted relatively little attention; the play *Die Seeschlacht am Skagerrak* (1935) and the story *Im Wald von Compiègne* (1939) were no more successful. With his intensely realistic and lengthy novel of the 1939–45 War, *Stalingrad* (1945), Plievier achieved an international best seller which was converted into a trilogy by the addition of *Moskau* (1952) and *Berlin* (1954). *Stalingrad* was dramatized in

1962. *Werke in Einzelbänden* began to appear in 1981.

Plimplamplasko, der hohe Geist, heut Genie, a satire by F. M. Klinger (q.v.), published in 1780. Purporting to be a MS. of the 16th c. ('eine Handschrift aus der Zeit Knipperdollings und Dr. Martin Luthers'), it derides the pretensions and follies of the age of genius ('Geniezeit', from which Klinger himself had just emerged), taking as its favourite butt C. Kaufmann (q.v.).

PLÖNNIES, LUISE VON (Hanau, 1803–72, Darmstadt), *née* Leisler, a physician's daughter, married August von Plönnies in 1824 and was widowed in 1843. Her first poems (*Gedichte*) appeared in 1844 and were followed by *Ein Kranz von Kindern* (1845). Her book on Belgium attracted attention by its account of Flemish literature (*Reiseerinnerungen aus Belgien, nebst einer Übersicht der flämischen Literatur*, 1845). She translated J. van Vondel's *Lucifer* (1654). Her numerous volumes of verse include the sonnet cycles *Abälard und Heloise* (1849) and *Oskar und Gianetta* (1851), and the religious poems *Lilien auf dem Felde* (1864).

POCCI, FRANZ, GRAF VON (Munich, 1807–76, Munich), a Bavarian civil servant who became court chamberlain, was a poet and humorist, publishing numerous deft and urbane sketches and poems which he illustrated himself (*Alte und neue Jägerlieder*, 1843; *Studentenlieder*, 1845; *Kinderlieder*, 1852; *Der Staatshämorrhoidarius*, 1857). He also wrote plays (*Gevatter Tod*, 1855; *Der Karfunkel*, 1860; *Der wahre Hort*, 1864) for the popular theatre and especially for marionettes (*Lustiges Komödienbüchlein*, 6 vols., 1859–77). *Sämtliche Kasperl-Komödien* (3 vols.) appeared in 1909 and a selection, ed. M. Kesting, in 1965. Pocci was a considerable artist in water-colour and black and white, and also a talented composer.

'Poesie ist tiefes schmerzen', first line of a short poem by J. Kerner (q.v.), stressing the association of poetry and grief.

POETHEN, JOHANNES (Wickrath/Lower Rhine, 1928–), settled in Hirschau near Tübingen. His poetry leans towards the mythical and has a sense of the horror of the contemporary world. His works include *Lorbeer über gestirntem Haupt* (1952), *Risse des Himmels* (1956), *Stille im trockenen Dorn* (1958), *Ankunft und Echo* (1961), *Im Namen der Trauer* (1969), and *Aus der unendlichen Kälte* (1970). His poetry was published in two volumes entitled *Gedichte 1946–1971* (1973) and *ach erde du alte—gedichte 1976–1980* (1980).

Poetischer Realismus (Poetic Realism), a term peculiar to German literature. There is no general agreement as to whether it should be used to characterize a period or to define a style. It is probably best considered as the generic term for narrative works written in an unsensational and unpretentious, yet consciously artistic, manner in a period roughly delimited by the years 1840 and 1880. The florescence of Poetic Realism after 1850 is often attributed to disillusionment with political activism after the Revolutions of 1848 (see REVOLUTIONEN, 1848–9). An important factor was the political structure of Germany, with its small units that bred provincialism. The same is also true of Swiss literature of the period. These factors probably explain, at any rate in some degree, why the German-speaking writers have a variety of 'realism' differing from that of the other important literatures. Conscious political or social engagement hardly occurs at all.

The most important writers usually classified as 'poetic realists' are A. von Droste-Hülshoff, Gotthelf, G. Keller, O. Ludwig, Stifter, Storm, C. F. Meyer, and Raabe (qq.v.). The general tone prevailing in their works suggests a desire to preserve the existing structure of society and an emphasis on moral duty and humane feeling. Class differences are accepted, including a distinction between upper and lower middle class. Where injustice or poverty is an issue the tone is mildly ironic or satirical rather than denunciatory. Keller epitomizes the spirit of the major writers when he commends the 'Liederbüchlein für Handwerksbursche', which the younger generation would replace with 'Arbeitermarseillaisen' (in *Das Sinngedicht*, q.v.). Lesser works often tend towards sentimentality.

The term 'Poetischer Realismus', adopted by O. Ludwig, was coined by F. W. J. Schelling (q.v.) in 1802. A somewhat analogous term 'poetische Materialisten' had also been used by JEAN PAUL (q.v.) in 1804. Ludwig defined it in the context of his *Shakespeare-Studien* and *Dramatische Aphorismen*, admiring Shakespeare's ability to create scenes which seem convincingly realistic 'ohne daß ein Wort darin naturgetreu wäre'. In his essay *Der poetische Realismus* he expresses his view that art should be neither 'naturalistic' nor 'idealistic'. His use of the term Naturalismus (q.v.) has nothing in common with later German Naturalism, but denotes a diffuse imitation of reality, which he considers to be as one-sided as idealistic representations. Ludwig's alternative term 'künstlerischer Realismus' expresses perhaps more obviously his approach to realism in literature as the coherent

artistic reproduction of reality. According to Ludwig, artistic realism avoids the monotony of idealism on the one hand, and on the other it disentangles the confusions of realistic detail. A wealth of shrewd and sensitive observations of the world, however small, is reproduced by artistic means and techniques of style without sacrificing the sense of permanence which characterizes human nature of all ages and environments (see also Stifter's *Vorrede zu Bunte Steine*).

Poetischer Trichter, see HARSDÖRFFER, G. P.

Poggenpuhls, Die, a short novel by Th. Fontane (q.v.), written between 1892 and 1895 and published in the periodical (Familienblatt) *Vom Fels zum Meer* (1895–6), and in book form in 1896. Set in Berlin, it is a subtle and detailed study of genteel poverty: an officer's widow, Frau von Poggenpuhl, and two sons and three daughters—all, except the widow, of noble descent—adapt themselves in various degrees to the new social world in which they have only a minor part to play. Of the two sons (grenadier officers at Thorn) only the younger makes an appearance in the novel, which is concentrated upon the widow, a woman of great goodness of heart, who has learned resignation in a hard school. Therese the eldest daughter regards herself as the upholder of the family tradition and is less than fair to her mother and the other members of the family. Sophie earns a modest supplement to the meagre family income by working hard at painting porcelain and drawing; Manon frequents the house of a rich Jewish banker, and in doing so obtains commissions for Sophie. There is scarcely a plot. Frau von Poggenpuhl's birthday is modestly celebrated, an occasion on which the younger son briefly visits Berlin. Sophie is then invited to Silesia, to the house of an uncle who is a general, and there she makes herself both agreeable and useful. The uncle dies, and his widow makes a modest provision for the Berlin family.

It is a highly sympathetic, as well as discreetly ironical, portrayal of the genteelly poor, but the characters who emerge with the truest humanity are the two widows, neither of whom is by birth a noblewoman.

Poland, Partitions of (Polnische Teilungen), the series of annexations which resulted in the total extinction of the state of Poland in 1795. Poland had flourished politically from the 14th c. to the 16th c. but in the 17th c. and 18th c. declined progressively into something approaching anarchy. Taking advantage of this, its neighbours Austria, Russia, and Prussia divided

Poland among themselves in what are known as the First, Second, and Third Partitions.

First Partition, 1772. Prussia acquired the north-west frontier portion together with Danzig (later Provinz Westpreußen). Austria annexed Galicia, including Lemberg, adding in 1775 the Bukovina. Russia took the eastern strip of White Russia (chief town Vitebsk).

Second Partition, 1793. A large tract of western Poland with the towns of Posen and Kalisch passed to Prussia (later Provinz Posen). The Russian share was the rest of White Russia and the western Ukraine, including Minsk and Pinsk. Austria was not a participant.

Third Partition, 1795. Prussia obtained the northern part of the rump (Neuostpreußen), including Warsaw. Austria took West Galicia with Cracow (Krakau) and Lublin; and Russia annexed an enormous area from the Baltic (Courland, Lithuania) southwards to Volhynia, acquiring the towns of Kovno, Wilna, Grodno, and Brest.

The Napoleonic Wars (q.v.) saw Poland resuscitated by Napoleon as the Grand Duchy of Warsaw, but after 1815 it was again partitioned. Prussia, however, lost all its acquisitions of the Third Partition and part of the second, the gainer being Russia. Austria also lost to Russia most of the spoils it had acquired in the Third Partition. Cracow was made a free republic in 1815, but was re-allocated to Austria in 1846.

POLENZ, Wilhelm von (Schloß Obercunewalde, Oberlausitz, 1861–1903, Bautzen), was at Berlin, Breslau, and Leipzig universities and then entered the civil service, from which he soon resigned. He lived on his estates as a country gentleman, inheriting eventually Obercunewalde. He devoted himself to writing and kept up literary contacts by winter visits to Berlin. His Naturalistic novels have as background either the great city or his own rural homeland. In the latter he has a link with Heimatkunst (q.v.). The general tone of his novels emphasizes the gloomier and harsher side of life.

Polenz began with *Sühne* (2 vols., 1890), and this was followed by the novels *Der Pfarrer von Breitendorf* (3 vols., 1893), *Der Büttnerbauer* (q.v., 1895), which is his masterpiece, *Der Grabenhäger* (2 vols., 1897), *Thekla Lüdekind* (2 vols., 1899), *Liebe ist ewig* (1900), and *Wurzellocker* (2 vols., 1902). His plays include *Heinrich von Kleist* (1891), *Andreas Bockholt* (1898), and *Heimatluft* (1900). He also wrote Novellen (including *Karline*, 1894, and *Reinheit*, 1896), and a volume of poems was published posthumously (*Erntezeit*, 1904). *Gesammelte Werke* (10 vols.) appeared in 1909–11.

Pole Poppenspäler, a story (Rahmennovelle, see Rahmen) by Th. Storm (q.v.), first published in vol. 4 of *Deutsche Jugend* in 1874. The author listens to the master turner, Paul Paulsen, who recalls his boyhood memories of the puppeteer Joseph Tendler who performed for a few months in Husum.

The boy Paul falls for the magic of the puppet theatre and wins the friendship of Tendler's daughter Lisei. After twelve years, when, upon the death of his father, he is about to return from his years of apprenticeship near Heiligenstadt to take over the family business, his and Lisei's paths meet again. Her father is in prison, a victim of an innkeeper's rash suspicion of itinerant players. His release follows the arrest of the real culprit. Paul marries his childhood love and the couple enjoy a happy married life, unperturbed by malicious gossip at their socially unequal match. Joseph Tendler shares their home, having retired from his life as a puppeteer. One day he makes the attempt to stage a performance in the town hall and meets with the ridicule of the city's youth headed by Paulsen's enemy, der schwarze Schmidt. The following morning Paul, who has met Tendler's longing to return to his art with understanding, finds the malicious nickname 'Pole Poppenspäler' written on his door. The broken-hearted puppeteer dies after selling his puppets. During his funeral his favourite puppet, Kasperle, is flung across the churchyard wall and falls into the open grave. Unwittingly der schwarze Schmidt has fulfilled the puppeteer's longing for the puppet which once delighted his audiences and which had united the families of two generations. Tendler, the dark-haired South German, is one of Storm's sensitive and vulnerable artist natures, whose naïve charm Lisei has inherited. (See also Puppenspiel.)

POLGAR, Alfred (Vienna, 1873–1955, Zurich), began as a parliamentary reporter and became a dramatic critic in Vienna and, from 1925 to 1933, in Berlin. Under National Socialism he returned to Vienna, whence he fled in 1938, reaching the U.S.A. in 1940, where he lived until 1947. He returned to Europe for his last years.

An ironist with a well-turned, elegant style, Polgar published several volumes of Novellen (*Der Quell des Übels*, 1908; *Bewegung ist alles*, 1909; *Hiob*, 1912; *Gestern und Heute*, 1922; *Geschichten ohne Moral*, 1943; *Begegnung im Zwielicht*, 1951; *Im Lauf der Zeit*, 1954). Two short satirical comedies, written in collaboration with E. Friedell (q.v.), were very successful: *Goethe im Examen. Eine Szene* (1908) and *Das Soldatenleben im Frieden* (1910). The comedy *Die Defraudanten* (1931) was his own work. Polgar was particularly gifted as a readable and perceptive critic of literature and drama and a commentator on cultural trends. In addition to his numerous contributions to newspapers, several collections of his essays were published in book form, including *Kleine Zeit* (1919), *An den Rand geschrieben* (1926), *Orchester von oben* (1926), *Ja und Nein* (4 vols. of criticism, 1926–7, and a selection, with other material, 1954), *Ich bin Zeuge* (1928), *Schwarz auf Weiß* (1929), *Hinterland* (1929), *Bei dieser Gelegenheit* (1930), *Ansichten* (1933), *In der Zwischenzeit* (1935), *Sekundenzeiger* (1937), *Handbuch des Kritikers* (1938), *Im Vorübergehen* (1947), *Standpunkte* (1953), and *Fensterplatz* (1959); the cliché-like titles are intentionally ironical. Selections, ed. B. Richter, appeared in 1968 and 1970 as *Alfred Polgar. Prosa aus 4 Jahrzehnten* and *Bei Lichte betrachtet, Texte aus vier Jahrzehnten.* Other posthumous editions include *Die Mission des Luftballons* (1975), *Die lila Wiese* (1977), and *Taschenspiegel* (1979). *Kleine Schriften* (4 vols.), ed. M. Reich-Ranicki, appeared in 1982, correspondence, *Lieber Freund!*, ed. E. Thanner, in 1981.

Polnische Teilungen, see POLAND, PARTITIONS OF.

Pommern (Pomerania), a former province (Provinz) of Prussia (see PREUSSEN), which ceased to exist in 1945. The greater part of Pomerania is now Polish. The territory was formerly divided into Vorderpommern, west of the River Oder, and the much larger area of Hinterpommern to the east of it, all of which now forms part of Poland. The greater part of Vorderpommern, except for a strip on the left bank stretching northwards from Stettin (now Szczecin), is in the DDR (see DEUTSCHE DEMOKRATISCHE REPUBLIK). In the Middle Ages Pomerania was a duchy. In the 17th c. and 18th c. it was a bone of contention between Prussia and Sweden; Prussia acquired it piecemeal between 1679 and 1815.

POMMERSCHE SAPPHO, DIE, appellation given to Sibylle Schwarz (q.v.), a poetess who died at 17 in 1638.

Ponce de Leon, a comedy written by C. Brentano (q.v.) in 1801 as an entry for a competition announced by Goethe and Schiller. For this it was unsuccessful, and it was published later, in 1804. It has only a slender plot, set in and near Seville. Ponce is loved by Valeria, but himself loves Isidora. Ponce and Isidora are united; Valeria pairs off with Porporino; and two other pairs of lovers are happily joined. The chief interest of the play is its pyrotechnic wit. The comedy became one of the sources of Büchner's *Leonce und Lena* (q.v.).

PONDO, GEORG, see PFUNDT, GEORG.

PONTANUS, JAKOB, real name Spanmüller (Brüx, 1542–1626, Prague), entered the Society of Jesus in 1564 and was later canon and dean in Prague. He was highly regarded for his schoolbooks (*Progymnasmata latinitatis sive dialogi,* 4 vols., 1588–94) and also wrote Latin plays for performance in schools on subjects taken from the Bible, the Apocrypha, and Ancient History (see JESUITENDRAMA). He was also the author of an influential treatise on rhetoric (*Poeticarum instutionum libri tres,* 1600).

PONTEN, JOSEF (Raeren nr. Eupen, 1883–1940, Munich), studied the history of art at several universities and colleges (Geneva, Bonn, Berlin, Aachen); a much-travelled man in Europe (including Russia) and in America, his trained eye, coupled with his descriptive powers, enabled him to produce landscape writing which is both evocative and accurate (*Griechische Landschaften,* 2 vols., 1915; *Europäisches Reisebuch,* 1928). He had also ambitions as an author of fiction, chiefly historical novels and Novellen. Among his earlier works should be mentioned the Novellen *Jungfräulichkeit* (1906), *Die Insel* (1918), *Die Bockreiter* (1919), *Der Gletscher* (1923), and *Der Urwald* (1924). The novels include *Siebenquellen* (1909), *Der babylonische Turm* (1918), *Salz* (2 vols., 1921–2), and *Die Studenten von Lyon* (1927). His work tended towards the sensational.

At his death Ponten left what he believed to be his principal work incomplete. Entitled *Volk auf dem Wege* (1931–42), it recounts in six separate novels (more were intended) the efforts of branches of a German family to settle in various foreign lands from the 17th c. onwards. Completed were: (1) *Die Väter zogen aus* (covering the period 1689–1808 and ranging from America to the Volga); (2) *Im Wolgaland, c.* 1910; (3) *Rheinisches Zwischenspiel,* which recounts the return of one of the family; (4) *Die Heiligen der letzten Tage,* which is set in *c.* 1820, partly in Germany and partly in eastern Hungary; (5) *Der Zug nach dem Kaukasus,* a sequel to the last; and (6) *Der Sprung ins Abenteuer,* which switches to Morocco in 1913.

Pontus und Sidonie, a German translation of the French prose romance *Pontus et la belle Sidoine.* The translator was the Archduchess Eleonore von Österreich (q.v.), who completed the work by 1456. *Pontus und Sidonie* tells of Pontus' battle against the heathen in Spain, and of his rescue and voyage to Britain, where he again fights against the heathen and delivers the land from them. He loves Sidonie and is rewarded with her hand. The story, first printed in 1483, achieved great popularity as a Volksbuch (q.v.), and was reprinted at intervals up to 1769. It was republished, ed. K. Schneider, 1961.

Porta Westphalica, see WESTFÄLISCHE PFORTE.

Portimunt, the title given to two fragments of a Middle High German narrative poem, probably written in the second half of the 13th c. In it a talking parrot plays an important part as messenger between two lovers. Portimunt is a kingdom, and the king's daughter is the heroine. The identity of the author is not known.

POSTEL, CHRISTIAN HEINRICH (Freyburg/Elbe, 1658–1705, Hamburg), a lawyer in Hamburg, provided extravagant and bombastic libretti for the baroque opera of his home city. His subjects included classical themes (*Die schöne und getreue Ariadne*, 1691; *Scipio Africanus*, 1694), biblical stories (*Kain und Abel*, 1689), and historical episodes (*Der tapfere Kaiser Carolus Magnus und dessen erste Gemahlin Hermingardis*, 1692). Postel also translated into German verse a part of Bk. XIV of the *Iliad* (*Die listige Juno*, 1700).

Postillon, Der, a poem written by N. Lenau (q.v.) in America in 1832–3. It tells of a postilion who invariably halts the coach in order to blow a salute to his dead comrade. Its first line runs 'Lieblich war die Maiennacht'.

POSTL, KARL ANTON, better known as Charles Sealsfield (q.v.).

Potsdam, small town in the German Democratic Republic (see DEUTSCHE DEMOKRATISCHE REPUBLIK), formerly Prussian (Provinz Brandenburg) and long serving as a royal residence. It has two palaces, the Stadtschloss built in the 17th c. and the 18th c. and Sanssouci (q.v.), the rococo building constructed for Friedrich II (q.v.) between 1745 and 1747. It is surrounded by a landscaped park. Potsdam was a garrison town with c. 7,500 troops of the Prussian Guard and was held to incorporate the spirit of Prussian militarism ('der Geist von Potsdam'). On 21 March 1933 Hitler and the aged President Hindenburg (qq.v.) walked through the streets lined by troops and SA (q.v.), and stood with bowed head before the sarcophagus of Friedrich II. This episode is known as Der Tag von Potsdam.

Pour le mérite, a Prussian order of chivalry created by Friedrich II (q.v.) in 1740 as the highest distinction for exceptional civil or military services. From 1810 to 1918 it was restricted to distinguished services in war.

Through the influence of A. von Humboldt,

Friedrich Wilhelm IV (qq.v.) created in 1842 a new civil class for services to the sciences and the arts, the Friedensklasse des Ordens Pour le mérite. In 1924 it was recognized as an independent association, and in 1952 it was reintroduced with the same designation by the Federal Republic (see BUNDESREPUBLIK DEUTSCHLAND) under the auspices of the Federal President (Bundespräsident).

Powenzbande, Die, a novel by E. Penzoldt (q.v.), published in 1930. It is an ironic story of a family of rogues (but not criminals) in a small German town in the first three decades of the 20th c. The aim of the head of the family, Baltus Powenz, is to build his own house; in doing so he encounters many difficulties, some municipal. On the eve of success, however, he is killed by a meteorite.

Prager Fenstersturz, see MAJESTÄTSBRIEF.

Pragmatische Armee, refers to the English and Dutch armies who, under the command of George II of England (who was also Elector of Hanover), supported the Empress Maria Theresia (q.v.) in the Austrian War of Succession (see ÖSTERREICHISCHER ERBFOLGEKRIEG) against France. The Pragmatic Army defeated the French on 27 June 1743 at Dettingen/Main. It was the last battle in which a British king commanded his army in the field.

Pragmatische Sanktion, Die, the Pragmatic Sanction was issued by the Emperor Karl VI (q.v.) on 19 April 1713 in order to secure, in the absence of a male heir, the succession of his daughters to the Habsburg crown and the indivisibility of the Habsburg dominions. In effect it aimed at securing the succession of his eldest daughter Maria Theresia (q.v.). By 1724 the Pragmatic Sanction was recognized by his own Estates, and, by 1738, by all principal European powers, though at the cost of the loss of Lorraine to Stanislas of Poland with reversion to France. Maria Theresia's husband, Franz of Lorraine (see FRANZ I), was compensated with Tuscany. When Maria Theresia succeeded her father in 1740, she nevertheless found her position seriously contested (see ÖSTERREICHISCHER ERBFOLGEKRIEG).

Prärie am Jacinto, Die, a Novelle by C. Sealsfield (q.v.), included in his novel *Das Kajütenbuch* (1841). It tells the story of Bob, the murderer who is pursued by his conscience and surrenders to justice. He is hanged, but at once cut down by the sheriff and so is preserved to continue the struggle to unite Texas with the U.S.A., in the course of which he loses his life. The story has often been separately reprinted.

PRASINUS, JOHANNES, real name PRASCH (b. Hallein, d. 1544), an Austrian humanist living in Vienna, wrote a Latin play *Philaemus* (printed 1548) and made a translation of four books of the *Odyssey* into Latin elegiacs.

Prater, park in Vienna (see JOSEPH II).

PRÄTORIUS, JOHANNES, real name Schulze (Zethlingen, Brandenburg, 1630–80, Leipzig), a scholar and poet laureate (see GEKRÖNTER DICHTER), published abundantly, especially on scientific and pseudo-scientific subjects such as magic and witchcraft. The following are a few characteristic titles from his prolific output: *De crotalistria oder Von des Storches Winterquartier,* 1656; *Daemonologia Rubinzalii Silesii* (in German as *Rübezahl*), 1662–5; *Saturnalia,* 1663; *Astrologia Germanica et Germana,* 1665; *Blocks-Berges-Verrichtung oder ausführlicher geographischer Bericht von dem Blocks-Berge,* 1668. The last-named work was one of the sources used by Goethe for the Walpurgisnacht scene of *Faust* (q.v.), Pt. One.

PRÄTORIUS, MICHAEL (Creuzburg, 1571–1621, Wolfenbüttel), master of music to the Duke of Brunswick from 1612, was one of the most eminent composers of the 17th c. His work consists largely of religious music, hymns, psalms, and motets. He is the composer of the well-known four-part setting of 'Es ist ein Ros' entsprungen' (q.v.).

PREHAUSER, GOTTFRIED (Vienna, 1699–1769, Vienna), became a comic actor in Vienna in 1725 under J. Stranitzky (q.v.), whose theatre he took over after Stranitzky's death in the following year. Prehauser, who played the stock character Hanswurst (q.v.), achieved his effects by witty improvisation of words and by facial expression.

PRERADOVIĆ, PAULA VON, married name Molden (Vienna, 1887–1951, Vienna), a Roman Catholic and a predominantly religious poet, published the volumes of verse *Südlicher Sommer* (1929), *Dalmatinische Sonette* (1933), *Lob Gottes im Gebirge* (1936), and *Ritter, Tod und Teufel* (1946). She wrote the words of the new Austrian national anthem, introduced in 1945. During the National Socialist regime she assisted the Austrian Resistance. Her work also includes a novel (*Pave und Pero,* 1940) and stories. *Gesammelte Werke,* ed. K. Eigl, appeared in 1967.

PREUSS, HUGO (Berlin, 1860–1925, Berlin), a distinguished constitutional lawyer, whose draft

was the basis of the constitution of the Weimar Republic (q.v.). In 1921 he published *Deutschlands republikanische Reichsverfassung.* A collection of his essays, *Staat, Recht und Freiheit,* was published by Th. Heuß (q.v.) in 1926.

PREUSS, JOHANN (Guben, 1620–96, Selchow, Brandenburg), an itinerant preacher, wrote religious poetry which was published in the volumes *Herzliches Saitenspiel* (1657), *Geistlicher Weihrauch* (1662), and *Fastspeise* (1678).

Preußen (Prussia) comprised until 1701 the future East Prussia (see OSTPREUSSEN). In that year the Elector Friedrich III of Brandenburg (see FRIEDRICH I, KÖNIG) succeeded in obtaining royal status and described himself as König in Preussen, so extending the name to his own electorate of Brandenburg and other territories. Under Friedrich II (q.v.) Silesia and a portion of Poland (1772) were annexed, and further Polish lands in 1793 and 1795 (see POLAND, PARTITIONS OF). This greatly enlarged Prussia became a serious rival of Austria within the Holy Roman Empire (see DEUTSCHES REICH, ALTES). In the Napoleonic Wars (q.v.) Prussia was reduced by the Treaty of Tilsit (1807) virtually to its extent at the beginning of the 18th c. The Congress of Vienna (1815, see WIENER KONGRESS) made ample amends by restoring lost territories (except for New East Prussia) and adding extensive lands in the west, including the Rhineland and parts of Westphalia. After the war of 1866 between Prussia and Austria (see DEUTSCHER KRIEG), those states north of the Main which had taken the Austrian side (Hanover, Hesse-Cassel, Nassau, Frankfurt) were annexed to Prussia, together with Schleswig-Holstein.

In the German Empire of 1871 Prussia was the most powerful state, the king of Prussia being hereditary German Emperor. Prussia's dominance in the late 19th c. and early 20th c. is shown by her population and area. In 1900 it had 60 per cent of both land and inhabitants. The remaining 40 per cent was made up of twenty-four sovereign states and the Reichsland of Alsace-Lorraine. The next largest state after Prussia was Bavaria with about 14 per cent. The same predominance continued in the Weimar Republic (q.v.).

Prussia was declared dissolved in the Potsdam decrees of 1945. Parts are now included in the Federal Republic (in West-Berlin, Bremen, Hamburg, Hessen, Niedersachsen, Nordrhein-Westfalen, Schleswig-Holstein, see BUNDESREPUBLIK DEUTSCHLAND), others in the German Democratic Republic (Ost-Berlin, Brandenburg in part, Provinz Sachsen, Mecklenburg, and a fragment of Pommern, see DEUTSCHE DEMO-

KRATISCHE REPUBLIK). Most of the rest is Polish (Pommern, Schlesien, eastern Brandenburg, the pre-1918 Provinzen Westpreußen and Posen), while East Prussia is divided between Poland and the U.S.S.R.

Preußische Union, see DREIKÖNIGSBÜNDNIS.

Priamel, a minor poetic form, cultivated in the 15th c. and 16th c., in which, after a preparatory cumulative build-up, a comic or witty *pointe* forms the final line. The word is derived from Latin *preambulum* (preamble). The chief exponent of Priameln is Hans Rosenplüt (q.v.).

PRINTZ, WOLFGANG CASPAR (Waldethurn, Upper Palatinate, 1641–1717, Sorau, Lausitz), a musician with experience in Italy and Hungary as well as at home, became director of music (Konzertmeister) to the counts of Promnitz. He is the author of the first German history of music (*Historische Beschreibung der edlen Sing- und Klingkunst*, 1690). Printz also wrote three novels in each of which the central figure is a musician, *Cotala* (1690), *Pancalus* (1691), and *Battalus* (1691). A 3-vol. selection, *Ausgewählte Werke*, ed. H. K. Krausse, began to appear in 1974.

Prinzessin Blandina, a fragment of a play (Ein romantisches Spiel in drei Aufzügen) written by E. T. A. Hoffmann (q.v.) in 1814, and published in 1815 in vol. 4 of *Fantasiestücke in Callots Manier* (q.v.). Only the first act was completed. Partly in verse, partly in prose, it is one of several works connected with Hoffmann's love for Julia Marc (q.v.).

Prinzessin Brambilla. Ein Capriccio nach Jakob Callot, a story by E. T. A. Hoffmann (q.v.), written in 1820 and published in 1821. Set in Rome at carnival time, it is a more than usually complex fairy-tale, in which two fairy-tale characters, Prince Cornelio and Princess Brambilla, and two characters in the real world, the actor Giglio and the sempstress Giacinta, are brought together. The two from the real world find that they are identical with the fairy creatures. The story, full of dazzling fantasy and eccentric invention, represents a kind of poetic allegory, justifying Romantic literature and symbolizing in Giacinta the poetic imagination. The magic land of Urdargarten, of which much is heard, is the land of poetry.

Prinzessin Fisch, a short novel written in 1881–2 by Wilhelm Raabe (q.v.) and published in 1883. The title, which alludes to 'die Prinzessin Fisch' in Goethe's early poem *Der junge Amadis*, hints at the hero's imaginative love for an unworthy object.

Theodor Rodburg, a straggler born late into a family in which the eldest (and runaway) brother was some twenty years older, is orphaned in early years and brought up by two homely figures, the widow Schubach and the journeyman in her book-binding workshop, der Bruseberger, while his formal education is watched over by the schoolmaster Dr. Drüding. Theodor falls in love at puberty with the wife of the next-door neighbour, Frau Romana Tieffenbacher, a striking, but idle and brainless Mexican, whom he admires from a distance. The situation becomes more serious with the return of Theodor's lost eldest brother Alexander, an alert and ingenious mind, selfish and unscrupulous, who in the intervening years has feathered his nest in Mexico. Alexander brings the boy into contact with Frau Romana (the Prinzessin Fisch of the title), and for a time Theodor's stability is threatened. But his own good nature, seconded by the efforts of Mutter Schubach and der Bruseberger, bring him safely through. In the end Alexander runs away with Romana, and Theodor resolves to complete his studies and some day to return to his home town with a maturity which can disregard the evil repute of Alexander. In the background, and faintly foreshadowed perhaps as Theodor's future wife, is Dr. Drüding's teenage daughter Florine, wholesome, sensible, and freshly attractive, a homely and genuine counterpart to the exotic false glitter of Romana.

The story is a form of Bildungsroman (q.v.), bringing its hero to the threshold of life, ending indeed with the words 'Auf der Schwelle!', which Raabe in the last sentence suggests could have been the title. The book is enriched by four original characters, Mutter Schubach, der Bruseberger, Dr. Drüding, and Romana's elderly husband Joseph Tieffenbacher. Its background is the little town of Ilmenthal, which is in process of conversion into a spa and caught up in the 'improvements' of civilization.

'Prinz Eugen der edle Ritter', first line of an Austrian patriotic song written after the capture and occupation of Belgrade in 1717 by Prince Eugene (see EUGEN, PRINZ). Its author is not known. The line is also used in a ballad by F. Freiligrath (q.v.) entitled *Prinz Eugen der edle Ritter* (published in *Gedichte*, 1838); it has been set to music by Carl Loewe (q.v.).

Prinz Friedrich von Homburg, a five-act play (Schauspiel) in blank verse by H. von Kleist (q.v.), completed in 1810, and published by L. Tieck (q.v.) in 1821. The historical event which is central to the play is the battle of Fehrbellin (1675), in which the Great Elector Friedrich Wilhelm of Brandenburg (see FRIEDRICH

WILHELM, DER GROSSE KURFÜRST) defeated the Swedes. Kleist was familiar with the *Mémoires pour servir à l'histoire de la Maison de Brandebourg* (1751) by Friedrich II (q.v.) of Prussia, as well as with a number of contemporary sources, of which he made selective use. He altered, however, the basic fact that the Elector took no action after Homburg's disobedience.

Homburg himself assumes an entirely new identity as he is introduced as a sleep-walker, unconsciously fashioning a wreath of laurels, which the Elector, in a mime, offers to him through his niece, Princess Natalie von Oranien. After the withdrawal of the party, Count Hohenzollern awakens Homburg to resume his duties and attend the communication of orders for battle. On the following day Homburg fails to obey his orders, impulsively ordering a charge, and he appears to be the victor and even commander, since the Elector is reported killed in battle. The Elector proves incorrect. The Elector emerges unharmed, arrests Homburg, has him tried by court martial, and sentenced to death. He makes preparations for the execution, but pardons Homburg after the Prince has unreservedly conceded the justification of the death sentence.

Although the Elector appears already to have undergone a self-scrutiny after his quick decision to deliver Homburg to death, he is also faced with the threat of revolt, instigated by Natalie and supported by the Prince's brother officers, foremost among them Obrist Kottwitz. In the final scene, in which the Elector exercises his prerogative of mercy, Homburg is celebrated as the victor and his military status and betrothal to Natalie are reaffirmed.

Kleist dedicated the play to Princess Amalie Marie Anne of Hesse-Homburg, the wife of Prince Wilhelm, brother of the King of Prussia. In his poem of dedication he expresses the hope that the play may meet with her approval, which it failed to do. The prostration which Homburg displays in the central act after seeing his open grave was to her unacceptable conduct in an officer. To Kleist it was an indispensable feature of his portrayal of the Prince, and a deliberate deviation from conventional historical and heroic drama. The fact that Kleist failed to win recognition for this great play contributed to the mental and financial distress that preceded his suicide. (For the opera *Der Prinz von Homburg*, see H. W. HENZE.)

Prinz Louis Ferdinand von Preußen, a historical tragedy (Ein Drama) in five acts by F. von Unruh (q.v.), published in 1913. It deals with the conflict in 1806 between the dynamic, impatient, warlike Louis Ferdinand and the hesitant King Friedrich Wilhelm III (qq.v.) and his timorous advisers, Haugwitz (q.v.) and Lombard, just before Jena (see NAPOLEONIC WARS). The Queen, whom Louis Ferdinand loves, persuades him to submit. The play belongs to Unruh's nationalistic phase before the experiences of the 1914–18 War and is epitomized in its motto, 'Wie über Sterne das Gesetz, erhebt sich über Menschen die Pflicht, groß und ernst'.

Problematische Naturen, a 4-vol. novel by F. Spielhagen (q.v.), published in 1861–2. It is divided into two sections (Abteilungen), each of 2 vols.; the second section originally appeared under the title *Durch Nacht zum Licht*, which was retained in parenthesis in subsequent editions after the words *Zweite Abteilung*. The title of the whole book alludes to a sentence in Goethe's *Maximen und Reflexionen* (later *Sprüche in Prosa*): 'Es gibt problematische Naturen, die keiner Lage gewachsen sind, in der sie sich befinden, und denen keine genug tut.'

Spielhagen's principal 'problematical nature' is Oswald Stein, a brilliant young academic, who is invited in June 1847 to become private tutor to the sons of the wealthy Baron Grenwitz in Pomerania. The action covers rather less than a year, ending in the revolutionary fighting in Berlin in March 1848 (see REVOLUTIONEN 1848–9). Stein has a deep hatred of the aristocracy, and his position as a tutor is marked by constant tension, as he holds his own in this world to which he is opposed. He is also a formidable lady-killer, taking as the successive objects of his attentions the young noblewomen Melitta von Berkow, Emilie von Breesen, and Helene von Grenwitz. In the course of these entanglements he crosses the paths of various noblemen and wounds his employer's son in a duel. It emerges that Stein is the illegitimate son of a former Baron Grenwitz, but the fact is kept from him by an intriguer, the retired officer Albert Timm. In the second part Stein reappears in an estimable burgher household. But he presently elopes to Paris with Emilie von Breesen, now the wife of Herr von Cloten, one of the Grenwitz circle. He returns to take his place on the barricades in Berlin, where he meets a heroic end.

Pro cadente morbo, see CONTRA CADUCUM MORBUM.

Pro commoedia commovente, title of the inaugural lecture delivered in Latin by C. F. Gellert (q.v.) in 1751 on taking his professorship at Leipzig University. It was translated into German by G. E. Lessing (q.v.) and published in *Die theatralische Bibliothek* (q.v.) in 1754 under the title *Des Hrn Prof. Gellerts Abhandlung für das* *rührende Lustspiel.* Gellert's essay puts the case for refined rather than for uproarious humour and reserves a place in comedy for tears.

Professor Bernhardi, a comedy in five acts by A. Schnitzler (q.v.), first performed in November 1912 at the Kleines Theater, Berlin, and published in the same year. In Austria performance was forbidden by the censorship until the dissolution of the Austro-Hungarian state in 1918.

Professor Bernhardi is a well-known Jewish consultant in charge of a ward in which a girl is dying of sepsis after an abortion. She is under sedation and unaware of the seriousness of her condition, and it is Bernhardi's intention that she should be allowed to die unperturbed under the analgesic. A priest, summoned by the ward sister, arrives to administer Extreme Unction but is refused admission by Bernhardi. While they argue the girl dies, having been told by the sister that the priest is there. Out of this episode arises a storm in the press, in the hospital administration, and in high society. As a Jew, Bernhardi becomes a special target for anti-Semitic demonstrations. He is tried for assault and obstruction of a priest in the performance of his duties, is convicted and sentenced to two months' imprisonment, which proves to be a far from rigorous experience. The fifth act is a relaxed discussion of the case between the released Bernhardi and Hofrat Winkler (a portrait of M. Burckhard, q.v.). The work is Schnitzler's best comedy, vivid in its dialogue, penetrating in its satire, and both serene and ironic in its ending.

Professorenroman, rather contemptuous term applied to certain historical novels or novels on archaeological subjects which display erudition rather than creative imagination or distinction of style. See DAHN, F., EBERS, G., and HAUSRATH, A., all of whom actually were university professors.

Professor Unrat oder Das Ende eines Tyrannen, a novel by H. Mann (q.v.), published in 1905. The setting is Lübeck. It is a bitter satire of the tyrant-pedant-schoolmaster, seen as a product of the Germany of Mann's day. Raat is his name, 'Unrat' (muck) the sarcastic nickname given him by his pupils. Raat discovers that three of the boys who are his particular targets (Kieselack, Graf Ertzum, and Lohmann) are visitors to a low cabaret, drawn by the attractions of Rosa Fröhlich, the singer. He goes to Der blaue Engel, as the cabaret is called, and so scares the boys off, but himself falls for Rosa, and becomes one of her lovers. The boys, in consequence of an act of silly vandalism, are expelled from school; and Raat, because of his *affaire* with Rosa, is dismissed. He marries her and turns his home into a gambling club, with Rosa as chief attraction, and in this, his act of revenge, some reputable citizens ruin themselves. When Raat finds Rosa in bed with Lohmann he attempts to strangle her and steals Lohmann's wallet. He is arrested, and we lose sight of him as he is carried off in the 'Black Maria' as 'ne Fuhre Unrat' (a load of muck). Mann's satire reaches monumental dimensions of caricature and grotesqueness. J. von Sternberg's film version (script by C. Zuckmayer, q.v.) had a sensational success as *Der blaue Engel* (1930). This title was substituted for *Professor Unrat* in the edition of 1947.

PROKOP VON TEMPLIN, also **PROCOPIUS** (Templin, Brandenburg, 1607–90, Linz/Danube), brought up a Protestant, was early converted to the Roman Catholic faith and became a Capuchin monk, spending most of his life in Austria and Bohemia. His works, which include numerous titles published between 1642 and 1679, consist principally of collections of sermons and hymns. A 6-vol. collection of sermons, *Catechismale,* was published in 1674–5.

Proletarisches Theater, see PISCATOR, E.

Prometheus, a poem written by Goethe in 1774. It was intended to be part of a projected drama about Prometheus, of which a substantial fragment (414 lines) survives. The poem was printed by F. H. Jacobi (q.v.) without permission in 1785 in *Über die Lehren des Spinoza* and thereupon published by Goethe in the next edition of his works (1789). The fragmentary drama was first published in 1830. *Prometheus,* one of the most powerful and dynamic of Goethe's poems, is a denunciation and defiance of the gods, hammered out in free rhythms. It has been set to music by F. Schubert and by Hugo Wolf (qq.v.).

Pro nessia, see WURMSEGEN.

Propyläen, Die, a periodical founded by Goethe in 1798 for the propagation of the aesthetic views of his classical phase, in particular with regard to the plastic arts. Goethe's principal collaborator was H. H. Meyer (q.v.). The *Propyläen* made little headway with the public and ceased publication in 1800.

Prosaroman, term used to denote prose versions of courtly verse romances, made for the most part in the 15th c. The distinction between the Prosaromane and the Volksbücher (see VOLKSBUCH), into which they developed at the end of the 15th c., lies in the persistence of the courtly and chivalric values in the Prosaroman, which are obscured in the Volksbuch.

Proserpina, see TRIUMPH DER EMPFINDSAM-KEIT, DER.

Protestantische Union, Die (also **Union von Auhausen** or **Deutsche Union**), was formed in May 1608 in Auhausen upon the initiative of Christian I, Fürst von Anhalt (1568–1630) after an attack by Protestants on Roman Catholics in Donauwörth (1607) had led to the occupation of this city by Maximilian I (q.v.) of Bavaria. The purpose of the Union was to defend Protestant, especially Calvinist, interests against the Habsburg dynasty. Its members were the Protestant states of the Rhineland, among them the Duchy of Württemberg and the Margravate of Baden, and the free cities of Strasburg, Nürnberg, and Ulm (1609), as well as the Electorate of Brandenburg and the Landgravate of Hesse (1610). The Union was headed by the Elector Palatine Friedrich V (q.v.), and Christian of Anhalt and the Margrave of Baden-Durlach were its generals. In 1612 it made an alliance with England and was dissolved in 1621, having failed effectively to rival the Catholic League (see KATHOLISCHE LIGA).

Protestantismus, Der, takes its name from the protesting princes at the Diet of Speyer (see REFORMATION) and produced in Germany various Churches and creeds of which only a brief outline can be given here. The success of Luther's Church lay in his insistence on vernacular services and lay communion. He initiated both in 1526 with his Deutsche Messe, soon making the sermon the central part of the service and giving less stress to the Eucharist.

Luther's collaborators Melanchthon and Bugenhagen (qq.v.) evolved the theological and administrative basis of the new Churches. The theological principles of the Protestant Churches, including the work of the Swiss reformers Zwingli (q.v.) and Calvin, emanated from the simple notion of justification by faith alone and the return to the word of the Scriptures. Each age since the Reformation has successively been preoccupied with revision as well as revival of the theological issues. More so than the Roman Catholic Church, Protestantism has always been in a state of flux, in Germany more intensely than in other countries to which it has spread since its first beginnings. Protestant theology of the 18th c. was dominated by Pietism (see PIETISMUS) on the one hand, and the rationalism of the Age of Enlightenment (see AUFKLÄRUNG) on the other. Pietistic Lutheranism under Spener and Francke (qq.v.) aimed at a deeper spiritual and moral scrutiny of the Bible, which was furthered by Francke's *collegia pietatis,* i.e. private discussion groups linked to the church service. Spener's work can best be understood as a reaction against sterile 17th-c. theological scholasticism. It was continued by Zinzendorf (q.v.) of Herrnhut, who founded the Brüdergemeinde (see HERRNHUTER GEMEINE), out of which a special and at times ecstatic kind of Pietism developed in the mid-18th c. The devotional emotionalism of Pietism produced a proliferation of Protestant church hymns (see KIRCHENLIED) and the literature of Empfindsamkeit (q.v.). Rationalistic approaches to religion were promoted notably by Bodmer and Füßli (qq.v.). Their optimistic reappraisal of God's creation provoked scepticism, free thought, and anti-clerical speculative thought (as, for instance, in Lessing's *Die Erziehung des Menschengeschlechts,* q.v.), which was, however, equally a reaction against hypocritical elements within the Church. The Swiss J. C. Lavater (q.v.) stood for the belief in a personal relationship with God established by irrational and sensuous perception, which he combined with the study of Mesmerism (see MESMER). With the moral philosophy of Kant (q.v.) the paradoxical concept of a secularized religion arose in conjunction with Fichte, Schelling, and Hegel, (qq.v.), none of whom was a philosopher of Protestantism; rather they prepared the ground for the negative assessments of D. F. Strauß and L. Feuerbach (qq.v.), despite Hegel's avowed aim of reconciling reason with religion. A 19th-c. Protestant revival began with F. Schleiermacher (q.v.), whose theology of feeling coincided with the subjective idealism nurturing the literary age of Romanticism (see ROMANTIK); all Christocentric orientation of modern theology has evolved from him. The influence of Kierkegaard (q.v.) on men of letters was considerable. Noted 20th-c. exponents of Protestant theology are Karl Barth and Rudolf Bultmann (qq.v.). See also LANDESKIRCHEN.

Prozeß, Der, a novel written by Franz Kafka (q.v.) in 1914–15 and published posthumously in 1925. Though the novel has a conclusion, it is unfinished and the order of the chapters is still disputed. K., a bank official, is arrested for reasons which are not apparent to him. He is subjected to protracted investigation and interrogation, in which the processes of law undergo dream-like distortions. All his efforts to penetrate the reasons for his arrest and trial fail. K. is finally marched off by two men in black who put him to death by knifing him. *Der Prozeß* is usually seen as an allegory of existential guilt. It contains the parable *Vor dem Gesetz,* which Kafka included in the collection *Ein Landarzt*

(1919). It is the subject of an opera by G. von Einem (q.v.).

Prozeß Belial, an early 15th-c. anonymous Middle High German translation of the Latin work, *Processus Belial* (1382) by Jacobus de Terramo, Bishop of Spoleto. It portrays the hearing of a suit brought by Lucifer against Christ. The parties do not appear in person, but are represented by advocates, Belial and Moses respectively. Solomon appears on behalf of God as judge.

Prussia, see PREUSSEN.

PRUTZ, ROBERT EDUARD (Stettin, 1816–72, Stettin), studied at Berlin, Breslau, and Halle universities with the aim of embarking on an academic career in literary studies, but, because of his Radical political activities, he failed to obtain appointment until 1846; he was promptly suspended in the following year. He next undertook theatre direction at Hamburg (1847–9), and in 1849 was appointed to a chair at Halle University, which he held until 1859.

Although Prutz published 2 vols. of lyrical poetry (*Gedichte*, 1841–3), his chief works were political. In addition to the satirical comedy in verse *Die politische Wochenstube* (1843), he published the politico-social novels *Das Engelchen* (3 vols., 1851, dealing with the poverty of village weavers and the new industrialization) and *Der Musikantenturm* (3 vols., 1855), and treated class and marriage in the Novelle *Die Schwägerin* (1851). The novels *Felix* (2 vols., 1851) and *Oberndorf* (1862) were also politically tinged. Towards the end of his life Prutz published two further volumes of poetry (*Herbstrosen*, 1864; *Buch der Lieder*, 1869). His Germanistic publications included a monograph on the Göttinger Hainbund (q.v., *Der Göttinger Dichterbund*, 1841) and works on the Danish writer L. Holberg (1684–1754).

Psalm 138, a free paraphrase in Old High German rhyming verse and Bavarian dialect of the psalm numbered 138 in the Vulgate (139 in the Authorized Version). It belongs to the 10th c. The original version was possibly made in St. Gall.

Pseudohrabanisches Glossar, see ABROGANS.

PÜCKLER-MUSKAU, HERMANN, FÜRST VON (Muskau, Prussia, 1785–1871, Branitz, Prussia), served in the Saxon Guards, became an officer in the Russian army in the War of Liberation (1813, see NAPOLEONIC WARS), and was for a short time in the suite of the Duke of Saxe-Weimar. In 1815 he withdrew to his extensive

estates, Muskau and Branitz, both near Cottbus. These he transformed into magnificent parks which attracted visitors from all over Europe. In his forties he travelled extensively in Africa and Asia Minor, and proved himself a travel writer of distinction (*Briefe eines Verstorbenen*, 4 vols., 1830–2, relating to a tour of the British Isles; *Vorletzter Weltgang von Semilasso*, 3 vols., 1835; *Semilasso in Afrika*, 5 vols., 1836; and *Südöstlicher Bildersaal*, 3 vols., 1840, reprint, ed. K. G. Just, 1968). A work on landscape gardening (*Andeutungen über die Landschaftsgärtnerei*) appeared in 1834. An early volume of poetry (*Gedichte*, 1811) may be regarded as a youthful indiscretion. In 1845 he sold Muskau, but he retained Branitz to the end of his life.

Allusions to Pückler-Muskau occur in Eichendorff's *Viel Lärmen um Nichts* and Brentano's *Die mehreren Wehmüller und die ungarischen Nationalgesichter* (qq.v., published together, 1833), and in Th. Fontane's *Frau Jenny Treibel* (q.v., 1892). In 1836 he secured permission for H. Laube (q.v.), condemned to a spell of detention (Festungshaft), to serve his term as a guest on the Muskau estate.

Pückler-Muskau's diaries and correspondence (9 vols.), ed. L. Assing, 1873–6, were reprinted in 1971.

PUFENDORF, SAMUEL, FREIHERR VON (Chemnitz, 1632–94, Berlin), German jurist and historian, became a professor at Heidelberg University in 1661 and at Lund in Sweden in 1670. In 1677 he was appointed Swedish historiographer in Stockholm and in 1686 received a similar appointment at the court of the Great Elector of Brandenburg (see FRIEDRICH WILHELM, DER GROSSE KURFÜRST). His (Swedish) patent of nobility was granted in the last year of his life.

Pufendorf's first important work, written in Latin, as were almost all his books, was *Elementa juris prudentiae universalis, libri duo* (1661). This was followed by a prudently pseudonymous, short, and trenchant essay *De statu imperii germanici, liber unus* (1667), which clear-sightedly and savagely analysed the constitutional shortcomings of the Holy Roman Empire (see DEUTSCHES REICH, ALTES). The pseudonym adopted was Severinus de Monzambano. His principal work, *De jure naturae et gentium, libri octo* (1672), the views of which are partly based on Grotius and Hobbes, made an important contribution to the development of the Aufklärung (q.v.). Its influence was greatly strengthened by the publication in 1673 of an abridged version, *De officio hominis et civis juxta legem naturalem*. After 1677, when his duties became primarily historiographical, Pufendorf produced several historical works, which, though drily written, are carefully based on the available documents. They are *Einleitung zu der Historie der vornehmsten Reiche und Staaten in Europa* (1682), *Commentarii rebus suecicis, libri XXVI, ab expeditione Gustavi Adolphi in Germaniam ad abdicationem usque Christinae* (1686), *De rebus gestis Friderici Wilhelmi Magni electoris Brandenburgici* (1695), and *De rebus a Carolo Gustavo Sveciae rege gestis* (1696).

Pufendorf advocated religious tolerance in *De habitu christianae religionis ad vitam civilem* (1687), and Protestant unity in *Jus feciale divinum* (1695). He corresponded with C. Thomasius (q.v.) between 1687 and 1693 (the letters were published in 1897).

Puntila, see HERR PUNTILA UND SEIN KNECHT MATTI.

Puppenspiel (also Figurentheater), puppet or marionette plays, which have provided a form of popular entertainment in Germany over many centuries. The best known is that alluded to by Goethe in *Aus meinem Leben. Dichtung und Wahrheit*, the *Puppenspiel vom Dr. Faust*, which he saw as a child in Frankfurt. A tradition approximating to Punch and Judy shows has long been established with Kasperl plays (see POCCI, F., GRAF VON), which are popular both with the marionette and with the hand-puppet theatre. A number of theatres devoted to puppet plays still exist in German-speaking countries, and of these the Salzburger Marionettentheater (q.v.) deserves special mention.

Several authors have used the puppet as a symbol for the human situation and for artistic aspiration, including H. von Kleist (*Über das Marionettentheater*, q.v., 1810), G. Büchner (to illustrate a fatalistic view, fate being the manipulator of the puppet, i.e. man), and A. Schnitzler (qq.v.), who published a cycle of one-act plays entitled *Marionetten* containing one true puppet play (*Der tapfere Kassian*) and one in which the marionettes perform a play within a play (*Zum großen Wurstel*). G. Hauptmann conceives the characters of *Festspiel in deutschen Reimen* (q.v.) as puppets and the wrapper of the first edition portrayed Napoleon as a manipulated figure. An essay by Rilke (q.v.) on dolls (*Einiges über Puppen*, 1914) contains this statement on marionettes: 'Es könnte ein Dichter unter die Herrschaft einer Marionette geraten, denn die Marionette hat nichts als Phantasie. Die Puppe hat keine und ist genau um so viel weniger als ein Ding, als die Marionette mehr ist.'

The traditional puppeteer originates from South rather than North Germany. One such itinerant artist and his daughter are the central figures in Storm's Novelle *Pole Poppenspäler* (q.v.,

Poppenspäler is Low German for Puppenspieler).

PUSCHMAN, ADAM (Görlitz, 1532–1600, Breslau), a friend of Hans Sachs (q.v.) in Nürnberg, was active as a Meistersinger in his native town and in Breslau. His *Gründtlicher Bericht des deudschen Meistergesangs, darinnen begriffen alles, was einem jeden, der sich Tichtens und Singens annemen will, zu wissen von nöten* (1571, repr. 1888) gives in homely terms a valuable contemporary account of the Meistergesang (q.v.). In 1576 he issued an obituary essay on Hans Sachs, *Elogium reverendi viri J. Sachs*. Revised and enlarged editions of his work of 1574 on theory appeared as *Singebuch* (1584) and as *Gründlicher Bericht der deutschen Reimen* (1596; ed. G. Münzer, 1907).

PUSTKUCHEN, JOHANN FRIEDRICH WILHELM (Detmold, 1793–1834, Wiebelskirchen), also known as Pustkuchen-Glanzow, was from 1830 until his death pastor in Wiebelskirchen. He attracted attention by a parody of *Wilhelm Meister*, published from 1821 to 1828 under the title *Wilhelm Meisters Wanderjahre*, which many readers assumed to be an authentic work by Goethe. The genuine *Wilhelm Meisters Wanderjahre* (q.v.) of Goethe also appeared in 1821.

PÜTERICH VON REICHERTSHAUSEN, JAKOB (Munich, *c.* 1400–69, Munich), a Munich patrician's son, Bavarian magistrate, and friend of Ulrich Füetrer (q.v.), is the author of an *Ehrenbrief*, addressed to the Archduchess Mechthild, consort of the Archduke Albrecht II of Outer Austria (Vorderösterreich, i.e. Breisgau). The *Ehrenbrief*, written in 1462, is a poem in 148 Titurelstrophen (see TITURELSTROPHE) and contains, apart from formal homage to the Archduchess, a list of Bavarian noble families and a catalogue of many of the books which the author owned. As Püterich was a bibliophile with a considerable library (164 books), this is of distinct historical interest. He was an ardent admirer of WOLFRAM von Eschenbach (q.v.), whose grave he visited.

PUTLITZ, GUSTAV HEINRICH GANS, EDLER HERR ZU (Retzin nr. Perleberg, 1821–90, Retzin), joined the Prussian civil service for a few years and resigned in 1848. From 1863 to 1867 he was Court Theatre director (Intendant) at Schwerin and from 1873 to 1889 was in charge (Generaldirektor) of the theatre at Karlsruhe. He wrote several comedies, some of which were successful in their day, including *Badekuren, Familienzwist und Frieden, Das Herz vergessen*, all published in *Lustspiele* (4 vols.,

1850–5), and *Das Schwert des Damokles* (1863). Of his historical plays (which include a play on English history, *Wilhelm von Oranien in Whitehall*, 1864), *Das Testament des großen Kurfürsten* (1859) and *Don Juan de Austria* (1863) were frequently performed. None of these works attained the popularity of his first publication, the verse fairy-tale *Was sich der Wald erzählt* (1850).

Ausgewählte Werke (6 vols.) were published in 1873–8 with an additional volume in 1888.

PUTTKAMER, ALBERTA VON (Groß-Glogau, 1849–1923, Baden-Baden), *née* Weise, married the civil servant M. von Puttkamer. Her married life was spent in Strasburg and Colmar. After her husband's death in 1906 she moved to Baden-Baden. She is the author of the volumes of poems *Dichtungen* (1885), *Akkorde und Gesänge* (1889), *Offenbarungen* (1894), *Aus Vergangenheiten* (1895), *Jenseits des Lärms* (1904), and *Mit vollem Saitenspiel* (1912). Her historical play *Kaiser Otto III* (1883) made little impression. She published an autobiography, *Mehr Wahrheit als Dichtung*, in 1920.

PYRA, JAKOB IMMANUEL (Kottbus, 1715–44, Berlin), studied at Halle University, where he became friendly with S. G. Lange (q.v.), and both together founded the Gesellschaft zur Förderung der deutschen Sprache, Poesie und Beredsamkeit in 1733. In Halle he wrote his first significant work, the didactic poem *Der Tempel der wahren Dichtkunst* (1737), which was written in unrhymed verse. Pyra's conception of true poetry is religious, and the heroes of his poem are David the psalmist and Milton, whilst the Protestant German poets of the 17th c., Opitz,

Dach, Gerhardt, Rist, Fleming, and Gryphius (qq.v.), also receive their share of praise. In 1742 Pyra became a schoolmaster in Berlin. *Thirsis und Damons freundschaftliche Lieder* (1745) was an exchange of poems of sentimental friendship with Lange (its co-author), in which Pyra is Thirsis and Lange Damon. Pyra was a notable early opponent of rhyme in German poetry and, in his insistence on religious subject-matter for great poetry, a forerunner of Klopstock (q.v.). He wrote a polemical work directed against Gottsched (q.v.) and his school (*Erweis, daß die Gottschedianische Sekte den Geschmack verderbe*, 1743–4). He was also active in furthering interest in and knowledge of English literature.

Pyramus und Thisbe, an anonymous short Middle High German verse tale of the 14th c. The story of the sad fate of two lovers, derived from Ovid, is best known to English readers from *A Midsummer Night's Dream*.

PYRKER, JOHANN LADISLAV (Lángh nr. Székesfehérrár, Hungary, 1772–1847, Vienna), was ordained priest in the Roman Catholic Church in 1796, appointed abbot at Lilienfeld in 1812, consecrated bishop of Zips in 1818, and archbishop of Erlau in 1827. He was ennobled as Pyrker von Felsö-Eör (or von Oberwart) towards the end of his life. Pyrker wrote patriotic plays (*Historische Schauspiele*, 1810), the epics *Tunisias* (1820) and *Rudolph von Habsburg* (1824), and lyric poetry (*Lieder der Sehnsucht nach den Alpen*, 1845). His collected works (3 vols.) were published in 1832–4, and his autobiography, ed. A. Czigler, in 1966.

Q

QUAD, MATTHIS (Deventer, Holland, 1557–1613, Eppingen), became a copperplate engraver in Heidelberg. He is the author of a verse chronicle and a poetic history of Germany (*Teutscher Nation Herrlichkeit*, 1609).

QUANTZ, JOHANN JOACHIM (Oberschweden, Hanover, 1697–1773, Potsdam), the son of a blacksmith, became a distinguished flautist in the court orchestra of Friedrich II (q.v.) of Prussia, for whom he composed nearly 300 flute concertos.

Quarantäne im Irrenhause, Eine, see EINE QUARANTÄNE IM IRRENHAUSE.

Quintus Fixlein, frequently used abbreviation for Jean Paul's novel *Des Quintus Fixlein Leben bis auf unsere Zeiten* (q.v.).

QUISTORP, THEODOR JOHANN (Rostock, 1772–(?)), was a student at Leipzig, where he came under the personal influence of Gottsched (q.v.) and was afterwards an administrator in Wismar. He wrote plays, among which the comedy *Der Hypochondrist*, included in Gottsched's *Deutsche Schaubühne* (q.v.), was particularly successful.

Quitzows, Die, a four-act play (Schauspiel) in mixed verse and prose by E. von Wildenbruch (q.v.), published in 1888 and first performed in the Berlin Opera House in November 1888. It was an immense success, achieving its hundredth performance in December 1890.

A patriotic Prussian play, it is set in the 15th c. Dietrich Quitzow, head of a powerful noble house, allies himself with the Pomeranian dukes to lay waste the March of Brandenburg. The cities join him and Margrave Jobst is powerless to resist. Jobst dies, and Friedrich von Hohenzollern, Burggraf von Nürnberg, is nominated to succeed him. His accession is opposed by the Quitzows, whose power is destroyed in the ensuing civil war; the just reign of Hohenzollern (q.v.) begins in Brandenburg-Prussia. A performance of this play figures in Fontane's novel Die Poggenpuhls (q.v.).

R

RAABE, Wilhelm (Eschershausen nr. Hildesheim, 1831–1910, Brunswick), German novelist of the second half of the 19th c., was the son of a civil servant, who was posted to Holzminden, where Raabe had his first schooling; in 1842 there was a further move to Stadtoldendorf. When the father died three years later, the widow settled in Wolfenbüttel, where Raabe completed an undistinguished school career. Leaving in 1849, he was apprenticed to a bookseller at Magdeburg, with whom he remained for four years, making full use of the opportunity to read voraciously. He abandoned his career in 1853, returned home for some months, and moved in 1854 to Berlin where he attended lectures at the university for four semesters. During this time he lived in the Spreegasse, now renamed Sperlingsgasse, because he there wrote his first novel, Die Chronik der Sperlingsgasse (q.v., 1856), which appeared under the pseudonym Jakob Corvinus (Lat. corvus means Rabe, 'crow' or 'raven').

In 1856 Raabe returned to Wolfenbüttel, trying for six years, with little success, to make a living by his pen. To this period belong the novels Ein Frühling (1857), Der heilige Born (1861), and Unsers Herrgotts Kanzlei (1862), the long tale Nach dem großen Kriege (q.v., 1861), and 3 vols. of shorter narratives, Halb Mär, halb mehr (1859), Die Kinder von Finkenrode (q.v., 1859), and Verworrenes Leben (1862). Whilst writing these works Raabe decided to travel, but the war of Italian Independence, which broke out in May 1859, obliged him to alter plans for an Italian journey and he made instead an educational tour of Germany and Austria, visiting Leipzig, Dresden, Prague, and Vienna, passing through the Salzkammergut and the Alps, to Munich and Stuttgart, on to Heidelberg and Cologne, and so home. In 1862 Raabe married, settling in Stuttgart, where he remained until 1870. During these years he achieved his first full success with Der Hungerpastor (q.v., 1864); and he continued his prolific activity with the novels Die Leute aus dem Walde (q.v., 1863), Drei Federn (1865), Abu Telfan (q.v., 1868), and Der Schüdderump (q.v., 1870), and the collections of stories Ferne Stimmen (1865) and Der Regenbogen (1869), which contains the well-known tale Else von der Tanne and also Im Siegeskranze. It was long customary to regard this as Raabe's greatest period and to conceive the three novels Der Hungerpastor, Abu Telfan, and Der Schüdderump as a trilogy conveying a message of sombre pessimism.

In fact the phase of Raabe's greatest originality was about to begin. In 1870 he settled in Brunswick, the capital of his native land, and there he remained for the rest of his life, at first obscure and solitary, but gradually developing into a well-known figure. Raabe abandoned the long book in these years, specializing in closely wrought and intricate shorter novels and collections of stories. Der Dräumling (1872) was followed by the tales Deutscher Mondschein and the novel Christoph Pechlin. Eine Internationale Liebesgeschichte (both 1873). The great works set in with Horacker (q.v., 1876), Wunnigel (q.v., 1879), Alte Nester (q.v., 1880), Das Horn von Wanza (q.v., 1881), Prinzessin Fisch (q.v., 1883), Villa Schönow, Pfisters Mühle (q.v., both 1884), Unruhige Gäste (q.v., 1886), Im alten Eisen (1887), Das Odfeld (q.v., 1889), Stopfkuchen (q.v., 1891), Gutmanns Reisen (1892), Die Akten des Vogelsangs (q.v., 1895), and Hastenbeck (q.v., 1898). An unfinished novel, Altershausen (q.v.) was published posthumously in 1912. Other later works are the novel Deutscher Adel (1880), a collection of stories, Krähenfelder Geschichten (1879), and the short novel Fabian und Sebastian (1882). Raabe's little-known poetry was collected in 1912 (posth.).

The whole of Raabe's work is impregnated with a sense of history, including an awareness of the present as a part of the historical process;

and his immense reading and retentive memory enabled him to bestrew his books with literary allusions. His consciousness of the omnipresence of death and his sensitiveness to suffering are complemented by a rich and compassionate humanity, which manifests itself in all the later works, reaching its highest points perhaps in the portrayal of such figures as Horacker, Magister Buchius (*Das Odfeld*), die Wackerhahnsche (*Hastenbeck*), and, above all, Stopfkuchen.

Sämtliche Werke, historisch-kritische Ausgabe (24 vols.), ed. K. Hoppe *et al.*, appeared in 1951 ff. and *Briefe 1862–1910*, ed. W. Fehse, in 1940.

RABE, JOHANN JAKOB (Strasburg, *c.* 1545–86, Strasburg), a polemical writer in the service of the Counter-Reformation, was the son of a Protestant pastor. In 1565 he was converted to the Roman Catholic faith and was ordained priest in 1571. After some years in Munich he returned to Strasburg. He is the butt of Fischart's early satire, *Nacht Rab oder Nebelkräh* (1570, see FISCHART, J.).

RABENER, GOTTLIEB WILHELM (Wachau nr. Leipzig, 1714–71, Dresden), was educated at St. Afra's, Meißen (see FÜRSTENSCHULEN), and at Leipzig University. Gärtner, Gellert, Cramer, and Giseke (qq.v.) were among his friends. In 1741 he was appointed a tax inspector in Leipzig. He contributed first to *Die Belustigungen des Verstandes und Witzes* (q.v., 1741–5) and then to the *Bremer Beiträge* (q.v., 1745–8), in which his prose satires first appeared. These were collected in 1751 (*Sammlung satyrischer Schriften*, 2 vols. in 1751, 2 further vols. in 1755). In 1753 he was promoted senior fiscal secretary (Obersteuersekretär) at Dresden. His house was destroyed in the Prussian bombardment of 1760 (see SIEBENJÄHRIGER KRIEG), but he obtained compensation and was honoured with the title Steuerrat. Rabener's satires ridicule the follies and vices of the rising middle class.

Sämtliche Schriften (6vols.), ed. C. F. Weiße (q.v.), were published in 1777, and *Ausgewählte Satiren* in 1884, re-ed. H. Kunze, 1968.

Rabenschlacht, Die, an anonymous Middle High German epic poem of nearly 7,000 lines, dealing with Dietrich von Bern (see THEODORIC). It is written in six-line stanzas and dates from the late 13th c. 'Raben', in the title, denotes Ravenna. The starting-point is Etzel's (Attila's) court. Dietrich sets out for Italy to defeat Ermenrich, who had formerly driven him from his possessions. He has with him two children of Attila and his own young brother. While Dietrich defeats Ermenrich in battle at Ravenna, the three boys perish in combat with Ermenrich's supporter Witege. Dietrich is overcome with grief, pursues Witege, and then returns to Etzel's court, where he receives forgiveness for his failure to protect the boys.

Die Rabenschlacht is accompanied in the MSS. by *Dietrichs Flucht* (q.v.).

Rache, Die, a short ballad of ten lines by L. Uhland (q.v.), written and published in 1810. The first line runs 'Der Knecht hat erstochen den edlen Herrn'. The yeoman stabs and despoils his lord and himself perishes because he cannot control the stolen horse and is borne down by the weight of the stolen armour as he falls from the saddle into the Rhine. It is one of Uhland's most concise and powerful poems.

RACHEL, JOACHIM (Lunden, Dithmarschen, 1618–69, Schleswig), satirist, was at school in Hamburg and afterwards studied at Rostock and Dorpat. On his return he became a schoolmaster, first in Heide, then in Norden, and finally in Schleswig. A disciple of Martin Opitz (q.v.), he is the author of a collection of verse satires (*Teutsche satyrische Gedichte*, 1664, revised 1668) in which his strictures are directed principally at women. These are in High German; up to 1659 Rachel wrote in Low German.

RADECKI, SIGISMUND VON, pronounced like the Austrian Radetzky (Riga, 1891–1970, Gladbeck), was by training an engineer, made the acquaintance of K. Kraus (q.v.) in Vienna, moved to Berlin in 1926, and became a Roman Catholic in 1931. Before the 1939–45 War he was a comparatively little-known writer of Novellen (the collections *Der eiserne Schraubendampfer Hurricane*, 1929, and *Nebenbei bemerkt*, 1936) and of essays (*Die Welt in der Tasche*, 1939, and *Alles Mögliche*, 1940). After the war he developed into a widely respected journalistic critic. His collected criticisms and essays include the following volumes: *Was ich sagen wollte* (1946), *Das Schwarze sind die Buchstaben* (1957), *Im Vorbeigehen* (1960), *Ein Zimmer mit Aussicht* (1961), *Gesichtspunkte* (1964), and *Rückblick auf meine Zukunft* (1965).

RADETZKY, JOSEPH, with full title Feldmarschall Joseph Graf Radetzky von Radetz (Trzebnitz, Bohemia, 1766–1858, Milan), was commissioned in the Austrian army in 1787, served as an instructor in Wels, and was Chief of Staff to Fürst K. P. Schwarzenberg (q.v.) at the battle of Leipzig in 1813 (see VÖLKERSCHLACHT). He served in various Austrian garrisons, including Ofen (q.v.), and was general commanding the Austrian forces in Lombardy and Venetia from 1831 onwards. He was a capable administrator and trainer of troops. After revolution broke out in 1848 at Milan and elsewhere, Radetzky put down the insurrection in two pitched battles, at Custozza in 1848 and Novara in 1849 (see REVOLUTIONEN 1848–9). J. Strauß (sen., 1804–49) composed a march in his honour (*Radetzkymarsch*) and Grillparzer (q.v.) a patriotic poem (*Feldmarschall Radetzky*); the victory march is used by J. Roth as the title for his novel *Radetzkymarsch* (q.v.).

Radetzkymarsch, a novel by J. Roth (q.v.), published in 1932. It is an ironical and gloomy portrayal of the decline and fall of the Dual Monarchy. The title, which refers to the most popular Austrian military march named after Field-Marshal J. von Radetzky (q.v.), is intended ironically.

Roth traces the decline through three generations, each composed of only one son. At Solferino (1859) the Slovene Lt. Trotta, a platoon commander, saves the life of the Emperor Franz Joseph (q.v.), then aged 29, and is rewarded with the Maria Theresienorden, the *von* of nobility, and promotion to captain. Irritated by the falsely embellished representation of his deed in a school-book, he resigns his commission, receiving at the same time promotion to major and a barony. His son, whom he forbids to take up a military career, enters the civil service in which he eventually becomes a district governor (Bezirkshauptmann). His son Carl Joseph, the Trotta of the third generation, is destined by his father for the cavalry; the story of his short career occupies the greater part of the novel.

Carl Joseph von Trotta is commissioned in a regiment of Uhlans in Moravia, though he detests horses and has no military ability. His men are Romanians and Ukrainians, for the Viennese War Office dare not have garrisons of local men. Mess life is shown at its worst with drink, cards, dirty stories, and mass visits to the local brothel. Carl Joseph's only friend is the Jewish regimental doctor, Demant. In a mess squabble Demant insults a captain, who retaliates with anti-Semitic abuse. A duel takes place and both men fall. Carl Joseph is involved, and asks for a transfer to the infantry. He is posted to a rifle (Feldjäger) battalion in Ruthenia near the Russian frontier. Mess life is no better, and Carl Joseph finds himself tricked into guaranteeing the gambling losses of a captain. At this time he is ordered to take his platoon out to deal with a political demonstration. Stones are thrown, soldiers fall, and Carl Joseph gives the order to fire. An inquiry which could threaten his career is hushed up because the Emperor is reminded that this officer is a Trotta, grandson of the hero of Solferino. Soon the gambling captain shoots himself, and Carl Joseph is unable to produce the sum he has guaranteed. Once more the Emperor has to come to the rescue of a Trotta. Carl Joseph's disillusionment with his career is further augmented when he finds that some of his brother officers are spies in Russian pay. The last straw for him is the assassination of the Archduke Franz Ferdinand (q.v.), over which some of the officers openly rejoice, calling the victim a swine. He resigns his commission in disgust, but a month later is recalled to the colours at the outbreak of the 1914–18 War. He is killed in September in an act of unostentatious bravery, trying to bring water to his men under fire. His father, whose personal life is now destroyed, dies in 1916, at approximately the same time as Franz Joseph dies at Schönbrunn.

The intense atmosphere of political and social disintegration in this novel is reinforced by recurring graveyard thoughts and meditations, all dwelling vividly on images of corruption. mortality, and decay. A relative of the Trottas of this novel is the central figure in Roth's *Die Kapuzinergruft* (q.v.).

Rädlein, Das, a Middle High German comic poem of erotic character. It is a story of seduction, and the title derives from a circle or wheel which the man draws with soot on the girl's body. The poem was written in the second half of the 13th c. by Johannes von Freiberg, who is not otherwise known.

RAFOLD, HEINRICH, author of *Der Nußberg*, a Middle High German poem, probably of the 13th c., dealing with the infidelity of a knight's wife, who elopes with a Saracen king captured by her husband. The poem, which is no longer complete, is closely related to the medieval Latin poem *Rudolf von Schlüsselberg* (q.v.). Heinrich Rafold was by trade a smith and could neither read nor write.

RAHEL, see VARNHAGEN VON ENSE, R.

Rahmen, a term mainly associated with the Novelle (q.v.). There are two basic forms of frame technique, the cyclic frame (zyklischer Rahmen) and the single frame (Einzelrahmen). The cyclic frame, modelled on the *Decameron* (q.v.) and introduced into German literature by Goethe (*Unterhaltungen deutscher Ausgewanderten*, q.v.), combines a series of stories into a unified entity, although it was used by some writers (e.g. Tieck) merely as a convenient way of presenting a collection of stories. The single frame embraces a story which is self-contained and yet linked with the frame by a narrator. The narrator can function in both first-person and third-person narratives. The frame can introduce the narrator's recollections (Erinnerungsnovelle); alternatively the narrator can function as observer. In either case he gives the narrative perspective. One type of frame technique uses a story which purports to be a historical chronicle (Chroniknovelle), the frame linking different ages.

In the original cyclic frame the stories were told to a group of persons who function as commentators. This technique has also been applied to narratives relating one or two connected stories. The term is also used of narratives which introduce the story and are not resumed at its close.

RAHNER, KARL (Freiburg/Breisgau, 1904–), a leading Roman Catholic theologian who in 1949 became professor of dogmatic theology in Innsbruck, and later in Munich. He was a member of the Papal Commission for the Second Vatican Council. His works, written with lucid elegance, include *Schriften zur Theologie* (9 vols. from 1954), *Kirche und Sakramente* (1961), and *Gnade als Freiheit* (1968). His elder brother Hugo (1900–68), a distinguished church historian and professor at Innsbruck from 1937 to 1966, was an authority on St. Ignatius Loyola. Both brothers were Jesuits.

RAIMUND, FERDINAND, real name Raimann (Vienna, 1790–1836, Pottenstein, Lower Austria), was a prominent exponent of the Volksstück (q.v.). The son of a turner who died when the boy was 14, Raimund was at first apprenticed to a confectioner, but his passion for the theatre led him to go on the stage at 18. His early appearances with small touring companies were failures. In 1814 he secured an engagement at the Theater in der Josefstadt in Vienna, but was unsuccessful in the tragic roles which it was his ambition to play. In his comic parts, however, he immediately won the applause of a large public. In 1817 he transferred to the better-known Theater in der Leopoldstadt, where he was the principal draw for the next decade. He married perforce the actress Luise Gleich, daughter of the comedy-writer J. A. Gleich (q.v.), in 1820. The marriage was a failure and a separation order was made in 1822. For the rest of his life Raimund maintained a liaison with Antonie Wagner, whom he was not free to marry.

During the years at the Theater in der Leopoldstadt Raimund began to write plays, beginning with *Der Barometermacher auf der Zauberinsel* (q.v.), performed in 1823. This magic farce (see VOLKSSTÜCK) was followed by similar, though more ambitious, works, *Der Diamant des Geisterkönigs*, performed in 1824, *Das Mädchen aus der Feenwelt oder Der Bauer als Millionär*, in 1826, *Moisasurs Zauberfluch*, in 1827, *Die gefesselte Phantasie*, and *Der Alpenkönig und der Menschen-*

feind (qq.v.), both performed in 1828. In that year Raimund was made director of the Theater in der Leopoldstadt, for which he wrote *Die unheilbringende Zauberkrone*, 1829, a tragi-comedy which ended his run of popular successes. He resigned as director in 1830 and thereafter won acclaim as a guest in various theatres. His last and most ambitious play, *Der Verschwender* (q.v.), was performed in 1834. Though Raimund was outwardly prosperous, living on the scale of a well-to-do burgher, he was a deeply unhappy man, hypersensitive and quick to suspect persecution. During a stay at his country villa in 1836 he was bitten by a dog and believed himself to have contracted hydrophobia. While on the way to Vienna to consult a doctor he shot himself.

The theatre which Raimund knew and wrote for was the popular theatre of Vienna, with its dialect, its local allusions, obligatory songs, and limited range of emotions. Though fully successful in this unpretentious type of play, he felt impelled to write works of wider purport and deeper feeling. In *Der Alpenkönig und der Menschenfeind* and especially in *Der Verschwender* he revealed new possibilities in the Volksstück, which he did not live to exploit. None of his plays were published in his lifetime. His collected plays were issued in 1837 by J. N. Vogl as *Sämtliche Werke* (4 vols.), the contents of which are: (1) *Der Diamant des Geisterkönigs, Der Alpenkönig und der Menschenfeind*; (2) *Moisasurs Zauberfluch, Das Mädchen aus der Feenwelt*; (3) *Der Barometermacher auf der Zauberinsel, Die gefesselte Phantasie*; and (4) *Die unheilbringende Zauberkrone* and *Der Verschwender*.

The *historisch-kritische Ausgabe* (6 vols.), ed. F. Brukner and E. Castle, appeared in 1924–34, and *Sämtliche Werke*, ed. F. Schreyvogl (q.v.), in 1960.

RAMBACH, FRIEDRICH EBERHARD (Quedlinburg, 1767–1826, Reval), a schoolmaster in Berlin and later professor at Dorpat University, was a prolific writer of medieval adventure and horror stories and plays. He exploited the precocious talent of the schoolboy Tieck (see TIECK, LUDWIG).

RAMBACH, JOHANN JAKOB (Halle/Saale, 1693–1735, Gießen), a theologian, was appointed a professor at Halle University in 1727, and in 1731 in Gießen. A Pietist (see PIETISMUS), he is the author of religious poems and hymns, collected in *Geistliche Poesien* (1720) and *Neueingerichtetes Hessen-Darmstädter Kirchengesangbuch* (1733).

RAMLER, KARL WILHELM (Kolberg, 1725–98, Berlin), after studying at Halle and Berlin universities became a private tutor in 1746, and in

1748 professor of logic at the Berlin military cadet school. A man of amiable character, he was befriended by J. W. L. Gleim (q.v.), and was himself a friend of G. E. Lessing, E. von Kleist, and F. Nicolai (qq.v.). A co-director at the royal theatre in Berlin (1786–96), he was favourably regarded by the court and received a crown pension under Friedrich Wilhelm II (q.v.).

Ramler was a skilful, competent poet, austere and precise in his handling of language and metre. He translated classical poets, including Horace, Martial, Catullus, and Anacreon, and in his occasional poetry showed an inclination for classical forms. His poetry appeared as *Oden* (1767), *Lyrische Gedichte* (1772), and *Kriegslieder für Josephs und Friedrichs Heere* (1778). Many of his odes were published singly. Ramler collaborated with Lessing in an edition of Logau's *Sinngedichte* (1759, see LOGAU, F. VON), and turned the prose *Idyllen* of S. Geßner (q.v.) into hexameters. Posthumous publications include *Horazens Oden* (2 vols., 1800), *Anakreons auserlesene Oden* (1801), and his own poetry, *Poetische Werke* (2 vols.), ed. L. F. G. von Göckingk. Ramler's collection of German songs, *Lieder der Deutschen* (4 vols.), of 1766 was reprinted, ed. A. Anger, in 1965.

Rangierbahnhof, Der, a novel by Helene Böhlau (q.v.), published in 1895. It is a story of a would-be artist, who marries Olly, a girl with artistic talent. She in turn is drawn to a more gifted painter, Köppert. She suffers from tuberculosis, but dies of an overdose of sleeping-draught. The book begins near a railway marshalling yard, the Rangierbahnhof, which is used as a symbol of dispersal and rearrangement.

RANK, JOSEPH (Friedrichsthal, Bohemian Forest, 1816–96, Vienna), became a private tutor in Vienna, and in 1848 was elected a member of the Frankfurt Parliament (see FRANKFURTER NATIONALVERSAMMLUNG). He worked as a journalist in Frankfurt and Weimar before being appointed secretary to the Vienna Court theatres (1865–79). He is the author of fiction, including the Bohemian stories *Aus dem Böhmerwalde* (1842) and *Florian* (1853) and the novel *Achtspännig* (1857). *Ausgewählte Werke* (11 vols.) were published in 1859–60.

RANKE, LEOPOLD VON (Wiehe, 1795–1886, Berlin), a scholar of the well-known school Schulpforta (see FÜRSTENSCHULEN), studied at Leipzig University, and in 1818 became a history master in Frankfurt/Oder. In 1825 he was appointed to a chair of history in Berlin, and began a career which led to his ennoblement (1865)

and to a prominence in his field which established him as the G.O.M. of German historians. Ranke was the founder in Germany of the school of objective historians who saw their task to be the investigation of the sources and the establishment of facts, rather than the furtherance of political aims. Ranke's speciality was the period of the 16th c. and 17th c.

Ranke's principal works are *Geschichte der romanischen und germanischen Völker von 1494 bis 1535* (1824), *Fürsten und Völker von Südeuropa im 16. und 17. Jahrhundert* (1827), *Die römischen Päpste, ihre Kirche und ihr Staat im 16. und 17. Jahrhundert* (3 vols., 1834–6), *Deutsche Geschichte im Zeitalter der Reformation* (6 vols., 1839–47), *Neun Bücher preußischer Geschichte* (3 vols., 1847–8), expanded to *Zwölf Bücher preußischer Geschichte* (5 vols., 1874), *Französische Geschichte, vornehmlich im 16. und 17. Jahrhundert* (5 vols., 1852–61), *Englische Geschichte vornehmlich im 16. und 17. Jahrhundert* (7 vols., 1859–68), *Geschichte Wallensteins* (1869), *Die Mächte und der Fürstenbund* (2 vols., 1871–2), and *Hardenberg und die Geschichte des preußischen Staates* (5 vols., 1879–81). In the 1870s, when Ranke's eyesight failed, he dictated a monumental history of the world, which was still unfinished at his death (*Weltgeschichte*, 16 vols., 1881–8). His own work on this only reached Otto the Great, but his assistant, R. Dove, used Ranke's notes to continue it to 1450.

Sämtliche Werke (54 vols.) appeared in 1867–90, and a selection (12 vols.), ed. W. Andreas, in 1957.

Rapallo, Vertrag von, treaty signed on 16 April 1922 between the Weimar Republic (q.v.) and Soviet Russia, resuming normal diplomatic relations, renouncing on both sides claims for compensation and reparation, and favouring closer relations. It was the first German independent act of foreign policy after 1918 and aroused misgiving and distrust in the west, most of all in France.

RASPE, RUDOLF ERICH (Hanover, 1737–94, Muckross, Ireland), was appointed librarian and professor at Kassel in 1767, whence he fled to England in 1775 to escape a criminal charge. He appears to have become a mining engineer in Great Britain. He published the first collection of narratives solely devoted to the tall stories of Baron von Münchhausen (see MÜNCHHAUSEN, K. F. H.), under the title *Baron Münchhausen's narrative of his marvellous travels and campaigns in Russia* (1785). This is partly translated from the *Vademecum für lustige Leute*, but contains additional material.

Rastbüchlein, a collection of Schwänke (see SCHWANK) by Michael Lindener (q.v.).

Rastlose Liebe, a poem written by Goethe in Ilmenau in May 1776. It reflects the restless energy accompanying his love for Charlotte von Stein (q.v.). It was published in 1789 in *Goethes Schriften.*

RATHENAU, WALTHER (Berlin, 1867–1922, Berlin), a wealthy industrialist, had charge of the department for military supplies in the 1914–18 War. After the war he advocated a policy of compliance with Allied demands for reparations. This, together with his Jewish descent, made him unpopular in nationalistic circles, and he was assassinated by two ex-officers. The nationalist writer E. von Salomon (q.v.) was also implicated in this murder. Rathenau is said to have been the model for the character Arnheim in Musil's novel *Der Mann ohne Eigenschaften* (q.v.). *Schriften und Reden,* ed. with a postscript by H. W. Richter (q.v.), appeared in 1964.

Rathenower, Die, colloquial designation of the red Zieten (q.v.) hussars—Husaren-Regiment von Zieten (Brandenburgisches) Nr. 3—garrisoned in Rathenow.

Ratisbon, see REGENSBURG.

Rat Krespel, a story by E. T. A. Hoffmann (q.v.), first published in 1816 in Fouqué's *Frauentaschenbuch* and then included in Vol. 1 of Hoffmann's *Die Serapionsbrüder* (q.v., 1818). Some editors prefer the title *Antonie,* by which Hoffmann refers to the story in a letter. Krespel, a man of extraordinary eccentricity, has a daughter with a beautiful voice. She is ill, and Krespel, in the belief that if she sings she will die, endeavours to isolate her from music. But a visitor persuades her and she dies in full-throated song. Antonie is a projection of Hoffmann's love for Julia Marc (q.v.), and Rat Krespel is a caricatured self-portrait. This story forms the basis of the third act of Offenbach's opera *Les Contes d'Hoffmann.*

Ratperts Lobgesang auf den Heiligen Gallus, designation of an Old High German poem in praise of St. Gall, written by the monk Ratpert who died at St. Gall abbey *c.* 890. The original poem is lost, but a Latin translation by Ekkehard IV of St. Gall is extant. This version, which was made approximately a century after Ratpert's death, is composed of 17 five-line stanzas of rhyming verse, recounting the life and death of the saint and the foundation of the abbey by him.

Ratschläge für Liebende, see HEIMLICHE BOTE, DER.

Ratsmädelgeschichten, a volume of seven stories by Helene Böhlau (q.v.), published in 1888. They are set in Weimar in 1805–15. The 'Ratsmädel' are two young sisters, who in after years were grandmother and great-aunt of the authoress. The stories of their tricks, truancies, and pranks were written up from anecdotes told by Helene Böhlau's grandmother. The seventh story, *Das Gomelchen,* is a portrait of the grandmother in old age.

Ratten, Die, a five-act play (Berliner Tragi-komödie) by G. Hauptmann (q.v.), first performed in Berlin, at the Lessingtheater, in January 1911, and published in the same year. Hauptmann's last success in the Naturalistic manner, it portrays a middle-aged couple named John, who live in rat-ridden former barracks, inhabited also by the dregs of society (the symbolical Ratten of the title). Frau John has an invincible longing for a baby, adopts one, and persuades her husband that she has given birth to it. But Pieperkarcka, the child's mother, makes trouble, John becomes suspicious, and Frau John's brother murders the troublesome Pieperkarcka. Exposed and despairing, Frau John takes her life. Parallel with this runs a study of an artistic character, the former theatre director Hassenreuter, who keeps his costumes in the same building and is marginally involved in the action. Said to be based on an actor named Alexander Heßler, Hassenreuter has also traits of Hauptmann himself.

RAU, HERIBERT (Frankfurt/Main, 1813–76, Offenbach), a clergyman, was the author of a series of novels in which poets and composers are the principal figures. They are mostly described by him as kulturhistorische Romane: *Mozart, ein Künstlerleben* (3 vols., 1858), *Alexander von Humboldt* (7 vols., 1860), *Jean Paul* (4 vols., 1861), *Hölderlin* (2 vols., 1862), *Shakespeare* (4 vols.), and *Karl Maria von Weber* (4 vols., both 1864). *Theodor Körner* (2 vols., 1863) is designated a vaterländischer Roman, and *Beethoven* (4 vols., 1859) a historischer Roman.

RAU, LUISE, for four years (1829–33) the betrothed of E. Mörike (q.v.), was the daughter of the pastor of Plattenhardt, Württemberg, whose death caused the vacancy which Mörike was appointed to fill in May 1829. She was three years younger than Mörike.

Räuber, Die, a play in five acts by Schiller, written between 1777 and 1780, published in 1781, and first performed at Mannheim on 13 January 1782. Schiller called it a Schauspiel, though the stage edition designated it a Trauerspiel. Its source was a story by C. F. D.

Schubart (q.v.), published in 1777. The play has two parallel actions which intersect towards the close.

A Franconian nobleman, Herr von Moor, has two sons; Karl, the elder, is at university where he lives riotously; Franz, the younger brother, a calculating materialist, uses forgery to persuade the reluctant father to disinherit Karl, and then, by playing on the old man's emotions, reduces him to a swoon, which Franz pretends is death. Franz tries unsuccessfully to win the affections of his cousin Amalia, who remains unswervingly devoted to Karl. Finding himself rejected by his father, Karl reacts violently against a civilization in which such injustice is possible, and accepts command of a robber band, which terrorizes the region of the Bohemian Forest. The atrocities which accompany these activities appal Karl, who nevertheless pledges himself to adhere to the gang. Anxious to obtain news of Amalia, he visits his home in disguise and discovers Franz's iniquity. The fullest revelation is still to come. His father is discovered in a dungeon in the park, naked, and starving. Karl dispatches a force to seize Franz, who avoids capture by suicide. Karl thinks for a moment of a happy life with Amalia, but the gang holds him to his pledge and the spectre of his offences bars return. He recognizes the monstrosity of his revolt and makes amends by deliberate surrender to the law.

The play is written in tense, impassioned, often violent, prose. It combines all the fury of the Sturm und Drang (q.v.), of which it is a late manifestation, with the affirmation of a stable and pre-eminent world order. In its day its success was immense. Later productions include two by E. Piscator (q.v.), in 1926 and in 1957 (the latter in Mannheim).

Raumbühne, the *théâtre en ronde*, is based on the anti-illusionary stage artistry of Gordon Craig (1872–1966) and was developed in Germany primarily by E. Piscator, C. Neher, and T. Otto (qq.v.).

RAUPACH, ERNST (Straupitz, Silesia, 1784–1852, Berlin), studied at Halle University and thereafter spent seventeen years in Russia (1805–22), first as a private tutor, and from 1816 to 1822 as a professor at a teacher training establishment in St. Petersburg. He next visited Italy, returning to Germany in 1824. He settled in Berlin and developed into the popular dramatist of the day, writing, it is said, 117 plays. Among those which deserve mention are his first play *Die Fürsten Chawansky* (1810), *Die Leibeigenen* (1826), which attacked serfdom in Russia, *Der Nibelungenhort* (1834), a five-act tragedy dealing with the Nibelungen story (first performed in

1828), and *Der Müller und sein Kind* (1835), a sentimental play (Volksstück) which survived on the stage in South Germany well into the 20th c. His cycle *Die Hohenstaufen* (1837) comprises sixteen plays.

Raupach, whose serious verse plays were influenced by Schiller, had all the gifts which a popular dramatist of the day required, great technical skill, an eye for dramatic situations, a facile rhetoric, and a feeling for the pulse of the public. His comedies were collected as *Dramatische Werke komischer Gattung* (4 vols., 1829–35) and his tragedies in *Dramatische Werke ernster Gattung* (16 vols., 1835–43).

Realismus, denotes in general the presentation of scenes, persons, actions, or speech in a recognizably lifelike and apparently undistorted manner. The difficulty of arriving at a precise definition in German literature has led to varied interpretation as bürgerlicher Realismus, sozialer Realismus, etc. Although works which can be termed realistic have been written from the Middle Ages onwards, the 19th c. has generally been recognized as the typical age of Realism, partly in contrast to German Idealism. See STURM UND DRANG, BIEDERMEIER, JUNGES DEUTSCHLAND, POETISCHER REALISMUS, and NATURALISMUS. For an approach to realism in aesthetics Schiller's essay *Über naive und sentimentalische Dichtung* (q.v.) provides an example.

For the official form of realism in the German Democratic Republic see SOZIALISTISCHER REALISMUS.

Reallexikon, a dictionary or encyclopedia confined to a particular subject and dealing with concepts, categories, historical phases, and general phenomena. The best-known example in German literature is the *Reallexikon der deutschen Literatur* by P. Merker and W. Stammler (3 vols., 1925), the second edition of which, revised by W. Kohlschmidt and W. Mohr (vol. 1, 1958, vol. 2, 1965, vol. 3, 1977), was completed by K. Kanzog and A. Masser (vol. 4, 1984).

Realpolitik, a term chiefly applied to foreign policy and indicating an alert opportunism. The word was coined by the political writer A. L. von Rochau (1810–73) in the title and text of his *Grundsätze der Realpolitik* (2 vols., 1853–69). Realpolitik was held in high esteem in Germany in the Bismarckian, Wilhelmine, and National Socialist eras.

REBHUN, PAUL (Waidhofen/Ybbs, Austria, *c.* 1505–46, Saxony, either Ölsnitz or Voigtsberg), a friend of Melanchthon (q.v.), studied in Wittenberg under Luther (q.v.), and became a schoolmaster at various places in Saxony, including Zwickau, Plauen, and Ölsnitz. He wrote biblical plays which are modelled on classical plays with chorus, and include *Ein Geystlich spiel von der Gottfürchtigen und Keuschen Frawen Susannen* (1536) and *Ein Hochzeitsspil auff die Hochzeit zu Cana Galileae* (1538). The edition of his plays by H. Palm (*Dramen*, 1859) was reprinted in 1969.

REBMANN, JOHANN RUDOLF (Berne, 1566–1605, Muri nr. Berne), a Protestant pastor in various towns and villages, including Thun and Muri, is the author of a didactic poem on Switzerland and Swiss affairs (*Poetisches Gastmahl und Gespräch zweier Bergen in der Wissenschaft des Niesen und Stockhorn*, 1606). As the title suggests, it is in the form of a dialogue between the two mountains and is allegedly in sonnets, actually in groups of seven pairs of couplets, though this grouping is not preserved throughout. *Das poetische Gastmahl* was a success in its day, and in 1620 Johann's son Valentin republished it with substantial additions.

REBORCH, a monk of Bordesholm Abbey who was the author, in 1475, of the *Bordesholmer Marienklage* (q.v.).

Rechte des Herzens, Die, a five-act tragedy (Trauerspiel) by O. Ludwig (q.v.), to which he gave the alternative title *Paul und Eugenie*. It is set in the early 19th c. The Fürst plans to marry his daughter Eugenie to the heir to another principality. But she has contracted a marriage to an exiled Polish count. The Fürst insults the count, who with his bride takes poison. The play is full of sensational scenes and improbable language.

RECKE, ELISA, GRÄFIN VON DER (Schloß Schönburg, Courland, 1756–1831, Dresden), originally Komtesse von Medem, was married at 15 to Baron von der Recke. The marriage was dissolved in 1777. She published a volume of hymns (*Geistliche Lieder*, 1780), but she became especially well known for her part in the exposure of Cagliostro (q.v., in *Nachricht von des berühmten Cagliostro Aufenthalt in Mitau im Jahr 1779*, 1787). She also published poems (*Gedichte*, 1806) and a diary (*Tagebuch einer Reise durch einen Teil Deutschlands und durch Italien in den Jahren 1804 bis 1806*, 4 vols., 1815–17). From 1804 onwards she lived with C. A. Tiedge (q.v.), who accompanied her on her Italian journey. From 1819 they lived in Dresden.

Rede des toten Christus vom Weltgebäude herab, daß kein Gott sei, title given to a passage of prose written by JEAN PAUL (q.v.) in the form of a dream, in which the dead rise to hear Christ's discovery that there is no God. It ends with an awakening which implies a return to belief. The *Rede*, which is appended as *1. Blumenstück* to his novel *Blumen-, Frucht- und Dornenstücke* (q.v., otherwise known as *Siebenkäs*, 1796–7), is a development of an earlier dream (of 1789) in which the sermon of godlessness is preached by Shakespeare.

Redentiner Osterspiel, designation of an Easter play in Middle Low German written at Redentin Abbey, Mecklenburg, in 1463 by a monk, Peter Kalff.

Reden über die Religion, a work of theological apologetics by F. D. E. Schleiermacher (q.v.).

Rede vom Glauben, see HARTMANN, DER ARME.

Rede von den fünfzehn Graden, a Middle High German religious poem written in the Rhineland, setting out the fifteen stages of the soul on its mystical progress to God. It belongs to the late 13th c., and some have thought that it was written by the author of *Die Lilie* (q.v.). The title is also given to a prose tract, which may have had the same author.

Rede zum Shakespeare-Tag, see ZUM SCHÄKESPEARS TAG.

REDING, JOSEF (Castrop-Rauxel, 1929–), after service during the 1939–45 War, took up university studies, which he continued in the U.S.A. where he graduated. In 1959 he was a visiting lecturer in New Orleans. His background differs greatly from that of Max von der Grün (q.v.) of the Gruppe 61 (q.v.). His engagement with civil rights and the Third World, based on personal experience and study, influenced his writing, in which he favoured the short story but also reportage (*Menschen im Ruhrgebiet*, 1974), diary (*Wir lassen ihre Wunden offen*, 1965), and satire (*Josef Redings Erfindungen für die Regierung*, 1962). In the 1950s and 1960s he wrote a number of radio plays (including *Nur ein Stück Seife*, 1956, *Das Amen der Partisanen*, 1966, and *Vereinzelt Störungen*, 1967). A selection of his short stories, diary notes, and essays from two decades is entitled *Mühsam stirbt der Schnee* (1980). Later publications include *Gold, Raureif und Möhren, Sprengt den Eisberg*, and *Die Stunde dazwischen* (all 1981), *Nennt sie beim Namen* (1982), and *Friedenstage sind gezählt* (1983).

REDWITZ, OSKAR, FREIHERR VON (Lichtenau, Franconia, 1823–91, St. Gilgenberg nr. Bayreuth), spent his early years in the Rhineland, studied law at Erlangen and Munich universities, and began to practise as a lawyer at Speyer. His sentimental narrative poem *Amaranth* (1849) was for a time widely read, but its overt Catholicism exposed it to rough handling by the predominantly Protestant literary historians of the 19th c. After its initial success he gave up the law and lived as a man of means, first near Kaiserslautern, then in Munich (where he was a Liberal Deputy from 1858 to 1863), and finally in Merano. He wrote the sentimental tragedy *Siglinde* (1853) and the dramas *Thomas Morus* (1856, Sir Thomas More), *Philippine Welser* (1859, see WELSER, PHILIPPINE), *Der Zunftmeister von Nürnberg* (1860), and *Der Doge von Venedig* (1863). Of these *Philippine Welser* was a considerable success on the stage. He also wrote *Hermann Stark, deutsches Leben* (3 vols., 1869), a novel of family life, and the epic *Odilo* (1878). *Das Lied vom neuen deutschen Reich* (1871), a cycle of 450 sonnets, owed its fleeting success to the mood of nationalistic euphoria after the war of 1870–1 (see DEUTSCH-FRANZÖSISCHER KRIEG).

Reformation, Die, the 16th-c. Reformation which affected the whole of Europe, emanated from Germany. Attempts at reforms within the Roman Church had already been made in the preceding centuries, and the age which became known as the Reformation was one of revolt as well as reform. It opened when Martin Luther (q.v.) publicly protested against the sale of indulgences (see ABLASSKRAM) by the Dominican monk Tetzel (c. 1465–1519), who was active near Wittenberg after having been barred from Saxon territory by Friedrich der Weise (q.v.). Luther expressed his protest in a manner traditionally adopted to open a dispute: on 31 October 1517 he pinned 95 theses to the door of the castle church. The conflict became inevitable when Luther refused to revoke his views at the Diet at Worms in January 1521 (see WORMSER REICHSTAGE). He had indeed, against his original intention, sparked off a radical reappraisal of the Church and the word of the Bible, which resulted in the division of Christendom. The close of the age is marked by the peace of Augsburg (1555, see AUGSBURGER RELIGIONSFRIEDE), which in turn opened the age of the Counter-Reformation (see GEGENREFORMATION) culminating in the Thirty Years War (see DREISSIGJÄHRIGER KRIEG).

The Reformation marks the end of the Middle Ages and the beginning of modern history. It may be divided into three stages, the first leading to the Peasants' War (1524–5, see BAUERNKRIEG), the second up to the peace of Nürnberg (see NÜRNBERGER RELIGIONSFRIEDE) in 1532, and the last up to the peace of Augsburg, which Luther did not live to see. He had in any case lost control over the course of events. By this time, too, the Emperor Karl V (q.v.), whose long reign was marked by largely frustrated efforts to check the Reformation, had retired into a Spanish monastery.

The non-Catholic states (and individuals) became known as Protestants (see PROTESTANTISMUS) after those secular heads who, at the Diet of Speyer in 1529 (see SPEYER, REICHSTAG ZU), protested against their dependence on Rome; they proclaimed their confession at the Diet at Augsburg (1530, see AUGSBURG, REICHSTAG ZU), and in 1531 formed the League of Schmalkalden (see SCHMALKALDISCHER BUND) in defence against Karl V. The Reformation established Luther's own Evangelical Church and paved the way for the Reformed Church, which was initiated by the German-speaking Swiss Zwingli (q.v.) and the French-speaking Swiss Calvin. The Reformed Church (see REFORMIERTE KIRCHE) was, however, not officially recognized in the empire until the peace of Westphalia (1648, see WESTFÄLISCHER FRIEDE).

This age of 'reform' would have been stillborn, and Luther might well have been burnt at the stake as a heretic like Jan Hus (q.v.), had it not drawn German rulers into action, who from the start were alert to the political implications of Luther's revolt of conscience against the Church of Rome. It is only since the end of the 1914–18 War that the Roman Catholic Church has recognized the Protestant Church by no longer viewing it as heretical, and since the second Vatican Council (1962–5) it has begun to reconsider Protestant theology in order to discover common ground.

The dark chapter in German history which resulted from the Reformation in the following century makes readily understandable the yearnings of some later spirits for the original unity of the medieval Church (e.g. Novalis's essay *Die Christenheit oder Europa*, q.v.), but in fact, for better or for worse, this was an age of change and of scientific, technical, and cultural progress and emancipation. The printing machine (see GUTENBERG) became a decisive feature of modern and popular communication and propaganda through pamphlets and broadsheets (though books remained for a long time to come the privilege of the wealthy). The Reformation prepared the ground for modern philosophy, for Rationalism and Idealism, and even promoted Erasmus's humanism (see HUMANISMUS) in spite of the dispute between Luther and Erasmus (q.v.) in 1524–5. Since religion determined trends in art and culture the Reformation inspired new styles in painting (notably Dürer, q.v.) and music. It has been said that Luther might in another time and position have embodied his spiritual vision in an epic instead of a new Church. No comment could pay greater tribute to Luther's poetic gifts, and it may be added that he did indeed create the epic in his translation of the Bible (see BIBLE, TRANSLATIONS OF), which sums up the maker of the Reformation as the maker of the modern German language.

Nürnberg (see NÜRNBERG, Dürer's birthplace) was the first important city to accept the new teaching. Others followed: Ulm, Nördlingen, Strasburg, Magdeburg, Hamburg, Bremen, and other north German cities, as well as Königsberg in East Prussia; among the first principalities were the members of the League of Schmalkalden, Württemberg, the Palatinate, Brandenburg, and Hesse.

Reformatio Sigismundi, see SIGISMUND, KAISER.

Reformierte Kirche, one of the two principal Protestant Churches in Germany, the other being the Lutheran Church (see PROTESTANTISMUS). The division originated in the dispute on the nature of the Eucharist (Abendmahlstreit) in the Marburger Religionsgespräch of 1529 and represented sharp dissent between Luther and Zwingli (qq.v.). From 1559 the Reformierte Kirche was in the main the work of J. Calvin. It remains the principal Protestant Church in Switzerland and is strongly represented in Germany, where it was officially recognized in the Treaty of Westphalia (1648, see WESTFÄLISCHER FRIEDE). It corresponds to the Presbyterian Church of Scotland and to the United Reformed Church in England. In Prussia the two Churches were arbitrarily united as the Evangelische Union in 1817 by decree of King Friedrich Wilhelm III (q.v.). This act was a source of controversy and dissent lasting for some decades. See also CALVINISM.

REGENBOGE, DER, an itinerant Middle High German poet of the late 13th c. Regenboge, a Swabian, perhaps from Ulm, was by trade a blacksmith. In later years a tradition arose that his first name was Barthel. He participated in a famous poetic contest with HEINRICH Frauenlob (q.v.). Regenboge is the recorded author of a number of Sprüche (see (SPRUCH), political and social, of verse riddles, and of a remarkable dialogue with Death contained in a cycle of five poems. The poetic 'ich' struggles with Death, to which he succumbs. He then faces the final ordeal of judgement, praying to the Virgin Mary for her intercession.

Regensburg, a German city situated in Bavaria at the confluence of the Danube and the River Regen. Pre-Roman in origin, it was occupied by the Romans c. A.D. 50, and by A.D. 179 was the headquarters of a legion. The Latin name Ratisbona survives in English Ratisbon. The see of Regensburg was instituted by St. BONIFACE (q.v.) in 739. The city emancipated itself from episcopal and Bavarian suzerainty, becoming a Free City (see FREIE STADT) by charters of 1207 and 1230. In 1663 it became the seat of the Perpetual Diet or Immerwährender Reichstag (see REICHSTAG), which was dissolved in 1803. In that year the city became the capital of the newly formed Principality (Fürstentum) of Regensburg, and in 1810 it passed with its territories definitively to Bavaria.

The site of the Roman *castrum* is still indicated by the layout of the principal streets. The Gothic cathedral (c. 1250–1618, with spires of 1859–69), the 14th-c. gatehouse (Brückentor), and the medieval Danube bridge (Steinerne Brücke) provide the characteristic view of Regensburg, best seen from the left bank of the river, but the city has a wealth of medieval churches and old buildings, including the ancient and important Benedictine Abbey of St. Emmeram, secularized in 1803 (Reichsdeputationshauptschluß, see REVOLUTIONSKRIEGE), the buildings of which were acquired by the family of Thurn und Taxis (q.v.).

REGENSBURG, BURGGRAF VON, a nobleman of the family of Rietenburg, was the author of Minnelieder, of which four stanzas are preserved. He is believed to have died soon after 1185. The Burggraf von Rietenburg (q.v.), also a poet, was probably his younger brother. Regensburg's poems are early examples of lyrics in which love is not yet a courtly ritual and is capable of fulfilment. In one of his stanzas, the woman, contrary to all courtly procedure, serves the man.

REGENSBURG, KONRAD VON, see KONRAD, PFAFFE.

REGER, ERIK, pseudonym of Heinrich Dannenberger (Bendorf/Rhine, 1893–1954, Vienna), was taken prisoner in the 1914–18 War and afterwards worked for Krupp at Essen. He emigrated in 1933, but returned to Germany in 1936. Reger is the author of several novels, of which the best are the first two, *Union der festen Hand* (1931), directed against high-powered industry as he had experienced it at Krupp's, and *Das wachsame Hähnchen* (1932), an attack on the bourgeoisie. Both portray the oppressed with Naturalistic precision, resorting to caricature for the middle and upper classes. After his return to Germany, Reger wrote novels which were acceptable to the National Socialist regime, to which he remained opposed. His essays *Vom künftigen Deutschland* and *Zwei Jahre nach Hitler* appeared in 1947.

REGIOMONTANUS, real name Johannes Müller (Königsberg nr. Haßfurt, Franconia, 1436–76, Rome), a mathematician and astronomer, was a priest who studied in Vienna and in Rome, acquiring in addition to mathematical skills a knowledge of Greek, an unusual accomplishment at that time. In 1471, with the support of a patrician, Bernhard Walther (1430–1504), he set up an observatory and a printing press in Nürnberg. He was consecrated bishop of Regensburg in 1475 and died the following year in Rome, whither he had been summoned to assist in the reform of the calendar. His writings include works on this subject, on mathematics, and on astronomy, all of which are in Latin, though his calendar also appeared in a German version (*Der deutsche Kalender, c.* 1474, ed. E. Zinner, 1937). His customary name derives from his birthplace.

REGNART, JAKOB (Douai, *c.* 1540–99, Prague), a singer in the imperial choir in Vienna and Prague, introduced part-songs in Italian style, composing the music and writing German words. His *Kurtzweilige Teutsche Lieder zu dreyen Stimmen* appeared in 3 vols., 1576–9.

Regulatoren in Arkansas, Die, a novel by F. Gerstäcker (q.v.), published in 3 vols. in 1845. It is a story of the Wild West by a man who knew it. The 'regulators' are a self-appointed group of vigilantes intent on stamping out horse rustling, which the law is powerless to stop. Of the principal horse-stealers, Johnson, Cotton, and Rowson, the last-named, outwardly a minister of religion, is the most dangerous. Rowson is engaged to Marion Harper, who knows only his apparently devout side. Horses are stolen by the gang, and Rowson murders Alipaha, the squaw of an Indian who assists the 'regulators'. The thieves are eventually tracked down and besieged in a farmhouse, where, however, they hold Marion and another girl, Ellen, as hostages. With the help of the Indian, an entry is forced, the girls freed, and the horse thieves seized, except for Cotton, who disappears and is presumed drowned. Rough justice is executed, Johnson being hanged and Rowson handed over to the Indian, who burns him alive in revenge for the murder of Alipaha. Marion willingly marries Brown, a young man who is one of the 'regulators'.

REHFISCH, HANS JOSÉ (Berlin, 1891–1960, Schuls, Switzerland), a lawyer, emigrated in 1936 to the U.S.A. A successful dramatist, originally with leanings to Expressionism, he revised some of his early plays after his return to Germany in 1950 when he published the novels *Die Hexen von Paris* (1951), *Lysistratas Hochzeit* (1959), and a narrative version of the play *Die Affäre Dreyfus* (1950) of 1929, which was written in collaboration with W. Herzog (1884–1960). His other plays include the tragedy *Der Chauffeur Martin* (1920), the comedy *Wer weint um Juckenack?* (1924, revised 1951), and *Dr. Semmelweis* (1934, revised and retitled *Der Dämon*, 1950); *Oberst Chabert* (1955) is an adaptation of Balzac's novel *Le Colonel Chabert. Ausgewählte Werke* (4 vols.) were published under the auspices of the German Academy in 1967.

REHFUES, PHILIPP JOSEPH VON (Tübingen, 1779–1843, Bonn), held important administrative offices under several sovereigns, residing at different times in Naples, Munich, Stuttgart, Koblenz, and Bonn, where he held from 1819 to 1842 the office of curator of the university, and in 1826 was ennobled. He is the author of a number of works on travel in Italy (*Gemälde von Neapel*, 3 vols., 1808; *Briefe aus Italien*, 4 vols., 1809–10) and Spain (*Spanien nach eigner Ansicht*, 4 vols., 1813) and of the patriotic *Reden an das deutsche Volk* (2 vols., 1814). In the 1830s he published historical novels modelled on Scott (*Scipio Cicala*, 4 vols., 1832; *Die Belagerung des Castells von Gozzo*, 2 vols., 1834).

REICHARDT, JOHANN FRIEDRICH (Königsberg, 1752–1814, Giebichenstein, Halle), was appointed master of the royal music to Friedrich II (q.v.) of Prussia in 1775, but was dismissed by his successor Friedrich Wilhelm II (q.v.) in 1794. He spent the rest of his life on his estate near Halle. He was on friendly terms with Goethe except in the years 1795 to 1801, when his pro-revolutionary opinions (which were also responsible for his difficulties at the Prussian court) led to an estrangement. He set to music 128 songs by Goethe as well as the Singspiele (see SINGSPIEL) *Claudine von Villa Bella, Erwin und Elmire,* and *Jery und Bätely* (qq.v.).

Reichardt also composed operas and orchestral and instrumental works. He was an active writer on music and is the author of a book on the young Handel, *G. F. Händels Jugend* (1785, reissued in 1959 in the *Händel-Jahrbuch*, vol. 5), and of readable works of criticism and comment including *Briefe eines aufmerksamen Reisenden, die Musik betreffend* (1774–6), *Vertraute Briefe aus Paris, geschrieben 1802 und 1803* (1804–5), and *Vertraute Briefe, geschrieben auf einer Reise nach Wien und den Österreichischen Staaten zu Ende des Jahres 1808 und zu Anfang 1809* (2 vols., 1810; ed. G. Gugitz, 1915). He edited at different times several musical journals, one of which included his autobiography; this was reprinted, ed. W. Zentner, in 1940.

Reichenau, a former abbey situated on the island of the same name in the western part of Lake Constance. Founded in 724, it was a great centre of learning and manuscript illumination in the 9th c., 10th c., and 11th c.; its most notable scholar was Abbot WALAHFRID Strabo (q.v.). Reichenau Abbey plays a part in J. V. von Scheffel's historical novel *Ekkehard* (q.v.).

Reichenau, Chronik von, see CHRONIK VON REICHENAU.

Reichenbach, Convention of, was concluded in July 1790 between Austria and Prussia. Prussia guaranteed the Austrian Netherlands and shelved its hopes of gaining Danzig and Thorn; Austria agreed to make peace with Turkey guaranteeing the *status quo*. Leopold II (q.v.) of Austria checked Prussian policy (see FRIEDRICH WILHELM II) in Poland which in the context of the 'Eastern Question', involving Russian and Austrian interests in Turkey, might have developed into a European war.

REICHENTAL, see ULRICH VON RICHENTAL.

Reichsdeputationshauptschluß, an edict issued in 1803 reorganizing German territory after the Revolutionary Wars (see REVOLUTIONSKRIEGE).

Reichskammergericht, the supreme court of the Holy Roman Empire (see DEUTSCHES REICH, ALTES). Instituted at the Imperial Diet at Worms in 1495, the court was concerned with the maintenance of the peace of the Empire, being empowered to pronounce sentence of outlawry (Reichsacht) on offenders; it heard suits involving parties who owed allegiance directly to the Emperor (Reichsunmittelbare); and it was the ultimate court of appeal for the Empire. The Reichskammergericht had its seat first at Frankfurt; in 1527 it was established in Speyer, and in 1693 was removed to Wetzlar. It was abolished in 1806 in consequence of the dissolution of the Holy Roman Empire. The building in which it functioned at Wetzlar still exists. In 1772 Goethe was sent to Wetzlar by his father to study the procedure of the court.

Reichsstadt, a city owing allegiance directly to the Emperor of the Holy Roman Empire (see DEUTSCHES REICH, ALTES) and not subject to any intermediate sovereign prince. The number of such cities was considerable and included cities of importance and tiny market towns.

Augsburg, Nürnberg, and Ulm, Dinkelsbühl, Memmingen, and Wetzlar serve as examples. The Reichsstadt was in a less privileged position from a military and financial standpoint than the Freie Stadt (q.v.) or Freie Reichsstadt. Representatives of the free cities attended imperial diets occasionally before 1489, and invariably after that year. The significance of the term Reichsstadt ceased with the dissolution of the Empire in 1806.

REICHSTADT, HERZOG VON (Paris, 1811–32, Schönbrunn), title conferred on Napoleon's only son, François Joseph Charles, after his father's abdication (see NAPOLEONIC WARS). His mother was the Empress Marie-Louise, daughter of the Emperor Franz I of Austria (see FRANZ II), and he was brought up at the Austrian court and commissioned in the Austrian army. At his birth he was proclaimed Roi de Rome, and the title Duke of Reichstadt was officially decreed in 1817. He died of consumption.

Reichstag. (1) A Diet or meeting of the Emperor of the Holy Roman Empire (see DEUTSCHES REICH, ALTES) and the rulers of the component territories. In the Middle Ages Diets were held at irregular intervals and in various cities. In time noblemen other than ruling princes were summoned to the Diets. Between the years 800 and 1663, Diets were held at Aachen, Arnstadt, Augsburg (7), Dortmund, Eger, Erfurt (3), Frankfurt (18), Goslar (4), Magdeburg, Mainz (2), Nürnberg (15), Regensburg (10), Strasburg, Worms (13), and Würzburg (9). The Diet also met at Besançon and Ravenna, and twice at Verona. From 1663 a commission sat as a permanent Reichstag (Immerwährender Reichstag) at Regensburg (q.v., Ratisbon). This ceased with the dissolution of the Empire in 1806.

(2) The elected parliament of the German Empire 1871–1945. The Reichstag building, completed in 1894, was destroyed by fire on 27 February 1933. The conflagration, attributed to Communists, was possibly arranged by a National Socialist group (see NSDAP), but the circumstances remain obscure.

Reichswehr, the official title of the German armed forces from 1919 to 1935. It comprised the Reichsheer and the Reichsmarine, though it is often used to refer to the army alone. The strength of the Reichsheer was fixed by the Treaty of Versailles (see VERSAILLES, TREATY OF) at 100,000 men, and the treaty also specified the relative proportions of the various arms of the service. Since conscription was abolished, it became a highly trained professional army, though it was not allowed by treaty to possess tanks or aircraft. In 1935 the name was changed from Reichswehr to Wehrmacht, with the divisions Heer, Marine, and Luftwaffe.

In the years 1919–33 the chiefs of the Reichswehr were unenthusiastic supporters of the Weimar Republic (q.v.). Under the National Socialist regime General von Blomberg as Reichswehrminister, by introducing the oath of personal fealty (Vereidigung auf den Führer) in 1934, placed the armed forces under Hitler's direct control.

Reigen, a cycle of ten dialogues by Schnitzler (q.v.), written in 1896–7 and privately printed for friends in 1900. Published in 1903, it was not performed in the original German until December 1920 (Kleines Schauspielhaus, Berlin), though a performance in Magyar had taken place in Budapest in 1912. These delays were caused by the sensational sexual frankness of the cycle, which, though capable of moral interpretation as a satire, appeared to many to be cynically immoral—a kind of *Così fan tutti e tutte*. Ten characters appear, five women and five men. Each has intercourse with two others on the pattern that the initial dialogue of A and B is followed by one between B and C, until in the tenth dialogue J is joined by A, thus completing the round dance (Reigen).

The characters involved in this circle have typical, not individual, designations: Die Dirne, Der Soldat, Das Stubenmädchen, Der junge Herr, Die junge Frau, Der Ehegatte, Das süße Mädel, Der Dichter, Die Schauspielerin, Der Graf. The flexible and apt dialogue and the acute perception of erotic ingenuity and dishonesty are remarkable. The sexual act (indicated in the text by a line of dashes) was implied in the original Berlin performance ‘by blacking out the stage lighting. The cycle is also known through a successful French film version, *La Ronde*.

Reim, see ENDREIM.

REIMANN, BRIGITTE (Burg nr. Magdeburg, 1933–1973, East Berlin), taught for two years, then after a variety of jobs settled as a writer in Neubrandenburg in 1966.

The bulk of her fiction is concerned with the task of reconstruction and personal adjustment in the young DDR. Titles include *Die Frau am Pranger, Kinder von Hellas* (both 1956), *Das Geständnis* (1960), *Ankunft im Alltag* (1961), and *Die Geschwister* (1963). In 1960 she wrote (in collaboration with S. Pitschmann) the radio plays *Ein Mann steht vor der Tür* and *Sieben Scheffel Salz*. In 1961 and again in 1962 the Freie Deutsche Gewerkschaftsbund awarded her its

prize for literature and in 1965 she received the Heinrich Mann Prize. In the same year *Das grüne Licht der Steppen* was published, the last work to appear during her lifetime.

The term Ankunftsliteratur, applied to literature representing the early Bitterfeld guidelines for writers (see SOZIALISTISCHER REALISMUS), originates from *Ankunft im Alltag*. Brigitte Reimann was herself for a time associated with the Kombinat Hoyerswerda, at which the young woman of the story undergoes her successful training. In 1963 Reimann began work on a novel, *Franziska Linkerhand*, which continued until shortly before her death from cancer. The substantial, unrevised fragment was published posthumously in 1974. After the initial, but quickly abandoned attempt to model her work on Zola, progress in the early years was slow. In the end, a Zeitroman (q.v.) emerged, covering the period of German defeat in 1945 until the emergence of the first teething troubles of the new socialist state (Aufbauperiode); of these Franziska, daughter of a publisher who leaves her the Cotta edition of Goethe's works as a parting present before reluctantly moving to the West, becomes increasingly aware in her career as an architect. *Brigitte Reimann in ihren Briefen und Tagebüchern*, ed. E. Elten-Krause and W. Lewerenz, published in 1983 (DDR) and 1984 (BRD), is a chronologically arranged selection; though the addressees' names are omitted from the letters (for reasons indicated in the Preface), the volume's three sections (Burg, Hoyerswerda, and Neubrandenburg) are an interesting documentation of a hard-working writer's life in the DDR and of her irrepressible individualism.

REIMARUS, HERMANN SAMUEL (Hamburg, 1694–1768, Hamburg), a noted Hebrew scholar, acquainted himself with English Theism on a study tour undertaken in 1720–1, and was from 1728 a professor at the Akademisches Gymnasium at Hamburg. Reimarus left in MS. a rationalistic work of theology with the reputed title *Schutzschrift für die vernünftigen Verehrer Gottes*. G. E. Lessing (q.v.), who had made Reimarus's acquaintance in Hamburg, published fragments from this work while he was ducal librarian of Brunswick at Wolfenbüttel. The first fragment appeared in 1774 as *Von Duldung der Deisten*, and five more were published in 1777 in Lessing's series *Zur Geschichte und Literatur* with the title *Ein Mehreres aus den Papieren eines Ungenannten, die Offenbarung betreffend*. This publication precipitated Lessing's theological controversy with Pastor J. M. Goeze (q.v.).

Apart from these works the cautious Reimarus released only the following titles for publication:

Abhandlungen von den vornehmsten Wahrheiten der natürlichen Religion (1754), *Vernunftlehre als Anweisung zum richtigen Gebrauch der Vernunft* (1756), and *Allgemeine Betrachtungen über die Triebe der Tiere* (1760). In 1862 D. F. Strauß (q.v.) published *Hermann Samuel Reimarus und seine 'Schutzschrift für die vernünftigen Verehrer Gottes'*; the full work, under its correct title *Apologie oder Schutzschrift für die vernünftigen Verehrer Gottes*, ed. G. Alexander, appeared in 1972.

Reimchronik von Preußen, Kurze, title used to designate a Middle High German verse chronicle of which only two short fragments survive. It is believed to have been written about the middle of the 14th c. Its subject is the history of the Teutonic Order (see DEUTSCHER ORDEN).

Reimrede, a form of poetry current in the 14th c. and 15th c. Reimreden were composed in rhyming couplets, were continuous and not in strophic form, and were spoken, not sung. The best-known poets writing in this form are Der KÖNIG vom Odenwald and HEINRICH der Teichner (qq.v.).

REINALT, also **REINOLT,** see HAIMONSKINDER, DIE.

REINBOT VON DÜRNE was the author of a Middle High German poem recounting the martyrdom of St. George (*Georgslegende*). Reinbot originated from the region of Wörth, on the Danube below Regensburg, or, less probably, from Walldürn. He appears to have been a court poet and composed his poem for Duke Otto II of Bavaria (reigned 1231–53) between 1231 and 1236.

The *Georgslegende* tells how George ventures into Cappadocia to witness for his faith against the heathens under King Dacian, submitting to terrible torments with equanimity. He succeeds in converting the heathen queen, who is also martyred. There is no mention of the slaying of the dragon, which is a later addition to the legend. Reinbot's St. George is a knight, skilled in courtly ways, and there is a sharp contrast between the courtly tone and the harrowing episodes. Reinbot includes passages of Christian apologetics, put in the mouth of the saint, and a brief allegory of the Castle of Virtue (Wunderburg der Tugend). He refers with approval to HEINRICH von Veldeke, WOLFRAM von Eschenbach, and HARTMANN von Aue (qq.v.), and is particularly indebted to Wolfram's *Titurel* (q.v.). The poem, which extends to some 6,000 lines, represents an adaptation of legend to the tone of court epic.

Reineke Fuchs, see REINKE DE VOS.

Reinfried von Braunschweig, an unfinished Middle High German verse romance by an unknown Swiss author. It comprises more than 27,000 lines and was written *c.* 1300. Its origin is a legend linking Heinrich der Löwe (q.v.) of Brunswick with a lion, after the manner of *Iwein* (q.v.), though the lion, in the extant portion of the romance, only appears in a dream. The poet by his own account is a commoner.

The first part of the poem narrates the wooing of the Danish princess Yrkane by Reinfried. He gains her love at a tournament and later successfully defends her in combat against the slanders of a rival lover. Reinfried's journey to the East follows, with a number of adventures which have their source in *Herzog Ernst* (q.v.), including the encounters with the Magnetic Mountain and with the Sirens. The romance breaks off at a point at which Reinfried's return is interrupted. Though the style is prolix, the psychological developments in the love story of the first part are acutely portrayed.

REINHARD, KARL FRIEDRICH VON (Schorndorf, 1761–1837, Paris), the son of a parson, studied theology, went to France as a tutor before the Revolution, and in 1791 commenced a remarkable career as a French diplomat, serving the Republic as foreign minister, and Napoleon as ambassador to Kassel (1808–13). He received the title Baron in 1808, and Graf in 1813. From 1816 to 1829 he represented France at the Federal Diet (see DEUTSCHER BUND) in Frankfurt. He spent the last two years in French service as ambassador in Dresden, was created *Pair* and left in the year of the death of Goethe (1832), with whom he had corresponded since they had met at Karlsbad in 1807. *Goethe und Reinhard, Briefwechsel in den Jahren 1807–32*, first published in 1850, appeared in an enlarged edition in 1957. Reinhard also conducted correspondence with Schiller and other men of letters, including J. K. Lavater, F. H. Jacobi, F. Schlegel (qq.v.) as well as H. von Gagern (q.v.).

REINHARD, KARL VON (Helmstedt, 1769–1840, Zossen), was editor of the annual *Göttinger Musenalmanach* (q.v.) from 1795 to 1804 (but not in 1803), and of the *Romanenkalender* from 1798 to 1803. He is the author of a once well-known poem beginning 'Jahre kommen, Jahre schwinden'.

REINHARDT, MAX, adopted name of M. Goldmann (Baden nr. Vienna, 1873–1943, New York), an actor who became one of the most prominent theatrical producers in Germany and Austria during the first four decades of the 20th c. From 1894 he was in Berlin, at first under O.

Brahm (q.v.), and later in the Neues Theater (1903–6). He turned away from Naturalistic techniques and developed an imaginative style of production for which he found added scope and facilities, including the revolving stage, at the Deutsches Theater, Berlin (1905–20 and 1924–33; he owned it until 1933, when he was forced to hand his Berlin theatres over 'to the German people'). He also founded the cabaret Schall und Rauch and, later, his own school of acting. His production of *A Midsummer Night's Dream* (1905) marked a climax in his public recognition. In 1906 he opened his Kammerspiele with *Ghosts* by Ibsen (q.v.) and in 1919 the Großes Schauspielhaus with the *Oresteia*; all houses served his wide repertory and a versatile style that drew from all the arts and was based on his belief in an inherent link between new and old. The Großes Schauspielhaus, the so-called 'Theatre of the Five Thousand', was built on the site of Circus Schumann, which he had used as an arena theatre (for *König Ödipus* and *Jedermann* by Hofmannsthal, in 1910 and 1911) that established his world-wide reputation.

In the Berlin of the 1920s Reinhardt's illusionary theatre formed a stark contrast to the anti-illusionary theatre of E. Piscator (q.v.), and the range of his productions was flexible and even unpredictable. In 1913 he collaborated with G. Hauptmann (q.v.) in the centenary celebrations of Germany's liberation from Napoleonic occupation, and in 1928 he surprised his Berlin audiences with a production of *Die Fledermaus* by J. Strauß (q.v.), in collaboration with E. W. Korngold (1897–1957). From 1917 Reinhardt was prominently associated with the planning and running of the Salzburger Festspiele (q.v.), which resulted in a close collaboration with H. von Hofmannsthal (q.v.), an admirer of Reinhardt's stage techniques in indoor and open-air performances. In 1924 Reinhardt took over the Theater in der Josefstadt in Vienna, which remained the centre of his activities and where, in 1928, he founded the Max Reinhardt Seminar. His emigration, in 1937, to the U.S.A. was accompanied by the loss of his considerable fortune. The Workshop for Stage, Screen and Radio and his film production of *A Midsummer Night's Dream* (1935, with music by F. Mendelssohn-Bartholdy, q.v., and E. W. Korngold) mark his exile. For some twenty-five years Helene Thimig (q.v.) collaborated with Reinhardt, who married her in 1932.

A volume of letters, speeches, and essays (*Ausgewählte Briefe, Reden, Schriften*), ed. F. Hadamowsky, appeared in 1963, and the correspondence of A. Schnitzler (q.v.) with Reinhardt (*Briefwechsel Arthur Schnitzlers mit Max Reinhardt*), ed. R. Wagner, in 1971.

Reinhart Fuchs, see HEINRICH DER GLÎCHEZAERE.

REINICK, ROBERT (Danzig, 1805–52, Dresden), painter and poet, studied painting in Berlin, was then at Düsseldorf, visited Italy from 1838 to 1841, and settled in 1844 in Dresden. His pictures are mainly historical and romantic. He was also a skilled artist in woodcuts, and in several works provided both illustrations and text (*Drei Umrisse nach Holzschnitten von Albrecht Dürer mit erläuterndem Text und Gesängen*, 1830, and *Lieder eines Malers mit Randzeichnungen seiner Freunde*, 1838). Several of his works are intended for the young, including *Lieder und Fabeln für die Jugend* (1844) and *Der deutsche Jugendkalender* (1849–52). He also wrote the verse for the series of woodcuts entitled *Auch ein Totentanz* by Alfred Rethel (q.v.). With his friend F. Th. Kugler (q.v.) he published a *Liederbuch für deutsche Künstler* (1833); L. Richter (q.v.) illustrated his *Johann Peter Hebels 'Alemannische Gedichte' ins Hochdeutsche übertragen* (1851).

REINIG, CHRISTA (Berlin, 1926–) studied history and Christian archaeology in East Berlin where she published her first stories and poetry. In 1964 she obtained permission to travel to the BRD in order to accept the award of the Bremer Literaturpreis, and subsequently settled in Munich. The author of terse, unemotive verse, she conveys a deep sense of humanity and revulsion against political and social injustice; her controlled verse contains brutal images of torture and death, and bitter irony. The Moritat (q.v.) song *Die Ballade vom blutigen Bomme* is an early work that helped establish her reputation. Her collections include *Die Steine von Finisterre* (1960 and, in revised form with other previously published poems, 1974), *Gedichte* (1963), *Schwalbe von Olevano* (1969), *Papanscha—Vielerlei* (1971), *Müßiggang ist aller Liebe Anfang* (1979), and *Prüfung des Lächelns* (1980); and *Schwabinger Marterln. Freche Grabsprüche für Huren, Gammler und Poeten* (1968). Her prose works, marked by humour and the short episodic form, began with *Ein Fischerdorf* (1951), followed by *Der Traum meiner Verkommenheit* (1961), *Drei Schiffe* (1965), the collection *Orion trat aus dem Haus—Neue Sternbilder* (1968), and *Das große Bechterew-Tantra* (1970). The novel *Die himmlische und die irdische Geometrie* (1975) was followed by *Entmannung. Die Geschichte Ottos und seiner vier Frauen* (1976). Written with wit and verve, it established her as the author of one of the best contemporary feminist novels, though feminism is not its sole concern. It was followed by the prose of *Der Wolf und die Witwen* (1980) and *Die ewige Schule* (1982). *Sämtliche Gedichte* appeared in 1984.

Reinke de Vos, a German beast epic narrating the adventures of Reynard the fox, was printed at Lübeck in 1498. It tells how the wicked and rapacious, but alert and resourceful fox is twice arraigned for his many crimes, and how he outwits by sheer effrontery and brazen deceit every attempt by his enemies, especially Ysegrim the wolf and Brun the bear, to secure his conviction and punishment. He finally triumphs over them all and is raised to favour and honour by the lion Nobel, who is their king. The poem, written in rhyming verse, parodies and criticizes social conditions in the later Middle Ages, showing cunning victorious over greed, brute force, and stupidity, and law disregarded by those who have power and influence. Lest its message should not be fully grasped, it is interspersed with moral interpretations, or glosses, in prose. Its anonymous author wrote in the Low German tongue. The immediate source for *Reinke de Vos* was a Dutch poem (1487) by Hinrek van Alkmaar.

The theme, elements of which reach back to Aesop, was a popular one in medieval times, flourishing particularly in Flanders. Its earliest occurrence is in the *Ecbasis captivi* (q.v.). Other versions are the French *Roman de Renart* of the 12th c. and the fragmentary German *Reinhart Fuchs* (*c.* 1180), written by HEINRICH der Glîchezaere (q.v.). A feature of this, which is absent from *Reinke de Vos*, is Reinhart's final crime, the poisoning of the lion-king. The Lübeck and Dutch versions are descended from *Von den vos Reinaerde* (*c.* 1250) by the Fleming Willem.

Reinke de Vos enjoyed great popularity, partly for its satire, but also on account of its comic qualities and the vigorous portrayal of its animal characters. It appeared in numerous editions (nine in the 16th c.) and was translated into High German (1544 and many later editions). Particularly noteworthy is the edition published by Gottsched (q.v.) in 1752 together with a prose translation, which formed the basis for Goethe's *Reineke Fuchs*, a version of the story in twelve books, written in hexameters. This epic treatment enhances the comic element, while imposing a certain discipline. Goethe was attracted to the story in 1793, at a time when his inspiration for original work was flagging. *Reineke Fuchs* (the bastard title was Gottsched's) offered him a fascinating parallel to the ruthless and egoistic world of Realpolitik (q.v.), which he had experienced in the political environment of the German anti-French coalition. He wrote it after his return from the campaign of 1792 and published it in *Goethes Neue Schriften* (vol. 2, 1794).

REINMAR DER ALTE, Middle High German Minnesänger. The epithet 'old' (alt), which occurs in the MSS., has nothing to do with his age, but distinguishes him from later bearers of the same name, notably REINMAR von Zweter (q.v.). He is often called Reinmar von Hagenau, in consequence of a reference by GOTTFRIED von Straßburg (q.v.) to 'diu [sc. nahtegal] von Hagenouwe', without, however, any mention of Reinmar's name. Nothing is known with certainty of Reinmar's life. The Hagenau from which, it is believed, he came, was probably the town in Alsace. It is generally thought that his poems were mostly, if not all, written at the Babenberger court in Vienna, where he seems long to have held a position equivalent to that of court poet. He may have taken part in a crusade in 1197-8, though the evidence for this is slender; he probably died between 1205 and 1210, some think as a relatively young man.

Reinmar was the acknowledged master of courtly Minnesang (q.v.), and many younger poets, including WALTHER von der Vogelweide (q.v.), regarded him as their preceptor. He took his aesthetic responsibilities seriously, and it has been widely held, though without proof, that he became involved in controversy, notably with Walther. Reinmar's poetry constantly repeats the motif of the knight's service (Minnedienst), unrewarded by his lady, and makes it clear that fulfilment would destroy a relationship which has an important function in educating the character to true nobility. The poems move within narrow limits, exhibiting refinement and tact and eschewing any utterance of passion. Their reticent perfection can appear monotonous, and their evident appreciation by Reinmar's contemporaries indicates a public schooled in poetic form and responsive to subtleties.

His great prestige is endorsed by his opponent Walther, who quotes in his elegy Reinmar's line 'Sô wol dir, wîp, wie reine ein nam'.

REINMAR DER FIEDLER (Videler), a minor Middle High German poet of the late 13th c. who is the author of some extant Sprüche (see SPRUCH), one of which contains an attack upon the poet LEUTHOLD von Seven (q.v.).

REINMAR VON HAGENAU, see REINMAR DER ALTE.

REINMAR VON ZWETER (*c.* 1200–*c.* 1260, Eßfeld nr. Ulm), a Middle High German poet of minor nobility, whose work consists almost exclusively of Sprüche (see SPRUCH). A few details of his life emerge from his poems. He was born in the Rhineland, went to Vienna as a boy or youth, and became a professional poet at the Viennese court. In *c.* 1234 he moved to the court of King Wenzel I of Bohemia (reigned 1230–53). In his later years he was again in the Rhineland at Cologne and Mainz.

About 300 of Reinmar's Sprüche survive, arranged systematically in an order probably determined by the poet himself. They range over religion, *minne*, morals, and politics. The central point of them all is honour, symbolized in the figure of Frau Ehre. Implicit in honour is the conception of *mâze*, the just mean in conduct. Honour is pleasing to God, the basis of *minne*, and the necessary fundamental quality of the knight. In his religious Sprüche, which reflect the orthodox faith of the god-fearing knight, a new element is manifest in the cult of the Blessed Virgin. The moral Sprüche cover not only the principles of ethics, but deal with specific problems, including the spread of drunkenness and gambling. The political poems, which are also primarily ethical, record a remarkable change of opinion. An early group of *c.* 1228 attacks Pope Gregory IX and supports the Emperor Friedrich II (q.v.). In *c.* 1240 the attitude to Friedrich is reversed and Reinmar goes so far as to pray God to oppose him and even to urge his deposition.

Reinmar's principal formal model is WALTHER von der Vogelweide (q.v.), of whom there are verbal reminiscences in his poems. The Sprüche are cast in a twelve-line strophe, and several are at times grouped together. This characteristic form was later appropriately termed the 'Frau-Ehren-Ton'. Reinmar also wrote a religious *leich* (q.v.).

Reisebilder, a 4-vol. collection in prose and verse by H. Heine (q.v.). Vol. 1 (1826) contained *Die Harzreise* (q.v.), the collection of poems *Die Heimkehr, Die Nordsee I*, and a few other poems. In vol. 2 (1827) appeared *Die Nordsee II* and (prose) *III, Ideen, Das Buch Le Grand*, and three letters, *Briefe aus Berlin*. The contents of vol. 3 (1830) were *Die Reise von München nach Genua* and *Die Bäder von Lucca*. The fourth volume (1831), originally called *Nachträge zu den Reisebildern*, comprised *Die Stadt Lucca* and *Englische Fragmente*.

Reise nach Braunschweig, Die, a novel by A. von Knigge (q.v.), published in 1792. Sub-titled *Ein komischer Roman*, this short work is something of a potpourri. It tells of a journey made by four worthies of Biesterberg (Amtmann Waumann and son, Pastor Schottenius and Förster Dornbusch) to Brunswick to see a balloon ascent. They encounter various comic misfortunes and by ill chance miss what they have come to see. Pastor and forester never get to Brunswick at all since they encounter the latter's niece apparently eloping. The elopement

turns out to be honourable, the forester is united with his long-lost brother, and all ends happily. Meanwhile an inserted section tells the story of the girl's father and suitor. The novel also contains a few pages of intrusive dramatic criticism strongly condemning the plays of Kotzebue (q.v.). An unexpected feature is some dialect speech.

Reise nach Tilsit, Die, see LITAUISCHE GESCHICHTEN.

Reise um die Welt mit der Romanzoffschen Entdeckungsexpedition in den Jahren 1815–1818 auf der Brigg Rurik, Kapitän Otto von Kotzebue, a travel book written by A. von Chamisso (q.v.) in the years 1834–5 on the basis of the diary which he kept on the voyage. It was published in 1836. Chamisso sailed as attached scientist with special botanical qualifications. The work consists of a rewritten diary (*Tagebuch*) and *Bemerkungen und Ansichten*. The latter had been previously published in 1821.

Reitergeschichte, a Novelle by H. von Hofmannsthal (q.v.) published in 1899. The story, reflecting Hofmannsthal's peace-time military experience, describes one day (it proves to be the last) in the life of Sergeant Anton Lerch of Captain Rofrano's squadron of the Austrian Wallmodenkürassiere. On 22 July 1848 the squadron successfully disposes of small remaining pockets of resistance in upper Italy, remnants of troops of irregulars fighting for national liberation. Only the Captain's procession through Austrian-occupied Milan to the midday chimes of the church bells, during which Lerch visits the house of Vuic, a woman of tempting beauty, is not part of the destructive military mission. Stimulated to a heightened sensual awareness through his encounter with Vuic, Lerch himself subsequently takes a path of his own on a ride through a village, in which he finds, instead of triumph and victory, only squalor and filth, blood and decay. Coming out of the village, he sees a fleeting vision of his double, horse and rider, moving towards him on a bridge which he crosses to capture a handsome grey horse, whose rider he kills in a brief skirmish. When the squadron reassembles at the end of the day, Captain Rofrano orders all enemy horses to be turned loose. As Lerch hesitates with a gesture of insubordination, Rofrano shoots him dead on the spot. Lerch's sudden execution is unexpected, and yet the condensed symbolism of this short narrative points to more fundamental motivation, including the contrast between the aggressive brutality of war and an aristocratic devotion to beauty.

The name Rofrano also occurs in Hofmannsthal's *Der Rosenkavalier* (q.v.) where it is the surname of Octavian.

RELLSTAB, LUDWIG (Berlin, 1799–1860, Berlin), at first an artillery officer, became an author and journalist. He wrote poems, some of which were set to music by Franz Schubert (q.v.), including the first seven songs of Schubert's *Schwanengesang*. His poems appeared in *Griechenlands Morgenröte* (1822), *Gedichte* (1827), and *Erzählungen, Skizzen und Gedichte* (1833). Rellstab provided the libretto for an opera by Bernhard Klein (*Dido*, 1823) and wrote the tragedies *Karl der Kühne* (1824) and *Eugen Aram* (1839); the latter is adapted from Bulwer-Lytton's novel *Eugene Aram*. He is the author of the historical novel *1812* (4 vols., 1834), his most successful work, and of the novel *Der Wildschütz* (1835). *Musikalische Beurteilungen* appeared in 1848, *Sommerfrüchte*, a collection of stories, in 1852, and his autobiography, *Aus meinem Leben* (2 vols.), in 1861. *Gesammelte Schriften* (20 vols.) were published in 1843–8, and posthumously (24 vols.) in 1860–1.

REMARQUE, ERICH MARIA, pseudonym of Erich Paul Remark (Osnabrück, 1898–1970, nr. Locarno), was twice wounded during the 1914–18 War, after which he had a variety of jobs before becoming a journalist in the 1920s and writing, within a short space of time, his war novel *Im Westen nichts Neues* (1929). A worldwide success as a novel and film, it portrays his experiences on the western front with great intensity and intentionally brutal realism. But the pacifism underlying Remarque's treatment of war also met with bitter criticism; in 1933 the novel was one of the books burnt publicly in Berlin by the National Socialist regime.

Remarque continued to make his living as a writer who confined himself to his contemporary background and always found a public. In 1938 he went to Switzerland and was deprived of his German citizenship, which he declined to resume after the 1939–45 War. In 1939 he went to the U.S.A., was granted American citizenship in 1947, and spent the remainder of his life partly in New York, but mainly in Porto Ronco. He wrote a sequel to his first novel, *Der Weg zurück* (1931); *Drei Kameraden* (1938) and *Der schwarze Obelisk* (1956) reflect conditions in the Weimar Republic (q.v.) and the period of inflation, and *Der Funke Leben* (1952) those in concentration camps (Konzentrationslager); *Zeit zu leben, Zeit zu sterben* (1954) depicts soldiers in retreat in Russia during the 1939–45 War; and *Der Himmel kennt keine Günstlinge* (1961), related to Heimkehrerliteratur (q.v.), resumes experiences following the 1914–18 War. His other novels deal with the problem of exile, *Liebe deinen Nächsten* (1941), *Die Nacht von Lissabon* (1963), and, above all, *Arc de Triomphe* (1946), which was his second great success. Its contrived and sensational plot deals with an act of revenge: Ravic, a German surgeon and a refugee in Paris without papers, manages to kill Haake, a Gestapo agent, who has cruelly murdered a woman with whom Ravic had once lived. The character of Ravic reappears in Remarque's last novel, completed shortly before his death and published posthumously in 1971, *Schatten im Paradies*. Three of the novels published between 1938 and 1952 appeared first in English translation, *Three Comrades* (1937), *Flotsam*, and *Spark of Life* (1941 and 1952 respectively, the year of their German publication).

Reminiscere, second Sunday in Lent and fifth before Easter. It takes its name from the introit, 'Remember, O Lord, thy tender mercies and thy lovingkindness' (Psalm 25:6, A.V., Vulgate 24:6).

Renate, a Novelle by Th. Storm (q.v.), published in the *Deutsche Rundschau* (q.v.) in 1878. The narrative conveys in the form of chronicles attributed to the years 1700–8 and 1778 the memories of Josias, a pastor, of his love for Renate, a farmer's daughter, of whom legend has it that she was a witch. Josias saves her from the hands of villagers who attempt to drown her. He is attracted to Renate, who in reality is a sensitive and courageous girl with a warm personality; yet he abstains from marrying her because he and his father misguidedly believe that to ask for her hand would be to yield to temptation. He sees his error as if by divine illumination towards the end of his life when Renate rides across the heath to visit him.

RENN, LUDWIG, pseudonym of Arnold Vieth von Golßenau (Dresden, 1889–1979, East Berlin), a Saxon nobleman. Renn is the name of the narrator and principal character of Vieth's first successful novel *Krieg* (q.v., 1928), and the name appeared as author on the title-page. Vieth's subsequent works appeared under this pseudonym. From 1911 an officer in the Saxon Guards, Renn served throughout the 1914–18 War, after which he studied law and economics and was for a short time a police officer. He became a communist, a step which the novel *Nachkrieg* (1930) reflects. Known for his political activities, he was arrested by the National Socialist regime in 1933 and imprisoned, but escaped abroad and served on the Republican side in the Spanish Civil War, of which he wrote an account in *Der spanische Krieg* (1955). From 1939 to 1947 he lived in Mexico, devoting himself to the Bewegung Freies Deutschland, of which he was president. In 1944 he published the novel *Adel im Untergang*. From 1947 to 1951 he held a chair at the Dresden Technical University. His later publications include the novels *Krieg ohne Schlacht* (1957) and *Auf den Trümmern des Kaiserreichs* (1961), and the autobiographical *Meine Kindheit und Jugend* (1957). His works, *Gesammelte Werke in Einzelausgaben* (10 vols.), were published in 1964–70.

RENNER, KARL (Untertannowitz, Southern Moravia, 1870–1950, Vienna), of country stock, studied law at Vienna University, and became leader of the Austrian Social Democratic Party (Sozialdemokratische Partei Österreichs, SPÖ, see SPD), leading the Austrian delegation at the peace negotiations after both the 1914–18 War and the 1939–45 War. The dissolution of the monarchy terminated his efforts to achieve a multi-national realm on a federal basis, and in 1918 as well as in 1938 he favoured Austrian unification with Germany. He was chancellor (Staatskanzler) from 1918–20, and again in the provisional government formed in April 1945. In December of that year he became federal president (Bundespräsident) and retained this office until his death.

Renner published his early political writings under various pseudonyms (O. W. Payer, Josef Karner Synopticus, Rudolf Springer). He dealt with Austria's complex political situation in the 20th c. as a moderate Socialist reformer. His principal works, published under his own name, are *Österreichs Erneuerung* (3 vols., 1916–17), *Marxismus, Krieg und Internationale* (1917), *Die Wirtschaft als Gesamtprozeß und die Sozialisierung* (1924), and *Staatswirtschaft, Weltwirtschaft und Sozialismus* (1929). *An der Wende zweier Zeiten* appeared in 1946, and *Österreich von der ersten zur zweiten Republik*, ed. A. Schärf, posthumously in 1953.

Renner, Der, a Middle High German didactic poem written in the last decade of the 13th c. by Hugo von Trimberg (q.v.). Hugo himself gives 1300 as the date of completion. It appears to be an elaboration and extension of one of his lost early works, which he entitles *Der Samener (Sammler)*. *Der Renner* is an immense moral sermon, running to nearly 25,000 lines and dealing in six long sections with the seven deadly sins (anger and envy are considered together).

For his examples Hugo draws chiefly upon the Old Testament, making use also of the fables of Aesop and of his own experience in a long life. He is skilled in the devices of rhetoric, but his interminable loquacity weakens the structure of his poem. He has not the power of the prophet or the satirist, but expresses with simple piety homely moral views. The curious title is not of Hugo's devising; it was given by the 14th-c. bibliophile Michael de Leone of Würzburg, who justifies it thus: 'Renner ist ditz bûch genant/ Wanne ez sol rennen durch diu lant./(*Renner* ist dieses Buch genannt,/Weil es durch die Lande rennen soll.)'

Though for a modern taste *Der Renner* is prolix and pedestrian, it was one of the most popular works of the Middle Ages. Over sixty MSS. are known, and the printed edition of 1549 proves that after more than 200 years it was still held in high esteem.

Renommist, Der, a comic epic poem in six cantos by J. F. W. Zachariä (q.v.), published in 1744 and revised in 1754. It is influenced by Pope's *Rape of the Lock*, to which Zachariä refers in Canto II ('Und eine Locke glänzt, die Popens Lied erhöht'). Zachariä's poem is a satire on the manners prevalent at two German universities, Jena, noted for the coarse, blustering behaviour of its duelling and drinking students, and Leipzig, the home of affectation, preciosity, and francophilia.

The plot of the poem deals with the brief sojourn in Leipzig of Raufbold, a student sent down from Jena, and his encounter with Sylvan, also formerly an alumnus of Jena, who has assumed Leipzig ways. Raufbold conducts himself in an uproarious fashion, until in Sylvan's company he meets the attractive Selinde. Persuaded to adorn himself suitably, he cuts a poor figure and is ridiculed. Thereupon he challenges Sylvan to a duel, is defeated, admits his discomfiture, and mounts his horse to ride to Halle, a less 'civilized' university. Zachariä surrounds his characters with a comic apparatus of gods, such as Die Galanterie and Der Putz. The poem bestows its gentle satire on both parties, though more frequently on Jena. It is written in deft Alexandrine verse and is attractively goodhumoured. It is Zachariä's first and best work.

Resignation, a poem written by Schiller, probably in 1784, and arising out of his attachment to Frau Charlotte von Kalb (q.v.). It was republished in abridged form in *Gedichte*, 1800.

Resistance Movements. (1) In the 1939–45 War hidden resistance arose in all German-occupied countries, and in many cases contact was established by sea or air with allied agents. Savage reprisals often stiffened resolve. The revolt of the Warsaw ghetto (January 1943) and the Warsaw rising of 1 August–4 October 1944 were large-scale resistance operations which ended tragically. In Russia fierce partisan warfare was conducted behind the German lines and large numbers of SS troops were used in savage attempts at suppression.

(2) The German Resistance Movements (Widerstandsbewegungen), since they opposed the legal, though inhuman, government of their own country, were liable to the stigma of treason. In the one-party state decreed on 14 July 1933 even dissent was treasonable. Opposition was first expressed by the Church (creation of the Evangelische Bekenntniskirche, see BEKENNENDE KIRCHE, May 1934). By preaching against the liquidation of the aged and of the incurably sick and against illegal acts of violence some pastors risked or incurred persecution. Notable among these were Bishop O. Dibelius (1880–1967), Bishop Th. Wurm (1868–1953), and Pastors M. Niemöller and D. Bonhoeffer (qq.v.). The Roman Catholic Church had also notable protesters, including Cardinal Graf Galen of Münster (1878–1946), Archbishop Gröber of Freiburg (1872–1948), and Graf Preysing, Bishop of Berlin (1880–1950).

Communist resistance crystallized around the Rote Kapelle (q.v.), organized in 1940 and discovered in 1942. Other groups were composed of individuals disenchanted with the inhumanities of the National Socialist regime and the recklessness of Hitler's foreign policy. In the armed forces opposition arose among highly placed officers, and in civilian life around C. Goerdeler (q.v.) and in the Kreisauer Kreis (q.v.), but it was not until 20 July 1944 that a carefully prepared attempt was made to assassinate Hitler and set up a new government in Berlin. Its failure was followed by many executions, including those of C. Schenck von Stauffenberg (q.v.), Dietrich Bonhoeffer, A. Delp, S.J. (1907–45), C. F. Goerdeler, Ulrich von Hassell (1881–1944), a former National Socialist ambassador in Rome, H. J. von Moltke (q.v.), A. von Trott zu Solz (1909–44), and Field-Marshal E. von Witzleben (1881–1944). Field-Marshal E. Rommel (q.v.) took poison as an alternative to execution.

Protest inspired by religious faith and idealism without hope of political success came from the Roman Catholic student group Weisse Rose, led by Professor K. Huber (1894–1944), to which Hans (1918–43) and Sophie Scholl (1921–43) belonged. Brother and sister, arrested while distributing leaflets, were executed.

Prosecutions for resistance, sedition, and subversion were conducted in the notorious people's court (Volksgerichtshof) set up in 1942 under the presidency of R. Freisler (q.v.).

Restitutionsedikt, see DREISSIGJÄHRIGER KRIEG.

RETHEL, ALFRED (nr. Aachen, 1816–59, Düsseldorf), a painter and graphic artist, studied in Düsseldorf and in 1840 made a series of woodcuts illustrating the *Nibelungenlied* (q.v.). In the same year he was commissioned to paint the large murals for the city hall of Aachen on the subject of Karl I (q.v.) der Grosse. Of these eight frescoes (Karlsfresken) he was able to execute five before losing his sanity in 1852; the other three were completed to Rethel's designs by Joseph Kehren. Some of the work was destroyed in the 1939–45 War. Rethel's best achievement is in the field of the woodcut, particularly *Auch ein Totentanz* (1849, reproduced 1957 with an introduction by Th. Heuß, q.v.), to which his fellow artist Robert Reinick (q.v.) provided a poetic text.

RETTENPACHER or **RETTENBACHER,** SIMON (Aigen nr. Salzburg, 1634–1706, Kremsmünster), became a Benedictine monk in 1661 at Kremsmünster and was in charge of the abbey school from 1668 to 1671. From 1671 he was professor of history and ethics at the Benedictine university at Salzburg; in 1675 he returned to Kremsmünster Abbey as librarian. Rettenpacher was a historian and a prolific poet and dramatist. Virtually all his works, except for about a hundred German poems, were written in Latin. He himself estimated the number of his Latin poems at more than 6,000. They provide a commentary on his own life and meditations, and reflect not only his religious experience but also political events, notably the siege of Vienna (1683) and the repulse of the Turks. His Latin plays were intended for school performance. A historical work in Latin (*Annales monasterii Cremifanensis*, 1677) was published in German translation in 1793.

A selection of his plays (*Selecta dramata*) was published in 1684. His Latin poems (*Lateinische Gedichte*, ed. Th. Lehner) appeared in 1905, and *Deutsche Gedichte* (ed. R. Newald) in 1930.

RETTICH, JULIE (Hamburg, 1809–66, Vienna), *née* Gley, a famous actress in the Viennese theatre of the mid-19th c. She made her début in Dresden in 1825 and appeared at the Burgtheater as guest in 1828, receiving a contract there in 1830. She married the actor Karl Rettich in 1833. Soon afterwards she left the Burgtheater, but she and her husband were permanently engaged there in 1835. She excelled in tragic and serious roles, including Goethe's Gretchen (see FAUST) and Iphigenie (see IPHIGENIE AUF TAURIS). In 1835 she ensured, by her performance in the title-part, the success of *Griseldis* by F. Halm (q.v.), and so began a lifelong friendship with the author, who wrote his principal female roles for her.

Rettungen, title given by G. E. Lessing (q.v.) to four substantial essays, published in 1754 in vol. 3 of his *Schriften* (1753–5). Each attempts the rehabilitation of a reputation. The first, *Die Rettungen des Horaz*, defends the moral character of the poet Quintus Horatius Flaccus. The other three are apologias for three supposedly heretical writers, Professor Hieronymus Cardanus of Pavia (1501–76), the anonymous author of *Ineptus Religiosus ad mores horum temporum descriptus* (1652), and J. Cochläus (q.v.).

REUCHLIN, JOHANNES (Pforzheim, 1455–1522, Bad Liebenzell), was, together with his friend Erasmus (q.v.), the driving force in early German humanism. He studied in Freiburg, Paris, and Basel, and was from 1484 to 1496 a lawyer prominent in the service of Eberhard im Bart, Duke of Württemberg (1450–90). From 1502 to 1513 he was a judge of the Swabian League (see SCHWÄBISCHER BUND). His main interests, however, were philological; in 1519 he became a professor at Ingolstadt University, and in 1521 at Tübingen.

Reuchlin contributed to the encouragement of Greek studies, and he is the founder of Hebraic studies in Germany with *Rudimenta hebraica* (1506) and *De accentibus et orthographia linguae hebraicae* (1518). His attainments in Hebrew, coupled with his judicious and tolerant personality, led to his involvement in a controversy with the anti-Semitic J. Pfefferkorn (q.v.) of Cologne, to whose abuse Reuchlin replied with *Der Augenspiegel* (1511). Pfefferkorn's backers, the Dominicans of Cologne University, sought to accuse him of heresy, but his high standing as a scholar and diplomat rendered him virtually immune to serious attack. In defence he published the *Epistolae clarorum virorum* (1514). An outcome of the controversy was the *Epistolae obscurorum virorum* (1515–17, q.v.), directed by Reuchlin's supporters against his opponents.

Reuchlin wrote two Latin comedies, *Sergius oder Capitis caput* (1496), a satire, and *Scena progymnasmata sive Henno* (1497), which is reckoned to be an important step in the development of German comedy; it resembles *La Farce de maistre Pierre Pathelin* (1470), but it is improbable that Reuchlin knew the French play.

Reuchlin also dabbled in cabbalistic philos-

ophy in *De arte cabbalistica* (1517). Like Erasmus an indefatigable correspondent, he spread his views by letters and personal contact, in which his integrity, tolerance, and intelligence influenced a wide circle of notable scholars and men of affairs. *Komödien. Ein Beitrag zur Geschichte des lateinischen Schuldramas*, ed. H. Holstein, appeared in 1888 (reissued 1973) and *Briefwechsel*, ed. L. Geiger, in 1875 (reissued 1962).

Reue nach der Tat, Die, a play by H. L. Wagner (q.v.), published in 1775 and first performed in that year by F. L. Schröder (q.v.) at Hamburg. It tells of the unhappy love of Langen and Friederike, which is thwarted by the snobbery of Langen's mother. Langen at one time proposes to marry without his mother's consent; but Friederike will have none of this. The mother intrigues against the lovers, Friederike poisons herself, and Langen goes mad, whereupon his mother goes mad too, from remorse. The play is in six acts, an implied protest by an author of the Sturm und Drang (q.v.) against the five of so-called Aristotelian orthodoxy.

REUENTAL, NEIDHART VON, see NEIDHART VON REUENTAL.

Reuner Relationen, title given to two Latin narrations, probably of the 12th c., found in Reun Abbey near Prague, the first of which is the source of the *Vorauer Novelle* (q.v.).

REUTER, CHRISTIAN (Kütten nr. Halle, 1665–?1712 or later) when a student at Leipzig fell out with his landlady, whom he lampooned in the comedies *L'Honnête Femme oder Die ehrliche Frau zu Plissine* (1695) and *La Maladie et la mort de l'honnête femme, das ist: Der ehrlichen Frau Schlampampe Krankheit und Tod* (1696). Sued by the landlady, Reuter was held in the university prison, and while there wrote another satire directed at her, *Letztes Denck- und Ehren-Mahl der weyland gewesenen Ehrlichen Frau Schlampampe* (1697). In consequence of this, Reuter, already under sentence of rustication, was sent down.

While under arrest he also wrote his best-known work, the satirical travel book *Schelmuffskys Wahrhafftige curiöse und sehr gefährliche Reisebeschreibung zu Wasser und Lande* (q.v., 1696 and, enlarged, 1697). In 1700 he was appointed secretary to Count Seyfferditz and while in his employment wrote the satirical comedy *Graf Ehrenfried* (1700), the hero of which wastes his substance in riotous living. From 1703 Reuter was in Berlin, and there he wrote a cantata, *Die frohlockende Spree* (1703). He also composed the text for an oratorio (*Passionsgedanken*, 1708). *Sämtliche Werke*, ed. G. Witkowski (2 vols.), appeared in 1916, and a select edition by W. Jäckel in 1965.

REUTER, Fritz (Stavenhagen, Mecklenburg, 1810–74, Eisenach), one of the most talented German dialect writers, was a son of the mayor of Stavenhagen. He studied law, first at Rostock University and then at Jena, where he became a member of one of the student corporations, which were then officially frowned upon as breeding-places of subversive views. When, after a student attack on a military guardhouse in Frankfurt (April 1833), new repressive measures were taken, Reuter, then staying in Berlin, was one of the victims. He was arrested by the Prussian police in October 1833, detained for more than three years on remand (Untersuchungshaft), and in January 1837 was condemned to death, a sentence which was simultaneously commuted to thirty years' imprisonment.

This astonishing sentence for treason, based on suspicion of membership of a non-existent conspiracy, was served in various Prussian prisons (Silberberg, Glogau, Magdeburg, and Graudenz) until in 1839 repeated pleas from the father, supported by the Mecklenburg authorities, led to Reuter's transfer to the Mecklenburg prison of Dömitz. In 1840 he was released. After an unsatisfactory attempt to resume his studies, he worked on the land in Mecklenburg until 1850, when he became a private tutor in Treptow. There in 1851 he married Luise Kuntze.

Reuter's first book, the dialect poetical anecdotes *Läuschen un Rimels* (q.v., 1853) proved a success, and a second volume (1858) was also well received. Reuter turned to dialect verse narratives with *De Reis nah Belligen* (1855), *Kein Hüsung* (q.v., 1857, a sombre tale of serfdom), and *Hanne Nüte* (q.v., 1860). In 1856 he moved to Neubrandenburg with the intention of continuing to give private tuition, but he abandoned this plan in order to devote himself entirely to literature. *Schurr-Murr*, a volume of short narratives, appeared in 1861.

Reuter's first novel, *Ut de Franzosentid* (q.v.), set in Mecklenburg in 1812 at the time of the Napoleonic Wars (q.v.), appeared in 1859; a remarkably detached, and in places even humorous, account of his years in prison followed in 1862 as *Ut mine Festungstid* (q.v.); an autobiographical novel based on his years on the land concluded this group of works (*Ut mine Stromtid*, q.v., 1862–4). He gave these works and others such as *Dörchläuchting* (1866) the generic title *Olle Kamellen* (q.v., 'Old Tales'). In 1863 Reuter moved to Eisenach, which remained his home for the rest of a life that was henceforth diversified only by a rather adventurous journey to the Mediterranean and the Near East (Corfu,

Smyrna, Constantinople, Venice). He wrote three comedies, all of which reached the stage (*Onkel Jakob und Onkel Jochen, Blücher in Teterow* (both 1857), *Die drei Langhänse*, 1858). Reuter, whose principal novels draw extensively upon his own experiences and those of his family, had a considerable gift for the creation of character in a small town or rural environment, a sense of fun, and a pervading good humour, all of which make him one of the most successful of German comic writers.

Reuter's collected works and posthumous papers (15 vols.) were published with a biography by A. Wilbrandt in 1861–75, *Lustspiele und Polterabendgedichte* (2 vols.) were added 1878. *Gesammelte Werke und Briefe* (9 vols., including biography), ed. K. Batt, appeared in 1967, selections, transposed into German by F. and B. Minssen, in 1977 (*Gezeiten des Lebens* and *Das Leben auf dem Lande*).

REUTER, Gabriele (Alexandria, 1859–1941, Weimar), spent her early childhood in Egypt and had her later schooling in Wolfenbüttel and Neuhaldensleben, near Magdeburg. From 1880 to 1895, when she moved to Munich and Berlin (1899), she lived in Weimar, where she also spent her last years. She began writing with the novels *Glück und Geld* (1888), set in modern Egypt, and *Kolonistenvolk* (1891), a story of Argentina. She achieved her first great success with her best-known novel *Aus guter Familie* (1895); Agathe Heidling, a young unmarried woman of good family, seeks contact with social reality, but is unable to break through the barriers of convention which limit the woman's sphere.

Most of Gabriele Reuter's novels deal with women seeking with greater or less success to emancipate themselves. They include *Ellen von der Weyden* (1900), *Liselotte von Reckling* (1903), *Das Tränenhaus* (1909), and *Benedikta* (1923). *Frau Bürgelin und ihre Söhne* (1899) deals with a loving but tyrannical mother. *Frühlingstaumel* (1911) and its sequel *Die Jugend eines Idealisten* (1917) are concerned with two phases in the private and family life of a great (fictional) actress, Elena Schneider. In 1905 she published biographies of Annette von Droste-Hülshoff and Marie von Ebner-Eschenbach (qq.v.). Her autobiography is entitled *Vom Kinde zum Menschen* (1921).

Revolution der Literatur, a polemical work by C. Bleibtreu (q.v.), published in 1886. It condemns polished writers such as Heyse (q.v.), rejects the Naturalists ('unreife Jünglinge, welche glauben, das Wesen des Realismus bestehe darin, gemeine Situationen und Konflikte zu pflegen'), and applauds the true realist who sees things 'sub specie aeterni'. His 'new poetry' ('Neue Poesie') was to consist in a fusion of the realistic and the Romantic. *Revolution der Literatur* was reprinted in 1973.

Revolutionen 1848-9. After the February Revolutions in France, similar revolutions broke out in central Europe, threatening the Habsburg Empire and the Prussian Monarchy. The German Confederation (see DEUTSCHER BUND) of 1815, which had remained intact under the stringent control of Metternich (q.v.), was discredited (see FRANKFURTER NATIONALVERSAMMLUNG). The heads and governments of the smaller German states refrained from active resistance to popular demands and so avoided civil war. Bavaria, where King Ludwig I (q.v.) abdicated in favour of Maximilian II (q.v.), also avoided internal strife.

(1) *The Revolution in Austria-Hungary.* The heterogeneous nature of the Austrian Empire explains the particularly serious course of the revolution there, which was not confined to popular demands for constitutional government but included demands for national independence. On 12 March Ferdinand I (q.v.) of Austria was confronted in Vienna with a protest march of university students and professors, and on the two following days clashes took place between the populace and the military; the latter proved unreliable, some soldiers joining the crowd. On 15 March the Emperor was sufficiently intimidated to issue edicts promising a liberal constitution, a parliament, and freedom of the press, which had been under severe censorship throughout the Metternich era. Risings in Budapest similarly succeeded by 17 March in inducing the Emperor, as King of Hungary, to give way to demands for a renewal of the old, more liberal, constitution. The concessions granted to the Austrians were extended to Hungary and Hungarian autonomy was recognized. Under their leader Kossuth (q.v.) the Hungarians provoked the Austrian government to an open declaration of war (October) with the aim of attaining complete independence. But the non-Hungarian subjects of Hungary, the Slavs, Serbs, and Croats, rightly feared the consequences of being a minority dependent upon Hungary. Under the leadership of Jellačić (q.v.) they declared war against Hungary, and at a later stage they joined forces with Austria. The sympathies of the Viennese revolutionaries, however, lay with Kossuth, who encouraged them by promising Hungarian support.

The position of the Emperor was already precarious after the concessions made in March. In May he lost still more ground by fresh demonstrations, and sought refuge in Innsbruck.

In June his General Windischgraetz (q.v.) succeeded in subduing the Czechs in Prague, and in August General Radetzky (q.v.) defeated the Piedmontese and occupied Milan. Thereupon Ferdinand returned to Vienna, only to flee again in October on the approach of Hungarian troops and fresh risings in Vienna. He abdicated in December in favour of Franz Joseph (q.v.), his young nephew. That Franz Joseph was able to succeed at all was due to the resolute military action of Windischgraetz, who, on his own initiative, bluntly insisted on the unconditional surrender of the rebellious army and populace, and in November appointed Prince Felix zu Schwarzenberg (q.v.) as Metternich's successor.

Despite the occupation of Budapest, the war against Hungary was by no means decided. The commander of the Hungarian army, A. von Görgey (1818-1916), defeated Windischgraetz (battle of Isaszet, April 1849), and would probably have won the war, had Görgey and Kossuth agreed on a common policy; but of greater importance was the humiliating, but effective, appeal by Franz Joseph to Tsar Nicholas I for military aid. Defeat, revenge, and atrocities ended the Hungarian bid for independence from the Habsburg crown. Italy fared no better. With Schwarzenberg's reactionary rule the Empire returned, for the time being, to its pre-revolutionary state. The most notable men of letters to witness the Viennese Revolution were Grillparzer, Bauernfeld, Nestroy, and Hebbel (qq.v.).

(2) *The Revolution in Prussia.* The Revolution in Prussia was the result of curious misunderstandings and confusions, for Friedrich Wilhelm IV (q.v.) was too irresolute to risk serious resistance to popular demands. On 18 March he issued a proclamation (Patent) promising to promote the cause of German unity, fulfil the promise of a constitution, and consent to the freedom of the press. Crowds gathered in the forecourt of the royal residence (Schloß) in a gesture of thanksgiving to their sovereign. But the army, which had been on the alert for some days, appears to have mistaken popular intentions and feared hostile demonstrations and violence. Clashes arose as the soldiers tried to clear the forecourt and shots were fired claiming civilian victims, who became martyrs in the eyes of an angry crowd. Believing themselves betrayed by the King, the masses expressed their resentment in street fighting, which threatened to let loose the civil war the King had sought to avoid. In the night of 18/19 March the deeply distressed Friedrich Wilhelm drafted his address to his 'beloved Berliners' (*An meine lieben Berliner*), in which he promised the prompt withdrawal of the troops if the people on their part would abandon the barricades. Friedrich Wilhelm was

now defenceless in Berlin. He was obliged to pay his respects in public to the bodies of the civilians which had been carried into the forecourt of the Palace. The events of the March Revolution ended with a fresh royal proclamation (19 March) after the King had ridden through his capital wearing a black, red, and gold armlet (see SCHWARZROTGOLD). But in his eyes this move and his address *An mein Volk und an die deutsche Nation* (21 March 1848) expressed a new turning-point in German history. His final assertion that Prussia would henceforth identify itself with the German nation ('Preußen geht fortan in Deutschland auf') was not meant as a concession of defeat, though it was so understood by his critics (headed by the conservative Royalists, who included Bismarck, q.v.); Friedrich Wilhelm's intention was to indicate that Prussia was to guide the German nation towards unity. He had to approve the formation of a liberal government and the re-election of the United Diet (Vereinigter Landtag) on the basis of universal and equal manhood suffrage.

On 22 May the duly elected Diet opened its session to discuss the new Prussian constitution. But the opposition of the recently formed conservative Junkerparlament grew as the radical left asserted itself in the Diet. Always afraid of popular moves and at last gladly yielding to pressure by his strong-minded counsellors, the King foiled the liberal constitution, which was modelled on the democratic Belgian constitution of 1831, by appointing a new conservative ministry under Count Brandenburg (5 July 1848). He took care to do this at a safe distance from Berlin, having taken up temporary residence at Brandenburg. Its first action was to dissolve the liberal ministry—without, however, withdrawing all the King's concessions. The Prussian constitution was ratified on 31 January 1850. The United Diet consisted of two chambers (Herrenhaus and Abgeordnetenhaus) chosen by indirect election. The Lower Chamber was granted the right of taxation, but the King retained the executive power, the right to appoint his ministers; he was head of the civil service and the army.

The political orientation of literature which was noteworthy before 1848 (see JUNGES DEUTSCHLAND) ceased after the failure of the revolutions. Realistic trends for the next decades become manifest in the writers of Poetic Realism (see POETISCHER REALISMUS) whose aesthetic aim is a balanced, politically unprovocative presentation of society.

Revolutionskriege, the wars resulting from the French Revolution (q.v.) of 1789.

The First Coalition. As the monarchy and the personal safety of Louis XVI and of his family were seriously endangered and many emigrant nobles (*émigrés*) fled to Germany pleading for support against revolutionary France, the German powers faced an increasingly difficult situation. But the Austrian Emperor Leopold II (q.v.) was anxious to avoid an open conflict with France, in spite of his concern for his sister Queen Marie-Antoinette (q.v.). In the declaration of Pillnitz (August 1791) Friedrich Wilhelm II (q.v.) of Prussia and Leopold refused to intervene on behalf of the *émigrés*; at the same time they expressed their concern for the French king and for the German princes in Alsace, for whom they demanded a reinstatement of their feudal rights. The Declaration of Pillnitz thus aggravated the danger of war, and the Girondins saw in the campaign against the enemies of the Revolution a means of consolidating their position at home.

In April 1792, a month after the death of Leopold II and the accession of Franz II (q.v.), France declared war on Austria, implicating Prussia as an ally of Austria. The war became an integral part of the Revolution; a manifesto of Duke Karl Wilhelm Ferdinand of Brunswick (reigned from 1780, killed in action 1806) on behalf of the militant *émigrés* provoked the storming of the Tuileries, and the advance of the Prussians across the Rhine to Longwy and Verdun (30 August 1792) determined the abolition of the French monarchy. Prussian progress was checked at Valmy (which Goethe witnessed, see CAMPAGNE IN FRANKREICH), and before the end of 1792 the French Republic occupied the Austrian Netherlands (Belgium), Speyer, Worms, and Mainz, as well as Savoy and Nice. For a short time Frankfurt was in French hands. In 1793 the coalition, now including Great Britain, gained successes by the recovery of the Austrian Netherlands and Mainz; the invasion of Alsace once more threatened an advance towards Paris. At this point the Jacobins reorganized national resistance under Carnot, and by the end of 1793 the French were well on the way to recovering their initial gains in Alsace and on the left bank of the Rhine. During 1794 French advances were made easier by the slackening of Prussian resistance. In 1795 Friedrich Wilhelm II, tempted by gains in Poland (see POLAND, PARTITIONS OF), deserted the coalition.

In the Peace of Basel (5 April 1795) Prussia recognized the French Republic and its claim on the left bank of the Rhine, and ceded Mörs, Kleve, and upper Geldern to France. In return France acknowledged Prussia's right to compensate herself for losses west of the Rhine by territorial acquisitions on the right bank. Other German powers likewise withdrew, including Baden, Bavaria, Hesse-Cassel, the Swabian

Circle (Schwäbischer Kreis), and Württemberg; Austria alone remained in the field against Napoleon, whose repeated successes in Italy culminated in the fall of the fortress of Mantua in February 1797. While Austria negotiated with France following a preliminary peace at Leoben, Napoleon invaded the republic of Venice.

In the Peace of Campo-Formio (17 October 1797) Austria recognized the Cisalpine Republic, created by Napoleon in the name of the French Republic, but was promised in return those parts of the republic of Venice which had not been incorporated into the Cisalpine state (the territory east of the Adige with Istria and Dalmatia). Austria ceded the Netherlands (Belgium) to France and, in secret clauses, acknowledged the French acquisition of the left bank of the Rhine, except for occupied Prussian territories. German princes were to be compensated for their losses with ecclesiastical principalities at a congress at Rastatt. France also agreed to the Austrian acquisition of the Bavarian Inn district (see INNVIERTEL) and of the bishopric of Salzburg.

The treaties of Basel and Campo-Formio ending the wars of the First Coalition realized French ambitions which had not been fulfilled in the Thirty Years War (see DREISSIGJÄHRIGER KRIEG). But apart from territorial losses they confirmed above all the dualism persisting in the relationship between Prussia and Austria since the Silesian Wars (see SCHLESISCHE KRIEGE) and exposed the Holy Roman Empire (see DEUTSCHES REICH, ALTES) as an ironic anachronism. At the Congress of Rastatt (November 1797–April 1799), which was supervised by envoys of the French Republic while the newly gained territories were being integrated into the French legal and administrative system, no agreement was reached over the embarrassing problem of territorial compensation.

The Second Coalition. In 1798 French conquests were resumed and the proclamation of the Roman Republic in the Papal States and the Helvetian Republic in Switzerland was the prelude to Napoleon's Egyptian campaign, which provoked the Second Coalition against the French Directory; Prussia abstained. Napoleon returned to France, deposed the Directory by a *coup d'état* (9 November 1799), and, as First Consul in the Consulate, took over the government of France; he immediately set out to repair French military disasters in Italy. His position was strengthened by Russia's withdrawal from the Coalition. On 14 June 1800 he won the battle of Marengo, to which Moreau added a victory at Hohenlinden in Bavaria (3 December 1800). These two battles decided the issue in favour of France. In the Treaty of

Lunéville (9 February 1801) Austria confirmed the conditions of the Treaty of Campo-Formio and recognized the French possessions on the Rhine as well as the Batavian, Helvetian, Cisalpine, and Ligurian (Genoa) republics. France and Great Britain concluded the Peace of Amiens in the following year (1802).

The question of compensation and redistribution of land had still to be solved, but to achieve results the conference was this time transferred to Paris, beginning its sessions in 1801. Its Resolution, the Reichsdeputationshauptschluß (Principal Resolution of the Imperial Deputation), was passed on 15 February 1803: more than half of the 360 odd petty German states existing since the Peace of Westphalia (see WESTFÄLISCHER FRIEDE) were eliminated; of the free cities (see FREIE STADT) only six survived (Hamburg, Lübeck, Bremen, Frankfurt, Nürnberg, and Augsburg); virtually all ecclesiastical states disappeared, notably Cologne, Trier, and Mainz (for the last of which Regensburg was substituted). The Reichsdeputationshauptschluß favoured above all Prussia and the secondary German states, thus balancing the German powers against Austria. Baden, Württemberg, and Bavaria especially emerged strengthened. Prussia gained five times more than it lost, absorbing a number of free cities and ecclesiastical states such as Hildesheim, Paderborn, Münster, Erfurt, and Goslar. The circumstances surrounding the preparations for the Resolution were in many respects humiliating, but the resultant map of Germany was a conspicuous inauguration of Napoleonic influence over Germany, abolishing the structure of the Holy Roman Empire which formally ceased to exist three years later (1806), after the formation of the Third Coalition (see NAPOLEONIC WARS).

Rezeptionsästhetik, term used for the study of the evaluation of a work of literature, or of art, by its 'recipient'. Proceeding from premises established by H. R. Jauß (*Literaturgeschichte als Provokation,* 1970) and W. Iser (*Der implizite Leser. Kommunikationsformen des Romans von Bunyan bis Beckett,* 1972), it aims at an assessment of socio-political factors underlying the judgement of the recipient. Neither the author's imagined reader, nor individual, subjective taste is the subject of enquiry, but the integration of a literary work and its reader into the objective historical process, the evaluation of which it is designed to promote. Rezeptionsästhetik has proved fruitful especially in respect of periods that yield detailed documentation bringing out a specific social structure, in the context of which literary work assumes renewed actuality. It has stimulated a reassessment of Trivialliteratur (q.v.), and it has served critics in their assessment of the reception of contemporary works of literature by other critics representing different societies.

REZZORI, Gregor von (Czernowitz, then Hungarian, 1914–), is the author of fiction of grotesque humour exploiting the oddities of some of the less sophisticated parts of south-east Europe: *Maghrebinische Geschichten* (1953), set in the imaginary 'Maghrebinien', was published with Rezzori's own trenchant illustrations. His novels include *Ödipus siegt bei Stalingrad* (1954), *Ein Hermelin in Tschernopol* (1958), *Der Tod meines Bruders Abel* (1976), and *Memoiren eines Antisemiten* (1979). A satirical work in 4 vols., *Idiotenführer durch die deutsche Gesellschaft*, appeared in 1962–5.

Rhein, Der, a hymnic poem written in 1801 by F. Hölderlin (q.v.). The Rhine's movement from the Alps to the plains is its basic motif, evoking a vision of a blessed, though precarious, harmony succeeding the fierce energy of youth. The poem is dedicated to Hölderlin's friend, Isaak von Sinclair.

Rheinauer Sündenklage, an anonymous Middle High German confession of sins. It was probably written in Rheinau Abbey in the second half of the 12th c.

Rheinauer Weltgerichtsspiel, a medieval religious play based on passages from St. Matthew 24 and 25 dealing with the Day of Judgement. The date of the MS. (of Rheinau Abbey near Schaffhausen) is 1467, and there are four other versions, all of the 15th c. All are derived from a lost 14th-c. original.

Rheinbund (*Confédération du Rhin*). The Confederation of the Rhine was set up by the Treaty of Paris, 7 July 1806, by Napoleon I with the object of extending French influence into Germany and neutralizing the power of Austria and Prussia (see Napoleonic Wars).

Napoleon appointed himself president (Protektor) of the Confederation and K. Th. von Dalberg (q.v.) primate (Fürst-Primas). The principal original members of the Confederation were the principality of Berg (newly created from Prussian and Bavarian territory and ruled by Napoleon's Marshal Murat), Bavaria and Württemberg (the rulers of which became kings), Baden and Hesse-Darmstadt (whose princes became Grand Dukes), the duchy of Nassau, and the newly created Grand Duchy of Frankfurt. A number of states of the Holy Roman Empire (see Deutsches Reich, Altes), which was in process of dissolution, were annexed in the scramble for territorial expansion;

they included the free cities of Frankfurt, Nürnberg, and Augsburg. The new Grand Duchy of Würzburg joined later in the year, and in 1807 Napoleon's newly formed Kingdom of Westphalia (see Westfalen) became a member state. The defeat of Prussia in October 1806 was followed by the accession of Saxony (now made a kingdom) and of Mecklenburg, together with a number of petty principalities. At its height the Confederation embraced thirty-six German states.

The states of the Confederation were a protectorate of France, and were bound by treaty to furnish substantial contingents of troops for Napoleon's campaigns (63,000 men in 1806). After the disaster of Napoleon's Russian campaign a *sauve qui peut* began among the princes of the Rheinbund. The Confederation ceased to exist in 1813.

Rheingold, Das, see Ring des Nibelungen, Der.

Rheinische Bund or **Rheinischer Städtebund.** (1) An alliance of four Rhenish cities, Mainz, Bingen, Oppenheim, and Worms, formed in 1254 in order to preserve order during the anarchic years of the Interregnum (q.v.). It grew rapidly to include other cities, but ceased to be effective after 1257.

(2) A similar alliance set up in 1381 and lasting until 1388.

Rheinische Merkur, Der, a political newspaper, founded and edited by J. J. Görres (q.v.), and appearing three times a week from 1814 to 1816. Devoted to the liberation and reunification of Germany, its influence is well summed up by Napoleon's comment that it represented the fifth of the Great Powers. Among its contributors were E. M. Arndt, L. J. (Achim) von Arnim, C. Brentano, the brothers J. and W. Grimm, the Freiherr vom Stein, and N. von Gneisenau (qq.v.), the two last being among the principal figures behind the political and military reorganization leading to the War of Liberation (see Napoleonic Wars).

Rheinisches Marienlob, an anonymous Middle High German poem in praise of the Virgin Mary, written c. 1220 for the use of a convent of nuns. It is notable for its warmth of feeling and for the rhetorical flexibility of its metrical structure.

Rheinische Thalia, Die, see Thalia, Die.

Rheinländische Hausfreund, Der, see Hebel, J. P.

Rheinsberg in der Mark, was bought in 1734 by Friedrich Wilhelm I (q.v.) of Prussia, who gave it to his heir Crown Prince Friedrich (see FRIEDRICH II, DER GROSSE). Friedrich altered Schloß Rheinsberg and spent there the last four years before his accession in 1740, frequently surrounded by friends in a style which he later resumed at Sanssouci (q.v.). He remembered it as a place in which he had spent days and nights in the study of Cicero, Plutarch, Lucretius, Tacitus, and Racine. In 1744 he gave it to his brother Heinrich, Prinz von Preußen (q.v.) as a reward for bravery. Heinrich, valuing his independence from both his brother and wife, used it after the Seven Years War (1763, see SIEBENJÄHRIGER KRIEG) until his death in 1802 as his permanent residence.

Rheinweinlied, a poem by M. Claudius (q.v.), better known by its first line 'Bekränzt mit Laub den lieben vollen Becher' (q.v.).

Rhenser Kurfürstentag or **Kurverein,** a meeting of the German King Ludwig IV (q.v.) and the electoral princes in 1338, at which a declaration was agreed affirming that a properly elected German King (see DEUTSCHER KÖNIG) did not need papal approbation for the exercise of his functions.

Richard Savage oder der Sohn einer Mutter, a tragedy (Trauerspiel) by K. Gutzkow (q.v.), first performed in July 1839 at Frankfurt/Main. The poet Richard Savage discovers that he is an illegitimate son of Lady Macclesfield. When he is in dire danger on a charge of murder, she does not stir a finger to help him, and it is the actress Miss Ellen who intercedes to secure his pardon. Savage is taken up by Lord Tyrconnel, who plans to use him to humiliate Lady Macclesfield, but he refuses to play such an ignoble role, turning his back on wealth to die in poverty. There is a touching reconciliation at the end with Lady Macclesfield. The play is conceived as a disguised attack on the German aristocracy, and to this end Gutzkow makes free use of his source, Dr. Johnson's account in *Lives of the Poets.* It was Gutzkow's first successful work for the stage.

RICHARDSON, SAMUEL (Derbyshire, 1689–1761, London), the author of the novels *Pamela or Virtue Rewarded* (2 vols., 1740–1), *Clarissa or The History of a Young Lady* (7 vols., 1747–8), and *The History of Sir Charles Grandison* (1753–4), which by their combination of sentiment and morality influenced German literature in the middle of the 18th c.

Gellert in his novel *Das Leben der schwedischen*

Gräfin von G . . . (q.v.) and Lessing in his play *Miß Sara Sampson* (q.v.) are clearly indebted to Richardson. *Pamela,* as a novel presented in letters (see BRIEFROMAN), influenced German writers both directly and indirectly, through *La Nouvelle Héloïse* of J.-J. Rousseau (q.v.). *Pamela* was translated into German in 1742, *Clarissa* 1748–51, and *Grandison* 1754–9. A parody of the last by J. K. A. Musäus (q.v., *Grandison der Zweite*) appeared in 1760–2.

RICHTER, HANS WERNER (Bansin, Usedom, 1908–), a fisherman's son from a Baltic island, worked in a bookshop in Swinemünde and later in Berlin. He served in the 1939–45 War, was taken prisoner in 1943, and after the war established himself as a writer. A co-editor of the periodical *Der Ruf* (q.v.), he became a founder of the Gruppe 47 (q.v.) and in 1962 edited the *Almanach der Gruppe 47. 1947–62.* His autobiographical novel *Spuren im Sand* (1953) conveys the impressions of the village school boy whose father served in the 1914–18 War, vividly conjuring up the background of Wilhelmine Germany and his hopes and disillusionments in the Weimar Republic. This and the 1939–45 War and further years of disillusionment remained the substance of his fiction, which includes the novels *Die Geschlagenen* (1949), *Sie fielen aus Gottes Hand* (1952, tracing the history of twelve refugees in a displaced persons' camp near Nürnberg, whose common problem is the future), *Du sollst nicht töten* (1954), *Linus Fleck oder Der Verlust der Würde* (1959), and *Rose weiß, Rose rot* (1971), a powerful return to the ideals and disillusionments of the 1914–18 War generation, represented mainly by young communists, during the Weimar Republic. *Blinder Alarm. Geschichten aus Bansin,* a collection of stories, appeared in 1970, and the story *Die Flucht nach Abanon* (1980) was followed by two more novels, *Die Stunde der falschen Triumphe* (1981) and *Ein Julitag* (1982). *Deine Söhne Europa* (1947) is a volume of poems, *Menschen in freundlicher Umgebung* (1965) a collection of six satires. His other writings include *Die Mauer oder der 13. August* (1961), *Bestandsaufnahme* (1962), and a volume for children, *Kinderfarm Ponyhof* (1976), which had a sequel in *Bärbel Hoppsala. Neue Abenteuer auf der Kinderfarm Ponyhof.*

RICHTER, JOHANN PAUL FRIEDRICH, real name of the writer universally known by his pseudonym JEAN Paul (q.v.).

RICHTER, JOSEPH (Vienna, 1749–1813, Vienna), was an Austrian journalist, who also wrote plays, stories, and poetry. He is best known as the pseudonymous author of the 'Eipeldauer letters' (*Briefe· eines Eipeldauers an seinen Herrn Vetter in Krakau über d'Wienerstadt*) which, except for a four-year spell (1798–1801), appeared from 1785 until his death. The 'Eipeldauer' is a simple-minded commentator on the Viennese scene, and the whole work is an exercise in satirical journalism. After Richter's death it was continued by other hands, including A. Bäuerle (q.v.). *Sämtliche Schriften* (12 vols.) were published in 1813.

RICHTER, LUDWIG (Dresden, 1803–84, Dresden), in full Adrian Ludwig, a draughtsman and painter, remained a Romantic all his life, long after the Romantic Movement (see ROMANTIK) had faded. He is at his best in water-colours and sets of illustrations with decorative detail for books. These include *Deutsche Volksbücher* (1838–46), *Volksmärchen von Musäus* (1842, see MUSÄUS), *Alemannische Gedichte von Hebel* (1851, see HEBEL, J. P.), *Andersens Märchen* (1851), *Bechsteins Märchenbuch* (1853), *Goethe-Album* (1853–6), and *Bilder zu Schillers Glocke* (1857). An autobiographical volume, *Lebenserinnerungen eines deutschen Malers*, was published posthumously by Richter's son (1885, re-ed. E. Marx, 1944).

Richterin, Die, a Novelle by C. F. Meyer (q.v.), published in 1885 in the *Deutsche Rundschau*. Divided into five chapters, it is set soon after the coronation of Charlemagne in 800 (see KARL I) and deals with the theme of justice, to which the murder of Count Wulf on the day of his marriage to Stemma is central. Sixteen years have passed since the Count's sudden death, and during this time Stemma has acted as judge (Judicatrix) in Rhaetia. Her position adds to the weight of conscience, for it is she who poisoned Count Wulf because she had given her love to another, Peregrin, the father of her daughter Palma. She confesses her guilt to the Emperor before poisoning herself as a judgement from God. The action is supported by the Count's family horn (Wulfenhorn), which is said to have the power to compel sinners to confess, and by Faustine, a woman who confesses to the murder of her husband despite the lapse of years which have taken her out of reach of secular justice. The narrative is made complex by the love of Wulfrin, the Count's son by a previous marriage, for his half-sister Palma, which intertwines with the main plot.

Richter und sein Henker, Der, a detective story by F. Dürrenmatt (q.v.), published in book form in 1952 after appearing in 1951 as a serial in *Der Schweizerische Beobachter*. Kommissär Bärlach, who makes a further appearance in *Der Verdacht* (q.v.), is the detective, an ageing man suffering from cancer. He is entrusted with the investigation of the murder of detective Schmied, a promising officer. As his assistant in this matter Bärlach chooses an ambitious officer named Tschanz. Suspicion falls on an old enemy of Bärlach, Gastmann, who has never been convicted of a crime, though Bärlach knows that he is a murderer. In a confrontation with Gastmann Bärlach calmly states that the executioner he has appointed will kill Gastmann before the day is over. He sends Tschanz to Gastmann's house. Gastmann, accompanied by his two bodyguards, is about to leave. When one of them draws a revolver, Tschanz, who is on his guard, is quicker and shoots all three dead. Bärlach invites him to dinner to 'celebrate', and reveals that he knows that Tschanz killed Schmied out of professional jealousy, and that he has now entrapped Tschanz into murdering again.

Acting as a judge (Richter), Bärlach has made the policeman his own executioner (Henker). Tschanz's body is found in his car, crushed by a train at a level-crossing, and Bärlach enters a nursing home for a major operation.

RIEDEL, FRIEDRICH JUSTUS (Vieselbach nr. Erfurt, 1742–85, Vienna), a pastor's son, studied at Jena, Leipzig, and Halle, where he became an associate of the dissolute professor C. A. Klotz (q.v.). He was a professor at Erfurt University in 1768 where he exercised some influence on J. J. W. Heinse (q.v.), and was given an appointment at Vienna in 1772. Of unstable character, Riedel eventually became insane. His *Theorie der schönen Künste und Wissenschaften* (1767) in some respects coincides with the views of Herder (q.v.).

RIEHL, WILHELM HEINRICH VON (Biebrich, 1823–97, Munich), after residence at Marburg, Göttingen, and Gießen universities became a journalist in 1844. He was active in various cities, including Wiesbaden and Augsburg, and in 1848 was a member of the Frankfurt Parliament (see FRANKFURTER NATIONALVERSAMMLUNG). In 1854 he accepted a chair as professor of politics in Munich, retiring in 1892. In 1885 he was additionally appointed director of the National Museum in Munich. He was raised to the nobility in 1883.

Riehl was a notable folk-lorist and cultural historian, publishing a *Naturgeschichte des deutschen Volkes als Grundlage einer deutschen Sozialpolitik* in the four volumes *Die bürgerliche Gesellschaft* (1851), *Land und Leute* (1853), *Die Familie* (1855), and *Das Wanderbuch* (1869). Other notable works in this field were *Die Pfälzer* (1857) and *Die deutsche Arbeit* (1861). He advocated fictional writing based on social history, but was unsuccessful with his own works of this kind. They include the novels *Geschichte vom Eisele und Beisele*

(1848) and *Ein ganzer Mann* (1897), and the *Kulturgeschichtliche Novellen* (1856). His historical sketches *Kulturstudien aus drei Jahrhunderten* (1859) are among his best works. *Geschichten und Novellen* (7 vols.) appeared in 1899–1900 and a selection of fifty Novellen, *Durch tausend Jahre* (4 vols.), of 1937 was reprinted in 1969.

RIEMENSCHNEIDER, TILMAN (Heiligenstadt nr. Erfurt, *c.* 1460–1531, Würzburg), one of the great German wood and stone carvers, presided as master over a busy workshop in Würzburg, where he became a councillor in 1505 and was burgomaster 1520–1. He took the peasant side in the Peasants' War in 1525 (see BAUERNKRIEG) and, after the rising was crushed, was tortured, imprisoned, and deprived of most of his considerable fortune. His greatest work is the Creglinger Altar (1505–10, at Creglingen nr. Rothenburg). Some other works remain in their original sites (New Minster and cathedral, Würzburg; St. Joseph's church, Rothenburg; and Bamberg Cathedral). Riemenschneider's wood carvings were intended to achieve their expression without the aid of paint.

RIEMER, FRIEDRICH WILHELM (Glatz, 1774–1845, Weimar), after completing his studies, became private tutor in the household of Wilhelm von Humboldt (q.v.), and in 1803 was engaged by Goethe as tutor to his son August. He became a schoolmaster at the Weimar grammar school in 1812 and librarian in 1814. With Eckermann (q.v.) he helped to edit the *Ausgabe letzter Hand* of Goethe's works. Riemer wrote poems published as *Blumen und Blätter* (1816–19, under the pseudonym Silvio Romano) and *Gedichte* (1826). He also edited Goethe's correspondence with Zelter (*Briefwechsel zwischen Goethe und Zelter in den Jahren 1796–1832*, 6 vols., 1833–6, and *Briefe von und an Goethe*, 1846) and published 2 vols. of recollections (*Mitteilungen über Goethe aus mündlichen und schriftlichen Quellen*, 2 vols., 1841).

RIEMER, JOHANNES (Halle, 1648–1714, Hamburg), studied in Jena and in 1678 succeeded Christian Weise (q.v.) as a school professor in Weißenfels. As a Protestant clergyman he held cures in Osterwieck and Hildesheim, and from 1704 was Pastor primarius at St. James's (Jakobikirche) in Hamburg. Riemer, a disciple of Weise, conscientiously composed plays as educational aids. They were collected as *Der Regenten bester Hoff-Meister, oder Lustiger Hoff-Parnassus* (1679). He also wrote didactic novels in Weise's manner: *Der politische Maulaffe* (1679), *Politische Colica* (1680), *Der politische Stockfisch* (1681), *Der politische Grillenfänger* (1682), and *Der untreue Ertz-Verläumbder oder böse Mann* (1682).

Rienzi, der letzte der Tribunen, a five-act opera by R. Wagner (q.v.), first performed in Dresden on 20 October 1842. It is based on Bulwer-Lytton's novel *Rienzi, last of the Tribunes* (1835). Cola di Rienzi, a commoner, becomes people's tribune in the 14th-c. Roman republic, but is attacked by the nobles, who are jealous of his success. He and his sister Irene die in the Capitol, which has been set on fire by nobles and people (1354). Wagner wrote his own words for this, his first operatic success.

RIESE, ADAM (Staffelstein, 1492–1559, Annaberg), also spelt Ries, German arithmetician whose name has become proverbial, often being used metonymously for 'arithmetic'. Adam Riese wrote a number of arithmetical manuals and devised the square-root sign.

RIETENBURG, BURGGRAF VON, a Middle High German lyric poet, was probably the younger brother of the Burggraf von Regensburg (q.v.), and is thought to have died not long after 1185. The seven extant stanzas attributed to him reveal the courtly conception of love, in which the man serves the lady without hope of reward.

RILKE, RENÉ KARL WILHELM JOSEF MARIA (Prague, 1875–1926, Val-Mont nr. Montreux), used the German form of his first name, styling himself Rainer Maria Rilke. He belonged to the German-speaking minority in Prague and was the son of an unsuccessful officer, turned railway employee, and of an emotional, ambitious, and possessive mother. His early years with her may well have produced in later life the exceptionally strong impulse to be mothered. Intended for a military career (in compensation for his father's failure), he was sent to the Military School (Militär-Akademie) at St. Pölten (Lower Austria) at the age of 11. He proved unsuited for the army and went in 1891 to the School of Commerce (Handelsakademie) at Linz, but was no better adapted to this course and returned home in 1892. The patronage of a well-to-do uncle enabled him to read for university entry, and in 1895 he began to study philosophy at Prague University. He did not care for this, however, and in 1896 migrated to Munich, ostensibly to study, but actually to devote himself to writing.

Rilke was brought up as a practising Roman Catholic, but in his adolescence he rebelled against the Christian faith, and between 1893 and 1898 wrote several anti-Christian poems which were included with other previously unpublished poetry in vol. 3 of *Sämtliche Werke* (1959). An early autobiographical novel, *Ewald Tragy*, written *c.* 1896, appeared in 1944. *Leben und Lieder* (1894) consists of somewhat sentimental and self-indulgent poetry, and *Larenopfer* (1896) largely of delicate impressions of old Prague. *Traumgekrönt* (1897) begins to show Rilke's evocative power; it comprises two cycles headed *Träumen* and *Lieben*. The collection *Advent* (1898), divided into *Gaben, Fahrten, Funde*, and *Mütter*, shows a deepening introspection. About this time Rilke was drawn into a liaison with Lou Andreas-Salomé (q.v.), whose influence confirmed him in his dedication to poetry. Together they undertook two journeys to Russia (1899 and 1900); deeply impressed by the piety and simplicity of the Russian people, he composed the poetry of *Das Stunden-Buch* (q.v., 1905), an extended cycle purporting to be written by a Russian monk; its haunting music has a truly inward tone reflecting Rilke's newly acquired cult, the religion of art. Contacts in Russia included meetings with Tolstoy, the painter L. O. Pasternak (father of the novelist), and the peasant Spiridon Droschin, author of poetry which Rilke rendered into German; Rilke's own attempt to write poetry in Russian is a further mark of the deep impression of the experiences in Russia. His new outlook also changed his mode of living when, on his return, he settled for a time in the Bremen fen country, attracted by a community of artists in Worpswede, on which he wrote the monograph *Worpswede* (1903). In 1901 he married the sculptress Clara Westhoff; a daughter, Ruth Rilke, was born and the couple lived in stringent financial circumstances in Rilke's farmhouse in Westerwede. They separated the following year; Clara Rilke died in 1954. In June 1902 Rilke had another kind of experience of North Germany, staying as guest of Prince Emil von Schoenaich-Carolath (q.v.) in Schloß Haseldorf in Holstein. While seeing himself as 'ein ziemlich Heimatloser', he was impressed by the sense of belonging emanating from the landscape descriptions of G. Frenssen (q.v.), with whose work he became acquainted through his host; yet he insisted on the supremacy of self-transcendence ('aber mehr, als in der Heimat stehen ist doch noch: weit aus ihr hinauswachsen in den Himmel. Das meine ich').

A few years earlier Rilke had been impressed by the monumental poetry of Stefan George (q.v.), though his own poetry continued to show the lush harmonies of *Mir zur Feier* (poems, 1899) and *Geschichten vom lieben Gott* (prose, with this title 1904, originally *Vom lieben Gott und Anderes*, 1900), and this soft richness persists in the *Stunden-Buch*, written between 1899 and 1903. To 1899 belongs also what was once Rilke's most popular work, *Die Weise vom Leben und Tod des Cornets Christoph Rilke* (1903), a narrative in rhythmic prose about a presumed ancestor in the Turkish war of 1663.

The transition from the old to the new stronger style first appears in the *Buch der Bilder* (q.v., 1902), which contains such diverse poems as the Impressionistic *Herbsttag* and *Herbst*, and the hard, ringing lines of *Ritter*. After the breakdown of his marriage Rilke travelled, and Paris proved, both in its sordid squalor and as a centre of art (especially as he saw it in Rodin's sculpture, *Auguste Rodin*, a monograph, 1903), to be a vital experience. For some months he was Rodin's secretary, and after his dismissal remained in Paris for most of the time until 1909. He composed *Die Aufzeichnungen des Malte Laurids Brigge* (q.v., 1910), and developed the new chiselled, precise, visual style which is largely associated with his name and characterizes the 2 vols. of the *Neue Gedichte* (q.v., 1907–8). They contain such notable poems as *Der Panther, Das Karussell, Orpheus. Eurydike. Hermes, Papageienpark*, and *Die Flamingos*. By the use of visual images and classical plasticity he was able to give his inwardly directed poems a firm structure (see DINGGEDICHT). He also wrote *Requiem* (1909), consisting of two remarkable threnodies, the *Requiem für eine Freundin* (Paula Modersohn-Becker, q.v.) and the shorter but even more beautiful *Requiem für Wolf Graf von Kalckreuth*, a young poet, not personally known to Rilke, who had taken his own life. In 1909 Rilke, overcome by restlessness, began to travel again, visiting North Africa and Spain and paying two visits to Duino Castle on the Dalmatian coast in 1911–12. Here he was the guest of Princess Marie von Thurn und Taxis-Hohenlohe, a lady of wealth, rank, and influence, who, assisted by the publisher Anton Kippenberg, enabled Rilke to fulfil in congenial conditions what he saw as his artistic mission.

In January 1912, left by arrangement alone in Duino, he began to compose the poems of the cycle which, when complete, was to be known as the *Duineser Elegien* (q.v., 1923). He also wrote during this time the poems of *Das Marien-Leben* (1913). Although his *Fünf Gesänge* of 1914 reflect the initial wave of patriotism they also anticipate the mood of profound doubt. Late in 1915 he was called up for the Austrian army. A less apt soldier could hardly be conceived, and Rilke was quickly transferred to the Military Records Office (Kriegsarchiv) and soon after was discharged. For the rest of the war he withdrew into himself, waiting numbly for the return of peace and civilization. The elegies still seemed to him incomplete, an important but fragmentary expression of his existential religion of art and its terribly exacting demands, but inspiration was dormant. In 1922 a congenial environment for writing was found for him by a Swiss patron, W. Reinhard, at the castle of Muzot (Valais, Switzerland). Here, in the astonishingly short time of three weeks, in February, he wrote the two groups of *Die Sonette an Orpheus* (q.v., 1923) and in between composed five more *Duineser Elegien*, so bringing them to the final canon of ten. With these two works, the uncompromising elegies and the relaxed and conciliatory sonnets, Rilke felt that he had fulfilled the demand made on him by his genius, though he continued to write elegant and delicate verse, some of which was in French (titles of their posthumous publications include *Les Fenêtres* and *Les Roses*, both 1927). Rilke translated poetry by Michelangelo, J. P. Jacobsen, Louïze Labé, Magallon, Elizabeth Barrett Browning, Mallarmé, Verlaine, Baudelaire, Leopardi, Paul Valéry; Russian authors include Tolstoy, Dostoevsky, and Chekhov. He died of leukaemia.

Probably no German poet is so well known in the English-speaking world as Rilke. In Germany he has often been seen as a systematic philosopher and a dispenser of wisdom, but the interpreters have mostly been at variance about the nature of the message. Beyond all doubt is his standing as one of the great poets of the 20th c., grappling in verse with the problems of his own sensitive soul, seeking rather to interpret himself than to communicate with readers, and willing to make and accept sacrifices to achieve the perfection of his art. Despite occasional preciosity and affectation his power to manipulate words, and the delicate precision of his workmanship are superlative, and since, for all his egocentricity, he was a sentient human being, communication is established, even if it is not willed. The striking speed with which he wrote his greatest works proves that he belonged to a small category of poets (including Goethe) with whom poems are not wrought, but are incubated, often over a long period in the unconscious mind, so that when they finally reach the surface they can be written down as it were to inner dictation. This process does not preclude subsequent careful revision. Rilke himself called the final euphoric stage of this process the Umschlag, which by some has been translated as the reversal.

Gesammelte Werke (6 vols.) was published in 1927 and *Sämtliche Werke* (6 vols.), ed. R. Sieber-Rilke and E. Zinn, in 1955–66 (and in 12 vols. 1975); editions of correspondence include *Gesammelte Briefe in sechs Bänden* (from the years 1899–1926), ed. R. Sieber-Rilke and C. Sieber, 1936–9, and correspondence with H. von Hofmannsthal (q.v.), 1978. Two vols. of *Selected Works*, containing notable translations by J. B. Leishman and C. Craig Houston, appeared in 1954 and 1960.

Rinaldo Rinaldini, der Räuberhauptmann,

a once popular novel about a noble brigand, published in 1798 by C. A. Vulpius (q.v.).

Ring, Der, see WITTENWEILER, HEINRICH.

Ring des Nibelungen, Der, a cyclical 'Bühnenfestspiel' by R. Wagner (q.v.), consisting of a Vorabend, *Das Rheingold*, and the three music dramas *Die Walküre, Siegfried,* and *Götterdämmerung*. In its first form the text of the cycle was written between 1848 and 1853, when it was privately printed. The musical composition of *Das Rheingold* was completed in 1854, and that of *Die Walküre* in 1856. A long interval ensued before Wagner again took up *Siegfried* (originally *Siegfrieds Tod*) in 1868, and this work was completed in 1871. The last work of the cycle, *Götterdämmerung*, was finished in 1874. *Das Rheingold* was performed at Munich on 22 September 1869 and *Die Walküre* at Munich on 26 June 1870. The first performance of the cycle, which was also the first performance of *Siegfried* and of *Götterdämmerung*, took place at Bayreuth on 13, 14, 16, and 17 August 1876.

Wagner's sources, freely used, were the Edda (q.v.) and the Völsungensaga. In *Das Rheingold*, the gold concealed in the Rhine and guarded by the daughters of the Rhine excites the cupidity of gnomes, giants, and gods alike. By abstaining from love Alberich, the dwarf, is enabled to carry it off; Wotan cheats Alberich of it in order to pay the giants Fasolt and Fafner for the palace of Walhalla, which they have built for him. The giants quarrel over the gold, and Fafner, having slain his brother Fasolt, makes off with it. Wotan and the other gods take possession of Walhalla. In the course of *Das Rheingold*, a ring giving supreme worldly power is fashioned from the gold, as well as the Tarnhelm of invisibility and transformation.

Die Walküre begins in the world of men. Siegmund, son of Wotan and of a mortal, is beset by the barbaric Hunding and his men and in dire need of weapons. Sieglinde, who proves to be his sister, shows him Wotan's sword embedded in a tree, and this he is able to draw forth and use. Wotan is persuaded by Fricka, his wife, to take sides against Siegmund, whom Brünnhilde, the Walküre, seeks to rescue. Siegmund is, however, slain through Wotan's intervention. Sieglinde is rescued by Brünnhilde and later bears Siegmund a son, named Siegfried. Meanwhile Wotan punishes Brünnhilde for opposing him by putting her into a sleep and surrounding her with flames.

When *Siegfried* opens, the young Siegfried, brought up by the dwarf Mime (for Sieglinde died at his birth), succeeds in fashioning afresh the shattered sword of Wotan which had failed Siegmund. Having slain the dwarf, who wished to poison him, he confronts and kills the guardian of the Rhine gold, Fafner, who has changed himself into a terrible dragon. Siegfried gains the treasure, then sets out, guided by the Wood Bird, for the fire-girt rock on which Brünnhilde sleeps. He awakens her and the two are united.

In *Götterdämmerung*, Siegfried comes to King Gunther's court. Hagen, Gunther's vassal, a son of Alberich, plans to rob Siegfried of the gold in order to gain worldly power. Siegfried is to be given a potion of oblivion, by which he will forget Brünnhilde, and is then to marry Gutrune, Gunther's sister. The plan succeeds. Siegfried, no longer recognizing Brünnhilde, aids Gunther to woo her. The outraged Brünnhilde is avenged by Hagen, who stabs Siegfried in the back in the one vulnerable spot revealed to him by Gutrune. Too late, Brünnhilde discovers the plot of the potion, of which she and the dead Siegfried are the victims. She gives orders for the erection of her funeral pyre, and rides her horse Grane into the flames; Hagen, striving vainly for the ring, is engulfed in the Rhine, the treasure returns to the Rhine daughters and the crashing beams and rafters of Walhalla mark the end of the gods.

This revival of Germanic myth is also a symbolic portrayal of the conflicting lusts for the flesh, for power, and for gold. A famous, if somewhat narrow, interpretation from the economic angle is given in G. B. Shaw's *The Perfect Wagnerite* (1898).

Ring des Polykrates, Der, a ballad by Schiller. It was written in June 1797, and published in *Der Musenalmanach* for 1798. It narrates a critical episode in the life of Polykrates, Tyrant of Samos. In the flush of his prosperity and power Polykrates is warned by a friend to beware of the envy of the gods. Polykrates heeds the advice and hurls into the sea a most precious ring. The following day the ring is restored to him, having been found in the belly of a fish. The friend departs, believing Polykrates to be destined for disaster—'Die Götter wollen dein Verderben'—an allusion to the cruel execution of Polykrates recounted by Herodotus. The sparse and taut poem has in its structure a resemblance to the opening of the Book of Job.

RINGELNATZ, JOACHIM, pseudonym of Hans Bötticher (Wurzen, 1883–1934, Berlin); the name is that of the harmless watersnake (*Natrix natrix*, common on the Continent, but not indigenous to Britain), and it is well suited to a humorist, satirist, and ironist, who was also a man of goodwill. Ringelnatz was restless in his youth, running away to sea, later becoming a librarian, then a cabaret singer with the cabarets Simplizissimus in Munich and Schall und Rauch

in Berlin. In the 1914–18 War he became a naval officer and an air pilot. He was highly successful in reciting and singing his own sarcastically satirical pieces, which guyed current intellectual and social trends and fashions. He published many volumes of comic verse and nonsense poetry, including *Schnupftabaksdose* (1912, with R. J. M. Seewald, b. 1889), *Kuttel Daddeldu* (1920, extended edn. 1923), *Nervosipopel* (1924), *Allerdings* (1928), *Kinder-Verwirr-Buch* (1931), *Gedichte dreier Jahre* (1932), and *Gedichte, Gedichte* (1934); posthumous publications include *Und auf einmal steht es neben dir* (1950) and *Kasperle-Verse* (1954). He was also the author of stories, collected as *Ein jeder lebt's* (1913) and *Die Woge* (1932), and of autobiographical writings, *Als Mariner im Krieg* (1928) and *Mein Leben bis zum Kriege* (1931, reissued 1966). Correspondence appeared posthumously, including an edition by Ringelnatz's wife whom he called Muschelkalk (*Reisebriefe an Muschelkalk*, 1964). *Das Gesamtwerk in sieben Bänden*, ed. W. Pape, appeared 1982 ff.

RINGWALDT, BARTHOLOMÄUS (Frankfurt/Oder, 1530–99, Langenfeld, Brandenburg), pastor in Langenfeld from 1567 until his death, was a prolific poet. Of his hymns, which are collected in *Evangelia* (1581), the best known is 'Es ist gewißlich an der Zeit'. *Die lauter Warheit* (1585) is a satire in which the spiritual and the worldly are compared, and *Die Warnung des Trewen Eckarts* (1588) is a long poem in which Eckart in a trance visits Heaven and Hell. *Speculum mundi* (in German in spite of the Latin title, 1590) is a drama, depicting realistically the persecution of Lutherans in Silesia and Moravia at the time of the Counter-Reformation (see GEGENREFORMATION).

RINKART or **RINCKART,** MARTIN (Eilenburg, Saxony, 1586–1649, Eilenburg), studied at Leipzig, was a deacon at Eisleben (1611) and from 1617 principal pastor in his native town. Rinkart's selfless devotion to the care of his parish through the tribulations of the Thirty Years War (see DREISSIGJÄHRIGER KRIEG) is recorded. His first literary works were plays, notably *Der Eislebische Ritter* (1613) in which Luther (q.v.) triumphs over the representatives of the other religions, the Pope and Calvin. All three are depicted as knights disputing an inheritance. He is the author of several collections of devout and deeply felt hymns, which include 'Nun danket alle Gott' (q.v.), known throughout the world. It was published in *Jesu-Hertz-Büchlein in geistlichen Oden* (1636).

RINSER, LUISE (Pitzling, Bavaria, 1911–), studied at Munich University, qualified as a teacher in 1934 and taught in a school for four years. In 1939 she married the opera conductor H. G. Schnell, who was killed on the Russian front in 1943. Her marriage, ten years later, to the composer C. Orff (q.v.) ended in 1959. She made Italy her home and she has travelled widely.

After the publication of her first story, *Die gläsernen Ringe* (1940, rev. 1948) she was officially banned from publishing further work. In 1944 she was arrested for alleged subversion, was condemned to death by the Volksgericht (see RESISTANCE MOVEMENTS, 2) but liberated through the German surrender. Her record of this grim period of detention is contained in *Gefängnis-Tagebuch* (1946). In the same year appeared the delicate and sensitive story *Erste Liebe* in a collection with that title, followed by the Novellen *Jan Lobel aus Warschau* (1948) and *Martins Reise* (1949), and the collections *Ein Bündel weißer Narzissen* (1956), *Vom Sinn der Traurigkeit* (1962), and *Septembertag* (1964). Her novels began with *Hochebene* and *Die Stärkeren* (both 1948); other titles include *Mitte des Lebens* (1950) and its sequel *Abenteuer der Tugend* (1957, both reissued in 1961 as *Nina* after the name of their principal character), *Daniela* (1953), *Der Sündenbock* (1955), *Die vollkommene Freude* (1962), *Ich bin Tobias* (1966), and *Bruder Feuer*, subtitled *Roman um Franz von Assisi* (1975). Her early studies in psychology and in the development of children have influenced her fiction, which, while not ignoring the evil forces in the world, portray love, marriage, adolescence, and the conflict of faith and scepticism. Luise Rinser's standpoint is Roman Catholic, and her theological interests are expressed in a number of tracts entitled *Hat Beten einen Sinn?* (1966), *Laie, nicht ferngesteuert, Zölibat und Frau* (both 1967), and *Von der Unmöglichkeit und der Möglichkeit heute Priester zu sein* (1968). Her autobiographical works contain personal information, accounts of her travels (including Russia), and assessments of the contemporary scene in Germany; *Baustelle. Eine Art Tagebuch*, covering the years 1967–70, appeared in 1970, and *Grenzübergänge. Tagebuch-Notizen* in 1972. To these were added in 1974 three works showing a continued commitment to the problems of the age, the novels *Der schwarze Esel* and *Wie, wenn wir ärmer würden oder Die Heimkehr des verlorenen Sohnes*, and the documentary tract *Dem Tode geweiht? Lepra ist heilbar*, discussing the plight of lepers. *Wenn die Wale kämpfen. Porträt eines Landes. Südkorea* (1976) and *Der verwundete Drache. Dialog über Leben und Werk des Komponisten Isang Yun* (1977, in collaboration with Isang Yun) were followed by *Kriegsspielzeug. Tagebuch 1972–1978* (1978) and *Nordkoreanisches Tagebuch* (1981), accounts of Rinser's Korean travels, first to South Korea

and later, following an official invitation by the government, to North Korea. An autobiography, *Den Wolf umarmen*, appeared in 1981.

RISSE, HEINZ (Düsseldorf, 1898–), served in the army 1914–18, studied at Marburg, Frankfurt, and Heidelberg universities, and spent the greater part of his life in business as an auditor. At the age of 50 he published his first story, *Irrfahrer* (1948), and followed this in rapid succession with eight novels and a number of shorter narratives. The novels are *Die Flucht hinter das Gitter* (1948), *Wenn die Erde bebt* (1950), *So frei von Schuld* (1951), *Dann kam der Tag* (1953), *Sören der Lump* (1955), *Große Fahrt und falsches Spiel* (1956), *Einer zuviel* (1957), and *Ringelreihen* (1963). The first story was followed by *Fledermäuse* (1951), *Die Grille* (1953), *Simson und die kleinen Leute* (1954), *Wuchernde Lianen* (1956), *Buchhalter Gottes* (1958), *Die Schiffsschaukel* (1960), *Fort geht's wie auf Samt* (1962), and *Macht und Schicksal einer Leiche* (1967).

As a writer, Risse has remained something of an outsider, ignored by most surveys of literature, even by those devoted to the decade (1950–60) in which he was most active. Four novels deserve particular attention. All of them combine realism of detail with a visionary quality linked with an idiosyncratic theology. In *Wenn die Erde bebt* the world, ruled by the industrious ants ('die betriebsamen Ameisen'), is shattered by an earthquake (which may be interpreted as the destructiveness of war). The order which it destroyed was entirely materialistic, devoted to competitiveness and gain. The earthquake is God's visitation. But when it is past the old materialistic society begins to reconstruct itself and the unnamed principal character, who is also the narrator of the story, murders his wife, Leonore, because she rejects the 'true' simple life that he offers her for the world of affluence and greed. The narrator purports to tell the story while awaiting trial. Haunting this novel and its successor *So frei von Schuld* (a quotation from Schiller's ballad *Die Kraniche des Ibykus*, q.v.) is the question of God's responsibility. Alexander Boethin rejects Rottmann's assertion of God's guilt, but when, at the end of his long sequence of misfortunes, he commits a murder under provocation, he gives a hint that the act may not be his fault but the fault of God, who made him as he is. *Dann kam der Tag* is another powerful indictment of materialistic society, in which the ruthless, 70-year-old industrialist Brocke is 'dethroned' by his own son, whom he has brought up to be as ruthless as himself. When he realizes this he destroys his own life's work in an act of grandiose arson. *Sören der Lump* depicts a character who devotes his life (ended by his death in the 1914–18 War) to parodying and ridiculing the deification of a mechanical and technological civilization and the conventions which screen it. Risse's tone is often explicitly didactic, and his outlook suggests a variant of the doctrine of predestination. His views are also expressed in his essays *Die Fackel des Prometheus* (1952), *Paul Cézanne und Gottfried Benn* (1957), and *Feiner Unfug auf Staatskosten* (1963). He has since published little, but an essay, *Wer denkt heute noch an Verdun?*, appeared in *PEN. Neue Texte deutscher Autoren* (1971).

RIST, JOHANN VON (Ottensen nr. Hamburg, 1607–67, Wedel nr. Hamburg), a pastor's son and a prolific poet, studied theology at Rinteln and Rostock universities. In 1633 he became a schoolmaster at Heide (Holstein) and moved to Wedel north-west of Hamburg as pastor in 1635. An ardent follower of Opitz (q.v.), Rist (already a member of the Hirten- und Blumenorden an der Pegnitz, q.v.) was received into Die Fruchtbringende Gesellschaft (q.v.) as 'Der Rüstige' in 1647; six years later he received a patent of nobility. In 1660 he founded in Hamburg the Elbschwanorden (q.v.).

Rist's earliest published work is a play, *Irenomachia* (1630). A volume of poems (*Musa Teutonica*) followed in 1634, *Poetischer Lust-Garte* in 1638, *Kriegs- und Friedensspiegel* in 1640. At this point Rist turned to religious poetry, though his earlier pastoral poems, *Galathee*, written some years before, were published against his wishes in 1642. His *Himmlische Lieder*, which appeared in 5 vols. between 1641 and 1643, contain some poems which have survived as hymns to the present day, including 'O Ewigkeit, du Donnerwort'. Rist wrote many plays, most of which have disappeared. Among the survivors, *Das Friedewünschende Teutschland* (1647) is the most notable. *Das Friedejauchzende Teutschland* (1653) is a Festspiel, which comes close to being an opera. His *Monats-Unterredungen* (vol. 1, 1663), intended to number twelve, were interrupted after the sixth by his death. They are urbane dialogues intended to educate; Rist himself participates as 'Der Rüstige' or 'Palatin'. They were completed in 1668 by E. Francisci. *Sämtliche Werke* (10 vols.), ed. E. Mannack, appeared in 1967 ff.

Ritter, a poem by Rilke (q.v.) included in *Das Buch der Bilder*.

Ritter Blaubart, a fairytale-play written in 1796 by L. Tieck (q.v.) and published in vol. 1 of his *Volksmärchen* (1797) and also separately in the same year. Tieck claims that it was the work of not much more than an evening—written 'fast nur an einem Abend'. Described as *Ein Ammenmärchen in vier Akten*, it tells the story of

Bluebeard, ending with the timely rescue of his last wife Agnes by her brothers, one of whom kills Bluebeard. It was produced by Immermann (q.v.) at Düsseldorf in 1835.

Ritterfahrt, Die, a short Middle High German allegorical poem, which recounts a symbolical campaign against a lady of Limburg who has transgressed the laws of *minne*. The principal instigator of the action is Irmgard, wife of Wilhelm I von Katzenellenbogen on the Rhine, and the poet himself was a Rhinelander, who wrote *c.* 1300. (See DER MINNEHOF, DAS TURNIER, and DER RITTERPREIS.)

Ritter Gluck, a story by E. T. A. Hoffmann (q.v.), first published in the *Allgemeine Musik-Zeitung* in 1809 and then in the first volume of his *Fantasiestücke in Callots Manier* in 1814. With remarkable realism of detail and verisimilitude of narration, Hoffmann exhibits the madness of a musician who believes himself to be the composer C. W. Gluck.

Ritterpreis, Der, a short Middle High German poem, which appears in MSS. in company with *Der Minnehof, Die Ritterfahrt,* and *Das Turnier* (qq.v.). It is the description of a courtly festival. Its author, who lived in the Rhineland, wrote it *c.* 1300. (See also DIE BÖHMERSCHLACHT.)

Ritterspiegel, see ROTHE, JOHANNES.

Ritter, Tod und Teufel, a copperplate engraving executed by A. Dürer (q.v.) in 1513. It plays a part in C. F. Meyer's *Huttens letzte Tage* (q.v.) and in F. Dürrenmatt's *Der Verdacht* (q.v.).

Rittertreue, an anonymous Middle High German verse tale of the 14th c. The knight Willekin von Montabaur on the way to a tournament spends all that he has on appropriate funeral rites for a dead knight, for whose interment no one has been willing to pay. He is then unable to pay for a horse, and obtains one from a knight by promising to give him half of any prize he wins at the tournament. His prize, however, is his lady's hand. Willekin reluctantly concedes a share to the knight and is immediately rewarded; for the whole procedure is a test of his truth to his word. The horse owner is the spirit of the knight whom Willekin had generously buried; he ascends to heaven leaving Willekin in sole possession of his bride.

'Ritter- und Räuberroman, term denoting contrived, extravagant, mostly long-winded stories, corresponding roughly to the 'Gothick' novel of late 18th-c. England. They exploit the historical interest fostered by J. G. Herder (q.v.)

and the violent and sentimental strains of Sturm und Drang (q.v.). See CRAMER, K. G., NAUBERT, BENEDIKTE, SCHLENKERT, F. C., SPIESS, C. H., and VULPIUS, C. A.

Plays of similar character were also popular, largely in imitation of Goethe's *Götz von Berlichingen* (q.v.) and Schiller's *Die Räuber* (q.v.).

Ritter vom Geiste, Die, a 9-vol. novel by K. Gutzkow (q.v.) first published in 1850-1 and issued in revised form in 1869. In spite of its enormous length (some 3,000 pages), Gutzkow took less than two years (1849-51) to write it. It is set in the years after the 1848 Revolution (see REVOLUTIONEN 1848-9) in an unnamed German state, the identity of which with Prussia is easily deduced. Gutzkow's novel is a remarkably comprehensive panorama of contemporary life, held together by an artificial plot of which the author himself wrote in contemptuous terms in the preface to the third edition (1854).

The young civil servant Dankmar Wildungen discovers a casket containing documents which give him and his brother Siegbert a claim to the wealth of the German Order of Knights Templar, and he resolves to devote this fortune, if his lawsuit succeeds, to the support of a new (fictitious) Order which he founds, the Order of the Knights of the Spirit. The casket and its contents are destroyed by fire at the end of the novel. Concealed illegitimate births, including that of the central figure, Fürst Egon von Hohenberg, also play a part in this insignificant framework. The climate of the novel reproduces the state of anxiety which followed the suppression of the Revolution. Prince Egon, who has spent his youth incognito in France as a journeyman cabinet-maker, returns to Germany full of idealism and sympathy for the common people. The ministry falls and he is invited to become premier (Chefminister), but instead of putting his idealism into practice he steadily becomes more and more dictatorial and is responsible for repressive measures, through which lives are lost and honourable men ruined. His character is redeemed at the end of the novel when he joins the Order of the Knights of the Spirit.

The real merit of the book (and Gutzkow's own interest) resides in the enormous gallery of characters ranging through the whole spectrum of society, generals, diplomats, prelates, civil servants, farmers, gamekeepers, livery stable keepers, innkeepers, humble workmen, police spies, etc., all with their womenfolk, and all portrayed with remarkable distinctness, vividness, acumen, and fairness. Gutzkow claimed in his first preface that he had created a new kind of novel, 'der Roman des Nebeneinander', which aims at simultaneous narrative, as distinct from a novel in which the action proceeds consecutively (Roman des Nacheinander); he certainly provided a broad conspectus of contemporary society and analysed its deep-seated *malaise*. (See ZEITROMAN.)

Ritter vom Thurn, Der, a German translation (1485) of the French collection of stories (in prose) entitled *Livre pour l'enseignement de ses filles* by the Chevalier de la Tour de Landry (hence the German title, 'Knight of the Tower'). The translator was Marquard vom Stein (d. 1559). The collection of ostensibly moral, though partially obscene, stories, was first printed in 1493 and was reprinted at intervals until 1682.

Rittmeister Brand, a story (Erzählung) by Marie von Ebner-Eschenbach (q.v.) published in 1896. Brand, who comes of a prosperous Viennese commercial family, has a passion for education and is convinced that the greatest exercise for his powers is to be found in the army and especially in the cavalry. Reluctantly his parents consent, and he is trained and commissioned in the 2nd Dragoons, rising in a few years to the command of a squadron (Rittmeister). After the death of his parents he falls in love with Sophie, the daughter of an extravagant and debt-ridden nobleman. But in the end he does not marry her, because he does not wish to abandon his career and disapproves of married officers unless they have wealth. The colonel of Brand's regiment is a malignant mediocrity who drives a brother officer of Brand's to suicide. Brand resigns his commission, challenges the colonel and disables him in the duel. Brand now gratifies his passion for education by taking a close interest in the children of friends and acquaintances, and in this way he comes again into contact with Sophie, who had married for the sake of her father's finances and is now a widow making a small income by trimming hats. The two are united through Sophie's son Georg, a delicate boy, whose death ends the story.

RITZ, WILHELMINE, see LICHTENAU, GRÄFIN VON.

RIVANDER, ZACHARIAS, real name Bachmann (Leisnig, Saxony, 1553-97, Bischofswerda, Saxony), a Lutheran theologian who wrote a polemical play with Luther (q.v.) as its hero, *Lutherus redivivus* (1593, in German in spite of the Latin title), and a Thuringian chronicle (1581).

ROBERT, LUDWIG (Berlin, 1778-1832, Baden-Baden), a member of a well-to-do Jewish family and a brother of Rahel VARNHAGEN von Ense (q.v.), wrote plays, including the tragedy *Die Macht der Verhältnisse* (1819) which deals with the position of Jews in society.

Robert Guiskard, Herzog der Normänner, a dramatic fragment (Fragment aus dem Trauerspiel) by H. von Kleist (q.v.) on Robert Guiscard (*c.* 1015–85). Its origin is associated with the first desperate crisis produced by Kleist's endeavour to prove himself as a dramatist of genius after the failures of his early career. He began the writing of the tragedy in 1802, but burned the MS., which was nearing completion, in 1803, while suffering a mental and physical breakdown in Paris. He believed that he had failed to produce, after 'half a thousand days, including most nights', a play worthy of his aspirations. C. M. Wieland (q.v.), to whom Kleist recited long passages from memory in 1804, was so impressed by Kleist's attempt to blend elements of Greek and modern tragedy in this play, that he assured him that his tragedy might fill a gap in German literature which even Goethe and Schiller had left open. Kleist abandoned the project in Dresden (1807–8) after a renewed promise to his publisher to complete the play, but in 1807 he wrote the 324 lines of the existing fragment for publication in his literary magazine *Phöbus* (q.v., 1808).

Kleist's source is an essay by K. W. F. von Funk (*Robert Guiscard, Herzog von Apulien und Calabrien*), published in 1797 by Schiller in *Die Horen* III (Stück 1–3). Kleist's Guiskard besieges Constantinople, the conquest of which is to seal his claim to the crown and untramelled power. But disaster befalls his camp when it is struck by the plague, and the people plead with Guiskard to abandon his campaign and save them from utter destruction by leading them back to their native country. The final lines suggest the inevitable catastrophe, as Guiskard himself can no longer conceal that he, too, has already been smitten by the disease. The problem of Guiskard's position as a usurper and that of his succession remains undeveloped. Nevertheless, the extant dramatic situation fully justifies, through the power of its language, its vision, and its balanced composition, the place it now holds among the great literary fragments.

ROBERTHIN, ROBERT (Saalfeld, East Prussia, 1600–48, Königsberg), also Robertin, a prominent personality in the Königsberg group of poets (see KÖNIGSBERGER DICHTERSCHULE), was an official in the courts of justice at Königsberg. Some poems by him were included by Heinrich Albert (q.v.) in the collections *Poetisch-Musikalisches Lust Wäldlein* (1648) and *Arien* (1638–50).

Robinsonade, term applied to novels of shipwreck and survival deriving from Defoe's *The Life and Strange Adventures of Robinson Crusoe* (1719). A German translation appeared in 1720 and numerous imitations followed. Of note are J. G. Schnabel's *Die Insel Felsenburg* (q.v., 1731 ff.), *Robinson der Jüngere* (1779–80) by J. H. Campe (q.v.), and *Der schweizerische Robinson* (1812–27) by J. D. Wyß (q.v.).

ROCHLITZ, JOHANN FRIEDRICH (Leipzig, 1770–1842, Leipzig), editor of the *Leipziger allgemeine musikalische Zeitung* from 1798 to 1818, wrote some comedies and a number of novels and stories, including *Amaliens Freuden und Leiden* (1798) and *Charaktere interessanter Menschen* (1799–1803). Rochlitz, whom Goethe knew and valued, was regarded as one of the better purveyors of light novels. His correspondence with Goethe, ed. W. von Biedermann, appeared in 1887.

RODA RODA, ALEXANDER, pseudonym of Sandór von Rosenfeld (Puszta Zdenci, Hungary, 1872–1945, New York), was from the age of 20 in the army of the old Austrian Empire, serving as an artillery officer. From 1902 he took to travel, journalism, and writing, and during the 1914–18 War was a reporter on the Austrian fronts. Pre-war Austria and his military experiences are the subject of much of his prolific work, which is full of humour and satire, and was mostly written before the end of the 1920s. In 1938 he emigrated to the U.S.A. His stories include *Der gemütskranke Husar* (1903), *Soldatengeschichten* (2 vols., 1904), *Schlangenbiß* (1931), and the novels *Der Schnaps, der Rauchtabak und die verfluchte Liebe* (1908) and *Die sieben Leidenschaften* (1921). A select edition of his works, *Ausgewählte Werke* (3 vols., 1932–4), was followed by another novel, *Die Panduren* (1935). His best-known work is the light comedy *Der Feldherrnhügel* (1910, in collaboration with C. Rößler, q.v.), a success on the stage and as a film. He entitled his first autobiographical work *Irrfahrten eines Humoristen 1914–19* (1920), and in 1925 published *Roda Rodas Roman* (reissued 1958). *Großmutter reitet und andere Kapriolen* appeared in 1981.

RODENBERG, JULIUS, real name Levy (Rodenberg, Hesse, 1831–1914, Berlin), a journalist in Berlin from 1862 onwards, wrote in many forms. His novels include *Die Straßensängerin von London* (1863), *Die neue Sündflut* (4 vols., 1865), and *Die Grandidiers* (3 vols., 1879), but he was best known for his verse romances *Dornröschen* (1852), *König Haralds Totenfeier* (1853), and the humorous epic *Der Majestäten Felsenbier und Rheinwein lustige Kriegshistorie* (1853). He travelled in England, France, and Italy and published *Pariserbilderbuch* (1856), *Ein Herbst in Wales* (1857), and *Alltagsleben in London*

(1860), as well as *Bilder aus dem Berliner Leben* (2 vols., 1885–7).

Rodenberg's correspondence with C. F. Meyer (q.v.) was published in 1918, and a select edition of his diaries (*Aus den Tagebüchern*), ed. E. Heilborn, in 1919.

RÖHM, ERNST (Munich, 1887–1934, Munich), a regular army officer, served in the infantry in the 1914–18 War, joining a Freikorps (q.v.) in 1919. Röhm, a man of ruthless brutality and drive, was an early member of the NSDAP (q.v.) and participated in the abortive National Socialist revolt in Munich in 1923 (see HITLER-PUTSCH). He fell out with Hitler (q.v.) after this episode and in 1929–30 took service as a mercenary in Bolivia. In 1930 he was recalled by Hitler and entrusted with the leadership of the SA and SS (qq.v.) with the title Chef des Stabes. After the National Socialists came to power in 1933, Röhm again diverged from Hitler, seeking a more radical revolution and aiming at a union of SA and Reichswehr under his own command. Alarmed at this threat to his position, Hitler ordered the assassination of Röhm and his lieutenants, which took place on 30 June 1934. Röhm's blatant homosexuality, long known to Hitler, was paraded as an excuse for this political murder.

Rokoko, as a literary term, is applied to types of poetry introduced in the early 1740s from France by the Anacreontics (see ANAKREON-TIKER) and culminating in Wieland's *Oberon* (q.v., 1780). It takes in poems by Klopstock, Hölty, and Goethe (see MIT EINEM GEMALTEN BAND, written 1771), idylls with pastoral themes like *Idyllen* by S. Geßner (q.v., 1756), and the Singspiel (q.v.). It has also been taken (by A. Anger, *Literarisches Rokoko*, 1962) to represent a distinct phase of transition from late Baroque (q.v.) to early Classicism (see KLASSIK, DEUTSCHE) on the grounds of the writers' predilection for Greek models, which offered a practicable philosophy of life, as opposed to French models which, unlike conditions in Germany, grew out of contemporary cultural, social, and moral conditions. In music a comparable ornamental style was introduced by Telemann (1681–1767), and developed by the early Haydn and his friend Mozart (qq.v.). Apart from its manifestation in the arts and in the Meißen porcelain industry, Rococo is more important as a style in Austrian and Bavarian architecture (Asamkirche, Munich), both ecclesiastical and secular, and especially in the interior ornamentation of buildings. It occurs sporadically in north Germany, notably at Sanssouci (q.v.).

Rokoko oder Die alten Herren, a five-act comedy (Lustspiel) by H. Laube (q.v.), first performed at Dresden in April 1842, and published in 1846. It is an effective comedy of intrigue set at Versailles about 1745 in the time of Louis XV and Madame de Pompadour, who is one of the characters. A marriage is arranged between Mélanie de Gérard and Prosper de Didier. The Pompadour, who wishes to have Mélanie available for the King's pleasure, sets an intriguer, the Abbé de la Sauce, to work to stop the match. He unearths family secrets, but his own lasciviousness (he tries, unsuccessfully, to abduct Mélanie) and the counter-intrigues of his enemies prove too much for him. Mélanie marries, not Gérard, but Viktor, a young nobleman, whom she loves, and the Abbé is consigned to the Bastille by a *lettre de cachet*. The play closes with a reference to the frankness of the new age replacing the corruption of the older generation (*Die alten Herren* of the title): 'denn der Jugend gehört die Zukunft'.

Roland, name given to certain medieval statues in a number of North German cities. They represent a bare-headed man holding a naked sword and are thought to symbolize the jurisdictional rights of the city. The best-known surviving example is in Bremen. The novel *Der Roland von Berlin* (q.v.) by W. Alexis (q.v.) records the destruction of the Roland in that city in 1448 on the orders of the Elector Friedrich II (q.v.) of Brandenburg.

Rolandslied des Pfaffen Konrad, a long Early Middle High German poem which is a translation, faithful as to incident, but free in style, of the French *Chanson de Roland* (*c.* 1100). The German rendering was undertaken by Konrad, a priest, probably of Regensburg, for Duke Heinrich der Löwe (q.v.) *c.* 1170. Konrad translated in two stages, first into Latin, then into German.

The historical basis of the *Chanson de Roland* is the campaign of Charlemagne (see KARL I) in Spain, which was interrupted by news of a revolt among the Saxons. In passing through the Pyrenees in 778, Charlemagne's rearguard was annihilated and several knights, including one Hruodland, were killed. Tradition made this warrior into the hero of the story. The *Chanson* shows Roland betrayed by the false knight Genelun (Ganelon) and waging a valiant battle against fearful odds. Shortly before his death he blows the horn to summon Charlemagne, who arrives too late to save his knights but wreaks a fearful vengeance on the infidels. In all this Konrad follows his model closely.

The underlying attitude of the poem is a

Christian heroism. The Emperor is the representative of God's kingdom; he and his knights fight to extend Christendom, and their death in battle is a martyrdom. An element of patriotism is also present, but the Christian element is stronger in the German poem than in the French. The *Rolandslied* is broader in presentation and laxer in style than the *Chanson*, but Konrad's poetic quality is nevertheless appreciable.

There are several fragmentary MSS., and one complete one of the late 12th c., which is at Heidelberg. An edition by C. Wesle (1928), revised by P. Wapnewski, was published in 1970.

Roland von Berlin, Der, a historical novel (Ein vaterländischer Roman) in 3 vols. by W. Alexis (q.v.), published in 1840. It is set in Berlin and its adjoining twin city Cölln and in the Mark Brandenburg between 1442 and 1448, and deals with the dissensions between the two cities and the drastic action to resolve them taken by the Elector Friedrich II (q.v.). Its principal figure is the burgomaster, Johannes Rathenow, a man of steadfast character and devotion to duty, who, in his rectitude, alienates his fellow citizens and also comes into collision with the Elector, who nevertheless respects his integrity. A personal plot, the love of the burgomaster's daughter Elsbeth and the lively Henning Mollner, who is eventually ennobled, runs through the historical events. The lovers, after many vicissitudes, are finally united. A short epilogue shows the weary Friedrich leaving Brandenburg after his abdication in 1470.

The novel is based on a careful study of the sources, and occasional quotations from documents are given in notes. The title alludes to the statue known as der Roland (q.v.), destroyed by the Elector in 1448. Its removal absolves the burgomaster from an uncomfortable oath.

ROLLENHAGEN, GABRIEL (Magdeburg, 1583–1619, Magdeburg), a son of Georg Rollenhagen (q.v.), was a lawyer in the service of the cathedral chapter of Magdeburg. His comedy *Amantes amentes, d. i. ein sehr anmutiges Spiel von der blinden Liebe, oder wie mans Deutsch nennet, von der Leffeley* (1609) is a dramatization of the story of Euryalus and Lucretia (as told by Enea Silvio PICCOLOMINI, q.v.) with comic scenes in Low German. He was also the author of a notable emblem book in Latin, *Nucleus emblematum selectissimorum* (2 vols., 1610–13).

ROLLENHAGEN, GEORG (Bernau nr. Berlin, 1542–1609, Magdeburg), studied at Wittenberg, was a schoolmaster at Halberstadt in 1563, a private tutor in Wittenberg in 1566, and pastor in Magdeburg in 1573. From 1575 he was also headmaster of the school at Magdeburg. Rollenhagen's principal work is the *Froschmeuseler* (1595), a beast epic after the manner of *Reinke de Vos* (q.v.). It is a free adaptation of the *Batrachomyomachy*, the pseudo-Homeric battle of the frogs and mice, to which Rollenhagen's attention was first drawn in his student days at Wittenberg. Clumsily constructed, the poem nevertheless has many attractive scenes of real observation. The aim is moral, a demonstration of the roles of the commoner, the cleric, and the ruler in the familiar terms of doing their duty in the station to which it has pleased God to call them. His standpoint is Lutheran. Rollenhagen also wrote or adapted biblical plays for school performance: *Des Ertzvaters Abraham Leben und Glauben* (1569), *Spiel vom Tobias* (1576), *Spiel vom reichen Manne und armen Lazaro* (q.v., 1590, critical edition by J. Bolte, 1929). He was the father of Gabriel Rollenhagen (q.v.).

ROLLETT, HERMANN (Baden by Vienna, 1819–1904, Baden by Vienna), an Austrian political writer of Radical views, was on the move for many years, but settled in his home town in 1855. His political literature included the poems *Frühlingsboten aus Österreich* (1845), *Wanderbuch eines Wiener Poeten* (1846), *Republikanisches Liederbuch* (1848), and *Kampflieder* (1848). In 1851 he published his plays (*Dramatische Dichtungen*) in 3 vols. His reminiscences (*Begegnungen*) appeared in 1903.

Rollwagenbüchlein, Das, a once-popular collection of anecdotes by Jörg Wickram (q.v.).

Roman Nr. 7, a tetralogy of novels projected by H. von Doderer (q.v.), but left incomplete at his death in 1966. The first part, entitled *Die Wasserfälle von Slunj* (q.v.), covers the years 1877 to 1941 and was published in 1963, the second, *Der Grenzwald* (q.v.), appeared posthumously (1967) as a substantial fragment, perhaps almost half the intended book. It is set partly in Vienna and partly in Siberia in the years 1915–20 approximately. The third volume was to have been set in Berlin in the late 1920s and the fourth was to deal with the '50s and '60s and would have included the volcanic eruption on Tristan da Cunha (1961). The whole cycle would have traced the evolution and interplay of forms of social life over a century.

Romantik, Die (The Romantic movement), the German phase of a general trend in Europe, which was, however, not synchronized in all countries. Though a cognate movement had existed in Germany in the 1770s (see STURM UND DRANG), the Romanticism beginning in the second half of the 1790s did not directly link up with it. German Romanticism is commonly classified into phases, both geographical and temporal, but the movement as a whole is fluid, even kaleidoscopic; prominent writers of classicism (see KLASSIZISMUS), such as Goethe and Schiller, are briefly associated with the Romantic movement. Biedermeier (q.v.) and Poetic Realism (see POETISCHER REALISMUS) are also indebted to Romanticism.

Early Romanticism (the Jenaer, Ältere Romantik, or Frühromantik) was until *c.* 1802 centred on Jena and Berlin; the main period of Romanticism (Jüngere Romantik or Hochromantik) falls into the years of the Napoleonic Wars (q.v.) and overlaps with the beginning of the restoration in 1815. It was centred on Heidelberg (Heidelberger Romantik) and Berlin, and later spread to the provinces, notably Württemberg (see SCHWÄBISCHER DICHTERKREIS).

Romanticism originated with a group of intellectuals of austerely classical education: in 1796 A. W. and F. von Schlegel (qq.v.), accompanied by the wife of the former (see SCHELLING, CAROLINE), arrived in Jena and found a congenial associate in J. G. Fichte (q.v.), since 1794 professor of philosophy at Jena University. Goethe's *Wilhelm Meisters Lehrjahre* (q.v., 1795) was in many respects a seminal work in this stage of Romanticism. In 1799 L. Tieck (q.v.) arrived, and through him contacts with his close friend Wackenroder (q.v., who never visited Jena) were established. Of particular importance was the association of Novalis (q.v.) with Jena. Both he and F. Schlegel made important contributions to *Das Athenäum* (q.v.), founded in 1798 by the Schlegel brothers as the organ for their reassessment of literature and aesthetics. Another contributor, forming one of the links of the Jena group with Berlin, was the Protestant theologian F. D. E. Schleiermacher (q.v.); his Christology was as influential as Fichte's subjective Idealism and the less systematized Nature philosophy of F. W. J. von Schelling (q.v.), Fichte's successor at Jena University from 1798.

In the first decade of the 19th c. L. J. von Arnim and C. Brentano (qq.v.) settled in Heidelberg. They were well acquainted with the works of the early Romantics, which tended to be exemplifications of theory. The fruit of the collaboration of Arnim and Brentano was the collection of German folk-songs, *Des Knaben Wunderhorn* (q.v., 1805–8), one of the landmarks of the German Romantic movement, and a collection which differs in its nationalistic stance from the cosmopolitan approach of Herder (q.v.). J. J. von Görres (q.v.), from 1806 in Heidelberg, published *Die teutschen Volksbücher* (1806–8) based on the chap-books (see VOLKSBUCH) of earlier centuries, especially the 16th c.

The interest in the folk-song and ballad, folktale and folk-lore was accompanied by a devotion to the fairy-tale (see MÄRCHEN). While in Kassel the brothers J. and W. Grimm (qq.v.) compiled *Kinder- und Hausmärchen* (2 vols., 1812–14), testifying to the timeless components of national consciousness. Fairy motifs stirred the imagination and provided an outlet for introspection. The fairy-tale (Kunstmärchen), frequently employing oriental motifs, became a favourite vehicle for allegory. The exploitation of the irrational, of the occult, and of animal magnetism (see MESMER, F. A.), accentuating the sinister and inexplicable, and culminating in the use of the grotesque (notably in E. T. A. Hoffmann, q.v.) accorded with the Romantic conception of irony (see ROMANTISCHE IRONIE), which found its last theorist in F. Solger (q.v.). The subconscious was further explored by Romantic preoccupation with dreams, influenced by G. H. Schubert (q.v., especially *Die Nachtseite der Naturwissenschaften*, 1808, and *Die Symbolik des Traumes*, 1814).

In 1807 J. von Eichendorff (q.v.) came to Heidelberg as a student; he became one of the greatest of German lyric poets, responsive to folk-songs and original in his use of their tradition. In him the Romantic passion for Nature is most fully developed. He later moved to Berlin with which, in addition to Fichte, were associated Arnim (a Berliner by birth), Brentano, A. W. Schlegel, Z. Werner, E. T. A. Hoffmann, F. de la Motte FOUQUÉ, and A. von Chamisso (qq.v.). The writers of provincial Romanticism are mainly Swabians and include L. Uhland, G. Schwab, J. Kerner, W. Hauff, and, more remotely, E. Mörike (qq.v.).

Two great names connected with Romanticism but not associated with any school of Romanticism are H. von Kleist and H. Heine (qq.v.). Hölderlin (q.v.), in spite of his classical preoccupations and aspirations, exhibits some Romantic features. An expression of a late Romantic attitude is also to be found in the Weltschmerz of N. Lenau (q.v.).

The Romantic movement gave a primacy to the artist; Tieck and Wackenroder collaborated in the perceptive essays of the *Herzensergießungen eines kunstliebenden Klosterbruders* (q.v., 1797), which demonstrate intuitive understanding of the creative process. Novalis invented the symbol of the blue flower (see BLAUE BLUME), expressing Romantic love and longing (Sehnsucht). Romanticism also had appreciable impacts upon painting (see FÜSSLI, J. H., RUNGE, P. O., FRIEDRICH, C. D., KERSTING, F. G., NAZARENER, RICHTER, L., SCHWIND, M., and SPITZWEG, C., and music (see BEETHOVEN, WEBER, C. M. VON, SCHUBERT, F., LOEWE, C., LORTZING, A., and SCHUMANN, R.).

Romantische Ironie denotes a method by which the Romantic writer can demonstrate his own superiority to his work, viz. by deliberately destroying or interrupting the illusion he has created. It was devised by F. Schlegel (q.v.) on the basis of the subjective Idealism of Fichte (q.v.). The writer, the ego (das Ich), refuses to allow his work (which is part of the non-ego, Nicht-Ich) to determine his course, and maintains it in subordination by this device. Examples of Romantic irony are common, especially in the work of Schlegel, Tieck, Brentano, and Heine (qq.v.). See also ROMANTIK, DIE.

Romantische Ödipus, Der, a satirical verse comedy (Ein Lustspiel in fünf Akten) written in 1827–8 in Sorrento and Capri by A. von Platen (q.v.), and published in 1829. It represents the story of Oedipus as a Romantic would tell it. Its satire is largely directed against Immermann (q.v.), who appears as 'Nimmermann'. The first, third, and fifth acts are followed by parabases addressed to the audience.

Roman von Vieren, an unwritten novel planned by A. von Chamisso, K. W. Contessa, J. E. Hitzig, and E. T. A. Hoffmann (qq.v.). Hoffmann's contribution appeared in 1821 as *Die Doppelgänger*, and Contessa completed his as *Das Bild der Mutter* (in *Sämtliche Schriften*, 1826).

Romanzen vom Rosenkranz, an unfinished poetic work of legendary religious character in the form of a succession of verse romances written 1804–12 by C. Brentano (q.v.) and published posthumously in 1852. The basis of the romance is a curse arising in the course of the Flight into Egypt. A young man has stolen Mary's wedding ring and Mary prophesies, according to Brentano's notes: 'Eure Schuld werden nur die drei Rosen retten, wenn sie endlich lebendig geworden und das Unglück der Ringe getilgt haben, wenn sie selbst ein Ring geworden ... Der Samen des Diebes wird hoffärtig und trostlos sein in alle Ewigkeit.' Generation after generation of the thief's descendants commit incest, and the poem deals with three girls of this line, born of a criminal and incestuous union (the mother is a seduced nun), who each overcome temptation and so defeat the curse and achieve redemption. Approximately two-thirds of the poem was written. The prologue is in *terza rima* (see TERZINEN), the romances are in four-line stanzas of rhyming trochaic verse. The nineteen extant romances comprise more than 2,600 such stanzas.

Romanzero, a volume of poems by H. Heine (q.v.), published in 1851. It is divided into three books, *Historien*, *Lamentationen*, and *Hebräische*

Melodien. The poems in this volume were written during Heine's long illness and represent both a renewal and an intensification of his poetic power. For inspiration he turns in part to the Jewish traditions into which he was born, and at the same time he grapples, with the aid of a courageous irony, with the realities of his slow death. Among the best-known poems of the *Historien* are *Der Dichter Firdusi* and *Vitzliputzli*, and of the *Hebräische Melodien*, *Jehuda ben Halevy*. The more personal poems of the second section of the *Lamentationen*, headed *Lazarus*, include the powerful and moving *Der Abgekühlte*, *Gedächtnisfeier*, *Frau Sorge*, *An die Engel*, and the final trio *Sie erlischt*, *Vermächtnis*, and *Enfant perdu*.

Romeo und Julia auf dem Dorfe, the best-known Novelle in G. Keller's *Die Leute von Seldwyla* (q.v.). It is the second story in vol. 1 (1856), and is based on a short item of news printed in the *Züricher Freitagszeitung* of 3 September 1847. Two neighbouring farmers, Manz and Marti, live in amity until they agree to misappropriate an orphan's field. They fall out over the boundary and ruin each other in a lawsuit. Manz's son Sali and Marti's daughter Vrenchen fall in love. The courtship is barred by the mutual enmity of the fathers. In protecting Vrenchen from her father's brutality, Sali strikes a blow which causes his mental derangement. After this tragedy the two young lovers can see no future. They resolve to spend one happy day, and at its close to end their lives. They have a meal at an inn, and dance in the 'Paradiesgärtchen'; a way out of their impasse by taking to vagrancy is offered to them by the Black Fiddler (Schwarzer Geiger). They turn down his suggestion, spend their bridal night on a barge, and at dawn slip into the water to drown.

The version of 1856 ended with two moralizing paragraphs, which were deleted in the definitive 2nd edn. of 1874. Keller's story is the basis of the libretto of Delius's opera *A Village Romeo and Juliet* (first performed, Berlin, 1907).

Römische Brunnen, Der, see SCHWEISSTUCH DER VERONIKA, DAS.

Römische Elegien, a cycle of twenty poems written by J. W. Goethe (q.v.) between the autumn of 1788 and the spring of 1790. They were first published as a whole in Schiller's periodical *Die Horen* (q.v.) in July 1795. *Elegie XIII* had been published separately in *Die deutsche Monatsschrift* in 1791. The designation *Elegien* refers to the elegiac metre in which they are written. The title in the MS. was *Erotica romana*, which was altered for publication to

Elegien, Rom 1788. The present title dates from 1806.

The elegies were written after Goethe's return from Italy to Weimar in the summer of 1788. They reflect his nostalgic appreciation of Italy, not only as the shrine of antique art, but as a vivid, pulsating reality which awakened him to a new vigorous life, and especially to an uninhibited physical enjoyment of love. They cannot be interpreted as detailed autobiography, though they are undoubtedly related to his life with Christiane Vulpius (see GOETHE, CHRISTIANE); they express with great frankness and precision a newly won and jealously guarded standpoint. Their publication in 1795 aroused much hostile criticism.

Römischer König, see DEUTSCHER KÖNIG.

ROMMEL, ERWIN, GENERALFELDMARSCHALL (Heidenheim, Württemberg, 1891–1944, nr. Ulm), served as an infantry officer in the 1914–18 War and was decorated for bravery. He was an officer in the Reichswehr, and in 1940 commanded an armoured division. In 1941 he became commander of the Afrika-Korps. After the end of German resistance in Africa in 1943 Rommel was given commands successively in Italy and in France. He was wounded in a low-flying air attack in 1944. Rommel joined the group of dissident officers in 1944 (see RESISTANCE MOVEMENTS, 2), was suspected of complicity, and on Hitler's orders committed suicide by taking poison. He was the author of a book, *Infanterie greift an* (q.v., 1937). Rommel's posthumous papers appeared as *Krieg ohne Haß* (1950).

ROON, ALBRECHT, GRAF VON (Pleushagen, Pomerania, 1803–79, Berlin), entered the Prussian army in 1821. Regimental duty convinced him of the army's inefficiency, and association on service with Prince Wilhelm (see WILHELM I) enabled him to air his views. In 1859, as a lieutenant-general, he was appointed to a commission to report on reorganization of the army. In 1859 Prince Wilhelm (since 1857 Prince Regent) made him War Minister, and he held this office until 1873.

Roon's reforms (military expansion, three-year service, and the creation of a Landwehr, q.v.), and their cost, provoked the crisis of 1862, which he survived through the skill of Bismarck (q.v.), whose appointment as Prussian Ministerpräsident he favoured. Roon's army organization was vindicated in the successful wars of 1864, 1866, and 1870 (see DEUTSCHER KRIEG and DEUTSCH-FRANZÖSISCHER KRIEG). The title Graf was conferred on him in 1871, and he was promoted field-marshal in 1873. Roon

was a man of harsh, intransigent character. In his early years he achieved some repute as a writer on military geography with *Grundzüge der Erd-, Völker- und Staatenkunde* (1832) and *Militärische Länderbeschreibung von Europa* (1837). *Denkwürdigkeiten* (2 vols.) and his speeches (*Kriegs-Minister von Roon als Redner*, 3 vols.), ed. W. von Roon, appeared in 1892 and 1895–6 respectively.

ROQUETTE, OTTO (Krotoschin, Posen, 1824–96, Darmstadt), first a schoolmaster and then a professor of literature at various institutions of higher education (but not at a university), lived mainly in Berlin until 1869 and thereafter at Darmstadt. Roquette was a facile poet, whose first light epic, *Waldmeisters Brautfahrt* (1851), was one of the best sellers of the middle of the century. Other, rather faded, romantic epics by him are *Der Tag von St. Jakob* (1852), *Hans Haide-Kuckuck* (1855), and *Cesario* (1888). He also wrote the novels *Heinrich Falk* (3 vols., 1858) and *Das Buchstabierbuch der Leidenschaft* (2 vols., 1878), as well as Novellen (*Erzählungen*, 1859; *Neue Erzählungen*, 1862; *Novellen*, 1870; *Neues Novellenbuch*, 1884). His plays include *Rudolf von Habsburg oder Die Sterner* (1856); a collection *Dramatische Dichtungen* (2 vols.) was published 1867–77, but Roquette has become an almost forgotten author. His *Geschichte der deutschen Literatur* (2 vols., 1862–3) appeared in 1878 as *Geschichte der deutschen Dichtung*, and his autobiography, *Siebzig Jahre* (2 vols.), in 1894.

ROSCHMANN-HÖRBURG, KASSIAN ANTON, RITTER VON (Innsbruck, 1739–1806, Vienna), a minor Austrian writer, was an official in the Viennese court record office. He was a dramatist (*Sirminde oder Die Afrikaner*, 1774) and is best known for his completion in 1764 of the unfinished tragedy *Olint und Sophronia* (q.v.) by J. F. von Cronegk (q.v.). Roschmann was also a historian, writing *Geschichte der Grafschaft Tirol* (1778) and *Geschichte von Tirol* (1792–1802).

Rose Bernd, a play (Schauspiel) in five acts by G. Hauptmann (q.v.), which was written, published, and first performed (October, Deutsches Theater, Berlin) in 1903. It is based on a court case in which Hauptmann participated as a member of the jury in the spring of 1903: a country girl was accused of perjury and infanticide, and was acquitted.

The action takes place within a typical country community and lasts from May to September. It focuses on the circumstances driving Rose, the daughter of the unemployed farm manager (Dominiumsverwalter) Bernd, to commit perjury and to murder the child to which she has given birth at night in the open after having concealed her pregnancy from all except Frau Flamm, the wife of her lover. Flamm, the village magistrate (Ortsvorsteher), has known Rose as a child and playmate of his deceased son Kurtel; his affection for her is genuine as well as a compensation in his childless marriage to an invalid wife. For her father's sake Rose is engaged to the bookbinder August Keil, who is of delicate health and physically unattractive but in faith and deed a dedicated man, upon whose financial support Bernd has become dependent. Torn by her conscience, Rose postpones the marriage for which her father presses, but is seduced by the agricultural mechanic Streckmann in an act of brutal jealousy, which also drives him to smear her name in public. In a clash between Streckmann and Keil, Keil loses an eye and Bernd files a lawsuit against Streckmann which, if Rose reveals the truth in court, can only bring ruin and heartbreak.

The play brings out above all Rose's victimization at the hands of a deterministically conceived fate. At the end of the play, when Rose, deserted by all, confesses to infanticide and is arrested, only Keil shows a full grasp of her suffering, which has brought her to the brink of insanity. The play combines Naturalistic techniques (see NATURALISMUS) with the social drama of domestic tragedy (see BÜRGERLICHES TRAUERSPIEL), giving this genre, which had reached a peak in the 1840s with Hebbel's *Maria Magdalene* (q.v.), a new and impressive place in 20th-c. drama.

ROSEGGER, PETER, in full Petri Kettenfeier Rosegger (Alpl, Styria, 1843–1918, Krieglach), the son of an impoverished Styrian farmer, grew up in humble rural conditions without formal schooling, learning only to read and write from an out-of-work schoolmaster, whom the parson had sacked. While apprenticed to an itinerant tailor (1858–62), he began to write and his early sketches and poems eventually found a home in the *Tagespost* in Graz. Its editor, Dr. Svoboda, befriended him and enabled him to attend the business college in Graz (1865–9). In 1869 his first book, *Zither und Hackbrett*, a collection of dialect poems, was published with a preface by R. Hamerling (q.v.). This led to a grant for travel in Europe (1870–2).

Rosegger married in 1873 a young admirer of his work, who died after the birth of their second child in 1875. His earliest and perhaps most popular novel, *Die Schriften des Waldschulmeisters* (q.v., 1875), was written during this brief marriage. In 1876 he founded the monthly magazine *Heimgarten*, which he edited until 1910, when one of his sons took over. Rosegger remarried in 1879. From 1877 to the end of his life he spent the summers in his native parish Krieglach and the winters in Graz.

A prolific writer, Rosegger was the author of several novels, including *Heidepeters Gabriel* (q.v., 1882), *Der Gottsucher* (q.v., 2 vols., 1883), *Jakob der Letzte* (q.v., 1888), *Peter Mayr, der Wirt an der Mahr* (1891), *Das ewige Licht* (1896), and *I.N.R.I.* (q.v., 1905). They mostly treat in simple terms the impact of an urban culture on a traditional rural society, taking sides with the latter. The numerous shorter stories, which appeared in several volumes, are accounts of country life which for the most part lack the social contrasts of the novels, but compensate for this by their freshness and life. The principal collections are *Sittenbilder aus dem steierischen Oberlande* (1870), revised as *Volksleben in Steiermark* (1875), *Geschichten aus Steiermark* (1871), *Die Älpler* (1872), *Aus Wäldern und Bergen* (1875), *Sonderlinge aus dem Volk der Alpen* (3 vols., 1875), *Lustige Geschichten* (1879), *Mann und Weib* (2 vols., 1879), *Dorfsünden* (1883), *Neue Waldgeschichten* (1883), *Das Geschichtenbuch des Wanderers* (1885), *Allerhand Leute* (1888), *Hoch vom Dachstein* (1892), and *Idyllen aus einer untergehenden Welt* (1899), the title of which sums up Rosegger's attitude. His collection *Als ich noch der Waldbauernbub war* (3 vols., 1902) was written for the young, whom his volume *Waldjugend* (1897) describes as people between the ages of 15 and 70. He also wrote in Styrian dialect (e.g. *Stoansteirisch*, 1885). Rosegger, who acknowledged his personal indebtedness to Svoboda, Hamerling, and others, and his literary indebtedness to Stifter, appears in his autobiographical writings (*Waldheimat*, 1877, and *Mein Weltleben*, 1888) as an engaging, modest, and sincere personality. His selected writings (*Ausgewählte Schriften*) appeared in 30 vols. from 1894 and his dialect works (*Schriften in steirischer Mundart*) in 4 vols. in 1894–6. *Gesammelte Werke* (40 vols.) was published in 1913–16 (reissued 1922–4) and correspondence (*Das Leben in seinen Briefen*), ed. O. Janda, in 1943.

Rosenband, Das, a short poem in four strophes of unrhymed verse written by F. G. Klopstock (q.v.) in 1752.

Rosengarten, an anonymous Middle High German heroic epic of the late 13th c. occurring in different versions. It is closely related to *Biterolf und Dietleib* (q.v.) and has analogies with Laurin (q.v.). It is one of the poems concerned with Dietrich von Bern (see DIETRICHSAGE).

Rosenkavalier, Der, an opera (Komödie für Musik) by R. Strauss (q.v.) with a libretto written for him by H. von Hofmannsthal (q.v.), and published in 1911. It is the second and the most popular of the six works in which Hofmannsthal and Strauss collaborated, and was first performed at Dresden in January 1911. Its setting is Vienna in the second half of the 18th c.

The Marschallin, married straight from school to a great nobleman who spends his time hunting, has as her lover the 18-year-old Octavian Count Rofrano, nicknamed Quinquin. While he is with her she is visited by a boorish country cousin, Baron Ochs von Lerchenau. Octavian disguises himself as a maid. Ochs has come to Vienna to marry a rich and pretty girl, Sophie, the daughter of a newly ennobled commoner, Edler von Faninal. Ochs asks the Marschallin's help in finding a suitable person to perform the indispensable ceremony of presenting to the betrothed girl a silver rose. She assures him that Count Rofrano will accept the duty. Ochs, who meanwhile takes a fancy to the supposed serving maid, is delighted. Left alone, the Marschallin, who foresees the end of Octavian's love for her, sings a moving lament for passing youth and beauty. In the second act Octavian duly performs the ceremony of the rose, and he and Sophie von Faninal at once fall deeply in love. An altercation occurs with Ochs, whom Octavian wounds in an impromptu duel. Ochs, who is as lecherous as he is cowardly, receives a note of assignation from 'Mariandel', the name Octavian used when disguised as a maid. In the third act Ochs falls into a trap set for him by Octavian, who has arranged the rendezvous in an inn of dubious propriety. Octavian attends in disguise as Mariandel, the Baron is frightened out of his wits by opening trapdoors and grinning faces, and his immoral intentions revealed to the Marschallin and Faninal, whom Octavian has summoned. Octavian, casting off his disguise, receives Sophie, and the Marschallin sadly but gracefully accepts the loss of his love.

Rosenkreuzer, Rosicrucians, members of the Orden der wahren Rosenkreuzer (alternatively styled Orden des goldenen Rosenkreuzes), a secret society with vague philosophical and religious aims, which included a fusion of a revived ideal classical antiquity with Christianity. It had its origins in the early 17th c. and was chiefly active in central and north-west Germany. A similar society (Orden der neuen Gold- und Rosenkreuzer) was founded in South Germany in the mid-18th c. and spread northwards and also north-eastwards into Russia. The Rosenkreuzer became notorious because the secrecy of their proceedings led to abuses. King Friedrich Wilhelm II (q.v.) of Prussia was connected with the order to the detriment of his reputation.

Rosenmontag, the day before Shrove Tuesday (see FASTNACHT). The word is a corruption of 'Rasenmontag', 'crazy Monday', and has nothing to do with roses.

Rosenmontag, eine Offiziers-Tragödie, a play by O. E. Hartleben (q.v.), published in 1900. The plot portrays the tragedy of an officer, Leutnant Hans Rudorff, resulting from his inability to subordinate his individuality to the rigid social code of the officer caste.

At the opening of the play Hans rejoins his Rhenish regiment after three months' sick leave caused by a nervous illness, but also intended as a cure for his love-affair with Gertrude Reimann, who is socially unacceptable to his family, which prides itself, through the services of his late grandfather as regimental commander, on a distinguished military tradition. In the course of the action, which is concentrated into a few days, Hans discovers that his well-intentioned cousins and brother officers, Peter and Paul von Ramberg, have betrayed his love and played Gertrude into the hands of the unscrupulous Oberleutnant von Grobitzsch. The urge to find out the whole truth drives him to arrange a meeting with Gertrude, thereby breaking the word of honour with which he had promised his commanding officer to terminate the association. Drawing the consequences of his actions, which have ruined his career as an officer, he shoots himself and, at her own wish, Gertrude.

The action is framed by the Rosenmontag (q.v.) festivities to which a performance, by the officers, of Schiller's poem *Der Handschuh* (q.v.) is to be a climax. Von Grobitzsch is detailed to act the tiger, and Peter and Paul the two leopards. Hartleben's skilful, if over-deliberate, handling of the serious, ironic, and burlesque ingredients of the play, including military music and bugle calls, as well as popular interest in the goings-on in an officers' mess made the play an immediate stage success. Hartleben was awarded the Grillparzer Prize for *Rosenmontag*.

ROSENPLÜT or ROSENBLUT, HANS (probably Nürnberg *c.* 1400–*c.* 1470, Nürnberg), also known as Der Schnepperer (Sneperer), was a brass-founder of Nürnberg, who in 1444 was appointed master-gunner (Büchsenmeister) in the city service. He wrote much poetry, though the authenticity of works ascribed to him is not always easy to determine. Some of his Sprüche (see SPRUCH) are political, referring to the Hussites, to the misgovernment of the Empire and the misdeeds of the princes, and to the rising political power of the Turks in south-eastern Europe. A Fastnachtspiel (*Des Türken Vasnachtspil,* 1456) which is attributed to him praises the orderly government of the Turks in contrast to oppression and exploitation in the Empire. In his hostility to the nobility Rosenplüt is a characteristic townsman of the 15th c. He is the first known author of Fastnachtspiele (see FASTNACHTSPIEL); he also wrote Sprüche of non-political character, such as *De hantwerger* (*Die Handwerker*) and *Die beycht* (*Die Beichte*), and Schwänke (see SCHWANK), of which the *Disputaz eines Freiheits* (a tramp triumphs over a Jew in a religious dispute) is the best known. He wrote further a number of drinking poems (Weinsegen and Weingrüße) and Priameln (q.v.), in which he was regarded as a specialist. Rosenplüt's work is liberally sprinkled with the crudities and obscenities in which the citizen-readers of the 15th c. delighted.

ROSMER, ERNST, pseudonym of Elsa Bernstein, *née* Porges (Vienna, 1866–1949, Hamburg), a dramatist; she is the author of the Naturalistic plays *Wir drei* (1889) and *Dämmerung* (1894), of the comedy *Tedeum* (1896), of the tragedies *Themistokles* (1897), *Mutter Maria* (1900), *Nausikaa* (1906), *Achill* (1910), and of the Schauspiele *Merete* (1902) and *Johannes Herkner* (1904). *Die Königskinder* (1894) provided the libretto for the opera of that title by E. Humperdinck (q.v.). She also published a volume of Novellen, *Madonna* (1894).

Roßbach, battle of (5 November 1757), in which Friedrich II (q.v.) of Prussia defeated the Franco-Imperialist army (see SIEBENJÄHRIGER KRIEG). Roßbach lies in Saxony, not far from Leipzig.

RÖSSLER, CARL (Vienna, 1864–1948, London), was an actor who wrote a number of light comedies, collaborating with A. Roda Roda (q.v.) in the extremely successful play *Der Feldherrnhügel* (1910). His fiction includes the novels *Die drei Niemandskinder* (1926) and *Wellen des Eros* (1927). In 1938 he emigrated to London.

ROST, JOHANN CHRISTOPH (Leipzig, 1717–65, Dresden), who studied at Leipzig, was for a time an adherent of Gottsched (q.v.), but went over to the opposition in 1742. He wrote anti-Gottschedian literary satires in verse (*Das Vorspiel*, 1743, and *Der Teufel*, 1755). His pastoral play *Die gelernte Liebe* (1742) had some success. His pastorals and poems were collected after his death (*Vermischte Gedichte*, 1769).

Rostocker Liederbuch, a MS. collection of 52 songs, mainly Middle Low German, made in the late 15th c. It contains the tunes as well as the words. It was edited by F. Ranke and J. M. Müller-Blattau in 1927.

ROSWITHA (also **HROTSWIT** or **HROTSWITHA**), a 10th-c. German nun who lived in the convent of Gandersheim near Brunswick and wrote six Latin comedies intended as a Christian substitute for the comedies of Terence. Their titles are *Gallicanus, Dulcitius, Callimachus, Abraham, Pafnutius,* and *Sapientia.* They appear to have received little attention in spite of their lively dialogue and skilful dramatic organization. They were rescued from obscurity by Conrad Celtis (q.v.) in 1501. Roswitha also wrote a life of the Virgin Mary, legends of the saints, and a historical poem on Otto I (q.v.), all in Latin.

Rote Hahn, Der, a four-act play (Tragikomödie) by G. Hauptmann (q.v.), first performed in the Deutsches Theater, Berlin, in November 1901, and published in the same year. It reintroduces several characters appearing in *Der Biberpelz* (q.v.), to which it is a sequel, notably Frau Fielitz (formerly Mutter Wolff and now the wife of a cobbler); Wehrhahn is still the magistrate, and the event of the play is Frau Fielitz's successful attempt to burn down her house and obtain the fire insurance money. Her health is subsequently so much eroded by her fear that the crime will be discovered that she dies. The dialogue, much of it in dialect, is as richly humorous as that of *Der Biberpelz,* and the discouragement and despair of Frau Fielitz contrasts grimly with its hilarity.

Rote Kapelle, Die, a Communist resistance group organized in Germany in 1940 by H. Schultze Boysen (1909–42), A. Harnack (1901–42), Professor W. Krauss (1900–), and A. Kuckhoff (1887–1943). It carried out sabotage and espionage for Russia until detected in August 1942. Most of its members were executed in 1942–3. Krauss, the only leader to survive, lived in East Germany after the war.

Rote Schanze, in W. Raabe's novel *Stopfkuchen* (q.v.), the farm of Farmer Quakatz, which later passes to Heinrich Schaumann.

Rote Signale, a collection of Communist poems published in 1931 in Berlin. Among the contributors were Erich Weinert, Kurt Tucholsky, and Erich Arendt (qq.v.), whose poems had previously been published in the *Arbeiter-Illustrierte-Zeitung.*

ROTH, EUGEN (Munich, 1895–1976, Munich), a Munich journalist and later an author, wrote many volumes of light satirical verse, including *Dinge, die unendlich uns umkreisen* (1918), *Erde, der Versöhnung Stern* (1920), *Rose und Nessel* (1951), *Heitere Verse* (1959), and *Ins Schwarze. Limericks* *und Schüttelreime* (1968). A selection (*Das neue Eugen-Roth-Buch*) appeared in 1970. He also wrote stories, including *Unter Brüdern* (1958) and *Lebenslauf in Anekdoten* (1962). An edition of his work (5 vols.) was published in 1977.

ROTH, JOSEPH (Brody, Galicia, 1894–1939, Paris), was born of Jewish parents in a polyglot region of Galicia close to the Russian frontier. Roth's father became insane before his birth and died in Russia in 1910. Biographical information on Roth is scarce and unreliable, partly because he himself distorted or transfigured his past life in recollecting it. It is often stated that he was baptized into the Roman Catholic Church, but this is uncertain. In his years of exile he professed Roman Catholicism and he was buried according to the Catholic rite, though the ceremony was declared provisional in view of the lack of evidence of baptism and confirmation.

Roth studied at Lemberg and Vienna and in 1916 joined the Austrian army in a rifle regiment (Feldjäger), and claimed to have spent some months in Russian captivity. His military experiences are, however, shrouded in uncertainty. After the war he took up journalism and was at different times in Vienna, Berlin, and Frankfurt. While working for the *Frankfurter Zeitung* he travelled much, and acquired the habit of living in hotels. He was married, but his wife was in a mental home. Roth grew up with an acute sense of a lost tradition and a threatened future. He mourned the stability of the vanished Habsburg Empire and he feared the coming of Nazism. During what was probably his best period he published seven novels: *Savoy Hotel* (1924), *Die Rebellion* (1924), *Die Flucht ohne Ende* (1927), *Zipper und sein Vater* (1928), *Rechts und Links* (1929), *Hiob* (q.v., 1930), and *Radetzkymarsch* (q.v., 1932) which is generally considered to be his best novel. All except the last are short works. In 1933 Roth emigrated to Paris, where he spent the remainder of his life, writing novels to earn a living, and drinking excessively as a form of painless suicide, or, as he himself expressed it, deferment of immediate death (by suicide). Through the last fifteen years of his life he regarded pre-1914 Austria with increasing nostalgia, though with a clear recognition of its shortcomings and its inevitable decay; and he sought to associate himself by his dress, company, and religious worship with the legitimist cause. The novels of this period of exile are *Tarabas, ein Gast auf dieser Erde* (1934), *Die hundert Tage* (1936), *Beichte eines Mörders* (1936), *Das falsche Gewicht* (1937), *Die Kapuzinergruft* (q.v., 1938) and *Die Geschichte der 1002. Nacht* (q.v., 1939). Of his minor fiction, *Die Büste des Kaisers* (q.v., in French 1934, in German 1964) is prob-

ably the most notable work. *Der stumme Prophet*, the story of the failure of a revolutionary, written in 1929, was published posthumously in 1966. Roth's novels have two principal settings, Vienna and the Galicia in which he grew up. Many novels embody both (e.g. *Radetzkymarsch*, *Geschichte der 1002. Nacht*).

Roth is noticeably preoccupied with family problems and especially with the relationship between father and son (*Zipper und sein Vater*, *Hiob*, *Radetzkymarsch*, *Die Kapuzinergruft*). The aged Emperor Franz Joseph is repeatedly represented as a paternal figure. On the other hand the atmosphere is predominantly one of decadence and *ennui*. Collections of essays published in Roth's lifetime are *Juden auf Wanderschaft* (1927) and *Panoptikum* (1930). His letters (*Briefe 1911–1939*) were published in 1970, and *Werke* (3 vols.), both ed. H. Kesten (q.v.), in 1956, and in extended form (4 vols.) in 1975–6.

ROTHE, HANS (Meißen, 1894–1978, Florence), worked as a dramatic adviser (Dramaturg) during the 1920s, first in Leipzig and then under Max Reinhardt (q.v.) at the Deutsches Theater in Berlin. In 1934 he emigrated to the U.S.A., settling later in Florence. He is the author of plays, including *Wen die Götter verderben wollen* (1939) and *Sainte Eugénie* (1944), of plays for radio (*Verwehte Spuren*, 1935, and *Die Vitrine*, 1964), and of the novels *Ankunft bei Nacht* and *Beweise das Gegenteil* (both 1949). In 1961 he published *Shakespeare als Provokation*, an essay on Shakespeare. He spent some forty years translating Shakespeare's plays into modern German; his edition appeared as *Der elisabethanische Shakespeare*, 1963–4. Rothe's *Gesammelte Dramen* were published in 1947.

ROTHE, JOHANNES (Kreuzburg, Thuringia, *c.* 1360–1434, Eisenach), a Thuringian priest, who was town clerk of Eisenach and later canon of the collegiate church of Our Lady there. Rothe wrote several works of primarily didactic character. The *Ritterspiegel* (after 1410) is an instructional work, setting forth the qualities a knight should cultivate and those he should avoid, together with descriptions of chivalric life, equipment, heraldry, and so on. *Von der stete ampten und von der fursten ratgeben* (a title given by A. F. C. Vilmar (1800–68), the first editor) deals with the organization of councils and the duties of counsellors. About 1420 Rothe wrote *Die Passion*, a poem recounting the legends of Judas and Pilate. His life of St. Elisabeth in verse (*Heilige Elisabeth*) was written at about the same time, and he also wrote a long poem on chastity (*Von der Keuschheit*).

Rothe is the author of a prose chronicle of Thuringia (*Düringische Chronik*), written *c.* 1421.

It exists in two versions, one dedicated to the Landgravine Anna, the other to Bruno von Teutleben, governor of the Wartburg (q.v.).

RÖTTGER, KARL (Lübbecke, 1877–1942, Düsseldorf) a religious poet, published *Wenn deine Seele einfach wird* (1909) and *Die Lieder von Gott und dem Tod*. He was associated with the periodical *Charon* (q.v.).

ROUSSEAU, JEAN-JACQUES (Geneva, 1712–78, Ermenonville), French–Swiss man of letters and novelist, who fostered the love of nature and the cult of emotion in the middle of the 18th c. Rousseau set primitive or rustic simplicity above the complexities and hypocrisy of civilization. His thought is expressed in a series of essays and treatises, including *Discours pour l'Académie de Dijon* (1750), *Discours sur l'origine et les fondements de l'inégalité parmi les hommes* (1755), *Lettre à d'Alembert sur les spectacles* (1758), and *Du contrat social ou principes de droit politique* (1762), to which must be added the novel *Émile ou de l'éducation* (1762). These works helped to prepare the way for the French Revolution (q.v.) and contributed to the rise of the Sturm und Drang (q.v.) and of the Romantic movement (see ROMANTIK) in Germany. J. G. Herder (q.v.) was possibly the writer most influenced by Rousseau. The novel *Julie ou la nouvelle Héloïse* had a direct impact on Goethe, manifested in *Die Leiden des jungen Werthers* (q.v., 1774), and also strongly influenced Sophie von La Roche's novel *Die Geschichte des Fräuleins von Sternheim* (q.v., 1771).

RUBIN, a minor Middle High German Minnesänger, was a native of Tyrol and, notwithstanding the absence of any prefix, is believed to have been of minor nobility. Rubin, whose home was possibly Merano, is represented by 22 songs, rather pale reflections of the manner of REINMAR der Alte (q.v.). His poems include a Tagelied and crusade songs, and it is thought that he may have participated in the crusade of 1228–9.

Rubin, Der, see HEBBEL, F.

Rückblick, a poem by D. von Liliencron (q.v.), published in *Gedichte* (1889). It is a recollection of battle.

RÜCKERT, FRIEDRICH (Schweinfurt, 1788–1866, Neuses nr. Coburg), son of a lawyer in government employment, was educated at Schweinfurt grammar school (1802–5) and at Würzburg (1805–8) and Heidelberg (1808–9) universities. Having decided on an academic career, he qualified at Jena in 1811, but in 1812 abandoned the idea. He took up a school-teaching post at Hanau, but soon resigned

and returned home. In the War of Liberation (see NAPOLEONIC WARS) he stayed at home at his parents' request, contributing *Geharnischte Sonette* (1814) to the war effort. For a time (1815–16) he was on the staff of Cotta's *Morgenblatt* at Stuttgart, but resigned and spent a year (1817–18) in Italy. On his return he met J. Hammer-Purgstall (q.v.) in Vienna and under his influence began to devote himself to oriental languages. He married in 1821 and subsisted on honoraria and the editorship of the *Frauentaschenbuch*. His impressive command of oriental philology led in 1826 to appointment as professor at Erlangen, where, however, he felt happy only for a short time. In 1841 he was appointed a professor at Berlin University, but his enthusiasm soon waned, and in 1848 he was given leave to retire on half pay. He settled at Neuses nr. Coburg, where he already owned a property, and devoted his remaining years to scholarship.

Rückert was a fertile and facile poet, and his large output of poetry is more notable for its neat workmanship in various verse forms than for vision or originality. His best work is probably in the *Liebesfrühling* (q.v., 1844), in which are collected the poems of his courtship. His principal poetic publications are *Deutsche Gedichte* (1814, including *Geharnischte Sonette*), *Kranz der Zeit* (1817), *Östliche Rosen* (1822), *Ghaselen* (1822), *Gesammelte Gedichte* (1834–8), and *Die Weisheit des Brahmanen* (q.v., 6 vols., 1836–9). The posthumously published *Kindertotenlieder* (1872) have been set by G. Mahler (q.v.). Rückert also wrote epic poems, Aristophanic comedies, and historical plays, none of which achieved success. *Gesammelte poetische Werke* (12 vols.), ed. by his son H. Rückert and D. Sauerländer, appeared in 1867–9.

Rückversicherungsvertrag, the secret Reinsurance Treaty between Germany and Russia, devised by Bismarck (q.v.), and signed on 18 June 1887. By it each power undertook to maintain a benevolent neutrality if the other were involved in war, except in the case of a German attack upon France or a Russian attack upon Austria-Hungary. The duration of the treaty, which was designed to avoid the encirclement of Germany (Einkreisung), was for three years in the first place. It sought to mitigate the military dangers inherent in the Dual Alliance (see ZWEIBUND), which was extended into a Triple Alliance (see DREIBUND) with Italy. The weakness of the Reinsurance Treaty, of which Bismarck was well aware, lay in the fact that Germany undertook obligations towards Russia as well as towards Austria and Italy, which it would have found difficult to fulfil in the event of an armed conflict between these powers.

Bismarck was dismissed in 1890 and his successor Caprivi (q.v.) did not renew the treaty.

RUDOLF I (1218–91, Speyer), the founder of the Habsburg (q.v.) dynasty, whose election, in 1273, as German King (see DEUTSCHER KÖNIG) ended the Interregnum (q.v.).

In his early life Rudolf supported the Emperor Friedrich II (q.v.) against the Papacy, and in 1254 was excommunicated. A man of greater wealth and power and of stronger character than it was thought by some, especially by his rival candidate Ottokar II (q.v.) of Bohemia, he owed his election mainly to the influence of the Elector of Mainz and Count Friedrich of Nürnberg. In October 1273 Rudolf was crowned German King at Aachen. He secured his position by renouncing rights in papal territories, so winning the Pope's recognition.

Ottokar of Bohemia remained opposed to Rudolf and, after reluctantly submitting in 1276, rose against him with Polish support. The campaign which followed was ended by Rudolf's triumph and Ottokar's death at the battle of Dürnkrut in 1278. In the following years Rudolf subdued other rebellious dukes and princes, checked unlawful oppression by predatory barons, and took the first step to establishing the hereditary claim of the Habsburgs to imperial power by investing his sons with the duchies of Austria and Styria (Steiermark). Rudolf was not able to enforce his authority as broadly as he wished, nor did he succeed in getting himself crowned emperor. His simple manners brought him considerable popularity among the common people; the territorial princes would have preferred him to take more interest in foreign ventures and leave them a free hand at home.

Rudolf is the principal character in Schiller's ballad *Der Graf von Habsburg* (1804) and the protagonist in Grillparzer's tragedy *König Ottokars Glück und Ende* (q.v., written in 1823). He is the subject of a ballad by J. Kerner (q.v.), *Kaiser Rudolfs Ritt zum Grabe*, and of plays by A. von Kotzebue (q.v., *Rudolf von Habsburg und Ottokar*, 1815), K. Pichler (q.v., *Rudolf von Habsburg*, 1818), and O. Roquette (q.v., *Rudolf von Habsburg oder Die Sterner*, 1856).

RUDOLF II (Vienna, 1552–1612, Prague), son of the Emperor Maximilian II (q.v.), was from 1572 king of Hungary, from 1575 king of Bohemia, and became emperor in 1576. He put much effort into seeking to prevent the nomination of his brother Matthias (q.v.) as his successor. Unmarried himself and without an heir, he distrusted the advice of his family even when political action, especially against the Turks and, in 1604, during the revolts in Hungary, was called for. In 1608 Rudolf was forced to transfer the government of Austria and Hungary to Matthias. Only Bohemia and Moravia remained faithful to him. But the Bohemians exploited his weakened position by forcing him to guarantee their religious freedom in the Letter of Majesty (see MAJESTÄTSBRIEF) of 1609. Difficulties arising out of the interpretation of the Letter of Majesty ended in his deposition in 1611 and in the recognition of Matthias as king of Bohemia.

Rudolf, who had been brought up at the Spanish court, was a highly gifted and educated man who devoted his voluntary isolation at the Hradschin in Prague to the collection and study of works of art and literature as well as to scientific and astronomical pursuits, inviting J. Kepler (1571–1630) and the astronomer Tycho Brahe (1546–1601) to his court. Towards the end of his life he became more and more eccentric and inefficient in the administration of government. The reign of Rudolf II is the subject of Grillparzer's tragedy *Ein Bruderzwist in Habsburg* (q.v.).

RUDOLF, ERZHERZOG UND KRONPRINZ (Laxenburg nr. Vienna, 1858–89, Mayerling), only son of the Emperor Franz Joseph and the Empress Elisabeth (qq.v.), and heir to the throne of Austria-Hungary. Prevented by his father from taking any part in political affairs, the Crown Prince became restless and dissatisfied. He is believed to have shot the 17-year-old Marie Vetsera and himself in a suicide pact, but the precise facts of his death were concealed and still remain obscure (see MAYERLING).

RUDOLF VON EMS, a Middle High German poet of the 12th c., belonged to the second generation of epic writers of the Blütezeit (q.v.). The date of his birth is not known, but his known works, six in number, were written between 1220 and 1250. He died in Italy in 1252 or 1253. Rudolf was a *ministeriale*, a nobleman in the train of a greater noble, the Count of Montfort. His name refers to the Montfort castle at Hohenems in what is now the western tip of Austria between Bregenz and Feldkirch. It has been suggested that he may have established the collection of MSS. formerly in the castle. His earliest known work is *Der gute Gerhard* (q.v.) a poem of charity and humility, the hero of which, surprisingly, is a merchant. This was followed by *Barlaam und Josaphat* (q.v.), which was in effect a renunciation of the conception of *minne* exalted by GOTTFRIED von Strassburg (q.v.), whom Rudolf had once revered. The poem *Eustachius* is lost, and the chronological order of the next two poems *Willehalm von Orlens* and *Alexander* (qq.v.) is uncertain; probably the former was the earlier. *Willehalm von Orlens* portrays a knight of ideal conduct in a world of reality, and *Alexander*, which is unfinished, represents an ideal ruler. Rudolf's last work, which also remained a torso, though a very substantial one, is the *Weltchronik* (q.v.), a vast history which proceeds as far as the Hebrew kings.

Rudolf was indebted stylistically to HART-MANN von Aue (q.v.) and especially to Gottfried von Straßburg, and *Willehalm von Orlens* and *Alexander* contain interesting passages of literary comment with tributes to his predecessors. His own style gravitated from the elegant literary to a sober historical manner, a development which corresponds to the shift of his own interests. Deeply rooted in his own class (in spite of *Der gute Gerhard*), Rudolf was a firm adherent of the Staufen emperors (see HOHENSTAUFEN) and met his death while accompanying Konrad IV (q.v.) to Italy.

RUDOLF VON FENIS, GRAF VON NEUENBURG (Neuchâtel), a Middle High German Minnesänger and great noble, who is documented between 1158 and 1192. His seven Minnelieder, conceived as a cycle, are strongly influenced by Provençal lyric poetry.

RUDOLF VON HABSBURG, see RUDOLF I.

RUDOLF VON ROTENBURG, a Middle High German poet of the mid-13th c., was probably a knight of Lucerne. His twelve Minnelieder continue the tradition of HART-MANN von Aue and WALTHER von der Vogelweide (qq.v.). He also wrote five leiche (see LEICH).

Rudolf von Schlüsselberg, a medieval Latin poem which recounts the infidelity of a knight's wife, her sadistic persecution of her husband, and the husband's revenge upon her and her lover, a Saracen king. The story has affinities with *Der Nußberg* by Heinrich Rafold (q.v.).

RÜEDEGER VON HÜNCHOVEN, author of a short Middle High German verse tale, *Der Schlegel*, is known to have lived at Regensburg at the end of the 13th c. In *Der Schlegel* an old man who resigns his property to his children finds himself disgracefully treated. He secures good treatment by a ruse, pretending that he still has treasure in a chest. After his death the chest is opened and found to contain only a mallet (Schlegel) with the instruction that it is to be used to brain a man foolish enough to give his property to his children in his lifetime. See also KOTZEMAERE, DAS.

RUF, JAKOB (Zurich, c. 1500–58, Zurich), a surgeon in Zurich, wrote anti-Catholic biblical plays, of which the principal titles are *Die Beschreibung Jobs* (1535), *Etter Heini* (1538, the title is that of an edition of 1847, the original was called *Vom wol- und übelstand einer löblichen Eidgenoschaft*), *Von des Herren Weingarten* (1539), *Spiel von Joseph* (1540), *Das Leiden des Herrn*, *Wilhelm Tell* (both 1545), *Spiel von der Erschaffung Adams und Eva* (1550), *Abraham*, and *Geistliches Spiel von der Geburt Christi* (both 1552).

Ruf, Der, a radical literary and political periodical founded in Munich in September 1946 by A. Andersch and H. W. Richter (qq.v.). It was closed down in April 1947 by the U.S. Military Government, which refused a licence to a projected successor, *Der Skorpion*, if Andersch and Richter were included on its board. Out of this incident arose the Gruppe 47 (q.v.).

Ruhe ist die erste Bürgerpflicht oder Vor fünfzig Jahren, a historical novel (Ein vaterländischer Roman) in 5 vols. by W. Alexis (q.v.), published in 1852. It is set in Prussia (and chiefly in Berlin) in the period preceding the Prussian defeat at Jena (q.v., October 1806). The title alludes to a proclamation issued on 17 October 1806 by the Prussian minister Count von der Schulenburg-Kehnert, containing the words, 'Der König hat eine Bataille verloren. Jetzt ist Ruhe die erste Bürgerpflicht. Ich fordere die Einwohner Berlins dazu auf.' The novel contrasts the corruption and decadence of an important element in the Prussian nobility with the robust and unpretentious integrity and patriotism of the citizen class. The hero, Louis Bovillard, dies serving the king in the war and is married on his deathbed to Adelheid Alltag, daughter of a civil servant. In 1854 Alexis published a sequel (*Isegrimm*, 3 vols.) dealing with the aftermath of Jena.

RÜHM, GERHARD (Vienna, 1930–   ), studied piano and composition at the Wiener Staatsakademie für Musik (1945–51) and, until 1952, oriental music in Beirut; he was a member of the Wiener Gruppe (q.v.) and is the author of concrete poetry (see KONKRETE POESIE) and of experimental prose. In 1958–9 he collaborated with H. C. Artmann and F. Achleitner (qq.v.) in the 'literarisches cabaret'. In 1964 he settled in West Berlin. The titles of his publications include *konstellationen* (1961); *farbengedicht*; *betrachtung des horizonts*; *lehrsätze über das weltall* (all 1965); *söbstmeadgroundz* (1966); *daheim. 10 Textmontagen* (1967); *rühms schablone für zeitungsleser*; and *fenster. Texte* (both 1968), which contains the best of his experimental writings. Rühm makes use of different type-founts as a visual feature of his work. *Gesammelte Gedichte und visuelle Texte* appeared in 1970. Rühm is the editor of the volume *Die Wiener Gruppe. Ach-*

leitner. Artmann. Bayer. Rühm. Wiener Texte. Gemeinschaftsarbeiten. Aktionen (1967, ext. edn. 1985). Other works include a collection of plays, *Ophelia und die Wörter* (1972), *Mann und Frau* (1972), *Wahnsinn. Litaneien* (1973), and *Comic* (1978). He has also staged exhibitions of his visual poetry and montage.

RÜHMKORF, PETER (Dortmund, 1929–), a radical critic of society and skilful manipulator of language in the *avant-garde* manner, began writing lyric poetry under the influence of G. Benn (q.v.), and plays influenced by Brecht (q.v.). His collections of poetry include *Ird. Vergnügen in g* (1959), a title alluding to poetry by B. H. Brockes (q.v.), and *Kunststücke* (1962); his play (Stück) *Lombard gibt den Letzten* (1972, performed in Dortmund in the same year) abounds in similar allusions serving his materialist point of view. A further play, *Die Handwerker kommen. Ein Familiendrama,* on the disintegration of a lower-middle-class family, appeared in 1974. *Walther von der Vogelweide. Klopstock und ich* (1975), on the subject of the writer, includes an essay on the medieval poet (see WALTHER VON DER VOGELWEIDE) and translations of some of his poems. The collection *Haltbar bis Ende 1999. Gedichte* was published in 1979, *Im Fahrtwind. Gedichte und Geschichte* in 1980, *Auf Wiedersehen in Kenilworth. Ein Märchen in dreizehn Kapiteln* in 1980, and *Der Hüter des Misthaufens. Aufgeklärte Märchen* in 1983. The volume *agar agar–zaurzaurim. Zur Naturgeschichte des Reims und der menschlichen Anklangsnerven* (1981) contains lectures delivered during his spell as visiting professor (Gastdozent für Poetik) at Frankfurt University. *Es muß doch noch einen zweiten Weg ums Gehirn rum geben* appeared in 1981, *Kleine Fleckenkunde* in 1982, and *Bleib erschütterbar und widersteh. Aufsätze. Reden* in 1984. In 1961 he published a biography of W. Borchert (q.v.). *Gesammelte Gedichte* appeared in 1976 and *Strömungslehre I. Poesie* in 1978.

Rührstücke, sentimental plays with tense emotional situations, popular in the late 18th c. and early 19th c. A. W. Iffland and A. von Kotzebue (qq.v.) were the best-known authors of these productions.

Ruinen des Campo Vaccino, Die, a poem by F. Grillparzer (q.v.), written in Rome on 20 April 1819, and published in the Taschenbuch *Aglaja* for 1820. The poem laments the passing of ancient Rome, evoking the great names associated with its rise and fall, and deploring in the last of the sixteen stanzas the emptiness of the present. The apparently anti-Catholic trend of the poem involved Grillparzer in difficulties with the Habsburg Court.

RULMANN, surname of two brothers, Anton (d. 1652) and Heinrich (1596–1651), natives of Lower Saxony. One of them, probably Anton, is believed to be the author of a volume of homely satirical poems in Low German, published anonymously about 1650 (*Etlike korte und Verstendlike Kling Gedikte van Allerhand Saken, so eine Tydt her hier und der under Olden Bekandten und Frunden vorgevallen*).

Rumänisches Tagebuch, see TAGEBUCH IM KRIEGE.

RUMELANT or **RUMZLANT VON SACHSEN,** a minor Middle High German poet, who was a native of Lower Saxony, lived in the second half of the 13th c., and wrote Sprüche (see SPRUCH). He was of a polemical turn of mind and readily attacked other poets.

RÜMELANT VON SCHWABEN, a minor Middle High German poet of the late 13th c., who is the author of four Sprüche (see SPRUCH), two of which are eulogies of noblemen, presumably his patrons.

RUMOHR, KARL FRIEDRICH VON (Reinhartsgrimma nr. Dresden, 1785–1843, Dresden), a distinguished art historian, published *Italienische Forschungen* (3 vols., 1826–31), *Hans Holbein der Jüngere in seinem Verhältnis zum deutschen Formschnittwesen* (1836), and other works on art, including a history of the Royal Danish collection of engravings. He wrote a 4-vol. novel entitled *Deutsche Denkwürdigkeiten aus alten Papieren* (1832), stories (*Novellen,* 1833–5), and a satirical poem, *Kynalopekomachie, der Hunde-Fuchsen-Streit* (1835). His talents extended also to culinary art (*Der Geist der Kochkunst,* 1823).

Rundköpfe und die Spitzköpfe, Die, oder Reich und Reich gesellt sich gern, a play with 11 scenes in mixed prose and blank verse by B. Brecht (q.v.), with music by H. Eisler (q.v.), written in 1931–4. Designated 'Ein Greuelmärchen' in its final version in 1938, it is described (*Versuche,* vol. 8, 1933) as a Schauspiel which was originally based on a stage adaptation of Shakespeare's *Measure for Measure.* The play's double title points to a double theme: racial conflict, in which the Tschuchen have round heads and the Tschichen pointed heads, and class division between the rich and the poor. The second title is a pun on the proverb 'Gleich und gleich gesellt sich gern.' The Tschuchen wear round masks and the minority people, the Tschichen, pointed masks which are 15 cm taller than the round masks. The setting is the imaginary city of Luma in Jahoo. The prologue explains that the action is inspired by the terrible quarrels dividing men of different colour and nationality, and the parable is intended to demonstrate that the principal division exists between the rich and the poor. Acting on racialist principles, Angelo Iberin, Statthalter of the Tschuchen, defeats the Tschichen. Callas, a tenant farmer, betrays the Sickle movement of his own and lower class, hoping that the new regime will introduce new justice. The various episodes illustrate the distortion of humanity by society and the power of money (e.g. *Lied von der belebenden Wirkung des Geldes*, scene 7). But the returning viceroy restores the old order, the power of the landlords, while the tenant peasants face a mass execution. They go to the gallows singing their revolutionary Sickle song (*Sichellied*: 'Bauer, steh auf!') as their final appeal.

Runen, a primitive Germanic alphabet in use from the early Christian era until the 11th c. It survives chiefly in inscriptions on stone, metal, or horn, the majority found in Scandinavia. It has been thought that the persistence of runes after the Latin alphabet had come into general use may be connected with a belief in the magic properties of runic letters. The word Rune to denote this lettering was first employed by J. G. Schottel (q.v.) in 1663.

Runenberg, Der, a Novelle by L. Tieck (q.v.), written in 1802 and published in 1804. Christian, drawn by the beauty of forest and mountain, sees at night a beautiful woman in a remote hut. He later marries and leads a respectable life as a successful farmer. But the once beautiful woman, now aged and ugly, reappears, and he goes off with her into the mountains in search of gold. He reappears briefly to his wife and then disappears for good. The most remarkable feature of this eerie story is Tieck's evocation of wild, romantic landscape.

RUNGE, Philipp Otto (Wolgast, Pomerania, 1777–1810, Hamburg), German painter, was at first destined for a commercial career, then studied art at the Copenhagen Academy (1799–1801) and was afterwards in Dresden. In 1804 he settled in Hamburg, where he died of tuberculosis at the age of 35. He painted a number of strong, sensitive, realistic portraits, chiefly of members of his family (notably the portraits of his parents and the self-portrait with his wife and brother, known as *Wir Drei*, which was destroyed by fire in 1931). He also produced a number of Romantic allegories, including an unfinished series on the phases of the day (*Tageszeiten*), and provided some illustrations for the *Kinder- und Hausmärchen*, the collection of fairy-tales of the brothers J. and W. Grimm (qq.v.),

to which he contributed *Von dem Fischer und syner Fru* and *Der Machandelboom* (1806). He is most strongly represented in the Hamburg Kunsthalle.

Runge's *Farbenkugel oder Konstruktion der Verhältnisse der Farben zueinander* appeared in 1810 (reissued 1959), posthumous papers (*Nachgelassene Schriften*, ed. D. Runge, 2 vols.) in 1840–1 (reissued 1965), *Briefwechsel mit Goethe*, ed. H. von Maltzahn, in 1940, and *Briefe und Schriften*, ed. P. Betthausen, in 1982.

Runkelstein, castle near Bolzano in South Tyrol, formerly belonging to the Vintler family, which has frescoes from the beginning of the 15th c. drawn from three medieval poems, Gottfried von Straßburg's *Tristan*, Rudolf von Ems's *Willehalm von Orlens*, and *Garel vom blühenden Tal* by Der Pleier (qq.v.).

Ruodlieb, a Latin poem which recounts the story of Ruodlieb, a youth of noble birth who goes out to seek his fortune and encounters diverse and sometimes miraculous adventures. It was written *c.* 1030–50 by a monk of Tegernsee Abbey (Bavaria). It is not complete. It has been regarded as an ancestor of the German novel, though this view is not universally accepted.

RUPERT OF THE RHINE or **OF THE PALATINATE,** see Ruprecht, Prinz.

RUPERT VON DEUTZ (d. 1135), a theological writer of the early 12th c., was a Benedictine monk and abbot of Deutz (on the Rhine opposite Cologne) from *c.* 1119 until his death. Rupert's writings, exegetical, allegorical, or mystical in conception, are in Latin. Modern editions exist of the following: *Liber de divinis officiis, Commentaria in Evangelium Sancti Johannis,* and *De victoria Verbi Dei* (ed. R. Haacke, 1967, 1969, and 1970 respectively), and *De operibus Spiritus sancti* (ed. E. de Solms and J. Gribomont, 1967). His works also include *De divinis officiis, De Trinitate et operibus ejus,* and a commentary on Exodus.

Ruppiner Bilderbogen, see Neuruppiner Bilderbogen.

RUPPIUS, Otto (Glochau, 1819–64, Berlin), after a varied early career became a left-wing journalist in 1845 and in 1848 fled to America in order to escape imprisonment. After a few years of music teaching he returned to journalism, first in New York and then in St. Louis, and also devoted himself to novel writing. *Der Pedlar* (1857) and *Das Vermächtnis des Pedlars* (1859), stories of American life, had a distinct success in Germany. Ruppius returned to Germany in

1861 and wrote for the magazine *Die Gartenlaube* (q.v.). Among his later stories *Eine Weberfamilie* (1862) deserves mention. *Gesammelte Werke* (6 vols.) appeared in 1857–64 and again in 1889 (15 vols.).

RUPRECHT, PRINZ (Prague, 1619–82, London), son of the unfortunate Elector Friedrich V (q.v.) of the Palatinate, was an able cavalry commander and fought with distinction in the Thirty Years War (see DREISSIGJÄHRIGER KRIEG) and on the Royalist side in the English Civil War from 1642. He was present at Marston Moor (1644) and Naseby (1645), defeats from which he and his troops emerged with credit. After the capitulation of Charles I, Rupert left England with a parliamentary safe-conduct. He

returned in 1660, became an admiral, and fought successfully against the Dutch.

He was an artist of note, working in mezzotint, and also a military technologist, who experimented with ordnance and explosives.

RUPRECHT VON WÜRZBURG, was the author of a 14th-c. verse tale, *Von zwein Kaufmannen*, celebrating the constancy of a wife. Bertram, a happily married merchant, while on a journey at Provins praises the fidelity of his wife. His host Hogier wagers his entire fortune that he will seduce her within six months. The wife, Irmingart, gives him an assignment in response to a substantial bribe, but substitutes her maid. So Bertram wins his wager and Hogier's fortune. He compensates Hogier with the maid's hand.

S

SA, abbreviation for Sturmabteilung, the paramilitary organization of the NSDAP (q.v.), founded in 1921. The storm troops (also referred to as brownshirts after the colour of their uniform, introduced in 1925) were involved in the abortive *coup d'état* in Munich (see HITLERPUTSCH) in 1923, after which the organization was forbidden for two years in Bavaria. In 1931 E. Röhm (q.v.) took over command; in April 1932 Brüning (q.v.) banned the SA for the whole of Germany, but in July the ban was lifted by Papen. The SA was notorious for its use of violence. After Röhm's assassination in 1934 its standing rapidly declined.

SAAR, FERDINAND VON (Vienna, 1833–1906, Döbling), descended from a line of civil servants, lost his father in infancy, and was brought up by his mother in his grandfather's house. At 16 he entered the Austrian Cadet School and was commissioned in the infantry in 1854. He saw active service in the campaign in North Italy in 1859, though his regiment was not engaged in the fighting. After the war he resigned in order to devote himself to literature, living in straitened circumstances for several years.

Saar's gifts are chiefly demonstrated in his numerous Novellen, some of which were published individually before being included in the two editions of *Novellen aus Österreich* (q.v., 1877 and 1897), which also contain the collections *Drei neue Novellen* (1883, see VAE VICTIS!), *Schicksale* (1889), and *Frauenbilder* (1892). Later volumes are *Herbstreigen* (1896), containing *Herr*

Fridolin und sein Glück, *Ninon*, and *Requiem der Liebe*; *Nachklänge* (1899) includes poems as well as the stories *Doktor Trojan*, *Conte Gasparo*, and *Sündenfall*; *Camera obscura* (1901) contains eight stories, *Die Brüder*, *Die Parzen*, *Der Burggraf*, *Der Brauer von Habrovan*, *Außer Dienst*, *Die Heirat des Herrn Stäubl*, *Der Hellene*, and *Dissonanzen*; the last collection, *Tragik des Lebens* (1906), consists of the four stories *Die Familie Worel*, *Sappho*, *Hymen*, and *Die Pfründner. Eine Wiener Geschichte* (q.v.).

Like many writers of the 19th c., Saar nourished a persistent ambition to write successfully for the stage, but his plays were repeatedly turned down by Viennese management, *Thassilo* (1885) being withdrawn by the Burgtheater after the first rehearsal. This and the earlier *Kaiser Heinrich IV* (1867), a play in two parts (*Hildebrand* and *Heinrichs Tod*, both consisting of five acts), are wholly, and *Die beiden de Witt* (1874) partly, in verse; *Tempesta* (1880), a five-act tragedy, is set in North Italy in the 17th c.; *Eine Wohltat* (1887), a four-act Volksdrama which has as its background the Austrian mountains was produced in the Burgtheater on 14 December 1903 to celebrate Saar's seventieth birthday, but it was not able to maintain a place in the repertoire. His lyric poetry was collected in 1881 as *Gedichte* (expanded 1888), and his elegies, *Wiener Elegien*, appeared in 1893. Saar wrote two verse epics, *Die Pincelliade* (1897), a comic erotic poem in *ottava rima* set in Vienna in the time of Maria Theresia (q.v.), and *Hermann und Dorothea* (1902), a sequel to

Goethe's poem of the same title. Although it employs the same classical metre, it is only a pale shadow of its model (see HERMANN UND DOROTHEA).

In his later life Saar enjoyed happier conditions, being given a permanent invitation to stay at the mansion of the princely family of Salm-Reifferscheidt in Blansko. Princess Marie zu Hohenlohe and the Princess von Sayn Wittgenstein were among his patrons. Saar married in 1881, but his ailing wife committed suicide in 1884. He spent his last years in and near Vienna; he put an end to his life with his revolver while suffering from cancer.

Sämtliche Werke (12 vols.), ed. J. Minor, appeared in 1909, *Gesamtausgabe des erzählerischen Werkes* (4 vols.), ed. J. F. Fuchs, in 1959–62.

SABINUS, GEORG, real name Schuler (Brandenburg, 1508–60, Frankfurt/Oder), a German humanist and author of Latin poetry, especially elegies, was a professor at Frankfurt/Oder, and became the first rector of Königsberg University, East Prussia (founded 1544), returning afterwards to Frankfurt. He was employed by the Elector of Brandenburg on diplomatic missions.

SACER, GEORG WILHELM (Naumburg, 1635–99, Wolfenbüttel), a lawyer by training, led a restless life, in the course of which he was by turns secretary, private tutor, and soldier. From 1670 he was a lawyer in the service of the Duke of Brunswick. He wrote religious poetry (*Der bluttriefende, siegende und triumphierende Jesus*, 1661), but his most remarkable work is *Reime dich oder ich fresse dich*, published under a pseudonym in 1673. It is a vigorous and combative treatise on poetry.

SACHER-MASOCH, LEOPOLD, RITTER VON (Lemberg, 1836–95, Lindheim), an Austrian subject, studied in Prague and Graz, served in the Italian war of 1859, and was for a short time a professor in Lemberg. Subsequently he devoted himself entirely to writing. His skilfully written short novels and Novellen are devoted almost exclusively to the perversion, which, since Krafft-Ebing (q.v.), is known as masochism. His heroes desire to be ill-treated and humiliated by the women they love. Among the best-known works are *Das Vermächtnis Kains*, *Venus im Pelz* (both 1870), and *Die Messalinen Wiens* (1874). He also edited a review, *Auf der Höhe* (1882–5). The revelations published by his widow, Wanda von Sacher-Masoch (*Meine Lebensbeichte*, 1906), are not universally accepted as true.

SACHS, HANS (Nürnberg, 1494–1576, Nürnberg), one of the most prolific writers of the

16th c., was by trade a shoemaker. After an education which included a grounding in Latin he was apprenticed in 1508. From 1511 to 1516 a travelling journeyman, he visited Vienna, Frankfurt, Lübeck, and Osnabrück, before settling permanently in Nürnberg. Trained in Meistergesang (q.v.) in his teens, Sachs became conspicuous in the Guild of Meistersinger, writing a large number of songs conforming to the strict rules of the form. In 1519 he married Kunigunde Kreutzer, who died in 1560, and he took as second wife in 1561 Barbara Harscher, a widow of 27. He was an adherent of Luther (q.v.), whom he celebrated in the poem *Die wittenbergisch Nachtigall* (q.v., 1523). Until checked by the city council, he supported the Protestant cause in prose dialogues, of which the most widely known was his *Disputation zwischen einem Chorherren und Schuchmacher* (1524).

The greater part of Sachs's literary output consists of Meisterlieder and Spruchgedichte, a designation which embraces a very large number of humorous fables, anecdotes, and tales in verse (Schwänke, see SCHWANK), as well as some 200 verse plays. Among the Schwänke are *Schlauraffenland* (1530), *Sanct Peter mit der Geiß* (q.v., 1555), *Gespräch Sanct Peters mit den Landsknechten* (1556), *Schwank von dem frommen Adel* (1562), and *Der Schneider mit dem Pannier* (1563). Sachs's plays include tragedies, comedies, and Fastnachtspiele (see FASTNACHTSPIEL). Their range of subjects is an indication of the extent and variety of his reading, and only a few titles can be mentioned. Among the tragedies are *Lucretia* (1527), *Tragödie von der Schöpfung* (1548), *Der Wüterich Herodes* (1552), *Die Maccabäer* (1552), *Die mörderisch Königin Klitemnestra* (1554), *Die getreu Fürstin Alcestis* (1555), *Tragödie König Sauls* (1557), *Der hörnen Siegfried* (1557), *Tragödie der ganz Passio* (1558), *Tragödie des jüngsten Gerichts* (1558), *Tragödie von Alexandro Magno* (1558), and *Andreas der ungerisch König mit Bancbano seinem getrewen Statthalter* (1561). The comedies (*Komödien*), which are often dramas rather than comic plays, include *Von dem Tobia und seinem Sohn* (1533), *Griselda* (1546), *Die Judith* (1551), *Die ungleichen Kinder Eve* (1553), *Komödie vom verlorenen Sohn* (1556), *David mit Batseba* (1557), *Die Komödie der Königin Esther* (1559), and *Die junge Witwe Franzisca* (1560). The following are among the better-known Fastnachtspiele: *Das Hofgesind Veneris* (1517), *Der Teufel mit dem alten Weib* (1545), *Der farent Schüler ins Paradeis* (1550), *Der böse Rauch* (1551), *Das heiß Eisen* (1551), *Der Bauer im Fegfeuer* (1552), and *Der Roßdieb zu Fünsing* (1553). In 1567, when he believed that he was about to die, he wrote an autobiographical poem, *Summa all meiner gedicht*, in which he expresses the religious purpose of his works and the hope that the common man may be the better for them. The poem gives the total of his poems (Spruchgedichte) as 1,700, including 208 plays; it also mentions that he composed 13 tunes (Meistertöne) for the Guild of Meistersinger.

Sachs's life was lived almost entirely within the confines of his native Nürnberg which combined the comfortable homeliness of a small compact city with considerable artistic and commercial activity. Dürer, Pirkheimer (qq.v.), and the sculptor in bronze Peter Vischer (c. 1460–1529) were his contemporaries and friends. He combines a naïve homespun simplicity with an awareness of the problems of his day, exhibiting in his better works shrewd observation and good-humoured tolerance.

Despised in the 17th c., he was restored to prominence by Goethe in his poem *Hans Sachsens poetische Sendung* (q.v., 1776), and, in Wagner's *Die Meistersinger von Nürnberg* (q.v., 1868), was converted into a German legend.

Sachs's own edition of his works (*Sehr herrliche schöne vnd warhaffte Gedichte*, 5 vols.) appeared in 1558–79. *Werke*, ed. A. von Keller and E. Götze, was published in 26 vols., 1870–1908, and reprinted in 1964.

SACHS, NELLY (Berlin, 1891–1970, Stockholm), wrote poetry for most of her life. A Jewess, she managed with the aid of foreign intervention to emigrate from Germany to Sweden in 1940. Her earliest book was *Legenden und Erzählungen* (1921), but all her finest poetry—and she is one of the most outstanding German poets of the mid-20th c.—belongs or refers to the period of the 1939–45 War and the persecution of the Jews. Her poems appeared in *In den Wohnungen des Todes* (1947), *Sternverdunkelung* (1949), *Und niemand weiß weiter* (1957), *Flucht und Wandlung* (1959), *Fahrt ins Staublose* (1961), *Glühende Rätsel* (1964), *Späte Gedichte* (1965), and *Die Suchende* (1966). *Teile dich Nacht. Die letzten Gedichte* appeared posthumously in 1971. She also translated Swedish poetry (*Schwedische Gedichte*, 1965). *Die Gedichte der Nelly Sachs* (2 vols.) appeared 1961–71 and hitherto unpublished letters, *Briefe der Nelly Sachs*, ed. R. Dinesen and H. Müssener, in 1984. A deep religious feeling and a sense of mourning pervade her work. She was awarded the Nobel Prize for literature in 1966.

Sachsen (Saxony), territorial designation which is no longer in official use, since the Länder of the DDR (see DEUTSCHE DEMOKRATISCHE REPUBLIK), which included Sachsen-Anhalt, were dissolved in 1952. Saxony, with Dresden as its capital and Leipzig, Meißen, and Chemnitz as its other principal cities, was a kingdom from 1806 (a creation by Napoleon) until 1918, then a Land until 1945. The early medieval dukes of

Saxony ruled over a different area (present Niedersachsen, q.v.) until the fall of Heinrich der Löwe (q.v.) in 1180. The Saxony of the early 20th c. was the surviving territory of an electorate created in 1423 (see FRIEDRICH DER STREITBARE). It played an important part in the Reformation (q.v.) under Friedrich der Weise (q.v.), Luther's supporter. In the early 18th c. the state enjoyed a period of cultural florescence with Leipzig University in high repute, and Dresden developing into a baroque and rococo city of great magnificence. The Electors, however, wasted their substance as unsuccessful kings of Poland (see AUGUST II, DER STARKE), and the land suffered repeatedly as a bufferstate between Austria and Prussia, being habitually on the losing side in wars (1763, see SIEBENJÄHRIGER KRIEG, 1813, see NAPOLEONIC WARS, 1866, see DEUTSCHER KRIEG), and having to forfeit territory, especially in 1815 and 1867.

Sachsenspiegel, a compendium of medieval law compiled in Middle Low German by EIKE von Repgau (q.v.) between 1220 and 1235. The laws set down are those valid in the region of Lüneburg and the Harz (Ostfalen), but the *Sachsenspiegel* exercised a considerable influence on legal writing and the drafting of laws throughout Germany. It exists in more than 200 manuscripts, some of which contain illustrations. It is preceded by two prefaces in verse, one of which reveals that it was originally written in Latin and was translated into German at the request of a Count Hoyer von Falkenstein.

Sachsen-Weimar, a petty Thuringian state (before 1815 Herzogtum, then Großherzogtum) which in the late 18th c. and early 19th c. became famous through the presence at the Weimar court of Goethe, Schiller, Herder, and Wieland (qq.v.) under the patronage of the Duke (later Grand Duke) Karl August (q.v.). The Grand Duchy later acquired Eisenach by inheritance, becoming Großherzogtum Sachsen-Weimar-Eisenach, simplified by law in 1877 as Großherzogtum Sachsen. In the Weimar Republic (q.v.) it became in 1920 part of Land Thüringen. (See also ANNA AMALIA, HERZOGIN.)

Sächsische Weltchronik, a history of the world in Middle Low German. The author (who compiled it from Latin sources) was probably EIKE von Repgau (q.v.), the compiler of the *Sachsenspiegel* (q.v.). It was written c. 1235 and outlines universal history from biblical times up to the author's own day. Later MSS. (known as B and C) expand and continue it.

The chronicle is written in simple, sober prose and has a preface in verse.

SACK, Gustav (Schermbeck nr. Wesel, 1885–1916, nr. Bucharest), led a restless life as a young man and was in Switzerland in 1914, where he remained for a time because of the war. Having become penniless, he returned to Germany, was called up, and was killed in the Romanian campaign. He was one of the earliest Expressionist writers (see Expressionismus), but except for poems in journals his works were published posthumously. He is the author of a play on conscientious objection to war (*Der Refraktair*, written 1914, published 1920) and of two violent and macabre novels, *Ein verbummelter Student* (1917) and *Ein Namenloser* (1919). His poems, collected as *Die drei Reiter*, appeared in 1958 and *Gesammelte Werke* (2 vols.) in 1920.

Sadduzäer von Amsterdam, Der, a Novelle by K. Gutzkow (q.v.), published in 1834. It is based on the autobiography of Uriel (or Gabriel) Acosta (*Exemplar humanae vitae*, 1687), who committed suicide in 1640. Uriel, son of a Jewish convert to Christianity, has returned to the Jewish faith, but is beset by renewed doubts. He also loves Judith, the daughter of the rich Jew Vanderstraten. For his unorthodoxy he is denounced and cursed by the rabbis and forsaken by Judith. He disappears; Judith seeks him out, they are reconciled, and she persuades him to make his peace with the synagogue. He is imprisoned for months and at last does humiliating penance. Finding Judith married to another, he shoots her and himself.

Sadowa, see Königgrätz.

Sage, term corresponding to the secular use of 'legend' in English; Legende in German is limited to religious connotations.

SAIKO, George Emmanuel (Seestadtle, then in the Kingdom of Bohemia, 1892–1962, Rekawinkel, Lower Austria), entered Vienna University, studying the history of art and archaeology. During the 1939–45 War he was employed at the Albertina in Vienna and from 1945 to 1950 was in charge of it. He did not begin creative writing until middle life. Preoccupied with the Freudian doctrine of the subconscious, he devised a quasi-surrealistic style of fictional presentation, which he termed magischer Realismus (*Die Wirklichkeit hat doppelten Boden. Gedanken zum magischen Realismus*, 1952).

In Saiko's narrative works the events are subordinated to the responses of the characters and, especially, to the undercurrents directing the responses. As with many of his generation of Austrians (he was a friend of H. Broch, q.v.), the old pre-1918 Austria-Hungary provides the background of much of his work. It includes the novels *Auf dem Floß* (1948, but completed in 1939), a treatment of social decay embodied in a weird story with subtle psychology, and *Der Mann im Schilf* (1955), a picture of what Saiko terms 'inner disintegration'. He also published the stories *Giraffe unter Palmen* and *Der Opferblock* (both 1962), a volume containing two Novellen of war, *In den Klauen des Doppeladlers* (dated October 1914) and *Die Badewanne* (dated April 1948). *Die erste und die letzte Erzählung* appeared posthumously in 1968. The Großer Österreichischer Staatspreis was awarded to him in the year of his death.

SAILER, Johann Valentin (Weißenborn, Swabia, 1714–77, Obermarchtal), a member of the Premonstratensians (White Canons) of Obermarchtal Abbey, was a noted preacher and the author of plays and poems in Swabian dialect; these were collected and published in 1819 (*Schriften im schwäbischen Dialekt*). The *Gesamt-Ausgabe* (41 vols.) appeared 1830–45. He is often referred to as Sebastian Sailer, since Sebastian was his monastic name.

SALAT, Hans (Sursee, Switzerland, 1498–1561, Freiburg), a Roman Catholic, led an adventurous life and fought against the Reformers in 1529. He became clerk of the law courts in Lucerne, but was dismissed in 1540. He attacked the Reformation (q.v.) in various historical and satirical works, including *Der Tanngrotz* (1531), *Triumphus Herculis Helvetici* (1532), *Chronik* (1536, of the Reformation), and *Des frommen Bruder Clausen Leben* (1536, see Nikolaus von der Flüe). He also wrote two anti-Protestant plays, *Judith* (1534) and *Der verlorene Sohn* (1535); the latter has been adapted by C. von Arx (q.v., 1935).

Salier, Die, (1) a branch of the Frankish people, originally inhabiting what is now Holland and Flemish-speaking Belgium.

(2) collective term for the Frankish emperors of the 11th c. and 12 c., Konrad II, Heinrich III, Heinrich IV, and Heinrich V (qq.v.), whose reigns cover the period 1024–1125.

Salisches Gesetz or **Lex Salica,** a body of common law valid in the Frankish territories of western Europe. The earliest Latin MSS. date from the 6th c. A fragment of a 9th-c. Old High German version is preserved in Trier. The *Lex Salica* was primarily a penal code, but contained a clause (59, 5) declaring that women may not inherit land. In the 14th c. this became the basis of the Salic Law denying succession in monarchies to female members of royal families and as such forms the background to the Austrian Wars of Succession (see ÖSTERREI-CHISCHER ERBFOLGEKRIEG) securing Maria Theresia's succession. The Salic Law, however, was enforced in other instances, debarring, for example, Queen Victoria from acceding to the throne of Hanover in 1837. It was a factor in the political crisis involving Schleswig and Holstein (see SCHLESWIG-HOLSTEINISCHE FRAGE). The Salic Law was published in three volumes, ed. K. A. Eckhardt, in 1953–6.

SALIS-SEEWIS, JOHANN GAUDENZ VON (Bothmar nr. Malans, Grisons, 1762–1834, Bothmar), a Swiss nobleman, who served from 1779 to 1789 as an officer (eventually captain) in the Swiss Guard Regiment at Versailles and Paris; he had literary inclinations and toured Germany in 1788–9, meeting Goethe, Schiller, Herder, and Wieland (qq.v.) in Weimar. He spent some time in revolutionary Paris as a civilian and served for a few more months as an officer, after which he returned to Switzerland and settled at Chur. There he became a civil servant, served later in Zurich as an officer, and then returned in an administrative capacity to Grisons. He resigned all his offices in 1817.

Salis possessed a gift for poetry, writing verse which was elegant and restrained in form, with a feeling for nature and friendship as the late 18th c. cultivated it. He sometimes used classical measures. His poems were published in 1793 by F. Matthisson (q.v.), with whom he corresponded (*Gedichte*). They have been twice reprinted in the 20th c. (1937 and 1964).

SALLET, FRIEDRICH VON (Neiße, 1812–43, Reichau nr. Nimptsch, Silesia), was educated at the Potsdam Cadet School and commissioned in 1829 in the 36th Infantry Regiment in Mainz. In 1830 he came before a court martial because of a satire he had written and was sentenced to be cashiered and to serve ten years' imprisonment. King Friedrich Wilhelm III (q.v.) quashed the sentence of cashiering and reduced the imprisonment to two months. Sallet resumed his service after serving his short sentence and was posted in 1834 to the Kriegsschule, where he remained for three years, returning to his regiment in 1837. He resigned his commission in 1838.

Sallet wrote two volumes of poems (*Gedichte*, 1835; *Funken*, 1837) and an epic (*Die wahnsinnige Flasche*, 1838); his best-known work was an interpretation of Christianity from a 19th-c. standpoint, written in blank verse (*Laienevangelium*, 1840). His collected poems (*Gesammelte Gedichte*) appeared in 1843, his complete

works (*Sämtliche Werke*, 5 vols.) were published posthumously, 1845–8.

Salman und Morolf, a long medieval poem in stanzas, presumed to have been written in the late 12th c., though the extant MSS. were made two to three hundred years later. It is a story of feminine infidelity, repetitive in form. Salman (Solomon) is married to a heathen princess, Salme. A heathen king, Fore (possibly Pharaoh), wins her love by a magic ring and elopes with her. Salman's brother Morolf sets out to find her, and, in doing so, encounters dangers from which his resourcefulness, expressing itself in crude and comic ruses, saves him. Salman sets out to win back Salme, makes a reconnaissance in disguise, is detected and condemned. He saves himself by blowing his horn. His forces defeat Fore, who is hanged, and Salme is brought back to Jerusalem.

The second part of the poem repeats the pattern of the first. Salme is again carried off, this time by Princian, who obtains her consent by means of a magic ring. Morolf undertakes a journey to find her, assuming various disguises. Salman undertakes a campaign against Princian, whom he defeats. This time Salme is put to death by Morolf. Salman marries Fore's sister, who had assisted him when he was in peril at Fore's court and had accepted baptism.

The vivid and entertaining quality of the poem derives from Morolf's alertness, his inexhaustible ruses, and the tricks he plays upon his adversaries. He disguises himself as a minstrel (Spielmann), and the roguery and drastic comedy which make up much of the poem are characteristic of Spielmannsdichtung (q.v.).

Salman and Morolf (or Markolf) appear also in other medieval works, notably in a Spruchgedicht of the 14th c. which consists of a disputation between the high-minded Salman and the vulgarly cynical Morolf, followed by a series of crude and droll adventures of Morolf. This poem is stated in the MS. to be a translation of a (medieval) Latin original (*Salomon et Marcolfus*).

SALOMON, ERNST VON (Kiel, 1902–72, Winsen/Luhe), a passionate nationalist, took part in the Baltic campaign of 1919, was a member of a Freikorps (q.v.) and was implicated in the assassination of Walther Rathenau (q.v.) in 1922. For this he was sentenced to five years' imprisonment. His autobiographical novels *Die Geächteten* (1930) and *Die Stadt* (1932) express his detestation of the Weimar Republic (q.v.). *Die Kadetten* (1933) refers to his early years as an officer's son at the Cadet School at Berlin-Lichterfelde. During the 1939–45 War he worked as a film propagandist. His most successful novel was the amusing and ironic *Boche in*

Frankreich (1950); his provocative autobiography *Der Fragebogen* (1951) is a purview of the post-war period based on the period spent in an allied internment camp. An unautobiographical, though still political, novel, *Das Schicksal des A. D.*, appeared in 1960, *Die schöne Wilhelmine* in 1965, *Glück in Frankreich* in 1966, and *Deutschland, deine Schleswig-Holsteiner* in 1971. *Die Kette der tausend Kraniche* (1972) is based on Salomon's activities as a pacifist in the course of which he attended a congress in Japan earning him the paper-chain of the title. It was followed by the novel *Der tote Preuße* (1973).

Salomonis Hus (Salomos Haus), a Middle High German tract of the 13th c. giving an allegorical interpretation of verses 9 and 10 of chapter 3 of The Song of Songs. Solomon's palace is conceived as a preparation for eternal love. The work begins in prose and passes eventually into a free rhyming verse or rhyming prose. The MS. also contains a description and interpretation of the mass under the heading 'ein bezeichenvnge der heiligen messen'.

SALTEN, FELIX, pseudonym of Siegmund Salzmann (Budapest, 1869–1947, Zurich), was a journalist in Vienna and then in Berlin, returning to Vienna to become dramatic critic to the *Neue Freie Presse.* In 1939 he emigrated to Switzerland. He wrote many works of fiction on historical or social themes, including the tale *Prinz Eugen* (1915) and the novel *Olga Frohmut* (1910), and also tried his hand at drama (*Kinder der Freude,* 1917). His chief success was with animal stories, of which that of the wild deer, *Bambi* (1923), toured the world as book and film. It was followed by *Bambis Kinder* (1940) and *Djibi das Kätzchen* (1946). His early writings on the theatre include *Schauen und Spielen. Studien zur Kritik des modernen Theaters* (2 vols., 1921) and *Das Burgtheater* (1922). *Gesammelte Werke* (6 vols.) appeared 1928–32.

SALUS, HUGO VON (Böhmisch-Leipa, 1866–1929, Prague), a doctor in Prague, made a name for himself with melodious melancholy verse, of which he published eleven volumes between 1898 and 1924. Probably the most widely read were *Ehefrühling* (1900) and *Neue Garben* (1904). His once popular poetry is clearly a derivative mixture of Romantic decadence and Austrian tradition. He also wrote Novellen closely related to the mood of his verse (*Novellen des Lyrikers,* 1904).

Salzburger Festspiele, an annual festival of music and drama originating in the Salzburger Festspielhausgemeinde, which was founded in August 1917 by H. von Hofmannsthal, R.

Strauss, and M. Reinhardt (qq.v.) who was until 1938 its director. Among other co-founders was the stage designer A. Roller (1864–1935). The first festival opened with the première of Hofmannsthal's *Jedermann* (q.v.) in 1920, and this work was performed annually from 1926 to 1937, and again from 1946. Other notable productions have been Goethe's *Faust,* Hofmannsthal's *Das Salzburger große Welttheater* (qq.v.), and a number of operas, among which are the modern works *Dantons Tod* (1947) and *Der Prozeß* (1953) by G. von Einem (q.v.), *Antigone* (1949) by C. Orff (q.v.), and *Die Bassariden* (1966) by H. W. Henze (q.v.). The festival is well known for its production of Mozart's operas; the first to be performed was *Don Giovanni* (q.v.) in 1926.

Salzburger große Welttheater, Das, a festival play written by H. von Hofmannsthal (q.v.) for performance at Salzburg in 1922 (see SALZBURGER FESTSPIELE). It is based on Calderón's play *El gran teatro del mundo.* The play is a baroque allegory, in which the stage is the world and the world a stage, and life is the play enacted on it; its actors are symbolically represented in stylized figures: König, Schönheit, Weisheit, Reicher, Bauer, Bettler. On the transcendent plane above these are the Meister (God), the two Angels, and beneath these Welt, Vorwitz, Tod, and Widersacher (the Devil). The central point of the action is the Bettler's intended onslaught on the König, Schönheit, and Weisheit, instigated as an act of levelling communism by the Widersacher. But the Bettler halts in mid-stroke, seized like St. Paul by divine illumination ('War das nicht/Des Saulus Blitz und redend Himmelslicht?'). As der Tod clears the stage, the Bettler gladly submits and is elevated by God.

Salzburger Marionettentheater, founded in 1913 by Anton Aicher (1859–1930), has made Mozart productions a special feature. Benefiting from the popularity of the Salzburger Festspiele (q.v.) it has established a reputation in its own right as a puppet theatre.

SALZMANN, JOSEPH DANIEL (Strasburg, 1722–1812, Strasburg) was a clerk of the courts in Strasburg from 1753. He was the accepted head of the pension table at which Goethe took his midday meal in Strasburg, and a brief appreciation of him is given in *Dichtung und Wahrheit* (q.v., Bk. 9).

Samanunga, see ABROGANS.

SAMAROW, GREGOR, pseudonym of Oskar Meding (Königsberg, 1828–1903, Charlottenburg, Berlin), who was first a Prussian then a Hanoverian civil servant. The historical novels of Samarow were popular for a time in the Germany of the Bismarckian era. They include *Um Szepter und Kronen*, an immense panorama of the period 1866–73 in 20 volumes.

SAMSON PINE, a Jew of Strasburg, who, according to Philipp Colin, explained the original French text for the German poem *Der neue Parzefal* (q.v.).

Samuel Henzi, see HENZI, S. and LESSING, G. E.

Sanctus, Das, a story by E. T. A. Hoffmann (q.v.), written in 1816 and published in the first volume of *Nachtstücke* (1816). Based on a real incident which befell Hoffmann's friend Betty Marcuse, it is the tale of a singer (Bettina) who loses her voice after leaving church in the middle of the Sanctus. The loss is psychosomatic, and in the end Bettina recovers her voice.

SAND, KARL LUDWIG (Wunsiedel, Bavaria, 1795–1820, Mannheim), a student of theology and a member of the Burschenschaft (q.v.) at Jena. He was one of the organizers of the Wartburgfest (q.v., 1817). His over-zealous nationalism drove him to assassinate A. von Kotzebue (q.v.) on 23 March 1819 in Mannheim, where he was subsequently condemned to death and executed. Metternich (q.v.) seized the opportunity to introduce stringent laws suppressing nationalism, drawing up the Karlsbad Decrees (see KARLSBADER BESCHLÜSSE). The majority of the members of the Burschenschaft disapproved of this political crime; yet, in the light of subsequent developments, Sand became in the minds of many a martyr of national liberation movements.

Sand's story was treated by a number of 20th-c. writers, among them Enrica von Handel-Mazetti, in a trilogy of novels, E. Penzoldt, in a play, and T. Dorst, in a play for television (qq.v.).

Sandmann, Der, a story by E. T. A. Hoffmann (q.v.), written in 1815 and published in 1816 in volume 1 of *Nachtstücke*. The central figure, Nathanael, is sensitive to a brooding presence of evil around him. In childhood he had feared the Sandmann (the traditional bringer of sleep to children) and had come to identify him with the sinister lawyer Coppelius, who frequently visited his father late at night. He discovers that his father and Coppelius perform chemical experiments, and in the end the father is killed by an explosion. Years after, Nathanael is pestered

by the barometer seller Coppola, whom he associates with Coppelius. He buys a pocket telescope from Coppola and with it observes Olimpia, the beautiful but secluded daughter of Professor Spalanzani. He falls in love with Olimpia at a ball, but in the end discovers to his horror that she is a clockwork automaton. Until his infatuation with Olimpia Nathanael had loved Clara, a balanced and harmonious character. He now returns to her, but falls into a fit of madness when enjoying the view from a tower with Clara and attempts to hurl her over the parapet. She is rescued by her brother, but Nathanael, tempted by the voice of Coppelius below, throws himself to his death. A brief postscript indicates that Clara married and found a happiness with her husband that she could never have enjoyed with the demented Nathanael.

The story provided Offenbach with the basis for the first act of *Les Contes de Hoffmann*.

SANDRUB, LAZARUS, pseudonym of Balthasar Schnurr (Lendsiedel, 1572–1644, Hengstfeld), under which appeared at Frankfurt in 1618 a collection of versified Schwänke (see SCHWANK), *Delitiae historicae et poeticae, historische und poetische Kurzweil*.

Sanfte Gesetz, Das, see BUNTE STEINE.

Sangbüchlein, common title for Luther's *Geystliche gesangk-Buchleyn*, published in 1524 by Johann Walter (q.v.) with Luther's consent. It contains four older German hymns, nine translations of Latin hymns, and thirteen hymns derived from paraphrases of psalms. It is regarded as the foundation of the German Protestant hymn.

Sänger, Der, a ballad written by Goethe in 1783 and first published in *Wilhelm Meisters Lehrjahre* (q.v., 1795, Bk. II, Ch. 11), where it is sung by the Harfenspieler. It is an expression of the artist's pleasure in his art; the minstrel declines all reward except a goblet of fine wine. *Der Sänger* has been set by F. Schubert, C. Loewe, R. Schumann, and Hugo Wolf (qq.v.).

Sanct Brandan, an anonymous Middle High German poem, recounting the remarkable maritime experiences of the Irish Saint Brandan. It was written at the end of the 13th c. or the beginning of the 14th c., but is believed to be based on a lost German poem of the middle of the 12th c. Part of the MS. is missing.

St. Emmeramer Gebet, a confession of sins and prayer for grace written in the Bavarian

Old High German dialect either in St. Emmeram Abbey, Regensburg, or in its offshoot Oberaltaich *c.* 820–30.

St. Gallen, the capital city of the Canton of St. Gall, grew up around the Benedictine abbey dedicated to the 7th-c. Irish missionary St. Gallus. The abbey was from the 9th c. to the 11th c. an outstanding centre of learning, numbering among its monks such scholars as the Ekkehards and Notkers (qq.v.). In the later Middle Ages the abbots became powerful feudal overlords, and tensions developed between abbey and city, which broke free from ecclesiastical rule in 1454. When, however, some thirty years later, it was proposed to remove the abbey to Rorschach, the inhabitants of St. Gall, threatened with the loss of their livelihood, vigorously opposed the move, even demolishing new buildings under construction. The city authorities were condemned in the courts, but the abbey stayed where it was. The most notable personality of St. Gall, after the scholars of the 10th c. and 11th c., was Burghermaster Joachim von Watt (latinized as Vadianus, q.v.), a humanistic scholar who fostered the Reformation in the city. The abbey church (since 1847 cathedral) was rebuilt in baroque style between 1756 and 1765. The abbey was secularized in 1805. The library is famous for both its architectural beauty and its contents, of which the most notable treasure is the MS. B of the *Nibelungenlied* (q.v.). St. Gall plays a part in J. V. von Scheffel's historical novel *Ekkehard* (q.v.).

St. Galler Chronik (*Nüwe Casus monasterii Sti Galli*), a chronicle of the Abbey of St. Gall written in Middle High German by a citizen of the town, Christian Kuchimeister. It was completed in 1335 and covers the period 1228–1329, forming a continuation of the Latin chronicles of the monks. Its interest is more than local, for it reflects the political events of the time.

St. Galler Himmelfahrtspiel, an anonymous Middle High German dialogue, in which Jesus bids farewell to his disciples. It probably dates from the 15th c.

St. Galler Passionsspiel, a religious play written in Middle High German in the 14th c. It is so called from the provenance of the MS., for the play was written far from St. Gall near the Middle Rhine (Rhenish Hesse or Nassau). Its subject goes beyond the scope of a passion play, covering the life of Jesus from the marriage at Cana to the Resurrection.

St. Galler Paternoster und Credo, an early clumsy translation of the Lord's Prayer and Apostles' Creed in Old High German, written in the same MS. as the *Abrogans* (q.v.) in St. Gall. It dates from the late 8th c.

St. Galler Spiel von der Kindheit Jesu, a Middle High German nativity play, written towards the end of the 13th c. and preserved at St. Gall in a MS. of the 14th c. It includes prophecies of Christ's coming, the Annunciation, the Nativity with the shepherds and the Magi, the Presentation in the Temple, the Slaughter of the Innocents, and the Flight into Egypt.

St. Galler Spottvers, a four-line satirical poem directed at 'Liubene', who married off his daughter but had her returned by the bridegroom. The lines have been entered on a blank page of a MS. of passages from the Vulgate and are thought to be a rare fragment of early popular poetry. They were probably written down in the 9th c.

Sankt Joseph der Zweite, a narration written in 1807 by Goethe, which forms Ch. 2, Bk. I in *Wilhelm Meisters Wanderjahre* (q.v.). Wilhelm Meister encounters a group of persons, a man and a woman with a baby riding upon an ass, who conjure up for him the picture of the Holy Family on the flight into Egypt. The following day he learns from the man, who bears the name Joseph, the story of their meeting and marriage, and is impressed by their robust integrity and simple piety.

Sante Margareten Marter, a Middle High German verse legend of St. Margaret, written in the 14th c.

ST. OSWALD, King of Northumbria (635–42) and martyr. He is the subject of two long Middle High German poems and a prose legend, all three written in the 14th c. They are versions of a single earlier poem, possibly of the 12th c., but their relation to it and to each other is not clear. The legend is in the form of a wooing. Oswald goes out to seek a heathen princess, who inclines to Christianity, as his bride. His adroit and resourceful messenger on this enterprise is a raven. Oswald is successful in his wooing and converts the bride's father and his warriors.

St. Pauler Predigten, a collection of Middle High German sermons, in a 13th-c. MS., discovered at St. Paul's Abbey in Carinthia. The MS. is a copy of a 12th-c. original.

Sanct Peter mit der Geiß, a Schwank (q.v.) by Hans Sachs (q.v., 1555). St. Peter, discontented with the evil in the world, deplores God's management of terrestrial affairs. God hands over the government to him for one day. St. Peter finds that he can scarcely cope with a goat entrusted to his care, let alone deal with more weighty matters. Chastened, he hands the power back to God with the words: 'O Herr, vergib mir mein Torheit. / Ich wil fort der Regierung dein, / Weil ich leb, nicht mehr reden ein.' The poem is written in Hans Sachs's characteristic Knittelverse (q.v.).

Sankt-Rochus-Fest zu Bingen, a description by Goethe of the Feast of St. Roch at Bingen on 16 August 1814. It was published in 1817 in the second instalment of *Über Kunst und Altertum* (q.v.), and is prefixed by a six-line stanza by Goethe himself, beginning 'Zu des Rheins gestreckten Hügeln'.

St. Trudperter Hohes Lied, a translation of The Song of Songs (*Hohes Lied*) and a commentary on it, the MS. of which was discovered in the Convent of St. Trudpert near Freiburg (Breisgau). Its author and the place of its composition are unknown, but modern scholarship inclines to Bavaria as the region from which it derives. The probable time of composition is the early 12th c. The *St. Trudperter Hohes Lied*, which was intended for the instruction and delectation of nuns, owes much to Williram (q.v.), but its interpretation differs significantly in exalting the Virgin Mary, who takes over from the Doctors of the Church the vital function of mediation. It is a work of early mysticism, a step towards the later cult of the Virgin.

Sansibar oder Der letzte Grund, a short novel published by A. Andersch (q.v.) in 1957. It is the story of the salvation in war-time of a piece of religious sculpture (it resembles Barlach's *Lesender Klosterschüler*, though Barlach, q.v., is not named), which the Nazi authorities wish to impound. Behind the story is the theme of the moral responsibility of individuals. Pastor Helander, Knudsen, a Communist fisherman, Gregor, a Communist organizer, Judith, a young Jewess on the run, and a boy (Der Junge) all participate, and the statue, together with Judith, is safely landed in Sweden. Pastor Helander resists the SS (q.v.) men coming to arrest him and is shot down after he has killed one of them. The story is narrated in short sections, each in turn seeing events from the standpoint of one of the participants.

Sanssouci, a palace at Potsdam built in 1745-7 by G. W. von Knobelsdorff (q.v.) for Friedrich II (q.v.) of Prussia, who prescribed the main lines of its design. Set in a handsome park, partly in the French and partly in the English style, it is a long single-storey building crowning a terraced slope. The interior is richly and tastefully decorated by various hands, and the whole ranks as a masterpiece of rococo architecture. Sanssouci was Friedrich's favourite residence and the house in which he entertained privileged guests, including Voltaire (q.v.). It was at Sanssouci that Friedrich died. (See Rococo.)

SAPHIR, Moritz Gottlieb (Lovas-Berény, Hungary, 1795-1858, Baden by Vienna), a Jew, studied theology at Prague, entered his father's business, and then studied the humanities at Pest (Budapest) University. In 1822 he took to journalism as a literary and dramatic critic. He was active in Vienna with the *Theaterzeitung*, and then in Berlin, where he edited the *Berliner Schnellpost* (1826-9) and the *Berliner Kurier* (1827). A severe critic of Berlin literary circles (notably the Mittwochsgesellschaft, q.v.), he founded in 1827 the Berliner Sonntagsverein, renamed Der Tunnel über der Spree (q.v.). He found himself unpopular in Berlin and moved to Paris in 1830, and in 1832 to Munich. In 1834 he returned to Vienna, where he settled permanently, editing *Der Humorist* from 1837 to 1858.

Saphir's wit was often malicious, and he was much feared in the 1830s and 1840s. His publications in book form include the *Humoristische Damenbibliothek* (6 vols., 1838-41), a volume of poems, *Wilde Rosen* (1847), and *Pariser Briefe* (1845). *Ausgewählte Schriften* (12 vols.) appeared posthumously in 1884.

'Saphire sind die Augen dein', first line of an untitled poem by H. Heine (q.v.), included as No. LVI in *Die Heimkehr, Buch der Lieder* (q.v.).

SAPPER, Agnes (Munich, 1852-1929, Würzburg), *née* Brater, was in her day a popular writer of stories and novels for young people. Her most widely read book was *Die Familie Pfäffling* (1906).

Sappho, a blank-verse tragedy in five acts by F. Grillparzer (q.v.), written in 1817. Its highly successful first performance took place in April 1818 at the Burgtheater, Vienna. It was published in 1819. The play's subject, first suggested by Dr. Felix Joël as a libretto for an opera, quickly provided the young playwright with the challenge he needed to emancipate himself from fate tragedy (see Schicksalstragödie). Grillparzer made free use of the vague legend attached to the Greek poetess Sappho of Lesbos (7th c.–6th c. B.C.) and her unhappy love for the handsome Phaon. The tragedy is written in classical style, with only three principal characters.

Sappho, at the height of her creative life as the celebrated poetess, returns, adorned with the laurel wreath, from the festivities at Olympia, with Phaon at her side. With him she hopes to enjoy happiness. But the youth can only admire her, and Sappho herself is the first to discover an affection between Phaon and her young maid Melitta, which her bitter jealousy self-destructively lures into full consciousness. Only when she has debased her humanity does she admit to her disloyalty to the gods, which she has incurred in abandoning her lyre. Seeking for Phaon and Melitta the blessing of the gods, and for herself forgiveness, she plunges to her death into the sea.

The irreconcilability of life and art is the theme of this tragedy of the artist. Grillparzer himself referred to this dilemma as 'le malheur d'être poète', an adaptation of a quotation from d'Alembert. He treats with considerable independence a theme prominent in German literature since Goethe's *Torquato Tasso* (q.v.).

Satyros oder Der vergötterte Waldteufel, a satirical play (Ein Drama) written by Goethe in 1773, but not published until 1817. It is in five acts, yet totals only 484 lines of Knittelverse (q.v.). Satyros is a vicious, lascivious satyr, who is befriended by a hermit. He preaches a new religion of Nature, yet grossly abuses the hermit's kindness, even bringing false accusations against him. Eventually he is caught in the act of molesting a priest's wife, whereupon he decamps unabashed, heaping insults on those he has deceived.

Much scholarship has been devoted to discovering the presumed original of this savage caricature, and Basedow, Herder, Klinger, Lavater, and Heinse (qq.v.) have all been suggested with more or less confidence. More recently suspicion has fallen on A. S. Goue (q.v.) of Wetzlar.

SAUTER, SAMUEL FRIEDRICH (Flehingen, Baden, 1766–1846, Flehingen), a village schoolmaster successively in Flehingen and in Zaisenhausen in Baden, who wrote simple and unpretentious verse which was collected as *Volkslieder und andere Reime* (1811) and *Die sämtlichen Gedichte des alten Dorfschulmeisters S. Fr. Sauter* (1845). Some of Sauter's more naïve poems were later published in the Munich *Fliegende Blätter*, as examples of unintentional humour, by L. Eichrodt and A. Kußmaul (qq.v.) and attributed to a Gottlieb Biedermaier. Sauter thus posthumously gave rise to the stylistic term Biedermeier (q.v.). Though Eichrodt's persiflage

has attached much ridicule to Sauter's name, the schoolmaster's religiously tinged nature poetry has at times a simple charm, and *Der Wachtelschlag* was set to music by both Beethoven and F. Schubert (qq.v.).

SAVIGNY, FRIEDRICH KARL VON (Frankfurt/ Main, 1779–1861, Berlin), German jurist, became a professor at Marburg University in 1802, and at Landshut in 1808. He was one of the group of distinguished professors appointed in 1810 to the newly founded University of Berlin. A brother-in-law of Clemens and his sister Bettina von Arnim (qq.v., he married their sister Kunigunde), he was in close touch and sympathy with the Romantic poets, including L. J. von Arnim and the brothers Grimm (qq.v.). His ability and eminence led to official appointments in 1817 as a member of the Prussian State Council (Staatsrat) and in 1842 as Minister for the Revision of Legislation, a portfolio which he held until 1848. His principal works are *Das Recht des Besitzes* (1806, reissued together with the first publication of an earlier work, *Juristische Methodenlehre*, ed. G. Wesenberg, 1951), *Vom Beruf unserer Zeit für Gesetzgebung und Rechtswissenschaft* (1815), *Geschichte des römischen Rechts im Mittelalter* (6 vols., 1815–31, reissued in the 7 vols. of a later edition, 1956), *System des heutigen römischen Rechts* (8 vols., 1840–9, reissued 1971), and *Das Obligationsrecht* (2 vols., 1851–3, reissued 1971).

Savigny was joint editor of the *Zeitschrift für geschichtliche Rechtswissenschaft* (1815–50) and is regarded in Germany as the father of historical jurisprudence. *Vermischte Schriften* (5 vols.) appeared in 1850 (reissued 1968).

Savonarola, an epic poem by N. Lenau (q.v.), published in 1837. It takes the form of a series of twenty-five romances in almost a thousand rhyming four-line stanzas and was originally conceived as the central work of a trilogy. The two outer poems, which were to deal respectively with J. Hus and Ulrich von Hutten (qq.v.), were never written.

SAXE, MAURICE DE, MARÉCHAL (Goslar, 1696–1750, Chambord), a distinguished French general of German birth and upbringing. He was an illegitimate son of Augustus the Strong of Saxony (see AUGUST II) and of his mistress Gräfin Aurora von KÖNIGSMARCK (q.v.). Baptized Moritz, he was created Graf von Sachsen. In 1720 he entered the French army and had an outstandingly successful career, reaching the rank of *maréchal* in 1744. He was the victor of Fontenoy (1745). He was noted for his attention to the welfare of the troops under his command, and of their horses.

Saxe-Weimar, see SACHSEN-WEIMAR.

Schaaf-Schur, eine pfälzische Idylle, Die, a short work in prose and verse published by F. Müller (q.v.) in 1775. His name is given on the title-page as Mahler Müller. The verse passages are intended to be sung. It is a rustic idyll. Farmer Walter and his family are busy sheep-shearing. They sing to pass the time, and the schoolmaster criticizes their literary taste. Walter's daughter and the young peasant Veitel, who is to leave next day, are in love. Their love is revealed to Walter, who approves it and invites Veitel to stay.

Schachbücher, designation for a number of medieval works which dilate upon chess as a parallel to human life and expound moral views by reference to the game. All the 'Schachbücher' are based upon *Solatium ludi scaccorum, scilicet, libellus de moribus hominum et de officiis nobilium,* a Latin dissertation on chess as an image of human affairs, written towards the end of the 13th c. by an Italian monk, Jacobus de Cessolis (Jacopo Dacciesole). The Latin prose was turned into German verse by HEINRICH von Beringen (c. 1295), KONRAD von Ammenhausen (1337), der PFARRER zu dem Hecht (1355), and Meister STEPHAN (between 1357 and 1375, qq.v.).

Schach von Wuthenow, a short historical novel by Th. Fontane (q.v.), written between 1878 and 1880 and published in 1882, though 1883 appears on the title-page. It is sub-titled *Erzählung aus der Zeit des Regiments Gendarmes* (the regiment was disbanded in 1807). The novel is set in the summer of 1806 in the sultry months of peace before the Prussian catastrophe in October (see NAPOLEONIC WARS). The politics of the day are the subject of animated discussion; but the novel is primarily social and psychological, concentrating on Captain of Horse (Rittmeister) Schach von Wuthenow of the Gendarmes and two ladies, Frau von Carayon and her daughter Victoire, in whose drawing-room he, with other officers, is a frequent visitor.

Frau von Carayon is an elegant widow in her late thirties, and some believe that Schach intends to marry her. Victoire, attractive in feature and in the freshness of youth, has suffered the tragedy of disfigurement by small-pox in her teens, and is only too well aware that marriage for her is improbable, though she feels drawn to Schach. Schach, a polished gentleman, conventional in his views and sensitive to the opinion of the world, shows little sign of coming to the point, and it becomes clear that he is reluctant to abandon his bachelor status. In an unguarded moment Victoire gives herself to him, only to find that after this episode he continues as formally polite and distant as ever. She confesses to her mother, who insists that Schach should regularize the relationship, which he reluctantly agrees to do. Schach finds himself publicly lampooned and leaves Berlin for his estate without notifying the Carayons. Frau von Carayon, believing that Victoire has been betrayed, complains to King Friedrich Wilhelm III (q.v.), who insists that Schach, an officer of his guards, shall keep his engagement. The marriage takes place and immediately after the wedding Schach shoots himself. A prominent feature of the background is the theatrical sensation of the day, Z. Werner's play *Die Weihe der Kraft* (q.v.).

SCHACK, ADOLF FRIEDRICH, GRAF VON (Brüsewitz nr. Schwerin, 1815–94, Rome), the son of a wealthy country gentleman, studied law at Bonn, Heidelberg, and Berlin universities and entered the Prussian civil service in 1838. In 1839 he was granted leave and made a tour of Italy and the Near East, afterwards spending some time in Spain on literary studies. His *Geschichte der dramatischen Literatur und Kunst in Spanien* appeared in three volumes, 1845–6, and was followed by translations of Spanish drama (*Spanisches Theater,* 2 vols., 1845). He translated Spanish and Portuguese poetry (jointly with E. Geibel, q.v., *Romanzero der Spanier und Portugiesen,* 1860), and oriental literature (*Epische Dichtungen des Firdusi,* 2 vols., 1853). A learned work on Arabic literature, *Poesie und Kunst der Araber in Spanien und Sicilien* (2 vols.), appeared in 1865. He transferred in 1840 from Prussian to Mecklenburg-Schwerin service and was until 1852 a representative of his state in Frankfurt. In 1855 he settled in Munich at the invitation of King Maximilian II (q.v.). He devoted much of his wealth to encouraging German painters and acquiring a collection of contemporary German works, which he left at his death to the German Emperor, who agreed to its being retained in Munich.

Towards the end of his life Schack published a historical work, *Geschichte der Normannen in Sicilien* (2 vols., 1889). He was one of the Munich school of poets (see MÜNCHNER DICHTERKREIS), and his poems, published in *Gedichte* (1867), are elegant and fluent. He also wrote three tragedies (*Pisaner,* 1872; *Timandra,* 1879; and *Atlantis,* 1880) and two epics, *Lothar* (1872) and *Die Plejaden* (1881). He published his recollections in *Ein halbes Jahrhundert* (3 vols., 1888). He was made Graf von Schack in 1876. *Gesammelte Werke* (10 vols.) appeared 1897–9.

SCHADE, JOHANN KASPAR (Kühndorf,

Thuringia, 1666–98, Berlin), was active as a lecturer in Leipzig and as a Lutheran cleric in Berlin. He wrote hymns which were collected in *Fasciculus cantionum* (1699). His *Geistreiche und erbauliche Schriften* appeared posthumously (5 vols., 1720).

SCHAEFFER, ALBRECHT (Elbing, 1885–1950, Munich), had a university education (Munich, Marburg, Berlin), was a journalist from 1911 to 1913, and then lived as an author in Bavaria. In 1939 he emigrated to the U.S.A., returning in 1950, and dying a few weeks after his arrival. He was an eclectic writer, mixing classical and medieval Christian ideals with an idiosyncratic mysticism. His output was prolific, and his aims were higher than his attainments. The works to which he attached most importance were the Bildungsroman (q.v.) *Helianth* (3 vols., 1920–1), sub-titled *Bilder aus dem Leben zweier Menschen von heute und aus der norddeutschen Tiefebene*, and the verse epics *Raub der Persephone* (1920), *Parzival* (1922), in which classical antiquity and Christianity are intermingled, and *Hölderlins Heimgang* (1923). His poetry, influenced by S. George (q.v.), includes the collection *Gedichte aus den Jahren 1915–30* (1931). Some of his works were republished in revised form and with new titles, including the novels *J. Montfort* (1918, *Das nie bewegte Herz*, 1931) and *Der Roßkamm von Lemgo* (1933, *Janna du Cœur*, 1949). His stories include *Knechte und Mägde*, *Das Opfertier* (both 1931), *Enak oder Das Auge Gottes* (1948), and *Der Auswanderer* (1950). One of his most attractive minor works is *Roß und Reiter* (1931), a perceptive commentary on a wide range of equestrian sculpture, copiously illustrated. Among his essays is one on Rilke (q.v.), published in 1916. A collection, ed. W. Ehlers, appeared in 1958 as *Mythos. Abhandlungen über die kulturellen Grundlagen der Menschheit*.

SCHÄFER, WILHELM (Ottrau, Hesse, 1868– 1952, Überlingen), of peasant stock, was trained as a teacher and taught in an elementary school (Volksschule) from 1888 to 1896, first in Vohwinkel, then in Elberfeld. He greatly admired Naturalistic writing and in 1894 published under its influence the stories collected in *Mannsleut*. He resigned his teaching post, travelled in France and Switzerland, and in 1900 became editor of the cultural journal *Die Rheinlande* (until 1918). His most successful works were collections of terse, mostly historical, anecdotes, inspired by the writings of J. P. Hebel (q.v.). The first volume appeared as *Anekdoten* (1907), the second as *33 Anekdoten* (1911), followed later by *Neue Anekdoten* (1926) and *Wendekreis neuer Anekdoten* (1937).

Schäfer was strongly nationalistic and much occupied with concepts such as Volkstum and Deutschheit, and he valued as his best work the pseudo-historical *Dreizehn Bücher der deutschen Seele* (1922). Of his many historical novels *Lebenstag eines Menschenfreundes* (1915, the story of Pestalozzi, q.v.), *Huldreich Zwingli* (1926), and *Der Hauptmann von Köpenick* (1930) deserve mention. Schäfer also wrote a story about Hölderlin, *Hölderlins Einkehr* (1930). A second work entitled *Huldreich Zwingli* (1927), termed 'Ein epischer Versuch', is written in classical elegiac verse. Schäfer was one of the fourteen founder members appointed in 1933 to the new Dichterakademie (see AKADEMIEN). His collected stories appeared as *Novellen* and his collected anecdotes as *Anekdoten* (3 vols.) in 1943. *Rechenschaft* (1948) was an extended edition of his autobiography, first published as *Mein Leben* in 1934.

Schäfers Sonntagslied, a short poem by L. Uhland (q.v.), the first line of which runs, 'Das ist der Tag des Herrn!'. Written in 1805 and published in 1806, it was set to music by K. Kreutzer (1780–1849) in 1821.

SCHAFFNER, JAKOB (Basel, 1875–1944, Strasburg), came of humble stock, and was apprenticed to a shoemaker. He then led a vagrant existence in western Europe. He absorbed the works of Nietzsche (q.v.) and Dostoevsky and also read a great deal of Gottfried Keller (q.v.), whose writings he emulated in some degree, though his stormy characters, alternating between exultation and despair, derive from the other two models. The greater part of Schaffner's mature life was spent in Berlin. His best-known early novel is *Konrad Pilater* (1910), in which a young man is torn between political radicalism and love. He abandons the girl the night before the wedding, but she follows him and dies on their wanderings. Pilater then submerges his identity in the anonymous world of modern industry. Schaffner's most ambitious work is a sequence of autobiographical novels tracing the development to maturity of the Swiss shoemaker's apprentice Johannes Schattenhold. The four novels, which form a protracted Bildungsroman (q.v.), are *Johannes* (1922), *Die Jünglingszeit des Johannes Schattenhold* (1930), *Eine deutsche Wanderschaft* (1933), and *Kampf und Reife* (1939). After many false starts and a period as a musician, Johannes finally becomes a writer.

Schaffner published two collections of poetry (*Der Kreislauf*, 1917, and *Bekenntnisse*, 1940) and a history of Switzerland (*Geschichte der Schweizer Eidgenossenschaft*, 1915). *Gesammelte Werke* (6 vols.) appeared in 1925. Schaffner was killed in an air raid.

Schafschur, Die, see Schaaf-Schur, Die.

SCHAIDENREISSER, Simon Minervius (*c.* 1500–*c.* 1573), a schoolmaster and town clerk in Munich, translated the *Odyssey* into homely German prose in 1537. It is the earliest German version of the work.

SCHALLENBERG, Christoph Dietrich von (Pieberstein, Lower Austria, 1561–97, Vienna), an Austrian officer, who died of a wound received in the Turkish wars, wrote Latin and German poems, which were edited in 1910 by H. Hurch and in 1935 by H. Cysarz. Educated in Austria, Württemberg, and Italy, Schallenberg was particularly susceptible to Italian influences.

SCHALLÜCK, Paul (Warendorf, Westphalia, 1922–76, Cologne), intended to enter the Roman Catholic missionary service. He was called up in 1941, wounded in 1944 and taken prisoner in Paris during the Allied advance. Released in 1946, he studied at Münster and Cologne universities. Instead of writing a doctoral thesis, he embarked on a novel, *Wenn man aufhören könnte zu lügen* (1951), a work which reflects the misery, drabness, and disillusionment of the first post-war years and especially the temptations besetting integrity. It draws copiously on the author's own experiences as a student at Cologne. From 1949 to 1952 Schallück was a dramatic critic; then he lived as a journalist and writer in Cologne. He followed his first novel with two others, *Ankunft null Uhr zwölf* (1953), a family novel, partially Expressionistic in technique, and *Die unsichtbare Pforte* (1954). The story *Weiße Fahnen im April* (1955), a title alluding to the visible signs of capitulation on the houses overtaken by the Allied advance, reverts to the bleak year of 1945. The novel *Engelbert Reineke,* covering the period 1935–55, is a critical review of the Nazi past and contemporary attitudes to it (including evasion), from which the eponymous central figure emerges with renewed determination. Later works are the stories *Sein frohes Gesicht* (1964), *Lakrizza und andere Erzählungen* (1966), and *Karlsbader Ponys* (1968) and the novel *Don Quichotte in Köln* (1967). Schallück has also written plays for radio and television (*Rund um den Ochsenkopf,* 1968; *Der Mann aus Casablanca,* 1969; *Beim Metzger,* 1971), for which he has also adapted *Karlsbader Ponys; Verurteilt. Szenenfolge* (1970) is a dramatic work. His essays include *Zum Beispiel* (1962) and *Deutschland. Kulturelle Entwicklungen seit 1945* and *Gegen Gewalt und Unmenschlichkeit* (both 1969), a notable plea for tolerance addressed to the Christian and Jewish

Society for Co-operation. Schallück was one of the early members of the Gruppe 47 (q.v.). His *Gesamtwerk* was published in 1977.

Schandfleck, Der, a novel by L. Anzengruber (q.v.), first serialized in 1876, published in book form in 1877, and reissued in revised form in 1884. The 'shameful stain' (Schandfleck) of the title is Leni, a girl born as a result of a single act of adultery by a farmer's wife. The novel, set in the Austrian countryside, begins with Leni's birth. The farmer, Joseph Reindorfer, who knows that the child's father is the dissolute son of Miller Herlinger, keeps up appearances, but at home cuts himself off from his wife and ignores the child. The miller dies, his son inherits, reforms, and marries. His boy Flori and Leni Reindorfer fall in love, only to learn that they are half-brother and half-sister. Flori goes to the bad and is killed in a brawl. Leni leaves home and becomes nurse to the ailing child of another farmer, the Grasbodenbauer, a widower. Meanwhile Reindorfer goes downhill and, Lear-like, is driven from home by his children. He finds refuge with Leni, who has married the Grasbodenbauer, and ends his days there.

SCHAPER, Edzard (Ostrovo, Posen, since 1919 Ostrów Wiekopolski, 1908–84, Berne), the son of a military official, began to study music, but, as one of eleven children, had to give up his ambition and go out to work. He was an assistant stage manager for a time and then went to sea as a trawler hand. In 1930 he settled in Esthonia and began to write. He later repudiated the three novels (*Der letzte Gast,* 1927; *Die Bekenntnisse des Försters Patrik Doyle,* 1928; and *Erde über dem Meer,* 1934) and reckoned the novel *Die Insel Tütarsaar* (1934) as his first work. The occupation of Esthonia by the Russians in June 1940 put him in a critical position, as sentence of death had been passed on him in absence by both Germans and Russians. He succeeded in escaping to Finland; four years later he was in danger of being handed over to the Russians and fled to Sweden, where he worked on the land and then became secretary to the organization concerned with the welfare of prisoners of war. In 1947 he established a home at Brigue in Switzerland, moving later to Cologne. In 1951 he left the Greek Orthodox and entered the Roman Catholic Church.

Though *Die sterbende Kirche* (q.v.) appeared in 1935 and *Der Henker* (later retitled *Sie mähten gewappnet die Saaten,* q.v., 1956) in 1940, most of Schaper's works of fiction belong to the post-war period. *Der letzte Advent,* a sequel to *Die sterbende Kirche,* appeared in 1949. The novels *Die Freiheit des Gefangenen* (1950) and *Die Macht der Ohnmächtigen* (1951, both set in the time of

Napoleon I and later published together as *Macht und Freiheit*, 1961) deal with the psychological problems of those who exercise power and of those who are victims of its abuse. *Der Gouverneur* (q.v., 1954), set in Reval in the 18th c., investigates the scope of conscience. There followed *Die letzte Welt* (1956), *Attentat auf den Mächtigen* (1957), *Das Tier oder die Geschichte eines Bären, der Oskar hieß* (1958), *Der vierte König* (1961), *Am Abend der Zeit* (1970), *Taurische Spiele* (1971), *Sperlingsschlacht* (1972), *Degenhall* (1975), and *Die Reise unter dem Abendstern* (1976).

Schaper also wrote shorter narrative works: *Der große offenbare Tag* (1950, in which the faith of a village is revived by the fate of a blaspheming Russian soldier), *Hinter den Linien* (1953, three stories set in the Russo-Finnish war), *Die Söhne Hiobs* (1962, two stories), *Der Aufruhr der Gerechten* (1963, described as a *Chronik*), *Dragonergeschichte* (1963, a story of the Thirty Years War), *Einer trage des anderen Last. Eine Elegie auf den letzten Gepäckträger* (1965), and *Schicksale und Abenteuer* (1968). *Gesammelte Erzählungen* (1965) contains all stories up to that date. Schaper's theme is almost always conscience in connection with power, responsibility, or freedom. The ubiquitous religious standpoint is unobtrusive. The terse style generates tension.

Schaper's slim dramatic output consists of *Der Gefangene der Botschaft* (1964), *Das Feuer Christi. Leben und Sterben des Johannes Hus in siebzehn dramatischen Szenen* (1965), and some radio plays. Other works are *Das Leben Jesu* (1936) and essays, including *Untergang und Verwandlung* (1952), *Bürger in Zeit und Ewigkeit* (1956), *Verhüllte Altäre* (1962), and *Auf der Brücke der Hoffnung* (1968). *Geschichten aus vielen Leben* appeared in 1977.

SCHARFFENSTEIN, GEORG FRIEDRICH, GENERAL (Montbéliard, 1760–1817, Ulm), was a fellow-pupil and friend of Schiller at the Ducal Academy at Stuttgart. Commissioned as an infantry officer, he served throughout the Napoleonic Wars (q.v.), in which Württemberg troops fought for the most part with the French. He was promoted general in 1809, and was commandant of Heilbronn in 1811 and of Ulm in 1817. Scharffenstein remained throughout his service an eminently humane man. His recollections of Schiller were published posthumously in 1837 (*Erinnerungen aus akademischen und Jugendjahren, vorzüglich in Bezug auf Schiller*). His name is particularly associated with a letter addressed to him by Schiller after a quarrel at school, probably in 1776, and with abetting Schiller's escape in 1782.

SCHARNHORST, GERHARD VON (Bordenau, Hanover, 1755–1813, Prague), Prussian general, was born of peasant stock and was the son of a sergeant. After a military education in Schaumburg-Lippe he entered the Hanoverian artillery, distinguishing himself in the Netherlands in 1793–4 against the French Revolutionary armies (see REVOLUTIONSKRIEGE). Various writings on military subjects led to an invitation to join the Prussian service, which he accepted in 1801 on condition that he was ennobled. He served as instructor in organization and tactics at the Berlin War Academy during the Napoleonic Wars (q.v.) and in 1806 was wounded at Auerstedt. He took part in Blücher's retreat to Lübeck and fought with distinction in the final engagement of the war at Preussisch-Eylau (1807).

After the Peace of Tilsit he was appointed Director of the War Department and carried out a thoroughgoing reform of the Prussian army. His aim was a national army, and he achieved this within the numerical limits imposed at Tilsit, introducing a system of short service, after which the soldiers were released to a reserve which he styled the Landwehr (q.v. and see KRÜMPERSYSTEM). Scharnhorst abolished corporal punishment, instituted promotion by merit, and opened commissioned rank to commoners. He revolutionized tactical drill, introducing individual training in musketry and flexible skirmishing, and created the army which fought in the Wars of Liberation. On the outbreak of war in 1813 he became chief of staff. At Großgörschen he received a wound from which he died at Prague two months later.

Scharnhorst's writings included *Handbuch für Offiziers in den anwendbaren Theilen der Kriegswissenschaften* (1781–90), *Militärisches Taschenbuch* (1793), *Die Ursachen des Glücks der Franzosen im Revolutionskrieg* (1803), and *Über die Wirkung des Feuergewehrs* (1813). From 1788 to 1805 he edited a military periodical (*Neues Militärisches Journal*, 1788–96; *Militairische Denkwürdigkeiten unserer Zeiten*, 1797–1805), in which appeared one of his best essays, *Die Verteidigung der Stadt Menin und die Selbstbefreiung der Garnison unter dem Königlich Großbritannisch-Kur-Hannoverschen Generalmajor von Hammerstein 1794*. This was particularly praised by Clausewitz (q.v.), Scharnhorst's pupil and later collaborator. See also GNEISENAU.

SCHARTENMEYER, P. V., a pseudonym of F. Th. Vischer (q.v.).

Schatten im Paradies, a novel by E. M. Remarque (q.v.), published posthumously in 1971. It is set in the U.S.A. in the 1939–45 War; its theme, exemplified in different ways by a number of characters, is the sadness, the strain, and the frustration of the refugees' life. One character, Dr. Ravic, had appeared in the novel *Arc de Triomphe* (1946).

Schatzgräber, Der, a ballad written by Goethe in Jena in May 1797 and published in the *Musenalmanach* for 1798. It is a mildly ironical moral poem, rejecting the treasureseeker's magic short-cuts and insisting on a balanced healthy life of alternating effort and relaxation.

Schatzkästlein des Rheinischen Hausfreundes, a collection of anecdotes, moral reflections, and practical suggestions taken from the various numbers of the annual *Der rheinländische Hausfreund* and published in 1811. Their author is J. P. Hebel (q.v.), who freely admits in the preface that many are not original, but lifted from earlier collections. Their simple homely form derives, however, from Hebel's direct, unaffected style.

Schaubühne, a volume of plays published in 1813 by L. J. von Arnim (q.v.), who devoted the proceeds to the purchase of munitions of war for the campaign against Napoleon. It contained ten plays, including the historical play *Vertreibung der Spanier aus Wesel* (q.v.), the comedy *Die Mißverständnisse,* and the epilogue (Nachspiel) *Das Frühlingsfest.*

Schaubühne als moralische Anstalt betrachtet, Die, Schiller's inaugural address to the Kurpfälzische Deutsche Gesellschaft, Mannheim, read on 26 June 1784. Schiller sees the theatre as an ally of religion and law, deriding folly, exposing vice, acquainting men with fate, and propagating wisdom. The address was published by Schiller in the first number of his periodical *Die Rheinische Thalia* (1785, see THALIA). When spoken, it was prefixed by an autobiographical apologia, which was omitted on publication.

SCHAUKAL, RICHARD VON (Brünn, now Brno, 1874–1942, Vienna), studied law at Vienna University and entered the Austrian civil service, retiring as Sektionsrat with a title of nobility in 1918. He was a lyric poet attached to the virtues of neat craftmanship, restrained elegance, and poetic mood (Stimmung). He published at least sixteen volumes of verse, beginning with *Gedichte* (1893) and ending with *Herbsthöhe* (1933), sifting and revising his work in the selections *Ausgewählte Gedichte* (1904), *Gedichte 1891–1918* (1918), and *Ausgewählte Gedichte* (1924). *Spätlese* (1943) is a posthumous collection.

Schaukal's best volumes of poetry are probably *Tristia* (1896) and *Die Gezeiten der Seele* (1926); his weakest is the derivative patriotic verse of

1914. Eherne Sonette (1914). He also wrote a number of stories, of which the five Novellen of *Eros Thanatos* should be especially mentioned, together with *Kapellmeister Kreisler* (both 1906). He is the author of books on literary subjects (*E. T. A. Hoffmann,* 1904, *Wilhelm Busch,* 1905, *Adalbert Stifter,* 1926, and *Karl Kraus,* 1933; *Richard Dehmels Lyrik,* 1907, is an essay in literary criticism).

Like many Austrians of his generation, Schaukal was hard hit by the collapse of the Dual Monarchy in 1918 and found it difficult to come to terms with the new age. He translated the work of several French Symbolists, including Gautier, Flaubert, and Duhamel. His *Werke in Einzelausgaben* (3 vols.), ed. L. von Schaukal and N. Langer, appeared in 1965–6.

SCHAUMANN, RUTH (Hamburg, 1899–1975), spent her childhood in Hagenau, Alsace, where her father was in garrison as a captain of dragoons. She developed not only poetic talents, but gifts for drawing and sculpture. In 1924 she became a Roman Catholic. Her style in poetry is traditional and its mood religious.

Ruth Schaumann's first published verse was *Die Kathedrale* (1920); further volumes, some of which have prose interludes and her own illustrations, include *Der Knospengrund, Die Rose* (both 1924), *Die Kinder und die Tiere* and *Der blühende Stab* (both 1929), *Der Krippenweg* (1932), *Ecce Homo* (1935), and *Klage und Trost* (1947). *Die Sternnacht* (1959) is a selection of her poetry. Her first work of fiction contains her own childhood recollections from the age of 4 in fortyeight stories (*Erzählungen*), which together make up a kind of autobiographical novel (*Amei,* 1933). She is the author of the novels *Yves* (1933), *Der Major* (1935), *Der schwarze Valentin und die weiße Osanna* (1938), *Die Übermacht* (1940), *Die Uhr* and *Elise* (both 1946), *Der Jagdhund* (1949), *Die Karlsbader Hochzeit* (1953), *Die Taube* (1955), and *Die Haarsträhne* (1959). She also published a single story, *Die Silberdistel* (1941), and two collections, *Der singende Fisch* (1934) and *Akazienblüte* (1959). *Das Arsenal* (1968) is autobiographical.

Schauspiel, term used for any public performance or any visual entertainment (spectacle) since the late 15th c. First used specifically for performance of a stage play by N. Manuel (q.v.), and as a generic term for all drama by G. P. Harsdörffer (q.v.). Analogous to it is the title *Deutsche Schaubühne* (1740–5) used by J. C. Gottsched (q.v.). From the late 18th c. the word came to be applied to a serious play which either does not end with the death of the hero or departs from the pattern of French Classical tragedy. Early examples of this use are found on the title-pages of Goethe's *Götz von Berlichingen, Stella, Die Geschwister, Iphigenie auf Tauris, Torquato Tasso*, F. M. Klinger's *Sturm und Drang*, Schiller's *Die Räuber* and *Wilhelm Tell*, and H. von Kleist's *Prinz Friedrich von Homburg* (qq.v.). Goethe's *Götz von Berlichingen* and Kleist's *Das Käthchen von Heilbronn* (q.v.) are notable examples of the Ritterschauspiel.

SCHEDE, PAUL, see **MELISSUS, PAULUS**.

SCHEDEL, HARTMANN (Nürnberg, 1440–1514, Nürnberg), an early humanist, was a physician at Nürnberg. His *Weltchronik* (1493), written in Latin and published in 1493 in a German translation by one Simon (or Georg) Alt, was richly illustrated with woodcuts by M. Wolgemut (q.v.) and W. Pleydenwurff (*c.* 1420–72).

SCHEFER, LEOPOLD (Muskau, 1784–1862, Muskau), was protected and assisted by Prince H. von Pückler-Muskau (q.v.), who in 1813 appointed him to administer his estates. Between 1816 and 1819 he travelled extensively in Europe, visiting England, the Mediterranean countries, and Asia Minor. His poems, first published anonymously by the prince (*Gedichte*, 1811), were augmented in a new edition in 1847. He also wrote widely read Novellen and novels, and a *Laienbrevier* (1834), a quasi-religious poetic work of pantheistic tendency, which was followed by two works of similar character, *Der Weltpriester* and *Hausreden*. Two volumes of pseudo-oriental poetry (*Hafis in Hellas*, 1853; *Koran der Liebe*, 1854) were ill received because of their sensual tone. Schefer also composed orchestral and piano music. *Ausgewählte Werke* (12 vols.) appeared in 1845–6.

SCHEFFEL, JOSEPH VIKTOR VON (Karlsruhe, 1826–86, Karlsruhe), was the son of an engineer officer in the Baden army who subsequently became a government architect (Oberbaurat). His mother, Josephine Scheffel, was a poetess of some talent. Scheffel's inclinations were for painting and poetry, but his father insisted on legal studies, and from 1843 to 1847 Scheffel led a pleasant life at Munich, Heidelberg, and Berlin universities. In 1848 he became secretary to a member of the Federal Diet and with him visited Schleswig-Holstein. His career in the legal branch of the Baden civil service had little appeal for him, and after service in Säckingen and Bruchsal he secured leave of absence to study painting in Italy, where his interests swung more and more to literature. In 1853 he resigned from the civil service and devoted himself to letters.

Scheffel's sentimentally humorous epic *Der Trompeter von Säckingen* (q.v., containing his best-

known poem 'Alt-Heidelberg, du feine') appeared in 1854, and was followed in 1857 by the historical novel *Ekkehard* (q.v.), for which he made meticulous studies at St. Gall. Both these works were immensely successful and made Scheffel one of the most widely read German authors for the next fifty years. He became librarian to Prince Fürstenberg at Donaueschingen (1857–9); in 1863 he was the guest of the Grand Duke of Saxe-Weimar at the Wartburg. A novel on the Wartburg, intended to be his principal work, was never completed. In 1865 the title Sachsen-Weimarischer Hofrat was conferred upon him and in 1876 he was given a patent of nobility. Two collections in verse followed in these years of honour, *Frau Aventiure, Lieder aus Heinrich von Ofterdingens Zeit* (1863) and *Gaudeamus, Lieder aus dem Engern und Weitern* (1868, a selection, ed. P. Wiesmann, 1963), which contains the poems 'Als die Römer frech geworden', 'Ein Römer stand in finstrer Nacht', 'Wohlauf, die Luft geht frisch und rein', and 'Das war der Zwerg Perkêo'. He is also the author of *Juniperus* (a fragment of the projected Wartburg novel, 1866), *Waldeinsamkeit* (poems, 1878), and a short Novelle, *Hugideo* (1878). His last years were darkened by illness and depressive neurosis.

Scheffel's success, which has since faded, rested partly on his nationalistic standpoint and on an outworn Romanticism, but he had a real and fluent talent and, in his verse, a certain panache. Towards the end of his life he was regarded by some as an exponent of Butzenscheibenpoesie (q.v.). *Werke* (10 vols.), ed. J. Franke, appeared in 1917, and correspondence (6 vols.), ed. W. Zentner, 1926–67.

SCHEFFLER, JOHANN, see **ANGELUS SILESIUS**.

Scheidekunst, sobriquet given to G. E. Lessing (q.v.) because of the distinction between the spheres of the plastic and literary arts which he established in his *Laokoon* (q.v.).

SCHEIDT, KASPAR (*c.* 1520–65, Worms), a schoolmaster of Worms, made in 1551 a free translation into German rhyming verse of the Latin satire *Grobianus* (q.v.) by Friedrich Dedekind (q.v.), giving it the sub-title *Von groben sitten und unhöflichen gebärden*. In this form the work enjoyed great popularity. He is also the author of a consolatory poem on the death of a wife, the *Frölich Heimfahrt* (1552). Scheidt was an uncle of Fischart (q.v.). His name is also spelt Scheit, Scheid, and Scheyt.

SCHEIN, JOHANN HERMANN (Grünhain, 1586–1630, Leipzig), German composer, was cantor at St. Thomas's Church in Leipzig from 1616 to his death. His compositions include choral works to liturgical texts, madrigals, and hymns (setting his own words) as well as instrumental music. A *Gesamtausgabe* of his works, ed. A. Prüfer, appeared in 1901–23 (7 vols.).

SCHELLING, CAROLINE VON (Göttingen, 1763–1809, Maulbronn), daughter of Professor J. D. Michaelis (1717–91), a distinguished Hebraist at Göttingen, married in 1784 a physician named Böhmer. A widow at 25, she moved in 1790 to Mainz and associated there with persons of revolutionary sympathies, including G. Forster (q.v.). On the capture of Mainz by Prussian forces in 1793 she was accused of collaboration with the revolutionary party and imprisoned for a time at Königstein near Frankfurt. In 1796 she married August Wilhelm Schlegel (q.v.) and lived with him in Jena, frequenting the Schiller household. She has been blamed for causing trouble between Schlegel and Schiller, though Schlegel was quite capable of falling out with Schiller on his own account. She encouraged and assisted Schlegel in his great translation of Shakespeare (q.v.). From 1800 she was often seen in public with F. W. J. Schelling (q.v.), and in 1801 she and Schlegel were divorced. She married Schelling in 1803. She was a fascinating and witty conversationalist and a brilliant letter-writer. A selection of her letters was published in 1871.

SCHELLING, FRIEDRICH WILHELM JOSEPH VON (Leonberg, 1775–1854, Ragaz), was educated at the Tübinger Stift (q.v.), where Hölderlin and Hegel (qq.v.) were among his friends. In 1797 he published *Ideen zur Philosophie der Natur* in which he established himself as an exponent of Nature philosophy, and in 1798, on the advice of Goethe and Fichte (q.v.), to whom his intellectual development was indebted, was appointed to the chair of philosophy at Jena University. He soon established contact with the early Romantics (see ROMANTIK) A. W. and F. Schlegel, Tieck, and Novalis (qq.v.), to whom his tract *Von der Weltseele* (1798) commended itself. In 1803 he married the divorced wife of A. W. Schlegel (see SCHELLING, CAROLINE VON), and moved to a chair at Würzburg University which he occupied until 1806.

Schelling attempted a systematic philosophy superseding that of Fichte. The *Erster Entwurf eines Systems der Naturphilosophie* (1799) was followed by *System des transzendentalen Idealismus* (1800), linking the duality between the empirical self (ego) with the non-ego, as conceived by Fichte, in synthesis in art. The consummate function of aesthetics is further discussed in *Philosophie der Kunst* (1809, containing lectures of 1802–3). Schelling's *Darstellung meines Systems*

im Ganzen (1801) remained unfinished, but its central idea that speculative and imaginative thought could form a basis for scientific study and thus yield objective knowledge compatible with scientific knowledge appealed to the Romantic frame of mind. In establishing the thesis of identity equating Nature and Spirit (termed by him Indifferenz von Natur und Geist) he laid the foundation of his 'objective Idealism'; it gave an appearance of detachment to views which essentially constituted a subjective pantheism. Nature, he considered, is visible Spirit, and Spirit is invisible Nature.

Before he was 30, Schelling was drawn into paths which were more mystical than philosophical. The Roman Catholic theologian F. X. Baader (q.v.), whom he met in 1806, and the writings of J. Böhme (q.v.) influenced his thought in *Philosophie und Religion* (1804) and *Philosophische Untersuchungen über die menschliche Freiheit* (1809, reissued 1950). This second phase of his development took place in Munich, where from 1806 until 1841 he was General Secretary to the Akademie der bildenden Künste before being invited by Friedrich Wilhelm IV (q.v.) of Prussia to Berlin University; the King hoped that Schelling's more clearly religious thought would change the tone of the philosophy taught in the Prussian capital (an expectation that was not realized). In Berlin Schelling completed his theosophic writings, *Philosophie der Mythologie* (1842) and *Philosophie der Offenbarung* (1854). Schelling's pantheism gave way to another form of theism, which had some influence on writers of the mid- and late 19th c. The dualism, so happily resolved, recurs in the distinct concepts of the existence of an Absolute Spirit and of a Being which is God. The Absolute Spirit is the Will and Creation its manifestation, evolution proceeding through the interdependence of God, man, and the inscrutable Will.

Schelling appears to have had an attractive personality and a persuasive and stimulating utterance. He is generally regarded as more poet than philosopher. He did in fact write poetry, which was collected and published by Erich Schmidt (q.v.) as *Gedichte* in 1913, and he was at one time believed to be the author of *Die Nachtwachen des Bonaventura* (q.v.). Schelling's *Sämtliche Werke* (14 vols.), ed. K. F. A. Schelling, appeared in 1856–61; a revised edition of 1927 ff. (12 vols.) was reissued in 1958–60.

Schelmenroman, picaresque novel, a form which traces the desultory and haphazard adventures of a quick-witted and usually amusing rogue. It is first represented in German literature early in the 17th c. by translations and adaptations from Spanish. The outstanding German Schelmenroman of the age is Grimmelshausen's *Der abenteurliche Simplicissimus* (q.v., 1669), and a typical example in the 18th c. is *Der im Lustgarten der Liebe herumtaumelnde Cavalier* (1738) by J. G. Schnabel (q.v.). Modern descendants are Th. Mann's *Bekenntnisse des Hochstaplers Felix Krull* (q.v., 1954) and G. Grass's *Die Blechtrommel* (q.v., 1959).

Schelmenzunft, Die, a satirical poem by Thomas Murner (q.v.).

Schelmuffskys Wahrhafftige curiöse und sehr gefährliche Reisebeschreibung zu Wasser und Lande, a comic narrative by Christian Reuter (q.v.), published in 1696 (enlarged 1697). Schelmuffsky, a ne'er-do-well, tells his own story; he sets out on a journey, first to Hamburg then to Sweden, in the course of which, by his own account, he distinguishes himself, being everywhere well received by the ladies and treated by everyone as a person of quality. It is clear from the outset that Schelmuffsky's story is full of extravagant hyperbole. He claims to have been shipwrecked on the way to Holland, fêted at Amsterdam, and in India to have become a friend of the Great Mogul. He also visits London, where he has successful amours, and Spain, and he has a fight with the French pirate Hans Barth (Jean Bart). When he reaches home his story is disbelieved by his family. Schelmuffsky sets out on a second journey, only to encounter disbelief a second time on his return home.

In this novel, in which literary parody, social satire, and whimsy intertwine, the characters of Schelmuffsky's mother, sisters, and brother are intended to lampoon Frau Müller, Reuter's landlady, and her family, with whom Reuter conducted a long feud. This forgotten novel was rediscovered in the early 19th c., and the author was given the sobriquet Der Tebelhohlmer from the hero's interminably repeated profane interjection 'Der Tebel hohl mer' (Devil take me). Both versions were reprinted by A. Schullerus (1885, re-edited W. Polenz, 1956; ed. I. M. Barth, 1964).

Schelm von Bergen, a ballad by H. Heine (q.v.), included in *Historien* in the *Romanzero* (q.v., 1851). At a public festival in Düsseldorf castle, the Duchess dances with a masked man, who proves to be the public executioner of Bergen. The stain of dishonour, which this entails, is erased by the action of the Duke, who ennobles him with the title Schelm von Bergen. C. Zuckmayer (q.v.) used the subject in his play *Der Schelm von Bergen* (1934).

SCHENCK VON STAUFFENBERG, CLAUS, GRAF VON (Jettingen, 1907–44, Berlin, by

execution), a senior officer (Oberst) who, from 1943, conspired against the National Socialist regime (see RESISTANCE MOVEMENTS, 2). He attempted unsuccessfully to assassinate Hitler on 20 July 1944 by placing a bomb at the H.Q. at Rastenburg, East Prussia.

SCHENK, EDUARD VON (Düsseldorf, 1788–1841, Munich), a Bavarian civil servant, who rose to be Minister of the Interior (1828), wrote in his spare time Schillerian tragedies which were collected in *Schauspiele* (3 vols., 1829–35). Among them were *Belisar* (1827) and *Henriette von England* (first published in the *Schauspiele*).

SCHENKENDORF, MAX VON (Tilsit, 1783–1834, Koblenz), baptized Gottlob Ferdinand Maximilian, used the abbreviation Max to express his admiration for a character in Schiller's *Wallenstein* (q.v.), Max Piccolomini. Of noble birth, he grew up on his father's estate in Lithuania and was sent in 1798 to Königsberg University. His parents, suspicious of his conduct, withdrew him and sent him to a tutor in the country. Visiting the Marienburg (q.v.) at this time, he was horrified at its dilapidation and misuse and wrote a letter of protest to *Der Freimütige* in Berlin, signing himself F. v. Sch.; these initials are said to have been mistaken for those of the influential Th. von Schön (q.v.), and in this way the young Schenkendorf gave the first impulse to the restoration of the castle. In 1804 he resumed study at Königsberg. In the summer of 1806, with Prussia isolated in Europe (see NAPOLEONIC WARS), Schenkendorf began to write the patriotic poetry for which he is best known. He continued to write soldiers' songs right up to 1815, and he supplemented them after the defeat of Prussia by articles in periodicals. Some of these were published in journals such as *Vesta*, edited by himself and quickly forbidden by the censor. In 1809 he was wounded in the right hand in a duel and permanently disabled.

Schenkendorf moved to Karlsruhe in 1812 and there married Elisabeth Barckley, widow of a former acquaintance in Königsberg. In 1813 he joined the forces against Napoleon in Schweidnitz. Unable to handle a weapon, he acted as a purveyor of poems, expressing a German rather than a Prussian patriotism. Baron vom Stein (q.v.) sent him to Frankfurt at the end of 1813 to assist in the raising and arming of new forces. After the restoration of peace Schenkendorf became a civil servant in the Rhineland. His death is said to have been a consequence of the old injury to his hand.

Patriotic poems predominate in his output; they are frequently in folk-song-like stanzas and have a chivalrous and manly character with no hint of ferocity or hatred. He also wrote some historical poems, mainly linked with places famous in German history, and a small body of love poetry. In his last years he turned to religious poems. His *Christliche Gedichte* were published in 1814, and his *Gedichte*, preceded by a characteristically modest preface, in 1815. *Sämtliche Gedichte* appeared in 1837.

SCHENK VON LIMBURG, a minor Middle High German poet of the middle of the 13th c., belonged to an important noble family in the service of the Hohenstaufen kings. He accompanied Konradin (q.v.) to Italy in 1268. Six of his charming Minnelieder, which include a winter and summer or May song, are preserved.

SCHENK VON LISSBERG, a minor Middle High German poet of the 13th c. He was not known until 1956, when Minnelieder from a MS. belonging to a canon of Mainz were published, one of which is by Schenk von Lißberg. The poem suggests the influence of ULRICH von Winterstetten (q.v.).

SCHENZINGER, KARL ALOIS (Neu-Ulm, 1886–1962, Priem, Bavaria), studied medicine and practised for a short time in Hanover. After 1928 he lived by his pen, developing a new form of fiction, the technological or industrial novel, such as *Anilin* (1936, a best seller), *Metall* (1939), *Atom* (1950), *Bei I.G.-Farben* (1953), 99% *Wasser* (1956), and *Magie der lebenden Zelle* (1957). The last two, it will be noted, delve into biochemistry.

SCHERENBERG, CHRISTIAN FRIEDRICH (Stettin, 1798–1881, Berlin-Zehlendorf), who began his life as a clerk, went on the stage in 1819 in Magdeburg, but gave up this career on marriage, returning to a business life. Having lost all his money, he set up in Berlin in 1837 as an author. His poems (*Vermischte Gedichte*) were published in 1845, and the epic poem *Ligny* in 1846. His fame was established in 1849 by the blank-verse epic *Waterloo*, which was followed by *Leuthen* (1852) and *Abukir, die Schlacht am Nil* (1856). He was a prominent and eccentric member of the literary club Der Tunnel über der Spree (q.v.), in which he met his biographer Theodor Fontane (q.v.), whose *Christian Friedrich Scherenberg und das literarische Berlin von 1840 bis 1860* was published in 1884. According to Fontane, the rising interest in the novel diverted attention from Scherenberg's narrative poetry, but he achieved a new success in 1869 with the epic *Hohenfriedberg*, which was written in stanzas. The patriotism of Scherenberg's work won him a reputation in influential circles and in 1855 he

received an appointment in the library of the War Ministry.
Ausgewählte Dichtungen, ed. H. Spiero, appeared in 1914.

SCHERER, WILHELM (Schönborn, Austria, 1841–86, Berlin), a scholar and historian of literature, was professor of German in Vienna (1868), Strasburg (1872), and Berlin (1877). He assisted K. Müllenhoff (1818–84) in *Denkmäler deutscher Poesie und Prosa aus dem VIII–XII Jahrhundert* (1864). His publications include *Jacob Grimm* (1865), *Leben Willirams* (1866), *Zur Geschichte der deutschen Sprache* (1868), *Geschichte des Elsasses* (with O. Lorenz, 2 vols., 1871), *Geschichte der deutschen Dichtung im 11. und 12. Jahrhundert* (1875), *Die Anfänge des deutschen Prosaromans und Jörg Wickram von Colmar* (1877), *Aus Goethes Frühzeit* (1879), and *Geschichte der deutschen Literatur von den Anfängen bis zu Goethes Tod* (1883). In 1918 O. Walzel (1864–1944) extended this to the early 20th c. (*Geschichte der deutschen Literatur*) and this edition was provided with an excellent bibliography by J. Körner (1888–1950). Scherer, who was influential over several decades, related biography and creative writing on a basis of cause and effect.

SCHERNBERG, DIETRICH, a cleric of the 15th c. who was imperial notary in Mühlhausen from 1483 to 1502. He wrote the *Spiel von Frau Jutten* (q.v.).

SCHERZ, JOHANN GEORG (Strasburg, 1678–1754, Halle), professor of moral philosophy (1702) in his native city and then professor of law (1711), was a scholar interested in older German literature. He published a collection of fables, including 51 by Ulrich Boner (q.v.), under the title *Philosophiae moralis Germanorum medii aevi specimina*, 1704–10.

Scherz, Satire, Ironie und tiefere Bedeutung, a three-act comedy (Lustspiel) by C. D. Grabbe (q.v.), written in 1822 and published in 1827 in *Dramatische Dichtungen*. It was not performed until 1907, in Munich. A succession of loosely connected scenes, the play is an expression of Grabbe's disillusionment with reality, presented in the form of satire and grotesque irony. The 'deeper meaning' of the title hints at Grabbe's ironical self-detachment. A principal character of the play is the Devil (Der Teufel), who describes the world as a mediocre comedy written by an angel during his school holidays. Many writers of Grabbe's time come under ruthless attack, and the characters of the play, among them a schoolmaster, the farmer Tobies, Freiherr von Mordax, the poet Rattengift, Herr von Wernthal, Baron von Haldungen, and his daughter Liddy, are caricatures. Liddy's marriage to the courageous though hideous Mollfels, the play's only constructive element, brings the slender action to a conclusion. Grabbe ultimately also ridicules himself, appearing as a character at the end of the last act.

SCHICKELE, René (Oberehnheim, Alsace, 1883–1940, Vence nr. Nice), was of mixed (German and French) parentage and suffered throughout his life from the national tensions of the Alsatian. He hoped for a diminution of national hostility, and his work reflects this and his consequent pacifism. He studied at Strasburg, Munich, the Sorbonne, and Berlin, and was afterwards a journalist and author. He evaded the problem of loyalties during the 1914–18 War by moving to Zurich. He returned to German journalism in 1919, but emigrated to France in 1932. His original manner was Expressionistic, but he later adopted a sober realism. He published the poetry of his student days in *Pan* and *Sommernächte* (both 1902); later verse is in *Der Ritt ins Leben* (1906), *Weiß und Rot* (1910), and *Die Leibwache* (1914). *Mein Herz, mein Land* (1915) is his own selection from his poems.

Schickele is better known for his fiction. The novels include *Der Fremde* (1909), *Benkal, der Frauentröster* (1914), *Das Erbe am Rhein* (q.v., a trilogy, 1925–31), *Symphonie für Jazz* (1929, the story of an Alsatian musician whose music and whose love are at cross purposes), *Die Witwe Bosca* (1933), and a humorous work, *Die Flaschenpost* (1937). He published two separate Novellen, *Meine Freundin Lo* (1911) and *Trimpopp und Manasse* (1914), and a collection, *Die Mädchen* (1920). Late in life he turned to drama and achieved a great success with his first play, *Hans im Schnakenloch* (1920), which dealt with the problems the war posed for Alsatians. He also wrote two Expressionist plays, *Am Glockenturm* (1920) and *Die neuen Kerle* (1921). The story *Le Retour* (1938) was first published in French, appearing in German as *Die Heimkehr* in 1938. His essays include *Liebe und Ärgernis D. H. Lawrence* (1934). *Werke* (3 vols.), ed. H. Kesten (q.v.), appeared in 1959–61.

Schicksalslied, a poem by F. Hölderlin (q.v.), first published in Bk. 2 of his novel *Hyperion, oder Der Eremit aus Griechenland* (q.v., 1797–9), and opening with the line 'Ihr wandelt droben im Licht'; in the novel it is addressed by Hyperion to the 'blessed genii', cosmic spirits who, unlike man, enjoy communion with the gods. The last of its three verses presents human life in the image of the cataract, dashed from rock to rock.

Schicksalstragödie, term applied to plays in vogue during the Romantic movement (see ROMANTIK), in which individuals or the entire members of a family perish, either on a fated day or through a fatal weapon (or both), usually as a consequence of some past crime. The first Schicksalstragödie is considered to be *Blunt, oder der Gast* by K. P. Moritz (q.v., 1781), and the best example is Z. Werner's *Der vierundzwanzigste Februar* (q.v., 1806), which was influenced by Schiller's *Die Braut von Messina* (q.v., 1803, a tragedy which has analogies with Sophocles' *Oedipus rex*). Others are *Der neunundzwanzigste Februar* (q.v., 1815) by A. Müllner (q.v.), who had acted in Werner's play and is also the author of *Die Schuld* (q.v., 1816), and *Das Bild* (1821), *Die Heimkehr* (1821), and *Der Leuchtturm* (q.v., 1821), all by Baron von Houwald (q.v.). These writers favour the trochaic verse with four accents, borrowed from Spain and widely used by the Romantics. Grillparzer's early play *Die Ahnfrau* (q.v., 1816) was branded as a Schicksalstragödie, much to his dismay, for the standard of plays associated with the term remained poor, and their actions were crude: the crime which begins the series of fateful repetitions is usually particularly horrifying, incest or parricide being the most frequent, and the perpetrator's awareness or ignorance of the relationship is immaterial.

Parodies of such fate tragedies include Houwald's *Seinem Schicksal kann niemand entgehen* (1819) and *Die verhängnisvolle Gabel* by A. von Platen (q.v., 1826).

SCHIEBELER, Daniel (Hamburg, 1741–71, Hamburg), studied law at Göttingen and Leipzig universities and became in 1768 a lay canon of the Hamburg Chapter. He is the author of a Singspiel, *Lisuart und Dariolette* (1766), which J. A. Hiller (1728–1804) set to music. *Romanzen* (1767 and 1771) were also set by Hiller. *Die Schule der Jünglinge* (1767) is a comedy without music.

'Schier dreißig Jahre bist du alt', first line of K. von Holtei's *Mantellied* (q.v.).

SCHIKANEDER, Emanuel Johann Jakob (Straubing, 1751–1812, Vienna), came to Vienna as actor and singer in 1785, and subsequently became an impresario. He wrote plays and also libretti for operas, of which the best known is Mozart's *Die Zauberflöte* (q.v., 1791). He sang the part of Papageno at the original performance. In his last year Schikaneder went out of his mind. *Sämtliche theatralische Werke* (2 vols.) appeared in 1792.

Schildbürger, Die, a Volksbuch printed in 1597. It is a collection of Schwänke (see

SCHWANK) centred upon the inhabitants of Schilda and exists in an alternative form as *Das Lalenbuch*. A variant printed in 1603 is entitled *Der Grillenvertreiber*.

Schilflieder, a group of five lyric poems written by N. Lenau (q.v.) in Stuttgart in 1831. They were addressed to Lotte Gmelin.

SCHILHER, or **SCHILLER,** JÖRG, a Swabian Meistersinger of the 15th c., some of whose poems are preserved in the Colmarer Liederhandschrift (q.v.) at Munich.

SCHILL, FERDINAND BAPTISTA VON (Wilmsdorf nr. Dresden, 1776–1809, Stralsund), a Prussian patriot who was elevated into a national hero. Schill was a regular officer of dragoons when war with France broke out in 1806 (see NAPOLEONIC WARS). He was wounded at Auerstedt but made his way to Colberg where he took a prominent part in the defence of the city. After the signing of the peace treaty at Tilsit in 1807 he was appointed to the command of a regiment of hussars. At the time of the war between France and Austria in 1809 he mistakenly judged the moment ripe for a national rising, and began hostilities with his regiment against the French and allied army of occupation. His action, however, was disavowed by the Prussian government, and his small force, after several engagements, was overwhelmed at Stralsund. Schill himself was killed in street fighting and eleven surviving officers were court-martialled and shot. The rank and file were sent to the French galleys.

SCHILLER, CHARLOTTE VON (Rudolstadt, 1766–1826, Bonn), wife of J. C. Friedrich Schiller (q.v.), was a Fräulein von Lengefeld and a cousin of Schiller's friend, Wilhelm von Wolzogen (1762–1809). Schiller and she first made acquaintance at Mannheim in 1784. In 1787 the acquaintance was renewed and in August 1789 Schiller and Charlotte were engaged. The marriage took place on 22 February 1790 at the village of Wenigenjena, near Jena. Schiller was for a short time equally attracted by the elder sister, Caroline von Beulwitz (see WOLZOGEN, CAROLINE VON). Charlotte von Schiller bore her husband four children.

SCHILLER, JOHANN CHRISTOPH FRIEDRICH (Marbach, 1759–1805, Weimar), generally referred to as Friedrich Schiller, who was elevated to the nobility as von Schiller in 1802, was the son of a commoner who had risen to commissioned rank in the army of the Duke of Württemberg. At 13 Schiller was separated

from his family and sent by the Duke to the Militär-Akademie, in which he remained until he was 21. He emerged from this environment, which he resented, as an army physician and something of a rebel. While at school he wrote a play, the unwieldy yet dynamic *Die Räuber* (q.v.), which he published at his own expense in 1781. From December 1780 to September 1782 Schiller served in Stuttgart, earning a reputation as an original and outspoken poet by his erotic and other poetry included in the *Anthologie auf das Jahr 1782* (q.v.) as well as by the successful performance of his play *Die Räuber* at Mannheim.

Finding the atmosphere of Stuttgart too repressive, Schiller fled in September 1782 to Mannheim, where he hoped mistakenly for encouragement. From December 1782 to July 1783 he enjoyed asylum on the estate of a patroness, Frau von Wolzogen. Here he completed *Fiesco* (see VERSCHWÖRUNG DES FIESCO ZU GENUA, DIE) and *Kabale und Liebe*, and began *Don Carlos* (qq.v.). He was next appointed house poet to the Mannheim Theatre for a year (1783–4), and there *Fiesco* was performed with moderate success, while *Kabale und Liebe*, a play in realistic idiom and with a strong element of contemporary social criticism, played to full houses. So far Schiller's writing was an impressive but belated echo of the Sturm und Drang (q.v.). An attempt at periodical journalism with *Die Rheinische Thalia* (see THALIA, DIE) was not successful. After a depressing winter spent without income and in a tormenting and hopeless love-affair with a married woman, Charlotte von Kalb (q.v.), Schiller accepted in the spring of 1785 an invitation to Leipzig to join two young couples who admired his work. This journey initiated a change in his life. The worst material cares were removed and he could devote himself to poetry. He made, however, only slow progress with *Don Carlos*, which did not appear until 1787. This historical drama in verse, intertwining the personal passions of individuals and the fate of nations, marks an entirely new phase in his development. During these years he also wrote a story expressing notable social concern, *Der Verbrecher aus verlorener Ehre* (q.v., 1786), and began an original novel, which remained a fragment, *Der Geisterseher* (q.v.).

In the summer of 1787 Schiller settled in Weimar, soon becoming attached to Charlotte von Lengefeld, whom he married in 1790. Meanwhile he engaged successfully in historical writing, publishing in 1788 a history of the conflict between Spain and the Netherlands in the 16th c. (*Geschichte des Abfalls der vereinigten Niederlande von der spanischen Regierung*, q.v.). This work gained him appointment in 1789 as professor of history at Jena. The university chair gave him a standing but little financial support and he contracted to write a popular history of the Thirty Years War (*Geschichte des dreißigjährigen Kriegs*, 1791–3). His health broke down through overwork early in 1791. In December 1791, however, he was freed from financial strain by a generous grant from Prince Friedrich Christian (q.v.) of Augustenburg and Count Heinrich Ernst von Schimmelmann (q.v.), both German Danes. After a visit to his homeland in 1793 he devoted himself in Jena to a study of the philosophy of Kant (q.v.). In 1794 he came into close contact and friendship with Goethe, with whom his relations had hitherto been cool. The results of Schiller's philosophical studies appeared in a series of works dealing with ethical and aesthetic problems. *Über Anmut und Würde* (q.v., 1793) championed instinctive rightdoing against Kant's rather joyless morality and established Schiller's aesthetic psychology. *Über die ästhetische Erziehung des Menschen in einer Reihe von Briefen* (q.v., 1795) set up the doctrine of psychological balance as the aesthetic state of mind, and in *Über naive und sentimentalische Dichtung* (q.v., 1795–6) he essayed a typology of poets, for which he and Goethe may be regarded as characteristic examples. He also wrote a series of short treatises on tragedy, including *Über den Grund des Vergnügens an tragischen Gegenständen* (1792), *Über das Pathetische* (q.v.), *Vom Erhabenen* (both 1793), and *Über das Erhabene* (q.v., 1801), which are significant for his later plays.

In 1795 Schiller began to edit with Goethe's support *Die Horen* (q.v.), a high-toned monthly, which was received coolly by the critics. Schiller and Goethe avenged themselves in satirical distichs, which they called *Xenien* (q.v., 1796) and published in the *Musenalmanach*. They followed up this destructive excursion by creative work in 1797 in the form of ballads, of which Schiller's share included *Der Taucher*, *Der Handschuh*, *Der Ring des Polykrates*, and *Die Kraniche des Ibykus* (qq.v.). It was now years since Schiller had written a play and, in spite of his poetry and various prose works, he was worried by a sense of sterility. In 1796 he at last embarked on a new tragedy, *Wallenstein* (q.v.), which he wrote principally in 1797–8. In its final form *Wallenstein* proved to be a trilogy, composed of *Wallensteins Lager*, a short prologue with deft comic touches, and the drama proper separated into *Die Piccolomini* and *Wallensteins Tod*. *Wallenstein*, which is perhaps Schiller's most impressive historical tragedy, was performed complete in 1799. The new work gave Schiller impetus, and play rapidly succeeded play. *Maria Stuart* (q.v.) was produced in 1800, *Die Jungfrau von Orleans* (q.v.), which Schiller termed 'eine romantische

Tragödie' and held in special affection, followed in 1801, and his most classical work, *Die Braut von Messina* (q.v.) with its chorus (which he justified in *Über den Gebrauch des Chors in der Tragödie*), in 1803. Schiller's next play, *Wilhelm Tell* (q.v., 1804), abandoned tragedy for Schauspiel, and achieved in its breadth and multiplicity of levels a universal success. He quickly reverted, however, to tragedy in *Demetrius* (q.v.); it remained unfinished, and work on it was resumed notably by Hebbel and Laube (qq.v.). Schiller, who had long struggled against illness, died in May 1805.

Schiller's character and poetic gifts are dominated by a powerfully developed will, which is embodied in many of his characters, is expressed in the vehemence of his style, and was the supporting element in his modest span of life. His early prose plays, though partially realistic, contain an important element of caricature. With *Don Carlos* he adopted verse and used it in all his subsequent plays, and in this later part of his œuvre he aimed at a harmonious manner, with a pronounced rhetorical style ('Schillersches Pathos'). His tragic heroes, he considered, achieve sublimity in the grandeur of their willed submission to an inescapable fate. In recent years increasing attention has been paid to Schiller's acute theoretical and critical writings and to the stylistic means by which he attained dramatic success.

Among older editions the *Säkular-Ausgabe* (ed. E. von der Hellen, 16 vols., 1904–5) should be mentioned. The *Nationalausgabe* (ed. J. Petersen *et al.*, 1943 ff.), planned for 43 volumes, includes correspondence and Gespräche (i.e. references to Schiller by contemporaries). The letters (*Schillers Briefe*) were edited by F. Jonas (7 vols., 1892–6).

SCHILLER, Jörg, see Schilher, Jörg.

Schillers Flucht von Stuttgart, see Streicher, Andreas.

SCHILLINGS, Max von (Düren, 1868–1933, Berlin), composer of music, was from 1919 to 1925 Intendant of the Berlin Opera House (Staatsoper). His works include the Melodrama (q.v.) *Das Hexenlied* (1902) to words by E. von Wildenbruch (q.v.), occasional music for Goethe's *Faust* (1908), and the opera *Mona Lisa* (1915). He also set to music poems from Goethe's *West-östlicher Divan* (q.v.).

SCHILTBERGER or **SCHILTPERGER,** Hans (nr. Freising, 1380–?), was taken prisoner by the Turks in 1396 or 1398 during a campaign in Bulgaria. He was later employed by them and travelled in the Near and Middle East, penetrating as far as Turkestan. He returned home in 1427 and took service with Duke Albrecht III (q.v.) of Bavaria. He wrote an account of his travels and experiences, which appeared as one of the early printed books (*History*, 1473). It was several times reprinted in the 16th c. and again in the 19th c. (*Hans Schiltbergers Reisen*, ed. V. Langmantel, 1885).

SCHIMMELMANN, HEINRICH ERNST, GRAF VON (Dresden, 1747–1831, Copenhagen), was from 1784 to 1813 Danish finance minister. In 1824 he became foreign minister. A humane man greatly interested in poetry and the arts, Schimmelmann, together with Prince Friedrich Christian (q.v.) of Augustenburg gave Schiller in 1791 a three-year pension, which was a turning-point in the poet's career.

Schimmelreiter, Der, a Novelle by Th. Storm (q.v.), written in 1886–8 and published in the *Deutsche Rundschau* in 1888. The story, set in a double frame (see RAHMEN), is based on the legend of the Schimmelreiter, the ghost of the grey horse and its rider, which appears on the Friesian shores when storm threatens the dikes. These consist of an old dike and of the Hauke-Haien-Deich, built by Hauke Haien, the Schimmelreiter, a hundred years ago in the middle of the 18th c.; of the two, Haien's dike provides the stronger protection against the sea. The character of its long-dead builder is still persistently maligned by the superstitious villagers, but a more detached view is secured by letting the enlightened schoolmaster narrate the Binnenerzählung containing episodes in the life of Hauke Haien, his rise to be Deichgraf, his achievement, and his tragic end.

The son of the clever but impecunious Tede Haien, Hauke, as a child, makes experiments in dike construction. As a young man he serves the Deichgraf Tede Volkerts, to whom he becomes indispensable. After Volkert's death he is appointed to this high office traditionally pertaining to a landowning villager. Hauke achieves both through his marriage to Volkert's daughter Elke. The jealous Ole Peters is quick to point this out, thus strengthening Hauke's determination to prove his own worth. He goes ahead with his original design for a new dike and with his plan to reclaim land from the sea, a motif prefigured in Goethe's *Faust* (q.v.), Pt. II. He succeeds because he combines supreme expertise with outstanding will-power. While supervising the work he rides a grey horse (Schimmel). The mystery of its origin and Hauke's refusal to sacrifice a living creature for the dike (a dog has been chosen for this superstitious ritual), and his rescue of the dog, all contribute to make the people' believe that

Hauke is in the hands of evil powers. Hauke's happiness is confined to Elke and their daughter Wienke, a mentally subnormal but much loved child. After an illness has deprived Hauke of his resolute spirit, he accepts, against his better judgement, Ole Peters's inadequate proposal for urgent repair to the old dike. With the October storms of 1756 disaster strikes: the old dike bursts as the Schimmelreiter fights his last battle against the elements. While the deluge pours into the low-lying village, and people and cattle escape to the safety of the higher land, Elke, with her child, drives her carriage out to meet Hauke. The sea takes them all.

Storm portrays in Hauke Haien the provocative and vulnerable daemonic force in man against the background of the sea, the force of destiny. His narrative technique, moreover, establishes a carefully gauged margin of ambiguity pervading the rational and mystic themes.

SCHINDERHANNES, nickname of Johann Bückler (1783–1803), the head of a gang of robbers active in the central Rhineland during the revolutionary disturbances in the first years of the 19th c. He was eventually caught, tried, and executed at Mainz. He is the subject of the novel *Unter dem Freiheitsbaum* by Clara Viebig (q.v.) and of the play *Schinderhannes* by C. Zuckmayer (q.v.).

SCHIRMBECK, HEINRICH (Recklinghausen, 1915–), a journalist, publicist, and author, who is preoccupied with the environmental problems raised by the developments of science and technology. His views are expressed in the essays *Vom Elend der Literatur im Zeitalter der Wissenschaft* (1967), *Die moderne Literatur und die Erziehung zum Frieden* (1970), *Ihr werdet sein wie die Götter. Der Mensch in der biologischen Revolution* (1972), and *Schönheit und Schrecken. Zum Humanismusproblem in der modernen Literatur* (1977). In 1980 he published *Die Pirouette des Elektrons*. His narrative works are related to his central preoccupations. They include the novel *Der junge Leutnant Nikolai* (1958, enlarged from the story *Gefährliche Täuschungen*, 1947), *Das Spiegellabyrinth* (1948), *Träume und Kristalle* (1968), and, by far the best known, the novel *Ärgert dich dein rechtes Auge* (1957), which is concerned with atomic physics and its implications for mankind.

SCHIRMER, DAVID (Pappendorf nr. Freiberg, Saxony, c. 1623–83), studied at Wittenberg and became librarian at Dresden. His poetry, which includes light and graceful love poems, was published in *Rosen-Gepüsche* (1650) and *Poetische Rautengepüsche* (1663). He was a member of Die

Teutschgesinnte Genossenschaft (see SPRACH-GESELLSCHAFTEN) under the name Der Beschirmende.

Schlacht bei Göllheim, Die, a Middle High German poem written *c*. 1300 by an unknown author. It deals with the battle, fought in 1298, in which Adolf von Nassau (q.v.), the German King, lost his crown and life to Albrecht of Austria. The poet, who was a Rhinelander, sides with the defeated Adolf. Attempts to attribute authorship to CILIES von Seyn (q.v.) are regarded as doubtful.

SCHLAF, JOHANNES (Querfurt nr. Merseburg, 1862–1941, Querfurt), was at school in Magdeburg and studied at Halle and Berlin universities. He formed a close friendship with A. Holz (q.v.), and the two collaborated in the pioneering sketches *Papa Hamlet* (q.v., 1889) and the drama *Die Familie Selicke* (q.v., 1890). Holz and Schlaf, however, soon fell out, and the drama *Meister Oelze* (q.v., 1892) was exclusively Schlaf's work. The idyllic sketches *In Dingsda* (q.v., 1892) are the beginning of an impressionistic and ultimately mystical trend in Schlaf's work. After a period of mental illness he settled in Weimar in 1904. Shortly before the 1939–45 War he returned to Querfurt, where he spent the last four years of his life.

Later works include Novellen (*Sommertod*, 1897; *Der Tod des Antichrist*, 1901; *Frühlingsblumen*, 1902; *Die Nonne*, 1905), novels (*Das dritte Reich*, 1900; *Peter Boies Freite*, 1903; *Der Kleine*, 1904; *Der Prinz*, 1908; *Am toten Punkt*, 1909; *Aufstieg*, 1910), and plays (*Gertrud*, 1898; *Der Weigand*, 1906). Towards the end of his life Schlaf's mystical tendencies took a political turn towards National Socialism. *Ausgewählte Werke* (2 vols.) appeared in 1934.

Schlafende Heer, Das, a realistic novel by Clara Viebig (q.v.), published in 1904. It is concerned with contemporary political tensions in the province of West Prussia, former Polish territory acquired by Prussia in the Second Partition of Poland (1793, see POLAND, PARTITIONS OF). The principal figures are, firstly, a family of German colonists from the Rhineland, whose son marries a Polish girl, who deceives him with one of her compatriots and, secondly, a patriotic German squire, Herr von Doleschal, who seeks to accelerate the assimilation of the province to German ways and thought. The young man is deserted by his wife and disappears in a bog. Doleschal, resented by the Poles for his German nationalism, is waylaid and beaten up. Conceiving himself to be dishonoured, he shoots himself. The Rhenish colonists give up and return to the Rhineland, but Frau von Doleschal determines to remain with her young sons.

The book gives a convincing portrayal of the Polish inhabitants, by turns warmly affectionate and brutally violent, and a wide range of characters is exhibited on both sides, including notably the scheming priest Gorka, and Dudek, the visionary shepherd who waits for the day when 'the sleeping army' of the title will rise up and drive the invader out.

Schlafwandler, Die, a trilogy of novels by H. Broch (q.v.), published in 1930–2. It is composed of *Pasenow oder die Romantik 1888, Esch oder die Anarchie 1903*, and *Huguenau oder die Sachlichkeit 1918*; the first two of these are intelligible as independent novels in their own right. The whole trilogy is a study of ethical and social decline.

In *Pasenow oder die Romantik 1888*, Joachim von Pasenow, of ancient nobility and an officer in the Guards, becomes unexpectedly, by the death of his elder brother in a duel, heir to a great estate. He abandons his chosen military career and the girl he loves, and satisfies convention by marrying in his own rank. He maintains a close friendship with Bertrand, the representative of the rising commercial world. *Esch oder die Anarchie 1903*, set fifteen years later, portrays a highly competitive and aggressive environment of anarchic disorganization. Esch makes his way in this world, dominating in turn Bertrand and the trade unionist Martin. In *Huguenau oder die Sachlichkeit 1918*, Huguenau, a deserter, triumphs alike over Pasenow and Esch, vindicating the complete amorality of the new world.

SCHLAGETER, ALBERT LEO (Schönau, Baden, 1894–1923, nr. Düsseldorf), took part in active resistance against the French occupation of the Ruhr, was arrested, tried by French court martial, and shot. He is the central figure of the nationalistic play *Schlageter* (1933) by H. Johst (q.v.), which was frequently performed under the National Socialist regime.

SCHLEGEL, AUGUST WILHELM VON (Hanover, 1767–1845, Bonn), elder brother of Friedrich Schlegel (q.v.), came of a family with literary traditions; J. E. Schlegel (q.v.) was his uncle, and his father, J. A. Schlegel (q.v.), was a noted hymn and fable writer. A. W. Schlegel was at school at Hanover, and studied at Göttingen University from 1786 to 1791. He then took up a post as private tutor in Amsterdam, which he retained until 1795, afterwards moving to Jena, where he established contact with Schiller, to whose *Die Horen* (q.v.) he contributed.

In 1796 Schlegel began to lecture at Jena University and in the same year he married Caroline Böhmer (see SCHELLING, CAROLINE VON), a young widow. In 1798 he was appointed a professor at Jena. He fell out with Schiller in 1797, but kept on the right side of Goethe. In 1798 he began to expound his Romantic conception of literature (see ROMANTIK) in a new journal, *Das Athenäum* (q.v.), which he edited with his brother Friedrich. A. W. Schlegel was a gifted translator, whose verse rendering of Shakespeare (17 plays, 1797–1810, see SHAKESPEARE) became a household book. His *Spanisches Theater* (1803–9) and *Blumensträuße italienischer, spanischer und portugiesischer Poesie* (1804) were almost as successful. In 1801, though still resident in Jena, he shifted the focus of his activities to Berlin, where he repeatedly stayed, delivering between 1801 and 1804 a series of lectures supporting the Romantic movement, later published (ed. J. Minor) under the title *Vorlesungen über schöne Literatur und Kunst* (3 vols., 1884, reissued 1966). In 1804 he joined Mme de Staël (q.v.) at Coppet and accompanied her in the next few years on extensive journeyings through Europe from Sweden to Italy, and France to Russia. In 1808 Schlegel gave a course of lectures in Vienna which constitute his most important critical work (*Vorlesungen über dramatische Kunst und Literatur*, q.v., 1809–11). In 1812 he took service with the Swedish prince and general Bernadotte, as a French-speaking propagandist, and after 1815 he again joined Mme de Staël. Following her death in 1817 he accepted a chair of oriental languages at Bonn University and devoted the rest of his life to Sanskrit and Indian studies, as both scholar and translator. His patent of nobility was granted in 1815.

Apart from his achievements as a translator, A. W. Schlegel is important as the propagandist of the first Romantic school, presenting his own thought and that of his brother Friedrich in systematic and coherent form. His original work, consisting mainly of poems, especially sonnets (*Gedichte*, 1800), made little impact. A classical play *Ion* (1803), produced by Goethe at Weimar, is an adaptation of Euripides. He is also the author of a bitter satire directed against A. von Kotzebue (q.v.), *Ehrenpforte und Triumphbogen für den Theaterpräsidenten von Kotzebue* (1801). *Kritische Schriften* appeared in 1828 and *Sämmtliche Werke* (26 vols.), ed. E. Böcking, in 1846–8 (reissued 1971–2), *Kritische Schriften und Briefe* (7 vols.), ed. E. Lohner, in 1962–74.

SCHLEGEL, DOROTHEA VON (Berlin, 1763–1839, Frankfurt/Main), the eldest daughter of Moses Mendelssohn (q.v.), married S. Veit, a banker, in 1783. The marriage was dissolved in 1798, and in 1804 she married Friedrich Schlegel (q.v.), with whom she had lived for some years. She actively supported Schlegel in advancing the new conception of Romantic literature (see ROMANTIK), and herself wrote an unfinished novel (*Florentin*, 1801). She also translated French medieval romances (*Romantische Dichtungen des Mittelalters*, 1804) and made a German version of Mme de Staël's novel *Corinne ou l'Italie* (*Corinne oder Italien*, 4 vols., 1807–8). Like her husband she was converted to Roman Catholicism in 1808, and in 1815, as a consequence of his ennoblement, she became Dorothea von Schlegel. Living in Vienna from 1808 to 1830, she devoted herself mainly to the service of the Roman Catholic Church. She lived her last years in Frankfurt in the house of her son, the painter Philipp Veit (q.v.).

SCHLEGEL, FRIEDRICH VON (Hanover, 1772–1829, Dresden), in full Karl Wilhelm Friedrich Schlegel, younger brother of A. W. Schlegel (q.v.), was the son of the pastor J. A. Schlegel (q.v.) and a nephew of the critic and dramatist J. E. Schlegel (q.v.).

Intended for a commercial career, Schlegel was apprenticed to a banker in Leipzig before receiving parental permission to study. He was at Göttingen and Leipzig universities from 1790 to 1794, reading law before turning to classical studies. In 1796 he joined his brother A. W. Schlegel in Jena; both hoped that Schiller, not only the most notable literary figure but also professor of history at the university, might favourably influence their careers. Instead they soon alienated Schiller by unfavourable reviews of *Die Horen* (q.v.) and *Der Musenalmanach*. Friedrich Schlegel next moved to Berlin and, as a leading spirit of the new Romantic school (see ROMANTIK), contributed to various journals, including Wieland's *Der Teutsche Merkur* and his brother's *Das Athenäum* (qq.v.). He applied the extreme subjectivism of J. G. Fichte (q.v.) to literature.

Schlegel's earliest publications were essays on classical antiquity, including *Über die Diotima* (published in *Biesters Berlinische Monatsschrift*, 1795), *Über das Studium der griechischen Poesie* (1797), *Die Griechen und Römer* (1797), *Geschichte der Poesie der Griechen und Römer* (1798), and a provocative essay, *Über Lessing* (q.v., 1797). His, in its day, sensational novel *Lucinde* (q.v., 1799) reflects on his love for Dorothea Veit (see SCHLEGEL, DOROTHEA), with whom he spent, after a brief spell as a lecturer in Jena (1801), two years in Paris; he married her in 1804. The *Charakteristik der Meisterischen Lehrjahre von Goethe* (1798) interprets Goethe's novel from a Romantic standpoint. His tragedy *Alarcos* (q.v., 1802), when performed on the Weimar stage, was a complete failure. In 1808 Schlegel became a Roman Catholic and took service with the Austrian government, spending much of his later life in administration (including official participation in the Congress of Vienna (see WIENER KONGRESS) and diplomatic employment in Austrian service at the Frankfurt headquarters of the German Confederation, 1815–18). He was ennobled in 1815. In 1819 he visited Italy in the suite of Prince Metternich (q.v.). He founded and edited the conservative journal *Concordia* (1818–23) which displays, along with *Europa* (1803–5) and the *Deutsches Museum* (1812–13), his journalistic acumen. He died of a stroke in Dresden, where he had gone to lecture on *Philosophie der Sprache und des Wortes*.

Schlegel's creative works are eccentric and negligible, but his critical writings, both the essays and the aphorisms and *aperçus* which he termed *Fragmente* (q.v., 1796–1801), are brilliant, provocative, and fertile, and he not only exercised a considerable influence on the Romantics, but used his knowledge of the literature of different ages and peoples to promote German approaches to literary history and criticism. He contributed to German oriental studies with his treatise *Über die Sprache und Weisheit der Indier* (1808).

F. Schlegel published his *Sämmtliche Werke* (10 vols., containing revisions) in 1822–3 and in extended form (15 vols.) in 1845. The *Kritische Friedrich-Schlegel-Ausgabe*, ed. E. Behler *et al.* appeared in 1958 ff. Correspondence includes *Friedrich Schlegels Briefe an seinen Bruder August Wilhelm*, ed. O. F. Walzel (1890), and *Krisenjahre der Frühromantik. Briefe aus dem Schlegelkreis* (3 vols.), ed. J. Körner, in 1936–58.

SCHLEGEL, JOHANN ADOLF (Meißen, 1721–93, Hanover), younger brother of J. E. Schlegel and father of A. W. and F. Schlegel (qq.v.), taught at Schulpforta (see FÜRSTENSCHULEN), became pastor at Zerbst (1754) and later moderator (Generalsuperintendent) at Hanover (1759). He wrote hymns (*Sammlung geistlicher Gesänge*, 1766–72) and didactic poetry (*Fabeln und Erzählungen*, 1769). He also translated *Les Beaux-arts réduits à un même principe* (1746) by the Abbé Batteux under the title *Einschränkung der schönen Künste auf einen einzigen Gegensatz* (1751).

SCHLEGEL, JOHANN ELIAS (Meißen, 1719–49, Sorø, Denmark), brother of J. A. Schlegel and uncle of A. W. and F. Schlegel (qq.v.), was at Schulpforta (see FÜRSTENSCHULEN) with Klopstock (q.v.), and then went up to Leipzig University, where he became an adherent of J. C. Gottsched (q.v.), contributing to *Die deutsche Schaubühne* (q.v.) and *Beyträge zur*

critischen Historie der deutschen Sprache, Poesie und Beredsamkeit. In 1743 he became secretary to the Saxon Minister in Copenhagen and in 1748 a professor at the noblemen's school (Ritterakademie) at Sorø. While in Denmark he contributed to the *Bremer Beiträge* (q.v.).

In his short life Schlegel was extremely active both as the author of plays and as a critic, writing at least fourteen plays, including eight tragedies in the French classical manner favoured by Gottsched, and six comedies, which, if they are more independent, still recall the French pattern. Schlegel's principal tragedies are *Canut* (q.v., 1746), in which the king is conceived as a benevolent despot, *Hekuba* (1736), after Euripides, later revised as *Die Trojanerinnen*, and the patriotic *Hermann* (1743). The outstanding comedies are *Der Geheimnisvolle* (1747), *Der Triumph der guten Frauen* (1746), and *Die stumme Schönheit* (1747). They were praised by Lessing in *Die Hamburgische Dramaturgie* (q.v.), where he describes the last as 'unstreitig unser bestes komisches Original' (13. Stück).

Schlegel is more notable for his critical and theoretical work. His *Vergleichung Shakespears und Andreas Gryphs* (1741, ed. H. Powell, 1964) is one of the first German endeavours to interpret Shakespeare (q.v.) sympathetically. The *Abhandlung von der Nachahmung* (1742–5) anticipates some of the views of the Abbé Batteux (1713–80), whose *Les beaux-arts* (1746) was translated in 1751 by Johann Adolf Schlegel. The *Gedanken zur Aufnahme des dänischen Theaters* (1747) reacts against Gottsched's didacticism, presses for national subjects, and rejects the class distinction between the characters of tragedy and those of comedy, accepted since the 17th c.

Theatralische Werke (5 vols.), ed. J. H. Schlegel, appeared 1761–70 (reissued 1971) and *Ästhetische und dramaturgische Schriften*, ed. J. von Antoniewicz, 1887 (reissued 1970).

Schleier, Der, a Novelle by E. Strauß (q.v.), published in 1920. In 1931 it was included in a collection entitled *Der Schleier*. Freiherr von Tettingen, a happy family man with a devoted wife, is bewitched by the beautiful young Countess Ittendorf; yielding to temptation the two meet regularly at his hunting lodge. Freifrau von Tettingen, suspecting unfaithfulness, visits the lodge at night and finds the lovers asleep together. The sight of the peaceful couple dispels jealousy and anger; she leaves quietly, having covered them with her veil. The Countess and Tettingen awake, recognize the situation, and part for ever. Tettingen finds forgiveness with his wife, to whom he returns the veil which had been his first present for her.

The theme of the veil resumes a motif from one of Bassompierre's tales adapted by Goethe in *Unterhaltungen deutscher Ausgewanderten* (q.v.). Strauß's delicate presentation is executed in the classical manner.

SCHLEIERMACHER, FRIEDRICH DANIEL ERNST (Breslau, 1768–1834, Breslau), ranks as the most important Protestant theologian of the Romantic movement (see ROMANTIK). Reacting against his strict Pietist upbringing (see PIETISMUS), Schleiermacher devoted himself during his stay at Halle University to theology and to the study of Plato, Spinoza (q.v.), and Kant (q.v.). He was ordained in 1794, and in 1796 moved to Berlin, where he enjoyed contacts with the early Romantics, among them F. Schlegel (q.v.) whose novel *Lucinde* (q.v.) inspired his *Vertraute Briefe über Lucinde* (1800).

In 1799 Schleiermacher published an ardent work of Christian apologetics, *Reden über die Religion*. While pastor in Stolpe in Pomerania (1802–4) he published his first philosophical work, *Grundlinien einer Kritik der bisherigen Sittenlehre* (1803), comparing favourably the systems of Plato and Spinoza with those of Kant and Fichte (q.v.). From 1804 Schleiermacher occupied the chair of theology at Halle University, but dedicated himself with equal vigour to preaching; in 1806 he published the dialogue *Weihnachtsfeier*. Political events, which brought about the dissolution of the university (see NAPOLEONIC WARS), caused him to move to Berlin in 1807.

He continued his dual activity as scholar and as practical reformer of the Church on an even larger scale as pastor of the Dreifaltigkeitskirche. He helped with the founding of Berlin University in 1810, became its first professor of theology, and was elected Secretary to the Academy of Sciences. His sermons were esteemed for their sincerity and religious fervour as well as, at this time of national depression, for their patriotism. In his work for reform of the Church his plea for a union of the Lutheran and Reformed Churches in Prussia ranked high. During the restoration he published his principal work, *Der christliche Glaube nach den Grundsätzen der evangelischen Kirche* (1821–2, revised 1830–1). He defended his Christology in letters to his friend Lücke in *Studien und Kritiken* (1829).

Schleiermacher influenced Romanticism principally by his definition of religion in the *Reden* of 1794 propounding the thesis that emotional experience (Gefühl) forms the basis of religion, a conviction to which he adhered despite modifications contained in his later works. He considered creeds to be the expression rather than the foundation of religious experience, which should aim at a union with the infinite. In the third of the five speeches he applies his principle of the primacy of feeling,

coupled with contemplation and intuition (Anschauung), to art and poetry. The last two speeches discuss the historical function of the Church in relation to natural religion and to non-Christian religions. *Der christliche Glaube nach den Grundsätzen der evangelischen Kirche* sums up Schleiermacher's Christocentric view linking the historical manifestation of Christianity with the conception that Christ is the centre of an inner religious consciousness. Schleiermacher's theology has not lost its influence, though it was opposed by the Swiss K. Barth (q.v.) and E. Brunner (1889–1966), and is evident in many works of literature. *Das Leben Schleiermachers* by W. Dilthey (q.v.) was published in 1870 (rev. edn. M. Redeker, 1970).

Sämtliche Werke (31 vols.) appeared 1835–64.

SCHLENKERT, FRIEDRICH CHRISTIAN (Dresden, 1757–1826, Tharandt nr. Dresden), a Saxon civil servant and teacher, was the author of plays (*Agathon und Psyche*, 1779; *Kein Faustrecht mehr*, 1798) and of a large number of extravagant historical novels (see RITTER- UND RÄUBERROMANE). Two of these works are novels in dialogue, *Friedrich mit der gebissenen Wange* (1785–8) and *Kaiser Heinrich IV* (1789–95).

Schlesische Dichterschulen, Erste und Zweite, designation for two groups of poets active in Silesia in the 17th c., which has been discarded because neither set of poets formed a school in the proper sense of the term. The First Silesian school included Martin Opitz, Andreas Gryphius, Johannes Heermann, Daniel Czepko, and Andreas Scultetus, together with the non-Silesians Paul Fleming, Andreas Tscherning, and Johann Peter Titz (qq.v.). It was customary to reckon that the Second Silesian school was composed of the following: Christian HOFFMANN von Hoffmannswaldau, Daniel Casper von Lohenstein, Johann von Besser, Friedrich von Canitz, and even Johann Christian Günther (qq.v.).

Schlesische Kriege, the wars resulting from the seizure of Silesia by Friedrich II (q.v.) of Prussia.

Erster schlesischer Krieg. Friedrich II, succeeding to the Prussian throne in the year that Maria Theresia (q.v.) succeeded in Austria (1740), exploited the disputed Pragmatic Sanction (see PRAGMATISCHE SANKTION), which guaranteed her succession, by offering his support in exchange for Lower Silesia, to which he had no title. He responded to her refusal by invading Silesia on 16 December 1740. After his victory at Mollwitz (April 1741), other powers (France, Spain, Bavaria, Saxony) joined him against Maria Theresia. Suspicious of French designs, Friedrich concluded behind their backs the secret Treaty of Klein-Schnellendorf with Maria Theresia, by which he hoped to secure Silesia without further military action. But when Karl of Bavaria (see KARL VII) invaded Bohemia, Friedrich went back on the treaty in order to occupy Moravia without Bavarian opposition. He captured Olmütz (December 1741), but was forced to abandon his Moravian campaign by an Austrian advance towards Prague endangering Silesia. He was, however, able to defeat the Austrians at Chotusitz (May 1742).

Through British intervention Maria Theresia and Friedrich concluded the Preliminaries of Breslau and the Peace of Berlin (July 1743), by which Friedrich, again deserting his allies, withdrew from the war, keeping Lower Silesia and the County of Glatz. August III (q.v.) of Saxony and Poland also concluded peace with Austria.

Zweiter schlesischer Krieg. The Second Silesian War was in effect fought to defend the gains Friedrich had made against fresh Austrian attempts to recover them. Alarmed at Maria Theresia's successes in the War of the Austrian Succession (see ÖSTERREICHISCHER ERBFOLGE-KRIEG), Friedrich entered into an agreement with France (Versailles, June 1744) and Bavaria, and invaded Bohemia, but was forced to retire owing to Saxon support of the Austrian invasion of Silesia. August III hoped to acquire a strip of Silesia which would link Saxony with Poland. Friedrich II repelled the invasion by his victory at Hohenfriedberg (q.v., June 1745); having failed to advance into Bohemia, he turned upon the Saxons before their Russian support arrived. After victories at Groß (or Katholisch) Hennersdorf (November 1745) and Kesselsdorf, he entered Dresden in December.

Through the mediation of George II (q.v.) in the Convention of Hanover, Friedrich and Maria Theresia concluded the Treaty of Dresden terminating the war (December 1745). Friedrich retained Silesia, but Maria Theresia did not resign herself to its permanent loss.

Dritter schlesischer Krieg. The Third Silesian War was fought on this issue twelve years later and is commonly referred to as the Seven Years War (see SIEBENJÄHRIGER KRIEG).

Schleswig-Holsteinische Frage. The political status of the two northern duchies came in the middle of the 19th c. to a crisis, which was resolved in 1867 by Prussian annexation. The following is an outline of their complex historical background.

Although Schleswig and Holstein were originally separate duchies, an agreement was made in the 14th c. that, while retaining their separate identities, they should remain permanently in union. Over the centuries they underwent a series of changes of sovereignty through inheritance, and the union at times was only nominal. In 1806 the Danes claimed the whole of both duchies. The situation was not helped by the language question (the northern part of Schleswig being Danish-speaking) and by the fact that the Danish court was almost as much German as it was Danish. In the settlement made in 1815, Holstein was included in the German Confederation (see DEUTSCHER BUND), but not Schleswig. Difficulties of succession arose in the 1830s and 1840s with the imminent extinction of the Danish male line, and the impossibility, in accordance with the Salic law (see SALISCHES GESETZ) of a woman ruling over a German state. In 1848, at the time of the Revolution (see REVOLUTIONEN, 1848-9), the Danish hold on the duchies was strengthened, and Schleswig virtually incorporated in Denmark. This led one of the German claimants to the ducal thrones, the Duke of Augustenburg (see AUGUSTENBURG, HERZOG VON), to ask for Prussian help. The German Confederation ordered federal execution (Bundesexecution) and Prussian troops marched in. The great powers, however, threatened war, and the Prussian troops marched out again. In the Punctation of Olmütz (see OLMÜTZER PUNKTA-TION) Prussia gave way to Austria. In the London protocols of 1850 and 1852 the integrity of Danish possessions was affirmed and the claim of Prince Christian von Glücksburg to the crown and adjoining lands recognized, in spite of his descent through the female line. The matter was now further bedevilled by the rapid rise in Europe of aggressive nationalism. In 1863 Frederick VII of Denmark published a new constitution which strengthened the Danish hold on Schleswig, without relinquishing his claims on Holstein. At this time Bismarck (q.v.) was in firm control of Prussian policy and he immediately perceived possibilities favourable to Prussia. Disregarding the machinery of the German Confederation, he made an agreement with Austria to take military action to settle the Schleswig-Holstein question, and an ultimatum was sent to Denmark on 16 January 1864, which was rejected. The ensuing campaign ended in the defeat of the Danes, though heavy losses were sustained by the Prussians at Düppel. Jutland was invaded and Denmark sued for terms.

In the Peace of Vienna, Denmark abandoned all claims to the duchies and, by the Convention of Gastein, Prussia and Austria set up a joint rule, Bismarck arranging for Prussian occupation of Schleswig and Austrian occupation of Holstein, so that in any future war the Austrian garrison of Holstein would have to fight on two fronts. War did ensue (see DEUTSCHER KRIEG), and the incorporation of Schleswig-Holstein in Prussia was one of Bismarck's aims.

The Austrian defeat in Bohemia in 1866 led to the Peace of Prague (23 August 1866), and included an article, effective from 24 January 1867, annexing Schleswig-Holstein to Prussia. The claim of the Duke of Augustenburg, to whom Prussia was committed, was rejected by Bismarck, in spite of remonstrance by his sovereign. Prussia, and eventually the German Empire of 1871, acquired a territory containing a Danish minority, which the treaty disregarded. By the Treaty of Versailles (1919) a plebiscite was held, and the northern part of Schleswig as far south as Tondern and Düppel was included in Denmark.

The stresses and conflicts involved in this matter are particularly perceptible in the life and work of Th. Storm (q.v.).

Schlieffen-Plan, plan of campaign on two fronts conceived by Alfred, Graf von Schlieffen (1833–1913), chief of the German General Staff from 1891 to 1906. The plan was for offensive war in the west, defensive war in the east, with a corresponding concentration on the western front. A massive advance on the right through Belgium and possibly Holland, while the left remained stationary, was to swing west and south in order to encircle the French armies in north-eastern France. In August 1914 the plan, put into action with modifications made by H. von Moltke (q.v., 'der jüngere Moltke'), failed.

'Schleswig-Holstein, meerumschlungen', see CHEMNITZ, M. F.

Schleswigsche Literaturbriefe, designation often used for H. W. von Gerstenberg's *Briefe über Merkwürdigkeiten der Literatur* (q.v., 1766–70).

SCHLIEMANN, HEINRICH (Neubukow, Mecklenburg, 1822–90, Naples), became an office boy at 14 and was employed in Amsterdam for several years. In 1847 he set up in business in St. Petersburg and in 1852 extended his activities to Moscow. Schliemann, who taught himself a number of languages, including classical Greek, used the substantial fortune he had made in trade to carry out archaeological excavations in Greece and Asia Minor. In 1868 he settled in Athens and over a period of years unearthed the site of Troy (1870–82). He also made important digs in Mycenae (1874) and Tiryns (1884). Though some of his conclusions were based on a too literal reading of Greek poets, Schliemann made some important dis-

coveries, retrieved valuable treasures, and by his enthusiasm, energy, and flair gave considerable impetus to archaeological exploration. His published works include *Ithaka, der Peloponnes und Troja* (1869), *Trojanische Altertümer* (1874), *Troja und seine Ruinen* (1875), *Mykenä* (1878), *Ilios* (1881), *Orchomenos* (1881), *Troja* (1884), and *Tiryns* (1886). An impression of contemporary reaction to Schliemann is given in Fontane's novel *Frau Jenny Treibel* (q.v., ch. VI). His autobiography (*Selbstbiographie*), completed by A. Brückner, ed. by S. Schliemann, appeared in 1892 (revised edn. by E. Meyer, 1968). Correspondence covering the years 1842–90 (2 vols.), ed. E. Meyer, was published in 1953 and 1958.

Schlimmen Monarchen, Die, a poem by Schiller which owes its inspiration to C. F. D. Schubart's *Die Fürstengruft* (q.v.). It was first published in *Die Anthologie auf das Jahr 1782.*

SCHLÖGL, FRIEDRICH (Vienna, 1821–92, Vienna), was an Austrian civil servant until 1870 and thereafter a Viennese journalist. He earned a reputation as a humorist by sketches such as *Wiener Blut* (1873), *Aus Alt- und Neu-Wien* (1882), and *Wien* (1886). Most of these originally appeared in weekly instalments. *Gesammelte Schriften* (3 vols.) appeared in 1893.

Schloß, Das, an unfinished novel written by Franz Kafka (q.v.) in 1922 and published by Max Brod (q.v.) in 1926. Brod prepared a dramatization in 1953.

K. comes in winter as surveyor into a mountain village dominated by a castle. He seeks to establish himself in the village and to enter into relations with the castle. But the village regards him with suspicion, and all efforts to make contact with the castle fail. His attempts at human relationships (e.g. liaison with Frieda, whom he believes to be the mistress of Klamm, an official of the castle) also miscarry. The castle remains intangible, possibly even unoccupied. Though the story has no conclusion, it seems certain that K. will never obtain entry to the castle. The surrealist, dream-like narrative, with abundant crass realistic detail and a compelling prose style, is a negative parable of human life and human aspiration, though in matters of detail interpretations diverge.

Schloß Boncourt, Das, a poem by Adalbert von Chamisso (q.v.), written in 1827 and first printed in an appendix to the second edition of *Peter Schlemihls wundersame Geschichte* (q.v.). It is a nostalgic recollection of the French château of his ancestors, in which he was born and spent his early childhood. Château Boncourt, destroyed in 1790, was situated in Champagne.

Schloß Dürande, Das, a story by J. von Eichendorff (q.v.), published in Brockhaus's *Urania* in 1837. Set in southern France at the time of the beginning of the French Revolution (q.v.), it is a story of seduction. The young Count Dürande wins the affections of Gabriele, sister of Renald, his huntsman. Renald pursues the Count to obtain satisfaction, but is at first unsuccessful. He leads an attack on the Count's castle, but his sister and the Count perish together. Renald sets fire to the castle, which blows up. An operatic version by Othmar Schoeck (1886–1957) was first performed in Berlin in 1943.

SCHLOSSER, CORNELIA, see GOETHE, CORNELIA.

SCHLOSSER, JOHANN GEORG (Frankfurt/Main, 1739–99, Frankfurt), practised as a lawyer in Frankfurt and then took a post as magistrate at Emmendingen in Baden. In 1773, immediately before taking up this appointment, he married Goethe's sister Cornelia (see GOETHE, CORNELIA), who died in 1777. Cornelia and her husband befriended J. M. R. Lenz (q.v.) in 1775. In 1778 Schlosser married Johanna Fahlmer (1744–1821). He obtained a higher appointment in Karlsruhe and retired in 1794.

Schluck und Jau, a comedy in prose and verse (Scherzspiel) and in six 'occurrences' (Vorgänge) by G. Hauptmann (q.v.), published in 1900. The sub-title *Ein Scherzspiel in sechs Vorgängen* was substituted by the author for the original *Spiel zu Scherz und Schimpf mit fünf Unterbrechungen.* The play is an adaptation of the Induction to Shakespeare's *Taming of the Shrew.*

SCHLÜSSELFELDER, HEINRICH, see ARIGO.

Schlüsselroman, a novel in which real persons are concealed under fictitious names. The form, which is usually denoted in English by the French *roman à clef,* is as old as the novel itself, and it is probable that almost all novels contain some characters closely based on the author's acquaintance. The Schlüsselroman attracts most attention when prominent personalities are involved, as in H. Mann's *Der Kopf* (q.v.). Many examples are to be found in the 17th c. (see ZESEN, LOHENSTEIN, and ANTON ULRICH, HERZOG VON BRAUNSCHWEIG). Much of the fiction of Th. Mann and Schnitzler (qq.v.) belongs to this category, though the key is not always available. Similarly, Schlüsselstücke exist in drama.

SCHLÜTER, ANDREAS (Homburg, 1664–1714, St. Petersburg), baroque sculptor and architect,

was active in Berlin and Potsdam (1694–1706); he was dismissed by Friedrich I (q.v.) of Prussia after the collapse of the Münzturm. He was the architect of the royal palace in Berlin (now destroyed).

Schmalkaldischer Bund, the League of Schmalkalden was formed in 1531 for the defence of the Protestant religion (see REFORMATION) on the basis of an agreement reached on 31 December 1530 at Schmalkalden near Meiningen between the Elector of Saxony, Johann der Beständige (reigned 1525–32) and his son Johann Friedrich (reigned 1532–47) and the Landgrave of Hesse, Philipp I, der Großmütige (reigned 1509–67, born 1504). Bremen and Magdeburg joined the League, which was renewed for twelve years in 1535, when it was reinforced by other Protestant states and cities, including Anhalt, Pomerania, Württemberg, Augsburg, Frankfurt, Hanover, Hamburg, and Kempten.

The League pursued a policy of resistance against the Emperor Karl V (q.v.) and his supporters, but a fatal weakness lay in its many-headed character. Karl found an opportunity, after the Peace of Crépy in 1544, to take action against the League, alleging as grounds disobedience to imperial decrees. At the Diet of Regensburg in 1546 Karl placed Saxony and Hesse under an interdict which resulted in war (see SCHMALKALDISCHER KRIEG).

Schmalkaldischer Krieg, the war between the Schmalkaldischer Bund (q.v.) and the Emperor Karl V (q.v.) began with cautious and desultory manœuvrings by both sides in the region of the middle Danube in 1546. It was decided by the victory of the imperial forces at the battle of Mühlberg in 1547. Karl defeated Johann Friedrich of Saxony, who was taken prisoner and deprived of his electorate, which he never recovered. Landgrave Philipp of Hesse surrendered and was also imprisoned. Karl V successfully exploited the ambition of Moritz of Saxony (see MORITZ VON SACHSEN), who deserted the League and yet failed to secure all the advantages for which he had hoped. The League, deprived of its leaders, ceased to exist.

SCHMELTZL, WOLFGANG (Kemnat, Palatinate, c. 1505–d. after 1557), became a schoolmaster in Vienna and later (c. 1550) parish priest of St. Lorenz, Steinfeld (south of Vienna). In middle life he seems for a time to have been a Lutheran. Schmeltzl wrote biblical plays of the kind performed in schools, which he combined with more popular elements. Seven are preserved: *Comoedia des verlorenen Sohnes* (1540, based on Gnapheus, q.v.), *Aussendung der* *zwelf poten, Judith* (both 1542), *Die Hochzeit zu Cana* and *Der blindgeborene Sohn* (both 1543), *David und Goliath* (1545), and *Samuel und Saul* (1551). He also wrote a poem in praise of Vienna (*Lobspruch der Stadt Wien*, 1548), a poetic account of the Archduke Ferdinand's campaign in 1556 against the Turks (*Der Zug in das Hungerland*, 1556), and songs. A selection of his works, ed. E. Triebnigg, appeared in 1915 as *Wolfgang Schmeltzl. Auswahl aus seinen Schriften.*

Schmetterlingsschlacht, Die, a satirical comedy (Komödie in vier Akten) by H. Sudermann (q.v.), published in 1895. Frau Hergentheim lives in genteel poverty with her three marriageable daughters. Else and Laura spend their time scheming for husbands, and the entire family is supported by Rosi, the youngest, who paints fans and sells them. One of these depicts a battle of butterflies, and from this the symbolical title is derived. Else and Laura overreach themselves and Cinderella-Rosi makes a satisfactory match.

SCHMID, CHRISTOPH VON (Dinkelsbühl, 1768–1854, Augsburg), a Roman Catholic priest and schoolmaster, wrote abundantly for children. His works include *Biblische Geschichte für die Kinder* (1801) and an unfinished autobiography (*Erinnerungen aus meinen Leben*, 1853–7). His poem 'Wie lieblich hallt durch Busch und Wald' (1816) has become well known as a song with a melody by F. Silcher (q.v.), and he is the author of the Christmas carol 'Ihr Kinderlein, kommet' (*c.* 1850). Schmid's preposition of nobility dates from 1837. *Gesammelte Schriften* (26 vols.) appeared 1841–56, *Erinnerungen und Briefe*, ed. H. Pörnbacher, 1968.

SCHMID, FERDINAND VON (Muri nr. Berne, 1823–88, Berne), pseudonym Dranmor, son of a Swiss banker, spent much of his life in commerce in Brazil, where he became consul-general for Austria-Hungary. His *Poetische Fragmente* appeared in 1860, his *Requiem* in 1869, and his *Gesammelte Dichtungen*, which first attracted public attention to him, in 1873. Schmid's unhappy and tormented character is reflected in his poetry.

SCHMID, HERMANN THEODOR (Weizenkirchen, Austria, 1815–80, Munich), became a civil servant in Munich, but was retired for political reasons after the 1848 Revolution. In middle and later life he was a prolific author, especially well known for his Bavarian stories (*Almenrausch und Edelweiß*, 1864; *Alte und neue Geschichten aus Bayern*, 1861; *Bayerische Geschichten aus Dorf und Stadt*, 1864). He also wrote novels (*Der Kanzler von Tirol*, 1862; *Im Morgenrot*, 1864; *Friedel und*

Oswald, 1866; *Mütze und Krone*, 1869), a number of plays, including *Camoens* (1843), *Karl Stuart* (1853), *Kolumbus* (1875), *Rose und Distel* (1876), and the Volksstücke (see VOLKSSTÜCK) *Der Tatzelwurm* (1873), *Die Z'widerwurz'n* (1878), *Der Stein der Weisen* (1880), and *Der Loder* (1880).

Schmid was a regular contributor to *Die Gartenlaube* (q.v.); he was ennobled in 1876. His *Gesammelte Schriften* (1867–71) and *Gesammelte Schriften Neue Folge* (1881–4) occupy 50 volumes.

SCHMIDT, ARNO (Hamburg, 1914–79, Celle), studied mathematics and astronomy, entered the textile business, and served in the German artillery, and was taken prisoner in 1940. After a short period as an interpreter after his release in 1945, he turned in 1947 to authorship, including translations of works by E. A. Poe, J. F. Cooper, P. Fleming, and others. A mind of originality and ingenuity, he wrote stories and novels which are exceptionally experimental in narrative technique, in orthography, in temporal concentration, and in the representation of consciousness.

Schmidt's work consists partly of short fiction of which the stories of *Brand's Haide, Schwarze Spiegel* (both 1951), and *Aus dem Leben eines Fauns* (1953) were published as the trilogy *Nobodaddys Kinder* (1963). Other titles include *Leviathan* (1949), *Die Umsiedler* (1953), *Das steinerne Herz* (1956), *Die Gelehrtenrepublik* (1957), *Kaff auch Marc Crisium* (1960), and the collection of short stories, *Sommermeteor* (1969). *Die Schule der Atheisten* (1972) is described as a six-act Novellen-Comödie. The many essays written for radio include *Dya na sore. Gespräche in einer Bibliothek* (1958), *Belphegor. Nachrichten von Büchern und Menschen* (1961), and *Der Titron mit dem Sonnenschirm. Großbritannische Gemütsergötzungen* (1969), Titron referring to James Joyce; other writers under discussion include the Brontë sisters, Cooper, Wilkie Collins, and Dickens. *Sitara und der Weg dorthin* (1963) is a study of Karl May (q.v.). Among his work on known and not so well known 18th- and 19th-c. writers special mention should be made of his thorough study of *Fouqué und einige seiner Zeitgenossen* (1958) and *Nachrichten von Büchern und Menschen* (1971).

In the late 1950s Schmidt settled in the seclusion of Bargfeld in the Lüneburger Heide. In his work on theory, *Berechnungen I und II* (1959), he deliberates on the function of the brain and its relevance to experimental writing, to which his essays in Novelle form, *Zettels Traum* (1970), are his most idiosyncratic contribution. This work occupies more than 1,300 pages in triple column pursuing simultaneously within the space of twenty-four hours three processes of thought; in the left-hand column Poe figures as the pacemaker, and the juxtaposition serves as a radical demonstration of the interaction of subconscious and conscious phenomena. Schmidt was a humanist as well as a man of wit.

SCHMIDT, ERICH (Jena, 1853–1913, Berlin), a noted Germanist, studied at Graz, Jena, Strasburg, and Würzburg universities and was successively professor at Strasburg (1877), and Vienna (1880), Director of the Goethe-Schiller-Archiv at Weimar (1885), and professor at Berlin in 1887. Schmidt is best known for his *Lessing* (1884–92) and his publication of *Urfaust* (*Goethes Faust in ursprünglicher Gestalt*, 1887), but he was active in many fields of German literature. His other principal publications are *Reinmar von Hagenau und Heinrich von Rugge* (1874), *Richardson, Rousseau und Goethe* (1875), *Goethes Jugendgenosse H. L. Wagner* (1875), *Lenz und Klinger* (1878), and *Charakteristiken* (2 vols., 1886–1901). His *Reden zur Literatur- und Universitätsgeschichte* appeared in 1911. Volumes of correspondence were published in 1963 (with W. Scherer, q.v.) and in 1972 (with Th. Storm, q.v.).

SCHMIDT, FRIEDRICH WILHELM (Schmidt von Werneuchen) (Fahrland nr. Potsdam, 1764–1838, Werneuchen nr. Berlin), a Prussian pastor, whose homespun poetry aroused Goethe's criticism in his poem *Musen und Grazien in der Mark.* Goethe later treated him more leniently in *Maximen und Reflexionen,* 972. There is a brief discussion of Schmidt's poetry in Fontane's *Vor dem Sturm* and also in Bk. 4 of *Wanderungen durch die Mark Brandenburg* (qq.v.). Between 1793 and 1797 Schmidt edited *Der neue Berliner Almanach* and in 1796–7 published *Kalender der Musen und Grazien.* His own poems appeared as *Gedichte* (1797) and *Neueste Gedichte* (1815).

SCHMIDT, GEORG PHILIPP (Lübeck, 1766–1849, Ottensen nr. Hamburg), usually known as Schmidt von Lübeck, studied first law then medicine. In 1794, while in Jena, he met Herder, Goethe, Schiller, and Wieland (qq.v.), through the mediation of Sophie Mereau (q.v.). He practised for a time as a physician in Lübeck, entered Danish service as a bank director, was appointed Danish Justizrat in 1816, and retired in 1829. His poetry, which was first collected in 1821 (*Gedichte*), includes the poem *Der Wanderer* (set to music by F. Schubert, q.v.), which in its last lines epitomizes one aspect of Romantic psychology: 'Dort, wo du nicht bist, / Dort ist das Glück.'

SCHMIDT, JAKOB FRIEDRICH (Zella St. Blasii nr. Gotha, 1730–96, Gotha), pastor in Gotha, wrote poetry which was published in *Sammlung vermischter Gedichte* (1758), *Kleine poetische Schriften*

(1766), and *Gedichte* (1786). A volume of religious poetry appeared in 1759 (*Poetische Gemälde und Empfindungen aus der Heiligen Schrift*) and a collection of hymns (*Sammlung einiger Kirchenlieder*) in 1779.

SCHMIDT, JULIAN (Marienwerder, East Prussia, 1818–86, Berlin), at first a schoolmaster, took to journalism in 1847, editing *Der Grenzbote,* in which he was joined in 1848 by G. Freytag (q.v.). In 1878 he received official recognition in the form of a royal pension. He was a literary historian of note (*Geschichte der deutschen Literatur von Leibniz bis auf unsere Zeit,* 5 vols., 1886–96) and also published *Bilder aus dem geistigen Leben unserer Zeit* (4 vols., 1870–5). An earlier version of his history of literature was attacked in 1862 by F. Lassalle (q.v.). G. Freytag placed a quotation from Schmidt as motto on the title-page of his novel *Soll und Haben* (q.v.).

SCHMIDT, MARIE SOPHIE (Langensalza, 1731–99, Eisenach), a cousin of the poet F. G. Klopstock (q.v.), who fell in love with her in 1748, when he was a private tutor in Langensalza. She is the Fanny of his odes. His love was not returned, and she married in 1754 a wealthy businessman, Johann Justinus Streiber of Eisenach.

SCHMIDTBONN, WILHELM, name assumed by Wilhelm Schmidt of Bonn (Bonn, 1876–1952, Bad Godesberg). He studied at the universities of Bonn, Göttingen, and Zurich, worked for a time as dramatic adviser (Dramaturg) to the municipal theatre at Düsseldorf, and then concentrated on authorship, living for a time in Bavaria, and later in his native district. During the 1914–18 War he was a war reporter.

Schmidtbonn began to write as a Naturalist (see NATURALISMUS), but quickly developed into a regional novelist and story-teller and a dramatist dealing with legendary as well as contemporary themes. The play *Mutter Landstraße* (1901) is a modern version of the story of the Prodigal Son, which, however, ends tragically. The novel *Der Heilsbringer* (1906) recounts the conversion of a young working-class radical to the belief that man's salavation lies in a return to rural life. The play *Der Graf von Gleichen* (1908) gives its legend a tragic conclusion. *Der Zorn des Achilles* (1909) was a venture into verse tragedy, and in *Lobgesang des Lebens* (1911) the author attempted rhapsodic poetry. Other plays are the comedies *Der spielende Eros* (1911) and *Der Schauspieler* (1921), and the Schauspiele *Stadt der Besessenen* (1915), *Der Geschlagene* (1920), and *Der Pfarrer von Mainz* (1923). His later work consists mainly of novels and stories of the Rhineland, written with knowledge and sympathy. The novels include *Mein Freund Dei* (1928), *Der dreieckige Marktplatz* (1935), *Hü Lü* (1936), and *Anna Brand* (1939); *Uferleute* (1903) and *Die unberührten Frauen* (1925) are collections of tales. The short narratives of *Der Wunderbaum* (1913) are described as legends (*Legenden*). *An einem Strom geboren* (1935) is his autobiography.

SCHMIDT-ROTTLUFF, KARL (Rottluff nr. Chemnitz, 1884–1976, West Berlin), studied architecture in Dresden from 1905, in which year he became a founder member of Die Brücke (q.v.). In 1911 he moved to Berlin, teaching at the Akademie from 1947 until his retirement. His woodcuts are a striking feature of his output, which includes the graphic arts as well as painting. In 1974, the year of his ninetieth birthday, he was made an honorary member of the American Academy of Arts and Letters and the American National Institute of Arts and Letters for his pioneering work in Expressionist art.

Schmied, Der, a short poem by L. Uhland (q.v.), written in 1809. It tells how the smith's forge glows as his love passes by. The poem has been set by Brahms (q.v.).

Schmied seines Glückes, Der, see LEUTE VON SELDWYLA, DIE.

SCHNABEL, ERNST (Zittau, 1913–), well known since the 1939–45 War as a leading figure in North German and West Berlin radio and television, went to sea at 18 and saw many parts of the world. His prolific writings, which include works of fiction, provide a conspectus of his adventures and their relevance to the contemporary scene.

Schnabel began with the novels *Die Reise nach Savannah* (1939) and *Nachtwind* (1942) and the stories *Schiffe und Sterne* (1943) and *Sie sehen den Marmor nicht* (1949), his first post-war work of fiction. The motif of Odysseus is central to the novel *Der sechste Gesang* (1956); *Ich und die Könige* (1958) was turned into a radio play in 1960. His stories include *Fremde ohne Souvenir* (1961), *Hurricane* (1972), and *Auf der Höhe der Messingstadt* (1979). He is the author of the libretto for *Das Floß der Medusa* by H. W. Henze (q.v., 1968).

SCHNABEL, JOHANN GOTTFRIED (Sandersdorf nr. Bitterfeld, 1692–c. 1752, Stolberg), was orphaned early and, after some medical study, led a restless life. He served as a surgeon in the army of Prince Eugen (q.v.) during the War

of the Spanish Succession (1708–12), and was later in the service of the Counts of Stolberg as surgeon and press officer. From 1731 to 1738 he edited an official newspaper, *Stolbergische Sammlung neuer und merkwürdiger Weltgeschichte.*

Schnabel's principal work is a novel in four parts in the tradition descending from *Robinson Crusoe* (see ROBINSONADE); its long descriptive title covering the title-page opens with *Wunderliche Fata einiger See-Fahrer, absonderlich Alberti Julii, eines gebohrnen Sachsens, Welcher in seinem 18den Jahre zu Schiffe gegangen* (1731–43), but the work, comprising some 2,300 pages, has long been known as *Die Insel Felsenburg* (q.v.). It has been published in abridged and revised versions several times, complete reprint, ed. E. Weber, 1973. *Der im Irrgarten der Liebe herumtaumelnde Cavalier* is an erotic novel of picaresque character (1738, ed. P. Ernst, q.v., 1907, and H. Mayer, 1968). Both these works were pseudonymously published under the name of Gisander. For the *Lebensgeschichte Prinz Eugens* (1736) Schnabel is said to have drawn extensively on his own diary. Shortly before his death appeared *Der aus dem Mond gefallene und nachher zur Sonne des Glücks gestiegene Prinz* (1750).

SCHNACK, ANTON (Rieneck, 1892–1973, Kahl/Main), a journalist in Darmstadt, Mannheim, and Frankfurt, served in both world wars. He began with poetry, *Strophen der Gier* (1919) and *Tier rang gewaltig mit Tier* (1920), containing 60 war sonnets, the quality of which has been compared to Wilfred Owen (P. Bridgwater). He later found a very different style in deftly written, light-hearted stories, which include *Kleines Lesebuch, Die fünfzehn Abenteuer* (both 1935), *Begegnungen am Abend* (1940), *Mädchenmedaillons* (1946), *Phantastische Geographie* (1949), *Die Reise aus Sehnsucht* (1954), and *Flirt mit dem Alltag* (1956). His early fiction contains the novels *Zugvögel der Liebe* (1936) and *Der finstere Franz. Eine Seeräubergeschichte* (1937). He was a brother of F. Schnack (q.v.).

SCHNACK, FRIEDRICH (Rieneck, 1888–1977, Munich), elder brother of A. Schnack (q.v.), was in the electrical industry before the 1914–18 War, during which he fought in Turkey. He worked as a journalist in Dresden before devoting himself to authorship. His close and sympathetic contact with nature is evident in his entire work. This includes the novels *Sebastian im Walde* (1926), *Die Orgel des Himmels*, *Beatus und Sabine* (both 1927), *Der Sternenbaum* (1929, retitled *Das Waldkind*, 1939), *Goldgräber in Franken* (1930), *Der erfrorene Engel* (1934), *Die brennende Liebe* (1935), *Die einsame Straße* (1936), and *Dorine vom Amselberg* (1955).

Among Schnack's excellent popular books on natural history *Das Leben der Schmetterlinge* (1928, reissued 1958), *Sibylle und die Feldblumen* (1937), and *Das Waldbuch* (1960) stand out. He addressed himself to children in *Klick aus dem Spielzeugladen* (1933) and its sequel *Klick und der Goldschatz* (1938). He also wrote on his travels to Madagascar (*Die große Insel von Madagaskar*, 1942, is a revised edition of *Auf ferner Insel*, 1931), and in 1951 published *Der Maler von Malaya*.

Gesamtausgabe des poetischen Werkes (8 vols.) appeared 1950–4, and *Gesammelte Werke* (2 vols.) 1961.

SCHNECKENBURGER, MAX (Thalheim nr. Tuttlingen, 1819–49, Burgdorf, Switzerland), who became an iron founder in Switzerland, wrote in 1840 the patriotic poem *Die Wacht am Rhein* (q.v.). His *Deutsche Lieder* were published posthumously in 1870.

Schneekind, Das, 13th-c. German version of a story recorded in a Latin poem of the 10th c., known, from its tune, as *Modus Liebinc*. It tells of an unfaithful wife who, on the return of her husband, explains away her illegitimate child as a consequence of drinking snow. After some years the husband sells the boy and declares to the wife that her son has melted in the sun.

SCHNEIDER, REINHOLD (Baden-Baden, 1903–58, Freiburg), after initial uncertainty, embarked in his twenties on a career of writing. Much influenced by Unamuno and Kierkegaard (q.v.), he was chiefly preoccupied with religious and ethical matters. A phase of scepticism ended in 1938 with his reaffirmation of the Roman Catholic faith in which he had been reared. His overtly historical works, as well as his plays and novels, are attempts to interpret history in terms of guilt and conscience, and are sombre productions, in which religion provides a hard-won solution.

In 1931 Schneider published a biography of Philip II (q.v.) of Spain, which discarded the traditional view of the oppressive tyrant, in 1932 a life of J. G. Fichte (q.v.), and in 1933 a book on the Hohenzollerns (q.v.). *Auf Wegen deutscher Geschichte* (1934) is a historical travelogue. A study of British history (*Das Inselreich*, 1936) was a turning-point in his understanding of the great crisis of conscience in the 16th c. Among his later historical works are *Kaiser Lothars Krone* (1937), *Theresia von Spanien* (1939), *Papst Gregor der Große* (1945), *Vom Geschichtsbewußtsein der Romantik* (1951), and the posthumously published *Innozenz III* (1960), which dates from 1930. His literary production comprised novels (*Las Casas vor Karl V; Szenen aus der Konquistadorenzeit*, 1938, and *Die silberne*

Ampel, 1956), stories (grouped in *Der fünfte Kelch* and *Das getilgte Antlitz*, both 1953), and plays (*Der Kronprinz*, 1948, *Der große Verzicht*, 1950, *Die Tarnkappe*, 1951, and *Innozenz und Franziskus*, 1952). He also wrote poetry, some of which was clandestinely circulated during the war (see INNERE EMIGRATION); the collections are *Sonette* (1939), *Apokalypse, Die neuen Türme*, and *Erscheinungen des Herrn* (all three 1946), *Herz am Erdsaume* (1947), and *Die Sonette von Leben und Zeit, dem Glauben und der Geschichte* (1954).

Schneider's numerous essays include the titles *Macht und Gnade* (1940), *Der Dichter vor der Geschichte* (1944), *Schriften zur Zeit* (1948), *Über Herrscher und Heilige* (1953), and *Macht und Herrschaft in der Geschichte* (1957). Schneider, as a man of integrity and compassion (cf. the story *Der Tröster*, 1943), was one of the best-known figures in the revived intellectual life of Germany immediately after the 1939–45 War. In 1956 he was awarded the Friedenspreis des Deutschen Buchhandels. In 1953 he published *Ausgewählte Werke* (4 vols.) and in his last years autobiographical writings, including *Der Balkon* (1957) and *Winter in Wien. Aus meinen Notizbüchern 1957–58*. Correspondence appeared posthumously, with B. von Heiseler (q.v.) in 1965, and with W. Bergengruen (q.v.) in 1966. *Gesammelte Werke* (10 vols.), ed. E. M. Landau, appeared in 1977–81.

SCHNEIDER, ROLF (Chemnitz, 1932–), after studying at Halle University and working as an editor, settled as a free-lance writer near Berlin in 1958. He was a member of the Schriftstellerverband until his expulsion in 1979. He later worked in the West. He is the author of fiction, plays (including the comedy *Einzug ins Schloß*, 1972), and radio plays, collections of which appeared in 1970 as *Stücke* (plays) and *Stimmen danach* (radio plays). His first story, *Brücken und Gitter*, and his first novel, *Die Tage in W.*, appeared in 1965. The latter, a detective novel, opens with the news of the murder of a young woman, allegedly by a grammar school teacher in W., Michailsky. Karsten, the most junior member on the staff of a Berlin daily, is sent by his boss to W. in order to collect material for the paper's story. Set in 1932 and the preceding years of mass unemployment, the novel introduces a great variety of figures representing the 'frenzied' era of the Weimar Republic at a time when the NSDAP and political murder are gaining ground. Michailsky dies, a victim of one for whom he cared, a wretched illiterate and unemployed man ('illiterate' is an official description). Michailsky, however, leaves a legacy which Karsten is likely to inherit for the socialist cause. Performed in both East (Berlin) and

West, the documentary *Prozeß in Nürnberg* was published in 1968. Other works include the parodistic collage *Der Tod des Nibelungen. Aufzeichnungen des deutschen Bildschöpfers Siegfried Amadeus Wruck, ediert von Freunden* (1970), the collection *Nekrolog. Unernste Geschichten* (1973, DDR; 1974, BRD as *Pilzomlett und andere Nekrologe*), *Die Reise nach Jaroslaw* (1974, DDR; 1975, BRD), *Von Paris nach Frankreich. Reisenotizen* (1975), *Das Glück* (1976), *Orphée oder Ich reise* (1977), and *Die Abenteuer des Herakles* (1978). In *November* (1979) the central figure, Natascha Roth, a writer assumed by those around her to be a niece of Joseph Roth (q.v.) and a member of an émigré family, is married to Rudolf Roth, art historian and son of a Jewish Berlin coal merchant who has survived National Socialism because his non-Jewish wife did not divorce him. Natascha, well-known in the DDR, is disgraced after having signed a public letter protesting against the expulsion of a fellow writer, Bodakov. The marriage, unlike that of Roth's parents, ends in divorce and Natascha leaves for Paris. Her son Stefan, who at the beginning of the novel, on a November morning, is involved in a serious road accident that leaves him permanently handicapped, is skilfully integrated into the novel's politically allusive themes. Set in 1976, its reference to W. Biermann (q.v.) was evident and publication in the DDR was prohibited. In the year of its appearance, Schneider published *Pfützen voll schwarzer Unvernunft. Zur Publikationsgeschichte von 'November'. Unerwartete Veränderung* was published in 1980. Later plays include adaptations of Nestroy (q.v.), *Die beiden Nachtwandler* (1975), and *Der alte Mann und die junge Frau* (1979). Schneider is also the author of a work on Musil (q.v.), *Die problematisierte Wirklichkeit. Leben und Werk Robert Musils. Versuch einer Interpretation. Kommentar zur Ausgabe des Romans 'Der Mann ohne Eigenschaften'* (1975).

SCHNEIDER, WILHELM MICHAEL (Altrip, Palatinate, 1891–1975, Frankfurt/Main), is the author of a book of war recollections, *Infantrist Perhobstler* (1929), written in the first person and combining earthy humour and crass realism. A Bavarian of the Palatinate, he served first in the ranks and was commissioned in 1916. The novel *Franzosen am Rhein* (1930) deals with the period of French occupation.

SCHNIFIS, see LAURENTIUS VON SCHNÜFFIS.

SCHNITZLER, ARTHUR (Vienna, 1862–1931, Vienna), the son of a prominent Jewish laryngologist, studied medicine at Vienna University and qualified, but showed no inclination to emulate his father, though for a time he held a

hospital appointment. For several years he led the life of a man about town, having liaisons of varying length, of which the most important were with the hotel-keeper's wife Olga Waissnix and the embroideress Jeanette Heger. From 1894 until 1899, his relationship with Marie Reinhard also provided him with much material which he exploited in his writing. He entered upon authorship as a boy, and in his mid-twenties had poems published in *An der schönen blauen Donau*, the literary supplement of the *Neue Freie Presse* (q.v.). In 1893 he left hospital work for private practice, but his list of patients did not extend far beyond his circle of friends. He was soon able to live by his pen, and his chief occupations were his writing and his erotic experiences. In 1901 he was removed from the list of reserve officers of the Austrian military medical corps because his story *Leutnant Gustl* (q.v.) was held to insult the honour of the army. He married in 1903 and the marriage was dissolved in 1921. In 1928 his daughter committed suicide at the age of 18.

Schnitzler's work, which attracted public attention from 1895 (première of *Liebelei*, q.v., in the Burgtheater), is concentrated on sex and death, and shows a remarkable capacity to create atmosphere and to pursue profound, ruthless, and often Freudian analysis of human motives, which he investigated most minutely in his own sceptical self. He was successful as a dramatist in both short and long forms. The seven one-act plays of *Anatol* (q.v., 1893) were followed by other groups, *Lebendige Stunden* (1902), *Marionetten* (1906), *Komödie der Worte* (1915), and, the last performed, the cycle of *Reigen* (q.v., 1900). His first full-length play, *Das Märchen*, appeared in 1894. An Ibsenist treatment of a 'fallen woman', it preceded *Liebelei*, which was followed by *Freiwild* (q.v.) in 1898, and *Das Vermächtnis* in 1899. All of these deal with illicit love, though *Freiwild* also raises the question of the code of honour and the duel. With *Paracelsus* (1899) and *Der Schleier der Beatrice* (1901) Schnitzler diverged for a short time unconvincingly into verse drama, though he adhered to prose for the 'Groteske' in one act, *Der grüne Kakadu* (1899), which has as background the French Revolution (q.v.). The play *Der einsame Weg* (q.v., 1904) marked his reversion to prose and was followed by the light-weight Freudian comedy *Zwischenspiel* (1906) and an implausible military-cum-erotic tragedy, *Der Ruf des Lebens* (1906). *Komtesse Mizzi oder der Familientag* (1909) was another light comedy, but *Der junge Medardus* (1910), set in Napoleonic times, represented a laboured attempt at historical drama. *Das weite Land* (q.v., 1911) struck many as Schnitzler's most cynical work, but in 1912 he achieved a remarkable and balanced comedy on the theme of anti-Semitism in *Professor Bernhardi* (q.v.). The war seems to have inhibited his dramatic vein, for the later plays, except for the verse drama *Der Gang zum Weiher* (1926), are superficial (*Fink und Fliederbusch*, 1917; *Die Schwestern oder Casanova im Spa*, 1919; *Komödie der Verführung*, 1924; *Im Spiel der Sommerlüfte*, 1930).

Schnitzler's greatest gifts emerge in his *Novellen*, which blend atmosphere and mood, so as to infuse into the almost invariably erotic material a bitter-sweet attractiveness, and occasionally invoke the uncanny. Outstanding among them are *Sterben* (q.v., 1893), *Die Frau des Weisen*, *Der Ehrentag* (both 1897), *Die Toten schweigen* (q.v., 1898), *Leutnant Gustl* (q.v.), *Frau Berta Garlan* (q.v., both 1901), *Der blinde Geronimo und sein Bruder* (1902, of fraternal love, undermined but restored), *Frau Beate und ihr Sohn* (q.v., 1913), *Doktor Gräsler, Badearzt* (q.v., 1917), *Casanovas Heimfahrt* (q.v., 1918; these last three were collected as *Die Alternden*, 1922), *Fräulein Else* (q.v., 1924), *Traumnovelle* (1926, a Freudian fantasy), *Spiel im Morgengrauen* (q.v., 1927), and *Flucht in die Finsternis* (q.v., 1931). The volume of early stories, *Die kleine Komödie*, appeared posthumously in 1932.

Schnitzler is the author of two novels, *Der Weg ins Freie* (q.v., 1908), a masterpiece in its portrayal of a stratum of Viennese society, and *Therese* (1928), a work in which the style and structure fail to sustain the long narrative; Therese, an orphan girl of good family, is seduced and goes steadily downhill, and in the end is murdered by her own illegitimate son.

Schnitzler's autobiographical *Jugend in Wien* (up to 1889), ed. by his son Heinrich Schnitzler and Th. Nickl, appeared in 1968, *Meisterdramen* in 1955, *Gesammelte Werke* (6 vols.), ed. R. O. Weiß, in 1961–7, *Meistererzählungen* in 1969, and correspondence with G. Brandes (q.v.) in 1956, with H. von Hofmannsthal (q.v.) in 1964, with Olga Waissnix, *Liebe, die starb vor der Zeit*, in 1970, with M. Reinhardt (q.v.) in 1971, and with Hedy Kempny, *Das Mädchen mit den dreizehn Seelen*, ed. H. P. Adamek, in 1984. *Briefe 1875–1912*, ed. Th. Nickl and H. Schnitzler, appeared in 1981, and *Briefe 1913–1931*, ed. P. M. Braunwarth, R. Miklin, S. Pertlik, and H. Schnitzler, in 1984; and an extensive edition of Schnitzler's diaries (*Tagebuch*) by W. Welzig *et al.* from 1981.

SCHNOOR, HEINRICH CHRISTIAN (Lübeck, *c.* 1760–after 1828, Breslau), an itinerant poet and musician, who probably wrote the well-known song 'Vom hoh'n Olymp herab ward uns die Freude' (q.v.).

SCHNÜFFIS, see LAURENTIUS VON SCHNÜFFIS.

SCHNURR, BALTHASAR, see SANDRUB, LAZARUS.

SCHNURRE, WOLFDIETRICH (Frankfurt/Main, 1920–), after serving throughout the 1939–45 War (in his own laconic phrase 'sechs sinnlose Jahre'), became a dramatic critic (1946–9) and an author. He was a co-founder of the Gruppe 47 (q.v.), which he left in 1951. A radical writer with a predilection for the grotesque, he has written a number of short stories and satires, published singly or in collections, and including *Die Rohrdommel ruft jeden Tag* (1951), *Sternstaub und Sänfte* (1953, republished as *Die Aufzeichnungen des Pudels Ali*, 1962), *Protest im Parterre* (1957), *Eine Rechnung, die nicht aufgeht* (1958), *Man sollte dagegen sein* (1960), and *Schnurre heiter* (1970); a collection, *Erzählungen 1945–1965*, appeared in 1977. Several of these have his own illustrations. In the late 1950s he established himself as the author of novels, *Als Vaters Bart noch rot war* (1958), *Das Los unserer Stadt* (1959, revised 1963), and *Richard kehrt zurück* (1970). The essays *Schreibtisch unter freiem Himmel* (1964) contain an early tract, *Auszug aus dem Elfenbeinturm* (1949). Schnurre's verse includes *Abendländer* (1957), *Kassiber* (1964), and *Kassiber und neue Gedichte* (1979). Numerous volumes for children include *Schnurren und Murren* (1974, with illustrations by the author) and *Ich frag ja bloß: Kinder und sich* (1973). Schnurre has continued to write plays for radio and television. *Klopfzeichen* appeared in 1978, sketches, *Der Schattenfotograf*, in 1978, and *Ein Unglücksfall*, described as a novel, in 1981, all volumes of an unnumbered paperback edn., 1976 ff..

SCHOBER, FRANZ VON (Torup nr. Malmö, Sweden, 1798–1882, Dresden), was a student in Vienna, where he met F. Schubert, E. Bauernfeld, and M. von Schwind (qq.v.). He is the author of twelve poems set to music by Schubert between 1815 and 1827, of which the best known is *An die Musik* ('Du holde Kunst'). In later life Schober was a secretary of legation (Legationsrat) in Weimar.

SCHOCH, JOHANN GEORG (Leipzig, 1634–*c.* 1690), a minor poet, spent his life as an official in various towns, Naumburg, Westerburg, Cölln/Spree, and Brunswick. His lyric poetry, though influenced by Opitz (q.v.), has homely touches, and his irony expresses itself in parody. His two collections of verse are *Poetischer Weihrauchbaum und Sonnenblume* (1656) and *Neu erbauter poetischer Lust- und Blumengarten* (1660). He published in 1657 a play, *Comoedia vom Studentenleben*, which is an adaptation of the Latin comedy *Cornelius relegatus* (1600) by Albert Wichgrevius.

SCHÖFFER, PETER (Gernsheim/Rhine, c. 1425–c. 1562, Mainz), an early printer in Mainz, took over the press of his father-in-law, Johann Fust (q.v.).

SCHOLL, HANS and SOPHIE, see RESISTANCE MOVEMENTS (2).

SCHOLTZ, ANDREAS, see SCULTETUS, ANDREAS.

SCHOLZ, HANS (Berlin, 1911–), studied history of art, worked as a teacher, lecturer, and musician in Berlin, and in later life was on the staff of a Berlin daily. He achieved a popular success with his novel *Am grünen Strand der Spree* (1955), which by means of a group of people, gathered in a Berlin bar, vividly evokes experiences of the 1939–45 War of the most varied kind and in different settings, among them Russia and Norway. A love story is woven into it, and a happy end provided. His other writings include the story *Schkola* (1956), *Berlin, jetzt freue Dich* (1960), *Berlin für Anfänger* (1961), *Der Prinz Kaspar Hauser* (1964). The travel guide *Südost und Zurück; Wanderungen und Fahrten in der Mark Brandenburg* (9 vols. inc. an Index), with illustrations by the author, appeared in 1973–83 and *Fontane* in 1978. *An Havel, Spree und Oder* (1962) is a radio play. Scholz, it need hardly be said, is a dyed-in-the-wool Berliner.

SCHOLZ, WILHELM VON (Berlin, 1874–1969, Constance), the son of one of Bismarck's ministers, entered the army in 1894, but soon resigned and devoted himself to poetry. He spent most of his life in South Germany, which is reflected in *Der Bodensee* (1921). His first three volumes of verse are *Frühlingsfeier* (1896), *Hohenklingen* (1898), and *Der Spiegel* (1892), all of which are influenced by Liliencron (q.v.). He became interested in the theatre, directing, and acting, as well as writing plays. These include *Der Besiegte* (1898), *Der Gast* (1900), *Der Jude von Konstanz* (1905), *Meroë* (1906), *Vertauschte Seelen* (1910), and *Gefährliche Liebe* (1913). Through his dramatic work he was associated with the short-lived movement of Neo-Classicism (see NEUKLASSIZISMUS and ERNST, P.), expressing his views in *Gedanken zum Drama* (1915).

In 1913 Scholz published *Neue Gedichte*, adding other volumes at intervals (*Die Häuser*, 1924; *Das Jahr*, 1927; *Lebensjahre*, 1939). He achieved great success with two further plays, *Der Wettlauf mit dem Schatten* (1921) and *Die gläserne Frau* (1924), but in later life his main interest lay in fiction. His Novellen include the volumes *Die Unwirklichen* (1915), *Die Beichte* (1919), *Vincenzo Trappola* (1922), *Zwischenreich*

(1922), and *Erzählungen* (1924). His historical novel *Perpetua* (1924) is set in Augsburg in the 16th c. and deals with two sisters, Perpetua and Katharina. Katharina is mistakenly burned as a witch, while Perpetua lives on to be revered as a saint. The theme of another historical novel, *Der Weg nach Ilok* (1930), is persecution and repentance; it was followed by *Unrecht der Liebe* (1931), a crime novel, which is of excellent workmanship and high intellectual quality. In the 1940s and 1950s Scholz returned to dramatic work (*Claudia Colonna*, 1941; *Ayatari*, 1944; *Ewige Jugend*, 1949; *Das Säckinger Trompeterspiel*, 1955) and theory (*Das Drama*, 1956).

Scholz's early biographical works are on A. von Droste-Hülshoff (1904) and on F. Hebbel (qq.v., 1905), whom he valued as a dramatist, writing later on Schiller (*Stürmische Jugend*, 1944), revised and retitled *Friedrich von Schiller*, 1949). From 1926 to 1928 he was President of the Berlin Akademie der Dichtung (see AKADEMIEN). His autobiographical writings began with the titles *Wanderungen* and *Lebensdeutung* (both 1924); *Mein Leben* appeared in 1934, and *Mein Theater* in 1964. *Gesammelte Werke* (5 vols.) were published in 1924, a selection of Novellen entitled *Das Inwendige* appeared in 1958, and a select edition of his plays in 1964.

SCHÖN, THEODOR VON (Schreitlauken in Lithuania, 1773–1856, Arnau nr. Königsberg), an eminent Prussian civil servant, entered the administration in 1800, and after the defeat of Prussia in 1806 assisted successively Stein and Hardenberg (qq.v.) in their reforms. In 1809 he was moved to East Prussia as governor of Gumbinnen. He was concerned with raising the Prussian militia (see LANDWEHR) for the War of Liberation in 1813 (see NAPOLEONIC WARS). After peace was concluded he was prefect (Oberpräsident) of West Prussia and his sphere was extended in 1824 to include also East Prussia. From 1815 onward he was the leading spirit behind the steps taken to restore the Marienburg (q.v.). It was at his request that J. von Eichendorff (q.v.) wrote the historical account *Die Wiederherstellung des Schlosses der deutschen Ordensritter in Marienburg* (1844).

SCHÖNAICH-CAROLATH, CHRISTOPH OTTO, FREIHERR VON (Amtitz nr. Guben, Saxony, 1725–1807, Amtitz), served as an officer in the Saxon army in the War of the Austrian Succession (see ÖSTERREICHISCHER ERBFOLGEKRIEG) and afterwards lived as a country gentleman. On the publication of his verse epic *Hermann oder Das befreyte Deutschland* (1751), Gottsched (q.v.), recognizing him as a supporter against Klopstock and Lessing (qq.v.), crowned him as laureate in 1752. Schönaich's satire *Die ganze Ästhetik in einer Nuß oder Neo-logisches Wörterbuch* (1754) ridicules Bodmer and Haller (qq.v.), as well as Klopstock. Schönaich is the author of another patriotic epic, *Heinrich der Vogler oder Die gedämpften Hunnen* (1757).

SCHÖNAICH-CAROLATH-SCHILDEN, EMIL, PRINZ VON (Breslau, 1852–1908, Haseldorf, Holstein), after short service as an officer of dragoons lived on his estates. He is the author of poetry (*Lieder an eine Verlorene*, 1878, and *Gedichte*, 1903) and of Novellen, mostly of regional character, including the collections *Geschichten aus Moll* (1884), *Der Freiherr*, and *Der Heiland der Tiere* (both 1896). His *Gesammelte Werke* (7 vols.) appeared in 1907.

Schönbartlaufen, properly 'Schembartlaufen', masquerades held in the streets of Nürnberg by the butchers' guild between 1348 and 1539, described in a poem by Hans Sachs (q.v.).

SCHÖNBERG, ARNOLD (Vienna, 1874–1951, Los Angeles), composer, was for a time with the cabaret Das Überbrettl (1901–3), and later took up painting. As a musician he was virtually self-taught. His early works show the influence of Brahms, R. Wagner, and R. Strauss (qq.v.), and an affinity to Jugendstil (q.v.) begins with the string sextet *Verklärte Nacht* (1899), based on *Zwei Menschen* by R. Dehmel (q.v.). The *Gurre-Lieder*, an opulent setting of poems by the Danish writer Jens Peter Jacobsen (1847–85) in German translation, followed in 1900–13, and *Das Buch der hängenden Gärten*, poems by S. George (q.v.), in 1908–9. Schönberg developed an atonal style, represented in his most famous work *Pierrot lunaire* (1912). In the next decade he elaborated a 12-note system (Zwölf-Töne-Theorie), characterized by note-rows (Tonreihen). The works of his last years show a partial return to tonality.

From 1933 Schönberg lived in the U.S.A. Alban Berg (q.v.) and Anton von Webern (1883–1945) were his pupils and friends. Schönberg published a *Harmonielehre* (1911) and, in English, *The Theory of Composition* (1940).

SCHONDOCH, a Middle High German Swiss poet of the 14th c., of whom nothing is known except his authorship of two poems, *Die Königin von Frankreich und der ungetreue Marschall* and *Der Littauer*. The former tells a story resembling that of St. Genevieve. The marshal slanders the queen, who is then taken into the forest by a knight bearing the king's order to put her to death. The marshal waylays the party and murders the knight. The knight's hound, as

his master's champion, savages the marshal before the court. The marshal confesses and the queen's honour is restored. *Der Littauer*, which has as its background the wars of the Teutonic Order (see DEUTSCHER ORDEN) against the heathen, recounts the conversion of a Lithuanian prince through the miracle of the Eucharist.

Schöne auserlesene Lieder, an early collection of folk-songs compiled by Heinrich Finck of Nürnberg in 1536.

Schöne Bernauerin, Die, the earliest preserved folk-song on the tragic tale of the love of a commoner, Agnes Bernauer (see BERNAUER, AGNES), for Duke Albrecht III, der Fromme (q.v.), of Bavaria-Munich. It was not printed until 1710.

Schöne Buche, Die, a short idyll in elegiacs written by E. Mörike (q.v.) in 1842. It consists of fifteen distichs praising a well-grown, majestic beech in the forest.

Schöne Fremde, a short poem by J. von Eichendorff (q.v.), expressing longing and spell-bound expectation. Its first line runs 'Es rauschen die Wipfel und schauern' and it occurs (untitled) in Eichendorff's novel *Dichter und ihre Gesellen* (q.v.). It has been set to music by R. Schumann (q.v.).

Schöne Magelone, Die, a legend which is the subject of a chap-book (see VOLKSBUCH) and occurs also in later adaptations. Magelone is the daughter of a king of Naples. She is already promised in marriage when she meets Peter, Count of Provence, with whom she immediately falls in love. The couple elope, but become separated when Peter sets out to pursue a bird which has made off with the rings of the sleeping Magelone, and he eventually falls into Turkish slavery. Magelone settles in Provence and tends the sick until she and Peter are eventually reunited.

This story of fidelity rewarded, derived from a French prose romance of 1457, was given a German version by Veit Warbeck (q.v.), whose book was written in 1527 and published posthumously in 1535. It bore the title *Die sehr lustige Histori vonn der schönen Magelona . . . und von einem Ritter genannt Peter mit den silberin schlüsseln*. An earlier German version is lost. In 1797 Tieck (q.v.) published, in *Volksmärchen herausgegeben von Peter Leberecht* (his pseudonym), a highly sentimentalized retelling of the story, with the title *Die wundersame Liebesgeschichte der schönen Magelone und des Grafen Peter aus der*

Provence; the poems included in this were set by Brahms (q.v., *Romanzen aus der 'schönen Magelone'*, Op. 33). The story was also told by Gustav Schwab (q.v.) in *Deutsche Volksbücher* (1836–7) and by Karl Simrock (q.v.) in *Deutsche Volksbücher*, (1845–66). The origins of the legend are in part oriental.

SCHÖNEMANN, LILI (really ANNE ELISABETH) (Frankfurt, 1758–1817, Krautergersheim), daughter of a Frankfurt patrician family, was engaged to Goethe for some months in 1775. The powerful mutual attraction was countered by social stresses, for Goethe both despised her social environment and felt ill at ease in it. She married in 1778 Baron Bernhard Friedrich von Türckheim, with whom she lived in Strasburg until the excesses of the French Revolution caused them to flee to Germany in 1794.

Goethe's poems *Neue Liebe, neues Leben, An Belinden, Lilis Park*, and *Auf dem See* (q.v.) reflect the ambivalence of their relationship. According to Eckermann (q.v.), Goethe spoke of Lili Schönemann in 1830 in these terms: 'Sie war in der Tat die Erste, die ich tief und wahrhaft liebte. Auch kann ich sagen, daß sie die letzte gewesen.' (5 March.)

Schöne Melusine, a beautiful mermaid, is the subject of a French legend of the Middle Ages, which was first given a German form, in prose, by Thüring von Ringoltingen, mayor (Schultheiß) of Berne, who died in 1483. Thüring made his version in 1456 for the Margrave of Hochberg.

Raimund, Count of Poitiers, marries Melusine, accepting the condition that he must not seek her out on a Saturday. One day he transgresses the condition, finds her in her bath, and discovers that her body ends in a fish's tail. Thereupon she returns to the sea from which she came. *Die schöne Melusine*, printed in Thüring's version in 1474, became the theme of a popular Volksbuch (q.v.) and was frequently reprinted in the 16th c., 17th c., and even 18th c. It was dramatically treated by Hans Sachs and Jakob Ayrer (qq.v.) and, with variants, by the Viennese K. F. Hensler (*Das Donauweibchen*, a comedy in two parts, 1798) and F. Grillparzer (see MELUSINA, 1833), and in the narratives *Sehr wunderbare Historie von der Melusina* (1800) by L. Tieck (q.v.), *Undine* (q.v., 1811) by Fouqué, and *Die neue Melusine* (q.v., published in two parts in 1817 and 1819) by Goethe, who included his story in *Wilhelm Meisters Wanderjahre* (q.v.).

Schöne Müllerin, Die, a cycle of 20 poems by Wilhelm Müller (q.v.), published in 1821 in

Siebenundsiebzig Gedichte aus den hinterlassenen Papieren eines reisenden Waldhornisten. They tell how a young miller falls in love with his master's daughter, believes his love to be returned, and then finds himself supplanted by a gamekeeper (Jäger), whereupon he drowns himself in the mill stream. The cycle was beautifully set for solo voice and piano accompaniment by F. Schubert (q.v.) in 1824.

SCHÖNHERR, KARL (Axams, Tyrol, 1867–1943, Vienna), practised as a doctor in Vienna from 1896 to 1905 before devoting himself to literature. His first publications were poems (*Innthaler Schnalzer*, 1895) and stories (*Allerhand Kreuzköpf*, 1895), but he was primarily a dramatist.

Die Bildschnitzer (1900) is a compassionate treatment of a triangular situation among poor wood carvers. *Sonnwendtag* (1902) derives its title from surviving pagan rites on Midsummer night, but its theme is an unsuccessful attempt at vicarious expiation of sin by bringing up a boy to be a priest (a subject treated earlier by Anzengruber in *Der Meineidbauer* and later by F. Werfel in *Der veruntreute Himmel*, qq.v.). The play was recast as a comedy entitled *Die Trenkwalder* (1914). In spite of a sympathetic portrait of a mild and pious priest, Schönherr is concerned to condemn what he sees as widespread abuse and exploitation of faith by clergy and laymen. *Karrnerleut'* (1905) is a purely human tragedy, the suicide of a boy who has unwittingly betrayed his father to the police. *Familie* (1906) is crudely theatrical and was omitted by the author from his collected works. *Erde. Eine Komödie des Lebens* (1908), treats grimly the refusal of an aged farmer to give up his farm to his son. All these works are realistic representations of Tyrolean peasant life. In *Das Königreich* (1908) Schönherr unsuccessfully attempted a symbolical play in the manner of Hofmannsthal (q.v.). *Glaube und Heimat* (1910) returns to realism and Tyrol, but is set in the 17th c., dealing with Protestants expelled from their homeland in the Counter-Reformation.

Although Schönherr continued to write plays until 1937, his later work, with the possible exception of the rural tragedies *Frau Suitner* (1916) and *Kindertragödie* (1919), represents a decline. *Gesammelte Werke* (4 vols.) appeared in 1927 and *Gesamtausgabe* (3 vols.), ed. V. K. Chiavacci *et al.*, in 1967–74.

SCHÖNHUTH, OTTMAR (Sindelfingen, 1806–64, Edelfingen), a Württemberg pastor and minor poet, was educated at the Tübinger Stift (q.v.) and was on friendly terms with L. Uhland (q.v.). At first curate at Hohentwiel, he was later pastor at Edelfingen.

Schönhuth collected a number of early chapbooks (see VOLKSBUCH). In 1830 he published MS. C of the *Nibelungenlied* (q.v.), which was then in the possession of Freiherr von Laßberg (q.v.), and in 1838 *Die Klage* (q.v.). A volume of poems (*Gedichte*) appeared in 1830. He is the author of several plays, including *Die Städterschlacht bei Döffingen*, *Die Ohrfeige* (both 1830), and *Käthchen von Engen* (1836). Later in life he wrote an account of the monuments of Württemberg (*Die Burgen, Klöster, Kirchen und Kapellen Württembergs und Hohenzollerns*, 1860).

SCHÖNKOPF, ANNA KATHARINA (Leipzig, 1746–1810, Leipzig), the daughter of an innkeeper, was Goethe's first love in his student days in Leipzig (1765–8). She is the Annette of the early poems and the Amine of *Die Laune des Verliebten* (q.v.). Goethe and Käthchen Schönkopf, as she is usually called, contemplated marriage but abandoned the intention early in 1768, though they corresponded until 1770. The relationship is referred to in *Dichtung und Wahrheit* (q.v., Bk. 7), in which Käthchen appears under the names Annette and Ännchen and is described as 'ein gar hübsches, nettes Mädchen' and 'jung, hübsch, munter, liebevoll und so angenehm, daß sie wohl verdiente in dem Schrein des Herzens eine Zeitlang als eine kleine Heilige aufgestellt zu werden'. She married in 1770 a Dr. Kanne, who later became deputy Bürgermeister of Leipzig.

Schön-Rohtraut, a ballad written by E. Mörike (q.v.) in 1838. It is sung by a page, and its climax is the revelation that he has kissed Princess Schön-Rohtraut. Each stanza finishes with the refrain 'Schweig stille, mein Herze!'

SCHÖNTHAN, EDLER VON PERNWALD, two brothers, FRANZ (Vienna, 1849–1913, Vienna) and PAUL (Vienna, 1853–1905, Vienna), who were active in the Viennese theatre in the last two decades of the 19th c. Both wrote light comedies and farces, of which the most successful was *Der Raub der Sabinerinnen* (1885), which they composed in collaboration. Franz also collaborated with G. von Moser (q.v.).

SCHÖNWIESE, ERNST (Vienna, 1905–), until 1970 an official in Austrian broadcasting, is the author of religious verse, mystical in tone and traditional in form. It includes the following volumes: *Der siebenfarbige Regenbogen, Ausfahrt und Wiederkehr* (both 1947), *Nacht und Verheißung, Das Bleibende* (both 1950), *Das unverlorene Paradies* (1951), *Ein Requiem in Versen* (1953), *Stufen des Herzens* (1956), *Traum und Verwandlung* (1959), *Baum und Träne* (1962), *Geheimnisvolles Ballspiel* (1964), and *Odysseus und*

der Alchimist (1968). A collection of essays, *Literatur in Wien zwischen 1930 und 1980*, appeared in 1980.

SCHOPENHAUER, ADELE (Hamburg, 1797–1848, Bonn), sister of the philosopher Arthur Schopenhauer (q.v.), lived from 1806 to 1828 with her mother Johanna Schopenhauer (q.v.) in Weimar, where she belonged to the circle around Goethe. She wrote *Haus-, Wald- und Feldmärchen* (2 vols., 1844), a novel (*Anna*, 2 vols., 1845), and *Eine dänische Geschichte* (1848). Her poems were published many years after her death (*Gedichte*, 1920), as was also her diary (*Tagebuch einer Einsamen*, 1921). She figures as a character in Th. Mann's novel *Lotte in Weimar* (q.v.).

SCHOPENHAUER, ARTHUR (Danzig, 1788–1860, Frankfurt/Main), the radical philosopher of pessimism, who described himself as the only worthy successor to Kant (q.v.), assimilated all the negative trends of a disillusioned age. Like Voltaire, he mocked at the optimism of Leibniz (q.v.), writing in a highly readable style, which enabled him to draw a Dantesque vision of suffering, demonstrating 'welcher Art dieser *meilleur des mondes possibles* ist'. He had other rare gifts which made him conscious, when speaking about the few men endowed with genius, that he was one of them. This explains his reference to the average product of the human species as 'Fabrikware der Natur'. He became known as a misanthropist (Menschenverächter), and as such is second only to Nietzsche (q.v.). Schopenhauer was a highly complex individualist. His personal background counted with him more than with most philosophers and encouraged a stubbornly introspective nature. He had a wealthy and cultured father, whose financial acumen led him, as bank director, to spend much time abroad, including a few months in England, which Schopenhauer used with such profit that he read *The Times* daily for the rest of his life. In 1805 his father committed suicide. His mother, Johanna Schopenhauer (q.v.), moved to Weimar. After studying science and philosophy at Göttingen and Berlin universities, Schopenhauer graduated in 1813 at Jena with his dissertation *Über die vierfache Wurzel des Satzes vom zureichenden Grunde*. A brief experience of the War of Liberation (see NAPOLEONIC WARS) left him still more disillusioned with human nature. His mother, of whose social excesses Schopenhauer already disapproved, provoked a final rift, which contributed to his lifelong dislike of women.

Contact with Goethe, and in particular the reading of Goethe's *Farbenlehre*, stimulated Schopenhauer's treatise *Über das Sehen und die*

Farben (1816), which he wrote in Dresden before he produced his principal work *Die Welt als Wille und Vorstellung* (1819). Years later it was followed by *Über den Willen der Natur* (1836), which was extended by further variants appearing in 1841 as *Die beiden Grundprobleme der Ethik*, containing two tracts, *Über die Freiheit des Willens* and *Über das Fundament der Moral*. Meanwhile he had qualified to lecture in Berlin (1820), where he hoped, by the force of his contrasting convictions, to deprive Hegel (q.v.) of his followers, an attempt which failed. He compensated himself by a ten-year stay in Italy before returning to Dresden and Berlin, which he left in haste for Frankfurt at the onset of the cholera epidemic which caused Hegel's death (1831). Thus he survived a great rival, but lived unnoticed and lonely, until the mid-century brought him recognition. His *Parerga und Paralipomena* of 1851 proved particularly popular, and contained the well-known *Aphorismen zur Lebensweisheit*. In the early Frankfurt period his considerable artistic and linguistic talents enabled him to translate, from the Spanish, a work of his favourite writer, *Balthasar Gracian's Hand-Orakel und Kunst der Weltklugheit*. It was published posthumously.

Die Welt als Wille und Vorstellung does not present a comprehensive philosophical system, but a view of life inspired by Kant and other unacknowledged influences, among them Fichte and Schelling (qq.v.), as well as by personal frustration. The work opens with the statement that the world is 'meine Vorstellung', Vorstellung being primarily conceived as sense perception. In defining the world (in analogy to Kant's thing-in-itself, Ding-an-sich) as Will ('als Wille'), he challenges the concept of the individual free will as found in Schiller's adaptation of Kant, but sees it as a blind will, which makes man slave to his nature, his emotions, and sexual drives; in stressing their superior function over the human intellect he anticipates the psychology of the subconscious. The will is the master, the intellect the servant. In another analogy Schopenhauer illustrates the will as the blind man who carries everything, including the intellect, the lame man, on his shoulders. He conceives the will as an absolute and irrational concept asserting itself independently of time and space over and above inorganic and organic phenomena; it includes everything 'was das Sein an sich jedes Dinges in der Welt und der alleinige Kern jeder Erscheinung ist.' Schopenhauer views happiness as an illusion and concedes that the most one can expect is boredom. The will thus constantly deceives and tricks humanity. Schopenhauer sees the most effective relief in art. But art can afford only temporary solace. The one fully effective means of coming to terms with life and the nothingness beyond death is to be found in the comprehension of the will. Schopenhauer was not the first to introduce Indian religious thought into German philosophy and literature, but he was the strongest advocate of the Buddhist principle of negation by contemplation. Self-denial and ascetic withdrawal effect the dissolution and extinction of the individual will, which in return instils into the mind a sense of perfect peace and serene happiness (Nirvāna). In German fiction H. Hesse's *Siddhartha* (q.v.) offers a concise example. Schopenhauer was convinced that humanity will go on exploiting life and accepting suffering until death puts an end to the painful process of living.

The appeal of Schopenhauer's work was mainly to sensitive, intelligent human beings seeking some explanation of the evil in life. Its considerable influence since the latter part of the 19th c. is particularly evident in the work of R. Wagner and Th. Mann (qq.v.). Schopenhauer's accent on Indian thought, moreover, harmonized with the advance of German Oriental studies. Perhaps the widest appeal emanated from his thesis that life was suffering and that the only consolation lay in compassion (Mitleid). But the subjective and introspective element prevails, reflected in a singularly lonely man who was at one neither with himself nor with the world. He bequeathed what material wealth he had to the relief of suffering.

Sämtliche Werke (6 vols.), ed. J. Frauenstädt, were published in 1873–4. Other editions include A. Hübscher (7 vols.), 1937–41, reissued 1972, by whom the historisch-kritische Ausgabe is being prepared (1966 ff.).

SCHOPENHAUER, JOHANNA (Danzig, 1766–1838, Weimar), *née* Trosiener, was the mother of the philosopher Arthur Schopenhauer (q.v.). Her husband committed suicide in 1805, and from 1806 to 1828 she lived in Weimar, associating with Goethe and his circle. From 1828 to 1837 she lived in Bonn. She wrote a number of novels and stories, including *Gabriele* (3 vols., 1819–20), *Die Tante* (2 vols., 1823), *Sidonia* (3 vols., 1827–8), *Richard Wood* (2 vols., 1837), *Erzählungen* (8 vols., 1825–8), *Novellen* (2 vols., 1830), and *Neue Novellen* (3 vols., 1832). Her collected works (*Sämtliche Schriften*, 24 vols.) appeared in 1830–1.

SCHOPPE, AMALIE (Burg, Fehmarn, 1791–1858, Schenectady, New York), *née* Weise, lived from early youth in Hamburg where, at the age of 20, she married a lawyer. She was widowed in 1829, and from 1851 lived with her son in the U.S.A. Her literary activities were confined to her Hamburg period, mainly the 1820s and 1830s. She is the author of novels and stories making an attempt at popular appeal, in which she was for a time successful. She founded a girls' grammar school, and much of her writing is addressed to the young. As editor of the *Pariser Modeblätter*, she published early poems and stories by F. Hebbel (q.v.), mainly in 1832 and 1833, and enabled Hebbel to move from Wesselburen to Hamburg. The death of one of her sons intensified her dedication to the advancement of Hebbel's education during the first period of Hebbel's stay in Hamburg, though her influence on him was short-lived.

Gesammelte Erzählungen und Novellen (3 vols.) appeared 1827–36, and *Erinnerungen aus meinem Leben* (2 vols.) in 1838. In 1846 she published a novel (3 vols.), *Der Prophet*.

SCHOPPE, also **SCHOPPIUS**, Kaspar (Neumarkt, Upper Palatinate, 1576–1649, Padua), philologist and Roman Catholic controversialist, was brought up as a Lutheran and converted to Catholicism in 1598. From 1599 he was in Rome and was from time to time employed on diplomatic missions by Pope Clement VII. Attacked by the Jesuits, he settled in 1636 in Padua. His anti-Protestant zeal is manifest in his learned Latin writings on the classics, as well as in his theological polemics, which were also written in Latin.

SCHÖPPER, Jakob (Dortmund, c. 1514–54, Dortmund), a Roman Catholic priest in his native city, was the author of Latin plays, including one on John the Baptist (*Ectrachelistes sive Joannes decollatus*, 1546) and another on David and Goliath (*Monomachia Davidis et Goliae*, 1550).

SCHORER, Christoph (Memmingen, 1618–71, Memmingen), physician and moralist, studied in Strasburg, and was a private tutor first at Polheim near Basel, then at Montbéliard. After further study at Padua he returned to his native town and was appointed city physician. His moral writings, which include *Der unartige deutsche Sprachverderber* (1643, reprinted 1921), *Der Mannverderber* (1644), and *Gespräch von dem Dantzen* (1645), are directed especially against foreign influences on language and manners; they also castigate the decline of morals in general. Over many years he produced annual almanachs or *Kalender*.

SCHORN, Henriette von (Nordheim, 1807–69, Weimar), by birth a Baroness (Freiin) von Stein, became a lady-in-waiting at the Weimar court and married in 1839 Ludwig von Schorn, who died in 1842. Under the pseudonym H. Nordheim, Henriette von Schorn published a volume of rural stories (*Ländliche Skizzen aus*

Franken, 1854), as well as contributing Novellen, etc. to magazines, including *Die Gartenlaube* (q.v.). Her collected stories were published in 1902 by her daughter Adelheid von Schorn (*Geschichten aus Franken*, 2 vols.).

SCHOTTEL, also **SCHOTTELIUS**, Justus Georg (Einbeck, Hanover, 1612–76, Wolfenbüttel), an eminent philologist and minor poet, studied law, became tutor to the sons of the Duke of Brunswick (see Anton Ulrich, Herzog von Braunschweig), and in 1646 was appointed a member of the Consistory Court. Schottel was a member of two poetic societies. In the Pegnesischer Blumenorden (see Hirten-und Blumenorden an der Pegnitz, Löblicher) he was known as Fontano, and in the Fruchtbringende Gesellschaft (q.v.) he went under the name Der Suchende. His principal poetic publications are *Lamentatio Germaniae expirantis*, *der nunmehr hinsterbenden Nymphe Germaniae elendeste Todesklage* (1640) and *Neu erfundenes Freuden-Spiel genanndt Friedens Sieg* (1648). In 1647 he published a collection, *Fruchtbringender Lustgarte* (reissued, ed. M. Burkhard, 1967). In 1669 appeared his *Ethica. Die Sittenkunst oder Wollebenskunst* (ed. J. J. Berns, 1979).

More important are Schottel's philological and prosodic writings, *Teutsche Sprachkunst* (1641), *Der teutschen Sprach Einleitung* (1643), *Teutsche Vers- oder Reim-Kunst* (1645), and *Ausführliche Arbeit von der teutschen Haubt-Sprache* (1663, reissued, ed. W. Hecht, 1967).

Schrätel und der Wasserbär, Das, a comic verse tale of the early 14th c., in which a goblin (Schrat) encounters unexpectedly and is mishandled by the polar bear belonging to an itinerant bear leader, and is driven by this experience to forsake the house it has beset. The anonymous tale was formerly wrongly ascribed to Heinrich von Freiberg (q.v.).

SCHRECKENBACH, Paul (Neumark, Thuringia, 1866–1922, Klitschen nr. Torgau), pastor at Klitschen, was the author of several carefully detailed historical novels, of which the best known is *Der König von Rothenburg* (1910). A collection (5 vols.) appeared in 1930.

Schreckenberger, a fictitious boastful soldier (*miles gloriosus*) of the Thirty Years War (see Dreissigjähriger Krieg), who is a minor character in Eichendorff's Novelle *Die Glücksritter* (q.v.). He sings a song, which in Eichendorff's collected poems bears the title *Der Schreckenberger*.

SCHREIBER, Alois Wilhelm (Bühl, Baden, 1763–1841, Baden-Baden), was successively a schoolmaster in Baden, a dramatic critic in Mainz and Frankfurt, and a professor at Heidelberg University. A prolific writer of novels, stories, and plays, he also published historical writings, topographical essays, and guide books (*Heidelberg und seine Umgebung*, 1811; *Handbuch für Reisende am Rhein*, 1816).

SCHREYER, LOTHAR (Blasewitz, 1886–1966, Hamburg), was a member of the Sturm group, co-editing *Der Sturm* (q.v.) from 1917 and publishing his programmatic tracts (notably *Die neue Kunst*, 1919). From 1912 to 1918 he was dramatic adviser (Dramaturg) at the Schauspielhaus, Hamburg; he was founder and director of the Kunstbühne, Berlin (1919–21), demonstrating his approach with his own plays *Jungfrau* (1917), *Meer. Sehnte. Mann* (1918), *Nacht* (1919), and *Kreuzigung* (1920). From 1921 to 1923 he was a professor at the Bauhaus (q.v.) in Weimar. His religious leanings showed in his essays on J. Böhme (q.v.); in 1933 he became a Roman Catholic. He is best known for his radical Expressionist theory on art, the theatre, and language (Das Wortwerk ist eine rhythmische Wortreihe). His reminiscences include *Expressionistisches Theater. Aus meinen Erinnerungen* (1948) and *Erinnerungen an Sturm und Bauhaus* (1956).

Schreyer's later work comprises religious legends (*Die Liebe der Heiligen Elisabeth*, 1933; *St. Christopherus*, 1936), novels (*Der Falkenschrei. Friedrich II. von Hohenstaufen*, 1940; *Der Untergang von Byzanz*, 1940; *Siegesfest in Karthago*, 1961), and plays (e.g. *Anbetung des göttlichen Kindes*, 1950; *Die Vogelpredigt*, 1951). An essayist (on topics related to religion and art), he also produced religious poetry (*Dichtungen*, 1928).

SCHREYVOGEL, JOSEPH (Vienna, 1768–1832, Vienna), left his native city in 1794, possibly because of alleged French Revolutionary sympathies, and lived for two years in Jena, where he met Schiller. From 1802 to 1804 he was secretary to the Burgtheater (q.v.) in Vienna. After a short period of journalistic activity with the *Sonntagsblatt oder Unterhaltungen des Thomas West* (a pseudonym which he afterwards changed to Karl August West), he was reappointed to the Burgtheater, this time as artistic director (Dramaturg), a post he held from 1815 to 1832. He translated and produced Calderón's *La vida es sueño* (*Das Leben ein Traum*, 1820) and *El médico de su honra* (*Don Gutierre*, 1818), and Moreto's *El desdén con el desdén* (*Donna Diana*, 1819).

Schreyvogel, who emphasized the classical tradition in repertoire and production, influenced Grillparzer (q.v.), whose *Die Ahnfrau* (q.v.) he helped to revise and several of whose plays he produced in the Burgtheater. His

writings were published as *Gesammelte Schriften von Thomas und Karl August West* (4 vols.) in 1829. Schreyvogel died of cholera. Posthumous publications include his diaries, *Tagebücher 1810–23*, ed. K. Glossy (2 vols.), 1903, and *Ausgewählte Werke*, 1910.

SCHREYVOGL, FRIEDRICH (Vienna, 1899–1976, Vienna), in spite of the difference of spelling, is a collateral descendant of J. Schreyvogel (q.v.), who was in charge of the Burgtheater from 1815 to 1832. F. Schreyvogl became in 1954 deputy and in 1959 principal dramatic adviser (Dramaturg) at his greatgreat-uncle's theatre.

Schreyvogl was a prolific author of poems, plays, and novels. His first volume of verse appeared in 1917 (*Singen und Sehnen*); several other collections followed, *Wir Kinder Gottes* in 1957. In drama he was successful in comedy (*Das Liebespaar*, 1940; *Die kluge Wienerin*, 1941; *Die weiße Dame*, 1942; *Der Liebhaber*, 1951; *Die Versuchung des Tasso*, 1955), but he also wrote serious plays (*Der zerrissene Vorhang*, 1920; *Das brennende Schiff*, 1926; *Habsburger Legende*, 1933). His numerous novels include *Der Antichrist* (1921), *Tristan und Isolde* (1929), *Grillparzer* (1935), *Brigitte und der Engel* (1936), *Die Nibelungen* (1938), *Eine Schicksalssymphonie* (1941, showing Austria 1900–14), *Der Friedländer* (1943, on Wallenstein, q.v.), *Das fremde Mädchen* (1954), *Die Dame in Gold* (1957), and *Venus im Skorpion* (1961). In 1960 Schreyvogl edited the works of F. Raimund (q.v.). In 1965 he published a work on the Burgtheater (*Das Burgtheater, Wirklichkeit und Illusion*). Select editions of his works appeared as *Bild und Sinnbild der Welt* (1959) and *Die große und die kleine Welt* (1970).

Schriften des Waldschulmeisters, Die, a novel by P. Rosegger (q.v.), written in 1874–5 and published in 1875. The novel is introduced by a narrator (see RAHMEN). He arrives in July in cold inclement weather at a Styrian hamlet from which he intends to climb the peak 'der graue Zahn'. He is lodged in the house of the village schoolmaster, who has disappeared. The narrator finds the schoolmaster's unfinished autobiography, which provides the story (Binnenerzählung).

Erdmann, the schoolmaster, in his youth served against Napoleon with Andreas Hofer (q.v.), in Napoleon's Grand Army in Russia, and on the Austrian side in the Battle of the Nations near Leipzig (see VÖLKERSCHLACHT). During the fighting he kills his friend. He is persuaded to begin a new life in the service of his fellow men in a remote Styrian village. Here he works selflessly and obscurely as schoolmaster, parish clerk, and counsellor for fifty years, earning the affection of the inhabitants. At Christmas 1863 he disappears.

The narrator climbs the mountain and discovers the frozen body of Erdmann; overcome by a longing to see the wider world once more, the old schoolmaster had reached a lofty viewpoint extending to the vast Adriatic Sea—and had died.

SCHRÖDER, FRIEDRICH LUDWIG (Schwerin, 1744–1816, Rellingen, Holstein), one of the outstanding actors and theatrical directors of the 18th c. He was in charge of the Hamburg theatre from 1771 to 1780 and from 1785 to 1800. In the years 1781–5 he was at the Burgtheater (q.v.), Vienna. His plays (*Dramatische Werke*, 4 vols., posth., 1831) are adaptations or translations.

SCHRÖDER, RUDOLF ALEXANDER (Bremen, 1878–1962, Bad Wiessee, Bavaria), of well-to-do Bremen patrician descent, had no university education. Being drawn to literature, he moved to Munich in 1897, where in 1899 he was a co-founder of the periodical *Die Insel* (q.v.), from which grew the Insel-Verlag. A religious (Protestant) and traditional poet, he was able to employ established forms to express his own originality. At different times he used classical metres, the strophe of the 17th c., odes, and sonnets. His predominantly poetic production includes up to 1933 *Unmut* (1899), *Empedokles, Lieder an eine Geliebte, Sprüche in Reimen* (all 1900), *An Belinde* (1902), *Sonette zum Andenken an eine Verstorbene* (1904), *Elysium* (1906), *Die Zwillingsbrüder* (1908), *Deutsche Oden* (1910), *Heilig Vaterland* (1914), *Audax omnia perpeti* (1919), *Der Herbst am Bodensee, Widmungen und Opfer* (both 1925), *Mitte des Lebens*, and *Jahreszeiten* (both 1930).

Schröder was also a man of considerable artistic talent and was active as an artist and interior designer before the 1914–18 War. He opposed National Socialism and joined the dissenting Confessing Church (see BEKENNENDE KIRCHE). During the 1939–45 War his poems were circulated clandestinely (see INNERE EMIGRATION). In 1946 he was appointed a member of the Protestant Church Synod, and he held many offices of distinction.

Schröder's post-1933 verse consists of *Ballade vom Wandersmann* (1937), *Kreuzgespräch* (1939), *Auf dem Heimweg* (1946), *Weihnachtslieder, Gute Nacht, Alten Mannes Sommer* (all 1947), *Stunden mit dem Wort* (1948), *Parabeln aus den Evangelien* (1951), *Das Sonntagsevangelium in Reimen* (1952). He published collected editions of his poems, first as *Elysium* (1912), then *Die weltlichen Gedichte* (1940) and *Die geistlichen Gedichte* (1949). He was also a distinguished essayist (*Aufsätze und Reden*, 2 vols., 1939) and a talented trans-

lator of Homer (*Odyssey*, 1910; *Iliad*, 1943), Horace (1935), and Virgil (complete, 1924–52). His post-1945 essays include *Goethe und Shakespeare* (1949) and *Meister der Sprache* (1953). *Aus meiner Kindheit* (1953) and *Abendstunde* (1960) are autobiographical. *Gesammelte Werke* (8 vols.) appeared 1952–65.

SCHRÖDER, SOPHIE (Paderborn, 1781–1868, Munich), née Bürger, the daughter of actor parents, began as a child actress in Hamburg in 1791 and became one of the leading tragédiennes in the period before Julie Rettich (q.v.). She first appeared in the Vienna Burgtheater through the influence of Kotzebue (q.v.), made her name in Hamburg, fell out with the French in 1813 (see NAPOLEONIC WARS), and, after restless travels, returned to Vienna (1815–29), where she created, in April 1818, the part of Sappho of Grillparzer's *Sappho* (q.v.) and, three years later, that of Medea (see GOLDENE VLIESS, DAS). Outstanding in classical parts, she spent the years from 1830 to 1835 in Munich. They virtually mark the end of her career. Her daughter Wilhelmine Schröder-Devrient (1804–60) became well known in Vienna as an actress and singer.

SCHRÖTER, CORONA (Guben, 1751–1802, Ilmenau), a gifted soprano singer and actress, who was appointed Kammersängerin in Weimar in 1776. In 1779 she took part in the original performance of Goethe's *Iphigenie auf Tauris* (q.v.), playing Iphigenie to Goethe's Orest. She also sang the first king in *Epiphanias* (q.v.). Goethe mentions her in *Auf Miedings Tod* (q.v.), and she is believed to be the original of the Amazone in *Wilhelm Meisters Lehrjahre* (q.v.).

SCHUBART, CHRISTIAN FRIEDRICH DANIEL (Obersontheim, Württemberg, 1739–91, Stuttgart), was brought up in Aalen. A pastor's son and intended for the Church, his reckless conduct at Erlangen University led his father to terminate his studies. Schubart had musical gifts and became organist at Geislingen (1764), and at the Ducal court at Ludwigsburg (1769). His bold tongue and dubious morals led Duke Karl Eugen (q.v.) to dismiss him in 1773. He moved to Augsburg and afterwards to Ulm, publishing an anti-Roman Catholic newspaper, *Deutsche Chronik* (1774–7, reissued by H. Krauss, 1975). In 1777 he was enticed on to Württemberg territory at Blaubeuren and imprisoned without trial for ten years in the fortress of Hohenasperg. The probable reason for this savage treatment was Schubart's satirical references in his newspaper to the Duke of Württemberg and his mistress, Franziska von Hohenheim (q.v.). During his incarceration the

Duke provided for the maintenance of Schubart's wife and for the education of his children. In 1787 Schubart was released and appointed Master of Music to the Court.

An enthusiastic admirer of Klopstock (q.v.), Schubart wrote religious poetry of pietistic character, vigorous patriotic and democratic political poems, others of folk-song type, and, during the years of his imprisonment, moving poems of anguished experience, such as *Der Gefangene* ('Gefangner Mann, ein armer Mann!'). His best-known poems are *Kaplied* (q.v.), a protest against the sale of citizens as mercenaries to foreign powers, and *Die Fürstengruft* (q.v.), which denounces the tyrant. Schubart was visited in prison by the young Schiller in 1781.

Sämmtliche Gedichte (2 vols.) were published in 1785–6 and *Gedichte, historisch-kritische Ausgabe*, ed. G. Hauff, in 1884 (a selection, ed. P. Härtling (q.v., 1968). *Schubarts des Patrioten gesammelte Schriften und Schicksale* (8 vols.) appeared in 1839–40 (reissued, 8 vols. in 4, 1972). Publications based on documents of Schubart's life include *Schubarts Leben in seinen Briefen* (2 vols., 1849, ext. edn. 1878 by D. F. Strauß, q.v.). *Schubart und seine Zeitgenossen* (1864, reissued 1926) is a novel by A. E. Brachvogel (q.v.).

SCHUBART, PAUL (Danzig, 1863–1915, Berlin), a poet humorist, published a volume of *Katerpoesie* (1909) and also wrote a fantastic novel, *Lesa béndio* (1913).

SCHUBERT, FRANZ (Vienna, 1797–1828, Vienna), the composer, was the son of a schoolmaster and became a choirboy of the Court Chapel. After assisting his father for a few years, Schubart was enabled by the assistance and encouragement of friends to devote himself from 1815 onwards to music. A modest and amiable man, he was fortunate in possessing a number of cultivated friends, including the painter M. von Schwind, the writers F. Grillparzer, E. von Bauernfeld, and J. Mayrhofer (q.v.), the singer J. M. Vogl (1768–1840), the musical amateur J. von Spaun (1788–1865), and the poet F. von Schober (1798–1883). Schubert's compositions include symphonies, of which those in C major (No. 9, formerly No. 7) and B minor ('The Unfinished') are the most notable, and much chamber music, and sonatas for solo piano and for four hands. He wrote several masses, including one (*Deutsche Messe*) in the vernacular, but his vocal music is especially notable for his Lieder (see LIED) for solo voice with piano accompaniment. In Schubert's intimate circle and in the musical evenings of high society these were first performed and popularized by his friend Vogl, whom Schubert accompanied.

From 1814 Schubert wrote more than 600 Lieder, among which over 90 poets are represented; they include 59 songs by Goethe, 12 by M. Claudius, 21 by Hölty, 13 by Klopstock, 26 by Matthisson, 42 by Schiller (qq.v.), and a number by lesser poets such as L. Kosegarten, L. Rellstab, J. F. Rochlitz, and E. K. F. Schulze (qq.v.), as well as his friend Mayrhofer.

Schubert's two song-cycles *Die schöne Müllerin* (q.v., 1824, 20 songs) and *Die Winterreise* (q.v., 1827, 24 songs) are to poems already arranged in cyclical form by W. Müller (q.v.). A third collection of 14 poems, *Schwanengesang* (q.v.), a posthumous publication (1828), is not an organized cycle. It marks Schubert's discovery of Heine's poems, six of which are included in the volume. *Der Hirt auf dem Felsen* (1828), a well-known song with clarinet obbligato, is compounded of verses taken from W. von Chézy (q.v.) and W. Müller. Though not the first writer of Lieder, Schubert is usually regarded as the creator of the form, a claim based on the extent and quality of his productions, his original use of accompaniment, and the powerful effect of his modulations. His œuvre also includes a number of part-songs.

SCHUBERT, GOTTHILF HEINRICH VON (Hohenstein, Saxony, 1780–1860, Laufzorn, Bavaria), abandoned the study of theology at Leipzig University and turned to medicine and the natural sciences, moving later to Jena. He practised medicine successively in Altenburg, Freiberg, and Dresden, was for a few years headmaster of a grammar school in Nürnberg, but in 1819 was appointed to the chair of natural sciences at Erlangen University. In 1827 he moved to Munich University, becoming a member of the Munich Academy. He was ennobled in 1853.

While at Jena, Schubert was influenced by the Nature philosophy of Schelling (q.v.) and reacted strongly against a purely materialistic approach to the sciences. He expressed this in his early work, *Ansichten von der Nachtseite der Naturwissenschaften* (1808). His interest in anthropology and in literature was stimulated by J. G. Herder (q.v.), with whose son Emil Herder he became acquainted as a student. His work *Die Symbolik des Traumes* (1814) investigates subconscious phenomena and their relationship to reality. It expresses attitudes nurtured by the Romantics (see ROMANTIK) and is based on the conviction that there is a distinction between dreams expressing anxiety and serving a premonitory function and dream experience as a communication of the soul with God; on this level the subconscious workings of the mind can lead to an understanding of the presence of God in Nature and in life, and so to a supreme awareness of the wholeness of existence. Schubert appended to the third edition of this work in 1840 posthumous papers of J. F. Oberlin (q.v.) under the heading *Nachlaß eines Visionärs, des J. Fr. Oberlin, gewesenen Pfarrers im Steinthale.* In later life he retreated somewhat from this mysticism (much influenced by J. Böhme, q.v.). Schubert's other important works include *Geschichte der Seele* (2 vols., 1830) and its sequel *Die Krankheiten und Störungen der menschlichen Seele* (1845), *Altes und Neues aus dem Gebiet der innern Seelenkunde* (5 vols., 1817–44), and *Biographien und Erzählungen* (3 vols., 1847–8). His *Allgemeine Naturgeschichte* (1826) appeared in revised form as *Die Geschichte der Natur* (3 vols., 1835–7, reissued 1961).

Schubert's travels in the mid-1830s are recorded in *Reise in das Morgenland* (3 vols., 1838–9); *Der Erwerb aus einem vergangenen und die Erwartungen von einem zukünftigen Leben* (3 vols., 1854–6) is autobiographical. *Vermischte Schriften* (2 vols.) appeared 1856–60.

SCHUBIN, OSSIP, pseudonym of Aloysia Kirschner (Prague, 1854–1934, Kosǎtek, Czechoslovakia), was brought up on an estate in Bohemia, lived for some years in western Europe (Paris, Brussels, and Rome) and then returned to her native country. Her pseudonym is a token of her admiration for Turgenev, whom she met in Paris. Her numerous novels, mostly dealing with high life, include *Schuldig* (1883), *Ehre* (1883), *Unter uns* (2 vols., 1884), *Ein Frühlingstraum* (1884), *Boris Lensky* (1889), *O du, mein Österreich* (3 vols., 1890), *Gräfin Erikas Lehr- und Wanderjahre* (1892), *Finis Poloniae* (1893), *Gebrochene Flügel* (1894), *Woher tönt dieser Mißklang durch die Welt* (1894), *Die Heimkehr* (1897), *Refugium peccatorum* (1903), and *Der arme Nicki* (2 vols., 1906). She continued in later life to pour out a succession of light works of fiction.

SCHÜCKING, LEVIN (Clemenswerth, Westphalia, 1814–83, Pyrmont), was befriended by Annette von Droste-Hülshoff (q.v.), who became deeply attached to him and secured for him in 1841 appointment as librarian to her wealthy brother-in-law, Freiherr von Laßberg (q.v.) in Meersburg castle. He resigned this post in 1842, and in 1843 married a Baroness von Gall, causing a breach with Annette von Droste. Schücking became a newspaper editor in Augsburg (1843–5), and then in Cologne (1845–52). He retired to Sassenberg, Westphalia, in 1852 and devoted himself to writing. His first publication, encouraged by Annette von Droste, and produced with the collaboration of F. Freiligrath (q.v.), was *Das malerische und romantische Westfalen* (1841). *Der Dom zu Köln und seine Vollendung* (1842) is an essay on a matter at that time in the public eye.

Schücking is the author of a number of novels,

mostly set in Westphalia and linked with recent history, rather after the manner of Sir Walter Scott's 18th-c. novels. They include *Die Ritterbürtigen* (3 vols., 1846), *Ein Sohn des Volkes* (2 vols., 1849), *Der Bauernfürst* (2 vols., 1851), *Paul Bronckhorst* (3 vols., 1858), and *Die Marketenderin von Köln* (1861), of which the last two were the most successful. He also wrote more fanciful historical novels (*Luther in Rom*, 3 vols., 1870) and a number of Novellen (*Historische Novellen*, 1862; *Krieg und Frieden*, 3 vols., 1872). He wrote a biography, *Annette von Droste* (1862), and his memoirs appeared as *Lebenserinnerungen* (2 vols., 1886). His collected tales were published as *Gesammelte Erzählungen und Novellen* (6 vols., 1859–66) and an edition of the novels as *Ausgewählte Romane* (24 vols., 1864–72).

Schüdderump, Der, a novel written by W. Raabe (q.v.) in the years 1867–9 and published in 1870. It tells the story of Tonie Häußler, an orphan, brought up in the poorhouse of 'Krodebeck' in the Harz. Tonie is the granddaughter of the village barber, a rasping egoist, who after unsuccessful years at Krodebeck is lost sight of. A few years later his daughter reappears, dying of consumption. She is lodged in the poorhouse and after her death the little girl Tonie is looked after by the aged permanent inmate Hanne Allmann. When Hanne dies, Tonie is taken into 'the big house' and brought up as a lady. The occupants of the manor are Frau Adelheid, the brisk, capable widow of the squire, her not very incisive son Hennig, and two ageing poor relatives, the gentle Ritter von Glaubigern and Fräulein Adelaïde de Trouin, a French *émigrée*, consumed with aristocratic pride. The Ritter and the Fräulein look after the education, first of Hennig, then of Tonie, who turns into a model of beauty and decorum.

At this point a blow falls from without. Tonie's grandfather, now a brilliantly successful speculator and a nobleman, 'der Edle von Haußenbleib', claims his granddaughter, whom he proposes to exploit as a speculation. No one at the manor seriously resists his claim, and Tonie is carried off to Vienna. When, some years later, after his mother's death, Hennig visits Vienna, he finds Tonie seriously ill and in danger of being driven by her grandfather into a commercial marriage with a rich Russian count. Hennig proposes to marry her, but Tonie rejects his offer because she realizes that his motive is not love but pity. She is saved by the aged Ritter von Glaubigern, who undertakes the journey to assert his moral rights over Tonie and so saves her from the threatened unwelcome marriage. But Tonie soon dies and the Ritter quickly declines into second childhood.

As in his two other long novels, Raabe inclines

to a somewhat artificial plot, and the real interest centres upon the characters and on the manner in which they face or fail to face the realities of life. Of these Hennig and Fräulein Adelaïde are inadequate in their response, by weakness or rigidity. But there remains a whole gallery of characters who gain Raabe's support: Frau Adelheid, brave, robust, and forthright; Hanne Allmann with her deep humility and feeling for 'das liebe Vieh'; Jane Warwolf, the pedlar with her tenacity and kindliness, and, finally, Raabe's obvious favourite, Ritter von Glaubigern, unselfish, gentle, chivalrous, and courageous. Ostensibly, the story is one of deep pessimism, but the assertion of human values in these figures counterbalances the certainty of death and the sufferings of life. And so the 'Schüdderump', a cart for the mass transport of the corpses of plague victims, is not a true symbol of the book, though Raabe parades it through his pages from time to time. The last sentence of *Der Schüdderump* has given rise to the contention that the three novels, *Der Hungerpastor*, *Abu Telfan*, and *Der Schüdderump*, constitute a trilogy: 'Wir sind am Schlusse—und es war ein langer und mühseliger Weg von der Hungerpfarre zu Grunzenow an der Ostsee über Abu Telfan im Tumurkielande und im Schatten des Mondgebirges, bis in dieses Siechenhaus zu Krodebeck am Fuße des alten germanischen Zauberberges.' The significance of this pronouncement has given rise to controversy.

Schuld, Die, a four-act fate tragedy (see SCHICKSALSTRAGÖDIE) in trochaic verse by Adolf Müllner (q.v.), written in 1812 and published in 1816. Graf Hugo has in the past murdered a man and married Elvire, his widow. He discovers that the dead man was his own brother. His wife requests him to kill her. Hugo does so and then kills himself. It is a concentrated play, the action of which is in the past and is gradually revealed, Oedipus-fashion, to the horrified participants. *Die Schuld* enjoyed a sensational success on the stage.

Schuldlosen, Die, a novel published in 1950 by H. Broch (q.v.), and described as a 'Roman mit 11 Erzählungen'. It consists of five existing stories which Broch has linked together and expanded with six new ones. The tangled story of involved sexual relations reflects a disintegrating morality and touches on the political decline of the years 1923–33, of which the final symbol is the swastika flag flying over the city square.

Schüler von Paris, Der, an anonymous short Middle High German verse tale of the 14th c. recounting a love affair in which the man dies at the climax of love, whereupon his partner dies also. The poem exists in other forms, including one known as *Frauentreue* (q.v.). It has affinities with the Middle High German *Hero und Leander* (see HERO UND LEANDER).

SCHULTHESS, BARBARA (Zurich, 1745–1818, Zurich), *née* Wolf, married David Schultheß of Zurich. Goethe met her in 1775 on his first Swiss journey, renewed the acquaintance in 1779, called on her on his return from Italy in 1788, and met her for the last time in 1797. The copy of Goethe's novel *Wilhelm Meisters Theatralische Sendung* (q.v.), which was discovered in December 1909, was among papers belonging to her and preserved by her descendants. It has been said that she was the model for Nachodine (also known as 'das nußbraune Mädchen', Susanne, and 'die Gute-Schöne') in *Wilhelm Meisters Wanderjahre* (q.v.).

SCHULZ or **SCHULTZ,** JOHANN ABRAHAM PETER (Lüneburg, 1747–1800, Schwedt), a minor composer, was a theatre music director in Berlin from 1776 to 1778 and from 1780 to 1787 conductor (Kapellmeister) of the orchestra of Prince Heinrich (q.v.) of Prussia in Rheinsberg. In the 1790s he was at the Danish court. Schulz had a gift for popular melody and wrote the tunes for 'Am Rhein, am Rhein', 'Des Jahres letzte Stunde', and Bürger's 'Herr Bacchus ist ein braver Mann' and 'Ich will einst bei Ja und Nein' (see BÜRGER, G. A.).

SCHULZE, ERNST KONRAD FRIEDRICH (Celle, 1789–1817, Celle), studied theology and philology at Göttingen University, where he became engaged to Cäcilie, daughter of Professor Tychsen, in 1811. She died in the following year. Schulze served with the Hanoverian Rifles in the War of Liberation, returned afterwards to Göttingen, and died within a year or two of pulmonary disease. His poetic activity in his short life was centred on the loss of his betrothed, projections of whom are the central figures in his principal works.

Cäcilie. Ein romantisches Gedicht appeared posthumously in 1818–19. It is an epic poem in twenty cantos dealing with the conquest and conversion of heathen Denmark by Otto I (q.v.) in the 10th c. Cäcilie, with her minstrel adorer Reinald, accompanies the army. When success crowns the expedition she dies and her soul flies up to Heaven, while Reinald remains below. A shorter work, *Die bezauberte Rose*, also described as a 'romantisches Gedicht' won a prize offered by the magazine *Urania* (q.v.), and news of this reached Schulze shortly before his death. It is in three cantos, comprising 107 stanzas of *ottava rima*. The enchanted rose is a princess (Klothilde) placed under a spell. After suitors of great pomp and power have failed to secure her release, the minstrel Alpino is successful and she is restored to her proper form. It was published in *Urania* (1818).

Schulze wrote with elegance and feeling, and his poems, especially *Die bezauberte Rose*, were widely read before epic poetry went out of fashion. His shorter poems (*Gedichte*, 1813), which contain a once well-known poem, *Cäcilie. Eine Geisterstimme*, show his skill in verse forms. including classical elegiacs.

SCHULZE, Friedrich August (Dresden, 1770–1849, Dresden), a prolific writer of popular novels and tales under the pseudonym Friedrich Laun, is now known only for the story *Der Mann auf Freiersfüßen* (1800). He published jointly with J. A. Apel (q.v.) a long series of ghost stories (*Gespensterbuch*, 6 vols., 1810–17), one of which is the basis of the plot of Weber's opera *Der Freischütz* (q.v.). He is the author of *Gedichte* (1828) and *Memoiren* (3 vols., 1837). His collected works were published in 1843 with an introduction by L. Tieck (q.v.).

SCHUMANN, Robert (Zwickau, 1810–56, Endenich nr. Bonn), studied music at Leipzig under F. Wieck, whose daughter Clara (1819–96) he married in 1840. From 1834 to 1844 Schumann edited the *Neue Zeitschrift für Musik*. In his musical writings his lively imagination conjured up a group of Romantic musicians (Davidsbündler), of whom the most important were Florestan and Eusebius, representing two aspects of his own romantic character,·the one stormy and enthusiastic, the other gentle, melancholy, and introspective. Before 1840 Schumann composed chiefly piano music, frequently with literary titles (*Papillons*, 1832; *Die Davidsbündler*, 1837; *Kinderszenen*, 1838; and *Kreisleriana*, an allusion to E. T. A. Hoffmann, q.v., also 1838). His marriage in 1840 was the signal for a remarkable outpouring of songs (see *Lied*).

A highly sensitive reader of poetry, Schumann set poems by Goethe, Schiller, Eichendorff, Heine, Chamisso, Kerner, Rückert, Mörike, and Lenau (qq.v.), as well as poems by lesser writers such as J. Mosen (q.v.), who is the author of *Der Nußbaum*. He also set a number of translated poems by Burns, Byron, Thomas Moore, and various Spanish poets in versions by Geibel (q.v.). Schumann was notably successful with cycles, *Liederkreis* (1840, Heine), *Liederkreis* (1842, Eichendorff), *Frauen-Liebe und Leben* (1840, Chamisso), and above all *Dichterliebe* (1840, Heine). One of his most ambitious works was the cantata *Szenen aus Goethes Faust* (1847–50).

From 1853 his mental powers declined and he had to resign his post as municipal director of music in Dresden, to which he was appointed in 1850. He died as a voluntary patient at an institution for nervous diseases, which he entered after an unsuccessful attempt at suicide in 1854. An incomplete edition of his collected works (31 vols.) was published by Clara Schumann and J. Brahms (q.v.), 1881–93. His diaries, ed. G. Eismann (3 vols.), appeared 1972 ff.

SCHUMANN, Valentin (Leipzig, *c.* 1520, d. after 1559), son of a bookseller, became an artisan, and later a mercenary in the Turkish wars, finally settling in Nürnberg as a typefounder. He is the author of a collection of Schwänke (see Schwank) entitled *Das Nachtbüchlein* (1558).

SCHUMMEL, Johann Gottlieb (Seidendorf, Silesia, 1748–1813, Breslau), a schoolmaster successively at Magdeburg, Liegnitz, and Breslau (as Professor), was the author of sentimental and satirical novels. His principal works are *Empfindsame Reisen durch Deutschland* (1770–2), *Spitzbart, eine komi-tragische Geschichte für unser pädagogisches Jahrhundert* (1779), a satire on the educational theories of J. B. Basedow (q.v.), and *Wilhem von Blumenthal oder das Kind der Natur* (1780–1).

Schundliteratur, a classification ranking much lower than Trivialliteratur (q.v.), which makes some pretensions to taste and form, and Kitsch (q.v.), a highly subjective noun of considerable range. Schundliteratur is written either for gain or to gratify perversion, and it appeals to baser impulses of various kinds, not necessarily sexual.

SCHUPP, or **SCHUPPIUS, Johann Balthasar** (Gießen, 1610–61, Hamburg), after studying at Marburg, travelled abroad and then returned to Marburg, becoming a professor in 1634 and a pastor in 1643. After losing his possessions on the capture of Marburg in 1646, he was employed on diplomatic missions by his Hessian sovereign. In 1649 he settled as pastor in Hamburg, where he became engaged in bitter controversy with pastor Johannes Müller. Schupp wrote poems (*Morgen- und Abendlieder*, 1643), a moral tract in the form of advice to a son, containing satirical anecdotes (*Freund in der Not*, 1657), *Der Regentenspiegel* (1657), and a social satire *Corinna. Die ehrbare Hure* (1660).

Schupp's *Schrifften* (posth. 1663) were edited by C. Vogt (*Streitschriften*, 2 vols.) and published in 1910–11, and *Der teutsche Lehrmeister* and *Vom Schulwesen* (posth. 1667?), ed. P. Stötzner (2 vols.), in 1891.

Schuß von der Kanzel, Der, a short Novelle, divided into eleven concise chapters, by C. F. Meyer (q.v.), published in 1878. Set in rural Switzerland near Lake Zurich with an antiquated 17th-c. flavour, it narrates an amusing and ingenious intrigue by General Wertmüller, designed to resolve an unfortunate dilemma of two young lovers. They are Rahel, the daughter of the General's cousin, Pastor Wertmüller of the parish of Mythikon, who has an unseemly addiction to firearms, and Pfannenstiel, a deacon without any such inclinations, who cannot marry until he has found a living. He therefore applies for a chaplaincy in one of the General's regiments. General Wertheim, alert to the unsuitability of the candidate, finds a solution benefiting the lovers as well as cousin Wertmüller. By a simple ruse he lures the Pastor into discharging a blank pistol-shot in the course of his sermon, obliging him to resign. But the General gladdens his cousin's heart by putting him in charge of the game and shooting on his estate. Pfannenstiel, who succeeds Wertmüller, gets the living for which he is suited and the hand of Rahel. The General dies suddenly in Germany not long after.

Schutt, a collection of political poetry by Anastasius Grün (q.v.), published in Hamburg (in order to circumvent the Viennese censorship) in 1835.

SCHÜTZ, HEINRICH (Köstritz nr. Gera, 1585–1672, Dresden), enjoyed the early patronage of the Elector Moritz (q.v.) of Hesse, and from 1617 until his death was Master of Music (Hofkapellmeister) to the Elector of Saxony, residing mostly in Dresden. Schütz, who was influenced by Gabrieli and Monteverdi, was the greatest German composer of the 17th c. and was widely recognized as such. In the 18th c. he was forgotten and was only rediscovered late in the 19th c. (edition by P. Spitta, 1885–94).

Though primarily a composer of religious and liturgical music, Schütz set a German translation (by M. Opitz, q.v.) of the opera *Daphne* by O. Rinuccini, which was performed at Torgau in 1627. This music has since been lost. Nearly all Schütz's religious works are vocal, and employ words from the Lutheran Bible. He composed three Passions (St. Matthew, St. Luke, and St. John, all 1665–6), early in life a Resurrection Oratorio (*Auferstehungshistorie*, 1623) and in old age a Christmas Oratorio (*Historia der freuden- und gnadenreichen Geburt Gottes und Marien Sohnes Jesus Christi*, 1664). He also made settings of some of the Psalms (*Die Psalmen Davids*, 1619), and late in life set the very long Psalm 119. He wrote funeral music (*Musikalische Exequien*) for his friend Prince Heinrich von Reuß and many motets. His last work was a *Deutsches Magnificat* (1671).

SCHÜTZ, WILHELM VON (Berlin, 1776–1847, Leipzig), a Prussian nobleman holding the office of Landrat at Ziebingen nr. Frankfurt/Oder (Neumark), was a minor dramatist of the Berlin Romantic generation (see ROMANTIK). His first tragedy, *Lacrimas* (1803), was followed by *Niobe* and *Der Graf und die Gräfin von Gleichen* (both 1807). Other plays include *Graf von Schwarzenberg* and *Karl der Kühne* (both 1819). Schütz resigned his office in 1811 and went to live in Dresden. He was converted to the Roman Catholic faith in 1833.

In later years Schütz wrote on such diverse matters as Roman Catholicism, banking, railways, wool production, ecclesiastical law, Goethe's *Faust*, and epic poetry. He also published the first (incomplete) German version of the memoirs of Casanova (q.v., *Aus den Memoiren des Venezianers J. Casanova*, 2 vols., 1822–8). The *Mémoires* did not appear in French until 1826–38.

SCHWAB, GUSTAV (Stuttgart, 1792–1850, Stuttgart), was educated at the Stuttgart grammar school (Gymnasium) and the Tübinger Stift (q.v.), entering the ministry of the Lutheran Church. As a student he was a close friend of L. Uhland, J. Kerner, and K. Mayer (qq.v.), and was associated with the Schwäbischer Dichterkreis (q.v.). In 1815 he undertook a journey through Germany, in the course of which he visited almost every reputable German writer, including Goethe as well as the younger Romantic authors such as A. von Chamisso, E. T. A. Hoffmann, and J. and W. Grimm (qq.v.). After a brief spell of teaching at the Tübinger Stift he became classics master (with the title professor) at his old school in Stuttgart (1817–37). From 1827 to 1837 he was also editor of the literary columns of the Stuttgart *Morgenblatt*, and from 1833 to 1838 he edited jointly with Chamisso the *Deutscher Musenalmanach*. From 1837 to 1841 he had a living at Gomaringen near Tübingen and then became pastor at St. Leonhard's in Stuttgart. He held additional offices in the ecclesiastical and educational administration of Württemberg.

Schwab was a prolific, if unoriginal, writer. His own poems (*Gedichte*, 1828–9, amplified in 1838) are largely forgotten, although the ballads *Das Gewitter, Der Reiter und der Bodensee*, and *Der Riese von Marbach* remained popular throughout the 19th c. In 1815 he began with the collection of student songs *Neues deutsches allgemeines Commers- und Liederbuch* (which included among his own contributions 'Bemooster Bursche zieh ich aus'), and followed this with *Romanzen aus dem Jugendleben Herzog Christophs von Württemberg* (1816). He found his most congenial vein with the collection *Buch der schönsten Geschichten und*

Sagen (2 vols., 1836–7), achieving his greatest success with *Die schönsten Sagen des klassischen Altertums* (3 vols., 1838–40, repeatedly reissued).

Schwab was an anthologist, an editor of chapbooks (*Deutsche Volksbücher*, 3 vols., 1836–7), a travel writer, and the author of *Schillers Leben* (1840). His own biography was written by his son C. Th. Schwab (1883). His correspondence with the brothers A. and D. E. Stöber (qq.v.), ed. K. Walter, appeared in 1930.

Schwabacher Artikel, Die, refer to a document drawn up in 1529 setting out the Lutheran confession of faith. The articles formed a basis of discussion between the followers of Luther and those of Zwingli (qq.v.) at the Marburger Religionsgespräch, arranged by Landgrave Philipp (q.v.) of Hesse. Agreement was reached on the first fourteen articles, but the fifteenth, relating to the interpretation of Holy Communion, proved an insurmountable stumbling-block.

SCHWABE, JOHANN JOACHIM (Magdeburg, 1714–84, Leipzig), a disciple of J. C. Gottsched (q.v.), studied at Leipzig and published between 1741 and 1745 the eight volumes of *Belustigungen des Verstandes und Witzes*, which supported Gottsched in his controversy with the Swiss critics J. J. Bodmer and J. J. Breitinger (qq.v.). Schwabe became university librarian at Leipzig in 1750 and a professor of philosophy in 1765.

Schwaben, (Swabia), historical duchy and now a common regional designation for south-west Germany (Baden-Württemberg). The original duchy also included Alsace and German-speaking Switzerland. The last duke of Swabia, Johannes Parricida, who appears as a character in Schiller's *Wilhelm Tell* (q.v.), died in 1313.

Schwabenkrieg, a war fought in 1499 between the Swiss cities on the one hand and the Swabian Confederation (Schwäbischer Bund, q.v.) and the Empire on the other. The root cause was Swiss fears that their practical (though not legally based) independence was endangered. The occasion was the decision of the city of Constance to enter the Swabian Confederation. The fiercely fought war ended in a Swiss victory which was confirmed at the Treaty of Basel (1499), by which imperial claims in Switzerland were withdrawn. The war established the effective independence of Switzerland, though its ties with the Empire were not officially severed until 1648, by the Peace of Westphalia. The war is also sometimes referred to as the 'Swiss War' (Schweizerkrieg).

SCHWABENMÄDLE, see HAHN, ELISE.

Schwabenspiegel, title in use since the 17th c. for the written laws of the German people originally known as the *Kaiserrecht*, the *Landrechtsbuch*, or the *Lehensrechtsbuch*. It is an adaptation of the *Deutschenspiegel* (q.v.) and is the work of a priest writing, possibly in Eastern Franconia, *c.* 1270–80. Its wide distribution is attested by more than 350 MSS.

Schwäbischer Bund also called der Schwäbische Städtebund, (1) a league of imperial cities founded in 1376 with Ulm at its head, and eventually including Franconian (Regensburg, Nürnberg) as well as Swabian cities. Its aim was the safeguarding of the rights, privileges, and, especially, independence of the cities against the expansionist policies of the territorial princes and the attacks of minor predatory nobles. The league was defeated by Eberhard II of Württemberg in 1388 and dissolved in 1389 under the terms of the Egerer Landfriede.

(2) An alliance made between the Archduke Sigmund of Tyrol, Count Eberhard V of Württemberg, a society of imperial knights, and twenty-two imperial cities, and inaugurated in Esslingen in 1488. It supported the Emperor Friedrich III (q.v.) against the house of Wittelsbach (q.v.). This Schwäbischer Bund was dissolved in 1533.

Schwäbischer Dichterkreis, a number of poets, writing chiefly between 1810 and 1840, not closely associated in a coterie, but all resident in Württemberg and most of them in touch by occasional or, in some cases, regular meetings, and by correspondence. They tend to appear in their poetry and occasional narrative works as stragglers of the Romantic movement (see ROMANTIK). The outstanding personalities are L. Uhland, J. Kerner, G. Schwab, and W. Hauff, together with W. F. Waiblinger, G. Pfizer, K. F. H. Mayer, J. G. Fischer, H. Kurz, and F. K. Gerok (qq.v.). Mörike (q.v.), though in contact at times with several of them, can hardly be called a member of any group.

Schwäbische Trauformel, also known as *Schwäbisches Verlöbnis*, is the designation of a Middle High German written version of Swabian marriage procedure. The document outlines the form of the ceremony and gives the words spoken by the bridegroom and the bride's father or guardian. The procedure laid down is not ecclesiastical, but an act of civil law. The MS. dates from the 12th c., but the legal process described is considerably older.

SCHWAMMERL, nickname of the musician Franz Schubert (q.v.). It is used by R. H. Bartsch (q.v.) as the title of a sentimental novel about Schubert.

SCHWAN, FRIEDRICH (1729–60), a robber active in the district round Stuttgart, who is the original of Christian Wolf in Schiller's story *Der Verbrecher aus verlorener Ehre* (q.v.).

Schwanengesang, a collection of fourteen songs composed by Franz Schubert (q.v.) and published in 1828. The first seven songs are by Ludwig Rellstab and the last is by J. G. Seidl (qq.v.). In between are six songs by H. Heine (q.v.), *Der Atlas, Ihr Bild, Das Fischermädchen, Die Stadt, Am Meer,* and *Der Doppelgänger.*

Schwangere Mönch, Der, a Middle High German erotic verse tale of the 13th c., in which a foolish and innocent monk is seduced and believes himself pregnant. The author gives his name as der Zwingäuer.

Schwank, denotes in medieval and 16th-c. literature a humorous narrative or play. The narratives, which are usually collections of anecdotes connected with a legendary practical joker, extend over the whole period and include *Der Pfaffe Amis* (13th c.) by Der Stricker (q.v.), *Neidhart Fuchs* (q.v., 15th c.), *Der Pfaffe vom Kahlenberg* (15th c.) by P. Frankfurter (q.v.), *Till Eulenspiegel* (q.v., 1515), *Schimpf und Ernst* (1522) by J. Pauli (q.v.), *Peter Leu* (c. 1550, see WIDMANN, A. J.), *Das Rollwagenbüchlein* (1555) by G. Wickram (q.v.), *Das Nachtbüchlein* (1558) by V. Schumann (q.v.), and *Das Lalenbuch* (q.v., 1597).

The dramatic Schwank is confined to the 16th c., and its principal exponent is Hans Sachs (q.v.). Schwänke occur in both verse and prose.

Schwanritter, Der, a Middle High German verse romance by KONRAD von Würzburg (q.v.). It is probably an early work and the MS. is not complete. The story is familiar under the name of Lohengrin, but the Knight of the Swan is not named in Konrad's poem. Konrad sets the poem in the time of Charlemagne (see KARL I).

The widow of Duke Godfrey of Brabant sues the Duke of Saxony, whom she accuses of usurping her inheritance, and Charlemagne sits in judgement. The proceedings are interrupted by the appearance of a swan drawing a boat containing a sleeping knight. The Knight of the Swan becomes the Duchess's champion and asserts her rights by defeating the Duke of Saxony. A lacuna in the MS. must have narrated the Knight's marriage and the forbidden question, asking his name; for, in the conclusion of the story, he sadly bids the Duchess farewell. From his children are descended the counts of Gueldres and Cleves. A notable feature is the elaborate account Konrad gives of the court proceedings.

SCHWARZ, GEORG, see NIGRINUS, GEORG.

SCHWARZ, SIBYLLE, whose name also occurs in the feminine form Sibylle Schwarzin (Greifswald, 1621–38, Greifswald), daughter of the mayor of her native town, wrote poems in the manner of Opitz (q.v.), which were published in 1650 by Samuel Gerlach (1615–54, *Deutsche poetische Gedichte*). In the hyperbolical fashion of the age, she was sometimes termed die pommersche Sappho.

SCHWARZBURG-RUDOLSTADT, see AEMILIA, GRÄFIN VON, and LUDAEMILIA, GRÄFIN VON.

SCHWARZENBERG, FELIX, FÜRST ZU (Krumau, 1800–52, Vienna), became prominent in Austrian and German politics during the last four years of his life. A field-marshal during the revolutionary wars (see REVOLUTIONEN 1848–9), Schwarzenberg was appointed by General Windischgrätz (q.v.) to be Austrian chief minister in succession to Metternich (q.v.). By his autocratic and reactionary policy he safeguarded the throne for Franz Joseph (q.v.). He repealed liberal concessions granted under popular pressure and suppressed national claims for greater independence in Austria-Hungary. He restored the German Confederation (see DEUTSCHER BUND) by putting pressure upon Prussia (see OLMÜTZER PUNKTATION).

SCHWARZENBERG, KARL PHILIPP, FÜRST ZU (Vienna, 1771–1820, Leipzig), an Austrian officer, first saw active service against the Turks in 1789. As a divisional commander at Ulm in 1805 he successfully broke out with his troops and escaped the general capitulation. He also fought at Wagram in 1809 and, in 1813, as Austrian commander-in-chief, at Leipzig (see NAPOLEONIC WARS).

Schwarze Spinne, Die, a Novelle by J. Gotthelf (q.v.), published in *Bilder und Sagen aus der Schweiz* (vol. 1, 1842). One of the most notable examples of the genre, it is set in a frame (see RAHMEN). Its basic theme is salvation through sacrifice linked with the ritual of christening. A Swiss village celebrates on Ascension Day the christening of a newly born child with customary lavish hospitality. In this contemporary, homely setting the grandfather functions as the narrator of the two stories (Binnenerzählungen) contained in the frame, of which the first is designed to dominate the entire narrative.

Going back some six hundred years, it provides a mythical treatment of suffering and tyranny, of the eruption of daemonic forces, and of their conquest through sacrifice. An overlord imposes upon his peasants, forebears of the villagers, the impossible task of creating within one month a beech avenue leading up to his castle. The tempter, exploiting their grim position, promises help if one of them surrenders his next born child to him. None dares, and the hour of doom approaches, when Christine, a stranger, encounters the devil in the guise of a green huntsman. By his kiss she concludes the pact which saves the peasants, but they still have to fulfil the condition of the pact which they are determined to resist. When the devil has been cheated for a third time, the third newly born child having been christened before he could grasp it, he revenges himself by causing Christine's metamorphosis into a black spider which causes plague and devastation and is the devil's tool in the fulfilment of the pact. The sacrifice of a mother, who, at the cost of her life, succeeds in arresting the spider by sealing it inside a hollow piece of wood, restores the God-fearing community to equanimity. The spider, however, remains ever present in its confinement, and its release will spell the renewed eruption of daemonic forces which exist in perpetuity and can be checked but not annihilated.

The second story, set in a less remote age, provides a contrasting illustration of this theme. Materialism leads to the disintegration of the spirit. When a farm-hand, resembling the green huntsman, one Christmas Day releases the spider from its confinement, the destruction is as devastating as it had been before, burning into ashes the house built by the wealthy forebears of the family. Only the action and self-sacrifice of Farmer Christen, a father (complementing a mother's capacity for sacrifice contained in the first story), successfully returns the spider to its captivity.

The old piece of wood has since been built into the family's new house as a constant reminder of the threat of evil which only the simplicity of faith can arrest.

Schwarzgelb, the Habsburg colours of black and yellow and so a term applied in the late 19th c. and early 20th c. to policies in support of the imperial house.

Schwarzrotgold denotes the black, red, and yellow flag of the German Federal Republic (see BUNDESREPUBLIK DEUTSCHLAND). These colours form also the basic flag of the German Democratic Republic (see DEUTSCHE DEMOKRATISCHE REPUBLIK). The 'black-red-gold' combination originated as the colours of the Lützowsches Freikorps in 1813 and were then adopted by the Burschenschaft (q.v.) in 1815. From 1919–33 the black, red, and yellow flag was the national flag of Germany and was readopted for West and East Germany in 1948. (See also SCHWARZWEISSROT.)

Schwarzwälder Dorfgeschichten, a collection of village stories by B. Auerbach (q.v.), published in four volumes 1843–53. The first volume contained *Der Tolpatsch* (q.v.), *Die Kriegspfeife, Des Schloßbauers Vefele, Tonele mit der gebissenen Wange* (q.v.), *Befehlerles* (q.v.), *Die feindlichen Brüder, Ivo der Hajrle, Florian und Kreszenz,* and *Der Lauterbacher. Sträflinge* (q.v.), *Die Frau Professorin* (q.v.), *Lucifer, Die Geschichte des Diethelm von Buchenberg* (q.v.), *Brosi und Moni, Der Viereckig,* and *Der Lehnhold* (q.v.) appeared in the subsequent volumes. A new edition, *Sämtliche Schwarzwälder Dorfgeschichten* (1871) added *Hopfen und Gerste, Ein eigen Haus, Erdmute,* and *Der Geigerlex.*

After Auerbach's death a further expanded edition, *Sämtliche Schwarzwälder Dorfgeschichten* (10 vols., 1884), included also the short novels *Barfüßele* (q.v., 1856), *Joseph im Schnee* (q.v., 1860), and *Edelweiß* (q.v., 1861), and (in its last two volumes) the sequels to some of the original stories which had been published in 1876 as *Nach dreißig Jahren.* These sequels are *Des Lorles Reinhard* (see DIE FRAU PROFESSORIN), *Der Tolpatsch aus Amerika* (see DER TOLPATSCH), *Das Nest an der Bahn* (see STRÄFLINGE), and *Brigitta.*

Schwarzwälder Prediger, a late 13th-c. collection of Middle High German sermons which exists in several MSS. and takes its name from the dialect in which it is written. The source is a Latin collection, *Sermones de Tempore,* by a Franciscan, Konrad von Sachsen (d. 1279); the style of the German version is lively and popular. The translator's name is not known.

Schwarzweißrot, colours of the German national flag from 1871 to 1919. The flag was reinstated in 1933 and was flown beside the swastika flag until 1935 (see HAKENKREUZ and SCHWARZROTGOLD).

Schwedische Gräfin, Die, see LEBEN DER SCHWEDISCHEN GRÄFIN VON G . . .

Schweidnitz, battle of (9 October 1762), a victory for Friedrich II (q.v.) of Prussia which finally crushed Austrian hopes for the recovery of Silesia (see SIEBENJÄHRIGER KRIEG).

Schweigen, a Novelle by Th. Storm (q.v.), published in *Westermanns Monatshefte* in 1883. A young forester, Rudolf von Schlitz, marries, after a brief courtship, Anna, the daughter of a village pastor. Under pressure from his domineering mother, he conceals from his young wife the fact that he has recently been having treatment for a mental illness. This concealment (Schweigen) weighs on his conscience and increases his fear of a recurrence of the disease. He decides to commit suicide, in order to make room for Anna's rival suitor Bernhard, a farmer, of whose affection for Anna he feels assured. The tragic end is, however, averted. In writing his farewell letter to his wife, Rudolf relieves his conscience and recovers from his depression at the very moment at which he prepares to shoot himself, and is found in the nick of time by his wife. The narrative structure of this psychological study is supported by motifs from Weber's *Der Freischütz* (q.v.).

SCHWEINICHEN, HANS VON (Gröditzberg, Silesia, 1552–1616, Liegnitz), a Silesian nobleman in the service of an extravagant duke of Liegnitz, Heinrich XI, wrote a diary (ending in 1602) which gives a forthright account of his and his lord's experiences and a picture of the times. Intended for private use as a warning to his descendants, it was first published 1820–3.

Schweißtuch der Veronika, Das, a novel by Gertrud von LE FORT (q.v.), first published. in 1928. In 1946 a sequel, *Der Kranz der Engel,* appeared, and in 1958 the two novels were combined into one work in two volumes. The title *Das Schweißtuch der Veronika* was transferred to the whole book, and the original volume was renamed *Der römische Brunnen.*

This novel, which was Gertrud von le Fort's preoccupation for so many years, examines, from a religious standpoint, the increasing secularization of mankind and the role in life of faith and sacrifice. The first volume is set in Rome, the meeting-place of Christianity and classical humanism. The young Veronika, an orphan, lives with her wealthy grandmother and her aunt. The grandmother, cultivated, tolerant, and generous, represents all that is best in the humanistic tradition; but her primarily aesthetic values are not sufficient to support her in all the vicissitudes of life. The aunt, an emotionally disappointed woman, inflicts much torment on herself and others, and at the last undergoes a deathbed conversion.

By her absent father's decree, Veronika is brought up without religion. She has a sensitive responsiveness which earns for her the nickname Spiegelchen, and she possesses the virtues of humility and love (*caritas*). Released by her father's testament from the anti-clerical ban, she is converted to the Roman Catholic faith, though this step is strongly disapproved by Enzio, a young protégé of Veronika's grandmother, who is for a time the girl's constant companion. The

Roman fountain of the title is in the courtyard of the grandmother's house, and serves as a symbol of purity. The kerchief of St. Veronica alludes to the suffering imprinted on Veronika's features by the disillusionment of her grandmother, the torments of her aunt, and the mockery of Enzio.

The 1914–18 War intervenes between the two volumes of the novel. In *Der Kranz der Engel* Veronika is persuaded by her guardian, a professor at Heidelberg, to postpone entry into a convent. She becomes a student, meets Enzio again (now returned from the war), and the two fall in love. Veronika conceives that her mission, entrusted to her by God, is to share her whole life, including her religion, with Enzio; he, however, rejects the sacrament and insists on a civil. marriage. She renounces her hope of a union blessed by the Church and with it all that is most sacred to her. But she still cherishes the belief that her self-sacrificing love will work for Enzio's redemption. The emotional strain of this situation leads to a serious illness, and the novel ends with a sign of her success; for Enzio, acting against his proclaimed opinions, summons a priest to her bedside.

SCHWEITZER, ALBERT (Kaysersberg, Alsace, 1875–1965, Lambaréné, Gaboon), began as a pastor (from 1899) and a New Testament scholar in Strasburg before qualifying as a physician and becoming an organist and interpreter of J. S. Bach (q.v.), on whom he wrote a monograph (in French 1905, in German 1908) and whose organ works he edited (*Bachs Orgelwerke*, jointly with C. M. Widor, 1912–14); he also became an authority on organ building (*Deutsche und französische Orgelbaukunst*, 1906, reissued 1962).

In 1913 Schweitzer founded a tropical hospital in Lambaréné, devoting his life to its development except for the years between 1917, when he was interned in France, and 1924. It was to a large extent financed by his organ concerts during his frequent visits to Europe. In 1927 the primitive old construction was replaced by a new and larger hospital. Schweitzer's perseverance and sense of mission were exceptional, and he expressed his strong convictions in lecture tours and in writings. He opened his contributions to the theological dispute on the life of Jesus with *Das Messianitäts- und Leidensgeheimnis* (1901) and *Von Reimarus zu Wrede* (1906); *Die Mystik des Apostels Paulus* appeared in 1930. His principal writings on ethics were published in 1966 (ed. H.-W.Bähr) as *Die Lehre von der Ehrfurcht vor dem Leben.* His respect for life was amply demonstrated by his work for the African natives, but it extended to all living creatures. He remained rooted in Christianity

(*Reich Gottes und Christentum*, ed. U. Neuenschwander, 1967), but was also influenced by Schopenhauer (q.v.) and published a work on Indian philosophy (*Die Weltanschauung der indischen Denker*, 1935). In 1939 appeared four lectures on Goethe (*Goethe*). His autobiographical works are *Aus meiner Kindheit und Jugendzeit* (1924) and *Aus meinem Leben und Denken* (1932, reissued 1960). In 1955 Schweitzer published *Das Problem des Friedens in der heutigen Welt* following the award of the Nobel Peace Prize in 1952 and the Friedensklasse des Ordens Pour le mérite (see POUR LE MÉRITE) in 1954.

Schweizer, Die, appellation frequently used in histories of literature and critical writings for the Swiss critics J. J. Bodmer and J. J. Breitinger (qq.v.).

Schweizerische Robinson, Der, a novel of shipwreck and adventure by Johann David Wyss (q.v.).

Schweizerkrieg, see SCHWABENKRIEG.

SCHWENCKFELD or **SCHWENKFELD,** KASPAR VON (Lüben, Silesia, 1489–1561, Ulm), originally a counsellor of the Duke of Liegnitz, went over to the Reformation (q.v.), but later diverged from Luther, who bitterly attacked him. Schwenckfeld abandoned literal belief in Holy Writ for a mystical attitude and is one of the forerunners of later German Pietism (see PIETISMUS). He enlisted many followers, especially in Swabia and Silesia. In 1734 the sect (Schwenckfeldianer) emigrated to Pennsylvania. Schwenckfeld's writings and correspondence are contained in *Corpus Schwenckfeldianorum* (19 vols.), 1907–61.

Schwertbrüderorden (Fratres militiae Christi), also known as Schwertritter, a religious order of knights founded in Livonia in 1201 by Bishop Albert of Riga in order to support the Christian community and to extend its influence and territory. The Order suffered a severe defeat in conflict with the Lithuanians in 1236, and soon afterwards its remnants were absorbed into the Teutonic Order (see DEUTSCHER ORDEN).

Schwestern von Lesbos, Die, a poem in six cantos written by Amalia von Imhoff (q.v.) and first published in *Der Musenalmanach für 1800*. It appeared as a separate work in the following year. *Die Schwestern von Lesbos*, which is written in the classical hexameters made familiar by Goethe's *Hermann und Dorothea* (q.v.) and the idylls of J. H. Voß (q.v.), recounts the story of two sisters who both love the same young man.

He is destined for the elder, Simaitha, but when she discovers that he loves Licoris, she renounces his hand and leaves him to her sister.

Schweyk im Zweiten Weltkrieg (1943), a play in eight scenes by B. Brecht (q.v.) based on the Czech novel *Osudy dobrého vojáka Švejka* (1920–3, *The Good Soldier Schweik*) by J. Hašek. The *Vorspiel in den höheren Regionen* introduces Hitler (with a blood-stained globe), Göring, Goebbels, and Himmler (qq.v.), all larger than life except for Goebbels, who is of very small stature. The central location of the action is the inn Zum Kelch in German-occupied Prague. It is run by the resolute Frau Kopecka, who helps those worthy of help and tries to keep politics outside.

It is at Frau Kopecka's that 'dog-dealer' Schweyk, one of her regulars, first encounters Brettschneider, the Gestapo spy. Brettschneider takes Schweyk to Headquarters, where he is interrogated by SS-Scharführer Bullinger. Schweyk escapes torture because Bullinger needs his help to provide his wife in Cologne with a Pomeranian dog. In a brief interlude, *Zwischenspiel in den niederen Regionen*, Schweyk praises Hitler to the SS man Müller for refraining from drinking alcohol, and for conducting his 'higher' politics in such a way that few would equal him when sober. The whole dialogue is conducted in an ambiguously witty vein, by which Schweyk convinces his interlocutors that he is a feeble-minded idiot who is a good Hitlerite. His failure to produce the Pomeranian, which he has killed for Frau Kopecka's kitchen, leads to his being sent to join a draft for Stalingrad. In the *Nachspiel* he walks round in circles through heavy snow, keeping at an equal distance from the city until he finally meets Hitler, who has lost his way. Schweyk convinces Hitler that there is no way out for him in any direction and enjoys the moment when he can speak the truth and throw Hitler into a frenzy of bewilderment.

SCHWIEGER or **SCHWIGER,** JAKOB (Altona, 1624–after 1667), was a pastor in Wittenberg and later in Hamburg. Schwieger wrote poetry (*Liebesgrillen*, 1654 and *Die adeliche Rosa*, 1659) and two novels (*Verlachte Venus aus Liebe der Tugend*, 1659, and *Die verführte Schäferin Cynthie*, 1660). *Die geharnschte Venus* (q.v.) has sometimes been attributed to him, but he was not its author. As Der Flüchtige he was a member of the Elbschwanenorden (q.v.).

Schwierige, Der, a comedy (Lustspiel) by H. von Hofmannsthal (q.v.), published serialized in the *Neue Freie Presse* in 1920 and revised in book form in 1921. Though set after the 1914–18 War, it is enacted in an aristocratic environment recalling pre-war Vienna. The 'difficult' central character, Graf Hans Karl Bühl, is not prickly or irritable, but hesitant and unwilling to commit himself, doubtful of his ability to communicate, though he can in fact be the most charming conversationalist in the affable Austrian manner. Nor does he lack courage, for a career of service at the front lies behind him. The action of the comedy involves the bringing-together of the diffident Hans Karl and the delightful Helene Altenwyl, in which she is obliged in the end to take the initiative. This refined comedy, Hofmannsthal's most engaging work, asserts the values of humanity, decency, decorum, and urbanity, which are presented as characteristic of Viennese society before the 1914–18 War. They are highlighted by contrast with the German Neuhoff in his angular brashness.

SCHWIND, MORITZ VON (Vienna, 1804–71, Munich), began as a Romantic painter (see ROMANTIK) and developed a grandiose style of historical and allegorical painting, which brought him commissions in Munich (with P. von Cornelius, q.v.), Karlsruhe, Frankfurt, the Wartburg, and Vienna. His abilities are seen to better advantage in small-scale work such as his illustrations for *Robinson Crusoe* (1821–3) and *Le nozze di Figaro* (q.v., 1825), and for magazines (*Münchner Bilderbogen, Fliegende Blätter*). He was always fascinated by Romantic and fairy-tale motifs (*Die sieben Raben*, 1857–8, *Die schöne Melusine*, q.v., 1869–70, both cycles; *Rübezahl, c.* 1860). His correspondence with E. Mörike (q.v.), ed. J. Bächtold, appeared in 1890.

SCHWITTERS, KURT (Hanover, 1887–1948, Ambleside, Westmorland), was primarily an artist. He contributed to *Der Sturm* (q.v.), advocating collage in his preface to *Anna Blume*. *Selbstbestimmungsrecht des Künstlers* (1919) (*Merzkunst, Merzmalerei* and *-dichtung*, 'Merz' being an abstraction from Kommerz). In 1920 he met H. Arp (q.v.) and became associated with Dada (see DADAISMUS) and from 1923 published his *Merz* series. In 1937 he emigrated to Norway and in 1940 to England, where he spent the remainder of his life as a portrait painter. His poetry, apart from *Anna Blume*, includes *Die Kathedrale*, 1920; *Memoiren Anna Blumes in Bleie*, 1922, reissued 1965; *Die Blume Anna*, 1923; *Veilchen*, 1931; *Ursonate*, 1932, considered to be a forerunner of concrete poetry (see KONKRETE POESIE); he was also influenced by A. Stramm (q.v.). Correspondence, *Wir spielen bis uns der Tod abholt. Briefe aus 5 Jahrzehnten*, ed. E. Nündel, appeared in 1975; *Das literarische Werk* (5 vols.) ed. E. Lach, in 1973–81.

Scopf, scoph, scop, designation of the court minstrel in early medieval times. The 'scopf'

belongs to the days of oral tradition and sung poetry, which he accompanied to the harp. The *Hildebrandslied* (q.v.) is a surviving relic of 'scopf' poetry. The word became obsolete in the 12th c.

Scopf von dem lône, Der, title given by the first editor to an Early Middle High German poem in Alsatian dialect. It is a moral poem of sermon-like character, inculcating virtues of which generosity (*milte*) is the highest. The author calls himself 'scoph', but it is now held that, in spite of the use of this obsolescent term, he was not a lay minstrel, but a cleric. It was probably written *c.* 1150. The MS. is in Colmar.

SCULTETUS, ANDREAS, originally Andreas Scholtz (Bunzlau, 1622/3–1647, Troppau), a Silesian religious poet; a Protestant, he was converted to Roman Catholicism in 1639 and accepted into the Society of Jesus. His principal poems are *Friedens Lob- und Leidgesang* (1641), *Österliche Triumphposaune* (1642), and *Blutschwitzender Todesringer Jesus* (1643). Scultetus's poems were published by G. E. Lessing (q.v.) in 1771 (*Gedichte von Andreas Scultetus*).

SEALSFIELD, CHARLES, adopted name and literary pseudonym of Karl Anton Postl (Poppitz, Moravia, 1793–1864, Solothurn, Switzerland), who entered a monastery in Prague in 1813, was ordained priest in the following year, and ran away in 1823. After a brief stay in Switzerland he emigrated to the U.S.A. (1823), taking the name Sealsfield, and concealing his identity so successfully that it was only discovered after his death.

Sealsfield published in 1827 *Die Vereinigten Staaten von Nordamerika, nach ihren politischen, religiösen und gesellschaftlichen Verhältnissen betrachtet* and followed this in 1828 with two works in English, the descriptive *Austria as it is or Sketches of Continental Courts by an Eye-Witness* (q.v., which was forbidden by the censorship in Austria because of its Liberal outlook) and the novel *Tokeah or the White Rose*. Except for a visit to Germany in 1826–7, he lived in America until 1832, being chiefly active as a journalist. Between 1832 and 1837 he was a newspaper correspondent in London and Paris, moving then to Switzerland, where he remained from 1837 until his death. He several times briefly revisited the United States. *Tokeah*, a novel of the Indian in conflict with an expanding colonial civilization, appeared in 1833 in revised form in German and with a new title, *Der Legitime und die Republikaner*. His German novels proper begin with *Der Virey und die Aristokraten* (q.v., 3 vols., 1834), followed between 1835 and 1837 by *Lebensbilder aus beiden Hemisphären*, a collective title for *Morton oder die große Tour, Ralph Doughbys Brautfahrt, Pflanzerleben und die Farbigen*, and *Nathan der Squatter-Regulator*. These works deal with crises of social change, the conflict between an effete or primitive tradition and a new aggressive capitalism.

Probably the most successful of Sealsfield's widely read novels was *Das Kajütenbuch* (2 vols., 1841), which has as its background the struggle for the union of Texas with the U.S.A. It is in the form of a frame novel (see RAHMEN), and *Die Prärie am Jacinto* (q.v.), the Novelle or short novel included in it, has been frequently reprinted. Sealsfield opened up the American scene for the German public, displaying not only his first-hand knowledge, but a command of character and a grasp of social and political forces.

Gesammelte Werke (15 vols.) appeared in 1842–3 (rev. edn. 1845–6), his American novels (5 vols.), ed. F. Riederer, in 1937, and *Briefe und Aktenstücke*, ed. J. A. von Bradish, in 1955. A reprint of *Werke*, ed. K. J. R. Arndt *et al.*, appeared in 1972–82 (23 vols.).

SECKENDORFF, KARL FRIEDRICH SIEGMUND VON (Erlangen, 1744–85, Ansbach), was a court chamberlain (Kammerherr) at Weimar 1775–84. He was musically gifted and composed the music for some of Goethe's minor dramatic works (*Jery und Bätely, Die Laune des Verliebten, Der Triumph der Empfindsamkeit*, qq.v.), as well as setting the poem *Der König in Thule* (q.v.). He wrote a tragedy, *Kalliste* (1782) and a novel, *Das Rad des Schicksals oder Die Geschichte des Thrangsi* (2 vols., 1783).

SED, abbreviation for the Sozialistische Einheitspartei Deutschlands, the official party of the Social Democratic Republic (see DEUTSCHE DEMOKRATISCHE REPUBLIK, KPD, and SPD).

Sedan, in northern France between Mézières and Montmédy close to the Belgian frontier, was the scene of the decisive German victory on 1 September 1870 (see DEUTSCH-FRANZÖSISCHER KRIEG). A French army under Macmahon marching to the relief of Metz was pushed back upon Sedan and encircled by the Prussian Third Army under the Crown Prince (see FRIEDRICH III) and the Army of the Meuse (Maasarmee) under Crown Prince Albert of Saxony. The outgeneralled French surrendered in the evening after fierce fighting and the capitulation was formally drawn up the following day. The Emperor Napoleon III was among the prisoners. The fighting was particularly severe in the south-east of the Bavarian sector at Bazeilles and at Givonne to the east where the Prussian Guard was engaged. The victory was regarded

as a great feat of arms and a triumph for the Chief of Staff Moltke (q.v.). A symbol for the new Germany, the Sedantag became a public holiday. In literature Sedan evoked poetry only of a meretricious kind. Sedan was the scene of further fighting in 1914 and in May 1940 during the German invasion of France (see WELTKRIEGE).

Seefahrt ist not!, a novel by G. Fock (q.v.), published in 1913. A story of life at sea in a North-Sea fishing cutter, it has little plot, but derives its quality from the vivid description of life on board, and of the sea and its moods and hazards. Klaus Mewes, the skipper, takes his boy to sea with him against his wife's wish. Eventually she prevails. Klaus perishes in a storm, but the sea is in the boy's blood and he insists on following his father's calling.

Seeschlacht, a tragedy by R. Goering (q.v., 1917). It is set in the turret of a German warship during the battle of Jutland.

SEGHERS, ANNA (Mainz, 1900–83, Berlin), pseudonym of Netti Radvanyi, *née* Reiling, and the name by which she is publicly known. She studied at Heidelberg University, and in 1929 joined the Communist Party. In 1933 she emigrated to France, and her books were banned by the National Socialist regime. In the Spanish Civil War she was an active helper of the Republicans. In 1940 she escaped from France to Mexico, returning to East Germany in 1947, where she became prominent as an author unswervingly supporting the official organs of the DDR and received high awards.

Anna Seghers's best novels are *Das siebte Kreuz* (q.v., 1942), *Transit* (1944, concerned with refugees), and *Die Toten bleiben jung* (1947), which weaves the story of individuals into a convincing picture of Germany between 1918 and 1945, emphasizing the rise of Fascism. Other novels include *Die Gefährten* (1932), *Der Kopflohn* (1933), *Der Weg durch den Februar* (1935), *Die Rettung* (1937), and *Der Abdecker* (1970). A skilful exponent of the Novelle and the short narrative, she produced with *Der Aufstand der Fischer von St. Barbara* (q.v.), published together with *Grubetsch* in 1928, one of her best stories. Others include *Auf dem Weg zur amerikanischen Botschaft* (1930, a collection, the first story dealing with the execution of Sacco and Vanzetti in the U.S.A. in 1927), *Die Stoppuhr* (1933), *Die drei Bäume* (1940), the collections *Der Ausflug der toten Mädchen* (1948), *Die Hochzeit von Haiti* (1949), *Die Linie* (1950), *Friedensgeschichten* (1950), and *Die Kinder* (1951). *Der Bienenstock* (1953) and *Karibische Geschichten* (1962) contain selections from her shorter fiction; in *Sonderbare Begegnungen* (1973) the story entitled *Treffpunkt* reflects her political commitment since the late 1920s in the context of the Communist Youth Movement. As an author concerned with social justice, she wrote with striking vitality, especially while in political opposition. Her essays include *Über Tolstoi* and *Über Dostojewski* (1962), and a collection of poetry, *Der Jagdherr liegt im Sterben*, was published in 1974. *Gesammelte Werke* (6 vols.) appeared 1951–3, *Erzählungen* (2 vols.), 1964, *Über Kunst und Wirklichkeit* (3 vols.), ed. and introduced by S. Bock, 1970–1, and *Gesammelte Werke in Einzelausgaben* (14 vols.) in 1975–80 (DDR) and (10 vols.) 1977 (BRD).

Segremors, title given to fragments of a Middle High German Arthurian romance in verse, based on a French poem. Segremors (i.e. Sagramore), a knight of the Round Table, sets out with a dwarf to search for Gawein. He sustains various adventures and eventually meets with success. The fragments, amounting to some 400 lines, were written in the first half of the 13th c. The author is unknown.

Sehnsucht, one of the most famous poems of J. von Eichendorff (q.v.), the first line of which runs 'Es schienen so golden die Sterne'. It unrolls a Romantic vision of a nocturnal landscape of bewitching beauty. The poem, without its title, occurs in Eichendorff's novel, *Dichter und ihre Gesellen* (q.v.). Eichendorff wrote a second and less well-known poem with the same title ('Selig, wer zur Kunst erlesen').

SEIDEL, HEINRICH (Perlin, Mecklenburg, 1842–1906, Berlin-Lichterfelde), a pastor's son, who trained as a civil engineer. In his 30s he began to write poetry and stories, and in 1880 soon after completing the iron-and-glass roof of the Anhalter Bahnhof, Berlin, his best-known engineering assignment, he retired to devote himself entirely to writing. His best work is found in his humorous Novellen *Vorstadtgeschichten* (1880) and especially the three series dealing with the amiable and contented eccentric, Leberecht Hühnchen (*Leberecht Hühnchen*, 1882; *Neues von Leberecht Hühnchen*, 1888; *Leberecht Hühnchen als Großvater*, 1890). All these were incorporated in *Leberecht Hühnchen* (1900). The setting of these stories is the Berlin Vorstadt, the drabness of which cannot dim Hühnchen's optimistic good humour. Seidel also wrote for children (*Kinderlieder und Geschichten*, 1903). H. W. Seidel (1876–1945) was his son and Ina Seidel (q.v.) his niece and daughter-in-law. *Von Perlin nach Berlin* (1894) is autobiographical. *Gesammelte Schriften* (20 vols.) appeared 1888–1907.

SEIDEL, INA (Halle, 1885–1974, Ebenhausen nr. Starnberg, Bavaria), daughter of a doctor, who committed suicide in 1897, and niece of Heinrich Seidel (q.v.), whose son Heinrich Wolfgang (1876–1945, a pastor in Berlin and also a writer) she married in 1907. Ina Seidel spent her childhood in Brunswick and Marburg; her formative years, which later had an important influence on her work, were spent in Munich before marriage led to residence in and near Berlin until her husband's retirement in 1934; during his nine-year pastorate (1914–23) in Eberswalde she lived near to and remained in touch with the capital, which, with Munich, was the main cultural centre influencing her work. The district of Starnberg, with which she had early associations, was an obvious choice for a permanent home after 1934, by which time her principal *œuvre* was nearing completion.

Ina Seidel's poetry includes ballads and rhapsodic odes, as well as lyric verse; it appeared in the collections *Gedichte* (1914), *Neben der Trommel her* (1915), *Weltinnigkeit* (1918), and *Die tröstliche Begegnung* (1932). Editions of *Gesammelte Gedichte* appeared in 1949, 1955, and 1958. It is, however, for her narrative work that she is primarily known. It is notable for an organic progression from novel to novel, a feature of which is the reappearance of characters in subsequent novels. Her first was *Das Haus zum Monde* (1916), which was complemented in 1923 by *Sterne der Heimkehr*; the two works form a diptych, in which decadence gives way to moral consciousness. These novels reappeared in 1952, combined as *Das Tor der Frühe*. *Das Labyrinth* (q.v., 1922) is a large-scale novel dealing with the life of the child prodigy, traveller, and geographer Georg Forster (q.v.). The social problems and the spiritual conflicts generated by the 1914–18 War are evoked in *Brömseshof* (1928), a sombre novel depicting the inability of the returning soldier, Conrad Brömses, to maintain the tradition of the farm Brömseshof, which is the family's long-standing inheritance. *Das Wunschkind* (q.v., 1930), Ina Seidel's best-known work, transfers similar preoccupations to the historical setting of the Napoleonic Wars (q.v.), and adds the important theme of motherhood and the stresses to which it is exposed in times of crisis and especially in war. In *Der Weg ohne Wahl* (1933) a brother and sister, Manno and Merula, are caught in the labyrinth of life, through which in the end compassion and idealism enable them to find their way. *Lennacker* (q.v., 1938), a novel of unusual construction, gives a fictional conspectus of the history of the Lutheran Church. It has a connection with *Das Wunschkind*.

Ina Seidel's next novel, *Unser Freund Peregrin. Aufzeichnungen des Jürgen Brook* (1940), introduces as narrator a character who was to play an important part in her last novel, *Michaela*. It is centred on the life of three children, two of them orphans, who build a world of fantasy around the figure of a poet long dead, the pseudonymous Peregrin, who is one of their ancestors; his real name purports to be Veit von Harvesthus, a character based on Novalis (q.v.). *Das unverwesliche Erbe* (q.v., 1954) has close links with *Das Wunschkind* and portrays the confessional division of Germany as a positive force in its cultural development. *Michaela* (1959) reflects the situation of a group of educated Germans under the National Socialist regime, though with strikingly limited success in form and structure.

Shorter works of fiction include the collection of Novellen *Hochwasser* (1921), *Der vergrabene Schatz* (1928), *Spuk in Wassermanns Haus* (1936), *Die Geschichte einer Frau Berngruber*, and *Die Versuchung des Briefträgers Federweiß* (both 1953), *Dresdner Pastorale* (1962), *Quartett* (1963), and *Die alte Dame und der Schmetterling* (1964). Her two outstanding Novellen are *Die Fürstin reitet* (1926), set in the Russia of Catherine the Great, and *Die Fahrt in den Abend* (1955), which employs the stream of consciousness (see INNERER MONOLOG) to portray the death of a doctor injured in a car accident.

The narrative *Renée und Reiner* (1928), midway between novel and Novelle, introduces two characters whose later life is pursued in *Michaela*.

Essays dealing with literature are contained in *Dichter, Volkstum und Sprache* (1934), and studies of three Romantic writers (C. Brentano, Bettina von Arnim, and L. J. von Arnim, qq.v.) were published in one volume as *Drei Dichter der Romantik* (1944). The essays *Die Vogelstube* (1946) and *Frau und Wort* (1965) are also worthy of note. Ina Seidel's autobiographical writings comprise *Meine Kindheit und Jugend* (1935), *Drei Städte meiner Jugend* (1960), *Vor Tau und Tag* (1962), and *Lebensbericht 1885–1923* (1970).

SEIDL, JOHANN GABRIEL (Vienna, 1804–75, Vienna), a schoolmaster, then a civil servant, was responsible for the modernized version of the Austrian national anthem ('Gott erhalte, Gott beschütze', based on the older 'Gott erhalte Franz den Kaiser'). In addition to poems in literary language (*Bifolien*, 1836) he wrote dialect poetry published as *Flinserln* (1828–37) and collected as *Gedichte in niederösterreichischer Mundart* (1844). *Gesammelte Schriften* (6 vols.), ed. H. Max, appeared 1877–81; a selection was republished in 1959.

Seifriet Helblinc, name by which a group of fifteen Middle High German poems of the late 13th c. is known. The author is anonymous, but he was evidently an Austrian of noble birth. Seifriet Helblinc is the name of a character who speaks in one poetic dialogue, and this name, which was once mistakenly believed to be that of the poet, has persisted as a convenient designation. Eight of the poems, called by the author *der kleine Lucidarius*, are in dialogue form, discussing morals and, in particular, the conduct of the poet's contemporaries (including reference to a conspiracy in 1295–6 against Duke Albrecht). The other seven dispense with dialogue; they survey the contemporary scene and deplore the decline in public and private morality since the poet's youth. One poem of this group takes the form of an allegorical combat between the virtues and the vices, presenting a so-called Psychomachie. The poet evidently lived in the Babenberg epoch of Austria and survived into Habsburg times, spanning approximately the period 1240–1300.

Sekundenstil, see NATURALISMUS.

SELADON, pseudonym of G. Greflinger (q.v.).

SELDEN, CAMILLE, pseudonym of Elise Krinitz. See MOUCHE.

SELDTE, FRANZ (Magdeburg, 1882–1947, Fürth), served in the 1914–18 War and in 1918 founded the Stahlhelm (q.v.), an association of ex-soldiers with right-wing tendencies. He took part in agitation against the Weimar Republic (q.v.) and during the National Socialist regime was employed in the Ministry of Labour. Arrested by the Allies in 1945, he died before trial.

Seldte, who claimed to have invented in 1914 the shield fitted to machine-guns to protect the detachment, published in 1929 a book of war recollections covering the period August–December 1914, *M.G.K.* (the abbreviation stands for Maschinengewehr-Kompanie). It was followed by other war books, *Dauerfeuer* and *Vor und hinter den Kulissen* (both 1931).

SELNECKER, NIKOLAUS, also Schnellenecker (Hersbrück nr. Nürnberg, 1530–92, Leipzig), a prominent Lutheran, was moderator (Generalsuperintendent) in Leipzig 1558–68, then court preacher in Brunswick. He returned to Leipzig in 1574. In addition to publishing writings on controversial theological matters, he composed hymns of merit, including 'Laß mich dein sein und bleiben', 'Ach Gott, wem soll ich klagen', and 'Wir danken dir, Herr Jesu Christ'. Selnecker was also an accomplished organist.

Seltzame Springinsfeld, Der, a novel published in 1670 by J. J. C. von Grimmelshausen (q.v.) under the pseudonym Philarchus Grossus von Tromersheim. Its hero is the soldier Springinsfeld, a character in Bk. 3 of *Der abenteuerliche Simplicissimus* (q.v.). It is set years later, and the story is told by Springinsfeld himself to the ageing Simplicius, so that the actual narration is contained within a frame (see RAHMEN). Springinsfeld recounts his life, including the discovery by his second wife of a magic bird's nest conferring invisibility. She makes dishonest use of it and comes to a bad end. Grimmelshausen published the novel under the pseudonym Philarchus Grossus von Tromersheim. See also DAS WUNDERBARLICHE VOGEL-NEST.

Sentimentalisch, term used by Schiller for one of the categories to which he assigned poets and poetry. See ÜBER NAIVE UND SENTIMENTALISCHE DICHTUNG.

Septembermorde, the massacres in which some 1,200 prisoners were murdered by the people of Paris between 2–5 September 1792 (see FRENCH REVOLUTION). Frequently referred to in German literature, they figure prominently in Büchner's play *Dantons Tod* (q.v.).

Septembermorgen, a poem of six short lines, composed by E. Mörike (q.v.) in 1827 while walking from Köngen to Nürtingen. It is an evocation of the mood of a landscape at a particular moment, achieved with an economy and precision recalling Goethe's 'Über allen Gipfeln' (q.v.). Mörike's poem, which begins 'Im Nebel ruhet noch die Welt', at first bore the title *Herbstfrühe*.

Septennat, the septennium operated for the military budget of the German Empire founded in 1871. In 1893 it was replaced by a quinquennium (Quinquennat). The septennial budget of 1887 met fierce parliamentary opposition and was passed only after a dissolution of the Reichstag and new elections. G. Hauptmann relates his comedy *Der Biberpelz* (q.v.) to the political climate of this crisis by the indication 'Zeit: Septennatskämpfe'.

Sequenz (*sequentia*), a form of liturgical song often believed to have arisen out of a coloratura-like extension of the last 'a' of the Alleluia in the Gradual of the Mass, the Sequenz being the substitution of words for the wordless singing. Created in France, the Sequenz reached St. Gall in the 9th c. There Notker Balbulus (q.v.) devised the first *sequentiae* in Germany.

Instead of being strophic, the Sequenz is divided into sections of unequal length, each of which is divided into two corresponding portions for antiphonal singing. It is sung between the

Gospel and the Gradual on certain days. In the 12th c. rhyme was introduced and vernacular examples of this originally purely Latin form began to occur (see MARIENSEQUENZ VON ST. LAMBRECHT and MARIENSEQUENZ VON MURI). In the 16th c. the number of *sequentiae* was reduced to four, *Victimae paschali* (Easter), *Veni, sancte Spiritus* (Whitsun), *Lauda Sion* (Corpus Christi), and *Dies irae* (All Souls'). In the 18th c. *Stabat mater* (Seven Sorrows of Our Lady) was added.

Seraphine, a novel by K. Gutzkow (q.v.) published in 1838. It is the tragedy of a governess whose capacity for love finds no suitable object.

Serapionsbrüder, Die, title given by E. T. A. Hoffmann (q.v.) to a collection of his stories in four volumes, published in 1819 (vols. 1 and 2), 1820 (vol. 3), and 1821 (vol. 4). The title refers to a group of friends consisting of Hoffmann, K. W. Contessa, A. von Chamisso, J. E. Hitzig, and D. F. Koreff (qq.v.), who styled themselves the Serapionsbrüder. The name was taken from St. Serapion Sindonita, whose feast is on 14 November, the day on which the group constituted itself.

The principal contents of the four volumes are *Serapion*, *Rat Krespel* (q.v., also known as *Antonie*), *Die Fermate*, *Der Dichter und der Komponist*, *Ein Fragment aus dem Leben dreier Freunde*, *Der Artushof* (q.v.), *Die Bergwerke zu Falun* (q.v.), *Nußknacker und Mausekönig* (q.v.), *Der Kampf der Sänger* (q.v.), *Der schwebende Teller*, *Die Automate* (q.v.), *Doge und Dogaressa* (q.v.), *Das fremde Kind*, *Nachricht aus dem Leben eines bekannten Mannes*, *Meister Martin, der Küfner, und seine Gesellen* (q.v.), *Die Brautwahl*, *Der unheimliche Gast* (q.v.), *Das Fräulein von Scuderi* (q.v.), *Spielerglück*, *Der Baron von B.* (q.v.), *Signor Formica*, *Erscheinungen*, *Der Zusammenhang der Dinge*, *Hyänen*, and *Die Königsbraut*.

The framing narration and discussions are published separately as *Unterhaltungen der Serapionsbrüder* (*Sämtliche Werke*, ed. W. Harich vol. 13, 1924).

Servatiuslegende, Oberdeutsche, a poem dealing with the life and martyrdom of St. Servatius, written probably between 1180 and 1190 by a cleric of Indersdorf in Bavaria. Its source is a Latin life of the saint. Servatius was the patron saint of Maastricht, and his adoption as subject of a Bavarian poem is linked with the marriage of Countess Agnes van Loon from Maastricht to Duke Otto I of Bavaria (b. *c.* 1120, in 1180 created first duke of Bavaria by the Emperor Friedrich I, q.v., d. 1183. See BAYERN).

Sesenheimer Lieder, customary designation of the poems linked with Goethe's love for Friederike Brion (q.v.). The more important among them are *Mit einem gemalten Band, Willkommen und Abschied,* and *Mailied* (qq.v.), all written in 1770–1.

SEUME, JOHANN GOTTFRIED (Poserna nr. Weißenfels, Saxony, 1763–1810, Teplitz), a peasant's son, received, through aristocratic patronage, a good education. In 1780, while on his way on foot to Paris, he was seized by a press-gang and forcibly enrolled in a Hessian contingent to be sent to reinforce the British in the American War of Independence. Seume's corps did not go into action and was repatriated in 1783. In Bremen he deserted, but was promptly impressed into the Prussian army. He deserted twice and on the second occasion was condemned to death, being saved only by the personal intervention of the Prussian General Courbière (1733–1811). He was allowed to return to Saxony, where he was active as a tutor and began to write. In 1793 he became secretary to the Russian General von Ingelström and was then commissioned in the Russian army, taking part in the suppression of the Polish revolution of 1794. In 1796 he resigned and returned home, working as proof corrector for the publisher G. J. Göschen (q.v.).

An indefatigable pedestrian, Seume walked in 1801–2 to Syracuse and back, publishing an account of his journey as *Spaziergang nach Syrakus im Jahre 1802* (1803, 3 vols.; 2nd rev. edn. 1805, ed. by A. Meier, 1985). In 1805 he made another great expedition, going first to St. Petersburg, then through Finland into Sweden and so home. He described this tour in *Mein Sommer im Jahre 1805* (1806, reissued 1968). Seume also wrote poetry (*Gedichte,* 1801), and two of his poems became very well known, 'Wo man singet, laß dich ruhig nieder' and *Der Huron,* the proverbial first line of which 'Ein Kanadier, der noch Europens übertünchte Höflichkeit nicht kannte' is quoted in Th. Fontane's *Frau Jenny Treibel* (q.v., 1892). Late in life he wrote a tragedy *(Miltiades,* 1808) and began an autobiography, which his friends Göschen and C. A. H. Clodius (1772–1836) completed (*Mein Leben,* 1813, reissued 1964). He is most notable as an alert and accurate writer with an eye for social conditions. *Sämmtliche Werke* (12 vols.) appeared in 1826–7, *Prosaische und poetische Werke* (10 vols.) in 1879.

SEUSE, HEINRICH (Constance, *c.* 1293–1366, Ulm), in Latinized form Suso, a Dominican mystic, entered the Order in 1306 and was in a monastery in his native city; in 1324 he studied at Cologne under Meister Eckhart (q.v.).

Among his duties was the pastoral supervision of the nuns in the convent at Töß near Winterthur. In 1347 he was transferred to Ulm, apparently in consequence of a calumny, which was subsequently proved false.

Seuse has left four books which he himself combined into a collected edition with the title *Exemplar Seuses.* It includes his 'life', *Das Büchlein der ewigen Weisheit, Das Büchlein der Wahrheit,* and *Das Briefbüchlein.* The earliest work is *Das Büchlein der Wahrheit* (1327), a discussion of metaphysical problems from a mystical standpoint, in which Meister Eckhart is defended and pantheistic interpretations of his doctrine reprobated. *Das Büchlein der ewigen Weisheit,* written before 1334, is a dialogue between eternal wisdom, i.e. God, and its servant (Diener), i.e. man; it aims at kindling the love of God by a contemplation of the Passion of Our Lord. It became one of the most widely read works of devotion, in its own time, in the 16th c. and 17th c., persisting even into the 19th c. Seuse himself translated it into Latin, with the title *Horologium Sapientiae.*

The life of Seuse (*Der Seuse* or *Vita*) was written down in his lifetime by his friend, the nun Elsbeth Stagel (q.v.), and subsequently revised and completed by Seuse himself. It is not intended as a reliable record of his outward life, but provides a portrait of the inner man. It is often regarded as the first German autobiography. *Das Briefbüchlein* contains eleven pastoral letters written to nuns (three of them to Elsbeth Stagel).

Seuse's deep piety and humility, purified by years of ascetic self-mortification, are enriched by a great warmth of feeling, and these qualities manifest themselves in his sensitive prose style, creating a vehicle for the expression of mystical thought.

Seven Years War, see SIEBENJÄHRIGER KRIEG.

Seydlitz, title of a cycle of three ballads by Th. Fontane (q.v.) dealing with F. W. von Seydlitz (q.v.). The first ballad of the cycle, *Herr Seydlitz auf dem Falben,* was published in the magazine *Der Soldatenfreund* in 1847, the second, *Seydlitz und der Bürgermeister von Ohlau* and third, *Und Calcar, das ist Sporn,* were included in Fontane's *Gedichte* (1889); *Calcar das ist Sporn* had been printed in 1878 in the novel *Vor dem Sturm* (q.v.).

SEYDLITZ, FRIEDRICH WILHELM VON (Calcar, 1721–73, Ohlau), of ancient noble descent, entered the Prussian cavalry in 1740, immediately saw active service, was promoted captain in 1742, and major in 1745. A successful cavalry charge at Kolin (1757) led to his being gazetted major-general. His handling of massed cavalry at Roßbach (1757), Zorndorf (1758), Kunersdorf (1759), and Freiberg (1762) was as skilful as earlier conduct with single squadrons and regiments. He was twice wounded. In 1767 he was promoted General der Kavallerie. He is the subject of three ballads by Th. Fontane (see above) and of a novel by E. von Naso (q.v.). See also SCHLESISCHE KRIEGE and SIEBENJÄHRIGER KRIEG.

Seylersche Truppe, a touring company of actors, founded in 1769 by Abel Seyler (1730–1800). Seyler's wife, Friederike Sophie (1738–89), née Sparmann, was a well-known actress.

Sezessionen, groups of protesters against an orthodox and lifeless 'establishment' in painting, which formed themselves by seceding from academies, etc. in the 1890s. The best known were in Munich (1892), in Berlin (1893, headed by M. Liebermann, q.v.), and in Vienna, where G. Klimt (q.v.) created a mode of *Art nouveau* (see JUGENDSTIL) which acquired the term Sezessionsstil.

SHAKESPEARE, WILLIAM, was known only uncertainly and indirectly in Germany in his lifetime, though some plays were performed by the Englische Komödianten (q.v.), probably in unauthentic versions and later in crude vernacular adaptations. He is first mentioned by name in Germany, in a short list, by D. G. Morhof (q.v., 1682), who candidly admits ignorance of his works. A further second-hand reference was made in 1708 by B. Feind (1678–1721) in *Gedancken von der Opera*. J. J. Bodmer (q.v.) in the preface to *Critische Abhandlung von dem Wunderbaren in der Poesie* (1740) couples him with Milton, but spells him 'Saspar'. J. C. Gottsched (q.v.), who wrote against him, appears to have known him little better, and in the first half of the 18th c. the general attitude (second-hand and influenced by French views) was that Shakespeare was ignorant of the rules of drama and was not worthy of serious consideration.

The first to take another view, and one based on knowledge, was J. E. Schlegel (q.v.) in *Vergleichung Shakespears und Andreas Gryphs* (1741), which was prompted by the appearance of the first German translation of a Shakespearian play *Julius Caesar* (in Alexandrine verse) by C. W. von Borcke (q.v.) in 1741. F. Nicolai (q.v.) in 1755, while adhering to the customary view of Shakespeare's ignorance of the rules, praised his creation of character (1755). G. E. Lessing (q.v.), in both the *Literaturbriefe* (1759) and *Die Hamburgische Dramaturgie* (1768), took Shakespeare seriously, and in the latter work discussed *Richard III* at some length. Meanwhile the first collection of Shakespeare's works in German

translation (8 vols., 1762–6) came from C. M. Wieland (q.v.), who rendered twenty-one into prose and one (*A Midsummer Night's Dream*) into verse. In this form Goethe and Schiller first met Shakespeare's work. The 1770s brought a wave of boundless admiration. Herder (q.v.) published his essay *Shakespeare* (written in 1771) in *Von deutscher Art und Kunst* (q.v., 1773) and Goethe composed his briefer panegyric *Rede zum Shakespeares-Tag* (1771, see ZUM SCHÄKESPEARS TAG). A new complete prose translation by J. J. Eschenburg appeared in 1775–7 (13 vols.). A perceptive and influential discussion of *Hamlet* is a feature of Goethe's *Wilhelm Meisters Lehrjahre* (q.v., Bk. 5). The enthusiasm for Shakespeare generated in the Sturm und Drang (q.v.) was sustained in the Romantic movement (see ROMANTIK), notably by A. W. and F. Schlegel and L. Tieck (qq.v.).

A. W. Schlegel, together with Dorothea Tieck (see SCHLEGEL, DOROTHEA VON) and W. von Baudissin (q.v.), produced the first verse translation of genius (9 vols., 1825–33). It was edited by Tieck and reissued (ed. A. Brandl) 1896–9 and by F. Gundolf (q.v.) in an edition which included his own translations (10 vols., 1908–23). At the time of its first publication it quickly superseded a translation by J. H. Voß (q.v., 9 vols., 1818–29). Shakespeare's position on the German stage was secure from this time on, and indeed he came to be played more frequently in Germany than in his native land. Essays on Shakespeare and studies of individual works abound among notable German dramatists, usually with particular relevance to their own work (*Über die Shakespearo-Manie*, q.v., 1827, by C. D. Grabbe and O. Ludwig's posthumously published *Shakespeare-Studien*, q.v., 1871, are examples of contrasting character). Not all are, of course, based on translations.

20th-c. translations include a bilingual edition by L. L. Schücking (20 vols., 1912–35), and others by M. J. Wolff, R. A. Schröder (q.v.), E. Fried (q.v.), R. Flatter, and H. Rothe (q.v., the most extreme attempt at modernization, begun early 1920s). Versions of the sonnets include those by K. Lachmann (q.v., 1820), G. Regis (1836), F. Bodenstedt (q.v., 1862), O. Gildemeister (q.v., 1871), M. J. Wolff (1903), S. George (q.v., 1909, revised 1931), Therese Robinson (1927), and P. Celan (q.v., 1967), and of epic poetry and poems those by F. Freiligrath (q.v., 1849), W. Jordan (q.v., 1861), and K. Simrock (q.v., 1867).

Shakespeare-Studien, a collection of more or less fragmentary essays and jottings by Otto Ludwig (q.v.) posthumously published in 1871. For the most part written down in the last ten years of Ludwig's life (1855–65), they include generalized observations on Shakespeare's technique, style, and characters, as well as a number of essays on individual plays, together with comparisons, notably with Schiller. Ludwig was an acute and perceptive critic, but his principal intention in these studies was to benefit his own dramatic writing. Instead, the work on Shakespeare seems to have inhibited his creative powers.

SIBOTE, a Thuringian poet of the first half of the 13th c., is the author of a short verse tale entitled *Frauenzucht*. A knight undertakes to marry a notorious shrew. He kills hawk, hound, and horse before her eyes, and then insists that he must ride home on her. Out of fear of being likewise killed she submits and soon collapses exhausted. Thereafter she is a good and obedient wife.

Sibyllen Weissagung, a Middle High German religious chronicle-cum-prophecy in prose, which extends from the last years of Adam to the Day of Judgement. It was probably written *c.* 1320, though it was later brought up to date to include Karl IV (q.v., 1347–78). *Sibyllen Weissagung* is in three parts. The first is the story of the Holy Cross, i.e. the tree from which the Cross is to be made, beginning with Eden and going on to the visit of the Queen of Sheba. The second contains the Sibyl's prophecies of the end of the world. The third part continues the story of the Cross to Christ and then passes again to the Day of Judgement.

'Sich hûb vor gotes trône', first line of an untitled anonymous 14th-c. Middle High German poem of nearly 500 lines, recounting with great brevity a history of man from the standpoint of his redemption by Jesus Christ. One section, in which the four Daughters of God consider man's sinfulness, and the Son of God proposes to redeem him, is worked out more fully in quasi-dramatic form.

SICKINGEN, Franz von (Ebernburg, 1481–1523, Landstuhl), was one of the best-known and most successful of the predatory barons of the late Middle Ages. A powerful nobleman owning large estates in the Palatinate, Sickingen made a practice of succouring the oppressed by buying out their claims and then himself exacting payment from reluctant payers by force of arms. He was outlawed in 1515, but was powerful enough to ignore the sentence. In 1518 he burned Metz and laid waste whole districts in the Palatinate and Hesse. For a time he put his military power at the service of the Emperor and was appointed Kaiserlicher Feldhaupt-

mann, but a defeat in northern France in 1521 terminated this phase of his career. Through the influence of Ulrich von Hutten (q.v.) he went over to the Lutheran Reformation. After the failure of an attack on Trier in 1522 Sickingen was again outlawed and was besieged in his castle of Landstuhl, which he surrendered after being mortally wounded.

Sickingen occurs as a character in Goethe's *Götz von Berlichingen* (q.v., 1773), and is the subject of plays by Julius von Soden (q.v., *Franz von Sickingen*, 1808), by E. Bauernfeld (q.v., *Franz von Sickingen*, 1849), and, remarkably, by F. Lassalle (q.v., *Franz von Sickingen*, 1859).

Siddhartha. Eine indische Dichtung, a narrative by H. Hesse (q.v.), published in 1922 and included as the first story of the collection *Weg nach Innen* (q.v., 1931). Inspired by Hesse's intimate study of Indian philosophy and religion, it takes the form of a Bildungsroman (q.v.) divided into two sections and numerous chapters with headings indicating the nature of the experience of the eponymous central figure, the Brahmin Siddhartha.

Siddhartha grows up with his friend Govinda destined for priesthood. He encounters Buddha and arrives at the conviction that Buddha does not owe his supreme spirit of equanimity to the philosophy of causality which he preaches but to experience incapable of learned formulation. The introductory section ends as Siddhartha leaves his parents' home and the Brahmin monk Govinda in search of self-fulfilment. The ultimate goal is the conquest of suffering and fear, the attainment of serene contentment fusing life and death and illuminating the wholeness of seeming contrasts, subsequently directly referred to as Sansāra and Nirvāna.

In the second section the beautiful courtesan Kamala teaches Siddhartha the art of love, and he acquires wealth by learning the skills of Kamaswami. But his luxurious life threatens to destroy his soul and he takes flight in despair. He is on the verge of death when he is found by Govinda, the wandering monk, who protects his sleep on snake-ridden ground, ignorant of the identity of the richly clad man to whom his watch is devoted. On waking they recognize one another, but Siddhartha bids his friend go and turns to the river. From the ferryman Vasudeva, with whom he stays, he learns to listen to the river. Unexpectedly Kamala arrives with her son, of whose existence he was ignorant. She is fatally wounded by a snake, but before she dies both recover a new spirit of love which Siddhartha hopes to impart to his son. It is not returned, the boy runs away, and Siddhartha must accept the separateness of individual existence which as a youth he had expected his own father to bear. The river helps him to overcome intense longing for his son and to attain the highest wisdom. Vasudeva, who had looked forward to Siddhartha's recognition, leaves at this stage, and their parting is one of serene harmony. The narrative ends with the reunion of Siddhartha, now the ferryman, and Govinda, to whom Siddhartha is able to impart his state of bliss.

Siebenjähriger Krieg, the Seven Years War between Prussia and Austria, was fought simultaneously with the colonial war between Great Britain and France, while all powers were committed by a new system of coalitions to the two German powers (see DIPLOMATIC REVOLUTION). The Seven Years War in Germany (also known as the Third Silesian War, see SCHLESISCHE KRIEGE) was a direct outcome of the Silesian wars and was opened by King Friedrich II (q.v.) of Prussia. Aware of the uneasy peace, he had by June 1756 reliable information about an alliance against him, aiming at the recovery of Silesia for Austria and the dismemberment of Prussia. After failing to receive satisfactory assurances about Austrian movements from Maria Theresia (q.v.), he invaded Saxony on 29 August 1756.

After the indecisive battle at Lobositz in Bohemia (1 October 1756) Saxony capitulated on 16 October 1756. In Dresden Friedrich found secret papers confirming hostile moves against him. August III (q.v.) fled to Warsaw and Friedrich incorporated his troops in the Prussian army. His aim to strike before the allies were ready was nevertheless only partially realized. By the following year he had to prepare to fight on all frontiers, against the Swedes in the north, against the Russians in the east, and in the south and west against the Austrians and the French. The Diet at Regensburg voted on 17 January 1757 for imperial execution (Reichsexecution) against Prussia. Prussia's only aid consisted of an Anglo-Hanoverian army, from 1757 under Ferdinand of Brunswick (Ferdinand von Braunschweig-Wolfenbüttel), and of British subsidies.

Although victorious at Prague (6 May 1757), Friedrich was soon defeated on Bohemian soil at Kolin (18 June 1757). His western frontier was weakened by the defeat of the Duke of Cumberland at Hastenbeck (26 July 1757) by the French, which was followed by the Convention of Kloster-Zeven (8 September 1757) necessitating the entire reorganization of the defence. Sweden invaded Pomerania from Stralsund, and the Russians defeated the Prussians in East Prussia at Groß-Jägersdorf (30 August 1757) and again in Silesia at Görlitz (7

September 1757). In Saxony Friedrich himself defeated the Franco-Imperial army at Roßbach (5 November 1757) and in Silesia at Leuthen (5 December 1757). Austria and France, however, remained unwilling to make peace. In 1758 Friedrich therefore advanced into Moravia, intending to take Olmütz on his way to Vienna. The town, however, resisted his siege for seven weeks, after which he had to withdraw, his failure going to the credit of the Austrian commander, Laudon (q.v.).

As East Prussia and Poland were by now in the hands of the Russians, Friedrich withdrew to defend Brandenburg at Zorndorf (25 August 1758). In the drawn battle heavy losses were sustained by both sides. From Brandenburg the King moved to Saxony, where the Austrian commander Daun (q.v.) defeated him at Hochkirch (14 October 1758); in spite of this defeat Friedrich succeeded in relieving Neiße. In the following year Duke Ferdinand endeavoured unsuccessfully to drive the French from Frankfurt. This setback was compensated by a victory won by the combined British and Hanoverian forces at Minden (1 August 1759). A few days later Friedrich was defeated by the Russians and the Austrians (under Laudon) at Kunersdorf (12 August 1759), in the south-east corner of Brandenburg. Of his 43,000 men he lost almost half. Only the fact that the allies did not follow up their victory saved Prussia from ruin. The Prussian army (under Fouquet) also suffered defeat in Silesia, at Landshut (23 June 1760) and Glatz (26 July 1760); but at Liegnitz (15 August 1760) and Torgau in Saxony (3 November 1760) Friedrich himself defeated the Austrian commanders Laudon and Daun. At Torgau the last-minute arrival of the veteran Zieten (q.v.) turned the battle into a decisive Prussian victory.

Marches and skirmishes continued, but victory was no longer, it seemed, to be decided on the battlefield. Yet Friedrich hung on, for as long as considerable Prussian territories (Pomerania, Neumark, and Silesia) were occupied, a peace effected by diplomacy would have reduced Prussia to a third-rate power. Friedrich was resigned to this after the loss of British subsidies, when the Tsarina Elizabeth died in January 1762 and her successor, Tsar Peter III, reversed the Russian alliance into support of Prussia. When he was deposed and succeeded by Catherine II in July 1762, the new Tsarina decided for neutrality. The war ended after all on the battlefield: Friedrich's victories at Burkersdorf (21 July 1762) and Schweidnitz (9 October 1762) freed Silesia, and the victory at Freiberg in Saxony (29 October 1762) by Prince Heinrich von Preußen (q.v.) over the combined imperial and Austrian army completely changed Friedrich's prestige in the negotiations for peace. It was a tragic blow to Maria Theresia and Kaunitz (q.v.).

Five days after the Peace of Paris terminating the colonial war was signed, Austria, Prussia, and Saxony concluded the Peace of Hubertusburg on 15 February 1763 (see HUBERTUSBURGER FRIEDE). As the treaties of Dresden and Breslau concluding the first two Silesian Wars were confirmed, the war, which had exhausted Austrian as well as Prussian resources, had altered no boundaries. Silesia and the County of Glatz remained with Prussia. Friedrich, however, promised to support the election of Maria Theresia's son Joseph (see JOSEPH II) as King of the Romans in preparation for his succession to the imperial title. The significance of the three Silesian wars for the future of Germany was nevertheless greater than the territorial changes wrought by Friedrich might suggest. Henceforth Austria and Prussia remained the two great rival powers in the struggle for supremacy in Germany, also referred to as the 'dualism' (Dualismus) affecting developments over the next hundred years.

The Seven Years War with its Prussian defeats (Kolin, Hochkirch, Kunersdorf) and victories (Rossbach, Leuthen, Liegnitz, Torgau) has, because of Friedrich's personal skill and bravery and, above all, his extraordinary endurance, exercised an unusual fascination and is remembered in literary works and popular songs, among them J. W. L. Gleim (q.v.), *Kriegs- und Siegeslieder der Preußen* (1758); C. A. Tiedge (q.v.), *Elegie auf dem Schlachtfelde bei Kunersdorf* (in *Elegien und vermischte Gedichte*, 1803–7); Lessing, *Minna von Barnhelm* (q.v.); Goethe, *Dichtung und Wahrheit* (Pt. I, Bk. 3); Johannes Gründler (1747–1845), *Friedrich der Große oder die Schlacht bei Kunersdorf* (1826); Otto Ludwig (q.v.), *Friedrich II von Preußen, Fragment: Die Torgauer Heide* (1844); C. F. Scherenberg (q.v.), *Leuthen*, *Schlachtengemälde* (1852); Gustav zu Putlitz (q.v.), *Die Schlacht bei Mollwitz* (1869); Martin Greif (q.v.), *Der Sieger von Torgau, Ballade* (1882); K. Bleibtreu (q.v.), *Friedrich der Große bei Kolin*, a story (1888); W. Raabe, *Das Odfeld* (1889) and *Hastenbeck* (qq.v., 1899), both novels.

Siebenkäs, commonly used title for the novel *Blumen-, Frucht- und Dornenstücke* (q.v., 1796–7) by JEAN Paul (q.v.). It is also the surname of the principal character, Firmian Stanislaus Siebenkäs.

Sieben Legenden, a volume of religious legends published by G. Keller (q.v.) in 1872. The stories, the writing of which was spread over twenty years, afforded him a relaxation from his rather bleak positivism. The tone of the legends varies from the sympathetic to the playful, and from the playful to the critical; it is apparent that they are written by an unbeliever, but mockery is absent from this graceful and engaging collection. The legends are: (1) *Eugenia*; (2) *Die Jungfrau und der Teufel*; (3) *Die Jungfrau und der Ritter*; (4) *Die Jungfrau und die Nonne* (these three form a trilogy on the B.V.M.); (5) *Der schlimm-heilige Vitalis*; (6) *Dorotheas Blumenkörbchen*; (7) *Tanzlegendchen.*

Siebente Ring, Der, a volume of poetry by S. George (q.v.), published in 1907.

Sieben vor Verdun, a war novel by J. M. Wehner (q.v.), published in 1930. Though accompanied by a chauvinistic preface, the novel is a moving account, from the standpoint of a group of German soldiers, of the long, bloody, and fruitless battle for Verdun in 1916. It begins with the preparations for the assault on Fort Douaumont and ends at the evacuation of the Fort de Vaux nine months later.

Sieben weisen Meister, Die, a Volksbuch (q.v.) of the 16th c. The story also occurs in various prose versions in the 15th c. See also HANS VON BÜHEL.

Siebte Kreuz, Das, a novel by Anna Seghers (q.v.), published in 1942. In 1935 seven men, Aldinger, Belloni, Beutler, Füllgraber, Georg, Pelzer, and Wallau, escape from a concentration camp. The commandant has seven crosses erected by nailing boards across tree-trunks; his orders are that the men are to be brought back and attached to these crosses, alive or dead. Beutler and Pelzer are caught almost at once, and later Wallau is recaptured. Füllgraber's nerves break down, he surrenders and is the fourth man to be returned alive. Belloni is shot dead, and Aldinger's corpse is found. Georg alone survives, helped by good luck and by many sympathizers, including a Jewish doctor. Finally the KPD secret organization provides him with false papers enabling him to escape into Holland. Consequently the seventh cross remains vacant.

Siebzigste Geburtstag, Der, an idyll by J. H. Voß (q.v.), published in *Idyllen* (1801). Written in classical hexameters, it depicts the reunion of a family on the seventieth birthday of the grandfather. The son, his young wife, whom the old couple have hitherto not met, and two young grandchildren arrive in the winter snow and in an atmosphere of warm affection.

Sie erlischt, a poem by H. Heine (q.v.), included in the section *Lamentationen* of the *Romanzero* (q.v.). No. 18 in the subdivision *Lazarus*, it begins 'Der Vorhang fällt, das Stück ist aus'.

Under the image of the close of a theatrical performance Heine likens his own death to the guttering-out of the last candle.

Sieg der Natur über die Schwärmerey, Der, oder Die Abentheuer des Don Sylvio von Rosalva, a novel by C. M. Wieland (q.v.), arranged in two parts, divided into four and three books respectively, and published in 1764. It is an educational novel, in which the hero, Don Sylvio, is cured of an obsessive belief in fairies, and brought to a balanced rational outlook. The setting is in Spain, and frequent references to *Don Quixote* suggest an affinity to Cervantes's work. Don Sylvio, a handsome, intelligent, and good-natured 18-year-old, grows up in seclusion, acquiring a belief in fairies; he is convinced he is in love with a fairy who has taken the form of a blue butterfly. To escape a marriage with a hideous and stupid heiress he takes to the road with his Sancho-like servant Pedrillo. They sustain various adventures, in the course of which Sylvio is seen asleep by Donna Felicia, a beautiful and virtuous 18-year-old widow, who falls in love with him. A little later he courageously rescues a young man and woman, who are assaulted by a group of kidnappers. Sylvio accompanies them to their destination, where the young man, Eugenio, proves to be Felicia's brother, and Jacinte, the young woman, turns out to be Sylvio's sister Serafina, kidnapped years before by a gipsy. Sylvio is cured of his fairy obsessions and both pairs are united.

SIEGFRIED DER DORFER, see FRAUENTROST.

Siegwart, eine Klostergeschichte, a two-volume novel by Johann Martin Miller (q.v.), published in 1776. It tells two parallel love stories, one of them happy, the other tragic. Xaver Siegwart is a highly strung, gifted youth, who, after visiting a monastery, opts for the monastic life and becomes a novice; but he later leaves the abbey and falls in love with Mariane Fischer. She is placed in a convent by a tyrannical father, and both she and Xaver die in their respective claustral establishments. The other pair of lovers, Xaver's friend Kronhelm and his sister Therese, overcome resistance and live happily. Their harmony and balance are clearly intended to provide a moral contrast to Xaver's instability.

Siegwart was one of the most popular novels of its day, accounting young Schiller and K. P. Moritz (qq.v.) among its enthusiastic readers. Since then it has been much maligned. In spite of its obvious sentimentality it has many robust passages presenting a fascinating and unideal-ized picture of important aspects of 18th-c. society, including some of the tougher sides of rural life. Many accounts of the novel imply that it is a watered down *Werther*, but it has in fact scarcely any real point of contact with Goethe's novel (see LEIDEN DES JUNGEN WERTHERS). *Siegwart* was reissued, ed. A. Sauer, in 1893.

Sie mähten gewappnet die Saaten, a novel by E. Schaper (q.v.), which received this title on republication in 1956, the original title being *Der Henker* (1940). It is set in Esthonia at the time of the Russian rebellion of 1905 and portrays the conflict between the indigenous Balts and the German governing and land-owning classes. It shows violence, both in the Esthonian revolt and in its suppression, but it is primarily concerned with conscience. Despite the destruction and desolation it ends on a note of conciliation.

'Sie sollen ihn nicht haben, den freien deutschen Rhein', first line of N. Becker's patriotic poem *Der freie deutsche Rhein* (q.v.).

SIGEHER, MEISTER, a medieval poet who was active in the 13th c. at the Bohemian court, where he successively supported in his Sprüche (see SPRUCH), the policies of Wenzel II and Ottokar (qq.v.). Sigeher, a commoner as the style Meister indicates, wrote in addition to Sprüche a hymn to the Virgin Mary.

Sigenot, a Middle High German epic poem of unknown authorship. It exists in two versions, one dating from the middle of the 13th c., the other from the middle of the 14th c. DIETRICH von Bern (q.v.) encounters the giant Sigenot, who throws him into a cave. Sigenot then encounters Hildebrand, whom he attacks. Hildebrand slays Sigenot and with the aid of a dwarf rescues Dietrich. The poem is written in the verse known as the Eckenstrophe (q.v.).

SIGISMUND, KAISER (1361-1437, Znaim), or Sigmund, son of the Emperor Karl IV (q.v.), became margrave of Brandenburg by inheritance in 1378 and acquired by marriage a claim to Poland and Hungary. Crowned king of Hungary in 1386, he sold Brandenburg to Jobst of Moravia (see JOBST VON MÄHREN) in order to finance hostilities against the Turks, and later disposed of the New March (Neumark) to the Teutonic Order (see DEUTSCHER ORDEN). In 1410, and again in 1411, he was elected German King and had a hand in calling the Council of Constance (see KONSTANZER KONZIL). The betrayal and burning of Hus (q.v.) in 1415 damaged Sigismund's prestige, and Bohemia subsequently defected. He was unable to re-establish himself there until 1436, and even then did so only by concessions to the offended Hussites. He was crowned Emperor in Rome in 1431.

Sigismund, who employed the restless and turbulent Tyrolean poet OSWALD von Wolkenstein (q.v.) on various missions, was an enlightened patron of letters and the arts. An anonymous tract proposing radical reforms of the constitution (*Reformatio Sigismundi*) has, though uncertainly, been attributed to him. An important occurrence in his reign was the investment of Friedrich von Hohenzollern, Burggraf of Nürnberg with the Margravate of Brandenburg in 1415 (see HOHENZOLLERN). Sigismund also founded the fortunes of the Wettin (q.v.) family in 1423 when he created Friedrich der Streitbare (q.v.) of Meißen Elector of Saxony.

Signor Formica, title of a story (Novelle) by E. T. A. Hoffmann (q.v.) written in 1819 and published in 1821 in vol. 4 of *Die Serapionsbrüder*. See ROMAN VON VIEREN.

SILCHER, FRIEDRICH, in full Philipp Friedrich (Schnaith, Württemberg, 1789–1860, Tübingen) was Director of Music at Tübingen University from 1817 until his death. His *Sammlung deutscher Volkslieder* (1827–40) included his settings of *Ännchen von Tharau*, *Die Lorelei*, *Morgen muß ich fort von hier* (qq.v.), and *Zu Straßburg auf der Schanz*, all of which have become familiar as the accepted versions for singing.

Silesian Wars, see SCHLESISCHE KRIEGE.

Silvester, see KONRAD VON WÜRZBURG and TRIERER SILVESTER.

Silvester, New Year's Eve, 31 December being St. Silvester's Day in the calendar.

Simone, a novel by L. Feuchtwanger (q.v.), published in Stockholm in 1944. It is the fourth and last of the novels designated as *Der Wartesaal* (q.v.). It draws a deliberate parallel with Joan of Arc in German-occupied France in 1940. Simone Machard, the new Joan, fifteen years of age, sets fire to her uncle's petrol depot in order to prevent it from falling into German hands. Under interrogation she denies, like Jeanne d'Arc, her idealistic motivation, only later to recover her courage, admit the truth, and submit to her punishment. This story was dramatized by B. Brecht and Feuchtwanger in 1956 as *Die Gesichte der Simone Machard* (q.v.).

Simplicianische Schriften, term used to denote the series of novels by J. J. C. von Grimmelshausen (q.v.), in which the characters Simplicius Simplicissimus, Springinsfeld, and the Landstörzerin Courage appear. They comprise *Der abenteuerliche Simplicissimus Teutsch* (1669), *Der seltzame Springinsfeld* (1669), *Trutz Simplex: Oder Ausführliche und wunderseltzame Lebensbeschreibung der Ertzbetrügerin und Landstörtzerin Courasche* (1670), and *Das wunderbarliche Vogel-Nest* (1672), qq.v.; they were published pseudonymously.

Simplicissimus, a satirical weekly, concerned mainly with political topics, which derives its title from the eponymous hero in the 17th-c. novel by Grimmelshausen (q.v.). It was founded in 1896 by the publisher A. Langen (1869–1909) and Th. Th. Heine (1867–1948) and appeared until 1944, and again from 1954 to 1967. Among the contributors were German journalists and writers (including Owlglass, L. Thoma, and F. Wedekind, qq.v.), and the Norwegian caricaturist O. Gulbransson (1873–1958, who lived from 1902 in Germany and was in 1929 appointed Professor at the Munich Academy of Arts).

SIMROCK, KARL (Bonn, 1802–76, Bonn), an adherent of the Romantic movement (see ROMANTIK) in his youth, studied at Bonn University and entered the Prussian civil service in Berlin in 1823. A political poem ('Drei Tage und drei Farben'), inspired by the July Revolution in Paris, caused his dismissal in 1830, whereupon, being a man of means, he withdrew into private life, pursuing literary and philological studies. In 1850 he was appointed to a newly founded chair of medieval literature at Bonn, which he retained until his death.

Simrock is chiefly known for his translations of medieval, and especially heroic, poetry. His version of the *Nibelungenlied* (q.v.) appeared in 1827 and of the *Edda* (q.v.) in 1851. *Das Heldenbuch*, including his translations of *Kudrun* (q.v.) and the *Amelungenlied*, as well as the *Nibelungenlied*, came out in six volumes between 1843 and 1849, and *Deutsche Volksbücher* in 13 volumes between 1845 and 1866. It might almost be said that Simrock translated the whole of medieval German literature: *Parzival* and *Titurel* (qq.v.), *König Orendel* (see ORENDEL), *Der gute Gerhard* (q.v.), *Tristan und Isolde* (see TRISTAN), and the *Heliand* (q.v.) are random examples. His successful policy of popularization extended also to Anglo-Saxon literature (*Beowulf*, 1859), to the translation of poems by Shakespeare (q.v., 1867), and to the 17th c. (Friedrich von LOGAU, 1875, Friedrich von SPEE, 1876, qq.v.). He published volumes of original poetry in 1844 and 1863, but only the folksong-like *Warnung vor dem Rhein* ('An den Rhein, an den Rhein, zieh nicht an den Rhein') has survived. *Ausgewählte Werke* (12 vols.), ed. G. Klee, appeared in 1907.

Simsone Grisaldo, a play (Schauspiel) in five acts published anonymously by F. M. Klinger (q.v.) in 1776. The eponymous hero is a man of great physical and moral strength, who bears his first name in allusion to the biblical Samson. Simsone, a general in the service of Castile, defeats the state's foreign enemies, first the Moors and then the Aragonese. His enemies at home conspire against him, seeking to exploit his weakness for women, and so to bind and blind him. Simsone, with the help of one of the women, Almerine, frustrates their machinations. The robust, manly, generous character of the hero is said to owe something to Goethe, whom Klinger greatly admired.

'**Sind wir vereint zur guten Stunde**', first line of the poem *Bundeslied* by E. M. Arndt (q.v.), written in 1815 as a poem of thanksgiving for victory.

SINED, inversion of Denis used as a pseudonym by J. N. C. M. Denis (q.v.).

Singspiel, a simple form of opera in which strophic songs are linked by spoken dialogue instead of accompanied recitative.

Though J. Ayrer (q.v.) wrote plays which he described as Singspiele, the form really begins in Germany in the 18th c. with imitations of John Gay's *The Beggar's Opera* (1728). The most successful German writer of Singspiele was C. F. Weiße (q.v.), whose *Der Teufel ist los* (q.v., 1752) began the vogue. C. M. Wieland (q.v.) and Goethe both wrote Singspiele, of which Wieland's *Alceste* (1773) and Goethe's *Erwin und Elmira* (q.v., 1775), *Claudine von Villa Bella* (q.v., 1776), and *Die Fischerin* (q.v., 1782) are well known.

With *Die Entführung aus dem Serail* (1782) Mozart raised the form to a higher musical level, introducing arias and vocal ensembles. His *Die Zauberflöte* (1791, q.v.) represents the apogee of the genre. In the earlier 19th c. the tradition was continued by A. Lortzing (q.v.), and in the second half of the century it developed into operetta.

Sinngedicht, Das, a Novelle with a cyclic frame (see RAHMEN) in 13 chapters by G. Keller (q.v.), first conceived 1851–5 and published in 1881–2. The title refers to a distich by Logau (q.v.) which Ludwig Reinhart, a scholar, finds in a volume of Lessing: 'Wie willst du weiße Lilien zu roten Rosen machen?/Küss' eine weiße Galatee: sie wird errötend lachen'. It is introduced at the opening of the Novelle and forms a leitmotiv throughout, leading up to the engagement between Reinhart and Lucie (or Lux). A seemingly simple device becomes fascinating through the skill of the ingenious author.

Reinhart's work has tired his eyes and, on chancing to find the epigram, he decides to ride forth and literally to test its truth. Three attempts fail before he arrives for the night at a mansion inhabited by a young woman named Lucie and her uncle, a retired colonel. Reinhart naïvely reveals his adventures, only to be mocked at by Lucie. Yet they stay together to illustrate, through the telling of stories, in which the Uncle, joins them, the attitudes of greatly differing characters to marriage, and the relationship of the sexes as it is shaped through environment and experience rather than in fantasy. Chapters 7 to 12 are devoted to them. As in Goethe's *Unterhaltungen deutscher Ausgewanderten* (q.v.), each story is briefly discussed by the listeners before the next narrator takes over, to tell a contrasting tale. Thus Lucie tells *Von einer törichten Jungfrau*, Reinhart *Regine* and *Die arme Baronin*. The Uncle next recounts a personal experience (*Die Geisterseher*), and Reinhart relates *Don Correa*. These are the only two stories stressing the imponderable and exotic.

Lucie's story *Die Berlocken* forms a transition to the last story in chapter 13. An account of her lonely childhood, it provides not only an omen of the happy end, but an unexpected revelation of character and maturity, which is not apparent in her defensive scepticism and irony in the early part of the Novelle. Left at the age of 15 in the care of a governess and a housekeeper, Lucie secretly becomes a Roman Catholic convert. She has taken this step because of her childhood love for her cousin, Leodegar who has teasingly called her his little wife and has later said that he could marry her only if she were a Catholic. When, a few years later, she learns that Leodegar has become a cardinal, she realizes to her chagrin that his endearments were meaningless. Her conversion, however, has left her with the pangs of a conflicting conscience. Reinhart is the sole listener to this story, and, in making him the witness of the secrets of her past and of her sense of guilt, she displays genuine trust in him. This sense of trust, not mere sensuous desire and the frustrations of an orphan, brings about their engagement. The forthcoming marriage between Reinhart and Lucie is clearly seen by Keller as based on an ideal relationship between partners in love. *Die arme Baronin*, the story of the baroness, impoverished and degraded by deceit, demonstrates that compassion can likewise result in a most happy relationship for the pure pleasure of giving. *Regine* is deliberately designed to form a counterpart to Auerbach's village story *Die Frau Professorin* (q.v.), examining the problem of class division in marriage. Regine's tragic end provides one answer.

While Logau's Sinngedicht brings, in the figure of Lucie, pervasive light (*lux*) into the Novelle, the illumination of this intimate product of Keller's maturity springs from the experience of existence *ante lucem* (the last words of the story). It reflects Keller's awareness of the plight of humanity, groping in the dark for lack of genuine communication, which alone can relieve the torment of isolation, irrespective of class or environment.

Sinn und Form. Beiträge zur Literatur, a bimonthly periodical, edited by the Berlin Academy of Art. Published since 1949, its first editor-in-chief was P. Huchel (q.v., until 1962); this office was held from 1964 by W. Girnus (1906–85). The most important literary journal of the DDR, its contributors also include authors of the BRD.

Sintflut, Die, collective title of a cycle of three novels by S. Andres (q.v.), made up of *Das Tier aus der Tiefe* (1949), *Die Arche* (1951), and *Der graue Regenbogen* (1959). The book is a pessimistic political parable. In *Das Tier aus der Tiefe* a theologian turned agitator takes over a movement for the 'normalization' of mankind. The 'Normer', as he is called, takes over Germany, ruthlessly suppressing all dissent and all individualism. The irresistible wave of the 'Normists' constitutes the Deluge. *Die Arche* represents the circles of resistance and secret opposition. In *Der graue Regenbogen*, war has overrun Germany, the 'Normer' and his supporters are dead, and the danger seems past. But adherents of 'Normism' begin to reappear and it becomes depressingly clear that for a second time the waters of the Deluge will cover the earth.

Sisto e Sesto, a historical novel by H. Federer (q.v.), published in 1913. It is set in the Abruzzi in the late 16th c. Sesto, driven to brigandage by the harsh rule of the local count, is arrested and condemned to death by Pope Sixtus V (Sisto). An eleventh-hour perception of the oppression and distress which have led Sesto to crime causes Sisto, who dies shortly afterwards, to pardon him.

Situation aus Fausts Leben, a fragment of a play by F. Müller (q.v.) dealing with Faust. It was published in 1776 and dedicated 'An Shakespeares Geist'. In the structure of Müller's unfinished play, it would have come after the published Erster Teil (see FAUSTS LEBEN DRA-

MATISIERT). At a critical phase in Faust's career, twelve years after the pact, he is at the Aragonese court, consumed with ambition and in love with the Queen. Mephistopheles offers him an escape: if he reverts to his humble self he will escape damnation. Faust refuses.

Skorpion, Der, (1) see RUF, DER; (2) a poem by S. Andres (q.v.).

Socialist Realism, see SOZIALISTISCHER REALISMUS.

SLEIDAN (SLEIDANUS), JOHANN BAPTIST, real name Philippi or Philippson (Schleiden, Eifel, 1506–56, Strasburg), studied at Cologne and Louvain and became a Protestant in 1541. His history of the Reformation (q.v., *Commentarii de statu religionis et reipublicae Carolo V Caesare,* 1555) is remarkable for its sober style and objectivity. His *De quatuor summis imperiis* (1556) was widely used as a manual of general history.

SMIDT, HEINRICH (Altona, 1798–1867, Berlin), went to sea in his youth, serving before the mast. In 1823 he left the sea, studied at Kiel and Berlin universities, and in 1834 joined the editorial staff of an official journal. In 1848 he was appointed a member of the Naval Commission (Marinekommission) and eventually became librarian in the Ministry of War.

A prolific writer, Smidt produced poetry (*Poetische Versuche,* 1825) and plays (*Kaufmann und Seefahrer,* 1844), but he is chiefly notable for his novels and stories of the sea (*Seegemälde,* 1828; *Die Belagerung von Glückstadt,* 1838; *Steuermann Johannes Smidt,* 1840; *Das Loggbuch,* 1844; *Berlin und Westafrika,* 1844; *Michael de Ruiter,* 1846; *Der Bergenfahrer,* 1850; *Hamburg und die Antillen,* 1860). He was a member of the literary club Der Tunnel über der Spree (q.v.).

SODEN, FRIEDRICH JULIUS HEINRICH, REICHSGRAF VON (Ansbach, 1754–1831, Nürnberg), was in Prussian service and spent some time as envoy at Bamberg, where he founded a theatre, in which E. T. A. Hoffmann (q.v.) was active. He was the author of many dramatic works, operas, tragedies, and historical plays, including *Leben und Tod Kaiser Heinrichs IV* (1784) and *Franz von Sickingen* (1808).

Sodoms Ende, a play (Drama in fünf Akten) by H. Sudermann (q.v.), first performed in the autumn of 1890 and published in 1890. It has as its hero an artist (Willy Janikow), who, corrupted by society women (notably Frau Adah Barczinowski), seduces his foster-sister and dies of pulmonary haemorrhage when he learns of her suicide. The play, intended as social criticism, failed in the theatre because of its sensationalism and theatrical artificiality. The symbolical title refers to Willy Janikow's 'great picture' of the destruction of Sodom, which is in the possession of Frau Adah.

Sohn, Der, an Expressionist play by W. Hasenclever (q.v.), written in 1913–14, published in 1914, and first performed in September 1916 in Prague. Banned in Germany, it was privately performed in Dresden in October 1916. The first public performance in Germany took place in March 1918. *Der Sohn* is based on Hasenclever's relationship with his domineering father. The principal characters are generically denoted (Der Sohn, Der Vater, Der Freund, etc.).

The father tyrannizes the son, seeking by brutal treatment to force him into an orthodox education and career. Encouraged by a mysterious friend, the son escapes, dons a mask, and makes, to a wildly applauding student audience, an impassioned speech, denouncing the tyranny exercised by fathers. He spends the night with a prostitute and is then picked up by the police and taken home. In the ensuing confrontation the father dies of a stroke when the son draws a revolver.

Sohn der Wildnis, Der, a five-act play in blank verse (dramatisches Gedicht), written in 1840–1 by F. Halm (q.v.), first performed at the Burgtheater, Vienna, in January 1842, and published in 1843. Set in and near Massalia (modern Marseilles) *c.* 500 B.C., it shows the raw barbarian Gaul Ingomar subdued and refined by the beautiful and brave Greek virgin Parthenia, so paying tribute to the pacifying influence of civilization and the civilizing influence of love.

The part of Parthenia was created by the actress Julie Rettich (q.v.). *Der Sohn der Wildnis* was a regular item in the repertoire of most German theatres up to the end of the 19th c.

Söhne des Senators, Die, a Novelle by Th. Storm (q.v.), published in the *Deutsche Rundschau* in 1880. Christian Albrecht and Friedrich Jovers, the sons and heirs of one of the last representatives of the flourishing merchant aristocracy in Husum, quarrel after their mother's death over the possession of the family garden. Friedrich, the younger, files a lawsuit against his brother and, to emphasize his alienation from Christian Albrecht, causes the wall separating the backyards of their adjoining properties to be built up so high that no one can look over it. But his family affection proves stronger than his obstinacy. One night, after hearing the family parrot cry the familiar jocular 'Come across!' ('Komm röwer!'), he orders the wall to be taken down to its original level, thus inaugurating a complete reconciliation. This simple tale is one of Storm's masterpieces in lighter vein.

Sokratische Denkwürdigkeiten, an essay on genius and knowledge by Johann Georg Hamann (q.v.), published in 1759. It is eccentrically dedicated 'An das Publikum oder niemand, den kundbaren'.

Soldat, Der, see 'ES GEHT BEI GEDÄMPFTER TROMMEL KLANG'.

Soldaten, a tragedy in free verse by R. Hochhuth (q.v.), published in 1967 and first performed in October 1967 in Berlin. It is subtitled *Nekrolog auf Genf*, referring to the Geneva Convention and simultaneously to the bombing of civilians in the 1939–45 War. Dedicated to E. Piscator (q.v.), the play begins within the ruined walls of the former medieval cathedral of Coventry in 1964 (the centenary of the Geneva Convention) and then reverts to the war. The play's three acts are headed *Das Schiff* (H.M.S. *Duke of York*), *Das Bett* (Churchill's bedroom), and *Der Park* (Chequers).

The central points of the indictment (the legal term is not out of place) are that Sir Winston Churchill engineered the death of General Sikorski in order to maintain the Russo-Anglo-American alliance, and that he rejected peace negotiations and supported the ruthless bombing of German cities against the arguments of the Bishop of Chichester. The suggestion of deliberate murder made almost immediately after the statesman's death and in the lifetime of his widow provoked prompt and violent controversy. Hochhuth has consistently maintained that he has evidence of the truth of his contentions. Disclosures made public in 1974 suggest that Hochhuth's imputations probably arose from a misreading of the documents. In 1972 the Czech pilot of Sikorski's aircraft brought a libel action in the High Court against Hochhuth and was awarded heavy damages.

Soldaten, Die, a play in five acts by J. M. R. Lenz (q.v.), written in 1774–5, and published in 1776. Lenz styled it *Eine Komödie*: the tragic action is based on situation rather than on character. The play was not performed until 1863 (Burgtheater, Vienna, under the title *Soldatenliebchen*). It is set in French Flanders, in Lille, Philippeville, and Armentières. Its theme is the seduction of a middle-class girl, Marie Wesener, by Desportes, an officer of noble birth, who abandons her. Marie, though conscious of her disloyalty to Stolzius, a draper whom she was to have married, becomes an easy prey to Mary, another officer of the garrison stationed in Lille, her home town. Concerned about the disgrace Marie has brought upon herself and her family, the Gräfin de la Roche takes her into her home to rehabilitate her; but the villain Mary interferes, and Marie runs away in despair.

Meanwhile Stolzius, who has not lost faith in Marie, has become Mary's batman. When he believes Marie to have perished, he takes his own life; but he also succeeds in poisoning Desportes at the same time. Wesener goes in search of his daughter. The sufferings of both have rendered them barely recognizable to one another when they meet in a poignant scene of reunion, in which Marie appears anonymously as Weibsperson. In the fifth act, which contains the tragedy of the middle-class characters, the last scene moves into the apartments of Graf von Spannheim, the colonel of the regiment. He and the Gräfin de la Roche sympathetically discuss the social implications of the action. The Colonel proposes a regiment of volunteer harlots, a suggestion which Lenz intended seriously, since he put it to Duke Karl August (q.v.) in a *pro memoria* entitled *Über die Soldatenehen* (first published 1914).

Although the play adopts features common to other plays of Sturm und Drang (q.v.), it is more radical in style and technique. Its 34 scenes, designed as a sequence of pictures, alternately characterize and parody a society in need of reform. The army chaplain Eisenhardt functions as a commentator who has no part in the action, of which many scenes (including those of seduction and of life in the officers' mess) are based on Lenz's own experience at Strasburg. The play influenced Büchner, especially in *Woyzeck* (q.v.). It achieved a new fame through the opera of the same title by Bernd-Alois Zimmermann (q.v.), which is based on Lenz's play and uses techniques of epic theatre (see EPISCHES THEATER).

SOLDATENKÖNIG, see FRIEDRICH WILHELM I.

Soldatenstücke, term sometimes used to denote plays written in the latter half of the 18th c., in which military and civilian life interlock. The first and classical example is Lessing's *Minna von Barnhelm* (q.v., 1767), the success of which inaugurated a spell of popularity for this type of play. Other Soldatenstücke are *Die Soldaten* (q.v., 1776) by J. M. R. Lenz; *Die abgedankten Offiziers* (1770) and others by G. Stephanie (q.v.); and *Der Graf von Walltron* (1776) by H. F. Möller (1745–98). Later plays of this type occur sporadically (e.g. *Rosenmontag* (q.v., 1900) by O. E. Hartleben (q.v.)).

SOLGER, KARL WILHELM FERDINAND (Schwedt, 1780–1819, Berlin), who died when

the Romantic movement had reached its peak, devoted himself to aesthetic dialectics and emerged as a principal exponent of Romantic Irony (see ROMANTISCHE IRONIE). At the age of 21 he attended lectures by F. W. J. von Schelling (q.v.) at Jena University, worked from 1803 to 1806 in the Prussian Kriegs- und Domänenkammer in Berlin, and in 1809 became a professor at the University of Frankfurt/Oder. In 1811 he moved, partly through the influence of Hegel (q.v.), to a chair at Berlin University. Alert to Romantic approaches to philosophy, ethics, and religion (F. D. E. Schleiermacher, q.v.), as well as literature, his many friends included L. Tieck (q.v.), and F. von Raumer (1781–1873), who acted as his literary executors.

Early studies of Greek philosophy and drama were a decisive factor in Solger's later development. Having translated Sophocles' *Oedipus* in 1804, he published in 1808 his interpretations of Greek tragedy in *Des Sophokles Tragödien*. In 1815 he formulated his views on beauty in art in his most widely read work, *Erwin. Vier Gespräche über das Schöne und die Kunst* (reissued, ed. W. Heckmann, 1971), consisting of four dialogues after the manner of Plato's dialogues. He resumed the dialogue form in *Philosophische Gespräche* (1817). Solger's dialectics are based on the paradox that beauty and irony form in synthesis the climax of tragedy, which is concerned with the ultimate comprehension of human suffering representing man's temporal existence against the background of universal forces responsible for the hero's plight in adversity. Tragedy builds up a mounting sense of incomprehension, which must be maintained until the hero's death. In death, however, the tragic hero perceives a fleeting vision of the true nature of the universe, which Solger terms the divine idea ('göttliche Idee', resembling Hegel's concept of a Weltgeist). This vision, evoking a sense of elevation and reconciliation with an otherwise incomprehensible divine spirit, constitutes the beauty justifying tragedy as the highest form of art. Solger's conception of irony is based on this assumption, for it is ironical that recognition can only be attained in death. The coincidence of beauty and irony at the tragic dénouement is the distinctive feature of Solger's contribution to Romantic aesthetics.

Solger's writings and correspondence were published by Tieck and Raumer as *Nachgelassene Schriften und Briefwechsel* (2 vols.) in 1826 (reissued, ed. H. Anton, 1973), and *Vorlesungen über Ästhetik*, ed. K. W. L. Heyse (1797–1855), in 1829 (reissued 1969).

SOLITAIRE, M., pseudonym of W. Nürnberger (q.v.).

Solitude, an elegant rococo palace of moderate size near Stuttgart, constructed between 1763 and 1767 to the order of Duke Karl Eugen (q.v.) of Württemberg on a hill and with a remarkable view. From 1773 to 1775 it was the home of the Herzogliche Militär-Akademie, of which Schiller was a pupil. The school occupied the buildings designed for the suite, behind the Schloß proper.

Soll und Haben, a three-volume novel by G. Freytag (q.v.), published in 1855. The title-page bears as a motto a quotation from Freytag's journalist friend Julian Schmidt (q.v.): 'Der Roman soll das deutsche Volk da suchen, wo es in seiner Tüchtigkeit zu finden ist, nämlich bei seiner Arbeit.'

Its central figure, Anton Wohlfart, enters, at his father's death, the firm T. O. Schröter in Breslau. There he is well treated and his intelligence and industry enable him to make rapid progress until he is working as Schröter's own right-hand man. A young nobleman, Herr von Fink, joins the firm and in his company Anton frequents aristocratic circles and begins to stray from the burgher environment. In particular, he is drawn to the daughter of a Baron von Rothsattel, apparently a man of wealth, but in reality near to ruin and in the hands of two competing Jews, Ehrenthal and Itzig. Anton shows his sterling worth by supporting his principal and even saving his life during a Polish rising. Freytag is clearly a supporter of East German colonization who distrusts the Slavs. When the Rothsattel family ask Anton to help liquidate the estate he accepts, giving up his post with Schröter. He carries out his duties with efficiency, but, when he finds that Rothsattel regards him as of a lower order, he returns to Schröter's at Breslau, marries Schröter's sister Sabine and becomes a partner. Lenore von Rothsattel marries Herr von Fink, and Anton is able to retrieve for them a proportion of the once great Rothsattel fortune. Itzig murders an accomplice and is drowned trying to make his escape.

The novel extols the integrity, industry, and efficiency of the German merchant, living up to the motto on the title-page. Well into the 20th c. it continued to be regarded by the general reading public as one of the great novels of the 19th c. It retains interest as a major Zeitroman (q.v.).

Somnium vitae humanae, a play in German verse (in spite of the Latin title) by Ludwig Hollonius (q.v.).

Sonett, in origin a Romance form, occurs in German abundantly in the 17th c., virtually disappears in the 18th c., and reappears with the Romantic movement (see ROMANTIK), since when it has been cultivated by many poets down to the present day. The structure and rhyme scheme in German have usually conformed to the so-called Petrarchan form, with two quatrains followed by two tercets and a rhyming pattern abba abba cdc dcd, though variations in the number of rhymes and the arrangement of rhymes in the tercets occur quite frequently. The Petrarchan hendecasyllables are rare in German, and the usual metres have been the Alexandrine iambic line of six accents with a central caesura (17th c.), or the iambic pentameter (in later sonnets).

Many poets of the 17th c. wrote sonnets, including Opitz, Weckherlin, Zesen, Lohenstein, Fleming, HOFFMANN von Hoffmannswaldau, Catharina von Greiffenberg, and, above all, A. Gryphius (qq.v.). A. W. Schlegel (q.v.), who was interested in metrical matters and Romance literature, reintroduced the sonnet. Among 19th-c. and 20th-c. authors of sonnets are Goethe, Rückert, Liliencron, Rilke, Eichendorff, Heine, J. Kerner, A. Grün, Mörike, C. F. Meyer, G. Keller, L. Uhland, Th. Däubler, P. Zech, F. Werfel (qq.v.). The modern writer most interested in developing the sonnet as a cyclical form is J. Weinheber (q.v.). Stefan George (q.v.) wrote a few sonnets and also translated those of Shakespeare (q.v.) and sonnets by Baudelaire and D. G. Rossetti.

(See PETRARKISMUS.)

Sonette an Orpheus, Die, two linked cycles of sonnets by Rilke (q.v.) published together in 1923. They bear the dedication 'Geschrieben als ein Grab-Mal für Wera Ouckama Knoop' (see KNOOP, WERA OUCKAMA). The two parts were written in Rilke's *mensis mirabilis*, February 1922, at Muzot. Part I (26 sonnets) was written between 2 and 5 February, then followed five of the *Duineser Elegien* (q.v.), and Pt. II (29 sonnets) was completed between the 11th and 20th. Such speed indicates that the poems must have been virtually complete in the poet's subconsciousness. The sonnets to Orpheus, as their title suggests, are euphonious and musical in tone and breathe an atmosphere of serenity in contrast to the challenging grandeur of the *Duineser Elegien*, which they may be said to complement. Though the dedication suggests a requiem, the sonnets are, in fact, concerned with life and especially with the role of poetry and art in life. Though a number of the sonnets are orthodox in form, Rilke included several bold and successful experiments with rhythm, rhyme, and length of line.

SONNENFELS, JOSEF VON, REICHSFREIHERR (Nikolsburg, 1733–1817, Vienna), was the son of a distinguished Jewish rabbi, who became a Roman Catholic convert and was ennobled in 1746. A professor of politics and an exponent of Enlightenment (see AUFKLÄRUNG), Sonnenfels attempted to introduce the theatrical reforms of Gottsched (q.v.) into Vienna, but his campaign against the Hanswurst (q.v.) in the 1760s failed in the face of determined local loyalty to established traditions. His *Briefe über die wienerische Schaubühne* appeared in 1768. Sonnenfels opposed judicial torture, writing *Über die Abschaffung der Tortur* (1722). In 1797 he was created Reichsfreiherr. *Gesammelte Schriften* (10 vols.) appeared in 1765.

Sonnenfinsternis, a tragedy by A. Holz (q.v.), published in 1908. An enormous play with a cast of only five, it deals with the failure of the gifted artist Hollrieder to achieve his ideal. It is the second in the series *Berlin. Die Wende einer Zeit in Dramen.* The figure of the 'Präsident der Sezession' is based on Max Liebermann (q.v.).

SONNLEITHNER, JOSEPH (Vienna, 1766–1835, Vienna), became secretary to the Emperor Joseph II (q.v.) in 1787, and in 1804 was appointed secretary to the Burgtheater. An accomplished musician, he wrote opera libretti; he was also active as a translator, and in 1815 founded a literary annual, *Aglaja* (q.v.). His best-known work is the libretto to Beethoven's opera *Fidelio* (q.v.). The dramatist F. Grillparzer (q.v.) was his nephew.

Sonnwendhof, Der, a play (Volksschauspiel) in five acts by S. Mosenthal (q.v.), published in 1857. It is written in High German prose with no attempt at dialect. The widow Monica has inherited the Sonnwendhof, loves her foreman Valentin, and intends to marry him. A girl (Anna) asks for a night's shelter, pleases Monica, and is engaged as a farm servant. Valentin is attracted to her, but she repulses him, revealing that she is under a curse because her father has committed arson. Monica is indignant and reproaches Valentin, but eventually relents and favours his love for Anna. Anna discovers that her father was innocent, and the real criminal falls to his death in a ravine. Anna and Valentin are united. The melodramatic and crudely motivated play had a great success on the stage.

SOPHIE DOROTHEA, KÖNIGIN VON PREUSSEN (Hanover, 1687–1757, Berlin), was a princess of Hanover (daughter of the future George I of England) when she married Crown Prince Friedrich Wilhelm (see FRIEDRICH WILHELM I) in 1706. She was the mother of Friedrich II (q.v.) of Prussia. Her plan to marry Friedrich's favourite sister Wilhelmine (q.v.) to the Prince

of Wales met with the blunt opposition of Friedrich Wilhelm.

Sophiens Reise von Memel nach Sachsen, a novel by J. T. Hermes (q.v.), published in five volumes in 1769–73. It is in epistolary form (see BRIEFROMAN), most of the letters purporting to be written by Sophie herself. The story is set in eastern Europe towards the end of the Seven Years War (see SIEBENJÄHRIGER KRIEG) in 1762, when wartime conditions make travel particularly difficult.

Sophie undertakes to go to Dresden in order to secure documents for her foster-mother and benefactress, Frau E. She is to be escorted on this journey by a brother she has not seen since early childhood. In fact, the brother is impersonated by a Russian agent who is to carry her off and hand her over as mistress to a Russian general. However, soon after they have left together he receives fresh orders and disappears, and Sophie continues her perilous journey unescorted. For some time her protector is a mysterious personality (Herr Less), whom she at first mistakes for a clergyman. He, too, turns out to be a Russian agent, but of higher character and on a harmless mission. By accident, he and Sophie find themselves in the same room at night, and Herr Less's behaviour is so virtuous and tactful that Sophie, already impressed by his looks and address, falls in love with him, as he does with her. But circumstances separate them, and Sophie, rescued by another virtuous man, Herr Puf, from an attempted seduction, gives to him a rather ambiguous promise of marriage. Her failure to be definite is her undoing, for Puf, a man of decision, turns from her and marries another; and Less, reappearing, renounces her because of her conduct to Herr Puf. In the end Sophie remains unmarried, devoting herself to bringing up the children of a friend.

The book, which contains descriptions of the devastation and hardships of war, was one of the more widely read novels of the late 18th c.

SORGE, REINHARD JOHANNES (Berlin, 1892–1916, killed in action, Ablaincourt/Somme), studied for a time at Jena University, gave up study for authorship, and in 1913 became a Roman Catholic. He wrote one of the earliest Expressionist works, the play *Der Bettler* (q.v., 1912). His next plays were religious, *Metanoeite. Drei Mysterien* (1915) and *König David* (1916). Several works were published posthumously, including an epic, *Mutter des Himmels* (1917), a play, *Der Sieg des Christos* (1924), and a 'vision' (Vision), *Gericht über Zarathustra,* but the complete works did not appear until 1962–7 (*Gesammelte Werke,* 3 vols., ed. H. G. Rötzer).

Sound Shifts (Lautverschiebungen), term denoting two shifts in consonantal pronunciation occurring early in the evolution of the German language. Certain anomalies in this sound shift are explained by Verner's Law, formulated in 1877 by the Danish philologist K. Verner (1846–96). The First Sound Shift (Erste or Germanische Lautverschiebung) took place before 500 B.C. and marks the change from Indo-European to Germanic, spreading over a long period. The Second Sound Shift (Zweite or Hochdeutsche Lautverschiebung) appears to have occurred progressively between the 5th c. and the 10th c. of the Christian era, affecting the language of central and southern Germany. The Second Sound Shift produced High German, the normal spoken and written language of Germany. Northern and western areas were unaffected, and their language, termed Low German, has survived only in dialect (Plattdeutsch); Low German languages also include Dutch, Flemish, and, up to a point, English.

Details of these consonantal changes, which are intricate and numerous and demand for their understanding a knowledge of phonetic symbols, are available in works such as *The German Language* by R. Priebsch and W. E. Collinson (6th edn. 1966) and *A Short History of the German Language* by W. W. Chambers and J. R. Wilkie (1970). See also GERMAN LANGUAGE, HISTORY OF.

Sozialaristokraten, a comedy by A. Holz (q.v.), published in 1896. Written in Berlin dialect, it treats amusingly of the difference in politics between words and actions, and of the subservience of protestations to interests. It is the first in the series *Berlin. Die Wende einer Zeit in Dramen.*

Sozialdemokratische Partei Deutschlands, see SPD.

Sozialistischer Realismus, the official form of literary realism of the German Democratic Republic (see DEUTSCHE DEMOKRATISCHE REPUBLIK). Its main objectives are contained in the terms Parteilichkeit, Volksverbundenheit, and Geschichtsbewußtsein, which must be seen in the context of re-education, the prime concern of DDR cultural policy in the 1950s and 1960s.

Socialist realism is designed to induce a spirit of optimism within the framework of SED ideology and social and economic developments. A focal notion is that of the 'positive hero', who serves the community and displays an affirmative attitude to the vicissitudes of life, the scope of which is narrowed down by excluding any morbid aspects. The official organ for writers in the DDR is the Schriftstellerverband, headed by a committee and a president (from 1978 Hermann Kant, q.v.). A notable variant is the so-called Bitterfeld experiment (abandoned by the 1970s), a policy adopted at the first writers' conference in 1959 at Bitterfeld, the aim of which was a closer co-ordination between writers and workers. Products of this experiment are known as Ankunftsliteratur (after a story by Brigitte Reimann, q.v.). (See BITTERFELDER WEG.)

DDR socialist realism is modelled on Soviet policy as practised in Russia and other communist countries. As propounded under the influence of M. Gorky in 1934 at the First Congress for Soviet writers, this policy attempted to define the role of 'realistic' literature in the context of the socialist revolution, which K. Marx (q.v.) had envisaged, but which even among his followers had lacked coherence. The then current debate on realism included reassessments of Heine (q.v., works written during the revolutionary period) and Th. Mann's *Buddenbrooks* (q.v.), as well as Brecht (q.v.), whose theatre was considered to be too experimental and indebted to the West. The aim was to draw a clear line between bourgeois literature, including that engaging in social and politically orientated criticism, and literature that was uniformly committed to communist objectives. Johannes R. Becher (q.v.) was among German writers, who, during their exile in Russia, adjusted to the new style, and whose approach to tradition corresponds to Soviet/DDR policy styled Erbe.

SPALATIN, GEORG, real name Burkhardt (Spalt, Franconia, 1484–1545, Altenburg, Saxony), studied at Erfurt and Wittenberg and was on friendly terms with the humanists Eobanus HESSUS and CROTUS Rubeanus (qq.v.), and with Luther (q.v.). Spalatin, who was in the service of the Elector of Saxony as tutor and librarian, was able to bring Luther and the Elector into contact. His writings include translations of Latin works by Luther, Melanchthon, and Erasmus (qq.v.), an autobiography, biographies of three electoral princes, and an important contemporary record, *Annales Reformationis* (first published in 1718). The customary name Spalatin is derived from his birthplace.

SPALDING, JOHANN JOACHIM (Tribsee, Pomerania, 1714–1804, Berlin), at first a pastor in Pomerania, became a member of the Berlin Consistory in 1764. He wrote widely read books of popular edification (*Über die Bestimmung des Menschen*, 1748; *Gedanken über den Wert der Gefühle im Christentum*, 1761), and was also a popular and effective preacher. His autobiography (*Lebensbeschreibung, von ihm selbst aufgesetzt*, 1804) was published by his son.

Spange, Die, a poem by S. George (q.v.), included in *Pilgerfahrten*.

SPANGENBERG, Cyriakus (Nordhausen, 1528–1604, Strasburg), son of the schoolmaster and religious poet Johann Spangenberg (q.v.), studied in Wittenberg under Luther and Melanchthon (qq.v.); he became rector of the school at Eisleben in 1546, and pastor in 1550. Later Moderator (Generaldekan) of Mansfeld, he fled from a civil commotion there in 1574 and spent his later years in Alsace, Hesse, and at last in Strasburg. Spangenberg, an ardent Lutheran, published Luther's sermons in *Cithara Lutheri* (1571–2). With his *Formularbüchlein der alten Adamssprache* (1555) he sought to rekindle religious zeal, and with his *Christliches Gesangsbüchlein* (1568) provided an expression of devotion. *Wider die Bösen Sieben ins Teuffels Karnöffelspiel* (1562) is a fierce onslaught on Pope Pius IV and other enemies of the Reformation. He also wrote a biblical play, *Von dem Cananeischen Weiblein* (1589).

SPANGENBERG, Johann (Hardegsen nr. Göttingen, 1484–1550, Eisleben), studied at Erfurt University, was a schoolmaster in Stolberg, then a parish clergyman in Nordhausen. He wrote *Alte und neue geistliche Lieder und Lobgesänge* (1543) and *Kirchengesänge, deutsch* (1545). Cyriakus Spangenberg (q.v.) was his son. *Sämtliche Werke* (7 vols.), ed. A. Vitzkelety, appeared in 1971–9.

SPANGENBERG, Wolfhart (Mansfeld, *c.* 1570–*c.* 1636, Buchenbach nr. Künzelsau), son of Cyriakus Spangenberg (q.v.), wrote the satirical beast poem (Tierdichtung) *Gans König* (1607), directed against the Roman Catholic Church, and later a beast tale, *Esel König* (1625), directed against the Rosicrucians (see Rosenkreuzer) and castigating political conditions and human folly. He was a pastor in Buchenbach after working for a few years as a teacher in Strasburg. He translated a number of Latin plays, both classical and modern, including *Der Brand Sodoms* (1607, *Conflagratio Sodomae* by Andreas Saurius) and was himself author of plays for school production (*Geist und Fleisch*, 1608; *Glückswechsel*, 1613; *Mammons Sold*, 1614).

SPANMÜLLER, Jakob, see Pontanus, Jakob.

Spartakusbund, an extreme left-wing political group originally formed at the end of 1916 by K. Liebknecht and Rosa Luxemburg (qq.v.). The name was adopted from articles published in the press by Liebknecht as *Spartakusbriefe*. Meeting in Berlin in the last days of December 1918, the Spartakusbund converted itself on 1 January 1919, in the presence of a Russian delegation, into the Revolutionäre Kommunistische Arbeiterpartei Deutschlands. Though a part of this name was to remain, summarized in the initials KPD (q.v.) in subsequent years, the members of the party continued to be known in January 1919 as Spartakisten. Between 16 and 21 December 1918 the Spartakisten clashed with local loyal troops. On 5 January 1919 the new Communist party attempted to seize power in an armed revolt known since as the Spartakusaufstand. The provisional government, having no regular troops on which it could depend, called in the assistance of Freikorps (q.v.), formations of determined, well-led, and well-armed right-wing groups including, as well as students and citizens, monarchist officers and soldiers with long front-line experience. After fierce fighting, which included the use of artillery and armoured cars as well as small arms, the Spartakus revolt was completely quelled on 12 January. Liebknecht and Rosa Luxemburg were taken and shot. From this point the term 'Spartakisten' fell out of use, being replaced by 'Kommunisten'.

Spätestens im November, a novel by H. E. Nossack (q.v.), published in 1955, has as its subject the love of a married woman, Marianne Helldegen, for the dramatist Berthold Möncken. Marianne herself functions as the narrator. The love between Marianne and Möncken is disturbed by Möncken's preoccupation with his play, which he hopes to complete at the latest in November. Meanwhile Marianne's husband persuades her to return to him. The theme of Möncken's play, which treats the tragedy of Dante's ill-starred lovers Paolo and Francesca, foreshadows the tragic ending of the novel. After the successful première, the elated Möncken meets Marianne, who at once rejoins him in the hope that their love will glow again as promised, 'spätestens im November'. Together they drive in a car which collides with a stone pier and both are killed.

Spaziergang, Der, a poem of 200 lines by Schiller, written in elegiac verse. Schiller wrote it in September 1795 and it was first published in No. 10 of *Die Horen* (q.v.) with the title *Elegie*. Using the walk as a thread for reflection, the poet meditates first on Nature, then on rural and city life, industry, commerce, and art, reflects on history, on war, and revolution, and returns finally to the stability of Nature, from which he had set out: 'Unter demselben Blau,

über dem nämlichen Grün/Wandeln die nahen
und wandeln vereint die fernen Geschlechter,/
Und die Sonne Homers, siehe! sie lächelt auch
uns.'

Spaziergänge eines Wiener Poeten, a
collection of political poetry published in 1831
by Anastasius Grün (q.v.), which touches
satirically and critically on the political climate
of Vienna in the period of repression during the
chancellorship of Metternich (q.v.). The poems,
written in the Nibelungenstrophe (q.v.), com-
pare the meanness of the present with former
Austrian greatness, and in particular attack the
oppressive censorship and the system of in-
formers and secret police. The best-known poem,
Salonszene, ends with the plea of the Austrian
people for freedom: 'Sieh, es fleht ganz artig:
Dürft' ich wohl so frei sein, frei zu sein?' In
order to avoid the censorship, the volume was
published in Hamburg.

SPD, abbreviation of Sozialdemokratische
Partei Deutschlands, the designation of the
German Social Democratic Party since 1890.
The various 19th-c. Socialist movements first
united in 1863 in Leipzig as Allgemeiner
Deutscher Arbeiterverein under F. Lassalle
(q.v.), whose programme of Socialist reform on
a national basis, with the exclusion of Austria
(see KLEINDEUTSCH), was adopted. In 1869 the
Marxist Socialists A. Bebel and W. Liebknecht
(qq.v.) formed in Eisenach (Eisenacher Programm) the Sozialdemokratische Arbeiterpartei,
which aimed at integration with the international Socialist movements.

In 1875 both parties decided in Gotha
(Gothaer Programm) to make common cause
in order to strengthen Socialist influence in the
Diet (Reichstag) of the Empire. As Sozialistische
Arbeiterpartei Deutschlands the party met from
1878, the year of the Socialist Laws, with
mounting antagonism and persecution, and its
leaders emigrated to Switzerland. The party
adopted its present designation on reorganization after Bismarck (q.v.) was dismissed in 1890.
In F. Ebert (q.v.) the party provided the first
president of the Weimar Republic. It developed
into a powerful political force during the early
years of the republic and supported Brüning
(q.v.) until his dismissal in 1932. It unanimously
rejected the Ermächtigungsgesetz (q.v.) of the
National Socialist regime; it continued to work
as an underground movement directed by its
leaders from Prague (until the German invasion
in 1938), from Paris (until 1940), and until the
end of the 1939–45 War from London. German
emigrants working for its principles in the
U.S.A. include P. Tillich (q.v.).

After the war the party was reorganized in
occupied Germany by K. Schumacher (1895–
1952). W. Brandt (b. 1913) was the first Social
Democratic chancellor (Bundeskanzler) of the
Federal Republic (see BUNDESREPUBLIK
DEUTSCHLAND). In East Germany the SPD
merged with the Communist Party (see KPD
and DEUTSCHE DEMOKRATISCHE REPUBLIK) to
form the SED (Sozialistische Einheitspartei
Deutschlands) on 21 April 1946.

In Austria (see ÖSTERREICH) the Socialists
first united in 1867, forming the Arbeiterbil-
dungsverein, and in 1888–9 the Sozialdemokratische Partei Österreichs (SPÖ) under V.
Adler (q.v.). Its influence declined after 1907.
After 1918–20 it came to prominence in the
Republic of Austria under its leader Karl
Renner (q.v.), and recovered its influence in the
late 1920s and early 1930s. The February revolt
of 1934 against the government of E. Dollfuß
(q.v.) resulted in its prohibition. After the 1939–
45 War the SPÖ was once more reorganized by
Renner and became a party to the coalition
government, having in Renner its first chancellor (Staatskanzler) and president (Bundespräsident).

Speculum humanae salvationis, a late Latin
poem written in 1324, covering the history of
mankind from the standpoint of redemption. It
begins with Eden and ends with the Day of
Judgement. The author is unknown. The poem
was translated into German by Konrad von
Helmsdorf in the 14th c. and by HEINRICH von
Laufberg (q.v.) in 1437.

SPEE VON LANGENFELD, FRIEDRICH
(Kaiserswert nr. Düsseldorf, 1591–1635, Trier),
a religious poet, came of a Rhenish noble
family, entered the Society of Jesus in 1610 and
taught as a professor of philosophy at Jesuit
colleges at Paderborn, Cologne, and Trier. The
members of the Society were engaged in the re-
establishment of Roman Catholicism in Westphalia and the Rhineland, and Spee, in the
course of these duties, was wounded when an
attempt was made on his life by a religious
opponent at Peine in 1629. His duties also
included the preparation for death of condemned witches, and as a result of his experiences (which convinced him that many were
condemned unjustly) he published (anonymously) an attack on the judicial procedures
observed (*Cautio criminalis,* 1631). He died of the
plague while tending the sick at Trier.

Spee is the most considerable Roman Catholic poet of the early part of the 17th c. His
verse (published posth. as *Trutznachtigall oder,*
Geistlichs-Poetisch Lustwäldlein in 1649, together
with his mainly prose *Güldenes Tugendbuch*) is
filled with a warm and gentle piety and with a

feeling for God's creatures, including especially the birds and flowers, which invites comparison with Franciscan rather than Jesuit tradition. As a Catholic in the north-west, he was largely exempt from the Protestant Silesian literary authority of Opitz (q.v.), and his verse is freer than that of many of his contemporaries. *Sämtliche Schriften. Historisch-kritische Ausgabe* ed. E. Rosenfeld, began to appear in 1968 with vol. 2, *Güldenes Tugendbuch. Trutznachtigall*, ed. O. Arlt, 1936, was reissued 1967; *Die anonymen geistlichen Lieder vor 1623*, ed. M. Härting, appeared in 1971.

SPENER, PHILIPP JAKOB (Rappoltsweiler, Alsace, 1635–1705, Berlin), a Lutheran divine and an early and influential Pietist (see PIETISMUS), was a preacher at Strasburg (1663), a parish priest in Frankfurt (1666), then court preacher at Dresden (1686), and finally in 1696 provost (Probst) at St. Nicholas's, Berlin. Spener lent his influence to stirrings of a new movement for reform, rejecting the pomp of ecclesiastical oratory and the rigidity of Lutheran dogma, and conducting in Strasburg and Frankfurt communal meditations (*Collegia pietatis*). His principal work is the *Pia desideria, oder herzliches Verlangen nach gottgefälliger Besserung der wahren evangelischen Kirche* (1675). He laid considerable stress on charitable works and encouraged institutions for the care of the aged poor and of orphans. Spener was the subject of polemical attack from the orthodox. Apart from his theological significance he was a heraldic expert, writing the *Opus heraldicum* (1680–90).

Spener's *Hauptschriften*, ed. P. Grünberg, appeared in 1889.

SPENGLER, LAZARUS (Nürnberg, 1479–1534, Nürnberg), town clerk of Nürnberg and an ardent supporter of Luther (q.v.), whom he knew personally, wrote religious tracts and poems. He was formerly held to be the author of *Die lutherische Strebkatz*, (q.v.).

SPENGLER, OSWALD (Blankenburg, 1880–1936, Munich), a historical philosopher, moved after a brief career as a teacher in Hamburg (1908–11) to Munich, where he spent the rest of his life as a writer and private scholar. He is chiefly known for *Der Untergang des Abendlandes* (2 vols., 1918–22, reissued 1969), which has affinities to attitudes associated with *fin de siècle*. Spengler prophesied the decline of the Western world and the corresponding rise of Asiatic and African powers. He was influenced in his theory of the history of civilization by J. G. Herder, Goethe, and Nietzsche (qq.v.), but K. F. Vollgraff (1792–1863), a professor of law and politics at Munich University, and Jacob

Burckhardt (q.v.) were forerunners of his pessimistic assessment of the future of the Western world. Adopting the principle of cyclic progression, Spengler asserted that growth (Blüte), maturity (Reife), and decline (Verfall) apply not only to nations, but also to whole continents. Spengler nevertheless saw himself as a spokesman of Germany, professing a firm belief in the principle of austerity, which he associated with the Prussian tradition. He predicted the disastrous results of the National Socialist regime, but retained the greatest faith in the Germanic race as being the least 'saturated' (saturiert) in Europe; he included the Jews living in Germanic countries, explicitly detaching himself from anti-Semitism.

Spengler's approach is biological, anthropological, and sociological with a strong emphasis on the significance of industrialism and technology as the prime forces in the destruction of culture. In *Der Mensch und die Technik* (1931) he claimed that only dreamers (Träumer) ignored this fact and defined optimism as cowardice. *Urfragen. Fragmente aus dem Nachlaß*, ed. A. M. Koktanek and M. Schröter, appeared in 1965, *Frühzeit der Weltgeschichte* followed in 1966, and correspondence (*Briefe 1913–36*) was published by the same editors in 1963.

SPERATUS, PAULUS (Rötlen nr. Ellwangen, 1484–1551, Marienwerder), a writer of hymns, was ordained priest in 1506 and became one of the cathedral clergy at Würzburg in 1519. He became a Lutheran and was court chaplain at Königsberg in 1524, later becoming bishop of Pomerania. He is the author of a number of Protestant hymns, including the well-known 'Es ist das Heil uns kommen her' (1524). His name is a Latinization of Paul Spret.

Sperber, Der, an erotic verse anecdote of the early 13th c., in which a nun, in her innocence, gives her 'love' for a hawk, and presently asks for and receives its 'return'.

SPERONTES, pseudonym of Johann Sigismund Scholze (1705–50), author of poems (*Singende Muse an der Pleiße*, 1736).

SPERR, MARTIN (Steinberg, Bavaria, 1944–   ), trained in Munich and, for two years, in Vienna as an actor and began his stage career in Bremen (1965–6). In the mid-sixties he also established himself as a playwright representing the 'critical' Volksstück (q.v.), notably with *Jagdszenen aus Niederbayern* (1966 in *Theater heute*), published in 1972 with *Landshuter Erzählungen* (1967) and *Münchner Freiheit* (1971) as *Bayerische Trilogie*. Other works include the play *Koralle Meier. Geschichte einer Privaten* (1970) and the television plays *Der Räuber Mathias Kneißl* (1971) and *Adele Spitzeder* (1972), which Sperr revised and adapted for the stage as *Die Spitzeder* (1977). Using *Pioniere in Ingolstadt* by Marieluise Fleißer (q.v.) as a model, he depicts in his first play the fate of a homosexual, an outsider (reminiscent of Büchner's *Woyzeck*, q.v.) in an intolerant village community, still capable of Nazi atrocities. Set in the late 1940s, the play's film version of 1967 was transposed into the 1960s. Sperr writes in stylized Bavarian, though in *Mathias Kneißl* he uses the more specific Dachau dialect and High German as a more definitive social demarcation. The subsequent plays aim at increasingly broader topical themes, including that of the Gastarbeiter. The volume '*Willst du Giraffen ohrfeigen, mußt du ihr Niveau haben*' was published in 1979.

SPERVOGEL, the name under which a number of Middle High German Sprüche (see SPRUCH) are recorded in the MSS. They are, however, the work of two distinct authors: a simple rugged poet generally referred to as Hergêr, whose poetry evinces piety and a tendency to lament his hard lot, and a younger man, for whom the name Spervogel (sometimes 'der jüngere Spervogel') has been retained. Spervogel probably belonged to the minor nobility and wrote about the end of the 12th c. His Sprüche are more sophisticated and less personal than Hergêr's, and they have no reference to religion. Their tone suggests a manly and accomplished personality.

Speyer, Reichstag zu, usually denotes, among the many Diets held in Speyer, that of 1529, known as the 'Protestation der evangelischen Stände', from which Protestantism derives its name. See PROTESTANTISMUS, REFORMATION, and LANDESKIRCHEN.

SPEYER, WILHELM (Berlin, 1887–1952, Riehen, Switzerland), had already published one good novel (*Das fürstliche Haus Herfürth*, 1914) before the 1914–18 War, in which he served. On resuming authorship, he wrote chiefly for the young, publishing in 1927 the school story *Der Kampf der Tertia*, which is concerned with the rescue by the 'Tertia' (i.e. Fourth Form) of cats and dogs condemned to destruction in consequence of a false report of rabies. He emigrated in 1933, returning to Europe in 1949. Of his large output the last two novels stand out, *Das Glück der Andernachs* (1947), the story of a Jewish family in Berlin in the late 19th c., and *Andrai und der Fisch* (1951), in which his religious outlook is expressed. He is also the author of Erzählungen and Novellen, the last of which is *Señorita Maria Teresa* (1951), and of plays,

mainly written after the 1914–18 War (including *Er kann nicht befehlen, Der Revolutionär, Karl V*, all 1919; *Napoleon*, 1930).

Spiegel das Kätzchen, see LEUTE VON SELDWYLA, DIE.

Spiegel der Natur, a Middle High German medical treatise in poetic form, setting out rules for the preservation of health. It was written by Eberhard von Wampen in 1325, probably while he was on a visit to Sweden. Eberhard (*fl. c.* 1330) was a physician of repute on the island of Rügen.

Spiegel des Regiments, see JOHANN VON MORSHEIM.

Spiegelmensch, a fantastic Expressionist play published by F. Werfel (q.v.) in 1920. Thamal, a Faust-like figure, who has sought peace in a monastery, tires of the contemplative life. He releases his base and sinful (Mephistophelean) self by shooting at his reflection in a mirror. Abetted by his second self, he commits a series of violent acts, including the murder of his father and the tyrannical oppression of a nation he has liberated. He brings himself to trial and condemns himself to death by poison. The evil mirror image returns to the mirror, which changes into a window giving a glimpse of a happy and tranquil world. The play contains elements of satire directed at K. Kraus and S. Freud (qq.v.).

Spielberg, castle at Brünn (Brno), Moravia, used to house political prisoners in Austria under the regime of Metternich (q.v.).

SPIELHAGEN, FRIEDRICH (Magdeburg, 1829–1911, Berlin), grew up in Stralsund, and studied first law and then philology at Berlin, Bonn, and Greifswald universities. In the 1850s he was a schoolmaster in Leipzig and then turned to journalism and writing. His first works were the Novellen *Clara Vere* (1857) and *Auf der Düne* (1858). From 1860 to 1862 he lived in Hanover, editing the *Zeitung für Norddeutschland*. The novel *Problematische Naturen* (q.v., 4 vols.) was published in 1861–2. After 1862 he lived in Berlin, devoting himself to the writing of novels and editing from 1878 to 1884 *Westermanns Monatshefte* (q.v.).

Spielhagen's novels followed each other in rapid succession: *In der zwölften Stunde* (1863), *Die von Hohenstein* (q.v., 4 vols., 1864), *In Reih und Glied* (q.v., 5 vols., 1867), *Hammer und Amboß* (q.v., 5 vols., 1869), *Allzeit voran!* (1871), *Sturmflut* (q.v., 3 vols., 1877), *Angela* (1881), *Uhlenhaus* (1884), *Was will das werden* (3 vols.,

1887), *Noblesse oblige* (1888), *Sonntagskind* (1893), *Selbstgerecht* (1896), *Opfer* (1899), and *Freigeboren* (1900). His best-known collection of Novellen was *Quisisana* (1885). His plays (*Liebe für Liebe*, 1875; *Gerettet*, 1884) are of little importance. In 1882 he published *Beiträge zur Theorie und Technik des Romans*. Spielhagen's novels present a curious mixture of realistic observation and conventional story-telling, and they reflect, though often in pretentious terms, the liberal trends of his day. For a time he was esteemed a great novelist, but in the last decade of the century a reaction set in, and his sincerity was called in question. His reputation has never recovered from this phase of depreciation, which has caused some real merits of characterization and description to be overlooked.

Sämtliche Romane were published in 1871 (10 vols.), and 1895–1904 (29 vols.), and *Ausgewählte Romane* (10 vols.) 1907–10.

Spiel im Morgengrauen, a Novelle by A. Schnitzler (q.v.), published in 1927. Its chief figure, the impecunious second Lieut. (Leutnant) Willy Kasda has prudently given up gambling. A former officer, now down on his luck, taps him for a loan, and Kasda goes out for the evening to win the money at cards. He manages it, but, seized by gambling fever, goes on playing, loses the lot and more than he can pay. As he is driven back to Vienna in the grey light of dawn by his creditor, Kasda finds that the latter insists on payment within twenty-four hours. The money is forthcoming from an uncle, however, but not until after the time-limit, at the end of which Kasda has shot himself. Woven into the story is a somewhat implausible erotic complication.

Spielmannsdichtung, formerly accepted as a generic term for a number of medieval German epics of adventurous and 'popular' character. The underlying assumption that these poems were composed and written down by itinerant minstrels is no longer regarded as tenable, since they were as a class vulgar and illiterate. Among the poems formerly classed as Spielmannsdichtung are *Herzog Ernst*, *König Rother*, *Orendel*, *Ortnit*, and *Salman und Morolf* (qq.v.). The expression is still sometimes used as a convenient means of grouping these poems.

Spiel vom reichen Manne und armen Lazaro, a biblical play by Georg Rollenhagen (q.v.), published in 1590. The play, which has a large cast, contrasts the worlds of rich and poor, treating the former satirically, and showing their roles reversed in the life to come.

Spiel von den zehn Jungfrauen, see LUDUS DE DECEM VIRGINIBUS.

Spiel von Frau Jutten, a late medieval play dealing with Jutta, who is tempted by the Devil, goes to Rome in male attire as Johan von Engelland, becomes cardinal, and then pope. She exorcizes a devil, who reveals her sex, is condemned, but saved from damnation through the intervention of the Virgin Mary. The figure of Jutta is identical with 'Pope Joan'. The author is Dietrich Schernberg, who lived in Thuringia and wrote the play *c.* 1490. It was first printed in 1565 and was reprinted by J. C. Gottsched (q.v.) in 1765.

SPIESS, CHRISTIAN HEINRICH (Helbigsdorf, Saxony, 1755–99, Schloß Bezdiekau, Bohemia), spent several restless years as an itinerant actor, before entering in 1788 the service of a Bohemian nobleman, Count von Künigl, whose steward and companion he became. His dramatic works include the tragedy *Maria Stuart* (1784) and the Ritterschauspiel *Klara von Hoheneichen* (1792). He was a prolific author of widely read Ritter- und Räuberromane (q.v.). They include *Das Petermännchen* (2 vols., 1791–2), in which the hero, like Faust, enters into a pact which destroys his soul, *Der alte Überall und Nirgends* (2 vols., 1792–4), in which a doubter is saved by repentance, *Die Löwenritter*, and *Die zwölf schlafenden Jungfrauen* (3 vols., both 1794–6). He compiled the collections of anecdotes, *Biographien der Selbstmörder* (1785) and *Biographien der Wahnsinnigen* (4 vols., 1795–6, reissued 1966). *Sämtliche Werke* (11 vols.) appeared 1840–1.

SPIESS or **SPIES,** JOHANN (d. *c.* 1607) a Frankfurt printer who published the chap-book of Dr. Faust (1587), now usually known as the *Spieß'sches Faustbuch* (see FAUSTBUCH, SPIESS'-SCHES) or *Das Volksbuch vom Dr. Faust*.

SPIESSHAYMER, HANS, see CUSPINIAN, JOHANNES.

SPINDLER, KARL (Breslau, 1796–1855, Bad Feiersbach, Baden), an actor for some years, was a prolific writer, who achieved success as a historical novelist with such works as *Der Bastard* (3 vols., 1826), *Der Jude* (3 vols., 1827), *Der Jesuit* (3 vols., 1829). His *Sämtliche Werke* appeared between 1830 and 1855, reaching a total of 101 volumes. *Ausgewählte Romane* (34 vols.) were published 1875–6.

SPINOZA, BENEDICTUS DE (Baruch Despinoza), (Amsterdam, 1632–77, The Hague), a philosopher of Jewish descent, was expelled from the Synagogue in 1656. In 1676 he rejected an invitation to teach philosophy at Heidelberg University, preferring to remain in obscurity.

He made his living by grinding and polishing lenses. Spinoza's writings are in Latin, and the anonymously published *Tractatus theologico-politicus* (1670) was the only one of his works to appear in his lifetime. Like his posthumous works, it was placed on the Roman Catholic Index Librorum Prohibitorum. His *Ethics*, for which he is most esteemed (*Ethica ordine geometrica demonstrata*), appeared in five parts in 1677 shortly after his death. In spite of the dryness of the Euclidean method, they express an attitude of love, kindness, and tolerance for his fellow-men. The ethical views, however, have not been specially influential in German literature, which has been attracted more by Spinoza's monistic view of God and Nature as one substance, existing in various modes.

The impact of Spinoza outside philosophical circles was first felt in Germany in the second half of the 18th c. In the time of the Sturm und Drang (q.v.) he came to be regarded as a kind of godfather to vague pantheistic ideas and sentiments. Largely at second hand he exercised considerable influence on the young Goethe and later on the writers of the Romantic movement (see ROMANTIK), who saw God in Nature and felt that Spinoza offered them a philosophical justification for doing so. Lessing (q.v.) showed a genuine interest in the philosophy of Spinoza and in a conversation with F. H. Jacobi (q.v.) in 1780 declared, 'Es gibt keine Philosophie als die Philosophie des Spinoza'. Spinoza is the subject of novels by B. Auerbach and by E. G. Kolbenheyer (qq.v.).

Spital von Jerusalem, Das, a Middle High German verse chronicle in praise of the Order of St. John of Jerusalem (see JOHANNITERORDEN). It was written in Strasburg in the late 13th c. and is partly legendary.

SPITTA, KARL JOHANN PHILIPP (Hanover, 1801–59, Burgdorf), pastor in various Hanoverian parishes, then moderator (Superintendent) in Wittingen, Peine, and Burgdorf, published an early collection of verse (*Sangbüchlein der Liebe für Handwerksburschen*, 1823), followed later by two volumes of hymns (*Psalter und Harfe*, 1833–43). Two posthumous publications were *Nachgelassene geistliche Lieder* (1861) and *Lieder aus der Jugendzeit* (1898).

SPITTELER, CARL (Liestal nr. Basel, 1845–1924, Lucerne), son of a Swiss civil servant, studied law and theology at Basel and Heidelberg universities. In 1871, through the mediation of his father's friend Colonel Sulzberger, he obtained a post as tutor in Finland (then under Russian government), where he remained until

1879. There followed six years as a schoolmaster, first in Berne, then in Neuville, after which he became a journalist, editing the literary pages of the *Neue Zürcher Zeitung* and becoming a major contributor to *Der Kunstwart* (q.v.). An inheritance enabled him to devote himself entirely to writing. He spent the rest of his life in Lucerne. In 1919 he received the Nobel Prize for Literature.

Spitteler's first work was *Prometheus und Epimetheus* (1880–1), an epic printed as prose, but with a persistent iambic rhythm. He used the figures of classical mythology to present the contrast of the human soul rising to its full grandeur (Prometheus) or falling into the drabness of subjection to habit and routine (Epimetheus). The work, which was ignored by his contemporaries, appeared over the pseudonym C. F. Tandem. Late in life Spitteler wrote a new version in iambic hexameters with the title *Prometheus der Dulder* (1924). A volume of poetry entitled *Schmetterlinge* appeared in 1889, and two stories followed at long intervals, *Conrad der Leutnant* in 1898 and *Gerold und Hansli, die Mädchenfeinde* in 1907. The novel *Imago* (q.v., 1906), which is autobiographical, has obvious connections with psycho-analysis and greatly impressed S. Freud (q.v.). The epic *Olympischer Frühling* (4 vols., 1900–6), long held to be Spitteler's best work, presents the human comedy in mythological terms. It is an attempt to revitalize mythology as a means of communication. Some see his sensitive short autobiographical study *Meine frühesten Erlebnisse* (1914) as his best work.

Spitteler's essays include *Unser Schweizer Standpunkt* (1915), *Meine Beziehungen zu Nietzsche* (1908), and *Rede über Gottfried Keller* (1919). *Gesammelte Werke* (11 vols.), ed. G. Bohnenblust, appeared 1945–58, *Kritische Schriften*, ed. W. Staufacher, 1965.

SPITZWEG, CARL (Munich, 1808–85, Munich), a self-taught painter, who started life as an apothecary. A representative of Biedermeier (q.v.), he painted odd, tumbledown townscapes, inhabited by quaint, old-fashioned men and women, in a spirit of amiable, almost affectionate, satire. After visiting Paris in 1851 he acquired a new vision and painted a number of pure landscapes without figures. He was successful as a cartoonist and illustrator for weeklies, including the *Fliegende Blätter*.

SPOHR, LUDWIG (LOUIS) (Braunschweig, 1784–1859, Kassel), was one of the greatest violinists of his time and a prolific composer. At the age of 6 he was able to participate in chamber music, and at 15 was appointed Kammermusikus by the Duke of Brunswick. He was a conductor in Gotha, Vienna, and Frankfurt before he settled in 1822 as Hofkapellmeister in Cassel, retiring in 1857. He was an admirer of Mozart and a friend of Mendelssohn (qq.v.). Although he owed his recommendation to the Elector of Hesse-Cassel to C. M. Weber (q.v.), he could not appreciate Weber's *Freischütz* (q.v.) or the last compositions of Beethoven (q.v.). Yet he produced Wagner's *Der fliegende Holländer* and *Tannhäuser* (qq.v.) in Kassel in 1842 and 1853 respectively.

Among Spohr's works, numbering more than 150, are nine symphonies, violin concertos, and oratorios, which met with great success; among them are *Die letzten Dinge*, *Des Heilands letzte Stunden* (*Calvary* or *The Crucifixion*), and *Der Fall Babylons*. Of the last two the former was produced at the Norwich Festival in 1839 and the latter in 1842. Spohr composed ten operas. In 1808 Goethe showed himself well disposed towards the second, *Alruna*. The third, *Faust*, completed in 1813 and produced by him in 1818 in Frankfurt, initiated his public success. Its plot, differing from that of Goethe's *Faust* (q.v.), Pt. I, was among the influences on Grabbe's *Don Juan und Faust* (q.v.).

Spohr's autobiography was published in two volumes in 1860–1 (ed. F. Göthel, 1968).

Sprachgesellschaften, societies founded in the 17th c. with the aim of purifying the German language and promoting its use in poetry. The most prominent were the Fruchtbringende Gesellschaft (q.v., 1617), the Tannengesellschaft (q.v., 1643), the Teutschgesinnte Genossenschaft (see ZESEN, P., 1643), the Hirten- und Blumenorden an der Pegnitz (q.v, 1644), and the Elbschwanenorden (q.v., 1660).

SPRENG, JOHANNES (Augsburg, 1524–1601, Augsburg), a lawyer of Augsburg, who was active as a Meistersinger (see MEISTERGESANG). His translations of the *Iliad* (1610) and the *Aeneid* (1616) into rhyming verse appeared posthumously.

SPRENG, JOHANN JAKOB (Basel, 1699–1768, Basel), a pastor in Heilbronn and later (1743) a professor at Basel, edited two short-lived Swiss moral weeklies, *Der Eidgenoß* (1749) and *Der Sintemal* (1759). He also retranslated the Psalms (*Neue Übersetzung der Psalmen Davids*, 1741). His hymns were published in *Geistreiche Kirchen- und Hausgesänge* (1741) and his collected poetry in *Geistliche und weltliche Gedichte* (1749).

SPRICKMANN, ANTON MATTHIAS ALOYSIUS (Münster, Westphalia, 1749–1833, Münster), studied at Göttingen, where he became friendly with G. A. Bürger (q.v.), and then settled as a lawyer in Münster. He is the author of an opera, *Die Wilddiebe* (1774), and of a play, *Die natürliche Tochter* (1774), and also wrote popular ballads with a tendency towards the horrifying. In his sixties he became a friend of Annette von Droste-Hülshoff (q.v.). He was appointed to a professorship at Breslau in 1812 and at Berlin in 1817.

Spruch, Spruchdichtung, name applied since the middle of the 19th c. to a genre of Middle High German poetry. The Spruch is closely related to the *liet* (see MINNESANG) and some authorities consider that the distinction is arbitrary. In general it can be said that the Spruch, like the *liet*, was sung. It consists of one or more strophes and deals with social or political matters, often in a personal context. The beginnings of the Spruch are found in Hergêr (see SPERVOGEL), its first great master is WALTHER von der Vogelweide, whom REINMAR von Zweter closely follows (qq.v.). The later authors of Sprüche are mostly commoners, many of them itinerant poets wandering from court to court. The most notable exponents of the form, apart from those already mentioned, are Der Marner and HEINRICH Frauenlob (qq.v.).

In the Middle Ages the term was applied primarily to short spoken poems, of moral or didactic character. In this sense, Freidank (q.v.) is the outstanding author of Sprüche. The term gnomische Dichtung (aphoristic or gnomic verse) is sometimes used as an alternative to Spruchdichtung.

In the 16th c. Hans Sachs (q.v.) used the term Spruchgedicht to include not only his moral poems, but also his plays.

Spuren im Sand, an autobiographical novel by H. W. Richter (q.v.).

SPYRI, JOHANNA (Hirzel nr. Zurich, 1829–1901, Zurich), née Heußer, married a Zurich lawyer. A successful and sensitive writer for children, she is the author of *Heidis Lehr- und Wanderjahre* and *Heidi kann brauchen, was es gelernt hat* (both 1881), books which in translation have been much read by children in other countries.

SS, abbreviation for Schutzstaffel, designation of a paramilitary force of the NSDAP (q.v.). The SS, the members of which wore black uniforms, originated in 1925, being formed within the SA (q.v.) to provide a personal bodyguard for Hitler. At first a hundred strong, it was rapidly augmented under H. Himmler (q.v.), its leader from 1929 on. After the murder of the SA leaders in June 1934 the influence of the élite SS increased at the expense of the mass SA. The SS absorbed the Gestapo (q.v.), which was originally separate and under H. Göring (q.v.).

As time passed, the SS, in a newly founded branch, the Waffen-SS, furnished the concentration camp guards and ran the systematic persecution of the Jews and eventually the extermination campaign. The SS provided the troops behind the lines who carried out numerous massacres in occupied territory. The Waffen-SS also formed a division of combat troops, Division Leibstandarte. This vast organization became in the end the executive force of National Socialist government.

Staberl, a comic figure invented by A. Bäuerle (q.v.) in *Die Bürger in Wien.* So great was Staberl's fascination for the public of the Viennese popular theatre that Bäuerle included him in three later plays, and imitators, notably Karl Carl (q.v.), kept the figure alive in new plays into the 1850s.

Stabreim, see ALLITERATION.

STADEN, HANS (Homburg, Hesse, *c.* 1525–76?) was in Spanish and Portuguese military service in South America between 1547 and 1555. He was captured by Indians and afterwards described his experiences and observations in his so-called 'Menschenfresserbuch' (*Warhafftige Historia und Beschreibung einer Landtschafft der Wilden, Nacketen, Grimmigen Menschenfresser,* 1557). A facsimile appeared in 1925, and an edition by K. Fouquet in 1963.

STADEN, SIGMUND GOTTLIEB (THEOPHIL) (Kulmbach, 1607–55, Nürnberg), was the organist of St. Lorenz, Nürnberg, and a member of the Hirten- und Blumenorden an der Pegnitz (q.v.). In 1644 he composed the music for the opera *Seelewig,* published in Harsdörffer's *Frauenzimmer-Gesprechspiele* (see HARSDÖRFFER, G. P.), which is reputed to be the first extant German opera (*Daphne,* composed by H. Schütz, q.v., in 1626, was a translation from the Italian by M. Opitz, q.v.).

STADION, PHILIPP (JOHANN PHILIPP), GRAF VON (Mainz, 1763–1824, Baden by Vienna), left Mainz to serve as Austrian foreign minister during the crucial years 1805–9 in the Napoleonic Wars (q.v.). His outstanding services for the Emperor Franz II (q.v.) ended with his dismissal after the Austrian defeat. He was replaced by Metternich (q.v.), but continued in the Austrian government as finance minister, and in 1816 founded the Austrian National Bank. It was in his capacity as finance minister that he helped F. Grillparzer (q.v.) by securing for him the position of Theaterdichter at the Burgtheater and by offering him a position in his finance department.

STADLER, ERNST (Colmar, Alsace, 1883–1914, killed in action nr. Ypres), studied at Strasburg and Munich universities, was a Rhodes Scholar at Magdalen College, Oxford 1906–8, where he later took a B.Litt., and was a lecturer at Brussels. A reserve officer, he was called up in August 1914. Stadler was one of the earliest Expressionist poets, writing powerful poetry with a sense of foreboding, though not without hope. The only collection made in his lifetime was *Der Aufbruch* (1914). The critical edition of his work, *Dichtungen, Schriften, Briefe,* ed. K. Hurlebusch and K. L. Schneider, appeared in 1983.

Stadt hinter dem Strom, Die, a novel by H. Kasack (q.v.), published in 1947. In realistic terms it describes the shattered city of the dead at which the living Dr. Robert Lindhoff alights after he has crossed the river by the railway. The city is ghostly, a simulacrum of the worst bombed towns of Germany, and its dead inhabitants repeat meaninglessly the routine motions of their lives until they disintegrate into nothingness. The city is ruled on a totalitarian model by the thirty-three *Weltenwächter* (in the symbolism the year 1933 is easily discernible and the link with the years of the National Socialist regime thus established). As the dead dissolve, all that remains is their spiritual achievement, which is sifted and categorized by the Archiv. 'Geist' survives when 'Ungeist' (Kasack's word) has perished.

Kasack has himself commented on his novel in an essay in *Mosaiksteine,* emphasizing his ironical portrayal of the age of ruin and ruins, in which the whole world, and especially Germany, existed while the book was being written (1942–6), and deprecating a more general nihilistic interpretation.

STAËL, ANNE LOUISE GERMAINE, BARONNE DE (Paris, 1766–1817, Paris), *née* Necker, was the daughter of the Swiss-born banker J. Necker (1732–1804) called in by Louis XVI to retrieve the financial situation of the French monarchy. In 1792 Madame de Staël took refuge from the French Revolution (q.v.) at Coppet, Switzerland. She published *De la littérature,* her first important work, in 1800. She toured Europe and in 1804 visited Weimar, meeting the celebrities Goethe, Schiller, and Wieland (qq.v.), and also Berlin, where she made the acquaintance of A. W. von Schlegel (q.v.), who thereafter accompanied her on her travels and lived with her at Coppet. In 1810 she published her book on Germany (*De l'Allemagne*) which was confiscated by order of Napoleon. It was reprinted in London in 1813 and translated into German in 1814 (*Über Deutschland*).

Madame de Staël was particularly impressed and influenced by German Romanticism (see ROMANTIK). Her novel *Corinne ou l'Italie* (1807), translated into German by D. von Schlegel (q.v.), was widely read throughout Europe.

STAGEL, ELSBETH (d. *c.* 1360, Töß nr. Winterthur), a Dominican nun, became a close friend of Heinrich Seuse (q.v.), her spiritual adviser, and collaborated with him, writing his life (which he subsequently revised) and collecting the pastoral letters written to various nuns; eleven of these letters were later incorporated in *Das Briefbüchlein.*

STÄGEMANN, FRIEDRICH AUGUST VON (Vierraden, Uckermark, 1763–1840, Berlin), a Prussian civil servant from 1785, collaborated with Baron vom Stein (q.v.) after Napoleon's victory at Jena and was a member of the Prussian mission at the Congress of Vienna (see WIENER KONGRESS). In 1819–20 he edited the official *Staatszeitung* in Berlin. In 1813 he published a volume of patriotic poetry (*Kriegsgesänge aus den Jahren 1806 bis 1813*). An expanded edition of this appeared in 1828 as *Historische Erinnerungen in lyrischen Gedichten.* Stägemann's patent of nobility dates from 1816.

Stahlhelm, Der, was founded by F. Seldte (q.v.) as an association of officers and soldiers who had seen front-line service in the 1914–18 War. It expressed its right-wing tendencies by supporting the formation of Freikorps (q.v.) and widened its membership by admitting all former soldiers as well as sympathizers among the young. It was opposed to the Treaty of Versailles (see VERSAILLES, TREATY OF, 1919), parliamentary government, pacifism, and socialism; it advocated rearmament and the recovery of territories and colonies lost in 1919. One of the three parties in the Harzburger Front (q.v., 11 October 1931), the Stahlhelm supported Hitler (q.v.) in January 1933, but was merged by him into the SA (q.v.) in July 1933.

STAHR, ADOLF (Prenzlau, 1805–76, Wiesbaden), a schoolmaster in Oldenburg, resigned in 1852, and in 1854 married the authoress Fanny Lewald (q.v.). He is the author of the historical novel *Die Republikaner in Neapel* (3 vols., 1849), of travel books, including *Ein Jahr in Italien* (3 vols., 1847–50), and of works on literature (*G. E. Lessing,* 2 vols., 1859; *Goethes Frauengestaltung,* 2 vols., 1865–8).

STAINHÖWEL, HEINRICH, see STEINHÖWEL, HEINRICH.

Stanzen, German term for the Italian verse form *ottava rima.* Goethe's *Zueignung* prefaced to *Faust* (q.v.) is a well-known example.

STARHEMBERG, RÜDIGER (ERNST RÜDIGER), GRAF VON (Graz, 1638–1701, Vienna), an Austrian general of ancient lineage, fought with distinction in the Turkish War of 1664. In the War of 1683, under his efficient and inspiring leadership as city commandant, Vienna withstood a Turkish siege from 15 July until it was relieved on 12 September. His letters, ed. V. von Renner, were published in two volumes, 1890–2 (*Vertrauliche Briefe 1682–1689*).

Stark, Herr Lorenz, see HERR LORENZ STARK.

Stationendrama, term denoting a type of Expressionist (see EXPRESSIONISMUS) play in which a succession of episodes replaces a continuously developing action. Implied in the term is an analogy to the depiction of the Stations of the Cross, and this analogy is sometimes made explicit, as in G. Kaiser's *Von morgens bis mitternachts* (q.v.). Characteristic exponents of Stationendrama are R. Sorge, G. Kaiser, and E. Toller (qq.v.). Originating in medieval drama, the immediate model was *Till Damaskus* (Pt. 1 and 2 in 1898, Pt. 3 in 1901; *Nach Damaskus, To Damascus*) by Strindberg (q.v.).

Staufer, see HOHENSTAUFEN.

Stechlin, Der, a novel by Th. Fontane (q.v.), written in 1895–6 and published in book form in 1898 (date on title-page 1899) after appearing in instalments in the magazine *Über Land und Meer* between October 1897 and March 1898. Although it is Fontane's second longest novel, it can scarcely be said to have a plot. Its central figure is the ageing Dubslav von Stechlin, widowed for thirty years and living alone in his somewhat dilapidated mansion near the shore of the Stechliner See, to be imagined in the neighbourhood of Fontane's native Neu-Ruppin.

Dubslav, a man of honest and humorous character, refuses to take himself seriously and lives modestly and contentedly in contact with his elderly valet Engelke, his politically progressive vicar Lorenzen, and Krippenstapel, the most Prussian of Prussian schoolmasters. Dubslav's only son, Woldemar, is a captain in a cavalry regiment of the guard in Berlin (1. Garde-Dragoner-Regt.). The novel opens with a visit paid by Woldemar with two friends to Schloß Stechlin. Woldemar next calls on his aunt, Dubslav's half-sister, the formidable Domina of a retreat (Stift) for elderly gentlewomen at nearby Wutz. The Domina, a target for discreet comedy, is the embodiment of an unimaginative narrow-mindedness typical of much of the provincial aristocracy. On his return to Berlin, Woldemar pursues with courtesy and without eagerness his courtship in the house of Count Barby. Indeed, so discreet is his approach that it is not at first clear which of the two daughters is intended. It proves to be the younger, Armgard, and the couple are married in Berlin and set out for a honeymoon in Italy. In the meantime Dubslav has allowed himself to be nominated as the local Conservative candidate for the Reichstag, but at the election he is defeated by his Social Democrat opponent Thorgelow. After the wedding festivities, Dubslav begins to ail and his condition soon worsens. His doctor goes on holiday and he does not take to the locum. Soon he declines and dies. Woldemar and Armgard only learn of the illness when death and burial have taken place. They return to Berlin, but within a few months Woldemar resigns his commission, and the couple take up residence in Schloß Stechlin.

Character is one of the most important aspects of this, Fontane's last novel, and the fullest attention is given to Dubslav and to those around him, all, in one way or another, eccentrics or oddities. Fontane clearly attaches great importance to the combination of integrity and goodwill which he thrice notes in the closing stages of the work under the word Gesinnung. *Der Stechlin* is also a social and political work. Dubslav and, in a lesser degree, Count Barby are seen as lovable anachronisms; but Dubslav, though he does not take to the new world that is coming, adopts no moral postures about it. Woldemar, a pupil of Lorenzen, has a broader political horizon; the termination of the work is, however, politically ambiguous. Fontane is most severe on the newly ennobled parvenu Herr von Gundermann. The dialogue, which is a prominent feature of the novel, shows Fontane at his best. A recurrent symbol, the turbulence of the Stechlin lake in times of earthquake or eruption, attempts to affirm the link between the rural quietude, which is the principal background of the novel, and the events of the world at large.

STEFFEN, ALBERT (Murgenthal nr. Berne, 1884–1963, Dornach nr. Basel), a Swiss writer with leanings towards mysticism, lived for a number of years in Berlin and Munich, becoming in 1920 a member and in 1925, after the death of R. Steiner (q.v.), president of the Anthroposophical Society; he was closely associated with the Goetheanum in Dornach. His novels include *Ott, Alois und Werelsche*

(1907), *Die Bestimmung der Roheit* (1912), *Die Erneuerung des Bundes* (1914), *Sibylla Marianna* (1917), *Lebensgeschichte eines jungen Menschen* (1927), *Wildeisen* (1929), *Sucher nach sich selbst* (1931), and *Wach auf, du Todesschläfer* (1941); Erzählungen and Novellen include the collections *Die Heilige mit dem Fische* (1919) and *Novellen* (1947). He wrote a number of plays (*Die Manichäer*, two plays, 1916; *Das Viergetier*, 1924; *Hieram und Salomo, Der Sturz des Antichrist*, both 1927; *Das Todeserlebnis des Mannes*, 1934; *Pestalozzi*, 1939; *Märtyrer*, 1944; *Alexanders Wandlung*, 1953; and *Lin*, 1957) and poetry (*Weg-Zehrung*, 1921; *Gedichte*, 1931; *Spätsaat*, 1947; and *Steig auf den Parnaß und schaue*, 1960, the last of his collections). His essays deal with the artist, mysticism, and religion (*Der Künstler zwischen Westen und Osten*, 1925; *Der Künstler und die Erfüllung der Mysterien*, 1927; *Mysterienflug*, 1948; *Geist—Erkenntnis—Gottes-Liebe*, 1949; *Dichtung als Weg zur Einweihung*, 1960); he wrote on Steiner, whom he first met in 1907 (*Begegnungen mit Rudolf Steiner*, 1926, and *Über den Keimgrund der Mysteriendramen Rudolf Steiners*, 1971), published a collection of essays on Goethe (*Goethes Geistgestalt*, 1932), and in 1971 published an essay on Christian Morgenstern (q.v., *Vom Geistesweg Christian Morgensterns*). From 1921 to 1950 he was editor of the periodical *Das Goetheanum*.

STEFFENS, HENRIK (Henrich) (Stavanger, Norway, 1773–1845, Berlin), Norwegian-born but of part German descent, became a lecturer at Kiel University (1796) and at Jena (1799), where he met the celebrities of Weimar and was especially influenced by the philosopher F. W. J. Schelling (q.v.), afterwards studying geology with A. G. Werner (1750–1817) in Freiberg (Hesse). In 1802 he lectured at Copenhagen University, and in 1804 became a professor at Halle, moving after the Napoleonic occupation of the city to Breslau University. Steffens felt himself a German and served with zeal in the War of Liberation (see NAPOLEONIC WARS). In 1832 he was appointed to a professorship at Berlin University.

Steffens sought to reconcile scientific investigation, philosophical speculation, and religion, and his influence on the Romantic movement (see ROMANTIK) is notable for his contact with A. G. Oehlenschläger (q.v.). His publications began with *Grundzüge der philosophischen Naturwissenschaft* (1806), which was followed by *Anthropologie* (1824) and *Christliche Religionsphilosophie* (1839). He also wrote on universities (*Über die Idee der Universitäten*, 1809) and published purely scientific works. His fiction appeared as *Novellen* (16 vols., 1837–8), and his remarkable memoirs as *Was ich erlebte* (10 vols.,

1840–4; a selection was edited by F. Gundolfinger in 1908 and by W. A. Koch in 1938).

Stegreif, extemporization, commonly practised in cabaret and night-club performances, and used at all times as an emergency procedure by competent actors to cover a 'dry-up', was widespread and popular in the 17th c. in Germany in the performances of the Englische Komödianten (q.v.). The heyday of extemporization was the Viennese popular theatre of the early 18th c., which possessed masters of irrepressible spontaneous wit in J. A. Stranitzky and G. Prehauser (qq.v.). Extemporization was, however, officially banned in 1737, but occasional gagging certainly continued for many years, and there exists a well known and relevant anecdote concerning Mozart and E. Schikaneder (qq.v.), when the latter was playing Papageno in *Die Zauberflöte* (q.v.) in 1791.

STEHR, HERMANN (Habelschwerdt, Silesia, 1864–1940, Oberschreiberhau, Silesia), grew up in straitened circumstances and became an elementary school teacher. The outspoken character of some of his early works caused offence and the education authorities posted him, as a punitive measure, to one remote village after another. In 1915 he abandoned teaching and devoted himself to writing. He had a deep awareness of suffering, and a tense and brooding religious mysticism underlies his works. He was one of the authors approved by the National Socialist regime, who appointed him to office in the Reichsschrifttumskammer. He published some gloomy volumes of Novellen (*Auf Leben und Tod*, 1898; *Das Abendrot*, 1916; *Die Krähen*, 1921; *Mythen und Mären*, 1929) and several novels (*Leonore Griebel*, 1900; *Der begrabene Gott*, 1905; *Drei Nächte*, 1909; *Geschichten aus dem Mandelhause*, 1913; *Der Heiligenhof*, q.v., 1918, his best-known work; *Peter Brindeisener*, 1924; *Das Geschlecht der Maechler*, q.v., a trilogy, 1929–44). An early play, *Meta Konegen* (1904), attracted some attention. He also wrote an engaging fairy-tale (*Wendelin Heinelt*, 1909) and some poems (*Das Lebensbuch*, 1920; *Der Mittelgarten*, 1936). A considerable number of unpublished poems were found in his posthumous papers. *Gesammelte Werke* (12 vols.) appeared 1927–36.

STEIN, CHARLOTTE, FREIFRAU VON (Weimar, 1742–1827, Weimar), *née* von Schardt, was married in 1764 to Josias Friedrich, Freiherr von Stein (1735–93), Master of the Horse (Hofstallmeister) to the Weimar court. There were seven children of the marriage, which was not a happy one. Frau von Stein was a lady-in-waiting at the Weimar court and from the time of Goethe's arrival exercised a powerful attraction, which

was more lasting than that of any other woman to whom he had so far been drawn. His utterances repeatedly emphasize the serenity of her temperament and the calming influence she had upon him. Goethe's sudden departure for Italy without leave-taking in 1786 led to a breach with Frau von Stein, which became complete when, on his return in 1788, Goethe took Christiane Vulpius into his house as his mistress. A measure of reconciliation between him and Frau von Stein took place in 1801. Frau von Stein and Goethe corresponded abundantly, but after the estrangement she asked for the return of her letters, which were then destroyed. Goethe's letters to her have been published (*Goethes Briefe an Frau von Stein*, 3 vols., 1848–51, in 2 vols., 1960).

Many of Goethe's poems are addressed to or refer to Frau von Stein, notably 'Warum gabst du uns die tiefen Blicke', *Jägers Abendlied*, *Rastlose Liebe*, *An den Mond* (qq.v.). She is also in greater or less degree the original of Marianne (*Die Geschwister*, q.v.), of Iphigenie (see IPHIGENIE AUF TAURIS), and of Leonore von Este (*Torquato Tasso*, q.v.). Goethe gave her the poetic name Lida.

STEIN, KARL, REICHSFREIHERR VOM UND ZUM (Nassau/Lahn, 1757–1831, Schloß Kappenberg, Westphalia), Prussian statesman, who planned far-reaching reforms of the Prussian administrative system during the Napoleonic occupation (see NAPOLEONIC WARS).

Having studied law and politics at Göttingen University, Stein began his career in the Prussian civil service in 1780. His experience in various administrative branches and a study tour to England strengthened his belief that the state should be a community of free citizens entrusted with responsibility as well as service. Unlike the French revolutionaries, he believed in an emancipation of the subjects within the existing class system. It needed the crisis of Jena (1806) to convince Friedrich Wilhelm III (q.v.) of the necessity of such radical changes as Stein envisaged. Even so it was not until after the Peace of Tilsit (July 1807) that the King entrusted Stein with hitherto unparalleled ministerial authority. Stein's three edicts laid the new foundations of the Prussian state: that of 9 October 1807 effected the emancipation of the peasants, which included the abolition of serfdom; that of 19 November 1808 established municipal self-government by elected councils; and that of 24 November introduced the centralization and reorganization of the machinery of government, aiming at a parliamentary constitution. On the same day, however, Friedrich Wilhelm yielded to Napoleon's demand to dismiss Stein. Napoleon's suspicions were well founded, for, like Scharnhorst (q.v.), Stein was preparing a move against the French, and a letter revealing his intentions had been intercepted by the French. Banned by Napoleon, Stein went into exile in Bohemia. He never returned to high office, but in the critical years to come he proved an important agent in the *rapprochement* between Prussia and Russia.

At the opening of Napoleon's Russian campaign, Alexander I called Stein as an unofficial adviser to St. Petersburg, where he worked out plans for the reorganization of Germany, and on the victorious advance of the Russian troops he won the Tsar's support for an alliance with Prussia and an all-out campaign against Napoleon. It was more difficult to win the support of the Prussian king. Both Hardenberg (q.v.), since 1810 Stein's successor as chancellor, and General von Yorck (q.v.) brought pressure upon the King, while public opinion, stirred up by Stein in East Prussia, became likewise a factor in the new alliance which led to the War of Liberation. But Stein's plans for the reorganization of Germany remained unrealized in spite of his presidency of the Central Administration (October 1813) responsible for occupied territory. And although he attended the Congress of Vienna (1815, see WIENER KONGRESS) as the unofficial adviser of Alexander, he had no authority to curb the overriding influence of the Austrian minister Metternich (q.v.). Stein spent the remainder of his years at Schloß Kappenberg. A few years before his death he accepted an appointment as marshal of the Provincial Estates (Provinziallandtag) in Westphalia. The Prussian king marked Stein's visit to Berlin in 1827, after eighteen years of absence, by naming him a member of the Council of State (Staatsrat).

In 1815 Stein met Goethe in Cologne Cathedral (which both happened to visit after the city's liberation) and in Nassau; he asked Goethe to submit (without reference to his name) a memorandum to Hardenberg urging Prussia to acquire Sulpiz Boisserée's collection of medieval German works of art, which, however, went to Munich. In 1819 Stein founded the Gesellschaft für Deutschlands ältere Geschichtskunde, for which he achieved recognition in 1827 by being elected honorary member of the Berlin Academy. The idea that led to the foundation of the *Monumenta Germaniae historica* (q.v.) arose from his desire to teach his daughter history based on facts and documents. The insistence on facts was a characteristic feature of this statesman, who had inspired his age with the idea of German unification while standing aloof from Romantic fantasy and idealism as well as from Prussian bureaucracy, and who (as Hardenberg stated) combined the rare qualities of learning, intellect, and common sense. Unlike Bismarck (q.v.), to whom he is comparable in stature, Stein was always at heart a German rather than a Prussian. In 1858 E. M. Arndt (q.v.) published a book of recollections.

Briefe und Schriften (6 vols.), ed. G. H. Pertz, appeared 1849–55, and *Briefe und sämtliche Schriften* (10 vols.), ed. W. Hubatsch, 1957–73.

STEINER, RUDOLF (Kraljević, Croatia, 1861–1925, Dornach nr. Basel), is the founder of Anthroposophy, of the Goetheanum (Hochschule für Geisteswissenschaften) in Dornach, and of schools bearing his name.

Having studied natural science at Vienna University, Steiner collaborated on the Weimar edition of Goethe's works, publishing *Einleitung zu den naturwissenschaftlichen Schriften Goethes* (4 vols.) during this period (1883–97) and *Goethes Weltanschauung* (1897). In 1902 he became head of the German Section of the Theosophical Society. But he dissociated himself from Theosophy by introducing the term Anthroposophy and in 1913 founded the Anthroposophical Society in order to promote his 'Geisteswissenschaft' (as distinct from 'religion'), which was based on man's own innate capacity to relate to cosmic forces (he described Christ as 'sun being' (Sonnenwesen) in his theories on evolution); Steiner's adherents were not confined to Germany. He lectured and wrote extensively on his ideas (including *Wahrheit und Wissenschaft*, 1892; *Die Philosophie der Freiheit*, 1894; *Die Rätsel der Philosophie*, 1900; *Die Mystik im Aufgange des neuzeitlichen Geisteslebens*, 1901; *Die Geheimwissenschaft im Umriß*, 1910; *Vom Menschenrätsel*, 1916; *Von Seelenrätseln*, 1917; *Geisteswissenschaft und Medizin*, 1920; and *Anthroposophie*, 1924). In 1913, the year of foundation, the building of the Goetheanum was begun according to plans designed by Steiner; after destruction by fire at the end of 1922 it was newly built in 1924, again according to his own plans (*Wege zu einem neuen Baustil*, 1914). An international cultural centre, it also served Steiner for vocal activities of speech and singing termed Eurhythmy (Eurhythmie), in which inner experience and 'curative' physical exercise collaborated in the attainment of artistic and therapeutic effects (*Eurhythmie als sichtbare Sprache*, 1927).

In 1919 the first school, of which Steiner was head until his death, was founded by a director of a cigarette-manufacturing firm (Waldorf) in Stuttgart, named after him Waldorfschule. It became the model for Steiner schools (Rudolf-Steiner-Schulen) in and outside Germany. The schools in Germany, which were prohibited in 1938 by the National Socialist regime and were refounded in the German Federal Republic, are centralized in Stuttgart (Bund der Freien

Waldorfschulen). A practical application of his intricate anthroposophical methodology, the schools combined individualism with community life (the evolution of mankind remaining central to his convictions) on the principle of dividing education from infancy into three stages, each of seven years, in which imitation, the need for leadership, and the capacity for individual conceptual judgement were successively encouraged; learning and the exercise of crafts according to individual inclination and aptitude were to promote self-awareness, in which Steiner saw a prerequisite for the progression of the human spirit towards cognition (*Die Erziehung des Kindes vom Standpunkt der Geisteswissenschaft*, 1907, *Allgemeine Menschenkunde als Grundlage der Pädagogik*, 1918). Steiner wrote four mystery plays, *Die Pforte der Einweihung* (1910), *Die Prüfung der Seele* (1911), *Der Hüter der Schwelle* (1912), and *Der Seelen Erwachen* (1913, collected 1935, repr. 1971).

Other reprints of Steiner's writings appeared in the 1950s and 1960s. A *Gesamtausgabe* appeared in 1935 (reissued 1962), and a new edition of his collected works, prepared by the Rudolf-Steiner-Archiv, from 1955. Diaries (14 vols.) were published 1961–7.

STEINHAUSEN, HEINRICH (Sorau, Lausitz, 1836–1917, Schöneiche, Prussia), a Protestant clergyman, who taught for a time at the Cadet School at Berlin-Lichterfelde and from 1868 was pastor in Blüthen (near Perleberg), is the author of mainly historical novels, of which the best known is the 14th-c. story, set in Maulbronn (q.v.), *Irmela, eine Geschichte aus alter Zeit* (1881). His *Memphis in Leipzig* (1880) satirizes G. Ebers (q.v.).

STEINHÖWEL, HEINRICH (Weil der Stadt, Württemberg, 1412–82, Ulm), a physician of Ulm, translated a number of Italian and late Latin works. His renderings are free, sometimes to the point of becoming adaptations. Among the works he translated were the *Historia Hierosolymitana* of Robertus Monachus (*Historia von der Kreuzfahrt Herzog Friedrichs*, 1461), the Latin version from the *Gesta Romanorum* of the lost Greek tale of Apollonius of Tyre (*Apollonius von Tyros*), and three stories of Boccaccio: *Griselda* (q.v.), *De claribus mulieribus* (*Von den sinnrychen erluchten Wyben*), and *Guiscardo und Sigismunda* (all 1473). Other works are *Das Büchlein von der Pestilenz* (1473), *Der Spiegel des menschlichen Lebens* (1475) after Rodericus de Arevalo, Poggio's *Facetiae* (1476 ff.), and the fables of Aesop (1477).

Steinklopfer, Die, a Novelle by F. von Saar (q.v.), published in 1874 and included in 1877 in *Novellen aus Österreich* (q.v.). Its setting is the railway over the Semmering south-west of Vienna at the time of its construction. A former soldier and a girl are among the workers, and both are tyrannized by the girl's stepfather, who is the foreman. The soldier, when attacked, strikes the other a fatal blow with a hammer. Handed over to the military authorities, he has the good fortune to come before a sympathetic colonel, who, on learning the truth, imposes a light sentence. The soldier and the girl marry, and the colonel arranges a job and a house for them as level-crossing keepers.

STEINMAR, a Middle High German poet of the second half of the 13th c., is the author of fourteen poems. Though he was formerly identified with Berthold Steinmar von Klingnau, it is not clear that he was of noble birth. All that is known of him is that he was present at the siege of Vienna by Rudolf von Habsburg (see RUDOLF I) in 1276 and participated in a campaign against Meißen, c. 1294–6. Steinmar wrote conventional Minnelieder of some freshness and vigour. He is even better known for his songs of so-called niedere Minne and his parody of the Tagelied (farm servants replace the knight and lady, and the cowherd's horn has the role of the watchman's warning). He also wrote an autumn song, in which the tippler sings of the pleasures of the full season of the year, and with this introduced a new minor poetic form.

Stella, a play written by J. W. Goethe (q.v.) in 1775 and first published in 1776 with the subscription 'Ein Schauspiel für Liebende'. The hero Fernando, who has loved and grown tired of two women (Cäcilie and Stella), returns to the latter, only to encounter the former as well. A situation, it would seem, for comedy or for tragedy is treated as neither. After an abortive attempt by Fernando to slip off with Cäcilie, a *ménage à trois* is proposed and accepted. For a performance in Weimar in 1806 (published as *Stella. Ein Trauerspiel*, 1816) Goethe altered the end: Stella takes poison and Fernando shoots himself.

'Stell auf den Tisch die duftenden Reseden', first line of a poem by Hermann von GILM (q.v.).

Stellvertreter, Der, a play in free verse by R. Hochhuth (q.v.), first performed in Berlin in 1963. It accuses and condemns Pope Pius XII, who is a prominent person in the play, for failure to condemn the racial policies of the National Socialist regime, which culminate in mass extermination in Auschwitz. The Pope, more concerned (in the play) with diplomacy and worldly goods, is portrayed as failing as Christ's representative on Earth, and the true exemplar of the faith is the obscure Jesuit Riccardo. The play, inspired by passionate indignation, aroused fierce controversy.

STELZHAMER, Franz (Großpiesenheim, Upper Austria, 1802–74, Henndorf nr. Salzburg), became an actor and later a recitalist of Austrian dialect poetry, being known in this capacity as Piesenhamer Franz. His volumes of verse include *Lieder in obderennsischer Volksmundart* (1837), *Neue Gesänge in obderennsischer Volksmundart* (1845), and *Neue Gedichte in obderennsischer Volksmundart* (1846). *Ausgewählte Dichtungen* (4 vols.), ed. P. Rosegger, q.v., appeared in 1884 and *Ausgewählte Werke* (2 vols.), ed. L. Hörmann, in 1913.

STEPHAN, Heinrich von (Stolp, Pomerania, 1831–97, Berlin), was in charge of the postal services (Generalpostmeister) of the new German Empire, introducing major reforms, including the postcard (1870, while serving the North German Confederation as Generalpostdirektor) and the telephone (1877). In 1885 he was ennobled. He is the author of works tracing early developments in postal communication in Prussia (*Geschichte der Preußischen Post von ihrem Ursprunge bis auf die Gegenwart*, 1859), in antiquity (*Das Verkehrsleben im Altertum*), in the Middle Ages (*Das Verkehrsleben im Mittelalter*, both 1869), and in his own time (*Weltpost und Luftschiffahrt*, 1874).

STEPHAN, Meister, a schoolmaster who wrote between 1357 and 1375 *Das mittelniederdeutsche Schachbuch*, a Middle High German verse translation of the *Solatium ludi scaccorum* of Jacopo Dacciesole (see SCHACHBÜCHER).

STEPHANIE, Christian Gottlob (Breslau, 1733–98, Vienna), usually known as Stephanie der Ältere, had a middle-class upbringing before he took to the stage. From 1760 on he was a prominent actor at the Vienna Court Theatre. He wrote plays, the majority of which are adaptations of foreign dramas, and he also published from 1766 to 1768 the periodical *Neue Sammlung zum Vergnügen und Unterricht*. He was the elder brother of Gottlieb Stephanie (q.v.). The brothers' surname was originally Stephan.

STEPHANIE, Gottlieb (Breslau, 1741–1800, Vienna), originally Stephan, usually known as Stephanie der Jüngere, was educated at a Gymnasium, served in a Prussian hussar regiment and was taken prisoner in 1760 by the Austrians. He transferred to the Austrian army, but in 1769 became, like his elder brother C. G. Stephanie (q.v.), an actor in Vienna. He wrote

numerous comedies, in which his practical experience of the stage is exploited, and made something of a speciality of plays about military life (see SOLDATENSTÜCKE), including *Die abgedankten Offiziers* (1770), *Die Kriegsgefangenen* (1771), and *Der Deserteur aus Kindesliebe* (1773). He wrote the libretto for Mozart's *Die Entführung aus dem Serail* (1782, see SINGSPIEL).

Stephanuslegende, a Middle High German verse legend written by Havich der Kelner, who is understood to be Hauch (i.e. Hug) von Köln. The poem, preserved in imperfect form, narrates briefly the death of St. Stephen, as given in Acts 7, and deals extensively with the transfer of his remains to Rome and the miracles wrought by him. It was possibly written at the beginning of the 14th c.

Steppenwolf, Der, a novel by H. Hesse (q.v.), published in 1927. An exceptionally bizarre product of Hesse's psycho-analytical preoccupations, it is a narrative in the surrealist manner. An unnamed person publishes the jottings and reflections of the central character, Harry Haller, as *Harry Hallers Aufzeichnungen*, introducing them with the motto 'Nur für Verrückte'. Harry Haller, the Steppenwolf of the title, is a man of sensitive artistic temperament, an outsider, who cannot come to terms with the world around him. On his solitary nocturnal walks through the city, in which he rents a garret, he repeatedly encounters advertisements for a Magic Theatre displayed as 'Magisches Theater Eintritt nicht für jedermann'. On one occasion he accosts the placard bearer, who hands him a 33-page booklet with the, for Harry, mysterious title *Tractat vom Steppenwolf*; once more confronted with the motto 'Nur für Verrückte', he finds that this Steppenwolf is called Harry. The *Tractat* suggests the cure of Harry Haller's misery. Harry does not, however, adopt the solution of the *Tractat*, which insists that he must abandon the rigid polarity of 'Wolf' and 'Mensch'. He continues to suffer his ordeals and disillusionments with the ordinary social world, and at the same time falls in with a group of sympathizers. Hermine, a woman of great goodness of heart, and her friend, the saxophone player Pablo, take him in hand and teach him that there are at least some others who do not fit into the respectable pattern and yet can communicate and be happy among themselves. Finally Harry visits the Magic Theatre, a surrealist world, in which he murders a Hermine who is not Hermine with a knife which is not a knife. The Magic Theatre has done its part, the inhibitions and tense psychological contortions can be released, if only momentarily, and Harry, in the last words of his *Aufzeichnungen*, speaks of hope, 'Einmal würde ich das Lachen lernen'.

Sterben, a Novelle by A. Schnitzler (q.v.), written in 1892 and published in 1893. It is almost a medical and psychiatric case history. Felix, suffering from tuberculosis, learns that he has only one year to live. He and his mistress Marie resolve to spend it together by a lake in the Austrian Alps, and Marie's love moves her to declare that she will not survive him. The initial harmony is threatened as the disease advances, and Felix becomes more and more jealous of her health and future. They move to Merano and there the end comes. Felix seizes her, and in a panic struggle for life she breaks away and flees. Felix dies alone. The barely perceptible, yet remorseless, progress of the disease and its psychological consequences are recorded with great subtlety.

Sterbende Cato, Der, a verse tragedy by J. C. Gottsched (q.v.), published in 1732. It is compounded out of two foreign tragedies, Joseph Addison's *Cato* (1713) and the *Caton d'Utique* (1715) of F. M. C. Deschamps; the first four acts are modelled on the French play, the end derives from Addison. The play is set in the time of the Roman Civil War. The stoic Cato, on the losing side, declines conciliatory propositions from Caesar, and ends his life rather than submit to a dictator. The work is intended as Gottsched's idea of a model German tragedy, classical in the French manner with confidantes, observing the unities, and written in Alexandrines.

Sterbende Kirche, Die, a novel by E. Schaper (q.v.), published in 1935. Set in ah Esthonian coastal town after 1918, it relates the efforts of Father Seraphim to revive and sustain the faith of a small congregation of the declining Orthodox Church. Though he perishes when the cupola of the church collapses, his death is not in vain, for his faith and encouragement live on in the survivors.

STERN, ADOLF, real name Adolf Ernst (Leipzig, 1835–1907, Dresden), a partly self-taught academic, became a professor at the Dresden Polytechnic in 1868. His principal academic works were a biography of his friend Ludwig, *Otto Ludwig. Ein Dichterleben* (1890), and two histories of literature, *Die Geschichte der neueren Literatur* (7 vols., 1883–5) and *Die deutsche Nationalliteratur von Goethes Tod bis zur Gegenwart* (1886). He was a prolific writer of novels and Novellen, among which *Die letzten Humanisten* (1881), *Camoëns* (1886), and the posthumous *Die Ausgestoßenen* (1911) deserve mention. In 1903 Stern received the title Hofrat from the king of Saxony.

Sternbald, see FRANZ STERNBALDS WANDE-RUNGEN.

Stern der Ungeborenen, the last novel by F. Werfel (q.v.), completed a few days before his death and published in 1946. It is set far in the future, in a world in which work and pain and natural death are eliminated. But this resolution of problems has deprived both persons and landscape of individuality and differentiation. The uniform people inhabit an endless plain. Reserves of 'jungle', however, still exist, in which varied landscape and human differences, good and bad, persist; and the 'jungle' threatens to encroach upon the Utopia. Moreover, the Utopia itself is threatened from within, discords arise, especially in the strange floral metamorphosis which is a substitute for death. Man's scientific perfection seems doomed, and he is saved by a redeeming sacrifice by a single youth, whose death suggests an analogy with the Messianic Christ. This world and its events are inspected by an easily identified narrator designated F. W. It is the final testament of Werfel's Messianic, eschatological, and quasi-christological belief.

STERNE, LAURENCE (Clonmel, 1713–68, London), English clergyman and novelist, whose mixture of whimsical humour and sentiment exhibited in *Tristram Shandy* (1760–7) and *A Sentimental Journey* (1768) influenced German writers from the Sturm und Drang (q.v.) to the Romantic movement (see ROMANTIK). C. M. Wieland, M. A. von Thümmel, Th. G. Hippel, and JEAN Paul (qq.v.) show his influence.

STERNHEIM, CARL (Leipzig, 1878–1942, Brussels), son of a Jewish banker, grew up in Berlin. From 1897 to 1902 he studied at Munich, Göttingen, Leipzig, and Berlin universities, reading philosophy, psychology, and law. He chose to do his military service as an officer cadet in a crack cavalry regiment (6th Kürassicre). With ample financial resources derived from his family and his first and second wives, he was not dependent on his writing for a living.

In a period of early experiment he wrote the comedy *Der Heiland* (1898), the tragedy *Judas Ischariot* (1901), and published a volume of verse, *Fanale* (1901), which showed the influence of Stefan George (q.v.). A group of plays dealing with relations between the sexes, *Vom König und der Königin* (1905), *Ulrich und Brigitte* (1907), and *Don Juan* (two parts, 1905–9), coincided with stresses in his first marriage. Sternheim lived in style near Munich, then in Belgium; after 1918 he moved to Switzerland, tried Berlin for a time, and settled definitively in Belgium in 1930, when he married, as his third wife, Pamela, daughter of F. Wedekind (q.v.). The marriage was brief. During the 1939–45 War Sternheim was allowed to remain in Belgium unmolested.

The most important creative period in Sternheim's life began in 1908. Encouraged by his friend F. Blei (q.v.) and by his own enthusiasm for Molière, he wrote *Die Hose* (q.v., 1911), which was banned by the censor, and followed it with a series of amusing and ruthlessly satirical comedies, lampooning middle-class society. Sternheim himself disclaimed satirical intention, asserting that he was simply realistic. Some comedies he grouped together under the ironic title *Aus dem bürgerlichen Heldenleben*; they are thought to include *Die Hose*, *Der Snob* (1914), *1913* (1915), *Das Fossil* (1925), *Die Kassette* (1911), and *Bürger Schippel* (q.v., 1913). Kaiser referred to the first four of these as the *Masketetralogie*, because they all have a prominent figure from a bourgeois family, which, conditioned by its class, is named Maske ('mask'). The comedies of 1911–16, which also include *Der Kandidat* (1914) and *Tabula rasa* (1916), mark the apogee of Sternheim's work. Among his other plays are *Der Stänker* (1917, originally *Perleberg*), *Die Marquise von Arcis* (1919), *Der entfesselte Zeitgenosse* (1920), *Oscar Wilde* (1925), and *John Pierpont Morgan* (1930).

Sternheim published a collection of ironic stories under the title *Chronik von des zwanzigsten Jahrhunderts Beginn* (2 vols., 1918), followed by *Vier Novellen. Neue Folge der Chronik vom Beginn des zwanzigsten Jahrhunderts* (also 1918). He issued some of these and other Novellen separately (*Busekow*, 1914; *Napoleon*, the story of a chef, 1915; *Schuhlin* and *Meta*, both 1916; and *Ulrike*, 1918). The collection *Mädchen* (1917) was reprinted in 1926 with alterations, and at about the same time he published collections of earlier Novellen as *Napoleon* (1927) and *Busekow* (1928). Later titles are *Fairfax* (1921) and *Libussa, des Kaisers Leibroß* (1922). Sternheim's only novel was *Europa* (1919–20). *Gauguin und Van Gogh* (1924) might, however, be considered a Künstlerroman.

Sternheim's miscellaneous prose includes *Die deutsche Revolution* (1919), *Berlin oder Juste milieu* (1920), and *Lutetia* (1926). *Vorkriegs-Europa im Gleichnis meines Lebens* (1936) is autobiographical. Throughout his work his style is noteworthy for its directness and its avoidance of metaphor.

A selection of Sternheim's writings appeared as *Werke* (4 vols.), 1947–8, and the *Gesamtausgabe* (10 vols.), ed. W. Emrich, 1963–76.

Sternsteinhof, Der, a novel by L. Anzengruber (q.v.), first published in 1884 and in book form in 1885. Set in the Austrian countryside, it tells the story of Zinsdorfer Helene's determination to become the mistress of the splendid farmhouse, Sternsteinhof. Helene early conceives the plan of marrying Toni, the son of the owner of the Sternsteinhof, but the latter has other ideas and intends, and eventually forces, a marriage between Toni and Sali, a rich farmer's daughter. Helene weds the ailing wood-carver Muckerl, but she does not give up. Muckerl, who learns of her continuing association with Toni, dies, and Helene is installed at the Sternsteinhof to look after Toni's child. Sali, conscious of being supplanted, dies also, and Helene achieves her ambition, marrying Toni, who has taken over the farm from his father. But war breaks out and Toni is called up and is later reported missing. Helene does not remarry and devotes herself to managing the farm, bringing up her own child and her stepdaughter and tending Toni's aged father.

Sterzinger Spiele, see TIROLER PASSIONSSPIELE.

StGB, abbreviation for *Strafgesetzbuch*, the German code of criminal law, first published in 1871.

STICH-CRELINGER, AUGUSTE (Berlin 1795–1865, Berlin), *née* Düring, an actress who became well known in the leading roles of Shakespearian and classical drama. She was from 1812 to 1863 at the Berlin Hoftheater and, having established her reputation, acted at other major theatres as well. Hebbel owed the acceptance of his first play, *Judith* (q.v.), in which she took the title role, to her influence (in 1840), and she promoted O. Ludwig's *Die Makkabäer* (q.v.) by acting Lea (in 1852).

STIEFEL, MICHAEL, see STYFEL, MICHAEL.

STIEGLITZ, CHARLOTTE (Hamburg, 1806–34, Berlin), *née* Willhöft, married H. Stieglitz (q.v.) in 1828. Her suicide by stabbing caused a public sensation. Her declared motive was to give her husband a tragic experience which would release his creative powers. Theodor Mundt (q.v.) wrote a sympathetic memoir entitled *Charlotte Stieglitz. Ein Denkmal* (1836).

STIEGLITZ, HEINRICH (Arolsen, 1801–49, Venice), of a Jewish family converted to Christianity, studied at Göttingen, Leipzig, and Berlin universities, and expressed his radical views in *Gedichte zum Besten der Griechen* (1823), written jointly with a friend. After appointment as a schoolmaster in Berlin he married in 1828 Charlotte Willhöft (see STIEGLITZ, C.). The marriage was unhappy and Charlotte committed suicide in 1834, with the avowed intention of stimulating her husband's genius.

Stieglitz, however, remained unproductive, moving restlessly about Europe, staying at different times in Italy and the Balkans as well as in various German cities. He died of cholera. His principal work is *Bilder des Orients* (1831–3), which contains, in addition to poems, two plays, *Ein Tag in Ispahan* and *Sultan Selim III*.

STIELER, KARL (Munich, 1842–85, Munich), an archivist under the Bavarian government, possessed a talent for attractive dialect poetry. His volumes of verse include *Bergbleamel* (1875), *Weil's mi freut!* (1876), *Habt's a Schneid!* (1877), *Um Sunnawend* (1878), and *In der Sommerfrisch'* (1883).

STIELER, KASPAR VON (Erfurt, 1632–1707, Erfurt), led a restless life as soldier and student and was later an administrator in princely service. He was ennobled in 1705. He is the author of *Die geharnschte Venus* (1660, reissued 1968), a collection of frankly erotic, even priapic poetry which appeared in Hamburg under the pseudonym Filidor der Dorfferer. He also wrote a comedy (*Willmut*, 1680), a tragedy (*Ballemperie*, 1680), and a treatise on language (*Der teutschen Sprache Stammbaum und Fortwachs, oder teutscher Sprachschatz*, 1691, reissued 1968) commissioned by the Fruchtbringende Gesellschaft (q.v.), of which he was a member (as 'Der Spate') from 1668.

STIFTER, ADALBERT (Oberplan, Bohemia, now Horné Planà, Czechoslovakia, 1805–68, Linz, Austria), by birth an Austrian citizen, was the son of a flax merchant, who died when the boy was 11. His childhood was spent in the Bohemian Forest, the landscape of which repeatedly enters into his novels and stories. In 1818 he was sent to school at the Benedictine monastery of Kremsmünster in Upper Austria, where his outstanding intelligence was recognized. He then studied at Vienna University, turning his attention especially to science. After a passionate attachment to Franziska Greipl (Fanni, also Fanny) in 1828–9, Stifter married in 1837 Amalia Mohaupt. He had hoped for an academic career as a scientist in Vienna, but was disappointed and for some years earned a meagre living by giving private lessons, in which he was extremely successful, gaining a well-disposed clientele in the upper strata of Viennese society.

Stifter's artistic interests were at first centred on landscape painting, but in 1840 he published, with some reluctance, his first story, *Der Condor*, which has as its central episode a balloon flight in which a young woman participates. This story was later included in the first volume of *Studien*

(q.v.). Other stories followed and were eventually incorporated in the six volumes of *Studien* (1844–50): *Feldblumen* (1840), *Das Haidedorf* (1840), *Der Hochwald* (q.v., 1842), which was Stifter's first masterpiece, *Die Narrenburg* (1843), *Die Mappe meines Urgroßvaters* (1841), *Abdias* (q.v., 1843), *Das alte Siegel* (1844), *Brigitta* (q.v., 1844), *Der Hagestolz* (q.v., 1844), *Der Waldsteig* (1845), *Schwestern* (1846, original version of 1845, *Zwei Schwestern*), and *Der beschriebene Tännling* (1845). The revolution of 1848 (see REVOLUTIONEN 1848–9) proved, in its later manifestations, a shock to Stifter, who withdrew to Linz, where for a short time he edited a newspaper. In 1850 he was appointed to a senior post in educational administration at Linz under the new dispensation which followed the revolution. For fifteen years he was inspector of schools in Upper Austria, but he found it difficult to realize the educational objectives he set himself and hard to reconcile his duties with his compulsion to write. The two volumes of *Bunte Steine* (q.v.), the preface to which contains the enunciation of Stifter's well-known 'Sanftes Gesetz', appeared in 1853; his novel *Der Nachsommer* (q.v.) followed in 1857.

In 1865 Stifter retired and devoted himself to completing his historical novel *Witiko* (q.v.), publication of which began in 1865 and was completed in 1867. Stifter's later years were darkened by the childlessness of his marriage and the suicide in 1859 of an adopted daughter. In 1867 he was seriously ill, suffering, it is thought, from cancer. During the night of 28 January 1868, while beset with agonizing pain he ended his life by cutting his throat with his razor. The nature of his end was concealed at the time, but the facts were revealed some thirty-five years later. A number of stories left in MS. were published posthumously in 1869 under the title *Erzählungen: Prokopus* (1848), *Die drei Schmiede ihres Schicksals* (1844), *Der Waldbrunnen* (1866), *Nachkommenschaften* (1864), *Der Waldgänger* (1847), *Der fromme Spruch* (1866), *Der Kuß von Sentze* (1866), *Zuversicht* (1846), *Zwei Witwen* (1860), *Die Barmherzigkeit* (1843), *Der späte Pfennig* (1843), and *Der Tod einer Jungfrau* (1847).

Stifter's novels enjoyed only a moderate popularity in his lifetime and after his death he was quickly forgotten. A revival of interest began in the 20th c., but his full stature was not recognized till after the 1914–18 War. The integrity of his vision, the recognition and simultaneous rejection of violence, the emphasis on the natural processes of growth and the sensitive perception of nature, especially in the form of landscape, are the essential features of his œuvre. He is by nature an educator, whose instrument is his art. His remarkably transparent, gentle, and deliberate style is a natural and apt expression of his outlook and aims.

Sämtliche Werke. Historisch-kritische Gesamtausgabe (24 vols.), ed. A. Sauer, F. Hüller, and G. Wilhelm, appeared 1904–60, *Gesammelte Werke* (14 vols.), ed. K. Steffen, 1962–72, and *Werke und Briefe. Historisch-kritische Gesamtausgabe*, ed. A. Doppler and W. Frühwald, from 1978.

Stille, Die, a poem by J. von Eichendorff (q.v.), the first line of which runs 'Es weiß und rät es doch keiner'. A poem of secret ecstasy, it is contained (untitled) in Eichendorff's novel *Ahnung und Gegenwart* (q.v.).

'Stille Nacht, heilige Nacht', first line of what is perhaps, outside Germany, the best-known German Christmas carol. It was written on Christmas Eve 1818 by Joseph Mohr (1792–1848), a young priest, at Oberndorf near Salzburg. The melody was composed by Franz Gruber (1787–1863), organist at the neighbouring village of Arnsdorf. The carol was not published until 1834.

Stiller, a novel by M. Frisch (q.v.), published in 1954. It deals with an attempt to shed an unwanted identity and to escape into a new one. A Swiss sculptor, Anatol Stiller, married to a dancer, Julika Tschudy, disappears and is not heard of for six years. An American, by name J. C. White, is stopped at the Swiss frontier, has an altercation with a passport official, and is arrested. At first there is a suspicion that he is connected with Soviet espionage; but he is then further detained because he is thought to be Stiller. The novel is largely told through the suspect's journal, kept while in custody, and beginning with the assertion, 'Ich bin nicht Stiller', a position which he persistently maintains. Stiller's wife instantly recognizes the man as her husband and his brother identifies a photograph. Still he insists that he is not Stiller, but a Swiss court decides that he is the missing man. In an epilogue (*Nachwort des Staatsanwalts*) the state prosecutor, who is a personal friend of Stiller and his wife, tells how the couple lead a miserable existence at Glion, which is ended by Julika's death after an operation. The last words confirm Stiller's identity, 'Stiller blieb in Glion und lebte allein.'

'Still ist die Nacht, es ruhen die Gassen', first line of an untitled poem by H. Heine (q.v.), which is often known by the apocryphal title *Der Doppelgänger.* It occurs in the *Buch der Lieder* (q.v.) as No. XX of *Die Heimkehr* and is one of the poems set by F. Schubert in *Schwanengesang* (q.v.).

Stilpo und seine Kinder, a tragedy in five acts by F. M. Klinger (q.v.). It was written in 1777 when Klinger was an itinerant actor with the Seylersche Truppe (q.v.). Klinger declared that it was only a stage work and that he would not publish it. It was, nevertheless, published in 1780. It is a story of conspiracy and revenge in Florence. Stilpo wishes to avenge his brother's death by execution. Through the courage of his wife, who assassinates his enemy, he triumphs, but his two sons are killed.

Stimme hebt an, Eine, see EINE STIMME HEBT AN.

Stimmen der Völker in Liedern, title of the second (1807) edition of a collection of folksongs made by J. G. Herder (q.v.), first published in 1778–9 as *Volkslieder* (see VOLKSLIED), and edited by J. von Müller (q.v.); subsequent editions have retained the title, which is based on posthumous papers.

STIMMER, TOBIAS (Schaffhausen, 1539–84, Strasburg), was by trade an artist producing woodcuts for book illustration, including the Bible (*Bilderbibel,* 1576). He is the author of a comedy (Fastnachtspiel) commonly called *Von zwei Eheleuten* (1580). It portrays the misunderstandings arising when a wife plans to seek a lover during her husband's absence and her subsequent discomfiture. Its full title is *Comoedia von zweien jungen eeleuten, wie sey sich in fürfallender reiss beiderseitz verhalten.*

STINDE, JULIUS (Kirch-Nüchel nr. Eutin, 1841–1905, Olsberg, Westphalia), a pastor's son, was a university-trained chemist (Kiel, Gießen, Jena), who held a position in a chemical factory in Hamburg from 1863 to 1866. Simultaneously he was active as a journalist, and after 1866 he devoted his time to writing and to travel (including a journey to Egypt). In 1872 he published a volume of stories (*Alltagsmärchen*) and followed this with the novel *In eiserner Faust* (1874) and the comedy *Tante Lotte* (1875). In 1876 he settled in Berlin. *Aus der Werkstatt der Natur* (3 vols., 1880) exploited, in the form of Novellen, some of his scientific knowledge. *Waldnovellen* appeared in 1881. Shortly afterwards he wrote the first of the Buchholz books, a series of stories and novels portraying in gently ironical form a typical middle-class family of Berlin. These comprise *Buchholzens in Italien* (1883), *Die Familie Buchholz* (3 vols., 1884–6), *Frau Buchholz im Orient* (1888), and *Frau Wilhelmine Buchholz' Memoiren* (1895). The series proved to be a great popular success.

Stine, a short novel or Novelle by Th. Fontane (q.v.), written between 1881 and 1888. It was first published in the magazine *Deutschland* in 1890 (January/March) and appeared in book form in April 1890. Stine is the sister of Pauline Pittelkow, a youngish widow and energetic and downright character, who is kept by a Graf von Haldern and views her dubious social circumstances with homely realism. Stine, however, while on the best of terms with her sister, lives a virtuous life and earns her living by her needle. Pauline's lover arranges a party at her flat. Stine is present, and so is the Count's young nephew, Graf Waldemar, who is condemned to ill-health as a consequence of wounds received as a young officer in August 1870. He takes to Stine, visits her in the afternoons throughout the summer, spending the hours in conversation, and in the end decides to marry her. His uncle is greatly put out and demands Pauline's help in putting an end to the threatened *mésalliance*. But their intervention is not needed. Stine realizes only too clearly the social, educational, and financial factors which would harshly destroy their happiness in a permanent union. Waldemar, having lost his last hope, takes an overdose of sleeping powder. He is buried with pomp in the family vault. Stine, uninvited, attends his funeral and is subsequently ill with shock. Fontane leaves the outcome for her open.

STIRNER, MAX, real name Johann Kaspar Schmidt (Bayreuth, 1806–56, Berlin), a schoolmaster, then a journalist, published in 1845 the philosophical work *Der Einzige und sein Eigentum* (reissued, ed. H. G. Helms, 1970). An exponent of uncompromising individualism, Stirner has been regarded by some as a forerunner of Nietzsche (q.v.). His other writings include *Die Geschichte der Reaction* (2 pts., 1852, reissued 1967).

STÖBER, ADOLF (Strasburg, 1810–92, Mülhausen), from 1840 Protestant pastor in Mülhausen, from 1860 president of the Reformed Consistory. The son of D. E. Stöber (q.v.), he edited jointly with his brother August Stöber (q.v.) the journal *Erwina* (1838–9) and contributed to a collection of legends and stories (*Alsabilder*, 1836). His publications include *Gedichte* (1845), *Reisebilder aus der Schweiz* (2 vols., 1850–7). He and his brother were friends of G. Büchner (q.v.).

STÖBER, AUGUST (Strasburg, 1808–84, Mülhausen), a teacher by profession, he edited together with his younger brother Adolf Stöber (q.v.) the journal *Erwina*, in which he published Oberlin's record of Lenz's stay at his rectory in Waldersbach early in 1778 (see OBERLIN, J. F.,

and LENZ, J. M. R.), which Stöber published for a second time in his monograph *Der Dichter Lenz und Friederike von Sesenheim* (1842), adding in a footnote that it formed the basis of his late friend Georg Büchner's Novelle *Lenz* (q.v.), which 'unfortunately had remained a fragment' and for which he had supplied all available material. Like his father D. E. Stöber (q.v.) and his brother, he contributed much towards the preservation of the cultural heritage of his native Alsace, notably *Die Sagen des Elsaß* (2 pts., 1851–2).

STÖBER, DANIEL EHRENFRIED (Strasburg, 1779–1835, Strasburg), contributed through various publications to the cultural traditions of his native Alsace, both in his own name and under the pseudonym 'Vetter Daniel' (in *Neujahrsbüchlein in Elsässer Mundart*, 1830). In 1807 he founded the *Alsatisches Taschenbuch* and the journal *Alsa*. He published a biography of J. F. Oberlin (q.v., 1831) and was co-editor of Oberlin's collected works (1843). His *Sämtliche Gedichte und kleinere prosaische Schriften* were published in three volumes in 1835–6. His sons Adolf and August Stöber (qq.v.) had similar interests.

STÖCKEN or **STÖKKEN**, CHRISTIAN VON (Rendsburg, 1633–84, Rendsburg), held high office in the Lutheran Church, being first court preacher at Eutin, then provost and moderator (Generalsuperintendent) of Holstein and Schleswig (1677). He is the author of a German version of the Psalms (*Neugestimmte Davidsharfe*, 1656) and of hymns (*Heilige Herzens-Seufzer*, 1668, *Heilige Nachtmahlsmusik*, 1676).

STOECKER, ADOLF, see CHRISTLICHSOZIALE PARTEI, (1).

STÖKKEN, CHRISTIAN VON, see STÖCKEN, CHRISTIAN VON.

STOLBERG, AUGUSTE, in full Auguste, Gräfin zu Stolberg-Stolberg (Bramstedt, Holstein, 1753–1835, Kiel), sister of Christian and Friedrich Leopold Stolberg (qq.v.), was, as a young woman, an enthusiastic admirer of Goethe's poetry. She entered (at first anonymously) into a correspondence with him, which lasted from January 1775 to 1782. Before her name was disclosed, Goethe addressed her as 'die teure Ungenannte'. She married in 1783 Andreas Peter, Graf Bernstorff (1735–97), Danish foreign minister from 1773 to 1780 and from 1784 to 1797. In 1822–3 she entered briefly into correspondence with Goethe again, seeking to reconvert him to Christianity.

STOLBERG, Christian, in full Christian, Graf zu Stolberg-Stolberg (Hamburg, 1748–1821, Windebye, nr. Eckernförde), a Danish subject, was very close to his brother Friedrich Leopold (Fritz), with whom he collaborated in poetry. The two brothers studied at Halle University and from 1772 to 1774 at Göttingen, where they became prominent members of the Göttinger Hainbund (q.v.). In 1775 they undertook with Graf Kurt von Haugwitz and, initially, with Goethe a prolonged Swiss tour, on which their well-meant enthusiastic naturism aroused hostility among the inhabitants. In 1777 Christian became magistrate (Amtmann) at Tremsbüttel in Holstein, and at this point the courses of the two brothers diverged. Christian held his office at Tremsbüttel until 1800 and then lived on his own estates at Windebye. Somewhat overshadowed by Friedrich Leopold, he contributed poems to their joint *Gedichte* (1779) and *Vaterländische Gedichte* (1815), and also participated in their classical *Schauspiele mit Chören* (1787). He was successful as a translator, with *Gedichte aus dem Griechischen übersetzt* (1782) and *Sophokles* (2 vols., 1787).

STOLBERG, Friedrich Leopold, in full Friedrich Leopold, Graf zu Stolberg-Stolberg (Bramstedt, Holstein, 1750–1819, Sondermühlen nr. Osnabrück), a Danish subject, early developed an enthusiasm for Klopstock (q.v.) and a hatred of tyranny. For years he collaborated closely with his less talented brother Christian (q.v.) and the two were inseparable into their middle twenties. They were together at Halle and at Göttingen universities, where they were prominent members of the Göttinger Hainbund (q.v.). In 1775, together with Graf Kurt von Haugwitz and, initially, Goethe, they undertook a tour in Switzerland, on which they met J. J. Bodmer (q.v.). In 1777 the brothers took up widely separated appointments, Friedrich Leopold (generally called Fritz) becoming envoy of the Bishop of Lübeck at Copenhagen. In 1779 the *Gedichte* of the two brothers were published.

Though Friedrich Stolberg stood close to the Sturm und Drang (q.v.) and had written poetry in the manner of Klopstock, he was also drawn to Homer and translated the *Iliad* (*Homers Ilias,* 1778). In 1781 he became cup-bearer (Oberschenk) to the episcopal court of Eutin and in the following year he married Agnes von Witzleben, who is the Agnes of his poetry. It was on their honeymoon that he wrote the well-known poem set by F. Schubert, *Auf dem Wasser zu singen* (q.v.). In the same year Friedrich Stolberg secured the appointment of J. H. Voß (q.v.) as headmaster of the school at Eutin. He published further poems (without his brother) in

1784 (*Jamben*) and a tragedy, *Timoleon*; but a new collaboration came with *Schauspiele mit Chören* (1787). He held diplomatic appointments for short periods in St. Petersburg (1785) and Berlin (1789). His first wife died in 1789 and later in the same year he married Sophie von Redern. He became increasingly religious and intolerant of what seemed to him an irreligious paganism, violently attacking Schiller's poem *Die Götter Griechenlands* (q.v., 1788). His creative writing slackened and he devoted himself to travel works (*Reise in Deutschland, der Schweiz, Italien und Sizilien,* 1794) and translation (*Auserlesene Gespräche des Platon,* 1796–7; *Vier Tragödien des Aeschylos,* 1802; *Die Gedichte von Ossian,* 1806).

After a long religious crisis he settled in Westphalia, lived for many years in Münster, and resigned all his offices in consequence of conversion to Roman Catholicism, a step which caused a considerable stir. His former protégé J. H. Voß turned against him, abusing his 'reactionary' religious and political views in *Wie ward Fritz Stolberg ein Unfreier?* Stolberg lived his later life in Westphalian Catholic circles, writing an immense *Geschichte der Religion Jesu Christi* reaching the year 430 (15 vols., 1806–18, continued by F. von Kerz and J. N. Brischar, 53 vols., 1825–64).

Gesammelte Werke of both brothers (20 vols.) appeared 1819–25 (reissued 1974), *Ausgewählte Werke* (2 vols.), ed. A. Sauer, 1891–5 (reissued 1969).

STOLLE, Meister, a Middle High German itinerant poet of the middle of the 13th c., who is the author of numerous Sprüche (see Spruch). Stolle's poems treat social and political themes with seriousness and irony. His poem conceding to Rudolf of Habsburg (see Rudolf I) every virtue except generosity has an unusual anaphoric structure.

Stollen, the term used by the Meistersinger for the two metrically identical stanzas of the Aufgesang in the structure of Minnesang and Meistergesang (qq.v.). The word means 'prop' and Paul/Betz, *Deutsches Wörterbuch,* 5th edn., 1966, suggests, that it came into use because the two Stollen are, in a sense, the two supports on which the following Abgesang rests.

STOLTZE, Friedrich (Frankfurt/Main, 1816–91, Frankfurt), took part in the 1848 Revolution (see Revolutionen 1848–9) and was afterwards a journalist in the city's dialect press. His paper *Frankfurter Latern* was suspended on political grounds in 1866, the year of the war with Prussia (see Deutscher Krieg), and Stoltze fled to

Switzerland, returning the following year after the proclamation of an amnesty. His *Gedichte in Frankfurter Mundart* (2 vols.) appeared 1864–71 and he also wrote *Novellen in Frankfurter Mundart* (2 vols., 1880–5). His *Gesammelte Werke* were published in four volumes in 1892.

Stopfkuchen, Eine See- und Mordge-schichte, a novel written by W. Raabe (q.v.) in the years 1888–90, and published in 1891. The unusual title is a nickname of the principal character Heinrich Schaumann, earned by his addiction to stuffing cake. The story-telling is peculiarly intricate, for the narrative purports to be written by Stopfkuchen's friend Eduard as he returns by ship to South Africa, whither he has immigrated. But what Eduard faithfully sets down is the tale of characters and events as told by Stopfkuchen, so that the story is enclosed by a double frame (see RAHMEN). At its centre is an unsolved murder and its consequences. Schaumann, alias Stopfkuchen, has married the daughter of Farmer Quakatz, who has spent much of his life under the cloud of an accusation and suspicion of murder.

Kienbaum, a cattle dealer, was found dead and Quakatz was known to have had an altercation with him not long before. The case was taken up and dropped three times for lack of evidence, but villagers and townspeople are convinced of Quakatz's guilt and make his and his daughter's life a misery. Stopfkuchen, who had conceived in childhood a passionate desire to own Quakatz's farm 'Die rote Schanze' (so called because the farm-buildings are within the embankments of a field fortification of the Seven Years War), defends Valentine Quakatz against her persecutors, assists her father, and on one occasion arrives in the nick of time to save them from violence at the hands of drunken farm servants. He marries Valentine and they live together in happiness and harmony, making Quakatz's last years tolerable. At the old man's funeral Stopfkuchen finds a clue to the murder of Kienbaum. He follows it up and solves the mystery. The murder was committed on impulse by Störzer, the postman, who all his life had been mocked and reviled by Kienbaum. But Stopfkuchen keeps his knowledge to himself, not even telling his wife. For what good will it do any one to invoke the principle of justice and have Störzer executed or imprisoned? Eduard's visit falls just after Störzer's death, and now at last Stopfkuchen can let out the secret. He does so in his own fashion, interminably, circuitously, and yet with his goal always clear in his mind.

The real focus of the novel is not the murder of Kienbaum, but the unfolding of the rich and balanced character of Heinrich Schaumann, self-confessed sloth and glutton, yet with an in-

fallible eye for the things which are truly important and a sovereign disregard for those which are not.

STOPPE, DANIEL (Hirschberg, 1697–1747, Hirschberg), a schoolmaster of Hirschberg, wrote poetry, including fables and hymns. His publications include *Erste Sammlung von des Silesischen teutschen Gedichten* (1728), which was followed by a *Zweite Sammlung* in 1729, *Der Parnaß im Sättler oder Scherz- und Ernsthafte Gedichte* (1735), *Sonntagsarbeit* (1737–42), and *Neue Fabeln* (1738–40).

STORCH, LUDWIG (Ruhla, 1803–81, Kreuzwertheim), was a prolific writer of novels, many with historical settings (*Kunz von Kauffungen,* 3 vols., 1828; *Der Freiknecht,* 3 vols., 1830; *Ein deutscher Leineweber,* 9 vols., 1846–50). *Vörwerts Häns* (1830) is a Thuringian regional novel. A select edition of his fiction, including Novellen, appeared 1855–62 (31 vols.).

Storchenbotschaft, a humorous poem written by E. Mörike (q.v.) in 1837. The shepherd is puzzled by the visit of two storks, until he realizes that it signifies twins. The poem has been set to music by Hugo Wolf (q.v.).

STORM, THEODOR WOLDSEN (Husum/Schleswig, 1817–88 Hademarschen/Holstein), studied law, which he practised for the greater part of his life in his native Husum before his retirement, in 1880, from his office as Amtsgerichtsrat. During the years of political crisis over the possession of Schleswig-Holstein (see SCHLESWIG-HOLSTEINISCHE FRAGE) Storm had to leave Husum with his family because his refusal to recognize the Danish occupation cost him his post. After an unpaid assignment with the Prussian civil service in Potsdam (1853–6), during which he depended upon his father's financial support, he settled as a stipendiary magistrate in Heiligenstadt, but, after the settlement of the political crisis, took the first opportunity to return to his native Husum (1864), which offered him the office of Landvogt.

During his student years, which (with the exception of three semesters in Berlin) were spent in Kiel, he became deeply attached to Berta von Buchau, an eleven-year-old girl, upon their first acquaintance. The courtship ended in frustration when she rejected his proposal of marriage in 1842. Soon after settling down as a lawyer (1843) Storm became engaged to his cousin Konstanze Esmarch, whom he married in 1846. A year after her death and after the family's return to Husum, he married Dorothea Jensen (1866), a friend of long standing. Neither marriage was initially easy, but both grew happy, though Storm remained a stern father and tutor to his sons, the oldest of whom predeceased him as a result of alcoholism. Storm's literary work sprang from personal experience rather than political attitudes. He was impatient both of the privileged nobility and of Prussia, and this antagonism did not diminish during the years spent in Prussian service; his contacts in Berlin with the literary club Der Tunnel über der Spree (q.v.), and notably with Th. Fontane (q.v.), sharpened his northern individualism, though they stimulated his development as a writer. His strong personality was nurtured by his profoundly emotional sensibility and an increasing scepticism, which inclined him to determinism.

Storm's early production was fostered by the influence of Romantic writers and of Mörike (q.v.), with whom he corresponded and whom he once visited. He commented (1882) that his prose narrative grew out of his poetry, to which he first devoted himself under the influence of Theodor Mommsen (q.v.). Storm formed a friendship with Theodor and his brother Tycho Mommsen at school, and the three were joint editors of the *Liederbuch dreier Freunde* (1843). Storm's own contributions reflect the frustrations of his love for Berta, from which he sought to free himself by writing *Immensee* (q.v., 1851), the Novelle which established his reputation as a writer and which has been the introduction to German literature for countless English readers. This lyrical Novelle aims at integrating song and narrative in a way which Storm admired in Mörike's *Maler Nolten* (q.v.). Storm set store by the therapeutic effect of reminiscences, which characterize, with at times excessive melancholy, his narrative technique, especially in stories involving a frame (see RAHMEN).

In a number of poems Storm's creative vein and consciousness of form achieve the high standard which he set for himself in comments to friends and in his introduction to the anthology *Hausbuch aus deutschen Dichtern seit Claudius* (1870). 'Schließe mir die Augen beide', *Trost*, *Auf einen Kirchturm* demonstrate his mastery of epigrammatic brevity, *Die Nachtigall* his reticent sensuality, and *Die Stadt* and *Meeresstrand* are examples of the local colouring indispensable to his peculiar portrayal of Husum and the North Sea coastal region. *Oktoberlied* is his most exuberant song; *Einer Toten* expresses grief at Konstanze's death, as well as illustrating his recurring treatment of the transience of life in association with the permanently ticking pendulum of the clock. Storm sought to write poems born of the immediate moment and appealing to the senses by their musical quality, sound, and rhythm, and the unobtrusive employment of imagery. His theories grew out of what he

recognized as the key to Goethe's intimate poetry. In his choice of images he confined himself to his personal and local heritage and environment, to which both his lyric poetry and his narrative work owe their plain but sincere authenticity. He was an admirer of H. Heine (q.v.), matching Heine's irony with his own brand of humour, which was to him a necessary antidote to the frailty of existence.

Particularly fruitful for his narrative work was his correspondence with G. Keller and P. Heyse (qq.v.). Periodicals such as *Westermanns Monatshefte* and the *Deutsche Rundschau* (qq.v.) ensured the quick circulation of his stories. The Novellen of his age (see POETISCHER REALISMUS) became almost a substitute for drama; Storm explicitly recognized this when calling the Novelle the sister of drama, since it presented human conflict in disciplined and closed form (1881). In Storm's historical Novellen (Chroniknovellen) *Aquis submersus* (1876), *Renate* (1878), *Eekenhof* (1879), *Zur Chronik von Griesshuus* (1884), *Ein Fest auf Haderslevhuus* (qq.v., 1885), the dramatic intensification is often motivated by inexplicable extraneous occurrences, rather than by a conflict within the characters themselves. Paul Heyse was one of the first to recognize the advantage of Storm's chronicle and framework technique for a narrative demanding shifting perspectives. In *Pole Poppenspäler* (1874), *Carsten Curator* (1878), *Hans und Heinz Kirch* (1882), *Bötjer Basch* (1886), *Ein Doppelgänger* (qq.v., 1887), Storm penetrates a seemingly secure narrow burgher (Kleinbürger) world to reveal human dependence on environment and heredity, for which he found little scope in patrician and professional family circles, in which he portrayed the delicacy of human relationships with humour (*Die Söhne des Senators*, 1880) and melancholy (*Viola tricolor*, qq.v., 1874). In one of his last Novellen, *Ein Bekenntnis* (q.v., 1887), he grapples with the problem of euthanasia. The warmth of his humanity for those who stand apart shows likewise consistently in his treatment of themes relating to the artist, *Eine Malerarbeit* (1867) and *Ein stiller Musikant* (1875); the undercurrents of an age of change are caught in *Auf dem Staatshof* (1859), *Im Schloß* (1862), and *Auf der Universität* (qq.v., 1863). In all Storm wrote more than 50 Novellen. None is constructed on the expansive scale adopted for the portrayal of the tragedy of the Deichgraf Hauke Haien in *Der Schimmelreiter* (q.v., 1888). Storm's work, especially in his later years, anticipates in some degree the Naturalistic manner (see NATURALISMUS).

Gesammelte Schriften (19 vols.) appeared 1868–89; *Sämtliche Werke* (8 vols.) appeared in 1898, (14 vols.), ed. A. Biese, 1919, (8 vols.), ed. K. M. Schiller, 1926 (reissued 1968), a critical edition by A. Köster, 1912–20 (8 vols., reissued 1939), and correspondence (4 vols.), ed. Gertrud Storm, 1915–17; other editions include Storm's correspondence with H. W. Seidel (see SEIDEL, H.), 1911, with E. Mörike, ed. H. W. Rath, 1919, with G. Keller, ed. A. Köster, 1904 (ed. P. Goldammer, 1960), with P. Heyse (2 vols.), ed. G. J. Plotke, 1917–18, with Th. Mommsen, ed. H.-E. Teitge, 1966, with E. Schmidt, ed. K.-E. Laage, 1972.

STRACHWITZ, MORITZ, GRAF VON (Peterwitz, Silesia, 1822–47, Vienna), a student at Breslau and later at Berlin universities, was a prominent and popular member of the literary club Der Tunnel über der Spree (q.v.) in Berlin. He returned to Silesia in 1843 and died four years later on his way back from a visit to Venice. He wrote lyrical and political poetry, but was most successful in patriotic and heroic ballads, of which *Das Herz von Douglas* survived in anthologies well into the 20th c. His poetry was collected in *Lieder eines Erwachenden* (1842) and in two posthumous volumes, *Neue Gedichte* (1848) and *Gedichte, Gesammelte Ausgabe* (1850). *Sämtliche Lieder und Balladen*, ed. H. M. Elster, appeared in 1912.

Sträflinge, a Novelle by B. Auerbach (q.v.), published in 1843 in *Schwarzwälder Dorfgeschichten* (q.v., vol. 1). It begins with the recent foundation of a society for rehabilitating released prisoners by finding them situations, and the story records the eventually successful struggle of two such characters to achieve a normal life. Jakob has been imprisoned for manslaughter, committed under provocation and while drunk, but Magdalena has been convicted of a theft of which she is innocent. They are in separate employment in the country, but are drawn together by their similar situations. A burglary is committed in circumstances suspicious for both and they are arrested, but a benevolent doctor is able to prove that they are innocent. They marry and settle as crossing-keepers on one of the new railways. They reappear in the story *Das Nest an der Bahn* included in *Nach dreißig Jahren* (q.v.), published in 1876.

Stralsund, a small Pomeranian harbour town which has a key position in the Baltic. In the Thirty Years War (see DREISSIGJÄHRIGER KRIEG) Wallenstein (q.v.), sweeping across North Germany in pursuit of the Danish invader, vainly laid siege to Stralsund. The resistance of the city, which was supported by Gustavus Adolphus of Sweden (q.v.), because he desired to have the harbour available for the landing of Swedish troops, was the first major set-back Wallenstein (q.v.) experienced in his early campaigns. After besieging Stralsund for a month

(July 1628), Wallenstein withdrew, abandoning his hopes for the formation of a fleet. Stralsund was also the scene of Schill's attempted rising against the French (1809, see SCHILL, F. B. VON). From 1648 to 1814, except for two brief intervals, Stralsund was Swedish. It came to Prussia by the settlement of 1815, and since 1952 belongs to the Bezirk Rostock of the DDR (see DEUTSCHE DEMOKRATISCHE REPUBLIK).

STRAMM, AUGUST (Münster, 1874–1915, killed in action, Horodec, Russia), after studying at Berlin and Halle, became a postal official. In 1913 he began to contribute to the periodical *Der Sturm* (q.v.). A reserve officer, he was recalled to the colours in 1914. His poetry, including notable war poems of 1914, is strikingly original in its terse formulation and original syntax, arrived at only after endless experiment. His plays are in the hectic Expressionist manner. Much of his work was published posthumously. It includes the poetry *Rudimentär* (1914), *Du* (1915), and *Tropfblut* (1919), and the plays *Sancta Susanne* (1914), *Die Haidebraut* (1914), *Die Menschheit* (1915), *Kräfte* (1915), *Geschehen* (1916), and *Die Unfruchtbaren* (1919). *Gesammelte Dichtungen* (2 vols.) appeared in 1920–1 and *Das Werk*, ed. R. Radrizzani, 1963.

STRANITZKY, JOSEF ANTON (Graz, 1676–1726, Vienna), was at first an itinerant actor, active in Munich and Salzburg. In 1706 he established himself in Vienna, and from 1712 until his death he directed a successful company in the Kärntnertor-Theater. He himself acted the comic part of Hanswurst (q.v.), which he created in the form of a local figure, wearing Salzburg peasant costume and speaking Viennese dialect (Wienerischer Hanswurst). In 1707 Stranitzky was registered as a medical practitioner in Vienna, but it is not known when or where he pursued his studies. His *Wiener Hauptund Staatsaktionen*, ed. F. Homeyer, appeared in 1907.

Straßburg (French Strasbourg), since 1919 an important French university city and Rhine port, belonged in the early Middle Ages to the bishop of the diocese. Among notable medieval inhabitants were GOTTFRIED von Straßburg, Meister Eckhart, and, rather later, the preacher GEILER von Kaisersberg (qq.v.). In the 15th c. S. Brant spent much of his life in the city, which was also the birthplace of J. Fischart (qq.v.). In the 14th c. Strasburg became a Free Imperial City (Freie Reichsstadt) in the Holy Roman Empire (see DEUTSCHES REICH, ALTES). It accepted the Reformation (q.v., in Calvinistic form) c. 1523, and was a member of the League of Schmalkalden (see SCHMALKALDISCHER BUND). Virtually unaffected by the Thirty Years War, it was in 1681 seized by Louis XIV in a sudden, but carefully prepared, *coup*. It retained its links with the Empire, however, and until the French Revolution (q.v.) its university was regarded as German.

Strasburg lost its special privileges during the Revolution, became a part of France, and the principal city of Alsace. With Alsace and a part of Lorraine it was annexed by Germany in 1871 (see DEUTSCH-FRANZÖSISCHER KRIEG), but forty-seven years later reverted to France by the Treaty of Versailles (see VERSAILLES, TREATY OF). In 1870 Strasburg resisted siege for seven weeks, suffering bombardment which destroyed not only buildings but important manuscripts.

Goethe was a student at the university in 1770–1, and Strasburg Münster (ascribed at the time to Erwin von Steinbach, d. 1318) opened his eyes to the qualities of Gothic architecture (see VON DEUTSCHER ART UND KUNST). While he was there the Archduchess Marie-Antoinette (q.v.) passed through the city on her way to be married to the Dauphin, later Louis XVI, an episode which is recorded in *Dichtung und Wahrheit* (q.v., Bk. 9). G. Büchner (q.v.) twice stayed in Strasburg, initially seeking refuge there as a political refugee.

Strasburg has remained an important centre for Franco-German cultural exchanges and European relations.

Straßburger Blutsegen, a magic spell formerly preserved in an 11th-c. MS. in Strasburg. The MS. was destroyed during bombardment in 1870. The spell has two parts, the first narrative, the second imperative. The latter is an adaptation of an ill-understood Latin spell. (See also ZAUBERSPRÜCHE.)

Straßburger Chronik, see CLOSENER, FRITSCHE.

Straßburger Eide, formal oaths sworn between Ludwig II, der Deutsche and Karl II, der Kahle (qq.v.) on 14 February 842 at Strasburg, and preserved in Nithart's *Historiarum libri quatuor* (see NITHART). After the death of Ludwig I der Fromme (q.v.) his three sons disputed the inheritance, Ludwig and Karl in alliance against their elder brother Lothar (see LOTHAR I, KAISER). They defeated Lothar at Fontenay in 841 and solemnly renewed their alliance at Strasburg in the presence of their armies. In order that the oaths should be understood by their allies, Ludwig swore in French (*romana lingua*), Karl in German; these oaths therefore are not only a historical document, but

one of philological interest for Old High German.

STRAUSS, BOTHO (Naumburg/Saale, 1944–), after studying Germanistik, drama, and sociology, worked as editor and theatre critic for the periodical *Theater heute*, in which some of his early work was first published. In 1970 he became producer at the Schaubühne am Halleschen Ufer in West Berlin, where he later settled as a free-lance writer.

The plays *Die Hypochonder* (1972) and *Bekannte Gesichter, gemischte Gefühle* (1974) were published together in book form in 1979, and *Trilogie des Wiedersehens* (1976, film version 1979) and *Groß und klein* (1978, film version 1980) in 1980; *Kalldewey. Farce* followed in 1981 (film version 1983), and *Der Park* in 1983. In 1972 Strauß produced (in collaboration with Peter Stein, who directed the film versions of 1975, 1979, and 1980) an adaptation of Kleist's *Prinz Friedrich von Homburg* (q.v.) under the title *Kleists Traum vom Prinzen Homburg*, and in 1974 an adaptation of a play by Gorky, *Sommergäste* (film version 1975). In the programme for the former play he explains that the characterization is based on montage, drawn both from Kleist's empirical reality (Real-Bild) and from 'the dreams of which (the characters) are made' (Ideal-Bild), which, applied to the Prince, reveals a nervous and unstable psychic disposition. This approach forms a key to his own figures, who live in the faceless contemporary world of his experience. His overriding interest lies in 'demasking' this reality in order to demonstrate man's 'struggle with the subconscious' (*Die Widmung*). The opposite of the documentary political theatre, Strauß's plays of the 1970s represent in this sense the 'Theater der Neuen Subjektivität und Sensibilität' (W. Hinck).

Groß und klein, a play reminiscent of Strindberg's and Kaiser's Stationendrama (q.v.), comprises ten scenes, held together by a woman in her mid-thirties, Lotte, whose husband has left her for another woman. Still professing to love her husband, and adopting a mission to preach love and charity, she finds at each location, each typifying modern life, different people, all of whom live in the same void. Even at a family reunion she fails to establish any contact and leaves, a witness to alcoholism. Mentally disturbed, she finally arrives in a physician's waiting room; she tells the doctor that nothing is wrong with her, he bids her leave, and the lights are switched off. The 'dream of which she is made' has led her into the darkness of her soul.

As a writer of fiction, Strauß established himself with a series of stories: *Schützenehre*, the volume *Marlenes Schwester*, which contains, apart from the title-story, *Theorie der Drohung* (both 1975), and *Die Widmung* (1977). In the latter, Richard Schroubek, whose partner Hannah Beyl has left him during the hot Berlin summer of 1976, writes in his increasingly neglected flat the 'biography' of their love during her absence ('die Biografie seiner leeren Stunden'), in the hope that she will return. As the story's title suggests, his script is dedicated to her. His at times macabre self-analysis has a therapeutic effect until, on seeing her, he meets with total disillusionment. His 'Liebesunglück' remains inexplicable.

Rumor appeared in 1980, *Paare Passanten* in 1981, and *Der junge Mann* in 1984. In the Introduction the narrator reflects with consternation and irony on contemporary readers and on his description of the novel as RomantischerReflektionsRoman, finishing with a quotation from Giordano Bruno's *De Immenso et Innumerabilis*, referring to the return of old things and unrecognized truths after a long night. In the first of the novel's five main sections, *Die Straße*, the narrator appears as the 'young man' of the title, Leon Pracht, who has turned his back on the theatre for a more fulfilling self-realization, which he has found in writing. Within this framework, the narrator's preoccupation with the nuclear age directs his intellect and fires his imagination. *Die Siedlung* suggests with ironic twists, discernible throughout, the emergence of a new society (the people of the Synkreas), in which man's creative drive (Spieltrieb) is reactivated. In *Der Wald* a business woman emerges from strange encounters and horrific nightmares (in a wood near Cologne) with the knowledge that boundless love and undeterred activity must replace the pervasive fear of extinction. In a prominent section of *Die Terrasse*, the medical orderly Rappenfries argues against the 'evolutionist' (representing society) and proves his worth as a follower of Blake. In *Der Turm* (here an ultra modern luxury hotel), Leon finds that the once admired comedian Ossia has himself, through his lavish lifestyle, become a comic figure and lost the ability to distance himself from his art. The novel ends with Leon's departure from him. He is alone, for Yossica, the 'perfect woman' whom he understands so well, has left him for an unspecified time in pursuit of her own artistic ambitions. She is the subject of the surrealist fairy-tale *Die beiden Talentsucher* contained in the last section.

In 1976 Strauß published seven poems under the title *Unüberwindliche Nähe* (in *Tintenfisch*, ed. M. Krüger) and in 1985 the volume *Diese Erinnerung an einen, der nur einen Tag zu Gast war. Gedicht*.

STRAUSS, DAVID FRIEDRICH (Ludwigsburg, 1808–74, Ludwigsburg), was at school at Blaubeuren and afterwards studied theology at Tübingen. In 1832 he began to teach at the Tübinger Stift (q.v.) and also lectured at the university. He was at that time at work on his life of Jesus, in which the gospel accounts were critically analysed, and he resigned in 1833 in order to complete the book, which was published in 1835–6 as *Das Leben Jesu, kritisch bearbeitet*, 2 vols. (q.v.). The work caused a considerable stir, and Strauß answered his critics in 1837 with *Streitschriften zur Verteidigung meiner Schrift über das Leben Jesu*.

For a short time Strauß was a schoolmaster in Ludwigsburg, his native town, and was then appointed a professor of theology at Zurich University, only to be pensioned even before he entered on his duties; for the population, indignant at his appointment, brought about the fall of the government. In 1840–1 he published the two volumes of *Die christliche Glaubenslehre in ihrer geschichtlichen Entwicklung und im Kampfe mit der modernen Wissenschaft*. In 1841 he married the opera singer Agnes Schebest, but a separation took place in 1846, Strauß spent the rest of his life as a writer, turning to historical and literary biography. *Schubarts Leben in seinen Briefen* appeared in 1849, *Christian Märklin, ein Lebens- und Charakterbild aus der Gegenwart* in 1851, *Nikodemus Frischlin* in 1855, *Ulrich von Hutten*, 1858–60, *Hermann Samuel Reimarus und seine Schutzschrift für die vernünftigen Verehrer Gottes* in 1862. Influenced by Reimarus (q.v.), he returned to theology with *Das Leben Jesu für das deutsche Volk bearbeitet* in 1864, and rounded on its critics with *Die Halben und die Ganzen* (1865). In *Der alte und der neue Glaube* (1872) he abandoned the Hegelian principles to which he had previously adhered and embraced materialism.

Gesammelte Schriften (12 vols., 1876–8) were published posthumously. The first volume of these was an autobiography. In 1846 George Eliot translated *Das Leben Jesu* into English.

STRAUSS, EMIL (Pforzheim, 1866–1960, Freiburg, Breisgau), studied at Berlin University 1886–90. He had friends among the adherents of Naturalism (see NATURALISMUS), including M. Halbe and G. Hauptmann (qq.v.). For a short time he farmed in southern Baden, and in 1892 emigrated to Brazil, where he remained until 1901. His first Novellen, collected in *Menschenwege* (1899), and a tragedy, *Don Pedro* (1899, revised 1914), appeared before his return to Europe. In 1903 he spent a short time in Switzerland, before settling at Überlingen, Württemberg, on Lake Constance. From 1911 to 1915 he lived at Hellerau near Dresden. From

1925 until his death he alternated between Freiburg and Badenweiler, both in Baden.

Strauß made his name primarily with the novels and Novellen written after his return to Europe at the age of 34. *Freund Hein. Eine Lebensgeschichte* (q.v., 1902), depicting the torments suffered by Heinrich Lindner, a schoolboy, nurtured a theme repeatedly treated by writers of the period (among them Hesse with *Unterm Rad* and Musil with *Die Verwirrungen des Zöglings Törleß*, qq.v., both 1906). *Der Engelwirt. Eine Schwabengeschichte* (1901), a shorter work, is a village tragicomedy, in which, after great tribulation, an erring husband (the Engelwirt) is reconciled with his wife. The novel *Kreuzungen* (1904) has as its principal characters a man (Hermann Anshelm) and two young women (Elfriede and Klara); their characters and the interplay of their relationships are traced with considerable insight. Later novels are *Der nackte Mann* (1912, set in Baden at the beginning of the 17th c.), *Das Riesenspielzeug* (1934, a long political parable), and *Lebenstanz* (1940).

Strauß is an acknowledged master of the shorter narrative, and the sensitive Novelle *Der Schleier* (q.v., 1920) is held by many to be his finest work. It was included in 1931 in a collection, to which Strauß gave the title *Der Schleier*; other stories included are *Das Grab zu Heidelberg*, *Liebe*, *Baptist*, *Gartenäre*, *Befund*, and *Der Skorpion*. Also noteworthy are *Hans und Grete*, *Der Laufen* (q.v., both in the collection *Hans und Grete*, 1909), *Der Spiegel* (1919), and the three stories of *Dreiklang* (1949), *Unterwegs*, *Frau Kampe*, and *Otta*, dealing with three women of different temperament and in different situations; each has to recognize her true being and come to terms with it (Einklang). Strauß wrote two more plays, the realistic drama *Hochzeit* (1908) and the patriotic play *Vaterland* (1923). *Ludens* (1955) contains recollections, essays, and poems.

STRAUSS, JOHANN (Vienna, 1825–99, Vienna), composer of waltzes and other dances and of operettas, is often called the Walzerkönig, king of waltzes. Among his best-known works are the waltzes *An der schönen blauen Donau*, *G'schichten aus dem Wiener Wald*, *Wiener Blut*, and *Rosen aus dem Süden*, and the *Tritsch-Tratsch-Polka*.

Encouraged by Offenbach's successes in Paris, Strauß composed a number of operettas, including *Die Fledermaus* (1874, libretto by C. Haffner and R. Genée), *Cagliostro in Wien* (1875, libretto by G. Quedenfeld), *Eine Nacht in Venedig* (1883, libretto by F. Zell and R. Genée), *Der Zigeunerbaron* (1885, libretto by J. Schnitzler), and *Wiener Blut* (1899, libretto by V. Léon and L. Stein).

Strauß is often referred to as Johann II, since his father (Johann I, 1804–49), brothers (Joseph

1827–70, and Eduard, 1835–1916), and nephew (Eduard's son, Johann III, 1866–1939) were all composers of waltzes.

STRAUSS, RICHARD (Munich, 1864–1949, Garmisch-Partenkirchen), had a distinguished career as a conductor, beginning with an appointment to the Meiningen court at 21 (1885), going on to Munich (1886–9), Weimar (1889–94), Munich again (1894–8), Berlin (Generalmusikdirektor, 1898–1919), and Vienna (Staatsoper, 1919–24). In 1933 he was appointed president of the Reichsmusikkammer, but he dissented from the racial policies of Nazism, and by 1935 was under suspicion.

Strauss wrote much orchestral music, beginning with a violin concerto in 1883. He became especially known for his illustrative orchestral works, which he called Tondichtungen. These were intended to convey visual (*Aus Italien*, (1887), or literary impressions *Don Juan*, 1889, based on Lenau's poem; *Macbeth*, 1890; *Till Eulenspiegels lustige Streiche*, 1895; *Also sprach Zarathustra*, after Nietzsche (q.v.), 1896; *Don Quichote*, 1898). *Ein Heldenleben* (1899) and *Sinfonia domestica* (1904) were intended as works of musical autobiography. *Eine Alpensymphonie* (1915) reverted to visual impression. Strauss was the most notable opera composer of the early 20th c. *Guntram* (to his own libretto, 1894) was noticeably influenced by R. Wagner (q.v.). Strauss's own individuality was first evident in the short opera *Feuersnot* (libretto by E. von Wolzogen, q.v., 1900).

In 1905 Strauss shocked the musical public with *Salome*, based on Oscar Wilde's play, and repeated the process in 1909 with *Elektra* (q.v.), which took H. von Hofmannsthal's tragedy as libretto. From this time until the death of Hofmannsthal (q.v.) in 1928 the two collaborated regularly and produced five more operas, which bore witness to their profound devotion to an ideal conception of opera. They are *Der Rosenkavalier* (q.v., 1911), *Ariadne auf Naxos* (q.v., 1912), *Die Frau ohne Schatten* (q.v., 1919), *Die ägyptische Helena* (q.v., 1928), and *Arabella* (q.v., 1933). Other operas are *Intermezzo* (libretto by Strauss himself, 1924), *Die schweigsame Frau* (1935, libretto by S. Zweig, q.v., based on Ben Jonson's *Epicene*, or *The Silent Woman*), two one-act operas, *Dafne* (1938, libretto by J. Gregor, 1888–1960) and *Capriccio* (1942, libretto by the conductor C. Krauß, 1893–1954, and Strauss himself), and a three-act work with Gregor as librettist, *Die Liebe der Danae* (1940).

Strauss wrote other vocal music, including motets, *Enoch Arden* after Tennyson (1899), and songs (see LIED, 2). Among the poets he set are O. J. Bierbaum, G. A. Bürger, R. Dehmel, H. von Gilm, Heine, and Weinheber (qq.v.).

His *Vier letzte Lieder*, written between 1946 and 1948, consist of three songs to poems by H. Hesse (*Frühling, September*, and *Beim Schlafengehen*) followed by *Im Abendrot* by Eichendorff (qq.v.). In these, as in a number of other songs, he used orchestral accompaniment in preference to the piano. His writings appeared as *Betrachtungen und Erinnerungen*, ed. W. Schuh, 1949 (reissued 1957), and editions of correspondence include that with H. von Hofmannsthal (1926, revised 1964), and with S. Zweig (1957).

STRAUSS UND TORNEY, LULU VON (Bückeburg, 1873–1956, Jena), a general's daughter, began in her twenties to write poetry and notable ballads (*Gedichte*, 1898, and *Balladen und Lieder*, 1902). Her first narrative work was a collection of Novellen, *Bauernstolz* (1901); most of her fiction consists of historical novels and stories with rural settings in north-west Germany and include *Ihres Vaters Tochter* (1905), *Der Hof am Brink, Das Meermineke* (both 1906), *Lucifer* (1907), *Sieger und Besiegte* (1909, a collection, of which *Auge um Auge* was published separately, 1933), *Der Judas* (1911, a novel restyled *Der Judashof*, 1937), and *Der jüngste Tag* (1922, a novel dealing with the extremist Anabaptists at Münster, q.v.), *Das Fenster* (1923, restyled *Das Kind am Fenster*, 1938), and *Schuld* (1940). She is best remembered for her ballads, of which another collection appeared as *Reif steht die Saat* (1919).

Vom Biedermeier zur Bismarckzeit (1932) belongs to her historical and biographical writings. In 1916 she married the publisher Eugen Diederichs (1867–1930); in 1936 she published *Eugen Diederichs Leben und Werk* (with correspondence) and in 1943 reminiscences entitled *Das verborgene Angesicht*. A selection of ballads and shorter fiction, *Tulipan. Balladen und Erzählungen*, appeared in 1966 and correspondence with Th. Heuß (q.v.) in 1965.

Stream of Consciousness, see INNERER MONOLOG and ERLEBTE REDE.

STREFF, E., pseudonym of E. E. Niebergall (q.v.).

STREHLENAU, NICOLAUS FRANZ NIEMBSCH, EDLER von, see LENAU, N.

STREICHER, ANDREAS (Stuttgart, 1761–1835, Vienna) became acquainted with Schiller, for whom he conceived a great admiration. He assisted Schiller in his escape from Stuttgart in September 1782, and was a loyal friend of the poet in the difficult months spent in and near Mannheim up to December 1782.

Streicher, who was a talented musician, settled in Vienna and achieved prosperity as a

piano-maker. Late in life he wrote an account of the months spent with Schiller, *Schillers Flucht von Stuttgart und Aufenthalt in Mannheim* (1836, repr. 1959), which is a conscious tribute to Schiller and, in its integrity and modesty, an unconscious one to Streicher himself.

Streit um den Sergeanten Grischa, Der, a novel by Arnold Zweig (q.v.), published in 1927. It is a war novel set on the eastern front in 1917.

The Russian sergeant Grischa Paprotkin, a prisoner of war and a man of honest, serene, and childlike character, plans flight to the east, in order to see wife and child again. He succeeds at first, concealing himself for several days in a goods wagon in a timber train. But he jumps off too soon and has to take refuge in a dense forest behind the German lines. Here he is enlisted by an armed gang of mixed-refugees and deserters led by Babka, a young woman with whom he falls in love. He takes the papers and identity of a dead Russian deserter, Bjuschew. Under this name he is captured by the Germans and sentenced to death as a spy under a German decree which treats Russian deserters who have not given themselves up in three days as spies. Grischa now declares his identity. His case is put to the divisional commander General Otto von Lychow, who orders an investigation, and it is confirmed that Grischa is not a deserter, but an escaped P.O.W. The politically minded, power-conscious quartermaster-general Schieffenzahn sees in the Bjuschew–Paprotkin case a political matter and demands that Grischa be handed over. Lychow refuses and Grischa's case seems hopeful. But Lychow goes on leave and Schieffenzahn orders the execution of Grischa in the divisional commander's absence. Babka bears the dead Grischa a child.

Though Zweig's sympathy with the common soldier is clear, he is fair to the elements of decency in the military hierarchy, General von Lychow, Advocate-General Posnansky, and Oberleutnant Winfried. His antagonism is directed at the military bureaucracy and at Schieffenzahn, in whom may be seen the figure of Ludendorff (q.v.), but he avoids caricature in this humane book. Zweig claims that the story is not invented ('dessen Fabel nicht erfunden ist') and indicates that it was conceived in 1917, drafted as a play in 1921, and converted into a novel in 1926–7. The stories *Erziehung vor Verdun* (1935) and *Einsetzung eines Königs* (1937) are intended to form a trilogy of novels with *Der Streit um den Sergeanten Grischa*.

STRESEMANN, GUSTAV (Berlin, 1878–1929, Berlin), a businessman elected to the Reichstag in 1907 as a National Liberal, played a prominent part in German politics during the Weimar Republic (q.v.).

In 1917 Stresemann became chairman of the National Liberal party, strongly supporting Hindenburg and Ludendorff (qq.v.) in their annexationist plans, and played a part in the intrigues which displaced von Bethmann Hollweg (q.v.). After the fall of the monarchy Stresemann founded and led the Deutsche Volkspartei, was a member of the Weimar National Assembly, and was elected to the new Reichstag in 1920. In August 1923 he became chancellor (Reichskanzler) and foreign minister. His government fell in November 1923, but Stresemann remained foreign minister until his death. During this period the Dawes Plan was accepted, the Locarno Pact signed, Germany was admitted to the League of Nations, and agreement for the French and Belgian evacuation of German territory in 1930 was secured (the British had left in 1926). Stresemann's policies were violently attacked by the German right wing, which opposed European co-operation, but he was widely respected abroad, as well as by some of his own countrymen. Since his death his policy and aims have been variously interpreted. Some have accepted him as a former nationalist converted to a European conception, others as a basically unwavering nationalist using the posture of Europeanism as a temporary measure to secure Germany's advancement in an adverse situation.

Stresemann's writings include *Von der Revolution bis zum Frieden von Versailles* (1919), *Reden und Schriften* (2 vols., 1926), and the posthumous *Vermächtnis. Aus dem Nachlaß Gustav Stresemanns*, ed. H. Bernard (3 vols., 1932–3). In 1926 he was awarded the Nobel Peace Prize.

STRICKER, DER, a Middle High German poet of the first half of the 13th c. He was by birth a commoner, a native of Franconia, and spent a considerable part of his life in Austria. Beyond these meagre details nothing is known of his life. He was a prolific poet and wrote in many forms. The chronology of his work is uncertain, but his two epics are probably early poems. Of these, *Karl*, a popular adaptation of the lay of Roland (see ROLANDSLIED), is believed to be the earlier. *Daniel vom blühenden Tal* is an Arthurian romance presenting a string of improbable adventures. Its unchivalric spirit provoked a retort by Der PLEIER (q.v.) in *Garel vom blühenden Tal*.

Der Stricker's principal field is the Schwank (q.v.) or humorous short story in verse. Many Schwänke attributed to Der Stricker in the MSS. are probably spurious, but the difficult task of determining the authentic poems has not so far been accomplished. Many of these verse

Schwänke throw vivid light on manners and conditions in the Middle Ages, others are fables depicting animals. The didactic purpose is conspicuous, and most of these realistic poems are in reality parables. The moralist Stricker condemns vice, but his counter to the evils of the world is rather in ingenuity and an alert and adaptable intelligence than in virtue.

A more ambitious work than the diverse series of Schwänke is *Der Pfaffe Amîs*, a poem of approximately 2,500 lines, in which a series of twelve exploits of clever deception and trickery are set in a serious frame. The English priest Amîs is beset by his bishop, whose inquisition he defeats by superior intelligence. Thereupon he sets out into the world and for a time lives, with great success, by his wits. Finally God turns him from deceit, he enters a monastery and becomes its respected abbot. Though the priestly image is a surprisingly worldly one, the story can be seen as a satire on human folly, and the end, though hardly consistent, is highly moral.

STRICKER, JOHANNES (Grube, Holstein, 1540–98, Lübeck), was a pastor in Cismar near Grube and, from 1584, in Lübeck. He translated Luther's catechism into Low German and wrote a realistic play on the theme of Everyman, *De Düdesche Schlömer* (*Der deutsche Schlemmer*, 1584) as well as a dramatic treatment of the Fall (*Geistliches Spiel von dem erbermlichen Falle Adams und Even*, 1570).

STRINDBERG, (Johan) AUGUST (Stockholm, 1849–1912, Stockholm), one of the most gifted writers and stormiest personalities of Swedish literature. He influenced German literature particularly through the Naturalistic plays *Fadren* (1887, *Der Vater, The Father*) and *Fröken Julie* (1888, *Fräulein Julie, Miss Julie*), and later through symbolical plays foreshadowing Expressionism, especially *Till Damaskus* (Pt. 1 and 2, 1898, Pt. 3, 1901, *Nach Damaskus, To Damascus*) and *Spöksonaten* (1907, *Gespenstersonate, The Ghost Sonata*). (See STATIONENDRAMA.)

STRITTMATTER, ERWIN (Spremberg, Lower Lusatia, 1912–   ), an official in the Soviet Zone, then a journalist in the German Democratic Republic. In 1959 he became First Secretary of the East German Writers Association. He has written novels of socialist realism (see SOZIALISTISCHER REALISMUS) which have a pronounced individualistic streak (*Ochsenkutscher*, 1950; *Tinko*, 1954; *Der Wundertäter*, vol. 1, 1957, vol. 2, 1973). *Ole Bienkopp* appeared in 1963. Shorter stories include *Eine Mauer fällt* (1953), *Paul und die Dame Daniel* (1956), *Pony Pedro* (1959), *Ein Dienstag im September. 16 Romane im Stenogramm* (1969, new edn. 1972), *Die blaue*

Nachtigall oder Der Anfang von etwas (1972, child-hood recollections, newly arranged in 1981). The collections *3/4 hundert Kleingeschichten* and *Schulzenhofer Krankenkalender* appeared in 1972, *Damals auf der Farm* in 1977, aphorisms written between 1966 and 1967, entitled *Selbstermunterungen*, in 1981. The novel *Der Wundertäter* was republished in three volumes in 1979–80 and a new novel, *Der Laden*, appeared in 1983.

Of Strittmatter's comedies (*Katzgraben*, 1954; *Die Holländerbraut*, 1961) *Katzgraben*, a village comedy in verse, received particular praise from B. Brecht (q.v.), who had it produced in 1953, before its publication, by the Berliner Ensemble (q.v.). He states in his tract '*Katzgraben*'-*Notate* (see KLEINES ORGANON FÜR DAS THEATER): '*Katzgraben* ist meines Wissens das erste Stück, das den modernen Klassenkampf auf dem Dorf auf die deutsche Bühne bringt'.

STRITTMATTER, EVA (Neuruppin, 1930–), after her university studies was active as a literary critic, a profession she publicly criticized. The wife of Erwin Strittmatter (q.v.), she has established herself as the author of lyric poetry, favouring poems on nature. Her collections include *Ich mach ein Lied aus Stille* (1973), *Mondschein liegt auf den Wiesen* (1975), *Die eine Rose überwältigt alles* (1977), *Zwiegespräche* (1980), *Eva Strittmatter. Gedichte* (1980, both DDR and BRD), and *Heliotrop* (1983). She is also the author of children's literature. Her edition *Briefe aus Schulzenhof* appeared in 1977 and a volume of essays, *Poesie und andre Nebendinge*, in 1983.

She was awarded the Heine Prize in 1975.

STRUBBERG, FRIEDRICH AUGUST (Kassel, 1808–89, Gelnhausen), pseudonym Armand, fled to America at 18 after being involved in a duel, returning in 1829. After a few years in his father's business a second duel led him once more to flee to America (1837), where he remained until 1854, becoming a physician. His novels of American life include *Sklaverei in Amerika* (3 vols., 1862) and *Der Krösus von Philadelphia* (4 vols., 1870). He also wrote sketches of American life (*Bis in die Wildnis*, 4 vols., 1858, and *Amerikanische Jagd- und Reiseabenteuer*, 1858) and an autobiographical work, *Aus Armands Frontierleben* (3 vols., 1868).

Strudlhofstiege, Die, a long novel by H. von Doderer (q.v.), published in 1951. Its sub-title *Melzer oder die Tiefe der Jahre* gives slight emphasis to one character in a work in which most are virtually of equal importance. *Die Strudlhofstiege* traces, both in parallel and in intersection, the lives of many persons drawn mainly from the lower nobility and the middle classes. It is an intensely Viennese novel and, except for two

episodes (Melzer's bear hunt in Bosnia and Etelka Grauermann's suicide in Budapest), is centred on the 9th District (IX. Bezirk, Alsergrund). It is here that the Strudlhofstiege is found, a flight of steps in the Strudlhofgasse, which connects the Liechtensteinstraße with the Waisenhausgasse (now Boltzmanngasse). The topographical detail is abundant.

The story opens in 1925 and looks back in recollection to 1910 and, to a lesser degree, to the years between. In 1910 Melzer, an infantry subaltern, wished to propose to Mary K., but feared a rebuff. In 1925 Mary K. is run over by a tram and loses a leg. First aid is expertly given by Melzer and Thea Rokitzer, who chance, independently, to be passing. Melzer and Thea fall in love and marry. Doderer uses these slender events as co-ordinates, by which he may plot the intertwining lives of a host of characters, including René, Etelka, and Asta von Stangeler, Rittmeister von Eulenfeld, Major Laska, Dr. Negria, Oskar K., and Grete Siebenschein. These intelligent and, for the most part, sympathetic, characters enable him to construct a picture of Vienna over the years, of almost photographic accuracy; and at the same time he succeeds in evoking the city's subtle personality. Doderer also provides an element of suspense and comedy in the Box and Cox act of the identical Pastré twins, whose near-criminal schemes are thwarted by Rittmeister von Eulenfeld and Melzer.

Struensee, a novel by R. Neumann (q.v.), published in 1935 and reissued in 1953 entitled *Der Favorit der Königin*.

Struensee, a historical tragedy in five acts by H. Laube (q.v.), first performed at Dresden in April 1845 and published in 1847. Its subject is the conspiracy by which the Danish dictator Count J. F. Struensee (b. 1731) was overthrown in Copenhagen in January 1772 and subsequently executed for treason.

The action is concentrated, occupying only a few hours, and the scene is limited to one room in the palace of Christiansborg. Struensee, the upstart and all-powerful minister, is represented by Laube as a disinterested humanitarian reformer, whose life work is jeopardized by political intrigues and ruined by private emotion; for he loves the Queen and refuses to give up his love. It is a play of intricate conspiracy in which Struensee's indiscretion plays into the hand of his reactionary opponents, by whom he is shot.

STRUWE or **STRUVE,** THERESE VON (Stuttgart, 1804–52, Java), married to Herr von Bacheracht, left him and devoted herself to K. Gutzkow (q.v.), who, however, when his first wife died, married another. She later married Baron Heinrich von Lützow, emigrating with him to Java in 1849. She is the author of several novels (*Falkenberg*, 1843; *Am Teetisch*, 1844; *Lydia*, 1844; *Weltglück*, 1845; *Heinrich Burkhart*, 1846; *Alma*, 1848).

Struwwelpeter, an illustrated book for children, published in 1845 with the full title *Lustige Geschichten und drollige Bilder mit 15 schön kolorierten Tafeln für Kinder von 3 bis 6 Jahren* by Heinrich Hoffmann (1809–94) under the pseudonym Reimerich Kinderlieb. Its author was a physician in Frankfurt/Main, who sought to teach children under school age in an amusing way the dire consequences of misconduct and disobedience. A nursery classic, it has been imitated and parodied, loved and scorned by adults in charge of children. The illustrations show Struwwelpeter as a grotesque-looking little boy, wearing the clothes a respectable mother with sufficient means would provide, but his fingernails are longer than his fingers because he won't have them cut, and his fair hair is long and fuzzy; on the title-page it stands up on end framing a somewhat bewildered-looking face.

STUCKEN, EDUARD (Moscow, 1865–1936, Berlin), an exotic and esoteric writer of neo-Romantic tendencies (see NEUROMANTIK) with a predilection for mythological themes. His four volumes of *Astralmythen* appeared over the years 1896–1907. He wrote several Arthurian plays in rhyming verse, which were produced by M. Reinhardt (q.v.), *Gawan* (1902), *Lanzelot* (1909), and *Merlins Geburt* (1912), and treated the story of Tristan in the play *Tristram und Ysolt* (1916). His most successful work was the trilogy of Aztec novels, *Die weißen Götter* (1918).

Studentische Verbindungen, associations of students, which carry out corporate activities according to a traditional ritual. They are divided into 'schlagende' and 'nicht-schlagende' Verbindungen. Social drinking and the singing of student songs and, for 'schlagende Verbindungen', formal duelling (see MENSUR) are the principal activities. The most exclusive Verbindungen are known as Corps. Each has its colours which are worn as a ribbon or on the cap, hence the term Couleurstudent. On ceremonial occasions the officers (Chargierte) parade in a braided tunic, white breeches, jackboots, and a pill-box cap, and carry a duelling sword.

The Verbindungen have their origin in the association of students from the same state, and were originally known as Landsmannschaften. They proliferated in the second half of the 19th c., in the wake of the Burschenschaft (q.v.). Names such as Rhenania, Borussia, Saxonia,

etc. recall their local origin. The National Socialist regime dissolved the Verbindungen in 1935. Their reconstitution was forbidden by the Allied Military Governments, but since 1949 the ban has been removed in the Federal Republic, though it continues in the DDR. Similar Verbindungen exist in Austria.

Studien, title used by A. Stifter (q.v.) for 6 vols. of his collected Novellen, published 1844–50. Some had previously appeared in magazines. The *Studien* comprise: *Der Kondor* (originally *Der Condor*), *Feldblumen*, *Das Heidedorf* (originally *Das Haidedorf*), *Der Hochwald* (q.v.), *Die Narrenburg*, *Die Mappe meines Urgroßvaters*, *Abdias* (q.v.), *Das alte Siegel*, *Brigitta* (q.v.), *Der Hagestolz* (q.v.), *Der Waldsteig*, *Zwei Schwestern*, and *Der beschriebene Tännling*.

Stunden-Buch, Das, a collection of poems by Rilke (q.v.), written 1899–1903 and published in 1905. It is linked with Rilke's visits to Russia in 1899 and 1900, and dedicated to his travelling companion Lou Andreas-Salomé (q.v.). The title alludes to medieval 'books of hours' and the book is based on the fiction that it is written by a Russian monk engaged in painting icons. The poems are organized in three sections, each rapidly written in a period of intense inspiration. *Das Buch vom mönchischen Leben* (1899) is the closest to Rilke's Russian experience. *Das Buch von der Pilgerschaft* belongs to 1901 and *Das Buch von der Armut und vom Tode* to 1903. All the poems are untitled and they form a continuous cycle. The god of Rilke's monk is not the Christian God, but the 'dark' god on whom the monk depends for his creation and existence. The poems embody Rilke's early admiration for the simple of heart. They are also an attempt to grapple with the fear of death and show the beginnings of his religion of art, which alone seems to offer him fulfilment.

STURM, JOHANNES (Schleiden, Eifel, 1507–89, Strasburg), founder and first head of the Protestant grammar school at Strasburg in 1538, from which he was later dismissed by the strictly Lutheran municipal inspector of schools. Sturm was educated at Liège and Louvain, and in his first educational work (*De literarum ludis recte aperiendis liber*, 1538) sought to embody the ideals of the humanistic age by encouraging the school performance of classical plays. His other writings include *Partitionum dialecticarum libri IV* (1571, reissued 1962) and *Classicae Epistolae* (ed. I. Rott, 1938).

STURM, JULIUS (Köstritz, 1816–96, Leipzig), studied theology at Jena University, after which he was for some years a private tutor in various families including that of the princely house of Reuß. He was appointed to a living in 1850. His mainly religious poetry appeared in collections with the titles *Gedichte* (1850), *Fromme Lieder* (3 vols., 1852–92), *Stilles Leben* (1865), and *Gott grüße dich* (1876). At the time of the Franco-Prussian War (see DEUTSCH-FRANZÖSISCHER KRIEG) he published a volume of patriotic poetry, *Kampf- und Siegesgedichte* (1870). A selection of his fables, *Spiegel der Zeit in Fabeln* (1872), was published in 1928 as *Ein deutsches Haus. Werke* and letters appeared in 1916.

Sturm, Der, a radical artistic periodical published and edited by H. Walden (q.v.) from 1910 to 1932, supporting Expressionism (see EXPRESSIONISMUS) in literature and painting. Among the writers associated with it are G. Benn, Döblin, K. Heynicke, K. Kraus, E. Lasker-Schüler, A. Stramm, L. Schreyer, P. Zech, W. Mehring, and K. Schwitters (qq.v.), whose preface to *Anna Blume. Selbstbestimmungsrecht des Künstlers* was published in it. Painters it supported included Marc (q.v.), Chagall, Kokoschka (q.v.), and W. Kandinsky (q.v.).

Originally a weekly, *Der Sturm* appeared from 1916 to 1932, when it ceased publication, as a monthly periodical. For a few years, until 1921, it sponsored the Kunstbühne in Berlin (also known as Sturm-Bühne), founded by L. Schreyer, who, with Nell Walden, published *Der 'Sturm'. Ein Erinnerungsbuch an H. Walden und die Künstler aus dem Sturmkreis* (1954).

Sturmflut, a 3-vol. novel by F. Spielhagen (q.v.), published in 1877. It is set in the years of speculation and financial disaster after 1870 (see GRÜNDERZEIT). The storm-driven sea, which in the last grandiose chapters inundates the land, is also a symbol of the financial flood and its calamities. The contrast between middle class and aristocracy is less harsh than in Spielhagen's earlier novels. Amidst all the vice and financial greed, Else von Werben, daughter of a general of sterling character, makes a love marriage with the young master mariner Reinhold Schmidt.

Sturm und Drang, a play (Schauspiel) in five acts by F. M. Klinger (q.v.). It was written at Weimar in 1776 and published in 1777, though the original title-page bears the date 1776. It was first performed at Dresden on 1 April 1777. Klinger's original title was *Der Wirrwarr*, and this was changed before publication to *Sturm und Drang* at the suggestion of C. Kaufmann (q.v.). This title later came to be used to designate the whole movement, otherwise called the Geniezeit (see STURM UND DRANG).

The play is set in America and deals with the hatred of two noble families, Berkley and Bushy. After five acts of threats and frantic talk of vengeance they are reconciled through the love of young Bushy (Wild) and Lord Berkley's daughter. The play contains a pair of eccentric exiles from Europe (La Feu and Blasius) and is written in Klinger's wildest exclamatory prose. Klinger himself wrote in 1777, 'Es ist meine Lieblings Arbeit'.

Sturm und Drang, 'Storm and Stress', denotes a literary movement and a group of writers of the 1770s, otherwise referred to as the Geniezeit. This short-lived phenomenon, which lasted from 1771 to 1778, has no exact parallel in other literatures. Its name, applied to it by later generations, derives from the title of a play, *Sturm und Drang* (q.v., 1777), by F. M. Klinger, and the phrase is also used about this time (for a state of mind, not a movement) by G. A. Bürger (q.v.).

Behind the Sturm und Drang are trends of European thought exemplified in the exaltation of freedom and nature by J.-J. Rousseau (q.v.), whom J. G. Herder (q.v.) followed, contributing notably a sense of historical evolution and a high estimation of folk poetry. Herder, whose *Von deutscher Art und Kunst* (q.v., 1773) operated as a manifesto, and J. G. Hamann (q.v.) supported the cult of genius, which Edward Young expressed in his widely read *Conjectures on Original Composition* (1759). Other factors in the growth of Sturm und Drang are the lyric vein of Klopstock and the Göttinger Hainbund (qq.v.), Young's *Night Thoughts* (1742–5), and Hamann's reaction against rationalism. The unmitigated suffering in Gerstenberg's *Ugolino* (q.v., 1768), the exaltation of Shakespeare (q.v.), and the attempts at realism in novel and drama by Lessing (see EMILIA GALOTTI, 1772) and Goethe are also significant. These contributed to the sense of social commitment which the movement evinced.

The Sturm und Drang is unthinkable without the dynamism of Goethe, who, in his Strasburg and Frankfurt days (1770–5), attracted a number of young men who, in different ways, responded to his promptings. His own works of this phase are *Götz von Berlichingen mit der eisernen Hand, Die Leiden des jungen Werthers, Clavigo, Stella* (qq.v.), the moving scenes of *Urfaust* (see FAUST), a handful of audacious verse satires, and many poems; *Prometheus, Ganymed, Wanderers Sturmlied*, and *Willkommen und Abschied* (qq.v.) are outstanding examples. J. M. R. Lenz, H. L. Wagner, F. Müller (qq.v.), and Klinger were Goethe's satellites. All these had spent their surplus energy by about 1776 and either died, run to seed, or, like Goethe and

Klinger, imposed discipline upon their early agitation.

Power and strength, conceived both in emotional and physical terms and designated Kraft, were expressed in extravagant terms, and this verbal explosion influenced the lives of a generation of readers, including Goethe's acquaintances, the Jacobis (see JACOBI, F. H. and J. G.) and the young K. P. Moritz (q.v.). Musicologists have implied a connection with J. Haydn and Mozart (qq.v.). The subsidence of the Sturm und Drang was followed by a belated manifestation in Schiller's early work (1780–5). This delay is to be ascribed to local conditions in Württemberg.

German literature sustained a decisive and long-enduring impulse from the Sturm und Drang, of which Goethe's own account in *Dichtung und Wahrheit* (q.v., Bk. 13) remains the most important record.

STURZ, HELFERICH (or HELFRICH) PETER (Darmstadt, 1736–79, Bremen), after studying at Jena, Göttingen, and Gießen universities, was appointed secretary to Adolf von Eyben in Glückstadt in 1760 and to Count Bernstorff in Copenhagen in 1762, rising to higher positions in the Danish foreign ministry. In 1768 he was in the suite of King Christian VII on his visit to London and Paris. When the Berhstorff regime was overthrown by Struensee (1731–72) in 1772, Sturz was dismissed and took service in the County of Oldenburg.

Sturz's first work was a tragedy (*Julie*, 1767), but he is chiefly notable as an essayist who wrote a remarkable tract on capital punishment (*Abhandlung über Todesstrafen*, 1776) and character sketches of Herder (1777) and Pitt (1778), as well as essays on many other subjects; most of these were published in H. C. Boie's *Das Deutsche Museum* (q.v.). Also noteworthy is his tribute to Count Bernstorff (*Erinnerungen aus dem Leben des Grafen von Bernstorff*, 1777). *Schriften* (2 vols.) appeared 1779–82 (reissued 1971).

Stuttgarter Handschrift, see WEINGARTNER LIEDERHANDSCHRIFT.

STYFEL, MICHAEL (Eßlingen, 1486 or 1487–1567, Jena), an Augustinian monk who became an early supporter of Luther (q.v.) and was later appointed professor of mathematics at Jena University. He is best known for his poem *Von der christfrömmigen rechtgegründten Lahr D. Martini Luthers* (1522).

SUCHENSINN, name (almost certainly assumed) of a strolling poet and singer of Bavarian origin, who is recorded in 1390 and 1392 at the court of Duke Albrecht II of Bavaria. His 22 songs are either religious, celebrating the Virgin Mary, or profane, praising human, especially wedded, love. Suchensinn is represented in the *Lieder der Klara Hätzlerin* (see HÄTZLERIN, CLARA).

SUCHENWIRT, PETER (d. Vienna, *c.* 1395), was a friend of HEINRICH der Teichner (q.v.), for whom he wrote an obituary poem. He also composed a number of moral and religious Sprüche (see SPRUCH) similar to those of his friend, and in addition wrote allegorical poetry. He became a man of some substance by providing the nobility with heraldic poetry (see HEROLDSDICHTUNG) in the form of eulogies of noble personages.

SUDERMANN, HERMANN (Matziken, East Prussia, 1857–1928, Berlin), went to school at Tilsit and afterwards studied at Königsberg and Berlin universities. The son of a small brewer, Sudermann possessed a gift for writing and began in the 1880s with literary journalism in Berlin, which, except for short periods spent in Königsberg and Dresden, was henceforth the focus of his activity. Sudermann's first publications were works of fiction, the volumes of Novellen *Im Zwielicht* (1887) and *Geschwister* (1888), and the novels *Frau Sorge* (q.v., 1887) and *Der Katzensteg* (q.v., 1889).

In 1889, with the first performance of *Die Ehre* (q.v.), Sudermann became at a stroke a celebrated dramatist. The ensuing plays *Sodoms Ende* (q.v., 1890) and *Heimat* (q.v., 1893) established his reputation as a skilled and daring exponent of the new Naturalistic manner (see NATURALISMUS). There followed other realistic plays including the comedy *Schmetterlingsschlacht* (q.v., 1895) and the Schauspiel *Das Glück im Winkel* (q.v., 1896). The cycle of one-act plays *Morituri* (q.v.) published in 1896, contains *Fritzchen*, which is perhaps his best dramatic work. As Naturalism lost ground, Sudermann turned to symbolism in *Die drei Reiherfedern* (q.v., 1898) and to biblical tragedy in *Johannes* (q.v., 1898). *Es lebe das Leben* (1902), *Stein unter Steinen* (1905), *Das Blumenboot* (1906), and the verse tragedy *Der Bettler von Syrakus* (1911) were less successful. He continued to write plays until the end of his life, though none achieved the impact of his early works (*Der gute Ruf*, 1913; *Die Lobgesänge des Claudian*, 1914; the cyclical *Die entgötterte Welt*, 1915; *Die Raschhoffer*, 1919; *Das deutsche Schicksal*, 1921, another cycle; and *Wie die Träumenden*, 1922). Though Sudermann's reputation as an advanced and original playwright quickly declined, his craftsmanship and theatrical economy were of high quality.

Sudermann published occasional novels (*Es war*, 1894; *Das hohe Lied*, 1908; *Der tolle Professor*, q.v., 1926; and *Die Frau des Steffen Tromholt*,

1927) and these have lasted better than the plays. They are more clearly involved in genuine experience and they strain less after effect. Many consider that his best work is found in the stories of *Litauische Geschichten* (q.v., 1917) centred on his home country by the Memel (see HEIMAT-KUNST). He published in 1922 a volume of recollections *Das Bilderbuch meiner Jugend*. His collected fiction was published in 1919 (*Romane und Novellen*, 6 vols.) and his plays in 1923 (*Dramatische Werke*, 6 vols.).

Sultanstochter im Blumengarten, an anonymous Middle High German verse legend which is preserved in a 15th-c. MS., but was possibly written as early as the first half of the 14th c.

It tells of the daughter of the Sultan of Babylon, who, in admiration of the beauty of a garden, is seized with adoration of Him who made the flowers. She is carried off by an angel (and so preserved from a detested marriage) and brought to a nunnery. She becomes the abbess and after thirty years dies and joins the heavenly bridegroom in Paradise. An injunction to chastity is contained in the poem, which, it is thought, may have been intended as reading matter for nuns.

SULZER, JOHANN GEORG (Winterthur, Switzerland, 1720–79, Berlin), originally intended for the ministry, became a private tutor, and in 1747 a teacher at the Joachimsthalsches Gymnasium, Berlin. Elected a member of the Prussian Academy in 1750, he enjoyed the special favour of Friedrich II (q.v.); in 1775 he was appointed director of the Philosophical Section of the Academy.

Sulzer acquired a considerable reputation as a writer on aesthetics, in which his outstanding work is *Allgemeine Theorie der schönen Künste und Wissenschaften* (1771–4, posthumously supplemented, 1792–9, 4 vols., reissued 1967–9). He saw the arts as a means of education, while conceiving beauty as a prerequisite of art; beauty, by its appeal to the senses and the emotions, aids the intellectual process of learning, the acquisition and comprehension of knowledge. He thus formulated views on aesthetics which were derived from his teacher Bodmer (q.v.). Published as the Sturm und Drang (q.v) burst upon Germany, they met with as much criticism as approval and emerged as a formative component of German classicism.

A story, *Damon oder die platonische Liebe,* was published by Bodmer in *Pygmalion* (1749), but Sulzer remained a theorist rather than a creative writer. Posthumous publications include *Nachträge oder Charakteristik der vornehmsten Dichter aller Nationen* (8 vols., 1792–1808). His autobiography, *Lebensbeschreibung von ihm selbst aufgesetzt,*

ed. J. B. Merian and F. Nicolai (q.v.), appeared in 1809.

Summa Theologiae, an Early Middle High German poem of some 300 lines, treating the doctrine of salvation from the creation of the angels to the Last Day. Its tone combines hymnlike and didactic elements; its matter is tightly packed. It is contained in the Vorauer Handschrift (q.v.) and was probably written *c.* 1125. The title was given to it by W. Scherer (q.v.).

Sündflut, Die, a play in five parts ('Drama in fünf Teilen') by Ernst Barlach (q.v.), published in 1924. It enacts a dialectical variation of the Old Testament story. God, appearing both as 'vornehmer Reisender' and 'Bettler', regrets the creation of mankind, which ignores his will. He decides to annihilate his creation by flooding the earth, but advises Noah, whom he visits as a beggar, to abandon his possessions and build an ark in the mountains. Noah obeys with some hesitation after Awah, who has been touched by the angels, has a vision confirming the truth of the prophecy. While Noah and his sons Japhet, Sem, and Ham carry out the work, Calan, defying Jehovah, rises to power in the valleys below and threatens to burn the ark. As the flood rises, Calan, dispossessed, can no longer implement his threat. In death he loses his human shape and experiences oneness with his own god, a god without form and voice who, by affirming his creation, represents beyond the concepts of good and evil a continuous process of rebirth. In the ark Noah depends for his survival on his blind trust in Jehovah. The play closes when the waters reach the ark. The open ending adds to the ambiguity of Barlach's treatment of the biblical subject, which is indebted to German mysticism.

SUONEGGE (SUNECK), DER VON, a minor Middle High German poet of the 13th c., who lived in Styria, is the author of three poems after the manner of GOTTFRIED von Neifen (q.v.).

Surrealismus (*surréalisme*), a common term in painting, owes its literary authentication to A. Breton's manifesto (1924); its approach to the irrational and its illogical and even absurd forms of expression are greatly indebted to S. Freud (q.v.). In German literature F. Kafka (q.v.) is, ahead of his time, an outstanding exponent of the new literature of dreams and nightmares, which is largely surrealist. Writers adopting this style in some of their works include A. Döblin, H. Hesse, H. Kasack, and H. E. Nossack (qq.v.). In drama, surrealism forms an aspect of Expressionism (see EXPRESSIONISMUS) and may be extended to later writers (e.g.

Frisch's *Biedermann und die Brandstifter*, q.v.); in poetry it is associated with symbolism (see SYMBOLISMUS) on the one hand, and with visual and concrete poetry (see KONKRETE POESIE), on the other; notable exponents are H. Arp, G. Benn, and A. Stramm (qq.v.).

SUSO, Latin name for Heinrich Seuse (q.v.).

SÜSSKIND VON TRIMBERG, a minor Middle High German poet of the 13th c., is the author of Sprüche (see SPRUCH) touching on the vanity of the world and the imminence of death, and also of a powerful poem portraying the misery of his own domestic life, beset with cares, in the course of which he speaks of setting out humbly like an old Jew. On the basis of this poem and the superscription 'Süßkind, der Jude von Trimberg', the poet was formerly believed to be a Jew, though this is now thought to be improbable. Süßkind was a strolling singer who came from the district of Bamberg, but beyond that nothing is known of him. He is the subject of a novel (*Süßkind von Trimberg*, 1972) by F. Torberg (q.v.).

SÜSS-OPPENHEIMER, JOSEF (Heidelberg, *c.* 1698–1738, Stuttgart, by execution), was financial aide to Duke Karl Alexander (reigned 1733–7) of Württemberg, whose extravagant mode of living, including especially the building of palaces (e.g. Ludwigsburg), led to a continually increasing need for money. Süß satisfied the Duke's financial needs, and simultaneously amassed a fortune, by the sale of offices of state, oppressive taxation, and extortion. After Karl Alexander's death he was arrested, convicted of treason, and hanged amid scenes of popular rejoicing. His story is told in narrative by W. Hauff (*Jud Süß*, q.v., 1828), L. Feuchtwanger (*Jud Süß*, q.v., 1925), who finds mitigating circumstances, and in dramatic form by P. Kornfeld (q.v., *Jud Süß*, 1930) and E. Ortner (q.v., *Jud Süß*, 1933). Other treatments exist.

SUTTNER, BERTA, FREIFRAU VON (Prague 1843–1914, Vienna), was the daughter of Count Kinsky, a field-marshal of ancient Bohemian descent. She married in 1876 Baron A. von Suttner, an engineer, whom she accompanied on his professional travels, spending ten years with him in Tiflis, Caucasus; he died in 1902. In 1891 she founded an Austrian pacifist movement and worked for this cause throughout her life. In 1905 she was awarded the Nobel Peace Prize, which she had encouraged Nobel to found thirteen years previously.

Berta von Suttner is the author of a number of novels, of which the best known is *Die Waffen nieder!* (2 vols., 1889). She used the same title for a pacifist journal which she published from 1894 to 1900. *Marthas Kinder* (1903) is a sequel to the novel. A collection of essays, *Der Kampf um die Vermeidung des Weltkrieges* (2 vols.), ed. A. H. Fried, was published in 1917; *Gesammelte Schriften* (12 vols.) appeared 1906–7.

SWIETEN, GOTTFRIED, FREIHERR VAN (Leyden, 1734–1803, Vienna), son of the physician to the Empress Maria Theresia (q.v.), was of Dutch descent, but lived his adolescent and adult years in Vienna. From 1777 on he was in charge of the Court (now National) Library and from 1781 also chief Censor of Books. Swieten wrote the libretti for *Die Schöpfung* (1798), based on Milton's *Paradise Lost*, Bk. VII, and *Die Jahreszeiten* (1801), which he invited Joseph Haydn (q.v.) to compose. Beethoven (q.v.) dedicated his First Symphony to him.

Swiss Family Robinson, The, see WYSS, JOHANN DAVID.

SYLVA, see CARMEN SYLVA.

SYLVIUS, PETRUS (Forst, Saxony, *c.* 1470–1547?), a priest who opposed Luther (q.v.) in a number of polemical tracts (*Büchlein*).

Symbolismus, denotes 'pure poetry' (*poésie pure*), consciously detached in style and content from reality. Commonly recognized German influences nurturing this initially French (S. Mallarmé) and English (E. A. Poe) approach to lyric poetry during the 1880s and 1890s are Novalis, E. T. A. Hoffmann, the aesthetic writings of Schopenhauer, C. F. Meyer, and R. Wagner (qq.v.). S. George cultivated symbolism in close contact with the French symbolists; the early style of H. von Hofmannsthal and Rilke (qq.v.) should similarly be noted; symbolism was an aspect of neo-Romanticism (see NEUROMANTIK) and merged with surrealism (see SURREALISMUS).

SZABO, WILHELM (Vienna, 1901–), for many years a schoolmaster in Lower Austria, lived by his pen from 1939 to 1945, resumed teaching, and held administrative posts in education in Weitra in Lower Austria. A lyric poet, he published volumes of verse reflecting village life and rural scenes, including *Das fremde Dorf* (1933), *Im Dunkel der Dörfer* (1940), *Der Unbefehligte* (1947), *Das Herz in der Kelter* (1954), and *Landnacht* (1966).

T

Tabulatur. The code of rules which rigidly governed the composition of Meistergesang (q.v.); it was so called because the rules were set out on a board or 'Tafel' (tabula).

TACITUS, see GERMANIA.

Tagebuch im Kriege, definitive title (1935) of the book of war reminiscences by H. Carossa (q.v.), first published in 1924 as *Rumänisches Tagebuch.* Its background is the Romanian campaign of 1916 and it covers the period October to December.

Tage der Kommune, Die, a play in 14 scenes with three songs by B. Brecht (q.v.) with music by Hanns Eisler (q.v.), written in Zurich in 1948–9 and based on *Die Niederlage (Nederlaget,* 1937) by the Norwegian playwright Nordahl Grieg, who was killed during the 1939–45 War (*Versuche,* vol. 15, Vorspruch). Two of the songs were written in 1934–5 and included in the *Svendborger Gedichte* (1939). The play was first performed in November 1956 at Karl Marx-Stadt (Chemnitz) and has undergone no final revision. It was published posthumously in 1957.

The central place of action is Paris from January to May 1871, the month in which the French Republican government under Thiers ratified the peace ending the Franco-Prussian War (see DEUTSCH-FRANZÖSISCHER KRIEG). The 'days of the Commune' are mainly depicted in March, the last two scenes referring to the 'blood-week' in May in which the government succeeded in crushing the revolt of the proletarians. Only scene 10 is set outside France. It takes place in the Frankfurt opera house. During the performance of Bellini's *Norma* Jules Favre and Bismarck (q.v.) discuss the war indemnities. Bismarck generously assures Favre that these need not be paid until the 'pacification' of Paris and reminds him that the first cheque should be paid into his private account. Scene 13 shows the heroic resistance of the few remaining members of the Commune at the street barricades, ending with the sacrifice of the teacher Geneviève Guéricault, holding the red flag, with the words 'Long live . . .' (i.e. the Commune).

Tagelied or dawn song, term applied to Middle High German poems which reflect in dialogue the parting of lovers at dawn. Tagelieder occur in Provençal poetry under the designation *alba,* and the German forms show some traces of derivation from this source. The dawn song was, however, current in European and oriental poetry. The earliest German Tagelieder are a poem attributed to DIETMAR von Eist and an untypical example by HEINRICH von Morungen (qq.v.). In the former the song of a bird leads to the awakening, in Morungen's poem it is the light of day. The characteristic figure of the Tagelied is later the watchman (Wächter), who by his horn or his voice rouses the lovers from their ecstasy and insists upon their parting. The most notable author of Tagelieder is WOLFRAM von Eschenbach (q.v.), with whom the watchman develops into a symbol of the forces opposing the illicit lovers, as in the poem beginning 'Sîne klâwen/durch die wolken sint geslagen' (Seine Klauen durch die Wolken haben geschlagen).

The form represents the only permitted intrusion of the sensual realization of love into the tenuous web of erotic stylization, which constitutes Minnedienst. Even this element of reality in the Tagelied is, however, a convention; the consummation is fictitious. The view has been expressed that the Tagelied represents a safety-valve for the stresses implicit in the Minnedienst relationship.

From the time of Wolfram to the 15th c. Tagelieder are a widely acknowledged form of lyric poetry. The social level of the form tends eventually to sink, and burlesque elements intrude. Among the more notable poets of Tagelieder after Wolfram are OTTO von Botenlauben, ULRICH von Singenberg, ULRICH von Winterstetten, KONRAD von Würzburg, Hadlaub, Steinmar, and OSWALD von Wolkenstein (qq.v.). An example also occurs in *Carmina Burana* (q.v.).

Tägliche Brot, Das, a 2-vol. novel of servant life by C. Viebig (q.v.), published in 1902. Mine, one of its characters, appears again in the later novel *Eine Handvoll Erde* (q.v.).

Tag- und Jahreshefte, autobiographical notes covering briefly Goethe's life, and especially his literary activity, up to 1822. Written mainly between 1822 and 1825, they were first published in vols. 31 and 32 of the *Ausgabe letzter Hand.* They are sometimes referred to as *Annalen,* a term occasionally used for them by Goethe himself.

TALANDER, pseudonym of August Bohse (q.v.).

Talisman, Der, a farce (Posse mit Gesang) by
J. N. Nestroy (q.v.), written in 1840, first performed in the Theater an der Wien, Vienna, in
December 1840, and published in 1843. The
plot turns on the mop of red hair of the hero,
Titus Feuerfuchs, which alienates the women he
meets. He comes into possession of a black wig
(the talisman of the title) and is immediately
courted, but loses his chances when the red hair
is rediscovered. In the end he marries an
attractive and homely redhead. The play, based
on a French *vaudeville*, is generally considered
Nestroy's masterpiece.

Tannenberg, a small town in the centre of East
Prussia (see OSTPREUSSEN), near which was
fought a battle (27–9 August 1914) resulting in
the destruction of the invading Russian army
(Samsonov) by the numerically inferior German
8th Army under its new commander Hindenburg with Ludendorff (qq.v.) as his chief of
staff. By this action and the battle of the
Masurian Lakes (see MASUREN) a few days later,
East Prussia was cleared of invaders. A grandiose
monument (Tannenberg-Nationaldenkmal) was
erected in 1927 and destroyed in 1945.

Tannengesellschaft, one of many language
societies (Sprachgesellschaften) formed on the
model of the Fruchtbringende Gesellschaft
(q.v.). Its exact title was Aufrichtige Gesellschaft
von der Tannen, and it was founded in 1633 in
Strasburg. Its most notable member was G. R.
Weckherlin (q.v.).

TANNHÄUSER, DER (*c.* 1200–*c.* 1270), was a
Middle High German lyric poet whose rank is
obscure. He is referred to in MSS. as 'der Tann-
hûser', which suggests a commoner, but some
have held him to be of petty nobility from
Tannhausen near Neumarkt in Franconia. He
took part in a crusade, probably in 1228–9 and
was later court poet with Friedrich der Streitbare (q.v.) of Austria, who rewarded him with
land. He appears to have squandered his assets
and after Friedrich's death led an unsettled life.

Tannhäuser's poetry marks historically a distinct phase in the decline of the concept of
minne, but his work has a freshness and vigour
which suggest rather an emancipation. His
poetry reveals humour and irony, together with
an alert sense of parody, which were necessarily
absent from Minnesang (q.v.) proper. His surviving poems probably represent only a small
proportion of his output. His love-songs are
Arcadian in tone, with a direct sensuality remote
from obscenity. He is one of the principal practitioners of the *leich* (q.v.) in the form of dance
songs, and he wrote a crusade song which dwells
with ironical humour on the hardships of a sea
crossing.

Tannhäuser, perhaps because of his years in
the Near East and the frankness of his love
poetry, became a legendary figure, as the knight
who visited Venus's grotto (see VENUSBERG),
repented, and sought absolution from the Pope.
In this form he is commemorated in the *Tannhäuserlied* (recorded in 1515), which reappears as
Der Tannhäuser in *Des Knaben Wunderhorn* (q.v.,
1805–8) and later, in combination with the
Wartburgkrieg (q.v.), in Wagner's *Tannhäuser und
der Sängerkrieg auf der Wartburg* (q.v.).

Tannhäuser's poems were edited by J. Siebert
(1934).

*Tannhäuser und der Sängerkrieg auf der
Wartburg,* an opera (Romantische Oper) in
three acts by R. Wagner (q.v.), the text of which
(in rhyming verse) he wrote in 1842–3. The
musical composition was completed in 1844. It
was first performed in Dresden on 19 October
1845. Wagner combines a legend of Tannhäuser
contained in a folk-song with the story of the
Wartburgkrieg (q.v.).

Tannhäuser is shown abandoned to sensual
enjoyment in the grotto of Venus (see VENUS-
BERG). He desires to return to normal life, and,
after leaving Venus, encounters Hermann
(q.v.), Landgrave of Thuringia, with minstrels,
including WOLFRAM von Eschenbach (q.v.). The
second act is occupied by the contest among the
singers, in which the Landgrave's daughter
Elisabeth is to reward the victor. Wolfram
praises spiritual love, but Tannhäuser allows
himself to be carried away in a passionate paean
in praise of 'Frau Venus', and the act ends
with his disgrace and banishment. Elisabeth,
who has loved Tannhäuser, dies. Tannhäuser
returns from a vain pilgrimage to Rome with his
sins still upon his head. Elisabeth's death not
only saves him from a return to the grotto of
Venus but brings him absolution, for messengers
arrive from Rome with news of a miracle (the
pilgrim's staff has put out leaves) proving that
Tannhäuser's sins are forgiven him.

Apart from the Venus music and Tannhäuser's song in praise of love, the most famous
numbers are Elisabeth's 'Dich, teure Halle,
grüße ich wieder' (Act II) and Wolfram's 'O du,
mein holder Abendstern' (Act III). Wagner
revised the opera for a performance in Paris in
1861 and preferred this version (Pariser Fassung). The original form of the opera (Dresdener
Fassung) is, however, still performed.

Tanz, Der, a poem in elegiac metre by Schiller,
first published in the *Musenalmanach* for 1796. It
was later substantially revised. The dance is
treated as a symbol of the order and complexity
of the universe.

Tarnhelm, the 'helmet of darkness' in Wagner's *Der Ring des Nibelungen* (q.v.). It is fashioned by Mime from the Rhine gold, seized by Alberich, and later used by Siegfried. It not only causes invisibility but also enables its wearer to change his shape.

Tarnkappe, the 'cloak of darkness' which, in the *Nibelungenlied* (q.v.), Siegfried obtains by force from the dwarf Alberich (3. Âventiure). He later uses it in assisting Gunther to win the contest of strength over Brünhild in Iceland (7. Âventiure) and in subduing her after Gunther has failed to tame her on the wedding night (10. Âventiure). Except for Wagner (see TARNHELM), all major treatments of the subject have adapted this device.

TARNOW, FRANZISKA (Güstrow, 1779–1862, Dessau), who wrote and was best known under the name Fanny Tarnow, was a governess but, after publishing her first novel *Natalie, ein Beitrag zur Geschichte des weiblichen Herzens* (1811), devoted herself to literature. Among her numerous novels *Lilien* (1821–3) and *Lebensbilder* (1824) were the most successful. Her selected works (*Ausgewählte Schriften,* 1830) occupy 15 vols.

Tat des Dietrich Stobäus, Die, a novel by M. Halbe (q.v.), published in 1911. The narrator of the frame (see RAHMEN) purports to have discovered in the 1890s the Stobäus papers of 1859–62. The story proper is told in the first person by Dietrich Stobäus, a Danzig patrician, and centres on his passion for the young actress Karola, who repeatedly deserts him and each time returns. She refuses to marry him and falls from the cliff edge into the sea. She drowns, and Stobäus is accused of her murder; he is acquitted, but the ending suggests that not all circumstances leading to her death have been revealed. A feature of the book is the atmospheric description of the landscape around Danzig.

TATIAN, 2nd c., a Mesopotamian, whose gospel harmony (see EVANGELIENHARMONIE), originally written in Syriac, was translated into Old High German from a Latin version in Fulda, which is said to have been brought from Italy by St. BONIFACE (q.v.). This fusion of the Gospels into one continuous narrative, made in Fulda *c.* 830, is one of the oldest examples of biblical translation into German. The MS. is in the St. Gall library.

TAU, MAX (Beuthen, 1897–), a publisher's reader, emigrated to Norway in 1938, taking refuge in 1940 in Sweden. He returned to Norway in 1945. He wrote mainly for the cause of peace, both in novels (*Glaube an Menschen,* 1948,

first published in Norwegian, and *Denn über uns ist der Himmel,* 1955) and in essays (*Albert Schweitzer und der Friede,* 1955, a collection). His recollections and convictions are recorded in the autobiographical works *Das Land, das ich verlassen mußte* (1961), *Ein Flüchtling findet sein Land* (1964), and *Auf dem Weg zur Versöhnung* (1968). In 1950 he received the Peace Prize (Friedenspreis des deutschen Buchhandels, q.v.).

TAUBE, OTTO, FREIHERR VON (Reval, 1879– 1973, Tutzing), poet, novelist, and art historian, travelled widely in his youth, and before 1914 was for a time attached to the Goethe-Museum at Weimar. He was a friend of H. von Hofmannsthal and of R. A. Schröder (qq.v.). In his first phase an aesthete, he expressed in his later work strong Protestant convictions.

Taube's many volumes of verse include *Verse* (1907), *Gedichte und Szenen* (1908), *Neue Gedichte* (1911), *Wanderlieder* (1937), *Vom Ufer, da wir abgestoßen* (1947), *Lob der Schöpfung* (1954), *Selig sind die Friedbereiter* (1956), and *Goldene Tage* (1959). His fiction embraced novels (*Der verborgene Herbst,* 1913; *Die Löwenpranke,* 1921; *Das Opferfest,* 1926; *Die Metzgerpost,* 1936; *Der Minotaurus,* 1959) and volumes of stories (*Adele und der Dichter,* 1919; *Baltischer Adel,* 1932; *Das Ende der Königsmarcks,* 1937; and *Dr. Alltags phantastische Aufzeichnungen,* 1951). Separately published stories are *Der Hausgeist* (1931) and *Die Wassermusik* (1948). One volume of *Ausgewählte Werke* appeared in 1959.

Tauben im Gras, a novel by W. Koeppen (q.v.), published in 1951. It is set in Munich in 1948 in the chaotic conditions which followed the currency reform. The action, occupying only twenty-four hours, intertwines strands of human relationships to emphasize the disorientation of a disintegrating society and to express a sympathetic view of the racial question posed by the presence of American Negro soldiers.

TAUBMANN, FRIEDRICH (Wonsee, 1565–1613, Bayreuth), the son of a shoemaker, was educated at the expense of the Margrave of Ansbach, and studied at Wittenberg, where he later became a professor. His ready wit made him welcome at the Saxon court in Dresden, where he filled the role of a learned 'court fool' or 'Kurtzweiliger Rath'. His anecdotes and witticisms appeared in 1579 as *Melodaesia* and were later renewed as *Taubmanniana* (1702).

Taucher, Der, Schiller's most famous ballad, written in 1797 and published in *Der Musenalmanach für das Jahr 1798.* It recounts the successful dive of a page into a whirlpool to retrieve a golden goblet. Challenged to repeat the feat for the hand of a king's daughter, he accepts, plunges in, and is not seen again. Much of the fame of the poem is due to the vivid rendering of the raging waters.

Taufgelöbnisse (baptismal vows), formal professions of faith and intention made in answer to a prescribed series of questions. They are divided into an abjuration of the Devil and his works (*Abrenuntiatio diaboli*) and a declaration of faith (*Confessio fidei*). The original language is Latin, but two Old High German versions (*Sächsisches Taufgelöbnis* and *Fränkisches Taufgelöbnis*) are important linguistic and historical documents. The Saxon formula belongs to Charlemagne's missionary wars against the North Germans (Saxons, see KARL I). It is intended for the conversion of adults and was made in Mainz not long after 775. The MS. is in the Vatican. The *Fränkisches Taufgelöbnis*, MS. in Merseburg, is a formula for the baptism of infants in an already converted community.

Taugenichts, see AUS DEM LEBEN EINES TAUGENICHTS.

TAULER, JOHANNES (Strasburg, *c.* 1300–61, Strasburg), Dominican monk and noted preacher, entered the monastery of his order in Strasburg. Evicted with his brother-monks in 1339, he spent eight years in Basel in a circle of the devout called Gottesfreunde. His last fourteen years were spent in his native city. Tauler, though an admirer of Meister ECKHART (q.v.), can hardly be called a disciple; for he had little interest in the intellectual processes of Eckhart's thought. He preached a simple piety, the love of God, and the need for humility. His religious interest is concentrated on the conversion of men to a better way of life rather than on metaphysical speculation. Some eighty sermons by Tauler survive, and though these are written down by hearers, chiefly nuns, it is possible that some have been checked by the preacher himself.

Tauler's sermons (*Predigten*) were edited by F. Vetter (1910) and by G. Hofmann (1961).

Tauroggen, see YORCK, H. D. L. VON.

Tausendundeine Nacht, the *Thousand and One Nights* or *Arabian Nights*, a collection of oriental tales, some of which date from the 8th c. Scheherazade, the King's wife for a night, tells the stories night by night in order to postpone her execution, fixed each day for the morrow. In the end the King is won over and grants her her life. It is one of the earliest examples of a series of stories with a narrative frame (see RAHMEN).

The work became known in Europe through the French translation of A. Galland (*Les mille et une nuits*, 12 vols., 1704–17). The standard German translation, which goes back to the original, is by E. Littmann (*Die Erzählungen aus den Tausendundein Nächten* (6 vols., 1921–8; 4th edn., 1960) with an introduction by H. von Hofmannsthal (q.v.).

TAVEL, RUDOLF VON (Berne, 1866–1934, Berne), of a family prominent in Berne for many generations, was a newspaper and periodical editor and the author of fiction written in Bernese dialect and dealing with the Bernese past, its people, and its customs. His collections of stories include *Ja gäll, so geit's* (1903) and *Geschichten aus dem Berner Land* (1934), and his stories *D'Frou Kätheli und ihri Buebe* (1910), *D'Haselmuus* (1922), and *Unspunne* (1924); his novels are *Veterane-Zyt* (1927), *Der Frondeur* (1929), and *Meischter und Ritter* (1933). Special mention should be made of his comedy *Di gfreutischti Frou* (1923). *Gedanken. Aus Werk und Werkstatt des Erzählers* (1937) is a posthumous publication.

TAYLOR, GEORGE, pseudonym of A. Hausrath (q.v.).

Technik des Dramas, Die, a treatise by G. Freytag (q.v.), published in 1863. An eclectic and unoriginal summary of views current in the middle of the 19th c., it was regarded for a time as an authoritative manual. After considering the Ancients and Shakespeare, Freytag concerns himself with elevated verse tragedy, which he regards as the norm.

Tegernseer Antichrist, see LUDUS DE ANTICHRISTO.

TELL, WILHELM, Swiss national hero, whose exploits, including the bowshot which cleaves the apple on his son's head and the near-miraculous escape from Geßler's barge, belong to the sphere of legend and are attributed to the 14th c. Tell's story is faithfully reproduced in Schiller's play *Wilhelm Tell* (q.v., 1804). The earliest source of the story is *Das Weiße Buch von Sarnen* (*c.* 1470), but the most celebrated of the older versions is the chronicle of Ägidius Tschudi (q.v.), which was used by Schiller. The story has often been poetically exploited. There are two early Tell ballads from the 15th c. and 16th c., the folk drama *Das Urner Tellenspiel* (*c.* 1511), Schiller's play, G. Rossini's opera *Guglielmo Tell* (1829), and *Wilhelm Tell für die Schule* by M. Frisch (q.v., 1971). The legend was given a historical blessing by Johannes von Müller (q.v.) in *Geschichten schweizerischer Eidgenossenschaft* (1786–1808), but its authenticity was soon questioned.

TEMME, Hubertus (Lette, Westphalia, 1798–1881, Zurich), a lawyer in Prussian government service, became in 1852 a professor of law at Zurich University. He wrote novels and stories of crime and detection, notably *Berliner Polizei- und Kriminalgeschichten* (1858–60).

TEMPLIN, Prokop von, see Prokop von Templin.

Tendenzdichtung, literature designed to further ideas, usually social or political.

Teppich des Lebens, Der, a volume of poetry by S. George (q.v.), published in 1899.

TERSTEEGEN, Gerhard, also ter Steegen (Moers, 1697–1769, Mülheim, Ruhr), a Pietist (see Pietismus) and poet of humble origin, worked his way up as a weaver and merchant before becoming an itinerant preacher in the Lower Rhine district. His publications include *Auserlesene Lebensbeschreibungen heiliger Seelen* (3 vols., 1733–53) and sermons, *Geistliche Brosamen* (2 vols., partly posthumous, 1769–73). He wrote more than a hundred hymns, some of which have remained in regular use in the Protestant Church; 'Gott ist gegenwärtig' and 'Ich bete an die Macht der Liebe' are notable examples contained in his collection *Geistliches Blumengärtlein inniger Seelen* (1729).

Terzinen, German term for the Italian verse form *terza rima*, the metre of Dante's *Divina Commedia*. Terzinen are composed of groups of three lines of eleven syllables with a simple rhyme scheme which permits their indefinite extension (aba, bcb, cdc, etc., closing at the desired moment with a single line rhyming with the middle line of the preceding group (pqp q). They were first introduced into German by Paul Melissus (q.v.) in his translation of the 37th Psalm (1572), but did not achieve frequent use until the Romantic movement (see Romantik), which inclined to exotic measures. 'Terzinen' were written by the brothers A. W. and F. Schlegel, L. Tieck, Chamisso, Rückert, and Platen (qq.v.). A passage of Goethe's *Faust* (q.v., Pt. 2, opening speech of Act I) is written in this form, and so, too, is his poem *Bei Betrachtung von Schillers Schädel* (see 'Im ernsten Beinhaus war's'). In the mid-19th c. *terza rima* was used by Herwegh and P. Heyse and later by S. George and H. von Hofmannsthal (qq.v.).

Terzinen über Vergänglichkeit, a cycle of three poems by H. von Hofmannsthal (q.v.). The third poem begins with Prospero's 'We are such stuff as dreams are made on' translated as 'Wir sind aus solchem Zeug wie das zu Träumen'. (See Terzinen.)

Teuerdank, an allegorical poem written by Melchior Pfintzing (q.v.) according to instructions and plans prescribed by the Emperor Maximilian I (q.v.), and published in 1517. Its basic theme is Maximilian's journey to the Netherlands to join his bride. Treated in terms of a chivalric romance, the poem sets out eighty adventures successfully withstood by the hero, Teuerdank; his three adversaries, Fürwittig, Unfalo, and Neydelhart, seek in various ways, especially on hunting expeditions, to lure Teuerdank to his death. In the end their treachery is exposed and they are executed.

The construction is repetitive and the verse wooden, but the work enjoyed considerable success, partly owing to Maximilian's popularity, and partly because it was illustrated with woodcuts by Hanns Schäufelein (c. 1480–c. 1540).

Teufel, Der, a historical novel by A. Neumann (q.v.), published in 1926. It is set in France and Flanders in the late 15th c. during the reign of Louis XI. 'The Devil' is the political intriguer Olivier Necker of Ghent, who becomes the king's principal partner and instrument in the manipulations and crimes, of which the murder of the heir apparent, Louis's brother Charles, is the culmination. Though this takes place by order of the King, it is done on Necker's advice. On Louis's death in 1483 Necker gives himself up and is condemned to death.

Teufel ist los, Der, a Singspiel (q.v.), based on C. Coffey's ballad opera *The Devil to pay or The wives metamorphosed* (1731), a work which owes much to Gay's *The Beggar's Opera* (1728). Translated into German in 1743 by C. W. von Borck (q.v.), it was successfully performed by the Schönemannsche Truppe, a company possessing the MS., which remained unpublished. In 1752 it was adapted by C. F. Weiße (q.v.) as *Die verschmähten Weiber oder Der Teufel ist los,* becoming generally known by its sub-title; this version, which was printed, drew full houses throughout Germany. In 1766 Weiße rewrote it, introducing new numbers and securing further success. The work contributed to establishing the Singspiel in Germany.

Teufelliteratur or **Teufelsliteratur,** term applied to tracts written mostly in the second half of the 16th c. in Protestant regions, attacking vices and foibles as the handiwork of a particular devil. The most important collection of Teufelliteratur is the *Theatrum Diabolorum*, published in 1569 by Sigmund Feyerabend (q.v.), and

Andreas Musculus (q.v.) is the best-known author.

Teufelsbeichte, Die, title given to a short, anonymous Middle High German poem of the 14th c. It tells how a devil seeks, like the human beings he sees, to confess and receive absolution. But he has to admit that he is incapable of repentance, and his yearning to recover lost goodness is therefore vain. The work is based on the *Dialogus de miraculis* of Cäsarius von Heisterbach (q.v.).

TEURE UNGENANNTE, DIE, Auguste Stolberg (q.v.), who in 1775 entered anonymously into correspondence with Goethe. His first reply is addressed to 'Der teuren Ungenannten'.

TEUTLEBEN, Kaspar von (?Jena, 1576–1629), at first a tutor and then master of ceremonies (Hofmarschall) at Weimar, was the moving spirit behind the foundation of the Fruchtbringende Gesellschaft (q.v.) in 1617. As a member of the society, he bore the name 'Der Mehlreiche'.

Teutoburger Wald, a range of wooded hills to the east of Münster, running north-west from Detmold towards Osnabrück. The southern portion was the scene of the defeat of the Romans under Varus in A.D. 9 (see Arminius).

Teutonic Order, see Deutscher Orden.

Teutsche Merkur, Der, a critical review founded by C. M. Wieland (q.v.) in 1773 and continuing until 1810. In its earlier years (1773–1800) under Wieland it had considerable and widespread influence, but after K. Bötticher (q.v.) took over the editorship in 1800 its reputation declined.

Teutschen Volkslieder, Die, an essay by J. J. von Görres (q.v.), published in 1807.

Teutschgesinnte Genossenschaft, see Zesen, P. and Sprachgesellschaften.

Text, (1) the generally accepted reference to an authentic text of a work of literature, letter, document, etc. It is a basic task of critical editions to establish the authenticity of texts.
(2) More frequently in the plural form of Texte, the word has been adopted by exponents of concrete and visual poetry (see Konkrete Poesie), extending its normal connotation to embrace an abstract approach to reading.

Thaddädl, a stock comic character in the Vien-

nese popular theatre at the end of the 18th c. created by A. Hasenhut (1766–1841).

Thaler, silver coin in common use from the 15th c. until the introduction of the Mark (1871–3). The value varied in the different German states, but the two principal coins were the Reichsthaler, which comprised 24 Groschen (q.v.), and the Speziesthaler, which comprised 32. Five Reichsthaler made one Friedrich d'or (q.v.) or Louis d'or. In rough reckoning the Reichsthaler was equivalent to 3 shillings in English currency.

Thalia, Die, a periodical founded by Schiller in the hope of financial reward and as a vehicle for his own publications. The intention was to publish every second month, but it appeared at irregular intervals. The first number, published in 1785 while Schiller was at Mannheim, bore the title *Die Rheinische Thalia*. This was altered in subsequent numbers to *Die Thalia* because Schiller had removed far from the Rhine. In 1792 the title was changed to *Die Neue Thalia*; the periodical ceased publication in 1793.
The first two acts of *Don Carlos* (q.v.) and a part of the third were published in instalments in *Die Rheinische Thalia* and *Die Thalia*. The fragment of a play by Schiller, *Der versöhnte Menschenfeind*, and the narratives *Der Verbrecher aus verlorener Ehre* and *Der Geisterseher* appeared in *Die Thalia*, and several of his philosophical works were first published in *Die Neue Thalia* (*Über den Grund des Vergnügens an tragischen Gegenständen*, *Über die tragische Kunst*, *Über Anmut und Würde*, and *Vom Erhabenen*). The fragment of Hölderlin's *Hyperion* (q.v.), published in 1794, is known as the 'Thalia-Fassung' (also 'Das Thalia-Fragment').

THARÄUS, Andreas, a Silesian pastor in the early 16th c., is the author of a didactic poem on the cultivation of flax and barley, in which the two crops engage in a dialogue (1609). He also wrote a comedy, *Der Weiberspiegel* (1628).

THEATERGRAF, Der see Hahn, K. F., Graf von.

Theatralische Bibliothek, Die, a periodical published in four issues by G. E. Lessing in 1754 with the collaboration of his friend F. Nicolai (qq.v.). Lessing's contributions consist of translations with brief introductions and essays; they include *Abhandlungen von dem weinerlichen oder rührenden Lustspiele*, *Leben des Herrn Jakob Thomson* (based on contemporary English sources), excerpts from the Spanish play *Virginia* by A. de Montiano y Luyando (based on French versions and connected with the genesis of *Emilia Galotti*, q.v.), and other essays on the Spanish, Italian, and English theatre (Dryden), as well as comments on Lessing's own play *Die Juden* (q.v.). It is an early instance of Lessing's preoccupation with original sources.

Théâtre Libre, a company inaugurated in Paris in 1887 by A.-L. Antoine (1858–1943) in order to perform Naturalistic plays and to develop appropriate acting techniques and settings. It continued until 1896. The Théâtre Libre visited Berlin in 1887 and its impact contributed to the founding of the Freie Bühne (q.v.).

Theodizee, a teleological term, revived by Leibniz (q.v., a tract published by him in 1710 bears the term as title) to demonstrate that the best of all worlds has an inherent tendency towards perfection. Thus he explains evil as a negative force, which, if seen in terms of imperfection, is instrumental in the recognition of the positive forces striving towards perfection. The term persisted in philosophical polemics (e.g. in Kant and Schopenhauer, qq.v.), and has also been adopted by literary criticism concerned with metaphysics (viz. the justification of God as opposed to his negation).

THEODORIC THE GREAT, Theoderich in German (Pannonia, *c.* 454–526), succeeded as king of the Ostrogoths (see OSTGOTEN) *c.* 473. Given command of armies by Zeno, the Byzantine emperor, he established himself as king of Italy by 493. He reigned for thirty-three years, during which his territories enjoyed prosperity and peace. Two crimes are imputed to Theodoric, the murder of Odoacer in 493, by which he established his rule in Italy, and the judicial murder of Boethius on suspicion of conspiracy *c.* 524.

His mausoleum still stands at Ravenna. He became a part of the early folklore of the Germanic lands, appearing in the guise of Dietrich von Bern (Verona, see DIETRICHSAGE) in medieval poetry.

Theologia Deutsch or **Teutsch,** commonly used designation of a Middle High German work written in the second half of the 14th c., probably by Heinrich von Bergen, a priest and warden (Kustos) of the house of the Teutonic Order (see DEUTSCHER ORDEN) in Frankfurt. The identity of the author was long unknown and the work was often styled *Der Frankforter*, because it was the work of a member of the Teutonic Order in Frankfurt. The attribution to Heinrich von Bergen dates from 1955. Luther (q.v.) thought highly of the book and published it in 1518 as *Eyn Deutsch Theologia*. It is not

systematic theology but a mystical work enjoining purity of heart and surrender to God's will.

Theses, 95, a series of statements condemning the sale of indulgences (see ABLASSKRAM) which Luther (q.v.) nailed to the door of the castle church in Wittenberg in 1517 in protest against the activities of the Dominican monk Tetzel, who was acting as agent in the trade in indulgences for the Archbishop of Mainz. Written in Latin, they take the form of a series of assertions and judgements rather than a coherent argument. They rapidly spread through Germany both in their original version and in a German translation.

Thidreksaga, also *Thidreks saga* or *Wilkina saga*, see DIETRICHSAGE.

THIESS, FRANK (Eluisenstein, Russian Livonia, 1890–1977, Darmstadt), served in the 1914–18 War, worked as a journalist, and from 1923 devoted himself to authorship. His early novels attracted attention by their preoccupation with delicate psychological problems; they include *Der Tod von Falern* (1921), *Angelika ten Swaart*, *Die Verdammten* (both 1923, dealing with incest), *Der Leibhaftige* (1924, linked with the inflation crisis of 1922–3), *Frauenraub* (1927, the eccentric story of a marriage, revised and retitled *Katharina Winter*, 1949), and *Johanna und Esther* (1933, revised as *Gäa*, 1957).

Thieß's later novels tend to be historical and include *Tsushima* (1936, treating the naval battle in the Russo-Japanese War), *Das Reich der Dämonen* (1941, sub-titled *Roman eines Jahrtausends*), *Die Straßen des Labyrinths* (1951) and its sequel *Geister werfen keinen Schatten* (1955), *Die griechischen Kaiser* (1959), and *Sturz nach oben* (1961). A novel in two parts, *Neapolitanische Legende* (1942) and *Caruso in Sorrent* (1946), is based on the life of the tenor Caruso. Erzählungen and Novellen include *Der Kampf mit dem Engel* (1925), *Narren* (1926), *Wir werden es nie wissen* (1949), *Tropische Dämmerung* (1951), and *Der schwarze Engel* (1966).

Thieß is the author of political and literary essays, contained in *Erziehung zur Freiheit* (1929) and *Die Zeit ist reif* (1932); *Der unbequeme Mensch* (1963), an essay on F. Hebbel (q.v.), was followed by *Dostojewskij: Realismus am Rande der Transzendenz* (1971). Thieß's recollections appeared as *Verbrannte Erde* (1963), *Freiheit bis Mitternacht* (1965), and *Jahre des Unheils* (1972). *Gesammelte Werke in Einzelausgaben* began to appear in 1956.

THIMIG, HELENE (Vienna, 1889–1974, Vienna), the wife and collaborator of Max Reinhardt (q.v.), was the daughter of the actor and director

Hugo Thimig (1854–1944), who was closely associated with the Burgtheater and the Theater in der Josephstadt in Vienna. Commencing her career as an actress in Meiningen (see MEI-NINGER), she came to prominence at the Deutsches Theater, Berlin, married Reinhardt in 1932, and worked from 1933 at the Theater in der Josephstadt. In 1938 she emigrated and collaborated with Reinhardt in his productions in the U.S.A. and in his foundation of a training centre for actors, first in Los Angeles and later in New York. Reinhardt died in exile and in 1946 she returned to Vienna, became a member of the Burgtheater, a professor at the Viennese Academy (Akademie für Musik und darstellende Kunst), and director of the Reinhardtseminar, re-establishing its reputation (1948–54 and again 1960). Her two younger brothers Hermann Thimig (1890–1982) and Hans Thimig (b. 1900) were also actors associated with the Burgtheater and with Reinhardt's school of acting, and in a number of plays all the Thimigs acted together; these included, among German plays, a farce by Nestroy (q.v.) and Schiller's *Kabale und Liebe* (q.v.), produced on a tour in the U.S.A. in 1927–8. Helene Thimig's own acting was exceptionally versatile; she made a considerable impression in the title role in Goethe's *Stella* (q.v.) in 1920, perfected prominent German classical, Shakespearian, and Shavian parts, and in 1958 performed in the German première of Rattigan's *Separate Tables* (1955).

The biography of Helene Thimig's father, ed. F. Hadamowsky (*Hugo Thimig erzählt von seinem Leben und dem Theater seiner Zeit*) appeared in 1962. As Helene Thimig-Reinhardt she is the author of *Wie Max Reinhardt lebte* (1973).

Third Reich, see DRITTES REICH.

THIRSIS, poetic name adopted by J. I. Pyra (q.v.) in *Thirsis und Damons freundschaftliche Lieder* (1745).

Thirty Years War, see DREISSIGJÄHRIGER KRIEG.

THOMA, LUDWIG (Oberammergau, 1867–1921, Rottach), a Bavarian lawyer, who turned to journalism, became in 1899 editor of *Simplicissimus* (q.v.). He wrote novels, stories, and plays with Bavarian rural settings. His serious novels include the anti-clerical *Andreas Vöst* (1906), *Der Wittiber* (1911) and *Der Ruepp* (1922), two works portraying the disruption and decline of a peasant family, and the story *Der Wilderer* (1903). The lighter rural novels *Altaich* (1918) and *Der Jägerloisl* (1921) exploit the contrast between town and country.

Thoma's most widely known work is the col-lection of humorous stories *Lausbubengeschichten* (1905), which has a scarcely less successful sequel *Tante Frieda* (1907); to this group also belongs *Kleinstadtgeschichten* (1908). Thoma was a popular writer of comedies, including *Witwen* (1900), *Die Medaille* (1901), *Moral* (1909), *Erster Klasse* (1910), *Lottens Geburtstag* (1911), and *Das Säuglingsheim* (1913); none, however, achieved more performances than *Die Lokalbahn* (1902). The tragic Volksstück *Magdalena* (1912) failed to evoke a comparable response. Thoma also published verse (*Grobheiten*, 1901, *Neue Grobheiten*, 1903, and *Kirchweih*, 1912). Memoirs include *Erinnerungen* (1919), *Stadelheimer Tagebuch*, and *Leute, die ich kannte* (both posth. 1923).

Thoma's *Gesammelte Werke* appeared in 7 vols. in 1922 and in extended form (8 vols.), ed. A. Knaus, in 1956.

THOMAS À KEMPIS, real name Thomas Hemerken (Kempen nr. Cologne, 1379 or 1380–1471, St. Agnetenberg, Holland), sometimes called Thomas von Kempen, was educated at a monastery at Deventer, Holland, and became a monk at the monastery at St. Agnetenberg near Zwolle, Holland, where he spent almost his whole life. He is internationally known through translations of the devotional book, written in medieval Latin, *De imitatione Christi*, which is generally, though not with complete certainty, attributed to him (published with German translation by P. Mons, 1959). He also wrote *De elevatione mentis* and a number of Latin hymns. *Opera omnia* (7 vols.), ed. M. J. Pohl, appeared in 1901–22.

THOMASIN VON CIRCLAERE or **ZER-CLAERE** (Cividale del Friuli, *c.* 1185–before 1238), was the Italian author of a Middle High German didactic poem, *Der Wälsche Gast*. Thomasin's appellation 'von Circlaere' is a transposition of his Italian style, Tommasino dei Cerchiari (a name which is variously spelt). He was a priest at Aquileia and belonged to the entourage of WOLFGER von Ellenbrechtskirchen (q.v.), the German bishop who became patriarch of Aquileia.

Thomasin wrote *Der Wälsche Gast* for an aristocratic German public as an exhortation and guide to a virtuous life. Its title refers to him, the Italian, whose German book makes him a guest among the Germans. He completed this poem of more than 14,000 lines in a (for him) foreign language in less than a year (1215–16). It is written in rhyming couplets and arranged in ten books. Of these the first has a special character, giving advice to children and young people, including hints on deportment for young women and on table manners for both sexes. In subsequent books Thomasin sets out his moral doctrine, of which *staete*, constancy of mind, is the centre, and condemns the evils resulting from *unstaete*. Linked with *staete* are *mâze*, which he conceives as moderation, *reht* (law-abidingness), and *milte* (generosity). His conception of society is aristocratic and stable; changes he views as deterioration. His political outlook, which is centred on religion, is a balanced one. An interesting passage of literary criticism in Bk. I admits the court epic as a means of education. In Bk. V he attacks the political Sprüche (see SPRUCH) of WALTHER von der Vogelweide (q.v.). The poem addresses itself directly and in homely tone to the reader, and, though not distinguished in style, is a remarkable achievement by a foreign writer.

Der Wälsche Gast (ed. H. Rückert, 1852) was reprinted in 1965 with additional apparatus by F. Neumann.

THOMASIUS, CHRISTIAN (Leipzig, 1655–1728, Halle/Saale), thinker and university teacher, was the son of a well-known professor of Leipzig, Jakob Thomasius, who in the 1660s had had Leibniz (q.v.) as a pupil. Christian Thomasius was appointed to lecture at Leipzig University in 1687 and was the first academic to use German instead of Latin as the teaching language in a lecture course. His unorthodox religious views led to persecution by senior theological colleagues and his licence to lecture was withdrawn in 1690. He was given asylum by the Elector of Brandenburg, and was appointed to teach in the Prussian Ritterakademie in Halle. When, in 1694, Halle University was founded, Thomasius was elected to the chair of law and he remained at Halle for the rest of his life. In 1709 he had the satisfaction of being offered a chair at Leipzig, but he declined it and in the following year became Rector of Halle University.

Thomasius is regarded as one of the progenitors of Enlightenment in Germany (see AUFKLÄRUNG), and he was certainly an enemy of prejudice and superstition, and an upholder of reason. He was a lifelong opponent of witch trials and burnings, against which he directed various tracts, consistently supporting the position of common sense. Some of his philosophical works were written in Latin (*Introductio ad philosophiam aulicam*, 1688, and *Introductio in philosophiam rationalem*, 1701), but most were written in German (*Einleitung zu der Vernunftlehre*; *Ausübung der Vernunftlehre*, both 1691; *Einleitung zur Sittenlehre*, 1692, reissued 1968; *Von der Arznei gegen die unvernünftige Liebe oder Ausübung der Sittenlehre*, 1696; and *Versuch vom Wesen des Geistes*, 1699). His principal legal treatise is *Fundamenta iuris naturae ac gentium* (1705, reissued 1963). Thomasius is one of the pioneers of the popular periodicals which became fashionable

in the early 18th c., founding and largely writing *Freimütige, lustige und ernsthafte, jedoch vernunft- und gesetzmäßige Gedanken oder Monatsgespräche über allerhand vornehmlich aber über neue Bücher* (1688–90).

Thomas von Kandelberg, a name disguising Thomas of Canterbury, i.e. St. Thomas à Becket; it is the title of a Middle High German poem. The saint is the hero of an anecdote, in which twelve scholars undertake each to bring a love token of his lady when next they meet. Thomas, the poorest and most pious, has no lady love but worships the Virgin Mary, who rewards him with a miraculous box which he exhibits as his token. In it is a handsome chasuble, symbolizing his future rise to episcopal authority. The author of the short poem (some 300 lines) is unknown; it is believed to have been written in the middle of the 13th c.

THOOR, JESSE (Berlin, 1905–52, Lienz, Austria), a political emigrant, wrote a volume of powerful religious sonnets (*Sonette*, 1948).

Throne stürzen, Die, title first given in 1951 to a group of three novels of recent history published by B. Brehm (q.v.): *Apis und Este* (1931), *Das war das Ende* (1932), and *Weder Kaiser noch König* (1933). The powerfully written and detailed novels cover the period 1903 to 1922.

Apis und Este begins with the assassination of King Alexander of Serbia and Queen Draga by members of the conspiracy led by Colonel Dimitrijević (known as Apis) and continues through the years of Austro-Serbian tension, culminating in the assassination of the Archduke Franz Ferdinand (q.v., von Österreich-Este, hence the Este of the title) and his consort at Sarajevo on 28 June 1914, and the ensuing hasty and undignified interment ('dritter Klasse') and mourning dictated by the view held in court circles that the Archduke had married beneath his rank. *Das war das Ende* opens with the death of the Emperor Franz Joseph (q.v.), and describes the attempts of his successor Karl to make a separate peace, the Treaty of Brest-Litovsk following the Russian Revolution, and the collapse of the central powers (involving the abdication of the German and Austrian emperors) before the allied offensives of the summer and autumn of 1918. *Weder Kaiser noch König* concerns the two unsuccessful attempts made in April and October 1921 by the ex-emperor Karl to regain his throne and his subsequent death in Madeira in 1922.

THÜMMEL, MORITZ AUGUST VON (Schönefeld nr. Leipzig, 1738–1817, Coburg), studied at Leipzig University, where he became friendly with C. F. Weiße, G. W. Rabener, and E. von Kleist (qq.v.). In 1761 he entered the administration of Saxe-Coburg, which he served for twenty-two years. His poetic narrative *Wilhelmine* (q.v., 1764) won him a reputation as an elegant and witty writer. In 1783 he retired and spent the rest of his life as a gentleman of means, travelling or on his estate. A journey to the south of France in 1785–6 provided him with material for a book after the manner of Sterne, *Reise in die mittäglichen Provinzen von Frankreich im Jahre 1785 bis 1786* (10 vols., 1791–1805). *Sämmtliche Werke* (7 vols.) appeared 1811–12.

THÜRING VON RINGOLTINGEN, see SCHÖNE MELUSINE.

THURN UND TAXIS, originally Taxis, a family of Spanish origin, ennobled in 1635 and granted the postal monopoly in the Holy Roman Empire (see DEUTSCHES REICH, ALTES) in 1649. The head of the house received the title Fürst in 1695. At first resident in Frankfurt, the family established itself in 1748 in Regensburg (q.v.). The postal monopoly continued in the German Confederation (see DEUTSCHER BUND), but was terminated in 1866. A Fürstin von Thurn und Taxis was prominent among the patrons of R. M. Rilke (q.v.).

TIAN, pseudonym of Karoline von Günderode (q.v.).

TIECK, LUDWIG (Berlin, 1773–1853, Berlin), a prominent and versatile representative of early Romanticism (see ROMANTIK), was the son of a master rope-maker. Educated in Berlin, he formed a close friendship with W. H. Wackenroder (q.v.). Tieck, who had wished to go on the stage, was prevailed upon to take the more respectable path of studying theology at Halle and Göttingen universities (1792–3), devoting himself, as time went on, more and more to the study of literature. In 1793 he spent a few months at Erlangen University with Wackenroder, after which he returned to Göttingen.

Tieck possessed from the beginning a facile pen, and wrote a number of stories while still a schoolboy. His first publication of note was a horror novel, *Abdallah, oder das furchtbare Opfer* (1795). This was quickly followed by the 3-vol. *Geschichte des Herrn William Lovell* (q.v., 1795–6), a story of the corruption of an innocent youth. *Peter Leberecht, eine Geschichte ohne Abenteuerlichkeiten* (q.v., 1795) is a partly humorous novel in the 18th-c. tradition. In the years 1795–8 Tieck contributed abundantly (for cash) to a kind of story magazine, *Die Straußfedern*, edited by F.

Nicolai (q.v.), demonstrating his almost fatal facility and fertility. His real interest at this stage was focused on the picturesque past, as he had seen it with Wackenroder in Nürnberg in the pictures of Dürer, and as it was manifested in fairy stories and chap-books. Taking from his own novel the pseudonym Peter Leberecht, he published *Volksmärchen herausgegeben von Peter Leberecht* (1797) in 3 vols.; it included versions of *Die Geschichte von den Heymonskindern* (see HAIMONSKINDER), *Die wundersame Liebesgeschichte von der schönen Magelone und des Grafen Peter aus der Provence* (see SCHÖNE MAGELONE), and the *Denkwürdige Geschichtschronik der Schildbürger* (see SCHILDBÜRGER), his early tragedy *Karl von Berneck* (1793–5), and his original tale *Der blonde Eckbert* (q.v., 1797, and see NOVELLE). The *Volksmärchen* contained two plays (Märchendramen), *Ritter Blaubart* and *Der gestiefelte Kater* (qq.v.), which appeared separately in the same year (1797). Together with Wackenroder, Tieck published the *Herzensergießungen eines kunstliebenden Klosterbruders* (q.v., 1797), and completed this productive year with a pot-boiler, the story *Die sieben Weiber des Blaubart*. In the following year he published the early idyll in dialogue form *Almansur*, the comedy *Die verkehrte Welt*, and the whimsical story *Merkwürdige Lebensgeschichte Sr. Majestät Abraham Tonelli*. These were overshadowed by his Künstlerroman (q.v.) *Franz Sternbalds Wanderungen* (q.v., 1798), which tells of a fictitious pupil of Dürer (q.v.).

Wackenroder died in December 1798, and Tieck commemorated him in a volume of essays partly drawn from Wackenroder's literary remains and partly written by himself, *Phantasien über die Kunst* (1799). In 1797 he had met Friedrich Schlegel (q.v.), and in 1799 he joined the Schlegels' circle in Jena. In this year he published *Der getreue Eckart und der Tannenhäuser*, *Prinz Zerbino* (another Märchendrama), and 2 vols. of *Romantische Dichtungen*, the first of which included *Leben und Tod des heiligen Genoveva* and the second *Leben und Tod des kleinen Rotkäppchens* and *Sehr wundersame Historie von der schönen Melusina* (see SCHÖNE MELUSINE). A translation of *Don Quixote* appeared in 1799–1800. In Jena he became a close friend of Novalis (q.v.). After a short stay in Berlin, Tieck moved to Dresden in 1801. There he pursued his artistic and medieval interests, publishing *Minnelieder aus dem Schwäbischen Zeitalter* (1803) and a satirical comedy *Anti-Faust oder Geschichte eines dummen Teufels* (1801), as well as the horror story *Der Runenberg* (1804). In Dresden also he began his immense, eccentric dramatic fantasy *Kaiser Oktavianus* (q.v., 1804). In 1805–6 Tieck travelled in Italy with his sister and a group of friends and on his return he moved restlessly from place to place. Up to this time he had been one of the most prominent of the Romantics, but his creativity had now spent itself, and, though he still produced some original works, he was active more and more as an editor.

In 1810 Tieck published *Frauendienst* by ULRICH von Lichtenstein, and in 1811 *Altenglisches Theater oder Supplement zum Shakespeare* as well as the works of F. Müller (q.v.). In *Phantasus* (1812–17) he republished early works, together with some new stories (*Die Elfen, Der Pokal*). *Deutsches Theater* (1817) included plays by Hans Sachs, M. Opitz, A. Gryphius, and Lohenstein (qq.v.). In the same year he visited England, where he pursued his study of Shakespeare (q.v.). In 1819 he settled in Dresden, where he was much lionized. He published the works of H. von Kleist (q.v.) in 1821 and (3 vols.) in 1826, and those of J. M. R. Lenz (q.v.) in 1828. In his studies of Shakespeare (*Shakespeares Vorschule*, 1823) he was assisted by his daughter Dorothea (1799–1841). Appointed dramatic adviser (Dramaturg) to the Dresden Theatre in 1825, he recorded his experiences in *Dramaturgische Blätter* (1826). Publication of his poems (*Gedichte*, 3 vols., 1821–3, reissued 1967), was followed by a collection of stories (*Novellen*, 7 vols., 1823–8). Of the works of his later years the Novelle *Der Aufruhr in den Cevennen* (1826) and the novel *Vittoria Accorombona* (q.v., 1840) stand out. In Dresden Tieck made a name for himself for his reading of plays and poetry, and in 1841 was appointed reader to King Friedrich Wilhelm IV (q.v.) of Prussia. In 1843 he published the works (*Sämtliche Schriften*) of F. A. Schulze (q.v., 6 vols.).

Tieck's definitive edition of his works appeared as *Schriften* (28 vols.) between 1828 and 1854 (reissued 1966 f.). Posthumous writings (*Nachgelassene Schriften*, 2 vols.), ed. E. Köpke, appeared in 1855 (reissued 1974). Various selections have been published, including those edited by E. Berend (6 vols., 1908) and M. Thalmann (4 vols., 1963–6).

TIECK, SOPHIE (Berlin, 1775–1833, Reval), younger sister of Ludwig Tieck (q.v.), married in 1799 the schoolmaster A. F. Bernhardi (q.v.), from whom she was divorced in 1804. She later married Ludwig von Knorring (1769–1837) and lived with him in Reval. She wrote *Wunderbilder und Träume* (1802), *Dramatische Phantasien* (1804), *Flore und Blanchefleur* (1822, an adaptation of the medieval *Floire und Blanscheflur*, q.v.), and a novel, *Evremont* (1836), all of which are contributions to the Romantic movement (see ROMANTIK).

TIEDGE, CHRISTOPH AUGUST (Gardelegen, 1752–1841, Dresden), after education at Magdeburg Gymnasium and Halle University, was be-

friended by J. W. L. Gleim (q.v.) in Halberstadt. In 1792 he became secretary to a canon of Stedern near Quedlinburg, spending the years 1799–1802 in Berlin. In 1804 he accompanied Countess Elisa von der RECKE (q.v.) on a journey to Italy, and from 1819 until her death he lived with her in Dresden.

Tiedge wrote several philosophical poems, of which *Urania über Gott, Unsterblichkeit und Freiheit* (1801), written in six cantos, had a considerable success and was frequently republished in the first half of the 19th c. His first work was the poem *Die Einsamkeit* (1792) and he also wrote *Elegien und vermischte Gedichte* (1803–7). His writings have fallen into oblivion, except for the lament in rhyming verse, *Elegie auf dem Schlachtfelde bei Kunersdorf.*

Gesammelte Werke (8 vols.), ed. A. G. Eberhard, appeared 1823–9 and posthumous writings (with biography, 4 vols.), ed. K. Falkenstein, in 1841.

Tiefurt, close to Weimar, has a park and former ducal mansion. In the park Goethe's *Die Fischerin* (q.v.) was given its first performance on 22 July 1782.

Till Eulenspiegel, a Volksbuch which enjoyed enormous popularity in the 16th c., especially among peasants and artisans. The oldest surviving edition is the High German version of 1515, but the lost original (which was written in Low German) probably dates from *c.* 1480. A relic of the low German origin is the occasional spelling 'Dyl Ulenspiegel'. Till's story takes the form of a series of unconnected comic episodes, so that it forms a collection of Schwänke (see SCHWANK). He is an adroit and inventive practical joker who takes in his victims, who belong to the governing classes, by his disarming air of simple innocence. The historical Till, around whom the stories have collected, was born at Kneitlingen in Brunswick, was a practical joker like his literary descendant, and died at Mölln near Lauenburg in 1350.

A modern Eulenspiegel is G. Hauptmann's *Des großen Kampffliegers, Landfahrers, Gauklers und Magiers Till Eulenspiegel Abenteuer, Streiche, Gaukeleien, Gesichte und Träume* (1928), written in hexameters and set out in eighteen adventures (Abenteuer). The composition (Tondichtung) by Richard Strauss (q.v.), *Till Eulenspiegels lustige Streiche*, Op. 28, was written in 1895.

TILLICH, PAUL (Starzeddel, 1886–1965, Chicago), a Protestant theologian and philosopher, was deprived of his chair of philosophy at Frankfurt University in 1933. He emigrated to the U.S.A., acquiring American nationality in 1940. He was a professor at Columbia University

(1938–55), at Harvard (1955–62), and at Chicago (1962–5). He was influenced by Existentialism (q.v.) and the theology of K. Barth (q.v.). In Germany he was a supporter of religious socialism, which he combined with 19th-c. historical conceptions and Christocentric theology.

Tillich's main thought is concerned with what he calls the 'frontier' areas between fields, as, for example, between religion and philosophy, between Church and State, between religion and art, between Idealism and Marxism. In matters of reason, knowledge, and belief, he distinguishes between 'technical reason', comprising science and technology, 'ontological reason', and 'ecstatic reason', covering faith. His wide-ranging views have some relevance to literature, since he maintains that culture, with all that is comprehended by that word in German, is the 'form' of religion. In 1962 Tillich was awarded the Friedenspreis des deutschen Buchhandels (q.v.). His work is most accessible in the selection *Auf der Grenze* (2 vols., 1962). Some of his books were first written in English. His *Systematische Theologie* (5 vols.), appeared in 1951–66, and his *Gesammelte Werke* (13 vols.) in 1959–72.

TILLY, Johann Tserklaes, Graf von (Château de Tilly, Brabant, 1559–1632, Ingolstadt), was intended for the priesthood, but preferred the profession of arms. He was at first in Spanish service, gaining quick promotion by bravery and skill. In the Thirty Years War (see Dreissigjähriger Krieg) he was in Austrian service. He greatly distinguished himself, although he was for a time (1625–30) overshadowed by Wallenstein (q.v.). Tilly was the victor in the decisive Battle of the White Mountain (see Weissen Berge, Schlacht am), and won numerous other victories. In 1631 he commanded the forces which besieged Magdeburg and stormed the city, putting more than half the inhabitants to the sword. Though Tilly, as commander, was responsible, he is said to have sought to restrain the soldiery and check the destruction. Tilly was defeated at Breitenfeld by King Gustavus Adolphus (q.v.) later that year, and in 1632 at Rain/Lech, in which battle he was mortally wounded.

Tilly figures in contemporary broadsheets, as an object of satire, and in a number of literary works dealing with the Thirty Years War; he is an important character in Gertrud von le Fort's novel *Die magdeburgische Hochzeit* (q.v., 1938).

TILO VON KULM, a Middle High German poet, who lived in the first half of the 14th c. and became a canon in the diocese of Samland in East Prussia. He is the author of the poem *Von den siben Ingesigeln*, which he completed in 1331

and dedicated to the High Master of the Teutonic Order (see Deutscher Orden), Luder von Braunschweig.

Von den siben Ingesigeln is a German abridgement of an unpublished Latin tract, *Libellus septem sigillorum*, formerly in Königsberg and now believed lost. The seven seals are interpreted as the symbols of the Incarnation, the Baptism, the Passion, the Resurrection, the Ascension, Pentecost, and the Day of Judgement, and the events of the life of Christ are treated in abstract fashion as symbols deprived of reality.

Tilsit, Treaty of, see Napoleonic Wars and Friedrich Wilhelm III, König.

Tiroler Passionsspiele, collective designation of a group of twelve passion plays which are all based on an original of the early 15th c. Written between 1480 and 1500, they are designed for performance on Maundy Thursday, Good Friday, and Easter Day. The first section comprises the story of the Passion up to Christ's appearance before Pilate, the second the Crucifixion, and the third the Resurrection. The best-known version is known as the *Sterzinger Spiele*.

TIROLFF, Hans, lived in the 16th c. and is the author of a didactic biblical play, *Von der Heirat Isaaks und seiner lieben Rebekken* (1539), and of a dramatization of the story of David and Goliath (1541). He also translated *Pammachius* by Thomas Naogeorgus (q.v.) into German in 1540.

Tirol und Fridebrant, title given to three Middle High German poems, the interrelationships of which are obscure. Two of them are included as one poem in the great Heidelberg MS. (see Heidelberger Liederhandschrift, Grosse).

The first, a poem in which Tirol sets riddles, which his son solves, is headed *Künik Tirol von Schotten unt Vridebrant sin sun*; the continuation, which is separately headed *Der künig tyrol leret sinen svn*, is a didactic poem dealing with the duties and conduct of a ruling prince. It was probably written *c.* 1250, whereas the poem containing the riddles is usually dated some twenty years later. The third poem appears to consist of slender fragments of an epic in which Tirol and his son are characters. Its date is probably between 1230 and 1250, and its strophic form suggests the influence of Wolfram's *Titurel* (q.v.); the fragments embody fabulous adventures.

TIRPITZ, Alfred von, Grossadmiral (Küstrin, 1849–1930, Ebenhausen nr. Munich), entered the small Prussian navy in 1865. In 1897 he became State Secretary of the Admiralty

(Reichsmarineamt) and prosecuted naval expansion with vigour. In the naval construction race with Great Britain the German Navy Laws of 1906 and 1908 were landmarks. During the 1914–18 War Tirpitz favoured aggressive use of the battle fleet, a view which was resisted by the Emperor Wilhelm II (q.v.). Tirpitz resigned in March 1916.

In 1917 Tirpitz and W. Kapp (q.v.) founded an extremist nationalist party (Deutsche Vaterlandspartei). Tirpitz published his memoirs (*Erinnerungen*, 1919) with two appendices, *Aus meinen Kriegsbriefen 1914–15* and *Bemerkungen zu unserer Schiffsbaupolitik*, justifying his policy. *Politische Dokumente* (2 vols.) appeared in 1924–6.

TISCHBEIN, JOHANN HEINRICH WILHELM (Hainau, Hesse, 1751–1829, Eutin), German painter, studied in Holland and then became a fashionable portrait painter in Berlin. He was in Rome from 1779 to 1781 and again from 1783 to 1789, meeting Goethe there in 1786, with whom he was on terms of close friendship for about a year. From this time date two of Tischbein's best-known works, the oil painting *Goethe in der Campagna* and the sepia drawing *Goethe am Fenster seiner Wohnung in Rom*.

In 1789 Tischbein became director of the Naples Academy, where he remained for ten years. After spells of work in Kassel and Hamburg he settled in Eutin, Holstein. In addition to portraits he painted historical pictures and landscapes. His autobiography, *Aus meinem Leben* (2 vols.), was published posthumously in 1861.

Titan, a novel by JEAN Paul (q.v.), published 1800–3 and divided into 35 'Jobelperioden', with 146 subdivisions ('Zykel'). The eccentric structure of the narrative is matched by a highly complex plot. The hero Albano is heir to the principality of Hohenflies, but is brought up to believe himself the son of a lesser nobleman. When the novel begins, he is on his way to meet his presumptive father on the island of Isolabella. With the librarian Schoppe he is sent to Pestiz, the princely residence of the state of Hohenflies. Here he meets the delicate and ethereal Liane, to whom he is attracted. Intrigues of state separate the lovers and Liane dies. At Ischia, where he visits his sister, he falls in love with Linda de Romeiro, who, though returning his love, is presently seduced by another (Roquairol). Linda proves to be Albano's sister. Albano marries a third beautiful young woman, Idoine, and takes over the government of his state.

The title alludes to the moral basis of the book, in which two 'titans' or geniuses ('Himmelsstürmer' in Jean Paul's phrase) devote themselves recklessly to freedom of thought (Schoppe) or to hedonism (Roquairol). Contrasted with them is the balanced figure of Albano. The style of the novel is involved and contemporary readers revelled in its effusive emotion. Its most lasting feature is its beautiful and evocative descriptions of landscape, especially by night.

Titurel, Arthurian hero, in Wolfram von Eschenbach's *Parzival* (q.v.) is the grandfather of Anfortas. *Titurel* is also the title given to an unfinished epic in four-line stanzas (the so-called Titurelstrophe, q.v.) by Wolfram, which deals, however, primarily with Schionatulander and Sigune. The fragments of Wolfram's *Titurel* were written between 1210 and 1220. One group of fragments, pertaining to the beginning of the poem, deals with the origins and early history of Schionatulander and Sigune. The second group sets the pair in a forest and introduces a hound whose lead carries an inscription. The hound escapes and Sigune insists on its recapture. Schionatulander was evidently destined to lose his life to the knight Orilus on this frivolous adventure.

Der jüngere Titurel is the title of a long poem based on Wolfram's fragments. It comprises more than 6,000 stanzas and was written c. 1272 by Albrecht, who has been identified (though without certainty) with Albrecht von Scharfenberg (q.v.). The love of Schionatulander and Sigune is portrayed, together with the latter's death on his quest for the lost hound; but the poem also contains a history of the Grail and the Grail family, originating in Troy, a detailed description of the Grail temple, chivalric festivities, a retelling of Parzival's story, and battles in the Orient. Though it seeks to unite religion and knighthood against a background of world history, the dualism between God and the world persists unreconciled. The style, based on Wolfram's fragments, is unusually obscure.

Der jüngere Titurel was one of the most widely read works of the Middle Ages and was held by many to be by Wolfram himself, an attribution at which Albrecht connived. Its popularity, attested by fifty-seven, surviving MSS., continued into the 15th c. and in 1477 it was printed. Its long-enduring repute probably derived from its didactic character; it presents a mirror of knightly virtue and conduct. The essence of this is contained in the long inscription on the hound's lead (occupying fifty-four stanzas), a summary of Christian and chivalric principles.

Titurelstrophe, a four-line stanza of special form used by Wolfram von Eschenbach in his *Titurel* (q.v.). The first line has 8 stresses, the second 10, the third 6, and the fourth resembles the second. All but the short third line have a caesura after the fourth stressed syllable. The stanza rhymes in the pattern aabb. In *Der jüngere Titurel* (see TITUREL), the pattern is more complex; the three long lines are each treated as two, so that a seven-line stanza is produced with a rhyme pattern ababcdc.

TITZ, also **TITIUS**, JOHANN PETER (Liegnitz, 1619–89, Danzig), a minor poet, lived in Danzig, Rostock, and again in Danzig, where in 1648 he became a professor at the grammar school. He is the author of two stories from the Dutch of Jakob Cats (*Leben aus dem Tode* and *Knemons Sendschreiben an Rhodopen*, 1644) and of a much-read Opitzian treatise on poetics (*Zwei Bücher von der Kunst Hochdeutsche Verse und Lieder zu machen*, 1642).

Tobias, see LAMPRECHT, PFAFFE.

Tochter Syon, Die, see LAMPRECHT VON REGENSBURG.

Tod des Empedokles, Der, the only tragedy attempted by F. Hölderlin (q.v.). It remained unfinished, preserved in three distinct handwritten MSS. and numerous variants, which were first published by L. Uhland and G. Schwab (qq.v.) in 1826 (*Gedichte*), and since in successive revised editions. The work was Hölderlin's major preoccupation before illness abruptly terminated his creative writing. Having initially (1794) intended to dramatize the death of Sophocles, he turned to the subject of Empedocles, setting out his original intentions in a plan known as the *Frankfurter Plan* (referred to by him in 1797) and wrote an ode *Empedokles*. Although Hölderlin used as his principal source Diogenes Laertius, other Greek and contemporary influences are known. They recede, however, as the plan shifts to become an explicitly classical and religious play, in which Empedocles figures, in exile, as the saviour, whose sacrifice will reconcile the gods with mankind, for his death will give permanence to the word of the spirit which it was given him to pronounce. All three versions are dominated by the monologues spoken by Empedokles.

The first version consists of two acts of over 2,000 lines. The priest Hermokrates, functioning as the antagonist, imputes blasphemy to Empedokles and so turns Kritias and his people against their idol. Pausanias figures as Empedokles' loyal disciple, sharing exile and banishment with him. The fragment ends with Empedokles' monologue, in which he prepares for the liberation of his spirit in the purifying fires of Mount Etna. Empedokles' insistence on death, conceived by him as atonement, as well

as serene union with the elements, is so timed that it immediately follows his pardon by his antagonists, thus removing the impression of external compulsion.

The second version, consisting of 724 lines, by introducing the sympathetic but hesitant figure of Mekades as a substitute for the archon Kritias, effects a new approach to the problem of Empedokles' guilt. The charge of blasphemy ('Wortschuld'), decisive in the first version, is replaced by the argument that Empedokles' visionary power endangers the country's stability. Empedokles is to be condemned, but at a stage when he has already condemned himself; for he feels himself deserted by the gods, who deny him the power of words to express his mystical visions.

The third version, known as *Empedokles auf dem Aetna*, is best understood in connection with Hölderlin's essay *Grund zum Empedokles*. The rigid blank verse and a different set of characters distinguish this fragment, which consists of three scenes with 507 lines, from the earlier versions. The version opens with a monologue. Empedokles, awaking from sleep, has overcome his sorrows; in the *Frankfurter Plan* these are defined as his hatred of civilization and of the dependence of life on a 'law of succession' ('Gesetz der Sukzession'), a fragmented existence denying man the full participation in the eternal mystery of Nature. Empedokles confesses his sins, which he defines as his abandonment of his fellow men while pursuing his own lonely path of destiny; he is convinced of his right to die and be liberated by the flames, for in eternity he will be reunited with all who curse him in this world.

The fragments are an important document of Hölderlin's development at the stage when his prophetic sense of mission had reached its zenith.

Tod des Vergil, Der, the best-known novel of H. Broch (q.v.), published in 1945. It is the inner monologue (expressed, however, in the third person) of the dying poet brought to shore at Brundisium and debating with himself in profoundly poetic language the value of his achievement in the face of imminent death. He is finally persuaded to refrain from destroying the *Aeneid*, but sees his true aim in love (*caritas*), exacting from Augustus a promise to liberate his slaves.

The mystical quality of this powerful work admits of widely varying interpretations.

Tod in Ähren, a short poem by D. von Liliencron (q.v.), written in 1877, and revised and reduced to twelve lines in 1880. It speaks of the agony, prolonged over two days, of a soldier mortally wounded in a cornfield. The title is a bitter pun. The poem was first published in 1883 in *Adjutantenritte* (q.v.).

Tod in Rom, Der, a novel by W. Koeppen (q.v.), published in 1954. A satire on contemporary conditions, it shows a Germany in which unrepentant officials of the Nazi regime have secured prominent positions, while a resentful younger generation is left to cope with the scars of the past. The central figure, the SS General Judejahn, who has escaped proceedings which would have led to his execution, embodies the intractable brutality of the Third Reich. He is obsessed with sex, which impels him to sleep with a Jewish barmaid, and with the 'final solution', which explains his decision to 'execute' her; he shoots the wrong woman, who happens to be Jewish and a near victim of his in the past. His death in Rome arises from his relentless sexual exertion. The finale includes an overt parody of Th. Mann's *Der Tod in Venedig* (q.v.).

Tod in Venedig, Der, a Novelle by Th. Mann (q.v.), published in 1912. Its central figure is Gustav von Aschenbach, a highly respected author, living in Munich and noted for the sobriety and discipline of his writings. As he ages, he decides to take a Mediterranean holiday, which culminates in his stay in Venice, where his attention is held by a 14-year-old Polish boy, Tadzio. The fascination of the boy's beauty causes him to prolong his stay even after he has received an intimation that a cholera epidemic is in progress.

This impending threat of death in Venice, the news of which is suppressed for the sake of the tourist trade, acts as an illumination for Aschenbach. He sacrifices his entire identity, his will, and his sense of dignity, which he has cultivated throughout his life, to the superior experience of beauty and art, which Tadzio reveals to him by his mere presence. The news of the epidemic spreads, and Aschenbach himself feels the onset of the disease. He learns that the Polish family is leaving. On the day of Tadzio's departure the German press reports the death of Gustav von Aschenbach.

The Novelle is the subject of Visconti's film *Morte a Venezia* (1971) and of Britten's opera *Death in Venice*, which was first performed at Aldeburgh in 1973.

Tod und das Mädchen, Der, a poem by M. Claudius (q.v.). It is a dialogue consisting of a four-line stanza spoken by the girl, to which Death replies in a similar stanza. It has been set to music by F. Schubert (q.v.), who also used part of the melody in his string quartet in D minor, which is often known by the title *Der Tod und das Mädchen (Death and the Maiden)*.

Tolle Invalide auf dem Fort Ratonneau, Der, a Novelle by L. J. von Arnim (q.v.), published in 1818. Sergeant Francoeur is wounded in the head in the Seven Years War, taken prisoner by the Prussians and nursed by Rosalie, who, upon his repatriation, follows him as his wife to Marseilles (whence the anecdote originates, as it is narrated by Grosson in the *Almanach historique de Marseille pour l'année bissextile 1772*).

Rosalie, however, bears the burden of her mother's curse for having gone over to the French; because of this she attributes the strangely erratic behaviour of her beloved husband to the workings of the devil. To isolate Francoeur, the commander of the invalided soldiers puts him in charge of Fort Ratonneau with Rosalie by his side, and with two soldiers under his command. Suspecting and banishing his wife, who barely escapes with her life, Francoeur, in a state of utter frenzy, defies the entire strength of the local troops, threatening to blow up the fort. To save the city, Rosalie, after the third day of the siege, sets out to the fort, ready to prove her loyalty to Francoeur by the sacrifice of her life. Francoeur trains a cannon on his wife, but, seized by conflicting emotions, tears his hair, thereby opening his old wound. A medical examination shows that an unextracted splinter was the cause of his mental disturbance. On the basis of these findings Francoeur receives a military pardon. A touch of mystery is added to the happy ending, for Rosalie learns that her mother, when dying, has lifted the curse.

Tolle Professor, Der, a novel by H. Sudermann (q.v.), published in 1926 with the subtitle *Roman aus der Bismarckzeit*.

Dr. Sieburth, appointed in his early 30s to an additional chair of philosophy at Königsberg University, has the almost certain prospect of succeeding to the principal chair. His academic brilliance is recognized, but in character and outlook he is a misfit. He has a great attraction for women, and a dislike both of the Bismarckian German nationalists and of the traditional Liberals. He becomes involved with three society ladies, the first an intelligent young woman, the second the discontented wife of a businessman, the third the wife of one of his professorial colleagues. With this last he becomes seriously compromised. His custom of entertaining an assortment of women at night in his lodgings also becomes known, to the detriment of his reputation. His colleagues seek to prevent his election to the principal chair when it falls vacant, and, partly to spite them, he throws in his lot with the governmental party which he has hitherto despised. He follows advice to consult the civil servant responsible for university patronage and learns that the government will press his appointment against the desires of the Faculty, but this information is accompanied by the discovery that the reason is not his quality as a philosopher, but his political support for the government together with a move on his behalf by a princely house with which he had formerly had a connection. After his last and most promising love-affair is brought to a close by the girl's mother and he is chillingly received by the Faculty, he shoots himself.

The best part of the novel is the portrayal of the political, social, and intellectual background of the 1880s.

TOLLER, ERNST (Samotschin nr. Bromberg, 1893–1939, New York), as a young Jew sensitive to the anti-Semitism around him, preferred to study at Grenoble University, but in 1914 voluntarily returned to Germany to enlist. His experiences at the front induced a breakdown, and he was invalided out in 1916. By this time a confirmed pacifist and a left-wing radical, he became involved in anti-war politics and agitation early in 1918, and was associated with the short-lived Communist government headed by K. Eisner (q.v.) in Bavaria in November.

In 1919 Toller was sentenced to five years' imprisonment, during which he wrote *Gedichte der Gefangenen* (1921) and Expressionist plays of the type of Stationendrama (q.v.). They bear the stamp of the strain of war and confinement, are hectic and strident in style, and filled with anger, scorn, and derision. *Die Wandlung* (1919), *Der Tag des Proletariats* and *Requiem den gemordeten Brüdern* (both 1920 and described as 'Chorwerke') were followed by the plays which achieved success, *Masse-Mensch* (q.v., 1921) and *Die Maschinenstürmer* (q.v., 1922, on the subject of the Luddites). *Der deutsche Hinkemann* (1923, republished as *Hinkemann*, 1924) caused consternation by the obtrusive, morbid details it introduces; its subject, the plight of the returning soldier (see HEIMKEHRERLITERATUR), is treated with glaring cynicism, but in realistic and not Expressionistic manner. *Der entfesselte Wotan* (1923) passed virtually unnoticed; *Hoppla, wir leben!* (1927) was produced by E. Piscator (q.v.) in 1927. Realism and resentment also permeate his later plays, *Feuer aus den Kesseln* (1930, on the subject of a naval mutiny in 1917) and *Die blinde Göttin* (1932). This play, intended to expose the gap between law and justice, deals with a wrongful conviction and the effect on the lovers who are its victims, all based on an actual trial.

In 1924 Toller published *Das Schwalbenbuch*, a work of tender humanity inspired by the efforts of swallows to nest in his cell. His sensitive nature is also revealed in his early autobiography *Eine Jugend in Deutschland* (1933). After 1933 Toller lived at first in Europe and later in the U.S.A. He published one more volume of poetry, *Weltliche Passion* (1934), and one more play, *Pastor Hall* (1939), but the knowledge of the increasing brutality of the National Socialist regime wore down his nerves; a few weeks after the German invasion of Czechoslovakia he committed suicide. A select edition of his works, *Ausgewählte Schriften*, appeared in 1959 and *Gesammelte Werke* (5 vols.), ed. J. Spalek and W. Frühwald, in 1978.

Tolpatsch, Der, the story which opens the first volume of *Schwarzwälder Dorfgeschichten* (q.v., 1843–53) by B. Auerbach (q.v.). 'Tolpatsch' (the equivalent of 'booby') is the nickname of the good-natured but simple-minded village boy Aloys. He loves Marannele, and his love seems to be returned. Determined to make his mark, he gets himself chosen for military service, but while he is away Marannele goes out with another, becomes pregnant, and marries the seducer. Aloys emigrates to America where he is successful, but he still sadly remembers Marannele.

Auerbach published a sequel, *Der Tolpatsch aus Amerika*, in *Nach dreißig Jahren* (1876).

Tonele mit der gebissenen Wange, a story in *Schwarzwälder Dorfgeschichten* by B. Auerbach (q.v., vol. 1, 1843). Sepper is Tonele's boyfriend; the gamekeeper (Jäger) of Mühringen, however, also takes a fancy to her. For a time, in spite of his jealousy, Sepper retains his footing with Tonele. But after a dance he inflicts on her a love bite (referred to in the title), which estranges her and drives her into the keeper's arms. Sepper is called up for manœuvres. On his return he quarrels with the keeper, who in the struggle is shot by his own gun. Sepper disappears and years later Tonele dies broken-hearted.

Toni, a three-act play (Ein Drama) in blank verse by Th. Körner (q.v.), published posthumously in 1815. It is an adaptation of H. von Kleist's story *Die Verlobung in St. Domingo* (q.v.), to which Körner has given a happy end. The play, written in about a week, was performed with acclaim in the Burgtheater, Vienna, in April 1812.

Tonio Kröger, a Novelle by Th. Mann (q.v.), published in 1903 and probably the most widely read of his shorter works. It also provided the title of a collection of stories published in 1914. Tonio, of mixed German and South American parentage, is a sensitive child, conscious that he is different from the rest of his school friends and their sisters. This normal world is represented by the uncomplicated extrovert Hans Hansen, who cannot share Tonio's enthusiasm for Schiller's *Don Carlos* (q.v.), and by Ingeborg Holm, to whose blonde good looks Tonio is irresistibly drawn without evoking any response.

Tonio grows up, leaves home, and settles in south Germany, where, after a period of dissipation, he develops into a writer of disciplined elegance. A discussion with a Russian woman friend, Lisaweta Iwanowna, epitomizes his problem in the phrases 'ein verirrter Bürger' and 'ein Bürger auf Irrwegen'. Two incidents illustrate Tonio's continued preoccupation with the world, in which he feels a stranger on lost paths. On a visit to his native northern city he is suspected by the hotel manager of being a swindler. All is cleared up with profuse apologies, but the sting of mistaken identity remains. In Denmark, Tonio chances to see, but does not meet, the two figures of his boyhood attachment. As Mann subtly indicates, they are not Hans and Inge, yet they are the norm, the Hans Hansen and the Ingeborg Holm for whom he has not ceased to yearn.

The work, which has obvious parallels with Mann's life, is a model presentation of his perpetual uneasiness over the dichotomy of Bürger and Künstler.

TÖPFER, KARL FRIEDRICH GUSTAV (Berlin 1792–1871, Hamburg), son of a civil servant, ran away from home in order to become an actor. In 1815 he was encouraged by J. Schreyvogel (q.v.) to write for the stage. After studying at Göttingen University (1820–2) he settled in Hamburg, where he wrote for various periodicals. He is the author of popular comedies (*Des Königs Befehl*, 1823; *Der beste Ton*, 1828; and *Rosenmüller und Finke*, 1830) and of fiction (*Novellen und Erzählungen*, 1842–3). His dramatization of Goethe's *Hermann und Dorothea* (q.v.) was performed in Vienna in 1820 and in Weimar in 1824.

Töpfer's *Lustspiele* (7 vols.) appeared 1830–51 and *Gesammelte Werke* (4 vols.) posthumously in 1873.

TORBERG, FRIEDRICH, pseudonym of Friedrich Kantor-Berg (Vienna, 1908–79, Vienna), a prominent Austrian critic skilled in the Viennese tradition of wit and parody. His works include poetry (*Der ewige Refrain*, 1929) and novels, of which the early ones are *Der Schüler Gerber hat absolviert* (1930, reissued as *Der Schüler Gerber*, 1954), the tragedy of the matriculation candidate Kurt Gerber, who commits suicide in the false belief that he has failed, ... *und glauben, es wäre die Liebe* (1932), and *Die Mannschaft* (1935). Forced to go into exile in 1938, Torberg did not return to Vienna until the early 1950s, editing from 1954 to 1965 the *Forum*. *Hier bin ich, mein Vater* (1948) deals with the fate of the German Jews, *Die zweite Begegnung* (1950) with Communist Czechoslovakia, and *Süßkind von Trimberg* (1972) with the medieval poet (q.v.). *Golems Wiederkehr* appeared in 1968, the collection *P. P. P. Pamphlete. Parodien. Post Scripta* (2 vols.) in 1964, and *Das fünfte Rad am Thespiskarren* (2 vols.) in 1966–7. He is noted for his translations of the satirical stories of the Israeli author Ephraim Kishon (b. 1924). *Gesammelte Werke in Einzelausgaben*, including correspondence (*In diesem Sinne*, 1981, *Kaffeehaus war überall*, 1982, and *Pegasus im Joch*, 1983), appeared 1962 ff. He promoted the works of F. von Herzmanovsky-Orlando (q.v.), of which he was the first editor. Shortly before his death he was awarded the Großer Österreichischer Staatspreis.

Torgau, battle of (3 November 1760), in which Friedrich II (q.v.) of Prussia defeated the Austrians under Daun (q.v.) in the last pitched battle of the Seven Years War (see SIEBENJÄHRIGER KRIEG). Torgau lies on the Elbe in Saxony. See TORGAUER HEIDE, DIE.

Torgauer Heide, Die, 'Vorspiel zum historischen Schauspiel: *Friedrich II von Preußen*', a scene in prose, all that remains of a projected patriotic play by O. Ludwig (q.v.). Published in 1844, it shows Prussian soldiers of all arms gathered round the bivouac fires on the evening after the victory of Torgau (q.v.).

Torquato Tasso, a verse play (Schauspiel) by J. W. Goethe, first published in vol. 6 of *Goethes Schriften*, 1790. Goethe first worked at the play in 1780 and 1781, at which time the first and part of the second act were written. This version is lost. The work in its present form was composed between February 1788 and the summer of 1790. The source was a biography of the poet Tasso by Giovanni Batista Manso (*Vita di Torquato Tasso*, 1621), which at a later stage was supplemented by a biography by Pierantonio Serassi published in 1785. The first performance did not take place until 16 February 1807 at Weimar.

The action takes place at Belriguardo in a highly cultivated court circle consisting of five persons, Duke Alfons, his sister Leonora von Este, referred to as Prinzessin, and her friend Leonora Sanvitale, the statesman Antonio Montecatino, and Tasso. The poet has just completed his great epic poem *La Gerusalemme liberata* for which he is fêted by the two ladies and by the Duke, who takes pride in his protégé. A wreath, taken from a statue of Virgil, is placed on the head of Tasso, who, however, shows himself humbly reluctant to accept it. At this stage Antonio arrives from Rome and, impatient with the apparently triumphant poet, adopts a hostile and disdainful attitude towards Tasso: the scene signifies an antagonism between the man of art and the man of practical life. In a confrontation between the two men (Act II) this antagonism comes fully into the open and Tasso draws against Antonio. Duke Alfons intervenes and, in view of this infraction of court etiquette, feels obliged to place Tasso under room arrest. Tasso reacts bitterly to this humiliation, seeing himself persecuted by those he has held most dear, for he is secretly in love with Leonora von Este. He begs leave to quit the court and, when this is granted, encounters Leonora, whom he passionately but imprudently embraces. Tasso's distress reaches its nadir when he watches the departure of the ducal party, an unmistakable omen that he has forfeited his position at the Ferrarese court. Antonio, however, has stayed behind; in him he finds an unexpected counsellor and friend. In his closing speech Tasso, desperately aware of the precariousness of his existence, glimpses the prospect of a recovery of his creative powers. The future is left indeterminate and has remained the subject of controversy.

Torquato Tasso embodies Goethe's regard for the Weimar court and its educative power, his sense of the beneficent influence of Charlotte von Stein (q.v.), and his admiration for Italy. In the figure of Tasso he shows his understanding for and criticism of a type of Romantic artist. To Eckermann he approved the description of the play as 'einen gesteigerten *Werther*' (3 May 1827, Pt. 3, see LEIDEN DES JUNGEN WERTHERS, DIE), adding three days later that it is 'Bein von meinem Bein und Fleisch von meinem Fleisch'.

TORRESANI, KARL VON (Milan, 1846–1907, Torbole, Lake Garda), was, in full designation, Karl Franz Ferdinand Torresani, Freiherr von Lanzenfeld und Componero. His father was the Austrian military police chief in Milan. Torresani was commissioned in the cavalry in 1865 and saw war service in 1866. He afterwards served on the General Staff, resigning in 1870. He wrote Novellen and novels chiefly drawn from the military life he knew. They include *Aus der schönen wilden Leutnantszeit* (3 vols., 1889), *Schwarzgelbe Reitergeschichten*, *Mit tausend Masten* (both 1890), *Die Juckerkomtesse* (1891), and *Ibi, ubi* (1893).

TÖRRING, JOSEF AUGUST, GRAF VON (Munich, 1753–1826, Munich), spent his life in the civil administration of Bavaria, rising from Hofkammerrat to president of the Council (Präsident des Staatsrats). In his early years he wrote two historical plays, *Agnes Bernauerin* (1780, see BERNAUER, AGNES) and *Caspar der Thorringer* (1785). Both deal with subjects from Bavarian history, which the lineage-conscious Törring treats from a dynastically loyal standpoint, and both touch his family history.

TORSVAN, TRAVEN, see TRAVEN, B.

Tor und der Tod, Der, a verse play in one short act by H. von Hofmannsthal (q.v.), written in 1893 and published in 1894. Claudio, whose life has been spent in perceiving the reflection of life in art, and has never given himself entirely to any person, is approached by Death, who, fiddle in hand, suggests a medieval dance of death, a motif which is accentuated by the dress of Death and those accompanying him, 'altmodisch nur von Tracht, / Wie Kupferstiche angezogen sind'. Claudio seeks to dismiss Death on the ground that he has not yet lived; Death produces his witnesses, Claudio's mother (Die Mutter), das junge Mädchen, who had loved him, and his friend (Der Mann). In his egocentric aestheticism, Claudio has aroused no devotion in others ('Der keinem etwas war und keiner ihm'). Only in the last, the parting moment, does it dawn upon Claudio what it means to live.

Toten an die Lebenden, Die, a political poem of 44 rhyming couplets, written in July 1848 by F. Freiligrath (q.v.) and first published in 1848. It was reprinted in *Neuere politische und soziale Gedichte* (vol. 1) in 1849. 'The dead' of the title were the 187 citizens killed in the fighting in Berlin on 18 March 1848 (see REVOLUTIONEN 1848–9), and the poem is a declamatory reminder to the German people of the cause for which they died.

Totenkopfhusaren, nickname for three regiments of hussars of the former German army. They were the Prussian Leib-Husaren-Regimenter Nr. 1 u. 2, both garrisoned at Danzig-Langfuhr and the Brunswick hussars (No. 17). All wore black uniforms with a skull badge on the front of the busby.

Toten schweigen, Die, a story by Schnitzler (q.v.), published in the collection entitled *Die Frau des Weisen* in 1898. Frau Emma and her lover Franz are on a pleasure ride at night just outside Vienna when the drunken driver loses control of his horses and the cab overturns in the ditch. Franz is killed and Emma runs away in panic from the scene of the accident in order to preserve her anonymity. She arrives home safely, but, in a trance-like state of exhaustion, reveals her illicit love to her husband. The story, in its closed structure, develops into a brief but concise psycho-analytical study. The effect of this Novelle is enhanced by a wealth of topographical detail.

Totentanz, a representation of dancing figures of dead persons, seeking to draw the living into their dance, and so to death. Its origin is a medieval superstition that the dead return at night to dance in the churchyard and attract the living. 'Totentänze' are primarily works of plastic art (e.g. *Der Tod von Basel, Lübecker Totentanz,* and *Totentanz* of Hans Holbein der Jüngere, q.v.); a literary by-product occurs in the form of verses, often a simple couplet, attached to each picture in a series.

Dances of death, which first arose in France, began to appear in Germany early in the 15th c., and were mostly executed between *c.* 1430 and 1520. The dance often takes the form of a review of social classes from the Pope downwards and readily becomes a vehicle of social satire. Holbein expanded the motif by introducing the figure of Death. At this point the actual dance disappears, though a dance-like element remains in the procession of figures each partnered by Death. In the 19th c. there was a revival of interest in the 'Totentanz', manifested in the series of woodcuts *Auch ein Totentanz* (1849) by Alfred Rethel (q.v.), which was provided with a verse text by Robert Reinick (q.v.).

Totenwald, Der, a report (Ein Bericht) by E. Wiechert (q.v.) on his horrifying experiences and the still worse experiences of many of his fellow inmates in a concentration camp in 1938. Published in 1945, it is dedicated 'den Toten zum Gedächtnis, den Lebenden zur Schande, den Kommenden zur Mahnung.'

Tote Tag, Der, a mythical play in five acts by E. Barlach (q.v.), published in 1912. It is partly inspired by the legend of *Parzival* (q.v.). Central to the action are the mother (die Mutter), the son (der Sohn), and Kule. Kule, a blind man, is in possession of a divine staff, which has guided him through his wanderings in the world to return to the mother when her son reaches manhood. It is Kule's mission to awaken in the son the awareness that he is destined to leave his mother to serve God, for he is a 'Göttersohn'; outside the cottage Herzhorn, a mythical horse, is waiting to take him away. In her anguish lest

she should lose her son, the mother kills the horse in the night.

The following day remains dark (the dead day of the title), for the sunlight has turned away from the earth. The lowly spirit Steißbart, the mother's tyrannized servant, exploits the dilemma and intensifies the son's yearning for his unknown father, which Kule has kindled in him. The mother feels impelled to confess to Herzhorn's murder. She stabs herself rather than lose her son's love, and the son, by likewise stabbing himself, follows the call of his mother's blood, for it proves stronger than his father's blood in him. Kule, who has lacked the strength to defy the mother's claim on her son, forfeits the divine staff, and is led away by Steißbart; he resumes his wanderings in order to help man recognize the call of the blood as the call of God, the invisible Father (the original title of the play was *Blutgeschrei*). Only by abandoning his physical origin ('Leibhaftigkeit') for the world of the spirit ('Geisthaftigkeit') can man move towards freedom and redemption.

Barlach created twenty-seven lithographs illustrating this, his first play.

TOUCEMENT, JEAN CHRETIEN, pseudonym of J. C. Trömer (q.v.).

TRAKL, GEORG (Salzburg, 1887–1914, Cracow), grew up in a comfortable middle-class environment, left grammar school early and trained as a pharmacist. He took to drugs and eventually became an addict; he was particularly attached to his similarly addicted pianist sister, who died in 1917. Trakl qualified in 1910, served a year as a volunteer (see EINJÄHRIGER) in the army medical service, but experienced considerable difficulty in finding an appointment. He was greatly helped by friends, especially L. von Ficker (1880–1967), who also published his early poems in his periodical *Der Brenner.*

Trakl's first volume of verse was *Gedichte* (1913); *Sebastian im Traum* (1915) was published posthumously. His poetry is at all times rich, heavily loaded with imagery of autumn and its associated colours, but is devoid of any affectation, such as commonly accompanies the *fin de siècle* mood. Trakl was acutely conscious that his world, both personal and external, was breaking apart ('entzweibricht' is his own word), and this gave rise to a state of suffering (*Leid*), which is the keynote of his poetry. The death of his sister, followed by the outbreak of war, his call-up as a reserve officer in the medical services, and the primitive and inadequate conditions in which he had to tend excessive numbers of men wounded in the battle of Grodek in Galicia overtaxed his resources, and he was sent to hospital at Cracow, where he died of an overdose of cocaine.

Whether he intended his death is uncertain. Trakl's poetry developed in his last year or two from the strophic form of the early poems to a hymnic mode, which, while owing something to Hölderlin (q.v.), is in its stark spareness, concentrated and elliptical syntax, and quivering personal tone, his own unique creation. His quality was early recognized by Rilke (q.v.), and his influence on later, especially Expressionist, poetry was considerable. Among poems which should be mentioned are *Seele des Lebens*, *Verklärter Herbst*, *An die Verstummten*, *Unterwegs*, *Klage*, and *Grodek*, one of the most celebrated poems of the 1914–18 War. Poems left in his papers were published in 1939 as *Aus goldenem Kelch*.

Gesammelte Werke (3 vols.), ed. W. Schneditz, appeared 1948–51. The historisch-kritische Ausgabe, *Dichtungen und Briefe* (2 vols., 1969) is edited by W. Killy and H. Szklenar. The volume *Der Wahrheit nachsinnen—viel Schmerz. Gedichte, Dramenfragmente, Briefe* was published in 1981.

Traum ein Leben, Der, a play (dramatisches Märchen) in four acts by F. Grillparzer (q.v.), written in four-foot trochaic verse. It was completed in 1831, performed in the Burgtheater in 1834, and published in 1839. Part of the first act appeared under the title *Des Lebens Schattenbild* in Lembert's *Taschenbuch für Schauspieler und Schauspielfreunde für das Jahr 1821*. The final title is adapted from Calderón's *La vida es sueño*. The play is influenced by the popular Viennese theatre (Besserungsstück, q.v.), the oriental settings of the *Arabian Nights*, and Mozart's *Die Zauberflöte* (q.v.). The plot is adapted from Voltaire's story *Le Blanc et le noir*, which also furnished the name of the principal character, Rustan.

The frame (see RAHMEN) introduces Rustan as a restless and ambitious young man yearning for adventure and reluctant to settle down in marriage with Mirza, the daughter of his uncle Massud, with whom he lives. In the central action Rustan experiences epic adventures worthy of a hero in search of love, conquest, and power. With the slave Zanga's help he wins the favour of the king of Samarkand, whose daughter Gülnare he marries after he has caused the king's death. But he has achieved his lofty ambitions by treachery and deceit; for this he has to pay with his life. As he plunges in despair into the river he awakes to realize that he has dreamed. The dream has the effect of stabilizing his personality. Grillparzer expresses subconscious impulses by means of the fanciful devices of the fairy world. The play was performed more often during Grillparzer's lifetime than any other of his plays.

Traumulus, a tragicomedy (Tragische Komödie) written by A. Holz in collaboration with O. Jerschke (qq.v.), and published in 1904. Its action deals with a sexual indiscretion by a schoolboy, which through small-town gossip and malice leads to his suicide. The play had considerable stage success.

Traurige Geschichte von Friedrich dem Großen, Die, an unfinished historical novel by H. Mann (q.v.) on Friedrich II (q.v.) of Prussia. Mann began to study the subject in 1940 and worked at the MS. until 1948. It was found among his papers and published in 1960 with the sub-title *Ein Fragment*. The second edition (1962) includes *Der König von Preußen. Ein Essay*, which had been published separately in 1949.

The work is written entirely in dialogue. About one-fifth is completed (up to the visit of the Prussian royal family to August II, q.v., of Saxony); but the complete plan of the novel (partly in English) shows its trend. Fragment, plan, and essay combine to underline Mann's harsh treatment of his subject. Writing at the time of the Third Reich, which he regarded as the ultimate expression of Friedrich's personality and policies, he drew a portrait of a crabbed yet vainglorious egoist with no redeeming virtue.

TRAVEN, B. (?–1969, Mexico City), is believed to have been identified as Otto Feige (b. 1882) of Schwiebus, E. Prussia, who worked under the pseudonyms Ret Marut and Richard Maurhut. An actor, writer, and journalist, he published the journal *Der Ziegelbrenner* (1917–21, from 1919 illegally) in Munich and was involved in the Communist revolution of 1919. Subsequently persecuted, he escaped in the early 1920s, adopted the name of Traven, and, as Traven Torsvan, became a Mexican citizen in 1951. He was himself responsible for the legend of adventure which surrounded his name.

Traven is the author of exotic novels and stories of North and South America; his concern for the underprivileged is central to his fiction, which includes the novels *Das Totenschiff* (1926), *Die Baumwollpflücker* (1929, as *Der Wobbly* in 1926), *Der Schatz der Sierra Madre* (1927), *Der Busch* (1928, as *Der Banditendoktor* in 1955), *Die Brücke im Dschungel* (1929), *Die weiße Rose* (1929), *Regierung* (1931), *Der Karren* (1931, revised as *Die Carreta*, 1953), *Der Marsch ins Reich der Caoba* (1933), *Die Rebellion der Gehenkten* (1936), *Ein General kommt aus dem Dschungel* (1940), *Trozas* (1959), and *Aslan Norval* (1960).

B. Traven in Einzelbänden, ed. E. Päßler, appeared in 1982 ff.; *The Man who was B. Traven* by Will Wyatt in 1980.

TREBITSCH, Siegfried (Vienna, 1869–1956, Zurich), a minor novelist, who is best known for his translation of G. B. Shaw's plays (1946–8).

Treibhaus, Das, a novel by W. Koeppen (q.v.), published in 1953. It is a harsh satire on the Germany of the 'economic miracle' (Wirtschaftswunder) and especially of life in Bonn, which is the hot-house of the title. The central figure, the member of parliament, Keetenheuve, is torn between the obligations of his respectable situation and a nihilism which believes neither in the present order nor in future revolution. He ends by taking his own life.

TREITSCHKE, Heinrich von (Dresden, 1834–96, Berlin), was the son of a Saxon general ennobled in 1821. Treitschke studied history and politics at Bonn, Leipzig, Tübingen, and Heidelberg universities, publishing two volumes of poems before qualifying to lecture (Habilitationsschrift: *Die Gesellschaftswissenschaft. Ein Versuch*, 1859). Deafness marred his ambition for a career in the civil service and in politics; instead he became one of the most influential historians of the 19th c.

In the early stages of his career he was a strong Liberal, advocating a united Germany under a parliamentary and constitutional monarchy. Unacceptable because of these views in Saxony, he obtained in 1863 a professorial chair in Freiburg. His strong Prussian sympathies during the Austro-Prussian War (see Deutscher Krieg) caused him to resign his chair in anti-Prussian Baden; he moved to Berlin, acquiring Prussian nationality. Treitschke's increasingly nationalistic writings used history as a means of furthering his own brand of patriotism. Soon after his arrival in Berlin he became editor of the *Preußische Jahrbücher*, in which he published his unswervingly nationalistic articles and studies until 1871. These included a fiercely polemical essay advocating the incorporation of Saxony in Prussia, which resulted in a breach with his family and with his native country. In spite of his deafness he was elected to the first Reichstag (1871–4) of the new Empire. His Liberalism vanished, and he became a leading propagandist for Prussian hegemony in Germany and one of the strongest supporters of Bismarck (q.v.), as well as an anti-Semite.

Treitschke possessed considerable literary powers; his works have many vivid and dramatic descriptions of historical events, and for some time his reputation as a historian was not confined to Germany. His contributions to the *Preußische Jahrbücher* were collected with other earlier essays as *Historische und politische Aufsätze* (3 vols., 1886, to which a fourth volume was added posthumously in 1897), and in *Zehn Jahre deutscher Kämpfe* (1874). Among the studies included in these volumes and in *Deutsche Kämpfe. Neue Folge* (1896) are *Die Grundlagen der englischen Freiheit* (1888), *Bundesstaat und Einheitsstaat* (1864), and *Das konstitutionelle Königtum in Deutschland* (1869–71). Treitschke's best-known work, *Deutsche Geschichte im 19. Jahrhundert* (5 vols., 1879–94), is unfinished, reaching only to the threshold of the 1848 Revolutions (see Revolutionen 1848–9). It extends back beyond the beginning of the 19th c. in order to provide a basis for the understanding of later events. A selection of stirring passages was published by the National Socialist A. Rosenberg, and a popular abridgement (2 vols.) in 1934. Treitschke wrote one study in which literature and political history intersect, *Heinrich von Kleist* (1858).

TREITZSAUERWEIN VON EHREN-TREITZ, Marx (nr. Innsbruck, c. 1450–1527, Wiener Neustadt), a commoner by birth, became the confidential secretary of the Emperor Maximilian I (q.v.) and assisted Melchior Pfinzing (q.v.) in the composition of *Teuerdank* and *Weißkunig* (qq.v.).

TRENCK, Friedrich, Freiherr von der (Königsberg, 1726–94, Paris), son of a Prussian general, entered the army in 1742 and was on the personal staff of Friedrich II (q.v.) of Prussia. He fell into disfavour and was imprisoned at Glatz for reasons that are not fully clear. He is said to have had a love affair with the King's sister, Princess Amalie, but he was also suspected of treasonable relations with Austria. He escaped and for a time took Russian and then Austrian service. In 1754, on a visit to Prussia, he was seized and imprisoned in Magdeburg under the most rigorous conditions until 1764, when his release, as an Austrian captain, was demanded by Maria Theresia (q.v.). While in prison Trenck made brave and ingenious attempts to escape. On his release he entered business and later served as a diplomatic agent. In 1791 he established himself in Paris and in 1794 was condemned as an agent of a hostile foreign power and guillotined.

Trenck achieved a European reputation through his autobiography (*Des Freiherrn von der Trenck merkwürdige Lebensgeschichte*, 3 vols., 1787, reissued as *Der Gefangene Friedrichs des Großen*, 1922, and under its original title, substituting *seltsame* for *merkwürdige*, 1925). He himself wrote a French version, and it was also translated into English. He is believed to be the author of an anonymous prose satire directed against the Roman Catholic Church (*P. Pavian, Voltaire und ich in der Unterwelt,* 1784). Trenck is the subject of a novel by Bruno Frank (q.v., *Trenck,* 1926). *Des Freiherrn von der Trenck sämmtliche Gedichte und Schriften* (8 vols.) were published in 1787.

TRENKER, LUIS (St. Ulrich, Tyrol, 1892–   ), trained as an architect, served as an Austrian officer during the 1914–18 War, and was an Alpine guide before establishing himself in Berlin (1927) as a writer, film producer, and actor. His first novel, *Berge in Flammen* (1931), an Alpine war story of the Austro-Italian front, was internationally successful as a film (1932). Later novels include *Der Feuerteufel* (1939), *Schicksal am Matterhorn* (1956), and *Sohn ohne Heimat* (1960); his other writings include *Berge im Schnee* (1932, reissued 1964). He entitled an autobiographical work *Alles gut gegangen. Geschichten aus meinem Leben* (1965).

TRESENREUTER, SOPHIE (Kiel, 1755–?), *née* von Thomson, wife of an actuary and the author of sentimental novels, of which *Lotte Wahlstein* (2 vols., 1791–2) and *Häusliches Glück oder Die rechtschaffene Witwe im Kreise ihrer Kinder* (1793) were very widely read.

Treuer Diener seines Herrn, Ein, see EIN TREUER DIENER SEINES HERRN.

Trier, ancient German city situated on the River Mosel at the point at which it enters German territory. It was founded by the Romans as Augusta Treverorum in 15 B.C., and the Porta Nigra, the ruins of the imperial baths, and the amphitheatre survive from Roman times. According to tradition, the bishopric of Trier was founded in the 1st c. A.D.; it was confirmed as an archbishopric under Charlemagne (see KARL I, DER GROSSE). The archbishop was an elector to the German crown and retained this privilege until 1801. Its most famous and most political archbishop was Balduin von Trier (q.v.) in the 14th c.

A part of the original Frankish Empire, the city became German in 925. In the 16th c. its archbishops successfully resisted the encroachment of the Reformation (q.v.). The Napoleonic Wars (q.v.) temporarily gave Trier to France (1801–14), after which period it became a part of Prussia. The most conspicuous ecclesiastical buildings are the Cathedral, built over many centuries, but primarily Romanesque, and the adjoining Gothic Church of Our Lady (Liebfrauenkirche). The reliques include the Holy Coat (Heiliger Rock Christi), first mentioned in the 11th c., which is a focus for pilgrimage. Trier is the birthplace of Karl Marx (q.v.).

Trierer Ägidius, a 12th-c. Middle High German poem recounting the story of St. Ägidius,

his life in the wilderness, his discovery by the Emperor, installation as abbot of a new monastery, and holy death. The poet stresses his humility and his idyllic existence in the wilderness. An atmosphere of peace and harmony pervades the work. The principal MS. is at Trier.

Trierer Blutsegen, see TRIERER ZAUBERSPRÜCHE.

Trierer Capitulare, a fragment of a legal document of 818 concerning Church property rights, translated later in the 9th c. into Old High German and philologically interesting as evidence for dialect (moselfränkisch). The MS., formerly in the Cathedral Library of Trier, is lost.

Trierer Marienklage, a medieval dialogue involving Mary, the Saviour on the Cross, St. John, and St. Peter. It probably originated in the 13th c. and is preserved in a MS. of the 14th/15th c. at Trier.

Trierer Osterspiel, a medieval Easter play, possibly of the 13th c., which belongs to the phase in which the 'Osterspiel' was still closely associated with the liturgy for Easter. The Latin text of the service is incorporated and also given in German translation. It is headed in the MS.: *Ludus de nocte paschae, de tribus Mariis et Maria Magdalena.*

Trierer Silvester, a Middle High German poem recounting the legend of St. Silvester. It is based on the *Kaiserchronik* (q.v.), but differs in its interpretation of the coronation. In the *Trierer Silvester* the Pope is the supreme temporal as well as spiritual authority. It was written in the second half of the 12th c.

Trierer Zaubersprüche, title given to two Low German magic spells preserved in a MS. of the 10th c. at Trier. The first, headed *Ad catarrum dic,* is a blood spell, designed to arrest haemorrhage. It is in Old Saxon verse. The second spell, which bears the superscription *Incantacio contra equorum egritudinem quam nos dicimus spurihalz,* is in Old Low German prose and was intended to prevent or cure lameness in horses.

TRILLER, DANIEL WILHELM (Erfurt, 1695–1782, Wittenberg), a professor of medicine at Wittenberg university, supported Gottsched (q.v.) in the literary disputes of 1740–50, mocking Klopstock (q.v.) in his comic epic poem *Der Wurmsamen* (1751). He is also the author of a didactic poem, *Gedicht von der Veränderung der Arzneikunst* (1768).

Trilogie der Leidenschaft, collective title of a group of three poems written by Goethe between August 1823 and March 1824, with the individual superscriptions *An Werther, Elegie,* and *Aussöhnung.* The biographical background to them was Goethe's love for a young girl, Ulrike von Levetzow (q.v.), which reached its climax in proposal, rejection, and parting at Marienbad in the late summer of 1823. *Aussöhnung,* though the third poem in the cycle, was written first, in the middle of August. Goethe himself recounts that the *Elegie* (often termed the *Marienbader Elegie*) was written down in the coach on the first stages of his journey home. *An Werther* was written in March 1824 for a jubilee edition of *Die Leiden des jungen Werthers* (q.v.) which, however, did not appear until 1825. The other two poems were first published in 1827, when they appeared in the *Ausgabe letzter Hand* (q.v.). *An Werther* and *Aussöhnung,* sad but consolatory, frame the *Elegie,* a sensitive and searing poem of suffering.

All three poems, though obviously autobiographical in genesis, speak of an experience which has a more than personal dimension. The *Elegie,* remarkable in the power of its expression of anguish, shows an unusual conservation of the lyrical faculty in a 74-year-old poet.

TRIMM, THOMAS, see WELK, EHM.

Tristan, the first fully developed Novelle by Th. Mann (q.v.), a Künstlernovelle (see KÜNST-LERROMAN), which was published in 1903. The action is confined to the sanatorium Einfried, which the writer Detlev Spinell uses as a retreat from reality. Imagining he has found in Gabriele Klöterjahn (*née* Eckhof), the wife of a prosperous businessman, a woman of like ideals, he urges her to play the piano while the majority of the patients are on a sledge outing. Gabriele, although in an advanced stage of tuberculosis, defies doctor's orders never to play, and yields to her old, but in recent years suppressed, artistic capabilities. Her rendering of the music from the second act of Wagner's opera *Tristan und Isolde* (q.v.), including Brangäne's 'Habet-Acht-Gesang' and the Todesmotif introducing the Liebestod, awakens in her under Spinell's ecstatic guidance emotions which induce a relapse. She dies soon after her husband has been called to her bedside.

A confrontation between Spinell and Klöterjahn accentuates irreconcilable extremes, the aesthetic world of the artist and the materialistic world of the man of business, both of which are the target of Mann's irony. While Spinell has practised on Gabriele his urge to illuminate unconscious aspects of the mind, regardless of her physical condition, the extrovert Klöterjahn is no less self-centred in his own way. Yet Gabriele

has borne him a healthy child—to Spinell's aesthetic sensibility excessively so—and it is this child, irradiated by the setting sun, which triumphs by its laughter.

Tristan or **Tristan und Isold,** accepted title of a Middle High German poem by GOTTFRIED von Strassburg (q.v.). It remains unfinished, in consequence, it is believed, of the author's death *c.* 1210. Gottfried's source was the *Roman de Tristan* by the Anglo-Norman poet Thomas (Thomas von Britanje), written in the middle of the 12th c. Thomas's poem survives only in fragments which mostly narrate the later part of the story, which Gottfried's poem does not reach. That the German poem conforms closely with its original can be assumed from Gottfried's insistence on fidelity to authority.

The story begins with the short-lived happiness of Tristan's parents, Riwalin and Blanscheflur. Riwalin falls in battle and Blanscheflur dies of a broken heart, leaving the infant Tristan, whose name is symbolical ('von triste Tristan was sîn nam'), to be brought up by noble foster-parents. The boy is spirited away by sea, finds himself in Cornwall, and is brought up by his uncle King Marke. When the mighty Morolt of Ireland comes to Cornwall to collect a tribute of maidens and youths, Tristan slays him, but sustains a wound which can only be cured by Morolt's sister, Isold of Ireland. To avoid identification and death Tristan journeys to Ireland as the minstrel Tantris, is cured, and returns to Cornwall. He is then sent to Ireland a second time by King Marke, for whom he is to woo the Irish queen's daughter. He kills a dragon which is oppressing the land, and so acquires the right of disposing of Isold's hand. She, however, recognizes him as the slayer of her uncle Morolt and is moved to kill him, but is dissuaded by her mother. Tristan woos Isold on Marke's behalf and carries her, with her woman Brangäne, by ship to Cornwall. On the voyage occurs the critical event, the accidental drinking of the love potion, destined for Marke and Isold, which now binds Tristan and Isold irrevocably. Their life henceforth is one of love and suffering. For a time they maintain secrecy. Brangäne, a virgin, is substituted for Isold on Marke's wedding night, and the lovers continue to meet for some time, suspected but undetected. When Marke at last sends them away, they live for a time an idyllic life of love in the *minnegrotte.* They return to the court and continue their clandestine relationship, but are finally discovered and Tristan is exiled by Marke. In Arundel he meets Isold with the White Hands and is drawn to her by the magic name. At this point, after some 19,000 lines, the poem breaks off. In the continuation Tristan would have sustained a wound which only Isold could cure, and deceived by a wrong signal, contrived by the rival Isold with the White Hands, would have died shortly before the true Isold's arrival, whereupon she too would have perished of a broken heart.

Though Gottfried has faithfully translated his original, he has interpolated passages which give the story an individual interpretation. The world for him is divided on the one hand into ordinary, sensual souls, and on the other into the elect who appreciate and experience love as a mystical union, crystallized in the chiastic formula 'ein man ein wîp, ein wîp ein man;/ Tristan Isolt, Isolt Tristan.' It is for the elect that he writes, and his poem celebrates two destined lovers, who maintain their love through suffering and adversity and are finally united in death. *Minne* appears as the central element in a new religious experience, without, however, displacing the accepted Christian creed. Since the two are doubtfully reconcilable, Gottfried's world exhibits an unresolved dualism. Gottfried is a master of form and his verse is the most flexible and mellifluous of any medieval German poet. The poem contains a famous passage on contemporary literature, in which WOLFRAM von Eschenbach is criticized (though his name is not mentioned) and HARTMANN von Aue, BLIGGER von Steinach, HEINRICH von Veldeke, Reinmar, and WALTHER von der Vogelweide (qq.v.) receive praise. The numerous MSS. (23, of which 11 are complete) attest the poem's popularity, and this is supported by the work of ULRICH von Türheim (q.v.) who completed the poem some twenty years after Gottfried's death.

Other medieval versions of the story are the earlier *Tristrant und Isalde* (q.v.) of Eilhart von Oberge, with its later prose form, and two works by successors of Gottfried, *Tristan als Mönch* (q.v.), and a fragment written in the late 13th c. and known as *Der niederfränkische Tristan*.

The standard text of *Tristan* (F. Ranke, 1930 and 1959) was reprinted in 1967 with explanatory material by G. Weber.

Tristan als Mönch, a short verse romance, of some 2,700 lines, which appears in two MSS. of Gottfried von Straßburg's *Tristan* (q.v.). It was probably written c. 1250 and the author was an Alsatian. Tristan, disguised as a monk, passes a dead knight off as Tristan's corpse and is able to enjoy Isolde's love undisturbed. The original of this crude and uncourtly poem was probably a lost French work.

Tristan und Isolde, an opera by Richard Wagner (q.v.), who completed the text in 1857 and the music in 1859. It was first performed in Munich on 10 June 1865. The plot follows the legend, which was chiefly known in Germany through the unfinished *Tristan* (q.v.) of GOTT-FRIED von Straßburg (q.v.).

The opera opens on the ship bearing Isolde under Tristan's escort to marriage with King Marke of Cornwall. Isolde reveals to her hand-maid Brangäne her unsettled score with Tristan, who, having slain her betrothed, Morold, dared to come to her in disguise, seeking cure for his wound. But Isolde expresses her resentment in terms which betray that her real, unacknowledged hatred of Tristan arises from frustrated love. She plans to poison Tristan and herself; but Brangäne mistakenly (yet divining Isolde's wish) brews a love potion instead. The two disembark as passionate lovers. Isolde is married to Marke, but she and Tristan satisfy their passion in clandestine meetings. They are betrayed by Melot, one of Marke's knights. Tristan, insensible from a wound received from Melot, is borne to Careol and tended by his faithful retainer Kurwenal. Isolde comes to him across the sea and Tristan dies in her arms. Marke follows in a second ship, but his aim, too late, is conciliation, not persecution. Isolde, now that Tristan is dead, wills her own death.

For many years the chromatic harmonies aroused the strongest musical objections in orthodox listeners and critics, and the languishing sensual music of the love night encountered equally strong reprobation.

Tristia ex Ponto, a collection of seventeen poems by F. Grillparzer (q.v.), published in *Vesta* in 1835. The title is adapted from Ovid's *Tristia* and *Epistulae ex Ponto*. It opens with *Böse Stunde*, the 'hour of anguish', self-scrutiny, and hope for creative strength, which indicates the theme underlying the flexible temper of the cycle. *Jugenderinnerungen im Grünen*, written 1824–6, includes in its reminiscences allusions to Grillparzer's relationship to Kathi Fröhlich. The last poem, *Schlußwort* (1830), passes the poet's sufferings into the sphere of dream.

Tristrant und Isalde, a Middle High German verse romance, some 9,400 lines in length, written by EILHART von Oberge (q.v.) probably c. 1170. Approximately 1,000 lines of something like its original form are preserved in three 13th-c. MSS. The complete poem is known in adaptations in MSS. from the 15th c. and a prose version in print, which was first published in 1484 (*Historie von Herrn Tristant und der schönen Isalden von Irlande*).

The story of Tristan and Isolde came from France and *Tristrant und Isalde* represents its earliest German form. The poem recounts Tristrant's education, which has made him into a perfect knight. His defeat of Morold follows and the cure of his wound by Isalde. Tristrant is once more sent to Isalde in Ireland to woo her on behalf of King Marke. On their journey to Cornwall both accidentally drink the love potion which binds them for four years. Their secret intercourse is discovered and they flee to the woods. Four years pass, the power of the potion lapses and Isalde returns to Marke, who takes her back but banishes Tristrant. There follows a section recounting various adventures and in particular Tristrant's unconsummated marriage with Isalde of the White Hand. Tristrant finally returns to Marke's court in disguise but is wounded. Isalde comes to heal him, yet a confusion of signals (black sail for white) leads to Tristrant's death. The pair are united in death, for Isalde, too, perishes.

Though there are many points of contact, Eilhart's poem is morally and linguistically cruder than Gottfried's *Tristan* (q.v.). Nevertheless, that the story was popular in this form is proved by the number of MSS. and the printed versions. F. Lichtenstein's edition of 1877 was reprinted in 1973.

TRITHEMIUS, Johannes von, real name Heidenberg (Trittenheim nr. Trier, 1462–1516, Würzburg), abbot of Sponheim near Kreuznach, 1485–1506, and thereafter of St. James's (St. Jakob) at Würzburg, was one of the outstanding German humanists of his age. His interests covered history, literature and its history, and secret codes of writing. Among his more important works are *Catalogus illustrium virorum Germaniae* (1491), *Catalogus scriptorum ecclesiae* (1494), and two works on secret handwriting, *Stenographia* (1500) and *Polygraphia* (1518). He wrote exclusively in Latin; Trithemius is the Latinized form of his birthplace.

Triumph der Empfindsamkeit, Der, a satirical prose play in six short acts written by Goethe probably in 1778 and published in *Goethes Schriften*, 1787, with the sub-title *Eine dramatische Grille*. It was performed at Weimar for the birthday of the Duchess Luise on 30 January 1778. It satirizes the sentimentality of the age, specifically ridiculing Rousseau's *La Nouvelle Héloïse*, J. M. Miller's *Siegwart*, and Goethe's own *Die Leiden des jungen Werthers* (q.v.).

The fourth act, in free verse, is a play within the play, a sad monologue spoken by Proserpina in the underworld, followed by a brief dialogue between her and the Parcae. This act is of separate origin and it has been conjectured that it may have been a threnody for Goethe's sister who died in 1777; but it is also possible that it was written in 1776 for performance by Corona Schröter (q.v.).

Trivialliteratur, term for long loosely applied to light literature appealing to popular taste. Since the 1960s a number of theories have suggested a more constructive use, the major trend being its dissociation from aesthetically subjective approaches (see Kitsch) and its integration into the historical assessment of literary periods, e.g. their changing communication structures and value judgements (H. Kreuzer and J. Schulte-Sasse). (See also Rezeptionsästhetik.) Historical examples include, for example, the Ritter- and Räuberroman (q.v.). The terms Trivialfilm and Trivialled have the same romantic and escapist connotations.

TROJAN, Johannes (Danzig, 1837–1915, Rostock), worked as a journalist in Berlin and was from 1886 to 1899 editor of the *Kladderadatsch* (q.v.). He published several volumes of light verse (*Beschauliches*, 1871; *Gedichte*, 1883; *Scherzgedichte*, 2 vols., 1883 and 1908, and *Hundert Kinderlieder*, 1899). He was also a nature lover, writing *Von Strand und Heide* (1887), *Aus dem Reich der deutschen Flora* (1910), and *Unsere deutschen Wälder* (1911).

Trojanerkrieg, Der, an immense unfinished Middle High German verse romance by Konrad von Würzburg (q.v.). It is the last of Konrad's three long verse narrations and reaches more than 40,000 lines before breaking off just before the death of Hector. Konrad's source is the *Roman de Troye* of Benoît de Sainte More (see also Herbort von Fritzlar), a very extensive work which Konrad has considerably expanded. The preparations for the Trojan War proper are reached after more than 23,000 lines. It is presumed that Konrad's death in 1287 was the reason for the unfinished state of the work. In *Der Trojanerkrieg* Konrad's tendency to expand speeches and descriptions reaches the point of downright verbosity. Helen, for instance, responds to Paris's declaration of love in a speech of more than 900 lines. Konrad names in the introduction his patron, Dietrich von Basel an dem Orte.

An unknown and less skilful hand completed the poem in some additional 9,000 lines. *Der Trojanerkrieg* was immensely popular and several subsequent treatments of the subject are known, of which one, *Der Göttweiger Trojanerkrieg* (q.v.), is in verse. One of these prose versions is anonymous, two others are by Hans Mair (*fl. c.* 1390) and Heinrich von Braunschweig (*fl. c.* 1400), respectively.

TRÖMER, Johann Christian (Dresden, before 1700?–56, Dresden), at first a bookseller's apprentice, became servant and then factotum to the Duke of Saxe-Weissenfels. He wrote a comic account of his own adventures in a barbarous gallicized German and in rhyming verse under the title *Jean Chretien Toucement des Deutsch Franços Schriften* (1736). A second edition in 1745 had a new title: *Die Avantures von Deutsch Franços mit all sein Scriptures*. 'Deutsch Franços' has been suggested as a source for Lessing's Riccaut in *Minna von Barnhelm* (q.v.).

TROMLITZ, AUGUST VON, pseudonym of K. A. F. von Witzleben (q.v.).

Trommeln in der Nacht, a five-act play by B. Brecht (q.v.), written in 1918–20, published in 1923 after its first performances in Munich (Kammerspiele) on 30 September, and Berlin (Deutsches Theater) on 20 December 1922. Against the background of the Spartacist revolt (see SPARTAKUSBUND) a returned soldier, Andreas Kragler, is drawn towards revolution when he finds himself deserted by his fiancée Anna, and abandons it as soon as she returns to him. The play was awarded the Kleist prize in 1922, but with characteristic self-scrutiny Brecht revised the final acts for publication as a comedy in 1927.

Trompeter von Säckingen, Der, a narrative poem by J. V. von Scheffel (q.v.). Begun in 1850 in Säkkingen, it was finished in Capri in 1853 and published in 1854. It was one of the best sellers of the 19th c. The poem is set in the second half of the 17th c.

Its hero, Jung-Werner, abandons his studies in Heidelberg and wanders southwards with his trumpet, on which he is a skilled performer. In Säkkingen he falls in love with a beautiful girl taking part in an ecclesiastical procession. Jung-Werner becomes master of music to her father, a nobleman (der Freiherr), and the daughter, Margareta, presently falls in love with him. Werner distinguishes himself in defence of the castle during a peasant revolt (der Hauensteiner Rummel), but his suit is rejected by the Baron because of his lower social station. He departs and journeys to Italy, and after some years becomes a musician in the Pope's service. Margareta and he meet in St. Peter's, the Pope ennobles him, and they return to be married in Säkkingen.

The sentimental tale is laced with rather arch humour. The poem, of some 6,000 lines, is written in unrhymed trochaic verse. It contains the famous poem 'Alt-Heidelberg, du feine' (q.v.) and a collection of songs, some of which are attributed to a cat, Hiddigeigi.

Trompete von Vionville, Die, also occurring as *Die Trompete von Gravelotte*, a patriotic war poem by F. Freiligrath (q.v.). It refers to an in-cident in the combined battle of Vionville and Mars-la-Tour on 16 August 1870, in which two regiments (7. Kürassiere and 16. Ulanen under General Bredow) made a desperate charge. According to the poem, the trumpet on which the rally was sounded was pierced by bullets and could give only a sad, mourning note.

Tröst-Einsamkeit, see ZEITUNG FÜR EIN-SIEDLER.

Trost in Verzweiflung, a fragment, 165 lines long, of an Early Middle High German poem. Beginning with praise of the ascetic life, it goes on to describe, in terms unusually personal for its age, the pursuit of the heart by passions, from which it is rescued by Our Lord. It was written in South Germany *c.* 1180.

TRUCHSESS VON ST. GALLEN, see ULRICH VON SINGENBERG.

TRUCHSESS VON WALDBURG, GEORG (1488–1531), a soldier of ancient family, commanded the forces of the Swabian League which in 1519 expelled Ulrich von Württemberg (q.v.) from his territory. He was also a successful commander against the peasants in 1525 in the Peasants' War (see BAUERNKRIEG); his victory at Böblingen was the turning-point of the war.

Trutznachtigal, title given to a collection of 51 poems by Friedrich von Spee (q.v.), published in 1649, fourteen years after his death, by Wilhelm Nakatenus (q.v.).

Trutz Simplex: oder Ausführliche und wunderseltzame Lebensbeschreibung der Ertzbetrügerin und Landstörtzerin Courasche, a novel by J. J. C. von Grimmelshausen (q.v.), published *c.*1669. Courasche (whose name is a hallmark of her sexuality, see Ch. 3) tells her life, which is a succession of love-affairs and marriages, in the course of which she has feathered her nest. A woman of remarkable beauty and no morals, she flits from man to man, and, as time goes on, contracts the pox and declines into crime. One of Courasche's lovers is Springinsfeld, a character in *Der abenteuerliche Simplicissimus* and *Der seltzsame Springinsfeld* (qq.v.). Courasche makes her first appearance in *Simplicissimus*, Bk. 5, and is the original of Brecht's play *Mutter Courage und ihre Kinder* (q.v.). Grimmelshausen published the *Trutz Simplex* under the pseudonym Philarchus Grossus von Trommenheim auf Griffsberg.

TSCHAMMER UND OSTEN, HIOB GOTT-HARD VON (Schloß Dromsdorff, Silesia, 1674–1735, Dromsdorff), a Silesian country squire, took to writing poetry, when increasing age debarred him from the usual rural pursuits. His poems were published after his death (*Geistliche und weltliche Gedichte*, 1737).

TSCHECH, JOHANN THEODOR VON (Voigtsdorf nr. Glatz, 1595–1649, Elbing), a Silesian nobleman and friend of Abraham von Franckenberg (q.v.), was an ardent disciple of Jakob Böhme (q.v.), in support of whose religious views he published *Das wahre Licht* (1627).

TSCHERNING, ANDREAS (Bunzlau, 1611–59, Rostock), a minor lyric poet, was a firm adherent of M. Opitz (q.v.), whom he knew personally. The son of a Protestant furrier, he moved to Görlitz in order to escape religious persecution. He studied in Rostock and was a tutor in Breslau. He returned to Rostock in 1642 and became professor of poetry there in 1644.

Tscherning's poetry makes no claim to originality, taking familiar themes and paraphrasing earlier poems. His *Deutsche und lateinische Gedichte* appeared in 1634, *Deutscher Getichte Frühling* in 1642, *Vortrab des Sommers deutscher Gedichte* in 1655. *Unvorgreiffliches Bedenken über etliche Missbräuche in der deutschen Schreib- und Sprachkunst, insonderheit der edlen Poeterey* (1658) is an amplification of Opitz's *Buch von der deutschen Poeterey* (q.v.).

TSCHUDI, ÄGIDIUS (Glarus, 1505–72, Glarus), a Swiss chronicler, was successively governor of Sargans (Landvogt, 1529), of Rorschach (Obervogt, 1532), and of Baden, Aargau (Landvogt, 1536). After travels in France and Italy, Tschudi was elected chief magistrate (Landamman) of Glarus in 1558. An ardent and intolerant Catholic, he attempted to suppress Protestantism in Glarus by force, but was unsuccessful and was for a time in exile. Throughout his life he pursued historical studies, but only his *Uralt wahrhafftig Alpisch Rhaetia* (1538) appeared in his lifetime. *Vom Fegfür*, ed. J. A. Knowles, was published in 1925. His best-known work, *Chronikon Helveticum*, used as a source by many Swiss historians and by Schiller in his *Wilhelm Tell* (q.v.), was published by J. R. Iselin (2 vols.) in 1734–6 (reissued, ed. P. Stadler and B. Stettler, 1968 ff.). It covers the period 1000 to 1470 and is the earliest work to present the story of Tell in its familiar form. Tschudi's historical reliability was successfully assailed in the late 19th c. and he is no longer accepted as an authority.

Tübinger Stift, a Protestant theological seminary founded in 1560 and housed in a former Augustinian monastery. Among famous alumni were Hegel, Hölderlin, Mörike, and Schelling (qq.v.). It is associated with the University of Tübingen.

TUCHOLSKY, KURT (Berlin, 1890–1935, nr. Göteborg, Sweden), a journalist and a writer of trenchant satires in the form of chansons for cabaret. He completed his law studies before the 1914–18 War, in which he served. He then lived in Berlin, from 1924 in Paris, and from 1929 in Sweden. He also used the pseudonyms Ignaz Wrobel, Peter Panter, Theobald Tiger, and Kaspar Hauser. He was a contributor to *Rote Signale* (q.v., 1931), a volume of left-wing verse, and was associated with the pacifist C. von Ossietzky (q.v.). In 1933 he was deprived of German nationality and his books were burned. He took his own life and released nothing for publication in the three years preceding his death.

Tucholsky's writings include the charming story *Rheinsberg. Ein Bilderbuch für Verliebte* (1912), *Zeitsparer* (1914, essays in the grotesque), *Fromme Gesänge* (1919), *Träumereien an preußischen Kaminen* (1920), and the novel *Schloß Gripsholm* (1931). With W. Hasenclever (q.v.) he wrote the comedy *Christoph Columbus* (1932). The edition by Mary Gerold-Tucholsky *Na und — ?* (1950) selects glossaries, aphorisms, and poems which highlight the Berlin wit underlying his satires in prose and in verse without concealing intense suffering. *Gesammelte Werke* (4 vols.), ed. M. Gerold-Tucholsky and F. J. Raddatz, appeared 1960–2 and a supplementary volume (*Deutsches Tempo*) in 1985. Volume 4 contains *Ausgewählte Briefe 1913–1935*. Other correspondence includes *Politische Briefe*, ed. F. J. Raddatz (1969), *Briefe aus dem Schweigen*, ed. M. Gerold-Tucholsky and G. Huonker (1977), and *Unser ungelebtes Leben. Briefe an Mary Gerold-Tucholsky*, ed. F. J. Raddatz (1982).

TÜGEL, LUDWIG (Hamburg, 1889–1972, Ludwigsburg), spent the greater part of his life in Ludwigsburg. He is the author of works of fiction mainly concerned with the people and countryside of North Germany, including the collections of stories *Die Freundschaft* (1939), *Auf der Felsentreppe* (1947), *Das alte Pulverfaß* (1948), and *Die Dinge hinter den Dingen* (1959), and the novels *Pferdemusik* (1935), *Die Charoniade* (1950, retitled *Auf dem Strom des Lebens*, 1961), and *Ein ewiges Feuer* (1963).

Tulifäntchen, a comic verse epic by K. L. Immermann (q.v.), published in 1830. Its satire is directed against A. von Platen and F. de la Motte FOUQUÉ (qq.v.).

Tundalus, a legendary Irish knight who died in 1149, passed through Hell and Heaven, and was restored to life at his burial. Two Early Middle High German poems, both based on the same Latin original, recount his experiences. The poem is a warning to the worldly. One version, in Hessian (Mittelfränkisch) dialect, is fragmentary. The other, in Bavarian, was written by a monk of Windberg monastery who gives his name as Alber. Both were written towards the end of the 12th c.

TÜNGER, AUGUSTIN (Endingen, Baden, 1455–d. after 1486), an administrative official in the episcopal see of Constance, is the first author of Facetien (see FAZETIE). His collection, in crude Latin with German translation, is entitled *Facetiae Latinae et Germanicae* (1486).

Tunnel über der Spree, Der, a literary club in Berlin founded in 1827 by M. G. Saphir (q.v., originally Berliner Sonntagsverein) and active particularly in the years 1840–60, long after Saphir had left Berlin (1830). The members met in a café or restaurant and original poems, stories, and extracts were read aloud and judged. Among the more notable literary and artist members were F. Th. Kugler, Th. Storm, E. Geibel, P. Heyse, Th. Fontane, C. F. Scherenberg, G. Hesekiel, H. Seidel, F. Dahn, and A. Menzel (qq.v.), and there were also a great number of dilettanti, who were by profession mostly civil servants or officers. G. Keller (q.v.) was once present as a guest. An account of the club is given by Th. Fontane in *Von Zwanzig bis Dreißig* (1898).

Turandot oder Der Kongreß der Weißwäscher, a play in 10 scenes with songs by B. Brecht (q.v.) and music by H. Eisler (q.v.), written in 1953–4, but not finally revised. The material, however, had interested Brecht since the 1930s, as a subject both for a play and for the novel *Das goldene Zeitalter der Tuis*. The play adapts facets of the oriental story, published in German translation in the collection *1001 Tag* by F. P. Greve (1879–1910). Princess Turandot will only marry the suitor who can solve a puzzle. If he fails to solve it he will be beheaded. In the end she does find an acceptable suitor. Apart from the operas based on this tale (Busoni, 1917, and Puccini, 1926), it was dramatized by Waldfried Burggraf in 1925. But already Carlo Gozzi had turned it into a *commedia dell'arte* which is the basis of the adaptation made by Schiller (q.v.). In both these versions the puzzle is replaced by a competition. Brecht adapted this variant and gave the action a Chinese setting.

The *Weißwäscher* of the sub-title are intellectuals, the Tuis (a bold corruption of the word *Intellektueller*), whom the Emperor invites to invent an explanation for the shortage of cotton, which he and his brother have monopolized. The most ingenious liar is to receive Turandot's hand in marriage. Several members of the Tui school, which specializes in teaching injustice, are beheaded for inventing useless explanations. Munka Du's head joins the others because he has ventured to give the true explanation. The heads find comfort, however, by enjoying free speech on the posts on which they have been stuck. The gangster Gogher Gogh, whom Turandot brings into the palace, believes neither in the conference of the Tuis which the Emperor has called nor in 'whitewashing', but suggests that questions about the whereabouts of the cotton should simply be prohibited. The Emperor is pleased and Gogher Gogh soon takes charge of the palace. The Emperor's brother Jan Jel is beheaded and half the cotton is burnt in the stores, so that the rest can be sold at a high price. At the wedding ceremony the bridegroom is to wear the ancient imperial cloak, which consists of rags symbolizing China's poverty in ancient times. As Turandot is, against her will, about to be married to Gogher Gogh, it is discovered that the cloak has disappeared from its place in the temple. Before the gangster can get away with his marriage and the Emperor with his betrayal of the people, the followers of the revolutionary Kai Ho storm the temple and triumphantly drive them all out, in headlong flight.

Turm, Der, a symbolical play by H. von Hofmannsthal (q.v.), published in 1925 and, in a revised form, in 1927. Hofmannsthal derived inspiration from Calderón's *La vida es sueño* and Grimmelshausen's *Der abenteuerliche Simplicissimus* (q.v.). The setting of the play suggests 'a past age, similar to the atmosphere of the 17th c.'.

Because of an oracle predicting that he will be a danger to the King, Sigismund, heir to the King of Poland, has been confined in a tower, where he barely exists in degrading conditions, ignorant of his own identity and rank. Magic potions play a part in the vicissitudes through which he passes. He emerges from the tower and is brought to court, where he attacks his father. Once more he is consigned to the tower, and is now, through the knowledge he has acquired of the world and of himself, satisfied to be alone. A revolt against the tyrannical king, whose principal agent is Olivier, the epitome of violent evil, is successful. Sigismund is set in his father's place and seeks to rule with mild humanity. Against his will, however, he is forced to violent action. Olivier is killed, but a gipsy in his pay stabs Sigismund with a poisoned dagger. As he dies, a

Child King (Kinderkönig) enters, and his boy followers sing a Latin chant ending with *renovabis faciem terrae* ('thou shalt renew the face of the earth'). It is on this conciliatory and hopeful note that the version of 1925 closes. In that of 1927 the episode with the children is omitted, Sigismund is shot by Olivier, and brutality prevails. The alterations are sometimes, though probably wrongly, regarded as foreshadowing the National Socialist regime in Germany. The difficulties which Hofmannsthal experienced in expressing his intentions in this play are reflected in the number of pregnant silences and the unusual quantity of stage directions. *Der Turm* remains a baffling work which invites divergent interpretations.

TURMAIER, JOHANNES, see AVENTINUS.

Turnier, Das, a short Middle High German poem the subject of which is a tournament. The author was a Rhinelander, writing *c.* 1300, and his poem contains a eulogy of a Rhenish nobleman, Adolf von Windhövel. (See DER MINNE-HOF, DIE RITTERFAHRT, and DER RITTERPREIS.)

Turnier von Nantes, Das (*Der Turnei von Nantheiz*), a poem of some 1,100 lines written by KONRAD von Würzburg (q.v.). It describes a tournament purporting to have been held at Nantes between German knights under King Richard of England and French knights under their own king. Much authentic detail is devoted to coats of arms, weapons, and procedure. The date of the poem is the subject of controversy. 'King Richard' is thought to be Richard of Cornwall, who was a candidate in the election of the German King in 1257. According to this view it is a political poem of this period and an early work of Konrad's. Others believe, on stylistic grounds, that it is a late work of *c.* 1280.

TWINGER VON KÖNIGSHOFEN, JAKOB (Königshofen, Alsace, 1346–1420, Strasburg), who was a canon of Strasburg, compiled a chronicle covering the history of mankind since the Creation. Three versions of the *Chronik von Königshofen* exist, of which the latest closes at 1415. Twinger drew substantially on the *Straßburger Chronik* of Fritsche Closener (q.v.).

U

'Über allen Gipfeln', first line of a poem by Goethe, written on 6 September 1780 in a hut on the Gickelhahn (q.v.) near Ilmenau, but not published until 1815. In Goethe's collected poems it is headed *Ein Gleiches*, that is to say it is intended to have the same title as the preceding poem, *Wanderers Nachtlied*. It consists of only eight short lines, comprising twenty-four words. In this compass Goethe has created an evocation of evening mood and harmony with nature which is matchless in its balance and precision. Goethe wrote it upon the wall of the hut, which was burnt down in 1870. The poem has been set to music by both F. Schubert and R. Schumann (qq.v.).

Über Anmut und Würde, a long philosophical essay written by Schiller in 1793 and published in the periodical *Die neue Thalia* (1793). It is the first fruit of the study of Kantian philosophy made possible for Schiller by the Danish benefaction of 1791 and reflects his preoccupation with the relationship between art and ethics. While accepting Kant's main principles, Schiller dissents from the view that the participation of pleasure in an action adversely affects its

moral meritoriousness. He sets up the conception of the 'beautiful soul' (schöne Seele), in which duty and inclination coincide, so that it does what is right by instinct rather than on principle. The expression of such a personality, he says, is 'grace' (Anmut). But this ideal harmony is rarely met and in practice duty and inclination are in collision. The character who, from motives of duty, subdues his personal impulses achieves dignity (Würde). Schiller's divergence from Kant is epitomized in the definition of virtue as 'eine Neigung zur Pflicht'.

Überbrettl, Das, a cabaret founded in Berlin in 1900 by E. von Wolzogen (q.v.).

Über das Erhabene, an essay on tragedy and the sublime published by Schiller in his *Kleinere prosaische Schriften* (vol. 3, 1801). When it was written is uncertain, but it was not before 1793. The essay puts with great clarity and eloquence the view that man, in the face of overwhelming disaster, can only retain his moral freedom by a willing acceptance. Moral education through art is the means by which man attains this sublimity ('Der moralisch gebildete Mensch, und nur dieser, ist ganz frei'). The essay contains Schiller's well-known assertion that the will is the specific characteristic of man: 'Alle andere Dinge müssen; der Mensch ist das Wesen, welches will'.

Über das Marionettentheater, a short essay by H. von Kleist (q.v.), published in December 1810 in the *Berliner Abendblätter.* It purports to record a discussion between the author and a male dancer in 1801 on the graceful and effortless perfection of movement achieved by marionettes. The puppet, only parts of which are controlled by the player, exhibits the unimpeded operation of the law of gravity. It demonstrates, by analogy, man's unselfconscious grace before the Fall, the first stage in his development. The second stage is represented by the human dancer striving to attain the same degree of perfection which he perceives in the soulless puppet. He fails because the art of dancing involves a reflective stage exemplified by self-scrutiny in a mirror and a resultant distortion of movement. A higher, third stage in the development of art, and of man's spiritual awareness, is suggested; self-reflection yields to self-transcendence effecting the fusion of man's finite being with the infinite consciousness of God and the perfect co-ordination and co-incidence of physical and spiritual planes. Its appearance, combining grace and beauty, is termed Grazie. The essay is widely held to illustrate principles underlying Kleist's creative works.

Über das Pathetische, an essay on tragedy by Schiller, published in *Die neue Thalia* in 1793. The title refers to suffering (πάθος) in tragedy. Schiller contends that the basic principles of tragic art are the depiction of suffering and the presentation of moral resistance to suffering.

Über den Gebrauch des Chors in der Tragödie, an essay by Schiller printed as a preface to *Die Braut von Messina* (q.v.) in 1803. Schiller defends his practice in introducing a chorus into a modern stage play, claiming that the chorus, acting as 'a living wall', separates the tragedy from the spectators and makes clear the non-real, idealistic nature of art.

Über den Umgang mit Menschen, a didactic work by Adolf, Freiherr von Knigge (q.v.), was first published in 1788. Its title is often quoted in conversation and it is widely believed to be a treatise on etiquette and manners. It is in reality a work of practical wisdom on the conduct of life and contains essays with such headings as *Über den Umgang mit sich selbst, Von dem*

Umgange mit Eheleuten, Über den Umgang mit Gelehrten und Künstlern, etc. Knigge is a man of broad humanity and good sense, who opts for moderation and the golden mean.

Über den Ursprung der Sprache. Von der Akademie der Wissenschaften zu Berlin im Jahr 1770 gekrönte Preisschrift, a prize-winning dissertation submitted by J. G. Herder (q.v.) for a competition and published in 1772. It was republished in revised form in 1789. The first part attacks various existing theories, the second puts forward the view that language arises with conscious thought ('So war auch das Moment der Besinnung, Moment zu innerer Erstehung der Sprache').

Über die ästhetische Erziehung des Menschen in einer Reihe von Briefen, an essay on aesthetics published by Schiller in 1795 in *Die Horen,* Nos. 1, 2, and 6. The work had its origin in philosophical letters which Schiller addressed in gratitude to his benefactor, Prince Friedrich Christian von Augustenburg (q.v.) in 1793 and 1794. These were then revised and published as an epistolary essay in aesthetics on a psychological basis. The letters champion the educative function of art and claim that the effect of successful poetry is a balance of the mind in which the potentialities are perfectly poised. In this sense 'das Schöne' (Erster Brief) is the essay's central motif.

Über die deutsche Litteratur, see FRIEDRICH II, DER GROSSE.

Über die Diotima, an essay published by Friedrich Schlegel (q.v.) in 1795 in J. E. Biester's *Berlinische Monatsschrift.* Its starting-point is the Diotima in Plato's *Symposium,* and Schlegel turns his essay into a justification of the intelligent, emancipated woman.

Über die Lehre des Spinoza, an essay on Spinoza (q.v.) in the form of letters published in 1785 by F. H. Jacobi (q.v.).

Über die neuere deutsche Literatur, a critical commentary on German literature published anonymously by J. G. Herder (q.v.) and often known as Herder's *Fragmente.* This title derives from the headings of the separate sections, (*Erste, Zwote, Dritte*) *Sammlung von Fragmenten.* All three collections were published in 1767, the first and second together, the third separately. The work links up with Lessing's *Literaturbriefe* (q.v.), as the sub-heading of the first section shows, *Eine Beilage zu den Briefen, die neueste Litteratur betreffend.* The first collection, *Von den Lebensaltern der Sprache,* applies Herder's idea of organic development to language, postulating phases of youth, maturity, and age; the second deals with Hebrew and Greek literature and their significance for German letters; and the third attacks a classicism based on Latin literature. The work, with its dynamic conception of original genius, was a potent influence on the Sturm und Drang (q.v.).

Über die Shakespearo-Manie, an essay by C. D. Grabbe (q.v.), written for inclusion in the second of his two volumes of *Dramatische Dichtungen* (1827), but backdated to 1822. Grabbe claims that the fashionable admiration of Shakespeare (q.v.) was based on a false image and tended to inhibit the independent development of German writers who had become 'Shakespearo-Manisten'. Goethe and Schiller had proved that they could achieve their aim without Shakespeare. The public blindly admired in Shakespeare features which they would have criticized in Schiller. Grabbe's claim that Shakespeare depended to a greater extent on other English models than was generally known had already been made by L. Tieck (q.v.).

Grabbe finds many faults in Shakespeare's plays, viz. the lack of an underlying central idea, excesses in language, false eloquence, poor expositions, unmotivated changes from prose to verse; only *A Midsummer Night's Dream* receives his unreserved praise. Shakespeare was neither the best model for tragedy (which was Sophocles), nor for comedy (which was Molière). Grabbe's criticism coincides with short-comings in his own plays, of which he was probably aware.

Über die Soldatenehen, a tract written by J. M. R. Lenz (q.v.) in 1776 (published in 1914) and submitted in manuscript to Duke Karl August (q.v.) of Saxe-Weimar. See SOLDATEN, DIE.

Über dramatische Kunst und Litteratur, see VORLESUNGEN ÜBER DRAMATISCHE KUNST UND LITTERATUR.

Über Kunst und Altertum, a series of volumes, six in all, edited by Goethe between 1816 and 1832. Their full title, rarely used, is *Über Kunst und Altertum in den Rhein- und Maingegenden.* Others beside Goethe contributed to these volumes. The series is best known for the first publication of the *Sankt-Rochus-Fest zu Bingen* (q.v.), in vol. 2 (1817).

Über Lessing, a critical essay by F. Schlegel (q.v.), first published in 1797 in Reichardt's *Lyzeum der schönen Künste.* Brilliant but erratic

and subjective, it advances the view that *Anti-Goeze* (q.v.) is Lessing's best work, and contains the famous lapidary judgement that *Emilia Galotti* is 'Unstreitig ein großes Exempel der dramatischen Algebra'.

Über naive und sentimentalische Dichtung, a critical essay by Schiller, published in *Die Horen* (Nos. 11 and 12 of 1795 and No. 1 of 1796). In the category of the 'naïve' the poet is at one with Nature. Such a state Schiller finds in the Greeks. The modern poet, who is conscious of his separation from Nature and longs to return to it, is 'sentimental' (Schiller uses 'sentimentalisch', not 'sentimental', and the sense is perhaps a compound of the reflective and the slightly sentimental). 'Naïve' poets tend to realism, 'sentimental' ones to the idealistic. The 'naïve' is in its nature indivisible. 'Sentimental' poetry can be satirical or elegiac, and the idyll is a subdivision of elegy. Schiller saw Goethe as a 'naïve' poet, whereas he himself pertained to the 'sentimental' category. The essay is both a stimulating adventure into general theory and an attempt to discover a typology which would classify both writers as autonomous poets in their own right.

Udo von Magdeburg, an anonymous Middle High German verse legend of some 800 lines, which was written in the 14th c. It recounts the life, death, and damnation of Archbishop Udo of Magdeburg, a fictitious prelate, whose story probably originated with Udo, Bishop of Hildesheim, in the late 11th c. and early 12th c.

Udo, a poor scholar, is given wisdom and understanding by the Virgin Mary, who promises him that he shall be archbishop of Magdeburg; but as archbishop he neglects his duties and lives in sin with nuns and an abbess. He is summoned to a divine judgement in the cathedral, which is accidentally witnessed by a canon. Udo is found guilty and executed. The same day a chaplain sees the Archbishop's soul seized by the Devil, tormented until he curses God, and consigned to everlasting torment. Finally, the disposal of his noisome body causes acute difficulties to the capitular authorities.

UECHTRITZ, FRIEDRICH VON (Görlitz, 1800–75, Görlitz), a Prussian civil servant, was in Berlin and Trier before becoming Landgerichtsrat in Düsseldorf (1833–58), where he was in friendly contact with K. L. Immermann (q.v.). Influenced by L. Tieck (q.v.), whom he met while a student at Leipzig University, he wrote unsuccessful verse plays in the Schillerian tradition, including *Alexander und Darius* (1826), *Die Babylonier in Jerusalem* (1836), and *Das Ehrenschwert* (1837). He commented on the cultural life in Düsseldorf during the 1830s in *Blicke in das Düsseldorfer Kunst- und Künstlerleben* (2 vols., 1839–40), and later turned to novels (*Albrecht Holm*, 7 vols., 1851–3; *Der Bruder der Braut*, 3 vols., 1860; and *Eleazar*, 3 vols., 1867). F. Hebbel (q.v.), who knew him as a dramatist, met him in 1854 in Marienbad, and corresponded with him until shortly before his death.

Ugolino, a tragedy (*Eine Tragödie*) by H. W. von Gerstenberg (q.v.), published in 1768. The story of Count Ugolino della Gherardesca is taken from Dante's *Inferno*, Canto XXXIII. Ugolino and his three sons are starving in the dungeon at Pisa. Hope rises as one son escapes through the roof, but his body is soon brought in in a coffin, together with the body of Ugolino's wife. As the pangs of hunger increase, Ugolino becomes insane for a spell and kills one of his sons. The last son dies of starvation and Ugolino kills himself.

It is a play of horror and agony, which is meant to wring the heart, and succeeded in doing so for its contemporaries; it was successfully performed, in both Berlin and Königsberg, in 1769. The play conforms to the discipline of the unities in its setting and in the duration of the action, which is limited to the last few hours of the Ugolinos, but it is written in prose; its style and its subject-matter appealed also to the Sturm und Drang (q.v.).

UHLAND, LUDWIG (Tübingen, 1787–1862, Tübingen), was the grandson of a professor of Tübingen University and a son of the University Secretary. He was educated at the grammar school and after matriculation at the early age of 14 began to read law at the university, completing his studies in 1808. During these years he formed friendships with J. Kerner and K. Mayer (qq.v.) and wrote ballads and Romantic poetry with folk-song affinities; he also became deeply interested in medieval literature and German legend and was subsequently regarded as a leading member of the Swabian School of the Romantic movement (see ROMANTIK).

In 1810–11 his father sponsored a journey to Paris, on which Uhland was to study French law (Code Napoléon), but he also devoted appreciable time to studying and copying medieval MSS. in the Bibliothèque impériale (now Bibliothèque nationale). In 1812 he was appointed to a civil service post in the Württemberg Ministry of Justice, but resigned in 1814 because of conscientious disagreement with policy and set up his own private practice of law in Stuttgart. In 1815 he published his *Gedichte* containing the bulk of his output, most of which was written in youth. The poems include *Die Kapelle*, *Der Schmied* (q.v.), 'Die linden Lüfte

sind erwacht', and the ballads *Die Rache* and *Des Sängers Fluch* (1814, qq.v.), his outstanding masterpieces in this form, for which he became well known. A few late poems written in 1829 and 1834 include *Tells Tod* and *Das Glück von Edenhall*; meanwhile he wrote political poetry (*Vaterländische Gedichte*, 1817) and completed two verse tragedies (*Ernst, Herzog von Schwaben*, 1818, and *Ludwig der Bayer*, 1819, qq.v.).

In 1820 Uhland married Emma Vischer (1799–1881), having the previous year become a member of parliament for Tübingen. A staunch Liberal, he was not well viewed in government circles, but he was re-elected in 1826, this time for Stuttgart. In 1829 he achieved an ambition by becoming a professor of German language and literature at his native university. He resigned, however, three years later, when the government refused him leave of absence for which he had asked to attend to his parliamentary duties. From 1839, when he left parliament, he devoted himself to private scholarship. In 1848 he was elected as a Liberal to the new German parliament (see FRANKFURTER NATIONALVERSAMMLUNG), resuming his literary and philological researches after its dissolution in 1849.

Uhland ranks as one of the founders of German literary and philological studies. His essay *Walther von der Vogelweide* appeared in 1822 and his *Sagenforschungen* in 1836. One of his most important publications was his scholarly *Alte hoch- und niederdeutsche Volkslieder* (1844–5). Other fruits of research appeared posthumously in *Schriften zur Geschichte der Dichtung und Sage* (8 vols., 1865–73), ed. A. von Keller, W. L. Holland, and F. Pfeiffer. *Werke* (6 vols.), ed. H. Fischer, appeared in 1892, the critical edition by E. Schmidt (q.v.) and J. Hartmann (2 vols.) in 1898, an edition by H. Fröschle and W. P. H. Scheffler 1980 ff. and (4 vols.) 1980–4.

UHSE, BODO (Rastatt, 1904–63, Berlin), the son of an army officer. After supporting the NSDAP (q.v., 1927–8), he devoted himself to anti-fascist and socialist activities, which mark the early part of his restless career as a journalist and author. A member of the Communist Party from the early 1930s, he emigrated to France in 1933 and, after participation in the Spanish Civil War, lived in Mexico. On his return to Germany in 1948 he settled in East Berlin, editing *Der Aufbau* and, in 1963, *Sinn und Form* (q.v.). He is the author of novels, including *Söldner und Soldat* (1935), *Die letzte Schlacht* (1938), *Leutnant Bertram* (1948), and *Die Patrioten* (1954–65). His stories include the collections *Die heilige Kunigunde im Schnee* and *Mexikanische Erzählungen* (1957); Käthe Kollwitz (q.v.) is the subject of *Die Aufgabe* (1958), which was followed by *Reise in einem blauen Schwan* (1959), *Das Wandbild* (1960), and *Sonntagsträumerei in der Alameda* (1961). *Tagebuch aus China* (1956) and *Im Rhythmus der Conga* (1962) are essays on his travels. *Gesammelte Werke in Einzelausgaben*, ed. G. Caspar, appeared 1974 ff. and, ed. E. E. Kisch, 1960 ff.

ULBRICHT, WALTER (Leipzig, 1893–1973, Berlin), was a Socialist from boyhood and a Communist from the inception of the KPD (see SPARTAKUSBUND) in 1919. In 1933 he fled to Moscow, returning to eastern Germany in 1945 to organize a Communist administration behind the Russian front. He was the effective head of the DDR (see DEUTSCHE DEMOKRATISCHE REPUBLIK) from its creation in 1949 until May 1971, when he resigned. He bore the principal responsibility for building the Berlin Wall in 1963. He was succeeded by E. Honecker. Ulbricht's work, *Zur Geschichte der Arbeiterbewegung* (8 vols.) appeared 1958 ff.

ULENHART, NIKLAS, possibly of Prague, was concerned with the translation of Spanish prose. He is known for his adaptation of Cervantes' *Rinconete y Cortadillo* (*Historia von Isaac Winckelfelder und Jobst von der Schneid*, 1617, ed. G. Hoffmeister, 1982–3). Ulenhart, Albertinus, Opitz, and Kuffstein (qq.v.) fashioned a prose style typical of the early Baroque (G. Hoffmeister).

Ulenspiegel or **Ulenspegel**, see TILL EULENSPIEGEL.

ULFILAS, Greek form of Wulfila (q.v.).

Uli der Knecht, title commonly used to refer to J. Gotthelf's novel *Wie Uli der Knecht glücklich wird* (q.v., 1841). The abbreviated title is borne by the edition of 1846.

Uli der Pächter, a novel written by J. Gotthelf (q.v.) in 1847 and published in 1849. It is a sequel to *Wie Uli der Knecht glücklich wird* (q.v., 1841), opening at the point at which the earlier novel stops.

Uli, married to Vreneli, takes over the farm in 'die Glungge' as a tenant. He works hard and is moderately successful, but he is consumed by a desire to get rich quickly. This leads to serious errors of judgement; he makes false economies by engaging cheap labour and listens to untrustworthy friends, the innkeeper and the miller. Though he sometimes loses patience with her, Vreneli supports him through thick and thin, and she herself finds help in her stepmother (the Base). The Base dies and Uli's situation is made worse by the foolish conduct of his landlord Joggeli and of Joggeli's grown-up children.

Uli is misled into making a questionable deal, is taken to court, and wins his case. Yet his conscience troubles him, and when, on his return, a terrible hailstorm lays all his crops, he sees it as a judgement of God. He is taken ill, but Vreneli, who nurses him, pulls him through. His avarice disappears and a happier life begins. Meanwhile Joggeli has got into financial difficulties, and when he dies the farm is sold. To the surprise of Uli and Vreneli, the new owner, Hagelhans, asks them to stay on, and presently reveals himself as the illegitimate Vreneli's father. Johannes the Bodenbauer, who was a principal character in *Uli der Knecht*, reappears from time to time as Uli's true friend and counsellor.

The novel is a Bildungsroman (q.v.) in the sense that Uli learns the right road by bitter experience; at the same time it demonstrates that the struggle of life continues, for the equilibrium achieved at the close of *Uli der Knecht* is soon disturbed.

ULITZ, ARNOLD (Breslau, 1888–1971, Tettnang), a schoolmaster, served in the 1914–18 War in Russia, and was expelled from Silesia in the transfer of population in 1945. He wrote a notable Expressionist novel, *Ararat* (1920), with the background of the Russian Revolution, in which the chief figure Daniel, caught up in horror and desolation, finds new meaning in life through the help of a Jewish rabbi. Other novels include *Testament* (1924, set in the time of inflation 1922–3), the Silesian novels *Der wunderbare Sommer*, *Der große Janja* (both 1939), and *Die Braut des Berühmten* (1942). Ulitz published two early volumes of poetry (*Der Arme und das Abenteuer*, 1919, and *Der Lotse*, 1924), and short works of fiction, most of which appeared in the 1940s as Novellen, of which *Hochzeit! Hochzeit!* (1940) should be mentioned.

ULLMANN, REGINA (St. Gall, Switzerland, 1884–1961, Munich), daughter of a well-to-do businessman, moved to Bavaria after her father's death, and in 1911 became a Roman Catholic. She was a friend of Rilke (q.v.), who wrote an introduction to her volume of poetic prose, *Von der Erde des Lebens. Dichtungen in Prosa* (1910). In 1945 she published *Erinnerungen an Rilke*. A volume of poetry (*Gedichte*) appeared in 1919 before she turned to the writing of stories which recall the outlook of Stifter (q.v.) and reflect her faith; they include the collections *Die Barockkirche* (1925), *Vier Erzählungen* (1930), *Vom Brot der Stillen* (1932), *Der Apfel in der Kirche* (1934), *Madonna auf Glas* (1944), *Von einem alten Wirtshausschild* (1949), and *Schwarzer Kerze* (1954). Her *Gesammelte Werke* (2 vols.) appeared in 1960.

ULRICH, HERZOG VON WÜRTTEMBERG (Reichenweier, Alsace, 1487–1550, Tübingen), came to the ducal throne in 1503. His extravagance led to the imposition of heavy taxes, which provoked a revolt known as 'der Arme Konrad' (see ARMER KONRAD). This he suppressed with the help of the cities, which exacted concessions from him. In 1516 Ulrich murdered Hans von Hutten, cousin of Ulrich von HUTTEN (q.v.), who thereupon became his bitter opponent. Duke Ulrich was attacked by the Swabian League (see SCHWÄBISCHER BUND), and fled abroad, living in exile for fifteen years. In 1534 he was restored, receiving his land as a fief from the Emperor Karl V (q.v.). He took part in the War of Schmalkalden (see SCHMALKALDISCHER KRIEG) and, having fallen out with the Emperor, was in danger of deposition; but he died before action could be taken. Ulrich made Württemberg a Protestant state. He is an important figure in W. Hauff's novel *Lichtenstein* (q.v.).

Ulrichlegende, an account in Middle High German verse of the life of St. Ulrich (Bishop of Augsburg, 923–73), translated from a Latin life of the saint by Berno von Reichenau (*c.* 1030). Its author gives his name as Albertus and he was probably a cleric of Augsburg, who wrote his poem towards the end of the 12th c.

ULRICH VON DEM TÜRLÎN, a Middle High German poet who originated in Carinthia, perhaps at St. Veit. He was a commoner and may have been of the kin of HEINRICH von dem Türlîn (q.v.). Ulrich wrote between 1261 and 1269 a *Willehalm,* which, though written later, is a kind of prelude to Wolfram von Eschenbach's *Willehalm* (q.v.); for it recounts the events before the beginning of the story, including Willehalm's wooing of Arabel and elopement with her, followed by Arabel's conversion and christening as Gyburg. The poem of approximately 10,000 lines is written in lengthy rhyming stanzas, each of 31 lines.

ULRICH VON ETZENBACH, formerly known as Ulrich von Eschenbach, a Bohemian by birth, was a learned commoner who wrote two substantial Middle High German verse works and was the probable author of a third. He was active at the courts of Ottokar II (q.v.), who was killed in 1278, and of his successor Wenzel II (q.v.).

Ulrich began his *Alexandreis,* a history of Alexander of Macedon, *c.* 1271, intending it as a glorification of Ottokar. The work, which contains 30,000 lines, was not, however, finished until *c.* 1287, long after Ottokar's death. Its source is the medieval Latin *Alexandreis* of Gualtherus de Castellion. Alexander is portrayed, in the main, as the ideal ruler, and the

elaborate depiction of court and military life is done in terms of the Middle Ages. A number of digressions recount Greek legends. The poem consists of eleven books, of which the last is thought to be by another hand.

Ulrich's *Wilhelm von Wenden,* a verse romance of some 8,000 lines, is designed to praise Wenzel II and his consort Guta, who appears in the story in the Latinized form of Bene. The Wenden of the title is Bohemia. The poem recounts the decision of Wenzel, on divine inspiration, to renounce wealth and rank, and his tribulations, including separation from his wife and loss of his children. It ends with reunion and reinstatement. It was probably written *c.* 1290.

The authorship of a version of *Herzog Ernst* (q.v.), often called the *Gothaer Herzog Ernst* from the location of the MS., is also often attributed to Ulrich.

ULRICH VON GUTENBURG, a Middle High German Minnesänger, belonged to an Alsatian noble family and is known to have been at the court of Friedrich I and Heinrich VI (qq.v.) between 1172 and *c.* 1200. Only a group of stanzas and an extended poem (see LEICH) are preserved. He was influenced by FRIEDRICH von Hausen (q.v.) and appears to have been primarily a formal poet.

ULRICH VON LIECHTENSTEIN (*c.* 1200–*c.* 1275), a Middle High German Minnesänger, belonged to an important Austrian noble family and was himself a man of mark, becoming high Steward (Truchseß) and, *c.* 1245, marshal of his native province of Styria. He fought in internal wars in Styria and in the campaign in Hungary in 1246, when he was present at the battle on the Leitha, in which his lord, Friedrich II, der Streitbare (q.v.), was killed.

Ulrich's Minnelieder are formally derived from WALTHER von der Vogelweide and REINMAR der Alte (qq.v.). They reflect conventional Minnedienst, and are classified by the poet under such headings as *tanzwîse, sincwîse, tagewîse,* or *ûzreise* (Ausreise). The most remarkable feature of Ulrich's poetry, however, is the framework in which he himself assembled it. Under his own title *Frauendienst* (*Vrowen dienst*) he collected some 60 songs and connected them with with an autobiographical narrative of *c.* 3,700 lines written in rhyming couplets. This purported autobiography (and there seems no doubt that the episodes are not entirely fictitious) is, however, partial, since it concerns only Ulrich's life of Minnedienst in the service, successively, of two different ladies. The devotion to the ideal mistress (which in Ulrich coexists satisfactorily with the claims of a wife) appears to have something of the role of a sport and is combined with a sporting activity to which Ulrich was passionately attached—the joust. Tournaments are his real-life substitute for Arthurian romance. In this way an element of realism, and with it an expression of sensuality, is infused into the stylized poetry of remote love. *Frauendienst* was completed in 1255; it was followed two years later by *Das Frauenbuch* (*Der frowen puoch*), a theoretical discussion of *minne* in rhyming dialogue between a knight and a lady.

ULRICH VON RICHENTAL or REICHEN-TAL (d. 1437), a citizen of Constance, compiled on a basis of observation and inquiry a history of the Council of Constance (1414–18, see KONSTANZER KONZIL). The original was in Latin and is lost; the German version is by Ulrich himself.

ULRICH VON SINGENBERG, a minor Middle High German Minnesänger, who is recorded as High Steward (Truchseß) of St. Gall between 1219 and 1228. More than thirty of his poems are preserved, including Minnelieder, Sprüche, Wechsel, Tagelieder, and a lament for WALTHER von der Vogelweide (q.v.), who was his avowed master in poetry. One of Singenberg's songs is an imitation of Walther's well-known alphabetical rhyming poem.

ULRICH VON TÜRHEIM, a Middle High German poet of the 13th c., belonged to the minor nobility, and was attached to the circle round the Staufen King Heinrich. Konrad von Winterstetten was his patron, for whom he wrote *c.* 1235 a completion of Gottfried von Straßburg's unfinished *Tristan* (q.v.). In this undertaking he drew on Eilhart's *Tristrant und Isalde* (q.v.). The work adds some 4,000 lines to Gottfried's poem. On a much larger scale he continued Wolfram von Eschenbach's *Willehalm* (q.v.), writing between 1240 and 1250 a gigantic work of more than 36,000 lines, which goes under the name of *Rennewart* and is mainly concerned with the history of this son of the heathen prince Terramer and of Rennewart's son Malefer. According to RUDOLF von Ems (q.v.), Ulrich wrote a poem entitled *Clies*, but it is generally assumed that this lost work is a continuation of a poem by Konrad Fleck (q.v.). Ulrich's extensions of the existing works of Gottfried and Wolfram fall far below their originals.

ULRICH VON WINTERSTETTEN, a Middle High German Minnesänger of the middle of the 13th c., is recorded between 1241 and 1280. He belonged to a noble family which bore the hereditary title Schenk (cupbearer). Ulrich was a canon of Augsburg in later life and it is assumed that his poetry belongs to his earlier years. His poems include Minnelieder and Tagelieder, but he is best known for his *tanzleiche* (see LEICH), flexible and frequently long poems with a pronounced, often jerky, rhythm.

Ulrich, though not the first to introduce refrain into Minnelieder, uses it much more frequently than any previous poet. He at times employs crass vocabulary in dialogue. One of his poems is an interesting lament on the decline of manners, in which the impertinent young are heard to refer to the 'Minner' of Ulrich's generation thus: 'Est ein argez minnerlîn' (Er ist ein arges Minnerlein).

ULRICH VON ZAZIKHOFEN (Zäzikon in Switzerland), author of the Middle High German Arthurian epic *Lanzelet*. Ulrich names himself as the author at the end of the poem and gives as his source a French work belonging to 'Hûc von Morville', who, as hostage for Richard I of England, was at the Emperor's court in 1194, and this is the approximate date of the poem. Nothing else is known of Ulrich and the French poem is lost. *Lanzelet* (as doubtless also the poem of which it is a translation) is compounded of diverse elements which have not been successfully integrated.

The poem, which is composed of more than 9,000 lines, tells the hero's life story from birth to death. After the death of his royal parents, Lanzelet is brought up by a sea fairy and is destined to avenge her against King Iweret. In due course he sets out on his adventures, defeats various adversaries and finally Iweret, whose daughter Iblis he marries. The second part of the poem concerns his quest for Arthur's queen, Ginover, who has been abducted. In the course of further adventures he liberates her and eventually settles and lives happily with Iblis until their death on the same day.

A curious feature of this poem about the bland insensitive Lanzelet ('er enweiz nicht waz trûren ist') is the series of his successes with women, whom he takes and unthinkingly discards. The work, which is devoid of .psychological insight and shows little trace of moral or religious feeling, is thought by some to have owed its success to emphasis on erotic adventure.

Umbehanc, mentioned by GOTTFRIED von Straßburg (q.v.) and by RUDOLF von Ems (q.v.), who, in reference to BLIGGER von Steinach (q.v.), writes the line, 'der den umbehanc gemalet hat'. The word has sometimes been seen as the title of a lost poem by Bligger; it is more probably an image referring to a conspicuous descriptive passage.

Umkehrende, Der, a cycle of five poems by J. von Eichendorff (q.v.). The best known is the last, 'Waldeinsamkeit!/Du grünes Revier', which occurs as an untitled song in Eichendorff's novel *Dichter und ihre Gesellen* (q.v.).

Um Mitternacht, a well-known poem by Goethe, written in February 1818, and first printed in Zelter's *Neue Liedersammlung* (1821). In three stanzas, each of which ends with the refrain 'Um Mitternacht', the poem reflects in economical symbolism man's youth, robust maturity, and old age. Employing the lyrical 'I', it is a masterly example of Goethe's late precise and discreet lyric style.

Um Mitternacht, a poem written in 1827 by E. Mörike (q.v.). In two eight-line stanzas of subtly varying rhythm, the poet writes of the serenity of the night, disturbed only by the streams' babbling recollection of the day. The poem, which begins 'Gelassen stieg die Nacht ans Land', has been set to music by Hugo Wolf (q.v.).·

Unauslöschliche Siegel, Das, a novel by E. Langgässer (q.v.), published in 1946. The 'seal' (Siegel) is the sacrament of baptism. The events take place partly in Hesse and partly in Senlis, north of Paris, between 1914 and 1926.

The work presents the conflict between the divine and the diabolical, in which the characters are symbols rather than individuals. The central figure, the Jew Belfontaine, accepts baptism in order to take to his wife the Roman Catholic Elisabeth; in the end he escapes from the works of the Devil and undergoes a true conversion.

Unbedeutende, Der, a comedy (Posse mit Gesang in drei Aufzügen) in Viennese dialect by J. N. Nestroy (q.v.), written in 1846, first performed in May 1846 and published in 1849. The good-natured young millionaire, Baron von Massengold, is dominated by his secretary, Puffmann. Fearing that Massengold is about to marry his ward Hermine, Puffmann, who sees in this a threat to his own influence, engineers an elopement and forges documents indicating that the girl is of age. Hard put to it to explain his activities on the evening in question, he contrives an alibi by alleging (and bribing a small boy as witness) a rendezvous with a girl, Klara Span, whom he has never even seen. The neighbours immediately and joyfully point the finger of scorn at Klara. Her carpenter brother Peter ('der Unbedeutende', the man of no account of the title) believes in her virtue and determines to restore her reputation. His persistence is rewarded, Puffmann is exposed and discredited (married off, in Volksstück

manner, to an elderly spinster as a punishment), and Klara fully rehabilitated.

The play has the integrity and courage of the honest man at its centre but it is not built on class opposition. Puffmann's opponent Herr von Packendorf and von Massengold himself are men of goodwill, and Klara's gossiping neighbours are comically despicable. *Der Unbedeutende* was one of Nestroy's greatest successes.

Unbekannte Größe, Die, a novel by H. Broch (q.v.), published in 1933. Its central figure is a brilliant young mathematician, Richard Hieck, who feels that all life is purposeless. He overcomes his depression when, faced with the accidental death of his brother, he realizes that logical thought and scientific investigation contribute only in part to a totality of knowledge, which is derived from both rational and irrational sources.

'Und Calcar, das ist Sporn', title and refrain of a ballad by Th. Fontane (q.v.) incorporated as No. 3 in his cycle *Seydlitz* (q.v.). It is printed under the title *General Seydlitz* in chapter 43 of *Vor dem Sturm* (q.v.), where it is read by Hansen-Grell.

Undine, a Romantic fairy-tale (Kunstmärchen, described as Erzählung) by F. de la Motte Fouqué (q.v.), published in 1811. Undine is a water-sprite, who is adopted and brought up by a poor fisherman and his wife. She falls in love with Huldbrand, a knight, whom she marries, and by this union she acquires a soul. Huldbrand imperceptibly falls in love with Bertalda, believed at first to be the daughter of a duke, but proving eventually to be the long-lost daughter of the fisher couple. The water spirit Kühleborn comes unasked to Undine's aid, and when Huldbrand rebukes her as they sail on the Danube, she is forced to return to her own watery element. Huldbrand marries Bertalda. Undine reappears and kills him with a kiss.

The story is Fouqué's most successful work, evoking a mysterious, enchanted landscape of forest and torrent. It has twice been used to provide a libretto for an opera, in 1816 by E. T. A. Hoffmann (q.v.) and in 1845 by A. Lortzing (q.v.).

Und Pippa tanzt, a four-act play in prose with some verse passages by G. Hauptmann (q.v.), first performed in the Lessingtheater, Berlin, in January 1906, and published in the same year. It is closely linked with the passionate love experienced by Hauptmann in 1905 for a young actress, Ida Orloff. Described as 'Ein Glashüttenmärchen', *Und Pippa tanzt* blends

Naturalistic detail with poetic symbolism. At its centre is the fascinating, ethereal Pippa, a young girl representing elusive beauty. Her exotic descent suggests also that she is intended to embody the traditional German yearning for the transalpine south, which Hauptmann himself vividly experienced.

The fragile crystal-like Pippa is transplanted from her natural Italian environment into the icy beauty of a moonlit winter's night in the remote mountains and forests of Silesia. In the tavern in which the play begins she is in the company of good-natured, but rugged and earthy glass-blowers. Her attraction is felt by all, but especially by two opposites, who are the chief contenders for her love—the uncouth, primitive giant Huhn and the blonde, idealistic visionary youth Hellriegel, in whom both Hauptmann and his vision of Hölderlin (q.v.) seem to be embodied. Pippa flees with Hellriegel and they wander through the ice-bound forest, escaping from Huhn. They reach the dwelling of Wann, the sage, and here the climax is enacted. When Huhn crushes a glass, Pippa, the glass-like creature, dies of shock. Huhn himself dies simultaneously as he performs this act, and Hellriegel loses his sight. He wanders out into the night, guided only by an inward vision. The wise and prescient Wann remains behind in resignation. There is an echo of Prospero in Wann, and more than an echo of *Faust* (q.v., Pt. II) in Hellriegel's blindness. Hauptmann's Glashüttenmärchen has a dreamlike incoherence.

Und sagte kein einziges Wort, a novel by H. Böll (q.v.), published in 1953. Its theme is a marriage in the immediate post-war period with its shortage of money, food, and housing. A first-person narrative, the story is told by Fred and Käthe Bogner, alternately.

Fred has separated from his family because the cramped conditions and the general frustration fray his temper to the point of maltreating his children. He drifts from job to job, and from time to time spends an hour with Käthe in a cheap hotel room. Some of the novel's satire is directed at the Roman Catholic Church. It ends with a faint ray of hope for Fred and Käthe. The title derives from a Negro spiritual: 'And he never said a mumbaling (*sic*) word'.

'Und wie zu Emmaus weiter ging's', beginning of the last four lines of the poem *Zwischen Lavater und Basedow* (q.v.) by J. W. Goethe. These lines are more widely known than the rest of the poem because they are quoted in isolation by Goethe in *Dichtung und Wahrheit* (q.v., Bk. 14).

Ungenannter, designation used by G. E. Lessing to conceal the identity of the author of

the fragments of a theological MS. which he published in 1777 as *Ein Mehreres aus den Papieren eines Ungenannten, die Offenbarung betreffend.* 'Der Ungenannte' was H. S. Reimarus (q.v.) and the work *Schutzschrift für die vernünftigen Verehrer Gottes.*

UNGER, Friederike Helene (Berlin, 1741–1813, Berlin), daughter of a General von Rothenburg, married the Berlin publisher J. F. Unger (1753–1804), who designed a new Gothic fount, less heavy than older types and known as Ungerfraktur. She published the story *Julchen Grünthal* in 1784 and wrote the novels *Gräfin Paulini* (1800) and *Die Franzosen in Berlin* (1809). She also wrote comedies and translated J.-J. Rousseau's *Confessions* (1782). She directed her husband's firm after his death.

UNGER, Hellmuth (Nordhausen, 1891–1953, Freiburg, Breisgau), an eye specialist, was the author of Expressionist plays (*Kentaurin, Verlorener Sohn,* both 1919, *Spiel der Schatten, Mammon,* and *Menschikow und Katharina,* all 1922), and of novels and biographies dealing with important physicians; they include *Robert Koch* (1936), *Vom Siegeszug der Heilkunde* (1936), *Röntgen* (1949), *Narkose* (1951), and *Virchow* (1953).

UNGERN-STERNBERG, Alexander, Freiherr von, also Alexander von Sternberg (Noistfer nr. Reval, 1806–68, Noistfer), a Baltic German, came to Germany in 1830 and lived chiefly in Dresden. He at first wrote Novellen and novels of social criticism, including *Der Missionär* (2 vols., 1842), *Diane* (3 vols., 1842), *Paul* (3 vols., 1845), and *Braune Märchen* (1850, reissued 1966). In 1848 he swung to the political right, publishing the novels *Royalisten* (2 vols., 1848) and *Die beiden Stützen* (1849). Late in life he produced biographical novels including one on Rubens (*Peter Paul Rubens,* 1862). His *Erinnerungsblätter* (6 vols.) appeared 1855–60.

Unheilbringende Zauberkrone, Die, a play by F. Raimund (q.v.), written in 1829 and first performed at the Theater in der Leopoldstadt, Vienna, in December 1829. Extravagantly described as a *tragisch-komisches Original-Zauberspiel* and paradoxically subtitled *König ohne Reich, Held ohne Mut, Schönheit ohne Jugend,* it was published in Raimund's *Sämtliche Werke* (1837, vol. 4). Its confused plot involves a conflict between Phalarius and Kreon, who has driven the former from his realm of Agrigento, a conflict between Hades and Latona, a pilgrimage of two comic Viennese (Zitternadel und Ewald) to discover the contradictory persons listed in the sub-title, avenging antics by the Eumenides, and a successful intervention by Lucina on behalf of

Kreon. The play, written in verse with comic Viennese scenes in prose, was one of the least successful of Raimund's plays.

Unheimliche Gast, Der, a story by E. T. A. Hoffmann (q.v.), written in 1819 and published in 1820 in *Die Serapionsbrüder* (q.v., vol. 3).

A young man and woman, he a cavalry captain and she a colonel's daughter, are attracted to each other and at length declare their love. But the colonel has destined Angelika for his friend, a middle-aged Italian count. The count, by the exercise of supernatural powers, has the captain detained in a French castle and subdues Angelika's mind to his will, inducing her to agree to marriage. On the day of the wedding he is found dead and leaves a letter revealing his plot and admitting its failure. Angelika rushes out of the house and throws herself into the arms of her approaching lover.

Universities. German universities existing before 1960 are included in the following list, which provides foundation dates; universities in the Deutsche Demokratische Republik (q.v.) are indicated by the abbreviation DDR.

Heidelberg (1386), Cologne (1388, dissolved 1798, refounded 1919), Leipzig (1409, DDR), Rostock (1419, DDR), Greifswald (1456, DDR), Freiburg (1457), Munich (1472 in Ingolstadt, 1800–26 in Landshut), Tübingen (1477), Mainz (1477, dissolved 1797, refounded 1946), Marburg (1527), Jena (1558, DDR), Würzburg (1582), Gießen (1607), Kiel (1665), Halle (1694, DDR, now Halle-Wittenberg), Göttingen (1737), Erlangen (1743, now Erlangen-Nürnberg) Münster (1780, dissolved 1818, refounded 1843 as academy, university in 1902), Bonn (1786, dissolved 1796, refounded 1818), Berlin (1810, DDR, Humboldt-Universität), Frankfurt/Main (1914), Hamburg (1919), Saarbrücken (1947), Berlin (1948, Freie Universität).

Universities founded since 1960 (or refounded, in which case dates are given) include Bochum, Regensburg, Bremen, Constance, Dortmund, Bielefeld, Ulm, Augsburg, Trier-Kaiserslautern, Bamberg (1648, dissolved 1803), Essen, Duisburg (1655, dissolved 1818), Paderborn (1614, dissolved 1808), Siegen-Hüttental, Wuppertal, Bayreuth, Eichstadt (1564), Lübeck (medical), Trier, Passau, Osnabrück, and Oldenburg. Hohenheim (1818), Gesamthochschule Kassel (1777), Stuttgart (1829), Düsseldorf (1907), and Mannheim (1907) are specialist colleges now raised to full university rank.

The six Austrian universities are Vienna (1365), Graz (1585), Salzburg (1622, dissolved 1810, refounded 1962), Innsbruck (1669), Klagenfurth (1970), and Linz (1975). The universities of the German-speaking part of Switzerland are Basel (1460), Berne (1528), and Zurich (1833). For Prague (1348) see KARL IV.

There are also a number of Technical Universities (Technische Universitäten) as well as Technische Hochschulen.

A number of early German foundations have disappeared, notably Erfurt (1392), Wittenberg (1502), Frankfurt/Oder (1506, amalgamated with Breslau in 1811), Königsberg (1544, now in Russia), Dillingen (1554), Braunsberg (1558), Helmstedt (1576), Herborn (1584), Strasburg (1621, in Germany 1871–1918), Rinteln (1621), Altdorf (1622), Osnabrück (1630), Dorpat (1632, now in Russia), Breslau (1702, now in Poland), and Fulda (1743). Most of the dissolutions took place during the French Revolutionary and the Napoleonic Wars (q.v., and see REVOLUTIONSKRIEGE).

Unordnung und frühes Leid, a story by Th. Mann published in 1926. It might be called a 'Familienbild'. Thinly disguised as Professor Cornelius, Mann portrays an afternoon and evening with his own family. The sketch is executed with masterly perception and gentle irony, and with sympathy for the young.

Unparteiische Kirchen- und Ketzer-historie, an epoch-making work of ecclesiastical history by Gottfried Arnold (q.v.), published 1699–1715.

UNRUH, FRIEDRICH FRANZ VON (Berlin 1893–), son of a general and brother of Fritz von Unruh (q.v.), was an officer who became a journalist and writer and, after the experience of the 1914–18 War, a pacifist. He is the author of stories, notably Novellen, including *Der Tod und Erika Ziska* (1937, retitled *Erika Ziska*, 1942), *Der Patriot wider Willen* (1944, retitled *Liebe wider Willen*, 1950), *Die Sohnesmutter* (1946), *Die jüngste Nacht, Vineta* (both 1948), *Nach langen Jahren* (1951), and *Sechs Novellen* (1958), and also of essays, including *Fichte, Friedrich Hölderlin* (both 1942), and *Wo aber Gefahr ist. Lebensdaten eines Novellisten* (1965).

UNRUH, FRITZ VON (Koblenz, 1885–1970, Diez/Lahn), son of a general, was destined for the army and was a pupil at the military school at Plön (Schleswig-Holstein), where the Emperor's sons were educated. He entered the cavalry, but did not take to peace-time soldiering, resigned his commission in 1911, and turned to literature with the plays *Offiziere* (q.v., 1911) and *Louis Ferdinand, Prinz von Preußen* (1913, see LOUIS FERDINAND, PRINZ VON PREUSSEN), which were produced by M. Reinhardt (q.v.). He joined a lancer (Ulan) regiment on the outbreak of the 1914–18 War, but was quickly disillusioned and disturbed by its horrors, as the symbolical play *Vor der Entscheidung* (written in October 1914) and the realistic narrative *Opfergang* (q.v., written near Verdun in 1916, both published 1919) show. Unruh's suffering, compassion, and pacifism are further reflected in the Expressionist play *Ein Geschlecht* (1917) and its successor *Platz* (1920).

Between the wars Unruh frequently spoke publicly in support of his humanitarian and pacifistic ideals (*Stirb und werde*, 1922; *Vaterland und Freiheit*, 1923; *Reden*, 1924; *Politeia*, 1933; and *Europa, erwache!*, 1936). He emigrated in 1932, escaping to the U.S.A. in 1940. Hitler (q.v.) is the subject of his novel *The end is not yet* (1947, in German as *Der nie verlor*, 1948), which was followed by *Die Heilige* (1952) and *Fürchte nichts* (1953), and the plays *Duell an der Havel* (a comedy), *Wilhelmus von Orleans* (both 1953), and *17. Juni* (1954).

After the 1939–45 War, Unruh continued to propagate his views in speeches, *Seid wachsam* (an address on Goethe), *Rede an die Deutschen*, *Friede auf Erden* (all 1948), *Mächtig seid ihr nicht in Waffen* (1957), and *Wir wollen Frieden* (1962). Notable late novels are the autobiographical *Der Sohn des Generals* (1957) and *Im Haus der Prinzen* (1967); set in a frame (see RAHMEN) in the 1950s, Unruh figures in the thin disguise of Uhle, the first-person narrator. *Sämtliche Werke* (20 vols.), ed. H. M. Elster, B. Rollka, *et al.*, began to appear in 1970.

Unruhige Gäste, a novel written by W. Raabe in 1884 and published in 1886 with the subtitle *Ein Roman aus dem Säkulum*. It is a concentrated work, turning on a single episode. Veit von Bielow, a good-natured and intelligent man of the world, is staying in the mountains, recognizably the Harz. On impulse, he decides to visit an old university friend, Prudens Hahnemeyer, who has become a village pastor. Hahnemeyer proves to be a grim fanatic. But with him is his sister Phoebe, a pure, dedicated soul, not of this world (the Säkulum of the subtitle). Before keeping house for her brother, Phoebe had worked for handicapped children.

Veit and Phoebe first meet in the forest when Phoebe is on an errand of mercy, bringing food and comforts to Anna, a young mother dying of typhus in her rough shelter on the Vierlingswiese, where the family has taken refuge after being outlawed by the community. When Anna dies, her husband, Volkmar Fuchs, refuses to have her buried in the churchyard of the hostile village. Veit breaks down Volkmar's resistance by engaging to isolate Anna from her neighbours; he buys a burial plot to right and left of her for him and Phoebe. It is clear that this is the expression of a love for Phoebe to

which she can respond and yet maintain her innocence and her integrity.

Veit leaves and returns to his friends and is at once caught up with Valerie, a handsome and able girl of his own class. He falls ill with typhus and is nursed through by Phoebe; on recovery, he is carried off and married by Valerie. Raabe implies that Veit will not recover fully from his illness. We leave Phoebe preparing to resume her task of looking after mentally subnormal children. Great as her suffering is, her integrity and her genuine piety surmount it. The forces of the Säkulum, strong though they are, have no power over her.

Unser Vrouwen Klage, an anonymous Middle High German poem of some 1,600 lines, probably written in the second half of the 13th c., which uses Mary's lament at the foot of the Cross (Marienklage) as a frame for a poem on man's love for Jesus. The poet gives as his source the *Interrogatio Sancti Anselmi de passione Domini*. Fifteen MSS. testify to its popularity.

Unsühnbar, a novel by Marie von Ebner-Eschenbach (q.v.), published in 1890. The characters are confined to the aristocratic circles, to which she herself belonged. The young Komtesse Maria Wolfsberg is married by her father to an entirely worthy and devoted young man, Count Dornach, whom she does not love. In an unguarded hour she is surprised and seduced by a former admirer, Felix Tessin, a notorious lady-killer. Overcome with horror and repentance she devotes herself to good works. She bears a second son, Erich, whom she knows to be Tessin's child. Over the years she learns to love Dornach, whose own devotion and love are unswervingly hers. He and her elder child are accidentally drowned and at the reading of the will she declares, to the shock and horror of those present, that Erich cannot inherit because he is not Dornach's son. She lives in retirement on her father's estate, and dies young, entrusting her son to a good-hearted and unselfish cousin of her husband.

The novel exposes dispassionately the superficiality, egoism, and pride of much of the aristocratic world of Vienna, balancing it with compassion for the suffering and sentient.

Unterhaltungen deutscher Ausgewanderten, a Novelle with a cyclic frame (see RAHMEN) by Goethe, published anonymously in 1795 in Schiller's *Die Horen* (q.v.). Modelled on Boccaccio's *Decameron*, the work is the first of its kind in modern German literature. It is set against the background of the Revolutionary Wars (see REVOLUTIONSKRIEGE).

The seven stories accommodated in the frame,

in which Baroness von C. is the leading figure, are narrated by three characters, the Geistlicher (priest), Fritz, and Karl; all belong to a small party of refugees, who are resting while waiting for news about the movements of the French armies in 1793. It soon proves that the party is very much in need of the stabilizing influence of the Baroness and of her firm yet gracious temperament; for Cousin Karl, the only fanatical defender of the French cause in their midst, has irresponsibly obliged the Geheimrat, a staunch adherent of the *ancien régime*, to leave the company with his wife. Thus alerted to the dangers of discussions growing out of stress and friction, the Baroness suggests that the company should engage in pleasing conversation ('freundliche Unterhaltungen') which may restore peace and composure in the minds of everyone: 'Laßt alle diese Unterhaltungen ... durch eine Verabredung, durch Vorsatz, durch ein Gesetz wieder bei uns eintreten, bietet alle eure Kräfte auf, lehrreich, nützlich und besonders gesellig zu sein.' The Geistlicher commands the attention of the company when he tells the unhappy story of the love of a young Genoese for the singer Antonelli (*Die Sängerin Antonelli*), which grows out of his long-standing and happy friendship with her. But doubts are expressed over the claim that the lover's prediction that, after his death, Antonelli would be haunted by his ghost, was fulfilled. Nevertheless, with his story of *Der Klopfgeist* Fritz adds to the supernatural phenomena.

Karl narrates two stories which Goethe has adapted almost literally from the *Mémoires* of the Marshal de Bassompierre (q.v.). The first relates an amorous exploit of the Marshal with a woman of lower station ('die schöne Krämerin') who, after a night of happiness, does not turn up to the second rendezvous. This story moves closer to the conditions of real life, the plague being the agent of destiny. The other story, also of an illicit love, relates a woman's discovery of two sleeping lovers, one of whom is her husband. In placing her veil across their feet, she offers forgiveness, a gesture to which the lovers, on waking, respond with the appropriate moral consciousness. In *Der Prokurator*, the Geistlicher takes up a more austere approach. The Prokurator resists the advances of a wealthy young lady during the prolonged absence of her husband, with the result that the lady overcomes her desire and admires his noble conduct. She emerges acquainted with the dangers of passion and the value of restraint. If the Prokurator could awaken in everyone 'die Kraft der Tugend', she says in conclusion, he would deserve the name 'Vater des Vaterlandes'. The Baroness values this story above all: 'Wirklich verdient die Erzählung vor vielen andern den

Ehrentitel einer moralischen Erzählung.' Encouraged by his audience, the Geistlicher now tells the story of *Ferdinand*, which maintains the moral tone of *Der Prokurator*, while presenting 'ein Familiengemälde' of the company's own environment. Ferdinand discovers by chance how to get access to a secret drawer in his father's desk containing money. Having grown up to detest debts, to admire, and at the age of maturity to envy his father, not least for his wealth, he cannot resist the temptation of taking some of his father's money in order to win the attractive but extravagant Ottilie. The incident turns out to be the turning-point in his life—the making of a man of exemplary character who accepts the trials following his repentance and takes no reward for granted. Nor does he marry Ottilie, for whom he had incurred guilt and repentance, but a lady worthy of his moral stature.

The frame closes with Karl's request to the Geistlicher to tell them a fairy-tale, a product of the imagination (Einbildungskraft). *Das Märchen* (q.v.), which is frequently published separately, concludes the *Unterhaltungen deutscher Ausgewanderten*, for which Goethe had planned a second part which did not materialize.

The stories vary considerably in length and structure, ranging between anecdote and Novelle. Only *Die Sängerin Antonelli* (using a report by Prinz August von Gotha), *Der Klopfgeist* (following an event reported to Frau von Stein, q.v., by von Pannewitz), and *Ferdinand* rank among Goethe's original contributions. Apart from the two adaptations from Bassompierre's *Mémoires*, *Der Prokurator* represents an adaptation of a story by the author of the *Cent nouvelles nouvelles* (Antoine de la Sales?).

Unterm Birnbaum, a Novelle by Th. Fontane (q.v.), published in 1885. It is the story of a crime set *c.* 1830 in a village near Frankfurt/Oder. The innkeeper Hradscheck, partly through his wife's extravagance and partly through his own inattention to business, is in financial difficulties. A commercial traveller is expected, who will press for the payment of a large outstanding debt. Knowing that he cannot pay, Hradscheck makes a plan, of which only a hint is provided. The traveller arrives and appears to leave early the next morning. His trap is found in the flooded river, but his body is not recovered. The occurrence is nowhere overtly explained, but the reader is allowed to infer that the traveller was murdered in the night and that he was impersonated in the early morning by Frau Hradscheck. Hradscheck is suspected and arrested, but he clears himself by proving that some freshly turned earth under his pear-tree conceals, not the traveller, but a French soldier from Napoleonic days. Frau

Hradscheck pines and dies. Hradscheck, feeling that suspicions still lurk, decides to move the body of the traveller which is in his cellar. He dies, apparently of a heart attack caused by the discovery that something has fallen and blocked the trapdoor. The reticent and low-toned story is less concerned with the events of the murder than with the question of whether the truth will emerge and how.

Unterm Rad, a novel by H. Hesse (q.v.), published in 1906. It is the story of an intellectually gifted boy, Hans Giebenrath, who collapses under the pressures of an uncomprehending father and a school (Maulbronn, q.v.), which drives him too hard. He dissents more and more from the attitudes of the school, gives up, and becomes an artisan. This proves no solution, and Hans loses his bearings in the real world. He is found drowned; it is left to the reader to infer accident or intent.

Untertan, Der, a satirical novel by H. Mann (q.v.), dealing with the Bürgertum and forming the first part of his *Kaiserreich-Trilogie*. Mann wrote it between 1906 and 1914, and extracts were published in *Simplicissimus* (q.v.) in 1911, 1912, and 1913. The novel began to appear as a serial in *Zeit im Bild* in 1914, and a whole chapter in *Der März* (1913), but the instalments were suspended on the outbreak of war. Apart from a limited edition in 1916 for private circulation, *Der Untertan* was not available until the end of the war (November 1918).

The novel is set in the period 1890–7, from the dismissal of Bismarck to the centenary of Wilhelm I (qq.v.), partly in the small town of 'Netzig', and partly in Berlin. Its main character is Diederich Heßling, of the lower middle class, a weakling who rises in the world by obsequiously deferring to those above him and brutally tyrannizing those of lower station. He evades military service, seduces a girl in Berlin, and deserts her without compunction; he rigs elections, and carries out shady business deals. He talks the hollow theatrical language of his Emperor Wilhelm II (q.v.), whom he twice encounters in (for himself) humiliating or ridiculous situations. The bitter satire is heightened by a dual parody: Wilhelm II speaks as if he were Heßling, and Heßling as if he were the Kaiser. The book ends with what should be Heßling's great day, the inauguration of the monumental statue to Wilhelm I, at which he makes the principal speech, only for the event to end in farce: torrential rain disperses the audience. (Film version by W. Staudte, 1951.)

Unverwesliche Erbe, Das, a novel by Ina Seidel (q.v.), published in 1954. It has a motto

taken from I Corinthians, 15 : 50, 'Das Verwesliche wird nicht erben das Unverwesliche'. The story is spread over three generations.

The forester Dornblüh and his wife Charlotte, who are Roman Catholics, have as their youngest child a daughter named Elisabeth. She marries a Protestant, Dr. Alves, and adopts his Lutheran allegiance, an act which her father refuses to forgive, even on his death-bed. Her mother, who comes to live with Elisabeth and Alves, breathes a serenity and spirituality which deeply influence her daughter and grandchildren. After Alves's death Elisabeth reverts to the Catholic faith; her youngest daughter, Maria, marries a Lutheran pastor, Hans Joachim Lennacker, who dies at 28. Their only child Hans Jakob Lennacker is the returning soldier of the earlier novel *Lennacker* (q.v.), and with him the two works intersect. Some of the characters of this novel are descendants of persons in Ina Seidel's much earlier novel *Das Wunschkind* (q.v.).

UNVERZAGTE, DER, a Middle High German poet of the 13th c., is the author of a number of Sprüche (see SPRUCH), many of which castigate the ungenerosity of patrons. Like Meister Stolle, (q.v.), he criticizes Rudolf I von Habsburg (q.v.) by name. One poem of Der Unverzagte's is an eloquent eulogy of song.

Unwiederbringlich, a novel by Th. Fontane (q.v.), written in 1887 and first published in *Die deutsche Rundschau* in 1890. Publication in book form followed in 1891. The novel is based on an anecdote referring to occurrences in Mecklenburg society told to Fontane by Geheimrätin Brunnemann. Fontane has transferred the scene to Copenhagen and Schleswig and set it in 1859, when Schleswig was under Danish rule.

Count Holk of Holkenäs near Glücksburg has been married for several years to Christine, daughter of a neighbouring baronial family. Outwardly happy, the marriage has developed tensions because Christine is the stronger character and adheres to narrow religious principles, which incline her to self-righteousness. The two children of the family are in their teens. Holk, a man of 45, is a lord-in-waiting to a princess of the Danish court and is summoned unexpectedly to duty in the autumn of 1859. He leaves home with relief, and his susceptibility in Copenhagen to the charms of his landlady's handsome daughter shows that his marriage is in some danger. But Brigitta Hansen, the landlady's daughter, is soon eclipsed by a witty and attractive lady-in-waiting, Ebba von Rosenberg, who, though she inwardly despises Holk, sets her cap at him. The Princess's court removes to Fredericksborg Castle, and Holk and Ebba are brought into closer contact, which, on the occasion of a skating party, glides into intimacy. When the castle catches fire (a historical occurrence on 17 December 1859) Holk helps to rescue Ebba. Convinced that she is in love with him, he returns to Holkenäs and asks his wife for his freedom, which is granted. But back in Copenhagen Ebba coolly rejects him. Holk goes away for eighteen months, visiting Italy and England, and is encouraged to think that a reconciliation with his wife is possible. The marriage is reconsecrated, but a cloud hangs over it, and in the autumn of 1861 Christine Holk takes her life by slipping into the sea.

The novel is a restrained study of a marriage which gradually loses its meaning, and of the foredoomed attempt of an ageing man to recapture youth by taking a young partner. Fontane knew the Danish environment from visits in 1864.

Unzeitgemäße Betrachtungen, a philosophical work by F. Nietzsche (q.v.).

Upsalaer Beichte, a formal confession of sins in rhyming verse written in Germany *c.* 1150 and preserved at Uppsala.

Urania, an annual publication of the publishing firm Brockhaus which appeared from 1810 to 1840. Described as a Taschenbuch, it included many works by Romantic authors.

Urbild des Tartüffe, Das, a five-act comedy (Lustspiel) by K. Gutzkow (q.v.), first performed in December 1845 at Oldenburg. Its subject is the web of intrigue which in 1667 almost succeeded in preventing the production of Molière's *Tartuffe.* Tartuffe is identified with President La Roquette, who, by appealing to the self-interest of a number of parties, persuades the chief of police to reverse his decision to permit the performance of the play. An appeal to King Louis XIV seems likely to fail because Molière proposes to marry Armande, an actress in whom the King is interested. But Armande manœuvres His Majesty into giving consent, the play is performed, and Roquette unmasked.

Urfaust, accepted title for an early version of Goethe's *Faust* (q.v.), which was first published in 1887 by Erich Schmidt (q.v.) under the title *Goethes Faust in ursprünglicher Gestalt.* The title *Urfaust* was applied by Schmidt himself to the text preserved in this copy.

Uriel Acosta, a five-act tragedy (Trauerspiel) in blank verse by K. Gutzkow (q.v.), first performed in December 1846 at Dresden. Gutzkow had already treated this story of the 17th c. in the Novelle *Der Sadduzäer von Amsterdam* (q.v., 1834). It deals with a free-thinking individual beset by the intolerance and bigotry of orthodox (in this case Judaic) religion.

Acosta's book is condemned by the synagogue and he is anathematized. Because he loves the Jewish girl Judith, and for the sake of his mother, he recants and does humiliating penance. But his mother dies and Judith, for her family's sake, abandons him and marries another. Acosta retracts his recantation, Judith poisons herself, and the play closes with Acosta shooting himself.

Urmeister, designation sometimes applied to *Wilhelm Meisters theatralische Sendung* (q.v.), an earlier version of *Wilhelm Meisters Lehrjahre* (q.v.).

Urstende, Diu, see KONRAD VON HEIMESFURT.

Ursula, a story by G. Keller (q.v.) included in *Züricher Novellen* (q.v., 1878). Set during the Reformation (q.v.), it recounts the love of Hansli Gyr for Ursula Schnurrenberger. Long separated by religious differences, they are reunited after the battle of Kappel, in which Zwingli (q.v.), who is a character in the story, was killed (1531). Keller's tale is directed against religious fanaticism.

Urteil, Das, a short story by F. Kafka (q.v.), written in a single night in 1912 and published in 1916. Georg Bendemann, a young businessman, is denounced by his aged father and condemned. Georg himself executes his father's death sentence by throwing himself from the bridge of his native city (recognizably Prague) into the river. The dream-like narrative, which demonstrates Kafka's familiarity with Freud (q.v.), is a parable on the theme of existential guilt.

Urworte. Orphisch, a philosophical poem written by Goethe in October 1817 and first published in *Zur Morphologie* in 1820. The 'Urworte' may be taken as a free translation of the orphic phrase ἱεροὶ λόγοι (holy words).

The poem consists of five stanzas, each with a Greek superscription to which Goethe adds a translation. The headings are as follows (in brackets are given variant translation found in one MS.): ΔΑΙΜΩΝ, *Dämon* (*Individualität, Charakter*); ΤΥΧΗ, *Das Zufällige* (*Zufälliges*); ΕΡΩΣ, *Liebe* (*Leidenschaft*); ΑΝΑΓΚΗ, *Nötigung* (*Beschränkung, Pflicht*); ΕΛΠΙΣ, *Hoffnung.* Thus the five stanzas are seen to set out the course and conditions of human life, as seen by Goethe in his maturity. In *Über Kunst und Altertum,* vol. 3, 1821, Goethe reprinted the poem, together with an extensive interpretation.

URZIDIL, JOHANNES (Prague, 1896–1970, Rome), a journalist and writer, worked during the 1920s and early 1930s for the German Embassy in Prague, where he came under the influence of Expressionism, counting F. Werfel and Kafka (qq.v.) among his friends. In 1939 he emigrated to England, moving in 1941 to the U.S.A.

Urzidil published three volumes of poetry, *Sturz der Verdammten* (1919), *Die Stimme* (1930), and *Die Memnonsäule* (1957), but he is better known for his stories, steeped in the tradition of his native Bohemia and of Prague, and presenting characters, whose struggle to come to terms with life reflects the deeper issues involved in the problem of emigration. *Der Trauermantel* (1945) deals freely with the life of A. Stifter (q.v.). The title refers to a butterfly, the Camberwell Beauty, *Nymphalis* (or *Vanessa*) *antiopa*, which occurs as a motif in Stifter's work. *Die verlorene Geliebte* (1956) is a collection of narratives; the lost beloved of the title is the city of Prague. In 1957 this work reappeared in abridged form as *Neujahrsrummel*. *Prager Triptychon* (1960) consists of sketches of life in Prague in the 1920s, to which the terms for the components of an altar-piece are applied; they are predella (*Relief der Stadt*), left panel (*Die Causa Wellner*), shrine (*Weißenstein Karl*), right panel (*Vermächtnis eines Jünglings*), and Gesprenge or decorative superstructure (*Die Zauberflöte*, which is concerned with a performance of Mozart's opera). Other stories include *Denkwürdigkeiten von Gibacht* (1958), *Das Elefantenblatt* (1962), *Entführung und sieben andere Ereignisse* (1964), *Die erbeuteten Frauen* (1966), *Bist du es, Ronald?* (1968), *Die letzte Tombola*, and *Morgen fahr' ich heim* (both 1971).

The biographical work *Goethe in Böhmen* (1932, extended 1962) was followed many years later by *Goethes Amerikabild* (1958). The essay *Da geht Kafka* appeared in 1965 (extended 1966). A posthumous selection of stories and essays with an introduction by H. Jacobi appeared in 1971 under the title *Bekenntnisse eines Pedanten*. Urzidil's only novel, *Das große Hallelujah* (1959), is concerned with the life of an emigrant in North America: it has links with his stories.

USINGER, FRITZ (Friedberg, Hesse, 1895–1982, Friedberg, Hesse), a schoolmaster after the 1914–18 War, is the author of a number of volumes of varied verse linking up with the hymnic and elegiac tradition of Hölderlin and Rilke (qq.v.). They include *Der ewige Kampf* (1918), *Große Elegie* (1920), *Irdisches Gedicht* (1927), *Das Wort* (1931), *Die Stimme* (1934), *Die Geheimnisse* (1937), *Hermes* (1942), *Hesperische Hymnen* (1948), and *Niemandsgesang* (1957). A selection, *Pentagramm*, ed. M. Roddewig, appeared in 1965, and *Canopus* in 1968.

Among his numerous essays are those on Schiller (*Friedrich Schiller und die Idee des Schönen*, 1955) and E. A. Poe (1959). *Tellurium*, a collection, appeared in 1966, *Das Ungeheuer Sprache* in 1965, and *Dichtung als Information* in 1970. The recipient of honours, he was awarded the Büchner Prize of 1946 and the Humboldt Plakette of 1980.

USTERI, JOHANN MARTIN (Zurich, 1763–1827, Rapperswil), son of a well-to-do merchant, made a grand tour in 1783–4, in the course of which he met Klopstock, M. Claudius, and Goethe (qq.v.). He afterwards entered his father's business, which he took over in 1803. He wrote poetry both in High German and in Swiss dialect. His best-known works are the dialect idyll *De Vikari* (1807–10) and the poem 'Freut euch des Lebens, weil noch das Lämpchen glüht' (q.v., 1793).

Usteri's *Dichtungen in Versen und Prosa* (3 vols.), ed. D. Hess, appeared in 1831, and posthumous writings, ed. K. Escher, in 1896.

Ut de Franzosentid, a short novel in Low German dialect (Mecklenburger Platt) by F. Reuter (q.v.), published in 1859 in the first volume of the narratives grouped by Reuter under the title *Olle Kamellen* (q.v.). 'De Franzosentid' is the time of the French occupation of Germany (see NAPOLEONIC WARS). The story is set in 1812 after the French defeat in Russia. In the Mecklenburg town of Stemhagen (Stavenhagen) a misunderstanding leads to the belief that a French soldier has been disposed of, arrests are made, but in the end all is satisfactorily resolved. The story then briefly recounts the national rising of 1813 and ends happily with the double marriage of Fiken Voss with her cousin Hinrich, and Fik Besserdich with Friedrich Schult.

The real substance of the novel is the gallery of rich characters; they include Amtshauptmann Wewer (Weber), Möller Voß, ruined by the French and by money-lenders, his man Friedrich Schult, Ratsherr Hesse, the 'Burmeister' (Bürgermeister), who was Reuter's father, Mamsell Westphalen, Fiken Voß, the miller's brave and sensible daughter, and Fik Besserdich. For these portrayals Reuter drew liberally on his own childhood recollections, and their originals have all been identified. It is a novel of great good humour and broad comedy, told in vigorous racy dialect.

Ut mine Festungstid, an autobiographical work by F. Reuter (q.v.), dealing with his last four years of detention. It appeared in 1862 as the second volume of *Olle Kamellen* (q.v.) and is written in the Low German dialect (Mecklenburger Platt). Of the three partially autobiographical novels (the others are *Ut de Franzosentid* and *Ut mine Stromtid*, qq.v.), this is the closest to reality.

It opens with Reuter's arrival at the fortress of Glogau; he is next transferred to Magdeburg, then to Graudenz, finally repatriated to Dömitz prison in Mecklenburg, and then released. Reuter portrays with sympathy and occasionally with severity the prison personnel, mostly military, and is able to describe his experiences with calm detachment and with touches of humour, though the moments of depression and despair are not concealed. The absence of *ressentiment* is remarkable.

Ut mine Stromtid, a three-volume novel in Low German dialect (Mecklenburger Platt) by F. Reuter (q.v.), published 1862–4 and constituting volumes III, IV, and V of *Olle Kamellen* (q.v.). The word 'Strom' in the title is a local dialect word meaning 'farm manager' or 'estate steward'; Reuter refers to his years on the land after his release from prison. In contrast to the episodic comedy of *Ut de Franzosentid* (q.v.) and the personal recollections of *Ut mine Festungstid* (q.v.), *Ut mine Stromtid* is a substantial well-developed novel with a wide range of fictitious characters. It begins in 1829 and finishes after the Revolution of 1848, passing from the depressed state of the Mecklenburg farmers early in the century to the eviction of oppressive landowners in the revolution (see REVOLUTIONEN 1848–9).

Havermann, driven from his farm tenancy and sold up by the landowner Pomuchelskopp, is befriended by an estate steward, Bräsig, and finds a new sphere as manager of the lands of Herr von Rambow at Pümpelhagen, where he demonstrates his efficiency and fidelity. Pomuchelskopp, however, has a financial lever over Rambow and establishes a hold on his estate. When Rambow dies, Pomuchelskopp poisons the mind of the heir, Axel von Rambow, against Havermann, who leaves under a cloud, though Bräsig disproves allegations of dishonesty made against his friend. In 1848, the year of revolution, Pomuchelskopp is chased away by his rebellious tenants and leaves the country. Havermann returns to Pümpelhagen and devotes himself to saving the estate for the sake of Axel von Rambow's wife. His loyal service is rewarded by consent to the marriage of his daughter Lowise to Franz von Rambow, who is in love with her and is a gentleman of a better stamp than his cousin Axel. The kind-hearted and eccentric Bräsig is a prominent and endearing character. Though it has a serious social background, *Ut mine Stromtid* is a good-humoured and heart-warming novel.

UZ, JOHANN PETER (Ansbach, 1720–96, Ansbach), a student friend of J. W. L. Gleim and J. N. Götz (qq.v.), studied law at Halle University (1739–43) and later held administrative appointments in his native town of Ansbach (1748–60) and in Nürnberg (Geheimer Justizrat). He is best known for his Anacreontic poetry (see ANAKREONTIKER), deft, light poems reflecting, in variations on wine, women, and song, an amiable hedonism. His *Der Sieg des Liebesgottes* (1753) is an imitation of *The Rape of the Lock*. He also wrote a didactic poem, *Versuch über die Kunst, fröhlich zu sein* (1760), in which his combination of hedonism and Christianity is apparent. His Anacreontic poems appeared in *Lyrische Gedichte* (1749) and his collected poetry in *Sämtliche Poetische Werke* (2 vols., 1768, reprinted 1964). In 1760 Uz published the works (*Schriften*) of J. F. von Cronegk (q.v.), who died in 1758.

V

Vademecum, commonly used abbreviated designation of a work of critical controversy by G. E. Lessing (q.v.), the full title of which runs *Ein Vademecum für den Herrn Sam. Gotth. Lange, Pastor in Laublingen, in diesem Taschenformate ausgefertiget von Gotth. Ephr. Lessing*. It was published in 1754. Lessing had reviewed Lange's translation of Horace in 1753 and had taken umbrage at a personal innuendo in Lange's reply. The *Vademecum*, which remorselessly analyses Lange's Latinity, established Lessing's reputation as a formidable literary controver-

sialist. The title 'Vademecum' alludes to a sneer by Lange at the small format (12mo) of Lessing's *Schriften* (1753). See also LANGE, S. G.

VADIAN or **VADIANUS,** JOACHIM, real name von Watt (St. Gall, 1484–1551, St. Gall), became a professor and in 1516 Rector of Vienna University. In 1518 he returned to St. Gall, devoting himself to Lutheran reform, which he supported in publications (*Karsthans und Kegelhans*, 1521, is attributed to him). He was a friend of Zwingli (q.v.). A humanist of note, he wrote a treatise on literature (*De poetica et carminis ratione liber*, 1518) and works on Swiss history, which were edited by E. Götzinger (3 vols.) in 1875–9. His Latin speeches, ed. M. Gabathuler, appeared in 1953.

Vae victis!, a Novelle by F. von Saar (q.v.) first published in 1879 as *Der General*. The present title dates from 1883, when it appeared in *Drei neue Novellen*. It was included in 1897 in *Novellen aus Österreich* (q.v.). 'The defeated' of the second title are the general, who shoots himself when he discovers the contempt in which his wife holds him, and his widow's second husband, a self-confident politician, whose performance in government does not come up to his promises.

VALENTIN, KARL, pseudonym of Valentin Ludwig Fey (Munich, 1882–1948, Munich), a Munich comedian, who wrote his own sketches for his popular cabaret. He had a gift for grotesque satire with a touch of the absurd; his style influenced the first short plays of Brecht (q.v.), with whom he collaborated in the early 1920s. He made a deft use of dialect and of masks. His publications include *Das Karl-Valentin-Buch* (1932), *Brillantfeuerwerk* (1938), *Valentinaden* (1941), and the posthumous *Das Lachkabinett* (1950), *Panoptikum* (1952), *Die Raubritter vor München. Szenen und Dialoge* (1963). The first volume of *Gesammelte Werke* (1961) was followed by a second volume entitled *Sturzflüge im Zuschauerraum. Der gesammelten Werke anderer Teil*, ed. M. Schulte (1970); *Das große Karl-Valentin-Buch* appeared in 1973, *Karl Valentins Filme* and correspondence, *Geschriebenes von und an Karl Valentin*, in 1978.

Valentin und Namelos, an anonymous Middle Low German epic poem of some 2,600 lines, written about the middle of the 15th c. It is a translation from the Flemish, but the original source is French. It is the story of twins who are saved by a servant from being exposed and left to die; one is brought up as a knight, the other (Namelos) grows up as a wild man in the woods. After various adventures each marries the woman of his choice.

Valentin und Orsus, a Volksbuch (q.v.) translated from the French by Wilhelm Ziely of Berne and first published in 1511. The content is virtually identical with that of the 15th-c. poem *Valentin und Namelos* (q.v.).

Valmy, Cannonade of, see REVOLUTIONS-KRIEGE and Goethe's CAMPAGNE IN FRANKREICH.

VARNHAGEN VON ENSE, KARL AUGUST (Düsseldorf, 1785–1858, Berlin), a physician's son, studied at various universities, interrupting his career to serve as an officer in the Austrian army against Napoleon. He was wounded in 1809 at Wagram and afterwards served as an adjutant to Prince Bentheim. He resigned in 1812 and entered Russian service. After the war he entered the Prussian diplomatic service, serving from 1816 to 1819 in Karlsruhe.

In 1814 Varnhagen married the well-known literary hostess Rahel Levin (see VARNHAGEN VON ENSE, R.), who was fourteen years older than he, and in 1819 the couple settled in Berlin, where their apartment became a centre of literary social life. Varnhagen, who was notable for his indiscretion, made enemies as well as friends in these years. Though acquainted with many poets, he was not himself a writer of distinction. He published poems (*Gedichte*, 1814; *Vermischte Gedichte*, 1816), stories (*Deutsche Erzählungen*, 1815), and a novel (*Die Sterner und die Psitticher*, 1831). He also compiled *Goethe in den Zeugnissen der Mitlebenden* (1823) and *Biographische Denkmale* (5 vols., 1824–30). His *Ausgewählte Schriften* (19 vols.) appeared 1870–6, diaries, ed. L. Assing (14 vols.), 1861–70 (reprinted 1971), and correspondence with his wife (6 vols.) 1874–5 (reprinted 1967).

VARNHAGEN VON ENSE, RAHEL (Berlin, 1771–1833, Berlin), originally Rahel Levin, was the sister of Ludwig Robert (q.v.) and came of a well-to-do Jewish family. At the beginning of the century she was a prominent literary hostess in Berlin, whose soirées were attended by many of the Romantics. The loss of her fortune in 1806 ended this way of life for a time. In 1808 she met K. A. Varnhagen von Ense (q.v.), whom she married in 1814 after being baptized. Varnhagen was a professional Prussian diplomat until his resignation in 1819, and the first years of the marriage were spent mainly in Karlsruhe. After 1819 the Varnhagens entertained in Berlin, and Rahel's *salon* was again the most notable social focus of Berlin literary life, frequented by H. Heine, Alexander von Humboldt, Bettina von Arnim, L. von Ranke, A. von Chamisso, H. Laube, and F. de la Motte FOUQUÉ (qq.v.).

Varnhagen published writings by her in *Rahel. Ein Buch des Andenkens für ihre Freunde* (1834, reprinted 1971). Correspondence, ed. F. Kemp, appeared in 4 vols. 1966–8 (rev. 1979), and *Gesammelte Werke* (10 vols.), ed. K. Feilchenfeldt, U. Schweikert, and R. E. Steiner, were published in 1983.

Vater, Der, a novel by J. Klepper (q.v.), published in 1937. It is sub-titled *Roman eines Königs* and attempts a justification of the much-vilified figure of King Friedrich Wilhelm I (q.v.) of Prussia, whom Klepper sees as convinced of his kingship by divine right. His aggressive paternalism is explained by his duty to act as God the Father to his country. In this position he is constantly a prey to a remorseless self-scrutiny. His most acute spiritual crisis is the flight of the heir apparent (the future Friedrich II, q.v., of Prussia), in which Klepper sides with the father.

Two minor works arose out of the historical studies for this book, *Der Soldatenkönig und die Stillen im Lande* (1938) and *In tormentis pinxit* (1939), which reproduces the strange pictures painted by the king during his sleepless nights.

Väterbuch, Das, a Middle High German collection of verse legends mainly concerning holy anchorites in the wilderness. It is in part a translation of the *Vitae patrum* but also draws on the *Legenda aurea* (q.v.). The author of this immense work of more than 40,000 lines was a cleric who wrote it *c.* 1280. He is believed to be identical with the author of *Das Passional* (q.v.).

Vaterunser, see PATERNOSTER.

VEGESACK, SIEGFRIED VON (nr. Wolmar, Livonia, 1888–1974, Burg Weißenstein, Bavaria), a Baltic German, lived after the 1914–18 War as a writer and farmed in the Bavarian Forest, making Weißenstein nr. Regen his home. In 1933 he emigrated to Sweden, and in 1936 moved to South America, returning to Germany in 1938; after the 1939–45 War he settled again in Weißenstein. A translator of Russian works, he is best known as the author of novels set in his native country, notably the trilogy *Baltische Tragödie*, the components of which are *Blumbergshof* (1933, the title bearing the name of his father's estate, revised as *Versunkene Welt*, 1949), *Herren ohne Heer* (1934), and *Totentanz in Livland* (1935). The trilogy depicts the life and decline of the German nobility in Livonia between 1890 and 1919. His other fiction includes the novels *Meerfeuer* (1936, reissued 1970), *Der letzte Akt* (1957), and *Überfahrt* (1967), and the stories *Zwischen Staub und Sternen* (1947), *Der Pastoratshase* (1958), and *Tanja* (1959). An early volume of poetry is entitled *Die kleine Welt vom Turm gesehen* (1925); it was followed by *Der Lebensstrom* (1943), *Das Unverlierbare* (1947), and *In dem Lande der Pygmäen* (1953); *Kleine Hausapotheke* (1944) is a collection of poems and stories. *Südamerikanisches Mosaik* (1962) is a record of his travels; family material is contained in *Vorfahren und Nachkommen, Aufzeichnungen aus einer altlivländischen Brieflade, 1689–1887* (1960). Vegesack, who during the war worked as an interpreter in Russia (*Als Dolmetscher im Osten,*

1965), was the recipient of a number of East German awards.

VEGHE, JOHANNES (Münster, *c.* 1431–1504, Münster), rector of the house of the 'Brüder vom gemeinsamen Leben' in Rostock and later (1475) in Münster, was an able preacher of mystical tendency, twenty-four of whose sermons in Middle Low German have been preserved. He is also in all probability the author of the Middle Low German allegorical tracts *Een bloemlich beddiken* (known in High German as *Das geistliche Blumenbeet*), edited by H. Rademacher in 1938, and *Wyngarde der zele* (*Weingarten der Seele* or *Geistlicher Weingarten*), published by the same editor in 1940.

VEHE, MICHAEL (Biberach, *c.* 1485–1593, Halle/Saale), a Dominican monk, was in the service of the Archbishop of Mainz and was later provost (Probst) in Halle. He opposed the Reformation in polemical writings and compiled a Catholic hymn-book (*Gesangbüchlein*, 1537) as a counterpart to Protestant hymnals.

Veilchen, Das, title given in 1800 to an originally untitled poem by J. W. Goethe, the first line of which is 'Ein Veilchen auf der Wiese stand' (q.v.).

VEIT, PHILIPP (Berlin, 1793–1877, Mainz), a Romantic painter, was a son of Dorothea von Schlegel (q.v.) by her first marriage, and hence a grandson of Moses Mendelssohn (q.v.). He was converted to Roman Catholicism at 16, shortly after the conversion of his mother and his stepfather, Friedrich von Schlegel (q.v.). He served as a volunteer in the Wars of Liberation (1814–15) and then studied painting in Rome, becoming one of the Nazarene school (see NAZARENER). He returned to Germany in 1830 and was director of the City Art Institute from 1830 to 1843. From 1853 until his death he was director of the Mainz Art Gallery.

VELDE, KARL FRANZ VON DER (Breslau, 1779–1824, Breslau), a Prussian Crown law officer, wrote plays and once-popular historical novels, including *Der Flibustier* (1818), *Die Lichtensteiner* (1821), *Der böhmische Magdekrieg* (1823), and *Christine und ihr Hof* (1823). His collected works (*Schriften*, 25 vols.) appeared between 1824 and 1827.

VELTEN, also **VELTHEN,** JOHANNES (Halle, 1640–92 or 1693, Hamburg), studied theology from 1657 to 1661 and then became an actor in the troupe of Andreas Paul or Paulsen, which toured not only in Germany but also in Scandinavia and the Baltic lands. He is thought to have married Paul's daughter *c.* 1672 and he took over the troupe in 1678. From 1685 he and his company were in the service of the Elector of Saxony. Velten took the actor's task seriously, raised the standard of acting and the quality of the repertoire, and sought to make the actor socially acceptable. After his death the company continued until 1711.

Veme, see FEMGERICHTE.

Venezianische Epigramme, title by which is known a collection of epigrams written by Goethe for the most part during his stay in Venice from March to May 1790 and in Silesia in the autumn of 1790. A small selection appeared in the *Deutsche Monatsschrift* in 1791; the complete set was published in Schiller's *Musenalmanach für das Jahr 1796.* In his collected works Goethe gave them the superscription *Epigramme Venedig 1790.*

There are 103 epigrams, all in elegiac metre, varying in length from one to eight distichs. Some relate to Goethe's relationship to Christiane Vulpius (see GOETHE, CHRISTIANE) and his enforced absence from her, others are critical of Italy (in sharp contrast to the *Römische Elegien,* q.v.) or make acid comments on human affairs. A note of frustration and irritation is perceptible, corresponding to Goethe's feeling of time mis-spent.

'Veni creator spiritus', first line of a Latin hymn for Whitsun believed to have been written in the 9th c. by Hrabanus Maurus (q.v.), Abbot of Fulda and later Archbishop of Mainz. Luther's translation begins 'Komm, Gott Schöpfer, Heiliger Geist', and English adaptations include Cosin's 'Come, Holy Ghost, our souls inspire' and Dryden's 'Creator Spirit by whose aid!' There is also a German translation by Goethe, dated 10 April 1820, beginning 'Komm, Heiliger Geist du Schaffender'.

The Latin text of the hymn provided Gustav Mahler (q.v.) with the words for the first movement of his 8th Symphony (1906, performed 1910).

Venusberg, a mountain containing a cave or grotto, where Venus holds sway in a realm of sensual pleasure. Venus's guests incur ultimate damnation. The legend is believed to date from the 13th c., and the name of Tannhäuser (q.v.) is connected with it from the 15th c., most notably in the ballad *Tannhäuserlied* of the early 16th c.

Venusgärtlein, a collection of poems, including folk-songs, published in Hamburg 1665-6. A later edition (1756) was reprinted in 1890 (ed. M. von Waldberg).

Venuswagen, Der, an early poem by Schiller, first published anonymously probably in 1781. It has a pronounced erotic character.

Verballhornen, see BALLHORN, JOHANN.

Verborgenheit, a poem written by E. Mörike (q.v.) in 1832. It is a poem of withdrawal and seclusion, beginning 'Laß, o Welt, o laß mich sein!' It has been set to music by Hugo Wolf (q.v.).

Verbrechen aus Ehrsucht, Das, a play by A. W. Iffland (q.v.), first performed in 1784. It is described as *Ein Familiengemälde in fünf Aufzügen.* Two families are about to be united by marriage, but the happiness of all is jeopardized by the bride's brother, who, having lived above his means and lost at cards, takes money from a fund administered by his father. Exposure is avoided by a loan provided by the rough-tongued but golden-hearted father of the other party to the marriage.

Verbrecher aus verlorener Ehre, Der, a story by Schiller, written in 1785 and published in his periodical *Die Thalia* in 1786, where its title ran *Der Verbrecher aus Infamie.* The revised title first appeared in Schiller's *Kleinere prosaische Schriften* (Pt. I, 1792).

Christian Wolf poaches in order to raise money to make presents to a girl in the village in which he keeps an inn. He is detected and fined, detected again and locked up, and on a third conviction receives a severe sentence. An outcast after each release, he takes to the woods, shoots the man who has betrayed him, and becomes the head of a gang of robbers. Utterly disillusioned, he tries in vain to obtain permission to return to normal life. In the end he gives himself up. The story, plainly and economically told, makes it clear that Wolf had good in him and, in more fortunate circumstances, would not have become a criminal; it also stresses the force of moral law. It is based on the story of Friedrich Schwan (1729-60), who surrendered to the father of Schiller's teacher J. F. Abel (q.v.), and is described as 'eine wahre Geschichte'. The impression of authenticity is reinforced by use of the first person for the narration of Wolf's crucial experience.

Verdacht, Der, a short detective novel by F. Dürrenmatt (q.v.), published in book form in 1953 after being serialized in *Der Schweizerische Beobachter* (1952). Its detective is Kommissär Bärlach of *Der Richter und sein Henker* (q.v.), and its action begins shortly after the end of the earlier novel.

Bärlach has survived an operation for

abdominal cancer and is unlikely to recover. He brings to book Dr. Emmenberger, a former Nazi concentration camp doctor, whose speciality was operating without an anaesthetic. Bärlach discovers that Emmenberger is the director of the nursing home Sonnenstein. Greatly daring, Bärlach takes a false name and has himself admitted to Sonnenstein as a patient. He nearly pays with his life, but is rescued in the nick of time from a terrible death on the operating table. His rescuer is the Jew Gulliver, a giant of formidable strength, and once a victim of Emmenberger. Gulliver's self-imposed task is to seek out and 'execute' Nazi criminals, and he forces Emmenberger to take poison a few minutes before the operation on Bärlach is due to begin.

Verdi. Roman der Oper, a novel by F. Werfel (q.v.), published in 1924. Its subject is the challenge which the apparently outmoded Italian opera (represented by Verdi) faces in the new Musikdrama of R. Wagner (q.v.). The novel is set in Venice at the time of Wagner's last visit to the city, 1882–3.

Wagner has completed in the preceding ten years *Siegfried, Götterdämmerung,* and *Parsifal* (qq.v.). In the same span of time Verdi has written nothing since *Aïda* (1871). Verdi, who arouses Werfel's admiration and sympathy, is conscious of defeat. Twice he and Wagner pass and exchange stern and inscrutable glances. On 14 February 1883 Verdi resolves to call on Wagner at his hotel, only to learn that the composer has died during the night. The discussion which Verdi sought is thus prevented. The period of tension, which covers two visits of Verdi to Venice, includes the intentional destruction of the score of his unfinished *King Lear.* In the 'Nachspiel' the reawakening of Verdi's genius in collaboration with Boito is indicated, the triumph of *Otello* (1887), representing the consummation of Verdi's style, is recorded, and a hint is given of the unexpected flowering of comedy in Verdi's last work *Falstaff* (1893). In the 'Nachspiel' Werfel takes care to emphasize, through the mouth of Boito, his view that Verdi's final style is not influenced by Wagner.

The novel, which is the result of careful study and intense enthusiasm, is a Künstlerroman (q.v.).

Verdun, Schlacht um, an offensive ordered by Field-Marshal E. von Falkenhayn (1861–1922) and carried out by the Crown Prince's Army Group in 1916. Verdun was the projecting bastion of the French lines, and since, for reasons of prestige and morale, it was unlikely that the French would abandon it, Falkenhayn believed that their army could be exhausted by massive and prolonged attacks (Ermattungsstrategie). These began on 2 February 1916 and were broken off in June. Verdun remained in French hands, though some of the outer forts fell; they were regained by the French later in the year. The appalling loss of life was about equal on both sides.

Works dealing with the battle for Verdun include *Sieben vor Verdun* (q.v.), by J. M. Wehner *Opfergang* (q.v.) by F. von Unruh, and *Erziehung vor Verdun* by A. Zweig (q.v.). In 1971 H. Risse (q.v.) published the essay *Wer denkt heute noch an Verdun?*

Verdun, Vertrag von, an agreement reached in 843 between Karl II, der Kahle, Lothar I, and Ludwig II, der Deutsche (qq.v.), by which they divided the Frankish Empire between them. The western portion (Neustrien) fell to Karl, a territory corresponding to modern France west of Verdun, Toul, and the Rhône. Austrien was split into two, Ludwig taking the eastern portion (modern Germany east of the Rhine and Aare and west of the Elbe and Bohemia). Lothar received a central strip from the North Sea to the Mediterranean, including Italy.

Though this partition did not finally divide the Empire, it is regarded as symbolizing the coming separation. From Karl's realm grew France, from Ludwig's Germany; Lothar's central kingdom proved less lasting. The agreement was a consequence of the defeat of Lothar at Fontenoy in 841.

Verein Freie Bühne, see FREIE BÜHNE, VEREIN.

Verfolgung und Ermordung Jean Paul Marats dargestellt durch die Schauspielgruppe des Hospizes zu Charenton unter Anleitung des Herrn de Sade, Die, a play (Drama in zwei Akten) in verse by P. Weiss (q.v.), written 1963–4, first performed in April 1964 at the Schiller-Theater, Berlin, and published in 1964 (fifth version in 1965).

A play-within-a-play, it is set in 1808 in an asylum at Charenton, whose inmates perform de Sade's play; this is set on 13 July 1793, the day of the murder of Marat by Charlotte Corday. The confinement of the Marquis de Sade at Charenton from 1801 to 1814 and his addiction to amateur theatricals, which were encouraged by the asylum director, M. Coulmier, are historical facts, but not the personal meeting between Marat and de Sade, though de Sade delivered Marat's funeral oration. The play resolves itself into an inconclusive argument between Marat, the revolutionary, and

Sade, the extreme individualist, aesthete, and nihilist, who seeks to expose Marat as a dangerous hypocrite. In incorporating authentic speeches made by Marat, the play represents the documentary theatre and resembles its forerunner, Büchner's *Dantons Tod* (q.v.), while clearly bearing the imprint of Brecht's epic theatre (for example *Der kaukasische Kreidekreis*, q.v.). There were three seminal productions of the play: Konrad Swinarski's at the Schiller-Theater (West Berlin, 1964), Peter Brook's with the Royal Shakespeare Company at the Aldwych Theatre (London, 1964), and Hanns-Anselm Perten's in Rostock (DDR, 1965). While Swinarski exploited the aesthetic attractions of the play-within-a-play, Brook saw it as an example of 'total theatre' and of Antonin Artaud's 'theatre of cruelty'; Perten stressed the dialectics of individuation and collectivism. Special features are the experimental style of Sade, who directs his own play, and the participation on stage of the audience, headed by Coulmier (who is described as a pompous bourgeois), to which the real audience adds a further dimension. Though initially not offering a solution to the play's complex dialectics, preferring to adopt a third, uncommitted standpoint, Weiss made it clear after experiencing Perten's version that authentic productions would present Marat as the moral victor and offer a rehabilitation of Marat, whom he sees as a forerunner of Marxism and a victim of 19th-c. bourgeois historians who have presented him as an abhorrently bloodthirsty figure (*Anmerkungen zum geschichtlichen Hintergrund unseres Stücks*, appended to the final version).

Verfremdungseffekt, also V-Effekt, a term deriving from Russian Formalism (ostranenie) and associated with the epic theatre (see EPISCHES THEATER) of B. Brecht (q.v.). The technique of Verfremdung aims at giving to familiar aspects of reality the appearance of being unfamiliar in order to arouse the spectator's critical judgement. It is used in the socialist theatre of Brecht, who adopted the term in order to dissociate himself from traditional German and Chinese models employing similar techniques for different ends, as well as from Aristotelian drama as he saw it.

An idiosyncratic terminology aids the V-effect, notably the use of Stückeschreiber (cf. *Lied des Stückeschreibers*), 'Thaeter' for theatre (Theater), and 'Misuk' for music (Musik), the latter being Brecht's signature tune for his criticism of traditional opera and especially Wagner's Gesamtkunstwerk, and for his own brand of songs. Brecht frequently commented on the function of the V-effect in his theoretical writings (e.g. § 42 of *Kleines Organon für das*

Theater, q.v.: 'Eine verfremdende Abbildung ist eine solche, die den Gegenstand zwar erkennen, ihn aber doch zugleich fremd erscheinen läßt'). He saw in Surrealism a 'primitive' (because non-political) forerunner since it produces disillusionment by means of shock-effect, and linked his own Verfremdungseffekt with the principle of revolution: 'Die echten V-Effekte haben kämpferischen Charakter' (*Nachträge zum 'Kleinen Organon'*, 1964).

Verhängnisvolle Gabel, Die, a satirical verse comedy written in Italy by A. von Platen (q.v.) in 1826 and published in that year. Platen referred to it as an 'Aristophanic comedy', but on the title-page it is simply described as *Ein Lustspiel in fünf Akten*. Parodying a fate tragedy (see SCHICKSALSTRAGÖDIE), Platen engineers the catastrophe of a whole family through the agency of a fateful fork. Each act is followed by an explanatory and satirical parabasis. Among the targets of Platen's wit are A. von Kotzebue and A. Müllner (qq.v.).

Verhör des Lukullus, Das, a radio play (Hörspiel) in 14 scenes with irregular verse by B. Brecht (q.v.), written in 1939 'before the outbreak of the Second World War' (*Versuche*, vol. 11, Vorspruch). General Lukullus is given a pompous funeral in Rome amid the comments of the soldiers, slaves, and women who have suffered through his wars, and of the businessmen who have gained by them. On entering the underworld he has to doff his imposing helmet and submit to trial before a jury introduced by the Sprecher des Totengerichts and consisting of a peasant, a slave, a fishwife, a baker, and a courtesan. The Totenrichter reminds the accused that in the underworld no one fears the names of 'the Great'. Thus humiliated, Lukullus finds that all the witnesses, including the figures stepping forth from his grand triumphal frieze, speak against him; only the peasant has something to say in his favour, since Lukullus introduced the cherry-tree into Europe.

The original version ended with the withdrawal of the jury. The judgement, committing him to annihilation, was later taken over from the opera *Die Verurteilung des Lukullus* (q.v.), but the original title was retained to distinguish the two versions.

Verlassene Mägdlein, Das, a poem written by E. Mörike (q.v.) in 1829 and included in the novel *Maler Nolten* (q.v.). It portrays, in the first person, the desolation of the forsaken servant girl. The first line runs 'Früh, wenn die Hähne krähn'. The best known of many musical settings is by Hugo Wolf (q.v.).

Verlobung in St. Domingo, Die, a Novelle by H. von Kleist (q.v.), which was first published in instalments in the Berlin journal *Der Freimütige* in March and April 1811 and included in the collected edition of his stories, *Erzählungen* (vol. 2, 1811). Based on the Negro revolt in Santo Domingo in 1803, the Novelle, concentrating its action into three days, narrates the tragic love of Gustav von der Ried for a 15-year-old half-caste, Toni, whom he kills when he mistakenly believes himself to have been betrayed by her.

Toni lives with her mother Babekan and her Negro foster-father Congo Hoango, who has trained her to trap white men. Gustav, who belongs to a small party of Swiss refugees, arrives unsuspecting, and seeks shelter in the house during Hoango's absence. Toni and Gustav instantly fall in love and become secretly betrothed. While Babekan schemes Gustav's destruction, Toni plans their flight. Hoango returns sooner than expected, and Toni, in order to save Gustav, who remains unaware of the plot against him, ties him to his bed, thus hoping to convince her father that she has not betrayed his cause. The Swiss, led by Herr Strömli, arrive at the moment of crisis and free Gustav, who impulsively shoots Toni. When he discovers the true motives of Toni's action, he shoots himself. Since the Swiss hold his two sons as hostages, Hoango refrains from further bloodshed and allows the Swiss free passage. They carry with them the bodies of Gustav and Toni for common burial. In 1807 Herr Strömli erects a monument to the lovers in his native Switzerland.

The tragedy arises out of Gustav's failure to trust his beloved when appearances belie the truth; this is a recurrent theme in works by Kleist, but the penetrating treatment of the problems of racial prejudice singles this Novelle out from his other stories.

Verlorene Handschrift, Die, a three-volume novel by G. Freytag (q.v.), published in 1864. Professor Felix Werner discovers a reference in a medieval MS., which leads him to infer the existence of a more ancient manuscript of a lost portion of Tacitus, and the pursuit of this becomes an *idée fixe* with him. His investigations lead him to a farmhouse, where he finds, not the MS., but a wife, the farmer's daughter Ilse Bauer. At the university, which is in the capital of the small state, Ilse Werner is for a time pursued by the reigning prince, who, for other reasons, is in the end declared insane and forced to abdicate. Werner at last abandons the search for the tantalizing MS., concluding that it was destroyed in the 17th c.

Freytag shows here, as elsewhere, the superi-

ority of the burgher to the nobleman, and also gives for the first time in a German novel a broad picture of university life.

Verlorene Lachen, Das, see LEUTE VON SELDWYLA, DIE.

Vermächtnis, a poem written by Goethe in 1829 and first printed with *Wilhelm Meisters Wanderjahre* in 1829. The poem takes as its point of departure the last two lines of *Eins und Alles* (q.v.), written eight years earlier: 'Denn alles muß in Nichts zerfallen,/Wenn es im Sein beharren will'. These lines had been lifted out of context at a scientific conference, and Goethe felt impelled to write a further poem beginning: 'Kein Wesen kann zu Nichts zerfallen!/Das Ew'ge regt sich fort in allen'.

Vermächtnis justifies the role of the individual in the scheme of things. The title is to be understood as a repository of wisdom *bequeathed* by the octogenarian poet.

Vermächtnis, a poem by H. Heine (q.v.), included in the section *Lamentationen* of the *Romanzero* (q.v.). No. 19 in the subdivision *Lazarus*, it begins 'Nun mein Leben geht zu End' and reveals in ironical guise Heine's still active capacity to hate.

Vernünftigen Tadlerinnen, Die, a moralizing weekly published by J. C. Gottsched (q.v.) and including some contributions by his wife, Luise A. V. Gottsched (q.v.).

Versailles, Treaty of, (1) effected an alliance between Austria, France, and Russia in 1756 (see DIPLOMATISCHE REVOLUTION). It was elaborated in the Second Treaty of Versailles in 1757.

(2) The document (Versailler Vertrag) of the Peace Conference held at Versailles from 18 January to 28 June 1919 (see WELTKRIEGE, 1). The treaty is usually described even by moderate German opinion as the Versailler Diktat, since the defeated were not admitted to the discussions. The text was handed to the German delegation on 7 May. After the resignation of one government in Germany and bitter debates in the Reichstag the German delegation signed the treaty on 28 June 1919.

The German text was a document of 229 pages, divided into fifteen parts and 440 Articles. Pt. I was the Covenant of the League of Nations (Völkerbund), to which Germany was not admitted. Pt. II concerned the frontiers of Germany. Alsace and Lorraine were returned to France, Eupen and Malmédy ceded to Belgium, the Prussian province of Posen became a part of the revived state of Poland, and a

Polish corridor was created providing access to the sea, so severing East Prussia from the rest of Germany. The northernmost strip of Schleswig, after a plebiscite, became Danish, the easternmost strip of East Prussia beyond the Memel passed to Lithuania. Pt. III contained the political clauses. The Saar District was to be administered by France until 1935, when a plebiscite would be held, Danzig became a Free State under the aegis of the League of Nations, and the independence of Austria was declared inalienable. Pt. IV deprived Germany of its colonies, which were later distributed as mandated territories by the League of Nations. Pt. V was military and naval, limiting the German army to 100,000 volunteers serving for twelve consecutive years, and banning tanks and aircraft. Naval building was limited to surface ships of not more than 10,000 tons.

Pt. VIII (Reparations) imposed repayments in kind or cash to cover not only damage inflicted on the Allies, but also the costs incurred by the Allies in conducting the war. The justification for these reparations is contained in Article 231, which is commonly referred to as the 'War Guilt' (Kriegsschuld) clause.

Pts. VI, VII, and IX were financial clauses, Pt. X economic clauses, Part XI covered aerial navigation, Pt. XII ports, waterways, and railways, and Pt. XIII labour. Pt. XIV provided for Allied occupation of the left bank of the Rhine for fifteen years.

Verschleierte Bild zu Sais, Das, a poem by Schiller first printed in 1795 in No. 9 of *Die Horen.* It recounts the fate of an Egyptian acolyte, who, in defiance of an express injunction, lifts the veil concealing the statue of truth, and perishes. The dying youth himself formulates the moral: 'Weh dem, der zu der Wahrheit geht durch Schuld.'

Verschmähten Weiber, Die, see TEUFEL IST LOS, DER.

Verschollene, Der, an unfinished novel written by F. Kafka (q.v.) between 1912 and 1914, and published by Max Brod (q.v.) in 1927 with the title *Amerika.* A dramatization appeared in 1957. The first chapter was published separately as *Der Heizer* (1913). The young emigrant to America, Karl Roßmann, is rejected by his relatives and by society, is robbed by tramps and roughly handled in his attempt to find a settled job until he seems at last likely to find acceptance in 'Das Naturtheater von Oklahoma'. At this point the story breaks off. *Der Verschollene,* ed. J. Schillemeit, appeared in 1982, *Der Verschollene. Apparatband* in 1983.

Verschwender, Der, a play (Original-Zaubermärchen in drei Aufzügen) by F. Raimund (q.v.), written in 1833, first performed at the Theater in der Josefstadt, Vienna, in February 1834, and published in *Sämtliche Werke* (1837, vol. 4).

The principal character, Julius von Flottwell, is a reckless spendthrift who squanders his immense wealth in extravagant hospitality and grandiose building projects. In the end he finds himself penniless, without possessions, family, or friends. But the play is a Zaubermärchen, and Flottwell is under the protection of the fairy Cheristane, who, unknown to him, has saved some of his wealth. In his need his faithful former servant Valentin willingly befriends him. And so Flottwell, justly saved because in his prosperity he had been generous to the poor and humble, is restored to a modest measure of wealth, and lives in harmony with Valentin and his wife Rosel.

Der Verschwender, though it contains the usual magic element and passages of conventional Viennese comedy, is Raimund's most serious play, written partly in verse. It contains several musical numbers, of which the best known is the 'plane song' (*Hobellied*), beginning 'Da streiten sich die Leut herum' and sung by Valentin, who after leaving Flottwell's service has become a joiner. Raimund himself played the part of Valentin as a guest artist in various productions.

Verschwörung des Fiesco zu Genua, Die, a historical tragedy written by Schiller in 1781–2, published in 1783, and first performed at Mannheim on 11 January 1784. It bears the subtitle *Ein republikanisches Trauerspiel* and a Latin motto from Sallust's *Ad Catilinam*: 'Nam id facinus inprimis ego memorabile existimo sceleris atque periculi novitate' (For I consider this outrage to be particularly noteworthy because of the novel nature of the crime and the danger).

The subject of the play is a revolt by Count Fiesco in Genoa in 1547, for which Schiller's principal source was *La Conjuration du comte Jean Louis de Fiesque* (1665) by the Cardinal de Retz. In the tragedy, various republicans plot the overthrow of the Genoese despot Doria and are disappointed by the apparent indifference of Fiesco. But Fiesco is biding his time. He lulls Doria's tyrannical nephew Gianettino into a false sense of security, and he makes love to Doria's niece Julia so convincingly that his own wife is in despair. When he is ready he reveals his hand to the conspirators and at once assumes the direction of the revolt. In two important monologues (end of Act II and III. ii) Fiesco, after hesitation, resolves to turn the revolution to his own profit and to succeed Doria as tyrant. The democrat Verrina perceives Fiesco's aim and contemplates an assassination if persuasion fails. The revolt is successful, but in the confusion Fiesco kills his own wife. He is nevertheless proclaimed prince. Verrina, failing to dissuade him from his despotic course, pushes Fiesco, as the latter goes aboard a galley, into the water, where he drowns.

Schiller himself is ambivalent in his attitude to his hero, poised between admiration and repulsion. The play is written in vigorous and sometimes strident prose. The acting version used in Mannheim was published in 1789, and that prepared for Leipzig in 1843 (reissued, as *Das Theater-Fiesco*, 1953).

Verschwörungen, Die, a collection of monographs on great conspiracies which Friedrich Schiller edited for the publisher S. L. Crusius from 1786 to 1788. Though Schiller had intended to contribute, the volume, published in 1788, contained nothing by his hand.

VERSHOFEN, Wilhelm (Bonn, 1878–1960, Tiefenbach, Bavaria), was for a time a schoolmaster at Jena. He turned to the study of economics and became prominent as an organizer in the commercial world (toy industry 1918, porcelain 1919–23). From 1923 he was a professor of economics at the Nürnberg Commercial College (Handelshochschule). His professional publications are numerous.

In 1904 Vershofen published with his friends J. Kneip and J. Winckler (qq.v.) the poems *Wir Drei!*, and in 1912 he participated in the founding of the literary group Werkleute auf Haus Nyland (q.v.). Vershofen's fiction, giving prominence to social and economic matters, includes the novels *Die Reisen Kunzens von der Rosen, des Optimisten* (1910), *Der Fenriswolf* (1914) *Das Weltreich und sein Kanzler* (1917), *Svennenbrügge. Das Schicksal einer Landschaft* (1928), *Poggeburg* (1934), *Zwischen Herbst und Winter* (1938), and the volumes of Novellen *Die Inflation* (1923) and *Der Ölkönig* (1927); *Rhein und Hudson* (1929) and *Seltsame Geschichten* (1939) are collections of grotesque tales. In 1948 appeared the story *William der Landedelmann*, and in 1951 *Das silberne Nixchen*. Vershofen wrote one play (Mysterienspiel), *Der hohe Dienst* (1921). His last work was the epic poem *Der große Webstuhl* (1954). In 1949 he published a volume of essays, *Erlebnis und Verklärung*.

Versprechen, Das. Requiem auf den Kriminalroman, a short detective novel by F. Dürrenmatt (q.v.), published in 1958. It began as a commission to write the script for a film, *Es geschah am hellichten Tage*, shown in

English as *Assault in Broad Daylight*. The story, told by a retired police chief of Zurich, concerns Matthäi, a former commissary of cantonal police; he and a friend have just seen Matthäi as the unkempt and apathetic owner of a petrol station. Matthäi's competence and single-minded determination as a detective had earned him the nickname 'Matthäi am Letzten' and secondment to Jordan.

A murder took place, and Matthäi went to the scene. The victim, discovered by a pedlar, was a little girl, Gritli Moser, who had been so brutally assaulted that no one could bring himself to break the news to her parents. Matthäi undertook the mission; in her distress, Gritli's mother asked him to swear that he would find the murderer. Matthäi gave the promise, which is the title of the novel.

In order to fulfil his promise Matthäi forfeited his appointment in the police by not going to Jordan. Single-handed, he constructed a theory, connecting the murder of Gritli with other sexual murders, deducing that the criminal had a black American car and travelled frequently between Graubünden and Zurich, luring the girls with a particular kind of chocolate. Following up his theory, he took over a filling station, acquired a housekeeper with a little girl like Gritli, Annemarie, whom he used as a bait. Annemarie stays out late, but remains mysterious about the affair, which has no sequel. Yet Matthäi goes on year by year, obsessed with the detection of the murderer. In the end he reaches the state of mental and physical deterioration in which he was seen at the petrol station by the narrator, the retired police chief. He knows that Matthäi's theory was correct, but an accident has denied him the confirmation; for on her death-bed, a Frau Schrott has made a confession to the police chief, revealing that the wanted criminal was Albertchen, her second husband, killed years before in a car crash on the day he set out to commit his fourth murder.

Versuch einer critischen Dichtkunst vor die Deutschen, a treatise on literature by Johann Christoph Gottsched (q.v.), published in 1730. It is largely a practical manual of poetry, based on French practice. The rules of poetry are established in accordance with reason, and poetry is good in so far as it accords with the rules. A moral aim, in Gottsched's opinion, is essential. He treats successively epic, tragedy, and comedy, extolling the French conception of these forms, and furnishing recipes for their confection.

Versuch einer deutschen Prosodie, an essay published in 1786 by K. P. Moritz (q.v.).

Versucher, Der, see BERGROMAN.

Versuch über den Roman, a treatise on the novel by C. F. von Blanckenburg (q.v.).

Versuchung des Pescara, Die, a novelle by C. F. Meyer (q.v.), published in the *Deutsche Rundschau* in 1887 and comprising five chapters and an epilogue. Set in Renaissance Italy, it centres on the abortive attempt of the Italian princes and Pope Clement to form a league against the Emperor Charles V (see KARL V). The Marquis of Pescara is to be the instrument of the betrayal; he is the Emperor's most successful field-marshal and the victor of Pavia. If he succeeds in defeating the imperial troops, he will be rewarded with the crown of Italy. Meyer used as his historical sources mainly the *Deutsche Geschichte im Zeitalter der Reformation* by L. von Ranke, the *Kultur der Renaissance in Italien* by J. Burckhardt, and the *Geschichte der Stadt Rom* (vol. 8) by F. Gregorovius (qq.v.), but he made free use of historical facts.

The first two chapters introduce the preparations for the plot against Charles. Morone, the chancellor of Duke Sforza, undertakes the temptation of Pescara, and the Pope (Ch. II) blesses Pescara's wife, the beautiful Vittoria, as the future queen, confident that the conspiracy will succeed. The central chapter presents the 'temptation scene'; Morone pleads with Pescara, while the conspirators Charles of Bourbon and Del Guasta listen from behind a curtain. All are left in doubt about Pescara's inscrutable attitude and none are aware that Pescara is no longer the man of action they knew.

A wound in his side, received in battle, has only superficially healed, and Pescara, in the presence of eternity, devises his own reckoning. This motif, forming the kernel of the story, adds poignancy and irony to Pescara's dismissal of the scene as a tragedy bearing the title 'Death and Fool'. During his last evening with Vittoria he confides no more than his conviction that loyalty is a virtue but that justice is the highest virtue of all, which echoes the argument with which the Pope had comforted Vittoria's uneasy conscience: the conflict between two supreme duties can only be resolved by the acknowledgement of the ethical order of man. Pescara, however, also reminds his wife that love does not seek reward before committing her to the convent of Heiligenwunden, where she is to stay until he has conquered Milan.

Before setting out, Pescara burns all papers indicating his identity, including a letter received from the Emperor in which Charles assures him of his trust in his field-marshal. Just before he leaves he is seen by the Emperor's agent Moncada. Since Moncada has in the past murdered Pescara's father, he is his personal enemy; thus confronted and asked for the last time to reveal his true intentions, Pescara responds with a gesture, which in Meyer's words signifies 'dust and ashes'. And, unperturbed by friend and foe, he sets forth in a litter, for he can no longer mount a horse, defeats the Italians, and enters Milan. He takes Duke Sforza prisoner and finds him to be a cringing weakling. With Charles of Bourbon and Leyva, the man of vengeance, he forms a triumvirate, successfully insisting that the Duke be pardoned and that Morone, though imprisoned, should not be tortured. Meanwhile the Pope makes an offer of peace to the Emperor. Although Pescara knows that Moncada and Leyva are plotting his arrest, he remains unmoved and dictates a farewell letter to the Emperor, assuring him of his loyalty and requesting protection for Sforza and pardon for Moncada. Ending the letter, he sinks back into Bourbon's arms. Vittoria arrives to find him lifeless. To the last, Pescara's face remains inscrutable, but he has demonstrated a principle of justice capable of forgiveness and moderation.

Versunkene Glocke, Die. Ein deutsches Märchendrama, a five-act verse play by G. Hauptmann (q.v.), first performed in the Deutsches Theater, Berlin, on 2 December 1896, and published in 1896. Except for a single scene in *Hanneles Himmelfahrt* (q.v.), it was Hauptmann's first breach with Naturalism and presents a conflict between the primal forces of Nature and the humdrum world of commonplace men. The former are embodied in the faun-like Waldschrat, the water goblin Nickelmann, the old crone Wittichen, and the touchingly beautiful elf Rautendelein; everyday life is represented by the schoolmaster, the barber, and the pastor. The artist Heinrich finds himself between these two worlds and perishes because he can reconcile himself with neither.

The play opens on the Bergwiese, a mountain meadow, with a dialogue between the creatures of Nature, who, except for Rautendelein, vanish when the bell-founder, Heinrich, appears injured and in need of help. In an accident, maliciously caused by the Waldschrat, Heinrich's new bell has slipped from the wagon transporting it, and plunged into the lake. Rautendelein, aided by Wittichen's potions, saves Heinrich from death. The schoolmaster, the barber, and the pastor arrive; they are ill received by the forest creatures, but they carry the still helpless Heinrich down into the valley. Reunited with his wife in his cottage (Act II), Heinrich rests, a sick man, wishing for death. Rautendelein enters and by her magic cures him; Heinrich succumbs to her spell and, deserting his wife, establishes himself near Rautendelein (Act III) in a derelict glass-blower's shack, which he turns into a bell foundry. He is at work on a new chime of bells, which is to hang in a temple devoted to the joy of life, and not to penitence and self-denial. The pastor arrives, hoping to persuade Heinrich to return to his wife, but, horrified by the bell-founder's blasphemous speech, he retires with the prophecy that the 'sunken bell will toll again'.

Heinrich works on his masterpiece of bell-founding (Act V), and is passionately in love with Rautendelein, when a vision of his children and the sound of the bell tolling from the bottom of the lake drive him from her arms; he curses his new love and work, and makes for the valley below, knowing that his wife has died. In the last act Rautendelein, abandoned and in sorrow, yields to Nickelmann and disappears into his well. As she does so, Heinrich returns, and, after a brief and sad reconciliation, he dies.

Stresses in Hauptmann's personal life are known to have had a part in the creation of this play.

Vertreibung der Spanier aus Wesel, 1629, a play (Schauspiel in drei Handlungen) by L. J. von Arnim (q.v.), published in his *Schaubühne* (q.v., 1813). It dramatizes a historical incident. An inhabitant of Wesel, Peter Mülder, plans a surprise attack with the support of Dutch troops, and the Spanish garrison under Count Lozan is defeated. This successful action simultaneously frees Mülder's sweetheart Susanna from Lozan's unwelcome attentions. The short play ends with the proclamation of Mülder as burgomaster.

Veruntreute Himmel, Der. Die Geschichte einer Magd, a novel by F. Werfel (q.v.), published in 1939. The pious Teta Linek spends her life as a servant in the home of the Argan family, enjoying good treatment. When she has been with them twenty-five years and is about 40, she is approached by her sister-in-law, who asks her to support Mojmir Linek, Teta's nephew, through school and university. She agrees, and it is her ambition for him to become a priest, an act which, she believes, will secure her entry into Heaven. As the years pass, Mojmir asks for more and more money, and provides her with only vague information about himself. In the end she finds that he is not and never has been a priest. She gives him the rest of her savings and sets out on a pilgrimage to Rome realizing at last that divine grace cannot be bought. Though suffering continuous severe pain, she reaches her destination, helped by a young priest, Johannes Seydel, who learns her story. She is received in audience by the Pope, but collapses in his pres-

ence. On her death-bed she receives the Holy Father's blessing.

The story has a frame (see RAHMEN). The first-person narrator, a friend of the Argans, learns of their sufferings after the National Socialist seizure of Austria and is instrumental in securing their release.

Verurteilung des Lukullus, Die, an opera by P. Dessau (q.v.) based on B. Brecht's radio play *Das Verhör des Lukullus* (q.v.), to which Brecht made minor alterations and added the final judgement. The opera was first produced in March 1951 in Berlin (Staatsoper).

Verwandlung, Die, a story by F. Kafka (q.v.) written in 1912, and first published in a magazine in 1915 and separately in 1917. It relates the final stage in the life of Gregor Samsa (whose name alludes to Kafka himself), after his metamorphosis into the shape of a beetle. The first of the three sections into which the story is divided describes the morning on which Gregor awakes from disturbed dreams to find that he has assumed the shape of a beetle. All that is left of his human characteristics is his consciousness.

This is brought into full play in his own adjustment to his tragic situation and to everyday reality, represented by the businessman (Prokurist) and his own family. Both suspect his integrity as he pleads through the closed door and, when confronted with the beetle, the Prokurist runs away, while the father drives Gregor back into his room behind the locked door, ignorant that, in pushing him, he has paralysed one of Gregor's many legs. In the second section his sister attends to his physical needs, soon finding in rotten vegetables and mouldy cheese what appears to be the right food. Only when she extends her care for him to the resolve to clear his room of his furniture does Gregor's suffering reach a new climax; he becomes more fully conscious of the deprivation of his individuality. This occasion gives Gregor the first opportunity to see his parents since he has, through his transformation, ceased to be their breadwinner and forced them to earn their own living. The change has affected the father most; he has taken a job as a uniformed porter at a bank. When Gregor leaves his captivity for the second time during the turmoil of the sister's rearrangements, it is again the father who drives Gregor back into his room by bombarding him with apples. One of them gets stuck in his body. It is the principal cause of his ensuing physical deterioration.

The third section shows the deterioration of the family's morale. Gregor's sufferings reach a climax when one evening his sister plays the long-neglected violin to entertain three tenants;

Gregor, responding to needs of which he had hitherto not been conscious, feels irresistibly drawn towards the music and unwittingly he comes into sight. This proves disastrous. The tenants quit and the parents are in a state of helpless paralysis. The sister solves the situation; she refuses to identify the beetle any longer with her brother and decides that it must be removed. Gregor succumbs and, amidst feelings of love towards his family, abandons himself peacefully to death.

The charwoman finds his remains the following morning and breaks the news to the family, while discreetly disposing of the beetle's body in the dustbin. For the first time since the metamorphosis, father, mother, and daughter enjoy the freedom of living and the promise of spring. The obviously psycho-analytical orientation of this tragic and grotesque tale allows a variety of interpretations.

Verwirrungen des Zöglings Törleß, Die, a short novel by R. Musil (q.v.), published in 1906. Young Törleß is sent, at his own request, to an exclusive boarding-school in a remote province of eastern Austria-Hungary. In compensation for his own diffidence, he chooses as his friends a coarse and brutal group of young noblemen. Despite the close supervision in the school he has his first sexual experiences with the middle-aged prostitute Božena. A situation arises in which one of the poorer pupils, Basini, is detected in theft by his comrades. Basini is a passive homosexual, and the group torment and humiliate him, treating him as a sexual object. Even Törleß, who maintains a certain remoteness, has homosexual contacts with him. Basini's virtual enslavement leads to his confession in order to escape from his tormentors, and he is in consequence expelled. Törleß, who has not persecuted Basini, but experiences contempt for him, leaves the school at his own request. Musil maintains that he has sustained a vital experience. Its external manifestation is a cool self-containment.

The film *Der junge Törleß,* 1966, for which H. W. Henze (q.v.) composed a Fantasy for string sextet, is a free adaptation by V. Schlöndorff.

VESPER, WILL (Barmen, 1882–1962, Gifhorn), was a literary journalist and littérateur, who, because of his nationalistic outlook found favour with the National Socialists and was one of the original fourteen members of the new Dichterakademie (see AKADEMIEN) in 1933. He wrote many poems dedicated to Der Führer and indulged in anti-Semitic vilification. He is best known as a popularizer of medieval poetry, with

modern versions of Wolfram's *Parzival* (1911), Gottfried's *Tristan* (1911), *Das Nibelungenlied* (1924), and *Kudrun* (1925), followed by *Die Historie von Reinecke dem Fuchs* (1928). His fiction consists mainly of historical stories (including *Martin Luthers Jugendjahre,* 1918, *Die Wanderung des Herrn Ulrich von Hutten,* 1922, and *Das harte Geschlecht,* 1931). A collection of Novellen appeared as *Geschichten von Liebe, Traum und Tod* (1937, reissued 1963) and his poetry as *Kranz des Lebens* (1934). *Gesammelte Werke* appeared 1963–7.

VIEBIG, CLARA (Trier, 1860–1952, Berlin-Zehlendorf), daughter of a highly placed civil servant, spent her childhood in the neighbourhood of Trier and part of her youth in the Polish-speaking province of West Prussia. Highly musical and endowed with a fine voice, she spent some time at the Conservatoire in Berlin. In 1896 she married a publisher (F. T. Cohn), and her first book, a volume of stories with a setting familiar to her from childhood (*Kinder der Eifel*), appeared in 1897. Throughout her life she used her maiden name in publication.

Clara Viebig's first novels were *Rheinlandstöchter* (1897) and *Dilettanten des Lebens* (1898), and her first considerable success was *Das Weiberdorf* (1900), a powerful novel about a village in the Eifel from which all the able-bodied men are absent working for the greater part of the year. *Das tägliche Brot* (q.v., 2 vols., 1902) is a compassionate novel of servant life in Berlin. These early works show her as an exemplar of the detailed realism cultivated by the Naturalistic movement (see NATURALISMUS). *Die Wacht am Rhein* (1902), though realistically written, turns from contemporary life to recent history, reflecting, in a Düsseldorf setting, the stresses in the 19th-c. Rhineland after its annexation by Prussia. One of her finest works is *Das schlafende Heer* (q.v., 1904), dealing with the Prussian attempt to assimilate West Prussia by the settlement of colonists and with the resentment of the indigenous Polish population. Numerous novels followed, including *Einer Mutter Sohn* (1907), *Absolvo te!* (1907), *Das Kreuz im Venn* (1908), *Die vor den Toren* (q.v., 1910), *Das Eisen im Feuer* (1913), *Eine Handvoll Erde* (q.v., 1915), *Töchter der Hekuba* (1917, a war novel), *Unter dem Freiheitsbaum* (1922), *Der einsame Mann* (q.v., 1924), *Die Passion* (1925), *Die mit den tausend Kindern* (1929), *Charlotte von Weiß* (1930), *Menschen unter Zwang* (1932), *Insel der Hoffnung* (1933), and *Der Vielgeliebte und die Vielgehaßte* (1935). She also published further collections of stories (*Vor Tau und Tag,* 1898; *Die Rosenkranzjungfer,* 1901; *Naturgewalten,* 1905; *Heimat,* q.v., 1914). Her plays *Barbara Hölzer* (1897) and *Das letzte Glück* (1909) were not successful. During the 1939–45 War she moved to Silesia, from which she was driven in the great expulsion carried out by the Polish government in 1945. Her old age was saddened before 1945 by persecution connected with her husband's Jewish descent, and after 1945, when she lived on the fringe of Berlin, by poverty.

Clara Viebig's novels and stories are noteworthy for the integrity of her vision and the imaginative realism of her nature descriptions. Although they show a keen insight into suffering female characters (e.g. in *Das tägliche Brot*), they also deal sympathetically with the problems and situations of men (in *Das schlafende Heer* or *Der einsame Mann*). Her *Berliner Novellen* appeared in 1952.

'Viel Feind, viel Ehr'!', an utterance attributed to Georg von Frundsberg (q.v.).

Viel Lärmen um Nichts, a Novelle by J. von Eichendorff (q.v.), published in 1832 in *Der Gesellschafter oder Blätter für Geist und Herz* and reissued in 1833 together with Brentano's Novelle *Die mehreren Wehmüller und die ungarischen Nationalgesichter* (q.v.). Its title is taken from the translation of Shakespeare's *Much Ado about Nothing* by A. W. Schlegel (q.v., modern German title *Viel Lärm um Nichts*); it owes its inspiration to Eichendorff's critical assessment of the literary scene in Germany during the decline of the Romantic movement (see ROMANTIK), and the social and literary climate in Berlin in the early 1830s. Its principal character, Prinz Romano, is modelled on Fürst Hermann von Pückler-Muskau (q.v.), the successful author of pseudo-Romantic poetry (*Gedichte*, 1811) and principal object of Brentano's satire. The characters Leontin and Faber from Eichendorff's novel *Ahnung und Gegenwart* (q.v.) reappear in this story; it contains a number of poems from the time of its writing, including *In der Fremde* (q.v.).

Vienna, see WIEN.

Viererzug, a short impressionistic poem by D. von Lilencron (q.v.). The title refers to driving four-in-hand.

Vierkönigsbündnis, a League of the Kings of Württemberg, Bavaria, Saxony, and Hanover formed under Austrian auspices to counter the formation of the Prussian Union (see DREIKÖNIGS-BÜNDNIS), which Prussia had prepared at a parliament at Erfurt (March 1850). It was dissolved upon the restoration of the German Confederation (Deutscher Bund) following the settlement of Olmütz (November 1850, see OLMÜTZER PUNKTATION) between Austria and Prussia.

Vierte Gebot, Das, a four-act play (Volksstück, q.v.) by L. Anzengruber (q.v.), first performed in the Josefstädter Theater, Vienna, in December 1877 and published in 1878. It is in a stylized Viennese dialect.

Three families are shown in parallel. The property-owner Anton ·Hutterer sacrifices his daughter's happiness in a loveless rich marriage. The criminal Schalanters applaud their son Martin's violence until, as a conscript, Martin shoots his sergeant and is condemned to death. Against these examples of wicked and perverse parents is set the Schön family with their devotion to the welfare of their son Eduard, who shrives Martin Schalanter in the condemned cell. The play is a practical sermon on the fourth commandment (the fifth of the Anglican catechism) 'Honour thy father and thy mother . . .', suggesting that the commandment imposes a duty upon the parents also.

Vierundzwanzigste Februar, Der, a fate tragedy (see SCHICKSALSTRAGÖDIE) in trochaic verse by Z. Werner (q.v.), written in 1809 and published with a *Prolog an deutsche Söhne und Töchter* (1814) in the periodical *Urania* (q.v.) in 1815. The writing of the play followed a challenge by Goethe to produce a one-act tragedy on the subject of the curse (Fluch). Goethe, who had promised to produce the play, postponed its performance until 1810. In dealing with the workings of conscience, the play reflects the earlier, unhappy, and unsteady phase of Werner's life. The date which he chose for the 'day of destiny', the 24th of February, is the day on which his mother and his closest friend died (cf. 'Und kam ein Unfall, der das Herz traf, war/Es stets am vierundzwanzigsten Februar!').

The action is set in a Swiss Alpine inn in Schwarrbach (its successor is the well-known Schwarrenbach Hotel) on the Gemmi Pass between Kandersteg and Leuk. It takes place about midnight on the 24th of February 1804 and involves three characters, the Swiss peasant Kunz Kuruth, his wife Trude, and their son Kurt. The play uses as a leitmotiv the theme of patricide in the Scottish ballad *Edward* (see VOLKSLIED) of which the opening lines are quoted in Trude's monologue in scene one and the final lines in the third and last scene.

Kunz suffers under his father's curse, inflicted upon him twenty-eight years before in the midnight hour of the 24th of February 1776. That night he had thrown a knife at his quarrelsome father. Although the father died moments later of a stroke, and not through the knife, which had failed to hit him, Kunz could never again free himself from a sense of guilt and from the curse of his dying father, predicting that he would become the 'murderer of the murderer'. Seven years later his 7-year-old son Kurt, while playing, killed his 2-year-old sister. In his grief Kunz placed a curse upon his son, who was sent away to be brought up by his uncle. On the twenty-fourth day in another February the son left his new home and was subsequently presumed to have become a victim of the French Revolution. In fact he has escaped with an officer of a Swiss regiment in whose service he was, and who soon made a fortune on a plantation in San Domingo. On another anniversary of the family's fateful day Kurt's master, who nursed him when he was struck down by the yellow fever, died himself of the disease. Kurt recovers and, having inherited his master's wealth, returns to Europe in the hope of being relieved from his father's curse. He arrives at his parents' home unrecognized. It is the day of the play's action, the past being revealed by analytical dialogue. Kunz and Trude have reached a crisis of despair. Their extreme poverty has driven them into debt, and Kunz has just been served with a writ committing him and his wife to the debtors' prison the following morning. He is determined to drown himself in the Daubensee rather than bear this disgrace. After Kunz has offered the stranger accommodation for the night, the relevance of the play's motto 'Führe uns nicht in Versuchung!' is finally demonstrated. Having seen the stranger's money, Kunz, in his desperation, feels driven to theft and murder. In dying, Kurt reveals his true identity, forgives, and lifts the curse from his parents.

The tragic irony pervading the plot is that the will to murder in a moment of extreme provocation has affected Kunz's sense of guilt as if he were a murderer in deed; and that he should (though not without a dramatic tour de force) in the end actually become a murderer. This is balanced by the regeneration of his spirit after he is freed from the curse and decides to submit to the executioner in just atonement for his deed.

The play is historically significant for having set the fashion of Schicksalstragödie, which became popular on the German stage. Werner reinforces the theme of the curse by the use of a number of motifs, among them the date, the scythe, the birthmark of the son denoting the sign of Cain, and the knife. Such devices were soon imitated. The tragedy stands out by its blending of classical and Romantic features with a strong element of social realism, by its mastery of atmosphere, and by its penetrating psychology producing claustrophobic tension.

Vierzig Tage des Musa Dagh, Die, a novel by F. Werfel (q.v.), published in 1933. Set in Syria in 1915, it recounts the heroic resistance of a remnant of the massacred Armenian people against more numerous and better equipped Young Turk forces. Led by Gabriel Bagradian, who has reluctantly accepted command, they entrench themselves on a mountain, the Musa Dagh. For forty days they hold out, and when they are about to be overwhelmed, the surviving Armenians are saved by the gunfire of an Anglo-French naval squadron, which takes them on board. Not, however, Bagradian, who chooses to remain, and falls to a Turkish bullet. The subject came to Werfel's notice during a visit to Syria in 1929.

VINTLER, Hans, a Tyrolean knight, documented between 1407 and 1419, was a member of the family which owned Runkelstein castle (near Bolzano), famous for its frescoes illustrating Tristan (q.v.) and other romances. Vintler completed in 1411 a moral poem of some 10,000 lines, entitled Die Pluemen der tugent (Die Blumen der Tugend) (printed 1486, ed. J. von Zingerle, 1874). The title is taken from its principal source, the prose Fiori di virtù (c. 1320) of Tommaso Leoni. The poem is only in part a eulogy of virtue; for it also fiercely castigates vice, and is particularly severe on the nobility, of which Vintler was a member.

Viola tricolor, a Novelle by Th. Storm (q.v.), published in Westermanns Monatshefte in 1874. A widowed professor, Rudolf, marries a young wife, Ines, whom his ten-year-old daughter Nesi refuses to accept as her mother. But the new marriage, too, bears the strain of the past. The portrait of the first wife, Marie, dominates the study and her parents' garden is locked in memory of her. The conflict between past and present is resolved when Ines gives birth to a daughter and becomes critically ill. She receives Nesi's spontaneous affection, but she also experiences in the face of death the urge to live on in the memory of her own child. Her recovery opens an entirely new and happy relationship.

The nadir of Ines's mental crisis is contained in her flight at night during the early stage of her pregnancy, when the dog's barking alarms Rudolf, who rescues his wife from her sleep-walk in the garden. This episode is outstanding for Storm's creative rendering of dream consciousness. The title, 'pansy' ('Stiefmütterchen'), points to the stepmother as the principal character; Storm portrays her with unusual sympathy.

VIRCHOW, Rudolf (Schivelbein, Pomerania 1821–1902, Berlin), an outstanding pathologist of humble origin, became a university lecturer in 1847 in Berlin, but as a Liberal had to flee abroad (i.e. to South Germany) in 1848 (see REVOLUTIONEN 1848-9). He obtained an appointment at Würzburg, and was recalled to Prussia in 1856 as director of the Pathologisches Institut and as professor of pathology at Berlin University. In his field of cellular pathology he made known important general principles as well as making significant discoveries of detail. His most widely known work was *Vorlesungen über Pathologie* (4 vols., 1858-67). His eminence was recognized abroad as well as at home and in 1892 he received the Copley medal of the Royal Society.

Virchow was also a noted anthropologist and visited the site of Troy with H. Schliemann (q.v.) in 1879, afterwards publishing *Zur Landeskunde des Troas* (1880) and *Alt-trojanische Gräber und Schädel* (1882). He remained all his life an active Liberal, founding the Progressive Party (Fortschrittspartei), and was a bitter opponent of Bismarck (q.v.). He was a member of the Reichstag from 1880 to 1893.

VIRDUNG, SEBASTIAN (Amberg, *c.* 1465-?), became a musician at Heidelberg and published in 1511 a dialogue on the instruments and music of his age entitled *Traktat Musica getutscht* (facsimile reprint, ed. L. Schrade, 1931).

Virey und die Aristokraten, Der, a novel by C. Sealsfield (q.v.) published in 1834 with the sub-title *Mexiko im Jahre 1812.* Its action is linked with the political tensions and conflicts of Mexico not long before the final revolt against Spain, and it demonstrates Sealsfield's talent for exotic description and evocation. 'Der Virey' is the Spanish viceroy.

Virginal, a Middle High German epic poem of unknown authorship, formerly ascribed to ALBRECHT von Kemnaten (q.v.). It exists in three versions, one of which has usually been called *Virginal,* while the other two have sometimes been known as *Dietrichs erste Ausfahrt* and *Dietrich und seine Gesellen.* All three derive from a lost original. Hildebrand urges the youthful Dietrich von Bern to rescue Queen Virginal of Tyrol from a heathen king. Dietrich slays the heathen and is then involved in many conflicts with giants and dragons before rescuing Virginal. Finally he is summoned back to Bern (Verona), which is threatened by his enemies.

The poem is written in the twelve-line stanza known as the Eckenstrophe (q.v.).

VISCHER, FRIEDRICH THEODOR (Ludwigsburg, 1807-87, Gmunden, Traunsee, Austria), was at school at Stuttgart and Blaubeuren and then studied for ordination at the Tübinger Stift. A curate at Horrheim in 1830, he taught at Maulbronn (1831) before moving to Tübingen. In 1837 he was appointed a supernumerary professor at Tübingen University and in 1844 was elected to an established chair; but the Liberal tone of his inaugural lecture resulted in his suspension for two years.

Vischer's first writings of note were aesthetic treatises based on Hegel's philosophy: *Über das Erhabene und Komische* (1837), *Kritische Gänge* (2 vols., 1844, in 6 vols., 1860-73), and *Ästhetik oder Wissenschaft des Schönen* (3 vols., 1846-57). Vischer was a moderate Liberal deputy at Frankfurt in 1848 (see FRANKFURTER NATIONAL-VERSAMMLUNG). He later became a professor at Zurich (1855) and then at Tübingen University (1866). Under the pseudonym Deutobold Symbolizetti Allegorowitsch Mystifizinsky he published in 1862 a parody of Goethe's *Faust* (*Faust. Der Tragödie dritter Theil,* q.v.) and, under another pseudonym (P. V. Schartenmeyer), a comic epic on the war of 1870 (*Der deutsche Krieg,* 1874). He is best remembered for his whimsical novel, *Auch Einer* (q.v., 2 vols., 1879), in which the hero A. E. wages unsuccessful war against a treacherous physical reality ('Tücke des Objekts'). A serious contribution to Faust literature (*Goethes Faust. Neue Beiträge zur Kritik des Gedichts*) appeared in 1875 (reprinted 1969). Two late works are a collection of poems, *Lyrische Gänge* (1882), the title of which is a counterpart to the earlier *Kritische Gänge,* and the Swabian comedy *Nicht Ia* (1884), which makes slight use of dialect.

Vischer was granted a patent of nobility as F. Th. von Vischer in 1870. His considerable impact upon German literature and thought in his lifetime depended in part on his dynamic and combative personality. A *Gesamtausgabe* (6 vols.), ed. R. Vischer, appeared in 1921-2; correspondence with Mörike in 1926, and with D. F. Strauß (qq.v.) in 1952-3 (2 vols.).

Visio Philiberti, a medieval Latin poem, the subject of which is the conflict between body and soul. A German verse translation of this is found in one of the MSS. of *Gottes Zukunft* (see HEIN-RICH VON NEUSTADT).

Visio Sti. Pauli, fragments of an Early Middle High German rhyming poem dealing with St. Paul's version of hell and based on a Latin original. It was written in Bavaria *c.* 1150.

Visuelle Poesie, see KONKRETE POESIE.

Vita Beatae Mariae Virginis et salvatoris rhythmica, medieval Latin versified life of the Virgin Mary, written early in the 13th c. It is the source of most of the German *Marienleben* of the next hundred years, including the *Grazer Marienleben* and the 'Lives' by WALTHER von Rheinau, Bruder Philipp der Karthäuser, and Bruder Wernher (qq.v.).

Vittoria Accorombona, a historical novel published by L. Tieck (q.v.) in 1840. Vittoria, its heroine, who is also the central figure of Webster's *The White Devil*, is treated by Tieck in a sympathetic light. A woman of outstanding intellectual quality, she allows herself for family reasons to be married to Perretti, a nonentity. She falls in love with Duke Bracciano, who murders Perretti and later marries Vittoria. Family enmities arise and Bracciano and Vittoria are successively murdered.

Vitzliputzli, a group of three poems by H. Heine (q.v.) included in the section *Historien* of the *Romanzero* (q.v.). The captured Spaniards are sacrificed to the Aztec god, Vitzliputzli, who foresees his own end, but foresees also a continuation of cruelty in the name of religion.

Vize-Mama, a story by E. von Wildenbruch (q.v.), published in 1902. Two boys, both officer's sons and cadets in Potsdam, form a close friendship. Georg von Drebkau, motherless and neglected by his father, a general, has wealth but no affection; Hans von Carstein is poor, but enjoys the love of his widowed mother. Carstein persuades his mother to invite Drebkau to stay, and she devotes herself to him, although years before she herself was jilted by his father. The boy becomes passionately devoted to her, but when his Sunday visits become known they are forbidden.

　　Georg von Drebkau encloses his emotions within himself until, at a confirmation service in Berlin, his affection for Frau von Carstein bursts out in the presence of the father, who is already an ailing man and is visibly shaken by the experience. The boy's state deteriorates, he begins to take morphine (to which his father had formerly been addicted) and falls seriously ill. He is sent on a tour with his aunt, but it becomes clear that his whole thoughts are with Frau von Carstein, who consents to join him and so to soothe his last weeks. After his death Frau von Carstein and the general, now so ill that he has had to resign his commission at the moment of mobilization against France in 1870, meet, and an unexpected reconciliation takes place. This perceptive psychological study has as its background the Cadet School in Potsdam in which Wildenbruch himself had been an inmate.

Vocabularius Sti. Galli, a Latin–German vocabulary, the MS. of which is in the library of

the monastery of St. Gall (see ST. GALLEN); the customary designation refers only to its location, not to authorship or place of origin. The *Vocabularius* is a Latin–German dictionary, the basis of which is the Latin element of the Latin–Greek vocabulary and dialogue 'Hermeneumata'. The MS. is not complete. Other fragments, originating in Fulda, are in Kassel (*Casseler Glossen* and *Casseler Gesprächsbüchlein*. The approximate date of the MSS. is 775.

VOGEL, JAKOB (Kornwestheim, Württemberg, 1584–?), a barber surgeon, wrote an epic (*Ungrische Schlacht*, 1626) in Knittelverse (q.v.), recounting in terms of his own day the battle of Merseburg (933). It contains the song 'Kein seelgr Tod ist in der Welt', which has become famous, not least through misuse, in the adaptation 'Kein schönrer Tod ist auf der Welt' (q.v.). Vogel also wrote *Wandersregeln* imitated from Hans Sachs (q.v.) and a prose satire, *Der diogenische Lasterbeller*.

VOGL, JOHANN NEPOMUK (Vienna, 1802–66, Vienna), an Austrian poet, was a civil servant, who wrote copiously in a belated Romantic manner. In 1837 he edited the complete edition of the works (*Sämtliche Werke*) of F. Raimund (q.v.). He is the author of the poem *Heinrich der Vogler* ('Herr Heinrich saß am Vogelherd'), well known in a setting by C. Loewe (q.v.). His *Balladen und Romanzen* (3 vols.) appeared 1835–41, and his once similarly popular *Bilder aus dem Soldatenleben* in 1852. A selection of his works, ed. R. Kleinecke, was published in 1911.

VOIGTLÄNDER, GABRIEL (d. 1643 Denmark), is recorded in 1626 as city trumpeter in Lübeck and in 1639 as court trumpeter to the Crown Prince of Denmark. He wrote songs to which he composed music, and others which he adapted to familiar tunes. His homely and sometimes humorous poems were collected in *Allerhand Oden und Lieder*, 1642.

VOIGTS, JENNY VON (Osnabrück, 1752–1814, Melle nr. Osnabrück), in full Johanna Wilhelmine Juliane von Voigts, was the daughter of the Osnabrück lawyer and writer, Justus Möser (q.v.). Her husband was also a jurist (Justizrat). She published her father's newspaper articles as *Patriotische Phantasien* (1774–8, q.v.).

VOIGT-DIEDERICHS, HELENE (Marienhoff nr. Eckernförde, 1875–1961, Jena), the fifth of ten children, grew up on the family estate, was twelve years old when her father died, and turned early to writing. She is best known as the author of regional fiction (see HEIMATKUNST) set in her native Schleswig-Holstein, presenting in

Naturalistic manner with the use of dialect and colloquialisms the life and hardships of simple country folk. Her stories include the collections *Schleswig-Holsteiner Landleute* (1898), *Abendrot* (1899), *Aus Kinderland* (1907, reissued 1955), and *Schleswig-Holsteiner Blut* (1928). Of her novels *Dreiviertel Stund vor Tag* (1905) deserves special mention; *Waage des Lebens* (1952) was her last.

Her marriage to the publisher Eugen Diederichs in 1898 was dissolved in 1911, and she travelled widely in Europe before settling in Jena in 1931. *Wandertage in England* (1912) and *Zwischen Himmel und Steinen* (1919) are works on her travels. Reminiscences of her parents' estate appeared as *Auf Marienhoff* (1925).

Völkerschlacht, the Battle of the Nations, fought 16–19 October 1813 in the neighbourhood of Leipzig between the French and supporting forces under Napoleon and the combined Austrian, Prussian, and Russian armies. Involving almost 500,000 men, it is said to be the largest battle known in history up to that time. It ended in the defeat of Napoleon (see NAPOLEONIC WARS), and is commemorated by a massive monument, the Völkerschlachtsdenkmal, built 1898–1913.

Völkerwanderung, the migration of peoples in Europe, which intensified from the 4th c. to the 6th c. A.D. Attributed by historians to climatic and economic factors, including the attraction of the prosperous Roman Empire, this migration begins for Germany with the crossing of the Danube by the West Goths in 376. For some two centuries unpredictable migrations of East Goths, West Goths, Vandals, and others took place in central Europe. The settling of the Langobards in North Italy in 568 is usually regarded as the close of the period of migrations.

Völkische Beobachter, Der, was originally a local Munich newspaper, *Der Münchner Beobachter*, founded in 1887. In December 1920 it was bought by the NSDAP (q.v.) and under the new title was the official newspaper of the party until its cessation in 1945.

Volksbuch, term applied to popular literature in prose narrative form, either continuous or in separate episodes. The heyday of Volksbücher is the 16th c., when they multiplied rapidly as printing-presses proliferated; they continued to be printed well into the 18th c. Their subject-matter came from a variety of sources, among which French and German verse romances of the 13th c. and 14th c. are conspicuous.

Among Volksbücher derived from the former are *Fierabras, Kaiser Octavianus* (qq.v.), *Olivier und Artus,* and *Von den vier Haimonskindern* (see

HAIMONSKINDER, DIE). Among German epics adapted are *Herzog Ernst, Gregorius, Tristrant und Isalde* (Eilhart von Oberge's form), *Willehalm,* and *Wigalois* (qq.v.). Shorter French and Italian works form the basis of *Die schöne Magelone, Die schöne Melusine,* and *Griselda* (qq.v.) and of the *Volksbuch von der Pfalzgräfin Genoveva* (1647). The source of *Fortunatus* (q.v.) is unknown, and the Volksbuch of *Dr. Faust* (see FAUSTBUCH, SPIESS'SCHES) brings new material. *Von dem gehörnten Siegfried,* based on *Der hürnen Seyfrid* (q.v.), was first printed as late as 1726. Animal stories such as *Reineke Fuchs* (see REINKE DE VOS and NEIDHART FUCHS) were popular; and so, too, were collections of Schwänke (see SCHWANK) such as *Till Eulenspiegel* (q.v.).

The stories of Elisabeth von Nassau-Saarbrücken and Eleonore von Österreich (qq.v.), both writing in the 15th c., are on a somewhat higher level and are usually designated Prosaromane.

The term Volksbuch was coined by J. J. Görres (q.v.), who published, in 1807 *Die teutschen Volksbücher.* The Romantics, especially Tieck (e.g. in *Kaiser Octavianus,* q.v.), attempted to give a new life to Volksbücher, and they were, for a time, regarded as a valuable expression of 'deutscher Volksgeist'. Gustav Schwab (1836) and Karl Simrock (qq.v., 1839) retold the stories for the public of the post-Romantic era. In reality Volksbücher, which are not all anonymous (e.g. Warbeck's *Die schöne Magelone* or Stainhöwel's *Grisardis*), represent a degenerate form of literature, expressed in the German phrase 'gesunkenes Kulturgut'. They were an attempt to satisfy the hunger of a new public, eager to read, but without knowledge or taste. An example is the collection *Das Buch der Liebe* by S. Feyerabend (q.v., 1578, ed. P. Ernst, q.v., 1911). An extensive collection of Volksbücher was published by R. Benz (q.v., *Die deutschen Volksbücher,* 6 vols.) in 1911–24 (reissued 1956).

Volksgerichtshof, see RESISTANCE MOVEMENTS (2) and FREISLER, R.

Volkslied, denotes a poem of unknown origin, handed down by oral tradition, in simple stanza form (usually four lines, rhyming either abab or abcb), and associated with an easily remembered and sung melody. Interest in poetry of this kind began with the reception in Germany of Bishop Percy's *Reliques of Ancient English Poetry* (1765). Herder (q.v.), an admirer of Percy's collection, himself collected German folk-songs, and invented the word Volkslied, which he first used in the essay *Auszug aus einem Briefwechsel über Ossian und die Lieder alter Völker,* published in 1773 in *Von deutscher Art und Kunst* (q.v.). Herder published in 1778–9 his collection of folk-songs of many lands under the title *Volkslieder* (q.v.). L. J. von Arnim and C. Brentano (qq.v.) published in 1805–8 an exclusively German collection, *Des Knaben Wunderhorn* (q.v.), and L. Uhland (q.v.) applied scholarship to the Volkslied in his anthology *Alte hoch- und niederdeutsche Volkslieder* (1844–5). In *Schlesische Lieder* (1842) A. H. Hoffmann von Fallersleben (q.v.) published the first regional collection.

In the early 19th c. it was commonly held that all Volkslieder were a spontaneous expression of the 'Volksseele' (emphasized by the brothers J. and W. Grimm, qq.v., in their theories on Naturpoesie and Kunstpoesie). It is, however, certain that all Volkslieder began as poems written by a specific person, that they have become, as it were, ownerless, and, being passed from one generation to another, have in many cases changed their wording, and sometimes their content. The process of mutation which they have undergone is often pejoratively termed 'Zersingen'.

Since the rise of interest in Volkslieder in the late 18th c. many poets have written poems of folk-song character, and to these the term 'volkstümliche Lieder' is often applied.

The research centre for the German folk-song is the Deutsches Volkslied-Archiv in Freiburg, Breisgau, founded in 1914. Its register catalogues some 320,000 versions of folk-songs from different ages, countries, and, within Germany, regions. The varieties include the Volksballade, Moritatenlieder, Liebeslieder, Tanzlieder, Handwerkslieder, Jägerlieder, Soldatenlieder, Seemannslieder, and Hirtenlieder. A collection *Die historischen Volkslieder der Deutschen vom 13. bis 16. Jahrhundert* (4 vols., 1865–9, reissued 1966) was compiled by the literary historian, musicologist, and writer Rochus von Liliencron (1820–1912), who was also the author of *Deutsches Leben im Volkslied um 1530* (1885, reissued 1966).

Volkslieder, a collection of folk-songs published by J. G. Herder (q.v.) in 2 volumes in 1778 and 1779. Herder's interest was aroused by Percy's *Reliques of Ancient English Poetry* (1765), and he himself collected folk-songs and encouraged others, including Goethe, to do so for him. His collection, divided into seven books, is international in scope and consists largely of translations into German.

The first book, *Lieder aus dem hohen Nord*, contains songs from Greenland and Lapland, from the southern Baltic (Esthonia, Latvia, Lithuania), and from Illyria. The second book (*Griechen und Römer*) includes odes of Horace as well as anonymous poems, and also a short section of medieval Latin poems. Italian, Spanish, and French ballads and songs are collected in Bk. 3 (*Romanische Lieder und*

Romanzen). Bk. 4 (*Nordische Lieder*) brings Ossian, and Icelandic and Danish songs. Bk. 5, the largest section, is English (*Englische Lieder*), which embraces Scots poems and includes *The Ballad of Chevy Chase* (*Die Chevy-Jagd*), *Fair Rosamund* (*Die schöne Rosemunde*), and *Edward*, as well as a number of songs from Shakespeare. The German section is in Bk. 6 (*Deutsche Lieder*), and the seventh book is devoted to 'tropical' poems, *Lieder aus dem heißen Erdstrich*. These originate from Madagascar, Brazil, and Peru.

After Herder's death a second edition was published in 1807 with the title *Stimmen der Völker in Liedern*.

Volksstück, a play written in local dialect and intended for popular audiences, often monopolizing a particular theatre (Volkstheater). Although Hamburger Volksstücke were written by F. Stavenhagen and Münchner Volksstücke by L. Thoma (q.v.), these are late 19th-c. offshoots of Naturalism (see NATURALISMUS) and their dialect speech is an aspect of Heimatkunst (q.v.). The Munich Volkstheater influenced B. Brecht (q.v.), and the 1960s produced the 'critical Volksstück'.

The real home of the Volksstück is Vienna (Wiener Volksstück), where it was indigenous from the early 18th c. The Viennese Volksstück makes frequent use of extravagant scenic display and stage machinery. The initiators of the tradition in the 18th c. were the actor and director of the Kärntnertor Theater J. A. Stranitzky and his protégé G. Prehauser (qq.v.), both of whom had the gift of comic extemporization in Viennese dialect and played the part of Hanswurst (q.v.). The site of this theatre is near where the Opera now stands. Though extemporization was officially banned in 1768, the popular tradition maintained itself with P. Hafner and J. von Kurz (qq.v.), and the Volksstück took on a new lease of life when K. von Marinelli (q.v.) opened the Leopoldstädter Theater in 1781 expressly for it and engaged J. Perinet (q.v.) as one of his principal playwrights. In 1788 the Josefstädter Theater and in 1801 the Theater an der Wien were also opened and used for Volksstücke.

The Volksstück took two main forms: the 'magic play' (Zauberstück), descended from baroque drama, offering lavish transformation scenes in a fairy world, through which a simple-minded or caustic-witted Viennese tradesman made his homely way, and the Lokalstück without magic, making comic use of realistic elements and poking fun at contemporary fashions or fads. An offshoot of the former was the moral Besserungsstück, in which a character was cured of some fault by the imagined experiences of a dream; not all Besserungsstücke were, however, intended as Volksstücke, Grillparzer's *Der Traum ein Leben* (q.v.) being a notable example. Most Viennese Volksstücke have inserted songs, many of them by the gifted Wenzel Müller (q.v.). These songs often possess great charm and humour. Among the writers at the end of the century were K. F. Hensler, J. F. Kringsteiner, and E. Schikaneder (qq.v.), who, as the librettist of Mozart's *Die Zauberflöte* (q.v.), collaborated in the only Zauberstück to achieve an international reputation.

At the beginning of the 19th c. J. A. Gleich, K. Meisl, and A. Bäuerle (qq.v.) were prolific in their output. F. Raimund (q.v.) raised the Volksstück to a higher level, though he was not able to realize his own aspirations. All his works were Zauberstücke, and these began to lose their attraction after 1830. The Volksstück reached its zenith with the plays of J. N. Nestroy (q.v.). He concentrated mainly on the Lokalposse without magic and wrote parodies. While he retained many traditional elements, Nestroy broke new ground by introducing trenchant satire into his Volksstücke. His major contribution to the form, however, is to be found in his verbal dexterity: dense imagistic wordplay and lively dialogue which exploits the tension between Viennese dialect and High German. F. Kaiser (q.v.) survived Nestroy and saw the disappearance of the Volksstück in its old form.

By 1870 the public had changed and the Volksstück had lost its impetus. L. Anzengruber (q.v.) turned for his Volksstücke to rural settings, and although he is the author of some Lokalstücke (at times 'mit Gesang'), the simple *bonhomie* was yielding to a more sophisticated and serious tone. The last representative of the Lokalstück was C. Karlweis (q.v.). Later local writing tended to merge into the general pattern of commercial comedy and farce, common to other European capitals, though the use of Viennese dialect did not die out. See also DIALEKTDICHTUNG. *Alt-Wiener Volkstheater* (7 vols.), ed. O. Rommel (no dates) contains plays by Hensler, Schikaneder, Kringsteiner, Meisl, Gleich, Bäuerle, and Kaiser. The collection *Barocktradition im österreichisch-bayrischen Volkstheater* (6 vols.), ed. also O. Rommel, was published 1935-9.

Both Carl Zuckmayer (q.v., *Der fröhliche Weinberg*, 1925) and Brecht (*Herr Puntila und sein Knecht Matti*, q.v., 1940) attempted to revive the form, alongside Ödön von Horváth and Marieluise Fleißer (qq.v.). It was not until the restless 1960s and early 1970s that Horváth's and Fleißer's works were revived on stage and championed by playwrights as representing a 'realistic' Volksstück, on which the contemporary 'critical' Volksstück should be modelled. Its main aim was to brandish social ills and any fascist tendencies, and its representatives included Franz Xaver Kroetz and the less wellknown Martin Sperr (qq.v.). It differed from traditional Dialektdichtung in that dialect became a more radical means of social identification. In general terms, Horváth's Volksstück is considered to mark a new phase in the development of the genre in both Austria and Germany.

Volkssturm, a *levée en masse* of all males from 16 to 60 decreed by Hitler (q.v.) on 18 April 1944.

Vollendung des Königs Henri Quatre, Die, see HENRI QUATRE-ROMANE.

VOLLMOELLER, KARL GUSTAV (Stuttgart, 1878–1948, Los Angeles), a wealthy businessman and amateur of literature, was in touch with S. George and his group of disciples (see GEORGE-KREIS). He translated poetry by G. d'Annunzio, and published a volume of his own verse (*Die frühen Gärten*, 1903). He is the author of the plays *Catherina, Gräfin von Armagnac und ihre beiden Liebhaber* (1903), *Assüs, Fitne und Sumurud* (1904), *Der deutsche Graf* (1906), *Wieland* (1910), *Venezianisches Abenteuer* (1912), *Das Mirakel* (1912), a Mysterienspiel produced by M. Reinhardt (q.v.) in Berlin in 1914, *Onkelchen hat geträumt* (1919), *Cocktail* (1930), and *La Paiva* (1931). A Novelle, *Die Geliebte*, appeared in 1920. Vollmoeller was also an amateur archaeologist and an aviator. A selection of his poetry, ed. H. Steiner, appeared in 1960.

VOLTAIRE, FRANÇOIS MARIE AROUET (Paris, 1694–1778, Paris), French poet, historian, and philosopher, whose humanitarian views on politics and religion influenced European thought. He ignored contemporary German thought as represented by G. E. Lessing (q.v.), who later attacked his views on drama in the *Hamburgische Dramaturgie* (q.v.), especially as represented in *Sémiramis* and its preface, and contributed to the decline of Voltaire's remarkable success on the German stage. During Voltaire's residence at the court of Friedrich II (q.v.) of Prussia, Lessing was commissioned to translate some of Voltaire's work. Through his contact with Voltaire's secretary he obtained in 1753 a copy of the (as yet unpublished) *Siècle de Louis XIV*, which he did not return upon his departure to Wittenberg. Voltaire, suspicious of his motives, made a complaint to the King who, remembering this unfortunate episode, refused Lessing the post of royal librarian some twelve years later.

Forty-eight years of correspondence with the

Prussian king testify to the constancy of the intellectual intercourse between both men. It began in August 1736 when the 24-year-old Crown Prince wrote his first letter admiring Voltaire's genius and hoping for his friendship. They met for the first time in 1740, and Friedrich cherished the idea that Voltaire would come to stay at his court. After the death of Mme du Châtelet in 1749, Voltaire decided on the move to Potsdam, resigning in 1750 his post as historiographer and Chamberlain, Gentilhomme Ordinaire, thus burning his boats at the court of Louis XV. Personal contact, however, cooled the friendship, and after Voltaire had got himself involved in a court action over illegal money transactions as well as attacking Friedrich's President of the Royal Academy (see AKADEMIEN), P.–L. M. de Maupertuis (q.v.), the King decided in 1753 that Voltaire must leave Prussia. He asked for the return of the order Pour le mérite (q.v.), with which he had honoured him, and of a book of his own verses. At Frankfurt, Voltaire was kept prisoner for several weeks before he could proceed to Geneva. This humiliating episode was the responsibility of Friedrich's over-zealous representative, but Voltaire never forgot the disgrace. Yet the correspondence of the following years resumed the exchange of ideas and verses, Friedrich realizing that Voltaire was 'only good to read', but read him he must, while Voltaire confessed his love for the Prussian king 'from a distance'. Their correspondence provides valuable insight into the minds of both men, whose extraordinary mutual fascination remains unique. Voltaire's defence of the reputations of Jean Calas, Sirven, and La Barre, who had been executed in the early 1760s, aroused the King's keen interest and sympathy.

In his last letter to Friedrich, Voltaire expressed the hope that the King might long survive him as the bulwark of Germanic liberty. Friedrich paid tribute to Voltaire's memory in a dignified and warm *Éloge* which was read to the Berlin Academy by Thiébault in November 1778. *Briefwechsel Friedrich des Großen mit Voltaire* was edited, in three volumes, by R. Koser and G. Droysen (1908–11) and followed by *Nachträge zu dem Briefwechsel Friedrich des Großen mit Maupertuis und Voltaire* (1917).

Vom armen Schüler, a Middle High German verse legend, in which a poor boy begs shoes of the Virgin Mary, a request which she refuses. She later offers him the choice of thirty years as a bishop or an early death and blessedness. He chooses the latter and is bidden to declare to the faithful the truth of her corporeal assumption. The author, who wrote in Bohemia *c.* 1280,

gives his name as Heinrich Clusener (Klausner), and states that he learned the subject-matter from Brother Pilgerim of Görlitz (Silesia).

Vom Geist der ebräischen Poesie, a two-volume treatise published by J. G. Herder (q.v.) in 1782–3. It is described in a sub-title as *Eine Anleitung für die Liebhaber derselben und der ältesten Geschichte des menschlichen Geistes.* The work, which remains incomplete, is a commentary on the Bible seen as literature. The first part is in the form of a dialogue between two friends, Alciphron and Euryphron; the second abandons dialogue for straightforward exposition. There are many poetic illustrations, scriptural and non-scriptural.

'Vom Himmel hoch da komm ich her', Christmas hymn by Luther, adapted from a folk-song and published in 1535. Luther is also believed to have composed the tune.

Vom Himmelreich (himelrîche), an Early Middle High German poem of 378 lines, beginning with a hymn praising God and continuing with a description of the Heavenly City, which makes free use of Revelation. It contains a famous passage of negative description phrased in crude and homely terms. The verse makes an original use of long lines. The poem was written by a monk of Windberg Monastery near Straubing, Bavaria, *c.* 1180.

'Vom hoh'n Olymp herab ward uns die Freude', first line of the song *Gesellschaftslied* probably written in 1795 or earlier by H. C. Schnoor (q.v.), who also composed the melody.

Vom Jüngsten Tage, an anonymous Middle High German poem some 750 lines in length, portraying the Day of Judgement. It was probably written in a Swiss monastery *c.* 1270. The poem begins with the signs and sounds of the approaching Day of Judgement, develops the conflict between body and soul, portrays Christ's court with the condemnation of the wicked and the hallowing of the good.

The vigorous and powerful poem is constructed as a verse sermon, vehemently denouncing sin and expatiating on the sufferings of the damned. The author's sympathies are especially engaged by the poor and oppressed.

Vom Rechte, title given to an Early Middle High German poem of moderate length (549 ll.), written *c.* 1150 probably by a country parish priest in Austria. It is in the form of a sermon and the 'Recht' with which it deals is understood in the medieval sense of 'duties'. Its subject is the order of society established by

God, and the mutual responsibilities of master and man, husband and wife, parents and children. Its tone is homely, its construction desultory. It was written away from the main stream of intellectual life and is an interesting document of rural social conditions. It occurs in the Milstätter Handschrift (q.v.), immediately before *Die Hochzeit* (q.v.).

Von dem Beginchen von Paris, a Middle High German poem recounting the legend of a béguine, who in her ardent devotion, neglects her routine duties, refuses all food and drink, and waits in her cell for Jesus, who at the end of seven years removes her to Heaven. The original, which has not survived, was probably written in the 15th c. The poem is known through two later versions, one of the 15th and one of the 16th c.

Von dem großen Lutherischen Narren, a polemical tract directed by Thomas Murner against Luther (qq.v.) in 1522.

Von dem Holte des hilligen Cruces (*Von dem Holz des heiligen Kreuzes*), an anonymous Middle Low German poem of 770 lines recounting the legend of the Holy Cross. It is a translation of a Dutch poem, *Boec van dem boute.* A further version exists with a slightly different title, *Van dem holte darane starf Marien Sone* (*Von dem Holz, an dem der Marie Sohn starb*).

Von dem Ritter und von dem Pfaffen, see HEINZELIN VON KONSTANZ.

Von den Falken, Pferden und Hunden, a Middle High German treatise on the care of the birds and animals which the nobleman required for sport and war. It was written by Heinrich Mynsinger, a doctor of medicine, for Count Ludwig of Württemberg between 1442 and 1450. One of the MSS. was written in 1473 by Clara Hätzlerin (q.v.) of Augsburg.

Von den fünfzehn Zeichen vor dem Jüngsten Tage, a short Middle High German verse paraphrase of Revelation. The poem, the author of which is unknown, was probably written towards the end of the 12th c.

Von den siben Ingesigeln, see TILO VON KULM.

Von den sieben Schläfern, a Middle High German poem of 935 lines, recounting the legend of the Seven Sleepers, young Christians of Ephesus immured in a cave, who awaken 248 years later and shortly after are received into Heaven. The poem, the authorship of which is unknown, was probably written in the late 13th c.

Von den sieben Todsünden, designation of a Middle Low German poem written probably in the first half of the 15th c. by Josep, an author not otherwise known.

Von den zehn Geboten, title given to fragments of an Early Middle High German poem explaining the Ten Commandments, which are divided into three concerning God and seven concerning man's neighbours. The fragments occur in the same defective MS. as *Von Esau und Jakob* (q.v.) and both poems are by the same author. It is not impossible, though it is unlikely, that both sets of fragments are part of one poem.

Von den zwein Sanct Johansen, see HEINZELIN VON KONSTANZ.

Von der babylonischen Gefangenschaft der Kirche, see DE CAPTIVITATE BABYLONICA.

Von der Freiheit eines Christenmenschen (*Von der Freyheyt eyniß Christen menschen*), one of the three important theological tracts by Luther (q.v.) from the year 1520. It discusses the problem of faith and works, reaching the conclusion: 'daß ein Christenmensch lebt nit in ihm selb, sondern in Christo und seinem Nähsten: in Christo durch den Glauben, im Nähsten durch die Liebe'.

Von der Keuschheit, see ROTHE, JOHANNES.

Von der schönen verlorenen Frau, an anonymous Middle High German poem, sometimes also known as *Frau Welt.* It tells how the poet is visited by a beautiful and regal woman who invites him to accompany her. She brings him to a courtly festival, which is interrupted by a pilgrim in grey garb escorted by twelve knights in red. It is Christ with his apostles. He bids the woman disclose herself. She removes her fine clothing, appears in loathsome form and announces her identity as Frau Welt. The tented field turns into a sea of fire in which the devotees of pleasure suffer torment.

The poem continues with a sermon representing the imminence of the Day of Judgement and denouncing fashionable immodest clothing. The poet was an Alsatian and wrote his work in Strasburg about the middle of the 14th c.

Von des tôdes gehugde, see HEINRICH VON MELK.

Von deutscher Art und Kunst, a volume of essays published by J. G. Herder (q.v.) in 1773. Of the five essays the first two are by Herder, the third by Goethe, the fourth by Paolo Frisi, and the last by J. Möser (q.v.). The volume is a manifesto of the new approach to art in the 1770s, represented by Herder and the writers of Sturm und Drang (q.v.). The essays are:

(1) *Auszug aus einem Briefwechsel über Ossian und die Lieder alter Völker* (Herder), a panegyric not only of Ossian (Herder accepts without hesitation the authenticity of Macpherson's translation) but of all folk-song and folk poetry.

(2) *Shakespeare* (Herder) an interpretation of Shakespeare as a historical rather than as a tragic dramatist.

(3) *Von deutscher Baukunst. D. M. Ervini a Steinbach* (Goethe), an essay on Gothic architecture which Goethe mistakenly sees as essentially German. His zestful appreciation of the harmonies and intricacies of Strasburg Minster finds expression in a paean to the presumed architect, Erwin von Steinbach (d. 1318).

(4) *Versuch über die gothische Baukunst* (Frisi), an anonymous translation of an Italian essay; it was apparently published in *Von deutscher Art und Kunst* without the author's knowledge. Its argument is anti-Gothic and in some degree it functions as a counter-blast to Goethe's essay. .

(5) *Deutsche Geschichte* (Möser) is a fragment from the preface to Möser's *Osnabrückische Geschichte*. It had previously been published in 1768. The focus of Möser's historical interest is the ownership and tenure of land, and he is led to deplore the change from small ownership by the many to large-scale ownership by the few.

Von Esau und Jakob, title given to fragments of an Early Middle High German theological poem interpreting allegorically the story of Esau and Jacob, Leah and Rachel, and Joseph and his brothers. The complete poem was probably an exposition of Genesis. The fragments are preserved in a defective MS., which contains also fragments of a poem on the Ten Commandments, usually referred to as *Von den zehn Geboten* (q.v.). It is possible that both sets of fragments form part of one poem, but more probable that they are distinct. The same author wrote both.

Von Gottes Zukunft, see HEINRICH VON NEUSTADT.

Von Meyer Betzen, see HEINRICH WITTENWEILER.

Von morgens bis mitternachts, a play (Stück in zwei Teilen) by G. Kaiser (q.v.), written in 1912, published in 1916, and first performed in 1917.

A Stationendrama (q.v.), it comprises the last day in the life of a bank clerk. The morning finds him as usual at his job as cashier, when the call

of a wealthy Italian lady awakens in him a lust for adventure. He makes off with 60,000 Mark deposited by a customer, but his wild intentions are shattered when he discovers that the Italian lady is a respectable mother accompanying her son on a study tour. He recovers his wits during his flight through the snow, where he meets the first vision of death: the branches of a tree take on the shape of a human skeleton. Defiantly he returns home at midday, posing as the family man to his wife, mother, and two daughters. When he casts off his disguise, revealing that he is on the run, his mother dies of shock; he escapes just before his bank manager arrives in search of him.

Intent on exploring the pleasures money can buy, he visits in turn the races at the Sportpalast and a hotel frequented by prostitutes (Ballhaus) before he turns up, late at night, in the crowded hall of the Salvation Army. Having witnessed the confessions of a number of people, whose experience reflects his own life, he stages his own confession: decrying the vanity of money, he flings the stolen cash, the best part of which is still intact, at the feet of the public who, far from regarding it with contempt, grab it instantly and disperse. He is left alone with the Salvation Army girl (Mädchen) who had promised to stand by him, but she has already alarmed the police in a bid for the reward put on his head. He has gone full cycle to return to the skeleton, shaped now by the wires of the dimly lit chandelier. As, at this last station, he shoots himself and sinks against the cross on the curtain, the stage direction suggests the words 'Ecce Homo'. The play is one of the most characteristic and best-known works of Expressionism (see EXPRESSIONISMUS).

Von zwein Kaufmannen, see RUPRECHT VON WÜRZBURG.

VORAGINE, JACOBUS DE, see LEGENDA AUREA.

Vorauer Bücher Mosis, Early Middle High German verse renderings of Genesis, Exodus, and part of Numbers, forming part of the Vorauer Handschrift (q.v.). Heavily weighted with theological interpretation of allegorical character, they were written, possibly by more than one monk, in an Austrian monastery *c.* 1130–40.

Vorauer Handschrift, an important MS. preserved in the library of the monastery of Vorau in Austria, probably written towards the end of the 12th c. It contains principally the *Kaiserchronik* (q.v.), the *Vorauer Bücher Mosis*, the *Alexanderlied* (see LAMPRECHT, PFAFFE), the

Ezzolied, and the religious poems of Frau Ava qq.v.). It presents, no doubt intentionally, a conspectus of history in the light of medieval Christianity.

Vorauer Marienlob, title given to a Middle High German poem, partly in praise of the Virgin Mary, but mainly an exposition of the beginning of Isaiah, chapter 11. It is a puzzling poem because it appears in the Vorauer Handschrift (q.v.) as an integral part of Genesis. It is possible that, though a separate poem, it was copied into the MS. through inadvertence.

Vorauer Novelle, title given to an anonymous Middle High German religious poem (the title derives from the Abbey of Vorau in Styria where the MS. was discovered). Two young inmates of a monastery run away from the severe discipline of the life, and give themselves up to necromancy. One falls ill and dies without having repented. His spirit appears and describes to the surviving youth the torments of Hell. The survivor repents and ends his life in a monastery.

 The latter part of the *Vorauer Novelle* is missing in the MS., but its course can be inferred from its source, one of the *Reuner Relationen* (q.v.). The poem, written in the 13th c., is the work of a highly gifted poet, who appears to have been a disciple of GOTTFRIED von Straßburg (q.v.).

Vorauer Pentateuch, see VORAUER BÜCHER MOSIS.

Vorauer Sündenklage, a poetic prayer to the Blessed Virgin Mary together with a confession of sins. Though based on the formula of confession, it is greatly amplified, extending to 850 lines. More than half of this is devoted to the prayer to the Virgin Mary, and the poem, written about the middle of the 12th c., constitutes one of the earliest examples in German literature of the cult of the Virgin. It forms part of the Vorauer Handschrift (q.v.).

Vor dem Ruhestand. Eine Komödie von deutscher Seele, a play by Th. Bernhard (q.v.), published in 1975. The former SS officer, protégé of Himmler (q.v.), and concentration camp commander Rudolf Höller ('hell') has hidden for ten years in the basement of the family home and then succeeded in becoming Chief Justice, an office which he is now due to relinquish, having reached retirement age. It is the anniversary of Himmler's birth, which he has celebrated for years, donning his SS uniform, medals, gun, and pistol, drinking, and treating his crippled sister Clara like a concentration camp inmate. His other sister, Vera, like her brother a Nazi, supports

the man's every whim. The orgy turns into a kind of exorcism, at the height of which he collapses and dies of a heart attack. After attempting to put her brother into civilian clothes, Vera telephones the family doctor, Dr. Fromm ('pious').

 Evil and hypocritical figures enact the family play, in which Clara, physically handicapped as a result of an American bomb hitting her school two days before the end of the war, takes the honest part of conscience, more through her silence than through speech. She thus intensifies the breathless self-analysis of Rudolf and Vera, which reverberates with phrases taught to them by their parents and the world in which they grew up. The treatment of a political topic is rare in Bernhard's work, but the play's motto hints at his gruesomely ironic approach: 'What is character but the determination of incident' (Henry James).

Vor dem Sturm, a historical novel by Th. Fontane (q.v.), begun in 1864 and published in 1878 in instalments in the magazine *Daheim* and later in 1878 in book form. It was Fontane's first novel and bore the sub-title *Roman aus dem Winter 1812 auf 13*. It is set partly in Berlin, but chiefly in the Prussian countryside to the west of Frankfurt/Oder and Küstrin, which Fontane knew at first hand. The 'storm' of the title is the War of Liberation in 1813 (see NAPOLEONIC WARS) and the action takes place between Christmas Eve 1812 and the end of March 1813.

 Berndt von Vitzewitz of Hohen-Vietz, a retired officer and a widower, chivalrous and honourable, is moved to take action against the French, partly from patriotic and partly from personal motives; for the French occupation has caused the death of his wife. For the time being the opportunity for action eludes him. His son, Lewin, is a student in Berlin, and his daughter Renate lives at home. Both are on the threshold of betrothal, Lewin to a distant cousin of Polish descent, Kathinka Ladalinska, and Renate to Kathinka's brother Tubal. But Kathinka runs away with the Polish count Bninski and Tubal's affections are clearly unstable. After the publication of the King's proclamation of national solidarity (see AN MEIN VOLK) Berndt and his friends plan a night assault on the French in Frankfurt with a local militia they have raised and with Russian support. But the preparations are inadequate and the Russians let them down. In consequence the attack is repulsed with loss, and Lewin is taken prisoner. Knowing that he is likely to be shot as a civilian taken in arms against the military, Berndt, his guests, and some villagers mount an urgent rescue operation which is successful; but Tubal Ladalinski is mortally wounded. Lewin discovers his love for the foundling Marie Kniehase, who has loved him for years, and Berndt, subduing his aristocratic pride, consents. Renate remains unmarried, devoting herself, as a perfunctory postscript tells us, to good works.

The quality of the work lies less in the plot than in the fascinating portrayal of the life of the period, and the gallery of figures, eccentric or orthodox, men and women of all shades of character, among whom, apart from the main characters, the poet Hansen-Grell (who has some features of Fontane himself), Schulze Kniehase, Pastor Seidentopf, the Pietist Tante Schorlemmer, and the grotesque female dwarf, Hoppenmarieken, stand out. Fluent and convincing dialogue plays an important part, though not to the same extent as in Fontane's later novels. The discursive structure represents an original contribution to the historical novel.

Vorlesungen über dramatische Kunst und Litteratur, a series of lectures delivered in Vienna in 1808 by A. W. Schlegel (q.v.). They were published in three volumes (1809–11). They praise Shakespeare, attack French classical tragedy, and exalt Romantic drama.

Vorlesungen über schöne Literatur und Kunst, published title of lectures given by A. W. Schlegel (q.v.) between 1801 and 1804 in Berlin, which were printed in 1884 long after Schlegel's death.

Vormärz, the period of political and intellectual unrest beginning *c.*1840 and culminating in the March Revolution of 1848 (see BIEDERMEIER, DAS, and REVOLUTIONEN 1848–9). The term is also applied to the years 1815–48 (or 1830–48) to mark a different approach to evaluations of this period associated with the terms Biedermeierzeit (F. Sengle) and Restaurationsepoche (J. Hermand).

Vorparlament, the preliminary Parliament which met upon the initiative and under the chairmanship of Heinrich von Gagern (see GAGERN, H. W., FREIHERR VON), Minister of Hesse-Darmstadt, on 31 March 1848 at Frankfurt to prepare for the National Assembly or Frankfurt Parliament (see FRANKFURTER NATIONALVERSAMMLUNG). While Austria sent only two representatives, Prussia sent 141, but South German representatives predominated.

Vorschule der Ästhetik, an essay on poetics by JEAN Paul (q.v.).

Vor Sonnenaufgang, a play by G. Hauptmann (q.v.), first performed privately by the Verein Freie Bühne (see FREIE BÜHNE) on 20 October 1889, and published in the same year. It bears the sub-title *Soziales Drama in fünf Akten* and was the first German Naturalistic play to be performed, though its public performance was for some time forbidden.

With crass realism Hauptmann exposes the degeneration of a Silesian farmer's family after it has become rich by the sale of mineral rights. Vulgar luxury and alcoholism pervade the home, and not only Farmer Krause, but his wife, his daughter Martha, and (so we finally learn) his young grandchild are all addicted to drink. Martha has married the mining engineer Hoffmann, an astute and crooked man of business. Into this household comes a friend from Hoffmann's student days, the socialist Alfred Loth, whose intention is to investigate the social conditions of the district. Hoffmann, fearing publicity, tries to persuade Loth to abandon his project and leave; but Helene, Krause's second daughter, who has been educated at boarding-school and is untouched by the prevailing demoralization, falls in love with Loth, and he returns her love. The local doctor, Dr. Schimmelpfennig, who happens to be another old acquaintance of Loth, opens the young man's eyes to the degenerate state of the family, whereupon the rather priggish Loth, acting on the theory of heredity, takes flight. Helene, her love rejected and her last hope gone, stabs herself with a hunting knife. In conformity with late 19th-c. Darwinian conceptions, alcoholism is considered to be a hereditary disease. The uneven, sensational, yet powerful play contains much dialect speech.

Vor Sonnenuntergang, a play (Schauspiel) by G. Hauptmann (q.v.), published in 1932. It attracts attention chiefly by the resemblance of the title to that of his first play, *Vor Sonnenaufgang* (q.v.), which had appeared forty-three years earlier. The theme is old age and sexual love. The widower Geheimrat Matthias Clausen, aged 70, is attracted by and wishes to marry a young girl, Inken Peters. His family, concerned as to what may happen to his fortune, take steps to have him declared insane. He denounces and dismisses them and then takes his own life.

VORWÄRTS, MARSCHALL, nickname of Blücher (q.v.).

Vorwärts, a German Social Democrat newspaper published from 1890 to 1933, when it was suppressed by National Socialist decree. The paper had previously existed under the title *Berliner Volksblatt* (1884–90). From 1890 to 1900 the editor of *Vorwärts* was W. Liebknecht (q.v.).

VOSS, JOHANN HEINRICH (Sommersdorf nr. Waren, Mecklenburg, 1751–1826, Heidelberg), of humble rural origin, was educated at the grammar school of Neubrandenburg, but was at first too poor to go to a university and became a private tutor. In 1772 he was helped to study at Göttingen by H. C. Boie (q.v.). He was one of the founder members of the Göttinger Hainbund (q.v.) in September 1772, and as its representative visited Klopstock in Hamburg in 1774.

In 1775 he took over the editorship of the *Göttinger Musenalmanach* (q.v.) and continued to publish an Almanach independently after L. F. G. von Goeckingk (q.v.) had succeeded him. Thereafter two Almanachs coexisted. Voß married Boie's sister Ernestine in 1777 and became headmaster (Rektor) of the school at Ottendorf in 1778. Since 1776 he had been working at a translation of the *Odyssey* into German hexameters and this appeared in 1781 as *Homers Odüssee*. Voß subsequently reverted to the spelling *Odyssee*. In 1782 he became headmaster at Eutin, where he remained for twenty-three years. An interest in Theocritus prompted him to attempt German idylls, some of which he wrote in dialect. The idylls, published as *Idyllen* in 1801, include *Der Frühlingsmorgen, De Winterawend (Der Winterabend)*, and, perhaps his finest work, *Der siebzigste Geburtstag* (q.v.). Some are in dialogue, and political criticism and indignation are woven into others, for Voß, recalling his oppressed childhood, remained a vivid hater of tyrants all his life (see LEIBEIGENEN, DIE). His best-known work in idyll form is the substantial poem *Luise. Ein ländliches Gedicht in drei Idyllen* (q.v., 1795). Many of his shorter poems are pastoral songs. In Eutin Voß continued his translation of Homer and published the *Iliad* and *Odyssey* together in 1793 as *Homers Werke*. In 1805 he was granted a pension by the Grand Duke of Baden and thereafter lived at Heidelberg.

As he grew older, Voß's hatred of old political and new literary ideas increased, and he aggressively denounced aristocrats and Romantic littérateurs. He conducted a specially venomous campaign against Friedrich von Stolberg (q.v.), who had once been his benefactor (*Wie ward Fritz Stolberg ein Unfreier?*, 1819). In addition to Homer, Voß translated Virgil (*Georgics*, 1789), Ovid (*Metamorphoses*, 1798), Horace (1806), Aristophanes (1821), Propertius (1830), and Shakespeare (q.v., 1818–29). *Sämtliche poetische Werke*, ed. A. Voß, appeared in 1835 and correspondence (4 vols.), ed. A. Voß, 1829–33, and a selection of his works, ed. A. Sauer, in 1885 (ed. H. Voegt, 1966).

VOSS, JULIUS VON (Brandenburg, 1768–1832, Berlin), an officer in the Prussian army from 1782 to 1792, was afterwards a prolific purveyor of popular literature. Some of his earlier works

draw on his military experiences (*Geschichte eines bei Jena gefangenen Offiziers*, 1807; *Geschichte einer Marketenderin*, 1808). His best-known novels were *Ignaz von Jalonsky* (1806), *Die Maitresse* (1808), and *Die Schildbürger* (1823). He wrote a tragedy on Faust (*Faust*, 1823) and parodies of Lessing's *Nathan der Weise* and Schiller's *Die Jungfrau von Orleans* (qq.v.). The lower middle-class of Berlin (Kleinbürgertum) is the subject of his comedies, published in nine volumes in 1807–18; they were followed by *Neuere Lustspiele* (7 vols.) in 1823–7.

VOSS, RICHARD (Neugrape nr. Pyritz, 1851–1918, Berchtesgaden), son of a country gentleman, volunteered for ambulance service in the war of 1870 (see DEUTSCH-FRANZÖSISCHER KRIEG), and was wounded. After studying at Jena and Rome, he devoted his time to writing plays and novels, living by turns in Berchtesgaden and Frascati. From 1884 to 1888 he was librarian at the Wartburg. A prolific writer, Voß tended to present characters who are misfits. Among his plays *Unfehlbar* (1874), *Regula Brand* (1884), and *Schuldig!* (1892) should be mentioned, and among the novels *Villa Falconieri* (1896), *Römisches Fieber* (1902), *Zwei Menschen* (1911), and *Mit Weinlaub im Haar* (1915). Two volumes of Novellen were entitled *Die Sabinerin* (1890) and *Südliches Blut* (1900). An early volume of poetry was characteristically styled *Die Scherben, gesammelt von einem müden Mann* (1874).

Voß's memoirs appeared posthumously as *Aus einem phantastischen Leben* (1920), and a selection of his works (5 vols.) as *Ausgewählte Werke* (1922–5).

Vossische Zeitung, a Berlin daily newspaper which bore this title from 1911 to its closure in 1934, but had a much longer history. Said to be descended from an untitled news-sheet first circulated in 1617, it became the *Berliner Privilegierte Zeitung* in 1721 and was published by J. A. Rüdiger (1683–1751). On Rüdiger's death it was taken over by C. F. Voß (1724–95), and for more than a century and a half it was commonly called the *Vossische Zeitung* after its owner, while preserving its old title. Th. Fontane (q.v.) often referred to it as 'die Vossin'. Its politics were liberal.

VRING, GEORG VON DER (Brake, Oldenburg, 1889–1968, Munich), trained as an art teacher, served as an officer on the Western Front in the 1914–18 War, and was taken prisoner. He incorporated his experience of war in the novels *Soldat Suhren* (1927, reissued 1980), *Camp Lafayette* (1929), and in *Der Goldhelm* (1938), a story. He is the author of numerous light novels including *Der Wettlauf mit der Rose* (1932), *Die Spur im Hafen* (1936), *Die kaukasische Flöte* (1939), and *Und wenn du willst, vergiß!* (1950). A translator of French and English poetry (*Angelsächsische Lyrik aus sechs Jahrhunderten*, 1962), he wrote unpretentious yet melodious verse of which a collection appeared in 1956 (*Die Lieder des Georg von der Vring 1906–1956*). In 1955 he published his autobiography, *Die Wege tausendundein*. In the 1960s he published more verse (*Der Schwan*, 1961; *Der Mann am Fenster*, 1964; and *Gesang im Schnee*, 1967) and the story *König Harlekin* (1966). A collection, *Gedichte und Lieder*, ed. B. Bondy and R. Goldschmit, appeared in 1979.

VRIOLSHEIMER, DER, the author of *Der entlaufene Hasenbraten*, a short Middle High German comic poem, in which a priest, accused of a liaison with a married woman, is cleared of suspicion. The poem was written in the 13th c.

Vrone Botschaft, an anonymous Middle High German poem of nearly 900 lines, in the form of a letter addressed by Christ to Christendom. It was probably written by a monk of Weihenstephan Abbey in Bavaria in the first half of the 13th c. The letter is said to appear graven on a marble tablet in St. Peter's in Jerusalem and contains reiterated complaints on failure to observe the Sabbath, together with the announcement of the imminent destruction of the world and consequent Day of Judgement.

VULPIUS, CHRISTIAN AUGUST (Weimar, 1762–1827, Weimar), after studying at Jena and Erlangen, became secretary to Baron von Soden in Munich. In 1790 he returned to Weimar, where in 1797 he became secretary of the library. He was the brother of Goethe's mistress Christiane Vulpius (see GOETHE, CHRISTIANE), and so after her marriage in 1806 was Goethe's brother-in-law.

Vulpius acquired a popular reputation as the author of robber novels, of which the best known was the outstandingly successful *Rinaldo Rinaldini, der Räuberhauptmann*, a three-volume work (1798, shortened version, 1971). He also had some poetic facility, and is the author of the songs 'In des Waldes düstern Gründen' (printed in *Rinaldo Rinaldini*) and 'Der Lenz ist angekommen'. See RITTER- UND RÄUBERROMAN.

VULPIUS, CHRISTIANE, see GOETHE, CHRISTIANE.

W

Wach, HENRIETTE, see PAALZOW, HENRIETTE.

Wacht am Rhein, Die, a patriotic poem by M. Schneckenburger (q.v.), written in 1840 at a time when the French were threatening war to annexe part of the Rhineland; it was set to music in 1854 by Karl Wilhelm (1815–73). Its popularity dates from 1870 (see DEUTSCH-FRANZÖSISCHER KRIEG).

WACKENRODER, WILHELM HEINRICH (Berlin, 1773–98, Berlin), a close friend from boyhood of L. Tieck (q.v.), was the son of a highly placed Prussian civil servant who became minister of justice. With Tieck he studied for a semester at Erlangen and was enraptured by Bamberg and Nürnberg and by Dürer's art. Sensitive and receptive, Wackenroder responded enthusiastically and intelligently to the great works of painting he saw on a visit to Dresden in 1796. He and Tieck were joint authors of the work by which he is principally known, *Herzensergießungen eines kunstliebenden Klosterbruders* (q.v., 1797, ed. R. Benz, 1961), but Wackenroder's share was much the larger.

Wackenroder, an outstandingly perceptive critic, rejected analysis, basing his work on empathy, and ranking inspiration and piety high in his assessment of artists. He was the author of three novels, *Die Unsichtbaren* (1794), *Der Demokrat* (1796), and *Das Schloß Montford* (1796). Tieck incorporated some writings from Wackenroder's posthumous papers in *Phantasien über die Kunst, für Freunde der Kunst* (1799) and published later a second edition composed entirely of Wackenroder's work (*Phantasien über die Kunst, von einem kunstliebenden Klosterbruder*, 1813). The sensitive narration *Das merkwürdige musikalische Leben des Tonkünstlers Joseph Berglinger* (see BERGLINGER, J.), portraying a composer torn between the inspiration of his art and the mundane ties of life, has affinities with Wackenroder's own situation. Wackenroder is thought to have collaborated with Tieck in the planning of Tieck's novel *Franz Sternbalds Wanderungen* (q.v.).

Wackenroder's works and correspondence

appeared in 2 vols., ed. F. von der Leyen, in 1910, and an edition by L. Schneider in 1938 (reprinted and extended 1967).

WACKERNAGEL, WILHELM (Berlin, 1806–69, Basel), an early Germanist and pupil of K. Lachmann (q.v.), was elected to a chair at Basel in 1835 and became a member of the city council in 1854. His works include editions of *Landrecht des Schwabenspiegels* (1840, see SCHWABENSPIEGEL), Hartmann's *Der arme Heinrich* (q.v., 1855), and the poems of WALTHER von der Vogelweide (q.v., 1862). He also wrote poetry, most of it light (*Gedichte eines fahrenden Schülers*, 1828; *Zeitgedichte*, 1843; *Weinbüchlein*, 1845).

Waffen für Amerika, a historical novel by L. Feuchtwanger (q.v.), published in 2 vols., 1947–8. Though concerned with Benjamin Franklin and the American War of Independence, it is set in France, portraying the *ancien régime* a few years before the French Revolution.

Waffen-SS, see SS.

WAGENFELD, KARL (Lüdinghausen, Westphalia, 1869–1939, Münster), a schoolmaster who spent his life in Westphalia, was a poet, dramatist, and fiction writer of some originality and stature. He first used his local Low German dialect in the story, *'n Öhm* (1905), and then made a new contribution to dialect poetry with his religious epics *Daud un Düwel* (1912) and *De Antichrist* (1916) and his verse drama *Luzifer* (1920). *Dat Gewitter* (1912) and *Hatt giegen Hatt* (1918) are serious rural plays in dialect, and he also wrote Volksstück (q.v.) comedies (*Altwestfälische Bauernhochzeit*, 1912; *Schützenfest*, 1922; and *In der Spinnstube*, 1934). His *Gesammelte Werke* (2 vols.) appeared 1954–6.

WAGENSEIL, JOHANN CHRISTOPH (Nürnberg, 1633–1705, Altdorf), a professor at Altdorf University from 1667, is the author of *Von der Meister-Singer Holdseligen Kunst* (contained in *De civitate Noribergensi commentatio . . . liber*, 1697), which served E. T. A. Hoffmann (q.v.) as source for his story *Der Kampf der Sänger* (q.v.).

WAGGERL, KARL HEINRICH (Bad Gastein, 1897–1973, Salzburg), trained to be a schoolmaster, and served in the Austrian army in the 1914–18 War as an officer on the Italian front. After the war he took to authorship, was at first influenced by the Norwegian Hamsun, and later developed into a popular writer of novels and stories of rural life in the Austrian Alps.

Waggerl's principal works are the novels *Brod* (1930), *Schweres Blut* (1931), *Das Jahr des Herrn* (1933), and *Mütter* (1935), and the collections

of stories *Du und Angela* (1934), *Kalendergeschichten* (1937), *Feierabend* (1944), *Die Pfingstreise* (1946), *Fröhliche Armut* (1948), and *Wagrainer Geschichtenbuch* (1951). A volume of verse, *Heiteres Herbarium*, appeared in 1950, and the autobiographical *Wanderung und Heimkehr* in 1959, followed by *Ein Mensch wie ich* in 1963. *Gesammelte Werke* (5 vols.) appeared 1948–52 and *Sämtliche Werke* (2 vols.) 1972.

WAGNER, HEINRICH LEOPOLD (Strasburg, 1747–79, Frankfurt/Main), met Goethe and Lenz as a student of law at Strasburg University and moved in 1774 to Frankfurt, where he practised law from 1776. His extravagant verse satire, *Prometheus, Deukalion und seine Recensenten*, attacking the critics of *Die Leiden des jungen Werthers* (q.v.), was published anonymously in 1775 and was widely believed to be the work of Goethe.

Wagner wrote two plays of note, *Die Reue nach der Tat* (1775) and *Die Kindermörderin* (qq.v., 1776), both of which owe much to Goethe and Lenz. Only one volume of *Leben und Tod des Sebastian Sillig*, a novel modelled on Sterne, was published (1776). His last work was an anti-Aufklärung (q.v.) sketch, *Voltaire am Abend seiner Apotheose*, published anonymously in 1778. He was also active as a translator, and in particular produced a German version of *Du théâtre ou nouvel essai sur l'art dramatique* by L.-S. Mercier, published under the title *Neuer Versuch über die Schauspielkunst aus dem Französischen* (1776).

Of the planned 5 vols. of *Gesammelte Werke*, ed. L. Hirschberg, 1 volume appeared in 1923.

WAGNER, RICHARD (Leipzig, 1813–83, Venice), whose father died when he was only a few months old, was at school first in Dresden, then in Leipzig, and entered Leipzig University to study music. Among his early compositions is his only symphony (C Major, 1832). In 1833 he was appointed choir master at Würzburg, where he wrote his first opera, *Die Feen* (performed 1888). His next post was as conductor at Magdeburg (1834–6), which was a failure. He met the actress Minna Planer (1809–66), whom he married in 1836 at Königsberg, where he obtained his next appointment.

In 1837 Wagner moved to Riga, again as conductor, and there he began the opera *Rienzi*. Finding himself in financial difficulties in 1839, he slipped away to London and from there to Paris, where he remained for three years, completing *Rienzi* (1840) and writing *Der fliegende Holländer* (q.v., 1841), which was partly inspired by the storms he experienced on his voyage to England. *Eine Faust-Ouvertüre* was also written during his first year in Paris (1840). *Rienzi*, based on Lord Lytton's novel *Rienzi, the Last of the Tribunes*, was performed in Dresden on 20 October 1842 and was an immediate success. This breakthrough was followed by an equally successful first night for *Der fliegende Holländer* (2 January 1843), with the result that Wagner was appointed conductor (Hofkapellmeister) at Dresden. Moral as well as musical objections were, however, soon advanced against his work (*Tannhäuser und der Sängerkrieg auf der Wartburg*, q.v., 19 October 1845). Wagner wrote in Dresden the text of *Siegfrieds Tod*, the first stage in the genesis of the *Ring*; he completed *Lohengrin* (q.v., first performed in Weimar under Liszt, q.v., 28 August 1850), and drafted the text of *Die Meistersinger*.

In May 1849, having taken part in the revolt in Saxony, Wagner had to flee abroad, settling in Zurich until 1859 and making in 1855 a successful visit to London, where he conducted concerts. While in Zurich he developed his theory of opera as a 'Gesamtkunstwerk' to which many arts contribute, expounding it in *Die Kunst und die Revolution* (1849), *Das Kunstwerk der Zukunft* (1850), and *Oper und Drama* (1851). In 1853 he completed the text of *Der Ring des Nibelungen* (q.v.), and wrote the music for *Das Rheingold* (1854), *Die Walküre* (1856), and the first two acts of *Siegfried* (1857). Meanwhile Wagner had fallen in love with Mathilde Wesendonck (q.v.), the gifted wife of a wealthy patron, and shifted his interest to a new subject, the love tragedy, *Tristan und Isolde* (q.v.). In an untenable situation he moved first to Venice, then to Lucerne, completing *Tristan und Isolde* in 1859. The composition of five poems by Frau Wesendonck was a further outcome of his (probably unrequited) passion (*Fünf Gedichte von Mathilde Wesendonck*, 1857–8). In 1861 he was in Paris for a carefully prepared and lavish production of *Tannhäuser*, which met, however, with organized opposition in the opera house.

A political amnesty made it possible for Wagner to return to the territories of the German Confederation. After a stay in Vienna he was invited by King Ludwig II (q.v.) to establish himself in Munich, and on 10 June 1865 the first performance of *Tristan und Isolde* took place there. However, local opposition, not entirely unprovoked, obliged Wagner to leave Munich in 1866 and set up house at Triebschen (or Tribschen) near Lucerne, where he wrote *Die Meistersinger von Nürnberg* (q.v., performed in Munich, 21 June 1868) and completed *Siegfried* (1871). Wagner's wife Minna died in 1866 and in 1868 Cosima von Bülow, wife of Wagner's friend the conductor Hans von Bülow (1830–94), and daughter of Liszt, settled with him in Triebschen, where they were married in 1870. The serenade *Siegfried-Idyll* (1870) celebrates the

birth of their son Siegfried Wagner (1869–1930). In 1871 a new opera house specially for Wagner's works was projected at Bayreuth, and Wagner himself laid the foundation-stone in 1872. This Festspielhaus was opened with the first performance of the *Ring* cycle (13–17 August 1876), at which the German Emperor was present. Wagner's last work, *Parsifal* (q.v.), was completed in 1882 and performed at Bayreuth on 26 July 1882. After 1872 Wagner lived at Bayreuth, from 1874 in the 'Villa Wahnfried', which he had built for himself. A friendship with Nietzsche (q.v.), begun in 1869, ended in 1878 in a complete breach.

The originality of Wagner's music, the most familiar feature of which is the Leitmotiv (q.v.), aroused widespread controversy and opposition, which has persisted into the 20th c. and is partly linked with his Germanic nationalism and the exaltation of strength. Among Wagner's critics was Eduard Hanslick (1825–1904), who wrote for the *Neue Freie Presse* and held a chair at Vienna University. He is satirized by Wagner in the figure of Beckmesser (*Die Meistersinger von Nürnberg*). Wagner prosecuted his far-reaching views on the Musikdrama with an intransigence betraying megalomania, and recent biographies have taken an increasingly unfavourable view of his character and conduct. His dramatic writings, regarded as literature, are derivative and trivial, reviving Old High German and Middle High German metres and affecting an archaic vocabulary, but in combination with his eloquent and persuasive music their defects disappear and even his detractors are obliged to admit their extraordinary power in this form.

A *Gesamtausgabe* (12 vols.), ed. M. Balling, was published 1907–23 and *Sämtliche Werke*, ed. C. Dahlhaus, 1970 ff.; correspondence includes *Richard Wagners Briefe in Originalausgaben* (17 vols.), 1910–12, and *Sämtliche Briefe* (2 vols. appeared 1967–70), ed. G. Strobel and W. Wolf.

Wagram, village in Lower Austria, *c.* 12 miles ENE. of Vienna. The decisive defeat of the Austrians by Napoleon at Wagram on 5–6 July 1809 marked the end of the campaign of that year and, for the time being, of Austrian resistance to the Napoleonic empire. See NAPOLEONIC WARS.

WAHL, RUDOLPH (Cologne, 1894–1961, Munich), is the author of historical novels on medieval emperors, including *Karl der Große* (1934, reissued 1956), *Der Gang nach Canossa* (1935, reissued 1951), *Kaiser Friedrich Barbarossa* (1941, reissued 1959), and *Wandler der Welt, Friedrich II., der sizilische Staufer* (1948); *Das Mittelalter endet erst jetzt* appeared in 1957.

Wahlverwandtschaften, Die, a novel in two parts written by Goethe in 1808–9 and published in 1809. First conceived as a Novelle, its compact plot concerns four people. Eduard and Charlotte, who had loved in youth, have been able to marry in mature years, as their respective partners have died. They are people of means, who devote their time to the embellishment of their estate. Two long-term guests, the Hauptmann and Ottilie, arrive, and the slightly monotonous life of Eduard and his wife is soon diversified. Charlotte is drawn to the Hauptmann and he to her, but they resist their passion. Eduard and the young girl Ottilie fall deeply in love and Eduard presses, unsuccessfully, for a divorce. Charlotte, in her message to Eduard, discloses that she is with child by him. Deeply shocked, Eduard leaves for war service, determined to seek death on the battlefield.

In the second part Charlotte and Ottilie are at first left alone in the mansion. When Charlotte's child is born, it is found to resemble both Ottilie and the Hauptmann, in token of the directions the thoughts of Charlotte and Eduard had taken at its begetting. Eduard, who has risked his life without finding death, returns and renews his passion for and his pleas to Ottilie. At the highest point of his expectations his child (and Charlotte's) falls from Ottilie's arms into the lake and is drowned. Charlotte now consents to a divorce, and Eduard sees no obstacle to his union with Ottilie. She, however, beset by a feeling of guilt, denies herself happiness and, by starving herself to death, reconciles her love with her conscience. Eduard dies soon after.

The title suggests and the novel expressly refers to a chemical process ('elective affinity'), but the complexities of human life refuse to conform to simple mutations of chemical formulae. *Die Wahlverwandtschaften*, at first sight a perfectly planned symmetrical story, proves to be full of ambiguity and irony; hardly any work of Goethe has been interpreted so variously and with such subtlety.

Wahnfried, 'peace from illusions', name given by R. Wagner (q.v.) to the house built for him at Bayreuth in 1873–4 by the architect Wölfel. It now contains a Wagner-Archiv. Wagner and his wife Cosima are buried in the grounds.

Wahrheit, Die, an Early Middle High German verse sermon of 183 lines urging repentance in order to achieve bliss and escape the pains of Hell. It is contained in the Vorauer Handschrift (q.v.) and was written c. 1150.

WAHSMUOT VON KÜNZICH, a minor Middle High German poet of the 13th c., was probably a native of Franconia. His pleasant and rather precious Minnelieder derive from REINMAR der Alte and WALTHER von der Vogelweide (qq.v.). Geltar (q.v.), a contemporary parodist, mocked his sentimentality.

WAIBLINGER, WILHELM FRIEDRICH (Heilbronn, 1804–30, Rome), was at school at Stuttgart and studied theology at the Tübinger Stift (q.v.). At Tübingen he regularly visited the deranged Hölderlin (q.v.), whose poems he admired, and his novel *Phaeton* (1823) has Hölderlin as its model. He settled in Italy in 1827, living in poverty since he had no income except from his writings. These include *Lieder der Griechen* (1823), *Vier Erzählungen aus der Geschichte des jetzigen Griechenlands* (1826), the poems *Blüten der Muse aus Rom* (1829), the tragedy *Anna Bullen* (Anne Boleyn, 1829) and a *Taschenbuch aus Italien und Griechenland* (1829–30). He also wrote a life of Hölderlin, published posthumously as *Friedrich Hölderlin, Dichtung und Wahnsinn* (1831, reissued 1951). Waiblinger died of pulmonary tuberculosis. E. Mörike (q.v.), a friend of his since his Tübingen days, collected and published his poetry, *Gedichte* (1844).

Gesammelte Werke (9 vols.) appeared 1839–42, and diaries, *Die Tagebücher 1821–26*, ed. H. Meyer, in 1956. A selection of his writings, ed. P. Friedrich, was published in 1922, and *Werke und Briefe*, ed. H. Königer, planned in 5 vols., 1980 ff.

Waise, a precious stone in the imperial crown, referred to by WALTHER von der Vogelweide (q.v.) metonymously for the crown itself: 'Philippe setze den weisen ûf' (Lachmann, 9, 15, 'Place the imperial crown on Philip's head'). The stone was known as the *lapis orphanus* and Waise is a translation of the epithet. It has been suggested that the *orphanus* indicated its singular perfection.

WALAHFRIED STRABO (809–49), was educated first at Reichenau Abbey (q.v.) and then, aged 17, under Hrabanus Maurus (q.v.) at Fulda. From 829 to 838 he was at the court of Ludwig I (q.v.), der Fromme, where he was tutor to Ludwig's son, the future Karl II (q.v.), der Kahle. In 838 he became abbot of Reichenau, was evicted for political reasons, and reinstated in 842. He was an elegant writer of Latin, in which his mainly theological works were composed.

WALDAU, MAX, real name Richard Georg Spiller von Hauenschild (Breslau, 1825–55, Tscheidt, Upper Silesia), wrote poetry and the novels *Nach der Natur* (3 vols.) and *Aus der Junkerwelt* (2 vols., both 1850). A selection appeared in 1926 as *Lug ins Land Oberschlesien*.

Waldbruder, Der, a short work of fiction written by J. M. R. Lenz (q.v.) in 1776. It bears the sub-title *Ein Pendant zu Werthers Leiden* and is told in the form of letters (see BRIEFROMAN). Its chief character, Herz, loves a lady above his station and, because of this unhappy affair, has taken to the woods. The story is based on Lenz's love for Henrietta von Waldner, a lady who was unacquainted with him. *Der Waldbruder* was published in Schiller's periodical *Die Horen* in 1797.

Waldeinsamkeit, a word beloved of the German Romantics, which was coined by L. Tieck (q.v.). It occurs at the beginning and end of a stanza which, in slightly varying form, is used three times in his story *Der blonde Eckbert* (q.v.).

'Waldeinsamkeit!/Du grünes Revier', opening lines of poem No. 5 in Eichendorff's cycle *Der Umkehrende* (q.v.).

WALDEN, HERWARTH, pseudonym of Georg Levin (Berlin, 1878–1941, Saratow, Wolga), was from 1901 to 1911 married to Else Lasker-Schüler (q.v.). In 1910 he founded and edited the periodical *Der Sturm* (q.v.), and was prominent in furthering the more extreme experiments of Expressionism. His creative work (he published three novels and ten plays 1916–31) was less important, but his essays include *Das Begriffliche in der Dichtung* and *Kritik der vorexpressionistischen Dichtung* (both in *Der Sturm*, 1919) and also *Die neue Malerei* (1920). In 1932 he went to Russia, where he subsequently remained in exile. He is noted for his contribution to the Expressionismusdebatte (also known as Realismusdebatte) through his tract *Vulgär-Expressionismus*, which appeared in February (Heft 2) 1938 in *Das Wort* (ed. by Brecht, L. Feuchtwanger, qq.v., and Willi Bredel, 1901–64), published 1936–9 in Moscow; he strongly rejected the notion that Expressionism prepared the way for fascism. (See LUKÁCS, G., and MANN, K.) He died during imprisonment in Saratow. *Gesammelte Schriften* (2 vols.) appeared in 1916 and 1923, *Gesammelte Tonwerke* in 1919.

Waldgespräch, a poem by J. von Eichendorff (q.v.), the first line of which runs 'Es ist schon spät, es wird schon kalt'. It is a version of the Lorelei (q.v.) legend and occurs as an untitled duet in the novel *Ahnung und Gegenwart* (q.v.). It has been set to music by Robert Schumann (q.v.).

Wald-Idylle, a poem in elegiacs written by E. Mörike (q.v.) in 1829, praising the genius of the fairy-tale, as expressed in the stories collected by the brothers J. and W. Grimm (qq.v.).

WALDINGER, ERNST (Vienna, 1896–1970, New York), a volunteer in the Austrian army, he was wounded in 1915. He studied at Vienna University. In 1938 he emigrated to the U.S.A. and in 1947 became a professor at the Skidmore College in Saratoga Springs. An author of lyric poetry rooted in the Viennese tradition, he published the collections *Die Kuppel* (1934) and *Der Gemmenschneider* (1937) before his emigration, and after the 1939–45 War, during which he worked in the U.S. civil service, *Die kühlen Bauernstuben, Musik für diese Zeit* (both 1946), and *Glück und Geduld* (1952); his last collections are *Gesang vor dem Abgrund* (1961) and *Ich kann mit meinem Bruder sprechen* (1965).

WALDIS, BURKARD (Allendorf, *c.* 1490–*c.* 1556, Abterode, Hesse), a Franciscan monk in Riga, who while on a journey to Rome in 1524 went over to the Reformation (q.v.). On his return to Riga he abandoned the monastic life and married. He was twice imprisoned, and when released in 1540 migrated to Hesse, where he became pastor in Abterode. He is best known for his version of the fables of Aesop (*Esopas*, 1548, reprinted 1862), which expand and elaborate the originals, setting them in localities with which he was familiar. Though his humour is often crude or even obscene, he is a sincere moralist.

His polemical Fastnachtspiel *De Parabell vam vorlorn Szohn* (*Vom verlorenen Sohn*, 1527, reprinted 1881 and 1935), written in Low German dialect, exhibits his fanatical Lutheranism, setting faith of the repentant prodigal high above the righteous son's adherence to the works of the law. The message is explicitly pointed and reinforced in a prologue (*De Vorrhede*) and an epilogue (*De Vorrede*). His polemical poetry against Duke Heinrich der Jüngere of Brunswick (reigned 1514–68) appeared in 1542 (reprinted 1883).

Waldplage, a humorous poem written by E. Mörike (q.v.) in 1841. It consists of 64 lines of antique trimeters. The poet, reading Klopstock (q.v.) in the woods, is tormented by gnats and takes his revenge, swatting as many as possible by snapping the book to.

Walküre, Die, see RING DES NIBELUNGEN, DER.

Walladmor, a historical novel in 3 vols. by W. Alexis (q.v.), published in 1824. It is a pastiche of the manner of Sir Walter Scott and purported to be a translation of one of his works ('frei nach dem Englischen des Walter Scott').

WALLENSTEIN, ALBRECHT EUSEBIUS WENZEL VON, Herzog von Friedland und Mecklenburg,

Fürst von Sagan (Hermanič near Königsgrätz, 1583–1634, Eger), was a general in the Thirty Years War (see DREISSIGJÄHRIGER KRIEG). Of noble origin, he became a devout Roman Catholic, but was nevertheless addicted to horoscopes. By marriage he acquired great wealth in Moravia. The Emperor Ferdinand I (q.v.) rewarded him with the title count and the rank of colonel for assistance against Venice.

After the suppression of the Bohemian revolt (see MAJESTÄTSBRIEF) Wallenstein bought confiscated estates, notably Friedland and Reichenberg, which he developed into a state, which Ferdinand acknowledged by granting Friedland the status of a principality in 1624. His second wife, Isabella Katharina von Harrach (m. 1623), had valuable contacts with the court at Vienna. Wallenstein advanced Ferdinand large sums of money and raised a formidable army, with which he proved his worth as imperial general against the Danish invasion. Although aided by Tilly (q.v.), the general of the Catholic League (see KATHOLISCHE LIGA), Wallenstein became the most successful commander on the imperial side. The secret of his popularity was that he offered the best pay and living conditions, by living off the country, whether friendly or hostile. By 1627 he had conquered Silesia, and only the successful resistance of Stralsund (q.v.) frustrated his plans to pursue the Danish king into his own land. Ferdinand rewarded him with the Silesian principality of Sagan and the Duchy of Mecklenburg. In May 1629 Wallenstein concluded the Peace of Lübeck with Denmark. The princes of the Catholic League, headed by Maximilian I of Bavaria (q.v.), persuaded Ferdinand to dismiss Wallenstein, whom they considered to be a dangerous rival to their dynastic interests (1630). Wallenstein retired for two years to his Bohemian estates, contemplating an alliance with the Swedish invader, Gustavus Adolphus (q.v.), whose successes, however, forced Ferdinand to recall his general.

Wallenstein was able to assume the supreme command of the imperial army on his own terms. He was defeated at Lützen (q.v., 16 November 1632), in which battle, however, Gustavus Adolphus was killed. Wallenstein, at the zenith of his power, desired peace, though his motives have been disputed. He was both a sick and an ambitious man, but his concern for his private fortune appears to have been greater than his care for the common good. By the time he took up winter quarters near Pilsen in the following year, he had become subject to criticism as a military leader, and rumours, both false and true, about his secret negotiations with the enemy were carried to the Viennese court. Wallenstein, alarmed at the increasing insecurity of his position, bound his officers by

personal oath. But his generals Gallas, Aldringen, and Piccolomini (q.v.) were won over to the Emperor, who twelve days later (24 January 1634) resolved to remove Wallenstein and his officers Ilow and Trčka. On 22 February Wallenstein was publicly declared a traitor. He withdrew to Eger, where he was taken by surprise and murdered in his sleeping quarters on 25 February 1634. The conspiracy against him had largely been in the hands of Piccolomini and, in the last stage, in those of colonels Butler and Lesley, and the Commandant of Eger, John Gordon. The assassin, the English Captain Devereux, and his helpers were richly rewarded.

Wallenstein, treated in a number of plays since the 17th c., is the subject of Schiller's trilogy *Wallenstein* (q.v.) and is a minor character in Grillparzer's tragedy *Ein Bruderzwist in Habsburg* (q.v.). 20th-c. German treatments include a novel, *Wallenstein*, by A. Döblin (q.v., 1920); the study *Wallenstein* by Ricarda Huch (q.v., 1915) and the extensive biography *Wallenstein. Sein Leben* by Golo Mann (q.v., 1971) take into account Schiller's play *Wallenstein*, as well as his historical studies on the subject, and the *Geschichte Wallensteins* by L. von Ranke (q.v., 1869).

Wallenstein, a tragedy by Schiller in the form of a trilogy and sub-titled *Ein dramatisches Gedicht*. Written mainly between 1797 and 1799, the complete cycle of plays was performed in Weimar between 15 and 20 April 1799. *Wallensteins Lager*, a prelude, was first performed on 12 October 1798, which coincided with the reopening of the Weimar Court Theatre, and *Die Piccolomini* in January 1799. *Die Piccolomini* and *Wallensteins Tod*, both written in blank verse, are two parts of one vast tragedy, divided by Schiller on Goethe's suggestion.

Wallensteins Lager sketches, in deliberately archaic Knittelverse (q.v.) and in the manner of comedy, the background to the success of the central figure, A. von Wallenstein (q.v.). It portrays the soldiers with their lusts and their diverse beliefs, their courage in battle, and their greedy violence in pillage; above all, it emphasizes their devotion to their leader and their trust in his superb generalship. In *Die Piccolomini* Wallenstein's plot to go over to the enemy with his army and the counterplot to overthrow Wallenstein are both developed, and Wallenstein's astrological confidence in his lieutenant Octavio Piccolomini is seen to be totally misplaced. Octavio's son Max and Wallenstein's daughter Thekla fall in love in innocence of the enmity to be generated between their houses. The most brilliant features of *Die Piccolomini* are the fourth act, in which officers faithful to Wallenstein seek signatures from the generals at a drunken carouse, and the fifth, in which Octavio reveals to Max his counter-conspiratorial plans.

In *Wallensteins Tod*, Wallenstein, urged on by Gräfin Terzky, overcomes his hesitation and, too late, opts for revolt. His plans to seize Prague are forestalled by Octavio, and the troops on which he had relied melt away into the darkness. Amid scenes of mutiny he removes from Pilsen to Eger with a handful of faithful regiments. Max Piccolomini falls, in an act of deliberate self-sacrifice, leading a charge against the Swedish forces. Wallenstein is assassinated by mercenaries under the orders of the rancorous Colonel Buttler, who, once faithful to Wallenstein, believes himself injured by him and therefore betrays him. Octavio Piccolomini can do no more than express horror at the catastrophe.

WALLOTH, WILHELM (Darmstadt, 1856–1932, Munich), a scientist by training, wrote historical novels of ancient Rome (*Oktavia*, 1885, *Der Gladiator*, 1888, *Tiberius*, 1889, and *Ovid*, 1890).

Wally die Zweiflerin, a short novel by Karl Gutzkow (q.v.), published in 1835. Wally, a spoilt and coquettish young woman, consents to marriage with a Sardinian nobleman whom she does not love, having previously entered into a spiritual marriage with Cäsar, a cynical man of the world, who undermines her vague and unreasoning faith. Her husband, for financial reasons, encourages the passion felt by his brother for Wally, an episode which culminates disastrously in the brother's suicide. Wally leaves her husband for Cäsar, who, however, abandons her for the rich Jewess Adolfine. Wally, deprived of her love and of her faith, stabs herself.

The book, written in a cool ironical style, was bitterly attacked, notably by W. Menzel (q.v.), for its supposed immorality, manifested in Wally's life and in Cäsar's strictures on marriage and on the Christian religion. The novel was later revised and was republished in 1852 under the title *Vergangene Tage*. Gutzkow indicates in the preface to this edition that, in portraying Wally's fate, he had in mind the suicide of Charlotte Stieglitz (q.v.).

Wälsche Gast, Der, see THOMASIN VON CIRCLAERE.

WALSER, MARTIN (Wasserburg, Lake Constance, 1927–   ), after war service (1944–5), studied German literature, history, and philosophy at Regensburg and Tübingen universities and was active in broadcasting before establishing himself as a writer of fiction and plays. He was associated with the Gruppe 47 (q.v.).

His strong criticism of established society and authority emerges particularly strikingly in his tantalizing analysis of crises of identity experienced by his characters. His early stories (reminiscent of Kafka, q.v., on whom he wrote his dissertation) are contained in the collections *Ein Flugzeug über dem Haus* (1955) and *Lügengeschichten* (1964). After *Ehen in Philippsburg* (q.v., 1957), his first novel, *Halbzeit* (q.v., 1960), together with *Das Einhorn* (q.v., 1966) and *Der Sturz* (1973), make up an ironical trilogy on modern life in West Germany, the central figure of which is the writer Anselm Kristlein. The first-person narrative of *Fiction* (1970) is an essay in self-analysis, and the novel *Die Gallistl'sche Krankheit* (1972) criticizes the attitude of society to the creative writer.

In *Jenseits der Liebe* (1976) the central figure, Franz Horn, has abandoned family life and love in his struggle for professional advancement; he fails in this and in his attempted suicide. The Novelle *Ein fliehendes Pferd* (1978) presents an ironic view of resignation involving the middle-aged grammar school teacher Helmut Halm, his wife, and an enviably young couple. *Seelenarbeit* (1979) centres on the chauffeur Xaver Zürn and the disintegration of his individuality in a society bent on achievement. The central figure of *Das Schwanenhaus* (1980) is the property agent Gottlieb Zürn, aged fifty and involved in a personal and professional crisis (his vain quest for the agency of the house of the title). A sequel to *Jenseits der Liebe*, *Brief an Lord Liszt* (1982), resumes Franz Horn's life four years after his suicide attempt. When he hears that a rival of his boss Thiele, to whom he had applied for another job, has taken his own life and burned his factory, Horn's prospects are finally shattered; the protracted letter he writes to the apparently successful Dr. Liszt (without dispatching it) reveals unexpected dimensions of his own inner conflict. The collection *Gesammelte Geschichten* was published in 1983, the novel *Brandung* in 1985.

Der Abstecher (1961, adapted for radio in 1963) was followed by Walser's first conspicuous stage success, *Eiche und Angora* (q.v., 1962, revised 1963), a sharp but amusing satire of post-war Germany and its connections with the Nazi period. It was followed by the deft political satires *Überlebensgroß Herr Krott* (1963) and *Der schwarze Schwan* (1964); *Die Zimmerschlacht* (1967, revised 1968) shows the influence of Albee's *Who's afraid of Virginia Woolf?*; *Der schwarze Flügel* (1968) attempts to shake off the conventions of the theatre. *Ein Kinderspiel* followed in 1970, *Aus dem Wortschatz unserer Kämpfe* in 1971, *Das Sauspiel. Szenen aus dem 16. Jahrhundert* in 1975, and *In Goethes Hand. Szenen aus dem 19. Jahrhundert* in 1982. *Gesammelte Stücke* (1971) is a collection of plays in revised form.

A Socialist to the left of the SPD (q.v.), Walser published *Gedanken zur Wahl* in 1972. His views on the writer's task are contained in *Wie und wovon handelt die Literatur* (1973). *Was zu bezweifeln war* (1976) is the title of a collection of essays and speeches covering the years 1958–75; the aphorisms *Der Grund zur Freude* appeared in 1978, *Meßmers Gedanken* in 1985, and *Wer ist ein Schriftsteller* in 1979. *Selbstbewußtsein und Ironie* (1981) is the title of a volume containing lectures on Kafka (irony and crises of identity) held at Frankfurt University as a visiting professor of Aesthetics (Gastdozent für Poetik). *Versuch, ein Gefühl zu verstehen und andere Versuche* (1982) was followed by *Liebeserklärungen* (1983), a collection of eleven essays and speeches on writers, among them Proust, Swift, and Robert Walser (q.v.).

Walser was awarded the Büchner Prize in 1981.

WALSER, ROBERT (Bienne, Switzerland, 1878–1956, Herisau, Switzerland), worked in a bank and later in a publisher's office. From 1906 he spent some years in Berlin with his brother Karl (1877–1943), a painter and stage designer. He returned to Bienne (or Biel) in 1913 and held no post except for a library appointment in 1921 in Berne, which he soon relinquished. From 1933 until his death he was in mental homes, latterly in Herisau.

Walser published three novels, which are partially autobiographical. *Geschwister Tanner* (1906) relates to his family; *Der Gehülfe* (1908) is a story of the decline of a family, based on his observations of one of his employers; and *Jakob von Gunten* (q.v., 1909), like the early story *Fritz Kochers Aufsätze* (1904), is written from the point of view of a schoolboy. Most of Walser's work consists of short essays, sketches, miniature stories, etc., many of which appeared in newspapers (including the *Neue Rundschau*). Collections of them were published as *Geschichten, Kleine Dichtungen* (both 1914), *Kleine Prosa, Der Spaziergang* (both 1917), *Seeland* (1921), and *Die Rose* (1924), but others remained unpublished until the appearance of the five volumes of *Dichtungen in Prosa*, edited by Walser's friend C. Seelig, 1953–62. Walser was a somewhat withdrawn individualist, and the prevalence of short pieces in his work corresponds to his dislike of anything pretentious. His writing is deceptively simple and has an appearance of spontaneity, but it operates with subtle irony and intentional absurdity. Although his work was praised by Kafka, Loerke, Morgenstern, and Musil (qq.v.), it fell into almost total neglect from the 1920s until the 1950s.

A comprehensive edition (*Das Gesamtwerk*) by K. J. W. Greven appeared in 12 vols. (incl. 12,1 and 12,2), 1966–75.

WALTER, or **WALTHER,** JOHANN (Thuringia, 1496–1571, Torgau), a musician of Luther's circle in Wittenberg, prepared Luther's *Sangbüchlein* (q.v.) for publication (*Geystlich gesangk-Buchleyn*, 1524). His later years were spent in Torgau and Dresden.

Waltharius, a Latin poem of roughly 1,500 lines, recounting a German heroic tale and believed to be founded on a lost German lay. It was written in the 9th or possibly the 10th c. It tells the story of Walther with the strong hand (*Waltharius manu fortis*), of Aquitaine, who with his bride Hiltgunt is detained by the Huns. Walther escapes with Hiltgunt. Near Worms they are attacked by Gunther and Hagen for the treasure they bear with them. In a series of single combats Walther defeats his lesser enemies and finally overcomes Gunther and Hagen. Walther and Hiltgunt continue their journey to Aquitaine, where they live happily.

The Latin hexameters in which the poem is written are based on Virgil, and the poem was formerly believed to be a school exercise done by Ekkehard I (see EKKEHARD) of St. Gall, who died in 973. This attribution is doubtful and recent scholarship inclines, though without certainty, to the view that the author was Geraldus, an otherwise unknown monk of the 9th c., possibly of the time of Charlemagne (see KARL I, DER GROSSE).

Walther und Hildegund, title given to two fragments of a Middle High German heroic epic written in Austria in the early 13th c. Its basis is an ancient Germanic legend (see WALTHARIUS). The one fragment (at Graz) shows Walther and Hildegund at Etzel's court, the other, preserved in Vienna, narrates their homecoming and marriage festival at Lengres (Langres) in France. The two fragments amount in all to 39 four-line stanzas.

WALTHER VON DER VOGELWEIDE (*c.* 1170–*c.* 1230), the greatest Middle High German lyric poet, was probably an Austrian by birth, but there is only one documentary reference to him (in the accounts of the Bishop of Passau in 1203), and his life is largely constructed from his poetry.

Walther may have been a *ministeriale*, a knight in the service of greater noblemen, but his rank is uncertain, and it is possible that he was a hanger-on of little social consequence, who undoubtedly spent his life in the service of the great. As a young man at the Viennese court of Leopold V (1157–94), he modelled himself on REINMAR der Alte (q.v.), with whom he later fell out. After 1198, when his patron Duke Friedrich von Österreich died, he attached himself for a time to PHILIPP von Schwaben (q.v.), and he was later at the courts of various lords, including Bishop Wolfger von Passau, Leopold von Österreich (1176–1230), Hermann, Landgraf von Thüringen (q.v.), and Dietrich von Meißen (d. 1221). About 1212 he was in the suite of Otto IV (q.v.), and from 1214 with King Friedrich II (see FRIEDRICH II, KAISER). In 1220 he was granted a fief or pension, an event which is the subject of one of his poems. He was probably buried in Würzburg.

Walther was exclusively a lyric poet, and his extant œuvre of more than 100 poems consists mainly of Minnelieder and political Sprüche (see SPRUCH). He also wrote two Kreuzlieder (crusaders' songs) and a leich (q.v.). Walther's poems are usually divided into three groups, which were formerly held to correspond to stages in his development. Those which were held to be the earliest are conventional expressions of Minnedienst, modelled on the poetry of Reinmar. The second group consists of poems in which courtly conventions are discarded, and man and woman love spontaneously and responsively on equal terms. These poems are technically expert and skilfully organized, but their emotional freshness gives them an appearance of naturalness, which is heightened by the references to landscape. These are sometimes aptly termed Mädchenlieder. A third (possibly later) group applies the spontaneousness of niedere Minne to the aristocratic sphere of hohe Minne. In these poems Walther transcends the courtly conventions and produces some of the finest examples of Minnesang.

The Sprüche are Walther's most individual creation. They represent his responses to political events and personal experiences, which are frequently combined in a single poem. His contact with men of power, especially at the Hohenstaufen (q.v.) court, gives much of his poetry a considerable documentary value. But it is also intensely poetic, the vivid record of the aspiration, exultation, indignation, and despondency of this sensitive and touchy man. It is not always possible to sympathize with his moods, for he can be both ungrateful and unjust, but the keenness of his vision and the power of his poetry are constant. The handful of late poems reveals the embittered, yet dignified and courageous response of an ageing man out of sympathy with the changing world around him.

Walther's personal flexibility, the vividness of his response, the freshness of his humour, and the vehemence of his anger are transmitted by an assured and brilliant technique, which enables him to perform tours de force of verbal manipulation.

In the larger Heidelberg manuscript collection of medieval poetry (see HEIDELBERGER LIEDERHANDSCHRIFT, GROSSE) Walther is represented seated with his legs crossed, a posture suggested by the Spruch beginning: 'Ich saz ûf eime steine/und dahte bein mit beine' (I sat on a stone and crossed one leg over the other).

K. Lachmann (q.v.) published a notable edition of Walther von der Vogelweide's poems (1827), which was revised by C. von Kraus (1936) and H. Kuhn (1965). F. Maurer's edition (2 vols., 1967–9) contains melodies as well as text and apparatus. P. Wapnewski's selection (1962, text based on Lachmann–Kraus) contains a modern German translation.

WALTHER VON KLINGEN (d. 1286, Basel), a minor Middle High German poet of the second half of the 13th c. He was a noted benefactor of the Church, founding the nunnery of Klingenthal in Basel. Eight of his formally attractive Minnelieder have been preserved.

WALTHER VON RHEINAU, a Middle High German poet, who, about the turn of the 13th to the 14th c., wrote a life of the Virgin Mary (*Marienleben*). The poem, which runs to 15,000 lines, suggests in its style the influence of RUDOLF von Ems and KONRAD von Würzburg (qq.v.). Its source was the *Vita Beatae Mariae Virginis et salvatoris rhythmica* (q.v.). Walther indicates that his home was Bremgarten in Aargau, Switzerland.

WALTRAM VON GRESTEN, see ALRAM VON GRESTEN.

Wanderer, Der, a poem written by G. P. Schmidt (q.v.), is well known as a song composed by F. Schubert (q.v.) in 1816.

Wanderer in der Sägemühle, Der, a poem on mortality by J. Kerner (q.v.), the first line of which runs 'Dort unten in der Mühle'.

Wanderers Nachtlied, title valid for two poems by Goethe. The actual superscription is borne by the poem beginning 'Der du von dem Himmel bist'. The other, immediately following in Goethe's arrangement, is styled *Ein Gleiches* and begins 'Über allen Gipfeln' (q.v.). 'Der du von dem Himmel bist' was written on the Ettersberg on 12 February 1776, was immediately sent to Charlotte von Stein with a letter, and was printed in 1780 in the *Christliches Magazin*. The poem reflects Goethe's restless energy and his longing for peace.

Wanderers Sturmlied, a poem in so-called Pindaric form written by Goethe probably in 1772. It was published without the poet's authority in 1810. The first publication approved by Goethe followed in 1815. The poet both asserts and mocks at the daemonic godlike quality which he feels within him. In *Dichtung und Wahrheit* (q.v.), some forty years later, Goethe refers to the poem as 'diesen Halbunsinn'.

Wanderlied, title of a poem (1809) by J. Kerner (q.v.), best known by its opening words 'Wohlauf! noch getrunken'.

Wanderungen durch die Mark Brandenburg, travel book by Th. Fontane (q.v.) in 5 vols. (1) *Grafschaft Ruppin* (1862); (2) *Oderland* (1863); (3) *Havelland* (1873); (4) *Spreeland* (1882); (5) *Fünf Schlösser* (1889). The fifth volume, which is not an actual Wanderung, was only included in the work after Fontane's death. The first publication of vols. 1–4 under the present title was in 1892.

The aim of the book was to bring home to the Prussian the fascination and the historical wealth of his own countryside, and the idea was first conceived by Fontane while visiting Scotland. The material was collected at first hand, and its interest is especially focused on relics and monuments of Prussian history in the 17th c., 18th c., and 19th c. It is a leisurely and richly informative guide-book, which successfully evokes the melancholy charm of the landscape and the historical anecdotes which it has framed.

Wandsbecker Bote, Der, a newspaper edited from 1771 to 1776 and largely written by M. Claudius (q.v.). It is also commonly used as a designation of Claudius himself.

Wappendichtung, see HEROLDSDICHTUNG.

WARBECK, VEIT (Schwäbisch-Gmünd, c. 1490–1534, Torgau), studied in Paris and was in the service of the Elector Friedrich der Weise (q.v.) of Saxony. He translated the story of Magelone (see SCHÖNE MAGELONE, DIE) from the French in 1527.

Warme Almosen, Das, a short Middle High German poem in which matrimonial infidelity on the part of a wife cures her husband of his miserliness. The poem was written in the 13th c. by an unknown author.

Warnung, Die a Middle High German poem of almost 4,000 lines written by an unknown Austrian author in the middle of the 13th c.

Doubt has been cast on the earlier belief that he was a priest or monk; he is thought instead to have been a nobleman who underwent an experience of conversion without subsequently taking holy orders.

Die Warnung records former delight in the pleasures of courtly life, which the author has come to reject in the thought of their nothingness in the presence of death, and laments the decline of civilized values. A satirical element probably refers to the anarchic conditions prevailing in Austria after the extinction of the Babenberg (q.v.) line in 1246. It is a powerful work in a style reflecting deep conviction. The end of the poem is lost.

War of the Austrian Succession, The, see ÖSTERREICHISCHER ERBFOLGEKRIEG.

Wars of Liberation, see NAPOLEONIC WARS.

Wartburg, 12th-c. castle near Eisenach in Thuringia. It was the seat of the Landgrafen von Thüringen for three centuries and the scene of the childhood and married life of St. Elizabeth (see ELISABETH, HEILIGE). The legendary contest of Minnesänger is said to have taken place in a still-existing hall of the castle (see WARTBURGKRIEG and TANNHÄUSER). From May 1521 to March 1522 Luther (q.v.) was given asylum in the Wartburg by the Elector of Saxony. In 1817 it was the scene of a Protestant tercentenary celebration, largely attended by students, which had important political consequences (see WARTBURGFEST). In the 19th c. it was extensively restored. It now stands in the German Democratic Republic, close to the western frontier.

Wartburgfest, a gathering of students held at the Wartburg (q.v.) on 18 October 1817. It was summoned by the students of Jena to celebrate simultaneously the tercentenary of the Reformation (Luther's 95 *Theses*, q.v.) and the fourth anniversary of the Battle of the Nations at Leipzig (Völkerschlacht, see NAPOLEONIC WARS).

Students were invited from all over Germany, and the assembly developed into a demonstration of patriotic and democratic fervour. The black, red, and gold flag of German unity was displayed, right-wing literature and symbols of despotic rule (a corset and a corporal's cane) were solemnly burned. The behaviour of the students excited apprehension in ruling circles, and the Wartburgfest led to increasing distrust of popular enthusiasm.

Wartburgkrieg, Der, a Middle High German poem of unknown authorship, the most important section of which deals with a contest of singers at the Wartburg (q.v.). Once thought to be a historical event which occurred between 1205 and 1208, the *Wartburgkrieg* is now recognized as legendary, and the poem is believed to have been written *c*. 1250.

The contest, which is couched in 24 eight-line strophes, is set in the form of a trial or ordeal. Heinrich von Ofterdingen, a poet not otherwise recorded and believed to be fictitious, appears as challenger at the court of Landgrave Hermann (q.v.) at the Wartburg, in order to prove in eulogistic song the superiority of the Duke of Austria; and he pledges his life in the contest. Known poets, including WOLFRAM von Eschenbach, REINMAR von Zweter, and Biterolf (qq.v.), appear for Landgrave Herman, and the contest is finally decided against Heinrich by WALTHER von der Vogelweide (q.v.). Heinrich von Ofterdingen's life is forfeit, but he is spared by the intervention of the Landgrave's consort. This episode is followed by the appearance of Klinsor (Clinschor from Wolfram von Eschenbach's *Parzival*, q.v.), who poses for Wolfram, in a different strophic form, a number of riddles of theological character which the poet successfully solves. So far worsted by his opponent, Klinsor summons a demon, Nasion, who puts astronomical questions. Wolfram is unable to answer them but puts the devil to flight by the sign of the cross.

The *Wartburgkrieg* contains a further, but extraneous poem, an attack upon simony, symbolized by 'Ausons Pfennig'. The version in the Jenaer Liederhandschrift (q.v.) also contains a threnody of a Henneberg and a Thuringian prince, generally supposed to be respectively Boppo VII, who died in 1245, and Heinrich Raspe (q.v.), who died two years later. The theme of the *Wartburgkrieg* was revived by E. T. A. Hoffmann in *Der Kampf der Sänger* (q.v.) and adapted by R. Wagner in *Tannhäuser und der Sängerkrieg auf der Wartburg* (q.v.).

The *Wartburgkrieg* was first printed in 1818. It was edited by T. A. Rompelmann in 1939.

Wartesaal, Der, collective title given to four novels by L. Feuchtwanger (q.v.), published between 1930 and 1944 and all dealing with National Socialism and its persecution of the Jews. Feuchtwanger described it as a *Zyklus aus dem Zeitgeschehen*.

'**Warum gabst du uns die tiefen Blicke**', first line of an untitled poem by Goethe, included in a letter to Charlotte von Stein (q.v.), dated 14 April 1776. It was first printed in the edition of Goethe's letters to Frau von Stein published in 1848. The intimate and sensitive poem reflects an early phase in Goethe's love for Charlotte.

Wärwolf, Der, a 3-vol. historical novel (Vaterländischer Roman) by W. Alexis (q.v.), published in 1848. A sequel to *Die Hosen des Herrn von Bredow* (q.v.), it deals with the later years of the reign of the Elector Joachim I of Brandenburg (1499–1535) and especially with his vain attempts to repress Lutheranism in his dominions. The climax to the story is Joachim's quarrel with his own consort, Elisabeth of Denmark, an adherent of the new faith, a quarrel which is followed by her flight. She takes refuge with the widowed Brigitte von Bredow, one of the chief characters of the preceding novel. Hans Jürgen von Bredow and his wife Eva are also in the Electress's service, but Alexis makes these fictional characters secondary to the principal historical figures, thus reversing the usual procedure of Sir Walter Scott and his followers.

The novel is based on a careful study of the historical sources, and from time to time a footnote reminds us that some of the less plausible anecdotes are 'historisch'. A comic element is derived from Joachim's obsessive belief, founded on astrological predictions, that a new Deluge is imminent. The title derives from the superstition of the Werwolf (q.v.), which plays a part in the opening chapters.

'**Was glänzt dort vom Walde im Sonnenschein?**', first line of the patriotic poem *Lützows wilde Jagd*, written in 1813 by Theodor Körner (q.v.) and included in *Leyer und Schwerdt* (q.v.).

'**Was ist des Deutschen Vaterland?**' first line of a patriotic poem written by E. M. Arndt (q.v.) in 1813.

Wasserfälle von Slunj, Die, a novel by H. von Doderer (q.v.), published in 1963. It is described as *Roman Nr. 7 Erster Teil*. Doderer's last completed work, it was intended to be the first novel of an Austrian tetralogy covering the period from 1877 to the 1960s.

Robert Clayton, son of a wealthy English manufacturer, spends his honeymoon in Austria-Hungary. The climax of the tour is a visit to the majestic waterfalls at Slunj in Croatia. The Claytons make agricultural machinery and decide to open a factory in Austria, which Robert is to manage. He and his wife Harriet make their home in Vienna. The prosperous Clayton firm becomes the focal point for a diverse collection of Viennese and Austrian characters, whose private lives ramify into all levels of Viennese society. The story extends over thirty years. Harriet Clayton never becomes acclimatized and dies in 1902 in her 40s.

Her son Donald, born in 1878, is educated in England, but joins the branch in Vienna when he grows up. A defective circulation and a phlegmatic temperament give the impression that he is scarcely younger than his exuberant father, and the two are often believed to be brothers. Donald becomes attached to Monica Bachler, a Diplomingenieur, who runs a technical publishing firm. He fails to notice her favourable response, and she transfers her affections to his father, whom she marries. This development is a blow to Donald, although he recognizes that it is due to his own inertia. The novel ends with Donald's death at Slunj. On a visit to the waterfalls he slips and dies of heart failure.

The plot provides the opportunity to bring to life a past generation of Viennese and Austrian characters, including doctors, diplomats, lawyers, businessmen, schoolboys, farmers, society ladies, and prostitutes, a conspectus which is made more vivid by Doderer's detailed knowledge of the topography of Vienna.

WASSERMANN, JAKOB (Fürth, Nürnberg, 1873–1934, Alt-Aussee, Austria), after beginning a career in business, turned to journalism, working for a time on the editorial staff of *Simplicissimus* (q.v.) in Munich. In 1898 he moved to Vienna. For nearly forty years he was a successful novelist, popular both at home and abroad.

Die Juden von Zirndorf (q.v., 1897) was followed by *Die Geschichte der jungen Renate Fuchs* (1900), a novel of self-emancipation; *Der Moloch* (1902, revised 1921), in which the central character, Arnold Ansorge, takes his own life because he cannot accommodate himself to the corruption of the metropolis; *Alexander in Babylon* (1905), a historical novel; *Caspar Hauser* (q.v., 1908); *Die Masken Erwin Reiners* (1910); *Der Mann von vierzig Jahren* (1913); *Das Gänsemännchen* (q.v., 1915); *Christian Wahnschaffe* (q.v., 1919, 2 vols.); *Laudin und die Seinen* (1925); *Der Fall Maurizius* (q.v., 1928); *Etzel Andergast* (q.v., 1931); and *Joseph Kerkhovens dritte Existenz* (q.v., 1934). He also wrote Novellen, publishing the collections *Schläfst du, Mutter* (1896), *Der geküßte Mund* (1903), *Die Schwestern* (1906), *Der goldene Spiegel* (1911), *Der Wendekreis* (1920, 3 vols.), and the separate story *Der Aufruhr um den Junker Ernst* (1926). A posthumous novel (*Olivia*) appeared in 1937.

Wassermann, who profited by the theories of S. Freud (q.v.), was an acute psychological observer. He wrote a number of essays (including *Die Kunst der Erzählung*, 1904, *Imaginäre Brücken*, 1921, and *Reden an die Jugend über das Leben im Geist*, 1932) and biographies of Columbus (1929) and H. M. Stanley (1932). By his

residence in Austria and his early death Wassermann escaped anti-Semitic persecution, but his autobiographical essay *Mein Weg als Deutscher und Jude* (1921) emphasizes his awareness of the problem. A penetrating self-analysis is contained in *Engelhardt oder Die zwei Welten*, the title given to a novel written in 1905 but withheld from publication during Wassermann's own lifetime and that of his second wife (died 1965); it was published after the discovery of the MS. in 1973.

Gesammelte Werke (7 vols.) appeared 1944–8.

'Was wär' ich ohne dich gewesen', first line of a hymn sung in the Lutheran Church. It is No. I in *Geistliche Lieder* (1802) by Novalis (q.v.).

Waterloo, Battle of (Schlacht von Waterloo or, alternatively, Schlacht bei Belle Alliance), fought on 18 June 1815 between the French under Napoleon and British (including Hanoverian and Dutch) and Prussian forces under Wellington and Blücher (q.v.).

After Blücher's defeat at Ligny on 16 June and Wellington's check at Quatre Bras on the same day, the allied forces were separated and in danger of being overwhelmed. Wellington withdrew to a battle position at the Mont St. Jean south of Waterloo and the Prussians withdrew northward with the intention of swinging to the west to link up with the British. In the night of 17/18 June Wellington received assurance of Prussian support and determined to accept battle. From 11.30 a.m. on 18 June the French launched fierce frontal attacks, which were repulsed. Prussian support on the British left began in the afternoon, increasing steadily. By 8 p.m. the combined action of the Allies had defeated the French, whose retreat under pursuit by the Prussian cavalry became a rout. From the German point of view Waterloo constituted the final stage of the Wars of Liberation (see NAPOLEONIC WARS).

WATT, BENEDIKT VON, a native of St. Gall and an ardent Lutheran, became a Meistersinger in Nürnberg in 1591.

WEBER, CARL MARIA, FREIHERR VON (Eutin, 1786–1826, London), son of an adventurer, whose patent of nobility was spurious, was a cousin once removed of Mozart's wife Constanze. A composer and conductor, he achieved success through his contributions to the Romantic movement (see ROMANTIK).

Weber's one-act Singspiel *Abu Hassan* (Komische Oper) was performed in Munich in 1811. The libretto was by his friend F. C. Hiemer (1768–1822), who also supplied the text for Weber's well-known cradle song 'Schlaf,

Herzenssöhnchen, mein Liebling bist du!'. In 1814 he composed a patriotic cantata based on poems from Th. Körner's *Leyer und Schwerdt* (q.v.). Weber made his name with the Romantic opera *Der Freischütz* (q.v.), which was performed on 18 June 1821 in Berlin; the libretto was by F. Kind (q.v.). It was followed by two more Romantic operas, *Euryanthe* (1823), which had an inadequate libretto by Wilhelmine von Chézy (q.v.), and *Oberon, König der Elfen* (1826), a work commissioned by Covent Garden; its English libretto by James Robinson Planché (1796–1880) was translated for German performance by Th. Hell. Weber wrote the music for *Oberon* while suffering from consumption; he was in London for the first performance on 12 April 1826 and died a few weeks later. *Oberon* fell into complete neglect, but was revised by G. Mahler (q.v.), whose version was performed with success in 1913, two years after Mahler's death.

Weber's *Gesammelte Schriften*, ed. G. Kaiser, appeared in 1908.

WEBER, FRIEDRICH WILHELM (Althausen, Westphalia, 1813–94, Nieheim nr. Höxter), studied philology, then medicine (Greifswald and Breslau), qualified as a physician, and practised in the Westphalian towns of Driburg and Lippspringe. A Roman Catholic, he was elected to the Prussian Parliament in 1861 as a member of the Zentrum (q.v.), and held his seat over a number of elections.

Weber first became known in letters as a translator of English and Scandinavian poets (Tennyson's *Enoch Arden*, 1868, *Aylmer's Field*, 1870, *Maud*, 1874, and *Schwedische Lieder*, 1872). His most successful original work was an epic on a religious subject, *Dreizehnlinden* (q.v., 1878). He published *Gedichte* in 1881, a volume of religious verse, *Marienblumen*, in 1885, and another with the title *Vaterunser* in 1887. A second epic, *Goliath*, and a further volume of religious verse (*Das Leiden unseres Heilandes*) followed in 1892. A last collection of verse appeared posthumously (*Herbstblätter*, 1895). *Gesammelte Dichtungen* (3 vols.) appeared in 1922.

WEBER, KARL JULIUS (Langenburg, Württemberg, 1767–1832, Kupferzell, Württemberg), a belated supporter of the Aufklärung (q.v.), attacked the medievalism of the Romantics. His works include *Die Möncherei* (1819–20), *Das Ritterwesen* (1822–4), and two satirical causeries, *Deutschland oder Briefe eines in Deutschland reisenden Deutschen* and *Demokritos oder Hinterlassene Papiere eines lachenden Philosophen* (12 vols., published posth. 1832–40), both replete with humorous anecdotes. *Sämtliche Werke* (30 vols.) appeared

1834–49 and a selection, ed. J. Hauser, in 1966.

WEBER, MAX (Erfurt, 1864–1920, Munich), a noted economist and sociologist who developed religious sociology; he published a series of essays including *Die protestantische Ethik und der Geist des Kapitalismus* of 1920 (*Gesammelte Aufsätze zur Religionssoziologie*, 1920, repeatedly reprinted). A democratic politician (*Gesammelte politische Schriften*, 1921, repr. 1971), he shared common ground with F. Naumann (q.v.).

Weber, Die, a Naturalistic play in five acts by G. Hauptmann (q.v.), published in 1892 and first performed by the Freie Bühne on 23 February 1893. First drafted in Silesian dialect, it was rewritten in High German with a strong flavour of dialect; both versions were published in 1892 (*De Waber* and *Die Weber*), the later version being the standard form used on the stage and printed in Hauptmann's works. Sub-titled *Schauspiel aus den vierziger Jahren*, it is dedicated to Hauptmann's father, whose oral account of the Silesian weavers' riots in 1844 was the germ of the play. Against this historical background it presents contemporary social tensions, and introduces a collective group, the weavers, instead of an individual, as protagonist.

The weavers, in the depths of poverty and hunger, are exploited by the businessman Dreißiger. In their despair they are egged on to revolt by the former soldier Moritz Jäger. Dreißiger's house is stormed and looted, the military are called in, and a company is driven back by the rioters. During the combat the old man Hilse, who is opposed to the rising, is killed by a stray bullet.

Wechsel, a form of indirect dialogue occurring in Minnesang (q.v.).

Wechselgesang, a song in the form of a dialogue between two persons. It occurs in folksongs and occasionally in later lyric poetry, e.g. in Goethe's *West-östlicher Divan* and Mörike's poem *Gesang zu zweien in der Nacht* (qq.v.).

WECKHERLIN, GEORG RUDOLF (Stuttgart, 1584–1653, London), German poet, was the son of a civil functionary in Württemberg. He studied at Tübingen, travelled in Germany, and visited Paris, after which he spent some time in London, marrying in 1616 an Englishwoman, Elizabeth Raworth. He held administrative office in Stuttgart and in 1620 joined the service of the exiled Elector Palatine, Friedrich V (q.v.), in London, where he settled for life, taking office in British administration. He became a 'Secretary for Foreign Tongues' in 1625 and was succeeded in 1649 by John Milton, whom Weckherlin later, because of Milton's increasing blindness, assisted from 1652 until his death.

Weckherlin was an early baroque poet, who for geographical reasons remained untouched by the influence of Opitz (q.v.). His poems include many festal and funeral odes with extensive mythological apparatus and a rhetorical technique exploiting accumulation and antithesis. His principal collections are *Das erste Buch Oden und Gesänge* (1618), followed in 1619 by *Das ander Buch Oden und Gesänge. Geistliche und Weltliche Gedichte* appeared in 1641. Weckherlin's poetry, perhaps because its author lived abroad, fell into obscurity, and was rediscovered in the late 18th c. by Herder (q.v.).

A Kritische Gesamtausgabe was edited by H. Fischer (3 vols., 1894–1907, reprinted 1968).

WEDEKIND, CHRISTOF FRIEDRICH, see KRAMBAMBULI, DER.

WEDEKIND, FRANK (Hanover, 1864–1918, Munich), was brought up in Switzerland. His father was a doctor, his mother had been an actress. Young Wedekind travelled and became familiar with modern art and ideas. He was in advertising for two years (1866–8) and for a time was a secretary with a circus (1888). He next turned to literature, to acting and producing plays, and was also a cabaret artist, performing with Die elf Scharfrichter in Munich. Wedekind was very much an individualist, and his plays, which constitute the greater part of his output, aroused much controversy. They can now be seen to anticipate the theatre of the absurd.

Influenced by Naturalism, Wedekind went beyond it in the direction of Expressionism (see NATURALISMUS and EXPRESSIONISMUS). His first play was *Frühlings Erwachen* (q.v., 1891), *Erdgeist* (q.v.) followed in 1895, its sequel *Die Büchse der Pandora* (q.v., and see BERG, A.) in 1904; other plays include *Der Liebestrank* (1899, also entitled *Fritz Schwigerling*), *Der Kammersänger* (q.v., 1899) and *Der Marquis von Keith* (q.v., 1901, two Shavian plays of argument), *So ist das Leben* (1902, retitled *König Nicolo oder So ist das Leben*, 1911), *Hidalla oder Sein und Haben* (1904, retitled *Carl Hetmann, der Zwergriese*, 1911), *Totentanz* (1906, retitled *Tod und Teufel*, 1909), *Oaha* (1908, retitled *Till Eulenspiegel*, 1916), the drama *Musik, Die Zensur* (both 1908), *Schloß Wetterstein* (1910), *Franziska* (1911), *Simson oder Scham und Eifersucht* (1914), *Bismarck* (1916), and *Herakles* (1917). Wedekind published a few narrative works, of which *Mine-Haha oder Über die körperliche Erziehung der jungen Mädchen* (1903, reprinted 1955) is the best known. *Die Fürstin*

Russalka (1897) is a novel and *Feuerwerk* (1906) a collection of stories. In 1905 Wedekind published a volume of poetry, *Die vier Jahreszeiten*; his *Lautenlieder*, ed. A. Kutscher and H. R. Weinhöppel, appeared in 1920.

Both as a cabaret artist and as a writer and contributor to the Munich *Simplicissimus* (q.v.), Wedekind indulged in *risqué* satire, and in 1899 was imprisoned for insulting the Crown. His works were attacked as immoral, although he was an author with a message, the morality which he advocated being a carnal one. The crassness with which he constructs his plays and conducts their dialogue is designed to shock his middle-class audience.

Gesammelte Werke (9 vols.), ed. A. Kutscher and R. Friedenthal, appeared 1912–21, *Gesammelte Briefe* (2 vols.), ed. F. Strich, 1924, and *Der vermummte Herr. Briefe Frank Wedekinds aus den Jahren 1881–1917*, ed. W. Rasch, 1967.

Weder Kaiser noch König, see DIE THRONE STÜRZEN.

WEERTH, GEORG (Detmold, 1822–56, Havanna), a political journalist, met F. Engels and K. Marx (qq.v.) on journeys abroad and worked on the communistic *Neue Rheinische Zeitung*. He wrote satirical fiction (*Humoristische Skizzen aus dem deutschen Handelsleben*, 1845–8; *Leben und Taten des berühmten Ritters Schnapphanski*, 1849). Weerth's *Sämtliche Werke* (5 vols.), ed. B. Kaiser, were published 1956–7.

Weg ins Freie, Der, a Viennese novel by A. Schnitzler (q.v.), published in 1908. It displays a society in which well-to-do members of a predominantly Jewish middle class mingle with the liberal-minded fringe of the aristocracy. The principal character is non-Jewish: Georg von Wergenthin, a composer without sufficient drive to make anything of his talent; the plot centres on his long and discreet love-affair with a non-Jewish middle-class girl, Anna Rosner, which, after Anna's pregnancy and a still-birth, ends in frustration and resignation.

The real substance of the novel is the vivid presentation of social functions and discussions, in the course of which a variety of characters are introduced; Georg's admirable brother Felician, Heinrich Bermann (largely a self-portrait of Schnitzler), the Zionist Leo Golowski and his sister Therese, the aristocratic-looking, impeccably dressed Willy Eißler, who reminds all comers of his Jewish descent, the hussar officer Demeter Stanzides (who also appears in the play *Das weite Land*, q.v.), the orthodox Jewish Ehrenbergs, whose son apes the nobility, the littérateur Edmund Nürnberger, and the touchy Jewish parliamentary deputy Dr. Berthold.

The topography, dialect, characters, and local problems create an authentic Viennese atmosphere in which the most urgent matters are anti-Semitism and the psychology of love.

Wegkürtzer, a collection of Schwänke (see SCHWANK) by Martin Montanus (q.v.).

Weg nach Innen, a collection of stories by H. Hesse (q.v.), all of which had been previously published. It appeared in 1931 and contains *Siddhartha. Eine indische Dichtung, Kinderseele, Klein und Wagner,* and *Klingsors letzter Sommer* (qq.v.). A brief *Nachwort* by Hesse on the genesis of his work includes the statement 'ich stehe zum *Steppenwolf* nicht minder als zum *Siddhartha*', and a tribute to Romain Rolland and Wilhelm Gundert 'der am tiefsten in das Denken des Ostens eingedrungen ist'.

Weh dem, der lügt!, a blank-verse comedy (Lustspiel) in five acts by F. Grillparzer (q.v.), which was first performed on 6 March 1838 in the Burgtheater in Vienna. The performance was a disastrous failure partly because of unsuitable casting, partly because the audience misunderstood Grillparzer's intentions; the nobility in particular, believing themselves caricatured, left the house noisily, slamming the doors of their boxes. The successful revival of the play under the direction of F. Dingelstedt (q.v.) in 1876 assured its future on the stage. But Grillparzer never recovered from the shocks of the spring of 1838, in which he felt himself deserted by both public and stage and ridiculed by the press. *Weh dem, der lügt!* remained the last play he offered for public performance. It was first published in 1840.

Grillparzer used as his main source the *Historia Francorum* by Gregory of Tours (6th c.) which he read either in the Latin original or in Augustin Thierry's French version (*Lettres sur l'histoire de France*). One of his modifications is the substitution for the Franks of the Gallo-Romans, who represent their superior and Christian civilization, and the invention of the heathen Rheingau barbarians of Germanic origin. The moral element of the play, indicated by the title, harmonizes with the tradition of the Viennese Volksstück (q.v.), but the title itself may be inspired by Lope de Vega and the predilection of Spanish comedy for using moral maxims as titles. As the title indicates serious reflection rather than light-hearted amusement, Grillparzer's insistence on using the designation Lustspiel on the programme in preference to Schauspiel was regretted by H. Laube (q.v.) as it may well have produced the wrong effect upon the first-night audience, which had come for a good laugh.

At the onset of the action, Leon, the kitchen-hand of Bishop Gregor at his palace at Dijon, asks for leave so that he may attempt to free the bishop's nephew Atalus, held as hostage by Kattwald, a count in the Rheingau. The bishop gladly accepts the courageous offer, but on condition that Leon achieves his mission without having recourse to lying. Leon finds his way to Trier and gets a pilgrim to sell him as a slave to Kattwald, whom he serves as an expert cook. He succeeds in his plan because his frank admission that he intends to flee is in effect taken as a lie. The whole play turns on this idea. Only Kattwald's daughter, Edrita, is not deceived by the truth and, by her own deceit, secures the key to the gates without which the flight cannot succeed. To escape her father's anger and marriage with her unloved, simple-minded kinsman Galomir, she joins Leon and Atalus. When they reach Metz the pursuers catch up with them, but, as if by God's own miracle, the town has been taken overnight by the Franks. Atalus is reunited with Gregor, and Leon, raised to higher rank, is to marry Edrita, who is converted to Christianity.

WEHNER, JOSEPH MAGNUS (Bermbach, Rhön, 1891–1973, Munich), the son of a schoolmaster, was at school in Fulda, and studied at Jena and Munich universities. He served in the 1914–18 War and was severely wounded at Verdun in 1916. He worked for most of his life as a journalist and writer in Munich, living during the 1939–45 War in Tutzing. His first work was a religious epic, *Der Weiler Gottes* (1920), which is written in hexameters and has a rural setting; it was followed by the play *Das Gewitter* (1926) and the novels *Der blaue Berg* (1922) and *Die Hochzeitskuh* (1928). Wehner achieved his first great success with the war novel *Sieben vor Verdun* (q.v., 1930), after which appeared another war novel, *Stadt und Festung Belgerad* (1936). The author of biographies, he published *Struensee* (1924, extended 1938), *Schlageter* (1934), *Hindenburg* (1936), and *Hebbel* (1938).

In the 1950s Wehner published three more novels, *Mohammed* (1952), *Der schwarze Kaiser* (1953), and *Der Kondottiere Gottes* (1956), and a series of religious plays, *Johannes der Täufer* (1953), *Das Fuldaer Bonifatiusspiel, Das Rosenwunder, Saul und David* (all 1954), and *Die aber ausharren bis ans Ende* (1956).

Wehrwolf, see WERWOLF.

Wehrwolf, Der, a novel by H. Löns (q.v.).

Weiber von Weinsberg, Die, title of two ballads, one by G. A. Bürger (published in J. H.

Voß's *Göttinger Musenalmanach 1777*), and the other by A. von Chamisso (published in *Deutscher Musenalmanach*, 1833 under its original title *Die Weiber von Winsperg*).

The ballads deal with the legendary rescue of the men of Weinsberg, who are carried out of their captured fortress on the backs of their wives, who have been given safeconduct. The legend refers to the siege of Weinsberg by Konrad III (q.v.) in 1140, and in allusion to it the ruined castle is known as Weibertreu. The story has also been treated by K. Immermann in *Münchhausen* (q.v., 1838–9) and by F. von Gaudy (q.v.).

WEIDIG, FRIEDRICH LUDWIG (Oberkleen nr. Wetzlar, 1791–1837, Darmstadt), a schoolmaster at Butzbach near Gießen and later pastor at Obergleen near Alsfeld, leader of the illegal Liberal party in Hesse. In the course of his subversive activities he had contacts with many movements, among them the Burschenschaften (q.v.) and the Frankfurter Gruppe, whose planned revolt came to the notice of the police and led to Weidig's arrest.

Weidig was the author of the clandestine pamphlet *Leuchter und Beleuchter für Hessen*. Early in 1834 G. Büchner (q.v.) joined his circle of conspirators, and Weidig helped him in the production of *Der Hessische Landbote* (q.v.). Betrayed by one from his own ranks, Weidig was arrested again in August 1834 and kept in prison without trial. He committed suicide in his cell in 1837. His poems were published posthumously in 1847.

WEIDMANN, PAUL (Vienna, 1744–1810, Vienna), a Viennese civil servant, wrote plays and poetry. His dramatic works include *Anna Boulen* (i.e. 'Anna Boleyn', 1771), *Dido* (1771), and *Adelhaid* (1772), as well as the historical play *Stefan Fädinger oder der Bauernkrieg* (1777). He wrote a serious epic, *Judith* (1773), and a comic poem, *Die Nonnenschlacht* (1782). Joseph Weidmann (1742–1810), who wrote a popular Singspiel, *Der Dorfbarbier* (1796), was his brother.

WEIGEL, HANS (Vienna, 1908–), a dramatic critic and a satirist, lived in Vienna and Maria-Enzersdorf, except for the period 1938–45, spent in political exile in Switzerland, and in his later years moved to Maria-Enzersdorf. He is the author of the tragedy *Barrabas und der fünfzigste Geburtstag* (1946), the comedy *Das wissen die Götter* (1947), the tragi-comedy *Hölle oder Fegefeuer* (1948), and of the novels *Der grüne Stern* (1946) and *Unvollendete Symphonie* (1951). In 1954 he published a Novelle, *Das himmlische Leben*. His essays, criticism, and other prose works include *Masken, Mimen und Mimo-*

sen, *O du mein Österreich* (both 1958), *Flucht vor der Größe* (1960), *Tausend und eine Première*. *Wiener Theater 1946–1961* (1961), *Lern dies Volk der Hirten kennen* (1962), *Apropos Musik* (1965), *Karl Kraus oder Die Macht der Ohnmacht* (1968), *Götterfunken mit Fehlzündung* (1971), *Die Leiden der jungen Wörter*. *Ein Antiwörterbuch* (1974), *Der exakte Schwindel oder Der Untergang des Abendlandes durch Zahlen und Ziffern* (1977), *Das Land der Deutschen mit der Seele suchend* (1978), *Gerichtstag vor 49 Leuten. Rückblick auf das Wiener Kabarett der dreißiger Jahre* (1981), *Große Mücken, kleine Elefanten*. *Vierzig Plädoyers für das Feuilleton* (1980), *Ad absurdum. Satiren, Attacken, Parodien aus drei Jahrzehnten* (1980), and *Das Schwarze sind die Buchstaben. Ein Buch über dieses Buch* (1980). *In memoriam* (1979) is a volume of reminiscences.

WEIGEL, HELENE (HELLI) (Vienna, 1900–71, East Berlin), the actress-wife and collaborator of B. Brecht (q.v.), to whose productions she devoted her remarkable talents after acting under Max Reinhardt (q.v.) at the Deutsches Theater, Berlin; her transition to the Epic Theatre (see EPISCHES THEATER) was referred to by Brecht as the 'Abstieg der Weigel in den Ruhm' (*Der Messingkauf, 3. Nacht*). Her co-operation with Brecht began with *Baal* (q.v.) in a production of 1926. She became particularly famous for her portrayal of Brecht's mother-figures, notably in *Die Mutter* (q.v.), performed in 1932 before she shared Brecht's years of exile, and in *Mutter Courage und ihre Kinder* (q.v.) in Berlin in 1949 and later in Paris and London. She last came into the limelight with the part of Volumnia in *Coriolan* (an adaptation by Brecht of Shakespeare's play) in 1964.

Helene Weigel was co-founder of the Berliner Ensemble (q.v.), which she directed after Brecht's death in 1956 until her own death.

WEIGEL, VALENTIN (Naundorf, Saxony, 1533–88, Zschopau, Erzgebirge), Lutheran pastor at Zschopau from 1567 to his death, wrote mystical books influenced by Paracelsus (q.v.). They were published long after his death and were condemned and publicly burned in 1624. They comprise *Von der seligmachenden Erkenntnis Gottes* (1613), *Nosce te ipsum* (1615), *Dialogus de Christianismo* (1616), *Der güldene Griff* (1616), and *Kirchen- oder Hauspostill* (1618).

A *Gesamtausgabe*, ed. W. E. Peuckert and W. Zeller, appeared 1962 ff.

Weihe der Kraft, Die, frequently used subtitle of Z. Werner's play *Martin Luther oder Die Weihe der Kraft* (q.v.).

Weihe der Unkraft, Die, a document written in 236 lines of Alexandrine verse by Z. Werner

(q.v.). It was published in 1814 as *Ein Ergän-
zungsblatt zur deutschen Haustafel*. Werner wrote
this verbose address to the German people as a
recantation of his tragedy *Martin Luther, oder Die
Weihe der Kraft* (q.v.), an act which had been
made a condition of his admission to Roman
Catholic holy orders. Not surprisingly, it set the
seal on Goethe's break with Werner.

Weiher, Der, a group of five poems by Annette
von Droste-Hülshoff (q.v.), included in *Heidebilder* (*Gedichte*, 1844).

WEILL, ALEXANDER (Schirrhofen, Alsace,
1811–99, Paris), wrote a volume of stories *Sittengemälde aus dem elsässischen Volksleben* (1843), followed by *Berliner Novellen* and *Knüttelreime eines
elsässer Propheten* (1843).

WEILL, KURT (Dessau, 1900–50, New York),
composed an opera to a libretto by G. Kaiser
(q.v.), *Der Protagonist* (1926), but his experimental style only came into its own with the
music for Brecht's *Dreigroschenoper* (q.v., 1928),
Aufstieg und Fall der Stadt Mahagonny (q.v., 1930),
Der Jasager und Der Neinsager (q.v., 1929 and
1930), and *Happy End* (1929). He also wrote, in
collaboration with P. Hindemith (q.v.), a cantata for Brecht's *Der Lindberghflug* (1929, see DER
OZEANFLUG). Kaiser wrote the libretto for *Der
Zar läßt sich photographieren* (1927), and Weill
composed the music to Kaiser's play *Silbersee*
(1933).

Weill's compositions for other German works
include *Die Bürgschaft* (1932), for which Caspar
Neher (q.v.) wrote the text, *Die sieben Todsünden*,
and a ballet written in collaboration with Brecht
and performed in 1933 under the title *Anna-Anna*
in London. Weill set poems by R. M. Rilke
(q.v.) to music, and composed incidental music
to works by F. Werfel, A. Strindberg, A. Bronnen, and L. Feuchtwanger (qq.v.). In 1935 he
settled in New York. He continued to work for
the two types of opera which were strongly
advocated by Brecht, the Volksoper, conceived
in reaction against traditional operas and the
Musikdrama of R. Wagner (q.v.), and the
Schuloper.

Weimar, small town, which was the capital of
the tiny state of Saxe-Weimar (see SACHSEN-
WEIMAR). In the late 18th c. the ruler Karl
August (q.v.) invited Goethe to stay indefinitely
in Weimar, and the town remained Goethe's
home for 57 years (1775–1832). Other notable
literary personalities invited to Weimar or
present there included Schiller, Wieland, and
Herder (qq.v.). This literary court is often
termed 'Der weimarische Musenhof' in German
histories of literature.

In the 19th c. Weimar was from 1844 to 1860
the residence of the composer Franz Liszt (q.v.).
In 1919 it was the meeting place of the National
Assembly (Nationalversammlung), which was
convened there in order to symbolize the links
between the new Germany of the republic about
to be created, and the idealism of Goethe and
Schiller. From these beginnings derives the term
Weimar Republic (q.v.). See also ANNA AMALIA;
EINSIEDEL, F. H. VON; GÖCHHAUSEN, LUISE VON;
KNEBEL, K. L. VON; LUISE, HERZOGIN; SCHRÖ-
TER, CORONA; and STEIN, CHARLOTTE VON.

Weimar Republic (Weimarer Republik),
usual designation of the German Empire
(Deutsches Reich) in its republican form 1919–
33. The preparatory National Assembly met in
January 1919, and a democratic constitution
drafted by H. Preuß (q.v.) was adopted. The
first president was F. Ebert (q.v.). The elections
produced no over-all majority for any party.

The republic began under unfavourable
auspices. The government had to sign the peace
treaty (see VERSAILLES, TREATY OF) and incurred opprobrium for its oppressive terms.
It had to defeat open insurrection in 1919, 1920,
and 1923 (see SPARTAKUSBUND, KAPP-PUTSCH,
HITLERPUTSCH), and to cope with reparations,
French occupation of the Ruhr (1923), and
disastrous inflation (1922–3). In 1924 the foreign
policy of G. Stresemann (q.v.) brought about an
improvement in the standing of the republic.
Ebert died in 1925 and was succeeded as president by Field-Marshal von Hindenburg (q.v.).
In 1926 Germany was admitted to the League
of Nations (Völkerbund). The multiplicity of
parties prevented the formation of a stable
government, and ministries fell in rapid succession. Finally parliamentary government became impossible, and the last democratic chancellor, H. Brüning (q.v.), ruled by invoking
Article 48 of the constitution by which the
President was enabled to issue decrees.

The Weimar Republic came to an end when
Hitler (q.v.) was appointed chancellor on 30
January 1933.

WEINERT, ERICH (Magdeburg, 1890–1953,
Berlin), originally a fitter, trained as an engineer
and machine draughtsman. Having served in
the 1914–18 War, he worked in a number of
jobs, as a teacher and as a cabaret artist and
actor, and in 1924 joined the Communist party;
he supported the party as a journalist (co-
founder of *Linkskurve*, 1929) and as a writer with
a gift for terse political poetry and prose (*Der
Gottesgnadenknecht und andere Abfälle*, 1923, *Affen-
theater*, 1925, and *Politische Gedichte*, 1928). He
was a contributor to the volume *Rote Signale*
(q.v.), published in 1931, the year in which the

Prussian government, by what became known as the 'Lex Weinert', barred him for a few months from speaking at public assemblies. *Alltägliche Balladen* appeared in 1933, when he emigrated to France, publishing *Es kommt der Tag* in 1934. In 1935 he went to Russia, fought in the Spanish Civil War, returned to Russia in 1939, and became president of the Nationalkomitee Freies Deutschland (q.v., 1943–5). A war reporter at Stalingrad, he published *Erziehung vor Stalingrad. Fronttagebuch eines Deutschen* in 1943.

In 1946 Weinert returned to East Germany and held office in the education department of the DDR. His poetry of this last phase opens with *Rufe in die Nacht* (1947), followed by a collection *Das Zwischenspiel. Deutsche Revue 1918–1933* (1950). His collections of essays and stories appeared as *Der Tod fürs Vaterland* (1942) and *Camaradas* (1951). The first posthumous publication was *Nachgelassene Lyrik aus drei Jahrzehnten* (1960). His *Gesammelte Werke* (9 vols.), ed. L. Weinert and A. Kantorowicz, appeared 1955–60, and *Gesammelte Gedichte*, ed. L. Weinert, E. Zenker *et al.*, 1970 ff.

Weingartner Liederhandschrift, a manuscript of courtly lyric poetry (Minnesang) with pictures of poets. It is shorter than the Heidelberg MS. (see HEIDELBERGER LIEDERHANDSCHRIFT, GROSSE), and contains poems attributed to 31 poets, 25 of whom are portrayed as the artist-monk imagined them to have looked. The MS. is early 14th-c. work, executed in the Abbey of Constance. Some poems were added later, including *Der Minne Lehre* (q.v.). It was already in the library of Weingarten Abbey in the 17th c., but the date of transfer is not known. In 1843 it was removed to Stuttgart, hence the occasional designation as Stuttgarter Handschrift. Facsimiles were published in 1931 and (limited edition) 1969.

Weingartner Reisesegen, a blessing for a journey, written in rhyming verse and resembling in its dual form Old High German magic spells (see ZAUBERSPRÜCHE). It was doubtless believed to have a magical protective effect. The MS., in Stuttgart, was formerly in Weingarten Abbey. It dates from the 12th c. That its origin is much earlier is testified by the alliterative features of the verse.

WEINHEBER, JOSEF (Vienna, 1892–1945, Kirchstetten nr. St. Pölten), having lost both parents in his early childhood, was sent to an orphanage at Mödling, an experience which forms the basis of his school novel *Das Waisenhaus* (1924). In 1911 he became an official in the Post Office. Originally a Roman Catholic, he adopted the Protestant faith in 1927.

Weinheber's poetry first attracted serious attention in 1934. Influenced by Hölderlin and Rilke (qq.v.), he achieved work of distinction in many lyric forms. Each volume of verse has an underlying unity, and he showed a predilection for cycles, in which the same form (sonnet, classical metre, *terza rima*, or strophic song) is repeated. His disciplined poetry reflects with sadness a declining world and yet maintains an optimistic humanism, which is most clearly seen in *Zwischen Göttern und Dämonen* (1938). The early verse included *Der einsame Mensch* (1920), *Von beiden Ufern* (1923), and *Boot in der Bucht* (1926). *Adel und Untergang* (q.v., 1934) established Weinheber's reputation. *Wien wörtlich* (1935) contains some remarkably effective poems in dialect. The climax of his later verse, which included *Späte Krone* (1936), was the volume *Kammermusik* (q.v., 1939). Posthumous verse (*Hier ist das Wort*) appeared in 1947. Weinheber, who had accepted the National Socialist regime, appears to have taken an overdose of sleeping tablets when Russia began to invade Austria.

Posthumous writings, including the novel *Gold außer Kurs*, were published in *Sämtliche Werke* (5 vols.), ed. J. Nadler and H. Weinheber, 1953–6 and, based on this edition, ed. F. Jenaczek, 1970 ff. (planned for 7 vols.).

WEININGER, OTTO (Vienna, 1880–1903, Vienna), published in 1903 *Geschlecht und Charakter*, a treatise asserting the intellectual superiority of man over woman. Though Weininger was himself of Jewish descent, the book, which was widely read, depreciated the character of the Jewish race. Weininger took his own life in the year of publication. *Über die letzten Dinge* (1904) and *Die Liebe und das Weib* (1917) appeared posthumously.

Weinschwelg, Der, a short Middle High German poem in the form of a monologue in praise of wine. Each section of discourse culminates in

> dô huob er ûf unde tranc,
> (Da hob er auf und trank)

and the compelling and imaginative poem proceeds by accumulation. *Der Weinschwelg* was written *c.* 1250 in Tyrol, probably by the poet who was the author of *Die böse Frau* (q.v.).

Weinspiel, Das, see MANUEL, HANS RUDOLF.

WEISE, CHRISTIAN (Zittau, 1642–1708, Zittau), a schoolmaster's son, studied at Leipzig, where he also lectured for a time. In 1668 he became secretary to the Count of Leiningen at Halle and in 1670 became a senior teacher (Professor) at the grammar school at Weißenfels. From 1678 until his death he was the headmaster of the grammar school in his native town, Zittau.

Weise's first publication was a collection of didactic poems, *Der grünenden Jugend überflüssige Gedanken* (2 vols., 1668–74); it was followed by *Nothwendige Gedanken der grünenden Jugend* (1675). He also turned his attention to the novel, writing a cycle of three novels, in which he reviews the follies and faults of the world from the standpoint of his own utilitarian ideal of the 'political' (that is worldly-wise yet honest) man. These novels are *Die drei Hauptverderber in Deutschland* (1671), *Die drei ärgsten Ertz-Narren in der ganzen Welt* (1672, reissued 1967), and *Die drei klügsten Leute in der ganzen Welt* (1675). The second and third are more closely linked, having certain characters in common. A fourth novel on the 'political' theme is *Der politische Näscher* (1678).

As headmaster in Zittau Weise took to playwriting. He introduced theatrical production into the curriculum, and between 1679 and 1705 himself wrote 61 plays for his top form. The plays were performed as a kind of festival on three successive days; one of the three plays was biblical, one historical, and the third 'free', and this usually took the form of a comedy. His best-known tragedy is the *Trauer-Spiel von dem Neapolitanischen Haupt-Rebellen Masaniello* (1683, ed. F. Martini, 1972), which was praised by Lessing. His comedies mock the peasantry (*Bäurischer Macchiavellus*, 1679, ed. W. Schubert, 1966; *Schauspiel vom niederländischen Bauer*, 1685) and are especially known for the extemporizing comic figure of Pickelhering (q.v.). Other plays deserving of mention are *Der Tochter-Mord* (1679), *Abraham und Isaac* (1682), and *Die böse Catharina* (1693, derived indirectly from *The Taming of the Shrew*). Weise also wrote a treatise on verse, *Curiöse Gedancken von Deutschen Versen* (1691–3).

Sämtliche Werke, ed. J. D. Lindberg, appeared from 1971.

WEISENBORN, GÜNTHER (Velbert, Rhineland, 1902–69, Berlin), studied medicine and German philology, lived for a time in Berlin before emigrating to Argentina in 1930. Keenly interested in politically orientated drama and the experimental stage, he first made his name as a playwright with the pacifist *U-Boot S4* (1928), followed by *SOS oder die Arbeiter von Jersey* (1929). A sympathizer of B. Brecht (q.v.), he collaborated in *Die Mutter* (q.v.) and published his first novel, *Barbaren*, in 1931. In 1933 his works were banned, and he published for some years under the pseudonyms Eberhard Foerster and Christian Munk. His play *Die Neuberin* (revised 1950) and the novel *Das Mädchen von Fanö* appeared in 1935.

In 1937 Weisenborn returned to Germany, joined Die rote Kapelle (q.v.) and, for his association with early plots against Hitler, was imprisoned in 1942, being released in 1945. In post-war Germany he resumed his work for the theatre, was co-founder of the Hebbel-Theater, Berlin, and in the early 1950s a producer in charge of the Kammerspiele, Hamburg. His play *Die Illegalen* (1946) is based on his experience during the National Socialist regime; its successors were *Die spanische Hochzeit* (1949), *Drei ehrenwerte Herren* (1953), *Fünfzehn Schnüre Geld* (1959), *Das Glück der Konkubinen* (1965), and the posthumously published *Die Familie von Makabah* (1970). In the early 1960s he was co-author of the script for the film version (1963) of *Die Dreigroschenoper* (q.v.).

After the autobiographical *Memorial* (1947), Weisenborn, using material collected by R. Huch (q.v.), wrote an account of the German Resistance Movements (q.v.), entitled *Der lautlose Aufstand* (1953). After *Der dritte Blick* (1956) he published *Der Verfolger*, his last novel, and *Am Yangtse steht ein Riese auf* (both 1961) and *Der gespaltene Horizont* (1964). The collection *Theater* (4 vols.) appeared 1964–7.

Weise vom Leben und Tod des Cornets Christoph Rilke, Die, a short narrative in poetic prose by R. M. Rilke (q.v.), published in 1903. It recounts the death in action of Cornet Rilke of Langenau, standard-bearer of an Austrian regiment of horse in Hungary during the Turkish war of 1663. Rilke implies that he and the Cornet, a historically authenticated person, belonged to the same ancestral line, but although the basis of the poem is historical, this relationship is improbable. The narrative was for many years one of Rilke's most popular works.

WEISFLOG, KARL (Sagan, 1770–1828, Warmbrunn), a magistrate in Sagan, was the author of ephemeral light literature once popular with the clients of circulating libraries. His principal collection is *Phantasiestücke und Historien* (12 vols., 1824–9). One of his stories (*Das große Loos*, 1827) was the source of J. N. Nestroy's *Der böse Geist Lumpazivagabundus* (q.v.).

WEISHAUPT, ADAM (Ingolstadt, 1748–1830, Gotha), became a professor at Ingolstadt University in 1772. In 1776 he founded the Illuminatenorden (q.v.), which was proscribed in 1784. Weishaupt, who was later in court employment at Gotha, defended his Order in *Vollständige Geschichte der Illuminaten in Bayern* and *Apologie der Illuminaten* (both 1786). References to Weishaupt and his Order occur in T. L. Peacock's *Nightmare Abbey* (1818).

Weisheit des Brahmanen, Die. Ein Lehr-gedicht in Bruchstücken, a philosophical poetical work by Friedrich Rückert (q.v.), published in 1836–9. It consists of a large number of poems of varying length, composed of from 2 to 72 distichs. These are arranged in twelve books termed *Stufen,* which attain a climax in the last, which is headed *Frieden.*

WEISMANTEL, LEO (Obersinn, Rhön, 1888–1964, Rodalben Palatinate), a prolific writer and an educationist, published his first novel, *Mari Madlen,* in 1918, and *Das unheilige Haus* in 1922. In 1925 appeared the work which gave the name to the school he founded in Marktbreit/ Main in 1928, *Die Schule der Volkschaft;* he explained his theories on education in further works, including *Der Geist als Sprache* (1927), *Vom Willen deutscher Kunsterziehung* (1929), and *Über die geistesbiologischen Grundlagen des Lesegutes der Kinder* (1931). A Roman Catholic, he was a member of the Reichstag for the Zentrum (q.v.) party.

In 1933 he completed the publication of a trilogy of novels, *Vom Leben und Sterben eines Volkes,* which was republished with new titles for each volume in 1943 (*Das alte Dorf,* 1928, re-titled *Das Jahr vom Sparbrot; Die Geschichte des Hauses Herkommer,* 1932, retitled *Die Leute von Sparbrot;* and *Das Sterben in den Gassen,* 1933, re-titled *Tertullian Wolf*). The trilogy was followed by *Rebellen in Herrgotts Namen* (1931, retitled *Der Vorläufer,* 1941) and *Gnade über Oberammergau* (1934). In 1935 Weismantel's school was closed, and in 1936 he published an autobiography (*Mein Leben,* followed in 1940 by *Jahre des Werdens*). At this period his interest in novels on artists became manifest in a number of publications; they make some pretensions to learning and are encumbered with additional informative material and include *Dill Riemenschneider* (1936, see RIEMENSCHNEIDER, T.), *Leonard da Vinci* (1938, earlier as a play, 1925), *Gericht über Veit Stoß* (1939), *Mathis-Nithart, der fälschlich Matthias Grünewald genannt wurde* (3 vols., 1940–3), and two works on A. Dürer (q.v.), *Albrecht Dürers Brautfahrt in die Welt* (1950) and *Albrecht Dürer, der junge Meister* (1950).

In the 1950s Weismantel resumed the publication of works on education (*Vom Wesen der Ganzheit in der Fibelfrage,* 1951, *Jugend und Schule in der Bundesrepublik,* 1958), and, having published a few Expressionist plays in the 1920s, wrote a similar work, *Das Interview* (1961).

WEISS, ERNST (Brünn, now Brno, 1884–1940, Paris), a doctor, served in the 1914–18 War and in 1938 emigrated to France. He committed suicide before the German entry into Paris. He is the author of fiction showing Expressionistic

and surrealist tendencies, owing something to Kafka (q.v.), whom he knew. His work frequently expresses violent perverted sexual impulses. His novels include *Die Galeere* (1913), *Der Kampf* (1916), *Tiere in Ketten* (1918), *Mensch gegen Mensch* (1919), *Stern der Dämonen* (1920), *Nahar* (1922), *Männer in der Nacht* (1925), *Boetius von Orlamünde* (1928, retitled *Der Aristokrat* in 1966), *Georg Letham. Arzt und Mörder* (1931), *Der Geisterseher* (1934), *Der Gefängnisarzt* (1934, reprinted 1969), *Der arme Verschwender* (1936, reprinted 1965), and *Der Verführer* (1937). *Der Augenzeuge* was published posthumously in 1963.

WEISS, KARL, see KARLWEIS, C.

WEISS, KONRAD (Gaildorf, Württemberg, 1880–1940, Munich), a Roman Catholic journalist and art critic, wrote mystical religious poetry and also, though with little success, tried his hand at drama. The poems appeared as *Tantum dic verbo* (1919), *Die cumäische Sibylle* (1921), *Die kleine Schöpfung* (1926), *Das Sinnreich der Erde* (1939). A collected edition was published posthumously (*Gedichte,* 2 vols., 1948–9). The plays are *Das kaiserliche Liebesgespräch* (1934) and *Konradin von Hohenstaufen* (1938). *Dichtungen und Schriften,* ed. F. Kamp, appeared in 1961.

WEISS, PETER, in full Peter Ulrich (Nowawes, Berlin, 1916–82, Stockholm), emigrated with his parents in 1934 and studied art in Prague (1936–8). After a year in Switzerland he settled in Sweden, taking Swedish nationality by naturalization in 1946. He temporarily returned to Berlin in 1962. For some years he worked primarily as a graphic artist and in the production of films. As a writer he began in middle life with idiosyncratic novels. *Der Schatten des Körpers des Kutschers* (1960) limits itself to recording observations, deliberately abstaining from suppositions on psychological occurrences. The title refers to the last episode, a coition observed in silhouette from outside the room in which it takes place. *Abschied von den Eltern* (1961) and *Fluchtpunkt* (1962) are autobiographical novels concerned with problems of emigration.

Weiss is best known for the plays *Die Verfolgung und Ermordung Jean Paul Marats dargestellt durch die Schauspielgruppe des Hospizes zu Charenton unter Anleitung des Herrn de Sade* (q.v., 1964, fifth version 1965), commonly referred to as *Marat/ Sade,* and *Die Ermittlung* (q.v., 1965), a documentary dealing with the mass murder of Jews in the 1939–45 War. The Marxist stance adopted by Weiss after completing the first version of *Marat/Sade* is also reflected in the play *Gesang vom lusitanischen Popanz* (1967), dealing with Portuguese colonial policy, and the documentary drama *Diskurs über die Vorgeschichte und den Verlauf des lang andauernden Freiheitskrieges in Viet Nam ...* (1968). His other plays include works written soon after the war but published later (*Der Turm*, 1963, and *Die Versicherung*, 1967), and *Nacht mit Gästen* (1963), *Wie dem Herrn Mockinpott das Leiden ausgetrieben wird* (1968, in collaboration with G. Palmstierna-Weiss), *Trotzki im Exil* (1970), and *Hölderlin* (1971, second version 1973), presenting the poet (see HÖLDERLIN, J. C. F.) as a victim of social conformity. The plays appeared in two volumes, *Stücke I* and *Stücke II/1* and *II/2*, in 1976–7, the last volume containing the two-act play *Der Prozeß*, based on Kafka's novel (see PROZESS, DER), which was published in entirely revised form as *Der neue Prozeß* in 1981. *Drei Stücke in der Übertragung von Peter Weiss* contain translations of *Fadren*, *Fröken Julie*, and *Ett drömspel* by Strindberg (q.v.).

Weiss published various prose writings, including *Dokument I* (1949), *Laokoon oder Über die Grenzen der Sprache* (1965), *Notizen zum kulturellen Leben in der Demokratischen Republik Viet Nam* (1968), *Rapporte* (1968), and *Rapporte 2* (1971). *Fluchtpunkt*, described as a novel, appeared in 1981, *Notizbücher 1960–1971* (2 vols.) in 1982, and *Notizbücher 1971–1980* (2 vols.) in 1981. The three volumes of the major work *Die Ästhetik des Widerstandes* appeared in 1975, 1978, and 1981. He also published poetry in Swedish.

Weiss had accepted the Büchner Prize of 1982 before his death. The customary Dankrede to this posthumous award was given by his wife Gunilla Palmstierna-Weiss and published in *Sprache im technischen Zeitalter* (85/1983).

WEISSE, CHRISTIAN FELIX (Annaberg, Saxony, 1726–1804, Leipzig), a headmaster's son, studied at Leipzig, where he met the men of letters of the day, including Lessing, Gottsched, and Gellert (qq.v.). After a period as a private tutor Weiße became a tax-collector in Leipzig, where he spent the rest of his life as a comfortably situated and well-respected citizen.

A facile writer, Weiße began with Anacreontic poetry and comedies in the French manner, of which the literary satire *Die Poeten nach der Mode* (1751) is the best known. His early poetry was published in 1758 as *Scherzhafte Lieder*. This was followed by a volume of patriotic poetry, *Amazonenlieder* (1760). In 1759 he edited for F. Nicolai (q.v.) the periodical *Allgemeine Deutsche Bibliothek*. About the same time he began to write tragedies in prose, including *Eduard III* (1758), *Richard III* (1759), *Krispus* (1764), and *Rosemunde* (1761), and later switched to verse (*Die Befreiung von Theben*, 1764, *Atreus und Thyest*, 1767). *Romeo und Julie* (1767) is an 'improved' Shakespearian tragedy, pruned and tidied, and *Die Flucht*

(1770) exploits the theme of hostile brothers. The tragedy *Jean Calas* (1774) borrows elements from the technique of *Götz von Berlichingen mit der eisernen Hand* (q.v.). His principal successes were as a writer of Singspiele (see SINGSPIEL), of which *Der Teufel ist los* (q.v., 1752, revised 1766), *Lottchen am Hofe* (1767), and *Die Jagd* (1770), all with music by J. A. Hiller (q.v.), are the best known; he also made a reputation as a writer for children. From 1775 to 1782 he produced a widely read periodical *Der Kinderfreund*, and this was followed by *Briefwechsel des Kinderfreundes* (1784–92). His *Selbstbiographie* was published in 1806.

Weiße's *Komische Opern* (2 vols.) appeared in 1768, *Kleine lyrische Gedichte* (3 vols.) in 1772, *Trauerspiele* (5 vols.) 1776–80, and *Lustspiele* (3 vols.) in 1783. A selection, ed. J. Minor, was published in 1883.

WEISSE, MICHAEL (d. 1534), cantor in Jungbunzlau, assembled a collection of hymns for the Moravian Brethren (see BÖHMISCHE BRÜDER) with the title *Ein neues Gesangbüchlein* (*Ein new Gesengbuchlein*, 1531). They are written for one voice only, blending of parts and instrumental accompaniment being regarded as immoral by the Brethren.

Weiße Buch von Sarnen, Das, title given to a collection of documents and historical narrations dating from *c.* 1476, which contains the legend of William Tell. See TELL, WILHELM.

Weiße Fächer, Der, a one-act play written in part prose, part verse by H. von Hofmannsthal (q.v.) in 1897, and published in *Kleine Dramen* (1906). Having lost his wife, the 24-year-old Fortunio is resolved to spend the remainder of his life mourning at her grave. His grandmother remonstrates with him in vain. Miranda comes to the cemetery to mourn her recently dead husband. Fortunio's and Miranda's eyes meet, and it becomes apparent that the vows of eternal mourning will soon be forgotten.

Weiße Frau, see ORLAMÜNDE, GRÄFIN VON.

Weißen Berge, Schlacht am (8 November 1620), the battle of the White Mountain (a hill —white through its chalk-pits—outside Prague), in which the imperial troops defeated the Bohemians led by Christian of Anhalt (1568–1630) in a surprise attack which put an end to the year-old reign of Friedrich V von der Pfalz (q.v.) as King of Bohemia (see DREISSIGJÄHRIGER KRIEG).

Weißenburger Katechismus, a collection of five theological passages in Old High German prose, written in the monastery of Weißenburg, present-day Wissembourg in Lorraine. Dating from *c.* 800, they comprise an interpretation of the Lord's Prayer, a list of sins, translations of the Apostles and Athanasian Creeds, and of the Gloria. The MS. is in Wolfenbüttel.

Weiße Rose, see RESISTANCE MOVEMENTS (2).

Weißkunig, Der, a quasi-autobiographical work, planned and partly dictated by the Emperor Maximilian I (q.v.). It tells of his youth, marriage, and campaigns. Its execution was entrusted to his secretary Marx Treitzsauerwein (1450–1527). The emperor died before the work was completed and the MS. was first printed in 1775. A further work, to be entitled *Freydal* and intended to deal with Maximilian's courtly pleasures, was planned but never executed. See also TEUERDANK.

Weite Land, Das, a play (Tragikomödie) by A. Schnitzler (q.v.), written in 1910 and first performed at the Lessing-Theater, Berlin, in October 1911. It portrays promiscuous middle-class society during the holiday months at Baden nr. Vienna.

Friedrich Hofreiter, a notorious *coureur de femmes,* deceives his wife with a young woman, Erna von Aigner. Frau Hofreiter, in turn, has an affair with a naval cadet, Otto Meinhold-Aigner, Erna's half-brother; when their liaison becomes public, a duel is arranged between Hofreiter and Otto. It is intended to be a mere formality, but Hofreiter, jealous of Otto's youth, shoots him dead. He feels no remorse, and is only concerned to escape the consequences.

'Welcher Unsterblichen/Soll der höchste Preis sein?', opening lines of Goethe's poem *Meine Göttin* (q.v.).

Welf, German dynastic house, from which the English Hanoverian kings were descended. The family originated in Carinthia in the 11th c., and ruled Bavaria in the 12th c. The most notable medieval Welf was Heinrich der Löwe (q.v.), who aspired to the imperial crown and was defeated and outlawed by Friedrich I (q.v.) in 1179. The Emperor Friedrich II (q.v.) conferred on Heinrich's son Otto the dukedom of Braunschweig-Lüneburg, from which the states of Brunswick and Hanover developed in the 18th c. Welf rulers held sway in Hanover until 1866 and in Brunswick until 1918. The Elector Georg Ludwig of Hanover became King George I of the United Kingdom in 1714.

WELK, EHM, pseudonym of Thomas Trimm (Biesenbrow, 1884–1966, Bad Doberan), a

Socialist, was imprisoned by the National Socialist regime 1934–7 (concentration camp Oranienburg). After 1945 he was prominent in founding and directing (Schwerin) Volkshoch-schulen in Mecklenburg (DDR), and in 1950 turned to authorship. Welk is the author of a number of novels, of which *Die Heiden von Kummerow* (1937) and its sequel *Die Gerechten von Kummerow* (1943) were the most successful. Other titles include *Der hohe Befehl* (1939), *Im Morgennebel* (1953), and *Der wackere Kühnemann aus Puttelfingen* (1959). His plays include the early *Gewitter über Gotland* (1926, produced by E. Piscator, q.v., in 1927) and *Kreuzabnahme* (1927).

WELLERSHOFF, DIETER (Neuß, 1925–), a publisher's reader, is the founder and the only well-known member of the literary group Kölner Schule des neuen Realismus (see NEUER REALISMUS). He is the author of the novels *Ein schöner Tag* (1966), *Die Schattengrenze* (1969), *Einladung an Alle* (1972), *Die Schönheit des Schimpansen* (1977), *Der Sieger nimmt alles* (1983), the volume *Doppelt belichtetes Seestück* (1974), which contains the story of the title and three other stories, three radio plays, and poetry written between 1969 and 1974, and *Die Sirene. Eine Novelle* (1980). *Das Schreien der Katze im Sack. Hörspiele, Stereostücke* (1970) also contains early plays, notably *Der Minotaurus* (1960). From the mid-1970s he turned to television plays, some of which are based on his own fiction.

His literary essays include *Literatur und Veränderung* (1969), *Literatur und Lustprinzip* (1973), and the collections *Das Verschwinden im Bild* (1980) and *'Von der Moral erwischt'. Aufsätze zur Trivialliteratur* (1983); *Die Arbeit des Lebens* (1985) consists of autobiographical texts. He is editor of the work of Gottfried Benn (q.v.).

Welsche Gast, Der, see THOMASIN VON CIRCLAERE.

Welschgattung, Die, a book published in 1513, opposing the gathering impetus of the Renaissance and advocating a return to established ways. The title refers to the Italian people. The unknown author was probably a priest from the region of the upper Rhine.

WELSER, name of an Augsburg family of merchant patricians, who acquired great wealth in the 16th c. They received a patent of nobility in the time of Bartholomäus Welser (1488–1561). In the later 16th c. the house was involved in colonial enterprises and in 1616 it went bankrupt. The most famous member of the family was Philippine Welser (q.v.).

WELSER, PHILIPPINE (Augsburg, 1527–80, Schloß Ambras, Innsbruck), the beautiful daughter of the Augsburg patrician Bartholomäus Welser (see WELSER), married in 1557 the Archduke Ferdinand of Tyrol. Her father-in-law, the Emperor Ferdinand I (q.v.), denied rights of succession to her children, who became Margraves of Burgau. Her story has been treated in a play by Emanuel Schikaneder (q.v., *Philippine Welser, die schöne Herzogin von Tirol*, 1792) and in a novel by Georg Hesekiel, q.v. (*Lux et umbra*, 1861), as well as by other writers.

Welt, Die, newspaper founded in Hamburg by British Military Government as a 'non-party' (unparteilich) newspaper for the British Zone of Occupation. Its editorial staff was German, and the direction lay in the hands of a small British military staff. The first number appeared on 2 April 1946. *Die Welt* passed into exclusively German ownership in 1950. It removed from Hamburg to Bonn in 1975.

Welt als Wille und Vorstellung, Die, see SCHOPENHAUER, A.

Weltchronik, an unfinished Middle High German historical poem by RUDOLF von Ems (q.v.), which breaks off amid the kings of Israel. Even so it reaches the prodigious length of 36,000 lines. In addition to the essential historical narrative, it includes digressions on geography and mythology.

WELTER, NIKOLAUS (Mersch, 1871–1951, Luxemburg), a Luxemburg philologist and author of lyric poetry, plays, and fiction, written in German. His collections of poetry include *Aus alten Tagen* (1900), *Hochöfen* (1913), *Über den Kämpfen* (1915), and *Mariensommer* (1929); his plays include *Siegfried und Melusine* (1900), *Die Söhne des Ösling* (1904, reprinted 1938), *Dantes Kaiser* (1922), and *Die Braut* (1931). He published *Das Luxemburgische und sein Schrifttum* (1914, reprinted 1938) and *Mundartliche und hochdeutsche Dichtung in Luxemburg* (1929). *Gesammelte Werke* (5 vols.) appeared in 1925.

Weltgerichtsspiele, term applied to medieval religious plays illustrating the Day of Judgement. The best-known example is the *Rheinauer Weltgerichtsspiel* (q.v.).

WELTI, ALBERT JAKOB (Höngg nr. Zurich, 1894–1965, Berne), son of the Swiss painter Albert Welti (1862–1912), trained as a painter before devoting himself to authorship. His works, in some of which he employs dialect, include the plays *Spiil ums Füür* (1928), *Servet in Genf* (1930), *Steibruch* (1939), and *Hiob der Sieger* (1954); *Summerfaart* (1942), *Inserat 82793* (1943), and *Sie aber hats nicht leicht* (1950) are comedies, and *Hie Schaffhausen* (1939), *Helfende Kräfte* (1945),

and *Der Paß* (1948) are Festspiele. *Die Heilige von Tenedo* (1943) is a Novelle; his novels include *Wenn Puritaner jung sind* (1941), *Martha und die Niemandssöhne* (1948), and *Der Dolch der Lukretia* (1958).

Weltkriege, I (1914–18 War). The war, the causes of which are still to some extent a matter of controversy, was occasioned by the assassination of the Austrian Archduke Franz Ferdinand (q.v.) on 28 June 1914. Hostilities began with the declaration of war by Austria-Hungary on 28 July. War between Russia and Germany with Austria-Hungary began on 1 August, between Germany and France on 3 August, and, after the German invasion of Belgium, between Germany and Great Britain on 4 August. Italy declared its neutrality on 1 August. Late accessions to the central powers were Turkey, on 1 November 1914, and Bulgaria, on 14 October 1915. Italy joined the Allies on 23 May 1915.

The German plan (see SCHLIEFFEN-PLAN) was concentration and quick victory by envelopment in the west, containment by small forces in the east with the support of the Austro-Hungarian army. In spite of great gains, the plan in the west failed, but the defending armies in the east won unexpected and important victories at Tannenberg and in Masuren (qq.v.). A stalemate in the form of trench warfare ensued in the west, which lasted for almost four years, interrupted from time to time by unsuccessful offensives, which entailed heavy losses of men and great expenditure of material (see VERDUN, SCHLACHT UM).

On the eastern front great gains were made against the Russians, but heavy losses seriously weakened the Austro-Hungarian armies. At sea the principal features were the British blockade and German submarine warfare, which began in February 1915 and was suspended in September 1915. Unrestricted submarine warfare, begun on 1 February 1917, led the U.S.A. to declare war on 6 April 1917. The Russian March and November Revolutions culminated in the virtual surrender of Russia in the Treaty of Brest-Litovsk (q.v., 3 March 1918).

The German attempt to break the deadlock in the west by a series of massive attacks began on 21 March 1918 (März-Offensive). These achieved initial successes, but were all halted. In July 1918 the Allies, reinforced by the Americans, attacked, and continued to advance slowly until the end. The Turkish front collapsed in the autumn of 1918, the Bulgarians surrendered in September, and the Austro-Hungarian monarchy disintegrated in October. The German request for an armistice was made on 3 October, the German navy mutinied on 28 October, the Emperor Wilhelm II (q.v.) fled to Holland on 10 November, and the armistice came into effect on 11 November 1918. (See also VERSAILLES, TREATY OF.)

Weltkriege, II (1939–45 War). (Passing reference only is made to Japanese operations.)

The war began with the invasion of Poland (1 September–6 October 1939), in which the U.S.S.R. joined with the Germans on 17 September. Great Britain and France declared war on Germany on 3 September 1939. After a winter of desultory naval warfare, Germany invaded Denmark and Norway (7 April–24 May 1940). On 10 May a German attack through Holland and Belgium into France isolated the British and some French forces, a large part of which were evacuated from Dunkirk (26 May–4 June). France was completely defeated (14–22 June) and occupied except for the south-east, which was governed by a puppet regime set up at Vichy under Pétain. When the French defeat was clear, Italy declared war. A projected invasion of Britain was prepared by air attack (Battle of Britain, 3 August–17 September), but failure to achieve air supremacy led to its indefinite postponement. Bombing of Britain continued until the spring of 1941. In February 1941 the Afrika-Korps came to the assistance of the Italians in North Africa, and remained there with varying fortunes until its surrender in May 1943. In April 1941 the Germans overran Yugoslavia and Greece, and in May captured Crete. On 22 June 1941 Germany invaded Russia, at first with success. On 11 December 1941, after the Japanese surprise attack on Pearl Harbor (7 December), Germany declared war on the U.S.A. While British and Anglo-American operations in North Africa captured the German forces there (23 October 1942–13 May 1943), the German 6th Army was liquidated at Stalingrad (8 January–2 February 1943).

The Anglo-Americans captured Sicily in July 1943, and invaded Italy. On 6 June 1944 the Anglo-American invasion of Normandy began, and Paris was occupied in August. By July the German armies on the Russian front were in retreat. The naval war, primarily a German submarine war, reached its climax in 1943; the air war led to Allied supremacy from July 1943. A German conspiracy to assassinate Hitler (q.v.) and seize power failed on 20 July 1944 (see RESISTANCE MOVEMENTS (2)). In August the British and American forces invaded southern France. On 11 October the Russians made their first crossing of the German frontier; they took Budapest on 13 February. On 23 March 1945 the final offensive from the west crossed the Rhine, and on 20 April the Russian investment of Berlin began. On 30 April Hitler committed

suicide, and on 7 May Germany capitulated unconditionally. The Japanese surrendered in August 1945 after the bombing of Hiroshima and Nagasaki.

Weltliche Klösterlein, Das, an anonymous poem of some 400 lines, describing an order of Minne after the manner of a religious order. The convent harbours eighteen couples who live in wedlock and occupy themselves with courtly pastimes. The poem was written in 1472.

WENDLER, OTTO BERNHARD (Frankenberg, Saxony, 1895–1958, Burg nr. Magdeburg), after serving in the 1914–18 War became a teacher and headmaster. A Socialist, he was dismissed in 1933, reinstated in 1945, but retired soon afterwards. His plays include *Theater eines Gesichts* (1925), *Liebe, Mord und Alkohol* (1931), the comedy *Pygmalia* (1942), and the tragicomedy *Die Glut in der Asche* (1950). His novels began with *Soldaten, Marieen* (1929), followed by *Drei Figuren aus einer Schießbude* (1932); later novels are *Rosenball* (1945) and *Als die Gewitter standen* (1954).

'Wenn alle untreu werden', first line of a hymn sung in the Lutheran Church. It is No. VI in *Geistliche Lieder* (1802) by Novalis (q.v.).

'Wenn ich ihn nur habe', first line of a hymn sung in the Lutheran Church. It is No. V in *Geistliche Lieder* (1802) by Novalis (q.v.).

WENTER, JOSEF (Merano, Tyrol, 1880–1947, Rattenberg/Inn), studied German philology and music, and during the 1914–18 War was an officer in the Tyrolese Kaiserjäger (q.v.) before making authorship his career. During the first phase of writing he devoted himself mainly to historical plays, including *Saul* (1908), *Canossa* (1916), *Der deutsche Heinrich* (1919), *Der sechste Heinrich* (1920), and *Johann Philipp Palm* (1923, see PALM, J. P.); in 1934 he resumed the subject of Heinrich IV (q.v.), begun with *Canossa*, with *Geschichte und Ende Kaiser Heinrichs IV*. Tyrolese history forms the background to *Der Kanzler von Tirol* (1925), and with his last play he turned to Maria Theresia (q.v., *Kaiserin Maria Theresia*, 1944). His contributions to drama earned him the position of Burgtheaterdichter at the Burgtheater, Vienna.

As the author of fiction, Wenter established a reputation for animal books based on astute observation: *Monsieur der Kuckuck* (1930), *Laikan, der Roman eines Lachses* (1931), *Mannsräuschlin, der Roman eines Wildpferdes* (1933, retitled *Situtunga*, 1938), *Tiergeschichten*, and *Tiere und Landschaften* (collections of stories, 1933 and 1937).

Im heiligen Land Tirol (1937) was Welter's last

novel; his autobiography is entitled *Leise, leise, liebe Quelle* (1941).

WENZEL II, KÖNIG VON BÖHMEN (1278–1305), the son of Ottokar II (q.v.) of Bohemia, encouraged German Minnesang (q.v.) at his court and himself composed poems, three of which survive: two Minnelieder and a Tagelied (q.v.). Though they show the influence of KONRAD von Würzburg (q.v.), they have original touches, such as the chivalrous renunciation in the line 'ich brach der rôsen niht und het ir doch gewalt' ('I did not pluck the rose although I had it in my power').

Among the poets protected by Wenzel was ULRICH von Etzenbach (q.v.).

WENZEL (Nürnberg, 1361–1419, Wenzelstein), German King (see DEUTSCHER KÖNIG) from 1378 to 1400 and also King Wenzel IV of Bohemia from 1378 until his death. Wenzel was elected in 1376 and succeeded his father Karl IV (q.v.) in 1378. In 1400 the four electors of the Rhineland conspired against him and deposed him, electing Rupert of the Palatinate (1352–1410).

WERDER, DIEDERICH VON DEM (Werdershausen, 1584–1657, Reinsdorf), spent his life in princely service, at Köthen as master of the horse and at Kassel as court chamberlain. As 'Der Vielgekrönte' he was a member of Die Fruchtbringende Gesellschaft (q.v.). He saw war service from 1631 to 1635 as a regimental commander under Gustavus Adolphus (q.v.) and in 1646 he became a civil servant in Brandenburg. He wrote sonnets, *Krieg und Sieg Christi in hundert Sonetten* (1631) and *Bußpsalmen* (1632), but was chiefly notable as a translator. His German version of Tasso's *La Gerusalemme liberata* (*Gottfried von Bulljon oder Das erlösete Jerusalem*) appeared in 1626, and his translation of Ariosto's *Orlando furioso* (*Der rasende Roland*) from 1632 to 1636.

Werder's *Andachten auf die Stunde des Todes* (1671) is a posthumous publication.

WERFEL, FRANZ (Prague, 1890–1945, Beverley Hills, California), grew up in the Jewish quarter of Prague and was a friend of Kafka (q.v.). He studied at Prague, Leipzig, and Hamburg universities, served a year as a volunteer (Einjährig-Freiwilliger) in the Austrian army, and became a publisher's reader. During the 1914–18 War he served on the Russian front until 1917 and after the war lived in Vienna.

Werfel first came into prominence as a writer during his Viennese years. In 1929 he married Alma Mahler (1879–1964), widow of the composer Gustav Mahler (q.v.). In 1938 Werfel and his wife emigrated to France, fled to Spain in 1940, and made their way with H. Mann (q.v.) to the U.S.A., where they settled in California. Werfel had a serious heart attack in 1943, from which he did not fully recover. The basis of Werfel's convictions and work was religious, and his consciousness of participation in the Jewish fate and mission was fundamental; but he was also strongly attracted to the Roman Catholic Church and yearned for the brotherhood of man.

Werfel's earliest Expressionist work was a volume of verse, *Der Weltfreund* (1911), followed by *Einander* (1915, containing his adaptation of the hymn 'Veni creator spiritus'), *Gesänge aus den drei Reichen* (1917), *Der Gerichtstag* (1919), and *Beschwörungen* (1923), his last important contribution to poetry. A play, *Besuch in Elysium*, appeared in a periodical in 1912 (in book form 1920); more significant was an adaptation of Euripides' *The Trojan Women*, *Die Troerinnen* (1915) preceding his post-war plays *Spiegelmensch* (q.v., 1920), *Bocksgesang* (1921), *Schweiger* (1922), *Juarez und Maximilian* (1924, dealing with the conflict between the deposed president and the emperor imposed by France and executed in 1867, see MAXIMILIAN, ERZHERZOG), *Paulus und die Juden* (1926), *Das Reich Gottes in Böhmen* (1930, treating the conflict between Cardinal Cesarini and Procop the Great in the 15th c.), and *Der Weg der Verheißung* (1935). *Jacobowsky und der Oberst* (1944), Werfel's last play, published as *Komödie einer Tragödie*, centres on the problems of race and emigration.

Best known as a novelist, Werfel expresses in *Verdi. Roman der Oper* (q.v., 1924) his love for music, especially opera, which is further borne out by his excellent translations of *La Forza del destino* (*Die Macht des Schicksals*, 1926), *Simone Boccanegra* (1929), and *Don Carlo* (*Don Carlos*, 1932). The novel *Der Abituriententag. Die Geschichte einer Jugendschuld* (1925) was followed by *Barbara oder Die Frömmigkeit* (q.v., 1929) and *Die Geschwister von Neapel* (q.v., 1931). *Die vierzig Tage des Musa Dagh* (q.v., 1933) recounts one of the overlooked horrors of the 1914–18 War; *Höret die Stimme* (q.v., 1937) was posthumously retitled *Jeremias* (1956), *Der veruntreute Himmel*. *Die Geschichte einer Magd* (q.v., 1939) has affinities with *Barbara*, and *Das Lied von Bernadette* (q.v., 1941), written to fulfil a vow made at Lourdes in 1940, commonly referred to as the 'Lourdes-Roman'. The posthumous novel *Stern der Ungeborenen* (1946), a Utopian work, expresses anxiety over the future. Of Werfel's few Novellen the most important are *Nicht der Mörder, der Ermordete ist schuldig* (1920), *Der Tod des Kleinbürgers* (1927), and *Geheimnis eines Menschen* (1927). *Zwischen Oben und Unten* is a collection of essays, published posthumously in 1946.

Werfel's *Gesammelte Werke* (8 vols.) appeared 1927–36, *Gesammelte Werke in Einzelausgaben* (15 vols.), ed. A. D. Klarmann, 1948–67 (14 vols.), and another edition 1975 f.

Wer ist der Verräter?, a story inserted into Goethe's *Wilhelm Meisters Wanderjahre* (q.v.) as Ch. 8 and part of Ch. 9 of Bk. I. It narrates the lawyer Lucidor's hesitant, but finally successful, wooing of Lucinde. Written in 1807, it bore in the first version of the *Wanderjahre* (1821) the title *Wo steckt der Verräter?*

Werkleute auf Haus Nyland, a group of poets including J. Kneip, H. Lersch, and J. Winckler (qq.v.), who first came together as admirers of the work of R. Dehmel (q.v.) in 1905. Their chief aim was to encourage a literature reflecting their own age with its growing use of technology. The group's name derives from a Westphalian farm which belonged to Winckler's family.

WERNER, E., pseudonym of Elisabeth Bürstenbinder (Berlin, 1838–1918, Merano), author of numerous light novels published in serial form. Her collected works appeared as *Gesammelte Romane und Novellen* (10 vols.) 1893–5.

WERNER, ZACHARIAS, in full Friedrich Ludwig Zacharias (Königsberg, 1768–1823, Vienna), the principal dramatist of the Romantic movement (see ROMANTIK), was the son of a professor at Königsberg University; his mother had a history of mental illness. Werner worked in the Prussian civil service in Warsaw and in Berlin. At the age of 31 he made his third marriage, two earlier unions having been dissolved. In 1807 this marriage also ended in divorce, and at the same time Werner resigned from the civil service and devoted himself to writing.

As a young man Werner had published poems (*Vermischte Gedichte*, 1789), but the influence of the Romantics unexpectedly sent him to the drama, in which he developed a new form of broadly presented historical play tinged with mysticism; *Die Söhne des Tals* (2 vols., 1803) was followed by *Das Kreuz an der Ostsee* (1806) and by *Martin Luther oder Die Weihe der Kraft* (q.v., 1807), which, when performed in 1806 in Berlin, had a *succès de scandale*. In the years 1807–10 Werner travelled in Europe, visiting Goethe in Weimar (1807–8) and Mme de Staël (q.v.) at Coppet. At this time he wrote his highly original and sensational long one-act fate tragedy (see SCHICKSALSTRAGÖDIE), *Der vierundzwanzigste Februar* (q.v., performed 1810 in Weimar, published 1815), which set a fashion in the theatre lasting for more than a decade. In

1810 Werner was converted in Rome to Roman Catholicism, and in 1813 he was ordained priest. *Die Weihe der Unkraft* (1814) is a recantation of the eulogy of Luther (q.v.) contained in the play of 1807.

Werner was appointed an honorary canon of the chapter of St. Stephen's Cathedral, Vienna, and was a fashionable preacher. The works of his later period (the plays *Cunegunde, die Heilige*, 1815, and *Die Mutter der Makkabäer*, 1820) are of little significance. Of his earlier works the tragedies *Attila* (1808) and *Wanda, Königin der Sarmaten* (1810) should be mentioned; his occasional poetry included *Klagen um seine Königin Luise von Preußen* (1810), *Kriegslied für die zum heiligen Kriege verbündeten deutschen Heere* (1813), and *Te Deum zur Einnahme von Paris* (1813).

An unauthorized edition of his plays (*Theater*, 6 vols.) appeared in 1818. *Ausgewählte Schriften* (15 vols.) were published posthumously in 1840–1 (reprinted with additions in 1969).

Werner oder Herz und Welt, a five-act play (Schauspiel) by K. Gutzkow (q.v.), first performed in February 1840 at Hamburg. Heinrich von Jordan, married to Julie, has, at a time when he bore the name Werner, loved Marie Winter. To his surprise, she turns out to be the newly engaged governess. When she discovers who her employer is she seeks to quit the house, but he insists on her staying. Julie leaves her husband, but Marie breaks out of the triangle by agreeing to marry another, a Dr. Fels. Julie returns and the marriage is re-established. Meanwhile Heinrich, who had been accused of peculation, is proved innocent. He resigns his post and, as plain Werner, accepts a professorship at a university.

WERNHER, BRUDER, a Middle High German poet, whose work consists of Sprüche (see SPRUCH). It is probable that Wernher was a layman, who became a friar or lay brother only in his later years, for he appears in his poems as a man versed in the courtly world. He was an Austrian who was at the court of Leopold VII and his successor and took part in the crusade of 1228–9. Later he was in Swabia. His poems, of which some seventy are known, were written between 1225 and 1250. They are based in the Christian faith and lament a decline in morals, which Wernher witnesses in the second quarter of the 13th c.

WERNHER, PRIESTER, author of a Marienleben (or life of the Virgin Mary) completed in 1172. Wernher himself indicates his name, his priestly calling, and the date of his work. Its dialect points to Augsburg as his residence. He entitles his poem *Drei Lieder von der Jungfrau* (*driu liet von der maget*).

The first song or section deals with the Virgin's parentage, conception, and birth; the second her childhood, marriage, the Annunciation, the Immaculate Conception, and the visit to Elizabeth; the third contains her early married life from the return of Joseph, including the birth of Jesus and the flight into Egypt. The poem closes with a brief section on the Crucifixion, Resurrection, and Ascension, and the Day of Judgement.

Wernher's source is the apocryphal *Pseudo-Evangelium Matthaei* up to the birth of Jesus. From this point he follows the canonical gospels. The poem is characterized by warm lyrical feeling and by vivid presentation of events in the costume and setting of Wernher's own day. In addition to its intrinsic merits, it is usually seen as a step in the direction of courtly poetry. The existence of several MSS., complete or fragmentary, indicates the popularity of the poem.

WERNHER DER GARTENAERE, see MEIER HELMBRECHT.

WERNHER VON ELMENDORF, chaplain in the 12th c. to the lords of Elmendorf near Oldenburg, wrote *c.* 1175 a rhymed German version of the *Moralium dogma philosophorum*, which is presumed to be of French authorship. Wernher's poem, which survives only in fragments, appears to abandon the systematic character of the original and to devote itself to practical counsel for a godly life in the material world.

WERNHER VOM NIEDERRHEIN, a 12th-c. priest, author of a poem dealing with the dogma of salvation in terms of a mystical interpretation of the number four: *Die vier schiven* (i.e. Die Vier Scheiben), a title which refers to the four wheels of Aminadab's chariot ('anima mea conturbavit me propter quadrigam Aminadab', Vulgate, Cant. 6. 11).

WERNHER VON HONBERG, GRAF (*c.* 1284–1320, Genoa), from the neighbourhood of Basel, had a distinguished career as a knight, taking part in an expedition to Lithuania in 1304 and in campaigns in Italy. Eight Minnelieder are attributed to him, some of which refer to his absence in foreign lands. The jealous hatred of a rival expressed in one of the poems is contrary to the tradition of Minnesang (q.v.).

WERNHER VON TEUFEN, a minor Middle High German Minnesänger from the neighbourhood of Zurich. Five poems are attributed to him. Formerly identified with a Wernher documented in 1220, he is believed to belong to the next generation.

WERNICKE, also **WERNIGKE,** CHRISTIAN (Elbing, 1661–1725, Copenhagen), spent most of his life in Danish service. He visited England and in 1714 became Danish minister in Paris. Poetry was only a by-product of a political and diplomatic life, but he found time for a literary dispute with C. H. Postel (q.v.), which brought him subsequently into controversy with C. F. Hunold (q.v.). Wernicke, whose style was at first inflated and bombastic in the 17th-c. fashion, eventually adopted a more chastened manner. The epigrams, which were his forte, appeared in two collections of *Überschriffte oder Epigrammata*, published in 1697 and 1701. The satirical poem he launched against Postel is entitled *Heldengedicht Hans Sachs genant* (1701).

A selection of Wernicke's work is contained in *Gegner der zweiten schlesischen Schule*, ed. L. Fulda, 1884. His *Epigramme* were published by W. Pechel in 1909 (reprinted 1970).

'Wer niemals einen Rausch gehabt', first line of a well-known song by J. Perinet (q.v.), included in the Viennese Volksstück *Das neue Sonntagskind* (1794).

'Wer nur den lieben Gott läßt walten', first line of the poem *Trostlied* by Georg Neumark (q.v.).

Werther, see DIE LEIDEN DES JUNGEN WERTHERS.

Wer weiß wo, a poem by D. von Liliencron (q.v.), published in *Adjutantenritte* (q.v.) in 1883. It bears the sub-title *Schlacht bei Kolin, 18 Juni 1757*. The fate of the young ensign 'Verscharrt im Sand zur ewgen Ruh./Wer weiß wo' is extended to all mankind.

Werwolf, early NHG Wärwolf (see DER WÄRWOLF by W. Alexis) originates from 'wer', a man who, for a specified time, is turned into a 'wolf', the symbols of which are a wolf's skin and a girdle or ring (in Greek and Roman mythology and other Western countries). The werewolf has persisted as a destructive creature of popular superstition. An exception is *Der Wehrwolf* by H. Löns (q.v.). Members of the secret National Socialist resistance movement of 1945 were called Werwölfe. (One of the homeless children in the novel *Die Kinder von Wien* by R. Neumann, q.v., emerged at the end of the war from a Werwolfausbildungslager.)

WESENDONCK, MATHILDE (Elberfeld, 1828–1902, Traunblick, Traunsee, Austria), *née* Luckemeyer, married Otto Wesendonck, a businessman with interests in Switzerland and America. When Wesendonck invited R. Wagner (q.v.) to stay on his estate in 1857, Wagner fell in love with his hostess, and after his departure in 1858 they corresponded. Mathilde wrote poetry, and at this time (1857–8) Wagner composed the *Fünf Lieder von Mathilde Wesendonck*.

Frau Wesendonck wrote industriously. Her works include *Märchen und Märchen-Spiele* (1864), *Deutsches Kinderbuch in Wort und Bild* (1869), *Gedichte, Volksweisen, Legenden und Sagen* (1875), and *Alte und neue Kinderlieder* (1900), as well as war poetry (*Patriotische Gedichte*, 1870) and plays (*Gudrun*, 1868; *Friedrich der Große*, 1871; *Edith oder Die Schlacht bei Hastings*, 1872; *Kalypso*, 1875; *Odysseus*, 1878; and *Alkestis*, 1881). Her diaries and correspondence with Wagner covering the years 1853–71 appeared in 1904, and correspondence conducted with J. Brahms (q.v.), *Ein Briefwechsel*, ed. E. H. Müller von Asow, in 1943.

WESSEL, Horst (Bielefeld, 1907–30, Berlin), a student who joined the NSDAP (q.v.) in 1926, abandoned his studies, and worked for the SA (q.v.). On 14 January 1930 he was attacked in his department and died six weeks later of sepsis. He was made a hero of National Socialism. A militant and anti-Semitic song which he had written (*Horst-Wessel-Lied*) was made a national anthem and sung after the *Deutschlandlied*. Wessel is the subject of a laudatory novel, *Horst Wessel. Ein deutsches Schicksal* (1932) by H. H. Ewers (q.v.).

Wessobrunner Gebet, one of the most important Old High German documents, is only in part a prayer. It consists firstly of an impressive poetic passage of nine lines, written in alliterative verse (Stabreim), describing primal chaos, followed by the introduction of God and his angels, doubtless as a prelude to the Creation. At this point the poem breaks off, and the second portion follows, a prose prayer for grace and power to withstand the devil and to do God's will.

The content and phrasing of the first section may possibly be derived from Anglo-Saxon poetry. The dialect is Bavarian and the MS., now in Munich, was originally located in the monastery of Wessobrunn in southern Bavaria. It is possible that the poem originated in the school of Fulda. It dates from the last thirty years of the 8th c.

Wessobrunner Glaube und Beichte, see BEICHTFORMELN.

Wessobrunner Predigten, translations into Old High German of sermons by Augustine, Gregory the Great, and others. They are popular sermons, simplifying, abridging, and explaining the originals, and were made in the first half of the 11th c. in the monastery of Wessobrunn (Bavaria).

WEST, THOMAS and CARL AUGUST, pseudonyms of J. Schreyvogel (q.v.).

Westermanns Monatshefte, one of the oldest German literary periodicals, was founded in 1856 by the publisher G. W. Westermann (1810–79) and in 1968 was retitled *westermann*.

Westfalen (Westphalia), a former Prussian Provinz, now part of Land Nordrhein-Westfalen in the Federal Republic (see BUNDESREPUBLIK DEUTSCHLAND). Until the beginning of the 19th c. a large part of it was a possession of the electoral archbishops of Cologne. It was acquired by Hesse-Darmstadt in 1803 and was part of the Kingdom of Westphalia created in 1807 by Napoleon, who made his brother Jérôme Bonaparte (1784–1860) its king. In addition to Westphalia, this kingdom included Hanover, Brunswick, Hesse-Darmstadt, and the Prussian territory west of the Elbe (later Provinz Sachsen). In 1813 Jérôme fled, and the kingdom was dissolved. See NAPOLEONIC WARS.

Westfälische Pforte or **Porta Westphalica,** a gap between high hills near Minden, through which the River Weser flows on its way to the sea.

Westfälischer Friede, the peace treaty concluding the Thirty Years War (see DREISSIG-JÄHRIGER KRIEG), was signed on 24 October 1648 after four years of negotiations, between the Holy Roman Empire (see DEUTSCHES REICH, ALTES) and France in Münster, and in Osnabrück between the Empire and Sweden. Among the numerous terms of the treaty the following may be mentioned:

France was confirmed in the possession of the bishoprics of Metz, Toul, and Verdun.

Austrian Alsace was ceded to France with Breisach, but without Strasburg.

Sweden received western Pomerania, which included the Oder estuary, and the strategically and economically important bishoprics of Bremen and Verden as imperial fiefs.

Of the principal German states Brandenburg received eastern Pomerania, the bishoprics of Halberstadt and Minden, and the promise of Magdeburg.

Brandenburg's possession of the duchies of Cleves, Mark, and Ravensberg was confirmed.

Saxony received Lusatia (Lausitz).

Maximilian I (q.v.) of Bavaria secured the hereditary succession of his dynasty (see

WITTELSBACH) and received the Upper Palatinate (Oberpfalz).

A new electoral vote (the eighth) was created for Karl Ludwig, son of the Elector Friedrich V (q.v.) of the Palatinate, who had been deprived of his electorate and territory in 1620. Lower (Rhenish) Palatinate was restored to his son.

As already agreed in the Peace of Prague (Prager Friede), all Roman Catholic possessions that had been Catholic on the first day of 1624 were to remain Catholic and, conversely, a similar proviso was applied to Protestant territory. The status of the predominantly Protestant north and the Catholic south was thus reaffirmed without major changes. A further modification of the terms of the Peace of Augsburg (see AUGSBURGER RELIGIONSFRIEDE, 1555) was the recognition of Calvinism and the equal division of Protestant and Catholic interests in the Imperial Diet. Religious tolerance or otherwise in individual states remained in the hands of the princes and the principle *cuius regio eius religio* (q.v.) was maintained. The rights of the Emperor were severely curtailed. The princes obtained the right to conduct their own affairs at home as well as in their foreign policy and to make alliances with other German or foreign states as long as such an alliance was not directed against the Empire. The free cities (Reichsstädte) were granted rights corresponding to those of the princes.

West-östlicher Divan, a collection of poetry of oriental character published by Goethe in 1819. He derived his inspiration from the Persian poet Hafiz (1320–89), whose verse he read in June 1814 in a translation by J. von Hammer-Purgstall (q.v.). Goethe experienced at the restoration of peace (see NAPOLEONIC WARS) a sense of rejuvenation, touring the Rhineland twice during the summer. At this time he wrote the first oriental poems. In September and October he was the guest of J. J. von Willemer and his attractive and gifted young bride (see WILLEMER, MARIANNE VON), and from this point his poetry, intended for the new volume, was intimately connected with Marianne. In the following winter Goethe assembled the poems so far completed chronologically under the provisional title *Versammlung deutscher Gedichte, mit stetem Bezug auf den Divan des persischen Sängers Mohammed Schemfeddin Hafis.* Goethe prepared the poems, a few of which had appeared separately, for publication in a cycle of twelve (unnumbered) books. Each book bears a transliterated title followed by its German equivalent, thus emphasizing the close collaboration of two kindred spirits.

The German titles are *Buch des Sängers, Buch Hafis, Buch der Liebe, Buch der Betrachtungen, Buch des Unmuts, Buch der Sprüche, Buch des Timus, Buch Suleika, Das Schenkenbuch, Buch der Parabeln, Buch des Parsen, Buch des Paradieses.* Goethe intended to add to these titles. He did not imitate the metrical form of the original (see GHASEL), but wrote rhyming verse, mostly arranged in four-line stanzas of deceptive simplicity. The collection is characterized by a mellifluous ease in the verse and by an attitude of mature contentment combined with gentle irony, and, in the *Buch des Paradieses,* an undogmatic reverence. The *Buch Suleika,* by far the longest, is closely connected with Goethe's relationship with Marianne (Suleika; to whom he plays Hatem); she contributed the poems 'Hochbeglückt in deiner Liebe . . .', 'Was bedeutet die Bewegung . . .', 'Ach, um deine feuchten Schwingen . . .', and, probably, 'Wie mit innigstem Behagen . . .'.

Goethe appended a lengthy explanatory essay on his oriental studies, first entitled *Zu besserem Verständnis,* and in subsequent editions appearing as *Noten und Abhandlungen zu besserem Verständnis des west-östlichen Divans.* The work is the culmination of Goethe's interest in oriental poetry and thought, first kindled by J. G. Herder (q.v.) in the 1770s; it represents his closest approximation to Romanticism (see ROMANTIK) and influenced a number of writers associated with the Romantic movement, among them A. von Chamisso, Julius Mosen, and H. Heine (qq.v.). *Ghaselen* by A. von Platen (q.v.) and *Östliche Rosen* by F. Rückert (q.v.) are directly indebted to Goethe's cycle.

Wettin, German dynastic house ruling in Saxony from 1423 to 1918. The most conspicuous Wettiner were Friedrich der Weise and Moritz von Sachsen in the 16th c. (qq.v.), and August II (der Starke) and August III in the 18th c. (qq.v.). Friedrich August of this family was the first king of Saxony (1806).

WETZEL, FRIEDRICH GOTTLOB (Bautzen, 1779–1819, Bamberg), lived mainly in Thuringia and Saxony and wrote poetry and plays. His works include *Strophen* (1803), *Gedichte* (1815), and the tragedies *Jeanne d'Arc* (1817) and *Hermannfried* (1818). All these have passed into virtual oblivion, but he remained generally known as the presumed author of *Nachtwachen. Von Bonaventura* (q.v.). This attribution was made in 1909 by F. Schultz, but it has been convincingly challenged in the 1960s and 1970s by J. Schillemeit and W. Kohlschmidt *et al.*

Wetzel's *Gesammelte Gedichte und Nachlaß,* ed. Z. Funck, appeared in 1838.

WEYRAUCH, WOLFGANG (Königsberg, 1907–80, Darmstadt), was an actor before working as a publisher's reader in Berlin (1933). Well known for his radio plays, written since the 1939–45 War, he was interested in the medium in its early stages, collaborating in *Anabasis* (1931, revised 1959). He published two novels, *Der Main. Legende* (1934) and *Strudel und Quell* (1938) before being called up in 1940 and becoming a prisoner of war in Russia. After his repatriation he worked as a newspaper editor in Berlin and from 1952–8 as a reader in Hamburg, devoting himself from 1959 to authorship.

An associate of the Gruppe 47 (q.v.), he was a strong supporter of the writers advocating the idea of a clean sweep in literature (see KAHLSCHLAG), and in 1949 published his prose anthology *Tausend Gramm*. Weyrauch, a politically committed humanitarian and a leading experimenter in literary techniques, published *Auf der bewegten Erde*, an amorphous story, in 1946 (reprinted 1967), *Die Liebenden* in 1947, *Die Davidsbündler* (consisting of 25 aphorisms) in 1948, *Bericht an die Regierung* in 1953, and in 1959 the collection *Mein Schiff das heißt Taifun*, containing thirteen tales. Later prose includes *Etwas geschieht* (1966), *Geschichten zum Weiterschreiben* (1969), *Ein Clown sagt*, and *Wie geht es Ihnen?* (both 1971). *Beinahe täglich* (1975) and *Hans Dumm. 111 Geschichten* are prose collections.

Although the distinction between Weyrauch's prose texts and his lyric poems is not always clear, he published his poetry in separate editions, including *Von des Glückes Barmherzigkeit* (1946), *Lerche und Sperber* (1948), *Ende und Anfang* (1949), *An die Wand geschrieben* (1950), *Die Minute des Negers* (1953), *Gesang, um nicht zu sterben* (1956), *Die Spur* (1963), *Judiths Strophen* (1973). *Das Komma danach* (1977), and *Fußgänger, B-Ebene Hauptwache Rolltreppe hinauf hinab* (1978). Weyrauch also edited prose and poetry anthologies.

In 1962 appeared a collection of Weyrauch's radio plays, *Dialog mit dem Unsichtbaren*, containing *Woher kennen wir uns bloß?* (1952), *Vor dem Schneegebirge* (1953), *Die japanischen Fischer* (1955), *Indianische Ballade* (1956), the revised *Anabasis* of 1959, *Jon und die großen Geister* (1961), and *Totentanz*, which, published in 1961, earned him a prize (Hörspielpreis der Kriegsblinden); he wrote a sequel, *Komm. Totentanz 2*, in 1963, in which year he published another radio play, *Das tapfere Schneiderlein*. His last radio play, *Hier wird Musik gemacht*, was broadcast in 1980, the year of his death.

A select edition, *Mit dem Kopf durch die Wand. Geschichten, Gedichte, Essays und ein Hörspiel. 1929–1977* appeared in 1977, *Dreimal geköpft. Unbekannte Gedichte* in 1983, and a collection of stories, *Proust beginnt zu brennen*, in 1985.

WEZEL, JOHANN KARL (Sondershausen, 1747–1819, Sondershausen), suffered in middle and later life from melancholia. A prolific writer of novels and stories, he owed something to C. F. Gellert (q.v.), whom he knew personally in his youth, and also to Sterne and Fielding.

Among a host of works, the following titles deserve mention: *Lebensgeschichte Tobias Knauts des Weisen, sonst der Stammler genannt* (4 vols., 1773–6, reprinted 1971), *Belphegor, oder die wahrscheinlichste Geschichte unter der Sonne* (2 vols., 1776, reprinted 1965), *Ehestandsgeschichte des Herrn Philipp Peter Marks* (1776), *Herrmann und Ulrike, ein komischer Roman* (4 vols., 1780, reprinted 1971), and *Wilhelmine Arend, oder die Gefahren der Empfindsamkeit* (1782). He also published *Versuch über die Kenntnis des Menschen* (2 vols., 1784–5, reprinted 1971). Critical writings, which included *Über Sprache, Wissenschaften und Geschmack der Deutschen* (1781), were published, ed. A. R. Schmitt, as *Kritische Schriften* (2 vols.) in 1971.

WICELIUS, see WITZEL, GEORG.

WICHERT, ERNST (Insterburg, East Prussia, 1831–1902, Berlin), not to be confused with Ernst Wiechert (q.v.), was a magistrate and judicial civil servant for many years in Königsberg and from 1888 in Berlin. Of his plays *Biegen oder Brechen* (1874), *An der Majorsecke* (1875), both comedies, and the historical plays *Aus eigenem Rechte* (1893) and *Im Dienst der Pflicht* (1894) enjoyed a passing esteem. He wrote a number of historical novels, including *Heinrich von Plauen* (1881) and *Der Große Kurfürst in Preußen* (3 vols., 1886–7), but his best work is generally reckoned to be the stories *Litauische Geschichten* (1881).

Gesammelte Werke (18 vols.) appeared 1896–1902.

WICKRAM, GEORG or JÖRG (Colmar, *c.* 1500–before 1562, Burgheim/Rhine), the illegitimate son of a municipal official in Colmar in Alsace, became a subordinate official in his native city and later town clerk of Burgheim. He was at the same time both goldsmith and painter.

An active and versatile writer, Wickram cultivated Meistergesang (q.v.), founding a school of Meistersinger in Colmar in 1549. His 'christliches Bürgerspiel' *Tobias* (1550) looks back to the Middle Ages, with its two-day duration, formal presentation, and audience participation in a hymn. His satirical dialogue *Von der Trunkenheit* and his *Der Irr Reitend Bilger* are didactic works touching in homely fashion on social and religious problems of the day; *Das Rollwagenbüchlin* (all 1555), a collection of anecdotes and Schwänke without moral pretensions, is intended as entertainment, well told and neatly turned, and (since Wickram aimed at a public of both sexes) with less obscenity than is customary in the story-telling of the age.

Wickram's principal achievements are in the novel. His *Ritter Galmy* (1539) and *Gabriotto und Reinhart* (1551) are echoes of the courtly romance. *Der Jungen Knaben Spiegel* (1554) contrasts a virtuous burgher's son with a prodigal young nobleman, to the latter's disadvantage. In *Der Goldtfaden* (1554, printed 1557) Wickram bridges the social gap. Leufried, a shepherd's son, is adopted by a merchant and becomes a menial in a count's castle. Here he falls in love with the count's daughter Angleana. He undertakes various adventures, is knighted and rewarded with Angleana's hand. *Der Goldtfaden*, in which a conscious stylistic sense is evinced, is often regarded as the foundation of the German novel. Wickram's last novel *Von Guten und Boesen Nachbaurn* (1556) draws an idealized picture of the burgher's life. Wickram also wrote Fastnachtspiele (*Die Zehen alter*, 1531, *Das Narrengießen*, 1538, *List der Weiber*, 1543) and in 1549 modernized the translation of Ovid's *Metamorphoses* by ALBRECHT von Halberstadt (q.v.).

Sämtliche Werke (12 vols.), ed. H. G. Roloff, appeared 1967 ff.

Widerstandsbewegungen, Deutsche, see RESISTANCE MOVEMENTS (2).

WIDMANN, ACHILLES JASON (d. *c.* 1590, Schwäbisch-Hall), published *c.* 1555 a collection of amusing anecdotes, concerning one Peter Lewen or Leu, as *Histori Peter Lewen*. The stories are traditional Schwänke (see SCHWANK); the central figure is linked with *Der Pfaffe vom Kalenberg* (see FRANKFURTER, P.) and has been referred to as 'der andere Kalenberger'.

WIDMANN, GEORG RUDOLF, of Schwäbisch-Hall, adapted and amplified the *Spieß'sches Faustbuch* (see FAUSTBUCH, SPIESS'SCHES) in 1599. His version with its didactic and theological additions attained a greater popularity than the original.

Wie Anne Bäbi Jowäger haushaltet und wie es ihm mit dem Doktern geht, a novel written in 1842–3 by J. Gotthelf (q.v.), and published in two volumes 1843–4. It originated in a proposal by a medical friend that he should write a tract against quacks, but the book quickly transcended the first intention and became one of Gotthelf's most substantial works.

Anne Bäbi, wife of Hansli Jowäger, a well-to-do farmer, is the embodiment of rural prejudice and superstition. A son, Jakobli, whom she idolizes, is born to her. He catches smallpox, survives with minimal medical attention, but is in danger of losing his sight. The doctor, summoned at last, saves one eye, but Anne Bäbi and her maid, Mädi, insist on following a quack's advice and Jakobli almost loses the sight of the other. He grows up a quiet and amenable, but somewhat melancholy young boy. Anne Bäbi is persuaded that he needs a wife and chooses Lisi, a robust, coarse, and vulgar farmer's daughter who is supposed to have money. But Jakobli falls in love at first sight with Meyeli, a poor girl of sunny temperament, to whom the Jowäger couple give a lift. After much resistance from Anne Bäbi a break is made with Lisi, and Jakobli and Meyeli are married. Their first child is completely annexed by Anne Bäbi, who spoils him and, when in the end he falls ill, goes to a quack rather than to a physician. The child dies and Anne Bäbi falls into a stupor. In this state she is visited by the curate, who persuades her that she is responsible for the child's death. The following night Anne Bäbi attempts suicide. A doctor is now summoned. He is the pastor's nephew, Rudi, a man impatient of religion but utterly devoted to his patients. He gradually undoes the harm done by the curate and also watches over Meyeli's health. The doctor sacrifices himself for his patients, is found by Jakobli on a winter's night sick and exhausted, rallies for a time, but succumbs to a second attack.

The curate and the doctor are both seen to be men whose view of God is obscured, but the doctor's self-sacrifice shows him to be the higher kind of man. The representative of true Christianity is the pastor, devout yet tolerant and full of charity, who preaches his nephew's funeral sermon. The narration is interspersed with persuasive and often vehement digressions on religion, quack medicine, medical morality, and contemporary politics.

WIECHERT, ERNST (Sensburg, 1887–1950, Uerikon, Switzerland), studied at Königsberg University, became a teacher in 1911, and during the 1914–18 War served as an officer. He returned to teaching, resigning in 1933 in order to devote himself to writing.

Wiechert spoke against National Socialism in addresses to students (*Der Dichter und die Jugend*, 1933, published 1936, and *Der Dichter und seine Zeit*, 1936, published 1945). Sent to a concentration camp for some months in 1938, he recorded his experiences in *Der Totenwald* (q.v., 1945). In *Der weiße Büffel, oder Von der großen Gerechtigkeit* (written 1937, published 1946) he advocated resistance to the National Socialist regime. Extracts from it were circulated in 1938. For a few years after the 1939–45 War, Wiechert was one of the most widely read German authors. His appeal for a new spirit (*Rede an die deutsche Jugend*, 1945) seemed to fall on deaf ears, and in 1948 he withdrew to Switzerland.

A recurring theme in Wiechert's fiction is contained in his first novel; written before 1914, *Die Flucht* (1916) contrasts the effects of civilization with the silence and beauty of the Masurian forest, the title's haunt of refuge. In *Der Totenwolf* (1924) a soldier returns to the forest formerly owned by his family and, in his hatred of the tendencies of the new world, kills the game and burns the woods rather than surrender the land he cherishes to the state. A destructive element is also evident in *Die blauen Schwingen* (1925) and *Der Knecht Gottes Andreas Nyland* (1926). *Die kleine Passion* (1929) and its sequel *Jedermann* (1932) trace the childhood and war experiences of the young man Johannes, contrasting stark evil and good: this antithetical manner of presentation was resumed in the Masurian novel *Die Magd des Jürgen Doskocil* (1932), in which the simple goodness of Jürgen and Marte is exposed to the treacherous local Mormon community. The novel, Wiechert's first real success, ends on an optimistic note. *Die Majorin* (1934) tells of a soldier, posted missing believed dead, returning to his East Prussian native region; the Majorin of the title, a major's widow, by establishing a mother–son relationship with him, helps him during the difficult phase of rehabilitation. *Das einfache Leben* (1939), which treats once more the problem of Wiechert's generation, ends in resignation.

Before the 1939–45 War several collections of stories appeared, including *Der silberne Wagen* (1928), *Die Flöte des Pan* (1930), *Der Todeskandidat* (1934), *Das heilige Jahr* (1936), and *Atli der Bestmann* (1938). In the separately published *Hirtennovelle* (1935) his sympathy is devoted to the obscure and inarticulate.

Wiechert published two novels directed against the National Socialist regime and dealing with the post-war period; *Die Jerominkinder* (q.v., 2 vols.), partly written during the war, appeared 1945–7, and *Missa sine nomine* (q.v.) in 1950. A further novel, *Der Exote*, written in 1932, was published posthumously in 1951. *Sämtliche Werke* (10 vols.) appeared in 1957 and *Werke* (5 vols.) in 1980.

WIED, MARTINA, pseudonym of Alexandra Martina Schnabl (Vienna, 1882–1957, Vienna), whose married name was Weisl. In 1938 she emigrated to England, where she became a teacher, returning to Vienna in 1950.

An undogmatic Christian, Martina Wied published an early volume of Impressionistic verse (*Bewegung*, 1919), but is chiefly known for her complex and panoramic novels, which seek to interpret the confusion and suffering of the human lot. They include *Rauch über Sanct Florian oder Die Welt der Mißverständnisse* (1936), *Kellingrath* (1950), *Das Krähennest* (1951), and *Die Geschichte des reichen Jünglings* (1952). She was also the author of Novellen (*Das Einhorn*, 1948, *Der Ehering*, 1954, *Das unvollendete Abenteuer*, 1955), and in 1952 published *Brücken ins Sichtbare*, another volume of poetry. *Jakobäa von Bayern* (1951) is autobiographical.

Wiedertäufer, Anabaptists, sectarian groups which broke away from the Lutheran and Calvinistic main streams of the Reformation (q.v.) and cultivated an individualistic biblical Christianity. Anabaptism originated in Zürich *c.* 1525, and because its adherents rejected secular interference in religious matters it became subject to political persecution. Most Anabaptists led quiet and humble lives, but in 1534 a revolutionary group of Wiedertäufer seized Münster by force and held it against siege for sixteen months (see MÜNSTER).

Wie die Alten den Tod gebildet, an essay published by Lessing in 1769. It attacks the macabre death symbolism of northern Europe and gives instances of antique representations of death as a youth bearing an inverted extinguished torch. The essay continues the polemics of the *Briefe antiquarischen Inhalts* (q.v.), adding more positive and constructive ideas.

WIEGAND VON MARBURG is the author of a Middle High German verse chronicle of the Teutonic Order (see DEUTSCHER ORDEN) from 1311 to 1394. His work, written at the end of the 14th c., survives only in relatively short fragments, but a Latin translation of 1464 proves that it was a substantial work of some 25,000 lines. Wiegand, who is known to have been still living in 1419, was the herald of the Order.

'Wie herrlich leuchtet/Mir die Natur!', first two lines of Goethe's poem *Mailied* (q.v.).

WIELAND, CHRISTOPH MARTIN (Oberholzheim nr. Biberach, 1733–1813, Weimar), spent his boyhood in Biberach until he was sent in 1747 to the school of Kloster Bergen near Magdeburg, which was conducted according to pietistic principles. After a short period of study at Erfurt he returned home and soon fell in love with his cousin Sophie von Gutermann (see LA ROCHE, SOPHIE VON), to whom he became engaged. His first published works were *Die Natur der Dinge* (1752), a didactic poem in six books of Alexandrine verse based on Lucretius' *De Rerum*

Natura, and *Lobgesang auf der Liebe* (1751, the year of its composition, just after the first title), in Klopstockian hexameters.

A MS., *Hermann* (see ARMINIUS), attracted the attention of J. J. Bodmer (q.v.), whose interest was further engaged by Wieland's flattering essay on the Swiss scholar's epic *Noah*, published under the title *Abhandlung von den Schönheiten des epischen Gedichtes 'Der Noah'* (1753). Bodmer thereupon invited Wieland to Switzerland, where he remained, first as Bodmer's guest (1752–4) in Zurich and then (until 1760) as a private tutor in Zurich and Berne. During this time, usually called Wieland's 'seraphic period', he wrote the sentimental *Briefe von Verstorbenen an hinterlassene Freunde* and a Bodmerian epic in hexameters *Der geprüfte Abraham* (both 1753). The pietistic vein of these works was continued in the two prose works *Sympathien* (1756) and *Empfindungen eines Christen* (1757). In these 'seraphic' years Wieland also wrote the very unseraphic satire against J. C. Gottsched (q.v.), *Ankündigung einer Dunciade für die Deutschen* (1755), and *Lady Johanna Gray* (1758), a tragedy in blank verse, which was performed in Winterthur. While he was in Switzerland his engagement with Sophie von Gutermann was broken off.

Towards the end of his years in Switzerland he was temporarily attracted to Julie von Bondeli of Berne, to whom he was betrothed. In 1760 he returned to Biberach, was elected a senator and appointed acting town clerk, being confirmed in this office in 1764. His second betrothal was revoked soon after because of a love-affair with Christine Hagel in Biberach in 1763.

In the 1760s Wieland came under the influence of Graf Stadion (1691–1768), a wealthy man of the world and an admirer of French literary style, who lived at Warthausen. Wieland, easily susceptible to the influence of mature personalities, swung rapidly over to a rationalistic philosophy and to light erotic poetry, which some resented as frivolous. *Der Sieg der Natur über die Schwärmerey oder Die Abentheuer des Don Sylvio von Rosalva* (q.v., 2 vols., 1764, reissued 1963), an ironical review of fantastic enthusiasms, was his first novel; it was soon followed by *Geschichte des Agathon* (q.v., 2 vols., 1766–7, reissued 1961), in which a young man, after errors and temptations, eventually adopts a balanced rationalism. The work is one of the most perfect expressions of enlightenment (see AUFKLÄRUNG). During these years Wieland also made his translation of 22 plays of Shakespeare (q.v.) into German prose (except for *A Midsummer Night's Dream*, which he turned into verse). Wieland's erotic playfulness appeared in his *Komische Erzählungen* (1765) in verse. A more serious view of love appears in the poem *Musarion oder Die Philosophie der Grazien* (q.v., 1768, reissued 1970).

In 1769 Wieland, who had married Dorothea von Hillenbrand in 1765, was elected to a professorship at Erfurt University. Here he wrote *Sokrates mainomenos oder Die Dialogen des Diogenes von Sinope* (1770) and a political novel, *Der goldene Spiegel* (q.v., 4 vols., 1772), which changed the course of his life, for it influenced the dowager Duchess Anna Amalia of Sachsen-Weimar to appoint him tutor to her sons (1772). Though Wieland was not a success as a pedagogue, he was well regarded as a man of letters and an ornament to the court circle, and he spent the rest of his life in or near Weimar. His Singspiel *Alceste* (1773) soon involved him in a brush with Goethe, in which his quiet tolerance showed up better than the young man's brash satire. In 1773 he launched *Der teutsche Merkur* (q.v.), which was a leading periodical in German intellectual life for 37 years. A satirical novel, *Die Abderiten. Eine sehr wahrscheinliche Geschichte* (q.v.), appeared in 1774 (reissued 1961), a verse romance, *Oberon* (q.v.), regarded by some as his best work, in 1780. Though Wieland continued to write original works in his later years (*Neue Götter-Gespräche*, 1791, *Aristipp*, 1800–1), much of his interest was given to translations, lovingly fashioned, of Horace and Lucian.

Versatile and protean, Wieland has often been berated by the more moralistic critics for lack of character. His early religious sentimentality was foreign to his real nature, and for the rest of his life he reflected in diverse original works and translations a *via media* of rational, sensual life. Without fully intending it, he was an exponent of classicism as well as, by his fantasy and orientalism, a forerunner of Romanticism (see ROMANTIK). Above all he wrote as a civilized man, having at his finger-tips a medium of sensitive, flexible, expressive prose, which brought to the German language qualities hitherto regarded as primarily French.

Wieland's *Sämtliche Werke, Ausgabe letzter Hand* (45 vols.) appeared 1794–1811 and was reissued (14 vols. with suppl. vol.), ed. H. Radspieler, in 1984; the *Historisch-kritische Ausgabe*, ed. B. Seuffert, from 1909, *Werke* (40 vols.), ed. H. Düntzer, 1867–79, and a selection by F. Martini and H. W. Seiffert (5 vols.) 1964–8. *Briefwechsel*, ed. H. W. Seiffert, appeared 1963 ff. (*c.*10–15 vols.).

'Wie lieblich hallt [schallt] durch Busch und Wald', first line of the song *Waldhornlied*, written in 1816 by C. von Schmid (q.v.) and set to music in the same year by F. Silcher (q.v.).

Wien (Vienna), capital of the Federal State of Austria since 1918 and residence of the emperors of Austria since 1806.

Vienna is on the site of the Roman frontier fortress of Vindobona, and is mentioned in 881 as Wenia. It became a city in 1137, and was from 1156 to 1246 the seat of the Babenberg (q.v.) dukes of Austria. After 1278 (see RUDOLF I) Vienna became the official seat of Habsburg and from 1438 to 1918 of Habsburg emperors. The university, the oldest university of a German-speaking land, was founded in 1365. The medieval city survives in important ecclesiastical buildings: the largest is St. Stephen's Cathedral (Stephansdom, 1304–1450), with south transeptal tower and spire (c. 1350–1433), affectionately known as 'der Steffl'. The Ruprechtskirche is the oldest, St. Maria am Gestade the quaintest of the smaller medieval churches. The Habsburg palace (Hofburg) is a huge agglomeration of buildings of disparate styles.

Vienna was besieged by the Turks in 1529 and again in 1683 (see STARHEMBERG, RÜDIGER, GRAF VON). The layout of the city within the ancient girdle of fortifications is largely preserved, together with many notable 17th-c. and 18th-c. palaces of the nobility. The most impressive 18th-c. baroque survivals are the Nationalbibliothek (formerly Hofbibliothek, Fischer von Erlach, 1723–6), Karlskirche (Fischer von Erlach, 1723–5), Prince Eugene's Belvedere (J. L. von Hildebrandt, q.v., 1696–1723), and the palace of Schönbrunn (Fischer von Erlach, then N. Pacassi, 1695–1749).

With the abolition of the Holy Roman Empire in 1806, Vienna became the capital of the Empire of Austria. The city walls, which provided agreeable promenades, were demolished in the 19th c. (a fragment, the Mölkerbastei, survives). In 1814–15 Vienna with its Congress (see WIENER KONGRESS) was the diplomatic centre of Europe. Its reputation for gaiety was sustained not only by the Congress, but also by the succession of Viennese waltz composers (see STRAUSS, J.) and the annual carnival with its balls (Redouten). The musical tradition of Vienna, from c. 1770 to almost 1914, was the most important in Europe, with the Austrians Haydn, Mozart, F. Schubert, A. Bruckner, and Mahler, and the 'adoptive Viennese' Beethoven, Brahms, and R. Strauss (qq.v.). The city's literary tradition is dual, a popular theatre (see VOLKSSTÜCK, LEOPOLDSTÄDTER THEATER) and a sequence of poets, novelists, and dramatists, including F. Grillparzer, A. Grün, J. N. Nestroy, E. von Bauernfeld, M. von Ebner-Eschenbach, A. Schnitzler, H. von Hofmannsthal, A. Wildgans, R. Musil, and H. von Doderer (qq.v.). The Burgtheater (q.v.), founded in 1776, was the home of serious drama. The university is especially famous for its medical school, and it was long the principal centre for psychoanalysis (see FREUD, S.).

WIENBARG, LUDOLF (Altona, 1802–72, Schleswig), studied at Kiel University, was then for a time a private tutor, and after further study qualified as a university teacher at Kiel, where he gave a series of lectures, published in 1834 as *Ästhetische Feldzüge*, which constitute his best-known work. They call for a rejection of the pseudo-medievalism of the Romantic age (see ROMANTIK) and for a renewal of literature by bringing it into touch with contemporary life. Almost inadvertently he gave the new political realism its title by dedicating it to 'Young Germany': 'Dir, junges Deutschland, widme ich diese Reden'. He was one of the authors named in the federal proclamation directed in December 1835 against Junges Deutschland (q.v.).

After publishing a travel book (*Holland in den Jahren 1831 und 1832*, 2 vols., 1834), written in the manner of Heine's *Reisebilder* (q.v.), Wienbarg looked in *Wanderungen durch den Tierkreis* (1835) at the social and political face of his time from the Liberal point of view, and defined in his essays *Zur neuesten Literatur* (1835) the political tendency of the new literature. After a review, planned together with Gutzkow (q.v.), was vetoed by the censorship, Wienbarg settled as a journalist in Altona. Of his *Die Dramatiker der Jetztzeit* (1839) only one instalment appeared, devoted to Uhland (q.v.). In 1848 he served as a volunteer in the Schleswig-Holstein campaign. His mental stamina began to fail and his next works (*Darstellungen aus den schleswig-holsteinischen Feldzügen*, 1850–1, and *Geschichte Schleswigs*, 1861–2) remained uncompleted. In 1868 his mind gave way and he was confined in the mental hospital at Schleswig.

Wienbarg's *Ästhetische Feldzüge* and *Zur neuesten Literatur* were reprinted, ed. W. Dietze, in 1964.

Wiener Exodus, title given to a Middle High German poetic narration of the first fifteen chapters of Exodus, ending with the destruction of the Egyptians in the Red Sea. It was written c. 1120 and the dialect is identical with that of the somewhat earlier *Wiener Genesis* (q.v.). The MS. is in Vienna. The poem also occurs in the Milstätter Handschrift (q.v.).

Wiener Genesis, so called from the present location of the MS., is a free rendering of the Book of Genesis in rhyming verse, interpreted from the standpoint of Christianity. Its unknown clerical author, a man of learning and experience and of no little poetic ability, was an Austrian and wrote the poem probably not long before 1075. The *Milstätter Genesis* is a later MS. version of the same work, written c. 1120, in which the style has been 'modernized', so that it matches with that of the *Wiener Exodus* (q.v.), which also occurs in the Milstätter Handschrift (q.v.).

Wiener Gruppe, a coterie of *avant-garde* Viennese writers, founded in 1952 by H. C. Artmann, F. Achleitner, G. Rühm (qq.v.), Konrad Bayer, and Oswald Wiener. Artmann left it in 1958, and the group ceased to meet shortly before Bayer's suicide in 1964. Rühm edited the volume *Die Wiener Gruppe. Achleitner. Artmann. Bayer. Rühm. Wiener Texte. Gemeinschaftsarbeiten. Aktionen* (1967, ext. edn. 1985). (See KONKRETE POESIE.)

Wiener Hundesegen, a magic spell for the protection and preservation of dogs, probably sheep-dogs. It preserves the dual structure of most spells, but the imperative second section is replaced by a prayer. It was written in the 10th c. in a MS. now in Vienna (Nationalbibliothek). It is likely that it is the Christianization of an ancient heathen spell. (See also ZAUBERSPRÜCHE).

Wiener Kongreß, 1814–15. Following the collapse of the Napoleonic Empire it was agreed by the signatories of the First Treaty of Paris (30 May 1814, see NAPOLEONIC WARS) that a representative gathering at Vienna should settle the European issues involving above all the future of Germany.

The Treaty of Vienna, comprising 121 Articles, was signed on 9 June 1815. The return of Napoleon from Elba and his defeat at Waterloo (q.v.) led to the Second Treaty of Paris (20 November 1815) imposing more stringent terms on France, although under Talleyrand France had recovered relatively high prestige at Vienna. In the European context the Great Powers recovered territories (or their equivalent) restoring the *status quo* of 1805, although Russia obtained more (at the expense of Poland, including Warsaw, which had been annexed by Prussia, see POLAND, PARTITIONS OF).

Against the nationalist aspirations for German unity which had inspired the War of Liberation and which was strongly advocated by the Prussian reformer Stein (q.v., acting as adviser to the Russian Tsar; Prussia was represented by Hardenberg, q.v.), the Austrian chancellor Metternich (q.v.) succeeded in moving the Congress to adopt his own principles based on a balance of powers in a loose confederation of states (see DEUTSCHER BUND). The settlements of Vienna were followed by a period of peace lasting until 1848. Yet the reactionary spirit of the principles underlying the Federal Acts of the German Confederation led, under the powerful influence of Metternich, to the ruthless suppression of progressive and liberal ideas (see KARLSBADER BESCHLÜSSE). The demand for constitutional government in the individual states, which the Federal Act had conceded, became a principal issue for internal unrest in the years to come. The phrase 'der Kongreß tanzt', first coined by Charles Joseph, Prince de Ligne ('Le congrès ne marche pas, il danse', *Souvenirs du Congrès de Vienne*, 1814–15), expresses irony and disillusionment at the fact that real work at the Congress was done by a few while the majority of the representatives were given little say and plenty of entertainment. Accounts of the open and secret goings on were recorded by Gentz (q.v.) who, at that time Metternich's personal secretary and publicity agent, acted as Secretary-General to the Congress.

Wiener Kreis, a group of neo-positivist philosophers formed in Vienna in 1929; it broke up in 1938. Its principal members were R. Carnap (1891–1970) and O. Neurath (1882–1945); the group was influenced by L. Wittgenstein (q.v.).

Wiener Meerfahrt, Die, a comic verse tale of the 13th c., in which drunken Viennese burghers in an inn fancy that they are on the way to the Holy Land and tossing on the sea. They throw a fellow tippler into the street (believing it to be the sea) and have to pay more in compensation than a real pilgrimage would have cost. The poet, probably an itinerant singer from Bohemia, gives his name as 'Der Freudenleere'.

Wiener Osterspiel, a medieval play originating in Rhenish Hesse in the second half of the 15th c. It is concerned with the events immediately before and after the Resurrection.

Wiener Passionsspiel, a passion play dating from the 13th c., though the MS. belongs to the 14th c. Only fragments amounting to *c.* 500 lines survive; they begin with the Fall of Lucifer. There are resemblances to the *Benediktbeurer Osterspiel* (see BENEDIKTBEURER SPIELE) and the *Alsfelder Passionsspiel* (q.v.).

Wiener Volksstück, see VOLKSSTÜCK.

Wie Uli der Knecht glücklich wird, a novel written by J. Gotthelf (q.v.) in 1840 and published in 1841. An abridged form, *Uli der Knecht*, was used for the edition of 1846 and the novel is often referred to by this title. The book bears as a sub-title *Eine Gabe für Dienstboten und Meisterleute*, a clear indication of its didactic intention.

Uli is a farm-hand working for farmer Johannes (der Bodenbauer) and wasting his money on drink, and spoiling his best clothes in tavern brawls. Johannes sees good qualities latent in Uli and gradually succeeds in diverting him to thrift and a sensible way of life. In spite of several back-slidings (the chief of which occurs when he is chosen to play at 'Hurnussen' (a ball game) against a neighbouring village), he justifies the trust placed in him and becomes a valued farm servant with a nest-egg of savings. Johannes is asked by a distant relative, the elderly farmer Joggeli, if he can recommend a good foreman, and Johannes puts forward Uli.

On the new farm at Glunggen Uli has to cope not only with disaffected farm servants, but also with Joggeli's mistrust. He makes his way, in spite of intrigues against him, of Joggeli's invincible suspicion, and of moments of despondency. His main support comes from the farmer's wife and her poor relation, the girl Vreneli, who is the mainstay of the farmhouse. Uli makes Joggeli dismiss the two worst rebels, but a new temptation besets him when the farmer's silly, affected, citified daughter Elisi makes a set at him. Uli is saved from a disastrous marriage by Elisi herself, who picks up a more impressive suitor at a spa. This man, a cotton dealer, plays his cards adroitly and secures parental consent, but he interferes so much on the farm that he overplays his hand. Joggeli's wife conceives the idea of saving the farm by Joggeli's retirement and a subsequent lease to Uli, who is to marry Vreneli. Vreneli, however, will not be treated as a chattel, and for a time, though she loves him, she does battle against Uli. The cotton merchant attempts to intervene against the leasing, but makes the mistake of making advances to Vreneli, who puts him to flight. Uli and Vreneli are married, with farmer Johannes and his wife as guests at the wedding, and they enter on their farming lease as a perfectly matched pair.

The novel, which has plentiful dialogue in Swiss dialect, is notable for its gallery of characters, ranging from the idealized harmony of Johannes and his wife, to the distrustful Joggeli; in the forefront are Uli, who develops through many relapses into a mature man, and Vreneli, a portrait of a woman as practical as she is warm-hearted; these are supplemented by satirical caricatures of the characters associated with urban life (Elisi and her brother and the cotton dealer). Gotthelf has provided a number of vivid and memorable scenes, including the epic struggle between the jealous maids Trini and Ursi after the former has fallen into the cesspit, the haymaking scene and the ensuing storm, Vreneli's victory over the cotton dealer, Uli's and Vreneli's walk through the blizzard to the parsonage, and their drive to the wedding.

Wigalois, see WIRNT VON GRAFENBERG.

Wigamur, a degenerate Arthurian romance of some 6,000 lines. Its authorship is unknown. The hero, Wigamur, carried off in childhood by a mermaid, rescues an eagle, which, after the manner of Iwein's lion, accompanies him henceforth. He rescues a maiden, becomes a Knight of the Round Table, and, after further adventures, confronts a knight, who proves to be his father. A recognition ensues and the father hands his realm to his son.

The poet, who was clearly not at home in the chivalric world, is an eclectic, drawing on HARTMANN von Aue, WOLFRAM von Eschenbach, *Lanzelet, Wigalois* (qq.v.), and many other models. Some authorities opt for 1250 as the approximate date, but it is possible that it was written some thirty or forty years later.

WILBRANDT, ADOLF VON (Rostock, 1837–1911, Rostock), son of a professor at Rostock University, studied at Rostock, Berlin, and Munich universities, and then settled in Munich where he was active as a journalist. After publishing in 1863 a monograph on H. von Kleist (q.v.), he wrote the novels *Geister und Menschen* (3 vols., 1864) and *Der Lizentiat* (3 vols., 1868) and two volumes of Novellen (*Novellen,* 1869, and *Neue Novellen,* 1870). A study of Hölderlin (q.v.) appeared in 1870.

In 1871 Wilbrandt moved to Vienna, where he married two years later Auguste Baudius, an actress of the Burgtheater (q.v.) company. In these years he devoted himself chiefly to verse tragedy and became a favourite author of the Burgtheater public. *Gracchus der Volkstribun* (1872) received the Schiller prize, and *Arria und Messalina* (1874), *Giordano Bruno* (1874), *Nero* (1876), and *Kriemhild* (1877) followed. He also wrote the comedies *Jugendliebe* (1873) and *Fridolins heimliche Ehe* (1876). With *Meister Amor* (2 vols., 1880) Wilbrandt returned to the novel, beginning a group of works in which personalities of the artistic world of Munich were recognizably depicted. For six years, 1881 to 1887, he was director of the Burgtheater, and in 1884 was ennobled by King Ludwig II (q.v.) of Bavaria. During his direction of the theatre he gave prominence to the classics and neglected modern plays. After resigning he returned to Rostock and soon after published the verse play (dramatisches Gedicht) *Der Meister von Palmyra* (q.v., 1889), perhaps his best-known work.

Wilbrandt's later fiction includes the novels *Adams Söhne* (1890), *Hermann Ifinger* (1892), *Der Dornenweg* (1894), *Die Osterinsel, Die Rothenburger* (both 1895), *Hildegard Mahlmann* (1897), *Franz* (1900), *Villa Maria* (1902), *Hiddensee* (1910), and *Die Tochter* (1911), and the collections of stories *Novellen aus der Heimat* (2 vols., 1882) and *Opus 23* (1909). Many of his works are conspicuously moralistic and have affinities with the Bildungsroman (q.v.). Wilbrandt's memoirs appeared as *Erinnerungen* (1905) and *Aus der Werdezeit* (1907).

WILD, SEBASTIAN, a Meistersinger of Augsburg in the 16th c., wrote a passion play in 1566, part of which is incorporated in the *Oberammergauer Passionsspiel* (q.v.). His poems are included in the Colmarer Liederhandschrift (q.v.).

WILDE ALEXANDER, DER, is the name attributed to a Middle High German poet of the late 13th c. He was of South German or more probably Swiss origin, and has usually been taken for a commoner, though some hold him to be of minor nobility. He was clearly, like WALTHER von der Vogelweide (q.v.), a wandering singer, who depended on his reception at courts and castles.

Der wilde Alexander sets a high value upon poetic art, insisting upon its descent from kings. The majority of his poems are Sprüche (see SPRUCH), and some of them deplore the decadence of the age, of which the disaster of the fall of Acre (to which one poem alludes) is a symptom. Though some of his poetry is obscure, Alexander can also write with great simplicity and apparent feeling. Yet even his most homely and direct poems are given a spiritual interpretation, as in the attractive poem beginning 'Hie vor, dô wir kinder waren', with its picture of the strawberry-picking children, which turns into a warning against sensual pleasures.

Der wilde Alexander also wrote a noble Christmas poem and a *minneleich* (see LEICH).

Wilde Jäger, Der, a ballad written by G. A. Bürger (q.v.) in 1777 and first published in the *Göttinger Musenalmanach 1786*. It is the story of an impious noble huntsman, cruel to man and beast, who is condemned to be himself pursued for ever by devils (das wilde Heer).

WILDENBRUCH, ERNST VON (Beirut, 1845–1909, Berlin), whose father, an illegitimate son of Prince Louis Ferdinand (q.v.) of Prussia, was in the Prussian consular service, was born within the Turkish Empire and spent his early years in south-east Europe (Constantinople, 1851, and Athens, 1852–7) as well as in Berlin.

In 1859 Wildenbruch was sent to the Cadet House at Potsdam, transferring in 1860 to the Cadet School at Berlin-Lichterfelde. Commissioned in 1863 in the 1st Regiment of Foot

Guards, he disliked Potsdam garrison life and resigned his commission in 1865, but rejoined in 1866 and 1870. He served with his regiment in Bohemia in 1866, but in 1870 his retention in a home battalion of another regiment caused him bitter resentment. Between the wars (see DEUTSCHER KRIEG and DEUTSCH-FRANZÖSISCHER KRIEG) he studied law in Berlin and was appointed to the judicial branch of the civil service in 1871, serving first in Frankfurt and then for a short time as magistrate in Eberswalde.

From 1877 to 1900 he was attached to the German Foreign Ministry, retiring with the rank of Counsellor of Legation (Geheimer Legationsrat).

Wildenbruch began to write poetry during his brief career as a regular officer, but his first publication was the satire *Die Philologen am Parnaß* (1868). A play, *Spartakus* (1873), was not performed. The short epic poems *Vionville* (3 cantos, 1874) and *Sedan* (3 cantos, 1875) were successful with the nationalistically minded public of the seventies. The once popular story *Der Meister von Tanagra* reflected the interest in the *objets d'art* (Tanagrafiguren) unearthed in the excavations at Tanagra in Boeotia in 1873. In the 1880s Wildenbruch turned his hand to play-writing. His dramas not only attracted attention, but were taken up by the Meiningen company (see MEININGER), which in 1881 performed his historical tragedy *Die Karolinger* (1882); it was followed by *Väter und Söhne* (1882), *Der Menonit* (q.v., 1882), *Christoph Marlow* (q.v., 1885), *Das neue Gebot* (1886), and *Die Quitzows* (q.v., 1888), the contemporary stage success of which is reflected in Th. Fontane's *Die Poggenpuhls* (q.v.). *Der Generalfeldoberst* (1889) and *Der neue Herr* (1891), plays about the Hohenzollerns, continued this trend. Wildenbruch turned to Naturalism with *Die Haubenlerche* (q.v., 1891) and *Meister Balzer* (1893), but reverted to historical plays, achieving another great theatrical success with a trilogy on the subject of the Emperor Heinrich IV (q.v.), *Heinrich und Heinrichs Geschlecht* (q.v., 1896). His last plays were *Die Tochter des Erasmus* (1899) and *Die Rabensteinerin* (1907).

Wildenbruch's Novellen include the collection *Kindertränen* (1884, *Der Letzte* and *Die Landpartei*), *Das edle Blut* (1892), *Claudias Garten* (1895), *Neid* (1900), and *Vize-Mama* (q.v., 1902). They retain more vitality than his plays and reveal a sensitiveness which is obscured in many of the plays by his nationalistic zeal. Wildenbruch's poetry appeared in *Dichtungen und Balladen* (1884, which in 1892 appeared as *Lieder und Balladen*) and in 1909 in *Letzte Gedichte*.

Gesammelte Werke (16 vols.), ed. B. Litzmann, were published 1911–24, and *Ausgewählte Werke* (4 vols.), ed. E. Elster, 1919.

WILDER MANN, the author of a poem, written *c.* 1170, dealing with the legend of Saint Veronica and the Emperor Vespasian. He was a priest in Cologne and the view has been advanced that the name is that of a house, which the owner's family assumed. He is also the author of a moral poem which preaches against avarice (*Von der girheit*).

WILDERMUTH, OTTILIE (Rottenburg/Nekkar, 1817–77), *née* Rooschütz, married in 1843 Professor J. D. Wildermuth of Tübingen. Described by a contemporary as an 'echte und rechte Hausfrau in glücklichster Ehe', she wrote many short stories and Novellen (*Bilder und Geschichten aus dem schwäbischen Leben*, 1862; *Schwäbische Pfarrhäuser*, not in book form until 1910), which had considerable popularity in the 19th c. Her stories appeared first in magazines. Her *Gesammelte Werke* (10 vols.) were published 1891–4.

WILDGANS, ANTON (Vienna, 1881–1932, Mödling nr. Vienna), a Viennese lawyer, gave up his legal practice for literature in 1912. After the 1914–18 War he had two short spells in 1921–2 and 1930–1 as director of the Burgtheater (q.v.). He was a fluent verse writer, evoking the quaint atmosphere of corners of old Vienna, or expressing the variations of love or the pangs of poverty with equal facility. His verse suggests, but does not equal, that of Baudelaire or Rilke.

Wildgans's collections of poetry are *Vom Wege* (1903), *Herbstfrühling* (1909), *Und hättet der Liebe nicht* (1911), *Die Sonette an Ead* (perhaps his best verse, 1913), *Österreichische Gedichte 1914–15* (1915), *Dreißig Gedichte* (1917), and *Wiener Gedichte* (1926). He published *Die sämtlichen Gedichte* (3 vols.) in 1923. He also wrote plays, some Naturalistic (*In Ewigkeit Amen*, 1913; *Armut*, 1914; *Liebe*, 1916, although in the last two, verse passages occur), some symbolical and Expressionistic (*Dies irae*, 1918; *Kain*, 1920). He published his reminiscences as *Musik der Kindheit* in 1928.

Ein Leben in Briefen, correspondence (3 vols.), ed. L. Wildgans, appeared in 1947, *Sämtliche Werke. Historisch-kritische Ausgabe* (8 vols.) 1948 ff., and a volume *Gedichte, Musik der Kindheit* in 1981. His correspondence with H. von Hofmannsthal (q.v.) appeared in 1935 and, in extended form, ed. N. Altenhofer, in 1971.

WILHELM I, DEUTSCHER KAISER UND KÖNIG VON PREUSSEN (Berlin, 1797–1888, Berlin), the second son of Friedrich Wilhelm III (q.v.), entered the Prussian army, which remained his lifelong interest. The year 1848 (see REVOLUTIONEN 1848–9) demonstrated his readiness to call out the military, and gave him the nickname Kartätschenprinz (Grapeshot Prince). Since he was thought to be in personal danger, he was persuaded to take temporary refuge in England. In 1849 he commanded the force which suppressed a revolt in Baden. In 1858 the illness of his brother King Friedrich Wilhelm IV (q.v.) obliged the Prince to undertake the regency, and in 1861 he succeeded to the throne.

The years of Wilhelm's regency and early reign were marked by conflict with the Prussian Diet, rising to a climax over a programme of military reform. The appointment of Bismarck (q.v.) as chancellor in 1862 solved Wilhelm's problems. Bismarck outmanœuvred the Diet, and was able to pursue a policy of Prussian assertion and German unification in a series of short and successful wars (see DEUTSCHER KRIEG and DEUTSCH-FRANZÖSISCHER KRIEG). The King was unwilling to make war, but reluctant to restrain his forces after victory was secured; Bismarck was able on each occasion to persuade him, and Wilhelm came to regard his chancellor as an indispensable collaborator.

In 1871 Wilhelm unwillingly accepted the title German Emperor. In later years this once unpopular figure became a symbol for the new Germany. Wilhelm died in his ninety-first year. His Queen, Augusta (1811–90), a princess of Saxe-Weimar and a Liberal in politics, was an opponent of Bismarck.

WILHELM II, DEUTSCHER KAISER UND KÖNIG VON PREUSSEN (Potsdam, 1859–1941, Doorn, Netherlands), was the son of the Crown Prince Friedrich Wilhelm (see FRIEDRICH III) and of his consort, Queen Victoria's eldest daughter the Princess Royal (see KAISERIN FRIEDRICH).

Under his tutor, G. Hinzpeter, Wilhelm had a Spartan upbringing. After coming of age he served, despite his crippled left arm, in the 1st Foot Guards and the Guard Hussars. He was on bad terms with his parents, especially with his mother. His father, a sick man when he succeeded Wilhelm I (q.v.) in March 1888, died after a reign of ninety-nine days, and in June 1888, at the age of 29, he became German Emperor as Wilhelm II.

'The Kaiser', as this Emperor is known in Britain, inherited from his grandfather the dynamic but aged Chancellor Bismarck (q.v.), who had created the German Empire. Though relations at first seemed normal, Wilhelm dismissed the Chancellor in 1890, replacing him with Caprivi (q.v.), a general with little inclination for, and no experience of, politics. He was followed in 1894 by the 75-year-old Prince Hohenlohe (q.v.). During Hohenlohe's chancellorship relations with foreign powers were adversely affected by two imperial indiscretions.

Wilhelm sent an anti-British telegram to President Krüger after the Jameson Raid (1896, see KRÜGERDEPESCHE), and in 1900 he made a speech to German troops about to sail for China inciting them to outdo the Huns in ferocity (Hunnenrede). In 1898 he enthusiastically endorsed Tirpitz's plan to build a large navy, which led to rivalry with Great Britain (see TIRPITZ). Hohenlohe was replaced in 1900 by Bülow (see BÜLOW, BERNHARD VON), an adroit politician of impeccable pedigree. In the following years the European situation deteriorated further, partly because of the Emperor's blustering speeches, which were an expression of his unstable personality and conflicted with his fundamentally peaceful intentions.

In 1908 an indiscreet interview was allowed to be published in an English newspaper (see DAILY TELEGRAPH-AFFÄRE). Bülow, the responsible minister, made no attempt to protect Wilhelm from the storm of protest at home and abroad, and the Emperor was obliged to make a humiliating retraction. In 1909 he seized an opportunity to dismiss Bülow, and appointed von Bethmann Hollweg (q.v.) in his place. In the crisis of July 1914 the Emperor, though filled with indignation against the Serbs, sought to avoid war, but was too irresolute to restrain his officials and the General Staff. During the 1914–18 War (see WELTKRIEGE, I) he lived much at G.H.Q. (OHL, q.v.) at Spa, but was a merely nominal military leader. Out of touch with his people, and living in an unreal world, he was suddenly faced in 1918 with defeat and revolution. Evading demands for his abdication, he took refuge in Holland and spent the rest of his life at Doorn. His final indiscretion was a congratulatory telegram sent to Hitler (q.v.) in 1940 after the defeat of France, bearing the signature 'Wilhelm, R. I.'.

His first wife was Auguste Viktoria, a princess of Augustenburg-Holstein (1858–1921). In 1922 he married Princess Hermine von Schönaich-Karolath (1887–1947).

His writings, which are of little historical importance, include *Ereignisse und Gestalten 1878–1918* (1922), *Aus meinem Leben 1859–88* (1927), and *Meine Vorfahren* (1929).

WILHELM, GRAF VON SCHAUMBURG-LIPPE (London, 1724–77, Bergleben), was educated partly in England and partly abroad, studying successively at Montpellier, Geneva, and Leyden, where he combined military and mathematical studies. In 1743 he joined his father, the commander-in-chief of the Dutch army, and fought beside the British at Dettingen on 17 June (see ÖSTERREICHISCHER ERBFOLGE-KRIEG and PRAGMATISCHE ARMEE). In 1748 he acceded to the throne of his tiny state, which he governed with skill and humanity. In the Seven Years War (see SIEBENJÄHRIGER KRIEG) he was again in the field, serving in Italy and commanding with success an Anglo-Portuguese force in Portugal in 1762. From 1764 he devoted himself entirely to the welfare of his state. He was an efficient administrator, and also an early advocate of national service as a substitute for the deception and brutality practised by recruiting officers. He sought to gather promising young men for service in the state, appointing Th. Abbt (q.v.) as Konsistorialrat (supervising schools and the Church) in 1765. Abbt unfortunately died in the following year; one of his successors (1771) was J. G. Herder (q.v.). Johann Christoph Friedrich Bach (1732–95) was appointed director of music by Count Wilhelm in 1756 and spent the rest of his life in Bückeburg (see BACH, J. S.).

Wilhelmine, a short prose narrative (Ein prosaisch-komisches Gedicht) by M. A. von Thümmel (q.v.), published in 1764 and consisting of six cantos (Gesänge). The first edition bore the alternative title *oder der vermählte Pedant.* It tells of a country girl who has been taken away and educated at court. The village pastor loves her and takes the opportunity, with the assistance of wine, to propose to her when she visits the village, and she accepts him. She is given leave by the court, a grand wedding takes place with many noble guests, and the marriage is consummated.

The slender story is told with gentle irony in poetic prose. The rococo impression is heightened by delicate erotic detail.

WILHELMINE FRIEDERIKE SOPHIE, MARKGRÄFIN VON BAYREUTH (Berlin, 1709–58, Bayreuth), the eldest daughter of Friedrich Wilhelm I (q.v.) of Prussia and the favourite sister of Friedrich II (q.v.), der Grosse. In 1731 her father married her to Friedrich, the future Margrave of Bayreuth.

Wilhelmine's close friendship with her brother Friedrich was only clouded in the first years of his reign, during which she composed her *Mémoires* (in French), which picture her father and the Prussian court, although not always accurately (2 vols., 1810; the 10th edn. includes her correspondence, 1899). After their reconciliation they corresponded, and Friedrich's letters to her bring out the most affectionate side of his personality. She shared Friedrich's friendship with Voltaire (q.v.), who turned to her in 1753 for a reconciliation with the King after his departure from Berlin. In 1757, after the defeat at Kolin (q.v.), Friedrich entrusted her with a secret mission to Mirabeau, the object of which was to intimate to Mme de

Pompadour the Prussian readiness for a negotiated peace. When she died, Friedrich lost the only woman whom he adored for her affection, loyalty, and intelligence. Voltaire wrote an ode to her memory. Her correspondence with Friedrich was edited by Volz and v. Oppeln-Bronikowski (2 vols., 1924–6).

Wilhelminisches Zeitalter, the period between the dismissal of Bismarck (q.v.) as chancellor by the young Emperor Wilhelm II (see WILHELM II, KAISER) in 1890 and the outbreak of the 1914–18 War. Some writers include the war when referring to the Wilhelmine era.

Wilhelm Meisters Lehrjahre, a novel by J. W. von Goethe, first published 1795–6 in four volumes divided into eight books.

Wilhelm, the son of well-to-do middle-class parents, is in love with the actress Mariane. Interwoven in his love is a passionate interest in the theatre. He is unexpectedly convinced that Mariane is unfaithful to him. When Bk. II opens, he is recovering from the illness into which the shock of Mariane's presumed deception has thrown him. He destroys his poetic writings and devotes himself to business. Some time later he sets out on a journey to visit clients. On the way he meets the carefree young woman Philine and the actor Laertes. He also discovers the mysterious child Mignon among a troupe of circus tight-rope dancers and buys her freedom for her, whereupon she follows him with submissive fidelity. An equally mysterious harper also attaches himself to Wilhelm, who is persuaded to support a troupe of actors led by Melina, a former acquaintance.

The third book sees the troupe invited to perform at a count's mansion, and Wilhelm himself accompanies them as young male lead. A vivid picture of the intrigues of actors and gentry follows. Wilhelm is introduced to Shakespeare's plays by Jarno, one of the guests. Bk. IV begins with the departure of the gentry. The actors, too, leave, are ambushed by robbers, and plundered. Wilhelm is wounded and is succoured by a beautiful lady (die schöne Amazone), who, in the few minutes of their acquaintance, makes the deepest impression upon him. On recovery Wilhelm journeys to the actor and director Serlo, in whose troupe he enrols at the beginning of Bk. V.

A performance of *Hamlet* is planned and the book contains an extensive critical discussion of Shakespeare's play. After the performance a fire breaks out. The mad harper suffers a fit of pyromania and Wilhelm feels obliged to put him in the care of a physician. During the fire Wilhelm first gathers, through the clue 'dein Felix', that he has a son, whom he is urged to

rescue; he has so far believed the 3-year-old Felix to be the child of Serlo's sister Aurelie. Bk. VI moves from these human complications to a world of renunciation and contentment. Headed *Bekenntnisse einer schönen Seele*, it purports to be the autobiography of a woman who devotes her life to God in serenity, and is based on the life and personality of Goethe's mother's friend, Susanna von Klettenberg (q.v.). Wilhelm next goes (Bk. VII) to the castle of Lothario, whose infidelity, he believes, has caused Aurelie's death. Lothario proves to be a responsible man who can justify himself, and Wilhelm learns to his chagrin that he himself is not guiltless: he has disturbed the life of the beautiful countess of Bk. III, and Felix proves to be his son by the now dead Mariane, whom he had rashly abandoned (Bk. I). He finds at the castle that he has been watched over by a secret society, 'die Gesellschaft vom Turm'; his apprenticeship (Lehrjahre) is at an end, and, matured by experience, he is admitted master (Meister). In Bk. VIII Mignon, who has secretly loved Wilhelm, dies, and it is discovered that she is the daughter, by an act of incest, of the harper, who kills himself. The 'schöne Amazone' of Bk. IV, now identified as Natalie, reappears and Wilhelm marries her. He is informed by the secret society that he must complete some years of travel (see WILHELM MEISTERS WANDERJAHRE).

The novel contains eight of Goethe's finest songs, 'Kennst du das Land', 'Nur wer die Sehnsucht kennt', 'Heiß mich nicht reden, heiß mich schweigen', and 'So laßt mich scheinen, bis ich werde' (sung by Mignon), 'Wer nie sein Brot mit Tränen aß', 'Wer sich der Einsamkeit ergibt', 'An die Türen will ich schleichen' (sung by the harper), and 'Singet nicht in Trauertönen' (sung by Philine). Also included are the ballad *Der Sänger* (q.v.) and the satirical poem 'Ich armer Teufel, Herr Baron'.

Wilhelm Meisters Lehrjahre grew out of *Wilhelm Meisters theatralische Sendung* (q.v.) and is in part an adaptation of it. The first four books approximately correspond to the six books of the *Sendung*. The latter was begun in 1777, dropped in 1779, and taken up again in 1780 and 1785; *Wilhelm Meisters Lehrjahre* was adapted from it with considerable alterations and additions in 1794–5. The new work shifts the emphasis from the theatre to the education of the personality by experience, and is an outstanding example of the German Bildungsroman (q.v.). It enjoyed great success with the new generation of Romantics (see ROMANTIK) and was translated into English by Thomas Carlyle (1824). Among early interpretations Schiller's letters to Goethe on the subject and *Charakteristik der Meisterischen Lehrjahre von Goethe* (1798) by F. Schlegel (q.v.) should be mentioned.

Wilhelm Meisters Theatralische Sendung, title of the original version of Goethe's novel about Wilhelm Meister. This title was mentioned by Goethe in a letter to Knebel in 1782. The composition of the *Sendung* seems to have extended from 1777 to 1785. Goethe's MS. is lost, but a copy, made by Barbara Schultheß of Zurich, was discovered in December 1909 and published in 1911 by Harry Maync. This version of the novel is in six books, which, together with certain elements from Bk. V (including the discussion of *Hamlet*), correspond to the first four books of the later *Wilhelm Meisters Lehrjahre* (q.v.).

In comparison with the later version of the novel, the *Sendung* shows a more positive attitude to the theatre, and it appears that, if completed, it would have been an artist novel (Künstlerroman, q.v.), rather than an educational one (see BILDUNGSROMAN).

Wilhelm Meisters Wanderjahre oder Die Entsagenden, Goethe's last novel published in 1821, and in its final, longer, form in 1829 as volumes 21, 22, and 23 in the *Ausgabe letzter Hand*. It was written between 1807 and 1828, with an interval between 1810 and 1821. The *Wanderjahre* is a sequel to *Wilhelm Meisters Lehrjahre* (q.v.) and shows Wilhelm's further progress after the completion of his apprenticeship.

The years of travel begin under the condition (imposed by the Gesellschaft vom Turm), that Wilhelm must not stay more than three days in any one place and, at each remove, must go to a point at least one league distant. As he passes from place to place with his son Felix, he encounters experiences and learns of others' experiences; these are all recounted in the form of stories or Novellen, many of which had been written earlier, chiefly in 1806 (see SANKT JOSEPH DER ZWEITE, DIE PILGERNDE TÖRIN, DIE NEUE MELUSINE, DER MANN VON FUNFZIG JAHREN, DAS NUSSBRAUNE MÄDCHEN, NICHT ZU WEIT). Wilhelm entertained well at the mansion of the Oheim, meets Hersilie, with whom he remains in correspondence, and pays a visit to her benevolent aunt Makarie. He encounters Makarie's wayward nephew Lenardo, who confesses his guilty conscience with regard to Nachodine, a farmer's daughter, whose plea for assistance he has not been able to satisfy. Wilhelm undertakes to search for her. Before setting out, he entrusts Felix to 'die pädagogische Provinz', an idealistic educational establishment in which the pupils learn by diverse experience and activity, aiming above all at the cultivation of reverence (Ehrfurcht), which, in its highest form, is self-reverence ('Ehrfurcht vor sich selbst').

After discovering Nachodine, Wilhelm obtains dispensation from the limitation of sojourn and learns the 'craft' of the surgeon. Trade or craft (Handwerk) plays a considerable part in the closing stages of the novel, in which characters from *Wilhelm Meisters Lehrjahre* reappear as worthy citizens instead of picturesque vagabonds. Wilhelm's practical devotion has its reward, for his surgical skill enables him to save the life of his son Felix. The book concludes with a series of philosophical and practical reflections headed *Aus Makariens Archiv*.

Wilhelm Meisters Wanderjahre did not readily find friends, and it was all too easy to regard it as a repository for a number of smaller works which the author had on hand. Goethe, however, insisted on its symbolic character and its fundamental unity ('ist es nicht aus Einem Stück, so ist es doch aus Einem Sinn'), and it is now easier for the reader to approach the work as a serious and consistent meditation on human life, in the course of which the coming of the industrial age is foreseen, and a vision of two Utopian societies developed.

Wilhelmstraße, metonymous term for the Prussian and, after 1871, German Foreign Ministry, which was situated in the Wilhelmstraße, Berlin.

Wilhelm Tell, a play (Ein Schauspiel) written by Schiller in 1803–4, first performed at Weimar on 17 March 1804, and published later in 1804. Except for the short allegory *Die Huldigung der Künste* (q.v.), it is his last completed play. Schiller expands his play on this now familiar theme (see TELL, W.) around a nucleus drawn from the *Chronikon Helveticum* of Ägidius Tschudi (q.v.). It is written in blank verse.

Two actions run parallel through the play. The tyranny and the outrages perpetrated by the Austrian intruders provoke a movement against them, shaping itself in the democratic assembly on the Rütli into a co-ordinated mass revolt, which, partly owing to the co-incident assassination of the Emperor by Johannes Parricida, is completely successful. The other thread traces the struggle forced upon Tell, who is arrested for failing to salute the Austrian hat set upon a pole, compelled by Geßler to shoot at the apple placed on his son's head, and, when he has successfully achieved this feat, tricked into an admission of murderous intentions towards the Austrian tyrant. Bound and taken on board ship, Tell is relieved of his bonds so that he may help in the navigation of the ship through stormy waters, and uses his freedom to steer a course close to the shore and jump clear. Straight away he sets out to ambush Geßler, whom he shoots dead, not from motives of vengeance, but in order to prevent future atrocities in the land. Tell's ethical position is clearly pointed in the scene in which Johannes Parricida seeks assistance from him in his flight.

The link between the two plots is provided firstly by the apple scene, which stimulates the conspirators to action, and secondly in the scene of general rejoicing at the close.

Wilhelm Tell für die Schule, see FRISCH, M.

WILHELM VON HOLLAND (1227–56) was elected German King in 1247 in succession to HEINRICH Raspe (q.v.). Both were, in fact, anti-kings, since the rightful king, Konrad IV, son of the Emperor Friedrich II (qq.v.), had been elected as early as 1237. From 1247 to 1251 Wilhelm was engaged in hostilities against Konrad, but after Konrad's departure for Italy his situation improved. He was killed fighting against the Frisians.

Wilhelm von Orlens, see WILLEHALM VON ORLENS.

Wilhelm von Österreich, see JOHANN VON WÜRZBURG.

Wilhelm von Wenden, see ULRICH VON ETZENBACH.

Willehalm, a Middle High German epic by WOLFRAM von Eschenbach (q.v.), comprising some 14,000 lines. It was written at the instigation of Landgraf Hermann (q.v.) of Thuringia, was begun *c.* 1210–12, and is unfinished. The source of *Willehalm* is a lost French *chanson de geste* dealing with a legendary battle with the Saracens at Aliscans (near Arles). The extant French poem on this subject was not Wolfram's model. The poem is set in the time of Ludwig I (q.v.), der Fromme.

Willehalm (St. William, Wilhelm, Guillaume) has been taken prisoner by the Saracens, and has escaped with Arabele, the daughter of the Saracen King Terramer. Arabele is baptized in the name Gyburg and is married to Willehalm. The poem falls into three sections, the two battles of Aliscans (Alischans) between which occurs a contrasting episode at the imperial court. The Saracens invade southern France to avenge and recover Arabele-Gyburg. In a fearful battle Willehalm's forces are annihilated and his nephew Vivianz, a model of knightly valour, is killed. Willehalm retires to Orange (Oransche), entrusts its defence to Gyburg, and sets out to enlist forces from the Emperor at Laon (Mun-leun). He meets with a tepid reception from the effete court, but in the end secures the promise of help. Accompanied by an army, and

notably by a youth of immense strength (Rennewart), Willehalm returns, relieves Orange, which has been successfully defended by the faithful Gyburg, and fights a second battle of Aliscans, in which the Saracens are totally defeated. Wolfram's poem ends at this point; in the legend Rennewart is revealed to be Gyburg's brother.

Wolfram freely adapts his source. The religious conflict between Christian and heathen is seen as necessary and the gulf between them as unbridgeable. War against the heathen is meritorious and death in such a conflict is martyrdom. But Wolfram conceives the Sara-cens as human beings, misguided but deserving of human sympathy, an attitude which is most clearly expressed by Gyburg. His outlook is profoundly humane, rather than tolerant. In Gyburg he creates an ideal of love in which deep devotion is allied with unswerving loyalty in adversity.

The poem exists in eight complete and a number of fragmentary MSS. and was evidently extremely popular. The principal MS. is at St. Gall. One MS., the Willehalm-Codex in Kassel, disappeared in mysterious circumstances in 1945, but was rediscovered at Philadelphia, U.S.A., in 1972 and returned to Kassel. A later poet, ULRICH von dem Türlin (q.v.), also wrote a *Willehalm,* recounting the events leading up to the beginning of Wolfram's poem; ULRICH von Türheim (q.v.) wrote a continuation. A prose version of *Willehalm,* with the introduction and continuation provided by the two Ulrichs, was made in the 15th c.

Willehalm von Orlens, a Middle High German poem of close on 16,000 lines by RUDOLF von Ems (q.v.). It was written, in rhyming couplets, after the lost *Eustachius* and probably before *Alexander,* between 1235 and 1240. A translation of the lost original French work was desired by a Count Johannes von Ravensburg, and on the advice of Konrad von Winterstetten the task was entrusted to Rudolf von Ems.

The hero of the poem, Willehalm, is the son of a knight who is killed by Jofrit (Godefroi) von Bouillon. The boy is adopted by the victor and brought up at the English court. He falls in love with Amelie, the king's daughter, and when the king plans for her a political marriage, Willehalm attempts to elope with her, is discovered, and wounded. He is pardoned by the king on condi-tion that he remains abroad and (literally) keeps silent until Amelie releases him from the promise. He keeps his word, conducts himself in knightly fashion, and eventually reconcilia-tion and marriage with Amelie take place.

The poem sets out an ideal of chivalrous

behaviour maintained in the hard conditions of the real world. Its extraordinary popularity is indicated by the existence of seventy-six MSS., and perhaps also by the choice of the subject for frescoes painted *c.* 1400 in the castle of Runkelstein (q.v.) near Bolzano.

Bk. II of *Willehalm von Orlens* contains an interesting review of contemporary and recent literature, in which are mentioned HEINRICH von Veldeke, HARTMANN von Aue (*Erec* and *Iwein*), WOLFRAM von Eschenbach (*Parzival* and *Willehalm*), GOTTFRIED von Straßburg (*Tristan*), BLIGGER von Steinach, ULRICH von Zazikhofen (*Lanzelet*), WIRNT von Grafenberg (*Wigalois*), Freidank, KONRAD von Fußesbrunnen, Konrad FLECK (*Floire und Blanscheflur*), der Stricker (*Daniel vom blühenden Tal*), ALBRECHT von Kemenaten, ULRICH von Türheim (*Clies*), and Meister HESSE (qq.v.). Rudolf also cites the names of two otherwise unknown poets, Heinrich von Linaue and Gottfried von Hohenlohe.

WILLEMER, MARIANNE VON (1784–1860, Frankfurt/Main), of uncertain origin, came to Frankfurt with a theatrical troupe in 1798 as Marianne Jung. The Frankfurt banker J. J. Willemer (1760–1838) took her into his house in 1800 and had her educated with his own children, and in 1814 he made her his third wife. A few weeks before the marriage (which took place on 27 September) she and Willemer visited Goethe at Wiesbaden.

Goethe was strongly attracted by Marianne, and she returned his affection. On 15 September he visited the couple at their summer residence, the Gerbermühle, and repeated the visit on 11 October, shortly after the wedding. He was there again in August 1815, and was last together with the Willemers at Heidelberg, 23–6 September 1815. Goethe continued to correspond with her for the rest of his life. His last letter to her, dated 10 February 1832, enclosed her letters to him with the request that the packet should be left unopened 'bis zu unbestimmter Stunde'; they were accompanied by a verse, which is a token of the warmth of their attachment:

> Vor die Augen meiner Lieben,
> Zu den Fingern die's geschrieben,—
> Einst mit heißestem Verlangen
> So erwartet, wie empfangen—
> Zu der Brust der sie entquollen
> Diese Blätter wandern sollen;
> Immer liebevoll bereit,
> Zeugen allerschönster Zeit.

During the fourteen months in which they were from time to time in personal contact, Goethe was studying oriental poetry and writing the poems of the *West-östlicher Divan*

(q.v.). Marianne von Willemer was the inspiration of one section, the *Buch Suleika*. So closely did their feelings coincide that Goethe was able to incorporate undetected in the book poems written by Marianne.

Wille zur Macht, Der, title given to a philosophical work, attributed to F. Nietzsche (q.v.) by its editors, E. Förster-Nietzsche (q.v.) and P. Gast (1854–1918), and published in 1906. It is sub-titled *Versuch einer Umwertung aller Werte*; the title was conceived by Nietzsche in 1885, when he devised for the work the sub-title *Versuch einer Welt-Auslegung*.

The work is based on notes and aphorisms begun in 1883, and presented in tendentious and misleading form by the editors. Nietzsche's notes have reappeared in more scholarly form in *Aus dem Nachlaß der Achtziger Jahre* (ed. K. Schlechta, 1960), but this arrangement is not universally accepted, and another version is included in *Friedrich Nietzsches Werke des Zusammenbruchs* (ed. E. F. Podach, 1961).

William Lovell, see GESCHICHTE DES HERRN WILLIAM LOVELL.

William Ratcliff, a one-act tragedy in blank verse by H. Heine (q.v.), published in *Tragödien, nebst einem lyrischen Intermezzo* in 1823. Consisting of five scenes, it is set in the Scottish Highlands 'in der neuesten Zeit'.

Mac-Gregor's daughter Maria has just been married to Douglas. The father recounts how she once rejected William Ratcliff, and how her first betrothed, Macdonald, and then her second, Lord Duncan, were murdered by Ratcliff. Douglas encounters Ratcliff and defeats him, but he spares his life because Ratcliff has once saved his. The wounded Ratcliff enters Maria's room. It now becomes apparent that Maria loves him. He kills Maria and her father and finally himself.

WILLIAM THE SILENT, also William of Orange, Regent of the Netherlands, Duke of Orange and Count of Nassau (Nassau, 1533–84, Delft). Although educated as a Roman Catholic at the court of Karl V (q.v.) in Spain, William became a leader of the Protestant cause and is regarded as the founder of the Dutch Republic. At the approach of Duke Alva to the Netherlands he went to Germany (1567). After raising forces and leading the resistance against the Spanish troops with temporary success, he was elected by the Dutch Estates to succeed Duke Alva as regent of the Netherlands. He strove vainly for an effective peace settlement with Spain during the continued unrest of the 1570s. In 1580 Philip II of Spain banned him and placed a price on his head. He was assassinated four years later.

William is the original of Oranien in Goethe's tragedy *Egmont* (q.v.).

WILLIRAM (*c.* 1010–85, Ebersberg), Abbot of Ebersberg in Bavaria from 1048 until his death, was earlier a monk at Fulda and at Bamberg. His paraphrase of The Song of Songs (*Hohes Lied*) is an important document of medieval theology, interpreting the text so as to emphasize its symbolical, hierarchical significance. Williram's *Hohes Lied* is divided into three sections: the text of the Vulgate; a Latin commentary in rhyming Latin hexameters; a commentary in German with an admixture of Latin. Its wide circulation, and therefore repute, is attested by the existence of eighteen MSS. It was first printed, entirely in Latin, in 1528; the German text first appeared in print in 1598. Williram's *Hohes Lied* is one of the earliest documents in Early Middle High German.

WILLKOMM, ERNST ADOLF (Herwigsdorf nr. Zittau, 1810–86, Zittau), studied at Leipzig University and was afterwards active as a journalist and editor, chiefly in Hamburg and Lübeck. A follower of Junges Deutschland (q.v.), he wrote social novels, one of which, *Die Europamüden* (2 vols., 1838), combines the motifs of disgust with Europe and longing for the New World, whilst another, *Weiße Sklaven* (3 vols., 1845), treats the conditions under which working men lived in the early stages of the Industrial Revolution, drawing a sensationally contrasting picture of a corrupt and luxury-loving upper class. Other titles include *Eisen, Gold und Geist* (1843), *Reeder und Matrose* (1857), *Männer der Tat* (4 vols., 1861), and *Wunde Herzen* (3 vols., 1875). *Sagen und Märchen aus der Oberlausitz* (3 vols.) appeared in 1845.

Willkommen und Abschied, a poem written by Goethe in Strasburg in the spring of 1771, and first printed in Jacobi's *Iris* (q.v.) in 1775. It portrays in exceptionally vivid verse the night ride, the lovers' meeting, and their parting. The underlying experiences are Goethe's love for Friederike Brion (q.v.) and his journeys between Strasburg and Sesenheim to be with her.

WIMPFELING, JAKOB (Schlettstadt, 1450–1528, Freiburg) (also spelt Wimpheling and Wympfeling), was an Alsatian scholar who was in general opposed to the New Learning, but was nevertheless influenced by it. At Heidelberg, where he was for a time Rector, he taught from 1471 to 1484 and again from 1498 to 1500. From 1484 to 1498 he was a canon of Speyer. Although he was critical of ecclesiastical abuses,

he remained faithful to the Roman Catholic Church.

In his polemical work *Germania* (1501) Wimpfeling asserted the German character of Alsace. He also wrote Latin comedies, of which *Stylpho* (1470) is the best known, and a history of Germany (*Epitoma rerum Germanicarum usque ad nostra tempora*, 1505). *Opera selecta* appeared in 1965 ff.

WINCKELMANN, JOHANN JOACHIM (Stendal, Prussia, 1717–68, Trieste), born into poverty as a cobbler's son, acquired as a boy, no one knows how, a love of Greek antiquity. He pressed for and obtained first a grammar school and then a university education. After some years (1743–8) as a schoolmaster, during which he frequently sat up reading Greek into the small hours, he became in 1748 librarian to a Count von Bünau. In 1754 he was converted to Roman Catholicism, allegedly in order to facilitate a journey to Rome.

Winckelmann's first, brilliant essay on Greek and Roman classicism, *Gedanken über die Nachahmung der griechischen Werke in der Malerei und Bildhauerkunst* (q.v., 1755), was written before this visit. Arriving in Rome in 1755, he quickly established himself as an authority on ancient art. He visited Naples, Herculaneum, and Paestum; and he was librarian successively to Cardinal Archinto and Cardinal Albani. The efforts of his early years in Rome were devoted to a monumental history of classical art (*Geschichte der Kunst des Altertums*, q.v., 1764, repr. 1934). He then annotated his own book (*Anmerkungen über die Geschichte der Kunst des Altertums*, 1767) and began work on a descriptive review of newly discovered works of classical sculpture (*Monumenti antichi inediti*, 1767–8).

In 1768 Winckelmann went to Munich, Regensburg, and Vienna, intending to travel further in Germany. He abandoned the journey and returned to Trieste. There he was murdered in his inn by Francesco Arcangeli for the gold he was carrying. Of his other works the *Anmerkungen über die Baukunst der alten Tempel zu Girgenti in Sizilien* (1762), the *Sendschreiben von den Herkulanischen Entdeckungen* (1762), and the *Versuch einer Allegorie, besonders für die Kunst* (1766) deserve mention. He was a particularly sensitive interpreter of the Greek ideal of male beauty.

Winckelmann's conception of the antique as noble, elevated, serene, and simple ('eine edle Einfalt und eine stille Größe'), which was present in his work from the outset, shaped for many decades the new classicism of Europe.

A *Gesamtausgabe* (11 vols.) appeared 1808–25, *Sämtliche Werke*, ed. J. Eiselein (12 vols.) 1825–9 (reissued, ed. O. Zeller, 1965), *Kleine Schriften.*

Vorreden. Entwürfe, ed. W. Rehm, in 1968, and *Briefe* (4 vols.), ed. W. Rehm *et al.*, 1952–7.

Winckelmann und sein Jahrhundert, a collection of essays edited by Goethe ('In Briefen und Aufsätzen herausgegeben von Goethe'), and published in 1805. The volume contains some letters written by J. J. Winckelmann (q.v.) to H. D. Berendis (1719–82), the essay *Entwurf einer Geschichte der Kunst des 18. Jahrhunderts* by H. H. Meyer (q.v.), *Bemerkungen eines Freundes* by C. L. Fernow (1753–1808), and three contributions on Winckelmann by Goethe, Meyer, and F. A. Wolf (q.v.) respectively.

WINCKLER, Josef (Rheine, Westphalia, 1881–1966), was a dental surgeon in Mörs until 1932, when he retired from practice in order to devote himself to writing. Primarily a regional novelist, Winckler founded in 1905 the group of similarly minded Westphalian writers Werkleute auf Haus Nyland (q.v.). An imaginative narrator, he wrote mainly about working men and women, differing from most social realists in his addiction to humour and in his Expressionistic use of language. His first publication was a volume of verse, written jointly with two of the Nyland group, J. Kneip and W. Vershofen (qq.v., *Wir drei!*, 1904). His *Eiserne Sonette* (1914) are concerned with technology. During the 1914–18 War he wrote the volume of verse *Mitten im Weltkrieg* (1915), turning after the war to fiction in *Der chiliastische Pilgerzug* (1923), *Der tolle Bomberg* (1924), *Dr. Eisenbart* (1929), and *Der Großschieber* (1933). *Pumpernickel* (1926) is a humorous, anecdotal book on the Westphalian scene and *Eiserne Welt* (1930) a collection of poetry.

Winckler's publications of the 1950s include *Gesammelte Gedichte* (1957). *Ausgewählte Werke* (4 vols.) appeared 1960–3.

WINDECKE, Eberhart (Mainz, *c.* 1380–*c.* 1440), spent the greater part of his life in Bohemia in royal service. He compiled a life of the Emperor Sigismund (q.v.) under the title *König Sigmunds Buch.* He spent his childhood and his last years in Mainz.

WINDISCHGRAETZ, Alfred, Fürst zu, also in the form Windisch-Graetz (Brussels, 1787–1862, Vienna), Austrian field-marshal, took an active part in suppressing the Revolutions of 1848–9 (see Revolutionen 1848–9). He checked the insurrection in Prague in June 1848, re-took Vienna in October, and marched against the Hungarians in 1849, but was relieved of his command in April 1849.

WINDTHORST, Ludwig (Kaldenhof nr. Osnabrück, 1812–91, Berlin), a lawyer and a Roman Catholic, became Minister of Justice in the Kingdom of Hanover. After the Franco-Prussian War (see Deutsch-Französischer Krieg) he entered the Reichstag and became the leader of the Zentrum (q.v.); he was a bitter opponent of Bismarck (q.v.) throughout the Kulturkampf (q.v.).

Windthorst's speeches were collected as *Ausgewählte Reden* (3 vols., 1901–2) and his correspondence appeared in *Stimmen aus Maria Laach 1882/83* (1912–13) and in *Briefe Windthorsts an F. Engelen* (ed. H. Schröter, 1954).

WINDTHORST, Margarete (Haus Hesseln nr. Halle, 1884–1958, Rothenfelde, Westphalia), a Westphalian and a Roman Catholic, strongly rooted in her native land and faith, is the author of fiction and of lyric poetry (*Gedichte*, 1911). Her novels include *Die Tau-Streicherin* (1922), *Der Basilisk* (1924), *Die Nacht der Erkenntnis* (1925), the trilogy *Die Sieben am Sandbach* (1937), *Mit Lust und Last* (1940), *Zu Erb und Eigen* and *Mit Leib und Leben* (both 1949), *Das lebendige Herz* (1951), and *Weizenkörner* (1954); *Die Verkündigung* (1924), her first story, was followed by *Das grüne Königreich* (1930), *Die Lichtboten* (1938), *Hoftöchter* (1947), and *Menschen und Mächte* (1949).

WINFRIED, see Boniface, Saint.

Winileot, an Old High German word found in various Glossen (q.v.) as a translation of the Latin *plebeios psalmos.* It occurs later as *wineliet* in the poetry of Neidhart von Reuental (q.v.). It is thought to denote folk-songs, sung in company. A passage in a Latin decree of Charlemagne (see Karl I) forbids nuns to write or send 'winileodos', and it seems likely that these were love-songs.

WINKELRIED, Arnold (von), a citizen soldier of Stans in Unterwalden, Switzerland, who is said to have heroically sacrificed himself at the battle of Sempach in 1386, gathering an armful of spear-points into his own body, and so making a gap by which his comrades poured through the enemy's line. Though the feat is probably legendary, Winkelried himself may well be authentic. The words attributed to him as he died ('Der Freiheit eine Gasse!') have become a common quotation.

WINKLER, Eugen Gottlob (Zurich, 1912–1936, Munich, by suicide), studied and worked in Munich as a painter and writer. His works, containing an essay on E. Jünger (q.v.), were published posthumously as *Gesammelte Schriften*

(2 vols.), ed. H. Rinn and J. Heitzmann, 1937, and as *Dichtungen, Gestalten und Probleme*, an enlarged edition, 1956.

WINKLER, Karl Gottlieb Theodor, see Hell, Theodor.

Winsbecke, Der, is the superscription of a short Middle High German didactic poem and denotes its author, a knight of Windsbach near Eschenbach (Wolfram's home) in Franconia.

The poem, which was written between 1210 and 1220 in 56 ten-line stanzas, is a code of knightly conduct set in the form of advice by a knight to his son, each stanza beginning with the word of address 'sun' (Sohn). The advice centres on the chivalric ideal, with its three essentials: fear of God, true *minne*, and the virtues of the soldier. A knight should live as he hopes to fare in the world to come. The poem breathes a sense of conviction and a warmth which have suggested to some scholars, including de Boor, the hypothesis that Winsbecke wrote it for his own son. Others regard it as merely conforming to a literary convention. An allusion to Wolfram's *Parzival* (q.v.) occurs in stanza 18.

The poem contains, in the Große Heidelberger Liederhandschrift a further twenty-four stanzas in which the son advises the father to found a hospital and the father confesses his sins. These stanzas, which are inconsistent with the tone of the father's advice, embody an unknightly, ecclesiastical attitude and are an addition, the work of an unknown cleric.

Winsbeckenparodie, a Middle High German parody of *Der Winsbecke*, in which the knight's good counsel is reversed and he advises his son to live a godless, riotous, and otherwise immoral life. It was probably written in the 14th c.

Winsbeckin, Die, a Middle High German didactic poem written early in the 13th c. which forms a parallel to *Der Winsbecke* (q.v.), giving the advice of a mother to her daughter. Its unknown author evidently knew *Der Winsbecke*, but has substituted dialogue for monologue. The advice is courtly, dealing with courtly behaviour and love.

WINTERKÖNIG, see Friedrich V von der Pfalz.

Winterreise, Die, a cycle of 24 poems by Wilhelm Müller (q.v.), first published in the periodical *Urania* in 1823 and then in *Zweites Bändchen der Gedichte aus den hinterlassenen Papieren eines reisenden Waldhornisten* in 1824. The poems, all in the first person, tell how a rejected lover

takes to the road in desolation, passing through stages of nostalgia, bitterness, defiance, and at last half-crazed despair. The cycle was set for a solo voice and piano accompaniment by F. Schubert (q.v.) in 1827 and is one of his finest achievements.

'Wir hatten gebauet ein stattliches Haus', first line of a student song by A. von Binzer (q.v.) written to be sung on the occasion of the dissolution of the Burschenschaft (q.v.) on 26 November 1819. It was first printed in 1821. The tune is quoted by Brahms (q.v.) in his *Akademische Fest-Ouvertüre*.

WIRNT VON GRAFENBERG, medieval poet, author of the Arthurian verse romance *Wigalois*. He was by rank a knight and by birth an East Franconian and his poem can be dated between 1204 and 1209. Other details of his life are not known, but he has the unusual distinction of appearing as a character in a later work (*Der Welt Lohn* by Konrad von Würzburg, q.v.).

Wigalois relates the life story of the hero in close on 12,000 lines. He is the son of Gawein and Florie. His father leaves and fails to return, and Wigalois, when he grows up, sets out to seek his father. He encounters many adventures, of which the most important is a battle with a heathen warrior who is in league with the Devil. Wigalois triumphs, is married, and lives happily. Gawein has already returned. Wirnt's poem has moral intentions and contains a number of wise maxims; but the extravagant adventures predominate. He was clearly influenced by Wolfram von Eschenbach, and the poem contains a tribute to Wolfram ('ein wise man von Eschenbach ... leien munt nie baz gesprach').

A 15th-c. prose version (*Wigoleysz*) of the romance was first printed in 1493. *Wigalois* was edited by J. Kapteyn in 1926.

Wir rufen Deutschland, a novel by E. E. Dwinger (q.v.), published in 1932. Sub-titled *Heimkehr und Vermächtnis*, it follows *Die Armee hinter Stacheldraht* and *Zwischen Weiß und Rot* (qq.v.). The narrator meets his comrades returned from Russian captivity. Disillusioned with post-war Germany, they migrate to an estate in East Prussia, where they work on the land. The strongly nationalistic book is mainly concerned with belief in the old Germany and criticism of the Weimar Republic (q.v.).

Wirrwarr, Der, originally intended title of F. M. Klinger's play *Sturm und Drang* (q.v.). *Der Wirrwarr* is also the title of a farce by A. von Kotzebue (q.v.).

'Wir saßen am Fischerhause', first line of an untitled poem by H. Heine (q.v.) included in the *Buch der Lieder* (q.v.) as No. VII of *Die Heimkehr*.

WIRSUNG, CHRISTOPH (Augsburg, ?1500–71, Heidelberg), a physician, who, while a student in Italy, translated the Spanish novel *Celestina* into German (*Ein hipsche Tragedia von zwaien liebhabenden mentschen Calixtus und Melibia*, 1520). Wirsung did not work from the original, using instead an Italian translation. He revised it in a second edition in 1534. He is also the author of a medical work, *Arcneybuch* (1568).

Wirt, Der, a short erotic verse tale of the late 13th c., in which three men enter into a wager to possess a woman undetected before the eyes of her husband.

Wirtshaus im Spessart, Das, a story by W. Hauff (q.v.) included in his *Märchen-Almanach* (1826). Two young craftsmen, one of whom is Felix, a goldsmith, walking through the forest known as the Spessart, take lodging in an inn and find there two others. None of them trusts the host, so they resolve to keep each other awake by telling stories. The four tales are *Die Sage vom Hirschgulden, Das kalte Herz, Saids Schicksale*, and *Die Höhle von Steenfoll*. In the night a count and countess arrive in a coach. Shortly afterwards robbers break in, intending to kidnap the countess and hold her to ransom. She and Felix change clothes, and she escapes. Felix is later rescued, and the countess proves to be his godmother, whom he was intending to visit. (Film version by K. Hoffmann, 1957.)

WIRZ, OTTO (Olten, Switzerland, 1877–1946, Gunten by Lake Thun), published a two-volume novel, *Gewalten eines Toren* (1923), and the *Novelle um Gott* (1925). In 1828 he abandoned a career as an engineer, for which he had qualified, and an appointment at the Zurich patent office, and devoted himself to authorship. The novel *Die gedruckte Kraft* appeared in 1928, *Prophet Müller-zwo* in 1933, *Späte Erfüllung* in 1936, *Rebellion der Liebe* in 1937, and *Maß für Maß* in 1944. The fragment of a novel, ed. W. Wirz, appeared in 1965 entitled *Rebellen und Geister*.

WISSE, CLAUS, see NEUE PARZEFAL, DER.

WISSOWATIUS, ANDREAS (1608–78), a Lithuanian theologian, whose arguments against the doctrine of the Trinity are discussed by Lessing (q.v.) in a contribution printed in *Zur Geschichte und Literatur* (*Des Andreas Wissowatius Einwürfe wider die Dreifältigkeit*, 1773). The name is a latinization of Wiszowaty.

WITEKIND, HERMANN (Neuenrade, Westphalia, 1522–1603, Heidelberg), a schoolmaster in Riga, became a professor at Heidelberg University (1561). His *Christlich bedenken und erinnerung von Zauberey* (1585) is an attack on the widely prevalent belief in witches and witchcraft. In later life Witekind wrote under the pseudonym Augustin Lercheimer von Steinfelden.

Witiko, a historical novel by A. Stifter (q.v.), published in three volumes, the first in 1865, the second and third in 1867, the year before his death. It is set in 12th-c. Bohemia, beginning in 1138 and closing in 1184.

Witiko, descended from a noble family fallen into obscurity, journeys from Passau into Bohemia, to participate in the events of the age. He assists Duke Soběslaw, distinguishing himself by courage matched with modesty. He settles in the southern region of Plan (Stifter's own homeland), and presently assists Soběslaw's successor, Wladislaw, to subdue two dangerous rebellions. When Wladislaw, as king of Bohemia, accompanies the Emperor Friedrich I (q.v.) to Italy to put down a Milanese revolt, Witiko is present and again acquits himself with distinction. He marries worthily and establishes himself as a great landlord and nobleman, whose feudal rule is marked by justice and compassion.

Although the book contains violent episodes, its whole tone is one of serenity and dignity. The character of Witiko, always marked by absolute integrity, grows by imperceptible degrees in power and awareness, a process of development which is reflected by the subsidiary figures. The style is deliberate, simple, and repetitive. It was Stifter's unfulfilled intention to extend this substantial Bildungsroman (q.v.) into a monumental trilogy.

Wittelsbach, a Bavarian dynastic house, founded by Otto von Wittelsbach, Duke of Bavaria (1180–3). He was a loyal supporter of Barbarossa, the Emperor Friedrich I (q.v.), hence the proverbial 'Wittelsbacher Treu'.

Wittembergisch Nachtigall, Die, by Hans Sachs (q.v.), a poem of 700 lines in Knittelverse (q.v.). Written in 1523, it praises the nightingale (Luther) and denounces the lion (the Pope). Its first lines have become famous, partly through Wagner's adaptation in *Die Meistersinger von Nürnberg* (q.v.):

> Wacht auf! es nahent gen dem Tag.
> Ich hör singen im grünen Hag
> Ein wunigliche Nachtigall,
> Ir Stim durchklinget Berg und Tal.

Wittenberg University was founded in 1502 by the charter of the Emperor Maximilian I (q.v.). In the 16th c. it was especially active in support of the Reformation (q.v.) and many of the most noted reformers, beginning with Luther and Melanchthon (qq.v.), taught or studied there. Its contemporary fame is further attested in *Hamlet* ('What make you from Wittenberg, Horatio?'). From the 17th c. its importance dwindled and in 1815 it was amalgamated with Halle. In the DDR (see DEUTSCHE DEMOKRATISCHE REPUBLIK) Halle University is now known as Halle-Wittenberg.

WITTENWEILER, HEINRICH, a citizen of Wyl in the Swiss district of St. Gall, wrote early in the 15th c. a comic epic of close on 10,000 lines, entitled *Der Ring*. It tells the story of the courtship and marriage of the peasant Bertschi Triefnas and his uncomely love Mätzli Rüerenzumpf, and of the village celebrations and feuds arising out of it. Coarseness and obscenity dominate the work which nevertheless has a moral aim, setting out to exemplify good conduct and to hold wrong up to ridicule. Wittenweiler's satire is focused on the peasantry, but the courtly way of life is not exempt from his criticism. His name derives from Wittenweil in Thurgau, whence his family moved to Wyl more than a century before his time. The story of *Der Ring* is derived from a short poem of 416 lines, *Von Meyer Betzen*, which Wittenweiler has enormously expanded.

The edition of E. Wiessner (1931) was reprinted 1964–70.

WITTGENSTEIN, LUDWIG (Vienna, 1889–1951, Cambridge), an Austrian-born philosopher, who became a naturalized British subject in 1938. Wittgenstein abandoned studies in engineering intended to prepare him for a career in his father's large industrial concern. At the suggestion of Gottlob Frege (1848–1925) of Jena university he went to Cambridge, studying philosophy from 1911 to 1913. During the 1914–18 War he served in the Austrian army and became an Italian prisoner-of-war. During his captivity he wrote the *Logisch-philosophisches Traktat*, which was first published in a German periodical (1921). This was recognized as a major contribution to philosophy on its appearance in 1922 under the title *Tractatus logico-philosophicus* with an English translation and an expository introduction by Bertrand Russell.

Abandoning both his career and his family fortune, Wittgenstein spent the years 1920–6 as an elementary school teacher in villages in lower Austria (Trattenbach, Otterthal, and Pachberg). He had a strong interest in the Austrian school reform movement. He worked as a gardener in a monastery and contemplated at one stage entering monastic life. In 1929 he returned to Cambridge, where he taught until 1947, succeeding G. E. Moore in his chair of philosophy in 1939. He continued to reside for prolonged periods in Austria, taking out Viennese citizenship as an architect. He designed a house for his sister in Vienna. For a time the *Tractatus* dominated the Vienna Circle (see WIENER KREIS) of which, despite his contact with M. Schlick, he was not a member.

Wittgenstein's argument in the *Tractatus*, which is presented in brief paragraphs numerically arranged according to a complex manipulation of decimals, rests on the assumption that the 'logical form' of language and reality can only be 'shown' in language, not 'stated'. Once language has fulfilled its function to 'picture' reality, from which he excluded the transcendental, silence is the only alternative to verbal expression. He states with concision in the last paragraph (7): 'Wovon man nicht sprechen kann, darüber muß man schweigen' ('Whereof one cannot speak, thereof one must be silent'). Before this conclusion he reiterates the tone of his introduction to the work, taking recourse to an image, the 'ladder', which he likens to his sentences 'through' and 'over' which the reader must climb (6.54): ('Er muß sozusagen die Leiter wegwerfen, nachdem er auf ihr hinaufgestiegen ist'); ('He must so to speak throw away the ladder, after he has climbed up on it'). His concept of 'silence' has a point of contact with the sense of crisis which H. von Hofmannsthal (q.v.) had experienced as a creative writer (see BRIEF DES LORD CHANDOS). Wittgenstein's interest in Tolstoy suggests one direct literary influence. His principal work, the *Philosophical Investigations* (1945 and 1949), was preceded by *The Blue and Brown Books* (1933–4 and 1934–5). These comprise lecture material, which first appeared in blue and brown wrappers. They form the basis of the *Preliminary Studies for the 'Philosophical Investigations'. Generally known as The Blue and Brown Books* (1958). *The Blue and Brown Books* were followed by *Remarks on the Foundation of Mathematics* (1937–44).

Wittgenstein revised *The Blue and Brown Books* in a German version with the significant title *Systeme menschlicher Verständigung*, 'systems of communication'. These 'systems' convey philosophical problems through 'Sprachspiele' ('language games'). Wittgenstein's influence is no longer confined to English philosophers. New assessments have corrected his reputation as a representative of neo-positivism, and his *œuvre*, which is not yet fully available in published form, is considered by some to stand in the same relationship to early 20th-c. philosophy as the *Critiques* of Kant (q.v.) to the philosophy of the mid-18th c., demolishing established concepts to clear the way for constructive formulations. These are based on Wittgenstein's view that the meaning of language is its 'use' and that language is a social phenomenon which constitutes 'forms of life'.

WITTMAACK, ADOLF (Itzehoe, 1878–1957, Hamburg), wrote novels of Hamburg life and of seafaring (*Butenbrink*, 1909; *Die kleine Lüge*, 1911; *Konsul Möllers Erben*, 1913; *Der Ozean*, 1937).

WITZEL, GEORG (Vacha, Hesse, 1501–73, Mainz), originally a supporter of Luther, became one of his most zealous opponents. In addition to writing polemical works he composed hymns for Catholic use (*Odae Christianae. Etliche Christliche Gesänge, Gebete und Reymen für die Gottesfürchtigen Layen*, 1541). Some of his writings appeared under a latinized form of his name, Wicelius.

WITZLEBEN, KARL AUGUST FRIEDRICH VON (Tromlitz, 1773–1839, Dresden), a Prussian regular officer, who fought in 1793 and 1806 and afterwards took part in the later Napoleonic Wars (q.v.), first on the French side in Spain and then, in 1813, on the Russian side. He retired after the war and, under the pseudonym August von Tromlitz, devoted himself to writing, living from 1821 to 1830 in Berlin, and then in Dresden.

An imitator of Scott, he wrote numerous historical novels, of which the best-known titles are *Die Pappenheimer* (1827), *Franz von Sickingen* (1828), *Der Fall von Missolunghi* (1828), and *Die Vierhundert von Pforzheim* (1830). These romances enjoyed considerable popularity for a time. Witzleben also wrote plays (*Die Entführung*, 1823, *Douglas*, 1825) and edited from 1827 to 1841 the annual (Taschenbuch) *Vielliebchen*. He was an unusually copious writer, whose collected works (*Sämtliche Schriften*, 1829–41) comprise 108 volumes.

WIZLAV, FÜRST VON RÜGEN (*c.* 1268–1325), gave encouragement to German Minnesang at his court, which became for a time an outpost of German literary culture on the Baltic. Wizlav himself learned the art of composing from a strolling singer, known as Der Ungelehrte, who settled in Stralsund. Of the prince's poetry fourteen somewhat involved Minnelieder and some Sprüche (see SPRUCH) are preserved. They are written in an interesting mixture of 'literary' German from more southerly regions with his own Low German dialect.

WOBESER, WILHELMINE KAROLINE VON (Berlin, 1769–1807, Wirschen nr. Stolpe), *née* von Rebeur, wife of a Prussian officer and country gentleman, was the author of one of the most widely read novels of the late 18th c., *Elisa, oder das Weib, wie es seyn sollte* (1795). A novel of moral and sentimental character, it was many times reprinted and was translated into English and French.

'Wohlauf! noch getrunken', first line of the poem *Wanderlied* by J. Kerner (q.v., 1809).

WOHMANN, GABRIELE (Darmstadt, 1932–   ), *née* Guyot, studied music and languages at Frankfurt University and became a teacher. A member of the Gruppe 47 (q.v.), she turned to authorship in 1956. Her experimental fiction, influenced by James Joyce, Virginia Woolf, and Katherine Mansfield, includes the stories *Mit einem Messer* (1958), *Sieg über die Dämmerung* (1960), *Trinken ist das Herrlichste* (1963), *Erzählungen* (1966), *Ländliches Fest* (1968), *Sonntag bei Kreisands*, *Treibjagd* (both 1970), *Der Fall Rufus* (1971), *Übersinnlich, Gegenangriff* (both 1972), and *Habgier* (1973). Some of these volumes comprise more than one story. Other volumes of stories include *Böse Streiche* (1977), *Das dicke Wilhelmchen* (1978), *Einsamkeit* (1982), and *Der kürzeste Tag des Jahres* (1983).

Gabriele Wohmann is also the author of the novels *Jetzt und nie* (1958), *Abschied für länger* (1965), *Die Bütows* (1967, described as 'Ein Mini-Roman'), *Ernste Absicht* (1970, a study of identity in illness), and *Paulinchen war allein zu Haus* (1974, the portrayal of an adopted child, for which a quotation from *Struwwelpeter*, q.v., furnishes the title). A change of attitude becomes increasingly perceptible in a preoccupation with the self. The novel *Schönes Gehege* (1975) was followed by the autobiographical *Ausflug mit der Mutter* (1976), dealing with a daughter's relationship with her widowed mother. In *Frühherbst in Badenweiler* (1978) the inner world of the artist Hubert Frey is explored. The next novel, *Ach wie gut, daß niemand weiß* (1980), the title of which adopts a line from the Grimms' fairy-tale *Rumpelstilzchen* (it runs on 'Daß ich Rumpelstilzchen heiß'), deals with the self-analysis of unresolved complexities within the central figure Marlene, a medical doctor. The volume *Selbstverteidigung* (1971) contains prose texts and lyric poetry. The collection of poetry *So ist die Lage* (1974) was followed by *Grund zur Aufregung* (1978). Her radio plays and plays for television include *Komm Donnerstag* (1964), *Die Gäste, Das Rendez-vous* (both 1965), *Grobe Liebe* (1966), *Norwegian Wood* (1967), *Der Fall Rufus* (1969), *Kurerfolg* (1970), *Der Geburtstag* (1971), *Tod in Basel* and *Die Witwen* (both 1972), and *Entziehung* (1974). A collection of three radio plays and one play for television appeared in 1978 under the title of one of the radio plays, *Heiratskandidaten* (1975). The collection *Ich lese. Ich schreibe. Autobiographische Essays* was published in 1984, *Ausgewählte Erzählungen 1957–1977* in 1978, *Ausgewählte Erzählungen aus zwanzig Jahren* (2 vols.) in 1979, *Gesammelte Gedichte 1964–1982* in 1983, and the volume of poetry *Passau, Gleis 3* in 1984.

'Wo ist des Sängers Vaterland?', first line of the patriotic poem *Mein Vaterland* written by Th. Körner (q.v.) in 1813 and included in *Leyer und Schwerdt* (q.v.).

WOLDEMAR, MARKGRAF VON BRANDENBURG (1281–1319, Bärwalde), became Margrave in 1303. He was a chivalrous ruler and his reign was recalled nostalgically during the decades that followed his death. He was succeeded in 1319 by his nephew Heinrich, who died in the following year, after which the Wittelsbachs (q.v.) of Bavaria became Margraves.

In 1347 a pretender appeared who claimed to be Woldemar and was briefly recognized by the Emperor Karl IV (q.v.). 'Der falsche Woldemar', as he is known, was dropped by the Emperor in 1350, but still retained some support in Brandenburg until his death in 1356. He is the subject of a novel, *Der falsche Woldemar* (1847), by W. Alexis (q.v.). The true Woldemar and his nephew Heinrich were the last Ascanians (see ASKANIER) to rule in the Mark Brandenburg.

WOLF, CHRISTA (Landsberg/Warthe, 1929–), after studying at Jena and Leipzig University, worked for several years as a critic and editor before becoming a free-lance writer and essayist in 1962. She joined the SED in 1949, married Gerhard Wolf in 1951, and settled near Berlin. She was on the committee of the DDR Schriftstellerverband from 1955 to 1977 and is a member of the Academy of Arts.

Having begun publication with *Moskauer Novelle* (1961), she established herself with her story *Der geteilte Himmel* (1963), the film version of which was directed by Konrad Wolf (1964). Set in Halle (DDR) and West Berlin, the action centres on Rita Seidel, a young student preparing to become a teacher, and the chemist Manfred Herrfurth. Their two-year love affair ends in separation as a result of Manfred's professional disillusionment and move to West Berlin. Rita, who decides against the move, adjusts to a new beginning. The story's climax is Rita's emotional crisis which causes a breakdown and almost fatal accident. The events of the two years from 1959 unfold through Rita's recollections during her recovery in hospital. The story's main substance lies in the depiction

of everyday life in the DDR, with a variety of figures in and around a coach-building factory and scenes in East and West Berlin in 1961, the year of the erection of the Wall. Following the official cultural policy known as the Bitterfelder Weg (q.v.), that has since become subject to modifications, Wolf herself had worked in a coach-building factory.

In *Nachdenken über Christa T.* (1968) a fictitious narrator reflects on the life of Christa T., who dies of leukaemia in her early thirties. Her experiences, preserved in various writings, reveal the disillusionment of an artistic personality, whose hopes for self-realization are shattered by her terminal illness, a psychosomatic response to her situation. *Kindheitsmuster* (1976, BRD 1977), an overtly autobiographical novel, links up with the periods covered in the earlier works, but depicts the two worlds in the life of the author through the introduction of Nelly Jordan, who grows up in Landsberg where her parents run a little shop. Childhood and early adolescence end in the family's flight towards the end of the war. The adult Nelly, now a DDR citizen, relates a visit with her husband, daughter, and brother Lutz to her native town (now Polish and renamed Gorzów Wielkopolski). The journey, undertaken at the height of summer in 1971, leads back to the grocer's shop and revives memories for Nelly, who over a period of three years leading up to 1975 puts her reminiscences on paper. They cover the entire era of the National Socialist regime. By far the longest of Wolf's novels, it combines references to major events with a depiction of life in a predominantly conformist provincial town and of hard-working parents, who relax over the newspaper with half-closed eyes before bedtime. Wolf's recurring preoccupation with reminiscence, reflection, and self-analysis characterizes her approach to writing. Her volume *Lesen und Schreiben* (1971, extended 1973) contains important essays clarifying her aesthetics. A major influence is Anna Seghers and another woman writer, Ingeborg Bachmann (qq.v.). In representing a younger generation of authors, Wolf aims at a new incisive style of 'subjective authenticity'. Her flexible narrative technique also uses dreams to point to disharmonies between inner experience and outer reality. The resultant trait of pessimism in her writing is balanced by her increasingly resourceful approach to individualism and to feminism.

The stories include *Juninachmittag* (1967 in *Neue Texte*), *Till Eulenspiegel. Erzählung für den Film* (1973, in collaboration with Gerhard Wolf), and the collection *Unter den Linden. Drei unwahrscheinliche Geschichten* (1974), which contains, apart from the title-story, *Neue Lebensansichten eines Katers* (using a quotation from

E. T. A. Hoffmann's novel *Lebensansichten des Kater Murr*, q.v., as an epigraph), and *Selbstversuch*. The story *Kein Ort. Nirgends* (1979) deals with an unproven meeting between Kleist and Karoline von Günderode (qq.v.), both of whom took their own lives. The emphasis is on the woman's urge for self-realization and on the solitary figure of Kleist, who is at that time in the care of his doctor, Wedekind. In the same year she edited *Karoline von Günderode: 'Der Schatten eines Traumes. Gedichte. Prosa. Briefe'*, with an essay; another essay, *Nun ja! Das nächste Leben geht aber heute an. Ein Brief über die Bettine*, is contained in an edition of Bettina von Arnim's epistolary novel *Die Günderode* (1981, see ARNIM, B. VON). A considerably extended edition of her essays appeared as *Fortgesetzter Versuch. Aufsätze, Gespräche, Essays* in 1979 (DDR), and as *Lesen und Schreiben. Neue Sammlung* in 1980 (BRD); other editions of this year include the volume *Geschlechtertausch. Drei Erzählungen* (in collaboration with Sarah Kirsch and Irmtraud Morgner, qq.v.) and *Gesammelte Erzählungen*. In 1983 another work appeared pertaining to the emancipation of women (and men), the story *Kassandra* (q.v.), and lectures, delivered at Frankfurt University (Gastdozentur für Poetik), *Voraussetzungen einer Erzählung: Kassandra. Poetik-Vorlesungen*, in which she again distances herself from established poetics and heroic mythology in order to demonstrate its unfamiliar side against the background of nuclear power.

The recipient of honours in East Germany in the 1960s and early 1970s, Wolf was awarded West German prizes in the 1970s and in 1980 the Büchner Prize.

WOLF, FRIEDRICH AUGUST (Hagenrode, 1759–1824, Marseille), a classical philologist, became a professor at Halle University in 1783, where he remained until the dissolution of the university in 1807 in the changes following the conquest of Prussia by Napoleon.

Modern classical studies, in the sense of a comprehensive approach to all aspects of antiquity, were created by Wolf, who was especially concerned with the genesis and authorship of the Homeric epics (*Prolegomena ad Homerum*, 1795). A dynamic and truculent man, he made an impact on learned circles throughout Europe. His most important work is the *Darstellung der Altertumswissenschaft* (1807). In 1810 Wolf became a professor at the new Berlin University. He died while undertaking a visit to the south of France for the sake of his health. *Gesammelte Werke in sechzehn Bänden*, ed. E. Wolf and W. Pollatschek, appeared in 1960–8.

WOLF, HUGO (Windischgrätz, 1860–1903, Vienna), one of the greatest composers of Lieder (see LIED), had a complex personality inhibiting composition for long periods; these were followed by bouts of a few days or weeks, in which he wrote numbers of Lieder of great individuality, sometimes several in a single day.

Wolf's greatness lay in his capacity to immerse himself in a poem and then to produce a musical setting which enhanced the words without introducing any tone or mood extraneous to the poem. His songs, written in groups according to the poet who interested him at the time, comprise *Mörike-Lieder* (53 songs, 1888), *Eichendorff-Lieder* (20 songs, 1888), *Goethe-Lieder* (51 songs, 1888–9), *Das spanische Liederbuch* (translations by E. Geibel and P. Heyse, qq.v., 1889–90), and *Das italienische Liederbuch* (translations by P. Heyse, 2 vols., 1891–1896). Wolf also set three poems by Michelangelo, several by other poets, and a few more poems by his favourites Mörike (q.v.) and Goethe. His few other works include an opera, *Der Corregidor* (performed Mannheim 1896, libretto by Rosa Mayreder, 1858–1938), a string quartet in D minor (1880), and a symphonic poem, *Penthesilea* (1883). In 1897 Wolf began to suffer from paralysis and mental abnormality and spent the rest of his life in a mental institution.

Wolfdietrich, a Middle High German epic written before 1250 and preserved incomplete in the Ambraser Handschrift (q.v.), where it forms a sequel to *Ortnit* (q.v.).

It tells of King Hugdietrich and three sons, the youngest of whom is said by the evil counsellor Sabene to be begotten by the devil. Hugdietrich decides to have the child killed and entrusts the task to the faithful lord Berchtung von Meran. Berchtung accepts the assignment reluctantly, and when he finds the child spared by a pack of wolves, determines, since it is obviously under God's protection, to save it. It is brought up under the name of Wolfdietrich. Sabene's wiles are exposed and Wolfdietrich restored; but, after Hugdietrich's death, Sabene returns and a battle ensues. Berchtung and Wolfdietrich flee. Wolfdietrich seeks help from the powerful king Ortnit, but finds his wife mourning her husband's death. Wolfdietrich sets out to kill the dragon responsible for Ortnit's death. At this point the MS., known as *Wolfdietrich A*, ends, but from a debased later version it is apparent that Wolfdietrich kills the dragon, encounters various fabulous adventures, marries Ortnit's widow, and finally retires to a monastery. The poet of the Ambras version, who was also the author of *Ortnit*, is a skilful story-teller whose interest is in narrative rather than character or ideas.

The poem also exists in another version known as *Wolfdietrich B* or *Wolfdietrich von*

Salnecke (Salonika), in which the basic thread is the same, but the adventures and detail differ. It is poetically inferior to A. Of two later versions, C exists only in fragments, D multiplies the adventures and adds an oriental element.

Wolfdietrich is connected in name and geography (Italy, Tyrol) with the legend of Theodoric (see DIETRICHSAGE).

Wolfdietrich A (ed. H. Schneider, 1931) was reprinted in 1968; *Wolfdietrich C* and *D* were edited by A. Holzmann (1865).

Wolfenbüttler Marienklage, title given to a quasi-dramatic poem of 464 lines written in Middle Low German possibly in the 14th c. and discovered at Wolfenbüttel. It expresses Mary's lamentations at the foot of the Cross, and represents a phase of transition from narrative poem to passion play.

WOLFENSTEIN, ALFRED (Halle/Saale, 1883–1945, Paris), an Expressionist writer, published poems (*Die gottlosen Jahre*, 1914; *Die Freundschaft*, 1917; *Menschlicher Kämpfer*, 1919; *Bewegungen*, 1928), plays (*Der Sturm auf den Tod*, 1921; *Der Mann*, 1922; *Mörder und Träumer*, 1923; *Der Narr der Insel*, 1925; *Die Nacht vor dem Beil*, 1929), collections of stories (*Der Lebendige*, 1918; *Die gefährlichen Engel*, 1936), and a single narrative (*Unter den Sternen*, 1924). He also translated poetry by Shelley, Verlaine, Gérard de Nerval, V. Hugo, and Emily Brontë.

In 1922 Wolfenstein published an essay, *Jüdisches Wesen und neue Dichtung*. In 1933 he emigrated to Prague and in 1938 to France; he was imprisoned for three months in 1940, living after his release with false papers in Paris. He took his own life.

WOLFF or **WOLF**, CHRISTIAN (Breslau, 1679–1754, Halle/Saale), the most influential philosopher in the early years of the age of Enlightenment (see AUFKLÄRUNG), became a professor at Halle University in 1706. Wolff's rationalistic views (though not anti-religious) aroused the antagonism of his theological colleagues. Some of these succeeded in 1723 in persuading King Friedrich Wilhelm I (q.v.) by a gross slander that Wolff was a subversive element, and he was thereupon dismissed, though he was immediately elected to a professorship at Marburg.

One of the first acts of Friedrich II (q.v.) on his accession in 1740 was to recall Wolff to Halle, and in 1745 he was raised to the peerage as Freiherr von Wolff. Wolff lucidly presented, in a clear logical order and system, views which in part took their origin from Leibniz (q.v.). His principal works, which bear the generic title *Vernünftige Gedanken* and are all written in German, range in succession over the whole field of philosophy. *Vernünftige Gedanken von Gott, der Welt und der Seele* (1719) presents his metaphysics. *Vernünftige Gedanken von der Menschen Tun und Lassen* (1720) is his treatise on ethics. These were followed by *Vernünftige Gedanken von dem gesellschaftlichen Leben der Menschen* (1721, reprinted 1971) and *Von den Wirkungen der Natur* (1723). His *Rede von der Sittenlehre der Chinesen* (1721), which first aroused the antagonism of the theologians, was published in Latin in 1726 and in German in *Kleine philosophische Schriften* (1740). Although Wolff's influence was for a time considerable, his views were too narrowly utilitarian to retain philosophical currency for long.

Gesammelte kleine philosophische Schriften (6 vols.), ed. G. F. Hagen, appeared 1736–40, Wolff's correspondence with Leibniz, ed. C. I. Gebhardt, was published 1860 (reissued 1963). *Gesammelte Werke*, ed. H. W. Arndt and J. Ecole, planned in two sections containing Wolff's German and Latin writings, began to appear in 1962.

WOLFF, EUGEN (Frankfurt/Oder, 1863–1929, Berlin), one of the founders of Durch (q.v.) in 1886, had trained as a Germanist and became professor of German at Kiel University in 1904. He wrote a history of modern German literature (*Geschichte der deutschen Literatur in der Gegenwart*, 1896), but his publications were mostly concerned with writers of the 18th c. (*J. E. Schlegel*, 1889; *Gottscheds Stellung im deutschen Bildungsleben*, 1895; *Entwicklungsgeschichtliche Goethekritik*, 1925). His lively interest in the theatre led him to found the Wissenschaftliche Gesellschaft für Literatur und Theater and the Theater Museum in Kiel.

WOLFF, JULIUS (Quedlinburg, 1834–1910, Berlin-Charlottenburg), was of comfortable middle-class origin, had a university education at Berlin, and took over for a time his father's business. In 1869 he founded the journal *Die Harz-Zeitung*. He served as an officer during the Franco-Prussian war (see DEUTSCH-FRANZÖSISCHER KRIEG), afterwards settling in Charlottenburg and devoting himself to writing. His first book publication was a volume of war poems (*Aus dem Felde*, 1871). This was followed by the verse epics *Till Eulenspiegel redivivus* (1874), *Der Rattenfänger von Hameln* (1876), *Der wilde Jäger* (1877), *Tannhäuser* (1880), *Der fliegende Holländer* (1892), *Assalide* (1896), *Der Landsknecht von Kochem* (1898), and *Der fahrende Schüler* (1900). The poem *Die Pappenheimer* dates from 1889. Wolff's works exemplify the Butzenscheibenpoesie (q.v.) derided by P. Heyse (q.v.).

Wolff is also the author of novels, including *Das schwarze Weib* (1894) and *Zweifel der Liebe*

(1904). His collected works, *Sämtliche Werke* (18 vols.), ed J. Lauff, appeared 1912–13.

WOLFF, Oskar Ludwig Bernhard (Altona, 1799–1851, Jena), turned a gift for instant poetry to commercial account by giving evenings of extemporization. He became a professor at Jena University in 1838. He was a prolific writer of fiction, for which he used the pseudonym Plinius der Jüngste, but was chiefly popular as a compiler of anthologies, the best known of which was *Poetischer Hausschatz des deutschen Volkes* (1839). He also wrote a history of the novel, *Geschichte des Romans von dessen Ursprüngen bis auf die neueste Zeit* (1841). His works appeared as *Schriften* (14 vols.), 1841–3.

WÖLFFLIN, Heinrich (Winterthur, 1864–1945, Zurich), son of the philologist Eduard Wölfflin (1831–1908), studied in Munich and was greatly influenced by J. Burckhardt (q.v.), on whose retirement in 1893 he became professor at Basel. In 1901 he moved to Berlin, becoming a member of the Prussian Academy in 1911; in the following year he took up a professorship in Munich and in 1924 a similar appointment in Zurich.

Wölfflin's works on art include *Prolegomena zu einer Psychologie der Architektur* (1886), *Renaissance und Barock* (1888, reissued 1961), *Die klassische Kunst* (1899, reissued 1953), two works on Dürer (q.v., *Die Kunst A. Dürers*, 1905, reissued 1943, and *Handzeichnungen von A. Dürer*, 1914), and *Das Erklären von Kunstwerken* (1922, ed. J. Gantner 1969). In 1915 he published *Kunstgeschichtliche Grundbegriffe. Das Problem der Stilentwicklung in der neueren Kunst* (14th edn. 1970). This stimulating analysis of stylistic features which characterize different ages evinces a refined sensitiveness to the variations of expression distinguishing individual artists. Its five sections deal with the principles underlying contrasts: the linear and the picturesque, space and depth, closed form and open form, multiplicity and unity, and the relationship between effects achieved in painting and architecture by clarity and by its diffuseness. The applicability of these principles to literary structures was first recognized by exponents of Geistesgeschichte (q.v.), with which Wölfflin expressed a kinship. Later works include *Italien und das deutsche Formgefühl* (1931) and *Gedanken zur Kunstgeschichte* (1941). *Kleine Schriften*, ed. J. Gantner, appeared posthumously in 1946, and correspondence with J. Burckhardt in 1948.

WOLFGER VON ELLENBRECHTS-KIRCHEN, German prelate and patron of literature, was bishop of Passau from 1194 to 1204. The unknown author of the *Nibelungenlied*

(q.v.) is thought to have stood in some relationship to the Bishop's court, and his reference to 'Bishop Pilgrim' (XXVI) is a compliment to his host or master. Albrecht von Johannsdorf and for a time Walther von der Vogelweide (qq.v.) were active at his court. The entry of a gift of a fur coat for Walther in the Bishop's accounts for 1203 is the only documentary attestation of Walther's life. In 1204 Wolfger became Patriarch of Aquileia in Italy, and here Thomasin von Circlaere (q.v.), author of *Der welsche Gast*, was a member of his entourage.

WOLFRAM VON ESCHENBACH (c. 1170–c. 1220), Middle High German poet, was the author of three epic poems, *Parzival*, *Willehalm*, and *Titurel*, and of eight lyric poems. Wolfram's life is known, rather sketchily, from allusions in his own poems. His home was at Eschenbach, probably the still existing village near Ansbach. He was a nobleman, but poor, and was dependent on patronage. Although he himself stated, perhaps ironically, that he was unlettered, it is no longer believed that he was truly illiterate. In spite of his insistence on the profession of arms, he may well have been a professional poet. He himself mentions as his patron Landgraf Hermann (q.v.) of Thuringia.

Parzival (q.v.), originating from the first decade of the 13th c., was followed by *Willehalm* (q.v.), which remained unfinished, and by two fragments of a third epic, *Titurel* (q.v.). Of Wolfram's eight extant poems only three are Minnelieder, and one of these is a rejection of Minnedienst. The remaining five are powerful and idiosyncratic Tagelieder (see Tagelied); the last, 'Ez ist nu tac', extols marital love, forsaking altogether the conventions of *minne*.

Wolfram's works were widely read and applauded, though perhaps not universally, since it has often been supposed that he is the subject of some scathing criticism directed by Gottfried von Straßburg (q.v.) at an unnamed poet. Wolfram was a highly individual poet with deep moral convictions, combining sincere religious belief, a sense of the worth of knighthood, and a profound humanity. His eccentric and difficult style corresponds to his idiosyncratic personality. In romanticized form Wolfram appears as a character in R. Wagner's *Tannhäuser* (q.v.).

Wolfram's *Gesamtwerk* was edited by K. Lachmann (1833, reprinted 1968, revised E. Hartl, 7th edn. 1952). *Parzival* and *Titurel* were edited by K. Bartsch (3 vols., 1870–1, 4th edn. M. Marti, 1927–32); *Willehalm* and *Titurel*, ed. W. J. Schröder and G. Hollandt, appeared in 1971 and *Die Lyrik von Wolfram von Eschenbach*, ed. P. Wapnewski, in 1972.

WOLFSKEHL, KARL (Darmstadt, 1869–1948, Auckland, N.Z.), of Jewish descent, studied German philology at Berlin, Leipzig, and Gießen universities. Associated with the circle of admirers of S. George (see GEORGE-KREIS), he developed into a poet of considerable power, his earliest collection being *Ulais* (1897). With George he edited *Deutsche Dichtung* (1900–3, anthology) and with Friedrich von der Leyen (1873–1966) the selection of texts and translations *Älteste deutsche Dichtung* (1909). In 1933 he emigrated to Italy and in 1938 to New Zealand, composing during years of agonizing isolation his greatest poetry; it reflects his consciousness of an intellectual heritage which is both Jewish and German, and comprises the volumes *Die Stimme spricht* (1947), *Sang aus dem Exil* (1950), and *Die vier Spiegel* (1951).

Gesammelte Werke (2 vols.), ed. M. Ruben and C. V. Bock, appeared in 1960.

WOLGEMUT, MICHAEL (Nürnberg, 1434–1519, Nürnberg), a master painter of Nürnberg who taught A. Dürer (q.v.). He was also notable for his woodcuts (see SCHEDEL, H.).

WOLKEN, KARL ALFRED (Wangeroog, 1929–), left his native Frisian island in the 1940s, worked as a cabinet maker in the 1950s before devoting himself to authorship in Rome and Stuttgart. His poetry includes the volumes *Halblaute Einfahrt* (1960), *Wortwechsel* (1964), and *Klare Verhältnisse* (1968). In 1961 he published his first novel, *Die Schnapsinsel*, and in 1964 *Zahltag*; a collection of stories (*Erzählungen*) appeared in 1967.

WOLTERS, FRIEDRICH (Werdingen nr. Krefeld, 1876–1930, Munich), professor of history successively at Marburg and at Kiel universities, was a member of the George-Kreis (q.v.), publishing in 1930 *Stefan George und die Blätter für die Kunst*. His other publications include speeches (*Vier Reden über das Vaterland*, 1927), a five-volume anthology *Der Deutsche. Ein Lesebuch* (1925–7), translations of Middle High German poetry, and a three-volume anthology *Hymnen und Lieder der christlichen Zeit* (1922–3). He entitled collections of his own poetry *Wandel und Glaube* (1911) and *Der Wanderer* (1924).

WOLZOGEN, CAROLINE VON (Rudolstadt, 1763–1847, Jena), Schiller's sister-in-law, *née* von Lengefeld, she married Wilhelm von Beulwitz in 1784, but soon parted from him. Schiller met Caroline and her sister Charlotte in 1787; a friendship sprang up in which Schiller's attention seemed equally divided, though in the end it was Charlotte to whom he proposed and, in 1789, married. Caroline found

it difficult to resign all claim to Schiller, but eventually in 1794 married Schiller's younger friend and her distant kinsman, Wilhelm von Wolzogen (1762–1809), having divorced Beulwitz in the preceding year.

She was the author of one of the earliest biographies of Schiller, *Schillers Leben, verfaßt aus Erinnerungen der Familie, seinen eigenen Briefen und den Nachrichten seines Freundes Körner*, 1830. Of her two novels *Agnes von Lilien* (1796–7) and *Cordelia* (1840), the former, published anonymously in *Die Horen*, was believed by many to be by Goethe. She also published *Erzählungen* (1826–7). Posthumous writings, *Literarischer Nachlaß* (2 vols.), were edited by K. Hase in 1848–9.

WOLZOGEN, ERNST, FREIHERR VON (Breslau, 1855–1934, Munich), of a Silesian noble family, studied in Strasburg and Leipzig, was in the service of the Grand Duke of Saxe-Weimar, and then took to journalism and authorship. In 1900 he founded the Berlin cabaret Das Überbrettl, for which he wrote social satire. He was a prolific author of humorous fiction (*Die tolle Komteß*, 2 vols., 1889; *Der Kraft-Mayr*, 2 vols., 1897; *Das dritte Geschlecht*, 2 vols., 1899) and also wrote works dealing with serious problems (*Fahnenflucht*, 1894, *Das Wunderbare*, 1898, and *Die arme Sünderin*, 1901). His comedies include *Die Kinder der Exzellenz* (1890, an adaptation of a novel, 1888), *Das Lumpengesindel* (1892), and *Ein unbeschriebenes Blatt* (1896). Wolzogen's style has wit and elegance.

WOLZOGEN, HANS PAUL, FREIHERR VON (Potsdam, 1848–1938, Bayreuth), German musicologist, founded in 1878 the periodical *Bayreuther Blätter* and wrote copiously and eulogistically on R. Wagner (q.v.). His standpoint in his Wagnerian and other writings was strongly nationalistic. Wolzogen coined the now universally accepted term Leitmotiv (q.v.).

Wonne der Wehmut, a poem written by Goethe in 1775 and printed in revised form in 1789. It has been set to music by F. Schubert (q.v.).

World Wars I and II, see WELTKRIEGE I and II.

Worms, city on the left bank of the Rhine, was a settlement in Celtic times and, as *civitas Vangionum*, a Roman city. At the time of the invasion of Attila and the Huns it was the capital of the ancient Burgundian kingdom. The bishopric, which lasted until 1797, was founded in Roman times. In the Middle Ages the city was one of the most important in the Empire and was the scene of numerous Imperial Diets (see WORMSER REICHSTAGE). Worms, which suffered repeatedly in the Thirty Years War (see DREIS-SIGJÄHRIGER KRIEG), was burned by the French in 1689 and heavily damaged in the 1939–45 War. Its political and economic influence declined in the 18th c. Its most important literary associations are with the *Nibelungenlied* (q.v.) and with Luther (q.v.).

Wormser Edikt, see WORMSER REICHSTAGE.

Wormser Konkordat, Das, an agreement between the Emperor Heinrich V (q.v.) and Pope Calixtus II which brought to an end by a compromise the long dispute known as the Investiture Contest (see INVESTITURSTREIT). It was proclaimed on 23 September 1122 after deliberations at Worms lasting a fortnight. The Emperor agreed to renounce the right of investiture of bishops with the spiritual insignia of ring and crozier; but elections of bishops were to be made in the Emperor's presence.

Wormser Reichstage, Diets of the Holy Roman Empire (see DEUTSCHES REICH, ALTES) held at Worms on many occasions from the 8th to the 16th c. The Diet of 1495, held under the Emperor Maximilian I (q.v.), sought to reform the administration of the Empire by establishing an independent judicial machinery and authorizing a special imperial tax. Its best-known measure was the proclamation of permanent public peace (see EWIGER LANDFRIEDE), which put an end to the right of private feud.

The most famous Diet of Worms was that held in 1521 (17–25 April) under Karl V (q.v.), to which Luther (q.v.) was summoned to make recantation. Luther complied with the summons, but declined to retract any of his writings. A graphic account of the occasion is given by L. von Ranke (q.v.) in his *Geschichte der Päpste*, in which Luther's famous words 'Hier stehe ich, ich kann nicht anders! Gott helfe mir! Amen' are quoted. After the Diet, which culminated in sentence of outlawry (Wormser Edikt), Luther was given asylum at the Wartburg by the Elector Friedrich der Weise (q.v.) of Saxony.

Wormser Vertrag, the agreement by which the Emperor Karl V (q.v.) transferred in 1521 to his brother Ferdinand of Austria (see FERDINAND I) the fiefs of Styria, Carinthia, and Krain.

Wörterbücher. For the principal German dictionaries see ADELUNG, J. C. (18th c.), GRIMM, J. (19th c. and 20th c.), and DUDEN (20th c.).

Also noteworthy are *Deutsches Wörterbuch* (1897, new edn. W. Betz, 1966) by H. Paul (1846–1921) and *Etymologisches Wörterbuch der deutschen Sprache* (1881–3) by F. Kluge (1856–1926).

Wo warst du, Adam?, a novel of the 1939–45 War by H. Böll (q.v.), published in 1951. The title's question, from Genesis, is answered by a quotation, 'Ich war im Weltkrieg'; the work is also prefaced by a short translated passage from de Saint-Exupéry ending '. . . Der Krieg ist eine Krankheit, wie der Typhus'. Consisting of a cycle of nine loosely connected episodes, it is mainly set in Rumania and Hungary between 1943 and 1945 and depicts wounds, death, mud, dysentery, and a brief interlude of extermination in a concentration camp. The only figure linking the episodes is the soldier Feinhals, an architect by profession; he arrives home in Germany to be killed, together with his parents, by German artillery firing senselessly into the village when the war is virtually at an end. The temper of the novel is one of futility and bitter irony.

Woyzeck, a dramatic fragment by G. Büchner (q.v.), written between 1835 and 1837 and, following its belated discovery, published in 1879 by K. E. Franzos (q.v.) under the title *Wozzek*. A revised version discussing variants of the MSS. and restoring the title *Woyzeck* was published by F. Bergemann in 1922. Uncertainties about textual detail as well as the scenic arrangement of the fragment, which consists of some 27 scenes with no act division, continued to be the subject of critical investigations. In 1967 W. R. Lehmann published a further revised edition containing scenic rearrangements, including the first and last scene of Bergemann's edition, and a reassessment of the phases of the composition of the fragment, the ending of which cannot be fully ascertained. The work was first performed on 8 November 1913 at the Residenztheater, Munich.

Büchner used as his source a report by Hofrat Clarus published in 1825 on the case of Johann Christian Woyzeck, an unemployed barber, wig-maker, and soldier, found guilty in Leipzig in 1821 of the murder of a widow, Frau Woost. He was sentenced to death, and executed in 1824 in the market-place in Leipzig. The execution, fixed for 1822, was delayed pending an appeal for a commutation of the sentence on the grounds of diminished responsibility. The appeal resulted in protracted investigations into Woyzeck's mental condition by more than one medical authority. Clarus's publication stirred up polemics on the case in a medical journal, *Henkes Zeitschrift für die Staatsarzneikunde*. From the analysis made by Clarus it may be inferred that Woyzeck was to a considerable degree victim of the political and social instability of the age. Orphaned at the age of 13, he served for want of other employment in various armies. Left in Stralsund in 1810 without identity papers, he was refused permission to marry the girl by whom he had a child. On his return to his native Leipzig, his application to enter the city's militia was turned down. At the time of the murder he was a beggar. (Other possible, but minor, sources have been suggested.)

Büchner uses his source freely and incorporates some of the prisoner's words to Clarus in his work. He concentrates on the simple humanity of Franz Woyzeck, whose sense of insecurity pervades all aspects of his life, including a mysterious sense of persecution by the Freemasons and visions of doom relating to the Bible. This aspect, revealing the extent of Woyzeck's mental disturbance, dominates the scene *Freies Feld. Die Stadt in der Ferne* with which the fragment opens in Lehmann's edition; it is followed by the scene *Die Stadt*, introducing the central figures of the dramatic complication: Marie, with whom Woyzeck has set up home and from whom he has a child, although for reasons of poverty they have remained unmarried, and the drum-major (Tambourmajor). The crucial ending of the scene reveals Marie's bewilderment at Woyzeck's increasingly strange behaviour.

Caricatured representatives of society responsible for Woyzeck's plight are the captain (Hauptmann) and the doctor (Doktor). Serving in the army as a barber, Woyzeck is the object of bitter humiliation, both in the derisive speeches of the captain in the shaving scene with which Bergemann's version opens, and in the inhuman experiments of the doctor, to whom Woyzeck's physical deterioration is of scientific interest. Woyzeck bears it all for the love of Marie and of his child, for whom he needs the extra money earned from the doctor in his experiments. On finding that Marie has allowed herself to be seduced by the drum-major, Woyzeck's world, meaningful only through his love, breaks down. He stabs Marie to death with a knife specially bought for the purpose and abandoned after the murder in the lake, into which he wades in the scene *Woyzeck an einem Teich* to cleanse himself of blood. Woyzeck's monologues in this and the preceding scene (*Abend. Die Stadt in der Ferne*) reflect his utter wretchedness, his fear of the consequences of his deed being accompanied by the sting of Marie's betrayal. His world finally collapses when Karl, the idiot, runs away with his little son, who is frightened at his father's attempt to hug him. In the Lehmann version the fragment breaks off with this scene (*Der Idiot. Das Kind. Woyzeck*). (Film version by W. Herzog, 1979.)

Alban Berg (q.v.) based his opera *Wozzeck* (1921) on the arrangement by K. E. Franzos in which the emphasis in the finale is laid on the orphaned boy riding his hobby-horse. Berg first planned the opera after having attended the Viennese performance at the Kammerspiele in May 1914.

Wozzeck, see WOYZECK.

WRONSKI, STEFAN, see HARDEKOPF, F.

WULFILA (b. in Danube basin in S.E. Europe, ?–383, Constantinople), a Christian missionary of half Gothic, half Cappadocian parentage, became bishop in 341 with the task of spreading Christianity among the West Goths in the Balkans. About 369 Wulfila, who had become the Primate of the Goths, began to translate the Bible into Gothic. He did not complete the work and the extent of the contribution of later translators is unknown. Wulfila's Bible is known principally in the Codex argenteus (q.v.), a MS. of the four Gospels, preserved at Uppsala. Fragments of the Gothic Bible are also preserved at Wolfenbüttel and Milan, and a further fragment was recently discovered in Speyer (1971).

Wulfila, who devised the first Gothic characters, used the Septuagint for the Old Testament, and a text similar to that used by St. Chrysostom for the New Testament. The Greek form of the name, Ulfilas, is also in use.

WULLENWEVER, JÜRGEN (Hamburg, before 1488–1537, Lübeck), settled in Lübeck and was prominent in the conflicts which led to the establishment there of the Reformation (q.v.). He became burgomaster in 1533 and sought to reaffirm the power of the Hanseatic League (see HANSE, DEUTSCHE). His policy failed, the former rulers of Lübeck rose against him, and he resigned in 1535. He was tried and executed under the Roman Catholic Duke of Brunswick-Wolfenbüttel. He is the subject of a tragedy by K. Gutzkow (q.v., *Jürgen Wullenweber*, 1848).

Wunderbarliche Vogel-Nest, Das, a novel in 2 parts published in 1672 by J. J. C. von Grimmelshausen (q.v.). The first part appears under the pseudonym Michael Rechulin von Sehmsdorf, the second is attributed to Samuel Griefnson von Hirschfeld.

The story links up with an episode near the end of *Der seltzame Springinsfeld* (q.v.); the bird's-nest, which makes its holder invisible, passes into the hands of a halberdier. He makes use of it for all kinds of tricks, but for the most part exposes dishonesty and deceit. In the end he repents and abandons the nest.

It is picked up by others and comes into the hands of a merchant who exploits it for dishonest and sinful purposes, including fornication. In the end he, too, repents and confesses, and the father confessor throws the nest into the Rhine. The moral basis of the story is somewhat impaired by the gusto devoted to the description of obscene incidents.

Wunderer, Der, a Middle High German poem, the original of which is lost. Its subject is the pursuit of a virgin (Saelde) by the fierce hunter, 'der Wunderer'; she seeks protection at Etzel's court, but Etzel and Rüdiger refuse her plea. She is rescued by Dietrich, who kills the 'Wunderer'. The poem exists in three late versions: in the *Dresdener Heldenbuch*, with the title *Etzels Hofhaltung*; in a poem entitled *Ain spruch von aim konig mit namen Etzell*; and in a Fastnachtspiel (q.v.), *Ein Spil von dem Perner* [Dietrich von Bern] *und Wunderer*. The original has been dated in the first half of the 13th c., but it is probably later.

Wunne Baum der minnenden Seele, Der, a prose allegory of heavenly love in the form of a tree which provides degrees of aspiration and approach to God. It was possibly written in the 13th c.

Wunnigel, a short novel written by W. Raabe (q.v.) in 1876 and published in 1879, and subtitled *Eine Erzählung*. Its plot is partly commonplace and partly absurd.

Dr. Weyland, a young physician, in comfortable circumstances, is summoned to a woman patient in a country inn. She is Anselma, the daughter of Regierungsrat a. D. Wunnigel, and in due course the Doctor marries her and they move into his ancient house, with its rich museum-like contents. Wunnigel, whose absorbing hobby is ancient bric-à-brac and oddments, has spent ·the time of the courtship in an exhaustive inspection of Weyland's treasures. After the marriage Wunnigel leaves for Italy, but presently reappears a changed man, worn and furtive. Evidently in need of refuge, he shifts his quarters to the tumbledown house of 90-year-old Uhrmachermeister Brüggemann and confesses that he has made a foolish second marriage in Italy. A Russian nobleman arrives, and therewith a further Italian *bêtise* of Wunnigel's is uncovered: he has offered to sell· the contents of Weyland's house to the Russian antiquary. Wunnigel takes to his bed and remains there despite the entreaties of his daughter, his son-in-law, and his Russian guest. The second wife arrives and departs in dudgeon, escorted by the Russian. Wunnigel dies without leaving his

bed, and nonagenarian Brüggemann closes the story by passing quietly away in his sleep.

The self-willed, impulsive, bristly Wunnigel expresses himself in tortuous arabesques with extraordinary verve and vigour. Hardly less important is the anachronistic figure of Brüggemann, who in his time has been useful enough and foolish enough and now in his old age achieves complete harmony and profound understanding of his fellow men.

Wünschelrute, a poem by J. von Eichendorff (q.v.) consisting of only four lines, the first of which runs 'Schläft ein Lied in allen Dingen'. Its title refers to the magic rod (analogous to the water dowser's) for the divining of buried treasure.

Wunschkind, Das, a novel by Ina Seidel (q.v.) published in 1930. The action spans the short life of Christoph von Echter from his conception at Mainz in 1792 during preparations for war against the French to his death in action in 1813 in the War of Liberation. The background of the Napoleonic Wars (q.v.) is used to show human emotions under stress and to scrutinize the cultural and national heritage, which is in danger of destruction.

The central figure is Christoph's Prussian mother, Cornelie von Echter, *née* von Tracht, widow of Hans Adam Echter von Mespelbrunn, of similarly ancient but South German lineage. Cornelie loves Christoph unselfishly and intensely, for he is 'das Wunschkind', conceived after the death of her first child, and in the very last night spent with her husband, who is killed in the campaign of 1792. Christoph is brought up with his orphan cousin, the half-French Delphine Loriot. After fighting as a boy of 16 at Jena in 1806, Christoph becomes a student. He falls passionately in love with Delphine, who goes off with an actor, though by his early death Christoph is spared the knowledge of her defection. Cornelie, the image of maternal care and solicitude, does not marry again despite the inclination to do so. She survives alone, looking after the older generation of her husband's and her own family and after children orphaned through the war.

The character of Delphine recurs in Ina Seidel's later novel, *Das unverwesliche Erbe* (q.v.), as the mother of Charlotte Dornblüh. Her death in early womanhood in the later novel is hinted at in *Das Wunschkind* by her name and character, representing the innocent and yet evil elemental spirit (Elementarwesen), a Romantic motif, described as 'schuldlos, wie auch die Natur schuldlos ist, . . . böse, wie die Natur böse ist'. She supports the novel's remarkable interplay of opposites. Cornelie, who disciplines the emotions by her spirituality which nurtures her compassion, illustrates through her conversion to the Roman Catholic Church for the sake of marriage, and her return to her Prussian home and to her Protestant heritage, the main theme to which the work is dedicated; for she emerges as a figure of reconciliation. This is based on hard experience and on the conviction that faith must be sustained and conceived in a heritage of conflict into which her son was born.

Würde der Frauen, a poem by Schiller first printed in the *Musenalmanach* for 1796. It is concerned with the different spiritual roles of man and woman, and reinforces its message by the use of two distinct metres. The text now generally printed is an abridgement made by Schiller for *Gedichte,* 1800.

WURM, THEOPHIL, Protestant Bishop of Württemberg, see RESISTANCE MOVEMENTS (2).

Wurmsegen, a spell for the cure of disease by the expulsion of the 'worm' causing it. The spell exists in two forms, Old Saxon and Old High German, which are entitled respectively *Contra vermes* and *Pro nessia.* They are recorded in MSS. of the 9th c. They represent the simplest and oldest type of spell, an imperative formula. (See also ZAUBERSPRÜCHE.)

Wurstelprater, an amusement park in the Prater in Vienna. See JOSEPH II.

Württemberg, former German county (Grafschaft), duchy (Herzogtum), 1495–1805, and kingdom, 1805–1918. It was a member of the German Confederation (see DEUTSCHER BUND) 1815–66 and a kingdom in the German Empire from 1871. After 1918 Württemberg became a Land (Freistaat) of the Weimar Republic (q.v.). In 1945 the northern half of Württemberg and of Baden were in the U.S. zone of occupation and were designated as the Land Württemberg-Baden. The southern half of Württemberg and the small state of Hohenzollern were in the French zone and were styled Südwürttemberg-Hohenzollern. These Länder continued under the Federal Republic (see BUNDESREPUBLIK DEUTSCHLAND) until 1952, when Baden, Württemberg, and Hohenzollern became the Land Baden-Württemberg.

WÜRTTEMBERGER, DER, see JAGD VON WÜRTTEMBERG, DIE.

Würzburger Kochbuch, designation of a Middle High German cookery book with the title *Das Buch von guter Speise.* The MS. was formerly at Würzburg, hence the designation; it dates from c. 1350.

Würzburger Liederhandschrift, a MS. important for the series of strophes it contains which are ascribed to WALTHER von der Vogelweide and REINMAR der Alte (qq.v.). In addition to lyrics it contains narrative, didactic, and religious poems. It was written some time before 1350, probably to the order of Michael de Leone (died 1355), the episcopal chancellor in Würzburg. It is now in Munich.

Wuz, see LEBEN DES VERGNÜGTEN SCHULMEISTERLEINS MARIA WUZ IN AUENTHAL.

WYLE, NIKLAS VON (Bremgarten/Aargau, c. 1410–78 or 1479, Stuttgart), early humanist and translator, became town clerk of Nürnberg in 1445 and of Esslingen in 1449, serving also as chancellor to Count Eberhard V of Württemberg. Wyle sought to create an elegant German prose style by direct imitation of Latin constructions. He translated a number of works by Italian authors, notably Enea Silvio Piccolomini (Pope Pius II) and Petrarch, and introduced into German literature the story of Griselda (q.v.). His versions were collected under the title *Translatzen* (1478).

WYMPFELING, JAKOB, see WIMPFELING, JAKOB.

WYNFRITH, see BONIFACE, SAINT.

WYSS, JOHANN DAVID (Berne, 1743–1818, Berne), a pastor attached to Berne Minster, wrote the moralizing novel of shipwreck and adventure, *Der schweizerische Robinson,* published between 1812 and 1827 (4 vols.) by his son J. R. Wyß (1782–1830). It enjoyed world-wide success, being translated into English as *The Swiss Family Robinson.*

X

Xanten, ancient cathedral city near the left bank of the Lower Rhine below Wesel. In the second Aventiure of the *Nibelungenlied* (q.v.) it is mentioned as the home of Siegfried (Santen).

Xenien, a collection of satirical epigrams, the joint work of Goethe and Schiller. Each epigram is a classical distich, composed of hexameter and pentameter. The *Xenien* were written in the winter of 1795–6 and published in October 1796 in the *Musenalmanach für das Jahr 1797*. The title, devised by Goethe, is borrowed from Martial (*Xenia*, presents made to departing guests) and is employed ironically both by Martial and the authors of the *Xenien*.

The occasion for this satirical campaign was the widespread unintelligent and hostile criticism to which Schiller's culturally ambitious periodical *Die Horen* (q.v.) was subjected during the year 1795. The first hint at retaliation came in October 1795 from Goethe, who suggested holding a brief 'trial' ('ein kurzes Gericht') of the criticisms. The idea of using a collection of epigrams for the purpose arose in December. Some of the distichs composed in 1796 were published in later years by Schiller and Goethe, and others, found among their papers, were published posthumously; but 414 were included by Schiller under the heading *Xenien* in the *Musenalmanach*. These were the aggressive distichs; others which they described as 'tame' ('zahm') were constructive generalizations on literature and art. The 'tame' distichs were included in the same volume of the *Almanach* under the headings *Tabulae votivae* (124 distichs arranged in 103 *tabulae*), *Vielen* (18 distichs), and *Einer* (19 distichs printed as one continuous poem). Although some can be allocated with certainty to Goethe or Schiller, a large number remain doubtful. The principal individual targets of the campaign were the critics L. H. Jakob (1759–1827), J. K. F. Manso (1759–1826), and F. Nicolai (q.v.), but the wider aim was to denounce the narrow-minded and platitudinous intellectual world of which they were representative. Among the many others who received 'gifts' in the *Almanach* were J. C. Adelung, J. B. von Alxinger, J. T. Hermes, Klopstock, A. von Kotzebue, J. K. Lavater, and F. L. von Stolberg (qq.v.). The *Xenien* were bitterly resented and savage retorts were made. Goethe and Schiller abstained from any further reply.

Y

YORCK, HANS DAVID LUDWIG VON (Potsdam, 1759–1830, Klein-Oels, Breslau), a Prussian officer of English descent, was created Graf Yorck von Wartenburg in 1814.

Yorck commanded an infantry brigade at Jena (q.v.) in 1806 and emerged from that disaster with an enhanced reputation. Later in the campaign he was wounded at Lübeck. In 1812 he was appointed to command the corps which Prussia was obliged to furnish for Napoleon's Grand Army against Russia. At first he fought ably against the Russians, but the defeat of Napoleon becoming evident, he found himself in a situation in which Prussian patriotism and obedience to orders conflicted. On 30 December 1812 he signed the Convention of Tauroggen by which the Prussian corps became neutral (see NAPOLEONIC WARS). Though his action was at first disavowed by the Prussian king, Yorck became in a short time a national hero. He distinguished himself in the campaigns of 1813–14. Yorck was promoted field-marshal (Generalfeldmarschall) in 1821.

YORCK VON WARTENBURG, PETER, GRAF VON, (1904–44), a descendant of the field-marshal (see YORCK, H. D. L. VON), at first a civil servant, later a subaltern in the German army, took part in the unsuccessful conspiracy to overthrow Hitler in July 1944, in consequence of which he was arraigned before the People's Court and executed. See RESISTANCE MOVEMENTS (2).

YOUNG, EDWARD (Winchester, 1681–1765, Welwyn), an Anglican divine, was the author of two works which influenced German literature. *The Complaint or Night Thoughts* (1742–4) was translated into German in 1751 by J. A. Ebert (q.v.) as *Nachtgedanken*. Ebert communicated his enthusiasm to his friend Klopstock (q.v.), and admiration of Young's poem is also expressed by J. G. Hamann, Herder, Wieland, Gerstenberg, and J. M. R. Lenz (qq.v.). G. E. Lessing (q.v.) regarded Young's poetry as maudlin and mocked its vogue.

Young's *Conjectures on Original Composition* (1759, in German 1760) influenced the view of genius developed in the Sturm und Drang (q.v.). Gerstenberg eloquently praised Young's essay in the literary periodical *Der nordische Aufseher*.

Z

Zabern, French Saverne, Alsatian town, which from 1871 to 1918 was in the German Empire. It became internationally known through the Zabern-Affäre in October 1913, a consequence of ill-feeling between the inhabitants and the German garrison. After a trivial incident a junior officer, Leutnant von Forstner, struck a crippled civilian. A commotion ensued, the military were called out, and on the orders of the regimental commander 28 people were illegally arrested. The Reichstag protested, but a court martial acquitted both Forstner and his C.O. The incident aroused much adverse comment abroad.

ZACCHI, FERDINAND (Wyk on the island of Föhr, 1884–1966), a writer of Heimatkunst (q.v.). His output is small, but, a genuine Frisian, he is the author of novels which are rich in peculiarly Frisian dialect and characterization. *Deus mare, Friso litora fecit*, which he translates 'Gott hat das Meer, der Friese das Land gemacht', is a motif in the novel *Klaar Kimming* (1922, the title meaning 'clear view' down to sea level). This work followed *Freerk Frendsens Blut* (1920) and preceded *Die liebe Not* (1924). *Freygeboren* (1927), perhaps his best novel, shows the development of Elert, whose forebears were serfs in Poland: despite the legal abolition of serfdom he only feels free when he returns to his ancestral Frisian lands and marries a woman of free descent. In the depiction of social themes Zacchi resembles G. Frenssen and W. Lobsien (qq.v.).

ZACHARIA or **ZACHARIÄ,** JUST FRIEDRICH WILHELM (Frankenhausen, 1726–77, Brunswick), German minor poet, studied at Leipzig and Göttingen universities. In 1748 he was appointed a master at the Carolinum (q.v.) in Brunswick, receiving the title of professor in 1761.

As a student Zachariä was an adherent of Gottsched (q.v.) and, as such, wrote his first and principal work, the verse satire *Der Renommist* (q.v., 1744, repr. 1909). He soon after transferred his allegiance to the Bremer Beiträger

(q.v.). His later, and less successful, works include *Das Schnupftuch* (a satire on the nobility) and *Murner in der Hölle*, a comic poem in 5 books, dealing with cats (1757). He also wrote short poems and translated Milton's *Paradise Lost* into German hexameters (2 vols., 1760). A serious epic poem on Hernando Cortés remained unfinished (*Cortes*, vol. 1, 1766).

Zachariä's *Poetische Schriften* (9 vols.) were published 1763–5, his *Hinterlassene Schriften*, ed. J. J. Eschenburg, 1781.

Zaches, see KLEIN ZACHES GENANNT ZINNOBER.

ZAHN, ERNST (Zurich, 1867–1952, Zurich), until 1920 a restaurateur at Göschenen, is the author of popular fiction of Swiss life, including the collections of Novellen *Bergvolk* (1897), *Schattenhalb* (1904), *Helden des Alltags* (1906), and the novels *Erni Behaim* (1898), *Albin Indergand* (1901), *Lukas Hochstraßers Haus* (1907), *Frau Sixta* (1926), *Gewalt über ihnen* (1929), *Der Weg hinauf* (1935), *Ins dritte Glied* (1937), *Die tausendjährige Straße* (1939), *Die große Lehre* (1943), and *Mütter* (1946).

In 1917 Zahn published *Bergland*, a collection of four Dichtungen: *Ein Blumenmärchen* (1903, in verse), *Mondelfe* (1910, in verse), *Der Schneegreis und die junge Anemone* (1911, a fairy-tale in prose), and *An mein Bergland. Eine Bekenntnisdichtung* (1916, in verse).

Zahn's *Gesammelte Werke* appeared in two series, each of 10 vols., 1909 and 1925.

ZAND, HERBERT (Bad Aussee, Austria, 1923–70, Vienna), a farmer's son, who after the 1939–45 War worked as a publisher's reader and latterly devoted himself entirely to authorship. He suffered a serious wound on the Russian front which he knew would eventually prove fatal. His novels and stories draw on his recollections of the Austrian Styria of his youth and on his wartime experiences (*Die Sonnenstadt*, 1947; *Letzte Ausfahrt*, 1953; *Der Weg nach Hassi el emel*, 1956; and *Erben des Feuers*, 1961, a critical review of post-war Austria). In 1953 he published a volume of poetry, *Die Glaskugel*.

Zand's *Werke* (5 vols.) appeared 1971–2 and in 1973 were supplemented by a posthumous collection of his essays, published as vol. 6 under the title *Träume im Spiegel*.

Zauberberg, Der, a novel by Th. Mann (q.v.), published in 1924. Its setting is modelled on a sanatorium at Davos, visited by Mann in 1912, when he first planned the work as a Novelle.

Hans Castorp, a young engineer from Hamburg, goes to see his friend and cousin Joachim Ziemssen, a young officer of the Hamburg garrison, who is a patient at the sanatorium on the magic mountain (Zauberberg). Castorp himself is found to have a spot on the lung (as happened to Mann, who had gone to Davos to visit his wife), prolongs his stay, and becomes engrossed by this self-contained world of the sick. Attracted to a Russian patient, Madame Clawdia Chauchat, he spends carnival night with her, but she disappears from the sanatorium the following day. Against medical advice Ziemssen also leaves to resume his commission; he returns after a short while, this time to die. Meanwhile Castorp becomes fascinated by two other patients, the 19th-c. Italian anti-clerical idealist Settembrini and the nihilistic anarchical Jew Naphta, a member of the Society of Jesus. The unequal company engages in philosophical discussions, in which Settembrini and Naphta fight for Castorp's soul. Madame Chauchat turns up again with a new companion, the wealthy and elderly Dutchman Mijnheer Peeperkorn, who dominates all around him before he commits suicide. The differences between Settembrini and Naphta culminate in a duel staged with grotesque irony; Settembrini deliberately shoots wide and Naphta turns his revolver on himself. Madame Chauchat vanishes once more, and Castorp remains in the company of Settembrini. On the outbreak of the 1914–18 War Castorp joins up as a volunteer and is glimpsed participating in an attack in Flanders.

In a sense *Der Zauberberg* is a Bildungsroman (q.v.), since it shows the maturing, through error, of Castorp's character; but it is even more a Zeitroman (q.v.), an eloquent and complex commentary on a futile and introverted civilization. By implication Mann supports a humane, enlightened democracy. Action, dialogues, and characters are set forth with unfailing irony.

Zauberei im Herbste, Die, a story (Märchen) by J. von Eichendorff (q.v.), written in 1809 and published posthumously in 1906. Set in a frame (see RAHMEN), in which two knights meet, it recounts how Raimund has spent his life in repentance of a crime (the murder of his friend and rival in love), which he has never committed. He is the victim of a hallucination.

Zauberer von Rom, Der, a nine-volume novel written by K. Gutzkow (q.v.) 1857–61 and published 1859–61. Its theme is the problem of Ultramontanism, the divisive effect on Germany of the Roman focus of the Catholic Church. The first volume has often been reprinted separately under the title *Lucindens Jugendgeschichte*.

Zauberflöte, Die, an opera in two acts with spoken dialogue (see SINGSPIEL), composed by W. A. Mozart (q.v.) in 1791 and first performed in Vienna on 30 September 1791. The libretto by E. Schikaneder (q.v.) is related to the Viennese Zauberstück (see VOLKSSTÜCK) and contains Masonic elements (librettist and composer were Freemasons) culminating in an allegory of the victory of humanitarianism, embodied in the wisdom of Sarastro, over evil, symbolized by the unbridled passion of the Queen of the Night. Schikaneder, whose line was robust pantomime, was not wholly successful in blending the ideas implicit in the libretto, but the sublimity of the music redeems all.

Among noted singers who have played Sarastro was the dramatist and actor J. N. Nestroy (q.v.). Goethe was present at a performance of *Die Zauberflöte* at Weimar in 1794; impressed by the work, he proposed to write a sequel, at which he worked in 1795–6 and again in 1799 without completing it. The fragment, *Der Zauberflöte zweiter Teil. Entwurf zu einem dramatischen Märchen* was published in *Taschenbuch auf das Jahr 1802*.

Zauberlehrling, Der, a ballad written by Goethe in July 1797 and published in the *Musenalmanach für 1798*. It tells, in monologue form, of the sorcerer's apprentice who attempts, in the absence of his master, to use the spells to do his work. He is unable to control what he has set in motion, but the situation is saved by the return of the sorcerer. The poem is the basis of the orchestral piece *L'Apprenti-sorcier* by P. Dukas (1865–1935).

Zauberring, Der, a three-volume novel by F. de la Motte FOUQUÉ (q.v.), published in 1813. It has a make-believe medieval setting, ranging from the Mediterranean to the Finnish forests. The end reveals unexpected family relationships between all the principal characters, valiant knights and fair ladies, who are repeatedly in danger of death. The peculiar mixture of the chivalric and the magic was popular with the general public of its day.

Zaubersprüche, magic spells, a number of which survive in MSS. of the 9th c. to the 12th c. They cover a number of important contingencies—rescue of a prisoner of war, stanching the flow of blood, cure of disease or of lameness in horses, preservation of sheep-dogs or bees, safe return from the perils of a journey. They go back into prehistoric times and persist into the Christian era. Only the figures invoked change with the change of religion.

The most primitive form of spell is the single imperative sentence (see WURMSEGEN). More common is the spell constructed in two stages, a narrative section describing a past instance of the emergency with a successful outcome, followed by an imperative or adhortative section, which constitutes the operative magic (see MERSEBURGER ZAUBERSPRÜCHE). In later Christian examples a prayer tends to replace the imperative stage (see WIENER HUNDESEGEN). Such spells must have been preserved mainly by oral tradition, going back far beyond the period of literary record, and the extant examples must be only a small proportion of those in use. In addition to those in German, a number of Latin spells are recorded in MSS. emanating from German monasteries. (See also LORSCHER BIENENSEGEN, STRASSBURGER BLUTSEGEN, WEINGARTNER REISESEGEN.)

Zauberstück, see VOLKSSTÜCK.

ZECH, PAUL (Briesen, West Prussia, now Poland, 1881–1946, Buenos Aires), a schoolmaster's son of Westphalian stock, studied at Bonn, Heidelberg, and Zurich universities, worked for two years in industry (as a stoker and miner) and, after a stay in Paris, was from 1910 a local official, librarian, and dramatic adviser in Berlin, and as co-editor of *Das neue Pathos* (1913–20), promoted Expressionist poetry. Arrested in 1933, he emigrated on his release in the following year, arriving in 1937 in South America. A worker poet without party political commitment but with strong religious convictions, he was concerned with brotherhood and the rebirth of man through contact with nature. This ideal was strengthened by personal experience of proletarian life in the cities, which marks his first collection of poetry, *Das schwarze Revier* (1909 and 1913, revised 1922), the title of which is contained in *Vom schwarzen Revier zur Neuen Welt* (ed. H. A. Smith, 1983), an edition of his poetry (notably sonnets) published over some thirty years; *Neue Welt* (1939) marks his emigration. *Das Grab der Welt. Eine Passion wider den Krieg auf Erden* (1919) is an example of his slow but strong reaction against the 1914–18 War. He is the

author of numerous plays and of fiction, including *Der schwarze Baal* (1917, revised 1919), *Die Reise um den Kummerberg* (1924), the novel *Die Geschichte einer armen Johanna* (1925), and posthumous publications: *Die Kinder von Paraná* (1952), *Die grüne Flöte vom Rio Beni* (1955), *Deutschland, dein Tänzer ist der Tod* (1981), *Menschen der Calle Tuyati. Erzählungen aus dem Exil* (ed. W. Kießling, 1982), *Das rote Messer. Begegnungen mit Tieren und seltsamen Menschen* (1983), and *Michael M. irrt durch Buenos Aires. Aufzeichnungen eines Emigranten* (ed. H. Nitzschke, 1985). He translated and adapted French poetry and prose. He acknowledged the influence of Rilke, George, and E. Lasker-Schüler (qq.v.) and published essays, e.g. on Rilke (*Ein Requiem*, 1927, also a biography, 1930), S. Zweig (q.v.), Rimbaud, and Verlaine. His correspondence with Zweig, *Briefe 1910–1942*, ed. D. G. Daviau, appeared in 1984.

ZEDLITZ, JOSEPH CHRISTIAN, FREIHERR VON (Johannesberg, Silesia, 1790–1862, Vienna), entered the Austrian army at 16 as an officer of hussars and from 1810 devoted himself to the management of his estates. In 1836 he entered the Austrian diplomatic service.

As a writer, Zedlitz began with the extravagant Romantic tragedy *Turturell* (1821), followed by *Zwei Nächte in Valladolid* (1825) and by poetry of Romantic character. He is mainly known for the poem *Die nächtliche Heerschau* (q.v., 1829). Its theme is the Napoleonic legend, comprehended in a ghostly nocturnal review of the fallen. *Totenkränze* (1828), a cycle of 134 poems in *canzone* form (13 lines with 6 rhymes in a complex pattern), is a review of some of the famous dead of history, including poets and philanthropists, as well as soldiers. His *Gedichte* appeared in 1832, the verse romance *Waldfräulein* in 1843, and the patriotic poems of his *Soldaten-Büchlein* (1849–50), praising Radetzky (q.v.) and his army, concluded his poetic output. Zedlitz also translated Byron's *Childe Harold* (*Ritter Harolds Pilgerfahrt*, 1836).

Zeichen des Jüngsten Gerichts, Die, an anonymous Middle High German poem, dealing with the disasters which, according to traditional belief, are to overtake the world in the fifteen days preceding the Day of Judgement. The MS. is corrupt and estimates of the date of the poem vary from 1180 to the middle of the 13th c.

ZEIDLER, JOHANN GOTTFRIED (1655–1711, Halle/Saale), a Protestant pastor, gave up his cure in 1700 and moved to Halle. He is the author of satires (*Neun Priesterteufel*, 1701; *Das verdeckte und entdeckte Carneval*, undated).

Zeit, Die, title of two periodicals concerned with political, economic, scientific, and cultural affairs. (1) A Viennese weekly founded in 1894 and continued from 1904 as *Österreichische Rundschau*. Publication ceased in 1924. (2) A weekly of international repute, published in Hamburg since 1946.

Zeit der Schuldlosen, a play in two parts (In zwei Teilen) by S. Lenz (q.v.), published in 1961. All the characters except one, Sason, are given general designations (Hotelier, Konsul, Student, Bauer, etc.).

In the first part, Sason, one of the instigators of an unsuccessful attempt to assassinate a dictator, is arrested and locked up with nine ordinary men chosen at random. They will be released as soon as Sason confesses the names of his accomplices. This he refuses to do. In the morning he is found strangled. By whom?

The second part takes place after the fall of the dictator. The nine former hostages are again confined together. This time the murderer of Sason must be found. The student of Pt. I takes over the role of judge, having been a secret member of the Resistance. In the end the consul shoots himself, and the hostages are freed. But it by no means follows that he committed the murder. The student, in his closing speech, uses the words 'die Tat ist gebüßt, aber die Schuld wird unter uns bleiben'. The message of complicity by silence, or by abstention, or by conformity, is one which Lenz expresses in a number of his works.

Zeitgeist und Berner Geist, a two-volume novel by J. Gotthelf (q.v.), written in 1850–1 and published in 1852. The title indicates the polemical nature of the book, which arose out of Gotthelf's concern for the old rural Switzerland and his hostility to the new Radical forces at work since 1848.

He takes two farmer families, united in friendship. Ankenbenz, solid, trustworthy, and cautious, declines to go with the times. He remains a God-fearing man, and his wife Lisi stands firm with him against the ridicule of the villagers. Hunghans, on the other hand, listens to the new political message, accepts office (as Amtsrichter), despises religion, and looks down on Benz. He neglects his farm, his wife grieves, declines, and dies; his elder son, a militia officer, cares nothing for his parents, lives high, contracts debts, and misappropriates public funds. This son meets a terrible end from apoplexy, and Hans turns from modern ways to the old friendship and support of Ankenbenz. Benz's elder daughter and Hans's younger son, it is implied, will marry and cement the alliance. The book finishes with a fully quoted sermon preached by the clergyman at the funeral of Hans's son, on

the text 'Als nun Maria kam an den Ort, da Jesus war, und sahe ihn, fiel sie zu seinen Füßen und sprach zu ihm: Herr wärest du hier gewesen, so wäre mein Bruder nicht gestorben!' (John 11: 32).

Zeitroman, term applied in German to novels which are primarily concerned with an author's critical analysis of the age in which he lives. Some Bildungsromane (q.v.) may be regarded as Zeitromane. Outstanding early examples of the genre are by Gutzkow (*Die Ritter vom Geiste*, q.v., 1850–1), Freytag (*Soll und Haben*, q.v., 1855), and Gotthelf (q.v.).

Zeitung für die elegante Welt, (1801–59), originally a Romantic journal, was edited in Leipzig by S. A. Mahlmann (q.v.) 1806–16, by H. Laube (q.v.) 1833–4 (reprint, ed. A. Estermann, 3 vols., 1971), by G. Kühne (q.v.), 1835–42, and again by Laube, 1843–4.

Zeitung für Einsiedler, a periodical founded jointly by L. J. von Arnim, C. Brentano, and J. Görres (qq.v.) in April 1808 and appearing, irregularly, until August 1808. Among its contributors were L. Uhland, J. Kerner, the brothers J. and W. Grimm, Z. Werner (qq.v.), and Bettina Brentano (see ARNIM, BETTINA VON). The editors aimed at publishing Volkspoesie, the literary heritage of past ages, fairy-tales, and legends, in order to promote a consciousness of German national culture. The organ of the Heidelberg Romantics (see ROMANTIK), it also appeared in book form in 1808, entitled *Tröst-Einsamkeit*.

Zeitungssänger, see BÄNKELSÄNGER.

ZELTER, KARL FRIEDRICH (Petzow-Werder/ Havel, 1758–1832, Berlin), German composer of vocal music. He was highly regarded by Goethe, who preferred Zelter's settings of his poems to those of any other composer. Zelter corresponded with Goethe from 1795 on, and the two men became lifelong friends. Both died in May 1832. The letters between them were published as *Briefwechsel zwischen Goethe und Zelter* (3 vols.), ed. W. Rintel, 1833–4, and in a new edition by L. Geiger (8 vols.) in 1904.

Zeno, a Middle Low German epic poem, relating how the bodies of the Three Kings are brought to Cologne. It was written c. 1300.

Zentrum, in full Zentrumspartei, the Roman Catholic party in the Prussian Diet and in the Reichstag after 1871. It was founded in Prussia shortly before the outbreak of war in 1870. It became conspicuous as the opposition party during the Kulturkampf (q.v.). It continued as a significant minority party, drawing support from all parts of Germany but especially the south, until the dissolution of all political parties except the NSDAP (q.v.) in 1933.

The first leader of the party was Ludwig Windthorst (1812–91). who in the 1860s had been minister of justice under George V, then king of Hanover. Windthorst became a formidable leader of the opposition against Bismarck (q.v.) in the Reichstag. The party's last leader was H. Brüning (q.v.).

Zerbrochene Ringlein, Das, one of J. von Eichendorff's best-known poems, first published in J. Kerner's *Deutscher Dichterwald* in 1813 and included (untitled) in *Ahnung und Gegenwart* (q.v.). Its first line runs 'In einem kühlen Grunde'. The broken ring of the title is the symbol of an untrue love; and the poem, which has virtually become a folk-song, ends with a death wish. The stimulus for the poem came from the folk-song 'Da droben auf jenem Berge', which Eichendorff read in *Des Knaben Wunderhorn* (q.v.).

Zerbrochne Krug, Der (Ein Lustpiel), a full-length one-act comedy in blank verse by H. von Kleist (q.v.), written between 1802/03 and 1807. Goethe produced his own stage version of Kleist's MS. (Lustspiel in drei Aufzügen) in March 1808 in Weimar with disastrous results. Kleist countered the misfortune of his rejected (and distorted) comedy by publishing parts of it in *Phöbus* (q.v.) in 1808; he published the whole work in 1811 after it had been rejected by Iffland (q.v.) for production in Berlin.

The situation on which the comedy is based was suggested to Kleist by a French engraving (by J. J. Le Veau) with the caption *La Cruche cassée*, which he saw in 1802 with his friends (among them H. D. Zschokke, q.v., who wrote a story with the same title) at Berne. The plot, set in the law-court of a Dutch village (a deliberate shift from the French environment), pursues the court proceedings dealing with the request for damages by Frau Marthe Rull against Ruprecht, her daughter Eve's lover, whom she accuses of having broken 'the most beautiful of all jugs'. The village judge, Adam, would gladly convict Ruprecht of the crime, were it not for the interference of Gerichtsrat Walter, who, attending the proceedings as a visiting inspector, exposes Judge Adam himself as the culprit. The jug was in fact broken as it was accidentally knocked over by Adam on his hasty retreat from a secret, but, in its amorous intentions, unsuccessful, midnight venture into Eve's room.

Kleist combines comedy of situation with that of character (concentrating overwhelmingly on Adam), and yet hints at the serious aspect of the fallibility of human feeling and the dilemma of human justice characterizing his work in general. The play stands out among German comedies by the deft realism of the simple peasant characters and their environment, and the ingenious play of words pervading the dialogue. The satire hits out above all against man in general, as the descendant of Adam, as is indicated through the name of the village judge.

An operatic version by an East German composer, F. Geißler, was first performed in Leipzig in 1971.

ZERNITZ, CHRISTIAN FRIEDRICH (Tangermünde, 1717–45, Klosterneuendorf nr. Gardelegen, Prussia), who became stipendiary magistrate in Klosterneuendorf in 1738, collaborated in *Belustigungen des Verstandes und Witzes* (q.v.) and wrote poetry, which was collected after his early death (*Versuch in moralischen und Schäfergedichten*, 1748).

Zerrissene, Der, a three-act farce (Posse mit Gesang in drei Aufzügen) by J. N. Nestroy (q.v.) written in 1844, first performed at the Theater an der Wien, Vienna, in April 1844, and published in 1845.

Herr Lips, 'der Zerrissene', is a man disillusioned and at odds with himself, as the refrain of his song indicates: 'Meiner Seel, 's is a fürchterlichs G'fühl,/Wenn man selber nicht weiß, was man will.' He makes a sudden decision to marry a worthless woman, but is involved in a struggle with her former suitor, the locksmith Gluthammer. They fall from the balcony into the river and are lamented as dead. In Act II both Lips and Gluthammer appear at a farm belonging to Lips. Each believes the other to be dead, and himself to be a fugitive from justice. For the time they do not meet. Lips recognizes the sterling worth and good nature of Kathi, niece of the farmer-tenant. The former friends of the supposedly deceased Lips arrive, believing themselves to be the heirs. But Kathi turns out to be the inheritor. In the final act Lips and Gluthammer discover each other to be alive and are reconciled. Lips dismisses his false friends, is cured of his neurosis, and declares his intention of marrying Kathi. The play is an adaptation of a French 'comédie-vaudeville', *L'homme blasé* by Duvert and Lauzanne.

ZESEN, PHILIPP VON (Priorau nr. Dessau, 1619–89, Hamburg), poet and novelist, studied in Wittenberg and Leyden, and visited Hamburg, where in 1643 he founded a society for the cultivation of the German language (Teutschgesinnte Genossenschaft, or Rosenzunft, see SPRACHGESELLSCHAFTEN) and encountered the enmity of J.

Rist (q.v.). In the same year he visited England and then settled in Holland, where he lived the greater part of his life, though he made many visits to towns in Germany and spent his last six years in Hamburg. He became a member of the Fruchtbringende Gesellschaft (q.v.) as 'Der Wohlsetzende' in 1648 and in 1653 was granted a patent of nobility.

Zesen held no office and, except for a short period in later life, engaged in no trade; he lived from the earnings of his pen and the patronage of great men. He undertook on his own account a campaign for the purification of the German language and the phonetic reform of its spelling; his proposals and practice were not free from pedantry. He was twice ridiculed in caricature by Johann Rist (as 'Sausewind' in *Das friedewünschende Deutschland*, 1647, and *Das friedejauchzende Deutschland*, 1649), and he seems to have been a humourless, determined, egoistic person. Yet his lyric poetry (in *Melpomene*, 1638, *Poetischer Rosen-Wälder Vorschmack*, 1642, and *Frühlingslust*, 1642) is technically inventive with a highly developed sense for words as poetic material. He translated three French novels, *Les Amours de Lysandre et de Caliste* by d'Andiguier, *Ibrahim* by Mme de Scudéry, and *Sophonisbe* by Gerzan, and wrote three original novels, *Die adriatische Rosemund* (1645), *Assenat* (1670), and *Simson* (1679). *Rosemund* is a tale of lovers separated by religion; Markhold is a Protestant, Rosemund a Venetian Catholic. *Assenat* treats the story of Joseph in Egypt and especially his temptation by Potiphar's wife.

Zesen's novels show originality of conception, which is not matched by imaginative power. He used the pseudonym Ritterhold der Blaue, and altered the spelling of his name from Caesius, his father's form, to Caesien and then to Zesen, at the same time phoneticizing his Christian name to Filip.

Sämtliche Werke (18 vols.), ed. F. van Ingen, appeared 1970 ff.

ZIEGLER, Hieronymus (Rothenburg ob der Tauber, *c.* 1514–62, Ingolstadt), was a Roman Catholic schoolmaster in Augsburg and later a professor at Ingolstadt. His Latin plays (including *Immolatio Isaac*, 1543, *Paedonothia*, 1543, and *Samson*, 1547) are a first step towards the Jesuit drama (see Jesuitendrama) of the 17th c.

ZIEGLER UND KLIPHAUSEN, Heinrich Anselm von (Radmeritz, Upper Luseatia, 1663–96, Liebertwolkwitz nr. Leipzig), was a man of means and property, to the management of which he devoted himself after completing his studies in 1684. A poetic work entitled *Helden-Liebe der Schrift des Alten Testaments* was published in 1691. It consists of imaginary love-letters in Alexandrine verse, purporting to be written by biblical personages. His best-known work is an oriental novel, *Die Asiatische Banise oder Das blutig, doch muthige Pegu* (1689, ed. W. Pfeiffer-Belli, 1965), which enjoyed an immense vogue well into the 18th c. Its principal figures are a pair of princely lovers, Balancin and Banise, whose happiness is threatened by the bloodthirsty tyrant Chaumigrem, who for a time holds Banise captive. Balancin rescues his princess and kills the tyrant. Contemporaries appreciated the information imparted about unknown regions, and the plentiful disguises and horrors were much to the taste of the age. *Die Asiatische Banise* has attracted censure for the quantity and quality of its atrocities, but many of these are present in the sources. A sequel by J. G. Hamann (q.v.) appeared in 1724.

Ziegler's surnames are also spelt Zigler and Klipphausen, and his second baptismal name is sometimes given as Anshelm.

ZIETEN, Hans Joachim von (Wiestrau nr. Ruppin, 1699–1786, Berlin), who became an almost legendary figure in Prussian military history as 'der alte Zieten', entered the Prussian army as an ensign of infantry in 1714. In 1724 he was transferred to the cavalry.

Promoted colonel in 1741, Zieten served with great distinction as a cavalry commander in the three principal wars of Friedrich II (q.v.) of Prussia, whose confidence he won and retained. Among his principal feats were a patrol in strength behind the Austrian lines known as the 'Zietenritt' (1745) and charges which decided the battles of Leuthen (1757) and Torgau (1760) He retired in the rank of general in 1763. He is the subject of a story by G. Hesekiel (q.v., *Hans Joachim*, 1871) and of a well-known ballad by Th. Fontane (q.v., *Der alte Zieten*, 1847).

Zimmerische Chronik, title given to a chronicle of the noble family of Zimmern of Rottweil (Neckar), written by Count Froben Christoph von Zimmern with the help of his clerk and including contributions by his uncle Baron (Freiherr) Wilhelm Werner von Zimmern. Completed in 1567, it is a valuable document of 16th-c. life. It was published in 1869. The family became extinct in 1594.

ZIMMERMANN, Bernd-Alois (Eifel, 1918–70, Groß-Königsdorf nr. Cologne), musicologist and composer, taught from 1953 at the Cologne Hochschule für Musik, at which he was appointed professor of composition in 1957. He was influenced in turn by Stravinsky and by Schönberg (q.v.). His orchestral works include a violin concerto (1952), a symphony (1952), and a cello concerto (1966). He also wrote ballet music. He is best known for the opera *Die Soldaten* (1960), which is based on the play of this title by J. M. R. Lenz (q.v.). It was performed with considerable success at Cologne in 1965. At the time of his death he was at work on an opera with a libretto taken from the *Medea* of H. H. Jahnn (q.v.). A Requiem, performed in 1969, employs experimental techniques, including electronic devices. Zimmermann devised his own epitaph: 'eine sehr rheinische Mischung von Mönch und Dionysus'.

ZIMMERMANN, JOHANN GEORG (Brugg, Switzerland, 1728–95, Hanover), studied medicine and was a favourite pupil of A. von Haller (q.v.) at Göttingen. He set up in practice at Berne in 1752, moving to Brugg in 1754. In 1768 he was appointed 'His Britannic Majesty's Physician' at Göttingen. He was called in to the last illness of Friedrich II (q.v.) of Prussia, but his subsequent account, *Über Friedrich den Großen und meine Unterredungen mit ihm kurz vor seinem Tode* (1788, reprinted 1920), was ill received. Insanity crept upon him in his later years.

Zimmermann was his master's biographer (*Leben des Herrn von Haller*, 1757), but he was best known as a popularizer of current philosophical and ethical ideas. *Betrachtungen über die Einsamkeit* appeared in 1756 and *Über die Einsamkeit* in 1784. His essay on patriotism, *Von dem Nationalstolze* (1758, reprinted 1937), written during the Seven Years War (see SIEBENJÄHRIGER KRIEG), distinguishes between true and false national pride. His *Von der Erfahrung in der Arzneikunst* (1763–4) departs from the established professional practice of the day in treating medical questions in the vernacular instead of in Latin.

ZINCGREF, JULIUS WILHELM (Heidelberg, 1591–1635, St. Goar), son of a counsellor in the service of the Elector Palatine, studied at Heidelberg and travelled in England and the Netherlands. He was caught up in the turmoil of the Thirty Years War (see DREISSIGJÄHRIGER KRIEG) and died of the plague while still a relatively young man. His satirical and lyrical poems were well received until the new mode imposed by M. Opitz (q.v.) made them unfashionable. His principal collections of verse are *Emblemata* (1619); *Der Teutschen scharpfsinnige kluge Sprüch* (1626), which are more often known by the title of the second edition *Apophthegmata*; and *Auserlesene Gedichte deutscher Poeten*, (1624), or *Teutsche Poemata*.

ZINK, BURCHARD (Memmingen, 1396–c. 1474, Augsburg), who travelled extensively in Europe in the service of a merchant of Augsburg, compiled a prose chronicle of his adopted city. The third book of this *Augsburger Chronik*, which was not printed until 1866, is autobiographical.

ZINZENDORF UND POTTENDORF, NIKOLAUS LUDWIG, GRAF VON (Dresden, 1700–60, Herrnhut), grew up in a pietistic religious atmosphere, becoming a man of great sanctity who devoted his life to missionary endeavours at home and abroad. He opened his estate at Herrnhut to the persecuted (see HERRNHUTER) and took part in a missionary expedition to America.

A prolific religious writer, Zinzendorf, in addition to works of devotion and edification, wrote over 2,000 hymns which were published in various collections beginning with *Sammlung geistlicher und lieblicher Lieder* (1725) and ending with *Sammlung von Liedern* (1755). Zinzendorf's *Hauptschriften* (6 vols.), ed. E. Beyreuther and G. Meyer, appeared 1962–4, to which were added *Ergänzungsbände* (12 vols.), 1964–72.

Zollverein, the Customs Union which promoted German national unity under Prussian hegemony. The idea that Prussia rather than the German Confederation (see DEUTSCHER BUND) should introduce economic reforms, such as the reorganization of fiscal barriers and other sources of revenue, was promoted by Karl Georg Maaßen (1769–1834) and Friedrich von Motz (1775–1830, from 1825 Prussian minister of finance). Maaßen was a follower of Adam Smith (1723–90), whose anti-mercantile views combining social welfare with liberal principles of productivity in the new industrial age coincided with those of F. List (q.v.).

The Tariff Reform Act of 28 May 1818 established safeguards for the commercial and economic unity of Prussia, among them above all the abolition of internal custom duties. Between 1819 and 1822 Prussia persuaded seven small states to join in a customs union by tariff treaties. For a time Bavaria and Württemberg maintained a separate union in which each member (including also smaller states) had equal fiscal rights. A third union between Saxony, Hesse-Cassel, Hanover, Brunswick, and the Free Cities broke up for lack of funds. But Prussia succeeded in 1828 in coming to an arrangement with Hesse-Darmstadt, modelled on the southern pattern, with the result that the southern union joined the Prussian union in the following year. Thus by 1829 there existed a formidable Customs Union from which Austria, miscalculating its political import, remained excluded. Baden joined in 1835 and Frankfurt in 1836. During the 1840s the teachings of List inspired the Union with a new national fervour which boosted its protectionist policy. To fortify itself against Austria, Prussia admitted Hanover and

Oldenburg on preferential terms (1852) and came to an arrangement with Austria by which it offered Austria some tariff concessions and the promise to review its admission into the Union in 1860.

The Zollverein was thus instrumental in preparing the political unification of Germany which, after the Austro-Prussian War (1866, see DEUTSCHER KRIEG), excluded Austria and commenced under Bismarck (q.v.) the formation of the North German Confederation (see NORDDEUTSCHER BUND). Its terms were revised for its members, while the Confederation established with the South German states a Customs Parliament (Zollparlament) which met in Berlin (1868–70). After the foundation of the Empire in 1871 few remained outside (in 1888 Bismarck succeeded in coercing Hamburg and Bremen into joining; Luxemburg, the only non-German member, on the other hand, stayed in the Union until 1919).

Zopf und Schwert, a five-act comedy (Lustspiel) by K. Gutzkow (q.v.), first performed in January 1844 at Dresden. Gutzkow's most popular play, it is set at the court of King Friedrich Wilhelm I (q.v.) of Prussia.

The dictatorial king proposes to marry his daughter Wilhelmine, first to the Prince of Wales, and, when that falls through, to the heir to the Habsburg Emperor. In the end he is prevailed upon to give Princess Wilhelmine's hand to the man she loves, the heir apparent of Baireuth. The King's frugality and his passion for running his palace and household like a barrack-room provide comic effect.

Zorndorf, battle of (25 August 1758). It was a drawn battle, although the Russian losses were heavier than those of the Prussian army (see SIEBENJÄHRIGER KRIEG). Zorndorf lies just to the north of Küstrin.

ZSCHOKKE, HEINRICH DANIEL (Magdeburg, 1771–1848, Aarau), was orphaned early, received a humanistic education at the Magdeburg Gymnasium, and for a short time joined an itinerant troupe of actors. From 1789 to 1792 he was at the University of Frankfurt/Oder and hoped to obtain a lecturing appointment. Being unsuccessful in this, he travelled through Germany and in 1796 settled in Switzerland, where he spent the rest of his life, occupying a variety of administrative posts in his adopted country and endeavouring in his official policy and in didactic writings to propagate enlightened political, economic, and social opinions. He retired from his offices in 1841.

A quick and facile writer, Zschokke began with plays (*Graf Monaldeschi*, 1790) and with novels, of which *Aballino, der große Bandit* (1794)

is the outstanding example. Zschokke adapted this tale of brigandage in the following year as a tragedy. It was his most successful work, but the novels *Kuno von Kyburg* (1795–9) and *Alamontade der Galeerensklave* (1802) were also widely read. He wrote many shorter narrative works which were published in various journals. The most famous of these are the stories *Der zerbrochene Krug*, written in friendly competition with Heinrich von Kleist (q.v.), and *Das Goldmacherdorf* (1817). Zschokke's first periodical was *Der Schweizer Bote* (1798–1800 and 1804–37), a work of popular edification. A similar aim was pursued by him in *Miszellen der neuesten Weltkunde* (1807–13) and in the *Erheiterungen* (1811–27). Among other didactic and educational works were *Anweisung für Schullehrer auf dem Lande* (1799), *Stunden der Andacht* (8 vols., 1809–16), *Des Schweizerlandes Geschichte für das Schweizervolk* (1822), and *Bilder aus der Schweiz* (5 vols., 1825–6, historical stories for the enlightenment of Swiss readers). He entitled his autobiography *Selbstschau* (1842).

Gesammelte Schriften (36 vols.), ed. E. Zschokke, appeared 1856–9 and *Sämtliche Novellen* (12 pts. in 4 vols.), ed. A. Vögtlin, in 1904.

'Zu Bacharach am Rheine', opening line of a ballad by C. Brentano (q.v.) first printed without title in vol. 2 of Brentano's novel *Godwi* in 1801. It usually bears the title *Die Lore Lay*.

ZUCCALMAGLIO, FLORENTIN VON (Waldbroel, 1803–69, Nachrodt), son of a lawyer who emigrated from Italy, spent part of his life in Russia, first as a private tutor and then as a professor, and was later a tutor in Germany. A poet and a collector of folk-songs, he is the author of the well-known poem 'Es fiel ein Reif in der Frühlingsnacht' (q.v.).

ZUCKMAYER, CARL (Nackenheim, 1896–1977, Visp), the son of a prosperous manufacturer, volunteered in 1914 and served throughout the war. After studying briefly at Heidelberg and Frankfurt universities (1918–19), he worked as an assistant producer at theatres in Berlin, Kiel, and Munich. By the late 1920s he had established his reputation as a dramatist. In 1926 he settled near Salzburg. In consequence of his plain-spoken opposition to National Socialism his plays were banned in Germany after 1933. In 1938 he escaped from Austria to Switzerland, and in 1939 settled in the U.S.A. He was in Germany in 1946–7 as an official of the American Military Government, and afterwards alternated for some years between Europe and America, settling in the end at Saas-Fee, Switzerland.

After the early Expressionist play *Kreuzweg* (1921) and the drama *Pankraz erwacht oder Die*

Hinterwäldler (1925, alternative title *Kiktahan*), Zuckmayer achieved a great stage success with the Volksstück (q.v.) *Der fröhliche Weinberg* (1925). This was followed by the historical play *Schinderhannes* (1927, dealing with the rebel Schinderhannes, q.v.), the drama of a tightrope dancer *Katharina Knie* (1929), and the first of his two international successes, the comedy *Der Hauptmann von Köpenick* (q.v., 1931, well known also in two film versions). *Der Schelm von Bergen* (1934) dramatizes a popular legend already treated by H. Heine (q.v., see SCHELM VON BERGEN); *Carl Michael Bellman* (1938), concerned with the Swedish poet Bellman (1740–95), was revised as *Ulla Winblad* (1953). *Des Teufels General* (q.v., 1946), a brilliant presentation of the higher levels of political and military life in Berlin during the 1939–45 War, marks the second peak of Zuckmayer's career as a dramatist. Later works are the historical play *Barbara Blomberg* (1949, see BLOMBERG, BARBARA), *Der Gesang im Feuerofen* (q.v., 1950), *Das kalte Licht* (q.v., 1955), the comedy *Der trunkene Herkules* (1958), *Die Uhr schlägt eins* (1961), *Das Leben des Horace A. W. Tabor* (1964), *Kranichtanz* (1961, first perf. 1967), and *Der Rattenfänger* (1975).

Although Zuckmayer is best known as a dramatist, he is the author of two novels, *Salwàre oder Die Magdalena von Bozen* (1936) and *Herr über Leben und Tod* (1938), and of a number of Novellen and Erzählungen. These include *Der Bauer aus dem Taunus* (1927, title of the volume and of one of the stories in it), *Die Affenhochzeit* (1932), *Eine Liebesgeschichte* (1934), *Ein Sommer in Österreich* (1937, filmed as *Frauensee*), *Der Seelenbräu* (1945), *Die Erzählungen* (1952), *Engele von Löwen* (1955, previously included in *Die Erzählungen*), and *Die Fastnachtsbeichte* (1959).

Zuckmayer's writings include essays on J. and W. Grimm (qq.v., *Die Brüder Grimm*, 1948), on Schiller (*Ein Weg zu Schiller*, 1959), and on G. Hauptmann (q.v., *Ein voller Erdentag*, 1962). His poetry appeared as *Der Baum* (1926), *Gedichte 1916–48*, (1948), and *Gedichte* (1960). Autobiographical works are *Second Wind* (1940, published in English only), *Die langen Wege. Ein Stück Rechenschaft* (1952), and the very successful *Als wär's ein Stück von mir* (1966).

Gesammelte Werke (4 vols.) appeared 1947–60 and *Werkausgabe in zehn Bänden* in 1976.

Zu ebener Erde und erster Stock oder die Launen des Glücks, a play (Lokalposse mit Gesang in drei Aufzügen) by J. N. Nestroy (q.v.), written in 1835, first performed at the Theater an der Wien, Vienna, in September 1835, and published in 1838. A strikingly original feature is the subdivision of the stage into floors, the wealthy family von Goldfuchs living upstairs and the miserably poor Schluckers down below, a juxtaposition which was a recognizable feature of 19th-c. Viennese housing conditions. A kind of counterpoint of rich and poor ensues.

The slender action of the play reverses the position of the two families, whose only contact is a clandestine love-affair between Schlucker's stepson Adolf and Goldfuchs's daughter Emilie. Financial disaster sends Goldfuchs down to the ground floor squalor, and the discovery that Adolf's real father has made a fortune in India and has sent him a large sum enables the Schluckers to move into the vacant first-floor apartment. Adolf remains faithful to the now impoverished Emilie, and the play thus reaches a conciliatory ending. In spite of the originality of its stagecraft, *Zu ebner Erde und erster Stock* is not an anticipation of modern social drama. The sharpest criticism is directed at Goldfuchs's dishonest servant Johann. Wealth and poverty have no consciousness of class, and both families realize that 'Fortune's wheel' can turn unexpectedly.

Zueignung, a poem written by Goethe in August 1784 as prelude to his fragmentary epic *Die Geheimnisse* (q.v.). It was published in 1787 as the introductory poem to vol. 1 of *Goethes Schriften*, and continued to occupy this place in later editions published under the poet's supervision. The poem mixes landscape portrayal with allegory, describing the poet's encounter with Truth and the gift of the veil of poetry: 'Aus Morgenduft gewebt und Sonnenklarheit,/Der Dichtung Schleier aus der Hand der Wahrheit.'

Zueignung is in 14 stanzas of *ottava rima*, which was also to be the metre of *Die Geheimnisse*.

Zug war pünktlich, Der, a story (Erzählung) by H. Böll (q.v.) which appeared in 1949 as his first published work. Partly by narration, partly by inner monologue, it relates the last days of Andreas, a soldier returning from leave in western Germany to the Galician front late in 1943. The conviction that he will die at the end of the journey grows on him as the train punctually travels across the breadth of Germany to Poland. He spends a *nuit blanche* in a brothel in Lemberg with a girl named Olina, and each discovers love for the other. Olina seeks to rescue him with two comrades making for Stryi. Their car is destroyed by an explosion, and all are killed.

Zum Schäkespears Tag, often referred to as *Rede zum Shakespeare-Tag*, title of a panegyric of Shakespeare written by Goethe in 1771 and sent to Strasburg for the celebration of Shakespeare's 'name day' on 14 October. Goethe apostrophizes Shakespeare as his friend, and praises the colossal grandeur of his figures. The enthusiastic speech contains the well-known exclamation, 'Natur, Natur! nichts so Natur als Schäkespears Menschen!'

ZUMSTEEG, JOHANN RUDOLPH (1760–1802), composer and boyhood friend of Schiller. He composed the original musical versions of the songs in *Die Räuber* (q.v.) and later set the *Reiterlied* in *Wallensteins Lager* (see WALLENSTEIN).

Zur Chronik von Grieshuus, a Novelle by Th. Storm (q.v.), published in *Westermanns Monatshefte* in 1884. The narrator purports to record the extinction of an ancient noble family in the 17th c., whose seat was Grieshuus in the North German heath. The older of twin brothers, Junker Heinrich, marries Bärbe, a girl of low station, and thus incurs the enmity of his brother Detlev. After losing his wife in childbirth, Heinrich kills his brother. Leaving his home and newly born daughter Henriette, he remains a fugitive until he returns in old age to serve his grandson Rolf at Grieshuus. He dies on the anniversary and at the scene of the fratricide while on a mission to warn Rolf, an officer in the Swedish army, of the approach of the Russians. Rolf is killed that night by the enemy. The story is divided into two books, the second book introducing a contemporary narrator, Junker Rolf's tutor, the Magister Caspar Bokenfeld. Storm treats the theme of fratricide as a variation of the Cain and Abel motif and in the manner of fate tragedy (see SCHICKSALSTRAGÖDIE).

Zur Diätetik der Seele, a popular work treating the border territory between philosophy and medicine, published by E. von Feuchtersleben (q.v.) in 1838. It asserts the power of the mind to dominate the body ('die Seelendiätetik, was ist sie sonst als eine Erziehung des Leibes durch die Seele?') and recommends an outlook in which the individual sees himself in proper proportion to his surroundings. The book was intended to be of therapeutic value against the hectic tempo of modern life, as Feuchtersleben saw it in the 1830s.

Zur Geschichte und Literatur. Aus den Schätzen der Herzoglichen Bibliothek zu Wolfenbüttel, title of a series of learned essays published by Lessing on the basis of hitherto unpublished papers in the Ducal Library of Brunswick. Six sets of contributions appeared, Nos. 1 and 2 in 1773, No. 3 in 1774, No. 4 in 1777, and Nos. 5 and 6 in 1781. They include such diverse matters as the identification of a

medieval fabulist as Ulrich Boner (q.v.), Marco Polo, the buildings, windows, and library of Hirschau Abbey (Hirsau in Württemberg), Andreas Wissowatius's (q.v.) views on the Trinity, the antiquity of oil-painting, and Adam Neuser (q.v.).

Züricher Novellen, a collection of five Novellen published in 1878–9 by G. Keller (q.v.). The first three stories are provided with a framing narrative (see RAHMEN), which provides an introduction and is reintroduced at the end of each of the three narratives. This frame is itself a story, the conversion of Jacques, an artistically minded young man, to a more sensible and prudent outlook by means of the three moral stories which his godfather provides, to which change of mind a personal disillusionment concerning a young artist also contributes. These first three Novellen are *Hadlaub, Der Narr auf Manegg,* and *Der Landvogt von Greifensee* (qq.v.), which, together with their frame, had previously been serialized in the *Deutsche Rundschau,* Nov. 1876–April 1877. The fourth Novelle, *Das Fähnlein der sieben Aufrechten* (q.v.), was first published in *Auerbachs Volkskalender* in 1860. The fifth, *Ursula* (q.v.), was entirely new.

ZUR LINDE, OTTO, see LINDE, OTTO ZUR.

Zwei Ansichten, a novel by U. Johnson (q.v.), published in 1965, four years after the erection of the Berlin Wall, a central motif of the narrative, which deals with the division of Germany by an arbitrary internal frontier. It is exemplified by the story of Herr B., a photographer in a town in Holstein, who intends to, but in the end does not, marry a nurse, referred to as die D., who works in a hospital in East Berlin. The use of initials for both characters underlines their scarcely determinable identities. The emphasis is placed on the problem of human communication in the two Germanies. Die D., adjusted to life in the DDR, sees in the building of the Wall a violation of the state's principles. She accepts B.'s aid to escape to the West, but rejects marriage and takes up nursing in a West Berlin hospital.

Zwei Beichten, Die, a short Middle High German poem of Schwank-like character (see SCHWANK). The partners of a marriage confess to each other their infidelities. The husband, who is the less guilty spouse, forgives the wife. She, however, beats the husband. The poem was written in the 13th c. by an unknown author.

Zweibund (Dual Alliance), a treaty agreed between the German Empire and Austria-Hungary in October 1879. It was negotiated (against the wishes of the Emperor Wilhelm I, q.v.) by Bismarck (q.v.) and the Austrian foreign minister Count G. Andrássy (1823–90). By its terms either party undertook to come to the assistance of the other if it became engaged in war with Russia. The treaty, which in 1882 was supplemented by the Triple Alliance (see DREIBUND), was devised by Bismarck in consequence of increasingly anti-German tendencies in Russian politics. In 1883 Romania was admitted to the Dual Alliance.

Zweifler, Der, title given to a legend of a monk who doubts the Psalmist's assertion that 'a thousand years are but a day in thy sight', hears the entrancing song of a bird, follows it, and returns to find that a thousand years are past. The legend occurs as a parable (Bispel) in a verse sermon. See also MÖNCH FELIX.

ZWEIG, ARNOLD (Groß-Glogau, Silesia, 1887–1968, Berlin), the son of a Jewish master saddler, studied from 1907 to 1911 at various universities and served in the 1914–18 War, latterly in a press unit on the Russian front. After the war he turned to authorship, and from 1923 lived in Berlin. In 1933 he fled to Czechoslovakia, moving from there to Palestine, where he worked as a journalist. From 1948 until his death he lived in East Berlin. During these last years he suffered from blindness, and his later works were dictated. Zweig was a recipient of various prizes for literature, and from 1950 to 1953 was president of the East German Academy of Arts. He was not related to Stefan Zweig (q.v.).

Zweig wrote two early tragedies, *Abigail und Nabal* (1913) and *Ritualmord in Ungarn* (1914), and the plays *Bonaparte in Jaffa* (1939) and *Soldatenspiele* (1956, a collection), but he is mainly the author of fiction. His collections of stories include *Die Novellen um Claudia* (1912), *Aufzeichnungen über eine Familie Klopfer* (1913), *Geschichtenbuch* (1916), *Söhne. Zweites Geschichtenbuch, Gerufene Schatten* (both 1923), *Frühe Fährten, Regenbogen* (both 1925), *Knaben und Männer, Mädchen und Frauen* (both 1931), *Spielzeug der Zeit* (1933), and *Allerlei Rauch* (1949); his Novellen *Der Spiegel des großen Kaisers* (1926) and *Pont und Anna* (1928, originally included in *Regenbogen*) were published separately.

In 1927 appeared Zweig's best novel, *Der Streit um den Sergeanten Grischa* (q.v., dramatized, *Das Spiel vom Sergeanten Grischa*, 1949), which was followed by *Junge Frau von 1914* (1931) and *De Vriendt kehrt heim* (1932). His next two novels, *Erziehung vor Verdun* (1935) and *Einsetzung eines Königs* (1937), were prefixed to *Der Streit um den Sergeanten Grischa* to form a trilogy; at their respective centres are Bertin and Winfried, two characters from the earlier novel. In his 60s

Zweig arranged his war novels to make a cycle of six works entitled *Der große Krieg der weißen Männer*. It consists of *Die Zeit ist reif* (1958), the earlier novels *Erziehung vor Verdun, Junge Frau von 1914, Einsetzung eines Königs*, and *Der Streit um den Sergeanten Grischa*, and *Feuerpause* (1954). Other novels are *Versunkene Tage* (1938, retitled *Verklungene Tage*, 1950), *Das Beil von Wandsbek* (1947), and *Traum ist teuer* (1962).

Zweig's essays, many of which are devoted to the Jewish problem, include *Lessing, Kleist, Büchner* (1925). A select edition of his works, *Ausgewählte Werke* (16 unnumbered vols.), appeared 1957–67.

ZWEIG, STEFAN (Vienna, 1881–1942, Petrópolis nr. Rio de Janeiro, Brazil, by suicide), of Jewish descent, studied at Berlin and Vienna universities and afterwards travelled extensively. In the 1914–18 War he developed pacifist views and moved to Zurich (where he met Romain Rolland) in order to be able to express them. Between the wars he lived chiefly in Salzburg, emigrating in 1938 to England. After a short period in New York he settled in Brazil.

Zweig's early collections of poems, *Silberne Saiten* (1901) and *Die frühen Kränze* (1906), are neo-Romantic (see NEUROMANTIK) and related to those of H. von Hofmannsthal (q.v.). His work was most deeply influenced, however, by the psychology of S. Freud (q.v.), notably the stories *Erstes Erlebnis* (1911), *Amok* (1922), *Verwirrung* (1927, all titles of collections), *Angst* (1920), and *Schachnovelle* (1942). In this highly intricate narrative the game of chess provides an analogy to the psychological deterioration of a Jewish intellectual under interrogation by the Gestapo. Zweig wrote several plays, including *Tereites* (1907), *Das Haus am Meer* (1912), and *Jeremias* (1917), and provided R. Strauss (q.v.) with the libretto for *Die schweigsame Frau* (1935). He also made a translation of Ben Jonson's *Volpone* (1926), which had a successful run on the stage.

Zweig found his most congenial form in perceptive biographical essays on poets and artists, which concentrate on their inner conflicts and on the creative process. *Drei Meister* (1920) contained studies of Balzac, Dickens, and Dostoevsky; *Der Kampf mit dem Dämon* (1925) has as its subjects Hölderlin, H. von Kleist, and Nietzsche (qq.v.). Essays on Casanova, Stendhal and Tolstoy make up *Drei Dichter ihres Lebens* (1928). Zweig is also the author of the longer biographical studies *Romain Rolland* (1921), *J. Fouché* (1929), *Marie Antoinette* (1932), *Maria Stuart* (1935), and *Triumph und Tragik des Erasmus von Rotterdam* (1935), which has an equal claim with the later *Schachnovelle* to be his best work. His poetry was collected in *Die gesammelten*

Gedichte (1924) and his shorter biographical essays in *Baumeister der Welt* (1936). He created a new minor genre in his 'historical miniatures' (Miniaturen), *Sternstunden der Menschheit* (1927, expanded in 1936, and again in a posthumous edition, 1943). Apart from the *biographies romancées* on Marie Antoinette and Mary Stuart, he wrote one novel, *Ungeduld des Herzens* (1939). *Die Welt von Gestern* (1942) is autobiographical. His correspondence with R. Strauss appeared in 1957 and with P. Zech (q.v.), *Briefe 1910–1942*, ed. D. G. Daviau, in 1984.

Gesammelte Werke (19 unnumbered vols.) appeared in 1946–67 and (25 unnumbered vols.), ed. K. Beck, in 1982–4.

Zwei Gesellen, Die, a poem by J. von Eichendorff (q.v.), the first line of which runs 'Es zogen zwei rüstge Gesellen'. It reflects the dangers of self-complacency on the one hand and of self-destructive indiscipline on the other. It was published in 1818 in the *Frauentaschenbuch*.

Zweikampf, Der, a Novelle by H. von Kleist (q.v.), published in *Erzählungen* (vol. 2) in 1811. Froissart's *Chronique de France* (15th c.) inspired this story which, set against the background of 13th-c. imperial Germany, opens with the murder of Duke Wilhelm von Breysach and closes with the confession of his half-brother, Count Jacob 'der Rotbart', that he was responsible for the crime by which he had hoped to secure his succession to the ducal crown.

In the course of legal proceedings against him, based on his proved ownership of the death-dealing arrow, Jacob has to establish, as the crux of his defence, an alibi for the night of the murder. This he convincingly accomplishes by revealing that he had kept a secret rendezvous with Littegarde von Auerstein, a widow of outstanding virtue. Thrust into deep distress by this assault upon her reputation, Littegarde finds in Friedrich von Trota a friend who is ready to prove her innocence before God and the world by challenging Count Jacob to single combat, at which the Emperor himself is present. The outcome, however, decides against him, for, while he is severely wounded, Count Jacob suffers only a slight injury. Friedrich and Littegarde are therefore duly imprisoned, sentenced to death, and are already tied to the stake when the strange 'Zweikampf', in which God appeared to have deserted the cause of innocence and justice, finds its explanation. While Friedrich has made a miraculous recovery from his seemingly fatal injuries, the Count's slight wound has rapidly developed into an incurable disease, prompting him to use his last breath to confess to the murder and to the blamelessness of Littegarde. The Emperor restores the condemned couple to honour, and their marriage enhances the triumph of truth.

Zwei Komtessen, a complementary pair of stories by Marie von Ebner-Eschenbach (q.v.), published in 1885. They take the form of character sketches. *Komtesse Muschi* consists of letters (see BRIEFROMAN) written by Muschi to an intimate friend, in which she unintentionally reveals her superficial mind and self-complacent egoistic character. Muschi is only concerned with horses, elegant clothes, and making a socially eligible and rich marriage. But the count who appears as a suitor proposes to another girl with less self-confidence and a kinder disposition.

The story of *Komtesse Paula* is given in diary form. Paula, an unpretentious but sensitive and warm-hearted girl, is almost married off to a wealthy and self-absorbed nobleman, is saved by the intervention of her elder sister, and becomes engaged to Baron Schwarzburg, whom she deeply loves.

Zwei Meilen Trab, a poem by D. von Liliencron (q.v.), published in *Gedichte* (1889).

Zwielicht, a poem by J. von Eichendorff (q.v.), the first line of which runs 'Dämmrung will die Flügel spreiten'. Its basic tone is an expression of Romantic dread.

Zwillinge, Die, a tragedy (Trauerspiel) in five acts by F. M. Klinger (q.v.), published in 1776. It was written in 1776 as an entry to a competition announced by the actor F. L. Schröder (q.v.) in Hamburg, and gained the prize. Schröder performed it in Hamburg in February 1776. There were two other entries, one of which was Leisewitz's *Julius von Tarent* (q.v.), which shares with *Die Zwillinge* the theme of hostile brothers. This seems not to have been entirely coincidental, for Klinger learned of Leisewitz's play from their mutual friend J. M. Miller (q.v.).

The scene is an Italian palace. Ferdinando and Guelpho are the twin sons of a prince. Ferdinando is the first born, but Guelpho, furiously jealous of his brother's rights, persuades himself that he is really the elder, works himself into a state of passionate anger, and murders Ferdinando. At first he maintains his innocence, but in the presence of Ferdinando's corpse he confesses the murder and is put to death by his father. Allusion is made in the last act to the murder of Abel by Cain. *Die Zwillinge* is a sombre, sultry play, which, in contrast to most plays of the Sturm und Drang (q.v.), adheres to the unities of time and place.

ZWINGÄUER, DER, names himself in the 13th-c. comic verse tale *Der Schwangere Mönch* (q.v.) as its author. He is not otherwise known.

ZWINGLI, ULRICH or (in his own spelling) HULDRYCH (Wildhaus, Toggenburg, 1484–1531, Kappel), German-Swiss reformer, became parish priest of Glarus in 1506. As an army chaplain he was present at the battles of Novara (1513) and Marignano (1515), becoming on his return a priest at the pilgrimage church of Maria Einsiedeln. His first thoughts on ecclesiastical reform came after contact with Erasmus (q.v.) had opened his mind to the new humanism.

From 1519 on, Zwingli was Leutpriester at the Großmünster in Zurich, where he succeeded, now also under the influence of Luther (q.v.), in establishing a reformed church in 1523. Zwingli, more politically minded and with fewer theological preoccupations than Luther, fell out with the latter over the doctrine of Transubstantiation, and a conference at Marburg, known as the Marburger Religionsgespräch, in 1529 failed to resolve their differences. It was at this point that the Lutheran and Reformed Churches (see REFORMIERTE KIRCHE) diverged, Zwingli preparing the way for Calvin (1509–64). Zwingli consistently maintained that salvation depended on faith alone and was granted only to those souls whom God had chosen. He was killed in the battle of Kappel in 1531. His principal work is doctrinal: *De vera ac falsa religione* (1525). He is the subject of an epic poem by W. Schäfer (q.v.). *Sämtliche Werke*, ed. E. Egli, G. Finsler *et al.*, appeared 1905 ff.

Zwischen Himmel und Erde, a narrative by O. Ludwig (q.v.) set in a frame (see RAHMEN) and usually referred to as a Novelle. The work stands foremost among those associated with Poetic Realism (see POETISCHER REALISMUS). It was published in 1856 (23 sections), and revised in 1858 (22 sections) and in 1862 (21 sections). The title, used as an epitaph closing the work, is expanded in the expository section by the factual statement 'Zwischen Himmel und Erde ist des Schieferdeckers Reich'; the tale gives the portrait of a family which combines the slater's and the steeplejack's craft in a small town.

The frame shows the principal character, Apollonius Nettenmair, at the beginning of the evening of life. The inner story (Binnenerzählung) opens with young Apollonius returning, thirty years before, from his apprenticeship in Cologne to join the family business, run by his father, who has gone blind. During Apollonius's absence, his older brother Fritz has, by deception, married Apollonius's intended bride, Christiane; the sense of guilt arising from this deceit determines the conflict, to which the relationship between the brothers and Christiane is central. The underlying painful consciousness of moral betrayal leads to a deterioration in the character of Fritz, who develops an obsessive hatred of his brother, which enfolds, with less conspicuous pathological force, the contrasting nature of his brother. Exercising tactful reticence, Apollonius devotes himself to duty and saves the business from bankruptcy.

A crisis arises when Fritz, accusing Christiane of infidelity, strikes her in the presence of their sick daughter Ännchen. Christiane, who is as innocent as is Apollonius, has, however, become estranged from Fritz, for she has found out his deceit, of which Apollonius is still unaware. The blow results in Ännchen's death and in Christiane's greater attraction to Apollonius, from whom she conceals it. Fritz reacts to the loss of their daughter by plotting the murder of Apollonius on the occasion of the repair of the roof and spire of the church of St. Georg, the story's central motif. He cuts a nick in the rope upon which the steeplejack's life depends when working suspended between heaven and earth. The attempt fails through the alertness of the family servant Valentin, who reveals to Nettenmair senior the stark reality behind the neat façade of the family residence. To the blind old man the threat to the family honour is foremost. To save it he is prepared to demand the life of his first born. Led by Valentin he climbs to the trapdoor leading to the roof, intending to force Fritz to jump to his death. The news that Apollonius is safe reaches him just before the time he allows Fritz is up. Fritz, his life thus saved, reacts by attacking Apollonius on the same church roof; but it is he who, in the dramatic encounter of the two brothers on the steep slope, loses his hold and plunges to death.

The action rises to another climax when Apollonius saves the church from destruction after lightning has set fire to it. Apollonius becomes the town's most admired and respected citizen and benefactor. The deeper importance of the incident, nevertheless, lies in its therapeutic effect on Apollonius, who in the hour of Fritz's revenge had not caused, but had willed, his death. Acting henceforth as guardian to Christiane's two sons, he declines to obey his father's order to marry her after the appropriate interval of family mourning, thus satisfying his own sense of integrity.

'Zwischen Lavater und Basedow', first line of the poem *Diné zu Koblenz* written in 1774 by Goethe and preserved in a copy made by Luise von Göchhausen (q.v.). The poem refers to a journey Goethe made in July 1774 in company with J. C. Lavater and J. B. Basedow (qq.v.). They went first to Ems, then to the

Rhine, and downstream as far as Neuwied. The last four lines of the poem (quoted in *Dichtung und Wahrheit*, Bk. 14) refer to their dinner at Coblence.

Zwischen Weiß und Rot (Die russische Tragödie), a novel of recollection by E. E. Dwinger (q.v.), published in 1930 as a sequel to his *Die Armee hinter Stacheldraht*, recounting experiences in Admiral Kolchak's White Russian army in 1919. The book is followed by *Wir rufen Deutschland* (qq.v.).

ZWÖLF ALTE MEISTER, twelve medieval poets whom the Meistersinger (see MEISTER-GESANG) acknowledged as the formal authorities and models for poetry and for the music to which it was set. Though the names varied, the earliest list, given in the 14th c. by Lupold Hornburg von Rotenburg, comprises Her Reimar (REIN-

MAR der Alte), Her Walther (von der Vogel-weide), Her Nithart (NEIDHART von Reuental), Her Wolfram (von Eschenbach), von Wirzeburg Cunrad (KONRAD von Würzburg), der Boppe, der Marner, der Regenboge, der Vrouwenlop (Heinrich Frauenlob), von Suneburg (Friedrich von Sonnenburg), Bruder Wernher (qq.v.), and Erenbote. A legend arose, recorded in a poem of the 16th c., that the twelve masters devised Meistergesang simultaneously and independently.

Zwölfjährige Mönchlein, Das, an anonymous Middle High German verse legend. It comprises roughly 300 lines and was written early in the 14th c. The author, a Swiss, was influenced by KONRAD von Würzburg (q.v.). The little monk is a boy of fervent piety, to whom the boy Jesus appears. The two play together and Jesus tells his companion that he shall be that day in Paradise. The boy dies and is carried off to Heaven.